Handbook of Radioactivity Analysis

Fourth Edition

Handbook of Radioactivity Analysis

Volume 1: Radiation Physics and Detectors

Fourth Edition

Edited by

Michael F. L'Annunziata

ELSEVIER

ACADEMIC PRESS
An imprint of Elsevier

Academic Press is an imprint of Elsevier
125 London Wall, London EC2Y 5AS, United Kingdom
525 B Street, Suite 1650, San Diego, CA 92101, United States
50 Hampshire Street, 5th Floor, Cambridge, MA 02139, United States
The Boulevard, Langford Lane, Kidlington, Oxford OX5 1GB, United Kingdom

Library of Congress Cataloging-in-Publication Data
A catalog record for this book is available from the Library of Congress

British Library Cataloguing-in-Publication Data
A catalogue record for this book is available from the British Library

ISBN: 978-0-12-814397-1

For information on all Academic Press publications visit our website at https://www.elsevier.com/books-and-journals

Publisher: Susan Dennis
Acquisitions Editor: Kathryn Eryilmaz
Editorial Project Manager: Hilary Carr
Production Project Manager: Prem Kumar Kaliamoorthi
Cover Designer: Matthew Limbert

Typeset by TNQ Technologies

Contents of Volume 1

3. Solid-state nuclear track detectors

Shi-Lun Guo, Bao-Liu Chen and S.A. Durrani

4. Semiconductor detectors

Ramkumar Venkataraman

5. Alpha spectrometry

Nóra Vajda, Roy Pöllänen, Paul Martin and Chang-Kyu Kim

6. Liquid scintillation analysis: principles and practice

Michael F. L'Annunziata, Alex Tarancón, Héctor Bagán and José F. García

8. Radioisotope mass spectrometry

Clemens Walther and Klaus Wendt

Contents of Volume 2

Contributors

Héctor Bagán, Department of Chemical Engineering and Analytical Chemistry, University of Barcelona, Barcelona, Spain

Karl Buchtela, Vienna University of Technology, Atominstitut, Vienna, Austria

Bao-Liu Chen, China Institute of Atomic Energy (CIAE), Beijing, China

S.A. Durrani, School of Physics and Astronomy, University of Birmingham, Birmingham, United Kingdom

José F. García, Department of Chemical Engineering and Analytical Chemistry, University of Barcelona, Barcelona, Spain

Shi-Lun Guo, China Institute of Atomic Energy (CIAE), Beijing, China

Chang-Kyu Kim, Department of Safeguards, International Atomic Energy Agency, Vienna International Centre, Vienna, Austria

Michael F. L'Annunziata, The Montague Group, Oceanside, CA, United States

Paul Martin, Australian Radiation Protection and Nuclear Safety Agency, Yallambie, VIC, Australia

Roy Pöllänen, Department of Physics, University of Helsinki, Helsinki, Finland

Georg Steinhauser, Leibniz Universität Hannover, Institute of Radioecology and Radiation Protection, Hannover, Germany

Alex Tarancón, Department of Chemical Engineering and Analytical Chemistry, University of Barcelona, Barcelona, Spain

Simon Temple, Meridian Biotechnologies Ltd., Chesterfield, United Kingdom

James Thomson, Meridian Biotechnologies Ltd., Chesterfield, United Kingdom

Nóra Vajda, Radiochemical Laboratory, RADANAL Ltd., Budapest, Hungary

Ramkumar Venkataraman, Oak Ridge National Laboratory, Oak Ridge, TN, United States

Clemens Walther, Institute of Radioecology and Radiation Protection, Leibniz University Hannover, Hannover, Germany

Klaus Wendt, Institute of Physics, Johannes Gutenberg-University Mainz, Mainz, Germany

About the Founding Editor

Michael F. L'Annunziata

Michael F. L'Annunziata, PhD, is the founding editor and coauthor of the *Handbook of Radioactivity Analysis*. He majored in chemistry with a BSc degree from St. Edward's University in 1965, and he was awarded MSc and PhD degrees from the University of Arizona, Tucson, in 1967 and 1970, respectively. His graduate thesis research in the 1960s, financed by the then US Atomic Energy Commission, dealt with the analysis of the radionuclides ^{89}Sr and ^{90}Sr and the remediation of soils contaminated with radiostrontium in the event of nuclear fallout, published as a thesis in 1967 (https://repository.arizona.edu/handle/10150/318640). After a short stint in the chemical industry (Amchem Products, Inc, Ambler, Pennsylvania) during 1970–71 as ^{14}C-tracer chemist, he joined the faculty at the Postgraduate College in Chapingo, Mexico, as a professor and thesis advisor during 1972–75; and during 1975–77, L'Annunziata was a senior research scientist at the Nuclear Center of the National Institute of Nuclear Research (ININ), Mexico, where he served also as a thesis research advisor to graduate students of chemistry of the Autonomous University of the State of Mexico in Toluca, Mexico, in the field of radionuclide analysis and applications. During 1977–91, he was a scientific officer in the Departments of Research and Isotopes and Technical Cooperation of the International Atomic Energy Agency (IAEA) in Vienna, Austria, where he served as IAEA Head of Fellowships and Training during 1987–91. From 1977 to 2007, he served as IAEA Expert in fact-finding, planning, and implementation assignments in peaceful applications of nuclear energy for development in more than 50 countries of the world in Asia, Africa, Europe, Latin America, and the Middle East. L'Annunziata was a member of the Board of Governors, International Science Programs at Uppsala University, between 1988 and 1991. His main research interests have been focused on the development of chemical and instrumental methods for the detection and measurement of radioactive nuclides as tracers in research. He was the first to postulate the soil microbial epimerization of myo-inositol to other inositol stereoisomers as a source of the stereoisomers of inositol phosphates in soils (PhD dissertation, 1970, https://dissexpress.proquest.com/dxweb/results.html?QryTxt=&By=L%27Annunziata&Title=&pubnum =) and in 1975 (*SSSA Journal* 39(2), 377–379) and first to demonstrate in 1977, with the use of the radioisotope carbon-14, the soil microbial epimerization of myo-inositol to chiro-inositol as a mechanism involved in the origin of the unique soil inositol phosphate stereoisomers (*SSSA Journal* 41(4), 733–736, https://dl.sciencesocieties.org/publications/sssaj/abstracts/41/4/SS0410040733). The first edition of the *Handbook of Radioactivity Analysis* was planned by L'Annunziata in 1995, and he edited and coauthored the subsequent editions, including the current fourth edition published by Elsevier in 2020. He has authored and coauthored 11 books since 1979 on radionuclide analysis and radiation physics among which his book entitled *Radioactivity: Introduction and History*, First Edition, published by Elsevier in 2007, was included on the Best Sellers List in Physics (*LibraryJournal Academic Newswire*) in 2008. His much expanded Second Edition entitled *Radioactivity: Introduction and History, From the Quantum to Quarks* (https://www.sciencedirect.com/book/9780444634894/radioactivity), published by Elsevier in 2016, was awarded an Honorary Mention in the 2017 PROSE AWARDS in the category of Chemistry and Physics.

Foreword

Radioactive sources play a significant role in promoting human development and health worldwide. Whether through its application to treat cancer, diagnose various diseases, develop new crop varieties, sterilize medical supplies, or provide clean energy, peaceful uses of radioactive sources are ubiquitous in our daily lives. These wide-ranging applications can only be implemented appropriately when radioactivity is measured precisely. Thus, the accurate measurement of nuclear radiation is indispensable for the peaceful applications of radioactive materials. For example, in fields such as nuclear medicine, whether for the treatment or diagnosis of disease, accurate measurements of radionuclides are essential. Dosimetric measurements are the cornerstone of safe and effective radiation therapy for the treatment of cancer whether for brachytherapy, proton beam therapy, or other sources of radiation therapy.

With more than 170 Member States in all continents of the world, the International Atomic Energy Agency (IAEA) serves as the global focal point for nuclear cooperation. The *Handbook of Radioactivity Analysis* will serve Member States as one of many tools available in the application of nuclear science and technology for peaceful purposes. The importance of this guidance is demonstrated by the wide range of areas in which the IAEA supports Member States to contribute to their well-being and development. Such examples include biological sciences research, insect pest control, health, fertilizer and water use efficiency, water resources and the environment including marine science and climate change, radiation technology, neutron diffraction, radiography and activation analysis, radiation processing in industry, radiation protection, nuclear power, nuclear safeguards, radiation preparedness and response, and research in the field of nuclear fusion, among others.

The *Handbook of Radioactivity Analysis* is now in its fourth edition since the successful first edition in 1998. Over the past two decades, this book has expanded in its scope from an initial 12 chapters to the current 22 chapters, encompassing the numerous modern applications and methods of radiation detection and measurement. The chapters in this book are written by experts from 16 countries around the world. This new edition will continue to serve as an important resource in our search to optimize radioactivity measurements both in research and in its applications, leading to the peaceful utilization of radioactive sources for health and development.

May Abdel-Wahab, MD, PhD, FACR
Director, Division of Human Health
Department of Nuclear Sciences and Applications
International Atomic Energy Agency, Vienna

Preface to the fourth edition

In 1996, I proposed to Academic Press the idea of a book that would provide readers with a reference source to state-of-the-art radiation detectors and methods of analysis of radionuclides and other sources of nuclear radiation. Thus, the first edition of this book was published in 1998 as a single volume with only 12 chapters, and the book has expanded in scope and depth over the past two decades with the current fourth edition and its 22 chapters in two volumes.

The numerous advances that have been made since the publication of the previous third edition warranted the partition of the *Handbook of Radioactivity Analysis* into two volumes. It was decided to separate the chapters into two categories, namely, Volume 1, *Radiation Physics and Detectors* and Volume 2, *Radioanalytical Applications*. The two volumes of this book were expanded greatly to provide material, which would serve as a valuable resource in teaching and a reference source to the researcher in his or her unique analytical requirements in the measurement of radioactive materials.

The first chapter in Volume 1, which was previously entitled *Radiation Physics and Radionuclide Decay*, was expanded to almost double in volume with a corresponding change in the chapter title to *The Atomic Nucleus, Nuclear Radiation, and the Interaction of Radiation with Matter*, which includes additional material helpful to supplement the academic curricula and aid in the decisions and calculations made by researchers in their measurement of nuclear radiation and radionuclide analysis. Current principles of operation of all classes of radiation detectors and their applications have been expanded and updated, including semiconductor detectors, gas ionization detectors, liquid and solid scintillation detectors, solid-state nuclear track detectors, Cherenkov detectors, calorimeters and bolometers, as well as advances in atom counting (i.e., mass spectrometry) for the measurement of radioactive and stable nuclides and radiation from other sources such as cosmic radiation, synchrotron radiation, and particle emissions from nuclear reactions.

In light of increased concern for radioactivity in the environment, a chapter was added on the *Analysis of Environmental Radionuclides* in Volume 2. Also, all chapters in Volume 2 have been expanded and updated with material required in the analysis of radionuclides and radiation in our land, air, and water resources, including the marine environment, as well as particle identification and measurement by Cherenkov counting, radiation counting statistics, radionuclide standardization, imaging techniques required in the applications of radionuclides in biological research and nuclear medicine, flow-cell analytical techniques, automation in radiochemical analysis together with analytical techniques required in the fields of nuclear safeguards and nuclear forensics.

Again, we have completed this book as an international effort by drawing upon the expertise of researchers and teachers from 16 countries of the world. Although coming from many branches of science, chapter authors all share one common objective, that being the most accurate measurement of radiation sources and radionuclides both natural and man-made, vital to all branches of science and human development. Readers interested in radiation physics, the applications of radionuclides and radiation sources, and how these have been vital to our well-being and development may refer to another text by the writer entitled "*Radioactivity: Introduction and History, From the Quantum to Quarks*" (ISBN: 978-0-444-63489-4), published in 2016 by Elsevier (https://www.elsevier.com/books/radioactivity/lannunziata/978-0-444-63489-4).

Women are the senior authors of three chapters in this new edition, which is evidence of the increasing role of women as leaders in this field of science. We may expect to see yet in the future an ever-increasing number of women, who will make great advances in this field of science following the pioneering examples of Marie Curie, Lise Meitner, Maria Goeppert-Mayer, Rosalind Franklin, Marietta Blau, and Chien-Shiung Wu, among others.

Mention of commercial products in this book does not imply recommendation or endorsement by the chapter authors or editor. Other or more suitable products may be available. Names of products are included for convenience or information purposes only.

I would like to thank the authors of each chapter, who have covered their fields of expertise with an unwavering commitment to meet the objectives of this book. Acknowledgment is extended to Kathryn Eryilmaz (nee Morrissey),

Aquisition Editor, at Elsevier in Cambridge for approaching me with the suggestion that we consider a fourth edition and for working with me during the planning stage of this book. Many thanks go to Hilary Carr, Elsevier Editorial Project Manager, for her constant support and advice throughout the writing and production of this book. I thank also Ashwathi P. Aravindakshan of Elsevier for her assistance in completing the legal requirements for the publication of this book. Appreciation is also extended to Prem Kumar Kaliamoorthi, Elsevier Production Project Manager, for his meticulous attention to every detail throughout the production process of this book. Thanks are also extended to Susan Dennis, Publisher of Elsevier Chemistry and Chemical Engineering Books, and Mona Zahir, Elsevier Editorial Project Manager, for their guidance and support during this project. Above all, I thank my wife Maria del Carmen (aka Reyna) for her understanding, encouragement, and unflagging patience.

Michael F. L'Annunziata, PhD

Acronyms, Abbreviations, and Symbols

A Mass number
A Ampere (1 A = 1 C/s), amplifier
a Year(s)
Å Angstrom (10^{-10} m = 0.1 nm)
AABW Antarctic Bottom Water
AAIW Antarctic Intermediate Water
AAS Atomic absorption spectrometry
AASI Advanced alpha-spectrometric simulation
ATTA Atom trap trace analysis
ABACC Brazilian–Argentine Agency for Accounting and Control of Nuclear Materials
ABEC aqueous biphasic extraction chromatography
AC Alternating current
ACC Antarctic Circumpolar Current
ACFM Actual cubic feet per minute (28.3 L/min.)
ADC Analog-to-digital converter
ADF Advanced digital filter
ADME Absorption, distribution, metabolism, and excretion
ADS Accelerator-driven subcritical reactor
AEC Automatic efficiency compensation, Atomic Energy Commission
AES Atomic emission spectrometry, Auger electron spectroscopy
AF Agulhas Front
AFM Atomic force microscope
AFS Atomic fluorescence spectrometry
α Alpha particle, internal conversion coefficient
$\propto$ Proportional to
ag Attogram = 10^{-18} g
AGeV GeV per nucleon
AkeV keV per nucleon
A2LA American Association for Laboratory Accreditation
AM β-artemether, arithmetic mean
AMAD Activity median aerodynamic diameter
AMANDA Antarctic Muon and Neutrino Detector Array, South Pole
AMANDE Accelerator for Metrology and Neutron Applications in External Dosimetry, IRSN, France
AMAP Arctic Monitoring and Assessment Programme
AMeV MeV per nucleon
AMP Adenosine monophosphate, ammonium molybdophosphate, amplifier
amp. Amplifier
AMS Accelerator mass spectrometry
amu Atomic mass units
ANDA 7-Amino-1,3-naphthalenedisulfonic acid
ANFESH Ferric potassium hexacyanoferrate on a cellulose carrier
ANITA ANtarctic Impulsive Transient Antenna
ANL Argonne National Laboratory
ANN Artificial neural network
ANSI American National Standards Institute
ANSTO Australian Nuclear Science and Technology Organisation
ANTARES ANTArctic RESearch, Astronomy with a Neutrino Telescope and Abyss Environmental RESearch, Mediterranean Sea
ANZECC Australian and New Zealand Environment Conservation Council
APCI Atmospheric pressure chemical ionization
APD Avalanche photodiode
APDC Ammonium pyrrolidine dithiocarbamate
APE Alkyl phenol ethoxylate
APMP Asia–Pacific Metrology Program
APS Advanced Photon Source, Argonne National Laboratory
AQC Automatic quench compensation
AQCS Analytical Quality Control Services (of IAEA)
AQP(I) Asymmetric quench parameter of the isotope
ARC Agulhas Return Current
ARMCANZ Agriculture and Resource Management Council of Australia and New Zealand
AS Alpha spectrometry
ASTAR Alpha stopping power and range
ASTM American Society for Testing and Materials
atm Atmosphere (standard) = 1.01325×10^5 Pa
at % Atom percent
ATP Adenosine triphosphate
ATSDR Agency for Toxic Substances and Disease Registry
AUV Autonomous underwater vehicle
AWCC Active Well Coincidence Counter
AWE United Kingdom Atomic Weapons Establishment
β Particle relative phase velocity, beta particle
$\beta\beta$ Double-beta decay
β^- Negatron, negative beta particle
β^+ Positron, positive beta particle
b Barn = 10^{-28} m^2 = 10^{-24} cm^2
BAC *N,N′*-bisacrylylcystamine
bar 10^5 N/m^2 = 100×10^3 Pa
BBD 2,5-Di-(4-biphenylyl)-1,3,4-oxadiazole
BBO 2,5-Di(4-biphenylyl)oxazole
BBOT 2,5-Bis-2-(5-t-butyl-benzoxazolyl) thiophene
BCC Burst counting circuitry, Bragg curve counter
BDs Bubble detectors
BDE Bond dissociation energy
BE Binding energy
BEAGLE Blue Ocean Global Expedition
BEGe Broad-energy germanium detector

BGO Bismuth germanate ($Bi_4Ge_3O_{12}$)
BIPM Bureau international des poids et mesures, Sèvres, France
bis-MSB *p*-Bis-(*o*-methylstyryl)benzene
B_K *K*-shell electron binding energy
bkg, BKG Background
BNCT boron neutron capture therapy
BNL Brookhaven National Laboratory, Upton, New York
BOD Biological oxygen demand
BOMARC Boeing Michigan Aeronautical Research Center
BOREXINO BOron EXperiment, solar neutrino detector, Italy
Bq Becquerel = 1 dps
BQM Bqmeter (Consortium BQM, Czech Republic)
BR Branching ratio
BS Backscatter
BSA Bovine serum albumin
BSI The British Standards Institute
BSO Bismuth silicate ($Bi_4Si_3O_{12}$)
BSS Bonner sphere spectrometer, Board of Safety Standards
BT Bound tritium
BTP Bistriazinylpyridine
butyl-PBD 2-(4-t-Butylphenyl)-5-(4-biphenylyl)1,3,4-oxadiazole
BWR Boiling water reactor
c Speed of light in vacuum (2.9979×10^8 m/s)
C Coulomb (1 C = 1 A s)
°C Degrees Celsius
CAI Calcium–aluminum-rich inclusions
CaF_2(Eu) Europium-activated calcium fluoride
CALEX Calorimetry Exchange Program
CAM Continuous air monitoring
CAMAC Computer-automated measurement and control
CANDLES CAlcium fluoride for the study of Neutrinos and Dark matter by Low Energy Spectrometer
CANDU Canadian deuterium uranium reactor
CART Classification and regression tree algorithm
CAVE Counting lAboratory for enVironmental radionuclidEs, Monaco
CC Charged current (interaction), charge comparison, carbonate carbon
CCD Charge-coupled device
CCRI Consultative Committee for Ionizing Radiation
CD ROM Compact disc read-only memory
CDW Circumpolar Deep Water
CE Chemical etching, capillary electrophoresis
CEA Commissariat à l'Energie Atomique
CEFAS Centre for Environment, Fisheries and Aquaculture Science (UK)
CE-ICP-MS Capillary electrophoresis–inductively coupled plasma mass spectrometry
CELLAR Collaboration of European Low-level Underground Laboratories
CENTA Centre for Nuclear and Accelerator Technologies, Bratislava
CERN European Organization for Nuclear Research, Geneva
CET Compton efficiency tracing method
C_F Feedback capacitor
CF Calibration factor, correction factor
CFD Constant fraction discriminator
cfm Cubic feet per minute
CFN Cross-flow nebulizer
CGE Chamber Gram Estimator
Ch Channel
CHEREN2 Anisotropy detection model for Cherenkov counting efficiency
CHU Centre hospitalier universitaire
Ci Curie = 2.22×10^{12} dpm = 3.7×10^{10} dps = 37 GBq
CIAE China Institute of Atomic Energy
CICM Conventional integral counting method
CID Collision-induced dissociation
CIEMAT Centro de Investigaciones Energéticas, Medioambientales y Technológicas, Madrid
CIRIA Construction Industry Research and Information Association
cm Centimeter
cm/d Unit of flux from cm^3/cm^2 per day
CMB Cosmic microwave background
CMOS Complementary metal-oxide-semiconductor
CMPO Octyl(phenyl)-N,N-di-isobutylcarbamoylmethylphosphine oxide
CMX-4 Collaborative Materials Exercise (fourth by the ITWG)
C/N CIEMAT/NIST (efficiency tracing method)
CN Cellulose nitrate
CN* Unstable compound nucleus
CNC Condensation nuclei counter
CNET CIEMAT/NIST efficiency tracing
CNRS Centre National de la Recherche Scientifique, France
CNS Central nervous system
COG Center of gravity
COMPASS Community Pentascale Project for Accelerator Science and Simulation
COTS Commercial off-the-shelf (system)
cph, CPH Counts per hour
cpm, CPM Counts per minute, channel photomultiplier
cps, CPS Counts per second
CR-39 Polyallyldiglycol carbonate plastic SSNTD
CRESST Cryogenic Rare Event Search with Superconducting Thermometers
CRL Compound refractive lens
CRM Certified reference material
CS Calibration source
CSDA Continuous Slowing Down Approximation range
CSIC Instituto de Física Fundamental, Madrid
CsI(Na) Sodium-activated cesium iodide
CsI(Tl) Thallium-activated cesium iodide
CT Computerized tomography
CTBT Comprehensive Nuclear-Test-Ban Treaty
CTBTO Comprehensive Nuclear-Test-Ban Treaty Organization
CTD Conductivity/temperature/density detector
CTF Contrast transfer function
CTFE Chlorotrifluoroethylene
CTR Controlled thermonuclear reactor
cts Counts
CV Core valence
cv Column volume
CWOSL Continuous wave optically stimulated luminescence
CZT Cadmium zinc telluride (semiconductor detectors)
D Deuterium

d Days, deuteron, down quark
$\bar{d}$ Antidown quark
2D Two-dimensional
DA Destructive analysis
Da Dalton (unified atomic mass unit, also abbreviated as u)
DAC Derived air concentration
DAP Diallyl phthalate
DASE Le Département analyse, surveillance, environnement, France
DATDA Diallyltartardiamide
DBD Double-beta decay
DC Direct current
DCC Digital coincidence counting
dc-GDMS Direct current−glow discharge mass spectrometry
DDCP Dibutyl-N,N-diethylcarbamylphosphonate
DDTC Diethyldithiocarbamate
DE Double escape
DEF Delayed ettringite formation
δ Delta rays
DEMO Demonstration Power Plant (fusion)
DESR Double external standard relation
DESY Deutsches Elektronen Synchrotron
Det. Detector
DF Decontamination factor
DF-ICP-MS Double focusing ICP-MS
DGA Diglycolamide
DIC Dissolved inorganic carbon
DIHEN Direct injection high-efficiency nebulizer
DIM Data interpretation module
dimethyl POPOP 1,4-Bis-2-(4-methyl-5-phenyloxazolyl)benzene
DiMF Decay in a magnetic field (method)
DIN Diisopropylnaphthalene
DIPE Diisopropyl ether
DIPEX Bis(2-ethylhexyl)methane-diphosphonic acid
DIRC Detector of internally reflected Cherenkov light
DJD Diffused junction detector
DLU Digital light units
DMCA Digital multichannel analyzer
DMF Digital microfluidics
DMG Dimethylglyoxime
DMM Direct matrices multiplication
DMSO Dimethyl sulfoxide
DNA Deoxyribonucleic acid
D_2O Heavy water
DOC Dissolved organic carbon
DOE US Department of Energy
DOELAP Department of Energy Laboratory Accreditation Program
DOM Digital optical module
DOP Dioctyl phthalate
DOT Digital overlay technique
dpm, DPM Disintegrations per minute
dps, DPS Disintegrations per second
DPSD Digital pulse shape discrimination
dpy, DPY Disintegrations per year
DQP Double quench parameter
DRAM Dynamic random access memory
DSA Defined solid angle
DSES Deep sea echo sounder
DSP Digital signal processing
DT Dead time
DTPA Diethylenetriamine pentaacetic acid
DTSA Desktop spectrum analyzer (software)
DU Depleted uranium
DWL Drinking water limit
DWPF Defense waste processing facility
E Counting efficiency, energy
E_b Binding energy
e^+ Positron
e^- Electron, negatron
e^-h^+ or e−h Electron−hole pair
EBq Exabecquerel (10^{18} Bq)
EC Electron capture, extraction chromatography, European Community, elemental carbon
ECD Effective cutoff diameter
ECDL Extended cavity diode laser
ECE Electrochemical etching
ECR Electron cyclotron resonance
ED Exponential decrease
EDS Energy dispersive spectrometer
EDTA Ethylenediamine tetraacetic acid
EDX Energy dispersive X-ray (spectrometer)
EDXRF Energy dispersive X-ray fluorescence
EESI-MS Extractive electrospray ionization tandem mass spectrometry
EeV Exaelectron volts (10^{18} eV)
E_F Fermi level
EF Enrichment factor
Eh Oxidation potential
EI Electron impact (e.g., in mass spectrometry)
EIA Enzyme immunoassay
EM Electromagnetic
EMA Extramural absorber
EMCCD Electron multiplier CCD
EML Environmental Measurement Laboratory, USA
EMPA Electron microprobe analysis
ENEA Italian National Agency for New Technologies, Energy and Sustainable Economic Development
ENSDF Evaluated Nuclear Structure Data File
EO Ethylene oxide
EPA US Environmental Protection Agency
EPCRA Emergency Planning and Community Right-to-Know Act
EPR Electron paramagnetic resonance
ERBSS Extended-range Bonner sphere spectrometer
erg Energy unit (1 erg = 6.2415×10^{11} eV = 10^{-7} J)
ES Elastic scattering, external standard
ESA European Space Agency, Paris; electrostatic analyzer
ESCR External standard channels ratio
ESI Electrospray ionization
ESIR WG Extended SIR Working Group
ESP External standard pulse
ESTAR Electron stopping power and range
esu Electrostatic unit
ET Efficiency tracing
ET-DPM Efficiency tracing disintegrations per minute (method)
ETH Eidgenössische Technische Hochschule, Zurich

ETV-ICP-MS Electrothermal vaporization−inductively coupled plasma mass spectrometry
E_{av} Average energy (beta particle)
E_{max} Maximum energy (beta particle), Compton electron energy maximum
E_{α} Alpha-particle energy
E_p Proton energy
E_{th} Threshold energy
EU European Union
EUChemS European Chemical Society
EURACHEM European organization for traceability of chemical measurements
EURADOS European Radiation Dosimetry Group
EURATOM European Atomic Energy Community
EUROMET European Collaboration in Measurement Standards
eV Electron volt $= 1.602176 \times 10^{-19}$ J $= 1.602176 \times 10^{-12}$ erg)
EXAFS X-ray absorption fine structure
°F Degrees Fahrenheit
FADC Fast analog digital converter
f_C Fraction of contemporary cabon
FDA US Food and Drug Administration
FDG Fluorodeoxyglucose
FDNPP Fukushima Dai-ichi Nuclear Power Plant, Japan
FDNPS Fukushima Dai-ichi Nuclear Power Station, Japan
FEP Full energy peak
FET Field effect transistor
FFF Field flow fractionation
fg Femtogram (10^{-15} g)
FGRM Flow-through gaseous radiochemical method
FI Flow injection
fm Fermi or femtometer (10^{-15} m)
f_M Fraction of modern carbon
fmol Femtomole (10^{-15} mol)
FNTD Fluorescent nuclear track detector
FOM Figure of merit
fov Field of view
fp Fission products
FPGA Field programmable gate array
FSA Flow scintillation analysis
FS-DPM Full-spectrum disintegrations per minute (method)
FT Fission track
FTD Fission track dating
FT-ICR Fourier transform−ion cyclotron resonance
FTIR Fourier transform infrared spectroscopy
FWHM Full width at half-maximum
FWT Free water tritium
FWTM Full width at 10th maximum
g Gram, gluon
G # G-number (Grau's-number, quench-indicating parameter)
γ Gamma radiation
G-8 Group of Eight Countries (IAEA Member States)
GBq Gigabecquerels (10^9 Bq)
GC Gas chromatography
GC/MS Gas chromatography/mass spectrometry
GCR Galactic cosmic rays
GD Glow discharge
GDMS Glow discharge mass spectrometry
GEANT Geometry ANd Tracking Monte Carlo code
Ge(Li) Lithium-compensated germanium
GEM Gas electron multiplier
GEOSECS Geochemical Ocean Sections Programme
GEOTRACES International Study on Marine Biogeochemical Cycling of Trace Elements and their Isotopes
GERDA GERmanium Detector Array
GeV Gigaelectron volts (10^9 eV)
GHz Gigahertz (10^9 Hz)
GICNT Global Initiative to Combat Nuclear Terrorism
GIS Geographical Information System
GISP Greenland Ice Sheet Projects
GLOMARD Global Marine Radioactivity Database
GLP Good laboratory practice
GM Geiger−Müller
GM-APD Geiger-mode avalanche photodiode
GPa Gigapascal
GPC Gas proportional counting (counter)
GPD Geometric progression decrease
CPG Coplanar grid
GPS Global positioning system
GRB Gamma ray burst
GS-20 Glass scintillator
GSD Geometric standard deviation
GSI Gesellschaft für Schwerionenforschung mbH, Darmstadt, Germany
GSO:Ce Cerium-activated gadolinium orthosilicate (Gd_2SiO_5:Ce)
GUM Guide to the Expression of Uncertainty in Measurement
GW Groundwater, gate width
GW_e Gigawatt electrical (10^9 W_e)
Gy Gray (1 Gy = 1 J/kg = 6.24×10^{12} MeV/kg)
GZK Greisen-Zatsepin-Kuz'min process of *proton-photon* interactions
h Hours
h Plank's constant (6.626×10^{-34} J s), hours
$\hbar$ Plank's constant reduced ($h/2\pi$)
***H* #** Horrock's number (quench indicating parameter)
HBT 2-(2-Hydroxyphenyl)-benzothiazole
HDE Heat distribution error
HDEHP Bis(2-ethylhexyl)phosphoric acid
HDPE High-density polyethylene (moderator)
HEDPA 1-Hydroxyethane-1,1-diphosphonic acid
HEN High efficiency nebulizer
HEP High-energy particle
HEPES N-2-hydroxyethylpiperazine-N′-2-ethanesulfonic acid
HERA-B RICH Particle detector of the Hadron-Elektron-Ring-anlage, Hamburg, Germany
HERM High-energy radio monitor
HEU Highly enriched uranium
HEX-ICP-MS Hexapole collision cell ICP-MS
HEX-ICP-QMS Hexapole collision cell quadrupole mass spectrometry
3HF 3-Hydroxy flavone
hg Hectograms (10^2 g)
h-index Hirsh index
HIBA Hydroxy-i-butyric acid
HKG Housekeeping gene
HLNC High-level neutron coincidence counter
HLW High-level waste

HPB High-pressure Bridgman
HPGe High-purity germanium
HPIC High-performance ionic chromatography
HPLC High-performance liquid chromatography
HPMT Hybrid photomultiplier tube
HRAS High-resolution alpha spectrometry
HRGS High-resolution gamma spectrometry
HR-ICP-MS High-resolution inductively coupled plasma mass spectrometry
HT High tension
HV High voltage
HWHM Half width at half-maximum
HWZPR Heavy water zero power reactor
Hz Hertz (cycles per second)
$\mathbf{i_{in}}$ Current pulse
IAEA International Atomic Energy Agency, Vienna
IAEA-EL IAEA Marine Environment Laboratory, Monaco
IC Internal conversion, ion chromatography
ICC Ice condenser chamber
IC-ICP-MS Ion chromatography–inductively coupled plasma mass spectrometry
IC# Isotope center number
IceCube Neutrino Observatory, South Pole
IceTop Surface array of stations for IceCube
ICF Inertial confinement fusion
ICP Inductively coupled plasma
ICP-CC-QMS Quadrupole inductively coupled plasma mass spectrometry with hexapole collision cell
ICP-FT-ICR-MS Inductively coupled plasma Fourier transform ion cyclotron resonance mass spectrometry
ICP-MS Inductively coupled plasma mass spectrometry
ICP-OES Inductively coupled plasma optical emission spectrometer (spectra)
ICP-QMS Inductively coupled plasma quadrupole mass spectrometry
ICP-SFMS Double-focusing sector field inductively coupled plasma mass spectrometry
ICRP International Commission on Radiological Protection
ICRU International Commission on Radiation Units and Measurements
ID or i.d. Inner diameter, inner detector
IDA Isotope dilution analysis
IDMS Isotope dilution mass spectrometry
ID-TIMS Isotope dilution thermal ionization mass spectrometry
IE Ion exchange
IEC International Electrotechnical Commission, inertial electrostatic confinement
IECF Inertial electrostatic confinement fusion
IEEE Institute of Electrical and Electronics Engineers
IEF Isoelectric focusing gel electrophoresis
IFIN-HH Horia Hulubei National Institute of Physics and Nuclear Engineering, Romania
IGPC Internal gas proportional counting
IL-5 Interleukin-5
IMS International Monitoring System of the CTBT
in. Inch = 2.54 cm = 25.4 mm
INES International Nuclear and Radiological Event Scale
INFN Instituto Nazionale di Fisica Nucleare (Italy)
INGE International Noble Gas Experiment
INP Institute of Nuclear Physics, Tirana, Albania
IN2P3 Institut National de Physique Nucléaire et de Physique des Particules, France
INSERM Institut national de la santé et de la recherché médicale. France
I/O Input/output
IPA Instrument performance assessment, isopropyl alcohol
IPRI Laboratoire Primaire des Rayonnements Ionisants, France
IPT Intramolecular proton transfer
IR Infrared (spectroscopy)
IRA Institut Universitaire de Radiophysique, Lausanne, Switzerland
IRMM Institute for Reference Materials and Measurements, Geel
IRMS Isotope ratio mass spectrometry
IRSN Institute of Radiation Protection and Nuclear Safety, France
IS Internal standard
ISH *In situ* hybridization
ISO International Organization for Standardization
ISOCS In-Situ object calibration software
IS-SCR Internal standard and sample channels ratio
IT Isomeric or internal transition
ITER International Thermonuclear Experimental Reactor
ITU Institute for Transuranium Elements, Europe
ITWG Nuclear Forensics International Technical Working Group
IUPAC International Union of Pure and Applied Chemistry
IUPAP International Union of Pure and Applied Physics
J Joule = 1 N m = 1 kg m^2/s^2 = 1 W s
JAERI Japan Atomic Energy Research Institute
JET Joint European Torus reactor
JFET Junction field effect transistor
JCGM Joint Committee for Guidelines in Metrology
JINR Joint Institute for Nuclear Research, Dubna, Moscow Oblast
JRC Joint Research Centre (of European Commission)
K particle kinetic energy
K^+, K^-, K^0 Kaons or K mesons
K Degrees Kelvin
ka Kiloannum (10^3 years)
KamLAND Kamioka Liquid Scintillator Anti-Neutrino Detector, Japan
KATRIN Karlsruhe TRItium Neutrino experiment
kBq Kilobecquerels (10^3 Bq)
KCFC Potassium cobalt ferrocyanide
kcps Kilocounts per second
KCRV Key comparison reference value
KEK The High Energy Accelerator Research Organization, Japan
keV Kiloelectron volts (10^3 eV)
kg Kilograms
kGy Kilogray
kHz Kilohertz
km.w.e km-water-equivalent
KNN k nearest neighbor algorithm
KRISS National Metrology Institute of Korea
KSTAR Korea Superconducting Tokamak Advanced Research fusion reactor
kt Kilotons
kV Kilovolts (10^3 V)
kW Kilowatts (10^3 W)
ky Kiloyears (10^3 y)

L, l Liters
LA Linear anode
LAAPD Large area avalanche photodiode
LAB Linear alkyl benzene, dodecylbenzene
LA-ICP-MS Laser ablation inductively coupled plasma mass spectrometry
LA-MC-ICP-MS Laser ablation multiple collector ICP-MS
λ Wavelength, decay constant, microliter (10^{-6} L), free parameter
λ_{nr} Nonrelativistic wavelength
λ_r Relativistic wavelength
LAMMA Laser microprobe mass analysis
LAN Local area network
LANL Los Alamos National Laboratory
LAr Liquid argon
LARA laser-assisted isotope ratio analysis
LAW Low-activity waste
LBD Ligand-binding domain
LBNL Lawrence Berkeley National Laboratory
LC Liquid chromatography
LCDW Lower circumpolar deep water
LCMS Liquid chromatography mass spectrometry
LD_{50} Median lethal dose
LED Light-emitting diode
LEGE Low-energy germanium detector
LENA Low-energy neutrino astrophysics detector
LET Linear energy transfer
LEU Low enriched uranium
LHCb RICH Large Hadron Collider beauty experiment detector at CERN
LHD Large Hadron Collider
LiI(Eu) Europium-activated lithium iodide
LIMS Laboratory Information Management System
LINAC or linac Linear accelerator
LIST Laser ion source trap
LL Lower level
LL-BSS Large ^{6}LiI(Eu) Bonner sphere spectrometer
LLC Liquid (mobile)—liquid (on solid phase) chromatography
LLCM Low-level count mode
LLD Lower limit of detection, lower level discriminator
LLE Liquid—liquid extraction
LLNL Lawrence Livermore National Laboratory
LLR Long-lived radionuclide
LMD Laser microdissection
LM-OSL Linear modulation optically stimulated luminescence
LN_2 Liquid nitrogen
LNE Laboratoire National de Métrologie et de E'ssais, France
LNGS Laboratori Nazionali del Gran Sasso, Italy
LNHB Laboratoire National Henri Becquerel, Saclay
LNMRI National Metrology Laboratory of Ionizing Radiation, Brazil
LOD Limit of detection
LOV lab-on-valve (system)
lp Line pairs
LPB Low-pressure Bridgman
LPI low-pressure cascade impactor
LPRI Laboratoire Primaire des Ionizants, Paris
LPS Lipopolysaccharide
LRAD Long-range alpha detector
LS Liquid scintillation, liquid scintillator, "linear-to-square" curve
LSA Liquid scintillation analysis (analyzer)
LSC Liquid scintillation counting (counter)
LSO Cerium-activated lutetium oxyorthosilicate ($Ce:Lu_2SiO_5$)
LSS Liquid scintillation spectrometer
LTC Live-time correction
LuAP Cerium-activated lutetium aluminum perovskite ($Ce:LuAlO_3$)
LY Light yield
LXe Liquid xenon
M Molar (solution concentration)
m Particle mass
m_0 Particle rest mass
m_r Speed-dependent particle mass
m Mass, meters, minutes
mA Milliampere (10^{-3} ampere)
Ma Megayear (10^6 years)
mAbs Monoclonal antibodies
MACS Magnetically assisted chemical separations
MALDI Matrix-assisted laser desorption/ionization
MAPD Micropixel avalanche photodiode
MAPMT Multianode photomultiplier tube
MARG Microautoradiography
MARIS Marine information system
MARSSIM Multi-Agency Radiation Survey and Site Investigation Manual
MATLAB MATrix LABoratory (numerical computing and programming language)
mb Millibarn (10^{-3} b)
mBq Millibecquerels (10^{-3} Bq)
MBq Megabecquerels (10^6 Bq)
mCi Millicurie (10^{-3} Ci) $= 2.22 \times 10^9$ dpm $= 3.7 \times 10^7$ dps $= 37$ MBq
MC Multiple ion counting
MCA Multichannel analyzer
MCF Moving curve fitting
MC-ICP-MS Multiple ion collector-ICP-MS
MCN Microconcentric nebulizer
MCNP Monte Carlo N-Particle code
MCNP-CP Monte Carlo N-Particle-Correlated Particle code
MCP Microchannel plate
MCP-PM Microchannel plate photomultiplier
MC-TIMS Multiple collector thermal ionization mass spectrometry
MD Molecular dynamics
MDA Minimal detectable activity
MDOA Methyldiooctylamine
METAS Federal Institute of Metrology, Berne-Wabern, Switzerland
METEPC Multielement tissue-equivalent proportional counter
MeV Megaelectron volts
MeVee Electron equivalent energy
MHSP Microhole and strip plate (imager)
MHz Megahertz (10^6 Hz)
MIBK Methyl isobutyl ketone
MICAD Microchannel Array Detector
MICM Modified integral counting method
MICROMEGAS Micromesh gas detector
mg Milligram (10^{-3} g)
mGy Milligray

MIBK Methyl isobutyl ketone
min Minutes
mK MilliKelvin (10^{-3} K)
mL, ml Milliliter (10^{-3} L)
MLR Multiple linear regression
mM Millimolar (10^{-3} M)
mm Millimeter (10^{-3} m)
MM Magnetic monopoles
MMAD Mass median aerodynamic diameter
MMC Metallic magnetic calorimeter
mmol Millimole (10^{-3} mol)
MNP Magnetic nanoparticle
mol Mole (gram-molecular weight)
MΩ Megaohm (10^6 Ω)
MOX Mixed oxide fuel
MP Multipurpose
M-P Mandel and Paule mean
MPa Megapascal (10^6 Pa)
MPGD Micropattern gas detector
MPPC Multipixel photon counter
mrad Millirad (1 mrad = 10 μGy)
MRI Magnetic resonance imaging
mRNA Messenger RNA
MS Mass spectrometry
ms, msec Milliseconds (10^{-3} s)
MSAP Microscale sample automation platform
MSB Methylstyrylbenzene
MSC Microplate scintillation counting
MSD Mean standard deviation
MSE Multisite events
MSGC Microstrip gas counter
MSI Mass spectrometry imaging
MS/MS Tandem mass spectrometry
mSv Millisievert
MW Megawatt (10^6 W)
Mt Megaton (10^6 t)
MTO Magnetooptical trap
μ^+, μ^- Muons
μ Attenuation coefficient
μA Microampere (10^{-6} A)
μCi Microcurie (10^{-6} Ci) = 2.22×10^6 dpm = 3.7×10^4 dps = 37 kBq
μg Microgram (10^{-6} g)
μL Microliter (10^{-6} L)
μm Micrometer (10^{-6} m)
μPIC Micropixel gas chamber
μs, μsec Microseconds (10^{-6} s)
μ-XANES Microfocused X-ray absorption near edge structure
μ-XRF Microfocused X-ray fluorescence
MEK Methyl ethyl ketone
MW Megawatt (10^6 W)
MWe Megawatt electrical
m.w.e. Meter water equivalent
MWPC Multiwire proportional chamber
MV Megavolts (10^6 V)
MVC Multivariate calibration
N Newton = 1 kg m/s^2
N Neutron number
n Neutron
n Index of refraction
NA Avogadro's constant (6.022×10^{23}/mol)
nA Nanoampere (10^{-9} A)
NAA Neutron activation analysis
NAC *N*-acetylcystein
NADW North Atlantic Deep Water
NaI(Tl) Thallium-activated sodium iodide
NARC Neutrino Array Radio Calibration
NASA National Aeronautics and Space Administration, Washington, D.C.
NBL New Brunswick Laboratory of the US DOE
NBR Natural background rejection
NBS National Bureau of Standards (now NIST)
NC Neutral current (interaction)
NCD Neutral current detector
nCi Nanocurie (10^{-9} Ci)
NCM Normal count mode
NCRP National Council on Radiation Protection and Measurements
NDA Nondestructive analysis
NEA Nuclear Energy Agency of the OECD
$N_{e/h}$ Number of electron−hole pairs
NEMO Nautic Environment Marine Observatoire
NE-OBT Nonexchangeable organically bound tritium
NF-LA-ICP-MS Near-field laser ablation inductively coupled plasma mass spectrometry
ng Nanograms (10^{-9} g)
NHMRC National Health and Medical Research Council, Australia
NIDW North Indian Deep Water
NIM Nuclear instrument module
NIMH Nickel metal hydride
NIST National Institute of Standards and Technology, Gaithersburg
nm Nanometer (10^{-9} m)
NMI National Metrology Institute
NMM Neutron moisture meter
NMR Nuclear magnetic resonance
NNDC National Nuclear Data Center
NOI Nuclide of interest
NORM Naturally occurring radioactive materials
NP Nanoparticle
NPD 2-(1-Naphthyl)-5-phenyl-1,3,4-oxadiazole
NPE Nonyl phenol ethoxylate
NPL National Physical Laboratory, UK
NPO 2-(1-Naphthyl)-5-phenyloxazole
NPP Nuclear power plant
NRC United States Nuclear Regulatory Commission
ν Neutrino, photon frequency, particle velocity
$\bar{\nu}$ Antineutrino
$0\nu\beta\beta$ Neutrinoless double-beta decay
$2\nu\beta\beta$ Two-neutrino double-beta decay
nM Nanomolar (10^{-9} M)
nm Nanometer (10^{-9} m)
NMM Neutron moisture meter
NMR Nuclear magnetic resonance
NNDC National Nuclear Data Center, BNL, Upton, New York
NNFL National nuclear forensics library
NORM Naturally occurring radioactive material

NPT Nonproliferation Treaty
NRC Nuclear Regulatory Commission
ns, nsec Nanosecond (10^{-9} s)
NSTAR Neutron sandwich transmitter/activation-γ radiator
NT200 Neutrino telescope, Lake Baikal, Siberia
NTD-Ge Neutron transmutation-doped Ge
N-TIMS Negative ion thermal ionization mass spectrometry
NTP Normal temperature and pressure
NTS Nevada test site
NU Natural uranium
NUDAT Nuclear Database of the NNDC
NWT Nuclear weapons test
N/Z Neutron/proton ratio
OC Organic carbon
OD or o.d. Outer detector, outer diameter
OECD Organization for Economic Cooperation and Development
OES Optical emission spectrometry
OFHC Oxygen-free high thermal conductivity
OGE Optogalvanic effect
OHM National Office of Measurement, Budapest
OLLSC Online liquid scintillation counting
OM Optical module
OSL Optically stimulated luminescence
OTPC Optical time projection chamber
P Parity quantum number
p Particle momentum
p, p^+ Proton
Pa Pascal = 1 N/m^2 = 1 kg/m$\cdot$s^2
PAC Pulse amplitude comparison (comparator)
PADC Polyallyldiglycol carbonate
PAGE Polyacrylamide gel electrophoresis
PAN Polyacrylonitrile
PANDA Particles and nondestructive analysis
PAW Physics Analysis Workstation
PAZ Partial annealing zone
PBBO 2-(4′-Biphenylyl)-6-phenylbenzoxazole
PBD 2-Phenyl-5-(4-biphenylyl)-1,3,4-oxadiazole
PBO 2-(4-Biphenylyl)-5-phenyloxazole
PBq Petabecquerel (10^{15} Bq)
PBS Phosphate buffered saline
PC Proportional counter(ing), personal computer, paper chromatogram, polycarbonate
PCA Principal component analysis
PCB Polychlorinated biphenyl
pCi Picocurie (10^{-12} Ci)
PCR Principle component regression
PD Photodiode
PDA Pulse decay analysis
PDB Pee Dee Belemnite (standard)
PDD Pulse decay discriminator
PE Phosphate ester, polyethylene
PEC Power and event controller
PENELOPE PENetration and Energy Loss of Positrons and Electrons Monte Carlo code
PERALS Photon Electron Rejecting Alpha Liquid Scintillation
PET Positron emission tomography, polyethylene terephthalate
PETAC Pentaerythritol tetrakis allyl carbonate
PeV Petaelectron volts (10^{15} eV)
pF Picofarad (10^{-12} F)
PF Polar front
PFA Perfluoroalkoxy
PFZ Polar frontal zone
pg Picogram (10^{-12} g)
PGA Pulse gradient analysis
ph Photons
PHA Pulse height analysis
PHITS Particle and heavy ion transport code system
PHOSWICH PHOSphor sandWICH (detector)
π Pi constant = 3.14159
π^+, π^-, π^0 Pions or pi mesons
PI Polyimide, pressurized injection
PID Particle identification
PIM Parallel ionization multiplier
PIMS Positive-ion mass spectrometry
PIPS, PIPSi Passivated implanted planar silicon
PIXE Proton-induced X-ray emission
PKC Protein kinase C
PLC Proportional long counter
PLI Pulse length index
PLS Partial least squares
PLS-DA Partial least squares discriminant analysis
PLSR Partial least squares regression
PM Photomultiplier, particulate matter
PMM Power-moderated weighted mean
PMBP 1-Phenyl-3-methyl-4-benzoylpyrazolone-5
pMC Percent modern carbon
PMMA Polymethylmethacrylate
PMP 1-Phenyl-3-mesityl-2-pyrazoline
PMT Photomultiplier tube
PN Pneumatic nebulizers
PNNL Pacific Northwest National Laboratory
PNX Pacific Northwest eXtraction system
POM Polyoxymethylene
POPOP 1,4-Bis-2-(5-phenyloxazolyl)benzene
PPAC Parallel plate avalanche chamber
ppb Parts per billion
PPC P-type point contact
PPD 2,5-Diphenyl-1,3,4-oxadiazole
PPE Personal protective equipment
ppm Parts per million
ppmw Parts per million by weight
PPO 2,5-Diphenyloxazole
PS Plastic scintillator, polystyrene
ps Picosecond (10^{-12} s)
PSA Pulse shape analysis
PSD Pulse shape discrimination
PSf Plastic scintillator foils
psi 6.895×10^3 Pa = 68.95×10^{-3} bar = 51.715 torr
PSL Photostimulable light (or luminescence)
PSm Plastic scintillator microspheres
PSPC Position-sensitive proportional counter
PSr Plastic scintillator resins
PSUP Photomultiplier SUPport structure
P/T Peak-to-total ratio
PTB Physikalisch-Technische Bundesanstalt, Braunschweig
PTBT Partial Test-Ban Treaty

PTFE Polytetrafluoroethylene
P-TIMS Positive ion thermal ionization mass spectrometry
PTP *p*-Terphenyl
PUR Pileup rejector
PUREX Plutonium URanium EXtraction
PVC Polyvinyl chloride
PVD Physical vapor deposition
PVDF Polyvinyldifluoride
PVT Polyvinyl toluene
PWR Pressurized water reactor
PXE Phenyl-ortho-xylylethane
Q Q value of nuclear reactions
QA Quality assurance
QC Quality control
QC-CPM Quench-corrected count rate
QCD Quantum chromodynamics
QD Quadrupole
QDC Charge-to-digital converter
QE Quantum efficiency
QIP Quench-indicating parameter
QWBA Quantitative whole-body autoradiography
R Roentgen ($1R = 2.58 \times 10^{-4}$ C/kg)
RAC Radon activity concentration
rad Radiation-absorbed dose (1 rad = 10 mGy = 100 erg/g)
RAD Radon-in-air monitor
RAST Radioallergosorbent test
RBE Relative biological effectiveness
RDC Remote detector chamber
RDD Radiological dispersal device ("dirty bomb")
RE Recovery efficiency
REE Rare earth elements
REFIT Radialelectron fluence around ion tracks
REGe Reverse-electrode coaxial Ge detector
REL Restricted energy loss
rem Roentgen equivalent mammal (1 rem = 10 mSv)
RF Radiofrequency
R_F Feedback resister
RFQ Radiofrequency quadruple
RH Relative humidity
ρ Density (g cm^{-3}), neutron absorption cross section, resistivity
RIA Radioimmunoassay
RICE Radio Ice Cherenkov Experiment
RICH Ring imaging Cherenkov (counters/detectors)
RIMS Resonance ionization mass spectrometry
RIS Resonant ionization
RM Reference material
RMS Rosette multibottle samplers
RMT Radiometric technique
RNA Ribonucleic acid
Ro5 Ring of Five (European radionuclide monitoring labs)
ROI Region of interest (spectral)
ROSEBUD The Rare Objects Search with Bolometers UndergrounD collaboration
ROV Remotely operating vehicle
RPC Resistive plate chamber
RPH Relative pulse height
RSC Renewable separation column, relative sensitivity coefficient
RSD Relative standard deviation
RSF Relative sensitivity factor
RST Reverse spectral transform
s Seconds
SAF Subantarcticfront
SAH *S*-adenosyl-homocysteine
SalSa Salt sensor array
SAM Standard analysis method, *S*-adenosyl-methionine
SAMAD Surface area mean aerodynamic diameter
SAS Semiconductor α-spectrometry
SBD Surface barrier detector
SCA Single channel analyzer
SCC Software coincidence counting, squamous cell carcinoma
SCI Science Citation Index
SCR Sample channels ratio, solar cosmic rays
SCX Strong cation exchange
SD Standard deviation
SDCC Simplified digital charge comparison
SDD Silicon drift detector
SDP Silicon drift photodiode
SDT Shared dead time
SE Single escape, secondary electron
sec Seconds
SEC Size exclusion chromatography
SEGe Standard electrode coaxial Ge detector
SEM Scanning electron microscopy
SF Spontaneous fission
SFC Supercritical fluid extraction
SFD Scintillation fiber detector
SF-ICP-MS Sector field−inductively coupled plasma mass spectrometry
SFU Stacked filter unit
SGD Submarine groundwater discharge
SHE Superheavy elements
SHOTS Southern Hemisphere Oceans Tracer Studies
SHRIMP Sensitive high mass resolution ion microprobe
SI International System of Units, sequential injection, spray ionization
SIA Sequential injection analysis
SIE Spectral index of the external standard
σ Reaction cross section, thermal neutron cross section
Si(Li) Lithium-compensated silicon
SIMS Secondary ionization mass spectrometry
Si PIN Silicon p-i-n diode
SiPM Silicon photomultiplier
SIR International Reference System (Système Internationale de Référence)
SI-RSC Sequential injection renewable separation column
SIS Spectral index of the sample
SJD Silicon junction detector
SLAC Stanford Linear Accelerator Center
SLIM System for Laboratory Information Management
SLM Standard laboratory module
SLSD Scintillator-Lucite sandwich detector
SMAD Surface median aerodynamic diameter
SMDA Specific minimum detectable activity
S/N Signal-to-noise
SNAP Systems Nuclear Auxiliary Power
SNICS Source of Negative Ions by Cesium Sputtering

SNF Spent nuclear fuel
SNM Special nuclear material
SNMS Secondary neutral mass spectrometry
SNO Sudbury Neutrino Observatory, Canada
SNR Signal-to-noise ratio
SNS Spallation neutron source
SNTS Semipalatinsk nuclear test site, Eastern Kazakhstan
SOA Secondary organic aerosol
SOI Silicon-on-insulator
SOP Standard operating procedure
SPA Scintillation proximity assay
SPC Single photon counting
SPD Self-powered detector
SPE Single photon event, solid phase extraction, solid polymer electrolyte
SPECT Single photon emission computed tomography
SPME Solid phase microextraction
SQM Strange quark matter
SQP(E) Spectral quench parameter of the external standard
SQP(I) Spectral quench parameter of the isotope
SQS Self-quenched streamer
SQUID Superconducting quantum interference device
SR Superresolution, synchrotron radiation
sr Steradian
SRAM Static random access memory
SRM Standard reference material
SRS Savannah River Site
SSB Silicon surface barrier detector
SSDD Segmented silicon drift detector
SSE Single site events
SSM Standard service module, selective scintillating microsphere
SSNTD Solid-state nuclear track detector
ST Supersensitive
STD Shared dead time concept
STE Self-trapped excitation
STF Subtropical front
STM Scanning tunneling microscope
STNTD Solid-state nuclear track detection (detectors)
STP Standard temperature and pressure
STS Semipalatinsk test site
STUK Radiation and Nuclear Safety Authority, Finland
Sv Sievert (1 Sv = 1 Gy = 100 rem = 1 J/kg)
SVOC Semivolatile organic carbon
t Ton(s)
$t_{½}$, $T_{½}$ Half-life
T Particle kinetic energy
T Tritium, tesla = 1 V s/m^2
TAEK Turkish Atomic Energy Authority
TALSPEAK Trivalent Actinide–Lanthanide Separation by Phosphorus Extractants and Aqueous Komplexants process
TAR Tissue–air ratio
TAT Targeted alpha therapy
TBP Tributyl phosphate
TBq Terabecquerel (10^{12} Bq)
TC Total carbon
TCA Trichloroacetic acid
TCS True coincidence summing
TD Time discriminator
TDCR Triple-to-double coincidence ratio (method)
TDS Total dissolved solids
TEA Triethylamine
TEM Transmission electron microscopy
TENORM Technologically enhanced naturally occurring radioactive materials
TEPC Tissue-equivalent proportional counter
TES Transition edge sensor
TBAB Tetrabutylammonium bromide
TeV Teraelectron volts (10^{12} eV)
Tf Transfer factor (radionuclide)
TFTR Tokamak fusion test reactor
TFWT Tissue-free water tritium
THGEM Thick gas electron multiplier
THM Traveling heater method
tHM y^{-1} Metric tons of heavy metal per year
TI Transfer instrument
~ Approximately
TIMS Thermal ionization mass spectrometry
TINCLE Track-in-cleavage (technique)
TINT Track-in-track (technique)
TIOA Triisooctylamine
TL Thermoluminescence
TLA Trilaurylamine
TLC Thin-layer chromatography (chromatogram)
TLD Thermoluminescence dosimeter
TMA Trimethylamine
TMI Three Mile Island
TMOS Tetramethoxysilane
TMS Tetramethylsilane
TNOA Tri-*n*-octylamine
TNSA Target normal sheath acceleration
TNT Trinitrotoluene
TOA Top of the atmosphere, trioctyl amine
TOF Time-of-flight
TOP Time-of-propagation
TOPO Trioctylphosphine oxide
torr 133.3224 Pa
TP *p*-Terphenyl
TPPS Triphenylphosphine sulfide
TR Tritium sensitive
TRACOS Automatic system for nuclear track evaluations
TRE 12-O-Tetradecanoyl phorbol-13-acetate responsive element
TRI Toxic release inventory
TR-LSC Time-resolved liquid scintillation counting
TR-PDA Time-resolved pulse decay analysis
TRPO Trialkyl phosphine oxide
TSC Task sequence controller
TSCA Toxic Substance Control Act
TSEE Thermally stimulated exoelectron emission
tSIE Transformed spectral index of the external standard
tSIS Transformed spectral index of the sample
TSP Total suspended particle
TTA Tenoyl-tri-fluoro acetone
TTL Transistor–transistor logic
TU Tritium unit (0.119 Bq ^3H kg^{-1} H_2O or 7.14 DPM of ^3H L^{-1} H_2O or ratio of 1 atom ^3H:10^{18} atoms of ^1H)

u Atomic mass unit (1/12 mass of $^{12}C = 1.66054 \times 10^{-27}$ kg), up quark
$\bar{u}$ Antiup quark
u Particle speed
u_{nr} Nonrelativistic particle speed
u_r Relativistic particle speed
UCN Ultracold neutrons
UHE Ultrahigh energy
UL Upper level
ULB Ultralow background
ULD Upper level discriminator
ULEGE Ultralow-energy Ge
UNSCEAR UN Scientific Committee on the Effects of Nuclear Radiation
UOC Uranium ore concentrate
U.S.A.E.C. US Atomic Energy Commission (now NRC)
U.S. DOE US Department of Energy
USEPA US Environmental Protection Agency
USN Ultrasonic nebulizers
UV Ultraviolet
V Volts
V_0 Step voltage
VAX Digital Equipment Corporation trade name
VCCI Variable configuration cascade impactor
VHPLC Very-high-pressure liquid chromatography
VMEbus Versa Module Europa bus
VSiPMT Vacuum silicon photomultiplier tube
VUV Vacuum ultraviolet (spectral region)
VYNS Vinyl acetate and vinyl chloride copolymer
W Watt (1 W = 1 J/s)
w/w Weight/weight
WAK Wiederaufarbeitungsanlage (nucleal fuel reprocessing plant), Karlruhe
WBA Whole-body autoradiography
WBEC Weak base extraction chromatography
WCVB Waste concentration vapor body
WDS Wavelength dispersive spectrometer
WDX Wavelength dispersive X-ray (analyzer)
WHO World Health Organization
WIMP Weakly interacting massive particle
WIPP Waste Isolation Power Plant
WM Weighted mean
WMO World Meteorological Organization, Geneva
WNO World Nuclear Organization, London
WOCE World Ocean Circulation Experiment
WOMARS Worldwide Marine Radioactivity Studies
WRA Warfare radioactive agent
WSF Wavelength shifting fiber
WSOC Water-soluble organic carbon
wt% Weight percent
XAF X-ray absorption spectroscopy
XANES X-ray absorption near edge structure
XRD X-ray diffraction
XRF X-ray fluorescence
XtRA Extended range
y Years
YAG:Yb Yb-doped $Y_3Al_5O_{12}$
YAP:Ce Cerium-activated yttrium aluminum perovskite ($Ce:YAlO_3$)
YG Yttrium glass
YSi(Ce) Cerium-activated yttrium silicate
Z Atomic number
Z_2 Average atomic number
Z_{ef} or Z_{eff} Effective atomic number
ZCH Central Analytical Laboratory, Jülich
ZnS(Ag) Silver-activated zinc sulfide

Chapter 1

The atomic nucleus, nuclear radiation, and the interaction of radiation with matter

Michael F. L'Annunziata
The Montague Group, Oceanside, CA, United States

Chapter outline

Handbook of Radioactivity Analysis: Volume 1. https://doi.org/10.1016/B978-0-12-814397-1.00001-7

I. Introduction

Radioactivity is the emission of radiation originating as a result of the spontaneous decay of unstable atomic nuclei or from a nuclear reaction. The term radioactive decay refers to the process whereby unstable atomic nuclei decay with the loss of energy by the emission of elementary particles (*e.g.*, alpha particles, beta particles, neutrons, gamma ray photons) directly from the nucleus or the atomic electron shells (*e.g.*, Auger electrons and X-ray photons). The rate of decay or disintegration rate of a radionuclide (*i.e.*, a specific radioisotope of an element), as we shall see in this chapter, is directly proportional to the mass of the radionuclide. Thus, radioactivity analysis is essentially the quantitative analysis of radionuclides. Methods of radioactivity analysis have, for the most part, two approaches, namely, the determination of the disintegration rate of a radionuclide by counting the radiation emissions from the atoms of a radionuclide disintegrating per unit time or by measuring the mass of a radionuclide, such as is done using mass spectrometry also referred to as atom counting. Thus, radioactivity analysis is synonymous to radionuclide analysis whereby we can calculate the mass of the radionuclide from its disintegration rate and vice versa.

The analysis of radioactivity is a challenging field. Both the sources of radioactivity (*i.e.*, radionuclides) and the media within which the radionuclides may be found can present themselves in a wide range of complexities. Also, nuclear radiation resulting from the decay of radionuclides can occur in various types, percent abundances or intensities, and energies. Furthermore, a given radionuclide may have more than one mode of decay. The presence of appreciable activities of more than one radionuclide in a sample can further complicate analysis. In addition, the different parent–daughter nuclide decay schemes, equilibria between parent and daughter radionuclides, and the rates of decay that radioactive nuclides undergo may facilitate or complicate the analysis of a given radionuclide. The problem of radioactivity analysis may be confounded further by the wide range of chemical and/or physical media (*i.e.* sample matrices) from which the nuclear radiation may emanate.

As we will find in this book, there are many modern methods of radioactivity analysis. The types of detectors available for the measurement of radioactivity are numerous, and they may be designed in the gaseous, liquid, or solid state. They will differ not only in their physical state but also in chemistry. The instrumentation and electronic circuitry associated with radiation detectors will also vary. As a result, the detectors and their associated electronic instrumentation will perform with varying efficiencies of radiation detection depending on many factors, including the characteristics of the instrumentation, the types and energies of the radiation, as well as sample properties.

The proper selection of a particular radiation detector or method of radioactivity analysis requires a good understanding of the properties of nuclear radiation, the mechanisms of interaction of radiation with matter, half-life, decay schemes, decay abundances, and energies of decay. This chapter will cover these concepts as a prelude to the various chapters that follow on radioactivity analysis. Throughout the book reference will be made to the concepts covered in this introductory chapter. For the experienced radioanalytical chemist, this chapter may serve only as a review. However, the newcomer to this field should find this introductory chapter essential to the understanding of the concepts of radiation detection and measurement. He or she will find that the concepts covered in this introductory chapter will facilitate the selection of the most suitable radiation detector and instrumentation required for any particular case.

The properties of nuclear radiation and the mechanisms whereby nuclear radiation dissipates its energy in matter, dealt with in this chapter, form the basis for the methods of detection and measurement of radionuclides.

II. Discovery and characterization of the atomic nucleus and radioactivity

A brief history of radioactivity and the pioneers, who have contributed much to our understanding of this fascinating field of science, is presented here. The history of the science is important to our understanding of how we have arrived to where we are today in this science, and it serves as a source of motivation to future pioneers in this field. For a more detailed treatment of the history of the discoveries that led to our current knowledge of radioactivity, the structure of the nucleus, and nuclear stability and decay, the reader is invited to peruse a previous work by the author (L'Annunziata, 2016).

Radioactivity was discovered in 1896 by Henri Becquerel. At the beginning of 1896, on the very day that news reached Paris of the discovery of X-rays, Henri Becquerel thought of carrying out research to see whether or not natural phosphorescent materials emitted similar rays. He was then Professor of the École Polytechnic in Paris where he went to work on some uranium salts that he had inherited from his father, who had previously studied phosphorescence as Professor of Applied Physics at the Polytechnic. Henri Becquerel placed samples of uranium sulfate onto photographic plates, which were enclosed in black paper or aluminum sheet to protect the plates from exposure to light. After developing the photographic plates, he discovered that the uranium salts emitted rays that could pass through the black paper and even a metal sheet or thin glass positioned between the uranium salts and the photographic plates. Becquerel reported his findings to the French Academy of Sciences in February and March of 1896 (Becquerel, 1896a,b) and summarized his discovery in 1901 in the journal *Nature* as follows:

> *At the commencement of the year 1896, in carrying out some experiments with the salts of uranium ..., I observed that these salts emitted an invisible radiation, which traversed metals and bodies opaque to light as well as glass and other transparent substances. This radiation impressed a photographic plate and discharged from a distance electrified bodies—properties giving two methods of studying the new rays.*

At first he thought the rays were a result of phosphorescence, that is, excitation of the crystals by sunlight forcing the crystals to give off their own rays. However, Henri Becquerel carried out further tests demonstrating that the rays emanating from the uranium salts were independent of any external source of excitation including light, electricity, or heat, and the intensity of the rays did not diminish appreciably with time. "We were thus faced with a spontaneous phenomenon of a new order", which were his words during his Nobel Lecture (Becquerel, 1903) given on December 11, 1903.

Becquerel provided evidence that all uranium salts emitted the same radiation, and that this was a property of the uranium atom particularly since uranium metal gave off much more intense radiation than the salts of that element. The new radiation produced ionization and the intensity of the radioactivity could be measured by this ionization. Not only did these rays produce ionization, but he was able to demonstrate that a large portion of these rays could be deflected by a magnetic field and were charged particles of property similar to cathode rays. It was J. J. Thomson, Cavendish Professor of Experimental Physics at Trinity College, Cambridge, who discovered in 1897 that the cathode rays were electrons (Thomson, 1897). Thus, Becquerel was the first to provide evidence that some of the radiation emitted by uranium and its salts were similar in properties to electrons. It would be years later that

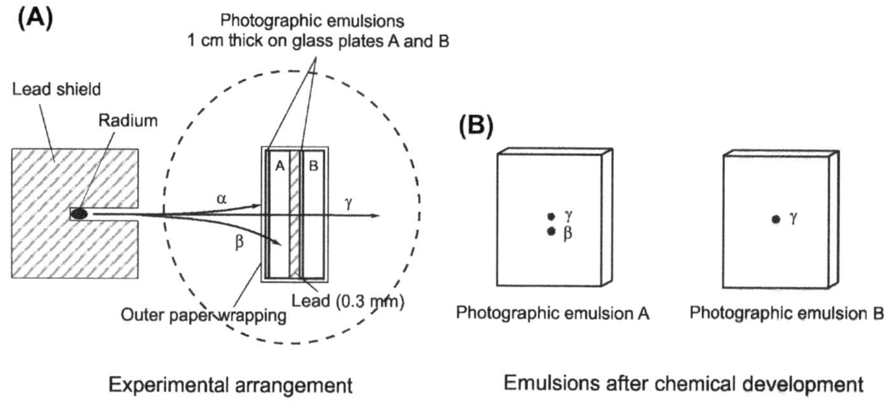

FIGURE 1.1 Paul Villard's (A) experimental arrangement and (B) experimental results that led to his discovery of gamma rays in 1900. Photographic plates A and B consisting of emulsion set on 1-cm-thick glass supports were separated from each other by a 0.3-mm-thick lead barrier and wrapped in an envelope of light-tight paper. The dashed circle represents the pole of a magnet from which the lines of force are directed into the plane of the page perpendicular to the path of radiation emitted by the radium source.

Rutherford (1903) would name the electrons originating from nuclear decay as beta particles.

Following Becquerel's discovery of spontaneous radiation from uranium, Marie Curie, who was born Maria Salomea Sklodowska in Warsaw, Poland, decided to study the mysterious rays emitted by uranium and to apply the work for a doctorate degree in the laboratory of her husband Pierre Curie, who was Professor of the Municipal School of Industrial Physics and Chemistry in Paris, France. In 1898 Marie Curie discovered that not only uranium gave off the mysterious rays discovered by Becquerel, but thorium did as well; this was independently discovered by Gerhard Schmidt in Germany the same year. Pierre and Marie Curie observed that the intensity of the spontaneous rays emitted by uranium or thorium increased as the amount of uranium or thorium increased. They concluded that these rays were a property of the atoms or uranium and thorium; thus, they decided to coin these substances as "radioactive". The emanation of such spontaneous rays from atoms would now be referred to as "radioactivity". Through tedious chemical separations and analyses, Marie and Pierre Curie worked as a team and found that another radioactive element with chemical properties similar to bismuth was present in pitchblende. She named this new element "polonium" in honor of her native country. They found yet a second new radioactive element in the pitchblende ore with chemical properties close to that of barium, and they named that new element "radium" from the Latin word *radius* meaning "ray" (Curie, 1905, 1911).

The discovery of a highly penetrating radiation that was nondeviable in an external magnetic field, which we now know to be gamma radiation, was discovered by Paul Villard at the Ecole Normal in Paris, France in 1900. Villard's discovery of gamma radiation was reported to the French Academy of Sciences (Villard, 1900a,b) and at the Meetings of the French Society of Physics (1900c). Villard did not provide any diagrams of his experimental arrangements, which led to the discovery of gamma rays; however, the writer sketched Fig. 1.1 to facilitate the description of his experiment. Villard placed a sample of barium chloride containing radium sealed in a glass ampoule within a lead shield that contained an opening, which essentially provided a collimated beam of the radiation from the radium source as illustrated in Fig. 1.1.

To the radiation beam he exposed two photographic plates wrapped in black light-tight protective paper. Between the two plates was sandwiched a 0.3-mm-thick lead barrier. A magnetic field was applied to the collimated beam to deflect the deviable rays. Alpha particles emitted by the radium are ignored because these are absorbed by the outer protective paper wrapping. The magnetic field caused a deviation of the beta particles whereas a very penetrating radiation remained unaffected or undeviable by the magnetic field as evidence from the images produced by the radium emanations on the developed photographic emulsions. The developed photographic emulsion A, which was the first to receive the nuclear radiations from the radium, showed two spots produced by two types of radiation, one deviable (marked β) and the other undeviable (marked γ) in the magnetic field. The second photographic emulsion B, which was placed behind a 0.3-mm-thick lead barrier, yielded only one spot produced by a highly penetrating radiation unaffected by the magnetic field. The intensity of the spot on emulsion B was the same as that on emulsion A, indicating that its intensity remained unaffected to any observable extent by the lead barrier. The spot was also more clearly discernible because it was not clouded by the deviable beta particles. Villard concluded that his experimental evidence demonstrated a radiation of property similar to X-rays, but with a greater penetrating power than X-radiation. In the journal *Nature* Becquerel

(1901) acknowledged Villard's discovery of gamma rays, which had not yet been named as such, with the following statement (with the writer's comments in brackets):

> *... there exists two kinds of radiations, one not capable of deviation [in a magnetic field] and of which the nature is still unknown, the other capable of deviation, which later experiments have identified with the cathode rays [later identified as beta-particles] ... I might add that recently Mr. Villard has proved the existence in the radium radiation of very penetrating rays, which are not capable of deviation.*

Villard did not venture to name this newly discovered highly penetrating radiation. He was a modest person possibly uninterested in fame or in the highly competitive pursuits of his fellow scientists in this new frontier.

Rutherford (1903), who is considered by many as the best experimental physicist of the 20th century, was first to name the highly penetrating radiation discovered by Villard, as gamma rays. A few years prior to Villard's discovery, Rutherford (1899) had already named two types of nuclear radiation as "alpha" and "beta", which he characterized on the basis of their relative penetrative power in matter, that is, alpha radiation would be more easily absorbed by matter than beta radiation. In harmony with this nomenclature, Rutherford assigned the term gamma rays to the yet more penetrating radiation. In the *Philosophical Magazine* Rutherford (1903) named and characterized the three types of nuclear radiation on the basis of their penetration power in matter as follows:

> *Radium gives out three distinct types of radiation:*
>
> 1. *The α rays, which are very easily absorbed by thin layers of matter, and which give rise to the greater portion of the ionization of the gas observed under the usual experimental conditions.*
> 2. *The β rays, which consist of negatively charged particles projected with high velocity, and which are similar in all respects to cathode rays produced in a vacuum tube.*
> 3. *The γ rays, which are nondeviable by a magnetic field, and which are of a very penetrating character.*
>
> *These rays differ very widely in their power of penetrating matter. The following approximate numbers, which show the thickness of aluminum traversed before the intensity is reduced to one-half, illustrate this difference:*

Radiation	Thickness of aluminum
α rays	.0005 cm
β rays	.05 cm
γ rays	8 cm

As noted by Rutherford in the above quotation, ionization in a gas was one of the principle properties of nuclear radiation, which helped distinguish the radiation type. It was well understood at the time that the three rays possessed different powers of ionization in matter. The number of ion pairs formed in a gas by α, β, and γ rays of given energy per path length of travel would differ according to the relationship α rays > β rays > γ rays. Consequently the property of ionization power of a given radiation was used to also measure the intensity of that radiation.

In addition to the different penetrating and ionization powers of α-, β-, and γ-radiation, other properties were used to identify these mysterious radiations, such as the differing deflections that the three radiations undergo in electric or magnetic fields. Rutherford (1903) contributed much to the understanding of these properties during the early years of research on radioactivity. Alpha radiation was known to possess a positive charge because it would be deflected toward the negative electrode in an electric field potential, while beta particles were known to be negatively charged due to their deflection in the opposite direction toward the anode or positive electrode. Whereas gamma radiation would not undergo any deflection whatsoever as illustrated in Fig. 1.2. Likewise, the alpha and beta radiations, when traveling in a path perpendicular to the lines of force of a magnetic field, will be deflected in opposite directions, which is a characteristic of charged particles. Gamma radiation, which carries no electric charge, would continue along a straight undeviating path in either electric or magnetic fields.

After 3 years at the Cavendish Laboratory, University of Cambridge, Ernest Rutherford, originally from New

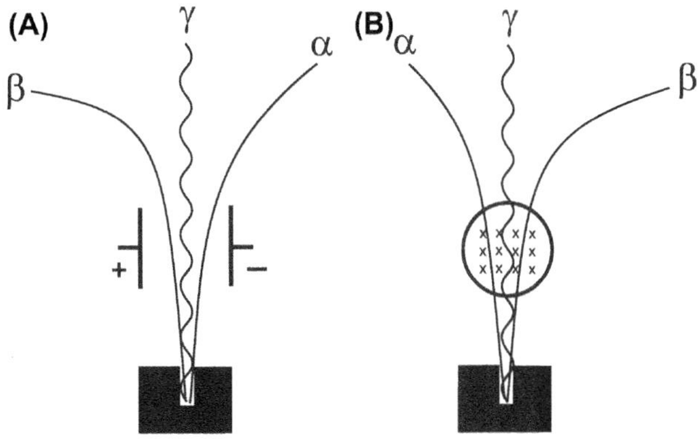

FIGURE 1.2 Paths of travel of collimated beams of alpha, beta, and gamma radiation in (A) electric and (B) magnetic fields. Beam collimation is provided by placing the radiation source in a hole drilled within shielded containers (e.g., lead) of which cross sections are illustrated. The electric field in (A) is illustrated by positively and negatively charged electrodes separated by a space through which the radiations pass, and the magnetic field in (B) is illustrated by the circle to depict the pole of a magnet through which the lines of force of magnetic flux (marked by the symbol x) are directed into the plane of the page (z-axis) perpendicular to the radiation paths in the xy-axis.

Zealand, moved in 1898 to McGill University in Montreal at the age of 27 to take on the position of Professor of Physics. It is at McGill University where he began to make his major discoveries in the field of nuclear physics. The first of these was the discovery that radioactive atoms emitting α-particles or β-particles disintegrate into atoms of lighter weight, in other words, atoms of an element, such as radium, that emit α-particles undergo transformations to atoms of a lighter and consequently different element. In 1907 Rutherford transferred to the University of Manchester in the UK where he and his assistant Thomas Royds (Rutherford and Royds, 1908a,b) and other coworkers demonstrated that alpha particles emanating from radium would turn into atoms of helium, and that helium gas would accumulate or be entrapped in minerals that contained radium. Thus, the alpha particle would later be determined to be equivalent to a nucleus of helium. Furthermore, Rutherford demonstrated that the lighter atom produced as a product of the decay of radium would likewise be radioactive, and in turn, decay to another even lighter atom, and so on until the final product atom was stable. It was for this work that Rutherford received the 1908 Nobel Prize in Chemistry (Rutherford, 1908). Together with another assistant Hans Geiger, who is best known for the development of the Geiger counter or Geiger–Muller tube, Rutherford (Rutherford and Geiger, 1908a,b) provided experimental evidence confirming the double positive charge on the alpha particle, thus confirming that the alpha particle was equivalent to a nucleus of helium.

Rutherford's work with numerous collaborators, including Frederick Soddy (Rutherford and Soddy, 1902), led to the conclusion that one chemical element can transform into other elements, which was previously only a centuries-old belief of alchemists, who tried to change lead into gold. Prof. Hasselberg (1908) in his speech of the 1908 Nobel Prize presentation, had foresight when he added further

> *[Rutherford's] disintegration theory [of atoms] and the experimental results upon which it is based, are synonymous with a new department of chemistry.*

We can thus give credit to Rutherford for giving birth to the field of radiochemistry. It was, however, Frederick Soddy, who established the "displacement law" or "periodic law" of radionuclides. While at Glasgow University in 1913, he published in the journal *Nature* (Soddy, 1913a) the law in the following words:

> *The successive expulsion of one α and two β particles in three radioactive changes in any order brings the interatomic charge of the element back to its initial value, and the element back to its original place in the (Periodic) Table, though its atomic mass is reduced by four units.*

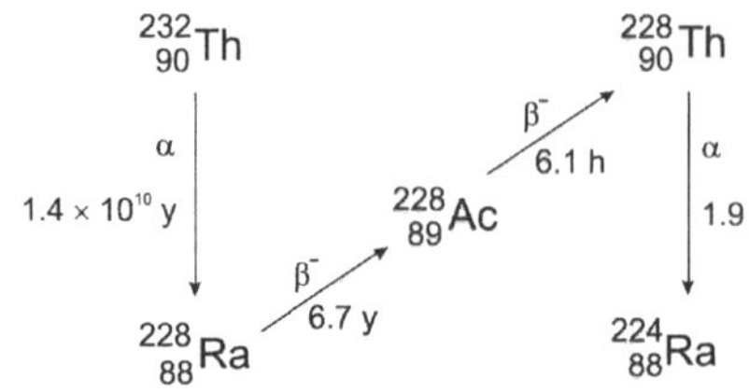

FIGURE 1.3 Soddy's displacement or periodic law as illustrated with thorium and radium isotopes.

An example of Soddy's displacement law can be taken from a part of Fig. 1.8 of the Thorium Decay Series, which is illustrated here in Fig. 1.3.

As illustrated in Fig. 1.3, Soddy demonstrated that the emission of an α-particle would produce an element that is two atomic numbers lower, while the emission of a negative β-particle would displace an atom one atomic number higher. In the illustration above we see that the radioactive atom of thorium-232 emits an α-particle to become an atom of radium-228, which in turn decays by the emission of a β-particle to become radioactive actinium-228. The actinium atom also decays by the emission of a negative β-particle to become thorium, the same element of atomic number 90 from which it had originated, but with a mass number of four units less. The emission of an α-particle from the thorium-228 brings the atom back to the element radium with four mass units less than the atom of radium-228. The parent and product atoms of atomic number 90 are of the same element thorium, but of different mass. These two distinct atoms of thorium ($^{232}_{90}$Th and $^{228}_{90}$Th) and radium ($^{228}_{88}$Ra and $^{224}_{88}$Ra) illustrated above are today known as "isotopes" because Soddy (1913a,b) coined the word "isotopes" to refer to such atoms that occupy the same place (from the Greek) in the periodic table. Isotopes have the same atomic number (*i.e.*, the number of protons in the nucleus), but different mass number (*i.e.*, number of protons plus neutrons in the nucleus).

One of Ernest Rutherford's greatest discoveries came in 1919 when he reported the first evidence of a man-made nuclear reaction, that is, the splitting of the atom. This he was able to demonstrate when a high-speed alpha particle would strike the nucleus of an atom and rearrange it into two different atoms. Rutherford observed that when alpha particles would strike air, he could detect scintillation on a zinc sulfide screen produced at a distance well beyond the distance of alpha-particle range of travel, which corresponded to the range of travel of hydrogen atoms (protons). He demonstrated that the production of high-speed protons, by collision of alpha particles with air, arose from the collision of the alpha particles with only the nitrogen atoms of the air because the effect would not occur with other constituents of air such as oxygen or carbon dioxide. Furthermore, when pure nitrogen was the target, the scintillations produced by the product hydrogen nuclei (protons) were greater than when air was bombarded with alpha particles (air contains 79% nitrogen). Rutherford was also able to show that the number of swift atoms of oxygen produced by the alpha particle collisions was about the same as the corresponding number of hydrogen nuclei (protons). This first man-made nuclear reaction is written as follows:

$$^{4}_{2}\text{He} + ^{14}_{7}\text{N} \rightarrow ^{17}_{8}\text{O} + ^{1}_{1}\text{H} \tag{1.1}$$

The historic nuclear reaction occurs when the colliding alpha particle, equivalent to a nucleus of helium ($^{4}_{2}He$), interacts with an atom of nitrogen-14 (the most abundant isotope of nitrogen, denoted as $^{14}_{7}N$ because its nucleus contains 7 protons and 7 neutrons). In the collision, a proton ($^{1}_{1}H$) is ejected and two protons and two neutrons from the alpha particle coalesce with the remaining nucleons of the original nitrogen to yield a nucleus having 8 protons and 9 neutrons, namely, the isotope oxygen-17 denoted above as $^{17}_{8}O$. This was the very first artificial transformation of one element into another, which was the age-old dream of alchemists.

As a result of the above man-made nuclear reaction where protons were emitted as a product, Rutherford has been given the honor of discovering the third elementary particle in matter, the proton. To put this discovery in perspective, the first elementary particle to be discovered was the electron from the work of J.J. Thomson in 1897, and the second elementary particle to be discovered was the photon from the work of Albert Einstein (Einstein, 1905a). Einsteins' discovery of the photon came on the heels of Max Planck's quantum theory of radiation (Planck, 1900), which was presented by Planck on 14 December 1900 at a meeting of the Physikalische Gesellschaft in Berlin. Thus, the world came to know that radiation is emitted at discrete energies corresponding to differences in energy states of a body. Planck defined the discrete energies of emitted radiation as the product of Planck's constant, h, and the radiation frequency, ν, or

$$E = h\nu \tag{1.2}$$

Planck referred to this discrete radiation energy as "the quantum" from the Latin *quantus* meaning "how great". Thus, Planck's discovery of the energy quantum in 1900 led to Einstein's interpretation of the photoelectric effect in 1905, where he explained that light not only traveled as waves of electromagnetic radiation but also as discrete packets of energy, namely "light quanta" or photons. This opened a new door to radiation physics, referred to as quantum physics, and Eq. (1.2) is now often referred to as the Planck–Einstein relation (Woan, 2000). If electromagnetic radiation, such as X-rays, gamma rays and visible light rays, had the properties of particles, that is, the energy quanta or photons, Louis Victor de Broglie at Paris University reasoned, in turn, that all particles in motion would have properties of a wave (de Broglie, 1923a,b,c, 1924, 1925). He reasoned that all particles of known rest mass would, in motion, have a dual nature, that of a particle and that of a wave. Thus the electron, neutron, proton, and alpha particle, etc could display this dual nature. He derived the fundamental relation for the wavelength of a particle in motion as

$$\lambda = h/p \tag{1.3}$$

where, λ is the wavelength of the particle, that is, the distance between two consecutive peaks of the particle wave, h is Planck's constant, and p is the particle momentum. This dual nature of particles (matter and wave) has been proven many times over and remains vital to applications in particle diffraction.

Another contribution, among many others made by Rutherford, deserves mention. This is the development, together with Hans Geiger, of the electronic means of detecting and counting individual alpha particle emissions from radioactive atoms. The alpha particle emissions were allowed to travel through a small opening or window into a vessel containing air or other gas exposed to an electric potential. The vessel is referred to today as an ionization chamber. Upon entering the vessel the alpha particle, which carries a double positive charge, would cause ionization of the gas, and the ions produced by the alpha particle would accelerate toward electrodes of the chamber thereby magnifying the ionization within the gas. The positive and negative ions produced by the alpha particle would be collected by their apposing electrodes and thereby produce a pulse that would cause a deflection of an electrometer needle. This instrument was the precursor of more modern Geiger counters, but it served its purpose, as Rutherford and Geiger were able to count each alpha particle emission from a radium sample and calculate its specific activity, in the words of Rutherford on his 1908 Nobel Lecture, "In this way it was shown that 3.4×10^{10} alpha-particles are expelled per second from 1 gram of radium." This was very close to the real value of 3.7×10^{10}, which is used today to define the unit of radioactivity known as the Curie (Ci), where $1 \text{ Ci} = 3.7 \times 10^{10}$ disintegrations per second.

Rutherford's best work was yet to come after his receipt of the Nobel Prize. In 1907 he moved to England to fill the position of Professor of Physics at Manchester University. It was at Manchester where Rutherford had a list of research topics to explore, and one of these was the deflection that alpha particles would undergo when passing through thin foils (approx. 5×10^{-5} cm thick) of materials such as mica, aluminum, gold, etc. He knew that the alpha particles, as they travel at high speed, would traverse the very thin foils as if the foil material was not even in the alpha particle path, and that the flux of alpha particle radiation would undergo only a very slight dispersion upon exiting the foils. This was understandable to him as strong electrical charges, expected to occur in atoms, could cause the slight dispersion of the positively charged alpha particles.

The story goes that Rutherford was approached by one of his students at Manchester, Dr. Hans Geiger (best known as the person who developed the Geiger counter still used today for the monitoring of radioactivity). Geiger had asked Rutherford "What do you suggest we give the new student Ernest Marsden to do?" Rutherford proposed that they try to see if any alpha particles would bounce back, that is, not traverse the foil but be deflected by over 90 degrees back toward the particle direction of travel. Rutherford did not expect to see any such deflection, but it had to be investigated. They could discern the alpha particle direction of travel and count the deflections by means of a zinc sulfide screen that would produce a microscopic fluorescence (flash of light scintillation) in the dark when each individual alpha particle hit the screen. They used radium as the source of the alpha particle beam to bombard a thin foil of gold. Geiger later informed Rutherford that they could see the occasional deflection by greater than 90 degrees of one alpha particle for every 8000 particles traversing the gold foil. In a presentation given by Rutherford years later, he described his reaction to this discovery by stating

> *It was quite the most incredible event that ever happened to me in my life. It was as incredible as if you fired a 15-inch shell at a piece of tissue paper and it came back and hit you.*

Rutherford grasped this discovery to conclude that there had to be a massive core or nucleus in the atoms of materials that would cause colliding alpha particles to bounce back. By using high-energy (7.7 MeV) alpha particles, that would travel at highest speeds available at that time, Rutherford was able to calculate the distance of closest approach and consequently the radius of the atomic nucleus to be approx. 5×10^{-15} m (Rutherford, 1906, 1913, 1919, 1920a,b).

In a Letter to the Editor of *Nature*, concerning the structure of the atom, published on December 11, 1913, Rutherford postulated correctly the atomic nucleus as the origin of α- and β-radiation. In his letter he stated the following:

> *There appears to me no doubt that the α particle does arise from the nucleus, and I have thought for some time that the evidence points to the conclusion that the β particle has a similar origin. This point has been discussed in some detail in a recent paper by Bohr (Philos. Mag., September 1913). The strongest evidence in support of this view is, to my mind, (1) that the β ray, like the α ray transformations, are independent of physical and chemical conditions, and (2) that the energy emitted in the form of β and γ rays by the transformation of an atom of radium C is much greater than could be expected to be stored up in the external electronic system.*

From here on Rutherford was able to begin to formulate the structure of atoms with a central massive nucleus (Rutherford, 1929, 1936). Rutherford's findings were the initial step that provided the foundation upon which other physicists such as Niels Bohr, Werner Heisenberg, and others were able to elaborate the structure of the atom.

Following Rutherford's alpha scattering experiment, which provided evidence for a massive central nucleus, Bohr (1913, 1914, 1921a,b, 1922) formulated the structure of the atom, known as the Bohr atom, consisting of a positively charged nucleus with electron orbitals about the nucleus, the number of which were sufficient to neutralize the positive charge on the nucleus (i.e., the number of electrons in each neutral atom equaled the number of protons in the nucleus or its atomic number). Also, the dimensions of the nucleus were very small in comparison to the size of the atom defined by the electron orbitals, and almost all of the mass of the atom was concentrated in the nucleus. By taking Rutherford's atom, Max Planck's quantum theory of energy, and the simplest of all atoms hydrogen, Niels Bohr was able to launch his quantum theory of atomic structure, referred to as the Bohr atom. With his quantum theory of atomic structure, Bohr was able to explain the spectra of radiation emitted by atomic electrons and the chemical properties of the elements based on electron groupings in quantum orbitals. Bohr proposed that electrons would exist in orbitals about the nucleus at discrete distances; and each orbital would represent specific quantized energies. Thus, when an atom absorbs energy, from an external form of excitation, an atomic electron would jump from one orbital to another further away from the atomic nucleus. Energy absorbed by the atom could be emitted from the atom as radiation when an electron falls from an outer to an inner orbital. Bohr explained that energy can be absorbed or emitted by an atom only as a single quantum of energy or light photon ($h\nu$) equal to the energy differences of specific orbitals. The difference between two such energy levels ($E_2 - E_1$), divided by Planck's constant h, would define the oscillation frequency of a spectral line that can be emitted by the atom, as described by the following

$$\nu = \frac{E_2 - E_1}{h} \tag{1.4}$$

With his quantum theory of atomic structure, Bohr derived and explained fully the Rydberg formula, which was a formula devised by the Swedish physicist Janne Rydberg (1854–1919) to calculate and predict the wavelengths of light photons emitted by hydrogen and later applied to the other elements of the periodic table to include, in addition to visible light, other types of electromagnetic radiation emitted by atoms.

The use of electron orbitals about a central atomic nucleus, as depicted by Bohr (*i.e.*, the Bohr atom), was not considered or intended as an accurate depiction of the atom; however, it remains to this day a very didactic method to illustrate the atom and explain radiation emission and absorption of electron origin and the chemical properties of the elements. Because electrons have properties of both particles and waves, that is, a dual nature as described by de Broglie, Heisenberg (1926a–c, 1929) demonstrated that we cannot always assign to an electron a position in space at a given time, nor follow it in its orbit, so therefore, we cannot assume that the planetary orbits postulated by Niels Bohr actually exist. Heisenberg thus established the Uncertainty Principle (Heisenberg, 1927), which states that it is not possible to accurately measure both the atomic particle (*e.g.*, electron) velocity or momentum and its location. Heisenberg's derivation from quantum mechanics showed that it is not possible to determine, in any given instant in time, both the velocity and the position of an atomic particle. It was demonstrated that the more accurately one tries to determine the velocity of the particle, the more uncertain becomes the determination of its position, and vice versa. Bohr (1928) presented his complementarity argument that the clearest analysis that can be derived from quantum mechanics must include both the corpuscular and wave nature of atomic particles, that is, both the particulate and wave nature of the electron are needed to complement each other in providing a complete description of the atomic particle.

Schrödinger (1926a–e) took the dual nature of the electron to describe the electron and its properties with wave equations. Schrödinger's wave mechanics provided scientists with a visual picture of the space coordinates occupied by the electron, which Heisenberg's mathematical matrix equations could not provide. Thus, we may also picture atomic electrons as possessing energy levels, rather than occupying definite orbits. Schrödinger had interpreted his wave function Ψ as the electron's matter–wave density distribution, that is, he wanted to dispense with the particulate properties of the electron and consider them to possess a continuous density distribution $|\Psi|^2$ meaning that the electron as a wave could have a high density in some regions of space and lower density in other regions. Max Born, however, interpreted the wave function Ψ as providing a picture of the electron with its coordinates in space, and $|\Psi|^2$ as the probability density of the electron (Born, 1926a,b,c,d; Born et al., 1926; Born and Oppenheimer, 1927; Born and Fock, 1928). Born's probability density interpretation is widely accepted today as that which gives the probability of finding the electron particle at particular coordinates in space.

At the time that Heisenberg, Schrödinger, and Born were studying the atomic electron, the Austrian physicist Wolfgang Pauli postulated the Exclusion Principle for which he received the 1945 Nobel Prize in Physics, as this proved to be a vital contribution to nuclear and atomic physics. The Exclusion Principle (Pauli, 1924, 1925, 1946) stated that no two electrons in an atom could exist in the same quantum or energy state. Pauli's Exclusion Principle enabled the assignment of electrons to different quantum levels *K*, *L*, *M*, *N*, etc., and our subsequent understanding of X-ray spectra that often follows nuclear decay. The Pauli Exclusion Principle also proved to be a vital tool decades later to explain the structure of protons and neutrons in the nucleus, particularly what is referred to as the shell structure of the nucleus, and the stability of particular nuclei to decay. Like the electrons, which are assigned to various quantum levels or orbitals on the basis of Pauli's Exclusion Principle, no two nucleons may possess the same quantum or energy state and are assigned to specific shells. Certain filled shells of protons and neutrons in the nucleus were found by Maria Goeppert-Mayer at the Argonne National Laboratory and J. Hans D. Jensen at the University of Heidelberg (Goeppert-Mayer and Jensen, 1955a,b) to be unusually stable. Maria Goeppert-Mayer and J. Hans D. Jensen were awarded the 1963 Nobel Prize for Physics for their work on nuclear structure, which they shared with Eugene Wigner for his work on the mechanics of proton–neutron interaction.

One of the greatest conundrums of modern physics was beta decay. It was understood that the kinetic energy of a particle emitted by a decaying nucleus was a manifest of the decay energy of that nucleus. However, it was observed that beta particles are emitted from decaying nuclei with a broad spectrum of energies over a range between zero and a maximum energy or E_{max}. In other words, it was observed that most of the beta particles emitted possessed an energy less than the nuclear decay energy. Only a very small number of the beta particles of the maximum energy of the beta-particle energy spectrum would correspond closely to the nuclear decay energy. In desperation for an answer, some physicists proposed that possibly the laws of conservation of energy did not apply in this particular case. However, Wolfgang Pauli could not agree to any exceptions to the laws of energy conservation, and he took a bold step. He proposed, that during beta decay, there should exist another particle of zero charge and very small rest mass that would be created at the time of beta decay, and this particle would share the decay energy with the beta particle. While he was Professor of Theoretical Physics at the Swiss Federal Institute of Technology in Zurich (Eidgenössische Technische Hochschule or ETH, Zurich) Pauli sent an Open Letter dated 4 December 1930 to the Radioactivity Group at their Regional Meeting at Tübingen, Germany. In this letter Pauli wrote

Dear Radioactive Ladies and Gentleman,

I beg you to most favorably listen to the carrier of this letter. He will tell you … of the continuous beta spectrum, I

have hit upon a desperate remedy to save ... the law of conservation of energy. There is the possibility that electrically neutral particles exist ... in nuclei, which have a spin ½ and obey the exclusion principle, and which differ from the photons also in that they do not move with the velocity of light. The continuous beta spectrum would then be understandable if one assumes that during beta decay with each electron a [neutrino][1] *is emitted in such a way that the sum of the energies of the [neutrino]*[1] *and electron (i.e., beta-particle) is constant.*

I admit that my remedy may seem incredible ... But only he who dares can win and the difficult situation caused by the continuous beta spectrum ... Henceforth every possible solution must be discussed. So, dear radioactive people, because I am indispensible here due to a ball which will take place in Zurich during the night from December 6 to 7. With my best regards to you...

Your humble servant,

W. Pauli

The writer believes that Wolfgang Pauli used the ball in Zurich as an excuse not to attend the scientific conference because he wanted his letter to be read to the participants of the conference, and it is very understandable that he would not want to listen to any ridiculous arguments against his theory, particularly any that would suggest an exception to the law of conservation of energy.

Wolfgang Pauli's prediction in 1930 of the neutrino as a particle, that would be produced in beta decay and share the decay energy with the beta particle, was only the beginning of the elucidation of a more complex scheme for beta decay. Two years earlier Paul A.M. Dirac, as a Fellow at St. John's College in Cambridge and a researcher at the Cavendish Laboratory, published the quantum theory of the electron (Dirac, 1928a,b, 1933). A result of Dirac's quantum theory of the electron was the prediction of the electron spin and the occurrence of a new electron, one of positive charge, known as the positron, the antiparticle of the negatron or negative electron. Dirac also predicted the phenomenon of annihilation, when a particle and its antiparticle (*e.g.*, negatron and positron) come into contact and are annihilated by the conversion of their electron masses into electromagnetic radiation. He also predicted pair production, which would be the converse of annihilation, that is, the creation of a positron and negatron pair from electromagnetic radiation. In his Nobel Lecture Dirac explained

From our theoretical picture, we should expect an ordinary electron, with positive energy, to be able to drop into a hole and fill up this hole, the energy being liberated in the form of electromagnetic radiation. This would mean a process in which an electron and a positron annihilate one another. The converse process, namely the creation of an electron and a positron from electromagnetic radiation, should also be able to take place. Such processes appear to have been found experimentally, and are at present being more closely investigated by experimenters.

The existence of the positron was experimentally discovered in 1932 by Carl D. Anderson while carrying out research on cosmic rays for which he was awarded the Nobel Prize in Physics 1936 (Anderson, 1932, 1933a,b). Anderson also demonstrated together with his graduate student, Seth Neddermeyer, that gamma rays from "ThC" (*i.e.*, ^{212}Bi) would generate positrons, by pair production, after passing through substances such as lead (Anderson and Neddermeyer, 1933). Pair production, as predicted by Dirac, is the creation of an electron and positron pair by the conversion of the energy of electromagnetic gamma radiation into the masses of two electrons of opposite charge. A cloud chamber photograph of positron–negatron pair production is illustrated in Fig. 1.4.

If negative and positive electrons (*i.e.*, beta particles) were known to exist and these were products of beta decay, the mystery remained how beta particles, which are

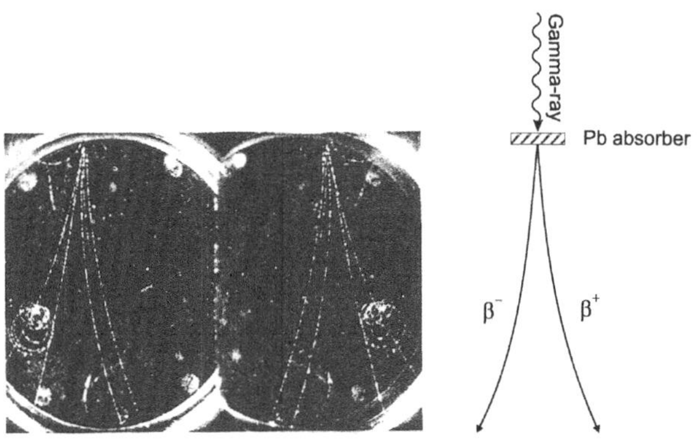

FIGURE 1.4 Photograph of cloud chamber tracks of positrons (curved tracks on the right) and negatrons (curved tracks on the left) originating from a common point of origin where gamma-ray photons of cosmic origin were absorbed by a lead barrier. The magnetic field was applied in the direction of the page of this book. The photograph was presented as a stereoscopic pair, and the direct image is at the left. A drawing to the right illustrates pair-production, that is, the production of a negatron–positron pair from the energy of a gamma ray photon in the region of the nucleus of an absorber material such as lead. The curved tracks illustrate the bending of the direction of travel of the negatron–positron pair as a result of the magnetic field. *From Anderson and Neddermeyer (1936). Reprinted with permission © 1936 American Physical Society.*

1. The actual naming of the particle as the neutrino, while not included in Pauli's letter, is included here for clarification with the reader as to the real particle referred to by Pauli. Enrico Fermi later coined the name of this neutral particle of near-zero rest mass as the "neutrino" from the Italian meaning "little neutral one". Due to the particle's neutral charge and near-zero rest mass, it was very difficult to detect the neutrinos, so difficult, that it was not until over 25 years later (June 1956) did technology permit Frederick Reines and Clyde Cowan, Jr. to finally achieve the detection of the neutrino.

equivalent to electrons, could originate and be expelled from an atomic nucleus. The beta decay process of radioisotopes, often referred to as "Fermi Beta Decay", was elaborated in 1934 by Enrico Fermi (1934a,b) only 2 years following the discovery of the neutron by Chadwick (1932a,b). Fermi proposed, that within the nucleus of a radioactive atom, the neutron may decay to a proton or vice versa, a proton into a neutron. He proposed the existence of a Weak Interaction as a new force, within the nucleus, that permits a neutron to decay into a proton and a negatively charged electron (*i.e.*, negatron) and antineutrino ($n \rightarrow p^+ + e^- + \bar{\nu}$) or vice versa, a proton could decay into a neutron and a positively charged electron (*i.e.*, positron) and neutrino ($p^+ \rightarrow n + e^+ + \nu$). The Weak Interaction would allow a nucleon (proton or neutron) to transform itself into the other to achieve nuclear stability. This force would be weaker than the Strong Nuclear Force or Strong Interaction that binds the nucleons together in the nucleus. The weak interaction proposed by Fermi advanced into a Unified Electroweak Theory for which the Nobel Prize in Physics 1979 was awarded to Sheldon L. Glashow, Abdus Salam, and Steven Weinberg "for their contributions to the theory of unified weak and electromagnetic interaction between elementary particles ...", as it has subsequently been discovered that the weak interactions and electromagnetic interactions are of similar strengths at short distances typical of nucleons (10^{-18} m). These two forces can this be combined into what is called the Electroweak Force.

Fermi's original manuscript on the transformations of nucleons within the nucleus was rejected by the editors of the journal *Nature* because of its theoretical nature, and therefore his first reports on the beta decay theory were published in the Italian and German languages (Fermi, 1934a,b). Evidence for these transformations was provided years later by Snell and Miller (1948), who demonstrated that the neutron when outside the protective confines of a nucleus would decay to a proton and an electron with a lifetime of ~15 minutes according to the scheme

$$n \rightarrow p^+ + e^- + \bar{\nu} + 0.782 \text{ MeV} \tag{1.5}$$

Although Fermi is well remembered for his beta decay theory, the Nobel Prize was awarded him in 1938 for his contribution to the application of neutron radiation to the production of many new radioactive isotopes and to his discovery that certain nuclear reactions are brought about by slow neutrons. In his Nobel Lecture (December 12, 1938) Fermi acknowledged that Rutherford was the first to start the technique of nuclear bombardments using high-energy alpha particles to enable the transmutation of one element into another in 1919. Fermi further noted that immediately after the discoveries of the artificial production of radioactive elements (radioisotopes) by the Joliot-Curies in 1934 via alpha particle–induced nuclear reactions (Joliot-Curie and Joliot (1934a–c), he thought that alpha particles might not be the only type of bombarding projectiles that could produce artificial radioactivity, particularly for atoms with atomic number greater than 20, and began then to investigate the effects of bombarding various elements with neutrons.

Fermi observed that neutrons having no electric charge should be able to reach the nuclei of all atoms without having to overcome the Coulombic barrier of the target nucleus. He knew that the neutron would not be hindered by the atomic electrons or the positive charge of the atomic nucleus, and that the probability of nuclear collisions by neutrons would be higher than that achievable with charged projectiles such as alpha particles or protons. The initial results obtained by Fermi in early 1934 (Fermi, 1934c,d; Fermi et al., 1934) and his coworkers during 1934–40 including H. Anderson, E. Amaldi, O. D'Agostino, F. Rasetti, E. Segré, and L. Szilard (Amaldi and Fermi, 1935, 1936a,b; Amaldi et al., 1935; Anderson et al., 1939a,b; Fermi and Amaldi, 1936; Fermi, 1940a,b; Fermi et al., 1934, 1938) yielded 63 artificial radioisotopes.

Only about 6 months after their first experiment with neutron bombardment Fermi made a great discovery. He found that the neutron irradiation of elements became more effective in producing new radioisotopes when the neutrons were made to slow down, by allowing the neutron beam to pass through a given barrier of water or paraffin. By so doing, the fast neutrons are slowed down to become thermal neutrons or in other words, neutrons corresponding to thermal agitation. In his Nobel Lecture, Fermi explained the properties of thermal neutrons

> *After their energy is reduced to a value corresponding to thermal agitation, the neutrons go on diffusing without further change of their average energy. All of the processes of capture of slow neutrons by any nucleus are generally accompanied by the emission of gamma rays: Immediately after the capture of the neutron, the nucleus remains in a state of high excitation and emits one or more gamma-quanta, before reaching the ground state.*

The idea of slowing down neutrons to improve the efficiency of neutron capture was contrary to what one would believe at that time, because previously only charged particles were used to bombard atomic nuclei and the higher the speed of the charged particle, the greater the probability of penetrating the Coulombic barrier of the target atom. Because neutrons, due to their neutral charge, have no difficulty diffusing into close proximity of atomic nuclei, Fermi and his coworkers were able to produce more than 400 new radioisotopes using slow neutron bombardment of all known elements with the exception of hydrogen and helium.

The work of Enrico Fermi on neutron reactions up to 1938 was the final precursor needed for the discovery of nuclear fission by Otto Hahn and Fritz Strassmann in 1939 with the collaboration and interpretation by Lise Meitner (Hahn and Strassman, 1939a,b; Meitner and Frisch, 1939; Meitner et al., 1938). Fermi's next great achievement would be the construction of the first nuclear reactor.

Enrico Fermi collaborated with Leo Szilard and a team of scientists in the construction of the first nuclear reactor. Leo Szilard was a renowned theoretical physicist and lifelong friend of Albert Einstein. Fermi's great knowledge of neutron physics and Szilard's determination to beat Germany to the production of a nuclear fission chain reaction was an excellent combination for the successful working reactor. In the words of Richard Rhodes (1999).

> *In 1939, still official enemy aliens, Fermi and Szilard co-invented the nuclear reactor at Columbia University, sketching out a three-dimensional lattice of uranium slugs dropped into holes in black, greasy blocks of graphite moderator, with sliding neutron-absorbing cadmium control rods to regulate the chain reaction.*

As soon as it became known that neutron-induced fission of Uranium-235 would yield more than one neutron per fission Fermi, Szilard and others knew that a fission chain reaction would be possible, which would release large amounts of energy. Walter Zinn and Leo Szilard (1939) determined that 2.4 neutrons were emitted on the average for neutron-induced ^{235}U fission, and it was understood that each fission reaction would yield a high amount of energy, ~200 MeV. The reactor, that was built under the direction of Enrico Fermi, was given the name of an "atomic pile", coined by Fermi because the unique apparatus had the appearance of a pile of uranium fuel, graphite moderator, and cadmium control elements. Fermi knew from his previous work with neutrons that cadmium has a very high efficiency for absorbing slow (thermal) neutrons and should serve to control the chain reaction, and that fission in ^{235}U is optimal at thermal incident neutron energies. They assembled the pile in such a way that neutrons would be produced with the correct energy needed to create fission in the uranium whereby the splitting of uranium nuclei would create additional neutrons and keep the chain reaction going to produce energy with the application of cadmium rods to control the reaction. The reactor "pile" was assembled by Fermi and his team on a squash court under the stands of the University of Chicago's Stagg (football) Field in late 1942. On December 2, 1942, under the direction of Enrico Fermi, the reactor went critical, that is, the chain reaction was sustained with the production of a neutron for every neutron absorbed by the uranium nucleus. Fermi allowed the reactor to continue as a carefully controlled chain reaction for 4.5 minutes at one-half Watt of power. A plaque, at the site of the first "atomic pile", commemorates this historic event with the words "On December 2, 1942, man achieved here the first self-sustaining chain reaction and thereby initiated the controlled release of nuclear energy."

On December 19, 1944 Enrico Fermi and Leo Szilard jointly filed for a highly classified patent as coinventors of the atomic pile, describing how the self-sustained nuclear chain reaction had been achieved. The patent (US Patent No. 2,708,656 entitled "Neutronic Reactor") was not issued until May 18, 1955. Detailed illustrations of the nuclear reactor patented by Fermi and Szilard are provided in a book by the writer (L'Annunziata, 2016) The invention of the nuclear reactor for the peaceful application of nuclear energy capable of providing electric power to cities was declared by some to be one of the most significant inventions of all time comparable to those of communication, the telegraph and telephone, by Samuel Morse and Alexander Graham Bell.

The first nuclear reactor was a giant step for mankind, as it has had since Fermi's "atomic pile" and continues to have a vital role in development with peaceful applications including (1) the production of isotopes for medical, biological, and other vital sciences, (2) the production of heat that can drive steam turbines and generate enormous amounts of electricity, and (3) the production of neutrons for research. A fourth ominous application of the nuclear reactor is, of course, for the production of plutonium for nuclear weapons.

The peaceful applications of nuclear power have an enormous impact on our daily lives. This is underscored in an article by Yukiya Amano (2013), IAEA Director General, where he noted

> *Many countries with existing nuclear power programmes plan to expand them. Many new countries, both developed and developing, plan to introduce nuclear power.*
>
> *The factors contributing to this growing interest include increasing global demand for energy, as well as concerns about climate change, volatile fossil fuel prices, and security of energy supply. It will be difficult for the world to achieve the twin goals of ensuring sustainable energy supplies and curbing greenhouse gases without nuclear power.*

As noted by Yukiya Amano, the number of nuclear power plants worldwide is expanding. The number of operating nuclear power reactors worldwide has increased to 448 in 2016, and by the end of 2016, there were 61 new power reactors under construction (IAEA, 2016).

About the same time that Enrico Fermi was producing radionuclides by nuclear reactions with neutrons, Ernest O. Lawrence at the University of California, Berkeley, invented the cyclotron, which could accelerate charged particles to high energies capable of penetrating the coulombic barrier of atoms resulting in nuclear reactions that produced numerous radionuclides. Lawrence visualized an accelerator, not consisting of a large number of accelerating electrodes in a straight line, but one that had only two electrodes, which would be used over and over again by sending the positive ions back and forth through the electrodes in a circular fashion in a magnetic field. He concluded that two hollow electrodes would be needed with a relatively low voltage (*e.g.*, 4000 V) and oscillating positive and negative potential so that each instant the accelerating ions exit one electrode and enter the other, the charges on the electrodes would change in accord with the proper frequency providing a "push" to the ions as they exit an electrode and thereby increase their energy and velocity.

The experimental work on the development of the cyclotron began in 1930 and, in the Spring of that year, Lawrence's first graduate student at Berkeley, Nels Edlefsen, constructed a crude model, which gave evidence of working. In September of 1930 they gave a preliminary report before a meeting of the National Academy of Sciences (Lawrence and Edlefsen, 1930). In the fall of the same year another graduate student, M. Stanley Livingston, improved on the model, which was only 4.5 inches (11.43 cm) in diameter. The diameter of the accelerator refers to the size of the chamber in which the ions move in a circular path of increasing radius. The first working cyclotron produced 0.08 MeV protons with less than 1000 V applied to the semi-circular accelerating electrode. This was reported before the American Physical Society (Lawrence and Livingston, 1931a) as a demonstration of the cyclotron as a practical apparatus for the acceleration of ions. The same year Lawrence and Livingston (1931b) reported the acceleration of protons to 0.5 MeV with only 5000 V potential applied to the electrodes.

Lawrence and Livingston (1932) described the basic design and principle of the cyclotron, with a simple diagram such as that illustrated in Fig. 1.5. The cyclotron utilizes the principle of repeated acceleration of ions by means of resonance with an oscillating field. The ions circulate from the interior of one electrode to the interior of another back and forth with increasing speed in a circular fashion. The hollow electrodes *A* and *B* of Fig. 1.5 are placed between the poles of a magnet, and the magnetic field normal to the plane of the hollow plates causes the ions to deflect in a circular fashion with a radius of curvature dependent on the mass, charge and velocity of the ions.

The oscillating electric field between the electrodes is vital to the operation of the cyclotron. For example, as Lawrence and Livingston (1932) explained, if at one instant there is an ion in the region between the electrodes, and if electrode *A* is negative with respect to electrode *B*, the ion will be "pushed" and accelerated to the interior of electrode (*A*). Within the hollow electrode the ion will travel along a circular path under the effect of the magnetic field. After completing a semi-circular path in electrode *A*, as illustrated by the arc *a* … *b* in Fig. 1.5, the ion emerges again between

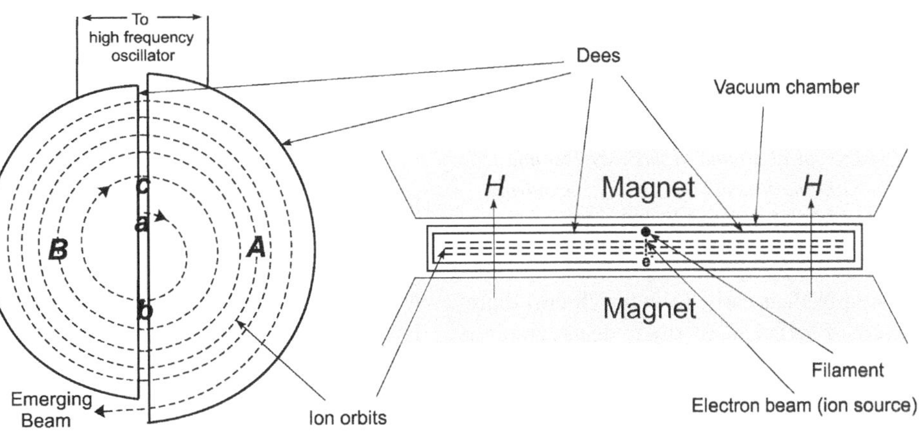

FIGURE 1.5 Diagrams illustrating the cyclotron method of multiple acceleration of ions. On the left is a drawing of the top or bottom of the cyclotron electrodes and on the right is a drawing of a view from the side with the electrodes between the poles of a magnet. Two electrodes *A* and *B* in the form of semi-circular hollow plates (called Dees because they have the shape of the letter D) are mounted in a vacuum chamber with their diametral edges adjacent. The electrodes of the early cyclotron of Lawrence and Livingston measured only 24 cm in diameter, and their hollow semi-circular chambers, through which the ions traveled, had a width of only 1 cm. The system is placed between the poles of a magnet whereby the magnetic field *H* is introduced normal to the plane of the plates. High frequency electric oscillations are applied to the plates so that there results an oscillating electric field in the diametral region between them. A filament placed above the diametral region creates a stream of electrons that produce the needed positive ions, which are pulled out sideways by the oscillating electric field. The electrons are not drawn out into the cyclotron electrodes because of their very small radii of curvature in the magnetic field. The positive ions (e.g., protons, deuterons, or helium nuclei) are accelerated by the oscillating electric field and travel with increasing speed and energy in a circular fashion of increasing radius in the magnetic field. The beam of positive ions emerges from the cyclotron to collide with specific targets creating new elements, stable isotopes, and radioisotopes, etc. *Adapted from Lawrence and Livingston (1932).*

the electrodes. If the time consumed in the travel of this semi-circular path (*e.g.*, from point *a* to point *b* of Fig. 1.5) is equal to the half-period of the electric oscillations, the electric field will have reversed and the ion will receive a second "push" and be accelerated into the interior of electrode *B* with higher velocity. The ion continues to travel in a circular path in electrode *B* and completes another semi-circular path with greater velocity and greater radius of curvature. Upon completion of the semi-circular path in electrode *B* at point *c*, the ion again emerges between the electrodes, and the ion is again accelerated by the oscillating electric field to travel at a greater speed into the interior of the opposing electrode with a path of greater radius of curvature. The process of acceleration repeats itself over and over until the ion achieves a radius of curvature whereby it emerges from the cyclotron.

A vital characteristic of the cyclotron principle is that the time of travel of an ion to complete a semi-circular path is independent of the ion's velocity and thus independent of the radius of the ion path. This was demonstrated and explained by Lawrence and Livingston (1932) following their demonstration of an 11-inch (28-cm) diameter cyclotron:

> *... the radius of the path is proportional to the velocity, so that the time required for the travel of a semi-circular path is independent of the ion's velocity. Therefore, if the ion travels its first half-circle in a half-cycle of the oscillations, it will do likewise on all succeeding paths. Hence it will circulate around on ever widening semi-circles from the interior of one electrode to the interior of the other, gaining an increment of energy on each crossing of the diametral region that corresponds to the momentary potential difference between the electrodes. Thus, if, as was done in the present experiments, high frequency oscillations having peak values of 4000 volts are applied to the electrodes, and protons are caused to spiral around in this way 150 times, they will receive 300 increments of energy, acquiring thereby a speed corresponding to 1,200,000 volts.*

The 1.2 million-volt hydrogen ions produced by Lawrence and Livingston in their early 11-inch (28-cm) diameter accelerator were of sufficient energy to disintegrate the light lithium nucleus; but much higher energies would be needed to penetrate larger atomic coulombic barriers and disintegrate the heavier elements. Larger diameter cyclotrons would require larger magnets, because the spiraling path taken by the ions in the cyclotron are produced by a uniform magnetic field normal to their plane of travel. Thus the pole faces of the magnets had to be at least as large as the diameter of the cyclotron vacuum chamber.

With a larger magnet at hand, Lawrence then proceeded to build a 27-inch (69 cm) diameter cyclotron with the electrodes ("dees") measuring 20 inches (51 cm) having a width of 1.12 inches (2.8 cm). With the new and larger cyclotron, Lawrence and Livingston (1934) were able to accelerate protons ($^{1}_{1}H$) to 5 MeV energy. With yet further modification by essentially increasing the size of the electrodes (*i.e.*, dees) to a diameter of 24.5 inches (62 cm) and width of 1.75 inches (4.4 cm) Lawrence and Cooksey (1936) produced deuterons (heavy hydrogen, $^{2}_{1}H$) ions and helium ions ($^{4}_{2}He$) of 5- and 10-MeV energy, respectively. The helium ions would be accelerated to twice the energy of the deuterons, explained by Lawrence and Cooksey (1936) as follows:

> *... The helium ions having twice the charge and mass relative to deuterons, receive twice the increments in energy each time they pass between the dees and arrive at the periphery with twice as much energy. Thus, apart from difficulties of producing doubly charged helium ions at the center, the cyclotron produces 10 MeV alpha particles as readily as 5 MeV deuterons.*

The most significant outcome of the new and improved cyclotron demonstrated by Lawrence and Cooksey (1936) was the production of neutrons and artificial radioactive isotopes. Neutron fluxes higher than ever achieved were reported. For example, the neutron flux by the deuteron bombardment of various elements including beryllium was 10^5 times greater than that produced by a mixture of 1 Ci of radon and beryllium. Lawrence began to produce many new radioisotopes as noted by Lawrence and Cooksey (1936).

> *With 5 MeV deuterons, it has been found possible (Lawrence, 1935) to produce radioactive isotopes of many of the elements throughout the periodic table. In many cases, the yields of the radioactive substances are quite large; as for example, a day's bombardment of sodium metal with 20 microampheres of 5 MeV deuterons produces more than 200 milligrams—equivalent of radio sodium.*

By 1939, the year Lawrence received the Nobel Prize, he and coworkers (Lawrence et al., 1939) at the Berkeley lab completed the construction of a larger 60-inch (1.5 m) cyclotron that produced 16 MeV deuterons. It had a magnet weighing 200 tons. In his Nobel Lecture at Stockholm, Lawrence (1951) commented about the larger cyclotron

> *... the cyclotron for the first time began to look like a well-engineered machine. It was with this machine that the discoveries of the transuranium elements were made, which have been rewarded this year by the award of the Nobel Prize in Chemistry [1951] to Edwin M. McMillan and Glenn T. Seaborg.*

Using the 60-inch cyclotron at the University of California, Berkeley, Edwin McMillan and Philip Abelson (McMillan and Abelson, 1940) created neptunium (Np) a new element of atomic number 93. McMillan moved to the Massachusetts Institute of Technology, and Glenn T.

Seaborg continued research on the creation of new radioactive elements with the 60-inch cyclotron. By bombarding U-238 with 16-MeV deuterons from the 60-inch cyclotron, Seaborg and coworkers (Seaborg et al., 1946a,b) discovered element 94 (*i.e.*, atomic number 94). The new element was named plutonium after the planet Pluto, because McMillan had suggested the name neptunium after the planet Neptune, the planet immediately beyond Uranus. Thus, Seaborg noted that it would be proper to name the new element plutonium (Pu) after what was then considered the next planet Pluto. In the summer of 1944, Seaborg and coworkers bombarded ^{239}Pu with 32 MeV helium ions (equivalent to alpha particles) in the Berkeley 60-inch cyclotron and created the new radioactive element 96, named curium, according to the reaction $^{239}_{94}Pu(\alpha, n)^{242}_{96}Cm$, which is an abbreviation for the nuclear reaction that reads "target nucleus(projectile, particle emitted)product nucleus". A few months later they identified a new element 95, named americium, created by the bombardment of ^{239}Pu with neutrons produced in a nuclear reactor. The resulting $^{241}_{94}Pu$ would decay by negative beta emission to yield $^{241}_{95}Am$. By December of 1949 they were able to identify new elements 97 and 98 created by bombarding americium and curium with 37 MeV helium ions in the 60-inch cyclotron according to the reactions $^{241}_{95}Am(\alpha, 2n)^{243}_{97}Bk$ and $^{242}_{96}Cm(\alpha, 2n)^{244}_{98}Cf$, respectively. The new elements berkelium (Bk) and californium (Cf) were named after the city and the university and state, respectively, where the discovery was made. This work laid the foundation for Seaborg's formulation of the "actinide" transition series of elements in the Periodic Table, which begins with the radioactive element actinium with a tripositive oxidation state. Included in the series are elements 99 to 103, namely, einsteinium, fermium, mendelevium, nobelium, and lawrencium. In recognition of this work, element 106 among the transuranium elements, was named "seaborgium" in 1997, which was the first time an element was named after a living person, less than 2 years before the passing of Glenn T. Seaborg. For a detailed description of this work, see Seaborg's Nobel Lecture (Seaborg, 1951) and a more recent paper published by him 4 years before his passing (Seaborg, 1995).

The cyclotron remains to this day an invaluable instrument for the production of radionuclides. Many short-lived nuclides are needed for the diagnosis and treatment of various types of cancer. Particularly the radionuclides used by medical physicians for positron emission tomography (PET) are short-lived, and only with the cyclotron is it possible to produce these radionuclides on-site at hospitals where patients can be diagnosed and treated. There are currently more than 350 cyclotrons in 39 countries of the world (Cantone and Hoeschen, 2011; IAEA, 2006, 2007), many of these are in hospitals, which produce short-lived radionuclides for medical use.

During the late 1950s, when the author of this chapter started to study chemistry, it was taught in all of the schools and widely accepted that the proton and neutron were elementary particles, that is, it was generally thought that the proton and neutron could not be broken down into smaller subatomic particles. However, by bombarding nucleons with high-energy electrons with an electron accelerator at the Stanford Linear Accelerator Center (SLAC), Robert Hofstadter and coworkers (Hoftsadter et al., 1953a,b; Hofstadter and McAllister; 1955; McAllister and Hofstadter, 1956; Chambers and Hofstadter, 1956; Yearian and Hofstadter, 1958a,b) determined that internal magnetic and charge forces of an individual proton or neutron would scatter the high-energy electrons of varying energies in different directions. From these experiments they could conclude that the proton and neutron are not elementary particles, but that they contain an inner structure that exhibit strong electric and magnetic forces. In his Nobel Lecture Hofstadter (1961) stated the following:

> *In concluding this discussion it may be appropriate to return to the theme introduced earlier in the paper and raise the question once again of the deeper, and possibly philosophical, meaning of the term "elementary" particle. As we have seen, the proton and neutron, which were once thought to be elementary particles are now seen to be highly complex bodies. It is almost certain that physicists will subsequently investigate the constituent parts of the proton and neutron ...*

Then in 1964 George Zweig, as a visiting theoretical physicist at CERN, wrote a lengthy treatise that contained a model for the internal structure of the nucleons to explain the strong interaction and its breaking (Zweig, 1964). His model depicted the proton and neutron to consist of three subatomic particles, which we now call quarks. He had called them aces. Zweig also proposed that the three quarks were held together within the proton or neutron by a force, which we now know to be mediated by gluons. He further proposed the yet unthinkable, that these subatomic particles (i.e., quarks) possess a fractional charge. At the time, it was a general belief that the smallest charge a particle could possess was unitary (+1 or −1). Zweig predicted the proton to consist of two up (u) quarks of charge +2/3 and a down (d) quark of charge −1/3 attributing the overall charge of +1 to the proton. In turn, the neutron would consist of two down (d) quarks and one up (u) quark giving the neutron its overall neutral charge. The same year in a very short paper, Gell-Mann (1964) also proposed the possibility that nucleons might be made up of three quarks similar to the model described by George Zweig.

The model presented by Zweig and Gell-Mann described above remained unconfirmed until the work of Jerome I. Friedman, Henry W. Kendall, Richard E. Taylor and their collaborators at MIT and SLAC, which yielded data demonstrating deep inelastic scattering of electrons on protons and bound neutrons obtained with a two-mile (3.2 km) electron accelerator buried underground at SLAC on the campus of Stanford University. Their deep inelastic scattering experiments with electrons demonstrated the presence of dense matter within the proton and neutron, which caused the scattering. More accurate measurements could be correlated to the fractional charges of the quark composition of the proton (with its two up quarks and one down quark, uud) and neutron (with its one up quark and two down quarks, udd). In addition, the electron deflections corresponded to only a fraction of the mass of the nucleons; the remaining fraction could be attributed to the mass of neutral gluons, which would not interact with the bombarding charged electrons. For this work, Friedman, Kendall, and Taylor were awarded the 1990 Nobel Prize in Physics. Most convincing was visible evidence for the presence of quarks and gluons, which was provided by high-energy positron–electron collisions produced by the large electron-positron (LEP) collider of 27 km circumference and build by CERN near Geneva, Switzerland. The LEP was operated as a synchrotron to accelerate positrons and electrons to travel close to the speed of light with energies as high as 209 GeV. Electrons and positrons colliding at such high energies would result in the annihilation of both particles, which can generate the emission of a quark and antiquark pair together with the radiation of a gluon traveling in three different directions terminating in the production of three jets of hadrons, as illustrated in Fig. 1.6. The LEP was dismantled in 2001 and replaced in 2012 with the Large Hadron Collider (LHD) capable of accelerating proton beams to energies of 4 TeV. A more complete description of these studies is provided in a book by the writer (L'Annunziata, 2016).

Our current understanding of the quark and gluon composition of protons and neutrons and the forces binding quarks have helped us understand the strong interaction or strong nuclear force between the nucleons that hold them together in the atomic nucleus. With this understanding, we can now explain both positron emission and electron capture (EC) when a proton changes to a neutron with the transformation of an up (u) quark into a down (d) quark. Likewise, we can now explain negatron emission when a neutron changes into a proton with a transformation of a down (d) quark into an up (u) quark within the decaying atomic nucleus, as illustrated in Fig. 1.7.

Since Max Planck's quantum theory of radiation (Planck, 1900) whereby he explained the emission of radiation in discreet packets of energy referred to as "the quantum" ($E = h\nu$) and Albert Einstein's explanation in 1905 of the photoelectric effect as the result of the photon quantum to consist of a distinct particle that can knock out atomic electrons from atoms to produce electricity, we have considered this as sufficient evidence to adhere to the concept of the particulate nature of photons. Finally in 2012, the Nobel Prize in Physics was awarded to Serge Haroche and David J. Wineland for their measurement and manipulation of individual quantum systems. As a result, science has taken us to the stage where an individual photon particle could be trapped and measured.

The previous paragraphs in this section provide a brief overview of a few key highlights of the historical

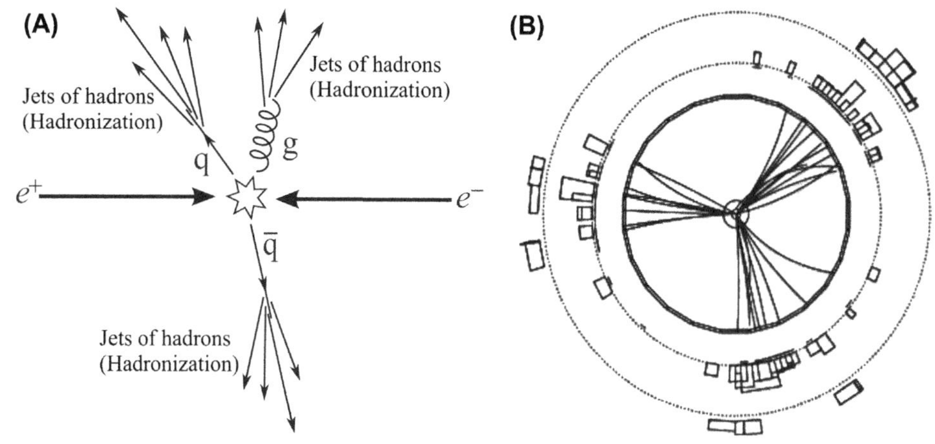

FIGURE 1.6 Hadron jets formed as a result of a high-energy collision and annihilation of a positron and electron produced in an electron–positron collider. (A) On the left is illustrated a collision of a high-energy positron and electron in a positron–electron collider producing a quark (q) and antiquark ($\bar{q}$) traveling apart in opposite direction, and a gluon (g) resulting in the production of three jets of hadrons traveling away from the point of collision. (B) On the right is an example of recordings of three hadron jets from a e^+e^- annihilation recorded in the jet chamber of a particle detector. The three jets include two jets produced by a quark–antiquark pair and the emission of a gluon by the separated quark–antiquark pair. The jets of hadrons are shown in the plane perpendicular to the direction of the e^+ and e^-, which annihilated in the center of the detector. Charged particles produce curved jets, bent by a solenoidal magnetic field, and uncharged particles produce straight lines. The boxes outside of the circular region represent recordings of the energy deposited by the neutral and charged particles in the outer components of the particle detector. *(A) From Driga (2012), reprinted with permission from Elsevier © 2012. (B) From Biebel (2001), reprinted with permission from Elsevier © 2001.*

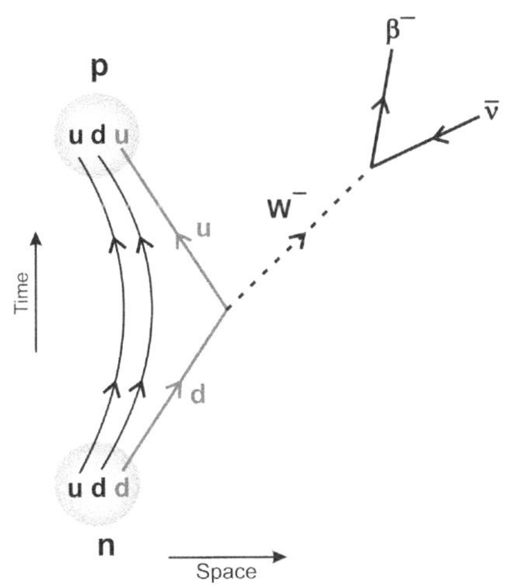

FIGURE 1.7 Feynman diagram illustrating negatron (β^-) emission in beta decay. A down quark (d) of a neutron (lower left of the figure) is transformed into an up quark (u) whereby the neutron changes into a proton (upper left of figure) with the emission of a W^- boson, which transforms into a negative electron (β^-) and antineutrino ($\overline{\nu}$). The W^- boson serves as a carrier of the weak interaction of elementary particles that occurs in beta decay. The quark transformation is depicted in red color, and the quarks that undergo no change are in black. The transformation of a neutron to a proton in the decaying nucleus results in an increase in atomic number of the nucleus; however, as in all forms of beta decay, the mass number of the nucleus does not change. For the color version of this figure the reader is referred to the electronic version of this book. *From L'Annunziata (2016); reprinted with permission from Elsevier © 2016.*

developments that have brought us to our current understanding of nuclear stability and decay. A more comprehensive coverage of the historical discoveries starting with the discovery of radioactivity to modern scientific findings that have helped us understand the structure of matter, and nuclear stability and decay can be obtained by the reader in a previous lengthy treatise by the writer (L'Annunziata, 2016).

III. Basic units and definitions

A. Properties of atomic constituents

The properties of the atomic constituents that are relevant to topics dealt with in this chapter are rest mass, charge, spin, magnetic moment, and stability. The atomic constituents in this section will be limited to the proton, neutron, and electron (see Table 1.1). The properties of particles that are found in the underlying structure of protons and neutrons, namely quarks and gluons, will be discussed further on in this chapter. Several units of rest mass are given in Table 1.1, as these are used throughout the chapter, including the expression of mass in terms of energy, the electron volt (eV) or joule (J) the unit of energy established by the International System of units (SI). A brief description of the terms used for the properties of the atomic constituents is as follows:

Rest Mass: The rest mass, which is the mass of the particle at rest relative to an observer, is expressed in four units, namely, kilograms (kg) or atomic mass units (u), and the energy equivalence of the rest mass ($E = m_0c^2$) in joules (J) or electron volts (eV). The atomic mass unit (u) is defined in terms of the stable carbon nuclide ^{12}C, which is assigned an atomic mass of exactly 12 u where 1 u $= 1.660540 \times 10^{-27}$ kg. The rest mass of the proton or neutron is approximately 1840 times that of the electron.

Charge: The charge on the particle is expressed on the basis of the unit electron charge $+e$ or $-e$ or in units of the Coulomb (C). The neutron is neutral or without charge. A **neutral atom** would be one that possesses a number of external electrons equal to the number of protons in the nucleus. It may be expressed as ^A_ZX, where Z is the **nuclear charge** or **atomic number**, i.e., the number of protons in the nucleus, A is the **mass number** of the nucleus or the sum of the number

TABLE 1.1 Properties of atomic constituents.

Property	Proton	Neutron	Electron
Rest mass	1.67262×10^{-27} kg 1.50327×10^{-10} J 938.271 MeV 1.007276466 u	1.67492×10^{-27} kg 1.50534×10^{-10} J 939.565 MeV 1.008664915 u	9.10938×10^{-31} kg 8.18710×10^{-14} J 0.510998 MeV 0.0005485799 u
Charge	$+e$ $+1.602 \times 10^{-19}$ C	Zero	$-e$ -1.602×10^{-19} C
Spin	½	½	½
Magnetic moment	1.410×10^{-26} J/T	-0.966×10^{-26} J/T	928.476×10^{-26} J/T
Stability	Stable	10.2 minutes half-life	Stable

of protons and neutrons in the nucleus. The number of electrons that the atom possesses is assumed to equal that of Z when the atom is a neutral atom.

Spin: In 1920, Otto Stern and Walter Gerlach observed two spin states of the electron, that is, the electron possessed a magnetic dipole moment, which may be quantized in either of two directions, namely with or against an external magnetic field, i.e., spin-up or spin-down. These spin states are assigned the values $+\frac{1}{2}\hbar$ or $-\frac{1}{2}\hbar$ where $\hbar$ is Planck's constant reduced or $h/2\pi$. The spin states are abbreviated as $+\frac{1}{2}$ and $-\frac{1}{2}$. The proton, neutron and electron possess half-integer spin states. Particles with half-integer spins belong to a class of particles called **fermions** (named after Enrico Fermi). Fermions obey the Pauli Exclusion Principle, which states that no two identical fermions can occupy the same quantum state (energy state) at any given time, as described further on in this chapter. The property of spin and the Pauli Exclusion Principle enable the assignment of protons and neutrons to specific quantum levels or shells within the nucleus and electrons to specific orbitals away from the nucleus.

Magnetic moment: The nuclear magnetic moments of the proton and neutron are about a thousand-fold smaller than that of the electron. Although the neutron is of neutral charge, it possesses a magnetic moment due to its underlying quark constituents that possess charge. The positive or negative signs designated to the magnetic moments indicate that the spin vectors and the magnetic moments are in opposite direction. The magnetic moments are important properties that establish the basis for magnetic resonance imaging and spectroscopy.

Stability: The neutron is unstable outside the confines of the nucleus. Once outside the nucleus, a neutron has a mean half-life of 10.2 minutes.

B. Nuclides, isotopes, isobars, isomers, and isotones

Ernest Rutherford (1911) discovered that the atom contains a massive center following experiments that revealed the backscattering of α-particles by atoms of gold, which was discussed previously in Section II of this chapter. Reference to the "nucleus" as a massive positively charged center of the atom was proposed by Rutherford (1913) and Niels Bohr (1913) in their classic papers on the structure of the atom. However, it was not until James Chadwick (1932a,b, 1935) discovered the neutron and determined its mass, could the structure of an atomic nucleus be explained to contain a definite number of protons and neutrons. Consequently the term nuclide and radionuclide could then be derived, which are defined as follows:

Nuclide: A nuclide is a species of an atom that is characterized by the number of protons (Z) and number of neutrons (N) in its nucleus. A symbol for a nuclide can be written in the form ${}^{A}_{Z}X_{N}$, where X is the symbol for the chemical element, Z is the atomic number or number of protons in the nucleus, A is the mass number or the number of protons plus neutrons in the nucleus, and N is the number of neutrons in the nucleus. Generally when A and Z are noted with the nuclide symbol, the neutron number N is not written, as it is deduced (i.e., $N = A-Z$). For example, the nuclide written as ${}^{90}_{38}Sr$ represents an atom of the element strontium with a nucleus containing 38 protons and 52 neutrons and a mass number of 90. The nuclide may also be written in a more abbreviated form as ${}^{90}Sr$, since the atomic number of the element strontium is known and an invariable number for that element.

There are 266 stable nuclides and 83 unstable or radioactive nuclides that exist naturally on earth. More than 3000 radioactive nuclides are known, that are artificially produced. The radioactive nuclides are referred to as **radionuclides**. All of the nuclides, both stable and radioactive, are organized in "The Chart of the Nuclides". The Chart provides a block containing the properties of each nuclide, which is plotted according to its Z versus N. A small part of the Chart of the Nuclides is illustrated in Fig. 1.50 in this chapter. The entire Chart of the Nuclides and the nuclear data for each nuclide can be viewed online on the website of the National Nuclear Data Center (NNDC, 2018a,b) of the Brookhaven National Laboratory (http://www.nndc.bnl.gov/nudat2/), or on the website of the International Atomic Energy Agency (http://www-nds.iaea.org/relnsd/vcharthtml/VChartHTML.html).

Isotopes: Nobel Laureate Frederick Soddy coined the word "isotopes" (Soddy, 1913a,b) to refer to such atoms that have the same atomic number (*i.e.*, the number of protons in the nucleus) but different mass number (*i.e.*, number of protons plus neutrons in the nucleus). Soddy recognized that these are atoms of the same element in the periodic table but of different mass. In his words Soddy (1913a) explained

> *I call (them) "isotopes" or "isotopic elements", because they occupy the same place (from the Greek) in the periodic table. They are chemically identical, and save only as regards the relatively few physical properties which depend on atomic mass directly, physically identical also.*

Soddy went further by alluding to the fact that many, if not most, of the stable elements actually consist of a mixture of isotopes. In his Nobel Lecture, Soddy (1922) gave credit to the Swedish scientists Strömholm and Svedberg (1909a,b) by quoting them as follows:

> *Perhaps, one can see, as an indication in this direction, the fact that the Mendeleev scheme (Periodic Table of Elements) is only an approximate rule as concerns the atomic weight, but does not possess the exactitude of a natural law; this would not be surprising if the elements of the scheme were mixtures of several homogeneous elements of similar but not completely identical atomic weight.*

Thus, **isotopes** are nuclides that have the same atomic number (Z), that is, the same number of protons; but they differ in their number of neutrons (N). Thus, they differ in their mass number (A), which is the sum of the number of protons and neutrons in the nucleus. Because isotopes have the same atomic number, they are nuclides of the same chemical element. For example, the following are isotopes of the chemical element carbon: ${}^{10}_{6}C$, ${}^{11}_{6}C$, ${}^{12}_{6}C$, ${}^{13}_{6}C$, ${}^{14}_{6}C$, where the subscript and superscript represent the atomic number Z and mass number A, respectively, whereby the number of neutrons in the nucleus, N, is calculated as $A-Z$. Among the five isotopes of carbon listed above ${}^{10}_{6}C$, ${}^{11}_{6}C$ and ${}^{14}_{6}C$ are radioactive, and thus referred to as **radioisotopes**. The other two isotopes of carbon listed above, namely, ${}^{12}_{6}C$ and ${}^{13}_{6}C$ are stable isotopes of carbon. The natural isotopes that make up the element of carbon are ${}^{12}_{6}C$, ${}^{13}_{6}C$, and ${}^{14}_{6}C$.

Isotope natural abundance: Isotopes of the elements, that are found in nature, occur at a particular **natural abundance**. The term natural abundance refers to the relative abundance of an isotope that occurs naturally on earth. For example, among the five isotopes of carbon listed above, ${}^{12}_{6}C$ and ${}^{13}_{6}C$ occur naturally at natural abundances of 98.90%, and 1.10%, respectively. The isotope ${}^{14}_{6}C$ is naturally occurring in only a small trace ($\sim 10^{-10}$% natural abundance), as it is formed continuously by the interaction of cosmic-ray protons with nuclides in the upper atmosphere. The isotope ${}^{14}C$ is radioactive, and thus a radioisotope of carbon. The percent natural abundances of the isotopes for each element should add up to 100%. Another example is the naturally occurring isotopes of potassium, namely, ${}^{39}_{19}K$, ${}^{40}_{19}K$, and ${}^{41}_{19}K$, which occur at natural abundances of 93.2581%, 0.0117% and 6.7302%, respectively. The isotopes ${}^{39}K$ and ${}^{41}K$ are stable isotopes; however, the isotope ${}^{40}K$ is a radioisotope that decays with a half-life of 1.26×10^9 years. Since the naturally occurring isotopes are remnants of the nucleosynthesis of the earth, which is 2.54×10^9 years of age (Dalrymple, 1991), the isotope ${}^{40}K$ has had time to decay since the earth's creation. When only one isotope of an element exists naturally on earth, its natural abundance is expressed as 100%. Examples are ${}^{19}_{9}F$, ${}^{27}_{13}Al$, ${}^{31}_{15}P$, ${}^{45}_{21}Sc$, ${}^{55}_{25}Mn$, ${}^{59}_{27}Co$, ${}^{89}_{39}Y$, ${}^{127}_{53}I$, ${}^{133}_{55}Cs$, ${}^{209}_{83}Bi$, and ${}^{232}_{90}Th$.

Further up the Chart of the Nuclides are found the last of the natural isotopes, namely, isotopes of uranium. Three of the isotopes of uranium are found naturally on earth, namely, ${}^{234}_{92}U$, ${}^{235}_{92}U$, and ${}^{238}_{92}U$, at abundances of 0.0055%, 0.720% and 99.2745%, respectively. The isotope ${}^{238}U$, which is the most abundant, has a half-life of 4.5×10^9 years. It has not undergone even a half-life of decay since the formation of the earth. However, the isotopes ${}^{234}U$ and ${}^{235}U$ with half-lives of 2.4×10^5 years and 7.1×10^8 years, respectively, have undergone considerable decay since the formation of the earth. All nuclides of atomic number greater than that of uranium do not occur naturally, and have been produced artificially by man as described in Section II of this chapter.

Isobars: Isobars are nuclides that possess the same mass number (A), but different atomic number (Z). For example, ${}^{56}_{28}Ni$, ${}^{56}_{27}Co$, ${}^{56}_{26}Fe$, ${}^{56}_{25}Mn$, and ${}^{56}_{24}Cr$ are isobars. In mass spectrometry where nuclides are separated and analyzed according to their mass, isobars can cause interference, and special techniques may be required to discriminate between nuclides of the same mass number (Kusnir et al., 2004; Litherland et al., 2007; Plaβ et al., 2008). In beta decay, discussed further on in this chapter, the parent and daughter nuclides are always isobars.

Isomers: Nuclear isomers are different energy states of the same nucleus, having different half-lives and even different modes of decay. A well known example is that of ${}^{99}_{43}Tc$, which may have two isomeric states. The isomers of ${}^{99}Tc$ are the following: ${}^{99}_{43}Tc$, which decays by negatron emission to ${}^{99}_{44}Ru$ with a half-life of 2.14×10^5 years, and ${}^{99m}_{43}Tc$, which decays by gamma emission to ${}^{99}_{43}Tc$ with a half-life of 6.00 hours. Isomeric states of nuclides other than the ground state are represented with a superscript "m" following the mass number. The isomer ${}^{99m}Tc$ is commonly used in the field of nuclear medicine as a diagnostic imaging source injected into the human body (L'Annunziata, 2016). When more than two excited isomeric states of a nuclide exist, the nuclear isomers are labeled as m_1, m_2, etc, in order of increasing excitation energy. For example, the isomers of ${}^{152}Eu$ are identified as follows: ${}^{152}_{63}Eu$ of 13.2 year half-life, ${}^{152m1}_{63}Eu$ of 9.3 hours half-life, and ${}^{152m2}_{63}Eu$ of 1.6 hours half-life. There are over 950 isomeric states known to exist among the nuclides. They are not considered as individual nuclides, but rather as isomers of individual nuclides.

Isotones: Nuclides that have the same number of neutrons, but different mass numbers are isotones. For example, ${}^{33}_{15}P$, ${}^{34}_{16}S$, ${}^{35}_{17}Cl$, ${}^{36}_{18}Ar$, and ${}^{37}_{19}K$ are isotones;

their nuclei all contain 18 neutrons. Isotones may be found in the Chart of the Nuclides, which plots the nuclides according to their Z versus N, as the column of nuclides having the same number N.

C. Mass and energy

The rest masses of the proton and neutron are of the order of 10^{-27} kg (see Table 1.1). Due to the small masses, most calculations involving atomic nuclei, such as nuclear decay energy, binding energy, and energies of nuclear reactions, make use of a more convenient scale of unit, namely, the atomic mass unit (u).

The atomic mass unit is defined in terms of the stable carbon nuclide ^{12}C, which is assigned an atomic mass of exactly 12.000000 u. Thus, the atomic mass unit is by definition exactly 1/12 the rest mass of ^{12}C or

$$1 \text{ u} = 1.66053873 \times 10^{-27} \text{ kg}. \tag{1.6}$$

The mass of all of the nuclides in atomic mass units are available in reference tables, such as those listed in the Handbook of Chemistry and Physics (Rumble, 2017) and the following NIST website: https://www.nist.gov/pml/atomic-weights-and-isotopic-compositions-relative-atomic-masses.

The International System of units (SI) for energy is the joule (J); however, the electron volt (eV) remains a very popular unit for expressing energy in nuclear decay, nuclear reactions, nuclear binding, etc. The electron volt (eV) is, by definition, the amount of kinetic energy gained by an electron when accelerated through a potential difference of 1 V. The electron volt is equivalent to the following units:

$$\begin{aligned} 1 \text{ eV} &= 1.602176462 \times 10^{-19} \text{ J} \\ &= 1.602176462 \times 10^{-12} \text{ erg} \end{aligned} \tag{1.7}$$

The erg is a unit of energy in the centimeter-gram-second (CGS) system of units, and less commonly used in nuclear energy calculations. From the above relation, the erg is equivalent to 10^{-7} J or 6.2415097×10^{11} eV.

Calculations concerning nuclear decay energy, nuclear reaction energy or nuclear binding energy, etc. generally require the conversion of mass to energy. A common factor that is used in such calculations is one that can convert atomic mass units (u) to the units of the electron volt (eV). The conversion factor is found using Einstein's equation of equivalency of energy and mass to convert the mass of 1 u to energy or

$$E = m_0 c^2 \tag{1.8}$$

where E is energy in joules, m_0 is the rest mass in kilograms, and c is the velocity of light in a vacuum or 2.99792458×10^8 m/s. If by definition, the mass of 1 u $= 1.66053873 \times 10^{-27}$ kg, the energy equivalent to 1 u is calculated as

$$\begin{aligned} E &= \left(1.66053873 \times 10^{-27} \text{kg/u}\right)\left(2.99792458 \times 10^{8} \text{m/s}\right)^2 \\ &= 14.92417783 \times 10^{-11} \text{ J/u} \end{aligned} \tag{1.9}$$

The energy in units of joules is then converted to units of electron volts, as 1 eV $= 1.602176462 \times 10^{-19}$ J or

$$\begin{aligned} E &= \frac{14.92417783 \times 10^{-11} \text{ J/u}}{1.602176462 \times 10^{-19} \text{ J/eV}} \\ &= 9.31494013 \times 10^{8} \text{ eV/u} \\ &= 931.494013 \text{ MeV/u} \end{aligned} \tag{1.10}$$

Thus 1 u $= 931.494013$ MeV or 931.494013 MeV/c^2. The latter units of MeV/c^2 are used conveniently when converting atomic mass units to MeV energy in Einstein's equation according to the following

$$\begin{aligned} E = mc^2 &= \left[(\text{atomic mass units}) \right. \\ &\quad \left. \times \left(931.494013 \text{ MeV} / c^2\right)\right] c^2 \end{aligned} \tag{1.11}$$

Eq. (1.11) is used to convert conveniently atomic mass units (u) to units of energy in MeV. Examples are given throughout this chapter in calculations of nuclear decay energy, nuclear reaction energy and nuclear binding energy.

D. Q value

The Q value of a nuclear reaction is a measure of the energy "released" or "absorbed" by the reaction. When the Q value is positive ($Q > 0$), energy is "released" by the reaction with a conversion of the rest mass of the reactants into kinetic energy of the products. When the Q value is negative in magnitude ($Q < 0$), an input of energy is needed for the reaction to proceed, as energy is consumed in the reaction. In other words, when Q is negative, kinetic energy must be added to one or more of the reactants (*i.e.*, bombarding particles) up to a threshold (*i.e.*, required) energy for the reaction to proceed. When the value of $Q = 0$, kinetic energy is neither consumed nor released. In this case, no reaction occurs with the exception of an elastic collision.

The energy released in a nuclear reaction is a function of the kinetic energies of the initial reactants K_i and final products K_f and, because the total energy (E) of the reactants or products is the sum of their rest energies (E_0) and kinetic energies (K), that is

$$E = E_0 + K \tag{1.12}$$

Because the rest energies and kinetic energies are conserved, the Q value is calculated as the difference between the initial and final rest masses

$$Q = E_{0,\ \mathrm{initial}} - E_{0,\ \mathrm{final}} = (m_\mathrm{i} - m_\mathrm{f})c^2 = (m_\mathrm{R} - m_\mathrm{P})c^2 \tag{1.13}$$

where $E_{0,\ \mathrm{initial}}$ and $E_{0,\ \mathrm{final}}$ are the initial and final rest energies, m_i and m_f are the initial and final rest masses, and m_R and m_P are the rest masses of the reactant and products, respectively.

Thus, the Q value of a nuclear reaction is calculated by the difference in rest masses of the nuclear reactants and nuclear products or

$$Q = E_{0,\ \mathrm{Reactants}} - E_{0,\ \mathrm{Products}} = (m_\mathrm{R} - m_\mathrm{P})c^2 \tag{1.14}$$

where m_R and m_P are the rest masses of the reactants and products of the nuclear reaction.

If we know the Q value of a nuclear reaction and we know the kinetic energies of the reactants, we can calculate the kinetic energies of the products according as

$$K_\mathrm{Products} = Q + K_\mathrm{Reactants} \tag{1.15}$$

Examples of the application of Eqs. (1.14) and (1.15) to a nuclear reaction is provided in the following paragraphs.

As nuclear reactions involve the reactions between nuclear particles and nuclei of atoms, we would need to know the rest masses of the nuclei or nuclear particles involved in the reaction to calculate a reaction Q value. In lieu of using rest masses of the atomic nuclei, it is common to use the atomic masses (u) which are available in reference tables, such as that of the Handbook of Chemistry and Physics or the Internet, such as data available from NIST, https://www.nist.gov/pml/atomic-weights-and-isotopic-compositions-relative-atomic-masses. However, it is important to keep in mind that atomic masses include the masses of the atomic electrons. To eliminate error that might occur from the added masses of the atomic electrons, it is important to assure that the number of electrons added to either side of the reaction is equal, whereby the difference of the electron masses of the reactants and products would be zero when utilizing Eq. (1.14). Let us take an example and calculate the Q value of the first man-made nuclear reaction whereby Earnest Rutherford discovered the proton described earlier in Part II of this Chapter. The reaction is as follows:

$${}^4_2\mathrm{He} + {}^{14}_{7}\mathrm{N} \rightarrow {}^{17}_{8}\mathrm{O} + {}^1_1\mathrm{H} \tag{1.16}$$

In this reaction the alpha particle interacts with an atom of ^{14}N to form an atom of ^{17}O with the ejection of a proton from the nucleus. The atomic masses may be used to calculate the Q value of the reaction as follows:

$$\begin{aligned} Q &= (M_{^4\mathrm{He}} + M_{^{14}\mathrm{N}}) - (M_{^{17}\mathrm{O}} + M_{^1\mathrm{H}}) \\ &= (4.002603254\ \mathrm{u} + 14.003074004\ \mathrm{u}) \\ &\quad - (16.999131756\ \mathrm{u} + 1.007825032\ \mathrm{u}) \\ &= -0.00127953\ \mathrm{u} \end{aligned} \tag{1.17}$$

By using atomic masses, we have added nine electron masses to each side of the reaction, and their difference is zero. Converting atomic mass units to energy according to Eq. (1.11) gives the Q value in units of MeV energy or

$$\begin{aligned} Q\ \text{value} &= (m_\mathrm{R} - m_\mathrm{P})c^2 \\ &= \left[(-0.00127953)(931.494013\ \mathrm{MeV}/c^2)\right]c^2 \\ Q\ \text{value} &= -1.19187\ \mathrm{MeV} = -1191.87\ \mathrm{keV} \end{aligned} \tag{1.18}$$

Atomic masses are conveniently used in such calculations, in lieu of nuclear masses, provided an equal number of electrons are added to either side of Eq. (1.14). The negative energy value above demonstrates that the reaction is endoergic or endothermic, that is, the reaction requires a net energy input of 1.192 MeV. When the Q value is positive, the nuclear reaction is exoergic or exothermic, resulting in a net release of energy. There are small differences in the binding energies of the electrons when using atomic masses; however, the differences in the binding energies are negligible and ignored.

The Q value for nuclear reactions or nuclear decays can be obtained using the QCALC computer program via the Internet on the website of the National Nuclear Data Center (NNDC, 2018a,b) of the Brookhaven National Laboratory in Upton, NY, USA (http://www.nndc.bnl.gov/qcalc/). In the case of a nuclear reaction, all that is required is to input on the website the symbols for the reactants and that of a product. For example, we can take the above reaction 1.16 and, on the NNDC website, we can input the symbol for the **target** nucleus as 14N (without superscript for the mass numbers), the symbol for the **projectile** 4He and the symbol for the **ejectile** 1H. The QCALC computer program can then calculate instantly the Q value to be −1191.87 keV, which is equivalent to the value calculated above by Eqs. (1.17) and (1.18).

When the reaction has a negative Q value, as in the example provided above, energy must be added for the reaction to proceed. According to the calculation of Eq. (1.18), a minimum energy of 1.19 MeV of kinetic energy must be added to the bombarding particle for the reaction to proceed. In the reaction described by Eq. (1.16), Ernest Rutherford (1919) used alpha-particles from ^{214}Po, which emits alpha particles with a kinetic energy of 7.69 MeV. This is well above the threshold energy for the reaction to proceed. The kinetic energy of the reaction products ($K_{Products}$) when 7.69 MeV alpha particles are bombarded onto ^{14}N nuclides at rest are calculated according to Eq. (1.15) as

$$\begin{aligned} K_{Products} &= Q + K_{Incident\ alpha-particle} \\ &= -1.19\ MeV + 7.69\ MeV \\ &= 6.5\ MeV \end{aligned} \tag{1.19}$$

The 6.5 MeV is shared between the product nuclides of reaction (1.16), namely the ^{17}O and ^{1}H.

IV. Naturally occuring radionuclides

If we take into account the age of the earth, which is 4.5×10^9 years, and the characteristic property of radionuclide half-life decay, one of the following conditions would have to be met for a natural radioactive nuclide to occur on earth: (1) the radionuclide would be produced continuously on earth or in its atmosphere by a natural phenomenon, (2) the radionuclide would be very long-lived, that is, it would have a half-life of the order of $\geq 10^9$ years, or (3) the radionuclide would be a short-lived radionuclide in equilibrium with a naturally occurring long-lived parent radionuclide. The above categories of naturally occurring radionuclides will be subsequently discussed.

A. Radionuclides of cosmogenic origin

A number of radionuclides are produced on a continuous basis by the interaction of cosmic-ray particles with nuclei in the earth's atmosphere. The cosmogenic isotopes are created by the interactions of high-energy primary and secondary particles of cosmic radiation with nuclei of gaseous molecules of the atmosphere (*e.g.*, N_2, O_2, Ar, etc.) resulting in the fragmentation of the target nuclei or by the capture of thermal neutrons of the cosmic radiation showers with target nuclei of the atmosphere (see Section XIV.B for additional information on cosmic-ray interactions in the earth's atmosphere). As reported by Lal (2009), most of the cosmic-ray energy (>98%) is dissipated in the earth's atmosphere. Secondary particles of cosmic radiation produce nuclear reactions at a much-reduced rate with the earth's superficial reservoirs including the hydrosphere, cryosphere and lithosphere. Cosmogenic radionuclides are produced, therefore, to the greatest extent in the earth's atmosphere. Table 1.2 provides a list of cosmogenic nuclides, their half-lives, rate of production in the atmosphere, and the global mass of these naturally occurring nuclides. Short-lived radionuclides are not included here. Once produced in the atmosphere, radionuclides, such as ^{3}H, ^{7}Be, ^{10}Be, ^{26}Al, and ^{36}Cl, will find their way to the earth's surface; whereas, radionuclides in gaseous form, such as $^{14}CO_2$, ^{37}Ar, ^{39}Ar, and ^{81}Kr find their way into oceans and rivers via air exchange.

B. Long-lived radionuclides

Based on the age of the earth (4.5×10^9 years), we can expect that all noncosmogenic radionuclides with a half-life $< 10^8$ years, that were formed during the formation of the earth and are not in equilibrium with a parent nuclide in a naturally occurring decay chain (see Section IV.C below), would have decayed to an undetectable level. Table 1.3 lists the naturally occurring long-lived radionuclides according to increasing atomic number (Z) together with their half-lives, relative isotope abundances, decay modes, and decay products. Two decay products are listed when there are two modes of decay. All of the decay products are stable with the exception of the decay products of ^{148}Sm and ^{152}Gd, which decay to the very long-lived ^{144}Nd and ^{148}Sm, respectively, and ^{232}Th, ^{235}U, and ^{238}U, which are long-lived parent radionuclides of the Thorium, Uranium and Actinium decay chains described in the following Section IV.C). Not listed in the table are 10 radionuclides that are exceptionally long-lived. These decay via a double-beta ($\beta\beta$) decay mode with half-lives ranging between 10^{18} and 10^{24} years. Double-beta decay is the rarest form of nuclear decay discussed further on in this chapter.

C. Natural radioactive decay chains

There are three naturally occurring decay chains, which begin with one of three long-lived parent nuclides. The three decay chains, namely, the ^{232}Th or Thorium Decay Chain (4n), the ^{238}U or Uranium Decay Chain (4n + 2), and the ^{235}U Decay Chain or also referred to as the Actinium Decay Chain (4n + 3) are illustrated in Figs. 1.8–1.10, respectively. The notations 4n, 4n + 2, and 4n + 3 refer to the mass number A for all nuclides in the respective decay chains where n is a positive integer between 50 and 60. The notations are derived from the fact that the mass number (A) for all radionuclides in a given decay chain are multiples of 4, because α-decay results in a mass reduction of 4, whereas the mass number between parent and daughter does not change in β-decay. Thus, the mass numbers in the respective series agree with the relations $4n = A$, $4n + 2 = A$, and $4n + 3 = A$ where n is an integer between 50 and 60. There are radionuclides in the decay chains with very short half-lives (of the order of minutes and seconds). These short-lived nuclides have survived the billions of years since the formation of the earth due to an equilibrium that develops between a long-lived parent and its shorter-lived daughter.

TABLE 1.2 Production rates of several isotopes in the earth's atmosphere, arranged in order of decreasing half-lives.[a]

		Production rate (atoms/cm^2 s)		
Isotope	Half-life	Troposphere	Total atmosphere	Global inventory
^{3}He	Stable	6.7×10^{-2}	0.2	3.2×10^{3} tons[b]
^{10}Be	1.5×10^{6} y	1.5×10^{-2}	4.5×10^{-2}	260 tons
^{26}Al	7.1×10^{5} y	3.8×10^{-5}	1.4×10^{-4}	1.1 tons
^{81}Kr[c]	2.3×10^{5} y	5.2×10^{-7}	1.2×10^{-6}	8.5 kg
^{36}Cl	3.0×10^{5} y	4×10^{-4}	1.1×10^{-3}	15 tons[d]
^{14}C	5730 y	1.1	2.5	75 tons
^{39}Ar[e]	268 y	4.5×10^{-3}	1.3×10^{-2}	52 kg
^{32}Si	~150 y	5.4×10^{-5}	1.6×10^{-4}	0.3 kg
^{3}H	12.3 y	8.4×10^{-2}	0.25	3.5 kg
^{22}Na	2.6 y	2.4×10^{-5}	8.6×10^{-5}	1.9 g
^{35}S	87 d	4.9×10^{-4}	1.4×10^{-3}	4.5 g
^{7}Be	53 d	2.7×10^{-2}	8.1×10^{-2}	3.2 g
^{37}Ar	35 d	2.8×10^{-4}	8.3×10^{-4}	1.1 g
^{33}P	25.3 d	2.2×10^{-4}	6.8×10^{-4}	0.6 g
^{32}P	14.3 d	2.7×10^{-4}	8.1×10^{-4}	0.4 g

[a]*Based on Lal and Peters (1967).*
[b]*The inventory of this stable nuclide is based on its atmospheric inventory, which includes an appreciable contribution from crustal degassing of ^{3}He.*
[c]*Based on atmospheric ^{81}Kr/Kr ratio of $(5.2 \pm 0.4) \times 10^{-13}$.*
[d]*Includes a rough estimate of ^{36}Cl produced by the capture of neutrons at the earth's surface.*
[e]*Based on atmospheric ^{39}Ar/Ar ratio of (0.107 ± 0.004) dpm/L Ar (STP).*
From Lal (2009), reprinted with permission from Elsevier Publishers.

Equilibrium between parent and daughter nuclides is discussed in detail further on in this chapter. All radionuclides in the Thorium, Uranium, and Actinium decay chains terminate with a stable isotope of lead, namely, ^{208}Pb, ^{206}Pb, and ^{207}Pb, respectively.

V. Artificially produced radionuclides

There are over 1300 artificially produced radionuclides. These are made by nuclear reactions often facilitated by a nuclear reactor or particle accelerator. A brief historical account of the development of man-made radionuclides including the synthetic elements, all of which are radioactive, will be provided here.

Frédéric Joliet and Irène Joliet-Curie (daughter of Pierre and Marie Curie) in 1934 were the first to produce artificially a radioactive nuclide, ^{30}P. The nuclide emits positrons, which are electrons of positive charge and the antiparticle of the negative electron. Positrons were discovered in cosmic radiation only 2 years earlier by Carl D. Anderson (1932, 1933a).

The discovery of the Joliot-Curies of artificially produced radioisotopes came about when they began their search for positrons. They decided to irradiate various metals with alpha particles of different energies to see if the result would be the production of a predominance of neutrons or positrons. They irradiated aluminum foil with alpha particles from a polonium source. The energy of the alpha particles colliding with the aluminum could be controlled by placing the polonium at various distances from the aluminum. The irradiation of aluminum foil by the polonium alpha particles produced neutrons and positrons. When the polonium source was moved away from the aluminum to a distance beyond the reach of the alpha particles (The alpha particles from polonium-216 have an energy of 6.78 MeV and these cannot travel beyond 5.3 cm in dry air as calculated according to Eq. (1.166).), the neutron emission from the alpha-irradiated aluminum would cease; however, to their amazement, the Joliot-Curies discovered that the positron emission would continue. They measured the positron emission with a Geiger counter and found that the number of positrons emitted would diminish with time with an average lifetime of less than 5 minutes. The interaction of the alpha particles with the aluminum foil produced a new radioactive isotope of short half-life, yet unknown to mankind.

TABLE 1.3 Long-lived naturally occurring radionuclides.[a]

Nuclide	Half-life (years)	Isotope abundance (%)	Decay mode(s)	Decay product(s)
$^{40}_{19}K$	1.26×10^{9}	0.0117	β^-, EC	$^{40}_{20}Ca$ (β^-), $^{40}_{18}Ar$ (EC)
$^{50}_{23}V$	1.4×10^{17}	0.250	β^-, EC	$^{50}_{24}Cr$ (β^-), $^{50}_{22}Ti$ (EC)
$^{87}_{37}Rb$	4.88×10^{10}	27.835	β^-	$^{87}_{38}Sr$
$^{113}_{48}Cd$	9×10^{15}	12.22	β^-	$^{113}_{49}In$
$^{115}_{49}In$	4.4×10^{14}	95.71	β^-	$^{115}_{50}Sn$
$^{123}_{52}Te$	1.3×10^{13}	0.908	EC	$^{123}_{51}Sb$
$^{138}_{57}La$	1.06×10^{11}	0.0902	β^-, EC	$^{138}_{58}Ce$ (β^-), $^{138}_{56}Ba$ (EC)
$^{142}_{58}Ce$	5×10^{16b}	11.08	α	$^{138}_{56}Ba$
$^{144}_{60}Nd$	2.1×10^{15}	23.80	α	$^{140}_{58}Ce$
$^{147}_{62}Sm$	1.06×10^{11}	15.0	α	$^{143}_{60}Nd$
$^{148}_{62}Sm$	7×10^{15}	11.3	α	$^{144}_{60}Nd$
$^{152}_{64}Gd$	1.1×10^{14}	0.20	α	$^{148}_{62}Sm$
$^{176}_{71}Lu$	3.8×10^{10}	2.59	β^-	$^{176}_{72}Hf$
$^{174}_{72}Hf$	2.0×10^{15}	0.162	α	$^{170}_{70}Yb$
$^{180}_{73}Ta$	$>1.2 \times 10^{15}$	0.012	β^-, EC	$^{180}_{74}W$ (β^-), $^{180}_{72}Hf$ (EC)
$^{187}_{75}Re$	4.4×10^{10}	62.60	β^-	$^{187}_{76}Os$
$^{186}_{76}Os$	2.0×10^{15}	1.58	α	$^{182}_{74}W$
$^{190}_{78}Pt$	6.5×10^{11}	0.01	α	$^{186}_{76}Os$
$^{232}_{90}Th$	1.4×10^{10}	100.0	α	$^{228}_{88}Ra$
$^{235}_{92}U$	7.04×10^{8}	0.720	α	$^{231}_{90}Th$
$^{238}_{92}U$	4.46×10^{9}	99.27	α	$^{234}_{90}Th$

[a] *Data from Lide (2010).*
[b] *Half-life from Macfarlane and Kohman (1961).*

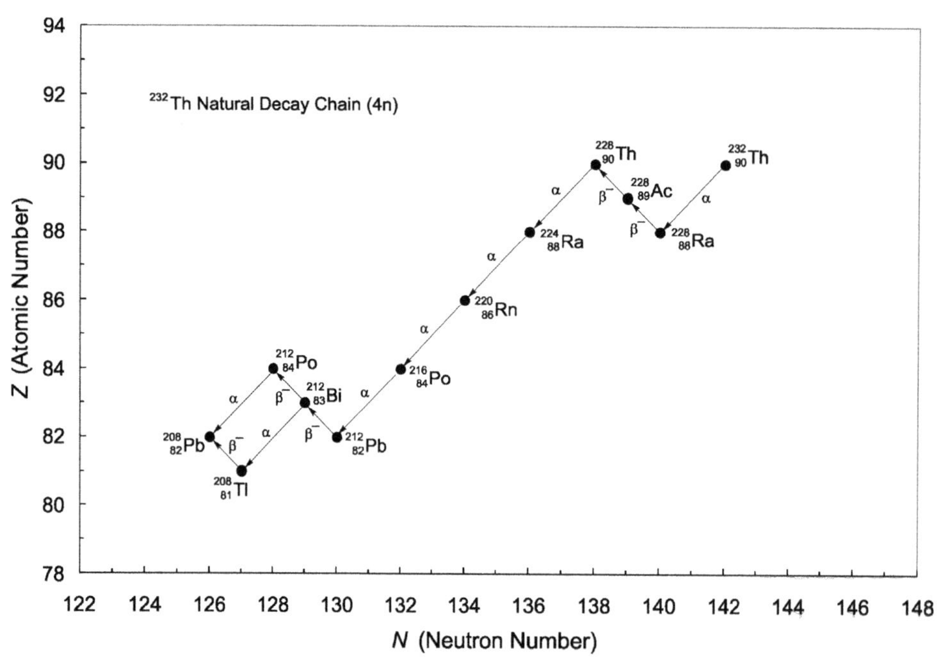

FIGURE 1.8 The ^{232}Th natural decay chain.

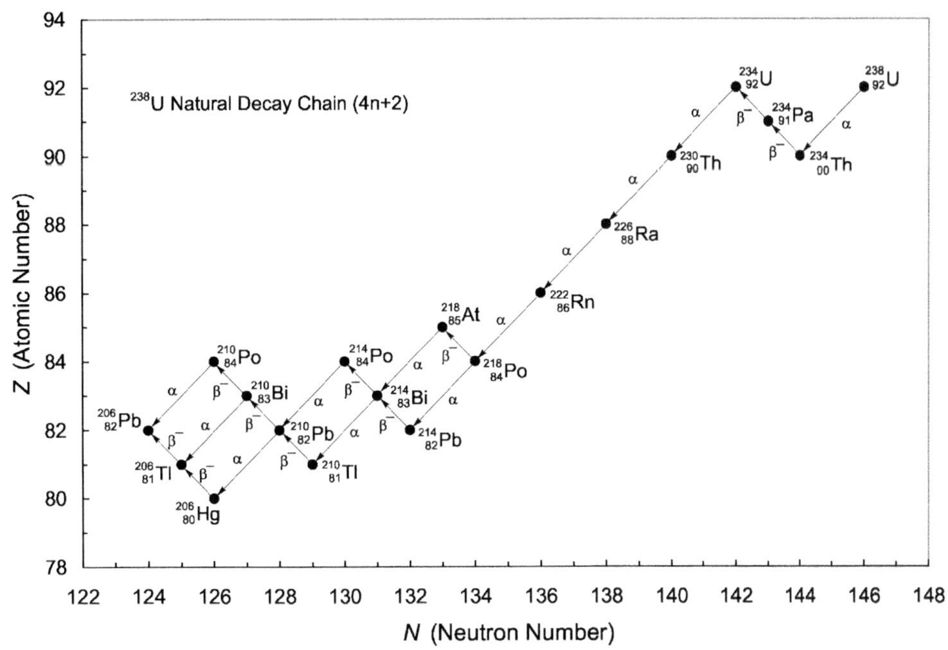

FIGURE 1.9 The ^{238}U natural decay chain.

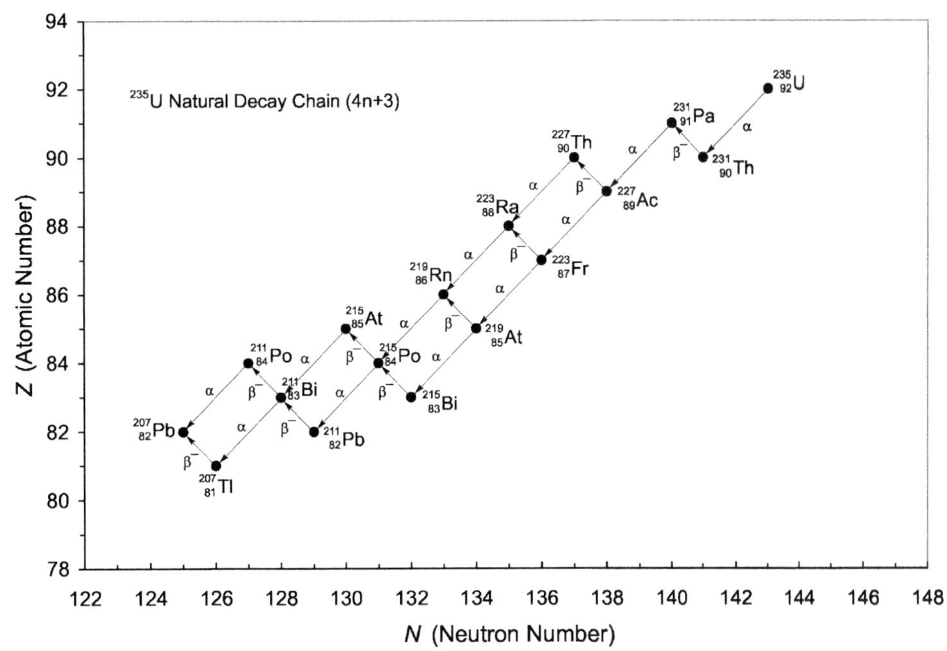

FIGURE 1.10 The ^{235}U natural decay chain also referred to as the Actinium decay chain.

The Joliot-Curies were able to demonstrate that the new isotope was chemically equivalent to elemental phosphorus, and it was radioactive phosphorus. By balancing the transformation of the nuclear reaction between alpha particles and aluminum yielding neutrons they could write the reaction

$$^{27}_{13}\text{Al} + ^{4}_{2}\text{He} = ^{30}_{15}\text{P} + ^{1}_{0}\text{n} \tag{1.20}$$

They used the equal sign rather than an arrow for the above reaction to demonstrate that the sum of the mass numbers and atomic numbers on one side of the nuclear reaction should equal those on the other side, *i.e.*, $27 + 4 = 30 + 1$ for the mass numbers and $13 + 2 = 15 + 0$ for the atomic numbers. Expressing nuclear reactions was not a practice at that time. The Joliot-Curies were one of the first to introduce the practice.

The Joliot-Curies demonstrated that the new radioactive isotope of phosphorus emitted positrons. The new radioisotope, phosphorus-30, decays to stable silicon-30 with positron emission and a short half-life of 2.5 minutes according to the following:

$$^{30}_{15}\text{P} \rightarrow {}^{30}_{14}\text{Si} + \beta^{+} + \nu \qquad (1.21)$$

Thus, the Joliot-Curies not only produced a previously nonexistent radioisotope, but they were the first to produce positrons in the laboratory. The neutrino included in the above decay equation accompanies positron emission; however, this particle was yet to be discovered at the time of the Joliet-Curies.

By irradiating other metals, such as boron and magnesium, with alpha particles the Joliot-Curies were able to produce radioactive nitrogen and radioactive silicon according to the following reactions (Joliot-Curie and Joliot, 1934a,b,c):

$$^{10}_{5}\text{B} + {}^{4}_{2}\text{He} \rightarrow {}^{13}_{7}\text{N} + {}^{1}_{0}\text{n} \qquad (1.22)$$

$$^{24}_{12}\text{Mg} + {}^{4}_{2}\text{He} \rightarrow {}^{27}_{14}\text{Si} + {}^{1}_{0}\text{n} \qquad (1.23)$$

The radioisotopes of nitrogen and silicon above decay by positron emission.

The Joliot-Curies were awarded the Nobel Prize in Chemistry the very year following the report of these discoveries. The rapid award of the Nobel Prize underscored the impact this discovery had on the scientific community and the advancement of science. Their work established firmly a new branch of chemistry called "radiochemistry". As explained by Frédéric Joliot in his part of the Nobel lecture

> *The succession of radioactive transformations provides numerous examples in which the quantities of radio-elements are extremely small and not capable of being weighed, yet nevertheless, by the methods of radiochemistry it has been possible to examine correctly their chemical properties, and identify some of them as being isotopes of elements ... This special kind of chemistry in which one handles unweighable quantities, sometimes of the order of 10^{-16} g, is made possible thanks to the fact that one can determine and follow by measuring the radiation emitted, infinitesimal traces of radioactive matter dispersed in the midst of other matter.*

While this work established firmly "radiochemistry" as a new branch of chemistry, the Joliot-Curies were the first to coin the terminology, which we are very accustomed to, when referring to radioactive elements. In their Nobel Lecture Frédéric Joliot stated

> *We have proposed that these new radio-elements (isotopes, not found in nature, of known elements) be called radio-nitrogen (for radioactive isotopes of nitrogen), radio-phosphorus (for radioactive isotopes of phosphorus), ...*

Only 2 years after the Joliot Curies reported the synthesis of radionuclides by the alpha particle bombardment of nuclei, Enrico Fermi and his coworkers reported the synthesis of new radionuclides by neutron bombardment of nuclei. This work, which was cited earlier in this chapter, yielded 63 previously unknown artificial radionuclides by the following three nuclear reactions:

$$(1)\quad ^{A}_{Z}\text{X} + {}^{1}_{0}\text{n} \rightarrow {}^{A-3}_{Z-2}\text{X} + {}^{4}_{2}\text{He} \qquad (1.24)$$

$$(2)\quad ^{A}_{Z}\text{X} + {}^{1}_{0}\text{n} \rightarrow {}_{Z-1}{}^{A}\text{X} + {}^{1}_{1}\text{H} \qquad (1.25)$$

$$(3)\quad ^{A}_{Z}\text{X} + {}^{1}_{0}\text{n} \rightarrow {}^{A+1}{}_{Z}\text{X} \qquad (1.26)$$

where $^{A}_{Z}\text{X}$ is a symbol for any element X with atomic number Z and mass number A, and n is the symbol for the neutron. Fermi found reactions (1) and (2) to occur chiefly among the light elements, and the reaction of type (3) was found to occur most often among the heavy elements. More than one of the above reactions would sometimes occur when elements were bombarded with neutrons. In Fermi's words at his Nobel Lecture

> *In many cases the three processes are found at the same time in a single element. For instance, neutron bombardment of aluminum that has a single isotope ^{27}Al, gives rise to three radioactive products: ^{24}Na, with a half-life of 15 hours by process (1); ^{27}Mg, with a period (half-life) of 10 minutes by process (2), and ^{28}Al with a period (half-life) of 2–3 minutes by process (3).*

In addition to the above three neutron reactions described by Fermi, he and his coworkers were also able to observe the production of artificial radioactive isotopes by a fourth process that he described:

> *The primary neutron does not remain bound in the nucleus, but knocks off instead, one of the nuclear neutrons out of the nucleus; the result is a new nucleus, that is isotopic with the original one and has an atomic weight less by one unit.*

This fourth neutron interaction, where the mass number is reduced by one (loss of a neutron) and the atomic number remains unaltered, can be written as follows:

$$(4)\quad ^{A}_{Z}\text{X} + {}^{1}_{0}\text{n} \rightarrow {}^{A-1}{}_{Z}\text{X} + 2{}^{1}_{0}\text{n} \qquad (1.27)$$

Fermi knew that neutrons, because of their neutral charge, did not have to overcome the Coulombic barrier to penetrate the nucleus of an atom. Furthermore, he discovered, that reducing the neutron energy or slowing the neutron down by placing a water or paraffin barrier between a neutron source and its target nuclei, the energy of the neutrons could be reduced to that corresponding to thermal agitation. At this energy state, the slow neutrons could diffuse through target atoms and be captured by the target nuclei. Thus, using slow neutron bombardment, Fermi and his coworkers were able to produce more than 400 new radioisotopes of all known elements with the exception of hydrogen and helium.

At the same time that Enrico Fermi was producing radionuclides by means of neutron bombardment of target nuclei, Ernest Lawrence was developing the cyclotron capable of accelerating protons, deuterons, and helium nuclei to very high energies. This work, discussed previously in this chapter, yielded many new artificial radionuclides, as the high energy of the accelerated ions could easily penetrate the Coulombic barrier of target nuclei.

The discovery of nuclear fission by Otto Hahn and Fritz Strassman with the interpretation of Lise Meitner in 1939 led to the discovery of the first nuclear reactor in 1942 by Enrico Fermi and coworkers. This work was discussed previously in more detail in Section II of this chapter. Today, the nuclear reactor and the cyclotron are the main instruments for the artificial production of radionuclides. There are now more than 1300 artificially produced radionuclides. Many of these radionuclides are vital for medical diagnosis and treatment, the sterilization of medical implements and supplies, and peaceful applications in industry and scientific research.

Elements with atomic number $Z > 92$, which are elements with atomic number higher than that of uranium, are artificially produced. They are all radioactive. By means of the cyclotron Glenn T. Seaborg pioneered the synthesis of many artificial elements. This work is described in detail in Section II of this chapter. There are no natural elements beyond $Z = 92$, because they all have half-lives too short to have survived since the formation of the earth. Also, there is the $4n + 1$ naturally occurring decay chain. The $4n + 1$ decay chain, illustrated in Fig. 1.11, consists of artificially produced radionuclides. It is called the Neptunium Decay Chain after ^{237}Np, which is the longest-lived nuclide of the chain with a half-life of 2.1×10^6 years. There are currently over 25 artificially produced elements. All of the elements are listed according to atomic number in Appendix B of this book.

VI. Properties of the nucleus

A. Nuclear radius and density

Rutherford (1911, 1919, 1920a,b) was the first to discover the atomic nucleus and to measure the size of the nucleus of an atom. A detailed accounting of this work is given in previous books (L'Annunziata, 2007, 2016). In brief, Rutherford bombarded very thin gold foil (4×10^{-5} cm thick) with alpha particles. Most of the alpha particles traversed the gold foil almost as if the foil was not in the path of the alpha particles; however, 1 alpha particle for every 20,000 particles would bounce back from the foil by more than 90 degrees from the direction of travel. From this work, Rutherford concluded that within much empty space of the atom there exists a massive central nucleus upon which the alpha particle would collide and consequently cause the alpha particle to ricochet back toward the direction from which it came. Rutherford expressed his reaction to the observed alpha particle back scattering with the following statement, related by Feather (1940) in a biographical essay a few years following Rutherford's death:

> *On consideration, I realized that the scattering backwards must be the result of a single collision, and when I made calculations I saw that it was impossible to get anything of that order of magnitude unless you took a system in which the greater part of the mass of the atom was concentrated in a minute nucleus.*

Rutherford went even further to make use of this interaction to determine the nuclear radius of aluminum. By selecting a metal foil of low Z (aluminum, $Z = 13$) and thus low Coulomb barrier to alpha penetration, and

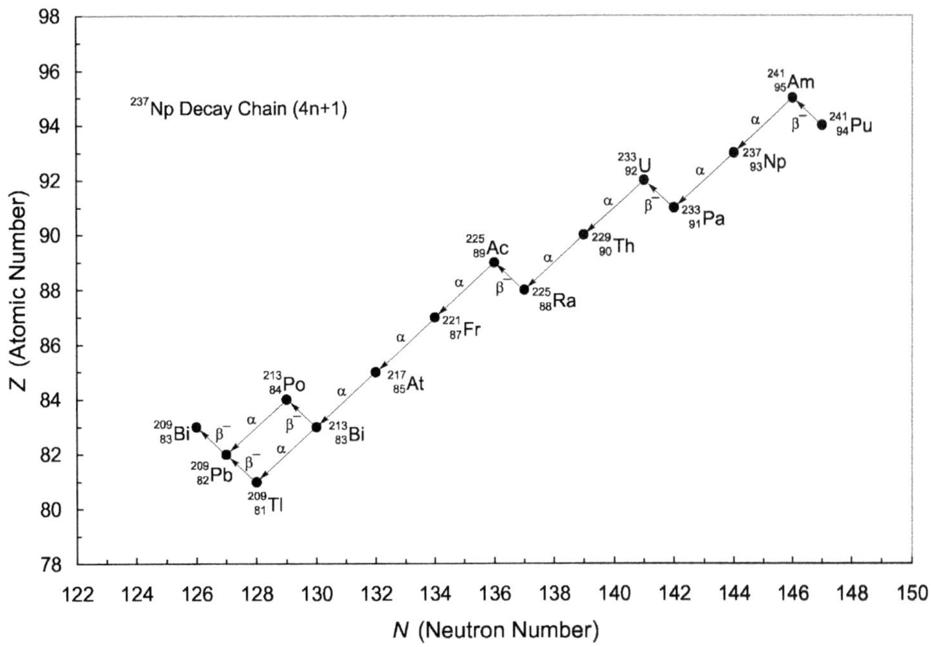

FIGURE 1.11 The Neptunium decay chain.

applying alpha particles of high energy (7.7 MeV) whereby defined alpha particle scattering at acute angle due to Coulombic repulsion would begin to fail, Rutherford (1919, 1920a,b) was able to demonstrate that the distance of closest approach of these alpha particles to the atom center according to Coulomb's law was equivalent to the nuclear radius of aluminum, ~ 5×10^{-15} m or five fm (*i.e.*, 5 fm or five fermi).

Since Rutherford's pioneering alpha-particle scattering experiment, numerous nuclear scattering experiments have been carried out to measure the size of atomic nuclei including electron scattering (Jansen et al., 1971; Sick, 1982), electron—proton scattering (Klarsfeld et al., 1986), electron—deuteron scattering (Sick and Trautmann, 1998), and high-energy beam interactions (Tanihata et al., 1985; Suzuki et al., 1999). The radius of an atomic nucleus obeys the general empirical formula

$$R = r_0 A^{1/3} \tag{1.28}$$

where r_0 is a constant, referred to as the radius parameter, and A is the mass number of the nucleus (Tolhoek and Brussaard, 1954; Elton, 1958; Angeli and Csatlós, 1977). A detailed derivation of Eq. (1.28) is provided by Elton (1958). The value of the radius parameter r_0, was calculated by Royer (2008) to be within the range of 1.22—1.23 fm. Obviously the radius of the nucleus will increase with mass number. For a small nucleus, such as the nucleus of ^4_2He (equivalent to an alpha particle), the radius is calculated to be 1.9×10^{-15} m or 1.9 fm. For larger nuclei, such as $^{27}_{13}Al$, $^{197}_{79}Au$ and $^{235}_{92}U$, the nuclear radii calculate to be 3.6, 6.9 and 7.4 fm, respectively.

Treating the nucleus as a sphere permits the calculation of the nuclear volume, and from the mass of the nucleus, we can calculate the nuclear density ρ_N as

$$\rho_N = \text{mass/volume} = m_n A / \frac{4}{3}\pi R^3 \tag{1.29}$$

where m_n is the mass of a nucleon (proton or neutron), A is the mass number, and R is the nuclear radius. The mass of either nucleon (the proton or neutron) may be used, as the two nucleons are very similar in mass and may be considered equal in mass, when only three significant figures are used, namely, 1.67×10^{-27} kg. Substituting $r_0 A^{1/3}$ for R in Eq. (1.29) gives

$$\rho_N = m_n / \frac{4}{3}\pi r_0^3 \tag{1.30}$$

The value of the nuclear density calculated according to Eq. 1.29 or 1.30, gives the value of $\rho_N = 2.3 \times 10^{17}$ kg/m^3. Treating the nucleus as a sphere facilitates the density calculation and provides a good estimate as provided in current textbooks; however, most nuclei are not spherical and some may exhibit oscillations such as depicted by the liquid-drop model of the nuclei. Some nuclei described by Gould et al. (2004) exhibit football-shaped quadrupole deformations and even pear-shape octupole deformations (Krappe and Wahsweiler, 1967; Vogel, 1967, 1968; Aponick et al., 1970; Cline, 1993). Gould et al. (2004) describe the interior of a large nucleus as uniform (0.17 nucleons/fm^3); but that the density falls off smoothly at the surface of the nucleus, dropping from 90% to 10% of the interior density over a distance of ~2.5 fm.

The density of the nucleus is extraordinarily high when compared with elements and materials consisting of atoms. For example, the atomic nucleus is 2.0×10^{13} times the density of lead (11.3×10^3 kg/m^3) or 1.2×10^{13} times the density of depleted uranium (19.1×10^3 kg/m^3). While the density of the nucleus is amazingly high, the nucleus consists of closely packed nucleons and thus, the nucleons themselves and even subnuclear particles, particularly leptons, would even have higher densities.

Subnuclear particles are classified into two groups, the leptons and hadrons. The word leptons comes from the Greek "lepto" meaning "refined" or "thin". **Leptons** are particles with no apparent substructure and thus considered to be truly elementary or point particles, such as, the electron, muon, and their respective neutrinos and quarks. Leptons, with the exception of quarks, participate in electromagnetic and weak interactions. **Hadrons**, on the other hand, are more complex in subnuclear structure and they participate in the strong nuclear force. Hadrons are classified into two groups, namely, the **mesons** containing two quarks (*e.g.*, pion and kaon) and the **baryons** (from the Greek "barys" meaning "heavy"), which are of higher mass and contain three quarks (*e.g.*, proton and neutron). The role of mesons as particles, that mediate the strong nuclear force between the nucleons, is discussed in the next section of this chapter.

The high density of the nucleus underscores the importance of a strong short-range nuclear force that exists between the nucleons (i.e., protons and neutrons) keeping them closely packed regardless of the repulsive forces of the positively charged protons or the *Z*/*N* ratios of the nucleus among the thousands of nuclides that exist.

B. Nuclear forces, quarks, gluons, and mesons

In the 1930s question arose concerning the force that was capable of holding the protons and neutrons together in the nucleus. It was unusual to find that the like positive charge of the protons did not force them to repel each other; but rather, they were held tightly together into a very dense nucleus in association with the neutrally charged neutrons. It was expected that the force holding the nucleons together would be a strong force. Nobel Laureate Hideki Yukawa (1935, 1949) first postulated the strong force between nucleons. Before doing this, Yukawa understood that charged particles would interact through the exchange of

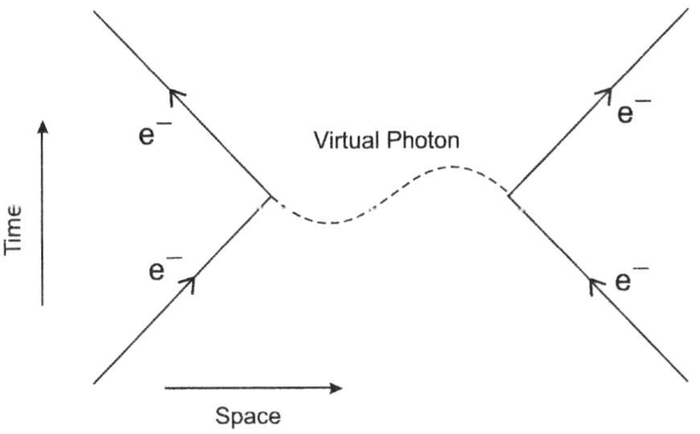

FIGURE 1.12 Feynman diagram illustrating the photon as a particle that mediates the electromagnetic force between two electrons. The electromagnetic interaction results in the absorption of the photon particle by an electron providing it with momentum to be repelled. The illustration can be read as follows: An electron (lower left) emits a photon and undergoes recoil (upper left). The photon, illustrated as a dashed wave, is absorbed by an electron (lower right) and it is repelled (upper right). The photon is referred to as a virtual photon, because it is emitted and absorbed without being detected. *From L'Annunziata (2016); reprinted with permission from Elsevier © 2016.*

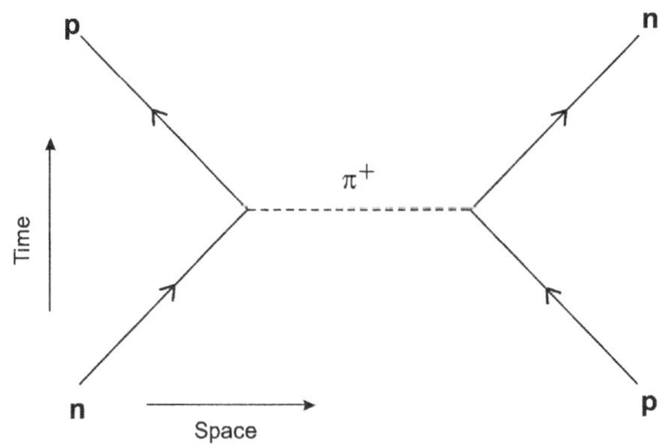

FIGURE 1.13 A feynman diagram illustrating the mediation of the strong nuclear force between nucleons in the nucleus. A neutron (lower left) is transformed into a proton (upper left) and a proton (lower right) is transformed into a neutron (upper right) via the intermediary pi meson. *From L'Annunziata (2016); reprinted with permission from Elsevier © 2016.*

photon particles. He noted that such a force, known as the electromagnetic force, would become manifest when the particles collide. For example, electrons would interact via the exchange of virtual photons as illustrated by the Feynman diagram[2] of Fig. 1.12. Yukawa explained the strong force between nucleons by proposing a new particle, which would mediate the strong force between the nucleons. He predicted that the new particle would have a mass inversely proportional to the force between the nucleons, and that the mass of the particle would be intermediate of the proton and electron masses ($\sim 200 \times$ the electron mass). The new particle was named the meson, which was taken from the Greek "meso" meaning "middle". The prediction of Yukawa was verified by the discovery of the π meson or pion by Occhialini and Powell (1947, 1948). The strong force keeping the nucleons together is mediated by the exchange of π mesons between the nucleons. The pion is the carrier of Yukawa's nuclear force. It is found in three charge states, π^+, π^-, and π^0. They have the mass predicted by Yukawa and are unstable, e.g., the π^+ has a mass of 139.5 MeV/c^2 and a mean lifetime of 2.6×10^{-8} seconds (Sundareson, 2001). The mediation of the strong force between a neutron and a proton in the nucleus is illustrated by the Feynman diagram in Fig. 1.13, which illustrates a neutron converting into a proton and a proton converting into a neutron and, in the process, a π^+ is exchanged between the two nucleons.

Pion exchange between nucleons, as the particle mediating the strong force, is explained on the basis of the subnuclear quark structure of the proton, neutron, and meson. George Zweig (1964) and Murray Gel-Mann (1964) independently proposed the idea of a substructure to mesons and baryons. This substructure is now universally referred to as quarks, which would possess quantum numbers and fractional charge, which are subsequently described. The mesons (*e.g.*, pion and kaon), which are of intermediate mass between that of the electron and proton, consist of two quarks; whereas, the baryons (*e.g.*, proton, neutron, lambda, etc.), which are of higher mass, contain three quarks. The quark compositions of pi mesons (pions) and of the nucleons (proton and neutron), and the properties of the quarks are provided in Table 1.4. In the table, it is seen that the π^+ is composed of an up quark (u) with charge of $+2/3$ e and an antidown ($\bar{d}$) quark with charge of $+1/3$ e yielding a total charge on the π^+ of $+1$. The π^- consists of a down quark (d) of charge $-1/3$ e and antiup quark ($\bar{u}$) of charge $-2/3$ e yielding an overall particle charge of -1. The π^0 is a mixture or superposition of $u\bar{u}, d\bar{d}$ quark pairs referred to as mixing of quark pairs with overall neutral charge.

The proton, neutron, and all of the baryons, which are particles of higher mass than the mesons, contain three quarks. The quark content of the proton, as listed in Table 1.4, is two up quarks and one down quark (uud) with respective charges of $+\frac{2}{3}$e, $+\frac{2}{3}$e and $-\frac{1}{3}$e or a total charge of $+1$; whereas, the neutron contains two down quarks and one up quark (ddu) with a total charge of zero.

2. A convenient way to illustrate the interactions and exchange forces between the elementary particles is with Feynman diagrams, which were used frequently by Nobel laureate Richard Feynman (1918–88). When he was a young scientist, he utilized the diagrams as a means of depicting, through sketches, the interactions among nuclear particles as shorthand for the equations that are used to put the interactions into more complex mathematical terms. Feynman diagrams show particle interactions in coordinates of space and time. The axis is not quantitative, that is, the length of the lines or spacing between particles is not a measure of relative time or distance. Particles are illustrated by solid lines with arrows giving the particle direction. Virtual particles are illustrated as dashed waves for particles with zero rest mass (*i.e.*, photons or gluons) and dashed straight lines for particles with a rest mass.

TABLE 1.4 Quark compositions and properties of quark components of pi mesons and nucleons.

Particle	Quark composition	Properties of quarks			
		Name[a]	Symbol	Charge	Mass
Mesons					
π^+	$u\bar{d}$	up	u	$+\frac{2}{3}e$	2.3 MeV/c^2
π^-	$d\bar{u}$	down	d	$-\frac{1}{3}e$	4.8 MeV/c^2
π^0	$u\bar{u}, d\bar{d}$	antiup	$\bar{u}$	$-\frac{2}{3}e$	$m_{\bar{u}} = m_u$
Nucleons/Baryons		antidown	$\bar{d}$	$+\frac{1}{3}e$	$m_{\bar{d}} = m_d$
p	uud				
n	udd				

[a]*Other quarks exist, such as the strange (s), charmed (c), bottom (b), and top (t) quarks, which are component quarks of other mesons and baryons.*

Quarks possess another property called color or color charge. The property of color or color charge has no relation to visible radiation as we see color; but rather, color charge is a quantum number and source of force that quarks and antiquarks possess. Color charge was proposed as a quantum number by Greenberg (1964) at the University of Maryland, and as a source of force between quarks by Mo-Young Han at Duke University and Nambu (1966) at the University of Chicago (Han and Nambu, 1965). To satisfy the Pauli Exclusion Principle, that states that no two fermions can exist simultaneously in the same energy or quantum state, the concept of color charge was introduced as a quantum state and force between quarks. Three color-charge quantum states carried by quarks are red, blue, and green, and those carried by antiquarks are antired, antiblue, and antigreen. In the baryons, all of which possess three quarks such as the proton (uud) and the neutron (ddu), there would exist three quarks of color charge, red, blue, and green. Quarks of different color charge possess an attractive force between each other, whereas quarks of like color would repel each other. Mesons, which possess only two quarks, a quark and antiquark (see Table 1.4), would display red and antired, blue and antiblue, or green and antigreen quantum states. There is an attractive force between quarks of a given color and its anticolor quantum state in mesons.

There are yet other particles in addition to quarks, called gluons, which exist in baryons and mesons. A gluon is a field particle like the electromagnetic photon, which has zero rest mass. Gluons differ from photons because they interact directly with quarks and with each other, and they carry, similar to quarks, a color charge quantum state. Quarks emit and absorb gluons. When a quark either emits or absorbs a gluon, the quantum state of the quark changes and thus, the color charge of the quark also changes. The process of emission and absorption of gluons by quarks, which is discussed subsequently, is part of the mechanism by which the strong nuclear forces keep the protons and neutrons closely packed in the atomic nucleus. The force that exists between quarks of different quantum color states is called color force, and the theory that these forces play in the strong interactions is called quantum chromodynamics (QCD).

In summary, mesons mediate the strong forces between the protons and neutrons in the nucleus; and it is the energy exchange or quantum state changes between quarks and gluons of the mesons and nucleons, that enable the strong nuclear forces to hold the protons and neutrons together in the nucleus. An example of a process, whereby the strong nuclear forces between the nucleons come into play, is illustrated by the Feynman diagram of Fig. 1.14. The figure illustrates the involvement of quarks and gluons in the transformation of a proton into a neutron and a neutron into a proton within an atomic nucleus via the mediation of a pi meson (π^+).

The mass of the nucleons as well as that of the mesons arise from the interaction of their quarks with the gluon field particles. Because gluons have a zero rest mass, the creation and absorption of gluons is a constant process in the nucleons attributing mass to the nucleons. As described by Gould et al. (2004), the mass of the nucleons comes from the interaction of the quarks with the nucleon field, and the nucleon is described as consisting of a core of three quarks confined within a sphere of radius ~ 0.5−1.0 fm. Surrounding the core is a cloud of pions (pi mesons), which are themselves composed of quark−antiquark pairs.

The range of the strong nuclear force is very close to that of the radius of the nucleon core. We can use Heisenberg's Uncertainty Principle to calculate the range of interaction of the strong nuclear force mediated by the pi meson or pion. Nobel Laureate Werner Heisenberg published the Uncertainty Principle (Heisenberg, 1927), which is a basic concept of quantum physics. In terms of energy and time variables, Heisenberg's Uncertainty principle can be expressed as

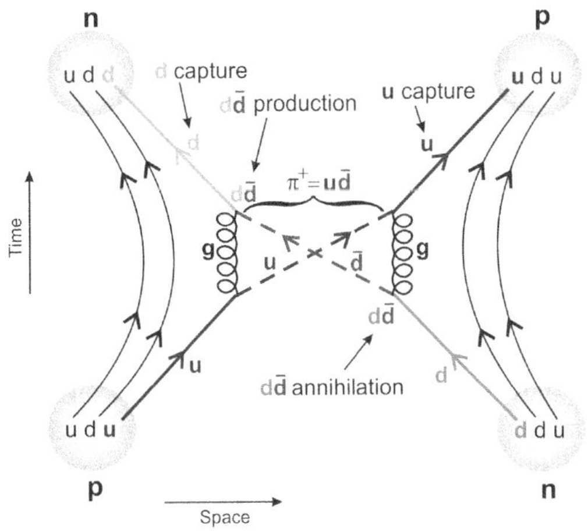

FIGURE 1.14 A Feynman diagram illustrating the mediation of a π^+ meson of the strong nuclear force between a proton and a neutron in the atomic nucleus with the interaction of quarks and gluons as follows: An up quark (u, colored red) in the proton on the lower left of the figure emits a gluon (g) illustrated by a helice in the left center of the figure resulting in pair production of a $d\bar{d}$ pair. The down quark (d, colored green) is captured (green line) changing the proton on the lower left to a neutron (upper left). The antidown quark ($\bar{d}$, colored purple) created in the pair production is annihilated by a down quark (d, colored brown) of the neutron at the lower right of the figure. The purple arrow of the antidown quark points in the opposite direction, because it is an antiparticle, and according to Feynman rules antiparticles are identified by arrows in the opposite direction. The $d\bar{d}$ annihilation results in the destruction of the down quark of the neutron on the lower right with the emission of a gluon (g, helice) in the right center of the figure, and the up quark (u, red colored) originating from the proton at the lower left is captured by the neutron on the lower right converting it into a proton on the upper right hand of the figure. For the color version of the figure, the reader is referred to the electronic version of the book. *From L'Annunziata (2016); reprinted with permission from Elsevier © 2016.*

$$\Delta E \Delta t \geq \frac{h}{4\pi} \geq \frac{\hbar}{2} \tag{1.31}$$

which states that the precision with which we measure the energy of an atomic particle (ΔE) is limited by the time available (Δt) to measure the particle energy, where h is Planck's constant ($h = 6.626 \times 10^{-34}$ J s) and $\hbar$ is a reduced form of Planck's constant ($\hbar = h/2\pi = 1.055 \times 10^{-34}$ J s $= 6.582 \times 10^{-22}$ MeV s). We can accurately calculate the energy of the pion from its mass, and from the Heisenberg Uncertainty Principle expressed by Eq. (1.31), we can state that the pion would violate the conservation of energy if it lasted for a period of time exceeding

$$\Delta t \approx \frac{\hbar}{2\Delta E} \tag{1.32}$$

where ΔE is the energy needed to create a pion of mass m_π defined by Einstein's equation of mass and energy as

$$\Delta E = m_\pi c^2 \tag{1.33}$$

And Δt is the time it would take for the pi meson to travel from one nucleon to the other (*e.g.*, from a proton to a neutron or vice versa). The mass m_π of the π^+ or $\pi^- = 139.5$ MeV/c^2. Since the pi meson cannot travel faster than the speed of light c, its distance (d) of travel is defined as

$$d = c\Delta t \tag{1.34}$$

Thus, the maximum distance, that the pi meson could travel, is calculated as

$$\begin{aligned} d &= c\Delta t = c\left(\frac{\hbar}{2m_\pi c^2}\right) \\ &= 2.99 \times 10^8 \text{ m/s}\left(\frac{6.582 \times 10^{-22} \text{ MeV s}}{2(139.5 \text{ MeV}/c^2)c^2}\right) \\ &= 7.0 \times 10^{-16} \text{ m} \approx 1 \times 10^{-15} \text{ m} \approx 1 \text{ fm} \end{aligned} \tag{1.35}$$

The range of the strong nuclear force mediated by the pion is thus ~ 1 fm similar to that of the radius of the proton or neutron.

For a more in-depth reading on subnuclear structure and its role in the strong nuclear force, see books by Serway et al. (2005), Martin (2011), and Taylor et al. (2015).

C. Binding energy

The difference in the total energy of the nucleus with its bound nucleons is always less than the total energy of the separate protons and neutrons, and this difference in energy is the binding energy (E_b) of the nucleus. The binding energy is equivalent to the energy required to pull a bound system apart, such as a bound nucleus, leaving its component protons and neutrons free of attractive forces. The binding energy is described by the equation

$$E_b = \sum_{i=1}^{n} m_i c^2 - Mc^2 \tag{1.36}$$

where m_i are the free component particles (protons and neutrons) of the nucleus numbered $i = 1$ to n, and M is the mass of the bound nucleus. We may use atomic mass units (u) to calculate the nuclear binding energies according to the above Eq. (1.36) for any nuclide, ${}^A_Z\text{X}$, of Z number of protons, A number of protons + neutrons, and $N = A - Z$ number of neutrons as follows:

$$E_b = \left[(Zm_{{}_1\text{H}} + Nm_n - M_{{}^A_Z\text{X}})931.494\text{MeV}/c^2\right]c^2 \tag{1.37}$$

where $m_{{}_1\text{H}}$ is the atomic mass of hydrogen, m_n is the neutron mass, and $M_{{}^A_Z\text{X}}$is the atomic mass of the nuclide of atomic number Z and mass number A, *i.e.*, ${}^A_Z\text{X}$, and by definition 1 u $= 931.494$ MeV$/c^2$. Atomic mass units are used in Eq. (1.37) because the number of electrons of the Z hydrogen atoms are the same as the electrons included in the atomic mass of the nuclide ${}^A_Z\text{X}$. The differences in electron binding energies of the hydrogen atoms and that of the nuclide are negligible and ignored. Let us take an example and use atomic mass units to calculate the binding energy for the helium nucleus, i.e., ${}^{4}_{2}He$, which is equivalent to an alpha particle, as follows:

$$\begin{aligned} E_b &= \left[(2m_{{}_1H} + 2m_n - M_{{}^4_2He})931.494\ \text{MeV}/c^2\right]c^2 \\ &= \left[((2)(1.007825032\ \text{u}) + (2)(1.008664915\ \text{u}) \right. \\ &\quad \left. - 4.00260325\ \text{u})931.494\ \text{MeV}/c^2\right]c^2 \\ &= 28.2956 \end{aligned} \tag{1.38}$$

Because nuclei vary in their number of nucleons or mass number (A), it is common to express the nuclear binding energy in terms of the number of nucleons in the nucleus, that is, the binding energy per nucleon or E_b/A. In the case of the nuclide, ${}^{4}_{2}He$, the binding energy per nucleon is

$$E_b/A = 28.2956\ \text{MeV}/4 = 7.0739\ \text{MeV} \tag{1.39}$$

The values of the binding energies per nucleon for all of the nuclides provides a measure of how tightly bound the nucleons are within a particular nucleus. Fig. 1.15 illustrates graphically the difference in the values of the binding energy per nucleon (E_b/A) as a function of the mass number (A) of the nucleus for nuclides over the entire range of mass numbers.

From Fig. 1.15 it is seen that the highest nuclear stability is found in the region of mass number $A = 50$ to 70, and the most stable nuclei, that is, those nuclei that are most tightly bound belong to the "iron group" including ${}^{62}Ni$, ${}^{58}Fe$, and ${}^{56}Fe$ with binding energies per nucleon at a maximum of approximately 8.8 MeV. On either side of this maximum, illustrated in Fig. 1.15, are (1) the smaller nuclei in the range of $A = 2$ to 20 with lower binding energies per nucleon, which could fuse to yield larger and more stable nuclei with the release of energy or (2) the very large nuclei in the range of about $A \geq 120$, which could break apart into approximately equal mass number by fission to form smaller and more stable nuclei with the release of energy. Consequently the size of the nucleus and its binding energy per nucleon will govern whether nuclear fusion or fission would be feasible, and the energy released in both cases can be estimated directly from Fig. 1.15 or by calculation. Let us take examples of both cases, fission and fusion.

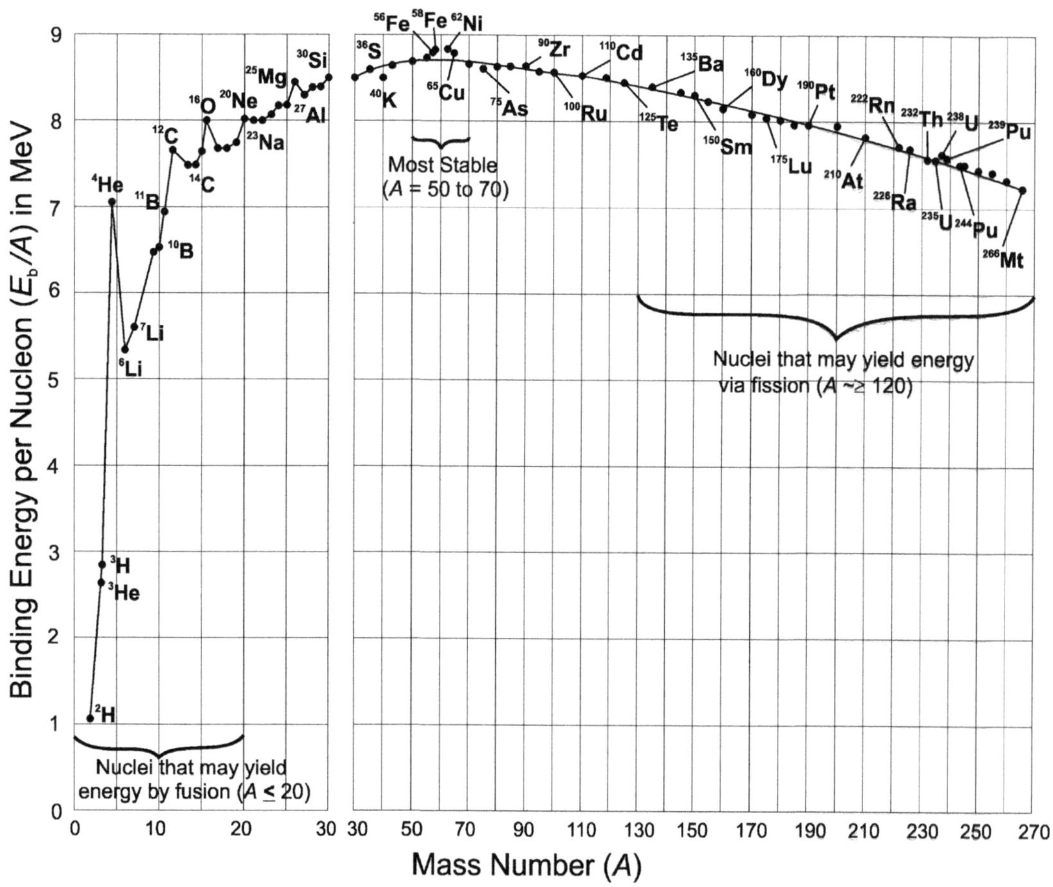

FIGURE 1.15 Binding energy per nucleon as a function of mass number. The scale between mass number 0 and 30 is broader than that illustrated for mass numbers 30–270. Several nuclides are identified along the curve. The most stable nuclei are ${}^{62}Ni$, ${}^{58}Fe$, and ${}^{56}Fe$ with E_b/A values of 8.794, 8.792 and 8.790 MeV, respectively. Notice the small peak of highest stability at the top of the curve.

1. Nuclear fission

The nuclide ^{236}U, which results from the capture of a slow neutron by ^{235}U, is unstable. It takes on the characteristic of an unstable oscillating droplet. The nuclear droplet, which oscillates with the opposing forces of two positively charged nuclides, splits into two fragments with the liberation of an average energy of 193.6 MeV (Koch, 1995). The mass numbers of the fission fragments vary between 70 and 160, and the following is just one example, illustrating the fission of ^{236}U into the nuclide fragments ^{90}Sr and ^{143}Xe with the emission of 3 neutrons:

$$^{236}_{92}\mathrm{U} \rightarrow {}^{90}_{38}\mathrm{Sr} + {}^{143}_{54}\mathrm{Xe} + 3^{1}_{0}\mathrm{n} \tag{1.40}$$

The binding energy (E_b) of the ^{236}U nucleus, *i.e.*, the difference in mass of the ^{236}U nucleus and the combined mass of its free component nucleons, is calculated as follows:

$$E_b = (M_{92\text{ protons}} + M_{144\text{ neutrons}}) - M_{^{236}\mathrm{U}} \tag{1.41}$$

where $M_{92\text{ protons}}$ is the mass of 92 protons, $M_{144\text{ neutrons}}$ is the mass of 144 neutrons, and $M_{^{236}\mathrm{U}}$ is the mass of the bound nucleus of ^{236}U. Using atomic mass units for the masses of the ^{236}U nucleus and its constituent nucleons, we calculate the binding energy (E_b) in atomic mass units (u) as

$$\begin{aligned} E_b &= [92(1.007825032\text{ u}) + 144(1.008664915\text{ u})] \\ &\quad - 236.0455682\text{ u} \\ &= 237.9676507\text{ u} - 236.0455682\text{ u} = 1.9220825\text{ u} \end{aligned} \tag{1.42}$$

Although the masses of the electrons are included in the atomic masses, the electron masses can be ignored since these are the same for the ^{236}U and the 92 hydrogen atoms. Values of the relative atomic masses are available from the National Institute of Standards database (https://www.nist.gov/pml/atomic-weights-and-isotopic-compositions-relative-atomic-masses). Using Einstein's equation for the equivalence of mass and energy we can convert the binding energy to MeV or

$$\begin{aligned} E = mc^2 &= \left[(1.9220825\text{ u})\left(931.494\text{ MeV}/c^2\right)\right] c^2 \\ &= 1790.41\text{ MeV} \end{aligned} \tag{1.43}$$

Dividing the binding energy by the number of nucleons in the nucleus (mass number, $A = 236$) will provide the binding energy per nucleon for ^{236}U or

$$(E_b/A)_{^{236}\mathrm{U}} = 1790.41\text{ MeV}/236 = 7.58\text{ MeV/nucleon} \tag{1.44}$$

Similar calculations for the binding energies of the fission fragments ^{90}Sr and ^{143}Xe yield $E_b/A = 8.69$ MeV/nucleon for ^{90}Sr and 8.19 MeV/nucleon for ^{143}Xe. The calculated binding energies per nucleon for all of the isotopes of atoms may be obtained from the website http://www.periodictable.com/Isotopes/054.132/index2.p.full.html. Estimated values for the binding energies per nucleon (E_b/A) as a function of mass number (A) can be obtained from Fig. 1.15. Because the binding energies per nucleon for the two fission products are slightly different, we could average the two values simply to get an estimated figure for the amount of energy released. We can now estimate the energy released from ^{236}U fission or the Q value, using as an example, the fission reaction of Eq. (1.40) as follows:

$$^{236}_{92}\mathrm{U} \rightarrow {}^{90}_{38}\mathrm{Sr} + {}^{143}_{54}\mathrm{Xe} + 3^{1}_{0}\mathrm{n}$$

$$\begin{aligned} Q &= \left[(E_b/A)_{\text{fission products}} - (E_b/A)_{^{236}\mathrm{U}}\right] \times 236\text{ nucleons} \\ Q &= \left[\left(\frac{(E_b/A)_{^{90}\mathrm{Sr}} + (E_b/A)_{^{143}\mathrm{Xe}}}{2}\right) - (E_b/A)_{^{236}\mathrm{U}}\right] \\ &\quad \times 236\text{ nucleons} \\ Q &= \left[\left(\frac{8.69\text{ MeV/nucleon} + 8.19\text{ MeV/nucleon}}{2}\right)\right. \\ &\quad \left. - 7.58\text{ MeV/nucleon}\right] \times 236\text{ nucleons} \\ Q &= 202\text{ MeV} \end{aligned} \tag{1.45}$$

Thus, approximately 200 MeV is the amount of energy released from ^{236}U fission, *i.e.*, neutron-induced fission of ^{235}U. The magnitude of the energy released from neutron-induced fission of ^{235}U would be an average figure because the actual amount of energy released would depend on the fission reaction, and there are many possible ways that the ^{236}U could break into two nuclides.

There are about 100 radionuclides of high mass number $A \geq 230$, which decay by spontaneous fission (SF), which involves the spontaneous noninduced splitting of the nucleus into two nuclides or fission fragments. A rough estimate of the energy released by fission reactions can be estimated quickly from Fig. 1.15. For example, ^{252}Cf undergoes SF in 3% of the decay events as an alternative to alpha decay. A rough estimate of the energy released in the SF is found by simply reading from Fig. 1.15 the binding energy per nucleon for a nuclide of mass number $A \approx 250$, which would be about 7.5 MeV, and the binding energy per nucleon for fission fragments that would have a mass number (A) in the range of 80–150, which varies between

8.5 and 8.3 or an average of ~8.4 MeV. The following calculation, in accord with Eq. (1.45), would give the approximate energy released as

$$Q \approx [(8.4\ \text{MeV/nucleon}) - (7.5\ \text{MeV/nucleon})] \times 252\ \text{nucleons} \approx 230\ \text{MeV} \quad (1.46)$$

Nuclear fission is discussed in more detail further on in this chapter.

2. Nuclear fusion

Nuclides of very low mass number ($A < 20$) have nuclei (see Fig. 1.15) with relatively low binding energies, and if these nuclei could coalesce, a larger more tightly bound nucleus would form, and the overall process would release energy as described by Eq. (1.36). For Example, if two nuclei of deuterium ($^{2}_{1}H$) were to fuse into a nucleus of helium ($^{4}_{2}He$), equivalent to an alpha-particle, the energy released by the nuclear fusion could be estimated from Fig. 1.15. Taking the values of binding energy per nucleon (E_b/A) from Fig. 1.15 for $^{2}_{1}H$ to be ~1 MeV and that for $^{4}_{2}He$ to be ~7 MeV, the energy released would be estimated from the following as

$$Q = \left[(E_b/A)_{^4He} - (E_b/A)_{^2H}\right] \times 4\ \text{nucleons}$$

$$Q \approx [(\sim 7\ \text{MeV/nucleon}) - (\sim 1\ \text{MeV/nucleon})] \times 4\ \text{nucleons} \approx 24\ \text{MeV} \quad (1.47)$$

The energy released is more precisely determined by calculating the binding energy, E_b, for the $^{4}_{2}He$ nucleus and for the $^{2}_{1}H$ nucleus according to Eq. (1.47). The nuclear binding energy for $^{4}_{2}He$ was previously calculated (see Eq. 1.38) to be 28.2956 MeV, and the precise binding energy per nucleon for $^{4}_{2}He$ would be $(E_b/A)_{^4He} = 28.2956\ \text{MeV}/4 = 7.0739\ \text{MeV}$. The nuclear binding energy for $^{2}_{1}H$ according to Eq. (1.37) is calculated as

$$\begin{aligned} E_b &= \left[(m_{{}_1H} + m_n - M_{{}^2_1H})931.494\ \text{MeV}/c^2\right]c^2 \\ &= \left[(1.007825032\,\text{u} + 1.008664915\,\text{u} - 2.014101778\,\text{u})\ 931.494\ \text{MeV}/c^2\right]c^2 \\ &= 2.2245\ \text{MeV} \end{aligned} \quad (1.48)$$

The binding energy per nucleon for $^{2}_{1}H$ would be $(E_b/A)_{^2H} = 2.2245\ \text{MeV}/2 = 1.11225\ \text{MeV}$. The energy released in the fusion of two nuclei of deuterium to a nucleus of helium would be calculated according to Eq. (1.47) as

$$\begin{aligned} Q &= [7.0739\ \text{MeV} - 1.11225\ \text{MeV}] \times 4\ \text{nucleons} \\ &= 23.846\ \text{MeV} \end{aligned} \quad (1.49)$$

The above calculated energy of 23.846 MeV for deuterium fusion agrees with the online computer calculated value of 23.846 MeV provided by the Brookhaven National Laboratory (BNL) of the National Nuclear Data Center (NNDC, 2018a,b): http://www.nndc.bnl.gov/qcalc. The deuterium fusion reaction is written as

$$^{2}_{1}H + ^{2}_{1}H \rightarrow ^{4}_{2}He + \gamma + 23.846\ \text{MeV} \quad (1.50)$$

The energy released in the deuterium fusion may be calculated according to Eq. (1.14) as

$$Q = (m_R - m_P)931.494\ \text{MeV/u} \quad (1.51)$$

where m_R and m_P are the rest masses of the reactants and products, respectively, in atomic mass units (u). Thus, the energy released in the fusion of two deuterium nuclei is calculated according to the above Eq. (1.51) as

$$\begin{aligned} Q &= [2(2.014101778\ \text{u}) - 4.00260325\ \text{u}]931.494\ \text{MeV/u} \\ &= 23.846\ \text{MeV} \end{aligned}$$

The nuclear fusion reaction described by Eq. (1.50) is one that releases a relatively large amount of energy; but it has a low reaction cross-section or low probability of occurrence, and the energy not easily captured because most of the energy would be carried off by the gamma radiation. A more practical source of energy by nuclear fusion would be the fusion of deuterium and tritium nuclei according to the reaction:

$$^{2}_{1}H + ^{3}_{1}H \rightarrow ^{4}_{2}He + ^{1}_{0}n + 17.589\ \text{MeV} \quad (1.52)$$

When comparing the energy released from neutron-induced fission of ^{235}U (~200 MeV, See Eq. 1.45) and that produced by the fusion of deuterium and tritium nuclei (17.6 MeV, See Eq. (1.52) above), we can immediately see that there is more than a 10-fold greater magnitude of energy released by the neutron-induced fission reaction. However, if we evaluate the reactions in terms of fuel mass, we will find that the above deuterium-tritium fusion reaction is the most efficient for energy production. For example, let us now calculate and compare the energy yields per kilogram of fuel for both nuclear fission and nuclear fusion.

Neutron-induced ^{235}U Fission energy yield:

$$E\ \text{yield}_{^{235}\text{U fission}} = \frac{\text{Reaction enegy release}}{\text{Fuel mass}}$$

$$= \frac{200\ \text{MeV}}{\left(M_{^{235}\text{U}} + M_{\text{neutron}}\right)\left(1.660 \times 10^{-27}\ \text{kg/u}\right)}$$

$$= \frac{200\ \text{MeV}}{(235.04392\ \text{u} \ + \ 1.00866\ \text{u})(1.660 \times 10^{-27}\ \text{kg/u})}$$

$$= 0.5104 \times 10^{27}\ \text{MeV/kg} \tag{1.53}$$

Deuterium-Tritium (D-T) Fusion:

$$E\ \text{yield}_{\text{D-T fusion}} = \frac{\text{Reaction energy release}}{\text{Fuel mass}}$$

$$= \frac{17.6\ \text{MeV}}{\left(M_{^{2}\text{H}} + M_{^{3}\text{H}}\right)\left(1.660 \times 10^{-27}\ \text{kg/u}\right)}$$

$$= \frac{17.6\ \text{MeV}}{(2.01410\ \text{u} \ + \ 3.016049\ \text{u})(1.660 \times 10^{-27}\ \text{kg/u})}$$

$$= 2.1077 \times 10^{27}\ \text{MeV/kg} \tag{1.54}$$

In the above Eqs. (1.53) and (1.54) the notation $M_{^{235}\text{U}}, M_{\text{neutron}}, M_{^{2}\text{H}}$, and $M_{^{3}\text{H}}$ refer to the masses of ^{235}U, the neutron, deuterium, and tritium in atomic mass units (u). Taking the ratio of energy yields of nuclear fusion over fission yields

$$\frac{E\ \text{yield}_{\text{D-T fusion}}}{E\ \text{yield}_{^{235}\text{U fission}}} = \frac{2.1077 \times 10^{27}\ \text{MeV/kg}}{0.5104 \times 10^{27}\ \text{MeV/kg}} = 4.13 \tag{1.55}$$

Thus, deuterium-tritium (D-T) nuclear fusion is more than four times as efficient than neutron-induced nuclear fission in terms of energy yield per mass of fuel consumption.

3. Nuclear fusion as an energy source

D-T fusion offers great promise as a future energy source, which would be more efficient and more environmentally friendly than neutron-induced fission. There are 448 nuclear power plants in 30 countries of the world operating on the basis of nuclear fission, which provides 16% of the energy needs of the world (IAEA, 2017). However, nuclear fusion is still in the developmental stage. The fusion of light nuclei is not possible under normal temperature and pressure because the repulsive Coulombic forces between atomic electrons and between positive charges of atomic nuclei prevent the nuclei from mingling into the required close proximity for them to coalesce into one nucleus. However, at temperatures of about 100×10^6 °C, the nuclei of atoms become plasmas in which nuclei and electrons move freely with high kinetic energy sufficient for the nuclei to overcome their repulsive forces and combine. Because of the high temperatures required, the process is referred to as thermonuclear fusion.

In summary, the fusion of the light nuclei, such as those of deuterium (^{2}H) and tritium (^{3}H), release an energy per nucleon (MeV/u) or energy per unit of material considerably greater than can be achieved by nuclear fission as calculated above. Thermonuclear fusion requires a large energy input to achieve high temperatures and maintain nuclear plasma; but once achieved, the energy output is greater than the input. An additional advantage of the nuclear fusion of deuterium and tritium is that no hazardous radioactive wastes are direct products of the reaction. The products of the thermonuclear fusion of deuterium and tritium nuclei are alpha particles and neutrons. The neutrons can produce some radioactive isotopes in the surrounding reactor shielding material, but most would be short-lived and there would be no need to store radioactive waste in geological deposits for long periods of time as is the custom with nuclear fission.

a. Advances in fusion reactor development

Nuclear fusion reactors or controlled thermonuclear reactors (CTRs) are under development to achieve nuclear fusion as a practical energy source. The reactors are based on maintaining plasmas at high temperature through magnetic confinement in a tokomak. The word "tokamak" is derived from the Russian "toroidal kamera ee magnetnaya katushka" meaning "torus-shaped magnetic chamber", which was initially designed in 1951 by Russian physicists and Nobel Laureates Andrei Sakharov and Igor Tamm (UIC, 2005). The reactors are based on maintaining plasmas through magnetic or inertial confinement. A deuterium-tritium (D-T) plasma burning experiment was performed with 0.2 g of tritium fuel with the Joint European Torus (JET) reactor in the UK in 1991 (JET Team, 1992), and a higher power D-T experimental program with 20–30 g of tritium was continued on the Tokamak Fusion Test Reactor (TFTR) at Princeton in 1993. The European device, JET, achieved 16 MW of fusion power over approximately 1s in 1997, and a long-duration plasma of 4.5 minutes was sustained with a heating power of 3 MW in the Tore Supra device at Cadarache, France in 2002 (Magaud et al., 2004). The Experimental Advanced Superconducting Tokamak (EAST) in China had a successful plasma test in 2006 (Wan et al., 2007; Xiao et al., 2008). The EAST tokamak was an advanced design for toroidal and poloidal magnetic confinement of plasma, which is a design currently undergoing construction in Cadarache, France by the international project ITER described in the following paragraphs. In addition, a plasma function with diagnostic tests was conducted on the Korea Superconducting Tokamak Advanced Research (KSTAR) fusion reactor in 2008 (Park et al., 2010).

Under the auspices of the International Atomic Energy Agency, the international project ITER, meaning "The Way" in Latin, to signify "the way to new energy" was given birth in 1988 with the purpose of designing and constructing a fully operational prototype fusion reactor. There are seven signatories of ITER including China, the European Union, India, Japan, Russian Federation, Republic of Korea, and the United States, all of which will provide the financial input and technical cooperation required for the construction and operation of the reactor. In addition, 35 nations are collaborating in the construction of this large-scale Tokamak. The design of the ITER Tokamak fusion reactor was approved by the signatory members in 2001, and the site where the reactor would be constructed was agreed in 2005 to be in Southern France at Cadarache near Aix-en-Provence. The fusion reactor now under construction will begin plasma experimentation in 2025, and the reactor is expected to be fully functional with deuterium-tritium fusion by the year 2035 (https://www.iter.org/). The prototype reactor has the purpose of demonstrating that fusion can produce relatively safe energy with a high net energy yield, that is, more energy output than energy put into the system. Fusion energy production via a large-scale commercial reactor with the capacity to produce electric power and with the capacity of producing its own tritium fuel will be the next step after ITER. This full-scale commercial fusion reactor is planned with the project DEMO or Demonstration Power Plant. DEMO is expected to begin operation with connection to the electrical grid in the year 2050 (https://www.euro-fusion.org/programme/).

b. Fusion power reactor

A schematic cutaway view of a fusion power reactor, based on magnetic confinement of the hot plasma, is illustrated in Fig. 1.16. Five major components of the reactor are identified by the encircled letters (A) through (E). Firstly, deuterium and tritium fuel (D + T) are introduced into an inner vacuum chamber at (A). This is a closed chamber having the shape of a torus. The deuterium and tritium are heated in excess of 150 million °C and enter into the plasma state (B). Heating, which is required to take the deuterium and tritium to the plasma state, illustrated at (A) in Fig. 1.16, is accomplished by three means, which are the following (1) Ohmic heating, which is the result of the changing magnetic fields used to confine the plasma within the center of the vacuum chamber. These magnetic fields create a very intense electrical current through the plasma causing the charged particles to travel and collide creating a resistance and heat, referred to as ohmic heating; (2) neutral beam injection and (3) high-frequency electromagnetic waves. As described by ITER (http://www.iter.org/sci/plasmaheating), neutral beam injection involves accelerating high-energy particles into the plasma. Deuterium particles are accelerated to the required energy level and subsequently are stripped of their electrons with an "ion beam neutralizer". Upon entering the plasma at high velocity, the high-energy deuterium nuclei undergo collisions and thus transfer their energy to the plasma particles. This process brings the temperature of the plasma close to that required for deuterium-tritium fusion. Irradiation with high-frequency electromagnetic radiation increases the temperature of the plasma to the desired ~150 million °C, which is more than sufficient for fusion to occur. The hot plasma is kept within the central region of the inner chamber by magnetic confinement by means of toroidal and poloidal field coils. The plasma produces energy from the fusion of deuterium and tritium with the emission of 3.52 MeV

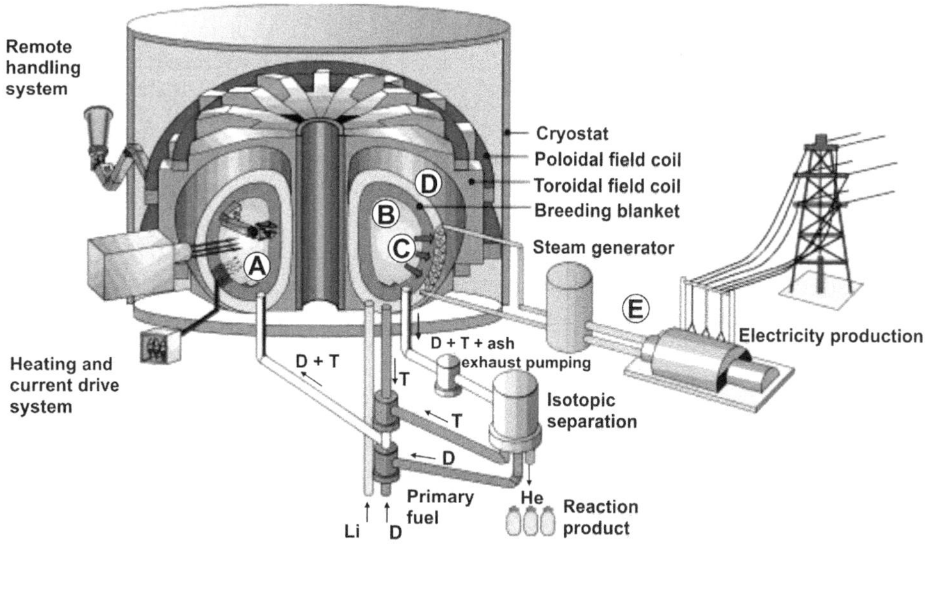

FIGURE 1.16 Schematic and cutaway view of a fusion reactor based on a magnetic confinement of plasma: (A) vacuum chamber, (B) plasma, (C) plasma radiation, (D) blanket, and (E) electricity production. For the color version of the figure, the reader is referred to the online or electronic version of the book. *From Magaud et al. (2004), printed with permission from Elsevier © 2004.*

helium nuclei and 14.06 MeV neutrons according to the fusion reaction

$$^{2}_{1}\mathrm{H} + ^{3}_{1}\mathrm{H} \rightarrow ^{4}_{2}\mathrm{He} + ^{1}_{0}\mathrm{n} + 17.58\ \mathrm{MeV} \quad (1.56)$$

The high-energy 14.06 neutrons account for 80% of the energy liberated from the above nuclear fusion. The high-energy neutrons possess no charge and thus can penetrate the magnetic confinement and deposit energy into the outer blanket of the chamber. The energy liberated is absorbed by an outer chamber blanket illustrated at (C) of Fig. 1.16, and the heat is utilized to produce steam to operate a turbine and generate electricity at (E). The neutrons thus lose energy and are absorbed by lithium nuclei in a breeding blanket at (D). The breeding blanket contains a high concentration of lithium to produce tritium according to the reaction

$$^{1}_{0}\mathrm{n} + ^{6}_{3}\mathrm{Li} \rightarrow ^{3}_{1}\mathrm{H} + ^{4}_{2}\mathrm{He} + 4.78\ \mathrm{MeV} \quad (1.57)$$

With the breeder blanket, The ITER fusion reactor will be equipped to produce more tritium fuel than it consumes to produce power. The tritium produced by the fusion reactor can be introduced together with deuterium as fuel when required. Ash, consisting of helium ions and unconsumed deuterium and tritium, are extracted from the inner chamber, as illustrated in Fig. 1.16 and separated to collect the helium; and the unconsumed deuterium and tritium are reintroduced into the inner chamber. The cryostat, illustrated in Fig. 1.16, is the cooling system and physical enclosure surrounding the vacuum chamber providing thermal insulation for the superconducting magnets required for plasma confinement.

c. Advantages of fusion over fission for power

Fusion is the energy of the future, and we may state optimistically that fusion is the energy of the relatively near future. The fusion of tritium and deuterium in a tokomak reactor has been demonstrated. As described previously, the ITER reactor will begin plasma experimentation in the year 2025, and the DEMO reactor, which will be the first full-scale fusion power plant, is planned to begin operation with connection to the electric power grid in year 2050. In 3 decades, we should witness a functioning full-scale fusion power plant producing relatively safe electric power and more power output than the energy consumed by the reactor. By the next century, we should expect and hope that fusion will be the energy of choice over power reactors based on nuclear fission that now produce considerable radioactive waste. Thus, fusion reactors together with solar and wind power would constitute sources of electric power that produce neither greenhouse gases nor long-lived radioactive waste.

Current optimism over fusion reactors for electric power is based on overwhelming advantages that deuterium-tritium fusion offers over current nuclear fission power reactors. These advantages are listed and described briefly as follows:

(i) **No danger of a runaway nuclear chain reaction.** Unlike nuclear fission, any misfortune such as a tsunami, earthquake, or simple negligence would not result in any out-of-control chain reaction that could spew radioactive contamination into the environment. Nuclear chain reactions are not involved in a fusion reactor. Any misfortune, that could break down the reactor, would automatically and immediately shut the fusion reactor down, because the continuous heating and feeding of deuterium and tritium into the plasma would be disrupted. As noted by Magaud et al. (2004) the slightest uncontrolled interference with the environment of a fusion reactor causes the rapid cooling and automatic shutdown of the fusion reaction.

(ii) **No long-lived radioactive waste.** Fusion reactors do not use or produce long-lived radioactive materials. Fusion reactors do not produce long-lived fission products. The fusion of deuterium and tritium produce alpha particles and high-energy neutrons. The neutrons would produce heat, more tritium fuel upon reaction with Li-6, and mostly short-lived radionuclides when neutrons are absorbed by nuclei of the materials with which the fusion reactor is constructed. These materials would have a relatively short half-life and would not require deep underground storage facilities for long-lived radioactive waste now necessitated by current fission reactor waste. In point of fact, if we compare an equal amount of energy produced by a fusion reactor and a coal-burning plant, the burning of coal would produce more long-term radioactive waste, as the ash from coal burning contains long-lived radioactive uranium and thorium (Magaud et al., 2004). Fig. 1.17 illustrates the relative amounts of long-lived radioactive waste that result from fission, coal, and fusion. Fission and coal burning leave behind significant radioactive waste beyond 500 years; whereas, after only 50 years of fusion reactor shutdown, there is no significant long-lived radioactive waste and no need for deep underground waste storage.

(iii) **No radioactive fuel is required.** The fuel required for fusion is deuterium and lithium, both of which are not radioactive. Although the fusion of deuterium with tritium is the reaction liberating energy for power, the elements considered as fuel components for fusion, are deuterium and lithium, which are the elements required in relatively large quantities. Tritium is needed only in relatively small quantities for the startup of the fusion reaction. The tritium, which is radioactive with a relatively short

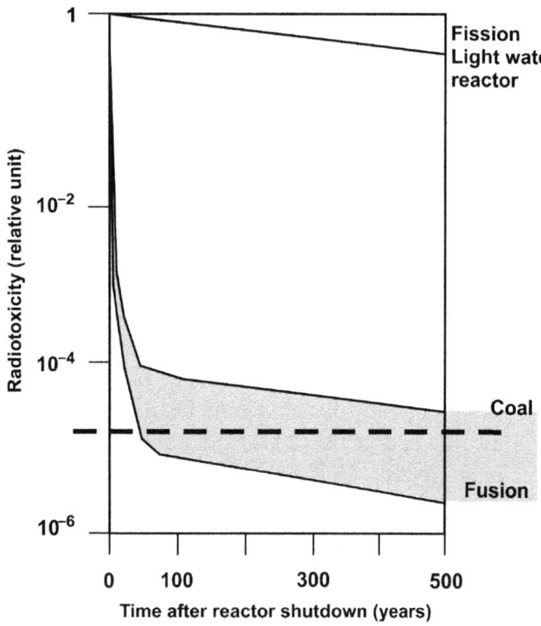

FIGURE 1.17 Radiotoxicity of fusion waste after reactor shutdown. The green shaded area shows the significant difference between coal and fusion with respect to long-term radioactive waste production. For the color version of the figure, the reader is referred to the online or electronic version of the book. *Figure from Magaud et al. (2004), reprinted with permission from Elsevier © 2004.*

half-life, would be produced by the neutrons emitted during fusion and subsequently fed into the fusion chamber when needed. All of the tritium needed for power would be produced by the fusion reactor.

(iv) **The only fuel elements deuterium and lithium, required for fusion, are abundantly available.** Deuterium is readily extracted from seawater, which contains inexhaustible amounts of deuterium (33 g of deuterium/m^3 of seawater), and lithium would be used in the reactor to breed the required tritium for the fusion reaction. Magaud et al. (2004) note that < 1 kg of deuterium and <10 kg of naturally occurring lithium would be required for a day's generation of 1000 MWe of power. Lithium deposits are massive with global deposits totaling 31.1 Mt (million tons) of Li (Kesler et al., 2012). In addition, the world's oceans contain 44.8 billion tons of recoverable lithium (Yaksic and Tilton, 2009).

(v) **Extraction of fuel for fusion will have minimal effect on the environment.** Obtaining deuterium and lithium fuel for fusion from seawater will cause minimal effect on the environment; whereas, the mining of uranium deposits for fission reactor fuel causes significant effect on the environment.

(vi) **The fusion reactor will yield more power than energy input.** The ITER prototype fusion reactor will require 50 MW of input power and it is designed to produce 500 MW of output power. The reactor will have a Q value ≥ 10, that is, it is designed to produce 10 times the power it will consume. This will be the first reactor to produce net energy (https://www.iter.org/proj/itermission).

(vii) **Not necessary to evacuate population in event of severe accident.** The overall radioactivity at a fusion reactor would be minimal. The only radioactive isotope involved in the production of fusion energy is tritium, and because tritium is not a fuel component but produced in the lithium breeding blanket to be fed into the plasma as needed, the tritium inventory on site would be minimal. Also, as noted by Magaud et al. (2004), to minimize emission there would be "three confinement barriers: the vacuum vessel, the cryostat, and the reactor building itself ... there is no accident sequence that could impugn reactor integrity. Even in the event of a severe accident, it would not be necessary to evacuate the population."

(viii) **Overall costs of fusion energy would be comparable to those of renewable resources.** Magaud et al. (2004) provides an excellent review of the overall costs of fusion power relative to other sources, such as fossil fuel, fission, and renewable resources such as solar and wind power. A brief summary will be provided here. The principle direct costs in the utilization of all energy sources are the investment, operation, maintenance, replacement of components, and fuel. Fuel costs of fusion account for less than 1% of the total kilowatt-hour costs; whereas, in the case of nuclear fission, fuel costs amount to ~35% of direct costs. Other costs, referred to as externalities, are measured in terms of the cost of effects on our environment and health, which includes fuel extraction, plant construction, operation, accident, dismantlement, and disposal of a dismantled power source. For example, the nuclear fuel cycle, required for nuclear fission, involves mining and nuclear enrichment, which have an effect on our environment. So does the mining and burning of coal. Fusion is predicted to have one of the lowest costs per kilowatt hour of energy produced similar to that of the renewable resources such as wind and solar, as determined by the European Union (ExternE Studies: http://www.externe.info/externe_d7/).

D. Nuclear models

Since Ernest Rutherford's alpha-scattering experiment that led to the discovery of the atomic nucleus (Rutherford, 1911), physicists have used experimental and empirical data to establish nuclear models to describe the nucleus. The nuclear models, that have been established, provide insight into the stability of the atomic nucleus and an understanding of the processes of nuclear decay.

1. Liquid drop model

During a meeting of the Royal Society in London on February 7, 1929, the Russian-born physicist George Gamow (1904–68) presented his thoughts on the properties of the nucleus and that the nucleus would possess forces similar to those of a liquid drop (Gamow et al., 1929). A very complete and detailed account of Gamow's conception of the liquid-drop model of the nucleus is given by Stuewer (1997). The writer will give here a small portion of this historical account, which provides the reasoning used by Gamow. At the time Gamow first presented his ideas on the liquid-drop characteristics of the nucleus, the neutron had not been discovered, and the alpha particle was thought to be the building block of the nucleus. In the discussion, Gamow et al. (1929) commented

> *Such an assembly of alpha particles with attractive forces between them, which vary rapidly with the distance may be treated somewhat as a small drop of water in which the particles are held together by surface tension.*

Gamow (1930) continued his development of the liquid-drop model, even before the neutron was discovered by Chadwick (1932a,b) to be a component particle of the nucleus. After the discovery of the neutron, Heisenberg (1934) presented his analysis of exchange forces and general theory of nuclear structure at the 7th Solvay Conference in Brussels in October 1933. At the meeting, Heisenberg reported his theory on nuclear structure, which could be "considered as corresponding to a form of Gamow's drop model made precise by the neutron hypothesis" (Stuewer, 1997). Only a year later, Heisenberg's student von Weizsäcker (1935) derived an empirical equation that would yield the binding energy of the nucleus as a function of its mass number (A), incorporating into the equation the various properties of the nucleus characteristic of a liquid drop. Von Weizsäcker's equation for the liquid-drop model of the nucleus gained notoriety when Hans Bethe and Robert Bacher (Bethe and Bacher, 1936; Bethe, 1937) from Cornell University reworked Weizsäcker's equation to include a pairing energy, which was referred to subsequently as the Bethe–Weizsäcker mass formula.

Gamow's original conception of the nucleus with the physical properties of a liquid drop soon became useful in the interpretation of nuclear fission. In December of 1938, Otto Hahn and Fritz Strassmann (Hahn and Strassman, 1939a,b) had confirmed that barium ($Z = 56$) was a product of neutron bombardment of uranium ($Z = 92$). Hahn could not arrive at an explanation of how a nuclide of lower atomic number would result from the reaction of a neutron with a nucleus of much higher atomic number. Hahn asked his former colleague and collaborator, Lise Meitner for an explanation. Meitner and Frish (1939) considered the recently developed concept of the nucleus having the properties of a liquid drop. They arrived at the conclusion that the capture of a neutron by the uranium nucleus attributes to the nucleus a degree of instability, whereby it would oscillate like a liquid drop forming a dumbbell shape (see Fig. 1.18) and that the apposing Coulomb forces of the fat sections of the dumbbell would cause the nucleus to split apart into two nuclides (fission fragments) of smaller atomic number, and that the sum of the atomic numbers of the two fission fragments would equal that of the original uranium ($Z = 92$). Meitner and Frisch concluded that if barium ($Z = 56$) was a product (fission fragment) of neutron bombardment of uranium ($Z = 92$), the other fission fragment would be krypton ($Z = 36$), as described by the following reaction:

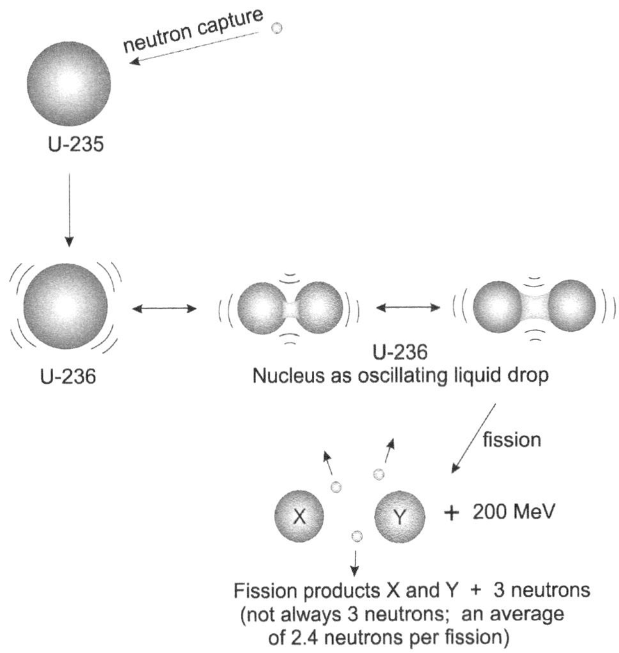

FIGURE 1.18 Neutron capture by uranium-235 and subsequent fission of unstable uranium-236. The capture of a slow neutron by uranium-235 yields the nuclide uranium-236, which is unstable, capable of oscillating like a liquid drop and breaking into two smaller nuclides X and Y generally of different masses and of atomic numbers that add up to $Z = 92$ (the atomic number of the parent uranium) plus the emission of an average of 2.4 neutrons and energy (approximately 200 MeV). The emission of three neutrons is illustrated. The uranium-236 nuclides break into two fission fragments in as many as 30 different possible ways producing, therefore, 60 different nuclide fission fragments.

$$^{235}_{92}U + ^{1}_{0}n \xrightarrow{\text{fission}} {}_{56}Ba + {}_{36}Kr + x^{1}_{0}n \tag{1.58}$$

They coined the word "fission" by borrowing the word from biology, a word used to describe the splitting of bacteria cells. The liquid-drop model remains to this day an excellent means of describing nuclear fission, and the liquid-drop model formula has been used to predict spontaneous fission half-lives (Ronen, 2004).

Today the liquid-drop model formula is referred to often as the Weizsäcker liquid-drop model formula or the semi-empirical mass formula. It has been refined by many researchers to include theory with empirical measurements including those described by Evans (1955), Friedlander et al. (1964), Ronen (2004), Serway et al. (2005), Mahboub (2008) and Royer (2008), and the formulas vary slightly among research reports; however, all have the basic mathematical structure as the equation first reported by von Weizsäcker. The liquid-drop formula described by Evans (1955), Friedlander et al. (1964), Choppin et al. (2002), Gould et al. (2004), and Serway et al. (2005) is basically the following:

$$E_b = a_v A - a_s A^{2/3} - a_c \frac{Z(Z-1)}{A^{1/3}} - a_{sy} \frac{(N-Z)^2}{A} \pm \delta_p(Z,N) \quad (1.59)$$

where E_b is the nuclear binding energy in MeV, A is the mass number, N is the neutron number, Z is the atomic number, a_v, a_s, a_c, and a_{sy} are coefficients for the volume energy, surface energy, Coulomb energy, and the symmetry energy, respectively, and $\delta(Z,N)$ is a pairing energy term described subsequently. The values of the coefficients a_v, a_s, a_c, and a_a reported by Woan (2000) are $a_v \approx 15.8$ MeV, $a_s \approx 18.0$ MeV, $a_c \approx 0.72$ MeV, and $a_{sy} \approx 23.5$ MeV. The coefficients vary slightly among research reports depending on the data used to obtain the mathematical best fit of the equation to measured values of nuclear binding energy, and the equation generally agrees with measured values to within a few percent for nuclides of $A > 40$. The various terms of the liquid-drop equation (Eq. 1.59) are described as follows:

Volume Energy, $a_v A$. The volume energy term assigns a direct proportionality of the nuclear volume to the total binding energy of the nucleus, i.e., the nuclear volume increases in proportion to the number of nucleons. The extension of the binding energy throughout the entire volume of the nucleus is a consequence of the short range and saturation characteristics of the nuclear forces. The strong nuclear forces binding neutrons and protons in the nucleus extend throughout the nuclear volume.

Surface Energy, $a_s A^{2/3}$. The surface energy term is negative, as the nuclear forces binding the nucleons are unsaturated at the surface similar to the surface tension effect that occurs with a liquid drop (see Fig. 1.19) and there is thus a reduction of binding at the surface. The nucleons at the surface have fewer neighboring nucleons with which to bind resulting in a reduction in binding energy as a function of surface area. The surface effect will increase with surface area, and surface area of a sphere is defined as πr^2. Since the radius of the nucleus is proportional to the mass number ($r \propto A^{1/3}$) as defined by Eq. (1.28), the surface area term can be written as being proportional to $A^{2/3}$.

Coulomb Energy, $a_c\left(Z(Z-1)/A^{1/3}\right)$. The Coulomb energy term takes into account the repulsive Coulomb forces of the protons in the nucleus. The nucleus contains Z protons and their long-range Coulomb forces interact with the remaining $Z-1$ protons. The Coulomb force therefore extends throughout the nucleus with radius proportional to $A^{1/3}$ (see Eq. 1.28), which is included in this Coulomb energy term of the liquid-drop model.

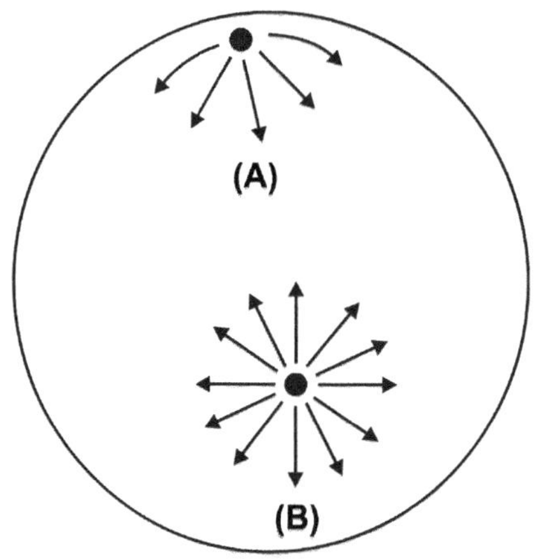

FIGURE 1.19 Nuclear forces (arrows) binding a nucleon at the nuclear surface (A) and a nucleon within the nucleus (B). The binding forces of the nucleon at the surface are unsaturated, as there are fewer neighboring nucleons with which to bind compared to nucleons within the nucleus. The phenomenon is similar to that which occurs in the binding of molecules of a liquid drop resulting in surface tension.

The strong nuclear binding forces between neutrons and protons in the nucleus, which are of short range, counteract the Coulomb repulsive forces. Thus, throughout the Chart of the Nuclides among the stable nuclei with $Z > 20$, we will find that their nuclei contain more neutrons than protons. If we look throughout The Chart of The Nuclides, we will notice that the stable nuclides of low atomic number ($Z \leq 20$) will have an N/Z ratio of ~ 1. However, as the atomic number increases ($Z > 20$), the N/Z ratio of the stable nuclides increases gradually and reaches as high as ~ 1.5 (*e.g.*, $^{209}_{83}$Bi, $Z = 83$, $N/Z = 1.518$). Furthermore, there are no stable nuclides of atomic number greater than 83.

In brief, the importance of the N/Z ratio to nuclear stability is explained on the basis of the existence of strong short-range nuclear binding forces, which are charge independent, and the long-range repelling Coulomb forces between protons. The short-range strong nuclear forces extend to a distance of ≈ 2 fm (2 fermi or 2×10^{-15} m), which is close to the radius of a small nucleus, such as that of $^4_2$He (see Section VI.A). These attractive strong nuclear forces are binding exchange forces that have charge independence (see Section VI.B). The binding exchange forces exist therefore, regardless of charge on the particles, between two protons, two neutrons, and a proton and neutron. Thus, at short distances within 2 fm, the strong forces between n–n, n–p, and p–p are equal and charge independent. However, at distances beyond 2 fm there exists repelling Coulombic forces between the positively charged protons that act to force them apart. For nuclides of low Z, the attractive nuclear forces exceed the repelling Coulombic forces when $N \approx Z$. However, increasing the number of protons (*e.g.*, $Z > 20$) further increases the strength of the repelling Coulombic forces over a larger nucleus, which will tend to force the nucleus apart. Therefore, additional neutrons, $N > Z$, provide additional strong attractive nuclear forces needed to overcome the repelling forces of the larger proton population. Fig. 1.20 illustrates a plot of the stable nuclides as a function of their *N* and *Z* numbers. The *N/Z* ratio of the stable nuclides (dots on the graph), illustrated in Fig, 1.20, increases significantly as a function of *Z* for nuclides of $Z > 20$. As the atomic number increases further, $Z > 83$, all nuclides are unstable. Nuclear stability is not achieved, when the number of protons in the nucleus exceeds 83.

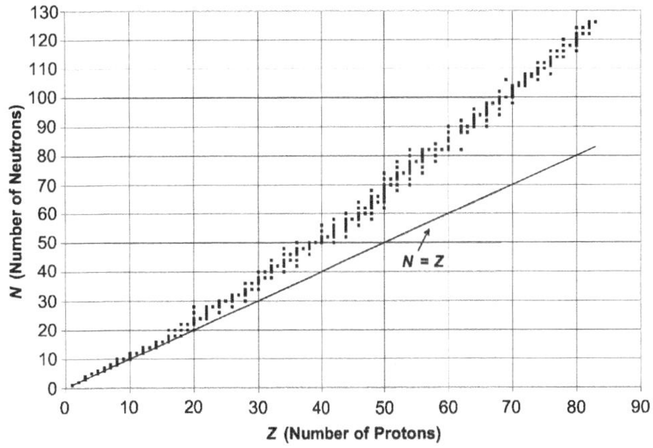

FIGURE 1.20 Plot of stable nuclides (dots). The number of neutrons (*N*) is plotted against the number of protons (*Z*) in the nuclei of the stable atoms. The straight line represents $N = Z$. Only 13 nuclides in the region $Z \leq 20$ are stable with an equal number of neutrons and protons ($N = Z$).

Volume Symmetry, $a_{sy}\left((N-Z)^2/A\right)$. The volume symmetry term quantifies the contribution of nuclear symmetry to the binding energy. Among the light nuclei, the unstable nuclei reside on either side of the symmetrical term $N = Z$. Thus, a negative sign is assigned to this symmetry term, which increases according to ($N - Z$). Nuclear stability among the larger nuclei is achieved by an increase in the *N/Z* ratio (see Fig. 1.20), as the strong short-range binding nuclear forces between neutrons and protons (see Section VI.B) are needed to hold the nucleus together over the entire volume of the nucleus. As these short-range n−p binding forces do not reach throughout the volume of the nucleus, the probability of occurrence of n−p binding within a nucleus would be inversely proportional to the nuclear volume or A^{-1}.

Pairing Energy, $\pm\delta_p(Z,N)$. The pairing energy term may have a positive sign whereby the term adds to the nuclear binding energy, or a negative sign diminishing the nuclear binding energy. Nuclei that have an even numbers of protons and even number of neutrons (even *Z*, even *N*) are known to be the most stable of all nuclei due to proton−proton spin coupling and neutron−neutron spin coupling (see Section VI.D.2 below). The pairing energy term for the binding energy of these nuclei are assigned a value of $+34/A^{3/4}$ MeV. Nuclides with even *Z*, odd *N* or odd *Z*, even *N* are generally less stable than even−even nuclides, and thus the even−odd and odd−even (i.e., odd *A*) nuclides are assigned a value of zero to the pairing energy term. The least stable of the nuclides are those with odd *Z* and odd *N* (i.e., odd−odd nuclides), and these are assigned a pairing energy term of $-34/A^{3/4}$ MeV. The negative value of this term would diminish the total binding energy of Eq. (1.59) for the liquid-drop model.

Thus, the pairing energy term would have the following values as summarized by Woan (2000) and Gould et al. (2004):

$$\pm\delta_p(Z,N) \approx \frac{34}{A^{3/4}} \begin{Bmatrix} +1 \text{ for even} - \text{even} \\ 0 \text{ for } A \text{ odd} \\ -1 \text{ for odd} - \text{odd} \end{Bmatrix} \quad (1.60)$$

The liquid-drop model is useful in describing certain properties that may attribute stability or instability to atomic nuclei, and it has proven successful in explaining nuclear fission in large nuclei. However, the liquid-drop model does not provide insight into the fine organized structure and spin coupling of protons and neutrons in the nucleus that attribute stability to certain nuclei as provided by the shell model described subsequently.

2. Shell model

Our knowledge of the shell structure of the atomic nucleus has its beginnings with the independent work of Maria Goeppert-Mayer while at the Argonne National Laboratory in the US (Goeppert-Mayer, 1948a; b, 1949) and J. Hans D. Jensen, from the University of Heidelberg, Germany (Haxel et al., 1949). They subsequently collaborated with a book on nuclear shell structure (Goeppert-Mayer and Jensen, 1955a,b). For this work they were awarded the 1963 Nobel Prize in physics, which they shared with Eugene Paul Wigner. This award made Maria Goeppert-Mayer the second woman to receive the Nobel Prize in physics since Marie Curie.

Maria Goeppert-Mayer's opportunity for discovery came while voluntarily collaborating at the University of Chicago's Institute for Nuclear Studies (now known as the Enrico Fermi Institute). At the institute she collaborated with Edward Teller in a search for the cosmological model of the origin of the universe. While compiling data she studied the abundance of the elements and found that there were certain elements that were highly abundant, and she found a correlation between the abundance of these elements and specific numbers of neutron and protons in their nuclei. J. Hans D. Jensen at the Institute of Theoretical Physics in Heidelberg together with his collaborators Otto Haxel at the Max Planck Institute in Göttingen and Hans E. Suess of the Institute of Physical Chemistry at Hamburg came to a similar conclusion that certain elements with particular numbers of neutrons and protons were particularly stable. It is interesting to note, that at both sides of the globe, the discoveries were made simultaneously and independently. When the stage is set, the discoveries will be made.

The numbers of protons and neutrons, correlated to particularly stable nuclei, were called the "magic numbers". In the words of Goeppert-Mayer in her Nobel Lecture

> *What makes a number magic is that a configuration of a magic number of neutrons, or of protons, is unusually stable whatever the associated number of the other nucleons. The magic numbers, as we know them today, are: 2, 8, 20, 28, 50, 82, 126 and most importantly, they are the same for neutrons and protons.*

Nuclides containing a magic number of neutrons and/or protons are listed in Table 1.5. Goeppert-Mayer (1963) pointed out that Sn ($Z = 50$) is the element with the largest number of stable isotopes among all of the elements in the Periodic Table. There are six stable nuclides with 50 neutrons and 7 stable nuclides with 82 neutrons, whereas normally one will find only 2 or 3 stable nuclei in the Chart of the Nuclides having identical numbers of neutrons.

Some nuclides are called "doubly magic" because they have equal numbers of protons and neutrons that are magic numbers. The nuclide $^{4}_{2}$He has a high binding energy, *i.e.*, it is very tightly bound (see Fig. 1.15), and an extra proton or neutron cannot be attached to its nucleus, as evidenced by the fact that $^{4}_{3}$Li and $^{5}_{2}$He do not exist. In addition, $^{16}_{8}O_8$, which is also "doubly magic", is very stable. A large amount of energy would be required to remove a neutron from the $^{16}_{8}O_8$ nucleus; whereas, the addition of an extra neutron to its nucleus to form $^{17}_{8}O_9$ would result in a weakly bound neutron, which is evidenced by the low natural abundance of ^{17}O (0.04%) compared to ^{16}O (99.76%).

The breakthrough in the development of the nuclear shell model came with the conclusion that the protons and neutrons in the nucleus exhibit nuclear spin coupling effects, which result in shifts in their energy levels within the nucleus, similar to what occurs with electrons in their orbitals about the atom. Pauli's Exclusion Principle, which was applied to atomic electrons, is also applied to the nucleons in the nucleus. The Exclusion Principle states that no two protons or two neutrons in a nucleus could exist in the same quantum or energy state. The proton or

TABLE 1.5 Magic number nuclides.

Number of protons	2	8	20	28	50	82	126
	^{4}He	^{16}O	^{40}Ca	^{58}Ni	^{112}Sn	^{204}Pb	
		^{17}O	^{42}Ca	^{60}Ni	^{114}Sn	^{206}Pb	
		^{18}O	^{43}Ca	^{61}Ni	^{115}Sn	^{207}Pb	
			^{44}Ca	^{62}Ni	^{116}Sn	^{208}Pb	
			^{46}Ca	^{64}Ni	^{117}Sn		
			^{48}Ca		^{118}Sn		
					^{119}Sn		
					^{120}Sn		
					^{122}Sn		
					^{124}Sn		
Number of neutrons	2	8	20	28	50	82	126
	^{4}He	^{15}N	^{36}S	^{48}Ca	^{86}Kr	^{136}Xe	^{208}Pb
		^{16}O	^{37}Cl	^{50}Ti	^{87}Rb	^{138}Ba	^{209}Bi
			^{38}Ar	^{51}V	^{88}Sr	^{139}La	
			^{39}K	^{52}Cr	^{89}Y	^{140}Ce	
			^{40}Ca	^{54}Fe	^{90}Zr	^{141}Pr	
					^{92}Mo	^{142}Nd	
						^{144}Sm	

Goeppert-Mayer (1963); reprinted with permission © The Nobel Foundation 1958.

neutron can exist in two spin quantum states, namely + 1/2 (i.e., spin-up) or −1/2 (i.e., spin-down). Goeppert-Mayer explained in her Nobel lecture that there is a strong interaction between spin and orbital angular momentum, whereby spin-up and spin-down refers to the spin being parallel or antiparallel to the orbital angular momentum. She provided evidence that this spin interaction raised considerably the energy levels of nuclei that contained 9, 21, 29, 51, or 83 protons or neutrons, or 127 or 128 neutrons, thereby attributing a high degree of stability to nuclei that contain 8, 20, 28, 50, 82 protons or neutrons, or 126 neutrons. The numbers 8, 20, 28, 50, 82, and 126 refer to closed shells of protons or neutrons attributing a high degree of stability to nuclei with these nucleon numbers. The closed shells of nucleons in the highly stable nuclei is analogous to the closed electron shells of the noble gases, which attribute to these gases a high degree of chemical inertness. Thus, protons and neutrons exist in the nucleus in different energy states in structured orbitals similar to the external electron orbitals of the atom. In Section VI.D.1 it was discussed how stability is achieved in the light nuclei ($Z < 20$) when $Z = N$, and this was illustrated previously in Fig. 1.20. In accord with the Exclusion Principle the nucleons are packed in the nucleus according to energy levels with two protons of opposite spin and two neutrons of opposite spin per energy level. As a proton or neutron is added to the nucleus, it will occupy the next highest energy state, and the increase in energy state of the nucleus can decrease its stability. Fig. 1.21 illustrates how the additional proton in ^{12}N compared to ^{12}C attributes a higher energy state to the ^{12}N nucleus, whereby it decays to the stable ^{12}C by the conversion of a proton to a neutron via positive beta decay. Also, an extra neutron in the nucleus of ^{12}B places that nucleus in a higher energy state compared to ^{12}C. The ^{12}B achieves stability by conversion of the excess neutron to a proton to form stable ^{12}C via negative beta decay.

In her Nobel lecture Maria Goeppert-Mayer provided evidence for the increased instability of nuclei when additional neutrons are added to nuclei above the magic numbers, which represent closed shells at 2, 8, 20, 28, 50, 82, and 126 neutrons. Fig. 1.22 illustrates the β^-decay energies of various nuclides with $N–Z = 3$. Maria Goeppert-Mayer explained the curve as follows:

> *One would expect to find a smooth curve, sloping downward. Except for one point it is indeed so, This point is ^{39}Ar with 21 neutrons and 18 protons. A smooth interpolation of the curve would predict ^{39}Ar to be stable … However, ^{39}Ar is unstable against β^- emission by about 0.5 MeV. The explanation of this anomaly is the low binding energy of the 21st neutron, while the 19th proton into which it is transformed has the higher binding energy of the proton shell which closes at 20. That the energies drop again sharply is due to the fact that now $Z = 20$ is involved. These types of discontinuity occur at all magic numbers.*

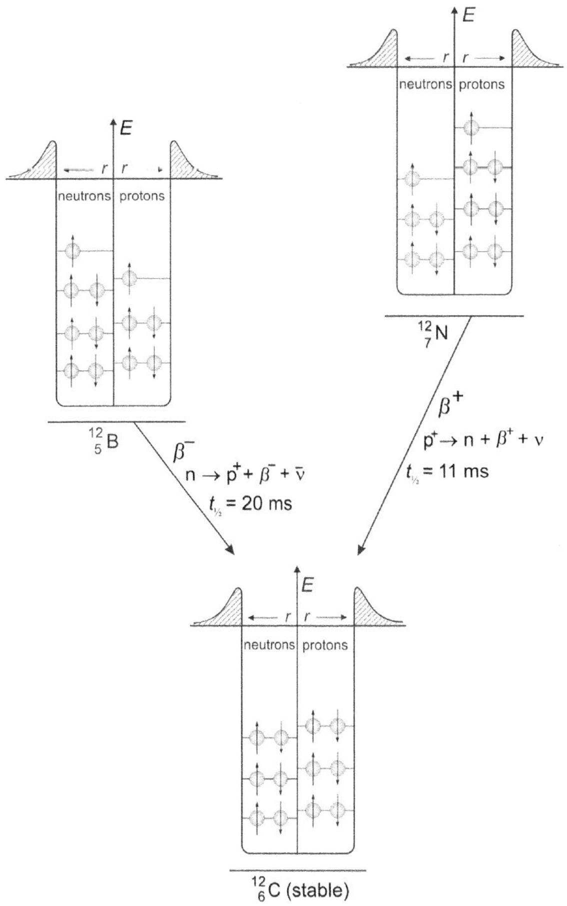

FIGURE 1.21 Nuclear potential energy-well illustrations of the pairing of nucleons in the isobars ^{12}N, ^{12}B, and ^{12}C. The neutrons and protons are illustrated with arrows to represent opposite spin states, spin up (↑) and spin down (↓). Only two nucleons of opposite spin state are allowed per energy level (E) in accord with the Pauli Exclusion Principle. The protons are illustrated at a higher energy level than the neutrons because of the added positive Coulomb potential of the protons. The ridge, at energy levels above those occupied by the protons, represents the Coulomb potential barrier to charged particles from entering or escaping the nucleus. The nuclear radius is represented by r.

A similar discontinuity of the β^-decay energies in the neighborhood of $N = 50$ is illustrated in Fig. 1.23. The addition of a single neutron to the nucleus of $N = 50$ increases the energy state of that nucleus illustrated by the higher decay energy of the nucleus of ^{89}Sr with $N = 51$ compared to ^{85}Kr with $N = 49$. This is only one of several examples given by Goeppert-Mayer (1972) of nuclear stability attributed to the closing of a neutron shell at $N = 50$.

The highest magic number of 126 occurs only for the closing of a neutron shell. Thus, it would be expected that the 127th and 128th neutrons would be less tightly bound, and that these nuclides would be at higher and less stable energy states compared to nuclides with 126 neutrons. This is just the case, as illustrated in Fig. 1.24. In the case of α-decay there is the emission of two neutrons together with two protons. As illustrated in Fig. 1.24, the α-particle decay energy is the highest for nuclides with 128 neutrons, and

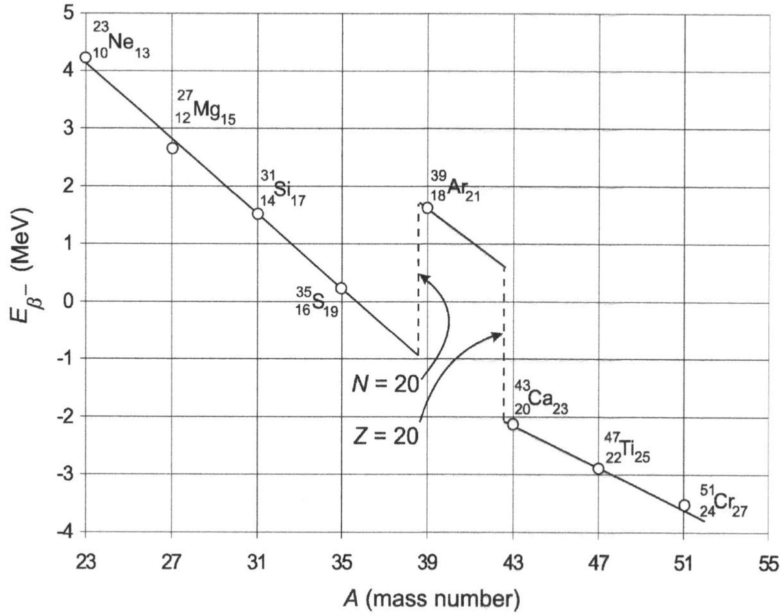

FIGURE 1.22 Beta decay energies in the neighborhood of $N = 20$. *Goeppert-Mayer (1963); reprinted with permission © The Nobel Foundation 1958.*

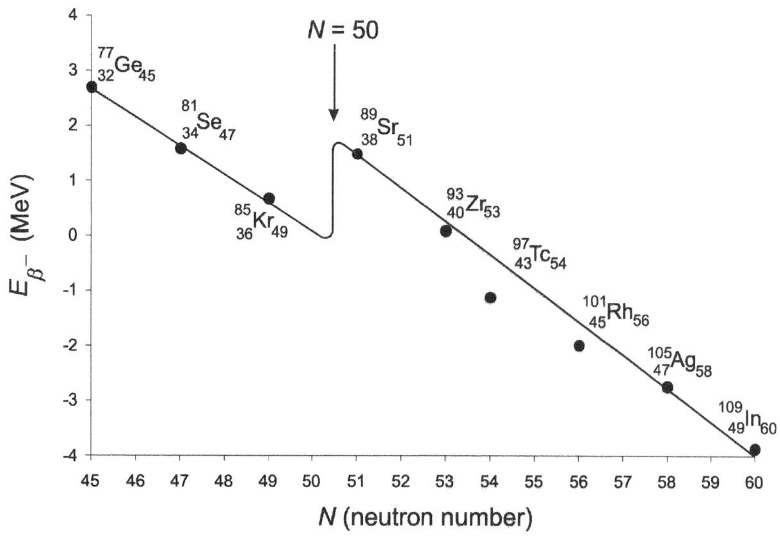

FIGURE 1.23 Beta decay energies in the neighborhood of $N = 50$. *Goeppert-Mayer (1963); reprinted with permission © The Nobel Foundation 1958.*

this drops sharply when the 128th and 127th neutron is removed from the nucleus.

For further information on this subject matter the reader is invited to peruse works by Goeppert-Mayer and Jensen (1955a,b, 1965), Goeppert-Mayer (1964), Choppin et al. (2002), and Rowe and Wood (2010).

3. Collective model

The collective model of the nucleus combines aspects of the liquid-drop and shell models, which provides a description of the collective properties of the nucleus. The oldest model was the liquid-drop model described previously. This model was very useful in describing certain nuclear reactions, such as nuclear fission. After the liquid-drop model came the shell model, which gave an organized picture of the nucleons in different energy states in structured orbitals. This provided an explanation for the stability of nuclei with certain "magic numbers". However, neither model could provide an explanation for excited states of the nucleus, which are a result or consequence of the collective action of the nucleons. The collective model, for which Aage Bohr (son of Niels Bohr), Ben Mottelson, and James Rainwater received the 1975 Nobel Prize in physics, depicts the nucleus with both the properties of a liquid drop and the ordered structure of the nucleons providing an explanation of excited states of the nucleus as a consequence of its rotations and vibrations.

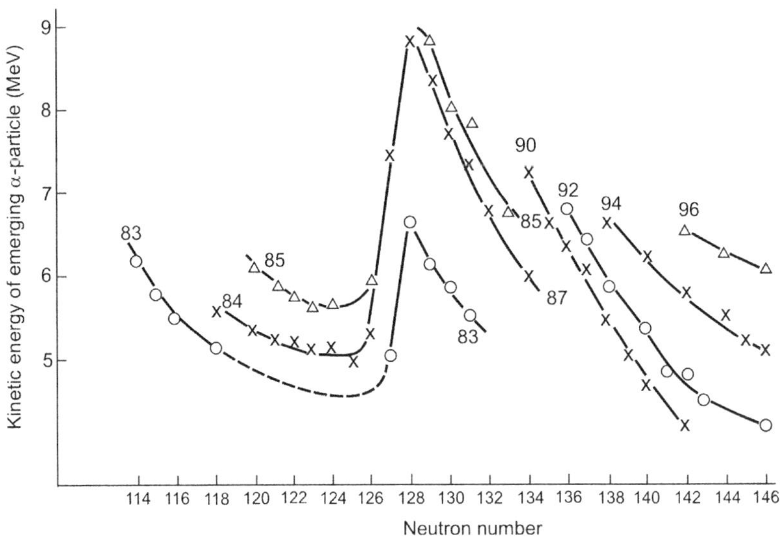

FIGURE 1.24 Energy release in alpha decays. *Goeppert-Mayer (1963); reprinted with permission © The Nobel Foundation 1958.*

Certain nuclei, particularly heavy nuclei ($A > 150$), exhibit large electric quadrupole moments. This property, which is a charge distribution that deviates from what one would encounter with a symmetrical spherical nucleus. This property clearly indicated that certain nuclei were not spherical in shape, but that they were deformed or ellipsoidal in shape. Rainwater (1950) provided a solution by considering that there was an interaction between an inner core and outer surface of valence nucleons whereby the valence nucleons could influence the shape of the nucleus. The movement of the valence nucleons with respect to the core would cause a polarizing effect resulting in a deformed nucleus as what would be expected in the liquid-drop model. Bohr (1951) published ideas similar to those of Rainwater in a paper that was submitted prior to the publication of Rainwater's paper demonstrating their independent discoveries. Bohr linked the movement of nucleons on the nuclear surface to oscillations of the nuclear shape, which would account for the large nuclear quadrupole moments. Bohr predicted collective excitations of the nucleus, namely vibrational excitations resulting in the periodic change in the shape of the nucleus and rotational excitations, which entails the rotation of the entire nucleus not as a single body, but that the motion consists of a surface wave of nucleons that propagates. In his words during his Nobel address, Aage Bohr stated "The rotational motion resembles a wave traveling across the nuclear surface and the moment of inertia is much smaller than for a rigid rotation." The moment of inertia of the nucleus will change as its shape changes. This is analogous to the actions of a figure skater, who with outstretched arms will pull in his or her arms thereby reducing their moment of inertia resulting in a faster spin. In his Nobel lecture Bohr (1975) illustrated the differences between the rotation of a rigid nucleus and one which has a nonrigid surface as depicted in Fig. 1.25.

Thus, nuclear excitations would be expected from the nucleus rotating not as rigid body, but resulting from the movement of a surface wave of nucleons propagating about and around the surface of the nucleus. The nuclear deformations (ellipsoidal shapes) would result from numerous nucleons acting in a collective manner as a wave about the surface of a more rigid inner nuclear core. The differing ellipsoidal shapes or rotational states of the nucleus are manifested as rotational bands in gamma emission spectra as subsequently discussed.

Bohr and Mottelson (1953a,b) were able to correlate the energy of rotational states with gamma ray emission spectra. The energy (E) of the rotational states for even–even nuclei would be defined by the equation

$$E = \hbar^2 \frac{I(I+1)}{2\vartheta} \tag{1.61}$$

where I is the nuclear spin and ϑ is the moment of inertia. Some of the first examples of energy spectra proportional to $I(I+1)$ are illustrated in Fig. 1.26. For example, the ratio of spin-state 4 to spin-state 2 is $4(4 + 1)/2(2 + 1) = 3.33$, and the product of this ratio and the observed energy at spin-state 2 yields (3.33) (93 keV) = 310 keV as given in parenthesis for the calculated energy at spin state 4 in Fig. 1.26.

Similar calculations can be made to arrive at the calculated energy levels of 651 and 1116 keV of ^{180}Hf illustrated in Fig. 1.26. In his Nobel Lecture Bohr stated

> *The spectrum of ^{238}Pu was established by Asaro and Perlman (1953) from measurements of the fine structure in the α-decay of ^{242}Cm. Subsequent evidence showed the spin-parity sequence to be 0+, 2+, 4+, and the energies are seen to be closely proportional to $I(I+1)$.*

These findings were the first to demonstrate that certain nuclear energy states can be explained on the basis that they

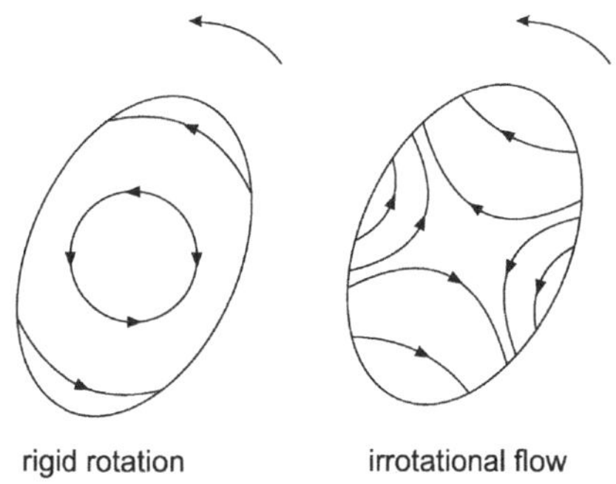

FIGURE 1.25 Velocity fields for rotational motion. For the rotation generated by irrotational flow, the velocity is proportional to the nuclear deformation (amplitude of the traveling wave). Thus for a spheroidal shape, the moment of inertia is $\vartheta = \Im_{rig}(\Delta R/R)^2$ where $\Im_{rig}$ is the moment for rigid rotation, while R is the mean radius and ΔR (assumed small compared to R) is the difference between major and minor semi-axes. *From Bohr (1992); reprinted with permission © The Nobel Foundation 1958.*

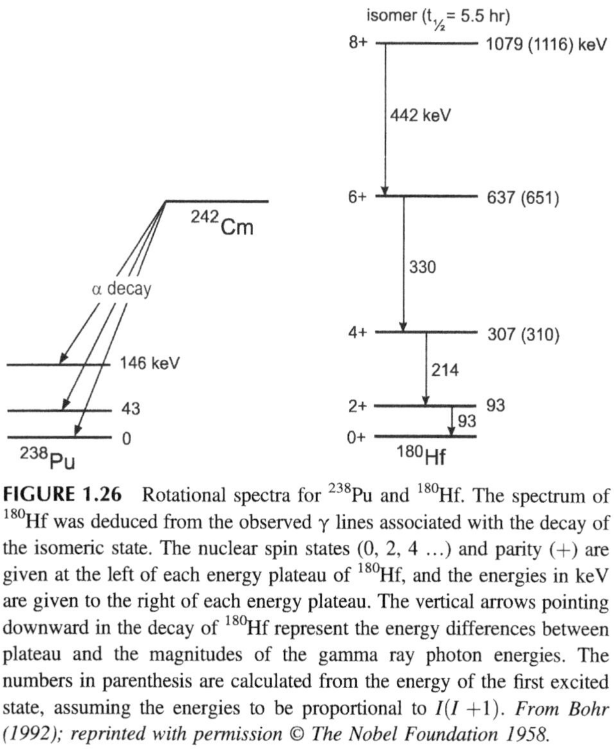

FIGURE 1.26 Rotational spectra for ^{238}Pu and ^{180}Hf. The spectrum of ^{180}Hf was deduced from the observed γ lines associated with the decay of the isomeric state. The nuclear spin states (0, 2, 4 ...) and parity (+) are given at the left of each energy plateau of ^{180}Hf, and the energies in keV are given to the right of each energy plateau. The vertical arrows pointing downward in the decay of ^{180}Hf represent the energy differences between plateau and the magnitudes of the gamma ray photon energies. The numbers in parenthesis are calculated from the energy of the first excited state, assuming the energies to be proportional to $I(I+1)$. *From Bohr (1992); reprinted with permission © The Nobel Foundation 1958.*

constitute a nuclear rotational spectrum; thus providing convincing evidence for the collective model of the nucleus.

E. Superheavy nuclei

The superheavy elements (SHE), also referred to as "superheavies" or SHEs, are those elements that contain nuclei with atomic number $Z > 104$. These do not occur naturally, and SHEs of atomic number $Z = 105$ to $Z = 118$ have been produced artificially. The quest to produce superheavy elements reached a seemingly scientific and political impasse during the period of 1965–74, which is described in an interesting historical account by Dunn (2013) in the *National Geographic*. During this period the Lawrence Berkeley National Laboratory in California and Flerov Laboratory of Nuclear Reactions of the Joint Institute for Nuclear Research in Dubna, Russia, both laid claim to the production of elements 102, 103, 104, 105, and 106. As Dunn (2013) describes, a dispute arose over which laboratory (at Berkeley or Dubna) was the first to produce these elements. This dispute got ugly as it was during the Cold War between the US and Soviet Union. Dunn (2013) related, compromise prevailed and "nuclear war" was averted by the naming of element 105 as dubnium (Db), after the city of Dubna, Russia, and element 106 seaborgium (Sg) after Glenn T. Seaborg, who was codiscoverer of the transuranium elements 94 through 103 and element 106.

The isotopes of the superheavy elements ($Z > 104$), which have been produced so far, are unstable with increasingly diminishing half-lives from elements 105 to 118. For example, the half-lives of these elements are measured in units of hours for element 105, minutes for elements 106 and 107, seconds for elements 108 to 114, and milliseconds for elements 115 to 118. While the superheavy elements, that have been produced so far (up to element 118), have very unstable nuclei, current theory of shell structure of the atomic nucleus predicts certain "magic numbers" of protons and neutrons that would attribute a relatively high stability to yet much heavier elements.

The shell structure of the atomic nucleus with its "magic numbers" of protons and neutrons in the nucleus, first discovered by Maria Goeppert-Mayer and Hans Jensen during 1948–49, was described earlier in this chapter. The shell structure described the stable atomic nuclei as having magic numbers of 2, 8, 20, 26, 28, 50, and 82 protons or neutrons or 126 neutrons, which comprised closed shells of protons or neutrons in the nucleus of an atom that attribute to the atom exceptional stability. Atomic nuclei that were "doubly magic" that is, those nuclei that contained magic numbers of closed shells for both protons and neutrons (*e.g.*, ^{4}He with $Z = 2$ and $N = 2$, ^{16}O with $Z = 8$ and $N = 8$, ^{40}Ca with $Z = 20$ and $N = 20$, ^{48}Ca with $Z = 20$ and $N = 28$, and ^{208}Pb with $Z = 82$ and $N = 126$), were shown to be most stable.

The concept of the shell structure and magic numbers of protons and neutrons in the nucleus has been extended to the theory and synthesis of the superheavy elements (SHE), also referred to as "the superheavies" or SHEs. New magic numbers are predicted that would constitute closed shells of protons and neutrons in an atomic nucleus that would attribute stability to superheavy nuclei with $Z = 114$ and $N = 184$, or $Z = 120$ (Bender et al., 2001) or $Z = 126$ (Ćwiok et al., 1996). These magic numbers of closed shells of protons and neutrons are predicted to yield one or more

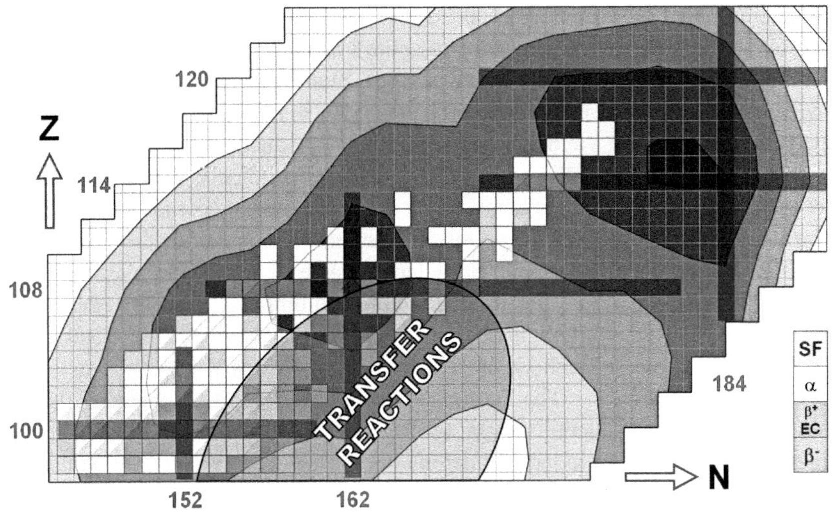

FIGURE 1.27 Cutout from the Chart of Nuclides in the region of the heaviest known elements. The underlying contour plot shows the magnitude of shell stabilization according to microscopic–macroscopic calculations (Patyk et al., 1997). The predicted Island of Stability is clearly visible as a dark area close to $N=184$ and $Z=114$. Note that other calculations favor a closed proton shell at $Z=120$ (Bender et al., 2001), or $Z=126$ (Ćwiok et al., 1996) instead. Known nuclei are indicated with color coded boxes according to their decay properties. The region of neutron-rich isotopes accessible via multinucleon transfer reactions is highlighted. For the color version of the figure, the reader is referred to the online or electronic version of the book. *From Dvorak et al. (2011); reprinted with permission from Elsevier © 2011.*

elements with unusual stability, which could be measured with half-lives in the thousands of years. These superheavy nuclides would have decayed, if they were produced at the time of the formation of the earth 4.5 billion years ago. Their predicted numbers of closed shells of protons and neutrons place them in the Chart of the Nuclides in a separate region referred to as the "Island of Stability", surrounded by a "Sea of Instability" composed of known and unstable radionuclides of very short half-life. Fig. 1.27 illustrates where the predicted "Island of Stability" would be found in the Chart of the Nuclides. The dark blue area and dark blue blocks of Fig. 1.27 (color visible in the electronic version of this book) in the region of $Z=114$ to 126 and $N=177$ to 186 identify the "Island of Stability" surrounded by light blue and yellow and green blocks representing the locations of very short-lived nuclides comprising a "Sea of Instability".

Currently, element 114, namely Flerovium (Fl), has been produced as isotopes with mass numbers in the range $A=285$ to 289. Thus, its heaviest known isotope (Fl-289) contains $A-Z=289-114=175$ neutrons, which is below the predicted magic number of $N=184$. Its measured half-life is reported to be 2.6 seconds. The difficulty lies in finding the mechanism of producing an isotope with the magic number of 184 neutrons, which would require adding an additional 9 neutrons to Flerovium-289, an isotope of Element 114.

The predicted magic number of $Z=120$, which would be Element 120 with its predicted filled shell of protons, is yet to be produced. In a review on the shell structure of superheavy nuclei, Greiner (2008) notes that research utilizing meson field theory and other approaches indicates that the proton number $Z=120$ should prove to be as much a magic number as $Z=114$ (Rutz et al., 1997; Papazoglou (1998); Papazoglou et al., (1999) Beckmann et al., 2002). The corresponding magic number of neutrons for Element 120 is predicted to be $N=172$ and to a lesser degree, $N=184$ (Greiner, 2008). Work by R. A. Gherghescu, Greiner, and coworkers (Poenaru et al., 1987, 1989) calculated the potential energy surface of the $Z=120$ nucleus; and the charge distribution of the $Z=120$, $N=184$ nucleus points to a hallow interior of that nucleus. This work indicates that such a nucleus would have a structure similar to the fullerene "buckyball" molecule that consists of 60 carbon atoms (C_{60}) in a spherical structure with hollow interior, as illustrated in Fig. 1.28. Greiner (2008) notes that these studies suggest that the nucleus with predicted doubly magic numbers $Z=120$, $N=184$ would consist of 60 α-particles (^4_2He nuclei) with one additional binding neutron per α-particle, as illustrated in Fig. 1.28. These studies point to a new frontier where elements beyond $Z=118$ will be produced, and some with an exceptional stability with long half-lives measured to have magnitudes in the thousands of years. When the writer was a university student in the early 1960s, the shell structure of

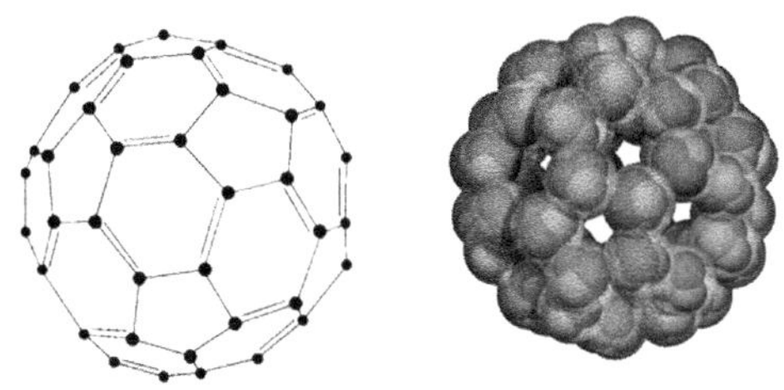

FIGURE 1.28 On the left is the typical molecular structure of the C_{60} carbon molecule fullerene. On the right is the nuclear case where the Carbon atoms are replaced by α-particles and the double bondings by the additional neutrons. Such a structure would immediately explain the semi-hollowness of that superheavy nucleus, which is revealed in the mean-field calculations within meson-field theories. For the color version of the figure, the reader is referred to the online or electronic version of the book. *From Greiner (2008); reprinted with permission from Elsevier © 2008.*

the atomic nucleus and the existence of magic numbers of protons and neutrons in the nucleus was an altogether new field. The buckyball structure of carbon molecules was unknown then. In years to come, amazing discoveries are expected with the production of new superheavy elements beyond Element 118.

The current goal is to find the ways and means of producing neutron-rich superheavy elements within the Island of Stability. Dean (2008) describes the methods used by Yuri Oganessian and his team of the Flerov Laboratory of Nuclear Research at the Joint Institute for Nuclear Research in Dubna, Russia, and methods utilized at the Lawrence Livermore National Laboratory (LLNL) in California. Also Hofmann (2009) describes the fusion reactions based on lead and bismuth targets, which were utilized by the Gesellschaft für Schwerionenforschung mbH (GSI) in Darmstadt, Germany, and/or at RIKEN in Wako, Japan, for the synthesis of neutron-deficient isotopes of the elements from 107 to 113, and the fusion reactions of ^{48}Ca projectiles with actinide targets whereby more neutron-rich isotopes of the elements from 112 to 116 and even 118 were produced at the Flerov Laboratory of Nuclear Reactions (FLNR) at the Joint Institute for Nuclear Research (JINR) in Dubna, Russia. Also, the Lawrence Livermore National Laboratory in California produced elements 113 through 118 in a joint collaboration between the Flerov Laboratory at the JINR in Dubna, Russia. Elements 114 and 116 were named flerovium (Fl) and livermorium (Lv) in honor of the two laboratories.

The production of the superheavy elements 113 through 118 involved fusion reactions of "doubly magic" neutron-rich ^{48}Ca with massive actinide target elements (Oganessian and Utyonkov, 2015). The ^{48}Ca stripped of its electrons is a highly stable nucleus containing both magic numbers of protons and neutrons ($Z = 20$ and $N = 28$). These nuclei are accelerated in a cyclotron against an actinide target resulting in the fusion of the ^{48}Ca projectile and actinide according to the following examples:

$$^{243}_{95}\text{Am} + ^{48}_{20}\text{Ca} \rightarrow ^{288}_{115}\text{Mc} + 3^1_0\text{n} \rightarrow ^{284}_{113}\text{Nh} + ^4_2\text{He} \quad (1.62)$$

$$^{243}_{94}\text{Pu} + ^{48}_{20}\text{Ca} \rightarrow ^{292}_{114}\text{Fl}* \rightarrow ^{289}_{114}\text{Fl} + 3^1_0\text{n} \quad (1.63)$$

$$^{243}_{95}\text{Am} + ^{48}_{20}\text{Ca} \rightarrow ^{288}_{115}\text{Mc} + 3^1_0\text{n} \quad (1.64)$$

$$^{248}_{96}\text{Cm} + ^{48}_{20}\text{Ca} \rightarrow ^{296}_{116}\text{Lv}* \rightarrow ^{293}_{116}\text{Lv} + 3^1_0\text{n} \quad (1.65)$$

$$^{249}_{97}\text{Bk} + ^{48}_{20}\text{Ca} \rightarrow ^{297}_{117}\text{Ts}* \rightarrow ^{293}_{117}\text{Ts} + 4^1_0\text{n} \quad (1.66)$$

$$^{249}_{98}\text{Cf} + ^{48}_{20}\text{Ca} \rightarrow ^{294}_{118}\text{Og} + 3^1_0\text{n} \quad (1.67)$$

A new approach to the production of neutron-rich elements that reside in the Island of Stability is the application of multinucleon transfer reactions reviewed by Dvorak et al. (2011) and Corradi et al. (2013). These reactions usually are carried out at lower projectile energies whereby one or more nucleons are transferred between the projectile and target nuclei (Götz et al., 2017). As described by Dvorak et al. (2011), these are deeply inelastic reactions whereby the projectile and target nuclei come into contact and exchange a large number of nucleons, before they separate. Dvorak et al. (2011) note that these reactions can yield a broad spectrum of products including the neutron-rich isotopes of the heaviest elements. This technique requires lower and more gentle projectile energies than those used in the fusion reactions, which can tend to "break up" target nuclei. A large source of neutron-rich isotopes that may be used in future multinucleon transfer reactions is highlighted in Fig. 1.27 in the region of the Chart of the Nuclides labeled "transfer reactions".

The acceleration of a beam of charged nuclei onto an actinide target for the production of superheavy elements can produce numerous nuclear reaction products in addition to the superheavy elements (Roberto et al., 2015). In addition, the superheavy elements produced to date all have short half-lives. The separation, detection, and measurement of newly produced superheavy elements require, after the beam irradiation of target nuclei, the separation of the beam nuclei and the scattered product ions, which emerge from the target, followed by their detection and measurement (Theisen et al., 2015; Popeko, 2016). An instrumental design used to separate, detect, and measure reaction products of this type, referred to as a quadrupole (QD) separator/spectrometer, is designed to contain a combination of magnetic quadrupoles (multipoles) and a magnetic dipole, as described by Dvorak et al. (2011) and illustrated in Fig. 1.29.

When the beam of projectile nuclei strikes the actinide target, reaction products (*e.g.*, transfer products) are scattered in many directions from the target material. These ions will enter the separator where quadrupole or multipole magnets will focus the product nuclei toward a central area in the first stage of the separator. The transfer products as well as nuclei of the original nuclear beam, which had not

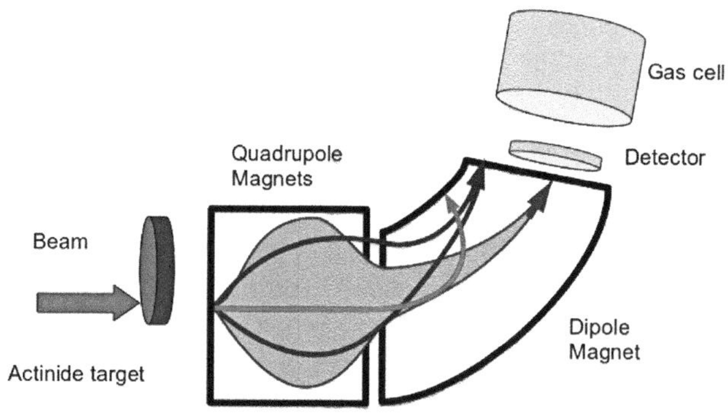

FIGURE 1.29 Schematic drawing of a separator based on the QD design. Ions are focused in a pair of quadrupole magnets before they enter the dipole magnet, where they are separated according to their magnetic rigidity. The beam, which leaves the target without a nuclear interaction has a low magnetic rigidity and is deflected to the inner wall of the dipole magnet. The intensity of the dipole field is tuned to guide the neutron-rich heavy transfer products into the focal plane detector (spectrometer mode) or into the gas cell (products collection mode). For the color version of the figure, the reader is referred to the online or electronic version of the book. *From Dvorak et al. (2011); reprinted with permission from Elsevier © 2011.*

undergone any nuclear reaction, enter into the field of the dipole magnet, which will separate the ions according to their momentum rigidity (i.e., their momentum-to-charge ratios). Ions of the nuclear beam, which had passed through the target material without reaction, are of low mass compared to the heavy transfer products. These ions thus have a lower magnetic rigidity and are deflected easily in the dipole magnetic field away from the detector and land onto the inner wall of the dipole magnet. These ions are, in a sense, discarded. The dipole magnetic field is tuned to focus the superheavy transfer products onto a detector, which serves as a spectrometer for the determination of the particle mass and proton number for particle identification (Dvorak et al., 2011). An alternative mode of the separator is to remove the detector and replace it with a gas cell, as illustrated in Fig. 1.29, for the collection of the neutron-rich isotopes of the heaviest elements. The heavy elements are stopped in the gas cell, that is, the gas medium serves as a collector of the neutron-rich transfer products. The neutron-rich heavy elements in the collection gas can then be transferred to another experimental arrangement for the measurement of their physical and chemical properties. Single or double charged ions are extracted from the gas via RF and DC electrical fields through a nozzle with extraction times as short as 0.1 s (Dvorak et al., 2011). Another type of ion separator/spectrometer used to collect, detect, and measure the heavy ions is a solenoid-based design, which utilizes a solenoid magnet to focus the transfer-product ions according to momentum and charge onto a detector and/or gas cell collector. This design is described by Dvorak et al. (2011).

The future offers great promise for discoveries in the ways and means of producing neutron-rich superheavy elements with very long half-lives. P. Roy Chowdhury of the Saha Institute of Nuclear Physics in Kolkata, India and coworkers (Chowdhury et al., 2008) have modeled the decay modes of numerous isotopes of the superheavy elements, which indicate that a few isotopes could have half-lives, which would measure from a few hundred years to as long as 100 million years. If these elements existed at the time of the creation of the earth, they would not be found, as such isotope lifetimes would be too short compared to the age of the earth (4.5 billion years). If such elements are produced eventually, Dvorak et al. (2011) note that our knowledge of the isotopes in this region would open new fields in nuclear chemistry, nuclear and atomic physics, and nuclear astrophysics of which the latter includes particularly nuclear processes that occur in supernovae.

For additional information the reader might be interested in perusing short and interesting articles by Dean (2008), Shiga (2011), Dunn (2013) and Gray (2017).

F. Cluster radioactivity

Cluster radioactivity is nuclear decay with the spontaneous emission of a particle heavier than an alpha particle with $Z \geq 6$. It is included in this part of the chapter, because cluster radioactivity is an important phenomenon that contributes to our understanding of the properties of the nucleus, in particular, nuclear stability and magic numbers.

Cluster radioactivity was first predicted by Sandulesco and Greiner (1977) and discovered by Rose and Jones (1984), who were the first to observe the emission of ^{14}C from the nucleus of ^{223}Ra. Since the discovery in 1984, intense experimental research at numerous institutions (see reviews by Price, 1989, 1994; Ronen, 1997; Bonetti and Guglielmetti, 1999; Santhosh et al., 2010; Santhosh and Biju, 2013; Soylu and Evlice, 2015; Mirea et al., 2017) have uncovered numerous radionuclides of high mass number ($A \geq 221$) that emit clusters including ^{14}C, ^{20}O, ^{23}F, ^{24}Ne, ^{28}Mg, or ^{32}Si. Radionuclides, that exhibit cluster radioactivity, are listed in Table 1.6 with their cluster decay modes and decay properties.

Cluster decay is a rare event that competes with alpha particle emission, i.e., alpha decay. The probabilities of cluster decay for observed decay modes (*i.e.*, branching ratios) range from 10^{-9} to 10^{-17} of the alpha decay probability. All of the decay modes listed in Table 1.6 yield product nuclei that possess magic numbers of protons ($Z = 82$) and/or neutrons ($N = 126$) with the exception of ^{211}Bi ($Z = 83$, $N = 128$), which is accompanied with the ^{14}C cluster in the decay of ^{225}Ac. However, in this exceptional case, the ^{14}C cluster has the highest binding energy (BE) per cluster. Ronen (1997) compared the magic numbers, binding energies per cluster and Q values of the cluster decays as listed in Table 1.6. From these comparisons, Ronen (1997) concluded the following:

1. *The most likely mode of cluster radioactivity is when the light cluster has the highest BE per cluster and the daughter product is the doubly magic nucleus ^{208}Pb.* We can see from Table 1.6 that, among the isotopes of Ra, which yield the ^{14}C cluster, the decay mode yielding the doubly magic ^{208}Pb has the highest Q value. Also, among the isotopes of U and Pu, the decay modes that yield the doubly magic ^{208}Pb also have high Q values.
2. Nuclei that decay by cluster radioactivity have either a daughter that is magic (N = 126 or Z = 82) or a light emitted cluster with the highest BE per cluster. Nuclei that obey both conditions (magic number and highest BE per cluster for the emitted cluster) have shorter half-lives).

Due to the rarity of cluster radioactivity in comparison to alpha decay, the detection of emitted clusters must be done with a relatively high alpha activity. Certain solid state nuclear track detectors insensitive to alpha particles (Durrani and Bull, 1987; Ilić and Durrani, 2003) serve as good means of detecting cluster radioactivity. However, Bonetti et al. (2001) point out that track detectors may not have enough A resolution to offer unambiguous identification. Thus, to decide in favor of one of the most probable Ne clusters, Bonetti et al. (2001) resorted to theoretical calculations.

TABLE 1.6 Cluster decay modes and properties.[a].

Decay mode[b]	Magic number(s)[c]	E Or Q value (MeV)	Half-life ($\log T_{1/2}$ (s))
$^{221}Fr \rightarrow ^{207}Tl + ^{14}C$	$N = 126$	29.28	14.5 ± 0.12
$^{221}Ra \rightarrow ^{207}Pb + ^{14}C$	$Z = 82$	30.34	13.0 ± 0.2
$^{222}Ra \rightarrow ^{208}Pb + ^{14}C$	$Z = 82$, $N = 126$	30.97	11.0 ± 0.06
$^{223}Ra \rightarrow ^{209}Pb + ^{14}C$	$Z = 82$	29.85	15.2 ± 0.05
$^{224}Ra \rightarrow ^{210}Pb + ^{14}C$	$Z = 82$	28.63	15.8 ± 0.12
$^{225}Ac \rightarrow ^{211}Bi + ^{14}C$	None	28.57	17.16 ± 0.06
$^{226}Ra \rightarrow ^{212}Pb + ^{14}C$	$Z = 82$	26.46	21.3 ± 0.2
$^{228}Th \rightarrow ^{208}Pb + ^{20}O$	$Z = 82$, $N = 126$	44.72	20.7 ± 0.08
$^{231}Pa \rightarrow ^{208}Pb + ^{23}F$	$Z = 82$, $N = 126$	46.68	26.0
$^{230}Th \rightarrow ^{206}Hg + ^{24}Ne$	$N = 126$	51.75	24.6 ± 0.07
$^{231}Pa \rightarrow ^{207}Tl + ^{24}Ne$	$N = 126$	54.14	22.9 ± 0.05
$^{232}U \rightarrow ^{208}Pb + ^{24}Ne$	$Z = 82$, $N = 126$	55.86	20.5 ± 0.03
$^{234}U \rightarrow ^{210}Pb + ^{24}Ne$	$Z = 82$	52.81	25.9 ± 0.2
$^{234}U \rightarrow ^{206}Hg + ^{28}Mg$	$N = 126$	65.26	25.7 ± 0.2
$^{236}Pu \rightarrow ^{208}Pb + ^{28}Mg$	$Z = 82$, $N = 126$	70.22	21.7 ± 0.3
$^{238}Pu \rightarrow ^{208}Pb + ^{30}Mg$	$Z = 82$, $N = 126$	67.00	25.7 ± 0.25
$^{238}Pu \rightarrow ^{206}Hg + ^{32}Si$	$N = 126$	78.95	25.3 ± 0.16

[a]*Decay mode, energy (E), and half-life data are from Price (1994) and Ronen (1997).*
[b]*Other cluster decays, not listed in the table, are* $^{236}U \rightarrow ^{206}Hg$ *(N = 126) +* ^{30}Mg, $^{236}Pu \rightarrow ^{212}Pb$ *(Z = 82) +* ^{24}Mg*, and* $^{242}Cm \rightarrow ^{208}Pb$ *(Z = 82, N = 126) +* ^{34}Si *reported by Tretyakova (1995) and* $^{230}U \rightarrow ^{208}Pb$ *(Z = 82, N = 126) +* ^{22}Ne *and* $^{226}Th \rightarrow ^{208}Pb$ *(Z = 82, N = 126) +* ^{18}O *reported by Bonetti et al. (2001). The magic numbers of the product nuclei are in parenthesis where the nucleus of* ^{208}Pb *is "doubly magic".*
[c]*The magic numbers are those of the cluster-decay product nuclei.*

Several of the cluster decays listed in Table 1.6 are illustrated in Fig. 1.30 together with sections of the Chart of the Nuclides (Z vs. N). This illustrates how cluster radioactivity competes with alpha particle emission and results in large leaps in the Chart of the Nuclides from parent to daughter nuclides.

An important outcome of the discovery of cluster radioactivity was the reconfirmation of the magic numbers of closed shells of protons and neutrons in the nucleus that yield exceptional stability to atomic nuclei discussed previously in this chapter. Numerous works have been devoted to explaining the cluster decay process. These are summarized by Santhosh et al. (2010). There are two approaches to explaining the cluster decay process, namely, that it is (1) fission-like and (2) alpha-like. In the fission model (Sandulescu et al., 1985; Greiner et al., 1985; Poenaru et al., 1985, 1986; Shi and Swiatecki, 1985; Shanmugam and Kamalaharan, 1988) the nucleus deforms continuously as the cluster penetrates the nuclear barrier. This would be somewhat similar to the nuclear fission process discussed previously in this chapter. In the alpha-like or cluster model of decay (Blendowske et al., 1987; Landowne and Dasso, 1986; Buck and Merchant, 1989; Malik and Gupta, 1989), the cluster is assembled in the nucleus in a similar fashion as the alpha particle is assembled (see alpha decay in this chapter.) before the cluster penetrates the nuclear potential barrier.

VII. Relativistic properties of nuclear radiation

A. Relativity

Einstein's theory of relativity is sometimes utilized in the measurement and detection of nuclear radiation and subatomic particles. In this part of the chapter, examples are provided on how relativity is taken into account in the measurement of muons traveling close to the speed of light in the cosmic radiation showers, the measurement of the lifetime of short-lived subatomic particles originating in the cosmic radiation showers, and the measurement of the mass and de Broglie wavelengths of particles, such as electrons and neutron radiation, traveling at relativistic speeds.

In the third of Einsteins' papers submitted to *Annalen der Physik* (Einstein, 1905a,b,c) he proposed his theory of special relativity. Einstein's *Principle of Relativity* is based

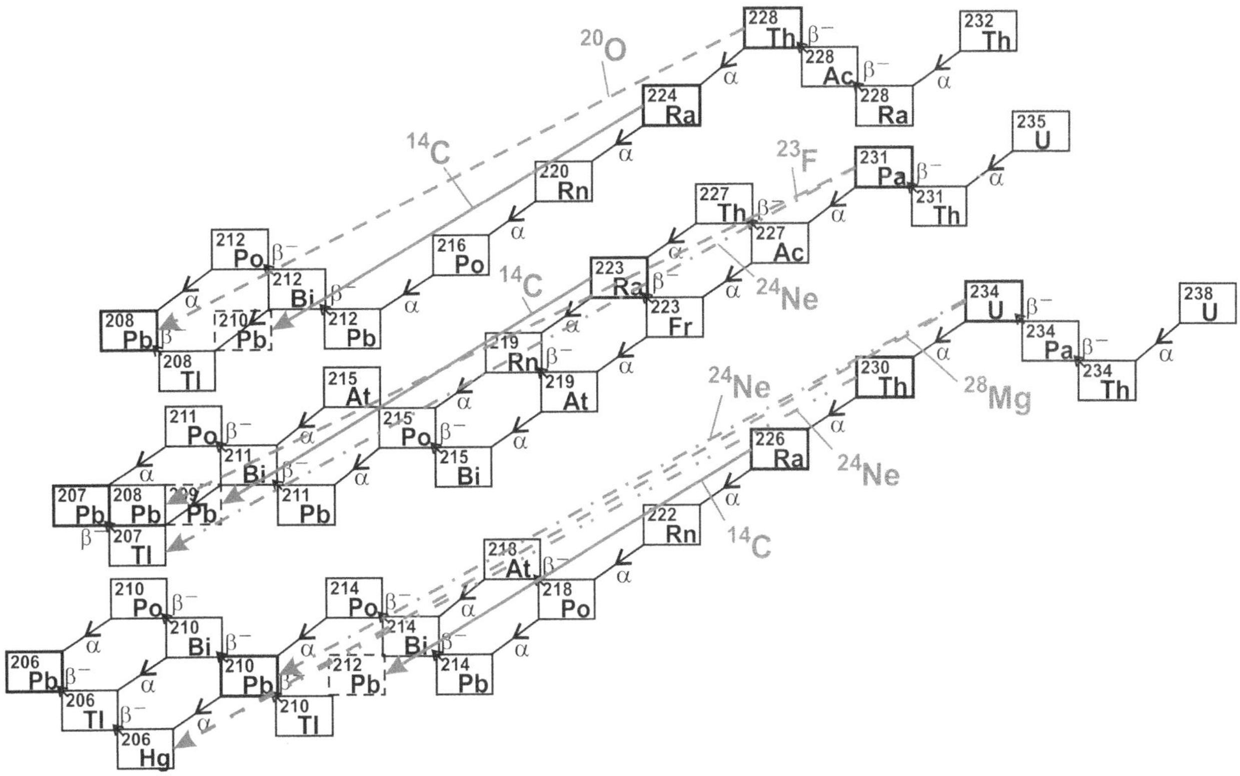

FIGURE 1.30 Cluster radioactivity of actinide nuclei. By emission of ^{14}C, ^{20}O, ... "big leaps" in the periodic system can occur, just contrary to the known α-, β-, γ-radioactivities, which are also partly shown in the figure. (Caption and figure are from Greiner, 2008 with red color added by the writer; reprinted with permission from Elsevier © 2008.) For the color version of the figure, the reader is referred to the online or electronic version of the book.

on two postulates, which are namely, (1) that in making observations or measurements in physics, there should be no preferred inertial (nonaccelerating) frame. In other words, for all inertial observers, the laws of physics are the same. (2) The speed of light in a vacuum is a constant ($c = 2.99 \times 10^8$ m/s) for all inertial observers, regardless of the velocity of travel of the source. Einstein developed his principle of relativity into what would be referred to later as the special theory of relativity that would provide an analysis of space and time. The special theory of relativity can be used to explain changes of length, time, and mass when an observer in one inertial frame makes measurements of length, time, and mass of another body or clock in another inertial frame. A person in one inertial frame of reference can make measurements of length, time, and mass on another inertial frame and vice versa. Space, time, mass, and speed are all relative depending on which inertial frame of reference is used. In his Nobel Lecture on the theory of relativity delivered to the Nordic Assembly of Naturalists at Gothenburg on July 11, 1923 (http://www.nobelprize.org), Einstein underscored the two principles upon which the special theory of relativity is based (Einstein, 1923) as follows:

> *The special theory of relativity is an adaption of physical principles to Maxwell–Lorentz electrodynamics. From earlier physics it takes the assumption that Euclidian geometry is valid for the laws governing the position of rigid bodies, the inertial frame and the law of inertia. The postulate of the equivalence of inertial frames for the formulation of the laws of Nature is assumed to be valid for the whole of physics (special relativity principle). From Maxwell–Lorentz electrodynamics it takes the postulate of invariance of the velocity of light in a vacuum (light principle).*

B. Relativistic length contraction and time dilation

In Einstein's treatise on relativity (Einstein, 1916a,b) he explained that, when making measurements of events from frames of reference that are in uniform motion, we must define the events by four coordinates, namely the space coordinates x, y, and z as well as the time coordinate t. Thus, any event wherever it takes place would be fixed in space by the perpendicular space coordinates x, y, and z and a fixed time t (see Fig. 1.31). The same event, relative to another frame of reference, is also fixed in space and time according to the space–time coordinates x', y', z', t'. In his treatise on relativity, Einstein (1916a,b) explained that the world we live in should be considered as a four-dimensional space–time continuum.

Einstein (1916a,b) demonstrated that, when events measured from one frame of reference (e.g., O, where the events are measured by an observer at O) on another frame of reference (e.g., O', where the same events are measured by an observer at O'), which is traveling at a velocity v relative to the observer at O, will yield different measurements of time (*e.g.*, time dilation) and length (e.g., length contraction), which is dependent on the frame of reference from which the measurements are taken and the velocity (v). The differences in measurement of time and length taken from the two frames of reference will be greater as the velocity v approaches the unattainable speed of light c (*i.e.*, speed of light in a vacuum = 2.9979×10^8 m/sec).

The problem of relativity was presented by Einstein (1916a,b) with a question, which may be worded as follows: What are the values of x', y', z', and t' of an event with respect to frame O', when we know the values of x, y, z, and t, of the same event with respect to reference frame O? He explained that the solution to the problem must satisfy a law of the transmission of light in a vacuum, abbreviated by the symbol c, as a constant. Einstein indicated that for the coordinate systems illustrated in Fig. 1.31, the problem is solved by the following Lorentz transformations (Eqs. 1.68 and 1.69), named after Nobel Laureate Hendrik A. Lorentz:

Length:

$$x' = \frac{x - vt}{\sqrt{1-(v^2/c^2)}}, \quad y' = y \quad \text{and} \quad z' = z \tag{1.68}$$

Time:

$$t' = \frac{t - \frac{v}{c^2}x}{\sqrt{1-(v^2/c^2)}} \tag{1.69}$$

where the variables are those defined in the previous paragraph. In his treatise, Einstein (1916a,b) provided a derivation of Eqs. (1.68) and (1.69). The detailed derivation is provided in the Appendix of a previous book by the writer (L'Annunziata, 2016).

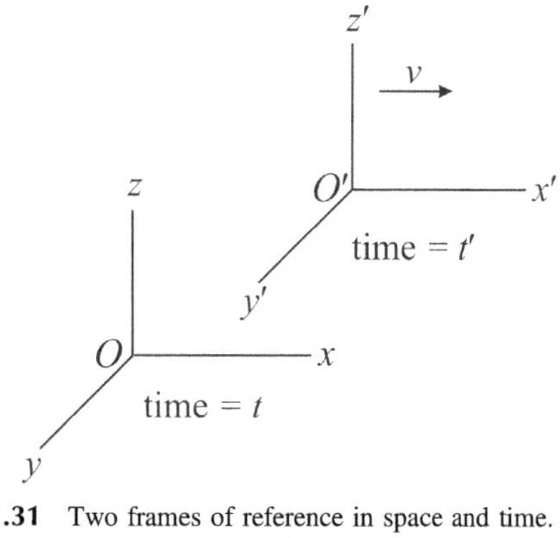

FIGURE 1.31 Two frames of reference in space and time. An event in space and time in reference frame O' is defined by space-time coordinates x', y', z', and time t', traveling at a velocity v relative to reference frame O, defined by coordinates x, y, z, and t.

1. Length contraction in relativity

Fig. 1.32 illustrates an observer (with waving hands) in reference frame O', who takes a measurement in his or her own frame of reference, and measures the length of an object at time t' along the x-axis to extend from x'_a to x'_b. Since the object is measured by an observer at O' in his or her own frame of reference, the object is at rest with respect to the observer at O'. At the same time, another observer in reference frame O takes a measurement of the same object in the reference frame O' to extend from x_a to x_b. Reference frame O' is traveling at a velocity v with respect to reference frame O. The measurement of the length of the object made by the observer in reference frame O will be smaller than that made by the observer in reference frame O' according to the Lorentz transformation defined in Eq. (1.68) above.

Let us look at Eq. (1.68) of the Lorentz transformation with respect to the specific example provided in the previous paragraph. The Lorentz transformation that defines changes in length along the x-axis (Eq. 1.68) is the following:

$$x' = \frac{x - vt}{\sqrt{1-(v^2/c^2)}} \tag{1.70}$$

where x' is a coordinate along the x axis measured at time t' in reference frame O' as measured by an observer in the

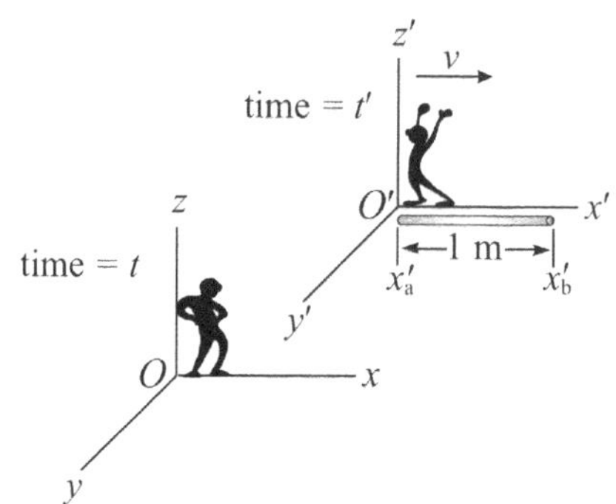

FIGURE 1.32 Two frames of reference in space and time. An event in space and time in reference frame O' is defined by space-time coordinates x', y', z', and time t', traveling at a velocity v relative to reference frame O, defined by coordinates x, y, z, and t. An observer at O (with hands on hips) may take measurements of length and time on the frame of reference at O' (where there is an observer with hands waving) and measure length to be contracted (*i.e.*, shorter) and time dilated (i.e., longer interval of time) than the length and time measured by the observer at O'. The observer at O', who makes measurements of an event in his or her own reference frame would experience different measurements, because the measurements would be made of an event that would be at rest with respect to the observer. A rod is illustrated in the reference frame O', which would measure 1 m by the observer at O', because the rod would be at rest with respect to the observer at O'. However, the rod measured by the observer in the reference frame O would measure the rod to be contracted or shorter than 1 m in length, and the amount of contraction would be a function of the velocity v at which the reference frame O' is traveling with respect to reference frame O. *From L'Annunziata (2016); reprinted with permission from Elsevier © 2016.*

same reference frame, which is traveling at the velocity v with respect to another reference frame O. The variable x is the coordinate along the x axis measured by an observer in reference frame O at time t, and c is the invariable speed of light in a vacuum. With respect to the example given in the previous paragraph, we can express the two measurements taken by the two observers in the two reference frames with the Lorentz transformation Eq. (1.68) as follows:

$$x'_b - x'_a = \frac{(x_b - x_a) - v(t' - t)}{\sqrt{1 - (v^2/c^2)}} \tag{1.71}$$

where $x'_b - x'_a$ are the measurements along the x-axis (length of an object) in reference frame O' taken by an observer in the same reference frame where the object would be at rest with respect to the observer, which we can now call L_r; and $x_b - x_a$ are the measurements of the same object made by an observer in reference frame O, which we may call L. If the measurements are made by the two observers in the two reference frames at the same time, $t' = t$ and $t' - t = 0$ and Eq. (1.71) becomes

$$L_r = \frac{L}{\sqrt{1 - (v^2/c^2)}} \tag{1.72}$$

where L_r is the length of the object that is at rest with respect to the observer (at O') traveling at the velocity v with respect to an observer at O in another reference frame, L is the length of the object in reference frame O' measured by an observer in reference frame O. Eq. (1.72) may be written as

$$L = L_r\sqrt{1 - (v^2/c^2)} \tag{1.73}$$

The above Eq. (1.73) states that an observer in a reference frame, who measures an object (L) in another reference frame that is traveling at a velocity v with respect to the observer will be contracted (*i.e.*, smaller) with respect to the measurement L_r taken by another observer, who make the measurement in the reference frame where the object is at rest with respect to that observer. Thus, L is always smaller than L_r (*i.e.*, $L < L_r$) by the factor $\sqrt{1 - (v^2/c^2)}$, and the degree of contraction is dependent on the velocity v. As the velocity v increases and approaches the unattainable speed of light c, the value of L decreases, that is, contraction increases.

Einstein (1916a,b) provided examples of how measurements from one frame of reference on another frame of reference will yield different measurements of time and length dependent on the frame of reference from which the measurements are taken and the velocity v at which one frame of reference is traveling with respect to the other. As an example, he chose a meter-rod situated in Fig. 1.32 in the coordinate system (reference frame) O' at $x' = 0$ at one end and $x' = 1$ m at the other end of the rod. He then proceeded to calculate the length of the meter-rod relative to system O at a particular time t. Using Eq. (1.70) of the Lorentz transformation reduced to Eq. (1.73), Einstein shows that at time $t = 0$, the value of $x = 0$ for the beginning of the rod, that is,

$$x = 0\left(\sqrt{1 - (v^2/c^2)}\right) \tag{1.74}$$

and the end of the rod would measure

$$x = 1\ \text{m}\left(\sqrt{1 - (v^2/c^2)}\right) \tag{1.75}$$

Thus, when the meter rod is traveling at the speed v relative to O, the meter-rod is shorter as judged from O according to the factor $\sqrt{1 - (v^2/c^2)}$, and the faster that the rod is traveling relative to O, the shorter will be the rod as measured from O. The degree of rod contraction is a function of the velocity v, and it is significant at relativistic speeds, that is, when v approaches c, the unattainable speed of light.

A specific example of relativistic length contraction in the detection of cosmic-ray muons on earth is provided subsequently in Part C.

2. Time dilation in relativity

Another example proposed by Einstein (1916a,b) in his treatise on relativity included a clock, which is permanently situated at the origin ($x = 0$) of the frame of reference O, which travels at the velocity v relative to the reference frame O' (see Fig. 1.33). He then assigned two successive ticks of the clock as $t = 0$ for the first tick and $t = 1$ second for the following tick. Using the Lorentz transformation provided by Eq. (1.69) above, Einstein (1916a,b) calculated that the time for the two ticks measured from the frame of reference O' to be $t = 0$ for the start and $t = 1/\sqrt{1 - (v^2/c^2)}$ for the following tick of the clock. Thus, from the reference frame of O', Einstein demonstrated that the time between two ticks of the clock is not one second, but rather the dilated time of $t = 1/\sqrt{1 - (v^2/c^2)}$ seconds. Thus, when measured from the frame of reference O', the time is enlarged, and the greater the velocity v relative to the observer at O', the longer will be the time measured from O'. As explained by Einstein (1916a,b), the value of v can never reach c, which remains an unattainable limiting velocity.

In the above example, Einstein (1916a,b) used the Lorentz transformation provided in Eq. (1.69), namely,

$$t' = \frac{t - \frac{v}{c^2}x}{\sqrt{1 - (v^2/c^2)}} \tag{1.76}$$

and

$$\Delta t' = \frac{(t_2 - t_1) - \frac{v}{c^2}(x_2 - x_1)}{\sqrt{1 - (v^2/c^2)}} \tag{1.77}$$

where x_2 and x_1 are the positions of the clock at the origin of reference frame O. The clocks remain at rest with respect to the observer in frame O, and thus $x_2 - x_1 = 0$. The above Eq. (1.77) is then written as

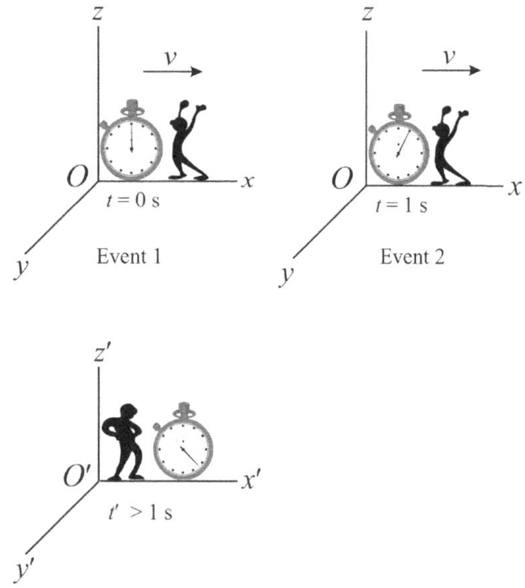

FIGURE 1.33 An observer in reference frame O measures two events with a clock situated at the origin. In reference frame O, the clock remains at the origin, that is, it does not move with respect to the observer in that reference frame. The reference frame O travels at a velocity v with respect to the frame of reference O'. However, the clock in reference frame O is at rest with respect to the observer in that reference frame, who measures the time interval between the two events to be 1 seconds. An observer in the frame of reference O' sees the two events as occurring in two locations and measures the time interval between these events to be $t' > 1$ seconds. *From L'Annunziata (2016); reprinted with permission from Elsevier © 2016.*

$$\Delta t' = \frac{\Delta t_r}{\sqrt{1-(v^2/c^2)}} \tag{1.78}$$

where $\Delta t_r = t_2 - t_1$, that is, the time interval measured by the observer in reference frame O where the clocks are at rest with respect to the observer. Thus, the time interval measured by the observer at O' is the time interval

$$\Delta t' = \frac{1\ \text{s}}{\sqrt{1-(v^2/c^2)}}, \tag{1.79}$$

which is a dilated or longer time interval than 1 seconds, that is $\Delta t' > \Delta t_r$. The length of the time interval is greater than 1 seconds by the factor $1\Big/\sqrt{1-(v^2/c^2)}$, and this dilated time with be greater as the velocity v increases and approaches c, the unattainable velocity of light in a vacuum.

The above results would be the same if we were to consider the reference frames O and O' in reverse, that is, if the reference frame O' would be moving with the velocity v with respect to O, the observer at O would measure the same results with the following Lorentz transformation with the expression Δt rather than $\Delta t'$:

$$\Delta t = \frac{\Delta t_r}{\sqrt{1-(v^2/c^2)}} \tag{1.80}$$

Real examples of relativistic calculations involving time dilation and length contraction with respect to the measurement of muons in the showers of cosmic radiation are provided in the following section.

C. Relativity in cosmic-ray muon detection and measurement

Muons are produced high in the atmosphere at an altitude of ~ 15 km where high-energy cosmic-ray nucleons encounter and undergo collision with nuclei of the atmospheric gases. The muons produced high in the atmosphere are components of the showers of cosmic radiation, and these muons travel toward the earth reaching the ground (sea level) with a mean energy of ≈ 4 GeV at speeds close to the speed of light ($>0.99c$).

One might ask at this point the following question: If muons of the cosmic radiation showers are produced high in the earth's atmosphere (at altitudes of $\sim 15{,}000$ m or higher), travel at speeds close to the speed of light, and have an average lifetime of only 2.2×10^{-6} seconds, how could such muons be detected to reach the earth's surface? The question is an obvious one, as it would appear with classical mathematics that, with such a short lifetime, muons would never be expected to reach the earth. Let's say, for example, a muon is produced at the altitude of 15,000 m and it travels at the speed v of $0.9993c$, which would correspond to a speed in units of m/s as

$$\begin{aligned} 0.9993c &= (0.9993)\left(2.9979 \times 10^8\ \text{m/s}\right) \\ &= 2.9958 \times 10^8\ \text{m/s}. \end{aligned} \tag{1.81}$$

The classical calculation of its average distance d of travel during its lifetime would be

$$\begin{aligned} d = (v)(t) &= \left(2.9958 \times 10^8\ \text{m/s}\right)\left(2.2 \times 10^{-6}\ \text{s}\right) \\ &= 659.1\ \text{m} \end{aligned} \tag{1.82}$$

Thus, the muon would be expected to decay, according to classical transformations, after traveling only 659 m on the average and, therefore, never be detected an earth. However, for such a muon viewed by an observer on earth, he or she would view the following two events: A, the creation of the muon and B, the decay of the muon, to occur in two locations in space–time coordinates. Einstein's theory of relativity utilizing the Lorentz time transformations demonstrate that the two events, A and B, separated by the time interval Δt, will be dilated to the time interval Δt_{earth}, when viewed by an observer on earth. The time dilation, as described by Einstein (1916a,b), is calculated according to the following:

$$\Delta t_{\text{earth}} = \frac{\Delta t_r}{\sqrt{1-(v^2/c^2)}} \tag{1.83}$$

where v is the velocity of the particle, and c is the invariable speed of light in a vacuum. For the muon traveling at

0.9993c, we can calculate Δt_{earth} or the observed dilated lifetime of the muon as follows:

$$\Delta t_{\text{earth}} = \frac{\Delta t_r}{\sqrt{1-(v^2/c^2)}} = \frac{2.2\times10^{-6}\ \text{s}}{\sqrt{1-(0.9993)^2}} = \frac{2.2\times10^{6}\ \text{s}}{0.0374}$$
$$= 58.8\times10^{-6}\ \text{seconds} \qquad (1.84)$$

According to the dilated time of 58.8 × 10^{-6} seconds, the observer on earth will determine that the muon would travel, before it decays, a distance of

$$d = (v)(\Delta_{\text{tearth}}) = (2.9958\times10^{8}\ \text{m/s})(58.8\times10^{-6}\ \text{s})$$
$$= 17615\ \text{meters}, \qquad (1.85)$$

which is well beyond the distance of 15,000 m elevation when it was created, permitting its detection on earth. The distance is only an average figure, because the muon lifetime of 2.2 × 10^{-6} seconds is an average lifetime. For thorough treatments of relativistic space–time measurements and the applications of Lorentz transformations to these measurements see books by Einstein (1916a,b), Wolfson (2003), Gautreau and Savin (1999) and Serway et al. (2005).

The example provided here is a realistic one, and the relativistic time dilation has been demonstrated by Bailey et al. (1977, 1979) based on the historical tests of Rossi and Hall (1941). The historical work of Rossi and Hall entailed the measurement of the muon flux on Mt. Washington in New Hampshire at the altitude of 1900 m and at the mountain base, as illustrated in Fig. 1.34. They found that the muon flux at both altitudes did not differ greatly, only by the ratio of 1.4, that is, the muon flux was 1.4 times as high at the altitude of 1900 m compared to the ground level. Calculations based on a muon half-life of 1.5 × 10^{-6} seconds (*i.e.*, average lifetime of 1.5 × 10^{-6} seconds/ln 2 = 2.2 × 10^{-6} seconds), the muon flux ratio for the two altitudes, should have been approximately 22 even when attributing the highest unrealistic speed of light to the muons. The actual low ratio of 1.4 could only be explained by applying the time dilation calculations for muons traveling at the speed of 0.994c, which are the following:

$$\Delta t_{\text{earth}} = \frac{\Delta t_r}{\sqrt{1-(v^2/c^2)}} = \frac{2.2\times10^{-6}\ \text{sec}}{\sqrt{1-(0.994)^2}}$$
$$= \frac{2.2\times10^{-6}\ \text{s}}{0.1094}$$
$$= 20.1\times10^{-6}\ \text{seconds} \qquad (1.86)$$

The dilated muon lifetime of 20.1 × 10^{-6} seconds measured by observers in the space–time reference frame of earth demonstrates that the muons measured at the 1900 m mountain peak altitude and at the base of the mountain, have a measured lifetime long enough to travel ~ 6000 m well beyond the base of the mountain, calculated as follows:

$$d = (v)(\Delta_{\text{tearth}})$$
$$- (0.994c)(\Delta_{\text{tearth}}) = (.994)(2.9979\times10^{8}\ \text{m/sec})$$
$$(20.1\times10^{-6}\ \text{sec}) = 5989\ \text{meters} \qquad (1.87)$$

The above calculation demonstrated the time dilation for the decay of the muon when observing the muon travel from the reference frames of observes on earth (Fig. 1.34), which explains why the muon of short lifetime (2.2 × 10^{-6} seconds) created high in the atmosphere can be observed (*i.e.*, detected) at sea level.

There is yet another aspect of relativity providing evidence for the observation of cosmic-ray muons at sea level. As noted by Einstein, when traveling at relativistic speeds (*i.e.*, speeds approaching the speed of light), measurements made in the space-time reference frame where they occur in different places will yield a contraction of objects compared to their sizes when measured in a reference frame where they occur at the same place. For example, an observer on the mountain would measure its altitude to be 1900 m; however, if an observer could travel with the muon at 0.994c, he or she would take two separate measurements, namely, one of the peak and one of the base of the mountain (two events) and find the height of the mountain to be contracted by the factor $\sqrt{1-(v^2/c^2)}$ previously described by Einstein (1916a,b)

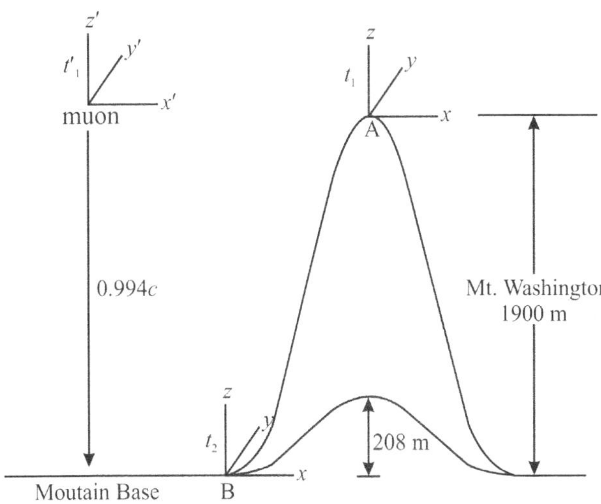

FIGURE 1.34 Observers at the peak of Mt. Washington measure the intensity of muons in the atmosphere at A at the altitude of 1900 m (mountain peak) and at (B) the base of the mountain. To the left of the figure a path of travel taken by a cosmic-ray muon is illustrated. Only one muon path of travel is illustrated for simplicity, although numerous muons are produced in the showers of cosmic radiation. The muon is illustrated traveling toward the earth at the relativistic speed of 0.994c. Relativistic calculations show that the mountain height would be squeezed or contracted in space to only 208 m when measured from the reference frame of the muon. *From L'Annunziata (2016); reprinted with permission from Elsevier © 2016.*

and discussed earlier in this section. The measured height of the mountain (x), from the reference frame of the muon, would be squeezed or contracted in space according to the following:

$$x = (1900\text{ meters})\left(\sqrt{1-(v^2/c^2)}\right)$$
$$= (1900\text{ m})\sqrt{1-\frac{(0.994c)^2}{c^2}} = 208\text{ meters} \quad (1.88)$$

Thus, the muon in the atmosphere traveling at $0.994c$ at the height of the mountain peak encounters a mountain squeezed in space to only 208 m from the mountain peak to its base (Fig. 1.34), which is more than enough distance for it to reach earth. One might question these calculations by asking "What is the true height of the mountain, 1900 m or 208 m?" Both calculations are correct relative to which inertial frame of reference the measurements are taken; the height of the mountain will depend on the reference frame (space–time coordinates) of an observer, who makes the measurement, namely, an observer located at the mountain, who measures the height to be 1900 m or an observer traveling close to the speed of light ($0.994c$) in the reference frame of the muon, who measures the height (208 m from peak to base) of the mountain while traveling past the mountain at this relativistic speed.

D. Relativistic measurements of particle lifetimes

1. Bubble chamber measurements

An interesting example of the application of relativistic time dilation can be taken from the work of Nobel Laureate Luis Alvarez. He was awarded the Nobel Prize in Physics in 1968. Luis Alvarez pioneered the development of the hydrogen bubble chamber, which enabled the discovery of subatomic particles produced by the collision of high-energy protons accelerated to energies reaching a billion electron volts in a proton synchrotron against hydrogen nuclei of the bubble chamber. The nuclear reactions that occurred would result in the production of very short-lived particles, and some would have lifetimes less than 10^{-10} seconds. Particles that possessed a positive or negative charge would produce tracks in the bubble chamber, which were photographed at the time of their production. Neutral particles would not produce any track; rather these would leave a gap in the bubble chamber from the time of their production to their decay into a charged particle. An example of the tracks produced by such short-lived particles can be seen in one of Alvarez's bubble chamber photographs illustrated in Fig. 1.35.

The particles produced in the bubble chamber would travel close to the speed of light ($0.99c$). Alvarez and his coworkers could measure the lengths of the tracks stored in photographs, and from the particle speed, they could calculate the particle lifetimes. The smallest track length, that Alvarez could measure accurately, was 1 μm or 10^{-6} m. In making such measurements of particle lifetimes in the laboratory, Alvarez would take into account that the particles travel at close to the speed of light, and from the reference frame of the laboratory, he would be measuring the dilated lifetime of the particle that travels in another reference frame. This is similar to the example presented in Fig. 1.33 where a measurement made by an observer in one frame of reference (e.g., the laboratory) on the time transpired for an event that occurs over time in another frame of reference traveling at a velocity v with respected to the observer. Serway et al. (2005) distinguishes the dilated measurement of time made by the observer in the lab as t_{lab}, according to the previously described Eq. (1.80) as follows

$$t_{\text{lab}} = \frac{t_{\text{proper}}}{\sqrt{1-(v^2/c^2)}}. \quad (1.89)$$

where t proper is the proper lifetime of the particle (i.e., the lifetime of the particle in the reference frame of the particle). Serway et al. (2005) calculated the dilated lifetime of a particle traveling at $0.99c$ for a distance of only 1 μm in the bubble chamber as

$$t_{\text{lab}} = \frac{d}{v} = \frac{10^{-6}\text{ m}}{(0.99)(2.99\times10^{8}\text{ m/s})} = 0.33\times10^{-14}\text{ s} \quad (1.90)$$

and the proper lifetime of the particle as

$$t_{\text{proper}} = (t_{\text{lab}})\sqrt{1-(v^2/c^2)}$$
$$= \left(0.33\times10^{-14}\text{ s}\right)\sqrt{1-\frac{(0.99c)^2}{c^2}} \quad (1.91)$$
$$= 4.6\times10^{-16}\text{ s}$$

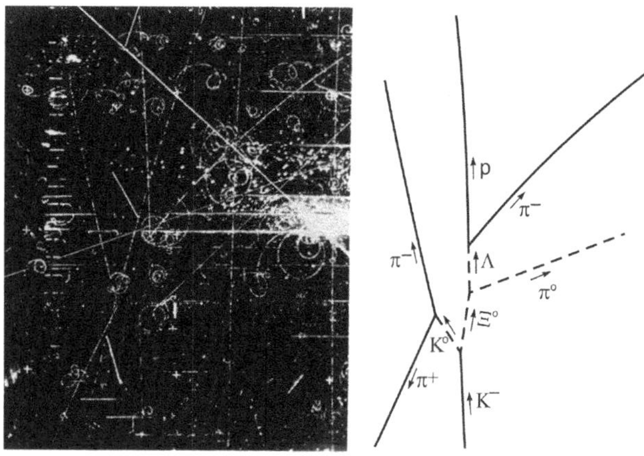

FIGURE 1.35 Left: photograph of particle tracks produced by high-energy proton collision with hydrogen nuclei in a hydrogen bubble chamber, https://doi.org/10.1103/PhysRevLett.2.215, reprinted with permission by the American Physical Society © 1959); right: sketch of specific particle tracks traced out of the photograph illustrating kaon (K^-, K^o), xi (Ξ^o), lamda (Λ^o), pion (π^+, π^o, π^-), and a proton (p) track. *From Alvarez et al. (1959).*

Thus, the proper lifetime of the particle traveling only 1 μm in the bubble chamber is much shorter than the lifetime measured in the reference frame of the laboratory (t_{lab}). If we consider a particle traveling at a yet faster speed of 0.9995c and apply the same calculations, we will arrive at the proper lifetime of the particle (t_{proper}) to be more than fourfold shorter or 1.0×10^{-16} seconds.

2. Measurements in CERN muon storage ring

Measurements made by Bailey et al. (1977, 1979) on relativistic muons in a large muon storage ring of the European Center for Nuclear Research (CERN) in Geneva provided very precise evidence for Einstein's special relativity. They found that the Einstein time dilation factor agreed with experiment with a fractional error of only 2×10^{-3} at a 95% confidence level. They calculated the proper lifetime (t_{proper}) of the negative muon (μ^{-1}) to be 2.2×10^{-6} seconds from a measured dilated lifetime (t_{lab}) of 64.4×10^{-6} seconds. The speed of the muon within the storage ring was found to be 0.9994c and the dilated lifetime of the muon as measured in the reference frame of the laboratory was found to be 64.4×10^{-6} seconds. This is the dilated lifetime according to Einstein's special relativity described by Eq. (1.89). The proper lifetime of the muon (i.e., its lifetime as measured in the reference frame of the muon or in other words the lifetime of the muon at rest) was calculated to be

$$\begin{aligned} t_{proper} &= (t_{lab})\sqrt{1-(v^2/c^2)} \\ &= \left(64.4 \times 10^{-6}\ \text{s}\right)\sqrt{1-\frac{(0.9994c)^2}{c^2}} \qquad (1.92) \\ &= 2.2 \times 10^{-6}\ \text{s} \end{aligned}$$

At the speed of 0.9994 times the speed of light, measurements of particle lifetimes made in the laboratory are highly dilated in accord with Einstein's special relativity. In the case of the muon, the measured lifetime at a muon speed of 0.9994c is close to 30-fold longer than its proper lifetime or

$$\frac{t_{lab}}{t_{proper}} = \frac{64.4 \times 10^{-6}\ \text{s}}{2.2 \times 10^{-6}\ \text{s}} = 29.3 \qquad (1.93)$$

This test of time dilation, provided by the muon lifetime measurements, is discussed in papers by Thompson (2004) and Tomaschitz (2004).

E. Energy and mass in relativity

In Einstein's treatise on relativity (1916a,b), he wrote on the relationship of energy and mass. He explained that, in the theory of relativity, the kinetic energy of a body of mass m is not given by the conventional expression

$$E = ½ mv^2, \qquad (1.94)$$

rather by the expression

$$E = \frac{mc^2}{\sqrt{1-(v^2/c^2)}} \qquad (1.95)$$

where the energy E approaches infinity, as the velocity v of the body approaches the speed of light c. Thus, the velocity v always remains below the unattainable speed of light c.

Einstein (1916a,b) elucidated that if a body, which he referred to as a material point (e.g., nuclear particle), takes up energy, its energy will increase by an amount

$$E_0 = mc^2 \qquad (1.96)$$

The increase in energy also represents an increase in mass. Thus, Einstein (1916a,b) wrote the increase in energy of the particle E_0 into the above Eq. (1.95) to give

$$E = \frac{mc^2 + E_o}{\sqrt{1-(v^2/c^2)}} \qquad (1.97)$$

or

$$E = \frac{(m + E_0/c^2)c^2}{\sqrt{1-(v^2/c^2)}}, \qquad (1.98)$$

which expresses that the body, to which energy is added, has the energy of a body of mass $(m + E_0/c^2)$ moving at a velocity v.

The above Eq. (1.98) may be written as

$$mc^2 = \frac{m_0 c^2}{\sqrt{1-(v^2/c^2)}} \quad \text{where} \quad m_0 = m + E_0/c^2 \qquad (1.99)$$

and

$$m = \frac{m_0}{\sqrt{1-(v^2/c^2)}} \qquad (1.100)$$

where m is the relativistic mass of the body, sometimes written as m_r; and m_0 is the rest mass of the body or, in other words, the mass of the body as measured in the coordinates of the observer, which would be at rest with respect to the observer, because the body and the observer in the same coordinates are traveling at the same velocity.

The expression of Eq. (1.100) is used further on in this chapter to define the energy of particles that travel at speeds approaching the speed of light (i.e., relativistic speeds). The equation is expressed as

$$E = \gamma mc^2 \quad \text{where} \quad \gamma = \frac{1}{\sqrt{1-(u^2/c^2)}} \quad \textit{and } u = v \qquad (1.101)$$

The symbol u is often used in place of v when particle speeds are relativistic. Einstein (1916a,b) explained in the above Eq. (1.101) that as the energy of a body (*e.g.*, nuclear particle) is increased, so is its mass increased. He expounded on this by clarifying that the law of conservation of energy must hold not only for one coordinate system O, but also for all coordinate systems, such as O', that are in uniform motion relative to O (see Fig. 1.31). Einstein (1916a,b) elaborated that when the energy of a body is increased by an amount E_0, its mass increases by the amount E_0/c^2. He stated that the inertial mass of a body is a variable; namely, the inertial mass of the body will vary according to the change in energy of the body. In turn, the inertial mass of a body is also a measure of its energy.

F. Relativistic mass calculations

In the above Eq. (1.100) Einstein described that the rest mass of a body (i.e., the mass of a body at rest with respect to the observer), m_0, would be measured to possess a higher mass, m, when the body is observed traveling at a velocity, v, relative to an observer, and the value of m is referred to the relativistic mass.

The relativistic speed of a 100 GeV cosmic-ray proton is calculated from its kinetic energy to be 0.999956c (see Eq. (1.493) in Part XIV.A of this chapter). As noted in Einstein's Eq. (1.100), we may calculate the relativistic mass, m, of the proton from its known rest mass (1.67262×10^{-27} kg) and its velocity, to be

$$m = \frac{m_0}{\sqrt{1-(v^2/c^2)}}$$

$$m = \frac{1.67262 \times 10^{-27}\ \text{kg}}{\sqrt{1-((0.999956c)^2/c^2)}}$$

$$m = \frac{1.67262 \times 10^{-27}\ \text{kg}}{\sqrt{1-(0.999956)^2}} = 178.3038 \times 10^{-27}\ \text{kg} \tag{1.102}$$

The ratio of the relativistic mass to the rest mass (m/m_0) of the proton traveling at 0.999956c is

$$m/m_0 = 178.3038 \times 10^{-27}\ \text{kg}/1.67262 \times 10^{-27}\ \text{kg} = 106.6 \tag{1.103}$$

which is calculated to be over 100-fold greater than its mass at rest.

Einstein's relativity calculations can be applied to photons of light, which are also particulate in nature. We know from Einstein's interpretation of the photoelectric effect, for which he was awarded the Nobel Prize, that the energy of a photoelectron is directly proportional to the energy of the photon, which had collided with the electron to cause its emission from an atom. The greater the energy of the photon, that is, the shorter its wavelength, the greater will be the energy of the photoelectron. Einstein demonstrated that the energy of the photoelectron depended on the energy of the photon that collided with the atomic electron or, the product of Planck's constant times the photon frequency according to the formula

$$E_{\text{photon}} = mc^2 = h\nu = \frac{hc}{\lambda} \tag{1.104}$$

and from the above equations, the effective mass of the photon can be written as

$$m = \frac{h}{\lambda c} \tag{1.105}$$

where m is the photon mass, c is the speed of light in a vacuum (2.9979×10^8 m/s), h is Planck's constant (6.626×10^{-34} J s), ν is the photon frequency in units of s^{-1}, and λ is the photon wavelength in meters. From Eq. (1.105), the effective mass of the photon as a function of its wavelength can be calculated. For example, the effective mass of a photon of 450 nm wavelength, which is in the visible blue wavelength of light, is calculated as

$$m = \frac{6.626 \times 10^{-34}\ \text{J}\cdot\text{s}}{(4.50 \times 10^{-7}\ \text{m})(2.9979 \times 10^{8}\ \text{m/s})} = 4.91 \times 10^{-36}\ \text{kg} \tag{1.106}$$

The above units of $\text{J}\cdot\text{s}^2/\text{m}^2$ convert to kilograms or kg as follows:

$$\frac{\text{J}\cdot\text{s}^2}{\text{m}^2} = \frac{\text{Nm}\cdot\text{s}^2}{\text{m}^2} = \frac{\text{N}\cdot\text{s}^2}{\text{m}} = \frac{\text{m kg/s}^2\cdot\text{s}^2}{\text{m}} = \text{kg} \tag{1.107}$$

where N is the Newton, a unit of force, which is equivalent to m kg/s^2. As seen from Eq. (1.105), the energy and consequently the effective mass of the photon is wavelength dependent.

G. Relativistic particle wavelength calculations

As described previously in accord with Einstein's theory of relativity, the mass of a particle will increase according to its energy. The change in mass and speed of a particle as a function of its energy must be accounted for in calculations of particle radiation wavelength. This is particularly relevant in studies of particle radiation diffraction.

The first documented application of relativistic particle wavelength calculations is that of Nobel Laureate George Thomson (1927, 1928, 1929), who demonstrated the diffraction of an electron beam thereby demonstrating the de Broglie matter wave theory as well as Einstein's theory of relativity. As a result of this work, George Thomson was awarded the 1937 Nobel prize in physics for his demonstration of the wave property of the electron, which he shared with Clinton Davisson. At the time of the 1937 Nobel prize award, the following joke spread throughout the scientific community: "Joseph Thomson, the father, was awarded the Nobel prize in physics in 1906 for demonstrating that the electron was a particle; and George Thomson, the son, won the Nobel prize in 1937 for demonstrating that the electron was a wave rather than a particle". This was only in jest because the scientific community finally had evidence for the de Broglie particle-wave duality, that is, particles not at rest will have a dual nature of particle and wave.

Thomson and coworkers (1927, 1928, 1929) calculated the wavelength of the electron beam according to the de Broglie matter-wave theory defined previously in Eq. (1.3) after correcting for the relativistic mass of the electron or

$$\lambda = \frac{h}{p} = \frac{h}{mv} \tag{1.108}$$

where m, the relativistic mass of the electron, is defined as

$$m = \frac{m_0}{\sqrt{1-(v^2/c^2)}} \tag{1.109}$$

and m_0 is the electron rest mass, v is the speed of the electron, and c is the speed of light in a vacuum. Eq. (1.108) can then be written as

$$\lambda = \frac{h}{\left[m_0/\sqrt{1-(v^2/c^2)}\right]v} \tag{1.110}$$

and

$$\lambda = \frac{h\sqrt{1-(v^2/c^2)}}{m_0 v} \tag{1.111}$$

Thomson (1928) used Eq. (1.111) from the de Broglie matter wave theory to calculate the wavelength of the electron beam produced in their electron diffraction apparatus. He only needed to know the speed of the electron, v, which was calculated from the electron kinetic energy. The following expression for the electron kinetic energy permits the calculation of the speed of the relativistic electron:

$$K = mc^2 - m_0c^2 \tag{1.112}$$

where m is the relativistic mass of the electron and m_0 its rest mass. From the expression of m of Eq. (1.109), the particle kinetic energy can be written as

$$K = \frac{m_0c^2}{\sqrt{1-(v^2/c^2)}} - m_0c^2 \tag{1.113}$$

The electron rest energy in units of MeV is calculated as

$$\begin{aligned} E = m_0c^2 &= \left(9.109\times10^{-31}\ \text{kg}\right)\left(2.9979\ \times\ 10^8\ \text{m/sec}\right)^2 \\ &= 8.186\ \times\ 10^{-14}\ \text{J} \end{aligned} \tag{1.114}$$

The electron rest energy in joules is converted to units of MeV as

$$8.186\times10^{-14}\ \text{J}/1.602\times10^{-19}\ \text{J/eV} = 0.511\ \text{MeV} \tag{1.115}$$

The known constant value of the electron rest energy, 0.511 MeV, can be inserted into Eq. (1.113) to calculate the electron velocity, v, from the measured value of its kinetic energy, K, as

$$K = \frac{0.511\ \text{MeV}}{\sqrt{1-(v^2/c^2)}} - 0.511\ \text{MeV} \tag{1.116}$$

Thomson measured the kinetic energy of the electron beam produced in his apparatus in units of eV from an accurate measurement of the potential difference used to accelerate the electron toward the anode. From the measured value of K, he calculated the velocity of the electron according to Eq. (1.116).

Thomson (1938) used electron beams over the range of 20,000 to 60,000 V energy, which he reports to have wavelengths varying according to energy over the respective range of 0.8×10^{-9} cm to 0.5×10^{-9} cm. For example, in his early detailed report of electron diffraction, Thomson (1928) used an electron beam of 25,000 V energy and calculated the wavelength of the electron beam according to Eq. (1.111) to be about 0.75×10^{-9} cm. To arrive at the value of the electron wavelength, the velocity of the 25 keV electron is first calculated according to Eq. (1.113) as

$$\begin{aligned} K &= \frac{m_0c^2}{\sqrt{1-(v^2/c^2)}} - m_0c^2 \quad \text{or} \\ 0.025\ \text{MeV} &= \frac{0.511\ \text{MeV}}{\sqrt{1-(v^2/c^2)}} - 0.511\ \text{MeV} \end{aligned} \tag{1.117}$$

which is transposed to solve for the electron velocity, v, as follows:

$$\sqrt{1-(v^2/c^2)} = 0.953 \tag{1.118}$$

and

$$1-\left(v^2/c^2\right) = (0.953)^2 \tag{1.119}$$

and

$$\frac{v^2}{c^2} = 1 - 0.90821 \tag{1.120}$$

and

$$v^2 = 0.09179c^2 \tag{1.121}$$

or

$$v = 0.3029\text{c} \tag{1.122}$$

Thus, the calculated value of $v = 0.3029c$ is the velocity of the electrons of 25,000 V energy, which is 0.3029 times the speed of light in a vacuum.

From the calculated value of v, the wavelength λ of the 25 keV electrons is calculated as

$$\lambda = \frac{h\sqrt{1-(v^2/c^2)}}{m_0 v}$$

$$= \frac{6.626 \times 10^{-34}\ \text{J sec}\sqrt{1-\frac{[(.3029)(2.9979 \times 10^8\ \text{m/sec}]^2}{(2.9979 \times 10^8\ \text{m/sec})^2}}}{(9.109 \times 10^{-31}\ \text{kg})(.3029)(2.9979 \times 10^8\ \text{m/sec})} \quad (1.123)$$

$$= 0.076 \times 10^{-10}\ \text{m} = 0.076\ \text{Å}$$

which is equivalent to the value calculated by Thompsom (1928) according to Eq. (1.123) in his discovery of the electron de Broglie matter-wave. The application of Eq. (1.124) to the calculation of the relativistic de Broglie wavelength of the neutron as a function of neutron energy is provided in Section X. E.2 of this chapter.

The above results agree closely with the calculated value of 0.75×10^{-9} cm reported by Thomson (1928).

Section X.E.2 provides a treatment of neutron velocity and wavelength as a function of neutron energy within the context of neutron classification. In that section, the following equation is derived for the calculation of the relativistic de Broglie wavelength:

$$\lambda_r = \frac{hc}{pc} = \frac{hc}{[2m_0c^2K(1+(K/2m_0c^2))]^{1/2}} \quad (1.124)$$

where the constant hc, Planck's constant times the speed of light in a vacuum, is calculated in Section XI.A to have the value

$$hc = 12.4\ \text{keV}\cdot\text{Å} = 12.4 \times 10^{-4}\ \text{MeV}\cdot\text{nm}, \quad (1.125)$$

m_0c^2 is the particle rest energy, and K is the particle kinetic energy. We can take the example of Thomson's electron accelerated to 0.025 MeV described in the previous paragraphs and use Eq. (1.124) to calculate the electron's relativistic wavelength as follows:

$$\lambda_r = \frac{hc}{pc} = \frac{12.4 \times 10^{-4}\ \text{MeV}\cdot\text{nm}}{\{\sqrt{2(0.51\ \text{MeV})(0.025\ \text{MeV})[1+(0.025\ \text{MeV}/2(0.51\ \text{MeV}))]}\}}$$

$$= \frac{12.4 \times 10^{-4}\ \text{MeV}\cdot\text{nm}}{0.162\ \text{MeV}}$$

$$= 7.6 \times 10^{-3}\ \text{nm} = 0.076 \times 10^{-10}\ \text{m} = 0.076\ \text{Å}$$

VIII. Nuclear decay modes

Throughout this book numerous decay modes of atomic nuclei are described. These decay modes are manifested in general through the emission of nuclear radiation, such as α-particles, β-particles, γ-ray photons, protons, neutrons, neutrinos, and heavy cluster emissions. We also refer to the modes of nuclear decay as radioactive decay modes. The modes of nuclear decay are given specific nomenclature for reason of identification, such as (1) alpha-decay (α), (2) beta decay including negatron emission (β^-), positron emission (β^+), and electron capture (ε or EC), (3) isomeric transition (IT resulting in γ emission), (4) proton emission (p), (5) neutron emission (n), (6) spontaneous fission (SF), and (7) heavy cluster emission (^{14}C, ^{20}O, ^{23}F, ...). Most nuclear decay modes result in a transformation (*i.e.*, transmutation) of an atom of one element into another, as illustrated in Fig. 1.36. This figure depicts nuclear decay transitions that occur between blocks occupied by the various radionuclides in the Chart of the Nuclides, which is the "Periodic Table" of the nuclides (both stable and radioactive) separated in blocks

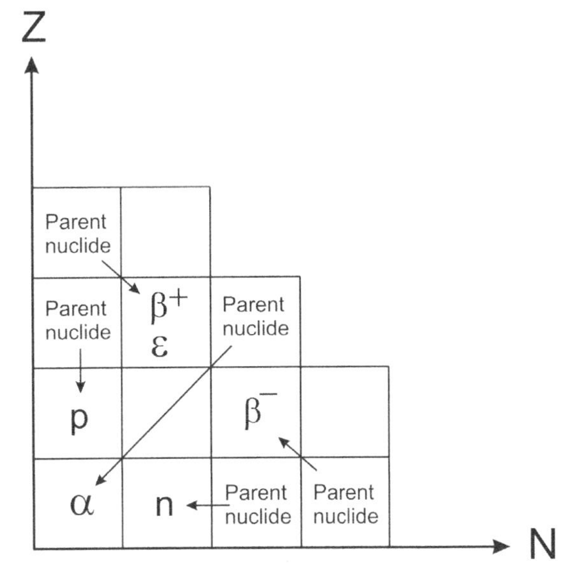

FIGURE 1.36 Nuclear decay mode transitions according to blocks occupied by the various nuclides in the Chart of the Nuclides plotted according to their number of protons (Z) and number of neutrons (N) in the nucleus. Various nuclear decay modes are illustrated that result in changes in the number of protons (atomic number, Z) or neutrons (neutron number, N) between parent and daughter nuclides. The radiation emitted or the decay modes are written in the blocks occupied by the daughter nuclides. For example, positron (β^+) emission or EC (ε) decay results in a reduction in proton number (Z − 1) and increase in neutron number (N + 1); whereas, negatron emission results in the opposite, that is, an increase in the number of protons in the nucleus (Z + 1) and reduction in the number of neutron (N − 1). Alpha (α) decay results in twofold reductions in the proton and neutron numbers in the nucleus (Z−2 and N−2). Proton (p) and neutron (n) radioactivity result in loss of only one or more protons or neutrons, respectively.

according to their atomic number (Z) and neutron number (N). The nuclear decays generally lead to a change in the atomic number (Z) and consequently a change from one element to another. The emission of a neutron (n) from the nucleus, also illustrated in Fig. 1.36, results in no change in atomic number and thus no change of element between parent and daughter nuclides; rather, the emission of one or more neutrons from the nucleus results in a change in the mass number of the same element resulting in an isotope of the same element. Also, another mode of nuclear decay, which does not result in a change in chemical elements between parent and daughter atoms, is the isomeric transition (IT) decay mode, not illustrated in Fig. 1.36. The IT decay results in a change from one energy state of an atomic nucleus to a lower energy state with the emission of gamma (γ) radiation. Heavy cluster emission is the emission of nuclei larger than the helium nucleus (*e.g.*, ^{14}C, ^{20}O, etc.) from a heavy nucleus, which for reasons of space, is not illustrated in Fig. 1.36.

Table 1.7 provides a list of nuclear decay modes together with their description and examples of each.

IX. Nuclear reactions

A nuclear reaction is an interaction between two nuclear particles or two nuclei that yields products which are different from the reactants, that is, new nuclei are formed. If the interacting particles or nuclei only collide and separate without change, the process is referred to as **elastic scattering**. When a nuclear particle collides with a nucleus, but the nucleus only undergoes excitation to a higher energy state, the process is referred to as an **inelastic collision**.

A. Reaction types

The term nuclear reaction generally refers to the interaction of a projectile particle and a target nucleus that results in the production of a residual (product) nucleus and an ejected particle.

An example of a nuclear reaction is the following:

$$^1_0\text{n} + {}^{10}_5\text{B} \rightarrow ({}^{11}_5\text{CN})^* \rightarrow {}^7_3\text{Li} + {}^4_2\text{He} + 2.8\text{ MeV} \quad (1.126)$$

In the example provided by Eq. (1.126) a neutron, $^1_0$n, collides with the target nucleus $^{10}_5$B, whereby they coalesce to form an excited and unstable compound nucleus (abbreviated as CN*). The compound nucleus generally is not written, as it is very short-lived ($\sim 10^{-16}$ seconds). In this example the compound nucleus decays to yield the product nucleus $^7_3$Li and an ejected alpha particle, $^4_2$He. The sum of the proton numbers and mass numbers must be equal on both sides of the reaction. For example, in Eq. (1.126) the proton numbers $0+5=3+2$ and mass numbers $1+10$ and $7+4$ are equal on both sides of the reaction. As illustrated in Eq. (1.126), there is kinetic energy released in this reaction to the amount of 2.8 MeV. This nuclear reaction is a very practical one because it is used for the detection of thermal neutrons. The nuclide ^{10}B has a high natural abundance of 20%, and when incorporated into the crystalline structure of a solid scintillation detector, the crystal becomes a sensitive detector for thermal neutrons. The alpha particles ejected as a product of the reaction have a very short range of travel in the crystalline detector. The kinetic energy of the alpha particles is in the MeV range, and the alpha particles dissipate their energy close to the point of neutron interaction. Solid scintillators of this type serve as good thermal neutron detectors.

Nuclear reactions between a specific projectile and target nucleus may result in more than one set of reaction products. For example, in his Nobel lecture, Enrico Fermi noted the following:

> Neutron bombardment of aluminum, that has a single isotope ^{27}Al, gives rise to three radioactive products: ^{24}Na, with a half-life of 15 h ...; ^{27}Mg, with a half-life of 10 min ...; and ^{28}Al with a half-life of 2 to 3 min ...

The three reactions described by Fermi (1938) may be written as follows:

$$^1_0\text{n} + {}^{27}_{13}\text{Al} \rightarrow ({}^{28}_{13}\text{CN})^* \rightarrow \begin{cases} {}^{24}_{11}\text{Na} + {}^4_2\text{He} & Q = -3.13\text{ MeV} \\ {}^{27}_{12}\text{Mg} + {}^1_1\text{H} & Q = -1.83\text{ MeV} \\ {}^{28}_{13}\text{Al} + \gamma & Q = 7.72\text{ MeV} \end{cases} \quad (1.127)$$

TABLE 1.7 Nuclear decay modes.

Mode	Designation and description	Example
α	**α emission.** A nucleus of helium ($^{4}_{2}He$) is emitted from the nucleus of the decaying atom.	$^{241}_{95}Am \rightarrow ^{137}_{93}Np + ^{4}_{2}He$
β^-	**β^- or negatron emission.** A neutron is converted to a proton with the emission of a negative electron and an antineutrino. There is no change in mass number.	$^{3}_{1}H \rightarrow ^{3}_{2}He + \beta^- + \overline{\nu}$
β^+	**β^+ or positron emission.** A proton is converted to a neutron with the emission of a positive electron and a neutrino. There is no change in mass number.	$^{11}_{6}C \rightarrow ^{11}_{5}B + \beta^+ + \nu$
ε or EC	**Electron capture.** An atomic electron (*e.g.*, K shell) is captured by a nuclear proton, and the proton is converted to a neutron with the emission of a neutrino. This is another form of beta decay, and there is no change in mass number. EC decay competes with positron emission.	$^{65}_{30}Zn \rightarrow ^{65}_{29}Cu + \nu$
IT	**Isomeric transition.** The transition of a nucleus from an excited elevated energy state to that of a lower energy state with the emission of gamma (γ) ray photons. There is no change in atomic number or mass number. The mass number of the parent nuclide is designated with the letter *m* to signify a metastable state when the gamma-ray emission is not immediate or a "prompt gamma".	$^{119m}_{50}Sn \rightarrow ^{119}_{50}Sn + \gamma$
SF	**Spontaneous fission.** The spontaneous splitting of a nucleus into two nuclear fragments with the emission of neutrons. The possible nuclear fission fragments vary, and there are numerous fragment combinations possible.	$^{252}_{98}Cf \rightarrow ^{101}_{40}Zr + ^{148}_{58}Ce + 3^{1}_{0}n$
p	**Proton emission.** The consecutive or simultaneous emission of one, two, or three protons from a proton-rich atomic nucleus. When two or three protons are emitted, the decay mode is identified as 2p and 3p, respectively.	$^{147}_{69}Tm \rightarrow ^{146}_{68}Er + p$
n	**Neutron emission.** The emission of one or two neutrons from a neutron-rich nucleus. The simultaneous emission of 2 neutrons is identified as a 2n decay mode.	$^{16}_{4}Be \rightarrow ^{14}_{4}Be + 2n$
β^+p	**Beta-delayed proton emission.** The emission of a proton from a populated elevated energy state of a proton-rich atomic nucleus following β^+ decay. The emission of two protons is designated as β^+2p.	$^{13}_{8}O \rightarrow ^{13}_{7}F + \beta^+ + \nu \rightarrow ^{12}_{6}C + p$
β^-n	**Beta-delayed neutron emission.** The emission of a neutron from a populated elevated energy state of a neutron-rich atomic nucleus following β^- decay. The emission of two neutrons is designated as β^-2n.	$^{94}_{35}Br \rightarrow ^{94}_{36}Kr + \beta^- + \overline{\nu} \rightarrow ^{93}_{36}Kr + n$
$2\beta^-$ or $\beta^-\beta^-$	**Double β^- decay.** Two neutrons decay to two protons in an atomic nucleus with the emission of two β^--particles and two antineutrinos.	$^{82}_{34}Se \rightarrow ^{82}_{36}Kr + 2\beta^- + 2\overline{\nu}$
$^{14}C, ^{20}O$ ...	**Cluster radioactivity.** The spontaneous emission of a particle heavier than an α-particle with $Z \geq 6$, such as the emission of $^{14}C, ^{20}O, ^{23}F, ^{24}Ne, ^{28}Mg, ^{30}Mg$, or ^{32}Si.	$^{221}_{87}Fr \rightarrow ^{207}_{81}Tl + ^{14}_{6}C$

From L'Annunziata (2016); reprinted with permission from Elsevier © 2016.

An intermediary compound nucleus is illustrated as decaying into three possible decay products. The above reactions, that leave behind the residual nuclei of ^{24}Na and ^{27}Mg, have negative Q values. These reactions are endoergic or endothermic. In these two cases, the neutron projectile must possess additional kinetic energy for the reaction to proceed. The reaction that produces the ^{28}Al residual nucleus has a positive Q value. This reaction is exoergic or exothermic resulting in a release of energy. Low-energy neutrons, classified as slow or thermal neutrons, are sufficient for this reaction to proceed. The capture of slow neutrons by nuclei is accompanied generally with the immediate emission of one or more gamma ray photons. The nuclei are left in a state of excitation after the neutron capture, and their ground state is reached after the emission of the gamma ray quanta. The Q value, as a measure of reaction energy, is discussed in Parts III.D and IX.B of this chapter.

The reactions were described in Eq. (1.127) as possessing a compound nucleus, and those reactions may be referred to as **compound nuclear reactions**. Two other types of nuclear reactions are classified as **pickup** and **stripping reactions**. When high-energy nuclear projectiles are used to bombard a target nucleus, the interaction can be more direct whereby the projectile does not interact with all of the nucleons of the target and thus, does not reside in a compound nucleus; but rather, the projectile may interact only with a few nucleons in specific nuclear shell-model orbits of the target. In such cases, the projectile may pick up or gain nucleons from the target nucleus in pickup reactions; or the projectile may lose nucleons to the target in stripping reactions. Some early studies of pickup and stripping reactions that may occur in specific nuclear shell-model orbits are the works of Kaschl et al. (1970) and Brien et al. (1972). Examples of pick-up and stripping reactions

TABLE 1.8 Examples of pick-up and stripping reactions.

Reaction	Condensed notation[a]	Projectile energy	Reference
Pick-up reactions:			
${}^{13}_{6}C + {}^{1}_{1}H \rightarrow {}^{12}_{6}C + {}^{2}_{1}H$	${}^{13}_{6}C(P,d){}^{12}_{6}C$	12 MeV	Glover and Jones (1966)
${}^{12}_{6}C + {}^{3}_{1}H \rightarrow {}^{11}_{5}B + {}^{4}_{2}He$	${}^{12}_{6}C(t,\alpha){}^{11}_{5}B$	10 MeV	Honda et al. (1965)
${}^{20}_{10}Ne + {}^{2}_{1}H \rightarrow {}^{19}_{9}F + {}^{3}_{2}He$	${}^{20}_{10}Ne(d,h){}^{19}_{9}F$	52 MeV	Kaschl et al. (1970)
${}^{89}_{39}Y + {}^{2}_{1}H \rightarrow {}^{87}_{38}Sr + {}^{4}_{2}He$	${}^{89}_{39}Y(d,\alpha){}^{87}_{38}Sr$	12 MeV	Brien et al. (1972)
Stripping reactions:			
${}^{12}_{6}C + {}^{2}_{1}H \rightarrow {}^{13}_{6}C + {}^{1}_{1}H$	${}^{12}_{6}C(d,p){}^{13}_{6}C$	12 MeV	Glover and Jones (1966)
${}^{12}_{6}C + {}^{3}_{1}H \rightarrow {}^{13}_{6}C + {}^{2}_{1}H$	${}^{12}_{6}C(t,d){}^{13}_{6}C$	12 MeV	Glover and Jones (1966)
${}^{27}_{13}Al + {}^{3}_{2}He \rightarrow {}^{28}_{14}Si + {}^{2}_{1}H$	${}^{27}_{13}Al(h,d){}^{28}_{14}Si$	41 MeV	Munger and Peterson (1978)
${}^{194}_{78}Pt + {}^{3}_{2}He \rightarrow {}^{195}_{79}Au + {}^{2}_{1}H$	${}^{194}_{78}Pt(h,d){}^{195}_{79}Au$	41 MeV	Munger and Peterson (1978)

[a] *The condensed form of writing nuclear reactions is discussed subsequently in this chapter.*

and the projectile energies at which these reactions were observed are listed in Table 1.8.

Spallation reactions are another class of nuclear reaction, which is gaining increased interest worldwide. Spallation occurs when a high-energy (>100 MeV) charged particle, such as a proton, strikes a nucleus whereby it initially "knocks-out" several nucleons from the target nucleus within a duration of about 10^{-22} seconds leaving the residual nucleus in an excited state. The excited residual nucleus may then immediately break up into numerous smaller components of nucleons such as helium and lithium nuclei, deuterons, mesons, and individual protons and neutrons, etc. In general, numerous neutrons are emitted from each spallation reaction making this reaction a good neutron source. Nobel Laureate Glenn T. Seaborg is credited with coining the term "nuclear spallation" in his doctoral dissertation in 1937 deriving the word from the verb "spall" meaning to chip or crumble.

Nuclear spallation is a common occurrence in our atmosphere when high-energy cosmic-ray particles of energies ranging from 10^8 to $>10^{20}$ eV strike nuclei in the atmosphere resulting in the fragmentation of nuclei into smaller nucleons constituting the cosmic-ray showers discussed further on in this chapter. Spallation reactions resulting from the interaction of high-energy particles in the cosmic-ray showers colliding with nuclei in mineral deposits on the earth's surface is attributed to the cosmogenic production of nuclides, such as, ${}^{36}Cl$ (Stone et al., 1996; Licciardi et al., 2008).

There is current worldwide interest in the artificial production of spallation reactions as neutron sources referred to as spallation neutron sources (SNS). Spallation neutron sources are accelerator driven systems whereby protons are accelerated to high-energies (1 GeV and above), which then collide with heavy metal targets resulting in nuclear spallation with the emission of neutrons useful in research (Slowiński, 2003; Mason et al., 2006; Thomsen, 2007; Zamani et al., 2008).

Two other nuclear reactions that are very important sources of energy are nuclear fission and nuclear fusion. These are discussed in much detail in other parts of this chapter.

B. Notation

The nuclear reaction, illustrated by the above in Eq. (1.126), may be abbreviated as

$$ {}^{10}_{5}B(n,\alpha){}^{7}_{3}Li \tag{1.128} $$

which is written in the following sequence:

$$ \text{target nucleus(projectile, ejected particle)product nucleus.} \tag{1.129} $$

This condensed form of writing the nuclear reaction, as written above in Eq. (1.128), reads as follows: a ${}^{10}B$ target nucleus bombarded by neutrons ejects an alpha-particle leaving the residual ${}^{7}Li$ nucleus. Particles, which are often projectiles or ejected particles, in nuclear reactions are the following in name and abbreviation in parenthesis: proton (${}^{1}_{1}H$ or p), neutron (n), deuteron (${}^{2}_{1}H$ or d), Triton (${}^{3}_{1}H$ or t), helium-3 (${}^{3}_{2}He$ or h), and helium-4, *i.e.*, the alpha particle (${}^{4}_{2}He$ or α). Neutrons as projectiles for nuclear reactions with target nuclei are easily available from alpha particle induced nuclear reactions, such as the alpha-particle bombardment of beryllium (*i.e.*, ${}^{9}_{4}Be(\alpha,n){}^{12}_{6}C$), which provide neutrons with average energies of 5.5 MeV. The nuclear fission reactor is another excellent source of neutrons available as projectile particles to induce nuclear reactions. Accelerators are also good sources of neutrons via spallation reactions induced by high-energy particle collisions with target nuclei. Various sources of neutrons are described further on in this chapter. The cyclotron is an excellent source of high-energy positively charged particles and nuclei, such as protons, deuterons, triton ions, and helium-3 and helium-4 ions. In the cyclotron, target material is placed in the path of the accelerating ion, and numerous radionuclides are made via the nuclear reactions initiated by the bombardment of the positively charged projectiles with the target nuclei.

C. Energy of reactions (Q value)

The Q value of a nuclear reaction is a measure of the reaction energy. It is the difference between the kinetic energy of the reaction products and the kinetic energy of the original particles of the nuclear reaction. The Q value for the reaction illustrated above in Eq. 1.126 is reported in the literature to be 2.8 MeV. The Q values for all nuclear reactions are readily available online at the following website of the National Nuclear Data Center, Brookhaven National Laboratory: http://www.nndc.bnl.gov/qcalc/ (NNDC, 2018a,b). At this website it is only needed to identify, the projectile, target nucleus, and the ejected particle. There is no need to input or identify the residual nucleus on this website, because the calculator can easily deduce the proton number and mass number of the residual nucleus by difference. The calculations made at this website are based on the data maintained by the Atomic Mass Data Center and on atomic mass evaluations reported by Wang et al. (2017). Q values for nuclear reactions are easy to calculate, and the calculation of Q values will be discussed subsequently in this section. The Q value is an important property of a nuclear reaction, because it will tell us whether a reaction will release kinetic energy ($Q > 0$) or lose kinetic energy ($Q < 0$). When $Q = 0$, the reaction is referred to as **elastic scattering** whereby the projectile and target nucleus collide and exchange energy; but they do not undergo any change. When the projectile and target nucleus collide, and part of the kinetic energy of the projectile is transferred to the target nucleus leaving the target at an excited energy state, the reaction is referred to as **inelastic scattering**. There is only energy transfer, but no change in atomic or mass number of the projectile or target nucleus in inelastic scattering reactions.

To arrive at the equation for the calculation of reaction Q values let us abbreviate the nuclear reaction of Eq. (1.126) in the form

$$A + B \rightarrow C + D \tag{1.130}$$

where projectile A bombards nucleus B, which leaves behind a residual nucleus C and ejected particle D. If the total energy of the reactants and products are conserved, we can write

$$E_A + E_B = E_C + E_D \tag{1.131}$$

Also, the total energy (E) of a particle or nucleus is the sum of its kinetic energy (K) and its rest energy (m_0c^2), that is

$$E = K + m_0c^2 \tag{1.132}$$

where m_0 is the particle or nuclear rest mass, and c is the speed of light in a vacuum. Furthermore, if we assume that the target nucleus is at rest, *i.e.*, it has zero kinetic energy, we can transform Eq. (1.131) to

$$K_A + m_Ac^2 + m_Bc^2 = K_C + m_Cc^2 + K_D + m_Dc^2 \tag{1.133}$$

The total kinetic energy released or absorbed in the reaction, Q, is the difference of the kinetic energies of the products and the reactants. In other words, it is the differences in the kinetic energies gotten out of the reaction versus what was put into the reaction, or

$$Q = K_{\text{out}} - K_{\text{in}} \tag{1.134}$$

Then from Eq. (1.133) we can write

$$Q = (K_C + K_D) - K_A = (m_A + m_B)c^2 - (m_C + m_D)c^2 \tag{1.135}$$

In accord with $E = mc^2$, Eq. (1.135) simple states

$$Q = E_{0,\text{ reactants}} - E_{0,\text{ products}} \tag{1.136}$$

where $E_{0,\text{ reactants}}$ and $E_{0,\text{ products}}$ are the rest energies of the nuclear reactants and products, respectively. Eqs. (1.135) and (1.136) may be simplified to read

$$Q = (m_{\text{reactants}})c^2 - (m_{\text{products}})c^2 \tag{1.137}$$

where $m_{\text{reactants}}$ and m_{products} are the combined masses of the nuclear reactants and products, respectively.

For example, in accord with Eqs. (1.135) and (1.137), the Q value for the nuclear reaction described previously in Eq. (1.126) and rewritten here in condensed form,

$$^1_0\text{n} + ^{10}_5\text{B} \rightarrow ^7_3\text{Li} + ^4_2\text{He}, \tag{1.138}$$

would be calculated as follows:

$$Q = (M_{^1\text{n}} + M_{^{10}\text{B}})c^2 - (M_{^7\text{Li}} + M_{^4\text{He}})c^2 \tag{1.139}$$

where $M_{^1\text{n}}$, $M_{^{10}\text{B}}$, $M_{^7\text{Li}}$ and $M_{^4\text{He}}$ are the masses of the neutron and ^{10}B reactants and the ^7Li and ^4He product nuclei, respectively, and c is the speed of light in a vacuum. We may use atomic mass units (u) taken from standard reference tables since the 5 atomic electron masses of the reactants and the product nuclides are included on both sides of the equation. Thus, using atomic mass units in Eq. (1.139) gives

$$Q = \left\{ \begin{array}{l} (1.00866492\text{ u} + 10.012937\text{ u})(931.494\text{ MeV/u}) \\ -(7.016004\text{ u} + 4.00260325\text{ u})(931.494\text{ MeV/u}) \end{array} \right\}$$

$$Q = 2.789\text{ MeV}$$

The Q value is positive. Thus, energy is released in the nuclear reaction. The above calculated Q value of 2.789 MeV agrees with the computer-calculated Q value provided via the QCalc website of the NNDC (http://www.nndc.bnl.gov/qcalc/ of the Brookhaven National Laboratory.

When Q is positive, the reaction is called **exoergic**. In such a case, the reaction releases kinetic energy by conversion of some of the rest mass to kinetic energy according to

$$E_{\text{Total}} = E_K + m_0c^2 \tag{1.140}$$

Reactions with a positive Q value ($Q > 0$) are also called **exothermic**, since kinetic energy is released. When the Q value is negative ($Q < 0$) the reaction is called **endoergic** or **endothermic**, as kinetic energy would be absorbed in the reaction. In such a case, the reaction could not take place unless energy is added to the reactant particle or projectile. For example, from atomic mass data of the Atomic Mass Data Center (AMDC) (http://amdc.impcas.ac.cn/web/masseval.html), let us calculate the Q value of the following nuclear reaction used by Ernest Rutherford (1919) whereby he discovered the proton:

$$^{4}_{2}\text{He} + ^{14}_{7}\text{N} \rightarrow ^{17}_{8}\text{O} + ^{1}_{1}\text{H} \tag{1.141}$$

$$Q = (M_{^4\text{He}} + M_{^{14}\text{N}})c^2 - (M_{^{17}\text{O}} + M_{^1\text{H}})c^2$$

$$Q = \left\{ \begin{array}{l} (4.002603254\ \text{u} + 14.003074004\ \text{u})(931.494\ \text{MeV/u}) \\ -(16.999131756\ \text{u} + 1.007825032\ \text{u})(931.494\ \text{MeV/u}) \end{array} \right\}$$

$$Q = -1.19187\ \text{MeV} = -1191.87\ \text{keV} \tag{1.142}$$

The National Nuclear Data Center (QCalc) computer-calculated Q value yields the same energy of -1191.87 keV for the above reaction (https://www.nndc.bnl.gov/qcalc/). The reaction is endoergic ($Q < 0$) or endothermic, that is, energy is absorbed in the reaction. The reaction would require at least 1.192 MeV of energy to occur. Since the target nucleus is at rest, it is the projectile particle or nucleus that would carry the kinetic energy needed for the reaction to take place. Furthermore, there is a threshold energy (E_{th}) energy, which is by definition the minimum bombarding energy that the projectile particle must possess when Q is negative ($Q < 0$), for the reaction to take place. The threshold energy is calculated by the following equation, derived by Evans (1955):

$$E_{\text{th}} = -Q\left(\frac{M_1 + M_2}{M_2}\right) \tag{1.143}$$

where M_1 and M_2 are the masses of the projectile particle and target nucleus, respectively, as illustrated in Fig. 1.37. Using the Q value of -1.192 MeV, we can calculate the threshold energy of the alpha-particle projectile for reaction 1.141 to take place, with atomic mass data of the AMDC used in Eq. (1.142), as follows:

$$E_{\text{th}} = -(-1.192\ \text{MeV})\left[\frac{(4.002603254\ \text{u} + 14.003074004\ \text{u})(931.494\ \text{MeV/u})}{(14.003074004\ \text{u})(931.494\ \text{MeV/u})}\right]$$
$$= 1.533\ \text{MeV} \tag{1.144}$$

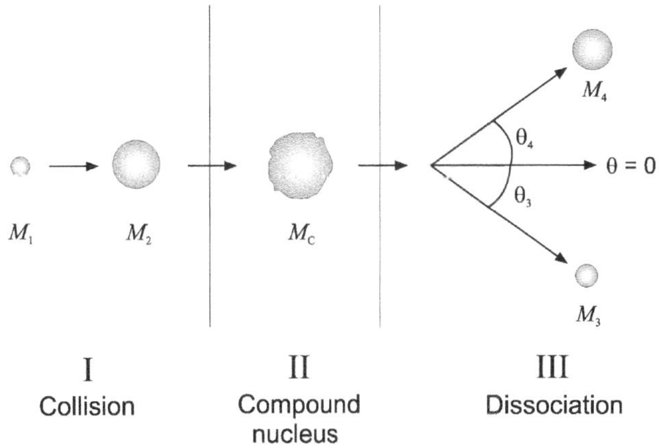

FIGURE 1.37 Three stages of a nuclear reaction, namely, (I) Collision of a projectile particle of mass M_1 with a target nucleus of mass M_2 forming a (II) Compound nucleus or intermediate nucleus M_C of mass equivalent to the combined masses of the projectile and target nucleus, and (III) Dissociation of the compound nucleus resulting in an ejected particle of mass M_3 and residual nucleus of mass M_4 expelled at angles θ_3 and θ_4, respectively.

The calculated threshold alpha-particle projectile energy of 1.533 MeV would be required for reaction 1.141 to take place. Rutherford (1919) used alpha particles from Radium C′, which is today known as ^{214}Po. The alpha particles from ^{214}Po have energies of 7.8 MeV well above the threshold energy for Reaction 1.141 to take place.

The projectile particle M_1 of Eq. (1.143) is illustrated in Fig. 1.37 traveling in the direction of the target nucleus M_2, which forms an intermediate compound nucleus M_C, which has a mass equivalent to the combined masses of the projectile particle and target nucleus. The compound nucleus then dissociates into an ejected particle M_3 and a residual nucleus M_4. Due to the very short lifetime of the combined nucleus ($\sim 10^{-22}$ seconds) or of an intermediate system that does not consist of the formation of a compound nucleus, Evans (1955) reports that M_C loses no appreciable energy due to ionization prior to dissociation into M_3 and M_4. When particle M_3 is ejected at an angle $\theta_3 = 0$, the energy of the projectile particle is at the minimum or threshold energy E_{th}. As the projectile particle energy is raised over the threshold energy, the ejected particles M_3 begin to appear at $\theta_3 > 0$.

D. Reaction cross section

The reaction cross section is, by definition, the measure of the probability that a nuclear reaction will occur. The cross section has units of area (cm^2), and the term reaction cross section is used, because it is a term that depicts the effective area of the target nucleus. It is not a measure of the true cross section of a target nucleus; rather, it is a measure of the effectiveness of the target nucleus to increase or reduce the probability of a nuclear reaction with a projectile particle. The larger the measured effective area or cross section of the target nucleus, the greater will be the probability that a projectile particle will hit or interact with the target nucleus. The probability, that a projectile particle collides with a target nucleus, will depend on the rate at which the projectile particles are bombarding the target material, that is, the particle beam intensity (I = projectile particles/second), the number of target nuclei in the target material or the target nuclei density (N = # of nuclei/cm^3), the effective area of the target nucleus, that is, the cross section (σ) of the target nucleus (cm^2), and the thickness (x) of the target material (cm). The variables affecting the probability of a nuclear reaction and consequently the rate of the reaction are illustrated in Fig. 1.38.

The rate of a nuclear reaction is a function of the effective target area exposed to the projectile particle beam. The effective target area is a product of the total number of target nuclei in the target material and the reaction cross section (σ) of the target nuclei. The number (#) of target nuclei in the target material is dependent on the density of the target nuclei and the target volume, that is,

$$\#\text{ of target nuclei} = NAx \tag{1.145}$$

where N in the target nuclei density (# of nuclei/cm^3), A is the target area (cm^2), and x is the target thickness (cm). The effective target area is then calculated as

$$\text{Effective target area} = (\#\text{ of target nuclei})(\sigma) \tag{1.146}$$

$$= NAx\sigma \tag{1.147}$$

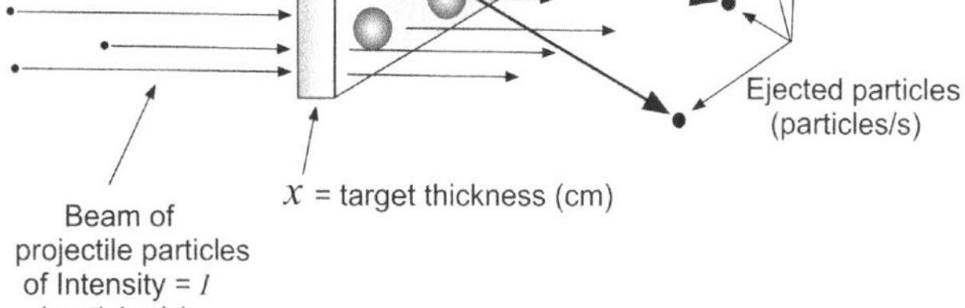

FIGURE 1.38 Reaction target area. A beam of projectile particles bombards a material containing target nuclei of cross section (σ). Some of the projectile particles react with the target nuclei resulting in the ejection of particles as products of the nuclear reaction. The reaction rate is measured by a detector sensitive to the ejected particles.

We can reason that the rate of the reaction is related to the particle beam intensity as the effective target area is to the total target area, or

$$\frac{\text{Rate of reaction (particles/s)}}{\text{Beam intensity (particles/s)}} = \frac{\text{Effective target area (cm}^2)}{\text{Total target area (cm}^2)} \tag{1.148}$$

or

$$\frac{R}{I} = \frac{N(Ax)\sigma}{A} = N\sigma x \tag{1.149}$$

and

$$R = IN\sigma x \tag{1.150}$$

where R in the rate of the reaction (ejected particles/s), I is the projectile beam intensity (projectile particles/s), N is the density of target nuclei (# of target nuclei/cm^3), σ is the reaction cross section (cm^2), and x is the target thickness (cm). The cross section may be determined experimentally from the reaction rate, projectile beam intensity, target nuclei density, and the target thickness. As described in the previous section on reaction Q value, the rate of a nuclear reaction will depend also on the energy of the projectile particles bombarding the target nuclei. Thus, reaction cross sections are reported together with the projectile particle energy (Zhou et al., 2010; Madani et al., 2010; Yigit, 2017; Ghosh et al., 2017a,b; Mosby et al., 2017). Thorough treatments of reaction cross section are given by Evans (1955), Friedlander et al. (1964), and Serway et al. (2005).

The common unit of measure for the reaction cross section is the barn where one b = 1×10^{-24} cm^2. This measure of cross section was first used by Baker et al. (1943) and proposed by Marshall P. Holloway and Charles P. Baker (1944) as the unit of measure for the cross section in secret technical reports of the Los Alamos National Laboratories during the Manhattan Project. Holloway and Baker (1944, 1972) reported that, while measuring reaction cross sections at Purdue University, they tried to come up with a suitable unit of measure. They thought of the word "Oppenheimer", but discarded the idea because of the word's length. They also considered the "Bethe", but discarded that, because of possible confusion with the widespread use of "beta". They also considered using the name "John" after John Manley, who headed the Purdue Project, but also discarded that because that word had widespread use for purposes other than the name of a person. One of the authors had a rural background, and on farms in the countryside, the broad side of a barn was always the largest building, which could be seen at a distance. There was then the very common colloquial term "big as a barn", and the term is still used today. It was a common saying then to sportsmen, who had poor aim such as poor marksman, archers, or golfers, that "they couldn't hit the side of a barn". Barns were built generally longer than they were wide, and thus the broad side or length of a barn was the largest part of a barn, which could be seen from a distance. They reported then that the unit "barn" was appropriate, because a cross section of 10^{-24} cm^2 for nuclear processes was very large and really "as big as a barn". Submultiples and multiples of the barn are also used, such as the millibarn (mb $= 10^{-3}$ barns) and the kilobarn (kb $= 10^3$ barns).

We may take an example of the practical application of Eq. (1.150) to calculate the rate of absorption of thermal neutrons via the $^{10}B(n,\alpha)^7Li$ reaction by crystalline boric acid (H_3BO_3). The nuclide ^{10}B has a high cross section for the absorption of thermal neutrons. The value of σ for the (n,α) reaction at the thermal neutron energy of 0.025 eV is 3.842×10^3 barns. According to Eq. (1.150), the rate of the (n,α) reaction as a function of the projectile beam intensity may be written as

$$R = IN\sigma x \quad (1.151)$$

The value of N, namely, the density of the ^{10}B target nuclei with units of # of target nuclei/cm^3 is calculated as

$$N = \frac{\delta N_A}{A} \quad (1.152)$$

where δ is the density of the absorbing material or compound, N_A is Avogadro's number, A is the molar mass of the absorbing compound, and the value of N is

$$N = \frac{(1.44\ g/cm^3)(6.022 \times 10^{23}\ \text{molecules}\ H_3BO_3/mol)(1\ \text{atom B/molecule})(0.199\ \text{atoms}^{10}B/\text{atom boron})}{61.8\ g\ H_3BO_3/mol}$$

$$= 2.79 \times 10^{21} \text{atoms of}\ ^{10}B/cm^3$$

The term 0.199 atoms ^{10}B/atom B is included here, because natural boron consists of 19.9% ^{10}B. Eq. (1.151) is solved for a boric acid thickness $x = 0.01$ cm and a neutron beam intensity of 6×10^{14} neutrons/cm^2•s as follows:

$$R = IN\sigma x$$

$$= (6 \times 10^{14}\ n/cm^2 \cdot s)(2.79 \times 10^{21}\ \text{atoms}^{10}B/cm^3)(3.842 \times 10^3\ \text{barns/atom}^{10}B)(1 \times 10^{-24}\ cm^2/\text{barn})(0.01\ cm)$$

$$= 6.43 \times 10^{13}\ \text{neutrons}/cm^2 \cdot s$$

Thus, the rate of neutron capture by the 1 cm^2 of the pure crystalline H_3BO_3 in this example is 6.43×10^{13} neutrons/s. The ratio of the rate of neutron capture and incident neutron intensity is

$$R/I = 6.43 \times 10^{13}\ n/cm^2 \cdot s / 6 \times 10^{14}\ n/cm^2 \cdot s = 0.107$$

which represents 10.7% of the incident neutron flux. Elemental boron, which contains a natural abundance of 19.1 atom % of ^{10}B has a high cross section for the absorption of thermal neutrons. In pressurized water reactors (PWRs) the boric acid may be dissolved in reactor coolant at a relatively low concentration (up to 2000 ppm or 0.2%) to reduce fission. Some reactors may use boric acid enriched in ^{10}B up to a 40% atom %.

Two very important nuclear reactions, not discussed in this section and vital energy sources, are nuclear fission and fusion. These are discussed in detail in other parts of this chapter.

X. Particulate radiation

All types of nuclear radiation over all ranges of energy will have a dual nature, that of a particle and a wave. Earlier in this chapter the work of Louis Victor de Broglie was described, whereby he demonstrated, that all particles of known rest mass (*e.g.*, alpha particles, beta particles, neutrons, etc.) would, in motion, display the properties of a wave with a wavelength inversely proportional to the particle momentum. For the purpose of classifying radiation in this chapter, particulate radiation will be all types of radiation that have a known mass at rest.

A. Alpha decay

Alpha decay is the nuclear decay process whereby the parent nucleus emits an alpha particle. The alpha particle, structurally equivalent to the nucleus of a helium atom and denoted by the Greek letter α, consists of two protons and two neutrons. Alpha particles are emitted as decay products of many radionuclides that consist of relatively large nuclei predominantly of atomic number greater than 72. For example, the radionuclide americium-241 decays by alpha particle emission to yield the daughter nuclide ^{237}Np according to the following equation:

$$^{241}_{95}Am \rightarrow {}^{237}_{93}Np + {}^4_2He + 5.638\ MeV \quad (1.153)$$

The loss of two protons and two neutrons from the americium nucleus results in a mass reduction of four and a charge reduction of two on the nucleus. In nuclear equations such as the preceding one, the subscript denotes the charge on the nucleus (*i.e.*, the number of protons or atomic number, also referred to as the Z number) and the superscript denotes the mass number (*i.e.*, the number of protons plus neutrons, also referred to as the A number). The 5.638 MeV of Eq. (1.153) is the decay energy, which is described subsequently.

1. Alpha decay energy

The energy liberated during nuclear decay is referred to as decay energy. Many reference sources report the precise decay energies of radioisotopes. The energy released in alpha decay will be reviewed in this section and the decay of ^{241}Am will be taken as an example. The value reported by the National Nuclear Data Center (NNDC, 2018a,b) of the Brookhaven National Laboratory (http://www.nndc.bnl.gov/qcalc) for the decay energy of ^{241}Am, illustrated in Eq. (1.153), is 5.638 megaelectron volts (MeV).

The alpha decay equation may be described in general terms as follows:

$$^{A}_{Z}P \rightarrow {}^{A-4}_{Z-2}D + {}^{4}_{2}\text{He} + K_T \tag{1.154}$$

where $^{A}_{Z}P$ is the parent nucleus of atomic number Z and mass number A, $^{A-4}_{Z-2}D$ is the daughter nucleus of atomic number Z–2 and mass number A–4, $^{4}_{2}$He is the alpha particle equivalent to a helium nucleus, and K_T is the total kinetic energy released in the alpha-decay. The total kinetic energy released, also referred to as the decay energy (Q_α) is shared between the alpha particle, recoil daughter nucleus, and gamma radiation or

$$Q_\alpha = K_T = E_\alpha + E_{\text{recoil}} + E_\gamma \tag{1.155}$$

where E_α is the kinetic energy of the alpha particle, E_{recoil} is the kinetic energy of the recoil daughter nucleus, and E_γ is the energy lost as gamma radiation from the daughter nucleus, when the daughter nucleus is left at an excited energy state and decays to its ground state. We will use the symbol "E" for kinetic energy, rather than "K", when referring to the kinetic energy of the radiation and recoil nuclei, as it is the symbol most commonly used, and the kinetic energy is inferred. The decay energy is represented by Q as the "Q value". The decay energy of 5.638 MeV of Eq. (1.153) above, may be written as Q_α for alpha-decay energy, which may be composed of the energies of the alpha-particle, recoil daughter nucleus, and gamma radiation described in Eq. (1.155). Energy and mass are conserved in the decay process; that is, the energy liberated in radioactive decay is equivalent to the loss of mass by the parent radionuclide (*e.g.*, ^{241}Am) or, in other words, the difference in masses between the parent radionuclide and the product nuclide and particle.

We can calculate the energy liberated in the decay of ^{241}Am, as well as for any radionuclide decay, by accounting for the mass loss in the decay equation. Using Einstein's equation for equivalence of mass and energy

$$E = mc^2 \tag{1.155}$$

we can write the general equation for the energy equivalence to mass loss as

$$Q = (\text{mass of parent nuclide} - \text{mass of decay products})c^2 \tag{1.157}$$

For the decay of ^{241}Am described in Eq. (1.153) above, we can write Eq. (1.157) as

$$Q_\alpha = (M_{^{241}\text{Am}} - M_{^{237}\text{Np}} - M_\alpha)c^2 \tag{1.158}$$

where Q_α is the disintegration or alpha decay energy, $M_{^{241}\text{Am}}$, $M_{^{237}\text{Np}}$ and M_α are the masses of ^{241}Am, ^{237}Np, and the alpha particle, and c is the speed of light in a vacuum. We may use atomic mass units (u) for the calculation, because the electron masses included with atomic mass units may be ignored. The number of electrons on both sides of the decay Eqs. (1.153) and (1.154) are equal, and we can ignore the negligible differences in the binding energies of the electrons. When the nuclide masses are expressed in the more convenient atomic mass units (u), the energy liberated in decay equations can be calculated in units of megaelectron volts according to the equation

$$Q_\alpha = (M_{^{241}\text{Am}} - M_{^{237}\text{Np}} - M_\alpha)(931.494\ \text{MeV/u}) \tag{1.159}$$

The precise atomic mass units obtained from reference tables (Rumble, 2017) or the National Institute of Standards and Technology (NIST, https://www.nist.gov/pml/atomic-weights-and-isotopic-compositions-relative-atomic-masses) can be inserted into Eq. (1.159) to obtain

$$\begin{aligned} Q_\alpha &= (241.0568293\ \text{u} - 237.0481736\ \text{u} - 4.0026032\ \text{u}) \\ (931.494\ \text{MeV/u}) &= (0.0060525\text{u})(931.494\ \text{MeV/u}) \\ &= 5.6378\ \text{MeV} \end{aligned}$$

which corresponds to the online computer calculated value of 5637.8 keV (National Nuclear Data Center, https://www.nndc.bnl.gov/nudat2/indx_dec.jsp). The energy liberated is shared between the daughter nucleus and the α-particle. If the parent nuclide (*e.g.*, ^{241}Am) is at rest when it decays, most of the decay energy will appear as kinetic energy of the liberated less-massive α-particle and only a small fraction of the kinetic energy remaining with the recoiling massive daughter nuclide (*e.g.*, ^{237}Np). The kinetic energy of the recoiling daughter nuclide is comparable to that of a recoiling canon after a shell is fired; the shell being analogous to that of the α-particle shooting out of the nucleus, as illustrated in Fig. 1.39.

Fig. 1.40 illustrates the transitions involved in the decay of ^{241}Am. The interpretation of this figure is given in the

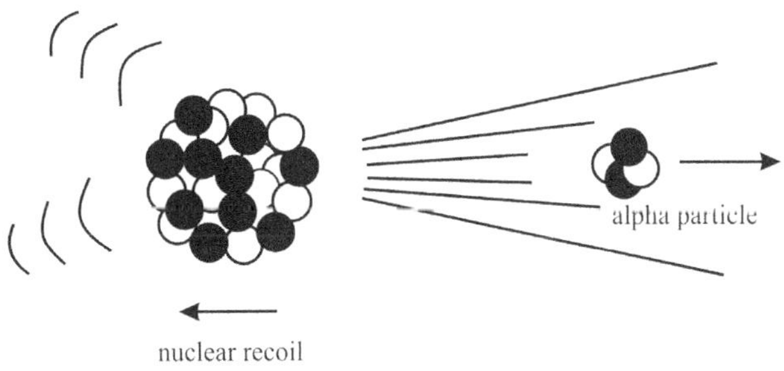

FIGURE 1.39 The alpha particle shoots out of a nucleus. In the decay of nuclides by alpha particle emission, the alpha particle will shoot out of the nucleus, and the daughter nucleus will recoil similar to that of a canon that recoils upon firing a shell from its barrel.

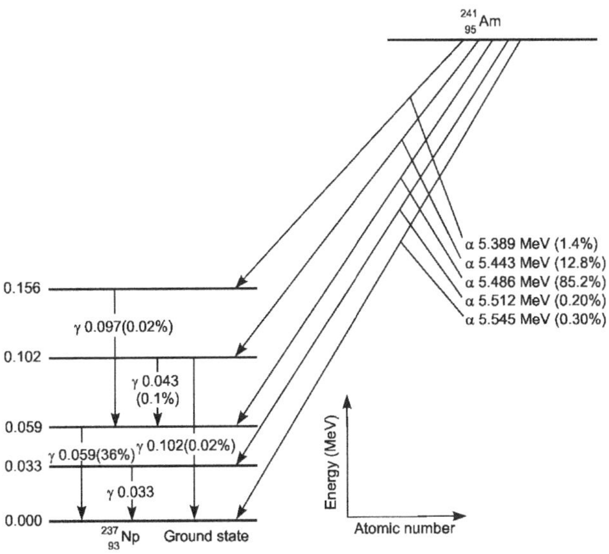

FIGURE 1.40 Decay scheme of ^{241}Am. The relative intensities of alpha-particle and gamma-ray emissions are expressed in percent beside the radiation energy values in MeV.

following paragraph. There are five major α-particle transitions in the decay of ^{241}Am, each involving an α-particle emission at different energies and relative intensities. These are illustrated in Fig. 1.40. The decay energy of 5.6378 MeV for ^{241}Am calculated above and reported in the literature is higher than any of the α-particle energies provided in Fig. 1.40. This is because there remains also the recoil energy of the daughter nucleus and any gamma ray energy that may be emitted by the daughter, when its nucleus remains at an excited state. The emission of gamma radiation often accompanies radionuclide decay processes that occur by alpha particle emission. Gamma radiation is described in Part XI.B of this chapter. The recoil energy, E_{recoil}, of the daughter nucleus can be calculated by the equation

$$E_{recoil} = (M_\alpha / M_{recoil})E_\alpha, \quad (1.160)$$

which is derived in Part XIII of this chapter, where M_α is the mass of the alpha particle as defined in Eqs. 1.159, M_{recoil} is the mass of the recoil nucleus, and E_α is the α-particle energy. For example, the recoil energy of the ^{237}Np daughter nucleus for the transition of the 5.545 MeV alpha particle (Fig. 1.40) can be calculated according to Eq. (1.160) as

$$\begin{aligned} E_{recoil} &= (4.00260325\text{u}/237.048166\text{u})5.545\text{ MeV} \\ &= (0.0168851897)(5.545\text{ MeV}) \\ &= 0.0936\text{ MeV} \end{aligned}$$

The transition energy, E_{trans}, for the above α-particle emission is the sum of the α-particle and recoil nuclear energies or

$$\begin{aligned} E_{trans} &= E_\alpha + E_{recoil} \\ &= 5.545\text{ MeV} + 0.0936\text{ MeV} \quad (1.161) \\ &= 5.638\text{ MeV} \end{aligned}$$

In the above case, the transition energy turns out to be equal to the reported and calculated decay energy because as illustrated in Fig. 1.40, the ^{241}Am radionuclides decay directly to the ground state whenever 5.545 MeV α-particles are emitted. This is not the case when α-particles of other energies are emitted from ^{241}Am. If we take, for example, the 5.486 MeV α-particle transition of Fig. 1.40, the decay energy, E_{decay}, would be the sum of the transition energy plus gamma ray energy, E_γ, emitted from the daughter nucleus or

$$\begin{aligned} E_{decay} &= E_{trans} + E_\gamma \\ &= E_\alpha + E_{recoil} + E_\gamma \\ &= E_\alpha + (M_\alpha/M_{recoil})E_\alpha + E_\gamma \\ &= 5.486\text{ MeV} + (4.00260325\text{ u}/237.048166\text{ u}) \\ &(5.486\text{ MeV}) + 0.059\text{ MeV} \\ &= 5.486\text{ MeV} + 0.0926\text{ MeV} + 0.059\text{ MeV} \\ &= 5.638\text{ MeV} \end{aligned} \quad (1.162)$$

The gamma ray energy emitted from the daughter nucleus for the 5.486 MeV α-particle transition in ^{241}Am decay is illustrated in Fig. 1.40. Gamma ray energy values of other radionuclides are available from reference tables (Browne et al., 1986; Firestone et al., 1996; Lederer and Shirley, 1978; Appendix A of this book and the Internet via the Interactive Chart of the Nuclides of the National Nuclear Data Center, Brookhaven National Laboratory, http://www.nndc.bnl.gov/nudat2/index.jsp).

As described in the previous paragraphs, alpha particles are emitted with a certain quantum of energy as the parent nuclide decays to a lower energy state. The energy emitted from radionuclides as nuclear radiation can be described by a decay scheme such as that given in Fig. 1.40. Decay schemes are written such that the energy levels of the nuclides are plateaus along the ordinate, and these energy plateaus are distributed along the abscissa according to atomic number. The alpha particles, as the example shows (Fig. 1.40), are emitted with certain magnitudes of kinetic energy, which is most often expressed in units of kiloelectron volts (keV) or megaelectron volts (MeV). The energies of alpha particles from most nuclear decay reactions fall within the range 4–10.5 MeV.

Alpha particles are emitted from unstable nuclei with discrete quanta of energy, often leaving the daughter nuclide at an excited energy state. In such cases, when the daughter nuclide occurs at an elevated energy state, it may reach the ground state via the emission of energy in the form of electromagnetic gamma radiation as illustrated in Fig. 1.40.

The nuclei of daughter atoms of alpha particle-emitting nuclides are often unstable themselves and may decay by further alpha or beta particle emission. Thus, alpha particle–emitting nuclides may consist of a mixture of radionuclides, all part of a decay chain, as illustrated in Figs. 1.8–1.11 described previously in this chapter.

2. Alpha decay energy and half-life relationship

Alpha particles are emitted by radionuclides with distinct energies that range between 4 and 10 MeV, and half-lives of the nuclides will vary over a wide range of time from 10^{10} years to microseconds. Radionuclides, emitting alpha particles of low energy, decay with long half-lives; whereas, those emitting alpha particles of high energy, have short half-lives.

A relationship between alpha decay energy and the half-life of the nuclide was recognized very early in the history of the study of radionuclide decay. German physicist Hans (Johannes Wilhelm) Geiger, best known for the Geiger–Müller counter, and British physicist John Mitchell Nuttall (Geiger and Nuttall, 1911) were the first to measure a relationship between alpha particle energy and half-life decay of even-*Z* even-*A* radionuclides, that is, radionuclides that had an even number of protons and even atomic number. By correlating half-life measurements with the range of alpha particles emitted by several radionuclides Geiger and Nuttall (1911) determined the best-fit linear relationship between the two variables and reported the following equation, which is known as the Geiger–Nuttall Law:

$$\log\lambda = a\log R + b \quad (1.163)$$

where λ is the radionuclide decay constant, R is the alpha particle range (cm) and a and b are constants that provided the best fit to the experimental data. Since $\lambda = 0.693/t_{1/2}$, where $t_{1/2}$ is the half-life, it is clear from Eq. (1.163), that for shorter values of $t_{1/2}$ there are correspondingly increasing magnitudes of the alpha-particle range (R). The range of travel of the alpha particle is proportional to its energy. Consequently, the Geiger–Nuttall Law provided a correlation between radionuclide decay half-life and alpha decay energy. Other variations of the best fit equation were reported subsequently to the work of Geiger and Nuttall, which provided best-fit empirical equations for specific radionuclide decay series (Beckerly, 1945), but the correlation of half-life decay and decay energy remained the same.

Alpha decay energy and half-life of heavy elements of even *Z* even *A* were studied by Perlman and Ypsilantis (1950) on the basis of the shell structure of nuclei. They found a direct and exponential relationship between half-life and alpha decay energy for many even *Z* even *A* radioisotopes of the heavy elements, which is illustrated in Fig. 1.41.

Radionuclides that are not even Z and even A (*i.e.*, even Z odd N, odd Z even N, or odd Z odd N) also obey the general rule whereby nuclides with low alpha decay energy exhibit long half-lives, and nuclides with high alpha decay energy exhibit short half-lives. However, these nuclides exhibit longer half-lives than even-even nuclides, because

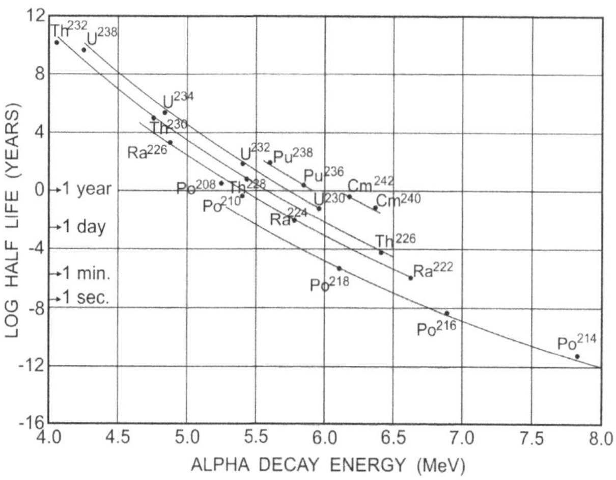

FIGURE 1.41 Observed values of half-life and total alpha-decay energy for each nuclide. The mass number of the radioactive parent nuclide is given at each point. The curves represent the theoretical values for radioactive parents having the same atomic number (isotopes). The range in half-life for the radionuclides, in the curves provided, is 1.6×10^{-4} seconds for ^{214}Po to 1.4×10^{10} y for ^{232}Th. The reader will note from the figure that, at the time of this report, the mass numbers were written as superscripts to the right of the element symbol. Currently mass numbers are written as superscripts to the left of the element symbol. *From Perlman and Ypsilantis (1950), United States Atomic Energy Commission Report UCRL-613, February 27, 1950.*

these nuclides must first transition to an excited state prior to the formation of the alpha particle, which is an even–even nucleus of two protons and two neutrons. The alpha particle must be assembled within the nucleus from proton and neutron nuclear shells, and the transition to the excited nuclear state will prolong the nuclear decay to yet longer half-lives (Perlman et al., 1950; Evans, 1955; Loveland, 2004).

The shorter half-lives exhibited by radionuclides with high alpha decay energies compared to the longer half-lives of nuclides with lower decay energies is explained on the basis of the nuclear potential barrier, that the alpha particle must overcome to penetrate or escape from the nucleus. By bombarding a thin foil of aluminum with 8.75 MeV alpha particles, Rutherford (1927) demonstrated that the high-energy alpha particles were scattered by the aluminum nuclei at a distance of 30×10^{-13} cm from the center of the nucleus, as illustrated in Fig. 1.42. A Coulomb potential barrier, which is inversely proportional to the distance from the center of the nucleus, repels incoming alpha particles as well as inhibits their escape from the nucleus. Question therefore arose as to how alpha particles, which are emitted from decaying nuclei with energies as low as 4 MeV, escape from the nucleus. Gamow (1928) and Gurney and Condon (1928) independently used Schrödinger's wave mechanics to demonstrate that, within the nucleus, the

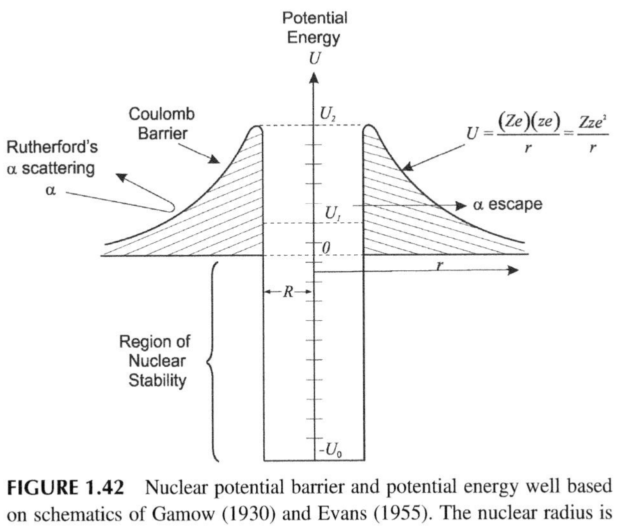

FIGURE 1.42 Nuclear potential barrier and potential energy well based on schematics of Gamow (1930) and Evans (1955). The nuclear radius is from 0 to R, and the distance from the center of the nucleus is r. The potential energy of the bound nucleus within the well is considered constant from $-U_0$ to 0. The Coulomb potential barrier U is inversely proportional to the distance from the center of the nucleus according to $U = (Ze)(ze)/r$ or Zze^2/r where Z is the atomic number or charge on the nucleus and z is the proton number of the alpha particle ($z = 2$). Excess energy of the unstable nucleus, required for alpha-particle emission, is represented by the energy level U_1 along the potential energy scale, and the height of the potential barrier occurs at U_2 where $r = R$. For a perspective on the nuclear dimensions, data from Evans (1955) for the uranium nucleus ($Z = 92$) are the following: $R \approx 9.3 \times 10^{-13}$ cm, $U_1 \approx$ 4 MeV, and $U_1 \approx$ 28 MeV.

alpha particle as a wave need not pass over the Coulomb barrier, but that the alpha particle wave could tunnel its way through the potential barrier and escape from the nucleus. The probability of the alpha-particle escape will be a function of the energy of the alpha particle and the amount of potential barrier that the alpha particle must overcome. Alpha decay energies, among all of the alpha-decaying radionuclides, vary between approximately 4 and 10 MeV. A high alpha decay energy yields alpha particles with less Coulomb barrier to overcome, which augments the probability for alpha particle escape resulting in a shorter half-life. The opposite would be the case for a low alpha decay energy, which would yield alpha particles with a greater Coulomb barrier or reduced penetrability and thus lower probability of escape, resulting in a longer decay time or half-life.

3. *Alpha-particle interactions with matter*

Now consider what happens to an alpha particle that dissipates its kinetic energy by interaction with matter. Alpha particles possess a double positive charge due to the two protons present. This permits ionization to occur within a given substance (solid, liquid or gas) by the formation of ion pairs due to coulombic attraction between a traversing alpha particle and atomic electrons of the atoms within the material the alpha particle travels. The two neutrons of the alpha particle give it additional mass, which further facilitates ionization by coulombic interaction or even direct collision of the alpha particle with atomic electrons. The much greater mass of the alpha particle, 4 atomic mass units (u), in comparison with the electron (5×10^{-4} u) facilitates the ejection of atomic electrons of atoms through which it passes, either by direct collision with the electron or by passing close enough to it to cause its ejection by coulombic attraction. The ion pairs formed consist of the positively charged atoms and the negatively charged ejected electrons. The alpha particle continues along its path suffering, for the most part, negligible deflection by these collisions or coulombic interactions because of the large difference in mass between the particle and the electron. An exception will be a less common (~one in a few thousand) deflection or scattering of the alpha particle due to a direct collision with the Coulombic barrier of an atomic nucleus. Thus, an alpha particle travels through matter producing thousands of ion pairs (see the following calculation) in such a fashion until its kinetic energy has been completely dissipated within the substance it traverses.

In air, an alpha particle dissipates an average of 35 eV (electron volts) of energy per ion pair formed. Before it stops, having lost most of its energy, an alpha particle produces many ion pairs. For example, as a rough estimate, a 5-MeV alpha particle will produce 1.4×10^5 ion pairs in air before coming to a stop:

$$\frac{5\times10^{6}\ \text{eV}}{35\ \text{eV/ion pair}} = 1.4\times10^{5}\ \text{ion pairs in air} \tag{1.164}$$

The thousands of interactions between a traveling alpha particle and atomic electrons can be abstractly compared with a rolling bowling ball colliding with stationary ping-pong balls. Because of the large mass difference of the two, it will take thousands of ping-pong balls to stop a bowling ball. The additional stopping force of electrons is the binding energy of the atomic electrons.

The amount of energy required to produce ion pairs is a function of the absorbing medium. For example, argon gas absorbs approximately 25 eV per ion pair formed and a semiconductor material requires only 2–3 eV to produce an ion pair. Ionization is one of the principal phenomena utilized to detect and measure radionuclides, and it is treated in more detail in subsequent chapters. The energy threshold for ion pair formation in semiconductor materials is approximately 10 times lower than in gases, which gives semiconductor materials an important advantage as radiation detectors (see Chapter 4) when energy resolution in radioactivity analysis is an important factor.

In addition to ionization, another principle mechanism by which alpha particles and charged particles, in general, may impart their energy in matter is via electron excitation. This occurs when the alpha particle fails to impart sufficient energy to an atomic electron to cause it to be ejected from the atom. Rather, the electrons of atoms of a given material may absorb a portion of the alpha particle energy and become elevated to a higher energy state. Depending on the absorbing material, the excited atoms of the material may immediately fall back to a lower energy state or ground state by dissipating the absorbed energy as photons of visible light. This process, referred to as fluorescence, was first observed by Sir William Crookes in London in 1903 (Crookes, 1903a,b) and soon confirmed by Julius Elster and Hans Geitel the same year (Elster and Geitel, 1903) in Wolfenbüttel, Germany. They observed fluorescence when alpha particles emitted from radium bombarded a zinc sulfide screen. In darkness, individual flashes of light were observed and counted on the screen with a magnifying glass with the screen positioned a few millimeters from the radium source. The phenomenon of fluorescence and its significance in the measurement of radionuclides are discussed in subsequent chapters. Thus, as described in the previous paragraphs, alpha particles as well as other types of charged particles, dissipate their energy in matter mainly by two mechanisms, ionization and electron excitation.

Because the atomic "radius" is much larger ($\approx 10^{-10}$ m) than the "radius" of the nucleus ($\approx 10^{-14}$ m), the interactions of alpha particles with matter via direct collision with an atomic nucleus are few and far between. However, when a direct collision of an alpha particle with an atomic nucleus does occur, the large mass of the nucleus causes deflection or ricocheting of the alpha particle via coulombic repulsion without generating any change within the atom. Such deflections were discovered in the early part of this century by Ernest Rutherford and his students Hans Geiger and Ernest Marsden in an experiment discussed previously in detail in Section II of this chapter.

Scattering of alpha particles at angles of less than 90 degrees may occur by coulombic repulsion between a nucleus and a particle that passes in close proximity to the nucleus. These deflected particles continue traveling until sufficient energy is lost via the formation of ion pairs. The formation of ion pairs remains, therefore, the principal interaction between alpha particles and matter.

The high mass and charge of the alpha particle in relation to other forms of nuclear radiation give it greater ionization power, but a poorer ability to penetrate matter. In air, alpha particles may travel only a few centimeters. This short range of travel varies depending on the initial energy of the particle. For example, a 5.5-MeV alpha particle, such as that emitted by the radionuclide ^{241}Am previously described, has a range of approximately 4 cm in dry air at standard temperature and pressure, as estimated by the empirical equation described by Lapp and Andrews (1948), namely Eq. (1.165).

$$R_{\text{air}} = (0.005E + 0.285)E^{3/2} \tag{1.165}$$

where R is the average linear range in cm of the alpha particle in air and E is the energy of the particle in MeV. The empirical formula is applied for alpha particles in the energy range 4–15 MeV. For example, Eq. (1.165) yields a value of 4.03 cm for the range of a 5.5 MeV alpha particle in dry air.

More accurate ranges are available from the following online ASTAR (Alpha Stopping Power and Range) computer program of the National Institute of Standards and Technology (NIST) https://physics.nist.gov/PhysRefData/Star/Text/programs.html, which utilizes the Bethe (1930) stopping-power formula for alpha-particle energies above 2 MeV and fitting formulas based on experimental stopping power data for alpha-particles below 2 MeV. The online ASTAR computer provides a projected range of 5.010×10^{-3} g/cm^2 for the 5.5. MeV alpha particle in dry air near sea level, which divided by the density of dry air near sea level (1.205×10^{-3} g/cm^3) yields a range of 4.15 cm, which agrees closely with the range calculated by the empirical formula of Eq. (1.165). The online ASTAR program provides the ranges of alpha particles or helium nuclei in dry air, other gases, chemical elements, glass, synthetic polymers, inorganic and organic compounds, and biological tissue, etc over a wide energy range from 0.001 to 1000 MeV. All that is needed is to select the medium and the energy of the alpha particle.

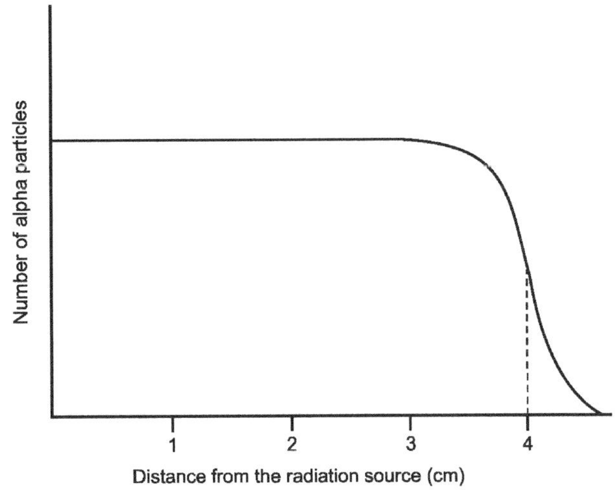

FIGURE 1.43 Range of 5.5 MeV alpha particles in air.

According to calculations of Fenyves and Haiman (1969), the ranges of alpha particles with energies between 4 and 7 MeV can be estimated by using a simplified version of Eq. (1.165) as follows:

$$R_{air} = 0.3E^{3/2} \tag{1.166}$$

Also, ranges of alpha particles in air over a wide range of alpha particle energy can be obtained from Figure B.1 of Appendix B of Volume 2 The online ASTAR program described above yields the more accurate calculations of alpha particles ranges.

The approximate 4-cm range of 5.5-MeV alpha particles in air is illustrated in Fig. 1.43. There is no abrupt drop in the number of alpha particles detected at the calculated range of 4 cm owing to statistical variations in the number of collisions that the particles may have with air molecules and to variations in the amount of energy loss by the particles for each ion pair formed. After being halted, an alpha particle acquires two free electrons through columbic attraction and is converted to helium gas.

In materials other than air, such as liquids and solids, the range of alpha particles is obviously much shorter owing to their higher densities, which enhance the number of collisions a particle may undergo per path length of travel. The range of alpha particles in liquids and solids may be approximated by comparison with ranges in air according to the empirical equation

$$R_{cm} = 0.00032\left(A^{1/2} / \rho\right)R_{air} \tag{1.167}$$

described in a previous text (L'Annunziata, 1987), where R_{cm} is the average range in cm of the alpha particle in an absorber other than air, A is the atomic weight of the absorber, ρ is the absorber density in g/cm^3, and R_{air} is the calculated average linear range of the alpha particle in air (from Eq. 1.165 or 1.166) or more accurate values of R_{air} from the online ASTAR computer program (https://physics.nist.gov/PhysRefData/Star/Text/programs.html). For example, from Eq. (1.167), the 5.5-MeV alpha particles emitted by ^{241}Am have a calculated range of 2.48×10^{-3} cm or 24.8 μm in aluminum ($A = 27$ and $\rho = 2.69$ g/cm^3 utilizing a value of R_{air} of 4.03 cm calculated previously from Eq. 1.165). However, if we utilize the more accurate online ASTAR program and compute the range of 5.5 MeV alpha particles in aluminum, we obtain the value of the range to be 6.6×10^{-3} g/cm^2, which divided by the density of aluminum (2.69 g/cm^3) yields a linear range of 2.45×10^{-3} cm or 24.5 μm, which corresponds closely to the value calculated by the empirical Eq. (1.167). Empirical equations should be used with caution, and it is best to compare empirical calculations to the computer calculated values such as the online ASTAR program.

The linear ranges, that is the depth to which charged particles, such as the alpha particle, will penetrate in liquids and solid materials can be too short to measure with conventional laboratory instrumentation. The alternative is to express range in units of weight of absorber material per unit area, such as mg/cm^2, which is a measure of milligrams of absorber per square centimeter in the absorption path, or in other words, a measure of absorber thickness. If we multiply the linear range of the alpha particle measured in cm of absorber material by the density of the absorber in units of mg/cm^3, the range of the alpha particle in an absorber will be expressed in terms of the weight of absorber per unit area (mg/cm^2) as described by Eq. (1.168), as follows

$$R_{mg/cm^2} = (\mathrm{R_{cm}})(\rho) \tag{1.168}$$

where R_{mg/cm^2} is the range of alpha particles of a given energy in units of mg/cm^2, also referred to as mass thickness units or material surface density, R_{cm} is the linear range of the alpha particles, and ρ is the absorber density. For example, the linear range of the 5.5 MeV alpha particles in aluminum, calculated above with Eq. (1.167) is converted to range in mass thickness units according to Eq. (1.168) as follows

$$R_{mg/cm^2} = \left(2.48 \times 10^{-3}\ \mathrm{cm}\right) \times \left(2690\ \mathrm{mg} / \mathrm{cm}^3\right) = 6.6\ \mathrm{mg/cm^2}$$

Therefore, the mass thickness of 6.6 mg/cm^2 of aluminum absorber is sufficient to absorb alpha particles of 5.5 MeV energy. The ASTAR (Alpha Stopping Power and Range Tables, https://physics.nist.gov/PhysRefData/Star/Text/ASTAR.html, provide a value of 6.60×10^{-3} g/cm^2

as the range of 5.5 MeV alpha particles in aluminum, which agrees with the calculated value of 6.6 mg/cm^2 calculated above.

Ranges of alpha particles as well as other charged particles such as protons and deuterons of a given energy in absorber elements of atomic number Z > 10 in units of absorber mass thickness can be calculated directly by comparison to the calculated range of the same charged particles of the same energy in air according to the following formula described by Friedlander et al. (1964).

$$\frac{R_Z}{R_{air}} = 0.90 + 0.0275Z + (0.06 - 0.0086Z)\log\frac{E}{M} \quad (1.169)$$

where R_Z is the range of the charged particle in mass thickness units, mg/cm^2, R_{air} is the range of the charged particle in air in the same mass thickness units, Z is the atomic number of the absorber element, E is the particle energy in MeV, and M is the mass number of the particle (*i.e.*, 1 for protons, 2 for deuterons, and 4 for alpha particles). For example, if we use the empirical formula provided above (Eq. 1.169) to calculate the range of 5.5 MeV alpha particles ($M = 4$) in aluminum ($Z = 13$) and a value of $R_{\text{air}} = 5.01 \times 10^{-3}$ g/cm^2, we obtain the value of $R_Z = 6.3 \times 10^{-3}$ g/cm^2, which is in close agreement to the mass thickness range calculated previously with Eqs. (1.167) and (1.168). In this example, Eq. (1.169) requires the value of R_{air} for 5.5 MeV alpha particles, which is determined according to Eq. (1.168) as the product of the 5.5 MeV alpha-particle linear range in air (previously calculated) and the density of air at STP ($\rho = 1.205$ mg/cm^3), that is, R_{air} = (4.15 cm) (1.205 mg/cm^3) = 5.01 mg/cm^2 or 5.01×10^{-3} g/cm^2. The value of $R_{\text{air}} = 5.01 \times 10^{-3}$ g/cm^2 is also available from the online ASTAR program (https://physics.nist.gov/PhysRefData/Star/Text/programs.html). The formula provided by Eq. (1.169) is applicable to charged particles over a wide range of energies (approximately over the range 0.001–1000 MeV) and for absorber elements of Z > 10. For lighter absorber elements, the term $0.90 + 0.0275Z$ is replaced by the value 1.00 with the exception of hydrogen and helium, where the value of 0.30 and 0.82 are used, respectively (Friedlander et al., 1964). Empirical equations, such as Eq. (1.169), should be used with caution and calculated results compared to the ASTAR computer program.

Where alpha particles alone are concerned, the range in mass thickness units can be calculated according to the following equation described by Ehman and Vance (1991), as follows

$$R_{\text{mg/cm}^2} = 0.173\text{E}^{3/2}\text{A}^{1/3} \quad (1.170)$$

where E is the energy of the alpha particle in MeV, and A is the atomic weight of the absorber. If we continue to use the 5.5 MeV alpha particles emitted from ^{241}Am as an example, we can calculate their range in mass thickness units in aluminum according to Eq. (1.170) as follows

$$R_{\text{mg/cm}^2} = 0.173(5.5)^{3/2}(27)^{1/3} = 6.6 \text{ mg/cm}^2$$

Ranges reported in mass thickness units (mg/cm^2) of absorber can be converted to linear range (cm) in that same absorber material from the absorber density (ρ) from the relationship described in Eq. (1.168) or

$$\text{R}_{\text{cm}} = R_{\text{mg/cm}^2}/\rho \quad (1.171)$$

For example, the linear range of the 5.5 MeV alpha particles in aluminum ($\rho = 2.69$ g/cm^3) is calculated as

$$R_{\text{cm}} = \frac{6.6 \text{ mg/cm}^2}{2690 \text{ mg/cm}^3} = 0.0024 \text{ cm} = 24 \text{ }\mu\text{m}$$

When the absorber material is not a pure element, but a molecular compound (*e.g.*, water, paper, polyethylene, etc) or mixture of elements, such as an alloy, the ranges of alpha particles in the absorber are calculated according to Eq. (1.172) on the basis of the atomic weights of the elements and their percent composition in the absorber material or, in other words, the weight fraction of each element in the complex material. Thus, the range in mass-thickness units for alpha particles in absorbers consisting of compounds or mixtures of elements is calculated according to the equation

$$\frac{1}{R_{\text{mg/cm}^2}} = \frac{w_1}{R_1} + \frac{w_2}{R_2} + \frac{w_3}{R_3} + \cdots \frac{w_n}{R_n}. \quad (1.172)$$

where $R_{\text{mg/cm}^2}$ is the range of the alpha-particles in mass-thickness of the complex absorber material, and w_1, w_2, w_3 ... w_n are the weight fractions of each element in the absorber, and R_1, R_2, R_3 ... R_n are the ranges in mg/cm^2 of the alpha particle of defined energy in each element of the absorber. For example, the range of 5.5 MeV alpha particles in Mylar (polyethylene terephthalate) in units of mass thickness is calculated as follows:

$$\frac{1}{R_{\text{mg/cm}^2}} = \frac{w_C}{R_C} + \frac{w_H}{R_H} + \frac{w_O}{R_O}$$

where w_C, w_H, and w_O are the weight fractions of carbon, hydrogen, and oxygen, respectively, in Mylar and R_C, R_H, and R_O are the mass thickness ranges of the alpha particles in pure carbon, hydrogen, and oxygen, respectively. The ranges of 5.5 MeV alpha particles in carbon, hydrogen, and oxygen are calculated according to Eq. (1.170) as

$$R_C = 0.173(5.5)^{3/2}(12)^{1/3} = 5.10 \text{ mg/cm}^2$$

$$R_H = 0.173(5.5)^{3/2}(1)^{1/3} = 2.23 \text{ mg/cm}^2$$

$$R_O = 0.173(5.5)^{3/2}(16)^{1/3} = 5.62 \text{ mg/cm}^2$$

The weight fractions of the carbon, hydrogen, and oxygen in Mylar [$-(C_{10}H_8O_4)_n-$] are calculated as

$$w_C = (12 \times 10)/192 = 0.625$$

$$w_H = (1 \times 8)/192 = 0.042$$

$$w_O = (16 \times 4)/192 = 0.333$$

The calculated ranges of the 5.5 MeV alpha particles in each element and the values of the weight fractions of each element in Mylar can now be used to calculate the alpha particle range in Mylar in mass-thickness units according to Eq. (1.172) as

$$\frac{1}{R_{\text{Mylar}}} = \frac{0.625}{5.10} + \frac{0.042}{2.23} + \frac{0.333}{5.62} = 0.201$$

$$R_{\text{Mylar}} = 4.9 \text{ mg/cm}^2$$

The more accurate online ASTAR program (https://physics.nist.gov/PhysRefData/Star/Text/ASTAR.html) provides a value of 4.7 mg/cm^2 for the alpha particle range in mass thickness units for Mylar (polyethylene terephthalate), which is close to that calculated above by the empirical formula. The linear range of these alpha particles in Mylar are obtained from the range in mass thickness units and the density of Mylar ($\rho = 1.38$ g/cm^3) as

$$R_{\text{cm}} = \frac{4.9 \text{ mg/cm}^2}{1380 \text{ mg/cm}^3} = 0.0036 \text{ cm} = 36 \ \mu\text{m}$$

To provide illustrative examples, the values of the ranges of 5.5 MeV alpha particles in units of mass thickness of various absorber materials are provided in Table 1.9. The data in this table was calculated by means of the empirical equations described earlier in this chapter and by the online ASTAR computer program. The empirical equations can provide good estimates of the alpha particle range; however, they may fall short for absorbers of high atomic number. These values in Table 1.9 represent the milligrams of absorber per square centimeter in the alpha particle absorption path. It can be difficult to envisage alpha particle distance of travel from the values of range when expressed in units of mass thickness. However, it is intuitively obvious that, the greater the charge on the nucleus of the absorber (*i.e.*, absorber atomic number, *Z*), the greater the atomic weight of the absorber (*A*), and the greater the absorber density (ρ), the shorter will be the path length of travel of the alpha particle through the absorber. This is more evident from the calculated values of linear range of 5.5 MeV alpha particles in various gaseous, liquid, and solid absorbers provided in Table 1.10. From the linear ranges, we can see that 5.5-MeV alpha particles could not pass through fine commercial aluminum foils 0.0025 cm thick. Also, the alpha particles of the same energy would not pass through a layer of Mylar only 0.0036 cm thick. Mylar is a polymer sometimes used as a window for gas ionization detectors. From our previous calculations in this chapter, we can see that a Mylar window of mass thickness 4.9 mg/cm^2 would not allow 5.5-MeV alpha particles to pass into the gas ionization chamber. A sample emitting such alpha particles would have to be placed directly into the chamber in a windowless fashion to be detected and counted.

From the above treatment, it is clear that the range of alpha particle travel depends on several variables including (1) the energy of the alpha particle, (2) the atomic number and atomic weight of the absorber, and (3) the density of the absorber. The higher the alpha particle energy, the greater will be its penetration power into or through a given substance as more coulombic interactions of the alpha particle with the electrons of the absorber will be required to dissipate its energy before coming to rest. Also, if we consider an alpha particle of given energy, its range will be shorter in absorbers of higher atomic number or atomic

TABLE 1.9 Ranges of 5.5 MeV alpha particles in various absorbers in units of surface density or mass thickness calculated by empirical equations and ASTAR program.

	Absorber				
Method of calculation	Water	Paper	Aluminum	Copper	Gold
Empirical equation	4.8[a] mg/cm^2	4.9[a,b] mg/cm^2	6.6[c] mg/cm^2	8.9[c] mg/cm^2	12.9[c] mg/cm^2
ASTAR[d]	4.3	4.9[e]	6.6	10.1	18.1

[a]*Calculated with empirical formula provided by Eq. (1.172) on the basis of the weight fraction of each element in the absorber.*
[b]*Cellulose, $(C_6H_{10}O_5)_n$ calculated on the basis of the weight fraction of each element in the monomer.*
[c]*Calculated with empirical formula provided by Eq.(1.170).*
[d]*Calculated by the online ASTAR (Alpha Stopping Power and Range) program https://physics.nist.gov/PhysRefData/Star/Text/programs.html.*
[e]*Calculated for cellulose nitrate.*

TABLE 1.10 Projected ranges[a] of 5.5-MeV alpha particles in various absorbers in units of cm and μm or 10^{-6} m.

Air[b]	Water[b]	Mylar[b]	Carbon[b]	Aluminum[b]	Copper[b]	Gold[b]
4.16 cm	0.0043 cm	0.0033 cm	0.0028 cm	0.0024 cm	0.00113 cm	0.00094
41,600 μm	43 μm	33 μm	28 μm	24 μm	11.3 μm	9.4 μm

[a]*Calculated by the online ASTAR computer program https://physics.nist.gov/PhysRefData/Star/Text/programs.html, which calculates the projected range of an alpha particle in an absorber in units of g/cm². The value of the projected range is then divided by the density of the absorber in units of g/cm³ to yield the range in units of cm. The projected range is defined by NIST as the average value of the depth to which a charged particle will penetrate in the course of slowing down to rest. The depth is measured along the initial direction of the particle.*
[b]*The absorbers are of increasing density from left to right, The densities of the absorbers are the following: Air (dry air at near sea level), $\rho = 1.205 \times 10^{-3}$ g/cm³; Water, liquid $\rho = 1.00$ g/cm³; Mylar (polyethylene terephthalate), $\rho = 1.40$ g/cm³; Carbon (graphite), $\rho = 1.70$ g/cm³; Aluminum, $\rho = 2.69$ g/cm³; Copper, $\rho = 8.96$ g/cm³; Gold, $\rho = 19.32$ g/cm³.*

weight and density, as the absorber atoms will contain a higher number of atomic electrons, which would consequently increase the number of columbic interactions of the alpha particle per path length of travel.

As the alpha particle travels through air and undergoes energy loss via numerous collisions, the velocity of the particle obviously diminishes. At reduced velocity and consequently reduced momentum, an alpha particle is more affected by columbic attraction within the vicinity of a given atom. Progressive reduction in the velocity of travel of the alpha particle therefore results in an increase in the number of ion pairs produced per millimeter of path length of travel. The increase in ionization per path length of travel of an alpha particle is illustrated in Fig. 1.44. The highest specific ionization (number of ion pairs formed per millimeter of path) occurs shortly before termination of the alpha particle's travel, some 2 or 3 mm before the end of its range.

B. Beta decay

A nuclear decay process, referred to as beta decay, results in some circumstances in the emission of beta radiation. Beta decay may be defined as any nuclear decay process whereby the mass number (A) of the nucleus remains the same and the atomic number (Z) changes. There are three main types of beta decay. These are

i. Negatron (β^-) emission, which involves the emission of a negative beta particle (i.e., negative electron) from the nucleus,
ii. Positron (β^+) emission whereby a positive beta particle (i.e., positively charged electron) is emitted from the nucleus, and
iii. EC, which does not result in the emission of any beta particle.

Each of the above beta decay processes will be treated in detail in the following sections of this chapter.

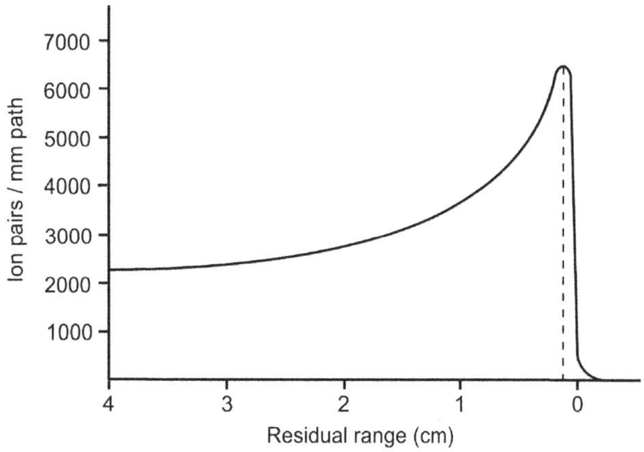

FIGURE 1.44 Specific ionization of an alpha particle in air along its range of travel.

1. Negatron (β^-) emission

Negatron emission can be described by the general equation

$$ {}^{A}_{Z}P \rightarrow {}_{Z+1}^{\ A}D + \beta^- + \bar{\nu} + K_T \qquad (1.173)$$

where ${}^{A}_{Z}P$ is the parent nuclide of atomic number Z and mass number A, ${}_{Z+1}^{\ A}D$ is the daughter nuclide of atomic number $Z+1$ and same mass number as the parent, β^- is the negatron or negative beta particle, $\bar{\nu}$ is the antineutrino, and K_T is the total kinetic energy released in the beta-decay process, The total kinetic energy released, also referred to as the decay energy (Q_β), is shared between the negatron, antineutrino, recoil daughter nucleus, and any gamma radiation that may be emitted by the daughter nucleus. Gamma ray emission can occur when the daughter nuclide is left in an excited energy state and decays to a lower energy state or ground (stable) state with the emission of energy as gamma ray photons. Gamma ray emission does not occur in all cases of negatron emission. There are some nuclides described further on in this section that decay with the emission of negatrons directly to the stable ground state without the emission of gamma radiation. The decay energy, Q_{β^-}, may be defined as

$$Q_{\beta^-} = K_T = E_{\beta^-} + E_{\bar{\nu}} + E_\gamma + E_{\text{recoil}} \qquad (1.174)$$

where E_{β^-} is the kinetic energy of the negatron, $E_{\overline{\nu}}$ is the kinetic energy of the antineutrino, E_γ is the gamma-ray energy, which is emitted when the daughter nucleus is at an elevated energy state and drops to a more stable or ground state, and E_{recoil} is the kinetic energy of the recoil daughter nucleus. The recoil energy may be ignored in most calculations of beta decay energy, because it may be only a few electron volts (eV) compared to the fractional or multiple magnitudes of megaelectron volts (MeV) energy possessed by the beta particle, antineutrino, and gamma radiation (see the Section on Nuclear Recoil in this chapter). We should keep in mind that there is a maximum energy shared by the beta particle and antineutrino ($E_{\beta^-}+E_{\overline{\nu}}$), which is often referred to as the E_{max} of the beta particle.

a. Basic principles

A negatron or negative beta particle (β^-) is an electron emitted from the nucleus of a decaying radionuclide that possesses an excess of neutrons or, in other words, a neutron/proton (n/p) imbalance. The nuclear instability, caused by the n/p imbalance, results in the conversion of a neutron to a proton within the nucleus. The balance of charge is conserved by the simultaneous formation of an electron (negatron) according to the equation

$$\mathrm{n} \rightarrow \mathrm{p}^+ + \beta^- + \overline{\nu}. \tag{1.175}$$

A neutrino, which is a particle of zero charge, accompanies beta-particle emission. The neutrino can be identified further as two types with opposite spin, namely, the antineutrino ($\overline{\nu}$), which accompanies negative beta particle (negatron) emission and the neutrino (ν), which accompanies positive beta particle (positron) emission. The explanation for the neutrino and its properties, also emitted from the decaying nucleus, is given further on in this section. The electron formed cannot remain within the nucleus, and it is thus ejected as a negatron or negative beta particle, β^-, with an energy equivalent to the mass difference between the parent and daughter atoms, less the energy of the antineutrino or neutrino in the case of positron emission, and any gamma ray energy that may be emitted by the daughter nucleus, if the nucleus is left in an excited energy state, and recoil energy of the daughter nucleus as described by Eq. (1.174). Tritium (^{3}H), for example, decays with β^- emission according to the following:

$$^3_1\mathrm{H} \rightarrow {}^3_2\mathrm{He} + \beta^- + \overline{\nu} + 0.0186\ \mathrm{MeV} \tag{1.176}$$

The ^{3}H nucleus decays directly to the ground state of ^{3}He. Thus, there is no gamma ray energy emission in this case. The value of 0.0186 MeV (megaelectron volts) is the maximum energy the beta particle may possess. The unstable tritium nucleus contains two neutrons and one proton. The transformation of a neutron to a proton within the tritium nucleus results in a charge transfer on the nucleus from +1 to +2 without any change in the mass number. Although there is no change in the mass number, the mass of the stable helium isotope produced is slightly less than that of its parent tritium atom. Eqs. (1.177) to (1.182) illustrate other examples of β^- decay, which decay to stable daughter nuclides whereby no gamma-ray emission occurs.

$$^{14}_{6}\mathrm{C} \rightarrow {}^{14}_{7}\mathrm{N} + \beta^- + \overline{\nu} + 0.156\ \mathrm{MeV} \tag{1.177}$$

$$^{32}_{15}\mathrm{P} \rightarrow {}^{32}_{16}\mathrm{S} + \beta^- + \overline{\nu} + 1.710\ \mathrm{MeV} \tag{1.178}$$

$$^{35}_{16}\mathrm{S} \rightarrow {}^{35}_{17}\mathrm{Cl} + \beta^- + \overline{\nu} + 0.167\ \mathrm{MeV} \tag{1.179}$$

$$^{36}_{17}\mathrm{Cl} \rightarrow {}^{36}_{18}\mathrm{Ar} + \beta^- + \overline{\nu} + 0.714\ \mathrm{MeV} \tag{1.180}$$

$$^{45}_{20}\mathrm{Ca} \rightarrow {}^{45}_{21}\mathrm{Sc} + \beta^- + \overline{\nu} + 0.258\ \mathrm{MeV} \tag{1.181}$$

$$^{89}_{38}\mathrm{Sr} \rightarrow {}^{89}_{39}\mathrm{Y} + \beta^- + \overline{\nu} + 1.490\ \mathrm{MeV} \tag{1.182}$$

The energies of beta particle decay processes are usually reported as the maximum energy, E_{max}, that the emitted beta particle or antineutrino may possess. The maximum energy is reported because beta particles are emitted from radionuclides with a broad spectrum of energies. A typical spectrum is illustrated in Fig. 1.45. The shape of the beta particle energy spectra will vary among the beta-emitting radionuclides, but all beta-emitting radionuclides will display a wide spectrum of energies between zero and E_{max}. Unlike alpha particles, which have a discrete energy, beta particles are emitted with a wide spectrum of energies ranging from zero to E_{max}.

The majority of the beta particles emitted have energies of approximately $\frac{1}{3}E_{max}$. Only a very small portion of the beta particles is emitted with the maximum possible energy from any radionuclide sample. In 1930 Wolfgang Pauli was the first to postulate why beta particles were not emitted with fixed quanta of energy, quite the contrary to what is observed in alpha particle emission. He proposed the

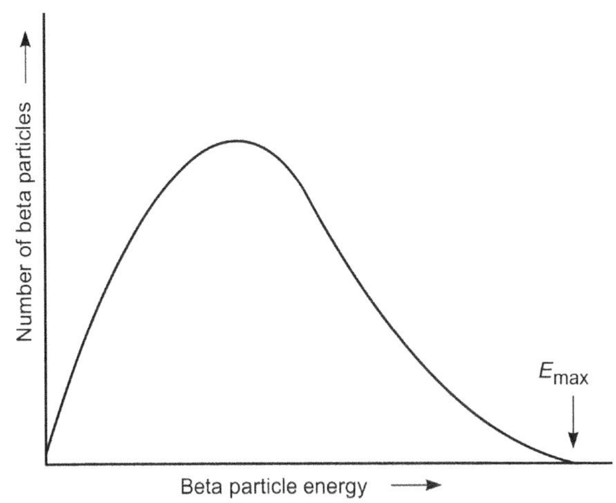

FIGURE 1.45 General beta particle energy spectrum.

existence of an elusive, neutral particle of near zero rest mass in a letter to the participants of a Regional Meeting on Radioactivity at Tübingen, Germany. The neutrino was considered elusive because if it existed, its zero charge and near zero rest mass would make the neutrino undetectable by conventional means and allow a neutrino to penetrate matter easily and even pass through the entire earth without causing a single interaction. The neutrino would be emitted simultaneously with the beta particle from the decaying nucleus and share the energy of decay with the beta particle. For example, if a beta particle was emitted from tritium (decay energy = 0.0186 MeV) with an energy of 0.0086 MeV, the accompanying neutrino would possess the remaining energy of 0.01 MeV, that is, the decay energy less the beta particle energy (0.0186 − 0.0086 MeV). Consequently, if we observe any number of beta particles emitted from a tritium sample or other beta-emitting nuclide sample (*e.g.*, ^{14}C, ^{32}P, ^{90}Sr), they would possess different energies and display an energy spectrum from zero to E_{max} as illustrated in Fig. 1.45.

b. Inverse beta decay

With Pauli's postulation of the neutral particle, Enrico Fermi elaborated the beta-decay theory in 1934 and coined the neutrino from the Italian language meaning "little neutral one". The particle remained elusive until the observation of the neutrino was demonstrated by Reines and Cowan in 1956 (see Reines and Cowan, 1956, 1957; Cowan et al., 1956; Reines, 1960, 1979, 1994). They confirmed the existence of the neutrino by demonstrating inverse beta decay where an antineutrino interacts with a proton to yield a neutron and positron

$$\bar{\nu} + p^+ \rightarrow n + e^+ \tag{1.183}$$

They used a tank of water containing a solution of $^{113}CdCl_2$ illustrated schematically in Fig. 1.46. Antineutrinos interacted with the protons of the water to produce neutrons and positrons. Some of the neutrons produced would be absorbed by the ^{113}Cd with the concomitant emission of characteristic gamma radiation. In coincidence, they observed two 511 keV gamma rays, which originate when a positron comes to rest in the vicinity of an electron, its antiparticle, which results in the annihilation of two electrons into two gamma ray photons of energy equivalent to the electron masses, 0.511 MeV. In the same year, Lee and Yang (1956) proposed that neutrinos and antineutrinos possessed left-handed and right-handed spins, respectively. Inverse beta decay remains an important nuclear process utilized in the measurement of solar neutrinos today (Zuber, 2003; Ianni et al., 2005; Fukuda et al., 2010; Ianni, 2017; Vannucci, 2017).

An inverse beta decay, first proposed by Raghaven (1976), which involves the interaction of a neutrino with a neutron of ^{115}In is the following:

$$\nu + {}^{115}_{49}In \rightarrow {}^{115}_{50}Sn^* + e^- \tag{1.184}$$

which can be written in general terms illustrating only the nucleons as

$$\nu + n \rightarrow p^+ + e^- \tag{1.185}$$

The superscript asterisk in Eq. (1.184) represents an excited state of ^{115}Sn, which decays into the ground state with a lifetime of 4.76 μs and emits two gamma ray photons of 116 and 497 keV. This provides a triple coincidence signal, namely the emitted electron together with the two gamma ray photons, that can be detected among huge backgrounds (Fukuda et al., 2010). Also, the neutrino energy may be calculated from the energy of the prompt electron according to the equation

$$E_{e^-} = E_\nu - (E_f - E_i) \tag{1.186}$$

where E_{e^-} is the energy of the electron produced, E_ν is the neutrino energy, and E_f and E_i are the final and initial energies of the nuclear state involved in the transition (Zuber, 2003). In the case of neutrino interactions with ^{115}In, Eq. (1.186) may be written, according to Fukuda et al. (2010) as

$$E_{e^-} = E_\nu - 118\ \text{keV} \tag{1.187}$$

Thus, neutrino spectroscopy is possible.

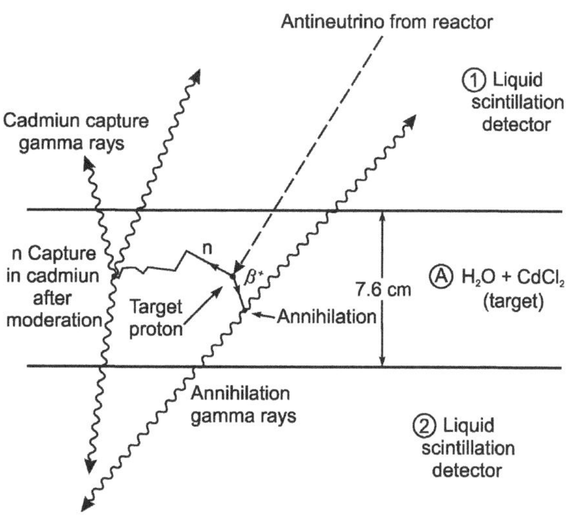

FIGURE 1.46 Detection scheme used by Reines and Cowan for the antineutrino signature signal. An antineutrino is illustrated entering the tank of aqueous $CdCl_2$ solution and striking a target proton. The proton converts to a neutron and positron. The positron annihilates on an electron with the emission of two 0.511 MeV gamma rays in opposite direction detected by the liquid scintillator in tanks above and below the water target tank. The neutron produced by the antineutrino interaction slows down in the water and is captured by a cadmium nucleus, and the resulting gamma rays are detected by the liquid scintillator in the adjacent tanks approximately 10 μs after the positron annihilation. *From Reines (1995) Nobel Lecture with permission © The Nobel Foundation 1995.*

In general, the detection principle of solar neutrinos involving the inverse beta decay reaction of Eq. (1.185), as described by Zuber (2003), may involve one of the following two reactions:

$$\nu + {}^{A}_{Z}\mathrm{A} \rightarrow {}_{Z+1}{}^{A}\mathrm{B}_{\mathrm{g.s.}} + e^{-} \rightarrow {}_{Z+2}{}^{A}\mathrm{C} + e^{-} + \overline{\nu} \quad (e - e \text{ tag}) \tag{1.188}$$

or

$$\nu + {}^{A}_{Z}\mathrm{X} \rightarrow {}_{Z+1}{}^{A}\mathrm{Y}^{*} + e^{-} \rightarrow {}_{Z+1}{}^{A}\mathrm{Y}_{\mathrm{g.s.}} + \gamma \quad (e - \gamma \text{ tag}) \tag{1.189}$$

The reaction described in Eq. (1.188) involves the interaction of a neutrino with a neutron of nuclide ${}^{A}_{Z}\mathrm{A}$ of atomic number Z and mass number A producing a nuclide ${}_{Z+1}{}^{A}\mathrm{B}_{\mathrm{g.s.}}$ at its ground state and a beta particle (e^{-}). The nuclide ${}_{Z+1}{}^{A}\mathrm{B}_{\mathrm{g.s.}}$ undergoes beta decay to nuclide ${}_{Z+2}{}^{A}\mathrm{C}$ with the emission of a beta particle (e^{-}) and antineutrino ($\overline{\nu}$). This reaction has an $e-e$ tag whereby a coincidence signal between the two electron emissions provides the signature for the neutrino detection. The following reaction described by Eq. (1.189) depicts a neutrino interacting with a neutron of nuclide ${}^{A}_{Z}\mathrm{X}$ producing nuclide ${}_{Z+1}{}^{A}\mathrm{Y}^{*}$ at an excited energy state indicated by the superscript asterisk plus an electron or beta particle (e^{-}). The nuclide ${}_{Z+1}{}^{A}\mathrm{Y}^{*}$ then undergoes deexcitation to its ground state ${}_{Z+1}{}^{A}\mathrm{Y}_{\mathrm{g.s.}}$ with the emission of a gamma ray photon. This reaction has an $e^{-}\gamma$ tag whereby a coincidence signal between the electron and one or more gamma ray photon emissions provides the signature for the neutrino detection. The neutrino energies are determined by Eq. (1.186). Zuber (2003) provides a review of these two approaches to neutrino detection and spectroscopy. The interaction of a neutrino with $^{115}\mathrm{In}$ previously discussed (Eq. 1.184), utilized by Fukuda et al. (2010), is an example of the generic reaction yielding an $e^{-}\gamma$ tag depicted by Eq. (1.189), which is specifically

$$\nu + {}^{115}_{49}\mathrm{In} \rightarrow {}^{115}_{50}\mathrm{Sn}^{*} + e^{-} \rightarrow {}^{115}_{50}\mathrm{Sn} + \gamma(116\text{ keV}) + \gamma(497\text{ keV}) \tag{1.190}$$

An example of the generic reaction depicted in Eq. (1.188) is the following:

$$\nu + {}^{13}_{6}\mathrm{C} \rightarrow {}^{13}_{7}\mathrm{N}_{\mathrm{g.s.}} + e^{-} \rightarrow {}^{13}_{6}\mathrm{C} + e^{+} \tag{1.191}$$

This reaction was utilized by Ianni et al. (2005) for the detection of neutrinos. The reaction provides an $e-e$ tag as the beta particle (e^{-}) produced in the reverse beta decay yields one signal while the positron (e^{+}) produced in the subsequent decay of $^{13}\mathrm{N}$ provides a time and space coincidence signal and signature for the neutrino detection.

c. Neutrino mass

Since its inception by Pauli in 1930 up to recent years, the neutrino or antineutrino had been thought to have no rest mass or to possess a near-zero rest mass. It was not until June 5, 1998 was it announced by the Super-Kamiokande Collaboration, including scientists from 23 institutions in Japan and the United States, at the "Neutrino 98" International Physics Conference in Takayama, Japan, that neutrinos possessed a definite mass (Gibbs, 1998; Kesterbaum, 1998; Kearns et al., 1999; Nakahata, 2000). The mass was not reported, but evidence was provided that the neutrino did possess mass although it was considered to be "very small", at least 0.07 eV, which would be less than a millionth of the electron mass. Evidence for the neutrino mass was provided by demonstrating that neutrinos can "oscillate" from one type into another (*i.e.*, electron-, muon-, and tau-neutrinos) as they travel through space and matter. Oscillation is the changing of neutrino types back and forth from one type to another, and this could occur only if the neutrino possessed mass. Experimental research on neutrino oscillation is reviewed by Messier (2006), and the Nobel prize in Physics 2015) was awarded jointly to Takaaki Kajita of the University of Tokyo, Kashiwa, Japan and Arthur B. McDonald of Queen's University, Kingston, Canada for, in the words of the Nobel Committee "the discovery of neutrino oscillations, which show that neutrinos have mass".

At the "Neutrino 2000" Conference held at Sudbury, Canada June 16–21, 2000, groups from the University of Mainz, Germany (Bonn et al., 2001) and Institute for Nuclear Research, Moscow (Lobashev et al., 2000) reported the mass of the neutrino to be between 2.2 and 2.5 eV/c^2, respectively, at 95% confidence levels. It is common to express subatomic particle mass in units of energy based on equivalence of mass and energy ($E = mc^2$), so that the particle mass m is measured in units of E/c^2, and with energy expressed in units of electron volts, eV/c^2. Neutrino mass experiments are reviewed by Kraus et al. (2005), Otten and Weinheimer (2008), Beck (2010), and Kristiansen et al. (2010), and Ferri (2016). Beck (2010) underscores that results from previous research reported by the Triosk (Russia) and Mainz (Germany) neutrino mass experiments (see Lobashev et al., 1999a,b,c, 2000; Kraus et al., 2005) have set the best upper limits of 2.3 eV/c^2 with a 95% confidence limit on the neutrino mass. Beck (1010) also reports that the KATRIN experiment (KArlsruhe TRItiun Neutrino experiment) is under preparation to search for the mass of the electron neutrino with a sensitivity of 0.2 eV/c^2. With this objective the KATRIN experiment will perform a precision measurement of the endpoint region of the β-decay spectrum of tritium, the shape of which depends highly on the neutrino mass. A windowless gaseous tritium source is used in a transport system that guides the beta particles of tritium without energy loss to PIN diode detectors described by Wüstling et al. (2006). A prespectrometer rejects all beta particles with energy <200 eV below the β-spectrum endpoint, and the beta particle energies can be measured with a resolution of 0.93 eV. Ferri (2016) reports that the KATRIN spectrometer will operate with the aim of reaching a sensitivity of 0.2 eV/c^2 in the determination of the neutrino mass during the next 5 years.

To put the mass of the neutrino in perspective, we can take the experimental value of the upper limit to the neutrino rest mass, $m_\nu = 2.3\ \text{eV}/c^2$, and convert this to kilograms as follows:

By definition $1\ \text{eV} = 1.60 \times 10^{-19}$ J and, from equation $E = mc^2$, we can calculate the neutrino rest mass in kilograms as

$$m_\nu = E/c^2$$
$$= (2.3\ \text{eV})(1.6 \times 10^{-19}\ \text{J/eV})/(3.0 \times 10^8\ \text{m/s})^2$$
$$= 4.1 \times 10^{-36}\ \text{kg} \tag{1.192}$$

If we compare the rest mass of the neutrino, m_ν, to that of the miniscule electron, m_e, we see that the neutrino rest mass is approximately 4 millionths that of the electron or

$$m_\nu/m_e = 4.1 \times 10^{-36}\ \text{kg}/9.1 \times 10^{-31}\ \text{kg} = 4.5 \times 10^{-6}$$

d. Negatron decay energy

The general reaction for negatron emission was described previously by Eq. (1.173). It is provided here again for a treatment on negatron decay energy calculations, and it is the following:

$${}^A_ZP \rightarrow {}_{Z+1}^{\ \ A}D + \beta^- + \overline{\nu} + K_T \tag{1.193}$$

where ${}^A_ZP$ is the parent nuclide of atomic number Z and mass number A, ${}_{Z+1}^{\ \ A}D$ is the daughter nuclide of atomic number $Z+1$ and same mass number as the parent, β^- is the negatron or negative beta particle, $\overline{\nu}$ is the antineutrino, and K_T is the total kinetic energy released in the beta decay process.

In terms of nuclear masses, we may write Eq. (1.193) as

$${}_ZM_P = {}_{Z+1}M_D + m_e + K_T \tag{1.194}$$

where ${}_ZM_P$ and ${}_{Z+1}M_D$ are the masses of the parent and daughter nuclei of atomic numbers Z and $Z+1$, respectively, and m_e is the mass of the beta particle (*i.e.*, one electron mass). We may use atomic mass units (u) to calculate the mass difference of the parent and daughter nuclei by adding firstly Z atomic electron masses to each side of Eq. (1.194) to give

$${}_ZM_P + Zm_e = {}_{Z+1}M_D + (Z+1)m_e + K_T \tag{1.195}$$

The mass difference of the parent and daughter nuclei is equivalent to the decay energy (Q_{β^-}). We can now calculate the decay energy Q_{β^-} by the difference of the parent and daughter nuclide atomic masses, after adding Z atomic electrons to each side of the equation, ignoring the small difference in the binding energies of the electrons in the two nuclides as

$$Q_{\beta^-} = ({}_ZM_P - {}_{Z+1}M_D)c^2 = K_T = E_{\beta^-} + E_{\overline{\nu}} + E_\gamma + E_{\text{recoil}} \tag{1.196}$$

where c is the velocity of light in a vacuum and the product Mc^2 yields the energy equivalence of the mass, E_{β^-} is the kinetic energy of the negatron, $E_{\overline{\nu}}$ is the kinetic energy of the antineutrino, E_γ is the gamma ray energy, which only appears with the decay of a daughter nucleus when left at an excited energy state, and E_{recoil} is the kinetic energy of the daughter recoil nucleus. As previously described, the recoil energy, E_{recoil}, is generally relatively very small (a few electron volts) and will be ignored here. Also, if there is no gamma radiation emitted by the daughter nucleus, we can exclude the term E_γ. The remaining decay energy ($E_{\beta^-} + E_{\overline{\nu}}$) is shared as kinetic energy between the beta particle and antineutrino, most often referred to as the E_{max}.

A simple example would be that of the radionuclide ^{14}C, which decays by negatron emission without the emission of gamma radiation to the stable daughter nuclide ^{14}N according to the following:

$${}^{14}_{6}\text{C} \rightarrow {}^{14}_{7}\text{N} + \beta^- + \overline{\nu} \quad (E_{\text{max}} = 0.156\ \text{MeV}) \tag{1.197}$$

The decay scheme is illustrated in Fig. 1.47.

The decay energy of ^{14}C can be calculated according to Eq. (1.196) using atomic mass units (u) of the parent and daughter atoms as follows:

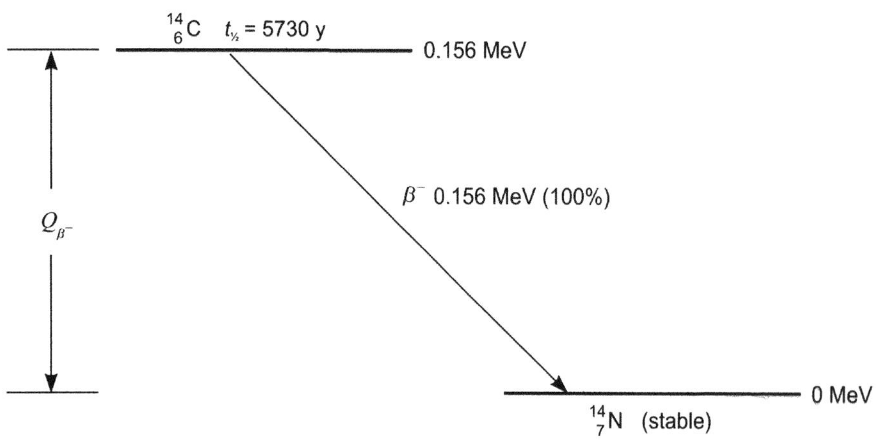

FIGURE 1.47 Negatron beta decay scheme of ^{14}C. The radionuclide ^{14}C decays to stable ^{14}N by negatron emission. Nuclear decay occurs entirely (100%) by emission of a negatron (β^-) and antineutrino ($\overline{\nu}$) with an energy maximum of 0.156 MeV, which is shared by the negatron and antineutrino.

$$\begin{aligned} Q_{\beta^-} &= (M_{^{14}C} - M_{^{14}N})c^2 \\ &= (14.00324199\ u - 14.00307401\ u)(931.494\ MeV/u) \\ &= 0.156\ MeV \end{aligned} \tag{1.198}$$

where 931.494 is the energy equivalence of 1 atomic mass unit as derived previously in this chapter. The energy of 0.156 MeV, calculated above, is shared between the beta particle and antineutrino. Thus, the maximum energy that the beta particle or antineutrino may possess is 0.156 MeV. The calculated decay energy may be found in Appendix A, which lists the decay energies of many radioactive nuclides. Also, one may utilize the online computer calculation (http://www.nndc.bnl.gov/qcalc), which agrees with the above beta-decay energy to be 156 keV or 0.156 MeV.

We may now look at yet a more complex example of ^{137}Cs decay where negatron β^- emission occurs with more than one energy maxima, or via what is referred to as a branching decay mode, and which decays with the emission of gamma radiation from the nucleus of the daughter nuclide ^{137}Ba. The decay equation is the following:

$$^{137}_{55}Cs \rightarrow {}^{137}_{56}Ba + \beta^- + \bar{\nu} + \gamma \tag{1.199}$$

The decay scheme of ^{137}Cs to stable ^{137}Ba is illustrated in Fig. 1.48.

The decay energy of ^{137}Cs is calculated as follows.

$$\begin{aligned} Q_{\beta^-} &= (M_{^{137}Cs} - M_{^{137}Ba})c^2 = E_{\beta^-} + E_{\bar{\nu}} + E_\gamma + E_{recoil} \\ &= (136.9070895\ u - 136.9058274\ u)(931.494\ MeV/u) \\ &= 1.1756\ MeV \end{aligned} \tag{1.200}$$

The 1.176 MeV decay energy is the energy liberated from ^{137}Cs decay to the ground state of stable ^{137}Ba. One branch of this decay occurs at a 94% intensity (abundance) via the emission of a negatron and antineutrino with an E_{max} of 0.514 MeV followed by the emission of gamma radiation of 0.662 MeV from the daughter nuclide ^{137m}Ba, which is metastable decaying to the stable ^{137}Ba (see Fig. 1.48). The remaining 6% of the negatron emissions (*i.e.*, 100% − 94%) arise from the decay of ^{137}Cs directly to the ground state of stable ^{137}Ba with the entire 1.176 MeV of decay energy shared by the negatron (β^-) and antineutrino ($\bar{\nu}$). The "QCalc" online (https://www.nndc.bnl.gov/qcalc/) computer-calculated decay energy is found to be 1175.6 keV or 1.1756 MeV as calculated above, which confirms the above calculated decay energy. The atomic mass units used in the above calculation are obtained from the Atomic Mass Data Center (AMDC) online via http://amdc.impcas.ac.cn/web/masseval.html). The recoil energy (E_{recoil}) in beta decay is relatively small (only several eV) and can be ignored in this case (see Part XIII of this chapter).

More complex negative beta decay schemes occur such as that of the radionuclide ^{131}I to stable ^{131}Xe, which is illustrated in Fig. 1.49.

The decay energy of ^{131}I to stable ^{131}Xe is calculated as.

$$\begin{aligned} Q_{\beta^-} &= (M_{^{131}I} - M_{^{131}Xe})c^2 = E_{\beta^-} + E_{\bar{\nu}} + E_\gamma + E_{recoil} \\ &= (130.9061264\ u - 130.9050841\ u)(931.494\ MeV/u) \\ &= 0.9708\ MeV \end{aligned} \tag{1.201}$$

The online computer-calculated decay energy (https://www.nndc.bnl.gov/qcalc/) is determined to be 970.8 keV or 0.9708 MeV as calculated above. The atomic mass units used in the above calculation are obtained from the Atomic Mass Data Center (AMDC) online via http://amdc.impcas.ac.cn/web/masseval.html).

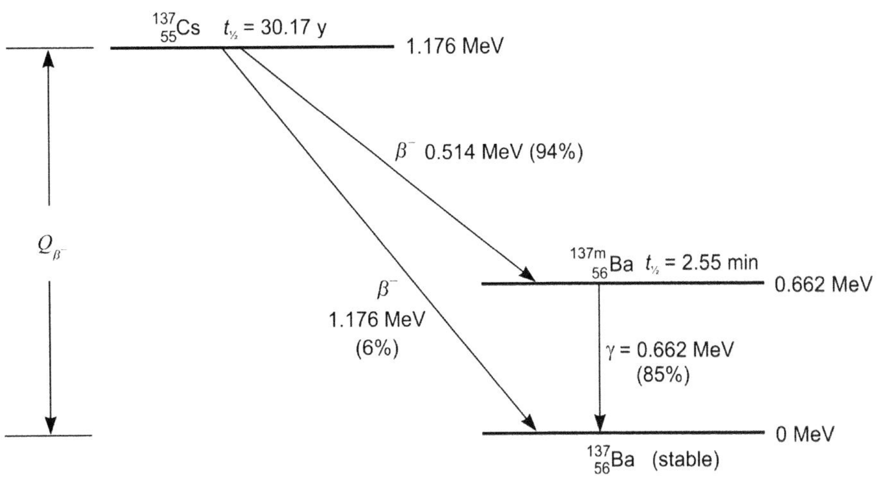

FIGURE 1.48 Negatron beta decay scheme of ^{137}Cs. The radionuclide ^{137}Cs decays to stable ^{137}Ba via two branching negatron emissions with energy maxima of 1.176 MeV at a relative intensity of 6% and 0.514 MeV at an intensity of 94%. Negatrons with energy maxima of 0.514 MeV produce metastable ^{137m}Ba daughter, which decays to stable ^{137}Ba with the emission of a gamma-ray photon of 0.662 MeV at an 85% intensity.

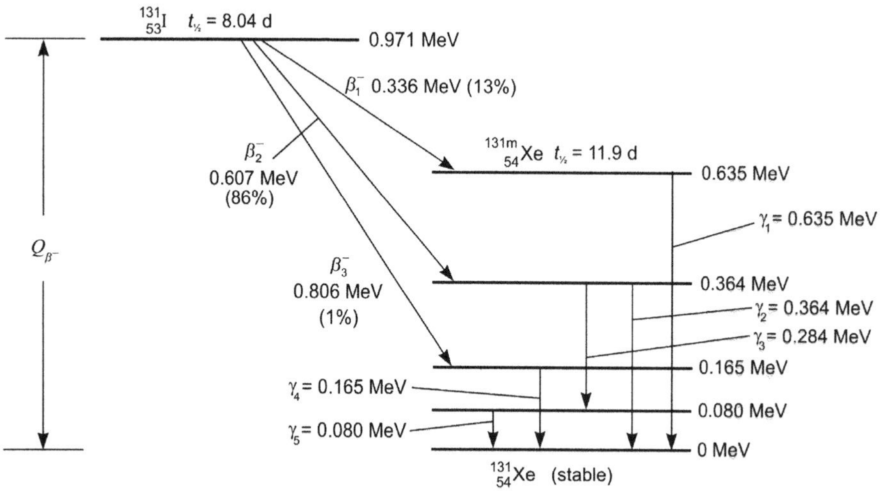

FIGURE 1.49 Negatron decay scheme of ^{131}I. The radionuclide ^{131}I decays to stable ^{131}Xe with three branching negatron emissions with energy maxima at 0.336 MeV, 0.607 MeV, and 0.806 MeV producing metastable ^{131m}Xe at three energy levels yielding several gamma ray lines of emissions, namely, $\gamma_1 = 0.635$ MeV, $\gamma_2 = 0.364$ MeV, $\gamma_3 = 0.284$ MeV, $\gamma_4 = 0.165$ MeV, and $\gamma_5 = 0.080$ MeV.

In summary, if we ignore the relatively very small nuclear recoil energy, we will find that the β^- energy maxima and the gamma-ray emission energies for the three beta decay branches of ^{131}I decay, illustrated in Fig. 1.49, yield the total decay energy of 0.971 MeV as follows:

β^- Branch 1 of 13% Intensity:

$$\begin{aligned} E_{\beta_1^-} + E_{\bar{\nu}} + \gamma_1 &= E_{\max} \text{ of } 0.336 \text{ MeV} + \gamma_1 \text{ of } 0.635 \text{ MeV} \\ &= 0.971 \text{ MeV} \end{aligned}$$

β^- Branch 2 of 86% Intensity:

$$\begin{aligned} E_{\beta_2^-} + E_{\bar{\nu}} + \gamma_2 &= E_{\max} \text{ of } 0.607 \text{ MeV} + \gamma_2 \text{ of } 0.364 \text{ MeV} \\ &= 0.971 \text{ MeV} \end{aligned}$$

β^- Branch 3 of 1% Intensity:

$$\begin{aligned} E_{\beta_3^-} + E_{\bar{\nu}} + \gamma_4 &= E_{\max} \text{ of } 0.806 \text{ MeV} + \gamma_4 \text{ of } 0.165 \text{ MeV} \\ &= 0.971 \text{ MeV} \end{aligned}$$

2. Positron (β^+) emission

Positron emission can be described by the general equation

$$^{A}_{Z}P \rightarrow {}_{Z-1}^{\ \ A}D + \beta^+ + \nu + K_T \qquad (1.202)$$

where $^{A}_{Z}P$ is the parent nuclide of atomic number Z and mass number A, ${}_{Z-1}^{\ \ A}D$ is the daughter nuclide of atomic number $Z-1$ and same mass number as the parent, β^+ is the positron or positive beta particle, ν is the neutrino, and K_T is the total kinetic energy released in the beta decay process.

a. Basic principles

In contrast to negatron emission from nuclei having neutron/proton (n/p) ratios too large for stability, positrons, which consist of positively charged electrons (positive beta particles), are emitted from nuclei having n/p ratios too small for stability, that is, those which have an excess of protons.

To attain nuclear stability, the n/p ratio is increased. This is realized by a transformation of a proton to a neutron within the nucleus. The previously discussed alteration of a neutron to a proton in a negatron-emitting nuclide (Eq. 1.175) may now be considered in reverse for the emission of positrons. Eq. (1.203) illustrates such a transformation

$$p^+ \rightarrow n + \beta^+ + \nu \qquad (1.203)$$

The decay of the radionuclide ^{11}C may be cited as an example of a nuclide that decays by positron emission:

$$^{11}_{6}C \rightarrow {}^{11}_{5}B + \beta^+ + \nu \qquad (1.204)$$

Note that the mass number does not change, but the charge on the nucleus (Z number) decreases by 1. A neutrino, ν, is emitted simultaneously with the positron (beta particle) and shares the decay energy with the positron. Thus, positrons, like negatrons emitted from a given radionuclide sample, may possess a broad spectrum of energies from near zero to $E_{\max}$, as illustrated in Fig. 1.45.

Decay by positron emission can occur only when the decay energy is significantly above 1.02 MeV. This is because two electrons of opposite charge are produced (e^+, e^-) within the nucleus, and the energy equivalence of the electron mass is 0.51 MeV (see Section XII.C of this chapter.). The positive electron, e^+, is ejected from the nucleus (i.e., β^+) and the negative electron, e^-, combines with a proton to form a neutron:

$$e^- + p^+ \rightarrow n \qquad (1.205)$$

From the Chart of the Nuclides, it is possible to cite specific examples of the n/p imbalance in relation to negatron and positron emission. Fig. 1.50 illustrates a segment of

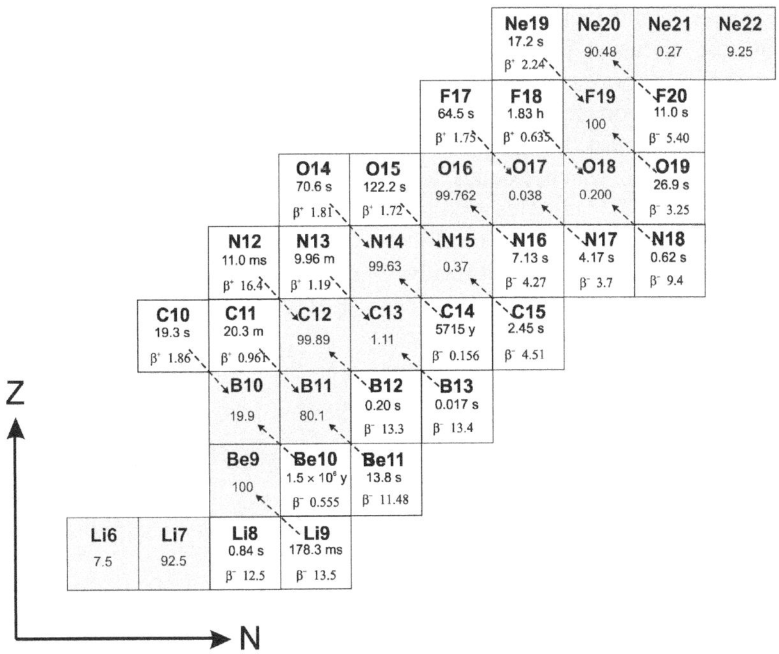

FIGURE 1.50 A Segment of the Chart of the Nuclides showing the relative positions of some stable (shaded) and unstable nuclides. The ordinate *Z* and abscissa *N* represent the number of protons (atomic number) and the number of neutrons (neutron number) within the nucleus, respectively. The mass number (number of protons + neutrons) in the nucleus of an isotope is provided alongside the abbreviation of the element. For example, the isotope ^{14}C (carbon-14) is written in the Chart as C14. Radionuclides that undergo positron emission (β^+) are illustrated with arrows pointing downward to stable nuclide products of lower atomic number, and radionuclides undergoing negatron (β^-) emission are illustrated with arrows pointing upward to stable nuclide products of higher atomic number. The maximum beta-particle energy in units of MeV is provided alongside the positron or negatron symbol. Numbers in shaded areas are the percent natural abundances of the stable isotopes, which for each element should add up to 100%. For example, the element lithium (Li) exists in nature as a combination of two stable isotopes, namely, ^{6}Li and ^{7}Li with abundances of 7.5% and 92.5%, respectively. When only one stable isotope of an element exists, its natural abundance will be expressed as 100%, such as is the case for ^{9}Be and ^{19}F illustrated above. The half-lives of the unstable nuclides are given in units of seconds (s), milliseconds (ms), minutes (m), hours (h) and years (y).

the Chart of the Nuclides and the relative positions of the stable nuclides (shaded areas) and of their neighboring radionuclides. The nuclides are positioned as a function of the number of protons, *Z*, and the number of neutrons, *N*, in their respective nuclei. Dashed arrows are placed through the blocks that segregate radionuclides interrelated with common daughter nuclides resulting from β^- or β^+ decay processes. For example, the stable nuclide ^{12}C of atomic number 6 has a nucleus with an n/p ratio of 6/6. However, the nuclide ^{12}N of atomic number 7 has an unstable n/p ratio of 5/7, an excess of protons. Thus, this radionuclide ^{12}N decays to ^{12}C via positron emission according to the equation

$$^{12}_{7}\mathrm{N} \rightarrow {}^{12}_{6}\mathrm{C} + \beta^+ + \nu \tag{1.206}$$

as indicated by a dashed arrow of Fig. 1.50.

The nuclide ^{12}B of atomic number 5 has the unstable n/p ratio of 7/5, an excess of neutrons. This nuclide thus decays to ^{12}C by negatron emission according to the equation

$$^{12}_{5}\mathrm{B} \rightarrow {}^{12}_{6}\mathrm{C} + \beta^- + \bar{\nu} \tag{1.207}$$

Similar reasoning may be used to explain positron and negatron decay of the unstable nuclides shown in Fig. 1.50 to the stable products ^{13}C, ^{14}N, and ^{15}N, etc. The interrelationship between β^- and β^+ decay leading to the formation of stable nuclides is to be found throughout the Chart of the Nuclides. Nuclear stability with n/p ratios close to unity only exists with nuclides of low atomic number ($Z \leq 20$), as discussed in Section VI.D.1 and in the following section of this chapter.

b. *N/Z* ratios and beta decay

Negatron and positron emission were discussed previously as processes whereby unstable nuclei may achieve stability via neutron or proton transformations, respectively. These processes in the nucleus of the radionuclide result in a change in the neutron/proton or *N/Z* ratio of the nucleus.

If we look throughout The Chart of The Nuclides, we will notice that the stable nuclides of low atomic number will have an *N/Z* ratio of approximately 1. However, as the atomic number increases ($Z > 20$), the *N/Z* ratio of the stable nuclides increases gradually and reaches as high as approximately 1.5 (*e.g.*, $^{209}_{83}$Bi, $Z = 83$, $N/Z = 1.518$). This was illustrated previously in Fig. 1.20. Furthermore, there are no stable nuclides of atomic number greater than 83.

The nature of nuclear forces and the relationship of *N/Z* ratio to nuclear stability are discussed in detail in Part VI.D.1 of this chapter, and by Serway et al. (2005) and Sundaresan (2001). In brief, the importance of *N/Z* ratio to nuclear stability is explained by the fact that there exists a short-range attractive nuclear force, which extends to a distance of $\approx$2 fm (2 fermi or 2×10^{-15} m). This attractive force has charge independence and is a consequence of the relative spins of the protons and neutrons and their relative positions in the nucleus. These binding exchange forces exist therefore, regardless of charge on the particles, between two protons, two neutrons, and a proton and neutron. While the attractive nuclear forces will tend to hold the nucleus together there exists, at the same time, repelling columbic forces between the positively charged protons that act to force them apart. For nuclides of low *Z*, the attractive nuclear forces exceed the repelling columbic forces when $N \approx Z$. However, increasing the number of protons (e.g., $Z > 20$) further increases the strength of the repelling columbic forces over a larger nucleus, which will tend to force the nucleus apart. Therefore, additional neutrons, $N > Z$, provide additional attractive nuclear forces needed to overcome the repelling forces of the larger proton population. As the atomic number increases further, $Z > 83$, all nuclides are unstable. Even though *N/Z* ratios reach 1.5, nuclear stability is not achieved when the number of protons in the nucleus exceeds 83.

c. Positron decay energy

The general reaction for positron emission was described previously by Eq. (1.202). It is provided here again for a treatment on positron decay energy, and it is the following:

$$ {}^{A}_{Z}P \rightarrow {}^{A}_{Z-1}D + \beta^{+} + \nu + K_T \qquad (1.208)$$

where ${}^{A}_{Z}P$ is the parent nuclide of atomic number *Z* and mass number *A*, ${}^{A}_{Z-1}D$ is the daughter nuclide of atomic number $Z-1$ and same mass number as the parent, β^{+} is the positron or positive beta particle, ν is the neutrino, and K_T is the total kinetic energy released in the beta decay process.

In terms of nuclear masses, Eq. (1.208) may be written as

$$ {}_{Z}M_P = {}_{Z-1}M_D + m_e + K_T \qquad (1.209)$$

where ${}_{Z}M_P$ and ${}_{Z-1}M_D$ are the masses of the parent and daughter nuclei of atomic number *Z* and $Z-1$, respectively, and m_e is the mass of the positron (*i.e.*, one electron mass). Atomic mass units (u) may be used to calculate conveniently the mass difference of the parent and daughter, if we add *Z* atomic electron masses to each side of Eq. (1.209) and ignore the negligible difference in electron binding energies of the parent and daughter atoms. Adding *Z* atomic electrons to each side of Eq. (1.209) gives

$$ {}_{Z}M_P + Zm_e = {}_{Z-1}M_D + (Z+1)m_e + K_T \qquad (1.210)$$

and

$$ {}_{Z}M_P = {}_{Z-1}M_D + 2m_e + K_T \qquad (1.211)$$

The term $2m_e$ of Eq. (1.211) arises from the creation of two electrons of opposite charge (e^{+}, e^{-}) within the nucleus. The positive electron, e^{+}, is ejected from the nucleus as the positron; and the negative electron, e^{-}, combines with a proton to form a neutron. The mass difference of the parent and daughter nuclei is equivalent to the decay energy, $Q_{\beta^{+}}$. The decay energy, $Q_{\beta^{+}}$, may be calculated now using atomic mass units and taking the product of mass and c^2 to determine the energy equivalence of mass using Eq. (1.211) to give

$$ Q_{\beta^{+}} = ({}_{Z}M_P - {}_{Z-1}M_D)c^2 = 2m_ec^2 + K_T \qquad (1.212)$$

or

$$ Q_{\beta^{+}} = ({}_{Z}M_P - {}_{Z-1}M_D)c^2 = 2m_ec^2 + E_{\beta^{+}} + E_\nu + E_\gamma + E_{\text{recoil}} \qquad (1.213)$$

where ${}_{Z}M_P - {}_{Z-1}M_D$ is the mass difference of the parent and daughter nuclides, *c* is the speed of light in a vacuum, and $2m_ec^2$ is the energy equivalence of two electron masses or 1.022 MeV, which is the energy required for the creation of the $e^{+}e^{-}$ pair in the positron decay process described previously in this section. The terms $E_{\beta^{+}} + E_\nu$ are the shared positron and neutrino kinetic energies, which combined is referred often to as the E_{max} of the beta decay. The term E_γ is gamma-ray energy that may be emitted by the daughter nucleus, and E_{recoil} is the nuclear recoil energy, which is relatively small (measured in eV) and can be ignored as negligible when calculating the overall decay energy (measured in MeV).

We may take an example of a simple case of positron emission in the decay of ^{13}N where there is no gamma radiation from the daughter nucleus, that is, the ^{13}N nucleus decays directly to ^{13}C at its stable ground state with the emission of a positron and neutrino with maximum energy 1.198 MeV. The scheme for the decay of ^{13}N is illustrated in Fig. 1.51.

The E_{max} is the combined energy of the positron and neutrino $(E_{\beta^{+}} + E_\nu)$, that is, the energy shared by the positron and neutrino. In other words, the E_{max} is the maximum energy of the positron when the neutrino is at its minimum energy or vice versa.

The decay energy, $Q_{\beta^{+}}$, for ^{13}N decay is calculated as

$$\begin{aligned} Q_{\beta^{+}} &= (M_{^{13}\text{N}} - M_{^{13}\text{C}})c^2 \\ &= (13.0057386\ \text{u} - 13.0033548\ \text{u})(931.494\ \text{MeV/u}) \\ &= 2.2204\ \text{MeV} \end{aligned} \qquad (1.214)$$

which is the magnitude of the *Q* value for ^{13}N listed in standard reference tables (*e.g.*, Table of the Isotopes in Rumble, 2017) and the National Nuclear Data Center online data base NuDat 2.7 (https://www.nndc.bnl.gov/

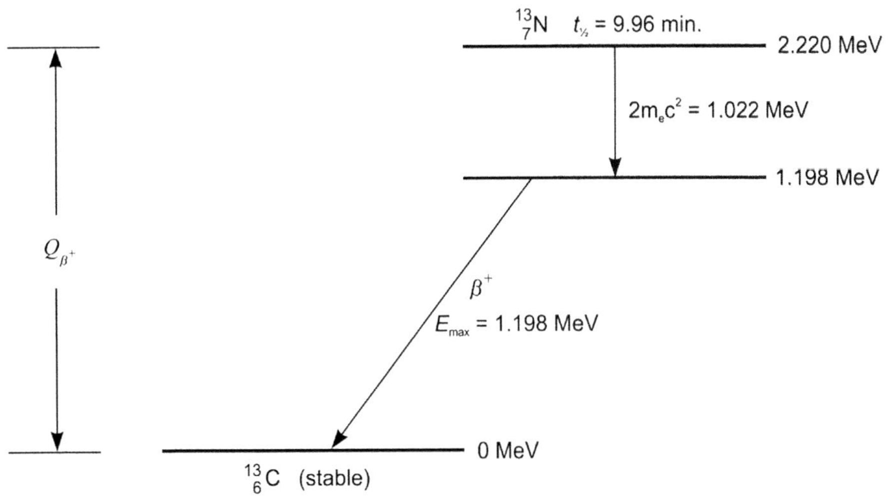

FIGURE 1.51 Beta decay of ^{13}N to stable ^{13}C by positron (β^+) emission. The scheme illustrates the energy equivalence (1.022 MeV) of two electron masses (e^+e^-) required for positron emission. The remaining 1.198 MeV energy is shared by the β^+ and ν.

nudat2/chartNuc.jsp). The maximum energy of the β^+ or neutrino in this case may be calculated easily according to Eq. (1.213) by ignoring the nuclear recoil energy (E_{recoil}), because of its small value and ignoring the gamma-ray energy (E_γ), because, in this example, no gamma-ray energy arises from the daughter nucleus following beta decay. Thus, the maximum energy of the β^+ or neutrino is

$$\begin{aligned} E_{max} &= E_{\beta^+} + E_\nu = Q_{\beta^+} - 2m_ec^2 \\ &= 2.220 \text{ MeV} - 1.022 \text{ MeV} \\ &= 1.198 \text{ MeV} \end{aligned} \quad (1.215)$$

The above value of 1198 keV or 1.198 MeV calculated above agrees with the maximum positron energy (End-point energy) provided the National Nuclear Data Center online database NuDat 2.7 (https://www.nndc.bnl.gov/nudat2/chartNuc.jsp).

A more complex example would be that of the decay of ^{22}Na, which decays by positron (β^+) emission and EC decay with the emission of gamma radiation according to the decay Eqs. (1.216) and (1.217) and the decay scheme illustrated in Fig. 1.52.

Positron emission (90.06%):

$$\begin{aligned} &^{22}_{11}Na \rightarrow ^{22}_{10}Ne + \beta^+ + \nu + \gamma(90.0\%) \text{ and} \\ &^{22}_{11}Na \rightarrow ^{22}_{10}Ne + \beta^+ + \nu(0.06\%) \end{aligned} \quad (1.216)$$

Electron capture (9.94%):

$$^{22}_{11}Na \rightarrow ^{22}_{10}Ne + \nu + \gamma(9.94\%) \quad (1.217)$$

As illustrated in Fig. 1.52, the radionuclide ^{22}Na may decay by EC or positron emission. EC is a decay mode that competes with positron emission as discussed in the following section of this chapter. In this case, 90.06% of

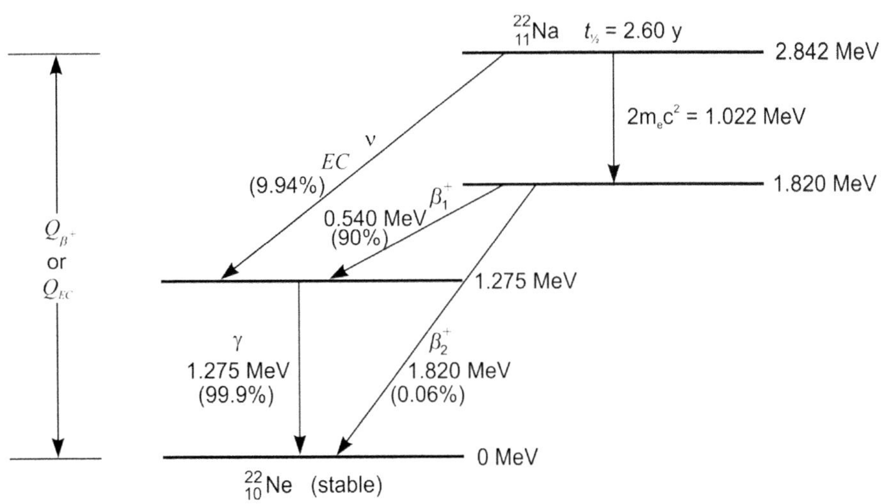

FIGURE 1.52 Beta decay scheme of ^{22}Na by competing positron (β^+) and electron capture (EC) decay. Positron emission occurs in 90.06% of the nuclear decays and EC in the remaining 9.94%. Decay by EC involves the emission of a mono-energetic neutrino of 1.567 MeV (*i.e.*, E_ν = 2.842 − 1.275 MeV) to metastable ^{22}Ne. Positron emission occurs with the consumption of the energy equivalence of two electron masses (*i.e.*, $2m_ec^2$) followed by two branching positron emissions with intensities of 90% and 0.06%.

the ^{22}Na radionuclides decay via positron emission and the remaining 9.94% decay via EC. When the decay energy, Q_{β^+}, is well above 1.022 MeV (the minimum energy required for e^+e^- pair production) positron emission will predominate over EC decay.

The decay energy for ^{22}Na may be calculated according to Eq. (1.213) as

$$\begin{aligned} Q_{\beta^+} &= (M_{^{22}\mathrm{Na}} - M_{^{22}\mathrm{Ne}})c^2 \\ &= (21.994437\ \mathrm{u} - 21.9913855\ \mathrm{u})(931.494\ \mathrm{MeV/u}) \\ &= 2.842\ \mathrm{MeV} \end{aligned} \tag{1.218}$$

which is in agreement with the magnitude of the Q value for ^{22}Na provided by the National Nuclear Data Center online database NuDat 2.7 (https://www.nndc.bnl.gov/nudat2/chartNuc.jsp). The calculated decay energy corresponds to the theoretical value described by Eq. (1.213), that is,

$$Q_{\beta^+} = 2m_ec^2 + E_{\beta^+} + E_\nu + E_\gamma + E_{\mathrm{recoil}} \tag{1.219}$$

where $E_{\beta^+} + E_\nu = \beta^+_{\max}$ and, if we ignore the relatively very small value of E_{recoil}, we calculate

$$\begin{aligned} Q_{\beta_1^+} &= 2m_ec^2 + \beta^+_{\max} + E_\gamma \\ &= 1.022\ \mathrm{MeV} + 0.540\ \mathrm{MeV} + 1.275\ \mathrm{MeV} \\ &= 2.84\ \mathrm{MeV} \end{aligned} \tag{1.220}$$

which is in accord the β_1^+ transition illustrated in Fig. 1.52, and

$$\begin{aligned} Q_{\beta_2^+} &= 2m_ec^2 + \beta^+_{\max} \\ &= 1.022\ \mathrm{MeV} + 1.820\ \mathrm{MeV} \\ &= 2.84\ \mathrm{MeV} \end{aligned} \tag{1.221}$$

which is in agreement with the β_2^+ transition illustrated in Fig. 1.52 whereby ^{22}Na decays directly to the stable ground state of ^{22}Ne without gamma emission in 0.06% of the nuclear decays.

3. Electron capture

The electron capture (EC) decay process is described by the following general equation.

$${}^A_ZP + e^- \xrightarrow{\mathrm{EC}} {}_{Z-1}^{\ A}D + \nu + K_T \tag{1.222}$$

where ${}^A_ZP$ is the parent nuclide of atomic number Z and mass number A, e^- is an orbital electron captured by the nucleus of the parent nuclide, ${}_{Z+1}^{\ A}D$ is the daughter nuclide of atomic number $Z-1$ and same mass number as the parent, ν is the neutrino, and K_T is the total kinetic energy released in the beta decay process.

a. Basic principles

EC is another form of beta decay. The change in atomic number is the same as occurs with positron (β^+) emission. EC competes with positron emission and most radionuclides that decay by positron emission also decay by EC to the same daughter nuclide.

The mechanism by which an unstable nucleus can increase its n/p ratio is via the capture by the nucleus of a proximate atomic electron (*e.g.*, *K*- or *L*-shell electron). The absorbed electron combines with a proton to yield a neutron within the nucleus as follows:

$$\begin{aligned} e^- + \mathrm{p}^+ \rightarrow \mathrm{n} + \nu &+ \text{ inner bremsstrahlung} \\ &+ \text{x-rays} + \text{Auger electrons} + \gamma \end{aligned} \tag{1.223}$$

The decay process is known as EC, or sometimes referred to as *K* capture, because most of the electrons are captured from the *K* shell, which is closest to the nucleus. A neutrino ν, is emitted and this is accompanied by the emission of internal bremsstrahlung, which is a continuous spectrum of electromagnetic radiation that originates from the atomic electron as it undergoes acceleration toward the nucleus. Unlike the beta decay process, which results in the emission of a neutrino from the nucleus with a broad spectrum of energies, the neutrino emitted from the EC decay process does not share the transition energy with another particle and, therefore, it is emitted with a single quantum of energy equal to the transition energy less the atomic electron binding energy. The capture of an atomic electron by the nucleus leaves a vacancy in an electron shell, and this is filled usually by an electron from an outer shell, resulting in the production of X-radiation (see Sections XI.D.2.c). The electron that fills the vacancy leaves yet another vacancy at a more distant shell. A cascade of electron vacancies and subsequent filling of vacancies from outer electron shells occurs with the production of X-rays characteristic of the daughter atom. The X-rays will either travel out of the atom or the energy emitted by the electron transition will be transferred to an outer orbital electron causing the electron to be ejected as an Auger electron. Gamma radiation is illustrated in the above Eq. (1.223) because it is emitted only when the daughter nuclide is left at an unstable elevated energy state (see Figs. 1.52–1.54).

The EC decay process may compete with β^+ emission. That is, some radionuclides may decay by either EC or, β^+-emission. As discussed previously, positron emission requires a transition energy of at least 1.022 MeV, the minimum energy required for pair production in the nucleus (*i.e.*, two electron rest mass energies or 2×0.511 MeV). Positron emission, therefore, will not compete with EC for decay transitions less than 1.022 MeV. In general, positron emission will predominate when the transition energy is high (well above 1.022 MeV) and for nuclides of low atomic number, while the EC decay

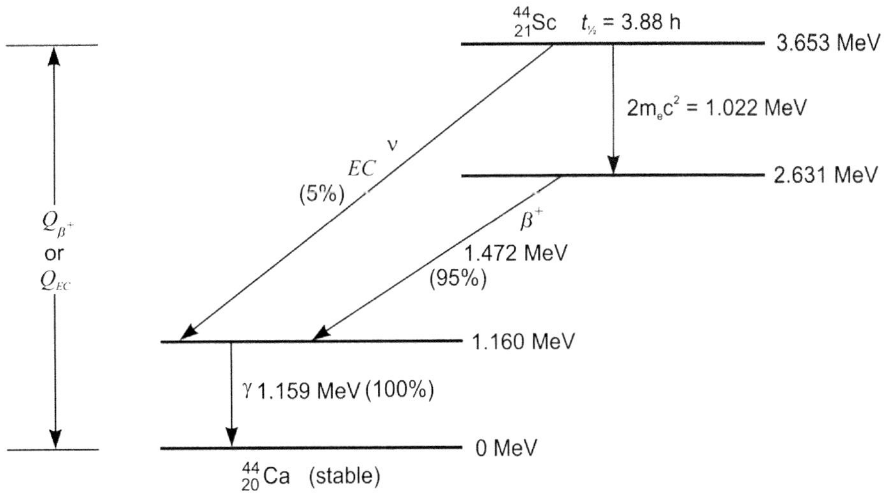

FIGURE 1.53 Beta decay scheme of ^{44}Sc by competing EC decay and positron (β^+) emission. The high decay energy of 3.653 MeV favors β^+ emission, as it is well above the minimum 1.022 MeV or $2m_ec^2$ required for positron emission. EC decay occurs in only 5% of the nuclear decays and β^+ emission in the remaining 95%. Both EC and β^+ decay yield metastable ^{44m}Ca, which decays to stable ^{44}Ca with gamma ray emission of 1.159 MeV.

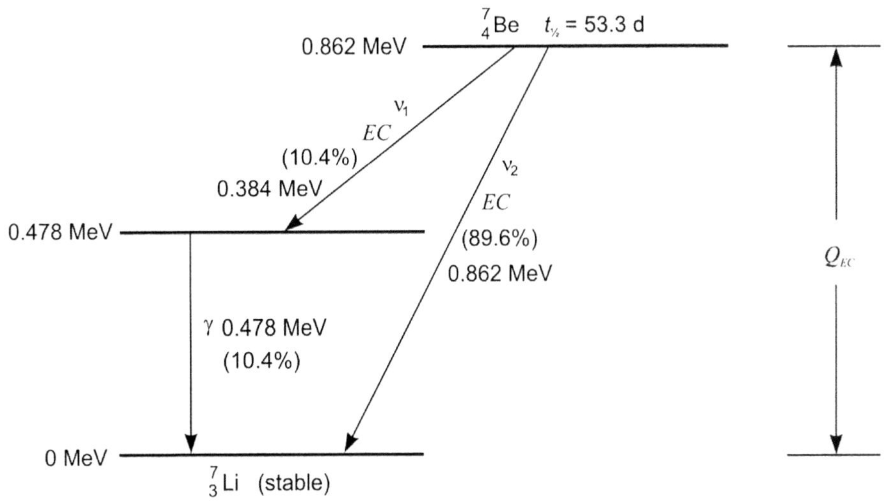

FIGURE 1.54 Decay scheme of ^{7}Be. The decay occurs exclusively by branching electron capture (EC). In 10.4% of the transitions, ^{7}Be emits a monoenergetic neutrino of 0.384 MeV (*i.e.*, 0.862 − 0.478 MeV) followed by the emission of a 0.478 MeV gamma-ray photon. The remaining EC decay events occur with the emission of a monoenergetic neutrino of 0.862 MeV directly to the ground state of stable ^{7}Li in 89.6% of the decay transitions.

process will predominate for low transition energies and nuclides of higher atomic number. The decay transitions of ^{22}Na and ^{65}Zn serve as examples. In the case of ^{22}Na, decay by β^+ emission predominates (90.06%) as compared with decay via EC (9.94%)

$$^{22}_{11}\text{Na} \rightarrow {}^{22}_{10}\text{Ne} + \beta^+ + \nu(90.06\%) \tag{1.224}$$

and

$$^{22}_{11}\text{Na} \overset{\text{EC}}{\rightarrow} {}^{22}_{10}\text{Ne} + \nu(9.94\%) \tag{1.225}$$

The transition energy of ^{22}Na is 2.842 MeV (Holden, 1997a and calculated previously, see Eq. 1.218), well above the 1.022 MeV minimum required for positron emission. Thus, positron emission predominates. On the other hand, taking the following example of the nuclide ^{65}Zn, we see that EC predominates over β^+ emission:

$$^{65}_{30}\text{Zn} \rightarrow {}^{65}_{29}\text{Cu} + \beta^+ + \nu(1.5\%) \tag{1.226}$$

and

$$^{65}_{30}\text{Zn} \overset{\text{EC}}{\rightarrow} {}^{65}_{29}\text{Cu} + \nu(98.5\%) \tag{1.227}$$

In the case of ^{65}Zn, the transition energy is only 1.35 MeV (Holden, 1997a), which is not much above the minimum energy of 1.02 MeV required for positron emission. Consequently, EC decay predominates.

b. EC decay energy

The general decay equation for EC was described previously by Eq. (1.222). It is provided here again for a treatment on EC decay calculations, and it is the following:

$$ {}^{A}_{Z}P + e^{-} \xrightarrow{EC} {}^{A}_{Z-1}D + \nu + K_T \quad (1.228) $$

where ${}^{A}_{Z}P$ is the parent nuclide of atomic number Z and mass number A, e^{-} is an orbital electron captured by the nucleus of the parent nuclide, ${}^{A}_{Z-1}D$ is the daughter nuclide of atomic number $Z-1$ and same mass number as the parent, ν is the neutrino, and K_T is the total kinetic energy released in the beta decay process.

In mass and energy terms, Eq. (1.228) can be written as

$$ {}_{Z}M_P + m_e = {}_{Z-1}M_D + K_T \quad (1.229) $$

where ${}_{Z}M_P$ is the mass of the parent nuclide of atomic number Z, m_e is the mass of the atomic electron captured by the nucleus, and ${}_{Z-1}M_D$ is the mass of the daughter nucleus of atomic number Z – 1, We may use atomic mass units (u) to calculate conveniently the mass difference of the parent and daughter nuclei and arrive at the decay energy by firstly adding $Z-1$ atomic electrons to each side of Eq. (1.229), which yields

$$ {}_{Z}M_P = {}_{Z-1}M_D + K_T \quad (1.230) $$

The mass difference of the parent and daughter nuclei is equivalent to the decay energy (Q_{EC}). We can now calculate the decay energy Q_{EC} by the difference of the parent and daughter nuclide atomic masses ignoring the small difference in the binding energies of the electrons in the two nuclides as

$$ Q_{EC} = ({}_{Z}M_P - {}_{Z-1}M_D)c^2 = K_T = E_\nu + E_\gamma + E_{recoil} \quad (1.231) $$

where c is the velocity of light in a vacuum and the product Mc^2 yields the energy equivalence of the mass, E_ν is the kinetic energy of the neutrino, E_γ is the gamma ray photon energy, which only appears with the decay of the daughter nucleus when it is left at a metastable or excited energy state, and E_{recoil} is the kinetic energy of the daughter recoil nucleus. As in previously described beta decay processes, we will ignore the relatively very small recoil energy when calculating the overall decay energy.

As an example, let us look at the decay scheme of ^{44}Sc, which decays by EC and positron emission. The decay equations are the following

$$ {}^{44}_{21}Sc \xrightarrow{EC} {}^{44}_{20}Ca + \nu + \gamma(5\%) \quad (1.232) $$

and

$$ {}^{44}_{21}Sc \rightarrow {}^{44}_{20}Ca + \beta^{+} + \nu + \gamma(95\%) \quad (1.233) $$

The decay scheme is illustrated in Fig. 1.53.

The scheme in Fig. 1.53 illustrates that 5% of the ^{44}Sc nuclides decay by EC and the remaining 95% of the nuclides decay with positron emission with a positron E_{max} of 1.472 MeV. The ^{44}Ca product nuclides then decay to a stable ground state with the emission of gamma radiation of 1.159 MeV energy.

The decay energy for either mode of decay (EC or β^{+} emission) is calculated according to Eqs. (1.231) and (1.218), respectively, as follows:

For electron capture

$$ \begin{aligned} Q_{EC} &= ({}_{Z}M_P - {}_{Z-1}M_D)c^2 = E_\nu + E_\gamma + E_{recoil} \\ &= (M_{^{44}Sc} - M_{^{44}Ca})c^2 \\ &= (43.9594029\ u - 43.9554815\ u)(931.494\ MeV/u) \\ &= 3.6527\ MeV \end{aligned} \quad (1.234) $$

which is in agreement with the decay energy of 3652.7 keV provided in the standard reference tables (*e.g.*, Holden, 1997a) and the online QCalc computation (https://www.nndc.bnl.gov/qcalc/). As we know, that the gamma ray energy $E_\gamma = 1.159$ MeV (see Fig. 1.53) and, if we disregard the very small magnitude of the recoil energy, we can calculate the neutrino energy in this case as

$$ \begin{aligned} E_\nu = Q_{EC} - E_\gamma &= 3.653\ MeV - 1.159\ MeV \\ &= 2.494\ MeV \end{aligned} \quad (1.235) $$

The neutrino energies are monoenergetic in the case of the EC decay mode.

For positron emission

$$ \begin{aligned} Q_{\beta^+} &= ({}_{Z}M_P - {}_{Z-1}M_D)c^2 \\ &= 2m_ec^2 + E_{\beta^+} + E_\nu + E_\gamma + E_{recoil} \\ &= (M_{^{44}Sc} - M_{^{44}Ca})c^2 = 3.653\ MeV \end{aligned} \quad (1.236) $$

If we again ignore the very small recoil energy of the daughter nucleus and combine the positron and neutrino energies as $E_{max} = E_{\beta^+} + E_\nu$, we can arrive at the decay energy taking the values from Fig. 1.53 as

$$ \begin{aligned} Q_{\beta^+} &= 2m_ec^2 + E_{max} + E_\gamma \\ &= 1.022\ MeV + 1.472\ MeV + 1.159\ MeV \\ &= 3.653\ MeV \end{aligned} \quad (1.237) $$

The decay energy of 3.653 MeV is well above the minimum energy for $e^{+}e^{-}$ pair production ($2m_ec^2$ or 1.022 MeV) required for positron emission. Thus, the EC decay does not compete well against positron emission in the case of ^{44}Sc, which decays via positron emission in 95% of the disintegrations and by EC decay in the remaining 5%.

Let us take another example, which is that of ^{7}Be decaying exclusively by EC according to the following equation:

$$^{7}_{4}\text{Be} \xrightarrow{\text{EC}} {}^{7}_{3}\text{Li} + \nu + \gamma \tag{1.238}$$

The decay energy is calculated as

$$\begin{aligned} Q_{\text{EC}} &= (M_{^7\text{Be}} - M_{^7\text{Li}})c^2 \\ &= (7.0169287\ \text{u} - 7.0160034\ \text{u})(931.494\ \text{MeV/u}) \\ &= 0.8619\ \text{MeV} \end{aligned} \tag{1.239}$$

which is in agreement with the magnitude of the decay energy reported in standard reference tables (*e.g.*, Holden, 1997a) and the online QCalc computation (https://www.nndc.bnl.gov/qcalc/). The decay energy is not sufficient to satisfy the minimum energy requirement of 1.022 MeV for positron emission in this case. Thus, there is no competition here between EC and β^+ decay modes, and the ^{7}Be nuclides decay exclusively by EC according to the decay scheme illustrated in Fig. 1.54.

The daughter nuclide, ^{7}Li, emits a gamma ray of 0.478 MeV in 10.4% of the transitions. The neutrino energy from the EC decay of ^{7}Be in these transitions would be derived from the difference of the decay energy and gamma ray energy while ignoring the relatively very small recoil energy or

$$\begin{aligned} E_\nu &= Q_{\text{EC}} - E_\gamma \\ &= 0.862\ \text{MeV} - 0.478\ \text{MeV} \\ &= 0.384\ \text{MeV} \end{aligned} \tag{1.240}$$

The remaining 89.6% of the transitions are due to a neutrino emission of 0.862 MeV directly to the stable ground state of the daughter nuclide without any gamma ray emission. Neutrinos emitted in the EC decay process are monoenergetic. They do not possess any broad spectrum of energies, contrary to the neutrinos emitted during β^+ or β^- emission, because the neutrinos emitted from the nucleus in EC decay do not share the decay energy with any other particle.

There is nuclear recoil of the parent nucleus during EC decay resulting from the emission of the neutrino. The nuclear recoils in EC decay have been measured by researchers confirming the mono-energetic character of the emitted neutrino. The calculation of nuclear recoil energies is discussed in detail in Section XIII of this chapter.

The ^{7}Li ion will recoil at two energy maxima corresponding to the neutrino emissions of 0.862 and 0.384 MeV calculated above. The maximum recoil energy corresponding to the neutrino emission of 0.862 MeV is calculated according to Eq. (1.487) of this chapter to be 56.84 eV. The calculation agrees with the experimental measurement of the ^{7}Li recoil energy reported by 2002 Nobel Laureate Raymond Davis Jr. (1952) to be 56 ± 1 eV. The recoil energy in this case is relatively high (>10 eV) because of the relatively low mass of the ^{7}Li nucleus. Nevertheless, the recoil energy here may be ignored as negligible when calculating the decay energy, which has units of MeV.

c. Chemical and pressure effects on EC decay rates

In general, the chemical and physical environment of radionuclides do not control nuclear decay processes. However, because the electron-capture decay process involves the capture of an orbital electron by the nucleus, atomic or molecular binding effects, which vary with chemical structure, can influence the electron-capture decay process. Also, external pressure that may alter electron densities in the vicinity of the nucleus could also have an altering effect on the EC decay rate.

Segré (1947) at the Los Alamos Laboratories in the US and Daudel (1947) at the Curie Laboratories of the Radium Institute in Paris, France, were the first to independently suggest that the rate of disintegration of nuclei decaying by EC or internal conversion (IC) would be subject to extranuclear environments. Numerous works were published from 1948 to the 1970s confirming the effect of chemical environment on the decay rate of ^{7}Be, and these are reviewed by Huh (1999). The effect of pressure on the decay constant of ^{99}Tc was first confirmed by Bainbridge (1952), and the mathematical expression for the shift in the binding energy of the atomic electron under pressure, that would affect EC decay rates, was derived by Mukoyama and Shimizu (1974).

During the last 2 decades or so much attention has been given to experimentation related to chemical and pressure effects on radionuclide EC decay rates due to the technological advances in detector instrumentation with improved detection efficiencies and energy resolution of semiconductor detectors that can now yield more precise decay-rate measurements. Utilizing a 100% efficiency high-purity germanium detector, Huh (1999) was able to determine the decay constant of ^{7}Be in three chemical environments, namely, $Be^{2+}(OH_2)_4$, $Be(OH)_2$, and BeO with a precision of 0.01% resulting in a ~1.5% difference in the decay rates of ^{7}Be in $Be(OH)_2$ and BeO. Also, Ray et al. (1999) found the decay rate of ^{7}Be implanted in Au was 0.72 ± 0.07% lower then ^{7}Be implanted in Al_2O_3. The variation in decay rates of the ^{7}Be was considered a result of the change in the electron density around the nucleus of ^{7}Be due to its association with different anions, and atomic electron affinity and geometry of the lattice within which the ^{7}Be was implanted. EC occurs mostly with electrons in closest proximity to the nucleus, such as electrons from the K shell, and any chemical bond or environment that would force atomic electrons closer to the nucleus or reduce the electron binding energy would increase the EC decay rate.

The combination of chemical environment and pressure can affect electron-capture decay rates. Liu and Huh (2000) found a steady increase in the decay constant of ^{7}Be in $Be(OH)_2$ as a function of pressure over the range of 1 atm–441 kbar. The decay constant increased by 1% at 441 kbar compared to the decay constant at 1 atm, which corresponded to a change in half-life from 53.414 ± 0.003 days at 1 atm to 52.884 ± 0.022 days at 441 kbar pressure. These results were consistent with calculations made by Tossell (2002) on the effects of Be–O bond compression in $Be(OH)_4^{2-}$ on the energy and the electron density at the nucleus of ^{7}Be, which indicates that several hundred kbar pressure could increase the electron density by 1% or more. This would expectedly facilitate the capture of an orbital electron by the nucleus and increase the decay rate. Lee and Steinle-Neumann (2008) demonstrated a 0.1–0.2% increase in the decay constant of ^{7}Be as a metal, chloride, or oxide under a pressure of 25 GPa. Ohtsuki et al. (2004) studied the half-life of ^{7}Be, which was found to be up to 1.1% shorter when the ^{7}Be was encapsulated in fullerene (C_{60}) molecules as compared to ^{7}Be embedded in metal at room temperature. The ^{7}Be half-life is reduced by close to 1.5%, when the ^{7}Be embedded in the C_{60} molecules is kept at a very low temperature of ~5K with liquid helium (Ohtsuki et al., 2007). A thorough study of the chemical environment of the ^{7}Be encapsulated in the C_{60} molecules by Ono et al. (2013) demonstrates the electron density at the Be nucleus to be proportional to the magnitude of the ^{7}Be decay rate.

The radionuclides ^{22}Na and ^{40}K in metallic form showed an increase of 0.06% and 0.025% in their decay constants, respectively, at 25 GPa pressure. Ray et al. (2009) also observed an increase in the orbital EC rates of ^{109}In and ^{110}Sn by $1.00 \pm 0.17\%$ and $0.48 \pm 0.25\%$, respectively, when implanted in the smaller Au lattice compared to implantation in a larger Pb lattice. They interpret the increased decay rates as due to an increase in the eigenstate energies of an atom under spatial confinement. This effect would increase the electron density of the orbital electrons at the nucleus and thus increase the decay rate of an electron capturing radioactive nucleus.

While the fluctuations in the observed EC decay rates may appear small, Zito and Schiferl (1987) indicate that the affects may be significant in terms of stellar energetics. They note that pressure-dependent decay may be an important source of heat generation in Jovian planets where central pressures may reach megabars. They give the example of Jupiter's rock-ice core, which contains as much potassium as Earth's continental crust, and that ^{40}K decay accounts for about 11.5% of the total excess energy output of Jupiter. This energy divided into two parts gives 9.3% due to normal (zero-pressure) beta and EC decay, and the remaining 2.2% would be due to pressure effects alone on the EC decay rate. Also, they surmise that pressure effects on the EC decay rate of ^{90m}Nb could account for 14% of the total energy output of Jupiter's core. The high pressures in Saturn, Uranus, and Neptune could have a role in pressure-accelerated EC decay as an important heat source in these planets. Zito and Schiferl (1987) also provide convincing arguments in support of the role of ^{50}V, ^{123}Te, ^{138}La, ^{40}K, ^{26}Al, and ^{36}Cl, which undergo EC decay, in stellar energetics. They note that peculiar A stars are enriched in La isotopes by a factor of 600 over normal stars; and other stars breed significant amounts of ^{26}Al and ^{36}Cl. They point out that sudden pressure changes in the interior of these stars could result in EC outbursts.

4. Branching β⁻, β⁺ and EC decay

It is a common practice to characterize radionuclide beta decay according to (1) negatron (β^-) emission, (2) positron (β^+) emission and/or EC. However, there exists also several radionuclides of odd atomic number Z and even mass number A that decay via all beta decay modes, namely β^-, β^+ and EC decay to varying degrees yielding two daughter nuclides that differ by two units of atomic number. Their decay may be described by the branching pathways illustrated in Fig. 1.55 where the parent nuclide $^{A}_{Z}P$ decays in two directions, namely, by β^- emission to a daughter nuclide $^{A}_{Z+1}D$ of atomic number $Z+1$ and by β^+ emission and by EC decay to a daughter nuclide $^{A}_{Z-1}D$ of atomic number $Z-1$. This decay phenomenon is also referred to as "dual beta decay"; however, it should not be confused with "double beta decay", which is an altogether different and more elusive decay process described in a subsequent section of this chapter.

Radionuclides that decay by means of this branching β^-, β^+ and EC decay are listed in Table 1.11 together with the relative intensity (%) to which these three beta decay modes occur, along with the decay energies of each mode, and their daughter nuclides. These radionuclides can be found in the Chart of the Nuclides together with their neighboring isobars that are stable. A segment of the Chart of the Nuclides illustrating the positions of ^{36}Cl and ^{40}K, which decay by the

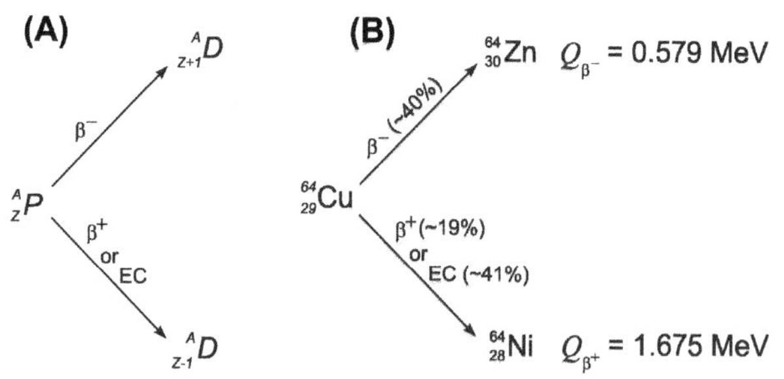

FIGURE 1.55 Branching β^-, β^+ and EC decay illustrated by (A) the general decay reaction of a parent nuclide $^{A}_{Z}P$ to daughter nuclides $^{A}_{Z+1}D$ and $^{A}_{Z-1}D$, and (B) the decay of $^{64}_{29}Cu$ to the daughter nuclide $^{64}_{30}Zn$ by negatron (β^-) emission and to the daughter nuclide $^{64}_{28}Ni$ by positron (β^+) emission or electron-capture (EC) decay. The intensities of the three modes of decay, that is, their branching ratios are ~40% for negatron emission, ~19% for positron emission and ~41% for EC decay.

TABLE 1.11 Radionuclides that decay via branching β^-, β^+, and EC.[a]

Radionuclide	Decay modes (%)	Decay energies (MeV)	Product nuclides
$^{36}_{17}Cl$	β^- (98%)	Q_{β^-}, 0.708	$^{36}_{18}Ar$ (β^-)
	β^+ (0.002%)	$Q_{\beta^+,EC}$, 1.421	$^{36}_{16}S$ (β^+, EC)
	EC (1.9%)		
$^{40}_{19}K$	β^- (89%)	Q_{β^-}, 1.311	$^{40}_{20}Ca$ (β^-)
	β^+ (0.3%)	$Q_{\beta^+,EC}$, 1.505	$^{40}_{18}Ar$ (β^+, EC)
	EC (10.7%)		
$^{64}_{29}Cu$	β^- (40%)	Q_{β^-}, 0.579	$^{64}_{30}Zn$ (β^-)
	β^+ (19%)	$Q_{\beta^+,EC}$, 1.675	$^{64}_{28}Ni$ (β^+, EC)
	EC (41%)		
$^{74}_{33}As$	β^- (32%)	Q_{β^-}, 1.353	$^{74}_{34}Se$ (β^-)
	β^+ (31%)	$Q_{\beta^+,EC}$, 2.562	$^{74}_{32}Ge$ (β^+, EC)
	EC (37%)		
$^{80}_{35}Br$	β^- (92%)	Q_{β^-}, 2.004	$^{80}_{36}Kr$ (β^-)
	β^+ (2.6%)	$Q_{\beta^+,EC}$, 1.871	$^{80}_{34}Se$ (β^+, EC)
	EC (5.7%)		
$^{84}_{37}Rb$	β^- (3%)	Q_{β^-}, 0.894	$^{84}_{38}Sr$ (β^-)
	β^+ (22%)	$Q_{\beta^+,EC}$, 2.681	$^{84}_{36}Kr$ (β^+, EC)
	EC (75%)		
$^{102}_{45}Rh$	β^- (22%)	Q_{β^-}, 1.150	$^{102}_{46}Pd$ (β^-)
	β^+ (14%)	$Q_{\beta^+,EC}$, 2.323	$^{102}_{44}Ru$ (β^+, EC)
	EC (64%)		
$^{108}_{47}Ag$	β^- (97%)	Q_{β^-}, 1.650	$^{108}_{48}Cd$ (β^-)
	β^+ (0.2%)	$Q_{\beta^+,EC}$, 1.922	$^{108}_{46}Pd$ (β^+, EC)
	EC (2.8%)		
$^{126}_{53}I$	β^- (47.3%)	Q_{β^-}, 1.258	$^{126}_{54}Xe$ (β^-)
	β^+ (1%)	$Q_{\beta^+,EC}$, 2.155	$^{126}_{52}Te$ (β^+, EC)
	EC (51.7%)		
$^{130}_{55}Cs$	β^- (1.6%)	Q_{β^-}, 0.370	$^{130}_{56}Ba$ (β^-)
	β^+ (43.6%)	$Q_{\beta^+,EC}$, 2.979	$^{130}_{54}Xe$ (β^+, EC)
	EC (54.8%)		

[a]*Data from Lide (2010), Firestone et al. (1996), and National Data Center, Brookhaven National Laboratory: https://www.nndc.bnl.gov/nudat2/).*

three β^-, β^+ and EC modes, and their neighboring stable daughter isobars is provided in Fig. 1.56.

The decay schemes of ^{36}Cl and ^{40}K exhibit very low relative occurrences of their β^+ decay modes, namely 0.002% and 0.3%, respectively (see Table 1.11), which may be due to their relatively low decay energies of (Q_{EC,β^+}) of 1.142 and 1.505 MeV, respectively (see Table 1.11). Positron decay will compete and predominate over the EC decay mode when the decay energy is well above 1.022 MeV, which is the energy required for the production of an e^+e^- pair needed for the positron decay mode. The following two nuclides listed in Table 1.11, namely ^{64}Cu and ^{74}As, have higher decay energies (Q_{EC,β^+}) of 1.675 and 2.562 MeV, respectively, yielding more significant intensities of 19% and 31%, respectively, for their β^+ decay modes as listed in Table 1.11.

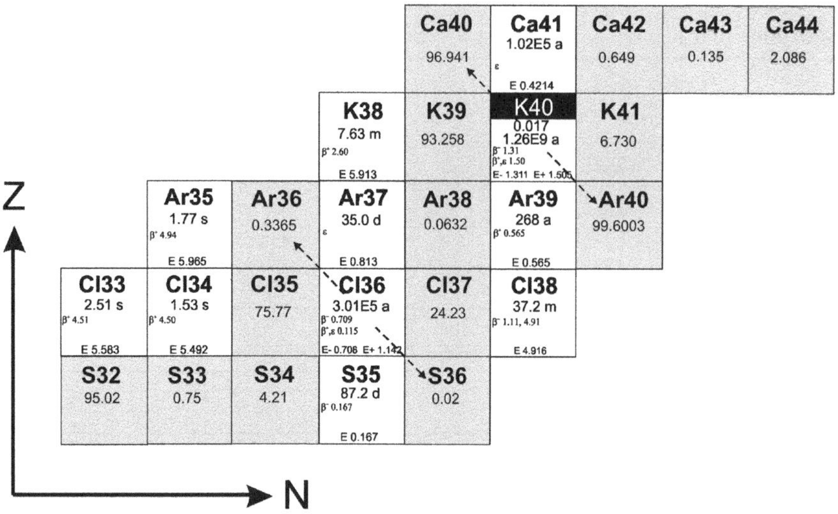

FIGURE 1.56 A segment of the Chart of the Nuclides illustrating the location of $^{36}_{17}$Cl and $^{40}_{19}$K, which decay to neighboring stable isobars (indicated by arrows via negatron (β^-) emission with increasing Z and via positron (β^+) emission and EC with decreasing Z. The stable and naturally occurring nuclides are shaded blocks, and the percent natural abundance of the isotope is included in the center of the shaded block. ^{40}K is a naturally occurring, but radioactive isotope of potassium, and it is thus displayed differently with a black bar over the symbol of the element and the mass number of the isotope. Units of half-life are abbreviated as "s", "m", "d", or "a" for seconds, minutes, days, or anni (years). Beta decay modes are abbreviated as β^-, β^+, and ε for negatron, positron, and EC, respectively, with decay energies in MeV. The decay energy of a radionuclide is in MeV at the bottom of each block alongside the letter E. Two decay energies are given for the radionuclides, ^{36}Cl and ^{40}K, because of their branching beta decay modes. The value after the negative sign (E−) is the decay energy for the negatron decay mode and the other with a positive sign (E+) is the decay energy for the positron and EC decay modes.

There are an additional 25 radionuclides that decay by branching negatron (β^-) and EC decay modes, whereas positron (β^+) emission does not occur. These radionuclides decay according to the general scheme provided in Fig. 1.55 with the exception that the positron (β^+) decay mode does not occur. The properties of these radionuclides, their decay modes, and product nuclides are included in the Table of Radioactive Isotopes listed in Appendix A. They can be easily identified by the fact that both β^- and EC decay modes are listed and there are two product nuclides rather than the customary one nuclide product. Although positron emission does not occur in these cases, the parent nuclide of atomic number Z decays to two daughter nuclides of atomic numbers $Z-1$ and $Z+1$. These nuclides, which do not decay by positron emission, have mostly low decay energies ($Q_{EC} < 1.0$ MeV). The decay energies are below the minimum decay energy equivalent to two electron masses (1.022 MeV) for the production of an e^+e^- pair required in the positron decay mode.

5. Double beta ($\beta\beta$) decay

Double-beta (ββ) decay is the rarest of all radioactive decay processes ever observed. Among these rare double beta decay events, the mode of decay that is the most common may be described by the following general equation:

$$^{A}_{Z}P \rightarrow {}^{A}_{Z+2}D + 2\beta^- + 2\bar{\nu} \tag{1.241}$$

which involves the simultaneous emission of two negatrons and two antineutrinos. This mode of double beta decay is abbreviated as ββ decay or 2νββ decay to read two-neutrino double beta decay. The parent nucleus transforms to a daughter nucleus of the same mass number A, but with an increase in atomic number Z by two. In accord with the beta-decay scheme, within the nucleus of the parent nuclide, two neutrons decay to two protons with the simultaneous emission of two negative beta particles (negatrons) and two antineutrinos, or

$$2n \rightarrow 2p^+ + 2\beta^- + 2\bar{\nu} \tag{1.242}$$

The National Nuclear Data Center (NNDC, 2018a,b) of the Brookhaven National Laboratory (https://www.nndc.bnl.gov/bbdecay/list2013.html) currently lists 11 nuclides that are known to undergo double-beta (ββ) decay according to the double negatron emission described above. These nuclides are the following: $^{48}_{20}$Ca, $^{76}_{32}$Ge, $^{82}_{34}$Se, $^{96}_{40}$Zr, $^{100}_{42}$Mo, $^{116}_{48}$Cd, $^{128}_{52}$Te, $^{130}_{52}$Te, $^{136}_{54}$Xe, $^{150}_{60}$Nd, and $^{238}_{92}$U. The NNDC lists an additional nuclide, ^{130}Ba, that decays by double electron capture (ECEC), which is another mode of double beta decay. These nuclides all have very long half-lives with orders of magnitude that range from 10^{19} to 10^{24} years.

As discussed previously, beta decay by definition includes nuclear decay processes where the mass number (A) of the nucleus remains constant and the atomic number (Z) changes. Thus, we must include EC decay, as a beta decay process, which does not result in the emission of any beta particle. Double beta decay, which involves exclusively EC, has been observed to this date to occur in the case of the decay of ^{130}Ba to ^{130}Xe (Meshik et al., 2001; Pujol et al., 2009). Double beta decay involving exclusively EC can be described by the following general equation:

$$^{A}_{Z}P + 2e^{-} \rightarrow {}^{A}_{Z-2}D + 2\nu \quad (1.243)$$

The parent nucleus captures two orbital electrons and transforms to a daughter nucleus of the same mass number A, but with a decrease in atomic number Z by two with the simultaneous emission of two neutrinos. In accord with the EC beta decay scheme, within the nucleus of the parent nuclide, two protons capture two orbital electrons converting the protons into two neutrons with the simultaneous emission of two neutrinos, or

$$2p^{+} + 2e^{-} \rightarrow 2n + 2\nu \quad (1.244)$$

This form of double beta decay is often abbreviated ECEC or 2 EC signifying the decay mechanism involving the simultaneous capture of two orbital electrons. It is in short referred to as two neutrino double EC.

Other possible modes of double beta decay whereby two nuclear protons transform to two neutrons resulting in the parent nuclide of atomic number Z transforming into a daughter nuclide of atomic number $Z-2$ may occur via (1) the simultaneous emission of two positrons ($\beta^{+}\beta^{+}$) and two neutrinos, (2) combined positron emission and electron capture (β^{+}EC) with the emission of two neutrinos, or the simultaneous double electron capture (ECEC) with the emission of two neutrinos. For example, the decay of ^{130}Ba to ^{130}Xe could occur by either of the three modes of decay, ^{130}Ba($\beta^{+}\beta^{+}$)^{130}Xe, ^{130}Ba(β^{+}EC)^{130}Xe, or ^{130}Ba(ECEC)^{130}Xe. Among the three possible modes of double beta decay, the latter mode ^{130}Ba(ECEC)^{130}Xe was calculated to have the shortest half-life by a magnitude of 10^{2}, and thus ECEC decay of ^{130}Ba would be the mode of decay in this case (Pujol et al., 2009). The measurement of double ECEC beta decay of ^{130}Ba cannot be accomplished by direct measurement of the radiation emissions, because the only emissions in this double EC decay process are the simultaneous emission of two neutrinos, which would be a technically impossible task for a rare decay event. Double ECEC decay of ^{130}Ba was measured by geochemical methods. These methods required the mass spectroscopic measurement of a strong isotope excess of the daughter nuclide ^{130}Xe that could be identified in mineral deposits close to a depth of ~90 m where the mineral deposits are shielded from cosmic radiation, and the observation of a Xe isotope spectrum was distinct from ancient or modern atmospheric Xe (Pujol et al., 2009).

a. Discovery and current research

Double-beta decay was first predicted by Nobel Laureate Maria Goeppert-Mayer (1935) at Johns Hopkins University, who calculated a predicted half-life of over 10^{17} years for this decay phenomenon. It was not until 1987 did Steven R. Elliott, Alan A. Hahn, and Michael K. Moe at the University of California, Irvine report the first observation of double-beta decay in a sample of ^{82}Se enriched from its natural abundance of 8.7% to an isotope abundance of 97% to provide more source material for the double beta decay as well as reduce contamination in the sample from other sources (Elliott et al., 1986, 1987; Moe, 1986; Moe and Rosen, 1989).

The rarity of this decay process is manifested by the very large half-lives (from 10^{18} to 10^{24} years) of the nuclides that decay by this process. Because of the rarity of double-beta decay, its observation in the laboratory is extremely difficult requiring highly enriched radionuclide sources and sophisticated coincidence (double-beta) counting instrumentation capable of extreme suppression or vetoing of background radiation (Elliott et al., 2006; Gómez et al., 2007; Campbell et al., 2008; Gehman et al., 2010; McGrath et al., 2010; Poda et al., 2013; Arnold et al., 2014; Belli et al., 2014; Barabash, 2015; Asakura et al., 2016). Thus, double-beta decay is difficult to detect and not measured routinely. The phenomenon is discussed briefly here, because our knowledge of its existence helps us understand more fully the beta decay process. More detailed treatments of double-beta decay are provided in various chapters of this book.

b. $2\nu\beta\beta$ decay energy

The double-beta decay of ^{82}Se, as observed by Elliott et al. (1987), can be written as follows:

$$^{82}_{34}\text{Se} \xrightarrow{t_{1/2}\;=\;9.2\times10^{19}\text{ years}} {}^{82}_{36}\text{Kr} + 2\beta^{-} + 2\bar{\nu} \quad (1.245)$$

The decay equation illustrated above would only be possible if the daughter nucleus, ^{82}Kr, is lower in mass than the nucleus of its parent ^{82}Se. As explained by Moe and Rosen (1989), the mass of the daughter nucleus is diminished by the mass equivalent to its binding energy, the energy needed to pull the nucleus apart. Thus, a more tightly bound nucleus is lighter than a more loosely bound nucleus containing the same number of nucleons. The number of nucleons or mass number (82) in the above equation remain the same; however, the mass of the nucleus of ^{82}Kr is less than that of its parent ^{82}Se, and the difference in mass of the two nuclei would yield the mass equivalent to the decay energy, $Q_{\beta\beta}$. In mass terms Eq. (1.245) may be written as

$$_{Z}M_{P} = {}_{Z+2}M_{D} + 2m_{e} \quad (1.246)$$

where $_{Z}M_{P}$ and $_{Z+2}M_{D}$ are the masses of the parent and daughter nuclei of atomic numbers Z and $Z+2$, respectively, and $2m_e$ is the mass of two beta particles (*i.e.*, two electron masses). We may use atomic mass units (u) to calculate the mass difference of the parent and daughter nuclei by adding firstly Z atomic electron masses to each side of Eq. (1.246) to give

$$_{Z}M_{P} + Zm_{e} = {}_{Z+2}M_{D} + (Z+2)m_{e} \quad (1.247)$$

The decay energy, $Q_{\beta\beta}$, may then be calculated by the difference in parent and daughter nuclide atomic mass units ignoring the small difference in the binding energies of the electrons in the two nuclides or

$$\begin{aligned} Q_{\beta\beta} &= (M_{^{82}\text{Se}} - M_{^{82}\text{Kr}})c^2 \\ &= (81.916699\ \text{u} - 81.913482\ \text{u})(931.494\ \text{MeV/u}) \\ &= 2.99662\ \text{MeV} \\ &= 2.997\ \text{MeV} \end{aligned} \tag{1.248}$$

The above calculated value is in agreement with the decay energy reported by the National Nuclear Data Center of the Brookhaven National Laboratory, which is 2.997 MeV (https://www.nndc.bnl.gov/qcalc/). The decay scheme of ^{82}Se is illustrated in Fig. 1.57. The scheme illustrates the decay of ^{82}Se of higher mass to ^{82}Kr of lower mass, and the mass difference of the two is equivalent to the decay energy, $Q_{\beta\beta}$. The decay energy is shared as energy of four particles, namely, two beta particles and two antineutrinos. Single beta decay of ^{82}Se to ^{82}Br, a nuclide of higher mass is energetically forbidden as depicted in Fig. 1.57.

All of the nuclides that undergo double beta (ββ) decay have an even number of protons and even number of neutrons. They exhibit a significant stability to beta decay with half-lives that vary over the range of $10^{19}-10^{24}$ years. The stability of these nuclei, as described by Moe and Rosen (1989), may be due to their "even—even" characteristic, that is, their nuclei consist of an even number of protons and neutrons versus a neighboring "odd—odd" nucleus in the Chart of the Nuclides with the same number of nucleons consisting of an odd number of protons and neutrons. The strong nuclear force that holds the nucleus together binds pairs of like particles more tightly than unlike particles. Thus, a nucleus with even number of protons and even number of neutrons can pair and be bound more tightly than a nucleus with the same mass number but with an odd number of protons and odd number of neutrons. For example, the nucleus of $^{82}_{34}$Se contains 34 protons and 48 neutrons, whereas, its neighboring isobar, $^{82}_{35}$Br contains the odd numbers of 35 protons and 47 neutrons. The mass of the ^{82}Br nucleus is slightly higher than that of ^{82}Se, as illustrated in Fig. 1.57. The diminished mass of the ^{82}Se is due to its higher binding energy. Thus, the ^{82}Se is a more tightly bound nucleus than the ^{82}Br, and the transition by beta decay from ^{82}Se to ^{82}Br is forbidden energetically, as illustrated in Fig. 1.57.

As discussed in previous parts of this chapter, beta decay is not confined to negatron emission. There also exist positron and EC decay processes, which are other forms of beta decay. Thus, there may exist double $\beta^+\beta^+$ and double electron capture (ECEC) as well as β^+EC double beta decay processes, and research is underway to observe these alternative double beta decay processes (Barabash et al., 2007, 2008; Kolhinen et al., 2010). The ECEC decay of ^{130}Ba to ^{130}Xe was confirmed by Pujol et al. (2009). In single positron emission a minimum decay energy of 1.022 MeV was discussed previously (see Section X.B.2). Double $\beta^+\beta^+$ would then require a minimum of 2.044 MeV decay energy.

c. Neutrinoless $\beta\beta$ decay

There is much research currently underway on double-beta decay particularly with respect to the search for neutrinoless double-beta decay by research groups around the world, and only a few of the research papers are cited here (Šimkovic et al., 2011; Giuliani, 2011; Pocar, 2012; Cremonesi, 2013; Dell'Oro, 2015; Maneschg, 2015; Henning, 2016). In accord with the Standard Model, which explains the continuous spectrum of beta energies emitted from decaying nuclei, the neutrino always accompanies beta emission, that is,

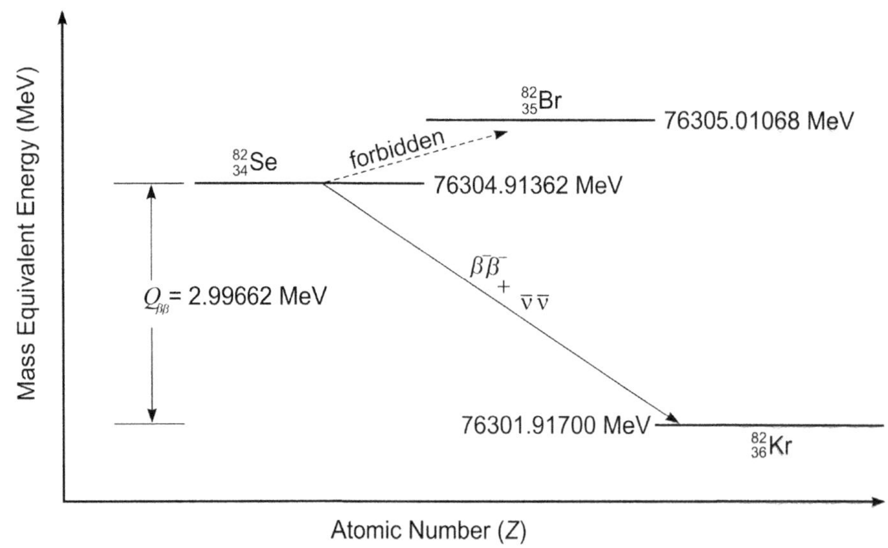

FIGURE 1.57 Double-beta decay of $^{82}_{34}$Se to $^{82}_{36}$Kr. The decay energy, $Q_{\beta\beta}$, of 2.99662 MeV (equivalent to the mass difference between the ^{82}Se and ^{82}Kr) is shared by the two beta particles and two antineutrinos created in the process. Transition via beta decay of ^{82}Se to ^{82}Br is forbidden energetically, as the mass of ^{82}Br is greater than that of ^{82}Se. The double-beta decay occurs with the simultaneous emission of two beta particles and two antineutrinos.

specifically the antineutrino accompanies negatron (β^-) emission and the neutrino accompanies positron (β^+) emission, and accordingly two antineutrinos would accompany double $\beta^-\beta^-$ decay and two neutrinos would accompany $\beta^+\beta^+$ emission or the ECEC decay mode. However, beyond this Standard Model there exists the possibility that the neutrino could be its own antiparticle referred to as a Majorana neutrino in honor of the exceptionally gifted Italian mathematician and theoretical physicist Ettore Majorana (1906–presumed year of death 1938), who first postulated the possibility. According to this theory, if the neutrino possesses a mass (now estimated to have a mass upper limit of 2.3 eV/c^2) and is its own antiparticle, a neutrino emitted in one beta decay could be reabsorbed in the second beta decay resulting in a double-beta decay in which no neutrinos are emitted, *i.e.*, a neutrinoless double-beta decay often abbreviated as $0\nu\beta\beta$. The search for the neutrinoless double-beta decay is important, because it would reaffirm the mass of the neutrino, as only a particle with a definite rest mass can oscillate from one form into another. Earlier in the chapter it was noted that the Nobel Prize in Physics 2015 was awarded jointly to Takaaki Kajita and Arthur B. McDonald for the discovery of neutrino oscillations, which show that neutrinos have mass. The current search and measurement of neutrinoless double beta decay would allow the measurement of the effective neutrino mass and define the absolute neutrino mass scale as described by Maneschg (2015).

Diagrams illustrating the double-beta decay processes that would occur in a decaying nucleus for the two neutrino double beta-decay ($2\nu\beta\beta$) and the, yet to be observed, neutrinoless double-beta decay ($0\nu\beta\beta$) are illustrated in Fig. 1.58. In both cases, two neutrons (n) transform into two protons (p) in the nucleus when two down quarks (d) in the two neutrons change into two up quarks (u) with the emission of 2 W^- bosons. The $W^{\pm}$ bosons were discovered at the Super Proton Synchrotron (SPS) proton-antiproton collider at CERN in 1983 by the UA1 and UA2 collaborations (Arnison et al., 1983a,b, Banner et al., 1983 and Bagnaia et al., 1983). These particles are the carriers of the weak interactions between elementary particles and are responsible for the weak decays of the particles (Sundaresan, 2001; L'Annunziata, 2016). As illustrated in Fig. 1.58, the 2 W^- bosons transform into two high-energy electrons (beta particles) and two antineutrinos. In the case of neutrinoless double-beta decay (yet to be confirmed) the Majorana neutrino emitted in one stage of the decay is reabsorbed in the second stage and two high-energy electrons are emitted without neutrinos.

In the $2\nu\beta\beta$ process, the two beta particles that are emitted simultaneously share the total decay energy, $Q_{\beta\beta}$, with the two neutrinos. Thus, the decay energy is shared between four particles, namely, the two neutrinos and two beta particles. In the $0\nu\beta\beta$ process, the two beta particles emitted share the entire decay energy, as there would be no neutrinos emitted with which to share the decay energy. The beta-particle energy spectra for $2\nu\beta\beta$ and $0\nu\beta\beta$ decay are illustrated in Fig. 1.59. The two beta-particle spectra are very distinct. In the $2\nu\beta\beta$ decay process, the numbers of beta particles (*N*) vary over a wide range of energy plotted as the sum of the two beta-particle energies divided by the decay energy, *i.e.*, $\sum(E_{\beta_1} + E_{\beta_2})/Q_{\beta\beta}$. The beta-particle energy spectrum for the $2\nu\beta\beta$ decay would encompass the continuous energy range with a spectral shape similar to that of the single beta-particle emission, because the beta particles share the decay energy with neutrinos. Notice in Fig. 1.59, that the endpoint of the continuous beta-particle spectrum of the $2\nu\beta\beta$ decay does not reach the value of 1.0. This is due to the fact that the two beta particles emitted in the $2\nu\beta\beta$ decay process must share decay energy with two neutrinos, and thus the sum of the two beta particle energies can never equal the decay energy, $Q_{\beta\beta}$. However, the beta-particle energy spectrum for the $0\nu\beta\beta$ decay, where neutrinos do not escape with any of the decay energy, would display a single energy line for the sum of the two beta particle energies, and this line would occur at $\sum(E_{\beta_1} + E_{\beta_2})/Q_{\beta\beta} = 1$. Current research is underway in the search for the sum of the two beta particles to equal the decay energy, $Q_{\beta\beta}$, or where $\sum(E_{\beta_1} + E_{\beta_2})/Q_{\beta\beta} = 1$, which would confirm $0\nu\beta\beta$ decay.

6. Parity violation in beta decay

Spatial parity involves the behavior of a physical system under the inversion of special coordinates (Sundaresan, 2001). This is equivalent to changing or transforming the coordinates (*e.g.*, x to $-x$ and vice versa) of a physical

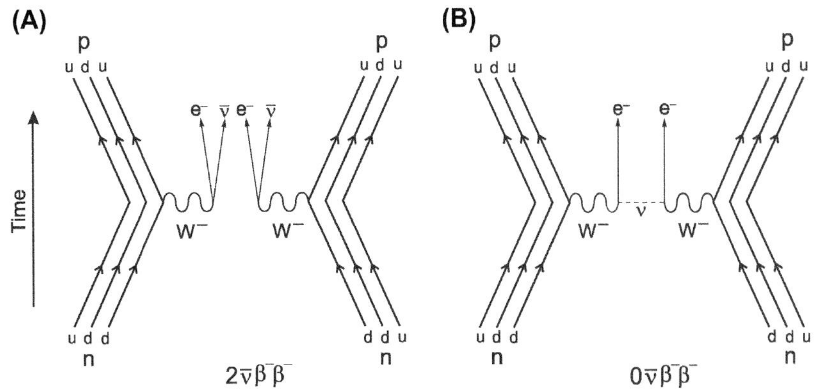

FIGURE 1.58 Feynman diagrams illustrating double-beta decay via (1) two antineutrino double-beta decay ($2\bar{\nu}\beta^-\beta^-$), and (B) neutrinoless double beta decay ($0\nu\beta^-\beta^-$). *From L'Annunziata (2016); reprinted with permission from Elsevier © 2016.*

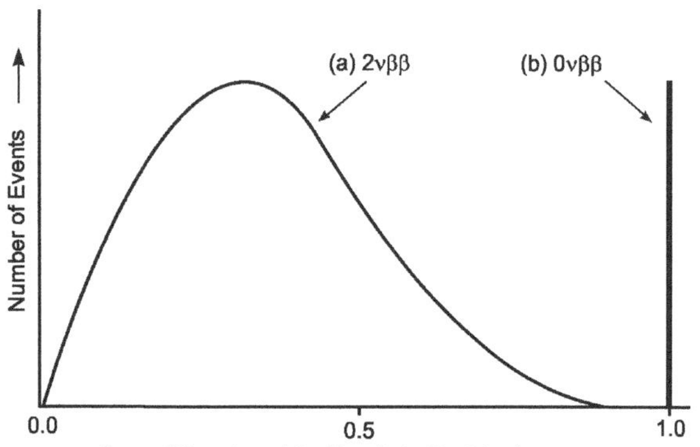

FIGURE 1.59 Beta-particle energy spectra for (A) two-neutrino double-beta decay ($2\nu\beta\beta$), and (B) neutrinoless double-beta decay ($0\nu\beta\beta$). In the $2\nu\beta\beta$ decay process, the sum of the two beta particle energies divided by the decay energy, $\sum(E_{\beta_1} + E_{\beta_2})/Q_{\beta\beta}$, displays a continuous spectrum, and the maximum will never reach 1.0. However, the spectrum for $0\nu\beta\beta$ would display a single line at $\sum(E_{\beta_1} + E_{\beta_2})/Q_{\beta\beta} = 1$.

system, as illustrated in Fig. 1.60. As described by Sundaresan (2001), the parity operation is equivalent to going from a right-handed system of coordinates to one that is left-handed, and parity is conserved if the system has "left-right symmetry". In Fig. 1.60, the clock and its mirror image represent a change in the sign of coordinates. Both clocks function equally and tell the same time. They are identical and they display symmetry and parity is conserved. In natural systems, we can find molecules in biological systems, which are mirror images of each other. Both molecules display the same physical and chemical properties, and thus display left-right symmetry.

Elementary particles possess the quantum number, parity (P), which may be either +1 or −1, which are referred to as even parity or odd parity, respectively. Particles with even parity (+1) have states which possess an even angular momentum (ℓ), and those with odd parity (−1) have states possessing an odd value for the angular momentum. The parity quantum number (P) is determined according to $P = (-1)^{\ell}$ where the angular momentum (ℓ) is an exponent. Parity is one of many quantum numbers that a particle may possess. A particle may have several quantum numbers, which will define the quantum state of the particle. For example, the quantum state of an atomic electron is defined by its principal number n, orbital quantum number ℓ, magnetic quantum number m_{ℓ}, and the spin quantum number m_s, which all together define the energy state and orbital of an atomic electron.

The principle of symmetry in nature was examined by Chen-Ning Yang in his Nobel Lecture given on 11 December 1957. He used the example of a person, who's heart is on the left side looking into a mirror, and the image in the mirror is that of a person with a heart on the right side. (Although the base of the human heart is really located along the midline of the body, the heart points to the left and about two-thirds of the mass of the heart are found on the left side of the body and the other one-third is on the right side of the body. Thus, it is common to refer to the heart being on the left side of the body). Chen-Ning Yang (1957) provided the following example in his Nobel Lecture [with the writer's comment in brackets] with:

> *One of the symmetry principles, the symmetry between the left and the right, is as old as human civilization ... Of course, in daily life, left and right are quite distinct from each other. Our hearts, for example, are on our left sides ... the laws of physics have always shown complete symmetry between the left and the right ... To illustrate the point, we mention that if there existed a mirror image man [from another world] with his heart on his right side, his internal organs reversed compared to ours, and in fact his body molecules, for example sugar molecules, the mirror image of ours, and if he ate the mirror image of the food that we eat, then according to the laws of physics, he should function as well as we do ... The law of right-left symmetry then leads also to a conservation law: the conservation of parity.*

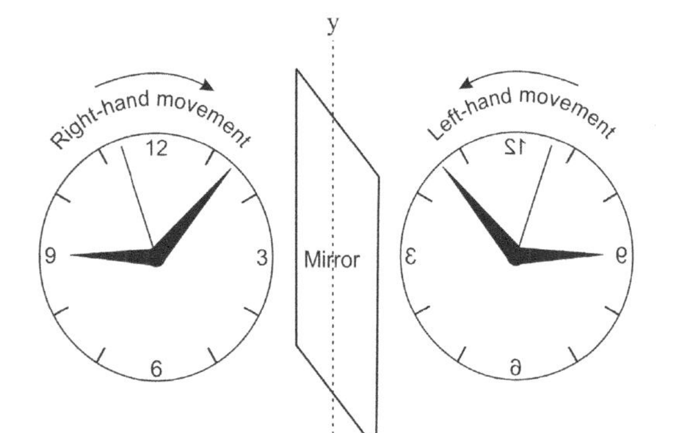

FIGURE 1.60 The inversion of the coordinates along the x axis ($-x$ to $+x$ and vice versa) resulting in a spatial parity transformation of a clock yielding the mirror image of each. Both clocks run identically and display symmetry, and parity is conserved. *From L'Annunziata (2016); reprinted with permission from Elsevier © 2016.*

Prior to the work of Nobel Laureates Lee and Yang (1956), and Chien-Shiung Wu and her collaborators (Wu et al., 1957), it was always assumed that right−left symmetry was conserved throughout all physical processes involving elementary particles and their reactions. This was explained clearly by Professor O. B. Klein, member of the Nobel Committee for Physics during his Award Ceremony Speech on the presentation of the Nobel Prize in Physics on 11 December 1957 when he stated the following:

> *But what has the question of right and left to do with elementary particle physics?.. it was assumed almost tacitly, that elementary particle reactions are symmetric with respect to right and left ... That this assumption was made was very natural,... it looked as if the electrons, the best known elementary particles, possessed no feature which would permit*

a distinction between right and left … In fact, most of us were inclined to regard the symmetry of elementary particles with respect to right and left as a necessary consequence of the general principle of right-left symmetry of Nature. Thanks to Lee and Yang and the experimental discoveries inspired by them we now know that this was a mistake. The starting-point of Lee and Yang in their revision of the whole question of right-left symmetry in elementary particle reactions were certain strange observations concerning a kind of new particles called K mesons, which looked as if they were in contrast with the assumption mentioned.

After a thorough study of the various forces in nature, Lee and Yang (1956) began to question if right−left symmetry is conserved in all interactions. They researched for evidence of the symmetry and the conservation of parity among the forces that act between subatomic particles, namely the nuclear forces, electromagnetic forces and the weak forces. They found that there was considerable evidence that parity was conserved in the strong nuclear and the electromagnetic forces between particles; however, it was only assumed by the scientific community that right-left symmetry and parity conservation was consistent in the weak forces that govern decay interactions, such as beta decay and the decay of subatomic particles. In his Nobel Lecture Yang (1957) explained these forces as follows [with the writer's comments in brackets]:[3]

We list the four classes of interactions below. The [relative] strength of these interactions is indicated in the column on the right.

1. Nuclear Forces	1
2. Electromagnetic Forces	10^{-2}
3. Weak Forces (Decay Interactions)[3]	10^{-13}
4. Gravitational Forces	10^{-38}

The strongest interactions are the nuclear interactions which include the forces that bind nuclei together and the interaction between the nuclei and the π mesons … The second class of interactions are the electromagnetic interactions of which physicists know a great deal. In fact, the crowning achievement of the physicists of the 19th century was a detailed understanding of the electromagnetic forces. [The electromagnetic forces hold electrons in orbitals about the nucleus of an atom; and these are the forces involved in the collision and scattering between charged particles via the exchange of photons between them.] … The third class of forces, the weak interactions, was first discovered around the beginning of this century in the β-radioactivity of nuclei,… About the last class of forces, the gravitational forces, we need only mention that in atomic and nuclear interactions they are so weak as to be completely negligible in all the observations with existing techniques

The fact that parity conservation was assumed by the scientific community and never demonstrated was emphasized by Lee and Yang (1956), and they proposed some experiments that might demonstrate right−left symmetry in the weak interactions including beta decay. The basic principles of these experiments, as described by Yang (1957) are the following:

One constructs two sets of experimental arrangements which are mirror images of each other, and which contain weak interactions. One then examines whether the two arrangements always give the same results in terms of the readings of their meters (or counters).

When applying such an experiment to β decay, the readings of the meters or counters described above refer to the counting of β-particles emitted by the decaying nuclei. If the results of two readings are the same for nuclei arranged as mirror images, then one can demonstrate right-left symmetry in the beta decay and the conservation of parity. Yang (1957) illustrated such as experiment in which Cobalt-60, which decays by the emission of β-particles, are polarized in a magnetic field so that the radioactive ^{60}Co nuclei spin in a given direction. The magnetic field can then be reversed so that the ^{60}Co nuclei spin in the opposite direction (the nuclei become a mirror image of those that spin in the opposite direction). If the radioactivity counters, that detect and count the emitted β-particles, yield the same count rate when the nuclei are spinning in either direction, then right−left symmetry and parity would be demonstrated. This is illustrated in Fig. 1.61.

Chien-Shiung Wu would become the physicist who would demonstrate whether or not there is conservation of parity in the weak interactions, such as in β-decay. As related in the biographical memoir of Chien-Shiung Wu by Benczer-Koller (2009), Nobel Laureate Tsung-Dao Lee consulted with Chien-Shiung Wu, because she was the leading experimentalist in β-decay. They confirmed that there was no evidence in the scientific literature for parity conservation or nonconservation in the weak interaction, and that an experiment would be necessary to find an answer. Benczer-Koller (2009) notes that Chien-Shiung Wu knew that the ^{60}Co, which undergoes β-decay, would be the isotope of choice for the experiment, because the spin of the ^{60}Co nucleus could be polarized in a certain direction with respect to emission of β-particles.

The design of the experiment to ultimately test the parity symmetry in the weak interaction (*i.e.*, β-decay) was left to Chien-Shiung Wu (Benczer-Kolle, 2009). She would design an arrangement whereby an external magnetic field could polarize ^{60}Co nuclei so that their nuclear spins would be directed uniformly according to the external magnetic field. This required embedding the radioactive ^{60}Co in a cerium magnesium (cobalt) nitrate crystal at the extremely low temperature of about 0.003K (Wu et al., 1957), which is very close to absolute zero. The very low temperature essentially eliminates thermal agitation and movement of the atomic nuclei, and thereby essentially freezing the polarized ^{60}Co nuclei in the orientation controlled by the external

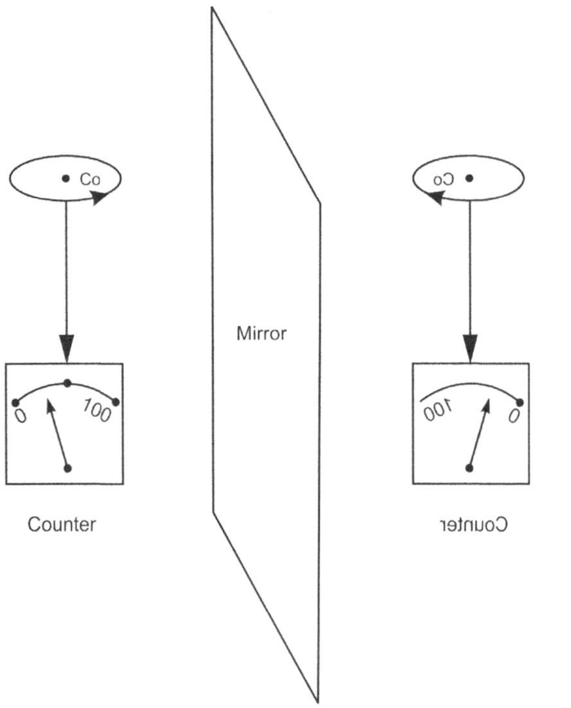

FIGURE 1.61 Experimental arrangement to test right–left symmetry and parity in β-decay of ^{60}Co. In the first experiment (on the left) nuclei of ^{60}Co are aligned in a magnetic field so that the nuclei exhibit spin in a fixed direction. The number of β-particles emitted by the nuclei is counted over a fixed period of time (count rate) by a radiation detector fixed at a given location relative to the ^{60}Co source. In the second experiment (on the right) the magnetic field is reversed so that the radioactive ^{60}Co nuclei spin in the opposite direction thereby creating the mirror image of the ^{60}Co nuclei of the first experiment. If the radiation detector yields the same count rate as in the first experiment than the β-decay will exhibit right–left symmetry and conservation of parity. If the count rates are significantly different in the two cases, the β-decay of ^{60}Co would demonstrate that parity is not conserved in the decay process. *Figure is from Yang (1957); printed with permission © The Nobel Foundation 1957.*

magnetic field. Such as experiment required expertise that was available at the low temperature laboratory of the National Bureau of Standards (NBS) in Washington, D.C., which is now the National Institute of Standards and Technology (NIST). The low temperature laboratory at the NBS had previous experience in the low-temperature alignment of atomic nuclei. Thus, the experiment that would demonstrate parity conservation would be a collaboration between Chien-Shiung Wu of Columbia University and Ernest Ambler, Raymond W. Hayward, Dale D. Hoppes, and Ralph P. Hudson of the low temperature laboratory of the NBS (Forman, 1982; http://physics.nist.gov/GenInt/Parity/parity.html). The specific experimental arrangement that they designed is illustrated in Fig. 1.62.

The apparatus used by Chien-Shiung Wu and collaborators to test parity symmetry in β-decay in ^{60}Co, illustrated in Fig. 1.62, consists of a cryostat housing the radioactive ^{60}Co (labeled as Specimen in the figure) as a thin source (0.005 cm thick) embedded in CeMg nitrate to facilitate the

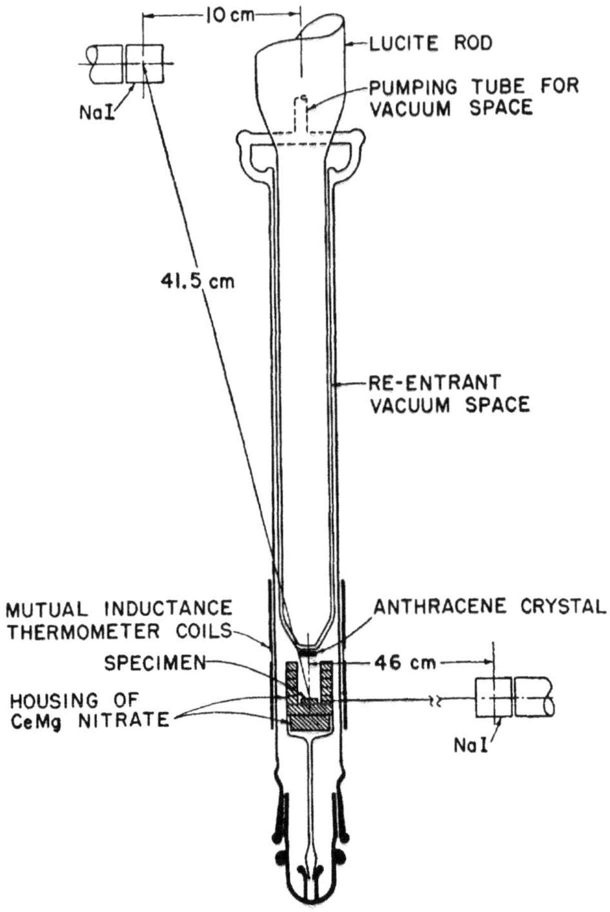

FIGURE 1.62 Schematic drawing of the ^{60}Co source, detector and lower part of the cryostat used by Chien-Shiung Wu and collaborators at NIST to test parity conservation in beta decay. *From Wu et al. (1957) © 1957 by the American Physical Society, https://doi.org/10.1103/PhysRev.105.1413.*

polarization of the ^{60}Co nuclei with the cryostat housing in a magnetic field at near absolute zero temperature. The anthracene crystal located about 1 cm above the ^{60}Co source is a scintillation detector sensitive to the β-particles emitted by the ^{60}Co. The anthracene scintillator emits a flash of light following the interaction of each β-particle with the crystal. The individual flashes of light are transmitted by the Lucite Rod, illustrated in Fig. 1.62, to an external photomultiplier tube (PMT), which will convert the light to voltage pulses. Each voltage pulse is then counted over a fixed period of time providing a counting rate (*e.g.*, counts per minute) of the β-particles emitted by the ^{60}Co. Beta-decay of ^{60}Co yields ^{60}Ni according to the following:

$$^{60}_{27}\mathrm{Co} \rightarrow ^{60}_{28}\mathrm{Ni} + \beta^- + \overline{\nu} + \gamma \tag{1.249}$$

The antineutrino emitted in the β-decay escapes without detection. The gamma radiation is emitted by the ^{60}Ni product nuclide, which is produced at an excited energy state resulting in the immediate emission of γ-rays before reaching a stable state. The γ-rays are detected by NaI solid

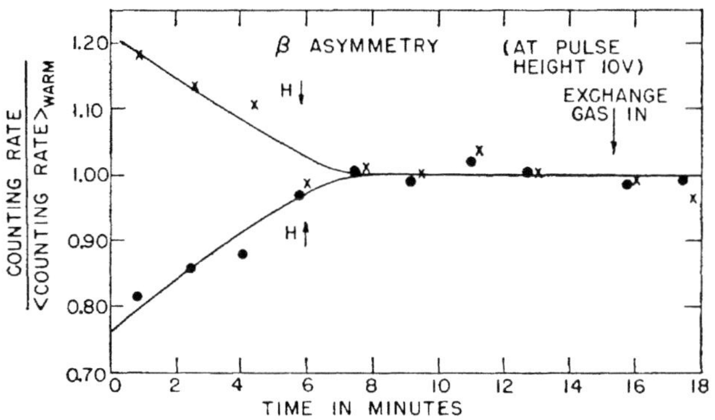

FIGURE 1.63 Beta particle asymmetry for polarizing fields pointing up and pointing down. *From Wu et al. (1957) © 1957 by the American Physical Society, https://doi.org/10.1103/PhysRev.105.1413.*

scintillation detectors located over 40 cm away from the apparatus at the equatorial and near the polar position of the ^{60}Co (see Fig. 1.62). The NaI detectors were used by Wu and collaborators to measure the γ-ray anisotropy to provide a measure of the ^{60}Co nuclear polarization.

Before cooling and before the application of an external magnetic field, the ^{60}Co nuclei exhibit thermal agitation and they are not polarized. Thus, the β-particles that they emit under these conditions would be isotropic, that is, they would be emitted in all directions. Under these warm (*i.e.*, room temperature) conditions, Wu and her collaborators would determine the count rate of the β-particles and label this count rate as "<COUNTING RATE>$_{warm}$" (see Fig. 1.63). When the ^{60}Co specimen is cooled and the magnetic field applied, the ^{60}Co nuclei become polarized and remain rigid in their polarized state. The number of β-particles detected by the anthracene crystal over a fixed period of time under these conditions were determined and labeled as the COUNTING RATE (see Fig. 1.63) The ratio COUNTING RATE/ <COUNTING RATE>$_{warm}$ was then used by Wu and coworkers to determine if the counting rate of the β-particles emitted by the ^{60}Co was higher or lower than the counting rate under warm conditions and in the absence of a magnetic field.

They discovered that when the magnetic field is applied in one direction (see H↓ in Fig. 1.63) with the ^{60}Co nuclei at near absolute zero temperature and the ^{60}Co nuclei polarized in one direction, the β-particle counting rate was at its highest, higher than when the sample is at room temperature and in the absence of a magnetic field. Reversing the direction of the magnetic field (see H↑ in Fig. 1.63) and thereby reversing the direction of the polarization of the ^{60}Co nuclei resulted in a reduction of the β-particle counting rate to magnitudes below that determined when the sample was at room temperature and in the absence of a magnetic field. When the ^{60}Co is warmed slowly to room temperature, the ^{60}Co nuclei would return to normal thermal agitation and lose their polarization after about 7 minutes of warming, as illustrated in Fig. 1.63.

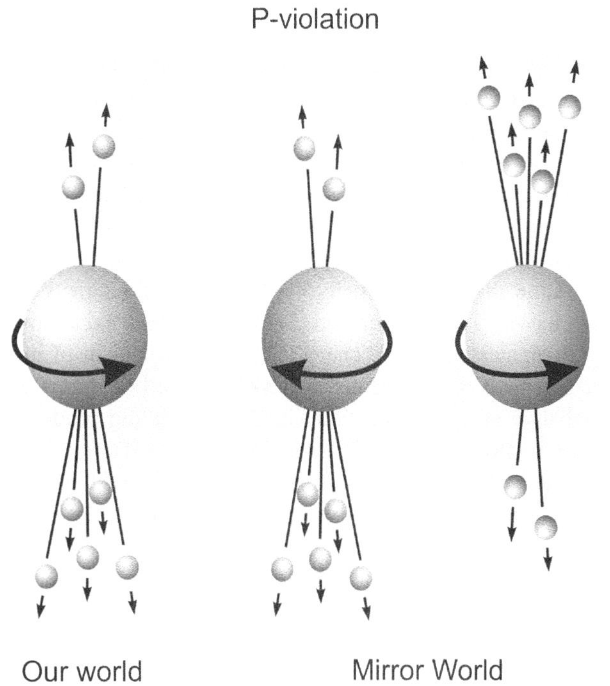

FIGURE 1.64 Parity violation in the decay of cobalt-60. The large spheres in the center represent groups of ^{60}Co nuclei. The β-particle emissions from the ^{60}Co nuclei are illustrated as smaller spheres, which are emitted predominantly in the direction opposed to that of the nuclear spin (curving arrow). The two orientations of the ^{60}Co nuclei in the "Mirror World" were created in the laboratory by Wu and coworkers, by polarization of the ^{60}Co nuclei in a magnetic field at near absolute zero temperature to fix the nuclei in one orientation and then flipping the nuclei by polarization in an opposite direction. The beta emissions of the two ^{60}Co nuclei are not identical in their mirror images and thus parity conservation is violated in the beta decay of ^{60}Co. *Figure is from Atkinson (2006); reprinted with permission from Elsevier © 2006.*

As described above, the findings of Wu and her collaborators demonstrated that the β-particle emissions were not the same for the two spin directions of the ^{60}Co nuclei; that is, the β-particle emissions differed for the two mirror images of the ^{60}Co nuclei. Thus, the β-decay process in ^{60}Co nuclei does not exhibit right-left symmetry or, in other words, this β-decay violates parity. This is illustrated in Fig. 1.64.

The experiment of Wu and her collaborators at the National Bureau of Standards (*i.e.*, National Institute of Standards and Technology) demonstrated that the β-particle emissions from decaying ^{60}Co nuclei is greatest in the direction of the south pole of the nucleus, which would be the pole that would point toward the north pole of an external magnetic field (Forman, 1982). This is illustrated in Fig. 1.64 labeled "Our World". Once the significance of Chien-Shiung Wu's work was ascertained, other researchers including Richard Garwin, Leon Lederman and Marcel Weinrich at the Nevis Cyclotron Laboratories at Columbia University, detected asymmetry in muon decay, which broadened the acceptance of parity violation in weak interactions other than β-decay (Garwin et al., 1957).

7. Beta-particle interactions with matter

Owing to the very low mass of the beta particle (i.e., electron) compared with the alpha particle, it travels at a much higher velocity than an alpha particle of equivalent energy. Because of its greater velocity, lower mass, and lower charge, the specific ionization produced in air by the traveling beta particle is much lower (by a factor of a 1000) than that of an alpha particle of equivalent energy.

Like the alpha particle, the beta particle interacts with matter via (1) ionization and (2) electron orbital excitation as it dissipates its kinetic energy. A third mechanism of interaction with matter, which distinguishes the beta particle, is radiative energy dissipation via Bremsstrahlung production (see Section XI.D.2). Thus, as described by Turner (1995) the stopping power for beta particles (β^- or β^+) is the sum of the collisional and radiative contributions or

$$\left(-\frac{dE}{dx}\right)^{\pm}_{tot} = \left(-\frac{dE}{dx}\right)^{\pm}_{col} + \left(-\frac{dE}{dx}\right)^{\pm}_{rad} \tag{1.250}$$

where the superscript $\pm$ refers to positively or negatively charged electrons. The radiative contribution, that is, the absorption of beta particle energy with the concomitant emission of Bremsstrahlung radiation is significant with high-energy beta particles (*e.g.*, ^{32}P and ^{80}Y beta-particle emissions) in absorbers of high atomic number (*e.g.*, Pb-glass). Bremsstrahlung radiation is discussed in Section XI.D.2 of this chapter.

Collisional interactions of beta particles are somewhat different from those that occur with alpha particles. A beta particle may collide with an orbital electron or come into close proximity to it and cause the electron to be ejected, resulting in the formation of an ion pair. Considerable scattering of beta particles occurs in such collisions, because the mass of the beta particle is equivalent to that of an atomic electron. This is in direct contrast to the alpha particle, which, for the most part, retains a relatively undeviating path while passing through matter and interacting with atomic electrons. The mass equivalence of beta particles and electrons is an important factor that gives bombarding beta particles the power to impart a major portion of their kinetic energy to atomic electrons in a single collision. The atomic electrons ejected upon beta particle collisions themselves cause ionization in a similar fashion. This is referred to as secondary ionization, and the ionization caused by initial beta particle-electron collisions is classified as primary ionization. Because the major portion of beta particle energy may be imparted to an atomic electron upon collision, secondary ionization may account for as much as 80% of the total ionization produced in a given material bombarded by beta particles.

The probability of beta particle interactions with atomic electrons increases with the density of the absorbing material. Beta particle absorption is consequently proportional to the density and thickness of an absorber. Figs. B.2 and B.3 of Appendix B provide graphs from which the range of negatrons (β^-) in units of surface density or mass thickness (g/cm^2) in air or water, respectively, can be estimated for particles of energies from 0.01 to 10 MeV. More precise ranges and graphs are available from the online computer program ESTAR "Electron Stopping Power and Range" (Berger et al., 2017a) available from the National Institute of Standards and Technology (NIST) at https://physics.nist.gov/PhysRefData/Star/Text/method.html. The range of negatrons expressed in terms of surface density or mass thickness (g/cm^2) of absorber can be converted to absorber thickness (cm) when the absorber density (g/cm^3) is known.

Several empirical equations exist for calculating negatron ranges based on the E_{max} of the beta particle and on experimental data collected by several researchers during the early years of research with nuclear radiation. These empirical equations may be used to obtain estimates of beta particle range, and should be used with caution. More accurate data is currently available from the online computer program ESTAR "Electron Stopping Power and Range" (Berger et al., 2017a) available from the National Institute of Standards and Technology (NIST) at https://physics.nist.gov/PhysRefData/Star/Text/method.html. The ESTAR computer program will provide Stopping Power and Ranges for electrons in 98 elements and in 200 compounds or materials. The data provided by ESTAR include collisional stopping power, radiative stopping power, and total stopping power and ranges for electrons over the electron energy range of 10^{-2} to 10^3 MeV. Empirical equations reported by Glendenin (1948) are

$$R = 0.542E - 0.133 \quad \text{for } E > 0.8 \text{ MeV} \tag{1.251}$$

and

$$R = 0.407E^{1.38} \quad \text{for } 0.15 \text{ MeV} < E < 0.8 \text{ MeV} \tag{1.252}$$

where R is the beta particle range in g/cm^2 and E is the energy of the beta particle (i.e., E_{max}) in MeV. In addition, the following is an empirical equation of Flammersfield (1946) described by Paul and Steinwedel (1955):

$$R = 0.11\left(\sqrt{1 + 22.4E^2} - 1\right) \quad \text{for } 0 < \text{E} < 3 \text{ MeV} \tag{1.253}$$

This equation provides calculated ranges in units of g/cm^2 in close agreement to those obtained from Eqs. (1.251) and (1.252).

For negatrons or electrons exceeding 2.5 MeV up to ~20 MeV, the following are empirical equations reported by Katz and Penfold (1952), the U.S. Public Health Service (1970) and Evans (1955):

$$R = 530E - 106 \quad \text{for } 2.5 \text{ MeV} < E < {\sim}20 \text{ MeV} \tag{1.254}$$

and

$$R = 412E^{1.265 - 0.0954 \ln E} \quad \text{for } 0.01 \text{ MeV to} {\sim} 3 \text{ MeV} \tag{1.255}$$

where R is the negatron or electron range in mg/cm^2. The above empirical equations have been used for low atomic number absorbers such as aluminum ($Z = 13$). Semi-empirical equations developed by Tabata et al. (1972) and described in detail by Tsoulfanidis (1995) yield beta-particle or electron ranges in absorber elements and compounds of a wide range of atomic number. Empirical equations are helpful in providing an estimate of the beta particle or electron ranges.

More accurate values and graphs on the ranges of beta particles or electrons with energies of 0.01–1000 MeV in 98 elements from hydrogen to californium and a wide range of media including air and natural and synthetic compounds are available from the NIST Electron Stopping Power and Range (ESTAR) data online via https://physics.nist.gov/PhysRefData/Star/Text/method.html. Beta particle energies general do not exceed 5.0 MeV. Using data from NIST, the ranges of beta particles with energies from 0.01 to 5.0 MeV in dry air, water, and aluminum are provided in Table 1.12 to provide a comparison of beta particle ranges in material of varying density and effective atomic number.

Table 1.12 illustrates that 1 cm of dry air has about the same stopping power as 0.001 cm of water or 0.0005 cm of aluminum.

The range of beta particles in matter is considerably greater than that of alpha particles of the same energy. Again, this is due to the lower mass, lower charge, and higher velocity of travel of the beta particle in comparison with an alpha particle of equivalent energy. The significance of this difference may be appreciated by reference to Table 1.13, in which the alpha particle and beta particle and/or electron ranges in air as a function of particle energy are compared.

To put the data of Table 1.13 into historical perspective, it is interesting to recall the origin of the names "alpha and beta radiation". Before alpha and beta particles were characterized fully, Rutherford (1899) carried out experiments that demonstrated two types of radiation existed; one radiation that was most easily absorbed by matter and another that possessed a greater penetrating power. Out of convenience, he named these radiations as "alpha" and "beta". Not much later, Villard (1900b,c) in France discovered a yet more penetrating radiation. In harmony with the nomenclature assigned to the "alpha" and "beta" radiation, Rutherford (1903) coined the term "gamma rays" to the yet more penetrating radiation. In the *Philosophical Magazine* Rutherford (1903) named and characterized the three types of nuclear radiation on the basis of their penetration power in matter as follows:

TABLE 1.12 CSDA[a] ranges of beta particles[b] or electrons of various energies in air, water, and aluminum.

	Dry air[c]		Water		Aluminum	
Energy (MeV)	(g/cm²)[d]	(cm)[e]	(g/cm²)[d]	(cm)[e]	(g/cm²)[d]	(cm)[e]
0.01	0.00028	0.02	0.00025	0.002	0.00035	0.00013
0.05	0.00491	4.07	0.00432	0.004	0.00057	0.00021
0.10	0.01623	13.4	0.01431	0.014	0.01872	0.007
0.15	0.03193	26.4	0.02817	0.028	0.03659	0.014
0.25	0.07213	59.8	0.06372	0.06	0.08217	0.030
0.50	0.1995	165	0.1766	0.17	0.2260	0.084
1.00	0.4912	408	0.4367	0.43	0.5546	0.206
1.50	0.7901	656	0.7075	0.70	0.8913	0.331
2.00	1.085	900	0.9785	0.97	1.224	0.455
2.50	1.374	1140	1.247	1.24	1.550	0.576
3.00	1.658	1375	1.514	1.51	1.869	0.695
3.50	1.935	1605	1.777	1.77	2.183	0.811
4.00	2.208	1832	2.037	2.03	2.491	0.926
4.50	2.476	2049	2.295	2.29	2.794	1.04
5.00	2.740	2274	2.550	2.55	3.092	1.15

[a]*Defined by NIST (https://physics.nist.gov/PhysRefData/Star/Text/appendix.html), the CSDA range is a very close approximation of the average path length traveled by a charged particle as it slows down to rest, calculated in the continuous-slowing-down approximation. In this approximation, the rate of energy loss at every point along the track is assumed to be equal to the total stopping power. Energy loss fluctuations are neglected. The CSDA range is obtained by integrating the reciprocal of the total stopping power with respect to energy.*
[b]*Beta particles are emitted with a broad range of energies from 0 to E_{max}. When calculating the maximum range of beta particles emitted by a given radionuclide, the energy of the emissions at E_{max} is used.*
[c]*The values are for dry air at near sea level.*
[d]*The values of range in units of surface density or mass thickness (g/cm²) are obtained from the NIST online ESTAR (Electron Stopping Power and Range) data (https://physics.nist.gov/PhysRefData/Star/Text/ESTAR.html (Berger et al. (2017a)).*
[e]*The values of range in units of path length (cm) are obtained by dividing the range in surface density (g/cm²) by the density (σ) of the absorber. The densities used are the following: $\sigma = 1.205 \times 10^{-3}$ g/cm³ for dry air at near seal level, $\sigma = 1.00$ g/cm³ for water and $\sigma = 2.69$ g/cm³ for aluminum.*

TABLE 1.13 CSDA[a] ranges of alpha and beta particles (or electrons) of various energies in air.

	Range (mg/cm^2 air)		Range (cm Air)[d]	
Energy (Mev)	Alpha particle[b]	Beta particle[c]	Alpha particle	Beta particle
0.1	0.166	16.2	0.14	13
0.5	0.419	199.5	0.35	165
1.0	0.670	491.2	0.56	407
1.5	0.952	790.1	0.79	656
2.0	1.30	1085	1.1	900
2.5	1.67	1374	1.4	1140
3.0	2.12	1658	1.8	1376
4.0	3.15	2208	2.6	1832
5.0	4.37	2740	3.6	2274
6.0	5.77	3255	4.8	2701
7.0	7.35	3757	6.1	3118
8.0	9.10	4246	7.5	3524

[a]*Defined by NIST (https://physics.nist.gov/PhysRefData/Star/Text/appendix.html), the CSDA range is a very close approximation of the average path length traveled by a charged particle as it slows down to rest, calculated in the continuous-slowing-down approximation. In this approximation, the rate of energy loss at every point along the track is assumed to be equal to the total stopping power. Energy loss fluctuations are neglected. The CSDA range is obtained by integrating the reciprocal of the total stopping power with respect to energy.*
[b]*Data is from the National Institute of Standards and Technology (NIST) Database ASTAR (Berger et al., 2017a) for CSDA (continuous slowing down approximation) ranges: http://www.nist.gov/pml/data/star/index.cfm).*
[c]*Data is from the National Institute of Standards and Technology (NIST) Database ESTAR (Berger et al., 2017a) for CSDA (continuous slowing down approximation) ranges: http://www.nist.gov/pml/data/star/index.cfm).*
[d]*Calculated from the range in mass thickness units (mg/cm^2) and the density of dry air at STP, $\rho_{air} = 1.205\ mg/cm^3$.*

Radium gives out three distinct types of radiation:

(1) *The α-rays, which are very easily absorbed by thin layers of matter, and which give rise to the greater portion of the ionization of the gas observed under the usual experimental conditions.*
(2) *The β rays, which consist of negatively charged particles projected with high velocity, and which are similar in all respects to cathode rays produced in a vacuum tube.*
(3) *The γ-rays, which are nondeviable by a magnetic field, and which are of a very penetrating character.*

These rays differ very widely in their power of penetrating matter. The following approximate numbers, which show the thickness of aluminum traversed before the intensity is reduced to one-half, illustrate this difference:

Radiation	Thickness of aluminum (cm)
α-rays	.0005
β-rays	.05
γ-rays	8

It is important to emphasize that, although all beta particles can be completely absorbed by matter, the shields we select can be of great consequence. Hazardous bremsstrahlung radiation can be significant when high-energy beta particles interact with shields of high atomic number. The phenomenon of bremsstrahlung production is discussed further in Section XI.D.2 of this chapter.

Positrons dissipate their energy in matter via the same mechanisms as previously described for negatrons, which is understandable, as both are electrons. When charged particles travel through matter they interact with the target atoms along their trajectory and dissipate energy in the absorber by two mechanisms, namely, (1) they may interact with the target nuclei by the transfer of momentum and (2) interact with the orbital electrons of the target/absorber material dissipating energy by means of electron excitation and ionization. The energy dissipated by positrons and electrons in the media with which they travel (i.e., stopping power) and their ranges in matter will differ and, as described in the following paragraphs, this difference is dependent on the positron and electron energy. The scattering of positrons by atomic nuclei (*i.e.*, elastic scattering) is less than that for negatrons of the same energy when traveling through matter of a given atomic number Z as demonstrated by Fowler and Oppenheimer (1938), Lipkin and White (1950) and Feshbach (1952). However, the positive charge on positrons would be expected to create Columbic forces of attraction between the positron and orbital electrons of the medium through which the

TABLE 1.14 Beta particle (β^+, β^-) or electron (e^+, e^-) collision stopping powers (S^+_{col}, S^-_{col}) and ranges (r^+, r^-) in dry air (near sea level).[a]

	Positrons (β^+)		Negatrons (β^-)			
Energy[b] (MeV)	S^+_{col}(MeV cm^2/g)	r^+ (g/cm^2)	S^-_{col}(MeV cm^2/g)	r^- (g/cm^2)	S^+_{col}/S^-_{col}	r^+/r^-
0.01	2.180E+01	2.571E−04	1.975E+01	2.883E−04	1.104	0.892
0.02	1.256E+01	8.885E−04	1.157E+01	9.781E−04	1.086	0.908
0.05	6.171E+00	4.566E−03	5.819E+00	4.912E−03	1.060	0.930
0.1	3.776E+00	1.537E−02	3.633E+00	1.623E−02	1.039	0.947
0.2	2.510E+00	4.908E−02	2.470E+00	5.082E−02	1.016	0.966
0.5	1.784E+00	1.978E−01	1.802E+00	1.995E−01	0.990	0.991
1	1.626E+00	4.945E−01	1.661E+00	4.912E−01	0.979	1.007
2	1.642E+00	1.103E+00	1.684E+00	1.085E+00	0.975	1.017
5	1.786E+00	2.800E+00	1.833E+00	2.740E+00	0.974	1.022
10	1.931E+00	5.311E+00	1.979E+00	5.192E+00	0.976	1.023

[a] *Data from Berger and Seltzer (1982), US Dept. of Commerce, National Bureau of Standards NBSIR 82–2550.*
[b] *Most beta particle emissions have an E_{max} < 5 MeV. Electrons (e^+, e^-) can be accelerated to very high energies; and data on electron and positron stopping powers and ranges in various materials for electron and positron energies up to 1000 MeV are available from Berger and Seltzer (1982) and online from NIST at https://www.nist.gov/publications/nist-7-nist-electron-and-positron-stopping-powers-materials. The work by Berger and Seltzer (1982) is available online via the NIST publications cite (https://nvlpubs.nist.gov/nistpubs/Legacy/IR/nbsir82-2550.pdf).*

positron travels. Negatrons, which carry a negative charge would not possess any attractive forces with atomic electrons of the target material.

In the early 1950s, Walter H. Barkas studied the tracks produced by high-energy pions in nuclear emulsions, as related in historical and technical accounts by Lindhard (1976) and Andersen (1985). Barkas and his coworkers discovered that the negatively charged pions (π^-) produced tracks that were recognizably longer in the emulsion than its antiparticle, the positively charged pions (π^+), of the same energy. Ten years later, Barkas and coworkers (Barkas et al., 1963) observed the same effect when measuring the tracks produced by the sigma hyperons, Σ^- and Σ^+. In this case, the difference in range was more definitive; whereby, the range of the negatively charged Σ^- hyperon was 3.6% longer than that of its antiparticle, positively charged Σ^+, which possessed the same energy or velocity. From then on, this effect was named "the Barkas effect" or the "Barkas-Andersen effect" after its original investigator Walther H. Barkas and Hans Hendrik Andersen, who undertook research into the effect (Andersen et al., 1966, 1967, 1969a,b). The Barkas effect may be defined as the difference in the stopping force on a charged particle compared to that of its antiparticle as it penetrates through matter (Sigmund and Schinner, 2003).

A few years following Barkas's confirmed observation, Heckman and Lindstrom (1969) made stopping power measurements of the π^- and π^+ at identical velocities in hydrogen, and they found that the differences in range between the π^- and π^+ would decrease with increasing energy. In other words, there was an energy level above which the particles would not exhibit a difference in their range. This observation agrees with electron and positron collision stopping powers and ranges in target matter or absorber media evidenced by data provided in Table 1.14, as an example. Similar examples of the Barkas effect of electrons and positrons in other absorber media (e.g., H_2O, C, Al, Cu, Ag, Pb, etc.) are available from data provided by Berger and Seltzer (1982) of the former National Bureau of Standards. Table 1.14 compares the negatron (β^-) and positron (β^+) (i.e., electron and positron) collision stopping powers and CSDA ranges.

Eq. (1.250) noted previously defines the total stopping power of beta particles or electrons in absorber media as the sum of collisional and radiative losses of energy by the beta particles as they travel through matter. Differences in radiative losses between negatrons and positrons are not included in Table 1.14, Berger and Seltzer (1982) assumed radiative losses to be the same for negatrons and positrons in their tabulations. Collision losses of energy, which are charge dependent, are higher for positrons at lower energies (~0.01–1.0 MeV), as illustrated in Table 1.14. From the data of Table 1.14, we can see that the atoms of the absorber have a higher stopping power for positrons of low energy (<1 MeV) than the negatrons of the same energy, which results in a shorter range for the positrons and longer range for the negatrons. As defined in part XVI.A of this chapter, stopping power is the average energy dissipated by ionizing radiation in a medium per unit path length of travel of the radiation in the medium. Taking into account the definition of stopping power, we can visualize in the case of positrons of low energy (<1 MeV), how a positron traveling through matter can interact with the Columbic field of the negatively charged electrons bound in their atomic orbitals. The attraction between the low-energy positron and atomic orbital electrons can create a drag on the positron as it travels through the atoms of the medium. This drag would cause the positron to dissipate more energy than a negatron of the same energy and consequently reduce the range of the low-energy positron compared to a negatron of the same energy. However, at higher energies (>1 MeV) and consequently higher speeds, the positrons would be more resistant to the negative charges of the atomic orbital electrons of the medium through which it travels and consequently dissipate less energy per path length of travel in the medium. As a result, the high-energy positrons (>1 MeV) would have ranges more similar to those of negatrons of the same energy.

Two equations (Eqs. 1.524 and Eq. 1.525) are cited in Section XVI.A of this chapter for calculating the ionization-excitation stopping powers for negatrons and positrons due to collision interactions with absorbers. Their difference, as noted by Tsoulfanadis (1995), is due only to the second term in the brackets of these two equations, which is much smaller than the logarithmic term, and consequently the differences between negatron and positron stopping powers are small as evidenced from the data in Table 1.14.

For further reading on the Barkas effect, the reader may peruse papers by Lindhard (1976), Mikkelsen (1992), Sigmund and Schinner (2003), Lodi Rizzini et al. (2004), Nersisyan and Das (2005), Sigmund et al. (2005), Fettouhi et al. (2006), and Archubi et al. (2013).

8. Beta particle absorption and transmission

Early research work on measuring the range of beta particles involved placing absorbers of increasing thickness between the radioactive source and the detector. The detector would measure the beta particles transmitted through the absorber. Increasing the absorber thickness would increasingly diminish the number of beta particles transmitted on to the detector. The transmission of beta particles was then plotted against absorber thickness in an attempt to determine the thickness of absorber required to stop all of the beta particles. Unfortunately, the plots could not be used directly to determine accurately beta particle ranges; rather they had to be compared to an absorption curve of a beta emitter of known range by what became known as Feather analysis (Feather, 1938; Glendenin, 1948). An auspicious outcome of this work was the observation that the plots of beta particle absorption had more or less an exponential character. When plotted logarithmically against absorber thickness the beta particle absorption and/or transmission through the absorber was linear or near linear at certain levels of transmission, as illustrated in Fig. 1.65 and described by Eq. (1.256). The linearity is a fortuitous outcome of the continuous energy spectrum of beta particles emitted from any given source. These findings are quite the contrast to the absorption curve of alpha particles discussed previously (Fig. 1.43), where the alpha particle is emitted from a radionuclide at a unique level of energy and comes to an almost abrupt stop at the end of its range in an absorber.

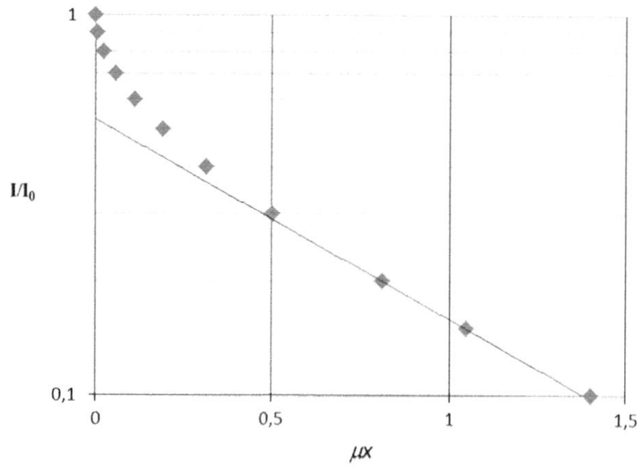

FIGURE 1.65 A typical beta transmission curve and its exponential approximation in an arbitrarily selected point at a low transmission value. *From Švec (2015), reprinted with permission from Elsevier © 2015.*

The curve illustrated in Fig. 1.65 is characteristic of beta particles. The linear segment of the semi-logarithmic plot of activity transmitted versus absorber thickness levels off horizontally (not illustrated in Fig. 1.65) due to a background of Bremsstrahlung radiation. Negatrons and positrons both display a linear semi-logarithmic plot with the exception that, in the case of positrons, the plot has an added background due to annihilation radiation (Glendenin, 1948). Because beta particles have a definite range in matter, beta particle transmission is not a purely logarithmic one as is the case for gamma radiation (see Section XII.D of this chapter.). The curves may not display a purely exponential character and the plots may have a degree of concavity to them depending on the distance of the source and detector to the absorber and on the shape of the beta particle continuous energy spectrum. The greater the atomic number of the beta particle emitter, and the more the beta spectrum is displaced toward the lower energies, the more nearly exponential (linear) will be the absorption curve (Glendenin, 1948). Studies by Thontadarya (1985) indicate that attenuation coefficient values of Eq. (1.256) or 1.257 depend on the part of the transmission curve selected and best fit (i.e., greatest linearity) is achieved at low transmission (i.e., low I/I_0) of <20%. This is explained by Švec (2015), who provides the following logic: (a) the beta spectra of various radionuclides differ in shape; however, their beta spectra resemble each other at the descending parts of their spectra close the maximum beta particle energies, (b) at a transmission of <20%, the absorber has removed 80% or more of the beta particles from the radiation field, and the remaining high-energy beta particles are partly collimated by passing the shortest paths (i.e., particles that have undergone less scattering) in the absorber representing a radiative "arrow head", and (c) high energy particles at the end of the beta spectrum encompass only a narrow band of energies providing them with an "almost monoenergetic" character.

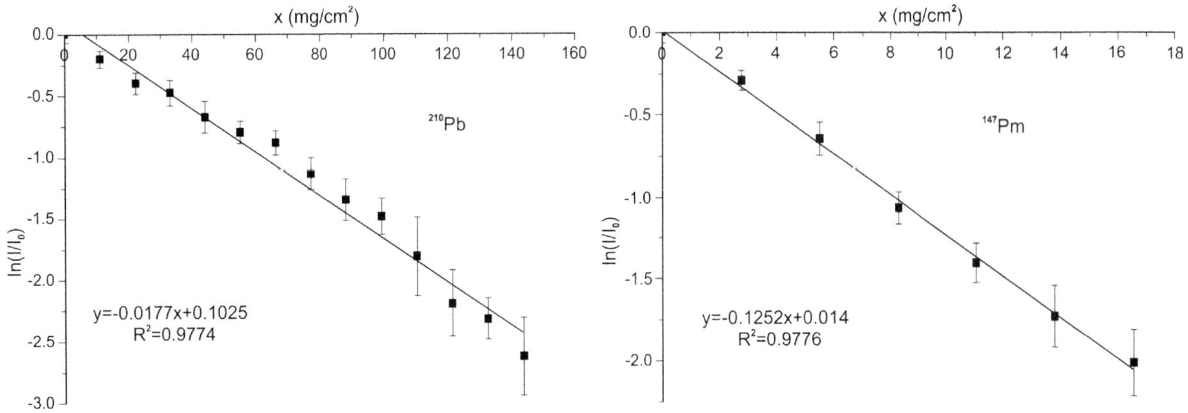

FIGURE 1.66 Beta transmission curves of $\ln(I/I_0)$ versus the thickness of Al absorber for beta emitters ^{210}Pb and ^{147}Pm. *From Şen et al. (2014), reprinted with permission from Elsevier © 2014.*

Based on the exponential character of beta-particle absorption we can describe the transmission of beta particles through the absorber as

$$I = I_0 e^{-\mu x} \tag{1.256}$$

where I is intensity of the beta-particles transmitted through the absorber (e.g., measured as disintegrations per minute, DPM, by a detector), I_0 is the initial intensity of the beta particles (DPM) incident on the absorber, μ is the linear absorption coefficient in units of cm^{-1} and x is the absorber thickness in cm. It is common to express the amount of absorber in mass thickness units, that is, mass per unit area (*e.g.*, g/cm^2), which is the product of absorber thickness and density, as it is easier to measure accurately very thin absorbers simply from their weight. If we express absorber thickness in mass thickness units (*e.g.*, g/cm^2 or mg/cm^2) we can rewrite Eq. (1.256) as

$$I = I_0 e^{-(\mu/\rho)(\rho x)} \tag{1.257}$$

or

$$\frac{I}{I_0} = e^{-(\mu/\rho)(\rho x)} \tag{1.258}$$

and

$$\ln\frac{I}{I_0} = -(\mu/\rho)(\rho x) \tag{1.259}$$

where μ/ρ is the mass absorption coefficient (also referred to as mass attenuation coefficient) in units of cm^2/g, that is, the linear absorption coefficient μ, with units of cm^{-1}, divided by the absorber density ρ with units of g/cm^3; and ρx is the absorber thickness in mass thickness units g/cm^2, that is, the product of the absorber density and absorber thickness. The μ/ρ value for each radionuclide source is determined by the least squares fit method for the linear portion of the beta transmission curve (Thontadarya, 1985).

Eq. (1.258) can be used to determine experimentally the unknown thickness of absorber materials. A standard curve is plotted with the ratio I/I_0 on a logarithmic scale versus mass thickness (ρx with units of mg/cm^2) of the absorber on a linear scale as illustrated in Fig. 1.66. A value for I in units of DPM are determined with a detector by measuring the beta-particle intensity transmitted through a given absorber thickness. This is repeated with absorbers of different thickness. The magnitude of the incident beta particle intensity, I_0, is a constant value and determined with the detector in the absence of absorber. The linear portion of the plot has a negative slope, such as that illustrated in Fig. 1.66, and from least squares analysis the mass attenuation coefficient μ/ρ is determined (Singh and Batra, 1987; Yi et al., 1999). Consequently, the thickness of an unknown similar material can be determined from the measured intensity, I, of the transmitted beta particle radiation after placing the material between the beta particle source and detector without altering the counting geometry. The sample thickness is calculated or determined directly from the aforementioned curve (Tumul'kan, 1991; Clapp et al., 1995).

The radionuclide utilized for a thickness gauge will depend on the maximum energy of the beta emissions and the thickness and atomic number of the absorber. Şen et al. (2014) demonstrate excellent beta transmission curves for the determination of Al absorber thickness, and two examples are provided in Fig. 1.66. Linear beta transmission curves are reported by Şen et al. (2014) with the following beta emitters and Al absorber thicknesses: ^{99}Tc (1–35 mg/cm^2), ^{36}Cl (10–120 mg/cm^2), ^{14}C (1–14 mg/cm^2), ^{210}Pb (10–150 mg/cm^2), and ^{147}Pm(1–17 mg/cm^2).

Jaklevic et al. (1983) describe the detailed schematic arrangement of a plastic scintillator detector, photomultiplier and electronics for a beta gauge system that can be easily set up in any laboratory. Timing measurements may also be used for the determination of linear (μ) or mass attenuation coefficients (μ/ρ) in beta transmission measurements. Ermis and Celiktas (2012) demonstrated timing measurements that focus on the determination of the time that the beta particle crosses the detector, which can be successfully used to measure the attenuation coefficients whereby the beta particle energy has a less significant importance or can even be disregarded.

Beta particle transmission has many practical applications today in industrial manufacturing. Beta particle sources and detectors are placed on the production line to test for thickness, uniformity, and defects in the manufacture of paper, metal, and plastic films (Yoshiyuki and Hiroshi, 1999; Gardner et al., 2004) as well as online inspection of sewn seams or fabric density in the textile industry (Ogando, 1993; Clapp et al., 1995; Mapleston, 1997; Titus et al., 1997; Kim et al., 2009) and in agronomic research to measure leaf water content (Mederski, 1961; Mederski and Alles, 1968; Nakayama and Ehrler, 1964; Jones, 1973; Obregewitsch et al., 1975; Barthakur, 1983) or to measure the biomass of a prairie (Knapp et al., 1985). These are commonly referred to as beta transmission thickness gauges. The beta particle sources used depend on the absorber thickness to be measured and the E_{max} of the beta particles. Three sources commonly used are ^{14}C ($E_{max} = 0.156$ MeV), ^{147}Pm ($E_{max} = 0.224$ MeV), ^{85}Kr ($E_{max} = 0.672$ MeV), ^{36}Cl ($E_{max} = 0.714$ Mev), ^{204}Tl ($E_{max} = 0.763$ MeV), and ^{90}Sr(^{90}Y) in secular equilibrium (E_{max} of ^{90}Sr and ^{90}Y $= 0.546$ and 2.280 MeV, respectively). A source of low beta particle E_{max} (*e.g.*, ^{14}C or ^{147}Pr) is used to measure the finest thickness of material (Jaklevic et al., 1983. Balasubramanian, 1997, 1998), and the sources are changed according to beta particle energy and material density and thickness to be tested. A practical reference for radioisotope thickness gauge measurements is provided by Johansen and Jackson (2004) and Wagner et al. (2014).

C. Internal conversion electrons

Decay by internal conversion (IC) results in the emission of an atomic electron. This electron, called the IC electron, is emitted from an atom after absorbing the excited energy of a nucleus. In this decay process, an unstable nucleus decaying to a lower more stable state transmits its decay energy to an atomic electron, which is emitted from the atom with an energy corresponding to the nuclear decay energy less the binding energy of the atomic electron. This mode of decay accompanies and even competes with gamma ray emission as a deexcitation process of unstable nuclei.

The kinetic energy of the electron emitted is equivalent to the energy lost by the nucleus (energy of transition of the excited nucleus to its ground or lower energy state) less the binding energy of the electron. This is illustrated by the following equation:

$$E_e = (E_i - E_f) - E_b \quad (1.260)$$

where E_e is the kinetic energy of the IC electron, $(E_i - E_f)$ is the energy of transition between the initial, E_i, and the final, E_f, nuclear energies normally associated with gamma ray emission, and E_b is the binding energy of the atomic electron. Lise Meitner and Otto Hahn in Germany (Meitner, 1924; Hahn and Mettner, 1924) were the first to interpret the origin of this mono-energetic electron energy as derived from the IC of the gamma decay energy. Nobel Laureate Luis W. Alvarez at the University of California, Berkeley was the first to provide experimental evidence of IC of artificially produced radioactive atoms (Alvarez, 1938).

An example of radionuclide decay by IC is found in Fig. 1.67, which illustrates the decay of the parent–daughter nuclides ^{109}Cd(^{109m}Ag). Note that the ^{109m}Ag daughter decays by IC with a 96% probability (*i.e.*, 45% for IC from the *K* shell + 48% from the *L* shell + 3% from higher electron shells – the latter is not illustrated in Fig. 1.67) and decay occurs also via gamma emission with the remaining 4% probability (Rachinhas et al., 2000).

Because the emission of internal conversion electrons competes with gamma ray emission as an alternative mode of nuclear deexcitation, many radioactive nuclei that emit gamma radiation will also emit IC electrons. The degree to which this competition occurs is expressed as the total internal conversion coefficient (α), which is the ratio of the

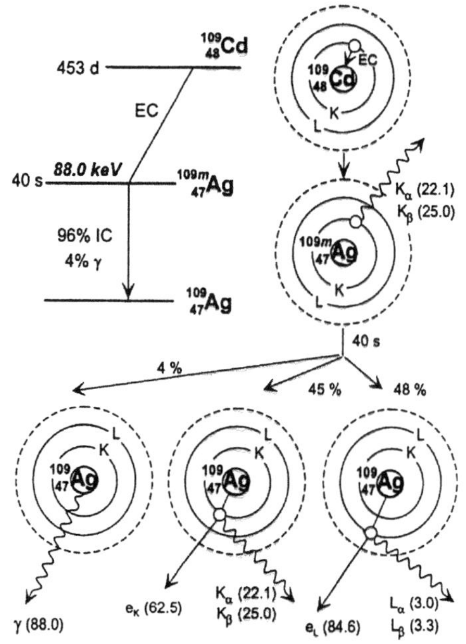

FIGURE 1.67 Decay scheme of ^{109}Cd(^{109m}Ag). The numbers in parenthesis indicate energy values in keV. The electron capture (EC) process occurs from *K*, *L* and outer shells with probabilities of 79%, 17% and 4%, respectively, but only K-capture is represented above. The ^{109m}Ag daughter decays by emission of 88.0 keV gamma rays with a 4% probability or by internal conversion (IC) with the probabilities of 45% and 48% for *K* and *L* shells. IC from shells higher than *L* contributes the remaining 3% (not illustrated). The *K* and *L* IC decay illustrated involve the ejection of a conversion electron with energy $e_K = 62.5$ keV or $e_L = 84.6$ keV, accompanied by the emission of an Ag *K*- or *L*-fluorescence X-ray photon ($K_\alpha = 22.1$, $K_\beta = 25.0$ keV, or $L_\alpha = 3.0$, $L_\beta = 3.3$ keV) or by the emission of Auger electrons (not represented) and X-ray photons following Auger electron emissions. *From Rachinhas et al. (2000), reprinted with permission from Elsevier Science.*

rate of emission of IC electrons to the rate of emission of gamma rays of equivalent energy or

$$\alpha = e/\gamma = N_e/N_\gamma \tag{1.261}$$

where N_e and N_γ are the number of electrons and number of gamma ray photons, respectively, emitted by a radionuclide sample in a given time interval for a given decay energy transition. In other words, the internal conversion coefficient is a quantitative measure of the number of internal conversion electrons divided by the number of gamma ray photons emitted from a radionuclide sample for a given radionuclide energy transition.

Internal conversion electrons may be emitted from specific electron shells of atoms and may be expressed in terms of internal conversion electrons and gamma rays of the same energy less the energy difference resulting from the binding energy of the electron as described by Eq. (1.260). When expressed in terms of electrons emitted from specific shells, the shell-conversion coefficients are written with a subscript denoting the electron shell of origin, for example, α_K or e_K/γ, α_L or e_L/γ, and α_i or e_i/γ, where $i = K, L, M$ ... electron shells, where the total internal conversion coefficient would be the sum of the shell-conversion coefficients or

$$\alpha = \alpha_K + \alpha_L + \alpha_M + \ldots \alpha_i \tag{1.262}$$

where $\alpha_K, \alpha_L, \alpha_M$ and α_i are the shell-conversion coefficients for the L, K, M electrons, etc.

As described by Evans (1955), theoretical values of the IC coefficient depend on the energy of the transition, the atomic number of the transforming nucleus, the atomic shell (K, L_1, L_2,...M_1, M_2, ...) in which conversion takes place, and the electric or multipole order of the transition. Values of internal conversion coefficients are provided in reference tables on isotope decay, such as those found in Lederer and Shirley (1978) and total theoretical IC coefficients based on gamma-ray energy lines are available from the online Nuclear Data Sheets via ScienceDirect (http://www.sciencedirect.com/science/journal/00903752/145?sdc=2). To utilize the data sheets, it is only necessary to specify a title and the mass number of the radionuclide of interest, such as Nuclear Data Sheets for A = (insert mass number) whereby it is necessary to specify a mass number. The Nuclear Data Sheets will provide the nuclear data including values of α of the total theoretical IC coefficients and corresponding gamma ray lines for all isobars of mass number A. For example, inserting the title Nuclear Data Sheets for A = 99 will provide the nuclear data for isobars of A = 99 compiled by Browne and Tuli (2017).

In general, internal conversion coefficients are small for gamma ray–emitting nuclides of low Z and high-energy transitions and larger for nuclides of high Z and low-energy transitions. This relationship is illustrated in Table 1.15, which lists a few radionuclides selected at random as examples in order of increasing Z. As can be seen, large internal conversion coefficients occur when internal conversion electrons are emitted with low-energy nuclear transitions as indicated by the large values of α associated with low gamma ray energies and high Z. Also, the internal conversion electron (e^-) energies listed are slightly lower than the gamma ray energies. This is because the energy of the IC electron is equal to the energy absorbed from the decaying nucleus (transition energy) less the binding energy of the atomic electron described previously in Eq. (1.260). On the other hand, gamma ray energies serve as a measure of the exact quanta of energies lost by a nucleus.

The loss of atomic electrons through the emission of internal conversion electrons leaves vacancies in atomic electron shells. The vacancies are filled by electrons from outer higher-energy shells, whereby there is a concomitant loss of electron energy as X-radiation. Emission of X-radiation resulting from electron filling of vacancies in electron shells (K, L, M ...) is also listed in Table 1.15. This is a process that occurs in the daughter atoms; the X-rays are a characteristic of the daughter rather than of the parent.

Internal conversion electrons are identical in their properties to beta particles. They differ, however, in their origin. Beta particles originate from the nucleus of an atom as a result of beta decay, whereas internal conversion electrons originate from atomic electron shells. A characteristic difference between these two types of electron is their energy spectra. Beta particles, as discussed previously, are emitted from nuclei with a broad spectrum or smear of energies ranging from near zero to E_{max}. However, internal conversion electrons are emitted from the atoms of decaying nuclei with discrete lines of energy of a magnitude equivalent to that of the energy lost by the nucleus less the electron binding energy. The energy of an internal conversion electron can be used to estimate the energy lost by a nucleus.

Like beta particles, internal conversion electrons dissipate their energy by ionization they cause in matter. The abundance of internal conversion electrons emitted from some nuclide samples can be significant and should not be ignored. In certain cases IC can play a significant role in radionuclide detection and measurement. Internal conversion electron energies are slightly lower than the true gamma decay energy because of the energy consumed in the ejection of the bound atomic electron (E_b in Eq. 1.260).

D. Auger and Coster-Kronig electrons

An Auger (pronounced OH-ZHAY) electron can be considered as the atomic analogue of the IC electron. Auger electrons are defined as atomic electrons that are emitted from atoms after acquiring energy from an atomic electron transition within the atom. For example, in the electron

TABLE 1.15 Relationship between gamma radiation and internal conversion electron radiation, e^-, associated with several nuclides listed in order of increasing Z.

Nuclide $^{A}_{Z}X$	Gamma radiation (MeV)[a]	e^- (MeV)	$\alpha = e/\gamma$	X-rays[a]
$^{7}_{4}Be$	0.477(10%)		7.0×10^{-7}	
$^{22}_{11}Na$	1.275(100%)		6.7×10^{-6}	
$^{44}_{22}Ti$	0.068(90%)	0.065	0.12	Sc K
	0.078(98%)	0.073	0.03	
$^{57}_{27}Co$	0.014(9%)	0.013	8.2	Fe K (55%)
	0.122(87%)	0.115	0.02	
	0.136(11%)	0.129	0.15	
$^{64}_{29}Cu$	1.34(0.6%)	1.33	1.3×10^{-4}	Ni K (14%)
$^{87m}_{38}Sr^{b}$	0.388(80%)	0.386	0.21	Sr K (9.4%)
$^{119m}_{50}Sn^{b}$	0.024(16%)	0.020	5.13	Sn K (28%)
$^{125}_{53}I$	0.035(7%)	0.030	13.6	Te K (138%)
$^{129}_{53}I$	0.040(9%)	0.034	22	Xe K (69%)
$^{169}_{68}Er$	0.008(0.3%)	0.006	220	Tm M
$^{181}_{74}W$	0.006(1%)	0.004	46	Ta K (65%)
$^{203}_{80}Hg$	0.279(82%)	0.275	0.23	Tl K (13%)
$^{239}_{94}Pu$	0.039(0.01%)	0.033	461	U K (0.012%)
	0.052(0.02%)	0.047	269	

[a]*Values in percent are radiation intensities or abundances.*
[b]*m denotes a metastable state.*

capture (EC) decay processes, vacancies are left in electron shells (K, L, ...) that can be filled by atomic electrons from higher energy levels. In the process of falling to a lower energy shell to fill a vacancy, electron energy is lost as a photon of X-radiation (see Section XI.D of this chapter). Alternatively the energy liberated in the shift of an electron from its higher energy state to a lower one can be transferred to an electron in an outer shell resulting in the emission of the electron as an Auger electron.

At the young age of 24 years Pierre Victor Auger at the Ecole Normale Supérieure de Paris discovered the emission of electrons from atoms excited by X-rays (Auger, 1923, 1925a,b). Pierre Auger irradiated krypton and argon gas with X-rays and discovered the emission of electrons. He reasoned that an X-ray photon causes the ejection of an atomic electron from a given shell or quantum level leaving a vacancy behind. He further reasoned that an electron from the next outer shell could fill the vacancy causing the emission of electromagnetic radiation that would correspond to the difference in the energy levels of the electrons in the outer and inner shells. For example, Auger noted that, if an electron in the K-shell was ejected by an artificially produced X-ray photon, an electron from the next outer L-shell could fill the vacancy resulting in the emission of energy, E, as electromagnetic radiation, which he described by the following equation:

$$E_{\text{x-ray}} = h\nu = E_L - E_K \tag{1.263}$$

where h is Planck's constant ($h = 4.14 \times 10^{-15}$ eV s $= 6.626 \times 10^{-34}$ J s), ν is the photon frequency in units of s^{-1}, and E_L and E_K are the energy levels of the electrons in the L and K shells, respectively. The transition energy $h\nu$ may be emitted as an X-ray photon characteristic of the atom or, alternatively, the transition energy may be absorbed by an atomic electron resulting in its emission from the atom. The electron emitted in this fashion is identified as an Auger electron in the name of Pierre Auger for his discovery and interpretation of this phenomenon. Lise Meitner (1923) also independently discovered Auger electrons the same year as Pierre Auger; however, the Auger electrons and the phenomenon that gives rise to these electrons, the Auger Effect, are named after Pierre Auger.

Electron transitions, that may result in the emission of Auger electrons, will occur when an atom becomes ionized by the loss of an electron from an inner shell. For example, an atom may become ionized by the ejection of an electron from an inner shell by one of several mechanisms such as (1) irradiation with artificial external X-rays or irradiation with particle beams, (2) the emission of an IC electron such as K-shell IC, or (3) by EC also known as K-capture. The vacancy left in the K-shell can be filled by an electron from an outer L-shell. In turn, the vacancy left in the L-shell could be filled by another electron from the outer M-shell, etc. The process of nonradiative rearrangement of atomic electrons as a result of the ionization of the atom in one of its inner shells is defined as the Auger effect (Borisenko and Ossicini, 2004). In the process of falling to a lower energy shell to fill a vacancy, electron energy may be lost as a photon of X-radiation. The energy of the photon radiation is equivalent to the differences in the energy levels of the electrons in the outer and inner shells as described by the above Eq. (1.263). The X-ray photon may be emitted from the atom, or alternatively the energy released in the electron transition will be transferred to an electron of an outer shell and cause its emission from the atom as an Auger electron. The energy of the Auger electron would be that defined by the above Eq. (1.263) less the binding energy, E_b, of the electron or

$$E_{\text{Auger}} = (E_L - E_K) - E_b \quad (1.264)$$

The energy of the Auger electron is thus equivalent to the energy of the X-ray photon less the binding energy of the electron or

$$E_{\text{Auger}} = E_{\text{x-ray}} - E_b \quad (1.265)$$

Values of the binding energies of electrons in various shells (K, L, M, ...) are found in references texts such as Rumble (2017). Either of two processes, the emission of an X-ray photon or the emission of an Auger electron, can occur as a result of electron energy-level transitions from higher to lower energies. Auger electron emission competes with X-ray emission. An example of the Auger effect and the resultant emission of an Auger electron as compared to X-ray emission are illustrated in Fig. 1.68. The emission of an Auger electron from an atom will occur as a consequence of the following transitions: (1) an atom losses an atomic electron from an inner shell leaving a hole or vacancy in that shell, (2) an electron from a higher energy level fills the vacancy, and (3) the energy emitted in the transition of the electron from the higher to lower energy levels is transferred to an outer atomic electron causing the electron to be emitted from the atom.

Whenever an X-ray photon causes the ejection of an atomic electron, another electron falls from an outer shell to a lower one to fill the vacancy, and there is a cascading effect of electrons falling from yet more distant shells to fill vacancies left behind until the atom reaches the ground or stable state. The downward transitions of electrons in this fashion produce additional X-ray photons of lower energy than the initial X-ray photon. The production of X-ray photons in this fashion is referred to as X-ray fluorescence.

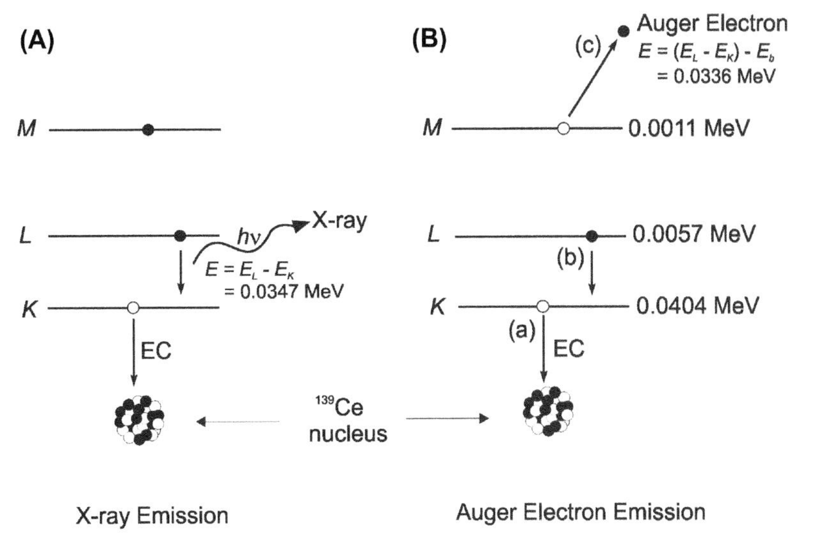

FIGURE 1.68 Decay of the radionuclide ^{139}Ce by electron capture (EC) resulting in (A) X-ray emission or (B) Auger electron emission. An electron vacancy in a shell is illustrated as an empty circle and an electron as a filled circle. Specific binding energies for Cerium atomic electrons in K, L and M shells were obtained from Lide (2010) and are provided along the right-hand side of lines illustrating the various electron energy levels. The Auger effect or process by which an Auger electron is emitted from an atom occurs according to the following sequence as noted in the figure: (a) an atom losses an atomic electron from an inner shell leaving a hole or vacancy, (b) an electron from a higher energy level fills the vacancy, and (c) the energy emitted in the electron transition in step (b) is transferred to an outer atomic electron causing the electron to be emitted from the atom. The energy of the K X-ray illustrated above is equal to the electron transition ($E_{\text{x-ray}} = h\nu = E_L - E_K = -0.0057 - (-0.0404) = 0.0347$ MeV. The energy of the Auger Electron is equal to the electron transition energy ($E_L - E_K$) less the binding energy of the electron (E_b) or, in other words, the X-ray energy less the binding energy of the electron ($E_{\text{Auger}} = 0.0347–0.0011 = 0.0336$ MeV. The daughter nucleus of the ^{139}Ce remains in an excited state after EC and emits a gamma-ray (not illustrated). Also the loss of an Auger electron will leave an electron vacancy, which can be filled by an electron of an outer shell. This will produce another X-ray, which is not illustrated here. The electron filling the vacancy will, in itself leave another vacancy, which could be filled by yet another electron in an outer shell. The production of X-rays by this cascading effect of electron transitions from outer to inner shells is known as X-ray fluorescence.

Auger electron emission competes with X-ray emission, and it can accompany any nuclear decay process that results in the production of X-rays, such as a consequence of EC decay or nuclides that have IC in their decay. Like internal-conversion electron emission described previously, the electron-capture decay process also results in the emission of appreciable quantities of X radiation. Auger electron emission can reduce appreciably the abundance of X-ray emission normally expected to accompany radionuclide decay processes. The two competing processes of Auger-electron emission and X-ray emission are important to consider in the detection and measurement of nuclides that decay by EC. This is measured by both the fluorescence yield and Auger yield. The fluorescence yield is the fraction of vacancies in a given electron shell that is filled with accompanying X-ray emission, and Auger yield is the fraction of vacancies that is filled resulting in the emission of Auger electrons (Friedlander et al., 1964). For example, as described by Burhop and Asaad (1972), an initial vacancy in the *K*-shell may be filled either by the emission of K series X-radiation (fluorescence yield ω_K) or *K* Auger electrons (Auger yield a_K) where

$$\omega_K + a_K = 1 \tag{1.266}$$

Fig. 1.69 illustrates the *K*-shell fluorescence yield as a function of nuclide atomic number. The *L*-shell, which consists of subshells, such as L_1, L_2, and L_3, can exhibit fluoresce yields characteristic of the subshells. The *L*-shell fluorescence yield also varies similarly with atomic number as the *K*-shell fluorescence yield, but it is several times lower in magnitude (Friedlander et al., 1964). When an Auger electron is emitted as a result of an electron transition from a higher subshell to a lower subshell of the same shell, it is referred to as a Coster–Kronig transition, which is a special type of Auger effect named in honor of Dirk Coster and Ralph Kronig from the Netherlands (Coster and Kronig, 1935). The ejected electrons in Coster–Kronig transitions are of very low energy ($\leq$1 keV), and in X-ray spectroscopy, they cause the broadening and lowering of intensities of corresponding X-ray lines.

Auger electron emission energies and intensities, including those arising from Coster–Kronig transitions are measured and accounted for in radionuclide standardization (Grau Carles and Grau Malonda, 2006; Grau Malonda et al., 2006; Bé et al., 2006; Kossert and Grau Carles, 2008, 2010; Grau Carles and Kossert, 2009; Tarancón Sanz and Kossert, 2011; Yunoki et al., 2014; García-Toraño et al., 2017). As described previously both Auger and IC electron emissions possess discrete energies, and these energy lines can be envisaged by the electron emission spectra illustrated in Fig. 1.70. The emission spectra illustrated in Fig. 1.70 provide an excellent picture of the energy magnitudes and monoenergetic character of Auger and IC electrons. Certain radionuclides, such as ^{123}I, ^{125}I, ^{111}In, and ^{119}Sb, which decay by EC with significant intensities of Auger electrons emissions, have received much attention for their applications in cancer therapy (Behr et al., 2000; Janson et al., 2000; Bodei et al., 2003; Capello et al., 2003; Michel et al., 2003; Chen et al., 2006; Constantini et al.,

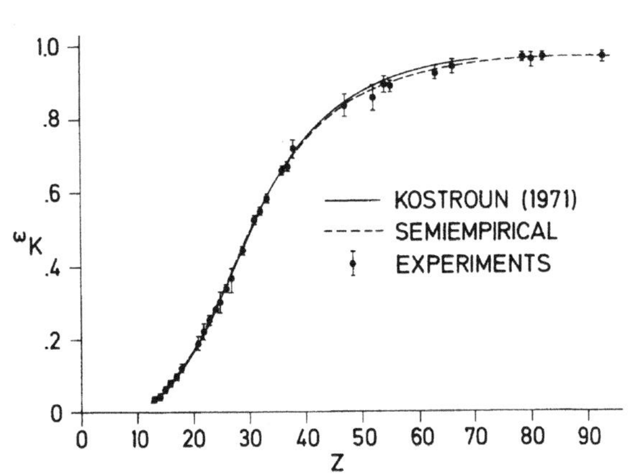

FIGURE 1.69 Fluorescence *K*-shell yields, ω_K, as a function of atomic number, Z: (A) according to Kostroun et al. (1971); (B) a best fit to selected experimental data; and (C) critically evaluated experimental results. *From Bambynek et al. (1972), https://doi.org/10.1103/RevModPhys.44.716, reprinted with permission ©1972 The American Physical Society.*

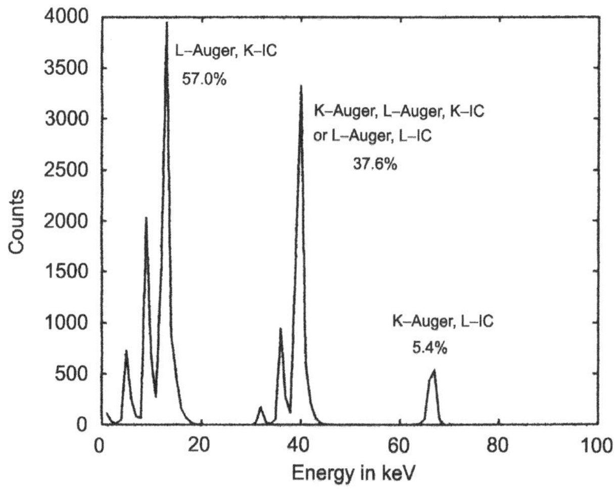

FIGURE 1.70 Simulation of the electron emission spectrum (Auger and IC) of ^{125}I. The 36-keV γ-transition of ^{125}I is highly converted with a probability of 93.3%. As described by Grau Carles and Kossert (2009), three regions characterize the spectrum. The first region below 20 keV comprises about 57% of the spectral events. The peaks are created by the emission of *K*-IC electrons with about 3.7 keV, which may occur in coincidence with one or two *L*-Auger electrons stemming from the subsequent rearrangement processes of the preceding EC decay. Thus, the region shows three sharp peaks. The second region at about 40 keV contains about 37.6% of the counts. It corresponds to the coincident interaction of *K*-IC and *K*-Auger electrons or the interaction of *L*-IC electrons. *K*-Auger electrons may have energies between ~21.8 and 31.8 keV. The energy of *L*-IC electrons is about 31 keV. Again, the emissions can be in coincidence with one or two *L*-Auger electrons yielding the subdivision into three peaks. The third region above 60 keV contains coincidences of *L*-IC and *K*-Auger electron emissions and encompasses ~5.4% of the total spectral events. *From Grau Carles and Kossert (2009) reprinted with permission from Elsevier © 2009.*

2007; Thisgaard and Jensen, 2009; Sedlacek et al., 2014; Zhang et al., 2014; Panosa et al., 2015; Cai et al., 2016; Aghevlian et al., 2017). The low energies of the Auger electrons emitted by these radionuclides, and consequently their short range in the human body, can increase the highly localized radiation-induced kill of cancer cells.

Another auspicious outcome of the discovery of the Auger Effect is the development of Auger electron spectroscopy (AES). The technique of AES involves irradiating the surface of a sample with an electron beam of energy sufficient to ionize the inner orbitals of atoms thereby inducing the concomitant emission of Auger electrons. The Auger-electron energy spectra drawn from the induced emissions serve as fingerprints of different atoms to enable their identification and quantitative analysis (Chourasia and Chopra, 1997; Melhorn, 1998; Vandendael, 2010; Ilyin, 2017).

E. Neutron radiation

The neutron is a neutral particle, which is stable only in the confines of the nucleus of the atom. Outside of the nucleus the neutron decays with a mean lifetime of about 15 minutes. Its mass, like that of the proton, is equivalent to 1 u (atomic mass unit). Unlike the particulate alpha and beta nuclear radiation previously discussed, neutron radiation is not emitted in any significant quantities from radionuclides that undergo the traditional nuclear decay processes with the exception of a few radionuclides such as ^{252}Cf and ^{248}Cm, which decay to a significant extent by spontaneous fission (see Section X.E.3.b). Significant quantities of neutron radiation occur when neutrons are ejected from the nuclei of atoms following reactions between the nuclei and particulate radiation. Due to its lack of charge, the neutron cannot produce directly any ionization in matter, again unlike alpha and beta radiation. The various sources, properties, and mechanisms of interaction of neutrons with matter are described subsequently.

1. Discovery of the neutron

The discovery of the neutron had eluded us until as late as 1932. The particle's neutral charge and high penetrating power when traveling through matter made its discovery more difficult during the early years of research on radioactivity. In 1932, James Chadwick provided evidence for the existence of the neutron (see Chadwick, 1932a,b). He placed a source of alpha particle radiation in close proximity to beryllium. It was known that bombarding beryllium with alpha radiation would produce another source of radiation, which had a penetration power through matter even greater than that of known gamma radiation. Chadwick observed that, when a sheet of paraffin (wax) was placed in the path of travel of this unknown radiation, he could detect a high degree of ionization in a gas ionization chamber caused by protons emitted from the paraffin. This phenomenon would not occur when other materials such as metals and even lead were placed in the path of this unknown radiation. Based on further measurements of the proton velocities and scattering intensities, it was concluded that the unknown radiation had a mass similar to that of the proton, but with a neutral charge. Only a particle with neutral charge would have a high penetration power through matter. A very detailed account of the experimental technique and calculations that Chadwick used to prove the existence of the neutron is given in a previous text by the writer (L'Annunziata, 2016). As noted in the previous discussion of beta particle decay, the neutron possesses a mass similar to that of the proton and, within the nucleus of an atom, the particle is a close union between a proton and an electron.

2. Neutron classification

Neutrons are generally classified according to their kinetic energies. There is no sharp division or energy line of demarcation between the various classes of neutrons; however, the following is an approximate categorization according to neutron energy:

*	Cold neutrons	<0.003 eV
*	Slow (thermal) neutrons	0.003–0.4 eV
*	Slow (epithermal) neutrons	0.4–100 eV
*	Intermediate neutrons	100 eV–200 keV
*	Fast neutrons	200 keV –10 MeV
*	High-energy (relativistic) neutrons	>10 MeV

The slow (thermal) neutrons are classified above within the encompassing energy range of 0.003–0.4 eV; however, many reference tables may provide the thermal neutron cross sections of the elements and their isotopes at the specific thermal neutron energy of 0.0253 eV and a neutron velocity of 2200 m per second, which are neutrons in thermal equilibrium with matter at room temperature (20.43°C) (Holden, 1997b).

The energies of neutrons are also expressed in terms of velocity (meters per second) as depicted in the terminology used to classify neutrons. A neutron of specific energy and velocity is also described in terms of wavelength, because particles in motion also have wave properties. It is the wavelength of the neutron that becomes important in studies of neutron diffraction. The values of energy, velocity, and wavelength of the neutron, as with all particles in motion, are interrelated. The velocity of neutrons increases according to the square root of the energy, and the wavelength of the neutron is inversely proportional to its velocity. Knowing only one of the properties, either the energy, velocity, or wavelength of a neutron, we can calculate the other two. We can relate the neutron energy and velocity using the kinetic energy equation

$$K = \frac{1}{2}mv^2 \quad \text{or} \quad v = \sqrt{2K/m} \tag{1.267}$$

where K is the particle kinetic energy in joules (1 eV = 1.602×10^{-19} J), m is the mass of the neutron (1.674×10^{-27} kg), and v is the particle velocity in meters per second. The wavelength is obtained from the particle mass and velocity according to

$$\lambda = \frac{h}{p} = \frac{h}{mv}, \tag{1.268}$$

where λ is the particle wavelength in meters, h is Planck's constant (6.626×10^{-34} J s), p is the particle momentum, and m and v are the particle mass and velocity as previously defined. The correlation between neutron energy, velocity, and wavelength is provided in Fig. 1.71, which is constructed from the classical Eqs. (1.267) and (1.268) relating particle mass, energy, velocity, and wavelength. However, calculations involving high-energy particles that approach the speed of light will contain a certain degree of error unless relativistic calculations are used, as the mass of the particle will increase according to the particle speed. In Section XII.C of this chapter we used the Einstein equation $E = mc^2$ to convert the rest mass of the positron or negatron to its rest energy (0.51 MeV). When gauging particles in motion, the *total energy* of the particle is the sum of its kinetic (K) and rest energies (mc^2) or

$$E = K + mc^2 = \gamma mc^2 \tag{1.269}$$

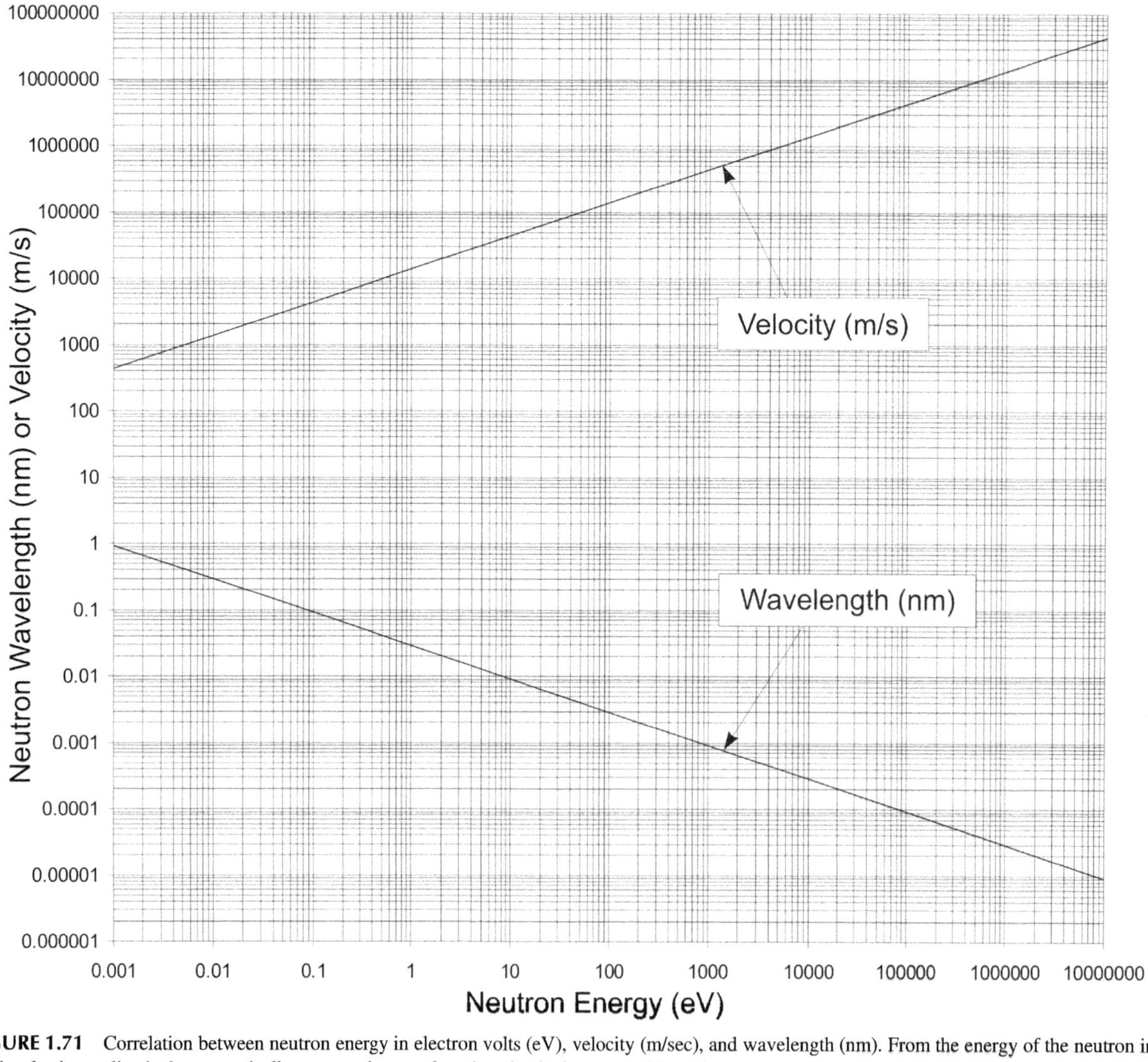

FIGURE 1.71 Correlation between neutron energy in electron volts (eV), velocity (m/sec), and wavelength (nm). From the energy of the neutron in eV on the abscissa, a line is drawn vertically to cross the wavelength and velocity curves. The values of neutron wavelength and velocity are obtained directly from the ordinate. For example, to determine the wavelength and velocity of 0.025 eV thermal neutrons, the value of 0.025 eV is found on the abscissa. A line is then drawn vertically from the point of 0.025 eV to cross the values of 0.18 nm wavelength and 2200 m/sec velocity.

And thus, the kinetic energy of the particle would be the difference between the total energy of the particle and its rest energy or

$$K = \gamma mc^2 - mc^2 \tag{1.270}$$

where

$$\Upsilon = \frac{1}{\sqrt{1-(u^2/c^2)}}, \tag{1.271}$$

u is the particle speed, and $u < c$, which was derived by Einstein (1916a,b). The complete derivation of Einstein is provided in the Appendix of a former book by the writer (L'Annunziata, 2016). If we call the particle rest mass m_0, then the relativistic mass, m_r, which is the speed-dependent mass of the particle is calculated as

$$m_r = \frac{m_0}{\sqrt{1-(u^2/c^2)}} \tag{1.272}$$

and from Eq. (1.269) the kinetic energy of the particle can be written as

$$K = \frac{m_0}{\sqrt{1-(u^2/c^2)}}c^2 - m_0c^2 \tag{1.273}$$

Eq. (1.273) can be transformed to read

$$K + m_0c^2 = \frac{m_0c^2}{\sqrt{1-(u^2/c^2)}} \tag{1.274}$$

and

$$\frac{K + m_0c^2}{m_0c^2} = \frac{1}{\sqrt{1-(u^2/c^2)}} \tag{1.275}$$

and

$$\left(\frac{K}{m_0c^2} + 1\right)^2 = \frac{1}{1-(u^2/c^2)} \tag{1.276}$$

which can be transformed to read

$$\frac{u^2}{c^2} = 1 - \frac{1}{(K/m_0c^2 + 1)^2} \tag{1.277}$$

From Eq. (1.277) the relativistic speed of the particle can be defined as

$$u_r = c\sqrt{1-(K/m_0c^2 + 1)^{-2}} \tag{1.278}$$

where K is the kinetic energy, and the particle speed u is always less than c (Serway et al., 2005). The nonrelativistic speed is that described by Eq. (1.267) or $u_{nr} = \sqrt{2K/m_0}$.

As particles increase in energy and approach the speed of light, the relativistic calculation of the particle speed should be used. To compare the use of nonrelativistic calculations of particle speed for the construction of Fig. 1.71, let us use Eqs. (1.267) and (1.278) to calculate the differences between the nonrelativistic and relativistic speeds of a neutron of 10 MeV kinetic energy. This energy was selected, because it is the highest neutron energy included in Fig. 1.71, and differences between nonrelativistic and relativistic calculations increase with particle energy. The difference between the two calculated speeds is defined by the ratio of the two or

$$\frac{u_{nr}}{u_r} = \frac{\sqrt{2K/m_0}}{c\sqrt{1-(K/m_0c^2 + 1)^{-2}}} \tag{1.279}$$

The rest energy of the neutron, mc^2, is first calculated as

$$\begin{aligned} mc^2 &= \left(1.6749 \times 10^{-27}\ \text{kg}\right)\left(2.9979 \times 10^8\ \text{m/s}\right)^2 \\ &= 1.505\text{x}10^{-10}\ \text{J} \end{aligned}$$

and

$$\frac{1.505 \times 10^{-10}\ \text{J}}{1.602 \times 10^{-19}\ \text{J/eV}} = 939.5\ \text{MeV}$$

since by definition, $1\ \text{eV} = 1.602 \times 10^{-19}$ J. From Eq. (1.279) the ratio of the nonrelativistic and relativistic speeds are calculated as

$$\begin{aligned} \frac{u_{nr}}{u_r} &= \frac{\sqrt{2(10\ \text{MeV})(1.602 \times 10^{-13}\ \text{J/MeV})/1.6749 \times 10^{-27}\ \text{kg}}}{c\sqrt{1-((10\ \text{MeV}/939.5\ \text{MeV})+1)^{-2}}} \\ &= \frac{4.3737 \times 10^7\ \text{m/s}}{0.1447751c} = \frac{4.3737 \times 10^7\ \text{m/s}}{(0.1447751)(2.9979 \times 10^8\ \text{m/s})} \\ &= \frac{4.3737 \times 10^7\ \text{m/s}}{4.340 \times 10^7\ \text{m/s}} = 1.0079 = 0.79\%\ \text{error} \end{aligned}$$

The error between the nonrelativistic and relativistic calculations is small at the 10 MeV neutron energy. However, if we consider higher neutron energies in excess of 10 MeV, the error of making nonrelativistic calculations increases.

As we observed above in the case of particle speed, we will also see that particle wavelength will also differ for nonrelativistic and relativistic calculations. In 1923 Louis Victor de Broglie first postulated that all particles or matter in motion should have wave characteristics just as photons display both a wave and particle character. We therefore attribute the wavelength of particles in motion as de Broglie wavelengths. Let us then compare calculated nonrelativistic and relativistic wavelengths. From Eq. (1.268), we can describe the nonrelativistic wavelength, λ_{nr}, as

$$\lambda_{nr} = \frac{h}{p} = \frac{hc}{pc} = \frac{hc}{cmv} = \frac{hc}{cm\sqrt{2K/m}} = \frac{hc}{\sqrt{2mc^2K}} \tag{1.280}$$

where $p = mv = m\sqrt{2K/m}$. For relativistic calculations the value of pc is calculated according to the following equation described by Halpern (1988):

$$pc = \left[2m_0c^2K\left(1+\frac{K}{2m_0c^2}\right)\right]^{1/2} \quad (1.281)$$

The above Eq. (1.281) is derived as follows:

From Eq. (1.374) of Part XI.A Dual Nature: Wave and Particle, the following expression was derived that defines the relationship between the total energy ($E = mc^2$) and rest energy ($E = m_0c^2$) of a particle:

$$E^2 = p^2c^2 + E_0^2 \quad (1.282)$$

where $p = \gamma mu$ is the relativistic momentum and $\gamma = 1/\sqrt{1-(u^2/c^2)}$, u is the particle speed and $u < c$ as described previously by Eq. (1.271). Because the total energy (E) of a particle is the sum of its kinetic (K) and rest energy (E_0), we can transform Eq. (1.282) to read

$$p^2c^2 = \left(K + m_0c^2\right)^2 - \left(m_0c^2\right)^2 \quad (1.283)$$

which is transposed to read

$$(pc)^2 = K^2 + 2Km_0c^2 + \left(m_0c^2\right)^2 - \left(m_0c^2\right)^2 \quad (1.284)$$

and

$$(pc)^2 = K^2 + 2m_0c^2K \quad (1.285)$$

and

$$(pc)^2 = 2m_0c^2K\left(1+\frac{K}{2m_0c^2}\right) \quad (1.286)$$

Finally Eq. (1.286) can be expressed as

$$pc = \left[2m_0c^2K\left(1+\frac{K}{2m_0c^2}\right)\right]^{1/2} \quad (1.287)$$

And, according to Eq. (1.280), the expression for the relativistic de Broglie wavelength, λ_r, then becomes

$$\lambda_r = \frac{hc}{pc} = \frac{hc}{[2m_0c^2K(1+(K/2m_0c^2))]^{1/2}} \quad (1.288)$$

With Eqs. (1.280) and (1.288) we can compare the difference between the calculated nonrelativistic and relativistic wavelengths for the 10 MeV neutron as follows:

$$\frac{\lambda_{nr}}{\lambda_r} = \frac{hc/\sqrt{2m_0c^2K}}{hc/\left[2m_0c^2K\left(1+\left(K/2m_0c^2\right)\right)\right]^{1/2}}$$

$$= \frac{\left\{\left[(6.626\times10^{-34}\text{ J s})(2.9979\times10^8\text{ m/s})/1.602\times10^{-13}\text{J/MeV}\right]/\sqrt{2(939.5\text{ Mev})(10\text{ MeV})}\right\}}{\left\{[(6.626\times10^{-34}\text{ J s})(2.9979\times10^8\text{ m/s})/1.602\times10^{-13}\text{ J/MeV}]/\sqrt{2(939.5\text{ MeV})(10\text{ MeV})[1+(10\text{ MeV}/2(939.5\text{ MeV}))]}\right\}}$$

$$= \frac{12.3958\times10^{-4}\text{ MeV nm}/\sqrt{18790\text{ MeV}^2}}{12.3958\times10^{-4}\text{ MeV nm}/\sqrt{18890\text{ MeV}^2}}$$

$$= \frac{9.0430\times10^{-6}\text{nm}}{9.0190\times10^{-6}\text{ nm}} = 1.0026 = 0.26\%\text{ error} \quad (1.290)$$

From the above comparison of nonrelativistic and relativistic calculations of neutron wavelength and velocity, we see that the data provided in Fig. 1.71 based on nonrelativistic calculations are valid with less than 1% error for the highest energy neutron included in that figure. However, if we consider higher energies beyond 10 MeV, where we classify the neutron as relativistic, the errors in making nonrelativistic calculations will increase with neutron energy. It will be clearly obvious to the reader that factors in Eq. (1.290) can be canceled out readily and the equation simplified to the following, which provides a quick evaluation of the effect of particle energy on the error in nonrelativistic calculation of the de Broglie wavelength:

$$\frac{\lambda_{nr}}{\lambda_r} = \sqrt{1+\frac{K}{2m_0c^2}} \quad (1.291)$$

where K is the particle kinetic energy in MeV and m_0c^2 is the rest energy of the particle (*e.g.*, 939.5 MeV for the neutron and 0.511 MeV for the beta particle). For example, a nonrelativistic calculation of the wavelength of a 50-MeV neutron would have the following error:

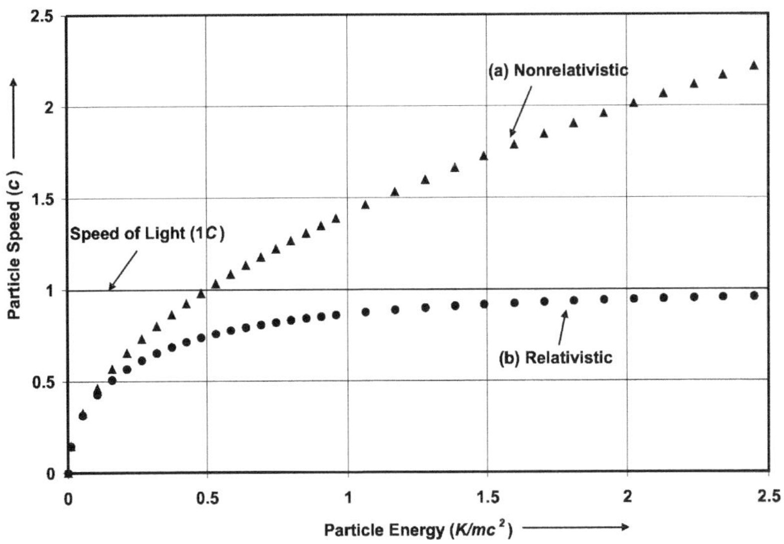

FIGURE 1.72 A graph comparing particle speeds derived from (a) nonrelativistic and (b) relativistic interpretations of the particle speed as a function of the particle kinetic energy. The particle speeds in units of the speed of light ($c = 2.9979 \times 10^8$ m/s) are plotted versus particle energy expressed as a ratio of its kinetic energy (K) over its rest energy (mc^2). In the nonrelativistic case, the particle kinetic energy is defined as $K = 1/2mv^2$ and its speed is calculated according to $v = \sqrt{2K/m}$; whereas in the relativistic case, the particle kinetic energy is the difference between the total energy of the particle and its rest energy ($K = \gamma mc^2 - mc^2$), where $\gamma = 1/\sqrt{1-(u^2/c^2)}$ and its speed $u = c\sqrt{1-(K/mc^2+1)^{-2}}$ is always less than c. *From L'Annunziata (2016); reprinted with permission from Elsevier © 2016.*

$$\frac{\lambda_{nr}}{\lambda_r} = \sqrt{1 + \frac{50 \text{ MeV}}{2(939.5 \text{ MeV})}} = 1.0132 = 1.32\% \text{ error.}$$

Note that the above-computed errors in nonrelativistic calculations of the de Broglie wavelength increased from 0.26% for a 10-MeV neutron to 1.32% for a 50-MeV neutron, and the error will increase with particle energy. Errors in nonrelativistic calculations are yet greater for particles of smaller mass (*e.g.*, beta particles or electrons) of a given energy compared to neutrons of the same energy. This is due obviously to the fact that particles of lower mass and a given energy will travel at higher speeds than particles of the same energy but higher mass. This is illustrated in Fig. 1.72 where the particle speed, u, is a function of the particle kinetic energy, K, and its mass or rest energy, mc^2. The particle energy in Fig. 1.72 is expressed as K/mc^2 to permit the reader to apply the curves for nonrelativistic and relativistic calculations to particles of different mass. For example, from the abscissa of Fig. 1.72, the values of K/mc^2 for a 2-MeV beta particle is 2 MeV/0.51 MeV = 3.9 and that for a 2-Mev neutron is 2 MeV/939.5 MeV = 0.0021. From Fig. 1.72 we see that the nonrelativistic calculation of the speed of a 2-MeV beta particle would be erroneously extreme (well beyond the speed of light), while there would be only a small error in the relativistic calculation of the speed of the massive neutron of the same energy.

TABLE 1.16 Processes for production of neutrons and short description of corresponding spectral fluence.

Process	Example	Neutron fluence energy distribution
Spontaneous fission	$^{252}Cf \rightarrow n + FP$	Maxwellian, $E_{mean} \sim 2$ MeV
Neutron-induced fission	$^{235}U(n,f)$	
Radioisotope (α,n) sources	$\alpha + {}^9Be \rightarrow {}^{12}C + n$	Mono[a] or polyenergetic neutrons
Photon reaction	${}^AX(\gamma,n){}^{A-1}X$	Mono[a] or polyenergetic
	$\gamma + {}^9Be \rightarrow {}^8Be + n$	neutrons (short half-life of γ-ray emitters)
Charged particle direct reaction	${}^2H(d,n){}^3He$	Monoenergetic[a] neutrons
Charged particle breakup reaction	${}^2H(d,np){}^2H$	Broad energy distribution $E_{mean} \sim 0.5\ E_d$
Spallation reaction	${}^{238}U(p,20n)$	Broad energy distribution E_n max $\sim E_p$

[a]*Definition of monoenergetic neutron source from ICRU Report 63 (2000): "A neutron source can be consider 'monoenergetic' if the energy spectrum consists of a single peak, which has an energy spread which is much less than the energy of the peak. The ideal source has a small energy spread and negligible neutron intensity at lower and higher energies."*
From Lacoste (2010) reprinted with Permission of Elsevier © 2010.

3. Neutron sources

Neutron sources are vital to our development and well being. They serve to produce the energy we need and for the production of many of the radionuclides vital to our well-being, such as radioactive sources used in medical diagnosis and treatment of many forms of cancer as well as numerous applications in industry including nondestructive testing and analysis. A review of current practical applications of neutrons and nuclear radiation is provided by the writer in a previous text (L'Annunziata, 2016). There are several sources of neutrons, including (1) alpha particle induced (α,n) nuclear reactions (2) radionuclides that decay by spontaneous fission, (3) neutron-induced fission, (4) photoneutron (γ,n) sources, (5) charged-particle accelerator sources, and (6) nuclear fusion. A basic list of these sources is provided in Table 1.16, and a more detailed description of the sources of neutrons will be provided subsequently. Also, the reader may peruse a review on neutron sources by Furrer (2016).

a. Alpha particle–induced nuclear reactions

It is interesting to note that the method used by Chadwick in 1932 in his discovery of the neutron, namely by an alpha particle–induced reaction, described in Section X.E.1, remains an important method of producing a neutron source, particularly when a relatively small or easily transportable neutron source is required. The source may be prepared by compressing an alpha particle–emitting radioisotope substance with beryllium metal. The nuclear reaction, which occurs between the alpha particle and the beryllium nucleus, terminates with the emission of a neutron and the production of stable carbon as follows

$$^{9}_{4}\mathrm{Be} + ^{4}_{2}\mathrm{He} \rightarrow ^{12}_{6}\mathrm{C} + ^{1}_{0}\mathrm{n} + 5.702\ \mathrm{MeV} \tag{1.292}$$

where 5.702 MeV is the energy released in the reaction. The kinetic energy released in the reaction is determined by calculating the Q value of the reaction according to Eq. (1.137), namely:

$$Q = (m_{\mathrm{reactants}})c^2 - (m_{\mathrm{products}})c^2 \tag{1.293}$$

which for the above reaction is written as

$$Q = (M_{^9\mathrm{Be}} + M_{^4\mathrm{He}})c^2 - (M_{^{12}\mathrm{C}} + M_{^1\mathrm{n}})c^2$$

$$= \begin{bmatrix} (9.01218306\ \mathrm{u} + 4.00260325\ \mathrm{u})931.494\ \mathrm{MeV/u} - \\ (12.000000\ \mathrm{u} + 1.00866492\ \mathrm{u})\ 931.494\ \mathrm{MeV/u} \end{bmatrix}$$

$$= 5.702\ \mathrm{MeV} \tag{1.294}$$

The calculated value agrees with the Q-value Calculator (QCalc) of the National Nuclear Data Center, Brookhaven National Laboratory (NNDC, 2010a,b: http://www.nndc.bnl.gov/qcalc/), which provides a computer-calculated value of 5702 keV. The neutrons emitted from the reaction will have an average energy of ~5.5 MeV, as the kinetic energy released in the reaction is shared between the neutron and ^{12}C nucleus, and the neutron may possess varying energies and velocities as a consequence of the various directions of emission of neutrons from the nucleus.

Several alpha particle sources are used to produce neutrons via the preceding (α,n) reaction. Among these are the alpha emitters ^{241}Am, ^{242}Cm, ^{210}Po, ^{239}Pu, and ^{226}Ra. The alpha radiation source selected may depend on its half-life as well as its gamma ray emissions. As noted previously in this chapter, gamma radiation often accompanies alpha decay. The use of an alpha source, which also emits abundant gamma radiation, requires additional protection for the user against penetrating gamma rays. For example, Am–Be sources may be preferred over the Ra–Be sources of neutrons used in soil moisture probes (Nielsen and Cassel, 1984; O'Leary and Incerti, 1993; Dwevedi et al., 2017), because the latter have a higher output of gamma radiation and require more shielding for operator protection.

The energies of the neutrons emitted from these sources will vary over the broad spectrum of 0–10 MeV. The neutrons produced by these sources vary in energy as a consequence of several factors, including the sharing of the liberated energy between the neutron and the daughter nucleus, the varying directions of emission of neutrons from the nucleus with consequent varying energies and velocities, and the variations in kinetic energies of the bombarding alpha particles.

The neutron activities available from these sources increase up to a maximum as a function of the amounts of alpha emitter and beryllium target material used (Didi, 2017). For example, as explained by Bacon (1969), the Ra–Be source, prepared by mixing and compressing radium bromide with beryllium powder, increases steadily in neutron activity (neutrons per second) for each gram of radium used as the amount of beryllium is increased to about 10 g; but no significant increase in neutron output is achieved if more beryllium is used. The maximum neutron output achieved is approximately 2×10^7 neutrons per second per gram of radium. Because alpha decay from any alpha particle–emitting source occurs by means of random events, the production of neutrons by (α,n) reactions is also a random event. Therefore, these reactions can be referred to as "not time correlated." This is contrary to the case of neutron sources provided through fission, discussed subsequently.

Beryllium is not the exclusive element that may be used for the alpha particle–induced neutron source. Other low Z elements may be packed with an alpha emitter (*e.g.*, boron or carbon); however, their neutron yields are about six times lower than Be (Lacoste, 2010). For example, Skidmore et al. (2009) tested oxygen enriched with ^{18}O packed

TABLE 1.17 **The total flux emitted by the neutron sources per second integrated over energy.**

Neutron source	Integrated flux ($s^{-1}g^{-1}$)	Approximate source Mass required for 100 W thermal power generation (g)
^{252}Cf	2.3×10^{12}	0.05
Am–Be	1.4×10^{6}	2000
Enriched AmO_2	3.7×10^{5}	2000
Pu–Be	7.2×10^{6}	2000

From Skidmore et al. (2009) reprinted with permission of Elsevier © 2009.

with ^{241}Am as a possible neutron energy source for planetary research. The neutron flux and energy yield from this source is compared to conventional sources of alpha emitters with Be as well as the ^{252}Cf spontaneous fission neutron source in Table 1.17.

Switchable neutron generators, that is generators that may be turned on or off, of the (α,n) type are possible when the alpha emitter and target are not packed intimately. The alpha emitter need not be packed intimately with the target material, but both must be close enough to allow the alpha particles to be absorbed by the target. Hertz et al. (2003) studied many combinations of alpha-emitter and target materials for switchable neutron sources, and they found the ^{9}Be(α,n) generator with an ^{241}Am alpha emitting source the most suitable as a switchable neutron source. In the "off" position a mechanical device is used to prevent the alpha particles from reaching the Be target. These switchable neutron sources are classified as micromachines because they can be constructed to have the dimensions of the same order of magnitude as the alpha particle range.

b. Spontaneous fission

About 100 radioisotopes are known to decay by spontaneous fission (SF) with the emission of neutrons (Karelin et al., 1997; Santhosh and Priyanka, 2015) as an alternative to another decay mode, such as alpha decay. SF involves the spontaneous noninduced splitting of the nucleus into two nuclides or fission fragments and the simultaneous emission of more than one neutron on the average. This phenomenon occurs with radionuclides of high mass number, $A \geq 230$. The radionuclide ^{252}Cf is a good example of a commercially available SF neutron source. It decays with a half-life of 2.65 years primarily by alpha emission (96.91% probability), and the remaining of the ^{252}Cf nuclides decay by SF with a probability of 3.09% (Martin et al., 2000, see also Appendix A). Decay of ^{252}Cf by SF produces an average number of 3.7 neutrons per fission. Because the sizes of the two fragments resulting from fission are not predictable, average sizes of the two fragments are determined. Consequently, the numbers of neutrons emitted from individual fissions are not the same, and an average number of neutrons produced per fission is determined. The fission rate of ^{252}Cf is 6.2×10^5 SF/s/μg (Isotope Products Laboratories, 1995). The neutron emission from ^{252}Cf in units of neutrons per second per unit mass is reported to be 2.314×10^6 neutrons/s/μg with a specific activity of 0.536 mCi/μg (Martin et al., 2000; Skidmore et al., 2009). Thus, 1 milligram of ^{242}Cf would emit 2.314×10^9 neutrons per second. If we know the radionuclide specific activity and the % probability of decay by SF, we can calculate the fission rate. For example, taking the specific activity and % probability of SF reported above for ^{252}Cf, we can calculate the fission rate as the product of decay rate and probability of SF per decay or

$$(0.536 \text{ mCi} / \mu\text{g})(3.7 \times 10^7 \text{dps} / \text{mCi})(0.0309)$$

$$= 6.13 \times 10^5 \text{ SF/s/}\mu g \tag{1.295}$$

which is in close agreement with the value cited above. See Section XVIII.A for a discussion of radioactivity units and calculations.

The variations in fission fragment sizes and number of neutrons emitted per fission provide variable neutron energies over the range 0–5.5 MeV with an average neutron energy from ^{252}Cf of approximately 2.3 MeV. Small sources of ^{252}Cf are commercially available for a wide range of applications such as prompt-gamma neutron activation analysis of coal, cement, minerals, interplanetary analysis, petroleum contamination analysis in soil (Mercer et al., 2007; Skidmore et al., 2009; Hadad et al., 2016), moisture distribution analysis in porous media (El Abd et al., 2013), detection of explosives and land mines (Elsheikh et al., 2012), and neutron radiography (Meshkian, 2016). There is an increasing interest in the use of ^{252}Cf neutron sources in radiation therapy or specifically neutron brachytherapy treatment of various types of cancer (Melhus and Rivard, 2006; Wang et al., 2008; Ghassoun et al., 2010; Lei, 2010; Samerdokiene et al., 2015; Ulinskas et al., 2016). Small commercial sources are described by Martin et al. (1997, 2000) among which include 50-mg sources of ^{252}Cf providing a neutron intensity > 10^{11}/s and measuring only 5 cm in length × 1 cm diameter. They report also larger sources of mass >100 mg of ^{252}Cf that approach reactor capabilities for neutrons.

Another standard nuclide source of neutrons is ^{248}Cm, which provides a SF intensity of only 4.12×10^4/s/mg and decays with a half-life of 3.6×10^5 years (Radchenko et al., 2000; Tudora, 2010). The lower neutron flux intensity of this source limits its application, although it has the advantage of a very long half-life providing invariability of sample intensity with time.

Some radionuclides of interest in nuclear energy and nuclear safeguards also decay by SF. The isotopes of plutonium of even mass number, namely ^{238}Pu, ^{240}Pu, and ^{242}Pu, decay principally by alpha particle emission but can also undergo SF to a lesser extent at rates of 1100, 471, and 800 SF/s/g, respectively (Esaka et al., 2017; Redd et al., 2017). The average number of neutrons emitted per fission is between 2.16 and 2.26 of broad energy spectrum (Canberra Nuclear, 1996). Because the neutrons produced with each fission occurrence are emitted simultaneously, we can refer to these emissions as "time correlated." Other isotopes of uranium and plutonium also undergo SF, but at a much lower rate.

c. Neutron-induced fission

When the naturally occurring isotope of uranium, ^{235}U, is exposed to slow neutrons, it can absorb the neutron to form the unstable nuclide ^{236}U. The newly formed nucleus may decay by alpha-particle and gamma ray emission with the long half-life of 2.4×10^7 years. This occurs in approximately 14% of the cases when ^{235}U absorbs a slow neutron. However, in the remaining 86% of the cases, the absorption of a slow neutron by ^{235}U results in the production of the unstable ^{236}U nuclide, which takes on the characteristics of an unstable oscillating droplet (see Fig. 1.18 of Section VI.D.1 of this chapter). This oscillating nuclear droplet with the opposing forces of two positively charged nuclides splits into two fragments, not necessarily of equal size, with the liberation of an average energy of 193.6 MeV. The general ^{235}U fission reaction may be written as

$$^{235}_{92}\text{U} + ^{1}_{0}\text{n} \rightarrow [^{236}_{92}\text{U}] \rightarrow fp \; + \; \nu^{1}_{0}\text{n} + E \qquad (1.296)$$

which represents the reaction of one atom of $^{235}_{92}$U with one thermal neutron $^{1}_{0}$n to yield the release of fission products fp of varying masses plus an average yield of $\nu = 2.42$ neutrons and an overall *average* release of energy $E = 193.6$ MeV (Koch, 1995). The intermediate nuclide ^{236}U is shown in brackets in the above reaction. Most of this energy (over 160 MeV) appears in the form of kinetic energy of the two fission fragments. The remaining energy is shared among the neutrons emitted, with prompt gamma radiation accompanying fission and also beta particles and gamma radiation from decaying fission fragments and neutrinos accompanying beta decay. When a sample of ^{235}U is bombarded with slow neutrons, the fission fragments produced are rarely of equal mass. The ^{236}U intermediate nuclide breaks into fragments in as many as 30 different possible ways, producing, therefore, 60 different nuclide fission fragments. In a review, Koch (1995) provides a list of the fission fragments and their relative abundances as produced in a typical PWR. The most common fission fragments have a mass difference in the ratio 3:2 (Bacon, 1969). Neutrons emitted from this fission process vary in energy over the range 0–10 MeV with an average neutron energy of 2 MeV, which are classified as fast neutrons.

Let us take one of many possible examples of ^{236}U fission to calculate the magnitude of energy liberated. One example of many possible fission reactions, that may be written conserving mass and charge of the ^{236}U nucleus with the emission of three neutrons, is the following:

$$^{235}_{92}\text{U} + ^{1}_{0}\text{n} \rightarrow [^{236}_{92}\text{U}] \rightarrow ^{90}_{38}\text{Sr} + ^{143}_{54}\text{Xe} + 3^{1}_{0}\text{n} \qquad (1.297)$$

The energy liberated in the above fission reaction can be determined from its Q value, which is calculated from Eqs. (1.136) and (1.137) of Section IX.C. as follows:

$$Q = E_{0,\,\text{reactants}} - E_{0,\,\text{products}} \qquad (1.298)$$

where $E_{0,\,\text{reactants}}$ and $E_{0,\,\text{products}}$ are the rest energies of the nuclear reactants and products, respectively. Eq. (1.298) may be written as

$$Q = (M_{\text{reactants}})c^2 - (M_{\text{products}})c^2 \qquad (1.299)$$

where $M_{\text{reactants}}$ and M_{products} are the combined masses of the nuclear reactants and products, respectively, and c is the velocity of light in a vacuum. For the fission reaction 1.297 we can calculate its Q value as

$$\begin{aligned} Q &= (M_{^{235}\text{U}} + M_{^{1}\text{n}})c^2 - (M_{^{90}\text{Sr}} + M_{^{143}\text{Xe}} + 3M_{^{1}\text{n}})c^2 \\ &= \begin{bmatrix} (235.0439282\text{ u} + 1.0086649158\text{ u}) - \\ (89.9077309\text{ u} + 142.935370\text{ u} + 3(1.0086649158\text{ u})) \end{bmatrix} \\ &\quad \times 931.494013\text{ MeV/u} = 170.927\text{ MeV}, \end{aligned} \qquad (1.300)$$

which is in agreement with the computer calculated Q value of 170927 keV determined with QCalc provided by the National Nuclear Data Center, Brookhaven National Laboratory (NNDC, 2018a,b: https://www.nndc.bnl.gov/qcalc/) with atomic mass units obtained from the Atomic Mass Data Center (http://amdc.impcas.ac.cn/web/masseval.html. As noted previously, most of the energy released appears as kinetic energy of the fission fragments (~165 MeV). The neutrons carry off about 2 MeV each on the average. Additional energy is released from fission via gamma radiation and beta decay of the fission products and the neutrino radiation that accompanies beta decay. The total energy released by the neutron-induced fission of ^{235}U, in the example given by Eq. (1.297), to the ultimate decay of the fission products ^{90}Sr and ^{143}Xe to their stable ground states is calculated by determining firstly the decay schemes of the two fission products as follows:

Decay of ^{90}Sr:

$$\begin{aligned} {}^{90}_{38}\text{Sr} &\rightarrow {}^{90}_{39}\text{Y} + \beta^- + \bar{\nu} \\ &\searrow {}^{90}_{40}\text{Zr (stable)} + \beta^- + \bar{\nu} \end{aligned} \quad (1.301)$$

Decay of ^{143}Xe:

$$\begin{aligned} {}^{143}_{54}\text{Xe} &\rightarrow {}^{143}_{55}\text{Cs} + \beta^- + \bar{\nu} + \gamma \\ &\searrow {}^{143}_{56}\text{Ba} + \beta^- + \bar{\nu} + \gamma \\ &\searrow {}^{143}_{57}\text{La} + \beta^- + \bar{\nu} \\ &\searrow {}^{143}_{58}\text{Ce} + \beta^- + \bar{\nu} + \gamma \\ &\searrow {}^{143}_{59}\text{Pr} + \beta^- + \bar{\nu} + \gamma \\ &\searrow {}^{143}_{60}\text{Nd (stable)} + \beta^- + \bar{\nu} + \gamma \end{aligned} \quad (1.302)$$

Thus, the final reaction to the stable nuclides can be written as

$${}^{235}_{92}\text{U} + {}^{1}_{0}\text{n} \rightarrow [{}^{236}_{92}\text{U}] \rightarrow {}^{90}_{40}\text{Zr} + {}^{143}_{60}\text{Nd} + 3{}^{1}_{0}\text{n} + 8\beta^- + 8\bar{\nu} + \gamma \quad (1.303)$$

If we use atomic mass units to calculate the Q value of the above reaction, the eight additional atomic electrons of the ${}^{90}_{40}\text{Zr} + {}^{143}_{60}\text{Nd}$ (100 atomic electrons) over the ${}^{235}_{92}\text{U}$ (92 atomic electrons) force the eight electrons represented by the eight beta particles ($8\beta^-$) to drop out of the equation. Thus, we can calculate the Q value as

$$Q = (M_{^{235}\text{U}} + M_{^1\text{n}})c^2 - (M_{^{90}\text{Zr}} + M_{^{143}\text{Nd}} + 3M_{^1\text{n}})c^2$$

$$= \begin{bmatrix} (235.0439282\text{ u} + 1.0086649158\text{ u}) - \\ (89.90469876\text{ u} + 142.9098199\text{ u} + 3 \times 1.0086649158\text{ u}) \end{bmatrix}$$

$$\times 931.494013\text{ MeV/u} = 197.55\text{ MeV} \quad (1.304)$$

The calculated Q value of 197.55 MeV represents the total energy released in the example of the neutron-induced fission of ^{235}U to the unstable ^{90}Sr and ^{143}Xe fission products, followed by the decay of these fission products, to the stable ^{90}Zr and ^{143}Nd nuclides. This calculated value is close to the estimated value of ~202 MeV using average binding energies of the parent and daughter nuclides found in Section VI.C of this chapter.

Let us take another example of many possible fission possibilities of the neutron-induced fission of ^{235}U. This example is one where the two fission products are of the same mass number, which is the following:

$${}^{235}_{92}\text{U} + {}^{1}_{0}\text{n} \rightarrow [{}^{236}_{92}\text{U}] \rightarrow 2{}^{117}_{46}\text{Pd} + 2{}^{1}_{0}\text{n} \quad (1.305)$$

The fission products in this case are two nuclides of ^{117}Pd and two neutrons. The energy released by the neutron-induced fission is calculated as

$$Q = (M_{^{235}\text{U}} + M_{^1\text{n}})c^2 - (2M_{^{117}\text{Pd}} + 2M_{^1\text{n}})c^2$$

$$= \begin{bmatrix} (235.0439282\text{ u} + 1.0086649158\text{ u}) \\ -(2 \times 116.917955\text{ u} + 2 \times 1.0086649158\text{ u}) \end{bmatrix}$$

$$\times 931.494013\text{ MeV/u} = 185.696\text{ MeV}, \quad (1.306)$$

which agrees with the value of 185696 keV calculated by the online QCalc provided by the National Nuclear Data Center, Brookhaven National Laboratory (NNDC, 2018a,b: https://www.nndc.bnl.gov/qcalc/) with atomic mass units obtained from the Atomic Mass Data Center (http://amdc.impcas.ac.cn/web/masseval.html. The total energy yield, that includes the decay of the radioactive ^{117}Pd, can be calculated after firstly determining the decay scheme of two nuclides of ^{117}Pd to final stable daughter nuclides. The decay scheme is the following:

$$\begin{aligned} 2{}^{117}_{46}\text{Pd} &\rightarrow 2{}^{117}_{47}\text{Ag} + 2\beta^- + 2\bar{\nu} + \gamma \\ &\searrow 2{}^{117}_{48}\text{Cd} + 2\beta^- + 2\bar{\nu} + \gamma \\ &\searrow 2{}^{117}_{49}\text{In} + 2\beta^- + 2\bar{\nu} + \gamma \\ &\searrow 2{}^{117}_{50}\text{Sn (stable)} + 2\beta^- + 2\bar{\nu} + \gamma \end{aligned} \quad (1.307)$$

From the above reaction schemes, the total reaction of the neutron-induced fission of ^{235}U to two nuclides of ^{117}Sn can be written as follows:

$${}^{235}_{92}\text{U} + {}^{1}_{0}\text{n} \rightarrow [{}^{236}_{92}\text{U}] \rightarrow 2{}^{117}_{50}\text{Sn} + 2{}^{1}_{0}\text{n} + 8\beta^- + 8\bar{\nu} + \gamma \quad (1.308)$$

The total energy released from this example of neutron-induced fission of ^{235}U is calculated as

$$Q = (M_{^{235}\text{U}} + M_{^1\text{n}})c^2 - (2M_{^{117}\text{Sn}} + 2M_{^1\text{n}})c^2$$

$$= \begin{bmatrix} (235.0439282\text{ u} + 1.0086649158\text{ u}) \\ -(2 \times 116.9029540\text{ u} + 2 \times 1.0086649158\text{ u}) \end{bmatrix}$$

$$\times 931.494013\text{ MeV/u} = 213.6\text{ MeV} \quad (1.309)$$

with atomic mass units obtained from the Atomic Mass Data Center (http://amdc.impcas.ac.cn/web/masseval.html.

TABLE 1.18 Approximate energy released from neutron-induced ^{235}U fission.

Type	Energy released (MeV)
Prompt energy	
Fission product kinetic energy	165
Neutrons (2.5 n/fission on average, ~2 MeV/n on average)	5
Gamma-ray emissions	7
Energy from fission product decay	
Gamma ray emissions	7
Beta radiation	6
Neutrinos	10
Total	200 MeV

Thus, from the two examples of possible fission reactions taken above (Eqs. 1.303 and 1.308), which include the decay of the fission products to stable nuclides, we calculated the energy released by the two fission processes as 197.5 MeV (Eq. 1.304) and 213.6 MeV (Eq. 1.309). We can see that approximately 200 MeV is the amount of energy released from the two examples taken of the neutron-induced fission of ^{235}U. The magnitude of the energy released from neutron-induced fission of ^{235}U would be an average figure, because the actual amount of energy released would depend on the fission reaction, and there are many possible ways that the ^{236}U could break into two nuclides. Some of the energy arises from fission product radioactivity including gamma ray emission as fission product nuclides decay to lower energy states, beta particle emission from fission product decay, and the emission of neutrinos that accompany beta decay. The energy released can be itemized into the various energy categories listed in Table 1.18, which are tallied to provide an approximate average figure for the energy released in neutron-induced fission of ^{235}U.

Because more than one neutron is released per fission, a self-sustaining chain reaction is possible with the liberation of considerable energy, forming the basis for the nuclear reactor as a principal source of neutrons and energy. In the case of ^{235}U, neutron capture and fission will occur with more efficiency when the neutrons are moderated to reduce their speed to slow neutrons. The nuclear reactor, therefore, will be equipped with a moderator such as heavy water (D_2O) or graphite, which can reduce the energies of the fast neutrons via elastic scattering of the neutrons with atoms of low atomic weight. The protons of water also serve as a good moderator of fast neutrons, provided the neutrons lost via the capture process $^1H(n,\gamma)^2H$ can be compensated by the use of a suitable enrichment of the ^{235}U in the nuclear reactor fuel (Bryne, 1994).

The previously described fission of ^{235}U represents the one and only fission of a naturally occurring radionuclide that can be induced by slow neutrons. The radionuclides ^{239}Pu and ^{233}U also undergo slow neutron-induced fission; however, these nuclides are man-made via the neutron irradiation and neutron capture of ^{238}U and ^{232}Th, as illustrated in the following equations (Murray, 2009). The preparation of ^{239}Pu occurs by means of neutron capture by ^{238}U followed by beta decay as follows:

$$^{238}_{92}U + ^{1}_{0}n \rightarrow ^{239}_{92}U + \gamma \tag{1.310}$$

$$^{239}_{92}U \xrightarrow{t_{1/2}=23.5\text{min}} ^{239}_{93}Np + \beta^- \tag{1.311}$$

$$^{239}_{93}Np \xrightarrow{t_{1/2}=2.35days} ^{239}_{94}Pu + \beta^- \tag{1.312}$$

The preparation of ^{233}U is carried out via neutron capture of ^{232}Th followed by beta decay according to the following:

$$^{232}_{90}Th + ^{1}_{0}n \rightarrow ^{233}_{90}Th + \gamma \tag{1.313}$$

$$^{233}_{90}Th \xrightarrow{t_{1/2}=22.4\text{min}} ^{233}_{91}Pa + \beta^- \tag{1.314}$$

$$^{233}_{91}Pa \xrightarrow{t_{1/2}=27.0days} ^{233}_{92}U + \beta^- \tag{1.315}$$

Nuclides that undergo slow neutron-induced fission are referred to as fissile materials. Although ^{235}U is the only naturally occurring fissile radionuclide, it stands to reason that, if an excess of neutrons is produced in a thermal reactor, it would be possible to produce fissile ^{239}Pu or ^{233}U fuel in a reactor in excess of the fuel actually consumed in the reactor. This is referred to as "breeding" fissile material, and it forms the basis for the new generation of breeder reactors (Murray, 2009).

Other heavy isotopes, such as ^{232}Th, ^{238}U, and ^{237}Np, undergo fission but require capture by fast neutrons of at least 1 MeV energy to provide sufficient energy to the nucleus for fission to occur. These radionuclides are referred to as fissionable isotopes.

d. Photoneutron (γ,n) sources

Many nuclides emit neutrons upon irradiation with gamma or X-radiation; however, most elements require high-energy electromagnetic radiation in the range of 10–19 MeV. The gamma or X-ray energy threshold for the production of neutrons varies with target element. Deuterium and beryllium metal are two exceptions, as they can yield appreciable levels of neutron radiation when bombarded by gamma radiation in the energy range of only 1.7–2.7 MeV. The target material of D_2O or beryllium metal is used to enclose a β^--emitting radionuclide, which also emits gamma rays. The gamma radiation bombards the targets deuterium or beryllium to produce neutrons according to the photonuclear reactions ^{2}H(γ,n)^{1}H and ^{9}Be(γ,n)^{8}Be, respectively. The photoneutron source ^{124}Sb + Be serves as a good example of a relatively high-yielding combination of gamma emitter with beryllium target. The ^{124}Sb gamma radiation of relevance in photoneutron production is emitted with an energy of 1.69 MeV at 50% abundance (*i.e.*, one-half of the ^{124}Sb radionuclides emit the 1.69-MeV gamma radiation following beta decay). A yield of 5.1 neutrons per 106 beta disintegrations per gram of target material has been reported (Bryne, 1994). The half-life ($t_{1/2}$) of ^{124}Sb is only 60.2 days, which limits the lifetime of the photoneutron generator; nevertheless, this isotope of antimony is easily prepared in the nuclear reactor by neutron irradiation of natural stable ^{123}Sb. An excellent example is that reported by Fantidis et al. (2009), whereby a SbBe neutron source was demonstrated to be very useful as a transportable neutron source applicable to a neutron radiography system. Also, Golshanian et al. (2016) demonstrate the use of an ^{124}Sb–Be source for Boron Neutron Capture Therapy (BNCT) for the therapeutic treatment of brain tumors. They demonstrate the use of 150 kCi of ^{124}Sb could yield an epithermal neutron flux of 0.23×10^9 n/cm^2 s, which would permit a treatment time of only 40 minutes for a patient with brain cancer. There is increasing research into the development of photoneutron sources for the treatment of cancer by Boron Neutron Capture Therapy such as the use of tungsten and D_2O as the photon converter and photoneutron target, respectively (Rahmani et al., 2015). The particular advantage of such a neutron source is the ability to effectively have an on/off switching capability because the production of neutrons can be stopped by simply removing the beryllium or deuterium target from the antimony gamma ray source. A unique application of the

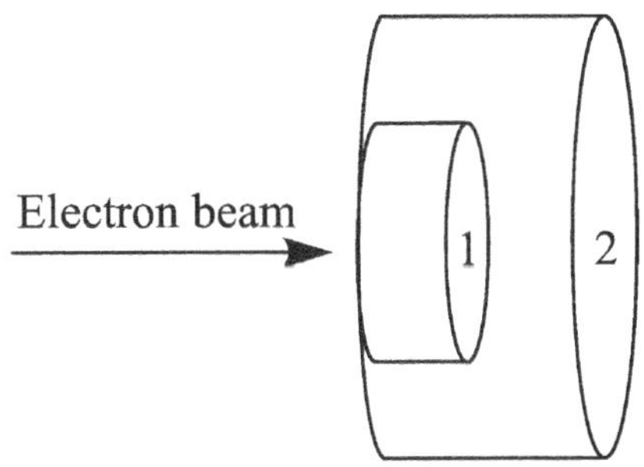

FIGURE 1.73 A photoneutron assembly consisting of (1) a disc of tungsten (W) target embedded within (2) a photoneutron converter consisting of Be or BeD_2. Typical dimensions are 2 × 8 cm (thickness × radius W target) and 10 × 12 cm or 60 × 60 cm Be or BeD_2 target. *From Rahmani and Shahriari (2010). Reprinted with permission from Elsevier © 2010.*

^{9}Be(γ,n)^{8}Be reaction is reported by Moussavi-Zarandi (2008), who used the gamma emissions from an ^{124}Sb source to survey for the Be content of minerals in the exploration of potential beryllium mines. The neutrons emitted from the Be nucleus via the (γ,n) reaction are detected by a BF_3 neutron counter providing a survey meter for Be exploration.

An alternative to a gamma source for the (γ,n) reaction is the use of electrostatic and linear accelerators (LINAC). Commercially available linear accelerators are capable of accelerating electrons against a high *Z* element, such as tungsten, to produce bremsstrahlung radiation. The bremsstrahlung photons interact with a converter material such as Be or BeD_2, which converts the photon beam to a neutron beam via the reactions ^{9}Be(γ,n)^{8}Be and ^{2}H(γ,n)^{1}H. The symbol γ is used here to denote the nuclear reaction projectile although bremsstrahlung photons (X-rays), rather than gamma rays, are the projectiles. This is done out of convenience, as bremsstrahlung and gamma rays are photons which differ only in their origins. The target and converter materials are arranged with respect to the electron beam in a fashion such as illustrated in Fig. 1.73. Alternatively the electron beam may be directed to impinge immediately onto a Be or BeD_2 target material, whereby the bremsstrahlung photons, that are produced, interact directly with the Be or BeD_2 target to yield neutrons (Auditore et al., 2005).

In such an arrangement, as illustrated in Fig. 1.73, neutrons are produced by means of the electron accelerator. The process, as described by Eshwarappa et al. (2007), entails the acceleration of electrons to impinge on a high *Z* target (*e.g.*, tungsten, W), which generates a cascade shower of bremsstrahlung photons with an energy spectrum having an end-point equal to the electron beam energy. The photons, produced in this fashion, travel through a suitable photoneutron target (*e.g.*, Be or BeD_2), producing excited compound nuclei of energy equal to the sum of the binding energy of the last neutron and the energy of the bremsstrahlung X-rays. When the excitation energy is larger than the binding energy of the last neutron in the compound nucleus, a neutron is emitted. Any excess energy is shared as kinetic energy of the neutron and the residual nucleus (Eshwarappa et al., 2007). Essentially the energy conversion follows the path of e−γ−n, that is, the accelerated electron energy is converted to bremsstrahlung photons, which subsequently yields neutrons.

LINACs producing high-energy (15−25 MeV) photon beams are becoming widespread (Yücel et al., 2016; Zolfaghari and Sedaghatizadeh, 2016; Gracanin et al., 2017). Low-energy LINACs (<10 MeV) are found useful for neutron radiography and radiation Boron Neutron Capture Therapy (BNCT) in radiation oncology. A 5 MeV electron LINAC with Be and BeD_2 targets can lead to thermal neutron beams of 8.48×10^7 n/cm^2/s/mA and 1.23×10^8 n/cm^2/s/mA, respectively (Auditore et al., 2005; Hodges et al., 2017).

The use of LINACs for e−γ−n conversions as neutron sources in cancer therapy using Boron Neutron Capture Therapy (BNCT) is under development (Martínez et al., 2010; Rahmani and Shahriari, 2010; Farhad Masoudi and Rasouli, 2015; Alikaniotis et al., 2016; Ghal−Eh et al., 2017). BNCT entails the use of thermal neutrons to kill individual cancer cells (*e.g.*, brain tumors) with minimal damage to cancer-free cells. The principle and technique of BNCT is firstly the administration of a ^{10}B-loaded pharmaceutical to a cancer patient followed by the irradiation of the tumor with thermal neutrons, produced as previously described with the LINAC e−γ−n conversion. The ^{10}B-loaded pharmaceutical concentrates in the cancer cells, and the neutrons react with the ^{10}B to yield ^{7}Li and alpha-particles within the cancer cell according to ^{10}B(n,α)^{7}Li, that is,

$$ {}^{1}_{0}\mathrm{n} + {}^{10}_{5}\mathrm{B}\text{-pharmaceutical} \rightarrow {}^{7}_{3}\mathrm{Li} + {}^{4}_{2}\mathrm{He} \quad Q = 2.79\ \mathrm{MeV} \tag{1.316} $$

The ^{7}Li and alpha particles share the kinetic energy liberated in the reaction. Their high mass attributes a high linear energy transfer (LET), high relative biological effectiveness (RBE) and short range of travel of 4.1 and 7.7 μm for the ^{7}Li and alpha particles, respectively (Eskandari and Kashian, 2009; Tanaka et al., 2009a; Rahmani and Shahriari, 2010). The total range of 11.8 μm is equivalent to the dimensions of a cell in the human body, thus providing BNCT with the advantage of cell selectivity over other methods of radiotherapy.

Accelerated electron-generated neutrons via the (γ,n) reactions have been reported to yield in a uranium target as many as 10^{-2} neutrons per accelerated electron at an electron energy of 30 MeV with a total yield of 2×10^{13} neutrons per second (Bryne, 1994). The use of a uranium target with a 20 MeV linear accelerator as a photoneutron source was demonstrated by Rahmani and Shahriari (2010) and Rahmani et al. (2010, 2015) yielding an epithermal neutron flux of 2.7×10^9 n/cm^{2}s/mA for Boron Neutron Capture Therapy. The accelerator is a good neutron source for the potential generation of nuclear fuels. Some other applications of photoneutrons produced by X-rays from electron accelerators are the detection of explosives by the measurement of photoneutron-induced gamma rays (Yang et al., 2007; McFee et al., 2013), photoneutron assay of highly enriched uranium via neutron-induced fission (Lakosi and Nguyen, 2008; Lakosi et al., 2011), and the elemental analysis of short-lived activation products (Patil et al., 2010a,b).

e. Charged-particle accelerator sources

The accelerator utilizes electric and magnetic fields to accelerate beams of charged particles such as protons, electrons, and deuterons into target materials. Nuclear reactions are made possible when the charged particles have sufficient kinetic energy to react with target nuclei. Some of the reactions between the accelerated charged particles and target material can be used to generate neutrons.

When electrons are accelerated, they gain kinetic energy as a function of the particle velocity. This kinetic energy is lost as Bremsstrahlung electromagnetic radiation when the accelerated electrons strike the target material. Bremsstrahlung radiation is described in Section XI.D.2 of this chapter. It is the Bremsstrahlung photons that interact with nuclei to produce neutrons according to the mechanisms described in the previous section under photoneutron (γ,n) sources.

Accelerated deuterons can be used to produce high neutron yields when deuterium and tritium are used as target materials according to the reactions ^{2}H(d,n)^{3}He and ^{3}H(d,n)^{4}He, respectively. In the deuterium energy range of 100−400 keV, it is possible to obtain neutron yields of the order of 10^{11} neutrons per second from these (d,n) reactions with relatively small electrostatic laboratory accelerators (Bryne, 1994; Kim et al., 2008; Eskandari and Kashian, 2009; Park et al., 2009; Zheng et al., 2016; Liu et al., 2016, 2018). The ^{2}H(d,n)^{3}He and ^{3}H(d,n)^{4}He reactions are referred to as the D-D and D-T neutron sources. When low-energy deuterons are used as projectiles, the D-D and D-T reactions yield 2.4 and 14.1 MeV fast monoenergetic neutrons, respectively (Eskandari and Kashian, 2009; Tracz et al., 2009; Sharma et al., 2010; Liu et al., 2016). The Q values for the D-D and D-T reactions are 3.27 and 17.58 MeV, respectively, which explains the energy of the neutrons liberated in these reactions. The remaining energy not carried away by the neutrons remains with the residual ^{3}He and ^{4}He nuclei of the D-D and D-T reactions. The two (d,n) reactions are the same as the fusion reactions discussed in the subsequent section of this chapter. The low energies of the deuteron projectiles required for the D-D and D-T reactions are advantageous when small accelerators are used in facilities that utilize neutrons for radiation therapy such as BNCT described in the previous sections of this chapter. Neutrons of higher energy (14–30 MeV) may be obtained from the D-T reactions by increasing the deuteron projectile energy, as the neutron energy is a function of the deuteron energy (Eskandari and Kashian, 2009). By means of a low-energy (350 keV) deuteron beam and a tritium target of 0.37 TBq of ^{3}H, Ochiai et al. (2007) report the D-T reaction yielding a neutron fluence of 1.5×10^{11} n/s, which was determined by counting the 3.5 MeV alpha-particles. The neutron flux from this generator could be increased 10-fold by increasing the thickness of the ^{3}H target (Csikai, 1987; Eskandari and Kashian, 2009).

By accelerating deuterons and protons to energies between 1.8 and 4 MeV, neutron generation using beryllium and lithium targets via the following reactions: ^{9}Be(d,n)^{10}B, ^{7}Li(p,n)^{7}Be, and ^{9}Be(p,n)^{9}B are routine (Bayanov et al., 2009a,b, 2010; Eskandari and Kashian, 2009; Feinberg et al., 2009; Iwashita et al., 2011). A higher energy 30 MeV proton beam against a beryllium target was used by Tanaka et al. (2009b), and the fast neutrons were moderated for boron-neutron capture therapy.

The acceleration of protons in the energy range of ~0.5–5 GeV against targets of high atomic number will result in nuclear spallation with the emission of typically 20–30 high-energy neutrons per incoming proton, depending on the energy of the proton beam and the nature of the target material (Samec et al., 2009; Krása et al., 2010). Even higher yields of 40–50 neutrons emitted per proton projectile are possible at the very high proton energies of 4 and 5 GeV (Krása et al., 2010). Nuclear spallation is the process that occurs when a nucleus of high atomic number emits a large number of neutrons following impact by a high-energy projectile. Such reactions are abbreviated often with the notation (p,*x*n) which signifies a proton projectile with the emission of an *x* number of neutrons (see Table 1.16). Such high-power neutron sources can yield a neutron flux of up to 10^{14} n/cm^{2} s at the surface of the source. Some typical targets of high mass number, which have favorable spallation-neutron production characteristics, are Hg, W, ^{238}U, Pb, and Th (Aydin et al., 2008; Fomin et al., 2015; Aguilar et al., 2017; Nowicki et al., 2017). The potential of utilizing neutron sources for the generation of nuclear fuels were noted previously in Section X.E.3.c. and research in this field is described by Kulikov et al. (2016).

When protons are accelerated at lower energies of 20–60 MeV against targets of intermediate mass number (*e.g.*, Cr and Cu), the spallation-neutron production characteristics are less favorable, that is, the yield of number of neutrons emitted per proton projectile is lower. Aydin et al. (2008) measured the spallation reaction cross sections for protons in the energy range of up to 100 MeV against target nuclei consisting of isotopes of Cr and Cu. The neutrons emitted per proton projectile ranged from (p,n) at the low 20–30 MeV proton energies to (p,4n) at the higher 50–60 MeV proton energies. Thus, a cyclotron capable of accelerating protons to 20–60 MeV was found suitable for the production of neutrons from these target materials.

f. Thermonuclear fusion

The fusion of two atomic nuclei into one nucleus is not possible under standard temperature and pressure. The repulsing coulombic forces between the positive charges of atomic nuclei prevent them from coming into the required close proximity of 10^{-15} m before they can coalesce into one. However, as described by Kudo (1995) in a review on nuclear fusion, if temperatures are raised to 100 million degrees, nuclei can become plasmas in which nuclei and electrons move independently at a speed of 1000 km/s, thereby overcoming the repulsing forces between nuclei. Nuclear fusion reactors or controlled thermonuclear reactors (CTRs) are under development to achieve nuclear fusion as a practical energy source. The reactors are based on maintaining plasmas through magnetic confinement as described previously in this chapter (Part VI.C.2). Some fusion reactions also produce neutrons.

The energy liberated during nuclear fusion is derived from the fact that the mass of any nucleus is less than the sum of its component protons and neutrons. This is because protons and neutrons in a nucleus are bound together by strong attractive nuclear forces discussed previously in Section VI.C. of this chapter. As described by Eq. (1.36), earlier in this chapter, the total energy of the nucleus with its bound nucleus is always less than the total energy of the separate protons and neutrons, and this difference is the binding energy of the nucleus. The binding energy is equivalent to the energy required to pull a bound system apart, such as a bound nucleus, leaving its component protons and neutrons free of attractive forces. From Eq. (1.36) we can see that, if it is possible to overcome the repulsive forces of protons in nuclei and fuse the nuclei into a new nucleus of lower mass, energy will be liberated.

Nuclear fusion reactions of two types emit neutrons, and these are of prime interest in man-made CTRs. The first type is fusion between deuterium and tritium nuclei according to

$$^{2}_{1}\mathrm{H} + ^{3}_{1}\mathrm{H} \rightarrow ^{4}_{2}\mathrm{He} + ^{1}_{0}\mathrm{n} + 17.58\ \mathrm{MeV} \tag{1.317}$$

and the other type involves fusion between two deuterium nuclei according to either of the following equations, which have approximately equal probabilities of occurring (Kudo, 1995):

$$^{2}_{1}\mathrm{H} + ^{2}_{1}\mathrm{H} \rightarrow ^{3}_{2}\mathrm{He} + ^{1}_{0}\mathrm{n} + 3.27\ \mathrm{MeV} \tag{1.318}$$

and

$$^{2}_{1}\mathrm{H} + ^{2}_{1}\mathrm{H} \rightarrow ^{3}_{1}\mathrm{H} + ^{1}_{1}\mathrm{H} + 4.04\ \mathrm{MeV} \tag{1.319}$$

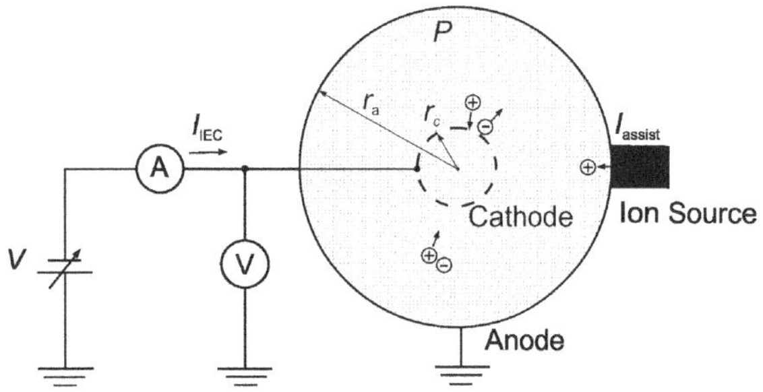

FIGURE 1.74 A schematic diagram of a spherical IECF device and its geometrical and electrical parameters, namely, r_a = anode radius (17.10 cm), r_c = cathode radius (3 cm), P = gas pressure (0.07–0.2 Pa), V = discharge voltage (60–120 kV), I_{IEC} = discharge current (2 mA–4 A), and I_{assist} = injected ion current (2–200 mA). The central inner cathode, illustrated as a dashed circle, consists as a spherical wire grid. *From Noborio et al. (2006) reprinted with permission of Elsevier @ 2006.*

The fusion reaction between deuterium and tritium or D-T reaction (Eq. 1.317) gives rise to a 14.06-MeV neutron and a 3.52-MeV alpha particle. The D-D fusion reaction of Eq. (1.318) gives rise to a 0.82 MeV ^{3}He nucleus and a 2.45 MeV neutron; while the D-D fusion reaction of Eq. (1.319) gives rise to a 1.01 MeV ^{3}H nucleus and 3.03 MeV proton (^{1}H).

g. Inertial electrostatic confinement fusion

Under development are compact neutron sources, which utilize either D-D or D-T fusion reactions as well as D-^{3}He fusion. One instrument described by Miley and Sved (1997) is the inertial electrostatic confinement fusion (IECF) device, which accelerates deuteron ions producing fusion reactions as the ions react with a pure deuterium or deuterium-tritium plasma target. The device is compact measuring 12 cm in diameter and provides a neutron flux of 10^6–10^7 2.5-MeV D-D n/s or 10^8–10^9 14-MeV D-T n/s. Advances in the development of IECF devices yield higher neutron fluxes greater than 10^{10} n/s range, and they present no or only a minimal safety concerns when turned off (Sztejnberg Gonçalves-Carralves and Miller, 2015; Ohnishi et al., 2016; Lang et al., 2018). Many designs have been developed for various practical applications as on/off neutron sources. Some of the basic designs will be described briefly in the following paragraphs.

As described by Noborio et al. (2006) IECF involves a basically simple device, which can cause the fusion reaction, and it can be used as a compact neutron-beam generator. A basic IECF device is illustrated in Fig. 1.74. The device consists of a spherical grounded anode that surrounds a much smaller inner spherical grid cathode. Ions are generated by glow discharge in low pressure deuterium gas (<~2 Pa), and the ions are accelerated toward the center region by a high electric field. Within the center cathode, concentrated energetic ions collide with each other or with background neutral gases resulting in the D–D fusion reaction or D–T fusion when tritium is also included as a fuel.

As described by Tomiyasu et al. (2010), ions are accelerated to fusion-relevant energies by the electric field between concentric electrodes and circulate inside the outer anode through the inner gridded cathode. The inner cathode consists of a spherical wire grid, which is highly transparent to ions, as illustrated in Fig. 1.75. The objective, as described by Khachan and Samarian (2007) is to create a deep potential well in order to trap ions and confine them for a long enough time, for which reason the concept was called inertial electrostatic confinement (IEC). The inner cathode spherical wire grid may be of many designs including a wire grid design such as that illustrated in Fig. 1.75 and even buckeyball-shaped spherical cathodes that resemble the geometry of the C-60 "buckeyball" fullerene (Wulfkühler and Tajmar (2016).

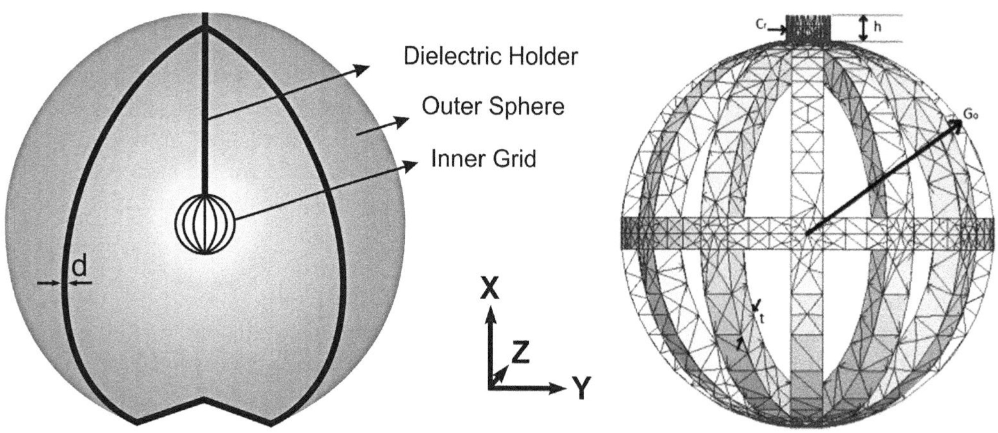

FIGURE 1.75 Basic design of a spherical IECF chamber. (Left) A cut-away view of a spherical outer chamber consisting of an anode sphere of 50 cm radius with an inner spherical cathode grid of 3.0 cm radius suspended in the center by a 7-mm-thick dielectric holder with inner cathode rod. The dielectric holder has a total length (h) of 50 cm, and the outer sphere has a skin depth (d) 5 mm. (Right) The mesh structure of the inner cathode sphere with six vertical ring structures. The grid radius G_o is 3 cm and the skin depth of the wire grid (t) is 5 mm. For the color version of the figure, the reader is referred to the online or electronic version of the book. *From Kurt and Arslan (2012); reprinted with permission from Elsevier © 2012.*

The first effective IECF device of the spherical design was reported by Hirsh (1967), who obtained a neutron output of ~ 10^8 D–D n/s and 10^{10} D–T n/s from a grided IECF device driven by six ion guns. In addition to the D–D and D–T fusion reactions as neutron sources described by Eqs. (1.317) to (1.319), the IECF device is also a portable proton source yielding 10^8 protons/s utilizing the D–^{3}He fusion reaction as described by Ashley et al. (2000), Masuda et al. (2006) and Buzarbaruah et al. (2015). In summary, the fusion reactions that may yield neutrons and/or protons are the following:

$$\text{D} - \text{D fusion:}\quad \begin{cases} {}^2_1\text{H} + {}^2_1\text{H} \rightarrow {}^3_2\text{He}(0.82\ \text{MeV}) + {}^1_0\text{n}(2.45\ \text{MeV})50\%\ Q = 3.27\ \text{MeV} \\ {}^2_1\text{H} + {}^2_1\text{H} \rightarrow {}^3_1\text{H}(1.01\ \text{MeV}) + {}^1_1\text{H}(3.03\ \text{MeV})50\%\ Q = 4.04\ \text{MeV} \end{cases} \tag{1.320}$$

$$\text{D} - \text{T fusion}: {}^2_1\text{H} + {}^3_1\text{H} \rightarrow {}^4_2\text{He}(3.52\ \text{MeV}) + {}^1_0\text{n}(14.06\ \text{MeV})\ \ Q = 17.58\ \text{MeV} \tag{1.321}$$

$$\text{D} - {}^3\text{He fusion}: {}^2_1\text{H} + {}^3_2\text{He} \rightarrow {}^4_2\text{He}(3.7\ \text{MeV}) + {}^1_1\text{H}(14.7\ \text{MeV})\ \ Q = 18.4\ \text{MeV} \tag{1.322}$$

The D–D fusion, as illustrated in Eq. (1.320), may occur via two possible reactions that have equal probabilities of occurrence yielding either 2.45 MeV neutrons or 3.03 MeV protons (Lang et al., 2018). The D–T and D–^{3}He fusion reactions yield high-energy (14 MeV) neutrons or protons respectively, according to Eqs. (1.321) and (1.322).

Although the concept of inertial electrostatic confinement fusion had an early beginning, only in the last several years has intense research been devoted to making this a compact portable neutron and proton source. In a review by Tomiyasu et al. (2010) two approaches have been taken to improve the IECF reaction rates, namely, (1) to increase the ion density by increasing the input current (Yamauchi et al., 2006; Ohnishi et al., 2007; Radel et al., 2007) and (2) to increase the ion energy by reducing the background gas pressure, which reduces the energy loss due to collisions with background neutral ions (Higashi et al., 2003; Piefer et al., 2005; Miley et al., 2005; Takamatsu et al., 2006; Buzarbaruah et al., 2015).

Tomiyasu et al. (2010), Dursun and Kurt (2016) and Dursun et al. (2017) proposed cylindrical chambers with inner cathode and outer anode with magnetic-assisted electrostatic confinement whereby the ion motion can be controlled, as illustrated in Fig. 1.76. With the application of an appropriate magnetic field, the ions are forced to an azimuthal direction by Lorentz force as they are accelerated toward the center of the electric field. Gridless cathode designs in IECF chambers are under research, as grided cathodes lead to heating, thermionic electron emission, power loss and even ultimate grid melting and these are reviewed by Knapp (2015).

In addition to efforts to increase the reaction rates achieved with the IECF device, research is conducted toward the practical applications of the device as a portable neutron or proton source in light of its simple configuration and compact size. Applications of such a portable neutron and proton source include the detection and measurement of highly enriched uranium (Radel et al., 2007), the detection of explosives including landmine detection (Yoshikawa et al., 2007a,b and Takahashi et al., 2010), isotope production (Cipiti and Kulcinski, 2005; Ohnishi et al., 2016), and medical applications including neutron irradiation facilities for Boron Neutron Capture Therapy in

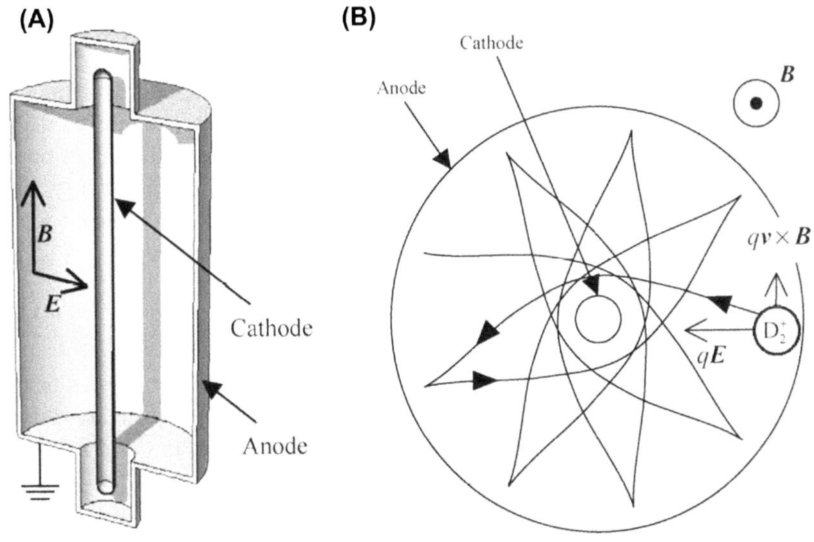

FIGURE 1.76 Schematic of (A) magnetic-assisted electrostatic confinement device, and (B) the motion of deuterium molecular ion under the influence of the applied axial magnetic field. *From Tomiyasu et al. (2010), Reprinted with permission of Elsevier © 2010.*

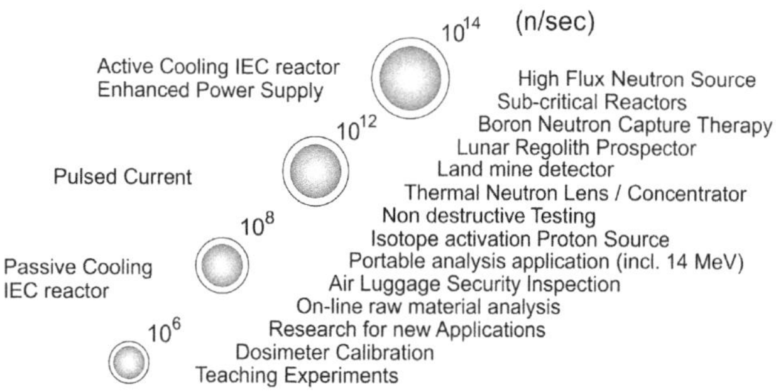

FIGURE 1.77 Applications of Inertial Electrostatic Confinement Fusion (IECF) devices. *From Yoshikawa et al. (2007a). Reprinted with permission from Elsevier © 2007.*

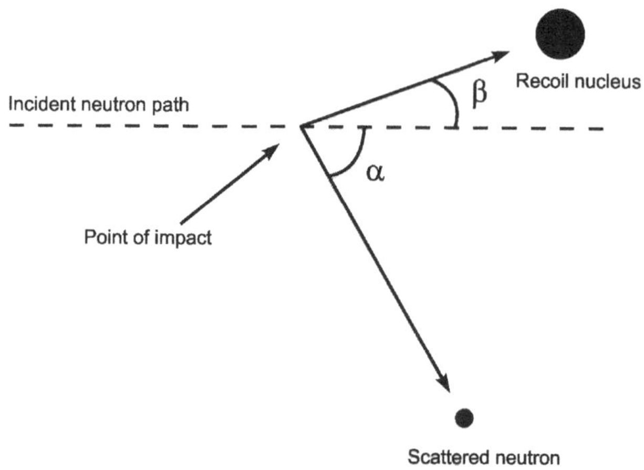

FIGURE 1.78 Elastic scattering of a neutron by collision of the neutron with an atomic nucleus. The neutron is scattered at an angle α and the nucleus recoils at an angle β to the direction of travel of the incident neutron.

the treatment of various cancers (Sztejnberg Gonçalves-Carralves and Miller, 2015). Because of the simplicity and compact size of the IECF device, Yoshikawa et al. (2007a) list numerous applications subject to the expected improvements in reactions rates achieved by this neutron/proton source. These applications are listed in Fig. 1.77.

4. Interactions of neutrons with matter

If a neutron possesses kinetic energy, it will travel through matter much more easily than other nuclear particles of similar energy, namely, alpha particles, negatrons, positrons, protons, or electrons. In great contrast to other nuclear particles, which carry charge, the neutron, because it lacks charge, can pass freely through the Coulombic barrier of the atomic electrons and actually collide with nuclei of atoms and be scattered in the process or be captured by the nucleus of an atom. Collision of neutrons with nuclei can result in scattering of the neutrons and recoil nuclei with conservation of momentum (elastic scattering) or loss of kinetic energy of the neutron as gamma radiation (inelastic scattering). The capture of a neutron by a nucleus of an atom may result in the emission of other nuclear particles from the nucleus (nonelastic reactions) or the fragmentation of the nucleus into two (nuclear fission). A brief treatment of the various types of neutron interactions, which are based on their scattering or capture by atomic nuclei, is provided next.

a. Elastic scattering

The elastic scattering of a neutron by collision with an atomic nucleus is similar to that of billiard-ball collisions. A portion of the kinetic energy of one particle is transferred to the other without loss of kinetic energy in the process. In other words, part of the kinetic energy of the neutron can be transferred to a nucleus via collision with the nucleus, and the sum of the kinetic energies of the scattered neutron and recoil nucleus will be equal to the original kinetic energy of the colliding neutron. This process of interaction of neutrons with matter results only in scattering of the neutron and recoil nucleus. It does not leave the recoil nucleus in an excited energy state. Elastic scattering is a common mechanism by which fast neutrons lose their energy when they interact with atomic nuclei of low atomic number, such as hydrogen (^{1}H) in light water or paraffin, deuterium (^{2}H) in heavy water, and ^{12}C in graphite, which may be encountered in nuclear reactor moderators. It is easy to conceptualize what would occur when particles of equal or similar mass collide; the event would result in energy transfer and scattering without any other secondary effects, similar to what occurs in billiard-ball collisions.

Neutron scattering is the principal mechanism for the slowing of fast neutrons, particularly in media with low atomic number. Let us consider what occurs when a neutron collides with a nucleus and undergoes elastic scattering. Fig. 1.78 illustrates the direction of travel of an incident neutron with given kinetic energy (dashed line). The neutron collides with the nucleus. The nucleus is illustrated as undergoing recoil at an angle β, while the neutron is scattered at an angle α to the direction of travel of the incident neutron. The kinetic energy (E_k) lost by the neutron in this collision is defined by the equation

$$E_k = \frac{4Mm_n}{(M+m_n)^2}\cos^2\beta \tag{1.323}$$

where M is the mass of the nucleus, m_n is the mass of the neutron, and β is the recoil angle of the nucleus. A derivation of Eq. (1.323) is provided by Bacon (1969). Let us look at two extreme examples of elastic collisions between a neutron and a nucleus. In the first example, it is intuitively obvious from Eq. (1.323), that for a recoil angle β = 90

TABLE 1.19 The maximum fraction of the kinetic energy (E_k) that a neutron can lose upon collision with the nucleus of various atoms listed in increasing mass in atomic mass units (u).

Nuclide	Nuclide mass, M	Neutron mass, m_n	$E_k = \left(4Mm_n / (M+m_n)^2\right)\cos^2\beta$
^{1}H	1.007825	1.008665	4.065566/4.066232 = 0.999 or 99.9%
^{2}H	2.014102	1.008665	8.126217/9.137120 = 0.89 or 89%
^{9}Be	9.012182	1.008665	36.36109/100.41737 = 0.362 or 36.2%
^{12}C	12.000000	1.008665	48.41592/169.22536 = 0.286 or 28.6%
^{16}O	15.994915	1.008665	64.53404/289.12173 = 0.223 or 22.3%
^{28}Si	27.976927	1.008665	112.87570/840.16454 = 0.134 or 13.4%
^{55}Mn	54.938047	1.008665	221.65633/3130.0329 = 0.071 or 7.1%
^{197}Au	196.96654	1.008665	787.86616/39194.175 = 0.020 or 2.0%

degrees, $\cos^2\beta = 0$, and consequently $E_k = 0$. Under such a circumstance, the neutron is undeflected by the nucleus and there is no energy transfer to the nucleus. The neutron continues along its path undeflected until it encounters another nucleus. For the second case, however, let us consider the other extreme in which the recoil angle, $\beta = 0°$, where we have a head-on collision of the neutron with the nucleus of an atom. In this case, the maximum possible energy of the neutron is imparted to the nucleus, where $\cos^2\beta = 1$. For example, Table 1.19 provides the maximum fraction of the kinetic energy, calculated according to Eq. (1.323), that a neutron can lose upon collision with various atomic nuclei. As illustrated in Table 1.19, the neutron can transfer more energy to the nuclei of atoms, which have a low mass, and the highest fraction of its energy can be transferred to the nucleus of the proton, which is almost equal in mass to the neutron. Nuclides of low mass number are, therefore, good moderators for the slowing down of fast neutrons. The substances often used are light water (H_2O), heavy water (D_2O), paraffin (C_nH_{2n+2}), and graphite (C).

b. Inelastic scattering

We may picture a fast neutron colliding with a nucleus. The neutron is scattered in another direction as described in the previous paragraph; however, part of the neutron's kinetic energy is lost to the recoil nucleus, leaving it in an excited metastable state. Inelastic scattering can occur when fast neutrons collide with nuclei of large atomic number. The recoil nucleus may lose this energy immediately as gamma radiation or remain for a period of time in the excited metastable state. Inelastic scattering occurs mainly with fast neutron collisions with nuclei of large atomic number.

Neutron scattering is a common mechanism by which fast and intermediate neutrons are slowed down to the thermal neutron energy levels. Thermal neutrons have an energy level at which they are in thermal equilibrium with the surrounding atoms at room temperature. There is an energy range for thermal neutrons, as described earlier in this chapter; however, the properties of thermal neutrons are often cited at an energy calculated to be the most probable thermal neutron energy of 0.0253 eV at 20.43°C corresponding to a velocity of 2200 m/s (Gibson and Piesch, 1985 and Holden, 1997a,b). Fig. 1.71, illustrated earlier in this chapter, may be used to find the velocity of the neutron at energy levels over the range 0.001 eV−10 MeV. For example, if we select the position 0.025 eV on the X axis and follow up the graph with a straight line to the upper curve, we find the value 2200 m/s. At the thermal energy state, the mechanisms of interaction of neutrons with matter change drastically as discussed in the following section.

c. Neutron capture

Because of the neutral charge on the neutron, it is relatively easy for slow neutrons, regardless of their low kinetic energy, to "find themselves" in the vicinity of the nucleus without having to hurdle the Columbic forces of the atom. Once in close proximity to nuclei, it is easy for slow neutrons to enter into and be captured by nuclei to cause nuclear reactions. The capture of thermal neutrons, therefore, is possible with most radionuclides, and neutron capture is the main reaction of slow neutrons with matter. The power of a nucleus to capture a neutron depends on the type of nucleus as well as the neutron energy. The neutron absorption cross section, σ, with units of 10^{-24} cm^2 or "barns," is used to measure the power of nuclides to absorb neutrons. A more detailed treatment of reaction cross section and its units and application was given earlier in Section IX.D, and this is provided further on in Section X.E.5 of this chapter. Because the capture of thermal neutrons is possible with most radionuclides, references

TABLE 1.20 Cross Sections σ in barns for thermal neutron capture reactions of selected nuclides in order of increasing magnitude.

Nuclide	σ (barns)
$^{3}_{1}H$	<0.000006
$^{16}_{8}O$	0.00019
$^{2}_{1}H$	0.00052
$^{12}_{6}C$	0.0035
$^{1}_{1}H$	0.332
$^{14}_{7}N$	1.8
$^{238}_{92}U$	2.7
$^{232}_{90}Th$	7.4
$^{55}_{25}Mn$	13.3
$^{233}_{92}U$	530
$^{235}_{92}U$	586
$^{239}_{94}Pu$	752
$^{6}_{3}Li$	940
$^{10}_{5}B$	3840
$^{3}_{2}He$	5330
$^{7}_{4}Be$	39,000
$^{155}_{64}Gd$	61,000
$^{157}_{64}Gd$	254,000

Data from Holden (1997a,b) and Lide (2010).

will cite the neutron capture cross sections of the nuclides for comparative purposes at the thermal neutron energy of 0.0253 eV equivalent to a neutron velocity of 2200 m/s. This is also the energy of the neutron, which is in thermal equilibrium with the surrounding atoms at room temperature. For comparative purposes, therefore, Table 1.20 lists the cross sections for thermal neutron capture reactions in barns (10^{-24} cm^2) for several nuclides. The nuclides selected for Table 1.20 show a broad range of power for thermal neutron capture. Some of the nuclides listed have practical applications, which are referred to in various sections of this book.

The capture of a slow neutron by a nucleus results in a compound nucleus, which finds itself in an excited energy state corresponding to an energy slightly higher than the binding energy of the neutron in the new compound nucleus. This energy of excitation is generally emitted as gamma radiation. Neutron capture reactions of this type are denoted as (n,γ) reactions. Two practical examples of (n,γ) neutron capture reactions were provided earlier in this chapter in the neutron irradiation of ^{238}U and ^{232}Th for the preparation of fissile ^{239}Pu and ^{233}U (Eqs. 1.310 to 1.315), respectively. Another interesting example of a (n,γ) reaction is the neutron capture by ^{235}U according to the following:

$$^{235}_{92}U + ^{1}_{0}n \rightarrow ^{236}_{92}U + \gamma \qquad (1.324)$$

This neutron capture reaction is interesting because the ^{236}U product nuclide decays by alpha emission with a very long half-life (2.4×10^7 y) in ~14% of the cases and decays by nuclear fission with emission of neutrons in the remaining 86% of the cases, as discussed previously in Section X.E.3.c. The subject of neutron capture is treated in more detail in Section X.E.5, which concerns the neutron cross section and neutron attenuation in matter.

d. Nonelastic reactions

Neutron capture can occur in nuclei resulting in nuclear reactions that entail the emission of nuclear particles such as protons (n,p), deuterons (n,d), alpha particles (n,α), and even neutrons (n,2n). These reactions may not occur in any specific energy range but may be prevalent at specific resonances, which are energy states of the excited compound nuclei that are specific to relatively narrow energies of the incident neutron. The effect of resonance in neutron capture by nuclei is discussed in more detail subsequently in Section X.E.5. The (n,2n) reactions occur at very high incident neutron energies, >10 MeV (Gibson and Piesch, 1985). The (n,p) and (n,α) reactions can occur in the slow neutron capture and reaction with nuclides of low atomic number (low Z), where the Coulomb forces of the electron shells are limited and present less a hurdle for the escape of charged particles from the confines of the atom. Some practical examples of these reactions are the (n,p) reaction used in the synthesis of ^{14}C by slow (thermal) neutron capture by ^{14}N

$$^{14}_{7}N + ^{1}_{0}n \rightarrow ^{14}_{6}C + ^{1}_{1}H \qquad (1.325)$$

and the (n,p) and (n,α) reactions used to detect neutrons by the interaction of slow neutrons with ^{3}He and ^{10}B, respectively, according to Eqs. (1.326) and (1.327).

$$^{3}_{2}He + ^{1}_{0}n \rightarrow ^{1}_{1}H + ^{3}_{1}H + 0.76\ \text{MeV} \qquad (1.326)$$

$$^{10}_{5}B + ^{1}_{0}n \rightarrow ^{7}_{3}Li + ^{4}_{2}He + 2.8\ \text{MeV} \qquad (1.327)$$

Either of these reactions is used to detect neutrons by using gas proportional detectors containing helium-3 gas or a gaseous form of boron (*e.g.*, boron trifluoride). Slow neutrons that penetrate these detectors produce either radioactive tritium (Eq. 1.326) or alpha particles (Eq. 1.327), which produce ionization in the gas. The ionization events or ion pairs formed can be collected and counted to determine a neutron count rate.

e. Nuclear fission

The reaction of neutron-induced fission occurs when a neutron interacts with a fissile or fissionable nucleus and the nucleus becomes unstable, taking on the characteristics of an oscillating liquid droplet, which then fragments into two nuclides (fission fragments). At the same time, there is the release of neutrons (2.4 neutrons on the average for neutron-induced ^{235}U fission) and a relatively high amount of energy ($\sim$194 MeV for neutron-induced ^{235}U fission). Neutron-induced fission of natural ^{235}U and man-made ^{233}U and ^{239}Pu is optimal at thermal incident neutron energies; whereas, neutron-induced fission in ^{232}Th, ^{238}U and ^{237}Np requires neutron energies of at least 1 MeV. A more detailed treatment of nuclear fission was provided previously in Sections VI.C, VI.D, and X.E.3.c of this chapter.

5. Neutron attenuation

The neutron, because of its neutral charge, has a relatively high penetration power in matter. Unlike charged particles, neutrons will not be impeded by the atomic electrons of the material through which the neutrons travel, and the neutron will consequently not lose energy via ionization. However, neutrons will interact with the nuclei of matter. As discussed previously in this chapter, there are several possible interactions of neutrons with nuclei. Among these are elastic scattering, inelastic scattering, neutron capture, nonelastic reactions, and neutron-induced nuclear fission. As we have seen in several examples, probabilities exist for any of these interactions to occur depending on the energy of the incident neutron and the mass of the nuclide with which the neutron interacts. Atoms of low mass (e.g., ^{1}H, ^{2}H, ^{9}Be, ^{12}C) have a higher probability of slowing-down neutrons than atoms of high mass, as seen previously

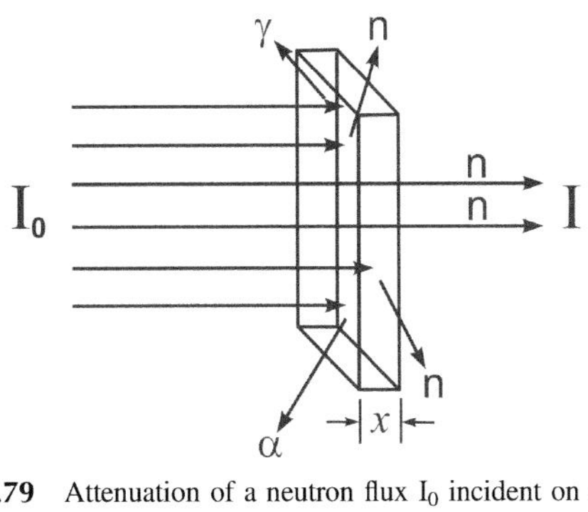

FIGURE 1.79 Attenuation of a neutron flux I_0 incident on a sample of thickness x. The neutron flux, after passing through the sample is reduced in intensity (I) as a result of interactions with atomic nuclei of the sample. Several interactions of neutrons with atomic nuclei are possible (e.g., elastic and inelastic scattering, neutron capture, and nonelastic reactions including (n, γ), (n, p), (n, α), and (n, d), among others. The reactions of neutrons with the nuclei will depend on the incident neutron energy, the magnitude of the cross section of a particular interaction of the nuclide with which the neutron interacts, and the density of the nuclide in the sample (nuclei/cm^3).

from the data of Table 1.19. We can define this probability of interaction of neutrons with specific nuclei by the magnitude of the cross section of the interaction, which is a measure of the power of interaction of a particular nucleus for neutrons of a specific energy.

The transmission of neutrons in matter is a function of the neutron energy and the cross section or interaction power of the atoms of the material through which the neutrons travel. The cross section then is a measure of the probability that a particular neutron interaction will occur, as described previously in Part IX.D of this chapter. Fig. 1.79 illustrates various types of interaction of neutrons with atomic nuclei as the neutrons travel through a target material. We can consider an incident beam of neutrons of given intensity or flux (I_0), which is defined in units of neutrons/cm^2•s). The incident neutrons impinge on a material of unit area (*e.g.*, cm^2) and thickness x, as illustrated in Fig. 1.79.

Earlier in Part IX.D of this chapter, Eqs. (1.148) and (1.149) were described whereby the rate of reaction of the particles of an incident particle beam with respect to the incident particle beam intensity is proportional to the effective target area (cm^2) and total target area that the nuclei present to the incident particles, which is expressed as

$$\frac{\text{Rate of reaction (particles/s)}}{\text{Beam intensity (particles/s)}} = \frac{\text{Effective target area (cm}^2\text{)}}{\text{Total target area (cm}^2\text{)}} \quad (1.328)$$

In accord with the Eq. (1.328) and the previously described Eq. (1.149) (Part IX.D of this chapter), we can express the rate of reaction as the change in the intensity of incident particles (dI) with respect to the incident particle intensity (I) to be proportional to the effective target area. i.e., effective nuclei and its numbers in the target area ($N\sigma Adx$) and the total target area (A) or or

$$-\frac{dI}{I} = \frac{N\sigma Adx}{A} = N\sigma dx \quad (1.329)$$

where the negative sign signifies the diminishing particle beam intensity. Thus, the rate of change of the neutron flux or intensity (dI) with respect to the incident neutron intensity (I) will be proportional to the number of effective nuclei (N) per unit volume in the material (# of nuclei/cm^3) and the "area of obstruction" (*e.g.*, cm^2) or nuclei cross section of obstruction that the nuclei present to the incident neutron flux. Eq. (1.329) can be integrated over the limits of beam intensity from I_0 to I and absorber thickness from 0 to x as follows:

$$\int_{I_0}^{I} dI/I = -N\sigma \int_0^x dx \quad (1.330)$$

to give the equation

$$\ln(I/I_0) = -N\sigma x \quad (1.331)$$

or

$$I = I_0 e^{-N\sigma x} \quad (1.332)$$

Eq. (1.332) is generally written as

$$I = I_0 e^{-\Sigma x} \quad (1.333)$$

where, as described in an early research paper by Vandecasteele et al. (1973), I is the neutron flux (n/cm^2 s) after passing through a sample of thickness x (cm), I_0 is the incident neutron flux (i.e., the neutron flux without the sample in its path), and Σ is the macroscopic cross section (cm^{-1}) for neutrons at a given neutron energy. The macroscopic cross section (Σ) embodies the effective target area of all of the nuclei in the target volume; whereas, the microscopic cross section (σ) embodies the effective target area of a single target nucleus. Because target areas generally contain different nuclides, Eq. (1.333) is written as

$$I = I_0 e^{-\Sigma_t x} \quad (1.334)$$

where $\Sigma_t = N\sigma_t$ is the total macroscopic cross section, as described by Tellili et al. (2014), with units of cm^{-1}, where N is the number of nuclei per unit volume (e.g., # of nuclei/cm^3) of target material, and σ_t is the total microscopic cross section with units in barns where 1 barn = 10^{-24} cm^2. The macroscopic cross section, Σ_t, has units of cm^{-1}, and it is calculated from the following

$$\Sigma_t = N\sigma_t = \left(\frac{\delta N_A}{A}\right)\sigma_t \quad (1.335)$$

where δ is the density of the target material (g/cm^3), N_A is Avogadroe's number (6.022×10^{23} molecules or nuclei/mol), A is the molar mass (g/mol), that is the nuclide mass in units of g/mol or gram molecular weight (g/mol) of a molecular compound, and σ_t is the total microscopic cross section of a nucleus in the target material at a particular neutron energy. The term

$$\frac{\delta N_A}{A} = N \quad (1.336)$$

is the number of nuclei of a given nuclide per unit volume (# of nuclei/cm^3), as described by Vandecasteele et al. (1973). The magnitude of σ_t is a measure of the probability that a neutron of a given energy interacts with the atoms of the material through which the neutrons travel, and it is defined as the sum of the microscopic cross section for neutron scattering (σ_s) and the microscopic cross section for neutron absorption (σ_a),

$$\sigma_t = \sigma_s + \sigma_a \quad (1.337)$$

as described by Tellili et al. (2014). In the case of a mixture of nuclides in the target material, the total macroscopic cross section, Σ_t, is the sum of the macroscopic cross sections of all of the different nuclides or

$$\Sigma_t = N_a\sigma_a + N_b\sigma_b + N_c\sigma_c + \ldots \quad (1.338)$$

where N_a, N_b, and N_c are the number of nuclei of each element per unit volume (# of nuclei/cm^3), σ_a, σ_b, and σ_c are the total microscopic cross sections of each of the different nuclides in the sample, as described in reviews by Tellili et al. (2014) and Elmahroug et al. (2015).

Microscopic cross sections (σ) in units of barns, where 1 barn = 10^{-24} cm^2, for specific nuclides as a function of neutron energy are available from tables or plots of cross section versus neutron energy online from the Brookhaven National Laboratory, National Nuclear Data Center (NNDC) (http://www.nndc.bnl.gov/sigma/index.jsp?as=1&lib=endfb7.1&nsub=10). The tables of microscopic cross sections for specific nuclides (e.g., ^{1}H, ^{2}H, ^{10}B, ^{12}C, etc.) and specific neutron energies in eV are separated into various types such as the total cross section or cross sections for specific neutron interactions or reactions with a nuclide such as elastic interactions, inelastic interactions, nonelastic reactions and specific reactions such as (n,γ), (n,p), (n,d), and (n,α), among others.

To take an example of a calculation of neutron attenuation, we can consider a case where the absorber material consists of only one pure nuclide, and calculate the attenuation of an incident neutron flux by pure aluminum. Elemental aluminum consists of 100 atom % of the nuclide ^{27}Al, and we will calculate the attenuation of thermal neutrons (0.0253 eV) by aluminum and determine the

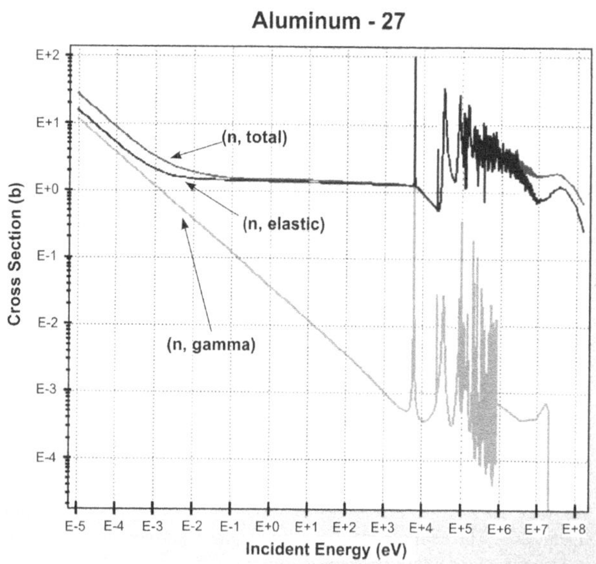

FIGURE 1.80 Cross section plots for neutron interactions in Aluminum-27 over the neutron energy range 10^{-5} to 10^8 eV. Three plots are illustrated from top to bottom, namely, total cross section (n, total), elastic interaction cross section (n, elastic), and neutron capture cross section (n, gamma). *From the Brookhaven National Laboratory (BNL), National Nuclear Data Center (NNDC) online data http://www.nndc.bnl.gov/sigma/index.jsp?as=Elemental&lib=endfb7.1&nsub=10.*

thickness (x) of aluminum required to attenuate an incident neutron flux by one-half or 50%. The total microscopic cross section in units of barns where 1 barn $= 1 \times 10^{-24}$ cm^2 for the most common nuclides and a wide range of neutron energies from 10^{-5}–10^{8} eV can be obtained from the Internet online tables of the Brookhaven National Laboratory (BNL) National Nuclear Data Center (NNDC) Website: http://www.nndc.bnl.gov/sigma/index.jsp?as=1&lib=endfb7.1&nsub=10). In the case of ^{27}Al, the total microscopic cross section (σ_t), identified as "n, total" on the BNL website, is 1.69 barns, which is equivalent to 1.69×10^{-24} cm^2. See also Fig. 1.80, which provides plots of the microscopic cross sections for neutron elastic scattering (n, elastic), neutron capture (n, gamma), and the total mictoscopic cross section (n, total). If we have an incident neutron flux of 5×10^{14} neutrons/cm^2•s, we can calulate the thickness, x, of aluminum necessary to attenuate the neutron flux as follows:

From Eq. (1.334) we can write

$$I = I_0 e^{-\Sigma_t x} = I_0 \exp-(N\sigma_t)x = I_0 \exp-\left(\frac{\delta N_A \sigma_t}{A}\right)x \tag{1.339}$$

Eq. (1.339) can be transposed to read

$$\ln\frac{I_0}{I} = \left(\frac{\delta N_A \sigma_t}{A}\right)x \tag{1.340}$$

If we assign to the incident unattenuated neutron flux I_0, the value of unity (i.e., $I_0 = 1$), we should assign the value of the attenuated neutron flux intensity $I = 0.5$ to determine the value of the thickness of the medium x when the incident neutron flux intensity is reduced by one-half. In this case, the ratio of $I_0/I = 1/0.5 = 2$, and we can then solve for the value of the thickness of aluminum required to attenuate the neutron flux by 50% according to Eq. (1.340) as

$$\ln\frac{I_0}{I} = \left(\frac{\delta N_A \sigma_t}{A}\right)x$$

$$\ln 2 = \left[\frac{(2.70\ \text{g/cm}^3)(6.022 \times 10^{23}\ {}^{27}\text{Al atoms/mol})\left(1.69 \times 10^{-24}\ \text{cm}^2/{}^{27}\text{Al atom}\right)}{26.98\ \text{g/mol}}\right]x$$

$$0.693 = \left[\left(6.026 \times 10^{22}\ {}^{27}\text{Al atoms/cm}^3\right)\left(1.69 \times 10^{-24}\ \text{cm}^2/{}^{27}\text{Al atom}\right)\right]x$$

$$0.693 = \left(0.1018\ \text{cm}^{-1}\right)x$$

$$x = 6.81\ \text{cm}$$

If we, in turn want to determine the magnitude of the attenuation of a certain incident neutron flux with respect to the thickness of the material through which the neutrons must travel, we can use Eq. (1.339). Taking the above calculated aluminum thickness of $x = 6.81$ cm, we can calculate the attenuation of an incident neutron flux $I_0 = 5 \times 10^{14}$ neutrons/cm^2•s as follows

$$I = I_0 \exp-\left(\frac{\delta N_A \sigma_t}{A}\right)x$$

$$I = \left(5 \times 10^{14}\ \text{n/cm}^2\cdot\text{s}\right)\exp-\left[\frac{(2.70\ \text{g/cm}^3)(6.022 \times 10^{23}\ {}^{27}\text{Al atoms/mol})\left(1.69 \times 10^{-24}\ \text{cm}^2/{}^{27}\text{Al atom}\right)}{26.98\ \text{g/mol}}\right]6.81\ \text{cm}$$

$$I = \left(5 \times 10^{14}\ \text{n/cm}^2\cdot\text{s}\right)(0.50)$$

$$I = 2.5 \times 10^{14}\ \text{n/cm}^2\cdot\text{s}$$

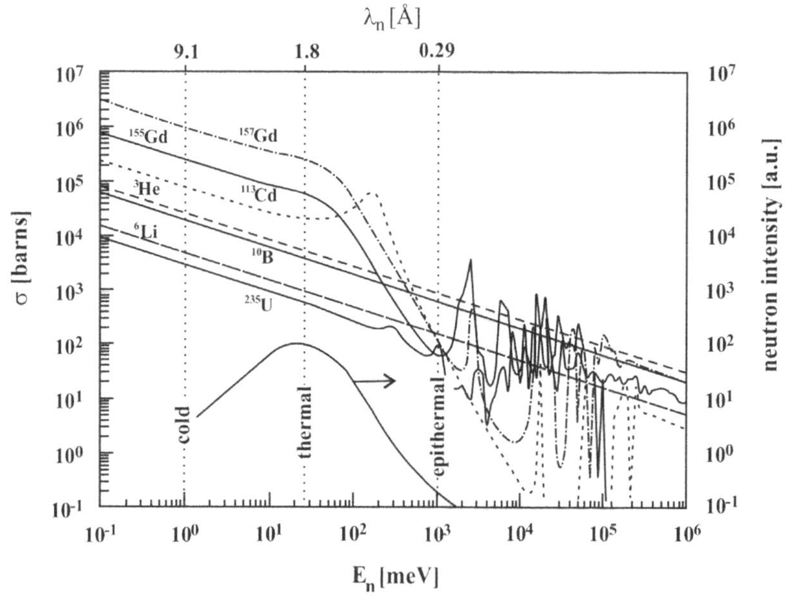

FIGURE 1.81 Neutron capture cross sections of some nuclides important for neutron detection as a function of neutron energy. The lowest curve shows a typical moderated neutron spectrum of a nuclear reactor (right y-axis).

If we were to consider other natural or synthetic materials that contain a mixture different nuclides, each nuclide would possess a different total cross section (σ_t) and different concentration per unit volume in the material. This would require the calculation of the total macroscopic cross section (Σ_t), which is the sum of the macroscopic cross sections of all of the different nuclides in the material as described by Eq. (1.338).

In the above example, the total cross section was chosen to calculate the neutron attenuation, because the total cross section represents all types of interaction that can result from the collision of a neutron with an atom within the material it travels (Anders and Briden, 1964; Vandecasteele et al., 1973). However, certain reaction cross sections are more relevant when it is our interest to optimize the detection efficiency of neutron radiation rather than the total neutron attenuation of neutron flux. For example, neutron capture cross sections for reactions of neutrons with nuclides that result in the production of easily detectable products or particles such as ^{3}H or protons from ^{6}Li, fission products from ^{235}U, alpha particles from ^{6}Li and ^{10}B, protons from ^{6}Li, conversion electrons and gamma radiation from ^{155}Gd, ^{157}Gd and ^{113}Cd would be more relevant than neutron scattering cross section interactions. The neutron capture cross sections of several nuclides used in the detection of neutrons are plotted as a function of neutron energy in Fig. 1.81.

The above treatment dealt with the attenuation of slow neutron energies where neutron absorption reactions can be relevant in the total neutron cross section σ_t used in the calculation of the macroscopic cross section Σ_t. During early research into neutron attenuation (Anders and Briden, 1964; Vandecasteele et al., 1973) it was recognized that at slow neutron energies the total neutron cross section would be the most appropriate, because σ_t "represents all interaction processes that may result from the collision of a neutron with an atom". However, early researchers also noted that for fast neutrons (2−12 MeV) elastic scattering with heavy nuclei would be significant, because the elastic scattering would change the direction of neutrons away from the beam of penetrating uncollided neutrons (Blizard and Abbot, 1962; Nargowalla et al., 1968, 1970; Adams et al., 1971; Vandecasteele et al., 1973). Thus, they concluded that another macroscopic cross section called the removal cross section, Σ_R with units of cm^{-1}, would provide the best measure of neutrons removed via scattering from the main penetrating neutron flux.

Treatments on the theory of fast neutron attenuation and the calculation of the macroscopic removal cross sections Σ_R of many natural and synthetic material are provided by El-Khayatt (2010, 2011), El-Khayatt and El-Sayed Abdo (2009), El-Khayatt and Akkurt (2013), Kurudirek et al. (2014), Tellili et al. (2014) and Elmahroug et al. (2015). Excellent descriptions of fast neutron attenuation is provided by Tellili et al. (2014) and Elmahroug et al. (2015) and a brief summary of their explanation will be provided here. In the case of fast neutrons, which are in the MeV energy range (e.g., 2−12 MeV), the most probable mechanism of removal of fast neutrons from the penetrating neutron beam is via elastic scattering, and all other known mechanisms of neutron interaction with atomic nuclei are negligible. An exception would be the case of fast neutrons penetrating water or other compound high in hydrogen atoms due to the similarity of mass between neutrons and protons resulting in efficient neutron attenuation via neutron-proton collision; however, for the heavy elements, elastic scattering of fast neutrons is the main mechanism for the removal of neutrons from the fast group of penetrating neutron flux. Tellini et al. (2014) and Elmahroug et al. (2015) note that fast neutron attenuation is described by the macroscopic removal cross section, Σ_R (cm^{-1}), which is different from the total macroscopic cross section, and it is a fraction of it. The Σ_R is approximately 2/3 of the total Σ_t between neutron energies of 6–8 MeV except for materials containing a lot of hydrogen for which the $\Sigma_R \cong \Sigma_t$. The removal cross section is a measure of the probability that a fast neutron will undergo a first elastic collision with an atomic nucleus, which removes the neutron of penetrating uncollided neutrons (Tellini et al., 2014). The neutrons that are removed by this initial elastic collision lose some of their initial kinetic energy and are then able to undergo additional elastic collisions with nuclei in the medium and until they fall into the thermal neutron energy range. At this low energy range, the neutrons will undergo neutron capture as described previously in this chapter.

Computer programs for the calculation of the macroscopic removal cross sections of natural and synthetic material are provided by El-Khayatt (2010, 2011), El-Khayatt and El-Sayed Abdo (2009), El-Khayatt and Akkurt (2013) and Elmahroug et al. (2015). The following empirical equations, which relate the macroscopic removal cross section to the atomic weight of an element have been developed based on experimental and calculated data based on measurements at 8 MeV neutron energies (Kurudirek et al., 2014):

$$\Sigma_R = 0.21A^{-m}\ \text{cm}^{-1} \tag{1.341}$$

where A is the atomic weight in g/mol and m is a constant. A value of $m = 0.56$ was found by Kurudirek et al. (2014) to provide the best agreement with experimental and calculated data. When more than one element is found in the material, Eq. (1.341) is modified with an effective atomic weight, A_{eff}, or

$$\Sigma_R = 0.21A_{\text{eff}}^{-m}\ \text{cm}^{-1} \tag{1.342}$$

and the effective atomic weight for a multielement material is obtained from the following:

$$A_{\text{eff}} = \left(\sum_i w_i A_i^{-m}\right)^{-1/m} \tag{1.343}$$

where w_i is the weight fraction and A_i is the atomic weight of the ith constituent element (Kurudirek et al. (2014).

As the examples illustrate in Figs. 1.80 and 1.81, the neutron cross section of nuclei decreases as the energy or velocity of the neutron increases. That is, the neutron reactions with nuclei obey the general rule of having some proportionality to $1/v$, where v is the velocity of the neutron. This inverse proportionality of cross section and neutron velocity is particularly pronounced in certain regions of energy as illustrated in the total neutron cross section curves illustrated in Figs. 1.80 and 1.81, respectively. However, this is not always the case with many nuclides at certain neutron energies where there exists a resonance between the neutron energy and the nucleus. At some specific or very narrow neutron energy ranges, certain nuclei have a high capacity for interaction with neutrons. The elevated neutron cross sections at specific neutron energies appear as sharp peaks in plots of neutron cross section versus energy, such as some of the cross section plots illustrated in Figs. 1.80 and 1.81. These peaks are called resonances and often occur with (n,γ) reactions. The high cross sections occur when the energy of the incident neutron corresponds exactly to the quantum state of the excited compound nucleus, which is the newly formed nucleus consisting of a compound between the incident neutron and the nucleus. Most nuclides display both the $1/v$ dependence on neutron cross section and the resonance effects over the entire possible neutron energy spectrum.

We should keep in mind that neutron cross sections can be specific and differ in value for certain reactions, such as proton (σ_p)- and alpha particle (σ_α)-producing reactions, fission reactions (σ_f), or neutron capture cross sections (σ_c). The total neutron cross section (σ_{tot}) would be the cross section representing the sum of all possible neutron reactions at that specific neutron energy. For example, the thermal neutron cross section for ^{235}U, which is the neutron cross section at 0.0253 eV neutron energy corresponding to a neutron velocity of 2200 m/s at room temperature, can be given as $\sigma_c = 95$ barns for the neutron capture cross section, $\sigma_f = 586$ barns for the fission cross section, and $\sigma_\alpha = 0.0001$ barns for the neutron cross section for the alpha particle-producing reaction. These neutron cross section values indicate that neutron fission would predominate at the thermal neutron energy of 0.0253 eV, although some neutron capture would also occur. The total neutron cross section, σ_{tot}, would be the total of the three possible reactions or $\sigma_{tot} = 95$ barns $+586$ barns $+0.0001$ barns $= 681$ barns. In our treatment of slow neutron capture by ^{235}U in Section X.E.4.c, illustrated by Eq. (1.324), we noted that about 14% of the slow neutron captures by ^{235}U nuclei result in the formation of a long-lived ^{236}U and gamma radiation, and the remaining 86% of the slow neutron captures result in nuclear fission. This is exactly what is predicted by the thermal neutron cross section values just provided; that is, for ^{235}U

$$\sigma_c/\sigma_{tot} = 95\text{barns}/681\text{barns} = 14\%\ \text{neutron capture}$$

and

$$\sigma_f/\sigma_{tot} = 586\ \text{barns}/681\ \text{barns} = 86\%\ \text{fission}$$

6. Neutron decay

We have seen that fast neutrons may lose their energy through elastic and inelastic collisions with other nuclei, and if these neutrons do not undergo other reactions with nuclei, they may lose sufficient energy to reach thermal equilibrium with surrounding atoms and possibly be captured by atomic nuclei. The question remains of what would happen to a free neutron that is not absorbed by any atomic nucleus.

Earlier in this chapter (Section X.B.1) we discussed the transformation of the neutron within nuclei of radioactive atoms, which have a neutron/proton ratio too high for stability. In these unstable nuclides the neutron breaks up into a proton, negatron (negative electron), and antineutrino (see Eq. 1.175). If the neutron can transform itself in unstable nuclei, it stands to reason that the neutron might be unstable outside the protective boundaries of the stable nucleus. This is just the case, which A. H. Snell and L. C. Miller demonstrated in 1948. This work was followed with further studies by Robson (1950a,b) and Snell et al. (1950), that demonstrated neutrons in free flight in a vacuum would indeed decay with a lifetime in the range of 9−25 minutes with a release of 0.782 MeV of energy. Later more accurate measurements of neutron decay by researchers including Andreev et al. (2017), Arzumanov (2015), Arzumanov, et al. (2000a,b), Pichlmaier et al. (2000, 2010) and Serebrov et al. (2008, 2013) among others, determined the neutron lifetime to be 878.5 ± 0.8 s, which is close to 15 minutes. Much research has been carried out over the past several years to obtain the most accurate measurement of the neutron lifetime (Andreev et al., 2017; Coakley, 2007; Wilkinson, 2007; Serebrov et al., 2008, 2013; Arzumanov, 2009, 2015; Dewey et al., 2009; Ezhov et al., 2009; Leung and Zimmer, 2009; Materne et al., 2009; Müller et al., 2009; O'Shaughnessy et al., 2009; Walstrom et al., 2009; Pichlmaier et al., 2010), which is an important physical parameter. The most accurate measurement of the neutron lifetime will help answer questions in particle and astrophysics. In particle physics, the neutron lifetime will help us address the weak interaction parameters within baryons, such as the transformation of a "d" quark into a "u" quark described in Section VI.B of this chapter (Walstrom et al., 2009). In astrophysics the neutron lifetime is a key process in establishing the model for the Big-Bang primordial nucleosynthesis whereby there was an equilibrium of neutrons and protons a few minutes after the Big-Bang described by Paul (2009). A detailed treatment of the Big-Bang primordial nucleosynthesis is given in previous books by the writer (L'Annunziata, 2007, 2016).

Various methods are used to measure the neutron lifetime including the trapping and magnetic storage of neutrons followed by the counting of decay products. Paul (2009) provides a very thorough review of the methods used worldwide to determine most accurately the lifetime of the neutron. The present world average of the neutron lifetime of $\tau_n = 878.5 \pm 0.8$ s was reported by the Petersburg Nuclear Physics Group Andreev (2017), Serebrov et al. (2008, 2013); however, research to achieve the measurement of the lifetime of the neutron to a precision of ± 0.1 s is currently under intensive research. The decay of elementary particles is characterized in terms of lifetime. The lifetime, usually symbolized as τ, is related to the term half-life, $t_{1/2}$, the mean time it takes for one-half of the particles to decay (Sundaresan, 2001) according to the relationship

$$t_{1/2} = (\ln 2)\tau = 0.693\tau \quad (1.344)$$

The free neutron decays according to the scheme

$$n \rightarrow p^+ + e^- + \overline{\nu} + 0.7823 \text{ MeV} \quad (1.345)$$

The 0.7823 MeV of energy released in the neutron decay corresponds to the difference in mass of the neutron and the sum of the masses of the products of the neutron decay, the proton plus the electron, or

$$\begin{aligned} Q &= m_n - (m_p + m_e) \\ &= [1.00866491588 \text{ u} - (1.007276466879 \text{ u} \\ &\quad +0.000548579909 \text{ u})](931.494013 \text{ MeV/u}) \\ &= 0.7823 \text{ MeV} \end{aligned} \quad (1.345)$$

The neutron, therefore, outside the protective confines of the stable nucleus, has a very short lifetime.

F. Proton and neutron radioactivity

In the previous sections of this chapter various radionuclide decay modes were described as mechanisms whereby unstable nuclei emit particles and energy in their transitions from an elevated energy state to a lower or stable state (ground state). Some of these transitions to lower energy states involve the transformation of a proton to a neutron or a neutron to a proton in the nucleus, as in beta decay, whereby a nucleus arrives at a more stable ratio of protons/neutrons in the nucleus. Some artificially produced radionuclides, that is, radionuclides that are not encountered naturally, but those produced through nuclear reactions such as reactions facilitated with an accelerator, may possess an excess number of protons or neutrons. Such nuclei that are unstable due to an excess number of protons or neutrons are said to have exceeded their nucleon "drip line". The nucleon drip, be it a proton drip line or neutron drip line, refers to the ratio of protons-to-neutrons in a nucleus (*i.e.*, the numbers of protons or neutrons in a nucleus) of a given element beyond which the isotopes of that element are unstable. Many isotopes produced by nuclear reactions via the bombardment of target nuclei with accelerated nuclei possess proton or neutron numbers beyond their drip lines and are unstable with very short half-lives, and these are referred to as exotic isotopes. Such isotopes may undergo nuclear decay to a more stable state by emitting a proton or a neutron directly from the unstable nucleus. The nucleus literally "spits out" a proton or a neutron to achieve a more stable or optimum proton/neutron ratio. The direct emission of a proton or neutron from the nucleus is a decay process referred to as "proton radioactivity" or "neutron radioactivity", respectively. The process is also referred to as "proton emission" or "neutron emission". Nuclides that decay by proton or neutron radioactivity generally have very short half-lives, which is measured most often in fractional seconds. As these radionuclides are produced artificially with very short half-lives, they are often referred to as "exotic nuclei". The decay process of direct "proton or neutron emission" may also be considered as exotic; however, much research is underway currently in this field, because it helps us understand the shell structure of the nucleus and its magic numbers of protons and neutrons in the stable nucleus.

1. Proton radioactivity

The phenomenon of proton radioactivity is eloquently introduced by Blank et al. (2003), and this introduction applies also to the phenomenon of neutron radioactivity. Their introduction is the following:

> *The atomic nucleus is made up of positively-charged protons and neutral neutrons, the nucleons. To form a stable atomic species, a balance between the number of neutrons and of protons has to be respected. For light nuclei, this requires an equal number of both nucleons. Beyond mass or nucleon number A = 40, stable atomic nuclei accommodate more neutrons than protons to outbalance the repulsion of the charged protons. The number of neutrons with respect to the number of protons increases until reaching a neutron number of N = 126 and a proton number Z = 82 for lead.*
>
> *Nuclei which do not respect this balance are unstable and decay by radioactive processes. For a minor disequilibrium, β decay occurs: an excess neutron is transformed into a proton and vice versa by emitting an electron/positron and an anti-neutrino/neutrino. If the imbalance between proton and neutron numbers is too large, the nuclear forces can no longer bind all nucleons and the atomic nucleus becomes particle-unstable. Thus the limits of nuclear stability are reached and the unstable nucleus will emit single excess nucleons or clusters of nucleons. The emission of clusters of two protons and two neutrons, a ^{4}He nucleus was the first radioactivity observed and was called α radioactivity. It occurs mainly for very heavy nuclei in the lead to uranium region (Z = 82–92), but also for very proton-rich lighter nuclei. For the heaviest nuclear species, fission may occur, where the heavy nucleus splits into two medium-heavy pieces.*
>
> *Theoretical investigations [Goldansky, 1960, 1961] in the 1960s predicted that for medium-mass (A = 50–100) nuclei, other new radioactive decays might occur. For very proton-rich odd-Z nuclei, one-proton radioactivity, i.e., the emission of one single excess proton, should be observable, whereas for even-Z nuclei with a large proton excess, two-proton radioactivity, the emission of a pair of protons, should occur. Although these one-proton or two-proton emitting nuclei are particle-unbound, they may exhibit a measurable half-life due to the Coulomb barrier the protons have to tunnel through.*

Reproduced from Blank et al. (2003), with permission

a. Beta-delayed proton emission

The first observation of proton radioactivity was a proton emission from an excited state (elevated energy state) of a nucleus following beta decay (Barton et al., 1963), which is referred to as "beta-delayed proton emission". This process is abbreviated as β^+(p) or β^+(2p) to signify the emission of one proton or two protons from a nucleus that was the product of positron emission. Silicon-25 was the radionuclide from which the beta-delayed proton emission was observed for the first time by Barton et al. (1963). The radionuclide was produced by bombarding aluminum with 97-MeV protons. The researchers also produced other radionuclides, namely ^{21}Mg, ^{17}Ne, and ^{13}O, that decayed by the same process, after bombarding various elements with accelerated protons. They observed that all of the beta-delayed proton emissions, where a parent nuclide with a proton and neutron number (*Z*, *N*) is of the form $(2k+2,\ 2k-1)$, exhibited the following decay sequence:

$$(2k+2, 2k-1) \xrightarrow{\beta^+} (2k+1, 2k)^* \xrightarrow{p} (2k, 2k) \qquad (1.347)$$

In the above reaction sequence, the notation above each arrow signifies the decay process (*i.e.*, β^+ for beta decay with positron emission and *p* for proton emission). The asterisk in Eq. (1.347) signifies an excited or elevated energy state of the daughter nuclide of the beta decay.

The beta-delayed proton emissions observed by Barton et al. (1963) in the decay of ^{25}Si, ^{21}Mg, ^{17}Ne, and ^{13}O, exhibited decay sequences in accord with Eq. (1.347), which are written as

$$^{25}_{14}\mathrm{Si}_{11} \xrightarrow[t_{1/2}=220\mathrm{ms}]{\beta^+} {}^{25}_{13}\mathrm{Al}_{12} + \beta^+ + \nu \xrightarrow{p} {}^{24}_{12}\mathrm{Mg}_{12} + p \qquad (1.348)$$

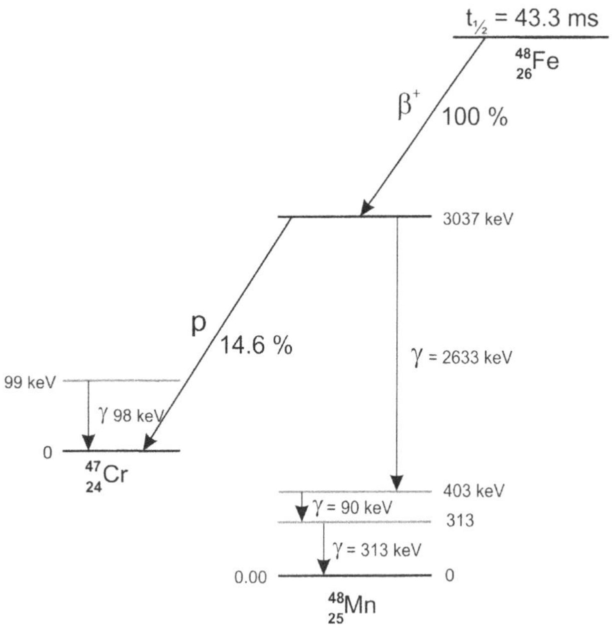

FIGURE 1.82 The scheme for beta-delayed proton emission in the decay of ^{48}Fe. The positron (β^+) and proton (p) emission intensities are from Audi et al. (2003). As illustrated, 14.6% of the elevated ^{48}Mn daughter nuclides undergo proton emission to ^{47}Cr. The remaining 85.4% decay by gamma-ray emission to the ground state of ^{48}Mn. *From Dossat et al. (2007); reprinted with permission from Elsevier © 2007..*

$$^{21}_{12}\text{Mg}_9 \xrightarrow[t_{1/2}=122\text{ms}]{\beta^+} {}^{21}_{11}\text{Na}_{10} + \beta^+ + \nu \xrightarrow{p} {}^{20}_{10}\text{Ne}_{10} + p \quad (1.349)$$

$$^{17}_{10}\text{Ne}_7 \xrightarrow[t_{1/2}=109.2\text{ms}]{\beta^+} {}^{17}_{9}\text{F}_8 + \beta^+ + \nu \xrightarrow{p} {}^{16}_{8}\text{Ne}_8 + p \quad (1.350)$$

$$^{13}_{8}\text{O}_5 \xrightarrow[t_{1/2}=8.6\text{ms}]{\beta^+} {}^{13}_{7}\text{F}_6 + \beta^+ + \nu \xrightarrow{p} {}^{12}_{6}\text{C}_6 + p \quad (1.351)$$

The decay process, beta decay with positron emission (β^+), is written above each arrow and the half-life of the radionuclide below an arrow.

A typical decay scheme of the beta-delayed proton emission, as an example, is illustrated in Fig. 1.82. The decay scheme illustrates ^{48}Fe, which decays with a half-life of 43.3 milliseconds by beta-decay with positron emission to a daughter nuclide ^{48}Mn. At an elevated energy state, 14.6% of the ^{48}Mn daughter nuclides decay by proton emission to the granddaughter nuclide ^{47}Cr. Gamma-ray emission accompanies daughter nuclide de-excitation. The decay equation is written as follows:

$$^{48}_{26}\text{Fe}_{22} \xrightarrow[t_{1/2}=43.3\text{ms}]{\beta^+} {}^{48}_{25}\text{Mn}_{23} + \beta^+ + \nu \xrightarrow{p} {}^{47}_{24}\text{Cr}_{23} + p \quad (1.352)$$

Beta-delayed proton emission may involve the sequential emission of two protons following beta decay, as in the decay of ^{26}P, described by Honkanen et al. (1983).

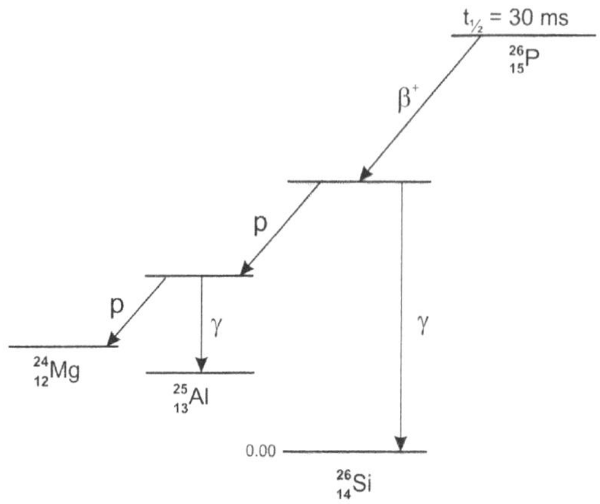

FIGURE 1.83 The scheme for the beta-delayed two-proton emission decay of ^{26}P according to the reaction sequence described by Eq. (1.353). *Adapted from Honkanen et al. (1983); © 1983 with permission from Elsevier.*

The decay of ^{26}P occurs according to the following sequence:

$$^{26}_{15}\text{P}_{11} \xrightarrow{\beta^+} {}^{26}_{14}\text{Si}_{12} + \beta^+ + \nu \xrightarrow{p} {}^{25}_{13}\text{Al}_{12} + p \xrightarrow{p} {}^{24}_{12}\text{Mg}_{12} + p \quad (1.353)$$

Such a decay scheme is abbreviated as β^+(2p), and its simplified decay scheme is illustrated in Fig. 1.83. As described by Honkanen et al. (1983), the two protons are not emitted simultaneously but in a sequential two-step process. As they explain and as illustrated in the decay scheme of Fig. 1.83, one proton is emitted to an energy level of an intermediate nucleus, which then becomes unbound to a subsequent emission of a second proton.

b. Direct proton emission

Proton emission from exotic nuclei may also occur directly from an unstable nucleus that possesses an excess number of protons without the initial beta decay. Direct proton radioactivity occurs by the following sequence:

$$^{A}_{Z}\text{X}_N \rightarrow {}^{A-1}_{Z-1}\text{Y}_N + p \quad (1.354)$$

where a radionuclide X of atomic number Z, mass number A, and neutron number N emits a proton p to transform into an isotope of element Y of atomic number $Z-1$, mass number $A-1$, and neutron number N.

The first direct proton emission as a radioactive decay process was observed by Jackson et al. (1970) and Cerny et al. (1970) in the decay of ^{53m}Co following the production of this exotic isotope of cobalt by the cyclotron bombardment of ^{40}Ca target with 81 MeV ^{16}O ions or the bombardment of an ^{54}Fe target with 32 MeV protons according to the reactions ^{40}Ca(^{16}O,2np)^{53m}Co and ^{54}Fe(p,2n)^{53m}Co, respectively. These abbreviated reaction notations with parenthesis read as follows: "target

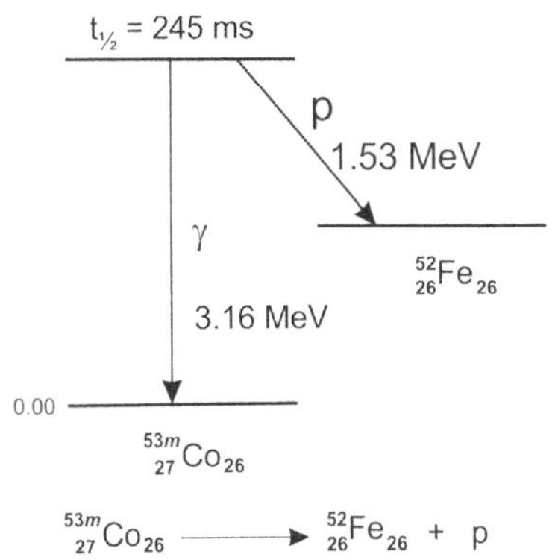

FIGURE 1.84 Direct proton emission decay scheme and decay equation of ^{53m}Co.

nucleus(projectile, emitted particles)product nucleus". The ^{53m}Co was observed by Jackson et al. (1970) to decay with the direct emission of a 1.53 MeV proton, and the decay scheme is illustrated in Fig. 1.84.

The direct emission of a proton as a disintegration mode of a nuclear ground-state was first observed by Hofmann et al. (1982) in the direct proton decay of ^{151}Lu and by Klepper et al. (1982) in the proton decay of ^{147}Tm. As noted by Klepper et al. (1982), this work demonstrated for the first time that the direct proton decay is a possible disintegration mode of nuclear ground-states similar to disintegration process that occur with alpha- and beta-decay as well as nuclear fission. The decay sequence, half-lives, and proton emission energies of the direct proton emission decay modes of the two radionuclides are the following:

$$^{151}_{71}\text{Lu}\xrightarrow{t_{1/2}=80.6\text{ms}}{}^{150}_{70}\text{Yb}+\text{p}(1.23\text{MeV}) \tag{1.355}$$

$$^{147}_{69}\text{Tm}\xrightarrow{t_{1/2}=580\text{ms}}{}^{146}_{68}\text{Er}+\text{p}(1.06\text{MeV}) \tag{1.356}$$

Two-proton radioactivity, that is, the direct emission of two protons from an atomic nucleus that has an excess number of protons, *i.e.* beyond their proton drip line, is also possible. The two-proton emission from such a nucleus could occur in a sequential process, described by Blank et al. (2003) as follows:

> ... *one-proton emission populates a well-defined quantal state in the one-proton daughter, before a second one-proton emission occurs to a two-proton daughter state.*

This sequential emission of two protons in nuclear decay is illustrated in the left-hand portion of Fig. 1.85. The term "mass excess" in the figure is a measure of the energy levels of each nuclear state, which is determined by the energy equivalence to mass of the nuclides. The mass excess, as defined by Audi et al. (2003), is the difference between the atomic mass (in mass units) and the mass number for each nuclear state. Thus, the mass excess is a measure of the energy levels, and the mass excess is expressed usually in units of keV or MeV. The simultaneous emission of two protons from the nucleus could occur when the one-proton daughter state is energetically not possible (*i.e.*, a forbidden transition) as illustrated in the right-hand portion of Fig. 1.85. The simultaneous emission of two protons is explained by Blank et al. (2003) as follows:

> ... *"the nuclear pairing force, which couples two nucleons of the same type to gain energy, is responsible for the fact that the masses of even-Z nuclei (in our case those of the two-proton emitter and the two-proton daughter) are smaller than the one of the one-proton daughter, thus rendering one-proton emission impossible (see Fig. 1.85). In such a case, the two protons have to be emitted simultaneously for energy conservation."*

An example of the simultaneous emission of two protons in a direct two-proton emission according to the sequence illustrated in the right-hand part of Fig. 1.85 is taken from the work of Mukha and Schneider (2001). They predicted the decay of ^{19}Mg to proceed by a simultaneous two-proton emission to ^{17}Ne, due to the higher energy state of the one-proton daughter, as illustrated in Fig. 1.86. The predicted two-proton daughter nuclide ^{17}Ne was found to decay by simultaneous two-proton emission by Chromik et al. (1998). It is

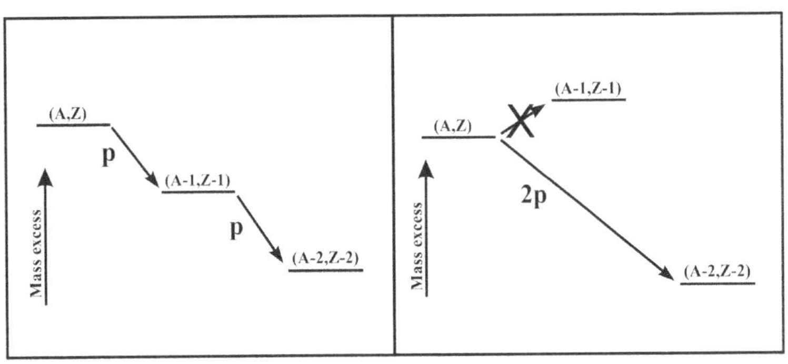

FIGURE 1.85 Schematic representation of sequential (left) and simultaneous two-proton emission (right). If the one-proton daughter state is energetically not accessible, the emission has to be simultaneous. *Reproduced from Blank et al. (2003), with permission*

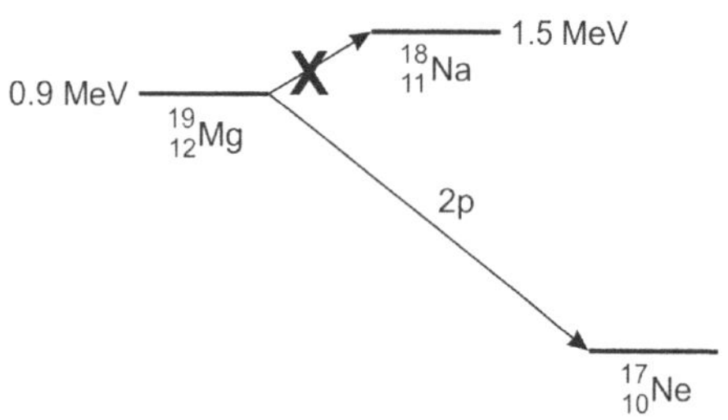

FIGURE 1.86 The predicted decay scheme of ^{19}Mg by the simultaneous emission of two protons.

interesting to note here that the decay sequence from ^{17}Ne would stop at the production of stable ^{15}N, which contains the magic number of eight neutrons according to the following:

$$^{17}_{10}\mathrm{Ne} \xrightarrow[t_{1/2}=109\mathrm{m/s}]{2p} {}^{15}_{8}\mathrm{O} + 2p \xrightarrow[122\mathrm{s}]{\beta^+} {}^{15}_{7}\mathrm{N(stable)} + \beta^+ + \nu \tag{1.357}$$

c. Detection and measurement of proton radioactivity

The detection and measurement of protons emitted simultaneously during radionuclide decay is accomplished with position-sensitive detectors, such as the Si semiconductor microstrip detector arrangement illustrated in Fig. 1.87.

A gas electron multiplier (GEM) detector is used often to detect and measure particles in nuclear reactions and decay. This is an advanced ionization detector, which can provide images of particle tracks with X- and Y-direction readout strips. GEM detectors were introduced by Sauli (1997), and the current state-of-the-art in the operation and applications of these detectors in particle physics is provided by Sauli (2014). An example of a two- and three-dimensional image of a 2-proton radioactivity event is illustrated in Fig. 1.88. This figure provides an image and the intensities of the two protons emitted simultaneously in the radioactive decay of ^{45}Fe. The radionuclide, ^{45}Fe, decays by beta-delayed proton emission; that is, the ^{45}Fe decays by positron emission to an elevated energy state of ^{45}Mn, which then emits simultaneously two protons to ^{43}V as described by the following decay sequence:

$$^{45}_{26}\mathrm{Fe} \xrightarrow[t_{1/2}=3.2\mathrm{ms}]{\beta^+} {}^{45}_{25}\mathrm{Mn} + \beta^+ + \nu \xrightarrow{2p} {}^{43}_{23}\mathrm{V} + 2p(1.1\mathrm{MeV}) \tag{1.358}$$

The two protons that are emitted simultaneously share 1.1 MeV (Pfützner et al., 2002; Giovinazzo et al., 2002), which is written into the decay sequence of Eq. (1.358). The relatively equal sharing of the 1.1 MeV energy by the two protons is evidenced in the three-dimensional Fig. 1.88 by the relatively equal length of travel of the two protons in the gas ionization detector.

Another experimental arrangement used for the measurement of proton radioactivity is the optical time-projection chamber (OTPC) developed by Miernik et al. (2007a,b) and utilized by Pomorski et al. (2011) for the beta-delayed emission of one, two, and three protons in the nuclear decay of ^{43}Cr. The OTPC, described by Pomorski et al. (2011), consists of an ionization counting chamber measuring $20 \times 20 \times 42$ cm^3 filled with a mixture of He, Ar, N_2, and methane gas. Charged particles (*e.g.*, protons) produce ionization (*i.e.*, ion pairs) in the gas, and the primary electrons drift in a uniform electric field toward a double-stage amplification structure formed by parallel-mesh flat electrodes. In this multiplication stage UV photons are emitted and the UV wavelengths are converted to the visual range by a thin luminescent foil providing an image of particle tracks, which are recorded by a CCD camera and photomultiplier tube. CCD camera images of the simultaneous emission of one, two, and three protons from the beta-delayed proton emission decay of ^{43}Cr are illustrated in Fig. 1.89. The decay schemes are the following:

$$^{43}_{24}\mathrm{Cr} \xrightarrow[t_{1/2}=21.6\mathrm{ms}]{\beta^+} {}^{43}_{23}\mathrm{V} + \beta^+ + \nu \xrightarrow{p} {}^{42}_{22}\mathrm{Ti} + p \tag{1.359}$$

$$^{43}_{24}\mathrm{Cr} \xrightarrow[t_{1/2}=21.6\mathrm{ms}]{\beta^+} {}^{43}_{23}\mathrm{V} + \beta^+ + \nu \xrightarrow{2p} {}^{41}_{21}\mathrm{Sc} + 2p \tag{1.360}$$

$$^{43}_{24}\mathrm{Cr} \xrightarrow[t_{1/2}=21.6\mathrm{ms}]{\beta^+} {}^{43}_{23}\mathrm{V} + \beta^+ + \nu \xrightarrow{3p} {}^{40}_{20}\mathrm{Ca} + 3p \tag{1.361}$$

The branching ratios for the emission of one, two, and three protons were determined by Pomorski et al. (2011) to be 81%, 7.1%, and 0.1%, respectively.

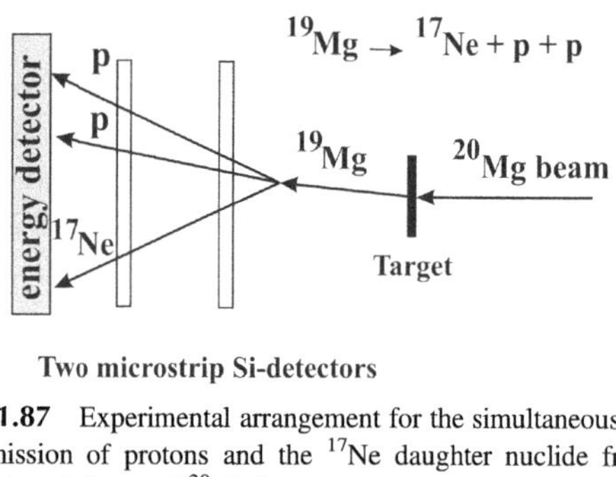

FIGURE 1.87 Experimental arrangement for the simultaneous detection of the emission of protons and the ^{17}Ne daughter nuclide from ^{19}Mg disintegration. A beam of ^{20}Mg ions are accelerated to bombard against a target resulting in a nuclear reaction causing a neutron "knock out" from the ^{20}Mg nucleus yielding ^{19}Mg. The ^{19}Mg disintegrates in flight by the emission of two protons producing the daughter nuclide ^{17}Ne. The directions of flight are determined by alignments of signals from the microstrip detectors, and the particle energies can be measured by scintillation detectors.

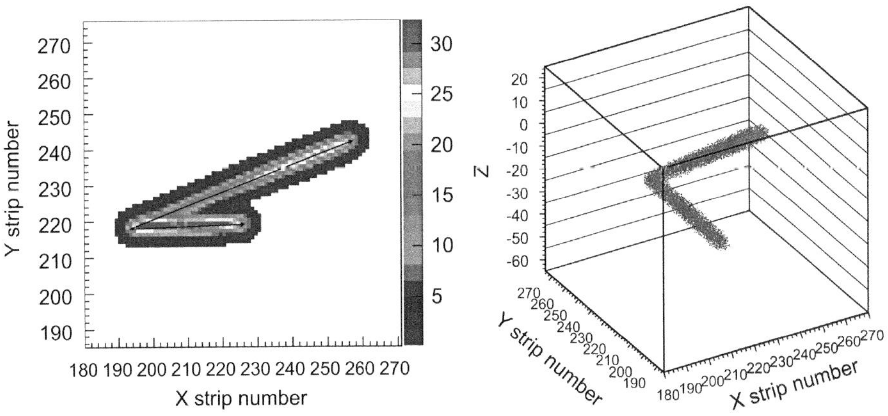

FIGURE 1.88 Two-dimensional (left) and three-dimensional (right) view of ^{45}Fe events as reconstructed from the strip information in X and Y. The color/gray code is indicative of the energy loss detected by the strips. For the color version of this figure, the reader is referred to the electronic version of the book.

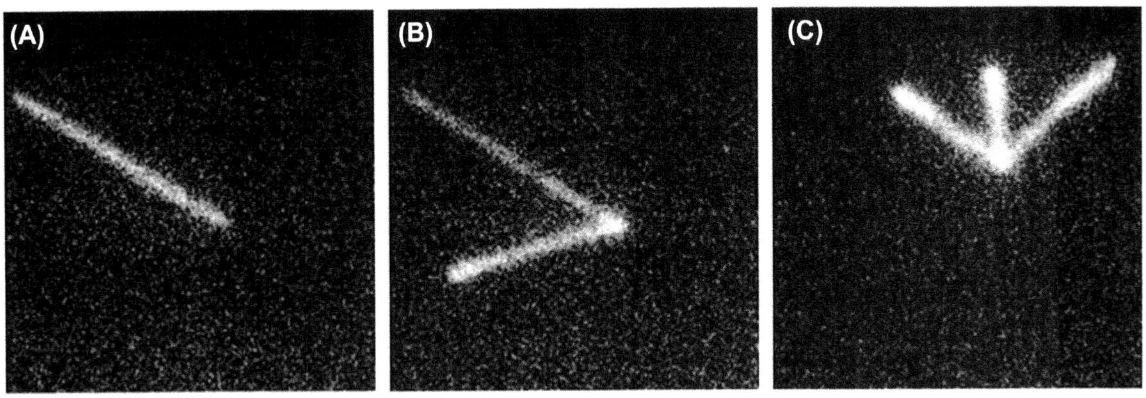

FIGURE 1.89 Example CCD images of ^{43}Cr decays by β-delayed proton emissions, recorded in the synchronous mode of an OTPC detector in which the track of an incoming ion is not seen. The cases of emission of (A) a single proton, (B) two protons, and (C) three protons are shown. The ionizing power of the positron is too small to see its trace. In the synchronous mode of the OTPC detector, the CCD camera image is started by a trigger, which avoids the image of the incoming heavy ions. For the color version of this figure, the reader is referred to the electronic version of the book. *From Pomorski et al. (2011); reprinted with permission © 2011 by the American Physical Society, https://doi.org/10.1103/PhysRevC.83.014306.*

Over 40 exotic nuclei have been identified that exhibit proton radioactivity. These are artificially produced via nuclear reactions, and they have short half-lives of the order of seconds or milliseconds. Their study has provided much insight into nuclear structure in terms of proton and neutron shells, nuclear stability, and proton and neutron drip lines. For further reading, reviews on proton radioactivity are available from Pfützner et al. (2012), Pfützner (2013), and Roeckl and Mukha (2013), Gonçalves et al. (2017), and Saxena et al. (2017).

2. Neutron radioactivity

The term "neutron radioactivity" should not be confused with the emission of neutrons from heavy nuclei following nuclear fission reactions or neutron emission as the result of the decay of nuclei by spontaneous fission. Somewhat analogous to proton radioactivity, which was described previously in this chapter, neutron radioactivity refers to the direct emission of neutrons from light nuclei that possess an excess number of neutrons, that is, a number of neutrons beyond their neutron drip line.

a. Beta-delayed neutron emisson

Neutron radioactivity most commonly occurs as a beta-delayed process, that is, neutron emission from the atomic nucleus occurs following beta (β^-) decay. Beta-delayed neutron emission by fission products is an important source of neutrons that contributes to the total number of neutrons responsible for the continuity of nuclear fission chain reactions in nuclear power reactors.

Beta-delayed neutron emission involves the emission of one or more neutrons from a neutron-rich nucleus when the neutrons exist in neutron-unbound states in a daughter nucleus at an elevated energy following beta decay (Birch et al., 2014). This decay process is abbreviated as β^- n, β^- $2n$, or β^- $3n$, for the emission of one, two, or three neutrons following beta decay. The emission of four neutrons from a beta-decay daughter nucleus at an excited state is possible; but not yet confirmed. The half-lives of nuclei that exhibit neutron radioactivity are considerably shorter than those that undergo proton radioactivity, because a neutron, due to its lack of charge, does not encounter any Coulomb barrier to hinder its emission from a nucleus. When half-lives are

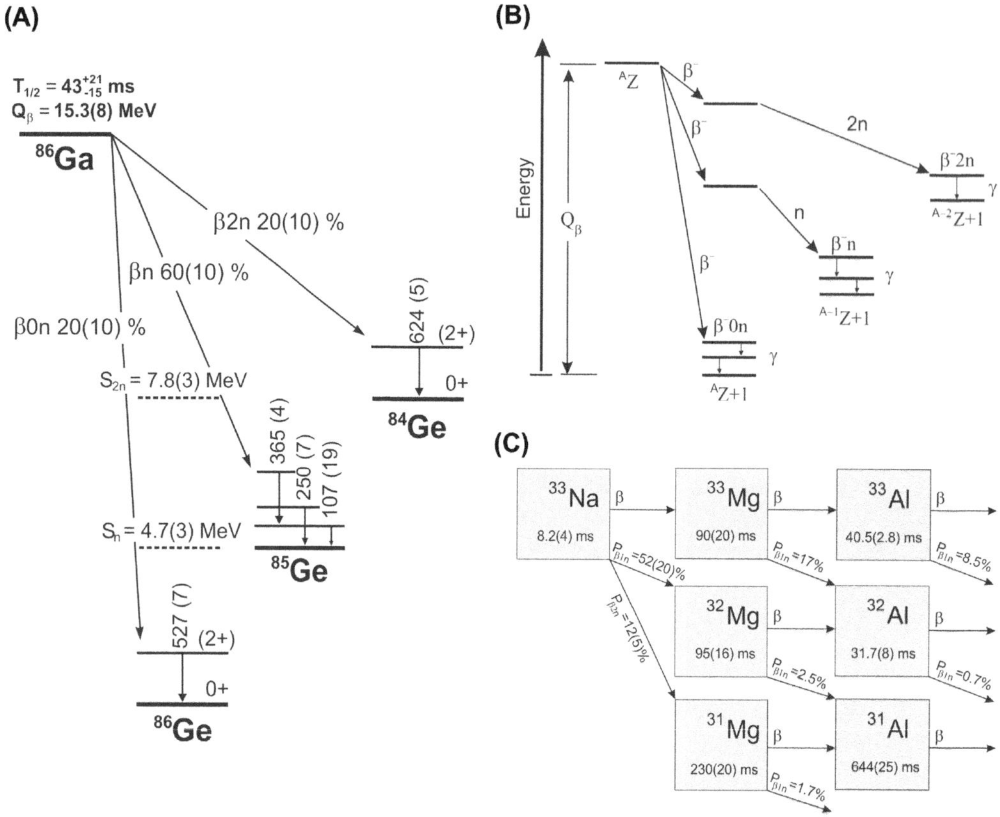

FIGURE 1.90 (A) The beta-delayed neutron emission decay scheme of neutron-rich ^{86}Ga (reprinted with permission from Miernik et al. (2013), https://doi.org/10.1103/PhysRevLett.111.132502 © 2013 American Physical Society; (B) basic beta-delayed neutron emission transitions of a neutron-rich radionuclide; and (C) beta-delayed neutron emission in the decay of neutron-rich ^{33}Na and its daughters and granddaughters. *From Radivojevic et al. (2002); reprinted with permission from Elsevier © 2002.*

very short ($<10^{-12}$ s), the process may be referred to as neutron emission, rather than neutron radioactivity, as described by Thoennessen et al. (2013).

The basic pathways of negatron emission of a neutron-rich radionuclide are illustrated in Fig. 1.90A. The figure illustrates the decay of a radionuclide of atomic number Z and mass number A by negatron emission to a nuclide of atomic number Z + 1 with the same mass number A with no neutron emission (β^-0n). Other possibilities are the delayed-emission of one or two neutrons following beta decay to nuclides of atomic number Z + 1 and mass numbers of A − 1 and A − 2 resulting from the emission of one neutron (β^-1n) or two neutrons (β^-2n), respectively. An example can be taken from a thorough study of the beta-delayed neutron emission in the decay of neutron-rich ^{86}Ga by Miernik et al. (2013). They measured the half-life and beta branching ratios of the three possible beta-decay modes (expressed in %) as follows:

$$\beta 0n(\sim 20\%):\quad {}^{86}_{31}\text{Ga} \xrightarrow[t_{1/2}=\ 43\text{ms}]{\beta^-} {}^{86}_{32}\text{Ge} + \beta^- + \gamma + \bar{\nu} \quad (1.362)$$

$$\beta 1n(\sim 60\%):\quad {}^{86}_{31}\text{Ga} \xrightarrow[t_{1/2}=\ 43\text{ms}]{\beta^-} {}^{86}_{32}\text{Ge} + \beta^- + \bar{\nu} \xrightarrow{n} {}^{85}_{32}\text{Ge} + n \quad (1.363)$$

$$\beta 2n(\sim 20\%):\quad {}^{86}_{31}\text{Ga} \xrightarrow[t_{1/2}=\ 43\text{ms}]{\beta^-} {}^{86}_{32}\text{Ge} + \beta^- + \bar{\nu} \xrightarrow{2n} {}^{84}_{32}\text{Ge} + 2n \quad (1.364)$$

The element gallium has an atomic number $Z = 31$, whereby the isotope ^{86}Ga with a neutron number of 55 ($N = 86 - 31 = 55$) is very rich in neutrons. The beta-decay of ^{86}Ga leaves neutrons of the germanium daughter nuclei ^{86}Ge at elevated unbound energy states permitting their emission from the ^{86}Ge nucleus to form ^{85}Ge or ^{84}Ge, respectively. Either one or two neutrons are emitted simultaneously from the ^{86}Ge nucleus according to the neutron-emission decay sequences described in the decay sequences 1.363 and 1.364. The beta-delayed neutron emission of neutron-rich ^{86}Ga, illustrated in Fig. 1.90A, can occur with branching ratios of ~60% and ~20% for the delayed emission of 1 neutron (βn) to ^{85}Ge or 2 neutrons (β2n) to ^{84}Ge, as illustrated in Fig. 1.90A. The remaining

~20% of the transitions occur without the emission of neutrons (β0n) to ^{86}Ge. All of the transitions yield daughter nuclides at excited states with the emission of gamma radiation at energy lines depicted in Fig. 1.90A.

The decay of neutron-rich nuclides by beta-delayed neutron emission are generally described by decay diagrams, as illustrated in Fig. 1.90B, where the atomic number Z increases by 1 (Z + 1) in negatron emission and the mass number (A) decreases by 1 (i.e., A − 1) and 2 (i.e., A − 2) in single neutron emission (β^-n) and double neutron emission (β^-2n), respectively, followed by gamma photon emission from daughter nuclides at excited energy states. Chains of beta-delayed neutron emission occur among neutron-rich parent, daughter, and granddaughter nuclides, as illustrated in the decay of neutron-rich ^{33}Na illustrated in Fig. 1.90C.

Among the light nuclei ($Z \leq 28$), there are numerous isotopes for which the beta-delayed neutron emission probabilities have been measured. By the year 2014, a total of 203 nuclei have been identified as potential precursors to beta-delayed neutron emission, and among these the probabilities of β^- n emission have been measured experimentally for over 109 nuclei; whereas, probabilities of β^- $2n$ emission have been measured experimentally for only 19; and the probabilities of β^- $3n$ emission measured in four nuclei (Birch et al., 2014). The beta-delayed emission of four neutrons (i.e., β^- $4n$) is yet to be measured. The nuclides that exhibit neutron radioactivity are found in the neutron-rich region of the Chart of the Nuclides, which is a Z versus N plot of the nuclides. These are illustrated in Fig. 1.91 for nuclides in the region of $Z \leq 28$. The radionuclides that decay by beta-delayed 1n emission (β^- n) are marked in red color. Nuclides decaying by β^- $2n$ and β^- $3n$ are fewer in number as mentioned previously. Also, as noted by Birch et al. (2014), all nuclides that decay by β^- 2n also decay by β^- 1n precursors, and all β^- 3n precursors are also β^- 2n and β^- 1n. Nuclides that exhibit negatron emission and neutron radioactivity reside in the neutron-rich side if the Chart of the Nuclides, *i.e.*, the region to the right of the black stable line of nuclides; whereas, nuclides that decay by positron emission and exhibit proton radioactivity are found in the region to the left of the black stable line of nuclides. Nuclides that decay by proton emission are not identified in Fig. 1.91; however, they reside among the yellow boxes of proton-rich nuclides to the left of the black boxes of stable nuclides.

b. Direct neutron emission

The direct emission of neutrons from neutron-rich nuclei can occur, and in such cases the half-life is generally much shorter than proton emission, because neutrons will not encounter any Coulombic barrier to their escape from the atomic nucleus. An example is the dineutron decay of

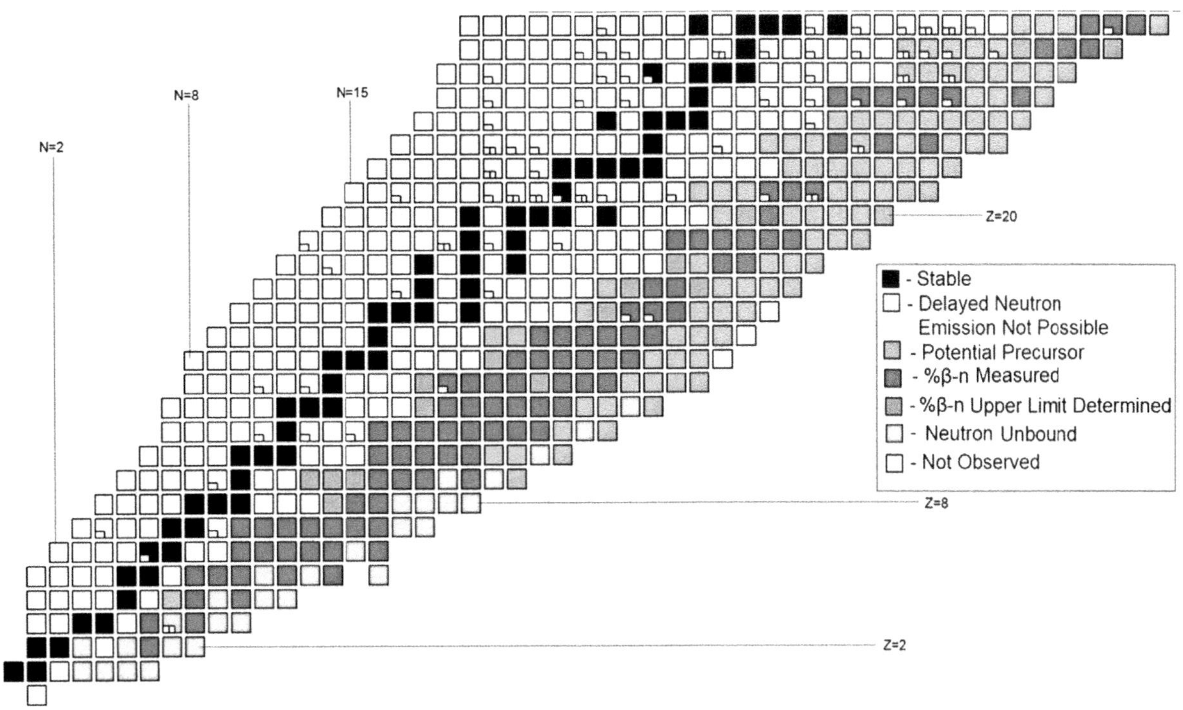

FIGURE 1.91 Chart of nuclides for $Z \leq 28$ nuclei. The coloring of the boxes indicates the measurement status of single neutron decay emission probability, P_{1n}. Small white boxes indicate the existence of an isomer for that isotope. This figure was adapted from the NUCLEUS display software (http://amdc.in2p3.fr/web/nubdisp_en.html) of the Atomic Mass Data Center. For the color version of the figure, the reader is referred to the online version of the book. *From Birch et al. (2014); reprinted with permission from Elsevier © 2014.*

neutron-rich ^{16}Be measured by Spyrou et al. (2012), which decays with a 200 ns half-life according to the following decay sequence:

$$^{16}_{4}\text{Be} \xrightarrow{t_{1/2}=200\text{nsec.}} {}^{14}_{4}\text{Be} + 2\text{n} \quad (1.365)$$

The emission of only one neutron or the sequential emission of two neutrons from ^{16}Be is not possible energetically, because the ground state of ^{15}Be is at a significantly higher energy than the parent ^{16}Be nuclide Spyrou et al. (2012).

c. Detection and measurement of neutron radioactivity

The detection and measurement of neutron radioactivity requires instrumental techniques that are different to the gas ionization detection and measurement techniques described previously for proton radioactivity. The neutral charge on the neutron will not permit its detection by gas ionization. However, certain inorganic and organic plastic scintillators are sensitive to neutron interactions such as proton recoil following elastic scattering in organic and plastic scintillators. The detection and measurement of neutrons is described in Chapters 6 and 9 of this volume.

A method proposed by Caesar et al. (2013) for the detection and measurement of neutron radioactivity and the very short half-lives involved is described in detail by Thoennessen et al. (2013). A diagram of the instrumental arrangement is illustrated in Fig. 1.92 with the decay scheme of ^{16}B as an example. The neutron-rich nucleus of ^{16}B emits a neutron with a half-life of <190 ps according to the decay sequence

$$^{16}_{5}\text{B} \xrightarrow{t_{1/2}<190\text{psec.}} {}^{15}_{5}\text{B} + \text{n} \quad (1.366)$$

The very short half-life requires the measurement of the neutron emissions from the onset of the production of the ^{16}B. A method of half-life measurement, illustrated in Fig. 1.92, is the measurement of the neutron distribution along position-sensitive scintillation detectors referred to as the Decay in a Magnetic Field (DiMF) method. As described by Thoennessen et al. (2013), the ^{16}B is produced by accelerating a beam of ^{17}C against a target material that knocks out a proton from the ^{17}C producing ^{16}B. The reaction ^{17}C(-p)^{16}B occurs at the location identified by a red dot in Fig. 1.92. The decay sequence $^{16}\text{B} \rightarrow {}^{15}\text{B} + \text{n}$ at various times in flight, which is illustrated by three blue dots along the flight path of the ^{16}B, bent by a deflecting magnetic field with a bend radius r and a deflecting angle θ after a possible drift distance d. The deflecting magnetic field will bend the path of the charged particles away from the paths of the emitted neutrons permitting the detection and counting of ^{15}B. The distribution of the neutrons measured along the distance x of the position-sensitive scintillation detectors (see Fig. 1.92) will provide a measure of the half-life of the nuclear decay. Decay times as short as 10 picoseconds (10 ps) will show a concentration of neutrons detected near the zero degree line (little deflection) at a distance l from the target. The longer the half-life of the order of 500 ps to 1 nanosecond (1 ns) will exhibit a distribution of neutrons further away from the zero degree line along the distance x. Longer decay times of up to 100 microseconds (100 μs) will exhibit a spread further away from the zero degree line.

For additional reading on neutron radioactivity the reader may peruse the papers by Spyrou et al. (2012), Thoennessen et al. (2013), Birch et al. (2014), Kahlbow et al. (2017), and Li et al. (2017).

G. Neutrino interactions with matter

Because neutrinos do not carry charge, they do not produce the Cherenkov Effect. However, neutrino interactions

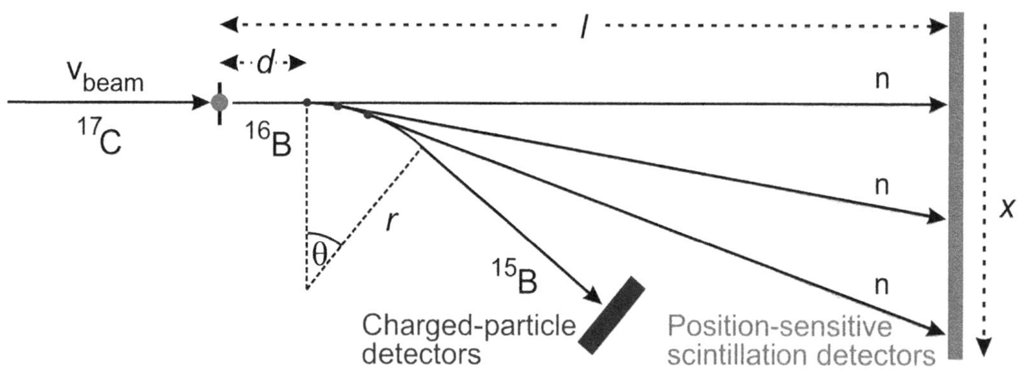

FIGURE 1.92 Schematics of the DiMF technique. The incoming ^{17}C beam produces ^{16}B in the target (red dot) which enters a magnetic field (bend radius r and deflection angle θ) after a drift distance d. The ^{16}B then can decay at different positions along the flight path (blue dots) into ^{15}B and a neutron. The ^{15}B fragments and neutrons are detected and identified in a set of charged-particle detectors and position sensitive scintillation detectors (at a distance l of 1500 cm from the target), respectively. For the color version of the figure, the reader is referred to the online or electronic version of the book. *Thoennessen et al. (2013); reprinted with permission from Elsevier © 2013.*

within very large bodies of transparent material can produce high-energy charged particles, which in turn produce Cherenkov photons. Photon detectors, namely photomultiplier tubes, are used to measure the Cherenkov photon intensity produced by the charged particles, which provides a measure of the neutrino flux.

The very large volumes of transparent material currently used for the detection of Cherenkov photons from neutrino interactions contain water or ice, and these have been designed to detect neutrinos from various sources such as solar neutrinos, atmospheric neutrinos of the cosmic ray showers, geo-neutrinos, artificially produced neutrinos arising from nuclear reactors and accelerators, and ultra-high energy (UHE) astrophysical neutrinos. The following large bodies of water or ice are utilized:

1. Heavy water (D_2O) weighing > 1 kiloton in an underground vessel, such as that utilized by the Sudbury Neutrino Observatory (SNO) near Sudbury, Canada (Boudjemline et al., 2010; Bellerive et al., 2016) or light water (H_2O) weighing > 50 kilotons contained in a large vessel underground, such as that utilized by the Super-Kamiokande and Hyper-Kamiokande Cherenkov detector under Mt. Ikenoyama, Japan (Fukuda et al., 2003; Abe et al., 2016 and F. Li et al., 2017). Both vessels are surrounded by photomultiplier detectors for the detection of the Cherenkov photons.
2. Lake or ocean waters with Cherenkov photon detectors (photomultiplier tubes) located deep within lake or ocean waters. In ocean waters, photon detectors have been placed at depths of up to 2.5 km. There are currently two large-scale neutrino telescopes in operation in lake or ocean waters, namely, the NT200 in Lake Baikal in Siberia (Aynutdinov et al., 2006, 2011; Avrorin et al., 2015, 2016), and the ANTARES (Astronomy with a Neutrino Telescope and Abyss Environmental RESearch) in the Mediterranean Sea (Carr, 2008; Aguilar et al., 2011; Brunner, 2011; De Bonis, 2016; Spurio et al., 2017).
3. Polar ice with Cherenkov photon detectors placed deep within holes drilled into the polar ice cap. The current large-scale neutrino telescope of this type is the Ice-Cube/AMANDA (Antartic Muon and Neutrino Detector Array) at the South Pole (Achterberg et al., 2006; Berghaus, 2009; Montaruli, 2009; Abbasi et al., 2010, 2011; Hultqvist, 2011; Desiati et al., 2014; Tosi et al., 2017; Xu, 2017 and Xu et al., 2017).

The very large bodies of water or ice are required to detect the neutrino interactions, because of the high penetrating power of neutrinos due to their lack of charge and very low mass.

The Cherenkov photons, that are detected, are those produced by charged particles resulting from neutrino interactions such as (i) elastic scattering (ES) of electrons by neutrinos in the detector medium, (ii) neutrino charged current (CC) interactions with deuterium nuclei in heavy water, and (iii) neutral current (NC) interactions of neutrinos with atomic nuclei of water or ice.

Electron Scattering. The elastic scattering (ES) of electrons by neutrinos, described by the following:

$$\nu_x + e^- \rightarrow \nu_x + e^- \tag{1.367}$$

is sensitive to all neutrino flavors (where $x = e$, μ, or τ), but with a reduced sensitivity to the muon neutrino (ν_μ) and the tau neutrino (ν_τ). In the ES interaction described by Eq. (1.367), neutrino energy is transferred to an atomic electron, and the electron will, in turn, travel as a scattered electron in light water, heavy water, or ice at relativistic speeds and produce Cherenkov photons as illustrated in Fig. 1.93. From this interaction and the anisotropic property of the Cherenkov radiation, the directional information of the neutrino can be determined.

Charged Current Interactions. The charged current (CC) interaction with a deuterium nucleus or deuteron (d) is described as follows:

$$\nu_e + d \rightarrow p + p + e^- \tag{1.368}$$

where the resultant energy of the electron (e^-) is well correlated with the energy of the electron neutrino (ν_e). The high-energy scattered electron produced from interactions 1.367 and the high-energy negatron from reaction 1.368 produce Cherenkov photons in transparent media. The electron neutrino ν_e is the only neutrino flavor that is able to convert the deuterium nucleus (*i.e.*, deuteron) into two protons and a negatron. The interaction of a neutrino with the deuteron described by Eq. (1.368) can be illustrated further whereby a neutron of the deuterium nucleus is transformed into a proton and a negatron as illustrated in Fig. 1.94.

The interaction described by Eq. (1.368) and Fig. 1.94 is called a charged current (CC) interaction, because a charged W boson, which can carry either a positive or negative charge ($W^{\pm}$), acts as an intermediate vector in the transformation process of a neutron in the nucleus of deuterium (d) into a proton (p). The charged W bosons act as exchange particles involved in the transformation of

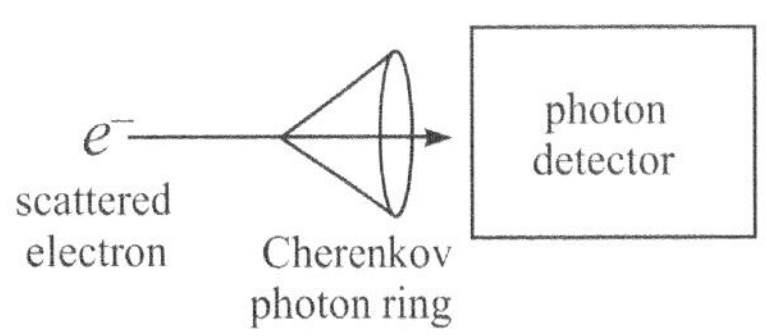

FIGURE 1.93 An electron with energy in excess of 262 keV traveling in H_2O, D_2O, or ice producing a Cherenkov photon ring, and the direction of travel of the Cherenkov photons are determined by the photon detectors (*e.g.*, photomultiplier tubes). The anisotropic property of Cherenkov photons permits the design and orientation of photon detectors to register only Cherenkov photons traveling in a certain direction. Thus, the directional information of the neutrino can be determined.

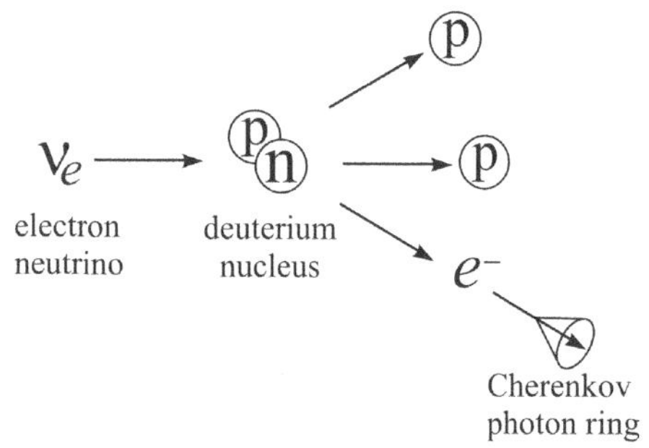

FIGURE 1.94 Charged current interaction of an electron neutrino with a deuterium nucleus resulting in the transformation of a neutron (n) into a proton (p) and negatron (e^-). The negatron or negative electron travels in the heavy water to produce Cherenkov photons illustrated with a Cherenkov photon ring. *From L'Annunziata (2016): reprinted with permission from Elsevier © 2016.*

quark flavor, which in this interaction, involves the transformation of a down quark (d) of the neutron into an up quark (u) transforming the neutron into a proton. This is illustrated by the Feynman diagram of Fig. 1.95.

Another **charged current** interaction that is currently employed to detect atmospheric neutrinos involves the creation of secondary muons by atmospheric neutrinos in large bodies of water, such as lakes, ocean floors, and artic ice. As described by Formaggio (2009), Katz (2009) and Abbasi et al. (2010), the following charged current reaction results in the creation of high-energy muons, which travel at relativistic speeds (0.8*c*) and produce Cherenkov photons detected by photon detectors located deep within lakes, ocean floors or artic ice:

$$\nu_\mu(\overline{\nu}_\mu) + N \rightarrow \mu^-(\mu^+) + X \tag{1.369}$$

where atmospheric muon neutrinos (ν_μ) or muon antineutrinos ($\overline{\nu}_\mu$) interact with nuclei *N* of elements that surround the Cherenkov photon detector to produce charge muons (μ^- or μ^+) and nuclei of element *X*. The nuclei change from *N* to *X*, because in the charge current reaction mediated by a W^+ boson, the quark flavor within a nucleon is changed as illustrated by the Feynman diagram of Fig. 1.96.

Neutral Current Interactions. The Cherenkov photon detection and measurement of neutrinos via neutral current interactions of neutrinos with deuterium nuclei in heavy water is described by the following:

$$\nu_x + d \rightarrow p + n + \nu_x \tag{1.370}$$

where ν_x is a neutrino of any flavor, that is $x = e$, μ or τ, to represent the electron, muon, or tau neutrino, *d* is the deuteron or nucleus of deuterium, and *p* and *n* are the proton and neutron, respectively, liberated when the deuterium nucleus breaks apart. The reaction is called a neutral current interaction, because a neutral Z boson (Z^0) acts as the intermediate exchange particle in this interaction of a neutrino

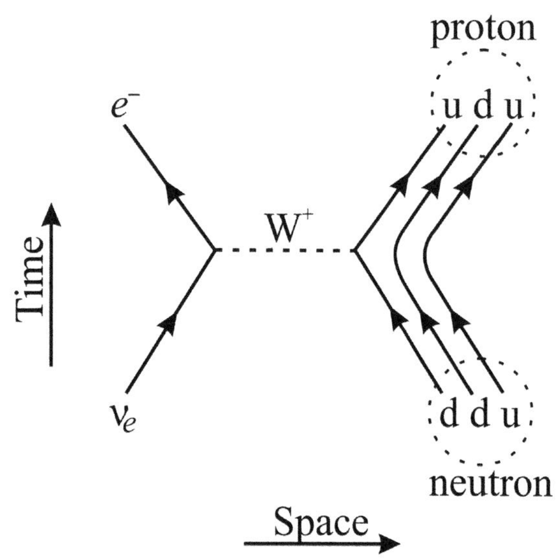

FIGURE 1.95 Feynman diagram illustrating the charged current (CC) interaction of an electron neutrino (ν_e) with a neutron of a deuterium nucleus via the mediation of a W^+ boson resulting in the transformation of a down quark (d) of the neutron into an up quark (u) resulting in the transformation of the neutron into a proton. The diagram illustrates on the lower left, an incoming electron neutrino (ν_e) interacting with a down quark (d) of a neutron (lower right) via a W^+ boson (dashed line). The down quark (d) of the neutron is transformed into an up quark (u) (upper right), thereby changing the neutron into a proton (upper right) with the creation of a negatron (e^-) illustrated with the arrow upward on the upper left of the diagram. *From L'Annunziata (2016): reprinted with permission from Elsevier © 2016.*

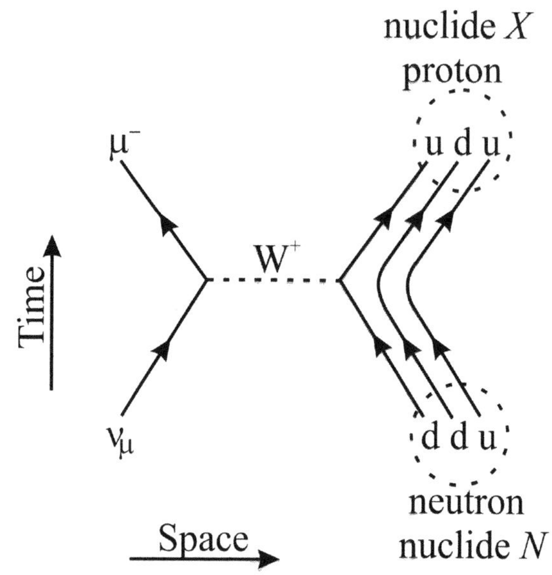

FIGURE 1.96 Feynman diagram describing the charge current interaction of a muon neutrino and a down quark of a nucleon in element *N*, mediated by a W^+ boson, converting the quark flavor from a down quark (d) to an up quark (u) with the production of a negative muon (μ^-). The conversion of the quark flavor from d to u results in the conversion of a neutron to a proton and consequently conversion of nuclide *N* to nuclide *X*. *From L'Annunziata (2016): reprinted with permission from Elsevier © 2016.*

with the deuterium nucleus. The Z^0 boson does not change the flavor of quarks in a nucleon; however, the energy transferred causes the nucleus to break up into its proton and neutron component nucleons. An illustration, by means of a Feynman diagram, of the neutral current interaction of a neutrino with a deuterium nucleus, is provided in Fig. 1.97.

All neutrino flavors react equally with deuterium in this neutral current reaction, which can provide a measure of the total neutrino flux for neutrino energies in excess of 2.2 MeV (Prior, 2009). The neutron produced in reaction 1.370 becomes quickly thermalized in the water, and the thermal neutron is subsequently captured by a deuterium (^2H) nucleus or by a nucleus of ^{35}Cl when the water is doped with a high concentration of NaCl. Neutron capture by ^2H or by ^{35}Cl results in the emission of gamma radiation. The gamma rays will interact with atomic electrons of the water to produce Compton electrons. All Compton electrons with energy in excess of the 262 keV threshold energy will produce Cherenkov photons in the water. The detection and measurement of the Cherenkov photons provide a measurement of the neutrino flux. The neutral current interaction in heavy water doped with NaCl is illustrated in Fig. 1.98.

An illustration of the layout of a neutral current detector, namely the SNO Cherenkov detector, which utilized heavy water (D_2O), described previously for the detection and measurement of neutrinos, is provided in Fig. 1.99. The SNO (Sudbury Neutrino Observatory) is located 2100 m underground in Sudbury, Ontario, Canada. Fig. 1.99 illustrates a person standing above the detector, which provides a perspective of the magnitude of the heavy water detector.

The Nobel Prize in Physics 2015 was awarded in one part to Takaaki Kajita of the University of Tokyo, Kashiwa, Japan for his work with the Super-Kamiokande Collaboration and in another part to Arthur B. McDonald of Queens University, Kingston, Canada for his work with the Sudbury Neutrino Observatory (SNO) Collaboration, which resulted, in the words of the Nobel Committee, in "the discovery of neutrino oscillations, which shows that neutrinos have mass." From the detection and analysis of neutrinos they were able to demonstrated that neutrinos can oscillate from one flavor to another, and this would be possible only if neutrinos have mass. The Super-Kamiokande detected atmospheric neutrinos. The muon neutrinos (ν_μ) that arrived at the Super-Kamiokande detector from the atmosphere above were more numerous than the muon neutrinos that traveled through the globe and collided with the detector from below. This demonstrated that the muon neutrinos that traveled longer (i.e., through the earth) had more time to change into a neutrino of another flavor (e.g., ν_e or ν_τ). In addition, the SNO measured the solar electron neutrino (ν_e) intensity, as only electron neutrinos are produced by the Sun. Neutrinos of all three flavors (i.e., ν_e, ν_μ, and ν_τ) could also be detected by

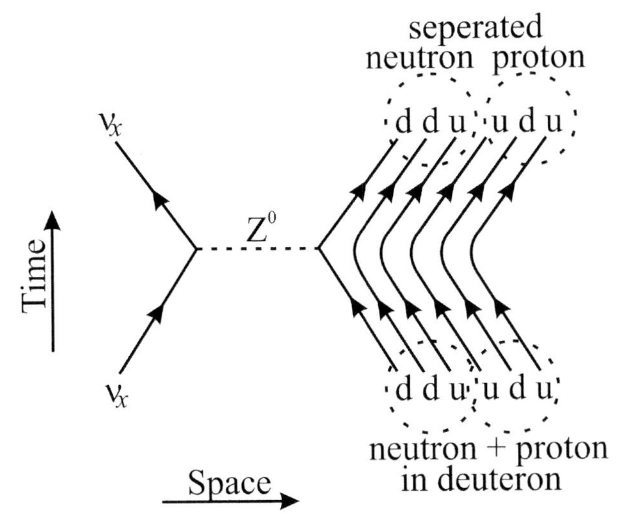

FIGURE 1.97 Feynman diagram describing a neutral current interaction of an electron, muon, or tau neutrino (where $x = e$, μ, or τ) with a down quark (d) of a neutron in a nucleus of deuterium (deuteron) via the neutral Z boson (Z^0) as an intermediate exchange particle. The flavor of the quark does not change, that is, the down quark (d) of the neutron remains a down quark, and thus the neutron does not change into a proton. However, energy of the neutrino is transferred, and the neutron and proton of the deuterium nucleus (lower right, dashed circles overlapping) break apart (upper right, dashed circles separated). *From L'Annunziata (2016): reprinted with permission from Elsevier © 2016.*

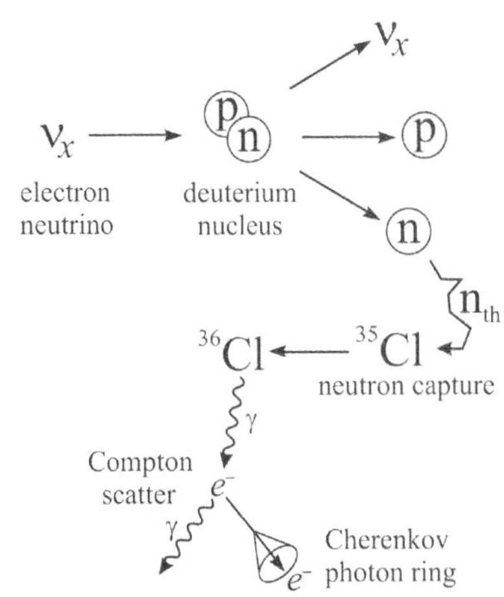

FIGURE 1.98 A neutral current interaction of a neutrino utilized for the detection of neutrinos of any flavor (electron, muon or tau) where $x = e$, μ, or τ, in heavy water (D_2O) doped with a high concentration of NaCl. The neutrino interacts with a deuterium nucleus causing the nucleus to break apart into its constituent proton and neutron. The neutron rapidly undergoes thermalization after several collisions (jagged path of arrow) with atomic nuclei, and the thermalized neutron (n_{th}) is captured by a ^{35}Cl nucleus. Gamma-ray emission from ^{36}Cl following neutron capture undergoes Compton scatter, and the Compton electrons, in excess of 262 keV energy, produce Cherenkov photons in the water. The Cherenkov photons are measured with suitable photon detectors (*e.g.*, photomultiplier tubes). *From L'Annunziata (2016): reprinted with permission from Elsevier © 2016.*

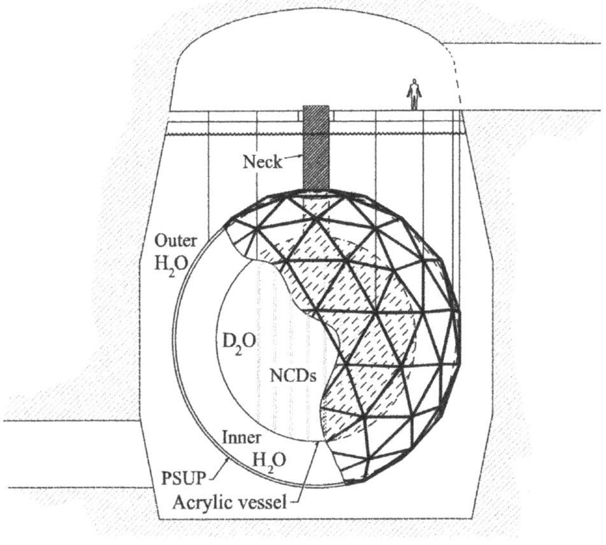

FIGURE 1.99 Layout of the SNO Cherenkov detector. The D_2O (1000 tons) is contained within an acrylic vessel. This volume could be accessed via the neck. The Neutral Current Detectors (NCDs), illustrated as vertical lines in the acrylic vessel housing the D_2O, were employed in the third phase of the experiment. The acrylic vessel was surrounded by water for shielding the radioactivity from the rock (outer H_2O) and photomultiplier tubes (inner H_2O). The photomultiplier SUPport Structure (PSUP) holds the photomultiplier tubes and light concentrators. A figure of a person is illustrated in the upper right to provide a perspective of the magnitude of the underground Cherenkov detector. *Reprinted from Boudjemline et al. (2010), with permission from Elsevier © 2010.*

the deuterium in the SNO. As described by the Nobel Committee, the electron neutrinos were smaller in number than expected, whereas the combined number of the three neutrino flavors corresponded to expectations, which indicated that the electron neutrinos had changed into a muon of another flavor (Rose, 2015).

The Sudbury Neutrino Observatory (SNO) is currently under modification to SNO+, which will involve reusing the SNO facility to investigate the Majorana nature of neutrinos (i.e., the ability of neutrinos to oscillate from one flavor to another via the search for neutrinoless double beta decay of ^{130}Te (Lozza et al., 2016). In the search for neutrinoless double beta decay the facility would be loaded with 0.3% natural tellurium in 780 tonnes of liquid scintillator (PPO fluor in linear alkyl benzene solvent), which would correspond to 800 kg of ^{130}Te. As described by Lozza et al. (2016) on behalf of the SNO+ Collaboration, the redesigned SNO+ will be equipped to measure reactor anti-neutrino oscillations, geo-antineutrinos, solar neutrinos, and supernova neutrinos. The solar neutrinos would be measured with the pure liquid scintillator (i.e., without the Te loading). Anti-neutrinos are detected by inverse beta decay as described previously in this chapter and via the basic reactions illustrated in Fig. 1.46 earlier in this chapter. The anti-neutrinos will be measured during the double beta-decay research on ^{130}Te. The measurement of anti-neutrinos originating from the Earth's crust and mantle (i.e., geo anti-neutrinos) could provide information on heat production of the Earth. Supernova neutrinos will be detected via charged current interactions and neutral current interactions on ^{12}C of the organic liquid scintillator solvent and via proton scattering (Lozza et al., 2016). Charged current and neutral current interactions are described earlier in this chapter via Eqs. (1.369) and (1.370) and Figs. 1.96 and 1.97. Neutral current interactons, as illustrated in Fig. 1.97, are possible with all neutrino flavors (i.e., electron, muon, and tau neutrinos), and thus the neutral current interaction can provide data on the combined flux of all neutrino flavors.

Ultra-high energy (UHE) astrophysical neutrino interactions with atomic nuclei can produce particles yielding Cherenkov radio waves, which are detected with antennas. For additional reading on the detection of neutrinos by the Cherenkov Effect, the reader may peruse a detailed review by the writer in a previous book (L'Annunziata, 2016).

XI. Electromagnetic radiation – photons

A. Dual nature: wave and particle

In the latter part of the 19th century, Hertz (1887) carried out a series of experiments demonstrating that an oscillating electric current sends out electromagnetic waves similar to light waves, but of different wavelength. Hertz proved, thereby, the earlier theory of James Clerk Maxwell, that electric current oscillations would create alternating electric and magnetic fields, and radiated electromagnetic waves would have the same physical properties of light. A subsequent discovery by Nobel Laureate Zeeman (1897) further linked the properties of light with electricity and magnetism when he discovered that a magnetic field would alter the frequency of light emitted by a glowing gas, which we now refer to as the Zeeman Effect. The magnetic field would alter the spectral line corresponding to the frequency of the light into a triplet or even a more complex multiplet of lines. This discovery was significant, because it was the very first to link the properties of light with electricity and magnetism.

Not long after the discoveries of Hertz and Zeeman came the work of Max Planck, who in 1900 proposed a formula to explain that the vibrating particles in the heated walls of a kiln could radiate light only at certain energies. These energies would be defined by the product of a constant having the units of energy × time and the radiation frequency. The constant, which Planck calculated became known as the universal Planck constant, $h = 6.626 \times 10^{-34}$ J s. Therefore, radiation would be emitted at discrete energies, which were multiples of Planck's constant and the radiation frequency, ν. Planck named the discrete radiation energy as the quantum from the Latin *quantus* meaning "how great".

In 1905 Einstein grasped the calculations of Planck to explain and provide evidence that light, not only traveled as waves, but also existed as discrete packets of energy or particles, which he named "energy quanta". Today we refer to these energy quanta as photons. Einstein demonstrated the existence of the photon in his explanation of the photoelectric effect (see Section XII.A of this chapter.). He demonstrated that the energy of an electron (photoelectron) ejected from its atomic orbital after being struck by light was not dependent on the light intensity, but rather on the wavelength or frequency of the light. In other words, increasing the light intensity would increase the number of photoelectrons, but not their energy. Whereas, altering the frequency, thus energy, of the light would alter the energy of the photoelectron. In summary, Einstein demonstrated that the energy of the photoelectron depended on the energy of the photon that collided with the electron or, the product of Planck's constant times the light frequency according to the formula

$$E = h\nu = \frac{hc}{\lambda} \tag{1.371}$$

Eq. (1.371) is referred to as the Planck–Einstein relation (Woan, 2000). From this observation Einstein is credited with discovering the photon as an elementary particle. Also, we may notice from Eq. 1.371 that the product of the photon frequency, ν, and wavelength, λ, always yields the velocity, c, the speed of light. The photon always travels at the constant speed in a vacuum, $c = 2.9979 \times 10^8$ m/s; it cannot travel at a speed less than c in a vacuum.

From our previous treatment we see that the photon behaves as a particle, which could knock out an electron from its atomic orbital provided it possessed sufficient energy to do so, that is, an energy in excess of the electron binding energy. Therefore, the photon is considered as another elementary particle. In his explanation of the photoelectric effect, Einstein was the first to demonstrate the particulate nature of light, and it is for this work he won the Nobel Prize. Since these findings of Einstein, electromagnetic radiation is known to have a dual nature as energy that travels as a wave and as a particle.

Electromagnetic radiation may be classified according to its wavelength or origin. For example, gamma rays and X-rays are similar, but have different origins. Gamma rays arise from energy-state transitions of the nucleus of an atom while X-rays come from energy-state transitions of extranuclear electrons. The classification of electromagnetic radiation according to wavelength and frequency is illustrated in Fig. 1.100.

Seeing as electromagnetic radiations or photons have properties of particles, they should also possess momentum. We calculate momentum as the product of mass and velocity. For relativistic conditions, the mass of a particle is a function of its speed according to Eq. (1.272) previously described or

$$m = \frac{m_0}{\sqrt{1 - (u^2/c^2)}} \tag{1.372}$$

where m and m_0 are the particle relativistic and rest masses, respectively, u is the particle speed and c, the speed of light. Gautreau and Savin (1999) transform Eq. (1.372) by squaring both sides and then multiplying each side by $c^4[1 - (u^2/c^2)]$ to yield the equation

$$m^2c^4 - m^2u^2c^2 = m_0^2c^4 \tag{1.373}$$

Using $E = mc^2$ and $E_0 = m_0c^2$ to define the relativistic and rest energies and $p = mu$ to define the particle momentum together with the fact that the rest energy of the photon is always zero, *i.e.*, $m_0 = 0$, Eq. (1.373) becomes

$$E^2 - p^2c^2 = \left(m_0c^2\right)^2 \tag{1.374}$$

$$E^2 - p^2c^2 = 0 \tag{1.375}$$

and

$$p = \frac{E}{c} \tag{1.376}$$

From Eqs. (1.371), (1.375), and (1.376) we can further describe the photon momentum as

$$p = \frac{E}{c} = \frac{h\nu}{c} = \frac{h}{\lambda} \tag{1.377}$$

To illustrate the use of the above equations defining the relationships of photon properties, we can calculate the wavelength, frequency, and momentum of a 2-MeV gamma-ray photon. From Eq. (1.371) we can write the equation for calculating the wavelength as

$$\lambda = \frac{hc}{E} \tag{1.378}$$

Planck's constant, h, can be converted from units of J·s to eV·s as

$$\begin{aligned} h &= \frac{6.626 \times 10^{-34}\ \text{J s}}{1.602 \times 10^{-19}\ \text{J/eV}} \\ &= 4.136 \times 10^{-15}\ \text{eV s} \end{aligned} \tag{1.379}$$

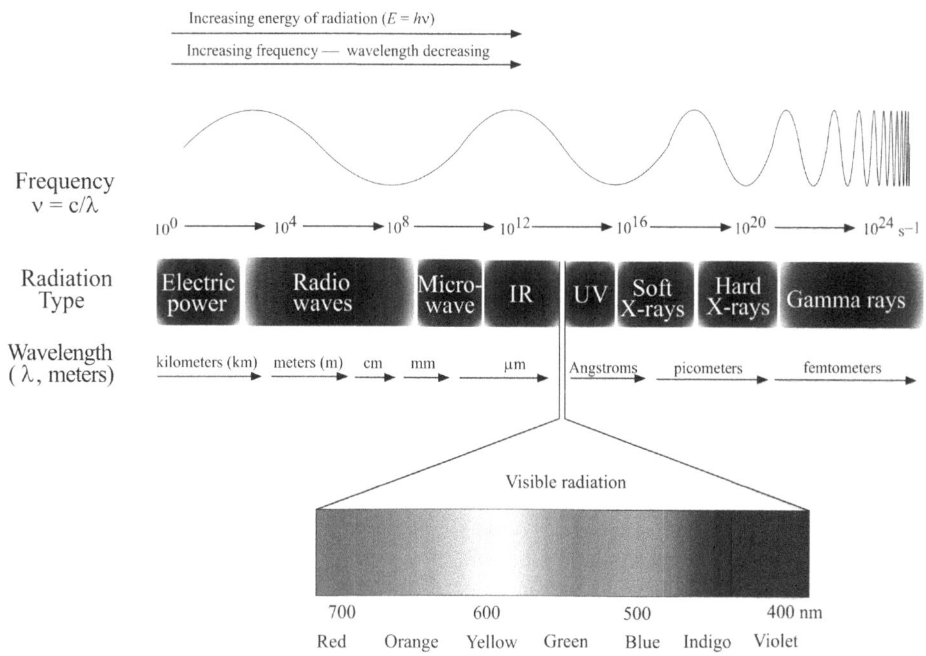

FIGURE 1.100 The electromagnetic radiation spectrum is illustrated above by classifying the types of radiation as a function (from left to right) of decreasing wavelength (λ) and increasing frequency (ν). As the wavelength of the radiation decreases, its frequency (ν) and its energy ($E = h\nu$) increase, where h is Planck's constant ($h = 6.626 \times 10^{-34}$ J s). There are no sharp lines of demarcation between the radiation types, as their wavelengths and frequencies can overlap between neighboring radiation types. Thus, the wavelengths and frequencies given above are only approximate. The uppermost wave in the figure is provided as an abstract illustration of how radiation frequency increases with decreasing wavelength. Radiation, visible to the human eye, is found in a very narrow region between the infrared (IR) and ultra violet radiation (UV) bands with decreasing wavelengths in the region of 740 to 380 nanometers (nm) yielding the colors of red, orange, yellow, green, blue, indigo, and violet. For the color version of this figure, the reader is referred to the electronic version of the book.

and hc is calculated as

$$\begin{aligned} hc &= \left(4.136 \times 10^{-15}\ \text{eV s}\right)\left(2.9979 \times 10^{8}\ \text{m/s}\right) \\ &= 12.4 \times 10^{-7}\ \text{eV m} \\ &= 12.4\ \text{keV Å} \end{aligned} \tag{1.380}$$

The wavelength of the 2.0 MeV gamma ray according to Eq. 1.378 becomes

$$\lambda = \frac{hc}{E} = \frac{12.4\ \text{keV Å}}{2 \times 10^{3}\ \text{keV}} = 0.0062\ \text{Å} \tag{1.381}$$

The frequency is calculated according to Eq. (1.371) as

$$\begin{aligned} \nu = \frac{c}{\lambda} &= \frac{2.9979 \times 10^{8}\ \text{m/s}}{0.0062 \times 10^{-10}\ \text{m}} = 484 \times 10^{18}\ \text{s}^{-1} \\ &= 4.84 \times 10^{20}\ \text{Hz} \end{aligned} \tag{1.382}$$

The momentum is expressed according to Eq. (1.376) as

$$p = \frac{E}{c} = 2.0\ \text{MeV}/c \tag{1.383}$$

Notice that relativistic calculations of momentum have units of MeV/c, while conventional units of momentum are derived from mass times velocity or kg m/s. Units of MeV/c can be converted to the conventional units with the conversion factor 1 MeV/c = 5.36×10^{-22} kg m/s (Fermi National lab: https://quarknet.fnal.gov/toolkits/ati/whatgevs.html).

B. Gamma radiation

Radionuclide decay processes often leave the product nuclide in an excited energy state. The product nuclide in such an excited state either falls directly to the ground state or descends in steps to lower energy states through the dissipation of energy as gamma radiation.

A nuclide in an excited energy state is referred to as a nuclear isomer, and the transition (or decay) from a higher to a lower energy state is referred to as isomeric transition. Gamma rays are emitted in discrete energies corresponding to the energy state transitions a nucleus may undergo when in an excited state. The energy, E_γ, of a gamma ray may be described as the difference in energy states of the nuclear isomers:

$$E_\gamma = h\nu = E_1 - E_2 \tag{1.384}$$

where $h\nu$ *is* the energy of the electromagnetic radiation described previously in Section XI.A, and E_1 and E_2 represent the energy levels of the nuclear isomers.

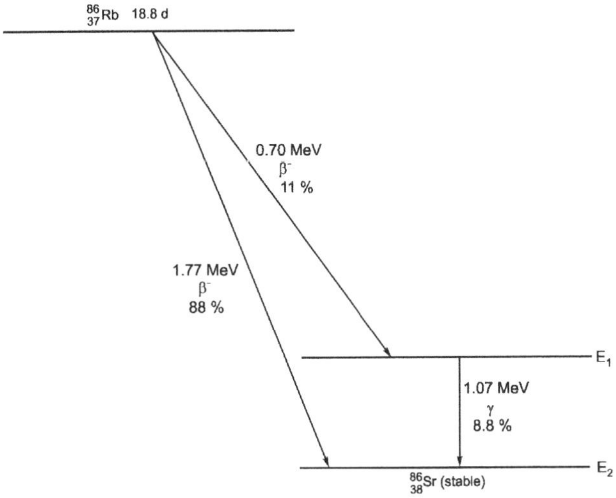

FIGURE 1.101 Decay scheme of $^{86}_{37}$Rb.

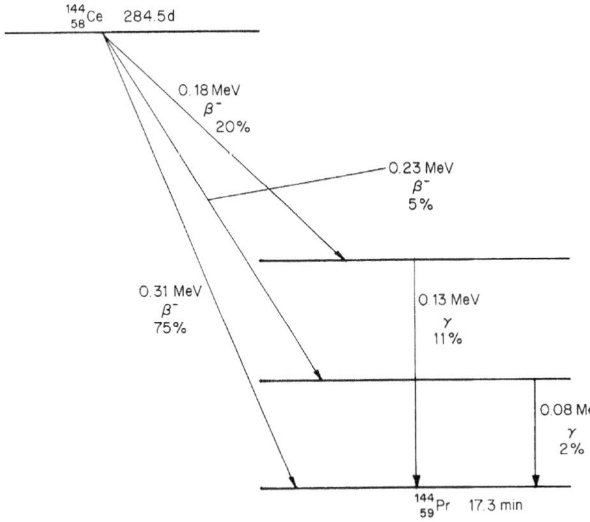

FIGURE 1.102 Decay scheme of $^{144}_{58}$Ce.

Let us consider the decay schemes of some radionuclides to illustrate the process in more detail.

Fig. 1.101 shows the decay scheme of $^{86}_{37}$Rb with a half-life of 18.8 days. This nuclide decays by β^- emission with an increase in atomic number to $^{86}_{38}$Sr. Eighty-eight percent of the beta particles emitted have a maximum energy of 1.77 MeV; the remaining 11% have a maximum energy of 0.70 MeV. The percentages cited and illustrated in the figure are referred to as transition probabilities or intensities. Obviously, a greater quantum of energy is released by the 1.77-MeV, β-decay process. As a consequence, the ^{86}Sr product nuclides that result from beta decay energy of 0.70 MeV (11%) are at a higher energy state than those that result from beta decay energy of 1.77 MeV. The energy difference of the two ^{86}Sr product nuclide isomers, $E_1 - E_2$, is equivalent to the difference of the two beta-decay energies, 1.77 MeV − 0.70 MeV = 1.07 MeV. Consequently, the ^{86}Sr nuclide isomers, which are products of the 0.70-MeV β-decay process, can emit the remaining energy as 1.07-MeV gamma-ray photons.

As illustrated in Fig. 1.101, 11 % of the parent ^{86}Rb nuclides decay to a ^{86}Sr nuclear isomer at an elevated energy state. Not all of these isomers immediately decay to the ground state. Only 8.8% of the ^{86}Rb → ^{86}Sr disintegrations result in the emission of a gamma-ray photon of 1.07 MeV in a given period of time. For example, a 37-kBq sample of ^{86}Rb by definition would emit 2.22×10^6 beta particles in 1 minute (37,000 dps × 60 s/min). However, only $(2.22 \times 10^6)(0.088) = 1.95 \times 10^5$ gamma-ray photons of 1.07 MeV can be expected to be emitted in 1 minutes from this sample.

Fig. 1.102 shows the somewhat more complicated decay scheme of $^{144}_{58}$Ce, which has a half-life of 284.5 days. This nuclide decays by β^- emission with an increase in atomic number to $^{144}_{59}$Pr. In this case, three distinct β-decay processes produce three nuclear isomers of the daughter ^{144}Pr. Seventy-five percent of the beta particles emitted have a maximum energy of 0.31 MeV, 20% have a maximum energy of 0.18 MeV, and the remaining 5% have a maximum energy of 0.23 MeV. Obviously, a greater amount of energy is released by the 0.31-MeV β-decay process. As a consequence, ^{144}Pr nuclides that result from beta-decay energy of 0.23 MeV can decay to the ground state with the emission of gamma-ray photons with an energy equivalent to 0.08 MeV, that is, 0.31 − 0.23 MeV. Likewise, ^{144}Pr isomers at an even higher energy state are products of the 0.18-MeV β-decay process. These can decay to the ground state with the emission of gamma-ray photons of energy 0.13 MeV, that is, 0.31 − 0.18 MeV. Not all of the product isomers decay with the immediate emission of gamma radiation, and the abundance of these transitions is given in Fig. 1.102.

It is also possible that essentially all of the product nuclides of a decay reaction will be at an excited or elevated energy state and subsequently fall to a lower energy state by the emission of gamma radiation.

The decay scheme of the nuclide $^{22}_{11}$Na with a 2.6-year half-life serves as an example (see Fig. 1.103). The $^{22}_{11}$Na nuclides decay by both electron capture (EC) and β^+ emission, at relative proportions of ~10% and ~90%, respectively, to yield immediate $^{22}_{10}$Ne product nuclides in an elevated energy state. Only a trace of the ^{22}Na nuclides (0.06%) decay directly to the ground state. All of the $^{22}_{10}$Ne isomers in the excited energy state decay immediately with the emission of gamma-ray photons of 1.28 MeV energy, which is equivalent to the difference of the energy levels of the two $^{22}_{10}$Ne isomers and also equivalent to the difference in energies released by the two β^+-decay processes (1.82 −0.54 MeV).

Isomeric transition, as described earlier, is a decay process in which γ emission is the sole process of

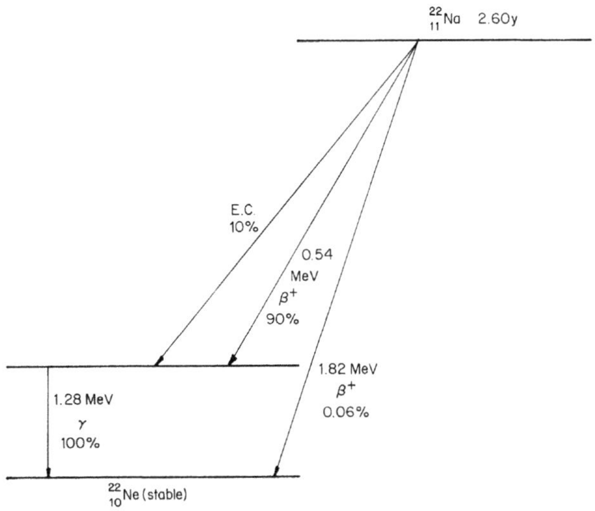

FIGURE 1.103 Decay scheme of $^{22}_{11}$Na.

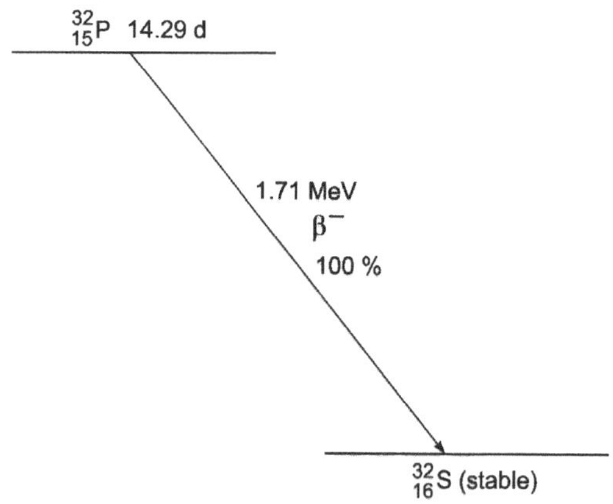

FIGURE 1.105 Decay scheme of $^{32}_{15}$P.

eliminating energy from an excited nucleus. This mode of decay is referred to as isomeric transition because neither the mass number, A, nor the atomic number, Z, of a nuclide ($^{A}_{Z}$X) changes in the decay process, and the nuclides are considered to be in isomeric energy states.

In the previous examples (Figs. 1.101−1.103) the isomeric energy state transitions are short-lived; that is, they occur virtually immediately after the other decay processes (*e.g.*, β^-, β^+, and EC); and the half-life of the parent nuclide is dependent on these initial processes. If, however, the isomeric transitions are long-lived, the nuclide is considered to be in a metastable state. These nuclides are denoted by a superscript m beside the mass number of the nuclide. The radionuclide $^{119m}_{50}$Sn with a 250-day half-life is an example. Its decay scheme, shown in Fig. 1.104, consists of the emission of two γ photons of 0.065 and 0.024 MeV energy falling from the 0.089-MeV excited state to the ground (stable) state.

Gamma radiation is not produced in all radionuclide decay processes. Instead, some radionuclides decay by emitting only particulate radiation to yield a product nuclide at an unexcited ground state. An example is the radionuclide ^{32}P, which decays by the scheme shown in Fig. 1.105.

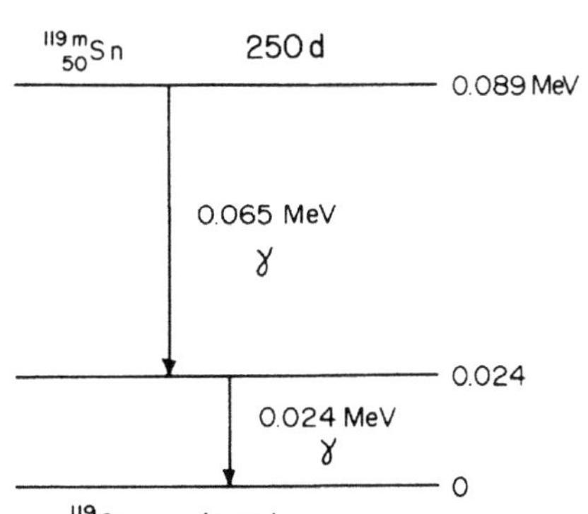

FIGURE 1.104 Decay scheme of $^{119m}_{50}$Sn.

C. Annihilation radiation

The negatron or negative beta particle, produced by β decay, will travel through matter until it has completely dissipated its kinetic energy via ionization, electron excitation, or bremsstrahlung. The negatron then at rest acts as an atomic or free electron in matter.

A positron or positive beta particle, however, is an "antiparticle" of an electron and consequently, in the electron environment of atoms, has a definite instability. A given positron (i.e., positive electron) emitted by pair production or by β^+ decay will also dissipate its kinetic energy in matter via interactions described previously for the case of the negatron. However, as the positron loses its kinetic energy and comes to a near stop, it makes contact with an electron (Fig. 1.106) with nearly simultaneous annihilation of the positron and the electron masses and their conversion into energy. The annihilation involves the formation of positronium, which is a short-lived association of the

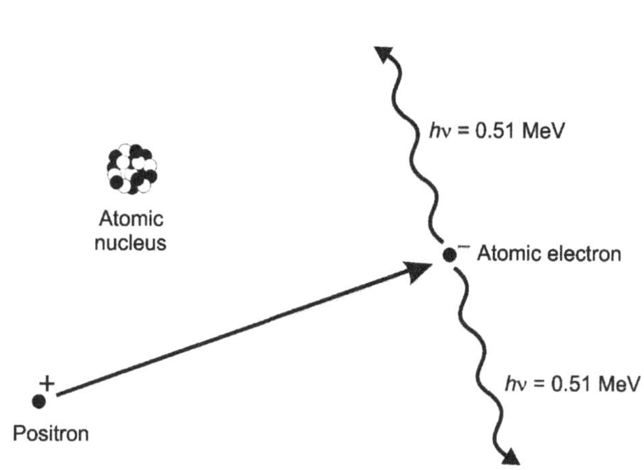

FIGURE 1.106 Annihilation. The interaction between a positron and an electron, and the conversion of their masses into two photons of 0.511 MeV energy.

positron and electron. Its lifetime is only approximately 10^{-10} or 10^{-7} seconds, depending on whether the spin states of the associated particles are parallel (ortho-positronium) or opposed (para-positronium). The para-positronium is the shorter-lived spin state. The energy released in this annihilation appears as two photons emitted in opposite directions. This transformation of mass into energy, considered as the reverse of pair production, is described as

$$e^{+} + e^{-} = 2h\nu = 2E_{\gamma} \tag{1.385}$$

where a positron, e^{+}, and negative electron, e^{-}, combine to form two gamma-ray photons of energy E_{γ}. To maintain the equivalence of mass and energy, the equivalent of two electron rest masses (0.511 MeV) must appear as photon energies. In agreement with Eq. (1.385), the annihilation results in the emission of two 0.511-MeV photons in opposite directions.

D. Line-spectra X-radiation and bremsstrahlung

X-radiation is electromagnetic radiation similar to gamma (γ) radiation. Both X-rays and γ-rays are identical in their properties. Both X-rays and γ-rays are photons that possess a quantum of energy defined by Eq. (1.371), namely, $E = h\nu$, which is the product of Planck's constant and the frequency of the radiation. X-rays and γ-rays differ only in their origin. X-rays originate from electron energy transitions, and γ-rays originate from the nucleus of an atom as a nucleus at an elevated energy state decays to a lower more stable energy state. There are two types of X-rays, which are the following:

(i) Electromagnetic radiation characterized by discrete spectral lines of energy resulting from electron transitions in shells of the atoms (e.g., *K*, *L*, *M*,... shells) from which they originate. These are X-ray emissions produced when an electron from a given shell is ejected from an atom and an electron from outer shell fills the vacancy left by the ejected electron. This is followed generally by a cascade of electron transitions from outer electron shells, since the filling of one electron vacancy by the transition of an electron from an outer to inner shell will leave another electron vacancy in the outer shell. For example, an electron vacancy in the *K*-shell created by the capture of an electron by the nucleus (e.g., EC or *K*-capture decay) is filled by an electron in the next outer *L*-shell creating an electron vacancy in that shell, which is followed by the filling of that electron vacancy by an electron in the next outer *M*-shell, etc. The spectra of these discrete energy lines of electromagnetic radiation are characteristic of the atoms from which they originate.

(ii) Electromagnetic radiation characterized as a broad smear of continuous energies referred to as **bremsstrahlung**, which is produced by electrons or other charged particles, such as protons and alpha-particles, as they are accelerated toward an atomic nucleus and are deflected by the coulomb field of the nucleus. This was the type of X-ray discovered by Wilhelm Röntgen in 1895, who was the first to discover X-rays. Röntgen discovered the mysterious invisible rays on November 8, 1895, when he was studying cathode rays, i.e., the current that would flow through a partially evacuated glass tube (cathode-ray tube then referred to as a Crooke's tube). With such a tube covered in black paper and in a dark room, he noticed that a paper plate covered with the chemical barium platinocyanide would become fluorescent (give off light) even at a distance of 2 meters from the cathode ray tube. He was able to demonstrate that the invisible rays came from the collision of the cathode rays (electrons) with the glass surface of the cathode-ray tube or from the collision of the cathode rays with other materials such as aluminum inserted into the glass of the tube. Further studies by Röntgen demonstrated that these rays could travel through various materials to varying extents, when these were placed in the path of the invisible rays, and the transmitted rays could be measured with photographic plates. He gave these mysterious rays the name X-rays, because of their unknown nature and "for the sake of brevity", as he stated in his original papers (Roentgen (1895, 1896).

1. X-rays characterized by discrete spectral lines

In Section X.B.3 we discussed the electron-capture (EC) or *K*-capture decay process whereby an electron from one of the atomic shells (generally the innermost *K* shell) is absorbed by the nucleus, where it combines with a proton to form a neutron. The vacancy left by the electron from the *K* shell is filled by an electron from an outer shell (generally the adjacent *L* shell). Transitions produced in electron shell energy levels result in the emission of energy as X-radiation. This radiation consists of photons of electromagnetic radiation similar to gamma radiation. X-radiation and gamma radiation differ only in their origin. X-rays arise from atomic electron energy transitions and gamma rays from energy state transitions of an atomic nucleus. Another nuclear decay process that results in the loss of atomic electrons is decay by internal conversion (IC), which competes with gamma-ray emission, discussed in Section X.C of this chapter. Also, Auger electron emission, that may follow EC decay, will leave a vacancy in an electron shell that would cause electron transitions from outer shells during the filling of that vacancy and the emission of X-

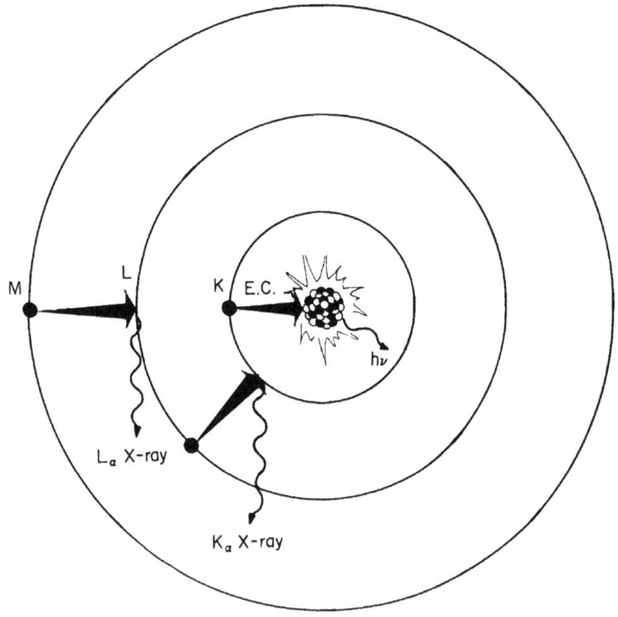

FIGURE 1.107 Electron capture (EC) decay and the accompanying gamma ($h\nu$) and X-radiation.

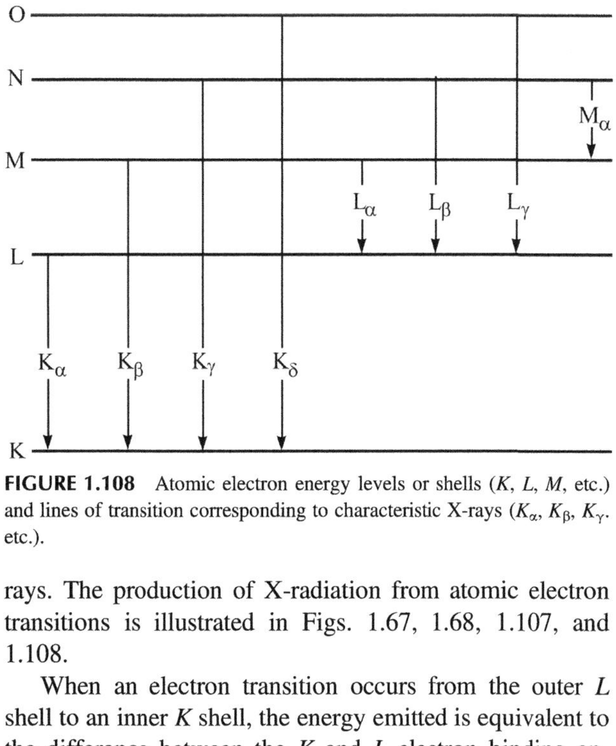

FIGURE 1.108 Atomic electron energy levels or shells (K, L, M, etc.) and lines of transition corresponding to characteristic X-rays (K_α, K_β, K_γ. etc.).

rays. The production of X-radiation from atomic electron transitions is illustrated in Figs. 1.67, 1.68, 1.107, and 1.108.

When an electron transition occurs from the outer L shell to an inner K shell, the energy emitted is equivalent to the difference between the K and L electron binding energies. The electron transitions that ensue in the filling of vacancies result in energy lost by the atom as X-radiation, which is equivalent to the difference of the electron binding energies of the outer, E_{outer}, and its new inner state, E_{inner}, as described by

$$h\nu = E_{outer} - E_{inner} \tag{1.386}$$

The radiation emitted consists of a discrete line of energy characteristic of the electron shell and, consequently, of the atom from which it arises.

The production of X-rays in radionuclide decay is, however, more complex. The filling of one electron vacancy in an inner shell is followed by a series of electron transitions in an overall adjustment of electrons in outer shells. This gives rise to further X-rays with lines characteristic of outer shells. Such electron transitions, each resulting in the emission of discrete lines of characteristic X-rays, are illustrated in Fig. 1.108. The transitions are identified by a letter corresponding to the shell (K, L, M, etc.) with vacancy giving rise to the X-ray photon and a subscript (α, β, γ, etc.) to identify, from among a series of outer electron shells of the atom, the shell from which the electron vacancy is filled. For example, an X-ray arising from an electron transition from the L to the K shell is denoted as K_α and that arising from a transition from the M to the K shell as K_β. Transitions involving the filling of electron vacancies in the L shell from outer M, N, and O shells are denoted by L_α, L_β, and L_γ, etc.

The complexity of X-ray lines emitted and their abundances of emission are compounded by the existence of other mechanisms of X-ray production in unstable atoms. One of these mechanisms is the production of Auger electrons. An electron transition involving the filling of an electron vacancy in an inner shell from an outer shell of an atom may produce an Auger electron (see Section X.D), which results in the emission of an atomic electron from a shell farther away from the nucleus. The vacancy left by the Auger electron gives rise to X-rays characteristic of outer shells following the electron readjustments that ensue. Auger electrons can be emitted from a variety of electron shells, followed by an equal variety of characteristic X-rays from subsequent electron adjustments in outer shells.

Any process that would cause the ejection of an atomic electron of an inner shell can result in the production of X-radiation. Other processes, such as radiation-induced ionization within inner electron shells of atoms, will result in the filling of the missing electron from an outer shell and a cascade of electron rearrangements from outer to inner shells. Any radiation source that can produce ionization, even X-rays that impinge on a target material, will produce additional X-rays as a consequence of the ionization of internal electron shells of atoms, and these X-ray emissions will be characteristic of the elements that were ionized by the external radiation. Nobel Laureate Charles Glover Barkla (Barkla, 1917) made this observation in 1917 in his Nobel lecture where he stated

Each element when traversed by x-rays emits x-radiations characteristic of the element; each characteristic radiation is unaffected by changes in the physical condition or

state of chemical combination of the radiating element, and its quality is independent of that of the exciting primary radiation. But only primary radiations of shorter wavelength are able to excite the characteristic x-radiations … all of the radiations hitherto definitely observed have fallen into three series, the K-, L-, and M-series.

Thus, each element can be characterized by its X-ray emission spectra. It was Henry Moseley who devised an apparatus whereby he could bombard individually all of the known and available elements with cathode rays and measure the wavelengths of the X-rays emitted by each element (Moseley, 1913, 1914). In his first report Moseley (1913) measured the X-ray emission spectra of 21 elements of increasing numbers of the Periodic Table of the Elements from aluminum to silver and determined the wavelengths of the α and lower intensity β lines that belong to Barkla's *K* series. In a subsequent paper Moseley (1914) measured the wavelengths of the *L* series of X-ray emissions of 24 elements of increasing numbers of the Periodic Table of the Elements from zirconium to gold. Of the *L*-series he measured the wavelengths of α, β, γ, and δ lines of decreasing wavelengths and decreasing intensities. Moseley then plotted the square root of the radiation frequencies against a characteristic integer (N) assigned to each element. Starting with aluminum, being the 13th element in the periodic table, he assigned it the number N = 13. The next element, silicon, was assigned the next highest integer N = 14, phosphorus N = 15, etc. A copy of his plots are illustrated in Fig. 1.109. The plots showed an incredibly straight line relationship between the assigned integer N and the X-radiation frequencies for each element. From this relationship he concluded the following with the writer's comments in brackets (Moseley, 1914):

Now if either the elements were not characterized by these integers or any mistake had been made in the order chosen, or in the number of places left for unknown elements [places for three unknown elements in the Periodic Table were found, See Fig. 1.109], these irregularities would at once disappear. We can therefore conclude from the evidence of the x-ray spectra alone, without using any theory of atomic structure, that these integers are really characteristic of the elements. Further, as it is improbable that two different stable elements should have the same integer, three, and only three, more elements are likely to exist between Al and Au. As the x-ray spectra of these elements can be confidently predicted, they should not be difficult to find … There is ever reason to suppose that the integer which controls the x-ray spectrum is the same number of electrical units in the nucleus …

In this work Moseley was able to demonstate that the number of electrons in the atoms of each element are equivalent to what we now refer to as the atomic number *Z* (Moseley's N) and that the X-ray emission spectra is a characteristic of each element.

Nobel Laureate Karl Manne Siegbahn provided insight into the energy quanta emitted by the X-rays and consequently the energy states of the electrons in the *K*-, *L*-, *M*-, *N*- and other electron energy levels of the elements. The impact of Siegbahn's findings was underscored in the Nobel Prize Presentation Speech (Gullstrand, 1925) with the statement

In an element that can emit both K-and L-rays, the former radiation has much shorter wavelengths and consequently greater frequencies than the latter. As the energy quanta are proportional to the frequencies, therefore, the K radiation involves a larger change in the energy of the atom than the L radiation; and in the atomic theory this is as much as to say that an orbit into which an electron falls on emission of a K-line must lie nearer the nucleus than an orbit to which an electron falls on emission of an L-line. In this way it was inferred that there is a K-level nearest the nucleus, outside that a L-level, and after that a M-level and a N-level (and so on …), all these four being experimentally determined [by Manne Siegbahn].

Because X-radiation is characteristic of the atom from which it arises, it is customary to identify the element along with the X-ray photon (*e.g.*, Cr *K* X-rays, Hg *L* X-rays, etc., and many examples are listed in Appendix A). In these examples, the fine structure of the X-ray emissions is not given in the Appendix, and the lines are grouped together as *K* and *L* X-rays.

Whenever, a nuclide undergoes alpha- or beta-decay including EC decay, the daughter nucleus may be left at an excited energy state whereby the daughter nucleus then decays to the ground state with the emission of gamma radiation. The emission of IC electrons often competes with gamma emission, which leaves vacancies in electron shells and the concomitant emission of X-rays. In such cases, the X-ray emissions are characteristic of the daughter atom. See examples throughout Appendix A in the Table of Radioactive Isotopes where the X-ray emission intensities are listed as those of the daughter nuclide rather than the parent.

2. Bremsstrahlung

Bremsstrahlung is a type of X-radiation that is characterized by a broad smear of continuous energies of electromagnetic radiation anywhere in the range between zero and an energy maximum. In general, any charged particle that undergoes acceleration will emit electromagnetic radiation;

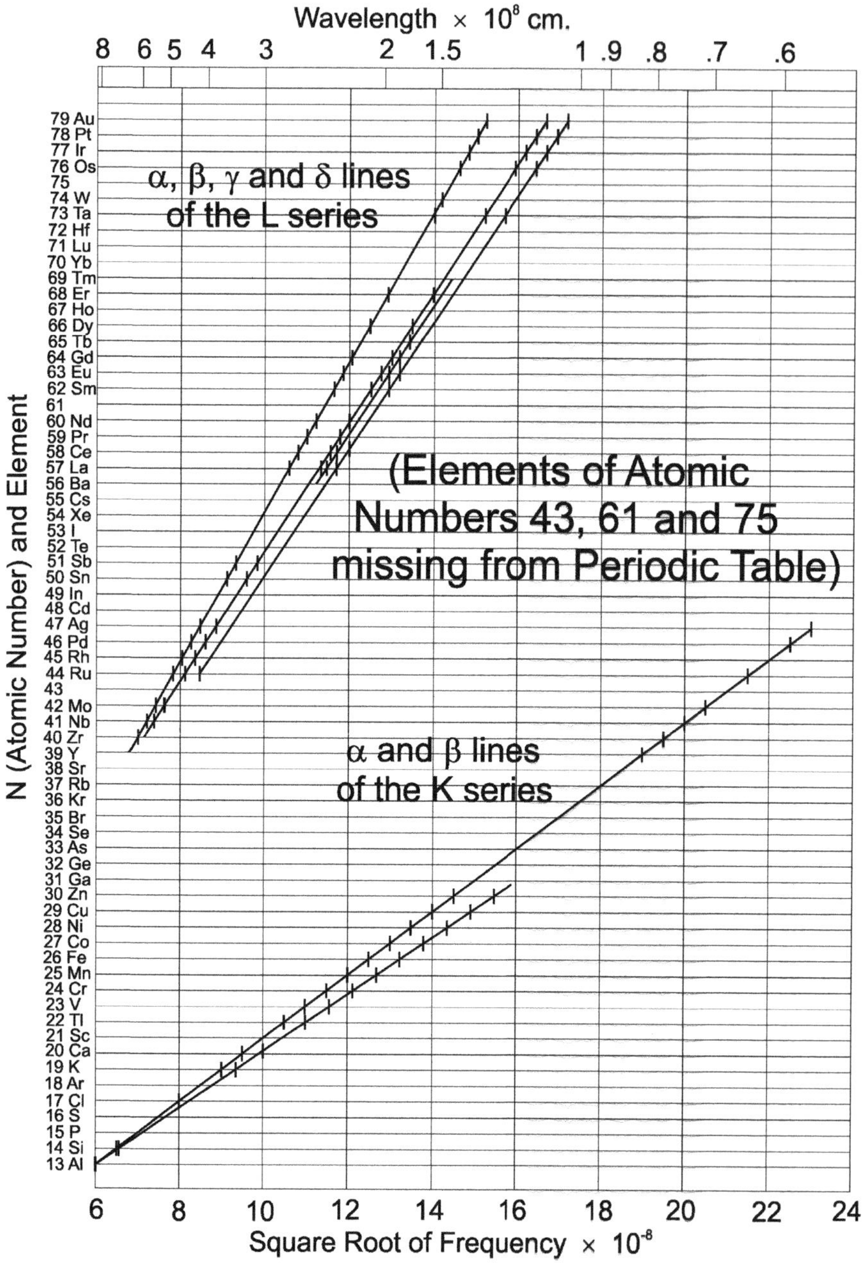

FIGURE 1.109 Artist's portrayal of Moseley's original 1914 graph of the frequencies and wavelengths of X-rays emitted by all of the then known elements plotted against each element and an integer N assigned to each element starting at aluminum (Al, assigned N = 13) to gold (Au, assigned N = 79). Three missing and yet unknown elements are listed with blank spaces adjacent to integers N = 43, 61, and 75. The two lines of the lower curves (elements N = 13 to 47) are X-rays resulting from α and β lines of the K series of electron transitions, and the four lines of the upper curves (elements N = 40 to 79) are X-rays resulting from α, β, γ, and δ lines of the L series of electron transitions. The original Moseley graph, drawn by his own hand, may be seen in the Moseley Room of the Cavendish Laboratory, University of Oxford). *From Moseley (1914).*

thus, an accelerating charged particle will emit bremsstrahlung when it is made to decelerate upon deflection from its path of travel by collision or near encounter with an atomic nucleus. Charged particles undergoing acceleration, such as beta particles, electrons, protons, deuterons, and alpha particles, can emit bremsstrahlung as these particles decelerate in a series of collisions with atomic nuclei. This mechanism is illustrated in Fig. 1.110, where a beta particle traveling through matter approaches a nucleus and is deflected by it. This deflection causes a deceleration of the beta particle and consequently a reduction in its kinetic energy with the emission of energy as a photon of bremsstrahlung, which is taken from the German language meaning "braking radiation." The phenomenon is described by

$$h\nu = E_i - E_f \tag{1.387}$$

where $h\nu$ is the energy of the photon of bremsstrahlung, E_i is the initial kinetic energy of the beta particle prior to collision or deflection, producing a final kinetic energy E_f of the electron.

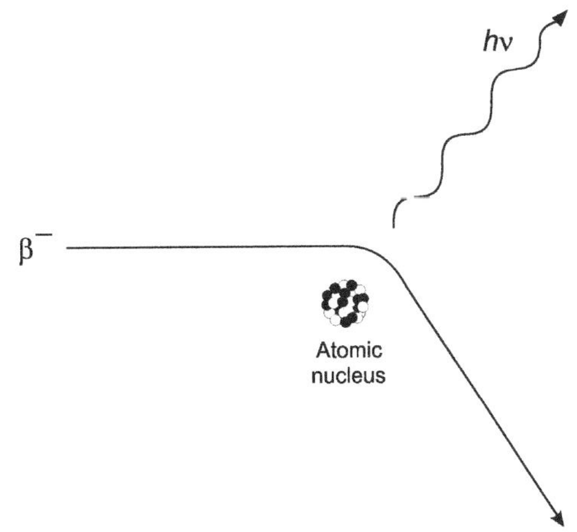

FIGURE 1.110 Bremsstrahlung production. A beta particle is deflected by an atomic nucleus and loses kinetic energy with the emission of a photon of X-radiation.

As described by Evans (1955), the acceleration produced by a nucleus of charge *Ze* on a particle of charge *ze* and mass *M* is proportional to Zze^2/M, where *Z* is the atomic number of the nucleus, *e* is the elementary charge $= 4.803 \times 10^{-10}$ esu, and *z* is the particle atomic number or 1 for electrons and protons, 2 for alpha particles. Thus, the bremsstrahlung intensity, which is proportional to the square of the product of the amplitude and the charge *ze*, will vary according to $Z^2z^4e^6/M^2$ (Evans, 1955). From this relation it is seen that the production of bremmstrahlung per atom of absorber is proportional to the square of the atomic number of the absorber material, and that the total bremsstrahlung is inversely proportional to the square of the mass *M* of the accelerating charged particle. It is intuitively obvious and it has been demonstrated through experiment, that an absorber material of higher atomic number will yield a greater production of bremsstrahlung per atom, because the size of the nucleus as well as its charge increases with atomic number, and thus the probabilities of interaction of the accelerating charged particle with a nucleus will be a function of the magnitude of the charge on the nucleus as well as the nuclear size. For example, a charged particle (*e.g.*, electron or beta particle) of given energy will yield more bremsstrahlung upon interaction with absorber material consisting of pure Pb than one of Cu or C (*e.g.*, graphite), that is, bremsstrahlung in Pb $(Z = 82) > \text{Cu}\ (Z = 29) > \text{C}\ (Z = 6)$. The inverse proportionality of bremsstrahlung to the square of the mass *M* of the charged particle, yields as an outcome that bremsstrahlung production by electrons ($M = 0.0005485$ u) is the most significant among all other charged particles, which are of much higher mass, *e.g.*, the muon ($M = 0.1134$ u), proton ($M = 1.007$ u), deuteron ($M = 2.014$ u), or the alpha particle ($M = 4.001$ u). For example the muon, which has a mass equivalent to 206.7 times that of the electron, would yield less bremsstrahlung than an electron of equal energy in the same absorber medium by a factor of more than 42,000, *i.e.*, $M_\mu^2/M_e^2 = 0.1134^2/0.0005485^2 = 42,700$. A proton of the same energy and in the same absorber material would yield yet less bremsstrahlung than that of the electron by a factor of $M_p^2/M_e^2 = 1.007^2/0.0005485^2 = 3.37 \times 10^6$. Nevertheless, bremsstrahlung from accelerated charged particles, such as accelerated muons and protons, are significant (Mahjour-Shafiei et al., 2006; Cohen et al., 2008; Johansson and Wilkin, 2009; Pérez et al., 2014; Sandrock et al., 2018).

The tremendous difference between the magnitudes of bremsstrahlung production by protons and electrons of the same energy in a given absorber material described in the previous paragraph (*i.e.*, $M_p^2/M_e^2 = 1.007^2/0.0005485^2 = 3.37 \times 10^6$) was utilized by Olzem et al. (2007) to measure cosmic-ray positron radiation through bremsstrahlung conversion. A major difficulty in the measurement of cosmic-ray positrons in the energy range of 50 GeV is the high background of cosmic-ray protons, which exceed the number of positrons by a factor of 10^4 (Aguilar et al., 2002; Beatty et al., 2004; L'Annunziata, 2007, 2016). By detecting the bremsstrahlung produced by the high-energy (1−50 GeV) cosmic-ray positrons, the proton background in this same energy range is suppressed by a factor of more than 3×10^6 with respect to the electron (*i.e.*, positrons), as described by Olzem et al. (2007). Bremsstrahlung by cosmic-ray pions are also significantly suppressed compared to electron bremsstrahlung by a factor of $M_\pi^2/M_e^2 = (0.1485\text{u})^2/(0.0005485\ \text{u})^2 = 7.3 \times 10^4$. A diagram illustrating the process of cosmic-ray positron bremsstrahlung conversion is illustrated in Fig. 1.111.

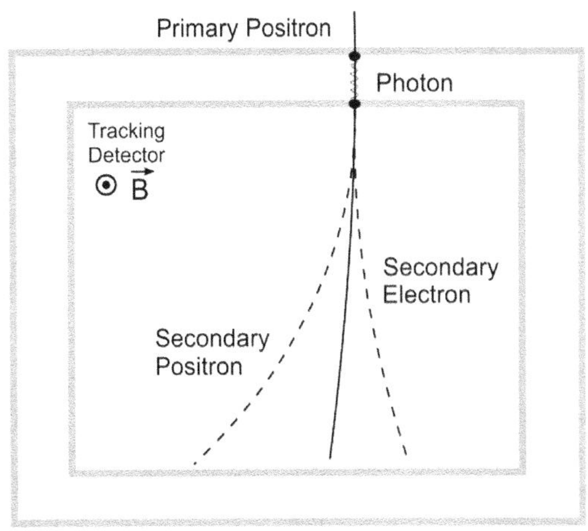

FIGURE 1.111 Signature of pair production by a bremsstrahlung photon. *From Olzem et al. (2007). Reprinted with permission of Elsevier © 2007.*

As illustrated in Fig. 1.111 a primary cosmic-ray positron enters into the detector material from above producing bremsstrahlung photons. The bremsstrahlung photon then converts into a positron—electron pair. Three particles are observed in the tracking detector, namely, the positron and negatron each of 0.511 MeV energy created in the pair production, which are highly deflected in a magnetic field, and the incident high-energy positron, which is only slightly deflected in the magnetic field. Olzem et al. (2007) report that additional background rejection of cosmic-ray mesons and protons are provided by the photon emission angles, which are essentially zero for high-energy positrons and significantly larger than zero for mesons and protons due to their higher mass.

a. Bremsstrahlung in beta-particle absorbers

When beta particles from a particular radionuclide source strike an absorber material, a wide spectrum of bremsstrahlung photon wavelengths (or energies) will be produced. The bremsstrahlung photon energies will range from zero to a maximum photon energy, $h\nu_{max}$. The broad spectrum of bremsstrahlung is due to the broad possibilities of different interactions, *i.e.*, deflections or collisions, that the beta particles can have with atomic nuclei of the absorber and the broad spectrum of beta-particle energies emitted from any given radionuclide. In a given spectrum of bremsstrahlung the shortest wavelength, λ_{min}, is observed when a beta particle or electron undergoes a direct collision with the nucleus of an atom and loses all of its kinetic energy, $h\nu_{max}$, as bremsstrahlung or X-radiation according to the relation

$$h\nu_{max} = \frac{hc}{\lambda_{min}}, \tag{1.388}$$

which follows the energy-wavelength Planck—Einstein relation described previously by Eq. (1.371).

Let us consider an example of a 1710 keV beta particle from ^{32}P ($E_{max} = 1.71$ MeV) striking a nucleus. If the beta particle loses all of its energy in the collision, the wavelength of the bremsstrahlung emitted from this interaction according to Eq. (1.388) would be

$$\lambda = \frac{hc}{h\nu_{max}} = \frac{12.4 \text{ keV Å}}{1710 \text{ keV}} = 0.00725 \text{ Å} \tag{1.389}$$

See Eq. (1.380) for the conversion of the constant hc to convenient units of eV m or eV Å. Bremsstrahlung production by high-energy beta particles in absorber material of high atomic number is significant (see Section XVI). Consequently to reduce the production of bremsstrahlung in radiation shielding against the harmful effects of high-energy beta particles, an absorber of low atomic number (*e.g.*, plastic) may be preferred over one of high atomic number (*e.g.*, Pb-glass).

Ionization and electron excitation were previously described as predominant mechanisms by which a traveling beta particle may lose its kinetic energy in matter. However, the production of bremsstrahlung may also be another significant mechanism for the dissipation of beta-particle energy, particularly as the beta-particle energy and the atomic number of the absorber increase (Kudo, 1995; Czarnecki et al., 2016, 2017). A more thorough treatment is found in Section XVI.A of this chapter, which includes examples of calculations involved to determine the degree of bremsstrahlung production as a function of beta-particle energy and absorber atomic number. In general terms we can state that for a high-energy beta particle such as the beta particle of highest-energy emitted from ^{32}P ($E_{max} = 1.71$ MeV) in a high-atomic-number absorber such as lead (Pb = 82), bremsstrahlung production is significant. In a substance of low atomic number such as aluminum (Al = 13) bremsstrahlung production by a 1.71 MeV beta particle is reduced to more than six-fold as compared to lead (see Table 1.21 in Section XI.D.3.a).

In view of the wide spectrum of beta-particle energies emitted from radionuclides and the wide variations of degree of beta-particle interactions with atomic particles, the production of a broad spectrum, or smear, of photon energies of bremsstrahlung is characteristic. This contrasts with X-radiation, which is emitted as a result of atomic electron rearrangement processes, resulting in the emission of discrete lines of photon energy.

b. Artificially produced bremsstrahlung

An apparatus used to produce artificially X-rays, such as those employed in medical diagnosis or X-ray diffraction, functions on a similar principle of bremsstrahlung described previously. The X-ray apparatus consists of an evacuated tube containing a cathode filament and a metal anode target such as tungsten ($A = 74$). A voltage potential is applied to the tube so that electrons emitted from the cathode accelerate toward the anode. Upon colliding with the tungsten anode the accelerated electrons lose energy as bremsstrahlung X-radiation. For example, an electron accelerated in an X-ray tube to an energy of 40 keV, which loses all of its energy upon impact with a tungsten nucleus would produce a single X-ray photon of wavelength calculated as

$$\lambda = \frac{hc}{h\nu} = \frac{12.4 \text{ keV} \cdot \text{Å}}{40 \text{ keV}} = 0.31 \text{ Å} = 0.031 \text{ nm} \tag{1.390}$$

A broad spectrum of bremsstrahlung is produced due to the various degrees of deflection or collision of the accelerated electrons with the target nuclei.

Electron accelerators are a good source of bremsstrahlung currently employed in food irradiation for the preservation and extension of the shelf life of food (Miller, 2003, 2005; Mehta et al., 2003; Auslender et al., 2004; Farkas, 2004, 2006; L'Annunziata, 2007, 2016; Farkas and Mohácsi-Farkas, 2010; Solanki et al., 2012; Sokovnin and Balezin, 2017). Food irradiation is the process where food is exposed to ionizing energy, utilizing gamma-ray photons emitted by ^{60}Co (or less frequently by ^{137}Cs), machine-generated X-rays (bremsstrahlung) of max. 5 MeV, or accelerated electrons of max. 10 MeV kinetic energy (Farkas, 2004). The use of accelerator sources is gradually replacing the use of radioisotope sources for the irradiation of food, mainly because the generation of the radiation can be controlled by a simple on-off switch (Cleland et al., 1991; Miller, 2003). The accelerators may be used to irradiate food directly with the electron beam or indirectly with the bremsstrahlung radiation produced after the electron beam strikes a converter material. To avoid the possibility of measurable food activation via photonuclear reactions, the kinetic energy of the electron beam is limited by regulation to 10 MeV, and the kinetic energy regulation for bremsstrahlung irradiation of foods is limited to 5 MeV.

Bremsstrahlung, electron beams, and gamma radiation have numerous industrial applications, which include, in addition to food preservation, the sterilization of health care products, the irradiation of blood to prevent transfusion-associated graft versus host disease, the sterilization of tissue for transplant surgery, the synthesis of radiation-cured rubber, and the manufacture of radiation-cured wood polymer composites, etc. These applications are reviewed in previous texts by the author (L'Annunziata, 2007, 2016).

c. Inner or internal bremsstrahlung

Bremsstrahlung of very low intensity, referred to as *inner* or *internal bremsstrahlung*, also occurs during the transforming nucleus in EC decay processes and beta particle emission. This is a continuous spectrum of X-ray photons that originates within the transforming atoms and can be attributed to the sudden change of nuclear charge when the beta particle is emitted or when an orbital electron is captured (Evans, 1955). Internal bremsstrahlung is explained by Cengiz and Almaz (2004) to occur during the emission of a β^- particle from the nucleus, as the β^- particle undergoes an acceleration at its birth and emits bremsstrahlung in the field of the emitting nucleus. The magnitude of inner bremsstrahlung was calculated to be approximately 1/137 quantum of internal bremsstrahlung per beta particle emitted from the nucleus (Knipp and Uhlenbeck, 1936; Evans, 1955). Internal bremsstrahlung spectra for radionuclides that emit β^- particles were calculated by Cengiz and Almaz (2004), Khalil (2011), and Singh and Dhaliwal (2014, 2015). This radiation was named inner or internal bremsstrahlung, because it originates from the nucleus the internal part of the atom in contrast to external bremsstrahlung that occurs when an external beta particle approaches an atom external to it from another source and is deflected by the nucleus of that atom.

In the EC decay process, the quantum of energy not carried away by the neutrino is emitted as internal bremsstrahlung. Thus, in EC decay, internal bremsstrahlung may possess energies between zero and the maximum, or transition energy of a radionuclide. When gamma radiation is also emitted, the internal bremsstrahlung may be masked by the more intense gamma rays and go undetected. However, in the absence of gamma radiation, the upper limit of the internal bremsstrahlung can be used to determine the transition energy of a nuclide in EC decay. Some examples of radionuclides that decay by EC without the emission of gamma radiation are as follows:

$$^{55}_{26}\mathrm{Fe} \rightarrow {}^{55}_{25}\mathrm{Mn} + \nu + h\nu \ (0.23\ \mathrm{MeV}) \tag{1.391}$$

$$^{37}_{18}\mathrm{Ar} \rightarrow {}^{37}_{17}\mathrm{Cl} + \nu + h\nu \ (0.81\mathrm{MeV}) \tag{1.392}$$

and

$$^{49}_{23}\mathrm{V} \rightarrow {}^{49}_{22}\mathrm{Ti} + \nu + h\nu \ (0.60\mathrm{MeV}) \tag{1.393}$$

where $h\nu$ *is* the internal Bremsstrahlung, the upper energy limits of which are expressed in MeV.

d. Nuclear bremsstrahlung (nuclear startstrahlung)

Nuclear bremsstrahlung was recently identified by Nie Luo, Magdi Ragheb, and George Miley at the University of Illinois (Luo et al., 2010). This class of bremsstrahlung refers to electromagnetic radiation of low energy (*i.e.*, soft X-rays of a few hundred eV to a few keV) emitted by a proton or deuteron or both as these undergo acceleration during the fusion of a proton with a neutron or the fusion of a proton with a deuteron. Luo et al. (2010) loosely assigned the name of nuclear bremsstrahlung to this phenomenon and have ventured to also call it nuclear startstrahlung from the German meaning "take-off radiation" depicting the acceleration of these nuclear particles as the origin of the soft X-rays. The identification of this class of bremsstrahlung opens the door to new plasma diagnostics such as in the identification of the parasitic D–D fusion reaction in future thermonuclear energy based on D–T fusion. The D–D fusion reaction, previously described by Eq. (1.319), produces protons according to

$$d + d \rightarrow {}^{3}\text{H} + p \quad Q = 4.04\ \text{MeV} \tag{1.394}$$

The D–T thermonuclear reaction is also an excellent source of neutrons, as described previously in this chapter (see Eq. 1.317), The protons and neutrons generated can, in turn, participate in the thermonuclear process via other reactions including the fusing of a proton with a neutron according to

$$p + n \rightarrow {}^{2}\text{H} + \gamma + x\text{-rays} \quad Q = 2.23\ \text{MeV} \tag{1.395}$$

and the fusion of a proton with unreacted deuterium fuel according to the reaction

$$p + d \rightarrow {}^{3}\text{He} + \gamma + x\text{-rays} \quad Q = 5.49\ \text{MeV} \tag{1.396}$$

Both of which give rise to the nuclear bremsstrahlung, which are included in the above two reactions.

The process of nuclear bremsstrahlung, in the simplest case of $p-n$ capture, is illustrated in Fig. 1.112. As described by Luo et al. (2010), the proton and neutron are attracted to each other by the strong nuclear force, and this attraction causes both particles to accelerate toward each other. Due to the charge on the accelerating proton, electromagnetic radiation is generated of the bremsstrahlung type. Luo et al. (2010) refer to this as a nuclear bremsstrahlung or nuclear startstrahlung to accentuate its origin in acceleration due to the strong nuclear force.

The fusion of the proton with deuterium yields the nuclear bremsstrahlung similar to the case of the proton and neutron fusion. However, Luo et al. (2010) explain that the Coulomb repulsion of proton and deuterium add a degree of complication to the process. They explain that the Coulomb barrier is overcome by quantum mechanical tunneling, and the strong nuclear force still causes both particles to accelerate toward each other, although the particles accelerate at different rates because of their mass difference. As both the proton and deuteron are charged they radiate quanta of electromagnetic radiation. Because the two charged particles are accelerating toward each other, they travel in opposite directions, the radiation they emit will tend to cancel each other; however, the cancellation is not complete, because the two particles accelerate at different rates due to their mass difference. Dhibar et al. (2018) describe large $LaBr_3$:Ce solid scintillation detectors, which can serve potentially in the measurement of nuclear bremsstrahlung.

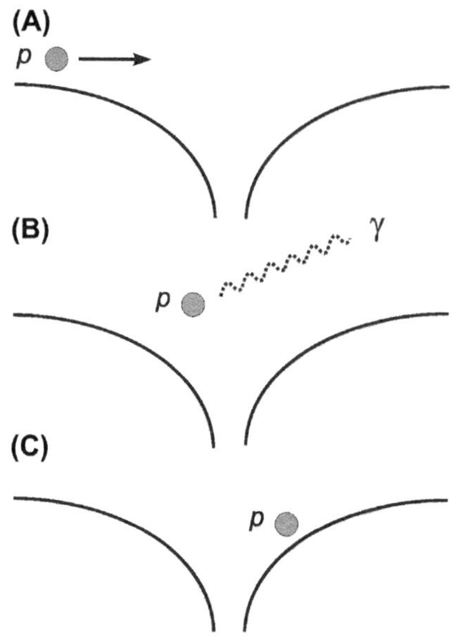

FIGURE 1.112 The nuclear bremsstrahlung process that releases soft X-rays in the proton-neutron capture and proton-deuteron fusion. The solid curves represent the attractive nuclear potential of a neutron (or deuteron) felt by a proton. The length dimension shown is ~10 fermi. (A) The proton, having an initial positive kinetic energy of several keV, is moving toward the neutron (or deuteron) due to nuclear attraction. (B) The proton accelerates in the process, and therefore starts radiating electromagnetic waves, *i.e.*, photons. (C) The proton, having lost all of its initial kinetic energy and maybe more, now has a total energy of negative value, *i.e.*, it is now a bound state around the neutron (or deuteron). Afterward it may emit a gamma of 2.23 MeV for proton-neutron fusion (or 5.49 MeV for proton-deuteron fusion) and then combine with the neutron (or deuteron) to finally form ^{2}H (or ^{3}He). The acceleration of the neutron is radiation-free, because the neutron is without charge. In $p-d$ fusion, the accelerating deuteron also radiates, but this is not illustrated in the figure for clarity. *From Luo et al. (2010) with permission of Elsevier © 2010.*

3. Bremsstrahlung and line spectra X-rays from beta-particle emitters

All beta-particle emissions from radionuclides, even those radionuclides classified as pure beta-emitters that are not the source of any gamma radiation, including ^{3}H, ^{14}C, ^{32}P, ^{35}S, ^{45}Ca, ^{63}Ni, ^{89}Sr, ^{90}Y, and ^{123}Sn, among others, will produce bremstrahlung X-ray photons. For example, the measurement of bremsstrahlung from the beta particles of ^{3}H has been demonstrated as a method for the monitoring of the fuel processing system in the development of thermonuclear fusion reactors (Shu et al., 2004, 2006; Röllig et al., 2013; Matsuyama and Abe, 2016). In addition, this section will describe X-ray production by beta particles in thick-target materials. Thick-target materials include any absorber within which a beta particle will come to a complete stop. When beta particles are emitted from decaying nuclei and escape the atom from which they originated, X-ray photons are created by several distinct physical phenomena, which are enumerated as follows:

a. ***External bremsstrahlung*** is a continuous spectrum of X-rays produced when the accelerating beta particle traverses atoms of absorber material and is deflected by the Coulomb field of atomic nuclei causing the beta particle to decelerate and emit electromagnetic radiation. This is the major source of X-ray energy that a beta particle will radiate. As discussed previously, the energy of the electromagnetic radiation emitted ($E = h\nu$) is proportional to the initial energy of the beta particle and the atomic number (Z) of the absorber. Thus, the bremsstrahlung energy loss (I) in MeV is written as

$$I \simeq kZE^2 \tag{1.397}$$

or

$$I/E \simeq kZE \tag{1.398}$$

where I/E is the fraction of incident beta-particle energy lost as bremsstrahlung, and $k \simeq 0.0007\ \text{MeV}^{-1}$ is a normalization constant (Evans, 1955; Faw and Shultis, 2004). Table 1.21 illustrates the effect of atomic number Z on the conversion of beta-particle energy to bremsstrahlung in a thick target according to the approximation given by Eq. (1.398).

The estimates of energy conversion in Table 1.21 for a 0.5 MeV electron absorbed in copper agrees with that reported by Evans (1955), and that of a 1 MeV electron in aluminum and a 2 MeV electron in lead listed in Table 1.21 agree with estimates quoted by Choppin et al. (2002) and Faw and Schultis (2004), respectively. Molybdenum (Mo) and tungsten (W), listed in Table 1.21, are relatively high-Z materials with heat stability used as target materials for the production of bremsstrahlung X-rays in commercial medical diagnostic X-ray tubes (Mavunda et al., 2004).

b. ***Internal bremsstrahlung*** occurs during the emission of the beta particle from the nucleus, as the beta particle accelerates at its birth and emits a continuous spectrum of X-rays in the field of the emitting nucleus (Cengiz and Almaz, 2004; Khalil, 2011; Singh and Dhaliwal, 2014, 2015). This mechanism of bremsstrahlung production is in direct contrast to that of outer bremsstrahlung whereby the beta particle approaches a nucleus from outside the atom described in the previous section.

c. ***X-rays corresponding to line spectra characteristic of the absorber material*** are produced when the beta particle traverses the absorber and produces ionization in inner electron shells (K, L, M,... shells). The electron vacancies in the inner shells are filled by electrons from outer shells with the concomitant emission of X-rays. These X-rays consist of discreet lines of energy characteristic of X-ray lines of the absorber material.

d. ***X-rays corresponding to line spectra characteristic of the emitting atom*** are produced when the beta particle produces ionization as it accelerates out of the atom from which is was created. Ionization of internal electron shells (K, L, M,... shells) of the emitting atom leaves electron vacancies, which are filled by electrons from outer shells, producing X-rays of discrete energy lines characteristic of the emitting atom. The probability of this internal ionization followed by X-ray emission by the β-emitting atom is of the order of 10^{-4} (Levinger, 1953; Boehm and Wu, 1954; Evans, 1955).

E. Cherenkov radiation

Charged particles, when they possess sufficient energy, may travel through a transparent medium (gas, liquid, or solid) at a speed greater than the speed of light in that medium. This occurrence, known as the Cherenkov effect, causes the emission of photons of light referred to as Cherenkov

TABLE 1.21 Approximate incident beta-particle energy loss (I/E) as bremsstrahlung in various thick targets.[a]

	Fraction of incident beta-particle energy loss (I/E)					
Incident energy (E)	H_2O ($Z = 6.6$)[b]	Al ($Z = 13$)	Cu ($Z = 29$)	Mo ($Z = 42$)	W ($Z = 74$)	Pb ($Z = 82$)
0.1 MeV	0.05%	0.09%	0.2%	0.3%	0.5%	0.6%
0.2	0.09	0.2	0.5	0.6	1.0	1.1
0.5	0.2	0.5	1.0	1.5	2.6	2.9
1.0	0.5	0.9	2.0	2.9	5.2	5.7
1.5	0.7	1.4	3.0	4.4	7.8	8.6
2.0	0.9	1.8	4.1	5.9	10.4	11.5
2.5	1.1	2.3	5.1	7.4	13.0	14.4

[a]*Calculated according to $I/E \simeq (0.0007\ \text{MeV}^{-1})ZE$ where I is the total bremsstrahlung energy (MeV), E is the incident electron or beta-particle energy (MeV), and Z is the atomic number of the thick-target absorber.*

[b]*For H_2O, a compound absorber, the effective or modified atomic number is used as calculated by Markowics and VanGriken (1984) and Manjunatha and Rudraswamy (2009, 2010), namely, $Z_{mod} = \left(\sum_i^I W_i Z_i^2 / A_i\right) / \left(\sum_i^I W_i Z_i / A_i\right)$ where W_i, Z_i, and A_i are the weight fraction, atomic number, and mass number, respectively, of the ith element.*

photons or Cherenkov radiation. These photons extend over a range of wavelengths from the ultraviolet into the visible portion of the electromagnetic radiation spectrum.

1. Origin and characteristics

Cherenkov photon emission is a result of a coherent disturbance of adjacent molecules in matter caused by the traveling charged particle, which must possess a certain threshold energy. This phenomenon has practical applications including (i) the measurement and detection of radionuclides that emit relatively high-energy beta particles, (ii) the measurement of gamma radiation via the Cherenkov effect produced by Compton electrons, and (iii) the identification of subatomic particles. The history of discovery, theory, and the applications of Cherenkov photons are discussed in detail in Volume 2, Chapter 6 of this book, and only a brief treatment is provided in this chapter.

Cherenkov radiation was discovered by Pavel A. Cherenkov (1934a,b, 1936) while working on his PhD dissertation under the direction of Sergei I. Vavilov, who was director of the P. N. Lebedev Institute of Physics in Moscow. For detailed accounts of the history of this discovery and Cherenkov's experimental technique, see papers by Cherenkov's daughter, Elena Cherenkova (2005, 2008) and previous books by the author (L'Annunziata, 2007, 2016). Once Cherenkov had discovered this unique visible radiation produced by high-energy Compton electrons or high-energy β-particles, and once Cherenkov found that this light was polarized and that the light emission was asymmetric, that is, it was emitted only in the direction of the fast electrons (Cherenkov, 1934a; b, 1936) did Cherenkov and his doctoral dissertation advisor Sergei I. Vavilov, know that they were dealing with a yet unknown phenomenon. Il'ja Frank and Igor Tamm of the Theoretical Physics Division of the same institute took notice and went to work on the interpretation of this new phenomenon. Their findings were published in a joint paper entitled "Coherent Visible Radiation of Fast Electrons Passing through Matter" (Frank and Tamm, 1937). They explained the phenomenon as caused by a charged particle traveling in a medium at a speed in excess of the speed of light in that medium. It was known from Einstein's theory of relativity that matter could not travel in excess of the speed of light in a vacuum ($c = 2.99 \times 10^8$ m/s); however, in gaseous, liquid, or solid media, the velocity of light will be less than its velocity in a vacuum and, an elementary charged particle with sufficient energy could travel in such media at speeds exceeding that of light. The charged particle in passing through the electron clouds of a transparent medium (liquid, solid or gaseous) would create an electromagnetic shock wave analogous to that of a "sonic boom" created by a jet airplane or projectile traveling in the atmosphere at a speed exceeding that of sound. In the words of Frank and Tamm (1937) with authors remarks in brackets:

> *We shall consider an electron moving with constant velocity v along [an] axis through a medium characterized by its index of refraction n. The field of the electron may be considered as the result of superposition of spherical waves of retarded potential, which are being continually emitted by the moving electron and are propagated with the velocity (c/n). [See Fig. 1.113] It is easy to see that all these consecutive waves emitted will be in phase along the directions making the angle θ with the axis of motion [of the particle], if only v, n, and θ do satisfy the condition*
>
> $$\cos\theta = \frac{1}{\beta n} \tag{1.399}$$
>
> *where β = v/c. Thus, there will be a radiation emitted in the direction θ, whereas the interference of waves will prevent radiation in any other direction. Now the condition [of radiation emission] can be fulfilled only if*
>
> $$\beta n > 1, \tag{1.400}$$
>
> *that is, only in the case of fast electrons in a medium, whose index of refraction n for frequencies in question is markedly larger than 1.*

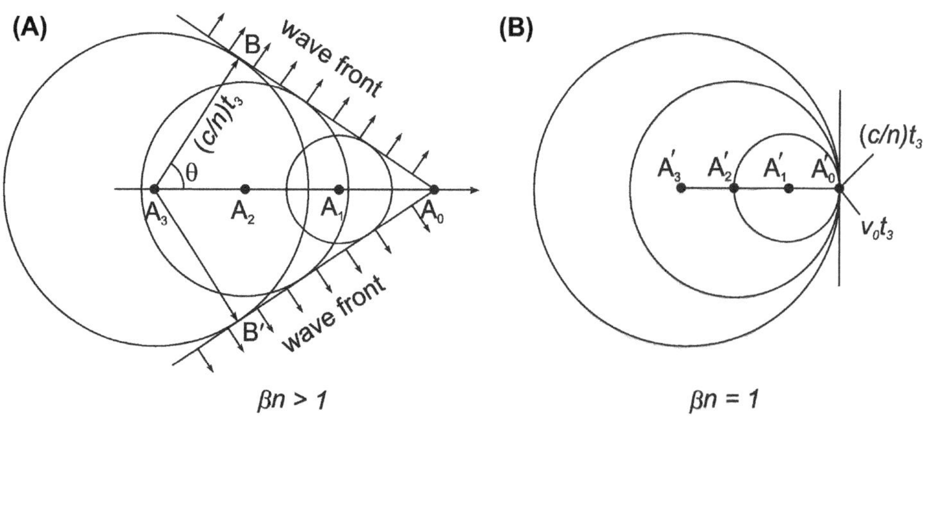

FIGURE 1.113 A Huygens construction of the spherical waves of retarded potential produced by a charged particle traveling along an axis in a refractive medium from points A_3 to A_0 at two velocities, namely, (A) $\beta n > 1$ and (B) $\beta n = 1$. *From Nobel Lecture of Il'ja Frank (1958) with permission © The Nobel Foundation 1958.*

The index of refraction, *n*, is by definition the ratio of the speed of light in a vacuum to its speed in a particular medium, and the value of *n* will vary from one medium to another. The term "fast electrons" used by Frank and Tamm (1937), quoted above, refers to electrons of such energy that they travel at a speed in excess of the speed of light in a particular medium. Also, the term β is called the relative phase velocity of the particle ($\beta = v/c$) or the velocity of the particle in a medium divided by the speed of light in a vacuum.

Eq. (1.399), derived by Frank and Tamm, can be found from the distance of travel of the spherical wave front relative to the distance traveled by the electron, as depicted in Fig. 1.113. In the words of Frank (1958) in his Nobel Lecture (© The Nobel Foundation):

> *... We have to consider each point of the particle trajectory as a source of waves. In this case the wave phase is determined by the instant of passage of the particle through a given point. Let us assume that at moment $t = -t_3$ the emitter was at point A_3, at moment $t = -t_2$ at point A_2, at moment $t = -t_1$ at A_1, and, finally, at the moment of observation $t = 0$ at A_0.*
>
> *... the velocity of the waves is equal to the phase velocity of light c/n ... The surfaces of the rays are simply spheres whose radii for points A_3, A_2, A_1, and A_0 are $(c/n)t_3$, $(c/n)t_2$, $(c/n)t_1$, and 0. respectively [See Fig. 1.113A]. The envelope of these spheres evidently represents a cone of circular cross section with the apex at A_0. Its generatrices in the plane of the drawing are A_0B and A_0B'.*
>
> *According to the Huygens' principle, the directions of the rays are defined by the radius vectors drawn from some centre of the waves to the point of tangency with the envelope. For example, in Fig. 1.113A, it is A_3B and A_3B' coinciding with the generatrices of the wave normal cone ...*
>
> *From Fig. 1.113 it is not difficult to determine the magnitude of the threshold velocity. When the velocity diminishes, the distances between points A decrease. The threshold case arises when point A, occupies the position A'_0 on the surface of the sphere (This case is depicted separately in Fig. 1.113B.) At lower velocities, one of the spheres lies completely within the other and they do not have a common envelope. In the threshold case [$\beta n = 1$], they have only a common point of tangency, A'_0. Thus evidently (c/n) $t_3 = v_0 t_3$, i.e. $v_0 = c/n$. The cone of wave normals is compressed in the direction of velocity v, and the wave cone transforms into a plane perpendicular to the axis of motion at point A'_0 [Fig. 1.113B].*

From Fig, 1.113A the equation defining the angle θ as a function of the particle phase velocity β and index of refraction *n*, *i.e.*, $\cos\theta = 1/\beta n$, is derived from the distances of travel of the wave front with respect to that traveled by the charged particle. The velocity *v* of travel of the wave front in a particular medium is a function of the index of refraction *n* of the medium, since by definition $n = c/v$. The distance traveled by the wave front from A_3 to B of Fig. 1.113A is the product of velocity and time *t* or $A_3B = vt_3$, and the distance traveled by the charged particle in the same time span from A_3 to A_0 is the product of the particle velocity in the medium v_p and time t_3 or $A_3A_0 = v_pt_3$, which may be written as $A_3A_0 = (v_p/c)ct_3$. Since the particle phase velocity β is defined as v_p/c, the distance A_3A_0 traveled by the charged particle becomes $A_3A_0 = \beta ct_3$. The angle of emission θ of the Cherenkov radiation with respect to the direction of travel of the charged particle is defined as

$$\cos\theta = \frac{AB}{A_3A_0} = \frac{(c/n)t_3}{\beta ct_3} = \frac{1}{\beta n} \tag{1.401}$$

Thus, the Cherenkov photons are anisotropic, that is, the photons are not emitted in all directions; rather, the photons are conical, *i.e.*, they are emitted as a cone at an angle θ. The angle of emission is a function of the phase velocity of the particle (β) and the index of refraction (n) of the medium within which the particle is traveling. If the particle is assigned a maximum velocity, where β approaches unity, there will be a maximum angle of emission θ of the Cherenkov photons in liquids, and the angle would be a function of the index of refraction of the medium or

$$\cos\theta_{\max} = \frac{1}{n} \quad \text{when } \beta = 1 \tag{1.402}$$

For example, if an electron traveling in water is assigned a maximum velocity, where $\beta = 1$, and the index of refraction of water at the sodium D line, $n_D^{20} = 1.3330$, the maximum angle of emission of Cherenkov photons in water would be calculated according to Eq. (1.402) as

$$\theta_{\max} = \cos^{-1}\frac{1}{1.3330} = 41.4^0 \tag{1.403}$$

In another medium such as benzene, where the index of refraction, $n_D^{20} = 1.50108$, the maximum angle of emission of the Cherenkov photons would be 48.2^0 calculated according to Eq. (1.401). Thus, Cherenkov photons are emitted as a cone at various angles between zero and a maximum value depending on the velocity of the particle and the index of refraction of the medium within which the particle travels. The conical properties of Cherenkov radiation, illustrated in Fig. 1.114, was described by Cherenkov in 1958 in his Nobel Lecture (Cherenkov, 1958), with the authors comments in brackets as follows:

> "*If we consider the picture* [of the radiation] *not in the* [2-dimensional] *plane but spatially, then the radiation must spread out along the surface of a cone whose axis forms the*

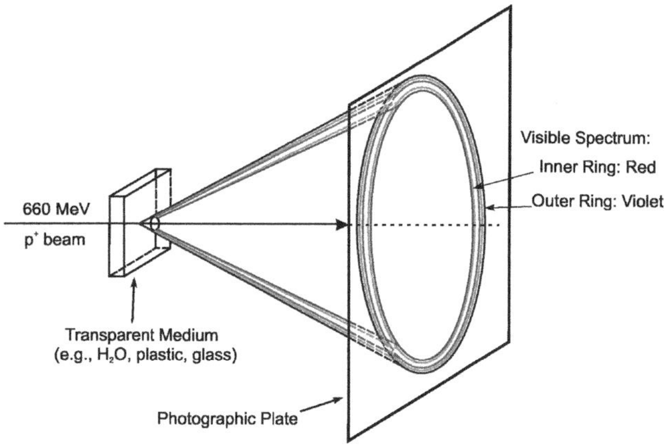

FIGURE 1.114 Diagram of an experimental arrangement for obtaining photographs of the cone section in the plane of a photographic plate placed perpendicular to the path of high-speed (600 MeV) protons in an accelerator beam. For the color version of this figure, the reader is referred to the electronic version of the book.

path of an electrically charged particle while the surface line forms with this axis the angle [θ].

If we place [a] *photographic plate perpendicular to the beam of high-speed particles* [See Fig. 1.114], *we shall obtain, in addition to an image of the track of the beam, also a photograph of the radiation in the form of a ring. This photograph was obtained with the aid of a fine beam of protons in the accelerator of the United Institute for Nuclear Research at Dubna.*

... we have in our considerations assumed some fixed frequency. In reality, however, the radiation spectrum is continuous. Since the medium exhibits dispersion, i.e., the refractive index is dependent on the frequency, this means that the light of different wavelengths is propagated at angles which, even with strictly constant velocity of the particles, differ somewhat from one another.

Thus, the radiation is broken up as in spectral analysis. The radiation cone will consequently show a definite intensity, and in the case of a medium with normal dispersion the spectral red will lie in the inner part of the cone while the violet is on the outside. That this is actually so was shown by a photograph showing part of the ring with a colour plate."

2. Threshold condition

As discussed in the previous section, the threshold condition for the production of Cherenkov radiation in a transparent medium is given by

$$\beta n = 1 \tag{1.404}$$

where β is the relative phase velocity of the particle, that is, the velocity of the particle divided by the speed of light in a vacuum, and n is the refractive index of the medium (*i.e.*, the ratio of the velocity of light in a vacuum to its velocity in the medium). Only charged particles that possess

$$\beta > 1/n \tag{1.405}$$

produce Cherenkov photons in transparent media (Frank and Tamm, 1937). The value of β of the charged particle is dependent on its kinetic energy, which is obtained from the equation describing the total energy of the particle. As described previously in this chapter, the total energy (E_{tot}) of a particle is the sum of its kinetic (E_K) and rest energies (m_0c^2) or

$$E_{tot} = E_K + m_0c^2 = \gamma m_0c^2 \tag{1.406}$$

and

$$E_K = \gamma m_0c^2 - m_0c^2 \tag{1.407}$$

where

$$\gamma = \frac{1}{\sqrt{1-(u^2/c^2)}} \quad \text{and } u/c^2 = \beta^2 \tag{1.408}$$

u is the particle speed and $u < c$. (The particle speed is often represented with the symbol v or u.) If m_0 is the particle rest mass, then the particle relativistic mass, m_r, which is the speed-dependent mass of the particle would be defined as

$$m_r = \frac{m_0}{\sqrt{1-(u^2/c^2)}} = \frac{m_0}{\sqrt{1-\beta^2}} \tag{1.409}$$

Thus, for a particle traveling at relativistic speed, Eq. (1.407) can be expressed as

$$E_K = m_0c^2\left[\frac{1}{(1-\beta^2)^{1/2}} - 1\right] \tag{1.410}$$

where m_0 is the particle rest mass in grams, c is the velocity of light in a vacuum (equal to 2.99×10^{10} cm/s), and E_K is the kinetic energy in erg (where 1.602×10^{-12} erg = 1 eV). Eq. (1.410) can be solved for β to give the expression for the phase velocity of the particle as a function of its energy or

$$\beta = \left[1 - \left(\frac{1}{(E_K/m_0c^2)+1}\right)^2\right]^{1/2} \tag{1.411}$$

The value of the rest energy for an electron is $m_0c^2 = (9.109 \times 10^{-28}$ g$)(2.9979 \times 10^{10}$ cm/s$)^2 = 8.19 \times 10^{-7}$ erg. The energy in ergs may be converted to electron volts as follows: $(8.19 \times 10^{-7}$ erg$)/(1.602 \times 10^{-12}$ erg/eV$) = 511$ keV. Substituting the value of 511 keV for m_0c^2 in Eq. 1.411 and solving for β gives

$$\beta = \left[1 - \left(\frac{1}{(E_K/511\text{keV})+1}\right)^2\right]^{1/2} \tag{1.412}$$

Thus, the value of β, where electrons or beta particles are concerned, is dependent upon the electron or beta-particle energy, E_K, in keV according to Eq. (1.412).

3. Threshold energies

As noted above, the particle phase velocity, β, is a critical factor in the production of Cherenkov photons in a medium of given index of refraction, n, that is, Cherenkov radiation will occur only when $\beta > 1/n$. Consequently the particle energy, upon which the particle phase velocity is dependent, is a critical factor. Thus, a minimum or threshold energy of the particle must be reached when traveling in a medium of given index of refraction before Cherenkov photons will be produced. If we apply the threshold value of β for the production of Cherenkov photons from Eq. (1.404), that is, $\beta = 1/n$, to Eq. (1.410), we will obtain the threshold energy that particles must possess for the production of Cherenkov radiation as a function of the index of refraction of the medium as

$$E_{th} = m_0c^2\left[\left(1-\frac{1}{n^2}\right)^{-1/2}-1\right] \quad (1.413)$$

where m_0 is the particle rest mass and c is the velocity of light in a vacuum (2.9979×10^{10} cm/s). Eq. 1.413 is dealt with in more detail in Volume 2, Chapter 6. Thus, the threshold energy for Cherenkov photon production will vary according to the particle rest mass and index of refraction of the medium. The energy required for the production of Cherenkov photons, E_{th}, increases with particle mass, and it will be lower for media of higher index of refraction.

The value of the rest energy for an electron is $m_0c^2 = (9.109 \times 10^{-28}\text{ g})(2.9979 \times 10^{10}\text{ cm/s})^2 = 8.19 \times 10^{-7}$ erg. The energy in ergs may be converted to electron volts as follows: $(8.19 \times 10^{-7}\text{ erg})/(1.602 \times 10^{-12}\text{ erg/eV}) = 511$ keV. Substituting the value of 511 keV for m_0c^2 in Eq. 1.413 will yield the equation for the calculation of the threshold energy that electrons or beta particles must possess for the production of Cherenkov radiation as a function of the index of refraction of the medium as

$$E_{th} = 511\text{keV}\left[\left(1-\frac{1}{n^2}\right)^{-1/2}-1\right] \quad (1.414)$$

The counting of Cherenkov photons for the analysis of radionuclides is a common technique, and Eq. (1.414) may be used to assess the potential of a given medium for the detection of beta particles that are emitted by particular radionuclides. For example, the radionuclides ^{33}P and ^{32}P are commonly used as radionuclide tracers in the biosciences, and aqueous solutions are generally the medium within which the radionuclides are analyzed. Eq. (1.414) may be used to determine the threshold energy for the production of Cherenkov photons by beta particles (*i.e.*, electrons) in water with the index of refraction of water ($n_D^{20} = 1.3330$) to give

$$\begin{aligned} E_{th} &= 511\text{ keV}\left[\frac{1}{\sqrt{1-\frac{1}{(1.333)^2}}}-1\right] \\ &= 511\text{ keV}\left[\frac{1}{\sqrt{0.43721}}-1\right] \\ &= 511\text{ keV}\left[\frac{1}{0.66121}-1\right] \\ &= (0.5110 \times 10^6\text{ eV})(0.51236) = 0.262\text{ MeV} \end{aligned} \quad (1.415)$$

Thus, the beta particle must possess a minimum of 0.262 MeV energy to produce Cherenkov photons in water. Comparing the maximum energies of the beta emissions of ^{32}P ($E_{max} = 1.710$ MeV and ^{33}P ($E_{max} = 0.249$ MeV) clearly indicate that the beta emissions from ^{33}P will not produce any Cherenkov photons in water, and only those beta-particle emission from ^{32}P in water with energies in excess of 0.262 MeV can be detected by Cherenkov photon counting. Thus, the Cherenkov effect may be used to discriminate between ^{32}P and ^{33}P in water by detecting only the ^{32}P fraction. Cherenkov photon counting is discussed in detail in Volume 2, Chapter 6 together with other applications of the Cherenkov effect.

4. Applications

There are numerous applications of the Cherenkov effect, and these are discussed in detail in Volume 2, Chapter 6. Among these applications are (i) the analysis of radionuclides that emit beta particles with sufficient energy to produce Cherenkov photons in the medium in which they are analyzed, (ii) the discrimination of beta-emitting radionuclides by the selection of a medium with an index of refraction that would allow or prevent the Cherenkov effect, (iii) the measurement of gamma radiation via the detection of Cherenkov photons produced by Compton electrons, and (iv) high-energy particle identification with ring-imaging Cherenkov (RICH) detectors.

Cherenkov Photon ring imaging has developed into a field of its own. Special ring-imaging Cherenkov (RICH) detectors have been developed, which are used for particle identification. Such detectors can provide information on the particle velocity, β, and discriminate between very high-energy relativistic particles of different mass, such as protons ($m = 938.27$ MeV/c^2), kaons

($m = 493.67$ MeV/c^2), pions ($m = 139.56$ MeV/c^2), and electrons ($m = 0.511$ MeV/c^2). The RICH detectors can be used to discriminate between kaons and pions of momentum of approximately 50 GeV/c. These detectors are discussed in detail in Volume 2, Chapter 6, and reviewed by Engelfried and Paic (2005), Bressan et al. (2008), Engelfried (2011), and Inami (2017).

F. Synchrotron radiation

Synchrotron radiation is electromagnetic radiation emitted by charged particles of low mass (*e.g.*, negatrons or positrons) traveling at relativistic speeds (*i.e.*, close to the speed of light) as the particles are caused to undergo radial acceleration when deflected in a magnetic field. The radiation emitted covers a broad spectrum from the long microwaves to the shorter wavelengths of hard X-rays. Synchrotron radiation is produced both naturally and artificially.

1. Synchrotron radiation from natural sources

A natural source of synchrotron radiation is observed in certain light emissions from astronomic bodies. An example is provided by the light emitted from the Crab Nebula, which is the remnant of a supernova explosion of a massive star. An excellent image of the Crab Nebula obtained by the Spitzer Space Telescope is illustrated in Fig. 1.115, which shows the bluish-white light originating from Synchrotron radiation caused by clouds of relativistic electrons deflected radially by the magnetic field of a rotating neutron star.

FIGURE 1.115 Image of the Crab Nebula photographed by the NASA Hubble Space Telescope's Wide Field and Planetary Camera released to the public on December 1, 2005. The bluish-white regions originate from synchrotron radiation produced by clouds of high energy electrons traveling radially at relativistic speeds under the influence of a magnetic field of a rapidly rotating neutron star. The Crab Nebula is located in the Constellation Taurus about 6500 light years from earth and the image along the bottom measures 12 light years. Courtesy of NASA, ESA, J. Hester and A. Loll, Arizona State University. For the color version of this figure, the reader is referred to the electronic version of the book. *From Owens (2012); reprinted with permission from Elsevier © 2012.*

2. Discovery of synchrotron radiation

The first artificial production of synchrotron radiation was reported in 1947 by Frank Elder, Anatole Gurewitsch, Robert Langmuir, and Herb Pollock in a letter entitled "Radiation from Electrons in a Synchrotron" in the journal *Physical Review* (Elder et al., 1947). They followed their initial observations with a paper on the properties of the radiation and it's origin the following year entitled "Radiation from Electrons Accelerated in a Synchrotron" (Elder et al., 1948). In this work they concluded that the radiation emissions were due to "high-energy electrons subjected to large radial accelerations". The radiation was named synchrotron radiation as it was discovered to occur when electrons are accelerated to relativistic speeds in a synchrotron where the electron path is diverted radially with a magnetic field. The phenomenon is illustrated in Fig. 1.116, which depicts electrons traveling near the speed of light in a storage ring. The storage ring is a ring-shaped chamber within which electrons traveling at near-light speed originating from a booster synchrotron, are stored for research. The synchrotron radiation is emitted when the path of travel of the electron is forced to bend in a magnetic field.

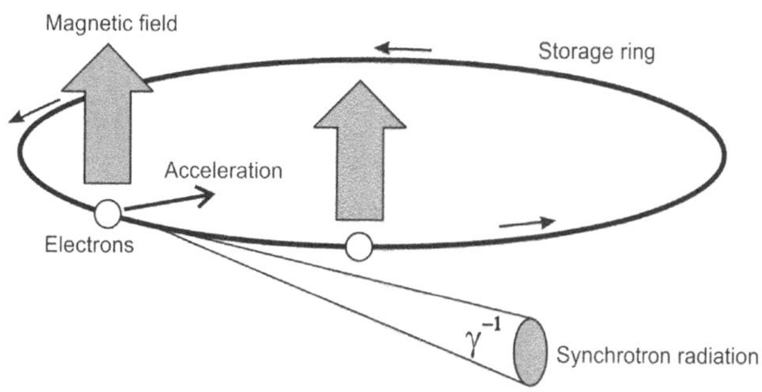

FIGURE 1.116 Synchrotron radiation from electrons traveling at relativistic speeds when the electrons are caused to travel radially by a bending magnet in a storage ring linked to a booster cyclotron. The high-energy electrons produce synchrotron radiation every time they are deflected by a magnetic field. The synchrotron radiation is concentrated into a cone with an angle of the order of γ^{-1}. *From Yagi (2014); reprinted with permission from Elsevier © 2014.*

The discovery of synchrotron radiation followed the construction of the synchrotron and the testing of the synchrotron principle. It was foreseen in the early 1940s that, as a charged particle is accelerated to relativistic speeds (*i.e.*, speeds approaching the speed of light), the relativistic mass of the particle would increase to the point that it would not be possible to maintain that particle in a constant orbit. Thus, current thought prevailed that there would be a limit to the speed to which the charged particle could be kept in a constant orbit. Wilson (1996) and Wei (2003) provide good historical accounts that led to the discovery of the synchrotron. In 1943, Sir Marcus Oliphant (1901–2000), an Australian physicist was working with Earnest Lawrence at the Oak Ridge National Laboratory on the Manhattan Project. As related by Wilson (1996), Marcus Oliphant, while at the Oak Ridge National Laboratory, was planning to return to his professorship post at the University of Birmingham, and in September of 1943, he wrote a memorandum to the Department of Scientific and Industrial Research (DSIR) in the UK (Oliphant, 1943) in which he proposed a new method of acceleration as follows:

> *Particles should be constrained to move in a circle of constant radius thus enabling the use of an annular ring of magnetic field ... which would be varied in such a way that the radius of curvature remains constant as the particles gain energy through successive accelerations by an alternating electric field applied between coaxial hollow electrodes.*

Oliphant is thus considered as the originator of the synchrotron concept (Wilson,1996; Wei, 2003). However, shortly thereafter the Russian Physicist Vladimir I. Veksler (1907–66) and Edwin M. McMillan (1907–91) at the University of California, Berkeley independently proposed the concept of phase stability upon which the operation of the synchrotron is based (Veksler, 1944, 1945; McMillan, 1945). As defined by Wei (2003), synchrotrons are circular accelerators in which charged particles are accelerated with electric fields and constrained to move in fixed orbits by magnets whose fields vary in time such that the radius of curvature of the particle beam remains constant as the particles gain energy. Increasing energy and acceleration are attributed to the particles, as described in Oliphant's memo, by alternating electric fields, which are produced by resonating radiofrequency (rf) cavities.

At the General Electric Company in Schenectady, New York, Frank Elder, Anatole Gurewitsch, Robert Langmuir, and Herb Pollock and their staff constructed a 70-MeV synchrotron to test the synchrotron principle of phase stability (Elder et al., 1947, 1948). During a test-run of the newly constructed synchrotron, synchrotron radiation was discovered. As related in a published memoir on "The Discovery of Synchrotron Radiation", Pollock (1983) recorded in his laboratory notebook the following accounts taken on 24 April 1947 (Reproduced with permission © 1983 American Association of Physics Teachers):

> *Frank Elder worked on the design of the power circuit and the laminated magnet, which was assembled by machinist Floyd Haber. Anatole Gurewitsch focused on the 163-Mc rf cavity resonator design ... Robert Langmuir and Pollock put the various components together and on 24 October 1946 we were able to tell Edwin McMillan in Berkeley that we had a synchrotron beam and that no problem had been encountered ... for synchrotron acceleration to 70 MeV ... On that day in April [24 April 1947] Langmuir and Pollock were running the machine and as usual were trying to push the electron gun and its associated pulse transformer to the limit. Some intermittent sparking had occurred, and Haber was asked to observe with a mirror around the protective concrete wall which separated the machine from the control room ... When Haber looked around the corner of the wall he noticed a very bright spot of light ... At first we thought it might be due to Cerenkov radiation, but it soon became clear that we were seeing Ivanenko and Pomeranchuk radiation ... We observed the bright spot with mirrors, looking tangent to the orbit at two or three points in the room ... The beam appeared stable and of small cross section (perhaps 1 mm square).*

The Ivanenko and Pomeranchuk radiation, referred to in Pollock's notes, is the prediction made in 1944 by the Russian physicists Dimitri Iwanenko (1904–94) and Isaak Pomeranchuk (1913–66) in their joint papers (see Iwanenko and Pomeranchuk, 1944a,b) on the emission of radiation by electrons traveling at relativistic speeds in a magnetic field. Iwanenko and Pomeranchuk (1944a,b) thought that this radiation emission might be a limiting factor in achieving maximal electron energies in the development of electron accelerators. However, the radiation emission (*i.e.*, synchrotron radiation) is named after the instrument within which it was first observed and which today remains the principle source of this radiation for experimental research. A very thorough review of the history of synchrotron radiation, including the theory on radiation from accelerated electrons from as early as 1912, is provided by Blewett (1988).

3. Synchrotron radiation and accelerated electron properties

Synchrotron radiation is currently the brightest source of X-rays available for numerous practical research applications, which will be discussed further on in this section. Suortti (2003), Owens (2012) and Eberhardt (2015) describe the very intense light emission properties of synchrotron radiation, and the brilliance is measured in the number of photons per second per mm^2 per $mrad^2$. Modern synchrotrons yield a brilliance of up to 10^{20} photons/s/mm^2/$mrad^2$ (Owens, 2012). The X-ray beam of synchrotron radiation is generally very highly collimated with an angular spread of only 1 mrad (Teresawa and Kihara, 1996). As illustrated previously in Fig. 1.116, the angular spread of the synchrotron radiation cone is

$$\gamma^{-1} = \sqrt{1-\beta^2} \quad \text{where } \beta = v/c \quad \text{and}$$
$$\gamma = \frac{1}{\sqrt{1-(v^2/c^2)}} \tag{1.416}$$

The variable β is the relative phase velocity of the electron, that is, the ratio of the velocity of the electron (v) to the speed of light in a vacuum (c), and γ is the Lorentz transformation factor described previously for relativistic calculations of mass, length and time. According to Eq. (1.269) described previously in this chapter, the total energy (E) of a particle is the sum of its kinetic (K) and rest energies (mc^2) or

$$E = K + mc^2 = \gamma mc^2 \tag{1.417}$$

According to Eq. (1.417) the synchrotron cone angle γ^{-1} can be expressed as

$$\frac{1}{\gamma} = \frac{mc^2}{E} \tag{1.418}$$

where the rest energy (mc^2) for the electron is 0.511 MeV. Thus, as the electron energy E increases, the synchrotron cone angle decreases. Teresawa and Kihara (1996) report $1/\gamma = 0.064$ mrad for $E = 8.0$ GeV electron, which is calculated according to Eq. (1.418) as

$$\frac{1}{\gamma} = \frac{mc^2}{E} = \frac{0.511 \text{ MeV}}{8 \times 10^3 \text{ MeV}} = 0.0000638 \text{ rad}$$
$$= 0.064 \text{ mrad} \tag{1.419}$$

which represents only a 0.64 mm spread of the synchrotron radiation at a distance of 10 m from the source or a 6.4 mm spread at a distance of 100 m. Synchrotron radiation is thus a very highly collimated beam. The radiation is highly polarized and emitted as pulsed light with short durations of 1 ns or less (Owens, 2012).

Teresawa and Kihara (1996) provide an example of electrons accelerated in a synchrotron to an energy of 1.0 GeV. They report the phase velocity $\beta = 0.999999868$, that is, the velocity (v) of the electron would be $0.999999868c$ and $\gamma = 1957$, which corresponds to an electron traveling at the relativistic speed, close to that of light and possessing a mass 1957 times that of its rest mass. This is found according to Eq. (1.418) or

$$E = \gamma mc^2 \quad \text{and} \quad \gamma = \frac{E}{mc^2} = \frac{1 \text{ GeV}}{0.511 \text{ MeV}}$$
$$= \frac{1 \times 10^3 \text{ MeV}}{0.511 \text{ MeV}} = 1956.95 \tag{1.420}$$

Also from the value of γ, the velocity of the electron (v) relative to the speed of light in a vacuum (c) can be calculated with the expression for the Lorentz transformation factor as

$$\gamma = \frac{1}{\sqrt{1-(v/c^2)}} \tag{1.421}$$

Substituting the value of 1957 for γ and calculating the velocity (v) of the electron according to Eq. (1.421) gives

$$1957 = \frac{1}{\sqrt{1-(v/c)^2}}$$

$$1-(v/c)^2 = \left(\frac{1}{1957}\right)^2$$

$$(v/c)^2 = 0.9999997389$$

$$v/c = \sqrt{0.9999997389}$$

$$v = 0.999999869c$$

Thus, the 1 GeV electron in the synchrotron is measured to have a relativistic mass of 1957 times that of its rest mass and travels at a velocity of 0.999999869 times the velocity of light in a vacuum.

The Advanced Photon Source at Argonne National Laboratory is a synchrotron radiation source, which can achieve an electron energy output of 7.0 GeV. The calculated mass of an electron of 7.0 GeV energy in the synchrotron according to Eq. (1.420) would possess a 7-fold increase in mass over the previously calculated 1.0 GeV electron, that is, a mass of 13698.6 times its rest mass. However, the velocity of the 7.0 GeV electron is calculated as above in Eq. (1.421) to be $0.9999999974c$, which is essentially the same as the 1.0 GeV electron, close to that of the unattainable speed of light in a vacuum.

4. Synchrotron radiation production and applications

Synchrotron radiation sources generally consist of the following basic elements: (i) an electron gun and linear accelerator (LINAC), (ii) a booster synchrotron, (iii) a storage ring, (iv) synchrotron radiation beam lines, and (v) experimental work stations at each synchrotron radiation beam line. An illustration of a modern synchrotron light source with an identification of its basic elements is illustrated in Fig. 1.117. The electron gun provides the initial source of electrons, which are emitted by thermionic emission from a heated cathode, which consists of a heated block of negatively charged metal. A linear accelerator or LINAC, which is labeled as the injector in Fig. 1.117, will accelerate the electrons to an energy sufficient for the electrons to be injected into a booster synchrotron ring. The energy output of the LINAC of the Advanced Photon Source (Synchrotron) of the Argonne National Laboratory will accelerate the electrons to 375 MeV (0.375 GeV).

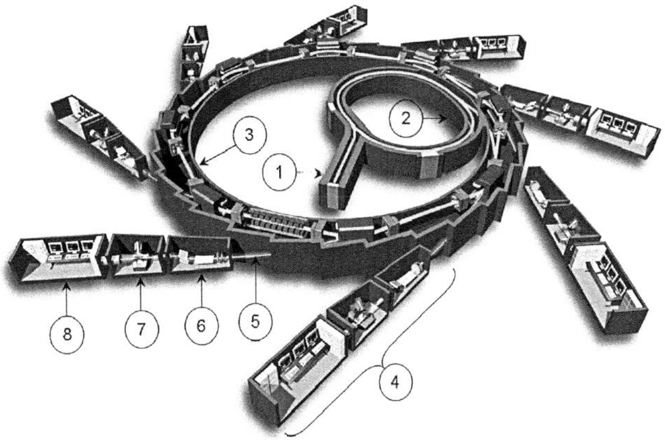

FIGURE 1.117 Essential elements of the third-generation 2.75-GeV Soleil Light Source at Saint-Aubin, France (http://www.synchrotron-soleil.fr/; printed with permission © Synchrotron SOLEIL–EPSIM/J.F. SANTARELLI). Its major components are (1) the injector, (2) booster ring, (3) storage ring, (4) experimental station consisting of (5) a beam-line, (6) optics and monochromator hutch, (7) experimental hutch, and (8) control room. Generally the experimental stations are custom designed for specific applications. For the color version of this figure, the reader is referred to the electronic version of the book. *From Owens (2012); reprinted with permission from Elsevier © 2012.*

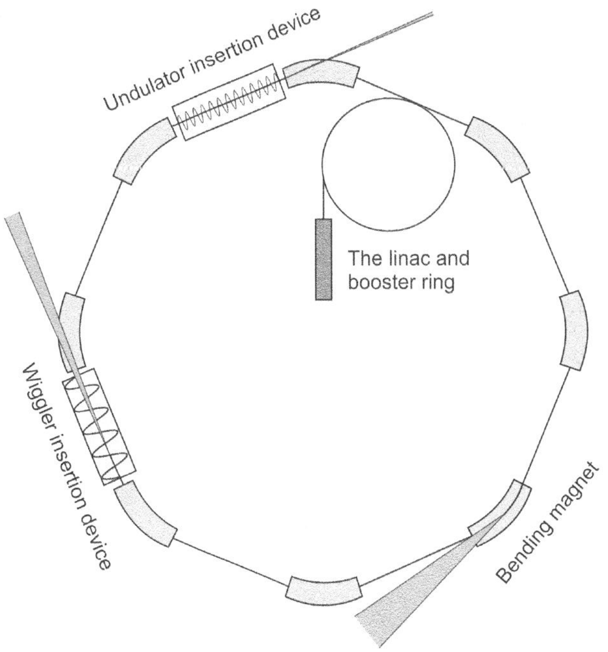

FIGURE 1.118 Schematic of a synchrotron radiation source showing the linac, inner synchrotron booster ring, outer storage ring, and the magnetic devices that produce synchrotron X-ray radiation. In the storage ring are illustrated bending magnets and wiggler and undulator insertion devices with synchrotron radiation emitting from each device. The properties of the source and resulting X-ray beam are described by a flux (photons/second/unit energy bandwidth), a brilliance (photons/second/mm^2/mrad^2/unit energy bandwidth) and a divergence (mrad). For the color version of this figure, the reader is referred to the electronic version of the book. *From Mitchell et al. (1999); reprinted with permission from Elsevier © 1999.*

These electrons travel at a speed of 0.9999990715c or 99.99990715% of the speed of light calculated according to Eqs. (1.420) and (1.421) described previously. The high-energy electrons are then fed into a booster synchrotron ring to accelerate the electrons further with radio frequency energy sources. Electron energies achieved at the booster synchrotron will vary from 2 to 8 GeV at different synchrotron facilities around the world (Clarke, 2005). The booster synchrotron at the Advanced Photon Source (Synchrotron) of the Argonne National Laboratory produces an electron energy output of 7.0 GeV with an electron velocity of 0.9999999974*c* or 99.99999974% of the velocity of light. The high-energy electrons produced by the booster synchrotron are then fed into a storage ring where the electrons will travel at relativistic speed in a circular path controlled by bending magnets, which also serve as the source of synchrotron radiation needed for practical research with X-rays in numerous fields of science. The synchrotron radiation beam-lines, illustrated in Fig. 1.117, lead into individual experimental stations where individualized experimentation with the synchrotron radiation, generally the X-radiation, is carried out. The circulating electrons will lose energy with the emission of synchrotron radiation, and the lost energy is replenished with surges of energy from radio frequency cavities as the electrons circulate in the ring. Eventually electrons are lost by interactions with molecules of gas that reside in the imperfect vacuum chamber within which the electron beam travels; however, beams can last in storage rings for many hours (*e.g.*, over 60 hours). Electron beams in storage rings are regularly replenished with the booster synchrotron to maintain the production of synchrotron radiation according to the needs of the research community.

The storage ring is not a perfect circle; rather it is considered as a polygon, as it will contain straight sections allowing for the insertion devices, referred to as wigglers and undulators, as illustrated in the simplified diagram of Fig. 1.118. The wigglers and undulators consist of varying poles of numerous magnets (see Fig. 1.119) to provide tightly curved transverse displacements of the particle beam (Clarke, 2005; Geloni et al., 2016). The wigglers use the poles of the various magnets to deflect the electron beam back and forth or from side to side, such as to "wiggle" the electron beam, and at each wiggle, synchrotron radiation is emitted as a wide horizontal fan of X-rays capable of feeding more than one work station with X-rays where each work station can take a portion of slice of the fan for research (Mitchell et al., 1999). An undulator, as described by Mitchell et al. (1999), operates in a similar fashion as the wiggler, but with a shorter period between each magnetic pole maintaining the deflected beam within the original beam envelope. The photons emitted at each pole result in partially coherent X-rays concentrated in a more narrow beam with enhanced

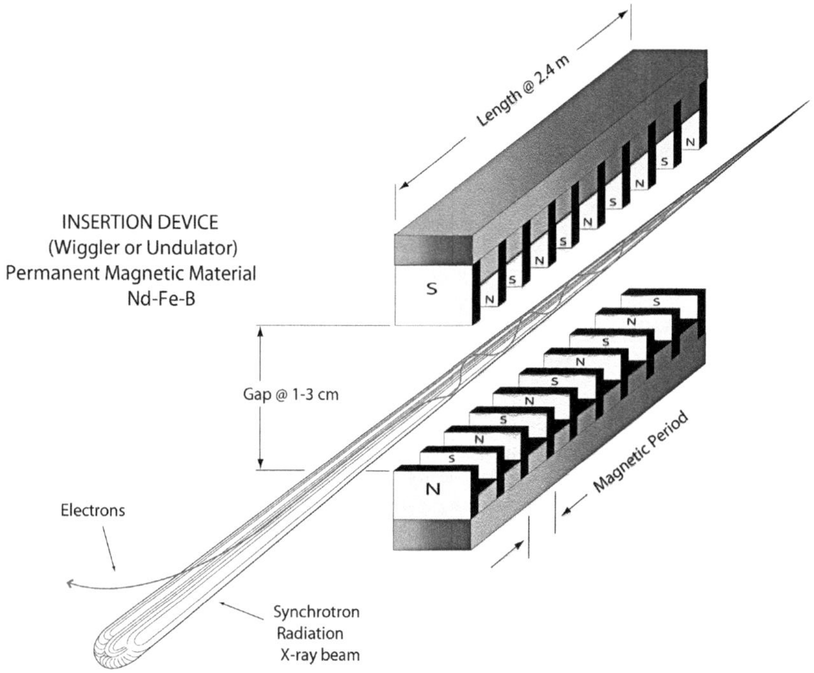

FIGURE 1.119 Basic structure and mechanism of a wiggler or undulator insertion device. For the color version of this figure, the reader is referred to the electronic version of the book. *Courtesy of Advanced Photon Source, Argonne National Laboratory, Argonne, IL, USA; http://www.aps.anl.gov/.*

brilliance. With N bending magnets, the brilliance is enhanced in indulators by a factor of N^2 emitted over a narrow bandwidth (Owens, 2012), whereas with wigglers, the brilliance is enhanced only by a factor of N bending magnets.

Many insertion devices are installed generally within the storage ring, which provide numerous synchrotron radiation beamlines for research and practical applications. Synchrotron radiation facilities can have large storage rings with a circumference exceeding a kilometer and with well over 40 beamlines and their associated workstations. Some storage rings of this magnitude include the European Synchrotron Radiation Facility (ESRF) in Grenoble, France, the Advanced Photon Source (APS) at Argonne, IL, USA, and the 8-GeV Super Photon Ring (Spring-8) at Nishi Harima, Japan. An example of the magnitude of such facilities is provided by an aerial photograph of the APS, and an illustration of its associated work stations in Figs. 1.120 and 1.121, respectively.

There are over 70 synchrotrons worldwide, and many of these are wholly or partially dedicated to the generation of synchrotron radiation for research (Sorensen et al., 2006). The IAEA maintains a database of synchrotron radiation facilities worldwide with links to the individual websites of these synchrotron radiation sources (http://www.iaea.org/). Synchrotron radiation facilities generally provide the opportunity for research groups to reserve a workstation coupled to a beamline for independent research.

Many applications of synchrotron radiation are available in almost every field of science including the

FIGURE 1.120 Aerial photograph of the Advanced Photon Source of the Argonne National Laboratory, Argonne, IL, USA, which houses a 1.1 km storage ring for the production and application of synchrotron radiation. For the color version of this figure, the reader is referred to the electronic version of the book. *Courtesy of Advanced Photon Source, Argonne National Laboratory, Argonne, IL, USA; http://www.aps.anl.gov/.*

biological and life sciences, geology and soil science, physics, chemistry, and materials science. Examples of synchrotron radiation applications in the scientific journal literature number in the thousands, and only a few examples will be cited here. Among these applications are

(i) the structural determinations of macromolecules by X-ray diffraction (Wasserman, 2012; Garino et al., 2014; Rao et al., 2016; Fischer et al., 2018), imaging by means of X-ray computed tomography (Meneses

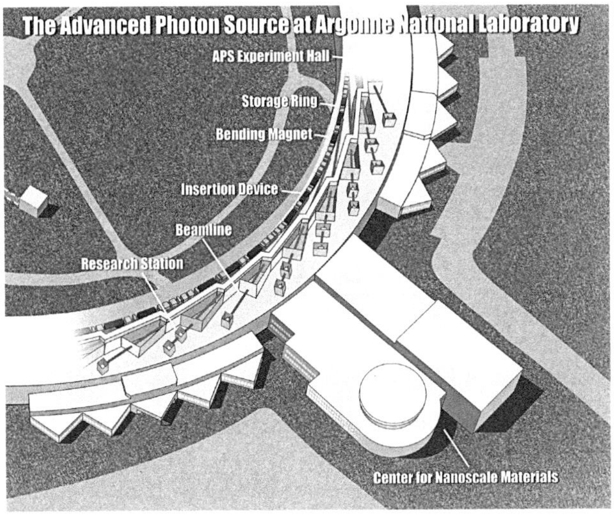

FIGURE 1.121 An artistic view of the layout within the experimental hall of the Advanced Photon Source storage ring illustrating the various workstations linked to synchrotron radiation beamlines and components of the storage ring. For the color version of this figure, the reader is referred to the electronic version of the book. *Courtesy of Advanced Photon Source, Argonne National Laboratory, Argonne, IL, USA; http://www.aps.anl.gov/.*

et al., 2012; Almeida et al., 2014; Basnayaka et al., 2014; Garcea et al., 2017),

(ii) X-ray spectroscopy (Sayen and Guillon, 2014; Young, 2014; Sindhupakorn et al., 2017),

(iii) infrared spectroscopy (McKellar, 2010; Miura et al., 2010; Marcelli et al., 2012; Li et al., 2017),

(iv) X-ray fluorescence analysis (Lis et al., 2014; Baldassini et al., 2015; Ghosh et al., 2017a,b), and

(v) nanotechnology (Wu and Becker, 2011; Zhu et al., 2014; Miao et al., 2017), among other fields where an X-ray or infrared or other suitable bandwidth of radiation source of high collimation and brilliance can be applied.

For a general review of the many applications of synchrotron radiation, the reader may peruse reviews by Abela et al. (2010) and Martins and Silva (2014), Grochulski et al. (2017), Sayers et al. (2017), and Zolotarev et al. (2016).

XII. Interaction of electromagnetic radiation with matter

The lack of charge and rest mass of electromagnetic gamma and X-radiation hinder its interaction with matter and the dissipation of its energy in matter. Consequently, gamma radiation and X-rays have greater penetration power and longer ranges in matter than the more massive and charged alpha and beta particles of the same energy. Nevertheless, gamma and X-radiation are absorbed by matter, and the principal mechanisms by which this type of radiation interacts with matter are discussed in this section.

A. Photoelectric effect

The energy of a photon may be completely absorbed by an atom in a single interaction. Under such circumstances, the entire absorbed photon energy is transferred to an electron of the atom and the electron is ejected, resulting in the formation of an ion pair. In such an interaction, known as the photoelectric effect, the energy of the emitted electron is equal to the energy of the impinging photon less the binding energy of the electron. This is described by the photoelectric equation of Einstein:

$$E_e = h\nu - \phi \tag{1.422}$$

where E_e is the energy of the ejected electron, $h\nu$ is the energy of the incident photon, and ϕ is the binding energy of the electron or the energy required to remove the electron from the atom. The ejected electron is identical in properties to a beta particle and produces ionization as it travels through matter as previously described for beta particles.

The photoelectric effect occurs in the inner electron shells, predominately in the *K*-electron shell, of the atom as described in Section XII.D of this chapter. When an electron from an inner atomic *K* or *L* shell is ejected, electrons from outer shells fall to inner electron shells to fill the resulting gap. These transitions in electron energy states require a release of energy by the atomic electrons, which appears as soft (low-energy) X-rays. X-radiation is identical in properties to gamma radiation. The essential difference lies in the origin of the two radiations. As previously described, gamma radiation originates from energy state transformations of the nucleus of an atom, whereas X-radiation originates from energy state transformations of electrons.

B. Compton effect

There is a second mechanism by which a photon (*e.g.*, X-ray or gamma ray) transfers its energy to an atomic orbital electron. In this interaction, illustrated in Fig. 1.122, the photon, E_γ, imparts only a fraction of its energy to the electron and in so doing is scattered with energy E'_λat an angle Θ to the initial trajectory of the photon, while the bombarded electron is ejected from the atom. This interaction is known as the Compton effect and also as Compton scattering. The result of this interaction is the formation of an ion pair, as in the case of the photoelectric effect. However, the deflected photon continues traveling through matter until it dissipates its entire kinetic energy by interacting with other electrons in a similar fashion or via other mechanisms of interaction with matter discussed in this section. The ejected electron, being identical in properties to a beta particle, loses its energy through the ionization it causes according to mechanisms previously described.

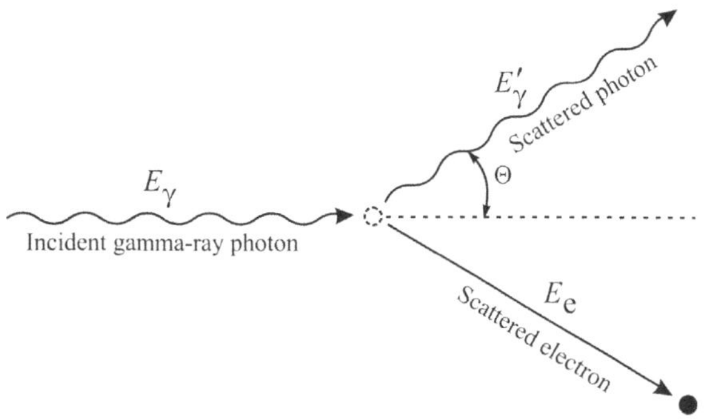

FIGURE 1.122 The Compton effect. An incident photon collides with an atomic electron and imparts energy to it, the photon is scattered at an angle Θ to the trajectory of the incident photon, and the electron is ejected from the atom.

Our understanding of the Compton effect comes from the original work of Arthur H. Compton (1923a,b) and the collaboration of his graduate student Wu Youxun (Woo, 1924, 1925a,b,c). They demonstrated that X-ray photons scattered by thin foils underwent a wavelength shift. The shift in wavelength of the scattered photon with respect to that of the incident photon was a function of the angle of scatter Θ. To interpret this effect Compton and Wu treated X-radiation as photon particles or quanta according to the Einstein–Plank relation $E = h\nu$ (see Eq. 1.371) and the scattering to occur as photon-electron collisions somewhat like billiard-ball collisions as illustrated in Fig. 1.122. Compton derived the equation, which describes the wavelength shift between the incident and scattered photons and angle of scatter as

$$\lambda' - \lambda = \frac{h}{m_0 c}(1 - \cos\Theta) \tag{1.423}$$

where λ' and λ are the wavelengths of the deflected and incident photons, respectively, h is Planck's constant, m_0 is the rest mass of the electron, c is the speed of light, and Θ is the angle of scatter of the photon relative to its original direction of travel.

The Compton-scatter photon will always be of longer wavelength (lower energy) than the incident photon, because of energy lost in the collision with the electron. For example, let us calculate the wavelength shift and energy loss by an incident photon of wavelength 0.300 nm that collides with an electron, and where the photon is scattered at an angle of 70 degrees. The wavelength of the scattered photon is calculated according to Eq. (1.423) as

$$\lambda' = \lambda + \frac{h}{m_0 c}(1 - \cos\Theta)$$

$$= 3.0 \times 10^{-10}\ \text{m} + \frac{6.626 \times 10^{-34}\ \text{J sec}}{(9.109 \times 10^{-31}\ \text{kg})(2.997 \times 10^{8}\ \text{m/sec})}(1 - \cos 70^\circ)$$

$$= 3.0 \times 10^{-10}\ \text{m} + 2.43 \times 10^{-12}\ \text{m}(1 - 0.342)$$

$$= 0.3016\ \text{nm}$$

The energy lost by the incident photon according to the Einstein–Planck relation is given by

$$\Delta E = E_\gamma - E'_\gamma$$

$$= \frac{hc}{\lambda} - \frac{hc}{\lambda'}$$

$$= \frac{12.4\ \text{keV Å}}{3.00\ \text{Å}} - \frac{12.4\ \text{keV Å}}{3.016\ \text{Å}}$$

$$= 4.133\ \text{keV} - 4.111\ \text{keV} = 0.022\ \text{keV}$$

and the fraction of photon energy lost becomes

$$\frac{\Delta E}{E} = \frac{0.022\ \text{keV}}{4.133\ \text{keV}} = 0.0053 = 0.53\%$$

We can calculate directly the energy of the Compton scatter photon, λ', if we know the incident X-ray or gamma-ray photon energy and angle of scatter of the photon according to the equation

$$E'_\gamma = \frac{E_\gamma}{1 + (E_\gamma/mc^2)(1 - \cos\Theta)} \tag{1.424}$$

where E'_γ is the energy of the Compton scatter photon, E_γ is the incident photon energy, mc^2 is the rest energy of the electron (511 keV or 0.511 MeV, see Section XII.C), and Θ is the Compton photon angle of scatter (Tait, 1980). If we take the data from the previous example, where the incident photon energy was 4.133 keV and the angle of scatter was 70 degrees, we can calculate the energy of the Compton photon according to Eq. (1.424) to be

$$E'_\gamma = \frac{4.133\ \text{keV}}{1 + (4.133\ \text{keV}/511\ \text{keV})(1 - \cos 70^\circ)}$$

$$= 4.111\ \text{keV}$$

The result is in agreement with the calculations above using Eq. (1.423) derived by Compton.

It has been shown by Compton that the angle of deflection of the photon is a function of the energy imparted to the electron. This angle may vary from just above $\Theta = 0$ degrees for low Compton electron energies to a maximum $\Theta = 180$ degrees for the highest Compton electron energy. Compton electrons are thus emitted with energies ranging between zero and a maximum energy referred to as the Compton edge. The Compton edge is the Compton electron energy corresponding to complete backscattering of the gamma ray photon. With $\Theta = 180$ degrees or $\cos\Theta = -1$, Eq. (1.424) is reduced to the following equation describing the energy, E'_γ, of the gamma ray photon at the Compton edge in MeV units:

$$E'_\gamma = \frac{E_\gamma}{1 + (E_\gamma/0.511\ \text{MeV})(1 - \cos 180^\circ)} \tag{1.425}$$

or

$$E'_\gamma = \frac{E_\gamma}{1 + (2E_\gamma/0.511)} \quad (1.426)$$

or

$$E'_\gamma = \frac{E_\gamma}{1 + 3.914E_\gamma} \quad (1.427)$$

As an example, the energy of the gamma-ray photon in MeV at the Compton edge for an incident gamma ray from ^{137}Cs (E_γ= 0.662 MeV) is calculated according to Eq. (1.427) to be

$$E_\gamma = \frac{0.662}{1 + 3.914(0.662)} = 0.184 \text{ MeV}$$

A Compton scatter photon is of longer wavelength and lower energy than the incident photon. Deflected Compton photons occur with a broad spectrum of energies. Spectra of Compton-scattered photon energies contain a peak known as the backscatter peak (see Chapter 9 of this volume). The backscatter peak arises from Compton scattering into a gamma photon detector [*e.g.*, NaI(Tl) crystal] from the surrounding detector shielding and housing materials. The backscatter peak occurs at increasing values of energy (MeV) in proportion to the incident photon energy and approaches a constant value of 0.25 MeV, according to Eq. (1.427), for incident photon energies greater than 1 MeV (Tait, 1980). The energy of the Compton electron, E_e, may be described by

$$E_e = E_\gamma - E'_\gamma - \phi \quad (1.428)$$

where E_γ and E'_γ are the energies of the incident and deflected photons, respectively, and ϕ is the binding energy of the electron. As the binding energy of the atomic electron is relatively small, the energy of the ejected electron is essentially the difference between the incident and deflected photon energies. Substituting the value of E'_γ from Eq. (1.424) into Eq. (1.428) and ignoring the electron binding energy, the Compton electron energy can be expressed as

$$E_e = E_\gamma - \frac{E_\gamma}{1 + (E_\gamma/mc^2)(1 - \cos\Theta)} \quad (1.429)$$

$$E_e = E_\gamma - \frac{E_\gamma}{1 + (E_\gamma/0.511 \text{ MeV})(1 - \cos\Theta)} \quad (1.430)$$

where the electron energies are given in MeV. For example, the energy of a Compton electron, E_e, scattered at 180 degrees (Compton edge: cosΘ = −1) and originating from an incident gamma ray photon from ^{137}Cs (E_γ= 0.662 MeV) is calculated according to Eq. (1.430) as

$$E_e = 0.662 - \frac{0.662}{1 + (0.662/0.511)(1 - \cos 180^o)}$$
$$= 0.478 \text{ MeV}$$

Alternatively, if we ignore the negligible electron binding energy and know the incident photon energy and Compton scatter photon energy, we can calculate the Compton electron energy by difference according to Eq. (1.428)

$$E_e = 0.662\text{MeV} - 0.184\text{MeV} = 0.478\text{MeV},$$

which is in agreement with the electron energy calculated above. The Compton edge and backscatter peak due to interactions of Compton electrons and Compton backscatter photons, respectively, in a scintillation crystal detector are described in Chapter 9 of this volume.

C. Pair production

The interactions of gamma radiation with matter, considered earlier, involve the transfer of γ energy, in whole or in part, to atomic electrons of the irradiated material. Pair production, as another mechanism of γ-energy dissipation in matter, results in the creation of matter (*i.e.*, electrons) from the γ energy. The particles produced are a negatron and a positron from an individual gamma-ray photon that interacts with the coulombic field of a nucleus (see Figs. 1.4 and 1.123). Consequently, this phenomenon involves the creation of mass from energy. Carl D. Anderson at the California Institute of Technology (Caltech) received the 1936 Nobel Prize in physics for his discovery of the positron, the first experimental evidence of antimatter. He and his graduate student Seth Neddermeyer also reported in their findings (Anderson and Neddermeyer, 1933) the first observation of pair production, which they described as follows:

> *One of the most striking phenomena, which have been observed in this experiment is the occasional simultaneous appearance of paired tracks consisting of one positive particle and one negative with a common point of origin … The process, which gives rise to the positrons is at present not known, but so far as these data go they are in accord with the view expressed by Blackett and Occhialini (1933) that the two particles may be formed by a process in which the energy of approximately 1 MeV required for the for-*

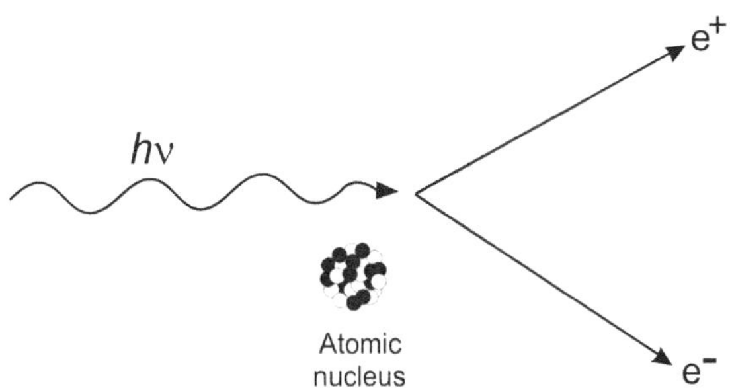

FIGURE 1.123 Pair production. The conversion of a gamma-ray photon into a negatron and positron pair.

mation of a positive and negative electron is supplied by the impinging radiation.

It was Nobel Laureate Patrick Blackett and his coworker G.P.S. Occhialini (Blackett and Occhialini, 1933) who recognized that the two particles, positron and negatron, are formed by the process known as "pair production" whereby the energy of 1.02 MeV required for the production of the two particles is provided by the gamma radiation. In his Nobel Lecture, Anderson (1936) explained the pair-production phenomena as follows:

> *... that the appearance of pairs of positive and negative electrons could be understood in terms of this theory [Dirac's theory] as the 'creation' of a positive-negative electron pair in the neighborhood of an atomic nucleus. The energy corresponding to the proper mass of both of the particles, as well as to their kinetic energies is supplied, according to this view, by the incident radiation. Since the energy corresponding to the proper mass of a pair of electrons is approximately 1 MeV, one should expect gamma rays of energy greater than this amount to produce positrons in their passage through matter, and further that the sum of the kinetic energies of the positive and negative electrons should be equal to the energy of the radiation producing them diminished by approximately 1 MeV.*

In the above statement Anderson described the process of pair-production, first theorized by Dirac (1928a,b), observed through cloud-chamber track photographs by Anderson and Neddermeyer (1933), and interpreted by Blackett and Occhialini (1933). The gamma-ray photon energy required for pair production is, as noted above by Anderson, equivalent to the sum of the masses of the electron and positron described by the equation

$$E_{pair} = m_{e^-}c^2 + m_{e^+}c^2 = 2mc^2 = 2(0.511\ \text{MeV}) = 1.022\ \text{MeV} \quad (1.431)$$

The gamma-ray photon energy in excess of 1.022 MeV appears as the kinetic energy of the electron and positron created, and the overall pair-production phenomena can be expressed as

$$h\nu = 2mc^2 + E_{e^-} + E_{e^+} \quad (1.432)$$

where $h\nu$ is the gamma-ray photon energy, $2mc^2$ is the energy equivalence of the mass of two electrons, and E_{e^-} and E_{e^+} are the kinetic energies of the electron and positron, respectively. In pair production, gamma-ray energy in excess of 1.02 MeV appears as kinetic energy of the negatron and positron produced. By the time of Anderson's Nobel Lecture in 1936, three sources of positrons were known, and Anderson took the opportunity of his lecture to describe these, which are the following: (1) cosmic ray showers of electron-positron pairs via pair production from cosmic gamma radiation, (2) pair production via the absorption of gamma radiation in excess of 1 MeV from radionuclide sources, and (3) positron emission via beta decay from radioactive elements, such as ^{30}P, ^{13}N, and ^{27}Si, artificially produced by Joliot-Curie and Joliot (1934a,b,c), which decay by positron emission.

The creation of an electron requires a certain quantum of energy of a gamma-ray photon, which may be calculated according to Einstein's equation for the equivalence of mass and energy

$$E = m_e c^2 \quad (1.433)$$

where E is energy, m_e is the electron rest mass, and c is the speed of light in a vacuum. Thus, the rest energy of the electron (negatron or positron) is calculated as

$$E = \left(9.109 \times 10^{-31}\ \text{kg}\right)\left(2.997 \times 10^{8}\ \text{m/sec}\right)^2 = 8.182 \times 10^{-14}\ \text{J}$$

Since by definition, 1 eV = 1.602×10^{-19} J, the electron rest energy in joules is converted to electron volts as

$$8.182 \times 10^{-14}\ \text{J}/1.602 \times 10^{-19}\ \text{J/eV} = 0.511\ \text{MeV}$$

Thus, the creation of an electron (negatron) requires a minimum energy of 0.511 MeV. However, a gamma ray of 0.511 MeV energy cannot alone create a negatron, as there must also be the simultaneous creation of its antiparticle, the positron of equal mass and opposite charge. The minimum gamma ray photon energy required for the creation of the negatron-positron pair is, as described by Eq. (1.431) above, 2(0.511) or 1.022 MeV. Thus, the absorption by matter of gamma radiation greater than 1.02 MeV may result in pair production. The probability of pair production increases in proportion to the magnitude of gamma-ray photon energy above 1.02 MeV, and pair production is the predominant mechanism of absorption of photons of energies of 10 MeV and above (see Figs. 1.124 and 1.125).

As discussed previously in Section XI.C, positrons will produce annihilation radiation when they come to rest in the proximity of a negative electron, *i.e.*, their antiparticle, resulting in the simultaneous conversion of two electron masses into two gamma-ray photons of 0.511 MeV energy.

Pair production does not occur only in the vicinity of atomic nuclei bombarded by gamma radiation. It may also originate from nuclei that emit gamma radiation with transition energies greater than 1.02 MeV. This is referred to as **internal pair production**, and the mechanism competes to a small extent with the emission of gamma radiation. The degree to which this competition occurs is measured by the ratio of intensities of positron−negatron pairs to gamma radiation or $(e^{\pm}/\gamma)$. Some examples of nuclides that emit such positron−negatron pairs and the

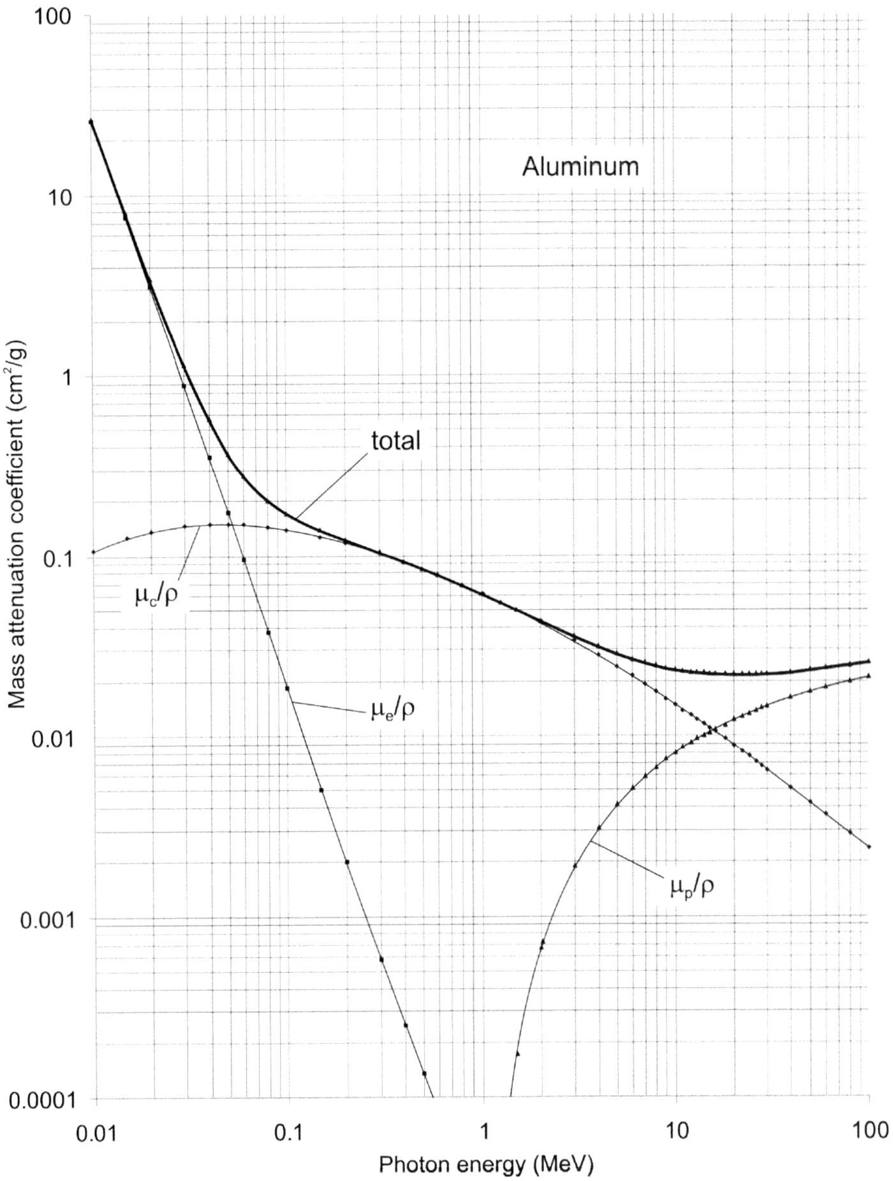

FIGURE 1.124 Mass attenuation coefficients for photons in aluminum. The total attenuation is given by the solid line, which is the sum of the partial attenuations due to the Compton effect, μ_c/ρ; the photoelectric effect, μ_e/ρ, and pair production, μ_p/ρ. XCOM: Photon Cross Sections Database available from the website of the National Institute of Standards and Technology (NIST): https://www.nist.gov/pml/xcom-photon-cross-sections-database). *Drawn from data of Berger et al. (2017b).*

intensities of these pairs relative to gamma radiation are given in Table 1.22.

The threshold nuclear transition energy $E_\gamma > 1.022$ MeV for internal pair production is intuitively obvious as electrons each of rest mass 0.511 MeV must be created from the transition energy. The decaying nucleus undergoes de-excitation with the creation of the positron–negatron pair which are ejected from the nucleus with a total kinetic energy, as described by Allan (1971) according to the equation

$$T = E_\gamma - 2m_0c^2 \quad (1.434)$$

where T is the total kinetic energy shared between the positron and negatron, E_γ in the nuclear transition energy, and $2m_0c^2$ is the energy equivalence of two electron rest masses (i.e., 1.02 MeV). Eq. (1.434) is described by Lombard et al. (1968) according to the shared positron and negatron energies or

$$W_+ + W_- = E_i - E_f - 2m_0c^2 \quad (1.435)$$

where W_+ and W_- are the kinetic energies of the positron and negatron, respectively, E_i and E_f are the initial and final energies of the nuclear states, respectively, and $2m_0c^2$ is the

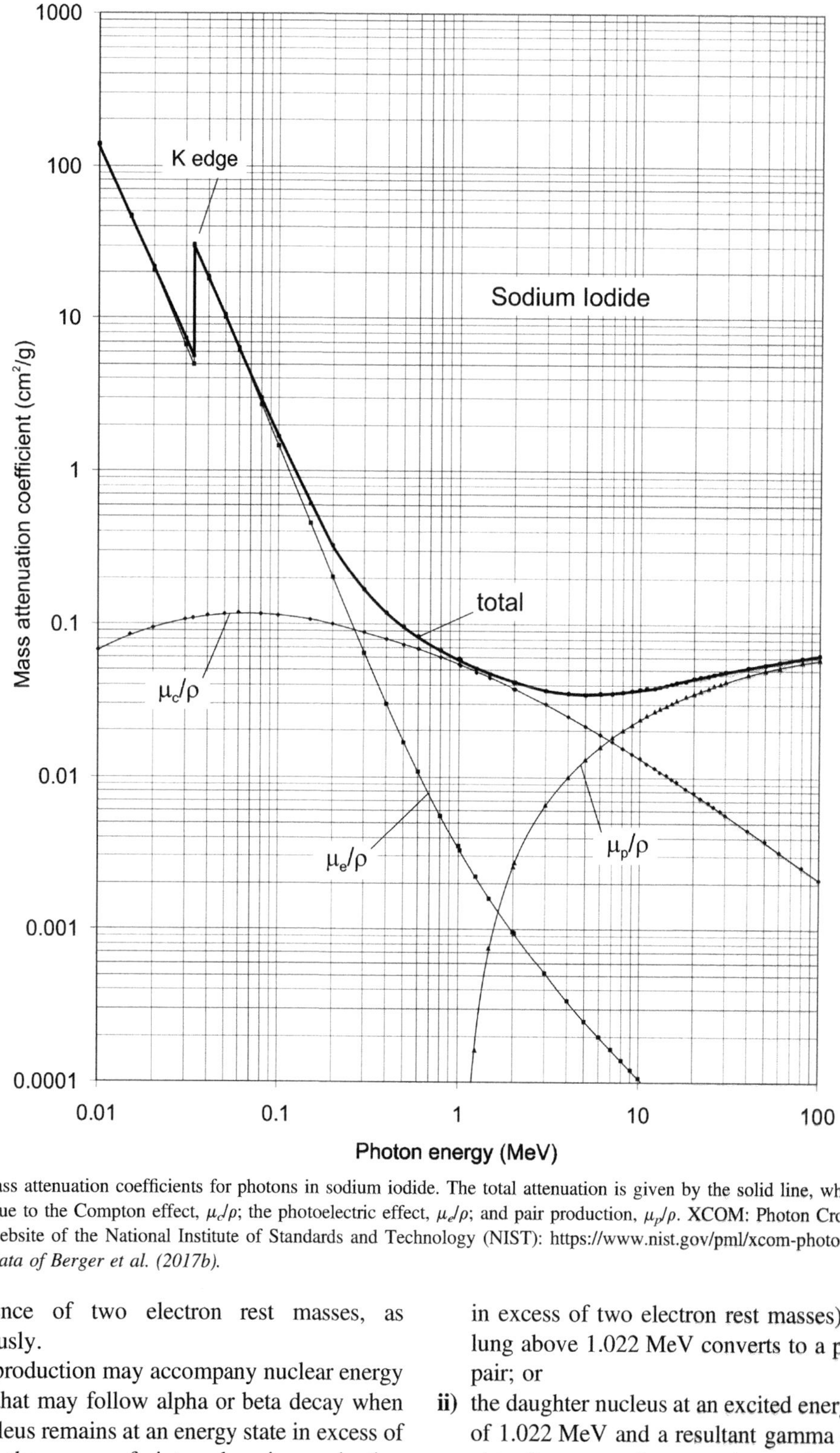

FIGURE 1.125 Mass attenuation coefficients for photons in sodium iodide. The total attenuation is given by the solid line, which is the sum of the partial attenuations due to the Compton effect, μ_c/ρ; the photoelectric effect, μ_e/ρ; and pair production, μ_p/ρ. XCOM: Photon Cross Sections Database available from the website of the National Institute of Standards and Technology (NIST): https://www.nist.gov/pml/xcom-photon-cross-sections-database). *Drawn from data of Berger et al. (2017b).*

energy equivalence of two electron rest masses, as described previously.

Internal pair production may accompany nuclear energy state transitions that may follow alpha or beta decay when the decaying nucleus remains at an energy state in excess of 1.022 MeV. In the case of internal pair production following alpha decay, Staníček and Povinec (1986) describe two possible mechanisms, which are the following:

i) The alpha particle accelerates in the Coulomb field of the daughter nucleus, which creates bremsstrahlung with energy above $2m_ec^2$ (i.e., $E > 1.022$ MeV, energy in excess of two electron rest masses). The bremstrahlung above 1.022 MeV converts to a positron–electron pair; or

ii) the daughter nucleus at an excited energy state in excess of 1.022 MeV and a resultant gamma quantum (real or virtual) converts into a positron–electron pair.

A practical application of internal pair production to the nondestructive activity analysis of ^{90}Y and ^{90}Sr was reported by Selwyn et al. (2007). The nuclide ^{90}Y decays by negatron emission with a Q value of 2280.1 keV as illustrated in Fig. 1.126. The activity of ^{90}Y is determined traditionally by liquid scintillation or Cherenkov counting

TABLE 1.22 Examples of nuclides that exhibit internal pair production, their gamma radiations and relative intensities of the Positron−Negatron pairs.

Nuclide	Gamma radiations		Pair/gamma ratio ($e^{\pm}/\gamma$)
	Energy (MeV)	Abundance (%)	
^{24}Na	1.369	100	6×10^{-5}
	2.754	100	7×10^{-4}
^{56}Mn	1.81	29	5.6×10^{-4}
	2.11	15	4.6×10^{-4}
^{59}Fe	1.099	57	1.4×10^{-4}
	1.292	43	1.1×10^{-4}
^{60}Co	1.17	100	3.7×10^{-5}
	1.33	100	Combined[a]
^{142}Pr	1.576	4	1.1×10^{-4}
^{144}Pr	1.489	0.3	1.9×10^{-4}
	2.186	0.7	6.7×10^{-4}
^{154}Eu	1.274	37	8.0×10^{-5}

[a]*The value of ($e^{\pm}/\gamma$)of 3.7×10^{-5} for ^{60}Co represents the pair/gamma ratio for both gamma emissions combined.*

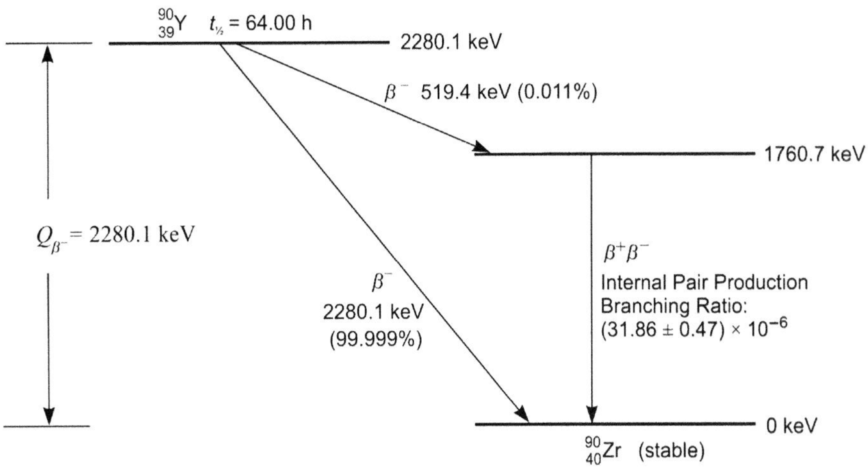

FIGURE 1.126 Decay scheme of ^{90}Y.

of the high-energy negatron emissions, which have an E_{max} of 2280.1 keV as described in Chapter 6, of this volume and Chapter 6 of Volume 2, as this beta transition occurs at an intensity of 99.999% to the ground state of stable ^{90}Zr. There is a very small branching beta transition with an E_{max} of 519.4 keV, which occurs at an intensity of 0.011%, to metastable ^{90}Zr at the energy level of 1760.7 keV (see Fig. 1.126). This is well above the 1022 keV required for internal pair production that can occur in the transition from 1760.7 keV to the ground state of ^{90}Zr. Selwyn et al. (2007) measured the internal pair production branching ratio for this transition to be $(31.86 \pm 0.47) \times 10^{-6}$ by measuring the count rate of the 511 keV peak with a high-purity germanium detector (HPGe) produced by the positron (β^+) annihilation that follows pair production. The high-energy beta emissions of ^{90}Y produce bremstrahlung background radiation, which had to be subtracted from the overall count rate to get an accurate measurement of the net count rate at the 511 keV peak. With the known branching ratio for the internal pair production, and measurement of the net count rate and detection efficiency of the 511 keV positron (β^+) annihilation photons, the ^{90}Y activity of unknown samples can be determined. The activity of ^{90}Sr can be calculated from the measured activity of ^{90}Y, as their activities are equal when in secular equilibrium (see Section XVII.C).

D. Combined photon interactions

Because of its zero rest mass and zero charge, gamma radiation has an extremely high penetration power in matter in comparison with alpha and beta particles.

Materials of high density and atomic number (such as lead) are used most often as absorbers to reduce X- or gamma radiation intensity. Radiation intensity, I, is defined here as the number of photons of a radiation beam that traverse a given area per second, the units of which can be photons/cm^2/s. Suppose a given absorber material of thickness x attenuates or reduces the intensity of incident gamma radiation by one-half. Placing a similar barrier of the same thickness along the path of the transmitted gamma radiation would reduce the intensity again by one-half. With three barriers each of thickness x and an initial gamma-ray intensity I_0, there is a progressive drop in the transmitted gamma-ray intensities: $I_1 = \frac{1}{2}I_0$, $I_2 = \frac{1}{2}I_1$, $I_3 = \frac{1}{2}I_2$, and $I_n = \frac{1}{2}I_{n-1}$. Obviously, incident X- or gamma radiation may be reduced from I_0 to I_3 by using a $3x$ thickness of the same material as an absorber. Consequently, the intensity of the transmitted electromagnetic radiation is proportional to the thickness of the absorber material and to the initial intensity of the radiation. An increasing absorber thickness increases the probability of photon removal, because there is a corresponding increase of absorber atoms that may attenuate the incident photons via the photoelectric effect, the Compton effect, and pair production mechanisms.

If gamma-ray attenuation with respect to absorber thickness is considered, the change in gamma-ray intensity, ΔI, with respect to the absorber thickness, Δx, is proportional to the initial gamma-ray photon intensity, I. This may be written as

$$\Delta I/\Delta x = -\mu I \tag{1.436}$$

where μ is the proportionality constant, referred to as the linear attenuation coefficient or linear absorption coefficient. Its value is dependent on the atomic composition and density of the absorber material. The change in intensity over an infinitely thin section of a given absorber material may be expressed as

$$dI/dx = -\mu I \tag{1.437}$$

or

$$dI/I = -\mu dx. \tag{1.438}$$

Integrating Eq. (1.438) over the limits defined by the initial intensity, I_0, to the transmitted intensity, I, and over the limits of absorber thickness from zero to a finite value x, such as

$$\int_{I_0}^{I} dI/I = -\mu\int_0^x dx \tag{1.439}$$

gives

$$\ln I - \ln I_0 = -\mu x \tag{1.440}$$

or

$$\ln(I/I_0) = -\mu x \tag{1.441}$$

Eq. (1.441) may be written in exponential form as

$$I = I_0 e^{-\mu x} \tag{1.442}$$

Because gamma ray absorption is exponential, the term half-value thickness, $x_{1/2}$, is used to define the attenuation of gamma radiation by matter. Half-value thickness is the thickness of a given material of defined density that can reduce the intensity of incident gamma radiation by one-half. The half-value thickness may also be defined according to Eq. (1.441), in which the initial gamma ray intensity, I_0, is given an arbitrary value of 1 and the transmitted intensity must, by definition, have a value of $1/2$, or

$$\ln(0.5/1) = -\mu x_{1/2} \tag{1.443}$$

where $x_{1/2}$ is the half-value thickness; and

$$\ln 0.5 = -\mu x_{1/2} \tag{1.444}$$

and

$$-0.693 = -\mu x_{1/2} \tag{1.445}$$

and

$$x_{1/2} = 0.693/\mu \tag{1.446}$$

From the linear attenuation coefficient, μ, of a given material and gamma-ray photon energy, it is possible to calculate the half-value thickness, $x_{1/2}$. The linear attenuation coefficient has units of cm^{-1}, so that the calculated half-value thickness is provided in units of material thickness (cm). Linear attenuation coefficients for some materials as a function of photon energy are provided in Table 1.23. The table refers to these as total linear attenuation coefficients, because they constitute the sum of coefficients due to Compton, photoelectric, and pair production interactions. Celiktas (2011) reports on the development of the experimental methods that may be used to determine gamma-ray linear attenuation coefficients. Calculated half-value thicknesses of various absorber materials as a function of gamma-ray energy are illustrated in Fig. 1.127 to illustrate some examples of the varying amounts of absorber material required to attenuate gamma-ray photons. The linear attenuation coefficient is a constant for a given absorber material and gamma-ray photon energy and has units of reciprocal length such as cm^{-1}. It is, however, dependent on the state of the absorber or the number of atoms per unit volume of absorber.

TABLE 1.23 Total linear attenuation coefficients (cm^{-1}) for gamma-ray photons in various materials[a].

Photon energy (MeV)	Water	Aluminum	Iron	Lead
0.1	0.167	0.435	2.704	59.99
0.2	0.136	0.324	1.085	10.16
0.4	0.106	0.2489	0.7223	2.359
0.8	0.0786	0.1844	0.5219	0.9480
1.0	0.0706	0.1658	0.4677	0.7757
1.5	0.0575	0.1350	0.3812	0.5806
2.0	0.0493	0.1166	0.3333	0.5182
4.0	0.0339	0.0837	0.2594	0.4763
8.0	0.0240	0.0651	0.2319	0.5205
10.0	0.0219	0.0618	0.2311	0.5545

[a] *Data from Argonne National Laboratory, ANL-5800 (1963), Hubbell (1969), and Berger et al. (2017b), and NIST (2017) database: https://physics.nist.gov/PhysRefData/FFast/html/form.html.*

A more popular coefficient is the mass attenuation coefficient, μ_m, which is independent of the physical state of the absorber material and is defined as

$$\mu_m = \mu/\rho \tag{1.447}$$

where ρ is the density of the absorber in units of g/cm^3, and μ_m has units of cm^2/g. Some examples of mass attenuation coefficients according to X- and gamma-ray photon energy are provided in Table 1.24. Using the mass attenuation coefficient, Eq. (1.442) transforms to

$$I = I_0 e^{-\mu_m \rho x} \tag{1.448}$$

and the half-value thickness is calculated according to Eq. (1.446) as

$$x_{1/2} = 0.693/\mu_m \rho \tag{1.449}$$

Mass attenuation coefficients for X- or gamma-ray photons over a wide range of energies from 1 keV to 1000 MeV in 100 elements are available from Berger and Hubbell (1997). In addition, mass attenuation coefficients for any element, chemical compounds, mixtures of elements, or mixtures of compounds over the X-ray- or gamma-ray photon energy range of 1 keV to 10^5 MeV are

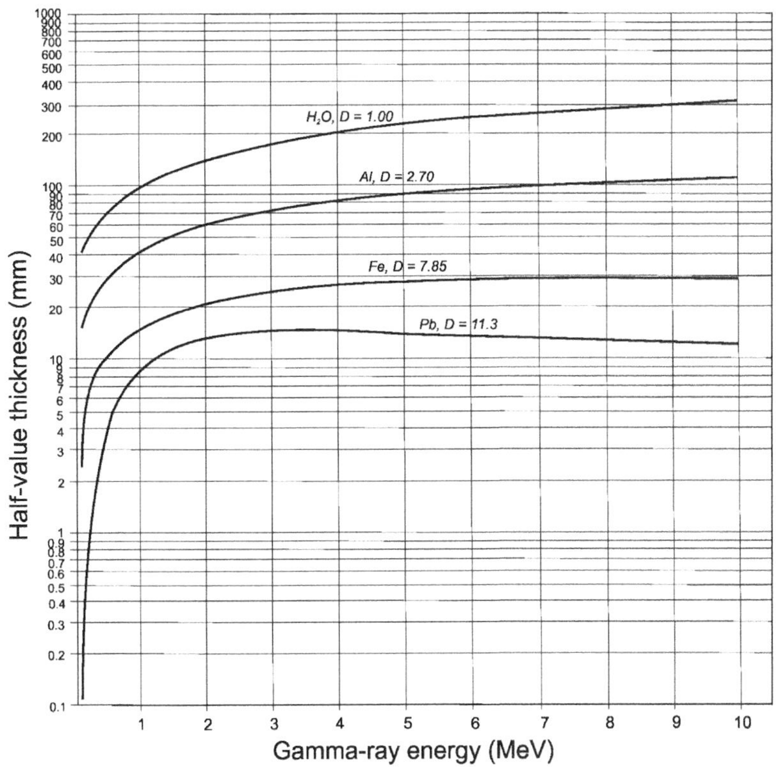

FIGURE 1.127 Half-value thicknesses of various materials as a function of gamma-ray energy. D is the density of each material. *From L'Annunziata (1987) reprinted with permission from Elsevier © 1987 Academic Press.*

available from a database, which can be found at the National Institute of Standards and Technology (NIST) Photon Cross Section Database XCOM: https://www.nist.gov/pml/xcom-photon-cross-sections-database (Accessed 12 December 2017).

A sample of mass attenuation coefficients over the range of 5 keV−10 MeV in a few materials are listed in Table 1.24. The following calculation illustrates the use of the data from Tables 1.23 and 1.24 to calculate half-value thickness and radiation attenuation:

Let us calculate the half-value thickness of lead (ρ = 11.3 g/cm^3) for 2.0 MeV gamma radiation, and further calculate what the reduction in radiation intensity would result if we positioned four times the half-value thickness of lead in the path of the radiation beam. First, the linear attenuation coefficient, μ, or mass attenuation coefficient, μ_m, for 1.0 MeV photons in lead are obtained from either Table 1.23 or 1.24, respectively, and the half-value thickness of lead for 2.0 MeV photons is calculated as

$$x_{1/2} = \frac{0.693}{\mu} \quad \text{or} \quad \frac{0.693}{\mu_m \rho} \tag{1.450}$$

or

$$x_{1/2} = \frac{0.693}{0.5182\ \text{cm}^{-1}} \quad \text{or} \quad \frac{0.693}{(0.0457\ \text{cm}^2/\text{g})(11.3\ \text{g/cm}^3)}$$

$$x_{1/2} = 1.34\ \text{cm}$$

Thus, a barrier of 1.34 cm thickness of lead is sufficient to reduce the radiation intensity of 2.0 MeV photons by 1/2 or 50%. According to Eq. (1.442) the relation between the initial radiation intensity, I_0, and the transmitted intensity, I is

$$\frac{I}{I_0} = e^{-\mu x} \tag{1.451}$$

and for x = 1.34 cm, if the initial radiation intensity is given an arbitrary value of 2, the transmitted intensity would be 50% of the initial intensity or equal to 1. We then can write

$$\frac{I}{I_0} = \frac{1}{2} = e^{-1.34\mu} \tag{1.452}$$

If we employ four times the half-value thickness of lead or 4 × 1.34 cm = 5.36 cm, we can calculate that the transmitted radiation would be reduced to the following:

$$\frac{I}{I_0} = \left(e^{-1.34\mu}\right)^4 = \left(\frac{1}{2}\right)^4$$

or

$$e^{-5.36\mu} = \frac{1}{16} = 0.0625 = 6.25\%\ \text{transmitted}$$

The remaining 15/16 or 93.75% of the initial radiation is attenuated by the 5.36 cm lead barrier. In general, we need not know the half-value thickness of the material or shield, but simply obtain the linear or mass attenuation coefficient for a given energy of X- or gamma radiation from reference tables and use Eqs. (1.442) or (1.448) to calculate the magnitude of radiation attenuation for any thickness of the absorber material. For example, if we used

TABLE 1.24 Total mass attenuation coefficients (cm^2/g) for X- or gamma-ray photons in various materials.[a]

Photon energy (MeV)	Air	Water	Aluminum	Iron	Lead
0.005	38.3	42.0	193	140	730
0.01	4.6	5.1	26.2	171	131
0.05	0.187	0.208	0.368	1.96	8.04
0.1	0.151	0.167	0.170	0.372	5.55
0.2	0.123	0.136	0.122	0.146	0.999
0.4	0.0953	0.106	0.0922	0.0919	0.208
0.8	0.0706	0.0786	0.0683	0.0664	0.0836
1.0	0.0655	0.0706	0.0614	0.0595	0.0684
1.5	0.0517	0.0575	0.0500	0.0485	0.0512
2.0	0.0445	0.0493	0.0432	0.0424	0.0457
4.0	0.0307	0.0339	0.0310	0.0330	0.0420
8.0	0.0220	0.0240	0.0241	0.0295	0.0459
10.0	0.0202	0.0219	0.0229	0.0294	0.0489

[a]*Data from Argonne National Laboratory, ANL-5800 (1963), Hubbell (1969), and Berger and Hubbell (1997), and Berger et al. (2017b), available at National Institute of Standards and Technology (NIST) Photon Cross Section Database XCOM: https://www.nist.gov/pml/xcom-photon-cross-sections-database (Accessed 12 December 2017).*

only 2.5 cm of lead barrier, the attenuation of 2.0 MeV gamma rays could be calculated as

$$\frac{I}{I_0} = e^{-\mu x} = e^{-\mu_m \rho x} \tag{1.453}$$

and

$$\frac{I}{I_0} = e^{-\left(0.5182\ \mathrm{cm}^{-1}\right)(2.5\ \mathrm{cm})}$$

$$= e^{-\left(0.0457\ \mathrm{cm}^2/\mathrm{g}\right)\left(11.3\ \mathrm{g/cm}^3\right)(2.5\ \mathrm{cm})}$$

$$= e^{-1.29} = 0.275 = 27.5\%$$

Thus, the 2.0 MeV radiation transmitted through a shield of 2.5 cm of lead would be 27.5% of the initial radiation intensity.

As previously discussed, the absorption of gamma radiation is a process that principally involves three mechanisms of attenuation: the Compton effect, the photoelectric effect, and pair production. The attenuation coefficients just discussed above are also referred to as total attenuation coefficients, because they consist of the sum of three independent coefficients or

$$\mu = \mu_c + \mu_e + \mu_p \tag{1.454}$$

where μ_c, μ_e, and μ_p are attenuation coefficients for Compton, photoelectric, and pair production processes. The attenuation coefficients are proportional to the probabilities of occurrence of these radiation attenuation processes and can be used as a measure of the relative roles these processes play in the absorption of gamma-ray photons.

Accordingly, the total and partial mass attenuation coefficients can be written as

$$\mu_m = \mu/\rho = \mu_c/\rho + \mu_e/\rho + \mu_p/\rho \tag{1.455}$$

Figs. 1.124 and 1.125 illustrated earlier in this chapter provide a graphic representation of the relative frequency of occurrence of the Compton, photoelectric, and pair production processes in aluminum and sodium iodide absorbers as a function of photon energy. From these curves, it is seen that the photoelectric effect plays an increasing role in total gamma-ray attenuation at lower gamma-ray energies and with absorber materials of higher atomic number. We can also note from these figures, that pair production becomes significant at photon energies above 1 MeV; as the threshold energy for pair production is 1.022 MeV. Data on the total and partial mass attenuation coefficients as a function of photon energy for any element, chemical compound, mixture of elements, or mixture of compounds over the photon energy range of 1 keV to 10^5 MeV are available from the National Institute of Standards and Technology (NIST) Photon Cross Section Database XCOM: https://www.nist.gov/pml/xcom-photon-cross-sections-database.

In some absorber materials of relatively high density, absorption edges can be measured for low photon energies, such as the *K* edge illustrated in Fig. 1.125. The absorption edge is a discontinuity in the attenuation coefficient curve for the photoelectric effect that is caused when photon energies are less than the binding energies of electrons of a certain shell (*e.g.*, *K* shell) and that reduces the number of electrons which may be ejected by the photoelectric effect. When photons possess the threshold binding energy of electrons of that shell, there is a sudden surge in attenuation owing to the ejection of electrons from that shell via photoelectric interactions. The photoelectric effect will occur primarily in the *K*-electron shell. At higher photon energies (>*K*-shell binding energy) the probability of the photoelectric effect occurring diminishes, for example, the photoelectric effect in the *L*-shell is only about 20% and even less probable for outer shells (Friedlander et al., 1964). Such a phenomenon may appear strange, considering that the binding energies of the electrons are lower in shells further away from the nucleus. If one considers only the binding energies of the electrons, it would appear, although incorrectly, that for photon energies in excess of the *K*-shell binding energy, electrons further away from the nucleus (e.g., *L*-, *M*-, *N*-shells) would be ejected as photoelectrons more easily. This is not the case, because energy and momentum must be conserved in the process, and the atom, because of its much higher mass, will carry off the excess momentum while taking away negligible energy in the process. The entire photon energy is transferred to the photoelectron less its binding energy. Consequently, the photoelectric effect cannot occur with a free electron, because there is no atom to carry off excess momentum, as demonstrated by Gautreau and Savin (1999). In the same sense, when the photon energy is much greater than the electron binding energy, the electrons, such as those in the outer shells with very low binding energies, act more like free electrons. These electrons contribute less to photon attenuation by the photoelectric effect. As explained by Lilley (2001), the conservation of energy and momentum is the reason why the most tightly bound electrons contribute most to photoelectric absorption for $E_\gamma > B_K$, where E_γ is the photon energy and B_K is the *K*-shell electron binding energy, and also why there is a marked dependence of photon attenuation on photon energy and absorber atomic number. The photon absorption in NaI, illustrated in Fig. 1.125, shows a *K* edge at ~0.033 MeV or 33 keV. This corresponds to the electron binding energy in the *K* shell of iodine, which from reference tables (Lide, 2010) is reported to be 33,169 eV or 33.1 keV. Photon attenuation by the photoelectric effect for photon energies in excess of 33 keV occur predominantly with the *K*-shell electrons. There is no *K* edge observed for sodium in Fig. 1.125 for the NaI absorber, because the binding energy of the *K*-shell of sodium is only 1070 eV or 1.07 keV, a factor of 10 lower than the photon energy scale provided in the figure. At much lower photon energies, corresponding to *L*- and *M*-shell electron binding energies, characteristic *L* and *M* edges can be observed. The binding energies of electrons in the various atomic electron shells of the elements are listed in reference tables, such as those provided by Lide (2010).

As illustrated in Figs. 1.124 and 1.125 the pair production process does not occur at gamma-ray energies below the threshold value of 1.02 MeV, as expected, in accord with the combined positron and negatron rest energies (*i.e.*, 2×0.511 MeV) required for pair production. A thorough treatment of the attenuation and absorption of gamma radiation in matter is available from Hubbell (1969), Krane (1988), Turner (1995), Lilley (2001), Serway et al. (2005), and Murray and Holbert (2015).

XIII. Radioactive nuclear recoil

When a decaying nucleus discharges or emits a particle such as an alpha particle, beta particle, gamma- or X-ray photon, or neutrino, the nucleus will recoil in a fashion somewhat like a cannon recoils when it shoots a cannon ball from its barrel. The energy of recoil is a function of the mass and energy of the particle that is emitted as well as the mass of the nucleus. This may appear intuitively obvious, if we relate the process to that of a cannon, that is, the heavier the cannon for a given mass and velocity of cannon ball, the less will be the recoil energy. Even radionuclides that emit electrons or photons from their atomic shell, such as Auger electrons and X-rays, respectively, will experience some recoil effect, albeit much less than would occur when a heavy particle, such as the alpha-particle, is emitted from a nucleus at high energy. In accorded with the principles of conservation of momentum, nuclear recoil will occur in all nuclear decay processes with the conservation of momentum between the emitted particle and the product nucleus, as illustrated in Fig. 1.128.

In various textbooks one can find reference to some equations for calculating nuclear recoil energies for some particular particle emissions. Van Rooyen et al. (2008) and Szucs et al. (2009) provide relativistic and nonrelativistic nuclear recoil expressions for all particle emissions including alpha particles, beta particles, Auger electrons, neutrinos, and gamma- and X-ray photons. Detailed derivations of the equations will be provided here.

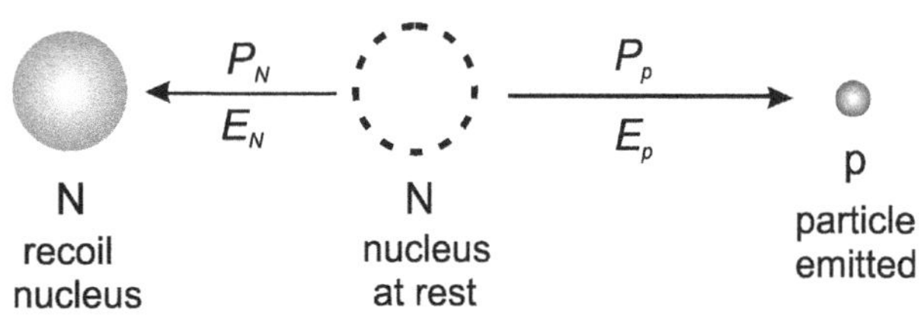

FIGURE 1.128 Nuclear recoil resulting from the emission of a particle from a nucleus similar to the shooting of a bullet from a gun. In the process there is conservation of energy shared between the emitted particle (E_p) and recoil nucleus (E_N) as well as conservation of momentum between the emitted particle (P_p) and recoil nucleus (P_N).

A. Relativistic expressions

When a particle is emitted by a decaying nucleus, it possesses a kinetic energy (K_p), which is the difference of the total particle energy (E_p) and its rest energy (m_pc^2) or

$$K_p = E_p - m_pc^2 \tag{1.456}$$

And if the expression of energy, momentum, and particle mass is written as (see Eq. 1.374)

$$E_p^2 = p_p^2c^2 + \left(m_pc^2\right)^2 \tag{1.457}$$

where p_p is the particle momentum and c is the speed of light in a vacuum. From Eqs. (1.456) and (1.457) we can write

$$\left(K_p + m_pc^2\right)^2 = p_p^2c^2 + \left(m_pc^2\right)^2 \tag{1.458}$$

which becomes

$$K_p^2 + 2m_pc^2K_p + \left(m_pc^2\right)^2 = p_p^2c^2 + \left(m_pc^2\right)^2 \tag{1.459}$$

and reduces to

$$K_p^2 + 2m_pc^2K_p = p_p^2c^2 \tag{1.460}$$

Solving Eq. (1.460) for the particle momentum, gives

$$p_p = \frac{\sqrt{K_p^2 + 2m_pc^2K_p}}{c} \tag{1.461}$$

As the momentum of the system is conserved whereby changes in momentum between the recoil nucleus and particle are equal in magnitude and opposite in direction, as in the nonrelativistic case where

$$m_N\Delta v_N = -m_p\Delta v_p \tag{1.462}$$

we can express the changes in nuclear recoil momentum and particle momentum according to Eq. (1.461) as

$$\frac{\sqrt{K_N^2 + 2m_Nc^2K_N}}{c} = -\frac{\sqrt{K_p^2 + 2m_pc^2K_p}}{c} \tag{1.463}$$

where K_N and m_N are the kinetic energy and mass, respectively, of the recoil nucleus. If the total energy of the nucleus (E_N), in terms of its momentum and mass, is written according to Eq. (1.457) as

$$E_N^2 = p_N^2c^2 + \left(m_Nc^2\right)^2 \tag{1.464}$$

we can input the expression of the particle momentum into Eq. (1.464) to give

$$E_N^2 = \left(\frac{\sqrt{K_p^2 + 2m_pc^2K_p}}{c}\right)^2 c^2 + \left(m_Nc^2\right)^2 \tag{1.465}$$

which becomes

$$E_N^2 = K_p^2 + 2m_pc^2K_p + m_N^2c^4 \tag{1.466}$$

Eq. (1.466) is reduced to the simple expression of the total energy of the nucleus as

$$E_N = \sqrt{K_p^2 + 2m_pc^2K_p + m_N^2c^4} \tag{1.467}$$

If by definition, the kinetic energy of the nucleus K_N, that is, it's recoil energy, is the difference of it's total energy, E_N, and it's rest energy, m_Nc^2, then

$$K_N = E_N - m_Nc^2 \tag{1.468}$$

From Eqs. (1.467) and (1.468) we can arrive at the final expression for the kinetic energy of the recoil nucleus in terms of the particle kinetic energy (K_p), particle mass (m_p), and mass of the recoil nucleus (m_N) to be

$$K_N = \sqrt{K_p^2 + 2m_pc^2K_P + m_N^2c^4} - m_Nc^2 \tag{1.469}$$

Eq. (1.469) is the fully relativistic expression for the calculation of the nuclear recoil energy for particle emissions that possess nonzero rest mass, such as alpha particles, beta particles, and Auger electrons.

For particle emissions consisting of gamma- and X-ray photons, which have a zero rest mass, the term $2m_pc^2K_p$ will fall out of Eq. (1.469), and in the case of the neutrino, which has a near-zero rest mass, the term $2m_pc^2K_p$ becomes negligible and can be dropped out of Eq. (1.469) so that it simplifies to

$$K_N = \sqrt{K_p^2 + m_N^2c^4} - m_Nc^2 \tag{1.470}$$

Eq. (1.470) is thus used to calculate the nuclear recoil energies for gamma-ray photon, X-ray photon, or neutrino emissions. The recoil nucleus is that of the daughter nucleus, because it is the daughter nucleus that recoils after the nucleus emits a particle.

Van Rooyen et al. (2008) and Szucs et al. (2009) utilized Eq. (1.469) for the calculation of the nuclear recoil energies arising from ^{103}Pd Auger electron emissions and Eq. (1.470) for nuclear recoil energies resulting from gamma-photon emissions of ^{103m}Rh.

B. Nonrelativistic expressions

1. Nuclear recoil energy from alpha-particle emissions

For relatively heavy particles, such as the alpha particle, the more simple nonrelativistic calculation of the nuclear recoil energy may also be used. The nonrelativistic expression of the nuclear recoil energy for alpha-particle emissions is derived subsequently. The nonrelativistic expression of the kinetic energy of the alpha particle is defined as

$$K_\alpha = \frac{1}{2}m_\alpha v^2 \tag{1.471}$$

where m_α and v are the mass and velocity of the particle, respectively. It's nonrelativistic momentum would be simply

$$p_\alpha = m_\alpha v \tag{1.472}$$

Multiplying Eq. (1.471) by $2m_\alpha$ gives

$$2m_\alpha K_\alpha = (m_\alpha v)^2 = p_\alpha^2 \tag{1.473}$$

And the momentum of the alpha particle can be expressed as

$$m_\alpha v = \sqrt{2m_\alpha K_\alpha} \tag{1.474}$$

which relates the alpha-particle mass, energy, and momentum. Conservation of momentum of the alpha particle (p_α) and recoil nucleus (p_N) dictates that changes in momentum between the recoil nucleus and particle are equal in magnitude and opposite in direction, we can express the momentum between the alpha particle and nucleus as

$$\sqrt{2m_\alpha K_\alpha} = -\sqrt{2m_NK_N} \tag{1.475}$$

where m_N and K_N are the mass and kinetic energies of the recoil nucleus. Squaring both sides of Eq. (1.475) and solving for K_N gives

$$K_N = \left(\frac{m_\alpha}{m_N}\right)K_\alpha \tag{1.476}$$

Eq. (1.476) was used in Section X.A.1 (Eq. 1.160) for the calculation of the recoil energy resulting from the 5.545 MeV alpha particle emitted from ^{241}Am. The relativistic expression, provided in Eq. (1.469), yields the same result.

2. Nuclear recoil energy from gamma-ray photon, X-ray photon, and neutrino emissions

For gamma-ray and X-ray photons, which have zero rest mass, and neutrinos, which have near-zero rest mass, we can express their momentum (see Eq. 1.376) as

$$p_p = \frac{E_p}{c} \tag{1.477}$$

where p_p is the momentum of the gamma-ray photon, X-ray photon, or neutrino, E_p is the particle energy, and c is the speed of light in a vacuum. If the nonrelativistic expression of the kinetic energy of the recoil nucleus is

$$K_N = \frac{1}{2}m_Nv^2 \tag{1.478}$$

where K_N is the kinetic energy of the recoil nucleus, m_N is the mass of the recoil nucleus and v is it's velocity. Multiplying both sides of Eq. (1.478) by $2m_N$ gives

$$2m_NK_N = (m_Nv)^2 = p_N^2 \tag{1.479}$$

where p_N is the momentum of the recoil nucleus. As momentum is conserved, we can write

$$2m_NK_N = \frac{E_p^2}{c^2} \tag{1.480}$$

where E_p is the energy of the emitted particle, albeit a gamma-ray photon, X-ray photon, or neutrino.

Solving Eq. (1.480) for the energy of the recoil nucleus gives

$$K_N = \frac{1}{2}\left(\frac{E_p^2}{m_N c^2}\right) \tag{1.481}$$

where E_p^2 is the square of the energy of the gamma-ray photon, X-ray photon, or neutrino, and m_N is the mass of the recoil nucleus.

The reader may peruse the works of Van Rooyen et al. (2008) and Szucs et al. (2009) for additional derivations and applications of the above equations for calculating recoil energies.

C. Sample calculations

1. Nuclear recoil from alpha emissions

Earlier in this chapter (Section X.A.1, Eq. 1.160), the nonrelativistic expression for the recoil energy was used to determine the recoil energy imparted to a nucleus as a result of alpha-particle emission of 5.545 MeV that occurs in the following decay of ^{241}Am:

$${}^{241}_{95}\mathrm{Am} \rightarrow {}^{237}_{93}\mathrm{Np} + {}^{4}_{2}\mathrm{He} \tag{1.482}$$

The relativistic expression provided by Eq. (1.469) will be used. Since nuclear recoil occurs after the nucleus emits a particle, it is the daughter nuclide that recoils. We will use atomic mass units (u) and the conversion factor of 931.494013 MeV/u to convert mass to energy units. Relevant data from nuclear tables (*e.g.*, Lide, 2010) are the mass of the alpha particle and that of ^{237}Np which are 4.00150617 u and 237.048166 u, respectively. Updated atomic mass data can be obtained online from the Atomic Mass data Center (AMDC) via the website http://amdc.impcas.ac.cn/web/masseval.html. Thus, from Eq. (1.469), the nuclear recoil energy is calculated as

$$\begin{aligned} K_N &= \sqrt{K_p^2 + 2m_p c^2 K_P + m_N^2 c^4} - m_N c^2 \\ &= \left\{\sqrt{\begin{array}{l}(5.545\ \mathrm{MeV})^2 + 2(4.00150617\ \mathrm{u}) \\ (931.494013\ \mathrm{MeV/u})(5.545\ \mathrm{MeV}) \\ +[(237.048166\ \mathrm{u})(931.494013\ \mathrm{MeV/u})]^2\end{array}}\right\} \\ &\quad - (237.048166\ \mathrm{u})(931.494013\ \mathrm{MeV/u}) \\ &= 0.0936\ \mathrm{MeV} \end{aligned} \tag{1.483}$$

The recoil energy of 0.0936 MeV agrees with the nonrelativistic calculation made earlier in Section X.A.1 (see Eq. 1.160) of this chapter. In general, the nuclear recoil energy resulting from alpha-particle emissions will be higher than that of other nuclear decay emissions for a nuclide of given mass and alpha-particle of given energy (see Table 1.25). This is characteristic of alpha-particle emissions, because the alpha particle is massive. We will notice much smaller nuclear recoil energies resulting from beta-particle emissions, if we compare emissions of the same energy from nuclides of similar mass. This is obviously due to the large difference in mass of the alpha particle compared to the electron or beta particle. The mass of the alpha particle is 7294 times that of the electron (*i.e.*, $\alpha/e = 4.00150617$ u/0.000548579 u). An analogy can be made by comparing the recoil imparted to a cannon upon firing a cannon ball compared to the kick imparted to a pistol upon firing a 0.45 caliber bullet.

2. Nuclear recoil from beta emissions

Let us calculate the recoil energy imparted to a nucleus by the emission of a beta particle. The following decay of ^{109}Pd by beta-particle emission may serve as an example:

$${}^{109}_{46}\mathrm{Pd} \rightarrow {}^{109m}_{47}\mathrm{Ag} + \beta^-(E_{\max} = 1.03\ \mathrm{MeV}) + \bar{\nu} + \gamma \tag{1.484}$$

Beta particles are emitted with a broad spectrum of energies from zero to $E_{\max}$, because the decay energies are shared with a neutrino. We will select the maximum energy (1.03 MeV) that the negative beta particle may possess when the antineutrino energy is at its minimum (near zero) in order to calculate the maximum recoil energy that the particle may impart to the nucleus. We will not calculate here any nuclear recoil energy that the gamma ray may impart to the nucleus. Again, as in the previous example, the nucleus that undergoes the recoil is that of the daughter nuclide, because it is the daughter that is produced when the beta particle is emitted from the nucleus. Also, we will use the relativistic expression of the recoil energy (Eq. 1.469) for the calculation. Relevant data from nuclear reference tables for the calculation are the mass of the beta-particle (*i.e.*, electron mass) and that of ^{109m}Ag, which are 0.000548579 u and 108.904756 u, respectively. The nuclear recoil energy is calculated as

$$\begin{aligned} K_N &= \sqrt{K_p^2 + 2m_p c^2 K_P + m_N^2 c^4} - m_N c^2 \\ &= \left\{\sqrt{\begin{array}{l}(1.03\times10^6\ \mathrm{eV})^2 + 2(5.48579\times10^{-4}\ \mathrm{u}) \\ (931494013\ \mathrm{eV/u})(1.03\times10^6\ \mathrm{eV}) \\ +[(108.904756\ \mathrm{u})(931494013\ \mathrm{eV/u})]^2\end{array}}\right\} \\ &\quad - (108.904756\ \mathrm{u})(931494013\ \mathrm{eV/u}) \\ &= 10.41\mathrm{eV} \end{aligned} \tag{1.485}$$

TABLE 1.25 Recoil energies in electron volts (eV) Imparted to nuclei of various masses by emitted particles of various energies in MeV.

Nuclear mass	Alpha particles[a]				
	$E_\alpha = 1.5$ MeV	$E_\alpha = 2$ MeV	$E_\alpha = 4$ MeV	$E_\alpha = 6$ MeV	$E_\alpha = 8$ MeV
150 u	4.0E4 eV	5.3E4 eV	10.7E4 eV	16.0E4 eV	21.3E4 eV
175	3.4E4	4.6E4	9.1E4	13.7E4	18.3E4
200	3.0E4	4.0E4	8.0E4	12.0E4	16.0E4
225	2.7E4	3.6E4	7.1E4	10.7E4	14.2E4
250	2.4E4	3.2E4	6.4E4	9.6E4	12.8E4
Nuclear mass	**Beta particles[b]**				
	$E_\beta = 0.5$ MeV	$E_\beta = 1$ MeV	$E_\beta = 2$ MeV	$E_\beta = 4$ MeV	$E_\beta = 6$ MeV
10 u	40.84 eV	108.35 eV	324.42 eV	1078.27 eV	2261.53 eV
20	20.42	54.27	162.21	539.13	1130.76
50	8.17	21.71	64.88	215.65	452.30
100	4.08	10.85	32.44	107.82	226.15
150	2.72	7.23	21.63	71.88	150.77
200	2.04	5.43	16.22	53.91	113.08
250	1.63	4.34	12.98	43.13	90.46
Nuclear mass	**Gamma-ray, X-ray photons, or neutrinos[c]**				
	$E_\gamma = 0.5$ MeV	$E_\gamma = 1$ MeV	$E_\gamma = 2$ MeV	$E_\gamma = 4$ MeV	$E_\gamma = 6$ MeV
10 u	13.42 eV	53.68 eV	214.71 eV	858.83 eV	1932.38 eV
20	6.71	26.84	107.35	429.42	966.19
50	2.68	10.74	42.94	171.77	386.47
100	1.34	5.37	21.47	85.88	193.24
150	0.89	3.58	14.32	57.25	128.82
200	0.67	2.68	10.74	42.94	96.62
250	0.54	2.15	8.59	34.35	77.29

[a] *Nuclear recoil energies resulting from alpha-particle emission are expressed in exponential form. For example, the nuclear recoil energy imparted to a nucleus of mass 150 u by an alpha-particle emitted with energy of 1.5 MeV is 4.0E4 or 4.0 × 10^4 eV. Also, nuclear recoil energies for nuclear masses <150 u are not listed for alpha particle emissions, because only nuclei of relatively high mass (A > 140 u) are unstable to alpha-particle emission. The recoil energies resulting from alpha-particle emissions were calculated by Eq. (1.476), i.e.,* $K_N = (m_\alpha/m_N)K_\alpha$.

[b] *The recoil energies resulting from beta-particle emissions were calculated according to Eq. (1.469), i.e.,* $K_N = \sqrt{K_p^2 + 2m_pc^2K_p + m_N^2c^4} - m_Nc^2$.

[c] *Recoil energies resulting from gamma-ray photon, X-ray photon or neutrino emissions were calculated according to Eq. (1.470), i.e.,* $K_N = \sqrt{K_p^2 + m_N^2c^4} - m_Nc^2$.

As we see from this example, the previous example, and from Table 1.25, the energies imparted by the massive alpha particles to recoil nuclei are measured in the tens of thousands of electron volts; whereas, energies imparted by the relatively miniscule beta particle range mostly from a few to several hundred electron volts, and energies imparted to nuclei by gamma-ray and X-ray photons and neutrino emissions are yet smaller.

3. Nuclear recoil from gamma-ray photon, X-ray photon, or neutrino emissions

Let us consider an example of a nuclide that decays exclusively by EC with the emission of neutrinos and gamma radiation. The nuclide ^{7}Be decays to ^{7}Li with the emission of neutrinos of 0.862 MeV at 89.6% relative intensity and 0.384 MeV at the remaining 10.4% intensity. The ^{7}Li daughter nuclides resulting from the lower 0.384 MeV neutrino emissions decay to the ground state with the emission of gamma radiation of 0.478 MeV. The decay scheme of ^{7}Be is illustrated in Fig. 1.54 and the decay equation is the following:

$$^{7}_{4}\text{Be} \xrightarrow{\text{EC}} {}^{7}_{3}\text{Li} + \nu + \gamma \qquad (1.486)$$

We can use Eq. (1.470) to calculate the nuclear recoil energy resulting from the emission of a neutrino or gamma-ray photon. Let us calculate the recoil energy imparted to the ^{7}Li nucleus resulting from the emission of a 0.862 MeV neutrino. The mass of ^{7}Li obtained from physical data tables is 7.016004 u and the constant for conversion of mass units to energy is 931.494013 MeV/u.

$$K_N = \sqrt{K_p^2 + m_N^2c^4} - m_Nc^2$$

$$K_N = \left\{ \sqrt{\begin{matrix}(8.62 \times 10^5\ \text{eV})^2 + \\ [(7.016004\ \text{u})(931494013\ \text{eV/u})]^2\end{matrix}} \right\}$$

$$- (7.016004\ \text{u})(931494013\ \text{eV/u})$$

$$= 56.84\ \text{eV} \tag{1.487}$$

This calculated value is in agreement with the experimental measurement of the ^{7}Li recoil energy of 56 ± 1 eV by 2002 Physics Nobel Laureate Raymond Davis Jr (see Davis, 1952). The recoil energy is higher than that calculated above for the recoil energy imparted by the beta-particle emission to a ^{109m}Ag nucleus (Eq. 1.485), because of the relatively small mass of the ^{7}Li nucleus. We can see from Table 1.25 that, when comparing particle emissions of the same mass and energy, higher recoil energies will result with nuclei of smaller mass.

For further examples of the practical application of the equations derived here for recoil energy, namely, Eq. (1.469), Eq. (1.470), Eq. (1.476) and Eq. (1.481), see Van Rooyen et al. (2008) and Szucs et al. (2009).

D. Radioactive recoil effects

The implications of nuclear recoil are numerous including applications of hot-atom chemistry where chemical bonds are broken by recoiling nuclei facilitating the preparation of high specific activity radionuclides by means of the Szilard–Chalmers Effect. Also, nuclear recoil during radioactive decay of naturally occurring radionuclides can create disequilibrium when the recoil liberates the nuclide from its mineral lattice. These recoil effects will be discussed subsequently.

1. Szilard–Chalmers process

Prior to discussing the Szilard–Chalmers process, which is important in the preparation of radionuclides with high specific activity, the historical background that led to its discovery is relevant and will be presented briefly. A more thorough account is given in previous books by the author (L'Annunziata, 2007, 2016). On 12 September 1933, Leo Szilard (1898–1964) read a news report in the *Times* where Ernest Rutherford was interviewed. The reporter had asked Rutherford what were his thoughts concerning nuclear transmutations and nuclear energy. Rutherford replied that charged particles, such as the proton, could eventually be accelerated with sufficient high voltage to eventually transform all of the elements. He added that the energy released would be too small to produce energy on a large scale for power and anyone who looked for transmutation as a source of power was talking "moonshine". Szilard agreed with Rutherford's comments concerning accelerated protons, that these would not be a useful source of energy. He knew that the proton had to overcome the Coulombic barrier of the repelling positive nuclear charge of atoms. However, he thought, that the neutron, because of its neutral charge, should be able to penetrate the Coulombic barrier of an atom and collide with the nucleus without any external force and cause a reaction or transmutation of the nucleus. Szilard's mind was restless. While standing at a street corner in London in the following month of October and during the time that it took a traffic light to change, the idea of the possibility of a neutron-induced nuclear chain reaction hit him. Szilard knew, that in chemistry, there existed chain reactions where one reaction could yield two products that could further interact with other reactants to produce four products and these could react to produce eight products, etc. resulting in an exponential increase in reactions. The chain reactions could be very numerous and limited only by the quantity of reactants available. He thought that if a chain reaction with neutrons could occur, that is, if a neutron would be able to interact with an atomic nucleus to produce more than one neutron and the additional neutrons created would interact with other atomic nuclei to produce yet more neutrons, a chain reaction would ensue and a tremendous amount of energy could be liberated. Szilard could foresee that the initial neutron reaction and the resultant multiplication of neutrons should occur in a very short period of time, in microseconds, as nuclear reactions would be fast, producing an explosion of unforeseeable magnitude. Leo Szilard was the first to conceive the possibility of obtaining large amounts of energy and even an explosion from nuclear transmutations.

Having formulated further his initial idea of the potential energy that could be released from neutron-induced chain reactions, on 12 March 1934, Szilard filed for a patent, which contained the basic concepts of a nuclear chain reaction, the concept of "critical mass", that is, the minimum mass of a material required for a sustained nuclear chain reaction, and the "explosion" that could result. He was awarded British Patent No. 630,726 entitled *Improvements in or Relating to the Transformation of Chemical Elements*. In his patent Szilard described the concept of critical mass in his statement "If the thickness is larger than the critical value ... I can produce an explosion." (Feld and Weiss-Szilard, 1972; Loeber, 2002). Otto

Hahn and Fritz Strassmann with the assistance of Lise Meitner had not discovered nuclear fission until 1939. Consequently, the fissile materials required to produce neutron-induced nuclear chain reactions had not yet been discovered and there is no record of any such nuclear chain reactions ever discussed prior to Szilard's patent. Szilard was clearly ahead of his time. The patent, as written by Szilard, stated "This invention has for its object the production of radioactive bodies and the liberation of nuclear energy for power production and other purposes through nuclear transmutation." (Feld and Weiss-Szilard, 1972). Szilard was a pacifist and idealist, and the objective of his patent was not personal gain but for control over the harmful use of nuclear weapons. After a year's time he convinced the British Government to accept the patent in order to keep the concepts of the patent secret and inaccessible to the world. Later in life, Leo Szilard joined the Manhattan Project, which was the secret American project for the production of the atomic bomb, out of fear that Germany could become the first to acquire the weapon. Rhodes (1986) and Loeber (2002) point out that Szilard tried to use his patent as a means to gain control of the decision-making process for the atomic bomb. He felt that the control of the bomb and its use should be in the hands of the wisest scientists and not in the hands of government officials. Rhodes (1986) noted that the US Government rejected Szilard's claim to the atomic bomb on simple legalistic grounds, because he did not disclose the existence of his patent before joining the Manhattan Project.

When Szilard first conceived of neutron-induced chain reactions in 1933, he did not know what element or isotope would undergo such reactions. His patent had made a slight mention of uranium and thorium, but his thoughts were mistakenly on beryllium. He therefore, started a search for the chain-reacting element. With this in mind, Szilard began a collaboration of nuclear research with T. H. Chalmers at the Physics Department of the Medical College of St. Bartholomew's Hospital, London. This research did not lead to the discovery of a neutron-induced chain reaction, but lead to other important discoveries including (i) the discovery of a photoneutron (γ,n) source, that is, the induction of neutron emission by gamma- or X-radiation, and (ii) the Szilard–Chalmers Effect, which is the rupture of a chemical bond between an atom and a molecule of which the atom is a constituent, as a result of nuclear recoil during radioactive decay of that atom. The Szilard–Chalmers Effect led to the Szilard–Chalmers Process or Reaction, which has enabled the isolation of high-specific activities of radionuclides. The two discoveries and their applications will be discussed subsequently.

In his search for neutron emission that might initiate a neutron-induced chain reaction, Leo Szilard and T.H. Chalmers irradiated beryllium with radium gamma-rays. They discovered a radiation from the beryllium that would induce radioactivity in iodine and concluded that neutrons were emitted from the beryllium by the gamma rays (Szilard and Chalmers, 1934a; b, 1935). Their report was the first where neutron emission was initiated by gamma radiation, and the reaction that they observed can be abbreviated as

$$^{9}\mathrm{Be}(\gamma,\mathrm{n})^{8}\mathrm{Be} \tag{1.488}$$

which reads:

^{9}Be target nuclide(gamma-ray projectile, neutron emitted)^{8}Be product nucleus.

This discovery was significant, because today photoneutron (γ,n) sources are useful portable sources of neutrons. A very common neutron source of this nature used today is the mixture of ^{124}Sb + Be where the gamma-rays from ^{124}Sb yield neutrons via a photonuclear reaction (see Table 1.16).

Szilard and Chalmers (1934b) used iodine as the indicator for neutrons as the neutrons would induce radioactivity in iodine by creating radioiodine via neutron capture. A subsequent work was reported shortly thereafter by Brasch and coworkers together with Szilard and Chalmers and assistance of Lise Meitner in Berlin (Brasch et al., 1934) whereby the neutron emission from beryllium was induced by X-ray photons and neutron-capture by bromine, and the isolation of the product nuclide, radiobromine, was used as the neutron indicator.

An important outcome of this work was the discovery of the Szilard–Chalmers process whereby radionuclides may be easily separated and isolated from their parent atoms and consequently high specific activities of radionuclides may be obtained which otherwise would be difficult when both the target and product atoms are isotopes. For example, Szilard and Chalmers (1934 a,b) surrounded 150 mg of radium (gamma-ray source) with 25 g of beryllium in a sealed container of 1-mm thick platinum. The Ra–Be neutron source was immersed in 100 mL of ethyl iodide. The neutrons produced radioactive ^{128}I from stable ^{127}I by neutron capture. The ^{128}I would precipitate as radioactive silver iodide (Ag^{128}I) after the addition of water containing silver ions (Ag^{+}). However, a control experiment without the neutron source produced no radioactivity over background in the silver iodide precipitate. This experiment indicated that the chemical bond between the carbon atom of the ethyl group and the ^{128}I was broken after ^{127}I neutron capture, because free unbound ionic $^{128}I^{-}$ would precipitate with the Ag^{+} to produce radioactive Ag^{128}I precipitate in the aqueous phase according to the sequence illustrated in Fig. 1.129.

The energy of bonding between the carbon and iodine atoms is about 2 eV, which is higher than the neutron recoil energy. Consequently, the rupture of this bond is due to the gamma-recoil when the ^{128}I undergoes gamma emission upon de-excitation after neutron capture. Some atoms of

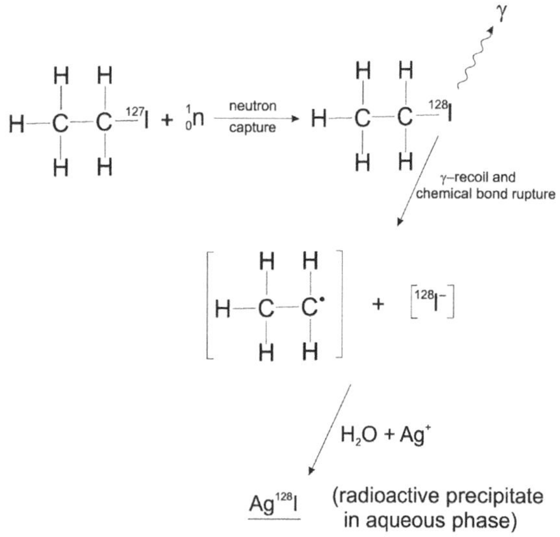

FIGURE 1.129 Szilard–Chalmers Process. High specific activity ^{128}I is separated from the stable ^{127}I as a result of the Szilard–Chalmers Effect. The capture of a neutron by the stable isotope ^{127}I in the ethyl iodide molecule created radioactive ^{128}I. The neutron capture process leaves the ^{128}I nucleus in an excited state, which emits a gamma-ray upon de-excitation. The gamma-recoil causes a rupture of the chemical bond between the carbon atom of the ethyl group and the ^{128}I thereby liberating the ^{128}I as an anion. The addition of water containing silver cations (Ag^+) to the organic solution enables the ionic $^{128}I^-$ to enter the aqueous phase and separate out as precipitated $Ag^{128}I$. The radioactivity of ^{128}I is detected by its beta-particle emissions, and it decays with a half-life of 25 minutes.

^{128}I could recombine with the free ethyl group and even exchange with some stable atoms of ^{127}I, but if these processes are slow and, if these processes are further reduced by the addition of water or even alcohol to dilute the organic phase, the precipitate of radioactive inorganic iodide proceeds yielding a highly enriched radioactive ^{128}I with minimal ^{127}I.

The Szilard–Chalmers process remains to this day a very practical application to the isolation of high specific activity radioactive sources from the medium in which the radionuclide sources were synthesized. A few examples that can be taken from the literature are an editorial review by Eckelman et al. (2008), the isolation of high specific activity radioisotopes after cyclotron production (Bonardi et al., 2004; Birattari et al., 2001), and a list of some radionuclides isolated with high specific activity listed in Table 1.26.

2. Radioactive disequilibrium

When chemical bonds are broken as a result of radioactive nuclear recoil, the recoiling nucleus may be displaced from the chemical structure (*e.g.*, mineral lattice) within which it resided. The liberated nuclide, which may be part of a natural decay chain, may then be leached from mineral deposits into ground water creating disequilibrium among the natural radionuclides. For example, Labidi et al. (2010) measured the isotopic levels of $^{234}U/^{238}U$ and $^{226}Ra/^{234}U$ in waters of Tunisia and found a disequilibrium between the two members of the ^{238}U decay series, which may be due to the direct action of nuclear recoil during radioactive decay and preferential leaching. This decay series was discussed and illustrated previously in Fig. 1.9. The researchers estimated the annual effective radiation doses due to the ingestion of the mineral waters to be well below the 0.1 mSv/year reference dose level. Fernandez et al. (2006) studied the radioecological characteristics of a uranium mining site in Brazil and point out that, as suggested by Ivanovich (1994) and Ivanovich and Harmon (1992) and Kobashi et al. (1979), disequilibrium between Ra and U isotopes may be due to (i) precipitation/dissolution reactions, (ii) alpha recoil, (iii) diffusion, and (iv) the Szilard–Chalmers effect. Almayahi et al. (2012) found elevated $^{226}Ra/^{238}U$ ratios in soil and sand systems, which may be the result of the Szilard–Chalmers effect causing disequilibrium in the decay series and elevated radon concentrations.

The activity ratios of $^{234}U/^{238}U$ were studied by Paces et al. (2002) for evidence of ground-water flow patterns in the United States. They found that ^{234}U enters solution preferentially from rocks as a result of mechanisms related to its origin by radioactive decay of ^{238}U including alpha recoil damage of crystal-lattice sites containing ^{234}U and the Szilard–Chalmers effect as well as radiation-induced oxidation of ^{234}U. Similar disequilibrium effects of natural radioactivity that may be due to nuclear recoil following alpha decay in the ^{238}U decay series was observed by Dowell and O'Dea (2002) and Davis and Krogh (2000). Disequilibrium among radionuclides in the natural decay series was underscored by Adloff (1989) to occur by "natural Szilard–Chalmers reactions" all along the radioactive decay series, and that this is triggered mostly by nuclear recoil effects after alpha decay. Adloff summarizes that these effects have found a wide range of applications in environmental studies such as geochronology, weathering, sedimental processes, and oceanic chemistry.

Kurosaki et al. (2014) provide evidence of isotopic disequilibrium between Pu in groundwater and sediment that may be explained by (i) ^{240}Pu produced by decay of ^{244}Cm at a higher oxidation state $Pu^{V}O_2^{+}$ and/or $Pu^{VI}O_2^{2+}$) as a result of the electronstripping Szilard–Chalmers process, and (ii) the oxidized forms of Pu have much lower tendencies to sorb onto sediments than the reduced forms of Pu. Their conclusions are supported by studies by Buesseler et al. (2009) and Dai et al. (2002) on Pu oxidation states and migration in groundwater.

XIV. Cosmic radiation

Stable charged particles and nuclei with lifetimes of 10^6 years or longer originating from space that strike the top of the atmosphere (TOA) from all directions constitute

TABLE 1.26 Some high specific activity radionuclides isolated by the Szilard–Chalmers process.

Radionuclide	Reference
Fluorine-18	Alfassi et al. (1986)
Phosphorus-32	Kichizo (1964), Tanaka et al., 1964
Chlorine-38	Shimojima and Lin (1971)
Chromium-51	Green and Maddock (1949) and Harbottle (1954), Iqbal and Ejaz (1978), Vimalnath et al., (2014)
Manganese-56	Zahn (1967) a,b
Cobalt-60	Matsuura (1967)
Copper-64	Ebihara (1966), Barnes et al., (1986), Hetherington et al. (1986), Sekine et al. (1986), and Gielow (1990), Van Elteren et al., (1999), Lin and Yeh (1966)
Copper-67	Mushtaq et al. (1990)
Bromine-82	Boyd et al., (1952), Collins and Willard (1962)
Yttrium-90	Tomar et al. (2010)
Molybdenum-99	Baba and Moki (1981), Tomar et al. (2010), van Dorp et al. (2018)
Tin-117m	Spano and Kahn (1952), Mausner et al. (1992)
Iodine-128	Svoboda (1963) (Nature Get ref)
Praseodymium-142g	Zeisler et al. (1999)
Samarium-153	Nassan et al. (2011)
Holmium-166	Zeisler, and Weber (1998), Nassan et al., (2011), Zhernosekov et al., (2012)
Hafnium-181	Abbe and Marques-Netto (1975), Marques-Netto and Abbe (1975)
Rhenium-186	Jia and Ehrhardt (1997), Zhang et al., (2000)
Rhenium-188	Zhang et al. (2000)
Osmium-191m	Mitchell and Martin, Jr. (1956)
Bismuth-210	Popplewell (1963)
Cyclotron-produced Radionuclides	Birattari et al., (2001), Bonardi et al., (2004)
Radioactive Lanthanides	Safavi-Tehrani et al. (2015)

what is known as the cosmic radiation. Air showers of the cosmic radiation consist of cascades of subatomic particles and electromagnetic radiation resulting from nucleon–nucleon collisions of high-energy cosmic-ray particles from space with atomic nuclei of the earths' atmosphere. Cosmic rays are classified according to their origin in space, and the air showers of the cosmic radiation are characterized according to the products of collisions of high-energy cosmic-ray particles and nuclei with atoms of the earth's atmosphere.

The discovery of cosmic radiation is attributed to Victor Hess, who made seven balloon ascents in Austria during 1911–13 into the atmosphere during the daytime, during the evening darkness, and even during a solar eclipse. On these ascents he took along three electroscopes that would measure the ionization caused by external radiation. He thought it best to make simultaneous measurements of ionization with more than one instrument to eliminate doubt in the event one instrument might give erroneous readings. Hess found that at an altitude of 500 m, the ionization dropped to about half that obtained from the soil surface. However, the readings would increase proportionally with altitude as he would ascend beyond 1000 m. He found the ionization at 1500 m would increase to be approximately equal to that at the soil surface, and further ascents to altitudes of 5000 m would provide ionization readings of several times that at the ground level (Hess, 1912, 1936, 1940). It was obvious to Hess that gamma rays, from radium in air, could not cause any increase in ionization with altitude where air got thinner, and he interpreted his results in the following words:

> *The only possible way to interpret my experimental findings was to conclude to the existence of a hitherto unknown and very penetrating radiation, coming mainly from above and being most probably of extra-terrestrial (cosmic) origin …*

Victor Hess was awarded the Nobel Prize in 1936 for his discovery of cosmic radiation.

The dose to the human body from cosmic radiation is of concern to astronauts, airline pilots, and persons living at high altitudes, as our atmosphere attenuates considerably cosmic radiation and the dose to the human body consequently increases with altitude. Cosmic radiation is also of much concern as background interference in the measurement of low levels of radioactivity. The accurate measurement of radionuclides in our environment requires the suppression and accurate measurement of the background interference from cosmic rays.

A. Classification and properties

Cosmic radiation incident on the earths' atmosphere is classified into "primary" or "secondary" cosmic rays. As defined by Patrignani et al. (2016), Beringer et al. (2012), and Gaisser and Stanev (2002), **primary cosmic rays** are stable charged particles and nuclei with lifetimes of 10^6 years or longer accelerated at astrophysical sources, and **secondary cosmic rays** are particles produced via the interaction of the primaries with interstellar gas. The nuclear interactions of the accelerated primary cosmic-ray particles with interstellar medium produce stable nuclei of the light elements (*e.g.*, Li, Be, and B), as well as many fundamental particles including pions or π mesons of zero charge (π^0), that decay into gamma rays. Galactic gamma radiation has been mapped by the Compton Gamma Ray Observatory satellite. The satellite has provided an image of the galaxy produced by gamma rays of approximately 100 MeV (Simpson, 2001). The gamma-ray image compares closely to the visible-light image of the galaxy of which we are most familiar. Galactic gamma rays can cover the full energy range from <100 MeV to >10 TeV.

Primary cosmic-ray particles include protons, helium nuclei, electrons, and nuclei of most elements of the periodic table (*e.g.*, carbon, iron, oxygen, etc) of stellar origin. Nuclei of the light elements lithium, beryllium, and boron ($Z = 3-5$) are classified as secondary radiation. The abundance of these light elements in cosmic radiation is highly enriched over the abundance of these elements in the universe, which is evidence that these are secondary nuclei created via collision (nuclear spallation interactions) with heavier primary particles during the interstellar propagation of primary nuclei. The radionuclide ^{10}Be ($t_{1/2} = 1.6 \times 10^6$ y) is found among other isotopes of the light elements in cosmic rays. From the measured abundance analysis of ^{7}Be, ^{9}Be and the radioactive ^{10}Be in cosmic radiation, it is found that cosmic-ray particles remain contained in galactic magnetic fields for approximately 10^7 years before escaping into our atmosphere or intergalactic space. The mean interstellar density for propagation of the cosmic radiation is ~0.2 atom/cm^3 (Simpson, 2001).

Nucleons of practically all elements of the periodic table will be found in cosmic radiation. Free protons account for about 80% of the primary nucleons, and approximately 15% are nucleons bound in helium nuclei (equivalent to alpha particles). Electrons constitute about 2% of the primaries. Nuclei of the elements of the periodic table other than the previously mentioned (H and He) make up the remaining components of cosmic radiation. Nuclei of the light elements (Li, Be, and B), which constitute secondary cosmic ray particles, account for a small fraction (~0.3%) of cosmic ray nuclei (see Table 1.27).

The composition of cosmic radiation will vary according to the 11-year solar cycle and the earth's magnetic latitude (Potgieter, 2008a,b; Patrignani et al., 2016). The earth's magnetic field, which extends well into space, affects the composition of the charged particles of cosmic

TABLE 1.27 Approximate composition of cosmic radiation of galactic origin incident on the top of the atmosphere (TOA)[a].

Radiation type	Approximate fraction (%)
H nuclei or protons (p^+ and p^-)[b]	80
He nuclei (equivalent to α particles)	15
Electrons (e^- and e^+)[c]	2
Heavier nuclei (*e.g.*, C, O, Mg, Fe, Si, etc)[d]	1
Lighter nuclei ($Z = 3-5$, *i.e.*, Li, Be, and (B)[e]	0.2

[a]*Cosmic radiation incident at the top of the terrestrial atmosphere includes all stable charged particles and nuclei with lifetimes of the order of 10^6 years or longer (Gaisser and Stanev, 2002).*
[b]*The ratio of antiprotons to protons ($\bar{p}/p$) is about 2×10^{-4} in the particle energy range of 10–20 GeV. Antipproton/proton ratios at lower energies are listed in this chapter.*
[c]*The positron fraction ($e^+/(e^+ + e^-)$) varies from ~0.2 below 1 GeV to ~ 0.1 around 2 GeV and to ~ 0.05 at higher electron energies of 5–20 GeV.*
[d]*Nucleons of all of the elements of the periodic table stripped of their atomic electrons with the exception of the lighter nucleons listed above.*
[e]*The nuclei of Li, Be, and B are classified as secondary cosmic radiation, that is, particles produced by the interaction of the primaries with interstellar gas.*

radiation. The geomagnetic latitude effect was first reported by Clay (1928) and confirmed by Compton (1932, 1933), who measured cosmic ray showers with ionization detectors while traversing latitudes onboard ocean vessels. Cosmic-ray particles will concentrate in the earth's Van Allen radiation belts, which extend from about 1000 to 60,000 km from the earth. The sea-level latitude effect of the electromagnetic and meson components of cosmic-ray interactions in the earth's atmosphere measured by Compton corresponded to an approximately 12–15% increase between the geometric equator and high latitudes. However, no information was obtained then on the latitude effect on the nucleonic component of cosmic radiation at sea level. Subsequent studies (Simpson, 1948, 1951, 2001) demonstrated a dramatic increase of 300–400% in the nucleonic component compared with a 10–15% increase in the meson component of cosmic radiation at sea level as one traveled from 0 degrees to 70 degrees latitude. This is particularly relevant to international airline pilots who are concerned about the cosmic radiation dose they accumulate when they travel frequently at high altitudes (12 km) and high latitudes (40–60 degrees) where the radiation intensities are higher.

Positrons and antiprotons are components of secondary cosmic radiation (Patrignani et al., 2016). Positrons are much lower in abundance than electrons (see Table 1.27 footnote). They are produced mainly via proton and nuclei interactions in the interstellar medium (Grimani, 2009). The differential flux of negatrons and positrons incident at the top of the atmosphere from galactic space is a function of particle energy. The positron fraction, measured as $e^+/(e^+ + e^-)$, decreases from ~0.2 below 1 GeV, to ~ 0.1 around 2 GeV and to ~ 0.05 at higher energies of 5–20 GeV (Grimani, 2009, 2005; Clem and Evenson, 2002; Gaisser and Stanev, 2002; Sundaresan, 2001). Antiprotons in the cosmic radiation were discovered by Golden et al. (1979, 1984). These are classified as secondaries, as antiprotons as well as positrons are produced by the interactions of primary cosmic radiation, principally protons with the interstellar matter, such as cosmic-ray nuclei (Adriani et al., 2010; Gaisser and Schaefer, 1992; Webber and Potgieter, 1989; Gaisser and Maurer, 1973). The proportions of antiprotons to protons ($\bar{p}/p$) is a function of energy with ratios varying from $\sim 0.3 \times 10^{-5}$ to 1×10^{-5} in the energy range of ~0.2–1 GeV, from $\sim 2 \times 10^{-5}$ to 1×10^{-4} in the energy range of ~1–10 GeV, and from $\sim 2 \times 10^{-3}$ to 3×10^{-3} in the energy range of 10–180 GeV (Adriani et al., 2010; Beach et al., 2001; Basini et al., 1999; Mitchell et al., 1996; Hof et al., 1996; Patrignani et al., 2016). In summary, the approximate composition of galactic cosmic ray particles is provided in Table 1.27.

Most galactic cosmic ray particles possess energies from about 0.5 to 50 GeV as illustrated in Fig. 1.130, which provides the fluxes of nuclei of primary cosmic radiation as a function of energy per nucleus. Protons (H) and He nuclei are the most abundant, as illustrated in Fig. 1.130. A series of very precise measurements of primary protons and helium nucleon intensities have been made by several groups including the AMS Collaboration (2000), Menn et al. (2000) and Sanuki et al. (2000), Bellotti et al. (1996, 1999), and Boezio et al. (1999), as described by Beringer et al. (2012) and Patrignani et al. (2016). Particle energies up to 10^{15} eV were first observed by Auger and Maze (1938, 1939), and a cosmic-ray particle of about 10^{20} eV energy was observed by Linsley (1963).

Protons in the kinetic energy range of 100 MeV to 10 GeV would have relativistic speeds ranging from 42.8% to 99.6% the speed of light (*i.e.*, 0.428*c* to 0.996*c*) calculated according to Eq. (1.278). For example, the relativistic speed of a proton with 10 GeV kinetic

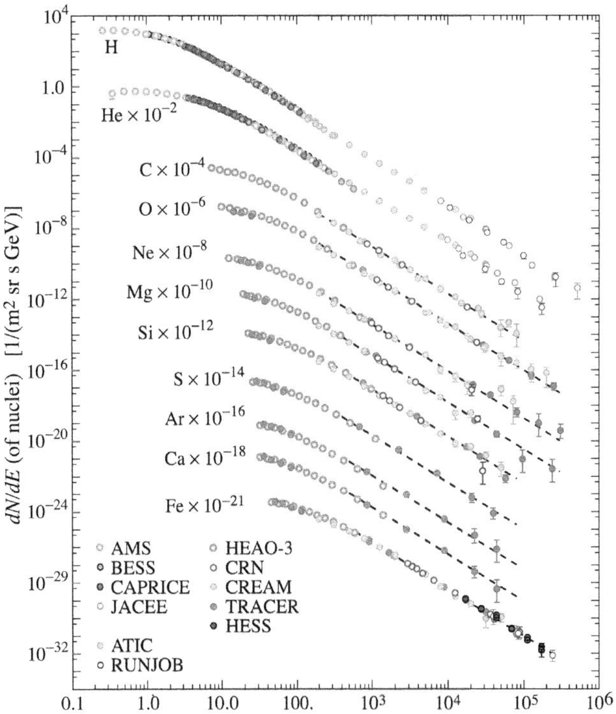

FIGURE 1.130 Fluxes of nuclei of the primary cosmic radiation in particles per energy-per-nucleus plotted versus energy-per nucleus. Acronyms in the lower left are those of research groups cited by Patrignani et al. (2016). The figure was created by P. Boyle and D. Muller. For the color version of the figure, the reader is referred to the online or electronic version of the book. *Reprinted with permission from and Beringer et al. (2012), © 2012 by the American Physical Society, https://journals.aps.org/prd/abstract/10.1103/PhysRevD.86.010001, and Patrigani et al. (2016), © Regents of the University of California, http://pdg.lbl.gov/2016/figures/figures.html.*

energy would be calculated according to the following equation:

$$u = c\sqrt{1 - \left(\frac{K}{mc^2} + 1\right)^{-2}} \tag{1.489}$$

The rest energy of the proton (mc^2) is calculated as

$$\begin{aligned} mc^2 &= \left(1.6726 \times 10^{-27}\ \text{kg}\right)\left(2.9979 \times 10^8\ \text{m/sec}\right)^2 \\ &= 1.503 \times 10^{-10}\ \text{J} \end{aligned} \tag{1.490}$$

which can be expressed in units of eV as follows:

$$\frac{1.503 \times 10^{-10}\ \text{J}}{1.602 \times 10^{-19}\ \text{J/eV}} = 0.9382 \times 10^9\ \text{eV} = 938.2\ \text{MeV} \tag{1.491}$$

The relativistic speed of the 10 GeV proton can then be calculated as

$$u = c\sqrt{1 - \frac{1}{\left(10^4\ \text{MeV}/938.2\ \text{MeV} + 1\right)^2}} = 0.996c \tag{1.492}$$

A higher kinetic energy 100-GeV (10^5 MeV) cosmic-ray proton would have a relativistic speed of

$$\begin{aligned} u &= c\sqrt{1 - \frac{1}{\left(10^5\ \text{MeV}/938.2\ \text{MeV} + 1\right)^2}} \\ &= 0.999956c \end{aligned} \tag{1.493}$$

The speed of the 100 GeV proton may also be calculated on the basis of the definition of the kinetic energy of the particle being the difference between the *total energy* of the particle and its rest energy (see Eq. 1.270) or

$$K = \gamma mc^2 - mc^2 = (\gamma m - m)c^2 \tag{1.494}$$

or

$$10^5\ \text{MeV} = (\gamma m - m)c^2$$

Converting the units of electron volts in the above equation to joules and imputing the speed of light provides

$$\begin{aligned} &\left(10^{11}\ \text{eV}\right)\left(1.602 \times 10^{-19}\ \text{J/eV}\right) \\ &= (\gamma m - m)\left(2.9979 \times 10^8\ \text{m/sec}\right)^2 \end{aligned} \tag{1.495}$$

from which the differences of the relativistic and rest masses of the proton can be calculated as

$$\begin{aligned} \gamma m - m &= \frac{1.602 \times 10^{-8}\ \text{J}}{8.9874 \times 10^{16}\ \text{m}^2/\text{sec}^2} \\ &= 0.1782495 \times 10^{-24}\ \text{kg} \end{aligned} \tag{1.496}$$

Inputting the known rest mass m of the proton (1.6726×10^{-27} kg) permits the calculation of the relativistic mass of the 100 GeV proton as

$$\begin{aligned} \gamma m &= 178.2495 \times 10^{-27}\ \text{kg} + 1.6726 \times 10^{-27}\ \text{kg} \\ &= 179.9221 \times 10^{-27}\ \text{kg} \end{aligned} \tag{1.497}$$

Since by definition, $\gamma m = m/\sqrt{1-(u^2/c^2)}$ (see Eq. 1.271), we can write

$$\frac{\gamma m}{m} = \frac{1}{\sqrt{1-(u^2/c^2)}} \tag{1.498}$$

which transforms to read

$$\left(\frac{m}{\gamma m}\right)^2 = 1 - \frac{u^2}{c^2} \tag{1.499}$$

or

$$\frac{u^2}{c^2} = 1 - \left(\frac{m}{\gamma m}\right)^2 \tag{1.500}$$

Inputting the values for the relativistic and rest masses of the proton yields

$$\begin{aligned} \frac{u^2}{c^2} &= 1 - \left(\frac{1.6726 \times 10^{-27}\ \text{kg}}{179.9221 \times 10^{-27}\ \text{kg}}\right)^2 \\ \frac{u^2}{c^2} &= 1 - (0.00929624)^2 \\ \frac{u^2}{c^2} &= 1 - 8.64200 \times 10^{-5} \\ u^2 &= (0.9999135)c^2 \end{aligned} \tag{1.501}$$

and

$$u = 0.999956c$$

where u is the relativistic speed of the 100 GeV proton exactly as calculated previously by Eq. (1.493).

High-energy cosmic-ray protons, as demonstrated above, approach the speed of light. Primary cosmic ray nucleon intensity drops rapidly according to energy beyond several GeV as illustrated in Fig. 1.130. The primary nucleon intensity from a few GeV to 100 TeV is described according to the power law

$$I_N(E) = 1.8E^{-\alpha}\text{nucleons/cm}^2\text{sec sr GeV} \tag{1.502}$$

where E is the energy per nucleon (including rest mass energy) and α is approximately 2.7.

As illustrated in Fig. 1.130, primary nucleon energies in excess of 1 TeV (*i.e.*, 10^{12} eV or 10^6 MeV) are relatively few in number, and primary cosmic-ray particles with energies in excess of 10^{20} eV or 10^{14} MeV) are very rare, but have been reported (Hörandel, 2010; Pierre Auger Collaboration, 2010; Scherini, 2011; Zavrtanik, 2011; Nagano and Watson, 2000; Bird et al., 1995; Hayashida et al., 1994; Linsley, 1963). The rarity of such extremely high-energy primary cosmic ray particles can be visualized with the graph of primary cosmic ray flux as a function of nucleon energy illustrated in Fig. 1.131.

The cosmic-ray scientific community is studying the knee and ankle regions of the cosmic-ray spectrum illustrated in Fig. 1.131. The knee region at particle energies of $10^{15}-10^{16}$ eV shows a decline in the particle flux while the ankle region of the spectrum occurring between 10^{18} to 10^{19} eV shows a rise in particle flux. The origins of the knee and ankle regions of the energy spectrum is not fully understood, and these are now under intense research as described by Patrignani et al. (2016), Beringer et al. (2012), Chiavassa (2010), Marcelli (2011), Blümer et al. (2009), Giller (2008), and Hörandel (2008a). The knee portion of the energy spectrum at about 4 PeV (*i.e.*, $\sim 4 \times 10^{15}$ eV), as described by Hörandel (2008a), is caused by a break in the energy spectra of the light elements. The mean mass of the cosmic rays is found to increase as a function of energy in the knee region. The knee in the energy spectrum is most likely caused by a combination of the maximum energy reached during acceleration and leakage from the Galaxy during propagation (Patrignani et al., 2016; Beringer et al., 2012; Hörandel, 2008a). A current thinking for the production of the ankle region of the spectrum is summarized by Patrignani et al. (2016). It appears that the ankle is caused by a higher energy cosmic-ray population mixed in with a lower energy population at energies a couple of orders of magnitude below the ankle, which could result from a higher-energy extragalatic flux dominating over a galatic flux. The spectrum below 10^{18} eV is considered to be of galactic origin, and the higher energy population in excess of 5×10^{18} eV is considered to be of extragalactic origin (Patrignani et al., 2016; Marcelli, 2011; Allard et al., 2007; Cronin, 1999; Sundaresan, 2001).

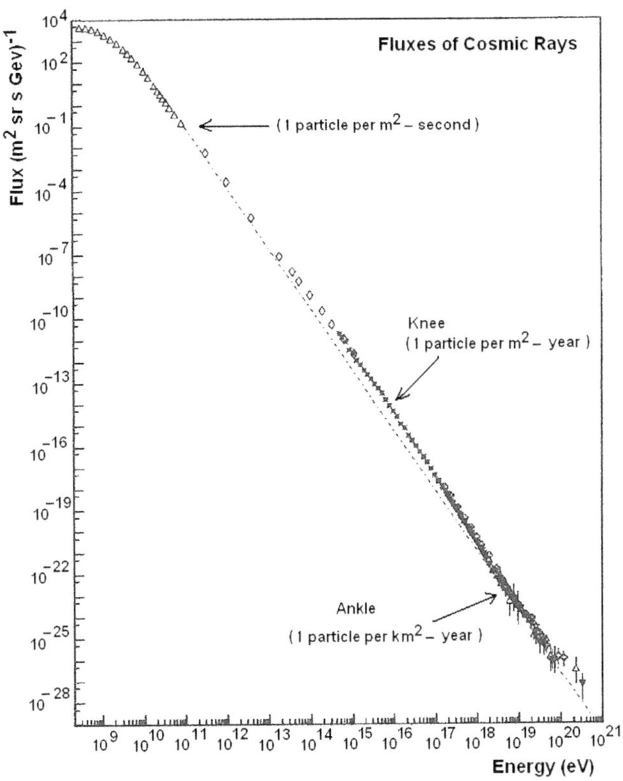

FIGURE 1.131 The all-particle spectrum of cosmic rays greater than 100 MeV. The cosmic rays consist predominantly of nucleons ranging in species from protons to iron nuclei with traces of nuclei of the heavier elements. Not included in the figures are notations for estimated cosmic ray fluxes of about 1 particle/cm^2/sec at the low energy end of 10^8 eV and only of the order of 1 particle/km^2/century at the high-energy end of 10^{20} eV of the spectrum (This figure was produced by S. Swordy, University of Chicago from published results of the LEAP, Proton, Akeno, AGASA. Fly's Eye, Haverah Park, and Yakutsk experiments. *Reprinted from Bhattacharjee and Sigl (2000), Marcelli (2011) with permission from Elsevier © 2000.*

Much attention in the cosmic-ray community is also focused on the mechanisms by which the cosmic-ray nucleons are accelerated in space to energies in excess of 10^{20} eV. Enrico Fermi (1949) proposed and calculated the acceleration of cosmic rays resulting from a series of collisions with magnetic fields or magnetic clouds traversing the universe whereby the cosmic ray particles gain energy each time they bounce off the magnetic fields. This is now referred to as Fermi acceleration, and Cronin (1999) explains that subsequent work has shown that multiple "bounces" off turbulent magnetic fields associated with supernova shock waves is the more efficient acceleration process (Drury, 1983). The energy spectrum of cosmic rays supports the "standard picture", as described by Patrignani et al. (2016) and Hörandel (2008a), that the bulk of galactic cosmic rays is accelerated in shocks of supernova remnants.

B. Showers of the cosmic radiation

Cosmic radiation consisting of high-energy nucleons striking the top of the atmosphere (TOA) collide with atoms of the air to produce a cascade of secondary subatomic particles and electromagnetic radiation referred to as **showers of the cosmic radiation**. During the years that preceded the development of man-made high-energy particle accelerators, cosmic rays provided much information to the field of high-energy particle physics. The collisions of high-energy nucleons with atomic nuclides of the atmosphere provided natural "atom-smashers" that led to the early discovery of subatomic particles.

When a cosmic-ray particle strikes an atomic nucleus of a gaseous molecule of the atmosphere (*e.g.*, N_2, O_2, etc.) a nuclear disintegration follows producing high-energy secondary nucleons and charged and neutral π mesons (pions).

These secondary nucleons collide with additional atomic nuclei of the atmosphere producing a cascade of particles including mesons, nucleons of various masses, neutrons, and products of meson-nuclear interactions and meson decay, namely, photons of electromagnetic radiation, positrons, negatrons, and neutrinos. Fig. 1.132 provides a good schematic of the various possible nucleon interactions that produce the cascade of secondary cosmic radiation in the atmosphere.

Mesons are strongly interacting particles of mass intermediate between that of the electron (0.511 MeV/c^2) and the proton (938 MeV/c^2). Among the mesons produced are (a) kaons (K-mesons), which have a mass of about 490 MeV/c^2, lifetimes of approximately 10^{-10} and 10^{-8} seconds for the neutral (K^0) and charged ($K^{\pm}$) kaon, respectively, decaying to pions and/or muons as illustrated in Fig. 1.133, and (b) the pions (π-mesons) that can decay into muons, such as described by Eqs. (1.507) and (1.508). A neutral pion has a mass of 135 MeV/c^2 and a lifetime of 8.3 $\times$ 10^{-17} seconds. It decays into two gamma-ray photons, as illustrated in Fig. 1.132, or into a positron–negatron pair and gamma ray with branching ratios of approximately 98.8% and 1.2%, respectively, for the two decay modes (Sundaresan, 2001; Glasser et al., 1961) as follows:

$$\pi^o \rightarrow \gamma + \gamma \ (98.798 \pm 0.032\%) \tag{1.503}$$

$$\pi^o \rightarrow e^+ + e^- + \gamma \ (1.198 \pm 0.032\%) \tag{1.504}$$

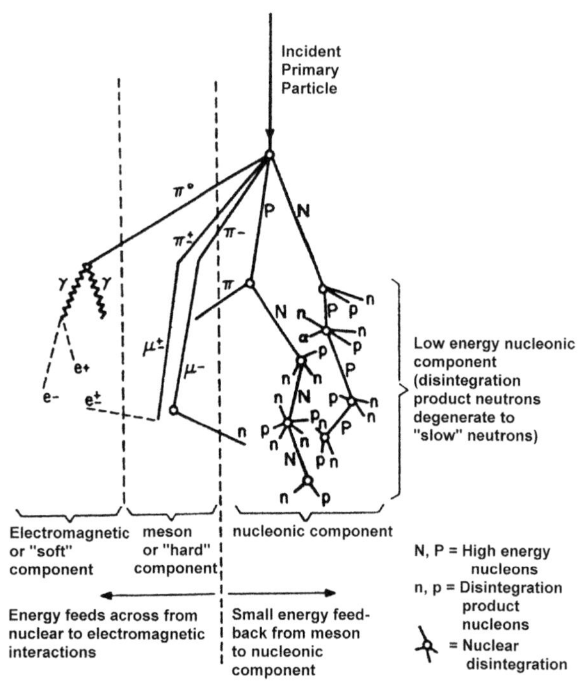

FIGURE 1.132 Schematic representation of the typical development of the secondary cosmic radiations within the atmosphere arising from an incident primary particle. *From Simpson et al. (1953), https://journals.aps.org/pr/abstract/10.1103/PhysRev.90.934, reprinted with permission © 1953 The American Physical Society.*

The positron and negatron pair (e^+e^-), as products of the neutral pion decay, illustrated by Eq. (1.504), is a result of internal conversion of a gamma-ray photon (i.e., internal pair production). The neutral pion decay may also manifest itself in yet a more rare form of dual internal conversion whereby two gamma-ray photons are converted to two positron–negatron pairs. The decay scheme is illustrated by Eq. (1.505), and the tracks from such a decay scheme produced in a bubble chamber are illustrated in Fig. 1.134.

$$\pi^o \rightarrow e^+ + e^- + e^+ + e^- \tag{1.505}$$

Out of eight million π^o decays in 836,000 bubble chamber pictures, Samios et al. (1962) were able to find only 206 neutral pion decays with double internal conversion. The numerous neutral pions were produced by Samios et al. (1962) by the interactions of π^- mesons from a cyclotron, with protons of the bubble chamber liquid according to the reaction

$$\pi^- + p^+ \rightarrow n + \pi^o \tag{1.506}$$

The π^- mesons were slowed down with a polyethylene absorber and then allowed to stop in the hydrogen bubble chamber. The numerous bubble chamber photographs could be analyzed for the identification of specific tracks by a digitized scanning machine.

Any of the gamma rays can, in turn, produce positron–negatron pairs via pair production, and the positron can undergo annihilation to gamma radiation. The positive and negative pions are antiparticles. Charged pions have a mass of 139.6 MeV/c^2 and a lifetime of 2.6 $\times$ 10^{-6} seconds, much longer than that of the neutral pion. The positive pion decays to a positive muon and muon neutrino, whereas the negative pion decays into a negative muon and muon antineutrino as illustrated by the following:

$$\pi^+ \rightarrow \mu^+ + \nu_\mu \tag{1.507}$$

$$\pi^- \rightarrow \mu^- + \overline{\nu_\mu} \tag{1.508}$$

The muon has a mass of 106 MeV/c^2 and lifetime of 2.2 $\times$ 10^{-6} seconds. The positive muon decays to a positron and muon neutrino, and the negative muon to a negatron and anti-muon neutrino as illustrated in Fig. 1.132 and by the following:

$$\mu^{+/-} \rightarrow e^{+/-} + \nu_\mu + \overline{\nu_\mu} \tag{1.509}$$

Slow negative muons undergo nuclear absorption by nucleons with sufficient high atomic number ($Z \approx 10$ or greater); however, for low Z nuclides such as carbon ($Z = 6$), slow negative muon decay occurs rather than absorption by the nucleus (Sundaresan, 2001).

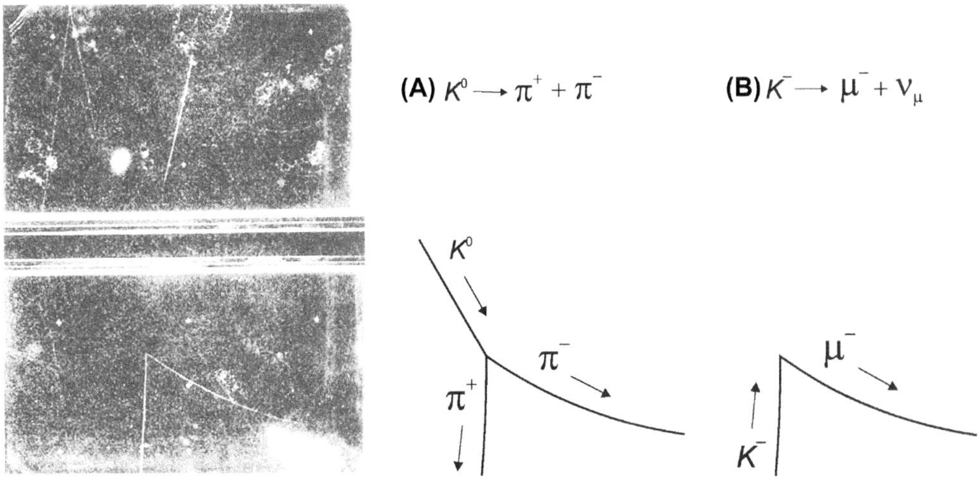

FIGURE 1.133 A cloud-chamber photograph of the tracks of a probable V^0 particle in a magnetic field of 5000 gauss. *V* particles got their name from the V-shapet tracks, that they formed from a given point in the cloud chamber. The cloud chamber consists of upper and lower chambers separated by the glass walls and copper tubing of a central Geiger counter. The V-shaped track appears left center in the lower-chamber. The track was the result of one of two possible modes of kaon decay, namely, (A) a neutral kaon (K^0) decaying into positive and negative pions where the neutral kaon does not leave a track because of its lack of charge, or (B) a negative kaon (K^-) entering the cloud chamber from below and decaying into a negative muon and neutrino. The neutrino obviously does not leave a track. The cloud chamber photograph is from the work of Nobel Laureate Carl D. Anderson and coworkers. *Reprinted with permission from Leighton et al. (1953), https://journals.aps.org/pr/abstract/10.1103/PhysRev.89.148, © 1953 by the American Physical Society.*

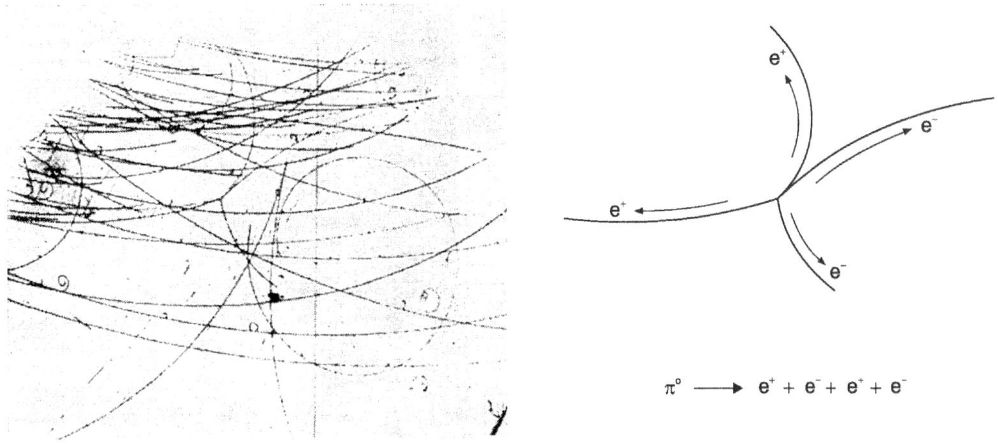

FIGURE 1.134 Photograph of a typical double internal conversion following π^0 decay on the left with a drawing of the specific tracks provided on the right. The tracks were produced in a hydrogen bubble chamber 30.5 cm in diameter and 15.25 cm in depth. A magnetic field of 5.5 gauss produced the curved deflection of the oppositely charged electrons. *Reprinted with permission, from Samios et al. (1962), https://journals.aps.org/pr/abstract/10.1103/PhysRev.126.1844, © 1962 by the American Physical Society.*

At sea level muons are the most numerous of the charged particles resulting from cosmic ray interactions with atomic nuclei of the atmosphere. These have been measured with an intensity of $100/m^2$ s at sea level. The relative abundances of the cosmic radiation shower components in the atmosphere at different altitudes with energies in excess of 1 GeV are illustrated in Fig. 1.135. As illustrated in the figure, most muons are produced high in the atmosphere at an altitude of ~15 km where high-energy cosmic-ray nucleons encounter and undergo collision with nuclei of the atmospheric gases. Pions likewise are produced high in the atmosphere; however, their number at sea level is highly diminished due to the prompt decay of the neutral pion to gamma radiation and the decay of the charged pions to muons. The muon produced high in the atmosphere loses about 2 GeV of energy to ionization before reaching the ground, and the mean energy of the muon at the ground is ≈ 4 GeV (Gaisser and Stanev, 2002), and these travel at speeds close to the speed of light (>0.99c).

The intensity of protons and neutrons in the atmosphere (i.e., cosmic ray showers) is the highest at the top of the atmosphere. The high flux of protons at the top of the atmosphere is intuitively obvious, because the proton is the most abundant component of the primary cosmic rays of galactic origin arriving from space, as depicted

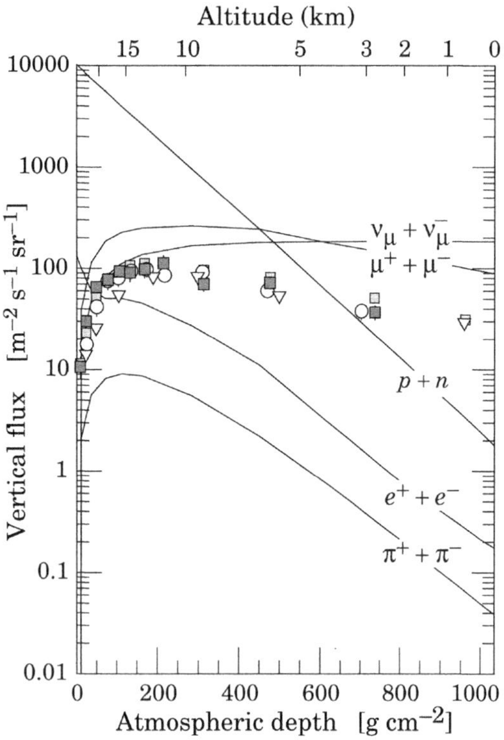

FIGURE 1.135 Vertical fluxes of cosmic rays in the atmosphere with $E > 1$ GeV estimated from the nucleon flux of Eq. (1.502). The points show measurements of negative muons with $E_\mu > 1$ GeV. For the color version of this figure, the reader is referred to the electronic version of the book. *Reprinted with permission from and Beringer et al. (2012), © 2012 by the American Physical Society, https://journals.aps.org/prd/abstract/10.1103/PhysRevD.86.010001, and Patrigani et al. (2016), © Regents of the University of California, http://pdg.lbl.gov/2016/figures/figures.html.*

previously in Table 1.27 and Fig. 1.130. These protons collide with nuclei in the atmosphere producing elementary particles and neutrons, as illustrated previously in Fig. 1.132. Thus the proton flux will diminish with atmospheric depth, and the integral intensity of protons with momentum >1 GeV/c at sea level is ≈ 0.9/m^2 s sr (Greider, 2001; Patrignani et al., 2016). The n/p ratio at the top of the atmosphere is ≈ 10%, and the neutron intensity increases to 1/3 of the nucleons at sea level (Patrignani et al., 2016).

C. Cosmic-ray muon detection and measurement

Muons in the cosmic-ray showers are produced high in the atmosphere at an altitude of ~ 15 km as a result of collisions of the galactic primary cosmic ray nuclei with nuclei of the atoms of our atmosphere. The muons have an average lifetime of only 2.2×10^{-6} seconds. A classical calculation of distance of travel in the earth's atmosphere for a particle with such a short lifetime would demonstrate that the muons should never be detected on earth at sea level. However, in Section VII.C of this chapter, calculations are made of time dilation in relativity whereby the muons traveling at relativistic speed are observed in the earth's time frame of reference as having a dilated lifetime, which would provide muons created at an altitude of 15 km sufficient time to reach earth's sea level. The calculations of time dilation in relativity made previously in this chapter (Section VII.C) also provide experimental evidence with respect to the muon lifetime measured on earth when the muons travel at relativistic speed. Section VII.D of this chapter also provides evidence of time dilation in relativity from bubble-chamber measurements of the dilated lifetimes of elementary particles produced by proton collisions with atomic nuclei.

D. Cosmic rays underground

The only significant cosmic radiation with energy sufficient to penetrate considerable depths of earth are charged muons and muon neutrinos. Thus, Patrignani et al. (2016) and Tueros and Sciutto (2010) describe muons as the "penetrating component" of the cosmic-ray-induced showers, capable of reaching deep underground without interacting in the atmosphere. High-energy muons will traverse rock and lose energy via either direct ionization or radiative processes. The radiative processes include (i) bremsstrahlung production with concomitant (ii) positron−electron pair production and (iii) photonuclear interactions. Muon-induced spallation is another mechanism of energy loss.

The muon cosmic-ray intensities diminish with depth, as illustrated in Fig. 1.136. This is a subject of concern in the measurement of very low levels of radiation. Consequently, underground laboratories have been built to achieve lowest possible background radiation interference in the counting of very low levels of natural radioactivity. Muon energy loss is described by Patrignani et al. (2016) and Beringer et al. (2012). The mechanisms of muon energy loss is via ionization and three radiative process, namely, (i) bremsstrahlung production followed by (ii) positron−electron pair production, and (iii) photonuclear interactions of which the latter two mechanisms originate from the bremsstrahlung produced by the muons. Patrignani et al. (2016) expressed the muon energy loss as a function of the amount of matter traversed as

$$-\frac{dE_\mu}{dx} = a + bE_\mu \tag{1.510}$$

where a is the ionization loss and b is the fractional energy loss by the three radiative process described earlier in this paragraph. They provide the parameters for a and b of the above equation for standard rock.

Muon ranges in rock are measured in units of km-water-equivalent (km.w.e) where 1 km.w.e. = 10^5 g/cm^2 of

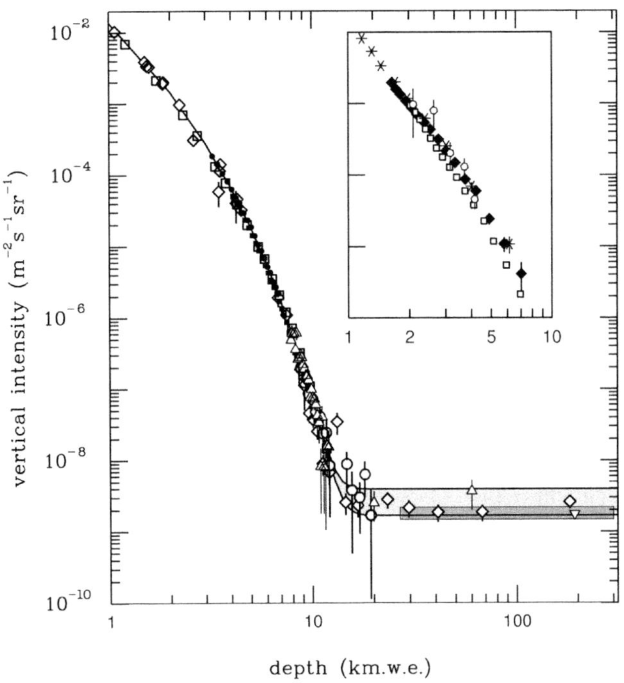

FIGURE 1.136 Vertical muon intensity versus depth measured in km-water-equivalent (km.w.e) where, by definition, 1 km.w.e = 10^5 g/cm^2. The experimental data are from several research collaborations, namely ◆: the compilations of Crouch (1987),: BAKSAN (Andreev et al., 1987), ○: LVD 9 (Aglietta et al., 1995), ●: MACRO (Ambrosio et al., 1995), ■: Frejus (Berger et al., 1989), and Δ: SNO (Waltham et al., 2001). The shaded area at large depths represents neutrino-induced muons of energy above 2 GeV. The upper line is for horizontal neutrino-induced muons, the lower one for vertically upward muons. Darker shading shows the muon flux measured by the Super-Kamiokande experiment. The inset shows the vertical intensity curve for water and ice. *Reprinted with permission from and Beringer et al. (2012), © 2012 by the American Physical Society, https://journals.aps.org/prd/abstract/10.1103/PhysRevD.86.010001, and Patrigani et al. (2016), © Regents of the University of California, http://pdg.lbl.gov/2016/figures/figures.html.*

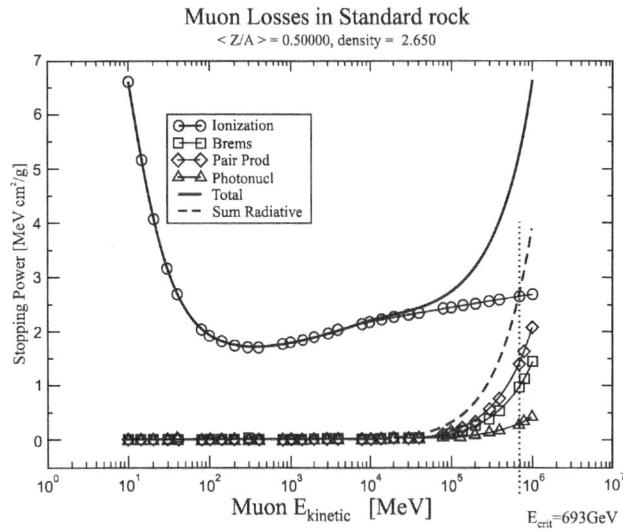

FIGURE 1.137 Muon energy loss versus muon energy in standard rock. *From Tueros and Sciutto (2010), reprinted with permission from Elsevier © 2010.*

standard rock. Average ranges in standard rock ($Z = 11$, $A = 22$, $Z/A = 0.5000$, $\rho = 2.650$ g/cm^3) for muons of 10, 100, 1000, and 10,000 GeV are 0.05, 0.41, 2.45, and 6.09 km.w.e (Patrignani et al., 2016). The vertical muon intensity versus underground depth of standard rock has been determined. The muon vertical intensity drops from 10^{-6} cm^{-2}sr^{-1}s^{-1} to 10^{-12} cm^{-2}sr^{-1}s^{-1} over the range of rock depths from 1 to 10 km.w.e., respectively (Gaisser and Stanev, 2002, see also Fig. 1.136).

Radiation from muons of the cosmic-ray showers is significant underground. The mechanisms of muon energy loss are dependent on the muon energy. Tueros and Sciotto (2010) report a Monte Carlo program that simulates high energy cosmic-ray showers underground. From nuclear data tables of Groom et al. (2001), they calculated the muon energy losses in standard rock by plotting the stopping power versus muon energy illustrated in Fig. 1.137 (Stopping power calculations are discussed in Section XVI.A of this chapter.) Muons traveling through matter will lose energy by several mechanisms including ionization, bremsstrahlung production, pair production, photonuclear reactions including spallation caused by muon collisions with nuclei. Tueros and Sciotto (2010) demonstate from Fig. 1.137 that ionization losses are dominate at low muon energies (<100 MeV) and ionization losses predominate among the other mechanisms of energy loss and remain fairly constant at 1.5–2.2 MeV cm^2/g below 10 GeV (10^4 MeV) muon energy in standard rock. At muon energies >10 GeV, radiative energy losses steadily increase and reach the critical energy at ~ 0.7 TeV (693 GeV, See Fig. 1.137), which is the energy at which the ionization loss is equal to the radiative loss. The critical energy for muons in air is a much higher 3.6 TeV as reported by Grieder (2001). Thus, Tueros and Sciotto (2010) point out that bremstrahlung and pair production energy losses for muons in air is negligible with the exception of very high muon energies; whereas it is increasingly important underground for muon energies > 50 GeV.

Another source of underground background interference from muons of the cosmic radiation showers is the production of gamma ray photons from radiative neutron capture reactions in hydrogen and elements encountered in rock and other soil materials. Neutrons are produced by muon spallation reactions as well as photonuclear reactions arising from bremsstrahlung radiative energy loss. Araújo et al. (2008) and Lindote et al. (2009) report a Monte Carlo simulation, which provides a very good estimate of the absolute neutron yield in lead as (1.31 ± 0.06) × 10^{-3} neutrons/muon/g/cm^2) for a mean muon energy of 260 GeV.

E. Origins of cosmic radiation

The origins of galactic and extragalactic cosmic rays remains an unsolved problem, and reviews on current research and thinking are provided by Tjus et al. (2016), Potgieter (2010), Blasi (2013, 2008), Hörandel (2008b), and Gaisser (2001). The subject of cosmic ray origin is somewhat out of scope of this chapter, and the reader is invited to peruse the review papers on this subject cited above. In brief, cosmic rays can originate from (i) energetic particles associated with mass ejections from solar flares and similar energetic solar events; (ii) anomalous cosmic rays, which are particles of interstellar origin accelerated at the edge of the heliopause and accelerated at the termination shock in the solar wind; (iii) high-energy particles of galactic origin far outside the heliosphere or our solar system; and (iv) extragalactic sources, which give rise to particles of the highest energies $\sim 10^{18}$ eV or higher.

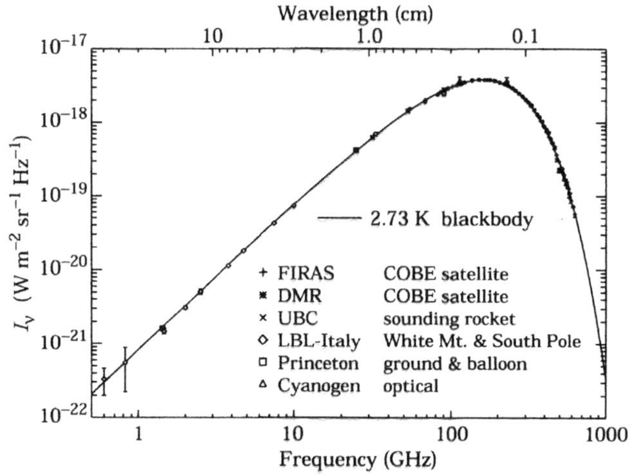

FIGURE 1.138 Measurement of the cosmic microwave background radiation (CMBR) spectrum. The line represents a 2.73 K blackbody. FIRAS, DMR, UBC, LBL-Italy, Princeton, and Cyanogen are sources of the experimental data points. *From Gawiser and Silk (2000). See also Hagiwara, K. et al. (2002). Reprinted with permission from Elsevier © 2000.*

F. Cosmic microwave background radiation

Cosmic microwave background (CMB) radiation does not have a direct relevance to the measurement of radioactivity; however, our knowledge of the CMB radiation helps us understand the hot beginning of the universe and the formation of atoms.

In 1965 Arno Penzias and Robert Wilson at Bell Telephone Laboratories in New Jersey accidently discovered a cosmic microwave background radiation. They were building and testing a new sensitive radio receiver, when they observed an excess noise that was of equal intensity regardless of the direction that the detector or antenna was pointing in the atmosphere. They concluded that the microwave radiation had originated from deep space even beyond our solar system and even our galaxy, because as the earth orbited the sun and rotated on its axis the antenna would be pointing to a different direction in space. At the same time, a team of physicists led by Robert Dicke at Princeton University was studying microwaves to devise a system that would demonstrate the cosmic microwave background predicted by Gamow (1948) and Alpher and Herman (1949). When Penzias and Wilson learned about this research work in search of the cosmic microwave background, they realized that they had discovered it. For this discovery, Penzias and Wilson were awarded the Nobel Prize in Physics 1978). In the Nobel Presentation Speech, Prof. Lamek Hulthén of the Royal Academy of Sciences on December 8, 1978 in Stockholm, Sweden, underscored the importance of the work and discovery of Penzias and Wilson. An excerpt of his speech is the following:

> "... *they* [Penzias and Wilson] *had found a background radiation, equally strong in all directions, independent of the time of the day and the year, so it could not come from the sun or our Galaxy ... Continued investigations have confirmed that this background radiation varies with wavelength in the way prescribed by well known laws for a space, kept at the temperature [of] 3K. Our Italian colleagues call it "la luce fredda" – the cold light ... Starting from the fact that the universe is now expanding uniformly, they concluded that it must have been very compact about 15 billion years ago and ventured to assume that the universe was born in a huge explosion, the 'Big Bang'. The temperature must then have been fabulous: 10 billion degrees [K], perhaps more. At such temperatures lighter chemical elements can be formed from existing elementary particles, and a tremendous amount of radiation of all wavelengths is released. In the ensuing expansion of the universe, the temperature of the radiation rapidly goes down ...*"

The precise determination of the CMB radiation spectrum involved large teams of researchers and decades of work, and the significance of this work was underscored by the Nobel Committee with the award of the Nobel Prize in Physics 2006) to John C. Mather (NASA Goddard Space Flight Center, USA) and George F. Smoot (University of California, Berkeley, USA) "for their discovery of the blackbody form and anisotropy of the cosmic microwave background radiation". In his Nobel address John C. Mather explained the impact of the determination of the precise cosmic microwave background radiation spectrum as follows:

> *In the beginning was the Big Bang, so we now say with great certainty. The COsmic Background Explorer (COBE) satellite, proposed to NASA in 1974 and launched in 1989,*

provided very strong evidence for it: the spectrum of the cosmic microwave background radiation (CMBR) has the spectrum of an almost-perfect blackbody emitter at 2.725 ± 0.001 K [See Fig. 1.138], and the radiation is isotropic … This radiation is interpreted as the relic of an incredibly hot and dense early phase of the universe. In such a hot and dense phase, the creation, destruction, and energy equilibration of photons with one another and with all other forms of matter and energy would occur very rapidly compared with the expansion time scale of the universe. Such a state would immediately produce a blackbody radiation field. The expanding universe should preserve this blackbody spectrum, so measurement of any significant deviation from a perfect blackbody spectrum would either invalidate the whole idea of the Big Bang, or show that energy (e.g., from decay of primordial particles) was added to the CMBR after the rapid equilibration ended …

The fact that the radiation is isotropic to such a high degree is key evidence for its origin in the Big Bang … Not only was the spectrum beautiful to look at, but at one stroke it banished the doubts of almost everyone about the Big Bang theory. For so many decades, the intense combat between the Big Bang and Steady State advocates had continued, and for so many years, a series of small discrepancies between theory and measurement had been explained by ingenious people. Now it was over … The Big Bang theory was safe, and the universe was simple, simple enough for theorists to go on to the next problem."

The precise spectrum of the cosmic microwave background radiation (CMBR) has been determined over 3 decades of frequency via many satellite, rocket, balloon, and ground measurements, and the cumulated data of radiant flux intensity as a function of radiation frequency is plotted in Fig. 1.138. The CMBR spectrum fits exactly to that predicted by a 2.73 K blackbody radiation spectrum whereby the brightness (B_ν) or intensity (I_ν) per unit of radiation frequency or wavelength is a function of the absolute temperature of the black body. The term "brightness" is used when measuring the radiant flux from a projected area of the surface of a blackbody, while the term "intensity" is used when reception of the radiant flux is measured (Woan, 2000). The Planck functions for the brightness or intensity are given by the following

$$I_\nu(T) = B_\nu(T) = \frac{2h\nu^3}{c^2}\left[\exp\left(\frac{h\nu}{kT}\right) - 1\right]^{-1} \tag{1.511}$$

and

$$I_\lambda(T) = B_\lambda(T) = B_\nu(T)\frac{d\nu}{d\lambda} = \frac{2hc^2}{\lambda^5}\left[\left(\frac{hc}{\lambda kT}\right) - 1\right]^{-1} \tag{1.512}$$

where I_ν and B_ν are the intensity or brightness, respectively, of blackbody radiation per unit frequency (W/m^2 Hz sr) and I_λ and B_λ are the intensity and brightness, respectively, of blackbody radiation per unit wavelength (W/m^2 m sr), h is Planck's constant (6.626×10^{-34} J s), c is the speed of light (2.99×10^8 m/sec), k is the Boltzmann constant (1.38×10^{-23} J/K), and T is the temperature Kelvin (Woan, 2000).

The hot Big Bang hypothesis describes the universe as having a very hot (~10^{10} K a second after the Big Bang) and small, dense beginning with an expansion and cooling of the universe ever since the Big Bang. The measured cosmic microwave background radiation spectrum illustrated in Fig. 1.138 provides evidence for the prediction of the Big Bang creation and the ever expansion and cooling of the universe. When the universe expands to twice its size, its temperature falls by half (Hawking, 1988). Consequently, when the universe was one-half its present size, its temperature would have been twice as hot (5.4 K) and the CMB radiation would have had a brightness peaking at a higher photon frequency. Likewise, if we go further back in time when the universe was 1/100th its current size, the universe would have been a hundred-fold hotter than now or 273 K equivalent to 0°C. The CMB radiation would have had a spectrum shifted to yet higher frequencies and shorter wavelengths.

XV. Radiation dose

Radiation dose, also referred to as absorbed radiation dose, is the amount of energy deposited in a given mass of a medium by ionizing radiation. The measurement of radiation dose, *i.e.*, radiation dosimetry, is not a subject of this book. Nevertheless, the basic units used in the measurement of radiation dose will be mentioned briefly here, as the monitoring of radiation workers for the dose received by the body is essential.

Radiation exposure was historically measured by the roentgen (R), which is a measure of the quantity of radiation deposited in air from the amount of charge or ionization produced by the radiation in air. By definition 1 $R = 2.58 \times 10^{-4}$ C/kg of air at STP, that is, 1 roentgen will produce 2.58×10^{-4} coulombs of ion pairs in 1 kilogram of air. The roentgen is a unit of exposure that is mostly historical and seldom used; it still occasionally appears on some dosimeter readings. Of more significance is the measure of absorbed dose, that is, the energy of radiation absorbed per unit mass of absorber. The original unit of absorbed dose is the rad, which is derived from the term "**r**adiation **a**bsorbed **d**ose". The rad has been replaced with the gray (Gy), which is the SI unit of absorbed dose. The use of SI units is recommended by the International Commission on Radiation Units and

Measurements (ICRU). The rad and gray have the following equivalents:

$$100 \text{ rad} = 10^4 \text{ erg/g} = 1 \text{ Gy} = 1 \text{ J/kg} \quad (1.513)$$

$$1 \text{ rad} = 10 \text{ mGy} = 100 \text{ erg/g} \quad (1.514)$$

$$1 \text{ mrad} = 10 \text{ } \mu\text{Gy} \quad (1.515)$$

As 1 eV = 1.602×10^{-19} J, we can convert the gray to units of electron-volt energy deposited in a kg of absorber or

$$1 \text{ Gy} = 6.24 \times 10^{12} \text{ MeV/kg} \quad (1.516)$$

The following example illustrates the calculation of radiation dose:

A 1.5-cm^2 beam of gamma radiation consisting of 1 MeV photons with an emission rate of 5×10^8 photons/sec is irradiated onto human tissue of density 1.1 g/cm^3. If 6.25% of the gamma radiation is absorbed by 1.25 cm thickness of the tissue (*i.e.*, 93.75% transmitted through the tissue), the rate of radiation dose administered to the tissue per second may be calculated as

$$\text{dose} = \frac{\text{energy}}{\text{mass}} \quad (1.517)$$

and

$$\text{dose/sec} = \frac{\text{energy/sec}}{\text{mass}}$$

$$= \frac{(5 \times 10^8 \text{ photons/sec})(0.0625)(1 \text{ MeV/photon})}{(1.1 \text{ g/cm}^3)(1.5 \text{ cm}^2)(1.25 \text{ cm})}$$

$$= \frac{3.12 \times 10^7 \text{ MeV/sec}}{2.0625 \text{ g}} = 1.51 \times 10^7 \text{ MeV/sec/kg} \quad (1.518)$$

If 1 Gray (Gy) is equivalent to 6.24×10^{12} MeV/kg (Eq. 1.516), the dose rate in MeV/sec/kg is converted to units of the Gray (Gy) as

$$\text{dose(Gy / sec)} = \frac{1.51 \times 10^7 \text{ MeV/sec/kg}}{6.24 \times 10^{12} \text{ MeV/kg/Gy}} = 2.42 \text{ } \mu\text{Gy/sec} \quad (1.519)$$

Another formerly very common and historical unit of radiation dose is the rem. The rem is a measure of absorbed dose in biological tissue. This unit of measure is derived from the term "**r**oentgen **e**quivalent for **m**an" or "**r**oentgen **e**quivalent **m**ammal". The rem was created as a measure of dose of ionizing radiation to body tissue in terms of its estimated biological effect; its SI unit is the sievert (Sv) and

$$100 \text{ rem} = 1 \text{ Sv} \quad \text{and } 1 \text{ rem} = 10\text{mSv} \quad (1.520)$$

The rem or Sv (sievert) are referred to as units of equivalent dose, because the dose is measured on the basis of a weighting factor (w_R), which defines the relative hazard of radiation on the basis of the types and energies of the radiations by placing all radiation classes on the same dose level or equivalent (L'Annunziata, 1987). The weighting factor, formerly known as the quality factor (*QF*), is defined as the ratio of the gamma-ray or X-ray dose to the dose required to produce the same biological effect by the radiation in question. Table 1.28 lists the radiation weighting factors (w_R) according to radiation type and energy. The weighting factor of a given radiation is a function of the radiation linear energy transfer (LET), that is, the radiation energy loss per path length of travel. As discussed in Section XVI of this chapter, radiations of high mass and charge, such as alpha particles or fission products will have a higher LET than electrons or beta particles, muons, or X-ray, or gamma-ray photons. Because photons will produce electrons as secondary particles, X- or gamma-ray photons are classified to have the same radiation weighting factor as electrons. According to the weighting factors provided in Table 1.28, alpha particles are considered to be 20 times more damaging to cells of the human body than beta particles, muons, and X- or gamma radiation. However, alpha particles and other heavy nuclei have short ranges in

TABLE 1.28 Radiation weighting factors.

Radiation type	w_R
Photons, electrons and muons	1
Neutrons, $E_n < 1$ MeV	$2.5 + 18.2 \times \exp[-(\ln E_n)^2/6]$
$1 \text{ MeV} \le E_n \le 50$ MeV	$5.0 + 17.0 \times \exp[-(\ln (2E_n))^2/6]$
$E_n > 50$ MeV	$2.5 + 3.25 \times \exp[-(\ln (0.04E_n))^2/6]$
Protons and charged pions	2
Alphas, fission fragments, heavy ions	20

Reprinted with permission. From Roesler and Silari (2018); Chapter 36 in Tanabashi et al. (2018) Particle Data Group © 2018 Regents of the University of California.

matter compared to other radiation types. Thus, distance provides the best protection from alpha particles and other heavy nuclei. Also, the weighting factors for neutrons in Table 1.28 classify the neutrally charged neutrons as more hazardous to biological tissue than the charged electrons or muons. This is because neutrons will produce recoil protons and ions of higher mass via neutron collisions in body tissue.

The equivalent dose in Sv can be calculated from the weighting factors as

$$\text{Sv} = \text{absorbed dose in grays} \times w_R \tag{1.521}$$

where the weighting factor expresses potential damage to cells, and long-term risk (primarily cancer and leukemia) from low-level chronic exposure, which is dependent on radiation type and other factors (ICRP, 1991 and Donahue and Fassò, 2002). Using the notation of Lilley (2001) for average absorbed dose in tissue Eq. (1.521) can be written as

$$\text{Sv} = D_{T,R} \times w_R \tag{1.522}$$

where Sv is the equivalent dose in sieverts and $D_{T,R}$ is the average absorbed dose in tissue T from a given type of radiation R. For example, from the weighting factors of Table 1.28 and Eq. (1.522) we can calculate that an average absorbed dose of 0.01 Gy from alpha particles ($w_R = 20$) will produce the same biological effect as a 0.2 Gy dose from X- or gamma radiation ($w_R = 1$). When more than one radiation type contributes to the absorbed dose, the equivalent dose is calculated according to the weighted sum of the contributions from each radiation type.

The ICRP (1991) and ICRU (2007) recommendations for limits of exposure for radiation workers (i.e., occupational exposure) for whole body dose is 20 mSv/yr (2 rem/yr) averaged over 5 years (i.e., 100 mSv in 5 years), with the effective dose in any 1 year not to exceed 50 mSv. It is estimated by the United States Environmental Protection Agency (EPA) that the average yearly radiation dose per person in the United States is only 6.2 mSv (0.62 rem or 620 mrem) of accumulated radiation dose per year from all sources including cosmic radiation, medical X-rays, and natural radioactivity, etc (https://www.epa.gov/radiation/calculate-your-radiation-dose). The Particle Data Group (Patrignani et al., 2016) and United Nations (2008) report a worldwide average for the annual whole-body dose equivalent from all sources of natural background radiation to range from 1.0 to 13 mSv (0.1−1.3 rem) with an annual average of 2.4 mSv; although in some areas annual averages of up to 50 mSv (5 rem) have been measured. Patrignani et al. (2016) report that a large fraction (>50%) of the exposure from natural background radiation originates from inhaled natural radioactivity, which includes mostly radon and radon daughters. Average dose from inhaled natural radioactivity will depend on the environment, namely, in open areas (0.1−0.2 mSv), within a house (2 mSv), or in a poorly ventilated mine (20 mSv).

As discussed previously in Section XIV.B, the atmosphere serves as an excellent shield against intense cosmic radiation, as cosmic-ray particles collide with nuclei of nitrogen, oxygen, and other gaseous atoms of the atmosphere and produce the less hazardous particles that constitute the cosmic-ray showers. Exposure to cosmic radiation is of some concern to intercontinental airline pilots, who fly at high altitudes (7−12 km above the earth) and high degrees of latitude and who compile large numbers of flight hours per year over long careers. Hammer et al. (2000) report a mean cosmic radiation dose estimate of 35 mSv for 509 pilots, who averaged 26.6 years of employment at an average of 481 flight hours per year. The accumulated cosmic radiation dose of 35 mSv over 26.6 years will yield an average dose rate of 1.3 mSv/y. The dose rates are well within the ICRP (1991) recommendations for limits of exposure for radiation workers for whole body dose set at 20 mSv/y. For pregnant women, however, the ICRP limit of dose is 2 mSv for the duration of pregnancy. Consequently, the study by NRL (1998) concludes that airline companies may apply a limit of 2 mSv for a pregnant member of the aircrew from the time of conception. Astronauts, who circle the earth in orbit for extended periods of time, those who have traveled and plan to travel to the moon and back, and those who may travel to Mars have reason for concern of exposure to cosmic radiation where there is no protection from the earth's partially protective atmospheric shield. Fujitaka (2005) reports that a 1-week mission on the space scuttle will result in a radiation dose of 3.5 mSv; while astronauts would accumulate a dose of 43 mSv during a trip to the moon. Travel from Earth to Mars would take 280 days and the return trip 256 days during which time astronauts would accumulate 1850 mSv, and a stay on Mars for 439 days would expose astronauts to an additional 410 mSv whereby each astronaut would accumulate a grand total of 2.26×10^3 mSv (Pachoa and Steinhäusler, 2010). A very interesting and stimulating note for thought is reported by Fujitaka (2005), that space travel to Pluto, which is at the boundaries of our solar system would expose astronauts to a cosmic radiation dose of 70 Sv or 70,000 mSv, which is enough radiation to kill every cell in the human body. This may be so at current speeds of space travel, as NASA's New Horizons Mission spacecraft to Pluto was launched in June 2006, and it reached Pluto after 9 years in July of 2015. However, with the proper thrust systems, high speeds of space travel in a manned spacecraft would be possible (Westmoreland, 2010; Walter, 2006), which would minimize space travel time well out of our solar system and thus keep cosmic radiation dose to acceptable limits.

For further information on radiation dosimetry, the reader is invited to peruse books dedicated to this subject matter such as those by Andreo and Burns (2017), Bevelacqua (2010), Stabin (2009), and Cember and Johnson (2008).

XVI. Stopping power and linear energy transfer

Previous sections in this chapter provide information on the mechanisms of interaction of radiation with matter. In summary, we can state that the principle mechanisms of interaction of charged particles with matter, which result in significant charged-particle energy loss are (i) ionization via coulombic interactions of the charged particles with atomic electrons of the absorbing medium; (ii) electron orbital excitation of the medium, which occurs when the energy transfer through coulombic interaction is not sufficient to actually eject an electron from an atom; and (iii) the radial emission of energy as bremsstrahlung (x radiation) when a charged particle decelerates as it is deflected by an atomic nucleus. Release of particle energy by bremsstrahlung radiation becomes increasingly significant as the charged particle (*e.g.*, beta-particle) energy and absorber atomic number increase. On the other hand, electromagnetic radiation dissipates its energy in matter via three mechanisms, namely, (i) the photoelectric effect, (ii) Compton scattering, and (iii) pair production. The photoelectric effect and Compton scattering generate ion pairs directly within the absorbing medium, whereas, pair production results in the creation of charged particles (positrons and negatrons) that will subsequently dissipate their energy via ionization, electron excitation and, in the case of positrons, the production of annihilation radiation. Also, we have seen that energetic neutrons will dissipate their energy in matter through elastic collisions with atomic nuclei of the absorbing medium. When hydrogen is present in the absorbing material, the bulk of fast-neutron energy is passed on to the hydrogen nuclei. In turn, the kinetic energy of these protons is absorbed in the medium via ionization and excitation processes. We have seen also that low- and high-energy neutrons are absorbed principally via inelastic neutron reactions, which can result in the production of charged particles and gamma radiation.

The radiation properties (*e.g.*, charge, mass, and energy) and mechanisms of interaction previously described govern the rate of dissipation of energy and consequently the range of travel of the nuclear radiation in the absorber. This brings to bare the concepts of stopping power and linear energy transfer (LET), which are described subsequently.

A. Stopping power

Stopping power is defined by The International Commission on Radiation Units and Measurements (ICRU, 2007) as the average energy dissipated by ionizing radiation in a medium per unit path length of travel of the radiation in the medium. It is, of course, impossible to predict how a given charged particle will interact with any given atom of the absorber medium. Also, when we consider that the coulombic forces of charged particles will interact simultaneously with many atoms as it travels through the absorbed medium, we can only predict an average effect of energy loss per particle distance of travel. Taking into account the charge, mass, and speed (energy) of the particle, and the density and atomic number of the absorbing medium, Bethe (1933) and Bethe and Ashkin, (1953) derived a formula for calculating the stopping power resulting from coulombic interactions of heavy charged particles (*e.g.*, alpha particles, protons, and deuterons) traveling through absorber media. Rohrlich and Carlson (1954) have refined the calculations to include energy losses via bremsstrahlung radiation, significant when high-energy electrons and beta particles interact with absorbers of high atomic number. Also, refinements to the stopping power formulae in the low energy ranges of heavy particles have been made by several researchers including Bohr and Lindhard (1954), Lindhard and Scharff (1960, 1961), Northcliffe (1963), and Mozumber et al. (1968). Derivations of stopping power formulas can be obtained from texts by Evans (1955), Friedlander et al. (1964), Roy and Reed (1968), and Segré (1968), Gümüs (2005, 2008), Nguyen-Truong (2015), and Taborda et al. (2015). Stopping power and range tables for electrons and protons over the energy range of 10^{-3} to 10^5 MeV and for helium ions (alpha particles) over the energy range of 10^{-3} to 10^3 MeV in 74 materials are available from Berger (1993) and Berger et al. (2017c); and these can be calculated online at the website of the National Institute of Standards and Technology (NIST, 2018), *i.e.*, https://www.nist.gov/pml/stopping-power-range-tables-electrons-protons-and-helium-ions, for electrons in any user-specified materials, and for protons and helium ions in 74 materials.

The formulas for the stopping power of charged particles due to coulombic or collisional interactions (i.e., ionization and electron orbital excitation) are clearly defined by Tsoulfanidis (1995) as the following:

(i) for heavy charged particles (e.g., protons, deuterons, and alpha particles),

$$\frac{dE}{dx} = 4\pi r_0^2 z^2 \frac{mc^2}{\beta^2} NZ \left[\ln\left(\frac{2mc^2}{I} \beta^2 \gamma^2 \right) - \beta^2 \right] \quad (1.523)$$

(ii) for electrons or negatrons (negative beta particles),

$$\frac{dE}{dx} = 4\pi r_0^2 \frac{mc^2}{\beta^2} NZ \left\{ \ln\left(\frac{\beta\gamma\sqrt{\gamma-1}}{I} mc^2 \right) + \frac{1}{2\gamma^2}\left[\frac{(\gamma-1)^2}{8} + 1 - \left(\gamma^2 + 2\gamma - 1\right)\ln 2 \right] \right\} \tag{1.524}$$

and **(iii)** for positrons (positive beta particles),

$$\frac{dE}{dx} = 4\pi r_0^2 \frac{mc^2}{\beta^2} NZ \left\{ \ln\left(\frac{\beta\gamma\sqrt{\gamma-1}}{I} mc^2 \right) - \frac{\beta^2}{24}\left[23 + \frac{14}{\gamma+1} + \frac{10}{(\gamma+1)^2} + \frac{4}{(\gamma+1)^3} \right] + \frac{\ln 2}{2} \right\} \tag{1.525}$$

where dE/dx is the particle stopping power in units of MeV/m, r_0 is the classical electron radius $= 2.818 \times 10^{-15}$ m, z is the charge on the particle ($z=1$ for p, d, β^-, β^+ and $z=2$ for α), mc^2 is the rest energy of the electron $= 0.511$ MeV (see Section XII.C of this chapter.), N is the number of atoms per m^3 in the absorber material through which the charged particle travels ($N = \rho(N_A/A)$ where ρ is the absorber density in units of g/cm^3, N_A is Avogadro's number $= 6.022 \times 10^{23}$ atoms per mol, A is the atomic weight (not the mass number) of the absorber material, and Z is the atomic number of the absorber, $\gamma = (T + mc^2)/mc^2 = 1/\sqrt{1-\beta^2}$ where T is the particle kinetic energy in MeV and m is the particle rest mass, (*e.g.*, proton $= 938.2$ MeV/c^2, deuteron $= 1875.6$ MeV/c^2, alpha particle $= 3727.3$ MeV/c^2, and β^- or $\beta^+ = 0.511$ MeV/c^2, and β the relative phase velocity of the particle $= v/c$, that is, the velocity of the particle in the medium divided by the speed of light in a vacuum $= \sqrt{1-(1/\gamma^2)}$ (see Section XI.E of this chapter and Chapter 6, Volume two for a treatment on β.), and I is the mean excitation potential of the absorber in units of eV approximated by the equation

$$I = \left(9.76 + 58.8Z^{-1.19}\right)Z, \quad \text{when } Z > 12 \tag{1.526}$$

where pure elements are involved as described by Tsoulfanidis (1995). However, when a compound or mixture of elements is concerned, a mean excitation energy, $\langle I \rangle$, must be calculated according to Bethe theory as follows

$$\langle I \rangle = \exp\left\{ \frac{\left[\sum_j w_j (Z_j/A_j) \ln I_j\right]}{\sum_j w_j (Z_j/A_j)} \right\} \tag{1.527}$$

where *wj*, *Zj*, *Aj*, and *Ij* are the weight fraction, atomic number, atomic weight, and mean excitation energy, respectively, of the *j*th element (Seltzer and Berger, 1982a). See Anderson et al. (1969a,b), Janni (1982), Seltzer and Berger (1982a,b, 1984), Berger and Seltzer (1982), and Tsoulfanidis (1995) for experimentally determined values of I for various elements and thorough treatments of stopping power calculations. Values of mean excitation potentials, I, for 100 elements and many inorganic and organic compounds are provided by Seltzer and Berger (1982a, 1984).

An example of the application of one of the above equations would be the following calculation of the stopping power for a 2.280 MeV beta particle (E_{max}) emitted from ^{90}Y traveling through a NaI solid scintillation crystal detector. This would be a practical example, as the NaI detector is used commonly for the measurement of ^{90}Y. The solution is as follows:

Firstly, the calculation of relevant variables are

$$\gamma = \frac{2.280\text{ MeV} + 0.511\text{ MeV}}{0.511\text{ MeV}} = 5.462$$

$$\beta = \sqrt{1 - \frac{1}{\gamma^2}} = 0.9831 \quad \text{and } \beta^2 = 0.9665$$

The atomic weight A for NaI would be the average atomic weight (A_{av}) based on the weight-fraction w_{Na} for Na (15.3%) and w_I for I (84.7%) in NaI or

$$A_{av} = (0.153)(A_{Na}) + (0.847)(A_I) = (0.153)(23) + (0.847)(127) = 111$$

Also, on the basis of the weight averages for Na and I, the atomic number Z would be the effective atomic number Z_{ef} calculated according to the following equation described by Tsoulfanidis (1995) and Andreo et al. (2005):

$$Z_{ef} = \frac{\sum_{i=1}^{L} (w_i/A_i) Z_i^2}{\sum_{i=1}^{L} (w_i/A_i) Z_i} \tag{1.528}$$

where L is the number of elements in the absorber, w_i is the weight fraction of the ith element, A_i is the atomic weight of the ith element, Z_i is the atomic number of the ith element, and $w_i = N_i A_i / M$ where N_i is the number of atoms of the ith element and M is the molecular weight of the absorber. If we apply Eq. (1.528) to the absorber NaI we find

$$Z_{ef} = \frac{(0.153/22.989)(11)^2 + (0.847/126.893)(53)^2}{(0.153/22.989)(11) + (0.847/126.893)(53)} = 45.798 \tag{1.529}$$

For pure elements the value of the mean excitation potential, I, can be calculated according to the empirical formula provided by Eq. (1.526). However, for the compound NaI, the mean excitation energy, $\langle I \rangle$, will be calculated according to Eq. (1.527) as follows

$$\langle I\rangle = \exp\left\{\begin{bmatrix}(0.153)(11/22.989)\ln 149+\\(0.847)(53/126.893)\ln 491\end{bmatrix}\right.$$

$$\left./\begin{bmatrix}(0.153)(11/22.989)+\\(0.847)(53/126.893)\end{bmatrix}\right\} \quad (1.530)$$

$$= 400 \text{ eV}$$

where the values of the mean excitation energy, I_j, for the elements Na and I of 149 and 941 eV, respectively, were obtained from Seltzer and Berger (1982a). From Eq. (1.524) the stopping power, due to collisional interactions, for the 2.280 MeV beta particle traveling through a NaI crystal is calculated as

This calculated value of stopping power is due to collision or coulombic interactions, and it agrees with the on-line computer-calculated value of 1.203 MeV cm^2/g for a 2.280 MeV beta particle or electron in NaI (see NIST, 2018, https://physics.nist.gov/PhysRefData/Star/Text/method.html). Slight variations on the calculated stopping power are expected, because of the statistical variations of charged particle interactions with matter and, as a consequence, there are small variations to the Bethe stopping power formula derived by researchers, described previously in this section.

Eq. (1.524) used above to calculate the stopping power for the 2.280 MeV beta particle from ^{90}Y in NaI accounts only for energy of the beta particle lost via collision interactions resulting in ionization and electron-orbital excitations. The equation does not account for radiative energy loss via the production of bremsstrahlung radiation, which can be very significant with beta particles of high energy and absorber materials of high atomic number. Thus, a complete calculation of the stopping power must include also the radiative energy loss via bremsstrahlung as described by Eq. 1.250 earlier in this chapter, that is,

$$\frac{dE}{dx} = 4(3.14)\left(2.818\times10^{-15}\text{ m}\right)^2\left(\frac{0.511\text{ MeV}}{0.9665}\right)\left(3.67\text{ g/cm}^3\right)\times\left(\frac{6.022\times10^{23}\text{ atoms/mol}}{111\text{ g/mol}}\right)\left(\frac{10^6\text{ cm}^3}{\text{m}^3}\right)$$

$$\times(45.798)\left\{\begin{array}{l}\ln\left(\dfrac{(0.9831)(5.462)\sqrt{4.462}}{400\text{ eV}}(0.511\text{ MeV})\left(10^6\text{ eV/MeV}\right)\right)\\ +\dfrac{1}{2(5.462)^2}\left[\dfrac{(4.462)^2}{8}+1-\left(5.462^2+2(5.462)-1\right)\ln 2\right]\end{array}\right\} \quad (1.531)$$

$$= 441.4\text{ MeV/m}$$

The term (10^6 cm^3/m^3) is included in the above calculation to maintain consistency of units, thus converting the units of density expressed in g/cm^3 to g/m^3. In SI units, the stopping power can be expressed in units of J/m or

$$\left(441.4\text{ MeV / m}\right)\left(1.602\times10^{-13}\text{ J / MeV}\right) = 7.07\times10^{-11}\text{ J/m} \quad (1.532)$$

The stopping power is often expressed in units of MeV cm^2/g or J m^2/kg, which provides values for stopping power without defining the density of the absorber medium (Taylor et al., 1970; Tsoulfanidis, 1995; NIST, 2018). In these units the above calculation can be expressed as

$$\frac{1}{\rho}\left(\frac{dE}{dx}\right)_{\text{col.}} = \frac{4.414\text{ MeV/cm}}{3.67\text{g/cm}^3} = 1.203\text{ MeV cm}^2/\text{g} \quad (1.533)$$

$$\left(\frac{dE}{dx}\right)_{\text{total}} = \left(\frac{dE}{dx}\right)_{\text{col.}} + \left(\frac{dE}{dx}\right)_{\text{rad.}} \quad (1.534)$$

where $(dE/dx)_{\text{total}}$ is the rate of total beta particle or electron energy loss in an absorber, $(dE/dx)_{\text{col.}}$ is the rate of energy loss due to collision or ionization interactions, and $(dE/dx)_{\text{rad.}}$ is the rate of radiative energy loss due to bremsstrahlung production. We must keep in mind that each stopping-power calculation, such as the above example, provides values for only one beta particle or electron energy. Beta particles, on the other hand, are emitted from decaying radionuclides with a broad spectrum of energies from zero to E_{max}, the majority of which may possess an average energy, E_{av}, of approximately one-third of E_{max}.

The ratio of beta-particle energy loss via radiative energy emission as bremsstrahlung to energy loss via collision interactions causing ionization and excitation is described by the approximation

$$\frac{E_{brems.}}{E_{ion.}} \simeq \frac{EZ}{750} \tag{1.535}$$

where E is the beta-particle energy in MeV and Z is the atomic number or effective atomic number Z_{ef} (e.g., 45.798 for NaI calculated by Eq. (1.529)) of the absorber material (Evans, 1955; Friedlander et al., 1964; Faw and Schultis, 2004). From Eqs. (1.534) and (1.535), we can approximate the radial energy loss for the 2.280 MeV beta particle in NaI using the previously calculated ionizational energy loss (Eq. 1.531) as

$$\begin{aligned}\left(\frac{dE}{dx}\right)_{rad.} &\simeq \frac{ZE}{750}\left(\frac{dE}{dx}\right)_{col.} \\ &\simeq \frac{(45.798)(2.280)}{750}(4.414 \text{ MeV/cm}) \\ &\simeq 0.614 \text{ MeV/cm}\end{aligned} \tag{1.536}$$

This radiative energy loss expressed in units of MeV cm^2/g would be

$$\frac{1}{\rho}\left(\frac{dE}{dx}\right) \simeq \frac{0.614 \text{ MeV/cm}}{3.67 \text{g/cm}^3} = 0.167 \text{ MeV cm}^2/\text{g} \tag{1.537}$$

This agrees closely to the online computer calculated value of 0.16 MeV cm^2/g for the radiative energy loss a 2.280 MeV beta particle or electron in NaI (see NIST, 2018, https://physics.nist.gov/PhysRefData/Star/Text/ESTAR.html). The total stopping power of the 2.280 MeV beta particle in NaI according to Eq. (1.534) is calculated as

$$\begin{aligned}\left(\frac{dE}{dx}\right)_{total} &= \left(\frac{dE}{dx}\right)_{col.} + \left(\frac{dE}{dx}\right)_{rad.} \\ &= 4.41 \text{ MeV/cm} + 0.61 \text{ MeV/cm} \\ &= 5.02 \text{ MeV/cm}\end{aligned} \tag{1.538}$$

The above total stopping power expressed in units of MeV cm^2/g is calculated as

$$\frac{1}{\rho}\left(\frac{dE}{dx}\right)_{total} = \frac{5.02 \text{ MeV/cm}}{3.67 \text{ g/cm}^3} = 1.37 \text{ MeVcm}^2/\text{g} \tag{1.539}$$

which agrees closely to the computed-calculated value of 1.36 MeV cm^2/g for the total stopping power of a 2.280 MeV beta particle in NaI calculated online via the NIST (2018) ESTAR computer program (https://physics.nist.gov/PhysRefData/Star/Text/ESTAR.html).

Beta-particle loss via bremsstrahlung radiation of the 2.280 MeV beta particles from ^{90}Y is significant in NaI, namely, 0.61/5.02 or 12% of the total energy loss. Consequently, NaI solid scintillation detectors are at times used for the analysis of ^{90}Y (Coursey et al., 1993). The actual detection efficiencies reported by Coursey et al. (1993) for the solid scintillation analysis of ^{90}Y fall in the range of 9.9–18% depending on sample and detector counting geometries. The detection efficiencies exceed the above-calculated 12% energy loss via bremsstrahlung production, because the NaI detector will also respond to collision-excitation energy of the beta-particle in addition to bremsstrahlung radiation excitation. Caution is warranted in making correlations between detector response to beta-particle radiation and stopping-power calculations, because we must keep in mind that each stopping-power calculation, such as the above example, provides values for only one beta-particle energy. Beta particles, on the other hand, are emitted with a broad spectrum of energies from zero to E_{max}.

Bremsstrahlung production will increase as the absorber atomic number (Z) and beta-particle or electron energy increase. Bremmstrahlung is highly penetrating electromagnetic radiation, and thus low atomic number absorbers such as aluminum or plastic (high in carbon and hydrogen) may be used as shields against beta-particle radiation. Shielding material containing high atomic number elements, such as lead glass, against high-energy beta particles can yield high levels of bremsstrahlung radiation. The effect of atomic number (Z) and beta-particle or electron energy (E) on the ratio of energy loss via bremsstrahlung production to ionization energy is illustrated in Fig. 1.139. The effective atomic numbers of compounds such as polyethylene ($(-C_2H_4)_n$, $Z_{ef} = 5.28$), water ($Z_{ef} = 6.60$), and air ($Z_{ef} = 7.3$), included as absorbers in Fig. 1.139, are a function of the weight fraction of each element in the compound or mixture calculated according to Eq. (1.528). As illustrated in Fig. 1.139, absorbers of low atomic number yield less bremsstrahlung radiative energy loss from high-energy beta-particle or electron interactions in matter.

B. Linear energy transfer

The International Commission on Radiation Units and Measurements defines linear energy transfer (L) of charged particles in a medium as

$$L = \frac{dE_L}{dl} \tag{1.540}$$

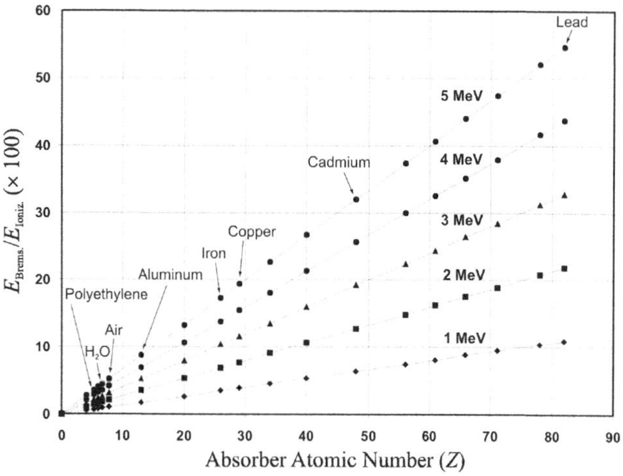

FIGURE 1.139 Ratio of energy loss via radiative bremsstrahlung production to energy loss via ionization or collision energy loss, expressed as a percent and calculated according to the approximation $E_{brems}/E_{ion.} \simeq EZ/750$, for beta-particles or electrons of energies E ranging from 1 to 5 MeV as a function of atomic number (Z) of the absorber material. Some common absorbers are identified above data points corresponding to their atomic number (Z) or effective atomic number (Z_{ef}) for absorbers that are compounds. *From L'Annunziata (2007, 2016), reprinted with permission of Elsevier © 2007.*

where dE_L is the average energy locally imparted to the medium by a charged particle of specified energy in traversing a distance dl (ICRU, 2007; Taylor et al., 1970). The term "locally imparted" refers either to a maximum distance from the particle track or to a maximum value of discrete energy loss by the particle beyond which losses are no longer considered as local. Linear energy transfer or LET is generally measured in units of keV/μm. The ICRU recommends, when a restricted form of LET is desired, that the energy cut-off form of LET be applied, because this can be evaluated using restricted stopping-power formulae (Taylor et al., 1970). The energy-restricted form of LET or L_Δ is therefore defined as that part of the total energy loss of a charged particle, which is due to energy transfers up to a specified energy cut-off value

$$L_\Delta = \left(\frac{dE}{dl}\right)_\Delta \tag{1.541}$$

where the cut-off energy (Δ) in eV units must be defined or stated. If no cut-off energy is applied, then the subscript ∞ is used in place of Δ, where L_∞ would signify the value of LET, which includes all energy losses and would therefore be equal to the total mass stopping power.

Fig. 1.140 illustrates charged particle interactions within an absorber involved in the measurement of LET. The possible types of energy loss, ΔE, of a charged particle of specified energy, E, traversing an absorber over a track length Δl is illustrated, where O represents a particle traversing the observer without any energy loss, U is the energy transferred to a localized interaction site, q is the energy transferred to a short-range secondary particle

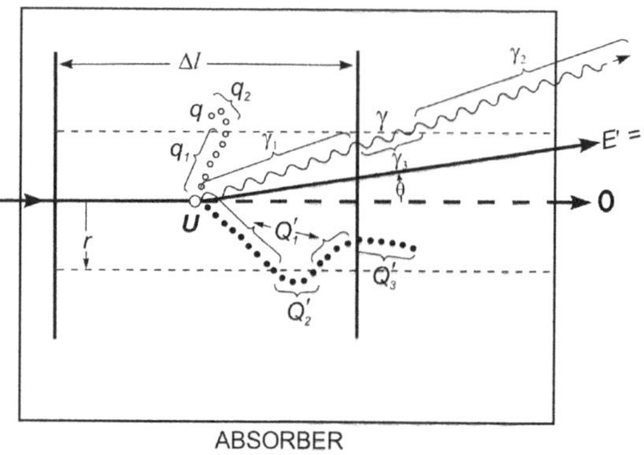

FIGURE 1.140 Diagram of the passage of a particle of energy E through a thickness Δl of material illustrating the several types of energy loss that may occur. *From Taylor et al. (1970) reprinted with permission from ICRU © 1970.*

when $q \leq \Delta$, and Δ is a selected cut-off energy level (*e.g.*, 100 eV), Q' is the energy transferred to a long-range secondary particle (*e.g.*, formation of delta rays) for which $Q' > \Delta$, γ is the energy transferred to photons (*e.g.*, excitation fluorescence, Cherenkov photons, etc.), r is a selected cut-off distance from the particle's initial trajectory or path of travel, and θ is the angle of particle scatter. The interactions q, Q, and γ are subdivided in Fig. 1.140 when these fall into different compartments of the absorber medium. See Taylor et al. (1970) and (ICRU, 2007) for methods used for the precise calculations of LET. Some examples of LET in water for various radiation types are given in Table 1.29. The table clearly illustrates that radiation of a given energy with shorter range in a medium will yield higher values of LET than radiations of the same energy with longer ranges in the same medium. This may be intuitively obvious because the shorter the range of the radiation, the greater is the energy dissipated per unit path length of travel. We can take this further and generalize that the following radiation types will yield LET values of decreasing orders of magnitude (the heavier charged particles are considered here to be of the same energy for purposes of comparison) according to the sequence:

Decreasing LET:

Fission Products
Alpha Particles
Deuterons
Protons
Low-energy x Rays and Beta Particles
High-energy x Rays and Beta Particles
Gamma Radiation and High-energy Beta Particles

Although the electromagnetic X- and gamma radiations are not charged particles, these radiations do have the characteristics of particles (photons) that produce ionization in matter. They are, therefore, included in the above sequence and among the radiations listed in Table 1.29.

TABLE 1.29 **Track-average values of LET (L_Δ) in water irradiated with various radiations.**[a]

Radiation	Cut-off energy, Δ (eV)	L_Δ (keV/μm)
^{60}Co gamma rays	Unrestricted	0.239
	10,000	0.232
	1000	0.230
	100	0.229
22-MeV X-rays	100	0.19
2-MeV electrons (whole track)	100	0.20
200-kV X-rays	100	1.7
^{3}H beta particles	100	4.7
50-kV X-rays	100	6.3
5.3 MeV alpha particles (whole track)	100	43

[a]*From Taylor et al. (1970).*

The term delta rays (δ − rays), referred to in the previous paragraph, is used to identify high-energy electrons that produce secondary ionization. When a charged particle, such as an alpha particle, travels through matter, ionization occurs principally through coulombic attraction of orbital electrons to the positive charge on the alpha particle with the ejection of electrons of such low energy that these electrons do not produce further ionization. However, direct head-on collisions of the primary ionizing particle with an electron does occur occasionally whereby a large amount of energy is transferred to the electron. These high-energy electrons are referred to as delta rays. Delta rays form ionization tracks away from the track produced by the primary ionizing particle (Anderson et al., 2017; Payne, 2015; Braby and Ford, 2006; Horowitz, 2006; Cucinotta et al., 1998; Laulainen and Bichsel, 1972). Also, most of the muon induced background in Ge gamma spectrometers is caused by bremsstrahlung radiation of delta electrons (Povinec et al., 2008).

When we compare particles of similar energy, we can state that, the ranges of particles of greater mass and charge will obviously be shorter and the magnitude of their LET values would be consequently higher in any given medium. The relationship between mass, charge, energy, and the range of particles and their corresponding LET values are compared in Table 1.30. The LET values in Table 1.30 are estimated by dividing the radiation energy by its range or path length in the medium. Such a calculation provides only an estimate of the LET, because the energy dissipated by the radiation will vary along its path of travel, particularly in the case of charged particles, more energy is released when the particle slows down before it comes to a stop as illustrated in Fig. 1.44 illustrated earlier in this chapter, when energy liberated in ion-pair formation is the highest. Nevertheless, the approximate LET values provided in Table 1.30 give good orders of magnitude for comparative purposes.

The concept of LET and the calculated values of LET for different radiation types and energies can help us interpret and sometimes even predict the effects of ionizing radiation on matter. For example, we can predict that heavy charged particles, such as alpha radiation, will dissipate their energy at shorter distances within a given absorber body than the more penetrating beta- or gamma radiations. Also, low-energy X-radiation can produce a similar effect as certain beta-radiations. The order of magnitude of the LET will help us predict the penetration power and degree of energy dissipation in an absorber body, which is critical information in studies of radiation chemistry, radiation therapy, and dosimetry, among others. For additional information on linear energy transfer and its applications, the reader is referred to works by Koshimizu et al. (2017, 2016), Yanagida et al. (2015), Liu et al. (2012), Jones et al. (2010), Pachoa and Steinhäusler (2010), Pickrell (2009), Choudhuri et al. (2007), Azab et al. (2006), Ehman and Vance (1991), Spinks and Woods (1990), and Farhataziz and Rodgers (1987).

XVII. Radionuclide decay, ingrowth, and equilibrium

The activity of a radioactive source or radionuclide sample is, by definition, its strength or intensity or, in other words, the number of nuclei decaying per unit time (*e.g.*, dpm or dps). The activity decreases with time as a result of radionuclide decay. A unit of time in which there is an

TABLE 1.30 Range and approximate LET values for various charged-particle radiations in water in order of decreasing mass and charge.[a]

Nuclide	Radiation, energy (MeV)	Range in water (mm)	Approximate LET in water (keV/μm)
Thorium-232	α, 4.0	0.029[b]	138
Americium-241	α, 5.5	0.048[b]	114
Thorium-227	α, 6.0	0.055[b]	109
Polonium-211	α, 7.4	0.075[b]	98
–	d, 4.0	0.123[c]	32.5
–	d, 5.5	0.214[c]	25.7
–	d, 6.0	0.249[c]	24.1
–	d, 7.5	0.369[c]	20.3
–	p, 4.0	0.246[d]	16.3
–	p, 5.5	0.428[d]	12.8
–	p, 6.0	0.498[d]	12.0
–	p, 7.5	0.738[d]	10.2
Tritium	β^-, 0.0186 (E_{max})	0.00575[e]	3.2
Carbon-14	β^-, 0.156 (E_{max})	0.280[e]	0.56
Phosphorus-32	β^-, 1.710 (E_{max})	7.92[e]	0.22
Yttrium-90	β^-, 2.280 (E_{max})	10.99[e]	0.21

[a] *The deuteron (d) and proton (p) energies were arbitrarily selected to correspond to the alpha-particle (α) energies to facilitate the comparison of the effects of particle mass and charge on range and LET.*
[b] *Calculated according to Eqs. (1.170) and (1.171).*
[c] *The deuteron range is calculated from the equation* $R_{Z,M,E} = (M/Z^2)R_{p,E/M}$. *The equation provides the range of a particle of charge Z, mass number M, and energy E, where* $R_{p,E/M}$ *is the range in the same absorber of a proton of energy E/M (Friedlander et al., 1964).*
[d] *CSDA ranges from Stopping Power and Range Tables for Protons (PSTAR) https://physics.nist.gov/PhysRefData/Star/Text/PSTAR.html.*
[e] *Calculations are based on the maximum energy* (E_{max}) *of the beta particles. The range was calculated according to the empirical formula* $R = 0.412E^{1.27-0.0954 \ln E}$ *available from the curve provided in Fig. B.4, Appendix B. Ranges are also available from the online data base Stopping Power and Range Tables for Electrons https://physics.nist.gov/PhysRefData/Star/Text/ESTAR.html.*

observable change in the activity of a given quantity of radionuclide may be very short, of the order of seconds, or very long, of the order of years. The rate of decay of some nuclides is so slow that it is impossible to measure any change in radioactivity during our lifetime.

A. Half-life

Rates of radionuclide decay are usually expressed in terms of half-life. This is the time, t, required for a given amount of radionuclide to lose 50% of its activity. In other words, it is the time required for one-half of a certain number of nuclei to decay. The decay curve of ^{32}P (Fig. 1.141) illustrates the concept of half-life. In Fig. 1.141, the activity of the ^{32}P is plotted against time in days. It can be seen that, after every interval of 14.3 days, the radioactivity of the ^{32}P is reduced by half. Thus, the half-life, $t_{1/2}$, of ^{32}P is 14.3 days. It is not possible to predict when one particular atom of ^{32}P will decay; however, it is possible to predict statistically for a large number of ^{32}P radionuclides that one-half of the atoms would decay in 14.3 days.

The phenomenon of radioactivity decay and its measurement in terms of half-life was first observed by Rutherford (1900). He published his observations on the rate of decay of a certain isotope of thorium in the *Philosophical Magazine* in January 1900 less than 4 years after Henri Becquerel first reported the existence of radioactivity to the French Academy of Sciences in February and March 1896. Rutherford (1900) measured the rate of decay by the level of ionization that the radiation would produce in air as a function of time, and he noted the following [with the writer's comments in brackets]:

> *... the intensity of the radiation has fallen to one-half its value after an interval of about one minute. The rate of leak [meaning rate of decay or rate of ionization caused by the radiation] due to the emanation was too small for measurement after an interval of 10 minutes [i.e., 10 half-lives] ... The current [produced by the ionization] reaches half its*

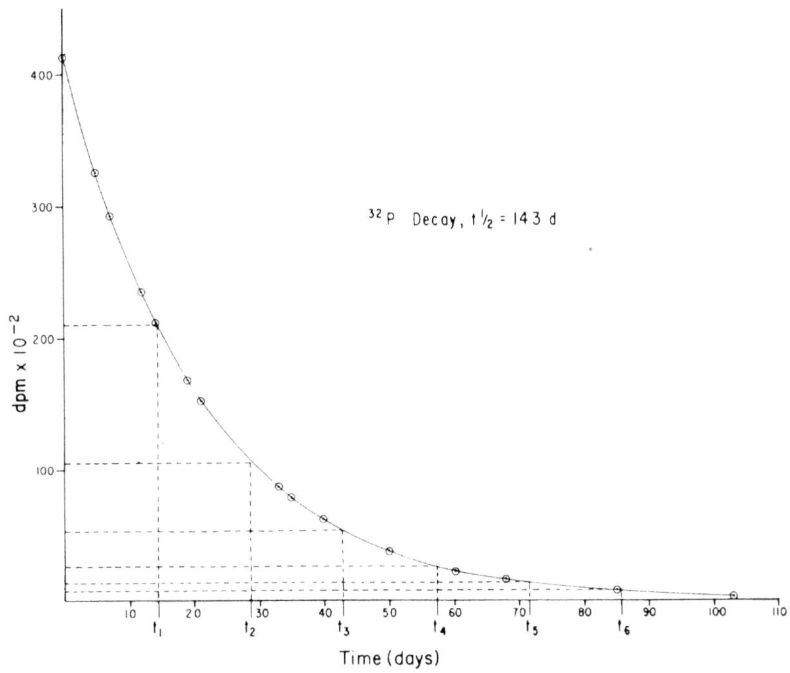

FIGURE 1.141 Decay of ^{32}P represented as a linear plot of activity in disintegrations per minute (dpm) against time in days. Horizontal and vertical lines between the ordinate and abscissa delineate ^{32}P activities (dpm) after six half-lives identified by the symbols t_1, t_2, t_3,...t_6. *From L'Annunziata (1965), unpublished work of the writer.*

value in about one minute — a result which agrees with the equation given, for $e^{-\lambda t} = 1/2$ when $t = 60$ seconds.

In his original paper Rutherford (1900) did not use the term "half-life"; but he defined mathematically the concept of half-life. As we shall see subsequently in Eq. (1.547) that the ratio of the number of radionuclides, N, in a sample at a given time, t, to the original number of nuclides, N_0, in that sample at time $t = 0$, is equal to $e^{-\lambda t}$ where e is the base to the natural logarithm, λ is a decay constant of that radionuclide, and t is the interval of time.

When radionuclide decay can be recorded within a reasonable period of time, the half-life of a nuclide can be determined by means of a semi-logarithmic plot of activity versus time, as shown in Fig. 1.142. Radionuclide decay is a logarithmic relation, and the straight line obtained on the semi-logarithmic plot permits a more accurate determination of the half-life.

Ernest Rutherford and Frederick Soddy (1902) in the *Philosophical Magazine* summarized their findings on radioactivity and radionuclide decay as follows [the writer's insertions in brackets]:

> *Turning from the experimental results to their theoretical interpretation, it is necessary to first consider the generally accepted view of the nature of radioactivity. It is well established that this property is the function of the atom and not of the molecule. Uranium and thorium, to take the most definite cases [example], possess the property in whatever molecular condition they occur ... So far as the radioactivity of different compounds of different density and states of division can be compared together, the intensity of the radiation appears to depend only on the quantity of active element [radioisotope] present.*

The above statement by Rutherford and Soddy in 1902 that "the intensity of the radiation appears to depend only on the quantity of active element present" is defined subsequently in mathematical terms.

The number, ΔN, of atoms disintegrating in a given time, Δt, is proportional to the number, N, of radioactive atoms present. This relationship may be written as

$$\Delta N/\Delta t = \lambda N \quad (1.542)$$

or

$$dN/dt = -\lambda N \quad (1.543)$$

where λ is a proportionality constant, commonly referred to as the decay constant, and the negative sign signifies a decreasing number of radionuclides with time.

One condition must be fulfilled for Eq. (1.543) to be rigorously applicable: the total number of radioactive atoms being considered must be large enough to make statistical methods valid. For example, in the case of a single isolated atom of ^{32}P there is no way to predict when the atom will decay. In fact, the atom might decay in the first second after $t = 0$ (the moment observations are initiated) or it might decay days or weeks later. The concept of half-life is a statistical one, which, when applied to a large number of atoms, as is usually the case, allows an accurate calculation of the activity of radionuclides after a given time interval.

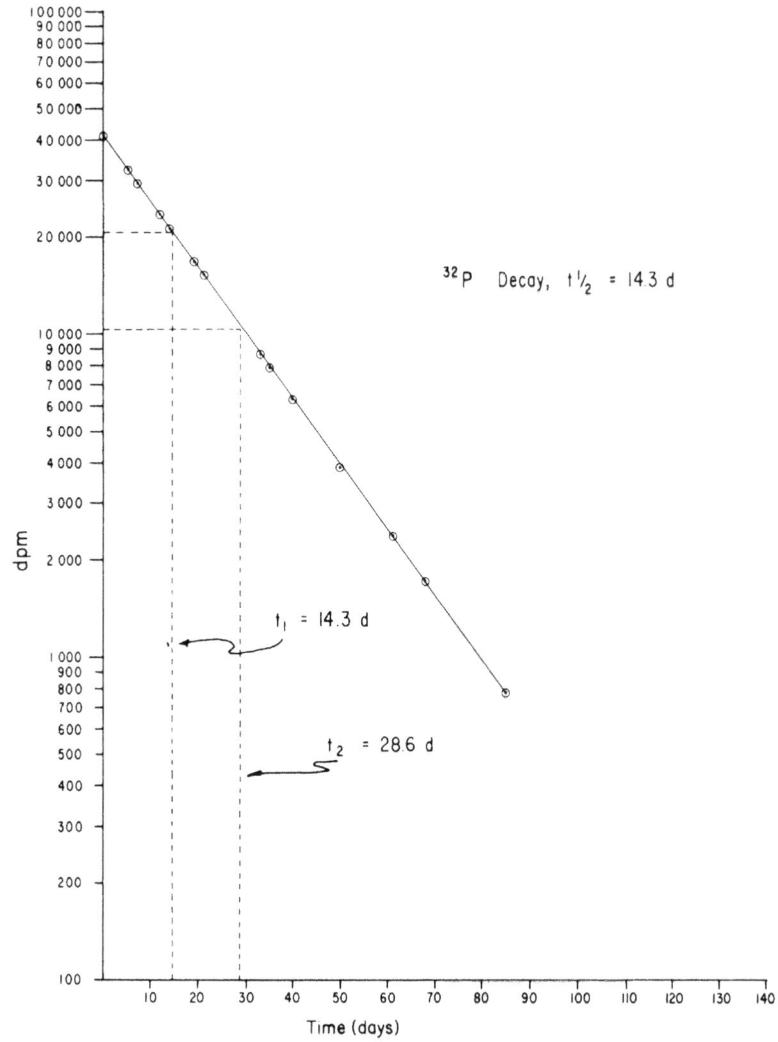

FIGURE 1.142 Semi-logarithmic plot of the decay of ^{32}P. Two half-lives (t_1 and t_2) are delineated by horizontal and vertical lines between the ordinate and abscissa. *From L'Annunziata, M. F. (1965), unpublished work of the writer.*

For radionuclide decay calculations, Eq. (1.543) may be transformed into the derivative as

$$dN/N = -\lambda dt, \quad (1.544)$$

which can be integrated between the limits N_0 and N and between t_0 and t, where t_0 is 0 (the moment observations are initiated), N_0 *is* the number of atoms originally present at time t_0, and N *is* the number of atoms remaining after time t:

$$\int_{N_0}^{N} dN/N = -\lambda \int_{t_0}^{t} dt \quad (1.545)$$

to give

$$\ln N/N_0 = -\lambda t \quad (1.546)$$

Eq. (1.546) may be written in exponential form as

$$N = N_0 e^{-\lambda t} \quad (1.547)$$

where e is the base of the natural logarithm, λ is the decay constant, and t is the interval of time. Eq. (1.547) is the form used to determine the decay of a radionuclide sample after a given time interval. To use Eq. (1.547), the value of the decay constant λ, must be known, and this is different for each radionuclide. To determine λ for a particular radionuclide, a relationship between the decay constant and the half-life may be derived from the decay Eq. (1.546), which may be transposed to

$$\ln N_0/N = \lambda t \quad (1.548)$$

By definition, we know that, after an interval of time corresponding to the half-life, half of the original activity remains. Therefore, we may assign the original activity N_0 as unity whereby after one-half life, where $t = t_{1/2}$, the remaining activity N would be one-half of unity, and Eq. (1.548) would become

$$\ln \frac{1}{1/2} = \lambda t_{1/2} \quad (1.549)$$

or

$$\ln 2 = \lambda t_{1/2} \quad (1.550)$$

and

$$0.693 = \lambda t_{1/2} \tag{1.551}$$

The decay constant can then be defined as

$$\lambda = 0.693/t_{1/2} \tag{1.552}$$

The value of λ can be calculated easily from the half-life of a radionuclide with Eq. (1.552). The units used for λ are expressed in reciprocal time, s^{-1}, m^{-1}, h^{-1}, d^{-1}, or y^{-1}, depending on the units used for the half-life of the radionuclide and also on the time interval t used in Eq. (1.552). For example, if ^{32}P, which has a half-life of 14.3 days, is used in an experiment, λ may be expressed in d^{-1}. The unit of the decay constant must agree with the time interval t of Eq. (1.547).

The following example illustrates the use of Eq. (1.547) to calculate the decay of a radionuclide sample within any time interval.

If a sample contained 3.7 MBq of ^{32}P on a given date and an investigator wished to determine the amount remaining after a 30-day period, he would first determine the decay constant for ^{32}P according to Eq. (1.552) and then calculate the activity after the specified time period using the decay Eq. (1.553) as follows. The decay constant in units of d^{-1} is determined by

$$\lambda = 0.693/t_{1/2} = 0.693/14.3\ \text{d} = 4.85 \times 10^{-2}\text{d}^{-1}$$

With the calculated value of λ and the known time interval t, the activity of the remaining ^{32}P is determined according to Eq. (1.547) expressed in terms of activity (A) or

$$A = A_0 e^{-\lambda t} \tag{1.553}$$

and

$$\begin{aligned} A &= 3.7 \times 10^6\ \text{dps}\left(e^{-\left[\left(4.85\times10^{-2}\text{day}^{-1}\right)30\ \text{days}\right]}\right) \\ &= 3.7 \times 10^6\ \text{dps}\left(e^{-1.455}\right) \\ &= 3.7 \times 10^6(0.2334) \\ &= 8.64 \times 10^5\ \text{dps} = 0.864\ \text{MBq} \end{aligned}$$

where A is the activity of the radionuclide in units of decay rate (e.g., dps) after time t, and A_0 is the initial activity at time t_0 where in this example, $A_0 = 3.7$ MBq $= 3.7 \times 10^6$ dps since by definition 1 MBq $= 1 \times 10^6$ dps. Thus, after the 30-day period, the activity of ^{32}P decayed from an initial activity of 3.7 to 0.864 MBq.

The decay equation has many practical applications, as it can be used also to calculate the time required for a given radionuclide sample to decay to a certain level of activity. Let us consider the following example:

A patient was administered intravenously 600 MBq of ^{99m}Tc-methylene diphosphate, which is a radiopharmaceutical administered for the purposes of carrying out a diagnostic bone scan. The doctor then wanted to know how much time would be required for the ^{99m}Tc radioactivity in the patient's body to be reduced to 0.6 MBq (0.1% of the original activity) from radionuclide decay alone ignoring any losses by bodily excretion. The half-life $t_{1/2}$ of ^{99m}Tc is 6.00 hours. To calculate the time required we can write Eq. (1.553) as

$$A/A_0 = e^{-\lambda t} \tag{1.554}$$

where A is the activity in dps (disintegrations per second) after time t and A_0 is the initial activity at time t_0. Eq. (1.554) can be transposed to

$$\ln A_0/A = \lambda t \tag{1.555}$$

or

$$t = \frac{1}{\lambda}\ln\frac{A_0}{A} \tag{1.556}$$

By definition (Eq. 1.552) the decay constant λ of ^{99m}Tc is $0.693/t_{1/2}$ or 0.693/6.00 h. Solving Eq. (1.556) after inserting the value of λ and the relevant activities of ^{99m}Tc gives

$$\begin{aligned} t &= (6.00\ \text{h} / 0.693)\ln(600\ \text{MBq} / 0.6\ \text{MBq}) \\ &= 59.8\ \text{hours} = 2.5\ \text{days} \end{aligned}$$

In the case of a mixture of independently decaying radionuclides, the rate of decay of each nuclide species does not change. However, the rate of decay of the overall sample is equal to the sum of the decay rates of the individual nuclide species. The cumulative decay of a mixture of independently decaying nuclides from the most simple case of a mixture of two nuclides to a more complex case of n number of nuclides is described by

$$N = N_1^0 e^{-\lambda t} + N_2^0 e^{-\lambda t} + \ldots N_n^0 e^{-\lambda t} \tag{1.557}$$

where N is the number of atoms remaining after time t, and N_1^0, N_2^0, and N_n^0 are the numbers of atoms originally present at time t_0 of 1, 2, and n number of nuclide species, respectively.

The semilogarithmic decay plot of a mixture of two independently decaying nuclides is not a straight line, contrary to pure radionuclide samples; rather, it is a composite plot, as in the case of a mixture of ^{32}P and ^{45}Ca (see Fig. 1.143). If the half-lives of the two nuclides are significantly different, the composite curve may be analyzed so that the decay curves of the two radionuclides may be determined. If the decay of the composite mixture can be observed over a reasonable period of time, the composite curve will eventually yield a straight line representing the decay of the longer-lived nuclide after the disappearance of the shorter-lived nuclide (depicted in Fig. 1.143). This straight line may be extrapolated to time $t = 0$ so that the activity (dpm) of this nuclide at $t = 0$ can be found. The difference between the activity at $t = 0$ of the longer-lived nuclide and the total activity of the sample at $t = 0$ gives the activity at $t = 0$ of the shorter-lived nuclide.

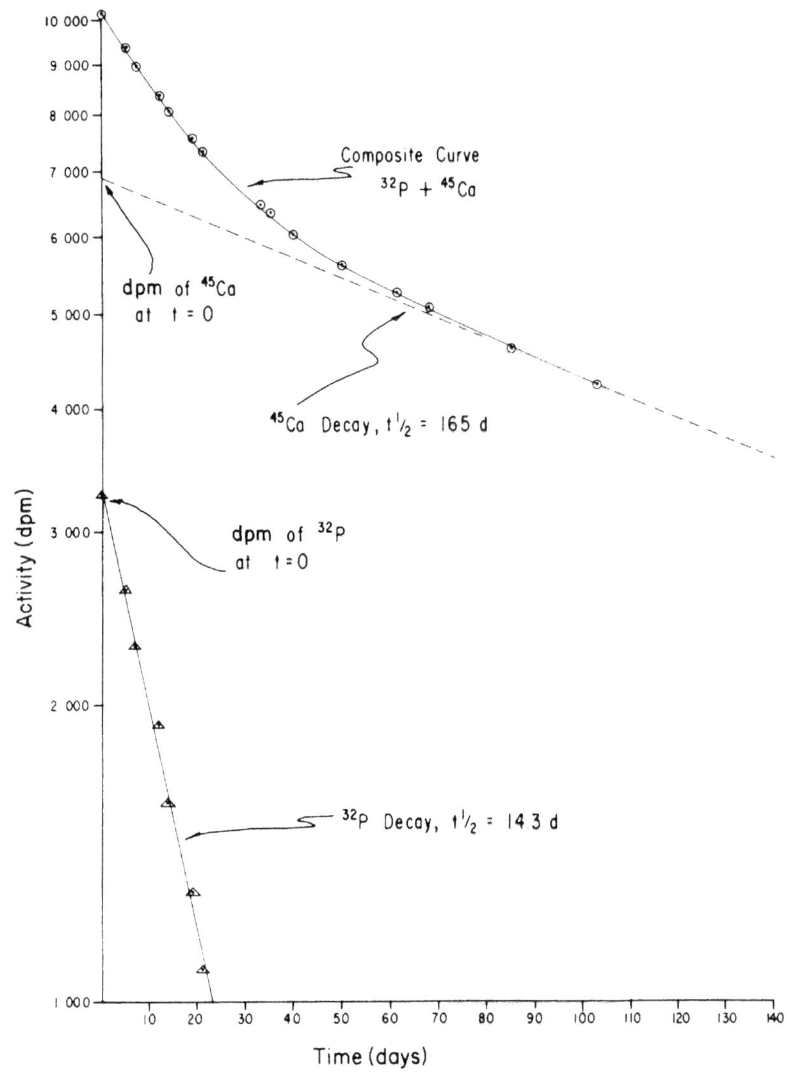

FIGURE 1.143 Semi-logarithmic decay curves of ^{32}P and ^{45}Ca isolated from a composite decay curve of a mixture of ^{32}P + ^{45}Ca. *From L'Annunziata, M. F. (1965), unpublished work of the writer.*

Likewise, further subtraction of points of the extrapolated decay curve from the composite curve yields the decay curve of the shorter-lived nuclide.

The half-lives of the two radionuclides are determined from the slopes of the two decay curves isolated from the composite curve. Eq. (1.548), which is expressed in natural logarithms, may be transformed to logarithms to the base 10 by

$$2.303 \log(N_1 / N_2) = \lambda(t_2 - t_1) \tag{1.558}$$

or

$$\log(N_1 / N_2) = \frac{\lambda}{2.303}(t_2 - t_1) \tag{1.559}$$

where N_1 and N_2 are the numbers of atoms or activity of the sample at times t_1 and t_2, respectively. As $\lambda/2.303$ of Eq. (1.559) is equal to the slope, the decay constant, λ, may be calculated from a graphical determination of the slope. With a calculated value of λ, the half-life of the nuclide is then calculated from Eq. (1.552).

Edwin M. McMillan (1907–91), who shared the Nobel Prize in Chemistry in 1951 with Glenn T. Seaborg (1912–99) for their discoveries in the chemistry of the transuranium isotopes, utilized the composite decay curve similar to that illustrated in Fig. 1.143 to determine graphically the half-lives of a mixture of ^{239}U ($t_{½} = 23.5$ minutes) and ^{239}Np ($t_{½} = 23.6$ days), which were prepared for the first time from the neutron capture following irradiation of ^{238}U with fast neutrons (McMillan, 1951). The detailed historical account and methods used by McMillan and Seaborg in their discovery of transuranium elemenys 94–103 and element 106 is provided by the writer in a previous text (L'Annunziata, 2016).

Many radionuclides have very long half-lives, which make the graphic representation of their decay impossible. Some examples are ^{3}H ($t_{1/2} = 12.3$ y), ^{14}C ($t_{1/2} = 5.73 \times 10^3$ y), ^{40}K ($t_{1/2} = 1.3 \times 10^9$ y), and ^{174}Hf ($t_{1/2} = 2 \times 10^{15}$ y) (see Appendix A). In such cases the half-lives can be calculated from Eqs. (1.543) and (1.552). The decay rate or activity, A, in disintegrations per year (DPY) of a given nuclide sample, defined by dN/dt of Eq. (1.543), is measured experimentally. The number of atoms of the radioassayed sample, defined by N of Eq. (1.543), must be known or determined. This is simple for pure samples. For example, the number of atoms of ^{40}K in a pure sample of KCl is easily calculated from Avogadro's number (6.022×10^{23} molecules/mol or atoms/mol) and the percentage natural abundance of ^{40}K (0.012%). Samples of unknown purity and isotopic abundance require a quantitative analysis of the element such as that provided by a mass spectral analysis of the isotopic abundance. The value of λ in y^{-1} is calculated as

$$\lambda = \frac{dN/dt}{N} = \frac{A}{N} = \frac{\text{CPM}/E}{N}(5.25 \times 10^5 \text{min / year}) \quad (1.560)$$

where A is the sample nuclide activity in DPY, N is the number of atoms of the nuclide in the sample, CPM is the sample count rate provided by the instrument radioactivity detector, E *is* the instrument counting efficiency, and 5.25×10^5 minutes/year is the factor used to convert counts per minute (CPM) to counts per year (CPY). The half-life can then be calculated according to Eqs. (1.552) and (1.560) both of which define the value of λ.

Let us look at a practical example of the use of the above equations to determine the half-life of ^{40}K taken from the work of Grau Malonda and Grau Carles (2002). The accurate determination of the half-life of ^{40}K has very practical implications, as it is currently used by geologists to date a rock's formation based on the measurement of the quantity of the stable daughter nuclide ^{40}Ar. Grau Malonda and Grau Carles (2002) report the accurate determination of the half-life of ^{40}K by measuring the activity of ^{40}K in a sample of pure KNO_3 and applying the relationships of half-life to λ according to Eqs. (1.552) and (1.560). They measured the ^{40}K specific activity in KNO_3 by the very accurate CIEMAT/NIST efficiency tracing liquid scintillation standardization method (see Chapter 6 of this volume) to be 12.24 ± 0.014 Bq/g. Also, applying the known isotopic concentration of ^{40}K in KNO_3 of 0.01167% and the value of Avogadro's number 6.022×10^{23} atoms per mole, they could calculate the number of atoms of ^{40}K in 1 g of KNO_3 as follows:

$$\left(\frac{6.022 \times 10^{23}\text{atoms K/mole KNO}_3}{(101.103\text{g KNO}_3/\text{mole KNO}_3)}\right) \times \left(\frac{0.01167 \text{ atoms } ^{40}\text{K}}{100 \text{ atoms K}}\right) = 6.951 \times 10^{17} \text{ atoms } ^{40}\text{K/g KNO}_3$$

From Eqs. (1.552) and (1.560), we can write

$$\frac{1}{\lambda} = \frac{t_{1/2}}{0.693} = \frac{N}{A} \quad (1.561)$$

or

$$t_{1/2} = 0.693\left(\frac{N}{A}\right) \quad (1.562)$$

From the specific radioactivity of ^{40}K in KNO_3 (*i.e.*, dps $^{40}K/g$ KNO_3) determined by the very accurate CIEMAT/NIST efficiency tracing liquid scintillation standardization method, and the calculated number of atoms of ^{40}K per gram of KNO_3, Grau Malonda and Grau Carles (2002) calculated the half-life of ^{40}K as

$$t_{1/2} = 0.693\left(\frac{6.951 \times 10^{17}\text{atoms } ^{40}\text{K/g KNO}_3}{(12.24\text{dps } ^{40}\text{K/g KNO}_3)(60\text{s/m})(5.25 \times 10^5\text{m/y})}\right)$$

and

$$t_{1/2} = 1.248 \times 10^9 \text{ y}$$

From the mean of nine determinations, Grau Malonda and Grau Carles (2002) were able to assign the value of the half-life ($t_{1/2}$) of ^{40}K to be $(1.248 \pm 0.004) \times 10^9$ y at a 95% confidence level.

Other radionuclides have very short half-lives such as ^{209}Ra ($t_{1/2} = 4.6$ seconds), ^{215}At ($t_{1/2} = 1.0 \times 10^{-4}$ seconds) and ^{212}Po ($t_{1/2} = 2.98 \times 10^{-7}$ seconds). Half-lives of such short duration can be determined by delayed-coincidence timing methods (Rudigier et al., 2017; Dias et al., 2014; Kojima et al., 2014; Suliman et al., 2013, 2012; D'Olympia et al., 2013; Takau et al. 2012; Jakubek et al., 2011; Kurtukian-Nieto et al., 2008; Morozov et al., 2006, 2002; 1998; Schwarzschild, 1963; Oms, 2004; Petkov et al., 2003; Ohm et al., 1990), which involve the use of multiple detectors (e.g., scintillation and semiconductor detectors) with detector response times as short as 10^{-11} seconds. The technique has also been applied to the measurement of the half-lives of excited nuclear states, such as the 8.7 and 98.1 ns half-lives of the 14.4 and 136.5 keV excited states of ^{57}Fe daughter nuclides, that emit gamma radiation, following the decay of ^{57}Co (Morozov et al., 2006, 2002). Radiation detectors with resolving times of fractions of a microsecond are set electronically so that a delay circuit will detect a radiation-induced pulse from the parent in coincidence with a radiation pulse produced from the daughter. Varying the delay time of the coincidence circuit results in a delay of the coincidence pulse rate from which a decay curve of the very short-lived daughter nuclide can be plotted and the half-life determined.

B. General decay equations

The simplest decay relationship between parent and daughter nuclides that can be considered is that of a parent nuclide which decays to form a stable daughter nuclide.

The decay of the radionuclide ^{33}P serves as an example. The parent nuclide ^{33}P decays with a half-life of 25 days with the production of the stable daughter ^{33}S, as indicated by

$$^{33}_{15}\mathrm{P} \rightarrow ^{33}_{16}\mathrm{S(stable)} + \beta^- + \bar{\nu} \quad (1.563)$$

Numerous radionuclides, such as ^{3}H, ^{14}C, ^{32}P, ^{35}S, ^{36}Cl, ^{45}Ca, and ^{131}I (see Appendix A), decay by this simple parent-daughter relationship.

However, numerous other radionuclides produce unstable daughter nuclides. The simplest case would be that in which the parent nuclide A decays to a daughter nuclide B, which in turn decays to a stable nuclide C:

$$\mathrm{A} \rightarrow \mathrm{B} \rightarrow \mathrm{C(stable)} \quad (1.564)$$

In such decay chains, the rate of decay and production of the daughter must be considered as well as the rate of decay of the parent. The decay of the parent is described by the simple rate equation

$$-dN_\mathrm{A}/dt = \lambda_\mathrm{A} N_\mathrm{A} \quad (1.565)$$

which is integrated to the form

$$N_\mathrm{A} = N_\mathrm{A}^0 e^{-\lambda_\mathrm{A} t} \quad (1.566)$$

where N_A^0 is the number of atoms of the parent at the time $t = 0$ and N_A is the number of atoms after a given period of time $t = t_1$.

The decay rate of the daughter is dependent on its own decay rate as well as the rate at which it is formed by the parent. It is written as

$$-dN_\mathrm{B}/dt = \lambda_\mathrm{B} N_\mathrm{B} - \lambda_\mathrm{A} N_\mathrm{A} \quad (1.567)$$

where $\lambda_\mathrm{B} N_\mathrm{B}$ is the rate of decay of the daughter alone and $\lambda_\mathrm{A} N_\mathrm{A}$ is the rate of decay of the parent or rate of formation of the daughter. Eqs. (1.566) and (1.567) may be transposed into the linear differential equation

$$dN_\mathrm{B}/dt + \lambda_\mathrm{B} N_\mathrm{B} - \lambda_\mathrm{A} N_\mathrm{A}^0 e^{-\lambda_\mathrm{A} t} = 0 \quad (1.568)$$

which is solved for the number of atoms of daughter, N_B , as a function of time to give

$$N_\mathrm{B} = \frac{\lambda_\mathrm{A}}{\lambda_\mathrm{B} - \lambda_\mathrm{A}} N_\mathrm{A}^0 \left(e^{-\lambda_\mathrm{A} t} - e^{-\lambda_\mathrm{B} t}\right) + N_\mathrm{B}^0 e^{-\lambda_\mathrm{B} t} \quad (1.569)$$

Although unnecessary in this treatment, the solution to Eq. (1.568) is given by Friedlander et al. (1964).

In decay schemes of this type, the following three conditions may predominate: (1) secular equilibrium, (2) transient equilibrium, and (3) the state of no equilibrium. Each of these cases will now be considered in detail.

C. Secular equilibrium

The phenomenon of radioisotope decay equilibrium was first observed by Ernest Rutherford and Frederick Soddy in 1902, which they reported in their classic paper on "The Cause and Nature of Radioactivity". They reported their observations as follows:

Radioactivity is shown to be accompanied by chemical changes in which new types of matter are being continuously produced. These reaction products are at first radioactive, the activity diminishing regularly from the moment of formation. Their continuous production maintains the radioactivity of the matter producing them at a definite equilibrium value.

Secular equilibrium is a steady-state condition of equal activities between a long-lived parent radionuclide and its short-lived daughter. The important criterion upon which secular equilibrium depends is the following: The parent must be long-lived compared to the life of the daughter nuclide; that is, the half-life of the daughter must be negligible compared to that of its parent (Evans, 1955). In other words, the activity of the parent does not decrease measurably during many half-lives of the daughter (Friedlander et al., 1964).that is,

$$\lambda_\mathrm{A} \ll \lambda_\mathrm{B} \quad (1.570)$$

where λ_A and λ_B are the respective decay constants of the parent and daughter nuclides. The importance of this criterion can be clearly seen, if the ^{90}Sr(^{90}Y) equilibrium is taken as an example.

The nuclide ^{90}Sr is the parent in the decay scheme

$$^{90}_{38}\mathrm{Sr} \xrightarrow{t_{1/2}=28.8\ \mathrm{y}} {}^{90}_{39}\mathrm{Y} \xrightarrow{t_{1/2}=2.7\ \mathrm{d}} {}^{90}_{40}\mathrm{Zr\ (stable)} \quad (1.571)$$

The long half-life of ^{90}Sr definitely satisfies the criterion for secular equilibrium, because over a quarter of a century is needed for it to lose 50% of its original activity. As will be seen, less than 3 weeks are required for secular equilibrium to be attained and, in this interim period, negligible decay of ^{90}Sr occurs. Also, the decay constants for ^{90}Sr and ^{90}Y, λ_A and λ_B, respectively, may be compared. The decay constants for ^{90}Sr and ^{90}Y are easily calculated from their half-lives, and the values are $6.60 \times 10^{-5}\ \mathrm{d}^{-1}$ and $2.57 \times 10^{-1}\ \mathrm{d}^{-1}$, respectively. In the comparison $\lambda_\mathrm{A}/\lambda_\mathrm{B} = 2.57 \times 10^{-4}$, and this is in agreement with the order of magnitude required for secular equilibrium where $\lambda_\mathrm{A} \ll \lambda_\mathrm{B}$.

An equation for the growth of daughter atoms from the parent can be obtained from Eq. (1.569) by consideration of the limiting requirements for secular equilibrium. Since $\lambda_\mathrm{A} \approx 0$ and $\lambda_\mathrm{A} \ll \lambda_\mathrm{B}$, $e^{-\lambda_\mathrm{A} t} = 1$ and λ_A falls out of the denominator in the first term of Eq. (1.569). If the daughter nuclide is separated physically from the parent (L'Annunziata, 1971), $N_B^0 = 0$ at time $t = 0$ (time of parent-daughter separation) and the last term would fall out of Eq. (1.569). Thus, in the case of secular equilibrium, the expression of the ingrowth of daughter atoms with parent can be written as

$$N_B = \frac{\lambda_A N_A^0}{\lambda_B}\left(1 - e^{-\lambda_B t}\right) \quad (1.572)$$

If the observation of the ingrowth of the daughter is made over many half-lives of the daughter, it is seen that the number of atoms of daughter approaches a maximum value $\lambda_A N_A^0/\lambda_B$, which is the rate of production of daughter divided by its decay constant. The final form of Eq. (1.572) to be used for the calculation of the ingrowth of daughter can be expressed as

$$N_B = (N_B)_{max}\left(1 - e^{-\lambda_B t}\right) \quad (1.573)$$

Since the activity of the daughter atoms, A_B, is proportional to the number of daughter atoms, or $A_B = k\lambda_B N_B$, where k is the coefficient of detection of the daughter atoms, Eq. (1.573) may also be written as

$$A_B = (A_B)_{max}\left(1 - e^{-\lambda_B t}\right) \quad (1.574)$$

Rutherford and Soddy (1902) were the first to write and interpret Eqs. (1.573) and (1.574) when they studied the equilibrium existing between radioactive thorium and a daughter radionuclide. They noted the following [with the author's comments in brackets]:

The radioactivity of thorium at any time is the resultant of two opposing processes: 1. The production of fresh radioactive material at a constant rate by the thorium compound, and 2. The decay of the radiating power of the active material with time. The normal or constant radioactivity possessed by thorium is an equilibrium value, where the rate of increase of radioactivity due to the production of fresh active material [daughter nuclide] is balanced by the rate of decay of radioactivity of that already formed … The experimental curve obtained with the hydroxide [This was the chemical form they used to separate the parent nuclide from the daughter.] for the rate of rise of its activity from a minimum to a maximum value will therefore be approximately expressed by the equation $I_t/I_0 = 1 - e^{-\lambda t}$, where I_0 represents the amount of activity recovered when the maximum is reached, and I_t the activity recovered after time t, λ being the same constant as before.

(Notice the similarity of Rutherford and Soddy's equation, which may be transposed to read $I_t = (I_0)\left(1 - e^{-\lambda t}\right)$ and Eqs. (1.573) and (1.574).)

Let us take an arbitrary example of equal activities of 100 dpm of parent ^{90}Sr and 100 dpm of daughter ^{90}Y, in secular equilibrium. From these activities we can calculate and graphically represent the ingrowth of ^{90}Y with its parent and also the decay of ^{90}Y subsequent to the separation of parent and daughter nuclides (L'Annunziata, 1971). Identical activities of ^{90}Sr and ^{90}Y are arbitrarily chosen, because their activities are equal while in secular equilibrium prior to their separation. Fig. 1.144 illustrates the calculated growth of ^{90}Y (curve B) as produced by ^{90}Sr using Eq. (1.574) with $(A_B)_{max} = 100$. The decay of separated ^{90}Y (curve A) is plotted by simple half-life decay ($t_{1/2} = 2.7$ d). The dashed line (line C) represents the decay of ^{90}Sr, which is negligible ($t_{1/2} = 28.8$ y) during the period of observation. The total activity (curve D) is the result of both ^{90}Sr decay and the ingrowth of ^{90}Y after the separation of the latter, and it is obtained by the addition of curve B to line C. It may be noted from Fig. 1.144, that after approximately six half-lives of ^{90}Y ($\sim$18 d), the activity of ^{90}Y has increased to essentially the same activity of ^{90}Sr, after which both nuclides decay with the same half-life, that of the parent ^{90}Sr (28.8 y).

Ernest Rutherford and Frederick Soddy in their classic paper on "The Cause and Nature of Radioactivity" in 1902 in the *Philosophical Magazine* made the first published observation of what is now known as secular equilibrium. They were studying the radioactivity of an isotope of thorium and a daughter isotope, which they called ThX. The term "daughter isotope" for an isotope decay product was not then established, because knowledge of radioactivity was only then at its infancy. They summarized their findings with the following:

The foregoing experimental results may be briefly summarized. The major part of the radioactivity of thorium – ordinarily about 54 percent – is due to a non-thorium type of matter, ThX, possessing distinct chemical properties, which is temporarily radioactive, its activity falling to half value in about four days. The constant radioactivity of thorium is maintained by the production of this material at a constant rate. Both the rate of production of the new material and the rate of decay of its activity appear to be independent of the physical and chemical condition of the system.

(We now know that the activity of parent and daughter nuclides are equal in secular equilibrium and the value of "about 54%" reported by Rutherford and Soddy was precisely 50%.)

As an example of the practical utility of this phenomenon, the application of secular equilibrium theory to the analysis of ^{90}Sr in the environment is discussed.

One method reported by the Los Alamos National Laboratory (see Gautier, 1995) entails the initial chelation (complex formation) of the sample strontium with the sodium salt of ethylenediaminetetraacetic acid (EDTA). The complexed strontium is then isolated by elution on an ion exchange column. The eluted strontium is then precipitated as a carbonate. The activity of radioactive strontium, which will include ^{89}Sr + ^{90}Sr in the sample, is determined by low-background counting. Low-background liquid scintillation counting is most often used for the total ^{89}Sr + ^{90}Sr analysis, as described by Passo and Cook (1994), Salonen et al. (2012) and Chapter 2 in Volume two of this book. The

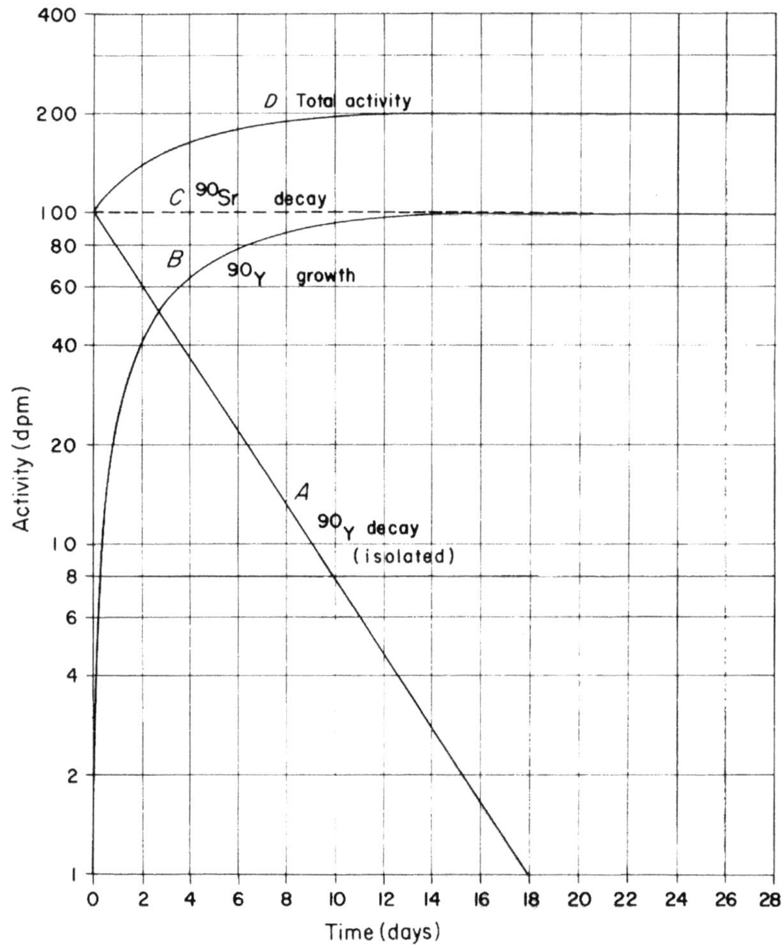

FIGURE 1.144 Growth and decay curves following the separation of ^{90}Sr(^{90}Y) in secular equilibrium. (A) Decay of isolated ^{90}Y. (B) Ingrowth of ^{90}Y with ^{90}Sr. (C) Decay of isolated ^{90}Sr. (D) Total activity from isolated ^{90}Sr, representing both ^{90}Sr decay and ^{90}Y growth until secular equilibrium is attained. *From L'Annunziata (1971); reprinted with permission © 1971 American Chemical Society.*

isolated radiostrontium is then allowed to remain in the sample without further treatment for a period of about 2 weeks to allow ingrowth of ^{90}Y. About 2 weeks are needed to ensure the parent and daughter radionuclides are in secular equilibrium before the chemical separation of yttrium from strontium. From Eq. (1.574) it is calculated, that after 2 weeks, the activity of ^{90}Y grows to 97.4% of its original level. Carrier yttrium is then added to the dissolved radiostrontium, and the yttrium is precipitated as the hydroxide, redissolved, and reprecipitated as an oxalate (see Section XVIII.C of this chapter for a discussion of the concepts of carrier and carrier-free radionuclides). The step involving the precipitation of yttrium from the sample results in the separation of ^{90}Y from the radiostrontium. The separated ^{90}Y can then be assayed by suitable low-background counting using liquid scintillation or Cherenkov counting (Passo and Cook, 1994; L'Annunziata and Passo, 2002). The ^{90}Sr activity in the sample is determined from the activity of ^{90}Y by calculating the ^{90}Y decay from the time of separation (precipitation) of yttrium from strontium. This is possible because the parent and daughter radionuclides were at secular equilibrium (i.e., ^{90}Sr dpm = ^{90}Y dpm) at time $t = t_0$ when the precipitation and separation of yttrium from strontium were carried out. The ^{89}Sr activity in the sample is determined from the difference between the total radiostrontium activity (^{89}Sr + ^{90}Sr) and the measured activity of ^{90}Sr. A review of radioanalytical methods for the analysis of radioactive isotopes of strontium including ^{90}Sr, which involves the concept of secular equilibrium, is provided by Vajda and Kim (2010) and Chapter 2 in Volume two of this book.

D. Transient equilibrium

Like secular equilibrium, transient equilibrium is a steady-state condition between the parent and daughter nuclides. However, in transient equilibrium the parent-daughter nuclides do not possess the same activities, but rather they decay at the same half-life, that of the parent nuclide.

The criterion upon which transient equilibrium rests is that the parent nuclide must be longer lived than its daughter, but not of the order of magnitude described by Eq. (1.570); that is, it is necessary that $\lambda_A < \lambda_B$; and, contrary to secular equilibrium, in transient equilibrium, the decay (i.e., half-life) of the daughter nuclide is not negligible compared to that of the parent. In transient equilibrium the ratio of half-life of the parent over the difference of half-lives of the parent (t_A) and daughter (t_B) should be clearly greater than unity (Evans, 1955) or

$$\frac{t_A}{t_A - t_B} > 1 \tag{1.575}$$

The decay chain of ^{100}Pd serves as an example of parent-daughter nuclides that may attain transient equilibrium. ^{100}Pd decays by electron capture to ^{100}Rh with a half-life of 96 hours. The daughter nuclide ^{100}Rh decays by electron capture and positron emission to the stable nuclide ^{100}Ru. The half-life of the daughter nuclide is 21 hours. The decay scheme may be represented as

$$^{100}_{46}\text{Pd} \xrightarrow{t_{1/2}=96\text{ h}} {}^{100}_{45}\text{Rh} \xrightarrow{t_{1/2}=21\text{ h}} {}^{100}_{44}\text{Ru (stable)} \tag{1.576}$$

The half-life of the parent nuclide is greater than that of the daughter, and the ratio

$$\frac{t_A}{t_A - t_B} = \frac{96\text{ h}}{96\text{ h} - 21\text{ h}} = 1.28 \tag{1.577}$$

is clearly greater than unity. Also, the value of λ_A, given by 0.693/96 hours, is $7.2 \times 10^{-3}\text{ h}^{-1}$, and that of λ_B, given by 0.693/21 hours, is $3.3 \times 10^{-2}\text{ h}^{-1}$ whereby $\lambda_A < \lambda_B$.

If the general decay Eq. (1.569) of the daughter nuclide is considered, the term $e^{-\lambda_B t}$ is negligible compared with $e^{-\lambda_A t}$ for sufficiently large values of t. Thus, the terms $e^{-\lambda_B t}$ and $N_B^0\, e^{-\lambda_B t}$ may be dropped from Eq. (1.569) to give

$$N_B = \frac{\lambda_A}{\lambda_B - \lambda_A}\left(N_A^0 e^{-\lambda_A t}\right) \tag{1.578}$$

for the ingrowth of the daughter nuclide as a function of time. Because $N_A = N_A^0 e^{-\lambda_A t}$, Eq. (1.578) may be written as

$$\frac{N_B}{N_A} = \frac{\lambda_A}{\lambda_B - \lambda_A} \tag{1.579}$$

From Eq. (1.579), it can be seen that the ratio of the number of atoms or the ratio of the activities of the parent and daughter nuclides is a constant in the case of transient equilibrium.

Since $A_A = k_A\lambda_A N_A$ and $A_B = k_B\lambda_B N_B$, where A_A and A_B are the activities of the parent and daughter nuclides, respectively, and k_A and k_B are the detection coefficients of these nuclides, Eq. (1.579) may be written in terms of activities as

$$\frac{A_B}{k_B\lambda_B}(\lambda_B - \lambda_A) = \frac{A_A}{k_A\lambda_A}\lambda_A \tag{1.580}$$

or

$$\frac{A_B}{A_A} = \frac{k_B\lambda_B}{k_A(\lambda_B - \lambda_A)} \tag{1.581}$$

If equal detection coefficients are assumed for the parent and daughter nuclides, Eq. (1.581) may be written as

$$\frac{A_B}{A_A} = \frac{\lambda_B}{(\lambda_B - \lambda_A)} \tag{1.582}$$

Thus, for transient equilibrium Eq. (1.582) indicates that the activity of the daughter is always greater than that of the parent by the factor $\lambda_B/(\lambda_B - \lambda_A)$. Eq. (1.582) may likewise be written as

$$\frac{A_A}{A_B} = 1 - \frac{\lambda_A}{\lambda_B} \tag{1.583}$$

whereby the ratio A_A/A_B falls within the limits $0 < A_A/A_B < 1$ in transient equilibrium.

If an activity of 100 dpm is arbitrarily chosen for the daughter nuclide ^{100}Rh in transient equilibrium with its parent ^{100}Pd, the activity of ^{100}Pd can be found using either Eq. 1.582 or 1.583. Eq. (1.582) gives

$$\frac{100\text{ dpm}}{A_A} = \frac{3.3\times10^{-2}\text{ h}^{-1}}{3.3\times10^{-2}\text{ h}^{-1} - 7.2\times10^{-3}\text{ h}^{-1}}$$

or

$$A_A = 78\text{ dpm}$$

With the use of Eq. (1.582) or 1.583, the decay of the daughter nuclide may be calculated as a function of parent decay in transient equilibrium. The ^{100}Pd$-^{100}$Rh parent-daughter decay in transient equilibrium is illustrated by curves A and B, respectively, of Fig. 1.145. The parent and daughter nuclides are shown to have respective activities of 78 dpm and 100 dpm at time $t = 0$. As curves A and B show, the parent and daughter nuclides in transient equilibrium decay with the same half-life, that corresponding to the half-life of the parent.

If the parent and daughter nuclides were to be separated, the daughter nuclide would decay according to its half-life as indicated by curve C. The isolated parent nuclide would, however, show an increase in activity with time owing to the ingrowth of daughter until transient equilibrium is attained. Curve D of Fig. 1.145 shows the ingrowth of daughter nuclide from a freshly isolated parent. Because $N_B^0 = 0$ at time $t = 0$ (time of separation of parent and daughter), the last term of Eq. (1.569) falls out to give

$$N_B = \frac{\lambda_A N_A^0}{\lambda_B - \lambda_A}\left(e^{-\lambda_A t} - e^{-\lambda_B t}\right) \tag{1.584}$$

The term $\lambda_A N_A^0/(\lambda_B - \lambda_A)$ approaches a maximum value, which is the rate of production of the daughter divided by the

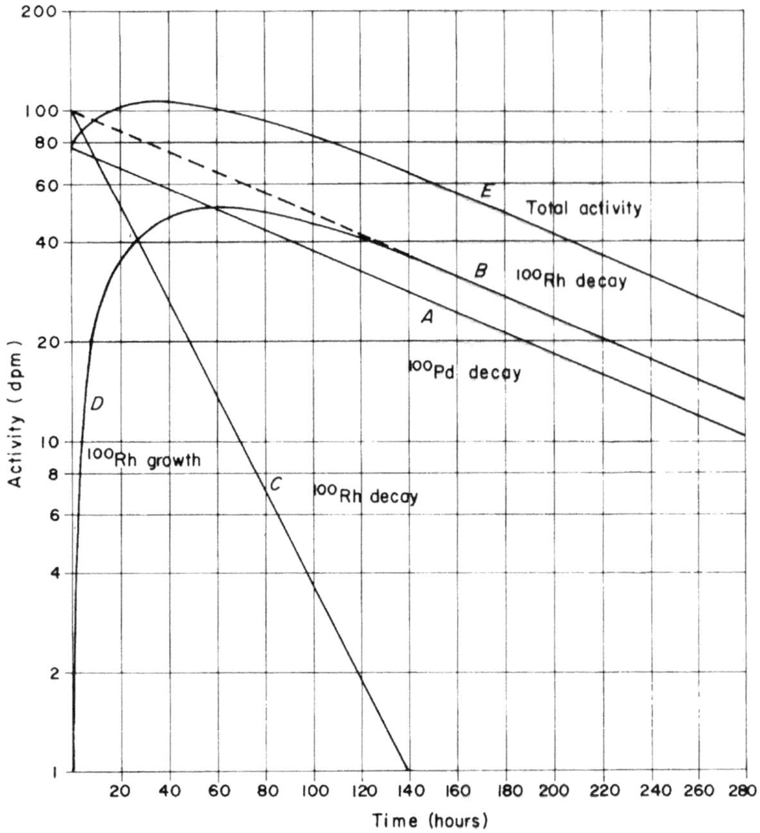

FIGURE 1.145 Growth and decay curves following the separation of ^{100}Pd(^{100}Rh) in transient equilibrium. (A) Decay of isolated parent nuclide ^{100}Pd. (B) Decay of ^{100}Rh daughter nuclide in transient equilibrium. The dashed portion of this curve represents ^{100}Rh decay if parent and daughter nuclides were not separated. (C) Decay of ^{100}Rh after separation from its parent. (D) The ingrowth of ^{100}Rh with the isolated parent ^{100}Pd. (E) Total activity from the isolated ^{100}Pd representing both ^{100}Pd decay and ^{100}Rh growth until transient equilibrium is attained. *From L'Annunziata (1987), original work of the writer.*

difference between the daughter and parent decay constants, and Eq. (1.548) may be written as

$$N_B = (N_B)_{max}\left(e^{-\lambda_A t} - e^{-\lambda_B t}\right) \quad (1.585)$$

similar to the case of Eq. (1.573). Since the activity, A_B, of the daughter atoms is proportional to the number of daughter atoms, or $A_B = k_B\lambda_B N_B$, where k is as defined previously, Eq. (1.585) may also be written as

$$A_B = (A_B)_{max}\left(e^{-\lambda_A t} - e^{-\lambda_B t}\right) \quad (1.586)$$

Because the maximum daughter activity in this sample is 100 dpm, Eq. (1.586) may be used to calculate the ingrowth of daughter nuclide with $(A_B)_{max} = 100$.

Curve E of Fig. 1.145 illustrates the total activity of the isolated parent and daughter nuclides. It is found by summing curves A and D and consequently accounts for the simultaneous decay of the parent nuclide and the ingrowth of the daughter. Notice that the slopes of curves A, B, and E are identical when transient equilibrium is attained, that is, the rates of decay of both the parent and daughter are identical.

Additional practical examples of decay and ingrowth calculations may be obtained from Harms and Jerome (2004).

E. No equilibrium

The cases of secular equilibrium and transient equilibrium, which involve decay schemes whereby the parent nuclide is longer lived than its daughter, were just considered. In other cases in which the daughter nuclide is longer lived than its parent, $\lambda_A > \lambda_B$, no equilibrium is attained. Instead, the parent nuclide of shorter half-life eventually decays to a negligible extent, leaving only the daughter nuclide, which decays by its own half-life. The following decay scheme of ^{56}Ni serves as an example:

$$^{56}_{28}\text{Ni} \xrightarrow{t_{1/2}=6.4\text{ d}} {}^{56}_{27}\text{Co} \xrightarrow{t_{1/2}=77.3\text{ d}} {}^{56}_{26}\text{Fe (stable)} \quad (1.587)$$

The parent nuclide ^{56}Ni decays by EC with a half-life of 6.4 d, whereas its daughter ^{56}Co decays with the longer half-life of 77.3 d by EC and β^+ emission. Curve A of Fig. 1.146 illustrates the decay of the ^{56}Ni parent nuclide. The decay of isolated ^{56}Ni is followed by the ingrowth (production) of the ^{56}Co daughter nuclide, shown by curve B. The ingrowth of daughter is calculated from Eq. (1.569) described earlier, of which the last term, $N_B^0 e^{-\lambda_B t}$, falls out because $N_B^0 = 0$ at time $t = 0$. The number of daughter atoms N_B of Eq. (1.569) may be converted to activity, A_B, by the term $A_B = k_B\lambda_B N_B$ as discussed previously. The total activity illustrated by curve C of Fig. 1.146 depicts

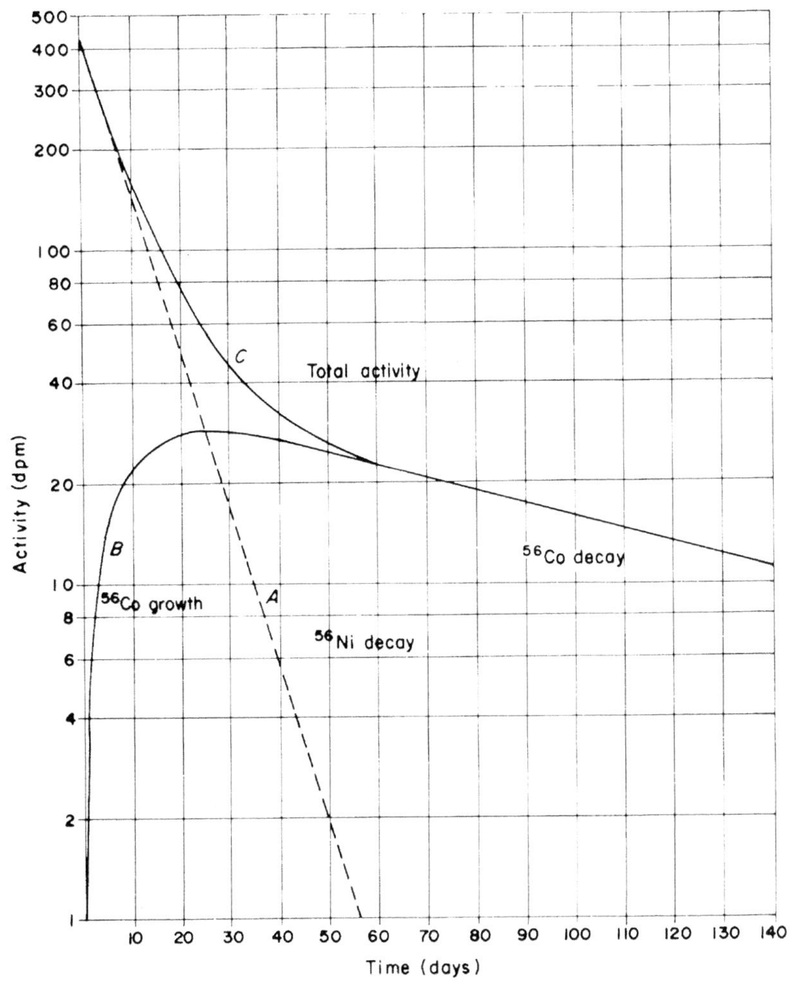

FIGURE 1.146 Growth and decay curves of the ^{56}Ni(^{56}Co) parent-daughter nuclides following the isolation or fresh preparation of the parent nuclide ^{56}Ni. (A) Decay of the parent nuclide ^{56}Ni. (B) Ingrowth of daughter nuclide ^{56}Co. (C) Total activity representing both ^{56}Ni decay and the simultaneous growth and decay of ^{56}Co daughter. *From L'Annunziata (1987), original work of the writer.*

both the simultaneous decay of parent nuclide and the ingrowth and decay of daughter determined by summing curves A and B. Notice from Fig. 1.146, that the parent nuclide activity in this example becomes negligible after around 55 d, after which the total activity, curve C, has a slope corresponding to the decay rate of the daughter nuclide.

F. More complex decay schemes

Other decay schemes exist that involve a chain of numerous nuclides such as

$$A \rightarrow B \rightarrow C \rightarrow \ldots N \tag{1.588}$$

where nuclides A, B, and C are followed by a chain of a number N of decaying nuclides. Long decay chains of this type may be observed in the complex decay schemes of high-atomic-number natural radionuclides such as the ^{232}Th, ^{238}U, and ^{235}U natural decay chains illustrated previously in Figs. 1.8–1.10. These decay schemes are described by the general Eq. (1.588). However, within each of the natural decay schemes, illustrated in Figs. 1.8–1.10, there are one or more branching decays of the type

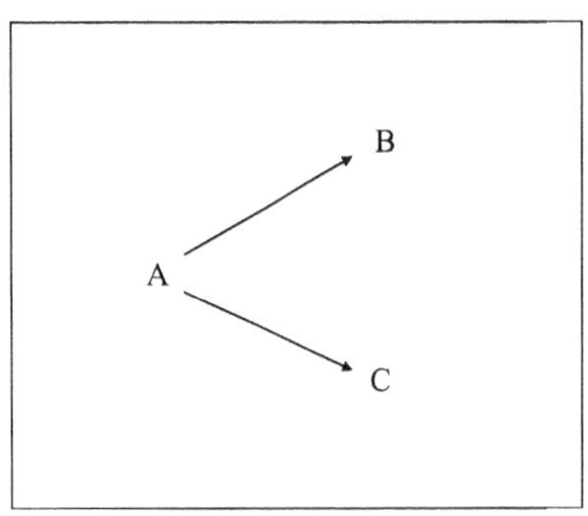

For example, in Fig. 1.8, $^{212}_{83}$Bi, within the ^{232}Th natural decay chain, is the parent of the two daughter nuclides $^{212}_{84}$Po and $^{208}_{81}$Tl. The half-life of the parent nuclide in branching decays is a function of the two decay processes and may be written as

$$t_{1/2} = 0.693/(\lambda_A + \lambda_B) \tag{1.589}$$

where λ_A and λ_B are the decay constants of the two separate decay processes.

XVIII. Radioactivity units and radionuclide mass

A. Units of radioactivity

The units used to define the level of radioactivity are written in terms of the number of atoms, N, disintegrating per unit of time, t. We can use Eq. (1.542), previously discussed in this chapter, to calculate the activity of any given mass of radionuclide. The equation, namely $\Delta N/\Delta t = \lambda N$, defines the proportionality between the rate of decay of a radionuclide and the number of atoms of the radionuclide in a sample. As an example, we may use Eq. (1.542) to calculate the activity of 1 g of ^{226}Ra as follows:

$$\Delta N/\Delta t = \lambda N \tag{1.590}$$

and

$$\Delta N/\Delta t = (0.693 / t_{1/2})(N) \tag{1.591}$$

where $\lambda = 0.693/t_{1/2}$ was derived previously. If we take the half-life, $t_{1/2}$, of ^{226}Ra to be 1600 years and substitute for N, in the preceding equation, the number of atoms per mol of ^{226}Ra, we can write

$$\Delta N/\Delta t = (0.693 / 1600 \text{ y})(6.022 \times 10^{23} \text{ atoms} / 226 \text{ g}) \tag{1.592}$$

where, according to Avogadro's number, there are 6.022×10^{23} atoms per gram mole of substance. If we now convert the half-life of ^{226}Ra from units of years to minutes, we can calculate the number of atoms of ^{226}Ra disintegrating per minute (dpm) per gram according to

$$\Delta N/\Delta t = \left[\frac{0.693}{(1600 \text{ y})(365 \text{ d/y})(24 \text{ h/d})(60 \text{ min/h})}\right] \times \left(\frac{6.022 \times 10^{23} \text{ atoms}}{226 \text{ g}}\right) \tag{1.593}$$

$$\Delta N/\Delta t = \left(\frac{0.693}{8.41 \times 10^8 \text{ min}}\right)(2.665 \times 10^{21} \text{ atoms/g})$$

$$= 2.19 \times 10^{12} \text{ atoms per minute per gram}$$

The activity of 1 g of ^{226}Ra is the basis of the unit of radioactivity known as the curie (Ci). One curie is almost equal to the activity of 1 g of ^{226}Ra or, by definition,

$$1 \text{ Ci} = 2.22 \times 10^{12} \text{ dpm} = 3.7 \times 10^{10} \text{ dps} \tag{1.594}$$

Therefore, 1 curie of activity or any multiple of the curie of any radionuclide defines the number of atoms disintegrating per unit of time in minutes or seconds.

The rate of decay in terms of time in seconds gives rise to the Système International d'Unités (SI) unit of activity, which is the Becquerel (Bq), where by definition

$$1 \text{ Bq} = 1 \text{ dps} \tag{1.595}$$

Therefore, we can interrelate the curie and Becquerel as follows:

$$1 \text{ Ci} = 2.22 \times 10^{12} \text{ dpm} = 3.7 \times 10^{10} \text{ dps} = 37 \text{ GBq} \tag{1.596}$$

Likewise, smaller units of the curie, namely the millicurie (mCi) and microcurie (μCi), may be interrelated with the becquerel as follows:

$$1 \text{ mCi} = 2.22 \times 10^{9} \text{ dpm} = 3.7 \times 10^{7} \text{ dps} = 37 \text{ MBq} \tag{1.597}$$

and

$$1 \text{ μCi} = 2.22 \times 10^{6} \text{ dpm} = 3.7 \times 10^{4} \text{ dps} = 37 \text{ kBq} \tag{1.598}$$

Another unit of activity recommended in the early 1960s by the International Union of Pure and Applied Physics, but less frequently used, is the rutherford, where 1 rutherford = 10^6 dps and 1 microrutherford would be equivalent to 1 dps or 1 Bq (Buttlar, 1968; Das and Ferbel, 1994).

B. Correlation of radioactivity and radionuclide mass

From Eq. (1.591) and calculations made in the previous Section XVIII.A, we can see that, for samples of a given level of activity, radionuclides of shorter half-life will contain a smaller number of radioactive atoms than radionuclides of longer half-life.

We can use Eq. (1.591) again to compare two radionuclides of relatively short and long half-lives to see the magnitude of the differences in radionuclide masses we would encounter for any given level of radioactivity. For example, we may take the radionuclide ^{32}P of 14.3-day half-life and the radionuclide ^{14}C of 5730-year half-life and calculate the activity per gram (*e.g.*, dpm/g) and grams per curie (*e.g.*, g/Ci) of each radionuclide for comparative purposes. These calculations are as follows.

1. ^{32}P, half-life = 14.3 days:

$$\Delta N/\Delta t = (0.693/t_{1/2})(N)$$

$$= \left[\frac{0.693}{(14.3 \text{ d})(24 \text{ h/d})(60 \text{ m/h})}\right]\left(\frac{6.022 \times 10^{23}}{32 \text{ g}}\right)$$

$$= 6.32 \times 10^{17} \text{ dpm/g}^{32}\text{P} \tag{1.599}$$

If, by definition, 1 curie $= 2.22 \times 10^{12}$ dpm, we can convert the above calculated activity per gram of ^{32}P to grams ^{32}P per curie as follows:

$$\frac{2.22 \times 10^{12}\ \text{dpm/Ci}}{6.32 \times 10^{17}\ \text{dpm/g}\ ^{32}\text{P}} = 3.51 \times 10^{-6}\ \text{g}\ ^{32}\text{P per Ci}$$

$$= 3.51 \times 10^{-6}\ \text{mg}\ ^{32}\text{P per mCi} \tag{1.600}$$

2. ^{14}C, half-life $= 5730$ years:

$$\Delta N/\Delta t = (0.693/t_{1/2})(N)$$

$$= \left[\frac{0.693}{(5730\ \text{y})(365\ \text{d/y})(24\ \text{h/d})(60\ \text{m/h})}\right]\left(\frac{6.022 \times 10^{23}}{14\ \text{g}}\right)$$

$$= 9.90 \times 10^{12}\ \text{dpm/g}\ ^{14}\text{C} \tag{1.601}$$

This activity per gram of ^{14}C is converted to grams ^{14}C per curie as follows:

$$\frac{2.22 \times 10^{12}\ \text{dpm/Ci}}{9.90 \times 10^{12}\ \text{dpm/g}\ ^{14}\text{C}} = 0.224\ \text{g}\ ^{14}\text{C per Ci}$$

$$= 0.224\ \text{mg}\ ^{14}\text{C per mCi} \tag{1.602}$$

The calculated mass of ^{32}P in 1 curie of activity is almost a million fold less than the calculated mass of ^{14}C in 1 curie of activity. In general, research with radionuclides often involves the applications and analysis of lower levels of radioactivity in microcuries, and picocuries, and so on. The masses of radioactive atoms in the micro-, and picocurie levels of radioactivity are obviously much smaller than encountered at the curie or millicurie level. It is important, therefore, to be aware of the order of magnitude of radioactive atom masses involved, which leads us to the concept of "carrier-free" samples of radionuclides, discussed subsequently.

C. Carrier-free radionuclides

A carrier-free radionuclide sample is, by definition, a sample in which all of the atoms of a particular element consist of the radioactive isotope; that is, no stable isotope of that element is present. A stable isotope of the particular element is referred to as carrier. Carrier-free radioisotopes are commonly used in research in the bio-, medical, and chemical sciences (Kobayashi, 2016; Chakraborty et al., 2014; Abbas et al., 2012; Aizatskiy et al., 2011; Elmore, 2009; Lin and Chao, 2009; Li and Deutscher, 2008; Steinman and Mulholland, 2007; Buchtela, 2005; Horányi, 2004; Lamson et al., 1975). Often chemical compounds are labeled as carrier-free, but are not literally carrier-free. Commercial suppliers of radionuclides and radioisotope-labeled compounds used in research will specify the specific activity of the radionuclide, which is the activity in disintegration rate per mass, *e.g.*, dpm/g. dps/g, Ci/g, Bq/g, or Bq/mol, etc. Organic compounds labeled with ^{14}C or ^{3}H are generally not carrier-free, and such compounds labeled with high specific-activity radioisotopes may contain >80% ^{14}C and >50% tritium (Elmore, 2009). In the previous section of this chapter Eqs. (1.599) and (1.601) were used to calculate the specific activities of carrier-free radioisotopes ^{32}P and ^{14}C, respectively. It is important, therefore, to be aware of the masses of radioactive isotope in a given sample and any consequences that may be involved when very small quantities (*e.g.*, 10^{-6} to 10^{-12} g or smaller) of radioactive nuclide may be involved.

For example, in Section XVIII.B we calculated that there was only 3.51×10^{-6} g of ^{32}P per curie of radioactivity. A millicurie of carrier-free ^{32}P, which is a level of activity commonly procured from a radioisotope supplier, would contain only 3.51×10^{-9} g of ^{32}P and possibly a small percentage of stable phosphorus. It is a common procedure to dilute the carrier-free ^{32}P to the microcurie level of activity prior to working with the radionuclide, such as in tracer studies. One microcurie of carrier-free ^{32}P would contain only 3.51×10^{-12} g of phosphorus. Obviously, therefore, we should consider the consequences of working with such small amounts of phosphorus in solution. Over the past 50 years of working with carrier-free radioactive nuclides sources, the author has experienced the absorption of significant quantities of radionuclides onto the surface of glassware. If we consider the ionic characteristics of the chemical forms of certain radionuclide sources and the minute quantities these may possess in the carrier-free form, significant quantities of certain radionuclides could be lost from solution by absorption onto the surface of glassware, onto the surface of precipitates, and so forth. The addition of carrier (*i.e.*, stable isotope or organic compound without the radioisotope label) may be added to a high specific activity sample to facilitate radioisotope tracer studies. For example, when working with sources of high specific activity, if a particular experiment calls for the addition of carrier, the author will add carrier to the radionuclide source during the dilution procedure. If carrier is not desired, the procedure recommended by Chase and Rabinowitz (1968) can be utilized. For example, if it is desired to dilute a carrier-free solution of $NaH_2{}^{32}PO_4$ in a volumetric flask, it is best to treat the flask first with a 1% solution of NaH_2PO_4 prior to the addition of the carrier-free solution. The volumetric flask and any other glassware used in the dilution may be rinsed with the 1% NaH_2PO_4. Alternatively, the volumetric flask may be filled with the 1% NaH_2PO_4 solution and allowed to sit for several hours.

The flask is then rinsed with deionized water to remove unabsorbed phosphorus. The flask can then be used to prepare a dilution of carrier-free or high specific activity $NaH_2{}^{32}PO_4$. It is important, however, to rinse the flask with a solution of the same chemical form as the radioisotope, if it is desirable to prevent contamination of the radioisotope with another chemical form.

References

Achterberg, A., et al., 2006. First year performance of the IceCube neutrino telescope. Astropart. Phys. 26, 155–173.

"Readers may view, browse, and/or download material for temporary copying purposes only, provided these uses are for noncommercial personal purposes. Except as provided by law, this material may not be further reproduced, distributed, transmitted, modified, adapted, performed, displayed, published, or sold in whole or in part, without prior permission from the American Physical Society or other organizations that own copyright to the material in the references listed herein."

Abbas, K., et al., 2012. Feasibility of ^{99}Mo production by proton-induced fission of ^{232}Th. Nucl. Instrum. Methods Phys. Res., Sect. B 278, 20–25.

Abbasi, R., et al., 2010. The energy spectrum of atmospheric neutrinos between 2 and 200 TeV with the AMANDA-II detector. Astropart. Phys. 34, 48–58.

Abbasi, R., et al., 2011. Search for neutrino-induced cascades with five years of AMANDA data. Astropart. Phys. 34, 420–430.

Abbe, J.C., Marques-Netto, A., 1975. Szilard-Chalmers effects in halfnium complexes. J. Inorg. Nucl. Chem. 37, 2239–2242.

Abe, K., et al., 2016. Real-time supernova neutrino burst monitor at Super-Kamiokande. Astropart. Phys. 81, 39–48.

Abela, R., et al., 2010. Analytical applications of synchrotron radiation. TrAC Trends Anal. Chem. 29 (6), 452.

Adams, F., et al., 1971. Instrumental neutron activation analysis. CRC Crit. Rev. Anal. Chem. 1 (4), 455–586.

Adloff, J.P., 1989. Radioactive disequilibrium in nature: origin, facts and applications. Int. J. Radiat. Appl. Instrum., Part C 33 (3), 257.

PAMELA Collaboration Adriani, O., et al., 2010. PAMELA results on the cosmic-ray antiproton flux from 60 MeV to 180 GeV in kinetic energy. Phys. Rev. Lett. 105, 121101.

Aghevlian, S., et al., 2017. Radioimmunotherapy of cancer with high linear energy transfer (LET) radiation delivered by radionuclides emitting α-particles or Auger electrons. Adv. Drug. Deliv. Rev. 109, 102–118.

(LVD Collaboration) Aglietta, M., et al., 1995. Neutrino-induced and atmospheric single-muon fluxes measured over five decades of intensity by LVD at Gran Sasso Laboratory. Astropart. Phys. 3, 311.

AMS Collaboration Aguilar, M., et al., 2002. The alpha magnetic spectrometer (AMS) on the international space station: Part I – results from the test flight on the space shuttle. Phys. Rep. 366 (6), 331–405.

Aguilar, J.A., et al., 2011. Time calibration of the ANTARES neutrino telescope. Astropart. Phys. 34, 539–549.

Aguilar, A., et al., 2017. Design specification for the European Spallation Source neutron generating target element. Nucl. Instrum. Methods Phys. Res., Sect. A 856, 99–108.

Aizatskiy, N.I., et al., 2011. Carrier-free production of ^{95}Tc at an electron accelerator. Nucl. Instrum. Methods Phys. Res., Sect. B 269, 3125–3128.

Alfassi, Z.B., et al., 1986. Optimation of the production of $Na^{18}F$ by use of the Szilard–Chalmers reaction following irradiation of fluorine compounds with 14 MeV neutrons. J. Radioanal. Nucl. Chem. 100 (2), 263–268.

Alikaniotis, K., et al., 2016. Radiotherapy dose enhancement using BNCT in conventional LINACs high-energy treatment: simulation and experiment. Rep. Pract. Oncol. Radiother. 21, 117–122.

Allan, C.J., 1971. Use of a trochoidal beta-ray spectrometer as an internal pair spectrometer. Nucl. Instrum. Methods 91, 117–123.

Allard, D., et al., 2007. On the transition from galactic to extragalactic cosmic-rays: spectral and composition features from two opposite scenarios. Astropart. Phys. 27 (1), 61–75.

Almayahi, B.A., et al., 2012. Effect of the natural radioactivity concentrations and ^{226}Ra/^{238}U disequilibrium on cancer diseases in Penang, Malaysia. Rad. Phys. Chem. 81, 1547–1558.

Almeida, A.P., et al., 2014. Phase contrast X-ray microtomography of the *Rhodnius prolixus* head: comparison of direct reconstruction and phase retrieval approach. Rad. Phys. Chem. 95, 243–246.

Alpher, R.A., Herman, R.C., 1949. Remarks on the evolution of an expanding universe. Phys. Rev. 75 (7), 1089–1095.

Alvarez, L.W., 1938. The capture of orbital electrons by nuclei. Phys. Rev. 54 (7), 486–497.

Alvarez, L.W., et al., 1959. Neutral cascade hyperon event. Phys. Rev. Lett. 2 (5), 215–219.

Amaldi, E., Fermi, E., 1935. Absorption of slow neutrons. Ricerca Sci. 6, 344–347.

Amaldi, E., Fermi, E., 1936a. Groups of slow neutrons. Ricerca Sci. 7 (1), 310–313.

Amaldi, E., Fermi, E., 1936b. Diffusion of slow neutrons. Ricerca Sci. 7 (1), 393–395.

Amaldi, E., D'Agostino, O., Fermi, E., Pontecorvo, B., Rasetti, F., Segre, E., 1935. Artificial radioactivity produced by neutron bombardment. II. Proc. R. Soc. London A149 (868), 522–558.

Amano, Y., 2013. Nuclear power in the 21st century. IAEA Bull. 54 (1), 2–3.

(MACRO Collaboration) Ambrosio, M., et al., 1995. Vertical muon intensity measured with MACRO at the Gran Sasso laboratory. Phys. Rev. D52, 3793–3802.

AMS Collaboration, 2000. Cosm. Protons Phys. Lett. B490, 27–35.

Anders, O.U., Briden, D.W., 1964. A rapid, nondestructive method of precision oxygen analysis by neutron activation. Anal. Chem. 36 (2), 287–292.

Andersen, H.H., 1985. The Barkas effect and other higher-order Z_1-contributions to the stopping power, pp. 409–429. In: Semiclassical Descriptions of Atomic and Nuclear Collisions. Elsevier, Amsterdam.

Andersen, H.H., et al., 1966. Stopping power of aluminium for 5-12 MeV protons and deuterons. Mat. Fys. Medd. Dan. Vid. Selsk. 35 (4).

Andersen, H.H., et al., 1967. Stopping power of Be, Al, Cu, Ag, Pt, and Au for 5-12 MeV protons and deuterons. Phys. Rev. 153, 338–342.

Andersen, H.H., et al., 1969a. An experimental investigation of charge-dependent deviations from the Bethe stopping power formula. Nucl. Phys. A 125, 171–175.

Andersen, H.H., et al., 1969b. Excitation potentials and shell corrections for the elements $Z_2 = 20$ to $Z_2 = 30$. Phys. Rev. 180, 373–380.

Anderson, C.D., 1932. The apparent existence of easily detectable positives. Science 76, 238.

Anderson, C.D., 1933a. The positive electron. Phys. Rev. 43, 491–498.

Anderson, C.D., 1933b. Free positive electrons resulting from the impacts upon atomic nuclei of the photons from ThC. Science 77, 432.

Anderson, C.D., 1936. The production and properties of positrons. Nobel Lecture, December 13, 1939. In: Nobel Lectures, Physics, 1922-1941 (1965). Elsevier Publishing, Amsterdam.

Anderson, C.D., Neddermeyer, S.H., 1933. Positrons from gamma rays. Phys. Rev. 43, 1034–1035.

Anderson, C.D., Neddermeyer, S.H., 1936. Cloud chamber observations of cosmic rays at 4300 meters elevation and near sea-level. Phys. Rev. 50, 263–271. http://prola.aps.org/toc/PR/v50/i4.

Anderson, H.L., Fermi, E., Hanstein, H.B., 1939a. Production of neutrons in uranium bombarded by neutrons. Phys. Rev. 55, 797–798.

Anderson, H.L., Fermi, E., Szilard, L., 1939b. Neutron production and absorption in uranium. Phys. Rev. 56, 284–286.

Anderson, C., et al., 2017. Delta-ray production in MCNP 6.2.0. Phys. Proc. 90, 229–236.

Andreev, V., 2017. Ultracold neutron detector for neutron lifetime measurements. Nucl. Instrum. Methods Phys. Res., Sect. A 845, 548–551.

Andreev, V., et al., 2017. Ultracold neutron detector for neutron lifetime measurements. Nucl. Instrum. Methods Phys. Res., Sect. A 845, 548–551.

Andreev, Y.M., et al., 1987. Muon intensity from the Baksan underground scintillation telescope. In: Proc. 20th Int. Cosmic Ray Conf., Moscow, vol. 6, p. 200. http://adsabs.harvard.edu/abs/1987ICRC....6..200A.

Andreo, P., Burns, D.T., 2017. Fundamentals of Ionizing Radiation Dosimetry. Wiley-VCH, p. 1000.

Andreo, P., et al., 2005. Calibration of photon and electron beams. In: Podgorsak, E.B. (Ed.), Review of Radiation Oncology Physics: A Handbook for Teachers and Students. International Atomic Energy Agency (IAEA), Vienna, pp. 301–355.

Angeli, I., Csatlós, M., 1977. Fine structure in the mass number dependence of rms charge radii. Nucl. Phys. A 288 (3), 480–492.

Aponick Jr., A.A., et al., 1970. Quadrupole and hexadecapole deformations in rare earth nuclei. Nucl. Phys. A 159 (2), 367–384.

Araújo, H.M., et al., 2008. Measurements of neutrons produced by high-energy muons at the Boulby Underground Laboratory. Astropart. Phys. 29, 471–481.

Archubi, C.D., et al., 2013. Barkas effect in the stopping power for ions with different ionization degrees. Nucl. Instrum. Methods Phys. Res., Sect. B 316, 88–93.

Argonne National Laboratory, 1963. Reactor Physics Constants. ANL-5800, second ed. United States Atomic Energy Commission, Washington, DC.

UA1 Collaboration Arnison, G., et al., 1983a. Experimental observation of isolated large transverse energy electrons with associated missing energy at $\sqrt{s} = 540$ GeV. Phys. Lett. 122B, 103–116.

UA1 Collaboration Arnison, G., et al., 1983b. Experimental observation of lepton pairs of invariant mass around 95 GeV/c^2 at the CERN SPS collider. Phys. Lett. 126B, 398–410.

Arnold, R., et al., 2014. Investigation of double beta decay of ^{100}Mo to excited states of ^{100}Ru. The NEMO-3 Collaboration. Nucl. Phys. A 925, 25–36.

Arzumanov, S., et al., 2000a. Neutron lifetime measured by monitored storing of ultra-cold neutrons. Nucl. Instrum. Methods Phys. Res., Sect. A 440, 511–516.

Arzumanov, S., et al., 2000b. Neutron lifetime measured by monitored storing of ultra-cold neutrons with detection of inelastically scattered neutrons. Phys. Lett. B 483, 15–22.

Arzumanov, S., 2009. A new project to measure the neutron lifetime using storage of ultracold neutrons and detection of inelastically scattered neutrons. Nucl. Instrum. Methods Phys. Res., Sect. A 611, 186–188.

Arzumanov, S., 2015. A measurement of the neutron lifetime using the method of storage of ultracold neutrons and detection of inelastically up-scattered neutrons. Phys. Lett. B 745, 79–89.

Asakura, K., et al., 2016. Search for double-beta decay of ^{136}Xe to excited states of ^{136}Ba with the KamLAND-Zen experiment. Nucl. Phys. A 946, 171–181.

Asaro, F., Perlman, I., 1953. Patterns in alpha spectra of even-even nuclei. Phys. Rev. 91, 763–764.

Ashley, R.P., et al., 2000. Steady-state D–^{3}He proton production in an IEC fusion device. Fusion Technology Institute, University of Wisconsin, Madison, p. 6. UWFDM-1144.

Atkinson, D., 2006. Does quantum electrodynamics have an arrow of time? Stud. Hist. Philos. Sci, Part B 37 (3), 528–541.

Audi, G., et al., 2003. The AME2003 atomic mass evaluation (II). Tables, graphs, and references. Nucl. Phys. A 729, 337–676.

Auditore, L., et al., 2005. Study of a 5 MeV electron linac based neutron source. Nucl. Instrum. Methods Phys. Res., Sect. B 229, 137–143.

Auger, P., 1923. Secondary β rays produced in a gas by X-rays. Compt. Rend. 177, 169–171.

Auger, P., 1925a. Sur les rayons β secondaires produits dans un gaz par des rayons-X. Compt. Rend. 180, 65–68.

Auger, P., 1925b. Sur l'effect photoélectrque compose. J. Phys. Rad. 6, 205–208.

Auger, P., Maze, R., 1938. Les grandes gerbes cosmiques de l'atmosphere. C. R. Acad. Sci. Ser. B 207, 228–230.

Auger, P., Maze, R., 1939. Extension et Pouvoir Pénétrant des Grandes Grebes de Rayons Cosmiques. Compt. Rend. 208, 1641–1643.

Auslender, V.L., et al., 2004. Bremsstrahlung converters for powerful industrial electron accelerators. Radiat. Phys. Chem. 71, 295–297.

Avrorin, A.D., et al., 2015. Search for neutrino emission from relic dark matter in the sun with the Baikal NT200 detector. Astropart. Phys. 62, 12–20.

Avrorin, A.D., et al., 2016. A search for neutrino signal from dark matter annihilation in the center of the Milky Way with Baikal NT200. Astropart. Phys. 81, 12–20.

Aydin, E.G., et al., 2008. Equilibrium and pre-equilibrium calculations of neutron production in medium–heavy targets irradiated by protons up to 100 MeV. An. Nucl. Energy 35, 2306–2312.

Aynutdinov, V., et al., 2006. Search for a diffuse flux of high-energy extraterrestrial neutrinos with the NT200 neutrino telescope. Astropart. Phys. 25 (2), 140–150.

Aynutdinov, V., et al., 2011. The Baikal Neutrino Project: present and perspective. Nucl. Instrum. Methods Phys. Res., Sect. A 628, 115–119.

Azab, A.K., et al., 2006. Boron neutron capture therapy. Stud. Inorg. Chem. 22, 337–366.

Baba, S., Moki, T., 1981. Preparation of ^{99}Mo by the use of the Szilard-Chalmers effect with oxymolybdenum. Radiochim. Acta 29 (2/3), 135–137.

Bacon, G.E., 1969. Neutron Physics. Wykeham Publications, London.

UA2 Collaboration Bagnaia, P., et al., 1983. Evidence for Z0→e+e− at the CERN $\bar{p}p$ collider. Phys. Lett. 129B, 130–140.

Bailey, J., et al., 1977. On muon lifetimes and time dilation. Nature 268 (5618), 301–305.

Bailey, J., et al., 1979. Final report of the CERN muon storage ring including the anomalous magnetic moment and the electric dipole moment of the muon, and the direct test of relativistic time dilation. Nucl. Phys. B 150, 1–75.

Bainbridge, K.T., 1952. Effect of chemical combination of radioactivity confirmed. Chem. Eng. News 30, 654.

Baker, C.P., et al., 1943. Cross section for the reaction 20(230,240)10. Los Alamos Technical Report, LAMS-2, p. 63.

Balasubramanian, P.S., 1997. Anodically oxidized aluminum layer as a useful substrate for the fabrication of ^{147}Pm sources for beta-ray thickness gauges. J. Radioanal. Nucl. Chem. 223 (1–2), 79–81.

Balasubramanian, P.S., 1998. A simple procedure for the fabrication of high activity beta-radiation sources of ^{147}Pm for use in beta-ray thickness gauges. J. Radioanal. Nucl. Chem. 229 (1–2), 157–160.

Baldassini, W.A., et al., 2015. Bioanalytical methods for the metalloproteomics study of bovine longissimus thoracis muscle tissue with different grades of meat tenderness in the Nellore breed (*Bos indicus*). Food Chem. 169, 65–72.

Bambynek, W., et al., 1972. X-ray fluorescence yields, Auger, and Coster-Kronig transition probabilities. Rev. Mod. Phys. 44, 716–813. http://rmp.aps.org/abstract/RMP/v44/i4/p716_1.

UA2 Collaboration Banner, M., et al., 1983. Observation of single isolated electrons of high transverse momentum in events with missing transverse energy at the CERN $\bar{p}p$collider. Phys. Lett. 122B, 476–485.

Barabash, A.S., 2015. Double beta decay experiments: present status and prospects for the future. Phys. Proc. 74, 416–422.

Barabash, A.S., et al., 2007. Search for β^{+}EC and ECEC processes in ^{74}Se. Nucl. Phys. A 785 (3–4), 371–380.

Barabash, A.S., et al., 2008. Search for β^{+}EC and ECEC processes in ^{112}Sn and $\beta^{-}\beta^{-}$ decay of ^{124}Sn to the excited states of ^{124}Te. Nucl. Phys. A 807 (3–4), 269–281.

Barkas, W.H., et al., 1963. Resolution of the Σ^{-} mass anomoly. Phys. Rev. Lett. 11, 26–29.

Barkla, C.G., 1917. Characteristic Röntgen radiation. In: Nobel Lectures, Physics 1901–1921. Elsevier, Amsterdam, 1967.

Barnes, R.K., et al., 1986. Polarographic quantification of the specific activity of copper-64 produced by Szilard–Chalmers reactions. Appl. Radiat. Isot. 37, 1241.

Barthakur, N.N., 1983. The β-ray gauge as a leaf surface wetness detector. Int. J. Appl. Radiat. Isot. 34 (11), 1549–1552.

Barton, R., et al., 1963. Observation of delayed proton radioactivity. Can. J. Phys. 41, 2007–2025.

Basini, G., et al., 1999. The flux of cosmic-ray antiprotons from 3.7 to 24 GeV. Proc. 26th Int. Cosmic Ray Conf., Salt Lake City 3, 101.

Basnayaka, U., et al., 2014. Diffraction-enhanced synchrotron imaging of bovine ovaries ex vivo. J. Med. Imag. Radiat. Sci. 45 (3), 307–315.

Bayanov, B., et al., 2009a. A neutron producing target for BINP accelerator-based neutron source. Appl. Radiat. Isot. 67, S282–S284.

Bayanov, B., et al., 2009b. First neutron generation in the BINP accelerator based neutron source. Appl. Radiat. Isot. 67, S285–S287.

Bayanov, B., et al., 2010. Dosimetry and spectrometry at accelerator based neutron source for boron neutron capture. Radiat. Meas. 45 (10), 1462–1464.

Bé, M.-M., et al., 2006. Detailed calculation of *K*- and *L*-Auger electron emission intensities following radioactive disintegration. Appl. Radiat. Isot. 64, 1435–1439.

Beach, A.S., et al., 2001. Measurement of the cosmic-ray antiproton-to-proton abundance ratio between 4 and 50 GeV. Phys. Rev. Lett. 87 (27), 271101.

Beatty, J.J., et al., 2004. New measurement of the cosmic-ray positron fraction from 5 to 15 GeV. Phys. Rev. Lett. 93, 241102.

Beck, M., 2010. The KATRIN experiment. J. Phys. Conf. Ser. 203 (1), 012097.

Beckerly, J.G., 1945. What is the "Geiger-Nuttall law"? Am. J. Phys. 13 (3), 158–159.

Beckmann, C., et al., 2002. Nuclei, superheavy nuclei, and hypermatter in a chiral SU(3) model. Phys. Rev. C 65 (2), 024301.

Becquerel, A.H., 1896a. On the rays emitted by phosphorescence. Compt. Rend. Acad. Sci., Paris 122, 420.

Becquerel, A.H., 1896b. On the invisible rays emitted by phosphorescent bodies. Compt. Rend. Acad. Sci., Paris 122, 501.

Becquerel, A.H., 1901. The radio-activity of matter. Nature 63, 396–398.

Becquerel, A.H., 1903. On radioactivity, a new property of matter. Nobel Lecture, December 11, 1903. In: Science Incorporated Elsevier (Ed.), Nobel Lectures in Physics, 1901–1970, Nobel Lectures, Physics, 1901–1921, vol. 1. Elsevier, Amsterdam, 1967.

Behr, T., et al., 2000. Therapeutic advantages of Auger electron-over beta emitting radiometals or radioiodine when conjugated to internalizing antibodies. Eur. J. Nucl. Med. 27 (7), 753–765.

for the SNO Collaboration Bellerive, A., et al., 2016. The Sudbury neutrino observatory. Nucl. Phys. B 908, 30–51.

Belli, P., et al., 2014. Search for double beta decay of ^{136}Ce and ^{138}Ce with HPGe gamma detector. Nucl. Phys. A 930, 195–208.

Bellotti, R., et al., 1996. Measurement of the negative muon spectrum between 0.3 and 40 GeV/*c* in the atmosphere. Phys. Rev. D53 (1), 35–43. http://prd.aps.org/abstract/PRD/v53/i1/p35_1.

Bellotti, R., et al., 1999. Balloon measurements of cosmic-ray muon spectra in the atmosphere along with those of primary protons and helium nuclei over midlatitude. Phys. Rev. D60 (5), 052002.

Benczer-Koller, N., 2009. Chien-Shiung Wu 1912–1997. National Academy of Sciences, Washington, D.C., pp. 1–17

Bender, M., et al., 2001. Shell stabilization of super- and hyperheavy nuclei without magic gaps. Phys. Lett. B 515 (1–2), 42–48.

Berger, M.J., 1993. Stopping-power and range tables for electrons, protons, and helium ions. NISTIR 4999. National Institute of Standards and Technology, Gaithersburg, MD.

Berger, M.J., Hubbell, J.H., 1997. Photon attenuation coefficients. In: Lide, D.R., Frederikse, H.P.R. (Eds.), Handbook of Chemictry and Physics, 77th ed. CRC Press, Boca Raton, pp. 10-250–10-254.

Berger, M.J., Seltzer, S.M., 1982. Stopping powers and ranges of electrons and positrons. Nat. Bureau Standards Publ. 168. NBSIR 82-2550. https://nvlpubs.nist.gov/nistpubs/Legacy/IR/nbsir82-2550.pdf.

(Fréjus Collaboration) Berger, C., et al., 1989. Experimental study of muon bundles observed in the Fréjus detector. Phys. Rev. D40, 2163–2171.

Berger, M.J., et al., 2017a. ESTAR, PSTAR, and ASTAR: Computer program: Stopping-power and Range Tables for Electrons, Protons, and Helium Ions. National Institute of Standards and Technology, Gaithersburg, MD [Online]. Available: https://www.nist.gov/pml/stopping-power-range-tables-electrons-protons-and-helium-ions.

Berger, M.J., et al., 2017b. NIST Standard Reference Database 8 (XGAM) NB SIR 87-3597. XCOM: Photon Cross Sections Database. The National Institute of Standards and Technology (NIST). https://www.nist.gov/pml/xcom-photon-cross-sections-database.

Berger, M.J., et al., 2017c. ESTAR, PSTAR, and ASTAR: computer programs for calculating stopping-power and range tables for electrons, protons, and helium ions version 1.21. http://physics.nist.gov/Star.

Berghaus, P., for the IceCube Collaboration, 2009. Muons in IceCube. Nucl. Phys. B (Proc. Suppl.) 196, 261–266.

(Particle Data Group) Beringer, J., et al., 2012. Review of particle physics. Phys. Rev. D 86, 010001.

Bethe, H.A., 1930. Zur Theorie des Durchgangs schneller Korpuskularstrahlen durch Materie. Ann. D. Physik 5, 325.

Bethe, H.A., 1933. Quantenmechanik der Ein- und Zwei-Electronen-Probleme. In: Geiger, H., Scheel, K. (Eds.), Handbuch der Physik, second ed., vol. 24. Springer, Berlin. Part I.

Bethe, H.A., 1937. Nuclear physics. B. Nuclear dynamics, theoretical. Rev. Mod. Phys. 9, 69–244.

Bethe, H.A., Bacher, R.F., 1936. Nuclear Physics A. Stationary states of nuclei. Rev. Mod. Phys. 8, 82–229.

Bethe, H.A., Ashkin, J., 1953. Passage of radiations through matter. In: Segré, E. (Ed.), Experimental Nuclear Physics, vol. 1. J. Wiley, New York.

Bevelacqua, J.J., 2010. Basic Health Physics: Problems and Solutions. Wiley-VCH, Hoboken, p. 768.

Bhattacharjee, P., Sigl, G., 2000. Origin and propagation of extremely high-energy cosmic rays. Phys. Rep. 327, 109–247.

Biebel, O., 2001. Experimental tests of the strong interaction and its energy dependence in electron-positron annihilation. Phys. Rep. 340, 165–289.

Birattari, C., et al., 2001. Review of cyclotron production and quality control of "high specific activity" radionuclides for biomedical, biological, industrial, and environmental applications at INFN-LASA. In: Marti, F. (Ed.), Cyclotrons and Their Applications 2001, Proceedings of Sixteenth International Conference. American Institute of Physics, Melville, New York.

Birch, M., et al., 2014. First compilation and evaluation of beta-delayed neutron emission probabilities and associated half-lives for $A \leq 72$ nuclei. Nucl. Data Sheets 120, 66–69.

Bird, D.J., et al., 1995. Detection of a cosmic ray with measured energy well beyond the expected spectral cutoff due to cosmic microwave radiation. Astrophys. J. 441, 144–150.

Blackett, P.M.S., Occhialini, G.P.S., 1933. Some photographs of the tracks of penetrating radiation. Proc. R. Soc. London 139A, 699–726.

Blank, B., et al., 2003. First observation of two-proton radioactivity from an atomic nucleus. Compt. Rend. Phys. 4 (4–5), 521–527.

Blank, B., et al., 2010. A time projection chamber for the three-dimensional reconstruction of two-proton radioactivity events. Nucl. Instrum. Methods Phys. Res., Sect. A 613, 65–78.

Blasi, P., 2008. Origin of high energy cosmic rays: a short review. Nucl. Instrum. Methods Phys. Res., Sect. A 588, 166–170.

Blasi, P., 2013. Origin of galactic cosmic rays. Nucl. Phys. B (Proc. Suppl.) 239–240, 140–147.

Blendowske, R., et al., 1987. Microscopic calculation of the ^{14}C decay of Ra nuclei. Nucl. Phys. A 464, 75–89.

Blewett, J.P., 1988. Synchrotron radiation – 1873 to 1947. Nucl. Instrum Methods Phys. Res., Sect. A 266, 1–9.

Blizard, E.P., Abbott, L.S., 1962. Reactor Handbook, vol. III. Interscience Publishers, New York.

Blümer, J., et al., 2009. Cosmic rays from the knee to the highest energies. Prog. Part. Nucl. Phys. 63, 293–338.

Bodei, L., et al., 2003. Radionuclide therapy with iodine-125 and other Auger-emitting radionuclides: experimental models and clinical applications. Cancer Radiother. Radiopharm. 18, 861–877.

Boehm, F., Wu, C.S., 1954. Internal bremsstrahlung and ionization accompanying beta decay. Phys. Rev. 93, 518–523.

Boezio, M., et al., 1999. The cosmic-ray proton and helium spectra between 0.2 and 200 GeV. Astrophys. J. 518 (1), 457–472.

Bohr, N., 1913. On the constitution of atoms and molecules. Philos. Mag. 26 (6), 1–25.

Bohr, N., 1914. On the spectrum of hydrogen (Address to the Physical Society of Copenhagen, Dec. 20, 1913). Physik Tidsskrift 12, 97, translated by A.D. Udden "The Theory of Spectra and Atomic Constitution – Three Essays 1922. In: Moulton, F.R., Schifferes, J.J. (Eds.), Autobiography of Science 1950. Doubleday, New York.

Bohr, N., 1921a. Atomic structure. Nature 107, 104–107.

Bohr, N., 1921b. Atomic structure. Nature 108, 208–209.

Bohr, N., 1922. The structure of the atom. Nobel Lecture, December 11, 1922. In: "Nobel Lectures, Physics 1922–1941" (1965). Elsevier Publishers, Amsterdam.

Bohr, N., 1928. The quantum postulate and the recent development of atomic theory. Nature 121, 580–591.

Bohr, A., 1951. On the quantization of angular momenta in heavy nuclei. Phys. Rev. 81, 134–138.

Bohr, A., 1975. Rotational motion in nuclei. Nobel Lecture. Dec. 11, 1975. http://www.nobelprize.org/nobel_prizes/physics/laureates/1975/bohr-lecture.pdf.

Bohr, A., 1992. Rotational motion in nuclei. In: Lundqvist, S. (Ed.), Nobel Lectures, Physics 1971-1980. World Scientific Publishing, Singapore.

Bohr, N., Lindhard, J., 1954. Electron capture and loss by heavy ions penetrating through matter. Kgl. Danske Videnskab Selskab, Mat.-Fys. Medd. 28 (7), 1–30.

Bohr, A., Mottelson, B.R., 1953a. Interpretation of isomeric transitions of electric quadrupole type. Phys. Rev. 89, 316–317.

Bohr, A., Mottelson, B.R., 1953b. Rotational states in even-even nuclei. Phys. Rev. 90, 717–719.

Bonardi, M.L., et al., 2004. Cyclotron production and quality control of "high specific activity" radionuclides in no-carrier added" form for radioanalytical applications in the life sciences. J. Radioanal. Nucl. Chem. 250 (3), 425-419.

Bonetti, R., Guglielmetti, A., 1999. Measurements of cluster radioactivity. In: Greiner, W., Gupta, R.K. (Eds.), Heavy Elements and Related New Phenomena, vol. II. World Scientific, Singapore, p. 1168.

Bonetti, R., et al., 2001. Cluster decay of ^{230}U via Ne emission. Nucl. Phys. A 686, 64–70.

Bonn, J., et al., 2001. The Mainz neutrino mass experiment. Nucl. Phys. B–Proc. Suppl. 91 (1–3), 273–279.

Borisenko, V.E., Ossicini, S., 2004. What is What in the Nanoworld: A Handbook on Nanoscience and Nanotechnology. Wiley-VCH Verlag GmbH & Co. KgaA, Weinheim.

Born, M., 1926a. The quantum mechanics of the impact process. Z. Phys. 37 (12), 863–867.

Born, M., 1926b. Quantum mechanics in impact process. Z. Phys. 38 (11/12), 803–840.

Born, M., 1926c. Zur Wellenmechanik der Stossvorgänge. Göttingen Nachr. Math. Phys. Kl. 146–160.

Born, M., 1926d. The adiabaten principle in the quantum mechanics. Z. Phys. 40 (3/4), 167–192.

Born, M., Fock, V., 1928. Bewas des Adiabatensatzes. Z. Phys. 51, 165–180.

Born, M., Oppenheimer, R., 1927. Quantum theory of molecules. Ann. Phys. 84 (20), 457–484.

Born, M., Heisenberg, W., Jordan, P., 1926. Quantum mechanics II. Z. Phys. 35 (8/9), 557–615.

Boudjemline, K., et al., 2010. The calibration of the Sudbury Neutrino Observatory using uniformly distributed radioactive sources. Nucl. Instrum. Methods Phys. Res., Sect. A 620, 171–181.

Boyd, G.E., et al., 1952. Recoil reactions with high intensity slow neutron sources. I. The Szilard-Chalmers enrichment of 35.9 h Br^{82}. J. Am. Chem. Soc. 74 (1), 237–240.

Braby, L.A., Ford, J.R., 2006. Energy deposition in different diameter sites can help characterize the incident radiation. Rad. Meas. 41, 1147–1151.

Brasch, A., Lange, F., Waly, A., Banks, T.E., Chalmers, T.A., Szilard, L., Hopwood, F.L., 1934. Liberation of neutrons from beryllium by x-rays: radioactivity induced by means of electron tubes. Nature 134, 880.

Bressan, A., et al., 2008. Proceedings of the sixth international workshop on ring imaging Cherenkov detectors. Nucl. Instrum. Methods Phys. Res., Sect. A 595 (1), 1–281.

Brien, M., et al., 1972. States of ^{87}Sr populated by deuteron pick-up from ^{89}Y. Nucl. Phys. A185, 289–302.

Browne, E., Tuli, J.K., 2017. Nuclear data sheets for A=99. Nucl. Data Sheets 145, 25–340.

Browne, E., Firestone, R.B., Shirley, V.S., 1986. Table of Radioactive Isotopes. John Wiley & Sons, New York.

Brunner, J., for the ANTARES Collaboration, 2011. The ANTARES neutrino telescope - status and first results. Nucl. Instrum. Methods Phys. Res., Sect. A 626–637, S19–S24.

Bryne, J., 1994. Neutrons, Nuclei and Matter, an Exploration of the Physics of Slow Neutrons. Institute of Physics, London.

Buchtela, K., 2005. Radiochemical methods: overview, pp. 1–8. In: Encyclopedia of Analytical Science, second ed. Elsevier, Amsterdam, p. 468.

Buck, B., Merchant, A.C., 1989. Cluster model calculations of exotic decays from heavy nuclei. Phys. Rev. C 39, 2097–2100.

Buesseler, K.O., et al., 2009. Source dependent and source-independent controls on plutonium oxidation state and colloid associations in groundwater. Environ. Sci. Technol. 43 (5), 1322–1328.

Burhop, E.H.S., Asaad, W.N., 1972. The Auger effect, pp. 163–184. In: Bates, D.R., Esterman, I. (Eds.), Advances in Atomic and Molecular Physics, vol. 8. Academic Press, New York, p. 299.

Buttlar, H.V., 1968. Nuclear Physics, an Introduction. Academic Press, New York.

Buzarbaruah, N., et al., 2015. Design of a linear neutron source. Fusion Eng. Des. 90, 97–104.

Caesar, C., et al., 2013. Beyond the neutron drip line: the unbound oxygen isotopes ^{25}O and ^{26}O. Phys. Rev. C 88, 034313.

Cai, Z., et al., 2016. ^{111}In-labeled trastuzumab-modified gold nanoparticles are cytotoxic in vitro to HER2-positive breast cancer cells and arrest tumor growth in vivo in athymic mice after intratumoral injection. Nucl. Med. Biol. 43, 818–826.

Campbell, D.B., et al., 2008. Evaluation of radioactive background rejection in ^{76}Ge neutrino-less double-beta decay experiments using a highly segmented HPGe detector. Nucl. Instrum. Methods Phys. Res., Sect. A 587, 60–67.

Canberra Nuclear, 1996. Neutron detection and counting. In: Canberra Nuclear Instruments Catalog, ninth ed. Canberra Industries, Meriden, CT, pp. 37–39.

Cantone, M.C., Hoeschen, C., 2011. Radiation Physics for Nuclear Medicine. Springer Science and Business Media, Berlin, p. 285.

Capello, A., et al., 2003. Peptide receptor radionuclide therapy in vitro using [In-111-DTPA(0)]octreotide. J. Nucl. Med. 44 (1), 98–104.

Carr, J., 2008. Data from the ANTARES Neutrino Telescope. Nucl. Instrum. Methods Phys. Res., Sect. A 588, 80–85.

Celiktas, C., 2011. A method to determine the gamma-ray linear attenuation coefficient. Ann. Nucl. Energy 38 (9), 2096–2100.

Cember, H., Johnson, T., 2008. Introduction to Health Physics, fourth ed. McGraw-Hill Medical, New York, p. 864.

Cengiz, A., Almaz, E., 2004. Internal bremsstrahlung spectra of β^- particle emitters using the Monte Carlo method. Radiat. Phys. Chem. 70, 661–668.

Cerny, J., et al., 1970. Confirmed proton radioactivity of ^{53m}Co. Phys. Lett. B 33 (4), 284–286.

Chadwick, J., 1932a. Possible existence of a neutron. Nature 129, 312.

Chadwick, J., 1932b. The existence of a neutron. Proc. R. Soc. A 136, 692–708.

Chadwick, J., 1935. The neutron and its properties. Nobel Lecture, December 12, 1935. In: Nobel Lectures, Physics 1922-1941 (1965). Elsevier Publishers, Amsterdam.

Chakraborty, R., et al., 2014. Nanostructured zirconium phosphate as ion exchanger: synthesis, size dependent property and analytical application in radiochemical separation. Appl. Radiat. Isot. 85, 34–38.

Chambers, E.E., Hofstadter, R., 1956. Structure of the proton. Phys. Rev. 103 (5), 1454–1463.

Chase, G.D., Rabinowitz, J.L., 1968. Principles of Radioisotope Methodology, third ed. Burgess Publishing Company, Minneapolis, pp. 140–143.

Chen, P., et al., 2006. Nuclear localizing sequences promote nuclear translocation and enhance the radiotoxicity of the anti-CD33 monoclonal antibody HuM195 labeled with In-111 in human myeloid leukemia cells. J. Nucl. Med. 47 (5), 827–836.

Cherenkov, P.A., 1934a. Visible light from clear liquids under the action of gamma radiation. C.R. Dokl. Akad. Nauk, SSSR 2 (8), 451–454.

Cherenkov, P.A., 1934b. Sichtbares Leichten von Reinen Flüssigkeiten unter der Einwirkung von γ-Strahlen. C.R. Dokl. Akad. Nauk, SSSR 2 (8), 455–457.

Cherenkov, P.A., 1936. Die Wirkung eines Magnetfeldes auf das durch Gamma-Strahlen Hervorgerfene sichtbare Leuchten der Flüssigkeiten. Dokl. Akad. Nauk, SSSR 3 (9), 413–416.

Cherenkov, P.A., 1958. Nobel Lecture: radiation of particles moving at a velocity exceeding that of light, and some of the possibilities for their use in experimental physics. Stockholm, December 11. In: Nobel Foundation Staff (Ed.), Nobel Lectures, Physics, 1942–1962. Elsevier Publishing, Amsterdam, 1964.

Cherenkova, E., 2005. Pavel Cherenkov – a person and a physicist – through the eyes of his daughter. Nucl. Instrum. Methods Phys. Res., Sect. A 553, 1–8.

Cherenkova, E.P., 2008. The discovery of the Cherenkov radiation. Nucl. Instrum. Methods Phys. Res., Sect. A 595, 8–11.

Chiavassa, A., 2010. Studies of the knee of the cosmic rays primary spectrum through measurements in the 10^{16}–10^{18} eV energy range. Nucl. Instrum. Methods Phys. Res., Sect. A 630 (1), 183–188.

Choppin, G.R., et al., 2002. Radiochemistry and Nuclear Chemistry, third ed. Elsevier, Amsterdam, p. 720.

Choudhuri, S., et al., 2007. Carcinogenesis: mechanisms and models, pp. 343–359. In: Gupta, R.C. (Ed.), Veterinary Toxicology: Basic and Clinical Principles. Academic Press, p. 1224.

Chourasia, A.R., Chopra, D.R., 1997. Auger electron spectroscopy. In: Settle, F.A. (Ed.), Handbook of Instrumental Techniques for Analytical Chemistry. Prentice Hall, Upper Saddle River, New Jersey, pp. 791–808.

Chowdhury, P.R., et al., 2008. Search for long lived heaviest nuclei beyond the valley of stability. Phys. Rev. C 77, 044603.

Chromik, M.J., et al., 1998. Two-proton decay of the first excited state of ^{17}Ne. In: AIP. Conf. Proc. 481, 187, Conf. 10-15 Aug 1998, Gatlinburg, Tennessee. https://doi.org/10.1063/1.59548.

Cipiti, B.B., Kulcinski, G.L., 2005. The production of ^{13}N using beam-target D$-^{3}$He fusion reaction. Fusion Sci. Techn. 47, 1245–1249.

Clapp, T.G., et al., 1995. The on-line inspection of sewn seams. National Textile Center Annual Report (August), pp. 221–230.

Clarke, R., 2005. Incoherent source/synchrotrons, pp. 217–224. In: Guenther, E.D. (Ed.), Encyclopedia of Modern Optics. Elsevier, Amsterdam.

Clay, J., 1928. Penetrating radiation. Proc. R. Acad. Amsterdam 31, 1091.

Cleland, M.R., et al., 1991. Evaluation of a new X-ray processing facility. Nucl. Instrum. Methods Phys. Res., Sect. B 56–57 (2), 1242–1245.

Clem, J.M., Evenson, P.A., 2002. Positron abundance in galactic cosmic rays. Astrophys. J. 568, 216–219.

Cline, D., 1993. Quadrupole and octupole shapes in nuclei. Nucl. Phys. A 557, 615–634.

Coakley, K.J., 2007. Optimal proton trapping strategy for a neutron life-time experiment. Nucl. Instrum. Methods Phys. Res., Sect. A 577, 702–707.

Cohen, D.D., et al., 2008. Experimental bremsstrahlung yields for MeV proton bombardment of beryllium and carbon. Nucl. Instrum. Methods Phys. Res., Sect. B 266, 1149–1153.

Collins, K.E., Willard, J.E., 1962. Szilard-Chalmers chemistry and radiation chemistry of solid hexabromoethane. Thermal annealing and radiation annealing effects. J. Chem. Phys. 37, 1908.

Compton, A.H., 1923a. A quantum theory of the scattering of x-rays by light elements. Phys. Rev. 21 (5), 483–502.

Compton, A.H., 1923b. The spectrum of scattered x-rays. Phys. Rev. 22 (5), 409–413.

Compton, A.H., 1932. Variation of the cosmic rays with latitude. Phys. Rev. 41, 111–113.

Compton, A.H., 1933. A geographic study of cosmic rays. Phys. Rev. 43, 387–404.

Constantini, D.L., et al., 2007. ^{111}In-labeled Trastuzumab (Herceptin) modified with nuclear localization sequences (NLS): an Auger electron-emitting radiotherapeutic agent for HER2/neu-amplified breast cancer. J. Nucl. Med. 48 (8), 1357–1368.

Corradi, L., et al., 2013. Multinucleon transfer reactions: present status and perspectives. Nucl. Instrum. Methods Phys. Res., Sect. B 317, 743–751.

Coster, D., Kronig, R. de L., 1935. A new type of Auger effect and its influence on the x-ray spectrum. Physica 2 (1–12), 13–24.

Coursey, B.M., et al., 1993. Radioassays of yttrium-90 used in nuclear medicine. Nucl. Med. Biol. 20 (5), 693–700.

Cowan Jr., C.L., Reines, F., Harrison, F.B., Kruse, H.W., McGuire, A.D., 1956. Detection of the free neutrino: a confirmation. Science 124, 103–104.

Cremonesi, O., 2013. Experimental searches of neutrinoless double beta decay. Nucl. Phys. B (Proc. Suppl.) 237–238, 7–12.

Cronin, J.W., 1999. Cosmic rays: the most energetic particles in the universe. Rev. Mod. Phys. 71 (2), S165–S172.

Crookes, W., 1903a. The emanation of radium. Proc. R. Soc. (London) A71, 405–408.

Crookes, W., 1903b. Certain properties of the emanation of radium. Chem. News 87, 241.

Crouch, M., 1987. An improved world survey expression for cosmic ray vertical intensity vs. depth in standard rock. In: Proc. 20th Int. Cosmic Ray Conf., Moscow, vol. 6, p. 165. http://adsabs.harvard.edu/abs/1987ICRC....6..165C.

Csikai, J., 1987. CRC Hand Book of Fast Neutron Generators, vol. 1. John Wiley and Sons, New York, p. 256.

Cucinotta, F.A., et al., 1998. The effects of delta rays on the number of particle-track traversals per cell in laboratory and space exposures. Rad. Res. 150, 115–119.

Curie, P., 1905. Radioactive substances, especially radium. Nobel Lecture on June 6, 1905. In: Science Incorporated Elsevier (Eds.), Nobel Lectures in Physics, 1901-1970. vol. 1. Nobel Lectures, Physics, 1901-1921. (1967). Elsevier Publishing, Amsterdam.

Curie, M., 1911. Radium and the new concepts in chemistry. Nobel Lecture on December 11, 1911. In: Nobel Foundation Staff (Eds.), Nobel Prize Lectures in Chemistry, 1901-1970. (1967). vol. 1, 1901-1921, vol. 2, 1922-1941, vol. 3, 1942-1962. Elsevier Science, Amsterdam.

Ćwiok, S., et al., 1996. Shell structure of the superheavy elements. Nucl. Phys. A 611 (2–3), 211–246.

Czarnecki, S., et al., 2016. Z-dependence of thick-target bremsstrahlung produced by monoenergetic low-energy electrons. Nucl. Instrum. Methods Phys. Res., Sect. B 378, 54–58.

Czarnecki, S., et al., 2017. Measurements of the effective atomic numbers of alloys using thick-target bremsstrahlung intensities. Phys. Proc. 90, 41–46.

D'Olympia, N., et al., 2013. Sub-nanosecond nuclear half-life and time-of-flight measurements with $CeBr_3$. Nucl. Instrum. Methods Phys. Res., Sect. A 728, 31–35.

Dai, M., et al., 2002. Sources and migration of plutonium in groundwater at the Savannah river site. Environ. Sci. Technol. 36, 3690–3699.

Dalrymple, G.B., 1991. The Age of the Earth. Stanford University Press, p. 474.

Das, A., Ferbel, T., 1994. Introduction to Nuclear and Particle Physics. John Wiley & Sons, New York.

Daudel, R., 1947. Alteration of radioactive periods of the elements with the aid of chemical methods. Rev. Sci. 85, 162.

Davis Jr., R., 1952. Nuclear recoil following neutrino emission from beryllium 7. Phys. Rev. 86 (6), 976–985.

Davis, D.W., Krogh, T.E., 2000. Preferential dissolution of ^{234}U and radiogenic Pb from α-recoil-damaged lattice sites in zircon: implication for thermal histories and Pb isotopic fractionation in the near surface environment. Chem. Geol. 172, 41–58.

De Bonis, J., for the ANTARES Collaboration, 2016. Recent results of the ANTARES neutrino telescope. Nucl. Part. Phys. Proc. 273–275, 419–424.

de Broglie, L., 1923a. Waves and quanta. Nature 112, 540.

de Broglie, L., 1923b. Ondes et quanta. Compt. Rend. 177, 507–510.

de Broglie, L., 1923c. Quanta de lumière, diffraction et interférence. Compt. Rend. 177, 548–550.

de Broglie, L., 1924. Rescherches sur la théorie des quantas. Thesis, Paris).

de Broglie, L., 1925. Rescherches sur la théorie des quantas. Ann. Phys. (Paris) 3 (10), 22–128.

Dean, T., 2008. Higher, higher! New Scientist, pp. 32–35.

Dell'Oro, S., 2015. Neutrinoless double beta decay: expectations and uncertainties. Nucl. Part. Phys. Proc. 256–266, 31–33.

Desiati, P., et al., 2014. Observation of TeV-PeV cosmic ray anisotropy with IceCube, IceTop and AMANDA. Nucl. Instrum. Methods Phys. Res., Sect. A 742, 199–202.

Dewey, M., et al., 2009. Prospects for a new cold neutron beam measurement of the neutron lifetime. Nucl. Instrum. Methods Phys. Res., Sect. A 611, 189–192.

Dhibar, M., et al., 2018. Characterization of a 2 × 2 array of large square bars of LaBr3:Ce detectors with γ-rays up to 22.5 MeV. Nucl. Instrum. Methods Phys. Res., Sect. A (in press).

Dias, M.S., et al., 2014. Disintegration rate, gamma-ray emission probabilities and metastable half-life measurements of ^{67}Ga. Appl. Radiat. Isot. 87, 126–131.

Didi, A., 2017. Neutron activation analysis: modelling studies to improve the neutron flux of Americium-Beryllium source. Nucl. Eng. Technol. 49, 787–791.

Dirac, P.A.M., 1928a. The quantum theory of the electron. Proc. R. Soc. Lond., A 117, 610–612.

Dirac, P.A.M., 1928b. The quantum theory of the electron, Part II. Proc. R. Soc. Lond., A 118, 351–361.

Dirac, P.A.M., 1933. Theory of electrons and positrons. Nobel lecture, December 12, 1933. In: Nobel Lectures, Physics, 1922-1941. (1965). Elsevier Publishing, Amsterdam.

Donahue, R.J., Fassó, A., 2002. Radioactivity and radiation protection. Phys. Rev. D66, 01001-1.

Dossat, C., et al., 2007. The decay of proton-rich nuclei in the mass A = 36 – 56 region. Nucl. Phys. A 792 (1–2), 18–86.

Dowdall, M., O'Dea, J., 2002. ^{226}Ra/^{238}U disequilibrium in an upland organic soil exhibiting elevated natural radioactivity. J. Environ. Radioact. 59, 91–104.

Driga, O., 2012. Soft hadron production at the LHC. Prog. Part. Nucl. Phys. 67, 434–439.

Drury, L.O.'C., 1983. An introduction to the theory of diffuse shock acceleration of energetic particles in tenuous plasmas. Rep. Prog. Phys. 46, 973–1027.

Dunn, R., 2013. Elements. Nat. Geographic May 2013, pp. 112–121.

Durrani, S.A., Bull, R.K., 1987. Solid State Nuclear Track Detection: Principles, Methods, and Applications. Pergamon Press, Oxford, p. 304.

Dursun, B., Kurt, E., 2016. Many-body solution to the D_2 gas filled inertial electrostatic confinement device. Int. J. Hydrogen Energy 41, 12546–12554.

Dursun, B., et al., 2017. Energy distributions and radiation emissions in an inertial electrostatic confinement (IEC) device under low and moderate magnetic fields. Int. J. Hydrogen Energy 42, 17874–17885.

Dvorak, J., et al., 2011. IRiS—exploring new frontiers in neutron-rich isotopes of the heaviest elements with a new Inelastic Reaction Isotope Separator. Nucl. Instrum. Methods Phys. Res., Sect. A 652, 687–691.

Dwevedi, A., et al., 2017. Soil sensors: detailed insight into research updates, significance, and future prospects, pp. 561–594. In: Grumezescu, A.M. (Ed.), New Pesticides and Soil Sensors. Elsevier, Academic Press), Amsterdam, p. 766.

Eberhardt, W., 2015. Synchrotron radiation: a continuing revolution in X-ray science—diffraction limited storage rings and beyond. J. Electron Spectrosc. Relat. Phenom. 200, 31–39.

Ebihara, H., 1966. Production of copper-64 in high specific activity by the Szilard-Chalmers Process with copper phthalocyanine. Radiochim. Acta 6 (3), 120–122.

Eckelman, W.C., et al., 2008. True radiotracers: are we approaching theoretical specific activity with Tc-99m and I-123? Nucl. Med. Biol. 35, 523–527.

Ehman, W.D., Vance, D.E., 1991. Radiochemistry and Nuclear Methods of Analysis. John Wiley & Sons, New York.

Einstein, A., 1905a. Über einen die Erzeugung und Verwandlung des Lichtes betreffenden heuristischen Gesichtspunkt, vol. 17. Annalen der Physik, Leipzig, pp. 132–148.

Einstein, A., 1905b. Zur Elektrodynamik bewegter Körper. Ann. Phys. 17, 891–921.

Einstein, A., 1905c. Ist die Träheit eines Körpers von seinem Energieinhalt abhängig? Ann. Phys. 18, 639–641.

Einstein, A., 1916a. Relativity, The Special and the General Theory, Translation by Robert W. Lawson, 1961. Three Rivers Press, New York, p. 188.

Einstein, A., 1916b. Die Grundlage der allgemeinen Relativitätstheorie. Annalen der Physik, Leipzig, pp. 769–822, 49.

Einstein, A., 1923. Fundamental ideas and problems of the theory of relativity. [Nobel] Lecture delivered to the Nordic Assembly of Naturalists at Gothenburg, July 11, 1923. In: Nobel Lectures, Physics, 1901–1921. Elsevier Publishing, Amsterdam, 1967.

El Abd, A., et al., 2013. Experimental determination of moisture distributions in fired clay brick using a ^{252}Cf source: a neutron transmission study. Appl. Radiat. Isot. 74, 78–85.

El-Khayatt, A.M., 2010. Calculation of fast neutron removal cross-sections for some compounds and materials. Ann. Nucl. Energy 37, 218–222.

El-Khayatt, A.M., 2011. NXcom – a program for calculating attenuation coefficients of fast neutrons and gamma-rays. Ann. Nucl. Energy 38, 128–132.

El-Khayatt, A.M., Akkurt, İ., 2013. Photon interaction, energy absorption and neutron removal cross section of concrete including marble. Ann. Nucl. Energy 60, 8–14.

El-Khayatt, A.M., El-Sayed Abdo, A., 2009. MERCSF-N: a program for the calculation of fast neutron removal cross sections in composite shields. Ann. Nucl. Energy 36, 832–836.

Elder, F.R., et al., 1947. Radiation from electrons in a cyclotron. Phys. Rev. 71, 829.

Elder, F.R., et al., 1948. Radiation from electrons accelerated in a cyclotron. Phys. Rev. 74, 52.

Elliott, S.R., Hahn, A.A., Moe, M.K., 1986. Experimental investigation of double-beta decay in ^{82}Se. Phys. Rev. Lett. 56, 2582–2585.

Elliott, S.R., Hahn, A.A., Moe, M.K., 1987. Direct evidence for two-neutrino double-beta decay in ^{82}Se. Phys. Rev. Lett. 59, 2020–2023.

Elliott, S.R., et al., 2006. Pulse shape analysis in segmented detectors as a technique for background reduction in Ge double-beta decay experiments. Nucl. Instrum. Methods Phys. Res., Sect. A 558, 504–510.

Elmahroug, Y., et al., 2015. ParShield: a computer program for calculating attenuation parameters of the gamma rays and the fast neutrons. Ann. Nucl. Energy 76, 94–99.

Elmore, C.S., 2009. The use of isotopically labeled compounds in drug discovery, pp. 515–524. In: Macor, J. (Ed.), Annual Reports in Medicinal Chemistry, vol. 44. Academic Press, p. 736.

Elsheikh, N., et al., 2012. On the use of a (^{252}Cf–^{3}He) assembly for landmine detection by the neutron back-scattering method. Appl. Radiat. Isot. 70, 643–649.

Elster, J., Geitel, H., 1903. Über die durch radioactive Emanation erregte scintillierende Phosphoreszenz der Sidot-Blende. Phys. Z. 4, 439–440.

Elton, L.R.B., 1958. A semi-empirical formula for the nuclear radius. Nucl. Phys. 5, 173–178.

Engelfried, J., 2011. Cherenkov light imaging—fundamentals and recent developments. Nucl. Instrum. Methods Phys. Res., Sect. A 639, 1–6.

Engelfried, J., Paic, G., 2005. Proceedings of the fifth international workshop on ring imaging Cherenkov detectors. Nucl. Instrum. Methods Phys. Res., Sect. A 553 (1–2), 1–380.

Ermis, E.E., Celiktas, C., 2012. Determination of beta attenuation coefficients by means of timing method. Ann. Nucl. Energy 41, 115–118.

Esaka, F., et al., 2017. Analysis of plutonium isotope ratios including $^{238}Pu/^{239}Pu$ in individual U–Pu mixed oxide particles by means of a combination of alpha spectrometry and ICP-MS. Talanta 165, 122–127.

Eshwarappa, K.M., et al., 2007. Comparison of photoneutron yield from beryllium irradiated with bremsstrahlung radiation of different peak energy. An. Nucl. Energy 34, 896–901.

Eskandari, M.R., Kashian, S., 2009. Design of moderator and multiplier systems for D-T neutron source in BNCT. An. Nucl. Energy 36, 1100–1102.

Evans, R.D., 1955. The Atomic Nucleus. McGraw-Hill, New York, 972 pp.

Ezhov, V.F., et al., 2009. Magnetic storage of UCN for a measurement of the neutron lifetime. Nucl. Instrum. Methods Phys. Res., Sect. A 611, 167–170.

Fantidis, J.G., et al., 2009. A transportable neutron radiography system based on a SbBe neutron source. Nucl. Instrum. Methods Phys. Res., Sect. A 606, 806–810.

Farhad Masoudi, S., Rasouli, F.S., 2015. Investigating a multi-purpose target for electron linac based photoneutron sources for BNCT of deep-seated tumors. Nucl. Instrum. Methods Phys. Res., Sect. B 356–357, 146–153.

Farhataziz, Rodgers, M.A.J., 1987. Radiation Chemistry, Principles and Applications. VCH Publishers, Inc., New York.

Farkas, J., 2004. Food irradiation, pp. 785–812. In: Mozumder, A., Hatano, Y. (Eds.), Charged Particle and Photon Interactions with Matter. Marcel Dekkar, New York.

Farkas, J., 2006. Irradiation for better foods. Trends Food Sci. Techn. 17, 148–152.

Farkas, J., Mohácsi-Farkas, C., 2010. History and future of food irradiation. Trends Food Sci. Techn. **22** (2–3), 121–126.

Faw, R.E., Shultis, J.K., 2004. Radiation sources, pp. 613–631. In: Meyers, R.A. (Ed.), Encyclopedia of Physical Science and Technology. Academic Press.

Feather, N., 1938. Further possibilities for the absorption method of investigating the primary β-particles from radioactive substances. Proc. Cambridge Philos. Soc. 34, 599–611.

Feather, N., 1940. Lord Rutherford. Blackie & Son, Ltd.

Feinberg, G., et al., 2009. LiLiT – a liquid lithium target as an intense neutron source for nuclear astrophysics at the Soreq Applied Research Accelerator Facility. Nucl. Phys. A 827, 590c–592c.

Feld, B.T., Weiss Szilard, G. (Eds.), 1972. The Collected Works of Leo Szilard: Scientific Papers. MIT Press, Cambridge, p. 642.

Fenyves, E., Haiman, O., 1969. The Physical Principles of Nuclear Radiation Measurements. Academic Press, New York.

Fermi, E., 1934a. Versuch einer theorie der β-Strahlen. Z. Phys. 88, 161–177.

Fermi, E., 1934b. Attempt at a theory of beta rays. Nuovo Cimento 11, 1–21.

Fermi, E., 1934c. Radioactivity induced by neutron bombardment. Nature 133, 757.

Fermi, E., 1934d. Possible production of elements of atomic number higher than 92. Nature 133, 898–899.

Fermi, E., 1938. Artificial radioactivity produced by neutron bombardment. Nobel lecture, December 12, 1938. In: Nobel Lectures, Physics, 1922–1941. Elsevier Publishing, Amsterdam, 1965.

Fermi, E., 1940a. Reactions produced by neutrons in heavy elements. Science 92, 269–271.

Fermi, E., 1940b. Reactions produced by neutrons in heavy elements. Nature 146, 640–642.

Fermi, E., 1949. On the origin of the cosmic radiation. Phys. Rev. 75 (8), 1169–1174.

Fermi, E., Amaldi, E., 1936. On the absorption and the diffusion of slow neutrons. Phys. Rev. 50, 899–928.

Fermi, E., Amaldi, E., D'Agostino, O., Rasetti, F., Segre, E., 1934. Artificial radioactivity produced by neutron bombardment. Proc. R. Soc. London A146 (857), 483–500.

Fermi, E., Amaldi, E., Wick, G.C., 1938. On the albedo of slow neutrons. Phys. Rev. 53, 493.

Fernandes, H.M., et al., 2006. Radioecological characterization of a uranium mining site located in a semi-arid region in Brazil. J. Environ. Radioact. 88, 140–157.

Ferri, E., 2016. Neutrino mass experiments with Ho. Nucl. Part. Phys. Proc. 273–275, 1922–1927.

Feshbach, H., 1952. The coulomb scattering of relativistic electrons and positrons by nuclei. Phys. Rev. 88 (2), 295–297.

Fettouhi, A., et al., 2006. Stopping of high-Z ions at intermediate velocities. Nucl. Instrum. Methods Phys. Res., Sect. B 245, 22–27.

Firestone, R.B., et al., 1996. Table of IsoPopes, eighth Fd., vols. I and II. JohD Wiley & Sons, New York.

Fischer, T.B., et al., 2018. Changes in the structure of birnessite during siderophore-promoted dissolution: a time-resolved synchrotron X-ray diffraction study. Chem. Geol. 476, 46–58.

Flammersfeld, A., 1946. Eine Beziehung zwischer Energie und Reichweite für Beta-Strahlen kleiner und mittlerer Energie. Naturwissenschaften 33, 280–281.

Fomin, N., et al., 2015. Fundamental neutron physics beamline at the spallation neutron source at ORNL. Nucl. Instrum. Methods Phys. Res., Sect. A 773, 45–51.

for the SNO Collaboration Formaggio, J.A., 2009. Measurement of atmospheric neutrinos at the Sudbury Neutrino Observatory. Nucl. Phys. A 827, 498c–500c.

Forman, P., 1982. The Fall of Parity. NIST, Washington, D.C. http://physics.nist.gov/GenInt/Parity/parity.html.

Fowler, W.A., Oppenheimer, J., 1938. Scattering and loss of energy of fast electrons and positrons in lead. Phys. Rev. 54 (5), 320–324.

Frank, I.M., 1958. Optics of light sources in refractive media. Nobel Lecture, December 11, 1958. In: Nobel Foundation Staff (Eds.), Nobel Lectures, Physics, 1942-62. (1964). Elsevier Publishing, Amsterdam.

Frank, I., Tamm, I., 1937. Coherent visible radiation of fast electrons passing through matter. Dokl. Akad. Nauk, SSSR 14 (3), 109–114.

Friedlander, G., et al., 1964. Nuclear and Radiochemistry, second ed. John Wiley & Sons, New York. 585 pp.

Fujitaka, K., 2005. High-level doses brought by cosmic rays. Int. Congr. Ser. 1276, 124–128.

Fukuda, S., et al., 2003. The Super-Kamiokande detector. Nucl. Instrum. Methods Phys. Res., Sect. A 501, 418–462.

Fukuda, Y., et al., 2010. InP solid state detector for measurement of low energy solar neutrinos. Nucl. Instrum. Methods Phys. Res. A 623 (1), 460–462.

Furrer, A., 2016. Neutron sources, pp. 1–6. In: Hashmi, M.S.J. (Ed.), Reference Module in Materials Science and Materials Engineering. Elsevier, Amsterdam.

Gaisser, T.K., 2001. Origin of cosmic radiation. Astron. AIP Conf. Proc. 558, 27–42.

Gaisser, T.K., Maurer, R.H., 1973. Cosmic $\bar{p}$ production in interstellar pp collisions. Phys. Rev. Lett. 30, 1264–1267.

Gaisser, T.K., Schaefer, R.K., 1992. Cosmic-ray secondary antiprotons: a closer look. Astrophys. J. 394, 174–184.

Gaisser, T.K., Stanev, T., 2002. Cosmic rays. In: Hagiwara, et al. (Eds.), The Review of Particle Physics, Phys. Rev., vol. D66. University of California, Lawrence Berkeley Laboratories, pp. 23.1–23.19.

Gamow, G., 1928. Zur Quantentheorie des Atomkernes. Zeit. F. Phys. 51, 204–212.

Gamow, G., 1930. Mass defect and nuclear constitution. Proc. R. Soc. Lond., A 126, 632–644.

Gamow, G., 1948. Evolution of the universe. Nature 162, 680–682.

Gamow, G., et al., 1929. Discussion on the Structure of Atomic Nuclei. Proc. R. Soc. Lond., A 6 (123), 373–390.

Garcea, S.C., et al., 2017. Mapping fibre failure in situ in carbon fibre reinforced polymers by fast synchrotron X-ray computed tomography. Compos. Sci. Technol. 149, 81–89.

García-Toraño, E., et al., 2017. Simulation of decay processes and radiation transport times in radioactivity measurements. Nucl. Instrum. Methods Phys. Res., Sect. B 396, 43–49.

Gardner, R.P., et al., 2004. A semi-empirical model for a ^{90}Sr beta-particle transmission thickness gauge for aluminum alloys. Nucl. Instrum. Methods Phys. Res. B 213, 357–363.

Garino, C., et al., 2014. Determination of the electronic and structural configuration of coordination compounds by synchrotron-radiation techniques. Coordin. Chem. Rev. 277–278, 130–186.

Garwin, R.L., et al., 1957. Observations of the failure of conservation of parity and charge conjugation in meson decays: the magnetic moment of the free muon. Phys. Rev. 105, 1413–1414.

Gautier, M.A. (Ed.), 1995. Health and Environmental Chemistry: Analytical Techniques, Data Management, and Quality Assurance. Los Alamos National Laboratory, Los Alamos, NM. Manual LA-10300-M, Vol. III, UC-907, pp. WR190-W191 -WR190-16.

Gautreau, R., Savin, W., 1999. Theory and Problems in Modern Physics. McGraw-Hill, New York.

Gawiser, E., Silk, J., 2000. The cosmic microwave background radiation. Phys. Rep. 333–334, 245–267.

Gehman, V.M., et al., 2010. Systematic effects in pulse shape analysis of HPGe detector signals for $0\nu\beta\beta$. Nucl. Instrum. Methods Phys. Res., Sect. A 615, 83–92.

Geiger, H., Nuttall, J.M., 1911. The range of alpha particles from various substances and a relationship between range and period of transformation. Philos. Mag. 22, 613–621.

Gell-Mann, M., 1964. A schematic model of baryons and mesons. Phys. Lett. 8, 214–215.

Geloni, G., et al., 2016. Brightness of synchrotron radiation from wigglers. Nucl. Instrum. Methods Phys. Res., Sect. A 807, 13–29.

Ghal–Eh, N., et al., 2017. FLUKA simulation studies on in–phantom dosimetric parameters of a LINAC–based BNCT. Radiat. Phys. Chem. 141, 36–40.

Ghassoun, J., et al., 2010. Detailed dose distribution prediction of Cf-252 brachytherapy source with boron loading dose enhancement. Appl. Radiat. Isot. 68 (2), 265–270.

Ghosh, M., et al., 2017a. Determination of impurities in graphite using synchrotron radiation based X-ray fluorescence spectrometry. Appl. Radiat. Isot. 128, 210–215.

Ghosh, R., et al., 2017b. Measurement of ^{58}Fe(p,n)^{58}Co reaction cross-section within the proton energy range of 3.38 to 19.63 MeV. Nucl. Phys. A 964, 86–92.

Gibbs, W.W., 1998. A massive discovery. Sci. Am. 279 (2), 18–20.

Gibson, J.A.B., Piesch, E., 1985. Neutron Monitoring for Radiological Protection. Technical Report Series No. 252. International Atomic Energy Agency, Vienna.

Gielow, P., 1990. Preparation of ^{64}Cu-labelled diethyl-HIDA. Appl. Radiat. Isot. 41 (2), 237–238.

Giller, M., 2008. Cosmic rays between the knee and the ankle. J. Phys. G 35 (2), 023201.

Giovinazzo, J., et al., 2002. Two-proton radioactivity of ^{45}Fe. Phys. Rev. Lett. 89 (10), 102501.

Giuliani, A., 2011. Advances in the detection of neutrinoless double beta decay. Nucl. Phys. B (Proc. Suppl.) 217, 9–14.

Glasser, R.G., et al., 1961. Mean lifetime of the neutral pion. Phys. Rev. 123, 1014–1020.

Glendenin, L.E., 1948. Determination of the energy of beta particles and photons by absorption. Nucleonics 2, 12–32.

Glover, R.N., Jones, A.D.W., 1966. Stripping and pick-up reactions on the carbon isotopes. Nucl. Phys. 84, 673–682.

Goeppert-Mayer, M., 1935. Double beta-disintegration. Phys. Rev. 48, 512–516.

Goeppert-Mayer, M., 1948a. On Closed Shells in Nuclei. United States Atomic Energy Commission, AECD – 1818. Argonne National Laboratory, p. 5.

Goeppert-Mayer, M., 1948b. On closed shells in nuclei. Phys. Rev. 74, 235–239.

Goeppert-Mayer, M., 1949. On closed shells in nuclei, II. Phys. Rev. 75, 1969–1970.

Goeppert-Mayer, M., Jensen, J.H.D., 1955a. Elementary Theory of Nuclear Shell Structure. John Wiley & Sons, New York, p. 269.

Goeppert-Mayer, M., Jensen, J.H.D., 1955b. The shell model. I. Shell closure and jj coupling. In: Siegbahn, K. (Ed.), Alpha, Beta- and Gamma-Ray Spectroscopy. North Holland Publishing, Amsterdam, p. 557.

Goeppert-Mayer, M., 1963. The Shell Model. Nobel Lecture, December 12, 1963. https://www.nobelprize.org/prizes/physics/1963/mayer/lecture/.

Goeppert-Mayer, M., 1964. The shell model. Science 145, 999–1006.

Goeppert-Mayer, M., 1972. The shell model. Nobel Lecture given on December 12, 1963. In: Nobel Lectures, Physics 1963-1970. Elsevier Publishing, Amsterdam.

Goeppert-Mayer, M., Jensen, J.H.D., 1965. Shell Closure and *jj* Coupling. In: Siegbahn, K. (Ed.), Alpha-, Beta- and Gamma-Ray Spectroscopy. North Holland Publishers, Amsterdam, pp. 557–581.

Golden, R.L., et al., 1979. Evidence for the existence of cosmic-ray antiprotons. Phys. Rev. Lett. 43, 1196–1199.

Golden, R.L., et al., 1984. Energy dependence of the $\overline{p}/p$ ratio in cosmic rays. Astrophys. Lett. 24, 75–83.

Golshanian, M., et al., 2016. ^{1ambo}–Be photo-neutron source for BNCT: is it possible? Nucl. Instrum. Methods Phys. Res., Sect. A 835, 182–185.

Gómez, H., et al., 2007. Background reduction and sensitivity for germanium double beta decay experiments. Astropart. Phys. 28, 435–447.

Gonçalves, M., et al., 2017. Two-proton emission half-lives in the effective liquid drop model. Phys. Lett. B 774, 14–19.

Götz, M., et al., 2017. Radiochemical study of the kinematics of multinucleon transfer reactions in ^{48}Ca + ^{248}Cm collisions 10% above the Coulomb barrier. Nucl. Phys. A 961, 1–21.

Gould, C.R., et al., 2004. Nuclear physics, pp. 721–738. In: Meyers, R.A. (Ed.), Encyclopedia of Physical Science and Technology. Academic Press, Orlando.

Gracanin, V., et al., 2017. A convenient verification method of the entrance photo-neutron dose for an 18 MV medical linac using silicon p-i-n diodes. Radiat. Meas. 106, 391–398.

Grau Carles, A., Grau Malonda, A., 2006. Computational aspects in modeling the interaction of low-energy X-rays with liquid scintillators. Appl. Radiat. Isot. 64, 1515–1519.

Grau Carles, A., Kossert, K., 2009. Monte Carlo simulation of Auger electron spectra. Appl. Radiat. Isot. 67, 192–196.

Grau Malonda, A., Grau Carles, A., 2002. Half-life determination of ^{40}K by LSC. Appl. Radiat. Isot. 56, 153–156.

Grau Malonda, A., Grau Carles, A., Garcia, G., 2006. Mean values of LMM Auger transition in a KLM model. Appl. Radiat. Isot. 64, 1485–1491.

Gray, R., 2017. Breaking the periodic table. New Scientist 234 (3121), 40–41.

Green, J.H., Maddock, A.G., 1949. (n, γ) recoil effects in potassium chromate and dichromate. Nature 164, 788–789.

Greenberg, O.W., 1964. Spin and unitary spin independence in a Paraquark model of baryons and mesons. Phys. Rev. Lett. 13, 598–602.

Greiner, W., 2008. Structure of vacuum and elementary matter: from superheavies via hypermatter to antimatter—the vacuum decay in supercritical fields, pp. 99–150. In: Salomonson, S., Lindroty, E. (Eds.), Advances in Quantum Chemistry, vol. 53. Elsevier.

Greiner, W., et al., 1985. An exotic nuclear decay of ^{223}Ra by emission of ^{14}C nuclei. Z. Phys. A 320, 347–348.

Grieder, P.K.F., 2001. "Cosmic Rays at Earth: Researcher's Reference Manual and Data Book". Elsevier, Amsterdam, p. 1093.

Grimani, C., 2005. Upper limit to the cosmic-ray positron flux generated at the pulsar polar cap. In: Proc. Of the 29th Int. Cosmic Ray Conf. (Pune, India, 3-10 August 2005), vol. 3, pp. 9–12.

Grimani, C., 2009. Origin of cosmic-ray positrons. In: Proc. Of the 31st Int. Cosmic Ray Conf., Lódź, Poland, 7-15 July 2009.

Grochulski, P., et al., 2017. Review of Canadian Light Source facilities for biological applications. Nucl. Instrum. Methods Phys. Res., Sect. B 411, 17–21.

Groom, D.E., et al., 2001. Muon stopping power and range tables 10 MeV to 100 TeV. Atom. Data Nucl. Data Tables 78, 183–356.

Gullstrand, A., 1925. Presentation Speech, The Nobel Prize in Physics 1924. In: Nobel Lectures, Physics 1922-1941. (1965). Elsevier Publishing, Amsterdam.

Gümüş, H., 2005. Simple stopping power formula for low and intermediate energy electrons. Rad. Phys. Chem. 72, 7–12.

Gümüş, H., 2008. New stopping power formula for intermediate energy electrons. Appl. Radiat. Isot. 66, 1886–1890.

Gurney, R.W., Condon, E.U., 1928. Wave mechanics and radioactive disintegration. Nature 122, 439.

Hadad, K., et al., 2016. Moderation and shielding optimization for a ^{252}Cf based prompt gamma neutron activation analyzer system. Int. J. Hydrogen Energy 41, 7221–7226.

Hagiwara, K., et al., (Particle Date Group), 2002. Physical Review D66, 010001-1, https://journals.aps.org/prd/abstract/10.1103/PhysRevD.66.010001.

Hahn, O., Meitner, L., 1924. Das β-strahlenspektrum von Radium und seine Deutung. Z. Phys. 26, 161–168.

Hahn, O., Strassmann, F., 1939a. Über den Nachweis und das Verhalten der bei Bestrahlung des Urans mittels Neutronen entstehenden Erdalkalimetalle. Naturwissenschaften 27, 11–15.

Hahn, O., Strassmann, F., 1939b. Nachweis der Entstehung activer bariumisotope aus Uran und Thorium durch Neutronenbestrahlung; nachweis weiterer aktiver Bruchtucke bei der Uranspaltung. Naturwissenschaften 27, 89–95.

Halpern, A., 1988. "Schaum's 3000 Solved Problems in Physics. McGraw-Hill, New York.

Hammer, G.P., et al., 2000. Comparing different methods of estimating cosmic radiation exposure of airline personnel. Radiat. Environ. Biophys. 39, 227–231.

Han, M.Y., Nambu, Y., 1965. Three-triplet model with double SU(3) symmetry. Phys. Rev. 139, B1006–B1010.

Harbottle, G., 1954. Szilard-Chalmers reaction in crystalline compounds of chromium, radiochemical analysis. J. Chem. Phys. 22, 1083.

Harms, A.V., Jerome, S.M., 2004. On the integrated decay and ingrowth equations used in the measurement of radioactive decay families: the general solution. Appl. Radiat. Isot. 61, 367–372.

Hasselberg, K.B., 1908. Presentation Speech of the Nobel Prize in Chemistry 1908. In: Nobel Lectures, Chemistry 1901-1921. Elsevier Publishing, Amsterdam (1966).

Hawking, S.W., 1988. A Brief History of Time. Bantam Press, London.

Haxel, O., Jensen, J.H.D., Suess, H.E., 1949. On the "Magic Numbers" in nuclear structure. Phys. Rev. 75, 1766.

Hayashida, N., et al., 1994. Observation of a very energetic cosmic ray well beyond the predicted 2.7 K cutoff of the primary energy spectrum. Phys. Rev. Lett. 73, 3491–3494.

Heckman, H.H., Lindstrom, P.J., 1969. Stopping-power differences between positive and negative pions at low velocities. Phys. Rev. Letts. 22 (17), 871–874.

Heisenberg, W., 1926a. Quantum mechanics. Naturwissenschaften 14, 989–994.

Heisenberg, W., 1926b. Multi-boby problem and resonance in the quantum mechanics. Z. Phys. 38 (6/7), 411–426.

Heisenberg, W., 1926c. Fluctuation appearances and quantum mechanics. Z. Phys. 40 (7), 501–506.

Heisenberg, W., 1927. Über den anschaulichen Inhalt der quantentheoretischen Kinematik und Mechanik. Z. Phys. 43, 172–198.

Heisenberg, W., 1929. The development of the quantum theory 1918-1928. Naturwissenschaften 17, 490–496.

Heisenberg, W., 1934. Considérations théoriques générales sur la structure du noyau, p. 316. In: Institut International de Physique Solvay, "Structure et Propriétés des Noyaux Atomiques: Rapports et

Discussions du Septième Conseil de Physique tenu a Bruxelles du 22 au Octobre 1933. Gauthier-Villars, Paris.

Henning, R., 2016. Current status of neutrinoless double-beta decay searches. Rev. Phys. 1, 29–35.

Hertz, H., 1887. Über einen Einfluss des ultravioleten Lichtes auf die elektrische Entladung. Sitz.-ber. Berl. Akad. 9 June, also Wied. Ann. Phys. Vol. 31.

Hertz, K.L., et al., 2003. Alpha-emitting radioisotopes for switchable neutron generators. Nucl. Instrum. Methods Phys. Res. A 505, 41–45.

Hess, V.F., 1912. Über Beobachtungen der durchdringenden Strahlung bei sieben Freiballonfahrten. Phys. Z. 13, 1084–1091.

Hess, V.F., 1936. Unsolved problems in physics: tasks for the immediate future in cosmic rays studies. Nobel Lecture, December 12, 1939. In: Nobel Lectures, Physics, 1922-1941. (1965). Elsevier Publishing, Amsterdam.

Hess, V.F., 1940. The discovery of cosmic radiation. Thought (Fordham Quarterly) 15, 225–236.

Hetherington, E.L., et al., 1986. The preparation of high specific activity copper-64 for medical diagnosis. Appl. Radiat. Isot. 37 (12), 1242–1243.

Higashi, T., et al., 2003. Preliminary results of low pressure discharge experiments of a cylindrical inertial electrostatic confinement fusion device aimed for a small neutron source. Fusion Sci. Techn. 44, 544–548.

Hirsh, R.L., 1967. Inertial-electrostatic confinement of ionized fusion gases. J. Appl. Phys. 38 (11), 4522–4535.

Hodges, M., et al., 2017. Characterization of a 6 MeV accelerator driven mixed neutron/photon source. Phys. Proc. 90, 164–169.

Hof, M., et al., 1996. Measurement of cosmic-ray antiprotons from 3.7 to 19 GeV. Astrophys. J. Lett. 467 (1), L33–L37.

Hofmann, S., 2009. Superheavy elements. Lect. Notes Phys. 764, 203–252.

Hofmann, S., et al., 1982. Proton radioactivity of ^{151}Lu. Z. Phys. A – Atoms and Nuclei 305, 111–123.

Hofstadter, R., et al., 1953a. Scattering of high-energy electrons and the method of nuclear recoil. Phys. Rev. 91 (2), 422.

Hofstadter, R., et al., 1953b. High-energy electron scattering and nuclear structure determinations. Phys. Rev. 92 (4), 978–987.

Hofstadter, R., 1961. The electron-scattering method and its application to the structure of nuclei and nucleons. In: Nobel Lectures, Physics 1942-1962. Elsevier Publishing Company, Amsterdam.

Hofstadter, R., McAllister, R.W., 1955. Electron scattering from the proton. Phys. Rev. 98 (1), 217.

Holden, N.E., 1997a. Table of isotopes. In: Lide, D.R. (Ed.), CRC Handbook of Chemistry and Physics, 77th ed. CRC Press, Boca Raton, FL, pp. 11-38–11-143.

Holden, N.E., 1997b. Neutron scattering and absorption properties. In: Lide, D.R. (Ed.), CRC Handbook of Chemistry and Physics, 77th ed. CRC Press, Boca Raton, FL, pp. 11-144–11-158.

Holloway, M.G., Baker, C.P., 1944. Note on the Origin of the Term "Barn". Los Alamos Technical Report LAMS-523, p. 4.

Holloway, M.G., Baker, C.P., 1972. How the barn was born. Phys. Today 25 (7), 9.

Honda, T., et al., 1965. Nuclear direct interaction in the $^{12}C(t,\alpha)^{11}B$ reaction. Nucl. Phys. 62, 561–574.

Honkanen, J., et al., 1983. Beta-delayed two-proton decay of ^{26}P. Phys. Lett. B 133 (304), 146–148.

Hörandel, J.R., 2008a. Cosmic-ray composition and its relation to shock acceleration by supernova remnants. Adv. Space Res. 41, 442–463.

Hörandel, J.R., 2008b. The origin of galactic cosmic rays. Nucl. Instrum. Methods Phys. Res., Sect. A 588, 181–188.

Hörandel, J.R., 2010. Cosmic rays at the highest energies. Prog. Part. Nucl. Phys. 64, 351–359.

Horányi, G., 2004. Advantages of the radiotracer technique, pp. 5–8. In: Horányi, G. (Ed.), Interface Science and Technology, Radiation Studies of Interfaces, vol. 3. Elsevier, Amsterdam, p. 435.

Horowitz, Y. (Ed.), 2006. Microdosimetric Response of Physical and Biological Systems to Low- and High-LET Radiations. " Elsevier, Amsterdam, p. 500.

Hubbell, J.H., 1969. "Photon Cross Sections, Attenuation Coefficients, and Energy Absorption Coefficients from 10 keV to 100 GeV." NSRDS-NBS 29, Natl. Stand. Ref. Data Ser. National Bureau of Standards (U.S.), p. 80.

Huh, C.-A., 1999. Dependence of the decay rate of ^{7}Be on chemical forms. Earth Planet. Sci. Lett. 171, 325–328.

Hultqvist, K., 2011. IceCube: Physics, status, and future. Nucl. Instrum. Methods Phys. Res., Sect. A 626–627, S6–S12.

IAEA, 2006. Directory of Cyclotrons used for Radionuclide Production in Member States. IAEA-DCRP/CD, International Atomic Energy Agency, Vienna.

IAEA, 2007. "Nuclear Technology Review". GC(951)/INF/3. International Atomic Energy Agency, Vienna, p. 32.

IAEA, 2016. IAEA Annual Report 2016. International Atomic Energy Agency, Vienna, p. 149.

IAEA, 2017. "Nuclear Technology Review". GC(61)/INF/4. International Atomic Energy Agency, Vienna, p. 45.

Ianni, A., 2017. Solar neutrinos. Prog. Part. Nucl. Phys. 94, 257–281.

Ianni, A., et al., 2005. How to observe ^{8}B solar neutrinos in liquid scintillator detectors. Phys. Lett. B 627, 38–48.

ICRP, 1991. Publication 60, "1990 Recommendation of the International Commission on Radiological Protection. Pergamon Press.

ICRU, 2000. Report 63. Nuclear Data for Neutron and proton Radiotherapy and for Radiation Protection, ISBN 0-913394-62-9.

ICRU, 2007. Report 16. Linear Energy Transfer. International Commission on Radiation Units and Measurements. https://icru.org/home/reports/linear-energy-transfer-report-16.

Ilić, R., Durrani, S.A., 2003. Solid state nuclear track detectors. pp.179–237. In: L'Annunziata, M.F. (Ed.), Handbook of Radioactivity Analysis, second ed. Elsevier, Amsterdam, p. 1273.

Ilyin, A.M., 2017. Auger electron spectroscopy, pp. 363–381. In: Thomas, S., et al. (Eds.), Microscopy Methods in Nanomaterials Characterization. Elsevier, Amsterdam, p. 414.

Inami, K., 2017. Cherenkov light imaging in particle and nuclear physics experiments. Nucl. Instrum. Methods Phys. Res., Sect. A 876, 278–281.

Iqbal, M., Ejaz, M., 1978. Enrichment of ^{51}Cr through Szilard-Chalmers effect. J. Radioanal. Nucl. Chem. 45 (1), 185–188.

Isotope Products Laboratories, 1995. Californium-252 fission foils and neutron sources. In: Radiation Sources for Research, Industry and Environmental Applications. Isotope Products Laboratories, Burbank, CA, p. 55.

Ivanovich, M., 1994. Uranium series disequilibrium: concepts and applications. Radiochim. Acta 64, 81–94.

Ivanovich, M., Harmon, R.S., 1992. Uranium Series Disequilibrium: Applications to Earth, Marine and Environmental Sciences. Clarendon Press, Oxford, p. 910.

Iwanenko, D.D., Pomeranchuk, I.Y., 1944a. On the maximal energy attainable in a betatron. Dokl. Akad. Nauk. SSSR 44, 364.

Iwanenko, D.D., Pomeranchuk, I.Y., 1944b. On the maximal energy attainable in a betatron. Phys. Rev. 65 (11–12), 343.

Iwashita, Y., et al., 2011. Satellite pulsed tiny neutron source at Kyoto University, Sakyo. Nucl. Instrum. Methods Phys. Res., Sect. A 634 (1, Suppl. 1), S97–S99.

Jackson, K.P., et al., 1970. ^{53m}Co: a proton-unstable isomer. Phys. Lett. B 33 (4), 281–283.

Jaklevic, J.M., et al., 1983. A precision beta gauge using a plastic scintillator and photomultiplier detector. Nucl. Instrum. Methods 214, 517–518.

Jakubek, J., et al., 2011. Direct observation of decay of radioactive nuclei with spatial and time coincidence technique. Nucl. Instrum. Methods Phys. Res., Sect. A 633 (Suppl. 1), S203–S205.

Janni, J.F., 1982. Proton range-energy tables, 1 keV – 10 GeV. Atomic Data Nucl. Date Tables 27, 147–339.

Jansen, J.A., et al., 1971. Nuclear charge radii of ^{12}C and ^{9}Be. Nucl. Phys. A 188 (2), 337–352.

Janson, E., et al., 2000. Nuclear localization of In-111 after intravenous injection of [In-111-DTPA-D-Phe(1)]-octreotide in patients with neuroendocrine tumors. J. Nucl. Med. 41 (9), 1514–1518.

Jia, W., Ehrhardt, G.J., 1997. Enhancing the specific activity of Re-186 using an inorganic Szilard-Chalmers process. Radiochim. Acta 79 (2), 131–136.

JET Team, 1992. Fusion energy production from a deuterium-tritium plasma in the JET tokamak. Nucl. Fusion 32, 187–201.

Johansson, A., Wilkin, C., 2009. Hard bremsstrahlung in the $pp \rightarrow pp\gamma$ reaction. Phys. Lett. B 673, 5–8.

Johansen, G.A., Jackson, P., 2004. Isotope Gauges for Industrial Process Measurements. John Wiley & Sons Ltd., Chichester, UK https://doi.org/10.1002/0470021098.fmatter.

Joliot-Curie, I., Joliot, F., 1934a. Artificial production of a new kind of radioelement. Nature 133, 201.

Joliot-Curie, I., Joliot, F., 1934b. Un nouveau type of radioactivité. Compt. Rend. 198S, 254–256.

Joliot-Curie, I., Joliot, F., 1934c. Séparation chimique des nouveaux radioéléments émetteurs d'électrons positifs. Compt. Rend. 198S, 559–561.

Jones, H.G., 1973. Estimation of plant water status with the beta-gauge. Agric. Meteorol. 11, 345–355.

Jones, J.A., et al., 2010. Ionizing radiation as a carcinogen, pp. 181–228. In: McQueen, C. (Ed.), Comprehensive Toxicology, second ed. Elsevier Science, Amsterdam, p. 250.

Kahlbow, J., et al., 2017. Neutron radioactivity—lifetime measurements of neutron-unbound states. Nucl. Instrum. Methods Phys. Res., Sect. A 866, 265–271.

Karelin, Y.A., et al., 1997. Californium-252 neutron sources. Appl. Radiat. Isot. 48 (10–12), 1563–1566.

Kaschl, G. Th., et al., 1970. Proton pick-up from ^{19}F, ^{20}Ne and ^{22}Ne. Nucl. Phys. A 155, 417–442.

Katz, U.F., 2009. Underwater neutrino telescopes. Nucl. Phys. B (Proc. Suppl.) 188, 245–251.

Katz, L., Penfold, A.S., 1952. Range-energy relations for electrons and the determination of beta-ray end-point energies by absorption. Rev. Mod. Phys. 24 (1), 28–44.

Kearns, E., et al., 1999. Detecting massive neutrinos. Sci. Am. 281 (2), 64–71.

Kesler, S.E., et al., 2012. Global lithium resources: relative importance of pegmatite, brine and other deposits. Ore Geol. Rev. 48, 55–69.

Kesterbaum, D., 1998. Neutrinos throw their weight around. Science 281 (5383), 1594–1595.

Khachan, J., Samarian, A., 2007. Dust diagnostics on an inertial electrostatic confinement device. Phys. Lett. A 363, 297–301.

Khalil, E.I., 2011. Investigation of the influence of nuclear charge and β-endpoint energy on the internal Bremsstrahlung of ^{89}Sr and ^{147}Pm. Radiat. Phys. Chem. 80, 669–672.

Kichizo, T., 1964. The preparation of phosphorus-32 in a high specific activity using the Szilard-Chalmers Effect. Bull. Chem. Soc. Jpn. 37 (12), 1730–1733.

Kim, I.J., et al., 2008. A D–D neutron generator using a titanium drive-in target. Nucl. Instrum. Methods Phys. Res., Sect. B 266, 829–833.

Kim, H.S., et al., 2009. Development of a beta gauge system for a fabric density measurement. Appl. Radiat. Isot. 67, 1213–1215.

Klarsfeld, S., et al., 1986. Determination of the deuteron mean square radius. Nucl. Phys. A 456 (3), 373–396.

Klepper, O., et al., 1982. Direct and beta-delayed proton decay of very neutron-deficient rare-earth isotopes produced in the reaction $^{58}Ni + {}^{92}Mo$. Z. Phys. A – Atoms Nucl. 305, 125–130.

Knapp, D.R., 2015. Planar geometry inertial electrostatic confinement fusion device. J. Phys.: Conf. Ser. 591, 012018. https://doi.org/10.1088/1742-6596/591/1/012018.

Knapp, A.K., et al., 1985. An evaluation of beta attenuation for estimating aboveground biomass in a tallgrass prairie. J. Range Manag. 38 (6), 556–558.

Knipp, J.K., Uhlenbeck, G.E., 1936. Emission of gamma radiation during the beta decay of nuclei. Physica III (6), 425–439.

Kobashi, J.A., et al., 1979. Radioactive disequilibrium with uranium, thorium and radium isotopes leached from euxenite. Radiochim. Acta 26 (2), 107–112.

Kobayashi, Y., 2016. Applications of nuclear technique to biological sciences labelled compounds, radioactive tracers, and x-ray tomography. Curr. Topics Med. Chem. 16 (24), 2694–2702.

Koch, L., 1995. Radioactivity and fission energy. Radiochim. Acta 70/71, 397–402.

Kojima, Y., et al., 2014. Half-life measurements of excited levels in fission products around mass number 150. Nucl. Data Sheets 120, 52–55.

Kolhinen, V.S., et al., 2010. Accurate Q value for the ^{74}Se double-electron-capture decay. Phys. Lett. B 684 (1), 17–21.

Koshimizu, M., et al., 2016. Influence of linear energy transfer on the scintillation decay behavior in a lithium glass scintillator. J. Luminesc. 169, 678–681.

Koshimizu, M., et al., 2017. Analysis of linear energy transfer effects on the scintillation properties of $Bi_4Ge_3O_{12}$ crystals. Nucl. Instrum. Methods Phys. Res., Sect. B 409, 19–22.

Kossert, K., Grau Carles, A., 2008. Study of a Monte Carlo rearrangement model for the activity determination of electron-capture nuclides by means of liquid scintillation counting. Appl. Radiat. Isot. 66, 998–1005.

Kossert, K., Grau Carles, A., 2010. Improved method for the calculation of the counting efficiency of electron-capture nuclides in liquid scintillation samples. Appl. Radiat. Isot. 68, 1482–1488.

Kostroun, V.O., et al., 1971. Atomic radiation transition probabilities to the 1s state and theoretical K-shell fluorescence yields. Phys. Rev. A3, 533–545.

Krane, K.S., 1988. Introductory Nuclear Physics. John Wiley & Sons, New York.

Krappe, H.J., Wahsweiler, H.G., 1967. Stability of nuclei against pear-shaped deformations. Nucl. Phys. A 104 (3), 633–641.

Krása, A., et al., 2010. Neutron production in a Pb/U-setup irradiated with 0.7–2.5 GeV protons and deuterons. Nucl. Instrum. Methods Phys. Res., Sect. A 615, 70–77.

Kraus, C., et al., 2005. Final results from phase II of the Mainz neutrino mass search in tritium decay. Eur. Phys. J. C 40 (4), 447–468.

Kristiansen, J.R., et al., 2010. Coupling between cold dark matter and dark energy from neutrino mass experiments. New Astron. 15, 609–613.

Kudo, H., 1995. Radioactivity and fusion energy. Radiochim. Acta 70/71, 403–412.

Kulikov, G.G., et al., 2016. Advanced nuclear fuel cycle for the RF using actinides breeding in thorium blankets of fusion neutron source. Nucl. Energy Technol. 2, 147–150.

Kurosaki, H., et al., 2014. Impact of environmental curium on plutonium migration and isotopic signatures. Environ. Sci. Technol. 48 (23), 13985–13991.

Kurt, E., Arslan, S., 2012. An inertial electrostatic confinement (IEC) device modeling and the effects of different cathode structures to the fields. Energy Convers. Manag. 63, 55–62.

Kurtukian-Nieto, T., et al., 2008. A new analysis method to determine β-decay half-lives in experiments with complex background. Nucl. Instrum. Methods Phys. Res., Sect. A 589, 472–483.

Kurudirek, M., et al., 2014. Remarks on the extension and validity of an empirical formula for the fast-neutron removal cross-section: the effective atomic weight. Ann. Nucl. Energy 70, 230–232.

Kushnir, M.M., et al., 2004. Simultaneous quantitative analysis of isobars by tandem mass spectrometry from unresolved chromatographic peaks. J. Mass Spectrom. 39 (5), 532–540.

Labidi, S., et al., 2010. Natural radioactivity levels in mineral, therapeutic and spring waters in Tunisia. Radiat. Phys. Chem. 79, 1196–1202.

Lacoste, V., 2010. Review of radiation sources, calibration facilities and simulated workplace fields. Radiat. Meas. 45 (10), 1083–1089.

Lakosi, L., Nguyen, C.T., 2008. Neutron interrogation of high-enriched uranium by a 4 MeV linac. Nucl. Instrum. Methods Phys. Res., Sect. B 266, 3295–3301.

Lakosi, L., et al., 2011. Neutron interrogation of shielded/unshielded uranium by a 4 MeV linac. Appl. Radiat. Isot. 69, 1251–1254.

Lal, D., 2009. Cosmogenic isotopes. In: Steele, J., Thorpe, S., Turekian, K. (Eds.), Encyclopedia of Ocean Sciences. Academic Press/Elsevier Science Publishers, Amsterdam, pp. 678–687.

Lal, D., Peters, B., 1967. Cosmic ray produced radioactivity on the earth. Handbuch der Physik 46 (2), 551–612.

Lamson III, M.L., et al., 1975. Generator-produced $^{99m}TcO_4^-$: carrier free? J. Nucl. Med. 16, 639–641.

Landowne, S., Dasso, C.H., 1986. Novel aspects of the carbon decay mode of radium. Phys. Rev. C 33, 387–389.

Lang, R.F., et al., 2018. Characterization of a deuteriumedeuterium plasma fusion neutron generator. Nucl. Instrum. Methods Phys. Res., Sect. A 879, 31–38.

L'Annunziata, M.F., 1971. Birth of a unique parent-daughter relation: secular equilibrium. J. Chem. Educ. 48, 700–703.

L'Annunziata, M.F., 1987. Radionuclide Tracers, Their Detection and Measurement. Academic Press, London, p. 505.

L'Annunziata, M.F., 2007. Radioactivity: Introduction and History, first ed. Elsevier Science, Amsterdam, Oxford, Cambridge, p. 609.

L'Annunziata, M.F., 2016. Radioactivity: Introduction and History, From the Quantum to Quarks, second ed. Elsevier Science, Amsterdam, Oxford, Cambridge, p. 902.

L'Annunziata, M.F., Passo Jr., C.J., 2002. Cherenkov counting of yttrium-90 in the dry state; correlations with phosphorus-32 Cherenkov counting data. Appl. Radiat. Isot. 56, 907–916.

Lapp, R.E., Andrews, H.L., 1948. Nuclear Radiation Physics. Prentice-Hall, Inc., p. 487

Laulainen, N., Bichsel, H., 1972. Energy removed by delta rays from finite volumes in passage of charged particles. Nucl. Instrum. Methods 104, 531–539.

Lawrence, E.O., 1935. Transmutation of sodium by deuterons. Phys. Rev. 47, 17–27.

Lawrence, E. O., 1951. "The Evolution of the Cyclotron", Nobel Lecture given on December 11, 1951 in Stockholm, Sweden. In: Nobel Lectures, Physics, 1922-1941. (1965). Elsevier Publishing, Amsterdam.

Lawrence, E.O., Cooksey, D., 1936. On the apparatus for the multiple acceleration of light ions to high speeds. Phys. Rev. 50, 1131–1140.

Lawrence, E.O., Edlefsen, N.E., 1930. The production of high-speed protons without the use of high voltages. Science 72, 376–377.

Lawrence, E.O., Livingston, M.S., 1931a. The production of high-speed protons without the use of high voltages. Phys. Rev. 37, 1707.

Lawrence, E.O., Livingston, M.S., 1931b. The production of high-speed protons without the use of high voltages. Phys. Rev. 38, 834.

Lawrence, E.O., Livingston, M.S., 1932. The production of high-speed light ions without the use of high voltages. Phys. Rev. 40, 19–35.

Lawrence, E.O., Livingston, M.S., 1934. The multiple acceleration of ions to very high speeds. Phys. Rev. 45, 608–612.

Lawrence, E.O., et al., 1939. Initial performance of the 60-inch cyclotron of the William H, Crocker Radiation Laboratory, University of California. Phys. Rev. 56, 124.

Lederer, M.C., Shirley, V.S., 1978. Table of Isotopes, seventh ed. John Wiley & Sons, New York.

Lee, K.K.M., Steinle-Neumann, G., 2008. Ab-initio study of the effects of pressure and chemistry on the electron-capture radioactive decay constants of ^{7}Be, ^{22}Na and ^{40}K. Earth Planet. Sci. Lett. 267, 628–636.

Lee, T.D., Yang, C.N., 1956. Question of parity conservation in weak interactions. Phys. Rev. 104, 254–258.

Lei, X., 2010. Treatment of low-lying rectal adenocarcinoma with californium-252 (^{252}Cf) neutron intracavitary brachytherapy (ICBT) and external beam radiotherapy (EBRT): the three years result for 130 patients (phase II trial). Int. J. Rad. Oncol. Biol. Phys. 78 (3 Suppl.), S331–S332.

Leighton, R.B., et al., 1953. The decay of V^0 particles. Phys. Rev. 89, 148–177. http://prola.aps.org/abstract/PR/v89/i1/p148_1.

Leung, K.K.H., Zimmer, O., 2009. Proposed neutron lifetime measurement using a hybrid magnetic trap for ultra-cold neutrons. Nucl. Instrum. Methods Phys. Res., Sect. A 611, 181–185.

Levinger, J.S., 1953. Effects of radioactive disintegrations on inner electrons of the atom. Phys. Rev. 90, 11–25.

Li, Z., Deutscher, M.P., 2008. Analyzing the decay of stable RNAs in *E. coli*. Ppp. 31–45. In: Maquat, L.E., Arraiano, C.M. (Eds.), Methods in Enzymology, vol. 447. Elsevier, Amsterdam, p. 564.

Li, F., et al., 2017. A study on growth and pyrolysis characteristics of microalgae using thermogravimetric analysis-infrared spectroscopy and synchrotron Fourier transform infrared spectroscopy. Bioresour. Technol. 229, 1–10.

Li, Z., et al., 2017. Atmospheric neutrinos and proton decay in Super-Kamiokande and Hyper-Kamiokande. Nucl. Part. Phys. Proc. 287–288, 147–150.

Licciardi, J.M., et al., 2008. Cosmogenic ^{36}Cl production rates from Ca spallation in Iceland. Earth Planet. Sci. Lett. 267, 365–377.

Lide, D.R., 2010. CRC Handbook of Chemistry and Physics, 81st ed. CRC Press, Boca Raton, pp. 10–220.

Lilley, J., 2001. Nuclear Physics, Principles and Applications. John Wiley & Sons, Ltd., West Sussex, p. 393.

Lin, C.C., Chao, J.H., 2009. Radiochemistry of iodine: relevance to health and disease, pp. 171–182. In: Preedy, V., Burrow, G., Watson, R. (Eds.), Comprehensive Handbook of Iodine. Academic Press, p. 1334.

Lindhard, J., 1976. The Barkas effect — or Z_1^3, Z_1^4- corrections to stopping of swift charged particles. Nucl. Instrum. Methods 132, 1–5.

Lindhard, J., Scharff, M., 1960. Recent developments in the theory of stopping power. I. Principles of the statistical method. In: Penetration of Charged Particles in Matter. National Academy of Sciences-National Research Council, Publication 752, p. 49.

Lindhard, J., Scharff, M., 1961. Energy dissipation by ions in the keV region. Phys. Rev. 124, 128–130.

Lindote, A., et al., 2009. Simulation of neutrons produced by high-energy muons underground. Astropart. Phys. 31, 366–375.

Linsley, J., 1963. Evidence for a primary cosmic-ray particle with energy 10^{20} eV. Phys. Rev. Lett. 10 (4), 146–148.

Lin, T.K., Yeh, S.J., 1966. Enrichment of copper-64 by the Szilard Chalmers Process. J. Nucl. Sci. Technol. 3 (7), 289–293.

Lipkin, H.J., White, M.G., 1950. Scattering of positrons and electrons by nuclei. Phys. Rev. 79 (5), 892L–893L erratum 80(4), 770L.

Lis, G.J., et al., 2014. Distribution of selected elements in calcific human aortic valves studied by microscopy combined with SR-μXRF: influence of lipids on progression of calcification. Micron 67, 141–148.

Litherland, A.E., et al., 2007. Isobar separation at very low energy for AMS. Nucl. Instrum. Methods Phys. Res., Sect. A 259 (1), 230–235.

Liu, L., Huh, C.-A., 2000. Effect of pressure on the decay rate of ^{7}Be. Earth Planet. Sci. Lett. 180, 163–167.

Liu, B., et al., 2012. A semi-empirical model of linear energy transfer for heavy ions and electron. Nucl. Eng. Design 245, 202–205.

Liu, J., et al., 2016. A method to improve the sensitivity of neutron porosity measurement based on D-T source. J. Nat. Gas Sci. Eng. 33, 879–884.

Liu, Z., et al., 2018. Measurement and analysis of ^{232}Th(n,2n)^{231}Th reaction rate in the thorium oxide cylinder with a D-T neutron source. Ann. Nucl. Energy 111, 660–665.

Lobashev, V.M., et al., 1999a. Direct search for mass of neutrino and anomaly in the tritium beta-spectrum. Phys. Lett. B. 460, 227–235.

Lobashev, V.M., et al., 1999b. Neutrino rest mass and anomaly in the tritium β spectrum. Nucl. Phys. A654, 982c–987c.

Lobashev, V.M., et al., 1999c. Neutrino mass and anomaly in the tritium beta-spectrum. Results of the "Troitsk ν-mass" experiment. Nucl. Phys. B (Proc. Suppl.) 77, 327–332.

Lobashev, V.M., et al., 2000. Neutrino mass and anomaly in the tritium beta-spectrum. In: Proceedings the XIX International Conference on Neutrino Physics and Astrophysics (Neutrino-2000), Sudbury, Canada, June 16-21, 2000.

Lodi Rizzini, E., et al., 2004. Antiproton stopping power in He in the energy range 1–900 keV and the Barkas effect. Phys. Lett. B 599, 190–196.

Loeber, C.R., 2002. Building the Bombs: A History of the Nuclear Weapons Complex. Sandia National Laboratories, Albuquerque, p. 262.

Lombard, R.L., et al., 1968. Internal pair formation and multipolarity of nuclear transitions. Nucl. Phys. A110, 41–55.

Loveland, W., 2004. Nuclear chemistry, pp. 597–615. In: Myers, R.A. (Ed.), Encyclopedia of Physical Science and Technology, third ed. Elsevier, Amsterdam.

Lozza, V., et al., 2016. On behalf of the SNO+ Collaboration. The SNO+ experiment for neutrinoless double-beta decay. Nucl. Part. Phys. Proc. 273–275, 1836–1841.

Luo, N., et al., 2010. Proton bremsstrahlung and its radiation effects in fusion reactors. Fusion Eng. Des. 85, 39–45.

Macfarlane, R.D., Kohman, T.P., 1961. Natural alpha radioactivity in medium-heavy elements. Phys. Rev. 121, 1758–1769.

Madani, J.H., et al., 2010. Microscopic study of deuteron–nucleus total reaction cross section at medium energies. Nucl. Phys. A 839 (1–4), 42–50.

Magaud, P., et al., 2004. Nuclear fusion reactors, pp. 365–381. In: Cleveland, C.J. (Ed.), Encyclopedia of Energy. Elsevier.

Mahboub, K., 2008. A simple theoretical approach to the liquid drop model. Ann. Nucl. Energy 35, 1381–1385.

Mahjour-Shafiei, M., et al., 2006. Proton–proton bremsstrahlung towards the elastic limit at 190 MeV incident beam energy. Phys. Lett. B 632, 480–484.

Malik, S.S., Gupta, R.K., 1989. Theory of cluster radioactive decay and of cluster formation in nuclei. Phys. Rev. C 39, 1992–2000.

Maneschg, W., 2015. Review of neutrinoless double beta decay experiments: present status and near future. Nucl. Part. Phys. Proc. 260, 188–193.

Manjunatha, H.C., Rudraswamy, B., 2009. Exposure of bremsstrahlung from beta-emitting therapeutic radionuclides. Rad. Meas. 44, 206–210.

Manjunatha, H.C., Rudraswamy, B., 2010. Bremstrahlung exposure of tissues from beta-therapeutic nuclides. Nucl. Instrum. Methods Phys. Res., Sect. A 621, 581–589.

Mapleston, P., 1997. Film thickness gauges meet market needs for quality, cost. Mod. Plastics 74, 73–76.

(On behalf of the Pamela Collaboration) Marcelli, L., 2011. Identification of the light nuclei component of cosmic rays with the PAMELA experiment. Nucl. Instrum. Methods Phys. Res., Sect. A 630 (1), 58–62.

Marcelli, A., et al., 2012. Biological applications of synchrotron radiation infrared spectromicroscopy. Biotechnol. Adv. 30 (6), 1390–1404.

Markowicz, A.A., VanGrikin, R.E., 1984. Composition dependence of Bremsstrahlung background in electron-probe X-ray microanalysis. Anal. Chem. 56, 2049–2051.

Marques-Netto, A., Abbe, J.C., 1975. Szilard-Chalmers effects in halfnium chelates. J. Inorg. Nucl. Chem. 37, 2235–2238.

Martin, B.R., 2011. Particle Physics. Oneworld Publications, Oxford, p. 201.

Martin, R.C., et al., 1997. Development of high-activity ^{252}Cf sources for neutron brachytherapy. Appl. Radiat. Isot. 48 (10–12), 1567–1570.

Martin, R.C., et al., 2000. Production, distribution and applications of Californium-252 neutron sources. Appl. Radiat. Isot. 53, 785–792.

Martínez, S.A., et al., 2010. Evaluation of neutron production in new accelerators for radiotherapy. Radiat. Meas. 45 (10), 1402–1405.

Martins, M.N., Silva, T.F., 2014. Electron accelerators: history, applications, and perspectives. Radiat. Phys. Chem. 95, 78–85.

Mason, T.E., et al., 2006. The spallation neutron source in Oak Ridge: a powerful tool for materials research. Physics B 385–386, 955–960.

Masuda, K., et al., 2006. Spatial distribution of D–D/D–^{3}He advanced fuels fusion reactions in an inertial electrostatic confinement device. In: Proceedings of the 21st IAEA Conference on Fusion Energy, Chengdu, 16-21 October 2006, IAEA-CN-149, pp. 1–8. IC/P7-9.

Materne, S., et al., 2009. PENELOPE—on the way towards a new neutron lifetime experiment with magnetic storage of ultra-cold neutrons and proton extraction. Nucl. Instrum. Methods Phys. Res., Sect. A 611, 176–180.

Matsuura, T., 1967. On the radiation-chemical interpretation of the specific activity values obtained by the Szilard-Chalmers reaction of some cobalt complex ions. Int. J. Appl. Radiat. Isot. 18, 697–708.

Matsuyama, M., Abe, S., 2016. Tracking of tritium charged into stainless steel by BIXS. Fusion Eng. Des. 113, 250–254.

Mausner, L.F., et al., 1992. Improved specific activity of reactor produced ^{117m}Sn with the Szilard–Chalmers process. Appl. Radiat. Isot. 43 (9), 1117–1122.

Mavunda, R.D., et al., 2004. Bremsstrahlung spectra from diagnostic X-rays. Radiat. Phys. Chem. 71, 991–992.

McAllister, R.W., Hofstadter, R., 1956. Elastic scattering of 188-Mev electrons from the proton and the alpha particle. Phys. Rev. 102 (3), 851–856.

McFee, J.E., et al., 2013. Photoneutron spectroscopy using monoenergetic gamma rays for bulk explosives detection. Nucl. Instrum. Methods Phys. Res., Sect. A 704, 131–139.

McGrath, J., et al., 2010. Detecting multi-hit events in a CdZnTe coplanar grid detector using pulse shape analysis: a method for improving background rejection in the COBRA $0\nu\beta\beta$ experiment. Nucl. Instrum. Methods Phys. Res., Sect. A 615, 57–61.

McKellar, A.R.W., 2010. High-resolution infrared spectroscopy with synchrotron sources. J. Mol. Spectrosc. 262 (1), 1–10.

McMillan, E.M., 1945. The synchrotron – a proposed high energy accelerator. Phys. Rev. 68, 143–144.

McMillan, E.M., 1951. The transuranium elements: early history. Nobel Lecture, December 12, 1951. The Nobel Foundation.

McMillan, E., Abelson, P.H., 1940. Radioactive element 93. Phys. Rev. 57, 1185–1186.

Mederski, H.J., 1961. Determination of internal water by beta gauging technique. Soil Sci. 92, 143–146.

Mederski, H.J., Alles, W., 1968. Beta gauging leaf water status: influence of changing leaf characteristics. Plant Physiol. 43, 470–472.

Mehta, K., et al., 2003. Applicability study on existing dosimetry systems to high-power Bremsstrahlung irradiation. Radiat. Phys. Chem. 68, 959–962.

Meitner, L., 1923. Das beta-Strahlenspektrum von UX_1 und seine Deutung. Z. Phys. 17, 54–66.

Meitner, L., 1924. Uber die Rolle der γ-strahlen beim Atomzerfall. Z. Phys. 26, 169–177.

Meitner, L., Frisch, O.R., 1939. Disintegration of uranium by neutrons: a new type of nuclear reaction. Nature 143, 239–240.

Meitner, L., Strassmann, F., Hahn, O., 1938. Künstliche Emwandlungsprozesse bei Bestrahlung des Thoriums mit Neutronen; Auftreten isomer Reihen durch Abspaltung von α-Strahlen. Z. Phys. 109, 538–552.

Melhorn, W., 1998. 70 years of Auger spectroscopy, a historical perspective. J. Electron Spectrosc. Relat. Phenom. 93 (1), 1–15.

Melhus, C.S., Rivard, M.J., 2006. Towards HDR ^{252}Cf brachytherapy: shielding assessment of a linear accelerator vault. Brachytherapy 5 (2), 83–84.

Meneses, A.A.M., et al., 2012. Segmentation of synchrotron radiation microcomputed tomography images using energy minimization via graph cuts. Appl. Radiat. Isot. 70 (7), 1284–1287.

Menn, W., et al., 2000. The absolute flux of protons and helium at the top of the atmosphere using IMAX. Astrophys. J. 533 (1), 281–297.

Mercer, J.A., et al., 2007. A non-intrusive neutron device for *in situ* detection of petroleum contamination in soil. Nucl. Instrum. Methods Phys. Res., Sect. B 263, 217–220.

Meshik, A.P., et al., 2001. Weak decay of ^{130}Ba and ^{132}Ba: geochemical measurements. Phys. Rev. C 64, 035205.

Meshkian, M., 2016. Monte Carlo simulation optimisation of zinc sulphide based fast-neutron detector for radiography using a ^{252}Cf source. Nucl. Instrum. Methods Phys. Res., Sect. A 808, 77–82.

Messier, M.D., 2006. Review of neutrino oscillation experiments. In: Flavor Phys. and CP Violation Conf. Vancouver 2002. hep-ex/0606013v1 5 Jun 2002.

Miao, Y., et al., 2017. *In situ* synchrotron investigation of grain growth behavior of nano-grained UO_2. Scr. Mater. 131, 29–32.

Michel, R., et al., 2003. A comparison of 4 radionuclides conjugated to antibodies for single-cell kill. J. Nucl. Med. 44 (4), 632–640.

Miernik, K., et al., 2007a. First observation of β-delayed three-proton emission in ^{45}Fe. Phys. Rev. C 76, 041304(R).

Miernik, K., et al., 2007b. Optical Time Projection Chamber for imaging nuclear decays. Nucl. Instrum. Methods Phys. Res., Sect. A 581 (1–2), 194–197.

Miernik, K., et al., 2013. Large β-delayed one and two neutron emission rates in the decay of ^{86}Ga. Phys. Rev. Lett. 111, 132502.

Mikkelsen, H.H., 1992. A note on the Barkas effect in gold. Nucl. Instrum. Methods Phys. Res., Sect. B 69, 22–23.

Miley, G.H., Sved, J., 1997. The IEC – A plasma-target-based neutron source. Appl. Radiat. Isot. 48 (10–12), 1557–1561.

Miley, G., et al., 2005. RF ion source-driven IEC design and operation. Fusion Sci. Techn. 47, 1233–1237.

Miller, R.B., 2003. Food irradiation using bremsstrahlung X-rays. Radiat. Phys. Chem. 68, 963–974.

Miller, R.B., 2005. Food Irradiation. Springer, Berlin, New York, London, p. 246.

Mirea, M., et al., 2017. Spontaneous fission, cluster emission and alpha decay of ^{222}Ra in a unified description. Ann. Phys. 380, 154–167.

Mitchell, R.F., Martin Jr., D.S., 1956. Szilard-Chalmers process for osmium from hexachloroosmate(IV) targets. J. Inorg. Nucl. Chem. 2, 286–289.

Mitchell, J.W., et al., 1996. Measurement of 0.25-3.2 GeV antiprotons in the cosmic radiation. Phys. Rev. Lett. 76 (17), 3057–3060.

Mitchell, E., et al., 1999. Demystifying the synchrotron trip: a first time user's guide. Structure 7 (5), R111–R121.

Miura, N., et al., 2010. Intermolecular vibrational study in liquid water and ice by using far infrared spectroscopy with synchrotron radiation of MIRRORCLE. Spectrochim. Acta, Part A 77 (5), 1048–1053.

Moe, M.K., 1986. Double-beta decay in ^{82}Se, ^{128}Te, and ^{130}Te. AIP Conf. Proc. 150, 1012–1016.

Moe, M.K., Rosen, S.P., 1989. Double-beta decay. Sci. Am. (November) 48–55.

Montaruli, T., 2009. Neutrino astronomy in the ice. Nucl. Phys. B (Proc. Suppl.) 188, 239–244.

Morozov, V.A., et al., 1998. Nuclear experimental techniques – three-dimensional delayed coincidence single-crystal scintillation time spectrometer. Instrum. Exp. Tech. 41 (5), 609.

Morozov, V.A., et al., 2002. High-sensitivity delayed-coincidence spectrometer to search for short-lived nuclear states. Nucl. Instrum. Methods Phys. Res., Sect. A 484, 225–232.

Morozov, V.A., et al., 2006. Measurement of the half-life for two ^{57}Fe excited states by a single-crystal scintillation time spectrometer. Nucl. Instrum. Methods Phys. Res., Sect. A 566, 448–451.

Mosby, M.A., et al., 2017. Cross sections for proton-induced reactions on ^{nat}Sb up to 68 MeV. Nucl. Instrum. Methods Phys. Res., Sect. B 412, 34–40.

Moseley, H.G.J., 1913. The high frequency spectra of the elements. Phil. Mag. 26 (1), 1024–1034.

Moseley, H.G.J., 1914. The high frequency spectra of the elements. Phil. Mag. 27 (2), 703–713.

Moussavi-Zarandi, A., 2008. Determination of beryllium by use of photonuclear activation techniques. Appl. Radiat. Isot. 66, 158–161.

Mozumber, A., et al., 1968. Theory of radiation chemistry. IX. Mode and structure of heavy particle tracks in water. Am. Chem. Soc. Ser. Adv. Chem. 1, 27.

Mukha, I., Schrieder, G., 2001. Two-proton radioactivity as a genuine three-body decay: the ^{19}Mg probe. Nucl. Phys. A 690, 280–283.

Mukoyama, T., Shimizu, S., 1974. The effect of pressure on orbital electron capture. Phys. Lett. 50A (4), 258–260.

Müller, A.R., et al., 2009. A low-temperature proton detector for a neutron lifetime experiment. Nucl. Instrum. Methods Phys. Res., Sect. A 611, 289–292.

Munger, M.L., Peterson, R.J., 1978. Proton stripping reactions on ^{194}Pt, ^{196}Pt and ^{198}Pt. Nucl. Phys. A303, 199–216.

Murray, R.L., 2009. Nuclear Energy. An Introduction to the Concepts, Systems, and Applications of Nuclear Processes, sixth ed. Elsevier Publishers, Amsterdam, p. 497.

Murray, R.L., Holbert, K.E., 2015. Nuclear Energy. An Introduction to the Concepts, Systems, and Applications of Nuclear Processes, seventh ed. Elsevier Publishers, Amsterdam, p. 531.

Mushtaq, A., et al., 1990. Production of no-carrier-added ^{64}Cu and ^{67}Cu in a reactor. J, Radioanal. Nucl. Chem. 141 (2), 261–269.

Nagano, M., Watson, A.A., 2000. Observations and implications of the ultrahigh-energy cosmic rays. Rev. Mod. Phys. 72, 689–732.

Nakahata, M., 2000. Neutrinos underground. Science 289 (5482), 1155–1156.

Nakayama, F.S., Ehrler, W.L., 1964. Beta ray gauging technique for measuring leaf water content changes and moisture status of plants. Plant Physiol. 39, 95–98.

Nambu, Y., 1966. A systematics of hardrons in subnuclear physics. In: de Shalit, A., Feshbach, H., Van Hove, L. (Eds.), Preludes in Theoretical Physics. North Holland, Amsterdam and John Wiley, New York, pp. 133–142.

Nargolwalla, S.S., et al., 1968. Correction. A technique for the evaluation of systematic errors in the activation analysis for oxygen with 14-MeV neutrons. Anal. Chem. 40 (7), 1134.

Nargolwalla, S.S., et al., 1970. Photon self-absorption corrections for the minimization of systematic errors in 14-mev neutron activation analysis. Anal. Chim. Acta 49 (30), 425–436.

Nassan, L., et al., 2011. Production of ^{166}Ho and ^{153}Sm using hot atom reactions in neutron irradiated tris(cyclopentadienyl) compounds. Nukleonika 56 (4), 263–267.

Nersisyan, H.B., Das, A.K., 2005. Proton and antiproton energy losses in a solid target: an exact trial potential approach. Nucl. Instrum. Methods Phys. Res., Sect. B 227, 455–460.

Nguyen-Truong, H.T., 2015. Modified Bethe formula for low-energy electron stopping power without fitting parameters. Ultramicros 149, 26–33.

Nielsen, D.R., Cassel, D.K., 1984. Soil water management. In: L'Annunziata, M.F., Legg, J.O. (Eds.), Isotopes and Radiation in Agricultural Sciences, vol. I. Academic Press, San Diego, pp. 37–43.

NIST, 2018. Stopping-Power and Range Tables for Electrons, Protons, and Helium Ions. National Institute of Standards and Technology, Gaithersburg, MD [Online]. https://www.nist.gov/pml/stopping-power-range-tables-electrons-protons-and-helium-ions.

NNDC, 2010a. National Nuclear Data Center, Brookhaven National Laboratory, "Q-value Calculator (QCalc)". http://www.nndc.bnl.gov/qcalc/.

NNDC, 2010b. National Nuclear Data Center, Brookhaven National Laboratory, "Nuclear Decay Data (NuDat)". Decay Radiation Data, NuDat 2.5. http://www.nndc.bnl.gov/nudat2/.

NNDC, 2018a. National Nuclear Data Center, Brookhaven National Laboratory, "Q-value Calculator (QCalc)". http://www.nndc.bnl.gov/qcalc/.

NNDC, 2018b. National Nuclear Data Center, Brookhaven National Laboratory, "Nuclear Decay Data (NuDat)". Decay Radiation Data. NuDat 2.5. http://www.nndc.bnl.gov/nudat2/.

Noborio, K., et al., 2006. Confinement of ions in an inertial electrostatic confinement fusion (IECF) device and its influence on neutron production rate. Fusion Eng. Des. 81 (8–14), 1701–1705.

Northcliffe, L.C., 1963. Passage of heavy ions through matter. Ann. Rev. Nucl. Sci. 13, 67–102.

Nowicki, S.F., et al., 2017. The Los Alamos Neutron Science Center spallation neutron sources. Phys. Proc. 90, 374–380.

NRL, 1998. The exposure of New Zealand aircrew to cosmic radiation. National Radiation Laboratory, Christchurch, New Zealand, p. 6. Information Sheet 19, February 1998.

Obregewitsch, R.P., et al., 1975. Estimating relative leaf water content with a simple beta gauge calibration. Agron. J. 67, 729–732.

Occhialini, G.P.S., Powell, C.F., 1947. Nuclear disintegrations produced by slow charged particles of small mass. Nature 159, 186–190.

Occhialini, G.P.S., Powell, C.F., 1948. Observations on the production of mesons by cosmic radiation. Nature 162, 168–173.

Ochiai, K., et al., 2007. Thin slit streaming experiment for ITER by using D-T neutron source. Fusion Eng. Des. 82, 2794–2798.

Ogando, J., 1993. Nuclear web gauging keeps pace with processor needs. Plast. Technol. 39, 46–49.

Oganessian, Y.T., Utyonkov, V.K., 2015. Superheavy nuclei from ^{48}Ca-induced reactions. Nucl. Phys. A 944, 62–98.

Ohm, H., et al., 1990. Delayed-coincidence measurement of subnanosecond lifetimes in fission fragments. In: The Spectroscopy of Heavy Nuclei 1989 Proceedings of the International Conference on the Spectroscopy of Heavy Nuclei, Agia Pelagia, Crete, June 25-July 1, 1989. Institute Physics Conference Series. No. 105. Adam Hilger, Ltd., Bristol, pp. 323–328.

Ohnishi, M., et al., 2007. Development of convergent D–D fusion neutron generator with large pulse current. Fusion Sci. Techn. 52, 1101–1104.

Ohnishi, M., et al., 2016. Tritium burning in inertial electrostatic confinement fusion facility. Fusion Eng. Des. 109–111, 1709–1713.

Ohtsuki, T., et al., 2004. Enhanced electron-capture decay rate of ^{7}Be encapsulated in C_{60} cages. Phys. Rev. Lett. 93, 112501.

Ohtsuki, T., et al., 2007. Radioactive decay speedup at $T = 5$ K: electron-capture decay rate of 7Be encapsulated in C_{60}. Phys. Rev. Lett. 98, 252501.

Oliphant, M.L., 1943. Classified memo submitted to the British Directorate of Atomic Energy. University of Birmingham Archive, UK).

Olzem, J., et al., 2007. Cosmic-ray positron identification through bremsstrahlung conversion. Nucl. Phys. B (Proc. Suppl.) 173, 51–55.

Oms, J., 2004. Lifetime extraction from delayed-coincidence experiments by a direct deconvolution method. Nucl. Instrum. Methods Phys. Res., Sect. A 527, 598–603.

Ono, S., et al., 2013. Sensitive electron capture decay rate of ^{7}Be encapsulated in carbon nanotubes: a density functional study. Chem. Phys. Lett. 561–562, 137–141.

Otten, E.W., Weinheiner, C., 2008. Neutrino mass limit from tritium β decay. Rep. Prog. Phys. 71 (8), 6201.

Owens, A., 2012. Synchrotron light sources and radiation detection metrology. Nucl. Instrum. Methods Phys. Res., Sect. A 695, 1–12.

O'Leary, G.J., Incerti, M., 1993. A field comparison of three neutron moisture meters. Aust. J. Exp. Agric. 33, 59–69.

O'Shaughnessy, C.M., et al., 2009. Measuring the neutron lifetime using magnetically trapped neutrons. Nucl. Instrum. Methods Phys. Res., Sect. A 611, 171–175.

Paces, J.B., et al., 2002. ^{234}U/^{238}U evidence for local recharge and patters of ground-water flow in the vicinity of Yucca Mountain, Nevada, USA. Appl. Geochem. 17, 751–779.

Pachoa, A.S., Steinhäusler, F., 2010. Cosmic Radiation, Including Its Effects on Airline Crew, Frequent Flyers, and Space Travel, pp. 87–121. In: Radioactivity in the Environment, vol. 17. Elsevier, Amsterdam, p. 244.

Panosa, C., et al., 2015. A comparison of non-biologically active truncated EGF (EGFt) and full-length hEGF for delivery of Auger electron-emitting ^{111}In to EGFR-positive breast cancer cells and tumor xenografts in athymic mice. Nucl. Med. Biol. 42, 931–938.

Papazoglou, P., 1998. PhD thesis, University of Frankfurt.

Papazoglou, P., et al., 1999. Nuclei in a chiral SU(3) model. Phys. Rev. C 59, 411.

Park, Y.J., et al., 2009. Performance characteristics of a prompt gamma-ray activation analysis (PGAA) system equipped with a new compact D–D neutron generator. Nucl. Instrum. Methods Phys. Res., Sect. A 606, 243–247.

Park, M., et al., 2010. Data acquisition system implemented with various platforms for KSTAR diagnostic systems. Fusion Eng. Des. 85, 350–355.

Passo Jr., C.J., Cook, G.T., 1994. Handbook of Environmental Liquid Scintillation Spectrometry. In: A Compilation of Theory and Methods. Packard Instrument Company, Meriden, CT.

Patil, B.J., et al., 2010a. Measurement of angulare distribution of neutron flux for the 6 MeV race-track microtron based pulsed neutron source. Appl. Radiat. Isot. 68, 1743–1745.

Patil, B.J., et al., 2010b. Simulation of e–γ–n targets by FLUKA and measurement of neutron flux at various angles for accelerator based neutron source. An. Nucl. Energy 37, 1369–1377.

(Particle Data Group) Patrignani, C., et al., 2016. Review of Particle Physics. Chin. Phys. C 40, 100001.

Patyk, Z., et al., 1997. Masses and shapes of heaviest nuclei. Nucl. Phys. A 626 (1–2), 337–340.

Paul, S., 2009. The puzzle of neutron lifetime. Nucl. Instrum. Methods Phys. Res., Sect. A 611, 157–166.

Paul, W., Steinwedel, H., 1955. Interaction of electrons with matter. In: Siegbahn, K. (Ed.), Beta- and Gamma-Ray Spectroscopy. North-Holland, Amsterdam.

Pauli, W., 1924. Zur Frage der theoretischen Deutung der Satelliten eineger Spektrallinien und ihrer Beeinflussung durch magnetische Felder. Naturwissenschaften 12, 741–743.

Pauli, W., 1925. Über den Zusammenhang des Abschlusses der Electronengruppen in Atom der Komplexstruktur der Spectren. Z. Phys. 31, 765–783.

Pauli, W., 1946. Exclusion principle and quantum mechanics. Nobel Lecture, December 13, 1946. In: Nobel Lectures, Physics, 1942-62.. Nobel Foundation Staff Elsevier Publishing, Amsterdam (1964).

Payne, S.A., 2015. Nonproportionality of scintillator detectors. IV. Resolution contribution from delta-rays. IEEE Trans. Nucl. Sci. 62 (1), 372–380.

Pérez, P.D., et al., 2014. Bremsstrahlung in carbon thick targets by proton incidence. Nucl. Instrum. Methods Phys. Res., Sect. B 318, 23–26.

Perlman, I., Ypsilantis, T.J., 1950. Consistency of nuclear radii of even-even nuclei from alpha-decay theory. Phys. Rev. 79, 30–34. http://prola.aps.org/abstract/PR/v79/i1/p30_1.

Perlman, I., Ghiorso, A., Seaborg, G.T., 1950. Systematics of alpha radioactivity. Phys. Rev. 77 (1), 26–50.

Petkov, P., et al., 2003. Lifetime determination in delayed-coincidence experiments using the differential decay-curve approach. Nucl. Instrum. Methods Phys. Res., Sect. A 500, 379–385.

Pfützner, M., 2013. Particle radioactivity of exotic nuclei. Phys. Scr. T152, 014014 (14pp).

Pfützner, M., et al., 2002. First evidence for the tw-proton decay of ^{45}Fe. Eur. Phys. J. A 14, 279–285.

Pfützner, M., et al., 2012. Radioactive decays at limits of nuclear stability. Rev. Mod. Phys. 84, 567–619.

Pichlmaier, A., et al., 2000. MAMBO II: neutron lifetime measurement with storage of ultra-cold neutrons. Nucl. Instrum. Methods Phys. Res., Sect. A 440, 517–521.

Pichlmaier, A., et al., 2010. Neutron lifetime measurement with the UCN trap-in-trap MAMBO II. Phys. Lett. B 693, 221–226.

Pickrell, J.A., 2009. Radiation and Health Effects, pp. 381–392. In: Gupta, R.C. (Ed.), Handbook of Toxicology of Chemical Warfare Agents. Academic Press, p. 1168.

Piefer, G.R., et al., 2005. Design of an ion source for ^{3}He fusion in a low pressure IEC device. Fusion Sci. Techn. 47, 1255–1259.

Pierre Auger Collaboration, 2010. A study of the effect of molecular and aerosol conditions in the atmosphere on air fluorescence measurements at the Pierre Auger Observatory. Astropart. Phys. 33, 108–129.

Planck, M., 1900. Zur Theorie des Gesetzes der Energieverteilung im Normalspectrum. In: Verhandlungen der Deutschen Physikalischen Gesellschaft im Jahre 1900, Jahrg. 2. Johann Ambrosius Barth, Leipzig, pp. 237–245, 17.

Plaß, W.R., et al., 2008. Isobar separation by time-of-flight mass spectrometry for loe-energy radioactive ion beam facilities. Nucl. Instrum. Methods Phys. Res., Sect. B 266 (19–20), 4560–4564.

Pocar, A., 2012. The quest for neutrino-less double beta decay. Phys. Proc. 37, 6–15.

Poda, D.V., et al., 2013. CdWO4 crystal scintillators from enriched isotopes for double beta decay experiments. Radiat. Meas. 56, 66–69.

Poenaru, D.N., et al., 1985. Atomic nuclei decay modes by spontaneous emission of heavy ions. Phys. Rev. C 32, 572–581.

Poenaru, D.N., et al., 1986. Calculated half-lives and kinetic energies for spontaneous emission of heavy ions from nuclei. Atom. Data Nucl. Data Tables 34, 423–538.

Poenaru, D.N., et al., 1987. Cold fission as heavy ion emission. Z. Phys. A 328 (3), 309–314.

Poenaru, D.N., et al., 1989. Inertia and fission paths in a wide range of mass asymmetry. Z. Phys. A 333 (3), 291–298.

Pollock, H.C., 1983. The discovery of synchrotron radiation. Am. J. Phys. 51 (3), 278–280. https://doi.org/10.1119/1.13289.

Pomorski, M., et al., 2011. β-delayed proton emission branches in ^{43}Cr. Phys. Rev. C 83, 014306.

Popeko, A.G., 2016. On-line separators for the Dubna Superheavy Element Factory. Nucl. Instrum. Methods Phys. Res., Sect. B 376, 144–149.

Popplewell, D.S., 1963. The preparation of ^{210}Bi by a Szilard-Chalmers reaction. J. Inorg. Nucl. Chem. 25 (3), 318–320.

Potgieter, M.S., 2008a. Solar cycle variations and cosmic rays. J. Atmos. Solar-Terr. Phys. 70, 207–218.

Potgieter, M.S., 2008b. Challenges to cosmic ray modeling: from beyond the solar wind termination shock. Adv. Space Res. 41, 245–258.

Potgieter, M.S., 2010. The dynamic heliosphere, solar activity, and cosmic rays. Adv. Space Res. 46, 402–412.

Povinec, P.P., et al., 2008. Monte Carlo simulation of background characteristics of gamma-ray spectrometers – a comparison with experiment. Radioact. Environ. 11, 163–208.

Price, P.B., 1989. Heavy particle radioactivity. Ann. Rev. Nucl. Part. Sci. 39, 19–42.

Price, P.B., 1994. Recent advances in cluster radioactivities. In: Scheid, W., Sandulesco, A. (Eds.), Frontier Topics in Nuclear Physics. Plenum Press, New York, p. 510.

the SNO Collaboration Prior, G., 2009. Results from the Sudbury Neutrino Observatory Phase III. Nucl. Phys. B (Proc. Suppl.). 188, 96–100.

Pujol, M., et al., 2009. Xenon in Archean barite: weak decay of ^{130}Ba, mass-dependent isotopic fractionation and implication for barite formation. Geochim. Cosmochim. Acta 73, 6834–6846.

Rachinhas, P.J.B.M., et al., 2000. Simulation and experimental results for the detection of conversion electrons with gas proportional scintillation counters. Nucl. Instrum. Methods Phys. Res., Sect. A 441, 468–478.

Radchenko, V.M., et al., 2000. Curium-248 standard neutron source. Appl. Radiat. Isot. 53, 833–835.

Radel, R.F., et al., 2007. Detection of highly enriched uranium using a pulsed D–D fusion source. Fusion Sci. Techn. 52, 1087–1091.

Radivojevic, Z., et al., 2002. Beta-delayed neutron decay of ^{33}Na. Nucl. Instrum. Methods Phys. Res., Sect. A 481, 464–474.

Raghaven, R.S., 1976. Inverse β decay of ^{115}In $\rightarrow$ ^{115}Sn*: a new possibility for detecting solar neutrinos from the proton-proton reaction. Phys. Rev. Lett. 37, 259–262.

Rahmani, F., Shahriari, M., 2010. Hybrid photoneutron source optimization for electron accelerator-based BNCT. Nucl. Instrum. Methods Phys. Res., Sect. A 618 (1–3), 48–53.

Rahmani, F., et al., 2010. Feasibility study on the use of uranium in photoneutron target and BSA optimization for Linac based BNCT. Nucl. Instrum. Methods Phys. Res., Sect. A 641, 136–140.

Rahmani, F., et al., 2015. Design of photon converter and photoneutron target for High power electron accelerator based BNCT. Appl. Radiat. Isot. 106, 45–48.

Rainwater, J., 1950. Nuclear energy level argument for a spheroidal nuclear model. Phys. Rev. 79, 432–434.

Rao, D.V., et al., 2016. Synchrotron-based crystal structure, associated morphology of snail and bivalve shells by X-ray diffraction. Rad. Phys. Chem. 127, 155–164.

Ray, A., et al., 1999. Observation of large change of ^{7}Be decay rate in Au and Al_2O_3 and its implications. Phys. Lett. B 455, 69–76.

Ray, A., et al., 2009. Observation of enhanced orbital electron-capture nuclear decay rate in a compact medium. Phys. Lett. B 679, 106–110.

Redd, E.M., et al., 2017. Computationally-generated nuclear forensic characteristics of early production reactors with an emphasis on sensitivity and uncertainty. Ann. Nucl. Energy 110, 941–947.

Reines, F., 1960. Neutrino interactions. Ann. Rev. Nucl. Sci. 10, 1–26.

Reines, F., 1979. The early days of experimental neutrino physics. Science 203, 11–16.

Reines, F., 1994. 40 years of neutrino physics. Prog. Part. Nucl. Phys. 32, 1–12.

Reines, F., 1995. The Neutrino: From Poltergeist to Particle, Nobel Lecture. https://www.nobelprize.org/prizes/physics/1995/reines/lecture/.

Reines, F., Cowen Jr., C.L., 1956. The neutrino. Nature 178, 446–449.

Reines, F., Cowen Jr., C.L., 1957. Neutrino physics. Phys. Today 10 (8), 12–18.

Rhodes, R., 1986. The Making of the Atomic Bomb. Touchstone/Simon & Schuster, New York, p. 886.

Rhodes, R., 1999. Atomic physicist, Enrico Fermi. Time Magazine. March. 28, 1999.

Roberto, J.B., et al., 2015. Actinide targets for the synthesis of super-heavy elements. Nucl. Phys. A 944, 99–116.

Robson, J.M., 1950a. Radioactive decay of the neutron. Phys. Rev. 77, 747A.

Robson, J.M., 1950b. Radioactive decay of the neutron. Phys. Rev. 78, 311–312.

Roeckl, E., Mukha, I., 2013. Q values of radioactive decay: examples from nuclear physics and related fields. Int. J. Mass Spectrom. 349–350, 47–56.

Roentgen, W.C., 1895. Über eine neue Art von Strahlen. Sitzungsberichte der Würzburger Physik-medic Gesellschaft. 9, 132-141.

Roentgen, W.C., 1896. On a new kind of rays. Nature 53 (1369), 274–276.

Roesler, S., Silari, M., 2018. Chapter 36. Radioactivity and Radiation Protection. In: Tanabashi, et al. (Eds.), Particle Data Group Phys. Rev D., vol. 98, p. 030001.

Rohrlich, F., Carlson, B.C., 1954. Positron-electron differences in energy loss and multiple scattering. Phys. Rev. 93, 38–44.

Röllig, M., et al., 2013. Activity monitoring of a gaseous tritium source by beta induced X-ray spectrometry. Fusion Eng. Des. 88, 1263–1266.

Ronen, Y., 1997. Systematic behaviour in cluster radioactivity. Ann. Nucl. Energy 24 (2), 161–164.

Ronen, Y., 2004. Indications of the validity of the liquid drop model for spontaneous fission half-lives. Ann. Nucl. Energy 31, 323–329.

Rose, J., 2015. The chameleons of space. The Royal Swedish Academy of Sciences, p. 6. https://www.nobelprize.org/nobel_prizes/physics/laureates/2015/popular-physicsprize2015.pdf.

Rose, H., Jones, G.A., 1984. A new kind of natural radioactivity. Nature 307, 245.

Rossi, B., Hall, D.B., 1941. Variation of the rate of decay of mesotrons with momentum. Phys. Rev. 59, 223–228.

Rowe, D.J., Wood, J.L., 2010. Fundamentals of Nuclear Models. World Scientific, Hackensack, NJ, p. 676.

Roy, R.R., Reed, R.D., 1968. Interactions of Photons and Leptons with Matter. Academic Press, New York.

Royer, G., 2008. On the coefficients of the liquid drop model mass formulae and nuclear radii. Nucl. Phys. A 807 (3–4), 105–118.

Rudigier, M., et al., 2017. Fast timing measurement using an $LaBr_3$(Ce) scintillator detector array coupled with Gammasphere. Acta Phys. Polonica B 48 (3), 351–357.

Rumble, J.R., 2017. Handbook of Chemistry and Physics, " 98th ed. CRC Press.

Rutherford, E., 1899. Uranium radiation and the electrical conduction produced by it. Philos. Mag. Ser. 5 47, 109–163.

Rutherford, E., 1900. A radioactive substance emitted from thorium compounds. Philos. Mag. Ser. 5 49, 1–14.

Rutherford, E., 1903. The magnetic and electric deviation of the easily absorbed rays from radium. Philos. Mag. Ser. 6, 177–187.

Rutherford, E., 1906. Retardation of the alpha particle from radium in passing through matter. Philos. Mag. Ser. 6 12, 134–146.

Rutherford, E., 1908. The chemical nature of the alpha particles from radioactive substances. Nobel Lecture. In: Nobel Lectures, Chemistry, 1901–1921. Elsevier Publishing, Amsterdam, 1966.

Rutherford, E., 1911. The scattering of α and β particles by matter and the structure of the atom. Philos. Mag. Ser. 6 21, 669–688.

Rutherford, E., 1913. The structure of the atom. Nature 92 (2302), 423.

Rutherford, E., 1919. Collision of α-particles with light atoms. Nature (London) 103, 415–418.

Rutherford, E., 1920a. Bakerian lecture: nuclear constitution of atoms. Proc. R. Soc. London 97A, 374–401.

Rutherford, E., 1920b. Nuclear constitution of atoms. Proc. R. Soc. London 97A, 374–401.

Rutherford, E., 1927. Structure of the radioactive atom and origin of the alpha rays. Philos. Mag. 4 (22), 580–605.

Rutherford, E., 1929. Annual address to the Royal Society. Proc. R. Soc. London 122A, 1–23.

Rutherford, E., 1936. The development of the theory of atomic structure. In: Needham, J., Pagel, W. (Eds.), Background to Modern Science. Macmillan Company, New York, pp. 61–74.

Rutherford, E., Geiger, H., 1908a. The charge and nature of the alpha particle. Proc. Roy. Soc. A 81, 162–173.

Rutherford, E., Geiger, H., 1908b. An electrical method of counting the number of α-particles from radioactive substances. Proc. Royal Soc. A 81, 141–161.

Rutherford, E., Royds, T., 1908a. Spectrum of the radium emanation. Nature 78, 220–221.

Rutherford, E., Royds, T., 1908b. Spectrum of the radium emanation. Philos. Mag. Ser. 6 16 (92), 313–317.

Rutherford, E., Soddy, F., 1902. The cause and nature of radioactivity. Philos. Mag. 4, 370–396.

Rutz, K., et al., 1997. Superheavy nuclei in self-consistent nuclear calculations. Phys. Rev. C 56, 238.

Safavi-Tehrani, L., et al., 2015. Production of high specific activity radiolanthanides for medical purposes using the UC Irvine TRIGA reactor. J. Radioanal. Nucl. Chem. 303 (2), 1099–1103.

Salonen, L., et al., 2012. Environmental liquid scintillation counting. In: L'Annunziata, M.F. (Ed.), Handbook of Radioactivity Analysis, third ed. Elsevier Science, Amsterdam, pp. 625–693.

Samec, K., et al., 2009. Design of a compact high-power neutron source—The EURISOL converter target. Nucl. Instrum. Methods Phys. Res., Sect. A 606, 281–290.

Samerdokiene, V., et al., 2015. Second primary malignancies after radiotherapy including HDR ^{252}Cf brachytherapy for cervical cancer. Brachytherapy 14, 898–904.

Samios, N.P., et al., 1962. Parity of the neutral pion and the decay $\pi^o \rightarrow 2e^+ + 2e^-$. Phys. Rev. 126, 1844–1849. http://prola.aps.org/abstract/PR/v126/i5/p1844_1.

Sandrock, A., et al., 2018. Radiative corrections to the average bremsstrahlung energy loss of high-energy muons. Phys. Lett. B 776, 350–354.

Sandulescu, A., Greiner, W., 1977. Mass asymmetry in fission, fusion and mass transfer due to the fragmentation in valleys. J. Phys. G 3 (8), L189.

Sandulescu, A., et al., 1985. Comment on "Exotic Nuclear Decay of ^{223}Ra by Emission of ^{14}C Nuclei". Phys. Rev. Lett. 54, 490.

Santhosh, K.P., Biju, R.K., 2013. Stability of $^{248-254}$Cf isotopes against alpha and cluster radioactivity. Ann. Phys. 334, 280–287.

Santhosh, K.P., Priyanka, B., 2015. The competition between alpha decay and spontaneous fission in odd–even and odd–odd nuclei in the range $99 \leq Z \leq 129$. Nucl. Phys. A 940, 21–52.

Santhosh, K.P., et al., 2010. Cluster formation probability in the trans-tin and trans-lead nuclei. Nucl. Phys. A 838, 38–49.

Sanuki, T., et al., 2000. Precise measurement of cosmic-ray proton and helium spectra with the BESS spectrometer. Astrophys. J. 545 (2), 1135–1142.

Sauli, F., 1997. GEM: a new concept for electron amplification in gas detectors. Nucl. Instrum. Methods Phys. Res., Sect. A 386, 531–534.

Sauli, F., 2014. Gas electron multiplier (GEM) detectors: principles of operation and applications, pp. 368–406. In: Brahme, A. (Ed.), Volume 8: Radiation Sources and Detectors" in "Comprehensive Biomedical Physics. Elsevier.

Saxena, G., et al., 2017. Two-proton radioactivity with 2p halo in light mass nuclei A=18–34. Phys. Lett. B. 775, 126–129.

Sayen, S., Guillon, E., 2014. Aging effect on Zn retention on a calcareous soil: column experiments and synchrotron X-ray micro-spectroscopic investigation. Sci. Total Environ. 487, 545–556.

Sayers, Z., et al., 2017. Application of advanced X-ray methods in life sciences. Biochim. Biophys. Acta 1861, 3671–3685.

Scherini, V., 2011. Search for ultra-high energy photons with the Pierre Auger Observatory. Nucl. Instrum. Methods Phys. Res., Sect. A 630 (1), 226–229.

Schrödinger, E., 1926a. Quantisierung als Eigenwertproblem I. Ann. Phys. 79 (4), 361–376.

Schrödinger, E., 1926b. Quantisierung als Eigenwertproblem II. Ann. Phys. 79 (6), 489–527.

Schrödinger, E., 1926c. Über das Verhaltnis der Heisenberg-Born-Jordan'schen Quantenmechanik zu der meinen. Ann. Phys. 79 (8), 734–756.

Schrödinger, E., 1926d. Quantisierung als Eigenwertproblem III. Ann. Phys. 80 (13), 437–490.

Schrödinger, E., 1926e. Quantisierung als Eigenwertproblem IV. Ann. Phys. 81 (18), 109–143.

Schwarzschild, A., 1963. A survey of the latest developments in delayed coincidence measurements. Nucl. Instrum. Methods 21, 1–16.

Seaborg, G.T., 1951. The transuranium elements: present status. In: Nobel Lectures, Chemistry 1942-1962. Elsevier Publishing Company, Amsterdam (1964).

Seaborg, G.T., 1995. Transuranium elements: the synthetic actinides. Radiochim. Acta 70/71, 69–90.

Seaborg, G.T., McMillan, E.M., Kennedy, J.W., Wahl, A.C., 1946a. Radioactive element 94 from deuterons on uranium. Phys. Rev. 69, 366–367 (submitted January 28, 1941).

Seaborg, G.T., Wahl, A.C., Kennedy, J.W., 1946b. Radioactive element 94 from deuterons on uranium. Phys. Rev. 69, 367 (submitted March 7, 1941).

Sedlacek, O., et al., 2014. Multistage-targeted pH-responsive polymer conjugate of Auger electron emitter: optimized design and in vivo activity. Eur. J. Pharm. Sci. 63, 216–225.

Segré, E., 1947. Possibility of altering the decay rate of a radioactive substance. Phys. Rev. 71, 274–275.

Segré, E., 1968. Nuclei and Particles. W. A. Benjamin, New York.

Sekine, T., et al., 1986. Application of nuclear recoil to radioisotope enrichment of copper-64. J. Nucl. Sci. Technol. 23 (12), 1064–1068.

Seltzer, S.M., Berger, M.J., 1982a. Evaluation of the collision stopping power of elements and compounds for electrons and positrons. Int. J. Appl. Radiat. Isot. 33, 1189–1218.

Seltzer, S.M., Berger, M.J., 1982b. Procedure for calculating the radiation stopping power for electrons. Int. J. Appl. Radiat. Isot. 33, 1219–1226.

Seltzer, S.M., Berger, M.J., 1984. Improved procedure for calculating the collision stopping power of elements and compounds for electrons and positrons. Int. J. Appl. Radiat. Isot. 35 (7), 665–676.

Selwyn, R.G., et al., 2007. A new internal pair production ratio of ^{90}Y: the development of a non-destructive assay for ^{90}Y and ^{90}Sr. Appl. Radiat. Isot. 65, 318–327.

Şen, M., et al., 2014. Determination of half-value thickness of aluminum foils for different beta sources by using fractional calculus. Nucl. Instrum. Methods Phys. Res., Sect. B 335, 78–84.

Serebrov, A., et al., 2008. Neutron lifetime measurements using gravitationally trapped ultracold neutrons. Phys. Rev. C78, 035505. https://doi.org/10.1103/PhysRevC.78.035505.

Serebrov, A., et al., 2013. New installation for measuring a neutron lifetime with a big gravitational trap of ultra cold neutrons. Tech. Phys. 58 (11), 1681–1687.

Serway, R.A., et al., 2005. Modern Physics, third ed. Brooks/Cole – Thomson, Belmont, CA. 600 pp.

Shanmugam, G., Kamalaharan, B., 1988. Application of a cubic barrier in exotic-decay studies. Phys. Rev. C 38, 1377–1381.

Sharma, S.K., et al., 2010. Explosive detection system using pulsed 14 MeV neutron source. Fusion Eng. Des. 85 (7–9), 1562–1564.

Shi, Y.J., Swiatecki, W.J., 1985. Estimate of radioactive decay by the emission of nuclei heavier than α-particles. Nucl. Phys. A 438, 450–460.

Shiga, D., 2011. Heaviest elements yet join the periodic table. New Scientist 11, 11.

Shimojima, H., Lin, T.-K., 1971. Enrichment of copper-64 by the Szilard Chalmers reactions of α-, β-, γ- and δ-benzene hexachlorides. J. Nucl. Sci. Technol. 8 (4), 179–183.

Shu, W.M., et al., 2004. Characteristics of a promising tritium process monitor detecting bremsstrahlung X-rays. Nucl. Instrum. Methods Phys. Res., Sect. A 521, 423–429.

Shu, W.M., et al., 2006. Monitoring of tritium in diluted gases by detecting bremsstrahlung X-rays. Fusion Eng. Des. 81, 803–808.

Sick, I., 1982. Precise nuclear radii from electron scattering. Phys. Lett. B 116 (4), 212–214.

Sick, I., Trautmann, D., 1998. On the rms radius of the deuteron. Nucl. Phys. A 637 (4), 559–575.

Sigmund, P., Schinner, A., 2003. Anatomy of the Barkas effect. Nucl. Instrum. Methods Phys. Res., Sect. B 212, 110–117.

Sigmund, P., et al., 2005. Valence structure effects in the stopping of swift ions. Nucl. Instrum. Methods Phys. Res., Sect. B 230, 1–6.

Šimkovic, F., et al., 2011. Neutrinoless double-beta decay and double-electron capture. Prog. Part. Nucl. Phys. 66, 446–451.

Simpson, J.A., 1948. The latitude dependence of neutron densities in the atmosphere as a function of altitude. Phys. Rev. 73, 1389–1391.

Simpson, J.A., 1951. Neutrons produced in the atmosphere by cosmic radiations. Phys. Rev. 83, 1175–1188.

Simpson, J.A., 2001. The cosmic radiation. In: Bleeker, J.A.M., Geiss, J., Huber, M.C.E. (Eds.), The Century of Space Science, vol. 1. Kluwer Academic Publishers, Dordrecht, The Netherlands, pp. 117–151.

Simpson, J.A., et al., 1953. Cosmic ray intensity-time variations and their origin: neutron intensity variation method and meteorological factors. Phys. Rev. 90, 934–950. http://prola.aps.org/abstract/PR/v90/i5/p934_1.

Sindhupakorn, B., et al., 2017. A structural study of bone changes in knee osteoarthritis by synchrotron-based X-ray fluorescence and X-ray absorption spectroscopy techniques. J. Mol. Struct. 1146, 254–258.

Singh, B., Batra, R.K., 1987. A method for calculating mass attenuation coefficients of beta particles. Appl. Radiat. Isot. 38 (12), 1027–1031.

Singh, A., Dhaliwal, A.S., 2014. Studies of internal bremsstrahlung spectrum of ^{35}S beta emitter in the photon energy region of 1–100 keV. Appl. Radiat. Isot. 94, 44–48.

Singh, A., Dhaliwal, A.S., 2015. Measurement of the internal bremsstrahlung spectrum of a ^{89}Sr beta emitter in the 1–100 keV photon energy regime. Appl. Radiat. Isot. 98, 113–116.

Skidmore, M.S., et al., 2009. Neutron sources for in-situ planetary science applications. Nucl. Instrum. Methods Phys. Res., Sect. A 608 (3), 403–409.

Slowiński, B., 2003. Spallation reactions and accelerator-driven systems. Appl. Energy 75, 129–136.

Snell, A.H., Miller, L.C., 1948. On the radioactive decay of the neutron. Phys. Rev. 74, 1217–1218.

Snell, A.H., et al., 1950. Radioactive decay of the neutron. Phys. Rev. 78, 310–311.

Soddy, F., 1913a. Inter-atomic charge. Nature 92, 399–400.

Soddy, F., 1913b. Radioactivity. Chem. Soc. Annu. Rep. 10, 262–288.

Soddy, F., 1922. The origins of the conceptions of isotopes. Nobel Lecture, December 12, 1922. In: Nobel Lectures, Chemistry 1901-1921 (1966), Elsevier Publishing Co., Amsterdam.

Sokovnin, S.Y., Balezin, M.E., 2017. Investigation of cold cathode for nanosecond electron accelerators. Vacuum 146, 79–86.

Solanki, R.B., et al., 2012. Probabilistic safety assessment for food irradiation facility. Ann. Nucl. Energy 43, 123–130.

Sorensen, T.L.-M., et al., 2006. New light for science: synchrotron radiation in structural medicine. Trends Biol. 24 (11), 500–508.

Soylu, A., Evlice, S., 2015. Deformation effects on cluster decays of radium isotopes. Nucl. Phys. A 936, 59–73.

Spano, H., Kahn, M., 1952. Enrichment of tin activity through the Szilard-Chalmers separation. J. Am. Chem. Soc. 74, 568–569.

Spinks, J.W.T., Woods, R.J., 1990. An Introduction to Radiation Chemistry, third ed. John Wiley & Sons, Inc., New York.

Spurio, M., et al., for the ANTARES Collaboration, 2017. Results from the ANTARES neutrino telescope. Nucl. Part. Phys. Proc. 291–293, 175–182.

Spyrou, A., et al., 2012. First observation of ground state dineutron decay: ^{16}Be. Phys. Rev. Lett. 108, 102501.

Stabin, M.G., 2009. Radiation Protection and Dosimetry" An Introduction to Health Physics. Springer, New York, p. 396.

Staníček, J., Povinec, P., 1986. Internal pair production in alpha-decaying nuclei and gamma-ray intensities of ^{241}Am. Nucl. Instrum. Methods Phys. Res., Sect. B 17, 462–466.

Steinman, A.D., Mulholland, P.J., 2007. Phosphorus limitation, uptake, and turnover in Benthic stream algae, pp. 187–212. In: Hauer, F.R., Lamberti, G.A. (Eds.), Methods in Stream Ecology, second ed. Elsevier, Amsterdam, p. 877.

Stone, J.O., et al., 1996. Cosmogenic chlorine-36 from calcium spallation. Geochim. Cosmochim. Acta 60 (4), 679–692.

Strömholm, D., Svedberg, T., 1909a. Untersuchungen über die Chemie der radioactiven Grundstoffe. I. Z. Anorg. Chem. 61, 338–346.

Strömholm, D., Svedberg, T., 1909b. Untersuchungen über die Chemie der radioactiven Grundstoffe. II. Z. Anorg. Chem. 63, 197–206.

Stuewer, R.H., 1997. Gamow, alpha decay, and the liquid-drop model of the nucleus, pp. 30–43. In: Harper, E., Parke, W.C., Anderson, D. (Eds.), George Gamow Symposium, ASP Conference Series, vol. 129. Astronomical Society of the Pacific, San Francisco.

Suliman, G., et al., 2012. Measurements of the half-life of ^{214}Po and ^{218}Rn using digital electronics. Appl. Radiat. Isot. 70, 1907–1912.

Suliman, G., et al., 2013. Half-lives of ^{221}Fr, ^{217}At, ^{213}Bi, ^{213}Po and ^{209}Pb from the ^{225}Ac decay series. Appl. Radiat. Isot. 77, 32–37.

Sundaresan, M.K., 2001. Handbook of Particle Physics. CRC Press, Boca Raton, FL, p. 446.

Suortti, P., 2003. X-ray, synchrotron radiation, and neutron diffraction, pp. 989–1023. In: Meyers, R.A. (Ed.), Encyclopedia of Physical Science and Technology, third ed. Elsevier, Amsterdam.

Suzuki, T., et al., 1999. Nuclear radii of ^{17,19}B and ^{14}Be. Nucl. Phys. A 658 (4), 313–326.

Švec, A., 2015. New analytical solution to calculate linear absorption coefficients of beta radiations. Appl. Radiat. Isot. 102, 63–69.

Svoboda, K., 1963. Szilard – Chalmers reaction on methyliodide with simultaneous extraction during irradiation. Nature 198, 986–987.

Szilard, L., Chalmers, T.A., 1934a. Chemical separation of the radioactive element from its bombarded isotope in the Fermi effect. Nature 134, 462.

Szilard, L., Chalmers, T.A., 1934b. Detection of neutrons liberated from beryllium by gamma rays: a new technique for inducing radioactivity. Nature 134, 494–495.

Szilard, L., Chalmers, T.A., 1935. Radioactivity induced by neutrons. Nature 135, 98.

Sztejnberg Gonçalves-Carralves, M.L., Miller, M.E., 2015. Neutron flux assessment of a neutron irradiation facility based on inertial electrostatic confinement fusion. Appl. Radiat. Isot. 106, 95–100.

Szucs, Z., et al., 2009. Recoil effect on β-decaying *in vivo* generators, interpreted for ^{103}Pd/^{103m}Rh. Appl. Radiat. Isot. 67, 1401–1404.

Tabata, T., Ito, R., Okabe, S., 1972. Generalized semiempirical equations for the extrapolated range of electrons. Nucl. Instrum. Methods 103, 85–91.

Taborda, A., et al., 2015. Simple polynomial approximation to modified Bethe formula low-energy electron stopping powers data. Nucl. Instrum. Methods Phys. Res., Sect. B 356–357, 172–175.

Tait, W.H., 1980. Radiation Detection. Butterworths, London.

Takahashi, Y., et al., 2010. Development of landmine detection system based on the measurement of radiation from landmines. Appl. Radiat. Isot. 68 (12), 2327–2334.

Takamatsu, T., et al., 2006. Inertial electrostatic fusion device with an ion source using a magnetron discharge. Nucl. Fusion 46, 142–148.

Takau, V.T., et al., 2012. A novel measurement of the b-decay half-life of ^{14}O, and its relevance to the standard model. Rad. Phys. Chem. 81, 1669–1672.

Tanabashi, M., et al., 2018. Review of particle physics. Phys. Rev. D 98 (3). Article No. 030001 (Open Access).

Tanaka, T., et al., 1964. The radiation annealing of phosphorus-32 with cobalt-60 gamma-rays for several neutron-irradiated inorganic phosphorus compounds. Bull. Chem. Soc. Jpn. 37 (7), 1031–1038.

Tanaka, H., et al., 2009a. Characteristics comparison between a cyclotron-based neutron source and KUR-HWNIF for boron neutron capture therapy. Nucl. Instrum. Methods Phys. Res., Sect. B 267, 1970–1977.

Tanaka, H., et al., 2009b. Improvement of dose distribution in phantom by using epithermal neutron source based on the Be(p,n) reaction using a 30 MeV proton cyclotron accelerator. Appl. Radiat. Isot. 67, S258–S261.

Tanihata, I., et al., 1985. Measurement of interaction cross sections and radii of He isotopes. Phys. Lett. B 160 (6), 380–384.

Tarancón Sanz, A., Kossert, K., 2011. Application of a free parameter model to plastic scintillation samples. Nucl. Instrum. Methods Phys. Res., Sect. A 648, 124–131.

Taylor, L.S., et al., 1970. Linear Energy Transfer. ICRU Report 16. International Commission on Radiation Units and Measurements, Washington, D.C.

Taylor, J.R., et al., 2015. Modern Physics for Scientists and Engineers. University Science Books, Mill Valley, California, p. 711.

Tellili, B., et al., 2014. Calculation of fast neutron removal cross sections for different lunar soils. Adv. Space Res. 53, 348–352.

Teresawa, M., Kihara, M., 1996. Basic characteristics of synchrotron radiation and its related facilities and instrumentation, pp. 1–78. In: Saisho, H., Gohshi, Y. (Eds.), Applications of Synchrotron Radiation to Materials Analysis. Elsevier, Amsterdam.

Theisen, C., et al., 2015. In-beam spectroscopy of heavy elements. Nucl. Phys. A 944, 333–375.

Thisgaard, H., Jensen, M., 2009. Production of the Auger emitter ^{119}Sb for targeted radionuclide therapy using a small PET-cyclotron. Appl. Radiat. Isot. 67, 34–38.

Thoennessen, M., et al., 2013. Novel techniques to search for neutron radioactivity. Nuc. Instrum. Methods Phys. Res., Sect. A 729, 207–211.

Thompson, K.A., 2004. Relativity, special, pp. 117–141. In: Meyers, R.A. (Ed.), Encyclopedia of Physical Science and Technology, third ed. Elsevier, Amsterdam.

Thomsen, K., 2007. A compound target concept for pulsed spallation sources. Nucl. Instrum. Methods Phys. Res. A 580, 1597–1599.

Thomson, J.J., 1897. Cathode rays. Philos. Mag. 44, 293–316.

Thomson, G.P., 1927. The diffraction of cathode rays by thin film of platinum. Nature 120, 802.

Thomson, G.P., 1928. Experiments in the diffraction of cathode rays. Proc. R. Soc. Lond., A 119, 600–609.

Thomson, G.P., 1929. The crystal structure of nickel films. Nature 123, 912.

Thomson, G.P., 1938. Nobel Lecture. "Electronic Waves", June 7, 1938. In: Nobel Lectures, Physics 1922–1941. Elsevier, Amsterdam, 1965.

Thontadarya, S.R., 1985. Determination of range of beta particles in aluminum from their mass attenuation coefficients. Int. J. Appl. Radiat. Isot. 36 (3), 251–252.

Titus, K.J., et al., 1997. A preliminary investigation of a beta-particle transmission gauge for seam quality determination. Textile Res. J. 67, 23–24.

Tjus, J.B., et al., 2016. Gamma-ray emitting supernova remnants as the origin of Galactic cosmic rays? Astropart. Phys. 81, 1–11.

Tolhoek, H.A., Brussaard, P.J., 1954. On the theory of alpha disintegration and the determination of nuclear radii. Physica 21 (1–5), 449–470.

Tomar, B.S., et al., 2010. Studies on production of high specific activity 99Mo and 90Y by Szilard Chalmers reaction. Radiochim. Acta 98, 499–506.

Tomaschitz, R., 2004. Cosmic rime dilation: the clock paradox revisited. Chaos Solut. Fract. 20, 713–717.

Tomiyasu, K., et al., 2010. Particle-in-cell simulation of magnetic-assisted electrostatic confinement device. Fusion Eng. Des. 85 (5), 728–733.

On behalf of the IceCube Collaboration Tosi, D., et al., 2017. Astrophysical neutrinos: ice cube highlights. Nucl. Part. Phys. Proc. 291–293, 167–174.

Tossell, J.A., 2002. Does the calculated decay constant for ^{7}Be vary significantly with chemical form and/or applied pressure? Earth Planet. Sci. Lett. 195, 131–139.

Tracz, G., et al., 2009. Pulsed thermal neutron source at the fast neutron generator. Appl. Radiat. Isot. 67, 1148–1155.

Tretyakova, S.P., 1995. Study of cluster decay of heavy nuclei. Radiat. Meas. 25 (1–4), 279–282.

Tsoulfanidis, N., 1995. Measurement and Detection of Radiation, second ed. Taylor and Francis, Washington, DC.

Tudora, A., 2010. "Point by Point" model calculation of prompt neutron emission data for ^{248}Cm(SF) and ^{244}Cm(SF). Ann. Nucl. Energy 37, 492–497.

Tueros, M., Sciutto, S., 2010. TIERRAS: a package to simulate high energy cosmic ray showers underground, underwater and under-ice. Comp. Phys. Commun. 181, 380–392.

Tumul'kan, A.D., 1991. Typical calibration curves for beta thickness gauges. Meas. Tech. 34 (1), 24.

Turner, J.E., 1995. Atoms, Radiation and Radiation Protection, second ed. John Wiley & Sons, New York.

UIC, 2005. Nuclear Fusion Power. UIC Nuclear Issues Briefing Paper No. 69, June 2005. Uranium Information Centre, Ltd., Melbourne, Australia, p. 5.

Ulinskas, K., et al., 2016. Long-term results for Stage IIIB cervical cancer patients receiving external beam radiotherapy combined with either HDR ^{252}Cf or HDR ^{60}Co intracavitary brachytherapy. Brachytherapy 15 (3), 353–360.

United Nations, 2008. Report of the United Nations Scientific Committee on the Effect of Atomic Radiation. General Assembly. Official Records A/63/46.

U.S. Public Health Service, 1970. Radiological Health Handbook. Publ. No. 2016. Bureau of Radiological Health, Rockville, MD.

Vajda, N., Kim, C.-K., 2010. Determination of radiostrontium isotopes: a review of analytical methodology. Appl. Radiat. Isot. 68, 2306–2326.

van Dorp, J.W.J., et al., 2018. Towards the production of carrier-free ^{99}Mo by neutron activation of ^{98}Mo in molybdenum hexacarbonyl – Szilard-Chalmers enrichment. Appl. Radiat. Isot. 140, 139–145.

Van Eijk, C.W.E., et al., 2004. Inorganic thermal-neutron scintillators. Nucl. Instrum. Methods Phys. Res., Sect. A 529, 260–267.

Van Elteren, J.T., et al., 1999. Voltammetry detection of copper in high specific activity ^{64}Cu. Appl. Radiat. Isot. 51, 15–19.

Van Rooyen, J., et al., 2008. A possible *in vivo* generator ^{103}Pd/^{103m}Rh – Recoil considerations. Appl. Radiat. Isot. 66, 1346–1349.

Vandecasteele, C., et al., 1973. Systematic errors in 14-MeV neutron activation analysis for oxygen. Part I. Neutron and γ-ray attenuation effects. Anal. Chim. Acta 64, 187–196.

Vandendael, I., 2010. Use of Auger Electron Spectroscopy to Study an Industrial Process. Lambert Academic Publishing, p. 228.

Vannucci, F., 2017. Interactions of neutrinos with matter. Prog. Part. Nucl. Phys. 95, 1–47.

Veksler, V.I., 1944. A new method of accelerating relativistic particles. C.R. Dokl. Akad. Nauk, SSSR 43 (8), 329–331.

Veksler, V.I., 1945. A new method of accelerating relativistic particles. J. Phys. USSR 9, 153–158.

Villard, P., 1900a. Sur la réflexion et la refraction des rayons cathodiques et des rayons déviables du radium. Compt. Rend. Acad. Sci., Paris 130, 1010–1012.

Villard, P., 1900b. Sur le rayonnement du radium. Compt. Rend. Acad. Sci., Paris 130, 1178–1179.

Villard, P., 1900c. Rayonnement du radium. Séances de la Société Française de Physique, pp. 45–46.

Vimalnath, K.V., et al., 2014. Large scale production of 51Cr for medical application in a medium flux research reactor: a comparative investigation of Szilard–Chalmers process and direct(n,γ) route. Appl. Radiat. Isot. 91, 104–108.

Vogel, P., 1967. On the nuclear octupole deformation in the $218 \leq A \leq 232$ region. Phys. Lett. B 25 (2), 65–67.

Vogel, P., 1968. On nuclear octupole deformations. Nucl. Phys. A 112 (3), 583–593.

von Weizsäcker, C.F., 1935. Zur Theorie der Kernmassen. Zeit. Phyik. 96, 431–458.

Wagner Jr., J.R., et al., 2014. Extrusion, second ed. Elsevier, Amsterdam, p. 620.

Walstrom, P.L., et al., 2009. A magneto-gravitational trap for absolute measurement of the ultra-cold neutron lifetime. Nucl. Instrum. Methods Phys. Res., Sect. A 599, 82–92.

Walter, U., 2006. Relativistic rocket and space flight. Acta Astronaut. 59, 453–461.

Waltham, C., et al., 2001. In: Proc. 27th Int. Cosmic Ray Conf., Hamburg, p. 991. http://adsabs.harvard.edu/abs/2001ICRC .... 3..991W.

Wan, B.N., et al., 2007. The first plasma of EAST. In: "21st IAEA Fusion Energy Conference", Chengdu, China, 16–21 October, 2007.

Wang, C., et al., 2008. Development of a new Cf-252 source for remote afterloading neutron brachytherapy: implications and applications. Brachytherapy 7 (2), 161–162.

Wang, M., et al., 2017. The AME2016 atomic mass evaluation (II). Tables, graphs and references. Chin. Phys. C 41 (3), 030003.

Wasserman, S.R., 2012. Rapid-access, high-throughput synchrotron crystallography for drug discovery. Trends Pharmacol. Sci. 33 (5), 261–267.

Webber, W.R., Potgieter, M.S., 1989. The artificial production of nuclear gamma-radiation. Astrophys. J. 344, 779.

Wei, J., 2003. Synchrotrons and accumulators for high-intensity proton beams. Rev. Mod. Phys. 75 (4), 1383–1432.

Westmoreland, S., 2010. A note on relativistic rocketry. Acta Astronaut. 67, 1248–1251.

Wilkinson, D.H., 2007. Super-allowed Fermi beta-decay: the neutron lifetime; G_A/G_V. Nucl. Instrum. Methods Phys. Res., Sect. A 574, 1–2.

Wilson, E.J.N., 1996. Fifty years of synchrotrons. In: CERN Accelerator Conference, 19 September 1996. https://accelconf.web.cern.ch/AccelConf/e96/PAPERS/ORALS/FRX04A.PDF.

Woan, G., 2000. The Cambridge Handbook of Physics Formulas. Cambridge University Press, Cambridge, p. 219.

Wolfson, R., 2003. Simply Einstein, Relativity Demystified. W. W. Norton & Co. Ltd., New York, p. 261.

Woo, Y.H., 1924. Note on absorption measurements of the x-rays reflected from a calcite crystal. Proc. Nat. Acad. Sci. 10 (4), 145–148.

Woo, Y.H., 1925a. The Compton Effect. PhD Dissertation. T-08048. The University of Chicago.

Woo, Y.H., 1925b. The intensity of the scattering of x-rays by recoiling electrons. Phys. Rev. 25 (4), 444–451.

Woo, Y.H., 1925c. The Compton Effect and tertiary x-radiation. Proc. Nat. Acad. Sci. 11 (2), 123–125.

Wu, B., Sabine Becker, J., 2011. Imaging of elements and molecules in biological tissues and cells in the low-micrometer and nanometer range. Int. J. Mass Spectrom. 307 (1–3), 112–122.

Wu, C.S., Ambler, E., Hayward, R.W., Hoppes, D.D., Hudson, R.P., 1957. Experimental test of parity conservation in beta decay. Phys. Rev. 105, 1413–1414.

Wulfkühler, J.-P., Tajmar, M., 2016. Novel inertial electrostatic confinement fusion with buckyball-shaped multi-grids. Am. Inst. Aeronaut. Astronaut. Conf. Paper. https://doi.org/10.2514/6.2016-4777.

Wüstling, S., et al., 2006. A large, 64-pixel PIN-diode detector for low-energy beta-electrons. Nucl. Instrum. Methods Phys. Res., Sect. A 568, 382–387.

Xiao, B.J., et al., 2008. EAST plasma control system. Fusion Eng. Des. 83, 181–187.

Xu, D., 2017. Search for astrophysical tau neutrinos with IceCube arXiv:1702.05238 [astro-ph.HE].

Xu, D., et al., on behalf of the IceCube Collaboration, 2017. Exploring the universe with neutrinos: recent results from IceCube. Nucl. Part. Phys. Proc. 287–288, 139–142.

Yagi, N., 2014. Synchrotron radiation, pp. 17–33. In: Brahme, A. (Ed.), Comprehensive Biomedical Physics, Radiation Sources and Detectors, vol. 8. Elsevier, Amsterdam, p. 418.

Yaksic, A., Tilton, J.E., 2009. Using the cumulative availability curve to assess the threat of mineral depletion: the case of lithium. Resour. Policy 34, 185–194.

Yamauchi, K., et al., 2006. Performance of neutron/proton source based on ion-source-assisted cylindrical radially convergent beam fusion. IEEJ Trans. Fund. Mat. 126, 1177–1182.

Yanagida, T., et al., 2015. Linear energy transfer effects on time profiles of scintillation of Ce-doped $LiCaAlF_6$ crystals. Nucl. Instrum. Methods Phys. Res., Sect. B 365, 529–532.

Yang, C.N., 1957. The law of parity conservation and other symmetry laws of physics. Nobel Lecture, December 11, 1957, In: Nobel Foundation Staff (Eds.), Nobel Lectures, Physics, 1942-1962. (1964). Elsevier Publishing, Amsterdam.

Yang, Y., et al., 2007. Explosives detection using photoneutrons produced by X-rays. Nucl. Instrum. Methods Phys. Res., Sect. A 579, 400–403.

Yearian, M.R., Hofstadter, R., 1958a. Magnetic form factor of the neutron. Phys. Rev. 110 (2), 552–564.

Yearian, M.R., Hofstadter, R., 1958b. Magnetic form factor of the neutron at 600 MeV. Phys. Rev. 111 (3), 934–939.

Yi, C.Y., et al., 1999. Mass attenuation coefficients of β^+-particles. Appl. Radiat. Isot. 51, 217–227.

Yigit, M., 2017. New empirical formulae for (n,t) cross sections at 14.6 MeV. Appl. Radiat. Isot. 128, 307–310.

Yoshikawa, K., et al., 2007a. Research and development of a compact discharge-driven D–D fusion neutron source for explosive detection. Nucl. Instrum. Methods Phys. Res., Sect. B 261 (1–2), 299–302.

Yoshikawa, K., et al., 2007b. Research and development on humanitarian landmine detection system by use of a compact D–D fusion neutron source. Fusion Sci. Techn. 52, 1092–1095.

Yoshiyuki, S., Hiroshi, T., 1999. Development of a small-sized painting thickness gauge based a a beta-ray transmission technique. Radioisotopes 48 (3), 163–171.

Young, N.A., 2014. The application of synchrotron radiation and in particular X-ray absorption spectroscopy to matrix isolated species. Coordin. Chem. Rev. 277–278, 224–274.

Yücel, H., et al., 2016. Measurement of photo-neutron dose from an 18-MV medical linac using a foil activation method in view of radiation protection of patients. Nucl. Eng. Technol. 48, 525–532.

Yukawa, H., 1935. On the interaction of elementary particles. I. Proc. Phys.-Math. Soc. Japan 17, 48–57.

Yukawa, H., 1949. Meson theory in its developments. Nobel Lecture, December 12, 1949. In: Nobel Lectures, Physics, 1942-62. (1964). Nobel Foundation Staff, Eds., Elsevier Publishing, Amsterdam.

Yunoki, A., et al., 2014. Observation of X-ray and Auger electron spectra in a 4π proportional counter for 4π(e, X)–γ coincidence measurements. Appl. Radiat. Isot. 87, 179–182.

Zahn, U., 1967a. A study of the recoil behavior of ^{56}Mn atoms in dilute solutions of manganese carbonyl compounds. Radiochim. Acta 7, 170–175.

Zahn, U., 1967b. Recoil reactions in crystalline ^{56}Mn carbonyls. Radiochim. Acta 8, 177–178.

Zamani, M., et al., 2008. A spallation neutron source based on Pb target surrounded by U blanket. Radiat. Meas. 43, S151–S155.

Zavrtanik, D., 2011. Results from the Pierre Auger Observatory. Nucl. Instrum. Methods Phys. Res., Sect. A 630 (7), 166–170.

Zeeman, P., 1897. The effect of magnetism on the nature of light emitted by a substance. Nature 55 (1424), 347.

Zeisler, S.K., Weber, K., 1998. Szilard-Chalmers effect in holmium complexes. J. Radioanal. Nucl. Chem. 227, 105–109.

Zeisler, S.K., et al., 1999. Szilard-Chalmers reaction in praseodymium compounds. J. Radioanal. Nucl. Chem. 240 (2), 637–641.

Zhang, Z., et al., 2000. Preparation of ^{186}Re and ^{188}Re with high specific activity by the Szilard–Chalmers effect. J. Label. Compounds Radiopharm. 43, 55–64.

Zhang, L., et al., 2014. CT-guided radioactive ^{125}I seed implantation treatment of multiple pulmonary metastases of hepatocellular carcinoma. Clin. Radiol. 69, 624–629.

Zheng, L., et al., 2016. Measurement and analysis of thorium fission rate in a polyethylene shell with a D-T neutron source. Fusion Eng. Des. 113, 177–182.

Zhernosekov, K.P., et al., 2012. The Szilard–Chalmers effect in macrocyclic ligands to increase the specific activity of reactor-produced radiolanthanides: experiments and explanations. Radiochim. Acta 100 (8–9), 669–674.

Zhou, F., et al., 2010. Measurements of the 89Y(n,γ)90mY cross-section in the neutron energy range of 13.5–14.6 MeV. Nucl. Instrum. Methods Phys. Res., Sect. B 268 (9), 1367–1369.

Zhu, Y., et al., 2014. Synchrotron-based X-ray microscopic studies for bioeffects of nanomaterials. Nanomed.: Nanotech. Biol. Med. 10 (3), 515–524.

Zinn, W.H., Szilard, L., 1939. Emission of neutrons by uranium. Phys. Rev. 56, 619–624.

Zito, R.R., Schiferl, D., 1987. Electron capture decay in Jovian planets. ICARUS 72, 647–649.

Zolfaghari, M., Sedaghatizadeh, M., 2016. Design of thermal neutron beam based on an electron linear accelerator for BNCT. Appl. Radiat. Isot. 118, 149–153.

Zuber, K., 2003. Spectroscopy of low energy solar neutrinos using CdTe detectors. Phys. Lett. B 571, 148–154.

Zolotarev, K., et al., 2016. Synchrotron radiation applications in the Siberian synchrotron and terahertz radiation center. Phys. Proc. 84, 4–12.

Zweig, G., 1964. *CERN Report No. 8419 TH 412* (unpublished), reprinted In: Lichtenberg, D.B., and Rosen, S.P. Eds. (1980). "Developments in the Quark Theory of Hadrons", Hadronic Press, Nonantum, MA.

Michael F. L'Annunziata

Michael F. L'Annunziata, Ph.D. is the Founding Editor and co-author of the *Handbook of Radioactivity Analysis* now in its Fourth Edition. His graduate thesis research at the University of Arizona, Tucson, AZ, USA, in the 1960s, financed by the then US Atomic Energy Commission, dealt with the analysis of ^{89}Sr and ^{90}Sr in the event of nuclear fallout. During 1970–71, he worked in the chemical industry (Amchem Products, Inc., Ambler, PA, USA) as ^{14}C-tracer chemist. Michael was then appointed Professor at the Postgraduate College, Chapingo, Mexico (1972–75) and Senior Research Scientist and graduate thesis advisor at the Nuclear Center of the National Institute of Nuclear Research (ININ), Mexico City during (1975–77). Michael served as Scientific Officer in the Department of Research and Isotopes and Department of Technical Co-operation of the International Atomic Energy Agency (IAEA) in Vienna, Austria during 1977–91 and Head of IAEA Fellowships and Training during 1987–91. Michael F. L'Annunziata was first to postulate the soil microbial epimerization of myo-inositol to other inositol stereoisomers as the chemical pathway and origin of the various inositol phosphate stereoisomers in soils (Ph.D. dissertation, 1970, https://dissexpress.proquest.com/dxweb/results.html?QryTxt=&By=L%27Annunziata&Title=&pubnum =) and in 1975 (*SSSA Journal* 39(2), 377–379), and first to demonstrate in 1977, with the use of the radioisotope carbon-14, the soil microbial epimerization of myo-inositol to chiro-inositol as a mechanism involved in the origin of the unique soil inositol phosphate stereoisomers (*SSSA Journal* 41(4), 733–736, https://dl.sciencesocieties.org/publications/sssaj/abstracts/41/4/SS0410040733). From 1977 to 2007, he served as IAEA Expert in peaceful applications of nuclear energy in over 50 countries of the world in Europe, Asia, Africa, Latin America and the Middle East. His book entitled *Radioactivity: Introduction and History*, First Edition, published by Elsevier, was on the Best Sellers List in Physics (*LibraryJournal Academic Newswire* in 2008), and his much expanded Second Edition entitled *Radioactivity: Introduction and History, From the Quantum to Quarks* published by Elsevier in 2016 (https://www.sciencedirect.com/book/9780444634894/radioactivity) was awarded an Honorary Mention in the 2017 PROSE AWARDS in the category of Chemistry & Physics.

Chapter 2

Gas ionization detectors

Georg Steinhauser[1] and Karl Buchtela[2]

[1]*Leibniz Universität Hannover, Institute of Radioecology and Radiation Protection, Hannover, Germany;* [2]*Vienna University of Technology, Atominstitut, Vienna, Austria*

Chapter outline

I. Introduction: principles of radiation detection by gas ionization

When radiation penetrates matter, energy of the radiation is passed on to the matter and the radiation is shielded or even stopped. The atoms or molecules of matter are brought to a state of higher energy, an excited state, or they are ionized if the energy of the radiation is high enough.

Alpha, beta, and gamma rays are known as ionizing radiation. On passing through a gas, these radiations create positive ions and electrons. Those charged particles either cause chemical reactions or recombine, finally producing

Handbook of Radioactivity Analysis: Volume 1. https://doi.org/10.1016/B978-0-12-814397-1.00002-9

neutral specimens again. However, if an electric field is applied, the positive ions start to migrate to the cathode, and the electrons are attracted by the anode. If the field strength, the applied voltage per unit length, is high enough to prevent recombination during migration of the ions and electrons, all of them arrive at the electrodes. They are collected at the electrodes, and by the detection of this electric charge using a suitable electric circuit, an indication of the presence of ionizing radiation is given.

Gas ionization detectors consist of a gas volume in an enclosure that is either sealed or constructed in such a way as to permit a continuous flow of the filling gas. Within that gas volume, an electric field is applied across the electrodes. The outer wall frequently serves as one of the electrodes, the cathode, whereas a wire rod, a grid, or a plate in the middle of the gas volume serves as the anode.

Although there are many different variations in the design of gas ionization counters, a cylindrical system with a central wire or rod, called a "counting tube," is very common. Many designs with different shapes and geometries have been realized. Some of them are suitable for a very wide range of useful applications, some were designed for a very special investigation, and others have been realized only to learn more about the operating principles of ionization detectors to improve the performance of this type of radiation detection device.

In this chapter, a selection is given from numerous developments in the field of gas ionization detectors. The principle of modern gas ionization detectors has been developed and used early in the "golden" era of radioactivity research in the early and mid-20th century. These detectors have made possible many discoveries, and their historic value cannot be overestimated. Although gas ionization detectors are still widely used in routine applications such as contamination monitors, α/β-gross counters, or dose rate instruments, it should be mentioned that radioactivity measurement methods today place emphasis mainly on radiation spectroscopy. Solid-state and scintillation detectors offer unique advantages in radiation spectroscopy and often outcompete classic gas ionization detector types/designs. The principle of gas ionization detectors also remains utilized in very special applications for high energy physics (Braccini et al., 2015; Siketic et al., 2015) or accelerator-based detectors, e.g., for accelerator mass spectrometry (AMS) facilities (Nadeau et al., 2015; Martschini et al., 2019; Suter et al., 2007; Zondervan et al., 2015) or in support of Mössbauer spectroscopy (Velasquez and Arroyave, 2014), in addition to the direct analysis of a radionuclide, where gas ionization detectors are increasingly replaced by other methods.

Many interesting developments have been made over the past decade or more in the field of tissue-equivalent proportional counter (TEPC), position-sensitive gaseous detectors such as the microstrip gas chamber (MSGC) and gas electron multiplier (GEM) detectors with good localization properties. These detectors are often referred to as micropattern gas detectors (MPGDs) because of their accurate position accuracy.

Although gas ionization detectors are extremely useful, problems and limitations have to be faced and careful planning of experiments to recognize and deal with those limitations is extremely important (Bateman et al., 1994). The suitability of gas ionization detector systems for a given kind and energy of radiation depends on the type (composition, pressure) of filling gas to be ionized; the applied field strength; the size, shape, and geometry of the detector volume and electrodes; and the type and thickness of the construction material that surrounds the detector gas volume. Also, environmental factors such as temperature should not be entirely ignored. Last but not least, the design of the electric circuit that handles the output signal plays a very important role.

The geometric design of a detector also depends mainly on its application. The size and shape have to be chosen appropriately if small or large areas have to be surveyed by the detector; if it has to be submerged in a liquid; or if, by use of a suitably thin wall, alpha and low-energy beta particles are permitted to enter the detector volume; finally, if radiation energy has to be determined or if the localization or distribution of the radioactive material in a given specimen is of primary importance.

There are three kinds of gas-filled detectors: ion chambers, proportional counters, and Geiger–Müller (GM) counters. They differ mainly in the strength of the electric field applied between their electrodes. Their common and different characteristics are discussed in this chapter.

II. Characterization of gas ionization detectors

A. Ion chambers

Gas ionization detectors can be characterized by the effects created by different field strengths between the charge-collecting electrodes. The relationship between the pulse size produced and the potential applied across the electrodes of a gas ionization detector is shown in Fig. 2.1. The pulse size depends on the field strength and also on the type of radiation that enters the detector volume and creates ions.

At low field strength, many slowly migrating ion pairs still have the opportunity to recombine. This recombination region is not used for radioactivity detectors. As more voltage is applied, more ions and electrons produced by the ionizing radiation are collected at the electrode. Finally, a field strength is reached at which the now rapidly migrating ions do not have a chance to recombine. Thus, a saturation region is reached where all the ions produced directly by

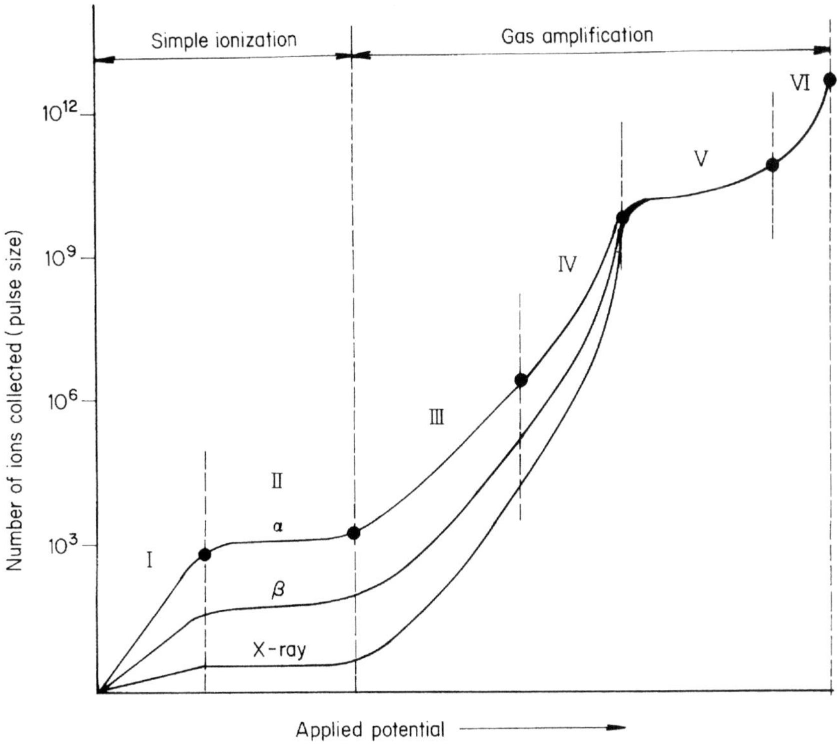

FIGURE 2.1 Relationship between the pulse size produced and the potential applied across the electrodes of a gas ionization chamber exposed to alpha, beta, and gamma radiation. Various regions are labeled by Roman numerals as follows: region I, recombination region; region II, simple ionization region; region III, proportional region; region IV, limited proportional region; region V, Geiger–Müller region; region VI, continuous discharge region. *From L'Annunziata (1987), reprinted with permission from Elsevier © 1987.*

the radiation event, the primary ions, are collected at the electrodes. A further increase of field strength cannot attract more ions, because all of them have already been collected. Ion chambers operate in this region. The amount of charge collected at the electrodes directly shows the ionization effects of the incident radiation.

The design of ion chambers can be tailored for a special type of radiation, and information about radiation energies can be provided. Since the output signal is directly related to the ionization effect, ion chambers are very useful in radiation dosimetry.

Alpha particles produce a great amount of ions along a short path length of travel (high linear energy transfer, LET). They are easily detected, because they provide a high output signal. Beta particles and gamma rays produce a very low signal, and rather sophisticated circuits are sometimes needed for amplification of such low-voltage signals.

The following short calculation provides an example of the signal requirement of electronic circuits combined with ion chambers:

Let us assume that a radioactive source emits one alpha particle per second (activity 1 Bq) with an energy of 5 MeV, and all the energy of the alpha particles is deposited in the gas volume (air) of the counting chamber. The average ionization energy of that gas is 32.5 eV/ion pair. The rate of ionpair formation in the gas is calculated as

$$\frac{5\ \text{MeV/alpha particle}}{32.5\ \text{eV/ion pair}} = 1.5 \times 10^5\ \text{ion pairs/alpha particle}$$

Thus, 1.5×10^5 ion pairs or 1.5×10^5 electrons are produced by one alpha particle per second, corresponding to an electric charge of

$(1.5 \times 10^5\ \text{electrons/s})(1.6 \times 10^{-19}\text{coulomb}) = 2.4 \times 10^{-14}\ \text{C/s}$

$= 2.4 \times 10^{-14}\ \text{A}.$

B. Proportional counters

If the field strength is increased further, additional ionization starts to occur, because of the higher kinetic energy of the migrating primary ions. These primary ions, now being accelerated to a higher energy than the ionization energy of the detector gas, produce secondary ions by impact. With increasing field strength, a great number of additionally produced ions are accelerated, the number still being proportional to the number of primary ions. This gas ionization detector region (see Fig. 2.1) is called the proportional region. In that region, radiation with different abilities to produce primary ions (alpha, beta, or gamma radiation) can still be discriminated, or they are registered by "gross counting" without separation. Also, radiation of the same type but with different energies can be discriminated (Garcia-Leon et al., 1984).

With further gas multiplication due to higher field strength, some nonlinearities will be observed. This effect marks the beginning of the limited proportional region.

C. Geiger–Müller counters

As the field strength is increased further, excitations of atoms and molecules are observed that, by the emission of ultraviolet (UV) light, can start additional ionization processes. In this region, referred to as the GM region, the total number of ions produced is independent of the number of primary ions and, therefore, also independent of the type and energy of radiation. A further increase of the field strength causes a continuous discharge (see Fig. 2.1).

In the GM region, all primary ionization effects produce the same maximum response in the detector. GM counting tubes operate in this region and thus provide no direct information about the type and energy of radiation. Information related to the type and energy of radiation can be provided only by observing shielding effects related to this radiation. Alpha particles are stopped by a thin layer of matter, beta particles show a maximum range in penetrating a shielding material before they enter the detector, and photons show a somehow logarithmic decrease in intensity with increasing thickness of the material. In the earlier days of radiation measurements, such experimental setups were frequently used for rough determination of radiation type and energy (Chase and Rabinowitz, 1967).

III. Definition of operating characteristics of gas ionization detectors

In the case of ionization detectors, as well as other detector types, some operating parameters are important for characterizing their capabilities: efficiency, resolution, and resolving time of the detector. For some special detector designs, such as position-sensitive detectors, the capability of determining precise information regarding the spatial distribution of particles or photons entering the detector volume is of importance.

A. Counting efficiency

The counting efficiency relates the number of particles or photons emitted by a radiation source to the number of interactions registered by the counting system. This is usually called the absolute efficiency.

$$\text{Absolute efficiency} = \frac{\text{number of signals recorded by the detector}}{\text{number of particles or photons emitted by the source}} \tag{2.1}$$

All particles striking the detector volume may not be registered. Therefore, another kind of efficiency is used, which is called the intrinsic efficiency, defined as:

$$\text{Intrinsic Efficiency} = \frac{\text{number of signals recorded by the detector}}{\text{number of particles or photons striking the detector}} \tag{2.2}$$

With ionization detectors, the absolute efficiency of charged particles can go up to nearly 100%. For gamma rays and X-rays, the efficiency is frequently much lower, because of the relatively poor interaction of the radiation with the gaseous detection volume of the ionization counters. Therefore, a higher density of the gaseous volume can sometimes be obtained by using a counting gas of high atomic number (xenon) and by increasing the gas pressure inside the ionization detector.

B. Energy resolution

The energy resolution characterizes the ability of the detector to discriminate between two types of radiation with energies that are different but rather close to each other. A characteristic figure is given by the full width at half-maximum (FWHM), the width of a peak in a radiation energy spectrum display halfway between the baseline and top of the peak. If E_0 is the energy at the peak maximum and ΔE is the FWHM, the resolution is given as

$$R = \Delta E / E_0, \tag{2.3}$$

which can also be recorded as a percentage.

The value of the FWHM is a measure of the resolution. The smaller its value for a given energy line, the greater the potential of a detector to provide individual information related to two radiations of approximate energy. Because of the statistical nature of any interaction of radiation with matter, resolution never can be perfect. In addition, electronic noise contributes to the deterioration of resolution.

Not all detectors can provide information about radiation energy.

C. Resolving time

The resolving time refers to the minimum time interval a detector needs to recover from the interaction with a radiation event and be able to register a following event. For many counting devices, it is not the resolving time of the detector but the resolving time of the electronic system (e.g., the data handling and processing steps) that sets the limits for dealing with high count rates. Counting losses induced by resolving time of a counting system can be a limiting factor in measurements. Several methods for resolving time determination and correction are presented in the literature (Gardner and Liu, 1997; Lee and Gardner, 2000; Vinagre and Conde, 2001).

D. Localization

Some detector designs can give information about the entrance region of particles or photons into the detector or about the distribution of radioactive material in a sample. They can give an image of a radioactive specimen by showing the longitudinal or even two-dimensional (2D) distribution of radioactivity.

Position-sensitive detectors based on proportional counting (PC) systems were developed by Nobel Laureate Georges Charpak in the 1960s, and these played a decisive role in many discoveries in particle physics. These types of detectors, providing the opportunity of "nuclear imaging," are nowadays very important also in many experiments in biology and medicine (Charpak, 1970; Charpak and Sauli, 1978; Geltenbort, 1994; Nickles et al., 2002; Shektman, 2004; van der Graaf, 2011; see also Chapter 9 of Volume 2).

IV. Ion chambers

Ionization chambers can be considered as one of the simplest devices for radioactivity measurements. They were used in the very early days of research dealing with the detection of ionizing radiation. However, even today, new designs for special purposes are being developed.

The chamber is made of a nonporous material, the electrodes are usually parallel plates, and the filling gas may have a pressure from a few tenths up to some tens of bars. When ionizing radiation passes through the gas, ion pairs are created. If a sufficiently high-voltage gradient prevents recombination, these ions drift toward the electrodes. The output signal registered by the electric circuit can be a flow of current, a charge or voltage pulse, or a total collected amount of electrical charge. Thus, three types of ion chambers are known: ion chambers operating in the current mode, ion chambers operating in the pulse mode, and electrostatic or charge integration ion chambers.

With ion chambers operating in the current mode, an electrical current flow is registered, which is initiated by the electrons and ions collected at the electrodes during the time of observation. With the pulse mode type of chamber, single signals, such as voltage pulses created by the ions arriving at the electrodes from a single ionization event, are registered by applying suitable electronic amplification circuits. Electrostatic or charge integration ion chambers are similar to electroscopes. A static electric charge is given to a system consisting of a thin foil or fiber that is suspended parallel to a solid support or to a second fiber or foil. Because of the repulsion of like charges, the fiber or foil will be bent to stay at some distance away from the support of the second foil or fiber. Ionizing radiation gradually discharges the system, and this causes the foils or fibers to move back to their original position.

Because of their simple construction and relatively low cost, ion chambers still have many applications. Information related to the type and energy of radiation can be obtained, and the ion chambers can be designed for the detection of low as well as high radioactivity levels. Many kinds of gases can be used to fill the detector volumes.

A. Operating modes of ion chambers

1. Ion chambers operating in the current mode

One of the most important applications of an ion chamber in everyday radiochemistry is as a portable survey instrument for radiation monitoring purposes. A volume of counting gas, mostly air, is enclosed within walls made of metal-lined plastic or aluminum. These types of walls are "air equivalent." Thus, accurate measurements can be made for gamma radiation if the energy of the gamma radiation is not only high enough to penetrate the walls without significant attenuation but also low enough to establish electronic equilibrium in these walls. Usually, for gamma radiation with energy lower than 50 keV, attenuation effects have a considerable impact on the efficiency of such detectors. With these instruments, the saturated ion current is measured by using an electrometer circuit that is battery powered. Converting the DC signal of an ion chamber to an AC signal provides a more stable amplification, such as with the vibrating-reed electrometer or dynamic capacitor.

2. Charge integration ionization chambers

A type of ionization counter, which is used frequently, is operated on the charge integration principle. This type of ionization chamber is charged initially. The drop in charge during exposure to a radiation field can be measured using a charger–reader mechanism and provides information regarding the dose from the radiation field to which the ionization chamber was exposed.

A familiar device is the ionization pocket chamber. These ionization chambers are also charged initially, but they are equipped with a small integral quartz fiber electroscope. An initial charging sets the scale of the electroscope to zero. The total integrated dose can be read periodically by observing the migration of the quartz fiber. This can be done very simply by optical observation, just by holding the pen-shaped pocket chamber up to a source of light and looking at the scale of the fiber electroscope through a small integrated magnifying glass. The accuracy and sensitivity of these devices are limited by leakage current across the insulator material of the ionization chamber.

3. Pulse mode ion chambers

Like other ionization detectors, such as proportional counters and GM tubes, ionization counters can also be used in pulse mode, in which each separate alpha particle, beta particle, or gamma quantum creates a distinguishable pulse signal. Advantages of pulse mode ionization chambers are their sensitivity and the ability to measure the energy of radiation and thus to be applicable in radiation spectroscopy. Today, such pulse mode ionization chambers have been mostly replaced by semiconductor detectors. Nevertheless, for special applications, such as neutron counting facilities, such chambers are still in use.

Pulse amplitudes from all types of ion chambers are relatively small. In theory, the maximum signal amplitude accumulated from the ion pairs produced by the interaction of, for example, an alpha particle in air along its track within the chamber is of the order of 10^{-5} V. Such a signal can be processed, but rather sophisticated electronic systems are required. Pulses from a single photon interaction are a hundred times smaller, and successful and accurate amplification is difficult and at times even impossible. Internal amplification within the detector volume, which is described in the section of this chapter dealing with PC tubes, helps to overcome these problems.

In any case, it has to be taken into account that for some filling gases, such as propane, the energy response to beams of heavy ions depends also on the atomic number of the observed ions and, as a consequence, deficits of the pulse heights are sometimes considerable. This is not only due to ion energy loss via elastic collisions in the filling gas (e.g., propane) but most probably also due to electron–ion recombination within the dense ionization column created by the incident ion beam. This may be of importance for many types of gas ionization detectors used for the spectroscopy of ion beams with higher atomic mass (Weijers et al., 2002).

B. Examples and applications of ion chambers

1. Calibration of radioactive sources

Standardization of gamma-emitting radionuclides (e.g., in nuclear medicine applications) is frequently carried out by comparing the ion current from a material with unknown activity with the ion current produced by a standard material of the same radionuclide (see also Chapter 7 of Volume 2). In that way, one takes advantage of the excellent long-term stability and reproducibility of the ion current produced from the same type of radiation. When operating in the saturation region, the current depends only on the geometry and the activity of a given radioactive material. Chamber volumes can be up to several 1000 cm^3, and the walls are made of solid materials, such as steel or brass. The collecting electrode in the inner part is made of a thin metal foil to avoid as much as possible attenuation of the radiation. An example is the production of fluorine-18 in clinical settings, where ion chambers are used as calibrators for the activity determination (Wurdiyanto et al., 2012). Using ion chambers as absolute methods of activity analysis allows for the (re)determination of half-lives, as shown by Rezende et al. (2012).

High sensitivity can be obtained if pressurized gas is used for the ionization chamber. Of course, this will cause the background current to increase but not be as great as that produced by radiation sources. Pressurized chambers are used for the measurement of gamma-emitting nuclides.

The ion chamber region is usually reached by adjusting the voltage for the electrodes. Saito and Suzuki (1999) used a multielectrode ion chamber for measuring absolute fluence rate of X-rays. They adjusted the ion chamber region by varying the gas pressure at a given voltage.

2. Measurement of gases

Many radioactive gases can be incorporated into the filling gas of ionization detectors. Also, in ionization chambers, a gas can be sampled on a continuous flow-through basis. The ionization current produced by a gas can be calculated simply and straightforwardly only if the radiation is fully absorbed in the gas volume of the ionization chamber. These types of flow-through ionization chambers are used for monitoring air that contains small amounts of radioactive gas. In any case, a number of difficulties arise if the air is subject to atmospheric changes. Such perturbations of air properties can be due to the content of aerosols, moisture, ions, etc. (Jalbert and Hiebert, 1971; Mustafa and Mahesh, 1978; Waters, 1974).

Some interesting examples of flow-through ionization chambers are those utilized in the activity measurements of ^{76}Kr and ^{85}Kr (Cooper et al., 2004; Yunoki et al., 2010). The flow-through gaseous radiochemical method was demonstrated by Lebedev et al. (2006) as a useful method for monitoring the release of radioactive inert gas molecules formed as a result of nuclear reactions. When a radioactive gas isotope is released from an ampoule-loaded target material, the radioactive gas, which is released, is transported by a carrier gas into a gas counter for flow-through measurements. Lebedev et al. (2006) demonstrated that the activity I in disintegrations per second (dps) of radioactive inert gas in the counter was related to the flux of ionizing radiation (e.g., proton beam)

$$I = \frac{V_C P_C N_0 \sigma \phi}{1.44tL} \tag{2.4}$$

where V_C is the active volume of the gaseous counting chamber in cm^3, P_C is the pressure of the counter in physical atmospheres, N_0 is the number of atoms of target material, t is the half-life of the radioactive inert gas in seconds, σ is the cross section of a nuclear reaction for the formation of a radioactive isotope of an inert gas averaging in a spectrum of ionizing radiation, ϕ is the flux of ionizing radiation in cm^2/s, and L is the consumption of transport gas in cm^3/s.

Yakushev et al. (2004) used a flow-through ionization chamber as a valuable tool in the identification and measurement of the decay properties of element 112. An Ar-CH_4 mixture was used as the carrier and counting gas used to transport element 112 into an ionization chamber $5000\ cm^3$ in volume. The IUPAC (2010) officially approved the name copernicium, with symbol Cn, for the element with $Z = 112$ on February 19, 2010.

The change of ionization current due to smoke particles is the operational basis for smoke detectors. In such smoke detectors, a built-in alpha source provides a constant ionization current under normal atmospheric conditions. A twin chamber with enclosed air without flow-through capability is used for the reference ion current.

The design of twin chambers can also be used for background compensation. A twin chamber filled with pure air records the background without flow-through of the air to be monitored. In that way, compensation for a changing background can easily be achieved, for example, in case of a changing gamma-ray background during air monitoring.

Current mode ion chambers have been very useful in the measurement of radon. The background is low, and the counting efficiency is high (practically 100%). Experiments have also been reported to provide data for the radon content of groundwater by placing an ion chamber together with a known amount of water for 3 hours in a leak proof container. The amount of radon in the air can be related to the concentration of radon in the water sample (Amrani et al., 2000).

Tritium measurements using ion chambers present a problem if elastomeric seals are applied. Those elastomeric materials are irreversibly contaminated, and the background of the ion chamber is increased. Colmenares (1974) constructed a chamber using ultra high-vacuum metallic seals, a metal construction of negligible water adsorption capacity and sapphire as insulator material. The chamber is bakeable up to 450°C, and the contamination problems are avoided.

3. Frisch grid ion chambers

Because of the slow ion mobility in gases and the slow drift of ions at the applied field strength in ion chambers, the use of pulse-type ion chambers is restricted to low pulse rates if signals are desired that are related accurately to the original charge of ions and electrons generated by the radiation. Pulse-shaping circuits designed for low frequencies also make these systems rather susceptible to interference from microphone signals produced by mechanical vibrations.

Therefore, pulse-type ion chambers are frequently operated in such a mode that they sense the collected electrons only, not the created positive ions, which migrate much slower than the electrons. In that case, the pulse amplitude is related to the drift of the electrons only. The signal therefore has a much faster rise time, and higher counting rates can be successfully registered. However, since the amplitude of the signal now depends also on the position of the interaction within the ion chamber gas, there is no well-defined information related to the total number of ions created, which means that no information about the energy of the radiation is provided. However, methods have been developed to overcome the problem of the dependence of the pulse amplitude on the position of the interaction within the chamber. The region of the chamber volume is divided into two parts by a grid. This grid is maintained at a potential between those of the cathode and anode. The mechanical construction of the grid should allow electrons to pass through; it should be as "transparent" to electrons as possible. By suitable positioning of the radiation source outside the chamber or by effective collimation of its radiation, the emitted particles or rays interact with the gas in the ion chamber in a well-defined region between this grid and the negative electrode of the chamber. Thus, positive ions simply migrate to the cathode. Electrons are attracted by the transparent grid initially but are further accelerated toward the anode, which is at a much more positive potential than the transparent grid.

Electronic circuits are designed in such a way that, with the electron migration from grid to anode, the voltage between grid and anode drops and a signal is created that depends only on the electron drift and not on the migration of both electrons and cations. Therefore, the slow rise related to ion drift is eliminated. Also, since all electrons are accelerated by the same potential difference, the amplitude of the pulse is independent of the position of the interaction. The amplitude is proportional only to the number of ion pairs, the number of electrons produced along the path of the interacting particle or ray. This type of ion chamber is called a Frisch grid chamber after Otto Robert Frisch who designed the detector (Knoll, 1989). Such Frisch grid chambers have been extremely useful in studies dealing with particle physics.

Gridded ionization chambers are ideally suitable for studies related to nuclear fission, because such detectors not only have a practically 100% detection efficiency, but they can also provide information about fission fragment properties such as energy, mass, charge, and emission angle. Tutin et al. (2001) designed an efficient device for such investigations, consisting of a combination of two parallel plate ionization chambers with grids and a common central cathode plate (Fig. 2.2). The central aluminum disk cathode has a hole of 78 mm diameter where two thin aluminum foils, covered with uniform layers of thorium-232 or

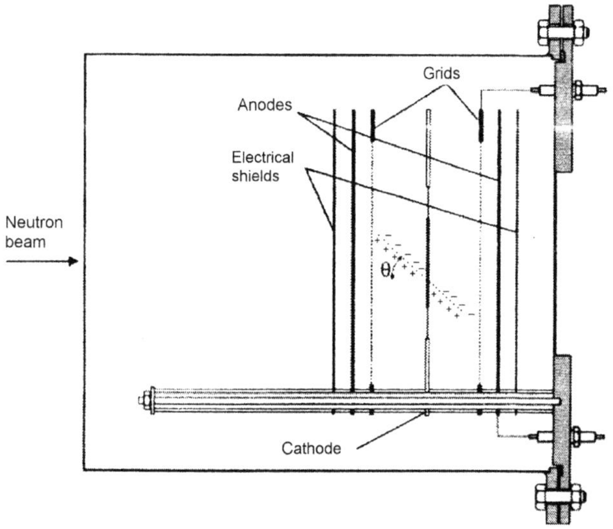

FIGURE 2.2 Schematic picture of a parallel gridded ionization chamber with one common cathode. *From Tutin et al. (2001), reprinted with permission from Elsevier © 2001.*

uranium-238 (fissile upon high-energy neutron capture), are mounted back to back. The grids are mounted on aluminum annular disks with an inner diameter of 160 mm and consist of tungsten wires 0.1 mm in diameter spaced by 1.25 mm. The chamber is filled with 90% argon and 10% methane at atmospheric pressure without continuous gas flow. Fission fragments emitted from the cathode are stopped in the space between the cathode and anode. Free electrons drift to the anode, whereas the slow ions can be treated as being static for a short interval of time. At the end of the electron drift, the collected charges can be related to the emission angle of the fission fragments. In a continuation of these studies, a similar multisection Frisch-gridded ionization chamber was designed by Ryzhov et al. (2006) to measure neutron-induced fission cross section of various radionuclides.

There are also applications in the life sciences. Lohmann et al. (1998, 2003) used a detector system of the Frisch grid chamber type in angiography for the determination of contrast agent (iodine) by "dichromography." According to this method, two images with monochromatic X-rays just below and above the absorption edge of the contrast agent are simultaneously obtained and subtracted. Although monochromatic X-rays having suitable intensity to visualize arteries of 1 mm diameter are only provided by synchrotron radiation, the authors concluded that the requirements for application of intravenous coronary angiography are fulfilled with the Frisch-gridded detector.

4. Radiation spectroscopy with ion chambers

Pulse-type ion chambers have attracted some interest again, after years, during which most of the interest was focused on semiconductor detectors. Ion chambers can be designed and constructed in any shape and size, and for charged particles, the pressure can be tailored to an optimum for a desired stopping power (Fulbright, 1979). Also, there is practically no deterioration or degradation due to irradiation, which may adversely affect the application of ion chambers in radiation fields, and ion chambers can be fabricated by using available workshop facilities without high expenses.

Pulse-type ion chambers have been applied in low-level alpha measurements, and good resolutions have been obtained that may even be comparable with the resolution of semiconductor detectors (Gruhn et al., 1982; Bertolini, 1984; Hoetzl and Winkler, 1984; Shenhav and Stelzer, 1985; Kotte et al., 1987; Nowack, 1987; Domnikov et al., 2001; Beck et al., 2009; Kada et al., 2010).

It was demonstrated that additional information regarding charged particle properties such as atomic number and charge state can be obtained by designing a chamber in such a way that particle pathways are parallel instead of perpendicular to the direction of the electric field. Thus, the drift time of electrons to the grid will be different for electrons created at the beginning of the track and those from the end of the path. The shape of the output pulse will therefore reflect the distribution of ion pairs along the track according to what is called a Bragg curve. With that technique, known as Bragg curve spectrometry, additional information such as atomic number and particle charge can be obtained. For that, a detailed analysis of the pulse shape is necessary.

Hagiwara et al. (2008) developed a method to extend the energy acceptance of a Bragg curve counter (BCC) at the high-energy end. They applied this method to the measurement of lithium and beryllium ions emitted from carbon in the course of the bombardment by 70 MeV protons. A schematic diagram of the BCC and its signal processing scheme, devised by Hagiwara et al. (2008), is illustrated in Fig. 2.3. A Bragg curve derived from the ionization intensity produced by an incident particle is illustrated in the figure by the shaded area above the particle path. From the Bragg curve, the Bragg peak signal PB and energy signal PE are obtained by the use of two pulse-shaping amplifiers with short (0.4 μs) and long (6 μs) shaping times, respectively. The long shaping time is longer than the drift time of the free electrons traveling from the cathode to the grid. A charge-sensitive preamplifier (PA) integrates and amplifies the original signals from the BCC before reaching the pulse-shaping amplifiers. Fig. 2.4 illustrates the 2D spectrum obtained when plotting PB versus PE (*i.e.*, Bragg peak pulse height vs. energy) illustrating clear particle discrimination for particles that deposit all of their energy within the BCC (left-hand spectrum); however, interferences are observed due to high-energy particles capable of passing through the BCC

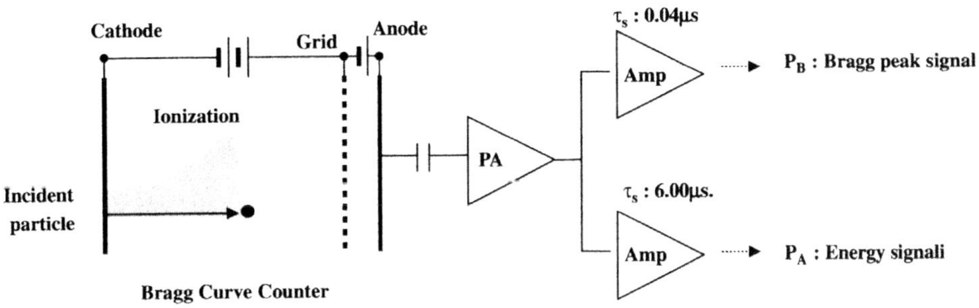

FIGURE 2.3 Schematic diagram of the electronic circuit to derive the incident energy and Bragg peak information from the anode of a BCC. The short and long shaping times are set to be 0.04 and 0.6 μs, respectively. *From Hagiwara et al. (2008), reprinted with permission from Elsevier © 2008.*

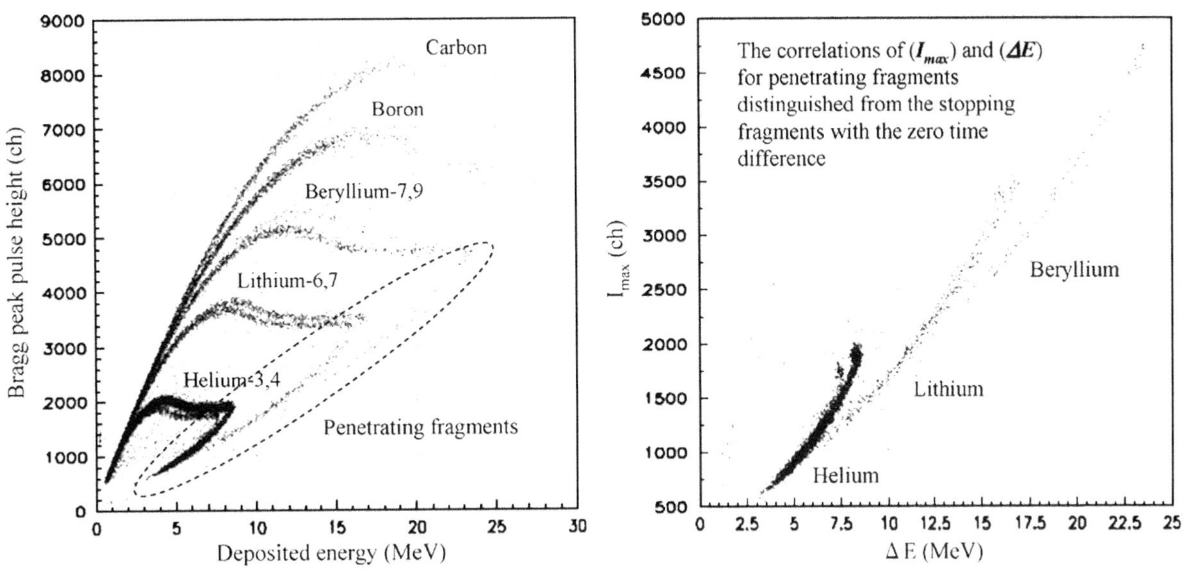

FIGURE 2.4 The two-dimensional spectrum of deposited energy versus Bragg peak of fragments emitted from a 4-μm-thick polypropylene sample at 30 degrees by 70-MeV proton-induced reaction. The dotted circle indicates the events of the penetrating fragments (i.e., those that pass through the BCC). The left-hand figure shows all events. The right-hand figure shows the correlations of I_{max} and ΔE for penetrating fragments from the fragments that stop within the active volume of the BCC. *From Hagiwara et al. (2008), reprinted with permission from Elsevier © 2008.*

(Fig. 2.4, dotted circle). Hagiwara et al. (2008) distinguished the high-energy particles (i.e., fragments) that penetrated the BCC counter by plotting I_{max} (i.e., maximum ionization) against deposited energy (ΔE) as illustrated in Fig. 2.4.

Khriachkov et al. (2000) used an alpha-particle spectrometer based on a Frisch grid chamber for studies of (n,α)-reactions induced by fast neutrons. Energy and emission angle of alphaparticles could be determined.

Combinations of ionization chambers with position-sensitive ionization detector devices were used by Menk et al. (2000) for small-angle X-ray scattering investigations. These systems are intended to be used for experiments in some European synchrotron centers.

Nariyama et al. (2004) developed a portable free-air ionization chamber for measurements of synchrotron radiation up to 150 keV. For high-energy X-rays, the plate separation becomes rather large, and therefore, the authors developed a special design to overcome that difficulty.

5. Electret detectors

Electret types of ion chambers make use of the drop of surface voltage on a plastic material. The plastic specimen is a dielectric material, usually Teflon, which is quasi-permanently charged. It is called an electret and usually has the shape of a disk about 1 mm thick and 10 mm in diameter. Electrets are prepared by being heated and simultaneously exposed to an electric field. In this process, many dipoles in the material become oriented in a preferred direction. After heating, the material is "frozen," which keeps the position of its electric dipoles for a long period of time. A voltage gradient of several hundred volts can be maintained between the surfaces of the electret disk.

One surface of the electret is kept in contact with the wall of an ion chamber, which builds up an electric field in the chamber. Ionizing radiation causes a decrease of charge in that system, resulting in a partial neutralization of the charge at the electret. Measurement of the electret voltage

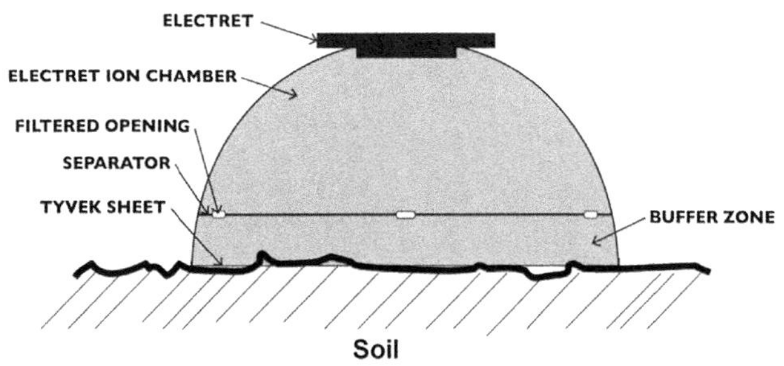

FIGURE 2.5 Scheme of an electret ion chamber for the measurement of radon flux from soil. The radon diffuses through the Tyvek window and into the electret ion chamber. The radon activity (Bq/m^3) is proportional to the potential drop due to radon decay on the positively charged electret located at the top of the ion chamber. *From Grossi et al. (2012), reprinted with permission from Elsevier © 2012.*

difference before and after irradiation provides a measurement of the amount of ionization. The system has to be calibrated and can be used for the determination of environmental radiation doses.

Amrani et al. (2000) used an electret ion chamber for the determination of the radon content of groundwater. They put an electret ion chamber together with a known amount of water in a leak proof container. The reading of the electret ion chamber provides the radon content in the air, and this value could be related to the concentration of radon in the water sample. Electret detectors are commonly used as radon monitors in the environment, including water and soil (Hamlat et al., 2003, Kitto et al., 2008, and Grossi et al., 2011, 2012), construction materials (Lavi et al., 2009), and in the public living and workplace (Denman et al., 2005; Righi et al., 2005; Papachristodoulou et al., 2010; Clouvas et al., 2011, and Okejiet al., 2012). The electret ion chamber is simple in design. An example of an electret detector used for the measurement of radon flux from soil is illustrated in Fig. 2.5. Radon and thoron gas from the soil surface diffuse through the Tyvek filter into a buffer zone. Tyvek is a brand of flashspun high-density polyethylene fibers, impermeable to water, but permeable to gases. The radon and thoron enter into a buffer zone specifically designed to reduce any thoron that may be present (Kotrappa and Steck, 2009; Grossi et al., 2012).

6. Fission chambers

For power monitoring purposes in nuclear reactors, miniaturized ion chambers have been constructed that are equipped with stainless steel walls lined with highly enriched uranium. Argon at a pressure of several bars is mainly used as a filling gas. Because of the high pressure, the dimensions of the detector volume can easily be kept larger than the range of the fission products created by the uranium-235 (n,f) reaction.

Long-term operation causes problems, because of the burnup of the fissile material (Böck and Balcar, 1975). To compensate for this, the so-called regenerative chambers have been designed. These chambers contain a combination of fertile (^{238}U, ^{234}U) and fissile (^{235}U, ^{239}Pu) material as a lining of the inner detector walls. Fission chambers may also show a memory effect after a prolonged period of operation in a reactor core. This is due to a buildup of fission products in the detector volume. Because of the fission product activity, some residual ionization still can be measured even without exposure to a flux of neutrons.

Because of the scarcity of conventional enriched uranium-235 material, fission detectors have been developed on the basis of uranium-233. Fig. 2.6 shows a schematic diagram of the uranium-233 fission chamber designed by Prasad and Balagi (1996). The chambers were filled with argon (97%) and nitrogen (3%) at 1 bar. Low and high sensitivities were obtained by using two kinds of electrode coatings. Low-sensitivity counters have a uranium-233 coating on the anode, and high-sensitivity counters have a coating on the cathode. The main disadvantage of uranium-233 is its high specific alpha activity. This can cause pileup effects and spurious counts, if the system is applied in pulse mode operation.

Filliatre et al. (2008, 2009) found ^{242}Pu to be the best suited fissile deposit for the fission chamber online measurement of the fast component of a high neutron flux ($\sim 10^{14}$ n/cm^2/s) with significant thermal component. This was confirmed by Cabellos et al. (2010) with tests on the use of various fissionable materials in fission chambers. They found fission chambers with deposits of ^{242}Pu to be stable with fluences as high as 10^{22} n/cm^2 while satisfying high sensitivity coefficients to fast neutrons.

A fission chamber of 3 mm diameter and 300 μg of ^{238}U deposit with argon filling gas at 5 atm pressure was used by Rapisarda et al. (2011) for the detection of fast neutrons in mixed neutron-gamma fields. The fission chamber was surrounded with a gadolinium (Gd) screen to

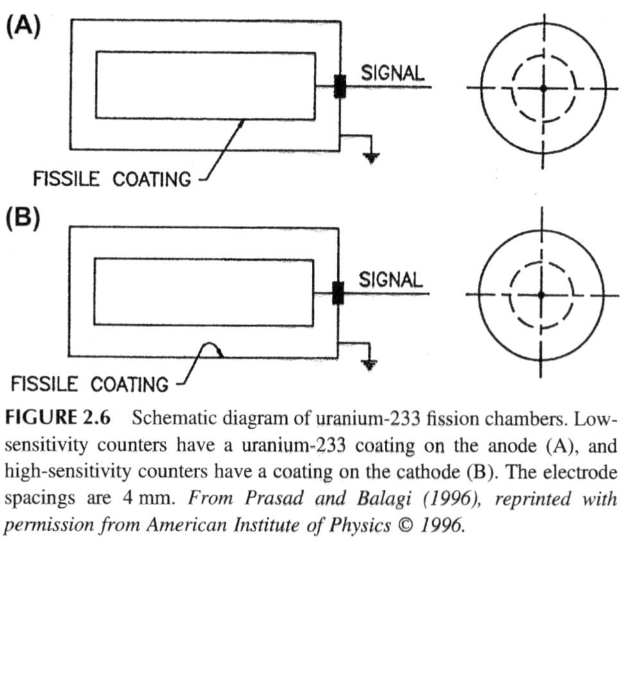

FIGURE 2.6 Schematic diagram of uranium-233 fission chambers. Low-sensitivity counters have a uranium-233 coating on the anode (A), and high-sensitivity counters have a coating on the cathode (B). The electrode spacings are 4 mm. *From Prasad and Balagi (1996), reprinted with permission from American Institute of Physics © 1996.*

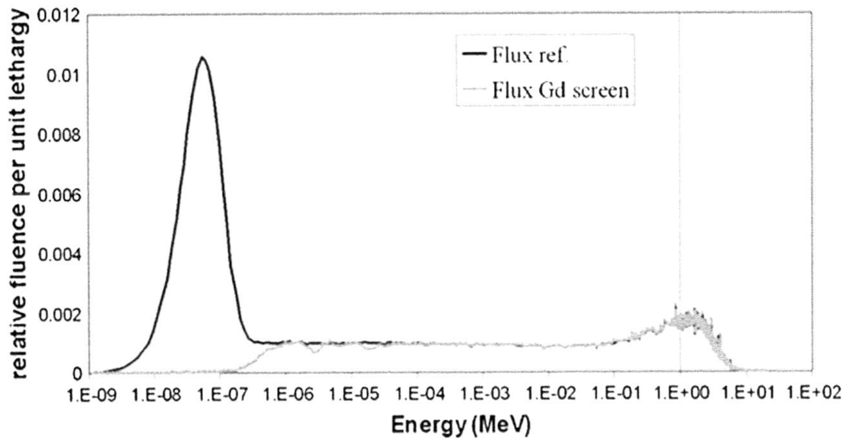

FIGURE 2.7 Calculated neutron spectra from a ^{238}U fission chamber with and without a Gd screen. *From Rapisarda et al. (2011), reprinted with permission from Elsevier © 2011.*

cut off the thermal neutron component. The typical calculated neutron spectrum at the position of the most sensitive part of the fission chamber with and without the Gd screen is illustrated in Fig. 2.7. The thermal neutron component is completely cut out by the Gd; however, the epithermal part (1 eV to 100 keV) is almost unaffected. The Gd screen obviously has a defined lifetime, due to complete burnup of the ^{155}Gd and ^{157}Gd isotopes, which are completely consumed for a 0.6-mm-thick Gd foil at a thermal neutron fluence of 2×10^{21} n/cm^2.

Transmutation of transuranium elements by fast neutron-induced fission could be a promising way to reduce the long-term radiotoxicity of these materials in radioactive waste. To measure online, the fission rate of actinide targets a new generation of microfission chambers that have been constructed by Fadil et al. (2002) for the use at the high flux reactor in Grenoble at a flux density of 10^{15} cm^2/s. To avoid pulse pileup, the chamber has to operate in current mode. Helium, a gas with high ionization potential, is used under such high flux conditions. Consequently, the problem of gas leakage during operation of the chamber at high temperatures has to be considered.

V. Proportional gas ionization detectors

Proportional gas ionization detectors operate at a higher voltage gradient than ion chambers. The primary ion pairs created by ionizing radiation are accelerated much more and acquire a high kinetic energy. By colliding with other atoms or molecules along their drift, these ions and electrons induce secondary ionization. More ions and electrons are released with energy for further ionization of the filling gas. This multiplication process is called a Townsend avalanche or Townsend cascade. The anode wire must be very thin to obtain a region of sufficient field strength by applying a reasonable voltage. The gas amplification takes place mainly in the region of high voltage gradient near the anode wire. Still, this amplification can be kept linearly proportional to the original ionization; the number of ions after gas amplification is proportional to the number of primary ions created by the ionizing radiation directly. Detailed explanations and descriptions of phenomena in gas ionization proportional counters are given by Charpak (1970) and Charpak and Sauli (1978).

PC tubes can be sealed, with the source of radiation kept outside the tube. A thin window permits radiation penetration into the detector volume. Another configuration is designed for flow-through of gas, and the sample can be inserted into the detector volume. These "windowless" counting systems are useful for the detection of alpha particles and low-energy beta particles. A maximum counting efficiency of 50%, theoretically for a 2π counting geometry, is achieved. A 4π geometry can be achieved by using two flow-through tubes with the sample mounted on a thin foil between the tubes.

Proportional counters usually operate in the pulse mode.

For proportional counters, special gases or mixtures of gases have to be used. The filling gas should not form anions and should not contain components that attract electrons. The noble gases meet this requirement optimally. The formation of secondary Townsend avalanches should also be avoided. Such secondary avalanches are created by the emission of UV photons. This light is produced in the course of deexcitation of molecules or atoms of the filling gas. To prevent this effect, a component is added to the filling gas that absorbs the energy from the excited species. This additive must get rid of the energy through nonradiative modes, such as dissociation. By this mechanism, the ion cascade is localized near its origin and propagates only along the electric field. No other secondary avalanches are created. A frequently used filling gas consists of 90% argon and 10% methane and is called P10 gas (Alkhazov et al., 1967; Marouli et al., 2007,2010; Phillips et al., 2010).

Other gas mixtures (Penning gas) consist of a noble gas (neon, argon), the parent gas, with a small amount of an additive (e.g., methane, acetylene, TMA) of lower ionization energy than the lowest excited state of the parent gas (Jarvinen and Sipila, 1984. Agrawal et al., 1989. Sakurai et al., 2003, and Tokanai et al., 2004).

Gas amplification factors of 10^4 can easily be obtained. Therefore, rather simple electric circuits can be used for pulse amplification and pulse handling. Also, the effects of electronic noise can easily be avoided, because the output pulses created by that phenomenon are small.

Gas gain in PC should be an exponential function of the applied high voltage. However, in proportional counters filled with mixtures of argon and a low amount of a molecular gas, secondary avalanches develop and, as a consequence, gas gain increases faster than exponentially with the applied high voltage (Bronic and Grosswendt, 2001; Curzio et al., 2005).

Proportional counters, using the fast pulses from electron collection, have a short resolving time of less than 1 μs. Proportional counters have a high intrinsic efficiency for alpha and beta particles. Photons are detected mainly by Compton effects produced in the walls of the counter. Thus, the intrinsic efficiency for gamma rays is rather low, especially for gamma photons with higher energies.

Counting losses with proportional counters are due to wall effects and nondetection of beta particles with very low energy. Stanga et al. (2002) proposed a calculation model for the correction of counting losses. By means of such calculations, the accuracy of internal gas counting methods can be improved, and tedious and time-consuming energy calibration procedures can be shortened or even avoided. PC is frequently applied to the preparation of reference sources by absolute activity measurements also referred to as radionuclide standardization. Such radionuclide standardization methods involving joint proportional and solid scintillation detector arrangements (i.e., $4\pi\beta$(PC) − γNaI(Tl)) counting are discussed in Chapter 7, Volume 2.

Aging effects in gas detectors under irradiation have been observed by Niebuhr (2006). The damage depends not only on the amount of radiation on the detector but also on many other parameters such as gas composition, gas quality, particle rate, and particle type.

Pulse shape analysis can be used to reduce background effects or any undesirable events (Pandola et al., 2004). Background pulses frequently show a longer rise time than the desired signals of events to be measured.

A. Examples and applications of proportional counters

1. Gross alpha−beta counting, alpha−beta discrimination, and radiation spectroscopy using proportional gas ionization counters

With gross alpha−beta counting, no attempt at any discrimination is made. Just the sum of all alpha and beta particles is detected. Gas PC is one of the methods frequently used for gross counting (Passo and Kessler, 1992; PerkinElmer, 1992). Proportional counters may be used in clinical settings, such as the determination of gross alpha and beta activity analysis in human urine (Chen et al., 2014).

Proportional counters are also frequently used to distinguish between alpha and beta particles from a mixed source (Waples et al., 2003; Vasile et al., 2010; Dueñas et al., 2011; Thakur and Mulholland, 2011). Alpha particles, with their high LET, produce a high number of interactions with the gas in the detector volume. A beta particle produces a much lower number of ions per centimeter along its track than an alpha particle. The gas amplification factor is constant at a given voltage, so the output pulse is much higher for interactions of alpha particles compared with beta particles. With a suitable discriminator level or gate, the pulses created by alpha particles can be detected at a rather low voltage setting. For the detection of beta particles, a higher voltage has to be used to overcome the discriminator level.

The alpha particles from a mixed source are registered at a lower voltage, the alpha plateau. At a higher voltage, alpha and beta particles are detected, and gross alpha−beta counting is accomplished (L'Annunziata, 1987).

Alpha and beta radiation can also be discriminated according to the pulse length. Alpha pulses have a different pulse shape than beta pulses. Semkow and Parekh (2001) could demonstrate that alpha-radioactivity can be measured accurately in the presence of beta-radioactivity, but the opposite is not always true due to alpha to beta cross-talk. This cross-talk depends mainly on the alpha-decay scheme and is due to the emission of conversion electrons, Auger electrons, and X-rays.

It is usually assumed that the counting efficiency of a 2π geometry alpha particle detector is 50%. Unfortunately, this is not true in practical measurements, because of self-absorption and backscattering. Several theories have been developed for the calculation of backscattering and self-absorption effects. Rodríguez et al. (1997) have presented a review on these topics and also developed new theories. Backscattering depends on the atomic mass of the backing material of the radiation source. Corrections can be found experimentally by preparing samples of various thickness and extrapolation to zero sample thickness. Such determinations are only possible with a radioactive material with suitable long half-life.

To a limited extent, proportional counters can also be used for radiation spectroscopy (Jarvinen and Sipila, 1984; Jahoda and McCammon, 1988). Pulse height analysis can be applied for radiation spectroscopy for a given type of radiation. To perform pulse height analysis properly, the particles or rays to be analyzed have to release their entire energy within the gas volume of the counter; that is, they must be totally absorbed within the counter. Proportional detectors are used for X-ray spectrometry in the field of X-ray fluorescence analysis, if high resolution is not required. Because of the gas amplification process, proportional counters have a poorer resolution than ion chambers. Today, mostly semiconductor detectors are used for classic X-ray spectroscopy.

Proportional counters are still widely used for contamination assessment in nuclear facilities. A recent example is the decommissioning of the Korean Research Reactor-1 (Min et al., 2017). Here, together with other techniques, proportional counters were used to find and assess contamination inside the facility.

Szaloki et al. (2000) have reviewed the essential progress in X-ray spectroscopy, and they point out that although the gas-filled proportional detectors are not superior to semiconductor detectors in resolution, microstrip proportional counters are applied for many investigations, including developments in the field of radioisotope excited X-ray fluorescence (XRF) analysis, especially at low energy regions (X-rays below 10 keV).

Xenon gas proportional scintillation counters are used for the detection of X-rays when high detection efficiency and good energy resolution are required (Lopes et al., 2000, 2001; Monteiro et al., 2001, 2004; Simões et al., 2001; Covita et al., 2004; Coelho et al., 2007). An excellent example is the detector used by Rachinhas et al. (2000) for the identification of internal conversion electrons produced by the decay of cadmium-109 and xenon-133m to investigate details of the decay scheme of these radionuclides. The main aim was to selectively detect and identify conversion electrons of cadmium-109. Fig. 2.8 shows the design of the gas proportional scintillation counting device. The stainless steel enclosure holds also a thin plastic window, which is aluminized on the inner side to provide a uniform field strength at the drift region. Drift and scintillations region are separated by a grid mesh with high electron transparency. A second grid is evaporated directly onto the photomultiplier; therefore, the scintillation region is in direct contact with the photomultiplier, and a high collection efficiency of the UV scintillation photons is guaranteed. Primary electron clouds are produced by ionizing radiation in the absorption region, and these drift under the influence of a low electric field toward and through the first grid into the scintillation region, where due to a much higher field strength, scintillations are produced in the xenon filling gas. The electric pulses of the photomultiplier are fed to an amplifier operating with very short shaping times, and as a result, pulse shapes resemble very closely the scintillation light bursts. This produces an efficient pulse shape discrimination and a very detailed interpretation of the pulse height spectra (see Fig. 2.9).

As reliable detectors, proportional counters are frequently used for the standardization of radionuclides. García-Toraño et al. (2002) compared three methods for the standardization of cesium-134: absolute counting with a 4π NaI(Tl) detector, liquid scintillation, and a proportional counter (argon and methane as counting gas at atmospheric pressure) in coincidence with a NaI(Tl) detector—system. It was shown that all the results of the standardization have been consistent and that any of the three methods that were applied was well suited for this type of application. Pommé

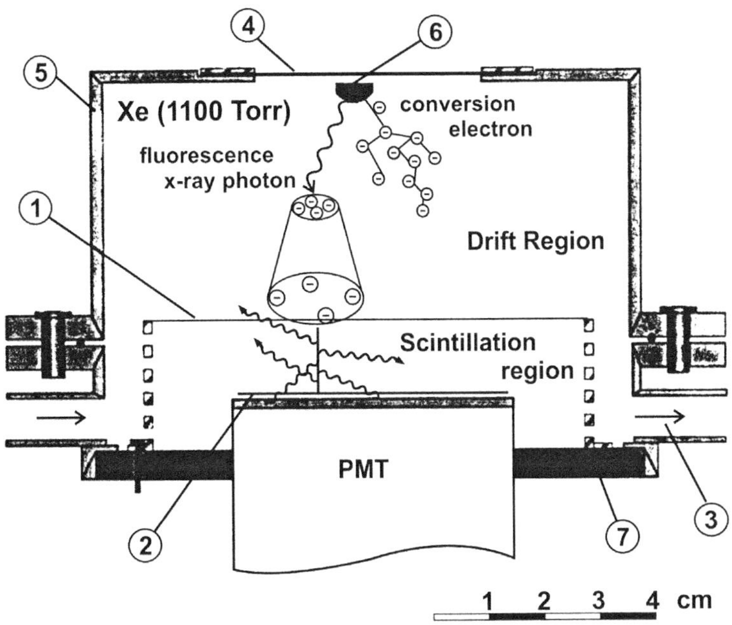

FIGURE 2.8 Design of a gas proportional scintillation counter. (1) grid mesh with high electron transparency; (2) grid evaporated onto the photomultiplier; (3) to xenon gas purification system; (4) thin aluminized plastic window; (5) stainless steel enclosure; (6) Cd-109 source; (7) insulating material. *From Rachinhas et al. (2000), reprinted with permission from Elsevier.*

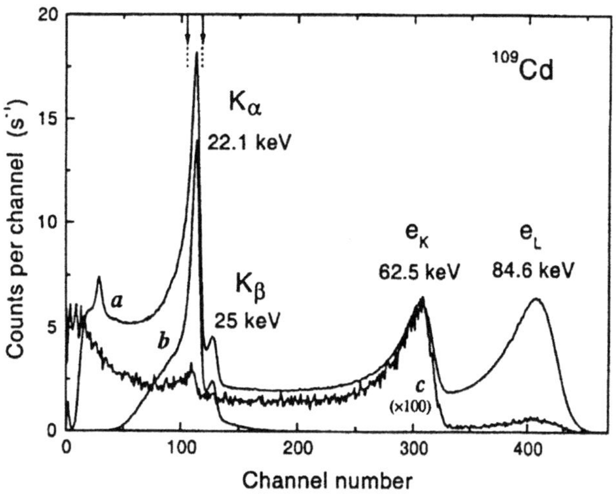

FIGURE 2.9 Pulse height spectrum obtained from a Cd-109 source placed inside the gas proportional scintillation counter. Curve *a* is raw pulse height distribution. Curve *b* is obtained when pulses with duration outside the range from 3.6 to 4.0 μs are rejected. Curve *c* (values multiplied by 100) shows pulses that appear within a 20-μs interval after a 22.1-keV pulse within the range determined by curve *b*. *From Rachinhas et al. (2000), reprinted with permission from Elsevier © 2000.*

et al. (2005) and Stagna et al. (2006) also compared successfully gas PC with other methods for the standardization of ^{125}I, which decays by electron capture with subsequent emission of conversion and Auger electrons, and the standardization of tritiated water, respectively.

The reliability of 4π pressurized gas proportional counters has been further demonstrated by Altzitzoglou et al. (2002) during their work dealing with the comparison of three methods to standardize a strontium-89 solution. Correction for self-absorption of the samples for gas PC was obtained by plotting the activity concentration of the solution against the mass of radioactive sample. A new half-life value for strontium-89 (50.61 ± 0.05 days) was determined in this work.

International comparison and standardization programs frequently result not only in getting more accurate data of radiation properties but also in improving measurement procedures. Self-absorption corrections for beta measurements of solid samples have to be applied depending on the thickness of the specimen. Johansson et al. (2002) demonstrated that the self-absorption of beta particles from thallium-204 shows a linear relation to the logarithm of the dry mass of the source. They described a way to minimize and correct for self-absorption in solid sources of thallium-204 of nuclides with similar decay properties. Also a special device for source drying is described. Warm dry nitrogen jets (60°C) are blown on the rotating source material, which is mixed with colloidal silica (Ludox) to decrease the crystal size of the solid deposit.

Improved gas PC techniques for the standardization of negatron emitters (e.g., ^{3}H, ^{85}Kr) and positron emitters (e.g., ^{11}C) are described by Stagna and Cassette (2006), Marouli et al. (2010), and Phillips et al. (2010). The theory and principles of 4πPC are provided in Section III.A of Chapter 7, Volume 2.

2. Position-sensitive proportional counters

a. Single-wire proportional counters

In a proportional counter, the position of the avalanche is limited to a small portion of the anode wire length. Some designs of proportional counters are capable of sensing the position of this avalanche and thus providing information about the position of an event taking place within the volume of the proportional counter. If the PC tube is cylindrical with a central wire, electrons drift along the radial field lines. Thus, the position of the avalanche indicates the axial position of the initial ion pairs and the position of the entering radiation to be detected. Of course, if the incident radiation extends for some distance along the counting tube, only an approximate region of the incident radiation can be determined. The principle of charge division is most frequently used to determine the position of the ion avalanche. For that purpose, the central anode wire is made of a material having a rather high electric resistance per unit length (Ohsawa et al., 2000). By that means, the charge that is collected at the wire electrode is divided between the amplifiers placed at both ends of the anode wire. The charges on those ends are collected in proportions related to the geometric position of the ion avalanche interacting with the wire electrode. A conventional output pulse is provided by summing up the response of the amplifiers and thus getting information about the total charge collected. A signal related to the position is provided by dividing the signal output of one amplifier by the output related to the total charge collected. The pulse height of this new signal indicates the relative position along the length of the central anode wire (Fischer, 1977; Westphal, 1976). Either analog signal handling or digital pulse processing techniques can be applied for this purpose.

Another approach to position sensing is the utilization of pulse rise time measurements. With this technique, the relative rise times of the output pulses of the preamplifiers placed on both ends of the anode wire are determined. Interactions that take place far from one of the preamplifiers result in pulses with a much longer rise time than events close to the preamplifier position. From the rise time difference of the two preamplifiers, a signal can be created that is related to the position of the ion avalanche along the electrode wire. Good results regarding spatial resolution are observed. For well-collimated alpha particles, the FWHM can be 0.15 mm for a tube 200 mm long. Such position-sensitive proportional detectors have been applied for X-rays and neutrons, for magnetic spectroscopy of charged particles, and for the localization of beta-emitting spots on thin-layer or paper chromatograms (Goulianos et al., 1980).

b. Multiwire proportional counters

For many purposes, proportional counters with a number of anode wires instead of one central anode wire offer advantages. A grid of anode wires can be placed between two flat cathode plates. Near the cathode plate, the field is nearly uniform and electrons drift in that homogeneous field toward the anode wire grid. Near the wires, the field strength increases and, as electrons approach this region, they are accelerated toward the nearest anode wire and an ion avalanche is created. Because of this, the signal appears only at a single anode wire and the position of the primary ionizing event can be localized in the dimension perpendicular to the direction of the anode wires. This multiwire proportional counter was developed by Nobel Laureate Georges Charpak in the 1960s and played a decisive role, not only in many discoveries in particle physics but also in many experiments in biology and medicine (Charpak, 1970; Charpak and Sauli, 1978; Geltenbort, 1994). The technique for position-sensitive counting by using cathode wires of high resistivity has already been discussed. This technique can be used in addition to the plate and multiwire design; and a 2D signal pattern can be obtained. Another technique uses a detector construction with the cathode plate divided into narrow strips perpendicular to the anode wires. The induced charge to the nearest strip is recorded. Such position-sensing detectors with large areas are applied in high-energy particle research (Uozumi et al., 1993; Hayakawa and Maeda, 1994).

The relatively low signal amplitude is a disadvantage of these detectors. Therefore, for some applications, a hybrid detector system, between proportional and GM detectors, may be useful because of the much higher signal amplitude achieved; these are referred to as self-quenched streamer (SQS) detectors (Knoll, 1989).

c. Microstrip and micropattern ionization counters

Wire proportional chambers were mostly developed at the European Organization for Nuclear Research (CERN) and have been a major step forward in particle detector technology. Even now the field of developing new varieties and improving available designs remains very active. MSGCs, micromesh designs, nonplanar variants of MSGCs, like the "compteur a trois (CAT)," secondary electron emission gas detectors, and some other varieties have been developed. A description of design with their special features and advantages is given by Fourme (1997). Christophel et al. (1998) present the development of a 2D microgap wire chamber. Such position-sensitive detectors can be used in other fields in addition to basic research in particle physics. Ortuño-Prados et al. (1999) described the use of a multiwire proportional counter as a potential detector for protein crystallography and other wide-angle diffraction experiments. Fried et al. (2002) describe a large curved 2D position-sensitive neutron detector, which had been constructed for the protein crystallography station at the Los Alamos National Laboratory. Babichev et al. (2001) report on their experience in medical radiography. The advantage of using multiwire proportional counters as high count rate detectors as well as their usefulness for producing dynamic images of high statistical quality is pointed out by Barr et al. (2002). A detailed summary on gas avalanche radiation detectors and their application in biomedical investigations is given by Breskin (2000) and Guillemot at al. (2007).

MSGCs are ionization counters in which anodes and cathodes are not single plates but are constructed as thin metal strips on a solid insulating support (Barbosa et al., 1992; Bouclier et al., 1992a, 1992b, 1992c, 1995, Oed, 1995, Pallares et al., 1995; Bellazzini et al., 2001, 2002; Clergeau et al., 2001; Sauli, 2001; Bateman et al., 2002; Mir et al., 2009). With such a system, the spot of the ionization track can be localized, because ion production and migration and current flow take place in a well-defined single electrode strip region. Thus, position-sensitive counting can be achieved. Such MSGCs can be obtained with very small spacing between the electrodes. A small pitch results in good resolution. Even at the early stages of their development, results were rather encouraging. At proportional gains above 10^4 with good energy resolution (12% for 5.9 keV), position accuracies around 30 μm, and high rate capabilities were obtained. An example is shown in Fig. 2.10. This microstrip chamber was constructed by Oed (1995) using photolithographic techniques. The small strips are the anodes, and the electric field lines between the electrode strips are plotted in Fig. 2.11. An electron, which is set free in the gas volume in front of the microstrip plate and reaches the microstrip plate, creates an avalanche in a very well-defined small region.

Bateman et al. (2002) presented results of investigations dealing with energy resolution measurements using an MSGC. The effects of gas filling, strip geometry, and other parameters have been examined in detail. A 2D position-sensitive detector was realized by Barbosa et al. (1992). Two sets of microstrips are orthogonally oriented, forming a 2D sensitive electrode, which is used in a multiwire proportional configuration as shown in Fig. 2.12. The two cathode systems are isolated by a silicon dioxide layer only 2 μm thick and are therefore at practically the same distance (3 mm) from the anode wires. Therefore, the signals induced in both orthogonal electrodes are of the same amplitude. The authors aimed to define a 2D X-ray detecting unit that also could be upgraded to a submillimeter spatial resolution detector.

There are some limitations to this detector design. One has to apply manufacturing techniques such as those used

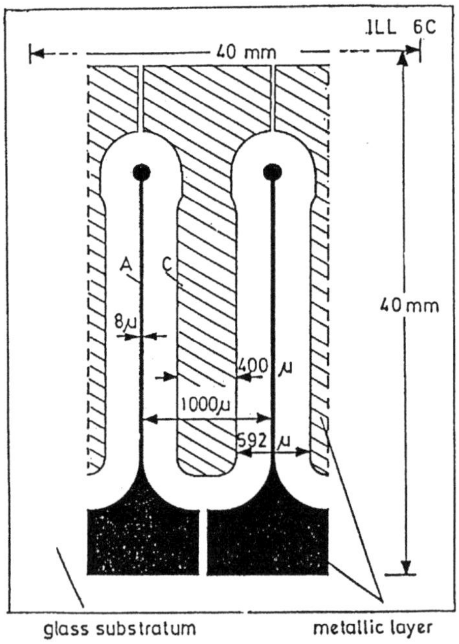

FIGURE 2.10 Structure of a microstrip plate. Electrode spacings are 400 μm, and the metallic layers on the glass support, the electrodes, have a thickness of 150 nm. The small strips are the anodes. *From Oed (1995), reprinted with permission from Elsevier © 1995.*

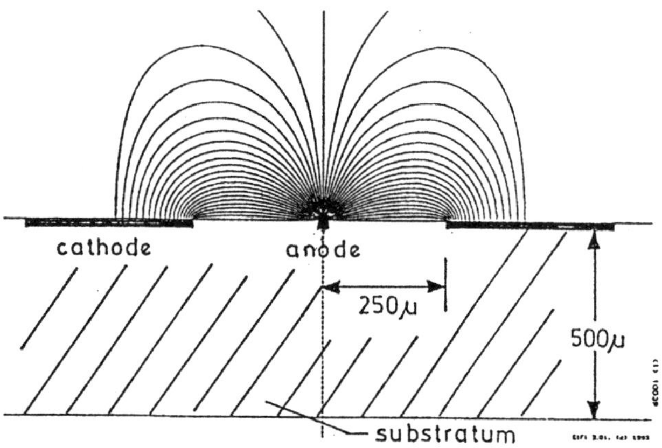

FIGURE 2.11 Plot of the electric field lines between the electrode strips of the microstrip plate. *From Oed (1995), reprinted with permission from Elsevier © 1995.*

in the field of microelectronics. The total sensitive area of such counters seems to be limited. Also, there are charge buildup effects of the supporting insulating materials. This can have a substantial influence on the gas gain at high fluxes. Ion avalanches can cause accumulation of electric charge on the insulating surface between the strips, which modifies the electric field around the electrodes and changes the gas multiplication characteristics. To avoid this, a surface conductivity of the insulating support can be created, for example, by ion implantation. However, the use of all these sophisticated manufacturing techniques imposes constraints on the size of such radiation detectors. In the beginning, glass and quartz were used for insulating support between the electrodes. Later developments dealt with the application of plastic supports. These materials offer some advantages. They are flexible, and therefore, nonplanar detectors can be designed. Cylindrical geometries with very small radii can be realized. Plastic materials not only have the advantage of lower atomic number of their constituents compared with glass but also can be made much thinner. Multiple scattering and photon conversion can be reduced. Plastic materials are also available with a wide range of electrical resistivities, and the design can be tailored to solve the problem of charge buildup. However, plastic materials have to meet the requirements of suitable mechanical stability. Bouclier et al. (1995) accomplished microstrip construction on plastic foils by applying a photolithographic etching technique on a layer of aluminum about 0.3 μm thick on plastic. The distance between the electrodes was about 400 μm. This is somehow wider than the usually applied 200 μm, and it is necessary because of the coarse optical quality of the plastic arrangements compared with glass support microstrips. Also, surface cleaning of plastic before vacuum evaporation of the aluminum cannot be done as perfectly as for glass supports.

Gains close to 10^3 could be reached with the equipment designed by Bouclier et al. (1995). Also, good energy resolution for low-energy X-rays was achieved.

The current tendency in the field of gaseous detectors is the replacement of wire chambers by advanced micropattern electron multipliers to obtain an improvement in spatial accuracy and counting rate capability. Electrode patterns are deposited by microlithographic techniques on insulating substrates. Due to the small distances between cathode and anode (50–200 μm), these multipliers offer localization accuracy around a few tens of micrometers. The rapid collection of the ion avalanches considerably reduces space charge buildup, which influences the counting rate limitations. Many types of detectors in this family provide 2D localization in a single detector element.

Many new(er) types of gas detectors with additional microstructures such as the GEM system and other designs continue under development (Horikawa et al., 2002; Sauli 2003, 2016; 2004a, 2007, 2010; Mörmann et al., 2004; Simon et al., 2009; Bucciantonio et al., 2013a, 2013b; Brunbauer et al., 2017; Veloso and Silva, 2018). These detectors are often referred to as micro pattern gaseous detectors (MPGDs) because of their position accuracy (Sauli, 2001; Bellazzini et al., 2001, 2002; Mir et al., 2009).

The GEM was introduced by Sauli (1997). A GEM detector consists of a thin polymer foil (approximately 25 μm), which is metal clad (approximately 18 μm) on both sides and perforated to yield a density of holes

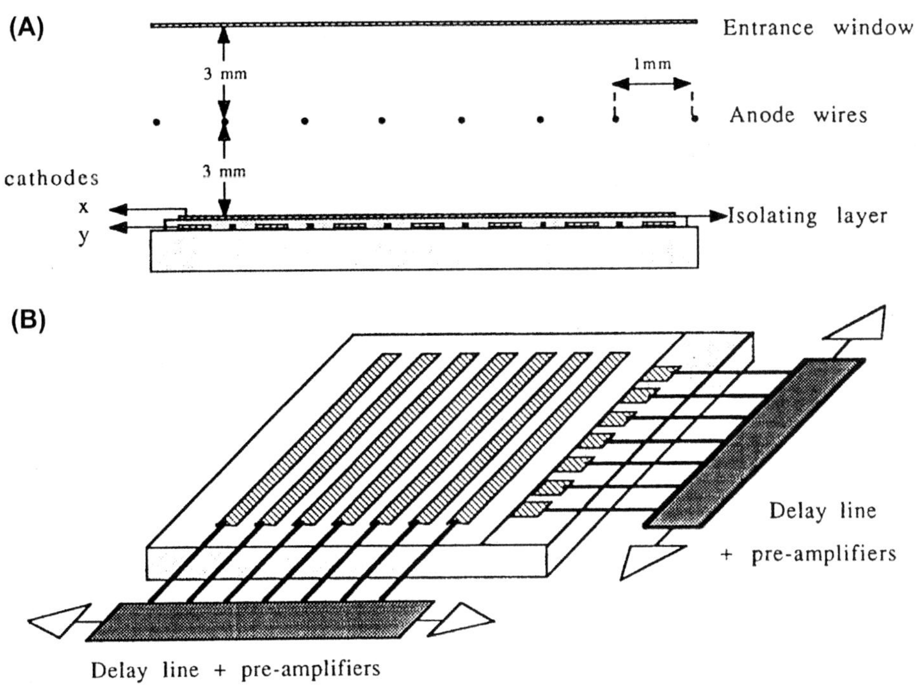

FIGURE 2.12 Two-dimensional position-sensitive detector arrangement (A). Two sets of microstrips at a distance of 2 μm are orthogonally oriented and connected to delay lines (B). The anode wires are arranged between the microstrips and the entrance window at a distance of 3 mm. *From Barbosa et al. (1992), reprinted with permission from Elsevier © 1992.*

(approximately 70 μm in diameter and approximately 100–150 μm apart) (see Fig. 2.13).

Photolithographic techniques have been used for manufacturing of these electrodes. A voltage is applied onto the two faces of the metal clad foil, causing a large difference of potential between the two sides of the foil, thus creating a very strong field inside the holes (Fig. 2.14).

The geometry of the holes and optimization of the device according to its purpose are still intensely researched (Benussi et al., 2016; Karadzhinova et al., 2015). The device is inserted in a gas detector on the path of drifting electrons (Fig. 2.15).

Primary electrons produced by ionization of the gas layer above the foil are accelerated into the holes where an avalanche process takes place. By that process, the charge drifting through the holes is amplified. Most of the secondary electrons produced in the avalanche are transferred

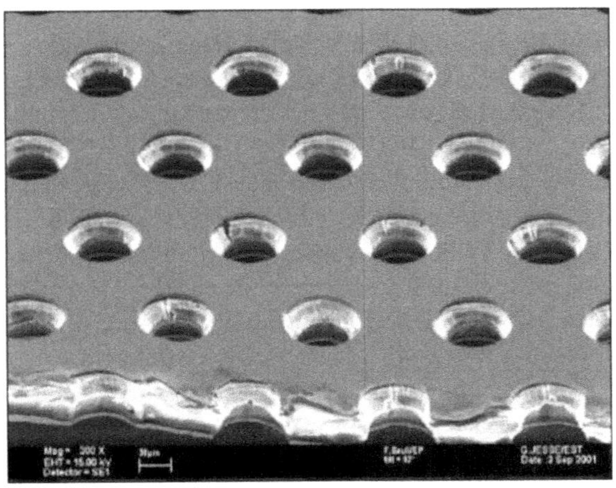

FIGURE 2.13 Electron microscopic image of a gas electron multiplier electrode (thickness 50 μm, pitch of the holes 140 μm, diameter of the holes 70 μm) (Sauli, 2016). *© CC by Sauli (2016).*

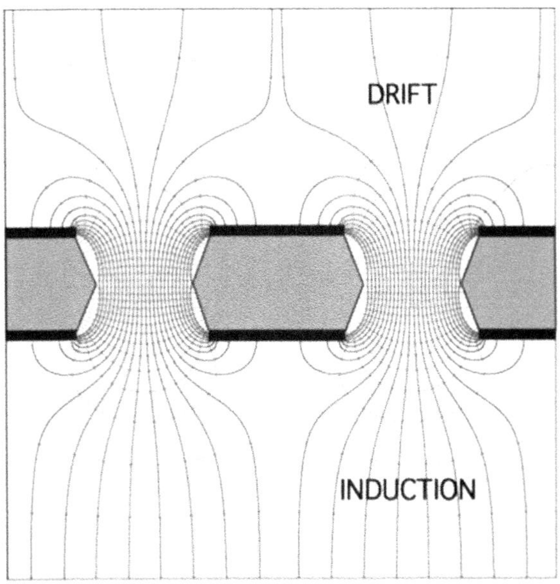

FIGURE 2.14 Electric field in the region of the holes of a gas electron multiplier electrode (Sauli, 2016). *© CC by Sauli (2016).*

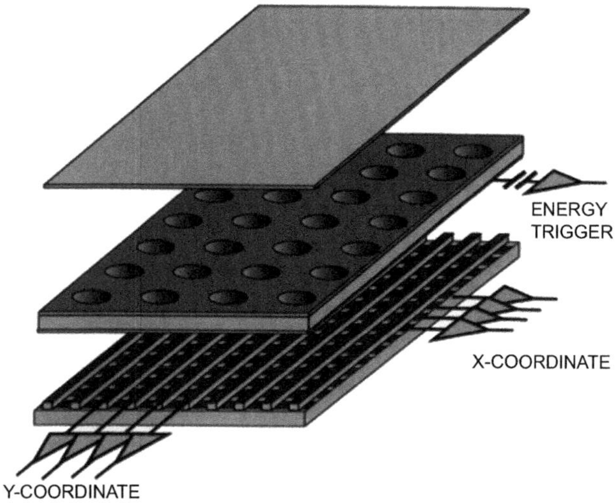

FIGURE 2.15 Schematic of a single gas electron multiplier detector with Cartesian two-dimensional strip readout (Sauli, 2016). © *CC by Sauli (2016).*

to the region below the foil where these electrons are collected by an anode and cause a detection signal. Coupled to other devices such as multiwire or micropattern chambers, higher gains are obtained or an operation in less critical field strength conditions is permitted. The fast response time generated by electrons is one of the main characteristic advantages of the GEM detector. The GEM detector has been originally developed for application in particle physics. It has also been applied successfully in other fields of research, though. 2D GEM detectors have been used to obtain X-ray absorption images to show their applicability in medical diagnostics. Fig. 2.16 shows an example of one of the earlier achievements of GEM technology in imaging in the form of an absorption radiography of a small mammal (bat) using 8 keV X-rays. The real size of the image is 3 × 6 cm. The position resolution depends on the photoelectron range in the gas.

The readout speed of the electronic system used to be the limiting factor in the application of GEM detectors. This challenge has been targeted, for example, by Bucciantonio et al. (2013a) by implementing a fast data acquisition system, thus allowing the application of GEM trackers for imaging and dosimetry in hadrontherapy (i.e., proton or carbon ion beams in cancer therapy). This type of therapy takes advantage of the Bragg peak (deposition of most of the energy of a heavy particle in the final section of its pathway when traveling through matter), which allows treatment of tumors in situ with utmost precision, while delivering only a small dose to the surrounding healthy tissue, as described by L'Annunziata (2016). GEMs have been used to monitor the ion beam in realtime (Aza et al., 2017).

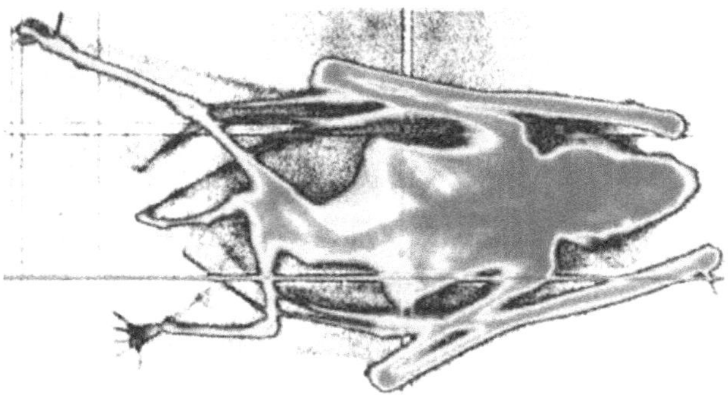

FIGURE 2.16 Absorption radiography of a small mammal recorded with a two-dimensionalgas electron multiplier detector (Sauli, 2004b). For the color version of the figure, the reader is referred to the online version of the book. *Reprinted with permission from Elsevier © 2004.*

Fraga et al. (2000) have shown that visible light emitted by the GEM avalanches can be successfully used for quality control of the material, to determine their uniformity, and to identify local defects. It is much more effective than the general practice of optical inspection. A modification of a GEM-type detector for medical purposes was presented by Nichiporov et al. (2016). The authors studied a GEM-based scintillation detector for proton beams in a clinical setting. The detector uses a mixture of He and CF_4 that allows (nearly) tissue-equivalent dose imaging. Mixtures of Ar/CF_4 had been studied previously, but the He/CF_4 mixture helps improve the detector's response in the Bragg peak area. The performance of this detector is similar to that of an ion chamber. A TEPC based on a single GEM foil has also been tested with tissue-equivalent gas mixtures and studied by De Nardo and Farahmand (2016).

GEM detectors have been used in fusion experiments to check the quality of the plasma. At the International Thermonuclear Experimental Reactor (ITER), a soft X-ray tomography diagnostic unit is based on GEM detectors. The GEM detectors allow for superb time and space resolution, as well as for the monitoring of the charge spectrum, from which the soft X-ray photon spectrum can be deconvoluted. Another advantage is that they are not prone to neutron damage compared with semiconductor detectors (Chernyshova et al., 2018; Wojenski et al., 2016). Such diagnostics require fast postprocessing computation systems (Krawczyk et al., 2018; Zabolotny et al., 2019).

Several designs of MPGDs are useful for (2D) imaging purposes (Veloso and Silva, 2018), including GEMs, thick gas electron multipliers (THGEMs), Micromegas, microhole and strip plates (MHSPs), and THCOBRA (Silva et al., 2013), not all of which shall be discussed here. Imaging is very useful for energy-dispersive X-ray fluorescence analysis (EDXRF), which can be pursued in the form of macro-XRF (using a 0D detector) or full-field scanning. The obvious drawback of macro-XRF (which has to scan over the surface of the object) is that, given the small X-ray beam diameter, scanning becomes time-consuming and complex. Data acquisition in full-field scanning mode using 2D detectors and a pinhole camera, however, permits simultaneous data acquisition (Dąbrowski et al., 2016).

The operating principle of MHSP in 2D imaging is shown in Fig. 2.17.

Beautiful examples of this technique in cultural heritage studies have been published by Zielińska et al. (2013) or Veloso and Silva (2018). Fig. 2.18 shows the 2D elemental distribution of a 17th century Portuguese tile. The distribution permits the position-sensitive imaging of Pb (glaze), Fe + Co (blue pigment), and Sb (yellow pigment) (Veloso and Silva, 2018).

Beyond these examples, there are continuing and promising developments of GEM-type detectors (Sauli, 2001, 2003, 2004a,b, 2010; Fraga et al., 2003; Tsyganov et al., 2008; Dalla Torre, 2011; Villa et al., 2011; Bucciantonio et al., 2013a; Bucciantonio et al., 2013b; Sauli, 2016; Brunbauer et al., 2017; Veloso and Silva, 2018).

GEMs have been included in rather novel designs of radiation detectors, which are constructed for applications in rare event experiments and, as the authors claim, will be of interest in neutrino-nucleus scattering, dark matter search, and probably also in nuclear imaging (Bondar et al., 2010, 2011). The authors describe experiments with a two-phase chamber, filled with liquid and gaseous argon and equipped with an assembly of THGEMs. Ionization signals as well as signals from an attached avalanche photodiode are recorded. Other investigations rely on the information given by the photodiode only (e.g., Renker, 2009; Kreslo et al., 2011). Photomultiplier tubes are frequently used in instrumentation for medical diagnosis such as with gamma cameras or CT equipment where light from large scintillator arrays has to be recorded. An alternative and probably more economic device for light detection and 2D recording would be the use of a thin solid photocathode combined with gas avalanche multipliers and a micropattern device (Fig. 2.19; Breskin, 2000). It is possible to include several GEMs to such a device in cascade (Sauli, 2001; Bachmann et al., 2002; Bondar et al., 2006; Aulchenko et al., 2009; Dalla Torre, 2011). Each GEM operates at a low gain whereby a high total gain is achieved. In addition, the photocathode is shielded from photon feedback induced by ion avalanches (Fig. 2.20; Breskin, 2000).

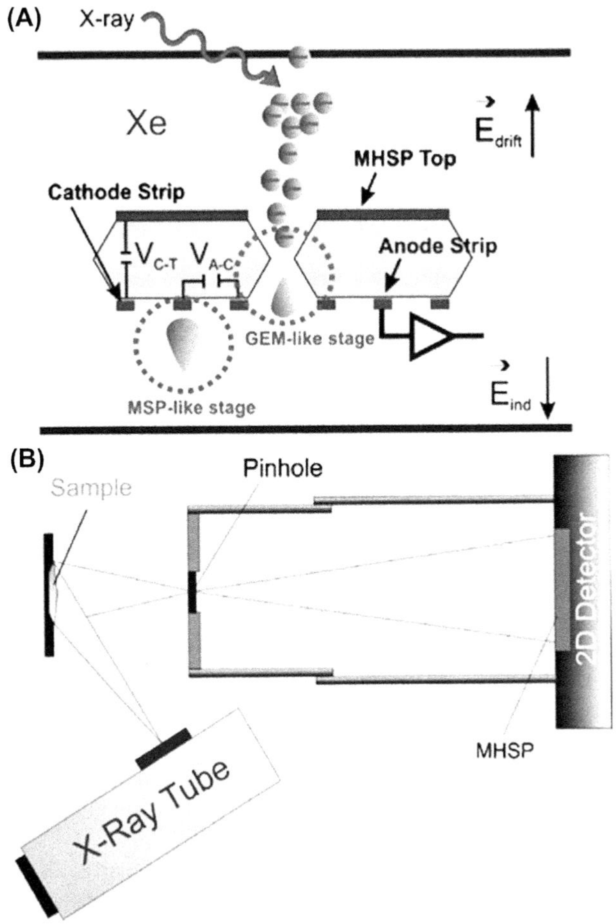

FIGURE 2.17 (A) Operating principle of microhole and strip plate (MHSP) and (B) application for X-ray fluorescence two-dimensional imaging (Silva et al., 2009). GEM, gas electron multiplier. For the color version of the figure, the reader is referred to the online version of the book. *Reprinted with permission from Springer © 2009.*

THGEMs can also be used as microdosimeters, as shown by Anjomani et al. (2017). These devices are sensitive for low-intensity neutron and gamma fields. The detector was found to exhibit enhanced neutron detection efficiency of three times the efficiency of a TEPC that was used as a benchmark. However, the low energy cutoff in gamma dosimetry is about 6 keV/μm, which is rather high and shall be improved in further studies.

A THGEM-based TEPC was presented by Darvish-Molla et al. (2018) with the purpose of dealing with high dose-rate fields. The microdosimetric response was found to be comparable with a commercially available TEPC.

Novel designs of THGEM-type detectors include the multilayer THGEM (M-THGEM) (Cortesi et al., 2017). This hole-type GEM detector is produced by multilayer printed circuit board technology and consists of a densely perforated assembly of multiple insulating substrate sheets, intercalated between thin metal electrode layers. The authors discuss several advantages of their novel detector, which includes a significant reduction of photon-mediated secondary effects, resulting in a stable high-gain operation in pure elemental gas, thus making the use of a quenching gas unnecessary. One possible disadvantage of the single-element M-THGEM is a higher ion flowback, which has to be accounted for.

Bellazzini et al. (1999) introduced the WELL detector as a rather new type of position-sensitive gas proportional counter. The basic design is similar to the GEM detector. The main difference between the GEM and the WELL detector is that the GEM alone acts only as an amplifying stage, whereas the WELL detector has readout strips directly placed onto the insulating foil providing a position-sensitive compact system. Printed circuit board technology

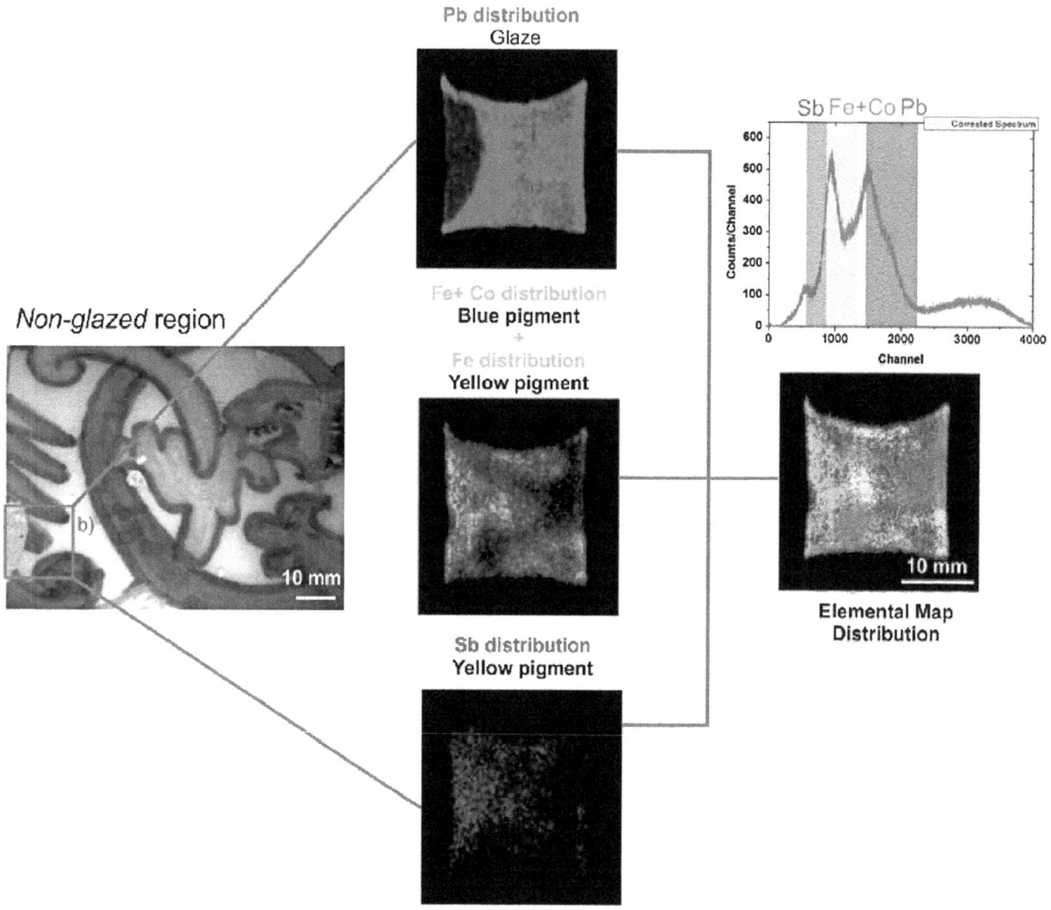

FIGURE 2.18 Pb *L*-lines (from the glaze), Fe + Co *K*-lines (from the blue and yellow pigment), and Sb *L*-lines elemental spatial distribution images, together with the total elemental map distribution of the analyzed region in the Portuguese tile from the 17th century (Veloso and Silva, 2018). For the color version of the figure, the reader is referred to the online version of the book. *Reprinted with permission from Elsevier © 2018.*

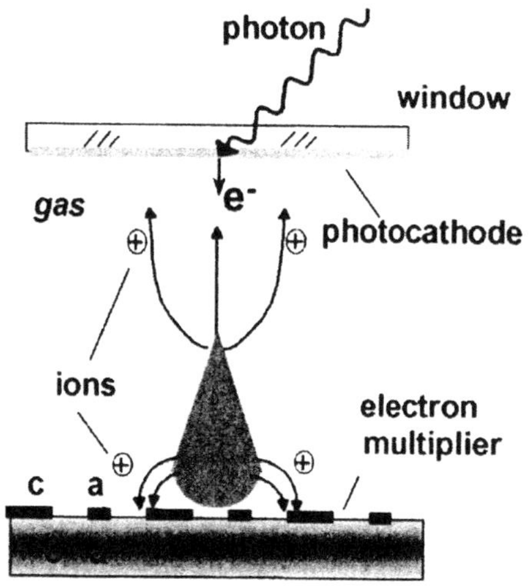

FIGURE 2.19 The principle of the gas avalanche photomultiplier: photons stimulate the emission of electrons from the photocathode into the gas, avalanche multiplication takes place near the anodes of the micropattern device, ions are collected on neighboring cathodes, and some ions may drift to the photocathode. *From Breskin (2000), reprinted with permission from Elsevier © 2000.*

was used to fabricate the amplifying structures (Bellazzini et al., 1999; Pitts and Martin, 2001).

Although the development of position-sensitive chambers is mainly dedicated for applications in high energy physics, these types of detectors are also instruments of choice for radiation detection and localization in other fields of basic and applied research. Breskin provides many examples for the application in biology and medicine. Among these are (1) the comparison of images obtained by autoradiographic techniques and ionization detectors, (2) images of ionizing particle track patterns demonstrated for applications in nanodosimetry, and (3) examples of the application of X-ray imaging and neutron imaging (Breskin, 2000).

Yu et al. (1999) designed a position-sensitive X-ray detector with curved electrodes for large angle X-ray diffraction experiments at a synchrotron at the Brookhaven National Laboratory. The detector can cover an angle of 45 degrees and has an arc length of 20 cm with a radius of curvature of 25 cm.

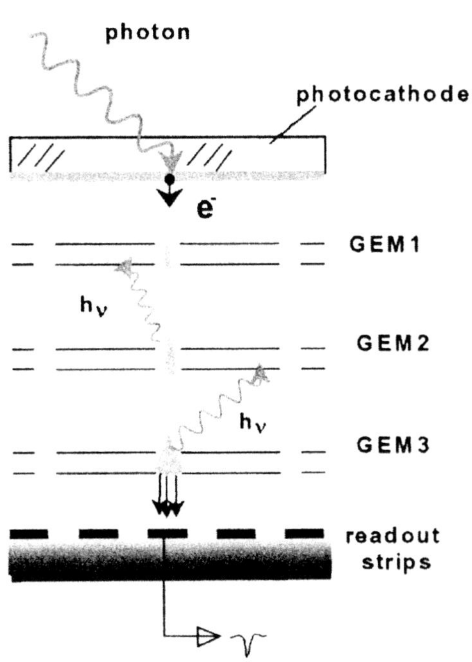

FIGURE 2.20 The multi-GEM photomultiplier concept, providing high total gain and two-dimensional recording by a micropattern device. GEM, gas electron multiplier. *From Breskin (2000), reprinted with permission from Elsevier © 2000.*

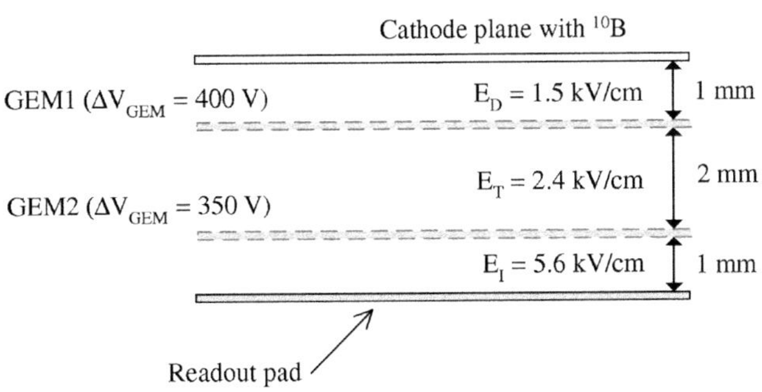

FIGURE 2.21 Schematic cross-sectional view of the double-GEM neutron detector. GEM, gas electron multiplier. *From Ohshita et al. (2010), reprinted with permission from Elsevier © 2010.*

Comprehensive reviews on the developments in gaseous detectors are presented by Hoch (2004), Tsyganov et al. (2008), Sauli (2010, 2016), and Dalla Torre (2011) and especially in the field of MPGDs by Sauli (1999, 2002, 2010, 2016), Oed (2001), Shektman (2002), Bellazzini et al. (2002), Titov (2007), and van der Graaf (2009, 2011).

Microstrip and micropattern gas chambers are filled with ^{3}He to be used as neutron counters (Iguchi et al., 1994; Hayakawa and Maeda, 1996, Radeka et al., 1998). The application of micropatterned devices for microdosimetry purposes was reviewed by Waker et al. (2009). By coating the cathode plane of a double-GEM detector with a 0.02-μm-thick layer of ^{10}B, Ohshita et al. (2010) demonstrated an efficient neutron detection, which could provide excellent 2D images of neutron-beam profiles. The neutron beam incident on the GEM was imaged according to the following neutron reactions:

$$^{10}\text{B} + \text{n} \rightarrow {}^{7}\text{Li} + \alpha + 2.79\ \text{MeV}\ (6\%)$$

$$^{10}\text{B} + \text{n} \rightarrow {}^{7}\text{Li}^{*} + \alpha + 2.31\ \text{MeV}\ (94\%)$$

where ^{7}Li* represents an excited particle, which spontaneously emits a 0.48 MeV gamma photon while returning to the ground state. Two GEMs were used for gas amplification as illustrated in Fig. 2.21 in a chamber gas mixture of Ar and CO_2 (70:30) providing a gas gain of approximately 400. The detector area was 100 × 100 mm with 120 × 120 readout channels. An example of 2D neutron beam images taken at various distances forming the beam collimator is illustrated in Fig. 2.22. Ohshita et al. (2010) measured the GEM detector position resolution at 1.2 mm in FWHM.

A triple-GEM neutron detector coupled to a CCD readout was devised by Fraga et al. (2002, 2003) for the detection of thermal neutrons. The detector was filled with ^{3}He and CF_4. The ^{3}He in the GEM is used for thermal neutron conversion according to the reaction

$$^{3}\text{He} + \text{n} \rightarrow {}^{3}\text{H} + \text{p} + 770\ \text{keV}$$

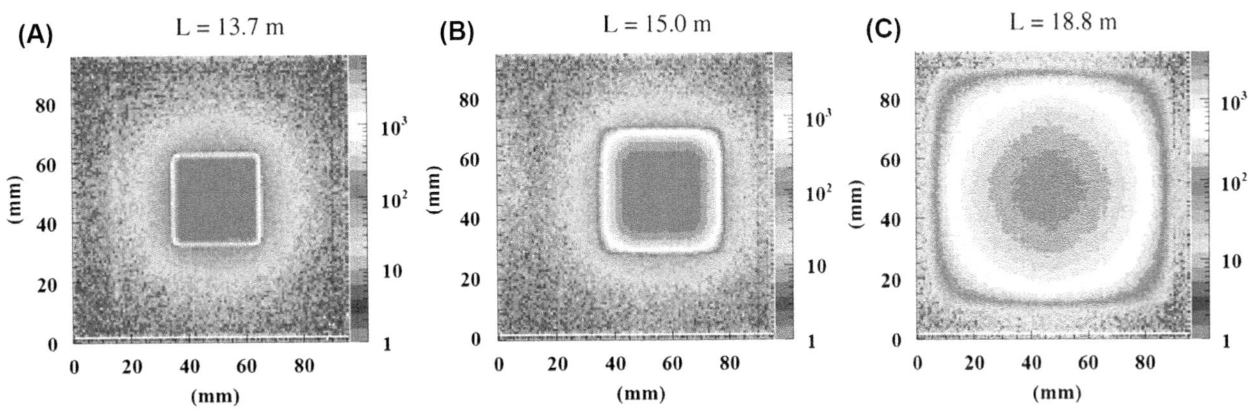

FIGURE 2.22 Neutron beam profiles at various distances: (A) 13.7 m, (B) 15.0 m, and (C) 18.8 m measured by the double-GEM neutron detector. GEM, gas electron multiplier. For the color version of the figure, the reader is referred to the online version of the book. *From Ohshita et al. (2010), reprinted with permission from Elsevier © 2010.*

FIGURE 2.23 Superimposed proton−triton tracks obtained with an exposure time of 1 s with a CCD readout of a gas electron multiplier−based neutron detector. *From Fraga et al. (2002), reprinted with permission from Elsevier © 2002.*

FIGURE 2.24 Image of three events with an electron multiplier CCD for 14.1 MeV neutrons from a neutron generator. The central event, appearing as an inverted Y, is considered the result of the inelastic reaction ^{12}C (n, n' + 3α), which has an energy threshold of 7.9 MeV. *From Roccaro et al. (2009), reprinted with permission from Elsevier © 2009.*

which releases a 589-keV proton and 193-keV triton, and the CF_4 is needed to decrease the range of the triton and proton in the GEM detector. The He-CF_4 gas mixture scintillates in the GEM with an emission maximum at approximately 625 nm, which is imaged with a CCD coupled to a glass window at the floor of the GEM. A typical image of the proton−triton tracks imaged in the GEM following neutron capture by ^3He is illustrated in Fig. 2.23.

Rocarro et al. (2009) devised a gaseous detector for the determination of the neutron flux, energy distribution, and direction of neutron motion for both fast and thermal neutrons. The detector is filled with CF_4 at low pressure with 1 bar or more of ^4He to provide a target for fast neutrons through elastic scattering and a few Torr of ^3He to detect the thermal neutrons via neutron capture according to the reaction described above. The charged particles leave a trail of electrons along their tracks, and an electron avalanche occurs accompanied by the emission of scintillation light from the CF_4. The light is then imaged by a lens coupled to a CCD camera, and the detector is referred to as an EMCCD camera (i.e., electron multiplier CCD camera). An example of images taken from events with 14.1 MeV neutrons in the detector is illustrated in Fig. 2.24 where an inelastic reaction with ^{12}C of the CF_4 gas is observed.

For additional information on the application of multiwire and multipattern proportional counters applied in the biological sciences, see Chapter 9, Volume 2.

In 2013, Takahashi et al. presented a novel glass GEM (G-GEM) type of detector that is based on the use of a photosensitive (photoetchable) glass substrate, which exhibits low electrical resistance (Fujiwara et al., 2016). G-GEMs were found to exhibit high gain and good energy resolution with a single plate. They allow for stable radiographic imaging under high X-ray count rates. The manufacturing process of this type of detector is shown in Fig. 2.25 (Fujiwara et al., 2018). G-GEMs take advantage of the photon yield in the course of the ionization and excitation processes, as illustrated in Fig. 2.26 (Fujiwara et al., 2017).

As stated above, the detection principle of G-GEM detectors is especially suited for imaging purposes (see Fig. 2.27). Photoelectrons created by X-ray excitation are multiplied by the G-GEM inside the gas-filled chamber. Scintillation light emitted during the avalanche process is detected with a mirror−lens−optical camera configuration and forms an X-ray transmission image. This method is applicable in industrial settings (Fig. 2.28) and can even be used for 3D computer tomography of objects (Fig. 2.29).

Multiwire proportional chambers originally have been designed for high rate applications; the applicability is mainly limited by the low ion drift velocity. Spatial resolutions are limited by wire spacing. To overcome these problems, the wires have been replaced by strips, printed on an insulating support. Developments include combinations and hybrids of the MSGC and GEM position-sensitive gaseous detectors such as the MHSP detector (Natal da Luz et al., 2007a, b, c), the micropixel gas chamber (μ-PIC), the MHSP, and Micromegas detectors (Titov, 2007; Homma et al., 2009; Veloso et al., 2010; Aune et al., 2014).

New structures of gas ionization detectors have been developed during the last years where the wires are replaced by a thin micromesh. Promising results were obtained with a very asymmetric two-stage parallel plate chamber with an arrangement that includes basically three electrodes: a large drift region (about 1 kV/cm) between cathode and a micromesh and a small multiplication region with high field strength between micromesh and anode. Electrons formed in the drift region with moderate electric field pass through the mesh into the multiplication region where avalanches are produced. Typical gains are 10^3−10^4, and due to high field strength, fast pulses of a few

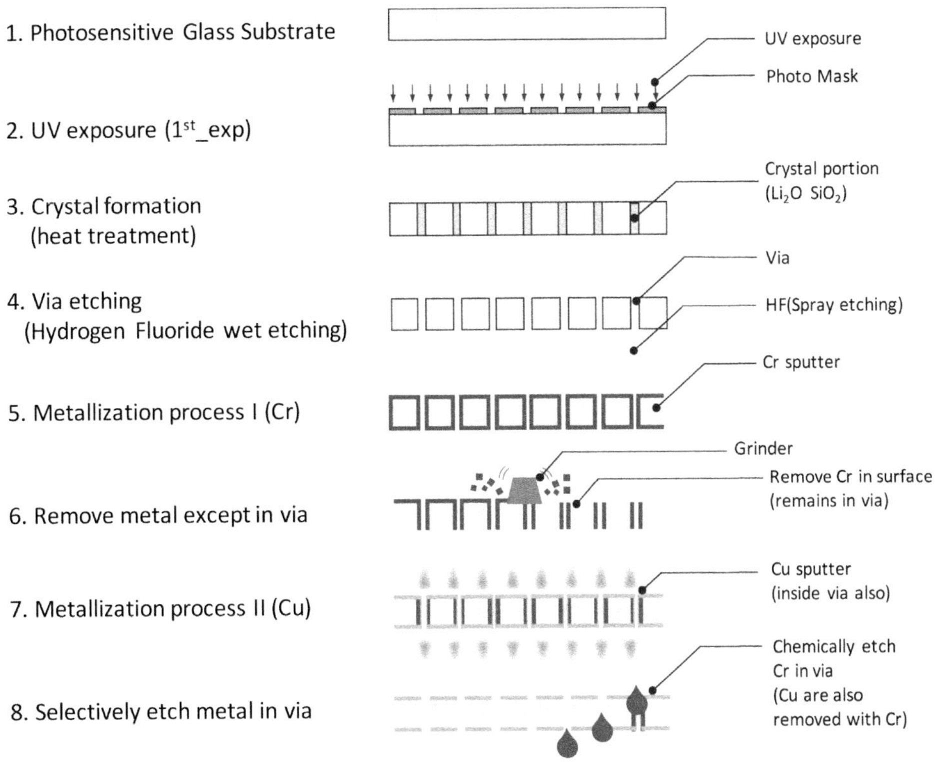

FIGURE 2.25 Manufacturing process of a glass gas electron multiplier on a photoetchable glass substrate. For the color version of the figure, the reader is referred to the online version of the book. *From Fujiwara et al. (2018), reprinted with permission from Elsevier © 2018.*

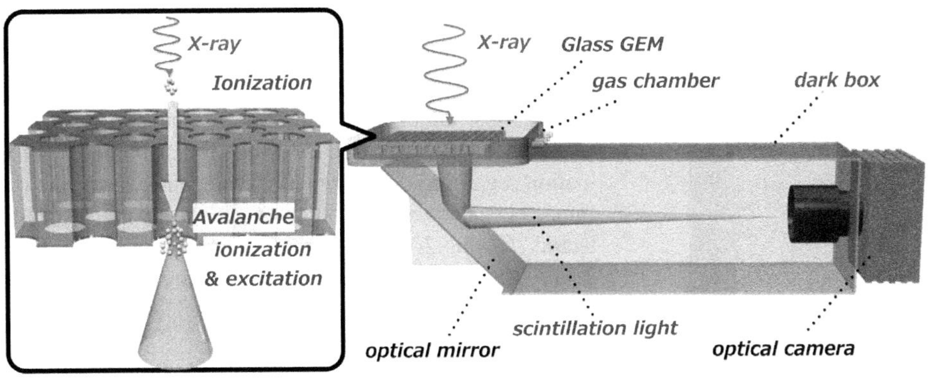

FIGURE 2.26 Scheme of a scintillating G-GEM (glass gas electron multiplier) detector with a mirror and optical camera readout. For the color version of the figure, the reader is referred to the online version of the book. *From Fujiwara et al. (2017), reprinted with permission from Elsevier © 2017.*

nanoseconds duration are produced and high count rates can be managed. Positive ions from the avalanche are collected by the mesh. The authors use the name Micromegas (micro−mesh gas−structure) for this detector arrangement (Giomataris et al., 1996). If the anode consists of a position-sensitive structure, the center, the avalanche can be spatially recorded. The Micromegas has been used mainly in high-energy particle detection and tracking, but other applications such as the measurement of neutron radiation and imaging in the biological and medical sciences are also described (Derréet al., 2001; Delbart et al., 2001; Andriamonje *et al.*, 2001; Pancin et al., 2007; Chefdeville et al., 2008; Donnard et al., 2009; Sauli, 2010; and Oger et al., 2012).

Another variant of the parallel plate counter is a detector device named resistive plate chamber (RPC) described by (Sauli, 2010). The electrodes of this detector are not made of a metallic but of a plastic material, which has a significant electrical resistivity (about $10^9-10^{13}\ \Omega\ cm^{-1}$). Avalanches do not grow too much, and spark formation is avoided. The time resolution is better than 100 ps. Until now, these devices have been used successfully in high energy physics, and further progress in developing this kind of detector can be expected (Santonico, 2003). Belli et al. (2007) studied possible applications of resistive plate counters for biomedical applications such as positron emission tomography. Couceiro et al. (2007) described the application of position-sensitive resistive plate counters for

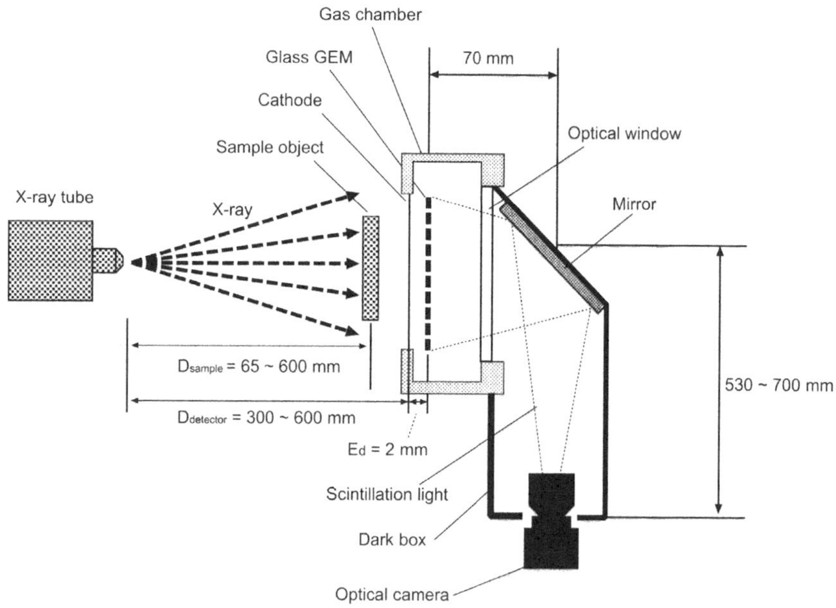

FIGURE 2.27 Scheme of a radiographic imaging of an object using a G-GEM (glass gas electron multiplier) detector. *From Fujiwara et al. (2017), reprinted with permission from Elsevier © 2017.*

FIGURE 2.28 Application of a glass gas electron multiplier detector in the imaging of a defective pouch. *From Fujiwara et al. (2018), reprinted with permission from Elsevier © 2018.*

animal PET investigations. Other applications (e.g., neutron counting) can be found in the reports of Jamil et al. (2010), Qian et al. (2015), and Qian et al. (2009).

For the first time, a Micromegas counter has been used for in-core measurements in a nuclear reactor by Pancin et al. (2008).

A comprehensive review discussing new developments and future perspectives of gas ionization detectors is presented by Titov (2007). New designs of GEM focus on Micromegas and RPC, such as the coupling of GEM and Micromegas structures to a new type of detector. The coupling of GEMs and Micromegas into hydride structures is being studied as a current topic of research (Zhang et al., 2017). Advances in photolithography and microprocessing techniques in the chip industry have triggered a major transition in the field of gas ionization detectors from wire structure to micropattern devices (Sauli, 2010). It has to be mentioned at this point that most applications are related to charged particle physics, UV detection, and X-ray investigations rather than classical radioactivity analysis. Other review articles available in journals provide information regarding developments, achievements, trends, and future perspectives of position-sensitive gas ionization detectors (Sauli, 2001, 2003, 2010; Agosteo, 2010; Dalla Torre, 2011; van der Graaf, 2011; Peskov et al., 2012).

3. Low-level counting techniques using proportional gas ionization detectors

For investigations involving low-level counting techniques, e.g., low-level radiocarbon dating experiments or forensic noble gas analysis, a low and stable background is a necessity (Aalseth et al., 2016). Today, this is achieved mostly by the application of "active shielding." The counting tube for the sample is surrounded by "guard tubes," which are combined with the sample counting tube by an anti-coincidence circuit. Only the counts due to the sample counting tube alone are counted and not those registered by both counting systems simultaneously, which are due to background radioactivity. In earlier times, this active shielding was a ring of sometimes more than 20 GM counting tubes. Later, umbrella-shaped guard tubes were designed. Those were, in some cases, displaced by liquid

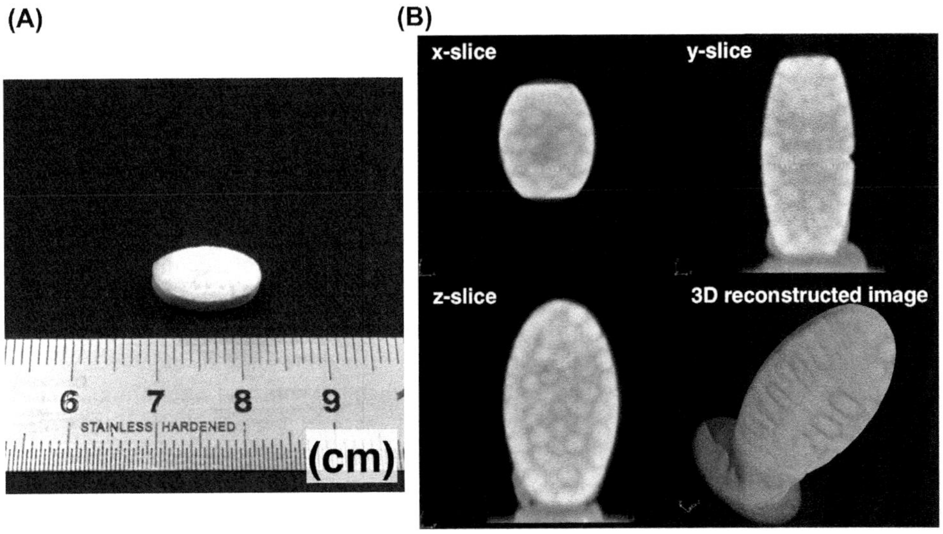

FIGURE 2.29 Photograph (A) and radiographic image of a microsphere-containing medicinal tablet (B), consisting of X, Y, and Z slices as well as the reconstructed 3D CT image. *From Fujiwara et al. (2017), reprinted with permission from Elsevier © 2017.*

scintillator guards, which were specially designed for low-level anticoincidence shielding. The liquid scintillation solution is frequently based on a mineral oil solvent and especially suitable for large tanks. Within the guard chamber, several counters based on proportional detectors are sometimes installed. Some systems are equipped with pressure transmitters and temperature sensors to ensure constant conditions for the counting gas. Also, measurement of the peak and median of the pulse height spectrum is used to obtain information about the purity of the counting gas.

Several guard counter designs are described in the literature, and a remarkable construction has been proposed and tested by Theodórsson and Heusser (1991). They suggest an arrangement of flat guard counters on the external sides of the main shield instead of the inner region of the shield as usual. In this way, the weight and space of the inner shield can be reduced. They also claim that the effects of secondary nuclear reactions causing background effects are considerably reduced.

A new detector type for low-level anticoincidence counting is designed and constructed by Zhang et al. (2002). A CdTe semiconductor counter is used as a guard detector forming also the wall of the low-level PC tube. This equipment is applied successfully to radiocarbon dating investigations, and the authors suggest also other fields of application.

Background reduction by electronic circuit design can be accomplished by pulse shape discrimination (Mäntynen et al., 1987; Äikkäaet al., 1992). By applying pulse shape discrimination, the background is reduced by more than 70%, and only 20% of the efficiency is lost. Figures of merit are improved by a factor of nearly 2. With a counting time of 44 hours, measurable ages up to 56,000 years are achieved.

Carbon dioxide, being a "slower" gas than, for example, methane, is better suited for pulse shape discrimination. On the other hand, purity requirements are much more severe for carbon dioxide. If pulse shape discrimination for background reduction is used, the total length of the rising pulses is measured. The accumulated rise time of an irregular (i.e., a background) pulse is much longer than the rise time of a beta pulse. Yet some background remains, for example, pulses arising from gamma-emitting radionuclides in the construction material. The factors necessary for all these improvements are provided by Äikää et al. (1992).

At present, three measurement methods for radiocarbon dating are available: AMS, low-level liquid scintillation counting, and low-level gas PC. During the past several years, gas PC methods had become less attractive for radiocarbon dating studies. Some authors are of the opinion that the application of gas PC for radiocarbon dating should be reconsidered, as multidetector gas PC systems offer some advantages.

A modern multidetector system has the advantage of parallel counting, which saves a great deal of time. Also, it takes less time to prepare carbon dioxide from a 1-g carbon sample than to carry out a benzene synthesis from the same amount of sample. The benzene is used as an additive to a liquid scintillation cocktail as discussed in Chapter 2, Volume 2. With parallel counting in a multidetector system based on ionization detectors, one of the samples is always a background sample, thus providing continuous monitoring of the background. Pulse rise discrimination techniques can be used in addition to reduce the number of background counts. If pulse rise analysis techniques are used to reject the slower rising background pulses, the counting efficiency is reduced by 18%, but at the same time the background is reduced by a factor of 3.3. A dramatic reduction of background counts is obtained by anticoincidence shielding. Like anticoincidence systems, liquid scintillation guard detectors are frequently used for active shielding.

According to the investigation of Theodórsson (1991), a multidetector gas PC system seems to be highly competitive. Of course, the AMS technique has clear superiority over radiometric methods (Hou and Roos, 2008), especially for very small samples, but considering the high price of AMS equipment, it seems likely that AMS systems and gas PC will be used in the future and these will complement each other very well. Because of the potential of AMS, scientists hesitated to apply and further improve gas PC.

PC devices still play an important role in radiocarbon dating investigations (e.g., Chiti et al., 2009).

Low-level liquid scintillation analyzers with active shielding can provide low background count rates of 0.3 cpm for ^{14}C measurements, making liquid scintillation an attractive method for ^{14}C dating. Chapter 6, Volume 1 and Chapter 2, Volume 2 of this book provide detailed information on low-level radiocarbon measurements by liquid scintillation analysis.

At the National Institute of Standards (NBS/NIST), tritium standards are calibrated regularly using liquid scintillation and gas PC methods. Using the available data from measurements over a period of 38 years, Unterweger and Lucas (2000) could obtain a more accurate and precise value for the half-life of that radionuclide (4504 $\pm$ 9 days).

The available data from international comparison projects have also been used to study the state-of-the-art of tritium low-level measurement techniques. The objective was to find a realistic value for the sensitivity, which could be demanded in ultra-low-level tritium investigations. Theodórsson (1999) reported that during intercomparison investigations, only two laboratories could reach a standard deviation of $\leq$0.03 TU for weak samples. The achievement of a good level of sensitivity and accuracy for tritium measurement is a requirement, because otherwise the possibility of obtaining reliable hydrological information that tritium can give as a natural tracer would be severely limited. Improved future counting systems are discussed. It is again mentioned that gas PC systems can be improved significantly by moving the guard counters to the outer surface of the shield as it had been already proposed by Theodórsson and Heusser (1991).

Measurements have been carried out also to verify theoretical aspects, such as the investigations of Kuzminov and Osetrova (2000) on the shape of the carbon-14 beta-spectrum. Their examinations yielded results, which are consistent with some of the theoretical predictions but which contradict the prediction of others. These findings may help researchers arrive at more accurate theories.

4. Application in environmental monitoring and health physics

a. Radon in water

Zikovsky and Roireau (1990) have developed a simple method for the measurement of radon in water using proportional counters. The method is based on the purging of radon from water with argon, which is bubbled through the water sample and then directed to the counting tube. Argon picks up the radon that was dissolved in the water. A gas purification system removes humidity and oxygen. The high voltage is set for the alpha plateau, and thus a very low background of less than 0.2 cpm and a counting efficiency of 25% are obtained, giving a detection limit of 0.02 Bq per liter. This detection limit compares favorably with that of other methods developed for the determination of radon in water.

Radon daughters (polonium-218, lead-214, polonium-214) contaminate the detector. Unless applying correction factors for the residual activities, a cooling period of at least 1 hour is essential to grant the decay of these daughter nuclides.

Beyond its relevance to health physics, the determination of radium and radon may also be essential for technical purposes. For example, the BOREXINO solar neutrino experiment at the Laboratory Nazionali des Gran Sasso (LNGS, Italy) requires an extremely low radioactivity background (Simgen et al., 2003). Special care has to be taken to control ubiquitous radon to resolve the low solar neutrino event rate properly. The BOREXINO experiment is discussed in detail in Chapter 6, Volume 1. Simgen et al. (2003) developed a system for the detection of radium and radon based on miniaturized proportional counters with sensitivities of ~ 1 Bq/L (for radium-226) and ~0.1 Bq/L (for radon-222), respectively.

b. Measurement of plutonium-241

Rosner et al. (1992) built a PC system that is especially suitable for the measurement of ^{241}Pu. Plutonium-241 is the only significant beta-emitting transuranium nuclide in the wastes from nuclear power plants. Quantitation of plutonium-241 in waste as well as in environmental samples is of interest because ^{241}Pu is a precursor of other transuranium nuclides that have longer half-lives, greater environmental mobility, and greater radiotoxicity. Alpha-emitting americium-241, with a half-life of 432 years, is the daughter product of plutonium-241 and has relatively high radiotoxicity.

Plutonium-241 can be determined indirectly by alpha spectroscopic measurements of its daughter nuclide americium-241. Measurements based on the ingrowth of the daughter radionuclide ^{241}Am can be done only after a long growth period. Even after 4 years, the activity ratio

^{241}Am:^{241}Pu is only 1:166. Thus a lower limit of detection for ^{241}Pu by direct measurement is desirable and has been achieved using PC in the range of about 10 mBq according to the work of Rosner et al. (1992), whereas via ^{241}Am buildup, about 200 mBq is needed for detection.

Some authors have applied liquid scintillation counting to directly measure plutonium-241. Because of the rather high background of commonly available liquid scintillation equipment, this method can be applied only for samples with a relatively high content of plutonium-241. Investigations of that type have been carried out in regions with elevated fallout levels such as Scandinavia or with samples from the nuclear industry or weapons test sites. Lower limits of detection of 35−65 mBq have been reported. However, a low-level liquid scintillation analyzer equipped with a BGO detector guard and time-resolved liquid scintillation counting (TR-LSC) background discrimination electronics is capable of counting environmental ^{241}Pu at a low background of 2.4 cpm (M. F. L'Annunziata, personal communication).

However, because of the nonspecific character (i.e., continuous energy spectrum) of beta radiation, liquid scintillation and PC require very pure samples for counting. Therefore, the chemical purity of the samples and the self-absorption due to the presence of matrix material in the counting sample are the critical points in the PC procedure. The effects of sample thickness, self-absorption, and energy loss in gas PC were studied in detail by Martin-Sánchez et al. (2009).

For PC, special equipment is needed. This equipment can be obtained by modification of commercially available systems.

c. Measurement of Iron-55

For some radionuclides that are difficult to detect during radioprotection measures, gas ionization detectors still offer good possibilities. Iron-55 is formed by neutron capture of iron-54 and disintegrates via electron capture with a half-life of 2.7 years. Consequently, it is formed by the action of neutrons on steel and hence occurs in and around nuclear reactors or accelerator facilities. During planned repairs, the suspension and dispersion of this radioisotope of iron have to be monitored to avoid uptake by workers. The low-energy X-rays of iron-55 (5.9 keV) are stopped by most detector windows, which makes its detection and quantification difficult. In the presence of other contaminating radionuclides, the measurement of iron-55 is further complicated. Surette and Waker (1994) have designed a monitoring system based on a sealed xenon-filled proportional counter with a thin beryllium window. The detector is combined with a single-channel analyzer and a shuttle mechanism that permits positioning of air filter or swipe media. For a counting time of 100 s, the detection limit is around 10 Bq. The thin window of the PC tube (0.05 mm) allows more than 90% of photons with an energy of 3 keV or greater to pass through. The monitoring system is sufficiently sensitive to detect well below the maximum permissible level of surface contamination and also below the maximum permissible concentration in air of the facility for which it was designed.

The low X-ray energy of iron-55 suggests its application for testing the performance of new detection instruments and methods in the lower energy region (see, e.g., the work of Simões et al. (2003) and Masuda et al. (2002)).

Alternatively, the 5 keV-Auger electrons emitted from iron-55 can be used for radioactivity analysis. Tosaki et al. (2013) mounted an iron-55 source at the inner cathode surface of a cylindrical proportional counter. By adjusting the gas mixtures (CH_4, dry air, N_2), a shift of the pulse height has been observed. The authors then studied the electron attenuation cross section of O_2 by evaluating the shifts of the pulse height caused by electron attachment to molecular oxygen.

Recent advances in GEM technology have raised interest in the measurement of iron-55 in activated steel by a small GEM detector coupled to a quad Timepix ADIC for readout (GEMPix) (Curioni et al., 2017). This method allows for a rapid assessment of compliance with the exemption limit. The technique is planned to be implemented as an operative test phase at CERN.

d. Tritium in air

Proportional counters can also be used for tritium monitoring in air, as demonstrated by Aoyama (1990). Monitoring of tritium in air is required in the environment of 14-MeV neutron generators, heavy-water reactors, and reprocessing plants and will also be necessary at nuclear fusion reactors. Tritium must be detected separately from other radioactive volatile noble gases and air activation products. For occupational radiation protection and emission control, a real-time measurement and high sensitivity are necessary to meet the legal requirements for radiation protection and emission control. To respond to an accidental release, a wide range of detection is essential. High sensitivity can be obtained by using systems equipped with anticoincidence shielding or pulse shape discrimination. Conventional proportional counters suffer from the disadvantage of requiring a counting gas and have a rather short operation range.

Aoyama (1990) described a method for tritium monitoring in air by the use of flow-through proportional counters with air as a counting gas. The counters need no counting gas other than the sampled air. The electronic equipment attached to the counting system comprises pulse height discrimination, anticoincidence shielding, and background compensation. In that way, it is possible to detect and measure tritium in an external gamma background and also

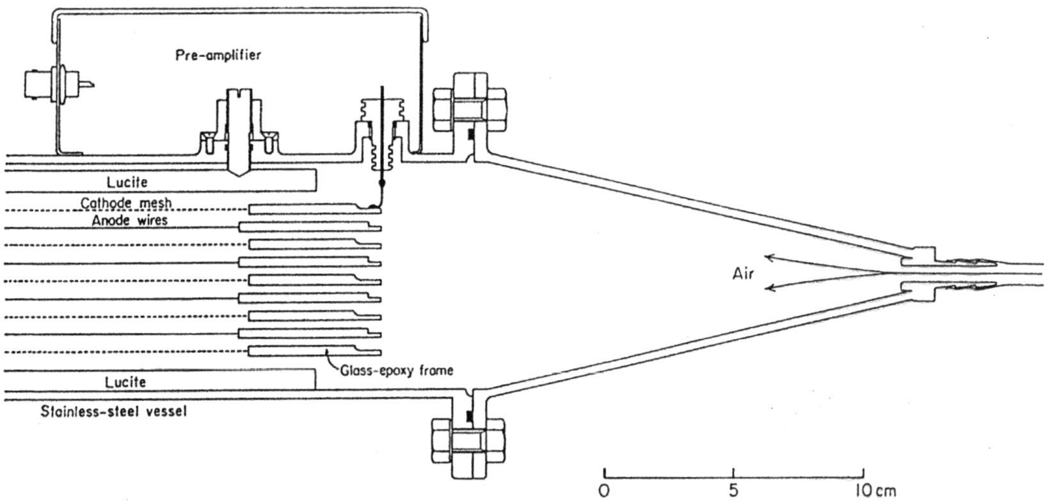

FIGURE 2.30 Cross section of a tritium monitor, which uses air as a counting gas. The detector consists of four layers of multiwire proportional counters. The air flows uniformly through the counter. *From Aoyama (1990), reprinted with permission from IEEE © 1990.*

in alpha and beta backgrounds originating from other gaseous radioactive materials in the air sample. It was reported that a lower detection limit of 0.005 Bq/cm^3 in the presence of natural background can be obtained in a counting time of 1 minute. Also, a wide range up to 5000 Bq/cm^3 (up to six orders of magnitude) can be managed by this system. The PC detector is rather complicated, consisting of an arrangement of anode wires and cathode meshes. A schematic illustration of this arrangement is shown in Fig. 2.30. Outer layers of the counter were used as guard counters to eliminate gamma background. Gaps between individual arrangements of anode and cathode were kept longer than the maximum range of tritium in air, thereby avoiding coincidence effects caused by tritium. Such coincidence effects were used to exclude other beta rays. The alpha component from radon and its daughter nuclides was eliminated by pulse height discrimination. Data derived from pressure, temperature, and humidity sensors were transferred to a computer and used to control the high voltage and to correct coincidence count rates. Loughlin and Lawrence (2007) carried out a thorough study of methods for the accurate monitoring of tritium in air in conjunction with the operation of the ITER. They conclude a gas proportional counter is best suited for the measurement of HT (i.e., a hydrogen molecule with one tritium atom bound to protium) and HTO (i.e., a water molecule incorporating one tritium atom), when discriminating samplers are used at ventilation outlets from fusion reactions.

e. Positron emitters in air

The increasing demand of positron emitters such as fluorine-18, carbon-11, and oxygen-15 for medical applications (such as positron emission tomography, PET) has resulted in the commissioning of new cyclotron facilities. Marouli et al. (2007) studied the feasibility of using proportional counters for the monitoring of the emissions from the exhaust ducts at these facilities. The positrons were found to deposit 1.5–2.5 times more energy than negatrons to the counting gas, which affects the minimum detectable activity concentration in a positive way.

f. Monitoring of gaseous fission products

Another important application of proportional counters is the verification of the Comprehensive Nuclear Test-Ban-Treaty (CTBT). The detection of fission products (most importantly ^{131}I and the xenon isotopes ^{131m}Xe, ^{133m}Xe, ^{133}Xe, and ^{135}Xe) or activation products (e.g., ^{37}Ar that can be produced from $^{40}Ca(n,\alpha)^{37}Ar$ or from $^{40}Ar(n,4n)^{37}Ar$) from a nuclear test remains the only safe way to verify the nuclear nature of any (underground) explosion. Argon-37 disintegrates solely by electron capture with the emission of Auger electrons. The measurement of its low decay energy (2.8 keV) is done with special low-level gas proportional counters (Saey, 2007) and can act as a smoking gun, especially in the course of on-site inspections (Aalseth et al., 2011; Seifert et al., 2013). Detection of radioxenon is also performed with the help of proportional counters, but usually in combination with other radioanalytical methods such as beta-gated gamma-coincidence counting or high-resolution gamma spectrometry (Saey, 2009, 2010; Xie et al., 2013, 2014).

g. Radiostrontium

Low-level proportional counters are applied for quantitative radionuclide measurement after radiochemical separation procedures. For high sensitivity measurement of low levels of radiostrontium, gas ionization detectors are still preferred over liquid scintillation counters (LSCs). The counting efficiency of LSC may be twice as high; however, the background level in LSC is approximately 10 times higher than that in gas ionization detectors (proportional counters or GM tubes). Rodriguez et al. (2012) reported on an automated strontium separation and preconcentration system that may achieve minimum detectable activities of strontium-90 as low as 8 mBq.

Mateos et al. (2000) designed and constructed a semi-automatic analysis system for the determination of $^{90}Sr/^{90}Y$ in aqueous samples using a sequential injection method. The beta measurements are made twice within 24 hours. From these results, the initial activity of strontium-90 and yttrium-90 is calculated. Thus, the time-consuming yttrium milking method can be overcome.

Vaca et al. (2001) compared strontium-90 measurement methods using a Berthold LB770 counter and a Quantulus 1220 liquid scintillation spectrometer. The proportional counter had a passive shield of 20-cm-thick lead and an active gas proportional guard counter. The samples can be measured simultaneously by that device, and a background from 0.3 to 0.6 cpm is obtained, depending on the detector location along the gas flow pathway. It is surprising that for gas PC, a minimum detectable activity of 0.13 Bq/kg is reported, for Cerenkov counting 0.37 Bq/kg.

Crown ether technologies were used by Scarpitta et al. (1999) to measure the strontium-90 content of Brookhaven National Laboratory groundwater samples. With gas proportional and liquid scintillation counting, minimum detectable levels of 37 Bq/m^3 were achieved using a processed sample of 1 L and a counting time of 1 hour.

PC is used also for the determination of strontium-90 in human bones and teeth in Greece. Measurement was performed on yttrium-90 after equilibration with strontium-90 and liquid extraction using bis (2-ethyl-hexyl) hydrogen phosphate. The analyses unfolded an average of 30 mBq strontium-90 per gram calcium, which was found with only a small variation with respect to age and sex. From the variation of the activity in teeth, it could be concluded that the contamination from atmospheric nuclear weapons tests exceeded by far that is caused by the Chernobyl accident (Stamoulis et al.,1999).

In an extensive study, Degteva et al. (2006) investigated the dose resulting from external and internal exposure to radioactivity in the Techa river region, which was affected by the operation of the Mayak Production Association in the former Soviet Union. The internal dose was dominated by radiostrontium. Hence, investigation of thousands of teeth and bones was the method of choice to assess dose aspects. The authors found that the median doses to the red bone marrow and the bone surface were 0.21 and 0.37 Gy, respectively. The maximum doses to the red bone marrow and bone surface 2.0 and 5.2 Gy, respectively, were reported.

In an analytical and environmental study, the groundwater from the site of the nuclear power plant A1 Jaslovske Bohunice (Slovakia) was investigated (Ometáková et al., 2011). Several separation methods for the preconcentration of the radionuclides were tested. The ^{90}Y activity was determined by a low-level alpha–beta proportional counter and by Cerenkov liquid scintillation counting.

Much work has also been done to develop faster radioanalytical protocols for emergency response. Authors from Savannah River Nuclear Solutions developed methods for the determination of strontium-90 and alpha-emitting actinides in urine (Maxwell III and Culligan, 2009a), air filters (2010a), vegetation samples (2010b), milk (Mawell III and Culligan, 2009b), and water (Mawell III and Culligan, 2009c). The authors used proportional counters for radiostrontium and alpha spectrometry for actinide determination.

Herranz et al. (2011) used a gas proportional counter to measure both ^{89}Sr and ^{90}Sr in environmental samples. The strontium is isolated firstly from the test sample, which also includes its isolation from the ^{90}Y daughter nuclide. Two counting times are required, one after the removal of ^{90}Y and a second count time selected after significant ingrowth of ^{90}Y. They used a proportional counter with a 5 cm diameter and 0.5 μm Mylar window. The detection efficiencies for ^{89}Sr, ^{90}Sr, and ^{90}Y were 43%, 36%, and 44%, respectively, with uncertainties between 0.4 and 0.5% and a typical background count rate of 0.8 cpm.

h. Health physics and dosimetry

Dosimetry ultimately aims at reflecting the radiation dose absorbed by the organism rather than by the dosimeter. To understand the damage induced by radiation, fundamental understanding of radiobiological processes is a prerequisite, which allows the description of radiation action at a microscopic level. Accordingly, much effort is put into bioequivalent dosimetry (Hajek, 2015).

TEPCs are a field with significant recent advances (Agosteo, 2010; Broughton and Waker, 2016). First introduced by Rossi and Zaider (1996), they can be used to evaluate the radiation dose and dose equivalent for gamma rays, charged particles, and neutrons. Agosteo (2010) reports on the efforts toward miniaturization of TEPCs, which is important in radiation therapy, because conventional TEPCs are too large in comparison with the beam in many specialized settings and hence suffer pulse pileup effects. TEPCs with cavities in the millimeter-range have been built to tackle this problem. An alternative approach consists in assessing the dose-mean linear energy through the variance method with a microdosimetric ion chamber operating in current mode. This method relies on the higher variance of the charge collected by an ionization chamber in a fixed time interval as the particle stopping power increases (Agosteo, 2010; ICRU, 1983; Hsu et al., 2008). In many cases, TEPCs use propane as a tissue-equivalent filling gas (Chiriotti et al., 2015; Moro et al., 2014).

A mini-TEPC was developed by Nardo et al. (2004a,b) for radiation therapy applications (Agosteo, 2010). Fig. 2.31 shows a sectional view of this detector. Its external diameter (2.7 mm) is the same as that of an 8-French (*i.e.*, 2.7 mm) cannula, which is used for mini-invasive surgery. The aim is to perform microdosimetric measurements directly inside the irradiated organ (for example, during the treatment of prostate tumors or proton radiotherapy of eye tumors; see DeNardo et al., 2010). The mini-TEPC sensitive volume is cylindrical (diameter and height 0.9 mm). The anode is a wire of gold-plated tungsten (diameter 10 μm). The cathode is made of Shonka A-150 plastic, 0.35 mm in thickness. The mini-TEPC is inserted in a titanium probe 2.7 mm in diameter, 170 mm in length, electrically grounded. The insulation between the conductive A-150 cathode and the titanium probe is ensured by a 0.35-mm-thick Rexolite cylinder. The propane-based tissue-equivalent gas flows continuously through the sensitive volume.

Another problem of conventionally sized TEPCs is their low sensitivity and energy response in low-dose rate applications (Waker et al., 2009). For low energy neutrons, the range of the secondary charged particles is lower than the simulated site size of a TEPC. The instrument hence loses its spectrometric properties with respect to LET (Agosteo, 2010). Waker et al. (2009) present and discuss the application of multielement and micropatterned devices to overcome these problems. Aslam et al. (2011) discuss their results on the investigation of the response of a commercially available spherical TEPC to mixed photon and neutron fields, which may affect the workers in nuclear power plants. They found that moderate dose rates, which do not result in dead times of more than 20%−25% due to either of the component radiation or due to both components of a mixed field, generate acceptable results for radiation monitoring.

A TEPC based on a THGEM has been proposed (Byun et al., 2009). The THGEM thickness ranges from the sub-millimeter to millimeter order, compared with standard GEM insulator foils with a thickness to 50 μm. This makes the THGEM more robust and practical for gas amplification in a 2D microdosimeter (Agosteo, 2010).

Multielement TEPCs based on GEMs have been developed by Farahmand et al. (2003a, 2003b, 2004) and were reviewed by Agosteo (2010). This millimetric device consists of five cylindrical cavities of 4.3 mm^3 in a sensitive volume sandwiched between two layers of A-150 plastic and filled with a tissue-equivalent gas. The GEM provides charge multiplication for the electrons from ion pair creation. The microdosimetric response of a TEPC-GEM to 14 MeV neutrons and a ^{252}Cf neutron source was measured by Farahmand et al. (2004) and found to be in good agreement with spectra available in the literature (Agosteo, 2010).

A twin miniaturized TEPC for application in boron neutron capture therapy was developed by Moro et al. (2006). Here, the wall of one TEPC is enriched in boron-10. For low energy neutron fields below 500 keV, Waker and Aslam (2011a,b) developed a TEPC using a graphite-walled proportional counter. Miniaturization of TEPCs for low-energy neutron dosimetry has also been achieved

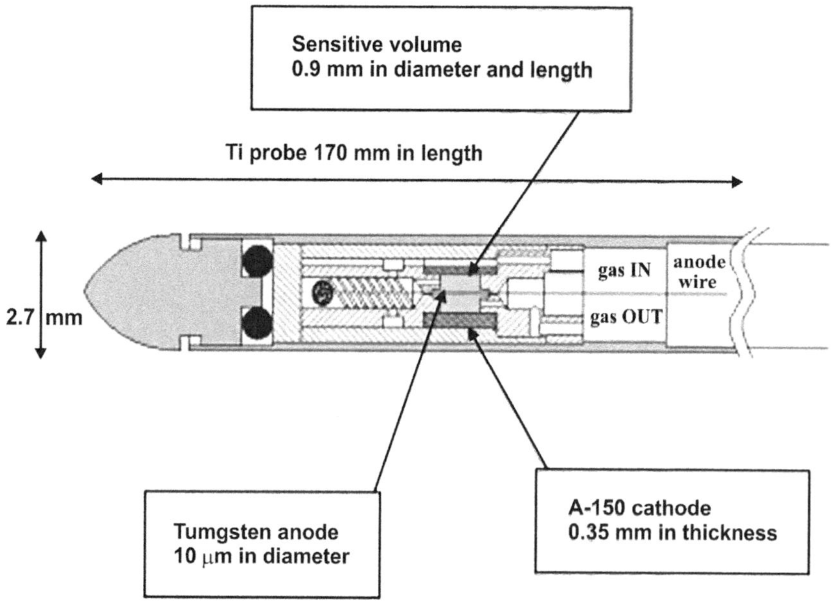

FIGURE 2.31 Cross-sectional view of a mini−tissue-equivalent proportional counter used in radiation therapy. The sensitive volume is the gray area in the center. *From DeNardo et al. (2004b), reprinted with permission from Elsevier © 2004.*

successfully with the development of compact multielement tissue-equivalent proportional counters (METEPCs) (Waker and Aslam, 2010).

More recently, TEPC-based technologies have received much attention in hadrotherapy (proton and carbon ion cancer therapy) for a realistic dose imaging in clinical settings, as they can measure the LET of the incident radiation. Nam et al. (2015) have developed a TEPC that can simulate a site diameter of 2 μm for microdosimetry (Fig. 2.32). The LET calibration factor of this detector is about 0.72 keV/μm. Other designs are GEM-based TEPCs that achieve simulated sites down to 140 nm for response to low-energy X-rays (Farahmand and De Nardo, 2016).

As stated above, research on TEPCs is nowadays much focused around the application at hadron therapy facilities. There is plenty of literature available on this field (that goes beyond classic radioactivity analysis). In particular, the application in proton therapy (Farah et al., 2015, 2017) and carbon ion therapy, both in experimental and in modeling studies (Burigo et al., 2013, 2014; Chiriotti et al., 2014; Galer et al., 2017; Mazzucconi et al., 2018; Taylor et al., 2015), has been researched. Novel detectors for application in ^{12}C therapy, such as the avalanche-confinement TEPC, have been presented (Bortot et al., 2017a), which can be calibrated using alpha emitters (Bortot et al., 2017b).

Techniques to separate out the dose and energy spectra of neutrons and charged particles are necessary for health physics investigations in space shuttle conditions. Braby and Badhwar (2001) used a combination of a tissue-equivalent and a hydrogen-free detector. Both have nearly the same response to photons, but the hydrogen-free detector is insensitive to neutrons below about 10 MeV. Thus, the neutron dose can be obtained by subtraction. Similar considerations are used also for the separation of charged particles and neutrons. Nunez-Perez and Braby (2011) report on the replacement of the TEPCs on the International Space Station and the improvements achieved with the new detectors.

A novel type of detector called European TEPC (EuTEPC) has been introduced recently (Moro and Chiriotti, 2015). The EuTEPC is applicable in gamma and neutron fields. Its distinguishing feature is a spherical cathode wall that is subdivided in nine sectors, which are separated by 0.5 mm Rexolite walls. Each sector is properly and differently biased, to obtain a uniform electric field along the anode wire, thus allowing good isotropic response and energy resolution. The authors suggest the application of the EuTEPC in space dosimetry.

Nanodosimetry will be of interest for investigations in microbiological radiation effects. Tamboul and Watt (2001) built a gridded parallel plate proportional counter, operating at low pressure (1 Torr). This corresponds to a mean chord diameter of 1.8 nm. The device is designed to have a response to radiation simulating that of a bimolecular target

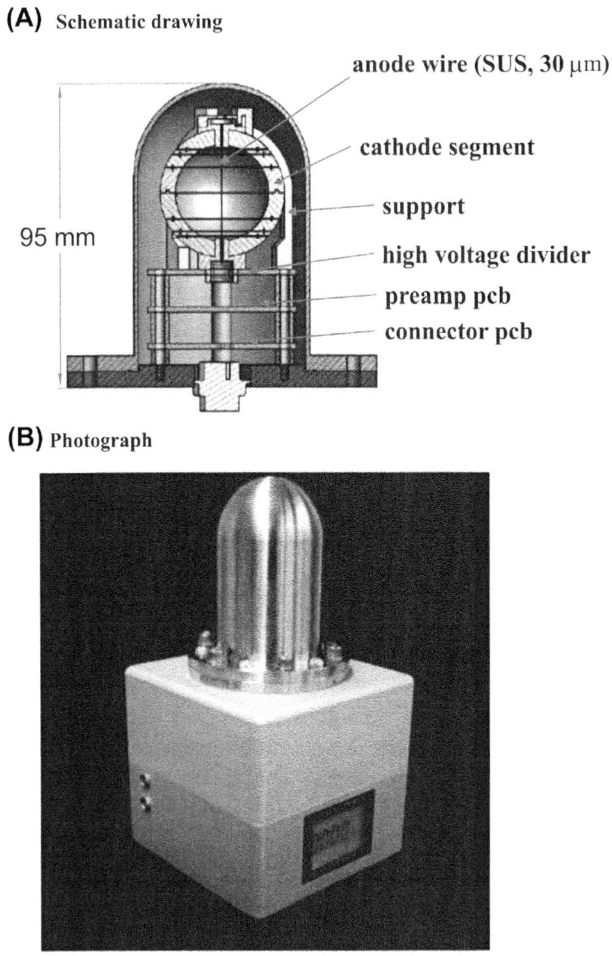

FIGURE 2.32 (A) Schematic drawing and (B) photograph of a tissue-equivalent proportional counter for proton therapy. For the color version of this figure, the reader is referred to the online version of the book. *From Nam et al. (2015), reprinted with permission from the Korean Physical Society © 2015.*

of about the same sensitive volume, for example, a double-stranded DNA molecule. A TEPC on nanometric level was presented by Cesari et al. (2002). This "avalanche confinement TEPC" is characterized by a high gas gain and good energy resolution. It was defined to study sites in the nanometer scale down to 35 nm (Agosteo, 2010). In many cases, TEPCs nowadays compete with silicon-based microdosimeters; however TEPC-based systems are often considered as a standard method (Zhang et al., 2014), which is why many current studies compare or cross-calibrate the performance of silicon-based microdosimeters (silicon on insulator, SOI) with TEPC-based microdosimeters (Tran et al., 2018; Bolst et al., 2017a, 2017b; Mukherjee et al., 2017; Rosenfeld, 2016). Comparative studies not only are limited to charged particle and photon radiation but also include neutron microdosimetry (Agosteo et al., 2011; Ali et al., 2014; De Saint-Hubert et al., 2016).

Thermoluminescence dating can be applied to objects that have been heated in the past such as bricks or pottery. In certain crystalline minerals, absorbed energy from ionizing radiation is reemitted as light upon heating of the material. This method takes advantage of the fact that the dose received is proportional to the age (past the initial heating). As a consequence, for thermoluminescence dating, the radiation dose the object had been exposed to must be known. Most frequently, this radiation dose is due to the content of alpha- and beta-emitting natural radionuclides within the material of the archeological specimen. PC techniques can be used to determine the activity of the material, and from this analytical result, the radiation dose can be calculated. Troja et al. (1995) give an example for this type of activity measurements and dose calculations.

VI. Geiger–Müller counters

As already mentioned, with GM tubes, much higher electric fields are applied than with ion chambers and proportional counters. Because of the high electric field, the intensity of an individual avalanche is enhanced. As a consequence of the emission of UV photons, which are released during deexcitation of atoms or molecules inside the tube, additional avalanches are created. One avalanche therefore can trigger another at a different position in the detector chamber volume. The number of avalanches grows exponentially. Also, the number of slowly migrating positive ions increases. The increasing number of positively charged ions near the electrode causes the field strength to decrease, and further creation of avalanches is stopped, because ion pair multiplication requires a sufficiently high electric field strength. The discharge in a GM tube is terminated at about the same total produced charge, regardless of the amount of ions initially created by the radiation event. Therefore, all output pulses from a GM tube are of about the same size. The output pulse amplitudes of GM tubes are very large compared with signals of ion chambers and proportional counters, usually of the order of several volts. Simple electronic circuits can be used to register GM output signals, but no information about the type and energy of the incident radiation can be obtained. Apart from this lack of information about the type and energy of radiation, GM tubes have a rather long resolving time compared with PC tubes. Therefore, their use is limited to relatively low count rates, only a few hundred counts per second. Geiger and Müller (1928) hence originally proposed their instrument for the detection of "weakest activities." Resolving time corrections can be applied, but the resolving time depends not only on the field strength but also on the observed count rate (Jones and Holford, 1981).

The decaying source method is probably the most general and accurate of the methods for measuring the observed and true counting rates over the entire counting rate range of interest. For that, a very pure radionuclide source of known half-life is essentially needed. True count rate and observed count rate differ considerably at high activity of the radioactive source; however, with time, the background corrected observed count rate will approach the true count rate. This type of experimental determination of dead time is frequently used to test the usefulness of mathematically based models for correction (e.g., Gardner and Liu, 1997; Lee and Gardner, 2000).

Counting losses induced by resolving time of a counting system can be a limiting factor in measurements. Vinagre and Conde (2001) presented an interesting method for the determination of resolving time of a counting system. They added an additional pulse to each pulse of the counting system and varied the delay time of this additional pulse. By observing the total count rate as a function of the delay time, good results for the resolving time could be obtained.

It warrants that GM tubes show a remarkable energy-dependent response for high-energy photons above 3 MeV. This was underscored by Neumann et al. (2002) to be a relevant factor in accurate dose determinations.

In summary, GM detectors are often used in industrial radiation gauges. They proved to be robust (even in harsh environments) and relatively cheap, and they only require simple readout electronics (Meric et al., 2012).

A. Designs and properties of Geiger–Müller counters

1. Fill gas

The fill gas for GM counting tubes has to meet requirements similar to those for the fill gas of proportional counters. Argon and helium are most frequently used. The gas pressure is in the order of tenths of bars, and depending on the size and shape of the tubes, a voltage in the order of hundreds of volts is applied. GM tubes are usually permanently sealed and operate at low gas pressure, although designs have been realized using atmospheric pressure and flow-through to replenish the fill gas and flush out impurities.

2. Quenching

After the termination of the discharge, the slowly migrating positive ions of the fill gas finally arrive at the cathode, which is usually the outer wall of the counting tube. At this electrode, the cations capture electrons from the cathode surface and a corresponding amount of energy is liberated. If this liberated energy exceeds the ionization energy of the cathode material, additional electrons are set free from this electrode. These newly generated free electrons migrate to the anode and create another avalanche. This finally results in a continuous output of pulses. The probability of this additional electron drift is rather low, but because of the high number of cations at the field strength conditions in a GM tube, this effect of multiple pulses is observed. With GM tubes, special precautions have to be taken to prevent the formation of additional avalanches. This can be done by reducing the bias voltage after the Geiger discharge. This external quenching can be achieved by using a suitable electronic circuit (resistor and capacitance) that determines the time of restoration of the high voltage following a Geiger discharge. The restoration time is usually on the order of milliseconds, and therefore, this design is suitable only for low count rates.

It is more common today to use internal quenching, which involves the addition of a suitable compound to the fill gas. The ionization energy for this additive to the fill gas (quench gas) must be lower than the ionization energy for the fill gas. Although confusing, the same expression, quench gas, is used for both the additive to a fill gas of proportional counters, which has to absorb UV photons, and the additive to a fill gas in the GM tube, which should be able to neutralize the drifting ions of the original filling gas by electron transfer. The ions of the quench gas migrate to the cathode and are also neutralized. However, the liberated ionization energy is now consumed by the quench gas and causes dissociation of the quench gas molecules. Some quench gases, such as halogens (e.g., chlorine or bromine), show spontaneous recombination; other quench gases, such as organic compounds (e.g., ethanol), are consumed, and therefore the lifetime of an organic-quenched GM tube is limited to about 10^9 counts. Quench gases are usually added at an amount of several percent to the fill gas of the GM tube.

A relatively long time is needed (100−500 μs) to clean the positive ions that are formed during the avalanche propagation. A mathematical approach for deadtime correction was published by Arkani et al. (2013).

Abilama et al. (2015) investigated the lifetime of bromine-quenched GM counters. The amount of quench gas needed is directly linked to the operational characteristics of the detector. Preserving the halogen is key in maintaining the operating lifetime of the GM tube. However, since halogens are highly corrosive, they will react with the materials they come into contact with and thus deplete themselves over time. Abilama et al. (2015) tested three corrosion-resistant mechanisms (surface treatments) to prevent the quench gas from reacting with the materials of the tube: oxygen-plasma bombarded ("raw"), a passivated surface (using nitric acid passivation followed by oxygen-plasma bombardment), and plating with a thin chromium layer of a few micrometers. The latter was found to be the most resistant surface.

The transition from proportional mode to GM mode takes place at increasing field strength. Golovatyuk and Grancagnolo (1999) could demonstrate that this transition also depends on the concentration of a quenching gas. This fact may be of relevance if pulse shape analysis is used for particle identification. If the concentration of quenching gas is low, gas amplification, as a function of high voltage, increases more rapidly and the boundary between the proportional region and the GM region may be crossed easily. Results of pulse shape analysis may not be interpreted correctly.

3. Plateau

For the simple electronic circuits that are usually designed for use with GM tubes, a minimum pulse amplitude is required for count registering. At a given voltage, this minimum pulse amplitude is exceeded by all signals, as soon as that voltage, the Geiger discharge region, is reached. Therefore, on increasing the voltage while exposing the GM tube to a radioactive source of constant activity, pulse registering starts rather abruptly and the counts per unit of time remain relatively constant (plateau of a GM counter).

GM tubes are frequently rated on the basis of the slope of the plateau region. The slope of the plateau region of halogen-quenched tubes is usually less flat than that of organic-quenched tubes (2−3% per 100 V). However, halogen-quenched tubes can usually be operated at a lower voltage than organic-quenched tubes.

4. Applications

The design of GM tubes is usually similar to that of proportional tubes. Most frequently, the end-window type is used. GM tubes can also have the shape of "needle tubes," in which the anode consists of a needle. In the vicinity of the needle point the field strength varies by $1/r^2$ instead of the $1/r$ variation near a wire or rod electrode. Therefore, counters with a very small active volume can be manufactured.

Since a Geiger discharge is created by a single ion pair, alpha and beta particles, once they penetrate the wall or window, are registered with very high efficiency. Gamma rays are detected by the electrons that are observed as a result of interaction of the gamma ray with the walls of the counting tube via the photoelectron effect or Compton effect. The efficiency of GM tubes for gamma rays is relatively low (in the order of a few percent over a wide energy range) and also depends on the atomic number of the material used to make the tubes. The range of medical or industrial applications of GM tubes could further be extended, provided that the stopping power of the tubes can be improved. In a first approach to this problem, Meric et al. (2011a, b) applied Monte Carlo methods for the simulation of the stopping efficiencies of GM counters and benchmarked the results experimentally.

Attempts have been made to enhance the interaction of GM detectors with gamma photons. Meric et al. (2012)

investigated the effect of placing metal disks made of high atomic number metals inside the tube and thus converted gamma photons to photo electrons or Compton electrons. The efficiency of the detection of gamma rays may thus be enhanced by a factor of 3.

Currently, GM tubes are used most frequently for radiation monitoring and contamination control in day-to-day radiochemistry work. For dose rate measurements, GM counters are likely being replaced in the future by novel, scintillating spectrometers based on $LaBr_3$, $CeBr_3$, SrI_2, and others (Kessler et al., 2018; Baeza et al., 2017). This could affect the European early warning network that consists of about 5000 stations and that are operated using mainly GM dose rate meters.

Photon doses in mixed fields (neutrons/gamma) are frequently measured with GM counters. However, it has to be mentioned that the response of GM detectors depends on photon energy, especially for photon energies above 3 MeV. Dealing with the analysis of neutron and photon components during calibration experiments,Neumann et al. (2002) point out that the knowledge of spectral distribution of the photons is essential for accurate dose determinations.

Beyond scientific value, GM counters are very suitable also for educational purposes. Blanco et al. (2009) and Perez et al. (2018) showed the application of GM counters for the detection of cosmic rays at different levels (from sea level, to mountains, to airplane cruise altitudes). In the Japanese Misho project (a "citizen science" project involving residents of Fukushima prefecture and Tohoku University), the earliest response to the Fukushima nuclear accident relied on the use of GM counters (and other radioanalytical techniques). The various data sets allowed for the estimation of the cumulative total external exposure as a function of time (Koike et al., 2014).

GM counters are easy to use and can be used for environmental monitoring even by laypersons after a short introduction (Jevtic and Drndarevic, 2015). They are very useful in screening for contaminations. Two recent examples include the screening of contamination after the Chernobyl and Fukushima nuclear accidents, respectively. In combination with other detectors, GM counters were used for the retrospective radiological assessment of personal items (sweaters) of two students from the United Kingdom who lived in Minsk and Kiev at the time of the Chernobyl accident (Alkhomashi and Monged, 2015); a study with a more anecdotal character. For Fukushima, screening of personal items is regarded as more relevant, because the accident happened in 2011 and the owners of the property may still be interested in using their property. Ogino and Hattori (2013) defined in their study the conditions and operational levels for unconditional release of contaminated personal items from the areas surrounding Fukushima Daiichi nuclear power plant. A level of 10 Bq/cm^2 is the limit for clearance of the contaminated property. By applying a factor for the conversion of the unit surface contamination to the count rate of a GM survey meter, which was widely used after the Fukushima accident, the operational level for the unconditional release of contaminated properties was calculated to be 2300 cpm on average and 23,000 cpm at the highest-contamination part (Ogino and Hattori, 2013).

a. Environmental radioassay

Radon. Bigu (1992) designed a fully automatic system for the unattended quantitation of radon-222 and radon-220 progeny. He used a GM beta particle detector with a pancake configuration. The instrument is a microprocessor-based system that consists of a sampling device, an electronic scaler, and a personal computer. The computer records all sampling and counting routines.

The sampling device consists of a filter about 5 cm in diameter facing the detector at a distance of about 0.5 cm. The air flow rate is 1.4 L/min. However, the measurement and data procedure is rather complex and requires a rather sophisticated computer program. Basically, the following steps are required:

- The sampling and counting for a given period provides results for the combined radon-222 and radon-220 progeny contribution.
- A counting period after the sampling records the beta particle activity versus time and thus permits assay of radon-222 daughter products.
- From the results of the preceding steps, the contribution of radon-220 progeny can be calculated.

Difficulties may arise if measurements have to be made under transient conditions, for example, when there are rapid changes in the concentration of radon-222 or changes in the aerosol concentration or aerosol size distribution. This effect is related to the half-life of the radionuclides of interest. For radon-220, it is particularly acute because of the long half-life of lead-212. The full effect of any of these changes (perturbations) is felt by the detectors after about one half-life of the dominating radionuclide of the decay chain.

Epstein et al. (2009) studied the health hazards caused by drinking water from "radium ore revigator," which was a pseudoscientific medical device developed in the early 20th century consisting of ceramic water jars lined with a carnotite layer. Direct radiation exposure was measured with a GM tube. Radon and its progeny in water and air from the revigator was measured by liquid scintillation counting. Trace elements were determined by application of inductively coupled plasma mass spectrometry. Interestingly, Epstein et al. (2009) conclude that the potential hazard from the revigator was not radiation but rather the chemically toxic elements such as arsenic and uranium that were extracted into the water.

A more curious and certainly very tragic case of the (failed) application of GM counters in the assessment of radon progeny was the polonium-210 poisoning of Mr. Alexander Litvinenko in London, United Kingdom, in 2006 (Nathwani et al., 2016). When the man was admitted to the hospital, the symptoms initially indicated gastroenteritis. Later poisoning with drugs or thallium was suspected. Only on the 22nd day after he first became ill, 1 day before his death, radiation poisoning was discovered by the characteristic (although extremely weak) 803 keV gamma peak in a gamma spectrometric measurement of his urine. Earlier body surface scans using GM counters failed, because the detector was not suited to detect the alpha emitter polonium-210 (combined with the minute increase of gamma dose rate).

Fluorine-18. Papp and Uray (2002) used a very simple experimental setup for the determination of fluorine-18 attached to aerosol particles in a laboratory where syntheses for positron emission tomography are carried out. Aerosol samples were collected by drawing the air through a glass−fiber filter using a mobile high-volume air sampler. The filter discs were counted under an end-window GM tube (mica window 2 mg/cm^2 thickness and 35 mm diameter, background about 32 cpm). Following this very simple experimental procedure, subsequent measurements and a rather complicated computation using Bateman-type differential equations have to be carried out to distinguish between the radioactivity of the airborne natural radionuclides such as ^{218}Po, ^{214}Pb, ^{214}Bi, ^{212}Pb, ^{212}Bi, ^{208}Tl, and fluorine-18. Therefore, the method cannot provide instantaneous results, but very low activity concentrations, around 1 Bq/m^3, corresponding to 160 atoms/m^3, can be detected. This method can be applied also to the determination of any other airborne beta-emitting radionuclide, if its half-life differs sufficiently from those of the progenies of radon and thoron.

Radiostrontium. The beta counting of yttrium-90 after ingrowth to equilibrium with strontium-90 had been used during an extensive and remarkable investigation carried out by Russian and Norwegian scientists in the South Ural region near the site of the first weapon grade plutonium production reactor complex in Russia. GM-counting tubes had been used for the determination of beta particles, and Strand et al. (1999) reported that they found 720 kBq/kg of strontium-90 in sediments and 8−14 kBq/L in water.

Cosma (2000) carried out strontium-90 determinations in Romania without previous chemical separation procedures. He used aluminum plates to absorb low-energy beta particles and thereby detect only the high energy beta radiation of yttrium-90. He obtained values between 40 and 75 kBq/kg in sediments and soil in Romania after the Chernobyl accident.

Chu et al. (1998) compared three methods for the determination of radiostrontium, the nitric acid precipitation method, ion exchange, and crown ether separation procedures. They analyzed soil, tea leaves, rice, and milk powder. Their main statement is that by application of the crown ether method, the time-consuming and hazardous nitric acid precipitation method is avoided. Measurements were carried out using gas flow GM tubes and Cerenkov counters. Data are given for strontium-89 and strontium-90.

Currently, attempts are being made to analyze strontium-90 in situ in aqueous solutions without chemical processing and by applying semiconductor and GM detectors (Samardzic et al., 2018). In this study, a methodology to simulate the spectral information obtained by Monte Carlo methods is presented.

VII. Special types of ionization detectors

A. Neutron detectors

Practically, every type of neutron detector consists of a target material that is designed to produce charged particles by interaction with neutrons. Those charged particles can be detected by any suitable detector, such as an ionization detector. The nuclear interactions resulting in the production of charged particles are governed by the reaction cross section. This cross section depends strongly on the energy of the neutrons as described in Chapter 1, Volume 1. In searching for such suitable nuclear reactions, one has to consider that the cross section should be as large as possible, thus allowing the design of detectors with high efficiency and small dimensions.

The most popular nuclear interaction for the measurement of neutrons is the ^{10}B(n,α)^{7}Li reaction. It can be used for the measurement of slow (thermal) neutrons. The cross section decreases rapidly with increasing neutron energy as illustrated in Fig. 1.81 of Chapter 1, Volume 1. This reaction is very useful because of the large cross section for capture of thermal neutrons (3840 b, see Table 1.20 in Chapter 1, Volume 1) and because of the relatively high isotopic abundance of the boron isotope with mass number 10 (19.8%). Usually, boron trifluoride is used as an additive to the host gas in PC tubes.

The reaction ^{3}He(n,p)^{3}H has a significantly higher cross section for thermal neutrons, but the relatively high cost of ^{3}He has limited the application of this target material for proportional neutron counting tubes. Helium-3 is a rare isotope, and the increasing demand due to its unique nuclear properties resulted in a shortage in its supply, which has caused a veritable "helium-3 crisis" (Kouzes, 2011; Shea and Morgan, 2010). Helium-3 counters can be found for what is usually called a hostile environment, and applications are found in well logging investigations. Glesius and Kniss (1988) provide a review of such applications for borehole measurements. Recent developments seek to replace ^3He by boron-based detector materials, which is not always a simple task. Helium-3 has some considerable advantages, that is, it is relatively insensitive to gamma radiation, nontoxic, and noncorrosive; can be operated at lower voltage; and can withstand extreme environments. Boron trifluoride (BF_3), as the obvious alternative, is not suitable for all situations and environments. It is corrosive, requires much higher voltage, and is more susceptible to interferences with gamma radiation (Lintereur et al., 2011).

The use of helium-3 is still justified for special applications. For example, it was used in a low-background thermal neutron detector for the Chinese Jinping Underground Laboratory (CJPL-II) (Zeng et al., 2017). Helium-3 was necessary to achieve the desired low sensitivity to other radiation. The authors used oxygen-free copper as the material of the tube to guarantee a minimal contribution to the count rate by radioactivity from the wall materials.

For experiments with spallation neutron sources, detectors are required with 2D response, good time resolution, and capability for neutron energy determination. Radeka et al. (1998) built multiwire chambers up to 50×50 cm with helium-3 and propane as filling gas mixture for the construction of a large curved detector for protein crystallography studies at a pulsed spallation source at the Los Alamos National Laboratory. The detector is designed to be placed 16 m from the neutron creation point, and thus a single neutron pulse time would act as a monochromator for neutrons. A review of neutron detector developments around particle accelerators was published by Nakamura (2012).

A new design of a boron-based gaseous neutron detector based on a boron-lined honeycomb neutron converter and a GEM was presented (Fang et al., 2017a, 2017b). Anode wires used in traditional detector designs were removed, and a GEM was used as the electron multiplier (Fig. 2.33). A boron layer on the honeycomb surface reacts with incoming neutrons in the nuclear reaction $^{10}B(n,\alpha)^7Li$. The emitted alpha particle or lithium-7 nucleus, respectively, will ionize the working gas. The electrons will migrate along the z-axis and will be multiplied by the GEM and collected by the anode. Various mixtures of working gases have been investigated (Fang et al., 2017a), showing that $Ar{:}iC_4H_{10}{:}CF_4 = 90{:}7{:}3$, $Ar{:}CO_2 = 95{:}5$, and Ar:DME = 95:5 show the most promising results and good performance in terms of stopping power of alpha particles and lithium-7 nuclei and limiting transverse diffusion of electrons. The boron lining is performed by plating a mixture of boron nanoparticles and epoxy resin onto the surface of the converter. This detector design helps overcoming the intrinsic problem of solid boron-based detectors: aiming at the use of a large amount of boron while keeping the boron layer as thin as possible to allow for the entering of the alpha particle or lithium nucleus (both of which have a short range in solid boron layers), respectively, into the working gas, by providing a large surface area.

Other variants of GEM detectors have also been introduced for neutron measurements. Lopes et al. (1999) combine the principles of proportional scintillation counters and GEMs, and Fraga et al. (2002, 2003) and Rocarro et al. (2009) applied helium-3 as a filling gas in GEM detectors to determine neutrons as described previously in this chapter. Ohshita et al. (2010) used a 0.02 μm-thick layer of ^{10}B on the entrance window of a GEM detector for the measurement and imaging of neutrons beams. The present helium-3 crisis further fuels efforts toward "^3He-free" detector systems. Albani et al. (2016) proposed several detector systems aiming at the development of a ^3He-free instrument for neutron diffraction experiments. They used $^{10}B_4C$ as the neutron absorber and developed three GEM-based detectors. Two of them have a planar; the third one has a three-dimensional (3D) neutron converter. The authors (Croci et al., 2014) found that the planar detectors exhibit lower efficiency, timing resolution, and signal/background ratio than conventional helium-3-based detectors. By using a 3D design and enriched boron-10, the authors improved the detector's efficiency from 7% of the 2D design to 25%, relative to the count rate obtained by ^3He-based detector.

The $^6Li(n,\alpha)^3H$ reaction cannot be used for gas ionization counters because a lithium-containing gas for proportional counters is not available. However, ^6Li counting scintillators are quite common as detectors for neutrons as described in Chapter 9, Volume 1.

The cross sections of uranium-233, uranium-235, and plutonium-239 for fission reaction with thermal neutrons are very large, and the fission products that form the "charged particles" to be detected in a PC tube have very high kinetic energy (about 160 MeV). This facilitates discrimination from the alpha emission of the fissile materials that are neutron targets of the counting system. Little success was achieved in attempts to produce these neutron targets as a gaseous additive to the host gas of PC tubes. Commonly the surfaces of the electrodes are covered with a deposit of the fissile material. This system is frequently applied, for example, for fission chambers that are used for reactor as well as nonreactor applications.

As mentioned previously, the BF_3 proportional tube is the most widely used detector for slow neutrons. Boron trifluoride can serve both purposes, as a target for slow neutrons and also as a PC gas for the reaction products of the $^{10}B(n,\alpha)^7Li$ reaction. Although other boron-containing gases have been investigated, BF_3 offers good properties as a proportional gas and also a high boron content compared with other gaseous boron compounds. Usually, boron-10 is enriched for use in boron trifluoride counting tubes. Tubes with enriched boron-10 have about five times

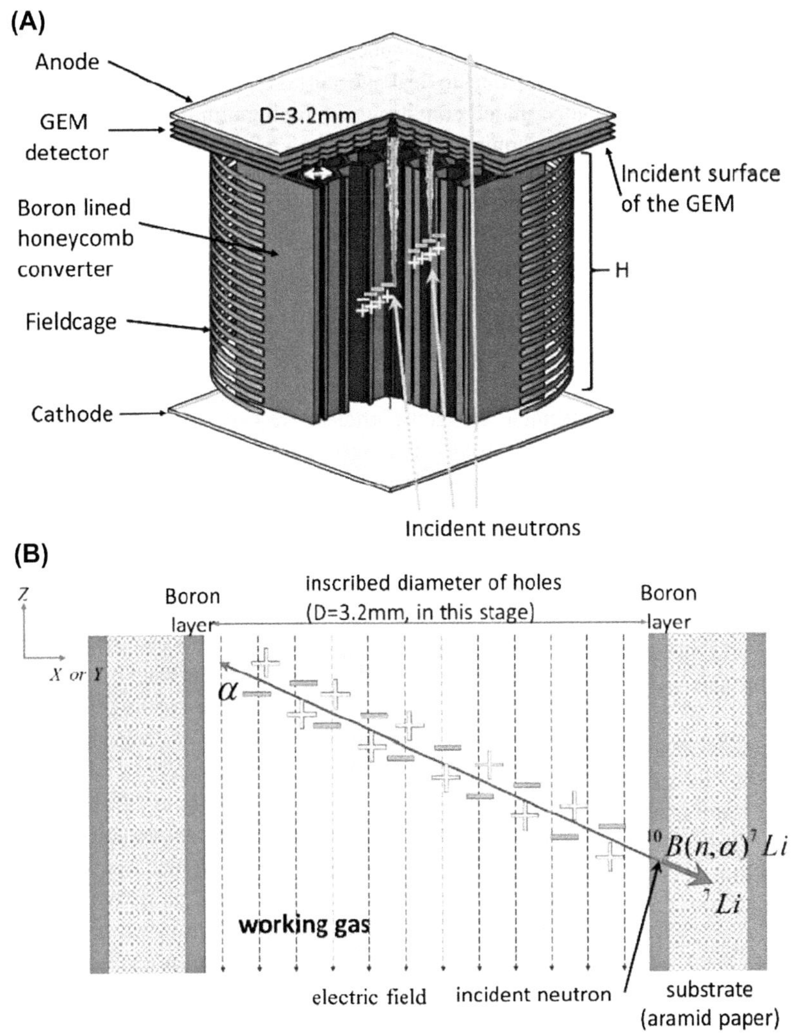

FIGURE 2.33 Physical structure (top) and detection principle (bottom) of the proposed honeycomb converter neutron detector. GEM, gas electron multiplier. For the color version of the figure, the reader is referred to the online version of the book. *From Fang et al. (2017a), reprinted with permission from IEEE © 2017.*

higher efficiency for thermal neutron counting than tubes filled with boron in its natural isotopic abundance.

According to the reaction $^{10}B(n,\alpha)^7Li$, the output signal handling may appear simple and straightforward for the application of boron-10 to the detection or even spectroscopy of neutrons. However, the energy spectrum and pulse processing for BF_3 tubes can be rather complicated in detail. Recoil ^7Li also contributes to the energy spectrum, and the nuclear reaction leads to either a ground state (94%) or excited state (6%) of ^7Li. Also, the volume of the counting tube, in general, is not sufficiently large compared with the range of the alpha particles or even the range of ^7Li recoil atoms. Therefore, not the entire energy of these reaction products is deposited in the gas volume, but interaction with the walls of the tube occurs instead. This results in distortion of the energy spectrum recorded from ionizing effects in the gaseous volume. Summarizing, one can say that the BF_3 tube is a detector from which, by differential pulse height analysis, little useful information is obtained about the energy spectrum of the incident radiation. The pulse height spectrum depends mainly on the size and shape of the detector. Therefore, counting is done only at a high voltage providing a flat region at a plateau, and a discriminator setting is used at which all neutrons are counted but all low-amplitude effects are rejected. Low-amplitude effects are due mainly to gamma rays producing secondary electrons from wall interactions. However, at very high gamma radiation fields, problems arise because of pileup effects. Also, BF_3 suffers from radiation decomposition in gamma fields of high intensity, thus making attempts necessary to absorb decomposition products of BF_3 by applying activated charcoal as an absorbent.

Position-sensitive neutron counters are essential for measuring the neutron flux distribution in critical assemblies. For that purpose, neutron counting tubes with unusual dimensions may be constructed, such as a 1.2-m-long and 8-mm-diameter tube designed by Uritani et al. (1995). With that device, nonuniformities in a critical assembly could be detected and correction measures are undertaken.

New structures of detectors have been designed, and these developments have also found their application in neutron counting. RPCs using highly neutron-absorbing gadolinium as converters have been constructed and investigated by Jamil et al. (2010), Qian et al. (2015), and Qian et al. (2009).

Although neutron detectors usually rely on the technical principles of proportional counters, modified GM counters can be applied for this purpose. Etaati et al. (2010), Lotfi et al. (2015), and others studied the performance of an activation counter. In this device, neutrons are moderated in a polyethylene packing and activate a silver foil that surrounds the GM tube. The activation products of silver (silver-108 and silver-110) disintegrate with short half-lives under emission of β^- particles, which can be detected by the counter. An alternative concept, where a silver-lined gas ionization neutron detector is operated in PC mode, was presented by Dighe and Das (2015). The authors claim that this concept shows enhanced performance in terms of operating range and dead time.

Moslehi and Raisali (2017, 2018) describe their development and evaluation of a multielement THGEM for use as a microdosimeter that can be applied from thermal to 14 MeV neutrons. The microdosimeter uses propane as a tissue-equivalent gas and has a small weight fraction of ^{3}He added.

1. BF_3 tube construction

With increasing dimension of BF_3 counting tubes, the detection efficiency is improved and wall effects are suppressed. To some extent, increasing the gas pressure inside the tube has the same effects. Some consideration has to be given to the materials used for tube construction to avoid radioactivation effects due to neutron capture by the materials used for tube wall construction. Because of its rather low neutron interaction, aluminum is frequently the material of choice; however, if a low background is essential, one has to keep in mind that aluminum contains a small amount of alpha-emitting nuclides. For such low-level investigations, stainless steel is preferred as a construction material for BF_3 tubes. Elevated temperature has some adverse effects on counting performance. Above 100–150°C pulse amplitude and pulse height resolutions are decreased, because of desorption of impurities from construction materials inside the tube. Extensive studies of the temperature dependence of BF_3 proportional counters were carried out by Sakamoto and Morioka (1994). Some phenomena that depend on temperature were related to impurities in the enclosed gas and also to construction details of the electrodes.

Usually, BF_3 tubes are operated at a rather higher voltage than ^{3}He tubes. Therefore, spurious pulses are possible due to leakage current through insulators, especially under conditions of high humidity. Also, detector microphonics have been observed, if the counting system is subject to shock or vibrations, which has been tackled by Czajkowski et al. (2012).

2. Fast neutron detectors

It has to be kept in mind that the gas ionization detectors previously described, namely BF_3 and ^{3}He detectors, which are based on the conversion of neutrons to directly detectable charged particles, are capable of detecting only slow neutrons. The cross section responsible for the $^{10}B(n,\alpha)^7Li$ and $^3He(n,p)^3H$ reactions decreases rapidly for neutrons with higher energies. To use these detectors for the determination of fast neutrons, the high-energy particles have to be slowed down, that is, to be moderated. The low detection efficiency for high-energy neutrons of slow neutron detectors can be greatly improved by surrounding the detector volume with a layer of moderating material, for example, hydrogen- and carbon-containing materials such as paraffin. Fast neutrons lose a fraction of their initial high kinetic energy by impacting the moderator molecules before reaching the sensitive volume of the detector. However, neutrons can escape from the moderator layer by scattering without reaching the detector volume or can be captured by moderator materials. Thus, an increase of the thickness of the moderator layer will not proportionally increase the number of thermalized neutrons counted by the detector. A maximum counting efficiency will be observed at a specific moderator thickness. This optimal thickness depends on the initial energy of the fast neutrons to be detected and varies from a few centimeters for neutrons with energies of keV up to several tens of centimeters for neutrons having energies in the MeV range.

There is no generally established method for neutron spectrometry, especially around the eV region. The "slowing-down time" method can be applied for such investigations, and conventional BF_3 tubes are used (Maekawa and Oyama, 1995a, 1995b; 1997).

Toyokawa et al. (1995) described a multipurpose neutron counter, applicable to the measurement of fluence, energy distribution, and radiation dose equivalent. This system consists of a spherical polyethylene moderator and three ^{3}He position-sensitive tubes inserted into the moderator orthogonally to each other. These three position-sensitive tubes provide information about the thermal neutron distribution in the spherical moderator, and from that information, the foregoing parameters can be evaluated.

For neutron spectrometry in the MeV range, ^{3}He ionization chambers can be used. Iguchi et al. (1994) carried out investigations dealing with the application of these detectors in neutron spectrometry. Their ^{3}He detector consists of a cylindrical gridded ionization chamber (Fig. 2.34). Monte Carlo simulation was applied to estimate the detector response. Four kinds of reactions in the detector gas were considered in the calculations: ^{3}He(n,p)t, ^{3}He(n,d)d, ^{3}He(n,n'), and ^{1}H(n,n') elastic scattering. Corresponding to these calculations, the response functions were measured with monoenergetic neutrons at various energy points.

An indirect attempt to achieve neutron spectroscopy can be approached by taking advantage of neutron-induced nuclear reactions of target neutrons with fast neutrons. For example, the ^{12}C(n,2n)^{11}C reaction has a threshold of 20 MeV. Bismuth detectors follow the reaction of ^{209}Bi(n,xn)$^{210-x}$Bi (x = 3−12), which are suitable for the energy range 20−150 MeV (Nakamura, 2012). The resulting radionuclide can be detected by gamma spectrometry, giving a hint at the neutron energies involved in a high-energy neutron source.

More recently, the prompt fission neutron spectrum of a californium-252 spontaneous fission neutron source was determined by EJ-301 liquid scintillators as well as an EJ-204 plastic scintillator (Blain et al., 2017). The plastic scintillator was used for the lower energy portion of the neutrons between 50 keV and 2 MeV, whereas the liquid scintillator was used for the higher energy neutron from 0.5 to 7 MeV. For more details on the principle of scintillators, see Chapter 6, Volume 1.

Pulse height and rise time distribution analysis of signals from neutron proportional counters were used to reject undesirable signals of hydrogen-filled proton recoil counters, ^{3}He-filled counters, and BF_3 counters. Gamma ray background and wall effect pulses can be reduced by that method (Sakamoto and Morioka, 1993).

Neutron measurements in an environment with high gamma radiation doses are of interest in the field of nuclear safeguards. Especially, neutron−gamma coincidence counting is of particular interest for spent-fuel measurements for burnup verification and in several steps of nuclear fuel reprocessing. The high gamma background has limited the selection of neutron detectors. Neutron fission chambers do not possess sufficient efficiency to be used in coincidence counting, and BF_3 tubes suffer from radiation damage. Beddingfield et al. (2000) have carried out comprehensive research to optimize the helium-3 neutron proportional counter performance in a high gamma ray dose environment. There are many parameters to be observed, such as tube size, gas pressure, gamma ray dose, gamma ray pileup, gamma ray energy, radiation damage to the gas mixture and to the preamplifier system, etc. There is no best option of counting tube design for all mixed field applications; however, from the presented amount of experimental data, a good choice for a useful special design can be made.

An inexpensive sealed ^{3}He drift-tube neutron detector was designed by Wang et al. (2009) at the Los Alamos National Laboratory. In light of the limited supply and high cost of ^{3}He, they developed a low-pressure ^{3}He drift-tube neutron detector that takes advantage of the fraction of thermal neutrons scattered off a polyethylene moderator surface. The basic design of the drift-tube neutron detector is illustrated in Fig. 2.35. Wang et al. (2009) surrounded the tube with approximately 25-mm-thick high-density polyethylene sheet moderator. With a detector diameter of 5 cm and a specific gas mixture and pressure, they could ensure that the full energy of the reaction ^{3}He(n,p)^{3}H would be absorbed by gas ionization, which results in a prominent *Q*-value peak at 0.764 MeV in the pulse height spectra and a detection efficiency of 5%. The detector was filled with

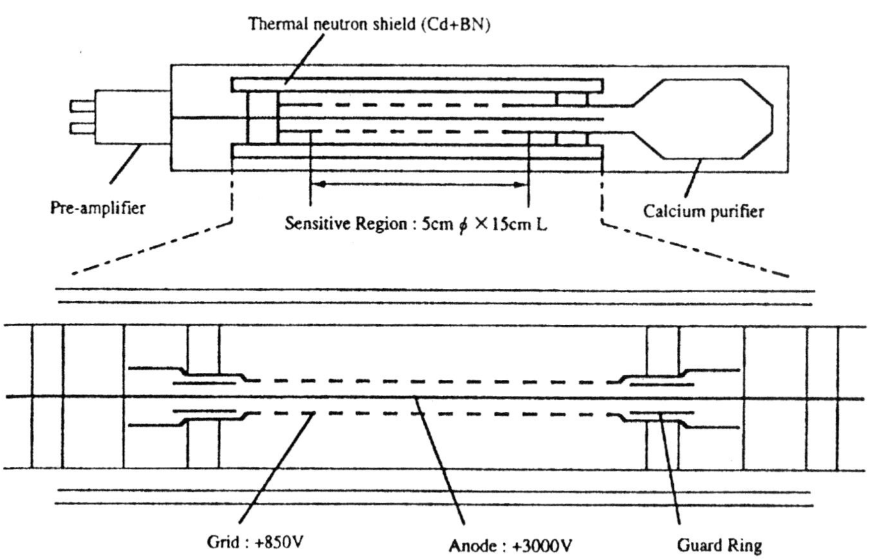

FIGURE 2.34 Schematic view of an ^{3}He gas ionization chamber. The detector consists of a cylindrical gridded ionization chamber. The sensitive volume is fixed to 5 cm in diameter and 15 cm in length by guard rings at both ends. The chamber is filled with helium-3, argon, and methane at a pressure of several bars. A calcium purifier in the chamber is used to remove hydrogen produced from the ^{3}He(n, p)^{3}H reaction from the detector gas. Thermal neutrons are shielded by a boron layer outside the tube. *From Iguchiet al. (1994), reprinted with permission from Elsevier © 1994.*

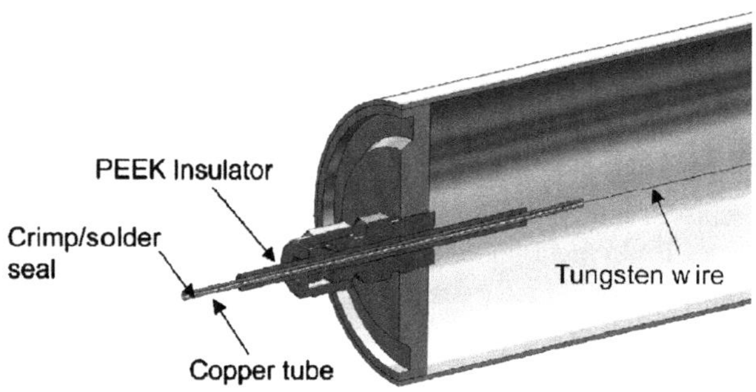

FIGURE 2.35 Schematic view of the drift-tube neutron detector construction. *From Wang et al. (2009), reprinted with permission from Elsevier © 2009.*

1 bar of total gas pressure with 25–300 mbar of ^3He with the remaining gas mixture consisting of C_2H_6, ^4He, CF_4, and Ar. Due to the relatively low pressure of ^3He, the cost of the ^3He is not more than the cost of the other materials used to construct the detector. The wall of the detector consists of 0.89-mm-thick aluminum, and the tube length can be varied from 30 cm up to 6.1 m in length. The detector length was kept at a minimum of 30 cm (12 in.) to prevent domination of the detector by end effects. The anode wire consisted of 50-μm-diameter gold-plated tungsten. The drift tube operates as a proportional counter at bias voltages ranging from 1400 V (300 mbar ^3He) to 1800 V (25 mbar ^3He). The detector was designed for fissile material detection, and the simple design permits the construction of 100 detectors by a two-person team in 1 work day of 8 h.

a. Long counter

Most neutron detector systems suffer from the disadvantage that the counting efficiency depends strongly on the energy of the neutrons to be detected. The so-called long counters try to avoid that disadvantage. A long counter consists of a neutron detector tube, most frequently a BF_3 tube, which is placed in the central region of a paraffin cylinder. The paraffin cylinder is covered with a layer of B_2O_3 and with an additional layer of paraffin. Only one end of the inner paraffin cylinder is not covered by the boron and additional paraffin. Thus, the device is sensitive only to neutrons coming from the direction of this end. Any neutron arriving from that direction is moderated and has a good chance of arriving at the central BF_3 tube. To give low-energy neutrons a better chance of reaching the tube, holes are drilled in the front end of the inner paraffin layer (Hunt and Mercer, 1978). Such a long counter is in use for the precise measurement of delayed neutrons in the French LOHENGRIN fission fragment separator (Mathieu et al., 2012; Serot et al., 2014) or for the study of the spectrum of cosmic ray–induced neutrons (Pazianotto et al., 2013).

Because of the nearly energy-independent response of this type of counting tube, the long counter is also called a "flat response" detector. Many variations of such flat response detectors have been designed and constructed, some of them using ^3He tubes, pressurized filling gas, multiple tube arrangements, polyethylene moderators, and so on (Tagziria and Thomas, 2000; Hu et al., 2009; Lacoste, 2010; Lacoste and Gressier, 2010; Roberts et al., 2010; Watanabe et al., 2011; Wang et al., 2015; Utsunomiya et al., 2017). One has to be aware that the counting efficiency of such neutron counting systems is rather low, sometimes much less than 1% (East and Walton, 1982, see Fig. 2.36). For high-energy neutrons (>6 MeV), the response was improved by introducing a tungsten irradiator, which acts as an energy converter through the (n,xn) reaction (Li et al., 2015).

As noted above, many variations of long counters may be used, and two examples of the most simple design and more complex design together with their response to a wide spectrum of neutron energies are illustrated in Figs. 2.36 and 2.37, respectively. The flat response neutron detector geometry illustrated in Fig. 2.36 is quite simple. The detector consists of a 20-cm-long ^3He position-sensitive proportional counter (PSPC) surrounded with polyethylene moderator. Polyethylene is often preferred over paraffin as moderator because it can be machined easily while also having a high content of hydrogen atoms. The neutrons enter from the left surface of the detector illustrated in Fig. 2.36A. The peak positions in the spatial response of the 20-cm-long ^3He PSPC counter shift to deeper positions from the detector surface with increasing neutron energy, as illustrated in the lower part of Fig. 2.36B.

A more complex design of a long counter, described by Lacoste and Gressier (2010) for the IRSN AMANDE facility in France, is illustrated in Fig. 2.37. It was modeled based on studies on the role and variation of different parameters, such as depth of holes in the front face and the position of the thermal neutron detector in the moderator. The long counter (PLC-IRSN), modeled with the Monte Carlo radiation transport code MCNPX, is illustrated in the upper portion of Fig. 2.37, and the response function of the counter over a wide range of neutron energies is illustrated in the lower portion of the figure. The detector measures approximately 40 cm in diameter and 45 cm in length. It is made of polyethylene and borated polyethylene, with a thin cadmium sheet in the front and at the rear. The front of the inner polyethylene layer is designed with an annulus of air and is 5 cm in depth and 3 cm wide. A polyethylene cylinder 6 cm in diameter and 7 cm long is placed in front of the thermal neutron detector. The ^3He gas was confined at a pressure of 10 atm. in an active volume 3.8 cm in diameter and 26.3 cm in length. The uniform response of the long counter to a wide range of neutron energies is illustrated in the lower portion of Fig. 2.37.

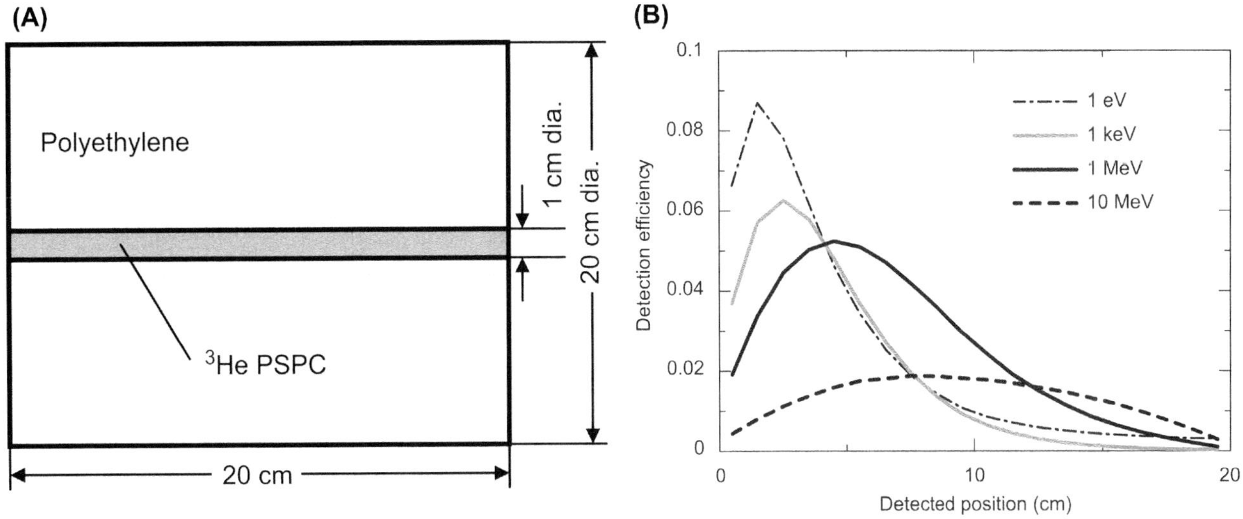

FIGURE 2.36 (A) Basic configuration of a flat response neutron detector. The neutrons enter the detector from the left side of the figure (B) Calculated spectral response distribution (i.e., number of detected neutrons per incident neutron) of neutron detection position in the ^{3}He position-sensitive proportional counter (PSPC) long counter for various incident neutron energies. *From Watanabe et al. (2011), reprinted with permission from Elsevier © 2011.*

Long counters are typically bulky because of the voluminous neutron moderator, which is required to achieve a flat detector response. This reduces portability and limits its use to a stationary instrument. A novel design of a compact flat response neutron detector was presented by Harano et al. (2011). Their detector design is based on two small spherical ^{3}He proportional counters (3.3 cm in diameter) with a reduced moderator volume, consisting of a polyethylene cylinder 20 cm in diameter and 35 cm in length (Fig. 2.38). The detector has shown (in MCNPX calculations) to yield a comparable sensitivity to long counters for californium-252 spontaneous fission and americium–beryllium (α,n) neutrons as well as monoenergetic neutrons, up to an energy range of 20 MeV, while weighing only 10.6 kg, which is 20% of conventional long counters.

3. Neutron counting in nuclear analysis of fissile materials and radioactive waste

Neutron counting tubes are also used in delayed neutron activation analysis. Some radionuclides follow a beta decay process; their beta emission is followed immediately by the emission of neutrons. Such nuclides are produced by fission of heavy elements. Therefore, this procedure can be used for the analysis of fissile materials. Using thermal neutrons for fission is a specific method for the determination of uranium-235 and plutonium-239. With fast neutrons, fission also occurs with uranium-238 and thorium-232.

Oxygen and calcium are interfering elements, creating nitrogen-17 and potassium-48 as activation products through fast neutron irradiation. However, since those radionuclides have short half-lives (nitrogen-17, 4.2 second; potassium-48, 5.8 second) compared with the neutron-emitting products from uranium and thorium, the interference can be avoided by counting after a decay period of at least 20 seconds.

Delayed neutron activation analysis is carried out using a pneumatic transfer system at a neutron source of sufficient flux density, usually a reactor. The samples are first positioned near the reactor core by the transfer system, and after a suitable irradiation period ($\sim$60 seconds) and decay period ($\sim$20 second), samples are counted ($\sim$60 second) at a neutron detector assembly. At the NIST research reactor at Gaithersburg, United States, the detection limit for natural uranium is in the range of 20 ng natural U, which corresponds to ca. 200 pg of ^{235}U (Eriksson et al., 2013). Kapsimalis et al. (2013) studied the simultaneous determination of both fissile nuclides uranium-235 and plutonium-239, which is possible by studying the different neutron emission spectra of both nuclides. The analysis, according to the authors, was successful within a precision of ca. 5%–10%.

Thorium interference due to fast neutron-induced fission is overcome by irradiation with and without cadmium shielding. Cadmium-113 has a large cross section for the capture of thermal (only) neutrons, whereas epithermal and fast neutrons will penetrate the cadmium shield. The delayed neutron activation analysis is used mainly for the determination of uranium and thorium at trace levels in minerals and for forensic and nonproliferation studies. Fully automatic systems are available, with detection limits in the range of 0.01 μg/g for uranium and 1 μg/g for thorium.

Neutron counters have been applied also to the determination of transuranium elements. A high-sensitivity

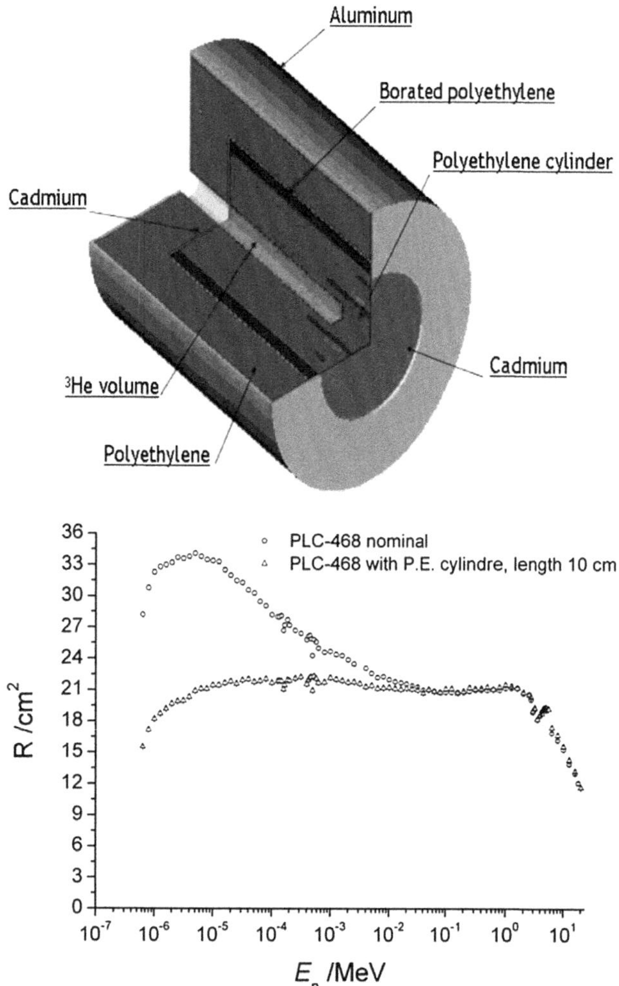

FIGURE 2.37 (Upper) PLC-468 long counter modeled with MCNPX. (Lower) Fluence response energy distributions of the PLC-468 long counter with and without a 10-cm-long cylinder of polyethylene inserted in front of the detector. The additional polyethylene cylinder decreases the sensitivity of the detector to the low-energy neutrons without affecting the detector response in the higher energy range. For the color version of the figure, the reader is referred to the online version of the book. *From Lacoste (2010), reprinted with permission from Elsevier © 2010.*

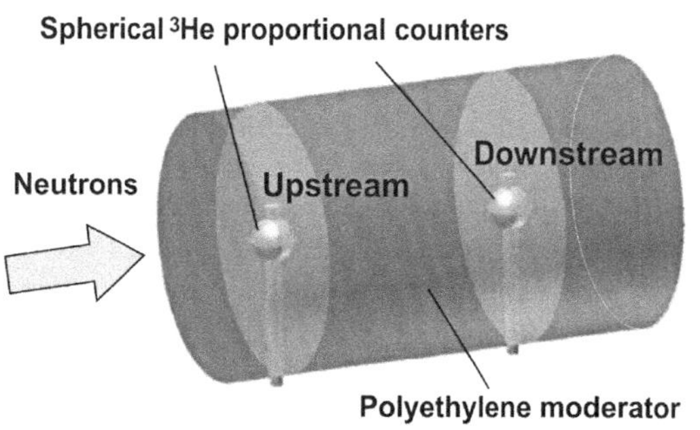

FIGURE 2.38 Structure of the compact flatresponse neutron detector (Harano et al., 2011). *Reprinted with permission from © 2011 IEEE.*

neutron counting tube arrangement was used successfully for the determination of plutonium in radioactive waste drums at Lawrence Livermore National Laboratory (Hankins and Thorngate, 1993a, 1993b). It was reported that the sensitivity of this equipment is about 10 times better than the sensitivity of X-ray and gamma-ray instruments that are normally used. Helium-3 counting tubes are arranged outside the waste package. These ^{3}He counters are covered with paraffin with an outside lining of cadmium. Fission neutrons passing the cadmium barrier are thermalized in the paraffin layer and detected by the ^{3}He tubes.

Another system uses a pulsed electron beam from a linear accelerator to produce high-energy photon bursts from a metallic converter. The photons induce fission in transuranium elements. When fission is induced in such material, delayed neutrons can be detected by a sensitive neutron counting system (Lyoussi et al., 1996).

A nanoparticle-doped micro-Geiger counter was designed to enable multispecies radiation detection, which is important for the detection of fissionable bomb-making materials (Whitney et al., 2009). In this forensic application, incoming radiation interacts with tailored nanoparticles, which release secondary charged particles into the fill gas. An array of multichanneled micro-GM tubes is formed to a counter platform. Each tube is doped with a nanoparticle from which the authors underscore the possibility of discriminating between alpha, beta, gamma/X-ray, and neutron radiation. Tungsten oxide nanoparticles convert beta particles to several electrons of lower energy; glass nanoparticles convert alphas to electrons; boron-10 converts neutrons to alphas; and lead oxide nanoparticles convert gammas/X-rays to photoelectrons. By analyzing the height of the resulting pulses of each individual channel under the same radioactive source, the type, location, and energy of the incoming radiation can be determined (Whitney et al., 2009).

Not only transuranium elements are determined in waste using neutron counting, but moisture measurements of the radioactive waste are also carried out. The thermalization of neutrons from an isotopic neutron source is detected by a proportional neutron counting tube. The moisture content of the waste is an important parameter that determines the combustibility of waste materials (Lentsch et al., 1996).

4. Moisture measurements

Moisture measurements are based on the principle of neutron moderation by hydrogen atoms. While various other methods have become available, the neutron-based moisture measurement in soil (neutron probe) is still regarded as one of the most accurate methods for the determination of moisture in soil (Fred Zhang, 2017). The method can even be used to analyze (significant) contaminations with hydrocarbons (in the percent range) in soil, provided a correction for moisture is being conducted (Naqvi et al., 2018).

In neutron water gauges, neutrons are most frequently produced through (α,n) reactions, for example, $^{9}Be(\alpha,n)^{12}C$. These neutrons have a spectrum of energies from 0 to about 10 MeV. The neutrons of high energies are moderated (slowed down) by elastic scattering with hydrogen atoms from water. These slow neutrons are detected by a counting device, which is only sensitive for slow neutrons, for example, a BF_3 counter. Modeling of the interactions of slow neutrons with different media, such as soils, is not easy, and therefore, a calibration is needed to convert the slow neutron counts to water content. Usually, a neutron moisture meter (NMM) device is combined with a density gauge.

Neutron moisture gauges are well-proven devices for soil water determination, or other difficult, bulky matrices such as asphalt (Huang et al., 2013). Tolk and Evett (2009) compared the performance of an NMM and a lysimeter with respect to evapotranspiration determination of soil. They found a deviation of not more than 4% when comparing both methods. Although the NMM has served the need for this purpose, increasing regulatory burdens, including the requirement that the NMM not be left unattended, limit the usefulness of the method. Newer methods, which respond to soil electromagnetic (EM) properties, typically allow data logging and unattended operation, but with uncertain precision, accuracy, and volume of sensitivity. One major advantage of the NMM over EM devices is its insensitivity to temperature (Evett et al., 2006). In a later study on the same problem, Evett et al. (2009) conclude that the EM methods poorly reproduced the spatial and temporal behavior of NMM and gravimetric sampling and implied spatial variability of profile water content that was not evident in either the NMM or gravimetric data. The authors conclude that EM sensors were influenced not only by the mean water content in the sampling volume but also by the smaller-scale structure of soil electrical properties.

The theory and practice of measuring the water content in large volumes of material by neutron thermalization and the measurement of thermal neutrons with BF_3 or ^{3}He detectors are reviewed by Nielsen and Cassel (1984) and Bacchi et al. (2002).

B. Multiple sample reading systems

In radioassay methods in biochemistry and medicine, a high sample number throughput is frequently essential. Radioactivity quantitation on solid supports, and radioimmune, dot blot, cell proliferation, and receptor binding assays require systems for counting a rather high number of samples in a given time. Simultaneous counting methods for a large number of samples are desirable. For these applications, multiple sample reading systems have been designed. Bateman (1994) has constructed a multipin detector. The pins are centered in holes in a metallic collimator system, and 60 beta-sensitive positions are obtained.

A system with 96 individual detectors working in the GM region has also been manufactured (Roessler et al., 1993; Hillman et al., 1993a). A high sample throughput is achieved, and the counting procedure is about 40 times faster than single-detector assay procedures. Of course, the counting efficiency for tritium is much lower than that achieved with liquid scintillation counting, but the background is reduced because the GM detectors are very small.

Roessler et al. (1993) compared several methods for receptor binding assays and compared the sample throughputs. Hillman et al. (1993b) applied the 96-sample measurement system for chromium-51 retention assays. Several other application examples can be found in the literature (Alteri, 1992; Hutchins, 1992).

Microplate assays related to investigations using radioactive tracers have attracted great interest during the past decade. For microplate assays, radioactivity has to be measured from samples on a solid support that may hold 96 samples in an area of 8 × 12 cm. Cells or tissues are incubated in the presence of a radiolabeled substrate simultaneously in all positions of such a microplate. After the incubation, the nonincorporated components must be separated from the incorporated radioactive substrate at each position of the microplate. Applying conventional techniques, this was usually done by filtration and washing one sample after the other. The radioactive residue on the filters was counted using liquid scintillation counting (LSC) techniques. All this was a rather time-consuming and expensive procedure. Great progress was achieved by developing a sample harvester that could harvest and wash 96 samples simultaneously, saving a great deal of time and work. This harvesting and washing procedure can be performed by a specially constructed fully automatic cell harvester from conventional microplates, or special filter bottom foils for these plates can be used. The application of a radioactivity reading system that can analyze 96 samples simultaneously greatly simplifies the microplate radioassay techniques.

Two different types of microplates can be chosen, those with and those without a removable bottom. The removable bottom consists of a membrane filter material that can be easily stripped from the bottom of the microplate.

These solid support samples can now be measured using either liquid scintillation counting (standard LSC or multidetector LSC) or ionization detector techniques, such as proportional ionization detector counting, PSPC scanning, or multidetector avalanche gas ionization detector quantitation. See Chapters 6 and 9 of Volume 1 for a detailed description of scintillation analysis in the microplate sample format.

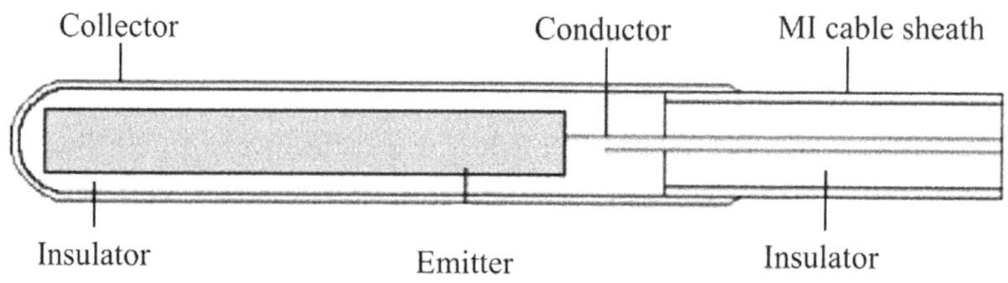

FIGURE 2.39 Schematic diagram of a self-powered neutron detector. *From Alex and Ghodgaonkar (2007), reprinted with permission from Elsevier © 2007.*

For PSPC scanning, systems similar to those used for scanning thin-layer chromatograms (TLCs) or paper chromatograms are used. With a position-sensitive wire detector, 12 samples in a single row can be counted simultaneously. This method suffers from some disadvantages. This type of detector has a very low counting efficiency for low-energy beta emitters, and it is subject to high amounts of cross talk when high-energy beta-emitting radionuclides such as phosphorus-32 are analyzed. Also, the efficiency is not uniform across the entire length of the wire. It seems that this technique is rather unsuitable for quantitative simultaneous multicounting applications. Therefore, systems with individual detectors in the format of the microplate were designed and manufactured. Open-end gas avalanche detectors are used, and the systems are capable of quantitating tritium, carbon-14, phophorus-32, sulfur-35, iodine-125, and many other beta emitters. Of course, the filter mat must be dry, but there is no addition of cocktail. The filter is not destroyed and can be used for further investigations. Also, the amount of waste is minimized.

A detailed description and examples of applications are given by Kessler (1991). This technique can be applied to the radioassay of dot blots and labeled cell proliferation assays. With conventional autoradiography and densitometry, the range of radioactivity measurements is much smaller than with a multidetector system, because an X-ray film shows a saturation effect in blackening. Also, the exposure time for X-ray films is much longer than the measurement time for ionization detectors. Other more quantitative imaging methods are described in Chapter 9, Volume 2. The ionization multidetector arrangement seems to be comparable to a liquid scintillation multidetector system (Kessler, 1991). However, commercially available high-sample-throughput multidetector microplate scintillation analyzers described in Chapters 6 and 9, Volume 1 provide higher counting efficiencies and higher sample throughput.

C. Self-powered detectors

Self-powered neutron detectors are fabricated with a material incorporated in the detector volume that has a high cross section for neutrons, for example, rhodium (Zhang et al., 2018), vanadium, cobalt, or silver (Khoshahval et al., 2018; Kong et al., 2017). By neutron capture, a beta-emitting radionuclide is formed. The detector operates by directly measuring the flow of current produced by the beta particles. No external bias voltage is needed. Other types of self-powered detectors (SPDs) for neutron counting are operated by the current that is produced by ionization due to gamma emission related to neutron capture during fission. The main advantages of these self-powered neutron detectors are their small size and the simple electronics necessary for this type of detector (Alex et al., 2004; Miller et al., 2004; Alex and Ghodgoankar, 2007).

The basic design of the SPD is illustrated in Fig. 2.39. The detector (emitter, Fig. 2.39) measures only 2 mm in diameter and 21 cm in length. The emitter is housed in a high purity alumina ceramic tube (2.2 mm ID × 2.8 mm OD) and enclosed in a 3 mm ID and 3.5 mm OD Inconel 600 tube. One end of the emitter is coupled to a 2-mm-diameter twin core mineral insulated cable with a length up to 12 m. The long cable permitted in-core neutron monitoring. The small size of the self-powered neutron detector allows for applications in nuclear medicine such as the implanting of the detector into patients for the online measurement of neutron flux during neutron capture therapy (Miller et al., 2004). Examples of some emitters, which can be used in SPDs, their neutron and/or gamma response, and applications, are listed in Table 2.1. From the table, it can be seen that not all SPDs are responsive to neutrons. For example, bismuth ($Z = 83$) has a neutron absorption cross section of 0.014 b exhibiting a negligible response to neutron flux, and it is essentially a pure gamma detector (Alex et al., 2004). Other emitters such as Pt yield nearly equal current contributions from neutrons and gamma rays.

Disadvantages are the low levels of the output signals, a slow response time, and sensitivity of the response to the neutron spectrum. SPDs have to be operated in the current mode, because the signal created by a neutron can be only a single electron.

D. Self-quenched streamer

Traditionally, gas ionization detectors are categorized as ion chambers, proportional counters, and GM tubes. A high-current gas-amplification mode in cylindrical detectors is considered as a time sequence of three amplification modes: a proportional mode, a mode of limited

TABLE 2.1 Response characteristics of self-powered detectors[a].

Emitter	Neutron response	Gamma response (γ, e)	Application
$^{103}_{45}$Rh	(n, β) (delayed)	Positive current	Flux mapping
$^{51}_{23}$V	Positive current	Negative current	
$^{59}_{27}$Co	(n, γ, e) (prompt)	Negative current	Reactor control
$^{195}_{78}$Pt	Positive current	Positive current	
$^{209}_{83}$Bi		Positive current	Gamma flux

[a]*From Alex et al. (2004), reprinted with permission from Elsevier.*

proportionality, and an SQS (or limited streamer) mode (Akopdzhanov, 2008). The latter is a type of gas multiplication detector that is used in position-sensitive multiwire detector systems. In conventional proportional and GM counters, UV photons play a significant role in the propagation of an ion avalanche. If the propagation of the avalanche is kept small by the field strength or by absorption of UV, the system works in the proportional mode. If UV photons are able to create additional avalanches that may spread through the entire length of the anode wire and the whole process is terminated only by the creation of a space charge around the anode, the system works in the GM mode and the output signal does not depend on the original ionization effect (e.g., on the number of primary ions produced by the radiation event). In the SQS mode, the ion avalanches are controlled in a special way. The counting tube is filled with a gas mixture that absorbs UV photons. Therefore, no additional avalanches far from the original avalanche pathway can be created through excitation by photon absorption. Avalanches, therefore, grow and propagate in the shape of a streamer. The streamers have a diameter of about 200 μm and extend a few millimeters from the anode. They terminate at low field strength at larger radii of the detector.

If the voltage is high enough, a single electron can create a streamer. The streamers have a final length that depends on the voltage applied. The formation of such streamers is supported by anode wires with relatively large diameters (0.1 mm).

SQS detectors have some properties of both proportional and GM detectors. The rather high internal gas amplification is useful for position-sensitive detectors. Position-sensitive detectors operating in the proportional region have much smaller signal amplitudes. However, as in the case of GM tubes, the signal amplitudes in SQS detectors no longer provide information about the energy of the primary radiation event. Spreading of the avalanche along the total length of the anode wire is prevented. Thus, the resolving time is much shorter than with GM tubes (Knoll, 1989).

E. Long-range alpha detectors

Traditional alpha detectors suffer from limitations related to the very short range of alpha particles in air. If sample and detector together are kept in a vacuum or sample and detector are operated in close proximity, reasonable efficiency is achieved. An alpha particle produces about 30,000 ions per 1 MeV of its particle energy (see Chapter 1, Volume 1). These ions can be transported over significant distances by a moving stream of air to a detector. For that purpose, a current of air can be generated by a small fan and the ions can be transported over a distance of several meters. The current of air is finally monitored by an ion chamber (Garner et al., 1994). By using air as the detector gas, alpha contamination on any complicated surface can be measured (MacArthur et al., 1992, 1993; Allander et al., 1994; Vu et al.,1994). Fig. 2.40 illustrates the principle of a long-range alpha detector (LRAD). It is shown that the detector is sensitive to the ionized air molecules produced by the passage of an alpha particle rather than to the alpha particle itself. The detector consists mainly of two grids (see Fig. 2.41) across which an electric field is applied. One type of ion is attracted by the high-voltage (HV) grid, the other by the sense grid. Both possible polarities for the grids have

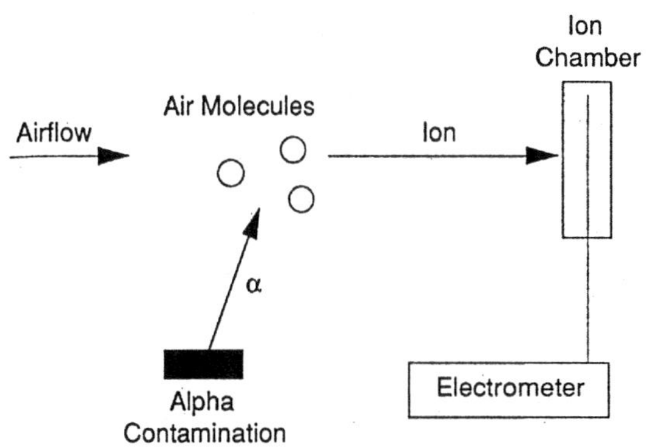

FIGURE 2.40 Principle of a long-range alpha detector operation. Ions created by alpha particles are transported to the detector by air flow. *From MacArthur et al. (1992). With permission © 1992 IEEE.*

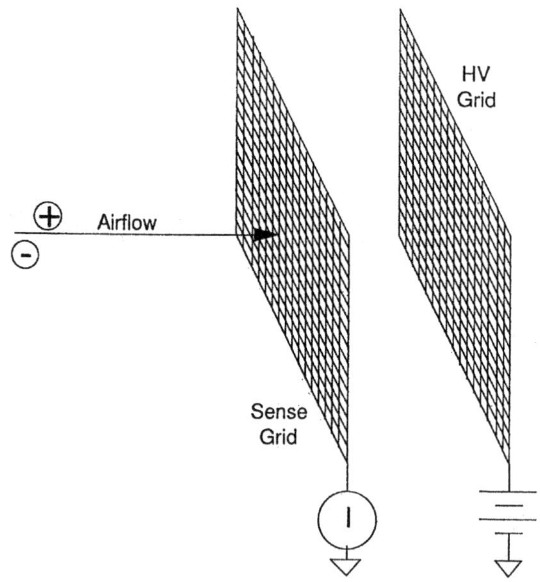

FIGURE 2.41 Construction detail of a long-range alpha detector. An electric field is applied across the grids. One type of ion is attracted by the high-voltage (HV) grid, the other one by the sense grid. *From After MacArthur et al. (1992). With permission © 1992 IEEE.*

been applied with equal sensitivity. The charge collected at the sense grid is measured by a suitable electronic circuit and used to determine the ionization, the number of alpha particles.

Instead of using air flow, the ions produced by alpha particles can be transported to the detector by an electrostatic field. At the Los Alamos National Laboratory, LRADs have been built for several applications, such as monitoring of soil surface and liquid effluents. A hand monitor has also been constructed. The method was applied to radon measurements (Bolton, 1994). Some effort has been made to use the long-range detectors for the measurement of beta contamination (Johnson et al.,1994a).

Real-time alpha activity monitoring is one of the applications for which the ionization detectors show several advantages. A monitoring system for real-time alpha monitoring was developed at Los Alamos and tested at the Radioactive Liquid Waste Treatment Facility as a means for real-time monitoring of liquid waste influent (Whitley et al., 1996). This system determines the alpha activity of the wastewater by measuring the ionization of ambient air above the surface at a rather long distance. The distance to the surface of the liquid described by Whitley et al. (1996) was about 4 in. Sometimes this type of design causes problems because of changing levels of the surface to be monitored, for example, with liquids.

The ionization counting system consists of a metal enclosure and a signal plate that is maintained at 300 V DC. The box is maintained at ground potential. A highly sensitive electronic circuit is used to detect changes in current to the plate. Changes in alpha activity in the contaminated liquid at the 370 Bq/L level could be detected.

The authors claim that this kind of measurement equipment can be useful for monitoring low-level liquid streams before discharge into the environment. However, a more sensitive design will be necessary to ensure regulatory compliance and offer the opportunity for field application.

LRADs are useful when a contamination is suspected in areas that are not easily accessible. Wu et al. (2013) showed that this measurement principle can be useful in the decommissioning of nuclear facilities, when the inside of a pipe needs to be checked for α-emitters. In their detection system, the screened pipe is placed inside a chamber, and a fan sucks air through the pipe (see Fig. 2.42). The authors showed that the time between switching on the fan and the detection of ions in the ion chamber is a decent indicator of the position of the contamination inside the pipe.

Ionization monitoring by LRAD also has a great advantage in simultaneous measurement of the entire body of a person. Air currents that are created by high capacity fans can be drawn from all surfaces of the body of a person who is positioned in a monitoring chamber (size 91 cm × 91 cm × 213 cm). High air flow rate is essential because of the rather short "ion lifetime."Koster et al. (1998) describe a contamination measurement facility for alpha particle radionuclides, which is used at the Los Alamos National Laboratory for test experiments.

In practical application, LRADs have been shown useful in monitoring swipe tests for rapid contamination control in the field (Johnson et al., 1994b). In their design, two electrometers are used to instantly subtract background radiation from radon buildup inside the chamber. Another example of this application was an LRAD-based portable swipe monitor, presented by Whitley et al., 1998. This

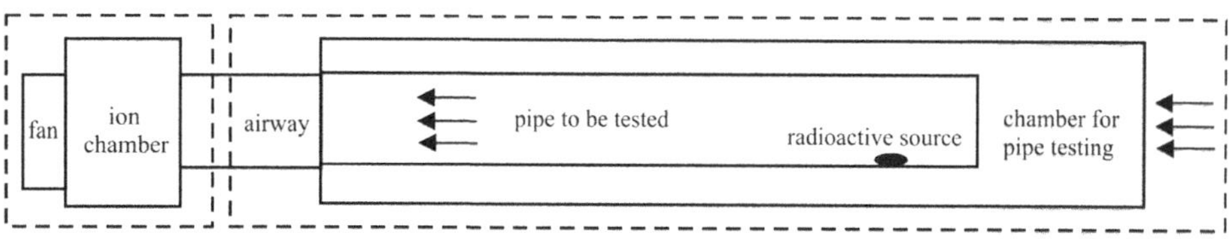

FIGURE 2.42 Setup of a long-range alpha detector that is used for the screening of contaminations inside a pipe. *From Wu et al. (2013). With permission © 2013. Chinese Physical Society and the Institute of High Energy Physics of the Chinese Academy of Sciences and the Institute of Modern Physics of the Chinese Academy of Sciences and IOP Publishing Ltd.*

TABLE 2.2 Comparison of instruments for the detection of alpha radiation (Gurkovskiy et al., 2015).

Parameter	Device proposed by Gurkovskiy et al. (2015)	Long-range alpha detector	Remote optical detection	Silicon ion–implanted alpha detector	ZnS alpha scintillation detector	Thin-window proportional counter	Thin-window Geiger–Müller counter
Principle of detection	Remote	Remote	Remote	Direct	Direct	Direct	Direct
Distance	40 cm and more	40 cm and more	40 cm[a]	<4 cm	<4 cm	<4 cm	<4 cm
Selectivity (β or γ)	Good (counting mode)	Bad (counting mode)	Good	Good	Good	Good	Bad
Problems	Power consumption (fan)	Power consumption (fan)	Discriminate the weak fluorescence signal from the background lighting	Direct detection	Direct detection, restricted area	Direct detection	Direct detection
Lower threshold of particle energy	~0.2 MeV			~0.02 MeV	~3 MeV	~3 MeV	~3 MeV
Power consumption	High	High	High	Low	Low	Low	Low

[a] *Literature suggests higher distances than given by Gurkovskiy et al. (2015) in this table, up to 30 m (Baschenko, 2004).*

facility consists of two independent detection chambers. The swipe is placed in one chamber for the detection of the alpha contamination, and the other chamber records signals due to the presence of radon or other background radionuclides. The response to beta contamination is about 100 times weaker compared with the same activity of alpha-emitting radionuclides. A unit applicable to rapid field measurements is available with dimensions of 28 cm × 13 cm × 14 cm and weight of 5 kg.

An unusual detector principle is based on alpha radiation–induced luminescence, which shall be discussed here briefly, although it is not based on gas ionization detectors. Baschenko (2004) reported that 95% of the intensity falls in the narrow range between 310 and 400 nm and that the spectrum corresponds to a number of narrow (2–3 nm) and broad bands, which are due to the transitions of ionized and excited nitrogen molecules. Each alpha disintegration is accompanied with an average of ca. 30 UV photons, which may be detected optically. This method may be useful for the remote detection of radiation sources, which has sparked some research in recent years, primarily in forensic (Kerst et al., 2018; Sand et al., 2015) and environmental sciences (Sand et al., 2016; Yamamoto et al., 2016, 2018), as well as in decommissioning of nuclear facilities (Crompton et al., 2017, 2018a, 2018b; Lamadie et al., 2005; Suran et al., 2018).

Gurkovskiy et al. (2015) proposed a device for long-distance detection of alpha radiation, by detecting negatively charged ions from the alpha particle tracks by a discharge wire counter open to the air. Ion clusters in this device are transferred from the particle tracks to the detector volume by an air flux. A comparison of LRAD detectors and other detection systems is given in Table 2.2 (Gurkovskiy et al., 2015).

F. Liquid ionization and proportional detectors

Detector materials of high density offer some advantages, particularly for the detection of radiation with low LET and high energy. Radiation spectroscopy in many cases can be carried out much more reliably using detector materials of higher density. Consequently, research related to liquid and solid-state ionization detectors is carried out. Noble gases in the liquid or solid phase are dielectric materials where created electrons remain free once all electronegative impurities have been removed.

Among the noble gases, xenon has attracted much interest as a filling medium for ionization-type detectors (Incicchitti et al., 1990), such as ion chambers and PC systems. The start of the ion multiplication phenomenon is observed at a field strength of 10^8 V/m. At 10^5 V/m, the electron drift velocity is about 3×10^3 m/s. Main obstacles to the construction of such detectors are the requirements for operation at a low temperature and for extensive purification of the detector medium. Liquid xenon ionization chambers compared with sodium iodide (NaI) detectors have a similar gamma efficiency and a higher energy resolution (L'Annunziata, 1987). Of course, the energy resolution of semiconductor gamma detectors is still better.

The size of useful liquid or solid noble gas ionization detectors depends on the purity of the filling material. Position sensing by large detectors can be carried out by measuring the electron drift time. Gridded versions of such ion chamber detectors have also been reported.

Liquid ionization chambers (He, Ar, Xe) are frequently used in basic nuclear physics, for example, for the search for weakly interacting massive particles, such as the neutrinos predicted by supersymmetric theories (Ovchinnikov and Parusov, 1999; Masaoka et al., 2000).

Liquid xenon is also used in proportional scintillation detectors (Aprile et al., 2014), which finds application in the search for dark matter. Arazi et al. (2013) observed proportional luminescence (secondary scintillation) in the holes of a THGEM immersed in liquid xenon, when bombarded with 662 keV gamma photons or 4 MeV alpha particles. This luminescence was found to be nearly linear to the applied voltage. The observation may result in novel detector systems.

Little information is available related to nonpolar liquids as ionization detectors at room temperature. Here, the purity that can be achieved and maintained for the applied material is extremely important. Research has been carried out using, for example, tetramethylsilane. This material was used for ion chambers working in pulse and current mode (Knoll, 1989).

References

Aalseth, C.E., Day, A.R., Haas, D.A., Hoppe, E.W., Hyronimus, B.J., Keillor, M.E., et al., 2011. Measurement of ^{37}Ar to support technology for on-site inspection under the comprehensive nuclear-test-ban treaty. Nucl. Instrum. Methods Phys. Res. Sect. A 652, 58–61.

Aalseth, C.E., Humble, P.H., Mace, E.K., Orrell, J.L., Seifert, A., Williams, R.M., 2016. Shielding concepts for low-background proportional counter arrays in surface laboratories. Appl. Radiat. Isot. 108, 92–99.

Abilama, M., Bates, M., Lohstroh, A., 2015. Investigating the lifetime of bromine-quenched G.M. Counters with temperature. Nucl. Instrum. Methods Phys. Res. Sect. A 795, 12–18.

Agosteo, S., 2010. Overview of novel techniques for radiation protection and dosimetry. Radiat. Meas. 45, 1171–1177.

Agosteo, S., Colautti, P., Fanton, I., Fazzi, A., Introini, M.V., Moro, D., et al., 2011. Study of a solid state microdosemeter based on a monolithic silicon telescope: irradiations with low-energy neutrons and direct comparison with a cylindrical TEPC. Radiat. Prot. Dosimetry 143, 432–435.

Agrawal, P.C., Ramsey, B.D., Weisskopf, M.C., 1989. Study of argon-based Penning gas mixtures for use in proportional counters. Nucl. Instrum. Methods Phys. Res. Sect. A 277, 557–564.

Äikää, O., Mäntynen, P., Kankainen, T., 1992. High performance ^{14}C gas proportional counting system applying pulse shape discrimination. Radiocarbon 34, 414–419.

Akopdzhanov, G.A., 2008. A postproportional gas-amplification mode in cylindrical counters. Instrum. Exp. Tech. 51, 371–380.

Alex, M., Ghodgaonkar, M.D., 2007. Development of an inconel self powered neutron detector for in-core reactor monitoring. Nucl. Instrum. Methods Phys. Res. Sect. A 574, 127–132.

Alex, M., Prasad, K.R., Kataria, S.K., 2004. Development of bismuth self-powered detector. Nucl. Instrum. Methods Phys. Res. Sect. A 523, 163–166.

Ali, F., Waker, A.J., Waller, E.J., 2014. Intercomparison of Monte Carlo radiation transport codes to model TEPC response in low-energy neutron and gamma-ray fields. Radiat. Prot. Dosim. 161, 257–260.

Alkhazov, G.D., Komar, A.P., Vorobev, A.A., 1967. Ionization fluctuation and resolution of ionization chambers and semiconductor detectors. Nucl. Instrum. Methods 48, 1–12.

Alkhomashi, N., Monged, M.H.E., 2015. Long-term assessment of contaminated articles from the Chernobyl reactor. J. Environ. Radioact. 144, 52–56.

Allander, K.S., Bounds, J.A., Mac Arthur, D.W., 1994. Application of the long-range alpha detector (LRAD) to the detection of natural-occurring radioactive materials (NORM). In: Proc. WM '94 Conf., Tucson, AZ.

Alteri, E., 1992. Measurement of Reverse Transcriptase Activity of an HIV-1 Virus Stock Prepared in A 3.01 Lymphblastiod Cells. Matrix Application Note. Packard Instrument Company, Meriden, CT.

Altzitzoglou, T., Denecke, B., Johansson, L., Sibbens, G., 2002. Standardisation of ^{89}Sr using different methods. Appl. Radiat. Isot. 56, 447–452.

Amrani, D., Cherouati, D.E., Cherchali, M.E.H., 2000. Groundwater radon measurements in Algeria. J. Environ. Radioact. 51, 173–180.

Anjomani, Z., Hanu, A.R., Prestwich, W.V., Byun, S.H., 2017. Development of a multi-element microdosimetric detector based on a thick gas electron multiplier. Nucl. Instrum. Methods Phys. Res. Sect. A 847, 117–124.

Aoyama, T., 1990. A tritium in-air monitor with compensation and additional recording of α, β and γ-backgrounds. IEEE Trans. Nucl. Sci. 37, 885–991.

Aprile, E., Contreras, H., Goetzke, L.W., Melgarejo Fernandez, A.J., Messina, M., Naganoma, J., et al., 2014. Measurements of proportional scintillation and electron multiplication in liquid xenon using thin wires. J. Instrum. 9, P11012/1–P11012/20, 20 pp.

Arazi, L., Coimbra, A.E.C., Itay, R., Landsman, H., Levinson, L., Pasmantirer, B., et al., 2013. First observation of liquid-xenon proportional electroluminescence in THGEM holes. J. Instrum. 8, C12004/1–C12004/17, 17 pp.

Arkani, M., Khalafi, H., Arkani, M., 2013. An improved formula for dead time correction of G-M detectors. Nukleonika 58, 533–536.

Aslam, Qashua, N., Waker, A.J., 2011. Study of the effect of high dose rate on tissue equivalent proportional counter microdosimetric measurements in mixed photon and neutron fields. Nucl. Instrum. Methods Phys. Res. Sect. A 652, 854–857.

Aulchenko, V.M., Bobrov, A.V., Bondar, A.E., Shekhtman, L.I., Usov, E.V., Zhilich, V.N., Zhulanov, V.V., 2009. Triple-GEM detectors for KEDR tagging system. Nucl. Instrum. Methods Phys. Res. Sect. A 598, 112–115.

Aune, S., Castel, J.F., Dafni, T., Davenport, M., Fanourakis, G., Ferrer-Ribas, E., et al., 2014. Low background x-ray detection with Micromegas for axion research. J. Instrum. 9, P01001/1–P01001/26, 26 pp.

Aza, E., Ciocca, M., Murtas, F., Puddu, S., Pullia, M., Silari, M., 2017. Preliminary results of the Gas Electron Multiplier (GEM) as real-time beam monitor in hadron therapy. Nucl. Instrum. Methods Phys. Res. Sect. A 841, 65–71.

Babichev, E.A., Baru, S.E., Groshev, V.R., Khabakhpashev, A.G., Krainov, G.S., Leonov, V.V., Neustroev, V.A., Porosev, V.V., Savinov, G.A., Shekhtman, L.I., 2001. Usage of two types of high-pressure xenon chambers for medical radiography. Nucl. Instrum. Methods Phys. Res. Sect. A 461, 430–434.

Bacchi, O.O., Reichardt, K., Calvache, M., Nielsen, D., Vachaud, G., Eaglesham, A., Chalk, P.M., Urquiaga, S., Zapata, F., Laurent, J.-P., Thony, J.L., Vauchlin, M., Moutonnet, P., 2002. Neutron and gamma probes: their use in agronomy. In: International Atomic Energy Agency Training Course Series, vol. 75. IAEA, Vienna. No. 16 (IAEA-TCS-16).

Bachmanna, S., Kapplera, S., Ketzer, B., Müller, T., Ropelewskia, L., Sauli, F., Schulte, E., 2002. High rate X-ray imaging using multi-GEM detectors with a novel readout design. Nucl. Instrum. Methods Phys. Res. Sect. A 478, 104–108.

Baeza, A., Corbacho, J.A., Caballero, J.M., Ontalba, M.A., Vasco, J., Valencia, D., 2017. Development of an advanced radioactive airborne particle monitoring system for use in early warning networks. J. Radiol. Prot. 37, 542–558.

Barbosa, A.F., Rickel, C., Wattecamps, P., 1992. Two dimensional X-ray detector based on microstrip and multiwire design. Nucl. Instrum. Methods A323, 247–251.

Barr, A., Bonaldi, L., Carugno, G., Charpak, G., Iannuzzi, D., Nicoletto, M., Pepato, A., Ventura, S., 2002. A high-speed, pressurised multi-wire gamma camera for dynamic imaging in nuclear medicine. Nucl. Instrum. Methods Phys. Res. Sect. A 477, 499–504.

Baschenko, S.M., 2004. Remote optical detection of alpha particle sources. J. Radiol. Prot. 24, 75–82.

Bateman, J.E., 1994. Fundamentals of gas counters. Med. Radiat. Detect. 1994, 89–99.

Bateman, J.E., Sore, J., Knight, S.C., Bedford, P., 1994. A new gas counter for radioimmunoassay. Nucl. Instrum. Methods Phys. Res. Sect. A 348, 288–292.

Bateman, J.,E., Conolly, J.,F., Derbyshire, G.,E., Duxbury, D.,M., Mir, J.,A., Spill, E.,J., Stephenson, R., 2002. Energy resolution in X-ray detection micro-strip gas counters. Nucl. Instrum. Methods Phys. Res. Sect. A 484, 384–395.

Beck, T.R., Buchröder, H., Schmidt, V., 2009. Performance tests for instruments measuring radon activity concentration. Appl. Radiat. Isot. 67, 876–880.

Beddingfield, D.H., Johnson, N.H., Menlove, H.O., 2000. ^{3}He neutron proportional counter performance in high gamma-ray dose environments. Nucl. Instrum. Methods Phys. Res. Sect. A 455, 670–682.

Bellazzini, R., Bozzo, M., Brez, A., Gariano, G., Latronico, L., Lumb, N., Massai, M.M., Papanestis, A., Raffo, R., Spandre, G., Spezziga, M.A., 1999a. A two-stage, high gain micro-strip detector. Nucl. Instrum. Methods Phys. Res. Sect. A 425, 218–227.

Bellazzini, R., Bozzo, M., Brez, A., Gariano, G., Latronico, L., Lumb, N., Papanestis, A., Spandre, G., Massa, M.M., Raffo, R., Spezziga, M.A., 1999b. The WELL detector. Nucl. Instrum. Methods Phys. Res. Sect. A 423, 125–134.

Bellazzini, R., Brez, A., Gariano, G., Latronico, L., Lumb, N., Moggi, A., Reale, S., Spandre, G., Massai, M.M., Spezziga, M.A., Toropin, A., Costa, E., Soffitta, P., Pacella, D., 2001. Micropattern gas detectors:

the CMS MSGC project and gaseous pixel detector application. Nucl. Instrum. Methods Phys. Res. Sect. A 471, 41–54.

Bellazzini, R., Spandre, G., Lumb, N., 2002. Progress with micro-pattern gas detectors. Nucl. Instrum. Methods Phys. Res. Sect. A 478, 13–25.

Belli, G., De Vecchi, C., Giroletti, E., Guida, R., Musitelli, G., Nardò, R., Necchi, M.,M., Pagano, D., Ratti, S.,P., Sani, G., Vicini, A., Vitulo, P., Viviani, C., 2007. RPCs in biomedical applications. Nucl. Instrum. Methods Phys. Res. Sect. B 158, 166–174.

Benussi, L., Bianco, S., Saviano, G., Muhammad, S., Piccolo, D., Suhaj, A., et al., 2016. Gas electron multiplier foil holes: a study of mechanical and deformation effects. J. Instrum. 11, P08002.

Bertolini, G., 1984. Alpha particle spectroscopy by gridded ionization chamber. Nucl. Instrum. Methods 223, 285–289.

Bigu, J., 1992. Design and operation of an automated beta-particle counting system for the measurement of ^{220}Rn (and ^{222}Rn) progeny. Appl. Radiat. Isot. 43, 443–448.

Blain, E., Daskalakis, A., Block, R.C., Danon, Y., 2017. Measurement of prompt fission neutron spectrum for spontaneous fission of ^{252}Cf using γ multiplicity tagging. Phys. Rev. C 95, 064615/1–064615/7.

Blanco, F., La Rocca, P., Riggi, F., 2009. Cosmic rays with portable Geiger counters : from sea level to airplane cruise altitudes. Eur. J. Phys. 30, 685–695.

Böck, H., Balcar, E., 1975. Long-time behaviour of regenerative in-core neutron detectors with ^{238}U-^{239}Pu electrodes during power cycling. Nucl. Instrum. Methods 124, 563–571.

Bolst, D., Guatelli, S., Tran, L.T., Chartier, L., Lerch, M.L.F., Matsufuji, N., et al., 2017a. Correction factors to convert microdosimetry measurements in silicon to tissue in 12C ion therapy. Phys. Med. Biol. 62, 2055–2069.

Bolst, D., Tran, L.T., Chartier, L., Prokopovich, D.A., Pogossov, A., Guatelli, S., et al., 2017b. RBE study using solid state microdosimetry in heavy ion therapy. Radiat. Meas. 106, 512–518.

Bolton, R.D., 1994. Radon Monitoring Using Long-Range Alpha Detector-Based Technology. Nuclear Science Symposium, Norfolk, VA.

Bondar, A., Buzulutskov, A., de Oliveira, R., Ropelewski, L., Sauli, F., Shekhtman, L., 2006. Nucl. Instrum. Methods Phys. Res. Sect. A 556, 495–497.

Bondar, A., Buzulutskov, A., Grebenuk, A., Sokolov, A., Akimov, D., Alexandrov, I., Breskin, A., 2010. Direct observation of avalanche scintillations in a THGEM-based two-phase Ar- avalanche detector using Geiger-mode APD. J. Instrum. 5. P08002 (20 pp.).

Bondar, A., Buzulutskov, A., Grebenuk, A., Sokolov, A., Akimov, D., Alexandrov, I., Breskin, A., 2011. Geiger mode APD performance in a cryogenic two-phase Ar avalanche detector based on THGEMs. Nucl. Instrum. Methods Phys. Res. Sect. A 628, 364–368.

Bortot, D., Pola, A., Agosteo, S., Pasquato, S., Introini, M.V., Colautti, P., et al., 2017a. A miniaturized alpha spectrometer for the calibration of an avalanche-confinement TEPC. Radiat. Meas. 106, 531–537.

Bortot, D., Pola, A., Agosteo, S., Pasquato, S., Mazzucconi, D., Fazzi, A., et al., 2017b. A novel avalanche-confinement TEPC for microdosimetry at nanometric level. Radiat. Meas. 103, 1–12.

Bouclier, R., Florent, J.J., Gaudaen, J., Million, G., Ropelewski, L., Sauli, F., 1992a. Microstrip gas chambers on thin plastic supports. IEEE Trans. Nucl. Sci. 39, 650–653.

Bouclier, R., Florent, J.J., Gaudaen, J., Sauli, F., Shekhtman, L., 1992b. Development of microstrip gas chambers on thin plastic supports. Nucl. Instrum. Methods Phys. Res. Sect. A 315, 521–528.

Bouclier, R., Florent, J.J., Gaudaen, J., Millon, G., Pasta, A., Ropelewski, L., Sauli, F., Shekhtman, L., 1992c. High flux operation of microstrip gas chambers on glass and plastic supports. Nucl. Instrum. Methods Phys. Res. Sect. A 323, 240–246.

Bouclier, R., Capeans, M., Evans, J., Garabatos, C., Manzin, G., Million, G., Ropelewski, L., Sauli, F., Shekhtman, L.I., Temmel, T., Fischer, G., 1995. Optimization of design and beam test of microstrip gas chambers. Nucl. Instrum. Methods Phys. Res. Sect. A 367, 163–167.

Braby, L.A., Badhwar, G.D., 2001. Proportional counter as neutron detector. Radiat. Meas. 33, 265–267.

Braccini, S., Ereditato, A., Nesteruk, K.P., Scampoli, P., Zihlmann, K., 2015. Study of the radioactivity induced in air by a 15-MeV proton beam. Radiat. Prot. Dosim. 163, 269–275.

Breskin, A., 2000. Advances in gas avalanche radiation detectors for biomedical applications. Nucl. Instrum. Methods Phys. Res. Sect. A 454, 26–39.

Bronic, I.K., Grosswendt, B., 2001. Experimental study of gas mixtures in strong non-uniform electric fields. Radiat. Phys. Chem. 61, 477–478.

Broughton, D., Waker, A.J., 2016. An analytical comparison of gas gain in spherical, cylindrical and hemispherical low-pressure proportional counters intended for use in experimental microdosimetry. Nucl. Instrum. Methods Phys. Res. Sect. A 840, 107–112.

Brunbauer, F.M., Oliveri, E., Resnati, F., Ropelewski, L., Sauli, F., Thuiner, P., et al., 2017. The planispherical chamber: a parallax-free gaseous X-ray detector for imaging applications. Nucl. Instrum. Methods Phys. Res. Sect. A 875, 16–20.

Bucciantonio, M., Amaldi, U., Kieffer, R., Malakhov, N., Sauli, F., Watts, D., 2013a. Fast readout of GEM detectors for medical imaging. Nucl. Instrum. Methods Phys. Res. Sect. A 718, 160–163.

Bucciantonio, M., Amaldi, U., Kieffer, R., Sauli, F., Watts, D., 2013b. Development of a fast proton range radiography system for quality assurance in hadrontherapy. Nucl. Instrum. Methods Phys. Res. Sect. A 732, 564–567.

Burigo, L., Pshenichnov, I., Mishustin, I., Bleicher, M., 2013. Microdosimetry of radiation field from a therapeutic 12C beam in water: a study with Geant4 toolkit. Nucl. Instrum. Methods Phys. Res. Sect. B 310, 37–53.

Burigo, L., Pshenichnov, I., Mishustin, I., Bleicher, M., 2014. Microdosimetry spectra and RBE of ^{1}H, ^{4}He, ^{7}Li and ^{12}C nuclei in water studied with Geant4. Nucl. Instrum. Methods Phys. Res. Sect. B 320, 89–99.

Byun, S.H., Spirou, G.M., Hanu, A., Prestwich, W.V., Waker, A.J., 2009. Simulation and first test of a microdosimetric detector based on a thick gas electron multiplier. IEEE Trans. Nucl. Sci. 56, 1108–1113.

Cabellos, O., Fernández, P., Rapisarda, D., García-Herranz, N., 2010. Assessment of fissionable material behaviour in fission chambers. Nucl. Instrum. Methods Phys. Res. Sect. A 618, 248–259.

Charpak, G., 1970. Evolution of the automatic spark chambers. An. Rev. Nucl. Sci. 20, 195–254.

Charpak, G., Sauli, F., 1978. Multiwire proportional chambers and drift chambers. Nucl. Instrum. Methods 162, 405–428.

Chase, G., Rabinowitz, J., 1967. Principles of Radioisotope Methodology. Burgess Publishing Company, Minneapolis.

Chefdeville, M., van der Graaf, H., Hartjes, F., Timmermans, J., Visschers, J., Blanco Carballo, V.,M., Salm, C., Schmitz, J., Smits, S., Colas, P., Giomataris, I., 2008. Pulse height fluctuations of integrated micromegas detectors. Nucl. Instrum. Methods Phys. Res. Sect. A 591, 147–150.

Chen, X., Zhao, L., Qin, H., Zhao, M., Zhou, Y., Yang, S., et al., 2014. Gross alpha and beta activity analyses in urine-A routine laboratory method for internal human radioactivity detection. Health Phys. 106, 545–550.

Chernyshova, M., Malinowski, K., Melikhov, Y., Kowalska-Strzeciwilk, E., Czarski, T., Wojeński, A., et al., 2018. Study of the optimal configuration for a Gas Electron Multiplier aimed at plasma impurity radiation monitoring. Fusion Eng. Des. 136 (Part A), 592–596.

Chiriotti, S., Moro, D., Colautti, P., Conte, V., Grosswendt, B., 2015. Equivalence of pure propane and propane TE gases for microdosimetric measurements. Radiat. Prot. Dosim. 166, 242–246.

Chiriotti, S., Moro, D., Conte, V., Colautti, P., D'Agostino, E., Sterpin, E., et al., 2014. Monte Carlo tools to supplement experimental microdosimetric spectra. Radiat. Prot. Dosim. 161, 454–458.

Chiti, T., Neubert, R.E.M., Janssens, I.A., Certini, G., Curiel Yuste, J., Sirignano, C., 2009. Radiocarbon dating reveals different past managements of adjacent forest soils in the Campine region, Belgium. Geoderma 149, 137–142.

Christophel, E., Dracos, M., Strub, R., 1998. The 2D-microgap wire chamber. Nucl. Instrum. Methods Phys. Res. Sect. A 419, 515–518.

Chu, T.,C., Wang, J.-J., Lin, Y.-M., 1998. Radiostrontium analytical method using crown - ether compound and Cerenkov counting and its application in environmental monitoring. Appl. Radiat. Isot. 49, 1671–1675.

Cesari, V., Colautti, P., Magrin, G., De Nardo, L., Baek, W.Y., Grosswendt, B., Alkaa, A., Khamphan, C., Ségur, P., Tornielli, G., 2002. Nanodosimetric measurements with an avalanche confinement TEPC. Radiat. Prot. Dosimetry 99, 337–342.

Clouvas, A., Xanthos, S., Takoudis, G., 2011. Indoor radon levels in Greek schools. J. Environ. Radioact. 102, 881–885.

Coelho, L.,C.,C., Lopes, J.,A.,M., Covita, D.,S., Conceicão, A.,S., Santos, J.,M.,F., 2007. Xenon GPSC high-pressure operation with large-area avalanche photodiode readout. Nucl. Instrum. Methods Phys. Res. Sect. A 575, 444–448.

Colmenares, C.A., 1974. Bakeable ionization chamber for low level tritium counting. Nucl. Instrum. Methods Phys. Res. 150, 549.

Coopera, J.R., Bernstein, L., McMahan, M.A., Powell, J., Wutte, D., Ahle, L., Benczer-Koller, N., Dashdorj, D., Kumbartzki, G., Mertzimekis, T.J., Schiller, A., Silver, C., Taylor, M.J., 2004. Production of a ^{76}Kr radioactive ion beam using a batch mode method. Nucl. Instrum. Methods Phys. Res. Sect. A 533, 287–294.

Cortesi, M., Rost, S., Mittig, W., Ayyad-Limonge, Y., Bazin, D., Yurkon, J., et al., 2017. Multi-layer thick gas electron multiplier (M-THGEM): a new MPGD structure for high-gain operation at low-pressure. Rev. Sci. Instrum. 88, 013303/1–013303/8.

Cosma, C., 2000. Strontium-90 measurement after the Chernobyl accident in Romanian samples without chemical separation. Spectrochim. Acta, Part B 55, 1165–1171.

Couceiro, M., Blanco, A., Ferreira, N.,C., Ferreira Marques, R., Fonte, P., Lopes, L., 2007. RPC- PET: Status and perspectives. Nucl. Instrum. Methods Phys. Res. Sect. A 580, 915–918.

Covita, D.,S., Simões, P.,C.P.,S., Fernandes, L.,M.,P., Freitas, E.,D.,C., Requicha Ferreira, L.,F., Veloso, J.,F.,C.,A., dos Santos, J.,M.,F., 2004. The X-ray performance of a driftless gas proportional scintillation counter using short shaping-time constants for pulse analysis. Nucl. Instrum. Methods Phys. Res. Sect. A 516, 134–142.

Croci, G., Albani, G., Cazzaniga, C., Cippo, E.P., Schooneveld, E., Claps, G., et al., 2014. Diffraction measurements with a boron-based GEM neutron detector. Europhys. Lett. 107, 12001/1–12001/4, 4 pp.

Crompton, A.J., Gamage, K.A.A., Bell, S., Wilson, A.P., Jenkins, A., Trivedi, D., 2017. First Results of Using a UVTron Flame Sensor to Detect Alpha-Induced Air Fluorescence in the UVC Wavelength Range. Sensors, Basel, p. 17.

Crompton, A.J., Gamage, K.A.A., Bell, S., Wilson, A.P., Jenkins, A.W., Trivedi, D., 2018a. Gas Flow to Enhance the Detection of Alpha-Induced Air Radioluminescence Based on a UVTron Flame Sensor. Sensors, Basel, p. 18.

Crompton, A.J., Taylor, C.J., Gamage, K.A.A., Jenkins, A., 2018b. Alpha Particle Detection Using Alpha-Induced Air Radioluminescence: A Review and Future Prospects for Preliminary Radiological Characterisation for Nuclear Facilities Decommissioning. Sensors, Basel, p. 18.

Curioni, A., Dinar, N., La Torre, F.P., Leidner, J., Murtas, F., Puddu, S., et al., 2017. Measurements of ^{55}Fe activity in activated steel samples with GEMPix. Nucl. Instrum. Methods Phys. Res. Sect. A 849, 60–71.

Curzio, G., Mazed, D., Ciolini, R., Del Gratta, A., Gentili, A., 2005. Effect of air on gas amplification characteristics in argon-propane (1%) – based proportional counters for airborne radon monitoring. Nucl. Instrum. Methods Phys. Res. Sect. A 537, 672–682.

Czajkowski, C., Finfrock, C., Philipsberg, P., Ghosh, V., 2012. Construction and development of a BF3 neutron detector at Brookhaven national laboratory. JNMM 40, 53–59.

Dabrowski, W., Fiutowski, T., Frączek, P., Koperny, S., Lankosz, M., Mendys, A., et al., 2016. Application of GEM-based detectors in full-field XRF imaging. J. Instrum. 11, C12025.

Dalla Torre, S., 2011. Status and perspectives of gaseous photon detectors. Nucl. Instrum. Methods Phys. Res. Sect. A 639, 111–116.

Darvish-Molla, S., Prestwich, W.V., Byun, S.H., 2018. Development of an advanced two-dimensional microdosimetric detector based on THick Gas Electron Multipliers. Med. Phys. 45, 1241–1254.

Degteva, M.O., Vorobiova, M.I., Tolstykh, E.I., Shagina, N.B., Shishkina, E.A., Anspaugh, L.R., Napier, B.A., Bougrov, N.G., Shved, V.A., Tokareva, E.E., 2006. Development of an improved dose reconstruction system for the Techa River population affected by the operation of the Mayak Production Association. Radiat. Res. 166, 255–270.

Delbart, A., De Oliveira, R., Derré, J., Giomataris, Y., Jeanneau, F., Papadopoulos, Y., Rabourgeard, P., 2001. New development of Micromegas detector. Nucl. Instrum. Methods Phys. Res. Sect. A 461, 84–87.

De Nardo, L., Cesari, V., Donà, G., Colautti, P., Conte, V., Tornielli, G., 2004a. Mini TEPCs for proton therapy. Radiat. Prot. Dosim. 108, 345–352.

De Nardo, L., Moro, D., Colautti, P., Conte, V., Tornielli, G., Cuttone, G., 2004b. Microdosimetric investigation at the therapeutic proton beam facility of CATANA. Radiat. Prot. Dosim. 110, 681–686.

De Nardo, L., Colautti, P., Hérault, J., Conte, V., Moro, D., 2010. Microdosimetric characterisation of a therapeutic proton beam used for conjunctival melanoma treatments. Radiat. Meas. 45, 1387–1390.

De Nardo, L., Farahmand, M., 2016. Operation of gas electron multiplier (GEM) with propane gas at low pressure and comparison with tissue-equivalent gas mixtures. Nucl. Instrum. Methods Phys. Res. Sect. A 819, 154–162.

Denman, A.R., Groves-Kirkby, C.J., Phillips, P.S., Crockett, R.G.M., Woolridge, A., Gillmore, G.K., 2005. The practical use of electrets in a public health radon remediation campaign. J. Environ. Radioact. 84, 375–391.

De Saint-Hubert, M., Vargas, C.S., Van Hoey, O., Schoonjans, W., De Smet, V., Mathot, G., et al., 2016. Secondary neutron doses in a proton therapy centre. Radiat. Prot. Dosim. 170, 336–341.

Dighe, P.M., Das, D., 2015. Annular shape silver lined proportional counter for on-line pulsed neutron yield measurement. Nucl. Instrum. Methods Phys. Res. Sect. A 778, 115–119.

Derré, J., Giomataris, Y., Zaccone, H., Bay, A., Perroud, J.-P., Ronga, F., 2001. Spatial resolution in Micromegas detectors. Nucl. Instrum. Methods Phys. Res. Sect. A 459, 523–531.

Domnikov, V.N., Saltykov, I.S., Slusarenko, L.I., Shevchenko, S.V., 2001. About the effectiveness of spectrometry in alpha-activity monitoring of industrial air-borne particles. Appl. Radiat. Isot. 55, 543–547.

Donnard, J., Berny, R., Carduner, H., Leray, P., Morteau, E., Provence, M., Servagent, N., Thers, D., 2009. The micro-pattern gas detector PIM: a multi-modality solution for novel investigations in functional imaging. Nucl. Instrum. Methods Phys. Res. Sect. A 610, 158–160.

Dueñas, C., Fernández, M.C., Gordo, E., Cañete, S., Pérez, M., 2011. Gross alpha, gross beta activities and gamma emitting radionuclides composition of rainwater samples and deposition to ground. Atmosph. Environ. 45, 1015–1024.

East, L.V., Walton, B.B., 1982. Polythene moderated ^{3}He neutron detectors. Nucl. Instrum. Methods Phys. Res. 199, 643.

Epstein, M.S., Miles Jr., D.G., Yu, L.L., 2009. What were they drinking? A critical study of the radium ore revigator. Appl. Spectrosc. 63, 1406–1409.

Eriksson, S.M., Mackey, E.A., Lindstrom, R.M., Lamaze, G.P., Grogan, K.P., Brady, D.E., 2013. Delayed-neutron activation analysis at NIST. J. Radioanal. Nucl. Chem. 298, 1819–1822.

Etaati, G., Amrollahi, R., Doust-Mohammadi, V., Ghal-Eh, N., Habibi, M., 2010. Calibration constant of a silver activated Geiger counter used for neutron efficiency measurements in plasma focus devices: theoretical vs. Experimental studies. J. Fusion Energ. 29, 381–386.

Evett, S.R., Tolk, J.A., Howell, T.A., 2006. Soil profile water content determination: Sensor accuracy, axial response, calibration, temperature dependence, and precision. Vadose Zone J. 5, 894–907.

Evett, S.R., Schwartz, R.C., Tolk, J.A., Howell, T.A., 2009. Soil profile water content determination: Spatiotemporal variability of electromagnetic and neutron probe sensors in access tubes. Vadose Zone J. 5, 926–941.

Fadil, M., Blandin, C., Christophe, S., Déruelle, O., Fioni, G., Marie, F., Mounier, C., Ridikas, D., Trapp, J.P., 2002. Development of fission micro-chambers for nuclear waste incineration studies. Nucl. Instrum. Methods Phys. Res. Sect. A 476, 313–317.

Fang, Z., Yang, Y., Li, Y., Wang, X., 2017a. Working gas selection of the honeycomb converter-based neutron detector. IEEE Trans. Nucl. Sci. 64, 1683–1688.

Fang, Z., Yang, Y., Li, Y., Zhang, Z., Wang, X., 2017b. Research on a neutron detector with a boron-lined honeycomb neutron converter. IEEE Trans. Nucl. Sci. 64, 1048–1055.

Farah, J., De Saint-Hubert, M., Mojzeszek, N., Chiriotti, S., Gryzinski, M., Ploc, O., et al., 2017. Performance tests and comparison of microdosimetric measurements with four tissue-equivalent proportional counters in scanning proton therapy. Radiat. Meas. 96, 42–52.

Farah, J., Mares, V., Romero-Exposito, M., Trinkl, S., Domingo, C., Dufek, V., et al., 2015. Measurement of stray radiation within a scanning proton therapy facility: EURADOS WG9 intercomparison exercise of active dosimetry systems. Med. Phys. 42, 2572–2584.

Farahmand, M., Bos, A.J.J., van Eijk, C.W.E., 2003a. Gas electron multiplier (GEM) operation with tissue-equivalent gases at various pressures. Nucl. Instrum. Methods Phys. Res. Sect. A 506, 160–165.

Farahmand, M., Bos, A.J.J., Huizenga, J., De Nardo, L., van Eijk, C.W.E., 2003b. Design of a new tissue equivalent proportional counter based on a gas electron multiplier. Nucl. Instrum. Methods Phys. Res. Sect. A 509, 262–267.

Farahmand, M., Bos, A.J.J., De Nardo, L., van Eijk, C.W.E., 2004. First microdosimetric measurements with a TEPC based on a GEM. Radiat. Prot. Dosim. 110, 839–843.

Farahmand, M., De Nardo, L., 2016. Microdosimetric measurements of a tissue-equivalent proportional counter based on a gas electron multiplier down to 140 nm simulated site sizes. Radiat. Prot. Dosim. 171, 304–312.

Filliatre, P., Oriol, L., Jammes, C., Vermeeren, L., 2008. Reasons why Plutonium 242 is the best fission chamber deposit to monitor the fast component of a high neutron flux. Nucl. Instrum. Methods Phys. Res. Sect. A 593, 510–518.

Filliatre, P., Oriol, L., Jammes, C., Vermeeren, L., 2009. Joint estimation of the fast and thermal components of a high neutron flux with a two on-line detector system. Nucl. Instrum. Methods Phys. Res. Sect. A 603, 415–420.

Fischer, B.E., 1977. A digital processor for position sensitive detectors. Nucl. Instrum. Methods 41, 173–181.

Fourme, R., 1997. Position sensitive gas detectors: MEPCs and their gifted descendants. Nucl. Instrum. Methods Phys. Res. Sect. A 392, 1–11.

Fraga, F.A.F., Fetal, S.T.G., Ferreira Maqes, R., Policarpo, A.P.L., 2000. Quality control of GEM detectors using scintillation techniques. Nucl. Instrum. Methods Phys. Res. Sect. A 442, 417–422.

Fraga, F.A.F., Margato, L.M.S., Fetal, S.T.G., Fraga, M.M., F. R., Maques, R.F., Policarpo, A.J.P.L., Guerard, B., Oed, A., Manzini, G., van Vuure, T., 2002. CCD readout of GEM - based detectors. Nucl. Instrum. Methods Phys. Res. Sect. A 478, 357–361.

Fraga, F.A.F., Margato, L.M.S., Fetal, S.T.G., Fraga, M.M.F.R., Ferreira Marques, R., Policarpo, A.J.P.L., 2003. Luminescence and imaging with gas electron multipliers. Nucl. Instrum. Methods Phys. Res. Sect. A 513, 379–387.

Fred Zhang, Z., 2017. Should the standard count be excluded from neutron probe calibration? Soil Sci. Soc. Am. J. 81, 1036–1044.

Fried, J., Harder, J.A., Mahler, G.J., Makowiecki, D.S., Mead, J.A., Radeka, V., Schaknowski, N.A., Smith, G.C., Yu, B., 2002. A large, high performance, curved 2D position-sensitive neutron detector. Nucl. Instrum. Methods Phys. Res. Sect. A 478, 415–419.

Fujiwara, T., Mitsuya, Y., Fushie, T., Murata, K., Kawamura, A., Koishikawa, A., et al., 2017. Gas scintillation glass GEM detector for high-resolution X-ray imaging and CT. Nucl. Instrum. Methods Phys. Res. Sect. A 850, 7–11.

Fujiwara, T., Mitsuya, Y., Takahashi, H., 2018. Radiation imaging with glass gas electron multipliers (G - GEMs). Nucl. Instrum. Methods Phys. Res. Sect. A 878, 40–49.

Fujiwara, T., Mitsuya, Y., Yanagida, T., Saito, T., Toyokawa, H., Takahashi, H., 2016. High-photon-yield scintillation detector with Ar/CF_4 and glass gas electron multiplier. Jpn. J. Appl. Phys. 55, 106401/1–106401/4.

Fulbright, H.W., 1979. Ionization chambers. Nucl. Instrum. Methods 62, 21–28.

Galer, S., Shipley, D.R., Palmans, H., Kirkby, K.J., Nisbet, A., 2017. Monte Carlo simulation of a TEPC for microdosimetry of carbon ions. Radiat. Phys. Chem. 140, 412–418.

García-León, M., García-Montaño, E., Madurga, G., 1984. Characterization of ^{99}Tc by shape of its plateau with a gas-flow proportional counter. Int. J. Appl. Radiat. Isot. 35, 195–200.

García-Toraño, E., Barquerro, R.L., Roteta, M., 2002. Standardization of ^{134}Cs by three methods. Appl. Radiat. Isot. 56, 211–214.

Gardner, R.P., Liu, L., 1997. On extending the accurate and useful counting rate of GM counter detector systems. Appl. Radiat. Isot. 48, 1605–1615.

Garner, S.E., Bounds, J.A., Allander, K.S., Caress, R.W., Johnson, J.D., MacArthur, D.W., 1994. A Compendium of Results from Long-Range Alpha Detector Soil Surface Monitoring: June 1992-May 1994. LA-12861-MS. Los Alamos National Laboratory Document.

Geiger, H., Müller, W., 1928. Elektronenzählrohr zur Messung schwächster Aktivitäten (Electron counting tube for the measurement of weakest activities). Naturwissenschaften 16, 617–618.

Geltenbort, P., 1994. Recent results with microstrip gas chambers. Nucl. Instrum. Methods Phys. Res. Sect. A 353, 168.

Giomataris, Y., Rebourgeard, P., Robert, J.,P., Charpak, G., 1996. MICROMEGAS: a high-granularity position-sensitive gaseous detector for high particle-flux environment. Nucl. Instrum. Methods Phys. Res. Sect. A 376, 29–35.

Glesius, F.L., Kniss, T.A., 1988. He-3 neutron detectors for hostile environments. IEEE Trans. Nucl. Sci. 35, 867–871.

Golovatyuk, V., Grancagnolo, F., 1999. Observation of transition between proportional and Geiger-Müller modes in helium-isobutane gas mixtures. Nucl. Instrum. Methods Phys. Res. Sect. A 428, 367–371.

Goulianos, K., Smith, K.K., White, S.N., 1980. A simple electronic apparatus for the analysis of radioactively labeled gel electrophoretograms. Anal. Biochem. 103, 64–69.

Grossi, C., Vargas, A., Camacho, A., López-Coto, I., Bolívar, J.P., Xia, Y., Conen, F., 2011. Inter-comparison of different direct and indirect methods to determine radon flux from soil. Radiat. Meas. 46, 112–118.

Grossi, C., Arnold, D., Adame, J.A., López-Coto, I., Bolívar, J.P., de la Morena, B.A., Vargas, A., 2012. Atmospheric ^{222}Rn concentration and source term at *El Arenosillo* 100 m meteorological tower in southwest Spain. Radiat. Meas. 47, 149–162.

Gruhn, C.R., Binimi, M., Legrain, R., Loveman, R., Pang, W., Loach, M., Scott, D.K., Shotter, A., Symons, T.J., Wouters, J., Zismon, M., Devier, R., 1982. Bragg curve spectroscopy. Nucl. Instrum. Methods 196, 33–40.

Guillemot, F., Brouillaud, B., Labrugère, C., Barthe, N., Durrieu, M.-C., 2007. High resolution β-imager: a new tool for characterizing 2D peptide distribution on biomimetic materials? IRBM 28, 86–92.

Gurkovskiy, B.V., Miroshnichenko, V.P., Onishchenko, E.M., Simakov, A.B., Streil, T.G., 2015. Long-distance detection of alpha-radioactivity: method and device. J. Environ. Radioact. 149, 150–157.

Hagiwara, M., Sanami, T., Oishi, T., Baba, M., Takada, M., 2008. Extension of energy acceptance of Bragg curve counter at the high-energy end. Nucl. Instrum. Methods Phys. Res. Sect. A 592, 73–79.

Hajek, M., 2015. Developments and trends in bioequivalent dosimetry. Radiat. Prot. Dosimetry 164, 65–69.

Hamlat, M.S., Kadi, H., Djeffal, S., Brahimi, H., 2003. Radon concentrations in Algerian oil and gas industry. Appl. Radiat. Isot. 58, 125–130.

Hankins, D.E., Thorngate, J.H., 1993a. A High Sensitive Neutron Counter and Waste Drum Counting with the High Sensitivity Neutron Instrument. UCRL-ID-111750. Document of the Lawrence Livemore National Laboratoy, University of California.

Hankins, D.E., Thorngate, J.H., 1993b. A Neutron Counting Instrument for Low Level Transuranic Waste. Document of the Lawrence Livermore National Laboratoy, University of California. UCRL-ID-115887.

Harano, H., Matsumoto, T., Nishiyama, J., Masuda, A., Uritani, A., Kudo, K., 2011. Development of a compact flat response neutron detector. IEEE Trans. Nucl. Sci. 58, 2421–2425.

Hayakawa, Y., Maeda, Y., 1994. Performance of a microstrip proportional counter. Annu. Rep. Res. Reactor Inst. Kyoto Univ. 27, 34–41.

Hayakawa, Y., Maeda, Y., 1996. Microstrip gas chamber for X-rays and neutrons. Jpn. J. Appl. Phys. 35, 123–125.

Herranz, M., Idoeta, R., Legarda, F., 2011. Analysis of uncertainties and detection limits for the double measurement method of ^{90}Sr and ^{89}Sr. Radiat. Meas. 46, 68–686.

Hillman, G.G., Roessler, N., Fulbright, R.S., Edson Pontes, J., Haas, G.P., 1993a. Application of the direct beta counter Matrix 96 for cytotoxic assays: simultaneous processing and reading of 96 wells using a ^{51}Cr retention assay. Cancer Immunol. Immunother. 36, 351–356.

Hillman, G.G., Roessler, N., Fulbright, R.S., Edson Pontes, J., Haas, G.P., 1993b. ^{51}Cr release assay adapted to a 96-well format sample reading. Biotechniques 15, 744–749.

Hoch, M., 2004. Trends and new developments in gaseous detectors. Nucl. Instrum. Methods Phys. Res. Sect. A 535, 1–15.

Hoetzl, H., Winkler, R., 1984. Experience with large-area Frisch grid chambers in low-level alpha spectrometry. Nucl. Instrum. Methods 223, 290–295.

Homma, Y., Ochi, A., Moriya, K., Matsuda, S., Yoshida, K., Kobayashi, S., 2009. Study of deposit associated with discharge in micro-pixel gas chamber. Nucl. Instrum. Methods Phys. Res. Sect. A 599, 47–52.

Horikawa, S., Inaba, S., Kawai, H., Matsumoto, T., Nakayama, H., Tajima, Y., Takamatsu, K., Tsurju, T., Yoshida, H.Y., 2002. Development of micro-gap wire chamber. Nucl. Instrum. Methods Phys. Res. Sect. A 481, 166–173.

Hou, X., Roos, P., 2008. Critical comparison of radiometric and mass spectrometric methods for the determination of radionuclides in environmental, biological and nuclear waste samples. Anal. Chim. Acta 608, 105–139.

Hsu, W.-H., Braby, L.A., Reece, W.D., 2008. Detection system built from commercial integrated circuits for real-time measurement of radiation dose and quality using the variance method. Radiat. Prot. Dosim. 128, 5–11.

Hu, Q.Y., Li, B.J., Zhang, D., Guo, H.S., Yang, G.Z., Si, F.N., Liu, J., 2009. Optimizing the design of a moderator-based neutron detector for a flat response curve in the 2–14 MeV energy range. Nucl. Instrum. Methods Phys. Res. Sect. A 609, 213–216.

Huang, B., Zhang, Y., Shu, X., Liu, Y., Penumadu, D., Ye, X.P., 2013. Neutron scattering for moisture detection in foamed asphalt. J. Mater. Civ. Eng. 25, 932–938.

Hunt, J.B., Mercer, R.A., 1978. The absolute calibration of a long counter by the associated activity techniques. Nucl. Instrum. Methods 156, 451–457.

Hutchins, D., 1992. The Bioassay of Cytokines. Matrix Application Note. Packard Instrument Company, Meriden, CT.
ICRU, 1983. International Commission on Radiation Units and Measurements. Microdosimetry. ICRU Report 36. ICRU, Bethesda, Maryland.
Iguchi, T., Nakayamada, N., Takahashi, H., Nakazawa, M., 1994. Neutron spectrometry using a ^{3}He gas ionization chamber. Nucl. Instrum. Methods Phys. Res. Sect. A 353, 152–155.
Incicchitti, A., Belli, P., Scafi, M., 1990. Liquid xenon as a detector medium. Nucl. Instrum. Methods Phys. Res. Sect. A A289, 236–242.
Jahoda, K., McCammon, D., 1988. Proportional counters as low energy photon detectors. Nucl. Instrum. Methods Phys. Res. Sect. A 272, 800–813.
Jalbert, R.A., Hiebert, R.D., 1971. Gamma insensitive air monitor for radioactive gases. Nucl. Instrum. Methods 96, 61–66.
Jamil, M., Jo, H.Y., Rhee, J.T., Jeon, Y.J., 2010. Simulation study using GEANT4 Monte Carlo code for a Gd-coated resistive plate chamber as a thermal neutron detector. Radiat. Meas. 45, 840–843.
Järvinen, M.L., Sipilä, H., 1984. Improved proportional counters for practical application. IEEE Trans. Nucl. Sci. NS-31, 356–359.
Jevtic, N., Drndarevic, V., 2015. Plug and play Geiger-Muller detector for environmental monitoring. Instrum. Sci. Technol. 43, 222–243.
Johansson, L., Sibbens, G., Altzitzoglou, T., Denecke, B., 2002. Self-absorption correction in standardisation of ^{204}Tl. Appl. Radiat. Isot. 56, 199–203.
Johnson, J.D., Allander, K.S., Bounds, J.A., Garner, S.E., Johnson, J.P., MacArthur, D.W., 1994a. Long range alpha detector (LRAD) sensitivity to beta contamination and soil moisture. IEEE Trans. Nucl. Sci. 41, 755–757.
Johnson, J.D., Allander, K.S., Bounds, J.A., Garner, S.E., Johnson, J.P., MacArthur, D.W., et al., 1994b. Long-range alpha detector sample monitoring. Nucl. Instrum. Methods Phys. Res. Sect. A 353, 486–488.
Jones, A.R., Holford, R.M., 1981. Application of Geiger-Mueller counters over a wide range of counting rates. Nucl. Instrum. Methods 189, 503–509.
Kada, W., Dwaikat, N., Datemichi, J., Sato, F., Murata, I., Kato, Y., Iida, T., 2010. A twin-type airflow pulse ionization chamber for continuous alpha-radioactivity monitoring in atmosphere. Radiat. Meas. 45, 1044–1048.
Karadzhinova, A., Nolvi, A., Veenhof, R., Tuominen, E., Haeggstrom, E., Kassamakov, I., 2015. Impact of GEM foil hole geometry on GEM detector gain. J. Instrum. 10, P12014.
Kerst, T., Sand, J., Ihantola, S., Perajarvi, K., Nicholl, A., Hrnecek, E., et al., 2018. Standoff alpha radiation detection for hot cell imaging and crime scene investigation. Opt. Rev. 25, 429–436.
Kessler, M.J., 1991. A new, rapid analysis technique for quantitation of radioactive samples isolated on a solid support. In: Proceedings of the International Conference on New Trends in Liquid Scintillation Counting and Organic Scintillators 1989. Lewis Publishers, Chelsea, MI.
Kessler, P., Behnke, B., Dabrowski, R., Dombrowski, H., Roettger, A., Neumaier, S., 2018. Novel spectrometers for environmental dose rate monitoring. J. Environ. Radioact. 187, 115–121.
Khoshahval, F., Park, M., Shin, H.C., Zhang, P., Lee, D., 2018. Vanadium, rhodium, silver and cobalt self-powered neutron detector calculations by RAST-K. Ann. Nucl. Energy 111, 644–659, v2.0.
Khriachkov, V.A., Ketlerov, V.V., Mitrofanov, V.F., Semenova, N.N., 2000. Low-background spectrometer for the study of fast neutron -induced (n,α) reactions. Nucl. Instrum. Methods Phys. Res. Sect. A 444, 614–621.
Kitto, M.E., Fielman, E.M., Haines, D.K., Menia, T.A., Bari, A., 2008. Performance of a commercial radon-in-water measurement kit. J. Environ. Radioact. 99, 1255–1257.
Knoll, G.F., 1989. Radiation Detection and Measurement. John Wiley & Sons, New York.
Koike, T., Suzuki, Y., Genyu, S., Kobayashi, I., Komori, H., Otsu, H., et al., 2014. Comprehensive data on ionising radiation from Fukushima Daiichi nuclear power plant in the town of Miharu, Fukushima prefecture: the Mishoproject. J. Radiol. Prot. 34, 675–698.
Kong, C., Lee, D., Shin, H.C., 2017. Lifetime extension of in-core self-powered neutron detector using new emitter materials. Int. J. Energy Res. 41, 2405–2412.
Koster, J.E., Bounds, J.A., Kerr, P.L., Steadman, P.A., Whitley, C.R., 1998. Whole body personnel monitoring via ionization detection. IEEE Trans. Nucl. Sci. 45, 976–980.
Kotrappa, P., Steck, D.J., 2009. Radon Flux Monitor for in Situ Measurement of Granite and Concrete Surfaces. Radon Symposium, 20–23 September. American Association of Radon Scientists and Technologists, Missouri.
Kotte, R., Keller, H.J., Ortlepp, H.G., Strary, F., 1987. Bragg peak spectroscopy of low-energy heavy ions. Nucl. Instrum. Methods Phys. Res. Sect. A 257, 244–252.
Kouzes, R.T., 2011. The ^{3}He Supply Problem. Technical Report 11–753. US Government Accountability Office.
Krawczyk, R.D., Czarski, T., Linczuk, P., Wojenski, A., Kolasinski, P., Gaska, M., et al., 2018. The software-defined fast post-processing for GEM soft x-ray diagnostics in the Tungsten Environment in Steady-state Tokamak thermal fusion reactor. Rev. Sci. Instrum. 89, 063504/1–063504/13.
Kreslo, I., Badhrees, I., Delaquis, S., Ereditato, A., Janos, S., Messina, M., Moser, U., Rossi, B., Zeller, M., 2011. Pulse-shape discrimination of scintillation from alpha and beta particles with liquid scintillator and Geiger-mode multipixel avalanche diodes. J. Instrum. 6 (P07009), 9.
Kuzminov, V.V., Osetrova, N.J., 2000. Precise measurement of ^{14}C beta Spectrum by using a wall-less proportional counter. Phys. At. Nucl. 63, 1292–1296.
Lacoste, V., 2010. Design of a new long counter for the determination of the neutron fluence reference values at the IRSN AMANDE facility. Radiat. Meas. 45, 1250–1253.
Lacoste, V., Gressier, V., 2010. Experimental characterization of the IRSN long counter for the determination of the neutron fluence reference values at the AMANDE facility. Radiat. Meas. 45, 1254–1257.
Lamadie, F., Delmas, F., Mahe, C., Girones, P., Le Goaller, C., Costes, J.R., 2005. Remote alpha imaging in nuclear installations: new results and prospects. IEEE Trans. Nucl. Sci. 52, 3035–3039.
L'Annunziata, M.F., 1987. "Radionuclide Tracers, Their Detection and Measurement". Academic Press, New York.
L'Annunziata, M.F., 2016. Radioactivity, Introduction and History: From the Quantum to Quarks. Elsevier, Oxford, p. 902.
Lavi, N., Steiner, V., Alfassi, Z.B., 2009. Measurement of radon emanation in construction materials. Radiat. Meas. 44, 396–400.
Lebedev, S.G., Akulinichev, S.V., Iljinov, A.S., Yants, V.E., 2006. A gaseous radiochemical method for registration of ionizing radiation

and its possible applications in science and economy. Nucl. Instrum. Methods Phys. Res. Sect. A 561, 90–99.

Lee, S.H., Gardner, R.P., 2000. A new G-M counter dead time model. Appl. Radiat. Isot. 53, 731–737.

Lentsch, J.W., Babad, H., Stokes, T.I., Hanson, C.E., Vargo, G.F., Boechler, G.N., 1996. New instruments for characterization of high level waste storage tanks at the Hanford site. In: Presented at WM'96 ,Tucson, AZ.

Li, Y., Li, T., Song, G., Mazunga, M., 2015. Response improved for neutron long counter. Radiat. Prot. Dosim. 164, 93–96.

Lintereur, A., Conlin, K., Ely, J., Erikson, L., Kouzes, R., Siciliano, E., et al., 2011. ^{3}He and BF_3 neutron detector pressure effect and model comparison. Nucl. Instrum. Methods Phys. Res. Sect. A 652, 347–350.

Lotfi, Y., Etaati, G.R., Doust-Mohammadi, V., Ghal-Eh, N., 2015. Design and fabrication of a data acquisition system for pulsed neutron flux measurement of plasma focus devices. J. Fusion Energy 34, 411–421.

Lohmann, M., Besch, H.J., Dix, W.-R., Dünger, O., Jung, M., Menk, R.H., Reime, B., Schildwächter, L., 1998. A high sensitive two-line detector with large dynamic range for intravenous coronary angiography. Nucl. Instrum. Methods Phys. Res. Sect. A 419, 276–283.

Lohmanna, M., Besch, H.J., Dix, W.-R., Metge, J., Reime, B., 2003. Demands on a detector for intravenous coronary angiography—experience after 379 patients. Nucl. Instrum. Methods Phys. Res. Sect. A 510, 126–137.

Lopes, J.A.M., dos Santos, J.M.F., Conde, C.A.N., Morgado, R.E., 1999. A new integrated photosensor for gas proportional scintillation counters based on the gas electron multiplier (GEM). Nucl. Instrum. Methods Phys. Res. Sect. A 426, 469–476.

Lopes, J.A.M., dos Santos, J.M.F., Conde, C.A.N., 2000. A large area avalanche photodiode as the VUV photosensor for gas proportional scintilllation counters. Nucl. Instrum. Methods Phys. Res. Sect. A 454, 421–425.

Lopes, J.A.M., dos Santos, J.M.F., Morgado, R.E., Conde, C.A.N., 2001. A Xenon gas proportional scintillation counter with a UV-sensitive large-area avalanche photodiode. IEEE Trans. Nucl. Sci. 48, 312–319.

Loughlin, M.J., Lawrence, G., 2007. Tritium monitoring in the ITER neutral beam test facility. Fus. Eng. Design 82, 646–651.

Lyoussi, A., Romeyer-Dherey, J., Buisson, A., 1996. Low Level Transuranic Waste Assay System Using Sequential Photon Interrogation and on Line Neutron Counting Signatures. Presented at WM'96, Tucson, AZ.

MacArthur, D.W., Allander, K.S., Bounds, J.A., Catlett, M.M., Mcatee, J.L., 1992. Long- range alpha detector (LRAD) for contamination monitoring. IEEE Trans. Nucl. Sci. 39, 952–957.

MacArthur, D.W., Allander, K.S., Bounds, J.A., Caress, R.W., Catlett, M.M., Rutherford, D.A., 1993. LRAD Surface Monitors. LA-12524-MS. Los Alamos National Laboratory Document.

Maekawa, F., Oyama, Y., 1995a. Neutron spectrum measurement in the energy region of eV with the slowing down time method. In: Miyajima, M., Sasaki, S., Iguchi, T., Nakazawa, N., Takebe, M. (Eds.), Proceedings of the 9th Workshop on Radiation Detectors and Their Uses (Nat. Inst. Phys. KEK Japan).

Maekawa, F., Oyama, Y., 1995b. Measurement of low energy neutron spectrum below 10 keV with the slowing down time method. Nucl. Instrum. Methods Phys. Res. Sect. A 372, 262–274.

Maekawa, F., Oyama, Y., 1997. Measurement of neutron energy spectrum below 10 keV in an iron shield bombarded by deuterium tritium neutrons and benchmark test of evaluated nuclear data from 14 MeV to 1 eV. Nucl. Sci. Eng. 125, 205–217.

Mäntynen, P., Äikää, O., Kankainen, T., Kaihola, L., 1987. Application of pulse shape discrimination to improve the precision of the carbon -14 gas proportional counting method. Appl. Radiat. Isot. 38, 869–873.

Marouli, M., Dean, J., Spyrou, N.M., 2007. Feasibility of using proportional gas counters as a primary standard for positron emitters in gas. Nucl. Instrum. Methods Phys. Res. Sect. A 580, 660–662.

Marouli, M., Dean, J.C.J., Sephton, J.P., Spyrou, N.M., 2010. Standardisation of positron-emitters in gas with the NPL primary gas counting system. Appl. Radiat. Isot. 68, 1378–1382.

Martín Sánchez, A., Sáenz García, G., Jurado Vargas, M., 2009. Study of self-absorption for the determination of gross alpha and beta activities in water and soil samples. Appl. Radiat. Isot. 67, 817–820.

Martschini, M., Fifield, L.K., Froehlich, M.B., Leckenby, G., Pavetich, S., Tims, S.G., et al., 2019. New and upgraded ionization chambers for AMS at the Australian National University. Nucl. Instrum. Methods Phys. Res. Sect. B. 438, 141–147.

Masaoka, S., Katano, R., Isozumi, Y., 2000. Helium-filled proportional counter ($\sim$4.2 K) and its new application. KEK Proc. 12–21, 2000-14.

Masuda, T., Sakurai, H., Inoue, Y., Gunji, S., Asamura, K., 2002. Optical imaging capillary gas proportional counter with Penning mixtures. IEEE Trans. Nucl. Sci. 49, 553–558.

Mateos, J.J., Gomez, E., Garcias, F., Casas, M., Cerdá, V., 2000. Rapid ^{90}Sr/^{90}Y determination in water samples using a sequential injection method. Appl. Radiat. Isot. 53, 139–144.

Mathieu, L., Serot, O., Materna, T., Bail, A., Koster, U., Faust, H., et al., 2012. New neutron long-counter for delayed neutron investigations with the LOHENGRIN fission fragment separator. J. Instrum. 7, P08029/1–P08029/31.

Maxwell III, S.L., Culligan, B.K., 2009a. New column separation method for emergency urine samples. J. Radioanal. Nucl. Chem. 279, 105–111.

Maxwell III, S.L., Culligan, B.K., 2009b. Rapid determination of radiostrontium in emergency milk samples. J. Radioanal. Nucl. Chem. 279, 757–760.

Maxwell III, S.L., Culligan, B.K., 2009c. Rapid separation method for emergency water and urine samples. J. Radioanal. Nucl. Chem. 279, 901–907.

Menk, R.H., Sarvestani, A., Besch, H.J., Walenta, A.H., Amenitsch, H., Bernstorff, 2000. Gas gain operations with single photon resolution using an integrating ionization chamber in small angle X-ray scattering experiment. Nucl. Instrum. Methods Phys. Res. Sect. A 440, 181–190.

Mazzucconi, D., Bortot, D., Pola, A., Agosteo, S., Pasquato, S., Fazzi, A., et al., 2018. Monte Carlo simulation of a new TEPC for microdosimetry at nanometric level: response against a carbon ion beam. Radiat. Meas. 113, 7–13.

Meric, I., Johansen, G.A., Holstad, M.B., Gardner, R.P., 2011a. Monte Carlo modelling of gamma-ray stopping efficiencies of Geiger-Müller counters. Nucl. Instrum. Methods Phys. Res. Sect. A 636, 61–66.

Meric, I., Johansen, G.A., Holstad, M.B., Lee, K.O., Calderon, A.F., Wang, J., Gardner, R.P., 2011b. A single scatter electron Monte Carlo approach for simulating gamma-ray stopping efficiencies of Geiger-Müller counters. Nucl. Instrum. Methods Phys. Res. Sect. A 654, 279–287.

Meric, I., Johansen, G.A., Holstad, M.B., Calderon, A.F., Gardner, R.P., 2012. Enhancement of the intrinsic gamma-ray stopping efficiency of Geiger-Mueller counters. Nucl. Instrum. Methods Phys. Res. Sect. A 696, 46–54.

Miller, M.E., Marianib, L.E., Sztejnberg Gonçalves-Carralves, M.L., Skumanicb, M., Thorp, S.I., 2004. Implantable self-powered detector for on-line determination of neutron flux in patients during NCT treatment. Appl. Radiat. Isot. 61, 1033–1037.

Min, J.S., Lee, K.W., Kim, H.R., Lee, C.W., 2017. Radiological assessment of the decontaminated and decommissioned Korea Research Reactor-1 building. Nucl. Eng. Des. 322, 492–496.

Mir, J.A., Conceição, A.S., Maia, J.M., Veloso, J.F.C.A., dos Santos, J, M.F., 2009. Comparative studies of MSGC and GEM for photon and electron detection at ambient conditions. Nucl. Instrum. Methods Phys. Res. Sect. A 598, 505–509.

Monteiro, C.M.B., Lopes, J.A.M., Simões, P.C.P.S., dos Santos, J.M.F., Conde, C.A.N., 2001. An argon gas proportional scintillation counter with UV avalanche photodiode scintillation readout. IEEE Trans. Nucl. Sci. 48, 1081–1086.

Monteiro, C.M.B., Morgado, R.E., dos Santosa, J.M.F., Conde, C.A.N., 2004. A curved-grid gas proportional scintillation counter instrumented with a 25-mm active-diameter photosensor. Nucl. Instrum. Methods Phys. Res. Sect. A 522, 407–412.

Moro, D., Colautti, P., Gualdrini, G., Masi, M., Conte, V., De Nardo, L., Tornielli, G., 2006. Two miniaturised TEPCs in a single detector for BNCT microdosimetry. Radiat. Prot. Dosim. 122, 396–400.

Moro, D., Chiriotti, S., 2015. EuTEPC: measurements in gamma and neutron fields. Radiat. Prot. Dosim. 166, 266–270.

Moro, D., Chiriotti, S., Colautti, P., Conte, V., 2014. TEPC gas gain measurements in propane. Radiat. Prot. Dosim. 161, 459–463.

Mörmann, D., Breskin, A., Chechik, R., Shalem, C., 2004. Operation principles and properties of the multi-GEM gaseous photomultiplier with reflective photocathode. Nucl. Instrum. Methods Phys. Res. Sect. A 530, 258–274.

Moslehi, A., Raisali, G., 2017. A multi-element thick gas electron multiplier-based microdosimeter for measurement of neutrons dose-equivalent: a Monte Carlo study. Radiat. Prot. Dosim. 176, 404–410.

Moslehi, A., Raisali, G., 2018. Simulated response of a multi-element thick gas electron multiplier-based microdosimeter to high energy neutrons. Appl. Radiat. Isot. 137, 236–240.

Mukherjee, B., Fuentes, C., Lambert, J., 2017. Out-of-field dosimetry and 2nd cancer risk assessment of child patients under proton therapy using a TLD-based microdosimeter. Radiat. Meas. 106, 336–340.

Mustafa, S.M., Mahesh, K., 1978. Criterion for determining saturation current in parallel plate ionization chambers. Nucl. Instrum. Methods 150, 549–553.

Nadeau, M.-J., Vaernes, E., Svarva, H.L., Larsen, E., Gulliksen, S., Klein, M., et al., 2015. Status of the "new" AMS facility in Trondheim. Nucl. Instrum. Methods Phys. Res. Sect. B 361, 149–155.

Nakamura, T., 2012. Neutron detector development and measurements around particle accelerators. Indian J. Pure Appl. Phys. 50, 427–438.

Nam, U.-W., Park, W.-K., Lee, J., Pyo, J., Moon, B., Kim, S., 2015. Proton beam dosimetry using a TEPC with a ^{252}Cf neutron calibration. J. Korean Phys. Soc. 67, 1459–1462.

Naqvi, A.A., Khiari, F.Z., Liadi, F.A., Khateeb ur, R., Raashid, M., Isab, A.A., 2018. Neutron moderation effects in phc contaminated soil samples. J. Radioanal. Nucl. Chem. 315, 475–480.

Nariyama, N., Kioshi, N., Ohnishi, S., 2004. Development of a portable free-air ionization chamber as an absolute intensity monitor for high-energy synchrotron radiation up to 150 keV. Nucl. Instrum. Methods Phys. Res. Sect. A 524, 324–331.

Natal da Luz, H., Veloso, J.F.C.A., Mendes, N.F.C., dos Santos, J.M.F., Mir, J.A., 2007a. MHSP with position detection capability. Nucl. Instrum. Methods Phys. Res. Sect. A 573, 191–194.

Natal da Luz, H., Mir, J.A., Veloso, J.F.C.A., dos Santos, J.M.F., Rhodes, N.J., Schooneveld, E.M., 2007b. Micro-hole and strip plate (MHSP) operation in CF_4. Nucl. Instrum. Methods Phys. Res. Sect. A 580, 286–288.

Natal da Luz, H., Veloso, J.F.C.A., dos Santos, J.M.F., Mir, J.A., 2007c. A simple X-ray position detection system based on a MHSP. Nucl. Instrum. Methods Phys. Res. Sect. A 580, 1083–1086.

Nathwani, A.C., Down, J.F., Goldstone, J., Yassin, J., Dargan, P.I., Virchis, A., et al., 2016. Polonium-210 poisoning: a first-hand account. Lancet 388, 1075–1080.

Neumann, S., Böttger, R., Guldbakke, S., Matzke, M., Sosaat, W., 2002. Neutron and photon spectrometry in mono energetic neutron fields. Nucl. Instrum. Methods Phys. Res. Sect. A 476, 353–357.

Nichiporov, D., Coutinho, L., Klyachko, A.V., 2016. Characterization of a GEM-based scintillation detector with He-CF_4 gas mixture in clinical proton beams. Phys. Med. Biol. 61, 2972–2990.

Nickles, J., Bräuning, H., Bräuning-Demian, A., Dangendorf, V., Breskin, A., Chechik, R., Rauschnabel, K., Schmidt Böcking, H., 2002. A gas scintillation counter with imaging optics and large area UV-detector. Nucl. Instrum. Methods Phys. Res. Sect. A 477, 59–63.

Niebuhr, C., 2006. Aging effects in gas detectors. Nucl. Instrum. Methods Phys. Res. Sect. A 566, 118–122.

Nielsen, D.R., Cassel, D.K., 1984. Soil water management. In: L'Annunziata, M.F., Legg, J.O. (Eds.), Isotopes and Radiation in Agricultural Sciences, vol. 1. Academic Press, London and New York, pp. 35–65.

Nowack, G.F., 1987. Electrical compensation method of mechanically induced disturbances in gas-filled radiation detectors. Nucl. Instrum. Methods Phys. Res. Sect. A 255, 217–221.

Oed, A., 1995. Properties of micro-strip gas chambers (MSGC) and recent developments. Nucl. Instrum. Methods Phys. Res. Sect. A 367, 34–40.

Oed, A., 2001. Micro pattern structures for gas detectors. Nucl. Instrum. Methods Phys. Res. Sect. A 471, 109–114.

Oger, T., Chen, W.-T., Cussonneau, J.-P., Donnard, J., Duval, S., Lamblin, J., Lemaire, O., Mohamad Hadi, A.F., Leray, P., Morteau, E., Scotto Lavina, L., Stutzmann, J.-S., Thers, D., 2012. A liquid xenon TPC for a medical imaging Compton telescope. Nucl. Instrum. Methods Phys. Res. Sect. A 695, 125–128.

Ogino, H., Hattori, T., 2013. Operational level for unconditional release of contaminated property from affected areas around Fukushima Daiichi nuclear power plant. Radiat. Prot. Dosim. 157, 446–454.

Ohsawa, D., Masaoka, S., Katano, R., Isozumi, Y., 2000. Resolution of a position sensitive proportional counter with a resistive anode wire of carbon fiber. Appl. Radiat. Isot. 52, 943–954.

Ohshita, H., Uno, S., Otomo, T., Koike, T., Murakami, T., Satoh, S., Sekimoto, M., Uchida, T., 2010. Development of a neutron detector with a GEM. Nucl. Instrum. Methods Phys. Res. Sect. A 623, 126–128.

Ometáková, J., Dulanská, S., Mátel, L., Remenec, B., 2011. A comparison of classical ^{90}Sr separation methods with selective separation using molecular recognition technology products AnaLig® SR-01 gel, 3M Empore™ Strontium Rad Disk and extraction chromatography Sr® Resin. J. Radioanal. Nucl. Chem. 290, 319–323.

Ortuño-Prados, F., Bazzano, A., Berry, A., Budtz-Jørgensen, C., Hall, C., Helsby, W., Lewis, R., Parker, B., Ubertini, P., 1999. A high-pressure MWPC detector for crystallography. Nucl. Instrum. Methods Phys. Res. Sect. A 420, 445–452.

Ovchinnikov, B.M., Parusov, V.V., 1999. A method for background reduction in an experiment for WIMP search with a Xe (Ar)-liquid ionization chamber. Astropart. Phys. 10, 129–132.

Pancin, J., Aune, S., Berthoumieux, E., Boyer, S., Delagnes, E., Macary, V., Poumarede, B., Safa, H., 2007. Neutron detection in high γ-background using a micomegas detector. Nucl. Instrum. Methods Phys. Res. Sect. A 572, 859–865.

Pancin, J., Andriamonje, S., Aune, S., Giganon, A., Giomataris, Y., Lecolley, J.,F., Riallot, M., Rosa, R., 2008. Piccolo Micromegas: first in-core measurement in a muclear reactor. Nucl. Instrum. Methods Phys. Res. Sect. A 592, 104–113.

Pandola, L., Cattadori, C., Ferrari, N., 2004. Neural network pulse shape analysis for proportional counters events. Nucl. Instrum. Methods Phys. Res. Sect. A 522, 521–528.

Papachristodoulou, C.A., Patiris, D.L., Ioannides, K.G., 2010. Exposure to indoor radon and natural gamma radiation in public workplaces in north-western Greece. Radiat. Meas. 45, 865–871.

Papp, Z., Uray, I., 2002. Sensitive method for the determination of ^{18}F attached to aerosol particles in a PET centre. Nucl. Instrum. Methods Phys. Res. Sect. A 480, 788–796.

Passo, C., Kessler, M., 1992. The Essentials of Alpha/beta Discrimination. Packard Instument Company, Meriden, CT.

Pazianotto, M.T., Goncalez, O.L., Federico, C.A., Carlson, B.V., 2013. Study of a long counter neutron detector for the cosmic-ray-induced neutron spectrum. IEEE Trans. Nucl. Sci. 60, 897–902.

Perez, M., Chavez, E., Echeverria, M., Cordova, R., Recalde, C., 2018. Assessment of natural background radiation in one of the highest regions of Ecuador. Radiat. Phys. Chem. 146, 73–76.

Peskov, V., Fonte, P., Martinengo, P., Nappi, E., Oliveira, R., Pietropaolo, F., Picchi, P., 2012. Advances in the development of micropattern gaseous detectors with resistive electrodes. Nucl. Instrum. Methods Phys. Res. Sect. A 661, S153–S155.

Phillips, H.C., Johansson, L.C., Sephton, J.P., 2010. Standardisationof^{85}Kr. Appl. Radiat. Isot. 68, 1335–1339.

Pitts, W.K., Martin, M.D., 2001. Experience with laser microfabricated detectors at the University of Louisville. Nucl. Instrum. Methods Phys. Res. Sect. A 471, 268–271.

Pommé, S., Altzitzoglou, T., Van Ammel, R., Sibbens, G., 2005. Standardisation of ^{125}I using seven techniques for radioactivity measurement. Nucl. Instrum. Methods Phys. Res. Sect. A 544, 584–592.

Prasad, K.R., Balagi, V., 1996. Uranium-233 fission detectors for neutron flux measurement in reactors. Rev. Sci. Instrum. 67, 2197–2201.

Qian, S., Ning, Z., Zhang, J., Zhang, Q., Han, R., 2015. The study and design of the large area neutron monitor with RPC-Gd. J. Instrum. 10, C02014/1–C02014/9.

Qian, S., Wang, Y.-F., Zhang, J.-W., Li, J., Chen, Y.-B., Chen, J., et al., 2009. Study of the RPC-Gd as thermal neutron detector. Chin. Phys. C 33, 769–773.

Rachinhas, P.J.B.M., Simões, P.C.P.S., Lopes, J.A.M., Dias, T.H.V.T., Morgado, R.E., dos Santos, J.M.F., Stauffer, A.D., Conde, C.A.N., 2000. Simulation and experimental results for the detection of conversion electrons with gas proportional scintillation counters. Nucl. Instrum. Methods Phys. Res. Sect. A 441, 468–478.

Radeka, V., Schaknowski, N.A., Smith, G.C., Yu, B., 1998. High performance, imaging, thermal neutron detectors. Nucl. Instrum. Methods Phys. Res. Sect. A 419, 642–647.

Rapisarda, D., Vermeerenc, L., García, Á., Cabellos, Ó., García, J.M., Ibarra, Á., Gómez-Rosa, J.M., Mota, F., Casal, N., Queral, V., 2011. Study on the response of IFMIF fission chambers to mixed neutron-gamma fields: PH-2 experimental tests. Fus. Eng. Design 86, 1232–1235.

Renker, D., 2009. Geiger-mode avalanche photodiodes for Cherenkov detectors. J. Instrum. 5 (P01001), 12.

Rezende, E.A., Correia, A.R., Iwahara, A., da Silva, C.J., Tauhata, L., Poledna, R., et al., 2012. Radioactivity measurements of ^{177}Lu, ^{111}In and ^{123}I by different absolute methods. Appl. Radiat. Isot. 70, 2081–2086.

Righia, S., Lucialli, P., Bruzzi, L., 2005. Health and environmental impacts of a fertilizer plant – Part II: assessment of radiation exposure. J. Environ. Radioact. 82, 183–198.

Roccaro, A., Tomita, H., Ahlen, S., Avery, D., Inglis, A., Battat, J., Dujmic, D., Fisher, P., Henderson, S., et al., 2009. A background-free direction-sensitive neutron detector. Nucl. Instrum. Methods Phys. Res. Sect. A 608, 305–309.

Rodríguez, P.B., Sánchez, A.M., Tomé, F.V., 1997. Experimental studies of self-absorption and backscattering in alpha-particle sources. Appl. Radiat. Isot. 48, 1215–1220.

Rodriguez, R., Avivar, J., Ferrer, L., Leal, L.O., Cerda, V., 2012. Automated total and radioactive strontium separation and preconcentration in samples of environmental interest exploiting a lab-on-valve system. Talanta 96, 96–101.

Roessler, N., Englert, D., Neumann, K., 1993. New instruments for high throughput receptor binding assays. J. Receptor Res. 13, 135–145.

Rosenfeld, A.B., 2016. Novel detectors for silicon based microdosimetry, their concepts and applications. Nucl. Instrum. Methods Phys. Res. Sect. A 809, 156–170.

Rosner, G., Hötzl, H., Winkler, R., 1992. Determination of ^{241}Pu by low level beta proportional counting, application to Chernobyl fallout samples and comparison with the ^{241}Am build-up method. J. Radioanal. Nucl. Chem. 163, 225–233.

Rossi, H.H., Zaider, M., 1996. Microdosimetry and its Applications. Springer, New York.

Ryzhov, I.V., Tutin, G.A., Mitryukhin, A.G., Oplavin, V.S., Soloviev, S.M., Blomgren, J., Renberg, P.-U., Meulders, J.P., El Masri, Y., Keutgen, T., Prieels, R., Nolte, R., 2006. Measurements of neutron-induced fission cross-sections of 205Tl, 204,206,207,208Pb and 209Bi with a multi-section Frisch-gridded ionization chamber. Nucl. Instrum. Methods Phys. Res. Sect. A 562, 439–448.

Saey, P.R.J., 2007. Ultra-low-level measurements of argon, Krypton and radioxenon for treaty verification purposes. ESARDA Bull. 36, 42–56.

Saey, P.R.J., 2009. The influence of radiopharmaceutical isotope production on the global radioxenon background. J. Environ. Radioact. 100, 396–406.

Saey, P.R.J., 2010. Xenon. In: Atwood, D.A. (Ed.), Radionuclides in the Environment. John Wiley & Sons, Chichester.

Saito, N., Suzuki, I.H., 1999. Absolute fluence rates of soft X-rays using a double ion chamber. J. Electron. Spectrosc. Relat. Phenom. 101–103, 33–37.

Sakamoto, S., Morioka, A., 1993. Pulse shape discrimination with proportional counters for neutron detection. In: Miyajima, M., Sasaki, S., Yoshimura, Y., Iguchi, T., Nakazawa, N. (Eds.), Proceedings of the Seventh Workshop on Radiation Detectors and Their Uses (Nat. Inst. Phys. KEK Japan).

Sakamoto, S., Morioka, A., 1994. Temperature dependence of BF_3 proportional counters. Nucl. Instrum. Methods Phys. Res. Sect. A 353, 160–163.

Sakuraia, H., Gunji, S., Tokanai, F., Maeda, T., Saitoh, N., Ujiie, N., 2003. Photoelectron track image of capillary gas proportional counter. Nucl. Instrum. Methods Phys. Res. Sect. A 505, 219–222.

Samardzic, S., Milosevic, M., Todorovic, N., Lakatos, R., 2018. Improved non-destructive method for 90Sr activity determination in aqueous solutions using Monte Carlo simulation. Appl. Radiat. Isot. 137, 199–204.

Sand, J., Ihantola, S., Perajarvi, K., Nicholl, A., Hrnecek, E., Toivonen, H., et al., 2015. Imaging of alpha emitters in a field environment. Nucl. Instrum. Methods Phys. Res. Sect. A 782, 13–19.

Sand, J., Ihantola, S., Perajarvi, K., Toivonen, H., Sand, J., Toivonen, J., 2016. Optical detection of radon decay in air. Sci. Rep. 6, 21532.

Santonico, R., 2003. RPCs: the challenges for the next two years. Nucl. Instrum. Methods Phys. Res. Sect. A 508, 1–5.

Sauli, F., 1997. GEM: a new concept for electron amplification in gas detectors. Nucl. Instrum. Methods Phys. Res. Sect. A 386, 531–534.

Sauli, F., 1999. Recent developments and applications of fast position-sensitive gas detectors. Nucl. Instrum. Methods Phys. Res. Sect. A 422, 257–262.

Sauli, F., 2001. Gas detectors: achievements and trends. Nucl. Instrum. Methods Phys. Res. Sect. A 461, 47–54.

Sauli, F., 2002. Micro-pattern gas detectors. Nucl. Instrum. Methods Phys. Res. Sect. A 477, 1–7.

Sauli, F., 2003. Development and applications of gas electron multiplier detecors. Nucl. Instrum. Methods Phys. Res. Sect. A 505, 195–198.

Sauli, F., 2004a. Progress with the gas electron multiplier. Nucl. Instrum. Methods Phys. Res. Sect. A 533, 93–98.

Sauli, F., 2004b. From bubble chambers to electronic systems: 25 years of evolution in particle detectors at CERN (1979–2004). Phys. Reports 403–407, 471–504.

Sauli, F., 2007. Imaging with the gas electron multiplier. Nucl. Instrum. Methods Phys. Res. Sect. A 580, 971–973.

Sauli, F., 2010. Recent topics on gaseous detectors. Nucl. Instrum. Methods Phys. Res. Sect. A 623, 29–34.

Sauli, F., 2016. The gas electron multiplier (GEM): operating principles and applications. Nucl. Instrum. Methods Phys. Res. Sect. A 805, 2–24.

Scarpitta, S., Odin-McCabe, J., Gaschott, R., Meier, A., Klug, E., 1999. Comparison of four 90Sr groundwater analytical methods. Health Phys. 76, 644–656.

Seifert, A., Aalseth, C.E., Bonicalzi, R.M., Bowyer, T.W., Day, A.R., Fuller, E.S., et al., 2013. Calibration of an ultra-low-background proportional counter for measuring 37Ar. AIP Conf. Proc. 1549, 26–29.

Semkow, T.M., Parekh, P.P., 2001. Principles of gross alpha and beta radioactivity detection in water. Health Phys. 81, 567–573.

Serot, O., Amouroux, C., Bidaud, A., Capellan, N., Chabod, S., Ebran, A., et al., 2014. Recent results from Lohengrin on fission yields and related decay properties. Nucl. Data Sheets 119, 320–323.

Shea, D.A., Morgan, D., 2010. The 3He Shortage: Supply, Demand, and Options for Congress. Congressional Research Service, The Library of Congress.

Shenhav, N.J., Stelzer, H., 1985. The mass dependence of the signal peak height of a Bragg-curve. Nucl. Instrum. Methods 228, 359–364.

Shekhtman, L., 2002. Micro-pattern gaseous detectors. Nucl. Instrum. Methods Phys. Res. Sect. A 494, 128–141.

Shekhtman, L., 2004. Novel position-sensitive gaseous detectors for X-ray imaging. Nucl. Instrum. Methods Phys. Res. Sect. A 522, 85–92.

Siketic, Z., Skukan, N., Bogdanovic, R.I., 2015. A gas ionisation detector in the axial (Bragg) geometry used for the time-of-flight elastic recoil detection analysis. Rev. Sci. Instrum. 86, 083301.

Silva, A.L.M., Oliveira, C.A.B., Gouvêa, A.L., dos Santos, J.M.F., Carvalho, M.L., Veloso, J.F.C.A., 2009. EDXRF imaging of Pb in glazed ceramics using a micropattern gas detector. Anal. Bioanal. Chem. 395, 2073–2080.

Silva, A.L.M., Azevedo, C.D.R., Carramate, L.F.N.D., Lopes, T., Castro, I.F., Oliveira, R.d., et al., 2013. X-ray imaging detector based on a position sensitive THCOBRA with resistive line. J. Instrum. 8, P05016.

Simgen, H., Buck, C., Heusser, G., Laubenstein, M., Rau, W., 2003. A new system fort he 222Rn and 226Ra assay of water and results in the BOREXINO project. Nucl. Instrum. Methods Phys. Res. Sect. A 497, 407–413.

Simões, P.C.P.S., dos Santos, J.M.F., Conde, C.A.N., 2001. Driftless gas proportional scintillation counter pulse analysis using digital processing techniques. X Ray Spectrom. 30, 342–347.

Simões, P.C.P.S., Covita, D.S., Veloso, J.F.C.A., dos Santos, J.M.F., Morgado, R.E., 2003. A new method for pulse analysis of driftless-gas proportional scintillation counters. Nucl. Instrum. Methods Phys. Res. Sect. A 505, 247–251.

Simon, F., Kelsey, J., Kohl, M., Majka, R., Plesko, M., Sakuma, T., Smirnov, N., Spinka, H., Surrow, B., Underwood, D, 2009. Beam performance of tracking detectors with industrially produced GEM foils. Nucl. Instrum. Methods Phys. Res. Sect. A 598, 432–438.

Stamoulis, K.C., Assimakopoulos, P.A., Ioannides, K.G., Johnson, E., Soucaco, P.N., 1999. Strontium-90 concentration measurements in human bones and teeth in Greece. Sci. Total Environ. 229, 165–182.

Stanga, D., Picolo, J.L., Coursol, N., Mitev, K., Moreau, I., 2002. Analytical calculations of counting losses in internal gas proportional counting. Appl. Radiat. Isot. 56, 231–236.

Stanga, D., Cassette, P., 2006. Improved method of measurement for tritiated water standardization by internal gas proportional counting. Appl. Radiat. Isot. 64, 160–162.

Stanga, D., Moreaub, I., Cassette, P., 2006. Standardization of tritiated water by two improved methods. Appl. Radiat. Isot. 64, 1203–1206.

Strand, P., Brown, J.E., Drozhko, E., Mokrov, Y., Salbu, B., Oughton, D., Christensen, G.C., Amundsen, I., 1999. Biogeochemical behaviour of 137Cs and 90Sr in the artificial reservoirs of Mayak PA, Russia. Sci. Total Environ. 241, 107–116.

Suran, J., Kovar, P., Smoldasova, J., Solc, J., Van Ammel, R., Garcia Miranda, M., et al., 2018. Metrology for decommissioning nuclear facilities: partial outcomes of joint research project within the European Metrology Research Program. Appl. Radiat. Isot. 134, 351–357.

Surette, R.A., Waker, A.J., 1994. Workplace monitoring of swipes and air filters for 55Fe. IEEE Trans. Nucl. Sci. 41, 1374–1378.

Suter, M., Doebeli, M., Grajcar, M., Mueller, A., Stocker, M., Sun, G., et al., 2007. Advances in particle identification in AMS at low energies. Nucl. Instrum. Methods Phys. Res. Sect. B 259, 165–172.

Szalóki, I., Török, S.B., Ro, C.U., Injuk, J., Van Grieken, R.E., 2000. X-ray spectrometry. Anal. Chem. 72, 211–233.

Tagziria, H., Thomas, D.J., 2000. Calibration and Monte Carlo modelling of neutron long counters. Nucl. Instrum. Methods Phys. Res. Sect. A 452, 470–483.

Takahashi, H., Mitsuya, Y., Fujiwara, T., Fushie, T., 2013. Development of a glass GEM. Nucl. Instrum. Methods Phys. Res. Sect. A 724, 1–4.

Tamboul, J.Y., Watt, D.E., 2001. A proportional counter for measurement of the bio-effectiveness of ionising radiations at the DNA level. Nucl. Instrum. Methods Phys. Res. Sect. B 184, 597–608.

Taylor, G.C., Hawkes, N.P., Shippen, A., 2015. Accurate simulations of TEPC neutron spectra using Geant4. Radiat. Phys. Chem. 116, 186–188.

Thakur, P., Mulholland, G.P., 2011. Monitoring of gross alpha, gross beta and actinides activities in exhaust air released from the waste isolation pilot plant. Appl. Radiat. Isot. 69, 1307–1312.

Theodórsson, P., 1991. Gas proportional versus liquid scintillation counting, radiometric versus AMS dating. Radiocarbon 33, 9–13.

Theodórsson, P., Heusser, G., 1991. External guard counters for low-level counting systems. Nucl. Instrum. Methods Phys. Res. Sect. B 53, 97–100.

Theodórsson, P., 1999. A review of low-level tritium system and sensitivity requirements. Appl. Radiat. Isot. 50, 311–316.

Titov, M., 2007. New developments and future perspectives of gaseous detectors. Nucl. Instrum. Methods Phys. Res. Sect. A 581, 25–37.

Tolk, J.A., Evett, S.R., 2009. Lysimetry versus neutron moisture meter for evapotranspiration determination in four soils. Soil Sci. Soc. Am. J. 73, 1693–1698.

Tosaki, M., Kawano, T., Isozumi, Y., 2013. Measurements of electron attachment by oxygen molecule in proportional counter. Nucl. Instrum. Methods Phys. Res. Sect. B 315, 72–75.

Toyokawa, H., Urotani, A., Mori, C., Takeda, N., Kudo, K., 1995. A multipurpose spherical neutron counter. IEEE Trans. Nucl. Sci. 42, 644–648.

Tran, L.T., Chartier, L., Bolst, D., Davis, J., Prokopovich, D.A., Pogossov, A., et al., 2018. In-field and out-of-file application in ^{12}C ion therapy using fully 3D silicon microdosimeters. Radiat. Meas. 115, 55–59.

Troja, S.O., Cro, A., Picouet, P., 1995. Alpha and beta dose-rate determination using a gas proportional counter. Radiat. Meas. 24, 297–308.

Tsyganov, E., Antich, P., Parkey, R., Seliounine, S., Golovatyuk, V., Lobastov, S., Zhezher, V., Buzulutskov, A., 2008. Gas electron Multiplying detectors for medical applications. Nucl. Instrum. Methods Phys. Res. Sect. A 597, 257–265.

Tutin, G.A., Ryzhov, I.V., Eismont, V.P., Kireev, A.V., Condé, H., Elmgren, K., Olsson, N., Renberg, P.U., 2001. An ionization chamber with Frisch grids for studies of high-energy neutron - induced fission. Nucl. Instrum. Methods Phys. Res. Sect. A 457, 646–652.

Unterweger, M.P., Lucas, L.L., 2000. Calibration of the national Institute of standards and technology tritiated-water standards. Appl. Radiat. Isot. 52, 527–431.

Uozumi, Y., Sakae, T., Matoba, M., Ijiri, H., Koori, N., 1993. Semi-microscopic formula for gas gain of proportional counters. Nucl. Instrum. Methods Phys. Res. Sect. A 324, 558–564.

Uritani, A., Kuniya, Y., Takenaka, Y., Toyokawa, H., Yamane, Y., Mori, S., Kobayashi, K., Shiroya, S., Ichihara, C., 1995. A long and slender position-sensitive helium-3 proportional counter with an anode wire supported by a ladder shaped solid insulator. J. Nucl. Sci. Technol. 32, 719–726.

Utsunomiya, H., Gheorghe, I., Filipescu, D.M., Glodariu, T., Belyshev, S., Stopani, K., et al., 2017. Direct neutron-multiplicity sorting with a flat-efficiency detector. Nucl. Instrum. Methods Phys. Res. Sect. A 871, 135–141.

Vaca, F., Manjón, G., Cuéllar, S., Garcia-Leon, M., 2001. Factor of merit and minimum detectable activity for ^{90}Sr determinations by gas-flow proportional counting or Cherenkov counting. Appl. Radiat. Isot. 55, 849–851.

van der Graaf, H., 2009. New developments in gaseous tracking and imaging detectors. Nucl. Instrum. Methods Phys. Res. Sect. A 607, 78–80.

van der Graaf, H., 2011. Gaseous detectors. Nucl. Instrum. Methods Phys. Res. Sect. A 628, 27–30.

Vasile, M., Benedik, L., Altzitzoglou, T., Spasova, Y., Wätjen, U., González de Orduña, R., Hult, M., Beyermann, M., Mihalcea, I., 2010. ^{226}Ra and ^{228}Ra determination in mineral waters—comparison of methods. Appl. Radiat. Isot. 68, 1236–1239.

Velasquez, A.A., Arroyave, M., 2014. Implementation of a preamplifier-amplifier system for radiation detectors used in Moessbauer spectroscopy. Hyperfine Interact. 224, 65–72.

Veloso, J.F.C.A., Silva, A.L.M., Oliveira, C.A.B., Gouvêa, A.L., Azevedo, C.D.R., Carramate, L., Natal da Luz, H., dos Santos, J.M.F., 2010. Energy resolved X-ray fluorescence imaging based on a micropattern gas detector. Spectrochim. Acta B 65, 241–247.

Veloso, J.F.C.A., Silva, A.L.M., 2018. Gaseous detectors for energy dispersive X-ray fluorescence analysis. Nucl. Instrum. Methods Phys. Res. Sect. A 878, 24–39.

Villa, M., Duarte Pinto, S., Alfonsi, M., Brock, I., Croci, G., David, E., de Oliveira, R., Ropelewski, L., Taureg, H., van Stenis, M., 2011. Progress on large area GEMs. Nucl. Instrum. Methods Phys. Res. Sect. A 628, 182–186.

Vinagre, F.L.R., Conde, C.A.N., 2001. Method for effective dead time measurement in counting systems. Nucl. Instrum. Methods Phys. Res. Sect. A 462, 555–560.

Vu, T.Q., Allander, K.S., Bolton, R.D., Bounds, J.A., Garner, S.E., Johnson, J.D., Johnson, J.P., MacArthur, D.W., 1994. Application of the long-range alpha detector for site-characterization technology. In: Proceedings of the WM '94 Conference. Tucson, AZ.

Waker, A.J., Dubeau, J., Surette, R.A., 2009. The application of micro-patterned devices for radiation protection dosimetry and monitoring. Nucl. Technol. 168, 202–206.

Waker, A.J., Aslam, 2010. A preliminary study of the performance of a novel design of multi-element tissue equivalent proportional counter for neutron monitoring. Radiat. Meas. 45, 1309–1312.

Waker, A.J., Aslam, 2011a. Study of microdosimetric energy deposition patterns in tissue-equivalent medium due to low-energy neutron fields using a graphite-walled proportional counter. Radiat. Res. 175, 806–813.

Waker, A.J., Aslam, 2011b. An experimental study of the microdosimetric response of a graphite walled proportional counter in low energy neutron fields. Nucl. Instrum. Methods Phys. Res. Sect. A. 652, 721–725.

Wang, Z., Morris, C.L., Makela, M.F., Bacon, J.D., Baer, E.E., Brockwell, M.I., Brooks, B.J., et al., 2009. Inexpensive and practical sealed drift-tube neutron detector. Nucl. Instrum. Methods Phys. Res. Sect. A. 605, 430–432.

Wang, G.-b., Qian, D.-z., Li, J.-j., Yang, X., Li, R.-d., Tang, B., 2015. Experimental and theoretical study of long counters on the departure of "point" assumption and scattering background influence. Radiat. Meas. 82, 146–153.

Waples, J.T., Orlandini, K.A., Weckerly, K.M., Edgington, D.N., Klump, J.V., 2003. Measuring low concentrations of ^{234}Th in water and sediment. Marine Chem. 80, 265–281.

Watanabe, K., Otsuka, J., Shigeyama, M., Suzuki, Y., Yamazaki, A., Unitani, A., 2011. Flat-response neutron detector using spatial distribution of thermal neutrons in a moderator. Nucl. Instrum. Methods Phys. Res. Sect. A. 652, 392–396.

Waters, J.R., 1974. Precautions in the measurement of tritium concentration in air when using flow-through chambers. Nucl. Instrum. Methods 117, 39–43.

Weijers, T.D.M., Ophel, T.R., Timmers, H., Elliman, R.G., 2002. A systematic study of the pulse height deficit in propane – filled gas ionization detectors. Nucl. Instrum. Methods Phys. Res. Sect. A 483, 676–688.

Westphal, G.P., 1976. A high precision pulse-ratio circuit. Nucl. Instrum. Methods 134, 387–390.

Whitley, C.R., Johnson, J.D., Rawool-Sullivan, M., 1996. Real-time Alpha Monitoring of a Radioactive Liquid Waste Stream at Los Alamos National Laboratorty. WM 1996 ,Tucson, AZ.

Whitley, C.R., Bounds, J.A., Steadman, P.A., 1998. A portable swipe monitor for alpha contamination. IEEE Trans. Nucl. Sci. 45, 533–535.

Whitney, C., Pellegrin, S.M., Wilson, C., 2009. A nanoparticle doped micro-Geiger counter for multispecies radiation detection. J. Microelectromech. Syst. 18, 998–1003.

Wojenski, A., Pozniak, K.T., Kasprowicz, G., Kolasinski, P., Krawczyk, R., Zabolotny, W., et al., 2016. FPGA-based GEM detector signal acquisition for SXR spectroscopy system. J. Instrum. 11, C11035.

Wu, X.-M., Tuo, X.-G., Li, Z., Liu, M.-Z., Zhang, J.-Z., Dong, X.-L., et al., 2013. Radioactive source localization inside pipes using a long-range alpha detector. Chin. Phys. C 37 (086201), 5 pp.

Wurdiyanto, G., Marsoem, P., Candra, H., Wijono, P., 2012. Radioactivity measurement of ^{18}F in 16 ml vials for calibration of radionuclide calibrators. Appl. Radiat. Isot. 70, 2237–2239.

Xie, F., He, X., Jiang, W., Zhang, X., Shi, Q., Wu, S., et al., 2014. Development of a radioxenon measurement system and its application in monitoring Fukushima nuclear accident. Radiat. Phys. Chem. 97, 85–89.

Xie, F., Jiang, W., Li, X., He, X., Zhang, J., Yu, G., 2013. Single channel beta-gamma coincidence system for radioxenon measurement using well-type HPGe and plastic scintillator detectors. Nucl. Instrum. Methods Phys. Res. Sect. A 729, 900–904.

Yakushev, A.B., Zvara, I., Oganessian, Y.T., Belozerov, A.V., Dmitriev, S.N., Eichler, B., Hübener, S., Sokol, E.A., Türler, A., Yeremin, A.V., et al., 2004. Chemical identification and properties of element 112. Nucl. Phys. A 734, 204–207.

Yamamoto, S., Kato, K., Kameyama, H., Abe, S., 2018. Imaging of Ra-223 solution using an optical method. Radiat. Meas. 108, 52–58.

Yamamoto, S., Komori, M., Koyama, S., Toshito, T., 2016. Luminescence imaging of water during alpha particle irradiation. Nucl. Instrum. Methods Phys. Res. Sect. A 819, 6–13.

Yu, B., Smith, G.C., Siddons, D.P., Pietraski, P.J., Zojceski, Z., 1999. Position sensitive gas proportional detectors with anode blades. IEEE Trans. Nucl. Sci. 46, 338–431.

Yunoki, A., Yamada, T., Kawada, Y., Unno, Y., Sato, Y., Hino, Y., 2010. Activity measurement of ^{85}Kr diluted by a large volume balloon technique. Appl. Radiat. Isot. 68, 1340–1343.

Zabolotny, W.M., Kasprowicz, G., Pozniak, K., Chernyshova, M., Czarski, T., Gaska, M., et al., 2019. FPGA and Embedded systems based fast data acquisition and processing for GEM detectors. J. Fusion Energy 38, 480–489.

Zeng, Z.-M., Gong, H., Li, J.-M., Yue, Q., Zeng, Z., Cheng, J.-P., 2017. Design of the thermal neutron detection system for CJPL-II. Chin. Phys. C 41, 056002/1–056002/5.

Zhang, L., Takahashi, H., Hinamoto, N., Nakazawa, M., Yoshida, K., 2002. Design of a hybrid gas proportional counter with CdTe guard counters for ^{14}C dating system. Nucl. Instrum. Methods Phys. Res. Sect. A 478, 431–434.

Zhang, Q., Liu, X., Deng, B., Cao, L., Tang, C., 2018. Numerical optimization of rhodium self-powered neutron detector. Ann. Nucl. Energy 113, 519–525.

Zhang, W., Wang, Z., Liu, Y., Li, C., Xiao, X., Luo, H., et al., 2014. TEPC performance for a reference standard. Radiat. Prot. Dosim. 158, 246–250.

Zhang, Y.-L., Qi, H.-R., Hu, B.-T., Wang, H.-Y., Ou-Yang, Q., Chen, Y.-B., et al., 2017. A hybrid structure gaseous detector for ion backflow suppression. Chin. Phys. C 41, 056003/1–056003/5.

Zielińska, A., Dąbrowski, W., Fiutowski, T., Mindur, B., Wiącek, P., Wróbel, P., 2013. X-ray fluorescence imaging system for fast mapping of pigment distributions in cultural heritage paintings. J. Instrum. 8, P10011.

Zikovsky, L., Roireau, N., 1990. Determination of radon in water by argon purging and alpha counting with a proportional counter. Appl. Radiat. Isot. 41, 679–681.

Zondervan, A., Hauser, T.M., Kaiser, J., Kitchen, R.L., Turnbull, J.C., West, J.G., 2015. XCAMS: the compact ^{14}C accelerator mass spectrometer extended for ^{10}Be and ^{26}Al at GNS Science, New Zealand. Nucl. Instrum. Methods Phys. Res. Sect. B 361, 25–33.

Further reading

Andriamonje, S., Aune, S., Bignan, G., Blandin, C., Ferrer, E., Giomataris, L., Jammes, C., Pancin, J., 2004. New neutron detectors based on Micromegas technology. Nucl. Instrum. Methods Phys. Res. Sect. A 525, 74–78.

Beddingfield, D.H., Menlove, H.O., Johnson, N.H., 1999. Neutron proportional counter design for high gamma-ray environments. Nucl. Instrum. Methods Phys. Res. Sect. A 422, 35–40.

dos Santos, J.M.F., Dias, T.H.V.T., Reyes Cortes, S.D.A., Conde, C.A.N., 1989. Novel techniques for designing gas proportional scintillation counters for X-ray spectrometry. Nucl. Instrum. Methods Phys. Res. Sect. A 280, 288–290.

dos Santos, J.M.F., Lopes, J.A.M., Veloso, J.F.C.A., Simões, P.C.P.S., Dias, T.H.V.T., dos Santos, F.P., Rachinhas, P.J.B.M., Ferreira, L.F.R., Conde, C.A.N., 2001. Development of portable gas proportional scintillation counters for X-ray spectrometry. XRay Spectrom. 30, 373–381.

Evans Jr., A.E., 1982. Energy dependence of the response of a ^{3}He long counter. Nucl. Instrum. Methods 199, 643–644.

Facorellis, Y., Kyparissi-Apostolika, N., Maniatis, Y., 2001. The cave of Theopatra, Kalambaka: radiocarbon evidence for 50,000 years of human presence. Radiocarbon 43, 1029–1048.

Fuitas, K., Takahashi, H., Siritiprussamee, P., Nikop, K.H., Nishi, K., Takada, Y., Oku, T., Suzuki, J., Ino, T., Shimizu, H.M., Furusaka, M., 2007. A high-resolution two-dimensional ^{3}He neutron MSGC with pads for neutron scattering experiments. Nucl. Instrum. Methods Phys. Res. Sect. A 580, 1027–1030.

Glergeau, J.,F., Convert, P., Feltin, D., Fischer, H.,E., Guerard, B., Hansen, T., Manin, G., Oed, A., Palleau, P., 2001. Operation of sealed microstrip gas chambers at the ILL. Nucl. Instrum. Methods Phys. Res. Sect. A 471, 60–68.

Levy, J.M., Mack, V., Schunck, J.P., Sigward, M.H., 1995. Microstrip gas chambers on implanted substrates. Nucl. Instrum. Methods Phys. Res. Sect. A 367, 185–188.

Maxwell III, S.L., Culligan, B.K., Noyes, G.W., 2010a. Rapid separation method for actinides in emergency air filters. Appl. Radiat. Isot. 68, 2125–2131.

Maxwell III, S.L., Culligan, B.K., Noyes, G.W., 2010b. Rapid separation of actinides and radiostrontium in vegetation samples. J. Radioanal. Nucl. Chem. 286, 273–282.

Okeji, M.C., Agwu, K.K., 2012. Assessment of indoor radon concentration in phosphate fertilizer warehouses in Nigeria. Radiat. Phys. Chem. 81, 253–255.

Pallares, A., Barthe, S., Bergtold, A.M., Brom, J.M., Cailleret, J., Christophel, E., Coffin, J., Eberle, H., Fang, R., Fontaine, J.C., Geist, W., Kachelhoffer, T., 1992. High Throughput Screening of Samples Containing Alpha &Beta Radionuclides: An Overview of Methods. Application note. PerkinElmer Life and Analytical Sciences, Boston.

Perez-Nunez, D., Braby, L.A., 2011. Replacement tissue-equivalent proportional counter for the international space station. Radiat. Prot. Dosim. 143, 394–397.

Roberts, N.J., Thomas, D.J., Lacoste, V., Böttger, R., Loeb, S., 2010. Comparison of long counter measurements of monoenergetic and radionuclide source-based neutron fluence. Radiat. Meas. 45, 1151–1153.

Sauli, F., 1998. Gas detectors: recent developments and future perspectives. Nucl. Instrum. Methods Phys. Res. Sect. A 419, 189–201.

Veloso, J.F.C.A., dos Santos, J.M.F., Conde, C.A.N., 2000. A proposed new microstructure for gas radiation detectors: the microhole and strip plate. Rev. Si. Instrum. 71, 2371–2376.

Veloso, J.F.C.A., dos Santos, J.M.F., Conde, C.A.N., 2001. Gas proportional scintillation counters with a CsI-covered microstrip plate UV photosensor for high-resolution X-ray spectrometry. Nucl. Instrum. Methods Phys. Res. Sect. A 457, 253–261.

Prof. Dr. Georg Steinhauser

Prof. Dr. Georg Steinhauser is alumnus of the University of Vienna (MSc in chemistry 2003) and the Vienna University of Technology (Atominstitut; PhD in radiochemistry 2005). After 10 years of research using the Atominstitut's TRIGA reactor, he joined Colorado State University's Health Physics Section in the Department of Environmental and Radiological Health Sciences in 2013 as an assistant professor of Radiochemistry. Since 2013, Prof. Steinhauser has been a member of the Radiation Protection Advisory Board of the Austrian Federal Ministry of Health. In October 2015, he was hired by Leibniz University Hannover (Institute of Radioecology and Radiation Protection) as a professor of Physical Radioecology. His main research focus is on environmental radioactivity studies in Chernobyl, Fukushima, and other release scenarios and environmental nuclear forensics as well as accident preparedness and response. He has (co) authored more than 100 publications. Since 2016, Prof. Steinhauser has been an editor of the Springer Nature Journal Environmental Science and Pollution Research.

Prof. Dr. Karl A. Buchtela

Prof. Dr. Karl A. Buchtela, born 1932, was the head of the Radiochemistry Unit at the Atominstitut of Vienna University of Technology from 1975 to 2000. Since then, he continued his participation in teaching and research projects as Professor emeritus at the University, as freelance collaborator and advisor. Prof. Buchtela received his scientific education in chemistry, physics, and economy at the Austrian universities, followed by professional scientific research and teaching positions at the universities, in collaboration with national and international organizations in the framework of planning and design of facilities for the production, application, and analysis of radioactive materials. He was hired as an expert by the International Atomic Energy Agency for special tasks in various member states. Prof. Buchtela has authored and coauthored about 150 scientific papers and chapters in books and has served in cooperation with editorial and peer-reviewing teams globally.

Chapter 3

Solid-state nuclear track detectors

Shi-Lun Guo[1], Bao-Liu Chen[1] and S.A. Durrani[2]

[1]China Institute of Atomic Energy (CIAE), Beijing, China; [2]School of Physics and Astronomy, University of Birmingham, Birmingham, United Kingdom

Chapter outline

Handbook of Radioactivity Analysis: Volume 1. https://doi.org/10.1016/B978-0-12-814397-1.00003-0

Part 1: Elements

I. Introduction

Solid-state nuclear track detector (SSNTD) is a special type of radiation detector, which records tracks of heavy charged particles ($Z \geq 1$) in insulating or dielectric solids.

In a broad sense, nuclear emulsions (Barkas, 1963), a special type of bubble detectors (BDs), etch track detectors, and fluorescent nuclear track detectors (FNTDs) (Akselrod et al., 2006) are all included in SSNTDs because these detector materials are all solid substances and the recorded signals are trajectories (or visible tracks) of charged particles. In this handbook, SSNTD refers to etch track detectors or dielectric track detectors. Track etching is the basic technique used to develop tracks in this type of track detector.

There are some other types of nuclear track detectors, such as cloud chamber, bubble chamber, and spark chamber (Wilson, 1951; Shutt, 1967). Their detector materials are saturated vapors, superheated liquids, and ionizable gases, respectively. These mediums are not solid substances. They do not belong to SSNTDs, even though they can record nuclear particle tracks. A special type of BD is able to record tracks of high-energy heavy ions in the whole Periodic Table of Elements ($Z \geqq 1$) (Guo et al., 2003, 2005, 2015). However, its principle for development of tracks is formation of bubbles from superheated droplets, not track etching. Therefore, it will be excluded from this handbook.

The historic stories of track etching to develop tracks in solid can be retraced to a very early time. H. Baumhauer in 1894 etched apatite with diluted sulfuric acid and saw etched figures (fission fragment tracks); but he did not know what they were because radioactivity was not discovered until two years later by Becquerel (1896a,b) and nuclear fission was discovered many years later (Hahn and Strassmann, 1939a,b; Meitner and Frisch, 1939). In recent books, L'Annunziata (2007, 2016) provides a detailed historical account of the discoveries and research into nuclear radiation. Baumhauer (1894) published his results in a treatise on etch methods. Since the end of the 19th century, etching techniques had become a traditional method to display dislocation in minerals. A. P. Hones in 1927, A. R. Patel, and S. Tolansky in 1957 and L. C. Lovell in 1958 also showed etched figures (tracks) in minerals. None of them recognized the etchings as tracks (Fleischer, 1998). D. A. Young was the first scientist who intentionally revealed fission fragment tracks in lithium fluoride (LiF) by etching in 1958 (Young, 1958). His discovery showed that radiation damage by heavy charged particles (fission fragments) could form tracks in crystal. After his early success he discontinued research in this field. E. C. H. Silk and R. S. Barnes in 1959 observed fission fragment tracks in muscovite mica with a transmission electron microscope (TEM) (Silk and Barnes, 1959). This was the first direct observation of straight latent (un-etched) tracks of fission fragments in mica. They also noticed that latent fission fragment tracks in muscovite mica disappeared (track fading) after a few seconds due to heating by the intensive electron beam of the microscope.

P. B. Price and R. M. Walker in 1961 found a way to circumvent the problem of the rapid fading of tracks in muscovite mica under the electron microscope. They stabilized and fixed the tracks by chemical etching (CE). They also discovered that CE could enlarge fission fragment tracks continuously and the enlarged tracks could be visible under the ordinary optical microscope when the diameter of tracks was enlarged to micrometers (Price and Walker, 1962a, 1962b; 1962c). Their discoveries made fission fragment tracks become sustainable records for continuous research, overcame the limitation of the electron microscope to survey only tiny volumes of solids, and made it possible for many difficult scientific problems to be easily tackled by very simple and popular instruments such as ordinary optical microscopes. Based on these discoveries, "a new track detector" was announced in 1962 by Price and Walker (Price and Walker, 1962d) and later was named the solid-state (nuclear) track detector (Walker, 1963; Fleischer et al., 1965a), which was similar to the cloud chamber, bubble chamber, and spark chamber, and superior to conventional nuclear emulsions in simplicity and ruggedness. At the same time, Price and Walker discovered fossil tracks in mica and invented fission track dating (FTD) (Price and Walker, 1962d, 1963a). They designed a low uranium determination technique (Price and Walker, 1963b) and suggested the idea to make nuclear track membrane (NTM).

R. L. Fleischer joined Price and Walker in 1962. By close cooperation they discovered that polymers and glasses could also record etchable tracks of heavy charged particles (Fleischer and Price, 1963a, 1963b). They extended nuclear track applications into nuclear physics (Price et al., 1963; Fleischer et al., 1964b), neutron detection (Walker et al., 1963), astrophysics (Price et al., 1968a, 1968b), meteorite (Fleischer et al., 1967c), cosmic rays (Price et al., 1967, 1968a, 1968b; Fleischer et al., 1967a), tektites (Fleischer and Price, 1964), archeology (Fleischer et al., 1965d), continental drift and ocean bottom spreading (Fleischer et al., 1968), Quaternary geology (Fleischer et al., 1969b), Crater dating (Fleischer et al., 1969e), dating of mechanical events and underground nuclear explosions (Fleischer et al., 1972, 1974), Lunar science (Fleischer and Hart, 1973), planetary science (Fleischer et al., 1965b), anthropology (Fleischer et al., 1965c, 1965d), search for magnetic monopoles (MMs) (Fleischer et al., 1969a, 1969c), mapping of uranium (Fleischer, 1966), and radon monitoring (Alter and Price, 1968). They made the first practical polymer track membrane (Fleischer et al., 1964a, 1964b).

Before 1967, etch track applications were basically limited to track counting. In 1967, P. B. Price and coworkers discovered a method of identification of charge (Z) of nuclear particles by measuring the etch rate along the trajectory of the particles. They found that track etch rate along the trajectory of the particles is a function of radiation damage density for a specific track detector. By measuring residual range and track etch rate one could determine the charge of the particle creating the track (Price et al., 1967; Price and Fleischer, 1971). This discovery greatly opened new fields of modern applications of SSNTDs to high energy physics, astrophysics, cosmic ray physics, nuclear fragmentation, exotic particles, and many other fields. These applications are discussed further on in this chapter.

Discovery of CR-39 as a track detector in 1978 was another important event (Cartwright et al., 1978). Up to now, CR-39 has been the most useful detector material compared with others.

Electrochemical etching was first reported by L. Tommasino in 1970, which is an alternative way to develop tracks and is used widely in radiation dosimetry (Tommasino, 1970). The nuclear track spark counter was first devised by W. G. Cross and L. Tommasino in 1970 (Cross and Tommasino, 1970), which is an often-used technique for rapid counting of nuclear tracks.

For track formation mechanism, Fleischer et al. (1964a, 1964b) first suggested total energy loss rate as a criterion to judge whether a particle could be recorded by a specific detector material, but soon they found that this criterion could not fit the experimental data and they suggested another: primary ionization rate criterion (Fleischer et al., 1967b, 1967d). Benton and Nix in 1969 suggested restricted energy loss criterion (Benton and Nix, 1969). Katz and Kobetich in 1968 suggested total energy deposition criterion to judge the formation of tracks (Katz and Kobetich, 1968). The last three criterions were almost satisfied by track research, but none of them are perfect theoretically and experimentally. The most often-used criteria are primary ionization rate and restricted energy loss rate. The calculation of total energy deposition criterion is more complicated, and fewer researchers used it in practice.

Due to the obvious advantage of etch track detectors, many researchers joined in the study of etch track detectors; especially many nuclear emulsion scientists fully or partially turned to etch track detectors. This new interest was reflected in the name and topics of the international conferences on nuclear track detectors. At first, the name of the series of the conferences was on nuclear emulsions ("Corpuscular Photography", "Nuclear Photography") from 1957 (the first international conference in Strasbourg) when the SSNTD was yet to be created. Since 1976 (the 9th international conference in Munchen), the name changed to "Solid State Nuclear Track Detector". This change reflected the rapid growth of the team of etch track detectors and the shift of scientific interest from emulsions to etch track detectors. Since 1992 (the 16th international conference in Beijing), the name changed to "Nuclear Tracks in Solids" for the reason that nuclear tracks can be used not only as nuclear particle detectors, but also as nano- and microstructures applied to nanotechnology.

The split-off of the FTD workshop from the international conference on SSNTDs also showed the rapid expansion of the field of etch track detectors. A group of scientists who paid more attention to dating (including measurements of reactor neutrons, induced fission in nuclear reactors and ancient spontaneous fission events) preferred to participate in FTD workshops.

Since 1978, a special journal entitled: "*Nuclear Track Detection*" and later "*Nuclear Tracks and Radiation Measurements*" and now "*Radiation Measurements*" was established to publish research papers on SSNTDs under the auspices of Elsevier Science publishers. All of the international conference proceedings on SSNTDs were published in the above journals or separately in the early years (1976, 1979, 1981) by Pergamon Press. The treatise works such as "Nuclear Tracks in Solids, Principles and Applications" by Fleischer et al. (1975), "Solid State Nuclear Track Detection" by Durrani and Bull (1987), "Ion Tracks and Microtechnology, Basic Principles and Applications" by Spohr (1990), "Fission Track Dating" by Wagner and Van den haute (1992), "Radon Measurements by Etch Track Detectors; Applications to Radiation Protection, Earth Sciences and Environment" by Durrani and Ilic (1997) and "Nuclear Track Detectors: Design, Methods and Applications" edited by Sidorov and Ivanov (2009) had given systematic details or specialized explanations on etch track detectors. The reader is referred to these reference books for additional information on nuclear track detectors.

II. Detector materials and classification of solid-state nuclear track detectors

Nearly all kinds of transparent insulating solids can be used as SSNTDs. These materials can be classified into natural and man-made solids. Natural solids include terrestrial and extraterrestrial ones. Terrestrial solids include all kinds of natural crystalline minerals and amorphous natural glasses. Tektites, obsidian, volcanic glasses, and impact glassed are examples of natural glasses. Extraterrestrial solids include meteorites and lunar rock. Most meteorites come from the asteroid belt between the orbits of Mars and Jupiter. Astronomers conjectured that there existed originally a planet in the asteroid belt. The planet broke at early time into many pieces or fragments. The fragments revolve round the sun on the original orbit of the planet but with some deviation. Some of the biased orbits intersect with the orbit of the earth. The fragments collide with the earth, forming meteorites. The lunar rocks were collected from the moon by astronauts of the United States in the early 1970s. Scientists from many countries participated in the research on lunar rocks with nuclear track techniques. The track lengths or diameters visible after etching under the optical microscope are usually several micrometers. Therefore, the lower limit of the sizes of the minerals or glasses for nuclear track studies is about 10 μm (μm). Very large areas of natural muscovite mica up to $10 \times 10\ cm^2$ can be found and applied to nuclear physics studies. Man-made solids applied to as SSNTDs include man-made crystals (such as synthesized fluoro-phlogopite mica), man-made glasses (such as soda-lime glass and phosphate glass) and plastics (such as cellulose nitrate (CN), polycarbonate, and polyallyldiglycol carbonate (CR-39)). Usually, man-made materials can be made very large for use as track detectors, especially in recording rare events such as cosmic ray particles.

According to track formation mechanisms and etching behaviors, etch track detectors are usually classified as follows with the inclusion of some examples of commonly used detector materials and newly developed materials:

- SSNTD
  - Crystals
    - Muscovite mica
    - Apatite
    - Zircon
    - Sphene
    - Olivine
    - Pyroxene
    - Whitlockite
  - Glasses
    - Man-made glasses
      - Soda-lime glass
      - Phosphate glass
      - BP-1 glass
    - Natural glasses
      - Tektite
      - Obsidian
      - Basaltic glass
  - Polymers
    - Polyallyldiglycol carbonate (PADC, CR-39)
    - Polycarbonate (PC, Lexan, Makrofol, Tuffak, Rodyne)
    - Cellulose nitrate (CN, LR-115, Daicell)
    - Polyethylene terephthalate (PET, Myar, Cronar, Lavsan)
    - Polyimide
    - CR-39-DAP Series
    - PETAC-ADC Copolymer detector
    - Silicon nitrate

In the following subsections, some basic knowledge on each detector material listed above will be given. For information about additional materials, the readers are referred to the monographs written by Fleischer et al. (1975), Durrani and Bull (1987) and Wagner and Van den haute (1992).

A. Crystalline solids

Most of the transparent natural and man-made crystalline solids can be used as solid-state nuclear track detectors. Crystalline solids widely exist in the earth crust as mineral compositions of rocks, soils and dust. Large-sized muscovite mica and quartz are examples of pure track detectors. The main compositions of granite are mica, feldspar, and quartz. All of them are transparent crystalline minerals and can be used as nuclear track detector materials. Many kinds of accessory minerals, such as apatite, zircon, and sphene are also contained in granite. They are also transparent crystals with small sizes from about tens to hundreds of micrometers. They constitute about several percent of the total weight of the granite. These accessory minerals are very useful track detectors for FTD. After separation from the main compositions of the granite, each single grain of the accessory minerals can be used as track detectors. In many other types of rocks on the earth's crust, similar situations exist, but the main and accessory minerals differ from rock to rock. One can select one or more special types of main minerals or accessory minerals as track detectors.

Each crystalline solid has a special crystal structure. The etch rates in different directions of the crystal usually vary. Therefore, etching proceeds anisotropically in crystals. The detection efficiencies of crystalline solid for charged particles (the ratio of recorded to incident particles) on different surfaces are usually diverse.

The most commonly applied crystalline solids and their properties are described below.

1. Muscovite mica

Muscovite mica is a natural resource excavated from a mica mine. Large slabs of muscovite mica are found in pegmatite. It can be easily cleaved into thin flat pieces by a sharp needle or knife. The atomic composition of muscovite mica is $KAl_3Si_3O_{10}(OH)_2$, density is $2.76-3.10\ g/cm^3$ (the commonly quoted density is $2.8\ g/cm^3$). It has three refractive indices $n_a = 1.560$, $n_b = 1.594$ and $n_c = 1.598$ in three directions, respectively. In track depth measurements under the optical microscope, the refractive index is taken as $n = 1.598$, that is n_c. Usually, the concentration of uranium in muscovite mica is very low ($\sim 10^{-9}$ g/g). It is about three orders of magnitude lower than that of the materials in the earth's crust ($\sim 2.7 \times 10^{-6}$ g/g). Therefore, muscovite mica is a good detector material with low concentration of uranium. It is suitable as a track detector for the determination of low uranium concentrations. The uranium concentration in muscovite mica varies. One must determine its uranium concentration before using it as a detector to determine the uranium concentration of other materials.

Muscovite mica is usually etched in hydrofluoric acid. For special purposes it can also be etched in hot concentrated NaOH solution (Khan et al., 1981; Guo et al., 1986). The etching condition of fission fragment tracks for optical microscope observation is listed in Table 3.1.

Fig. 3.1A shows a microphotograph of fission fragment tracks in muscovite mica taken with the ordinary optical microscope.

The etching condition is in 40% HF at 25°C for 500 minutes (Guo et al., 1982). Such an etching condition for muscovite mica was selected in order that the tracks were easily visible under optical microscope and the observed number of tracks was at the middle point of the flat response with respect to the change of etching time. Under this etching condition, the detection efficiency of muscovite mica for fission fragments from thermal neutron induced fission of ^{235}U in 2π geometry is $(93.6 \pm 0.3)\%$, critical angle is 3°41′ (Guo et al., 1982). Muscovite mica has been widely applied to fission rate measurements (Li et al., 2001), neutron dose and energy-spectrum determination (Walker et al., 1963), FTD as external detectors (Fleischer et al., 1975; Durrani and Bull, 1987; Wagner and

TABLE 3.1 Etching conditions of common nuclear track detectors for optical microscope observation.

Material	Etching condition	Reference
1. Crystalline solids		
Muscovite mica	48% HF, 23°C, 100 minutes–2 hours	Price and Walker (1962c)
Apatite	0.1–5% HNO_3, 20–25°C, 30 seconds–3 minutes	Fleischer and Price (1964)
Zircon	20 g NaOH + 5 g H_2O, 220°C, 15 minutes–4 hours	Wagner and Van den haute (1992)
Sphene	37% HCl, 80°C, 80 minutes	Guo et al. (1980)
Olivine	WN solution, boiling point, 5 hours (1 g $H_2C_2O_4 \cdot 2H_2O$ + 1 mL H_3PO_4 + 40 g Na_2EDTA + 100 mL H_2O, + NaOH until pH = 8.0 ± 0.3)	Liu et al. (1979)
Pyroxene	60% NaOH, boiling point, 80 minutes	Liu et al. (1979)
Whitlockite	0.25% HNO_3, room temperature, 60 seconds	Liu et al. (1979)
2. Glasses		
Soda-lime glass	48% HF, 20–25°C, 5–10 seconds	Fleischer and Price (1963b)
Phosphate glass	48% HF, 20–25°C, 5–20 seconds	Fleischer and Price (1963b)
BP-1 glass	49% HBF_4, 50°C, 58–100 hours	Wang et al. (1988)
Tektite	48% HF, 20–25°C, 30 seconds	Fleischer and Price (1964)
Obsidian	48% HF, 23°C, 60 seconds	Fleischer and Price (1964)
Basaltic glass	5–20% HF, 23°C, 1–5 minutes	Fleischer et al. (1969)
3. Plastics (high polymers)		
CR-39	6.25N NaOH, 75°C, 6 hours	Cartwright et al. (1978).
Polycarbonate (Lexan, Makrofol)	6.25N NaOH, 50°C, 20 minutes	Fleischer and Price (1963a)
Cellulose nitrate (LR-115)	6.25N NaOH, 23°C, 2–4 hours	Fleischer et al. (1975)
Polyethylene terephthalate	6.25N NaOH, 70°C, 10 minutes	Fleischer and Price (1963a)
Polyimide (PI, Kapton, Upilex)	5–15% NaClO, 5–70°C, 3–25 hours	Vater et al., 1988
4. New detector materials		
CR-39-DAP Series	7N NaOH, 70–90°C, 1–18 hours	Kodairo et al. (2008)
PETAC-ADC Copolymer	6N NaOH, 70°C, 60 minutes (α) 30 minutes (FT)	Manderekar et al. (2010)
Silicon nitrate (Si_3N_4)	HF	Vlasukova et al. (2010)

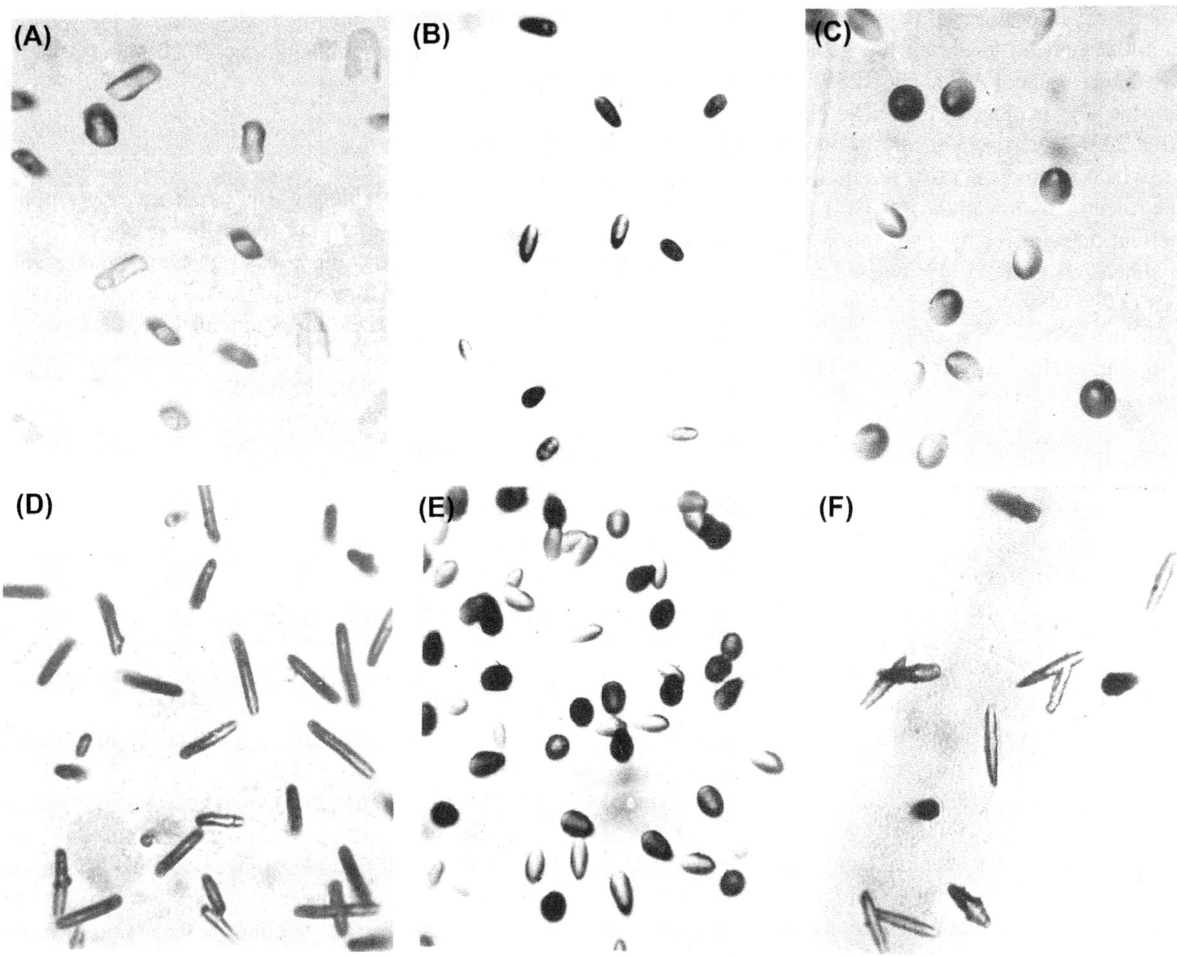

FIG. 3.1 Microphotographs of fission-fragment tracks in (A) muscovite mica, (B) phosphate glass, (C) soda-lime glass, (D) polycarbonate, (E) quartz, and (F) polyester taken under an ordinary optical microscope. The etching conditions were selected such that the tracks were most clear and easy to observe under the optical microscope and the number of tracks was at the middle point of the constant part of the response curve for each type of detector. For the color version of the figure, the reader is referred to the online version of the book.

Van den haute, 1992), uranium and plutonium concentration determinations (Price and Walker, 1963b), and multi-prong sequential fission studies of heavy targets with heavy projectiles (Brandt, 1980; Vater et al., 1977, 1986; Gottschalk et al., 1983; Khan et al., 1987, 2001; Qureshi et al., 2005; Shahzad, 1995; Shahzad et al., 1999; Nasir et al., 2008; Baluch et al., 2006).

2. Apatite

Apatite is a sort of widespread accessory mineral, which exists ubiquitously in various rocks, such as plutonic-, volcanic-, metamorphic-, and sedimentary rocks. It is the most frequently used mineral in FTD and geothermal chronology. Its atomic composition is $Ca_5(F,Cl)(PO_4)_3$, density 3.16–3.22 (usually 3.18) g/cm^3, refractive index $n_0 = 1.629-1.667$ in the direction of axis with average 1.648, $n_e = 1.624-1.666$ in the direction perpendicular to the axis with average 1.645. Its hardness is five (5), quite easy to be ground and polished.

The track-etching of apatite is very easy at room temperature and very fast (several seconds to several minutes) in concentrated or diluted nitric acid (See Table 3.1). Track etch rate V_T, bulk etch rate V_B, and detection efficiency for fission fragment tracks vary according to the orientation of apatite surfaces. The average detection efficiency of apatite for fission fragments of $^{252}C_f$ in 2π geometry is $(87.8 \pm 5.2)\%$ (Surinder Singh et al., 1986).

The uranium concentration in apatite is usually very high from about 1 to 100 ppm (parts per million $= 10^{-6}$ g/g) (For comparison, the average concentration of uranium in the earth's crust is about 2.7 ppm).

Apatite has a relatively low thermal stability of fission tracks. Its partial annealing zone (PAZ) of fission tracks is in the range of about 60–125°C (Gleadow and Duddy, 1981). The lower limit of temperature about 60°C means that fission tracks in apatite can be treated as stable at temperature lower than 60°C during the geologic time. Over the upper limit of the PAZ about 125°C, fission tracks will disappear after geological time. The closure temperature is (100 ± 20)°C (Harrison et al., 1979). The relatively low PAZ, which is almost consistent with the formation window of petroleum (hydrocarbon) 60–130°C, makes apatite an important detector for the exploration of petroleum formation (Gleadow et al., 1983; Naeser and McCulloh, 1989; Green et al., 1989; Kang and Wang, 1991).

All the above advantages make apatite one of most useful mineral materials in FTD and geothermal chronology.

3. Zircon

Zircon, similar to apatite, as an accessory mineral, exists ubiquitously in most types of rocks; and it is rich enough in uranium to permit age determinations down to million years or so. Its atomic composition is $ZrSiO_4$, molecular weight is 183.304 g/mol, $A/Z = 2.128$, and ionization potential is $I = 209.5$ eV. The colorless zircon has a density of 4.6–4.8 (usually 4.7) g/cm^3, refractive index of 1.925–1.984, and hardness of 7–7.5. The green or dark brown zircon has a density of 3.9–4.1 g/cm^3, refractive index of 1.78–1.815, and hardness of 6. Zircon can be etched by H_3PO_4, or NaOH, KOH, and their mixture at high temperature. The etching condition 20 g NaOH:5 g H_2O (≈ 100 N) at 220°C for 15 min to 4 h (See Table 3.1) is often adopted. For zircon etching, the grains of zircon are mounted on one surface of polyvinyldifluoride (PVDF) sheet by pressing with a glass plate at high temperature. The grains of zircon will be adhered and fixed onto the PVDF sheet. After grinding and polishing, the adhered zircon is etched in a tightly closed container made of Teflon. The Teflon container is held by a steel container, which is placed in an electric oven for heating to 220°C.

The uranium concentration of zircon is generally very high, between 300 and 600 ppm (Wagner and Van den haute, 1992), which is well-suited for FTD and even for archeological dating. The latent fission tracks in zircon are more thermal resistant than that in apatite. The track retention temperature of zircon is 210 ± 40°C (Wagner and Van den haute, 1992).

Zircon and apatite often exist in the same geological body. The apparent age (age calculated by the simple age equation without taking track fading into account, See Section III.A.1 of Part 2) of zircon and apatite reflect the years the geological body passed through their closure temperatures, respectively. From the depth–temperature relationship in the earth and the age difference between zircon and apatite, the uplift rate of the geological body can be derived. This technique of uplift rate determination is referred to as the mineral pair—closure temperature method or simply mineral pair method. Zircon is the second most commonly used mineral in FTD and geothermal chronology.

4. Sphene

Sphene (titanite) widely exists as an accessory mineral in plutonic and metamorphic rocks; but it is less common in pegmatite and volcanic rocks. Its atomic composition is $CaTiO[SiO_4]$, molecular weight 196.04 g/mol, $A/Z = 2.042$, ionization potential $I = 144.4$ eV, density 3.4–3.6 (usually 3.54) g/cm^3, hardness 5.0–5.5, and refractive index 1.888–2.054.

The etching of sphene is easy to carry out with concentrated HCl at 90°C for 30–90 minutes. The etching behavior of sphene exhibits a very strong anisotropic nature. The shapes of etched fission tracks are very different on different surfaces, depending on the orientation of the surfaces with respect to the crystal structure of the sphene and on the annealing before irradiation to fission fragments (Gleadow, 1978). The detection efficiency for fission tracks is also dependent on the orientation of the surface. The detection efficiency of annealed sphene can be down to one-half or even one-third compared with muscovite mica for spontaneous fission tracks. A calibration with annealed sphenes with random orientations showed that the ratio of detection efficiencies of the sphenes to muscovite mica is 0.516 ± 0.035 for fission fragments of ^{235}U fission induced by thermal neutrons (Guo et al., 1980, 1982a). This value is much less than 1. It means that the detection efficiency of annealed sphene for fission fragments is reduced.

Uranium concentration in sphene is usually rather high, between 100 and 1000 ppm. It is suitable for FTD and geothermal chronology. Thermal resistivity of fission tracks in sphene is higher than apatite and zircon. The effective track retention temperature (closure temperature) is 250 ± 50°C (Wagner and Van den haute, 1992). Its PAZ is 200–280°C (Harrison et al., 1979).

Sphene has provided much data on geological dating (Naeser, 1967; Wagner and Van den haute, 1992) and also archeological dating (Guo, 1982a; Guo et al., 1980, 1991).

5. Olivine

Olivine is a general composition of stone meteorites and stony-iron meteorites and exists in the earth's crust. It contains plenty of information about radiation in space. It is a space radiation detector. It has been recording heavy charged particle tracks since its formation billions of years ago. Fossil tracks in extraterrestrial minerals were first found in olivine in a stony-iron meteorite (Fleischer et al., 1975).

The atomic composition of olivine is $(Mg,Fe)_2SiO_4$, in which the ratio of Mg to Fe can be any value. The molecular weight is 140.708 g/mol for Mg_2SiO_4 and

203.778 g/mol for Fe_2SiO_4, the average value 172.2335 g/mol is usually taken. Its density increases with increasing Fe content from 3.3 to 3.5 g/cm^3, usually 3.35 g/cm^3 is taken. Its hardness is 6.5–7.0, and refractive index 1.654–1.690. The A/Z = 2.050, and its ionization potential I = 158.9 eV.

The etching condition used by many workers is the following:

1 mL H_3PO_4:1 g oxalic acidic acid: 40 g disodium salt of EDTA: 100 g H_2O: ~4.5 g NaOH, boiling temperature (~125°C) for 5 hour. Some amount of NaOH is added to make the pH value equal to 8.0 ± 0.3 of the etchant (Liu et al., 1979).

The uranium concentration of olivine is usually very low. It is $(0.046 \pm 0.004) \times 10^{-9}$ g/g in Jilin meteorite (Liu et al., 1979).

The tracks in olivine from meteorites or lunar rocks can provide information in the following fields:

1. Spontaneous fission of ^{238}U and ^{244}Pu. The latter nuclide is extinct in nature since its half life is 8.08×10^7 a (year), which is much less than the age of the earth $(4.6 \pm 0.1) \times 10^9$ a. It could exist only in the early era of the solar system. The tracks of spontaneous fission from both ^{238}U and ^{244}Pu in meteorites can provide information on the age of meteorites (Liu et al., 1979) as well as of lunar rocks.
2. Space radiation particles, including cosmic ray particles and possible magnetic monopoles.
3. Records of interactions of comic ray particles with nuclei in the meteorites and lunar rocks, such as high-energy spallation reactions and space neutron-induced fission.

From the above information, the origin of cosmic ray particles and the phenomena occurring in cosmic ray sources can be derived (Fleischer et al., 1975).

6. Pyroxene

Pyroxene is one of the main minerals in stone meteorites, stony-iron meteorites, and lunar rocks as well as in the earth's crust. Its atomic composition is $R_2[Si_2O_6]$, where R represents Mg, Fe or Ca; Mg and Fe are exchangeable. In some other pyroxenes, R represents Na, Li, Al, Fe, Ti, and the other elements. The basic structure of the above varieties of pyroxene are nearly the same, their properties are also very similar. For diapside $CaMg[Si_2O_6]$, its density is 3.28–3.40 (usually 3.29) g/cm^3, refractive index 1.675–1.701, hardness 5.5–6, A/Z = 2.000, and ionization potential I = 138.3 eV. The etching condition for varieties of pyroxene (diapside, pigeonite, augite, enstatite, bronzite, hypersthene and ferrohypersthene) is 3 g NaOH: 2 g H_2O, boiling temperature, 35–90 minutes (Lal et al., 1968). The uranium concentration in pyroxene is usually very low, for example, $(1.50 \pm 0.13) \times 10^{-9}$ g/g in Jilin meteorite (Liu et al., 1979), $(0.95-15.0) \times 10^{-9}$ g/g in lunar mare basaltic fragments 70017–291 (Hu et al., 1982).

Pyroxene, similar to olivine, is an important mineral detector for retrieving space radiation information as described in II.E of Part 2 and in the classic book by Fleischer et al. (1975).

7. Whitlockite

Whitlockite (or pyrophosphorite) is an accessory mineral in chondrite, achondrite, and aerosiderolite meteorites and a composition in enclosures of aerosiderite meteorites. It was formed in the early stage of condensation of nebula in the solar system. Whitlockite takes up to 0.31–0.57% by weight in chondritic meteorites (Ouyang, 1988). The sizes of the whitlockite crystslline grains are usually < 0.5 mm, among which the size from 0.1 to 0.4 mm are easy to be used as track detectors for FTD and determination of uranium concentration.

The atomic composition of whitlockite is $Ca_3(PO_4)_2$. Its molecular weight is 310.18 g/mol, and density 3.1 g/cm^3. The structure of whitlockite belongs to trigonal system in crystollography. Fission tracks in whitlockites can be etched in 0.25% HNO_3 at room temperature for 60s. The annealing condition in the laboratory for erasing all fission tracks (100% annealing degree) is at 584°C for 1 hour. The annealing temperatures are 274°C and 212°C for 100% reduction of fission track density for 106 and 109a of geological time, respectively. The uranium concentration in crystalline whitlockite in meteorites is usually relatively high. For example, $(80 \pm 16) \times 10^{-9}$ g/g U is contained in whitlockites in the Jilin meteorite, which fell in Northeast part of China in 1976 (Liu et al., 1979); whereas, the coexisted essential minerals of olivine and pyroxene contain only $(0.046 \pm 0.004) \times 10^{-9}$ and $(1.50 \pm 0.13) \times 10^{-9}$ g/g U, respectively.

The fission tracks in whitlockite in meteorites may not only come from spontaneous fission of ^{238}U, but also from spontaneous fission of ^{244}Pu. The half-life of ^{244}Pu is $(8.26 \pm 0.09) \times 10^7$a, which is much shorter than the age of chondritic meteorites 4.55×10^9a (Ouyang, 1988). The percentage of ^{244}Pu atoms, survived now in chondrites, inferred from the decay law is only 2.6×10^{-17}, which actually is zero. Therefore, ^{244}Pu is an extinct element in the solar system, but it created a large number of spontaneous fission tracks in the early stage of formation of whitlockites in meteorites. For example, the density of ^{244}Pu spontaneous fission tracks is about 7.9×10^5 cm^{-2} on the internal surface of whitlockite, while the density of ^{238}U spontaneous fission tracks is about 3.6×10^5 cm^{-2}. The ratio of ^{244}Pu to ^{238}U spontaneous fission is about 2.2. It means that there are more spontaneous fission tracks from ^{244}Pu than from ^{238}U in whitlockites in the Jilin meteorite. The tracks of extinct ^{244}Pu in whitlockites are valuable historical remains of the early solar system contributing to our understanding of the history and evolution of the universe.

8. Other crystalline solids

Up to now, about 100 crystalline minerals have been successfully etched to show fission fragment tracks (Fleischer et al., 1975; Wagner and Van den haute, 1992). Besides the most commonly used minerals described above, quartz, feldspar, garnet and epidote are also used frequently.

Quartz is one of the crystalline detectors with low uranium concentration. By using quartz as a track detector and irradiating samples in nuclear reactors, an atomic concentration of uranium as low as 10^{-12} can be determined (Fleischer et al., 1965a).

Feldspar is one of the main mineral compositions in meteorites and lunar rocks. Feldspar has recorded a large number of tracks of space radiation particles, which should undergo further detailed research.

For more information about the etching and applications of all of the crystalline minerals, the readers are referred to the monographs by Fleischer et al. (1975) and by Wagner and Van den haute (1992).

B. Glasses

Glasses, as nuclear track detectors, are a type of amorphous noncrystalline (uncrystallized) inorganic solid. It can be divided into natural glasses and man-made glasses. Natural glasses include terrestrial glasses and extraterrestrial glasses. Terrestrial glasses are formed in volcanic eruption, rock-melts solidification, tectonic frictional fusion, and meteoritic impaction. In other words, in all geological events in which the temperature dropped rapidly, the fused substance could not form crystalline structures, resulting in the formation of glasses.

Volcanic glasses and tektites are examples of this type of glass. Extraterrestrial glasses exist in meteorites and lunar rocks. When the meteoritic or lunar substances underwent cooling, some kinds of main mineral crystals, such as olivine, pyroxene, and feldspar were first formed. The remaining substance with low crystallization temperature would form glassy substances by the rapid cooling process (Hu et al., 1982). Owing to the larger ionic radius of the uranium atom than that of other metallic ions, uranium atoms were pushed aside from the forming crystals into the gap substance, which were later formed in glassy interstitial materials. Therefore, glassy substance, distributed in the gaps of crystalline minerals, have a higher concentration of uranium than crystalline minerals.

Man-made glasses include different varieties. Such as soda-lime glass, phosphate glass, BP-1 glass, and others.

The common feature of glass track detectors is isotropic etching behavior. This feature results from the amorphous structure of glassy materials. Detailed explanation of etching of particle tracks in amorphous solids can be found in more recent publications (Nikezic and Kostic, 1997; Nikezic, 2000; Nikezic and Yu, 2003, 2004).

1. Man-made glasses

a. Soda-lime glass

The atomic compositions of most man-made glasses vary according to their manufacture from place to place. For example, the compositions of soda-lime glasses are $23SiO_2$:$5Na_2O$: $5CaO$:Al_2O_3 in the USA, (Fleischer et al., 1975), $67SiO_2$:$14Na_2O$:$14CaO$:$5Al_2O_3$ in the U. K (Durrani and Bull, 1987), and $71SiO_2$:$15Na_2O$: $9CaO$:$3MgO$: $2Al_2O_3$ in China (Guo et al., 1976b, 1982). All of these are commonly used as window glass, microscope slides, and cover slips, etc. It is very convenient to place a slip of window glass or microscope slide in the beam of particles ($Z \geq 10$) at an accelerator to make an exposure for a certain time. Then, by etching it in HF for a few seconds at room temperature and washing it immediately in water, the tracks of the particles as well as the flux and radial distribution of the particles from the accelerator can be determined easily. Similar determinations can be carried out within several minutes at nuclear reactors for flux and radial distributions of thermal neutron beams by folding a uranium foil with glass plate.

A microphotograph of fission fragment tracks in soda-lime glass is shown in Fig. 3.1C. The critical angle of soda-lime glass for detecting fission fragments of ^{235}U is $\theta_c = 37°21'$; and its detection efficiency in 2π geometry for isotropic emitted fission fragments from thin ^{235}U foil is $(39.3 \pm 0.4)\%$ (Guo et al., 1982).

The advantage of soda-lime glass as a track detector is insensitivity to particles lighter than or equal to C ($Z = 6$). It can overcome intense background of α-particles in fission studies and against intense accelerator beam particles in studies of nuclear reactions in the synthesis of transuranium elements (Flerov et al., 1964).

b. Phosphate glass

The composition of phosphate glasses is also varied. For example, $10P_2O_5$:$1.6BaO$:Ag_2O:$2K_2O$:$2Al_2O_3$ in the United States (Fleischer et al., 1975), $34P_2O_5$:$7ZnO$:$4Al_2O_3$:$4BrO_3$:SiO_2 in the United Kingdom (Durrani and Bull, 1987), and $50LiPO_3$:$50AL(PO_3)_3$:$7AgPO_3$:$3B_2O_3$ in China (Guo et al., 1976b, 1982).

The critical angle of Chinese phosphate glass for recording fission fragment tracks is $7°6'$ while the 2π detection efficiency for isotropic ^{235}U fission source is

87.6 ± 0.5%, which is higher than soda-lime glasses (Guo et al., 1976b, 1982).

BP-1 glass is a special type of phosphate glass, whose composition is 65wt% P_2O_5, 25wt% BaO, 5wt% Na_2O, and 5wt% SiO_2 (Wang et al., 1988). BP-1 glass has been used in discoveries of several cluster radioactivities, studies of high-energy nuclear reactions, and measurements of isotopic compositions of manganese and nickel in galactic cosmic rays (GCRs). The latter measurements were carried out by using a stack of plates of BP-1 glass attached outside the Mir station for 5 years. The charge resolution was obtained being $\sigma_z = 0.08e$ for $Z = 69-79$, $E = 10.6$ AGeV for three individual measurements over three sheets of glass. The mass resolution is $\sigma_m = 0.43-0.58$ amu for $Z = 25-28$, $A = 53-62$, and $E = 150-400$ AMeV of GCRs. These are the highest charge and mass (isotope) resolutions yet obtained (Westphal et al., 1996).

c. BP-1 glass

BP-1 glass has been used in the identification of the emission of ^{23}F and ^{24}Na clusters from ^{231}Pa with branch ratios of 1×10^{-14} and 1.3×10^{-11}, respectively. In this study, BP-1 glass only recorded the emitted nuclei ^{23}F and ^{24}Na without recording α (background), whose intensity is 10^{14} or 10^{11} times higher than that of the emitted nuclei (Price et al., 1992a).

BP-1 glass has also been used in the measurement of the fragmentation cross section of 160 AGeV Pb. Charge resolution is ~0.14 charge unit for one single cone measurement. For successive 5 sheets of glass, the $\sigma_z = \sim 0.06e$. This charge resolution is more than adequate to identify fragments and hence to measure cross-sections for charge-changing interactions. BP-1 glass is expected to be able to detect magnetic monopoles (MMs) with charge n = 2 and $\beta > 0.2$. No tracks of MMs with n = 2 were found.

2. Natural glasses

Natural glasses were formed in the earth's crust or in meteorites or lunar rocks. Most important natural glasses are tektite and obsidian. Some other types are basaltic deep-sea glass and frictionite glass.

a. Tektite

Tektite is a special type of glassy object, which distributes in four restricted areas in the world: (1) South-East Asia and Australia (0.732 ± 0.039 Ma) (Guo et al., 1997), (2) Ivory Coast (1.08 ± 0.10 Ma) (Wagner, 1966), (3) Czech Republic and Slovakia (Moldova) (14.7 ± 0.4 Ma) (Wagner, 1966; Storzer and Wagner, 1977), and (4) North America (34.6 ± 0.7 Ma) (Storzer and Wagner, 1977). Many authors have dated the tektites and obtained results consistent with each other for each restricted area. The above ages in the brackets are given in order to show the time of their formation. The most acceptable hypothesis of the origin of tektites is that they were formed by impacts of meteoroids on the earth. During impact, a large kinetic energy of the meteoroid was transferred to the substance in the earth's crust and subsequently melted and sputtered the impacted substance to the sky; then the substance cooled down quickly to form glassy objects in sizes from several centimeters to less than one millimeter. The glassy objects drop down to the earth, forming tektites or micro-tektites. The chemical compositions of tektites are similar to that of the earth's crust. For example, 74.63% SiO_2, 12.19% Al_2O_3, 4.16% FeO, 2.32% K_2O, 1.87% CaO, 1.82% MgO, 1.36% Na_2O, 0.77% TiO_2, 0.54% Fe_2O_3, 0.09% MnO (Li, 1982). The uranium concentration of tektites is also consistent with that of the average earth's crust, for example, 1.74 ± 0.18 and 1.39 ± 0.15 ppm in the tektites found in Hainan island, China (Collaboration Group of China Institute of Atomic Energy and Guiyang Institute of Geochemistry, 1976).

The density of tektites is 2.32–2.5 g/cm^3, and refractive index: 1.48–1.52. Tektite is a suitable detector for fission track registration. FTD of tektites can provide a variety of information in space research, stratigraphy, and paleoanthropology (Guo et al., 1997).

b. Obsidian

Obsidian is a special type of natural glassy detector to record fission fragment tracks. Obsidian was formed during the young volcanism. When magma on the earth's surface met with water in lakes or rivers, it rapidly cooled down at normal atmospheric pressure, forming obsidian glasses. Obsidian exists worldwide with volcanoes, such as around the Pacific and Mediterranean Oceans, west part of North and South America, New Zealand, Japan, Near East, East Africa, and Iceland. The sizes of obsidians are varied from several hundred meters to centimeters or even as tephra. The refractive index of obsidian is 1.48–1.52 (usually 1.49), density 2.33–2.41 g/cm^3; and hardness: 5–5.5; Its uranium concentration is 5–15 ppm, which is suitable for FTD. Obsidian was used as sharp tools by ancient man, such as stone knifes, saws, and arrow heads. It was a sort of object for commercial trade. By fission track techniques, the following information may be obtained: age and geographic provenance of obsidian raw materials, age of heating for manufacture of obsidian artifacts, and the age and uranium concentration of the obsidian, which are specific for each source of obsidian. Both can be criterions for judging the source (Bigazzi et al., 1990).

c. Basaltic glass

Basaltic glass was formed during the extrusion of fresh lava from the upper mantle through a fracture of the earth's crust to deep-sea water and quenched by the cold seawater. The substance of the extruded lava did not have enough time to differentiate to grow into crystalline minerals, forming an amorphous glassy substance—basaltic glass. According to the concept of seafloor spreading, basaltic glass may exist in the whole surface layer of the Mid-Atlantic Ocean floor, where lava has been extruding outward sequentially and the surface layer of lava was cooled by deep-sea water. Basaltic glass is suitable to record fission tracks for dating of geological events.

Due to the fact that basaltic glass was formed directly from mantle lava, the uranium concentration in basaltic glass would be lower than that of zircon, apatite, and sphene, which are accessory minerals enriched with uranium. The uranium concentration in basaltic glass is $(0.1-0.3) \times 10^{-6}$ g/g. Two special techniques have been adopted to locate an adequate number of spontaneous fission tracks in the low uranium samples. The first technique is that following the original scan of spontaneous fission tracks, the glass is re-etched (removing a 20 μm layer on the surface of the glass) to reveal tracks at a different depth, then to scan, re-etch, scan, re-etch, until enough spontaneous fission tracks were found. Finally, the piece of glass was irradiated in a nuclear reactor for exposure to thermal neutrons for the determination of uranium concentration in the glass and dating of the glass (Fleischer et al., 1975). Another technique used to increase the scanning area of spontaneous fission tracks, one can repeat re-grinding (>20 μm off), re-polishing and re-etching of the sample, until one can count a large enough number of spontaneous fission tracks. Then, one undertakes an irradiation of the sample in a nuclear reactor for the determination of uranium concentration and dating (Wagner and Van den haute, 1992). The etching condition of basaltic glass is 5−20% HF, 23°C,1−5 m.

Fading of spontaneous fission tracks in basaltic glass for a geological time cannot be ignored even in deep-sea water temperature (4°C). A correction for track fading in basaltic glass must be carried out (Selo and Storzer, 1981). The activation energy for total fading of fission tracks in basaltic glass is 1.9 eV and for start of fading is 1.3 eV. The 1-hour annealing temperature in the laboratory for total fading is 240°C and for start of track loss is 190°C (MacDougall, 1973; Fleischer et al., 1975; Wagner and Van den haute, 1992).

Basaltic glass had been used to provide a strong proof for seafloor spreading hypothesis by FTD of Mid-Atlantic Ridge, combined with KAr dating (Fleischer et al., 1968, 1975).

Other natural glasses, such as frictional glasses formed during fault displacement, impact glasses formed during the formation of a crater by meteorite bombardment, are all useful for fission track measurements for many different purposes (Fleischer et al., 1975; Wagner and Van den haute, 1992).

C. Plastics

Since the discovery that polymers (plastics) can record etchable tracks of heavy charged particles (Fleischer and Price, 1963a), more than 30 types of polymer detectors have been successfully etched to reveal tracks. Compared with inorganic crystalline solids and glasses, plastic track detectors are more sensitive to heavy charged particles, such as the charge as low as $Z = 1$ (proton). The sensitivities of different polymers are different. CR-39 (polyallyldiglycol carbonate, PADC for short) is the most sensitive detector in common use (the recordable charge $Z \geq 1$); amber is the most insensitive plastic detector (the lowest recordable charge is fission fragments, $Z > 30$).

1. CR-39 (polyallyldiglycol carbonate, PADC, PM-355, PM-500, PM-600)

CR-39, as a unique sensitivity and high-charge resolution detector, was discovered by Cartwright et al. (1978). CR-39 is the trade name of allyl diglycol carbonate. CR stands for "Columbia Resin". CR-39 plate (sheet) is produced by casting CR-39 monomer with some kinds of initiators in a frame. Kinoshita and Price (1980) found that IPP (di-isopropyl peroxydicarbonate) as initiator could produce a smoother surface than the other initiators. Tarle et al. (1981) found that adding DOP (dioctyl phthalate) as plasticizer into the monomer before casting could reduce roughness after long etching. In order to meet the need of scientific research, many firms have made CR-39 sheets with special care. The most successful firms include: Intercast Europe in Italy,[1] Page Molding in the United Kingdom,[2] American Acrylic and Plastics in the United States,[3] Fukuvi Chemical Industry in Japan,[4] Tastrak in the United Kingdom,[5] and others (Pang and Ma, 1988).

The Chemical composition of the monomer of CR-39 is $C_{12}H_{18}O_7$, with its molecular weight being 274.2707 g/mol. Its molecular structure is:

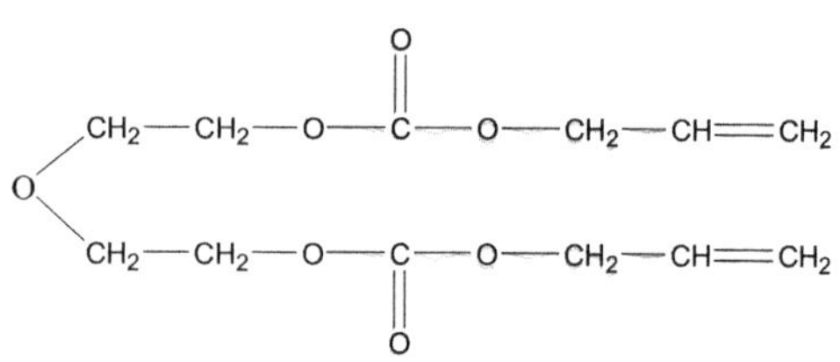

1. Intercast Europe Srl., 10/A Via G. Natta, Area SPIP, 43,122 Parma, Italy. Tel: (+39) (0)521 607555; Fax: (+39)(0)521 607924. Web: http://www.intercastit. E-mail: intercast@intercast.it
2. Pershore Mouldings Limited, Trade Estate, Poreshore, Worcestershire, WR10 2DH, England, Tel.: (+44)(0)386 553425.
3. American Acrylics and Plastics, Stratford, CT, USA.
4. BARYOTRAK, Fukuvi Chemical Industry, Tokyo, Japan.
5. Tastrak, Track Analysis Systems Ltd., c/o H. H. Wills Physics Laboratory, Bristol, UK.

It has a chain-like structure. In the polymerization of CR-39 monomers, the double bonds of allyl groups ($CH_2CH{=}CH_2$) in the structure of the monomers open and form a cross-linked net structure in three dimensional directions with other molecules, forming polyallyldiglycol carbonate (PADC).

The chemical structure of polyallyldiglycol carbonate (PADC) is shown below:

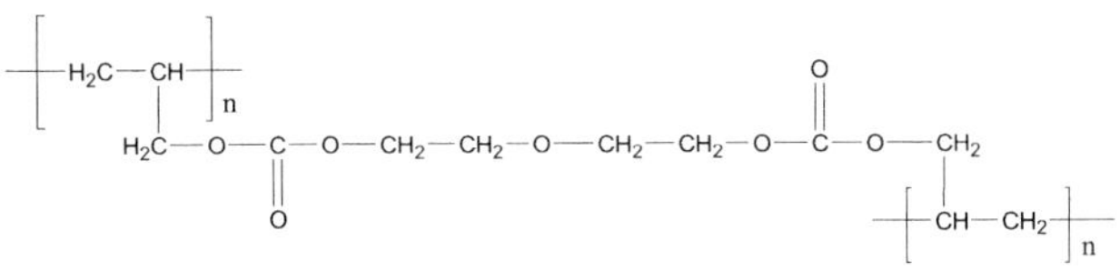

In general, chain-like structures are more sensitive to radiation than the other structures, such as a benzene ring in polycarbonate. It is thus understandable why CR-39 is more sensitive than polycarbonate (Tsuruta et al., 2008).

The density of CR-39 is 1.32 g/cm^3, refractive index is n = 1.504, A/Z = 1.877, and its mean ionization potential I = 70.2 eV (Durrani and Bull, 1987). CR-39 track detectors can be etched with NaOH and KOH solutions. The recordable minimum ionizing particle is a 20 keV proton by etching with NaOH solution (Duan et al., 2009, 2010). The etched tracks of protons from 20 keV to 1.02 MeV are shown in Fig. 3.2 (Duan et al., 2010).

The infrared (IR) absorption spectrum of virgin CR-39 is shown in Fig. 3.3. Every peak in the spectrum corresponds to a chemical group in the CR-39 structure and its type of vibration.

The major infrared absorption frequencies (wavenumber, cm^{-1}) measured with Fourier transform infrared absorption spectroscopy (FT-IR) in virgin CR-39 and irradiated one with 22.5 MeV protons are listed in Table 3.2 (Lounis-Mokrani et al., 2003).

The CR-39 detector can not only record track number N, positions (X, Y, Z), directions (zenith angle θ, azimuth angle ψ), but also it can derive particle charge Z, mass M, and energy E from the track parameters. It is the most widely used SSNTD.

2. Polycarbonate (PC, Lexan, Makrofol, Taffak)

Before the discovery of the CR-39 track detector in 1978, polycarbonate (PC) was the most commonly used track detector. Since CR-39 is more sensitive than polycarbonate, many of the measurements previously carried out with polycarbonate have been replaced by CR-39. But polycarbonate is still used frequently when proton tracks may be a strong background. For example, during nuclear reactor irradiation fast neutrons will produce background tracks of recoil protons in CR-39, but in polycarbonate the recoil protons will not be recorded as background tracks. In this case, polycarbonate is better than CR-39. Up to now, polycarbonate is preferred as an external detector in FTD. The flux of protons in cosmic rays at the top of the atmosphere is over 10^2 times more than the sum of the other nuclei, as described in Chapter 1, Volume 1 of this book. Protons will produce a large background in CR-39, but will have no effect on polycarbonate. Therefore, polycarbonate is still used in satellites and balloons to collect data on heavy cosmic rays.

The composition of polycarbonate is $C_{16}H_{14}O_3$, molecular weight 254.2855 g/mol, and its molecular structure is

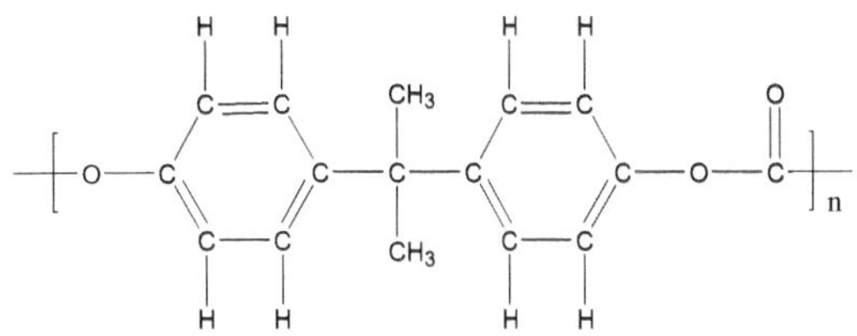

n ≈ 100

The benzene ring in the structure of polycarbonate has a tendency to absorb more radiation energy than the chain structure of CR-39 (Tsuruta et al., 2008; Chadderton et al., 1993). Therefore, polycarbonate is less sensitive to recording tracks than CR-39. The minimum recordable charge (atomic number) Z of particles by polycarbonate is 2 (α-particle). The ionization potential I = 69.5 eV, A/Z = 1.896, refractive index, n = 1.586, and density = 1.29 g/cm^3. Lexan is a commercial name of the product of Bisphenol-A polycarbonate foil made by General Electric Company. Lexan is one of the most widely used detectors in the polycarbonate series. Makrofol is a commercial name of Bisphenol-A polycarbonate foil produced by Bayer AG, Germany (Note: Bayer AG has stopped the production of Makrofol foil).

The properties of polycarbonate foils, produced by different firms, are similar but with noticeable differences. For example, the detection efficiency of polycarbonate foil produced in China for fission fragments of ^{235}U in 2π geometry is (95.5 ± 0.4)%, the critical angle is 2°24′; while the detection efficiency of Makrofol K from Bayer AG for the same source of fission fragments is (93.6 ± 0.7)%, the critical angle is 3°40′ (Guo et al., 1976b, 1982). The two results are close to each other, but not within statistical uncertainties. Similar situations happen for other types of plastic track detectors. For this reason, the parameters calibrated for a particular type of detector material cannot be used for another type of the detector when very precise results are required.

Experiences also show that the properties of different batches of the detector materials produced by the same firm

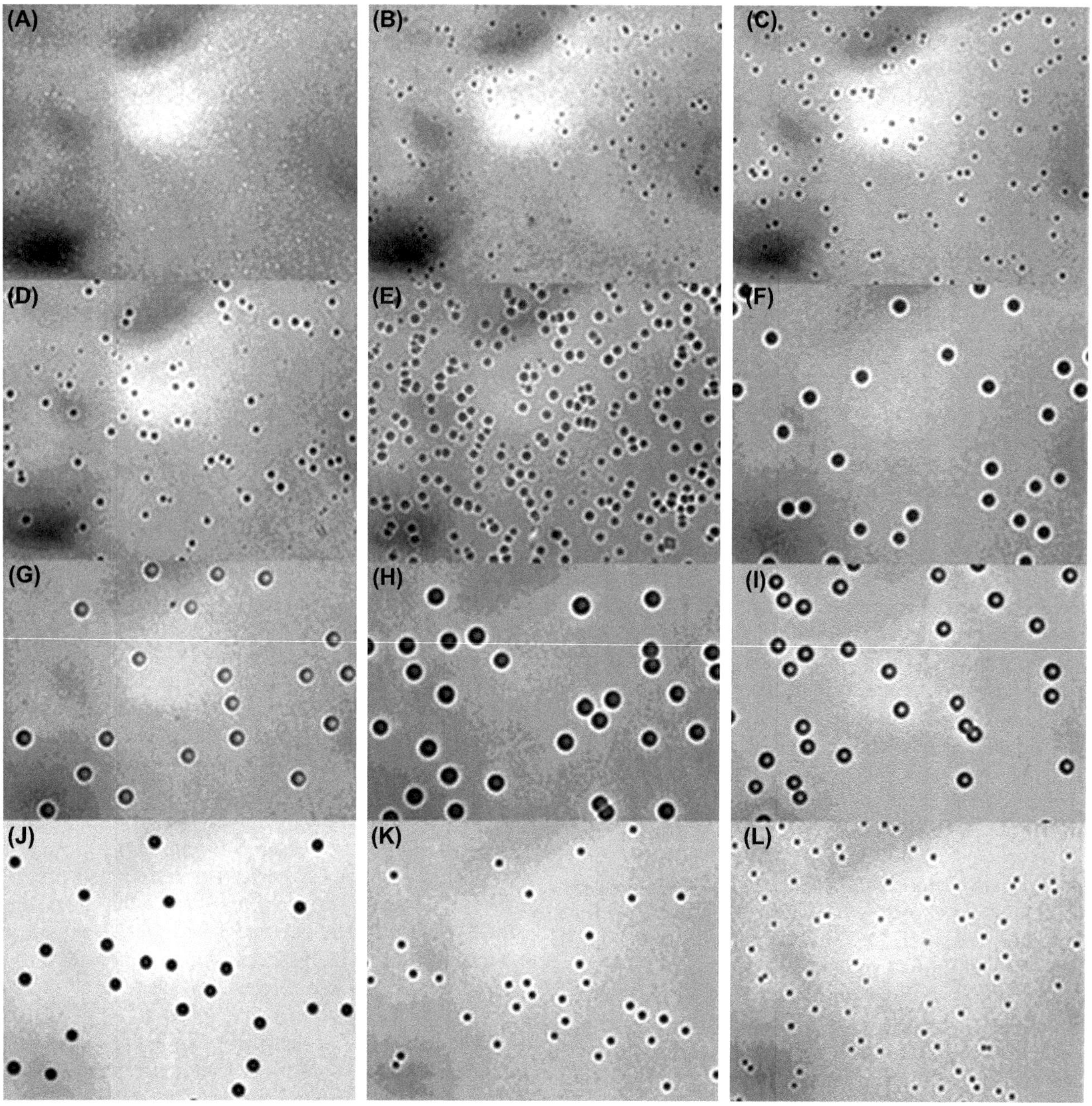

FIG. 3.2 Microphotographs of tracks of monoenergetic protons of (A) 20, (B) 40, (C) 60, (D) 80, (E) 100, (F) 220, (G) 320, (H) 420, (I) 520, (J) 620, (K) 820 keV and (L) 1.2 MeV, respectively, etched in 6 mol/L NaOH at 70°C for 9 h. The minimum detectable energy of protons with CR-39 is ~ 20 keV. *From Duan et al. (2010). Reprinted with kind permission from Acta Physica Sinica © 2010.*

with the same raw material are usually different. One must calibrate each batch of detectors before use. It does not mean one cannot use the results obtained by other workers. We just want to note that the best way to get reliable results is to calibrate the detector for every batch and for each situation, if one needs high precision.

Tuffak is a commercial name of the polycarbonate foil produced by Rohm & Haas Company, USA. The density of Tuffak polycarbonate is 1.249 g/cm^3, refractive index is 1.586. The smoothness of the surface of Tuffak is much better than that of Lexan. A comparative study between Lexan and Tuffak showed that $\sigma_D/\overline{D}$ are 0.015 and 0.0034 for Lexan and Tuffak, respectively, for 955 MeV/u U tracks. $\overline{D}$ is the average diameter of U tracks in Lexan or Tuffak; and σ_D is the standard deviation of the track diameters. The value $\sigma_D/\overline{D} = 0.0034$ infers that by

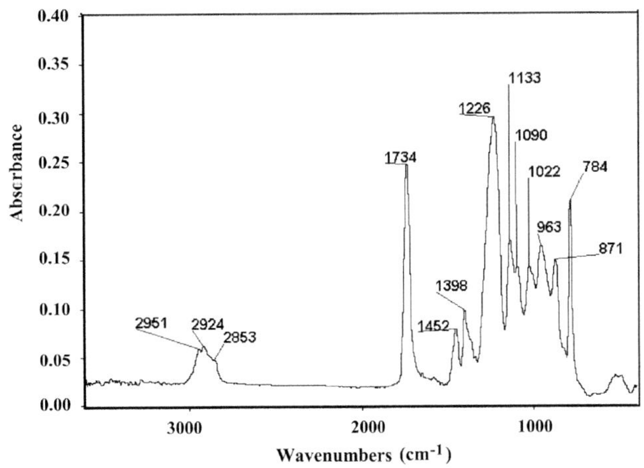

FIG. 3.3 Infrared (IR) spectrum of virgin CR-39 (PADC). *From Lounis-Mokrani et al. (2003). Reprinted with kind permission from Elsevier © 2003.*

measuring a single track diameter, one can determine the charge Z with an uncertainly $\sigma_Z = 0.9e$ near the charge Z around 92 (U). In other words, by measuring the diameter of a single cone of the track in Tuffak, one can resolve an uranium ion from the neighboring elements (O'Sullivan and Thomson, 1982; Ahlen et al., 1984; Salamon et al., 1985; Guo et al., 1984a).

The peaks of the infrared absorption spectrum of polycarbonate (PC) foil, the corresponding chemical functional groups for the peaks, the rules of the changes of the absorption as a function of the fluence, and energy of the heavy ions irradiated to the foils have been studied by many authors. Interested readers may refer to the following articles: Barillon et al. (2013); Dehaye et al. (2003); Gagnadre et al. (1993); and Yamauchi et al. (2010).

Polycarbonate track detectors have been used extensively in α radiography, neutron physics, heavy ion studies, cosmic ray research, and NTM production (Fleischer et al., 1975; Durrani and Bull, 1987).

3. Cellulose nitrate (CN, LR-115, Daicell)

CN is one of the most sensitive nuclear track detector materials. Its sensitivity is just less than CR-39. The least ionizing ion is 0.55 MeV protons (Daicell). Fleischer (1981) showed that the least ionizing ion seen was 1.2 MeV protons for CN (VEB Eilenburg). In this case, the CN is more sensitive than CR-39.

CN is made from absorbent cotton (cellulose) reacted with concentrated nitric acid. The three —OH groups in the molecular structure of cellulose partly (1, 2, or all the 3 —OH groups) are replaced by $-O-NO_2$ groups in the reaction.

TABLE 3.2 Absorption bands of CR-39 in the infrared (IR) region measured by FT-IR spectrometer and the corresponding chemical bonds and vibration types for the absorption bonds.

	Absorption band		
Chemical bond	Wavenumber, cm^{-1}	λ, μm	Vibration type
CH_3	2951	3.389	C—H stretching
CH_2	2924	3.420	Doublet H stretching
	2853	3.505	
> C=O	1734	5.767	Carbonate stretching
CH_3, CH_2	1452	6.887	Bending vibration
	1398	7.153	
—C—O—C—	1226	8.157	Ether bond vibration
	1133	8.826	Ether bond stretching
	1090	9.174	
	1022	9.785	
—CH=CH—	963	10.784	
C—H	871	11.481	Double bond wagging vibration
	784	12.755	CH rocking out-of-plane

More information about the IR absorption peaks and their changes after irradiation by heavy charged particles could be found in the following publications: Barillon et al., (2013, 2015); Hassan et al. (2013); Kusumoto (2017), Kusumoto et al. (2015, 2016, 2018) ; Lounis-Mokrani et al. (2003); Mori et al. (2009, 2011, 2012, 2013); Yamauchi et al. (2003, 2008, 2010, 2012).

The structure of CN formed with 2 $-O-NO_2$ groups is the following:

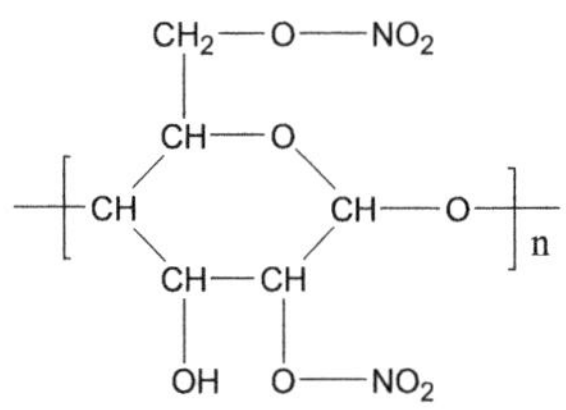

with formula $[C_6H_8O_9N_2]_n$ or $C_6H_8O_9N_2$ for short.

The density of CN is 1.4 g/cm^3, A/Z = 1.939, and ionization potential I = 81.1 eV.

LR-115 is a commercial name of CN detector made by Kodak-Pathe, or DOSIRAD, France. LR-115 type-II has an active layer of 12 μm red-dyed CN film coated on inert polyester base of 100 μm thickness. The thickness of 12 μm is designed for spark counting of α-particle tracks recorded in the film. LR-115 type-II is welcomed by radon scientists for the reason that α-particles emitted from radon progenies attached onto the surface of the detector cannot be developed by CE for short etching times. The energies of the α-particles from the short-lived progenies are all higher than 6.0 MeV, which are beyond the energy window 0.06–6 MeV of LR-115. In contrast, all the α-particles from the progenies can form etchable tracks in CR-39 by a short etching time. This is an advantage of LR-115 over CR-39 (Eappan and Mayya, 2004; Nikolaev and Ilic, 1999; Barillon, 2005; Yamauchi et al., 2008a, 2008b, 2008c; Durrani and Bull, 1987; Durrani and Ilic, 1997; Narula et al., 2009a, 2009b; Nikezic et al., 2009; Sajo-Bohus et al., 2008, 2009; Bochicchio et al., 2003, 2009).

The peaks of the infrared absorption spectrum of CN foil, the corresponding chemical functional groups for the peaks, the rules of the changes of the absorption as a function of the fluence, and the energy of the heavy ions irradiated to the foils have been studied. Interested readers may refer to the following articles: Barillon et al. (2002); Barillon and Katz (2000); and Kovalenko et al. (1993).

4. Polyethylene terephthalate (PET, Mylar, Cronar, Melinex, Lavsan, Terphane, Hostphan)

Polyethylene terephthalate (PET) is another commonly used track detector. Its molecular structure is:

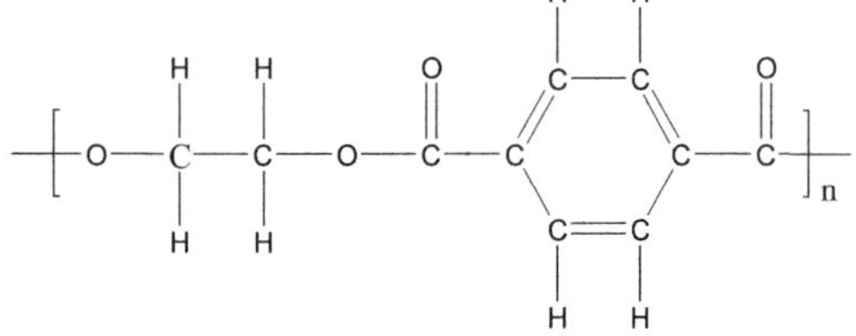

The molecular formula is: $(C_{10}H_8O_4)_n$ or $C_5H_4O_2$ for short.

The density of PET is 1.397 g/cm^3, refractive index n = 1.640, A/Z = 1.915, and ionization potential I = 73.2 eV. The most commonly used etching condition is 6.25 N NaOH, 70°C, 10–40 min for fission fragment tracks.

PET is less sensitive than polycarbonate: PET cannot record tracks of α-particles. Therefore, PET is often used to selectively record heavy ion tracks in the environments where intensive α-particles exist. Its charge and mass resolutions for several MeV/amu Ne (Z = 10) to Si (Z = 14) ions are good enough for the measurement of track parameters; thus, it was one of the detectors used in the discovery of cluster radioactivity (Barwick et al., 1985; Price, 1987; Tretyakova et al., 1985). See also Chapter 1, Volume 1 of this book for a more detailed treatment of cluster radioactivity.

The peaks of the infrared absorption spectrum of polyethylene terephthalate (PET) foil, the corresponding chemical functional groups for the peaks, the rules of the changes of the absorption as a function of the fluence,and energy of the heavy ions irradiated to the foils have been studied. Interested readers may refer to the following articles: Yamauchi et al. (2012, 2018); Barillon et al. (2013).

PET track detectors have also been used widely in many fields, such as:

- heavy cosmic ray analyses (Drach et al., 1987; Price and Drach, 1987),
- NTMs (Apel et al., 2008; Husaini et al., 2008; Zhu et al., 2004; Guo et al., 2008; Cao and Wang, 2009; Gomez Alvarez-Arenas et al., 2009),
- nano- and micro-structure device manufacture (Spohr, 1990, 2008; Fink et al., 2008; Chakarvarti, 2009; Wang et al., 2009; Zagorskiy et al., 2009) and
- anti-counterfeiting techniques (Wang et al., 2008).

5. CR-39–DAP series

Tsuruta (1999, 2000) found that the mixture of monomers of CR-39 and diallyl phthalate (DAP) can be copolymerized into a new type of track detector CR-39–DAP, whose property is in between that of CR-39 and DAP.

The structure of DAP monomer is:

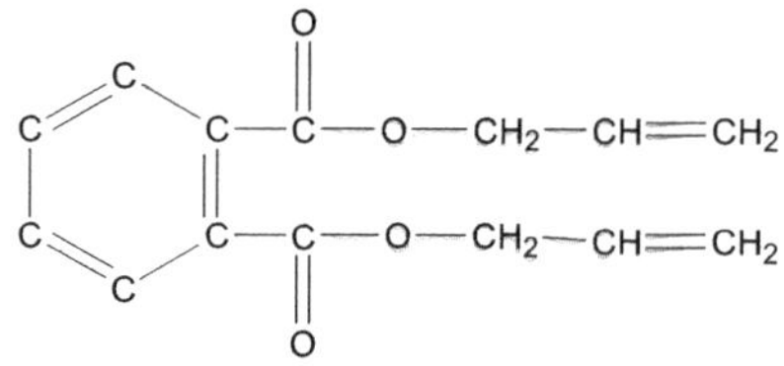

For comparison, the structure of CR-39 monomer is copied here:

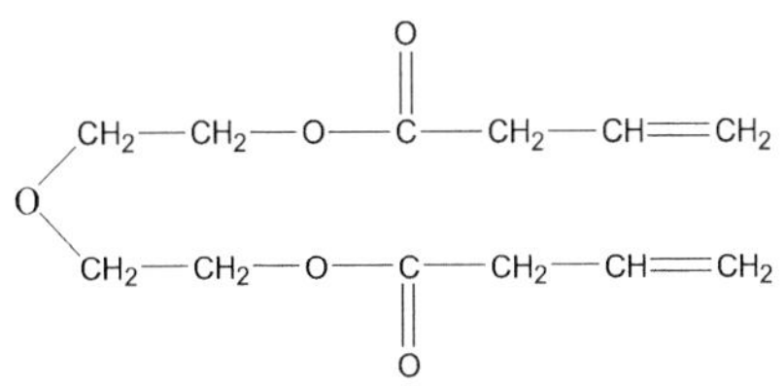

Similar to the CR-39 monomer, DAP monomer has also two allyl groups ($-CH_2CH{=}CH_2$), each of which has a double bond. When copolymerization starts, the double bond opens and connects with the opened bonds of CR-39 to form a three-dimensional structured copolymer CR-39–DAP.

In the DAP monomer there is a benzene ring, which has a tendency to adsorb radiation energy. For this reason, CR-39–DAP and DAP has a higher threshold than CR-39. Tsuruta et al. (2008) showed that the detection threshold of CR-39–DAP is lower than that of DAP and higher than CR-39, and the threshold increases as the percentage of DAP increases as follows:

2 keV/μm (water) (CR-39, DAP:CR-39 = 0:100), 40 keV/μm (50:50), 88 keV/μm (60:40), 108 keV/μm (70:30), 125 keV/μm (80:20), and ~ 250 keV/μm (DAP 100:0). They express the threshold as the total rate of energy loss in water on the right-hand side of the Bragg peak of the *dE/dX* curve. By the same expression, the thresholds of CN and polycarbonate (Lexan) are 110 and 330 keV/μm (water), respectively. This means that the sensitivity of CR-39–DAP is higher than Lexan and lower than CR-39; CN is nearly equal to the copolymer (70:30). These thresholds rank up as 2, 40, 88, 108, 125 and 250 keV/μm (water), which look like a spectrum of thresholds. This study opens up a new road to the development of nuclear track detectors, that is, to make a series of detectors with a rank of thresholds.

Doke (2008), Kodaira et al. (2008) and Ogura et al. (2006) confirmed the above development and suggested the application of these new types of detectors to the study of trans-iron nuclei in GCRs at a space station or on long-duration balloon experiments.

6. Polyimide (PI, Kapton, Upilex)

Polyimide (PI) has been used to make NTMs and as track template to synthesize nanostructures. During recent years, it has also been used to study track formation mechanisms. Polyimide has excellent temperature resistivity of 310°C for short exposures. It can be used at lower temperature down to −269°C. Kapton is a trade name of polyimide film produced by Du Pont, USA. It is also produced by Nilaco Corporation in Japan (Yamauchi et al., 2013). Upilex is a trade name of polyimide film produced by Ube Industries Ltd., Japan.

The molecular formula of polyimide is $[C_{22}H_{10}O_5N_2]_n$, or $C_{22}H_{10}O_5N_2$ for short. Its molecular structure is

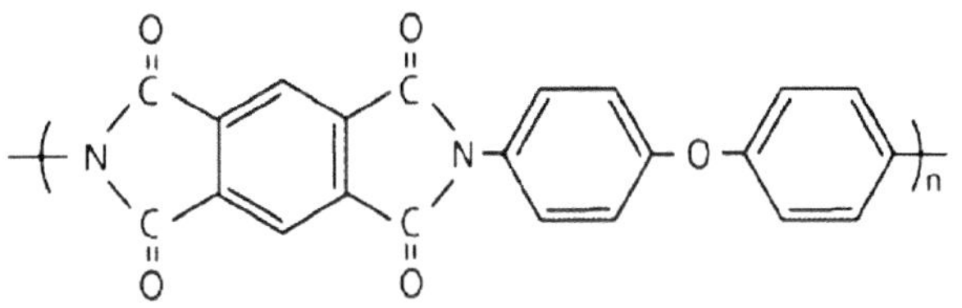

Its molecular weight (a repeat unit in the brackets) is 382.3321 g/mol. Its density is 1.40 g/cm^3.

The tracks in polyimide can be etched in $KMnO_4$ in water (Monnin and Isabelle, 1970). More authors used sodium hypochlorite solution NaClO with various concentrations 5–15% of free chlorine as etchant at temperature 50–70°C for 3–25 hours (Vater 1988; Zhu et al., 1988; Guo et al., 1990; Chakarvarti and Mahna, 1992; Vonderau et al., 1995; Ensinger et al., 2003; Yamauchi et al., 2013a, 2013b).

Gas separation has been tested by using Upilex nuclear track microfilters aimed at the separation of tritium from deuterium for fusion energy generation. The earlier stage of research was carried out by the separation of nonradioactive carbon monoxide CO from carbon oxide CO_2. The thickness of Upilex microfilter was 40 μm, the pores are cylindrical in diameter in the range of 100 nm. A pressure pushed a gas mixture of CO + CO_2 to pass through the pores with an aspect ratio of 400 (length/diameter). It shows that the light molecules CO pass through the microfilter faster than the heavy molecules CO_2, the Upilex microfilter can be used to separate CO from CO_2. Considering that tritium gas has a high β radioactivity, a test was carried out to show the resistivity of Upilex microfilter to β radioactivity. The test demonstrated that the microfilter turned out to be resistant (Ensinger et al., 2003). For fusion energy generation, a practical tritium/deuterium separation with nuclear track microfilters would be of vital need in the future.

Fourier transform infrared absorption spectroscopy (FT-IR) and other instruments can be adopted to analyze the molecular structures of track detecting materials and reveal the damages in the detectors.

Fig. 3.4 shows the absorption spectra of a Kapton film with thickness of 12 μm before and after the exposure to 5.4 MeV/u Mo ions with a fluence of 2×10^{12} ions/cm^2 (Severin et al., 2005). The broken curve indicates the spectrum of the pristine film; the solid curve is that of irradiated film.

The assignment of each peak in Fig. 3.4 has been well investigated (Guenther et al., 2001; Mishra et al., 2003; Sun et al., 2004; Severin et al., 2005). The absorption bands in

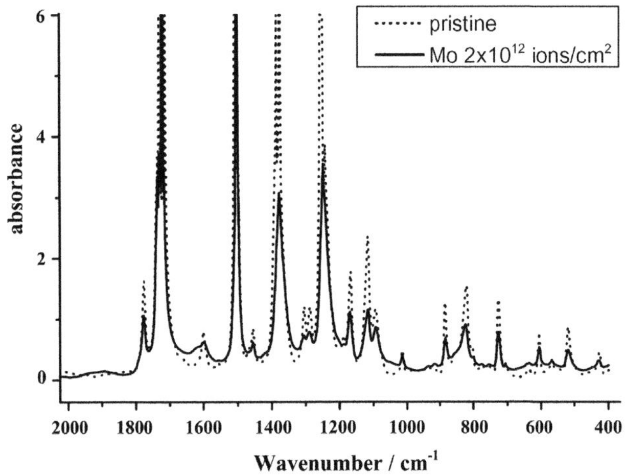

FIG. 3.4 FTIR spectra of 12 μm thick Kapton foils, before and after irradiation with Mo ions (5.4 MeV/u, mean $dE/dX = 8.4$ keV/nm, 2×10^{12} ions/cm^2, 1.9×10^7 MGy). *From Severin et al. (2005). Reprinted with kind permission from Elsevier © 2005.*

polyimide at some selected frequencies are listed in Table 3.3.

The sensitivity of polyimide for the registration of heavy charged particles is relatively low compared with CR-39, CN and polycarbonate (PC). In other words, the threshold of polyimide is relatively high. Fleischer (1981) showed that the least ionizing ion for forming etchable track in polyimide is 36 MeV ^{16}O; but recently Yamauchi et al. (2013a, 2013b) did not see any etched tracks of ^{20}Ne with energy from very low energy to 79 MeV. The damage density of ^{20}Ne ions in polyimide at the Bragg peak should be higher than that of ^{16}O ions. If ^{16}O can be recorded by polyimide, the ^{20}Ne should be easer than ^{16}O to be recorded in polyimide. One needs to do more work to clarify the true value of threshold of polyimide.

7. Other new track detector materials

One of the important tasks of nuclear track studies is to look for or synthesize new detector materials which have specific properties, such as a sensitivity higher than CR-39, or a threshold higher than that of the existing detector materials, or anti-annealing temperature of tracks higher than 1000°C for days.

a. PETAC-ADC copolymer detector

Mandrekar et al. (2010) claimed that a copolymer of pentaerythritol tetrakis (allyl carbonate) (PETAC) and allyl diglycol carbonate (ADC) in weight percent 4:6 shows that its maximum alpha sensitivity is almost double that of CR-39 under identical experimental conditions. The polymers can be conveniently etched by aqueous sodium hydroxide. In order to further verify the above conclusion, one should irradiate the copolymer and CR-39 sheets with heavy ions from accelerators to see at which energy the heavy ions can be recorded as tracks in each of the detectors.

TABLE 3.3 Chemical function groups, IR absorption bands and their corresponding wavelengths, and the vibration types of the bonds in polyimide.

Chemical bond	Absorption band: Wavenumber, cm^{-1}	Absorption band: λ, μm	Vibration type
C=O	1777	5.627	Stretching vibration of carbonyl group
	1732	5.774	
	1725	5.797	
C–C	1598	6.258	Skeletal in-plane phenyl ring vibration (C–C stretching of phenyl rings)
	1504	6.649	
C–N–C	1383	7.231	C–N–C asymmetric stretching
C–O–C	1289	7.758	C–O–C group vibration between two aromatic cycles
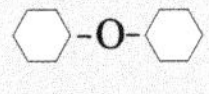	1250	8.000	Diphenyl ether symmetric stretching
	1169	8.554	
	1118	8.945	
	1094	9.141	
	883	11.325	C–H deformation of phenyl ring
C–H	823	12.151	

b. Silicon nitride (Si_3N_4) as track detector

Vlasukova et al. (2010) found that amorphous silicon nitride Si_3N_4 can record discontinuous tracks of W ($Z=74$) with electron energy loss of 20.4 $keVnm^{-1}$ being maximum with etchant HF solution; Fe ($Z=26$) and Kr ($Z=36$) ions in the electron energy loss regime cannot form tracks in it. From the above results one can see that the threshold of Si_3N_4 is quite high, but it is difficult to be used as a track detector at this stage due to its discontinuous tracks.

III. Recordable particles with solid state nuclear track detectors

When a heavy charged particle enters an insulating solid, it produces radiation damage to materials in the solid along the trajectory of the particle. If the density of the damaged materials is high enough, the damaged materials can be selectively etched by a chemical reagent resulting in an etched channel (hole) along the trajectory. The chemical reagent will continue to attack the bulk material of the solid in all directions. The inner diameter of the channel (hole) will increase in size. When the diameter of the channel increases to the size comparable with the wavelength of light, the channel becomes visible with an optical microscope. The visible channel is called an etched track of the particle. The original radiation damage region along the trajectory before etching is called the latent track of the particle. In some types of very thin solid layers, latent tracks of particles have been observed under transition electron microscopes (TEMs), such as muscovite mica. The diameter of latent tracks of fission fragments is several nanometers. If the radiation damaged materials are not dense enough, the damaged materials cannot be etched to form an etched track. In this case, the solid is not sensitive to the particle. The particle cannot be recorded by the solid. Conversely, when the solid is sensitive to the particle, it is recordable with the solid.

The density of radiation damaged materials is related not only to the species of particles (Z), but also to the velocity V (or energy E) of the particles. When the velocity (or energy) of a particle is very high, the time of interaction between the particle and the atomic electrons of the atom in the solid is very short. The energy transferred from the particle to the electrons is very little. In such a case, the resultant radiation damage is very low. The radiation damaged region cannot be etched to become a track. As the velocity of the particle slows down, the interaction time increases gradually as described in Chapter 1, Volume 1 of this book. The density of the radiation damaged materials increases. Up to a certain density, the damaged materials become etchable to form a continuous track. As the charged particle is further slowed down, the particle will catch more and more electrons from the solid. The effective charge (Z_{eff}) of the particle gradually reduces. Its ability to produce damage also reduces. At some velocity, the density of damaged materials drops down to the least value required for etching. Afterward, the charged particle cannot produce an etchable track. The charged particle can produce an etchable track only in a limited region of velocity or energy. Above or below this region, the particle cannot leave etchable tracks in the solid. Here, we conclude that as long as the radiation damage density, even in a small region, can reach the etchable level, we consider the charged particle recordable with the solid (SSNTD).

Generally speaking, selecting appropriate detector materials, SSNTDs can record tracks of protons, α-particles, fission fragments, heavy ions, neutrons via its charged reaction-products, α-recoil particles, exotic particles such as MMs, dark matter, and fractional charge particles if they exist. In contrast, γ-rays, β-rays and X-rays cannot be recorded by any kind of SSNTDs.

A. Protons

Proton measurements with SSNTDs has suddenly become an urgent need since the advent of laser acceleration and the latest advances in the study of laser inertial confinement fusion (ICF). In these studies, the duration of a proton burst is about 10^{-12} seconds or less for laser acceleration and about 10^{-9} seconds for a particle blast in laser fusion. In these cases, electronic detectors are blocked by dead time, but solid-state track detectors are superior in light of their advantages, which are the following: no dead time limitation, high detection efficiency, large dynamic range, good spatial-, charge-, mass- and energy resolutions, insensitivity to background radiation (neutrons, X- and γ rays), and insensitivity to electromagnetic disturbances. Therefore, CR-39 has become the key material for the detection of protons and other light nuclei produced in these studies (Duan et al., 2009, 2010; Szydlowski et al., 1999; Seguin et al., 2003).

The sources of protons, which are currently studied with SSNTDs include:

(1) Proton beams from laser acceleration;
(2) Protons produced from an ICF assembly and other plasma facilities;
(3) Proton beams from particle accelerators;
(4) Recoil protons from foil-radiators for neutron spectrometry and dosimetry;
(5) Recoil protons in hydrogen-containing plastic track detectors (CR-39); and
(6) Protons from other nuclear reactions.

1. Suitable detectors for proton detection

CR-39 is the most sensitive detector material in SSNTDs. It is the most suitable detector for recording proton tracks

(Baiocchi et al., 1995; Bernardi et al., 1991; Cross, 1986; Cross et al., 1986; Jager et al., 1985; Sadowski et al., 1994; Szydlowski et al., 1999; Wang et al., 2002). The lowest recordable energy of protons in CR-39 is ~20 keV (Duan et al., 2009, 2010). The upper limit of energy of protons is ~6 MeV (Seguin et al., 2003). The tracks of protons from 20 keV to 1.020 MeV in CR-39 are shown in Fig. 3.2.

CN detector is less sensitive than CR-39. It can also record proton tracks. The minimum energy of protons recordable in Daicell (CN) is 0.55 MeV (Fleischer et al., 1980).

2. Proton intensity measurements

Intensity usually refers to the number of specific particles that are produced from a source in unit time (second). To measure the intensity of a source, one needs to record the time (t) and to count the number of the particles (N) emitted from the source. The intensity is N/t. For proton intensity measurements with a SSNTD, one needs to record the time t, count the tracks N in a small area S of the detector, measure the distance X from the proton source to the detector, measure the angle θ between the proton direction and the normal of the detector, know the detection efficiency ε of the detector to protons and the anisotropy $\Phi(\varphi)$ of the source. The proton intensity can be calculated according to following equation:

$$I = \frac{4\pi X^2 N}{S\cos\theta\Phi(\varphi)\varepsilon t} \tag{3.1}$$

where φ is the angle of the detector position with respect to the proton source. For an isotropic source, $\Phi(\varphi) = 1$.

If particles (protons) from a source are collimated by apertures to become a beam such as that from particle accelerators, the particle (proton) intensity in these cases means particle flux, that is, the number of particles passed through a unit area (cm^2) in unit time (second), also referred to as beam intensity.

Proton sources may be a point source, or an extended source. Any particle source has a definite size (extended source). A so-called point source is a simple approximation, that the size of the source is negligible. Eq. (3.1) is derived by an assumption that the particle source is a point source.

Placing a CR-39 sheet in the path of the protons and after counting the proton tracks, one can easily measure the intensity of the proton source or proton beam flux.

3. Proton energy measurements

a. Track diameter method

In order to determine proton energy with a CR-39 or LR-115 detector, one must first calibrate the detector with mono-energetic protons. A detailed explanation of the calibration techniques with real monoenergetic protons (not by scattered protons through foil) can be found in the paper of Duan et al. (2009, 2010).

Fig. 3.5 illustrates the results of a typical calibration with proton energies of 20, 40, 60, 80, and 100 keV:

In this energy region, it shows that the diameter of proton tracks increases monotonically with increasing proton energy in a defined etching condition. By measuring the track diameter of protons of unknown energy, one can immediately find the energy E_p of the protons from the calibration curve as illustrated in Fig. 3.5. The uncertainty $\sigma(Ep)$ can be calculated from the following formula:

$$\sigma(E_p) = \frac{\sigma(D)}{\Delta D/\Delta E_p} \tag{3.2}$$

where $\sigma(D)$ is the statistical deviation of the measurements of the track diameter, and $\Delta(D)/\Delta E_p$ is the slope of D versus E_p curve at the point of track diameter D.

b. Track contrast (gray level) method

Fig. 3.6 illustrates the variation of track diameter D versus proton energy E_p in an extended range of energy.

There are two proton energies corresponding to a given track diameter. In this case, the two proton energies can be distinguished by the track contrast or track gray levels. The tracks of the lower energy protons are generally shallower and less dark due to the shallow and blunt etch pits. The tracks of the higher energy protons are deeper and darker due to the deep and sharp etch pits. This method of energy measurement of protons is practical with the use of the optical microscope and track image analyzers (Duan et al., 2009, 2010; Seguin et al., 2003; Nikezic and Yu, 2009).

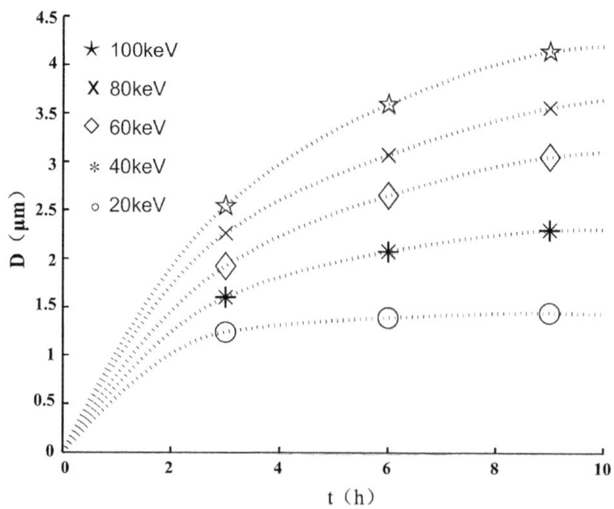

FIG. 3.5 Relation between track diameter and proton energy in CR-39 etched in 6N NaOH, 70°C for 3, 6 and 9 hours. *From Duan et al. (2009); printed with permission from Elsevier © 2009.*

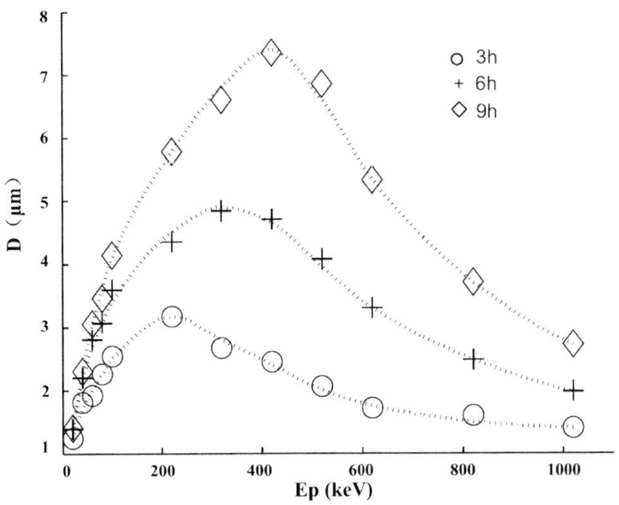

FIG. 3.6 Relation between track diameter and energy of protons in CR-39 etched in 6N NaOH at 70°C for 3, 6 and 9 h for proton energies from 20 keV to 1.2 MeV (Duan et al., 2009). Each track diameter corresponds to two proton energies. Track contrast or track gray level must be adopted in order to distinguish the energies of the protons. *From Duan et al. (2009). Reprinted with permission from Elsevier © 2009.*

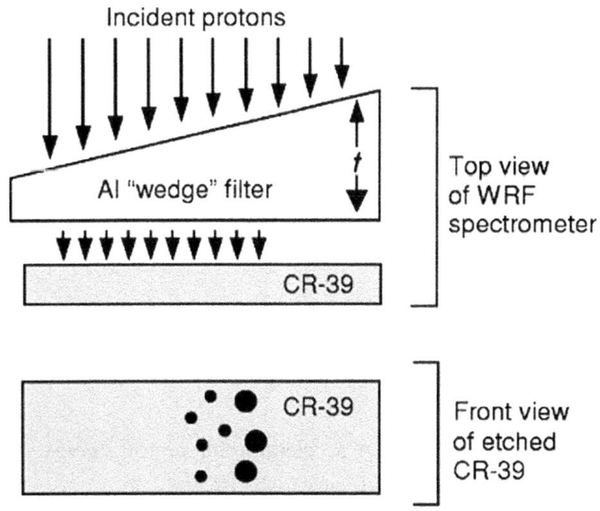

FIG. 3.7 Schematic illustration of a wedge-shaped-filter spectrometer. In front of a CR-39 there is a wedge filter (Al). The thickness of the wedge changes gradually. Some of the protons incident on the filter are energetic enough to pass through it, and they impinge on the CR-39, forming tracks within the CR-39. The energy of the protons can be determined by the diameters of the tracks and the thickness of the filter at the track positions. A two-dimension histogram of track number versus track diameter and versus the thickness of the filter can be converted to a histogram of proton number versus energy of the protons. *Seguin et al. (2003). Reprinted with permission from American Institute of Physics © 2003.*

c. Step filter method

The energy and energy spectrum of protons can be retrieved by the step filter method. If we let the protons pass through a plate of stopping material (filter of Al, Cu, or plastic sheet) and allow the protons to reach a CR-39 detector; and if the energy of the protons is in the energy window of the CR-39 registration (~20 keV–6 MeV), the protons will be recorded as tracks in the CR-39. The energy of protons on reaching the CR-39 can be obtained by measurement of track diameters. From the thickness of the stopping material (the filter) and the range–energy relation of protons in the material, one can further derive the energy of the protons arriving at the front surface of the filter. On the other hand, if a wedge-shaped filter is used to replace the plate filter, a wedge-range-filter (WRF) spectrometer is composed. This kind of filter can provide information of the proton energy spectrum of the source. Fig. 3.7 is a schematic illustration of a wedge-filter spectrometer used for measuring the proton energy spectrum.

The wedge can be composed of a series of films overlapping partly together. Different thicknesses of stopping material in the composed areas can be obtained. The wedge looks like a terraced field. The thickness of each part of the stopping material can be known precisely. This is the reason it is called a "step filter".

4. Obtaining mono-energetic proton beam for track detector calibration

a. Defocusing proton beam

Defocusing of heavy ion beams including protons for irradiation of nuclear track detectors is a special requirement. Most of the nuclear physics researches require very well-focused strong beams of heavy ions. On the contrary, nuclear a track detector as a tool of nuclear physics research needs a reduced and defocused beam of heavy ions. The reason for the special requirement is that nuclear track analysis needs to separate each track from the others. The track density usually should be less than about 10^4–10^5 cm^{-2} and should be kept mono-energetically. Therefore, one needs a specially designed defocusing beam facility to reduce the intensity of the beam. Usually the beam should be uniform in a quite large area perpendicular to the beam direction.

b. Deflection by high voltage electric pulse

If the particle accelerator cannot defocus to the required level, a deflection high voltage electric pulse may be used to deflect the beam in very short time. Fig. 3.8 shows a sketch of a high voltage pulse generator system to deflect

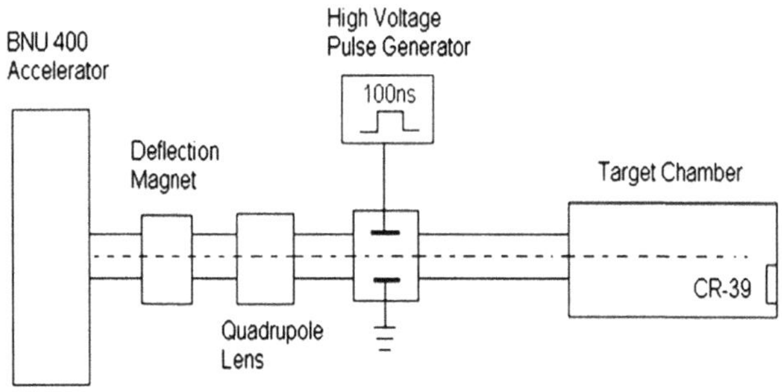

FIG. 3.8 Sketch of the calibration system. *From Duan et al. (2009). Reprinted with kind permission from Elsevier © 2009.*

the proton beam at an accelerator for proton energy calibration in the studies of laser acceleration.

The output of the high voltage waveform of the high-voltage pulse generator is shown in Fig. 3.9.

In this method of irradiation with protons, two requirements need to be satisfied. First, the pulse duration should be set according to the requirement of the heavy ion energy. The adjustable duration of the pulse of the generator is from 0 to larger than 100 ns. Second, the magnitude of the voltage must be high enough to ensure enough deflection of the beam direction. The adjustable magnitude of the generator ranges from 0 to 5000 V (Duan et al., 2009). The track detector CR-39 sheet is placed at the angle where the beam of protons is to be deflected. By counting the pulse, one can obtain the required number of proton tracks.

c. High-speed rotation wheel

A reduced density of tracks of mono-energetic protons in CR-39 sheet for calibration purposes can be irradiated at an accelerator by using a high-speed rotation wheel. CR-39 sheets are fixed on a wheel with a certain distance r from the center of the wheel. The wheel is then arranged in the

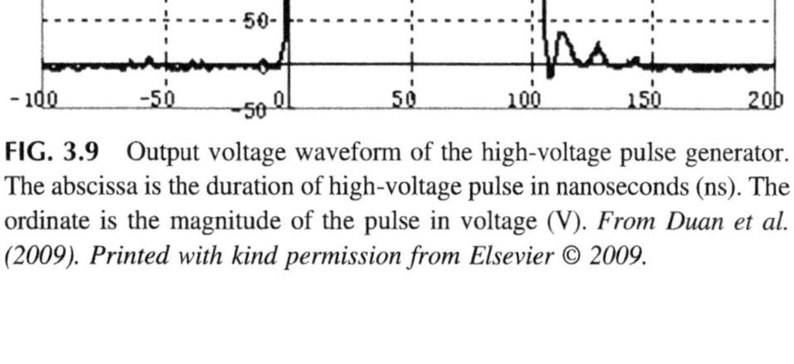

FIG. 3.9 Output voltage waveform of the high-voltage pulse generator. The abscissa is the duration of high-voltage pulse in nanoseconds (ns). The ordinate is the magnitude of the pulse in voltage (V). *From Duan et al. (2009). Printed with kind permission from Elsevier © 2009.*

target chamber and the chamber is pumped to vacuum. The CR-39 sheets will be faced toward the beam of protons. High-speed rotation of the wheel brings the CR-39 sheets to the beam, and the CR-39 sheets are exposed to the beam and left without stopping; but the wheel stops immediately after the exposure. Duan et al. (2009) made proton irradiation successfully at the accelerator in Beijing Normal University for calibration of CR-39 used in laser acceleration.

All the above techniques are able to obtain mono-energetic protons for calibration of track detectors.

5. Proton spatial distribution measurements

The SSNTD is a type of position sensitive detector. Its track diameter after etching is about several micrometers. Therefore, its spatial resolution is in the same range (several micrometers).

In some cases, the information of emission angles of protons is essential. For example, recoil protons from a thin hydrogen-containing foil, such as polyethylene $(CH)_n$, produced by elastic scattering of fast neutrons, can be recorded by a flat or curved CR-39 detector. The track coordinate on the detector reflects the recoil angle θ with respect to the direction of the incident neutron. The diameter of the track reflects the energy of the proton E_p (See Section III.A.3). The energy of the incident fast neutron can be calculated by the equation:

$$E_n = E_p/\cos^2\theta \tag{3.3}$$

This technique is often called the recoil proton method for neutron spectrum determination (Wan et al., 2003).

Another example of the measurement of proton spatial distribution is in laser acceleration. In order to determine the energy spectrum of protons accelerated in laser–plasma interaction, an electromagnetic spectrometer (Thomson spectrometer) is used to deflect the energetic protons produced in the interaction. Different energies of protons will be deflected to different angles. A flat or curved CR-39 sheet is placed behind the spectrometer to record the protons at different angles. Counting the proton tracks at different angles provides the energy spectrum of the protons after acceleration (Lan, 2009).

6. Applications of proton detection

The SSNTD is a kind of passive detector. It exists simultaneously with active detectors such as electronic types of detectors. Electronic types of detectors can instantaneously acquire data of measurements; but suffers from their dead time, which is in the level of ns (10^{-9} s). If there are many particles incident to the electronic type of detector at the same time, only one particle can be recorded, the other particles will be lost. More over, the signal of the first recorded particle will be distorted. Under these circumstances, electronic types of detectors are incompetent; but the SSNTD is almost unique.

a. Laser acceleration

Proton registration has become a budding and expanding field since the advent of laser acceleration. When an ultrashort and ultraintense laser hits on a target, the target substances become plasma. The electrons with much higher speed in the plasma fly out toward the direction nearly perpendicularly to the original target substances and forming a shell (sheath). In the gap of the high-speed shell of electrons and the positive ions forms a very strong electric field, which accelerates hydrogen nuclei (protons) contained in this region to very high velocity. This process accelerates the protons to very high energy (up to now about 100 MeV). This is the so-called laser acceleration. Large numbers of protons are produced and accelerated within a very short time, which is in the range of about pico-second (ps, 10^{-12} s) to femto-second (fs, 10^{-15} s). Electronic types of detectors cannot record and analyze the protons, but SSNTD can. Therefore, SSNTD is an indispensable tool in laser acceleration studies (Kanasaki et al., 2013; Yogo et al., 2011).

b. Laser confinement fusion

Laser confinement fusion is also a process produced within a short time from $\sim 10^{-12}$ to 10^{-15} s by laser on bullet target of fuel of deuterium and tritium. At this stage, CR-39, polycarbonate (PC) and the other SSNTD play important roles in these studies.

The application of proton detection in these fields will be described in more detail in Section II.D.1 and II D.2 of Part 2 of this chapter. See also Chapter 1, Volume 1 of this book for a description of current studies on laser and inertial electrostatic confinement fusion.

B. Alpha particles

Alpha particles (α-particle) are equivalent to a helium (He) nucleus with kinetic energy. The following sources of alpha particles are usually analyzed with SSNTDs:

(1) Alpha particle sources used in laboratories for calibration purposes;
(2) U-bearing aerosol particles for nuclear safeguards;
(3) Hot particles and fallout from nuclear accidents;
(4) Radon (Rn) and thoron dosimetry in dwellings, mines, and environments;
(5) Radon (Rn) monitoring for exploration of uranium (U) and petroleum (oil), and prediction of earthquake and landslip;
(6) $^{10}B(n,\alpha)^{7}Li$ reaction α-particle mapping for boron determination in metallurgy, agriculture, and cancer therapy (BNCT).

α-Particle registration is the most popular work in developing countries because it does not require expensive equipment, such as nuclear reactors and particle accelerators. One needs only CR-39, LR-115, an optical microscope, etching bath, and chemicals; and it provides valuable research results for natural resources, human health, and scientific training.

1. *Suitable detectors for alpha-particle detection*

CR-39 (Polyallyldiglycol carbonate, PADC), LR-115 (CN) and Lexan or Makrofol (Polycarbonate, PC) are suitable detectors for recording α-particles.

a. CR-39

CR-39 is the most sensitive and popular detector for recording α-particles. Its energy window for recording α-particles is from ~ 0.1 to > 20 MeV, or approximately from 0 to infinite (Durrani and Bull, 1987). This means that no matter what is the medium between the source and the detector (CR-39), gas, liquid, or solid, the α-particles will create tracks in the detector as long as the α-particle can reach the detector, with the exception of the limitation of critical angle θ_c (See Section V.A.3).

b. LR-115

The sensitivity of CN LR-115 is less than CR-39. Its energy window for α-particles is about 1.2–3.9 MeV (Yu and Nikezic, 2009; Benton, 1968; Eappan and Mayya, 2004; Ilic and Durrani, 2003).

c. Lexan or Makrofol

Polycarbonate (Lexan, Makrofol and Tuffak) is less sensitive than CN for recording α-particle tracks. In order to reveal α-particles clearly, it is better to irradiate the detector with ultraviolet light before etching (Stern and Price, 1972; Benton and Henke, 1969a; Crawford et al., 1968; DeSorbo and Humphrey, 1970; DeSorbo, 1979; Fleischer et al., 1975; Durrani and Bull, 1987). The energy window of polycarbonate for recording α-particles is in between ~ 0.2 and 3 MeV (Durrani and Bull, 1987).

2. *Alpha-particle intensity measurements*

α-Particle sources mainly have three states: solid, liquid, and gaseous. Solid α-sources are prepared on smooth plane surfaces of metals, such as stainless steel, or aluminum. The plane sources can be divided into thin sources and thick sources. If the thicknesses of α-emitting material on the surfaces are less than about 100 $\mu g/cm^2$, the sources may be considered thin sources. Because the ranges of α-particles of natural radioactivities in the source materials are in the range of about several to over ten mg/cm^2 (see Chapter 1, Volume 1), the absorption loss or energy losses of α-particles in the thin source are negligible. Conversely, if the thicknesses are over ~100 μg/cm^2, the absorption losses or energy losses cannot be ignored.

Some α-emitting materials are dissolved or suspended in water or other solutions. Determination of α-concentration in the liquids is often required. Radon in air is an example of gaseous α-source.

a. Determination techniques of α-particle intensity on plane surfaces

α Intensity is defined as the number of α particles emitted from a source in unit time (s^{-1}). Two types of techniques can be used in the determination of α intensity of plane sources: small sterad technique and 2π geometrical measurement.

Small Sterad Technique. The small sterad technique can give the most accurate and absolute determination of the intensity of α-particle sources. The α-plane source to be determined and a CR-39 sheet are put face to face, in a vacuum chamber with a distance 10–30 cm from each other, both surfaces being parallel with common axis. After pumping to remove air from the vacuum chamber, a shutter between the source and CR-39 sheet is opened, and α recording starts. After some time, the shutter is closed, and α recording stops. The α tracks on the CR-39 sheet are etched in NaOH solution and counted with a microscope. The α intensity I of the α-plane source can be calculated by the equations

$$I = \frac{N_T}{g\varepsilon t} \tag{3.4}$$

and

$$g = \frac{1}{2}\left[1 - \frac{h}{\sqrt{h^2 + r^2}}\right]\left\{1 - \frac{3}{8}R^2\left[\frac{h\left(h + \sqrt{h^2 + r^2}\right)}{(h^2 + r^2)}\right]\right\}, \quad (R << h,\ r << h) \tag{3.5}$$

where N_T is the number of α-tracks on the CR-39 sheet, t is the α recording time in seconds (s), g is the geometric factor for recording α particles with CR-39 sheet, h is the distance between the α source and CR-39 detector (cm), R is the radius of the α source (cm), r is the radius of the CR-39 detector (cm), and ε is the revealing efficiency of the α-particles in the CR-39 sheet.

If the CR-39 is shaped as a square with side a and all the tracks are counted as N_T, then r^2 in Eq. (3.5) should be replaced by a^2/π.

If one counts only a part of tracks in a small area S of CR-39 sheet, such as in several fields of view, one should replace r^2 in Eq. (3.5) by S/π. If the energy of the α particles is in the energy window of the detector, $\varepsilon = 1$. Otherwise, $\varepsilon = 0$. If the energy of the α particles is higher than the energy window of the detector, one can put a thin degrader foil in front of the detector to degrade the α energy. The thickness of the degrader can be calculated by an empirical Bragg–Kleeman formula (L'Annunziata, 2003; Eappan and Mayya, 2004):

$$R_s(\text{cm}) = 3.2 \times 10^{-4}\, R_a \text{A}^{1/2}/\rho \tag{3.6}$$

where A is the mean atomic weight of the solid (g), ρ is density of the solid in g/cm^3, R_s is the range of α particle in a given solid (the degrader foil), and R_a is the range of the α particle in air at an ambient temperature of 15°C under pressure of 760 mm Hg. R_a may be estimated through the relation:

$$R_a(\text{cm}) = 0.324E^{3/2} \tag{3.7}$$

where E is the energy of the α particle (MeV). The above equations for the calculation of α particle ranges are the same as Eqs. (1.166) and (1.167) in Chapter 1, Volume 1, where example calculations are provided.

When one has calculated the α particle range, one can select a thickness of the degrader foil, which is thinner than the calculated range. The selected foil can be used as degrader. The thickness of α sources for small sterad measurement should be thin, that is, at least less than half of the range of the α particles in the source material to ensure the accuracy of the measurement. In measurements utilizing the small sterad technique, other detectors can also be used, such as a surface barrier silicon detector or a scintillation detector. The advantage of the SSNTD (CR-39) is that it does not need the complicated electronic instrumentation.

2π geometrical measurement. The 2π geometrical measurement is a simple technique for determining the intensity of an α plane source. In this measurement, it is best that it be carried out as a relative measurement.

First, one should make a thin plane source of α particles as a standard source, determine its intensity I_s by using other techniques, such as small sterad, surface barrier detector, ionization chamber, or chemical analysis. Then the standard source is covered with a CR-39 detector (2π geometry), for a time t_s. After etching the CR-39 in NaOH solution, the α tracks are counted with an optical microscope (N_{Ts}). Afterward, an unknown α source is covered with another CR-39 sheet of the same batch as that used for the standard source for a time t. The tracks are etched and counted (N_T) in the same manner, as above. An approximation of the intensity I of the unknown source can be calculated by the following equation:

$$I = \frac{N_T}{N_{Ts}}\frac{t_s}{t}I_s \tag{3.8}$$

In Eq. (3.8), the difference of detection efficiencies for the standard and unknown sources due to possible difference in thicknesses of the sources are ignored.

From Eq. (3.8) one can see that the intensity of the unknown source I can be calculated from the measured data N_T, N_{Ts}, t, t_s and the know intensity I_s. The advantage of this relative measurement for plane sources is that one does not need to calibrate the detection efficiency of the CR-39 detector for the α particles produced in the sources.

b. Determination technique of α-particle intensity in liquids

In the process of uranium production and nuclear disposal as well as in laboratories, various liquids having α radioactivity have to be managed. The α radioactivities in these liquids can be easily determined by using SSNTDs (CR-39 or LR-115) through the relative measurement. Here, we will give an example to illustrate how to determine the α intensity of uranium in liquids. Firstly, some amount of uranium nitrate of natural uranium is dissolved in water. The uranium concentration in the solution can be calculated from the formula of uranium nitrate. The α intensity I_s in a unit volume of solution is obtained by the isotopic ratio of ^{238}U (99.2745%), ^{235}U (0.720%) and ^{234}U (0.0055%) and the half-life of ^{238}U (4.468×10^9 a), ^{235}U (7.038×10^8 a) and ^{234}U (2.455×10^5 a). This solution can be used as a standard solution for the α intensity determination. A piece of SSNTD, such as CR-39 is immersed into the solution for a certain time t_s. The CR-39 sheet can be cleaned by water. After etching the CR-39 sheet in NaOH solution (see Table 3.1), the track density N_{Ts} of α particles is measured with an optical microscope. Afterward, another sheet of CR-39 of the same batch is immersed in the solution of unknown α intensity for a time duration t_x. The track density N_{Tx} is measured. The α intensity I_x in the unknown solution can be calculated by the equation

$$I_x = \frac{N_{Tx}}{N_{Ts}} \frac{t_s}{t_x} I_s \tag{3.9}$$

Eq. (3.9) is strictly valid only in the case when the composition and concentrations of the chemicals and α energy are the same for the standard and unknown solutions. The variation of concentration will not affect the calculated value of I_x very seriously. The effect of α energy can be corrected as inversely proportional to the range of the α particles, that is,

$$I_x(\text{corrected}) = I_x \frac{R_s}{R_x} \tag{3.10}$$

The calibration result of CR-39 with standard solution can be treated as a constant c where

$$c = \frac{I_s t_s}{N_{Ts}} \tag{3.11}$$

Once the constant c is obtained, it can be used for follow-up determinations of unknown solutions with the same batch of detector.

The advantage of this relative determination of α intensity in liquids is that one does not need to worry about the critical angles of α particles in the CR-39. The effect of the critical angles will be canceled in the calculation. The distribution of tracks on the surface of the CR-39 is homogeneous in the case of a liquid source. One can select any place on the detector surface to measure the density of tracks and derive the α intensity of the liquid source.

c. Determination of α-particle intensity in air (gaseous substances)

The determination of the activities of radon (^{222}Rn), thoron (^{220}Rn), and ^{219}Rn in air is a typical example for α intensity measurement in gaseous substances. This subject is important in earth science, environmental studies, health physics, and resource prospecting. LR-115 and CR-35 are traditionally used to carry out the research. It has become a special field of Radon (Rn) monitoring (Jonsson, 1995; Ilic, 2005; Monnin et al., 1993; Segovia et al., 2005; Virk and Valia, 2001; Saad, 2008; Dwivedi et al., 2005; Mishra et al., 2005; Font, 2009; Font et al., 2008; Baixeras et al., 2005). More detailed explanation can be found in the Subsection Radon Measurement (Section IV.A.1 in Part 2 of this chapter).

3. *α-particle energy measurements*

Three methods can be adopted in α particle energy determinations, namely, the track diameter method, the residual range method, and the stopping-foil method.

a. Track diameter method

The experimental determination of energy of α particles with a SSNTD by the track diameter method includes two steps. The first step is to calibrate the detector, such as CR-39 with mono-energetic α particles from a particle accelerator or from an α source, such as an ^{241}Am thin source (5.486 MeV (85.2%), 5.443 MeV (12.8%), other α energies are negligible). The energy of α particles may be degraded by air or foils of various thicknesses. The α particles impinge on the CR-39 detector perpendicularly. After etching of the CR-39 under certain etching conditions, the diameter of the α tracks are measured with an optical microscope. A calibration curve will be obtained, which is the relationship between track diameter and the energy of the α particles. The calibration curve is similar to that of protons (Figs. 3.5 and 3.6). The second step is to record the α particles of unknown energy in normal incidence with the calibrated CR-39 detector (the detector produced in the same batch if possible). After etching under the same etching conditions, the track diameter is measured. By comparison of the track diameter with the calibration curve, one can obtain the energy of the α particles. On the calibration curve, one diameter may correspond to two energies of α particles (Similar to Fig. 3.6). In this case, the gray level and the depth of track can be used to distinguish which energy is correct (See Section III.A.3, Proton Energy Measurements); on the low-energy side, the track is shallower and the gray level is lower than that on the high-energy side.

b. Residual range method

The residual range method makes use of the range–energy relation (R–E relation) of α particles in the detector materials. Measuring the residual range (R_0) allows one to retrieve the energy (E) of the particles. The range–energy relation can be calculated by the empirical Bragg–Kleeman formula, or by theoretical formulas (Bethe, 1930; Fano, 1964; Ahlen, 1980; Barkas and Berger, 1964; Benton and Henke, 1969b; Weaver, 2001; Henke and Benton, 1966), or by more popular computational codes SRIM (Ziegler and Biersack, 2003; Ziegler et al., 1985). The residual range of a particle is the distance from the point considered to the end point of the range. The coordinates of the end point of the range can be obtained by etching the track to a round blunt tip. The actual end point is at the point with a distance r to the etched tip where r is the radius of the blunt tip. When the residual range R_0 is measured, one can immediately obtain the energy from the range–energy relation (Fews and Henshaw, 1982; Seguin et al., 2003).

c. Stopping-foil or range-filter method

A flat foil or wedge-like stopping material (range filter) and a CR-39 detector are used in this method, which is based on the range–energy relation of α particles in stopping material (Range filter) to determine the energy of the α particles. Only the α particles whose range is larger than the thickness of the range filter can be recorded as etchable tracks in the CR-39 detector behind the filter. The track diameter on the CR-39 sheet can give the energy of α particles. From this energy and the thickness of the filter one can derive the incident energy of the α particle to the filter.

If one uses several filters of different thicknesses and CR-39 sheet, a series of energies (energy spectrum) can be obtained. If one uses a WRF of appropriate thickness intervals, an accurate energy spectrum of α particles may be obtained (Seguin et al., 2003; Kacenjar et al., 1982).

4. α-particle spatial distribution measurements

The SSNTD is a position sensitive detector. Its spatial resolution is normally about several micrometers for α particles. α spatial distribution measurements have two meanings. The first meaning relates to the measurement of the distribution of α emitting nuclei on a surface or in a volume. This is a homogeneity measurement of an α source. The second meaning relates to the measurement of the directions of an α emission from a source. This is the angular distribution measurement of an α source. For the first measurement, one needs to put a CR-39 or LR-115 sheet on the area or in the volume of the α source for a certain time to allow enough numbers of α particles to bombard the detector sheet. Afterward, the detector is etched, whereby one can see the distribution of tracks produced by the α particles. Subsequently, one will be aware of the distribution of the source nuclei. The following measurements belong to the first type of measurements: distribution of α contaminations on surface of working places; the distribution of α emitting nuclei, such as ^{239}Pu, ^{235}U, and ^{237}Np in organs of animals and human bodies; distributions of α hot particles in the environments; uranium distribution in minerals; boron distribution measurements in steels, leaves, and soils through (n,α) reaction-produced α particles.

For the second type of spatial distribution measurement, one needs to deploy CR-39 or LR-115 sheets around the α source in different directions. In laser ICF studies, the CR-39 detector is the most useful detector to determine the yield and spatial distribution of α particles from D + T, T + T, D + ^{3}He, and T + ^{3}He reactions (Seguin et al., 2003). By counting the tracks in unit area on the detectors in different directions, one will know the spatial distribution of the α particles.

5. Applications of α-particle detection

α-particle detection is one of the commonly applied domains of SSNTDs. Nearly all the fields where α particles are produced can be studied by the track detectors. The particular fields requiring α detection and where SSNTDs are particularly useful are as follows:

- **a.** Long-term radon and thoron monitoring,
- **b.** Nuclear fallout and hot particles,
- **c.** Nuclear contamination around nuclear plants,
- **d.** Neutron spectrometry and dosimetry,
- **e.** Laser and plasma interactions,
- **f.** Uranium and petroleum exploration and earthquake prediction,

g. Agriculture and metallurgy, and
h. α-particle radiograph (Somogyi and Srivastava, 1970, 1971; Su, 1992, 1993).

The details of applications in these fields will be explained in the related subsections in Part 2 of this Chapter.

C. Fission fragments

A heavy nucleus can split into two nuclei, which is called nuclear fission. This process may occur automatically by the nucleus itself without any disturbance by other particles. This fission is called spontaneous fission. For example, the ^{238}U nucleus can split into two nuclei (the two new nuclei are called fission fragments, nuclear fission into three or more fission fragments is very rare (See Chapter 1, Volume 1 for a detailed treatment of nuclear fission). The two fission fragments have different masses. The masses of light fragments of ^{238}U spontaneous fission range from $A = 85-105$ with an average $A = 97$. The heavy fragments have masses $A = 130-150$ with an average $A = 139$. The light fragments have an average kinetic energy of 97.6±0.7 MeV. The heavy fragments have an average kinetic energy of 67.0±1.2 MeV. The total kinetic energy is 164.6 MeV (Kase et al., 1978).

The rate of spontaneous fission of heavy nuclei is expressed by the spontaneous fission decay constant λ_f. For ^{238}U spontaneous fission, one value of the decay constant $\lambda_f = 7.03 \times 10^{-17}\ a^{-1}$ (Roberts et al., 1968) may be recommended. This value corresponds to the half life of ^{238}U spontaneous fission $T_{1/2} = 9.86 \times 10^{15}$ a ($T_{1/2} = \ln 2/\lambda_f \approx 0.693/\lambda_f$).

It should be noted that more than 30 measurements on the decay constant of spontaneous fission of ^{238}U have been published (See the review articles: Bigazzi, 1981; Durrani and Bull, 1987; Wagner and Van den haute, 1992; Holden and Hoffman, 2000). A wide spread of the results is apparent (λ_f from $1.7 \times 10^{-17}\ a^{-1}$ to $11.8 \times 10^{-17}\ a^{-1}$) from different measuring techniques. Most of the results are gathered into two groups. The largest group is peaked around $7 \times 10^{-17}\ a^{-1}$, all of them were measured by SSNTDs. Another group is peaked around $8.46 \times 10^{-17}\ a^{-1}$, which are measured mostly by the other techniques. A technical report on spontaneous fission half-lives for ground-state nuclides was published in 2000 by the International Union of Pure and Applied Chemistry, Analytical Chemistry Division, Commission on Radiochemistry and Nuclear Techniques (Holden and Hoffman, 2000), in which a value of ^{238}U spontaneous fission half-life $T_{1/2} = (8.2 \pm 0.1) \times 10^{15}$ a was recommended, which corresponds to $\lambda_f = 8.45 \times 10^{-17}\ a^{-1}$.

To obtain the recommended value, all the data of λ_f measured by SSNTDs were ruled out (16 out of 26 data in total). The reasons for ruling out the track data were explained in detail in the report. The basic reason is that most of the results of the λ_f measured by etch track detectors are lower than those by other detectors. The simplest way to get an average value is to throw away all the data from etch track measurements. The true reason for causing the large differences of λ_f values determined by different detectors might not have been exposed. Durrani and Bull (1987) suggested that if λ_f of ^{238}U spontaneous fission has been explicitly used in FTD, the value λ_f assumed must be stated in reporting results, so that valid comparisons can be made with other age determinations.

In addition to spontaneous fission, a heavy nucleus can split into two fission fragments upon bombardment with an incident particle, such as neutron, proton, heavy ion, γ-ray photon, and other elementary particle. For example, the ^{235}U nucleus bombarded by a thermal neutron will split into two fission fragments. This type of nuclear fission is called induced fission, as described in Chapter 1, Volume 1.

The mass distributions of ^{235}U fission induced by thermal neutrons are similar to those in ^{238}U spontaneous fission. The masses of light fragments are in the range of $A = 78-107$, with a peak value at $A = 95$. The heavy fragments are in the range of $A = 118-146$, with an average value $A = 140$ and a peak value $A = 134$. The charges of the light fragments are in the range $Z = 31-43$ with a peak value at $Z = 38$. The charges of the heavy fragments are $Z = 47-57$ with a peak at $Z = 52$. The charge Z and mass A of the fission fragments are in the range suitable for detection with SSNTDs.

The probability of nuclear reaction between a target nucleus and incident particle is described by the quantity reaction cross section. For example, the fission cross section σ_f of ^{235}U by thermal neutrons ($v = 2200$ m/s, $E_n = 0.0253$ eV) is 582 b (1 b $= 10^{-24}$ cm^2). The fission cross section of ^{239}Pu by thermal neutrons is 738 b (Garber and Kinsey, 1976a,b). The fission cross section of ^{239}Pu is larger than that of ^{235}U for thermal neutrons. Both spontaneous and induced fission of heavy nuclei are important fields of research and application of SSNTDs.

1. Suitable detectors for fission fragments

The energies of fission fragments of ^{235}U induced fission are about 1 and 0.5 MeV/amu for light and heavy fragments, respectively (MeV/amu = energy per atomic mass unit, nearly equal to energy per nucleon). These energies are around the peak regions of energy-loss rate (dE/dx), or restricted energy-loss rate (REL) or primary ionization (J) of nuclear particles in solids. Therefore, fission fragments are very easily recorded by SSNTDs. Even the most insensitive track detector of meteoritic minerals can record tracks of fission fragments. In other words, all of the known etch track detectors can be used to record tracks of fission fragments.

The track detectors for recording fission fragments can be classified into natural and man-made detectors. The most often used natural detectors include minerals (such as muscovite mica, apatite, zircon, sphene, olivine, pyroxene, whitlockite) and natural glasses (such as tektite and obsidian). These detectors not only can record currently produced fission fragments, but also have retained large number of fission fragment tracks recorded since the formation of the detectors millions or billions of years ago. These detectors have been used in FTD, geothermal history and paleoanthropology studies. If one wants to use these detectors to record newly produced particles in the laboratory, one should first anneal out the fossil tracks of fission fragments and other tracks to produce clean detectors (free from the backgrounds of the fossil tracks). If the annealing process had been made by ancient mankind by making fires, the newly recorded fission fragment tracks after firing can provide the information of the age of the ancient man.

Man-made detectors for fission fragments can be classified into inorganic detectors and organic detectors. Inorganic detectors include synthetic mineral (such as synthetic phlogopite) and man-made glasses (such as soda-lime glass or window glass and phosphate glass). The advantage of these detectors is that these are free from background tracks. Organic detectors include polycarbonate, polyethylene terephthalate, CR-39 and cellulose nitrate (LR-115).

a. Muscovite mica

Muscovite mica is an ideal track detector for the study of nuclear fission at the low-energy region. It can be cleaved into suitable thickness with a perfect smooth surface and cut to the required area. The lowest atomic number Z of particles recordable by muscovite mica is 10 (Ne). α particles emitted from fission sources cannot be recorded as background tracks in mica. During irradiation with neutrons in nuclear reactors, the recoil nuclei and most of nuclear reaction products from the compositions of mica and from other materials cannot form background tracks in mica. Therefore, muscovite mica is a clean detector for the study of nuclear fission. In this respect, muscovite mica is much better than plastic track detectors, such as polycarbonate, CN, cellulose acetate, polyethylene terephthalate, and CR-39.

In nuclear fission rate measurements, 2π geometry is often used. An amount of fissile material such as natural uranium (99.2745% ^{238}U, 0.720% ^{235}U and 0.0055% ^{234}U) is electrodeposited onto a stainless-steel plate. A sheet of muscovite mica overlaps on it, constituting a 2π geometry. If one places this combination on the beam of thermal neutrons from a nuclear reactor channel, one can record fission fragment tracks with the muscovite mica. Because the angular distribution of fission fragments is isotropic in thermal neutron induced fission and every fission event releases two fission fragments in opposite directions, the detection efficiency of the muscovite mica for the fission fragments can be written as follows (Guo et al., 1976b, 1982).

$$\varepsilon = \begin{cases} 1 - \sin\theta_c & 0 \le t \le t_c \\ 1 - \sin\theta_c - \dfrac{(t - R\sin\theta_c)^2}{2Rt} & t_c \le t \le R \\ \dfrac{R}{2t}\cos^2\theta_c & t \ge R \end{cases} \tag{3.12}$$

where ε is the detection efficiency of muscovite mica for fission fragments, θ_c is the critical angle of track etching of fission fragments in the detector (See Section V.A.2), R is the average range of fission fragments in the fission source (mg/cm^2), t is the thickness of the fission source (mg/cm^2), and t_c is the critical thickness of the fissile material, $t_c = R\sin\theta_c$.

For other track detectors and isotropic distribution of fission fragments, Eq. (3.12) can also be useful.

Many authors have determined the detection efficiency ε and the critical angle θ_c of muscovite mica for fission fragments. For example, $\varepsilon = (93.6 \pm 0.3)\%$ and $\theta_c = 3°41'$ for fission fragments of ^{235}U fission induced by thermal neutrons (Guo et al., 1976b, 1982), $\varepsilon = 91.8\%$ and $\theta_c = 4°31'$ for fission fragments of ^{252}Cf spontaneous fission (Khan and Durrani, 1972), and $\varepsilon = (94.8 \pm 0.53)\%$ for ^{244}Cm spontaneous fission (Gold et al., 1968).

In fission rate measurements with SSNTDs in 2π geometry, the backscattering effect of fission fragments from the substrate of fission source must be taken into consideration. For a thin fissile source attached to the surface of the detector, one fission fragment from each fission event will reach the surface of the detector; another fission fragment will go in the opposite direction and hit the substrate of the source. But there is a possibility that the fission fragment undergoes elastic scattering on the atoms in the substrate and changes its direction and enters into the detector. This fragment may also create a track in the detector. For this reason, one fission event may create two fission fragment tracks in the detector. In other words, the number of tracks recorded in the detector is more than expected from the 2π geometry. For an accurate determination of fission rate, one must make a backscattering correction for the number of tracks recorded by track detectors. A careful calibration showed that for gold–silicon substrate (Au–Si surface barrier detector), 2.1% more tracks will be recorded by muscovite mica detectors and 2.1−2.7% more tracks will be recorded by polycarbonate track detectors (Guo et al., 1976b, 1982).

Muscovite mica has been used in measurements of fission cross sections, angular distribution of fission fragments, neutron spectrometry and dosimetry, fission and neutron distribution in nuclear reactors, sequential fission in medium energy nuclear interaction and FTD as external detectors. More details will be explained in Part 2 of this chapter.

b. Polycarbonate (Lexan, Makrofol, Tuffak)

The polycarbonate track detector is more sensitive than muscovite mica for fission fragments. Its least recordable particle is helium ($Z = 2$). However, it does not mean that full energy α-particles emitted from uranium and plutonium can be recorded by polycarbonate. The α particles can be recorded by polycarbonate detector only when the α particles are slowed down within the detector to a very low energy ($\leq$0.75 MeV/u) (Fleischer et al., 1975; Fleischer, 1981). In other words, by short etching, polycarbonate cannot show tracks of α particles from natural α radioactivity of uranium and plutonium and more heavier elements. For this reason, polycarbonate is also a clean detector in fission track measurements. It is comparable with muscovite mica. Therefore, the polycarbonate track detector is used wherever muscovite mica can be applied except for fossil tracks, or at high temperature. In addition, polycarbonate is much cheaper ($\sim$100 times) than muscovite mica, is more flexible, and has a higher charge resolution.

Currently, polycarbonate is more popular than muscovite mica in fission track recording.

c. Polyethylene terephthalate (PET for short; Mylar, Chronar, Melinex, Terphane, Lavsan)

The sensitivity of polyethylene terephthalate is less than polycarbonate. The lowest recordable particles vary from Li ($Z = 3$) to B ($Z = 5$). α particles and protons cannot be recorded as tracks in this detector. Therefore, PET is a clean detector for recording fission fragments of ^{235}U and ^{239}Pu.

d. CR-39

Even though CR-39 is sensitive to low Z heavy charged particles, it can be used safely to detect fission fragments induced by high-energy projectiles. At very high energies over several hundreds of MeV/amu, CR-39 can identify the charges of projectiles and distinguish the tracks of projectiles and fission fragments.

By measuring track diameters or track etched lengths, the charge resolution achieves the separation of each element in the periodic table of elements. Therefore, CR-39 is one of the best detectors in studies of relativistic projectile fission and fragmentations.

e. Glasses

One of the advantages of glass track detectors is their short etching time for fission fragment tracks. For example, several seconds to minutes are enough to develop fission fragment tracks in soda lime glass (e.g., window glass, microscope slide, or cover glass) with 48% HF at room temperature. This advantage makes glass a convenient detector for the rapid testing of whether or not a fission source, such as ^{252}Cf meets, the requirements. For this purpose, one needs to put a piece of glass over the source at a distance for a certain time period, then etch the glass in HF for several seconds, wash it in water, dry it in air, and scan it with microscope, after which one will know the intensity of the fission source. In total, one needs less than 10 minutes for the test. If one wants to judge the flux or shape or distribution of neutrons in a beam, one needs to cover a plane of natural-uranium source with a glass plate and put the uranium-glass combination in the beam to irradiate it for a certain time. The track of fission fragments of ^{235}U induced by thermal neutrons will be recorded in the glass. Then the glass detector plate is etched, washed, dried, and scanned, after which one will see the shape of the neutron beam, the distribution of beam neutrons, and possibly measure the flux of neutrons from the nuclear reactor.

The detection efficiency of soda lime glass ($71SiO_2$:$15Na_2O$: 9CaO:3MgO: $2Al_2O_3$) in 2π geometry for fission fragments from a thin layer of uranium source is (39.3 ± 0.4) %, which is less than that of muscovite mica and polycarbonate detectors (Guo et al., 1976b, 1982).

f. Geological minerals

The substances which can be used as SSNTD to record fission fragments not only include large pieces of plastics, such as CR-39, polycarbonate (PC), polyethylene terephthalate (PET), cellulose nitrate (CN, LR-115), and glasses, but also include many unusual substances, such as stones, meteorites, tektites, lunar rocks, and lunar soils. The latter substances include essential (main) minerals and accessory minerals. For example, in granite, the essential minerals are quartz, feldspar and mica, which are composed of the main compositions of granite and with large sizes, such as in mm size or larger sizes. The accessory minerals in granite are apatite, zircon and some other minerals, which are contained in very small amount and very small grain size, such as several to hundreds of micrometers (μm). All these minerals (essential and accessory) can be used as a SSNTD. One can use any one of the minerals as your track detector for fission fragment studies. Up to now, more than 100 kinds of minerals contained in various stones or geological substances have been tested and shown that they are candidates for selection to be track detectors to study fission fragments. The recorded fission tracks can provide information about the age and thermal history of the geological objects. For more details, the readers are referred to the monographs written by Fleischer et al. (1975), Durrani and Bull (1987) and Wagner and Van den haute (1992).

g. Space minerals

Space objects such as various types of meteorites, lunar rocks, and lunar soils are composed of various kinds of minerals similar to the geological substances. The main compositions are olivine and pyroxene that depend on the real substances. These minerals have recorded tracks of heavy charged particles from space in their history and fission fragments from ^{238}U and ^{244}Pu spontaneous fission. These tracks can provide information of the history and evolution of the objects.

2. Fission rate determination

a. Thin, thick, and asymptotic fission sources

The SSNTD is a convenient detector for the determination of fission rates in a fission source. Fission sources can be classified into solid sources and liquid sources. Solid sources include plane sources and point sources. Plane sources include thin sources, thick sources, and infinite-thickness sources (i.e., asymptotic sources). A plane (fission) source exposes a plane surface, which can be covered with a sheet of track detector.

Thin, thick, and infinite-thickness sources are differentiated by the critical thickness t_c of the source and the average range R of fission fragments in the source material. t_c is a constant for a given fissile material, where $t_c = R\sin\theta_c$ (Guo et al., 1976b, 1982). If $0 \le t \le t_c$, the detection efficiency ε of the detector exposed or placed in contact with the fissile material will be equal to $1 - \sin\theta_c$, independent of the thickness t of the source. In other words, the detection efficiency is a constant, which is equal to that of a zero-thickness source. Therefore, the source with thickness less than t_c is referred to as a thin source. If $t_c \le t \le R$, some fission fragments will stop in the fissile material, so that they cannot reach the detector. The detection efficiency changes as the thickness changes. The source in this thickness interval is referred to as a thick source. If $t \ge R$, only the fission fragments originating from the surface layer of the source material have the possibility of reaching the detector. Beyond the thickness R, no fission fragments can reach the detector. In this case, the thickness has reached an infinite thickness, or so-called asymptotic thickness. If one uses muscovite mica as fission detector for ^{235}U foil source, $\theta_c = 3°41'$. If the fission source material is uranium, $R \approx 7.5$ mg/cm^2, then $t_c \approx 482$ μg/cm.^{2} This means that the uranium fission source with thickness less than about 482 μg/cm^2 is a thin source. The source with a thickness ranging from about 482 μg/cm^2 to about 7.5 mg/cm^2 is a thick source. The source with thickness larger than about 7.5 mg/cm^2 is an asymptotic source.

b. Point fission source

A point source is defined as a source whereby the size of the source in all directions is smaller than the range of fission fragments in the source material. In this case, all the fission fragments emitted toward the detector in a 2π geometry can reach the detector, and no self absorption loss occurs in such source material. Fine dust such as aerosol particles bearing uranium or plutonium are examples of point sources.

c. Liquid fission source

A liquid fission source is a source material in which the fissile material is dissolved or dispersed in a liquid. The rate of spontaneous or induced fission events are calculated according to the equation

$$A_f = \frac{N_T}{t\varepsilon} \tag{3.13}$$

where A_f is the number of spontaneous or induced fission events that have occurred in unit time (s), N_T is the number of tracks recorded by the detector, ε is the detection efficiency of the detector for fission fragments (see Eq. 3.12), and t is the time of irradiation of the detector to the fission fragments.

For a liquid fission source, a strip of track detector is immersed into the liquid to record the tracks of fission fragments during irradiation. The fission rate can be calculated by Eq. (3.13). In this case, the thickness of the fission source represented by "t" in Eq. (3.12) is unit thickness (1 cm). Compared with the average range of fission fragments in liquid ($R = 15.6$ μm in water), the thickness "t" is larger than the range of fission fragments (R). Therefore, $\varepsilon = \frac{R}{2t}\cos^2\theta_c$ with $t = 1$ cm. The distribution of fission tracks on the surface of the detector is uniform. One does not need to count all the tracks on the detector. N_t should be considered as track density (number of tracks in unit area of the detector, cm^{-2}). The meaning of fission rate in a liquid is the fission number occurring in unit volume of the liquid (cm^{-3}).

For point fission sources, such as dusts or aerosol particles, or particles on swipes, one can move them to the surface of a track detector by micromanipulator or other means and stick them with diluted glue and cover them with another track detector for recording fission fragments from the source particles. The fission tracks gather like stars. If the tracks are not too crowded together, one can count the total tracks to know the number of fissions that had occurred.

3. Determination of detection efficiency of fission fragments

The definition and significance of detection efficiency of fission fragments are as follows:

The detection efficiency ε of SSNTDs for fission fragments is defined as the ratio of the number of tracks N_T recorded on SSNTD to the number of nuclear fission N_f occurred in the fission source:

$$\varepsilon = \frac{N_T}{N_f} \tag{3.14}$$

The significance of the determination of the detection efficiency of a SSNTD for fission fragments is exemplified in the following cases:

a. Determination of the number of atoms of ^{235}U, ^{238}U, ^{239}Pu, and ^{233}U

For the quantitative determination of the number of atoms of ^{235}U, ^{238}U, ^{239}Pu, ^{233}U, as well as the other fissile nuclides in a sample, one needs to know the detection efficiencies ε. The easiest way to determine the number of the atoms (or nuclei) in the sample is to irradiate the sample to neutrons at nuclear reactors or accelerators. As long as one counts the number of tracks of fission fragments N_T recorded by the SSNTD, knowing the neutron energy E_n, the fluence of neutrons F_n, the fission cross section σ_f and the detection efficiency ε_T, one can immediately calculate the number of the atoms N in the sample by the following equation:

$$N = \frac{N_T}{F_n \sigma_f \varepsilon_T} \tag{3.15}$$

This is a type of traditional work carried out in nuclear research.

b. Determination of neutron number, fluence, and energy spectrum

The SSNTD is often used to determine the number N_n, fluence F_n and energy spectrum N_n (E_n) of neutrons from nuclear reactors or accelerators. Fission foils (^{235}U, ^{237}Np, ^{238}U, and/or ^{232}Th) overlapped with muscovite mica are placed at certain distances and in ε_T certain directions from the source of neutrons. The number, fluence, and energy spectrum of neutrons irradiated to the samples can be determined by the SSNTD. These types of determinations are widely applied in neutron physics and neutron dosimetry. In these determinations, the detection efficiency of the SSNTD for fission fragments ε_T is required.

c. Fission cross-section σ_f measurements

For systematic study of fission cross-sections σ_f of heavy nuclei induced by neutrons, heavy ions, γ-rays, elementary particles such as the pion (π), or other particles, one needs to know the detection efficiencies of the recording SSNTD for fission fragments.

High precision of fission cross-sections σ_f of ^{235}U, ^{238}U, ^{239}Pu, as well as the other fissile nuclides is of vital importance in the designing of any fission energy facilities. Therefore, one needs a high precision of detection efficiencies, especially for determination of σ_f of the nuclei with high α-ray radioactivity. For samples with high α-ray radioactivity, electronic types of detectors cannot record fission events accurately due to the existence of dead time (~μs to ns) of the detectors, or due to pileup of two or more electric pulses of α particles into one pulse. Therefore, the N_f measured by electronic detectors will be less than that of the real fission events. On the contrary, the number of tracks recorded in SSNTD for short-time exposure is not affected by the high radioactivity of α particles.

From the above discussions, one can see that accurate determination of the detection efficiency ε_T of SSNTD for fission fragments is very important in nuclear science.

Several methods have been devised for determination of detection efficiencies of SSNTD for fission fragments. They are: statistical counting method, critical angle method, twin fragment method, and horizontal projected confined track-length method, which are discussed in the following paragraphs.

4. Statistical counting method for determination of detection efficiency

The statistical counting method can be classified into two types: The first type utilizes spontaneous fission sources; and the second type exploits induced fission sources.

a. Statistical counting method using spontaneous fission sources

Statistical counting method by using spontaneous fission sources for determination of detection efficiency of SSNTD for fission fragments is carried out as follows: Spontaneous fission sources, such as ^{244}Cm or ^{252}Cf evaporated or electrodeposited on stainless steel plates can be used to determine the detection efficiencies of various track detectors. The sheet of track detector such as muscovite mica is put over the surface of the fission source for a period of time. The spontaneous fission fragments from the fission source hit on the detector and form tracks in the detector. The detector is taken away from the source. Then, the detector is etched and the tracks are counted. One can immediately obtain the detection efficiency ε of the detector for fission fragments (Gold et al., 1968), which is the ratio of the number of tracks N_T recorded in the detector to the number of fission fragments N_f hit on the detector. In this determination, the number of spontaneous fission fragments N_f hit on the detector and the number of tracks N_T recorded in the detector both occur statistically.

b. Statistical counting method using induced fission sources

At nuclear reactors or accelerators, one can use induced fission fragments to determine the detection efficiency of a SSNTD for fission fragments. The simple way to carry out this determination is to use a fission chamber and a SSNTD sandwich. In the fission chamber, a natural U source deposited on a stainless steel or Al foil is used as cathode attached to the bottom surface inside the chamber. Outside the chamber, the sandwich of SSNTD and another natural U source is attached to the opposite surface of the U foil inside the chamber. The two U foils are set back to back to each other, but between the two backs of the U foils, there is only a thin window of the chamber. One is often called the measurement as back-to-back calibration of the detection efficiency of SSNTD for fission fragments. The detailed procedures are as follows:

(a) Preparing two layers of natural U ($\leq$482 μg/cm^2, thin sources, defined as in Section I.C.2.a) on two thin plates of substrates (aluminum <0.3 mm, or stainless steel, 0.1–0.2 mm in thickness) by vacuum evaporation, or electric deposition, or electrochemical deposition.

(b) A cylindrical fission chamber is prepared. The bottom outer surface of the chamber is machined from outside to form a depressed trough with wall being very thin (<1 mm in thickness). The diameter of the trough is about 3 cm. One can use the thin circular depressed part as the window of the chamber. On the inner surface of the window, a circle is drawn just opposite to the depression (the trough) to show the correct position of the U foil to be placed. A parallel metal plate is arranged in the chamber to act as anode of the fission chamber. After filling in argon (Ar) gas and plugging in the direct current, the fission chamber can give pulses of α particles. One should select a proper threshold value of the electronic circuit system to discriminate (against) the α particles produced from the U foil; only allow the pulses of fission fragments to be recorded. When one has selected the threshold value, one can obtain the detection efficiency ε_c of the fission chamber for fission fragments. This detection efficiency must be used to correct the number of the fission fragments N_C recorded by the fission chamber.

(c) In the trough outside the chamber, another U foil is attached to the window of the chamber. A SSNTD sheet is overlapped to the U foil to form a sandwich U foil + SSNTD.

(d) The fission chamber is again placed on the neutron beam at a nuclear reactor or accelerator (with moderating substances usually paraffin). The thermal neutrons induce fission of ^{235}U in the two foils. The fission fragments from the foil inside the chamber are recorded by the fission Chamber. The fission fragments from the foil outside the chamber are recorded by the SSNTD. The detection efficiency of fission chamber ε_c is determined by the method mentioned above. The detection efficiency of SSNTD, ε_T, for fission fragments is calculated from the following equation

$$\varepsilon_T = \frac{N_T}{N_C}\frac{N_{UI}}{N_{UO}}\varepsilon_C \tag{3.16}$$

where N_T is the number of tracks of fission fragments recorded in the SSNTD, N_C is the number of pulses recorded by the fission chamber, N_{UI} and N_{UO} are the numbers of ^{235}U nuclei on the foils of U placed inside and outside of the chamber, respectively.

(e) In order to compensate for the difference in the numbers of thermal neutrons penetrating through the two U foils, which is caused by the absorption of the window and the substrates of the U foils for neutrons, the fission chamber with the sandwich (a composite) must be turned over (the front end of the composite turns to the back end of the composite) to have the neutrons first penetrating through the U foil with SSNTD to first penetrating through the U foil in the chamber. Counting the electron pulses of fission fragments from the chamber to the same number as that counted before turning the fission chamber. The total number of electron pulses before and after turning the chamber must be used in Eq. (3.14) to calculate the detection efficiency of the SSNTD for fission fragments. The total number of tracks N_T of fission fragments are counted after irradiation in two directions of the composite.

(f) The fission chamber can be replaced by a surface barrier Si detector (Au–Si detector) for the determination of the detection efficiency of the SSNTD for fission fragments. But comparatively, a fission chamber is more convenient than a surface barrier Si detector.

(g) The angular distribution of fission fragments emitted from the fission of ^{235}U induced by thermal neutrons is isotropic. The above determination of detection efficiency of a SSNTD in 2π geometry is based on the isotropic distribution of fission fragments. Anisotropic distribution of fission fragments may occur in fissions induced by high energy particle (fast neutrons, high-energy heavy ions and so on). In these cases, Eq. (3.14) may give the wrong detection efficiencies for fission fragments.

The statistical counting method is simple to carry out, but generates many sources of uncertainties: such as the uncertainties from N_T, N_C, N_{UI}, N_{UO}, and ε_C. All the uncertainties add together to form the uncertainty of ε_T, which are reducible by adopting other methods of calibration of detection efficiency. For this reason, several other methods for the determination of detection efficiency of SSNTD for fission fragments have been proposed and used (see below).

5. Critical angle method

The dominant factor for track etching is the threshold property of SSNTDs. Only when the damage density created by heavy charged particles in the detector is higher than a certain value (threshold value) can the damaged tracks be etched. However, not all latent tracks formed in the detector, whose damage density is higher than the threshold, can be etched to form visible tracks. The second factor to influence the appearance of tracks is the critical angle of etching. If the angle θ between the direction of incident particle and the surface of the detector is less than a certain value θ_c, the latent track created by the particle cannot be etched to form a visible track. Only when the incident angle θ is larger than the value θ_c, that is $\theta > \theta_c$, can the track be etched to form an observable track. The specific value θ_c of incident angle is one of the critical conditions to show the etched track. The angle θ_c is called the critical value for etching the track (see Section V.A.3).

The relationship between detection efficiency ε and critical angle θ_c is as follows:

$$\varepsilon = 1 - \sin \theta_c \tag{3.17}$$

From the determination of the critical angle θ_c, one can derive the detection efficiencies ε of a SSNTD for the particle, or in this case, for fission fragments.

Khan and Durrani (1972) developed a technique to determine the detection efficiencies of a SSNTD for fission fragments. They used a beam of fission fragments from a ^{252}Cf spontaneous fission source to irradiate a SSNTD at different angles (θ) to find the critical angle θ_c; less than or equal to θ_c, no tracks can be seen on the surface of the detector.

This method converts the measurements of track numbers into observation of the occurrence of tracks.

This method is not concerned with the number N_T of tracks recorded in the detector as well as their statistics. However, the uncertainty of angle measurements arises, and its absolute value is relatively large compared to the other methods.

6. Twin fragment method for determination of detection efficiency

a. Principle of twin fragment method

Usually, each fission event of ^{235}U or ^{239}Pu induced by thermal neutrons produces only two fission fragments (binary fission). Scientists have found that some fission events may release three or more fission fragments with about the same masses or relatively large fragments, but the ratio of this type of fission to binary fission is only about 10^{-5} to 10^{-6}. Therefore, in the discussion of detection efficiency of a SSNTD for fission fragments, one can consider that every fission event produces only two fission fragments (twin fragments).

In order to achieve higher accuracy in the determination of detection efficiency of fission fragments, the twin fragment method was suggested and adopted (Guo et al., 1976a,b, 1982). The principle of twin fragment method in the determination of the detection efficiency is that one records the twin fragments from the same event with two types of detectors, such as a Au–Si surface barrier detector and a SSNTD. The fission source (^{235}U) is prepared on the surface of one of the detectors, such as the Au–Si detector, and the source is very thin (<10 μg/cm^2) and uniform, so that the thickness of the source is negligible. Another detector such as a SSNTD is overlapped onto the surface of the Au–Si detector, forming a sandwich Au–Si detector-fission source-SSNTD.

Fission fragments of ^{235}U induced by thermal neutrons are recorded by the SSNTD and Au–Si surface barrier detector. Since the two correlated fission fragments from one event are emitted in opposite directions, each of the two fragments should be emitted into one of the detectors, respectively. Therefore, the number of fission fragments entering the SSNTD should be the same as that entering the Au–Si detector. However, the number of recorded tracks N_T of fission fragments is less than the number of fission fragments N_S entering into the Au–Si detector. The reasons for $N_T < N_S$ are: (A) the SSNTD has a critical angle θ_c in track etching. For the fission fragments with incident angle $\theta < \theta_c$, the tracks are removed out from the surface of the SSNTD. (B) Eyesight is limited. For very shallow etch pits of tracks, the eyes of mankind even with the help of a microscope, cannot see them clearly. Finally, they are lost in counting tracks. In the determination of detection efficiency of a SSNTD, one cannot eliminate these two types of losses. It should be noted that for determining N_S with Au–Si detector, one must take into account the loss by the thickness of the Au layer (dead layer) on the surface of Au–Si detector (Guo et al., 1976a,b; 1982).

b. Statistical uncertainty of tracks in the twin fragment method

The aim of adopting the twin fragment method is to reduce statistical uncertainty in counting the number of tracks (N_T). Even though the numbers of fission fragments emitted to the Au–Si detector and SSNTD are both Ns, the number of tracks N_T on SSNTDs may not be equal to N_S. In other words, the number of tracks on the surface of SSNTDs should follow the rule of statistical fluctuation. In the twin fragment method, the way of statistical fluctuation takes place as follows: due to the negligible thickness of the fission source t ($t \ll R$), when N_S of fragments travel to the Au–Si detector, the same number N_S of fragments in the opposite direction will be certain to travel to the SSNTD. However, the latter fragments will enter into two regions due to the existence of a critical angle of the SSNTD. A part of fragments enter into a region where $\theta > \theta_c$, the tracks of these fragments can be etched visibly. Another part of fragments falls into the region $\theta \leq \theta_c$, their tracks disappeared after etching. The Au–Si detector records only a total number of fragments Ns, but cannot indicate which fragment enters into the region of $\theta > \theta_c$, or which one inters into the region of $\theta \leq \theta_c$. Therefore, the probability for a fragment to fall into the region $\theta > \theta_c$ and form a visible track should follow a binomial distribution. Supposing that the detection efficiency of a SSNTD for fission fragments in 2π geometry is ε, then, the probability for entered N_S fragments to form N_T tracks in SSNTDs can be written as:

$$P(N_T) = \frac{N_S!}{N_T!(N_S - N_T)!}\varepsilon^{N_T}(1-\varepsilon)^{N_S - N_T} \tag{3.18}$$

The standard uncertainty of N_T is

$$\sigma(N_T) = \sqrt{N_T(1-\varepsilon)} \tag{3.19}$$

From Eq. (3.19) one can see that due to the fission fragments impinging on the SSNTD and Au–Si detector are corresponding to each other one by one, the statistical uncertainty $\sigma(N_T)$ of the number of tracks recorded on the SSNTD being $\sqrt{N_T(1-\varepsilon)}$ is less than the uncertainty $\sqrt{N_T}$ of the total number of N_T formed on SSNTD by a statistical source. If the detection efficiency ε of the SSNTD is close to 1 (it is true for most of SSNTD), then the uncertainty $\sigma(N_T)$ for recording twin fragments with the aid of a Au–Si detector would be much less than that of recording statistical tracks of ^{235}U source:

$$\sqrt{N_T(1-\varepsilon)} < \sqrt{N_T} \tag{3.20}$$

This is the advantage of the twin fragment method over the other methods for the determination of detection efficiency of SSNTDs for fission fragments.

Table 3.4 shows a comparison of the results between the twin fragment method and statistical counting method as well as critical angle method.

c. Effect of backscattering on detection efficiency of fission fragments

The fission fragments originally flying toward the substrate of the fission source may be scattered by the nuclei in the substrate and turn its direction to fly out of the substrate and finally enter into the detector as shown in Fig. 3.10.

Suppose a fission event occurs at point N in a thin fission source between a nuclear track detector and the substrate of the fission source (See Fig. 3.10). One of the fission fragments (fragment 1) impinges on the track detector directly and forms a track (the first track) in the detector. Another fission fragment (fragment 2) enters into the substrate and then is scattered at point B by a nucleus. Fragment 2 changes its direction and also enters into the track detector, creating a track (the second track) in the detector. Therefore, two tracks can be formed by one fission event due to backscattering of a fission fragment. This phenomenon may result in extra tracks in an SSNTD and make the detection efficiency of SSNTD look higher than its intrinsic value. Therefore, one must make a correction for the backscattering of fission fragments from the substrate. In the following section, an example will be shown on how backscattering occurs and how to correct it.

A thin layer of enriched uranium (U) was sprayed by high electric voltage onto the surface of a Au–Si surface barrier detector. An SSNTD detector sheet was put on the U layer. The composition of the two detectors was placed at the nuclear reactor for irradiation with thermal neutrons. The fission fragments were recorded by the two detectors. However, as mentioned above, the fragment entering the Au–Si detector (the substrate of the U layer) may be backscattered by a nucleus or nuclei and turn its direction to the SSNTD covering on the Au–Si detector. If the incident angle θ of the fragment to the surface of SSNTD is larger than the critical angle θ_c of the SSNTD, the backscattered fragment will create a visible track in the SSNTD. Therefore, one event of fission, that occurred in the thin layer of ^{235}U, may create two tracks in the SSNTD. The extra track will result in an erroneous (larger) value of detection efficiency of the SSNTD for fission fragments. Alternatively, the fragment entering the SSNTD may be backscattered by nuclei in the SSNTD and turn its direction into the Au–Si detector, whereby the two fragments that entered into Au–Si detector will produce only one electronic pulse in the Au–Si detector because the dead time of Au–Si detector (~ns) is much longer than the gap of time between the two fragments entering the Au–Si detector. Therefore,

TABLE 3.4 Comparison of detection efficiencies of muscovite mica and polycarbonate for fission fragments in 2π geometry determined by the twin fragment method, statistical counting method, and critical angle method.

SSNTD	Detection efficiency, ε (2π geometry) %	Calibration method	Fission source	Substrate	Reference
Muscovite mica	93.6 ± 0.3	Twin fragments	^{235}U (n_{Th},f)	Au–Si	Guo et al., 1976a,b, 1982
	94.8 ± 0.5	Statistical Counting	^{244}Cm spont. fission	Pt	Gold et al. (1968)
	91.9 ± 1.9	Statistical counting			Wall (1970)
	96.1 ± 1.0	Statistical counting	^{235}U (n_{Th},f)	Si	Rumiazev et al. (1968)
	92.2 ± 0.9^a	Critical angle	^{252}Cf spont. fission	_	Khan and Durrani (1972)
Polycarbonate	95.5 ± 0.4	Twin fragments	^{235}U(n_{Th},f)	Au–Si	Guo et al., 1976a,b, 1982
	95.2 ± 0.5	Statistical counting	^{244}Cm Spont. fission	Pt	Gold et al. (1968)
	94.8 ± 0.9^b	Critical angle	^{252}Cf Spont. fission	_	Khan and Durrani (1972)
	95.6 ± 0.9^c	Critical angle	^{252}Cf Spont. fission	_	Khan and Durrani (1972)

[a]*Calculated from* $\theta_c = 4°30' \pm 30'$ *and* $\varepsilon = 1 - \sin\theta_c$, *(Khan and Durrani, 1972).*
[b]*Calculated from* $\theta_c = 3° \pm 30'$ *and* $\varepsilon = 1 - \sin\theta_c$, *for polycarbonate (Makrofol, Bayer) (Khan and Durrani, 1972).*
[c]*Calculated from* $\theta_c = 2°30' \pm 30'$ *and* $\varepsilon = 1 - \sin\theta_c$, *for polycarbonate (Lexan) (Khan and Durrani, 1972).*

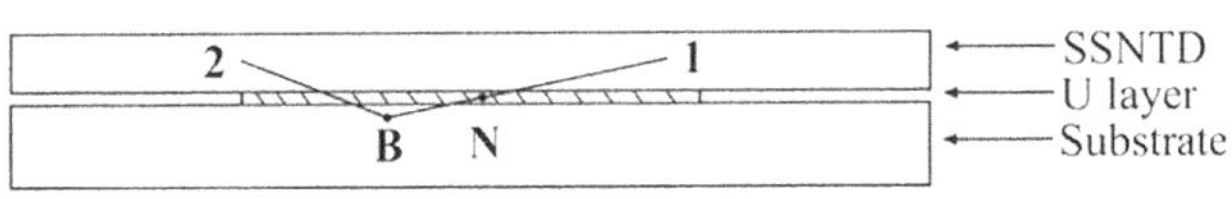

FIG. 3.10 Sketch of backscattering of a fission fragment by a nucleus in the substrate of fission source. A nuclear fission event occurs at point N in the fission source (U layer). A fission fragment (1) enters the SSNTD forming a track in the detector. The other fission fragment (2) flying to the opposite direction of fragment (1) is scattered by a nucleus at point B of the substrate causing the fragment to change its direction and create a second track in the SSNTD.

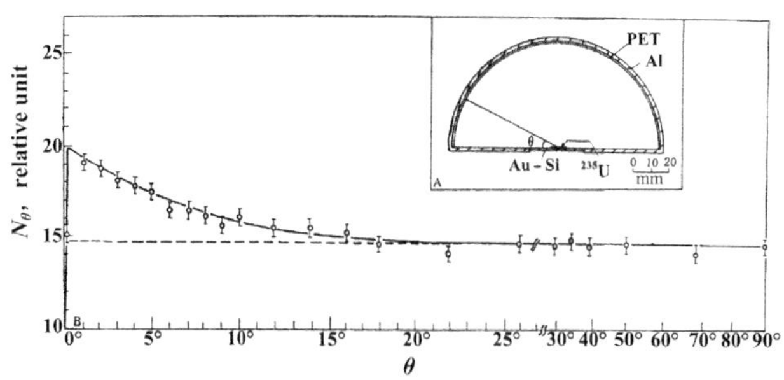

FIG. 3.11 Experimental device and measured curve of backscattering correction coefficient. (A) Experimental device (inset). (B) Number of tracks in unit area as a function of θ (lower curve). Φ — measured value. — Smoothed curve. - - - Average value measured is in the range from $\theta = 20°$ to $90°$.

backscattering that occurs in the SSNTD will not increase the number of pulses of fission fragments in the Au–Si detector, but the backscattering that occurs in the substrate may increase the number of tracks in the SSNTD. For this reason, a correction of the track number recorded by the SSNTD must be made in a determination of the detection efficiency of SSNTDs for fission fragments.

A device for the determination of the value of backscattering is shown in Fig. 3.11 (Guo et al., 1976b; 1982).

A frame was made in a half circular shape by an Al plate with 1 mm in thickness and 20 mm in width as shown in Fig. 3.11A. The Au–Si detector which was used to determine the detection efficiency was placed at the center of the frame. The center of the U (90% ^{235}U) fission source was precisely coincident with the center of the frame. The normal line of the surface of the ^{235}U source through the center of the source was precisely penetrating through the midpoint of the Al plate. A polyester detector strip with the same width and same length as the Al plate was attached to the inner surface of the Al plate and used to record the tracks of fission fragments emitted to different angles from the fission source. The whole device was put inside a vacuum chamber with pressure 0.1 mm height of Hg column. The ^{235}U fission source was irradiated with thermal neutrons from a nuclear reactor. Thermal neutrons induce fission of ^{235}U. The fission fragments from the source were recorded by the strip of polyester detector at different angles θ between the direction of the fission fragment and the surface of the source. The end point of the strip corresponds to $\theta = 0°$. The midpoint of the strip corresponds to $\theta = 90°$. The total number of ^{235}U fissions for the irradiation was 1.8×10^8. After etching the polyester track detector, the numbers of fission fragment tracks N_θ in unit area at different angles θ on the polyester were measured with an optical microscope. The numbers of tracks at different angles are shown in Fig. 3.11B. From Fig. 3.11B one can see that in the interval of $\theta > 40°$, N_θ is not changed at different angles, $N_\theta = N_D$, where N_D is the track density directly incident to the circular strip of polyester detector. In the interval of $\theta < 40°$, $N_\theta > N_D$. The extra tracks of N_θ over N_D, that is $N_\theta - N_D$, are backscattered from the Au–Si detector and recorded by the polyester detector.

The backscattering coefficient b for the tracks N_T observed on the surface of SSNTD can be written as the following equation:

$$b = \frac{1}{1+F} \tag{3.21}$$

where F is represented by the following equation:

$$F = \int_{\theta_c}^{\pi/2} \left[\frac{N_\theta}{N_D} - 1\right] \cos\theta d\theta \tag{3.22}$$

The meaning of b is that for recorded N_T tracks on the surface of the SSNTD, one must multiply N_T by a faction b which is less than 1, that is bN_T being the corrected number of tracks of the direct fission fragments.

The backscattering coefficients b and backscattering ratio F from Au–Si detector to several SSNTDs are listed in Table 3.5.

From the above results one can see that the backscattering of fission fragments by the substrate (here Au and Si) of the Au–Si detector contributes 2.6% more tracks in the detector polycarbonate (PC, China). When one determines the detection efficiency of PC for fission fragments, one must deduct the 2.6% tracks from N_T recorded by the PC detector. Furthermore, when one counts the fission number, that occurred in a fission source prepared on a Au–Si detector, one must deduct 2.6% from the track number N_T. The true fission number, that occurred in the source, is $N_f = N_T\,(1-2.6\%)/\varepsilon$.

The value of the backscattering coefficient b is related to the atomic compositions of the substrate of the fission source and the critical angle θ_c of the detector. Fortunately, by using a simple device similar to that in Fig. 3.11, one can easily determine the backscattering coefficient for the source and the detector.

From the result (Table 3.5) of the backscattering coefficient of muscovite mica one can see that the number of tracks N_T recorded on the surface of muscovite mica must be deducted (2.1%) from N_T. That is, the true number of fission event is $N_f = N_T(1-2.1\%)/\varepsilon$.

Many authors did not take into account the backscattering phenomenon in determining the fission number and the detection efficiency. This omission will result in about 2.6% bias error for work with polycarbonate and about 2.1% error with muscovite mica.

The equation for the calculation of the detection efficiency ε of SSNTDs for fission fragments is as follows:

$$\varepsilon = \frac{bN_T\varepsilon_S}{N_S} \tag{3.23}$$

where N_S is the number of fission fragments counted by the Au–Si detector, ε_S is the detection efficiency of the Au–Si detector for fission fragments.

TABLE 3.5 Backscattering coefficients b and backscattering ratio F of fission fragments from a ^{235}U source on a Au–Si detector to several types of SSNTDs.[a]

SSNTD	Backscattering coefficient, b	Backscattering ratio F
Polycarbonate (Chaoyang, China)	0.974 ± 0.002	2.6%
Polycarbonate (Makrofol K, Bayer, Germany)	0.979 ± 0.002	2.1%
Muscovite mica (Sichuan, China)	0.979 ± 0.002	2.1%
Polyester (Changzhou, China)	0.979 ± 0.002	2.1%
Phosphate glass (Beijing, China)	0.989 ± 0.002	1.1%
Quartz (China)	0.999 ± 0.002	0.1%
Silicate glass (China)	1.000 ± 0.000	0%

[a] *θ_c in Eq. (3.22) is derived from $\varepsilon = 1-\sin\theta_c$; ε is determined by twin-fragment method (Guo et al., 1976a,b, 1982).*

The ε_S is determined with a fission chamber. The Au–Si detector was placed inside the fission chamber to record one of the twin fragments. Another fragment was recorded by the fission chamber with an Al plate as anode inside the chamber. The detection efficiency ε_S of the Au–Si detector for fission fragments can be obtained by equation:

$$\varepsilon_S = \frac{N_S'}{N_F'}\left(1 - \frac{t}{2R}\right) \tag{3.24}$$

where N_S' and N_F' are the numbers of fragments recorded for the twin fission fragments in the Au–Si detector and fission chamber, respectively; and $\left(1 - \frac{t}{2R}\right)$ is the self-absorption factor for the detection efficiency of the fission chamber due to the thickness of the ^{235}U source. Here, t is the thickness of the ^{235}U source, and R is the range of fission fragments in the source. From Eqs. (3.23) and (3.24), one can see that by means of Au–Si detector and fission chamber, one can determine the detection efficiency ε of a SSNTD for fission fragments by the twin-fragment method.

7. Projected track-length method for determination of detection efficiency

The detection efficiency of internal and external surfaces of a mineral and external detector for fission fragments can be determined by the distribution of horizontal projected (confined) track lengths. Confined tracks are those that are entirely localized within a detector (usually mineral). The etchant for developing confined tracks comes through another track which intersects with the confined tracks and reaches the outer surface of the detector. The etchant first etches the outer track and reaches the confined tracks and etches them. The etched outer track is a man-made or natural tunnel for transporting etchant to the confined tracks. The etched confined tracks are called "track in track" or TINT. Confined tracks can also be etched by the etchant that passed through a crack or a cleavage of the detector, if the confined tracks penetrate through the crack or cleavage. The etched confined tracks by this way are called "track in cleavage" or TINCLE (Lal, 1969; Lal et al., 1968). For internal surface of a mineral, the maximum length of the confined tracks is the sum of the lengths of the two fission fragments from one fission event. For external detector, the maximum length of tracks is equal to the range of one fission fragment from the mineral. The "internal surface", "external surface", and "external detector" are technical terms for FTD. It means that each of them is a flat surface produced by cutting a mineral or on the attached track detector, and the boundary of the flat surface has a gap distance of one range of fission fragment to the outer edge of the mineral. The requirement of a gap distance from the edge of the mineral is to ensure that no fission fragments from outside the mineral could reach the detecting surfaces. To further describe the definitions and uses of the technical terms, a sketch is shown in Fig. 3.12.

For the components illustrated in Fig. 3.12, the following explanations according to the numbering in the figure are given in more detail:

1 Mineral, including crystalline minerals and amorphous glassy substances.
2 Cutting plane. Large-sized crystalline minerals or glassy substances can be cut by various saws or cutting wheel with diamond. Small-sized minerals may be mounted in epoxy resin or EFP (Tetrafluoroethylene-hexafluoro-propylene copolymer), and then ground by emery and polished.
3 Internal surface of mineral, which (1) may be used to record tracks of spontaneous fission occurred in the mineral before cutting and (2) may be used to record tracks of induced fission of ^{235}U in the mineral exposed to thermal neutrons after annealing and before cutting.
4 G, the gap distance from the border line of the internal surface to the outer surface of the mineral. G ≧ R, where R is the range of a fission fragment.
5 External surface of mineral, which may be used to record tracks of (1) spontaneous fission or (2) of induced ^{235}U fission by thermal neutrons in the mineral (after or without annealing), that occurred in the mineral after cutting or grinding and polishing of the mineral.
6 External detector, free from uranium (U) or with very low concentration of U, which may be used to record tracks of (1) spontaneous fission occurred in the mineral after overlapping with the mineral and (2) induced fission of ^{235}U in the mineral by thermal neutrons.
7 External detector surface (same application as component 6 above).
8 G′, the gap distance from the border line of the external surface to the outer surface of the mineral. G′≧R, where R is the range of a fission fragment.

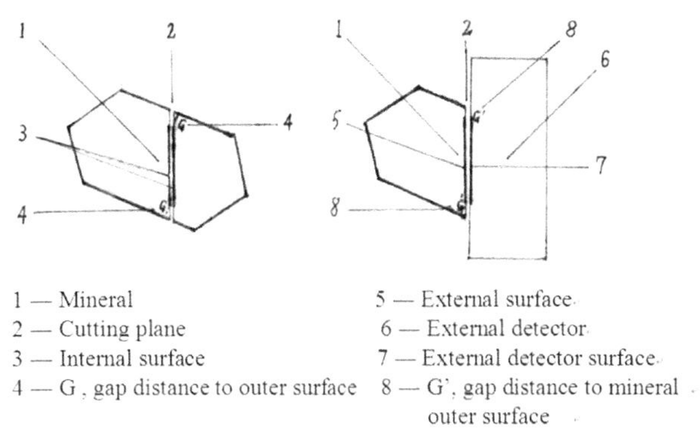

FIG. 3.12 Definitions and uses of technical terms on internal surface and external surface of a mineral and an external detector in FTD.

A new technique was suggested by Jonckheere and Van den haute (1999, 2002) for calibration of the detection efficiency of a SSNTD for fission fragments. It measures the distribution of horizontal projected confined fission-track lengths intersected with the mineral internal surface or external surface or with the surface of external detector. This method can be used to measure the detection efficiency of internal surface of the mineral for spontaneous fission tracks in the mineral. It can also determine the detection efficiency of the external surface of the mineral for induced ^{235}U fission tracks in the mineral. By covering with an external detector, the detection efficiency of the external detector for induced fission in the mineral can also be determined. In this method, the number of tracks N can be counted in an area. The number of fission events N_f ($\varepsilon = N/N_f$), that occurred in the corresponding area in the mineral, can be obtained by integration of the area below a straight line from $p = 0$ to $p = 2R$, where p is the projected track length, $2R$ is the maximum length of the confined tracks, the straight line is a function of the number N of confined tracks versus the projected track length p as shown in Fig. 3.13.

In Fig. 3.13, the abscissa is the horizontal projected track length. The ordinate is the number of tracks in the unit track length. A straight line can be found from counting the tracks. The intercept on the ordinate by the straight line is the number of tracks with zero length. The intercept on the abscissa by the straight line is the mean length of the confined tracks. The observed number of spontaneous fission tracks is $N = 2611$ in this specific example. The number of spontaneous fission N_f of ^{238}U, that occurred in the apatite was obtained by integration as mentioned above. The detection efficiency $\varepsilon = N/N_f$ being 0.91% was obtained with the apatite internal surface for spontaneous fission tracks that occurred in the apatite (Jonckheere and Van den haute, 2002).

From similar measurements, one can determine the detection efficiency of the external surface of a mineral for induced fission of ^{235}U in the mineral (The external surface of the mineral must be covered with a substance having very low U concentration such as muscovite mica). An example of this type of measurements is shown in Fig. 3.14.

One can also determine the detection efficiency of an external detector for fission events that occur in a mineral (apatite here). Fig. 3.15 shows an example.

From Fig. 3.15 one can determine the detection efficiency of the external detector (muscovite mica) for induced fission of ^{235}U in the apatite, which, in this example, was 0.91% as the result.

From the above discussion, one can see that this method does not require the measurement of the number of fission events N_f in the sample. The number N_f is derived from integration of the data in the measurements. This is the advantage of the projected track length method for the determination of detection efficiency of fission fragments.

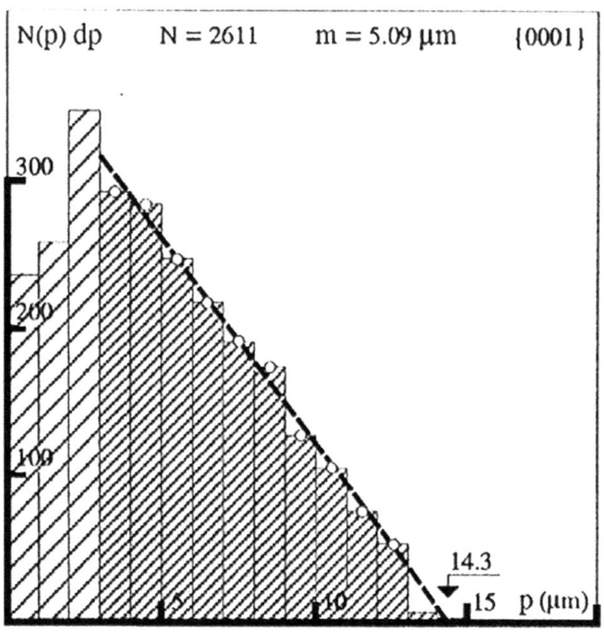

FIG. 3.13 Distribution of horizontal projected confined track lengths of spontaneous fission fragments of ^{238}U in Durango apatite. *From Jonckheere and Van den haute (2002). Printed with permission from Elsevier © 2002.*

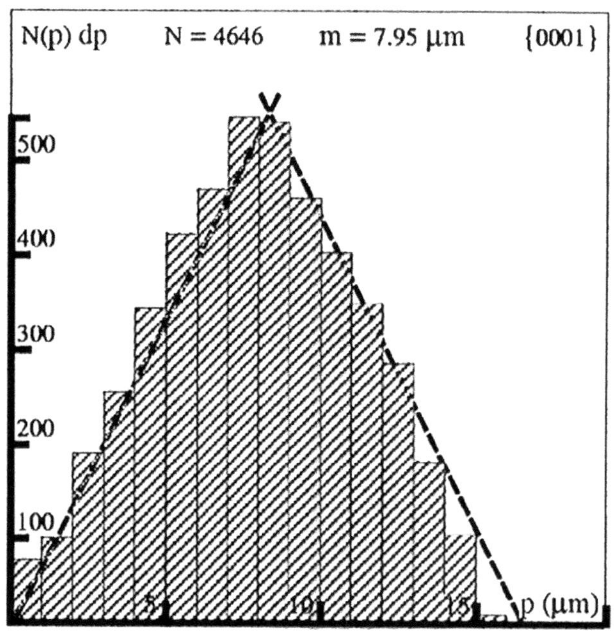

FIG. 3.14 Distribution of horizontal projected confined track lengths on the external surface of a Durango apatite from induced fission occurred in the apatite. *From Jonckheere and Van den haute (2002). Printed with permission from Elsevier © 2002.*

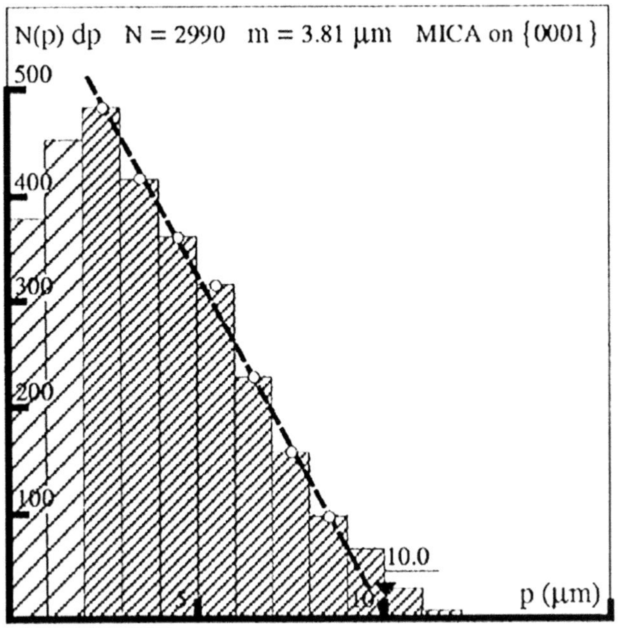

FIG. 3.15 Distribution of horizontal projected induced fission track lengths on a muscovite external detector surface. *From Jonckheere and Van den haute (2002). Reprinted with permission of Elsevier © 2002.*

8. Backscattering effect of fission fragments from substrate and fission source

As discussed in Section III.C.3.d, the phenomenon of backscattering of fission fragments from the substrate and from the fission source may result in an increase of more than 2% tracks in the number of real fission events. This bias is obviously not negligible in physics research. On the other hand, many workers are not concerned with the effect of backscattering in their measurements with track detectors. Their results of measurements must be given a correction before using them.

The backscattering coefficient (See Eqs. 3.21 and 3.22) is easily determined by a device similar to that shown in Fig. 3.11. One should make a correction measurement at the same time as one counts the number of fission fragments in the track detector with the same fission source.

9. Spatial distribution of fission and angular distribution of fission fragments

a. Spatial distribution of fission events

The SSNTD is a passive position sensitive detector. The spatial resolution is about several micrometers. It is suitable for the measurement of the distribution of fission events in a certain direction (one dimension) or on an area (two dimensions) or in a volume (three dimensions). The techniques of measurements are similar to that described in the previous subsections for protons. More detailed techniques and applications can be found in Section II.C of Part 2 of this chapter.

b. Angular distribution of fission fragments

The fission fragments released in thermal neutron induced fission of ^{233}U, ^{235}U, and ^{239}Pu are isotropically emitted to the 4π direction. On the contrary, the fragments emitted from fission induced by fast or high energy neutrons as well as by high-energy particles are anisotropic (Vandenbosch and Huizenga, 1973; Zhou et al., 1981; Zhang et al., 1990a,b, 1994). Therefore, the detection efficiencies for fission fragments determined in Section III.C.3 are correct only for thermal neutron induced fission or for spontaneous fission. They cannot be used for calculations of high-energy fission. However, it does not decrease the importance of thermal-energy fission because up to now, most of the work on nuclear energy is related to fission by thermal neutrons. On the other hand, one must be careful if one is working on high-energy fission. In this case, one must clarify the anisotropy of the fission fragments in one's study.

10. Application of fission detection

a. Nuclear fission study

Nuclear energy mainly has two sources: fission energy and fusion energy. The utilization of large amounts of fusion energy is now only in the promising stage, as described in Chapter 1, Volume 1. The application of fission energy is the main activity in the present era. Therefore, the detection and quantitative analysis of fission events are the main tasks of nuclear scientists and engineers.

SSNTDs are vastly used in fission studies. As far as fission reactions are concerned, the following measurements have been carried out:

(1) Incident particles: n, p, γ, π, α, heavy ion;
(2) Target nuclei: U, Pu, Th, Bi, Pb, Tl, Hg, Au, Ho;
(3) Energy range: Spontaneous fission; Low-energy fission ($E < 50$ MeV); Medium-energy fission (50 MeV $< E <$ 1000 MeV); High and relativistic energy fission ($E >$ 1000 MeV).

SSNTDs have the following advantages for studying fission:

(1) they can withstand intense low ionizing beam particles,
(2) they can undergo a long duration of irradiation without the need for a power supply,
(3) they have a high efficiency for fission fragments,
(4) they are position sensitive for fission events, and
(5) they exhibit a high charge resolution at high energy.

A very broad range of research fields utilize track detectors. Among these research fields, several areas of fission detection have been developed, which belong to the most important branches of nuclear science and technology, such as:

a. neutron physics and dosimetry,
b. nuclear reactor physics,

c. nuclear forensic analysis and safeguards,
d. FTD and geothermal chronology, and
e. uranium exploration.

For additional information, readers are referred to books by Fleischer et al. (1975); Durrani and Bull (1987); and Wagner and Van den haute (1992) and the related Sections III.C, II.F.1, II.F.2, II.F.3, II.G, III.A, and III.B in Part 2 of this chapter.

D. Heavy ions (Z ≥ 3)

The origins of heavy ions are mainly the following:

(1) Low-, medium- and high-energy nuclear-interaction products, (2) Cosmic ray nuclei, (3) Cluster radioactivity emissions, and (4) Heavy-ion accelerator beams.

1. Suitable detectors for heavy ions $Z \geq 3$

The following are suitable detectors for heavy ions with $Z \geq 3$:

a. CR-39 (PADC)
b. Cellulose nitrate (CN, LR-115)
c. Polycarbonate (PC, Lexan, Makrofol)
d. Polyethylene terephthalate (PET, Mylar, Lavsan, Melinex)
e. Polyimide (PI, Kapton)
f. BP-1 glass
g. Phosphate glass

Some basic points on the selection of detectors are as follows:

a. Plastic track detectors, such as CR-39, CN, PC, PET, and PI, are the most sensitive detectors for recording heavy ions ($Z \geqq 3$).
b. Glass detectors, such as BP-1 glass and phosphate, are the most resistant to high temperature and radiation.
c. Muscovite mica is superior for the discrimination of less ionizing particles.
d. The criterion to select the most suitable detector is the capability to discriminate against background—;less ionizing particles, if they exist. Otherwise, the detector should have the best potential to identify the charge Z and mass A of the beam and product heavy ions.

2. Identification of charge Z

In the first several years after the discovery of SSNTDs, scientists were very disappointed that this type of detector could not distinguish particles of different charges (Z). In those early years, these track detectors were mainly used to count tracks of particles. It was Price et al. (1967), who first discovered that plastic track detectors could distinguish charge Z of particles. The principles of identification of charge Z is similar to that used for silicon surface-barrier detectors. This discovery opened up very broad applications of etch track detectors.

The techniques of identification of charge Z by etch track detectors can be classified as: maximum etchable track-length (L_{max}) method, track etch rate versus radiation damage density method, track etch rate versus residual rage method, track diameter method for relativistic energy, and track length method for relativistic energy.

Detailed explanations of these methods will be provided in Section VI of Part 1 of this chapter.

3. Identification of mass A of isotopes

Isotope identification is an extension of identification of charge Z (atomic number). It is based on the fact that the residual range R_0 of nuclei in a uniform substance is related to their mass A if their charge Z and phase velocity β ($= v/c$) are the same. By measuring the etch rate V_t (or etch track length L) and residual range R_0, drawing a figure $V_t - R_0$, one can separate nuclei with different A (isotopes). The reader can refer to the monographs on track detectors (Fleischer et al., 1975; Durrani and Bull, 1987) for additional information on this subject.

4. Heavy-ion energy determination

Energy determination of heavy ions is based on charge identification and range-energy relation. In order to determine the energy of heavy ions, one must first identify what kind of nuclei (Z) created the tracks, then solve the problem about what energy the heavy ions possess.

The following techniques may be used in determination of energy of the heavy ions:

a. Range-Energy Relation for Charge Z
b. Track Diameter (D) versus Energy (E) Calibration Curve Method
c. Etch Rate (V or S) versus Residual Range (R) Method
d. Two Points Etch Rates versus Gap Distance Method.

The techniques used to identify charge Z of the heavy ions, which created the tracks, have been explained in Section III.D.2. The range—energy relation of heavy ions in a nuclear track detector will be discussed in the following Section IV.D.2. When one has obtained the value of charge Z of the heavy ion, one can find the energy value E on the curve of range—energy relation of the special ion in the given detector material based on the residual range R_0 measured in the process of identification of charge Z.

5. Applications of heavy-ion detection

Since the 1980s, great achievements have been made by using SSNTDs. The most important achievements are in the following two fields:

a. Heavy ion emission from heavy nuclei

Heavy nuclei can release not only α-, β-, γ-rays, but also heavy ions. Even though the first discovery of heavy ion emission, that is ^{14}C emission from the nucleus ^{223}Ra was achieved by a silicon $\Delta E - E$ detector by Rose and Jones (1984), the similar emission from other nuclei are not feasible to detect by the same detector, because of the more intense α-ray background. The solid-state track detector has taken the role of measuring the ions emitted from the other nuclei. Up to now, about 20 parent nuclei of heavy-ion emission have been discovered and measured (^{221}Fr, ^{221,222,223,224,226}Ra, ^{225}Ac, ^{228,230,232}Th, ^{231}Pa, ^{232,233,234,235}U, ^{237}Np, ^{236,238}Pu, ^{241}Am, ^{242}Cm, and ^{114}Ba) by nuclear track detectors. No other detectors can take the role in the subsequent discoveries. (See Chapter 1, Volume 1 for a more detailed treatment of cluster radioactivity).

b. Relativistic projectile fragmentations and the their products

Since the first acceleration of heavy nuclei to relativistic energy at Bevalac at the Lawrence Berkeley National Laboratory in California in the early 1980s, the capability of charge resolution $\sigma_z < 1$ was achieved for heavy nuclei ($Z > 26$) with the use of a nuclear track detector and detector stacks. This capability made possible the charge determination of the products from relativistic fragmentation. A series of studies emerged, such as:

a. Searching for fractional charge particles,
b. Measurements of charge pickup cross sections in nuclear interactions,
c. Projectile fission studies,
d. Electron capture and stripping process studies of ions in matters,
e. Charge-changing cross-section measurements of heavy ions in nuclear interactions,
f. Measurements of the composition of ultraheavy cosmic rays,
g. Laser Acceleration and Laser Confinement Fusion,
h. Deep-Inelastic Collision and Multi-Fragment Sequential Fission

More detailed information can be found in the relevant sections in Part 2 of this chapter.

For other applications of heavy ion detection with track detectors, the readers are referred to the book of Fleischer et al. (1975); Durrani and Bull (1987); and Benton et al. (1973); Hashemi-Nezhad and Durrani (1986); Trautmann et al. (1998); Hou et al. (1990); and Dunlop et al. (2007).

E. Neutrons

1. Principles of neutron detection

The neutron is a neutral particle. It cannot directly produce ionization and excitation of atoms. Therefore, it cannot be recorded directly by any kind of radiation detector including etch track detectors. Neutrons are recorded through their recoil nuclei and reaction products with the atoms in or in front of the detector materials. Recoil nuclei H, C, O in CR-39 ($C_{12}H_{18}O_7$) can be recorded directly in the detectors. Polycarbonate ($C_{16}H_{14}O_3$) sheet can only record recoil C and O in itself. Recoil nuclei may come from a radiator, such as a thin foil polyethylene $(CH_2)_n$. The downstream detector such as CR-39 will record the recoil protons and carbon nuclei. The energy E of recoil nuclei follows equation $E = \alpha E_n \cos^2\varphi$, where E_n is the energy of incident fast neutrons, φ is the angle of the recoil nucleus with respect to the direction of the incident neutrons in the laboratory system. $\alpha = 4A/(A+1)^2$, where A is the atomic number of the recoil nucleus. The nuclear reactions often used to record neutrons are (n,α) and (n,f) reactions, such as ^{10}B(n,α)^{7}Li, ^{6}Li(n,α)^{3}H, ^{235}U(n,f), ^{237}Np(n,f), ^{238}U(n,f), and ^{232}Th(n,f). Their threshold energies are $E = 0$ (^{10}B, ^{6}Li, ^{235}U), 0.4 MeV (^{237}Np), 1.1 MeV (^{238}U), and 1.17 MeV (^{232}Th).

To record neutron reaction product α particles, usually CR-39 or LR-115 (CN) sheet is used. For recording neutron produced fission fragments, muscovite mica, Lexan or Makrofol polycarbonate foil is used.

2. Suitable detectors for neutron detection

a. CR-39

CR-39 is suitable for recording neutrons through (1) recoil nuclei, *i.e.*, p, C, O in its own compositions (CR-39 sheet is used as both radiator and detector); (2) recoil nuclei, *i.e.*, p and C from radiator polyethylene $(CH_2)_n$. In this case, CR-39 is used only as detector, but fast neutrons will produce nuclear recoils in the CR-39 as background; (3) neutron reaction products, namely, α−particles and ^{7}Li from the ^{10}B(n,α)^{7}Li reaction; and α-particles and tritium from the ^{6}Li(n,α)^{3}H reaction.

b. LR-115

LR-115 CN is nearly the same as CR-39 in neutron detection, but it is less sensitive than CR-39.

c. Lexan, Makrofol (polycarbonate)

Polycarbonate detector can be used as a nuclear recoil detector without radiator for fast neutron detection. Bare polycarbonate sheets will record recoil nuclei of oxygen and nitrogen produced in air by neutrons. Usually, an aluminum plate is placed in contact and in front of the detector to shield off the recoil nuclei from the air; or alternatively, to treat air as a radiator.

Polycarbonate sheets sandwiched with fissile foils can record fission fragments (fission rate measurement) and consequently derive the neutron fluences and neutron spectrum.

d. Muscovite mica

Muscovite mica is the most ideal detector of neutrons with the help of fissile foils. It is insensitive to α particles and β- and γ-rays. It is suitable for the measurement of neutrons in mixed radiation fields, where neutrons coexist with γ-rays.

3. Neutron intensity measurements

The advantages of the SSNTD in neutron intensity measurements are (1) its small size and (2) insensitivity to β- and γ-rays and to less ionizing nuclear particles. It can be inserted into small gaps to measure neutron intensities in nuclear facilities. A modern example is the measurement of the neutron production yield, neutron spatial distribution, and neutron spectrum in the study of accelerator driven subcritical reactors (ADSs) for the disposal of long-lived radioactive wastes (Brandt et al., 2005, 2008; Zamani et al., 2008; Fragopoulou et al., 2008; Wan et al., 2001, 2003). A variety of fields of neutron applications and measurements have been studied with SSNTDs at different facilities (Zaki-Dizaji et al., 2008; Belafrites et al., 2008; Esposito et al., 2008; Bedogni et al., 2008; Al-Ghamdi et al., 2008; Al-Jarallah et al., 2002; Ohguchi et al., 2008; Abu-Jarad et al., 2002; Palfalvi, 2009; Guo et al., 1976a).

4. Neutron energy measurements

Early developments of neutron energy determination by track detectors have been reviewed and explained in books (Fleischer et al., 1975; Durrani and Bull, 1987). Here, only the basic techniques and new developments are pointed out.

a. Fission foil and track detector sandwich techniques

The commonly used fission foils are ^{235}U, ^{237}Np, ^{238}U, and ^{232}Th with muscovite mica and polycarbonate track detectors.

b. Recoil carbon and oxygen nuclei method

By measuring the etched track parameters: the mouth and depth, one can calculate the REL spectrum. By computer simulation with the composition of the detector, the reaction cross sections of carbon and oxygen with fast neutrons and the REL spectrum, one can obtain the neutron spectrum (Wang et al., 1998).

c. Radiator-Degrader-CR-39 method

The radiator is hydrogen containing material, such as polyethylene $(CH_2)_n$ thin foil. The degraders are Pb, Au, or Al in different thicknesses. CR-39 is the track detector to record the recoil protons with energy in the energy window of CR-39 (Aslam et al., 1988; Durrani et al., 1988; Dajko, 1990; Wan et al., 2003; Oda et al., 2009).

d. Bonner sphere spectrometers

More detailed explanation about Bonner sphere spectrometers (BSSs) will be given in **IV**.A.2 of Part 2 of this chapter, and Chapter 9, Volume 1 of this book.

5. Neutron dosimetry

Neutron dosimeters are used in measurements of neutron dose. From the readings of neutron dosimeters, one can directly get the information of the dose received by a person. Accurate measurements of neutron dose are very difficult. Theoretically, the process of energy deposition of neutrons and its effects on the human body are very complicated. Recent renewal of the recommendations of the ICRP (International Commission on Radiological Protection) from publication 60 (1991) to Publication 103 (ICRP 103, 2007) reflects the progressive deepening of the understanding of neutron energy deposition and its effects on human health. Practically, ideal neutron dosimeters have not been found at the moment. All the existing neutron dosimeters cannot give accurate measurements of neutron dose owing to their intrinsic deficiencies.

The principle of neutron dose measurements includes two steps. The first step is to measure the fluence of neutrons Φ irradiated to the human body. The second step is to convert the neutron fluence Φ to dose equivalent H(d) by using the documented conversion coefficients $H(d)/\Phi$ (ICRP Publication 74, 1996). New replacements for the ICRP 74 conversion coefficients will be published after having the ICRP recommendation 103. The procedure for the calibration of a neutron detector to become a neutron dosimeter is as follows: (1) the neutron detector is put in the field of neutrons of known energy E_n; (2) the detector is irradiated with the neutrons to a certain fluence Φ; and readings M_c of the detector and the fluence Φ of neutrons with their energy E_n are recorded. (3) Then, the response of neutrons M_c/Φ is obtained at the energy E_n of neutrons. (4) Dividing M_c/Φ by the conversion coefficient $H(d)/\Phi$, the response of dose equivalent $M_c/H(d)$ is obtained. The conversion factor $H(d)/\Phi$ was documented in ICRP Publication 74 (1996) and will be replaced with a new one. The reciprocal of the response of dose equivalent is $H(d)/M_c$, the dose equivalent corresponding to one reading of the dosimeter. The above discussion is correct for monoenergetic neutrons of known energy. For a neutron spectrum of known energy distribution, the average conversion coefficient $H(d)/\Phi$ must be calculated from the neutron spectrum. For example, with a AmBe neutron source, the dose equivalent received can be calculated from $H(d)/\Phi$ and Φ, where Φ is calculated from the distance between the detector (dosimeter) and the source and the intensity of the source.

An ideal neutron dosimeter should satisfy the following equations at all energies:

$$\frac{(M_c/\phi)}{(H(d)/\phi)} = c \quad (3.25)$$

or

$$\frac{M_c}{H(d)} = c \quad (3.26)$$

where c is a constant.

In other words, the two curves of M_c/Φ versus E_n and $H(d)/\Phi$ versus E_n can be overlapped if one of the curves is multiplied by a constant (c or $1/c$). In this case, one does not need to consider neutron energy E_n. One can directly obtain the dose equivalent $H(d)$ from the reading M_c. Up to now, a neutron dosimeter of this type has not been found. In other words, none of the existing neutron dosimeters is ideal.

CR-39 is a better neutron dosimeter than a nuclear emulsion plate (e.g., NTA nuclear emulsion) and thermoluminescence detector, but it is less efficient than a bubble detector (Guo et al., 2000) as a neutron dosimeter. A rem-counter was designed by Agosteo et al. (2010) to measure neutron dose with CR-39 and ^{10}B converter by recording tracks of (n,α) reactions as a passive detector and with a ^{3}He proportional counter as active detector in the rem-counter. It shows that both active and passive detectors in the rem-counter can give relatively good agreement with the ideal $H(d)/\Phi$ curve; and the CR-39 + ^{10}B detector does not undergo deleterious signal saturation, pileup effects, and the influence of intensive γ-rays as the active detector.

6. Applications of neutron detection

The neutron is one of the most important particles in modern science and technology. Nuclear energy and nuclear weapons are both most important issues in the current world. One can say that neutrons are the key to world development and security. Therefore, one cannot make a complete list to show how many fields are related to neutron detection; but one must give some examples to show how important neutron detection is in modern science and technology, including the following:

a. Neutron Physics,
b. Reactor Physics,
c. Neutron Dosimetry,
d. FTD and Geothermal Chronology (Yuan et al., 2013; Guo et al., 1996, 1999, 2013; Chen et al., 2001),
e. Nuclear Transmutation by Neutrons and Accelerator-Driven Subcritical Reactor (ADS),
f. Nuclear Forensics and Safeguards (Chen et al., 2013, 2015; Vlasova et al., 2015; Wang et al., 2018),
g. Neutron Radiography,
h. Neutron Cancer Therapy, such as BNCT (See Chapter 1, Volume 1), and
i. Element Mapping.

F. Exotic particle detection

Exotic particles include several hypothetical particles including the following:

(1) Magnetic monopoles (MMs), which was predicted theoretically by Dirac (1931) and is required by the Grand Unification theories in theoretical physics.
(2) Q-balls, which was predicted by supersymmetry theories to be a coherent state of squarks, leptons and Higgs field. Q-balls could be a candidate, among others, of cold dark matter which is proposed to solve the problem of the missing mass in the universe.
(3) Nuclearites are known as Strange Quark Matter (SQM). SQM is made up of approximately equal numbers of up, down, and strange quarks, and is surrounded by an electron cloud, forming a sort of atom. SQM could contribute to the cold dark matter in the universe.

1. Suitable detectors for exotic particle detection

CR-39, polycarbonate, phosphate glasses, and muscovite mica have been used to detect exotic particles.

2. Magnetic monopole detection

MMs are particles theoretically predicted by Dirac (1931). These particles are similar to protons or electrons or other charged particles, which can be separated into isolated particles from their opposite charge state. The magnetic charge of a monopole can be $g = ng_D = 68.5ne$, where $n = 1, 2, 3, \ldots$, and g_D is the Dirac minimum charge: $g_D = hc/2e$.

From the relationship between magnetic charge g and electronic charge e, one can calculate the energy loss rate of a MM passing through the stopping materials.

$$(dE/dx)_m \approx (g\beta/e)^2 (dE/dx)_e \text{ if } \beta << \alpha = \frac{1}{137}$$

$$(dE/dx)_m \approx \left(n^2/4\right)(dE/dx)_e = K\beta \text{ if } \beta \leq \alpha = \frac{1}{137}$$

$$(dE/dx)_m \approx 33 \text{ GeV/cm for plastic if } n = 1 \ (g = g_D)$$

A very large value dE/dx (equivalent to the dE/dx of an heavy nucleus with charge $Z \approx 68.5e$ for $g = g_D$, and equivalent to that of a superheavy elemental ion with $Z \approx 137e$ for $g = 2g_D$) of monopoles classifies these as heavily ionizing particles, if they exist and consequently, these would be easy to detect with the use of nuclear track detectors, such as CR-39, polycarbonate (Lexan, Makrofol), BP-1 glasses, and UG-5 glasses (Derkaoui et al., 1999; Ahlen, 1980; Pinfold, 2009; Giacomelli and Patrizii, 2005; Kinoshita et al., 1992; Cecchini et al., 2005). Another important feature of energy loss of MMs is that the electronic energy loss *dE/dx* decreases as the energy *E* of the MM decreases in the entire electronic energy loss region, which is contrary to the feature of electrically charged particles. In the latter case (electrically charged particles), *dE/dx* gradually increases up to the Bragg peak as the energy of the particle decreases from $E \approx 3$ GeV/amu.

The most sensitive etch track detector CR-39 has a threshold $Z/\beta \approx 5$. It allows the detection of a MM with one unit Dirac charge ($g = g_D$) for a velocity β ($= v/c$) of around 10^{-4} and $\beta > 10^{-3}$. The detector can record $g \geqq 2g_D$ MM in all the velocity regions from $4 \times 10^{-5} < \beta < 1$ (Derkaoui et al., 1998, 1999). Lexan and Makrofol polycarbonate detectors have thresholds at $Z/\beta > 50$. They can record only relativistic MMs.

The phosphate glass detector (UG-5) is less sensitive. It can record only multiply charged MMs.

The detection of MMs can also be carried out by recording tracks of bound systems (dyons) of MMs with a nucleus having a large nuclear magnetic moment, such as the proton, aluminum nucleus, and Mn nucleus. The energy loss of the bound system (MM + p, MM + Al, MM + Mn) is composed of the energy loss contributions of MM and the nucleus (Price et al., 1984; Derkaoui et al., 1999).

Up to now, no MMs have been found (Pinfold, 2009).

3. Dark matter particle detection

Great efforts have been made to detect various hypothetical dark matter particles by using solid-state track detectors. The solution to the problem of missing mass in the universe is of significance in theoretical physics, astrophysics, and cosmology. More detailed explanation of this subject is beyond the main scope of this handbook. The readers are referred to articles written by Price (2005) and Cecchini and Patrizii (2008), from which more information can be obtained.

IV. Track formation mechanisms and criterions

A. Introduction

Track formation mechanisms are studied to determine how etchable tracks of charged particles are formed in solids and what is the criterion or standard to expect which solids can record what particles.

The inferences we derive from the track formation mechanism and the criterion we use for the selection of SSNTDs must fit with the following facts:

(1) Insulating solids and poor semiconductors with electrical resistivity higher than about 2000 ohm-cm are basically track-recording materials;
(2) Metals and good semiconductors are not track-recording solids;
(3) Electrons (β-rays), x-rays and γ-rays cannot create etchable tracks in solids;
(4) The sensitivities of different insulating solids are different, such as muscovite mica, which can record the nucleus of Ne ($Z = 10$) and heavier nuclei, but not a lighter nucleus of $Z < 10$. Polycarbonate can record α-particles, but not protons; CR-39 can record protons ($Z = 1$) and heavier particles. In other words, the SSNTD is a type of threshold detector.
(5) Generally speaking, plastic or polymer detectors are more sensitive than inorganic solids;
(6) Before etching, the recorded tracks in solids may be annealed out (*i.e.*, track fading) at higher temperature.

What mechanisms can cause the above phenomena and what criterion can be used to judge the differences are essential questions for our understanding and application of nuclear track detectors.

B. Track formation mechanisms

1. Ion explosion spike for inorganic solids

The incident charged particle ionizes the atoms along its trajectory. Subsequently, the ionized ions with positive charges repel each other and are rejected into the region around the trajectory by coulomb repulsion, forming a vacant region along the central line of the particle trajectory and interstitial atoms in the surrounding region as shown in Fig. 3.16.

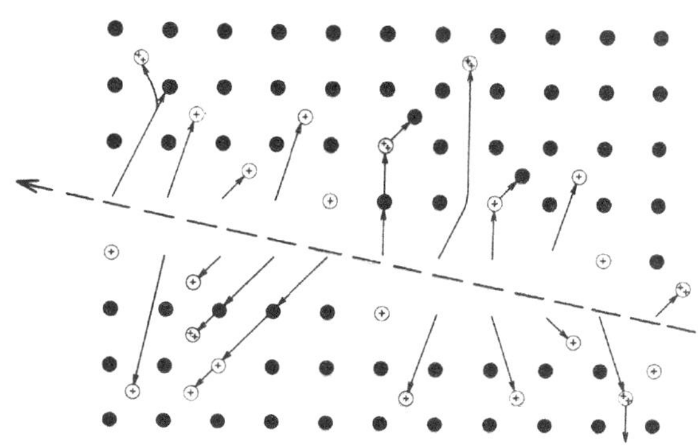

FIG. 3.16 The ion explosion spike mechanism for track formation in inorganic solids. The incident heavy charged particle ionizes the atoms in the detector along its trajectory and forms a positively charged region. The positively charged ions repel each other by coulomb force and are rejected to the surrounding region, forming a long vacant line along the trajectory and interstitial atoms around the vacant line. These damaged materials easily react with appropriate chemicals and are dissolved in a chemical reagent, forming an etched track. *From Fleischer (1998). Reprinted with kind permission from Springer Science + Business Media © 1998.*

The formed interstitial and vacant region relaxes elastically, straining the undamaged matrix. The long range of the strained region can be observed under a TEM as a latent track and selectively etched by special etchant to form an etched track of the incident particle. The electrons in metals can diffuse very fast to reach the electrically positive ions and neutralize them. Therefore, no ion explosions occur in metal after the passage of a heavy charged particle. For this reason, metal cannot be used as a SSNTD.

The positive ions formed by a charged particle in good semiconductors act as holes which move relatively fast and dissipate their charges before explosion of the ions. Therefore, good semiconductor material cannot record tracks of heavy charged particles.

2. Chain breaking mechanism in high polymers

In a high polymer, molecular chains are broken by the passage of a heavy charged particle. This process needs only about 2 eV of energy. The process occurs more readily in high polymers. The open ends of the chains are easily attacked by appropriate chemical reagents. This mechanism is shown in Fig. 3.17 (Fleischer, 1981).

The energy of about 2 eV, required to break a chain of polymers, is much less than that required to ionize an atom (at least 5−17 eV) for inorganic solids. Therefore, it needs less energy to create continuous damage sites in polymers than that in inorganic solids. This is the reason that higher polymers have a higher sensitivity than inorganic solids.

C. Criteria of track formation

The study of track formation criteria involves the search for a parameter or a quantity by which one can predict whether a particle can be recorded by a specific solid and describe quantitatively the threshold of the solid. This quantity is the property of the solid and has no relationship with any kind to charged particles. Several parameters have been proposed for the criteria, but none of these satisfactorily fit all data. Every parameter is successful for certain properties but fail with respect to other properties of the solids.

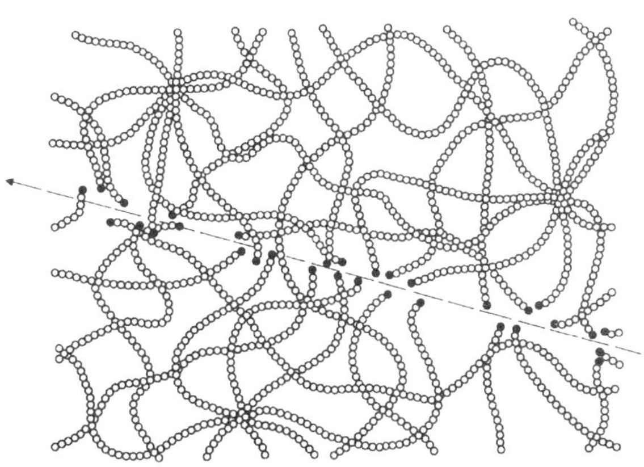

FIG. 3.17 Chain breaking mechanism in the formation of etchable tracks in higher polymers. Chain breaks by passage of a heavy charged particle allow preferential etching at a lower radiation damage density. For this reason, organic solids have a higher sensitivity to record tracks of heavy ions. *From Fleischer (1998a,b). Reprinted with kind permission from Springer Science + Business Media © 1998.*

1. Primary ionization rate criterion

This criterion was suggested by Fleischer et al. (1967b). According to this criterion, the formation of etchable tracks is related to the number of primary ionizations produced along the trajectory of the particles.

a. Formulation

The formulation for calculation of primary ionization rate is given as follows:

$$J = \frac{2\pi n_e Z_{eff}^2 e^4}{mv^2}\frac{f_e}{I_0}\left[\ln\frac{2mv^2}{I_0} - \ln\left(1-\beta^2\right) - \beta^2 - \delta - K\right] \tag{3.27}$$

where n_e is the number of electrons in a unit volume of the detector, v is the velocity of incident particle, m is the mass of the electron, f_e is the effective fraction of the electrons in the detector in the most loosely bound state, I_0 is the ionization potential of the most loosely bound electrons in the detector, β is the particle phase velocity, *i.e.*, v/c where c is the speed of light in a vacuum, δ is a correction term for the effect of polarization of a medium at relativistic velocities, K is a constant depending on the composition of the stopping medium (detector), and Z_{eff} is the effective charge of the incident particle where $Z_{eff} = Z\left[1-\exp\left(-130\beta/Z^{2/3}\right)\right]$ and Z is the atomic number of the incident particle.

The value of the adjustable constant K in Eq. (3.27) is selected by measuring track etch rate V_T for at least 2 types of incident particles and adjusting the K value in Eq. (3.27) in order that for the same V_T, the corresponding J values are the same for all of the incident particles. This means that V_T is related only to J but not to the types of particles. By adopting $I_0 = 13$ eV for muscovite mica and $I_0 = 2$ eV for polycarbonate and CN, Fleischer et al. (1967b) adopted $K = 3.04$ to calculate J to fit the data well.

Fig. 3.18 shows the relationships between J values and β or energy E of various incident charged particles as well as the thresholds of different types of detector materials.

b. Successes in explanation of thresholds

The criterion of primary ionization rate can give a good fit to the experimental results both for inorganic solids and organic polymers. By adjusting the parameter K, a horizontal line (threshold) can be obtained to separate the data of track formation from that of nontrack formation.

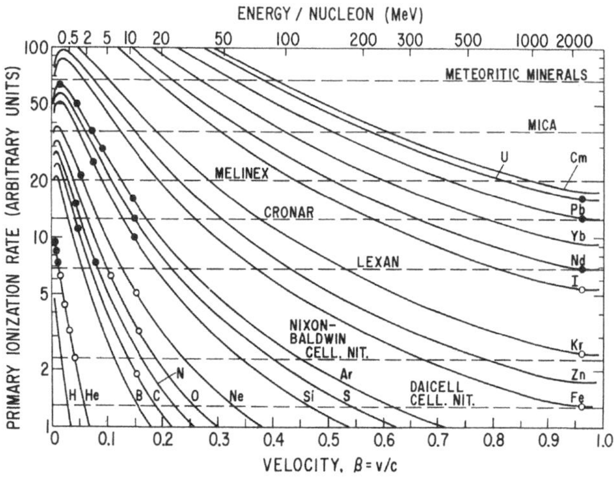

FIG. 3.18 Relationship between primary ionization rate J and velocity β or energy E of various projectiles. The horizontal lines represent the thresholds of different types of track detectors. The solid circles denote etchable track formation in Lexan. The open circles denote that no etchable tracks were formed in Lexan. The minimum charge Z, which can form tracks in muscovite mica, is Ne ($Z = 10$). The registration threshold of CR-39 lies below the x-axis of the figure. *Figure from Fleischer (1998). Reprinted with kind permission from Springer Science + Business Media © 1998.*

c. Existing problems

A number of problems exist with respect to primary ionization rate criterion, which are the following:

(1) The main problem is that it does not consider the ionization induced by δ-rays.

(2) Another problem is that the calculated primary ionization rate is only a relative value, and the formulation has not been approved scientifically.

(3) The third problem is that in organic polymers, the parameter I_0 must be taken as small as about 2 eV. This energy is not enough to ionize any atoms. Even though a good fit can be obtained for an organic polymer. $I_0 \approx 2$ eV indicates that the tracks are formed by molecular chain breaking, not by primary ionization.

2. *Restricted energy loss for plastic track detectors*

The restricted energy loss (*REL*) criterion was suggested by Benton (1967) and Benton and Nix (1969). The restricted energy loss is the portion of the total energy loss that produces δ-rays of less than some specified energy ω_0. Not all of energy loss in solids contributes to track formation. A significant fraction of the energy transferred to electrons in a detector goes to electrons with sufficiently high energy and with range much larger than the scale of latent tracks. These higher-energy δ-rays will not contribute to the track formation. Only low-energy δ-rays with energy lower than ω_0 will contribute to the formation of etchable tracks.

a. Formulation

$$REL_{\omega_0} = \frac{2\pi n_e Z_{eff}^2 e^4}{mv^2}\left[\ln\frac{2mv^2}{I}\frac{\omega_0}{I} - \ln\left(1-\beta^2\right) - \beta^2 - \delta\right] \tag{3.28}$$

where I is the mean ionization potential of the detector material, and ω_0 is the upper limit of the δ-ray energy below which the deposited energy contributes to the formation of tracks.

The value of ω_0 has been taken as 200, 350, and 1000 eV by different authors. It plays the role of an adjustable parameter in the calculation of *REL*.

Fig. 3.19 illustrates the relationships between *REL* and energy E of incident particles.

b. Successes in explanation of thresholds

The criterion of restricted energy loss rate has taken into account the insignificance of δ-rays with higher energy such as higher than ω_0. Ideally, it is correct and easy to be accepted. For this reason, the horizontal line of the threshold fits to the data for the most part. Only in the low-energy region some deviation occurs.

c. Existing problems

No horizontal line quite fits the measured data as shown in Fig. 3.19. Ahlen (1980) showed that *REL* cannot fit the V_T values in a straight line for particles with different charge Z.

3. *Energy deposition model* (e_v)

The energy deposition model (E_v) criterion was suggested by Katz and Kobetich (1968). It suggests that a charged

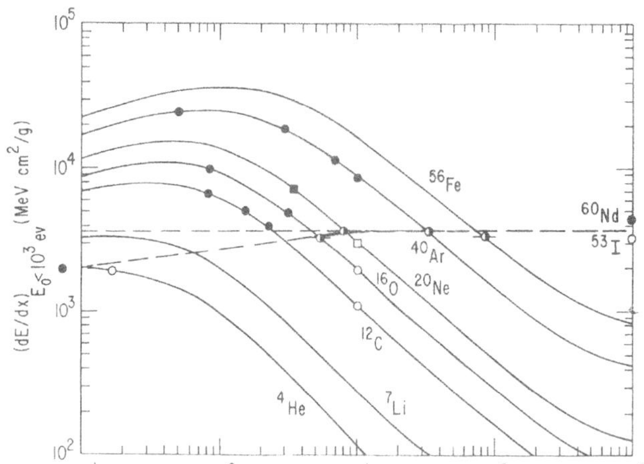

FIG. 3.19 Restricted energy loss rate versus energy of the heavy charged particles in a Lexan polycarbonate track detector. The upper limit of the energy of δ-rays $\omega_0 = 1000$ eV was taken. Solid circles denote the formation of etchable tracks. Open circles denote no track formation. No horizontal line quite fits the measured data. *From Fleischer (1981). Reprinted with permission from Elsevier © 1981.*

particle will create an etchable track in a dielectric solid if the particle deposits energy E_v, which reaches a critical dose (volume density) by δ-rays at a critical distance r from the path of the particle. The distance r from the path is 19Å ($1Å = 10^{-8}$ cm $= 10^{-10}$ m) for muscovite mica, 15Å for CN and 17Å for Lexan polycarbonate. The critical energy density E_v deposited at the critical distance r of muscovite mica is $\sim 3.5 \times 10^9$ erg/g.

a. Formulation

The calculations for obtaining E_v at a distance r by δ-rays are quite complicated. The reader is referred to the original articles by Katz and his coauthors (Katz and Kobetich, 1968; Kobetich and Katz, 1968).

Fig. 3.20 shows the relationships between E_v and E of incident particles.

b. Successes in explanation of thresholds

This criterion has taken into account the energy density deposited by δ-rays and the critical radius from the trajectory of the particle, and the results can approximately predict the thresholds both for inorganic and organic solids.

c. Existing problems

The obvious shortfall of this criterion is that it ignores the energy deposition of the primary incident particles. The second problem is that it cannot obtain a horizontal line to denote the existing threshold. The third problem is the sharp cutoff at low energy of incident particles, and it does not hold true for failure to detect very low-energy incident particles.

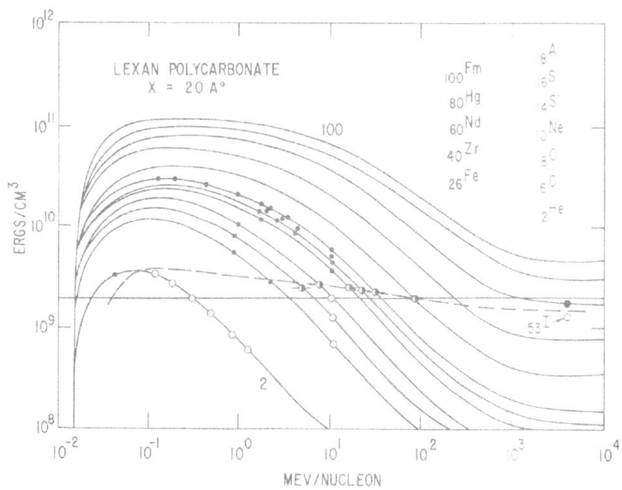

FIG. 3.20 Energy deposition density E_v versus energy E of heavy charged particles (nuclei) in Lexan polycarbonate. Solid circles represent etchability in Lexan, open circles denote zero registration. The horizontal line denoting the threshold of Lexan is tilted toward the low energy region. *From Fleischer (1981). Reprinted with permission from Elsevier © 1981.*

D. Extended and transitional criterions

Besides the criteria explained in the previous paragraphs, namely, J, REL, and E_v, there are two criteria often used by others, which are Z_{eff}/β and dE/dx. The value of Z_{eff}/β may be used as a criterion for track formation, but dE/dx can be used only as a transition parameter.

1. Z_{eff}/β

Ahlen (1980) compared the measured results of V_T of ^{28}Si, ^{56}Fe, and ^{20}Ne in a Lexan detector with REL and Z_{eff}/β. He concluded that Z_{eff}/β is preferred as a universal parameter over REL, and the experimental data are sufficiently good to rule out the REL model. The reason for Z_{eff}/β being a good parameter is that when the constant K in equation for J approaches infinity (∞), it is equivalent to, i.e., $J \propto (Z_{eff}/\beta)^2$. Therefore, $(Z_{eff}/\beta) \propto J^{1/2}$ will be another parameter as J to describe formation of etchable tracks.

Fig. 3.21 illustrates the response (V_T/V_T-1) versus Z_{eff}/β of BP-1 glass etched by 48% HBF_4 at 50°C, 49% HF at

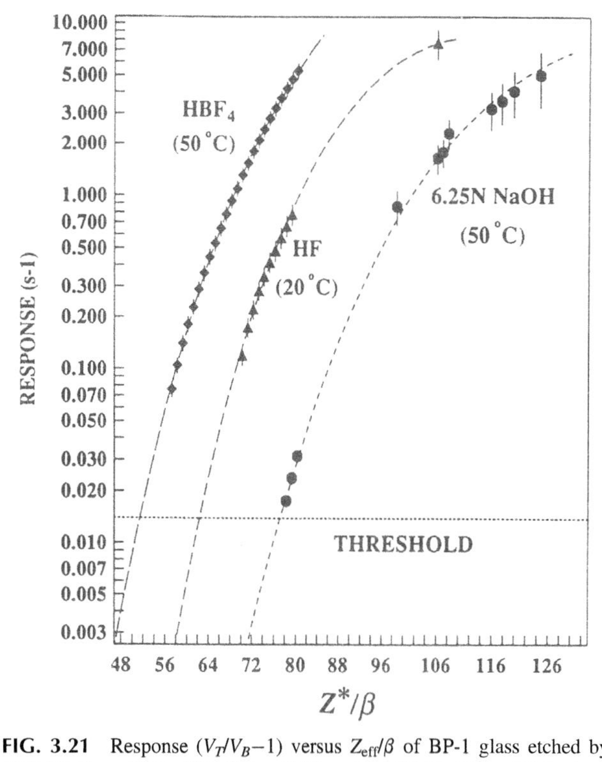

FIG. 3.21 Response (V_T/V_B-1) versus Z_{eff}/β of BP-1 glass etched by 48% HBF_4 at 50°C, 49% HF at 20°C and 6.25N NaOH at 50°C where $s = V_T/V_B$, $Z^* = Z_{eff}$. The solid points are measured data with an 11.4 AGeV ^{197}Au beam and its fragments in the three etching conditions. The horizontal dotted line denotes the threshold ($s-1 \leq 0.015$) of track registration at normal incidence. The threshold $(Z_{eff}/\beta)_{th}$ are given according to the lowest value of (Z_{eff}/β) in the figure, which are the lower limit of the instrument used. *From He et al. (1994); printed with permission from Elsevier © 1994.*

20°C, and 6.25N NaOH at 50°C (He et al., 1994). From the measured data, the authors found that the thresholds for formation of etchable tracks are related to the etching condition for the same detecting material. The threshold Z_{eff}/β for BP-1 glass are 57 ± 1, 68 ± 1 and 78 ± 1 respectively for the above three etching conditions.

a. Successes and advantages of Z_{eff}/β as a criterion of track formation

The Z_{eff}/β criterion of track formation has exhibited the following success and advantages:

(1) It is easy to convert to the charge *Z* of relativistic heavy charged particles which can be recorded by the detector. For relativistic energy, $\beta \rightarrow 1$, $Z_{eff} \rightarrow Z$, and $Z_{eff}/\beta \rightarrow Z$. Therefore, $(Z_{eff}/\beta)_{th} = 26$ can be considered whereby relativistic Fe ions ($Z = 26$) in cosmic rays may be recorded by the detector.

(2) It is easy to convert the registration threshold $(Z_{eff}/\beta)_{th}$ to primary ionization *J* and restricted energy loss (*REL*), since in all equations of *J*, *REL* as well as *dE/dx*, $(Z_{eff}/\beta)^2$ are included. Other parameters in the equations can be obtained from the compositions of the detectors. Knowing the value of $(Z_{eff}/\beta)_{th}$, the thresholds of *J*, *REL* as well as the transitional parameter *dE/dx* (See below), can be obtained for the detectors.

(3) It is easy to determine the threshold of Z_{eff}/β of a detector by using different energies of at least one type of heavy ion, or different heavy ions.

(4) The determined value of the threshold of Z_{eff}/β for each detector is absolute and fixed, not adjustable.

From the above advantages, one can see that the criterion Z_{eff}/β for track formation is convenient, straightforward, and easy to assigning a quantity, compared with other criteria suggested.

b. Existing problems

How to derive the threshold of track formation Z_{eff}/β for each type of detector remains to be solved.

2. *dE/dx transitional parameter*

Some authors still use total energy loss rate *dE/dx* to denote the thresholds of etch track detectors. This parameter has been demonstrated by extensive experiments that it cannot give a satisfied description of the thresholds. No horizontal line separates the ions that register from those that do not register. The lines that separate the experimental data for recording tracks from the none-recording of tracks tilt up at the high-energy region. One cannot find a threshold value, which is independent of projectile charges (*Z*). However, the *dE/dx* value can be treated as a conditional reference mark, such as the "threshold" for recording He ions are about 40 (50% CR-39), 88 (40%), 108 (30%), 125 (20%), and ~250 (0%) keV/μm equivalent in water for CR-39-DAP plastics, where 50% CR-39 means the ratio of CR-39/DAP in the composition of CR-39-DAP is 50/50 (Tsuruta et al., 2008). Since the *dE/dx* value of a particle of charge (*Z*) and energy (*E*) in a specific detector material can be determined or calculated very well. From the *dE/dx* value one can trace back the energy *E* and to calculate *REL* and *J*.

The formula used to calculate total energy loss rate is.

$$\frac{dE}{dx} = \frac{4\pi n_e Z_{eff}^2 e^4}{mv^2}\left[\ln\frac{2mv^2}{I} - \ln\left(1-\beta^2\right) - \beta^2 - \frac{C}{Z_2} - \frac{\delta}{2}\right] \tag{3.29}$$

where C/Z_2 is a shell correction term.

The other parameters in Eq. (3.29) are the same as in Eqs. (3.27) and (3.28). The calculation methods of C/Z_2 and δ can be found in the review articles by Ahlen (1980), Fano (1964) and Barkas and Berger (1964).

E. Incapability of the former adopted criteria

The criteria of track formation proposed in the1960s have played important roles in our understanding of the formation of etchable tracks in insulating solids, which include (1) criterion of total energy loss rate $(dE/dx)_c$ (1964), (2) criterion of primary ionization rate $(dJ/dx)_c$ or, J_c (1967), (3) criterion of restricted energy loss rate $(dE/dx)_{\omega<\omega_0}$ or REL_{ω_0} (1967), and (4) criterion of energy deposition model E_V (1968), as described in the paragraph above.

Ideally, the success of the above criteria is very limited. It cannot help scientists to do much work in the following subjects:

1. *Incapability to estimate the threshold values of existing track detectors*

Threshold is a general and basic property of SSNTDs. A successful theory on track formation should be able to derive the threshold value of a detector from its chemical compositions and molecular structure. Unfortunately, none of the above criterion can provide accurate values of the thresholds.

2. *Incapability to design a new material possessing the expected threshold value*

Since the 1980s, track workers hoped to synthesize new chemical materials which are more sensitive than CR-39 to record etchable tracks of heavy charged particles; but none of them meet with success (O'Sullivan et al., 1981).

F. Conflict between track formation criteria and chain breaking mechanism

Researchers in this field are familiar with the curves drawn in Fig. 3.18 (J vs. $\beta = (v/c)$, Fig. 3.19 (REL_{ω_0} vs. E (MeV)), and Fig. 3.20 (E_v(ergs/cm^3) versus E (MeV/N)) and their corresponding Eqs. (3.27), (3.28) as well as (3.29) for dE/dx. All these criteria of track formation rely on the mean ionization energy I or its derived quantities. On the other hand, chain-breaking mechanism in Fig. 3.17 is also accepted by all. The chain-breaking mechanism is dependent on chemical bond dissociation energy (BDE, or called bond breaking energy) (Luo, 2002). One should have noticed that the above four track criteria and the track formation mechanism of polymer track detectors are in conflict with each other. The former is based on ionization energy; the latter is based on bond breaking energy. The quantitative values of I for polymer track detectors are in the range from about 57 to about 80 eV, that is, polyethylene (57.4 eV), polypropylene (59.2 eV), cellulose acetate butyrate (63.4 eV), polystyrene (68.7 eV), polycarbonate (Lexan, 69.5 eV), polyallyl diclycol carbonate (PADC, CR-39, 70.2 eV), polymethyl methacrylate (PMMA, Lucite, 74.0 eV), polyethylene terephthalate (PET, 73.2 eV), polyimide (Capton, 79.34 eV), and CN (81.1 eV) (Sternheimer et al., 1982; Durrani and Bull, 1987). The quantitative values of bond breaking energy for polymer track detector materials is in the range from a few tenths of eV to about 4 eV (Luo, 2002; Fleischer, 1998). The difference between the two parameters amounts to tenfold. If one accepts the chain-breaking track formation mechanism, one should develop theoretical formulae for the criteria based on bond breaking energy instead of ionization potential for polymer track detectors.

G. Latent track structures

Infrared (IR) absorption spectroscopy, among some other methods, such as UV and visible spectroscopy, has been extensively used to investigate the track formation mechanism in polymer track detectors since the early 1980s (Chambaudet and Roncin, 1981). In the very early stage, IR absorption spectroscopy was used by track workers to investigate the constituents of radiation damage. Later, many authors joined in the studies on modification of polymer materials (El-Shahawy et al., 1992; Gagnadre et al., 1993; Darraud et al., 1994; Balanzat et al., 1995; Chong et al., 1997; Saad et al., 2001; Costantini et al., 2002; Malek and Chong, 2002; Lounis-Mokrani et al., 2003; Phukan et al., 2003; Barillon et al., 1999; Barillon, 2005a,b; Barillon et al., 2002; Sun et al., 2004; Severin et al., 2005; Dehaye et al., 2003; Mishra et al., 2003).

1. *IR absorption spectrometery for polymer track detectors*

In recent years, research advances have been carried out on track formation in polymer track detectors with IR absorption spectrometry, and rapid progress has been made.

Fig. 3.22 shows main IR absorption spectra and their related chemical bonds in polycarbonate track detector.

The major peaks of the IR absorption spectra marked with CH_3, C=O, C=C, and C−O−C are corresponding to methyl groups, carbonyl bonds, double bonds in phenyl rings, and carbonate ester bonds, respectively, in the polycarbonate structure (See Section II.C.2).

The peaks shown by dotted lines are for unirradiated (pristine) film of polycarbonate. The solid lines are for the irradiated film. The height of the peak of absorbance in Fig. 3.22 should be proportional to the number of bonds of the responsible types in the film if the film is thin enough. During irradiation by heavy ions, some bonds are broken by the irradiation of ions. Therefore, the heights of the peaks are reduced.

2. *Cross-section of bond breaking by heavy ions*

It was noticed that after irradiation of polymer film by heavy ions, the number of chemical bonds decreases with the fluence of heavy ions irradiated onto the film. Fig. 3.23 shows an example of reduction of peak intensities of the film of CN LR-115 of 12 μm in thickness as a function of proton fluence.

The straight line in Fig. 3.23 can be simply described by an equation

$$N / N_0 = 1 - \sigma_i F \tag{3.30}$$

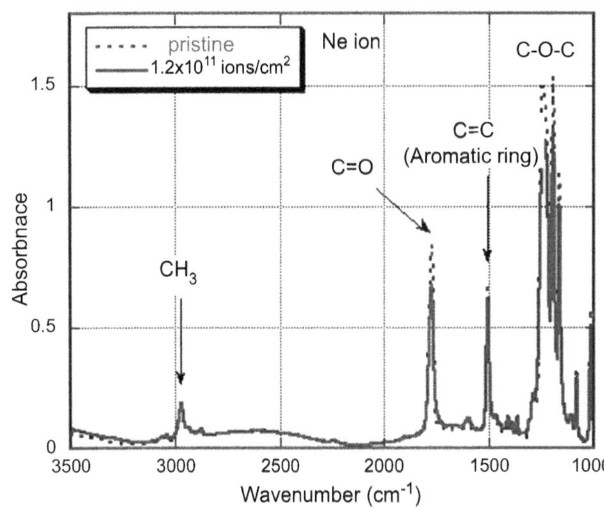

FIG. 3.22 IR absorption spectra of pristine and irradiated PC with 24 MeV Ne ions.

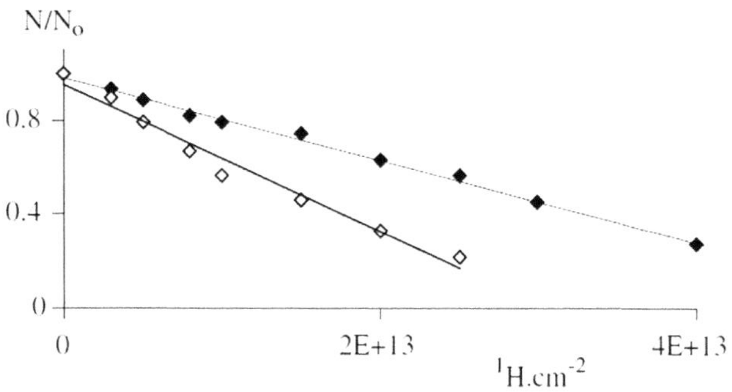

FIG. 3.23 Relative bond decrease (N/N_0) versus proton fluences. Average proton energy in the film: 3.1 MeV (LET = 14 keV/μm). Black point: nitrate functions; white points: glycosidic bonds. *From Barillon and Yamauchi (2003). Reprinted with permission from Elsevier © 2003.*

where N_0 represents the initial number of bonds in unit volume (cm^{-3}) of the film, N the number of bonds (cm^{-3}) left after irradiation of the film by heavy ions, σ_i a constant for each type of bond scission, F ion fluence (cm^{-2}) irradiated to the film.

If the film is thin enough and the overlapping of latent tracks of heavy ions in the film can be ignored, then $N/N_0 = A/A_0$, where A_0 is the absorbance of the pristine film for the responsible wave number of IR spectra, A is that of irradiated film. A and A_0 can be measured by Fourier transforms infrared (FT-IR) spectrometer.

Similar figures and equations as Fig. 3.23 and Eq. (3.30) were obtained for the other types of thin polymer detectors (Dehaye et al., 2003; Yamauchi et al., 2008a,b,c, 2010, 2013a,b; Mori et al., 2009, 2011; Kusumoto et al., 2016).

The meaning of σ_i is the cross-section (probability) of breaking particular bonds by one heavy ion. The unit of σ_i is cm^2, which is similar to that in nuclear reactions.

Fig. 3.24 shows that the cross-sections of breaking glycosidic groups (σ_{CO}) and nitrate groups (σ_{NO}) in CN film as a function of the energy of oxygen ions (Barillon and Yamauchi, 2003).

Fig. 3.25 illustrates the cross-sections of ester bond scission in polyethylene terephthalate (PET) as a function

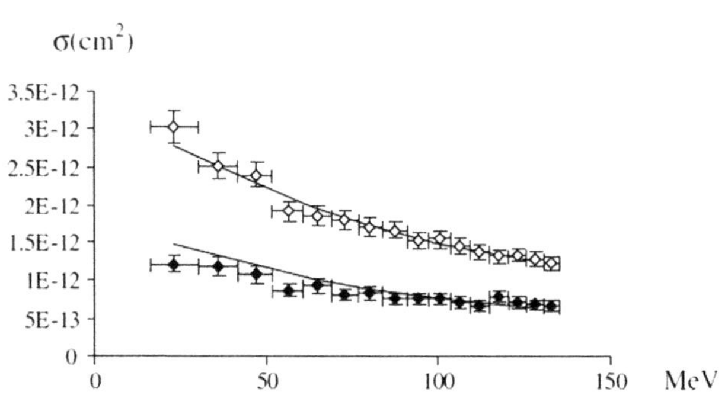

FIG. 3.24 Cross-sections of glycosidic bonds (white points) and nitrate groups (black points) in CN detector as a function of the energy of oxygen ions. *From Barillon and Yamauchi (2003). Reprinted with permission from Elsevier © 2003.*

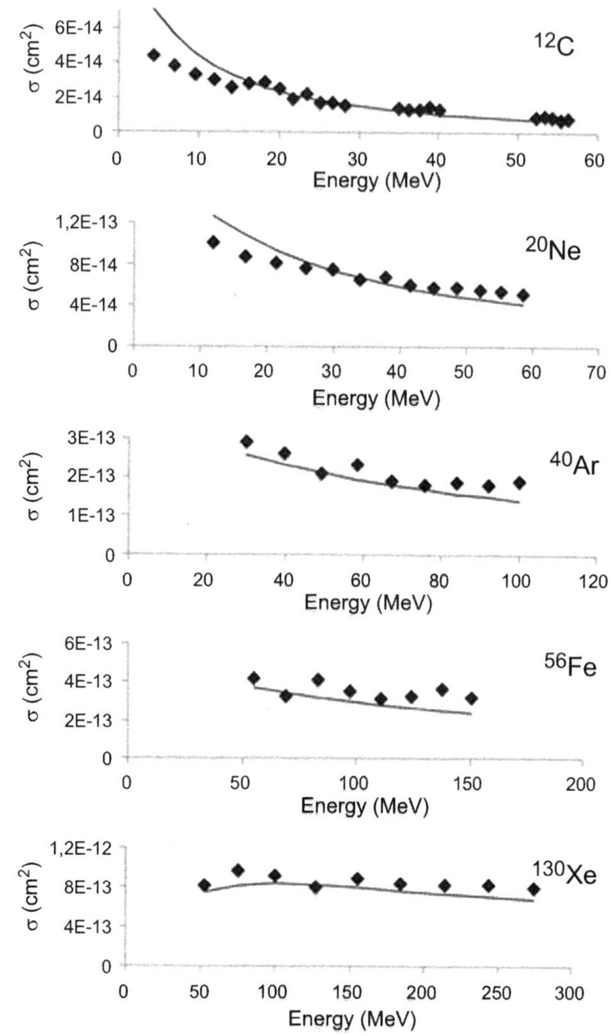

FIG. 3.25 Experimental (points) and simulated (lines) cross-sections (σ) for ester bond scission in PET according to ion energy. From top to bottom: ^{12}C, ^{20}Ne, ^{40}Ar, ^{56}Fe, ^{131}Xe. Simulation parameters:$P(r) = \left[1 - e^{-D(r)/D_0}\right]^m$, $m = 2$ and $D_0 = 1.8$ MGy. *Barillon et al. (2013). Reprinted with permission from Elsevier © 2015.*

of the energy E of heavy ions: ^{12}C, ^{20}Ne, ^{40}Ar, ^{56}Fe, and ^{131}Xe.

From Figs. 3.24 and 3.25, one can see that the cross-sections of bond breaking in polymer track detectors are similar to the cross-sections of nuclear reactions induced by neutrons (Garber and Kinsey, 1976a,b) and by heavy ions (www.nndc.bnl.gov).

3. Effective track core radius

The radius r_i of a cylinder in which the responsible bonds are effectively broken in the film can be calculated by the equation:

$$r_i = \sqrt{\sigma_i/\pi} \tag{3.31}$$

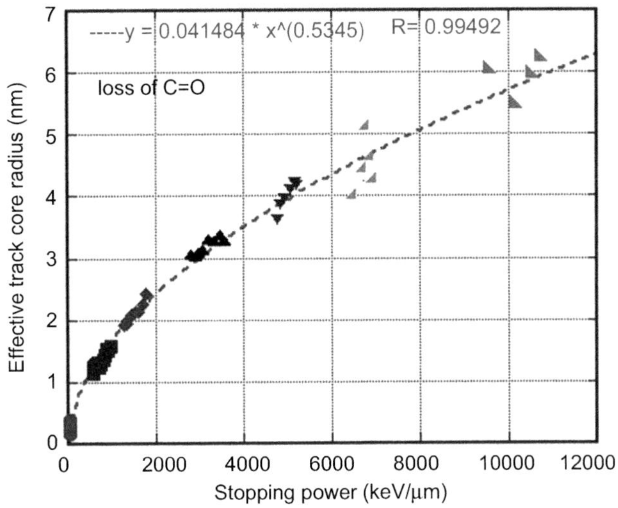

FIG. 3.26 Effective track core radius for loss of carbonyl in PC. For the color version of the figure, the reader is referred to the online version of the book. *From Yamauchi et al. (2010). Reprinted with permission from the author © 2018.*

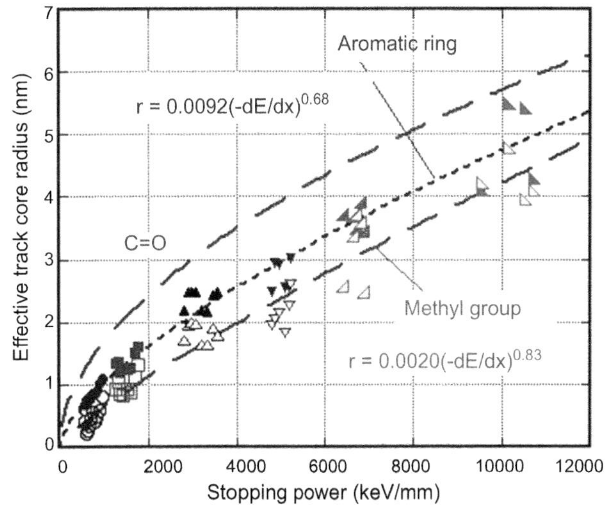

FIG. 3.27 Effective track core radius for loss of bonds C = O, phenyl ring and methyl group in polycarbonate (PC) track detectors. For the color version of the figure, the reader is referred to the online version of the book. *From Yamauchi et al. (2010). Reprinted with permission from the author © 2018.*

Fig. 3.26 shows the effective track core radius (r_i) for breaking carbonyl (C = 0) bonds in polycarbonate (PC) by heavy ions of H, He, C, Ne, Ar, Fe, Kr, and Xe.

From Fig. 3.26, one can see that the effective track core radius for breaking bonds of carbonyl (C = 0) by heavy ions is in the range from several tenths to several nm and the core radius is related to the stopping power of the heavy ions. In other words, the radius is related to the energies of heavy ions.

Fig. 3.27 shows a comparison of the effective track core radius from breaking bonds of carbonyl (C = 0), phenyl rings and methyl groups (CH_3).

From Fig. 3.27, one can see that the bond (C = 0) is easier to be broken than that of phenyl ring, and the bond in methyl group (CH_3) is the most stable among the three.

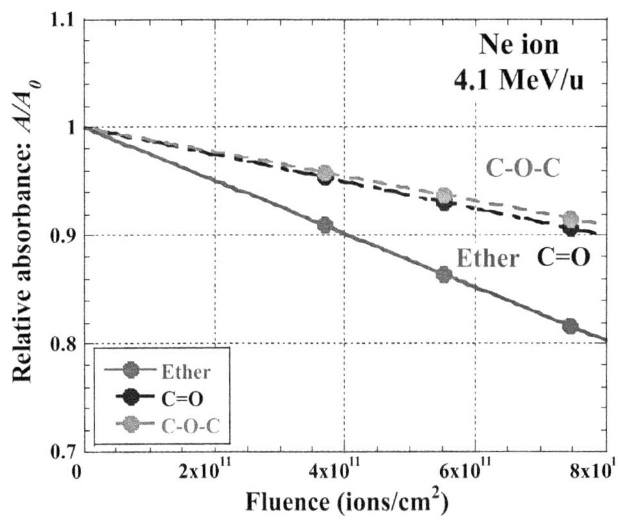

FIG. 3.28 Reduction of the relative absorbance A/A_0 of ether, carbonyl, and C−O−C in PADC against the fluence of Ne ions. *From Kusumoto (2017). Reprinted with permission from the author © 2018.*

4. Layered structure of latent tracks

Different chemical bond breaking needs different energy from the bombarded particles. If the energy required is in small amount, the positions of the bonds to be broken can be farther from the trajectory of the particles. On the contrary, if the bond breaking energy required is in large amount, the positions of the bonds must be close to the trajectory of the particles. Therefore, the diameters of the damaged regions for different bonds are different. Fig. 3.28 shows the reduction of the relative absorbance A/A_0 of ether, carbonyl, and C−O−C in PADC against the fluence of Ne ions.

From Fig. 3.28 one can see that the ether bonds are easier to be broken than C═O, and further than C−O−C. The diameter of the region for breaking bonds for ether should be larger than that of C═O, and the diameter of the region for C−O−C should be the smallest among the three.

Fig. 3.29 shows the diameters of the damaged regions for the three types of bonds in PADC.

From Fig. 3.29 one can see that the lateral distributions of breaking bonds are different for different bonds.

5. Chemical etching and OH groups in polymers

Yamauchi et al. (2005) point out that we must rewrite our overview on the nuclear track formation process in CR-39 as

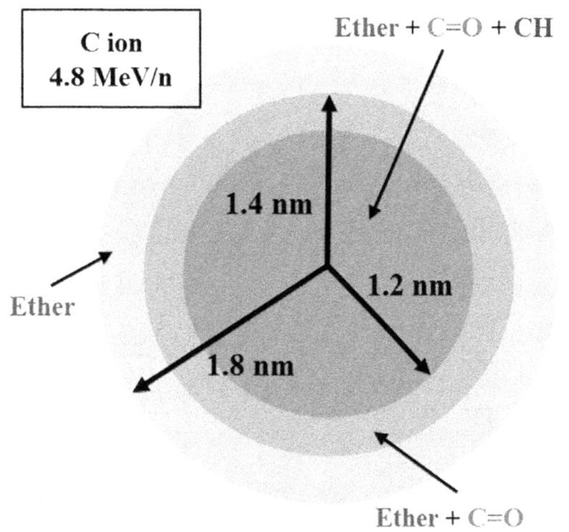

FIG. 3.29 Layered structure of damaged region by C ions with an energy of 4.8 MeV/u in PADC. *From Kusumoto (2017). Reprinted with permission of the author © 2018.*

follows: (A) The parts between the two carbonate ester bonds should be segmented into small molecules, including CO_2 gases, along the particle trajectory; (B) after the segmentation, the CO_2 gases are diffused away and a lower density region with chemically active end points is formed simultaneously; (C) subsequent chemical modifications, including a reaction with dissolved oxygen, resulting in OH groups as new end points in the polymer network. They also conclude that inside the track core, all of the ether bonds and most of the carbonate ester bonds will break. This means that the parts of the polymer chain between two carbonate ester bonds (−O−(C=O)−O−CH_2−CH_2−O−CH_2−CH_2−O−(C=O)−O−) should be segmented into small molecules composed by one or two carbon atoms including CO_2. The unique sensitivity of CR-39 may be attributable to such a relatively long part segmentation.

Fig. 3.30 shows the processes of bond breaking of ether and forming OH, and bond breaking of a carbonate ester and forming another OH group. The first formed OH group was lost due to breaking of the ester.

Fig. 3.31 shows that the number of OH groups formed in PADC is proportional to the number of ether bonds broken by the incident charged particle.

Fig. 3.32 shows the network of CR-39 (PADC) (restricted to two dimensions for our treatment of this subject matter).

The active end point OH groups are easy to react with the attacking reagent (etchant). The etching starts from the end point of OH groups, forming a hole (etched track) in the PADC detector.

6. Radial Electron Fluence around ion tracks

Kusumoto (2017) proposed a new criterion for track formation, which is named as Radial Electron Fluence around Ion Tracks (REFIT). A schematic drawing is shown in Fig. 3.33.

In Fig. 3.33A, the secondary electron trajectories are indicated as the lines in red. The purple points show the interaction points. Each secondary electron starts from each original point on the ion trajectory and reaches to the terminal after its travel with a certain distance, producing the other multielectrons. As a result, the number density of secondary electrons will decrease with increasing distance from the ion trajectory. Kusumoto (2017) counted in Monte Carlo simulation the secondary electrons (fluence, nm^{-2}), which passed through the surface of the cylinder with a certain radius from the trajectory. The direction of electrons includes outward and inward through the surface. The fluence of the secondary electrons at certain distance was taken as the criterion for track formation.

This criterion is based on the study of IR absorption spectra for PADC (CR-39).

It is a new advance in the study of track formation. At this moment, it is evidenced in some aspects by using CR-39. There is a need to acquire additional evidence from

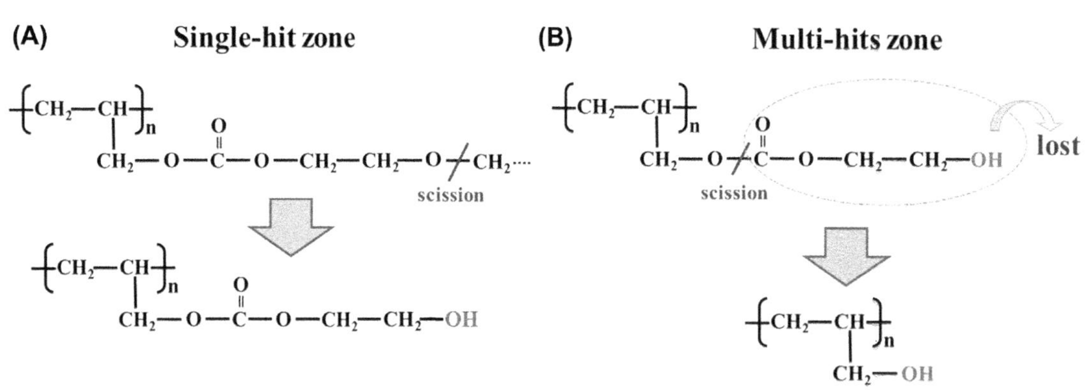

FIG. 3.30 Two steps in the damage formation process in PADC. The ether is broken in a single-hit zone (A) and a carbonate ester is lost with the OH groups in multi-hits zone (B). For the color version of the figure, the reader is referred to the online version of the book. *From Kusumoto (2017). Reprinted with permission from the author © 2018.*

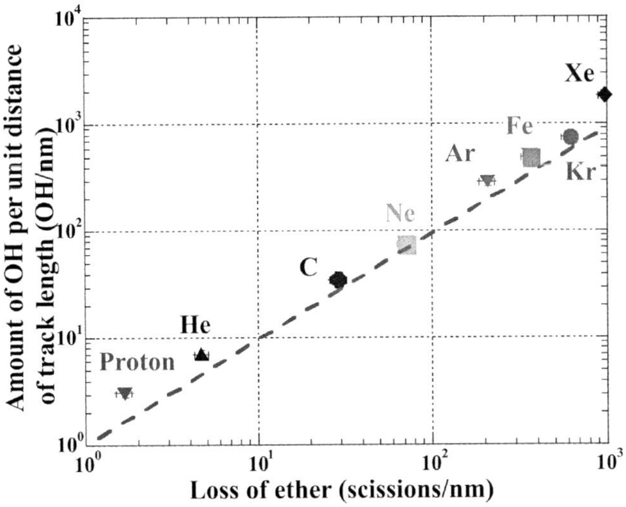

FIG. 3.31 Correlation between the generation density of OH groups and damage density of ether. The line is guided visually. For the color version of the figure, the reader is referred to the online version of the book. *From Kusumoto (2017). Reprinted with permission from the author © 2018.*

other types of polymer track detectors, such PC, PET, CN, and PI, and from heavy ions with higher charge Z.

The final purpose of the study of track formation mechanisms is to find theories, by which one can calculate the values of thresholds of all types of SSNTDs, and to design materials which meet the requirements for expected values of thresholds.

V. Track revelation

Track revelation refers to the techniques used to reveal charged particle tracks in solids. The most successful and popularly used technique is track etching, which is the subject of this chapter. For other techniques of track revelation, the reader is referred to the treatise by Fleischer et al. (1975).

A. Chemical etching

1. Etching condition

Chemical etching (CE) is the basic technique used to reveal nuclear particle tracks in solids. The indispensable conditions for CE are two, namely, etchant and etching device.

a. Etchant

An etchant is a particular chemical formulation, which preferentially reacts with the damaged materials in the particle track resulting from the passage of a heavy charged particle in the SSNTD. One can find the etchants for different detector materials in the lists of etching conditions in the book by Fleischer et al. (1975) or by Wagner and Van den haute (1992). The etching conditions include three aspects: (1) compositions of the chemicals used as etchant; (2) temperature during etching; and (3) duration of the etching. The etchant is particular for each type of detector material. Selection of the wrong etchant will never reveal tracks. Usually, a higher concentration of etchant and higher temperature will etch tracks more rapidly; longer etching times will make the etched track larger in diameter.

b. Etching device

Etching devices hold etchants and ensure the etching conditions. The etchant must be held in a closed container to

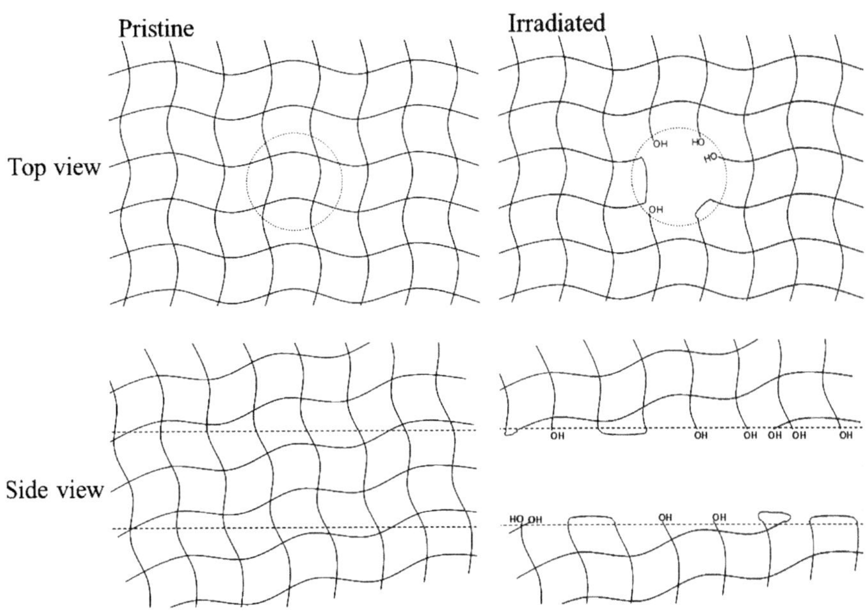

FIG. 3.32 Schematic view of the latent track formed in PADC. *From Kusumoto (2017). Reprinted with permission from the author © 2018.*

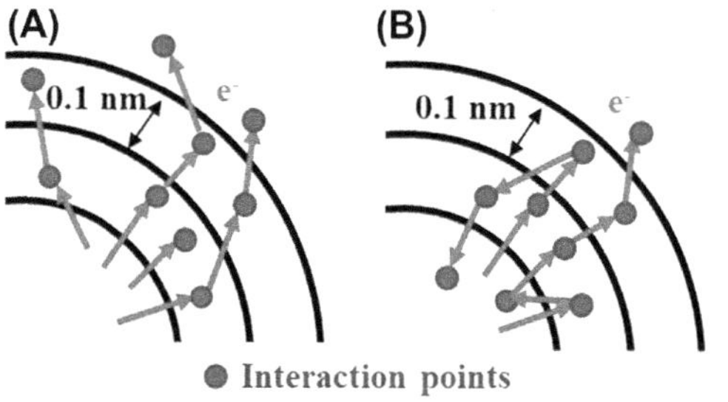

FIG. 3.33 Schematic description of the criterion for track formation in polymer detector—Radial Electron Fluence around Ion Tracks. For the color version of the figure, the reader is referred to the online version of the book. *From Kusumoto (2017). Reprinted with permission from the author © 2018.*

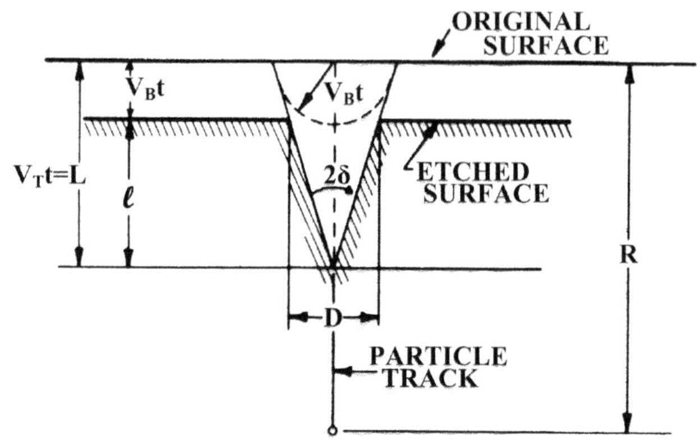

FIG. 3.34 Track etching geometry with constant V_T and V_B in an amorphous solid. The particle perpendicularly entered the solid and stopped at depth R below the surface. The etchant preferentially reacts with the damaged materials formed along the trajectory of the particle with a constant velocity V_T for a time t. The etchant attacks the bulk material from the original surface of the detector with a constant velocity V_B. After etching for a time t, an etch cone is formed along the trajectory with pit at depth $L = V_Tt$ below the original surface. During time t, the surface layer with thickness V_Tt of the detector is removed by the etchant. The original surface moves to the new surface that is the etched surface. The visible track length is $l = (V_T - V_B)t$. *From Price and Fleischer (1971). Republished with permission of Annual Review of Nuclear Science © 1971; permission conveyed through Copyright Clearance Center.*

prevent exchange of substances from inside or outside to ensure that the concentration of etchant remains constant.

c. Etching temperature

Homogeneous and constant temperature in the whole volume of etchant are required in many studies. The etch rate increases exponentially with an increase in temperature. A small change in temperature will influence considerably the etch rate. In very precise measurements, such as in the identification of the charges of cosmic ray particles, in the identification of masses of isotopes, and in NTM production, the temperature must be controlled within $\pm$ 0.01°C. When using the SSNTD for the counting of tracks with the optical microscope, a precision of temperature of $\pm$ 1°C is sufficient in the temperature range of 10−100°C. One or more stirrers are necessary to keep the etchant homogeneous in temperature (Guo, 1988).

2. Track etching geometry

The etching process includes two types of chemical reactions: (1) etchant with damaged material and (2) etchant with bulk material of the detector. The first proceeds with velocity V_T along the trajectory, called track etch rate. The second proceeds perpendicularly to the surface of the detector with velocity V_B called bulk etch rate. The etching process is a combined reaction of V_T and V_B.

Fig. 3.34 shows the process of CE in an amorphous solid having a track of particle perpendicularly incident to the detector.

3. Critical angle of etching

The track etch rate V_T and bulk etch rate V_B sometimes are competitive with each other. The critical angle θ_c of etching is a result of the competition of V_B with V_T, which is shown in Fig. 3.35. The first particle entered the detector with incident angle θ (the angle between the direction of incident particle and the surface of the detector).

V_Tt is the etched track length at time t. V_Bt is the removed thickness from the surface of the detector. θ is the incident angle of the particle (the angle between the particle direction and the detector surface). If $V_Tt \sin\theta < V_Bt$, the track will be removed away by the etchant. No etched track is left in the detector after etching for the example illustrated in part (A) of Fig. 3.35. If $V_Tt \sin\theta > V_Bt$, a track cone will be left after etching (parts c and d). If $V_Tt \sin\theta = V_Bt$, it is the critical situation for formation of an etched track (part b). In this situation, the incident angle θ is denoted as θ_c. θ_c is the critical angle of etching. Then, one has

$$\theta_c = \arcsin\frac{V_B}{V_T} \tag{3.32}$$

4. Techniques of critical angle measurements

a. Direct measurement of half cone angle

From Fig. 3.35B one can see $\theta_c = \delta$ where δ is the half angle of the etched track cone. This signifies that if one can measure the half angle δ of the etched cone, one can determine the critical angle θ_c, and V_T, provided one knows the value of V_B.

b. Direct measurement of critical angle

From Fig. 3.35B one can see that changing the beam angle θ of the incident particles, the etched tracks will start to

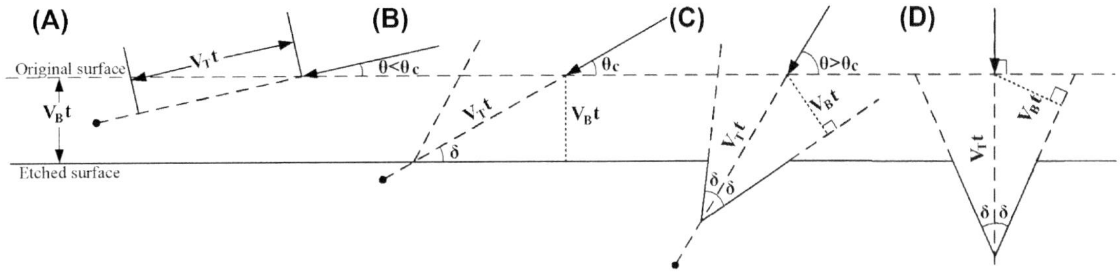

FIG. 3.35 Existence of critical angle of etching for observation of etched tracks.

appear at $\theta = \theta_c$. By this technique Khan and Durrani (1972) measured the critical angle θ_c of ^{252}Cf fission fragments in soda-lime glass (35°30′), U-2 Reference glass (31°45′), obsidian (26°00′), tektite (25°45′), quartz (7°15′), muscovite mica (4°30′), Makrofol polycarbonate (3°00′), and Lexan polycarbonate (2°31′).

c. Direct measurement of detection efficiency

For a point source or thin plane source of given particles, one can directly measure the detection efficiency ε of the detector for the particles, that is, the ratio of the number of recorded tracks to the number of the emitted particles, to derive the critical angle θ_c by Eq. (3.33) for an isotropic source in 2π geometry (Fleischer et al., 1975).

$$\varepsilon = 1 - \sin\theta_c \tag{3.33}$$

By this technique Guo et al. (1976b) (See also Guo et al., 1982) measured the critical angle θ_c of fission fragments of ^{235}U induced fission by thermal neutrons in muscovite mica (3°41′), phosphate glass (7°6′), silicate glass (37°21′), quartz (16°2′), polycarbonate (Chaoyang, China, 2°24′), Makrofol K polycarbonate (3°40′), and polyester (3°29′).

5. Track etching geometry

From Fig. 3.34 one can easily obtain the diameter of a track that perpendicularly enters the surface of detector ($\theta = 90°$) according to the equation

$$d = 2V_B t\sqrt{\frac{V_T - V_B}{V_T + V_B}} \tag{3.34}$$

or

$$d = 2V_B t\sqrt{\frac{V - 1}{V + 1}} \tag{3.35}$$

where $V = (V_T/V_B)$. V is called the etch rate ratio.

For an incident angle $\theta \neq 90°$ and constant track etch rate V_T and bulk etch rate V_B, the major axis D of the opening of etched track is

$$D = \frac{2V_B t\sqrt{V^2 - 1}}{V \sin\theta + 1} \tag{3.36}$$

The minor axis of opening is

$$d = 2V_B t\sqrt{\frac{V\sin\theta - 1}{V\sin\theta + 1}} \tag{3.37}$$

For other parameters of etched tracks, one can find the formulas in the book written by Durrani and Bull (1987).

The equations of etched track geometry for constant V_T are only approximations to the true situation because only for a very few cases is V_T constant. Generally, V_T varies along the trajectory of a particle. Calculations based on a constant V_T often give results which are good enough for the studies of particles.

For track etching geometry with varying V_T and for track etching geometry in anisotropic solids, such as all crystalline minerals, the formulations can be found in the book by Durrani and Bull (1987).

6. Progress in track etching geometry

The study on track etching geometry has made tremendous progress. Nikezic and Kostic (1997) derived an analytical two-dimensional equation of etch pit wall in amorphous SSNTDs. Nikezic (2000) derived an analytical three-dimensional equation of track wall (for variable V_T, conical track phase) described by the equation

$$\sqrt{x^2 + y^2} = \int_Z^L \frac{d\xi}{\sqrt{v^2(\xi) - 1}} \tag{3.38}$$

where Z is the axis along the particle track; (x, y) are the coordinates of a point on the track wall; v is the etch rate ratio V_T/V_B; L is the depth to which the chemical solution penetrates the detector material along the particle trajectory; and ξ is the parameter upon which v depends, such as *REL* or others.

Nikezic and Yu (2003, 2004) extended the above track wall equation to overetched tracks (or round tip tracks). The whole set of equations can be used to calculate the track wall, track opening diameter, major axis, miner axis, and other parameters for constant V_T, variable V_T, normal incidence, oblique incidence, sharp tip etched tracks, and round tip etched tracks. In summary, all parameters of tracks in amorphous solids can be calculated except the tracks in anisotropic crystalline minerals. For more information, the reader is referred to the original papers mentioned above.

B. Electrochemical etching

Electrochemical etching was first suggested by Tommasino (1970). A plastic detector foil with tracks of particles is used to divide a cell containing a suitable etchant, such as NaOH solution. A stainless-steel electrode is placed in each half of the cell. A high-frequency oscillating voltage is connected to the two electrodes. The voltage is typically equivalent to ~30−50 kVcm^{-1} field strength. The frequency is ranged from several KHz to several tens of KHz. A 50 Hz 220 V sinusoidal supply can also work. Under the joint reactions of etchant and electric voltage, the tracks start to be etched. The forming track tip has a strong field, which makes the foil break down near the tip as show in Fig. 3.36. The breakdown looks like a tree. The tree is a consequence of the charged particle track. Fig. 3.37 shows microphotographs of ^{252}Cf fission fragment tracks in CR-39 sheets of 485 μm in thickness.

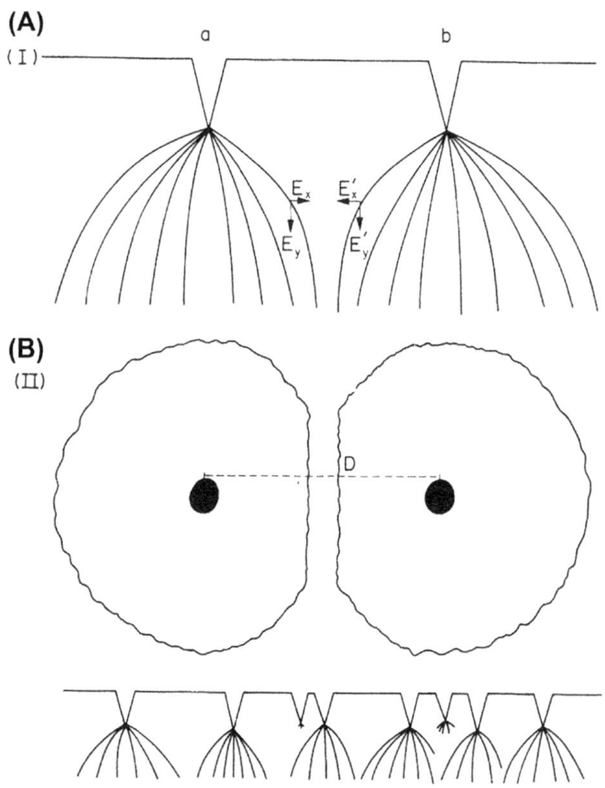

FIG. 3.36 "Treeing" phenomenon of electrochemical etching (ECE) of charged particle tracks in plastic foil. The sharp tips of etched cones have very strong field, which makes the foil beak down, forming treelike tracks. The breakdown spreads up to several hundred micrometers in width, which is easy to observe with a low magnification optical microscope or even by the naked eye. *From Al-Najjar et al. (1979); printed with permission from Elsevier © 1979.*

Many researchers have studied the "treeing" phenomenon (Tommasino et al., 1981; Sohrabi, 1981; Sohrabi and Mahdi, 1993; Li, 1993, 2008a, 2008b).

The advantage of electrochemical etching is that the size of the etched tracks is very large up to several hundreds of micrometers, which are easy to scan with an optical microscope or track image analyzer, or even be seen with the naked eye. The disadvantage is that the track density must be lower than ~10^5 cm^{-2}.

The applications of electrochemical etching are mainly in the detection of fission fragments, α particles and neutron-recoil tracks. The main fields of application are neutron dosimetry (Tanner et al., 2005; Dhairyawan et al., 2003; Zainali and Afkar, 2005; Hankins et al., 1989) and Rn monitoring (Zainali and Afkar, 2005).

C. Track etching kinetics

1. Objectives and required parameters

The ultimate purpose of the study of track etching kinetics is to deduce the species of the particles from the tracks created by the particles.

For this purpose, two lines of calculations must be carried out.

(1) First line of calculation—Calculation of track parameters from the parameters of the particles (Forward calculation);
(2) Second line of calculation—Derivation of the parameters of the particles from the parameters of the etched tracks (Inverse calculation).

In the above calculation and derivation, the following parameters are required:

(a) Particle parameters including: atomic number Z_1, mass M_1, and energy E_1 (or velocity v_1);
(B) Detector parameters including: average atomic number Z_2, average atomic mass M_2, density ρ, and mean ionization potential *I*.
(c) Radiation damage density parameter or threshold parameter, such as primary ionization rate *J*, or restricted energy loss rate *REL*, or Z_{eff}/β, or others;
(d) Track etching parameter under special etching condition, such as track etch rate V_T and bulk etch rate V_B, or track etch rate ratio $V = V_T/V_B$, or reduced etch rate $S = V_T/V_B - 1$, as a function of the radiation damage density, such as $V = f\,(REL)$, as well as the etching time *t* or the thickness *h* of removal on a single surface of the detector.
(e) Track parameters, such as track diameter *d* on the etched surface of the detector, major axis *D* and minor axis *d* of the mouth of tracks, and etched track length *L*, which are represented by equations as functions of track etching parameters.

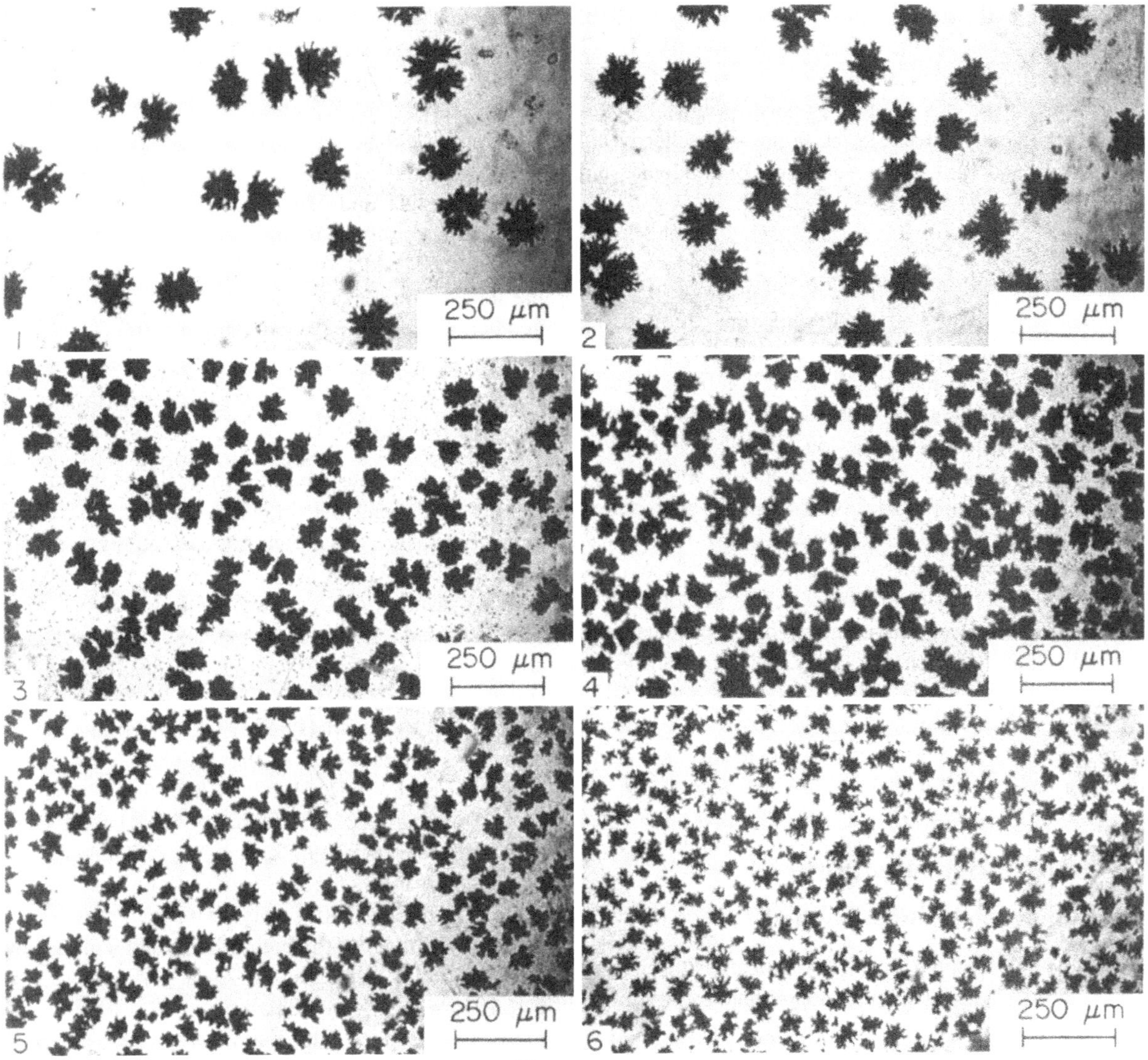

FIG. 3.37 Microphotographs of electrochemically etched tracks of ^{252}Cf fission fragments recorded in CR-39 plates. The thickness of the CR-39 plate is 485 μm. The track densities are (1) 1.5×10^3, (2) 3.0×10^3, (3) 1.2×10^4, (4) 2.4×10^4, (5) 3.6×10^4 and (6) 1.2×10^5 cm^{-2}, respectively. The etching condition was 30 wt% KOH, 60°C, 90 min. The electrical field condition was 15 $kVcm^{-1}$ and 40 KHz. The track densities after ECE are the same as CE. The track size changes diminish as the density increases. *From Al-Najjar et al. (1979); printed with permission from Elsevier © 1979.*

2. Forward calculation

For the first line of calculation (forward calculation), Z_1, M_1, E_1, in (a), Z_2, M_2, ρ, and I in (b) are known. One should select one of the threshold parameters in (c), such as *REL*, and know the relationship between the track etch parameter and threshold parameter in (d), such as $V = f\ (REL)$, and use the equations of *d*, *D* and *d*, and *L* versus track etching parameter in (e), such as $d = \varphi_1\ (V)$, $D = \varphi_2\ (V)$, and $L = \varphi_3\ (V)$. After calculation through the first line, a group of curves will be obtained of *d*, or *D* (*d*), or *L*, versus Z_1, M_1 or E_1 in the condition *t* or *h*. These curves can be used to identify the particles that created the tracks in the detector. These curves represent the relations between the tracks and particles.

The calculation through the first line is relatively easy and much progress has been made: Much work has been done along the first line on threshold parameters in (c) (Fleischer et al., 1975; Fleischer, 1981; Benton, 1967; Benton and Nix, 1969; Katz and Kobetich, 1968; Kobetich and Katz, 1968; Ahlen, 1980. See Section IV above). A large number of papers have been published on track etching parameters in (d) (Hermsdorf, 2009; Hermsdorf and Reichelt, 2010; Rana, 2008; Fromm et al., 1988; also, the articles cited by these papers). Equations for calculating track parameters in (e) have been derived in amorphous solids for constant and variable *V*, for conic tracks, over-etched tracks, and round-tip tracks (Nikezic and Kostic, 1997; Nikezic, 2000; Nikezic and Yu, 2003, 2004, 2006,

2008; Nikezic et al., 2008a; Ditlov, 1995; Somogyi, 1980; Somogyi and Szalay, 1973; Henke and Benton, 1971).

Track evolution in anisotropic crystals is very complicated and less work has been done on this subject. It remains to be studied in detail in the future (Fleischer et al., 1975; Durrani and Bull, 1987; Somogyi, 1980; Henke and Benton, 1971).

3. Inverse calculation

Inverse calculation by analytic formulations through the second line from the parameters of etched tracks to derive the parameters of the particles is more complicated than that of forward calculation. It should be possible, but no paper has been seen in scientific journals.

In practice, experimental calibration of detectors by using accelerator beam particles is the current way to assign the correct charge Z_I, mass M_I, and energy E_I of particles. These kinds of techniques have been used in the discoveries of cluster radioactivity of heavy nuclei (See Section II.A of Part 2 of this chapter) and in many other physical and cosmic ray studies (Fleischer et al., 1975; Durrani and Bull, 1987).

Experimental calibration is a more basic and straight way to retrace the particle parameters than the theoretical calculation through the track etching kinetics.

VI. Particle identification

A. Maximum track length method

The first method adopted for the identification of charges (atomic number) Z is that after an appropriate etching time, the entire portion L_{max} of the trajectory is etched out. L_{max} is called the maximum etchable track length, in which the density of radiation damage is higher than the critical (threshold) value. Therefore, the maximum etchable track length L_{max} is the whole etchable length of the track. By using an appropriate model of radiation damage density (*dE/dX, J, REL, or* E_v), a curve (relationship) between L_{max} and Z can be calculated (Fleischer et al., 1965a). By using the curve of L_{max} versus Z, one can determine the unknown Z from the measured maximum track length L_{max}. The "track-in-track" (TINT) and "track-in-cleavage" (TINCLE) techniques proposed by Lal et al. (1968) and Lal (1969) are two simple techniques to measure L_{max}. If a track (the first track) within a detector intersects another track (the second track), and the second track intersects the outer surface of the detector, the etchant will etch the second track and then etch the first track. The first track after etching is called "track-in-track" or TINT. A second track may be formed in the past (e.g., fossil track) or by irradiation with heavy ion at accelerator. If a track within a detector intersects a crack or cleavage in the detector, the etchant will pass through the crack or cleavage and etch the track. The track after etching is called "track-in-cleavage" or TINCLE. The maximum etchable track length L_{max} can be achieved after a certain etching time. The L_{max} and Z values in a specific track detector should be one to one, if the Z value is low (by restriction of the uncertainty of measurement). The techniques of TINT and TINCLE are often used in the identification of cosmic ray particles, identification of superheavy elements, FTD (determinations of detection efficiencies of the minerals and external detectors for fission events) and geothermal chronology (geothermal history).

B. Track etch rate versus radiation damage density method

This track etch rate method is based on the fact that the etching rate along a track is a function of the radiation damage density. Track etch rate can be represented by *Vt* along the track (Price et al., 1967), or by etch cone length $L = V_T t$, where t is etching time (Price et al., 1967; Fleischer et al., 1975), or by the etch rate ratio $V = V_T/V_B$, where V_B is the bulk etch rate of the detector material (Fleischer et al., 1969d, 1975), or by the reduced track-etch rate $V-1 = V_T/V_B - 1$ (Price, 1982). The radiation damage density is represented by the primary ionization rate J (Fleischer et al., 1967a, 1967b, 1967c, 1967d), or restricted energy loss rate (*REL*) (Benton, 1967), or energy density E_v (Katz and Kobetich, 1968), or by Z_{eff}/β (Ahlen, 1980).

Fig. 3.38 shows the curves of $V_T/V_B - 1$ versus Z_{eff}/β for CR-39 (DOP) and Tuffak polycarbonate. From these curves one can see that V_T for both CR-39 and Tuffak increases steeply in the range $10 \leq Z_{eff}/\beta \leq 99$ and $55 \leq Z_{eff}/\beta \leq 105$, respectively (Salamon et al., 1985). CR-39 is more sensitive than Tuffak polycarbonate.

C. Track etch rate versus residual range method

Fig. 3.39 shows the relationship between track reduced etch rates and residual ranges of some nuclei with different charge Z and A (isotopes) in phosphate glass (PSK-50, Schott Glass Technologies, Inc.).

The lines in Fig. 3.39 are fits to data for heavy ions of ^{28}Si, ^{24}Mg, and ^{20}Ne from the accelerator Superhilac in Berkeley, USA. The curves show that the phosphate glass has excellent charge resolution in the region of Ne and Mg. The glass was used to record and identify Mg and Ne isotopes as well as spontaneous fission from a source of ^{234}U. The points in Fig. 3.39 are the measured results (Wang et al., 1987; Price, 1987). A similar identification technique of charge Z of heavy ions emitted from cluster radioactivity has been adopted by using polycarbonate (Price et al., 1985), polyethylene terephthalate (Tretyakova et al., 1985; Barwick et al., 1985), barium phosphate

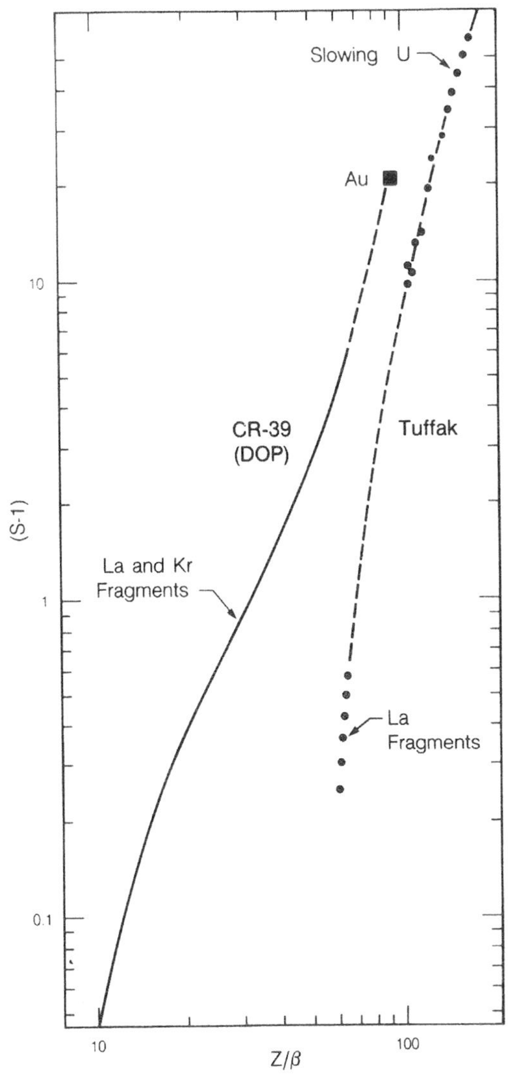

FIG. 3.38 The curves of $V_T/V_B - 1$ versus Z_{eff}/β for CR-39 (DOP) and Tuffak polycarbonate. From these curves one can see that V_T for both CR-39 and Tuffak increases steeply in the range $10 \leq Z_{eff}/\beta \leq 92$ and $55 \leq Z_{eff}/\beta \leq 105$, respectively. Here $s = V_T/V_B$. *From Salamon et al. (1985); printed with permission from Elsevier © 1985.*

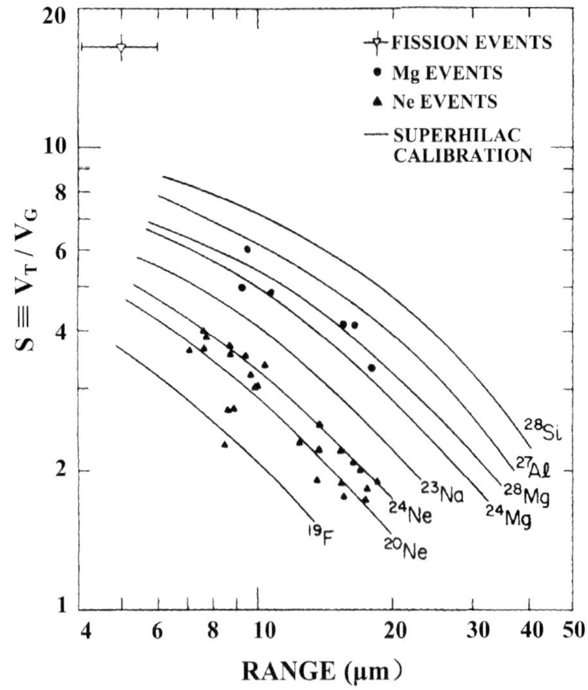

FIG. 3.39 The relations between track reduced etch rates and residual ranges of some nuclei with different charge Z and A (isotopes) in phosphate glass (PSK-50, Schott Glass Technologies, Inc.). *From Wang et al. (1987), https://doi.org/10.1103/PhysRevC.36.2717; printed with permission from Phys. Rev. C36, 2717–2720 (1987). Copyright 1987 The American Physical Society.*

glasses (Bonetti et al., 1993a, 1993b), BP-1 glass (Price et al., 1992), PSK-50 phosphate glass (Wang et al., 1987; Price et al., 1991; Moody et al., 1989, 1992; Bonetti et al., 1991), LG-750 phosphate glass (Wang et al., 1989; Moody et al., 1987), and phosphate glass (Tretyakova et al., 2001).

D. Track diameter method for identification of charge *Z* at high and relativistic energy

Experiments show that heavy ions (projectiles) with charge *Z* at high and relativistic energy will interact with nuclei in the targets and detectors and form a series of projectile fragments with charge $Z-1$, $Z-2$,.... These fragments have nearly the same velocity as the original projectiles. These interactions and their fragments can be used to calibrate the track detector, such as CR-39, to obtain its charge resolution σ_Z (standard deviation of charge *Z* measured) in charge identification.

The procedure of calibration of the track detector is as follows:

(1) A thick stack of nuclear track detectors (detector stack or stack detector), such as CR-39, is irradiated with projectiles perpendicularly to the surface of the detectors to a dose of $\sim 10^4$ cm^{-2}. The CR-39 stack may consist of several or even several 10s of sheets (depending on the energy of the heavy ions to be detected).
(2) After etching the detectors at appropriate etching conditions, tracks of projectiles and fragments are examined under an optical microscope.
(3) The diameters of the tracks are scanned with an optical microscope or a track image analyzer on one surface of the detector. The track diameter distribution is drawn as a histogram. If the scanned surface belongs to the sheet in the middle of the stack, the histogram of the track

diameter distribution will show a series of peaks. The highest peak on the right hand (on larger diameter side) belongs to the projectiles (when matching the tracks with the tracks on the front surface of the first sheet, one will understand this conclusion). The charge corresponding to the highest peak is Z (the charge of the projectiles). The charge of the second peak (on the left side of the highest peak) is $Z-1$, then the third $Z-2$, and so on.

(4) The average diameter $\overline{D}$ and standard deviation σ_D are calculated from all the diameters of each peak.

(5) The slope δ_D/δ_z is calculated in the figure of $\overline{D}$ versus Z at each charge.

(6) The charge resolution $\sigma_Z(1)$ is calculated from equation $\sigma_Z(1) = \sigma_D/(\delta_D/\delta_z)$, where $\sigma_Z(1)$ is the charge resolution on a single surface for a given charge Z.

(7) One can follow the same projectile to the next and the third and more surfaces to measure the track diameters of the same projectile. Then, one can obtain a mean of the diameters in successive surfaces for the same projectile. When all the means for all the projectiles have been obtained, one can draw a new histogram of the distribution of the mean for all the projectiles averaged over all the surfaces. For example, in n surfaces, the peaks in the new histogram are narrower than that obtained on a single surface. One can calculate a new average diameter and its standard deviation $\sigma_D(n)$ for each peak. This $\sigma_D(n)$ is for n surfaces for each charge Z (each peak). A new δ_D/δ_z at each charge (or peak) can be obtained. The final charge resolution $\sigma_z(n) = \sigma_D(n)/(\delta_D/\delta_z)$ can be calculated, which is for all projectiles in each peak (each charge Z) averaged over all the surfaces. It has been shown that $\sigma_Z(n) = \sigma_Z(1)/\sqrt{n}$ is correct. The correctness shows that the energy loss in each sheet of the detector and in all the detectors for measuring the n track diameters is negligible compared with the total energy of the projectile. This means that the velocity ($\beta = v/c$) of the projectile approximates no change when the projectile is traveling through the stack which is measured.

(8) One should point out that the correctness of the $1/\sqrt{n}$ relationship is conditional, which is the following: When a heavy ion (projectile) is passing through the detectors (or stopping materials), it might pick up one or more electrons from the detector atoms to fill in its own inner shell orbits. The effective charge of the projectile changes to $Z-1$ or $Z-2$. This phenomenon disrupts the original charge state and makes the charge resolution become poorer than the $1/\sqrt{n}$ relation. This problem can be overcome by using stripping foils interleaved in the detector stack. (Ahlen et al., 1984; Salamon et al., 1984; Guo et al., 1984a).

Fig. 3.40 shows a histogram of charge distribution of nuclear interaction fragments from a 1.85 GeV/u^{40}Ar beam in CR-39 (DOP) stack by the track diameter measurement, after averaging over 16 cones on surfaces of 8 sheets of the detector, the charge resolution $\sigma_z(16) = 0.06e$ (Price et al., 1983; Salamon et al., 1984).

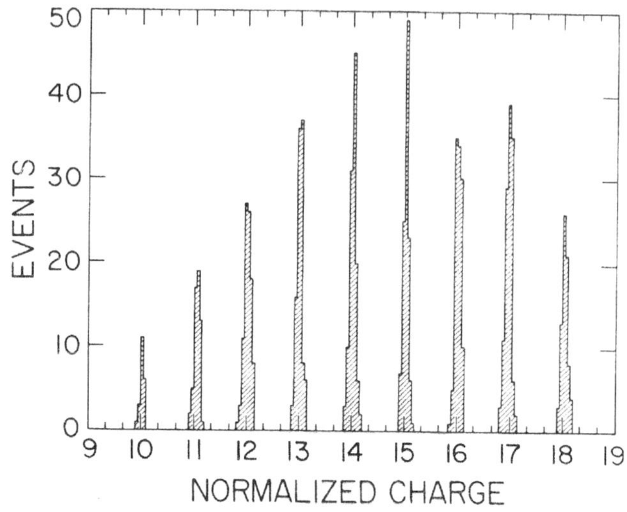

FIG. 3.40 A histogram of charge distribution of nuclear interaction fragments from a 1.85 GeV/u^{40}Ar beam in CR-39 (DOP) stack by track diameter measurement, after averaging over 16 cones on surfaces of 8 sheets of the detector, the charge resolution $\sigma_z(16) = 0.06e$. *From Price et al. (1983); Salamon et al. (1984); Reprinted with permission © 1983 the American Physical Society, https://doi.org/10.1103/PhysRevLett.50.566. CR-39 (DOP) is a special type of CR-39 (See Section II.C.1 above).*

It should be pointed out that the charge resolution of CR-39 is much higher than that of a Si semiconductor detector with equivalent thickness. Tarle et al. (1981) was the first to demonstrate that the etch track detector CR-39 has a charge resolution superior to semiconductor detectors of equivalent thickness. Ahlen (1980) theoretically predicted the superiority of the charge resolution of plastic track detectors to those of scintillator, nuclear emulsion, ionization chamber and semiconductor detectors for high- and relativistic-energy heavy ions. The reason for the higher charge resolution of plastic track detectors is as follows:

According to the restricted-energy-loss or the *REL* model (Benton, 1970), it is assumed that knock-on electrons with energy greater than ω_0 (350−1000 eV) are ineffective in causing the permanent radiation damage. The high-energy δ-rays deposit most of their energy quite far from the particle track. They do not contribute significantly to the primary track. Only the low-energy δ-rays deposit energy in the core region of the track, which is effective for etching. Ionization chambers and solid-state Si detectors correspond to total energy loss, including δ-rays of all energies. Scintillators and nuclear emulsions correspond to the δ-rays with energy higher than 1500 eV. Due to Poisson fluctuations in the collision frequency between the charged projectile and electrons within the detector medium, a large number of low-energy δ-rays are produced. The number of δ-rays then falls off rapidly as the δ-ray energy increases. The fluctuation of the number of high-energy δ-rays is the largest, which results in poor charge resolution of the scintillation detectors and nuclear emulsions. The fluctuation of the effective δ-rays is smaller in ionization chambers and Si detectors, which results in better charge resolution. The plastic track detectors correspond to δ-rays of low energy, the fluctuation is the least, thus the charge resolution is the best for high and relativistic energy heavy ions.

From a calibration with La, Kr, Fe, and Ar, Salamon et al. (1985) obtained the following empirical relation between charge resolution σ_z and Z/β of heavy ions for a CR-39 (DOP) detector.

$$\sigma_z = \frac{1}{\sqrt{n}}\left[0.914 + 1.94 \times 10^{-3}\left(\frac{Z}{\beta}\right) + 4.35 \times 10^{-5}\left(\frac{Z}{\beta}\right)^2\right]e \tag{3.39}$$

Similar calibrations have been made by several groups with similar results (Heinrich et al., 1995; Cecchini and Patrizii, 2008; Patrizii, 2001). From a calibration with Pb projectiles, Huntrup et al. (1997) achieved charge resolutions for a single etch cone $\sigma_z = 0.16e$ at $Z = 6$, $\sigma_z = 0.4e$ at $Z = 50$, and $\sigma_z = 1.0e$ at $Z = 82$ for a CR-39 detector.

He et al. (1994) found that BP-1 glass (Wang et al., 1989) has charge resolution $\sigma_z(1) = 0.19e$ with only a single ($n = 1$) measurement of an etch-pit of ions with $Z = 79$ and $\beta = 0.997$. This charge resolution is equivalent to that of ~16 successive cone measurements in a CR-39 detector stack according to the law $\sigma_z(n) = \sigma_z(1)n^{-1/2}$. Better charge resolution of BP-1 glass can be achieved by measuring many successive track cones along the same particle trajectory. The unprecedented charge resolution of BP-1 glass opens up several fields of applications in nuclear physics, high-energy heavy-ion physics and cosmic-ray astrophysics.

Track diameter identification of charges Z at relativistic energy has been used in many projects including the following:

(1) Searching for fractionally charged particles such as charges of $Z+\frac{1}{3}$, $Z+\frac{2}{3}$ in projectile fragmentations. (Price et al., 1983, Price, 2005; He and Price, 1991; Cecchini et al., 1993; Heinrich et al., 1995).
(2) Measurements of charge pickup cross-section, in which the projectile pickups one or two charges (Ren et al., 1989; Jin et al., 1990; Price, 2008; Westphal et al., 1991).
(3) Relativistic projectile fission studies (Guo et al., 1984b, 1986a).
(4) Measurements of compositions of ultraheavy cosmic rays (Westphal et al., 1998).
(5) Cross-sections for orbital electron capture and stripping of ions in stopping materials (Westphal and He, 1993).
(6) Charge changing fragmentation cross-section (Cecchini and Patrizii, 2008, Cecchini et al., 2002; He and Price, 1994).

Compared with other types of radiation detectors, such as Au–Si surface barrier detectors ($\Delta E-E$ detectors), nuclear emulsions and others, the solid-state track detector has also the following advantages (Price and Fleischer, 1970) for the identification of charge Z of particles.

(1) Identification of heavy particles recorded during ancient times.
(2) Identification of very rare heavy particles such as those that occur at an extremely low rate as 1 per cm^2 per year (See Chapter 1, Volume 1).
(3) Identification of heavy particles in a high background of less heavily ionizing particles.
(4) Identification of particles of very high ionization rate for which resolution by solid-state track detectors exceeds that of other detectors.
(5) Cross-sections for orbital electron capture and stripping of ions in stopping materials (Westphal and He, 1993).
(6) Charge changing fragmentation cross-sections (Cecchini and Patrizii, 2008, 2002; He and Price, 1994).

E. Track length method for identification of charge Z at high and relativistic energy

This method is similar to the track diameter (D) method, whereby the length (L) of an etched cone at high and relativistic energy can also be used to identify charges Z (Giacomelli et al., 1998). The study shows that with a CR-39 sheet for one cone height (length) measurement, the charge resolution can reach $\sigma_z \approx 0.19-0.29e$ for $Z = 74-82$. In other words, one can separate nuclear fragments of high Z ($Z \geq 74$) by one cone length (L) measurement in a CR-39 detector. The track cone length (L) method has a better charge resolution in the charge region $74 \leq Z \leq 82$ than the track diameter (D) method (Giacomelli et al., 1998).

The track length method has been used to measure the charge-changing cross sections of 158 AGeV ^{207}Pb with Bi, Al, Cu and Pb targets. The data can be described by strong electromagnetic dissociation (Sher et al., 2007; Manzoor et al., 2000).

VII. Track fading and annealing

This section describes the phenomenon of track fading and how to use this phenomenon to increase precision in the analyses of radioactivities.

A. Track fading and annealing

Latent tracks or radiation-damaged materials at higher temperature will gradually disappear. This process is called track fading. Track fading is an intrinsic behavior of latent tracks or radiation-damaged materials. Track annealing describes the same behavior of latent tracks but stresses the existence of an external factor, such as laboratory heating or heated by magma intrusion in the earth's crust.

Track fading appears as (1) shortening of the etched track length; (2) reduction of the etched track opening; (3) reduction of the number of tracks observable on internal surface compared with the fresh tracks of the same kind of particles.

B. Mechanisms of track fading

Track fading occurs presumably by the diffusion of atoms in the damaged region of inorganic solids or movements of molecular fragments within a polymer. Interstitial atoms can then recombine with lattice vacancies, and broken molecular chains may rejoin and the existed active species recombine. The activation energies of ~1−2 eV for track annealing support these conclusions.

From thermal dynamics, one can write

$$D = D_0 e^{-E_a/kT} \tag{3.40}$$

where D is the diffusion coefficient; E_a is the activation energy for annealing; k is the Boltzmann constant; and T is the absolute temperature.

The time to diffuse to some extent is inversely proportional to the diffusion coefficient D:

$$t \sim \frac{1}{D} \sim e^{E_a/kT} \tag{3.41}$$

From Eq. (3.41) one can see that the larger the activation energy E_a, the longer time it takes to remove the latent tracks; and the higher the temperature in the annealing device, the shorter will be the time it takes to erase the latent tracks.

C. Arrhenius diagram

Eq. (3.41) can be written as

$$\log t = c + \frac{E_a}{k}\frac{1}{T} \tag{3.42}$$

The relation between log t and $1/T$ can be drawn as a straight line as in Fig. 3.41.

From the slope of the straight line, one can derive the activation energy E_a for track annealing.

It should be noted that the activation energy E_a is related to the degree of annealing, that is, how large a percentage of the tracks are annealed away, or how many

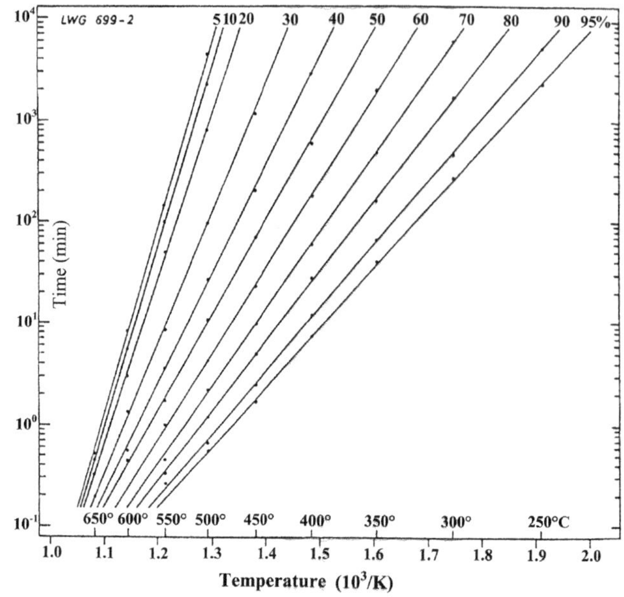

FIG. 3.41 Arrhenius diagram between log(t) and $1/T$. The slope is E_a/k, from which one can determine the activation energy E_a for track annealing. *From Wagner and Van den haute (1992). Reprinted with kind permission from Springer Science + Business Media © 1992.*

of the tracks (percent) are lost in the annealing process on the internal surface of a mineral. Therefore, annealing time t, annealing temperature T, and annealing degree are three parameters; from any two, one can determine the third.

D. Application of track fading and annealing

The track fading phenomenon has two sides or characteristics. In one respect, it can cause problems; and on the other hand, it can provide new information or improve analytic accuracy.

1. Problems resulting from track fading

The problems that may arise from track fading are the following:

a. Track loss increasing with etching temperature and time

Particle tracks would be lost or partly lost, if one etches the detector at an excessively high temperature and an etching time that lasts too long. Usually, for plastic track detectors, the etching temperature should not exceed 70°C in order that the annealing of tracks remains negligible.

b. Track loss increasing with temperature in the experimental environments

The temperature in the experimental environments should not be higher than that allowed by the detector. For example, the temperature for irradiation of plastic track detectors should not be higher than 70°C. If the temperature of the environment is higher than 70°C, one should use muscovite mica track detectors instead of plastic track detectors. Muscovite mica can retain 100% of its tracks up to 450°C for 1 hour (Fleischer et al., 1975).

2. Improving analysis with the aid of track annealing

Track annealing can be used to provide an experimental advantage, as described in the following:

a. Muscovite mica is a fission detector, but it contains spontaneous fission tracks of ^{238}U. By annealing muscovite sheets over 600°C, for more than 1 hour, one can remove the spontaneous tracks from mica. Thus, one can produce very clean detectors without track background.

b. In FTD, the detection efficiency of an external detector is usually different from that of the minerals to be dated. In this situation, one can use the same type of minerals as detectors for uranium determination. The detection efficiency will be equal to the minerals for dating. In this case, one can anneal out all the latent spontaneous fission tracks in one aliquot of the minerals. This aliquot will be used to determine the uranium concentration of the minerals in a nuclear reactor by neutron induced fission measurements. Another aliquot is left for the ^{238}U spontaneous fission number measurement. Both aliquots are mounted, ground, polished and etched in the same condition. The detection efficiencies of the minerals for dating and for uranium determination will be equal to each other, which can be cancelled in the age equation. In this way, the dating precision can be improved.

3. Apparent fission track age and its corrections

FTD is based on the number of ^{238}U spontaneous fission tracks accumulated in a mineral from the formation of the mineral. If the temperature of the geological body containing the mineral was too high to retain fission tracks, the tracks formed in this period would be completely lost. After the geological body cooling down to some temperature (upper limit temperature of partial annealing zone (PAZ) of the mineral), ^{238}U spontaneous fission tracks would be retained, but they would be annealed partially by any higher temperature that existed later on. Only when the temperature cooled down to the lower limit temperature of the PAZ, can the spontaneous fission track be retained completely. The tracks formed in the PAZ would be shorter than the original tracks. This shortening of the tracks would reduce the number of tracks etched on the internal surface of the mineral, consequently reducing the measured age by FTD. The reduced fission track age is named apparent fission track age.

Apparent fission track age can be corrected to true age. There are three methods to make this correction: (1) track length correction method; (2) track diameter correction method; and (3) Plateau-age correction method. For a detailed treatment on these methods, the reader is referred to the books written by Fleischer et al. (1975), Durrani and Bull (1987) and Wagner and Van den haute (1992).

4. Geothermal chronology

^{238}U spontaneous fission-fragment tracks are prints formed in minerals since the fission events occurred. Track annealing by follow-up thermal events make secondary prints on fission tracks. Every track contains plentiful records of the thermal history which the mineral experienced since it was registered. The tracks were registered at different times (year or age). Comparisons, subtractions and differentiations of the records contained in different tracks can retrieve the whole geothermal history from the time when the first spontaneous fission track was created in the minerals. Chambaudet et al. (1993) suggested a model coined as the "Smoke Convection Model" to retrieve the thermal history experienced by a mineral, such as apatite. The authors claimed that it can retrieve the thermal history uniquely. Geothermal chronology has become a hot spot in geology and nuclear tracks in solids. For more information on this field of measurements, the reader is referred to review articles written by Reiners and Brandon (2006), Reiners and Ehlers (2005) and Bernet and Garver (2005).

VIII. Instrumentation

A. Size of latent tracks and etched tracks

The lower limit of the diameter of etchable latent tracks is about 30–40Å (3–4 nm); and the etched track diameter can range from this size to more than 100 μm (Fleischer, 1981). The first instrument to study latent tracks was the transmission electron microscope (TEM) (Silk and Barnes, 1959; Price and Walker, 1962a). The first instrument to study etched tracks was the optical microscope (Young, 1958; Price and Walker, 1962c).

Before the 1980s, there was no way to see the structures of latent and etched tracks in the range below the nanometer (nm). The dream of imaging subnanometer-size tracks was realized since 1982 by the discovery of the scanning tunneling microscope (STM) (Binning et al.,

1982) and since 1986 after the discovery of the atomic force microscope (AFM) (Binning et al., 1986).

B. Optical microscope

The optical microscope can be used to analyze etched tracks in the size range of ~0.5 μm to several hundreds of μm. Optical microscopes are composed of (1) eyepiece (one or two), (2) objective (one or more), (3) stage, (4) condenser (transmitted light and reflected light), and (5) stand.

The accessaries should include: (1) standard ruler, used to calibrate the scales, (2) reticule, which can be put in the eyepiece for measurement of track or field size, (3) reticulate, which is used as a standard area for the measurement of track density (tracks/cm^2), (4) displacement transducer (one, two or three), at least one for the measurement of depth (z axis). Another two are for x and y axis measurements.

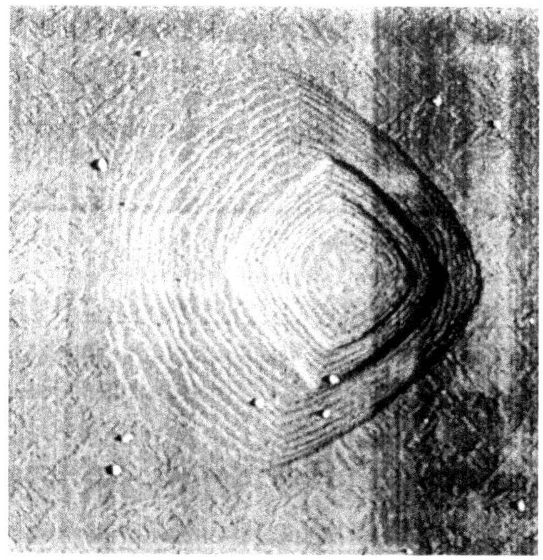

FIG. 3.42 Photograph of an α-recoil track taken by an AFM. This track can be seen with three segments. Each segment was produced by one α-emission. Three sections are consecutively produced by three α-emissions. The ^{238}U nucleus and its progeny can have 8 α-emissions. At most, it may leave 8 segments in the α-recoil track. ^{232}Th nucleus and its progeny can emit 6 α-particles. Their progeny nuclei are ^{228}Ra, ^{224}Ra, ^{220}Rn, ^{216}Po, ^{212}Pb, ^{208}Pb, respectively. At most, it may leave 6 segments in the α-recoil track. *From Price (1993); printed with permission from Elsevier © 1993.*

C. Track image analyser

The composition of a track image analyzer includes (1) optical microscope, with x, y and z controllers; (2) CCD camera; (3) image plate; and (4) computer.

The parameters which can be reported by a track image analyzer are (1) track number; (2) track width, major axis and minor axis; (3) track length; (4) track coordinates (x, y, z); (5) track orientation (Zenith angle θ, azimuth angle φ); (6) area of measurement; (7) areal density of tracks; and (8) gray level of tracks.

D. Electron microscope

The electron microscope is operated in a vacuum. It can be classified into two types. One type is the TEM, which requires a very think solid as a track detector, such as a mica flake. Its resolution usually is about 1 nm−10 μm. Very high voltage and antivibration TEMs can even resolve individual atoms. Another type is the scanning electron microscope (SEM), which can resolve about 2 nm−50 μm. When using a SEM to measure tracks, it is best to have a sputter coater or a vacuum evaporation chamber to prepare to conductive layer on the surface of the etched detector so that the detector surface changes to be conductive to electrons.

E. Scanning tunneling microscope (STM) and atomic force microscope

The STM is used to scan electronic conductive substances; and the atomic force microscope (AFM) can measure insulating solids, which is characteristic of SSNTDs. The above microscopes are capable of resolving individual atoms. Fig. 3.42 shows a photograph of an etched α-recoil etchpit taken by an AFM. In this photograph one can see three segments of track. Each segment represents one emission of α particle. The ^{238}U nucleus and its progeny may have 8 α particle emissions to decay to ^{208}Pb and at the most can create 8 segments in an α-recoil track (Price, 1993). Each segment has several steps. Each step in the photograph is 2 nm in height.

Fig. 3.43 shows a photograph taken by an AFM of a 200 keV Ag etchpit showing 1 nm steps joining pairwise. One nm is the distance between the two neighboring layers of potassium (K) atoms in the structure of muscovite mica. The layers of K atoms are known as easily cleavage planes. These layers separate the tightly bound Si−O−Al structure, i.e., the TOT structure. Here T means tetrahedron the structure within which one Si atom is in the center and three oxygen atoms at the four pinnacles of the tetrahedron. O means octahedron which is composed of Si−O or Al−O. The TOT structure is very rigid. This structure forms the step during etching. The pairing of two steps reflects the etching behavior of mica as explained by Price (1993).

From Figs. 3.42 and 3.43, one can see that the AFM is a unique tool to study the etched track structure and track-etch kinetics.

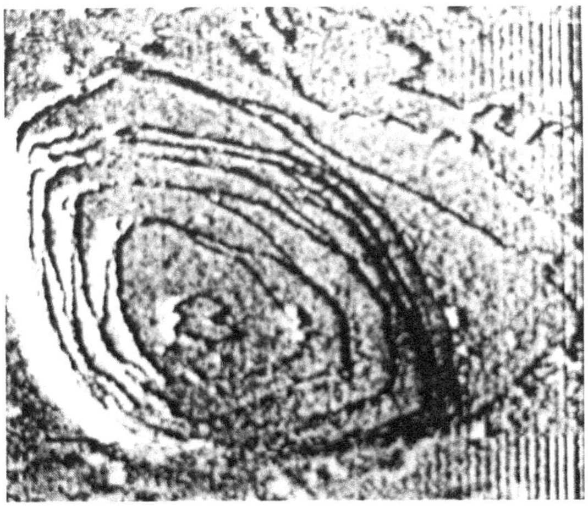

FIG. 3.43 Photograph of an etchpit of a 200 keV Ag track in muscovite mica. The black lines are etched steps bit by the HF etchant. The central part of the etchpit is the tip of the track. In certain directions two steps join together. The height of one single step is 1 nm. The height of the joining step is 2 nm. The pairing of two steps reflects the etching property of muscovite mica by HF etchant. *From Price (1993); printed with permission from Elsevier © 1993.*

F. Spark counter

The spark counter was first devised by Cross and Tommasino (1968), which is a very simple device; and it is a very fast counter of tracks recorded in thin foils of track detectors. The advantages of this device are: (1) It is quite accurate in counting tracks; (2) The track spots sparked in the thin foil can be seen by the naked eye. The spot distribution reflects the particle distribution entered into the detector foil. (3) It is a fast counting device. Usually, it takes several to 10 s to count all the tracks in a piece of track detector. This device is widely used in dosimetry of Rn and neutrons and in autoradiography (Eappen and Mayya, 2004).

For more information, the reader is referred to the original article by Cross and Tommasino (1968), and a book by Durrani and Bull (1987).

Part 2: Applications

I. Introduction

Applications of nuclear track detectors are widespread in science and technology, among which the most important fields are as follows:

(1) Physical sciences including low and high energy physics, astrophysics, cosmic-ray physics, plasma physics, nuclear reactor physics, and elemental analysis.
(2) Earth and planetary sciences including FTD, geothermal chronology, meteorite studies, lunar and solar particles, uranium and petroleum exploration and earthquake prediction.
(3) Life and environmental sciences, which include radiation protection dosimetry, radon and thoron monitoring, neutron dosimetry, and radioactive contamination.
(4) Nanotechnology and material sciences.

This chapter focuses on the most common and newly emerged fields of application.

II. Physical sciences and nuclear technology

Nuclear science is the first field of applications of etch track detectors (Price and Walker, 1962e). A large number of examples of research in the physical sciences with etch track detectors were reviewed in the journal articles and monographs (Fleischer et al., 1965a, 1975; Durrani and Bull, 1987). Some new research and results will be introduced below.

A. Cluster radioactivities

Since the 1980s, the application of SSNTDs to heavy ion detection has gained great success. One of the extremely exciting achievements is a series of discoveries in heavy ion emission from nuclei called cluster radioactivity, or heavy ion decay of nuclei. The first discovery of this type of nuclear decay ($^{223}Ra \rightarrow {}^{14}C + {}^{209}Pb$) was made with a silicon $\Delta E{-}E$ detector (Rose and Jones, 1984); however, for further discoveries in this field, the silicon detector cannot work well because the branching ratio of heavy ion decay to α decay is too low ($<10^{-10}$). The SSNTD can be used to find and identify the tracks of the rare heavy ions in a very intensive α track background. The branching ratio of ^{14}C decay from ^{223}Ra is $B = 3.98 \times 10^{-10}$, which already reaches the bare limit of detection of the silicon detector. For other nuclei with branching ratios below this limit, the nuclear track detector is the only capable detector for the discovery of this type of decay.

Up to now, more than 20 parent nuclei with $Z = 87{-}96$ and $Z = 56$ were discovered having cluster radioactivity. They are $Z = 87$ (^{221}Fr), $Z = 88$ ($^{221,\ 222,\ 223,\ 224,\ 226}Ra$), $Z = 89$ (^{225}Ac), $Z = 90$ ($^{228,\ 230,\ 232}Th$), $Z = 91$ (^{231}Pa), $Z = 92$ ($^{232,\ 233,\ 234,\ 235}U$), $Z = 93$ (^{237}Np), $Z = 94$ ($^{236,\ 238}Pu$), $Z = 95$ (^{241}Am), $Z = 96$ (^{242}Cm), and $Z = 56$ (^{114}Ba).

Up to now, 11 species of emitted heavy ions with $Z = 6$, 8, 9, 10, 12 and 14 were discovered in this type of nuclear decay. The emitted heavy ions are: $Z = 6$($^{12,\ 14}C$), $Z = 8$ (^{20}O), $Z = 9$ (^{23}F), $Z = 10$ ($^{24,\ 25,\ 26}Ne$), $Z = 12$ ($^{28,\ 30}Mg$) and $Z = 14$ ($^{32,\ 34}Si$).

In total, 11 species of daughter nuclei with $Z = 80$, 81, 82, 83 and 50 were formed in the decay. They are: $Z = 80$ ($^{205,\ 206}Hg$), $Z = 81$ (^{207}Tl), $Z = 82$ ($^{207,\ 208,\ 209,\ 210,\ 211,\ 212}Pb$), $Z = 83$ (^{211}Bi) and $Z = 50$ (^{102}Sn). The daughters are the nuclei with a magic number in Z (82, 50) or close to a magic number in Z (80, 81 and 83 are close to the magic number 82) (Price, 1993, 1994, 2005; Guglielmetti et al., 1995; Tretyakova et al., 1992). For a more detailed treatment of cluster radioactivity and magic numbers, the reader may refer to Chapter 1, Volume 1 of this book.

The half-life $T_{1/2}$ and branching ratio B measured with nuclear track detectors for heavy ion emission modes are listed in Table 3.6.

B. Heavy ion interactions

1. Relativistic projectile fragmentation

Since the 1980s, another exciting application of SSNTDs in nuclear physics is the identification and study of heavy ions resulting from high- and relativistic-energy fragmentation of nuclei carried out in Berkeley (Bevalac), Brookhaven and CERN.

These studies have opened a new branch of application in high-energy physics, astrophysics and cosmic rays (He and Price, 1995a,b; Price, 2005, 2008; Cecchini and Patrizii, 2008; Manzoor et al., 2000; Grabez and Dragic, 2008).

It has been calibrated that the CR-39 has a very high-charge resolution $\sigma_Z \leq 0.5e/\sqrt{n}$ for relativistic nuclei of $10 \leq Z \leq 65$ by measuring etch pit diameters, where n is the number of successive etch pits measured along the trajectory of a nucleus, as illustrated in Fig. 3.44 (Salamon et al., 1985). The value of n can be larger than 10 for CR-39 with a detector thickness of 0.72 mm for each sheet. The charge resolution is high enough to assign the charges of fission fragments.

Owing to the high charge resolution of CR-39, relativistic projectile fission of 1 AGeV Au nuclei in CR-39 has been studied (Guo et al., 1984b). They observed that some Au projectiles were broken into two fragments in a CR-39 stack composed of 80 sheets each in size $15 \times 15 \times 0.072\ cm^3$ doped with 1% dioctyl phthalate. After a calibration of track diameter D to charge Z, they could determine the charge Z of each fragment in the fission events. By measuring the zenith and azimuth angles of each fission fragment with respect to the direction of the projectiles, they could determine the total energy release of each event and the interaction nature between the projectile and the target nucleus C, H or O in the CR-39. They showed that the energy release in projectile fission is consistent with that of low-energy fission and the transverse momentum transferred to fissioning species is consistent with a peripheral interaction. The mass distribution of fission fragments is very broad as that of fission of nuclei with $28 \leq Z \leq 83$ by 600 MeV and 1 GeV protons (Guo et al., 1986a).

The above example shows that CR-39 is one of the best detectors to possess the highest charge resolution for fission fragments in the relativistic-energy region.

2. Sequential fission after inelastic collisions

The crystal structure of muscovite mica is anisotropic. The bulk etch rate by hydrofluoric acid parallel to the cleavage surface $V_{//}$ is much larger than that perpendicular to the cleavage $V_{\perp}$, that is $V_{//} \gg V_{\perp}$, and the track etch rate V_T for fission fragments is several thousand times larger than $V_{//}$. These properties make the ranges of fission fragments stable after etching. Taking this advantage of muscovite mica, Brandt and co-workers opened up a new field of research on sequential fission of heavy nuclei, such as U + U, U + Pb, breaking into 3, 4, 5 pronged fission events in the low-energy region (Brandt, 1980; Vater et al., 1977, 1986; Gottschalk et al., 1979, 1983; Khan et al., 1980; Qureshi et al., 1994, 1998a, 1998b, 2005; Khan, 1985; Shahzad et al., 1999).

In this field of research, the energy and mass of fission fragments cannot be derived from the shapes of tips, but by the ranges (lengths), zenith and azimuth angles of the fragment tracks and by mass and energy conservations (Gottschalk et al., 1983).

C. Nuclear fission and neutron physics

1. Nuclear fission

In nuclear fission, two fission fragments are emitted in opposite directions (A small number of fission events (<1/300) release a third particle, called ternary fission or even release a fourth particle, quaternary fission, which is outside the scope of this chapter). To which directions (orientations) the two fission fragments fly away from the point of fission is a primitive question. Experiments show that fission events induced by thermal neutrons are isotropic. Other fission events are all anisotropic, even in the center of mass system. The emitted fission fragments have some kind of preference in directions. In order to clarify the phenomenon of the anisotropic distribution, many experiments and theoretical studies have been carried out.

The SSNTD has been used in the measurements of angular distribution of fission fragments emitted from fission. The angular distribution of fission fragments is dominated by the properties of the transit state at fission barrier of fissioning nuclei. In the transit state, fission occurs only through one or a few channels at low energy. The characteristics of the transit state will be reflected in the angular distribution of fission fragments. SSNTDs, such as muscovite mica and polycarbonate (Lexan and Makrofol) can cover very large angles from ~0° to ~180° and can withstand the bombardments of intense light beam particles, such as γ-rays, neutrons, protons, α particles and nuclei even heavier than neon with higher energies. These detectors can clearly show the tracks of fission fragments

TABLE 3.6 Cluster radioactivity nuclei discovered or confirmed by solid-state nuclear track detectors.

Parent nucleus	Heavy ion emitted	Daughter nucleus	Ek (MeV)	Measured half life logT (sec)	Branch ratio -logB
^{221}Fr	^{14}C	^{207}Tl	29.28	14.5 ± 0.12	12.00 ± 0.12
^{221}Ra	^{14}C	^{207}Pb	30.34	13.0 ± 0.2	11.7 ± 0.2
^{222}Ra	^{14}C	^{208}Pb	30.97	11.0 ± 0.06	9.4 ± 0.06
^{223}Ra	^{14}C	^{209}Pb	29.85	15.2 ± 0.05	9.2 ± 0.05
^{224}Ra	^{14}C	^{210}Pb	28.63	15.8 ± 0.12	10.3 ± 0.12
^{225}Ac	^{14}C	^{211}Bi	28.57	17.16 ± 0.06	11.2 ± 0.06
^{226}Ra	^{14}C	^{212}Pb	26.46	21.3 ± 0.2	10.6 ± 0.2
^{228}Th	^{20}O	^{208}Pb	44.72	20.7 ± 0.08	12.9 ± 0.08
^{231}Pa	^{23}F	^{208}Pb	46.68	26	14
^{230}Th	^{24}Ne	^{206}Hg	51.75	24.6 ± 0.07	12.3 ± 0.07
^{232}Th	^{26}Ne	^{206}Hg	49.7	> 27.9	> 10.3
^{231}Pa	^{24}Ne	^{207}Tl	54.14	22.9 ± 0.05	10.9 ± 0.05
^{232}U	^{24}Ne	^{208}Pb	55.86	20.5 ± 0.03	11.1 ± 0.03
^{233}U ^{233}U	^{24}Ne ^{25}Ne	^{209}Pb ^{208}Pb	54.27 54.32	24.8 ± 0.03	12.1 ± 0.03
^{234}U ^{234}U	^{24}Ne ^{26}Ne	^{210}Pb ^{208}Pb	52.81 52.87	25.9 ± 0.2	13.0 ± 0.2
^{235}U ^{235}U	^{24}Ne ^{25}Ne	^{211}Pb ^{210}Pb	51.50 51.68	> 27.4	> 11.1
^{233}U	^{28}Mg	^{205}Hg	65.32	> 27.8	> 15.1
^{234}U	^{28}Mg	^{206}Hg	65.26	25.7 ± 0.2	12.8 ± 0.2
^{237}Np	^{30}Mg	^{207}Tl	65.52	> 27.4	> 13.6
^{236}Pu	^{28}Mg	^{208}Pb	70.22	21.7 ± 0.3	13.7 ± 0.3
^{238}Pu ^{238}Pu	^{30}Mg ^{28}Mg	^{208}Pb ^{210}Pb	67.00 67.32	25.7 ± 0.25	16.3 ± 0.25
^{238}Pu	^{32}Si	^{206}Hg	78.95	25.3 ± 0.16	15.9 ± 0.16
^{241}Am	^{34}Si	^{207}Tl	80.60	> 25.3	> 15.1
^{242}Cm	^{34}Si	^{208}Pb	82.88	> 21.5	> 14.4
^{114}Ba	^{12}C	^{102}Sn	~14.4	$T_c \geqslant 1.1 \times 10^3$ s	$T_\alpha \geqslant 1.2 \times 10^2$ s

and without background from the light particles of the beams. Because of the importance to fission physics, researchers have carried out this research with SSNTDs (Vandenbosch and Huizenga, 1973).

A thin fission source in a small area supported by a thin foil is placed on the beam of incident particles in a vacuum chamber. A mica or polycarbonate strip is bent into a circular arc and supported by a frame of aluminum and placed around the source at a distance and stretched to a large angle from ~0° to 90° with respect to the beam direction. Fission fragments emitted from the source are recorded by the track detector. Measuring the track density in the

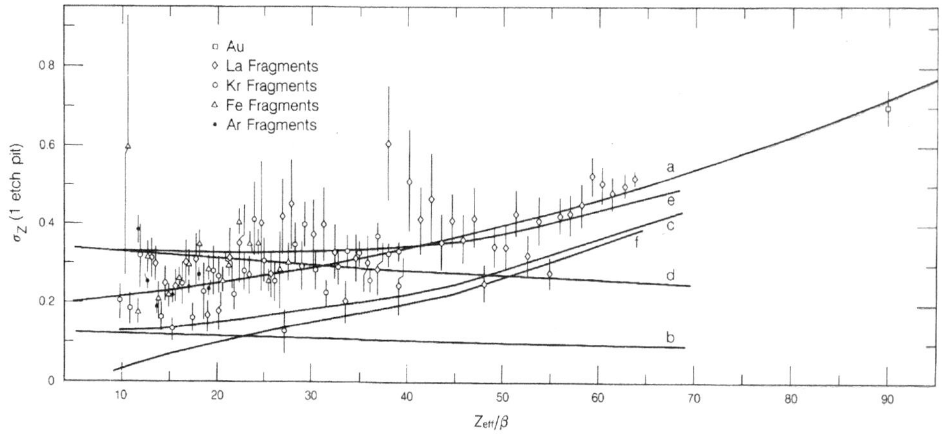

FIG. 3.44 Measured charge resolution σ_z for a single etch pit ($n = 1$) as a function of Z/β in CR-39. The curve a is a weighted least squares quadratic fit to the data. *From Salamon et al. (1985); printed with permission from Elsevier © 1985.*

detector for a certain range of angles can provide the angular distribution of fission fragments in this special condition (Zhou et al., 1981; Guo, 1982b).

2. Neutron physics

The neutron is an uncharged particle. It cannot be recorded by direct ionization in detecting material, but it can be detected through its recoil nuclei or nuclear reaction products, which are energetic charged particles. Nuclear fission is the first reaction used for the detection of neutrons by SSNTD (Walker et al., 1963). ^{235}U, ^{237}Np, ^{238}U and ^{232}Th are often used as the fissile materials. Fig. 3.45 shows their relationships between fission cross sections and neutron energy.

The threshold energies for fission are, respectively: ^{235}U (0), ^{237}Np (~0.1 MeV), ^{238}U (1.0 MeV), and ^{232}Th

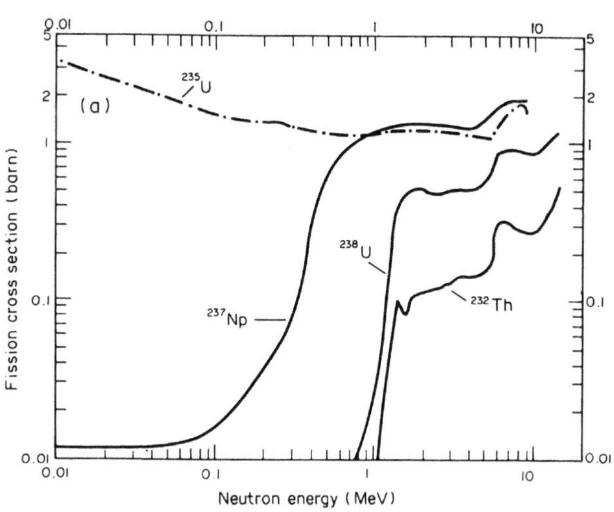

FIG. 3.45 Fission cross-sections versus neutron energy of ^{235}U, ^{237}Np, ^{238}U and ^{232}Th. *From Durrani and Bull (1987), reprinted with permission from Elsevier © 1987.*

(1.17 MeV). The accurate data of the cross-sections can be found in BNL 325, Third Edition, Volume II (1976). The fissile sources of ^{235}U, ^{237}Np, and ^{238}U can be electrolytically deposited on metal foils. The ^{232}Th source can be made by pressure in a mold from ThO_2 powder (Guo et al., 1976a). The effective thickness of the pressed ThO_2 tablet for recording fission fragments with muscovite mica covering on its surface has been calibrated (Guo et al., 1976a) to be: $\frac{1}{2}mR\cos^2\theta_c = 2.32 \pm 0.07$ mgTh/cm^2, where m is the mass of ThO_2 in unit volume of tablet, R is the average range of fission fragments in the tablet, and θ_c is the critical angle of muscovite mica for recording fission fragments (3°41′, see Guo et al., 1982).

The above fissile material mica sandwiches can be used separately to measure the flux or fluence of monoenergetic fast neutrons from accelerator neutron sources, or measure the neutron dose. Four sandwiches of detectors can be put together to measure the neutron spectrum by dividing the whole range of energy into intervals (Durrani and Bull, 1987).

D. Plasma physics

1. Laser acceleration

Since the year 2000, table-top accelerators of proton and heavy ion beams has become a focus of interest in scientific research. It is expected that table-top accelerators based on laser-plasma interactions might replace the current heavy-ion accelerators, which are large in size and length up to more than several tens of meters or even more than several kilometers. A high-intensity laser beam could produce a > 50 MeV proton beam. This result has attracted worldwide researchers to join in this new research field. The SSNTD, CR-39, has become the most important tool to study this phenomenon. The possible mechanism of proton acceleration by laser-plasma interaction is called Target Normal

Sheath Acceleration (TNSA) whereby a high intensity laser beam hits on a solid target (foil), the prepulse of the laser (the small pulse before the main pulse of laser shoot) creates a thin plasma layer at the surface of the foil. Then, the intense part of the pulse (the main pulse) of the laser shoot interacts with the thin plasma layer, which accelerates the electrons toward the foil by the ponderomotive force. Finally, the electron beam reaches the rear surface (the opposite surface to the laser irradiation) and creates a strong electrostatic field which first ionizes and then accelerates protons to high energy. When the electrons have passed the foil, they act as a sheath with negative charge. The electron sheath moves very quickly. It attracts the positive protons to go to the same direction (normal to the surface of the foil). Following the electron sheath, the protons can reach a very high energy depending on the parameters of the laser and the foil. The target foil may be varied, and most accelerated particles are protons. These protons come from H_2O or a hydrocarbon adsorbed on the surface of the foil.

The proton beam accelerated by laser-plasma interaction has the following properties: (1) The direction of the proton beam is normal to the surface of the target. The beam spread is less than 20°. It is nearly a collimated beam of protons. (2) The maximum proton energy is related to the laser intensity. Therefore, the proton energy is controllable by the laser beam. (3) The intensity of proton beams can reach 10^9–10^{13} protons per pulse. The usual frequency is about 10 pulses per second. It is strong enough for many applications. (4) The size of the beam source is about 10 μm, which is much smaller than that of the current heavy ion accelerators.

Due to the above properties, laser acceleration is expected to feasibly replace some types of current big accelerators, to work as proton beam medical accelerators or spallation neutron sources.

In laser acceleration, the proton beam duration is less than 10^{-12} seconds. In this very short time, more than 10^9 protons will reach the detector. Electronic detectors have a dead time. It is difficult to record a single proton and analyze its energy. In this case, the CR-39 is the best detector for recording a single proton and obtaining its energy.

Many laboratories are now carrying out laser acceleration studies with CR-39 track detectors (Snavely et al., 2000; Ceccotti et al., 2007; Duan et al., 2009, 2010; Li et al., 2007; McKenna et al., 2005; Hicks et al., 2001; Fukuda et al., 2013; Kanasaki et al., 2012, 2013, 2015, 2016; Macchi et al., 2013).

2. *Inertial confinement fusion*

In inertial confinement fusion (ICF) induced by pulsed laser with high intensity and short time, the following primary fusion reactions in the capsules will take place, in which α particles are produced:

$$\mathrm{D} + \mathrm{T} \rightarrow \alpha\ [3.5\ \mathrm{MeV}] + \mathrm{n}\ [14.1\ \mathrm{MeV}]$$

$$\mathrm{T} + \mathrm{T} \rightarrow \alpha\ [\leq 6.6\ \mathrm{MeV}] + 2\mathrm{n}\ [\leq 10.6\ \mathrm{MeV}]$$

$$\mathrm{D} + {}^3\mathrm{He} \rightarrow \alpha\ [3.6\ \mathrm{MeV}] + \mathrm{p}\ [14.7\ \mathrm{MeV}]$$

$$\mathrm{T} + {}^3\mathrm{He} \rightarrow \alpha\ [4.8\ \mathrm{MeV}] + \mathrm{D}\ [9.5\ \mathrm{MeV}]$$

$$\mathrm{T} + {}^3\mathrm{He} \rightarrow \alpha + \mathrm{p} + \mathrm{n} + 12.1\ \mathrm{MeV}$$

where D and T are the original compositions of the fuel in the capsule, ^{3}He is the direct product of D + D → ^{3}He [0.82 MeV] + n [2.45 MeV] reaction.

The best detector for recording α particles in these reactions is CR-39 nuclear track detector (Seguin et al., 2003). The first actual instrument for measuring the charge-particle spectra in these fusion reactions was designed with electronic detectors; but it was eventually implemented with CR-39 detectors. The reasons for the big change of detector are due to the advantages of CR-39 detector, which are the following: (1) The detection efficiency for the formed particles is high (about 100% for normal incidence); (2) The area coverage of CR-39 detector can be large and at various angles (~2π geometry being possible if necessary); (3) The energy resolution is higher by CR-39 than electronic detectors; (4) The immunity of CR-39 is superior to electromagnetic radiations (*x*-ray, γ-ray and electric transients accompanied by shooting of the laser); (5) There is no overloading for less than ~10^7cm^{-2} particles and no dead time compared with electronic detectors; and (6) The ability to distinguish the types and energies of individual particles. CR-39 detector is used in many laser-plasma fusion laboratories throughout the world (Kacenjar et al., 1982; Szydlowski et al., 2009).

E. Astrophysics and cosmic rays

Astrophysics and cosmic ray research is the oldest field of application since the discovery of SSNTDs (Price and Walker, 1962e). Great achievements were made in the past 50 years with SSNTDs in research on the following:

(1) Galactic cosmic rays
(2) Solar particles
(3) Ancient cosmic rays
(4) Search for super-heavy elements in cosmic rays
(5) Elementary and exotic particles

These studies with etch track detectors have been reviewed in the monographs of Fleischer et al. (1975) and Durrani and Bull (1987). More recent achievements are conveyed by the following authors: O'Sullivan (1995); Zhou et al. (2009); Doke (2008); Westphal et al. (1998);

Hasebe et al. (2009); Yasuda et al. (2008); Weaver, 2001; Weaver and Westphal, 2002; Price (2005, 2008).

F. Nuclear technology

1. Nuclear reactor physics

The basic physical processes in nuclear reactors are neutron-induced fission and the consequent fission release of neutrons. These processes continue as chain reactions. To study neutron number, flux, spatial distribution, and energy spectrum is one part of the work in reactor physics. SSNTDs are used in these studies.

a. Determination of neutron temperature

In a steady-state operation of a nuclear reactor, neutrons and other molecules in the reactor are in thermal equilibrium. This state is described by temperature as in thermodynamics. The temperature is a decisive factor for the neutron energy spectrum in the thermal energy region. The neutron energy spectrum is considered to follow Maxwell energy distribution.

The principle of the determination of neutron temperature in a nuclear reactor is based on the Westcott equation:

$$\widehat{\sigma} = \sigma_0\left(g + r\sqrt{\frac{T_n}{T_0}}S_0\right) \tag{3.43}$$

where $\widehat{\sigma}$ is the effective fission cross-section of a given fissile nucleus for neutrons with velocity of 2200 m/s; σ_0 is the fission cross section of the given nucleus for neutrons with velocity of 2200 m/s (For ^{235}U, $\sigma_0 = 582.6$ b; for ^{239}Pu, $\sigma_0 = 748.1$ b); g is a correction factor for non $1/v$ fission cross section of the given nucleus in Maxwell neutron spectrum; $r\sqrt{T_n/T_0}$ is the relative ratio of epithermal neutrons (epithermal index); S_0 is the correction factor for resonance absorption of the nucleus (resonance parameter). The variables g and S_0 are functions of neutron temperature T_n.

The track number N_T recorded on a track detector is described by equation

$$N_T = N\widehat{\sigma}\phi_0\varepsilon \tag{3.44}$$

where N is the number of the specific nuclei of fissile species; ϕ_0 is the integrated neutron flux (fluence) equivalent to 2200 m/s; and ε is the detection efficiency of the detector for fission fragments (see Eq. 3.12);

Substituting Eq. (3.43) into Eq. (3.44), gives

$$N_T = N\sigma_0\left(g + r\sqrt{\frac{T_n}{T_0}}S_0\right)\phi_0\varepsilon \tag{3.45}$$

In Eq. (3.45) for a combination of fissile species and a track detector, such as ^{235}U and mica, N, σ_0 and ε are known, and N_T can be measured by track counting; only ϕ_0 and the variable T_n of g and S_0 are unknown. If one uses a second fissile species, such as ^{239}Pu plus mica, one has two equations and two unknowns. One can solve the equations to find ϕ_0 and T_n where T_n is the neutron temperature in a nuclear reactor. From T_n one can calculate the neutron energy distribution from Maxwell equation (Liu and Su, 1971).

The advantages of the track detector for the measurement of the temperature of neutrons in nuclear reactor are the following:

(1) The measurements are simple and convenient to perform.
(2) One can determine small disturbances to the nuclear reactor and to the neutron energy spectrum.
(3) The neutron fluence irradiated to the fissile materials can be obtained as a by-product of the measurements. One does not need to measure the neutron fluence with other techniques.
(4) If one uses thin sources of both fissile materials, the detection efficiency ε for both sources will be canceled in the calculation. The uncertainty of the final result T_n will not be affected by ε.
(5) No background effects come from α-, β- or γ-rays.

The main sources of uncertainty are track counting, numbers of the atoms of the two fissile materials and the parameters of g, r, and S_0 as functions of the two species of nuclei.

The determination shows that the neutron temperatures at different positions in the nuclear reactor are different. The temperature in the core region is higher than that in the thermal column.

b. Determination of fast fission factor in nuclear reactors

In thermal neutron reactors, the ^{235}U nucleus absorbs a thermal neutron and goes to fission. On the average, ν_5 neutrons will be produced. These neutrons in the beginning are fast neutrons with an average energy about 2 MeV. The fast neutrons have a probability to be absorbed by ^{238}U nuclei and the formed compound nuclei probably go to fission and release on the average ν_8 neutrons, and the new neutrons may induce ^{238}U nuclei to go to fission again. As a result of the above process, there will be in total $\varepsilon\ \nu_5$ neutrons produced ($\varepsilon > 1$) where ε is called the fast neutron fission factor or fast fission factor. In the design of thermal neutron reactors, the knowledge of the fast fission factor is of practical importance. For this purpose, Shi and Li (2001) have developed a method for the determination of the fast fission factor ε with SSNTDs. The procedure is the following:

Two fission sources are electrolytically deposited on two Ni disks, one of which is natural uranium, another is enriched uranium. The thicknesses of uranium are 200 μg/cm^2 and 20 μg/cm^2 for the natural and enriched uranium, respectively; both are thinner than the critical thickness (See Section III.C.2). The Ni disk is 30 μm in thickness. The two uranium fission sources were covered with SSNTD sheets. The sizes of the sources and track detectors were the same as the cross section of the fuel element. The two fission sources and their detectors were packed together back to back, and inserted into the fuel element assembly. After irradiation in a heavy water zero power reactor (HWZPR), the track numbers recorded on the track detectors were counted. From the two track numbers and the atom numbers of ^{235}U and ^{238}U in the natural uranium source, enriched uranium source and fuel element, they obtained the fast fission factor ε (= 1.050) which is in agreement with the theoretical calculation (1.071). The detailed formulations to calculate ε from the above-mentioned data of measurements are provided by Shi and Li, 2001.

c. Measurement of reactor fission rate and reactor power by track detector

The decisive factor of nuclear reactor power is how many fission events occur in the whole reactor in 1 s, that is, the fission rate in the whole reactor. One way of measuring nuclear reactor power is to determine the fission rate in the whole nuclear reactor. Basically, it is difficult to measure fission rate in the whole reactor with conventional detectors. However, with the advent of the SSNTD, the measurement of fission rate in the whole reactor becomes quite easy. Li et al. (2001) sandwiched track detector foils on the fuel elements, which are regularly at different heights and different radius in the core of a miniature neutron source reactor. In total, 80 track foils were collocated. The distribution of fission rate at each place as well as in the whole volume of reactor is obtained. By this method, they obtained the reactor power (29.9 KW), which was in agreement with that of the thermal-hydraulic method (30.9 KW) and Gold foil activation method (27.7 KW) (Li et al., 2001). The conversion factor from the whole fission rate F to reactor power P is P_0, $P = P_0F$ where $P_0 = 0.321 \times 10^{-10}$ Ws/fission = 200 MeV/fission. The advantages of the track detector technique for measuring reactor fission rate are by its absolute nature, simplicity, and applicability at low reactor power.

2. Accelerator-driven subcritical reactors

The accelerator-driven subcritical reactor (ADS) is a new type of proposed nuclear reactor. It has the following characteristics:

(1) A subcritical nuclear reactor. It is stable and without any obvious nuclear reactions in it.
(2) Reaching critical condition only when an accelerator provides enough neutrons to it. When the accelerator stops, the criticality vanishes. It is a controllable reactor without any potential of a Chernobyl-type accident.
(3) Generation of electricity from nuclear fission. It will be a nuclear power plant.
(4) A breeder reactor. ^{238}U nuclei after capture of fast neutrons may turn to ^{239}Pu nuclei.
(5) Transmuting long-lived isotopes into stable or short-lived nuclides.

The ADS was first proposed by Bowman et al. (1992); and Rubbia et al. (1997) gave a clearer explanation of the reactor. Up to now, nobody knows what target materials and beam particles of the accelerator should be used to produce neutrons, what are the adequate fuel elements to release fission energy, how to exchange heat from the core region to the coolant, what is the appropriate structure for the whole reactor. A few prototypes have been built. Some of them are quite simple, including special target material and size, special beam and energy, special reactor core and blanket (Brandt et al., 2008; Hashemi-Nezhad et al., 2008).

The SSNTD, being thin pieces, small in size, nearly no disturbance to neutron fields, easy to insert into gaps in the testing reactors, being able to measure neutron flux, energy and distributions in three dimensions, has been applied to studies on accelerator-driven critical reactors. Zamani et al. (2008) used $^6Li_2B_4O_7$ + LR-115, CR-39 and CR-39 covered with Cd foil to measure thermal-epithermal and intermediate-fast neutrons. Adloff et al. (1999) investigated the production of thermal neutrons with thin ^{235}U targets on Lexan track detectors and ^{232}Th targets on Lexan detectors for neutron energies $E_n > 2$ MeV. Dwivedi et al. (1993) used gold foils and track detectors to measure neutrons of energy $E_n > 30$ MeV (Wan et al., 2001). Zhuk et al. (2008) and Perelygin et al. (2001) used Pb foils and mica and PET foil to measure proton beam profiles and neutrons of energy $E_n > 30$ MeV. Hashemi-Nezhad et al. (2008) used Pb foils and mica to measure neutrons and compared the results with Monte Carlo predictions using MCNPX code. Guo et al. (2001) used CR-39 sheets to measure neutrons and come to conclusions that the Hg target is disadvantageous compared to U(Pb) and Pb targets for neutron yield. In order to obtain a uniform field of neutrons in an adequate target size, the energy of protons from accelerators should not be lower than 1 GeV for ADS developments.

3. Nuclear forensic analysis and nuclear safeguards

a. Nuclear forensics

For the requirements of arms control and nonproliferation of nuclear weapons, a new research field gave birth in the last decade of the 20th century. It was named as strengthened nuclear safeguards (Donohue, 2002). In this field of science, environmental samples within and outside nuclear facilities are collected to disclose undeclared nuclear material and expose undeclared nuclear activities.

Nuclear materials for weapons are mainly ^{235}U and ^{239}Pu. The abundance of $^{235}U \geq 90\%$ and the ratio of $^{240}Pu/^{239}Pu \leq 6\%$ are requirements for nuclear weapons. Lower abundances of $^{235}U \leq 5\%$ and the ratio of $^{240}Pu/^{239}Pu \geq 20\%$ are for commercial uses. Enriched ^{234}U comes from a uranium enriching plant. The existence of enriched ^{234}U particles indicates that there is a uranium enriching plant. Existence of ^{236}U indicates the existence of a nuclear fuel reprocessing plant. The analysis of isotopic compositions of uranium and plutonium in microscopic single particles has become a decisive method in nuclear safeguards and nuclear forensic analysis to disclose the undeclared activities. See Chapter 13, Volume 2 of this book for a more comprehensive treatment of nuclear forensics.

b. Fission track technique

The fission track (FT) technique has been regarded as a feasible method to search for, locate, and transfer uranium-bearing particles from swipe samples. These particles are transferred to a thermal ionization mass spectrometer (TIMS), secondary ion mass spectrometer (SIMS), or inductively coupled plasma mass spectrometer (ICP-MS) for analysis of isotopic abundances.

The fission track (FT) technique includes the following steps:

(1) **Picking up uranium-bearing particles**. Swipe particles are transferred onto a Lexan polycarbonate detector of size $2 \times 2\ cm^2$. The Lexan detector with the particles is covered with a newly cleaved clean surface of mica detector. The sandwich formed of Lexan-mica is marked with a needle at three points (A, B, and C) **coincident** to each other on the outer surfaces of the Lexan and mica. The marked sandwich is tightly pressed together.

(2) **Irradiation**. The sandwich of Lexan-mica is irradiated with thermal neutrons in a nuclear reactor to a neutron fluence about $10^{15}-10^{16}\ cm^{-2}$. The neutrons induce fission of ^{235}U in the swipe particles. The fission fragments are recorded by the Lexan and mica track detectors.

(3) **Etching.** After irradiation in the nuclear reactor, the mica detector is etched in HF and track stars will be seen under an optical microscope.

(4) **Searching for and locating uranium-bearing particles**. Put the Lexan sheet on the mica back to back with the particles on the upper surface and the marks A, B, and C on both detectors overlapping each other. The particle on the upper surface of the Lexan coincident with the track star on the bottom surface of mica will be the uranium particle which produces the star. In this way, uranium-bearing particles are found. The coordinates (X_U, Y_U) of each particle as well as the coordinates (X_A, Y_A), (X_B, Y_B) and (X_C, Y_C) of the marks are recorded by the instruments connected with the stage of the microscope.

(5) **Relocating particles and isotope analysis**. After coating with carbon or aluminum on the Lexan sheet by sputtering or vacuum evaporation, the Lexan sheet is placed under a scanning electron microscope (SEM) or SIMS. By a mathematic formulation of relocation (see below) one can easily and quickly find the uranium-bearing particle. Further analyses on the elemental and isotopic composition of the particle can be carried out.

(6) If TIMS is used to analyze the isotopic composition of the particle, one should use a micromanipulator (see Fig. 3.46) to pick up the particle and transfer it to the TIMS rhenium filament for analysis.

(7) If ICP-MS is used for isotopic analysis, one should prepare a proper solution dissolving the particle for the measurement. A micromanipulator is also needed to transfer the particle to the solvent.

c. Mathematical formula of relocation

For relocation of a uranium-bearing microparticle on the surface of a track detector or on other substrate such as a graphite planchet, one needs two or three reference points such as the marks made on the Lexan or mica. The reference points must be very easy to be found in the instruments. For locating and relocating the particle, one should first record the coordinates of the reference points A (X_A, Y_A), B (X_B,

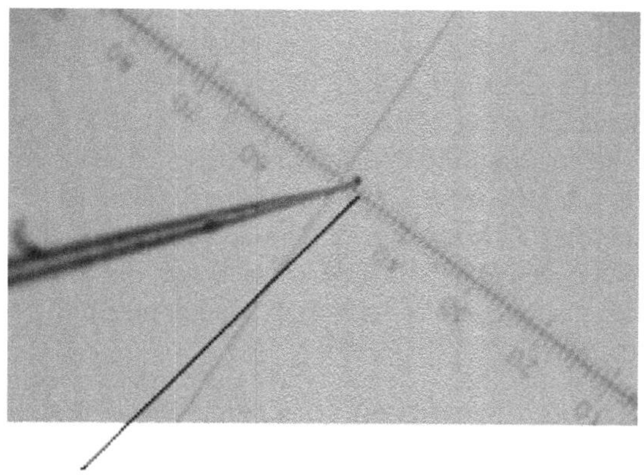

FIG. 3.46 Microphotograph of glass needle sticking a uranium-bearing particle at its tip by static electricity. The tip diameter of the needle is about 1 μm. The movement of the needle is controlled by micromanipulator (NARISHIGE, TYPE MO-202D). The precision of movement of the micromanipulator is 0.5 μm. The glass needle was made by using an instrument (Puller, type PC-10, NARISHIGE).

Y_B), and C (X_C, Y_C) and the coordinates of the uranium-bearing particle U (X_U, Y_U) in the first instrument. After putting the sample or planchet in the second instrument, one should first find the reference points and record the coordinates A (X'_A, Y'_A) and B (X'_B, Y'_B). The uranium-bearing particle can be found easily at the position U (X'_U, Y'_U) calculated by the following formula (Guo et al., 2008):

$$X'_U = \overline{AU} \times R \times \cos\phi' + X'_A \quad (3.46)$$

$$Y'_U = \overline{AU} \times R \times \sin\phi' + Y'_A \quad (3.47)$$

where $\overline{AU}$ is the distance between A and U, R is the ratio of the readings of the same distance in the second instrument to that in the first instrument:

$$R = \frac{\sqrt{(X'_B - X'_A)^2 + (Y'_B - Y'_A)^2}}{\sqrt{(X_B - X_A)^2 + (Y_B - Y_A)^2}} \quad (3.48)$$

ϕ' is the angle between AU and the X'-axis in the second instrument:

$$\varphi' = \alpha' + \theta' \quad (3.49)$$

where α' and θ' are the angles between AU and AB and between AB and the X'-axis in the second instrument, respectively.

$$\alpha' = \alpha \quad (3.50)$$

where α is the angle between AU and AB in the first instrument.

$$\theta' = \begin{cases} Arctg\dfrac{Y'_B - Y'_A}{X'_B - X'_A}, & X'_B > X'_A \\ Arctg\dfrac{Y'_B - Y'_A}{X'_B - X'_A} + \pi, & X'_B < X'_A \\ \dfrac{\pi}{2}, & X'_B = X'_A, Y'_B > Y'_A \\ -\dfrac{\pi}{2}, & X'_B = X'_A, Y'_B < Y'_A. \end{cases} \quad (3.52)$$

In other words, recording the coordinate of two reference points and the particle in the first instrument and the coordinates of the two reference points in the second instrument, one can immediately calculate the coordinates of the particle in the second instrument. If one uses three or more reference points, such as A, B, and C, every two reference points give a pair of coordinates (X'_U, Y'_U) of the particle. Averaging all the three coordinates X'_U as well as the three Y'_U, one can achieve better accuracy of the coordinates in the second instrument.

Guo et al. (2008) made reference points on a graphite planchet by a using NTM having a pore diameter of 10 μm and simulated particles having the same diameter. A relocation measurement in the second instrument (SEM, JEOL 6360 LV) showed that the relocation accuracy was (5.8 ± 2.6) μm compared with the coordinates calculated from the formula for 87 simulated particles. It is accurate enough to find the uranium-bearing particle in the second

$$\alpha = \begin{cases} Arctg\dfrac{Y_U - Y_A}{X_U - X_A} - Arctg\dfrac{Y_B - Y_A}{X_B - X_A}, & X_U > X_A, X_B > X_A, or, X_U < X_A, X_U < X_A \\ Arctg\dfrac{Y_U - Y_A}{X_U - X_A} - Arctg\dfrac{Y_B - Y_A}{X_B - X_A} + \pi, & X_U < X_A, X_B > X_A, \\ Arctg\dfrac{Y_U - Y_A}{X_U - X_A} - Arctg\dfrac{Y_B - Y_A}{X_B - X_A} - \pi, & X_U > X_A, X_B < X_A, \\ \dfrac{\pi}{2} - Arctg\dfrac{Y_B - Y_A}{X_B - X_A}, & X_U = X_A, Y_U > Y_A, \\ -\dfrac{\pi}{2} - Arctg\dfrac{Y_B - Y_A}{X_B - X_A}, & X_U = X_A, Y_U < Y_A, \\ Arctg\dfrac{Y_U - Y_A}{X_U - X_A} - \dfrac{\pi}{2}, & X_B = X_A, Y_B > Y_A, \\ Arctg\dfrac{Y_U - Y_A}{X_U - X_A} + \dfrac{\pi}{2}, & X_B = X_A, Y_B < Y_A, \end{cases} \quad (3.51)$$

instrument for further analysis. More detailed information can be found in Guo et al. (2008).

In all the above analyses, the SSNTD plays important roles (Shen et al., 2008; Guo et al., 2008, 2017; Lipponen and Zilliacus, 2004, 2003, 2002; Esaka et al., 2004; Baude and Chiappini, 2002; Donohue, 2002; Chen et al., 2013, 2015, 2018; Lee et al., 2011; Vlasova et al., 2015; Wang et al., 2018; Wang, 2018).

d. α-track technique

Research papers on plutonium-bearing particles are not common. The reason for scarce research of plutonium particles is mainly that the identification of plutonium-bearing particles is difficult by the fission track (FT) technique, which is commonly used in the study of uranium-bearing particles because one cannot efficiently distinguish between uranium fission and plutonium fission by tracks in SSNTD. The kinetic energy of fission fragments of ^{235}U induced by thermal neutrons in the nuclear reactor is 168.3 MeV on the average, while the average kinetic energy for ^{239}Pu fission fragments is 175.0 MeV. Both values of kinetic energy are too close to distinguish plutonium fission by parameters of tracks. The masses of the fission fragments of plutonium and uranium are also close to each other, which make the identification of Pu among U particles twofold difficult.

In contrast, the energies of α particles emitted from plutonium isotopes and that of uranium have relatively large differences: ~5.2–5.8 MeV and ~4.2–4.8 MeV for Pu and U, respectively. This large difference may provide more feasibility to distinguish plutonium particles from that of uranium. After a simple test, it was found that by the α track diameter alone, one still cannot efficiently identify plutonium particles in the mixture of uranium particles and plutonium particles. It was found that with more track parameters, one can distinguish α tracks more easily than with the use of only the track diameter.

Fig. 3.47 shows the difference of α-tracks between highly enriched uranium (HEU) and plutonium. The α-particles are emitted from highly enriched U ($^{235}U/(^{235}U+^{238}U) \geqq 90\%$) or weapons-grade Pu. The sizes of the U-bearing solid particles and that of Pu are in micron (μm) level. The α tracks recorded on CR-39 detector from one U or Pu particle distributed like a track star. R in Fig. 3.47 is the radius of the circular α tracks in the center part of the star. Rc is the radius of curvature of the tapered α tracks which are located about 130 μm away from the center of the star. The ratio Rc/R for HEU and Pu are different obviously at etching time from 10 hours to 16 hours. The large difference makes it very easy to distinguish Pu-bearing particles from HEU-bearing particles. After selecting the Pu particles, the Pu age is determined by ICP-MS. This technique with a micro-sized Pu particle has been successful in the determination of the age of Pu (the production year of Pu) (Wang, 2018; Wang et al., 2018).

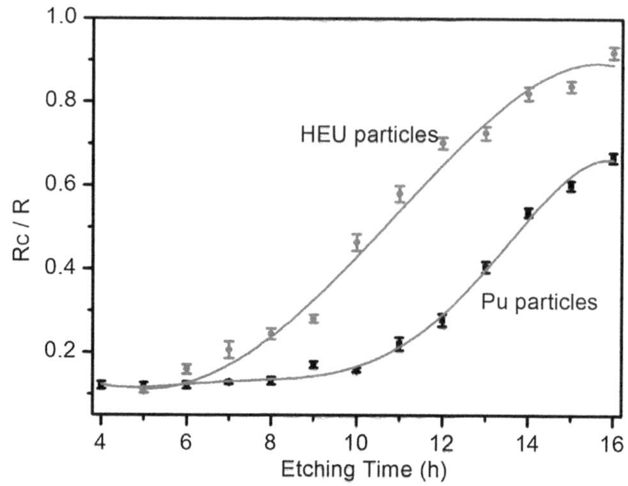

FIG. 3.47 Principle to distinguish microscopic Pu particles from U particles by α tracks on CR-39 sheet. The selected Pu single particle can be used to determine the year when the Pu was produced (the age of Pu) (Wang, 2018; Wang et al., 2018).

G. Elemental analysis and mapping

Elemental mapping has been discussed in detail in the book of Fleischer et al. (1975). Here only an example will be presented on boron mapping in metallurgy.

Natural boron is composed of ^{11}B (80.1%) and ^{10}B (19.9%). ^{10}B has a large cross-section (3840 ± 11 b) for thermal neutrons in nuclear reactors. The reaction products of the reaction $^{10}B(n,\alpha)^7Li$ consist of α particles with energies of 1.47 MeV (93.7%) and 1.77 MeV (6.3%). Both of these are easy to be recorded by SSNTDs (CR-39, LR-115, and cellulose acetate).

In order to observe the boron distribution in steel, a small block of steel is ground and polished to expose a smooth surface. Then, the smooth surface is covered with a thick solution of cellulose acetate (in acetone or cyclohexanone). After drying the solution, the steel block with the cellulose acetate layer is placed outside the channel of thermal column of nuclear reactor for irradiation with thermal neutrons. The reaction products α and Li particles will create tracks in the cellulose acetate. The microphotographs of tracks after etching and the microphotograph of the steel of the same area are shown in Fig. 3.48.

Boron is concentrated on the crystal boundaries and uniformly distributed within the crystal cells. The recipes, smelting temperature and quench condition of the steel, can be selected and improved through these microphotographs (Guo, 1995). A large quantity of special steels has been smelted.

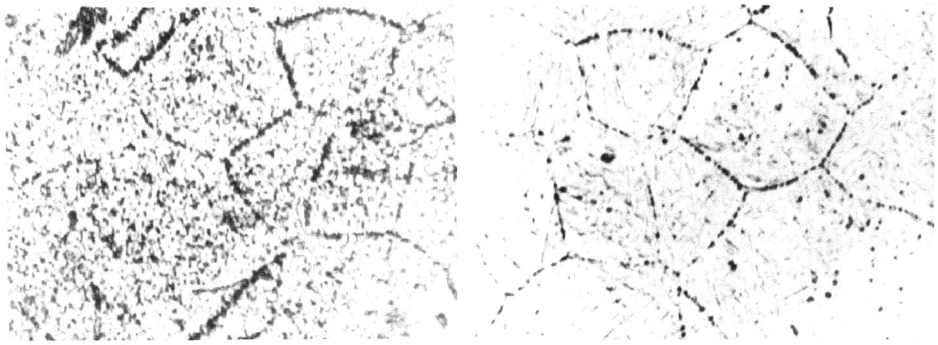

FIG. 3.48 Track microphotograph of MnTiB steel (A) and its corresponding microphotograph of the steel (B). After turning 180° of (A), the two microphotographs can be matched together.

III. Earth and planetary sciences

A. Fission track dating

1. Absolute approach

Fission track dating (FDT) was first devised by Price and Walker in 1962 (Price and Walker, 1963a,b). It has become one of the important dating techniques in geology and archeology (Fleischer et al., 1975; Durrani and Bull, 1987; Wagner and Van den haute, 1992; Dumitru, 2000).

a. Age equation

The principle of FTD is the following:

The fragments of spontaneous fission of ^{238}U nuclei in minerals create radiation damage tracks which can be developed as etched tracks in the mineral and can be observed with an optical microscope. The number of spontaneous fission tracks (N_s) is proportional to the age (T) of the mineral and the uranium content (N_u) in the mineral. After counting the spontaneous fission tracks and measuring the uranium content in the mineral, one can determine the age of the mineral. The easiest way to determine the uranium content in the mineral is to irradiate the mineral with thermal neutrons in a nuclear reactor. Thermal neutrons induce fission of ^{235}U in the mineral. The induced fission will be recorded in the mineral. Counting the induced fission tracks (N_i) and measuring the neutron fluence (ϕ_0) irradiated to the mineral, one can derive the uranium content from the fission cross section ($\widehat{\sigma}$) of ^{235}U and N_i and ϕ_0. The equation of age of the mineral can be written as follows (Collaboration Group of Institute of Atomic Energy and Guiyang Institute of Geochemistry, Academia Sinica, 1976):

$$T = \frac{1}{\lambda_D}\ln\left\{\frac{C_{235}}{C_{238}}\cdot\frac{\lambda_D}{\lambda_F}\cdot\frac{\rho_s}{\rho_i}\cdot\phi_0\cdot\widehat{\sigma}\,\frac{[GR_i\cos^2\theta_{ci}+(1-\sin\theta_{ci})l_i]}{R_s\cos^2\theta_{cs}+(1-\sin\theta_{cs})l_s}+1\right\} \tag{3.53}$$

where C_{235} is the relative abundance of ^{235}U (0.720%), C_{238} is the relative abundance of ^{238}U (99.2745%), λ_D is the total decay constant of ^{238}U ($1.5514\times10^{-10}\text{a}^{-1}$), λ_F is the spontaneous fission decay constant of ^{238}U, ρ_s is the spontaneous fission track density of ^{238}U (cm^{-2}), ρ_i is the induced fission track density of ^{235}U (cm^{-2}), R is the average etchable range of fission fragments in the mineral (cm), θ_c is the critical angle of fission fragments in the mineral, l is the thickness of the single surface layer removed by etchant (cm), s denotes that the quantity is for spontaneous fission fragments, i denotes that the quantity is for induced fission fragments, G is a geometry factor where $G=1$, when an internal surface is used to record induced fission fragment tracks and $G=1/2$, when an external surface is used to record induced fission fragment tracks.

In a special case, one can assume that $R_i=R_s$, $\theta_{ci}=\theta_{cs}$ and $l_i=l_s=0$. Then, Eq. (3.53) is simplified to the equation

$$T = \frac{1}{\lambda_D}\ln\left\{\frac{C_{235}}{C_{238}}\cdot\frac{\lambda_D}{\lambda_F}\cdot\frac{\rho_s}{\rho_i}\cdot\phi_0\cdot\widehat{\sigma}\cdot G+1\right\} \tag{3.54}$$

It should be pointed out that if the detector used for recording induced fission tracks is not the same kind of mineral as that to be dated (by spontaneous fission tracks), then $R_i \neq R_s$, $\theta_{ci} \neq \theta_{cs}$ and $l_i \neq l_s$. In this case, Eq. (3.54) cannot be used to calculate age T.

It is always necessary to calibrate the ratio of the detection efficiency of an external detector for induced fission fragments to that of the mineral for spontaneous fission fragments (Guo, 1982a; Guo et al., 1980; Durrani and Bull, 1987; Holden and Hoffman, 2000). Otherwise, the resulting age value of T calculated from the simplified Eq. (3.54) would be erroneous.

If $\lambda_D T \ll 1$ ($T \leq 2 \times 10^8$ a), then Eq. (3.54) can be further simplified as follows:

$$T = \frac{1}{\lambda_f} \frac{C_{235}}{C_{238}} \frac{\rho_s}{\rho_i} \cdot \phi_0 \cdot \hat{\sigma} \cdot G \quad (3.55)$$

Eq. (3.55) has the same limitations as Eq. (3.54) in addition to condition $T \leq 2 \times 10^8$ a.

The technique of FTD by using Eqs. (3.53), (3.54) and (3.55) is called the absolute approach (Wagner and Van den haute, 1992; Hurford, 1990).

Two problems have been met with the absolute approach, which are described subsequently.

b. ^{238}U spontaneous fission decay constant λ_f

Up to now, more than 30 data of λ_f have been published in the literature (Wagner and Van den haute, 1992; Bigazzi, 1981; Durrani and Bull, 1987). The values of λ_f spread over a very broad range. No one can recommend an integrated value from them. In the year 2000, the International Union of Pure and Applied Chemistry (IUPAC) published a Technical Report on "Spontaneous fission half-lives for ground-state nuclides" (Holden and Hoffman, 2000) and gave a special treatment on the ^{238}U spontaneous fission decay constant λ_f.

The IUPAC did not include the results of all λ_f values of ^{238}U spontaneous fission measured by SSNTDs. They averaged the results measured by the other detectors (ionization chamber, double ionization chamber, rotating bubble chamber, fission products from ^{238}U, third-order coincidence, and multiple neutron coincidence) and give a recommended half-life value $t_{1/2} = (8.2 \pm 0.1) \times 10^{15}$ a of ^{238}U spontaneous fission, which is equivalent to a $\lambda_f = 8.45 \times 10^{-17}$ a^{-1}. They considered that the recommended value of $t_{1/2}$ is the correct spontaneous fission half-life of ^{238}U.

The reasons for not including the results measured by SSNTDs are the following:

(1) The values of half-life derived from track detectors do not agree with that derived from other techniques. The specific activity (λ_f), in general, is lower than that determined from other detectors.

(2) The lower λ_f values are related to special properties of track detectors. The lower λ_f values are suitable to FTD, in which the factors that cause the lower values can be compensated in age determination. For a detailed explanation, the readers are referred to the abovementioned Technical Report.

It has been demonstrated that no trend can be found with respect to whether the zeta calibration (see below) or the decay constant of 7×10^{-17} a^{-1} was used in FTD. The results based on a decay constant of 8.46×10^{-17} a^{-1} tend to be low compared with the mean age obtained in an international comparison of fission track ages for the 1988 fission track workshop (Miller et al., 1990). It means that by using $\lambda_f = 6.85 \times 10^{-17}$ a^{-1} (Fleischer and Price, 1964) or 7.03×10^{-17} a^{-1} (Roberts et al., 1968), the age values obtained are in agreement with that of age standard, or in agreement with the mean value obtained by different techniques in FTD (population method, external detector method and zeta calibration approach). By using $\lambda_f = 8.46 \times 10^{-17}$ a^{-1} the results of ages are generally lower than those obtained by other techniques.

In the absolute approach of FTD, an important point is to state which decay constant λ_f of ^{238}U spontaneous fission was used in calculation of age, so that other workers can make comparison with other age determination by substituting a new λ_f value (Durrani and Bull, 1987).

In dating with the absolute approach, the effect of track fading on age must be corrected. The correction techniques include: track diameter method (relation of track diameter d/d_0 to track density ρ/ρ_0, where the subscript "0" signifies that for no fading of tracks), track length method (relation of track length l/l_0 to track density ρ/ρ_0) and plateau-age correction method (annealing spontaneous and induced tracks simultaneously before etching until ρ/ρ_i reaches a constant value. The age calculated by the constant ρ_s/ρ_i is the true age of the sample.)

c. Neutron fluence ϕ_0

Neutron fluence determination in nuclear reactors is a mature technique for reactor physicists. For thermal neutron determination, the often-used technique is activation foil measurements. Co, Au, and Cu foils are most suitable for thermal neutron fluences of 1.5×10^{15}–6×10^{15}, 1×10^{11}–1×10^{14}, and 1×10^{12}–1×10^{15} cm^{-2}, respectively. Co foil is most suitable for FTD.

In the irradiation of dating samples in nuclear reactors, one must pay attention to the total thickness of the samples. If the sample package is too thick, the outlayer of the sample will absorb neutrons and reduce the expected fluence of neutrons for the central part of sample. In this case, one should insert more than one (at least two) activation foils to get an average value of neutron fluence. Similar attention must be paid to the thickness of the activation foils. If the thickness of the activation foil is too large, the surface layer of the foil will absorb a significant amount of neutrons and a reduced fluence of neutrons reaches the central part of the foil (self-shielding). In this case, the fluence of thermal neutrons can be written as following equations:

$$\phi_0 = n v_0 t = \frac{A_0 t}{N\widehat{\sigma}(1 - e^{-\lambda t})} \tag{3.56}$$

$$\widehat{\sigma} = \sigma_0 \left(K_{th} g + r\sqrt{\frac{T_n}{T_0}} S_0 K_r \right) \tag{3.57}$$

where n is the volume density of thermal neutrons in the nuclear reactor (cm^{-3}), v_0 is the neutron velocity (2200 m/s), K_{th} is the self-shielding factor of activation foil for thermal neutrons, K_r is the self-shielding factor of activation foil for epithermal neutrons, A_0 is the activity of the activation foil, t is the exposure time of the activation foil to the reactor thermal neutrons in seconds (s), and λ is the decay constant of the nuclei formed in the exposure (s^{-1}).

The nuclear parameters of the activation foils of ^{59}Co, ^{63}Cu, and ^{197}Au, as well as the ^{235}U (n,f) reaction for thermal and epithermal neutrons in nuclear reactors are listed in Table 3.7. For example, 50 mg/cm^2 of Co foil will have ~3% ($K_{th} = 0.97$) and ~11% ($K_r = 0.885$) effects on thermal and epithermal neutrons, respectively. These effects will be brought to the fluence value. If one does not give a correction for these effects, the final value of fluence of thermal neutrons (ϕ_0) calculated from the Co foil activity A_0 will be ~3.6% smaller than the real fluence. This reduction of calculated ϕ_0 would result in a younger age of the sample as calculated from Eqs. (3.53), (3.54) or (3.55).

2. Zeta approach

In order to avoid direct measurements of fluence of thermal neutrons ϕ_0 in nuclear reactors and avert adoption of the decay constant λ_f of ^{238}U spontaneous fission in the calculation of age, the Zeta approach was proposed (Fleischer et al., 1975) and advocated (Hurford and Green, 1982, 1983) and extensively accepted and adopted (Miller and Duddy, 1985; Miller et al., 1990; Van den haute and Chambaudet, 1990; Wagner and Van den haute, 1992; Durrani and Bull, 1987; Wang and Kang, 1993; Wang et al., 1993; Yang et al., 2003).

a. Age equation

The Zeta approach of FTD is based on the adoption of standard uranium glass to measure fluence of thermal neutrons in nuclear reactors and the utilization of age standards to replace the decay constant of ^{238}U spontaneous fission.

When adopting standard uranium glass, one has the following equation:

$$\phi_0 = B\rho_D \tag{3.58}$$

where ρ_D is the areal track density on the external detector attached to the standard uranium glass, and B is a proportionality constant.

If the age standard mineral is employed in dating, whose age is known to be T_{st}, the ratio of track density for spontaneous to induced fission is $(\rho_s/\rho_i)_{st}$, and assuming that $l_i \approx l_s \approx 0$, $\theta_{ci} \approx \theta_{cs}$, $R_i \approx R_s$, $C_{235}/C_{238} = I$, Eq. (3.53) becomes:

$$T_{st} = \frac{1}{\lambda_D} \ln\left\{ 1 + \frac{\lambda_D B \rho_D \widehat{\sigma} I G}{\lambda_f} \left(\frac{\rho_s}{\rho_i} \right)_{st} \right\} \tag{3.59}$$

If we let

$$\zeta = \frac{B\widehat{\sigma} I}{\lambda_f} \tag{3.60}$$

we have

$$T_{st} = \frac{1}{\lambda_D} \ln\left\{ 1 + \lambda_D \rho_D \zeta G \left(\frac{\rho_s}{\rho_i} \right)_{st} \right\} \tag{3.61}$$

$$\zeta = \frac{(e^{\lambda_D T_{st}} - 1)}{\lambda_D (\rho_s/\rho_i)_{st} G \rho_d} \tag{3.62}$$

b. ζ value

For a given age standard, ζ is approximately a constant. The minor difference would come from: (1) different external detectors used, whose $\theta_{ci} \neq \theta_{cs}$, (2) different etching conditions adopted where $l_s \neq 0$, and (3) differences in subjective judgments by different workers for the numbers of spontaneous and induced tracks.

Once the ζ value is obtained, the age of unknown sample can be determined after co-irradiation with the standard uranium glass from Eq. (3.61), but all the quantities in Eq. (3.61) change to unknown quantities: $Tu = 1/\lambda D.\ \ln\{1 + \lambda D\rho Du\zeta G(\rho_s/\rho_i)u\}$, where the letter "$u$" indicates that the quantities are for unknown sample and the track density for uranium glass is for the irradiation with unknown sample.

The Zeta approach has been demonstrated to be a good way to train newcomers into the FTD field. It forces them to count tracks correctly so that the determined age is close to the real age. Zeta approach is a relative method of determination of age and it relies on the age value of the age standard mineral, which was dated by other isotopic techniques. Therefore, it is not an independent dating method.

The age standards proposed and adopted by the FTD community are listed in Table 3.8.

It should be noted that the age obtained by ζ approach is an apparent age of the mineral. In the case where track

TABLE 3.7 Parameters of activation foils of ^{59}Co, ^{63}Cu, and ^{197}Au and ^{235}U (n,f) in a thermal neutron field in nuclear reactors.

	^{59}Co		^{63}Cu		^{197}Au		$^{235}U(n, f)$	
Temperature, °C	g	S_o	g	S_o	g	S_o	g	S_o
20	1.000	1.736	1.000	0.795	1.0053	17.30	0.9759	−0.0502
60	1.000	1.851	1.000	0.848	1.0075	18.44	0.9665	−0.0324
100	1.000	1.959	1.000	0.897	1.0097	19.51	0.9581	−0.0169
140	1.000	2.061	1.000	0.943	1.0119	20.52	0.9507	−0.0039
180	1.000	2.159	1.000	0.992	1.0141	21.49	0.9441	+0.0066
$\sigma_0(\times 10^{-24} cm^2)$	37.40 ± 0.03		4.3 ± 0.2		98.8 ± 0.3		582.6	
$\lambda(s^{-1})$(activated nucleus)	4.1673×10^{-9}		1.5161×10^{-5}		2.976634×10^{-6}			

TABLE 3.8 Prominent age standards for fission track dating.

Mineral	Geological origin	Age(Ma)	Dated by	Contact
Apatite:	Fish Canyon tuff	27.8 ± 0.2	$^{40}A_r/^{39}A_r$	C. W. Naeser
	Durango	31.4 ± 0.6	KA_r	F. McDowell
Zircon:	Fish Canyon tuff	27.8 ± 0.2	$^{40}A_r/^{39}A_r$	C. W. Naeser
	Buluk Member tuff (FTBM)	16.4 ± 0.2	KA_r	A. J. Hurford
	Tardree Rhyolite	58.7 ± 1.1	KA_r, $^{40}A_r/^{39}A_r$	A. J. Hurford
Sphene:	Mount Dromedary Banatite	98.8 ± 0.6	R_b–S_r	A. J. Gleadow
Glass:	Moldavite (but only splits from a single Moldavite)	15.21 ± 0.15	KA_r	(there is no contact)

annealing occurred in the history, age correction for track annealing must be made.

3. LA-ICP-MS-based fission track dating

FTD has provided a great amount of age data in geology, archeology, and thermochronology. However, it has two problems. The first problem is the need of irradiation with nuclear reactor. After irradiation, the samples and the devices used for irradiation are heavily activated. The second problem is that the dating cycle is quite long. Several weeks even several months are needed to cool down the radioactivity. Even if one waited for the cooling down, the samples and the devices will still have residual radioactivity, which is also harmful to workers. The main purpose of reactor irradiation is to determine the number of atoms of ^{238}U in the sample (mineral). Another advantage of reactor irradiation is to cancel the two detection efficiencies of the mineral and the external detector for fission fragments. However, the irradiation is troublesome for dating practice.

In order to solve the problems above, some scientists intended to use laser ablation inductively coupled plasma spectrometry (LA-ICP-MS) to determine the number of atoms of ^{238}U in the mineral to avoid nuclear reactor irradiation of samples.

a. Age equation

The basic equation for age determination is simple.

$$N_S = N_{238}\lambda_F T\varepsilon \tag{3.63}$$

where N_S is the number of spontaneous fission tracks of ^{238}U, N_{238} the number of atoms of ^{238}U, λ_F the decay constant of spontaneous fission of ^{238}U, T the age of the mineral, ε the detection efficiency of internal surface of the mineral for spontaneous fission fragments of ^{238}U.

If one adopts the absolute approach for dating as explained in the previous section, one can write the following equation (Soares et al., 2014):

$$T = \frac{1}{\lambda_D}\ln\left[1 + \frac{\lambda_D}{\lambda_F}\frac{1}{g_{4\pi}[\eta q]_{iS} r_S}\frac{\rho_S}{N_{238}}\right] \tag{3.64}$$

where $g_{4\pi}$ is the geometry factor for an internal surface of the mineral ($g_{4\pi} = 1$), $[\eta q]_{iS}$ is the detection efficiency of the internal surface of the mineral for the spontaneous fission tracks of ^{238}U in the mineral, $[\eta q]_{iS} = \varepsilon$ for the special geometry, r_S is the mean length of horizontal confined spontaneous fission tracks, and ρ_S is the spontaneous fission track density in the polished surface. The other parameters are the same as previously defined.

If one adopts the *Zeta* approach, one can write the following equation:

$$T = \frac{1}{\lambda_D}\ln\left[1 + \zeta_{ICPMS}\frac{1}{[\eta q]_{iS} r_S}\frac{\rho_S}{238_U}\right] \tag{3.65}$$

where

$$\zeta_{ICOMS} = \frac{\lambda_D}{\lambda_F}\frac{M}{g_{4\pi}\cdot N_A\cdot 10^{-6}\cdot d} \tag{3.66}$$

where ζ_{ICOMS} is a constant for the mineral, ρ_S is the spontaneous fission track density in the internal surface of the mineral, M is the isotopic weight of ^{238}U, N_A is the Avogadro constant, and d is the average density of the mineral ($d = 3.19$ g/cm^3 for apatite).

b. Determination of ^{238}U content in mineral by LA-ICP-MS

It is not easy to determine the content of ^{238}U in minerals by LA-ICP-MS. There are two difficulties for this determination:

(1) U occurs in minerals in a minute amount of the element. For apatite, $Ca_5(F,Cl)(PO_4)_3$, the often adopted mineral in FTD, the content of U is about 10^{-5} g/g. Within one instrument, such as a LA-ICP-MS, the measurement of two compositions simultaneously with the number of the other elements which differ greatly in amount, is difficult. In other words, to count accurately the number of atoms of uranium (^{238}U) with the counting of the huge number of the other atoms (Ca, F, Cl, P, and O) in the MS, is very difficult.

(2) The mass of the atoms of ^{238}U is much heavier than that of the atoms of the main compositions of the minerals (Ca, F, Cl, P, and O). It is difficult to collect different ions of the atoms of the elements with so big differences in masses (238−16 for O). In order to reduce the differences, one must find some way to overcome the obstacles.

c. Determination of ^{43}Ca to derive the effective mass of minerals

Soares et al. (2014) tried to use ^{43}Ca as a representative of the atoms of minerals and a bridge between U and the mineral atoms to determine ^{238}U content. Fortunately, they dated the mineral (apatite) by using LA-ICP-MS.

d. Determination of detection efficiency of internal surface for spontaneous fission fragments

Soares et al. (2014) adopted the technique horizontal projected confined track method to determine the detection efficiency of the internal surface of apatite for spontaneous fission fragments of ^{238}U (Jonckheere and Ven den haute, 1999, 2002). Their results of dating are in agreement with that by other methods. This work shows that LA-ICP-MS may be used in FTD to avoid using nuclear reactors in the future. Certainly, one should make more improvements in dating processes to ensure that every step is reliable.

4. Continental drift and ocean-bottom speading

One of the most prominent achievements of FTD is the direct demonstration of the conjectures of continental drift and ocean bottom spreading by dating the ages of the basalts taken at different distances from the Median Valley in the Atlantic Ocean where lava rose out from the interior of the earth and cooled down forming the basalts. The fission track ages of basalts increase as the distance increases from the Median Valley, which is consistent with the K−Ar age, but ruled out some of the spurious ages of K−Ar dating. Both fission track and K−Ar dating showed that the American and African continents are moving apart from each other and the ocean bottom is spreading about 3 cm/a in more recent time since about 2 Ma ago. The old spreading rate is much slower, which was about 0.6 cm/a before 2 Ma ago (Fleischer et al., 1968, 1971 and 1975).

5. Archeology and anthropology

One of the advantages of FTD is the wide range of age from about 10^5 to 10^9 years which can be dated, especially in the range from 10^5 to several million years. In this very young range of age, other dating techniques have some deficiencies. FTD is almost unique in some cases.

a. Fission track dating of ancient man in Bed I, Olduvai Gorge, East Africa

One of the best examples of dating ancient man is in East Africa. FTD of volcanic pumice from Bed I, Olduvai Gorge showed that ancient man Zinjanthropus lived 2.0 ($\pm$0.3) Ma ago (Fleischer et al., 1965c, 1965e), which is in agreement with the age 1.75 ($\pm$0.05) Ma by K−Ar dating (Fleischer et al., 1975). The dating by K−Ar was originally treated as unbelievable, but the agreement by FTD made this site to be regarded as one of the best-dated of the anthropological sites. Consequently, it was established that the earliest man up to now is believed to be Zinjanthropus.

b. Dating of Peking Man

The first skull of Peking Man Site at Zhoukoudian near Beijing, China, was discovered in 1929. The model of Peking Man is shown in Fig. 3.49. About 50 years transpired since the skull was discovered before the first dating of the site by fission tracks in 1979. The difficulty of dating the site was caused by the lack of suitable materials in the deposited cave. Guo et al. (1980) separated sphenes, a kind of fine minerals of 50–300 μm size from the ashes left by Peking Man for FTD. The sphene minerals were burnt to high temperature in the fire made by Peking Man. The original ^{238}U spontaneous fission tracks recorded in the sphenes before firing were completely annealed out. New spontaneous fission tracks formed after firing were used to date Peking Man. The age of the Peking Man Site at layer 10 of Locality 1 is 0.462 (±0.045) Ma (Guo et al., 1980, 1982a) and layer 4 of Locality 1 is 0.299 (±0.055) Ma. It should be noted that the annealing degree of fission tracks in each grain of the sphene was judged individually (Guo et al., 1980, 1982a). This is one of the advantages of FTD over other dating techniques.

c. Dating of ancient man in Baise, China

More than 50 years ago Harvard University anthropologist Hallam Movius divided the peoples of the Early Stone Age into two cultures: those who could make sophisticated, two-sided stone hand axes, known as Acheulean or mode 2 technology and those who could not. This invisible technological barrier, which came to be known as Movius's line, separated handy *Homo erectus* in Africa, the Middle East, and Europe from their less-adept cousins in Asia, who left only mode I technology such as, simple stone flakes and cobbles used for chopping. Movius therefore wrote off the entire Asian continent as "a marginal region of cultural retardation". Because the climate and terrain in the forests of Asia remained stable for the past 2 million years, Movius wrote, humans there did not advance in culture but stayed backward for eons (Gibbons, 1998). In the 1980–90s, archeologists discovered a large number of bifacial hand axes in the Baise basin in South China. Tektites were also found in the undisturbed primitive layer of deposits (red soil or laterite) within which bifacial hand axes were found nearby tektites.

Fig. 3.50A shows two photographs of the two sides of a bifacial hand axe and Fig. 3.50B shows a photograph of tektite found in the same layer of the red soil.

Guo et al. (1996, 1997) used the fission track method to date the tektite. Their result after correction for track fading was 0.732 ± 0.039 Ma, about the same age as the Australia–Indochina tektite fall. This age is also similar to the reversal time of the geomagnetic field from the Matuyama epoch to the Brunhes epoch (0.789 ± 0.008 Ma), which led Guo and coworkers to conjecture that impacts of space objects on the Earth might cause at least some reversals of the geomagnetic field (Price, 2008).

The dating result also shows that at least a few early Asians were also making two-sided, or bifacial, stone tools as much as 730,000 years ago. Gibbons (1998) wrote a paper in "Science" announcing that "the Movius line is breached" by the discovery and dating of the bifacial hand axes and tektites in the Baise basin.

For more information about FTD, the reader is referred to the series of proceedings of Fission Track Dating Symposiums and the articles by Yuan et al. (2009), Hadler et al. (2009), Shen et al. (2009) and the book by (Wagner and Van den haute, 1992).

FIG. 3.49 Restored figure of Peking Man. For the color version of the figure, the reader is referred to the online version of the book.

6. *Tectonic up-Lift rate determination*

a. Retention temperature of minerals

Fission tracks are very sensitive to temperature. If the temperature is higher than the effective retention temperature T_r of the mineral, the tracks will be annealed away. Only when the temperature drops to lower than T_r, the tracks can be retained. Different minerals have different T_r values. For apatite, $T_r = 100 \pm 20°C$; Zircon, $210 \pm 40°C$; epidote, $240 \pm 40°C$; sphene, $250 \pm 40°C$; garnet, $270 \pm 30°C$ (Wagner and Van den haute, 1992).

b. Height difference method

For geological reasons, some mountains are rising up. For example, Himalayan mountains are rising up as a result of the thrusting of the Indian continental block toward the Tibet–Qinghai crust. The Qinling mountains in central China are rising by the clash between the Southern China and Northern China blocks.

FIG. 3.50 Photograph of (A) bifacial hand axe (two sided, the two photographs in (A) were taken from two sides of the same hand axe) excavated from Baise basin, South China, and (B) a tektite found within the same layer as the axe. For the color version of the figure, the reader is referred to the online version of the book.

The rising of the Himalayan mountains stops the humid air from the Indian plain from reaching the Tibet–Qinghai plateau, resulting in a dry and cold climate in the Tibet–Qinghai plateau.

Similarly, the Qinling mountains stop humid air from the south from reaching the north, resulting in a dry climate and the loess (yellow soil) in North China. At the same time, the Qinling mountains prevent the south from attack by the dry wind from the north.

Chen et al. (2001) measured the uplift rate of the Taibai mountains (the western part of the Qinling mountains) by fission tracks in apatites and determined that the Taibai mountains were formed around 40 Ma ago (according to apatite apparent age). The uplift rate by height difference over age difference was about 130 m/Ma in the first stage from 40 Ma to about 35.5 Ma. Then, the overburden on the mountains started a denudation accompanying the uplift. The sum of the uplift rate plus denudation rate was 430 m/Ma.

c. Mineral pair method

Yin et al. (2001) determined the uplift rate of the highest peak of Hua mountains (the east part of the Qinling mountains) by the height difference method (apatites) and mineral pair method (apatite and zircon), that is:

$$V = \frac{T_r(\text{Zircon}) - T_r(\text{Apatite})}{t_{\text{Zircon}} - t_{\text{Apatite}}} / \text{geothermal gradient}$$

where V is the uplift rate, T_r is the effective retention temperature, t is the age of the mineral by FTD, and the geothermal gradient was taken as 35°C/km. The effective retention temperatures were taken as 110°C and 220°C for apatite and zircon, respectively. It shows that the Hua mountains started rising 68.2 Ma ago according to the zircon apparent age, which is equivalent to about 41 Ma of the apatite apparent age.

The average uplift rates were determined to be 120 m/Ma and 140 m/Ma, respectively by the height difference method and the mineral pair method.

The whole Qinling mountains including Taibai and Hua mountains are growing year by year. It is bad for North China but good for South China.

B. Geothermal chronology

The temperature in a room can be measured by a thermometer. What thermometer can measure the temperature in the deep underground, especially in ancient time and the temperature as a function of time (year)? The answer is ^{238}U spontaneous fission tracks.

Latent fission tracks are sensitive to temperature. At normal temperature, fission tracks are stable for a long time. But at higher temperature track fading occurs, latent fission tracks shrink in etchable lengths and gradually disappear.

The degree of fading depends on two factors: temperature and time. The higher is the temperature, the faster is the fading. The longer the time at higher temperature, the greater is the shrinkage of the track length.

In minerals, ^{238}U spontaneous fission tracks are continually generated with time. At the same time, track fading accompanies every track. The first track suffered the longest history of heating. Its length is the shortest. The last track was created just a moment ago. It suffered the shortest history of heating. Its length remains intact. Every track records the thermal history of its own. An analysis of the track length distribution in the mineral will provide information on the whole thermal history experienced by the mineral (Lutz and Omar, 1991; Chambaudet et al., 1993; Wang et al., 1993).

In a retrieving test of geothermal history (Guo et al., 1999), a mono constant activation energy E_0 was used for the isothermal annealing process:

$$l = l_s \exp\{-\alpha_0 \Delta\tau \exp[-E_0/(k\theta)]\} \tag{3.67}$$

or

$$\theta = \frac{E_0}{k \ln\left[\frac{\alpha_0 \Delta\tau}{\ln(l_s/l)}\right]} \tag{3.68}$$

where l is the length of track at the end of the isothermal annealing interval, l_s is the length of track at the beginning of isothermal annealing interval, $\Delta\tau$ is the length of time of the isothermal annealing interval, θ is the temperature of isothermal annealing, α_0 is the coefficient of annealing, and k is the Boltzmann constant.

Before making the calculation (inverse modeling), the track length distribution is obtained by careful measurements. The whole range of distribution is divided into many small intervals. In each interval, the temperature is regarded as a constant (isothermal annealing process). The corresponding time length of the small interval can be calculated as

$$d\tau = \omega(l)dl/b(\tau) \tag{3.69}$$

where $d\tau$ is the differential form of $\Delta\tau$, dl is the differential form of Δl, $\Delta l = l_s - l$, the length of the small interval, $b(\tau)$ is the effective generation rate of ^{238}U spontaneous fission at time τ, $\omega(l)$ is the track density in the small interval having track length l, and $b(\tau)$ is calculated by the equation:

$$\int_0^T b(\tau)d\tau = \int_0^\infty \omega(l)dl \tag{3.70}$$

where T is the age of the mineral for retaining tracks.

From Eqs. (3.68), (3.69), and (3.70) one can calculate the temperature θ and the time τ of each small interval. Thus, one determines the ancient temperature θ as a function of time (age) τ.

The results of retrieved (inverse modeling) temperature is consistent with that by the smoke convection model developed by Chambaudet et al. (1993) and in agreement with that by a direct modeling from thermal history to track length distribution (Guo et al., 1999).

C. Uranium and oil exploration and earthquake prediction

Radon is harmful to human health, but sometimes, it provides valuable information about the underground resources of uranium, petroleum, and natural gas.

The technique of radon measurement used for geological purpose is similar to that for radon dosimetry. Additional schemes are required for geological investigations. For uranium exploration, first one must select a promising uranium-bearing area designated by geologists. Then, a grid pattern is set with distance about 50 meters (or closer if necessary) between neighboring grid-points. At each grid-point, plastic track detectors (CR-39 or LR-115) inside cups are buried at a small distance below the surface of the soil. After several weeks the CR-39 sheets are collected, etched and the tracks counted. The track densities on the detectors over uranium ores should be higher than those in surrounding areas deficient in ore deposits. In uranium ores, more radon (^{222}Rn) is produced from ^{226}Ra, which is the progeny of ^{238}U. The radon gaseous atoms will diffuse upwards to the surface of the soil. According to measurements, the diffusion length of ^{222}Rn in soil is about 1.2 m. If a uranium ore contains 10^4 times more uranium than that of common earth's crust rocks, the radioactivity of the ^{222}Rn from the ore will reach the background level of the crust rocks after diffusion to a distance ~12 m. It means that uranium ore at more than 12 m depth will not be detected. However, Fleischer et al. (1980) located uranium ore at 100 m depth. How can ^{222}Rn move such a long distance (100 meters in depth)?

There is evidence that the radon transport is faster than that expected by simple diffusion of radon gas in the earth. Some upward bulk flow of radon-bearing air must be occurring underground. The possible source of upward flow motion is fluid convection caused by nonuniformity in geothermal distribution. The ore body has a higher temperature and produces more radon gas. The air above the ore increases in buoyance than the surrounding area (Fleischer, 1998; Durrani and Bull, 1987; Durrani and Ilic, 1997).

For petroleum or gas exploration, the method of the deployment of a nuclear track detector (CR-39 or LR-115) is similar to that for uranium exploration and the selection of the area must be guided by geologists. The upward air flow around the oil-bearing structure is caused by the leakage of oil vapor from the edge of the structure shaped like a big bowl placed upside down similar to a dome (Fleischer, 1988, 1998; Durrani and Bull, 1987).

The detection of α particles from radon can also serve as a tool for the prediction of an earthquake. The basic technique is similar to that used in uranium exploration, but the deployment of CR-39 detectors is on the seismic belt. Unusual subterranean activity and possibly attendant fissures and strains result in release of larger quantities of radon gas from the U and Th content of the earth's crust than at normal conditions. The radon intensities, once measured in the underground soil at three prediction stations, exhibited a significant increase starting 6 weeks prior to a $M_s = 6.3$ magnitude of earthquake even at a 178 km distance from the epicenter (Fleischer and Mogro-Campero, 1985; Fleischer, 1988).

Fission track uranium exploration was developed based on hydrogeochemical prospecting of uranium. The original prospecting adopted a fluoro-colorimeter and a laser-fluorometer for the determination of uranium. After testing the application of the fission track technique, it was found that the fission track measurement of uranium is more sensitive than the other two instruments.

Uranium readily forms compounds dissolvable in water. Due to groundwater movement, uranium in ores disperses outward to form uranium-rich halos around the ores. Nearly all uranium ores have uranium-rich halos around them. The uranium-rich halos can be found in the field by uranium determination in ground and surface water.

A small droplet (5 μm) of ground or surface water is picked at each point in the field and dropped onto a small piece of polycarbonate (PC) sheet. After drying the water, the uranium contained in the water is deposited in the residue on the surface of the PC sheet. It is covered with another PC sheet to form a specimen. Many specimens are made at many points which are distributed in the field as a grid net. These specimens are packed together and put into a nuclear reactor core for neutron irradiation. Thermal neutrons induce fissions of ^{235}U in the specimens. The fission fragments are recorded in the PC sheets. After etching and counting the tracks, the distribution of uranium concentration is mapped and uranium-rich halos can be found on the map.

The fission track technique has been found to be the most sensitive method in the determination of uranium in natural water. Its sensitivity reaches 10^{-12} g/g, while the sensitivity is $\sim 10^{-11}$ g/g for laser-fluorometry and $\sim 5 \times 10^{-11}$ g/g for fluoro-colorimetry (Guo et al., 1986c). Fission track detectors are considered among the best sensors for uranium ore.

A test exploration of uranium ore with fission track technique showed that several regions have been found to be rich in uranium. The distinction of different regions is clearer by fission tracks than by laser-fluorometry, while fluoro-colorimetry was senseless to the difference. A region outlined by fission tracks was excavated and identified to have a uranium ore in underground deposits (Guo, 1993; Guo et al., 1992).

IV. Life and environmental sciences

A. Radiation protection dosimetry

1. Radon and thoron monitoring and dosimetry

Radon (^{222}Rn) and thoron (^{220}Rn) are produced in the decay series of ^{238}U and ^{232}Th, respectively. The latter nuclear species ^{238}U and ^{232}Th exist ubiquitously on the earth in rocks, soils, building materials, and waters. ^{222}Rn and ^{220}Rn are inert gases, which diffuse through all kinds of materials: solids, liquids, and gases, and may undergo rapid convection under the influence of external flows or pressure. It has been estimated that α decays of radon and thoron and their progeny contribute more than 55% of the annual effective dose equivalent (μSv) to the general population in the world from all sources of ionizing radiation exposure (ATSDR, 1999; Lowder, 1990, 1993). It is a big hazard to mankind. For this reason, a worldwide investigation is underway to monitor radon and thoron concentrations in houses, mines, caves, and all places where mankind can reach. There are two types of radon monitoring devices: active and passive detectors. The active detector, such as the silicon solid detector, can get results of radon concentrations very quickly, but for long-term monitoring, it is not so easy as the SSNTD, such as LR-115 and CR-39. The quantities, which can be measured for radon α detection with the track detector, include radon concentration in indoor air, open air, mines, working places, caves, and in natural water; radon emanation rate, or exhalation rate from various substances, such as rocks, soils, all kinds of building materials; radon diffusion rates through various materials; and radon dosimetry evaluation (Baixeras et al., 2005; Font, 2009; Font et al., 2008; Narula et al., 2009a, 2009b; Chauhan et al., 2006, 2008; Eappen and Mayya, 2004; Shweikani and Raja, 2009; Kumar et al., 2008; Chauhan and Chakarrarti, 2002; Nain et al., 2006, 2008; Mahur et al., 2005; Mayya, 2004; Kant et al., 2006; Mujahid et al., 2005; Abumurad and Al-Omari, 2008; Urosevic and Nikezic, 2003; Tian and Xiao, 1994; Xiao et al., 2001; Yang et al., 2002; Yang, 2007; Chen et al., 2006; Shang et al., 2006; Nikolaev, 2015; Nikezic et al., 2009).

2. Neutron dosimetry

a. Bonner neutron rem counter

Alpha-particle recording from the $^{10}B(n,\alpha)^{7}Li$ reaction with CR-39 detector has been used in the neutron Bonner rem counter for neutron dosimetry (Agosteo et al., 2010, 2009, 2007). Its sensitivity ranges from thermal energies up to 1 GeV of neutrons. The Bonner rem counter is composed of outer polyethylene sphere with lead shell, seven cadmium buttons and inner shell of polyethylene. Passive CR-39 and enriched ^{10}B converter are plugged inside the inner polyethylene shell. Adjusting the thicknesses of the outer and inner shells and the size of the cadmium buttons, Agosteo et al. (2010) has optimized the response function close to the $H^*(10)/\Phi$ fluence to ambient dose equivalent conversion coefficients in the whole range from thermal to 1 GeV neutrons. The calibration factor of the rem counter is 6.91 ± 0.45 tracks/(cm^2 μSv). The rem counter has been demonstrated to be able to measure the cosmic neutron background and to give nonzero background when used in environmental monitoring applications. This rem counter is suitable for routine radiation safety monitoring in public areas around particle accelerators. This counter does not suffer from α pulse pile-up and γ-ray background in neutron-γ-ray mixed fields.

Kralik et al. (2008, 2010) used CR-39 sheets sandwiched with ^{10}B radiators in 10 Bonner sphere spectrometers (BSSs) with different diameters to measure the response of each BSS and unfold the responses from all the BSS. They obtained the energy spectra of photoneutrons at the therapeutic accelerator. See Chapter 9, Volume 1 of this book for additional information on BSSs.

b. Recoiling carbon and oxygen nuclei method

Wang et al. (1998) have suggested a method by measuring the major and minor axes of the track opening and the depth of the track tips of recoiled carbon and oxygen nuclei in CR-39 by fast neutrons to determine V (= V_t/V_B and to obtain a REL spectrum. They combine the REL spectrum with the differential recoiling cross sections of neutrons with carbon and oxygen nuclei to separate the REL spectrum to carbon and oxygen and use computer simulation to find the true neutron energy spectrum.

c. Radiator-Degrader-CR-39 method

Aslam et al. (1988), Dajko (1990), Durrani et al. (1988), Wan et al. (2003) and Oda et al. (2009) have studied the Radiator-Degrader-CR-39 assembly as a neutron energy spectrometer. Fast neutrons induce proton recoils in the radiator (polyethylene). If the energy of the recoil protons is high enough, the protons will pass through the degrader (a foil of Pb, Au or Al) and reach the CR-39 detector. If the energy of the protons after the degrader is in the range of the energy window of CR-39 detector, the protons will be recorded as tracks in the CR-39. Therefore, each thickness of the degrader corresponds to an energy interval of protons. By using a group of thickness of degrader, one can compose a neutron energy spectrum. Wan et al. (2003) have made a Monte Carlo program to simulate this process and reached an agreement with the experiments by Durrani et al. (1988).

d. Bonner spheres spectrometer filled with boron radiators plus CR-39

An upsurge in research on Bonner spheres filled with boron radiators plus CR-39 detectors is evident in laboratories. Kralik et al. (2010) used a series of diameters of Bonner Spheres: 3, 3.5, 4, 4.5, 5, 6, 7, 8, 10, and 12 inches (1 inch = 2.54 cm) to compose the spectrometer. They used the GRAVEL Code from the UMC3.3 Package (Reginatto, 2004) to unfold the spectrum.

Similar Bonner spheres have been built in Barcelona (Fernandez et al., 2007) in Frascati (Bedogni et al., 2010), in which CR-39, TLD. Au foil and other types of passive detectors are used. (Domingo et al., 2010).

B. Environmental sciences

1. Radioactive fallout from nuclear accidents

In the processes of production and transportation of nuclear fuel (U and Pu), or in serious nuclear accidents, or in the detonation of nuclear weapons, small amounts or a large number of radioactive solid particles containing U or Pu will be produced, which mostly have sizes from several tens of nanometers (nm) to several hundreds of micrometers (μm). The particles less than several micrometers can be suspended in air for very long time, which are referred to as aerosols. These particles fall out onto the surface of the earth due to their gravity or rainfall. The larger particles are deposited more easily onto the earth's surface. Because all of these particles carry radioactivity, they are often called radioactive fallout as a whole, or hot particles as individuals. These particles, such as those exploded from the Chernobyl nuclear reactor, from uranium mining combinations, or from testing of nuclear weapons, results in serious environmental contamination and are harmful to human health. One of the main species of radioactivities of these hot particles is alpha radioactivity. SSNTDs, such as CR-39 and LR-115, are used to record tracks of α particles emitted from the hot particles. If one covers a sheet of track detector on a sample containing the hot particles, a specific feature of the tracks formed by a hot particle shapes like a sunburst or star of tracks on the surface of the track detector. The α particles are emitted isotropically from the hot particle. If one extrapolates all the tracks, they will approximately intersect at one point where the hot particle is located. This feature makes it easy to find and to locate a hot particle among a huge number of common dust or common aerosol particles. Counting the α tracks allows one to derive the α radioactivity of the hot particle. If the isotopic composition is known, one can derive the amount of radioactive nuclei in the hot particle. In a brief summary, with the use of a SSNTD, one can find, locate and estimate the intensity of radioactivity and roughly derive the amount of radionuclides in the aerosol particle. The problem is that one must wait for a relative long time to make the exposure of α particles, usually several days depending on the intensity of the α particle emissions.

Another scheme involves combining α track counting with fission track counting. In other words, after α exposure of one sheet of track detector to the hot particles, another sheet of track detector is put on the same sample having the hot particles. The whole sandwich is inserted into the core of a nuclear reactor for irradiation with thermal neutrons. The thermal neutrons induce fission of ^{235}U or ^{239}Pu in the hot particles. The fission fragments are recorded by the track detector. After etching the detector, the tracks of fission fragments emitted from the same hot particle are counted, whose distribution is similar to the α tracks like a star, but the tracks are much longer than the α tracks.

Measuring the fluence of thermal neutrons irradiated to the hot particle and counting the tracks of fission fragments, one can derive the number of atoms of uranium (U) or plutonium (Pu) in the hot particle.

For a given hot particle with special composition of isotopes, the number of fission tracks irradiated at constant neutron fluence and the number of α particle tracks at constant exposure time are both constants. All of the hot particles with the same composition of isotopes will fall on a straight line in the figure of the number of fission tracks versus the number of α tracks, if their irradiation in the nuclear reactor and exposure to α particles are both the same, respectively.

Fig. 3.51 shows the straight lines of natural uranium (U) and ^{239}Pu under the conditions of thermal neutrons (1×10^{15} n/cm^2) and α exposure time (1 h).

Comparing the results of measurements for an unknown hot particle with the straight lines, one can roughly infer the composition, origin and migration information of the hot particle (Vlasova et al., 2008). For more exact information about the composition, origin and migration of the hot particle, one can use a SEM with energy disperse x-ray spectrometry (EDX) to measure the elemental composition, and secondary ion mass-spectrometr (SIMS) to analyze the isotopic composition, and silicon semiconductor detector to measure the α particle energy (Vlasova et al., 2008). All of the above measurements must use the same hot particle, which can be located and relocated in each instrument by its α tracks or fission tracks in the track detectors. The movements of the hot particle from one instrument to anther can be performed with a micromanipulator.

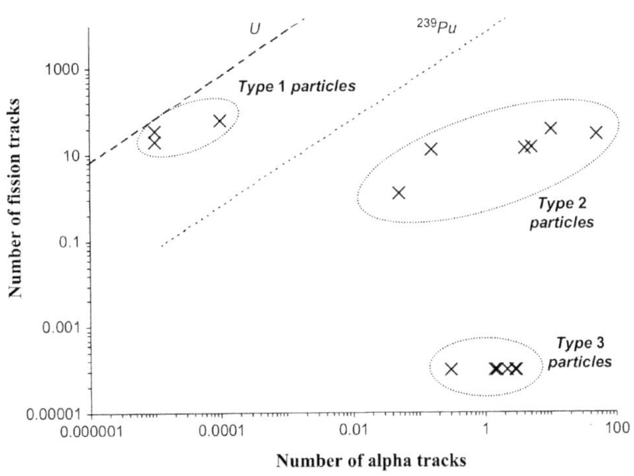

FIG. 3.51 The straight lines of natural ^{238}U and ^{239}Pu under the conditions of thermal neutrons (1×10^{15} n/cm^2) and α exposure time (1 h). *From Vlasova et al. (2008); printed with permission from Elsevier © 2008.*

An α particle track investigation in soil samples from the region of the Chernobyl nuclear power plant accident was carried out (Kashkarov et al., 1995, 2003; Bondarenko et al., 1995). The investigation showed that at least four groups of α-hot particles can be identified. The first group consists of the smallest ($\leq$5 μm) hot particles with a very high α activity ($\leq$10 Bq/particle); the second group is characterized by size 8–12 μm and activity 1–5 Bq/particle; the third and fourth groups are aggregates of hot particles with soil dust. The results showed that the major process of the formation of the hot particles was high-temperature melting and explosion of reactor core substance in the Chernobyl nuclear power plant and subsequent fast cooling of the hot particles, which formed the first and second groups of high α-intensity hot particles with small sizes. Afterward, aggregates of the hot particles with soil dust were formed. The hot particles (aerosols) fell out gradually from the air and deposited onto the soils. The number of the fallout hot particles in the soil decreases with increase of the distance from the nuclear power plant (Dushin et al., 1995).

Nuclear weapons testing at the Nevada test site (Mercury, Nevada, USA) formed a tremendous amount of hot particles with α-radioactivity. CR-39 detectors pressed tightly onto the surface of the soil sample for several days had provided the information of α radioactivities in the soils in the area. It showed that α radioactivity reached a maximum value in the depth between 0 and 5 cm beneath the surface of the soils (Espinosa et al., 1995).

2. *Drainage contamination of nuclear plants*

Hot particles with α radioactivity are always more or less produced in nuclear plants. A solid-state track detector sensitive to α particles can be used to measure the density of the hot particles in the working places and surrounding environments. The techniques used in this type of measurement are similar to those used in the investigation for the fallout of hot particles from nuclear disasters like Chernobyl. Detailed investigations have been carried out with track techniques among other techniques for the measurement of contamination of the Yenisey River bottom sediments near the Krasnoyarsk Mining and Chemical Combine in Siberia, Russia (Vlasova et al., 2008, 2015; Pavlotskaya et al., 2003; Baskaran et al., 1995; Bolsunovsky and Tcherkezian, 2001; Linnik et al., 2006).

V. Nanotechnology and radiation induced material modifications

This field of science includes: (1) nuclear track membranes, (2) microstructures, (3) nanostructures, and (4) material modifications. All these fields are derivatives from

SSNTDs. These fields developed rapidly since the 1980s. In light of these developments, the nuclear track community changed the name of the series of international conferences from Solid State Nuclear Track Detectors to Nuclear Tracks in Solids in 1992, because the micro- or nanodevice is no longer a detector for recording radiation. Micro- or nanostructures might be used as detectors in the future, but not in the present. The interested reader is referred to the treatise "Ion Tracks and Microtechnology, Principles and Applications" by Reimar Spohr (1990) and the review papers by Chakarvarti (2009), Fink et al. (2008), Spohr (2008), Apel et al. (2008), and articles: Liu et al. (2006); He and Guo (2015); Yang et al. (2018); Lounis-Mokrani et al. (2003); Mishra et al. (2003); Severin et al. (2005); Sun et al. (2004); Vater, 1988; Vonderau et al. (1995); Zhu et al. (1988); Guo (1998); Wang et al. (2018); Shakeri Jooybari et al. (2015); Shakeri Jooybari et al. (2015a); Zagorskiy et al. (2015); Gumirova et al. (2015); Guo and Chen (2014); Chauhan and Rana (2015).

Acknowledgments

The authors wish to thank Michael F. L'Annunziata for his careful correction of the manuscript sentence by sentence and giving valuable help and suggestions at every stage in finishing the manuscript. The authors also thank Academician Prof. Wang Yannai and Prof. Guo Bing in China Institute of Atomic Energy (CIAE) for their help and support in drafting the manuscript. The authors are also grateful to Liu Senlin, Wang Chen, Shen Yan, Chen Yan, Liu Fulong, Tian Baoxian, Huang Haiyan, and Guo Li in CIAE and Zhang Yanxia in National Library of China for their generous help in preparation of the manuscript.

References

Abu-Jarad, F., Fazal-ur-Rehman, M.N., Al-Jarallah, M.N., 2002. Thermal neutron equivalent dose assessment around the KFUPM neutron source storage area using NTDs. Radiat. Prot. Dosim. 102 (4), 349–354.

Abumurad, K.M., Al-Omari, R.A., 2008. Indoor radon levels in irbid and health risk from internal doses. Radiat. Meas. 43, S389–S391.

Adloff, J.C., Brandt, R., Debeauvais, M., Fernandez, F., Krivopustov, M., Kulakov, B.A., Sosnin, A., Zamani, M., 1999. Secondary neutron production from thick Pb target by light particle irradiation. Radiat. Meas. 31, 551–554.

Agosteo, S., Campi, F., Caresana, M., Ferrarini, M., Porta, A., Silari, M., 2007. Sensitivity study of CR39 track detector in a system of extended range Bonner spectrometer. Radiat. Prot. Dosim. 126, 310–313.

Agosteo, S., Caresana, M., Ferrarini, M., Silari, M., 2009. A passive rem counter based on CR39 SSNTD coupled with a boron converter. Radiat. Meas. 44 (9–10), 985–987.

Agosteo, S., Caresana, M., Ferrarini, M., Silari, M., 2010. A dual-detector extended range rem-counter. Radiat. Meas. 45, 1217–1219.

Ahlen, S.P., 1980. Theoretical and experimental aspects of the energy loss of relativistic heavily ioning particles. Rev. Mod. Phys. 52, 121–173.

Ahlen, S.P., Coan, T.E., Drach, J., Guo, S.-L., Price, P.B., Salamon, M.H., Tarle, G., Tincknell, M.L., 1984. Identification of relativistic nuclei with $10 \leq Z \leq 92$. Nucl. Tracks Radiat. Meas. 8, 571–574.

Akselrod, G.M., Akselrod, M.S., Benton, E.R., Yasuda, N., 2006. A novel Al_2O_3 fluorescent nuclear track detector for heavy charged particles and neutrons. Nucl. Instrum. Meth. B247, 295–306.

Al-Ghamdi, H., Fazal-ur-Rehman, Al-Jarallah, M.I., Maalej, N., 2008. Photoneutron intensity variation with field size around radiotherapy linear accelerator 18-MeV X-ray beam. Radiat. Meas. 43, S495–S499.

Al-Jarallah, M.I., Naqvi, A.A., Fazal-ur-Rehman, Abu-Jarad, F., 2002. Fast and thermal neutron intensity measurements at the KFUPM PGNAA setup. Nucl. Instrum. Meth. Phys. Res. Sect. B 195, 435–441.

Al-Najjar, S.A.R., Bull, R.K., Durrani, S.A., 1979. Electrochemical etching of CR-39 plastic: applications to radiation dosimetry. Nucl. Tracks 3, 169–183.

Alter, H.W., Price, P.B., 1968. Radon Detection Using Track Registration Material. U.S. Patent 3,665,194, filed (1968), issued (1972).

Apel, P.Y., Blonskaya, I.V., Dmitriev, S.N., Mamonova, T.I., Orelovitch, O.L., Sartowska, B., Yamauchi, Y., 2008. Surfactant-controlled etching of ion track nanopores and its practical applications in membrane technology. Radiat. Meas. 43, S552–S559.

Aslam, A.A., Matiullah, Durrani, S.A., 1988. Fast neutron spectrometry based on the registration of proton recoils in chemically etched CR-39 detectors: part 1. Theoretical. Nucl. Tracks Radiat. Meas. 15, 503–506.

ATSDR, 1999. Agency for Toxic Substances and Disease Registry. Toxicological Profile BEIR VI. Report of the Committee on the Biological Effects of Ionzing Radiation. Natl. Res. Council. Natl. Acad. Press, Washington, DC.

Baiocchi, P., Cecchini, S., Dekhissi, H., Garuth, V., Giacomelli, G., Giani, G.G., Katsavounidis, E., Iori, G., Patrizii, L., Popa, I.V., et al., 1995. Calibration with relativistic and low velocity ions of a CR-39 nuclear track detector. Radiat. Meas. 25, 145–150.

Baixeras, C., Bach, J., Amgarou, K., Moreno, V., Font, L., 2005. Radon level in the volcanic region of La Garrotxa, Spain. Radiat. Meas. 40, 509–512.

Balanzat, E., Beth, N., Bouffard, S., 1995. Swift heavy ion modification of polymers. Nucl. Instrum. Methods B 105, 46–54.

Baluch, J.J., Khan, E.U., Tahseen, R., Qureshi, I.E., Nasir, T., Hassan, N., 2006. Study of deep inelastic collisions in the heavy-ion interaction of (14.0MeV/u) $^{208}Pb + ^{238}U$. Radiat. Meas. 41, 217–221.

Barillon, R., 2005. Track etch velocity and chemical damages induced by ions in a cellulose nitrate detector. Radiat. Meas. 40, 214–217.

Barillon, R., Fromm, M., Chambaudet, A., Katz, R., Stoquert, J.P., Pape, A., 1999. Radiat. Meas. 31, 71.

Barillon, R., Fromm, M., Katz, R., Chambaudet, A., 2002. Chemical bonds broken in latent tracks of light ions in plastic track detectors. Rad. Prot. Dosim. 99 (1–4), 359–362.

Barillon, R., Katz, R., 2000. Tokyo, November 6-8. Proceedings of the 1st International Symposium on Supercritical Water Cooled Reactor, 406, p. 1.

Barillon, R., Mori, Y., Kanasaki, M., Yamauchi, T., Yasuda, N., 2013. Chemical cross sections induced by ions in solid organic detectors: experimentation and simulation. Radiat. Meas. 38–42.

Barillon, R., Yamauchi, T., 2003. Chemical bond scission induced by $^{1}H^{+}$, $^{16}O^{8+}$ and γ-rays in a cellulose nitrate detector. Nucl. Instrum. Meth. B 208, 336–339.

Barillon, R., Yamauchi, T., Mori, Y., Raffy, Q., 2015. A first attempt to simulate oxidation effects on latent track structure in PADC combining the radial dose theory and a radio-oxidation kinetic model. Radiat. Meas. 83, 1–4.

Barkas, W.H., 1963. Nuclear Research Emulsions. Academic Press, New York, p. 481.

Barkas, W.H., Berger, M.J., 1964. Tables of energy losses and ranges of heavy charged particles. In: Studies in Penetration of Charged Particles in Matter. Nuclear Science Series, Report Number 39, Committee on Nuclear Science. Publication 1133. National Academy of Sciences – Natural Research Council, Washington, D. C., pp. 103–172

Barwick, S.W., Price, P.B., Stevenson, J.D., 1985. Radioactive decay of ^{232}U by ^{24}Ne emission. Phys. Rev. C31, 1984–1986.

Baskaran, M., Asbill, S., Santschi, P., Davis, T., Brooks, J., Champ, M., Makeyev, V., Khlebovich, V., 1995. Distribution of ^{239,240}Pu and ^{238}Pu concentrations in sediments from the Ob and Yenisey rivers and Kara sea. Appl. Radiat. Isot. 46, 1109–1119.

Baude, S., Chiappini, R., 2002. Isotopic Measurements on Micrometric Particles: The French Experience to Detect Fission Material. IAEA-SM-367/10/05.

Baumhauer, H., 1894. Die Resultate der Aetzmethode. Leipzig, pp).

Becquerel, A.H., 1896a. On the rays emitted by phosphorescence. Comptes Rendus Acad. Sci. Paris 122, 420.

Becquerel, A.H., 1896b. On the invisible rays emitted by phosphorescent bodies. Comptes Rendus Acad. Sci. Paris 122, 501.

Bedogni, R., Esposito, A., Lovestam, G., Garcia, M.J., Virgolici, M., 2008. The new PADC fast neutron dosimetry system of the INFN-LNF. Radiat. Meas. 43, S491–S494.

Bedogni, R., Ferrari, P., Gualdrini, G., Esposito, A., 2010. Design and experimental validation of a Bonner Sphere Spectrometer based on Dysprosium activation foils. Radiat. Meas. 45, 1201–1204.

Belafrites, A., Nourreddine, A., Higueret, S., Le, T., Trocme, M., 2008. Dependence of PN3 response to Am-Be neutrons on etching and reading process. Radiat. Meas. 43, S482–S486.

Benton, E.V., 1967. Charged Particle Tracks in Polymers No. 4: Criterion for Track Registration. USNRDL-TR-67-80. U. S. Nav. Rad. Def. Lab., San Francisco, Calif.

Benton, E.V., 1968. A Study of Charge Particle Tracksin Cellulose Nitrate. U. S. Naval Radiological Defense Laboratory. Report TR-68-14.

Benton, E.V., 1970. On latent track formation in organic nuclear charged particle track detectors. Radiat. Eff. 2, 273–280.

Benton, E.V., Henke, R.P., 1969a. Sensitivity enhancement of Lexan nuclear track detector. Nucl. Instrum. Meth. 70, 183–184.

Benton, E.V., Henke, R.P., 1969b. Heavy particle range-energy relations for dielectric nuclear track detectors. Nucl. Instrum. Methods 67, 87–92.

Benton, E.V., Nix, W.D., 1969. The restricted energy loss criterion for registration of charged particles in plastics. Nucl. Instrum. Meth. 67, 343–347.

Benton, E.V., Henke, R.P., Tobias, C.A., 1973. Heavy-particle radiography. Science 182, 474–476.

Bernardi, L., Cecchi, A., Gori, C., Lucarelli, F., Renzi, R., 1991. Studies of the response of CR-39 track detectors to protons from a 3 MeV Vab de Graaff accelerator. Nucl. Instrum. Methods Phys. Res. Sect. B 53, 61–66.

Bernet, M., Garver, J.I., 2005. Fission-track analysis of detrital zircon. Rev. Miner. Geochem. 58, 205–238.

Bethe, H.A., 1930. Theory of the passage of rapid corpuscular rays through matter. Ann. Phys. 5, 325–400.

Bigazzi, G., 1981. The problem of the decay constant lf of ^{238}U. Nucl. Tracks 5, 35–44.

Bigazzi, G., Marton, P., Norelli, P., Rozloznik, L., 1990. Fission track dating of Carpathian obsidians and provenance identification. Nucl. Tracks Radiat. Meas. 17, 391–396.

Binnig, G., Roher, H., Gerber, C., Weibe, E., 1982. Surface studies by scanning tunneling microscope. Phys. Rev. Lett. 49, 57–61.

Binning, G., Quate, C.F., Gerber, C., 1986. Atomic force microscope. Phys. Rev. Lett. 56, 930–933.

Bochicchio, F., Forastiere, F., Farchi, S., Marocco, D., Quarto, M., Sera, F., 2003. Quality assurance program for LR-115 based radon concentration measurements in a case-control study: description and results. Radiat. Meas. 36, 205–210.

Bochicchio, F., Ampollini, M., Tommasino, L., Sorimachi, A., Tokonami, S., 2009. Sensitivity to thoron of an SSNTD-based passive radon measuring device: experimental evaluation and implications for radon concentration measurements and risk assessment. Radiat. Meas. 44, 1024–1027.

Bolsunovsky, A.Y., Tcherkezian, V.O., 2001. Hot-particles of the Yenisey River flood plain, Russia. J. Environ. Radioact. 57, 167–174.

Bondarenko, O.A., Henshaw, D.L., Salmon, P.L., Ross, A.N., 1995. The method of simultaneous size and activity measurement of alpha emitting hot particles using multiple track analysis of solid state nuclear track detectors. Radiat. Meas. 25 (1–4), 373–376.

Bonetti, R., Chiesa, C., Guglielmetti, A., Migliorino, C., Cesana, A., Terrani, M., Price, P.B., 1991. Neon radioactivity of uranium isotopes. Phys. Rev. C 44, 888–890.

Bonetti, R., Chiesa, C., Guglielmetti, A., Migliorino, C., Monti, P., 1993a. In: Proceedings of Second International Conference on Cluster Phenomena in Atoms and Nuclei (Santorini, Greece).

Bonetti, R., Chiesa, C., Guglielmetti, A., Matheoud, R., Migliorino, C., 1993b. Nuclear structure effects in the exotic decay of 225Ac via 14C emission. Nucl. Phys. A 562, 32–40.

Bowman, C.D., Arthur, E.D., Lisowski, P.W., Lawrence, G.P., Jensen, R.J., Anderson, J.L., Blind, B., Cappiello, M., Davidson, J.W., England, T.R., Engel, L.N., Haight, R.C., Hughes III, H.G., Ireland, J.R., Krakowski, R.A., Labauve, R.J., Letellier, B.C., Perry, R.T., Russell, G.J., Staudhammer, K.P., Versamis, G., Wilson, W.B., 1992. Nuclear energy generation and waste transmutation using an accelerator-driven intense thermal neutron source. Nucl. Instrum. Meth. A320, 336–367.

Brandt, R., 1980. Overall review of the applications of SSNTD's in fission physics. Nucl. Instrum. Meth. 173, 147–153.

Brandt, R., Ditlov, V.A., Dwivedi, K.K., Ensinger, W., Ganssauge, E., Shi-Lun, G., Haiduc, M., Hashemi-Nezhad, S.R., Khan, H.A., Krivopustov, M.I., Odoj, R., Pozharova, E.A., Smirnitzky, V.A., Sosnin, A.N., Westmeier, W., Zamani-Valasiadou, M., 2005. Interactions of relativistic heavy ions in thick element targets e and some unsolved problems. JINR Preprint El-2005-167, Dubna, pp. 1–44.

Brandt, R., Ditlov, V.A., Dwivedi, K.K., Ensinger, W., Ganssauge, E., Shi-Lun, G., Haiduc, M., Hashemi-Nezhad, S.R., Khan, H.A., Krivopustov, M.I., Odoj, R., Pozharova, E.A., Smirnitzky, V.A., Sosnin, A.N., Westmeier, W., Zamani-Valasiadou, M., 2008. Studies with SSNTD and nuclear chemistry on nuclear reactions induced by relativistic heavy ions in thick targets: a review. Radiat. Meas. 43, S132–S138.

Cao, L., Wang, Y., 2009. Fabrication and investigation of single track-etched nanopore and its applications. Radiat. Meas. 44, 1093–1099.

Cartwright, B.G., Shirk, E.K., Price, P.B., 1978. A nuclear-track-recording polymer of unique sensitivity and resolution. Nucl. Instrum. Meth. 153, 457–460.

Cecchini, S., Patrizii, L., 2008. Nuclear track detectors in astroparticle and nuclear physics. Radiat. Meas. 43, S144–S150.

Cecchini, S., Dekhissi, H., Giacomelli, G., Mantrioli, G., Margiotta, A.R., Patrizii, L., Predieri, F., Serra, P., Spurio, M., 1993. Fragmentation cross sections and search for nuclear fragments with fractional charge in relativistic heavy ion collisions. Astropart. Phys. 1, 369–376.

Cecchini, S., Chiarusi, T., Di Ferdinando, D., Cozzi, M., Frutti, M., Giacomelli, G., Kumar, A., Manzoor, S., McDonald, J., Medinaceli, E., Nogales, J., Patrizii, L., Pinfold, J., Popa, V., Qureshi, I.E., Saavedra, O., Sher, G., Shahzad, M.I., Spurio, M., Ticona, R., Togo, V., Velarde, A., 2005. Search for intermediate mass magnetic monopoles and nuclearites with the SLIM experiment. Radiat. Meas. 40, 405–409.

Cecchini, S., Giacomelli, G., Giorgini, M., Mandrioli, G., Patrizii, L., Pora, V., Serra, P., Sirri, G., Spurio, M., 2002. Fragmentation cross sections of 158 A GeV Pb ions in various targets measured with CR39 nuclear track detectors. Nucl. Phys. A 707, 513–524.

Ceccotti, T., Levy, A., Popescu, H., Reau, F., D'Oliveira, P., Monot, P., Geindre, J.P., Lefebvre, E., Martin, P., 2007. Proton acceleration with high-intensity ultrahigh-contrast laser pulses. Phys. Rev. Lett. 99 (185002), 1–4.

Chadderton, L.T., Cruz, S.A., Fink, D.W., 1993. Theory for latent particle tracks in polymers. Nucl. Tracks Radiat. Meas. 22, 29–38.

Chakarvarti, S.K., 2009. Track-etch membranes enabled nano-/micro-technology: a review. Radiat. Meas. 44, 1085–1092.

Chakarvarti, S.K., Mahna, S.K., 1992. Bulk-etching behavior of Kapton and thermalimide track detectors. Nucl. Tracks Radiat. Meas. 20 (4), 589–594.

Chambaudet, A., Miellou, J.C., Igli, H., Rebetez, M., Grivet, M., 1993. Thermochronology by fission tracks: an exact inverse method associated with the resolution of a single ordinary differential equation (ode). Nucl. Tracks Radiat. Meas. 22 (1–4), 763–772.

Chambaudet, A., Roncin, J., 1981. I. R., U. V. and E. S. R. studies on plastic track detectors irradiated by heavy ions. In: Proceedings of the 11th International Conference on Solid State Nuclear Track Detectors, Bristol, 7-12 September, 1981. Pergamon Press, Oxford.

Chauhan, R.P., Chakarvarti, S.K., 2002. Indian J. Pure Appl. Phys. 39, 491.

Chauhan, R.P., Rana, P., 2015. Nickel ion induced modification in the electrical conductivity of Cu nanowires. Radiat. Meas. 83, 43–46.

Chauhan, R.P., Upadhyay, S.B., Narula, A.K., Chakarvarti, S.K., 2006. Effect of grain size on radon diffusion through some building construction materials. Chem. Environ. Res. 15 (3&4), 223–227.

Chauhan, R.P., Nain, M., Kant, K., 2008. Radon diffusion studies through some building materials: effect of grain size. Radiat. Meas. 43, S445–S448.

Chen, B.-L., Yin, G.M., Li, W.-L., Guo, L., Li, L., Lu, Y.-C., Guo, S.-L., 2001. Determination of tectonic uplift rates of Qinling Mountains in Central China by fission tracks. Radiat. Meas. 34, 405–408.

Chen, G., Li, S., Zhang, H., Yang, F., Ding, C., Lei, P., Hu, Y., 2013. Fluid inclusion analysis for constraining the hydrocarbon accumulation periods of the Permian reservoirs in Northeast ordos basin. J. Earth Sci. 24, 589–598.

Chen, L., Pan, Z., Liu, S., Ma, J., Wu, Y., Liu, F., Wang, C., 2006. Investigation of the levels of ^{222}Rn and ^{220}Rn in Coal mines in China. Radiat. Protect. 26, 193.

Chen, Y., Shen, Y., Chang, Z.-Y., Zhao, Y.-G., Guo, S.-L., Cui, J.-Y., Liu, Y.-A., 2013. Studies on the analyzing single uranium-bearing particle by FT-TIMS. Radiat. Meas. 50, 43–45.

Chen, Y., Wang, F., Zhao, Y.-G., Li, L.-L., Zhang, Y., Shen, Y., Chang, Z.-Y., Guo, Shi-Lun, Wang, Xiao-ming, Cui, J.-Y., Liu, Y.-A., 2015. An improved FT-TIMS method of measuring uranium isotope ratios in the uranium-bearing particles. Radiat. Meas. 83, 63–67.

Chen, Y., Xu, C.-K., Zhao, Y.-G., Guo, S.-L., Li, L.-L., Chang, Z.-Y., Liu, G.-R., 2018. Laser Microsissection System Applied in Improving FT-TIMS. Perspective in Sciences (in press).

Chong, C.S., Ishak, I., Mahat, R.H., Amin, Y.M., 1997. Radiat. Meas. 28, 119.

Collaboration Group of China Institute of Atomic Energy and Guiyang Institute of Geochemistry, 1976. Fission track dating of muscovite and tektite. Geochimica 2, 148–156.

Costantini, J.–M., Couvreur, F., Salvetat, J.–P., Bouffard, S., 2002. Micro-Raman Study of the carbonization of polyimide induced by swift heavy ion irradiations. Nucl. Instrum. Meth. B 194, 132–140.

Crawford, W.T., DeSorbo, W., Humphrey, J.S., 1968. Enhancement of track etching rates in charged particle-irradiated plastics by a photo-oxidation effect. Nature 220, 1313–1314.

Cross, W.G., 1986. Characteristics of track detectors for personnel neutron dosimetry. Nucl. Tracks 12, 533–542.

Cross, W.G., Tommasino, L., 1968. Electrical detection of fission fragment tracks for fast neutron dosimetry. Health Phys. 15, 196.

Cross, W.G., Tommasino, L., 1970. A rapid reading technique for nuclear particle damage tracks in thin foils. Radiat. Effects 5, 85–89.

Cross, W.G., Arneja, A., Ing, H., 1986. The response of electrochemically etched CR-39 to protons of 10 keV to 3 MeV. Nucl. Tracks 12, 649–652.

Dajko, G., 1990. Fast neutron spectrometry using CR-39 track detectors. Radiat. Prot. Dosim. 34, 9–12.

Darraud, C., Bennamane, B., Gagnadre, C., Decossas, J.L., Vareille, J.C., 1994. Optical modifications of polymers by ion beam irradiation. Polymers 35 (11), 2447–2451.

Dehaye, F., Balanzat, E., Ferain, E., Legras, R., 2003. Chemical modifications induced in bisphenol A polycarbonate by swift heavy ions. Nucl. Instrum. Meth. B 209, 103–113.

Derkaoui, J., Giacomelli, G., Lari, T., Margiotta, A., Ouchrif, M., Patrizii, L., Popa, V., Togo, V., 1998. Energy losses of magnetic monopoles and dyons in the earth. Astropart. Phys. 9, 173–183.

Derkaoui, J., Giacomelli, G., Lari, T., Mantrioli, G., Ouchrif, M., Patrizii, L., Popa, V., 1999. Energy loss of magnetic monopoles and dyons in scintillators, streamer tubes and nuclear track detectors. Astropart. Phys. 10, 339–352.

DeSorbo, W., 1979. Ultraviolet effects and aging effects on etching characteristics of fission tracks in polycarbonate film. Nucl. Tracks 3, 13–32.

Dhairyawan, M.P., Marathe, P.K., Massand, O.P., 2003. Use of CR-39 solid state nuclear track detector in neutron personnel monitoring. Radiat. Meas. 36, 435–438.

Dirac, P.A.M., 1931. Quantised singularities in the electromagnetic field. Proc. Roy. Soc. London A 133, 60–72.

Ditlov, V., 1995. Calculated tracks in plastics and crystals. Radiat. Meas. 25 (1–4), 89–94.

Doke, T., 2008. Development of an automatic system for precisisely analyzing massive data in track detectors and its application to Space Station or longduration balloon experiments. Radiat. Meas. 43, S156–S162.

Domingo, C., Garcia-Fuste, M.J., Morales, E., Amgarou, K., Terron, J.A., Rosello, J., Brualla, L., Nunez, L., Colmenares, R., Gomez, F., Hartmann, G.H., Sanchez-Doblado, F., Fernandez, F., 2010. Neutron spectrometry and determination of neutron ambient dose equivalents in different LINAC radiotherapy rooms. Radiat. Meas. 45, 1391–1397.

Donohue, D.L., 2002. Strengthened nuclear safeguards. Anal. Chem. 74, 28A.

Drach, J., Price, P.B., Salamon, M.H., 1987. Characteristics of Cronar polyethylene terephthalate track detectors. Nucl. Instrum. Meth. Phys. Res. Sect. B 28, 49–52.

Duan, X., Lan, X., Tan, Z., Huang, Y., Guo, S., Yang, D., Wang, N., 2009. Calibration of CR-39 with monoenergetic protons. Nucl. Instrum. Meth. Phys. Res. Sect. A 609, 190–193.

Duan, X.-J., Tan, Z.-X., Lan, X.-F., Huang, Y.-S., Guo, S.-L., Yang, D.-W., Tan, X.-Z., Wang, N.-Y., 2010. Calibration of solid state nuclear track detector CR-39 with monoenergetic protons. Acta Phys. Sin. 59, 221–227.

Dumitru, T.A., 2000. Fission-Track geochronology. In: AGU Reference Shelf 4. Quaternary Geochronology: Method and Application, pp. 131–155. Copyright 2000 by the American Geophysical Union.

Dunlop, A., Jaskierowicz, G., Ossi, P.M., Della-Negra, S., 2007. Transformation of graphite into nanodiamond following extreme electronic excitations. Phys. Rev. B 76, 155403 (14 pages).

Durrani, S.A., Bull, R.K., 1987. Solid State Nuclear Track Detection. Pergamon Press, Oxford, p. 304.

Durrani, S.A., Ilic, R., 1997. Radon Measurements by Etch Track Detectors: Applications to Radiation Protection, Earth Sciences and Environment. World Scientific, Singapore, p. 387.

Durrani, S.A., Matiullah, Aslum, A.A., 1988. Fast neutron spectrometry based on the registration of proton tracks in electrochemically etched CR-39 detectors: Part 2. Exp. Nucl. Tracks Radiat. Meas. 15, 499–502.

Dushin, V.N., Gedeonov, A.D., Petrov, B.F., Pleskachevsky, L.A., Shigaev, O.E., 1995. Application of polycarbonate track detectors for the technogenic alpha-radioactive contamination analysis. Radiat. Meas. 25 (1–4), 409–412.

Dwivedi, K.K., Raju, J., Vater, P., Brandt, R., 1993. Study on the fission of heavy ions and particle evaporation using nuclear track detectors. Nucl. Tracks Radiat. Meas. 22, 577–582.

Dwivedi, K.K., Mishra, R., Tripathy, S.P., 2005. An extensive indoor ^{222}Rn/^{220}Rn monitoring in north-east India. Radiat. Meas. 40, 621–624.

Eappan, K.P., Mayya, Y.S., 2004. Calibration factors for LR-115 (type-II) based radon thoron discriminating dosimeter. Radiat. Meas. 38, 5–17.

El-Shahawy, A., Hussein, A., Tawansi, J., 1992. J. Mater. Sci. 27, 6605.

Ensinger, W., Vater, P., Happel, S., Sudowe, R., Brandt, R., 2003. Nuclear track microfilters: gas separation ability and β-radiation stability. Radiat. Meas. 36, 707–711.

Esaka, K.T., Esaka, F., Inagawa, J., Iguchi, K., Lee, C.G., Sakurai, S., Watanabe, K., Usuda, K., 2004. Application of fission track technique for the analysis of individual particles containing uranium in safeguard swipe samples. Jpn. J. Appl. Phys. 43, L915–L916.

Espinosa, G., Meyer, K., Gammage, R.B., 1995. Soil measurement by nuclear track detectors. Radiat. Meas. 25 (1–4), 401–404.

Esposito, A., Bedogni, R., Gentile, A., 2008. Monitoring the ambient neutron radiation around a high energy electron/positron collider using PADC dosimeters. Radiat. Meas. 43, S487–S490.

Fano, U., 1964. Penetration of proton, alpha particles, and Mesons. In: Studies in Penetration of Charged Particles in Matter. Nuclear Science Series, Report Number 39, Committee on Nuclear Science. Publication 1133. National Academy of Sciences - National Research Council, Washington, D. C., pp. 287–352. Appendix A.

Fernandez, F., Bouassoule, T., Amgarou, K., Domingo, C., Garcia, M.J., Lacoste, V., Gressier, V., Muller, H., 2007. Monte Carlo calculations and validation of a gold-foil based Bonner sphere system. Radiat. Prot. Dosim. 126, 366–370.

Fews, A.P., Henshaw, D.L., 1982. High resolution alpha particle spectroscopy using CR-39 plastic track detector. Nucl. Instrum. Methods 197, 517–529.

Fink, D., Saad, A., Dhamodaran, S., Chandra, A., Fahrner, W.R., Hoppe, K., Chadderton, L.T., 2008. Multiparametric electronic devices based on nuclear tracks. Radiat. Meas. 43, S546–S551.

Fleischer, R.L., 1966. Uranium micromaps: techniques for in situ mapping of distributions of fissionable impurities. Rev. Sci. Instr. 37, 1738–1739.

Fleischer, R.L., 1981. Nuclear track production in solids. In: Christian, J.W., Haasen, P., Massalski, T.B. (Eds.), Progress in Materials Science, Chalmers Anniv, Vol. Pergamon Press, Oxford, pp. 97–123.

Fleischer, R.L., 1988. Radon in the environment-opportunities and hazards. Nucl. Tracks Radiat. Meas. 14 (4), 421–435.

Fleischer, R.L., 1998. Tracks to Innovation, Nuclear Tracks in Science and Technology. Springer, New York.

Fleischer, R.L., Hart Jr., R.H., 1973. Mechanical erasure of particle tracks, a tool for lunar microstratigraphic chronology. J. Geophys. Res. 78, 4841–4851.

Fleischer, R.L., Mogro-Campero, A., 1985. Association of subsurface radon charges in Alaska and the northeastern United States with earthquakes. Geochim. Cosmochim. Acta 49, 1061–1071.

Fleischer, R.L., Price, P.B., 1963a. Tracks of charged particles in high polymers. Science 140, 1221–1222.

Fleischer, R.L., Price, P.B., 1963b. Charge particle tracks in glass. J. Appl. Phys. 34, 2903–2904.

Fleischer, R.L., Price, P.B., 1964. Fission track evidence for the simultaneous origin of tektites and other natural glasses. Geochim. Cosmochim. Acta 28, 755–760.

Fleischer, R.L., Price, P.B., Symes, E.M., 1964a. A novel filter for biological studies. Science 143, 249–250.

Fleischer, R.L., Price, P.B., Walker, R.M., Hubbard, E., 1964b. Track registration in various solid state nuclear track detectors. Phys. Rev. 133A, 1443–1449.

Fleischer, R.L., Price, P.B., Walker, R.M., 1965a. Solid-state track detectors: applications to nuclear science and geophysics. Ann. Rev. Nucl. Sci. 15, 1–28.

Fleischer, R.L., Price, P.B., Walker, R.M., 1965b. Spontaneous fission tracks from extinct Pu244 in meteorites and the early history of the solar system. J. Geophys. Res. 70, 2703–2707.

Fleischer, R.L., Price, P.B., Walker, R.M., 1965c. Applications of fission tracks and fission track dating to anthropology. Proc. Seventh Int. Congress on Glass, Brussels 224, 1–7.

Fleischer, R.L., Price, P.B., Walker, R.M., Leakey, L.S.B., 1965d. Fission track dating of a Mesolithic knife. Nature 205, 1138.

Fleischer, R.L., Price, P.B., Walker, R.M., Leakey, L.S.B., 1965e. Fission track dating of BedI, Olduvai Gorge. Science 148, 72–74.

Fleischer, R.L., Price, P.B., Walker, R.M., Filz, R.C., Fukui, K., Holeman, E., Friedlander, M.W., Rajan, R.S., Tamhane, A.S., 1967a. Tracks of cosmic rays in plastics. Science 155, 187–189.

Fleischer, R.L., Price, P.B., Walker, R.M., Hubbard, E., 1967b. Criterion for registration in dielectric track detectors. Phys. Rev. 156, 353–355.

Fleischer, R.L., Price, P.B., Walker, R.M., Maurette, M., 1967c. Origin of fossil charged-particle tracks in meteorites. J. Geophys. Res. 72, 331–353.

Fleischer, R.L., Price, P.B., Walker, R.M., 1967d. Criterion for Registration in dielectric track detectors. Phys. Rev. 156, 353.

Fleischer, R.L., Viertl, J.R.M., Price, P.B., Aumento, F., 1968. Mid-Atlantic Ridge: age and spreading rates. Science 161, 1339–1342.

Fleischer, R.L., Hart Jr., R.H., Jacobs, I.S., Price, P.B., Schwarz, W.M., Aumento, F., 1969a. Search for magnetic monopoles in deep ocean deposits. Phys. Rev. 184, 1393–1397.

Fleischer, R.L., Price, P.B., Walker, R.M., 1969b. Quaternary dating by the fission-track techniques. In: Brothwell, D.R., Higgs, H. (Eds.), Science in Archaeology, second ed. Thames and Hudson, London.

Fleischer, R.L., Price, P.B., Woods, R.T., 1969c. Search for tracks of massive, multiple charged magnetic poles. Phys. Rev. 184, 1398–1401.

Fleischer, R.L., Price, P.B., Woods, R.T., 1969d. Nuclear particle track identification in inorganic solids. Phys. Rev. 188, 563–567.

Fleischer, R.L., Viertl, J.R.M., Price, P.B., 1969e. Age of the Manicouagan and clearwater lakes craters. Geochim. Cosmochim. Acta 33, 523–527.

Fleischer, R.L., Viertl, J.R.M., Price, P.B., Aumento, F., 1971. A chronological test of ocean bottom spreading in the North Atlantic. Radiat. Effects 11, 193–194.

Fleischer, R.L., Comstock, G.M., Hart Jr., H.R., 1972. Dating of mechanical events by deformation-induced erasure of particle tracks. J. Geophys. Res. 77, 5050–5053.

Fleischer, R.L., Woods, R.T., Hart Jr., H.R., Price, P.B., Short, N.M., 1974. Effect of shock on fission track dating of apatite and sphene crystals from the Hardhat and Sedan Underground nuclear explosions. J. Geophys. Res. 79, 339–342.

Fleischer, R.L., Price, P.B., Walker, R.M., 1975. Nuclear Tracks in Solids: Principles and Applications. University of California Press, Berkeley, p. 605.

Fleischer, R.L., Hart Jr., H.R., Mogro - Camparo, A., 1980. Radon emanation over an ore body: search for long-distance transport of radon. Nucl. Instrum. Meth. 173, 169–181.

Flerov, G.N., Oganesyan, Y.T., Lobanov, Y.V., Kuznetsov, V.I., Druin, V.A., Perelygin, V.P., Gavrilov, K.A., Tretiakova, S.P., Plotko, V.M., 1964. Synthesis and physical identification of the isotope of element 104 with mass number 260. Phys. Lett. 13, 73–75.

Font, L., 2009. On radon Surveys: design and data interpretation. Radiat. Meas. 44, 964–968.

Font, L., Baixeras, C., Moreno, V., 2008. Indoor radon levels in underground workplaces of Catalonia, Spain. Radiat. Meas. 43, S467–S470.

Fragopoulou, M., Manolopoulou, M., Jokic, S., Zamani, M., Krivopustov, M., Sosnin, A., Stoulos, S., 2008. Preliminary results on neutron production from a Pb/U target irradiated by deuteron beam at 1.25 GeV/amu. Radiat. Meas. 43, S215–S218.

Fromm, M., Chambaudet, A., Membrey, F., 1988. Data bank for alpha particle tracks in CR-39 with energies ranging from 0.5 to 5 MeV recording for various incident angles. NTRM 15, 115–118.

Fukuda, Y., Sakaki, H., Kanasaki, M., Yogo, A., Jinno, S., Tampo, M., Faenov, A.Y., Pikuz, T.A., Hayashi, Y., Kando, M., Pirozhkov, A.S., Shimomura, T., Kiriyama, H., Kurashima, S., Kamiya, T., Oda, K., Yamauchi, T., Kondo, K., Bulanov, S.V., 2013. Identification of high energy ions using backscattered particles in laser-driven ion acceleration with cluster-gas targets. Radiat. Meas. 50, 92–96.

Gagnadre, C., Decossas, J.L., Vareille, J.C., 1993. IR spectroscopy studies of polycarbonate irradiated by H^+ and Li^+ ions. Nucl. Instrum. Meth. B 73, 48–52.

Garber, D.I., Kinsey, R.R., 1976a. Neutron Cross Sections. BNL 325, third ed., vol. II. Brookhaven National Laboratory, New York, p. 491.

Garber, D.I., Kinsey, R.R., 1976b. Neutron Cross Sections. Volume II, Curves. BNL-325, third ed. National Neutron Cross Section Center, Broohaven National Laboratory Associated Universities, Inc.

Giacomelli, G., Giorgini, M., Mandrioli, G., Manzoor, S., Patrizii, L., Popa, V., Serra, P., Togo, V., Vilela, E.C., 1998. Extended calibration of a CR39 nuclear track detector with 158 A GeV 207Pb ions. Nucl. Instrum. Meth. Phys. Res. Sect. A 411, 41–45.

Giacomelli, G., Patrizii, L., 2005. Magnetic Monopole Searches Hep-Ex/0506014.

Gibbons, A., 1998. In China, a handier Homo Erectus. Science 279, 1636. www.sciencemag.org.

Gleadow, A.J.W., 1978. Anisotropic and variable track etching characteristics in natural sphenes. Nucl. Track Detect. 2, 105–117.

Gleadow, A.J.W., Duddy, I.R., 1981. A natural long-term track annealing experiment for apatite. Nucl. Tracks 5, 169–174.

Gleadow, A.J.W., Duddy, I.R., Lovering, J.F., 1983. Fission track analysis: a new tool for the evaluation of thermal histories and hydrocarbon potential. Austr. Petrol. Exp. Ass. J. 23, 93–102.

Gold, R., Armani, R.J., Roberts, J.H., 1968. Absolute fission rate measurements with solid-state track recorders. Nucl. Sci. Eng. 34, 13–32.

Gomez Alvarez-Arenas, T.E., Apel, P.Y., Orelovitch, O.L., Munoz, M., 2009. New ultrasonic technique for the study of the pore shape of track-etched pores in polymer films. Radiat. Meas. 44, 1114–1118.

Gottschalk, P.A., Vater, P., Becker, H.J., Brandt, R., Grawert, G., Fiedler, G., Haag, R., Rautenber, T., 1979. Direct evidence for multiple sequential fission in the interaction of ^{208}Pb and ^{238}U with uranium. Phys. Rev. Lett. 42, 1728–1732.

Gottschalk, P.A., Grawert, G., Vater, P., Brandt, R., 1983. Two-three- and fourparticle exit channels in the reaction (806 MeV) Kr t U. Phys. Rev. C 27, 2703–2719.

Grabez, B., Dragic, A., 2008. Largest fragment as an order parameter in thermal multifragmentation. Radiat. Meas. 43, S184–S187.

Green, P.F., Duddy, I.R., Gleadow, A.J.W., Lovering, J.F., 1989. Apatite fissiontrack analysis as a paleotemperature indicator for hydrocarbon exploration. In: Naeser, N.D., McCulloh, T.H. (Eds.), Thermal History of Sedimentary Basins. Springer-Verlag, New York.

Guenther, M., Sahre, K., Suchaneck, G., Gerlach, G., Eichhorn, K.-J., 2001. Influence of ion-beam induced chemical and structural modification of polymers on moisture uptake. Surf. Coat. Tech. 142–144, 482–488.

Guglielmetti, A., Bonetti, R., Poli, G., Price, P.B., Westphal, A.J., Janas, Z., Keller, H., Kirchner, R., Klepper, O., Piechaczek, A., Roeckl, E., Schmidt, K., Plochocki, A., Szerypo, J., Blank, B., 1995. Identification of the new isotope 114Ba and search for its a and cluster radioactivity. Phys. Rev. C 52, 740–743.

Gumirova, V.N., Bedin, S.A., Abdulashidov, G.S., Razumovskaya, I.V., 2015. The influence of pores in track etch membranes and prepared on their base polymer/metal composites on their fracture strength. Phys. Procedia 80, 148–150.

Guo, H., Guo, S., Wang, Y., Li, L., Sun, Z., Qiu, N., 1999. Thermal history retracement of geological bodies. Nucl. Tech. 22 (7), 422–424.

Guo, L., Chen, B.-L., 2014. Prospects of nano- and micro nuclear electronics and semi-conductor nuclear detectors based on track templates. Nul. Sci. Tech. Inform. (2), 56–60.

Guo, S., Deng, X., Sun, S., Meng, W., Zhang, P., Hao, X., 1986. Uranium exploration by fission track method. Nucl. Tech. (12), 11–26.

Guo, S., Sun, S., Meng, W., Zhang, P., 1982. Determination of uranium content in natural water by means of fission track method. Nucl. Tech. (1), 61–62.

Guo, S., Yuan, W., Chen, B., Huang, W., Liu, S., 2013. Track fading and its applications in archaeology, tectonics and geothermal chronology in China. J. Earth Sci. 24 (4), 645–651.

Guo, S., Zhou, S., Meng, W., Zhang, P., Sun, S., Hao, X., Liu, S., Zhang, F., Hu, R., Liu, J., 1980. Age determination of Beijing man (Peking man) by fission track dating. Chin. J. Nucl. Phys. 2, 183–188.

Guo, S.-L., 1998. Basic nature of nuclear track membranes for liquid media. New Astron. Rev. 42, 205–216.

Guo, S.-L., 1982a. Some methods in fission track dating of Peking Man. In: Proc. 11th Int. Conf. Solid State Nuclear Track Detectors, Bristol. Pergamon Press, Oxford, pp. 265–269.

Guo, S.-L., 1982b. Study of solid State nuclear track detectors of atomic energy of the academy of sciences of China. In: Proc. 11th Int. Conf. Solid State Nuclear Track Detectors, Bristol. Pergamon Press, Oxford, pp. 189–192.

Guo, S.-L., 1988. Principles, techniques and applications of solid state nuclear track detectors. In: Neutron Physics and Nuclear Data Measurements with Accelerators and Research Reactors. IAEA-TECDOC-469, pp. 142–168. A technical document issued by the international atomic energy agency, Vienna.

Guo, S.-L., 1993. Latest development of nuclear track study in China Institute of atomic energy. Radiat. Meas. 22, 659–668.

Guo, S.-L., 1995. Nuclear track studies and applications relying on reactor neutrons. Radiat. Meas. 25 (1–4), 689–694.

Guo, S.-L., Ganz, M., Fuest, M., Vater, P., Brandt, R., 1990. Further studies on the filtration of liquids using Kapton nuclear track microfilters. Isotopenpraxis 26, 272–275.

Guo, S.-L., Li, Z., Zhou, S.-H., Li, X.-H., 1976a. Preparation of ThO_2 plate threshold neutron detectors. At. Energy Sci. Technol. 1, 77–79.

Guo, S.-L., Zhang, D.-H., Wang, Y.-G., 2017. Advances in studies of nuclear tracks in solids. J. Shanxi Normal University, Natural Science Edition 31, 40–49.

Guo, S.-L., Zhou, S.-H., Meng, W., Sun, S.-F., Shi, R.-F., 1976b. Detection efficiencies of solid state nuclear track detectors for fission fragments. At. Energy Sci. Technol. 1, 65–73.

Guo, S.-L., Zhou, S.-H., Meng, W., Sun, S.-F., Shi, R.-fa, 1982. Detection efficiencies of solid state nuclear track detectors for fission fragments. In: Proc. 11th Int. Conf. Solid State Nuclear Track Detectors, Bristol. Pergamon Press, Oxford, pp. 265–269.

Guo, S.-L., Drach, J., Price, P.B., Salamon, M.L., Tincknell, M.L., Ahlen, S.P., Tarle, G., 1984a. Calibration of Tuffak polycarbonate track detector for identification of relativistic nuclei. Nucl. Tracks 9, 183–188.

Guo, S.-L., Tincknell, M.L., Price, P.B., 1984b. Fission of relativistic intermediate-mass nuclei. Phys. Rev. C 30, 1737–1739.

Guo, S.-L., Price, P.B., Tincknell, M.L., 1986a. Study of relativistic projectile fission with CR-39 stack. Nucl. Tracks 12, 379–382.

Guo, S.-L., Tress, G., Vater, P., Khan, E.U., Dersch, R., Plachky, M., Brandt, R., 1986b. New approach to increase the gas throughput through mica track microfilters by changing their pore structure. Nucl. Tracks Radiat. Meas. 11, 1–4.

Guo, S.-L., Deng, X.-L., Sun, S.-F., Meng, W., Zhang, P.-F., Hao, X.-H., 1986c. Comparison between fission track method and laser-fluorometry and fluorocolorimetry for determination of uranium concentration in natural water. Nucl. Tracks 12, 801–804.

Guo, S.-L., Liu, S.-S., Sun, S.-F., Zhang, Feng, Zhou, S.-H., Hao, X.-H., Hu, R.-Y., Meng, W., Zhang, P.-F., Liu, J.-F., 1991. Age and duration of Peking Man Site by fission track method. Nucl. Tracks Radiat. Meas. 19, 719–724.

Guo, Shilun, Hao, Xiuhong, Sun, Shengfen, Meng, Wu, Deng, Xinlu, Lu, Qi, 1992. Fission track method for large area exploration of uranium. Nucl. Tech. 15 (4), 236–238.

Guo, S.L., Hao, X.H., Chen, B.L., 1996. Fission track dating of Paleolithic site at bose in Guangxi, South China. Acta Anthropol. Sin. 15 (4), 347–350.

Guo, S.-L., Huang, W., Hao, X.-hong, Chen, B.-L., 1997. Fission track dating of ancient man site in Baise, China, and its significances in space research, paleomagnetism and stratigraphy. Radiat. Meas. 28, 565–570.

Guo, S.-L., Li, L., Shen, Q.-B., Shi, Y.Q., Brandt, R., Vater, P., Wan, J.-S., Ensinger, W., Kulakov, B.A., Krivopustov, M.I., Sosnin, A.N., Bradnova, V., 2001. Measurements of neutron yields and spatial distributions in U/Pb, Pb and Hg thick targets bombarded by 0.5 and 1.0 GeV protons. Radiat. Meas 34, 301–304.

Guo, S.-L., Tu, C.-qing, Rong, C.-F., Zhai, G.-N., Wang, Z.-Q., Wang, Y.-L., 2000. Calibration of neutron detection efficiency of bubble detectors. At. Energy Sci. Technol. 34, 212–215.

Guo, S.L., Zhao, Y.G., Shen, Y., Liu, G.R., Li, J.H., Wang, C., 2008. Mathematical and technical procedures of relocation of microscopic aerosol particles for chemical and isotopic analysis. Radiat. Meas. 43, S309–S312.

Guo, S.-L., Li, L., Chen, B.-L., Doke, T., Kikuchi, J., Terasawa, K., Komiyama, M., Hara, K., Fuse, T., Murakami, T., 2003. Status of bubble detectors for high-energy heavy ions. Radiat. Meas. 36, 183–187.

Guo, S.-L., Doke, T., Li, L., Chen, B.-L., Zhang, D.-H., Kikuchi, J., Terasawa, K., Komiyama, M., Hara, K., Fuse, T., Yasuda, N., Murakami, T., 2005. Comparison between theoretical model and experimental calibrations and its inference for track formation in bubble detectors. Radiat. Meas. 40, 229–233.

Guo, S.-L., Doke, T., Zhang, D.-H., Chen, B.-L., Li, L., Hasebe, N., Kikuchi, J., Yasuda, N., Murakami, T., 2015. Parameters of 500 MeV/u ^{56}Fe tracks in bubble detector (BD) T-15 – a new technique to estimate the number and diameter of superheated droplets in bubble detectors. Radiat. Meas. 83, 5–11.

Guo, W., Xue, J.M., Zhang, W.M., Zou, X.Q., Wang, Y.G., 2008. Electrolytic conduction properties of single conical nanopores. Radiat. Meas. S623–S626.

Hadler, J.C., Iunes, P.J., Tello, C.A., Chemale Jr., F., Kawashita, K., Curvo, E.A.C., Santos, F.G.S., Moreira, P.A.F.P., Guedes, S., 2009. Experimental study of a methodology for fission-track dating without neutron irradiation. Radiat. Meas. 44, 955–957.

Hahn, O., Strassmann, F., 1939a. Ü ber den Nachweis und das Verhalten der bei Bestrahlung des Urans mittels Neutronen entstehenden Erdalkalimetalle. Naturwissenschaften 27, 11–15.

Hahn, O., Strassmann, F., 1939b. Nachweis der Entstehung activer bariumisotope aus Uran und Thorium durch Neutronenbestrahlung; nachweis weiterer aktiver Bruchtucke bei der Uranspaltung. Naturwissenschaften 27, 89–95.

Hankins, D.E., Homann, S.G., Buddemeier, B., 1989. Personnel Neutron Dosimetry Using Electrochemically Etched CR-39 Foils. Revision 1.

Report UCRL-53833 Rev. 1. LLNL, Livermore. Lawrence Livermore National Laboratory.

Harrison, T.M., Armstrong, R.I., Naeser, C.W., Harakal, J.E., 1979. Geochronology and thermal history of the Coast plutonic complex, near Prince Rupert, British Columbia. Can. J. Earth Sci. 16, 400–410.

Hasebe, N., Kodaira, S., Hareyama, M., Yasuda, N., 2009. A future plan in observing ultra-heavy nuclei (Z = 30 - 110) of cosmic rays with large-scale collector at the lunar base. Radiat. Meas. 44, 913–916.

Hashemi-Nezhad, S.R., Durrani, S.A., 1986. Charged-particle radiography of insects using accelerated alpha particles. Nucl. Tracks 12, 393–397.

Hashemi-Nezhad, S.R., Zhuk, I.V., Potapenko, A.S., Krivopustov, M.I., Westmeier, W., Brandt, R., 2008. Determination of uranium fission rate in an arbitrary neutron field using fission track detectors. Radiat. Meas. 43, S204–S209.

Hassan, N.M., Matai, Y., Kusumoto, T., Mori, Y., Nakasaki, M., Oda, K., Kitamura, H., Konishi, T., Kodaira, S., Yasuda, N., Yamauchi, T., 2013. On the mechanism of the sensitization of PADC (poly (allyl diglycol carbonate)) track detectors by carbon dioxide treatment. Radiat. Meas. 59, 23–29.

He, Z.-B., Guo, S.-L., 2015. Application of nuclear track membranes to filtration of medical injections and various transfusions to remove solid particles. Phys. Procedia 80, 131–134.

He, Y.D., Price, P.B., 1995a. Fragmentation of 160 A GeV Pb in various targets. In: Iucci, N., Lamanna, E. (Eds.), 24th International Cosmic Ray Conference, Vol. 1, August 28-September 8, 1995 in Rome, Italy. International Union of Pure and Applied Physics, p. 119.

He, Y.D., Price, P.B., 1991. Search for fractional charge states in high-energy heavy fragments produced in collisions of 14.5 A GeV ^{28}Si with Pb and Cu target. Phys. Rev. C 44, 1672–1674.

He, Y.D., Price, P.B., 1994. First measurement of charge-changing cross sections for 11.4 A GeV ^{197}Au in various targets. Nucl. Phys. A566, 363c–366c.

He, Y.D., Price, P.B., 1995b. Search for Dirac magnetic monopole production in high energy heavy ion collisions. In: Iucci, N., Lamanna, E. (Eds.), 24th International Cosmic Ray Conference, Vol. 1, August 28-September 8, 1995 in Rome, Italy. International Union of Pure and Applied Physics, p. 845.

He, Y.D., Westphal, A.J., Price, P.B., 1994. Response of the BP-1 phosphate glass detector to relativistic heavy ions. Nucl. Instrum. Meth. Phys. Res. Sect. B 84, 67–76.

Heinrich, W., Becker, E., Dreute, J., Hirzebruch, S.E., Huntrup, G., Kurth, M., Rocher, H., Rusch, G., Schmit, M., Streibel, T., Winkel, E., 1995. High energy heavy ion interactions studied with SSNTDs. Radiat. Meas. 25, 203–218.

Henke, R.P., Benton, E.V., 1966. Charged Particle Tracks in Polymers: No.3 - Range and Energy Loss Tables. US Naval Radiological Defense Laboratory, San Francisco. Report USNRDL-TR-1102.

Henke, R.P., Benton, E.V., 1971. On geometry of tracks in dielectric nuclear track detectors. Nucl. Instrum. Meth. 97, 483–489.

Hermsdorf, D., 2009. Evaluation of the sensitivity V for registration of α-particles in PADC CR-39 solid state nuclear track detector material. Radiat. Meas. 44, 283–288.

Hermsdorf, D., Reichelt, U., 2010. Measurement of track opening contours of oblique incident ^{4}He and ^{7}Li-ions in CR-39: relevance for calculation of track formation parameters. Radiat. Meas. 45, 1000–1013.

Hicks, D.G., Li, C.K., Seguin, F.H., Schnittman, J.D., Ram, A.K., Frenje, J.A., 2001. Observations of fast protons above 1 MeV produced in direct-drive laser-fusion experiments. Phys. Plasmas 8 (2), 606–610.

Holden, N.E., Hoffman, D.C., 2000. Spontaneous fission half-lives for ground state nuclei. Pure Appl. Chem. 72, 1525–1562.

Hou, M.-D., Klaumunzer, S., Schumacher, G., 1990. Dimensional changes of metallic glasses during bombardment with fast heavy ions. Phys. Rev. B 41, 1144–1157.

Hu, R.-Y., Liu, S.-S., Sun, S.-F., Guo, Shi-Lun, Meng, W., Deng, X.-L.U., 1982. Studies of the content and the distribution of uranium of lunar mare basaltic fragments taken by Apollo 17. Chin. J. Space Sci. 2, 205–208.

Huntrup, G., Streibel, T., Winkel, E., Kurth, M., Rocher, H., Heinrich, W., 1997. Study of fragmentation interactions at beam energies from 1 to 158 A GeV. Radiat. Meas. 28, 323–328.

Hurford, A.J., 1990. International union of geological sciences subcommission on geochronology recommendation for the standardization of fission track dating calibration and data reporting. Nucl. Tracks Radiat. Meas. 17 (3), 233–236.

Hurford, A.J., Green, P.F., 1982. A user's guide to fission track dating caliberation. Earth Planet. Sci. Lett. 59, 343–354.

Hurford, A.J., Green, P.F., 1983. The Zeta age caliberation of fission track dating. Isotope Geosci. 1, 285–307.

Husaini, S.N., Zaidi, J.H., Malik, F., Arif, M., 2008. Application of nuclear track membrane for reduction of pollutants in the industrial effluent. Radiat. Meas. 43, S607–S611.

ICRP, 1991. Recommendations of the International Commission on Radiological Protection. ICRP Publication 60, Ann. ICRP 21 (1-3). Elsevier Science, Oxford.

ICRP, 1996. Conversion Coefficients for Use in Radiological Protection against External Radiation. Elsevier Science, Oxford. ICRP Publication 74. International Commission on Radiological Protection. Ann. ICRP 26 (3-4).

ICRP, 2007. The 2007 Recommendations of the International Commission on Radiological Protection. Elsevier Science, Oxford. ICRP Publication 103, International Commission on Radiological Protection.

Ilic, R., 2005. Radon in Antarctica. Radiat. Meas. 40, 415–422.

Ilic, R., Durrani, S.A., 2003. Solid state nuclear track detectors (Chapter 3). In: L'Annunziata, M.F. (Ed.), Handbook of Radioactivity Analysis, second ed. Elsevier Science, USA. pp. 179-237, p. 1273.

Jager, U., Bertalet, L., herold, H., 1985. Energy spectra and space resolved measurements of fusion reaction protons from plasma focus devices. Rev. Sci. Instrum. 56, 77–80.

Jin, G., Williams, W.T., Price, P.B., 1990. Energy dependence of cross section for charge pickup of relativistic heavy ions. Phys. Rev. C 42, 769–770.

Jonckheere, R., Van den haute, P., 1999. On the frequency distributions per unit area of the projected and etchable lengths of surface-intersecting fission tracks: influences of track revelations, observation and measurement. Radiat. Meas. 30, 155–157.

Jonckheere, R., Van den haute, P., 2002. On the efficiency of fission-track counts in an internal and external apatite surface and in a muscovite external detector. Radia. Meas. 35, 29–40.

Jonsson, G., 1995. Radon gas-where from and what to do? Radiat. Meas. 25, 537–546.

Kacenjar, S., Skupsky, S., Entenberg, A., Goldman, L., Richardson, M., 1982. Direct measurement of fuel density - radius product in laser – fusion experiments. Phys. Rev. Lett. 49, 463–467.

Kanasaki, M., et al., 2012. The diagnosis method for high energy ion beams using backscattered particles for laser driven ion acceleration experiments. Jpn. J. Appl. Phys. 51, 056401.

Kanasaki, M., Hattori, A., Sakaki, H., Fukuda, Y., Yogo, A., Jinno, S., Nishiuchi, M., Ogura, K., Kondo, K., Oda, K., Yamauchi, T., 2013. A high energy component of the intense laser-accelerated proton beams detected by stacked CR-39. Radiat. Meas. 50, 46–49.

Kanasaki, M., Jinno, S., Sakaki, H., Faenov, A.Y., Pikuz, T.A., Nishiuchi, M., Kiriyama, H., Kando, M., Sugiyama, A., Kondo, K., Matsui, R., kishimoto, Y., Morishima, K., Watanabe, Y., Scullion, C., Smyth, A.G., Alejo, A., Doria, D., Kar, S., Borghesi, M., Oda, K., Yamauchi, T., Fukuda, Y., 2015. Observation of the inhomogeneous spatial distribution of MeV ions accelerated by the hydrodynamic ambipolar expansion of clusters. Radiat. Meas. 83, 12–14.

Kanasaki, M., Jinno, S., Sakaki, H., Kondo, K., Oda, K., Yamauchi, T., Fukuda, Y., 2016. The precise energy spectra measurement of laser-accelerated MeV/n-class high-Z ions and protons using CR-39 detectors. Plasma Phys. Control. Fusion 58, 034013.

Kang, T.-S., Wang, S.-C., 1991. Fission Track Methods for Study of Geothermal History. Science Press, Beijing, p. 112.

Kant, K., Upadhyay, S.B., Sharma, G.S., Chakarvarti, S.K., 2006. Indoor Built Environ. 15, 177.

Kase, M., Kikuchi, J., Doke, T., 1978. Half-life of 238U spontaneous fission and its fragment kinetic energies. Nucl. Instrum. Meth. 154, 335–341.

Kashkarov, L.L., Kalinina, G.V., Ivliev, A.I., Cherkisyan, V.O., 1995. "HOT" particles in soil from Chernobyl aes region. Radiat. Meas. 25 (1–4), 413–414.

Kashkarov, L.L., Kalinina, G.V., Perelygin, V.P., 2003. a-particle track investigation of the Chernobyl Nuclear Power Plant accident region soil samples. Radiat. Meas. 36, 529–532.

Katz, R., Kobetich, E.J., 1968. Formation of etchable tracks in dielectrics. Phys. Rev. 170, 401–405.

Khan, E.U., 1985. Ph.D. thesis, Philipps Universitat, Marburg, FR Germany.

Khan, H.A., Durrani, S.A., 1972. Efficiency calibration of solid state nuclear track detectors. Nucl. Instrum. Meth. 98, 229–236.

Khan, H.A., Vater, P., Brandt, R., 1980. Multiprong fission events produced by 1477 MeV ^{208}Pb ions in natural uranium. Lyon, and Suppl. 2, Nuclear Tracks. In: Proc. 10th Int. Conf. Solid State Nucl. Track Detectors. Pergamon, Oxford, pp. 915–920.

Khan, H.A., Khan, N.A., Spohr, R., 1981. Scanning electron microscope analysis of etch pits obtained in a muscovite mica track detector by etching in hydrofluoric acid and aqueous solution of NaOH, and KOH. Nucl. Instrum. Meth. 189, 577–581.

Khan, H.A., Qureshi, I.E., Jamil, K., Gottschalk, P.A., Vater, P., Brandt, R., 1987. On the determination of reaction product parameters using muscovite mica detectors. Nucl. Instrum. Meth. Phys. Res. Sect. B 28, 41–48.

Khan, E.U., Qureshi, I.E., Baluch, J.J., Shahzad, M.I., Karim, S., Sajid, M., Malik, F., Husaini, S.N., Khan, H.A., 2001. Heavy ion interactions of (14.0 MeV/u) Pb with U. Radiat. Meas. 34, 227–230.

Kinoshita, K., Price, P.B., 1980. Method for producing thin sheets of protonsensitive CR-39 plastic track detectors. Rev. Sci. Instrum. 51, 63–65.

Kinoshita, K., Du, R., Giacomelli, G., Patrizii, L., Predieli, F., Serra, p., Spurio, M., Pinfold, J.L., 1992. Search for highly ionizing particles in e e annihilations at $\sqrt{s} = 91.1$ GeV. Phys. Rev. D 46, R881–R884.

Kobetich, E.J., Katz, R., 1968. Energy deposition by electron beams and d rays. Phys. Rev. 170, 391–396.

Kodaira, S., Aasaeda, M., Doke, T., Hareyama, M., Hasebe, N., Ogura, K., Yasuda, N., Tsuruta, T., Kori, Y., 2008. Track detector of CR-39-DAPcopolymer with variable threshold to detect trans-iron nuclei in galactic cosmic rays. Radiat. Meas. 43, S52–S55.

Kovalenko, V.I., Mukhamadeeva, R.M., Maklakova, L.N., Gustova, N.G., 1993. J. Struct. Chem. 31 (4), 540.

Kralik, M., Turek, K., Vondracek, V., 2008. Characterisation of neutron fields around high-energy radiotherapy machines. Radiat. Prot. Dosim. 132, 13–17.

Kralik, M., Turek, K., Vondracek, V., Krasa, J., Velyhan, A., Scholz, M., Ivanova-Stanik, I.M., 2010. Measurement with Bonner spheres spectrometer in pulsed neutron fields. Radiat. Meas. 45, 1245–1249.

Kumar, R., Mahur, A.K., Rao, N.S., Sengupta, D., Prasad, R., 2008. Radon exhalation rate from sand samples from the newly discovered high background radiation area at Erasama beach placer deposit of Orissa, India. Radiat. Meas. 43, S508–S511.

Kusumoto, T., 2017. Radial Electron Fluence Around Ion Tracks as a New Physical Concept for the Detection Threshold of PADC Detector (Ph.D. thesis). Kobe University and Universite de Strasbourg.

Kusumoto, Tamon, Bitar, Ziad, E.L., Okada, S., Gillet, P., Arbor, N., Kanasaki, M., Mori, Y., Oda, K., Nourreddine, A.-M., Kurashige, H., Fromm, M., Cloutier, P., Bass, A.D., Sanche, L., Kodaira, S., Barillon, R., Yamauchi, T., 2018. Radial electron fluence around ion tracks as a new physical parameter for the detection threshold of PADC using Geant4-DNA toolkit. Radiat. Meas. 118, 50–53.

Kusumoto, T., Mori, Y., Kanasaki, M., Ikenaga, R., Oda, K., Kodaira, S., Kitamura, H., Barillon, R., Yamauchi, T., 2016. Radiation chemical yields for the loss of typical functional groups in PADC films for high energy protons registered as unetchable tracks. Radiat. Meas. 87, 35–42.

Kusumoto, T., Mori, Y., Kanasaki, M., Ueno, T., Kameda, Y., Oda, K., Kodaira, S., Kitamura, H., Barillon, R., Yamauchi, T., 2015. Yields on the formation of OH groups and the loss of CH groups along nuclear tracks in PADC films. Radiat. Meas. 83, 59–62.

Lal, D., 1969. Recent advances in the study of fossil tracks in meteorites due to heavy nuclei of the cosmic radiation. Space Sci. Rev. 9, 623–650.

Lal, D., Muralli, A.V., Rajan, R.S., Tamhane, A.S., Lorin, J.C., Pellas, P., 1968. Techniques for proper revelation of etch-tracks in meteoritic and terrestrial minerals. Earth Planet. Sci. Lett. 5, 111–119.

Lan, X., 2009. The Experimental Study of the Fast Protons Generation from Ultrashort Intense Laser Interaction with Metallic Foil Target (Ph.D. thesis). Tsinghua University, Beijing, China.

L'Annunziata, M.F., 2003. Nuclear radiation, its interaction with matter and radioisotope decay. In: L'Annunziata, M.F. (Ed.), Handbook of Radioactivity Analysis, second ed. Elsevier, Amsterdam and Oxford. pp. 1-121, p. 1273.

L'Annunziata, M.F., 2007. Radioactivity: Introduction and History, vol. 609. Elsevier, Amsterdam, ISBN 978-0-444-52715-8.

L'Annunziata, M.F., 2016. Radioactivity: Introduction and History, from the Quantum to Quarks. Elsevier, Amsterdam, ISBN 978-0-444-63489-4, p. 902.

Lee, C.-G., Suzuki, D., Esaka, F., et al., 2011. Combined application of alpha-track and fission-track techniques for detection of plutonium particles in environmental samples prior to isotopic measurement using thermal-ionization mass spectrometry. Talanta 85, 644–649.

Li, B., 1982. Chemical Compositions and Origin of Tektites (Ph.D thesis). Institute of Geochemistry, Acadimia Sinica, Guiyang, China.

Li, B.Y., 1993. Track treeing mechanism and its application. Nucl. Tracks Radiat. Meas. 22, 131–142.

Li, B.Y., 2008a. Track treeing mechanism and plastic zone in solid. Part 1: initial development of plastic zone. Radiat. Meas. 43, S87–S90.

Li, B.Y., 2008b. Track treeing mechanism and plastic zone in solid. Part 2: saturation of plastic zone. Radiat. Meas. 43, S91–S94.

Li, Y.G., Shi, Y.Q., Zhang, Z.B., Xia, P., 2001. Reactor fission rate measurement for miniature neutron source reactor by solid state nuclear track detector. Radiat. Meas. 34, 589–591.

Li, Y.-T., Xu, M.-H., Zhang, J., 2007. Generation of high energy ions in ultraintense laser-plasma interactions. Physics 36, 39–45.

Linnik, V.G., Brown, J.E., Dowdall, M., Potapov, V.N., Nosov, A.V., Surkov, V.V., Sokolov, A.V., Wright, S.M., Borghuis, S., 2006. Patterns and inventories of radioactive contamination of island sites of the Yenisey River, Russia. J. Environ. Radioact. 87, 188–208.

Lipponen, M., Zilliacus, R., 2002. Development of track-etch method for for exracting uranium containing particles from swipes. In: Final Report on Task FIN A 1318 of the Finnish Support Programme to IAEA Safeguards. STUK-YTO-TR 188.

Lipponen, M., Zilliacus, R., 2003. Track-etch method for exracting uranium containing particles from swipes and their analysis with ICP-MS. In: Final Report on Task FIN A 1318 of the Finnish Support Programme to IAEA Safeguards. STUK-YTO-TR 200.

Lipponen, M., Zilliacus, R., 2004. Track-etch method for exracting uranium containing glass particles from swipes and their analysis with ICP-MS. In: Final Report on Task FIN A 1318 of the Finnish Support Programme to IAEA Safeguards. STUK-YTO-TR 212.

Liu, J., Duan, J.L., Toimil-Molares, M.E., Karim, S., Cornelius, T.W., Dobrev, D., Yao, H.J., Sun, Y.M., Hou, M.D., Mo, D., Wang, Z.G., Neumann, R., 2006. Electrochemical fabrication of single-crystalline and polycrystalline Au nanowires: the influence of deposition parameters. Nanotechnology 17, 1922–1926.

Liu, T.C., Su, C.S., 1971. The application of fission track detectors as reactor neutron temperature monitor. Int. J. Appl. Radiat. Isotopes 22 (4), 227.

Liu, S., Zhang, F., Zhou, S., Sun, S., Meng, W., Hao, X., Guo, S., 1979. Fission track age of Jilin meteorite (editor group of the theses). In: Collected Papers of Research on Jilin Meteorite Shower. Science Press, Beijing.

Lounis-Mokrani, Z., Fromm, M., Barillon, R., Chambaudet, A., Allab, M., 2003. Characterization of chemical and optical modifications induced by 22.5 MeV proton beams in CR-39 detectors. Radiat. Meas. 36, 615–620.

Lowder, W.M., 1990. Natural environmental radioactivity and radon gas. In: Tommasino, L., Furlan, G., Khan, H.A., Monnin, M. (Eds.), Proceedings of the International Workshop on Radon Monitoring in Radioprotection, Environmental Radioactivity and Earth Sciences, ICTP, Trieste, Italy, April 3-14, 1989. World Scientific Publishing Co. Pte. Ltd., ISBN 981-02-0187-7, pp. 1–17. Copyright © 1990.

Lowder, W.M., 1993. Natural environmental radioactivity with particular regard to radon gas and cosmic radiation. In: Furlan, G., Tommasino, L. (Eds.), Proceedings of the Second Workshop on Radon Monitoring in Radioprotection, Environmental And/or Earth Sciences, ICTP, Trieste, Italy, November 21- December 6, 1991. World Scientific Publishing Co. Pte. Ltd, ISBN 9810212267, pp. 12–23. Copyright © 1993.

Luo, Y.R., 2002. Handbook of Bond Dissociation Energies in Organic Compounds. CRC Press.

Lutz, T.M., Omar, G., 1991. An inverse method of modeling thermal histories from apatite fission-track length data. Earth Planet. Sci. Lett. 104, 181–195.

Macchi, A., Borghesi, M., Passoni, M., 2013. Ion acceleration by superintense laser-plasma interaction. Rev. Mod. Phys. 85, 751–793.

MacDougall, D., 1973. Fission track dating oceanic basalts. Trans. Amer. Geophys. Union 54, 987.

Mahur, A.K., Kumar, R., Parsad, R., 2005. Radon exhalation rate from different types of buildings construction materials through SSNTDs. Environ. Geochem. J. 8, 300–305.

Malek, M.A., Chong, C.S., 2002. Generation of CO2 in γ-ray-irradiated CR-39 plastic. Radiat. Meas. 35, 109–112.

Mandrekar, V.K., Chourasiya, G., Kalsi, P.C., Tilve, S.G., Nadkarni, V.S., 2010. Nuclear track detection using thermoset polycarbonates derived from pentaerythritol. Nucl. Instrum. Meth. Phys. Res. Sect. B 268, 537–542.

Manzoor, S., Qureshi, I.E., Rana, M.A., Shahzad, M.I., Sher, G., Sajid, M., Khan, H.A., Giacomelli, G., Giorgini, M., Mandrioli, G., Patrizii, L., Popa, V., Serra, P., Togo, V., 2000. Charge identification in CR-39 nuclear track detector using relativistic lead ion fragmentation. Nucl. Instrum. Methods Phys. Res. Sect. A 453, 525–529.

Mayya, Y.S., 2004. Theory of radon exhalation into accumulators placed at the soil-atmosphere interface. Radiat. Prot. Dosim. 111, 305.

McKenna, P., Ledingham, K.W.D., Shimizu, S., Yang, J.M., Robson, L., McCanny, T., Galy, J., Magill, J., Clarke, R.J., Neely, D., Norreys, P.A., Singhal, R.P., Krushelnick, K., Wei, M.S., 2005. Broad energy spectrum of laser-accelerated protons for spallation-related physics. Phys. Rev. Lett. 94 (084801), 1–4.

Meitner, L., Frisch, O.R., 1939. Disintegration of uranium by neutrons: a new type of nuclear reaction. Nature 143, 239–240.

Miller, D.S., Duddy, I.R., 1985. Results of interlaboratory comparison of fission-track age standards: fission-track workshop e 1984. Nucl. Tracks. 3, 383–391.

Miller, D.S., Eby, N., Mc Corkell, R., Rosenberg, P.E., Suzuki, M., 1990. Results of interlaboratory comparison of fission track ages for the 1988 fission track workshop. Nucl. Tracks Radiat. Meas. 17 (3), 237–245.

Mishra, R., Orlando, C., Tommasino, L., Tonnarini, S., Trevisi, R., 2005. A better understanding of the background of CR-39 detectors. Radiat. Meas. 40, 325–328.

Mishra, R., Tripathy, S.P., Dvivedi, K.K., Khathing, D.T., Ghosh, S., Muller, M., Fink, D., 2003. Spectroscopic and thermal studies of electron irradiated polyimide. Radiat. Meas. 36, 621–624.

Monnin, M., Isabelle, D.B., 1970. Solid track detectors and their applications in biology. Ann. Phys. Biol. Med. 4, 95–113.

Monnin, M.M., Morin, J.-P., Seidel, J.-L., 1993. A comprehensive approach of radon measurements for Geophysical studies. Nucl. Tracks Radiat. Meas. 22, 403–411.

Moody, K.J., Hulet, E.K., Wang, S., Price, P.B., Barwick, S.W., 1987. Search for heavy-ion radioactivity of ^{241}Am. Phys. Rev. C 36, 2710–2712.

Moody, K.J., Hulet, E.K., Shicheng, Wang, Price, P.B., 1989. Heavy-fragment radioactivity of ^{234}U. Phys. Rev. C 39, 2445–2447.

Moody, K.J., Hulet, E.K., Price, P.B., 1992. Search for cluster radioactivity of ^{237}Np. Phys. Rev. C 45, 1392–1393.

Mori, Y., Ikeda, T., Yamauchi, T., Sakamoto, A., Chikada, H., Honda, Y., Oda, K., 2009. Radiation chemical yields for loss of carbonate ester bonds in PADC films exposed to gamma ray. Radiat. Meas. 44, 211–213.

Mori, Y., Yamauchi, T., Kanasaki, M., Yusuke, O., Keiji, K., Satoshi, T., Yasuda, N., Barillon, R., 2011. Radiation Chemical yields for loss of ether and carbonate ester bonds in PADC films exposed to proton and heavy ion beams. Radiat. Meas. 46, 1147–1153.

Mori, Y., Yamauchi, T., Kanasaki, M., Hattori, A., Matai, Y., Matukawa, K., Oda, K., Kodaira, S., Kitamura, H., Konishi, T., Yasuda, N., Tojo, S., Honda, Y., Barillon, R., 2012. Greater radiation chemical yields for losses of ether and carbonate ester bonds at low stopping powers along heavy ion tracks in poly (allyl diglycol carbonate) films. Appl. Phys. Express 5, 086401.

Mori, Y., Yamauchi, T., Kanasaki, M., Hattori, A., Oda, K., Kodaira, S., Konishi, T., Yasuda, N., Tojo, S., Honda, Y., Barillon, R., 2013. Vacuum effects on the radiation chemical yields in PADC films exposed to gamma rays and heavy ions. Radiat. Meas. 50, 97–102.

Mujahid, S.A., Hussain, S., Dogar, A.H., Karim, S., 2005. Determination of porosity of different materials by radon diffusion. Radiat. Meas. 40, 106–109.

Naeser, C.W., 1967. The use of apatite and sphene for fission track age determinations. Bull. Geol. Soc. Am. 78, 1523–1526.

Naeser, N.D., McCulloh, T.H. (Eds.), 1989. Thermal History of Sedimentary Basins. Springer-Verlag, New York.

Nain, M., Chauhan, R.P., Chakarvati, S.K., 2008. Alpha radioactivity in tobacco leaves: effect of fertilizers. Radiat. Meas. 43, S515–S519.

Nain, M., Chauhan, R.P., Chakarvati, S.K., 2006. Radium concentration and radon exhalation measurements in various brands of cement in India. Iran. J. Radiat. Res. 3 (4), 183–188.

Narula, A.K., Goyal, S.K., Saini, S., Chakarvarti, S.K., 2009a. Calculation of radon diffusion coefficient and diffusion length for different building construction materials. Indian J. Phys. 83, 1171–1175.

Narula, A.K., Saini, R.S., Goyal, S.K., Chauhan, R.P., Chakarvarti, S.K., 2009b. Indoor radiation levels enhanced by underground radon diffusion. Asian J. Chem. 21, S275–S278.

Nasir, T., Khan, E.U., Baluch, J.J., Qureshi, I.E., Sajid, M., Shahzad, M.I., 2008. Analysis of the intermediate stage in the heavy ion interactions of ^{208}Pb + ^{197}Au and ^{197}Au + ^{197}Au. Radiat. Meas. 43, S274–S278.

Nikezic, D., 2000. Three dimensional analytic determination of the track parameters. Radiat. Meas. 32, 277–282.

Nikezic, D., Kostic, D., 1997. Simulation of the track growth and determining the track parameters. Radiat. Meas. 28, 185–190.

Nikezic, D., Yu, K.N., 2003. Three-dimensional analytical determination of the track parameters: over-etched tracks. Radiat. Meas. 37, 39–45.

Nikezic, D., Yu, K.N., 2004. Formation and growth of tracks in nuclear track materials. Mater. Sci. Eng. R 46, 51–123.

Nikezic, D., Yu, K.N., 2006. Computer program TRACK-TEST for calculating parameters and plotting profiles for pits in nuclear track materials. Comput. Phys. Commun. 174, 160–165.

Nikezic, D., Yu, K.N., 2008. Computer program TRACK-VISION for simulating optical appearance of etched tracks in CR-39 nuclear track detectors. Radiat. Prot. Dosim. 178, 591–595.

Nikezic, D., Yu, K.N., 2009. Optical characteristics of tracks in solid state nuclear track detectors studied with ray tracing method. In: Sidrov, M., Ivanov, O. (Eds.), Nuclear Track Detectors: Design, Methods, and Applications. Nova Science Publishers, Inc.

Nikezic, D., Stevanovic, N., Kostic, D., Savovic, S., Tse, R.C.C., Yu, K.N., 2008. Solving the track wall equation by the finite difference method. Radiat. Meas. 43, S76–S78.

Nikezic, D., Krstic, D., Savovic, S., 2009. Response of diffusion chamber with LR115 detector and electret. Radiat. Meas. 44, 783–786.

Nikolaev, V.A., 2015. Solid State Nuclear Track Detectors in Radiation Research. Publishing House of SPbPU.

Nikolaev, V.A., Ilic, R., 1999. Etched track radiometers in radon measurements: a review. Radiat. Meas. 30, 1–13.

Oda, K., Hayano, D., Yamauchi, T., Ohguchi, H., Yamamoto, T., 2009. Several techniques for improving energy dependence of a commercial personal neutron dosemeter package based on PADC track detector. Radiat. Meas. 44, 977–980.

Ogura, K., Kodairo, S., Asaeda, M., Doke, T., Hasebe, N., Harayama, M., Yasuda, N., Tsuruta, T., Kori, Y., 2006. Relation between the track registration threshold and DAP concentration of CR-39/DAP copolymer. Space Radiat. 5, 77–86.

Ohguchi, H., Oda, K., Yamauchi, T., Nakamura, T., Maki, D., 2008. New presoaking technique for PADC and application to wide-range personal neutron dosimeter. Radiat. Meas. 43, S500–S503.

O'Sullivan, D., 1995. Some results from the investigation of cosmic rays using SSNTD at Dublin. Radiat. Meas. 25, 295–300.

O' Sullivan, D., Price, P.B., Kinoshita, K., Wilson, C.G., 1981. Correlative studies of track-etch behavior and chemical development of lithographic polymer resists. In: Proc. 11th Int. Conf. Solid State Nuclear Track Detectors, Bristol. Pergamon Press, Oxford, pp. 81–84.

O'Sullivan, D., Thomson, A., 1982. Tuffak polycarbonate as a track detector. In: Proc. 11th Inter. Conf. SSNTD, Bristol, p. 85.

Ouyang, Z., 1988. Space Chemistry. Science Press, Beijing.

Palfalvi, J.K., 2009. Fluence and dose of mixed space radiation by SSNTDS achievements and constraints. Radiat. Meas. 44, 724–728.

Pang, D.-L., Ma, G.-C., 1988. Preparation and etching properties of CR-39 SSNTD. Nucl. Tech. 11, 4–5.

Patrizii, L., 2001. Search for massive rare particles with the MACRO track-etch detector at Gran Sasso. Radiat. Meas. 34, 259–263.

Pavlotskaya, F.I., Goryachenkova, T.A., Kazinskaya, I.E., Novikov, A.P., Myasoedov, B.F., Kuznetsov, Y.V., Legin, V.K., Shishkunova, L.V., 2003. Forms of occurring and migration behavior of Pu and Am in flood land soils and bottom sediments of Yenisey River. Russ. Radiochem. 45, 524–532.

Perelygin, V.P., Zaveriukha, O.S., Kulakov, B.N., Krivopustov, M.I., Abdullaev, I.G., Knjazeva, G.P., Petrova, R.I., Brandt, R., Ochs, M., Wan, J.S., Vater, P., 2001. Fission of Pb nuclei induced by 0.5, 1.0, 1.5, 3.7 and 7.4 GeV protons in the volume of massive U/Pb and Pb targets. Radiat. Meas. 34, 287–291.

Phukan, T., Kanjilal, D., Goswami, T.D., Das, H.L., 2003. Study of optical properties of swift heavy ion irradiated PADC polymer. Radiat. Meas. 36, 611–614.

Pinfold, J.L., 2009. Searching for the Magnetic monopole and other highly ionizing particles at accelerators using nuclear track detectors. Radiat. Meas. 44, 834–839.

Price, P.B., 1982. Application of plastic track detectors to atomic, nuclear, particle, and cosmic ray physics. In: Proc. 11th Int. Conf. Solid State Nuclear Track Detectors. Pergamon Press, Oxford, Bristol, pp. 265–269.

Price, P.B., 1987. Heavy fragment radioactivities. In: Towner, I.S. (Ed.), American Institute of Physics (AIP) Conference Proceedings 164. Fifth International Conference on Nuclei Far from Stability, Rosseau Lake, Ontario (Canada), 1987. American Institute of Physics, pp. 800–809.

Price, P.B., 1993. Advances in solid state nuclear track detectors. Nucl. Tracks Radiat. Meas. 22, 9–21.

Price, P.B., 1994. Cluster radioactivity. In: Scheid, W., Sandulescu, A. (Eds.), Frontier Topics in Nuclear Physics. Plenum Press, New York.

Price, P.B., 2005. Science and technology with nuclear tracks in solids. Radiat. Meas. 40, 146–159.

Price, P.B., 2008. Recent applications of nuclear tracks in solids. Radiat. Meas. 43, S13–S25.

Price, P.B., Drach, J., 1987. Dependence of response of plastic track detectors on post-irradiation aging time, temperature, and atmosphere. Nucl. Instrum. Meth. Phys. Res. Sect. B 28, 275–279.

Price, P.B., Fleischer, R.L., 1970. Particle identification by dielectric track detectors. Radiat. Effects 2, 291–298.

Price, P.B., Fleischer, R.L., 1971. Identification of energetic heavy nuclei with solid dielectric track detectors: applications to astrophysical and planetary studies. Ann. Rev. Nucl. Sci. 21, 295–334.

Price, P.B., Walker, R.M., 1962a. Electron microscope observation of a radiation- nucleated phase transformation in mica. J. Appl. Phys. 33, 2625–2628.

Price, P.B., Walker, R.M., 1962b. Observations of charged-particle tracks in solids. J. Appl. Phys. 33, 3400–3406.

Price, P.B., Walker, R.M., 1962c. Chemical etching of charged-particle tracks in solids. J. Appl. Phys. 33, 3407–3412.

Price, P.B., Walker, R.M., 1962d. A new track detector for heavy particle studies. Phys. Lett. 3, 113–115.

Price, P.B., Walker, R.M., 1962e. Electron microscope observation of etched tracks from spallation recoils in mica. Phys. Rev. Lett. 8, 217–219.

Price, P.B., Walker, R.M., 1963a. Fossil tracks of charged particles in mica and the age of minerals. J. Geophys. Res. 68, 4847–4862.

Price, P.B., Walker, R.M., 1963b. A simple method of measuring low uranium concentrations in natural crystals. Appl. Phys. Lett. 2, 23–25.

Price, P.B., Fleischer, R.L., Walker, R.M., Hubbard, E.L., 1963. Ternary fission of heavy compound nuclei. In: Ghiorso, A., Diamond, R.M., Conzett, H.E. (Eds.), Reactions between Complex Nuclei. U. C. Press, Berkeley.

Price, P.B., Fleischer, R.L., Peterson, D.D., 1967. Identification of Isotopes of energetic particles with dielectric track detectors. Phys. Rev. 164, 1618–1620.

Price, P.B., Fleischer, R.L., Walker, R.M., 1968a. The utilization of nuclear particle tracks in solids to study the distribution of certain elements in nature and in the cosmic radiation. In: Ahrens, L.H. (Ed.), Origin and Distribution of the Elements. Pergamon Press, Oxford.

Price, P.B., Rajan, R.S., Tamhane, A.S., 1968b. The abundance of nuclei heavier than iron in the cosmic radiation in the geological past. Astrophys. J. 151, L109–L116.

Price, P.B., Tincknell, M.L., Tarle, G., Ahlen, S.P., Frankel, K.A., Permutter, S., 1983. Search for nonintegrally charged projectile fragments in relativistic nucleus-nucleus collisions. Phys. Rev. Lett. 50, 566–569.

Price, P.B., Guo, S.L., Ahlen, S.P., Fleischer, R.L., 1984. Search for grandunified- theory magnetic monopoles at a level below the Parker limit. Phys. Rev. Lett. 52, 1265–1268.

Price, P.B., Stevenson, J.D., Barwick, S.W., Ravn, H.L., 1985. Discovery of radioactive decay of ^{222}Ra and ^{234}Ra by ^{14}C emission. Phys. Rev. Lett. 54, 297–299.

Price, P.B., Moody, K.J., Hulet, E.K., Bonetti, R., Migliorino, C., 1991. Highstatistics study of cluster radioactivity from ^{233}U. Phys. Rev. 43, 1781–1788.

Price, P.B., Bonetti, R., Guglielmetti, A., Chiesa, C., Matheoud, R., Migliorino, C., Moody, K.J., 1992. Emission of ^{23}F and ^{23}Ne in cluster radioactivity of ^{231}Pa. Phys. Rev. C 46, 1939–1945.

Qureshi, I.E., Khan, H.A., Rashid, K., Gottschalk, P.A., Vater, P., Brandt, R., 1998a. Four-fragment exit channel in the interaction of 1050 MeV ^{84}Kr with U studied with mica detectors. Phys. Rev. C 37, 393.

Qureshi, I.E., Khan, H.A., Rashid, K., Vater, P., Brandt, R., Gottschalk, P.A., 1998b. Kinematical analysis of 2-, 3- and 4- body channels in the reaction (12.5 MeV/u) ^{84}Kr + $^{nat.}$U observed with mica track detectors. Nucl. Phys. A 477, 510–522.

Qureshi, I.E., Shahzad, M.I., Khan, E.U., Sher, G., Khan, H.A., 2005. Study of the reaction step preceeding sequential fission in the nuclear interaction (16.7 MeV/u) ^{238}U + $^{nat.}$Au. Ridiat. Meas. 40, 464–469.

Qureshi, I.E., Shahzad, M.I., Manzoor, S., Khan, H.A., 1994. Systematics of nuclear cross sections observed with different dielectric detectors. Nucleus 31, 25–33.

Rana, M.A., 2008. Microscopic model for chemical etchability along radiation damage paths in solids. Nucl. Sci. Tech. 19, 174–177.

Ren, G., Price, P.B., Williams, W.T., 1989. Systematics of charge-pickup reactions by GeV/nucleon heavy nuclei. Phys. Rev. C 39, 1351–1358.

Reginatto, M., March 1, 2004. The "Few Channels" Unfolding Programs in the UMG Package. UMG Package, Version 3.3. PTB, Braunschweig.

Reiners, P.W., Brandon, M.T., 2006. Using thermochronolgy to understand orogenic erosion. Ann. Rev. Earth Planet. Sci. 34, 419–466.

Reiners, P.W., Ehlers, T.A., 2005. Low-temperature thermochronology. Rev. Miner. Geochem. 58, 1–620.

Roberts, J.H., Gold, R., Armani, R.J., 1968. Spontaneous-fission decay constant ^{238}U. Phys. Rev. 174, 1482–1484.

Rose, H.J., Jones, G.A., 1984. A new kind of natural radioactivity. Nature 307, 245–247.

Rubbia, C., Buono, S., Kadi, Y., Rubio, J.A., 1997. Fast Neutron Incineration in the Energy Amplifier as Alternative to Geological Storage: The Case of Spain. CERN/LHC/97-01.

Rumiazev, O.V., Selizky, Y.A., Funshtein, V.B., 1968. Registration of tracks of fission fragments in mica under irradiation of high intensity of charged particles. Exp. Instrum. Technol. 1, 51 (in Russian).

Saad, A.F., 2008. Radium activity and radon exhalation rates from phosphate ores using CR-39 on-line with an electronic radon gas analyzer "Alpha GUARD". Radiat. Meas. 43, S463–S466.

Saad, F.A., Atwa, S.T., Yasuda, N., Fujii, M., 2001. FT-IR spectroscopy of carbon dioxide in CR-39 and SR-90 track detectors irradiated with ions and gamma-rays at different energies and fluencies. Radiat. Meas. 34, 51–54.

Sadowski, M., Al-Mashhadani, E.M., Szydlowski, A., Czyzewski, T., Glowacka, L., Jaskola, M., Wielunski, M., 1994. Investigation on the response of CR-39 and PM-355 track detectors to fast protons in the energy range 0.2e4.5 MeV. Nucl. Instrum. Meth. Phys. Res. Sect. B 86, 311–316.

Sajo-Bohus, L., Palasios, D., Barros, H., Greaves, E.D., Nemeth, P., Liendo, J., Bermudez, J., 2008. (n,α) Reaction study with LR-115 for binary glass metal boron distribution. Radiat. Meas. 43, S656–S658.

Sajo-Bohus, L., Simon, A., Csako, T., Nemeth, P., Palasios, D., Espinosa, G., Greaves, E.D., Szorenyi, T., Barros, H., 2009. 10Boron distribution measurement in laser ablated B4C thin films using (n,α) reaction and LR- 115 passive detector. Radiat. Meas. 44, 795–797.

Salamon, M.H., Drach, J., Guo, S.-L., Price, P.B., Tarle, G., Ahlen, S.P., 1984. Effect of electron capture and loss on the response of polycarbonate track detectors to relativistic uranium ions. Nucl. Instrum. Meth. 224, 217–224.

Salamon, M.H., Price, P.B., Tincknell, M., Guo, S.-L., Tarle, G., 1985. Charge resolution of plastic track detectors used to identify relativistic nuclei. Nucl. Instrum. Meth. Phys. Res. Sect. B 6, 504–512.

Segovia, N., Pulinets, S.A., Leyva, A., Mena, M., Monnin, M., Camacho, M.E., Ponciano, M.G., Fernandez, V., 2005. Group radon exhalation, an electrostatic contribution for upper atmospheric layers processes. Radiat. Meas. 40, 670–672.

Seguin, F.H., Frenje, J.A., Li, C.K., Hicks, D.G., Kurebayashi, S., Rygg, J.R., Schwartz, B.-E., Petrasso, R.D., Roberts, S., Soures, J.M., et al., 2003. Spectrometry of charged particles from inertial-confinement-fusion plasma. Rev. Sci. Instrum. 74, 975–995.

Selo, M., Storzer, D., 1981. Uranium distribution and age pattern of some deep-sea basalts from the Entrecasteaux area, south-western Pacific: a fission track analysis. Nucl. Tracks 5, 137–145.

Severin, D., Ensinger, W., Neumann, R., Ttrautmann, C., Walter, G., Alig, I., Dudkin, S., 2005. Degradation of polyimide under irradiation with swift heavy ions. Nucl. Instrum. Meth. B 236, 456–460.

Shahzad, M.I., 1995. Study of Heavy Ions Interactions in Low and Relativistic Energy Range with Solid State Nuclear Track Detectors Using Automatic Track Analysis System (Ph.D. thesis). University of the Punjab, Lahore.

Shahzad, M.I., Qureshi, I.E., Manzoor, S., Khan, H.A., 1999. Sequential fission process observed in the reaction (16.7 MeV/u) $^{238}U + ^{nat}.Au$ using mica as dielectric track detector. Nucl. Phys. A 645, 92–106.

Shakeri Jooybari, B., Afarideh, H., Lamehi-Racti, M., Moghimi, R., Ghergherehchi, M., 2015. Copper nano- and micro wires electrodeposited in etched cellulose nitrate and Makrofol KG nuclear track detector. Phys. Procedia 80, 41–45.

Shakeri Jooybari, B., Moghimi, R., Gorgiri, D., Afarideh, H., Lamehi-Rachti, M., Ghergherehchi, M., 2015a. Design and construction of optimized electrochemical cell and data analysis system for etching of ion tracks and electro deposition of nano and micro wires in porous ion tracks foils. Phys. Procedia 80, 57–61.

Shang, B., Chen, L., Liu, F., Wu, Y., 2006. Determination of ^{222}Rn and ^{220}Rn and their progeny in two coal mines of Fangshan district in Beijing. Radiat. Protect. 26, 279.

Shen, Y., Zhao, Y.-G., Guo, S.-L., Cui, J.-Y., Liu, Y., Li, J.-H., Xu, J., Zhang, H.-Z., 2008. Study on analysis of isotopic ratio of uraniumbearing particle in swipe samples by FT-TIMS. Radiat. Meas. 43, S299–S302.

Shen, C.-B., Mei, L.-F., Xu, S.-H., 2009. Fission track dating of Mesozoic sandstones and its tectonic significance in the Eastern Sichuan Basin, China. Radiat. Meas. 44, 945–949.

Sher, G., Shahzad, M.I., Hussain, M., 2007. Fragmentation of (158 A Ge v) Pb ions with Bi target. Radiat. Meas. 42, 1692–1695.

Shi, Y.Q., Li, Y.G., 2001. Measurement of fast fission factor for heavy water zero power reactor (HWZPR) by solid state nuclear track detector. Radiat. Meas. 34, 605–607.

Shutt, R.P., 1967. Bubble and Spark Chambers, Principles and Uses, vols. I and II. Academic Press, New York, p. 656.

Shweikani, R., Raja, G., 2009. Radon exhalation from some finishing materials frequently used in Syria. Radiat. Meas. 44, 1019–1023.

Sidorov, M., Ivanov, O., 2009. Nuclear Track Detectors: Design, Methods and Applications. Nova Science Publishers, Inc.

Silk, E.C.H., Barnes, R.S., 1959. Examination of fission fragment tracks with electron microscope. Phil. Mag. 4, 970–971.

Singh, S., Singh, D., Sandhu, A.S., Singh, G., Virk, H.S., 1986. A study of track etch anisotropy in apatite. Nucl. Tracks 12, 927–930.

Snavely, R.A., Key, M.H., Hatchett, S.P., Cowan, T.E., Roth, M., Phillips, T.W., Stoyer, M.A., Henry, E.A., Sangster, T.C., Singh, M.S., Wilks, S.C., MacKinnon, A., Offenberger, A., Pennington, D.M., Yasuike, K., Langdon, A.B., Lasinski, B.F., Johnson, J., Perry, M.D., Campbell, E.M., 2000. Intense high-energy proton beams from Petawatt-laser irradiation of solids. Phys. Rev. Lett. 85 (4), 2945–2948.

Soares, C.J., Guedes, S., Hadler, J.C., Mertz-Kraus, R., Zack, T., Iunes, P.J., 2014. Phys. Chem. Minerals 41, 65–73.

Sohrabi, M., 1981. Electrochemical etching of fast-neutron-induced recoil tracks: the effects of field strength and frequency. Nucl. Tracks 4, 131–140.

Sohrabi, M., Mahdi, S., 1993. A "quick dyecet" method for ece particle tracks in polymer detectors. Nucl. Tracks Radiat. Meas. 22, 165–169.

Somogyi, G., 1980. Development of etched nuclear tracks. NIM 173, 21–42.

Somogyi, G., Srivastava, D.S., 1970. Investigations on Alpha Radiography with Plastic Track Detectors, vol. 12. Atomki Kozlem, pp. 101–117.

Somogyi, G., Srivastava, D.S., 1971. Alpha radiography with plastic track detectors. Int. J. Appl. Radiat. Isotopes 22, 289–299.

Somogyi, G., Szalay, A.S., 1973. Track-diameter kinetics in dielectric track detectors. Nucl. Instrum. Meth. 109, 211–232.

Spohr, R., 1990. Ion Tracks and Microtechnology: Principles and Applications. Vieweg, Braunschweig.

Spohr, R., 2008. Real-time control of track etching and recent experiments relevant to micro and nano fabrication. Radiat. Meas. 43, S560–S570.

Stern, R.A., Price, P.B., 1972. Charge and energy information from heavy ion tracks in Lexan. Nat. Phys. Sci. 240, 83–85.

Sternheimer, R.M., Seltzer, S.M., Berger, M.J., 1982. Density effect for the ionization loss of charged particles in various substances. Phys. Rev. B 26 (11), 6067–6076.

Storzer, D., Wagner, G.A., 1977. Fission track dating of meteorite impacts. Meteoritics 12, 368–369.

Su, C.-S., 1992. Alpha particle radiography of ants using a 244Cm alpha source. Nucl. Tracks Radiat. Meas. 20, 335–340.

Su, C.-S., 1993. Alpha particle radiography of small insects. Nucl. Tracks Radiat. Meas. 22 (1–4), 877–883.

Sun, Y., Zhang, C., Zhu, Z., Wang, Z., Jin, Y., Liu, J., Wang, Y., 2004. The thermal-spike model description of the ion-irradiated polyimide. Nucl. Instrum. Meth. B 218, 318–322.

Szydlowski, A., Sadowski, M., Czyzewski, T., Jaskola, M., Korman, A., 1999. Comparison of responses of CR-39, PM-355, and PM-600 track detectors to low-energy hydrogen, and helium ions. Nucl. Instrum. Meth. Phys. Res. Sect. B 149, 113–118.

Szydlowski, A., Badziak, J., Fuchs, J., Kubkowska, M., Parys, P., Suchanska, R., Wolowski, J., Antici, P., Mancic, A., 2009. Application of solid-state nuclear track detectors of the CR-39/PM-355 type for measurements of energetic protons emitted from plasma produced by an ultra-intense laser. Radiat. Meas. 44, 881–884.

Tanner, R.J., Bartlett, D.T., Hager, L.G., 2005. Operational and dosimetric characteristics of etched-track neutron detectors in routine neutron radiation protection dosimetry. Radiat. Meas. 40, 549–559.

Tarle, G., Ahlen, S.P., Price, P.B., 1981. Energy straggling eliminated as a limitation to charge resolution of transmission detectors. Nature 297 (5833), 556–558.

Tian, Z., Xiao, D., et al., 1994. Passive radon monitoring techniques with and without electret collection. J. Radiol. Prot. 14, 241–249.

Tommasino, L., 1970. Electrochemical etching of damaged track detectors by H. V. pulse and sinusoidal waveform. In: Internal Rept. Lab. Dosimetria e Standardizzazione. CNEN Casaccia, Rome.

Tommasino, L., Zapparoli, G., Griffith, R.V., 1981. Short Communication: electrochemical etching - I. Nucl. Tracks 4, 191–196.

Trautmann, C., Schwartz, K., Geiss, O., 1998. Chemical etching of ion tracks in Li F crystals. J. Appl. Phys. 83, 3560–3564.

Tretyakova, S.P., Sandulescu, A., Micheev, V.L., Hasegan, D., Lebedev, I.A., Zamyatnin, Y.S., Korotkin, Y.S., Myasoedov, B.F., 1985. On the spontaneous emission of clusters by the ^{230}Th, ^{237}Np and ^{241}Am nuclei. JINR Rapid Commun. 13, 34.

Tretyakova, S.P., Ogloblin, A.A., Mikheev, V.L., Zamyatnin, Y.S., 1992. Observation of nuclear clusters in the spontaneous decay of heavy clusters. In: Brenner, M., Lonnroth, T., Malik, F.B. (Eds.), Clustering Phenomena in Atoms and Nuclei. Springer-Verlag, Berlin, pp. 283–292.

Tretyakova, S.P., Bonetti, R., Golovchenko, A., Gugliemetti, A., Ilic, R., Mazzocchi, C., Mikheev, V., Ogloblin, A., Ponomarenko, Shigin, V., Skvar, F., 2001. Study of cluster decay of 242Cm using SSNTD. Radiat. Meas. 34, 241–243.

Tsuruta, T., 1999. Characteristics of allyl phthalate resin as a fission track detector. Radiat. Meas. 31, 99–102.

Tsuruta, T., 2000. Diallyl phthalate resin and its copolymers containing allyl diglycol carbonate as nuclear track detectors. Radiat. Meas. 32, 289–297.

Tsuruta, T., Koguchi, Y., Yasuda, N., 2008. Discrimination of heavy ions using copolymers of CR-39 and DAP. Radiat. Meas. 43, S48–S51.

Urosevic, V., Nikezic, D., 2003. Radon transport through concrete and determination of its diffusion coefficient. Radiat. Prot. Dosim. 104 (1), 65.

Van den haute, P., Chamaudet, A., 1990. Results of an interlaboratory experiment for the 1988 fission track workshop on a putative apatite standard for internal calibration. Nucl. Tracks Radiat. Meas. 17 (3), 247–252.

Vandenbosch, R., Huizenga, J.R., 1973. Nuclear Fission. Academic Press, New York.

Vater, P., 1988. Production and applications of nuclear track microfilters. Nucl. Tracks Radiat. Meas. 15, 743–749.

Vater, P., Becker, H.J., Brandt, R., Freiesleben, H., 1977. Study of heavy-ion induced reactions on uranium with use of mica detectors. Phys. Rev. Lett. 39, 594–598.

Vater, P., Khan, E.U., Backmann, R., Gottschalk, P.A., Brandt, R., 1986. Sequential fission studies in the interaction of 9.03 MeV/N ^{238}U with $^{nat.}$U using mica track detectors. Nucl. Tracks Radiat. Meas. 11, 5–16.

Virk, H.S., Valia, V., 2001. Helium/radon precursory signals of Chamoli earthquake, India. Radiat. Meas. 34, 379–384.

Vlasova, I.E., Kalmykov, S.N., Konevnik, Y.V., Simakin, S.G., Simakin, I.S., Anokhin, A.Y., Sapozhnikov, Y.A., 2008. Alpha track analysis and fission track analysis for localizing actinide-bearing micro-particles in the Yenisey River bottom sediments. Radiat. Meas. 43, S303–S308.

Vlasova, I., Shiryaev, A., Ogorodnikov, B., Burakov, B., Dolgopolova, E., Senin, R., Averin, A., Zubavichus, Y., Kalmykov, S., 2015. Radioactivity distribution in fuel-containing materials (Chernobyl "lava") and aerosols from the Chernolyl "Shelter". Radiat. Meas. 83, 20–25.

Vlasukova, L.A., et al., 2010. Ion tracks in amorphous silicon nitride. Bull. Russ. Acad. Sci. Phys. 74, 206–208.

Vonderau, S., Vater, P., Brandt, R., 1995. Upilex as material for nuclear track filters. Radiat. Meas. 25, 727–728.

Wagner, G.A., 1966. Altersbestimmungen an Tektiten und anderen naturlichen Glasern mittels Spuren der spontanen Kernspaltung des Uran^{238} ("fissiontrack"- Methode). Z. Naturforschung 21a, 733–745.

Wagner, G., Van den haute, P., 1992. Fission Track Dating. Kluwer Academic publishers, Dordrecht, p. 285.

Walker, R.M., 1963. Characteristics and applications of solid state track detectors. In: Proc. Strasbourg Conf. On New Methods of Track Detection. Centre de Recherches Nucleaires, Strasbourg, France.

Walker, R.M., Price, P.B., Fleischer, R.M., 1963. A versatile disposable dosimeter for low and fast neutrons. Appl. Phys. Lett. 3, 28–29.

Wall, T., 1970. Measurement of the fission track recording efficiency of mica. Nucl. Sci. Abstracts 24, 50542.

Wan, J.-S., Schmidt, T., Langrock, E.-J., Vater, P., Brandt, R., Adam, J., Bradnova, V., Bamblevski, V.P., Gelovani, L., Gridnev, T.D., Kalinnikov, V.G., Krivopustov, M.I., Kulakov, B.A., Sosnin, A.N., Perelygin, V.P., Pronskikh, V.S., Stegailov, V.I., Tsoupko-Sitnikov, V.M., Modolo, G., Odoj, R., Phippen, P.-W., Zamani-Valasiadou, M., Adloff, J.C., Debeauvai, M., Hashemi-Nezhad, S.R., Guo, S.-L., Li, L., Wang, Y.-L., Dwivedi, K.K., Zhuk, I.V., Boulyga, S.F., Lomonossova, E.M., Kievitskaja, A.F., Rakhno, I.L., Chigrinov, S.E., Wilson, W.B., 2001. Transmutation of ^{129}I and ^{237}Np using spallation neutrons produced by 1.5, 3.7 and 7.4 GeV protons. Nucl. Instrum. Meth. Phys. Res. Sect. A 463, 634–652.

Wan, J.-S., Zhu, G.-N., Zhao, Y., Liu, N.-N., Zhuang, Q.-C., Zhang, L.-X., Guo, S.-L., 2003. Computer studies of detection efficiency of fast-neutron spectrum based on PADC using the Monte Carlo method. Radiat. Meas. 36, 193–197.

Wang, C., 2018. Study on the Application of Solid State Nuclear Tracks in Pu-Bearing Particles Analysis (Ph.D. thesis). China Institute of Atomic Energy.

Wang, C., Guo, S.-L., Chang, Z.-Y., Guo-Rong, L., Zhao, Y.-G., 2018. Detection and identification of Pu particles and HEU particles by nuclear detector CR-39. Radiat. Meas (in press).

Wang, P., Wang, X., Ling, Y., Wang, M., Ding, S., Shen, W., Wang, Z., Wang, Y., Liu, F., 2018. Ultrafast selective ionic transport through heat-treated polyethylene terephthate track membranes with sub-nanometer pores. Radiat. Meas. 119, 80–84.

Wang, S.C., Kang, T.S., 1993. A study on standardization of fission track dating. Nucl. Tracks Radiat. Meas. 22 (1–4), 843–846.

Wang, S., Price, P.B., Barwick, S.W., Moody, K.J., Hulet, E.K., 1987. Radioactive decay of ^{234}U via Ne and Mg emission. Phys. Rev. 36, 2717–2720.

Wang, S., Barwick, S.W., Ifft, D., Price, P.B., Westphal, A.J., Day, D.E., 1988. Phosphate glass detectors with high sensitivity to nuclear particles. Nucl. Instrum. Meth. Phys. Res. Sect. B 35, 43–49.

Wang, S., Snowden-Ifft, D., Price, P.B., Moody, K.J., Hulet, E.K., 1989. Heavy-fragment radioactivity of ^{238}Pu: Si and Mg emission. Phys. Rev. C 39, 1647–1650.

Wang, S.C., Kang, T.S., Jing, G.R., 1993. Recent progress in fission track analysis and its applications in China. Nucl. Tracks Radiat. Meas. 22 (1–4), 835–842.

Wang, Y., Guo, S.-L., Li, L., Tu, C., 1998. New methods for measurement of neutron energy spectrum with CR-39 nuclear track detectors. Jpn. J. Appl. Phys. 37, 4164–4168.

Wang, Y., Du, G.hua, Xue, J., Liu, F., Wang, S., Yan, S., Zhao, W., 2002. The primary target model of energetic ions penetration in thin botanic samples. Phys. Lett. A 300, 611–618.

Wang, Y., Xu, S., Lin, J., 2008. Nuclear track anti-counterfeiting technique and its application in secure ID. Radiat. Meas. 43, S659–S661.

Wang, L., Guo, W., Xie, Y.B., Wang, X.W., Xue, J.M., Wang, Y.G., 2009. Nanofluidic diode generated by pH gradient inside track-etched conical nanopore. Radiat. Meas. 44, 1119–1122.

Weaver, B.A., 2001. Actinides in the Cosmic Rays and Their Detection (Ph.D. thesis). University of California at Berkeley, USA.

Weaver, B.A., Westphal, A.J., 2002. Extended analysis of the TREK ultraheavy collector. Astrophys. J. 569, 493–500.

Westphal, A., He, Y.D., 1993. Measurement of cross sections for electron capture and stripping by highly relativistic ions. Phy. Rev. Lett. 71, 1160–1163.

Westphal, A.J., Jin, G., Price, P.B., 1991. Measurement of cross sections for charge pickup by relativistic holmium ions on heavy targets. Phys. Rev. C 44, 1687–1690.

Westphal, J.A., Afanasyev, V.G., Price, P.B., Solarz, M., Akimov, V.V., Rodin, V.G., Shvets, N.I., 1996. Measurement of the isotopic composition of manganese, iron, and nickel in the galactic cosmic rays. Astrophys. J. 468, 679–685.

Westphal, A.J., Price, P.B., Weaver, B.A., Afanasiev, V.G., 1998. Evidence against stellar chromospheric origin of galactic cosmic rays. Nature 396, 50–52.

Wilson, J.G., 1951. The Principles of Cloud-Chamber Technique. Cambridge.

Xiao, D., Pan, Z., Qiu, L., Dong, L., 2001. The research on passive and integrated thoron exposure monitoring techniques. At. Energy Sci. Technol. 35, 325–330.

Yamauchi, T., Barillon, R., Balanzat, E., Asuka, T., Izumi, K., Masutani, T., Oda, K., 2005. Yields of CO_2 formation and scissions at ether bonds along nuclear tracks in CR-39. Radiat. Meas. 40, 224–228.

Yamauchi, T., Mineyama, D., Nakai, H., Oda, K., Yasuda, N., 2003. Track core size estimation in CR-39 track detector using atomic force microscope and UV-Visible spectrophotometer. Nucl. Instrum. Meth. B 208, 149–154.

Yamauchi, T., Mori, Y., Oda, K., Yasuda, N., Kitamura, H., Barillon, R., 2008a. Structure modification along heavy ion tracks in poly (allyl diglycol carbonate) films. Jpn. J. Appl. Phys. 47, 3606–3609.

Yamauchi, T., Mori, Y., Oda, K., Kodaira, S., Yasuda, N., Barillon, R., 2010. On the tracks of proton and heavy ions in PC and PADC Plastics detectors. In: Proceedings of the 24th Workshop on Radiation Detectors and Their Uses. KEK Proceedings. Tsukuba, Japan, vol. 10, pp. 1–11.

Yamauchi, T., Kaifu, S., Mori, Y., Kanasaki, M., Oda, K., Kodaira, S., Konishi, T., Yasuda, N., Barillon, R., 2013a. Applicability of the polyimide films as SSNTD material. Radiat. Meas. 50, 16–21.

Yamauchi, T., Kusumoto, T., Ueno, T., Mori, Y., Kanasaki, M., Oda, K., Kodaira, S., Barillon, R., 2018. Distinct step-like changes in G values for the losses of typical functional groups in poly(ethylene terephthalate) along boron ion tracks around the detection threshold. Radiat. Meas. 116, 51–54.

Yamauchi, T., Matsukawa, K., Mori, Y., Kanasaki, M., Hattori, A., Matai, Y., Kusumoto, T., Tao, A., Oda, K., Kodaira, S., Konishi, T., Kitamura, H., Yasuda, N., Barillon, R., 2013b. Applicability of polyimide films as etched-track detectors for ultra-heavy cosmic ray components. Appl. Phys. Express 6 (1–4), 046401.

Yamauchi, T., Mori, Y., Morimoto, A., Kanasaki, M., Oda, K., Kodaira, S., Konishi, T., Yasuda, N., Tojo, S., Honda, Y., Barillon, R., 2012. Threshold of etchable track formation and chemically damaged parameters in poly(ethylene terephthalate), bisphenol A polycarbonate and poly(allyl diglycol carbonate) films at the stopping powers ranging from 10 to 12,000 keV/um. Jpn. J. Appl. Phys. 51 (05631), 1–5.

Yamauchi, T., Watanabe, S., Oda, K., Yasuda, N., Barillon, R., 2008b. An evaluation of radial track etch rate in LR-115 detectors exposed to Fe ions by means of FT-IR spectrometry. Radiat. Meas. 43, S116–S119.

Yamauchi, T., Watanabe, S., Seto, A., Oda, K., Yasuda, N., Barillon, R., 2008c. Loss of carbonate ester bonds along Fe ion tracks in thin CR-39 films. Radiat. Meas. 43, S106–S110.

Yang, J.-X., He, Z.-B., Guo, S.-L., 2018. Identification and harmfulness analysis of solid particles contained in medical injections and their removal by nuclear track membranes. Perspect. Sci. (in press).

Yang, T.F., Chen, C.-H., Tian, R.L., Song, S.R., Liu, T.K., 2003. Remnant magmatic activity in the Coastal Range of East Taiwan after arc-continent collision: fission-track data and $^3He/^4He$ ratio evidence. Radiat. Meas. 36, 343–349.

Yang, M., 2007. KF 606Rn/g personal dosimeter. Radiat. Prot. 22, 1–8.

Yang, M., Shinji, T., Cai, Z., Li, X., He, W., 2002. Rn survey in the surrounding region of an uranium mill tailing impoundment. Nucl. Tech. 25, 1–6.

Yasuda, N., Zhang, D.H., Kodaira, S., Koguchi, Y., Takebayashi, S., Shinozaki, W., Fujisaki, S., Juto, N., Kobayashi, I., Kurano, M., Shu, D., Kawashima, H., 2008. Verification of angular dependence for track sensitivity on several types of CR-39. Radiat. Meas. 43, S269–S273.

Yin, G.-M., Lu, Y.-C., Zhao, H., Li, W.-L., Li, L., Guo, S.-L., 2001. Tectonic uplift rate of cenozoic Hua Mountains. Sci. Bull. 46, 1121–1123.

Yogo, A., Sato, K., Nishikino, M., Maeda, T., Sakaki, H., Hori, T., Ogura, K., Nishiuchi, M., Teshima, T., Nishimura, H., Kondo, K., Bolton, P.R., Kawanishi, S., 2011. Measurement of DNA double-strand break yield in human cancer cells by high-current, short-duration bunches of laser-accelerated protons. Jpn. J. Appl. Phys. 50, 1–7, 106401.

Young, D.A., 1958. Etching of radiation damage in lithium fluoride. Nature 182, 375–377.

Yu, K.N., Nikezic, D., 2009. Long-term measurements of radon progeny concentrations with solid state nuclear track detectors. In: Sidorov, M., Ivanov, O. (Eds.), Nuclear Track Detectors: Design, Methods and Applications. Nova Science Publishers, Inc (Chapter 2).

Yuan, W., Mo, X., Zhang, A., Chen, X., Duan, H., Li, X., Hao, N., Wang, X., 2013. Fission track thermochronology evidence for multiple periods of mineralization in the Wulongguo gold deposits, Eastern Kunlun Mountains, Qinghai Province. J. Earth Sci. 24 (4), 471–478.

Yuan, W., Zheng, Q.-G., Bao, Z.-K., Dong, J.-Q., Carter, A., An, Y.-C., Deng, J., 2009. Radiat. Meas. 44, 950–954.

Zagorskiy, D.L., Bedin, S.A., Oleinikov, V.A., Polyakov, N.B., Rybalko, O.G., Mchedlishvili, B.V., 2009. Metallic nanowires obtained as replicas of etched ion tracks in polymer matrixes: microscopy and emission properties. Radiat. Meas. 44, 1123–1129.

Zagorskiy, D.L., Korotkov, V.V., Frolov, K.V., Sulyanov, S.N., Kudryavtsev, V.N., Kruglikov, S.S., Bedin, S.A., 2015. Track pore Matrixes for the preparation of Co, Ni and Fe nanowires: electrodeposition and their properties. Phys. Procedia 80, 144–147.

Zainali, G., Afkar, A., 2005. Study of current drain during electrochemical etching of polycarbonate detectors. Radiat. Meas. 40, 337–342.

Zaki-Dizaji, H., Shahriari, M., Etaati, G.R., 2008. Calculation of CR-39 detection efficiency for fast neutrons using the MCNP and SRIM codes. Radiat. Meas. 43, S283–S285.

Zamani, M., Fragopoulou, M., Stoulos, S., Krivopustov, M.I., Sosnin, A.N., Brandt, R., Westmeier, W., Manolopoulou, M., 2008. A spallation source based on Pb target surrounded by U blanket. Radiat. Meas. 43, S151–S155.

Zhang, H., Liu, Z., Xu, J., Xu, K., Lu, J., Ruan, M., 1990a. Fission fragment angular distributions for the $^{19}F+^{208}Pb$ near- and sub-barrier fusion-fission reaction. Nucl. Phys. A512, 531–540.

Zhang, H., Liu, Z., Xu, J., Qian, X., Qiao, Y., Lin, C., Xu, K., 1994. Anomalous anisotropy of fission fragments in near- and sub-barrier complete fusion-fission reactions of $^{16}O+^{232}Th$, $^{19}F+^{232}Th$, and $^{16}O+^{238}U$. Phys. Rev. C 49 (2), 926–931.

Zhang, H., Xu, J., Liu, Z., Lu, J., Ruan, M., Xu, K., 1990b. Anomalous anisotropies of fission fragments for the $^{16}O+^{232}Th$ sub-barrier fusion-fission reaction. Phys. Rev. C 42 (3), 1086–1091.

Zhou, S.H., Guo, S.L., Meng, W., Sun, H.C., Hao, X.H., 1981. Angular distributions of fission fragments at the ^{238}U (n, 2n'f) threshold. Chin. J. Nucl. Phys. 3, 39–46.

Zhou, D., O'Sullivan, D., Semones, E., Zapp, N., Benton, E.R., 2009. Radiat. Meas. 44, 909–912.

Zhu, T.-C., Brandt, R., Vater, P., Vetter, J., 1988. Kapton nuclear track microfilter. Nucl. Tracks Radiat. Meas. 15, 771–774.

Zhu, Z., Maekawa, Y., Koshikawa, H., Suzuki, y., Yonezawa, Y., Yoshida, M., 2004. Role of UV light illumination and DMF soaking in production of PET ion track membranes. Nucl. Instrum. Meth. Phys. Res. Sect. B 217, 449–456.

Zhuk, I., Potapenko, A., Hashemi-Nezhad, S.R., Krivopustov, M.I., 2008. Determination of high-energy deuteron beam profile and spatial distribution of high-energy neutrons in a U/Pb assembly under 1.26 GeV/nucleon deuteron irradiation. Radiat. Meas. 43, S199–S203.

Ziegler, J.F., Biersack, J.P., 2003. SRIM-2003 Stopping and Range of Ions in Matter. http://www.SRIM.org.

Ziegler, J.F., Biersack, J.P., Littmark, U., 1985. The Stopping and Range of Ions in Solids, vol. 1. Pergamon, New York.

Further reading

Barillon, R., Yamauchi, T., Mori, Y., Raffy, Q., 2015. A first attempt to simulate oxidization effects on latent track structure in PADC combining the radial dose theory and a radio-oxidation kinetic model. Radiat. Meas. 83, 1–4.

DeSorbo, W., Humphrey Jr., J.S., 1970. Studies of environmental effects upon track etching rates in charged particle irradiated polycarbonate film. Radiat. Effects 3, 281–282.

Fleischer, R.L., Delany, A.C., 1976. Determination of suspended and dissolved uranium in water. Anal. Chem. 48, 642–645.

Liu, J., Wang, Y., 2004. The thermal-spike model description of the ion-irradiated polyimide. Nucl. Instrum. Meth. B 218, 318–322.

Nakashiro, 2010. On the tracks of proton and heavy ions in PC and PADC plastic detectors. In: Proceedings of the 24th Workshop on Radiation Detectors and Their Uses. KEK Proceedings, Tsukuba, Japan, vol. 10, pp. 1–11.

Nourreddine, A., Salem, Y.O., Nachab, A., Roy, C., 2015. Study of a new neutron dosimeter incorporating RPL detectors. Radiat. Meas. 83, 47–50.

Paretzke, G.H., Benton, E., Henke, P.R., 1973. On particle track evolution in dielutric track detectors and charge identification through track radius measurement. Nucl. Instrum. Methods 108, 73–80.

Stajic, J.M., Milenkovic, B., Nikezic, D., 2018. Study of CR-39 and Makrofol efficiency for radon measurements. Radiat. Meas. 117, 19–23.

Yamauchi, T., Watanabe, S., Oda, K., Yasuda, N., Barillon, R., 2009. An evaluation of radial track etch rate in LR-115 detectors exposed to Fe ions by means of FT-IR spectrometry. Radiat. Meas. 43, S116–S119.

Yuan, W.M., Mo, X.X., Yu, X.H., et al., 2000. Gold mineralization belt and prospecting direction in eastern Kunlun area. Geol. Prospect 36 (5), 20–23.

Zuo, Y., Qiu, N., Pang, X., Chang, jian, Hao, Q., Gao, X., 2013. Geothermal evidence of the Mesozoic and Cenozoic lithospheric thinning in the Liaohe depression. J. Earth Sci. 24 (4), 529–540.

Shi-Lun Guo

Shi-Lun Guo is a professor in the China Institute of Atomic Energy (CIAE), former President of the International Nuclear Track Society (INTS) and honorary member of the INTS. He was born in Hebei, China in 1937 and graduated in Zhengzhou University in China in 1961 and then joined the CIAE. He initiated studes on solid state nuclear track detectors (SSNTD) in the CIAE in 1965 and continued in the University of California at Berkeley in USA in 1982–4 and Kernchemie, Universitat Marburg, Germany in 1984, 1985, 1988 and 1990. He established the first laboratory on fission track dating (FTD) in China in 1975 and determined the age of Peking Man by ancient ashes from which grains of mineral sphene annealed completely by Peking Man were identified and used. He worked on nuclear track membranes (NTMs) and applied it to the determination of solute concentrations in various solutions and removal of bacteria from water and removal of solid particles from liquid medical injections. He created a new field of bubble detectors (BD) for recording tracks of high energy heavy ions. He is working on nuclear forensics and nuclear nonproliferation with fission- and α-tracks and on laser acceleration with CR-39 detectors. He has published more than 200 scientific articles in journals and won 2 awards by the National Congress of Science and Technology of China in 1978 and 5 awards by the China National Nuclear Corporation and the People's Liberation Army of China.

Bao-Liu Chen

Bao-Liu Chen is an associate professor in the China Institute of Atomic Energy (CIAE). He was born in Fujian, China in October 1968 and graduated from East China University of Geology in 1989. In the early 1990s he participated in dating of the Baise (Bose) archaeological site in South China by fission tracks in tektites imbedded in undisturbed stratum containing bifacial stone artifacts (bifacial hand axe) made by ancient man with the measured age 0.732 Ma. This result breached "the Movius Line" theory in paleoanthropology. He was in charge of dating by fission tracks and determining the uplift rate of the Qinling mountains in Central China and demonstrated that the Qinling mountains are still uplifting till now, which will exert an increasing influence upon climates in South and North China. He undertook the preparation of bubble detectors for detecting and identifying high-energy heavy ion tracks in collaboration between China and Japan. He is also specialized in oil field logging, radioisotope tracing and γ-ray tower scanning. He has quite a number of scientific publications, patents and is a winner of awards in science and technology from the China National Nuclear Corporation.

Chapter 4

Semiconductor detectors[1]

Ramkumar Venkataraman
Oak Ridge National Laboratory, Oak Ridge, TN, United States

Chapter outline

1. Contributors to this chapter in previous editions are Jan Verplancke, Paul F. Fettweis, Ramkumar Venkataraman, Brian M. Young, and Harold Schwenn.

Handbook of Radioactivity Analysis: Volume 1. https://doi.org/10.1016/B978-0-12-814397-1.00004-2

I. Introduction

A. The gas-filled ionization chamber

A semiconductor detector can be best compared with a classical ionization chamber described elsewhere in this book (Chapter 3). A schematic diagram of such an ionization chamber is given in Fig. 4.1. It consists essentially of a gas-filled (Kr, Xe, ...) capacitor to which a bias (H.T.) is applied. An ionizing particle (alpha, p, d, beta, ...) will create a certain number N of pairs of positive ions and electrons, where N is given by

$$N = \frac{E}{\varepsilon}. \tag{4.1}$$

where E represents the kinetic energy of the particle and ε the energy necessary to create one ion−electron pair.

To act as a spectrometer (i.e., an instrument able to count the number of entering particles and to measure their energy), an ionization chamber must fulfill three basic conditions:

1. The ionizing particle must lose all its kinetic energy inside the sensitive volume of the detector.
2. All created charges must be collected by the applied bias and contribute to the pulse formation.
3. In the absence of any ionizing particles, no charges may be collected by the electric field.

B. The semiconductor detector

A reverse-biased p-n or p-i-n Ge or Si diode fulfills all three of these basic conditions to function as a solid-state ionization chamber (Fig. 4.2). Indeed, the intrinsic or depleted region of the junction acts as the sensitive volume, and the whole may be regarded as a capacitor having a (small) leakage current between the p+ and n+ contacts in the absence of any ionizing radiation. From an electronic point of view, it may be regarded as a capacitor in parallel with a direct current (DC) source.

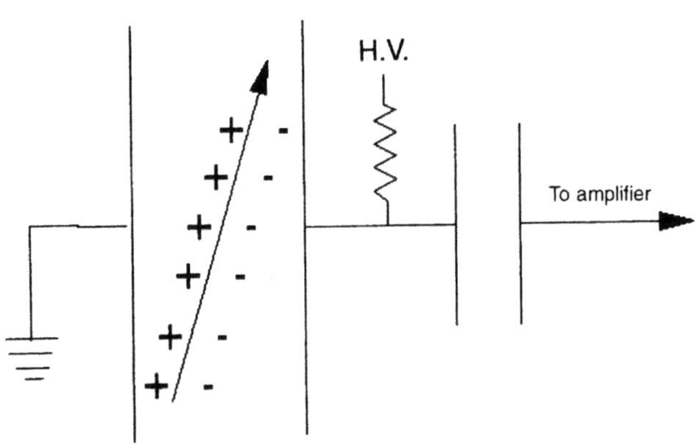

FIGURE 4.1 Schematic view of a gas-ionization chamber. The positive and negative charges formed by the ionizing particle are integrated, and the resulting pulse, whose height is proportional to the deposited energy, is sent to the amplifier for further treatment.

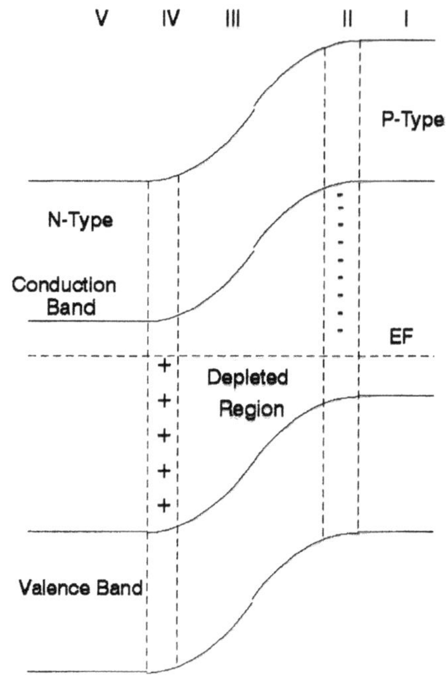

FIGURE 4.2 Band structure of a p-n junction. The probability $P(E)$ of occupation of an electronic level E in a solid is given by $P(E) = I/\{[\exp(E - E_F)/kT + 1\}$, where E_F represents the Fermi level, k the Boltzmann constant, and T the temperature. Note that $P(E_F) = 0.5$. In a p-type semiconductor, E_F lies close to the valence band, in an n-type semiconductor close to the conduction band, and in an intrinsic semiconductor approximately halfway between both bands. In an unbiased p-n junction, the height of the Fermi level E_F depends only on the temperature. Five regions are distinguished in a p-n junction: the p-region, the negative space charge region, the intrinsic region, the positive space charge region, and the n-region. In a reversed biased p-n junction, the potential barrier is enhanced, and the p-n junction acts as a diode. In a p-i-n junction, E_F is no longer constant, and the intrinsic region is increased.

The detector capacitance depends on the detector dimensions. Its magnitude is determined by the area of the p+ and n+ contacts, their separation, and the dielectric constant of the semiconductor. The p+ contact carries a negative space charge, and the n+ contact carries a positive space charge. In the intrinsic region, an electrical field exists due to both the space charges and the applied reverse bias. In Ge detectors, this intrinsic region may be very large (up to 60 mm), typical values for silicon detectors are 150−1000 μm, and Si(Li) detectors have a thickness of 3−4 mm.

An ionizing particle entering (or created in) the intrinsic region will excite a certain number of electrons from the valence band, into the conduction band, generating a certain number of pairs consisting of positive holes and negative electrons swept away to the p+ and n+ contacts, respectively, by the existing electric field.

Variations in shape and rise time make the amplitude of the current pulse unsuitable for spectroscopic aims, as its

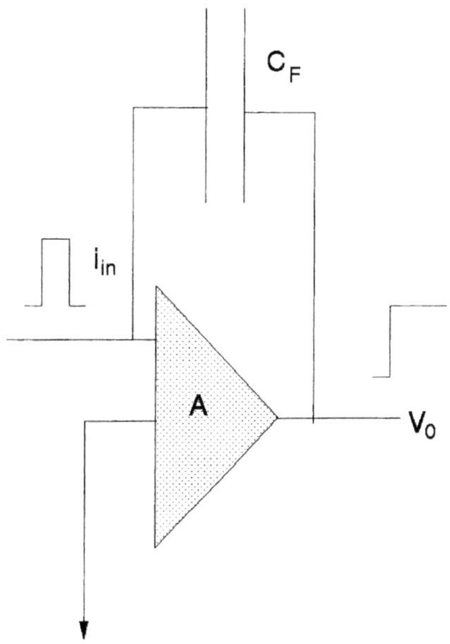

FIGURE 4.3 Charge-sensitive preamplifier.

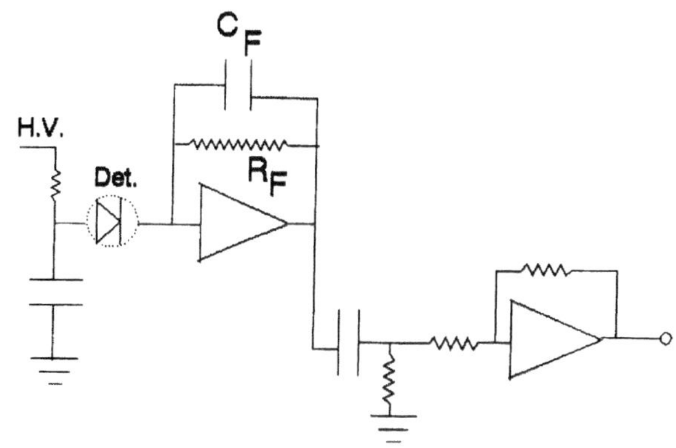

FIGURE 4.4 Schematic drawing of a resistive feedback preamplifier.

intensity is not proportional to the deposited energy (Kröll et al., 1996). What is important for spectroscopic applications is the integral of the current pulse. Therefore, a charge-sensitive (integrating) preamplifier (Fig. 4.3) has to be used, which transforms the current pulse, i_{in}, into a step voltage V_o. The latter is proportional to the incident energy, if the amplification factor, A, is very large:

$$V_0 = \int \mathrm{i_{in}}\ \mathrm{dt}/C_F = Q/C_F = Nq/C_F = Eq/(\varepsilon C_F) \tag{4.2}$$

where Q represents the total charge Nq and ε the energy necessary to excite an electron–hole pair. This energy may not be confused with the forbidden energy gap, which is much smaller (Table 4.1). It means that about 33% of the available energy is actually converted into electron–hole pairs. The rest serves to excite lattice vibrations and is lost in the pulse formation (Leo, 1987; Goulding and Landis, 1982). Protracted accumulation of charges on the feedback capacitor C_F must be avoided. Therefore, C_F has to be discharged in time, most commonly by a resistor R_F or by pulsed reset techniques (see Fig. 4.4).

TABLE 4.1 Some important Ge and Si properties.

	Ge (at 77 K)	Si (at 300 K)
Electron mobility: μ_e *in* $cm^2/V \cdot s$	3.6E4	1350
Hole mobility: μ_h *in* $cm^2/V \cdot s$	4.2E4	480
Energy ε needed to create 1 $e^- - h^+$ pair	2.96 eV	3.62 eV
Atomic number Z	32	14
Forbidden energy gap	0.746 eV	1.115

Apart from the integrator stage, a (resistive feedback) preamplifier may have a second stage. A differentiation and a pole-zero cancellation circuit couple the two stages (Fig. 4.4). The rise time of the signal is determined by the output signal of the detector along with the preamplifier response time. A typical fall time is of the order of 50 μs. For digitization, the signal has to be further transformed by a shaping amplifier. The task of the shaping amplifier is complex. It transforms the shape and amplitude from the preamplifier signal to do the following:

1. Improve the signal to noise ratio via adjustable shape form and width.
2. Make it suitable for digitization in an analog-to-digital converter (ADC).
3. Make the output independent of the signal rise time.
4. Facilitate calibration of the spectrum.

A well-chosen peaking time, that is, the time needed for the signal to reach its maximum amplitude, is important to reduce the electronic noise and thus to improve the detector resolution. The output pulse should not be too long to prevent spurious summation of independent pulses separated by very small time intervals. On the contrary, for very short peaking times (~1 μs), the peak shaping may be ended before completion of the integration, which would mean important loss of information. This so-called ballistic deficit is particularly important in large γ detectors. Analog ("gated integrator") or digital techniques may be incorporated in the pulse processing to minimize spectrum broadening due to ballistic deficit. A comparative study of different ballistic deficit correction methods versus input count rate has been carried out (Duchêne. and Moszynski, 1995). Finally, the signal is transferred to an ADC and a multichannel analyzer (MCA), which measures the pulse height and constructs a spectrum, which is a histogram of pulses classified as a function of their pulse height. The analog amplifier and ADC can be replaced by a digital signal processing (DSP) module. DSP is a technique whereby the detector signal is digitized directly as it comes from the preamplifier, with only some minor preconditioning. The digitized data are then filtered and optimized using digital processing algorithms and finally transferred to the MCA for storage, view, and analysis. DSP allows implementation of signal filtering functions that are not possible through traditional analog signal processing. Benefits include higher throughput, reduced sensitivity to ballistic deficit, adaptive processing, improved resolution, and improved temperature stability for repeatable performance.

C. Fundamental differences between Ge and Si detectors

In Table 4.1, three important differences between Ge and Si are given. These are the energy gap, the atomic number Z, and the mobilities μ_e and μ_h of the majority carriers. Together with the purity and charge-carrier lifetime, they influence the thickness of the depletion region of a biased p-n junction.

1. The energy gap

There is a 50% difference between the energies needed to create an electron–hole pair in germanium and silicon. A Si detector may be used at room temperature for the spectroscopy of charged particles. A Ge detector has to be cooled below 100 K to reduce the leakage current due to thermal generation of charge carriers to an acceptable level. This has important consequences: a Ge detector has to be operated inside a vacuum chamber. The sensitive detector surfaces are thus protected from moisture and other condensable contaminants. Consequently, independent of the junction itself, traditional Ge detector cryostats have an entrance window that makes Ge detectors less suited for the detection of charged particles and also affects the efficiency for low-energy photons.

2. The atomic number

In Chapter 1, Volume 1, the three typical interactions of electromagnetic radiation with matter have been detailed. The electrons scattered (photoelectric effect or Compton scattering) or generated (pair production) by one of the three basic interactions excite a certain number of electron–hole pairs and are responsible for the peak formation. For γ spectroscopy, the photoelectric effect contributes directly to the full-energy peak (FEP). Indeed, as the total energy of a γ ray is transferred to an electron, the kinetic energy of the electron will be proportional to the energy of the incoming γ ray. For the efficiency of a γ spectrometer, preference should thus be given to a semiconductor material having a high photoelectric cross section. Fig. 4.5 shows the photoelectric cross section of Si and Ge as a function of energy. One sees immediately that Ge cross section is greater than that of Si by one to two orders of magnitude. This is expected, as the photoelectric cross section depends roughly on the fifth power of the atomic number Z. Fig. 4.6 shows the Compton, photoelectric, and pair production cross section of Ge for γ ray energies up to 10 MeV.

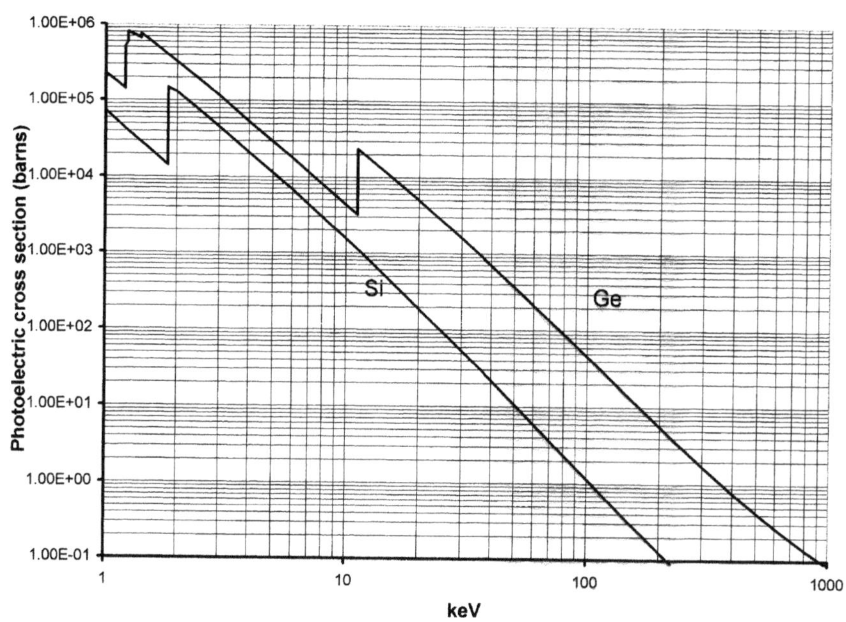

FIGURE 4.5 Photoelectric cross section (barns) of Si (lower curve) and Ge (higher curve) as a function of energy (keV).

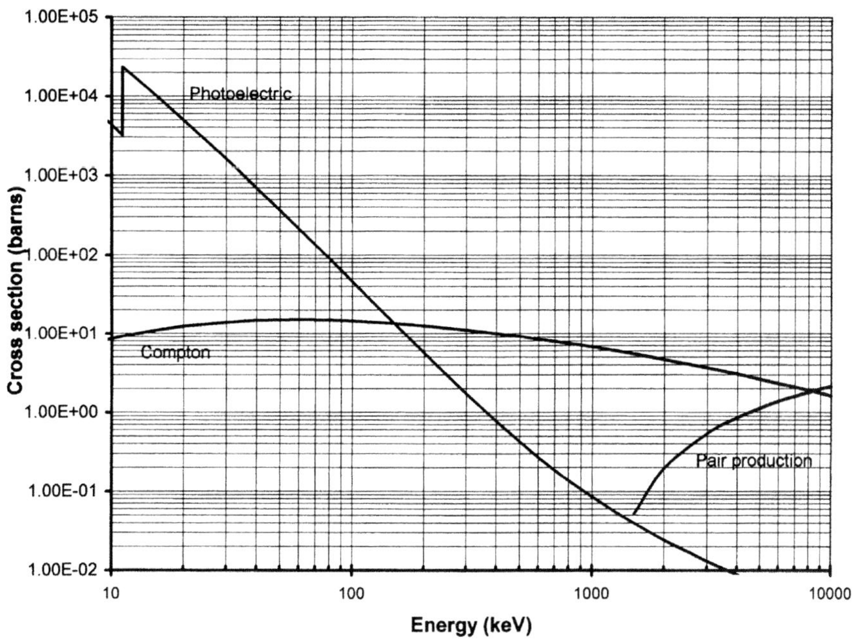

FIGURE 4.6 Compton scattering, photoelectric, and pair production cross section of Ge for high-energy γ rays.

The Compton scattering (or inelastic scattering) cross section is the dominant one for all energies except the very lowest ($E_\gamma \leq 150$ keV) and the very highest (E_γ 8.5 MeV). In this energy range, most full-energy depositions occur via several Compton scattering interactions—or multisite events inside the germanium crystal under the condition that the last interaction is a photoelectric one and that all the preceding Compton interactions take place in the Ge crystal. In large-volume detectors, the probability of multiple Compton scattering increases. If the last interaction does not occur by the photoelectric effect or if one of the multiple Compton scattering interactions takes place outside the sensitive volume of the detector, the pulse will contribute to the Compton continuum. Depending on the size and shape of the crystal and the energy of the interacting gamma quantum, in about half of the interactions, there is only one interaction point inside the crystal. The continuum background in this energy range has therefore a high probability for single site interactions (single site events, SSE). This observation has led to techniques to lower the continuum background by rejecting SSE using pulse-shape discrimination (Budjas et al., 2009).

The threshold of 1022 keV for the pair production process (see Chapter 1) is clearly seen. It is remarkable that the pair production cross section of a 10 MeV γ ray equals that of the photoelectric cross section at about 300 keV. It plays an important role in the spectroscopy of high-energy γ rays. Full absorption of two, one, or none of the 511 keV annihilation lines will contribute to the "full energy," the "single escape," or the "double escape" peak. All three peaks carry full spectroscopic information and are discussed in some detail in Section II.B.2.

3. The purity or resistivity of the semiconductor material

It is known (Knoll, 1999) that the thickness of the depletion region of a planar semiconductor is given by

$$d = \sqrt{\frac{2\varepsilon V}{qN}} \tag{4.3}$$

where V represents the applied bias, N the net concentration level of electrically active impurities in the bulk, q the electronic charge, and ε the energy needed to excite one e^-h^+ pair. For Ge, net active impurity concentration levels as low as 10^{10} atoms/cm^3 of either p or n type can be achieved. This corresponds roughly to 1 impurity atom per 10^{12} atoms! The application of a reversed bias of up to 5000 V thus leads to a depletion thickness of several centimeters. This is not the case in Si. Indeed, the resistivity ρ of the semiconductor material can be expressed as

$$\rho = \frac{1}{q\mu N} \tag{4.4}$$

where μ represents the mobility of the majority carrier. Eq. (4.4) may thus be written as

$$d = \sqrt{2\varepsilon V \rho \mu} \tag{4.5}$$

From Table 4.1 one sees that μ is much smaller for Si than for Ge. If d is expressed in μm, V in volts and ρ in ohm-cm, Eq. (4.5) reduces for Si to

$$d = 0.562\sqrt{\rho V} \tag{4.6}$$

A thickness of up to 315 μm can be obtained for typical resistivities of ≈3000 Ω-cm and a bias of ~100 V. It is thus not possible to realize high-volume detectors with Si. Except for X-rays or low-energy γ rays, Si detectors are used mainly for charged particles. Since Si detectors may be used at room temperature, they may be placed in a vacuum chamber together with the source. The absence of any supplementary entrance window allows the particles to reach the sensitive volume of the detector.

In Si(Li) detectors, the excess acceptor ions in p-type Si may be compensated by Li donor ions. This way, a thickness of up to 5 mm of the active p-i-n region can be obtained. These detectors are predominantly used in X-ray spectroscopy.

4. Charge carrier lifetime τ

The charge carrier lifetime τ is the time that the carriers (electrons in the conduction and holes in the valence band) remain free. Trapping centers reduce this lifetime. The maximum signal height V_0 (Eq. 4.2) from the preamplifier after interaction of the detector with the ionizing radiation is given by

$$V_0 = \frac{Eq}{\varepsilon C_F}\left(1 - \frac{d}{\mu \mathrm{E}\tau}\right) \tag{4.7}$$

where d is the distance traveled by the charge, E the energy deposited in the detector, E the electric field, μ the mobility of the charges, q the elementary charge, ε the energy needed to excite one e^-h^+ pair, τ the charge carrier lifetime, and C_F the feedback capacitor value.

To have good charge collection and thus to avoid tailing, $\mu \mathrm{E}\ \tau >> d$ where the minimum value for τ for detector grade semiconductor material is 5 ms for Si at 300 K and 20 μs for Ge at 77 K.

II. Ge detectors

A. High-purity Ge detectors: merits and limitations

The depletion layer of a gamma detector must have a thickness of several centimeters to enhance the probability of an interaction of a γ ray with the sensitive detector material and thus be useful as a γ ray spectrometer.

Today, Ge crystals of either p or n type are grown with the low-impurity levels needed with dimensions up to 10–12 cm. The detectors fabricated from these crystals are called intrinsic or high-purity detectors. They can be stored at room temperature; however, the detector response will decrease if stored at room temperature for a prolonged time, especially with respect to p-type coaxial detectors. The effect of storing a germanium detector for a prolonged time in warm condition is discussed in Section II.C.3.f entitled "Stability of the Germanium Detector Efficiency."

The facts that germanium can be made in large sizes, that it can be purified to the levels needed for full depletion, and that it needs only a very small energy to create one electron–hole pair make it a unique gamma ray detection material. A Ge detector produces more electron–hole pairs for a given gamma ray that is totally absorbed in its active volume than any other detector material. By consequence, energy resolutions are obtained that are more than 10 times better than by competing gamma detector technologies. The advantage that Ge has with regard to the small bandgap also represents the operational challenge for these detectors because they must be cooled to very low temperatures. Historically, this cooling has been achieved using liquid nitrogen baths, but, today, electromechanical coolers are becoming the standard (see commercial information Canberra, Ortec, PHDs Inc., and others). This also requires that, for most applications, the Ge detector crystal be stored in a cryostat having a window that stops (partly) low-energy gamma rays (see Fig. 4.15 further on in this chapter).

Because of these unique advantages, the applications in which Ge detectors are used span a very wide range from fundamental scientific research to nuclear material safeguards and security, environmental protection, and human health and safety.

Yet, for some applications, gamma spectroscopy alone does not suffice to quantify and qualify gamma emitters to a level that is acceptable for safety and security. Particular examples of these are the transuranic isotopes. Some of these isotopes are only weak gamma emitters that could be hidden in dense waste matrices or behind thick, absorbing walls and/or masked by very strong gamma emitters such as fission and activation products. For some applications, it may be necessary to use more detectable signals emitted by these special nuclear materials, besides their gamma signature. These detectable signals that are emitted per gram of this isotope are summarized in Table 4.2. The associated nondestructive assay techniques that can be used to make best use of these signals are well described in Reilly et al. (1991).

B. Analysis of typical γ spectra

Ge detectors of different size or geometry, such as planar detectors, coaxial detectors, and well-type detectors, are available. Others differ in the choice of contacts, of the choice of the entrance window, the selection of the cryostat construction materials, and so on. In Section II.E, these will be briefly described together with their main applications. However, before doing so, it is important to analyze the main features of a γ spectrum, to understand the influence of the parameters that are used to characterize a germanium detector, and to know the different sources of background. Only a clear understanding of these features will allow the user to choose the right detector for a specific application.

TABLE 4.2 List of transuranic isotopes with characteristic signals that could be used to detect these.

Isotope	Gamma (keV)	γ/g/s	n/g/s (for metal)	α/g/s	Watt/g
Th-228	239	1.3E+13		3.0E+13	2.7E+01
	583	9.4E+12			
Th-229				7.8E+09	
Th-230				7.6E+08	5.7E−04
Th-231	84	1.1E+15			
	108	2.6E+14			
	163	2.7E+13			
Th-232			4.1E−05	4.1E+03	2.7E−09
U-232	58	1.6E+09	1.3E+00	8.0E+11	6.9E−01
	129	5.4E+08			
U-233			8.6E−04	3.6E+08	2.8E−04
U-234			5.5E−03	2.3E+08	1.8E−04
U-235	144	8.5E+03	2.7E−04	8.0E+04	6.0E−08
	186	4.4E+04			
U-236			5.5E−03	2.4E+06	1.8E−06
U-237	60	1.0E+15			1.6E+02
	208	6.6E+14			
U-238	258	9.0E+00	1.3E−02	1.2E+04	8.6E−09
	569	2.5E+00			
	1001	1.0E+02			
Np-237	86	3.3E+06	1.1E−04	2.6E+07	2.1E−05
	95	2.2E+05			
Pu-236	47	6.1E+09	1.6E+04	2.0E+13	1.9E+01
	110	2.4E+09			
Pu-238	100	4.1E+07	2.6E+03	6.4E+11	5.7E−01
	153	6.4E+06			
Pu-239	129	1.4E+05	2.2E−02	2.3E+09	1.9E−03
	414	3.4E+04			
Pu-240	104	5.8E+05	1.0E+03	8.4E+09	7.2E−03
	160	3.4E+04			
Pu-241	104	3.8E+06	2.5E−02	9.4E+07	3.3E−03
	149	7.1E+06			
	208	2.0E+07			
Pu-242	103	3.8E+03	1.7E+03	1.4E+08	1.2E−04
	159	4.4E+02			
Am-241	60	4.6E+10	1.2E+00	1.3E+11	1.2E−01
	99	2.6E+07			
Cm-242	102	3.1E+09	2.3E+07	1.2E+14	1.2E+02
	561	1.9E+08			

Continued

TABLE 4.2 List of transuranic isotopes with characteristic signals that could be used to detect these.—cont'd

Isotope	Gamma (keV)	γ/g/s	n/g/s (for metal)	α/g/s	Watt/g
Cm-243	277	6.4E+11		1.9E+12	1.9E+00
	228	5.2E+11			
Cm-244	43	7.8E+08	1.1E+07	3.0E+12	2.9E+00
	555	2.8E+06			
Cm-245	173	2.9E+08		6.4E+09	5.8E−03
	133	2.9E+08			
Cm-246	45	2.9E+06	7.5E+06	1.1E+10	1.0E−02
Bk-249	327	1.5E+02	9.5E+04	8.8E+08	3.2E−01
Cf-252	43	3.0E+09	2.3E+12	2.0E+13	4.1E+01
	160	4.0E+08			

1. Spectrum of a source emitting a single γ ray with Eγ < 1022 keV

Fig. 4.7 shows the decay scheme of ^{137}Cs, one of the important long-lived ($T_{1/2} = 30.17$ y) fission products and a common contaminant. It emits two β-rays of 1176 (6%) and 514 keV (94%) exciting a 2.55-min isomeric level of ^{137}Ba. This isomeric level deexcites itself by the emission of a single γ ray of 661.66 keV. The M4 isomeric transition is highly converted ($\alpha_{total} = 0.11$); that is, the deexcitation can take place through the emission of a γ ray but also by the ejection of an atomic electron (a conversion electron) with subsequent delayed emission of the characteristic ^{137}Ba X-rays. Even though ^{137}Cs generates one of the simplest spectra possible (Fig. 4.8), it is worthwhile to take a closer look at it. The spectrum was taken with a 25% n-type Ge ("REGe") detector placed in an RDC low-background cryostat and a ULB Pb castle. The most striking is the FEP at 661.66 keV carrying the full spectroscopic information. The X-rays of the daughter element ^{137m}Ba are clearly seen: two doublets at 31.82−32.19 keV and 36.4−37.3 keV. For most other γ transitions, the intensity of the X-rays of the daughter element will be less pronounced, as most γ transitions have a much smaller total conversion coefficient. The small peak at 693.9 keV corresponds to the random sum peak between the Ba X-rays and the 661.66-keV photopeak. The weak peaks at 651.8, 22.31, and 26.52 keV correspond to the Ge escape peaks (see Section II.B.3) of the 661.66-keV line and of the Ba X-rays.

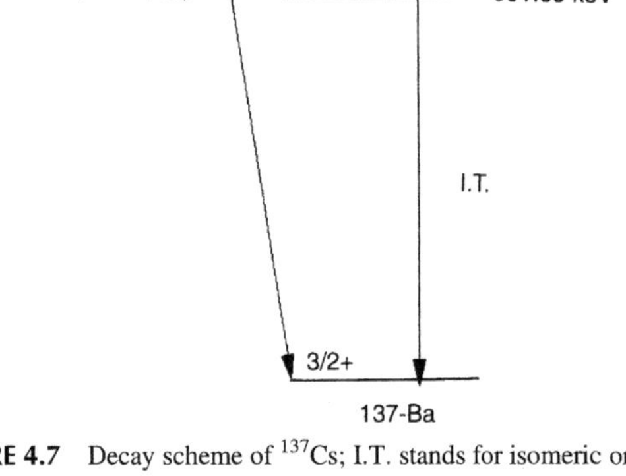

FIGURE 4.7 Decay scheme of ^{137}Cs; I.T. stands for isomeric or internal transition.

In addition to these well-defined peaks, two broad peaks are seen, namely, the Compton edge and the backscatter edge. The first has an energy of about 478 keV. It is due to a 180 degrees Compton scattering inside the active volume of the detector with subsequent escape of the Compton γ ray from the detector's active volume. The second is due to a 180 degrees Compton scattering in the detector surroundings with subsequent detection in the detector of the escaped Compton-scattered γ ray having an energy of about 184 keV. The broadness of these peaks is due to the fact that the scattering angle of 180 degrees is only approximately fulfilled. Finally, the broad elevation in the continuum around 845 keV is due to the summation of the backscatter edge with the photopeak. In Fig. 4.9, the energies of the Compton edge and the backscatter edge are given as a function of energy of the primary gamma ray. Notice that the backscatter edge tends toward a saturation value of about 200 keV. The energy of the Compton edge is given by

$$E_{CE} = E_\gamma - E_{BS}. \tag{4.8}$$

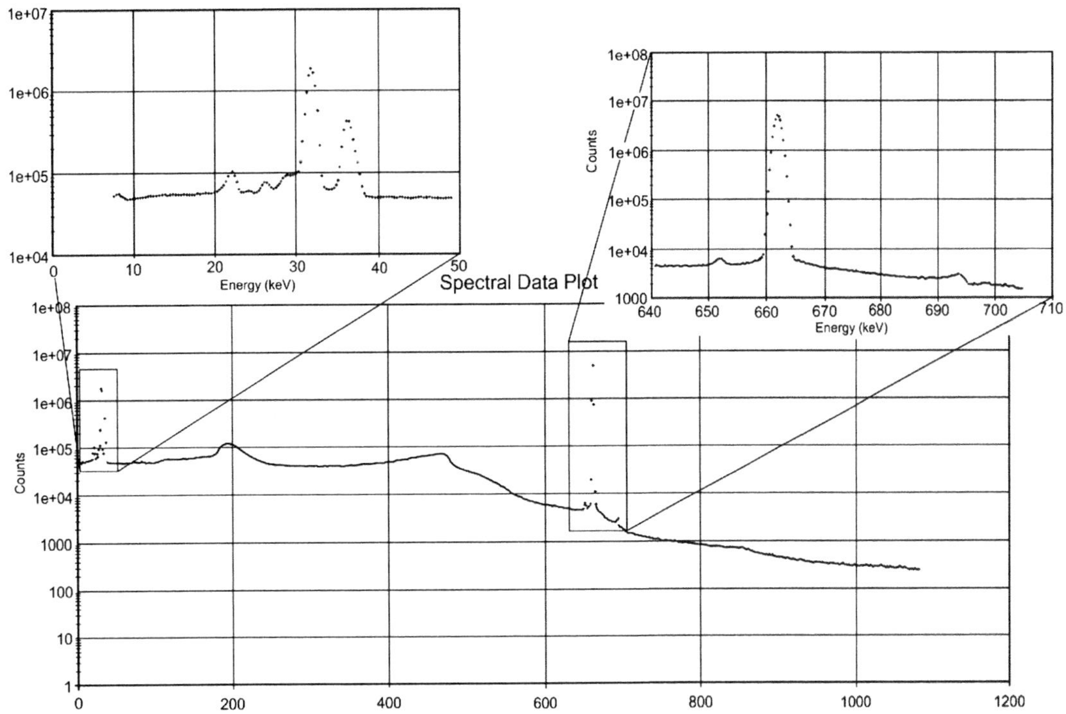

FIGURE 4.8 Gamma spectrum of ^{137}Cs emitting a single γ-ray at 661.66 keV taken with a 25% n-type Ge detector placed in an RDC low-background cryostat and a ULB Pb castle. Besides the photopeak at 661.66 keV, the X-rays of the daughter element ^{137m}Ba are seen: two doublets at 31.82–32.19 keV and 36.4–37.3 keV. The weaker lines at 22.11, 26.3, and 651.8 keV correspond to the Ge escape peaks, and the 693.9 keV peak is a random sum peak between the Ba X-rays and the 661.66 keV main peak.

Note that both curves cross at about 250 keV. For γ rays with $E_\gamma < 250$ keV, the positions of the Compton edge and the backscatter edge are thus reversed. The continuum at the lower energy side from the Compton edge is due to Compton scattering inside the active volume of the detector with subsequent escape of the Compton-scattered γ ray and to the bremsstrahlung emitted during the interaction of the betas and electrons with the detector surroundings. The maximum of this bremsstrahlung continuum is equal to that of the emitted beta, 1176 keV in the case of ^{137}Cs.

The continuum between the Compton edge and the FEP is due to multiple Compton scattering, where the last interaction is a Compton event rather than a photoelectric effect. This leads to the fact that the continuum to the left of the FEP is generally higher than to its right.

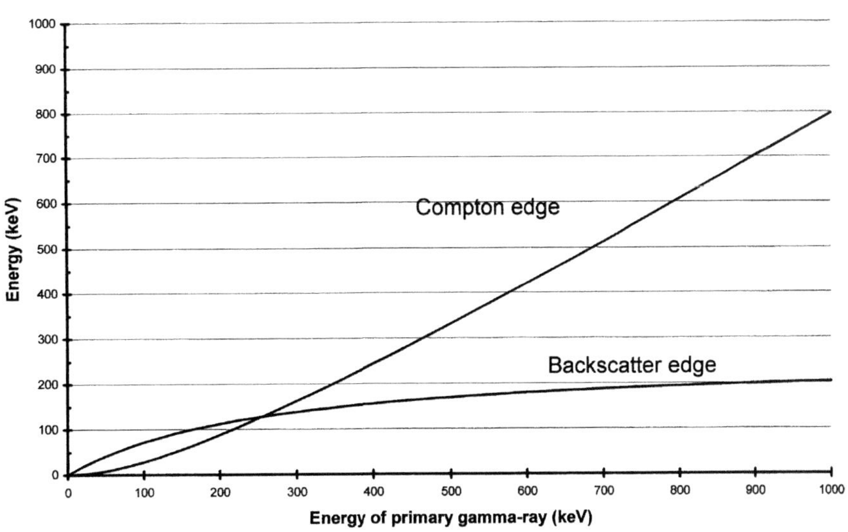

FIGURE 4.9 Backscatter edge and Compton edge as a function of primary γ ray energy.

The background above the FEP is due to the bremsstrahlung of the 1176-keV β transition and to origins not related to the source, as discussed in greater detail later in Section D.

2. Spectrum of a multiple-γ-ray source emitting at least one γ ray with an energy ≥1022 keV

The spectrum can be complicated even when only a small number of γ rays are emitted during the radioactive decay. The case in which one or several γ rays surpass the energy of 1022 keV is especially interesting. This will be illustrated with the help of the γ spectrum of ^{24}Na formed, for example, by the ^{23}Na(n,γ)^{24}Na reaction and decaying with a half-life of 15.03 hours to ^{24}Mg. This decay takes place in >99% of all cases by a β transition of 1.389 MeV and in 0.06% by a β transition of 276 keV. From the decay scheme shown in Fig. 4.10, one sees that two strong (>99%) coincident γ rays of 1368.9 and 2755.2 keV exist, as well as a weak γ transition of 3867.2 keV (0.06%), also in coincidence with the 1368.9-keV line. The total of only three γ rays, all surpassing the threshold for a possible pair production, leads to the quite complex spectrum of Fig. 4.11. It shows, besides the backscatter peak at about 200 keV, a total of 13 well-defined peaks.

The three FEPs at 1368.9, 2755.2, and 3867.2 keV are clearly seen. The intense first two are accompanied by a well-pronounced Compton edge at approximately 1100 and 2400 keV, whereas the Compton edge of the weak 3867.2-keV line is almost lost in the general background.

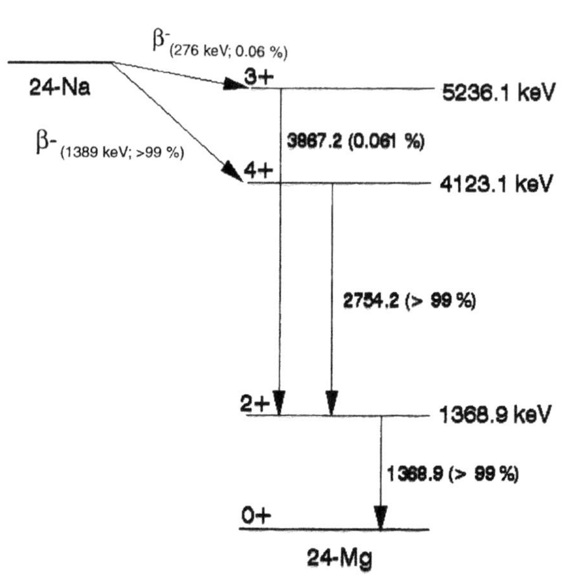

FIGURE 4.10 Decay scheme of ^{24}Na.

If pair production takes place, two annihilation quanta of 511 keV are emitted at 180 degrees. When the two are fully absorbed, they contribute to the FEP. When one escapes from the detector without interaction, a discrete single escape (SE) peak is generated having an energy of

$$E_{SE} = E_\gamma - 511\text{keV}. \quad (4.9)$$

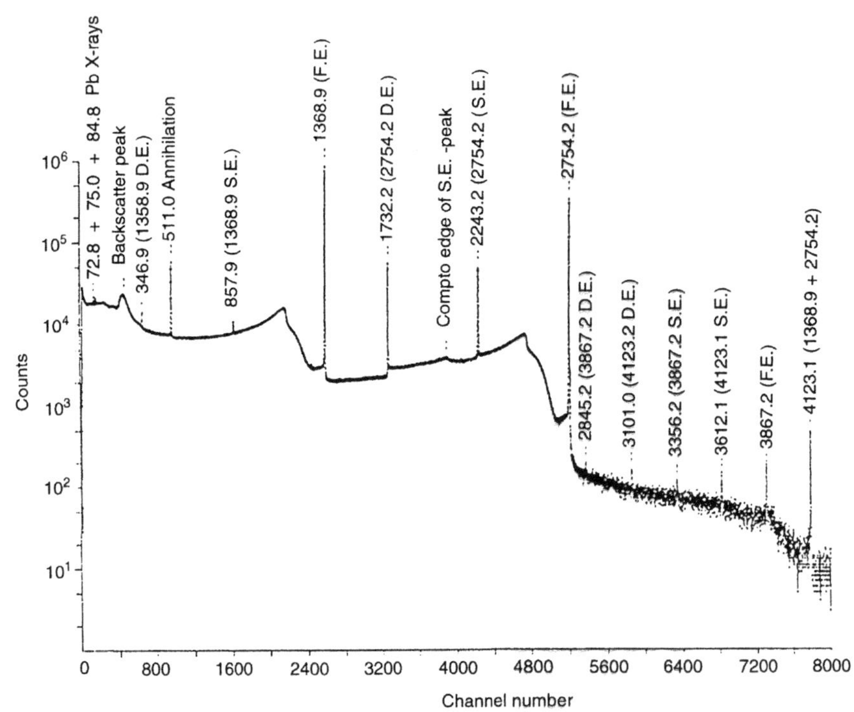

FIGURE 4.11 γ spectrum of ^{24}Na-emitting two strong (>99%) coincident γ rays at 1368.9 and 2754.2 keV and a weak γ ray at 3867.2 keV (0.06%). Thirteen well-defined peaks are observed. Their origin is explained in the text.

When both annihilation quanta escape, the double escape peak (DE) is generated with an energy of

$$E_{DE} = E_\gamma - 1022\text{keV}. \tag{4.10}$$

Besides the Doppler-broadened 511-keV line (see Section C.1.c), all six escape peaks can be recognized in the spectrum. Those of the 1368.9-keV line are weak, as the energy is too close to the threshold energy of 1022 keV (Fig. 4.6). It is worthwhile to take a closer look at the strong DE peak at 2755.2−1022 = 1732.2 keV and the corresponding SE peak at 2243.2 keV. The peak shape of the first one is a mirror image of the FEP. The background to the right of the peak is higher than that to the left! This is due to multiple Compton scattering of one or both annihilation quanta (the last interaction not being a photoelectric effect), whereby the energy of the Compton electron adds to the energy of the DE peak, increasing the continuum to its right. On the contrary, the SE peak is perfectly symmetric, as the continuum to the left and the right of the peak is increased by multiple Compton scattering. These shapes are characteristic for the escape peaks. For weaker peaks, however, they are often masked by the general continuum. Note also that the SE peak is accompanied by a Compton edge but that the gap between the SE peak and its Compton edge does not correspond to that of a 2243.2-keV γ line but to that of a Compton scattered 511-keV γ line (Fig. 4.9). Indeed, one of the two annihilation quanta escaped from the detector, whereas the other was Compton scattered. Note also that the SE peak is Doppler-broadened. Indeed, this is due to the summation of a sharp DE peak with a Doppler-broadened 511-keV quantum. To a lesser amount, the same is also true for the FEP of high-energy γ rays due, partially, to the full absorption of the two annihilation quanta.

Finally, one recognizes the sum peak at 4123.1 keV of the two strong coincident γ rays at 1368.9 and 2755.2 keV as well as the two escape peaks corresponding to this energy.

It is not always easy to distinguish these different peaks in an unknown spectrum. The best criterion for recognizing the different escape peaks is the exact energy difference of 511 or 1022 keV. For complicated spectra, a comparison of their relative intensities with the expected ones from the relative efficiency curves for the three peak types can give further confirmation.

3. Peak summation

In Fig. 4.11, different sum peaks are discernible. They merit further attention. Real sum peaks have to be distinguished from random sum peaks. Real sum peaks are due to coincident γ rays simultaneously detected. Their energies equal the sum of two individual energies. The interpretation can be confirmed by their intensities if measured with the same detector at a different source-to-detector distances. Indeed, the probability P of a real sum peak is given by:

$$P = I \cdot p \cdot \varepsilon_1 \cdot \varepsilon_2 \tag{4.11}$$

where I is the intensity (Bq) of the source, ε_1, ε_2 are the counting efficiencies for γ_1 and γ_2, respectively, and p is the intensity of the less abundant of the two coincident γ rays summing up.

If $\varepsilon_1 \approx \varepsilon_2$, one sees that the intensity of the sum peak varies roughly as the square of the efficiency. The phenomenon of true coincidence summing, also referred to as cascade summing, and its impact on gamma ray FEPs is discussed in Section II.B.4 of this chapter.

In addition to real sum peaks, spurious sum peaks due to the finite time resolution can occur. Their probability is given by

$$P = 2 \cdot \tau \cdot I^2 \cdot \varepsilon_1 \cdot p_1 \cdot \varepsilon_2 \cdot p_2 \tag{4.12}$$

where τ is the time resolution of the detection system and p_1 and p_2 are the branching ratios of the two γ rays summed up accidentally.

The intensity of random sum peaks depends, therefore, on the square of the source intensity and on the time resolution. An illustration of a random sum peak is given in Fig. 4.8. Avoid, if possible, the use of intense sources to minimize random summation. If the radioactive source decays with a certain transition probability λ, the intensity I is given by

$$I = I_0 \cdot e^{-\lambda \cdot t} = I_0 \cdot e^{-\frac{0.693}{T_{1/2}} \cdot t} \tag{4.13}$$

where I_0 represents the initial intensity and $T_{1/2}$ the half-life. Inserting Eq. (4.13) into Eq. (4.12), one sees that the probability P of occurrence of a spurious sum peak depends on $I^2 = I_0^2 \cdot e^{-2 \cdot \lambda \cdot t}$. The probability of occurrence of a spurious sum peak decays thus with a transition probability of $2 \cdot \lambda$ rather than λ, or a half-life $T_{1/2}/2$ rather than $T_{1/2}$, i.e., twice as fast as the isotope itself. This is a firm criterion for their recognition. Pulse pileup rejection in modern processing electronics can reduce random summing to a great degree.

4. True coincidence summing effects

In most cases of radioactive decay, a parent nuclide decays to an excited energy level of a daughter nuclide by emitting an alpha or a beta particle, or via electron capture. The transition from the excited state to the ground state of the daughter nuclide may then occur by the emission of two or more gamma rays in a cascade. Since the excited states have life times on the order of picoseconds, it is highly probable that the γ rays emitted in a cascade are detected within the resolving time of a gamma ray spectrometer. The γ rays are then said to be detected in true coincidence. In the case of a nuclear decay occurring via electron capture,

X-rays will be emitted, which may also be detected in true coincidence with a gamma ray. As a result of true coincidence summing or cascade summing, the detector accumulates the sum total of the energy deposited by the cascading gammas from a given nuclear decay. Therefore, events are lost (summing-out) or gained (summing-in) from the FEP of the gamma ray of interest, and any activity determination based on the FEP will be in error. It is, therefore, necessary to correct for true coincidence effects.

Fig. 4.12 gives an example of a radioactive decay where cascade summing occurs. In the above example, the parent nuclide undergoes a beta decay to the excited energy state E_1 of the daughter nucleus. The deexcitation to the ground state of the daughter nuclide occurs via the emission of gamma rays γ_1 and γ_2 in a cascade or via the emission of gamma ray γ_3 directly to the ground state. Assuming that the gamma rays γ_1 and γ_2 are detected in true coincidence, an FEP measurement of γ_1 or γ_2 suffers from cascade summing losses and the FEP measurement of γ_3 suffers from cascade summing gains. It must be noted that cascade summing losses are not just limited to the counts appearing in the sum peak. Rather, the detector may accumulate the full-energy deposition from one of the gamma rays (say γ_1) and a partial energy deposition from the second gamma ray (say γ_2), resulting in a count being lost from the FEP of γ_1. In the pulse height spectrum, these events will appear in the continuum between the energy of γ_1 and the energy of the sum peak. The same argument can be made for γ_2 as well. Therefore, the cascade summing losses depend on *the total efficiency* of the detector for the gamma rays in the cascade. In the case of cascade summing gains, the two gamma rays γ_1 and γ_2 in the above example deposit their full energy in the detector. The resulting event appears at the same energy as that of γ_3. Cascade summing gains are dependent on the peak efficiencies of the detector at the gamma ray energies of interest. Besides the detection efficiencies, cascade summing losses or gains also depend on the gamma ray emission probabilities and any angular correlations involved in the gamma ray emission. It is important to note that the magnitude of cascade or *true* coincidence summing losses or gains is dependent on the counting geometry, and not on the source activity.

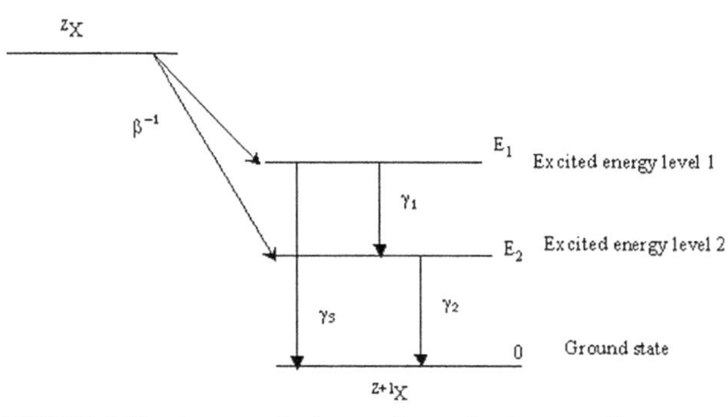

FIGURE 4.12 An example decay scheme showing cascading gamma rays.

Angular correlation between two gamma rays emitted in a cascade is defined as the relative yield of γ_2 about the 0° direction defined by the detector position, given that γ_1 is emitted in the same direction. Angular correlations arise because the direction of emission of the first gamma ray is related to the orientation of the angular momentum of the intermediate state. If the lifetime of the intermediate state is short, the orientation of the angular momentum will persist. The direction of the second gamma ray will be related to the angular momentum of the intermediate state, and hence to the direction of the first gamma ray (Evans, 1955). Angular correlation effects, in general, are not very significant when correcting for cascade summing effects. But for measurements requiring a high degree of accuracy (few tenths of a percent), it is indeed necessary to take angular correlation effects into account.

Detailed discussions on the subject of true coincidence summing can be found in several standard text books (Debertin and Helmer, 1988; Knoll, 1999).

a. True coincidence correction for a simple case

For a simple decay scheme such as the one shown in Fig. 4.12, it is straightforward to derive a correction factor for cascade summing losses or gains. For γ_1, the FEP rate in the absence of cascade summing can be written as follows:

$$\dot{N}_{10} = A \cdot p_1 \cdot \varepsilon_1 \tag{4.14}$$

The quantity A is the source activity, p_1 is the emission probability of γ_1, and ε_1 is the FEP efficiency at the energy of γ_1. Since γ_1 and γ_2 are emitted and detected in true coincidence, the energy deposited in the detector may be the sum of the full energy from γ_1 and a part of the energy (up to the full energy) from γ_2. This results in events being lost from the FEP of γ_1. Since any type of interaction involving γ_2 will result in a loss of count from the FEP of γ_1, the *total detection efficiency* of γ_2 is used in determining the cascade summing loss.

The peak rate of γ_1 in the presence of cascade summing is written as follows:

$$\dot{N}_1 = A \cdot p_1 \cdot \varepsilon_1 - A \cdot p_1 \cdot \varepsilon_1 \cdot \varepsilon_{t2} \tag{4.15}$$

The correction factor is derived from Eqs. (4.14) and (4.15):

$$COI = \frac{\dot{N}_1}{\dot{N}_{10}} = 1 - \varepsilon_{t2} \tag{4.16}$$

In deriving a correction factor for γ_2, one has to keep in mind that not all emissions of γ_2 are preceded by γ_1. A fraction of γ_2 is preceded by beta decay. The peak rate of γ_2 in the presence of cascade summing is given in Eq. (4.17).

$$\dot{N}_2 = A \cdot p_2 \cdot \varepsilon_2 - A \cdot p_2 \cdot \varepsilon_2 \cdot \left(\frac{p_1}{p_2}\right) \varepsilon_{t1} \tag{4.17}$$

The cascade summing correction factor for γ_2 is therefore,

$$COI = \frac{\dot{N}_2}{\dot{N}_{20}} = 1 - \left(p_1/p_2\right)\varepsilon_{t1} \tag{4.18}$$

From Eqs. (4.16) and (4.18), it is evident that total efficiency ε_t should be known to determine the correction factor for cascade summing losses.

In the case of γ_3, one has to correct the FEP for cascade summing gains. The peak count rate in the absence of cascade summing is written as

$$\dot{N}_{30} = A \bullet p_3 \bullet \varepsilon_3 \tag{4.19}$$

The FEP of γ_3 will gain events only when γ_1 and γ_2 deposit their full energies in the detector.

$$\dot{N}_3 = A \bullet p_3 \bullet \varepsilon_3 + A \bullet p_1 \bullet \varepsilon_1 \bullet \varepsilon_2 \tag{4.20}$$

The correction factor is therefore

$$COI = \frac{\dot{N}_3}{\dot{N}_{30}} = 1 + \left(p_1\varepsilon_1\varepsilon_2/p_3\varepsilon_3\right) \tag{4.21}$$

For decay schemes involving three or more gamma rays in a cascade, the analytical formulae for summing out probabilities especially become quite cumbersome to calculate. Over the past two decades, several authors have generalized these formulae for complex decay schemes and have reported them in the literature (Andreev et al., 1972; Moens et al., 1982; De Corte, 1987).

b. True coincidence correction using Canberra's Genie2000 software

Cascade summing losses could be as high as 30%–40% at close-in geometries, depending on the type of detector used and the specific nuclide that is being measured. If the detector is calibrated with a standard source identical in shape and size to that of the sample, and the nuclide(s) under study are the same in the standard and sample, then no correction needs to be applied for true coincidence summing. In all other cases, correction factors must be applied if measurements are required to be performed at close-in geometries. Canberra Industries has developed and patented a technique for calculating correction factors for true coincidence or cascade summing losses and gains (U.S. Patent 6,225,634). The algorithms that perform the calculations have been incorporated into Canberra's Genie2000 Gamma Analysis software package (version 2.0 and later). In version 3.2 of Genie2000 and later, the capability to compute true coincidence summing losses due to X-ray and gamma ray summing or annihilation photon (511 keV) and gamma ray summing has been incorporated (Zhu et al., 2009a,b). The coincidence library is created using nuclear data directly extracted from the internationally recognized database ENSDF. This provides direct traceability to the ENSDF database. Using Genie2000, one can thus calculate the true coincidence correction factors for a wide variety of counting geometries and for an exhaustive list of nuclides and gamma ray lines. To compute the correction factors, Canberra's method requires a single intrinsic peak-to-total efficiency curve or alternatively, the total efficiencies directly calculated from Canberra's ISOCS (In Situ Object Calibration Software) method. In addition, the true coincidence correction algorithm also uses the so-called spatial response characterization or ISOCS characterization for the specific detector. Canberra's ISOCS is a powerful mathematical tool to calculate HPGe FEP efficiencies for practically any source geometry (Bronson and Young, 1997; Venkataraman et al., 1999). For thin window HPGe detectors, the capability to compute correction factors to include X-ray and gamma ray coincidences was made possible by extending the lower energy limit of the ISOCS methodology down to 10 keV (Mueller et al., 2009).

The Genie2000 algorithms for calculating the true coincidence correction factors for voluminous sources are based on the work done by Kolotov et al. (1996). In this method, the voluminous source is first divided into a large number of equal volume subsources. A point location is selected within each subsource using a pseudorandom sequence. The true coincidence correction factor at each of these point locations is calculated and then integrated to determine the overall correction factor for the entire source.

It was previously noted that the total efficiency of the detector, ε_t, is required to compute the correction factor for true coincidence losses. For a point source at a location "I," the total efficiency at a given gamma ray energy may be determined, provided the FEP efficiency ε_p and the peak-to-total ratio (P/T) are known at the given energy.

$$\varepsilon_{t,i} = \varepsilon_{p,i}/(P/T) \tag{4.22}$$

The FEP efficiency is calculated using the ISOCS characterization for the given detector. The P/T ratio is obtained from the intrinsic peak-to-total efficiency curve determined for the detector. It is possible to determine the total efficiency directly by using Canberra's ISOCS methodology. However, this capability currently exists for HPGe detector measurements only.

In general, the TCS correction factor (COI_A) for the analytic γ ray (denoted by subscript "A") due to γ-γ and γ-X (γ-511 keV) TCS can be written as

$$COI_A = \left(1 - L_A^{\gamma-\gamma}\right) \bullet \left(1 + S_A^{\gamma-\gamma}\right) \bullet \left(1 - L_A^{\gamma-X,511}\right) \bullet \left(1 + S_A^{\gamma-X,511}\right) \tag{4.23}$$

where $L_A^{\gamma-\gamma}$ ($S_A^{\gamma-\gamma}$) is the loss (gain) probability due to γ-γ and γ-X (IC) TCS, and $L_A^{\gamma-X,511}$ ($S_A^{\gamma-X,511}$) is the loss (gain) probability due to γ-X (EC) and γ-511 keV TCS; where L_A is the probability of summing out and S_A is the probability of summing in. These probabilities are the sum of the partial probabilities calculated for individual decay chains involving the gamma line of interest.

$$L_A = \sum_{i=1}^{N} L_{A,i} \tag{4.24}$$

$$S_A = \sum_{j=1}^{M} S_{A,j} \tag{4.25}$$

The calculation of summing out probability L_A requires the knowledge of nuclear data such as the gamma ray yields, branching ratios, and internal conversion coefficients, as well as total detection efficiencies. Summing in probability S_A requires the knowledge of nuclear data and FEP efficiencies. F. De Corte has worked out the general formulae for the γ-γ true coincidence loss (gain) probability in a 5-γ cascade chain, where the loss probability depends on the location of the analytic γ ray in the decay chain. The generalized formulae reported in the literature to compute the summing out and summing in probabilities for complex decay schemes have been incorporated into the methodology developed by Kolotov et al. (1996). The extensions to the γ-γ true coincidence loss or gain due to γ-X-ray true coincidence are given by Zhu et al. (2009a,b). By calculating the coincidence correction factors ($COI_{A,i}$) for a large number of infinitesimally small subsources and then integrating, the correction factor for the whole voluminous source is obtained.

It is desirable to use the spatial characterization for the specific HPGe detector, if available. However, it is not a necessary condition. Koskelo et al. (2001) have shown that it is sufficient to use an approximate detector characterization to obtain good cascade summing results with Genie2000. Venkataraman and Moeslinger have demonstrated the feasibility of employing a discrete number of generic detector response characterizations for carrying out cascade summing corrections on gamma ray spectra obtained with noncharacterized HPGe detectors (Venkataraman and Moeslinger, 2001). A set of generic detector characterizations has therefore been made available within Genie2000.

c. Monte Carlo methods to compute true coincidence summing correction factors

Berlizov (2006) has developed an upgraded patch for the general-purpose Monte Carlo code MCNP (Monte Carlo N-Particle) code to allow computations with a source of correlated nuclear particles. The upgraded version of the code, MCNP-CP, performs statistical simulation of processes accompanying radioactive decay of a specified radionuclide, yielding characteristics of emitted correlated nuclear particles, which are then tracked through the problem geometry within the same history. When radioactive decay is modeled, MCNP-CP considers the emission of cascade gamma rays; continuum energy electrons at β^- decay; annihilation photons accompanying β^+ decay; monoenergetic electrons at the internal conversion on the atomic K-shell and L1,2,3-subshells; K- and L X-rays due to single and double fluorescence occurred at both electron capture and internal conversion processes; the K-L X and K-M X Auger electrons. Quantity, types, energies, and times of emitted particles are sampled according to the properties of a decay scheme of a particular radionuclide, which are taken from the evaluated nuclear structure data file ENSDF. γ-γ angular correlations are taken into account at sampling emission directions of cascade γ rays. As a basic source of information about the decay properties of a radionuclide of interest, MCNP-CP uses ENSDF. When necessary, the ENSDF data are supplemented by the calculations with application of known theoretical models and approaches.

To estimate true coincidence summing losses or gains using MCNP-CP, an accurate model of the detector and the source geometry must be developed. To ensure that the computed correction factors are accurate, it will be necessary to validate the Monte Carlo model of the detector using sources of known activities, located in geometries where true coincidence effects are negligible. In general, it is a good practice to validate and benchmark computational methods with real measurements before applying the computational method to solve other problems. Besides modeling the detector, the source geometry for which true coincidence summing losses or gains are needed must be accurately modeled. Two runs are performed using MCNP-CP; first with the correlated particle (CP) emission turned on, and a second computation with the correlated particle emission turned off. The ratio of pulse height tally with CP turned on to the pulse height tally with CP turned off gives the fraction of counts lost or gained due to true coincidence summing.

Besides MCNP-CP, other Monte Carlo codes such as GEANT can also be used to compute true coincidence summing losses and gains. These codes provide an independent method to verify cascade summing corrections obtained using analytical methods, e.g., in Genie2000. True coincidence correction factors computed using the analytical and ray tracing methods used in Genie2000 have been compared with those computed using MCNP-CP (Zhu et al., 2008, 2009a,b).

d. True coincidence correction using Ortec's GammaVision software

The technical details given in this section are from a paper published by Keyser et al. (2001). The method for true coincidence correction implemented in Ortec's GammaVision software is based on the work by Blaauw (1993). In this method, the probability of recording a count in the FEP is given by,

$$P_{Ei} = g_i \varepsilon_{fullenergy,Ei} \prod_{j \neq i} (1 - g_j \varepsilon_{total,Ej}) \quad (4.26)$$

where P_{Ei} is the probability of a count in the FEP, $\varepsilon_{fullenergy,Ei}$ is the full energy efficiency at an energy E_i, $\varepsilon_{total,Ej}$ is the total efficiency at an energy E_j, and g_i and g_j are the transition probabilities for gamma rays with energies E_i and E_j, respectively. Thus, the determination of the correction factor is reduced to knowing the full energy efficiency, the total efficiency, and the decay scheme of the nuclides in question. In addition to the full energy efficiency, the total efficiency includes the peak-to-total ratio, an absorption correction, and terms that correct the efficiency for an extended source.

5. Ge-escape peaks

For low-energy γ rays or for extremely thin detectors, when the interaction takes place close to the detector border, a certain probability exists that a Ge X-ray escapes from the detector. This probability is thus particularly important in detectors having thin windows (see Fig. 4.20). The parasitic peaks are observed at energies of

E_γ − 9.88 keV (escape of the K_α line)

and

E_γ −10.98 keV (escape of the K_β line)

The latter is five times less probable than the former. An illustration of several Ge escape peaks can be found in Fig. 4.8.

With these general aspects of a γ spectrum in mind, it is time now to take a closer look at the characteristics of a Ge detector such as resolution and efficiency, which play an important role in the choice of an appropriate detector.

C. Standard characteristics of Ge detectors

1. Energy resolution

From the spectra discussed in Section B, it is clear that the observed peaks have a finite width. Peak broadening is due to the statistical fluctuations in the number of electron−hole pairs created in the active detector volume *$(FWHM)_{det}$* and to the electronic noise of the different elements of the amplification chain. The resolution is expressed by full width at half-maximum *(FWHM)*, and it can be readily obtained from the spectra. The different noise contributions add quadratically according to the equation

$$FWHM = \sqrt{(FWHM)^2_{\det} + (FWHM)^2_{elect}}. \quad (4.27)$$

$(FWHM)_{det}$ and $(FWHM)_{elect}$ represent the detector and the electronic contribution in the total *FWHM*.

The energy E released in the detector is shared by two processes, namely, direct ionization and lattice vibrations. Both processes may lead to the generation of

$$N = \frac{E}{\varepsilon} \quad (4.28)$$

electron−hole pairs according to Eq. (4.1) described in Section I.A. The second process obeys a Gaussian distribution and, if direct ionization would be negligible, the variance σ_N of the number of charge carriers N would be given by the equation

$$\sigma_N = \sqrt{N} = \sqrt{\frac{E}{\varepsilon}}. \quad (4.29)$$

When the variance σ is expressed in energy units (eV), Eq. (4.29) becomes

$$\sigma = \varepsilon\sqrt{N} = \sqrt{E\varepsilon} \quad (4.30)$$

and the intrinsic $FWHM_{det}$ is calculated as

$$(FWHM)_{\det} = 2.35\sqrt{E\varepsilon} \quad (4.31)$$

where the factor 2.35 is a statistical property of the Gaussian distribution and gives the ratio between FWHM and the variance of a Gaussian distribution. In practice, however, direct ionization is not negligible at all, justifying the introduction of a correction factor F, the so-called Fano factor:

$$FWHM_{\det} = 2.35\sqrt{FE\varepsilon} \quad (4.32)$$

Fitting this formula at various energies and with various Ge and Si detectors shows that the effective Fano factor may vary between 0.08 and 0.15. The average value is 0.1 for Ge and Si. In Fig. 4.13, the approximate intrinsic FWHM is given as a function of γ ray energy (Fig. 4.13).

a. The electronic noise contribution $(FWHM)_{elect}$ and its time behavior

Depending on the detector type, resolutions (FWHM) lower than 1.8 keV at 1332 keV, 0.50 keV at 122 keV, and 0.15 keV at 5.9 keV are common. This implies an electronic noise contribution of <0.8, 0.22, and 0.10 keV, respectively. The electronic noise depends mainly on two factors: the capacitance of the detector and any current flowing over the detector. The capacitance of the detector depends on the detector dimensions and geometry.

Electronic noise is any undesired fluctuation that is superimposed on the signal. It contributes to the finite resolution of the detector. In electrical circuits, it stems from random processes such as the random collection of electrons or the arbitrary thermal movement of electrons in a resistor (Goulding and Landis, 1982). It can be represented as a voltage or current generator with zero average value and random positive and negative peaks. Noise is a

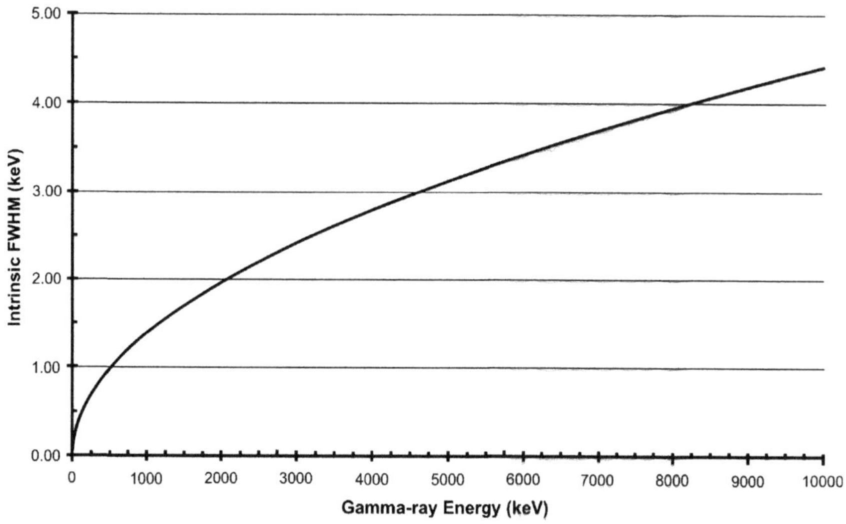

FIGURE 4.13 Approximate intrinsic full width at half-maximum as a function of γ-ray energy.

statistical phenomenon and can be described as a time average of the squares of all positive and negative values. One has to realize that a counting rate of one 1-MeV γ ray per second losing its complete energy in the active volume of the detector generates a current of only 5.41×10^{-14} C/s, and this has to be registered with a precision of better than 0.2% if a resolution (FWHM) of 1.8 keV is desired. This is a very difficult task for any electronic measuring chain.

In a detector amplifier system, three different noise contributions may be distinguished as function of their time behavior.

The Step Noise or Parallel Noise $(FWHM)_S$ arises from the discrete character of any current i_n flowing in the input circuit of the preamplifier. This current is integrated on the capacitor C_f (see Figs. 4.3 and 4.4). The two main sources of step noise are the detector leakage current and the thermal noise of the feedback resistor. It can be represented by a current generator, generating current pulses at the input of the preamplifier. It is proportional to

$$(FWHM)_s \propto \sqrt{\left(I_l + 2\frac{kT}{R_f}\right)\tau} \tag{4.33}$$

where I_l represents the total current of the detector (leakage current plus current generated by the detected radiation), k the Boltzmann constant, T the temperature of the feedback resistor, and τ the shaping time (measuring time) of the amplifier.

Step noise can be reduced by

1. measuring at shorter shaping times;
2. reducing the current through the detector (e.g., by measuring at a lower counting rate); and
3. choosing a feedback resistor with a high resistance, or avoiding it by using a different reset mechanism.

The delta noise or series noise $(FWHM)_D$ is mainly associated with the shot noise in the first stage of the preamplifier (FET). Delta noise is proportional to

$$(FWHM)_D \propto C\sqrt{\frac{T}{g_m \tau}} \tag{4.34}$$

where g_m represents the transconductance of the FET and C is the total capacitance at the input of the preamplifier.

Delta noise can thus be reduced by

1. measuring at longer shaping time;
2. minimizing the detector and stray capacitance; and
3. selecting a low-noise FET with large transconductance.

The Flicker noise or 1/f noise $(FWHM)_F$ is independent of the detector capacitance and exists only in association with a direct current. It is independent of the shaping time τ and is thus less relevant for the present discussion.

All these different noise contributions sum up quadratically with the intrinsic noise discussed in Section C.1. The total noise is thus given by

$$(FWHM)_{tot} = \sqrt{(FWHM)^2_{\det} + (FWHM)^2_D + (FWHM)^2_S + (FWHM)^2_F} \tag{4.35}$$

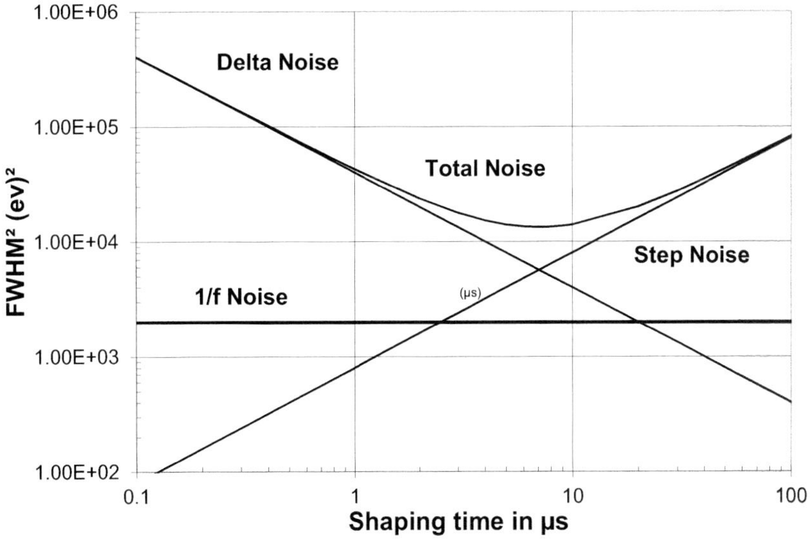

FIGURE 4.14 Noise as a function of the shaping time τ.

It is particularly instructive to look at the dependence of the noise on the shaping time τ. Fig. 4.14 gives a schematic view of the square of the total FWHM. For most detectors, the optimum resolution is obtained with a shaping time between 3 and 12 μs (if a Gaussian, analog shaper is used) or a signal rise time between 10 and 28 μs (if a digital amplifier is used). It is important to realize that, at high counting rates, the average DC current through the detector will increase and consequently also the step noise. The optimum shaping time will thus tend to lower values at high counting rate! In either case, the optimum shaping/peaking time for a given measurement condition should be determined experimentally.

The importance of step noise and delta noise also depends on the actual shape of the amplifier signal. A semi-Gaussian shaper gives a good compromise between both step noise and delta noise. For high-count-rate measurements with large coaxial detectors (see Section E), a gated integrator (e.g., the Canberra model 2024 Spectroscopy Amplifier) or longer "Rise-" and "Flat Top-" times with DSP processors may be used to minimize the ballistic deficit.

b. Interference with mechanical vibrations and with external radio frequency noise

Vibrations of the detector cryostat, or even audible noise, may also lead to spectrum broadening. This has to do with the fact that the germanium detector crystal and some leads are at high voltage, whereas they are closely surrounded by conductors on ground potential. This way, effective capacitors are formed. Their value can change when the crystal, the leads, or cryostat parts vibrate. Since a capacity, C, can be written as the ratio of a charge, Q, over a voltage, V, a changing capacity can be interpreted by the detector's electronics circuit as being due to a changing charge, in the same way as a detected photon gives rise to a change of charges. Provided that the frequency of the mechanical vibration or noise is not filtered out by the RC filter network of the amplifier, this noise can sum up with real photon events and show up as peak broadening.

To minimize this sensitivity for "microphonics," the user should avoid excessive audible noise and vibrations in the vicinity of the detector, e.g., by placing the detector on some damping material. Detectors are also less sensitive for microphonics at lower shaping times. For extreme applications, e.g., for use on board of helicopters or air planes, detector manufacturers can change the mechanical construction of the detector so that its eigen frequencies do not correspond with the characteristic noise or vibration frequencies of the plane or helicopter.

The spectroscopy system can also behave as an effective antenna for strong RF signals from the environment. Pick up of these may also lead to peak broadenings. Sensitivity for pickup depends strongly on details such as orientation, grounding and bundling of cables, and contact resistance between the various components of the spectroscopy system components. For extreme RF noise, detector manufacturers can change the cryostat and preamplifier hardware to render them virtually immune for pickup of RF signals.

Finally, peak degradation resulting from more fundamental physical processes can also occur, including detector temperature change, Doppler broadening, recoil shift, and recoil broadening. These will be discussed briefly.

c. Other sources of peak degradation

Temperature change of the detector. A small temperature dependence of the energy gap and the energy ε necessary for creating an electron–hole pair (Table 4.1) of Si and Ge exists and is given by

$$\frac{\Delta\varepsilon}{\varepsilon} \approx \frac{0.00015}{K} \quad (4.36)$$

The peak position of a 1.3-MeV transition thus changes by about 0.2 keV per kelvin. This effect can cause some peak broadening, especially at higher γ ray energies, when the cryostat temperature is not stable.

Doppler broadening. This is observed when the γ ray is emitted by an "object" that is not at rest. The most commonly observed Doppler-broadened gamma line is the 511-keV annihilation line. When a positron (e^+) comes to rest, it combines ("annihilates") with an electron (e^-) to form a positronium that decays almost immediately into two quanta of 511 keV. Due to momentum conservation, these are emitted at 180 degrees (see Chapter 1). As the annihilating electrons may have a rest kinetic energy, the annihilation lines are Doppler-broadened.

Doppler broadening can also occur when a γ ray is emitted by an isotope "in flight" after a nuclear reaction. An example is provided by the well-known 479.9-keV γ line emitted in the reaction

$$^{10}_{5}\mathrm{B}(\mathrm{n}, 2\alpha)\mathrm{t}.$$

As the cross section for this reaction is very high (3837 b), boron constitutes a very effective thermal neutron shield but adds an intense Doppler-broadened background line if no special shielding is used.

Note that because of the almost isotropic emission of γ rays, a Doppler broadened line is always symmetric.

Recoil broadening. This must not be confused with Doppler broadening. A fast neutron can transfer a large amount of its energy to a recoiling nucleus (see Chapter 1). This recoil energy is generally not seen by the detector as it takes place in the target. However, Stelson et al. (1972) showed that, in the case in which the target is the detector itself, the supplementary energy from the recoil contributes to the formation of electron–hole pairs in the detector. These supplementary charge carriers add to those due to the γ transition, resulting in an odd-shaped peak, having a normal low-energy and a long high-energy slope. Bunting and Kraushaar (1974) detected this phenomenon. It was further described by Verplancke (1992) and Heusser (1993). In particular, it is seen for certain background lines induced by (n,n′) reactions in the Ge crystal itself, such as the 691.0-keV line due to the ^{72}Ge(n,n′)^{72}Ge (see Table 4.4 further on in this chapter). Recoil broadening always results in a right-side asymmetric peak.

Recoil Energy Shift. A γ ray is emitted by a nucleus when it passes from an excited state to a lower energy state. The latter can be an excited state or the ground state. However, depending on the nature (multipolarity) of the transition, the excited state can deexcite alternatively through the emission of a conversion electron. Furthermore, it shares its energy with the energy of the recoiling nucleus. This recoil energy E_R is given by

$$E_R = \frac{E_\gamma^2}{2Mc^2} \quad (4.37)$$

where E_γ represents the energy of the emitted γ ray, M the mass of the nucleus, and c the speed of light (see Chapter 1 for the derivation of Eq. 4.37). If the energy is given in keV and the mass in atomic units (M of ^{12}C = 12), the corresponding energy difference E_g of the nuclear states involved is given by

$$E_g = E_\gamma + E_R = E_\gamma\left(1 + \frac{E_\gamma}{1{,}862{,}300M}\right). \quad (4.38)$$

This effect is completely negligible, in general, and therefore, $E_g \approx E_\gamma$. However, if M is small and E_γ is high, the difference between E_g and E_γ can be significant, as pointed out by Greenwood and Chrien (1980), among others. For instance, the γ ray of 10,829.1 keV produced in the ^{14}N(n,γ)^{15}N reaction populates the ground state and is issued from an excited level at 10,833.3 keV. This energy corresponds to the binding energy of the neutron. Note that medium- and high-efficiency Ge detectors perform excellently even at this very high energy, the most useful peak being the double escape peak, which suffers no Doppler broadening as outlined in Section II.B.2.

Radiation damage. Low-energy tailing can be due to electronics but also to the presence of trapping centers or "deep levels" in the detector. These trap electrons or holes for periods longer than the time needed for pulse formation (Eq. 4.7). Trapping centers may be created by radiation damage in the detector induced by fast charged particles and/or fast neutrons. A 16-MeV neutron creates four times more trapping centers than a 1.6-MeV neutron. Charged particles are easy to shield. This is not true for fast neutrons. They have to be thermalized by a large, hydrogen-rich layer and must be subsequently absorbed by a high-cross section material. The effects of deep-level defects in a high radiation environment have been studied by Lutz (1996).

d. The Gaussian peak shape

The peak shape is closely related to the resolution. In principle, the peak shape follows Poisson statistics. If the number of counts is ≥ 20, the shape of the FEP is given by a Gaussian distribution of the values x around the energy channel E according to the equation

$$f(x) = e^{-(x-E)^2/2\sigma^2} \quad (4.39)$$

The maximum of the distribution lies at $x = E$. Table 4.3 gives the width of the distribution for a few points. Few detectors, if any, exhibit the theoretical peak shape. Most modern detectors have a ratio FWTM (full

TABLE 4.3 Theoretical Gaussian peak widths.

A = fraction from the maximum	B = width at fraction	B/FWHM
1/2 (FWHM)	2.35σ	1
1/10 (FW1/10M)	4.29σ	1.82
1/20	4.9σ	2.08
1/50	5.59σ	2.37

FWHM, full width at half-maximum.

width at tenth-maximum) to FWHM (full width at half-maximum) of better than 1.9, but a ratio of 2 is common for larger detectors. Mainly, detectors made of n-type germanium may show higher FWTM/FWHM ratios.

An FWTM/FWHM ratio smaller than 1.82 may indicate that the peak has shifted during the accumulation of the spectrum or that the peak is actually a doublet.

2. The peak-to-compton ratio

Following the IEEE standards (ANSI N42.14, 1991), the peak-to-Compton ratio is defined as the ratio between the maximum number of counts in the channel at the top of the 1332.5-eV peak of ^{60}Co and the average channel count between 1040 and 1096 keV. It depends not only on the resolution and efficiency but also on the presence of material in the vicinity of the active detector region, as these materials may backscatter γ rays into the detector. It plays a role in the "background due to the presence of the source," as will be discussed in Section II.D.

3. The detector efficiency

The efficiency ε_γ is a measure of the probability (expressed in absolute values or in %) that a γ ray of energy E_γ is fully absorbed in the active volume of the detector or, in other words, the probability that it contributes to the FEP. It depends basically on the solid angle Ω under which the source is seen by the detector and on intrinsic factors characteristic of the detector.

a. Geometrical efficiency factor

In the case of a point source situated on the axis of a circular detector with a flat surface facing the source, the geometric efficiency η is given by a simple analytical formula

$$\eta = \frac{\Omega}{4\pi} = \frac{1 - \cos[arc\ \tan(r/d)]}{2} \tag{4.40}$$

where r represents the radius of the detector, d the distance between detector and source including the distance between detector and endcap, and Ω the solid angle under which the source sees the detector. Moens et al. (1981) and Moens and Hoste (1983) proposed an extension of this formula for the case of an extended source or a nonaxial source. More realistic computations are based on numerical approaches (e.g., Monte Carlo, Canberra's ISOCS/LabSOCS software) (see Section C.3.e below and references Mueller et al., 2009, Bronson and Atrashkevitch, 2001 and Venkataraman et al., 1999).

b. The intrinsic efficiency ε_i and the transmission T_γ

The overall efficiency may be given by

$$\varepsilon_\gamma = \eta \varepsilon_i T_\gamma \tag{4.41}$$

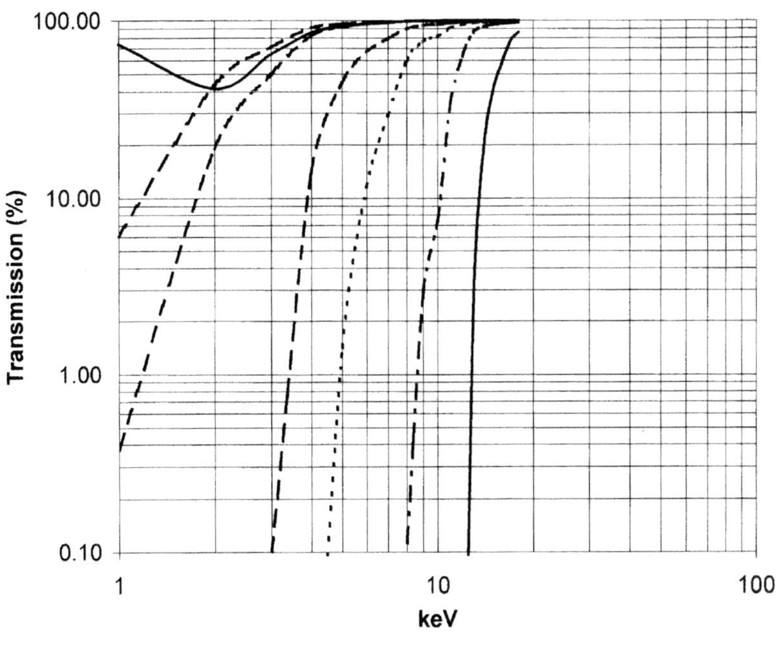

FIGURE 4.15 Some typical transmission curves. Plain curves: Ge dead layer of 0.3 μm (implanted window for *REGe* detector or "thin window" for Canberra's *XtRa* or *BEGe* detectors) and 0.5 mm (Li diffused layer). Dashed curves: cryostat Be window of 0.05, 0.1, and 0.5 mm. Dotted curve: cryostat Carbon window of 0.5 mm. Dashed-dotted curve: cryostat Al window of 0.5 mm.

Fundamental effects such as the photoelectric effect and multiple Compton scattering discussed in Section II.B.1 are included in an intrinsic factor ε_i. Other factors such as the thickness of the different entrance windows, the p^+ or n^+ layers and the encapsulation of the source itself are included in the transmission factor T_γ, and η represents the geometric efficiency (Eq. 4.40). Fig. 4.15 gives the transmission through different endcaps and dead Ge layers. The importance of T_γ is illustrated in Fig. 4.16, which shows two experimental efficiency curves for the same low-energy germanium detector (LEGE) of 200 mm^2 surface and 10 mm thickness obtained with a mixed ^{241}Am ^{137}Cs ^{60}Co source placed at 5 cm. The only difference is the entrance window, 0.15-mm Be in the first case and 0.5-mm Al in the second. A big difference in efficiency is observed below 20 keV. The increase in efficiency at the very low-energy side is due to the beginning of the influence of the K-absorption edge of Ge. It is clearly seen in the upper curve (Be window) but is strongly reduced by the higher absorption of Al as shown by the lower curve.

The transmission T_γ is given by

$$T_\gamma = e^{-\mu d}. \tag{4.42}$$

Here, μ represents the total absorption coefficient for the γ ray and d the thickness of the specific absorber. The absorption coefficient μ can be expressed in g/cm^2 (the mass absorption coefficient μ_m) or in cm^{-1} (the linear absorption coefficient μ_l) whether d is expressed in g/cm^2 or in cm. The following relation exists between both:

$$\mu_m = \frac{\mu_l}{\rho} \tag{4.43}$$

where ρ represents the density of the absorber in g/cm^3. It is sometimes convenient to express the thickness d as half-thickness $d_{1/2}$. The relation between both is given by

$$d_{1/2} = \frac{\ln 2}{\mu} \tag{4.44}$$

where μ stands for either μ_l or μ_m.

c. Relative efficiency

The efficiency cited by the manufacturer of a Ge detector following the IEEE standards (ANSI/IEEE Standard 325-1986) does not represent the absolute efficiency of the detector. It represents the ratio of the absolute detector efficiency at 1332.5 keV (^{60}Co) to that of the same γ ray obtained with a 3 × 3 inch NaI(Tl) scintillation detector, the point source being placed at 25 cm on the axis of the endcap (measured from the center of the source to the front of the endcap). The absolute efficiency at 1.3 MeV of such a 3 × 3 inch NaI(Tl) scintillation detector measured at the same distance of 25 cm is 1.2×10^{-3}. The relative efficiencies offer some means to compare detectors. In Section E, however, it will be shown that the notion of "relative efficiency" could lead to completely false expectations.

The relative efficiency can be helpful to construct a very crude absolute efficiency curve when, besides the detector's relative efficiency, the diameter of its active volume is also given. Indeed, two points of this curve for a point source at 25 cm from the endcap may be roughly estimated. The first point, at 1.3 MeV, can be calculated from the relative efficiency (absolute efficiency at 1.3 MeV and at 25 cm = relative efficiency × 1.2×10^{-3}). The second

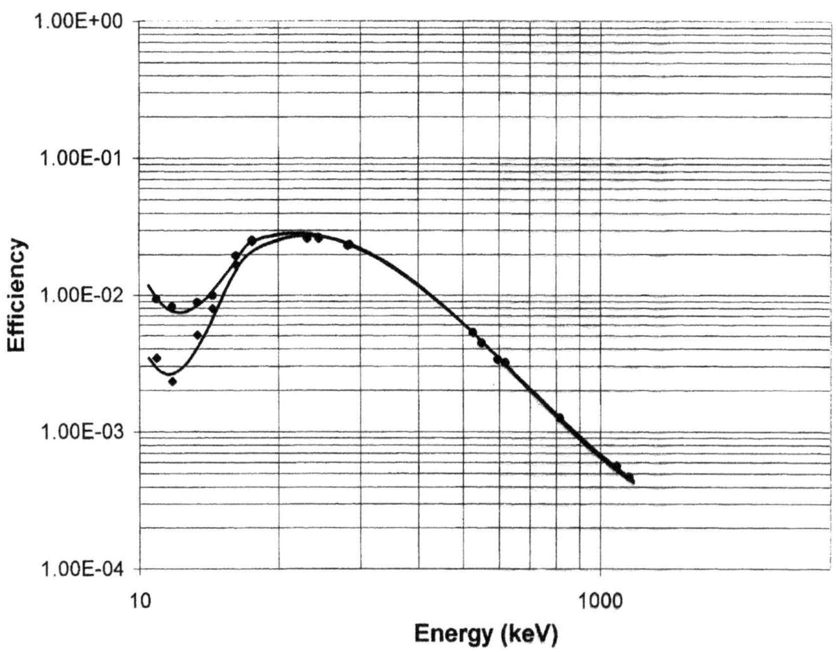

FIGURE 4.16 Superposition of two efficiency curves for the same low-energy detector mounted with two different endcaps: 0.15 mm Be (upper curve) and 0.5 mm Al (lower curve). The source-to-detector distance was 5 cm. The curves are polynomial fits through the experimental points.

point, at 100−150 keV, can be approximated as $\varepsilon = \eta$ with η calculated from Eq. (4.25) using the information about the active diameter. This approximation is based on the assumption that, around 100−150 keV, ε_i is near 100% and the transmission correction negligible (at this precision). These two points may be joined by a straight line on a log-log scale. It is evident that such a rough estimate may not be used for actual measurements, as large errors are introduced by the application of the oversimplified assumptions.

d. The experimental efficiency curve

To analyze a gamma ray spectrum to obtain a source activity or gamma emission rate, it is necessary to know the detection efficiency for each peak observed in the spectrum. This can be accomplished by mapping the detection efficiency curve versus gamma ray energy over a range of energies. Such a curve can be established by the use of one or more calibration sources. If N_0 represents the number of radioactive atoms present in the calibration source at the starting moment of the measurement and $\lambda = 0.693/T_{1/2}$, its transition probability, where $T_{1/2}$ represents the half-life,

$$N_d = N_0\left(1 - e^{-\lambda\,\Delta\,t}\right) \tag{4.45}$$

atoms decay during the measuring time Δt. If Δt is small with respect to $T_{1/2}$ as is generally the case, Eq. (4.45) reduces to

$$N_d = \mathrm{N}_0\lambda\Delta t = I_0\Delta t \tag{4.46}$$

I_0 being the activity in Bq. The number N_r of registered counts in the FEP is thus given by

$$N_r = I_0\Delta t\varepsilon p \tag{4.47}$$

where ε contains all efficiency factors discussed and p represents the branching ratio of the γ ray measured. The efficiency can thus be readily calculated. Evidently, I_0 has to be corrected for the decay during the time elapsed when the calibration source was certified, according to the equation

$$I_0 = I_{cert}e^{-\lambda\,t} \tag{4.48}$$

where I_{cert} represents the certified intensity of the source for a given day and hour and t the time elapsed until the actual measurement.

To establish the complete efficiency curve, a great number of efficiencies ε_γ for various energies should be obtained experimentally, either by the use of several calibrated standard monoenergetic γ sources such as ^{137}Cs or, depending on the energy range desired, by the use of one or several multigamma sources such as ^{241}Am (11.9 59 keV), ^{60}Co (1173−1333 keV), ^{56}Co (1−3 MeV), ^{152}Eu (121−1408 keV), or ^{133}Ba (53−383 keV). For still lower energies, X-ray sources such as the Mn X-rays [5.88,765 keV (50.5%), 5.89,875 keV (100%), and 6.49 keV (20.3%)] emitted during the EC decay of ^{55}Fe can be useful. This calibration work is straightforward using modern software such as GENIE (Canberra Industries). Nevertheless, attention should be paid to the following points:

1. Ensure that all standard sources have the same form and are placed at the same distance from the detector. These must be identical to those of the samples to be measured.
2. Ensure that the encapsulations used for all calibration sources and the samples to be measured are the same, especially when (very)-low-energy measurements have to be performed.
3. Do not use intense sources if multigamma ray sources are used emitting two or more coincident gammas. This will lead to losses due to random summing (see Section II.B.3).
4. If the standard source used in the calibration emits multiple gamma rays in true coincidence (^{152}Eu, ^{60}Co, ^{88}Y, etc.), then one has to be cognizant of true coincidence summing (or cascade summing) losses or gains affecting the FEPs. True coincidence summing losses or gains lead to an underestimation of the measured efficiencies, whereas the summing gains lead to an overestimation. These effects become worse with high efficiencies (small source-to-detector distances and/or large detectors). A correction factor may have to be used to correct for the effects of true coincidence summing (see Section II.B.4) before the efficiency calibration curve can be used in the analysis.

Fig. 4.17 shows a typical efficiency curve of a 25% p-type coaxial detector. A mixed ^{133}Ba ^{137}Cs ^{60}Co source was placed at a distance of 5 cm. The full line represents a fourth-order polynomial fit in $1/E$, E being the energy in keV of the experimental points.

e. Mathematical efficiency calculations

As described in the previous section, the detection efficiency curve may be obtained by measuring one or more calibration standards. These standards should emit gamma rays that span the range of energies expected to be present in the actual samples to be measured. In addition, it is essential that the calibration measurements have the same geometry as the actual samples. Specifically, this means that the source dimensions, source material and density, container wall thickness, container wall thickness and density, and source-to-detector positioning must all be same for the calibration standards versus the samples. In many cases it is not practical to obtain calibration standards that match the samples to be counted. Examples of this situation are as follows:

- The samples to be counted span a large variety of shapes and sizes and densities (e.g., decommissioning

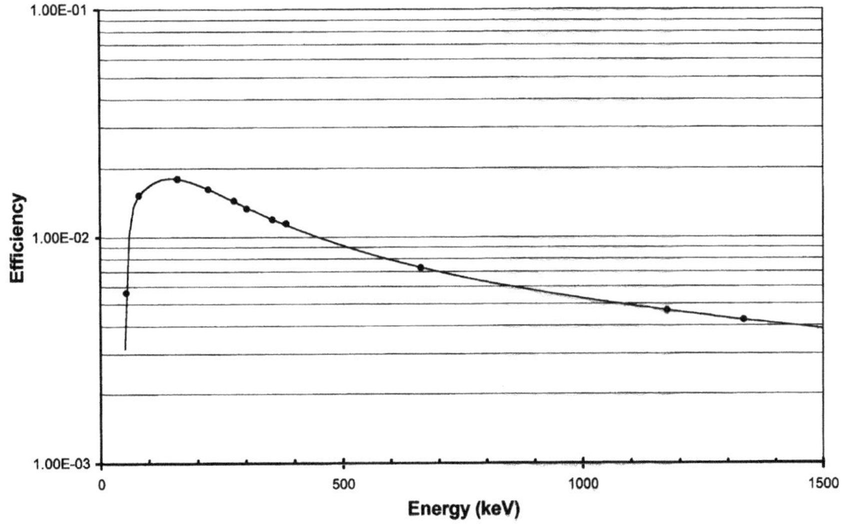

FIGURE 4.17 Experimental efficiency curve of a 25% p-type coaxial detector. A mixed ^{133}Ba$-^{137}$Cs$-^{60}$Co source was placed at a distance of 5 cm. The full line represents a fourth-order polynomial fit in $1/E$, E being the energy in keV of the experimental points.

activities), thereby requiring an unacceptably large number of calibration standards to buy, count, and dispose of.

- The samples to be counted are too large to fabricate equivalent calibration standards (e.g., railroad cars full of soil and debris).

In such cases, it becomes necessary to utilize mathematical methods to obtain detection efficiency curves. One common approach is to use Monte Carlo computer modeling techniques. These techniques derive their name from the use of computer-generated random numbers to mimic the random processes that take place in real-life gamma ray emission, scattering, absorption, and detection events. Three computer packages that are in common use are as follows:

- MCNP (Monte Carlo N-Particle). See Briesmeister, 2002.
- EGS (Electron Gamma Shower). See Nelson et al. (1985).
- GEANT. See CERN Applications Software Group, 1994.

These codes allow for description of the counting geometry as well as features of the source gamma ray emission spectrum and features of the detector. These packages are extremely flexible and, consequently, extremely complex. To obtain accurate results, it is particularly important to provide detailed information about the structure of the detector. The best results are obtained by benchmarking the calculations from a detector model against measurements with calibration standards; in effect, calibrating the mathematical model. An example of this sort of approach, including rather sophisticated details of the detector structure, is given in Friedman et al. (2001). After developing an accurate model of the detector (i.e., a model proven to be able to reproduce measured efficiencies), it can be used to calculate efficiencies for other source geometries. Clearly, this is a very complex process. Development of the geometry model and execution of the calculation take a lot of time and require a high level of sophistication and experience on the part of the user. It is well rewarded; however, a good Monte Carlo model can reproduce detection efficiencies with accuracies that rival those obtainable with calibration standards.

To reduce the amount of time and expertise necessary to obtain reliable efficiencies from Monte Carlo techniques, other approaches have been developed. Typically, these are simplified discrete-ordinated calculations that divide the source region into small volume pixels ("voxels") that can be approximated as point sources. The detection efficiency for a given voxel is obtained by approximating the efficiency for the equivalent naked point source and then accounting for the attenuation losses through any absorbing materials between the point source and the detector. The detection efficiency for the entire source is simply obtained by summing the efficiencies for all the voxels.

Examples of such computer codes are Canberra's ISOCS and LabSOCS software packages (Venkataraman et al., 1999). At the Canberra factory, a given detector is "characterized" by developing an MCNP model that reproduces several standard source measurements made with the detector. From there, the MCNP code is used to map out the efficiency of the detector for a naked (i.e., unattenuated) point source at any location within 500 m of the detector. This map of efficiency versus position and energy (referred to as the "characterization" for that detector) is then provided in the form of a lookup table to be used with the ISOCS and LabSOCS software. The software itself provides a simple interface for users to select a basic geometry template (e.g., box, cylinder, planar source, etc.)

and to specify the details of the source and possible passive absorbers (e.g., dimensions, densities, etc.) as well as the details of the source-to-detector vector. Utilizing the characterization to obtain the naked point source efficiencies, the software divides the source into point like voxels to calculate the detection efficiency as described in the previous paragraph. Typical calculation times for common geometries are on the order of seconds and rarely more than 1 minute; the calculated results are typically accurate to better than 10% depending on the complexity of the geometry. For detectors that have been characterized by Canberra, this is an extremely flexible and powerful technique.

f. Stability of the germanium detector efficiency

Users of HPGe detectors must know that the efficiency of their detector may change over time. The probability that this happens and the degree of change depends on the type of detector used as well as on the temperature at which this detector is stored.

As explained in Section E below, all germanium detectors have surface "dead layers" that are necessary for operation. These dead layers do not have charge collection and therefore are not part of the active detector. P^+-contacts as well as the thin N^+-contacts that Canberra applies on their BEGe and XtRa detectors are both very thin and stable with time and temperature. All germanium detectors available today still have Li-diffused contacts. This N^+-contact is both thick (several hundred microns) and unstable at elevated temperatures. The diffused lithium layer is unstable because it continues to diffuse over time. The rate of diffusion is a function of temperature and lithium concentration gradient, with warmer temperatures, resulting in higher diffusion rates. Dead layer thickness growth is affected by the amount of time spent at higher (room) temperatures. When kept at liquid nitrogen temperatures, the dead layer growth rate is insignificant!

The effect of this basic property of Li-diffused contacts on the stability of the germanium detector efficiency depends on the precise geometry of each type of germanium detector. Fig. 4.20 shows the various detector models offered by Canberra and the locations of the problematic thicker, unstable n-type dead layer. From this table, it is clear that the detection efficiency at low energies with a p-type coaxial detector (SEGe) is most affected. With the growth of the Li-diffused contact, the transmission of low energies may decrease strongly. Also the detection efficiency at low energies of an XtRa or BEGe detector model is affected when the source is located in a Marinelli beaker wrapped around the germanium detector. The effect on other detector models is limited to the loss of some active detector volume when the thickness of the inactive Li layer increases, which, in practice, is negligible.

Comparing the Li layer thickness of new germanium detectors with germanium detectors that have been used in the field for a number of years shows that the typical growth rate of the Li layer is 0.2 mm/y. Diffusion theory predicts that the growth rate will decrease over time, so this linear approximation is conservative. In practice, the typical dead layer thickness of "fresh" germanium detectors is between 0.3 and 0.6 mm. Detectors that have been in the field for some years have typical Li layer thicknesses between 0.7 and 1 mm, occasionally up to 2 mm. Consequently, the detector efficiency may degrade with 5%−10% in less than a year for 60 keV and in 2−4 years at higher energies measured with a standard p-type coaxial detector (SEGe). Similar efficiency losses are observed with BEGe or XtRa detectors in 4−5 years' time of "typical" use. N-type coaxial detectors (REGes) hardly show any efficiency degradation with time.

The gamma spectroscopist must be aware of these phenomena. He or she can minimize the effects on measured data by taking some basic precautions, such as the following:

- Use the proper detector for a given application. If an important measurement includes ^{241}Am, then a standard coaxial p-type detector, or a BEGe or XtRa detector in combination of a Marinelli geometry, should not be used.
- Check the detector calibration periodically with a QC check source. These data can then also be used to correct the mathematically generated efficiency curves in a simple way if needed.
- Keep the detector at liquid nitrogen temperatures as much as possible and practical. If not, store it in a cool place. Storage in an air-conditioned room rather than in a hot area may reduce the dead layer growth rate by 20%−40%. Storing it in a deep freezer is not a good alternative since this may freeze the O-ring seal used and ruin the detector vacuum.

D. Background and background reduction

It seems natural to define the background as all pulses registered by the detection system when no source is present. However, it was seen earlier (see Section II.B) that the Compton effect and bremsstrahlung give rise to an important continuum that also has to be regarded as a real background. It is thus useful to distinguish between the background with and without a source.

1. *Background in the presence of a source*

The background due to the source itself is essentially the continuum generated by Compton scattering and by bremsstrahlung. These effects have been analyzed in Section II.B, and no further discussion is needed. However, it is this background that very often governs the detection limits.

2. Background in the absence of the source

Background in the absence of a source has three different origins: man-made isotopes, natural isotopes, and cosmic radiation. In contrast to the background from a source, it contains, besides a continuum due to cosmic interactions in the crystal (Verplancke, 1992), many discrete γ lines. The most common discrete lines are summarized in Table 4.4. The first column of Table 4.4 gives the energy, the second column gives the isotope in which the nuclear transition responsible for the emission of a γ ray takes place, and the third column gives the decaying isotope and/or the reaction responsible for the formation of the isotope. The fourth column gives the intensity of the γ ray. A fifth column is reserved for various remarks, such as the origin of the γ ray, the principal decay mode, the half-life, and, when possible, the intensity of the prompt reaction of γ rays.

a. Man-made isotopes

Here, we find essentially fission isotopes such as ^{137}Cs due to fallout from the former bomb testing in the atmosphere, nuclear accidents, or isotopes formed by man-made nuclear reactions such as those of ^{60}Co.

b. Natural isotopes

Here, we find ^{40}K and the isotopes belonging to the natural decay chains: ^{238}U (Table 4.5), ^{235}U (Table 4.6), and ^{232}Th (Table 4.7). These tables give the decay mode, the half-life, and the main γ rays of the various isotopes. The parent nuclei, ^{238}U, ^{235}U, and ^{232}Th, are very long lived. Their half-lives are several orders of magnitude longer than those of the longest-lived daughter elements. They reach a secular equilibrium meaning that the intensities of the various γ rays may be compared directly with each other after correction for α or β branching (see Chapter 1). However, the equilibrium may be disturbed if physical or chemical separation took place. Just two examples:

1. The "emanation" of noble gases (^{222}Rn, ^{220}Rn, ^{219}Rn), daughters from natural U and Th. In particular, U is often found underground or in the construction materials of buildings. Consequently, ^{222}Rn (radon) may concentrate in closed rooms. The characteristic γ rays of its daughters ^{214}Pb and ^{214}Bi are very common background lines. The intensity of Rn lines in the background spectrum may fluctuate a lot with the weather conditions.
2. Separation in geological time due to different solubilities of the various elements. This can be used for dating of geological formations based on the $^{234}U/^{230}Th$ ratio.

When there is a chance that the equilibrium of the daughter isotope with its parent is disturbed, only the intensities of the γ rays belonging to the partial decay chain of this daughter can be compared directly. These long-lived isotopes, whose equilibrium with the parent can be disturbed, are marked by a frame in Tables 4.5, 4.6, and 4.7.

3. Background of cosmic origin

Cosmic rays comprise primarily very high-energy (up to $10^8 - 10^9$ GeV) protons and α particles originating from stellar processes in supernovas with a mean energy between 5 and 10 GeV per nucleon. These particles undergo collisions in the stratosphere, where they give rise to various π and K mesons as well as to muons, neutrinos, electrons, neutrons, and photons. Typical fluxes at sea level are 10^{-2} particles/cm^2 $\cdot$s$\cdot$steradian distributed according to a $\cos^2\theta$ law, θ being the polar angle. About 75% of all particles at sea level are π-mesons, and the absolute proton flux is of the order of 0.1% of all particles. These cosmic rays constitute a very important part of the background in the absence of a source and contribute to the continuum as well as to the activation of various nuclei. These effects continue to attract the attention of various researchers, such as Wordel et al. (1996) Heusser (1996), Laubenstein and Heusser (2009), and Jovancevic et al. (2010). A comprehensive overview of the origin of cosmic rays has been given by Celnikier (1996).

a. "Prompt," continuously distributed background

Charged particles can penetrate the sensitive volume of the detector, giving rise to a continuous background in coincidence with the primary particle. The energy loss per collision of 10-MeV electrons, 100-MeV mesons, or 1000-MeV protons is approximately 1.8 MeV/g$\cdot$cm^2, generating in a Ge detector a signal of about 10 MeV per centimeter traversed. Cosmic particles also produce showers of secondary particles (p, e^-, e^+), mainly in the detector shielding. In turn, these secondary charged particles produce bremsstrahlung and annihilation lines. This secondary radiation contributes more specifically to the background in the lower energy region of a shielded detector.

b. Neutron-induced "prompt" discrete γ rays

Fast neutrons can induce prompt γ rays by the (n,n′) reaction. This is particularly important when the reaction takes place in the Ge itself or in other materials in the vicinity of the detector such as Cu, Fe, Pb, and Cd. These lines are summarized in Table 4.4. The Compton scattering of these γ rays also adds to the continuum.

c. "Delayed" γ rays

Delayed γ rays are due to the deexcitation of isotopes formed either in the Ge itself or in the material surrounding the detector. They are due to capture of thermalized neutrons or to more exotic nuclear reactions also mentioned in Table 4.4. These isotopes also contribute to the continuum by bremsstrahlung and Compton scattering.

TABLE 4.4 Background lines observed in Ge spectra (This list is neither complete nor should all lines be present in each spectrum).

γ line (keV)	Isotope[a)]	Reaction[b)]	$I_\gamma^{c)}$ (%)	Remarks
13.26	^{73m}Ge	^{72}Ge(n,γ)^{73m}Ge	0.09	$T_{1/2} = 0.5$ s: isomeric transition produced continuously by thermalized neutrons from cosmic origin, (see also 66.7 keV line).
14.41	^{57}Fe	^{57}Fe(p,n)^{57}Co ^{56}Fe(p,γ)^{57}Co ^{56}Fe(d,n)^{57}Co	8.8	EC-decay ($T_{1/2} = 271.3$ d): Particles from cosmic origin.
46.5	^{210}Bi	^{210}Pb	3.65	β$^-$-decay ($T_{1/2} = 22.28.3$ h): ^{238}U series.
49.9 50.1	^{223}Ra	^{227}Th	0.52 7.28	α-Decay ($T_{1/2} = 11.43$ d): ^{235}U series.
53.2	^{230}Th	^{234}U	0.12	α-decay ($T_{1/2} = 1.47$E5 y): ^{238}U series.
53.4	^{73m}Ge	^{72}Ge(n,γ)^{73m}Ge	10.5	$T_{1/2} = 0.5$ s is produced continuously by thermalized neutrons from cosmic origin.
63.32	^{234}Pa	^{234}Th	4.49	β$^-$-decay ($T_{1/2} = 24.1$ d): ^{238}U series.
66.7	^{73m}Ge	^{72}Ge(n,γ)^{73m}Ge	0.5	$T_{1/2} = 0.5$ s is produced continuously by thermalized neutrons from cosmic origin. Sum peak 53.4 + 13.26 and individual line. As the lines are produced inside the detector, the probability for summation is almost 100%.
67.7	^{226}Ra	^{230}Th	0.38	α-decay ($T_{1/2} = 8$E4 y): ^{238}U series.
68.7	^{73}Ge	^{73}Ge(n,n')^{73}Ge		Prompt γ-line produced by inelastic scattering of fast neutrons from cosmic origin.
72.80 74.97 84.45 84.94 87.3	Pb	Pb X-ray		Mainly due to external conversion in the Pb shield.
81.23	^{231}Pa	^{231}Th	0.89	β$^-$-decay ($T_{1/2} = 25.5$ h): ^{235}U series.
82.09	^{231}Pa	^{231}Th	0.4	β$^-$-decay ($T_{1/2} = 25.5$ h): ^{235}U series.
84.21	^{231}Pa	^{231}Th	6.6	β$^-$-decay ($T_{1/2} = 25.5$ h): ^{235}U series.
84.37	^{224}Ra	^{228}Th	1.9	α-decay ($T_{1/2} = 1.91$ y): ^{232}Th series.
92.6	^{234}Pa	^{234}Th	5.16	β$^-$-decay ($T_{1/2} = 24.1$ d): ^{238}U series.
93.32	^{67}Zn	^{65}Cu(α,2n)^{67}Ga	48.0	EC-decay ($T_{1/2} = 78.3$ h): α-particles from cosmic origin. See also 184.5 and 194.25 keV lines.
99.6	^{228}Th	^{228}Ac	1.37	β$^-$-decay ($T_{1/2} = 6.15$ h): ^{232}Th series.
109.89	^{19}F	^{19}F(n,n')^{19}F		Prompt γ-line produced by inelastic scattering of fast neutrons from cosmic origin.
122.4	^{57}Fe	^{57}Fe(p,n)^{57}Co ^{56}Fe(d,n)^{57}Co ^{56}Fe(p,γ)^{57}Co ^{57}Co(n,n')^{57}Co		EC-decay ($T_{1/2} = 271.3$ d): particles from cosmic origin
122.4	^{219}Rn	^{223}Ra	1.19	α-decay ($T_{1/2} = 11.43$ d): ^{235}U series.
129.6	^{228}Th	^{228}Ac	2.45	β$^-$-decay ($T_{1/2} = 6.15$ h): ^{232}Th series.
131.2	^{234}U	^{234}Pa	20	β$^-$-decay ($T_{1/2} = 6.7$ h): ^{238}U series.

Continued

TABLE 4.4 Background lines observed in Ge spectra (This list is neither complete nor should all lines be present in each spectrum).—cont'd

γ line (keV)	Isotope[a)]	Reaction[b)]	$I_\gamma^{c)}$ (%)	Remarks
136.47	^{57}Fe	^{57}Fe(p,n)^{57}Co ^{56}Fe(d,n)^{57}Co ^{56}Fe(p,γ)^{57}Co ^{57}Co(n,n′)^{57}Co	11.0	EC-decay ($T_{1/2}$ = 271.3 d): particles from cosmic origin.
139.7	^{75m}Ge	^{74}Ge(n,γ)^{75m}Ge	39.0	$T_{1/2}$ = 48 s: isomeric transition produced continuously by thermalized neutrons from cosmic origin.
143.58	^{57}Fe	^{57}Fe(p,n)^{57}Co ^{56}Fe(d,n)^{57}Co ^{56}Fe(p,γ)^{57}Co ^{57}Co(n,n′)^{57}Co	1.0	See also 14.12, 122.4 and 136.47 eV lines.
143.8	^{231}Th	^{235}U	10.9	α-decay ($T_{1/2}$ = 7.05E8 y): ^{235}U series.
143.9	^{226}Ra	^{230}Th	0.05	α-decay ($T_{1/2}$ = 8E4 y): ^{238}U series.
144.2	^{219}Rn	^{223}Ra	3.26	α-decay ($T_{1/2}$ = 11.43 d): ^{235}U series.
154.1	^{219}Rn	^{223}Ra	3.26	α-decay ($T_{1/2}$ = 11.43 d): ^{235}U series.
159.7	^{77m}Ge	^{76}Ge(n,γ)^{77m}Ge	11.0	$T_{1/2}$ = 52.9 s: isomeric transition produced continuously by thermalized neutrons from cosmic origin.
163.3	^{231}Th	^{235}U	5.0	α-Decay ($T_{1/2}$ = 7.05E8 y); ^{235}U series.
174.9	^{71m1}Ge	^{70}Ge(n,γ)^{71m1}Ge	1.0	$T_{1/2}$ = 73 ns: isomeric transition produced continuously by thermalized neutrons from cosmic origin.
184.59	^{67}Zn	^{65}Cu(α,2n)^{67}Ga	62.0	EC-decay ($T_{1/2}$ = 78.3 h): isomeric transition; α-particles from cosmic origin. See also 93.32 and 194.24 keV lines.
185.7	^{231}Th	^{235}U	57.5	α-decay ($T_{1/2}$ = 7.05E8 y): ^{235}U series.
185.91	^{66}Cu	^{65}Cu(n,γ)^{66}Cu		Prompt neutron capture γ-line produced by thermalized neutrons from cosmic origin.
186.1	^{222}Rn	^{226}Ra	3.57	α-decay ($T_{1/2}$ = 1601 y): ^{238}U series.
194.25	^{67}Zn	^{65}Cu(α,2n)^{67}Ga	1.0	β$^+$-decay ($T_{1/2}$ = 78.3 h): α-particles from cosmic origin.
198.4	^{71m2}Ge	^{70}Ge(n,γ)^{71m2}Ge	99.0	$T_{1/2}$ = 22 ms is produced continuously by thermal neutrons from cosmic origin. Sum peak 23.5 + 174.
203.1	Cu	^{63}Cu(n,γ)^{64}Cu		Prompt neutron capture γ-ray, I = 6.64% in nat. isotope mixture; is produced continuously by thermalized neutrons from cosmic origin.
205.3	^{231}Th	^{235}U	5.0	α-decay ($T_{1/2}$ = 7.05E8 y): ^{235}U series.
209.3	^{228}Th	^{228}Ac	3.88	β$^-$-decay ($T_{1/2}$ = 6.15 h): ^{232}Th series.
215.5	^{77}As	^{76}Ge(n,γ)^{77m}Ge	21.0	$T_{1/2}$ = 52.9 s: β$^-$-decay of isomeric level excited continuously by thermal neutrons from cosmic origin.
215.99	^{224}Ra	^{228}Th	0.3	α-decay ($T_{1/2}$ = 1.91 y): ^{232}Th series.
226.4	^{234}U	^{234}Pa	5.9	β$^-$-decay ($T_{1/2}$ = 6.7 h): ^{238}U series.
227.2	^{234}U	^{234}Pa	5.5	β$^-$-decay ($T_{1/2}$ = 6.7 h): ^{238}U series.
236.0	^{223}Ra	^{227}Th	11.2	α-decay ($T_{1/2}$ = 11.43 d): ^{235}U series.
238.6	^{212}Bi	^{212}Pb	43.6	β$^-$-decay ($T_{1/2}$ = 10.64 h): ^{232}Th series.
241.0	^{220}Rn	^{224}Ra	3.97	α-decay ($T_{1/2}$ = 11.43 d): ^{235}U series.
241.98	^{214}Bi	^{214}Pb	7.5	β$^-$-decay ($T_{1/2}$ = 26.8 m): ^{238}U series.

Continued

TABLE 4.4 Background lines observed in Ge spectra (This list is neither complete nor should all lines be present in each spectrum).—cont'd

γ line (keV)	Isotope[a)]	Reaction[b)]	$I_\gamma^{c)}$ (%)	Remarks
256.0	^{223}Ra	^{227}Th	7.6	α-decay ($T_{1/2}$ = 11.43 d): ^{235}U series.
269.2	^{219}Rn	^{223}Ra	13.6	α-decay ($T_{1/2}$ = 11.43 d): ^{235}U series.
270.2	^{228}Th	^{228}Ac	3.43	β^--decay ($T_{1/2}$ = 6.15 h): ^{232}Th series.
271.2	^{215}Po	^{219}Rn	9.9	α-decay ($T_{1/2}$ = 3.96 s): ^{235}U series.
277.4	^{208}Pb	^{208}Tl	6.31	β^--decay ($T_{1/2}$ = 3.05 m): ^{232}Th series.
278.3	^{64}Cu	^{63}Cu(n,γ)^{64}Cu		Prompt neutron capture γ-ray, I = 30.12% in nat. isotope mixture; is produced continuously by thermalized neutrons from cosmic origin.
283.7	^{227}Ac	^{231}Pa	1.6	α-decay ($T_{1/2}$ = 4243 y): ^{235}U series.
288.1	^{208}Tl	^{212}Bi	0.34	α-decay ($T_{1/2}$ = 1.01 h): ^{232}Th series.
295.2	^{214}Bi	^{214}Pb	18.5	β^--decay ($T_{1/2}$ = 26.8 m): ^{238}U series.
300.0	^{227}Ac	^{231}Pa	2.39	α-decay ($T_{1/2}$ = 4243 y) ^{235}U series.
300.1	^{212}Bi	^{212}Pb	3.34	β^--decay ($T_{1/2}$ = 10.64 h): ^{232}Th series.
302.7	^{227}Ac	^{231}Pa	2.24	α-decay ($T_{1/2}$ = 4243 y): ^{235}U series.
323.3	^{219}Rn	^{223}Ra	3.9	α-decay ($T_{1/2}$ = 11.43 d): ^{235}U series.
328.3	^{228}Th	^{228}Ac	2.95	β^--decay ($T_{1/2}$ = 6.15 h): ^{232}Th series.
330.1	^{227}Ac	^{231}Pa	1.31	α-decay ($T_{1/2}$ = 4243 y): ^{235}U series.
338.3	^{219}Rn	^{223}Ra	2.789	α-decay ($T_{1/2}$ = 11.43 d): ^{235}U series
338.3	^{228}Th	^{228}Ac	1.25	β^--decay ($T_{1/2}$ = 6.15 h): ^{232}Th series.
351.0	207 Tl	^{211}Bi	2.76	α-decay ($T_{1/2}$ = 2.14 m): ^{235}U series.
351.92	^{214}Bi	^{214}Pb	38.5	β^--decay ($T_{1/2}$ = 19.9 m): ^{238}U series.
367.94	^{200}Hg	^{199}Hg(n,γ)^{200}Hg		Prompt neutron capture γ-ray, I = 81.35% in nat. isotope mixture; is produced continuously by thermalized neutrons from cosmic origin. Its observation is mainly due to the high reaction yield and the enormous thermal cross section of ^{199}Hg of 2000 b.
401.7	^{215}Po	^{219}Rn	6.64	α-decay ($T_{1/2}$ = 3.96 s): ^{235}U series.
404.8	^{211}Bi	^{211}Pb	3.83	β^--decay ($T_{1/2}$ = 36.1 m): ^{235}U series
409.5	^{228}Th	^{228}Ac	1.94	β^--decay ($T_{1/2}$ = 6.15 h): ^{232}Th series.
426.99	^{211}Bi	^{211}Pb	1.72	β^--decay ($T_{1/2}$ = 36.1 m): ^{235}U series.
427.89	^{125}Te	^{124}Sn(p,γ)^{125}Sb	29.4	β^--decay ($T_{1/2}$ = 2.77 a): protons from cosmic origin.
444.9	^{219}Rn	^{223}Ra	1.27	α-decay ($T_{1/2}$ = 11.43 d): ^{235}U series.
452.83	^{208}Tl	^{212}Bi	0.31	α-decay ($T_{1/2}$ = 1.01 h): ^{232}Th series.
463.0	^{228}Th	^{228}Ac	4.44	β^--decay ($T_{1/2}$ = 6.15 h): ^{232}Th series.
463.38	^{125}Te	^{124}Sn(p,γ)^{125}Sb	0.15	β^--decay ($T_{1/2}$ = 2.77 a): protons from cosmic origin.
510.8	^{208}Pb	^{208}Tl	22.6	β^--decay ($T_{1/2}$ = 3.05 m): ^{232}Th series.
511.0	Anni.			This very common Doppler broadened line finds its origin in the annihilation of β^+-particles occurring in the β^+-decay or the pair production process induced by high-energy γ rays (E_γ>1022 keV) of cosmic origin and/or due to nuclear decay or various nuclear reactions. Many possible origins allow no prediction of its intensity. It may not be used to estimate the intensity of a β^+-decay branching. Is also produced by muon-induced pair production.

Continued

TABLE 4.4 Background lines observed in Ge spectra (This list is neither complete nor should all lines be present in each spectrum).—cont'd

γ line (keV)	Isotope[a)]	Reaction[b)]	$I_\gamma^{c)}$ (%)	Remarks
549.7	^{216}Po	^{220}Rn	0.1	α-decay ($T_{1/2}=55.6$ s): ^{232}Th series.
558.2	^{114}Cd	$^{113}Cd(n,\gamma)^{114}Cd$		Prompt neutron capture γ-ray, I = 79.71% in nat. isotope mixture; is produced continuously by thermalized neutrons of cosmic origin.
562.9	^{76}Ge	$^{76}Ge(n,n')^{76}Ge$		Prompt γ-line produced by inelastic scattering of fast neutrons from cosmic origin. Right asymmetric line-shape due to recoil of the Ge-atoms induced by (n,n')-reaction.
563.3	^{134}Ba	$^{133}Cs(n,\gamma)^{134}Cs$	8.38	β^--decay ($T_{1/2}=2.06$ a). This isotope is found in reactor waste (Chernobyl fallout) but not in the fall-out of bomb testing. This is due to the fact that it is no fission product, as it is screened by the stable ^{134}Xe. It is however found among the reactor fission products, as ^{133}Cs is the stable end product of the A = 133 fission chain having a yield of 7.87%.
568.7	^{234}U	^{234}Pa	3.3	β^--decay ($T_{1/2}=6.7$ h): ^{238}U series.
569.5	^{234}U	^{234}Pa	10.0	β^--decay ($T_{1/2}=6.7$ h): ^{238}U series.
569.79	^{207}Pb	$^{207}Pb(n,n')^{207}Pb$ $^{206}Pb(n,\gamma)^{207}Pb$		Prompt γ-line produced by inelastic scattering of fast neutrons from cosmic origin or by thermal neutron capture.
583.2	^{208}Pb	^{208}Tl	84.5	β^--decay ($T_{1/2}=3.05$ m): ^{232}Th series.
595.9	^{74}Ge	$^{73}Ge(n,\gamma)^{74}Ge$ $^{74}Ge(n,n')^{74}Ge$		Prompt neutron capture γ-ray, I = 34.65% in nat. isotope mixture; is produced continuously by thermalized neutrons from cosmic origin. Prompt γ-line produced by inelastic scattering of fast neutrons from cosmic origin. Right asymmetric line shape due to recoil of the Ge atoms induced by (n,n') reaction.
604.7	134Ba	$^{133}Cs(n,\gamma)^{134}Cs$	97.6	See comments 563.3 keV line.
651.0	^{114}Cd	$^{113}Cd(n,\gamma)^{114}Cd$		Prompt neutron capture γ-ray, I = 15.23% in nat. isotope mixture; is produced continuously by thermalized neutrons from cosmic origin.
600.55		$^{124}Sn(p,\gamma)^{125}Sb$	17.78	β^--decay ($T_{1/2}=2.77$ y): protons from cosmic origin.
606.64	^{125}Te	$^{124}Sn(p,\gamma)$	5.02	β^--decay ($T_{1/2}=2.77$ y): protons from cosmic origin
609.3	^{214}Po	^{214}Bi	44.8	β^-decay ($T_{1/2}=19.9$ m): ^{238}U series.
635.9	^{125}Te	$^{124}Sn(p,\gamma)^{125}Sb$	11.32	β^--decay ($T_{1/2}=2.77$ y): protons from cosmic origin.
661.66	^{137m}Ba	^{137}Cs	85.0	Fission isotope β^--decay ($T_{1/2}=30.17$ y): bomb testing + - Chernobyl fallout.
669.6	^{63}Cu	$^{63}Cu(n,n')^{63}Cu$		Prompt γ-line produced by inelastic scattering of fast neutrons from cosmic origin
671.40	^{125}Te	$^{124}Sn(p,\gamma)^{125}Sb$	1.8	β^--decay ($T_{1/2}=2.77$ y): protons from cosmic origin.
691.0	^{72}Ge	$^{72}Ge(n,n')^{72}Ge$		Prompt γ-line produced by inelastic scattering of fast neutrons from cosmic origin. This line is a 0+ to 0+ and can thus only take place by internal conversion as electrical monopol transitions are strictly forbidden. The asymmetric right-side shape is due to imperfect transformation of the recoil energy due to the neutron scattering and is observed—in contrary to other (n,n') reactions, due to the fact that the recoil takes place in the Ge and thus inside to the detector.

Continued

TABLE 4.4 Background lines observed in Ge spectra (This list is neither complete nor should all lines be present in each spectrum).—cont'd

γ line (keV)	Isotope[a)]	Reaction[b)]	$I_\gamma^{c)}$ (%)	Remarks
727.3	^{212}Po	^{212}Bi	6.25	β^--decay ($T_{1/2}$ = 1.01 h): ^{232}Th series.
751.8	^{65}Zn	^{63}Cu(α,2n)^{65}Ga	50.7	β^+-decay ($T_{1/2}$ = 15 m): continuously formed by α-particles from cosmic origin.
766.0	^{234}U	^{234m}Pa	0.21	β^--decay ($T_{1/2}$ = 1.17 m): ^{238}U series.
768.4	^{214}Po	^{214}Bi	4.88	β^--decay ($T_{1/2}$ = 19.9 m): ^{238}U series.
769.7	^{73}As	^{73}Ge(p,nγ)^{73}As		Prompt γ-line produced by p,n-reaction with protons from cosmic origin.
770.8	^{65}Cu	^{65}Cu(n,n')^{65}Cu		Prompt γ-line produced by inelastic scattering of fast neutrons from cosmic origin.
772.4	^{228}Th	^{228}Ac	1.58	β^--decay ($T_{1/2}$ = 6.15 h): ^{232}Th series.
785.6	^{212}Po	^{212}Bi	1.11	β^--decay ($T_{1/2}$ = 1.01 h): ^{232}Th series.
794.9	^{228}Th	^{228}Ac	4.34	β^--decay ($T_{1/2}$ = 6.15 h): ^{232}Th series.
795.8	^{134}Ba	^{133}Cs(n,γ)^{134}Cs	85.4	See 563.3 keV line.
801.9	^{134}Ba	^{133}Cs(n,γ)^{134}Cs	8.73	See 563.3 keV line
803.3	^{206}Pb	^{206}Pb(n,n')^{206}Pb ^{210}Po	0.001	Prompt γ-line produced by inelastic scattering of fast neutrons from cosmic origin. α-decay ($T_{1/2}$ = 138.4 d): ^{238}U series.
805.7	^{114}Cd	^{113}Cd(n,γ)^{114}Cd		Prompt neutron capture γ-ray, I = 5.1% in nat. isotope mixture; is produced continuously by thermalized neutrons from cosmic origin.
810.80	^{58}Fe	^{59}Co(γ,n)^{58}Co 59Co(n,2n)^{58}Co ^{58}Fe(p,n)^{58}Co ^{57}Fe(p,γ)^{58}Co ^{57}Fe(d,n)^{58}Co ^{58}Fe(n,p)^{58}Mn	≈100 82.2	β^--decay ($T_{1/2}$ = 63 s) is produced continuously by fast γ's and particles of cosmic origin.
831.8	^{211}Bi	^{211}Pb	3.83	β^--decay ($T_{1/2}$ = 36.1 m): ^{235}U series.
833.95	^{77}Ge	^{72}Ge(n,n')^{72}Ge		Prompt γ-line produced by inelastic scattering of fast neutrons from cosmic origin. Right asymmetric line-shape due to recoil of the Ge atoms induced by (n,n')-reaction.
834.6	^{54}Cr	^{54}Cr(p,n)^{54}Mn ^{53}Cr(d,n)^{54}Mn ^{53}Cr(p,γ)^{54}Mn	100	EC-decay ($T_{1/2}$ = 312.2 d): protons from cosmic origin.
835.7	^{228}Th	^{228}Ac	1.68	β^--decay ($T_{1/2}$ = 6.15 h): ^{232}Th series.
846.8	^{56}Fe	^{56}Fe(n,n')^{76}Fe	19.0	Prompt γ-line produced by inelastic scattering of fast neutrons from cosmic origin. The absence of the 1282.6 keV line allows to distinguish it from the same line excited in the decay of ^{56}Co.
846.8	^{56}Fe	^{56}Fe(p,n)^{56}Co		β^+-decay ($T_{1/2}$ = 78.76 d): the presence of the 1238.2 keV line allows to distinguish it from the same line excited in the ^{56}Fe(n,n')^{76}Fe reaction.
860.6	^{208}Pb	^{208}Tl	12.42	β^--decay ($T_{1/2}$ = 3.05 m): ^{232}Th series.

Continued

TABLE 4.4 Background lines observed in Ge spectra (This list is neither complete nor should all lines be present in each spectrum).—cont'd

γ line (keV)	Isotope[a)]	Reaction[b)]	$I_\gamma^{c)}$ (%)	Remarks
868.1	^{73}Ge	^{72}Ge(n,γ)^{73}Ge		Prompt neutron capture γ-ray, I = 30.12% in nat. isotope mixture; is produced continuously by thermalized neutrons from cosmic origin.
880.51	^{234}U	^{234}Pa	9	β^--decay ($T_{1/2}$ = 6.7 h): ^{238}U series.
883.24	^{234}U	^{234}Pa	15	β^--decay ($T_{1/2}$ = 6.7 h): ^{238}U series.
897.6	^{207}Bp	^{207}Tl	0.24	β^--decay ($T_{1/2}$ = 4.79 m): ^{235}U series.
911.2	^{228}Th	^{228}Ac	26.6	β^--decay ($T_{1/2}$ = 6.15 h): ^{232}Th series.
925.0	^{234}U	^{234}Pa	2.9	β^--decay ($T_{1/2}$ = 6.7 h): ^{238}U series.
926.0	^{234}U	^{234}Pa	11.0	β^--decay ($T_{1/2}$ = 6.7 h): ^{238}U series.
927.1	^{234}U	^{234}Pa	11.0	β^--decay ($T_{1/2}$ = 6.7 h): ^{238}U series.
934.1	^{214}Po	^{214}Bi	3.03	β^--decay ($T_{1/2}$ = 19.9 m): ^{238}U series.
946.0	^{234}U	^{234}Pa	12	β^--decay ($T_{1/2}$ = 6.7 h): ^{238}U series.
962.1	^{65}Cu	^{63}Cu(nn′)^{63}Cu		Prompt γ-line produced by inelastic scattering of fast neutrons from cosmic origin.
964.8	^{228}Th	^{228}Ac	5.11	β^--decay ($T_{1/2}$ = 6.15 h): ^{232}Ht series.
969.0	^{228}Th	^{228}Ac	16.20	β^--decay ($T_{1/2}$ = 6.15 h): ^{232}Ht series
1001.0	^{234}U	^{234m}Pa	0.59	β^--decay ($T_{1/2}$ = 1.17 m): ^{238}U series.
1039.5	^{70}Ge	^{70}Ge(n,n′)^{70}Ge		Prompt γ-line produced by inelastic scattering of fast neutrons from cosmic origin. Right asymmetric line-shape due to recoil of the Ge atoms induced by (n,n′)-reaction
1063.64	^{207}Pb	^{207}Pb(n,n′)^{207}Pb ^{206}Pb(n,γ)^{207}Pb		Prompt γ-line produced by inelastic scattering of fast neutrons from cosmic origin or by thermal neutron capture. See also 569.79 keV line.
1077.41	^{68}Zn	^{65}Cu(α,n)′^{68}Ga	3.0	β^+-decay ($T_{1/2}$ = 68.3 m): α-particles of cosmic origin.
1097.3	^{116}Sn	^{115}In(n,γ)^{116m1}	55.7	β^--decay ($T_{1/2}$ = 54.1 m): formed by thermalized neutrons from cosmic origin.
1115.5	^{65}Cu	^{65}Cu(n,n′)^{65}Cu ^{65}Cu(p,n)^{65}Zn	50.75	Prompt γ-line produced by inelastic scattering of fast neutrons from cosmic origin. EC + β^+-decay ($T_{1/2}$ = 244 d) formed by fast neutrons or protons from cosmic origin.
1120.4	^{214}Po	^{214}Bi	14.8	β^--decay ($T_{1/2}$ = 19.9 m): ^{238}U series.
1124.51	^{65}Cu	^{70}Ge(n,α2n)^{65}Zn	50.75	EC + β^+-decay ($T_{1/2}$ = 244 d).formed by fast neutrons from cosmic origin. Note that it is the same line as the above mentioned 1115.5 keV transition. As the reaction takes place inside the Ge-detector itself, its energy sums up with the Kα–X-ray of Cu. It is thus possible to distinguish the formation reaction of ^{65}Zn.
1173.2	^{60}Ni	^{59}Co(n,γ)^{60}Co	100	β^--decay ($T_{1/2}$ = 5.172 y): this isotope is a common contamination in modern steel and is introduced at the high furnace level.

Continued

TABLE 4.4 Background lines observed in Ge spectra (This list is neither complete nor should all lines be present in each spectrum).—cont'd

γ line (keV)	Isotope[a)]	Reaction[b)]	$I_\gamma^{c)}$ (%)	Remarks
1204.1	^{74}Ge	$^{74}Ge(n,n')^{74}Ge$		Fast neutrons from cosmic origin. Right asymmetric line-shape due to recoil of the Ge-atoms induced by (n,n′)-reaction.
1238.26	^{56}Fe	$^{56}Fe(p,n)^{56}Co$	13.4	β^+-decay ($T_{1/2} = 78.76$ d): see 846.8 keV line.
1238.8	^{214}Po	^{214}Bi	5.86	β^--decay ($T_{1/2} = 19.9$ m): ^{238}U series.
1291.65	^{59}Co	$^{58}Fe(n,\gamma)^{59}Fe$	57.0	β^--decay ($T_{1/2} = 45.1$d). Is produced continuously by thermalized neutrons of cosmic origin.
1293.5	^{116}Sn	$^{115}In(n,\gamma)^{116m1}$	85.0	In β^--decay ($T_{1/2} = 54.1$ m): formed by thermalized neutrons from cosmic origin.
1293.64	^{41}K	$^{40}Ar(n,\gamma)^{41}Ar$	99.16	β^--decay ($T_{1/2} = 1.83$ h); is produced continuously by thermalized neutrons from cosmic origin. Is a common B.G. line near air-cooled fission reactors.
1327.0	^{63}Cu	$^{63}Cu(n,n')^{63}Cu$		Prompt γ-line produced by inelastic scattering of fast neutrons from cosmic origin.
1332.5	^{60}Ni	$^{59}Co(n,\gamma)^{60}Co$	100	See 1173.2 keV-line.
1377.6	^{57}Co	$^{58}Ni(\gamma,n)^{57}Ni$ $^{58}Ni(n,2n)^{57}Ni$	30.0	β^+ + EC-decay ($T_{1/2} = 36.0$ h)
1377.6	^{214}Po	^{214}Bi	3.92	β^--decay ($T_{1/2} = 19.9$ m): ^{238}U series.
1408.0	^{214}Po	^{214}Bi	2.48	β^--decay ($T_{1/2} = 19.9$ m): ^{238}U series.
1412.1	^{63}Cu	$^{63}Cu(n,n')^{63}Cu$		Prompt γ-line produced by inelastic scattering of fast neutrons from cosmic origin.
1460.8	^{40}Ar	^{40}K	99.16	EC and β^{+-}decay. ($T_{1/2} = 1.277E+8$ Y) widespread natural radioactive isotope. The modal human body contains about 4000 Bq of this isotope.
1481.7	^{65}Cu	$^{65}Cu(n,n')^{65}Cu$		Prompt γ-line produced by inelastic scattering of fast neutrons from cosmic origin.
1547.0	^{63}Cu	$^{63}Cu(n,n')^{63}Cu$		Prompt γ-line produced by inelastic scattering of fast neutrons from cosmic origin.
1588.2	^{228}Th	^{228}Ac	3.27	β^--decay ($T_{1/2} = 6.15$ h): ^{232}Th series.
1620.6	^{212}Po	^{212}Bi	1.6	β^--decay ($T_{1/2} = 1.01$ h): ^{232}Th series.
1729.6	^{214}Po	^{214}Bi	2.88	β^--decay ($T_{1/2} = 19.9$ m): ^{238}U series.
1764.5	^{214}Po	^{214}Bi	15.96	β^--decay ($T_{1/2} = 19.9$ m): ^{238}U series.
2204.1	^{214}Po	^{214}Bi	4.86	β^--decay ($T_{1/2} = 19.9$ m): ^{238}U series.
2223.2	^{2}H	$^{1}H(n,\gamma)\ ^{2}H$		Prompt neutron capture γ-ray, I = 100% in nat. isotope mixture; is produced continuously by thermalized neutrons from cosmic origin.
2614.6	^{208}Pb	$^{208}Pb(n,n')^{208}Pb$ ^{208}Tl	99.2	Prompt γ-line produced by inelastic scattering of fast neutrons from cosmic origin. β^--decay ($T_{1/2} = 3.05$ m): ^{232}Th series.

TABLE 4.5 ^{238}U natural decay chain.

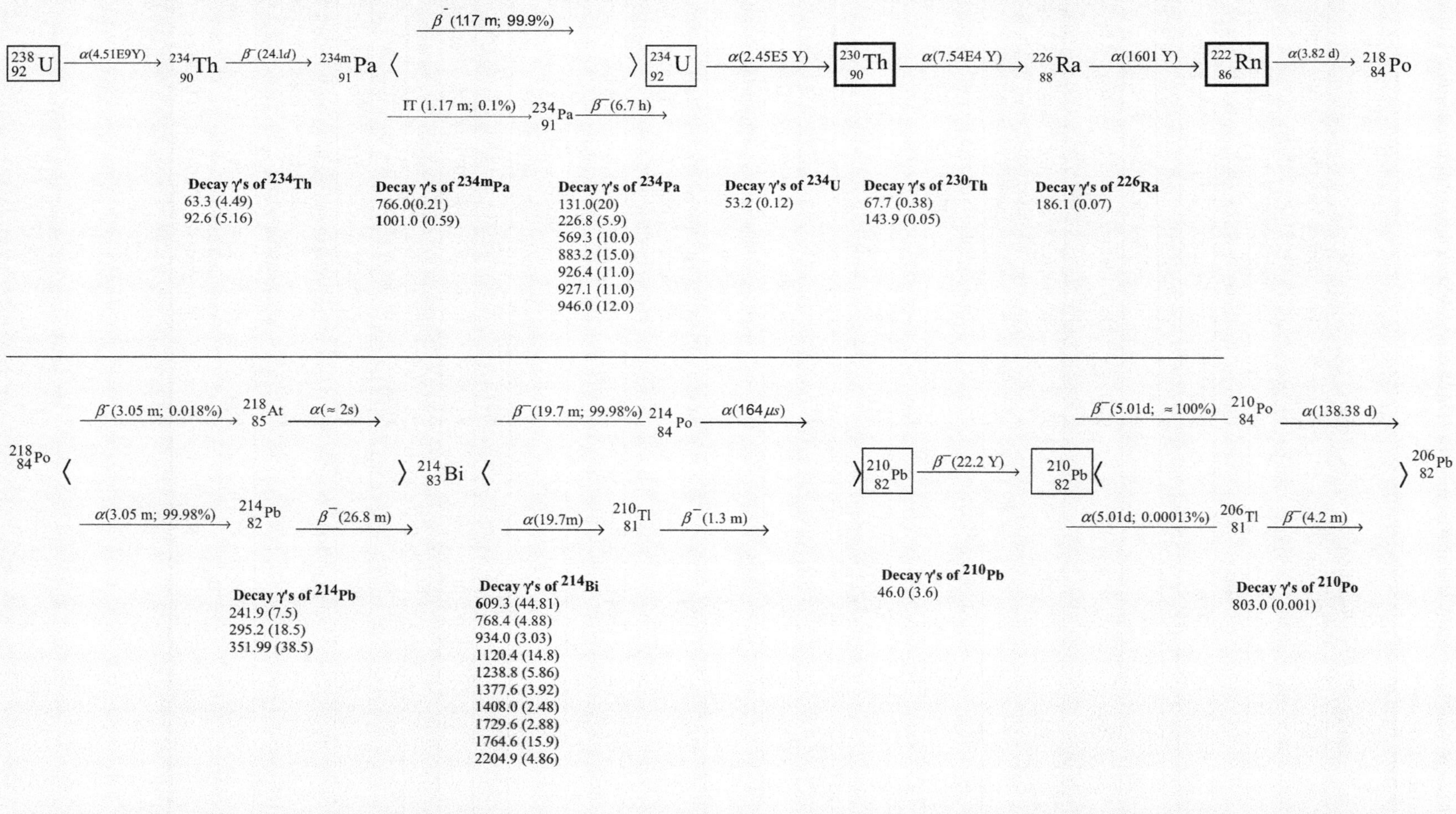

TABLE 4.6 ^{235}U natural decay chain.

$$\boxed{^{235}_{92}U} \xrightarrow{\alpha(7.05E8\ Y)} {}^{231}_{90}Th \xrightarrow{\beta^-(25.5\ h)} {}^{231}_{91}Pa \xrightarrow{\alpha(3.28E4\ Y)} {}^{227}_{89}Ac \left\langle \begin{array}{l} \xrightarrow{\beta^-(21.733\ Y;\ 98.62)} {}^{227}_{90}Th \xrightarrow{\alpha(18.7\ d)} \\ \xrightarrow{\alpha(21.733\ Y;\ 1.38)} {}^{223}_{87}Fr \xrightarrow{\beta^-(21.8\ m)} \end{array} \right\rangle {}^{223}_{88}Ra \xrightarrow{\alpha(11.43\ d)} \boxed{^{219}_{86}Rn} \xrightarrow{\alpha(3.96\ s)} {}^{215}_{84}Po \left\langle \begin{array}{l} \xrightarrow{\beta^-(1.78\ m\ s;\ 0.00023)} {}^{215}_{85}At \xrightarrow{\alpha(0.1\ ms)} \\ \xrightarrow{\alpha(1.78\ ms;\ \approx 100)} {}^{211}_{82}Pb \xrightarrow{\beta^-(36.1\ m)} \end{array} \right\rangle \boxed{^{211}_{83}Bi}$$

Decay γ's of ^{235}U	Decay γ's of ^{231}Th	Decay γ's of ^{231}Pa	Decay γ's of ^{227}Th	Decay γ's of ^{223}Ra	Decay γ's of ^{219}Rn	Decay γ's of ^{211}Pb
143.8 (10.9)	84.2 (6.6)	283.7 (1.6)	49.9 (0.52)	122.4 (1.19)	271.2 (9.9)	404.8 (3.83)
163.3 (5.0)		300.0 (2.39)	50.1 (7.28)	144.2 (3.26)	401.7 (6.64)	427.0 (1.72)
185.7 (57.5)		330.1 (1.31)	236.0 (11.65)	154.2 (5.59)		831.8 (3.8)
205.3 (5.0)			256.0 (7.6)	269.4 (13.6)		
				323.3 (2.78)		
				444.9 (1.27)		

$$\boxed{^{211}_{83}Bi} \left\langle \begin{array}{l} \xrightarrow{\beta^-(2.14\ m;\ 0.28\%)} {}^{211}_{84}Po \xrightarrow{\alpha(0.516\ s)} \\ \xrightarrow{\alpha(2.14\ m;\ 99.72\%)} {}^{207}_{81}Tl \xrightarrow{\beta^-(4.79\ m)} \end{array} \right\rangle {}^{207}_{82}Pb$$

Decay γ's of ^{207}Tl

TABLE 4.7 ^{232}Th natural decay chain.

$$\boxed{^{232}_{90}\mathrm{Th}} \xrightarrow{\alpha(4.1\mathrm{E}10\mathrm{Y})} {}^{228}_{88}\mathrm{Ra} \xrightarrow{\beta^-\ (6.79\ \mathrm{Y})} {}^{228}_{89}\mathrm{Ac} \xrightarrow{\beta^-\ (6.15\ \mathrm{h})} {}^{228}_{90}\mathrm{Th} \xrightarrow{\alpha\ (1.91\ \mathrm{Y})} {}^{224}_{88}\mathrm{Ra} \xrightarrow{\alpha\ (3.66\ d)} \boxed{^{220}_{86}\mathrm{Rn}} \xrightarrow{\beta^-\ (55.6\ \mathrm{s})} {}^{216}_{84}\mathrm{Po} \xrightarrow{\alpha(0.156s)} {}^{212}_{82}\mathrm{Pb}$$

Decay γ's of ^{228}Ac
129.6(2.45)
209.3(3.88)
270.24 (3.43)
328.0(2.95)
338.3(11.3)
463.0(4.44)
772.4(1.5)
794.9(4.34)
835.7(1.68)
911.2(26.6)
964.8(5.11)
969.0(16.2)
1588.2(3.27)
1630.6(1.6)

Decay γ's of ^{228}Th
84.37 (1.27)
216.0 (0.26

Decay γ's of ^{224}Ra
241.0 (3.97)

Dcay γ ^{220}Rn
549.7 (0.1)

$$\boxed{^{212}_{82}\mathrm{Pb}} \xrightarrow{\beta^-\ (10.64\ \mathrm{h})} {}^{212}_{83}\mathrm{Bi} \xrightarrow{(60.55s)} \left\langle \begin{array}{l} \xrightarrow{\beta^-(64.0\%)} {}^{212}_{84}\mathrm{Po} \xrightarrow{\alpha(304ns)} \\ \xrightarrow{\alpha(36.0\%)} {}^{208}_{81}\mathrm{Tl} \xrightarrow{\beta^-\ (3.05\ \mathrm{m})} \end{array} \right\rangle {}^{208}_{82}\mathrm{Pb}$$

Decay γ's of ^{212}Pb
238.6 (43.6)
300.1 (3.34)

Decay γ's of ^{212}Bi
288.1 (0.31)
452.83 (0.31)
727.2 (6.65)
785.4 (1.11)
893.39 (0.37)
1512.8 (0.54)
1620.6 (1.51)

Decay γ's of ^{208}Tl
277.4 (6.31)
510.8 (22.6)
583.2 (84.5)
763.1 (1.81)
860.6 (12.42)
2614.5 (99.2)

4. Background reduction

Background reduction is a difficult and delicate operation. The optimum shielding should take the isotopes to be measured into account as well as the energy range and the lower limit of detection desired. But as local conditions can vary strongly, no off-the-shelf solution can be given for all cases. Some general rules remain valid under all conditions and will be discussed subsequently.

a. Passive background reduction

Passive background reduction is based on the absorption of undesired γ rays by an absorber placed between the detector and the source of the background. The transmitted intensity is given by Eq. (4.49). If the absorption coefficient is expressed in half-thickness (cm), the transmitted intensity I is given by

$$I = I_0 e^{-0.693d/d_{1/2}} \tag{4.49}$$

where I_0 is the initial flux, $d_{1/2}$ the half-thickness, and d the actual thickness of the shielding. In Fig. 4.18, the half-thickness for Cu, Sn, Pb, and Si is given as a function of energy. Good shielding should be sufficiently thick; for example, 10 times the half-thickness, to reduce the background by a factor of 1000. This would translate into a Pb thickness of 8.8 cm for a 1000-keV γ ray. In practice, a thickness of 10 or 15 cm is often chosen. It must be remembered here that lead contains ^{210}Pb ($T_{1/2} = 21$ y), as the Pb ores and the coke used in the melting process contain U traces that continuously form ^{210}Pb. The ^{210}Pb content in lead varies according to its origin and age. Specific lead is available for ultralow background shielding (Fig. 4.18).

On the other hand, the Pb shield should not be too thick, to reduce the production of fast neutrons by cosmic particles with the subsequent production of n-induced background lines summarized in Table 4.4. Fast neutrons are difficult to stop. Several tens of centimeters of hydrogen-rich material is needed to thermalize them. Once thermalized, they can be stopped by high-cross section materials such as B or Cd. However, the absorption process generates new γ rays, including a Doppler-broadened 480 keV for B and a whole spectrum of neutron capture γ rays for Cd. Shields have been built to minimize the effect of cosmic neutrons on the background spectrum. These consist of a "sandwich," combining materials that slow down fast neutrons and absorb these neutrons and materials that stop environmental gamma rays and gamma rays produced in the process of eliminating the (n,γ) signals (Heusser, 1996).

Fluorescent Pb X-rays can be reduced by a supplementary lining of the Pb shield by lower Z material such as Cd or Sn. A 1-mm thickness of Sn stops 95% of all Pb X-rays, and a supplementary lining of 1.5 mm of Cu stops most of the Sn X-rays and raises the total absorption of lead X-rays to 98.5%. Once again, thick linings should not be used; otherwise, the continuous background due to the backscattering of the source γ rays will increase. This is due to the fact that the Compton effect, responsible for the backscattering, varies with the atomic number Z, whereas the photoelectric effect is proportional to Z^5. Also, the plastic inner layer often used to prevent contamination of the shield should be as thin as possible. Lining a 10-cm Pb shield with

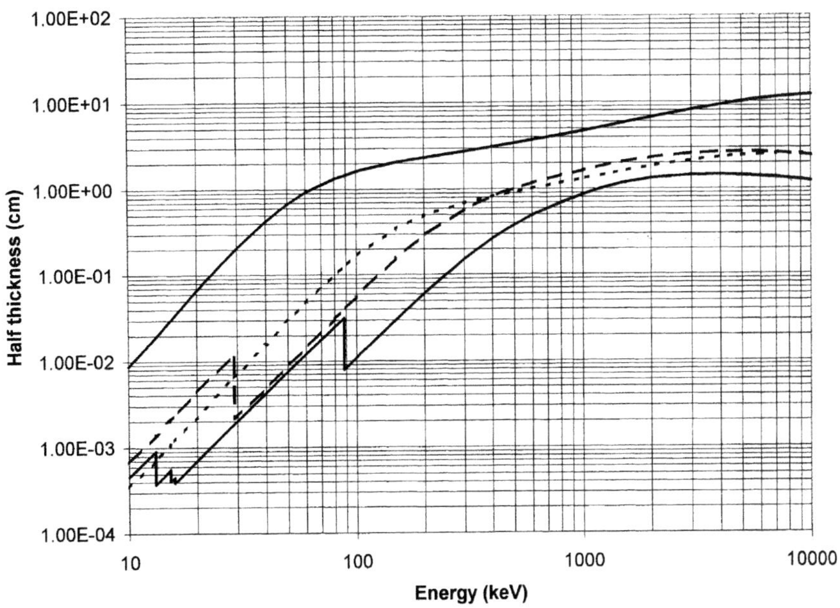

FIGURE 4.18 Half-thickness in centimeter for Pb (lower plain curve), Sn (dashed curve), Cu (dotted curve), and Si (upper plain curve) as a function of energy.

1-mm Cd, 2-mm Cu, and 10-mm Plexiglas increases the background by as much as 30% at 25 keV and 15% at 1000 keV.

Cosmic background and the associated activation and reaction lines can be adequately reduced by placing the detector deep underground. At a depth of 1000-m water equivalent, the neutron flux induced by cosmic particles is less than 1% of the flux observed at sea level.

b. Active background reduction

In active background reduction, we cover all measures that are not based on the absorption of the undesired γ rays. Active background reduction techniques eliminate their causes or limit their effect.

Venting. It has been seen that ^{222}Rn and its daughters may accumulate in a closed area and, in particular, inside the shielding of the detector. Venting with an Rn-free gas such as N_2 or Ar may help to reduce their presence strongly. This can easily be achieved by using the nitrogen gas boiling off from the liquid nitrogen tank used to cool the germanium detector.

Choice of construction materials and cryostat design. Materials used in the detector surroundings and especially for the construction of the cryostat as well as the electronic components of the preamplifier can contain elements such as Al, Be, and Sn, which may contain traces of U or Th, constituting an undesired source of background. The industry offers different types of cryostats whose design and selection of construction materials minimize these effects. Figure 4.19 shows the experimental background per keV per hour (cph/keV) for the same detector element mounted in different Canberra cryostats: the classical vertical dipstick cryostat and the same cryostat with the materials close to the detector element selected for low background. The third spectrum is taken with the same cryostat but with an additional low-background lead disk between the crystal and preamplifier. It is seen that this additional lead disk reduces the background from the preamplifier and from the floor at higher energies only. The lowest background at all energies is obtained with the 7500Sl-RDC-ULB cryostat. In this cryostat, the low-background detector chamber is separated from the preamplifier chamber and from the rest of the cryostat with a thin tube holding the cold finger. This part of the cold finger is off-set from the lower part in the dipstick cryostat, preventing a direct line of view between the detector element and the floor (Ceuppens et al., 1996) (Fig. 4.19).

The Compton suppression Spectrometer. The Compton continuum is observed when the Compton-scattered γ ray escapes from the detector. When a large scintillation detector [NaI(Tl), plastic or BGO) surrounds the Ge detector and the source, a coincidence signal between this shield and the Ge detector can be used to suppress the Compton pulse. However, the following two important aspects must be noted.:

1. Above 200 keV Compton scattering occurs predominantly in the forward direction. The optimum active shield should be designed with this property and the actual source-detector geometry in mind. By Compton suppression, the remaining continuum after adequate passive shielding, can be reduced by a factor of 5 or more.

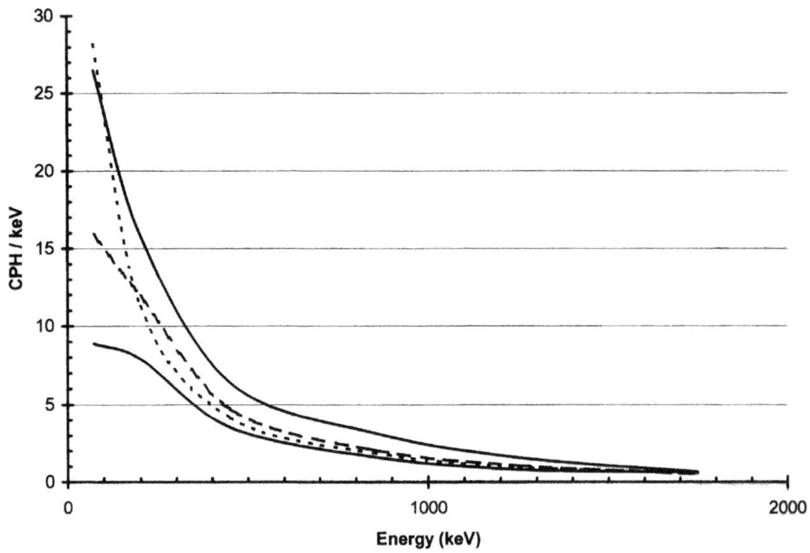

FIGURE 4.19 Background in counts per keV and per hour for the same detector placed in different cryostats: Upper plain curve: standard Canberra vertical dipstick cryostat; dashed curve: same cryostat with the building materials close to the crystal replaced by low background alternatives; dotted curve: same cryostat but with an additional low-background lead disk between the crystal and preamplifier; lower plain curve: ultralow-background Canberra cryostat model 7500SL RDC-ULB.

2. A Compton suppression active shield also rejects coincident lines such as the 1173.2- and 1332.5-keV lines of ^{60}Co. For isotopes having two or more coincident lines, strong spectrum deformation will occur, and the previously established efficiency curve will no longer be valid. The application of this technique is limited, therefore, to specific applications. This effect is similar to that observed in a well-type detector with one important difference. In both cases, the intensity of coincident γ lines is reduced, but in the case of a well-type detector, the intensity of the coincident lines is recovered back in the sum peaks, as will be discussed in Section E.

The cosmic veto shield. High-energy charged cosmic particles contribute to the continuum in the spectrum. This background can be drastically reduced by an active veto detector, generally a plastic scintillator, placed above, or surrounding, the Ge detector's passive shield (see, for instance, Semkow et al., 2002). Any cosmic particle entering the Ge detector will also interact with the veto detector generating two coincident signals that can be used to suppress the Ge pulse. Long dead times of a few tens of microseconds will also reduce the "delayed" cosmic background. With the help of a veto shield, background reductions of 99% have been obtained in the 10-MeV region, where the background is solely due to cosmic events (Müller et al., 1990).

E. The choice of a detector

1. General criteria

The industry offers a large variety of germanium detector models each of which is tailored for a particular application or energy range. Fig. 4.20, for instance, summarizes the various models offered by Canberra. This figure is self-explanatory. More details of each of the models can be found in the catalogs from the various manufacturers.

2. Increasing the size of high-purity germanium detectors

High-purity germanium (HPGe) became industrially available in the late 1970s and the beginning of 1980s. To satisfy the demand for devices that provide continually better measurement statistics in less counting time, optimum detector geometries as well as the enabling contacting technologies have continually developed since the beginning of the 1980s. This also led to the race to grow the biggest germanium ingot between the few industrial Ge crystal manufacturers. This resulted in the first 100% detector (i.e., a germanium detector with roughly 80 mm diameter and 80 mm length that has the same efficiency at 1.3 MeV as a 3″ × 3″ NaI detector) in 1988, the first 150% detector in 1992 (diameter 90 mm, length 86 mm), and the first 200% detector (diameter 100 mm and length 126 mm) built in 1998. This latter achievement (produced in Canberra Oak Ridge) ended the race for ever bigger germanium detectors. Development of larger detectors ceased because the costs to grow such huge crystals turn out to be unacceptably high and, secondly, the extra detector dimensions do not lead to a sizable improvement in "counting efficiency" for an industrial lab, over the energy domain between 100 and 2 MeV, which is their primary range of interest. Indeed, these gamma rays do not penetrate germanium more than about 7–8 cm. Anything beyond this length only sees background radiation from the environment. This long size also increases the detector capacitance and, thus, the noise and energy resolution. This explains why these very large crystals are almost exclusively used to make well-type Ge detectors. In this type of detector, the source is located in a well drilled in the detector crystal.

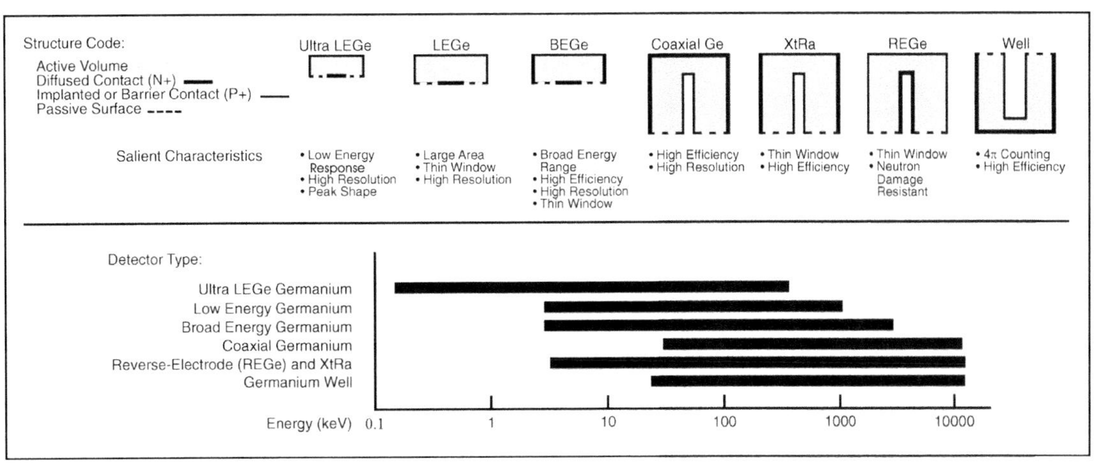

FIGURE 4.20 Summary of the various Ge detector models offered by Canberra, with the energy range they cover and their salient performance characteristics. *Image courtesy of Mirion Technologies.*

3. The germanium well-type detector

A well-type detector is designed to surround the sample so that close to 4π geometry is obtained. The source sees the negligible thin ion-implanted p^+ contact. It is thus the ideal detector when small samples (test tube sized) have to be measured routinely. It must be emphasized, however, that with this detector type, coincident γ rays are subject to intense summation, leading to strong spectrum deformation, namely a reduction of the individual peaks and an enlargement of the sum peaks. Consequently, for each sample (or at least for each isotope mixture), a specific and carefully established efficiency calibration is needed.

Also, it should be noted that, when large quantities of source materials are available that can be put on the detector window, or around the detector chamber, lower minimum detectable activities (MDAs) per gram of source material can be obtained than when only a very small portion is measured in the favorable well-type geometry.

4. Limitations to the "relative efficiency" quoted for coaxial detectors

Traditionally, the shapes and geometries of most HPGe coaxial detectors that are offered on the market today are designed to optimize resolution and relative efficiency as defined in Section II.C.3. The notion of "relative efficiency," however, does not tell the spectroscopist anything about the real behavior of this detector at energies other than 1.3 MeV or in real measurement situations with sources different from a point source at 25 cm distance. Fig. 4.21, for instance, compares the absolute efficiencies of the two detectors from Fig. 4.22. Both detectors have the same "relative efficiency" of 35%, but the absolute efficiencies, even for a point source at 25 cm, are very different for all energies different from 1.3 MeV. It is seen that the detector with the best absolute efficiency in the energy range considered has a large diameter, a shorter length, and sharper edges at the side of the entrance window (this window is facing down in this picture) than the other detector.

A goal that is pursued by most environmental and low-level gamma spectroscopists is to lower the MDA of their detection system, i.e., to obtain more statistical evidence in less time. It has been shown that the MDA depends in the first place on the detection efficiency and in the second place on number of background counts and peak width. It thus appears that it pays the most to increase the detection efficiency. This, however, does not always mean that one needs to choose a bigger detector or a detector with a higher relative efficiency. More important is to select a detector that is better matched with the source to be measured.

This principle is dramatically demonstrated with the following example, summarized in Table 4.8. It shows the net count rates, the number of background counts, energy resolutions, and relative MDAs obtained with a cylindrical multigamma source positioned on the endcaps of two different detectors. The "50% XtRa" detector is a Canberra thin window p-type coaxial detector with relatively sharp edges at the window side, a diameter of 65.5 mm and a length of 65 mm, similar to the shape of the crystal on the left side in Fig. 4.22. The "70% n-type" or "REGe-type" detector has strongly rounded edges—beyond the diameter of the source, a diameter of 69.7 mm and a length of 80.3 mm. It is seen in Table 4.8 that the "smaller" detector for this particular detector-source geometry yields a higher counting rate at all energies below 1.2 MeV, a lower background, better energy resolutions, and thus lower MDAs than the "bigger" detector!

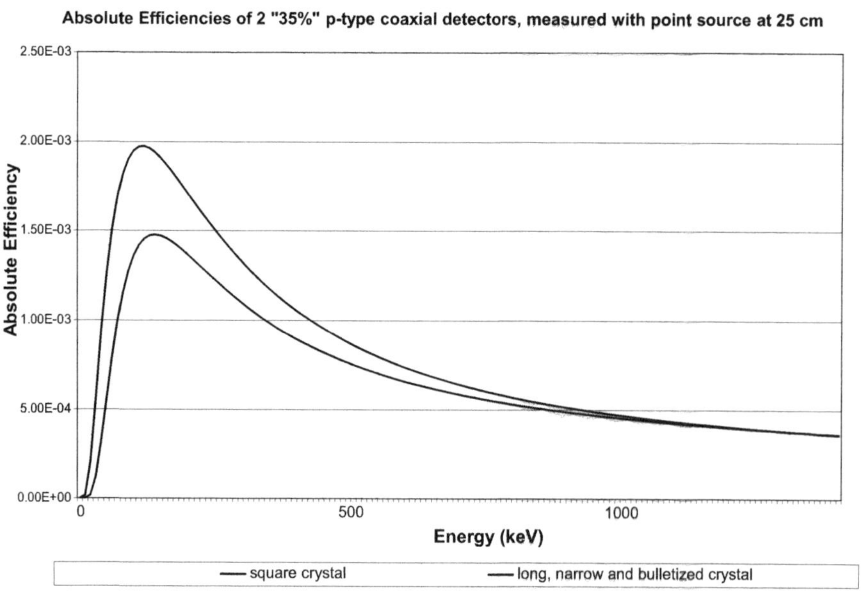

FIGURE 4.21 Absolute efficiency curves for the two "35%" detectors shown in Fig. 4.22. Point sources at 25 cm from the endcap were used to obtain these curves.

FIGURE 4.22 Two germanium crystals yielding a relative efficiency of 35%. The entrance windows are facing down. The right crystal has rounded edges at the window side.

5. The Broad-Energy Germanium, or "BEGe" detector

Observations like those described in Section 3 above led some detector manufacturers to build detectors that are optimized for certain specific applications (Verplancke, 1999; Keyser et al., 1998). The broad-energy germanium or "BEGe" detector from Canberra is developed to give a detector that is best adopted for low-level applications with extended sources and energies ranging from 5 keV to >2 MeV. It makes use of the best available and selected germanium material (generally of p-type) and has relatively sharp edges, a very thin window that is 100% transparent for energies of 3 keV and up, a low capacitance (and thus very low electronic noise), very high resolutions (low FWHM) at lower energies, large active surfaces (up to 5000 mm^2), and fixed dimensions. The cryostat is equipped with a carbon window that has a transmission of more than 85% at 10 keV and close to 100% for energies higher than 20 keV.

It is interesting to note that detectors with either small p^+-contact (like this BEGe detector) or small n^+-contacts (like the LEGe geometry—see Fig. 4.20) have long been rated as less suitable for physics experiments involving precise timing information (such as time-of-flight experiments) because the time of interaction between the impinging gamma ray on the detector cannot be determined as precisely as with a coaxial-type detector. Indeed, the time to collect the signal generated in such a small-contact detector depends very strongly on the place where the signal is generated. The collection time is very short when the gamma interaction takes place in the vicinity of the central small contact, and can be much longer if it takes place near the rim of the detector. More recently, it has been recognized that this same phenomenon can be used to discriminate "single-site events (SSEs)" from "multisite events (MSEs)" (where the observed signal is due to multiple Compton interactions in the crystal), using pulse shape analysis. Distinction between SSE and MSE can be used to suppress Compton backgrounds and increase the signal-to-noise ratio when searching for full-energy gamma lines.

TABLE 4.8 Performance of a long, 70% n-type detector with rounded edges versus a 50% thin window (XtRa) p-type Ge detector.

	Net count rate (cps)		Background (cps)		FWHM (keV)		Relative MDA	
Energy	70%	50%	70%	50%	70%	50%	70%	50%
(keV)	n-type	XtRa	n-type	XtRa	n-type	XtRa	n-type	XtRa
59	16.3	18.9	2.06	1.81	1.04	0.82	1	0.72
88	16.3	18.6	0.82	0.65				
122	16.2	17.5	1.11	0.77				
165	11.7	13.1	1.08	0.79				
392	6.84	7.42	0.74	0.56				
514	5.45	5.87	0.78	0.34				
662	4.63	4.89	0.61	0.46	1.63	1.34	1	0.75
898	3.24	3.33	0.56	0.45				
1173	2.63	2.66	0.33	0.26				
1333	2.38	2.36	0.16	0.14	2.23	1.81	1	0.85
1836	1.87	1.83	0.11	0.06				

The source is a disk source on the detector window.

This technique is also proposed to look for a very rare SSE, namely, the neutrinoless double beta decay. For that purpose, more "point-contact" (or "small-contact") detector types have been built (Budjas et al., 2009; Agostini et al., 2011; Barbeau et al., 2007). See also Chapters 1, 6, and 9 of Volume 1 for the theory and detection methods applied in the search for neutrinoless double beta decay.

III. Si detectors

A. Si(Li) X-ray detectors

Si(Li) detectors are made by compensating the excess acceptor ions of a p-type crystal with Li donor ions by a process called lithium drifting. The nominal thickness varies between 2 and 5 mm, and the active area between 12 and 80 mm^2 and resolutions between 140 and 190 eV are achieved. Like Ge detectors, they are operated in a liquid nitrogen cryostat. They find their application mainly in X-ray analysis. From this point of view, they should be compared with low-energy or ultralow-energy Ge detectors. The latter have better resolutions for reasons explained in Section I.C. All generalities mentioned in Section II related to Ge detectors can be transposed readily to Si(Li) detectors. As with low-energy Ge detectors, the efficiency for low-energy γ rays is governed by the various entrance windows. The efficiency for high-energy γ rays drops drastically above 20 keV and reaches nearly zero at 100 keV, whereas a low-energy Ge detector still has appreciable efficiency at 1000 keV as illustrated in Fig. 4.16. What seems to be a disadvantage may turn out to be an advantage in many applications. Indeed, the low efficiency for higher energy γ rays reduces not only the FEPs but also the continuous background due to the presence of the source and more particularly to Compton scattering decreasing the lower limit of detection in the X-ray region. Finally, γ spectra or X-ray spectra taken with Si(Li) detectors are less disturbed by the escape of Si X-rays than Ge detectors by the escape of Ge X-rays (see Section II.B.3). Indeed Si K_α X-rays have an energy of only 1.74 keV. The choice between a low-energy Ge detector and a Si(Li) detector is thus governed solely by the projected application. At room temperature, Si(Li) detectors are sometimes used as high-energy particle detectors.

B. Si-charged particle detectors

Silicon-charged particle detectors, such as diffused junction detectors (DJDs) or silicon surface barrier detectors (SSBs), have served the scientific and industrial community for several decades (Knoll, 1999). In the gold-silicon detector, the n-type silicon has a gold surface barrier as the front contact and deposited aluminum at the back of the detector as the ohmic contact. Current applications, however, require detectors having lower noise, better resolution, higher efficiency, greater reliability, more ruggedness, and higher stability than older technologies could produce. Modern ion-implanted detectors such as the passivated implanted planar silicon (PIPS) detectors are now recommended as charged particle detectors. They surpass the older detector types in almost every respect.

Salient advantages of PIPS technology include the following:

- Buried ion-implanted junctions. No epoxy edge sealant is needed or used. This increases the detector stability. Ion implantation ensures thin, abrupt junctions for good α resolution.
- SiO_2 passivation. It allows long-term stability and low leakage currents.
- Low leakage current, typically 1/10 to 1/100 of an SSB (surface barrier detector).
- Low noise.
- Thin windows ($\leq$500 Å equivalent of Si). This results in less straggling in the entrance windows and thus yields better α resolution.
- Ruggedness (cleanable surface).
- Bakeable at high temperatures.
- Long lifetime.

In the detection process, the particle is stopped in the depletion region, forming electron–hole pairs. The energy necessary to form a single electron–hole pair depends on the energy gap ε (Table 4.1) of the detector material, but it is essentially independent of the energy of the incoming particle. Consequently, the number N of electron–hole pairs ultimately formed is directly proportional to the energy of the stopped particle as expressed in Eq. (4.1). This eventually results in a pulse proportional to the energy of the charged particle. The thickness d (Eq. 4.6) of the depletion region depends on the applied bias voltage. Partial or full depletion with or without overvoltage is possible as illustrated in Fig. 4.23. The capacitance in pF for a PIN diode structure is given by

$$C = \frac{1.05A}{d} \tag{4.50}$$

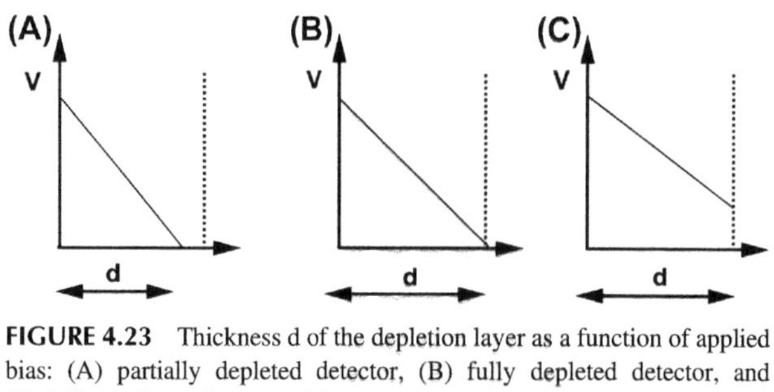

FIGURE 4.23 Thickness d of the depletion layer as a function of applied bias: (A) partially depleted detector, (B) fully depleted detector, and (C) fully depleted detector with overvoltage.

where A represents the surface area of the junction in cm^2 and d its thickness in cm. The surface seen by the charged particles is called the active area of the detector. It is required for the calculation of the efficiency. The junction area is typically 20% larger than the active area.

The noise level of charge-sensitive preamplifiers is usually given by the manufacturer for zero input capacitance. It increases with capacitance (Eq. 4.34), and the rate of increase is also specified by the manufacturer. The detector capacitance is reduced at higher bias voltages as long as the detector is not fully depleted. The lowest noise and best resolution are thus obtained at higher voltages within the recommended range. At voltages above that recommended by the manufacturer, the reverse leakage current is likely to increase, causing excessive noise and loss of resolution.

1. Alpha detectors

Alpha spectroscopy finds applications in widely different disciplines such as follows:

- Radiochemical analysis
- Environmental studies and surveys
- Health physics
- Survey of nuclear sites through the off-line detection of emitted actinides
- Geological and geomorphologic studies (such as U–Th dating)

It requires high resolution, high sensitivity, and low background.

1. High resolution is ensured by the thin entrance window over the detector surface. It reduces energy straggling in the entrance window. Energy straggling is due to the random nature of the interaction of a charged particle with the detector material. This leads to a spread in energy if a beam of charged particles passes through a certain thickness of absorber and, consequently results in an increase of the peak width (Knoll, 1999). Since the alpha radiations generally penetrate the entrance window with variable angles of incidence, alphas will undergo different travel lengths in this layer, leading to variable energy losses and, thus, to a broadening of the peak. A thin window means less straggling and better resolution. Furthermore, the low leakage current ensures a low electronic noise contribution. Both properties together allow high α resolution. Values ≤ 16 keV (FWHM) are routinely achieved for a detector with an active area of 450 mm^2. Note, however, that the obtainable resolution depends not only on the detector but also on external factors such as vacuum and source preparation described later in this chapter. Table 4.9 shows some typical specifications and operating characteristics for modern, ion-implanted α detectors.
2. High sensitivity is enhanced by good resolution, which reduces the background below the peak. A depletion depth of 140 μm is enough to absorb α particles of up to 15 MeV covering the complete range of all α-emitting radionuclides. For larger detector diameters (1200 mm^2), absolute efficiencies $\geq 40\%$ can be achieved. This is illustrated in Fig. 4.24 and discussed in more detail later in this section. Packaging and mounting materials have to be carefully selected to avoid possible contaminants. Low background is further ensured by clean manufacturing and testing procedures. Backgrounds of $\leq$0.05 cts/(h cm^2) in the energy range 3–8 MeV are achieved routinely.

a. Factors influencing resolution and efficiency

Detector–Source Distance. All α particles reaching the active area of the detector will be counted. The counting

TABLE 4.9 Some examples of operating characteristics for α detectors.

Active area (mm^2)	300	450	600	900	1200
Thickness (μm min/max)	150/315	150/315	150/315	150/315	150/315
Recommended bias (V)	+20/80	+20/80	+20/80	+20/80	+20/80
Si resistivity (min Ω-cm)	2000	2000	2000	2000	2000
Operating temperature (min/max)	−20/+40	−20/+40	−20/+40	−20/+40	−20/+40
Leakage current (at 25 °C) (typical/max in nA)[a]	15/70	25/100	30/120	40/200	60/300
α resolution (keV)[b]	17/19	18/20	23/25	25/30	30/37
Absolute efficiency (%)[c]					
At 2 mm spacing	36.8	40.2	41.0	43.6	44.5
At 5 mm spacing	23.5	28.3	31.2	34.8	36.9
At 15 mm spacing	7.3	10.1	12.4	16.1	18.9

FIGURE 4.24 Calculated efficiencies for a 1200 mm^2 (upper curve), a 900 mm^2 (middle curve), and a 300 mm^2 (lower curve) detector as a function of the source–detector distance h; the source diameters were respectively 35, 32, and 15 mm.

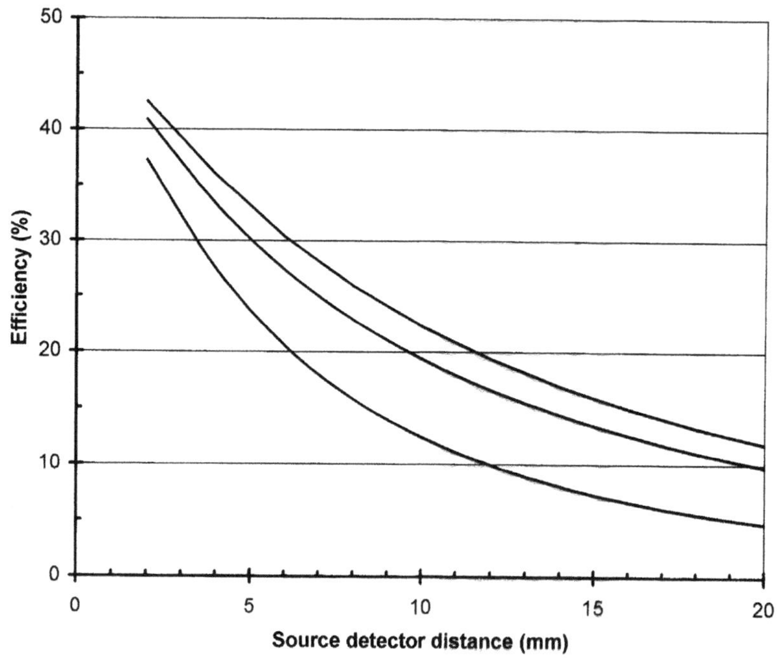

efficiency is thus given by the geometric efficiency, $\eta = \Omega/4\pi$, where Ω is the solid angle under which the detector subtends the source. For the case of a circular detector on axis with a circular isotropic source disk, this solid angle can be computed by Monte Carlo calculations (Williams, 1966; Carchon et al., 1975) and is available in tabulated form (Gardner et al., 1980). Fig. 4.24 gives the calculated efficiencies for 1200-, 900-, and 300-mm^2 detectors as a function of source-to-detector distance. The source diameters are 35, 32, and 15 mm. Actual efficiencies may be slightly different, especially at small source–detector distances, because of factors such as self-absorption in the source. Efficiencies of $\geq$40% are obtainable.

In Table 4.9, alpha resolutions (FWHM) for the 5.486-MeV alpha line of ^{241}Am are given in the case of a detector–source distance of $d = 15$ mm, using standard Canberra electronics. When the source approaches the detector, line broadening is expected, as the mean slope of the α particles entering the detector is increased, resulting in an effectively increased maximal path in the entrance window and subsequent increase of the spread of incident angles and therefore higher energy straggling (Aggarwal et al., 1988). For ion-implanted detectors, this energy straggling is minimized because of the very thin entrance window of 500 Å. For comparison, the entrance window in equivalent Si is ≅ 800 Å for an SSB with a gold window and >2000 Å with an aluminum window. Empirically, it has been proven that for a 300- to 600-mm^2 detector the increase in FWHM stays below 50% for distances as small as 2 mm. Consequently, for a 300-mm^2 detector, the increase of the α resolution at a source–detector distance of 2 mm with respect to that at 15 mm is thus expected to be $\leq 17\times$ 0.50 or $\leq$8.5 keV. This results in an FWHM $\leq$26 keV. The increase in FWHM decreases to 10% at d = 8 mm and is practically negligible for distances >10 mm.

Source radius. It is interesting to take a closer look at the influence of the source diameter on the efficiency. Fig. 4.25 shows the geometric efficiency of a 450-mm^2 and a 1700-mm^2 detector as a function of the source radius for a source to detector distance of 5 mm. One sees immediately that the efficiency of the bigger detector is much greater, whatever source radius is chosen. Note, however, the existence of an inflection point for $R_s = R_d$ as well as the sharp decrease in efficiency beyond this point. R_s and R_d represent the source and detector radii. The diameter of the source should thus never exceed the diameter of the detector. If a uniform specific source activity A_s (Bq/cm^2) is assumed, the total number of counts registered in a time t is proportional not only to the efficiency but also to the total activity of the source deposited on the surface area or, in other words, the efficiency multiplied by A_s. Fig. 4.26 gives this number as a function of the source radius in arbitrary units. Note that when the source radius exceeds that of the detector, the gain in source surface is exactly compensated by the loss in efficiency. The optimum source radius thus equals the radius of the detector. This general rule is independent of the source-to-detector distance.

Source thickness. Sources must be homogeneous and thin to avoid energy straggling due to self-absorption

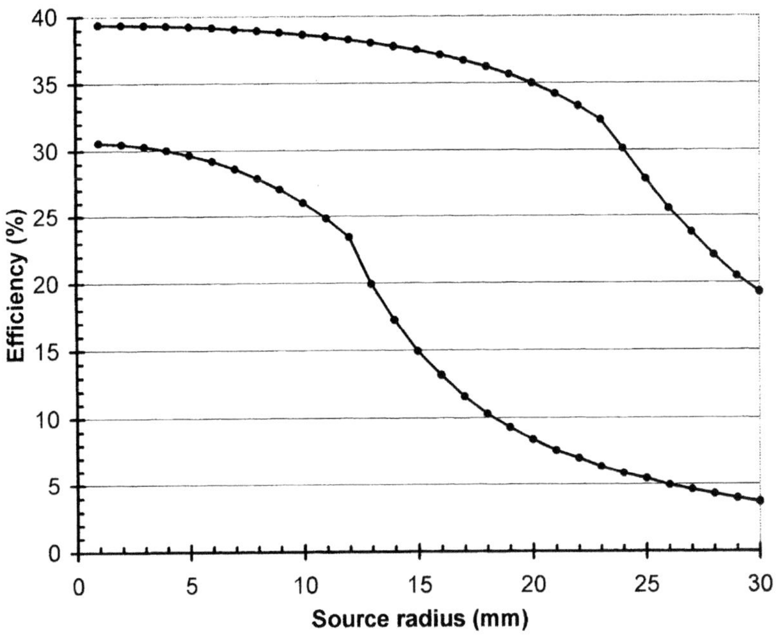

FIGURE 4.25 Geometrical efficiency of a 1700 mm^2 (upper curve) and a 450 mm^2 (lower curve) α detector as a function of the source diameter given in mm for a source detector distance of 5 mm.

(Burger et al., 1985). Self-absorption is proportional to the thickness of the source and inversely proportional to the specific activity. For typical values of specific activities on the order of 100 Bq/cm^2, the self-absorption is generally negligible for carrier-free sources. However, the effect of thickness of the carrier-free source depends on the transition probability of the isotope in question, which increases with increasing half-life. Expressed in energy loss, it is on the order of 0.03 keV for "short"-lived isotopes such as ^{239}Pu ($T_{1/2} = 2.4 \times 10^4$ y) and ^{230}Th ($T_{1/2} = 7.5 \times 10^4$ y), whereas for "long"-lived isotopes such as ^{238}U ($T_{1/2} = 5.7 \times 10^9$ y), it is on the order of 5 keV. Indeed, a 10^5 times smaller transition probability requires the presence of 10^5 times more source material to reach the same activity (see Chapter 1). When estimating the source thickness of a non–carrier-free source, all isotopes deposited together with the isotope of interest must be considered. This can be due either to a different isotope of the same element or to the simultaneous deposition of other elements during source preparation.

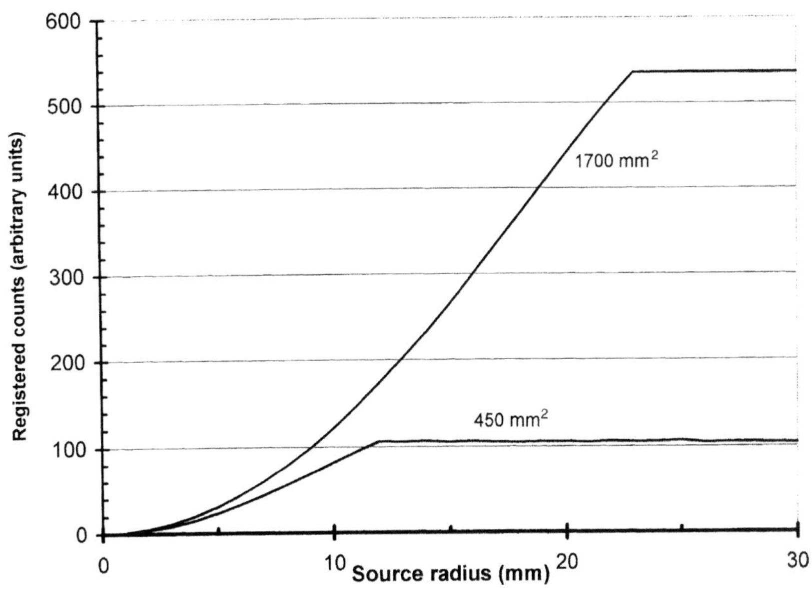

FIGURE 4.26 Number of counts registered during a certain time Δt (arbitrary units) for a 1700 mm^2 and a 450 mm^2 detector, as function of the source radius in mm.

Problems can also arise with very intense sources, as the source thickness and, therefore, the self-absorption is proportional to the total source activity. For a given total activity, the specific activity can be reduced by choosing a larger source diameter. In this case, preference should be given to a detector with a diameter about equal to that of the source to increase the efficiency (Fig. 4.26) and to reduce the energy straggling, as relatively fewer α particles will strike the detector at an acute angle.

b. Factors influencing contamination and stability

Oil contamination. Alpha sources have to be placed together with the detector in a vacuum chamber to avoid any energy loss in the air gap. Typical α spectroscopy systems use a rotary vacuum pump to evacuate the α spectrometer(s). When static conditions are established in the vacuum system (the ultimate pressure has been reached) and there is no substantial gas flow toward the pump, oil particles can back-stream toward the spectrometer and deposit on the detector and the source surfaces. The same can happen in a more dramatic fashion if the pump is disabled and the spectrometer draws air backward toward the manifold connecting the two. For this reason, a dry pump or a back-streaming filter is recommended between the pump and the detector source vacuum chamber to prevent oil contamination.

Particulate and recoil contamination. Contamination of detectors can take place when particles from sources gravitate to the detector surface and stick there or are splattered, sputtered, or splashed onto the detector surface by the recoil energy imparted to the nucleus of an α-emitting atom. In the latter case, the energy of the particles may be sufficient to implant themselves in the detector so that they cannot be removed nondestructively. Much of the casual contamination can be removed from PIPS detectors by cleaning with a cotton ball saturated with isopropanol. Vigorous scrubbing will not harm the PIPS detector. Recoil contamination is almost never 100% removable. It is best avoided by careful sample preparation, avoiding hot samples, or using the techniques reported by Sill and Olson (1970), which involve operating the spectrometer with an air barrier and/or a bias voltage between the detector and source. They show that recoil contamination can be reduced by a factor of up to 1000 if an air layer of about 12 mg/cm^2 exists between the detector and source and if the source is negatively biased by a few volts. By straggling, the air gap will increase the FWHM of α peaks by a few keV, which is probably acceptable in all but the most demanding of applications.

c. Stability of the detection system

Both long-term stability and temperature stability are important in detectors used for α spectroscopy because count times are often many hours or days and gain shifts during data accumulation lead to erroneous or unusable spectra.

Long-term stability. Long-term stability is affected by the impact of the environment on the detector junctions. SSB detectors sometimes fail with prolonged exposure to room atmosphere and at other times fail when operated for prolonged periods under high vacuum. This instability is caused by the epoxy edge encapsulation that is required for this type of detector. The PIPS detector has junctions that are buried in the silicon bulk, and no epoxy encapsulation is needed or used, which ensures intrinsic long-term stability.

Temperature stability. The leakage current of silicon diodes doubles for every 5.5–7.5°C change in ambient temperature. Since the preamplifier HV bias resistor is a noise contributor, it is necessarily of high value, typically 100 MΩ. With an SSB detector having a leakage current of 0.5 μA, the change in bias voltage at the detector for a 2°C change in ambient temperature can be as much as 13 V. This is enough bias change to affect the overall gain of the preamplifier by a substantial amount. Modern PIPS detectors have a typical leakage current of less than one-tenth that of SSB detectors or DJD. Consequently, system gain change as a function of temperature is proportionally less, so that for operational temperatures of up to 35°C, no significant peak shifts are observed.

d. The minimum detectable activity

The MDA at the 95% confidence level is given by

$$MDA = \frac{2.71 + 4.65\sqrt{b}}{t\eta P} \tag{4.51}$$

where t is the counting time, η the counting efficiency, P the yield of the α measured, and b the background counts. The numerator on the right-hand side of Eq. (4.51) corresponds to the detection limit at the 95% confidence limit based on the Currie formalism (Currie 1968)

The two detector-bound parameters, background (B) and efficiency (η), are particularly favorable in the case of an α-PIPS detector. For a 450-mm^2 detector ($\eta = 0.40$, $b = 6$ counts/d) and for an overnight run ($t = 15$ hours $= 54{,}000$ seconds), one has thus $MDA = 0.54$ mBq if a 100% yield for the α ray is assumed, as well as the worst-case condition that all background counts are in the peak or region of interest. The limiting factor is often not the absolute MDA expressed in Bq, but rather the specific minimum detectable activity (SMDA) expressed in Bq/cm^2:

$$SMDA = \frac{MDA}{S_s} \tag{4.52}$$

where S_s represents the area of the source in cm^2.

The background in practical applications is often compromised by the presence of higher-energy α lines that produce counts in the spectrum at lower energies. PIPS detectors are notably free of these tailing effects in comparison with SSB detectors of equivalent efficiency, in part because of their thin entrance window. Comparisons between the two types of detectors have shown a difference of as much as a factor of 3 in this background tailing or continuum. This translates into an improvement in MDA by a factor $\sqrt{3}$.

2. The Si drift detector technology

For applications where low energies must be measured, such as electron spectroscopy and low-energy X-ray spectroscopy, and for the read-out of scintillators, the electronic noise of the Si detector becomes more important. The drift detector geometry is very useful here because the capacitance of this detector is much lower than with conventional Si detector structures.

Silicon drift detectors are multielectrode photodiodes realized with double-sided technology. These photodiodes were introduced by Gatti and Rehak (Castoldi et al., 2000). The main differentiator in comparison with the well-known PIN technology is that the collection of the charge in the SDD is realized onto a very small readout anode (∅ 100 μm) using a lateral drift field within the detector. Radiations (such as X-rays or visual photons) absorbed in the detector generate electron–hole pairs. The path of electrons migrating toward the anode is shown in Fig. 4.27.

The complexity of the structure depicted (a double-sided multielectrode diode) is justified by the drastic reduction of capacitance of this device. In a PIN diode of similar size, this capacitance will typically be 90 pF, whereas with SDDs, the capacitance ranges from 0.1 pF for a traditional SDD to 3 pF for a linear anode segmented silicon drift detector (LA-SSDD—see further). This fact leads to detector benefits that explain the popularity of this detector type for X-ray spectroscopy. Indeed, low capacitance diodes create less noise and thus better energy resolution. They can be operated at lower processing times, which allows higher counting rates, without resolution loss. An FWHM of 130 eV at the 5.9 keV K_α X-ray line of ^{55}Mn has been measured.

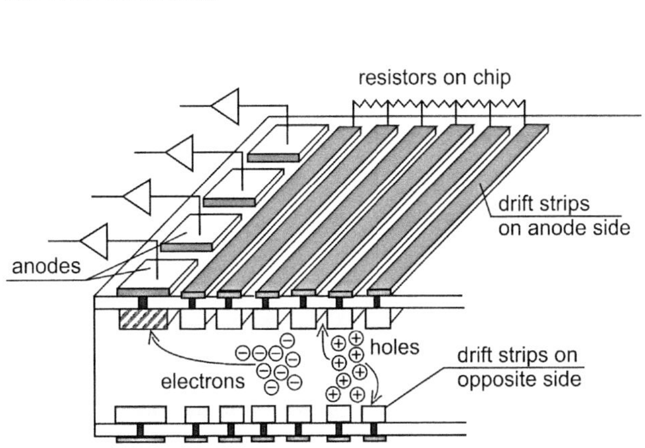

FIGURE 4.27 The Si drift detector.

3. Electron spectroscopy and β counting

Traditional PIPS detectors and SDDs can be used for electron spectroscopy and β counting. The thin entrance window of these detectors provides little attenuation even for weak beta particles. In the β ray and (conversion) electron energy region (<2000 keV), the resolution (FWHM) is approximated by the pulser line width. Due to the limited thickness of these detectors (0.3–0.5 mm), the absorption of X-rays is only efficient in the 0–30 keV range. In such applications, the SDD or PIN diode is directly absorbing X-rays.

Canberra provides special β PIPS detectors fabricated from higher ohmic material, having a minimum thickness of 475 μm and allowing full absorption of electrons of up to 400 keV. Note, however, that higher-energy electrons can be fully absorbed too. This is due to the fact that high-energy electrons do not follow a straight path inside the detector but rather change direction, so the real path of the electrons inside the detector is much greater than the detector thickness *d*. For example, the conversion electrons of the 661-keV γ line of ^{137}Cs at 625.8 and 629.7 keV are clearly seen. If only β counting is needed, the efficiency is uncompromised as long as the detector absorbs enough energy from the β ray to exceed the noise level.

As in the case of α-spectroscopy, the main factor influencing detector efficiency for electron spectroscopy (e.g., spectroscopy of low-energy conversion electrons) is governed by the geometric efficiency η. Note, however, that in the calculations the junction area and not the active area has to be taken into account, as the detector mount is partially transparent for electrons. Furthermore, backscattering of low-energy electrons from the detector surface may cause significant loss of efficiency. By analogy with the experimental values of the fraction of normally incident electrons backscattered from thick slabs of aluminum (see, e.g., Knoll, 1999), it can be inferred that between 10% and 13% of the electrons of energies between 50 and 700 keV are backscattered by thick slabs of Si, and the backscattered fraction drops sharply for higher energies. On the other hand, if backscattering occurs in the source, it may increase the apparent number of β particles, as electrons emitted outside the solid angle, sustained by the detector, can be scattered inside this solid angle. Efficiency calibration for electron spectroscopy must be done, therefore, with multienergy standards, prepared in the same way as the unknowns. Source backings should be of low-Z materials to minimize source backscattering effects. Conversion electrons show up most clearly if they are not in coincidence with β particles. This is the case if the decay takes place through an isomeric level of the daughter such as in the decay of ^{137}Cs (Fig. 4.7) or if it takes place by almost pure electron capture (such as with ^{207}Bi, often used as standard). If the conversion electrons are in coincidence with the β particles, they can sum up with the nuclear electrons (Eq. 4.11). The resulting sum peak will be continuously distributed as the β particles. If β activities have to be measured, these conversion electrons can furnish supplementary counts. This is the case, for example, if the sum peak surpasses the energy of the β threshold. This effect is isotope specific. The β threshold is not given by the thickness of the entrance window, which is negligible for all practical cases, but by the noise of the detector and electronics. In practice, a value of three times the electronic noise (FWHM) is taken. As β rays have a continuous energy distribution, this effect has to be taken into account when source intensities have to be measured. Indeed, part of the emitted electrons can lie under the threshold. This depends essentially on the form of the β spectrum and has to be considered individually.

Beta particles can "channel" between crystal planes of the detector and lose energy at a lesser rate than if they cross planes. To minimize this effect, β-PIPS detectors [(as well as continuous air monitoring (CAM)] are made from silicon wafers that are cutoff axis. Small errors in calculated efficiency, however, remain possible.

Finally, it has to be noted that β detectors with an active thickness of 475 μm have small sensitivity for γ rays. Indeed, from Fig. 5.18, it follows that the half-thickness of Si for the total absorption of γ rays of 50 and 100 keV is 0.631 and 1.63 cm, respectively, so that for these energies 5.72% or 2.00% of all γ rays falling on the detector undergo an interaction. This can lead to a supplementary pulse or a sum pulse (Eq. 4.11).

4. Readout of scintillators

SDD diodes can also be used as readout electrode for scintillators (Fiorini et al., 2006) and contribute to the detection of gamma rays of much higher energies (60 keV−3 MeV range). As shown in Fig. 4.28, the gamma ray is converted into tens of thousands of visual photons reaching the SDD in a continuous distribution from the center to the border of the SDD. The upper limit of the number of scintillated photons generated by a gamma photon is the ratio of the gamma ray energy and the product of the beta factor times the bandgap of the scintillator:

$$N_{photons} = \frac{E_y}{\beta \cdot E_g} \tag{4.53}$$

where E_γ is the gamma photon energy, E_g is the bandgap of the scintillating material, β is the beta factor having a value of 2.5 eV.

Table 4.10 lists the bandgap and the number of photons generated by a 1 MeV photon for three of the most luminous scintillators.

The very broad distribution of visual photons leads to the generation of electron−hole pairs across the whole radius of the SDD as illustrated in Fig. 4.28. This situation is therefore different than in the case of direct X-ray absorption. Directly absorbed X-rays lead to the generation of electron−hole pairs in a single point, which all arrive with a quasisimultaneity to the anode. In the case of scintillated

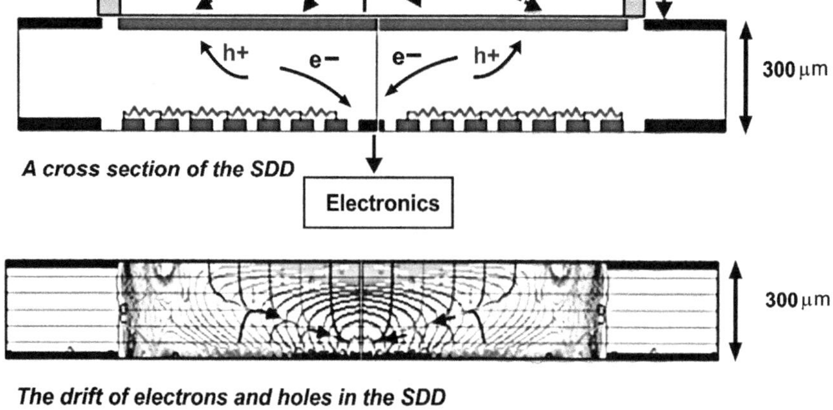

FIGURE 4.28 The conversion of a gamma ray into visual photons and the generation of electron−hole pairs in the SDD detector. For the color version of the figure, the reader is referred to the electronic version of the book.

TABLE 4.10 Three of the most luminous scintillators.

	Bandgap E_g	Number of photons scintillated by a 1 MeV photon
LuI_3	4.2 eV	100,000
SrI_2	4.8 eV	90,000
$LaBr_3$	5.3 eV	60,000

events, a continuous signal will be generated, its width being the longest drift time.

The key detector parameters to reach optimum resolutions are a good quantum efficiency, a low leakage current density, small drift lengths, and a small detector capacitance.

While the low capacitance of the diodes allows SDD X-ray detectors to be used at low shaping times, the peaking times used for scintillation applications must be higher than the largest drift time seen by electrons generated at the border of the SDD.

To collect the complete charge, the shaping time of the shaper amplifier following the charge-sensitive amplifier needs to be higher than the pulse signal width and thus the longest electron drift time. The use of lower shaping times will lead to incomplete charge collection.

SDDs are attractive at low shaping times because of their small series noise, but are not as good as PIN diodes at high shaping times, because of their higher leakage current density. So, SDDs having a large radius are not attractive for scintillation applications because of the need of using long shaping times.

LA-SSDDs have been developed and optimized for scintillation applications to overcome this problem (Morichi et al., 2011). These detectors are circular shaped and have areas of 250 and 300 mm^2 and a thickness of 300 μm. The drift length of electrons is minimized by the implementation of a series of concentric ring−shaped anodes. This permits minimization of the number of anodes while keeping very small drift distances, thus allowing the use of short shaping times.

5. Continuous air monitoring

The increasing demand for safety of nuclear installations calls for continuous survey of airborne radioactive particles inside and around nuclear sites, and the potential for nuclear accidents calls for a worldwide survey of the atmosphere. In particular, it is important to know whether, instantaneously or over a certain time, β and/or α activities remain below imposed limits. For a judicious choice of a continuous air monitoring system, the influence of the detector on the system performance should be understood.

Airborne radioactive particle concentration limits are expressed in derived air concentration (DAC) units and are isotope specific. One DAC corresponds to an isotope concentration of 1 Bq/m^3. For certain α emitters, these limits are extremely low. For example, for ^{239}Pu in soluble form, the DAC limit corresponds to a value of 0.08 Bq/m^3. The exposure is expressed in DAC hours, that is, the concentration in Bq/m^3 multiplied by the exposure time in hours. To detect these activities, air is pumped through a filter at a speed of about 1 m^3/h. A detector continuously measures the accumulated activity. An instrument should be able to detect an activity concentration of 8 DAC-hours, that is, 1 DAC in 8 hours, 2 DAC in 4 hours, and so on.

This requirement is further complicated by the fact that the α background varies due to simultaneous collection and counting of the α activity from ^{222}Rn progeny, which can be significantly higher than the desired MDA. The β background also varies but, unlike the cause of the α background, this is mainly due to cosmic events.

For off-line measurements of filter samples, standard α or β detectors can be used under certain conditions. Online measurements, however, require special characteristics, in particular, light-tightness, moisture resistance, and corrosion protection. Fig. 4.29 shows an exploded view of a Canberra CAM PIPS detector. Depletion layers between 120 and 325 μm are possible. Their main characteristics are as follows:

1. Operable in light to 5000 lumens
2. Corrosion-resistant varnish coated
3. Moisture-resistant varnish coated
4. Low bias voltage (10−90 V)
5. β and α discriminated by energy
6. Wide temperature range and low leakage current
7. High β sensitivity, 300 μm active thickness

a. Light-tightness and resistance to harmful environments

Silicon detectors are fundamentally light sensitive. In continuous air monitoring, the detector is not protected by a vacuum chamber, and light may reach the detector in some cases. CAM PIPS detectors are made with a front surface coating of 0.5-μm-thick aluminum, which blocks the light. Furthermore, because of the nature of continuous air

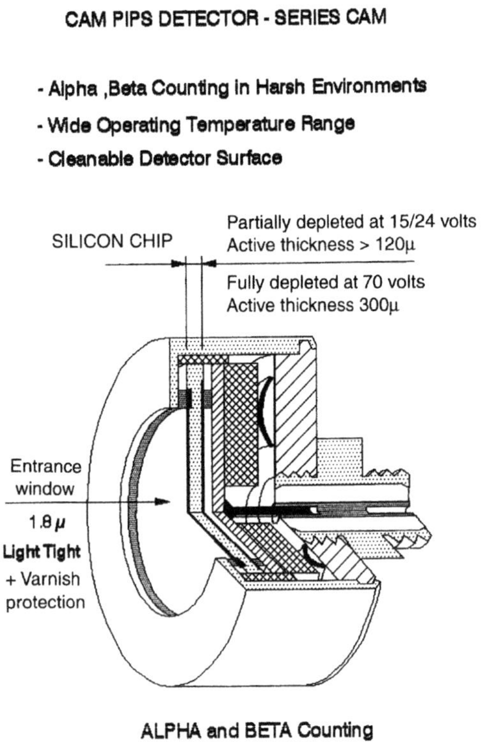

FIGURE 4.29 Exploded view of a Canberra CAM detector. *Image courtesy of Mirion Technologies.*

monitoring, detectors are often used in a harmful environment, such as a humid and/or dusty atmosphere charged with corrosive gases. To extend the useable lifetime of the detectors, CAM detectors are covered with a 1 μm varnish coating, providing mechanical and chemical resistance against abrasion, solvents, and corrosion. This varnish corresponds to a supplementary absorption layer of about 0.6 μm silicon equivalent. In vacuum, these supplementary windows cause roughly a doubling of the α resolution. However, one has to take into account the energy straggling in the air gap between filter and detector and in the filter itself, which makes straggling in the entrance window relatively unimportant. This is illustrated by Fig. 4.30, showing the empirical resolution (FWHM) of a CAM450 and a CAM1700 detector for the 5499.2-keV α line of ^{238}Pu as a function of the source–detector distance. The FWHM decreases with the distance, contrary to the situation for a detector in a vacuum. Note the quasilinear increase of the FWHM for distances of up to 10 mm. This degraded resolution is, however, still good enough to separate completely the α and β activity as illustrated in Fig. 4.31, showing the β spectrum of ^{137}Cs (decay scheme illustrated in Fig. 4.7) in the presence of a ^{239}Pu alpha source taken with a CAM1700 detector with a source–detector distance of 5.3 mm.

b. Efficiency

In normal continuous air measurements, no efficiency loss is expected due to the air gap between the source and the detector. Indeed, the range in air of typical α particles of about 5 MeV is several centimeters, and the air gap is normally <1 cm. All earlier remarks on the efficiency remain valid, in particular that the optimum source diameter equals the detector diameter. The advantage of a big detector and a large source radius is evident, as the resolution is dominated by the air gap and not by the detector radius as in normal α spectroscopy. However, the source diameter should never exceed the detector diameter as seen in

FIGURE 4.30 Empirical resolution (FWHM) of a CAM450 and CAM1700 detector for the 5499.2 keV α line of ^{238}Pu as a function of the source–detector distance.

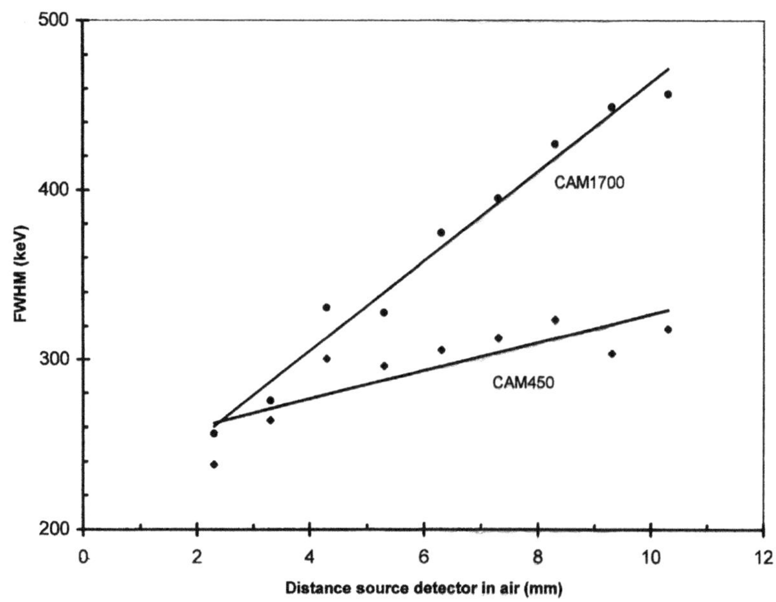

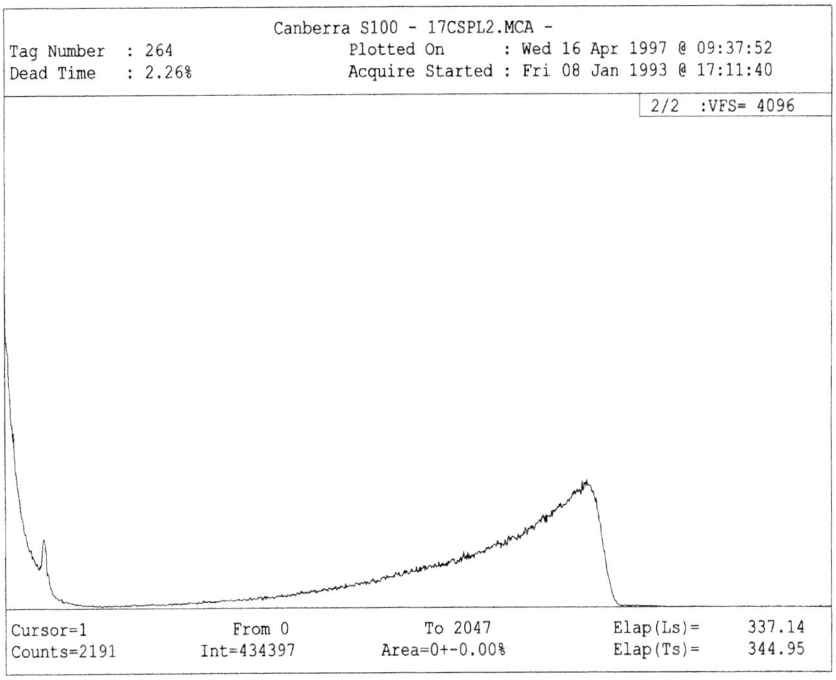

FIGURE 4.31 Beta spectrum of ^{137}Cs in the presence of an alpha source of ^{239}Pu, taken in air with a CAM1700 detector.

Fig. 4.25. Furthermore, the total activity deposited on the filter depends on the pumping speed, which in turn is limited by the pressure drop through the filter. The pressure drop increases linearly with the pumping speed. For a given throughput, the required pumping speed decreases with the square of the filter diameter. A large detector, therefore, permits the use of a large filter and, as a consequence, higher air flow for the same pressure drop, permitting larger total activities to be deposited on the filter in less time.

c. Background and minimum detectable activity problems in continuous air monitoring

In the case of continuous air monitoring, Eq. (4.51) can be written in the form:

$$MDA = \frac{2.71 + 65\sigma_b}{t * \eta * 3600} \tag{4.54}$$

where t is the pumping and measuring time expressed in hours, σ_b the standard deviation of the background, and η the fractional counting efficiency. Besides the measuring time t, the most important parameter is the standard deviation of the background, which is quite different in the α and β regions. The MDA, therefore, must be examined separately for α and β emitters.

For α emitters, the background b is no longer given by the proper background of the detector but rather by the activity of the ^{222}Rn progeny accumulated simultaneously on the filter, which can be higher than the α activity of concern. Whether or not the air in the laboratory is filtered, values of 4—40 Bq/m^3 can be regarded as quite normal, and DAC values of 0.08 Bq/m^3 have to be detected for soluble ^{239}Pu. Furthermore, the concentration of the Rn progeny in air varies with time. Therefore, the standard deviation σ_b is determined not only by the square root of the registered number of background counts but also by the concentration fluctuations. Indeed, all α lines due to ^{222}Rn and its progeny lie above the α energies of ^{239}Pu. Consequently, due to tailing effects, these peaks contribute to the background beneath the ^{239}Pu peaks. The energy discrimination shown in Fig. 4.30 is good enough to ensure complete α and β separation despite the tailing effects inherent in continuous air measurements.

If a counting efficiency of $\eta = 40\%$, a pumping speed of 1 m^3/h, a pumping time of 8 hours, and a constant background of 40 Bq/m^3 are assumed, a total number of $0.5 \times 8 \times 3600 \times 40$ disintegrations occur due to the background accumulated on the filter. This leads to an MDA (Eq. 4.51) of 0.3 Bq in 4 m^3 of air or 0.08 Bq/m^3. Up to four times better results can be obtained by using background subtraction based on stripping methods, that is, by subtracting the independently determined contribution of higher-energy background peaks under the peak of interest.

The background in the β region (2.1 ct/min cm^2) is largely of cosmic and γ ray origin. Let us assume that a 450-mm^2 detector is used close to a filter of almost equal size. The background in an 8-hour run is thus $8 \times 60 \times 2.1 \times 5.5 = 4536$ counts. Assuming an 8-hour accumulation on the filter, this leads to an MDA of 0.03 Bq for the mean 4 m^3 counted during 8 hours or to 0.01 Bq/m. We must emphasize that the actual MDA will depend on the experimental setup.

IV. Cadmium zinc telluride detectors

A. Characteristics of cadmium zinc telluride detectors

In recent years, cadmium zinc telluride (CZT) semiconductor detectors are being increasingly used in applications involving X-ray and gamma ray detection. CZT has a "tunable" bandgap from 1.5 to 2.2 eV, depending on the concentration, x, of zinc present in the $Cd_{1-x}Zn_xTe$ a relatively high bandgap energy that makes room temperature operation possible. The effective Z of CZT is relatively high, which results in a high stopping power for photons of energies of several hundred keV. CZT has many of the desirable properties for detector applications, such as a high resistivity ($\sim 10^{11}$ Ω-cm) for low leakage current operation and absence of significant polarization effects. However, until the last 10 years or so, limitations due to charge transport properties (poor hole mobility) and availability of large volume crystals have restricted the use of CZT detectors to photon energies in the 10–200 keV energy range. The typical thickness of CZT for detecting energies less than 200 keV is around 6 mm. While this is adequate for applications such as X-ray astronomy or gamma imaging in nuclear medicine, gamma spectrometry applications involving the detection of higher energy gamma rays (200–20,000 keV) such as those in homeland security require the use of thicker CZT crystals. Advances in crystal growth techniques and the development of methods to alleviate limitations due to poor hole mobility have resulted in CZT crystal sizes of several cubic centimeters. CZT detectors with energy resolutions on the order of 1% at 662 keV are commercially available. The widespread use of CZT detectors (along with $LaBr_3$ scintillation detectors) has resulted in the emergence of a class of instruments known as medium-resolution gamma spectrometers (MRGS). The MRGS instruments fill the gap in the application space between low-energy resolution detectors such as NaI(Tl) and high-energy resolution detectors such as HPGe. The material properties of CZT are presented in Table 4.11.

B. Crystal growth techniques

Detector-grade CZT is typically grown by melt growth techniques or solution growth techniques. Melt growth techniques involve melting the material and then crystallizing the material using changing pressure, temperature, or a combination of the two. CdTe and CZT single crystals have historically been grown using the Bridgman method and the traveling heater method (Del Sordo et al., 2009).

1. The Bridgman process

The Bridgman technique uses a "hot" and "cold" zone to create a temperature difference within a furnace as shown in Fig. 4.32 (Dutta, 2010, Dhanraj et al., 2010). The hot zone is kept at a temperature above the melting point of the material. The precursor material used to create the product material is sealed within a crucible or an ampoule and placed in the furnace along with a seed material. The seed crystal is of a known orientation and is placed at the base of the crucible or the ampoule used for crystal growth and is used to facilitate crystal growth in the same orientation as

TABLE 4.11 Physical properties of CdZnTe at 25°C.[a,b]

CdZnTe material property	Specification
Growth method	HPB, THM[a]
Bandgap energy	Tunable from 1.5 to 2.2 eV at 300 K[b]
Atomic number	Cd = 48, Zn = 30, and Te = 52[a,b]
Density	5.78 g/cm^3[a]
Electrical resistivity	10^{10} Ω-cm[a], 10^{11} Ω-cm[b]
Electron mobility × lifetime	$\mu_e\tau = 10^{-2}$ to 10^{-3} cm^2/V[a,b]
Electron mobility	≥1000 cm^2/V·s[b]
Pair creation energy (ionization potential)	4.6 eV[a]
Poor hole transport	$\mu_h\tau_h = 10^{-5}$ cm^2/V[a,b], μ_h = 50 V s[b]
Operational temperature	Room temperature[b]

[a]*From Del Sordo et al. (2009).*
[b]*From Krishna (2013).*

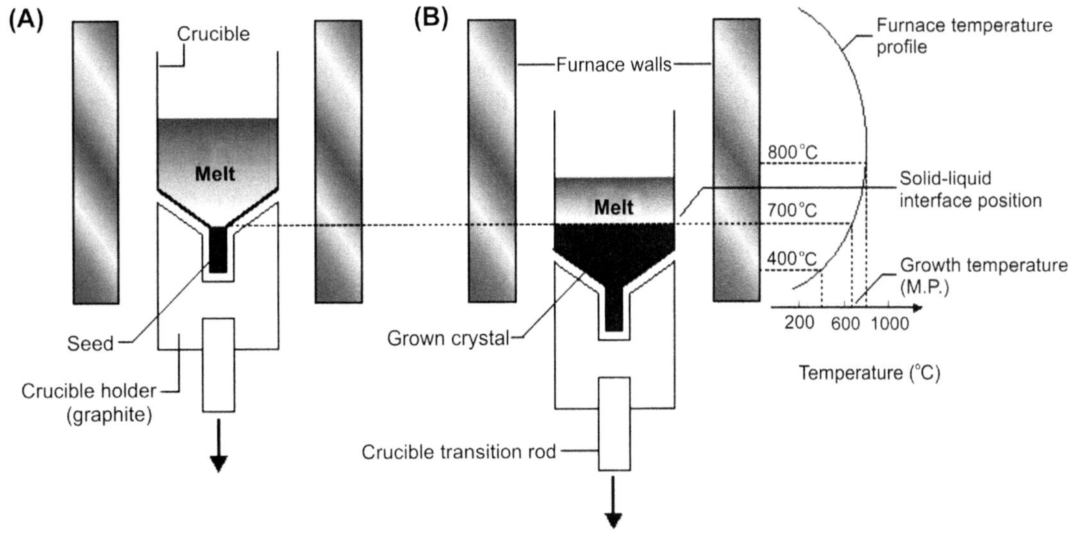

FIGURE 4.32 Schematic diagram of a vertical Bridgman crystal growth process in a single-zone furnace: (A) in the hot zone at the beginning of the experiment, and (B) while moving through the cold zone with partially grown crystal. *From Dutta (2010); reprinted with permission from Springer Nature © 2010.*

the seed crystal. The precursor material is then melted in the hot zone and translated into the cold zone by motion. The material solidifies as it moves through the temperature gradient in the furnace. Bridgman method can be performed in the vertical or horizontal configurations.

Two types of methods have been employed for growing CZT crystals: high-pressure Bridgman (HPB) and low-pressure Bridgman (LPB) methods. In both HPB and LPB processes, CZT is grown using a stoichiometry of $Cd_{0.9}Zn_{0.1}Te$ with a hot zone temperature of 1100°C (Schlesinger et al., 2001). Quartz and graphite are commonly used materials for crucibles due to their high temperature and pressure-resistant properties. Typical growth rates are in the 1 mm/h with a temperature gradient of about 1.5°C/cm (Schlesinger et al., 2001). HPB methods typically employ 5 MPa of argon pressure (Kolesnikov et al., 1997), whereas LPB methods evacuate the crucible/ampoule to an ultrahigh vacuum (approximately 10^{-6} torr). CZT crystals grown using either HPB or LPB methods suffer from a large amount of defects due to impurities introduced because of the high growth temperature. CZT crystals grown using the HPB method are inhomogeneous on the scale of a few centimeters, due to macro defects such as small grain sizes. Also, there are significant variations in the concentration of zinc present from one end of the ingot to another (Del Sordo et al., 2009). These issues limit the size of the CZT crystals that can be grown; HPB results in 25% of an ingot that yields sizable single crystals and only 10% of the ingot yielding detector-grade CZT (Schlesinger et al., 2001). The LPB method requires a simpler furnace setup than the HPB, but the yields are similar to the HPB method.

2. *Traveling heater method*

The traveling heater method (THM) requires that the material to be grown be dissolved in a compatible solution. In this method, a molten zone of material slowly migrates through a solid, homogeneous source material. Matter transport occurs through convection and diffusion across the solvent zone due to the temperature gradient (Grille. 1991). This method also purifies the source materials due to the zone melting nature of this technique. CZT growth using the THM involves placing the presynthesized $Cd_{0.9}Zn_{0.1}Te$, the dopant In, and additional tellurium as the solvent material within a quartz ampoule (Roy et al., 2009). The ampoules are sealed under ultrahigh vacuum and placed in a furnace with a hot zone. Before the THM is initiated, another furnace heats and evenly distributes the material to ensure all of the material is dissolved in Te. The material is then placed in the THM furnace, and multiple passes of the heater moving through the material collect all of the excess Te to one end of the ingot. The THM has several advantages over the Bridgman method. THM combines the low growth temperatures of a solution growth method with the steady-state conditions present between the dissolution of the source material before the heater and the crystal growth section after the heater. This ensures that Zn does not segregate and is uniform throughout the ingot (Roy et al., 2010). However, THM requires a homogeneous ingot be used.

C. Correction schemes to mitigate poor hole mobility in cadmium zinc telluride detectors

Poor hole mobility results in significant tailing on the low-energy side of the gamma ray peaks recorded using CZT detectors. Correction schemes have been developed to mitigate the impact of poor hole mobility and improve the energy resolution of CZT detectors. Two such schemes, the Frisch grid and coplanar grid (CPG) electrode configurations are discussed in the following two subsections.

1. Frisch grid device configuration

A good way to address the hole trapping issue in room temperature semiconductor detectors is to utilize single-polarity charge sensing. This term is derived from the manner that only one polarity of charge (electrons) is sensed to infer the energy deposited by the incident gamma ray. This is advantageous because the effect of poor hole mobility can be mitigated, since the resulting signal will only be sensitive to the fast-moving electrons. This method was first carried out by Frisch in 1944 for gas ionization chambers (Frisch 1944). Frisch employed a gridded electrode, now known as a Frisch grid, placed inside the ionization chamber surrounding the anode. The Frisch grid acts as an electrostatic shield to the region between the cathode and the grid, while making the region between the grid and the anode highly sensitive to the movement of charge. Therefore, events that occur between the cathode and the grid, which constitutes most of the detector volume, will result in the fast-moving electrons passing through the grid and inducing all of their charge on the anode, whereas the slow-moving positive charges will travel in the opposite direction and induce no charge on the anode.

In the recent years, many improved Frisch grid designs have been developed, including the so-called virtual Frisch grid. A schematic drawing of the virtual Frisch grid is given in Fig. 4.33. This is from the report by Cui et al. (2009) from Brookhaven National Laboratory (BNL).

The design includes the capture device (CZT) and the Frisch ring (or capacitive Frisch grid) detector. The shielding electrode (Frisch ring), separated from the cathode by a layer of insulator and grounded in the final assembly, overlaps the cathode and extends up to the far end near the anode. This shielding, together with the cathode cap, ensures that the side surfaces of the crystals are shielded properly. When the cathode is biased negatively, this configuration (cathode, anode, and shielding) forms a focusing electrical field in the crystal's bulk.

The focusing electrical field defines an active region in which the device's response changes only a little (i.e., <5%) from its maximum. The boundary of this region is defined as the virtual Frisch grid; since the cathode is isolated from the shielding on the side surfaces, the signal readout from cathode also can serve as a depth-sensing technique to correct the charge loss caused by electron trapping and further improve the spectral response.

Frisch grid detectors offer many advantages such as ease of device fabrication (and hence high yields) and a low power consumption compared with pixelated three-dimensional CZT detectors. The low power requirements enable the implementation of Frisch grid CZT detectors in portable device configurations. Energy resolution of less than 1.4% at 662 keV can be achieved by Frisch grid detectors.

2. Coplanar grid device configuration

The coplanar grid (CPG) is another single-polarity charge-sensing device configuration used in CZT. The CPG consists of a series of narrow strips, which are connected together in an alternate fashion producing two sets of interdigitated grid electrodes (Luke 1994). By biasing these grid electrodes properly and taking the difference between the two grid signals, one can achieve a resulting signal that is only sensitive to the movement of electrons. Hence, the CPG structure achieves single polarity charge sensing much like the Frisch grid. The primary advantage of this electrode design is that it can be constructed on the surface of the crystal.

CPG detectors use a subtractive method to compensate for tailing due to hole trapping. The anode electrodes take the form of interdigitated grids that are connected to separate charge-sensitive preamplifiers (Figure XX). An electric field is established in the detector bulk by applying bias to the cathode, which is a full-area contact located on the side opposite the grid electrodes. The two grid preamplifiers are connected to a subtraction circuit to produce a difference signal. Bias is applied between the grid electrodes so that one of the grids preferentially collects charge. Charge motion within the detector is sensed equally by the grid electrodes. Consequently, the difference signal is insensitive to charge motion within the bulk of the detector. When charge approaches the anode, the grid signals begin to differ and a signal is registered at the output of the difference circuit. In a well-designed detector and in the absence of electron trapping, the magnitude of the difference signal is the same no matter where the charge is generated in the device. This results in a large improvement in performance for gamma ray spectroscopy when compared with conventional planar device technology (Prettyman et al., 1999) (Fig. 4.34).

V. Spectroscopic analyses with semiconductor detectors

Semiconductor detectors (e.g., SSB, PIPS, Ge(Li), HPGe, and Si(Li) detectors) are typically operated in a pulse mode, and the pulse amplitude is taken as a measure of the energy deposited in the detector. Typically, the pulse amplitude data are presented as a differential pulse height spectrum. Because of real-world effects (such as electronic noise and the various interactions that can occur within the sample, between the sample and the detector, and within the detector itself), the peaks that result in alpha spectra have a

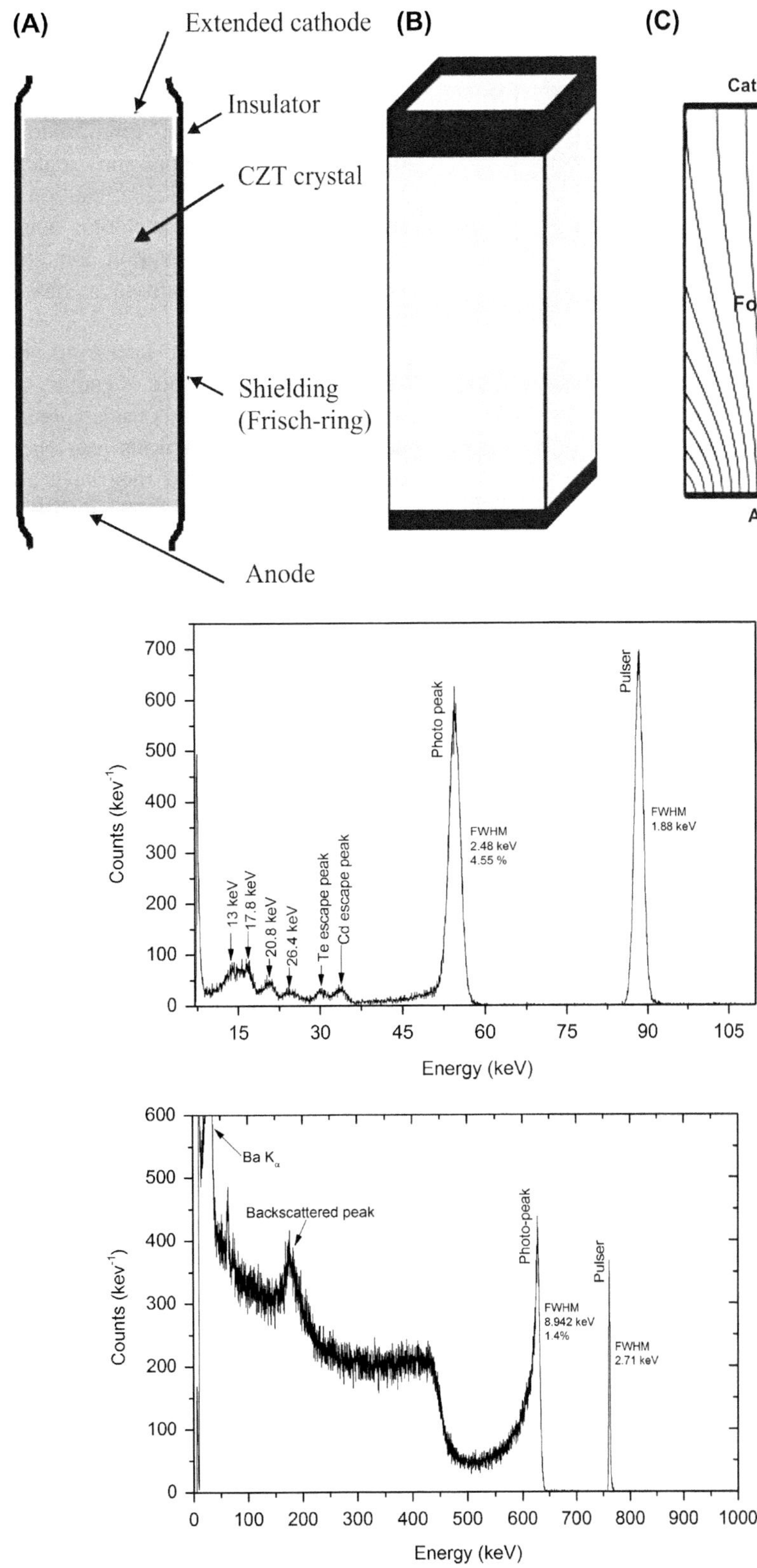

FIGURE 4.33 (Top) Schematic of virtual Frisch grid CZT detector; from left to right, (A) cross-sectional view, (B) three-dimensional view, and (C) focusing electrical field. (Middle and bottom) ^{241}Am spectrum obtained using the virtual Frisch grid detector at an applied bias of 600 V, and ^{137}Cs spectrum obtained using the virtual Frisch grid detector at an applied bias of 1100 V. *(Top) From Cui et al. (2009) Brookhaven National Laboratory Report BNL-82,046-2009-CP, U.S. Department of Energy. (Middle and Bottom) From Krishna et al. (2013); reprinted with permission from Elsevier © 2013.*

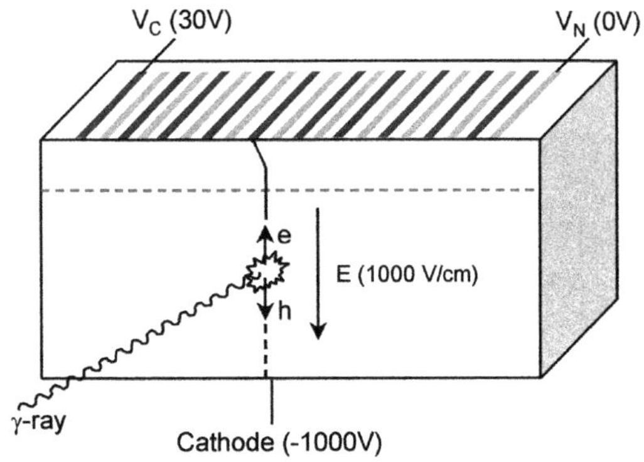

FIGURE 4.34 Coplanar grid electrode configuration. *From Prettyman et al. (1999); with permission from the SPIE, International Society for Optics and Photonics.*

very different shape from those that result in gamma spectra. The peak shapes that occur in gamma ray spectra have been studied and described extensively in the literature (e.g., Gunnink R., and Niday, J.B., 1972; Helmer and Lee, 1980). Fig. 4.35 shows a detailed analysis of a gamma ray peak and the shape components proposed by Gunnink and Niday (1972) to model the peak. The shape of a peak in an alpha particle spectrum has been described and modeled by Wätzig and Westmeier (1978). Other models of alpha particle peaks have been proposed by Garcia-Toraño and Aceña (1981), Amoudry and Burger (1984), and Kirby and Sheehan (1984). Representative examples of a gamma ray peak and alpha particle peaks are presented in Figs. 4.36 and 4.37, respectively.

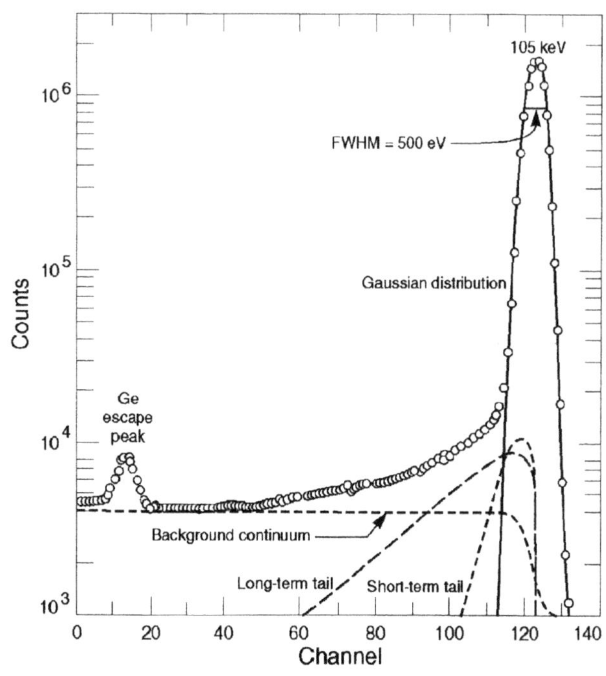

FIGURE 4.35 The detailed shape of an observed peak from a Ge(Li) detector with the principal shape components indicated. *Adapted from Gunnick (1990).*

A. Sample preparation

Warning: The chemical procedures discussed in this section involve the use of strong acids, caustic solutions, and very high temperatures. Appropriate precautions should be observed when handling such materials or working with such conditions. Particular caution should be exercised when working with perchloric acid, as the addition of perchloric acid to a solution containing any organic (carbon) material can result in a very vigorous reaction or *explosion*!

The very different interaction mechanisms and thus attenuation characteristics of gamma rays and alpha particles demand very different considerations in their sample preparations. As alpha particles lose energy virtually continuously along their track, they have a distinct range. In the energy range of interest (typically 4–7 MeV), alpha particles can be stopped by a sheet of paper or approximately 2–8 cm of air (at STP). Thus, encapsulating the sample is out of the question. In fact, even minimal amounts of material between the emitting nuclide and the detector can degrade the energy of the alpha particles to the point that spectroscopic identification becomes difficult, if not impossible. On the other hand, gamma rays can penetrate relatively long distances in a material without interaction or loss of energy, so containment of volumetric (thick) samples of γ emitters is not only possible but routinely employed.

1. Sample preparation for alpha spectrometry

Sample preparation must convert the raw sample into a form that is suitable for alpha spectrometry. This implies two requirements for the preparation:

1. Produce a thin sample.
2. Chemically separate elements that would produce chemical or radiochemical interferences.

In addition, the final form of the source should be rugged enough to be handled safely, chemically stable, and free of all traces of acid and solvent to prevent damage to the counting chambers and detectors. Proper sample preparation is essential to ensure an accurate quantitative assay as well as high resolution. In general, sample preparation requires three steps:

1. Preliminary treatment
2. Chemical separation
3. Sample mounting

These three steps will now be reviewed, starting with sample mounting and ending with preliminary treatment. This order has been chosen because it is easier to understand why certain conditions are required in the early steps after which one understands the requirements of the later steps.

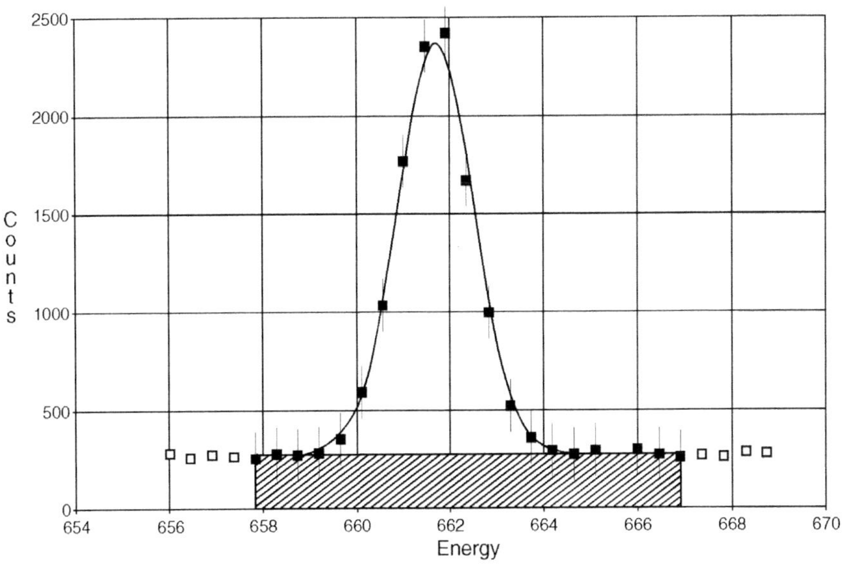

FIGURE 4.36 A 661.6 keV peak from ^{137}Cs as observed at approximately 0.5 keV/channel.

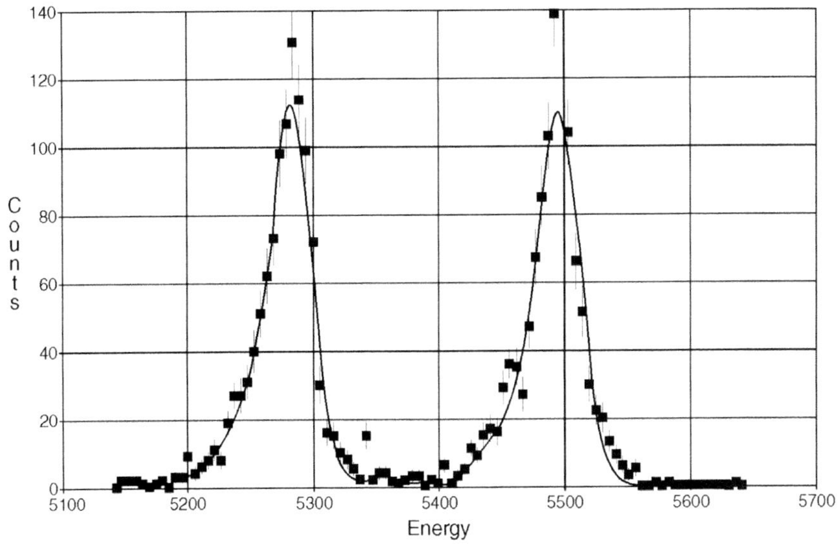

FIGURE 4.37 The 5.276 and 5.485 MeV alpha peaks of ^{243}Am and ^{241}Am, respectively. Note the asymmetry (tailing) of the peaks.

a. Sample mounting

In addition to energy straggling, there are geometric effects that alter the energy resolution of alpha spectra. The need for a thin sample is demonstrated in Figs. 4.38 and 4.36. Fig. 4.35 demonstrates that the variation in the energy of (originally) monoenergetic particles escaping the sample is proportional to the sample thickness, and Fig. 4.36 illustrates the fact that particles leaving the sample or entering the detector at angles other than perpendicular have a longer path length in the energy-degrading materials of the sample matrix and detector dead layer. This variation of the track length in energy-degrading material causes a variation in the observed energy, which contributes to increased line breadth and tailing. Referring to the notation of Fig. 4.39, the difference in the track length of a particle traveling along path a versus path b is given by

$$\Delta = \text{difference in track length} = (d+t)\left(\frac{1}{\cos\theta} - 1\right) \tag{4.55}$$

Thus, one can reduce the line breadth by

1. minimizing d, the dead layer (window) on the detector;
2. minimizing t, the thickness of the sample; and
3. minimizing θ, the acceptance angle of the detector

These items were introduced in a general sense in Sections III.B.1.a and III.B.l.b.

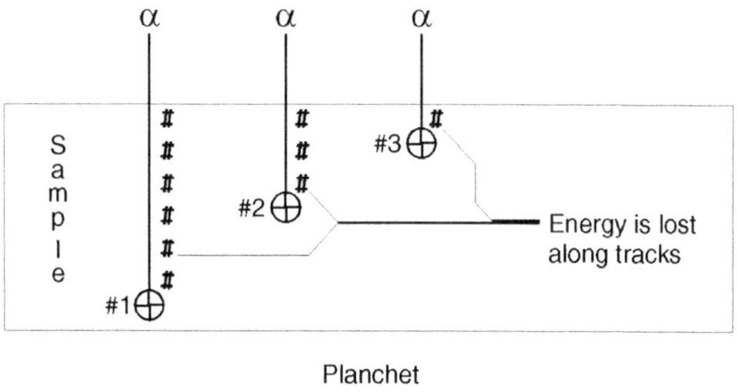

FIGURE 4.38 Three atoms of an alpha-emitting nuclide (labeled 1, 2, and 3) are deposited at different depths within the thickness of the sample. The energy of the alpha particle from the atom labeled #1 will be degraded more than that of #2, which in turn is degraded more than that of #3. Thus, the observed energy of the alpha particles from a thick (monoenergetic) sample will have a distribution of energies reflecting the thickness of the sample (as well as due to straggling).

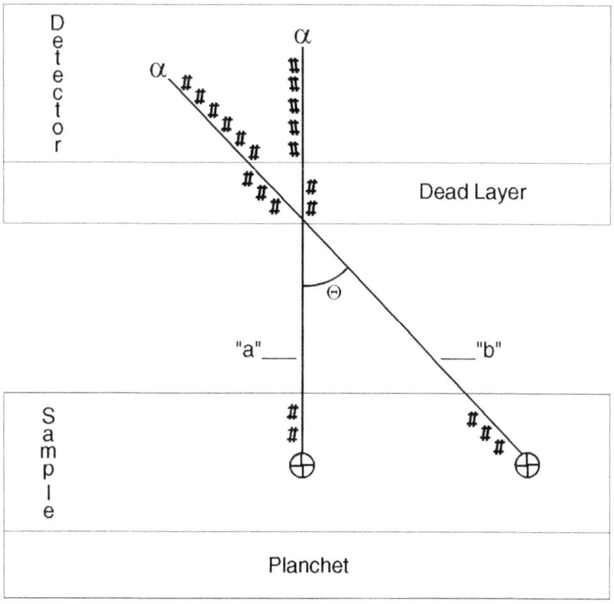

FIGURE 4.39 Since the sample (may) and the detector (definitely does) have a finite radius, alpha particles can leave the sample and enter the detector at angles other than perpendicular. Track *a* leaves the sample and enters the detector at right angles, wherea track *b* leaves the sample and enters the detector at an angle θ from perpendicular. Track *b* has a path length (through the sample and through the detector dead layer) that is greater than track *a* by a factor of 1/cos θ. Thus, an alpha particle emitted along track *b* will have a greater energy degradation than a particle emitted along track *a*.

From the preceding discussion, it is clear that (all else being equal) the thinner the sample, the better the resolution. Thus, the optimum sample mount in terms of resolution would be a monatomic layer of sample atoms; however, in practice, thicker mounts are typical. Methods that have been used to mount sources for alpha spectrometric measurements using semiconductor detectors include the following:

1. Vacuum sublimation
2. Electrospraying
3. Electrodeposition (A) from an aqueous solution and (B) from an organic solution—also referred to as molecular plating
4. Hydroxide or fluoride coprecipitation and filtration as a thin source
5. Evaporation from an organic solvent
6. Evaporation from an aqueous solution

An excellent review of the various sample mounting methods is given by Lally and Glover (1984).

Vacuum sublimation. If the overriding concern is to achieve the highest possible resolution, one should consider mounting the sample by vacuum sublimation. Although the method is capable of producing very good resolution, it is not quantitative, and it is more appropriate to metrology applications (such as the precise measurement of alpha particle energies) than to general radiochemical assay. Vacuum sublimation requires an apparatus in which the sample is heated to a sufficiently high temperature in a vacuum that the sample is vaporized and then sublimed onto a substrate. Samples mounted by this method have produced resolutions of 4–5 keV with magnetic spectrographs and approximately 11 keV (FWHM) with a surface barrier detector.

Electrospraying. Sample mounting by electrospraying can produce extremely thin sources as well as deposits of up to 1 mg/cm^2 with high efficiency. The method requires an apparatus in which the sample is dissolved in an organic solvent and sprayed from a fine capillary tube or hypodermic needle (with the tip squared off) against a substrate that forms the cathode of the apparatus. An electrode may be placed in the solution, or the needle itself can be made the anode of this device. With a potential of up to 8 kV applied between the needle and substrate, the organic solution is ejected as a fine spray so that the organic solvent evaporates before reaching the cathode. In this manner, only solid particles reach the cathode. To ensure a uniform deposit, the cathode is typically rotated during the spraying. Electrosprayed sources have produced resolutions of approximately 17 keV (FWHM).

Electrodeposition. Samples may be mounted by electrodeposition from an aqueous solution or an organic solution, in which case the method is generally referred to as molecular plating. The method produces rugged sources that may be kept in the laboratory indefinitely and is frequently used for preparing rugged alpha sources. Electrodeposition is applicable to a wide range of work from metrology measurements to radiochemical assays. Resolutions of <20 keV are possible with semiconductor detectors. In addition to the production of thick sample deposits, impurities can affect the yield of the technique. Thus, steps to chemically separate the element of interest and place this element in an appropriate solution for electrodeposition must precede the electrodeposition.

Electrodeposition from organic solutions. This technique involves passing a low current at high voltage through an organic solution. It is reasonably rapid and virtually quantitative; near-quantitative recoveries of many of the actinides have been reported in about 1 hour. The method requires the use of reasonably pure solutions. As little as 100 μg of iron or aluminum (which deposits on the cathode along with the actinides) in solution can cause the deposit to be thick and produce degraded resolution. One precaution concerning molecular plating, which should be noted, is the use of high voltages and volatile organic solvents. This combination can present a hazard, particularly in confined areas such as a glove box.

Electrodeposition from aqueous solutions. In contrast to molecular plating, electrodeposition from aqueous solutions is usually performed at voltages of approximately 12–20 V with sufficient current capacity to provide a few hundred mA/cm^2. The method can produce quantitative yields from pure actinide solutions; however, impure solutions may produce less than quantitative yields. The use of a complexing agent, such as hydrofluoric acid, sodium bisulfate, tri-/diethylenetriaminepentaacetic acid (DTPA), or ethylenediaminetetraacetic acid (EDTA), can make the electrolyte more tolerant of impurities. One drawback of electrodeposition from an aqueous solution is that it is somewhat time-consuming, taking up to several hours to complete a deposition. Fig. 4.40 shows the amount of Pu remaining in the plating solution as a function of time for electrodeposition of Pu from a 1 M H_2SO_4 solution. It is apparent from this figure that, to achieve high recovery, one must commit a substantial amount of time to the electrodeposition step.

Electrodeposition is applicable to many elements, including the actinides (Talvitie, 1972). Procedures for electrodepositing radium (Roman, 1984), thorium (Roman, 1980), and uranium, thorium, and protactinium (McCabe et al., 1979; Ditchburn and McCabe, 1984) have been presented. For alpha spectrometry, Cable et al. (1994) have investigated the optimization of the chemical and physical parameters affecting the electrodeposition of uranium, thorium, protactinium, americium, and plutonium using a custom-designed electrodeposition unit.

Electrodeposition cells vary from very simple to rather elaborate. Schematic drawings of two designs are presented in Figs. 4.41 and 4.42. The sample is electrodeposited on a metallic substrate, typically a disk of stainless steel, nickel, or copper (although other materials have been used successfully), which functions as the cathode of the electrodeposition cell. Only one side of the disk should be exposed to the plating solution. The anode is normally made of platinum.

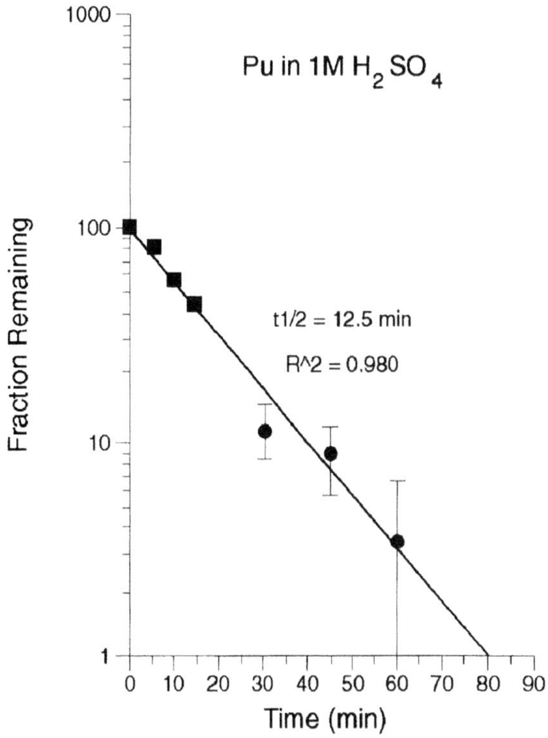

FIGURE 4.40 The results of an experiment designed to evaluate the rate of electrodeposition of Pu from 1 M H_2SO_4 (pH = 3.2). Such experiments are used to optimize conditions and evaluate plating times. *From Burnett (1992), image courtesy of Mirion Technologies.*

In general, the actinide elements thorium through curium can be electrodeposited as hydrous oxides from a buffered, slightly acidic aqueous solution without prior oxidation. Following electrodeposition, the cathode disk is often heated to convert the deposited actinide compound to the anhydrous state or flamed to convert it to an oxide. The high temperature will also volatilize the spontaneously volatile component of any polonium that may have inadvertently deposited on the disk. Sill and Olson (1970)

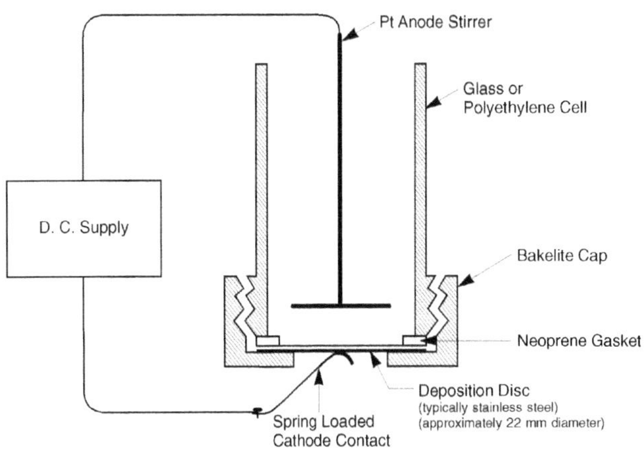

FIGURE 4.41 Schematic drawing of a simple electrodeposition cell.

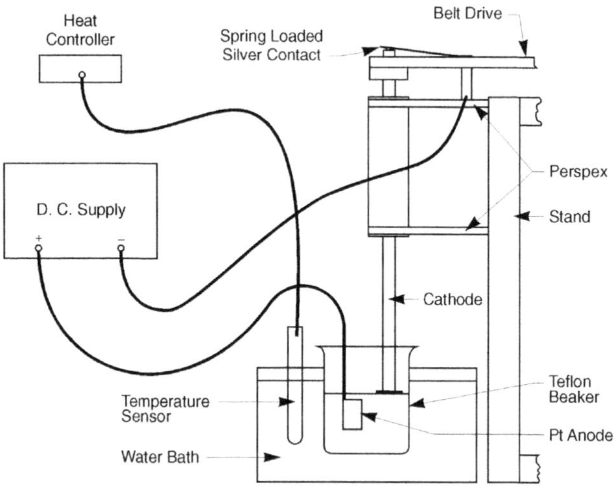

FIGURE 4.42 Schematic drawing of a rotating disc electrodeposition unit. The disc upon which the sample is to be deposited (cathode) is mounted on the end of the spindle, which rotates at 3600 rpm. *From Burnett (1990), image courtesy of Mirion Technologies.*

report that heating the disk on an uncovered hot plate for 5 minute reduces the spontaneously volatile component of polonium to a generally acceptable level without loss of lead or polonium. As the volatility of polonium produces a pseudorecoil effect, by which the detector can become contaminated, it is desirable to eliminate the spontaneously volatile component of polonium to prevent contamination of the detector (see Section IV.B.1) and counting interferences. Care should be taken in heating the disk, as ignition at red heat can volatilize lead (if present), which can carry other nonvolatile components with it, resulting in a loss of material. The volatility of polonium is highly dependent on the disk material and the conditions of the deposition. Contamination of samples by polonium has been shown to occur via a variety of pathways, including spontaneous deposition from the air and from acid baths used to clean recycled deposition disks. For a more complete discussion of the role of polonium as a contaminant and interference, see Sill and Olson (1970) and Sill (1995).

Coprecipitation and filtration as a thin source. Coprecipitation and filtration is a fast, inexpensive method used to mount samples for alpha spectrometry. Basically, the method provides for the coprecipitation of the nuclides of interest as either a hydroxide or a fluoride using either cerium or neodymium[2] as a carrier to produce an extremely finely divided precipitate, which is deposited by filtration over a substrate of ceric hydroxide, cerous fluoride, or neodymium fluoride. The substrate is prepared by filtering the substrate solution (typically ceric hydroxide or neodymium fluoride) through a 0.1 μm membrane filter. These very finely divided precipitates plug the filter and provide a very smooth and nearly impenetrable surface upon which the coprecipitated (with cerous hydroxide or neodymium fluoride) nuclides of interest lie.

The method as presented by Sill and Williams (1981) uses cerium carrier and substrate (in both the oxide and fluoride forms), and it was proposed that lanthanum and neodymium could be used equally well. Subsequent extensions of this procedure tended to focus on the chemical separations that allow the elements of interest to be separated from each other and placed in a chemical form that permits them to be coprecipitated (typically as a fluoride or hydroxide). Hindman (1986) presented a method by which the actinides (thorium, uranium, plutonium, and americium) are separated from each other by coprecipitation and mounted as fluorides on a neodymium fluoride substrate, and Sill (1987a) presented methods for the precipitation of actinides as fluorides or hydroxides for high-resolution alpha spectrometry.

The method continues to evolve, being combined with many separation procedures to handle a wide variety of sample types (Sill, 1987b; Sill and Sill, 1989). More recently, it has been demonstrated that satisfactory resolutions can be obtained even with the sample mounted directly upon the filter, that is, without first depositing a substrate on the filter (Sill and Sill, 1994).

The method can produce excellent resolution[3] provided the total mass of the sample layer (nuclides of interest, carrier, and any impurities) is kept below approximately 100 µg.[4] Using an SSB detector, Sill and Williams (1981) found an FWHM for ^{239}Pu of about 65 keV when mounted by this method compared with an FWHM of about 50 keV for a ^{239}Pu source electrodeposited on polished stainless steel. Noting the limitation on the size of the mounted sample, some care should be exercised in selecting the initial size of the sample. If the sample contains more than 100 µg of the nuclide of interest, the resolution will suffer. Obviously, a smaller initial sample size should be chosen.

In addition, it should be noted that certain sample types (e.g., soils) frequently contain trace quantities of elements

2. Lanthanum has also been proposed as a carrier (Sill and Williams, 1981); however, later work (Hindman, 1986) indicated that there are certain disadvantages associated with the use of lanthanum: the purity of available lanthanum reagents is a problem, lanthanum is not as soluble as neodymium in the small pyrosulfate fusions of this procedure, and the precipitation characteristics of lanthanum are not as advantageous as those of neodymium.

3. Today, possibly because of improved filters and detectors, one can expect to achieve routinely a resolution of 40–50 keV with samples mounted by coprecipitation and filtration, while electrodeposited samples typically produce a resolution of 20–40 keV.

4. The 100-µg limit applies when deposited in a 7/8-inch-diameter circle (on a 25-mm filter) producing a thickness of ~25 -µg/cm^2. Sill and Williams (1981) warn against attempting to distribute the sample over an area greater than that of the detector in an attempt to decrease the sample thickness, as the large entry angle of alpha particles into the detector produces unacceptable amounts of tailing in the spectrum.

that produce chemical interferences with the elements of interest. This can lead to degraded resolution if the total mounted mass exceeds approximately 100 μg. In this case, one has two choices:

1. If the mass of the interfering elements exceeds 100 μg, then a more specific separation is required.
2. If the mass of the interfering elements is less than approximately 75 μg, one might be able to use them in place of the carrier in the coprecipitation of the sample fraction in which these interfering elements occur. For example, 1 g of an average soil contains approximately 75 μg of the light lanthanides (lanthanum, praseodymium, neodymium, etc.), which can be used in place of the cerium carrier to coprecipitate the (actinide) element of interest that occurs in the same fraction as the light lanthanides. (As the light lanthanides are typically trivalent, they typically end up in the americium fraction.)

This method of sample mounting is not limited to chemical separations by coprecipitation. Any separation scheme that produces purified fractions of the elements of interest (e.g., ion exchange, extraction) can precede this method of sample mounting.

Direct evaporation of an organic solution. Direct evaporation of an organic solution produces sources with reasonable resolution by generating nearly solid-free deposits of some alpha-emitting elements. Basically, the method requires extraction of the elements of interest into an organic solution followed by the evaporation of this solution on a stainless steel disk. Examples of organic solutions that may be used include the noyltrifluoroacetone (TTA) in benzene or xylene to complex uranium and thorium and TTA in toluene to chelate plutonium.

The method typically starts with reasonably pure fractions of the elements of interest obtained by ion exchange or solvent extraction. This solution is then evaporated to dryness and treated with a small volume[5] of perchloric and nitric acids to oxidize any residual organic matter. Following the dissolution of the sample, the pH is adjusted to about 3.0 by the addition of 1.0 M NaOH, and the elements of interest are extracted into approximately 1 mL of an approximately 0.4 M TTA solution. Small stoppered centrifuge tubes may be used to avoid the introduction of excessive amounts of air during the mixing of the organic and aqueous phases. A small Pasteur pipette may then be used to transfer the separated organic phase dropwise onto the stainless steel disk for evaporation. To promote uniform drying, the disk can be placed on a heated brass cylinder or common iron washer. As a final step, the disk may be flamed to a dull red to ensure the removal of all residual organic material.

Direct evaporation of an aqueous solution. This method is typically not used for the preparation of high-resolution sources as the material does not deposit uniformly. Any salts in the aqueous solution, including the active material itself, tend to deposit as crystals and aggregates. The resulting self-absorption causes a decrease in resolution. Although spreading agents (such as tetraethylene glycol) can be added to the solution to reduce the crystallization problem during the evaporative deposition, they tend to leave substantial quantities of organic material in the deposit that must later be burned off, causing poor adherence of the nuclide to the disk.

b. Chemical separation

As the initial sample may be rather large (on the order of 1 g or more) and the mounted sample needs to be very small (in the microgram range), it is necessary to separate the elements of interest from the bulk of the sample. Once the alpha-emitting elements are separated from the bulk sample, it may not be necessary to separate the various alpha-emitting elements from each other before counting; see, for instance, Sill and Sill (1994). However, as the alpha particle energies of many nuclides differ by as little as 15–30 keV (which is comparable with the energy resolution of the detectors used in alpha spectrometry), chemical separation of such nuclides is required to eliminate these radiochemical interferences and make quantitative analysis possible.

Unlike cold chemistry, in which standard methods abound, there are no standard (prescribed) methods for radiochemical procedures other than for drinking water as given in the EPA 900 series. The trend in the United States in recent years has been for the acceptability of a radiochemical procedure to be performance based. That is, there is no one mandatory procedure with which to perform a given analysis. Rather, a procedure is considered acceptable if one can demonstrate acceptable performance in cross-checks, analysis of knowns, and so forth.

To perform the necessary chemical separations, one must get the elements of interest into solution. This will be discussed in Section IV.A.1.c. Assuming the elements of interest have been dissolved, numerous separation procedures are available. A brief overview of the various methods is presented in the following with references to the scientific and commercial literature from which the detailed procedures may be obtained.

Separation by precipitation/coprecipitation. This technique has been documented extensively in the literature (Sill, 1969, 1977; 1980; Sill and Williams, 1969; Sill et al.,

5. As the organic solution will eventually have to be evaporated, it is expedient to keep the volume to a minimum. Since the chemical yield of the extraction increases as the ratio of the volume of the aqueous phase to that of the organic phase decreases, it follows that the extraction should be carried out from small volumes (~5 mL) to maximize the recovery.

1974). The method is frequently used in conjunction with sample mounting by the method of coprecipitation and filtration as a thin source (Sill and Williams, 1981; Hindman, 1986; Sill and Sill, 1994), but it can also be used preparatory to electrodeposition.

Separation by ion exchange. This is probably still the most common method of chemical separation for the preparation of samples for alpha spectrometry. The method depends on the selective adsorption and desorption of ionic species on ion exchange resins and thus requires that the element of interest be in a form that may be adsorbed by the resin.

Numerous procedures for chemical separations by ion exchange have been presented. Quantitative separation of uranium, thorium, and protactinium by ion exchange has been demonstrated by McCabe et al. (1979). An improved method for the purification of protactinium was later presented by Ditchburn and McCabe (1984). Numerous other procedures for chemical separations by ion exchange have also been presented in the literature. In addition to the numerous texts on the subject (e.g., Small, 1989), manufacturers of ion exchange resins[6] are often excellent sources of resource material.

Chemical extraction. Chemical extraction is a separation technique that relies on the difference in the solubility of the element of interest in an organic solvent versus an aqueous solution. Traditionally, the two components of the system were maintained in the liquid phase, and the method was referred to as liquid–liquid extraction. However, an innovative application of solvent extraction has been developed at the Argonne National Laboratory in which the solvent extraction system is adsorbed on a macroporous polymeric support that immobilizes the extractant and diluent to form the stationary phase of an extraction chromatographic system.

Separation by liquid–liquid extraction. This method of extraction requires that the element of interest be in true ionic solution in an aqueous medium and not complexed (chelated or bound) in any manner. That is, liquid–liquid extraction will not extract the element of interest from suspended solid or colloidal material. In addition, the presence of organic (and in some cases inorganic) complexing materials in the aqueous phase will, in many cases, cause the extraction to be unsuccessful.

The difference in the solubility of the element of interest in the organic solvent versus the aqueous solution is expressed in terms of the distribution coefficient, K_d, which is defined as

$$K_d = \frac{C_{org}}{C_{aq}} \tag{4.56}$$

From this definition, it follows that the percent recovery of an extraction is given by

$$\%\ \text{recovery} = \frac{K_d V_{org}}{K_d V_{org} + V_{aq}} \times 100 \tag{4.57}$$

In general, organic acids, ketones, ethers, esters, alcohols, and organic derivatives of phosphoric acid have all been used for extraction. The Purex process, which is generally used for the reprocessing of nuclear fuel, makes use of tributyl phosphate (TBP) in an inert hydrocarbon diluent to extract both uranium and plutonium. Methyl isobutyl ketone (MIBK) has also been used for the extraction of U and Pu from spent fuel. The noyltrifluoroacetone (TTA) can be used to extract some actinides. Sill et al. (1974) have presented a procedure by which the actinides are extracted into Aliquat 336, followed by stripping of these elements from the organic extracts. Although liquid–liquid extraction can be used as a precursor to further separations, samples are frequently mounted directly from the organic phase by evaporation of the organic solvent.

Extraction chromatographic systems. This extraction system is used much like ion exchange resins. One advantage of these materials is their high specificity. They are marketed by Eichrom Industries, Inc.[7] Procedures are available for the separation of americium, plutonium, and uranium in water (Eichrom Industries, 1995a); uranium and thorium in water (Eichrom Industries, 1995b); uranium and thorium in soil (Eichrom Industries, 1994); and thorium and neptunium in water (Eichrom Industries, 1995c). A method for screening urine samples for the presence of actinides using these extraction chromatographic materials has been presented by Horwitz et al. (1990).

c. Preliminary treatments

Preliminary treatments typically vary with the objectives of the experiment and the sample matrix. Basically, they are performed to attain one or more of the following objectives:

1. To separate the component(s) of interest from the remainder of the sample
2. To ensure that the sample is representative of the bulk sample
3. To ensure that the sample remains representative of the bulk sample
4. To preconcentrate the component(s) of interest
5. To introduce chemical tracers and ensure equilibration with analyte isotopes

6. For example, Bio-Rad Laboratories, Inc., 2000 Alfred Nobel Drive, Hercules, CA 94547 and The Dow Chemical Company, P.O. Box 1206, Midland, MI 48641-1206.

7. Eichrom Industries, Inc., 8205 S. Cass Avenue, Suite 107, Darien, IL 60559.

6. To prepare the sample for the chemical procedures that are to follow, that is, dissolve the sample

Variable and/or incomplete sample dissolution is a major cause of inaccurate radiochemical analyses. To ensure accurate and reproducible results, it is essential that all of the element of interest be brought into solution. A variety of methods have been suggested and used to prepare samples for alpha spectrometry, including high-temperature fusions, acid leaching, and a variety of "digestions" typically involving acid bombs at elevated temperature and pressure. Sill and Sill (1995) provide convincing arguments for the use of high-temperature fusions, citing examples of the failure of other methods to place selected elements into solution. For a more complete discussion of decomposition methods, see Bock (1979) or Sulcek and Povondra (1989).

For liquid samples, one must first decide what is of concern. Is it the dissolved material, the particulate material, or the total (both together)? If the sample is to be separated into soluble and particulate components, the first step should be to filter the sample. Following filtration, the liquid portion should be acidified to prevent biological growth as well as to keep trace elements in solution (at $6 < pH < 8$, many metallic elements form insoluble hydroxides, which can then "plate" onto the walls of the sample container). Acidification of a liquid sample before filtration can introduce a bias in the individual components, as the acidification of the sample will leach the particulate matter. Radiochemical tracers, if they are to be used, should be added immediately following acidification. Sufficient time for isotopic equilibration should be allowed before any further chemical procedures are performed. Burnett (1990) suggests that "24 h appears to be sufficient for equilibration of most radiotracers with uranium-series isotopes in natural waters."

Following isotopic equilibration, one can perform a preconcentration step if desired. Preconcentration is frequently used to obtain sufficient material when the concentration of the material of interest is very low. Common methods of preconcentration include ion exchange, coprecipitation, and the use of adsorptive filters such as manganese-coated acrylic fibers, which have high adsorptive capacities and can be used to preconcentrate elements such as radium, thorium, protactinium, and actinium.

To ensure total dissolution of the element(s) of interest in a total water sample or even the liquid phase when there is a possibility that the element(s) of interest is chelated with organic material or otherwise bound in a form that would interfere with its separation, a high-temperature fusion may be employed. Such a procedure is described by Sill and Sill (1994).

A simple flowchart for the preliminary treatment of liquid samples might appear as follows:

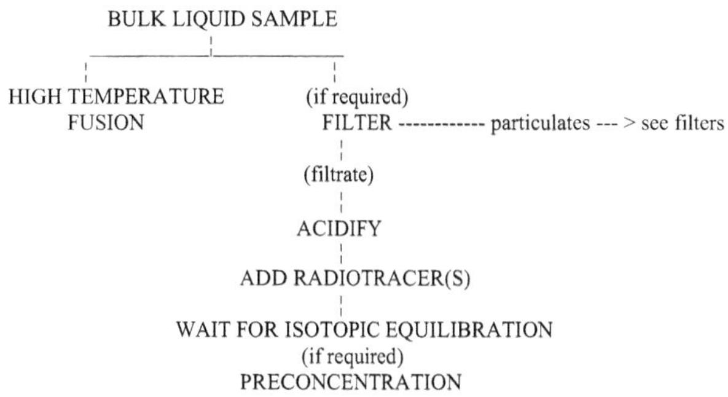

The particulate fraction can be treated in the same way as solid samples (soils, etc.) once the presence of the filter is addressed. Typically, the filter is "digested" either by ashing or by dissolving in strong acid and treating the residue as one would a solid sample; however, polycarbonate (membrane) filters are resistant to acids and do not submit to acid dissolution.

A simple flowchart for the preliminary treatment of filtrates might appear as follows:

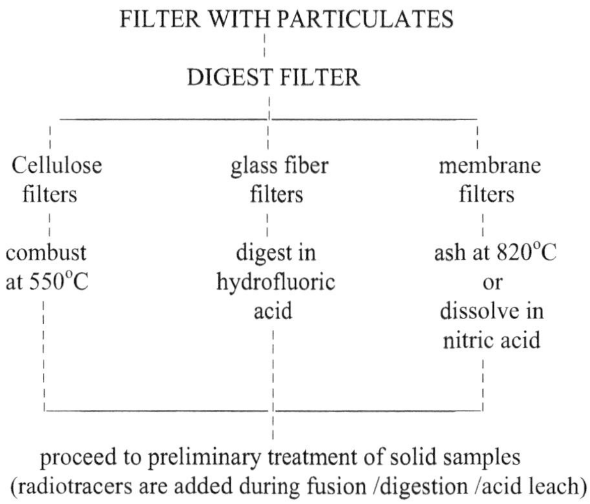

For solid samples, one must first decide what is of concern. Is it the total sample or some fraction thereof? Typically, for soil samples, one is concerned with the sample in total; however, if the sample is to be separated according to particle size, then the first step should be to fractionate the sample according to particle size. Following fractionation, the sample should be ground and mixed well to ensure homogeneity. Finally, the subsample, on which the chemical separations will be performed, should be measured.

As with any sample being prepared for alpha spectrometry, the elements of interest need to be brought into solution before their separation. Although in some cases, it may be possible to remove the element(s) of interest from the bulk of the sample by leaching in strong acid and separating the liquid and solid phases by centrifuging or filtration, it is generally recommended that a total dissolution of the sample be performed to ensure that the element(s) of interest is indeed brought into solution.

Typically, a high-temperature fusion (e.g., pyrosulfate or potassium fluoride fusion) is used to ensure the total dissolution of a solid sample (Sill and Williams, 1981; Hindman, 1986; Sill and Sill, 1994). Detailed procedures for this technique have been presented by Hindman (1984) and Sill and Sill (1994). One drawback of this method is the expense of the required platinum dish and its limitations in terms of compatibility with certain chemicals and processes. Acknowledging this drawback, Sill and Sill (1995) have presented a procedure for performing a pyrosulfate fusion in borosilicate glassware.

A simple flowchart for the preliminary treatment of solid samples might appear as follows:

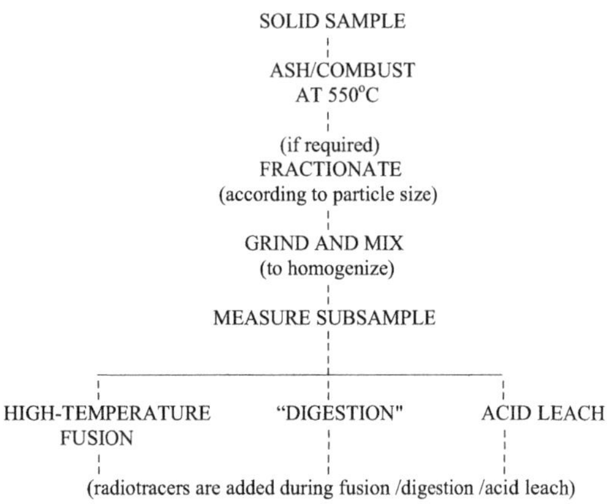

2. Sample preparation for gamma spectrometry

Before the advent of high-resolution spectrometers, radiochemical separations were often required prior to counting. Although such procedures are still useful in some cases, they are not covered in this section. Rather, this section focuses on the preparation of samples that do not require extensive chemical preparation.

The first step in sample preparation is to collect the sample. Care should be taken during sample collection to ensure that the sample is representative of the bulk material. For example, air sampling for particles should employ isokinetic sampling. For soil sampling, care should be taken to prevent cross-contamination of samples by the collection tools.

Assuming one is analyzing bulk samples (e.g., there is no chemical separation or preconcentration), the basic function of the sample preparation is to make the sample look like the standard that was assumed for the efficiency calibration. Whether the calibration standard is an actual source or a mathematical model such as that used for Monte Carlo calibrations, the standard is prepared with or assumed to have certain properties (e.g., dimensions, density, distribution). The sample must be prepared in a manner that reproduces these properties. For example, if it is assumed that the active material is uniformly distributed in a liquid sample, then plating of the active material on the walls of the container must be avoided. To this end, liquid samples may be acidified.

A significant difference between alpha and gamma spectrometry is that in gamma spectrometry, the nuclides of interest are not removed from the bulk sample, so the properties of the bulk sample (density, homogeneity, etc.) become important.

In other ways, the sample preparation considerations for gamma spectrometry are similar to those for alpha spectrometry. For example, for liquid samples, one must still decide which component is of concern. Is it the dissolved material, the particulate material, or the total (both together)? If the sample is to be separated into soluble and particulate components, the first step should be to filter the sample. Following filtration, the filtrate should be acidified both to prevent biological growth and to keep trace elements in solution (at $6 < pH < 8$, many metallic elements form insoluble hydroxides, which can plate onto the walls of the sample container). In alpha spectrometry, one wants to keep the active material in solution so that it might be chemically separated; however, in gamma spectrometry, one tries to keep the active material in solution to ensure that the geometric distribution of the active material is consistent with the assumed distribution of the calibration standard. Acidification of a liquid sample before filtration can introduce a bias in the individual components, as the acidification of the sample will leach the particulate matter. If the sample is to be analyzed in total (without regard to which fraction contains what activity), acidification as the first step in sample preparation is appropriate as it keeps trace elements in solution. The fact that acidification also leaches the particulate matter is not critical in this case, as the dissolved material will then be uniformly distributed in the liquid, which is most likely the distribution assumed for the efficiency calibration. In fact, if a sample is to be analyzed in total and it contains particulate matter, one should pay particular attention to ensuring that the material is distributed as assumed. That is, if one prepares the calibration standard assuming the active material is uniformly distributed, then one should attempt to ensure that the active material in the sample is also uniformly distributed. In other words, shake it up. Particulate matter in liquid samples can present a difficulty, particularly with long sample counts, as the particulate matter can settle during the counting period, causing a bias to develop.

B. Analysis—analytical considerations

1. Analytical considerations in alpha spectrometry

One can use a peak search program to identify peaks in an alpha spectrum, but it is more typical to use a library-driven and/or user-defined search, as the separations that are typically performed in the preparation of the sample severely limit the nuclides that could be found in any given fraction. Thus, one simply analyzes the regions of the spectrum where the nuclides of interest could be. In addition, library-driven routines are more suited to the analysis of small, poorly defined peaks that are frequently encountered in low-level (environmental) alpha spectrometry.

If the peaks are fully resolved from one another, a simple summation of the counts in each peak provides an accurate value for the peak area. If there is any overlap of peaks, one should use an algorithm (typically implemented in a computer program) that is capable of calculating the areas of peaks that overlap. The algorithm should use a peak model that includes a low-energy tail, which is typical of alpha peaks. A variety of mathematical models and methods have been used in various computer codes designed to analyze complex alpha spectra. Examples of these include ALFUN (Wätzig and Westmeier, 1978), NOLIN (Garcia-Toraño and Aceña, 1981), DEMO (Amoudry and Burger, 1984), and GENIE-PC (Koskelo et al., 1996).

These programs assume no continuum distribution under the peaks, as alpha particle interactions do not provide a mechanism by which a continuum distribution could be generated. Rather, any alpha particle interaction in the detector would necessarily contribute to the peak (or tail of the peak). This should not be interpreted as meaning that there is no "background" contribution, but rather that the background contributions also form or contribute to a peak (or tail). Background contributions may result from

1. contamination of the counting chamber and/or detector, which can be determined by counting the empty chamber
2. contaminants in the process reagents and/or mounting materials, which can be determined by counting a method blank

For radiochemical analyses (assays), these background contributions must be subtracted from the observed spectrum to determine the sample (only) count rate to determine accurately the sample activity. Some commercially available alpha spectrometry software packages differentiate between these two contributions as an area correction (item I above) and a reagent correction (item II above). The reagent correction is often implemented by scaling the contribution of one (reagent) nuclide to another (reagent nuclide), in which case the area correction should be implemented before the reagent correction. For example, if a ^{242}Pu tracer is used that contains trace levels of ^{239}Pu, one would need to subtract the tracer's contribution to the ^{239}Pu peak from the observed spectrum to determine the sample's contribution to the ^{239}Pu peak. The tracer's contribution to the ^{239}Pu peak may be determined as a fraction of the ^{242}Pu that is present due to the tracer. However, if ^{242}Pu contamination is present in the counting chamber, the chamber's contribution to the observed ^{242}Pu peak must be subtracted from the observed ^{242}Pu peak before the ^{242}Pu peak can be used to determine the quantity of reagent present and thereby the reagent's contribution to the ^{239}Pu peak.

As discussed in Section III.B.1.a, the efficiency for the detection of alpha particles is independent of energy or emitter and is strictly a function of the geometric efficiency of the source–detector configuration. Thus, the question becomes, what is the ideal source–detector configuration?

As discussed in Section III.B.1.a, for a given specific source activity, A_s (Bq/cm^2), the optimum source diameter (from efficiency considerations only) is equal to the detector diameter. However, for a fixed amount of activity (as one would obtain from a given sample), the count rate depends only on the geometric efficiency, which, as Fig. 4.25 shows, increases with decreasing source diameter. The practical ramification of this is that one can increase the counting efficiency by depositing the sample in a smaller diameter. However, as the sample diameter decreases, the sample thickness increases and can cause a decrease in resolution. Thus, the minimum sample diameter is constrained by the effect of sample thickness on resolution. For example, a 100-μg sample deposited in a diameter of 1 inch ($\sim$5 cm^2) results in a sample thickness of 20 μg/cm^2, as does a 25-μg sample deposited in a diameter of 1/2 inch ($\sim$1.25 cm^2). With approximately equal sample thicknesses, the resolution of the two samples will be roughly equivalent,[8] but the smaller diameter sample will have a higher counting efficiency. Thus, the counting efficiency for low-activity (actually low-mass) samples can be increased by depositing them in a smaller diameter within the constraints imposed by the effects of sample thickness on resolution. The resolution of samples mounted by the fluoride precipitation method is generally acceptable if the sample thickness is kept at 25 μg/cm^2.

However, in practice, the diameter of the sample deposition is not adjusted from sample to sample but rather specified by the sample mounting procedure. The size of the sample deposition, as well as the initial mass of the sample, called for in a procedure should be based on the anticipated sample quantities (concentrations), maximum desired sample thickness, diameter of the detector, and available sizes of commercially available filters, filter holders, electrodeposition disks, and so on. Thus, instead of altering the diameter of the deposition to increase the count

8. Actually, the smaller diameter sample should produce a slight advantage in terms of spectral tailing, as the maximum entry angle (of alpha particles into the detector) is less for the smaller diameter sample.

rate of low-activity (low-mass) samples, one typically increases the size of the initial sample. Applying these considerations to the practical problem of achieving the lowest possible MDA (in terms of Bq/g) for a given sample analysis produces a protocol that requires

1. The diameter of the sample mount should be approximately equal to the diameter of the detector.
2. The amount of initial raw sample to be used in the analysis should be maximized within the constraint that the final mounted sample thickness does not exceed 25 μg/cm^2 (or whatever thickness is demanded by resolution considerations).

Although it has been recommended by Sill and Olson (1970) that "sources should be placed at least 1.5 diameters from the detector to obtain optimum resolution," one needs to appreciate that there is a trade-off between efficiency and resolution. While placing the source closer to the detector causes a decrease in resolution (larger FWHM), it also increases the counting efficiency. As discussed in Section III.B.1.a, the increase in the FWHM at distances as close as 2 mm can be expected to be no greater than 50% (for 300- to 600-mm^2 detectors). Such a decrease in resolution may be deemed tolerable in light of the increase in efficiency so obtained. For low-level counting, it is not unusual to sacrifice resolution to increase the counting efficiency.

Another important consideration in alpha counting is the problem of recoil contamination, which can occur when the progeny of the alpha-emitting nuclides being observed is ejected from the sample (due to the kinetic energy of recoil from the initial alpha emission) and becomes attached to the detector. The short-lived alpha-emitting progeny then contributes to the alpha spectrum.

Sill and Olson (1970) have demonstrated a reduction in recoil contamination "by a factor of at least 10^3 with a loss in resolution of only 1 or 2 keV by leaving enough air in the counting chamber to produce 12 μg/cm^2 of absorber between the source and detector, and applying a negative potential of 6 V to the source plate." Fig. 4.43 summarizes the relationship between distance and pressure required to stop recoiling ^{221}Fr atoms as described by Sill and Olson (1970), who observed that "the range of the recoiling atoms was between 12 and 16 μg/cm^2 for all distances checked."

Neither the air layer nor the negative bias on the sample plate individually is sufficient to prevent recoil contamination of the detector. Both the air layer and the negative bias together are required to prevent recoiling daughter atoms from reaching the detector. Today, most commercially available alpha spectrometers provide a readout of the chamber pressure and a negative bias on the sample plate relative to the detector.[9]

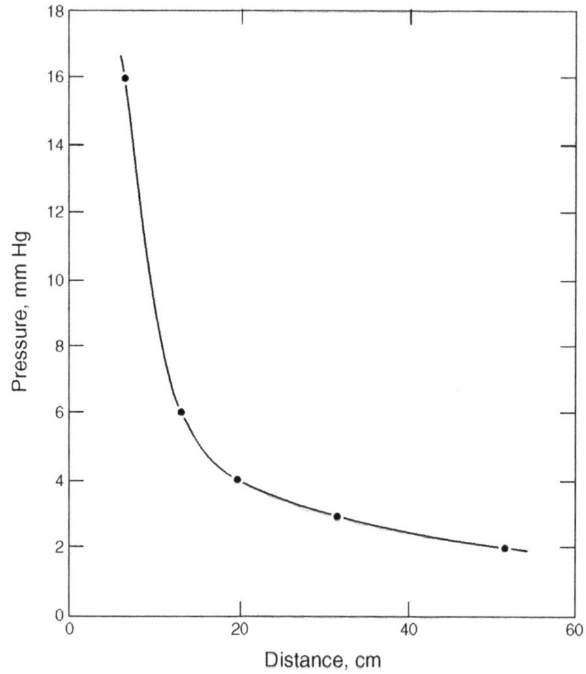

FIGURE 4.43 Relationship between distance and pressure required to stop recoiling ^{221}Fr atoms. *From Sill and Olson (1970), reprinted with permission from the American Chemical Society, © 1970.*

In summary, the resolution is improved with the following:

1 . A thinner entrance window on the detector.
   Trade-off: None.
2. Thinner sample deposition.
   Trade-off: Larger diameter samples (of the same total activity) have a lower counting efficiency and greater low-energy tailing due to large angle entry of the alpha particles into the detector.
3. A thinner air column (absorber) as measured in g/cm^2 between the sample and detector.
   Trade-off: Increased recoil contamination if the air column is too thin (<12 μg/cm^2).
4. A longer (straighter) path between the sample and detector.
   Trade-off: Lower counting efficiency.

2. Analytical considerations in gamma spectrometry

The American National Standards Institute has published a standard that provides guidance in the calibration and use of germanium spectrometers:

Methods for the calibration and use of germanium spectrometers for the measurement of gamma ray energies and emission rates over the energy range from 59 keV to approximately 3000 keV and for the calculation of source

9. Equivalently, the detector can be biased positive relative to the sample plate.

activities from these measurements are established. Minimum requirements for automated peak finding are stated. Methods for measuring the FEP efficiency with calibrated sources are given. Performance tests that ascertain the proper functioning of the Ge spectrometer and evaluate the limitations of the algorithms used for locating and fitting single and multiple peaks are described. Methods for the measurement of and the correction for pulse pileup are suggested. Techniques are recommended for the inspection of spectral analysis results for large errors resulting from summing of cascade gamma rays in the detector. Suggestions are provided for the establishment of data libraries for radionuclide identification, decay corrections, and the conversion of gamma ray rates to decay rates.[10]

Typically, an automated gamma spectral analysis requires the following steps:

1. Peak location
2. Calculation of peak areas
3. Correction of peak areas (if required; e.g., subtraction of system background or reference peak correction for random summing losses)
4. Calculation of the efficiency at the peak energies
5. Calculation of activity

Generally, the activity of a sample is calculated in gamma spectrometry from the following equation:

$$\text{Activity (in Bq at t = 0)} = \frac{\text{Net Peak Area}(E)}{T_{\text{L}}\text{Efficiency}(E)\text{Intensity}_{\gamma}(E)} \frac{\lambda T_{\text{R}}}{1 - e^{-\lambda T_R}} \quad (4.58)$$

where $t = 0$ is the starting time of data acquisition, T_{L} the live time of the count, T_{R} the real time of the count, and λ the decay constant:

$$\lambda = \frac{\ln(2)}{T_{1/2}} \quad (4.59)$$

where $T_{1/2}$ is the radionuclide half-life and[11]

$$\frac{\lambda T_{\text{R}}}{1 - e^{-\lambda T_R}} = \text{decay correction for decay during the counting period} \quad (4.60)$$

$$\text{Intensity}_{\gamma}\ (\text{E}) = \frac{\text{gamma emission rate (at energy } E)}{\text{disintegration rate}} \quad (4.61)$$

and

$$\text{Efficiency(E)} = \frac{\text{full-energy deposition rate (at energy } E)}{\text{gamma emission rate (at energy } E)} \quad (4.62)$$

The efficiency is supposed to express the relationship between the full-energy deposition rate and the emission rate of a sample. Typically, the full-energy deposition rate is approximated by the net count rate, so that the emission rate of a sample is determined by a simple proportional scaling of the observed count rate from a standard of known emission rate as follows:

$$\frac{\text{Emission rate of sample}}{\text{Count rate of sample}} = \frac{\text{(known) emission rate of standard}}{\text{observed count rate of known standard}} \quad (4.63)$$

so that

$$\text{Emission rate of sample} = \frac{\text{count rate of sample}}{\left(\dfrac{\text{observed count rate of known standard}}{\text{(known) emission rate of standard}}\right)} = \frac{\text{count rate of sample}}{\text{efficiency}} \quad (4.64)$$

In other words, the efficiency is supposed to specify the number of full-energy depositions (of energy E) in the detector per gamma ray (of energy E) emitted by a source of a given geometry.[12] The efficiency may be determined by mathematical methods such as Monte Carlo calculations.[13] More traditionally, it is determined by dividing an observed count rate for a calibration standard of a given geometry by the known emission rate of the standard as indicated above. The assumption inherent in this methodology is that the observed count rate (or total net counts in a given time period) is equal to the full-energy deposition rate of the standard (or total number of full-energy depositions of energy E in a given time period).

Most problems in the quantification of gamma spectra can be traced to a discrepancy between the observed count rate and the full-energy deposition rate. There are many

10. ANSI N42.14-1991, copyright © 1991, IEEE. All rights reserved.

11. This correction factor [which essentially converts the nominal count rate (peak area/T_{L}) to the count rate at time $t = 0$] is derived assuming the dead time is constant during the counting period. As such, it is an approximation that is valid only for materials whose half-life is long relative to the count time.

12. The term geometry is used to indicate the geometric distribution of a source (or sample) relative to the detector, the materials between the source and detector, the size and configuration of the detector, and so forth.

13. With the increase in computer power that has become available in recent years, mathematical calibrations have become more practical and more widely available. Mathematical calibrations are particularly well suited to in situ counting and other geometries for which the production of a calibration standard would be impractical if not impossible.

potential causes for such a discrepancy, among which are the following:

- Failure to correct for dead time causes full-energy depositions to be unrecognized.
- Pulse pileup causes full-energy depositions to be unrecognized.
- Random summing causes full-energy depositions to be unrecognized.
- Coincidence summing causes full-energy depositions to be unrecognized.
- Incorrect assessment of peak area produces an incorrect count rate that does not properly represent the full-energy deposition rate.

The basic steps involved in a gamma spectral analysis will now be reviewed.

a. Peak location

The first step in the analysis is to locate the peaks in the spectrum. This can be accomplished by either a library-driven routine or a search-driven routine.

The library-driven routine uses a list of energies (a library[14]) of peaks for which one wishes to search. It then calculates the net area of the region over which each listed peak, if present, would exist. The area so calculated may then be reported (even if it is negative), whereas other programs first determine whether the net area is statistically significant. The region over which the peak is assumed to exist is usually determined from a "shape calibration," that is, a relationship of the FWHM versus energy and possibly a tailing parameter versus energy.

The ability of a library-driven routine to identify peaks (and nuclides) is limited to the entries in the library. One should also be aware that spectral artifacts (such as the backscatter peak) can produce false-positive peak identifications.

On the positive side, library-driven routines provide the following advantages:

1. The ability to identify small peaks
2. The ability to identify poorly shaped peaks
3. The ability to unfold complex multiplets

This last strength is used to great advantage in a program known as MGA (Gunnink, 1990), which quite arguably epitomizes the capabilities of library-driven routines. It was designed to determine the isotopic abundances of a plutonium sample. A mixture of plutonium isotopes (238–242) produces a spectrum that is too complex to analyze by traditional means. However, knowing that the sample was pure plutonium at one time, all of the potential component nuclides are known and a library of all their energy lines and relative abundances can be specified. By knowing the detailed line shape of gamma and X-ray peaks, one can generate the envelope function for a given mixture of the component nuclides. Essentially, a least-squares fit of the envelope function to the observed spectrum using (among other parameters) the relative abundances of the plutonium isotopes as independent variables then yields the relative abundances of the plutonium isotopes.

A search-driven routine applies some mathematical methodology to the spectral data to distinguish peaks from the continuum distribution. A method that is often employed is to apply a symmetric zero-area transform (often referred to as a sliding transform, sliding filter, digital filter, or filter) to the spectral data. The method was proposed by Mariscotti (1967) and employed in the programs SAMPO (Routti and Prussin, 1969) and HYPERMET (Phillips and Marlow, 1976) and several commercially available programs that followed (e.g., Canberra Industries' GENIE family of spectrometry systems). The transformed spectrum (which can be thought of as a response function) will be zero where the spectrum is constant, nearly zero where the spectrum is slowly varying, and large (either positive or negative, depending on the definition of the transform) in the region of a peak. Thus, one merely needs to scan the response for regions that exceed some threshold value to find peak locations in the spectrum. The response will be strongest when the width of the feature in the spectrum (ideally a peak) most closely matches the width of the filter. This has two implications:

1. The width of the filter should be chosen to match the expected width of the peaks. This is typically accomplished by use of a shape calibration.
2. This algorithm tends to discriminate against features that are both wider and narrower than the filter width. Hence, spectral artifacts in Ge and Si(Li) detector spectra whose width differs significantly from the expected peak width (such as the backscatter peak) tend to be filtered out.

A common error in the use of search-driven routines is failure to match the sensitivity of the peak search routine to the detection limit assumed in the calculation of the MDA or the lower limit of detection (LLD). These calculations assume that peaks of a given size (relative to the background) can be detected at a given confidence level. This is not necessarily true if the sensitivity (response threshold) is not selected appropriately. This has been recognized and addressed in ANSI standard N42.14 (1991), which states in Section 5.

If an automated peak-finding routine is used in the spectral analysis, it should be able to find small well-formed single peaks whose areas are statistically significant (above background)

14. As the library is typically used again during the analysis process for nuclide identification and activity calculation, it generally includes the nuclide name, the nuclide half-life, and the gamma ray intensities, in addition to the gamma ray energies.

and provides a test for automatic peak-finding algorithms in Section 8.1, which

> *… has been designed to determine how well singlet peaks on a flat baseline that are at or above an "observable" level can be found (i.e., detected) with the peak-finding algorithm.*

The standard goes on to state that

> *the peak-finding algorithm is expected to find a peak in a spectrum whose area, A, = $L_P\sqrt{[(2.55)(FWHM)y_i]}$, where 2.55 is based on $\pm 3\sigma$ for a Gaussian peak, FWHM is the full width in channels at half maximum of the peak, y_i are the average counts in each baseline channel, and $L_P = 2.33$ corresponds to the value L_P initially suggested for this test.*[15]

The value of 2.33 for L_P was chosen to correspond to the critical level, L_C as defined by Currie (1968) to be the decision limit at which a count is assumed to be detectable with 95% confidence (i.e., $\alpha =$ probability of a false positive = 5%).[16]

Advantages of search-driven routines include the following:

1. The ability to locate peaks even if one did not anticipate their presence
2. The ability to differentiate between peaks and other spectral features

A search-driven algorithm should also be able to locate the individual components of a multiplet (two or more peaks that overlap). Depending on the methodology employed, this functionality can be incorporated in the peak-locating routine; however, some other methods can be implemented only during the peak (area) analysis routine. Therefore, the discussion of multiplet deconvolution will be taken up in the next section on peak area analysis.

15. ANSI N42.14-1991, copyright © 1991, IEEE. All rights reserved.

16. Currie's derivation of the limit of detection was based on single-channel (gross counting) considerations and, as such, is not strictly applicable to multichannel analysis; however, it has become common practice to apply the equations and concepts from his derivation to multichannel analysis even though there are additional considerations and uncertainties in multichannel analysis that are not incorporated in these equations. For instance, the probability of a false-positive identification is not strictly a function of the size of the background, as the peak-locating algorithm will not (falsely) identify a peak if the region of the null spectrum under consideration is reasonably flat (regardless of size). Furthermore, the uncertainty associated with the ability to detect a peak as a function of peak shape is not included. As an example, consider two peaks with equal net area at (or slightly above) the critical level. A peak-locating algorithm may detect one and not the other, because its ability to recognize a peak is dependent on the shape of the spectral distribution.

b. Peak area analysis

The next step in the analysis is to calculate the net area of the observed peaks. This is typically accomplished either by a summation method or by fitting a function that represents the assumed peak shape to the observed data and reporting the area under the peak function as the net area:

$$\text{Summation: Area} = \sum_{i=\text{left}}^{\text{right}} y_i - b_i \qquad (4.65)$$

$$\text{Fit: Area} = \int P(\alpha_1, \alpha_2, \alpha_3, \ldots, x)dx \qquad (4.66)$$

where i = channel number, left = the leftmost channel of peak region (to be fitted), right = the rightmost channel of peak region (to be fitted), y_j = number of (gross) counts in channel i, b_i = continuum contribution to channel i, and $P(\alpha_1, \alpha_2, \alpha_3, \ldots x)$ = the "best fit" mathematical function that models the assumed peak shape. The best fit is typically determined by the method of least squares, which requires that χ^2 be minimized, where χ^2 is defined as

$$\chi^2 = \sum_i w_i[y_i - b_i - P(\alpha_1, \alpha_2, \alpha_3, \ldots, x_i]^2 \qquad (4.67)$$

where w_i is the the weighting applied to the ith point and the α_k are the free parameters of the model.

The fit method is applicable to both singlets and multiplets; the summation method (by itself) cannot assess the contributions from the individual components of a multiplet. Thus, multiplet analysis requires some sort of fit to be performed. Notice that both the summation and fit methods require that the continuum contribution (background) under the peak, b_i, be specified. One way to estimate this contribution is to assume a particular mathematical model for the background and determine its parameters from the channels immediately to either side of the peak.

Two commonly used background models are the linear background and step background (Gunnink, 1979),[17] which may be determined from the spectral data as follows:

$$\text{Linear model: } b_i = \frac{B_L}{n} + \frac{B_R - B_L}{n(N+1)}i \qquad (4.68)$$

$$\text{Step model: } b_i = \frac{B_L}{n} + \frac{B_R - B_L}{nG}\sum_{i=\text{left}}^{i} y_i \qquad (4.69)$$

where

$$B_L = \sum_{i=\text{left}-n}^{\text{left}-1} y_i \qquad (4.70)$$

17. Mathematically, a step function can be expressed by a variety of functions. The function presented here is the one proposed by Gunnink (1979) and used in Canberra Industries' GENIE software.

$$B_R = \sum_{i=\text{right}+1}^{\text{right}+n} y_i \quad (4.71)$$

where n is the number of channels to be averaged on each side of the peak to determine the background, N is the number of channels in the peak region, and

$$G = \text{integral of the peak region} = \sum_{i=\text{left}}^{\text{right}} y_i \quad (4.72)$$

For single well-resolved peaks, the linear and step backgrounds produce approximately equivalent results for the area calculation. However, for multiplets, the linear and step backgrounds can produce different results. As shown in Figs. 4.44 and 4.45, the step background is greater than the linear background on the left side of the multiplet and less than the linear background on the right side. For multiplets containing a large component and a small component, the step background places the major portion of the background change under the major component, and the linear background changes linearly across the peak region. This results in the linear approximation understating (relative to the step approximation) the background under the minor component (thus overstating the net area) when the minor component is on the left side of the multiplet and overstating the background (relative to the step approximation) under the minor component (thus understating the net area) when the minor component is on the right side of the multiplet.

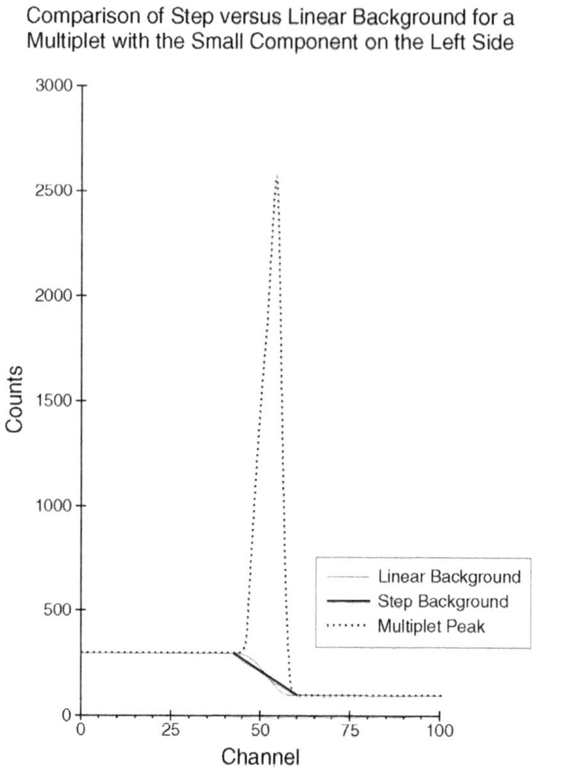

FIGURE 4.45 A comparison of a step background versus a linear background for a multiplet with the small component on left side.

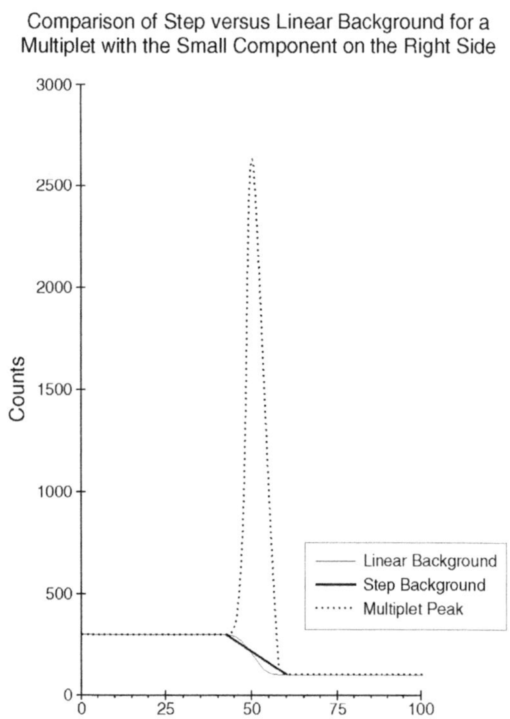

FIGURE 4.44 A comparison of a step background versus a linear background for a multiplet with the small component on right side.

The accuracy of the analysis of multiplet components and the ability to detect the presence of these components are also addressed in ANSI standard N42.14 (1991), as Section 5 states:

> *... The peak fitting routine should be able to find multiplet peaks that meet the peak area criteria for a singlet, are approximately the same intensity, and are separated by 1 FWHM. (see Section 8 for the test procedures). Optimization of the peak search parameters of the peak-finding algorithm is left to the user or software vendor. These should be adjusted so that statistically significant peaks are found with a minimum number of false peaks being reported (see performance tests in Section 8).*[18]

Section 8.3 of the standard (ANSI N42.14, 1991) provides performance tests for "the Doublet-Peak Finding and Fitting Algorithms." These tests can be used to determine which background function provides the most accurate multiplet analysis as well as to evaluate an algorithm's overall accuracy in determining the areas of the

18. ANSI N42.14-1991, copyright © 1991, IEEE. All rights reserved.

components of a multiplet. Investigations using such tests have been presented by Koskelo and Mercier (1990) and Mercier and Koskelo (1992), in which several commercially available analysis programs were evaluated.

As stated previously, the ability to detect a multiplet can be incorporated in the peak locate routine or the peak (area) analysis routine. Peak search algorithms that employ a symmetric zero-area transform (and determine the presence of the peak from a large response in the transformed spectrum) to detect the individual components of a multiplet at the peak search stage can generally resolve components of equal size separated by one FWHM or more. As the difference in size of the individual components increases, the separation required to be able to detect the smaller component becomes greater. The individual components of a multiplet can also be detected during the peak fit stage by inspecting the residuals of the fit. After the initial fit, the residuals are examined. The presence of an unresolved component causes the residuals to deviate significantly from zero. Thus, one can add another component (peak) to the fitting function and fit the data again. This algorithm (particularly when used in conjunction with shape information) tends to be more sensitive to the detection of multiplet components than the symmetric zero-area transform; however, some care should be taken in the application of this method, as one needs to distinguish between normal statistical fluctuations and significant deviations.

c. Peak area corrections

The next step in the analysis is to correct the observed net area of the peaks for any systematic errors. Such errors can be caused by (but are not limited to)

- Environmental background
- Pulse pileup
- Random summing
- Coincidence summing

Correcting for the presence of environmental background can be done by subtracting the count rate of the environmental background peak from the observed count rate of the sample peak. The count rate of the environmental background peak should be established by counting a sample blank (as opposed to an empty shield), as the sample itself can shield the detector from the source of the environmental background. Counting the empty shield rather than a sample blank biases the environmental background count rate high, resulting in a low bias for the corrected sample count rate.

Pulse pileup, random summing, and coincidence summing all cause an event of energy E to increment a channel that corresponds to an energy $E' > E$. Thus, the net area of the peak at energy E does not accurately represent the number of full-energy events of energy E.

For purposes of this discussion, random summing is defined as two independent depositions occurring within ΔT of one another, where ΔT is less than the discrimination time of the fast discriminator used by the pileup rejector (PUR) (typically, approximately 500 ns), so that the amplifier pulse appears to be a single event.

Pulse pileup is defined as two depositions occurring within ΔT of one another where ΔT is greater than the discrimination time of the fast discriminator used by the pileup rejector but less than the amplifier pulse width, so that one amplifier pulse starts before the preceding pulse has ended. The resulting pulse is thus distorted in that it has an amplitude and/or width that differs from the first event alone. The distinction between pulse pileup and random summing is that the pileup rejector can tell that the pileup pulse is the result of two events (and thus discriminate against these pulses, whereas it cannot discriminate against pulses resulting from random summing).[19]

Coincidence summing is defined as two depositions originating from a single event within ΔT of one another, where ΔT is less than the discrimination time of the fast discriminator used by the pileup rejector, so that the amplifier pulse appears to be a single event. The distinction between coincidence summing and random summing is that coincidence summing originates from a single event. Examples are cascade gammas (such as the 1332-keV gamma that follows the 1173-keV gamma following the decay of ^{60}Co), coincident x and γ rays following electron capture (such as the 14-keV X-ray and 1836-keV γ ray following the electron capture decay of ^{88}Y), and coincident emission of γ rays along with the 511-keV annihilation photons following positron emission (such as in the decay of ^{58}Co in which the daughter emits the 810-keV gamma in coincidence with the annihilation photons). A further distinction between random summing and coincident summing is that coincident summing is independent of count rate, but random summing is a function of count rate.

The discrepancy (bias) caused by pulse pileup between the number of full-energy counts (net peak area) and the number of full-energy events can be reduced by the use of a pileup rejector or corrected for in software by use of a reference peak correction. PURs can keep the observed full-energy count rate (net area divided by live time) within approximately 1% of the true full-energy deposition rate for input count rates[20] below approximately 20,000 counts/s. A PUR is ineffective against random summing, as the summed pulse appears to be a single event, against which the

19. With a further distinction that $\Delta T <$ the linear gate time (LGT) of the ADC in the case of "leading-edge pileup," and LGT $< \Delta T <$ pulse width in the case of "trailing-edge pileup."

20. Where the input count rate must be defined as the full-spectrum input rate, because an event of any size summing with an event of interest will remove the count from the peak of interest.

PUR cannot discriminate. With a discrimination time of 500 ns, the probability of random summing approaches 1%[21] at rates of approximately 20,000 counts/s.

An alternative to PUR is reference peak correction, which can be used to correct for both pileup and random summing (even dead time if the count times are measured in real time rather than live time). The corrected peak area is given by

$$A_C = A_O \frac{P_{NOM}}{P_O} \tag{4.73}$$

where A_O is the observed net peak area, P_{NOM} the nominal (net) count rate of the reference peak, and P_O, the observed (net) count rate of the reference peak.

The reference peak can be produced by either a pulser or a radioactive source in a fixed position. It should be noted that when a pulser is used to inject a signal into the preamplifier, the preamplifier output may have a shape slightly different from that of the output that results from charge injection by the detector (i.e., the detection of a gamma ray). The difference in preamplifier pulse shapes makes it impossible to pole zero the amplifier properly for both pulse types. This will cause a slight degradation in the system resolution.

Both pileup rejection and reference peak corrections are ineffective for correcting for coincidence summing. However, correction factors of the form

$$C = \frac{\text{full-energy events (of energy } E)}{\text{full-energy counts (at energy } E) \text{ (i.e., net peak area)}} \tag{4.74}$$

can be calculated as presented by Andreev et al. (1972, 1973) and McCallum and Coote (1975). Debertin and Schötzig (1979) have extended these equations and incorporated them in the computer program KORSUM, which allows calculation of true coincidence summing corrections for arbitrary decay schemes. This program does not include effects due to angular correlations or coincidence with β rays or bremsstrahlung, as the authors considered contributions from these effects to the total summing correction to be low and smaller than the uncertainty of the correction. The authors used the program to calculate correction factors for a point source geometry and a beaker geometry and obtained good agreement with experimental values.

Note that to correct for true coincidence summing, the correction factors must be calculated and applied on a line-by-line basis, and their calculation requires knowledge of the peak and total efficiencies for the particular detector–geometry combination. The specificity of the correction factors to be determined made it difficult to incorporate the techniques into a commercial analysis package, until now. Within the past few years, however, manufacturers such as Canberra Industries, and Ortec, have developed their own analysis packages that include algorithms to correct for true coincidence summing (or cascade summing) effects. The effects of true coincidence summing and the techniques employed to correct for it are discussed in Section II.B.4. of this chapter.

Virtually, all commercially available spectrometers include live-time correction and pileup rejection, and some systems also include reference peak correction capability in their software.

d. Efficiency calculation

The next step in the analysis is to calculate the efficiency at the energy of each observed peak. Because it would not be practical to try to measure the efficiency at every possible energy that might be encountered in a sample, most software uses an efficiency calibration to calculate the efficiency for any given energy. Typically, these calibrations consist of a functional expression of efficiency as a function of energy:

$$\text{Efficiency}(E) = f(E) \tag{4.75}$$

The functional dependence is typically determined by the method of least-squares fitting or sometimes spline fitting. The energy efficiency coordinates from which these functions are calculated can be determined by measuring a source of known emission rate or by mathematical calculation, such as with the Monte Carlo method. A typical efficiency calibration from a commercial spectroscopy package is shown in Fig. 4.46.

Care should be taken to ensure that the calibration standard represents the samples to be counted. That is, the calibration standard and samples should be identical in size, shape, density, spatial distribution of active material, and so on. Source position is relatively more critical for close geometries. These precautions apply to mathematical calibrations as well as to those performed with radioactive standards.

e. Nuclide identification and activity calculation

The next step in the analysis is to identify the nuclides that are present in the sample and to calculate their activity. Some simple schemes allow the nuclide identification to be independent of the activity calculation, whereas other schemes depend on the activity calculation to identify a nuclide positively. This will be clarified subsequently by example.

A variety of nuclide identification algorithms have been used in commercial software packages. Following is a list

21. Note that this application of a PUR does not imply an overall accuracy or precision of 1%, but rather a limitation on the *bias* caused by just pulse pileup. All of the normal considerations associated with the measurement of a nuclear count rate still apply.

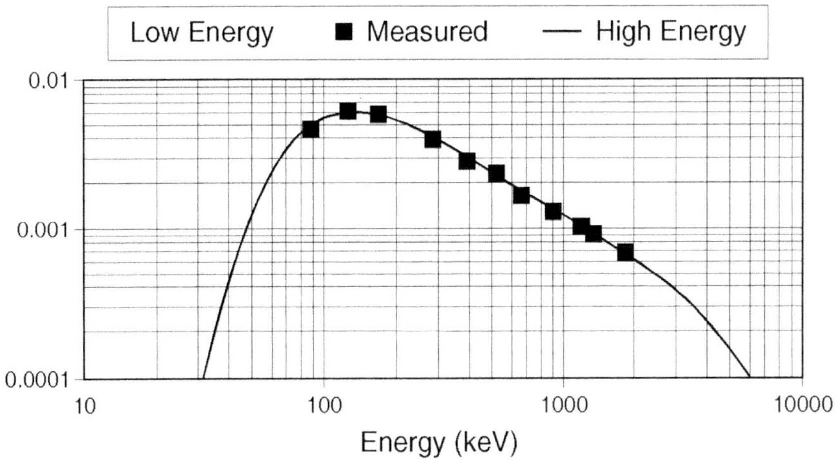

FIGURE 4.46 A typical efficiency calibration as displayed by a Canberra GENIE-PC system.

of some of the nuclide identification algorithms that have been used in commercial packages.

1. The simplest algorithm requires that for a particular nuclide to be identified, a peak must exist in the spectrum (within some user-defined energy tolerance) at every gamma ray energy listed in the library for that nuclide.
2. Another algorithm requires that only a given (user-defined) fraction of the listed gamma rays of a particular nuclide be observed (as peaks within some user-defined energy tolerance) to identify that nuclide positively.
3. Yet another algorithm requires that only a given (user-defined) fraction of the listed gamma rays of a particular nuclide be observed (as peaks within some user-defined energy tolerance) to identify the nuclide potentially, its positive identification being dependent on a nonzero activity being calculated for it.

The equation for activity is obtained from the following expression:

Observed net peak area (E)

$$= \int_{t=0}^{t=T_R} A_O e^{-\lambda t} \text{Intensity}_\gamma(E)\ \text{Efficiency}(E)(1 - DT)dt \tag{4.76}$$

where DT is the fractional dead time. If the dead time is assumed to be constant for the duration of the acquisition, one can solve this equation for A_O to obtain Eq. (4.77) (as stated previously and restated here):

Activity (in Bq at t = 0)

$$= \frac{\text{Net Peak Area } (E)}{T_L \text{Efficiency } (E)\ \text{Intensity}_\gamma\ (E)} \frac{\lambda T_R}{1 - e^{-\lambda T_R}} \tag{4.77}$$

where $T_L = T_R\ (1 - DT)$.

This equation has been applied in a variety of ways in commercial packages. One of the simplest nuclide identification–activity calculation methodologies calculates the activity of a nuclide from a single "key line." Following an independent nuclide identification routine (typically, algorithm 1 or 2 from the preceding list of nuclide identification algorithms), the activities of the identified nuclides are calculated from Eq. (4.77).

The most serious deficiency of these simple independent nuclide identification algorithms occurs in spectra that have an interference (i.e., two or more nuclides contributing to a single peak). In particular, these algorithms tend to produce both false-positive identifications and grossly inaccurate quantifications due to the presence of interfering nuclides. This is easily demonstrated by considering the case of ^{75}Se. If one were to analyze a sample containing only ^{75}Se, there would be (at least) five peaks in the spectrum at 121, 136, 264, 279, and 400 keV. With a 1-keV energy tolerance, these peaks would also be identified as ^{57}Co and ^{203}Hg in addition to ^{75}Se.

If one were to calculate the activities of the "identified nuclides" by assigning the entire area of a peak to each nuclide without regard to the interferences, then for every 100 Bq of ^{75}Se, the following activities (within statistical fluctuations) would be reported:

^{57}Co: 20.0 Bq based on the 121-keV peak
545.6 Bq based on the 136-keV peak
X Bq where $10.0 < X < 545.6$

If a weighted mean activity is employed

^{203}Hg: 30.4 Bq based on the 279-keV peak
^{75}Se: 100.0 Bq based on any of the lines

Of course, for mixtures, the situation is even worse, as none of the nuclides are calculated correctly. To remedy this situation, one of two algorithms is generally employed. The first makes use of an interference library in which one explicitly identifies which lines of which nuclides interfere with what other nuclides. The activity of the nuclide that

^{75}Se		^{57}Co		^{203}Hg	
Energy	Intensity	Energy	Intensity	Energy	Intensity
121.117	17.14%	122.061	85.60%		
136.001	58.27%	136.474	10.68%		
264.658	58.5%				
279.544	25.79%			279.197	81.46%
400.660	11.37%				

has lines without interference is then calculated from the peaks for which there is no interference (e.g., the 264- and 400-keV lines of ^{75}Se in the preceding case). This nuclide's contribution to the peaks with which it interferes is then calculated and subtracted from the observed area of these peaks. In the preceding example, the contribution of ^{75}Se to the 121- and 136-keV peaks is calculated from the calculated activity of ^{75}Se and subtracted from the observed areas of the 121- and 136-keV peaks. The balance of the peak areas is then attributed to ^{57}Co, and its activity is calculated from these "corrected" peak areas. Similarly, the contribution of ^{75}Se to the 279-keV peak is calculated from the calculated activity of ^{75}Se and subtracted from the observed area of the 279-keV peak. The remaining area is then attributed to ^{203}Hg and used to calculate the activity of ^{203}Hg. One limitation of this algorithm is that the user must explicitly identify all of the potential interferences before the analysis for this method to identify the interferences that are present.

Another algorithm, originally proposed by Gunnink and Niday (1972) and implemented in SAMP080 (Koskelo et al., 198 1) and GENIE-PC (Koskelo and Mercier, 1995), that can resolve this situation involves setting up a simultaneous set of equations that expresses each potentially identified nuclide's contribution to each observed peak. Using the previous example, in which the observed peaks (at 121-, 136-, 264-, 279-, and 400-keV) cause ^{75}Se, ^{57}Co, and ^{203}Hg to be identified, one would obtain the following set of equations:

$$\text{Int}_{121-\text{Se}} * \text{A}_{\text{Se}} + \text{Int}_{122-\text{Co}} * \text{A}_{\text{Co}} = \frac{\text{Peak Area}(121)}{\varepsilon(121)^{*} T_{\text{L}}} \tag{4.78a}$$

$$\text{Int}_{136-\text{Se}} * \text{A}_{\text{Se}} + \text{Int}_{136-\text{Co}} * \text{A}_{\text{Co}} = \frac{\text{Peak Area}(136)}{\varepsilon(136)^{*} T_{\text{L}}} \tag{4.787b}$$

$$\text{Int}_{264-\text{Se}} * \text{A}_{\text{Se}} = \frac{\text{Peak Area}(264)}{\varepsilon(264)^{*} T_{\text{L}}} \tag{4.78c}$$

$$\text{Int}_{279-\text{Se}} * \text{A}_{\text{Se}} + \text{Int}_{279-\text{Hg}} * \text{A}_{\text{Hg}} = \frac{\text{Peak Area}(279)}{\varepsilon(279)^{*} T_{\text{L}}} \tag{4.78d}$$

$$\text{Int}_{400-\text{Se}} * \text{A}_{\text{Se}} = \frac{\text{Peak Area}(400)}{\varepsilon(400)^{*} T_{\text{L}}} \tag{4.78e}$$

in which A_{Se} is the (unknown) activity of ^{75}Se; A_{Co} the (unknown) activity of ^{57}Co; A_{Hg} the (unknown) activity of ^{203}Hg; $\text{Int}_{121\text{-Se}}$ the gamma intensity of the 121-keV emission from ^{75}Se, and so on; peak area(121) is the observed peak area of the 121-keV peak; $\varepsilon(121)$ is the peak efficiency at 121-keV, and T_{L} is the live time of the data acquisition.

This set of five equations in only three unknowns is obviously overdefined; however, a best fit solution can be obtained by minimizing chi squared (the sum of the squares of the residuals[22]). Because each observed emission rate has some uncertainty associated with its measurement, we might prefer to place more weight on the lines that have the smallest uncertainty and less weight on the lines that have the greatest uncertainty. This weighting can be accomplished by multiplying the residual (for each line) by a weighting factor, which is the inverse of the variance of the measured emission rate (for that line) so that a weighted chi squared is given by

$$\chi^2 = \sum_{i=\text{lines}} \sum_{j=\text{nuclides}} w_i \left[\frac{\text{Peak Area}(i)}{\varepsilon(i) T_{\text{L}}} - \text{Int}_{ij} A_j \right]^2 \tag{4.79}$$

This quantity is minimized when the activities, A_j, satisfy the condition

$$\frac{\partial \chi^2}{\partial A_j} = 0 \text{ for all } \text{A}_\text{j} \tag{4.80}$$

The solutions obtained with this formalism are the weighted average activities for each nuclide. One advantage of this algorithm is that the user is not required to

22. The residuals being defined as the differences between the observed emission rate and the emission rate implied by the solution.

identify explicitly all of the potential interferences before the analysis for this method to identify the interferences that are present. As long as the nuclides that need to be included in the interference set satisfy the identification criteria of the nuclide identification algorithm, their interferences will be recognized by this method.

To take fullest advantage of this algorithm, one should have a very complete nuclide library—in terms of both nuclides and lines of each nuclide—to ensure that all potential nuclides and interferences can be identified. To illustrate the need for a complete library, consider the analysis of a natural soil containing ^{228}Ac, which has a gamma emission of 835.7 keV with an intensity of approximately 1.7%. Because of this low intensity, people often omit it from their library. As a result, ^{54}Mn (which has a gamma emission of 835.8 keV) is reported because there is nothing else to which the 835-keV peak can be attributed. If the 835.7-keV line is included in the library for ^{228}Ac, the 835 keV peak can be (correctly) attributed to the ^{228}Ac, and the simultaneous solution can eliminate the (false) identification of the ^{54}Mn.

Another precaution that should be taken when setting up libraries is related to the half-life that is entered for each nuclide. Most commercial software implement corrections for decay during data acquisition decay back to some sample time and, if the sampling occurred over an extended period, decay during sample collection. These corrections produce erroneous (high) values for short-lived material in equilibrium with a long-lived parent if the true half-life of the short-lived material is used for the decay correction. If the short-lived material is in equilibrium with the long-lived parent, one can substitute the half-life of the long-lived parent in the decay corrections by using the parent's half-life in the short-lived daughter's library entry.

VI. Advances in HPGe detector technology: gamma-ray imaging with HPGe detectors

With modern low-noise electronics and the quality of Ge material available, the energy resolution that can be achieved using HPGe detectors is very close to the fundamental limits imposed by charge carrier generation statistics. So, further improvement in energy resolution is not expected. The attention has shifted to manufacturing larger volume crystals that can yield higher detection efficiencies. Another area of focus for HPGe applications is gamma ray imaging (Luke et al., 2005).

There are different methods to image gamma rays, including the use of parallel-hole collimators, pinholes, coded apertures (Fennimore and Cannon, 1978), and Compton cameras (Phillip 1995). In almost all cases, position-sensitive detectors are needed to implement an imaging system. A common method to provide position sensitivity is to segment the contacts of a detector. Conventional contacts used in Ge detectors are lithium (Li)-diffused contacts (n-type) and boron-(B) implanted contacts (p-type). The Li contacts are typically several hundred microns thick, making it difficult to segment using standard etching processes. Often times, mechanical means such as cutting using diamond saws or ultrasonic machining were employed, although segmentation using a photolithography technique has also been developed. Whatever the segmentation methods employed, the relatively large thickness of the Li contact limits the minimum feature size of contact elements to >1 mm. In addition, the diffusion of Li, which occurs even at room temperature, could lead to electrical shorts between contact elements when detectors are stored at room temperature for long periods of time, or when radiation damage annealing becomes necessary. Another difficulty in making multisegment detectors is the instability of the Ge surface. When contacts of a detector are segmented, the intrinsic Ge surface between the contact elements can often become unstable with time or with changes in ambient conditions, which may lead to increased leakage current, increased noise, or inadequate isolation between contact elements. Some type of surface passivation is desirable, but it must be able to be applied to the surface between contacts without affecting the ability to make electrical connections to each contact element. In recent years, improved contact segmentation techniques have been developed, contributing to the increasing use of segmented Ge coaxial and planar detectors (Vetter et al., 2000; Yang et al., 2001; Philips et al., 2002).

In recent years, improved contact segmentation techniques have been developed, contributing to the increasing use of segmented Ge coaxial and planar detectors. As an example, the design developed at Lawrence Berkley National Laboratory (LBNL) is highlighted (Luke et al., 2005). The LBNL design employs contacts made of an amorphous-Ge (a-Ge) thin film to create an electrically blocking junction between it and the bulk Ge crystal. The blocking junction permits the application of bias voltage to fully deplete the detector while maintaining a low leakage current. The structure of a multisegment detector using such contacts is shown schematically in Fig. 4.47. The fabrication process involves the sputter deposition of an a-Ge layer onto the surface of the Ge crystal. Selected areas of the a-Ge film are then metalized to form the contact elements. The resistivity of the a-Ge film is sufficiently high to provide good electrical isolation between contact elements, whereas the small thickness of the film ensures that it does not affect the detector signals or the passage of signal currents to the metal contacts. An interesting property of the a-Ge contact is its ability to block carrier injection under either polarity of bias voltage. Therefore, the same process can be used on both sides of a detector to

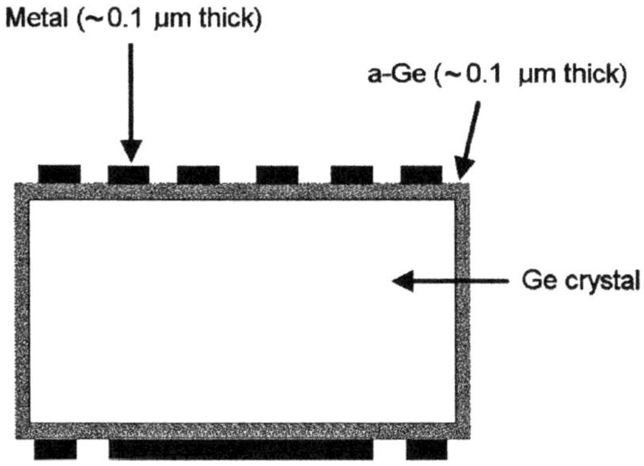

FIGURE 4.47 Structure of a-Ge contact. *From Luke et al. (2005); reprinted with permission from Springer Nature © 2005.*

FIGURE 4.48 Compton image of highly enriched uranium sample, taken using a GeGI Spectrometer (Dreyer, 2014).

realize double-sided segmented detectors. One advantage of this contact technology is its simplicity, which involves very few processing steps and can be carried out entirely with the detectors at room temperature. Another advantage is that the a-Ge film covers the full surface of the Ge crystal, forming a passivation layer. There is no need for additional processes to passivate the surface between contact elements afterward. The amorphous contacts are thin and stable with annealing up to 150 °C. Very small-scale contact structures, on the order of ~100 μm or less, can be realized.

A number of orthogonal strip detectors have been successfully produced using the a-Ge contact technology (Amman and Luke, 2000a; Hull et al., 2001). These detectors are amenable to the development of practical gamma ray imaging systems.

In addition to the two-dimensional position information obtained from the orthogonal strips, the depth of gamma ray interaction in the detector can also be determined by measuring the time difference between signals from strips on opposite sides of the detector (Amman and Luke, 2000b). The ability to localize events within the detector in three dimensions improves the accuracy of reconstructed images.

Dreyer et al. (2014) describe the application of a portable, mechanically cooled gamma ray imaging spectrometer referred to as the GeGI, which has become commercially available (PHDs Co). The imaging capability is largely due to the advances made in the segmentation techniques (Hull and Pehl, 2005; Hull et al., 2011). The GeGI is a 16 × 16 strip detector with 5-mm pitch strips on an 8 cm diameter, 1-cm-thick germanium crystal. As a spectrometer, the instrument has a useable energy range from 30 keV to several MeV. Two imaging modalities are available, the choice of which depends on the energy of interest. For gamma-ray energies ranging from 40 to 500 keV, a collimator can be placed in front of the detector for pinhole imaging. This modality offers high contrast and spatial resolution; however, it is limited to a 60 degrees forward field of view. Compton imaging can be used for higher energy radiation from ~150 keV to <1 MeV and provides a 4π field of view with a spatial resolution of ~6 degrees.

Fig. 4.48 shows the Compton image of highly enrichment uranium (~100 g of material) located in a building approximately 28 m away, taken using a GeGI spectrometer. The 186 keV gamma ray from ^{235}U was detected spectrally in 15 minutes and localized within 40 minutes (Dreyer, 2014).

VII. Segmented Ge detectors and their applications in nuclear physics research

The structure of the atomic nucleus has been studied to a large extent by gamma ray spectroscopy using germanium detectors. Ge detectors were developed in the 1960s and still offer the best compromise between energy resolution and efficiency for high-resolution gamma-ray spectroscopy. In the 1970s and 1980s, the main advances were in the purity and volume of the Ge crystals. More recently, the development of highly segmented Ge detectors has significantly increased the performance and power of detection systems. In a segmented Ge detector, the crystal is divided into sections by isolating the contacts. This enables the position of interaction to be determined. The accuracy of the position determination can be improved by the analysis of the shape of the charge signals and of the transient signals in adjacent segments. This technique has led to the concept of gamma ray tracking in a segmented Ge detector whereby the energy, time, and position of all interactions are recorded and the full interaction is reconstructed. This enables a spectrometer system with unprecedented efficiency and energy resolution to be realized.

A. Segmented HPGe detectors

Gamma ray imaging using HPGe detectors has become possible as a result of the segmentation technology that was developed in the 1990s. In his paper titled "Photomask Technique for Fabricating High Purity Germanium Strip Detectors," D. Gutknecht (Gutknecht 1990) describes a new technology for realizing germanium strip detectors. The technique uses a photomask process that needs no growing and etching and allows various kinds of segmented electrodes. Using his technique on planar detectors, Gutknecht successfully produced a double-sided X−Y HPGe strip detector with a thickness of 30 mm, and a pitch of 200 μm. The segmentation techniques that Gutknecht pioneered have been implemented on coaxial HPGe detectors. Large-volume segmented HPGe detectors have been available for a couple of decades now (Shepherd et al., 1999; Boston 2000; Boston et al., 2007, 2009; Akkoyun et al., 2012).

The use of internal and external contacts of a segmented detector provides information on interaction position: (1) vertically and transversally by analyzing signals induced by mirror charges, and (2) radially by performing a pulse shape analysis. The external contact of a detector can be longitudinally or transversally segmented without a dead zone generation. Segmentations on the front side of the detector (like a checker-board pattern) and laterally (in one or two directions) can be realized. Segmentation separation is such that cross-talk between adjacent segments is avoided. Fig. 4.49 shows the picture of a segmented germanium detector.

B. Neutrinoless double beta decay

Double beta decay is a type of radioactive decay in which two protons are simultaneously transformed into two neutrons, or vice versa, inside an atomic nucleus. Unstable nuclei may undergo $2\nu\beta\beta$ if single β-decay is energetically forbidden or highly suppressed by conservation of angular momentum. This process has been observed in many nuclei and has a half-life on the order of $10^{19}-10^{21}$ years. As in single beta decay, this process allows the atom to move closer to the optimal ratio of protons and neutrons. As a result of this transformation, the nucleus emits two detectable beta particles, which are electrons or positrons.

The literature distinguishes between two types of double beta decay: ordinary double beta decay and neutrinoless double beta decay. In ordinary double beta decay, which has been observed in several isotopes, two electrons and two electron antineutrinos are emitted from the decaying nucleus. In neutrinoless double beta decay, a hypothesized process that has never been observed, only electrons would be emitted. This is possible only if neutrinos are their own antiparticles (called Majorana particles). In this case, neutrinoless double-β ($0\nu\beta\beta$) decay could be observed, a process in which lepton number is not conserved. Several experiments are underway or are under preparation searching for this decay using a variety of suitable isotopes. The sum of the kinetic energies of the two electrons emitted in the $0\nu\beta\beta$ decay is equal to the mass difference or (Q value) $Q_{\beta\beta}$ (equal to 2.039 MeV) of the two nuclei. A sharp peak in the energy spectrum is the prime signature for all $0\nu\beta\beta$ experiments. Key parameters of these rare event searches are large mass M and long measuring time t on the one hand, and high energy resolution and low background on the other. ^{76}Ge is one of the isotopes that undergoes double beta decay. Therefore, using an HPGe detector, enriched in ^{76}Ge, would be very convenient; the sample and the detector will be one and the same! There are two major collaborations, Majorana and the GERmanium Detector Array (GERDA) that are currently in progress, based on the approach of using HPGe detectors, enriched in ^{76}Ge.

FIGURE 4.49 55% n-type coaxial HPGe of tapered shape segmented in 36 folds. *Image courtesy of Mirion Technologies, "Segmented Coaxial Detectors".*

C. Majorana collaboration

The Majorana project is an international effort to search for neutrinoless double-beta (0νββ) decay in ^{76}Ge. The project builds upon the work of previous experiments, notably those performed by the Heidelberg–Moscow (Klapdor-Kleingrothaus et al., 2001) and IGEX (Aalserth et al., 2000) collaborations, which used high-purity germanium (HPGe) detectors, to study neutrinoless double-beta decay (Klapdor-Kleingrothaus et al., 2004).

The first stage of the project is the Majorana Demonstrator (Fig. 4.50, designed to demonstrate the backgrounds low enough to justify building a ton-scale experiment, and to establish feasibility to construct and field modular arrays of Ge detectors. Cryostats housing up to 40 kg of natural and enriched germanium detectors are being deployed in low-background vacuum cryostats, underground at the Sanford Underground Laboratory in Lead, South Dakota. Following the Demonstrator, the collaboration intends to merge with the GERDA collaboration to build a much larger experiment.

The principal goal of the Majorana Demonstrator is to demonstrate the feasibility of achieving the background required in a ton-scale experiment. This corresponds to four counts/ton/yr in a 4 keV window around the 0νββ Q value of 2.039 MeV, which scales to one count/ton/yr in a ton-scale experiment. The experiment will use a mixture of detectors made with natural germanium and enriched germanium, allowing it to confirm or refute the controversial claim for 0νββ observation in ^{76}Ge. Electroformed copper and lead bricks protect the cryostats. Polyethylene shields the setup and includes PMTs to act as a veto. Nitrogen flushing removes trace radon. If low enough electronic noise is achieved, the Demonstrator may also be used to search for weakly interacting massive particles (WIMPs)

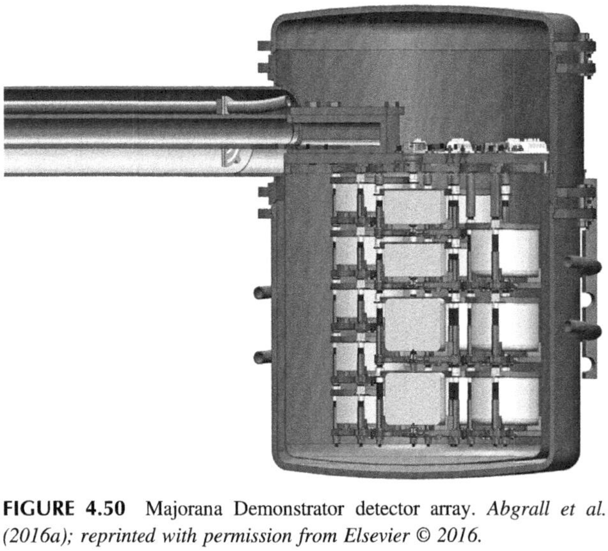

FIGURE 4.50 Majorana Demonstrator detector array. *Abgrall et al. (2016a); reprinted with permission from Elsevier © 2016.*

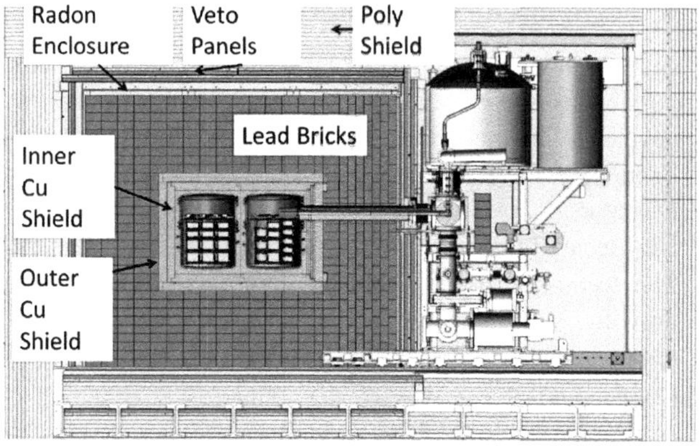

FIGURE 4.51 Schematic drawing of Majorana Demonstrator setup. *From Abgrall et al. (2016b); reprinted with permission from Elsevier © 2016.*

in dark matter research. The Majorana Demonstrator measurement setup is illustrated in Fig. 4.51.

The detector array consists of "strings" of P-type point-contact (PPC) detectors (Luke et al., 1989), with four to five detectors stacked in a string. The modular cryostat has the cooling capacity for seven strings of detectors (up to 22.5 kg of Ge). The Majorana Demonstrator detector array is illustrated in Fig. 4.52. The insets in Fig. 4.52 show an individual PPC detector and a "string" of detectors.

PPC detectors allow efficient discrimination of multiple scattering gamma backgrounds. This results from the weighting potential being strongly peaked close to the small electrode, meaning that as charge drifts toward the electrode, there is a high probability of seeing distinct signals from each energy deposition, thus being able to reject events corresponding to these signals. Other advantages include the low capacitance due to the small contacts, reducing electronic noise and thresholds; and shielding surface alpha decays by the thick outer n-type contact.

The projected background rates from Majorana Demonstrator experiment is shown in Fig. 4.53.

D. GERDA collaboration

The GERDA experiment is located at the Gran Sasso underground laboratory (LNGS) of Instituto Nazionale di Fisica Nucleare (INFN) or "National Institute of Nuclear Physics" in Italy. HPGe detectors made from material with enriched ^{76}Ge fraction of $\sim$87% are operated in a 64 m^3 liquid argon (LAr) bath. The argon cryostat is located inside a tank filled with 590 m^3 of high-purity water. LAr and water shield against the external radioactivity. The water tank is instrumented with photomultipliers and operates as a Cherenkov detector to veto residual muon-induced events. Material for structural support of the detectors and for cabling is minimized to limit the background from nearby radioactive sources.

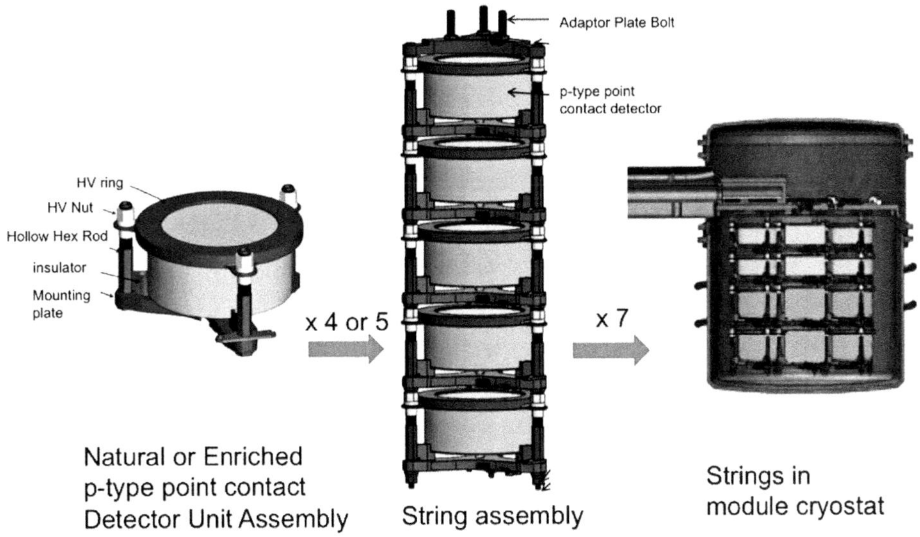

FIGURE 4.52 Majorana Demonstrator detector array. *From Xu et al. (2015); reprinted with permission from Elsevier © 2915.*

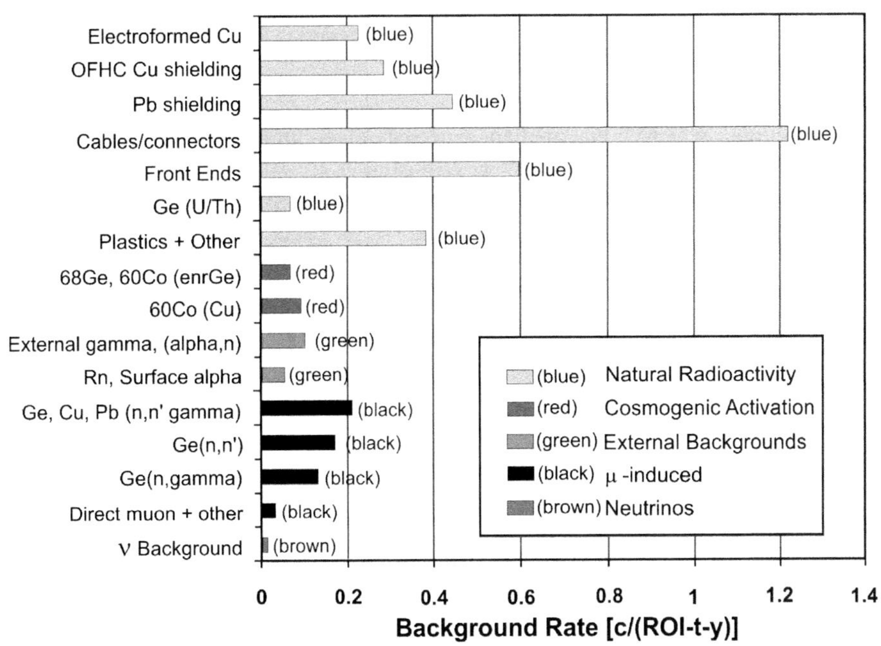

FIGURE 4.53 Projected background rates from Majorana Demonstrator. *From Green et al. (2015); reprinted with permission © 2015.*

A schematic drawing of the GERDA experiment is illustrated in Fig. 4.54.

A first phase of data taking ended in 2013 with no indication of a signal (Agostini et al., 2013). The background index achieved at the ^{76}Ge Qββ value of 2.039 MeV was 10^{-2} counts/(keV Kg Year). For the second phase, a new component has been installed to detect argon scintillation light (Agostini, to be published). The enriched germanium mass was doubled in the form of small readout electrode detectors (the Canberra Broad Energy Germanium or BEGe detector model) supplementing the previously used coaxial detectors. A schematic drawing of BEGe detectors is shown in Fig. 4.55.

The detector is made of p-type HPGe with Li^- drifted by n^+ contact (0.7 mm specified thickness) covering almost the entire outer surface of the detector. The small p^+ contact is located in the middle of the bottom surface.

Both enhancements allow for a more efficient rejection of background events, which can be characterized by their energy deposition in the LAr, in several detectors or in several locations (including the surface) of a single detector. In contrast,

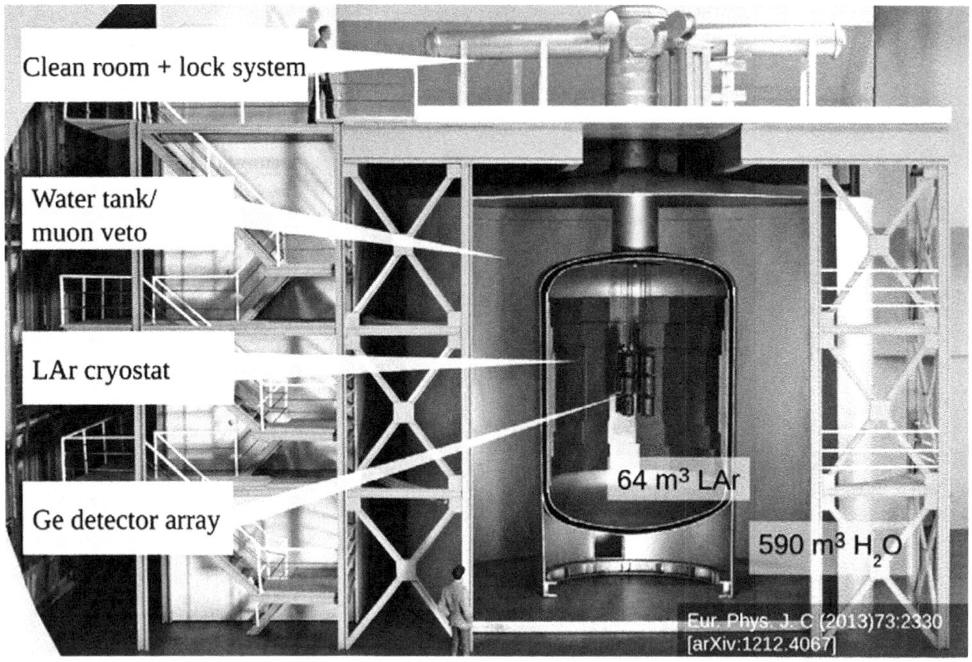

FIGURE 4.54 Section of the GERDA experiment. Human figures are illustrated in the upper left and lower center of the figure to illustrate the relative dimensions of the experimental setup. *From D'Andrea et al. (2016).*

0νββ energy deposits are made by two electrons, which typically release all their energy in a small volume of a single detector. Localized and delocalized energy deposits are distinguished by pulse shape discrimination (PSD) based on the time profile of the detector signal.

Phase II data taking started in December 2015 with a target background index of 10^{-3} counts/(keV Kg Year), a tenfold reduction of background with respect to Phase I. Thirty BEGe detectors (20.0 kg total mass) and seven coaxial detectors (15.6 kg) are deployed, whose energy resolution at $Q_{\beta\beta}$ is typically better than 3 and 4 keV FWHM, respectively.

As in Phase I, a 25-keV window around Qββ was blinded: events with an energy in one detector within this window were hidden until the entire data selection was finalized. The first unblinding of Phase II took place in June 2016, and no 0νββ signal was found. A lower limit of half-life for neutrinoless decay $T_{1/2}^{0\nu} > 5.3 \times 10^{25}$ yr (90% confidence level) was extracted with a sensitivity, defined as

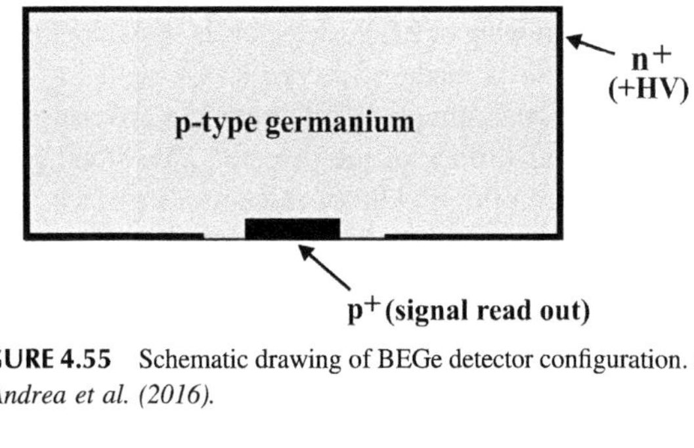

FIGURE 4.55 Schematic drawing of BEGe detector configuration. *From D'Andrea et al. (2016).*

the median expected lower limit assuming no signal, of 4.0×10^{25} yr (Agostini et al., 2017).

The low background index (BI) of BEGe detectors, previously based on one single event, has been confirmed with a more than threefold exposure to be 1×10^{-3} counts/(keV Kg Year). If normalized according to the energy resolution and total signal efficiency ε, i.e., BI × FWHM/ε, this value corresponds to 4.9 counts/(ton year). Hence the average background in the energy interval 1 × FWHM at $Q_{\beta\beta}$ is expected to be less than 1 for the entire design exposure of 100 kg yr.

With the same normalization for BI, the Majorana collaboration has reported a background is 5.7 counts/(ton year) (Aalserth et al., 2018).

References

Aalserth, C.E., et al., 2000. Recent results of the IGEX 76Ge double-beta decay experiment. Phys. At. Nucl. 63 (7), 1225–1228.

Aalserth, C.E., et al., 2018. (Majorana Collaboration), preceding letter. Phys. Rev. Lett. 120, 132502.

Abgrall, N., et al., 2016a. The Majorana demonstrator radioassay program. Nucl. Instrum. Methods Phys. Res., Sect. A 828, 22–36.

Abgrall, N., et al., 2016b. High voltage testing for Majorana demonstrator. Nucl. Instrum. Methods Phys. Res., Sect. A 823, 83–90.

Aggarwal, S.K., Duggal, R.K., Shah, P.M., Rao, R., Jain, H.C., 1988. Experimental evaluation of the characteristic features of passivated ion implanted and surface barrier detectors for alpha spectrometry of plutonium. J. Radioanal. Nucl. Chem. 120, 29.

Agostini, M., Ur, C.A., Budjas, D., Bellotti, E., Brugnera, R., Cattadori, C.M., di Vacri, E., Garfagnini, A., Pandola, L., Schönert, S., 2011. Signal Modeling of High-Purity Ge Detectors with a Small Read-Out Electrode and Application to Neutrinoless Double Beta Decay Search in Ge-76. IOP Publishing Ltd and SISSA.

Agostini, M., et al., 2013. GERDA collaboration. Phys. Rev. Lett. 111, 122503.

Agostini, M., et al., 2017. GERDA collaboration. Nature (London) 544, 47.

Akkoyun, S., et al., 2012. AGATA-advanced – advanced gamma tracking array. Nucl. Instrum. Methods Phys. Res. A668, 26–58.

Amoudry, F., Burger, P., 1984. Determination of the 239Pu/240Pu isotopic ratio by high resolution alpha spectrometry. Nucl. Instrum. Methods Phys. Res. 223, 360–367.

Amman, M., Luke, P.N., 2000a. Nucl. Instrum. Methods A452, 155.

Amman, M., Luke, P.N., 2000b. Proc. SPIE 4141, 144.

Andreev, D.S., Erokhina, K.I., Zvonov, V.S., Lemberg, 1 K., 1972. Instrum. Exp. Tech. 25, 1358.

Andreev, D.S., Erokhina, K.L., Zvonov, V.S., Lemberg, 1 K., 1973. Izv. Akad. Nauk. SSR Ser. Fiz. 37 (No. 8), 1609.

ANSI N42.14, 1991. American National Standard Calibration and Use of Germanium Spectrometers for the Measurement of Gamma Ray Emission Rates of Radionuclides. Copyright © 1991 by the Institute of Electrical and Electronics Engineers, Inc (IEEE). The IEEE disclaims any responsibility or liability resulting from the placement and use in the described manner. Information is reprinted with the permission of the IEEE.

Barbeau, P.S., Collar, J.I., Tench, O., 2007. Large-mass ultralow noise germanium detectors: performance and applications in neutrino and astroparticle physics. J. Cosmol. Astropartcile Phys. 2007 (9). Art. No. 009.

Berlizov, A., 2006. MCNP-CP - a correlated particle radiation source extension of a general purpose Monte Carlo N-particle transport code. In: Semkow, T.M., et al. (Eds.), Applied Modeling and Computations in Nuclear Science. ACS Symposium Series 945. American Chemical Society, Washington, DC, pp. 183–194.

Blaauw, M., 1993. The use of sources emitting coincident γ-rays for determination of absolute efficiency curves of highly efficient Ge detectors. Nucl. Instrum. Methods Phys. Res., Sect. A 332, 493–500.

Bock, R., 1979. A Handbook of Decomposition Methods in Analytical Chemistry. International Textbook Company.

Boston, A.J., 2000. "Recent developments in large gamma-ray arrays: a European perspective", Capture gamma-ray spectroscopy and related topics. AIP Conf. Proc. 529, 357–363.

Boston, A.J., Boston, H.C., Cresswell, J.R., Dimmock, M.R., Nelson, L., Nolan, P.J., Rigby, S., Lazarus, I., Simpson, J., Medina, P., Santos, C., Parisel, C., 2007. Gamma-ray tracking: characterisation of the AGATA symmetric prototype detectors. Nucl. Instrum. Methods Phys. Res. B261, 1098–1102.

Boston, A.J., Dimmock, M.R., Unsworth, C., Boston, H.C., Cooper, R.J., Grint, A.N., Harkness, L.J., Lazarus, I.H., Jones, M., Nolan, P.J., Oxley, D.C., Simpson, J., Slee, M., 2009. Performance of an AGATA asymmetric detector. Nucl. Instrum. Methods Phys. Res. A604, 48–52.

Briesmeister, J., 2002. MCNP – A General Monte Carlo N-Particle Transport Code. Los Alamos National Laboratory Publication LA-13709-M. Version 4C.

Bronson, F., Atrashkevitch, V., 2001. Calibration Method for Radiation Spectroscopy. U.S. Patent 6.228.664 B1.

Bronson, F.L., Young, B.M., 1997. Mathematical calibrations of germanium detectors and instruments that use them. In: Proceedings of the 5th Annual NDA/NDE Waste Characterization Conference, Salt Lake City, Utah, January, 1997.

Budjas, D., Heider, M.B., Chkvorets, O., Khanbekov, N., Schönert, S., 2009. Pulse Shape discrimination studies with a Broad-Energy Germanium detector for signal identification and background suppression in the GERDA double beta decay experiment. J. Instrum. 4 (10), 24. Article No. P10007.

Bunting, R.L., Kraushaar, J.J., 1974. Short lived radioactivity induced in Ge(Li) gamma ray detectors by neutrons. Nucl. Instrum. Methods 118, 565–572.

Burger, P., De Backer, K., Schoemnaeckers, W., 1985. 2nd International Technical Symposium on Optical and Electro Optical Science and Engineering. Cannes, France.

Burnett, W.C., 1990. Alpha Spectrometry: A Short Course Emphasizing the Practical Applications of Alpha Spectrometry. Canberra Industries, Meriden, CT.

Burnett, W.C., 1992. Advanced Alpha Spectrometry: A Short Course Emphasizing Advanced Techniques in Alpha Spectrometry. Canberra Industries, Meriden, CT.

Cable, P., Burnett, W.C., Hunley, D., Winnie, J., McCabe, W., Ditchburn, R., 1994. Investi¬gating the chemical and physical controls on electrodeposition for alpha spectrometry. In: Pro-ceedings, 40th Conference on Bioassay, Analytical and Environmental Radiochemistry, Nov. 13 17,1994, Cincinnati, OH.

Carchon, R., Van Camp, E., Knuyt, G., Van De Vijver, R., Devos, J., Ferdinande, H., 1975. A general solid angle calculation by a Monte Carlo method. Nucl. Instrum. Methods 128, 195–199.

Castoldi, A., Chen, W., Gatti, E., Holl, P., Rehak, P., 2000. Fast silicon drift photodiodes free from bias connections on the light entering side. Nucl. Instrum. Methods A439, 483–496.

Celnikier, L.M., 1996. Cherche source des rayons cosmiques … Désespérément. Bull. Soc. Fr. Phys. 108, 6–10.

CERN Applications Software Group, 1994. GEANT: Detector Description and Simulation Tool. CERN Program Library Long Writeup W5013.

Ceuppens, M., Verplancke, J., Tench, 0., 1996. Low background germanium detectors; envi¬ronmental laboratory to underground counting facility. In: Presented at the Workshop on Meth-Ods and Applications of Low Level Radioactivity Measurements, Nov. 7 8. D Rossenclorf–Dresden.

Y. Cui, A. Bolotnikov, G. Camarda, A. Hossain, G. Yang and R.B. James, "CZT Virtual Frisch-Grid Detector: Principles and Applications", 2009 Brookhaven National Laboratory Report BNL-82046-2009-CP.

Currie, L.A., 1968. Limits for qualitative detection and quantitative determination: application to radiochemistry. Anal. Chem. 40, 586–593.

D'Andrea, V., 2016. "Status report of GERDA phase II startup", 8th international conference on neutrino physics and astrophysics. Int. J. Math. Comput. Phys. Electr. Comput. Eng. 10, 3.

De Corte, F., 1987. The K0-Standardiation Method: A Move to the Optimization of Reactor Neutron Activation Analysis (Doctoral thesis). University of Gent, Belgium.

Debertin, K., Helmer, R.G., 1988. Gamma and X Ray Spectrometry with Semiconductor Detectors. North Holland, Amsterdam.

Debertin, K., Schötzig, U., 1979. Coincidence summing corrections in Ge (Li) spectrometry at low source to detector distances. Nucl. Instrum. Methods 158, 471–477.

Del Sordo, S., Abbene, L., Caroli, E., Mancini, A.M., Zappettini, A., Ubertini, P., 2009. Progress in the development of CdTe and CdZnTe semiconfuctor radiation detectors for astrophysical and medical applications. Sensors 9 (5), 3491–3526.

Dhanraj, G., Byrappa, K., Prasad, V., Dudley, M. (Eds.), 2010. Springer Handbook of Crystal Growth, New York.

Ditchburn, R.G., McCabe, W.J., 1984. An Improved Method for the Purification and Electro¬deposition of Protactinium for Application to the INS Uranium Series Dating Project. Institute of Nuclear Sciences (New Zealand) R 325.

J.G. Dreyer, M.T. Burks, E.L. Hull, "Next generation germanium systems for safeguards applications", U.S. Department of energy, LLNF-CONF-660195. Proc. of Symposium on International Safeguards: Linking Strategy, Implementation and People, October 20-24, 2014, Vienna, Austria.

Duchêne, G., Moszynski, M., 1995. Ballistic deficit correction method for large Ge detectors. High counting rate study. Nucl. Instrum. Methods Phys. Res., Sect. A 357, 546–558.

Dutta, P.S., 2010. Bulk crystal growth of Ternary III-semiconductors. In: Dhanaraj, G., Byrappa, K., Prasad, V., Dudley, M. (Eds.), Springer Handbook of Crystal Growth. Springer Nature, pp. 281–325, p. 1451.

Eichrom Industries, Inc, 1995a. ACW03: Americium, Plutonium and Uranium in Water.

Eichrom Industries, Inc, 1995b. ACW01: Uranium and Thorium in Water. Illinois.

Eichrom Industries, Inc, 1994. ACS06: Uranium and Thorium in Soil (Single Column).

Eichrom Industries, Inc, 1995. ACW08: Thorium and Neptunium in Water. Illinois.

Evans, R.D., 1955. The Atomic Nucleus. McGraw-Hill, New York.

Fennimore, E.E., Cannon, T.M., 1978. Coded aperture imaging with uniformly redundant arrays. Appl. Opt. 17, 337.

Fiorini, C., Gola, A., Zanchi, M., 2006. Gamma-ray spectrocopy with LaBr3:Ce scintillator readout by a silicon drift detector. IEEE Trans. Nucl. Sci. 53 (4), 2392–2397.

Friedman, R.J., Reichard, M.C., Blue, T.E., Brown, A.S., 2001. Evaluation of scatter contribution from shielding materials used in scatter measurements for calibration range characterization. Health Phys. 80, 54–61.

Frisch, O., 1944. British atomic energy report, BR-49.

Garcia Toraño, E., Aceña, M.L., 1981. NOLIN: nonlinear analysis of complex alpha spectra. Nucl. Instrum. Methods 185, 261–269.

Gardner, R., Verghese, K., Lee, H.M., 1980. The average solid angle subtended by a circular detector coaxial to a isotopic source. Nucl. Instrum. Methods 176, 615–617.

Goulding, F.S., Landis, D.A., 1982. Signal processing for semiconductor detectors. IEEE Trans. Nucl. Sci. 29, 1125–1141.

Green, M.P., et al., 2015. The Majorana Demonstrator for 0nbb: Current Status and Future Plans. Physics Procedia 61, 232–240.

Greenwood, R.C., Chrien, R.E., 1980. Precise γ ray energies from the 14N(n,γ)15N and 23Na(n,γ)24Na reactions. Nucl. Instrum. Methods 175, 515–519.

Gunnink, R., 1979. Computer techniques for analysis of gamma ray spectra. In: Proceedings, ANS Topical Conference, Computers in Activation Analysis for Gamma Ray Spectroscopy. CONF 780421, Mayaguez, Puerto Rico, pp. 109–138.

Gunnink, R., 1990. MGA: a gamma ray spectrum code for determining plutonium isotopic abundances. In: Methods and Algorithms, vol. 1. UCRL LR 103220, Lawrence Livermore National Laboratory.

Gunnink, R., Niday, J.B., 1972. Computerized Quantitative Analysis by Gamma Ray Spectrometry, vol. 1. Lawrence Livermore National Laboratory Rept. UCRL 51061.

Gutknecht, D., 1990. Photomask technique for fabricating high purity germanium strip detectors. Nucl. Instrum. Methods Phys. Res. A288, 13–18.

Helmer, R.G., Lee, M.A., 1980. Analytical functions for fitting peaks from Ge semiconductor detectors. Nucl. Instrum. Methods 178, 499–512.

Heusser, G., 1993. Cosmic ray induced background in Ge spectrometry. Nucl. Instrum. Phys. Res., Sect B 83, 223–228.

Heusser, G., 1996. Cosmic ray interaction study with low level Ge spectrometry. Nucl. Instrum. Phys. Res., Sect A 369, 539–543.

Hindman, F.D., 1986. Actinide separations for alpha spectrometry using neodymium fluoride coprecipitation. Anal. Chem. 58, 1238–1241.

Horwitz, E.P., Dietz, M.L., Nelson, D.M., LaRosa, J.J., Fairman, W.D., 1990. Concentration and separation of actinides from urine using a supported bifunctional organophosphorus extractant. Anal. Chim. Acta 238, 263–271.

Hull, E.L., Pehl, R.H., 2005. Amorphous germanium contacts on germanium detectors. Nucl. Instrum. Methods 538 (1–3), 651–656.

Hull, E.L., Burks, M.T., Cork, C.P., Craig, W., Eckels, D., Fabris, L., Lavietes, A., Luke, P.N., Madden, N.W., Pehl, R.H., Ziock, K.P., 2001. Proc. SPIE 4507, 132.

Hull, E.L., Pehl, R.H., Lathrop, J.R., Suttle, B.S., 2011. Yttrium hole-barrier contacts for germanium semiconductor detectors. Nucl. Instrum. Methods A 626–627, 39–42.

Jovancevic, N., Krnar, M., Mrda, D., Slivka, J., Bikit, I., 2010. Neutron induced gamma activity in low-level Ge spectroscopy systems. Nucl. Instrum. Methods Phys. Res., Sect. A 612 (2), 303–308.

Keyser, R.M., Twomey, T.R., Sangsingkeow, P., 1998. Matching Ge detector element geometry to sample size and shape: one does not fit it all. In: Proceedings of the 1998 Winter Meeting of the ANS.

Keyser, R.M., Haywood, S.E., Upp, D.L., 2001. Performance of the true coincidence correction method in gamma vision. In: Proceedings of American Nuclear Society 2001 Annual Meeting, Milwaukee, WI.

Kirby, H.W., Sheehan, W.E., 1984. Determination of 238Pu and 241Pu in 239Pu by alpha-spectrometry. Nucl. Instrum. Methods Phys. Res. 223, 356–359.

Klapdor-Kleingrothaus, H.V., et al., 2001. Latest results from the Heidelberg–Moscow double beta decay experiment. Eur. Phys. J. A 12 (2), 147–154.

Klapdor-Kleingrothaus, H.V., et al., 2004. Search for neutrinoless double beta decay with enriched 76Ge in Gran Sasso 1990-2003. Phys. Lett. B 586 (3–4), 198–212.

Knoll, G.F., 1999. Radiation Detection and Measurement. John Wiley and Sons, New York.

Kolesnikov, N., Kolchin, A., Alov, D., Ivanov, Y., Chernov, A., Schieber, M., Hermon, H., James, R., Goorsky, M., Yoon, H., Toney, J., Brunett, B., Schlesinger, T., 1997. Growth and Characterization of p-type Cd(1-x)Zn(x)Te (x=0.2, 0.3, 0.4). J. Cryst. Growth 174, 256–262.

Kolotov, V.P., Atrashkevich, V.V., Gelsema, S.J., 1996. Estimation of true coincidence corrections for voluminous sources. J. Radioanal. Nucl. Chem. 210 (1), 183–196.

Koskelo, M.J., Mercier, M.T., 1990. Verification of gamma spectroscopy programs: a standardized approach. Nucl. Instrum. Methods Phys. Res., Sect. A 299, 318–321.

Koskelo, M.J., Mercier, M.T., 1995. Verification of gamma spectroscopy programs: multiple area problems and solutions. J. Radioanal. Nucl. Chem. 193, 211–217.

Koskelo, M.J., Aarnio, P.A., Routti, J.T., 1981. SAMPOSO: an accurate gamma spectrum analysis method for microcomputers. Nucl. Instrum. Methods 190, 89–90.

Koskelo, M.J., Burnett, W.C., Cable, P.H., 1996. An advanced analysis program for alpha-¬particle spectrometry. Radioact. Radiochem. 7 (1), 18–27.

Koskelo, M.J., Venkataraman, R., Kolotov, V.P., 2001. Coincidence summing corrections using alternative detector characterization data. J. Radioanal. Nucl. Chem. 248 (2), 333–337.

Krishna, R.M., 2013. Crystal Growth, Characterization, and Fabrication of CdZnTe-Based Nuclear Detectors (Doctoral dissertation). University of South Carolina.

Kröll, T., Peter, L., Elze, T.W., Gerl, J., Happ, T., Kaspar, M., Schaffner, H., Schremmer, S., Schubert, R., Vetter, K., Wollerrsheim, H.J., 1996. Analysis of simulated and measured pulse shapes of closed ended HPGE detectors. Nucl. Instrum. Methods Phys. Res., Sect. A 371, 489–496.

Lally, A.E., Glover, K.M., 1984. Source preparation in alpha spectrometry. Nucl. Instrum. Methods Phys. Res. 223, 259–265.

Laubenstein, M., Heusser, G., 2009. Cosmogenic radionuclides in metals as indicator for sea level exposure history. Appl. Radiat. Isot. 67 (5), 750–754.

Leo, W.R., 1987. Techniques for Nuclear and Particle Physics Experiments. Springer Verlag, New York.

Luke, P.N., 1994. Single-polarity charge sensing in ionization detectors and coplanar electrodes. Appl. Phys. Lett. 65 (22), 2884–2886.

Luke, P., et al., 1989. Low capacitance large volume shaped-field germanium detector. IEEE Trans. Nucl. Sci. 36 (1), 926–930.

Luke, P.N., Amman, M., Tindall, C., Lee, J.S., 2005. Recent developments in semiconductor gamma-ray detectors. J. Radioanal. Nucl. Chem. 264 (No.1), 145–153.

Lutz, G., 1996. Effects of deep level defects in semiconductor detectors. Nucl. Instrum. Methods Phys. Res., Sect. A 377, 234–243.

Mariscotti, M., 1967. A method for automatic identification of peaks in the presence of back-ground and its application to spectrum analysis. Nucl. Instrum. Methods 50, 309–320.

McCabe, W.J., Ditchburn, R.G., Whitehead, N.E., 1979. The quantitative separation, electrodeposition and alpha spectrometry of uranium, thorium and protactinium in silicates and carbonates. R 262, DSIR. Institute of Nuclear Sciences, New Zealand, 29 pp.

McCallum, G.J., Coote, G.E., 1975. Influence of source detector distance on relative intensity and angular correlation measurements with Ge (Li) spectrometers. Nucl. Instrum. Methods 130, 189–197.

Mercier, M.T., Koskelo, M.J., 1992. Verification of gamma spectroscopy programs: accuracy and detectability. J. Radioanal. Nucl. Chem. 160 (No. 1), 23 3 243.

Moens, L., Hoste, J., 1983. Calculation of the peak efficiency of high purity germanium detectors. Int. J. Appl. Radiat. Isot. 34, 1085–1095.

Moens, L., De Donder, J., Lin, X.lei, De Corte, F., De Wispelaere, A., Simonits, A., Hoste, J., 1981. Calculation of the absolute peak efficiency of gamma ray detectors for different count-ing geometries. Nucl. Instrum. Methods 187, 451–472.

Moens, L., De Corte, F., Simontis, A., Lin, X., De Wispelaere, A., De Donder, J., Hoste, J., 1982. Calculation of the absolute peak efficiency of Ge and Ge(Li) detectors for different counting geometries. J. Radioanal. Nucl. Chem. 70, 539–550.

Morichi, M., Evrard, O., Keters, M., 2011. Simplified Silicon Drift Detecttor and Waraparound Neutron Detector. Patent filed, Appl# 12/ 838.267.

Mueller, W.F., Bronson, F., Field, M., Morris, K., Nakazawa, D., Venkataraman, R., Atrashkevitch, V., 2009. Challenges and techniques to effectively characterize the efficiency of broad-energy germanium detectors at energies less than 45 keV. J. Radioanal. Nucl. Chem. 282 (1), 217–221.

Müller, G., Wissmann, F., Schröder, F., Mondry, G., Brinkmann, H.J., Smend, F., Schumacher, M., Fettweis, P., Carchon, R., 1990. Low background counting using Ge(Li) detectors with anti muon shield. Nucl. Instrum. Methods Phys. Res., Sect. A 295, 133–139.

Nelson, W.R., Hirayama, H., Rogers, D.W.O., 1985. The EGS4 Code System. Stanford Linear Accelerator, Stanford University. SLAC-265.

Philips, G.W., Marlow, K.W., 1976. Automatic analysis of gamma ray spectra from germanium detectors. Nucl. Instrum. Methods 137, 526–536.

Philips, B.F., Johnson, W.N., Kroeger, R.A., Kurfess, J.D., Phillips, G., Wulf, E.A., Luke, P.N., 2002. IEEE Trans. Nucl. Sci. 49, 597.

Phillips, G.W., 1995. Gamma-ray imaging with Compton cameras. Nucl. Instrum. Methods B99, 674 (and references therein).

Prettyman, T.H., Smith, M.K., Soldner, S.E., 1999. Design and characterization of cylindrical CdZnTe detectors with coplanar grids. In: Proceedings of SPIE. Los Alamos National Laboratory document LA-UR-99-3117.

Reilly, D., Ensslin, N., Smith, H., Krelner, S., 1991. Passive Nondestructie Assay of Nuclear Materials. NUREG/CR-5550, Washington.

Roman, D., 1980. The electrodeposition of thoriurn in natural materials for alpha spectrometry. J. Radioanal. Chem. 60, 317–322.

Roman, D., 1984. Electrodeposition of radium on stainless steel from aqueous solutions. Appl. Radiat. Isot. 35, 990–992.

Routti, J.T., Prussin, S.G., 1969. Photopeak method for the computer analysis of gamma ray spectra from semiconductor detectors. Nucl. Instrum. Methods 72, 125–142.

Roy, U., Gueorguiev, A., Weiller, S., Stein, J., 2009. Growth of spectroscopic grade Cd(0.9)Zn(0.1)Te:in by THM technique. J. Cryst. Growth 312, 33–36.

Roy, U., Weiller, S., Stein, J., 2010. Growth and interface study of 2 in diameter CdZnTe by THM technique. J. Cryst. Growth 312, 2840–2845.

Schlesinger, T., Toney, J., Yoon, H., Lee, E., Brunett, B., Franks, L., James, R., 2001. Cadmium Zinc Telluride and its use as a nuclear radiation detector material. Mater. Sci. Eng. 32, 103–189.

Semkow, T.M., Parekh, P.P., Schwenker, C.D., Khan, A.J., Bari, A., Colaresi, J.F., Tench, O.K., David, G., Guryn, W., 2002. Low background gamma spectrometry for environmental radioactivity. Appl. Radiat. Isot. 57, 213–223.

Shepherd, S.L., Nolan, P.J., Cuellen, D.M., Applebe, D.E., Simpson, J., Gerl, J., Kaspar, M., Kleinboehl, A., Peter, I., Rejmund, M., Schaffner, H., Shlegel, C., de France, G., 1999. Measurements on a prototype segmented Clover detector. Nucl. Instrum. Methods Phys. Res. A434, 373–386.

Sill, C.W., 1969. Separation and radiochemical determination of uranium and the transuranium elements using barium sulfate. Health Phys. 17, 89–107.

Sill, C.W., 1977. Determination of thorium and uranium isotopes in ores and mill tailings by alpha spectrometry. Anal. Chem. 49, 618–621 (See Anal. Chem. 49,1648, for correction.).

Sill, C.W., 1980. Determination of gross alpha strontium, neptunium and/ or uranium by gross alpha counting on barium sulfate. Anal. Chem. 52, 1452–1459.

Sill, C.W., 1987a. Precipitation of actinides as fluorides or hydroxides for high resolution alpha spectrometry. Nucl. Chem. Waste Manage. 7, 201–215.

Sill, C.W., 1987b. Determination of radium 226 in ores, nuclear wastes and environmental samples by high resolution alpha spectrometry. Nucl. Chem. Waste Manage. 7, 239–256.

Sill, C.W., 1995. Rapid monitoring of soil, water, and air dusts by direct large area alpha spectrometry. Health Phys. 69, 21–33.

Sill, C.W., Olson, D.G., 1970. Sources and prevention of recoil contamination of solid state alpha detectors. Anal. Chem. 42, 1596–1607.

Sill, C.W., Sill, D.S., 1989. Determination of actinides in nuclear wastes and reference mate¬rials for ores and mill tailings. Waste Manage 9, 219–229.

Sill, C.W., Sill, D.S., 1994. Simultaneous determination of actinides in small environmental samples. Radioact. Radiochem. 5 (2), 8–19.

Sill, C.W., Sill, D.S., 1995. Sample dissolution. Radioact. Radiochem. 6 (2), 8–14.

Sill, C.W., Williams, R.L., 1969. Radiochemical determination of uranium and the transuranium elements in process solutions and environmental samples. Anal. Chem. 41, 1624–1632.

Sill, C.W., Williams, R.L., 1981. Preparation of actinides for alpha spectrometry without electrodeposition. Anal. Chem. 53, 412–415.

Sill, C.W., Puphal, K.W., Hindman, F.D., 1974. Simultaneous determination of alpha¬-emitting nuclides of radium through californium in soil. Anal. Chem. 46, 1725–1737.

Small, H., 1989. Ion Chromatography. Plenum, New York.

Stelson, P.H., Dickens, J.K., Raman, S., Tramell, R.C., 1972. Deterioration of large Ge(Li) diodes caused by fast neutrons. Nucl. Instrum. Methods 98, 481–484.

Sulcek, Z., Povondra, P., 1989. Methods of Decomposition in Inorganic Analysis. CRC Press, Boca Raton, FL.

Talvitie, N.A., 1972. Electrodeposition of actinides for alpha spectrometric determination. Anal. Chem. 44, 280–283.

Venkataraman, R., Moeslinger, M., 2001. Using generic detector characterization templates for Cascade Summing Correction. In: Proceedings of American Nuclear Society 2001 Annual Meeting, Milwaukee, WI, June 2001.

Venkataraman, R., Bronson, F., Atrashkevich, V., Young, B.M., Field, M., 1999. Validation of in situ object counting system (ISOCS) mathematical efficiency calibration software. Nucl. Instrum. Methods Phys. Res., Sect. A 442, 450–454.

Verplancke, J., 1992. Low level gamma spectroscopy: low, lower, lowest. Nucl. Instrum. Methods Phys. Res., Sect. A 312, 174–182.

Verplancke, J., 1999. About shapes and geometries of high purity germanium detectors. In: Presented at the Nuclear Physics Conference in Madrid, Aug. 1999.

Vetter, K., Kuhna, A., Deleplanque, M.A., Lee, I.Y., Stephens, F.S., Schmid, G.J., Beckedahl, D., Blair, J.J., Clark, R.M., Cromaz, M., Diamond, R.M., Fallon, P., Lane, G.J., Kammeraad, J.E., Macchiavelli, A.O., Svensson, C.E., 2000. Nucl. Instrum. Methods A452, 223.

Wätzig, W., Westmeier, W., 1978. ALFUN a program for the evaluation of complex alpha¬-spectra. Nucl. Instrum. Methods 153, 517–524.

Williams, I.R., 1966. Monte Carlo calculation of source to detector geometry. Nucl. Instrum. Methods 44, 160–162.

Wordel, R., Mouchel, D., Altzitzoglou, T., Heusser, G., Quintana, A.B., Meynendonckx, P., 1996. Study of neutron and muon background in low level germanium gamma ray spectrometry. Nucl. Instrum. Methods Phys. Res., Sect. A 369, 557–562.

Xu, W., et al., 2015. Testing the Ge Detector for Majorana Demonstrator. Physics Procedia 61, 807–815.

Yang, Y.F., Gono, Y., Motomura, S., Enomoto, S., Yano, Y., 2001. IEEE Trans. Nucl. Sci. 48, 656.

Zhu, H., Venkataraman, R., Menaa, N., Mueller, W., Croft, S., Berlizov, A., 2008. Validation of gamma-ray true coincidence summing effects modeled by the Monte Carlo code MCNP-CP. J. Radioanal. Nucl. Chem. 278 (2), 359–363.

Zhu, H., Venkataraman, R., Mueller, W., Lamontagne, J., Bronson, F., Morris, K., Berlizov, A., 2009a. X-ray true coincidence summing correction in Genie 2000. Appl. Radiat. Isotopes 67, 696.

Zhu, H., Morris, K., Mueller, W., Field, M., Venkataraman, R., Lamontagne, J., Bronson, F., Berlizov, A., 2009b. Validation of true coincidence summing correction in Genie2000 V3.2. J. Radioanal. Nucl. Chem. 282, 205–209.

Further reading

Abt, I., Caldwell, A., Gutknecht, D., Kroninger, K., Lampert, M., Liu, X., Majororvits, B., Quirion, D., Stelzer, F., Wendling, P., 2007. Characterization of the first true coaxial 18-fold segmented n-type prototype HPGe detector for the GERDA project. Nucl. Instrum. Methods Phys. Res. A577, 574–584.

Agostini, M., et al. GERDA Collaboration, in publication.

De Corte, F., Freitas, C., 1992. The correction for γ-γ, γ-KX and γ-LX true-coincidences in ko- standardized NAA with counting in a LEPD. J. Radioanal. Nucl. Chem. 160, 253–267.

Ejiri, H., de Voight, M.J.A., 1989. Gamma ray electron spectroscopy in nuclear physics. In: Oxford Studies in Nuclear Physics. Clarendon Press, Oxford.

Gimore, G., Hemingway, J.D., 1995. Practical Gamma Ray Spectrometry. John Wiley and Sons, New York.

Grille, P., Kiessling, F., Burkert, M., 1991. A new approach to crystal growth of Hg(1-x)Cd(x)Te by the travelling heater method (THM). J. Cryst. Growth 114, 77–86.

Laborie, J.M., Le Petit, G., Abt, D., Girard, M., 2002. Monte Carlo calibration of the efficiency response of a low-background well-type HPGe detector. Nucl. Instrum. Methods Phys. Res., Sect. A 479, 618–630.

Turner, J.E., 1986. Atoms, Radiation and Radiation Protection. Pergamon Press, New York.

Wulf, E.A., Ampe, J., Johnson, W.N., Kroeger, R.A., Kurfess, J.D., Philips, B.E., 2002. IEEE Trans. Nucl. Sci. 49, 1876.

Ramkumar ("Ram") Venkataraman

Dr. Ramkumar ("Ram") Venkataraman has been in the field of radiation detection and measurements for more than 25 years. His areas of interest include gamma spectroscopy, gamma- and neutron-based nondestructive assay methods for nuclear safeguards applications, new detector materials, neutron activation analysis, and mathematical modeling of radiation detectors and systems. Ram is a staff member in the Safeguards and Security Technology group at Oak Ridge National Laboratory (ORNL). At ORNL, Ram is developing new methods for accurate identification and quantification of special nuclear materials of interest in nuclear safeguards and nonproliferation. Ram has authored more than 75 papers in the field of nuclear science and technology and has made numerous presentations at international scientific conferences. Ram actively participates in standard committees. He has been serving as the Chairman of the of ASTM Subcommittee C26.10 (Nuclear Fuel Cycle: Non-Destructive Assay Techniques) since 2008.

Ram holds a master's degree in Physics from the University of Madras, India, and a doctoral degree in Nuclear Engineering from the University of Michigan, Ann Arbor, MI.

Chapter 5

Alpha spectrometry

Nóra Vajda[1], Roy Pöllänen[2], Paul Martin[3] and Chang-Kyu Kim[4]

[1]*Radiochemical Laboratory, RADANAL Ltd., Budapest, Hungary;* [2]*Department of Physics, University of Helsinki, Helsinki, Finland;* [3]*Australian Radiation Protection and Nuclear Safety Agency, Yallambie, VIC, Australia;* [4]*Department of Safeguards, International Atomic Energy Agency, Vienna International Centre, Vienna, Austria*

Chapter outline

Handbook of Radioactivity Analysis: Volume 1. https://doi.org/10.1016/B978-0-12-814397-1.00005-4

I. Introduction

Alpha particle spectrometry (in short alpha spectrometry) is a widely applied radioanalytical technique, primarily due to its high counting efficiency and low intrinsic background (resulting in a high sensitivity), its versatility both in terms of the range of radionuclides that can be determined and sample types that can be analyzed, and the reliability of the technique due to the possibility to use an alpha particle–emitting isotope of the element of interest as an internal tracer.

For some special applications, it is possible to carry out alpha spectrometry using only minimal sample preparation or no sample processing at all (Surbeck, 2000; Röttger et al., 2001; Ranebo et al., 2010; Samuelsson, 2011). However, the short range of alpha particles through matter means that, in general, separation of the radionuclides of interest from the bulk sample matrix is required. This involves dissolution of the sample and chemical separation of the radionuclides prior to source preparation.

In practice, a separate source is normally prepared for each element for which radioisotopes are to be determined. For each of these sources, the chemical separation should fulfill three aims. Firstly, it should remove elements that would interfere in the final source preparation step in such a way as to reduce the recovery of the element of interest. Secondly, it should remove elements that would deposit on the final source, resulting in a thick source and hence degraded spectral resolution. Thirdly, it should remove other alpha particle–emitting radionuclides that might interfere with the peaks of interest in the spectrum. A major part of the effort expended in development of alpha spectrometry relates to improvements in these chemical separation procedures, and in particular minimizing the time required.

The number of radionuclides that can be reliably determined using alpha spectrometry is quite large. It includes almost all of the alpha particle–emitting radionuclides with half-lives long enough relative to the time required for source preparation. Beta particle–emitting radionuclides can also be determined if they have alpha particle–emitting progeny, which may be measured on the source following a suitable ingrowth period. For example, ^{228}Ra may be determined using its progeny ^{228}Th and ^{224}Ra (Hancock and Martin, 1991), and ^{227}Ac may be determined using ^{227}Th and ^{223}Ra (Bojanowski et al.,1987).

Another major advantage of alpha spectrometry is its high sensitivity, with detection limits as low as 1 mBq per sample being easily achievable. This arises due to several factors. The first of these is the high yield of the alpha decay process. For most alpha-particle emitting radionuclides there are only a few peaks within a relatively small energy range, representing a total intensity of or close to 100% of decays. Secondly, semiconductor alpha detectors have a low intrinsic background, being typically 10^{-5} to 10^{-6} counts per second for an uncontaminated detector. Thirdly, the intrinsic efficiency of the detector system is close to 100% for incident alpha particles. Consequently, the counting efficiency is dominated by the geometric efficiency, which is typically about 20%–40% for sources placed close to semiconductor detectors. A fourth factor is the elimination of competing radiation by chemical separation, and the last factor is the low sensitivity of the semiconductor alpha detector to beta and gamma radiation.

An outcome of the sensitivity of the technique is the ability to analyze relatively small sample sizes. For example, for the determination of uranium and thorium series radionuclides in environmental samples, common sample sizes are of the order of 1–10 g for soils, sediments, and biological samples and 0.1–1 L for water samples.

The use of an alpha particle–emitting isotope as an internal tracer makes alpha spectrometry a highly reliable analytical technique, provided no losses occur before the sample and tracer isotopes have been equilibrated. The internal tracer compensates not only for losses during the chemical separations but also for the counting efficiency of the detection system. This is because the result is obtained directly from the ratio of sample to tracer peak areas. Consequently, a highly accurate knowledge of the counting efficiency is not required to achieve accurate determinations of the target radionuclides.

A further advantage is that the capital cost for alpha spectrometry equipment, including the detector, counting chamber and associated electronics, is significantly lower than that for a number of other systems capable of isotope discrimination. Counterbalancing this is the requirement for chemical laboratory facilities and reagents.

The major disadvantage of alpha spectrometry is that time-consuming chemical separation procedures are normally required to remove the bulk matrix, elements that interfere with recovery of the element of interest, and other alpha emitters. Alpha spectrometry without chemical separation is possible only in some special applications (such as the determination of airborne concentrations of radon progeny). Typically, the full chemical digestion, separation, and source preparation procedure require several steps over several hours. Some automated systems have been developed to reduce the need for operator time (Kim et al., 2008). In recent years, various efforts have been made to develop rapid procedures for sample preparation by combining the chemical separation with the source preparation. Preparation of good-quality alpha sources has been reported by treating (coating, drafting) polymer films with element-specific reagents. For example, Ra was selectively retained on MnO_2 disks; Pu, Am, and U nuclides were retained by various ligands anchored on polymer films. The disks and the polymer ligand films (PLFs) were directly counted by alpha spectrometry (Aggarwal, 2016).

Unlike many other radioanalytical techniques, alpha spectrometry is element based in the sense that each spectrum consists of the alpha-emitting radionuclides of typically one element (plus those alpha particle–emitting progeny, which grows in on the source during data acquisition and in the interval between preparation and counting). This means that a separate source must be prepared and counted for each element of interest in the sample. However, it also gives rise to the advantage that the spectrum is in a sense "predictable." Given the element for which the source has been prepared, the analyst knows which peaks are expected in the spectrum, with only the relative areas under each peak varying for different sources. This simplifies spectrum analysis. It also enables an important quality control check, which is the examination of the spectrum for peaks, which would not be expected for a source of the given element. The presence of such unexpected peaks may indicate that the chemical separation was not sufficiently effective in removing other alpha-emitting radionuclides from the sample extract.

In some cases, alpha spectrometry has disadvantages in comparison with other techniques for case-specific reasons. One important example is the impossibility of distinguishing ^{239}Pu and ^{240}Pu with conventional alpha spectrometry due to their overlapping peaks.

Alpha spectrometry has applications in diverse fields, from the analysis of alpha particle–emitting radionuclides in nuclear materials of high activity to environmental samples of low and ultralow activity (Aggarwal, 2016). An example of its widespread use in environmental studies is the measurement of ^{210}Po in soil and sediment samples, as an indirect measure of ^{210}Pb for erosion and sedimentation investigations (Robbins, 1978). Other examples include measurement of U and Ra radionuclides to investigate groundwater, surface water, and seawater processes (Hancock et al., 2000; Porcelli, 2008) and that of ^{210}Po as an atmospheric tracer (Baskaran, 2011).

Alpha spectrometry is commonly used for monitoring of environmental samples for the impact of nuclear activities. Examples include actinide determination on soil samples and analysis of water samples downstream of uranium mining operations for uranium series radionuclides. The technique is also used for internal dosimetry for workers, by means of analysis of biological samples such as urine (Kehagia et al., 2007; Manickam et al., 2010). Alpha spectrometry is a basic technique for the determination of man-made radionuclides in nuclear power production, nuclear fuel processing and reprocessing, management and storage of radioactive wastes, decommissioning of nuclear facilities, nuclear medicine, nuclear safeguards and forensics, etc.

As a result of development in detector fabrication and source preparation procedures, as well as progress in the technology of multichannel analyzers (MCAs) and spectrum unfolding programs, high-resolution alpha spectrometry (HRAS) became available in the 1980s. Thus, a precise and sensitive technique was developed that is suitable for measurement of nuclear data such as half-life, cross section (Wiltshire, 1984), and emission probabilities (Bortels and Collaer, 1987). Based on development in spectrum interpretation and deconvolution, in situ (nondestructive) alpha spectrometry has become available and has been applied for in-field measurements (Pöllanen, 2016). Recently, attempts have been made to prepare the alpha source directly on the detector surface, thus combining source preparation and detection occasionally as well as chemical separation into a single-step process (Krupp and Scherer, 2018). More recently, cryogenic detectors have dramatically improved the achievable peak resolution, and the application of magnetic spectrometers brought this highly specialized technique closer to routine alpha spectrometry (Hoover et al., 2015).

Some mass spectrometry techniques, which detect the radioactive atom itself rather than the emitted radiation, have shown significant advances in sensitivity for the determination of long-lived alpha particle–emitting radionuclides such as ^{238}U and ^{235}U. These developments have made these techniques rivals to, and in some applications superior to, alpha spectrometry in terms of sensitivity and/or sample throughput. Nevertheless, the advantages of alpha spectrometry, especially for shorter-lived radionuclides, mean that it will remain an important, standard technique for the foreseeable future.

II. Alpha decay and alpha particle–emitting radionuclides

Alpha decay occurs when a nucleus of helium $^4_2He^{2+}$ called an "alpha particle" is emitted from a heavy nucleus. After the decay, the mass number (A) and the atomic number (Z) of the new nucleus are lower by 4 and 2, respectively.

$$^A_ZX \rightarrow {}^{A-4}_{Z-2}Y + {}^4_2He + Q \tag{5.1}$$

During the spontaneous decay, the decay energy Q is liberated and shared between the reaction products $^{A-4}_{Z-2}Y$ and $^4_2He^{2+}$ according to the law of conservation of momentum. The kinetic energy of the alpha particle (E_α) can be calculated from:

$$E_\alpha = Q\left(1 - \frac{1}{1 + m_Y/m_\alpha}\right) \quad (5.2)$$

where Q is the disintegration energy, m_Y is the mass of the daughter nuclide Y (also called the recoil atom), and m_α is the mass of the alpha particle (α).

Since the masses of the decay products are significantly different, the kinetic energy of the alpha particle is almost equal to the total decay energy Q. The decay energy Q and therefore the energy of the alpha particle E_α is in the range of 2–8 MeV. (A more detailed description of alpha decay is given in Chapter 1, Volume 1.)

Alpha decay is the major decay mode of proton-rich and high atomic number nuclides because in heavy nuclides the electrostatic repulsive forces increase more rapidly with atomic number than do the cohesive nuclear forces. Hence, alpha decay is characteristic for heavy nuclides, predominantly for those with atomic numbers greater than 83 (i.e., beyond Bi). The radionuclide of lowest atomic number that emits alpha particles is $^{144}_{60}$Nd. There are in total 815 alpha emitters in the Nuclides.net database (NUDAT 2, 2009) (see in Fig. 5.1).

There are relatively few long-lived alpha particle–emitting radionuclides of relevance in the measurement of environmental and nuclear samples. A list of some important long-lived alpha particle–emitting radionuclides and their short-lived progeny is presented in Table 5.1, together with the half life, alpha particle energy (E_α), and emission probability (I_α) of the most intensive alpha emissions.

A unique feature of the alpha decay process is the strong correlation between alpha particle energy and the half-life of the parent nuclide. It can be clearly seen from Table 5.1 that isotopes of the highest alpha particle energies

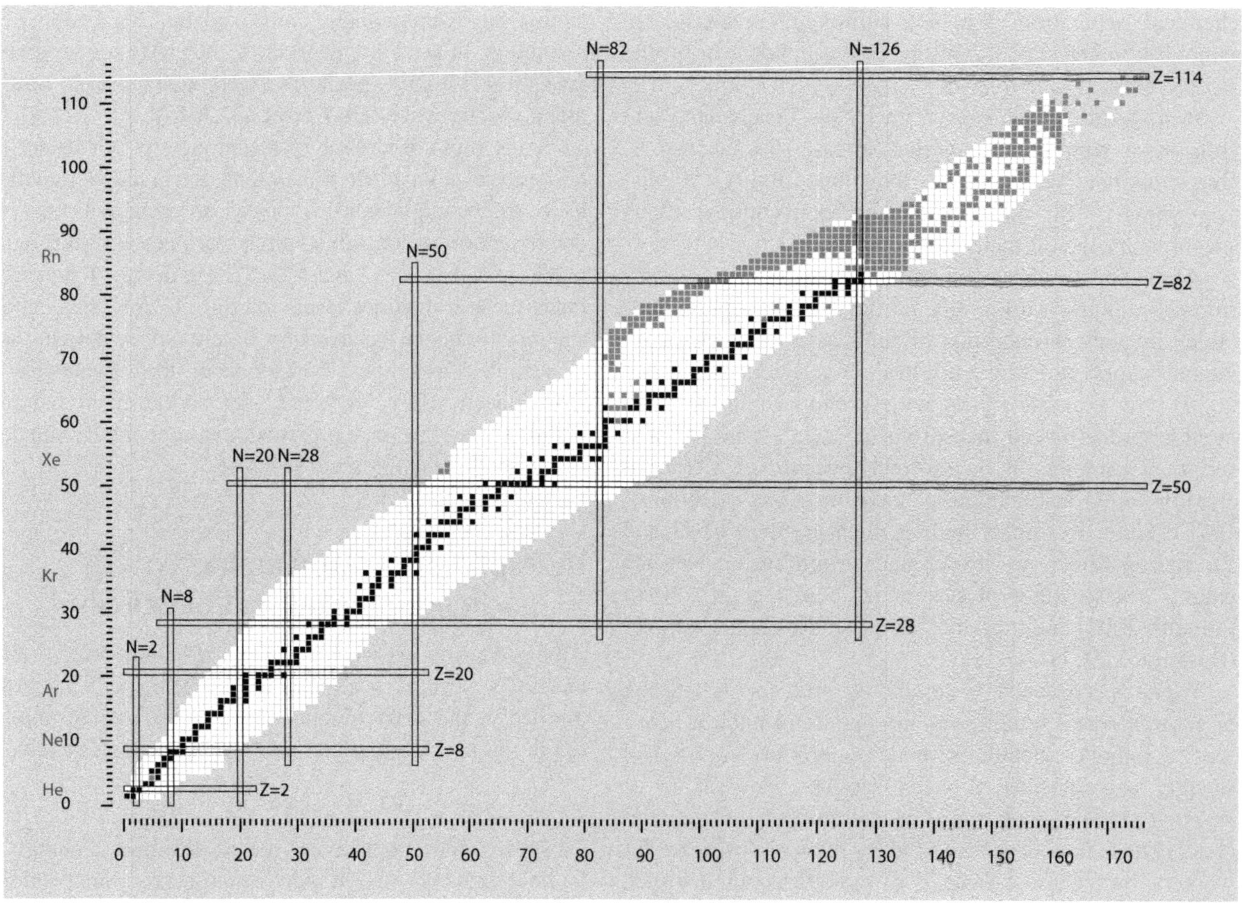

FIGURE 5.1 Alpha emitters in the upper right region of the concise chart of the nuclides indicated by yellow (online version) or gray color (printed version). Nuclides are presented by their atomic numbers as a function of neutron number. *From Magill and Galy (2005); reprinted by permission of Springer-Verlag © 2005.*

TABLE 5.1 Basic decay data of alpha emitting radionuclides of interest (DDEP 2017, *NUDAT2 2016).

Nuclide	Half life	E_α (keV)	I_α (%)	Nuclide	Half life	E_α (keV)	I_α (%)
^{208}Po*	2.898 a	5114.9	100	^{226}Ra	1600 a	4684	5.95
^{209}Po	115 a	4977	79			4870.54	94.04
		4979	20	^{227}Th*	18.68 d	5668.0	2.06
^{210}Po	138.376 d	5407.45	100			5693.0	1.50
^{211}At	7.216 h	5982.4	41.78			5700.8	3.63
^{211}Bi	2.15 m	6399.8	16.16			5708.8	8.3
		6750.4	83.56			5713.2	4.89
^{212}Bi	60.54 m	6167.40	25.1			5756.87	20.4
		6207.26	9.7			5807.5	1.27
^{212}Po	0.300 μs	8954.12	100			5866.6	2.42
^{213}Po	3.70 μs	8536.2	100			5959.7	3.00
^{214}Po	162.3 μs	7833.46	99.99			5977.72	23.5
^{215}Po	1.781 ms	7526.3	99.93			6008.8	2.90
^{216}Po	0.148 s	6906.3	100			6038.01	24.2
^{217}At	32.3 ms	7199.6	99.93	^{228}Th	1.9126 a	5435.71	26.0
^{218}Po	3.071 m	6114.68	99.98			5520.08	73.4
^{220}Rn	55.8 s	6404.67	99.88	^{229}Th*	7340 a	4761	1.0
^{222}Rn	3.8232 d	5590.3	99.92			4797.8	1.5
^{223}Ra	11.43 d	5533.96	1.60			4814.6	9.30
		5640.72	10.6			4838	5.0
		5709.50	25.8			4845.3	56.2
		5820.35	49.6			4901.0	10.20
		5852.22	10.0			4967.5	5.97
^{224}Ra	3.631 d	5547.86	5.25			4978.5	3.17
		5788.85	94.73			5053	6.6
^{225}Ac	10.0 d	5711	1.09	^{230}Th*	7.538E4 a	4620.5	23.4
		5739.3	4.16			4687.0	76.3
		5785.0	1.31	^{230}U*	20.8 d	5817.5	32.0
		5826.7	2.03			5888.4	67.4
		5834.2	1.6	^{231}Pa	32670 a	4762.6	1.8
		5835.3	1.24			4795.5	1.2
		5835.6	9.0			4819.8	8,4
		5896.5	6.2			4939.1	1,4
		5898.0	18.9			5023.0	2.9
		5935.1	52.4			5039.9	22.5
						5075.7	1.6
						5103.5	25.3
						5119.9	20
						5122.5	2.8
						5149.9.6	11.7

Continued

TABLE 5.1 Basic decay data of alpha emitting radionuclides of interest (DDEP 2017, *NUDAT2 2016).—cont'd

Nuclide	Half life	E_α (keV)	I_α (%)	Nuclide	Half life	E_α (keV)	I_α (%)
^{232}Th	1.402E10 a	4017.8	21.0	^{238}U	4.468E9 a	4220.2	22.33
		4081.6	78.9			4269.7	77.54
^{232}U	70.6 a	5355.87	30.6	^{238}Pu	87.74 a	5549.70	28.85
		5413.63	69.1			5593.20	71.04
^{233}U*	1.592E5 a	4729	1.61	^{239}Pu	24100 a	5191.81	11.87
		4783.5	13.2			5231.47	17.14
		4824.2	84.3			5244.43	70.79
^{234}U	2.455E5 a	4804.5	28.42	^{240}Pu	6561 a	5210.54	27.16
		4857.6	71.37			5255.75	72.74
^{235}U	7.04E8 a	4287.7	5.95	^{241}Am	432.2 a	5388.23	1.6
		4397	3.33			5442.80	13.0
		4441.7	18.80			5485.56	84.5
		4474.0	57.19	^{242}Pu	3.73E5 a	4939.6	23.44
		4491.3	3.01			4984.5	76.53
		4580.4	1.28	^{242}Cm	162.8 d	6069.42	25.0
		4635.0	3.79			6112.72	74.0
		4676.0	4.74	^{243}Am	7367 a	5268	1.383
^{236}U	2.343E7 a	4522	26.1			5320.9	11.46
		4571	73.8			5363.6	86.74
^{236}Pu*	2.858 a	5720.87	30.8	^{243}Cm	28.9 a	5781.4	1.6
		5767.53	69.1			5838.7	11.3
^{237}Np	2.144E6 a	4720.4	6.43			5883.3	73.4
		4745.9	3.46			6093.1	5.7
		4795.0	1.17			6111.5	1.05
		4849.3	9.5			6160.9	4.4
		4854.7	23.0			6168.8	1.3
		4871.8	47.64	^{244}Pu*	8.11E7 a	4546	19.4
		4887.8	2.02			4589	80.5
		4901.2	2.43	^{244}Cm	18.11 a	5858.92	23.3
		4958.3	2.41			5901.74	76.7

have the shortest half-lives (and therefore highest decay constants). For example, within the series of ^{232}Th, ^{230}Th, and ^{228}Th, the half-life is reduced from 10^{10} years to about 2 years, whereas the alpha particle energy is increased from 4 to 5.4 MeV. This illustrates the validity of the law recognized by Geiger–Nuttall in 1911, which can be expressed by the following empirical equation (see Fig. 5.2):

$$\log\lambda = A + B \log E_a \tag{5.3}$$

where λ is the decay constant, E_α is the energy of the alpha particle, and A and B are constants.

Later this process was explained by the quantum mechanical barrier penetration theory of Gamow (see Chapter 1, Volume 1). Typically, alpha particles have energies in a

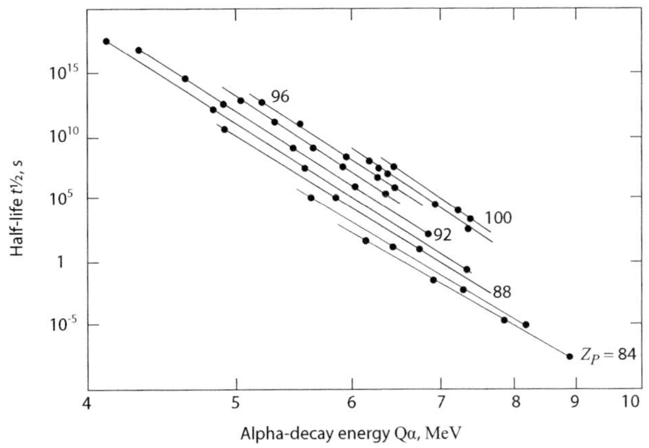

FIGURE 5.2 Geiger–Nuttall plot: half-life as a function of atomic number $Z = 84-100$ (Loveland et al., 2006). *Reprinted by permission of John Wiley and Sons © 2006.*

charge carriers: electron–cation pairs, electron–hole pairs, or electrons, respectively. The collection of these charge carriers at the electrodes gives a current, which flows through a resistor producing a voltage pulse. These interactions between the alpha particle and the detector can be utilized in alpha spectrometry where the height of the generated pulse is proportional to the kinetic energy of the alpha particle, and the pulse frequency or count rate is proportional to the activity of the alpha source. Alpha spectra are commonly depicted as the number of counts collected during a given counting time against the channel number, where channel number is a linear function of the pulse height and thus of the energy of the alpha particle.

An example of a typical alpha decay scheme and corresponding alpha particle energy spectrum are presented in Figs. 5.3 and 5.4.

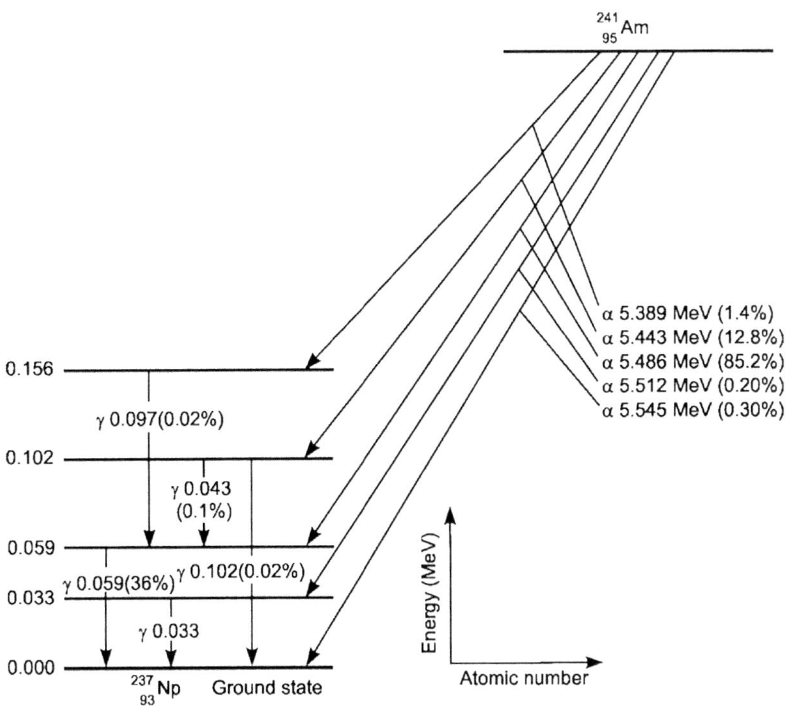

FIGURE 5.3 Simplified decay scheme of ^{241}Am. The relative emission probabilities of alpha particles and the relevant gamma ray emissions are expressed in % (in parenthesis) beside the radiation energy values in MeV. Energy levels of the nuclides are shown as plateaus along the ordinate and are distributed along the abscissa, representing the atomic number. *From Chapter 1, Volume 1. The full scheme can be obtained from the website of the Decay Data Evaluation Project, DDEP, http://www.nucleide.org/DDEP.htm.*

narrow range, between 4 and 8 MeV. If the alpha energy is beyond these values, the half-lives of the nuclides become either too short for the radionuclides to be easily detected or so long that they can be regarded as practically stable. Both cases require special skills and highly sophisticated analytical techniques for successful measurement.

In alpha spectrometry, alpha particles of different kinetic energies are "detected" using various types of nuclear detectors, the most common being gas ionization, semiconductor, and scintillation detectors. When an alpha particle enters the detector, it produces (directly or indirectly)

Alpha decay is characterized by the similarity of the decay schemes of many radionuclides (in particular the even mass number nuclides) in the following sense: Several radionuclides have only a few possible alpha particle transitions involving different alpha particle energies and emission probabilities (see Table 5.1). Direct transformation to the ground state of the daughter results in the emission of alpha particles of the highest energy. For almost all even-mass number and for many odd-mass number alpha particle–emitting radionuclides, this is the emission with the highest intensity. Thus, a series of alpha

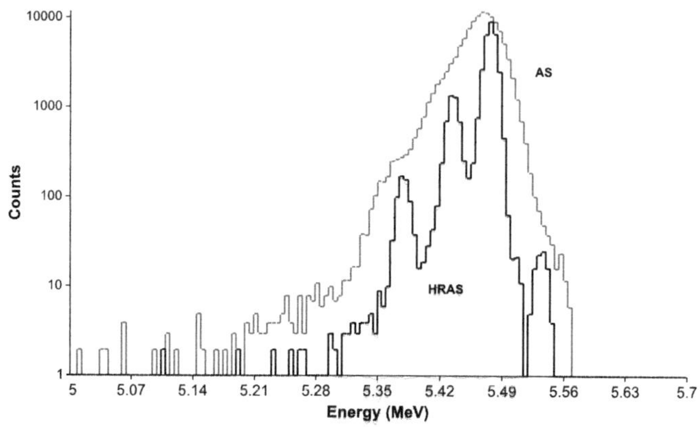

FIGURE 5.4 Simulated alpha spectrum of a ^{241}Am source as it appears in a high resolution (full width at half maximum smaller than approximately 20 keV—blue curve) and a low-resolution alpha spectrum (green curve). In high-resolution spectrometry, the alpha peaks at 5.389, 5.443, 5.486, and 5.545 MeV can be identified as partially overlapping peaks, whereas in low-resolution spectrometry, the peaks are strongly overlapping appearing as a single broad peak. Spectra were generated by the AASI simulation software (Siiskonen and Pöllänen, 2004) with estimated resolutions of Si semiconductor detectors of 14 and 40 keV.

particles of slightly different energies and of increasing emission yield with increasing energy can be assigned usually to a single radionuclide.

Each alpha particle energy results in an alpha peak in the spectrum; however, due to the relatively small energy difference between the peaks of the radionuclide, in question, they are more or less overlapping, resulting typically in a single broad asymmetric alpha peak similar to the one shown in Fig. 5.4. Thus, instead of individual peaks, the spectrum may consist of a group of overlapping peaks where the individual peaks can be distinguished only by using HRAS with full width at half maximum (FWHM) of the alpha peaks in the order of 20 keV or less.

For normal analytical purposes, if the group of alpha peaks belonging to a single radionuclide is well separated from those of the other nuclides, then there is no need to analyze the individual peaks for nuclide identification and activity calculation. Rather, each group of peaks is evaluated as a unit, and the total counts assigned to the appropriate radionuclide. This approach may be referred to as conventional alpha spectrometry.

III. Detection systems

The core part of an alpha spectrometer is the detector, where the interaction of the alpha particles with the matter of the detector material takes place. The energy of the alpha particle is transformed to a measurable physical quantity, e.g., electric charge, current, or voltage. In the process of detection, the charge carriers are collected and converted to electric pulses.

In spectrometry, pulses are processed in so-called "pulse mode" where each interacting particle is detected independently. Electric pulse processing units such as preamplifiers and amplifiers attached to the detector output both amplify and form (shape) the pulses. Pulses are then digitalized by an analog-to-digital converter (ADC). From a series of digitalized pulses of different height at the ADC output, the MCA generates a pulse height spectrum by sorting the pulses according to their heights and assigning them to a channel number proportional to the pulse height as illustrated in Fig. 5.5.

A schematic drawing of an alpha spectrometer is shown in Fig. 5.6. An alpha spectrometer usually contains a

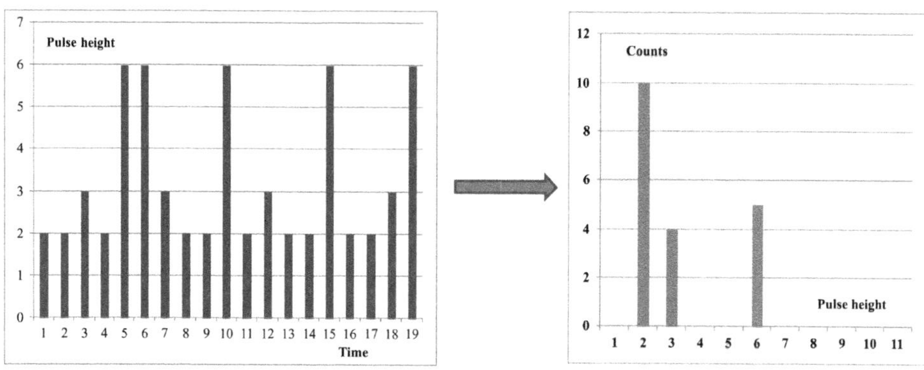

FIGURE 5.5 Transformation of digital pulse data into a pulse height spectrum using a multichannel analyzer. The left side of the figure shows the pulses appearing at the detector as a function of time. The right side shows the generated spectrum with the number of pulses as a function of the pulse height.

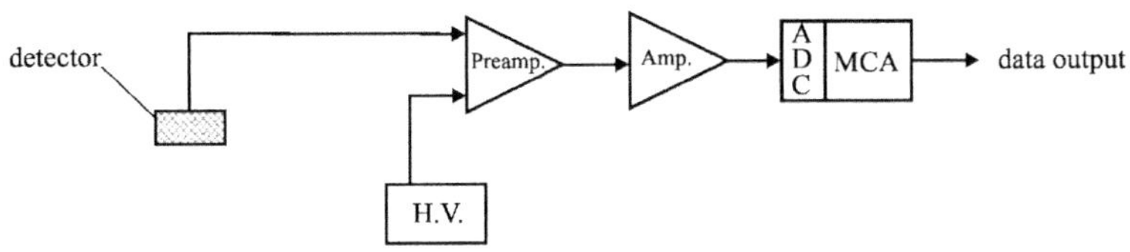

FIGURE 5.6 Schematic drawing of an alpha spectrometer. H.V. and Amp. refer to the high-voltage supply unit and the amplifier, respectively. ADC, analog-to-digital converter.

vacuum chamber surrounding the detector and the source. The vacuum supply unit contains a vacuum pump such as an oil rotation pump, a vacuum gauge, and occasionally a vacuum control system to adjust to the desired pressure.

In spectrometric applications, the basic information of the radiation (i.e., the type of radiation, the energy of the particles, and the activity of the source) is conserved in the form of the shape and height of the electric pulses, and the pulse frequency or count rate.

The components of the alpha spectrometer are discussed in more detail below.

A. Detectors

1. *Interaction of alpha radiation with detector materials*

Only a few detector types are suitable for the accurate determination of the energy distribution of alpha particles. Of these, silicon semiconductor detectors are the most frequently used at present. Due to instrumental improvements and developments in spectrum analysis software, HRAS using Si detectors became available in the 1980s.

Among gas ionization detectors, ionization chambers and proportional counters have been used in alpha spectrometry, and a special type of gas detector, the gridded ionization chamber, has been developed for the determination of pulse height distribution of the alpha particles.

Although scintillation detectors have, in general, relatively poor energy resolution, the development of "high-resolution" alpha liquid scintillation (LS) spectrometers in combination with selective chemical procedures offers some advantages in alpha spectrometry.

The process of detection of alpha particles by various nuclear detectors is determined by the interactions between the particle and the detector. The principal characteristics of the interactions have been described in Chapter 1, Volume 1. The most important aspects influencing the detection process in alpha spectrometry are summarized briefly below.

When alpha particles enter the detector, they immediately interact with the atomic electrons of the detector material. The main results of this interaction are **excitation** and **ionization**. In the latter case, electron and positive ion (cation/hole) pairs are formed. Along its path through the detector, the high-energy alpha particle will create a large number (N) of electron−cation pairs. N can be expressed as

$$N = \frac{E_\alpha}{\varepsilon} \tag{5.4}$$

where E_α is the energy of the alpha particle and ε is the energy needed to form one electron−cation pair.

Since E_α is in the range of a few MeV, and ε is—depending on the type of detector—in the range of $1-100$ eV, a single alpha particle can produce 10^4–10^6 charge carriers, whereas it loses its energy before it is stopped.

The track of the alpha particles in the detector is almost **straight** because the heavy alpha particles are not significantly deflected by the interacting electrons. The specific energy loss of the alpha particle, also called the **stopping**

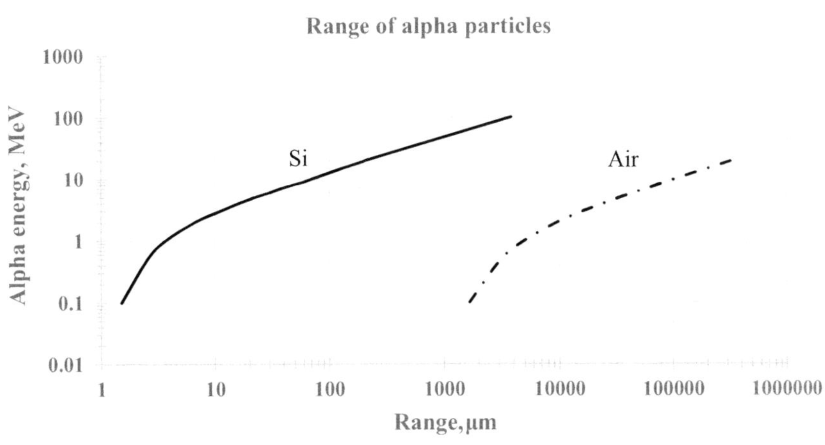

FIGURE 5.7 The range of alpha particles in silicon and air.

power, is described by the Bethe formula (see in Chapter 1, Volume 1). The alpha particle interaction with the detector is characterized by a **range** or distance beyond which no particles penetrate (Fig. 5.7). This range depends on the energy of the alpha particle and the type of the detector, i.e., its atomic number and density.

Although alpha particle energy loss in the detector material or in other media usually occurs with electromagnetic interactions of the electrons, interactions with the nuclei are also possible. In these cases, the track of the alpha particles is not straightforward and may lead to backscattering from the detector and, thus, may slightly reduce the detection efficiency.

Since alpha energies of radionuclides are not higher than 10 MeV, they are completely absorbed in a Si layer of 100 μm in thickness. Thus, Si detectors should not be significantly thicker, if they are used for the measurement of alpha-emitting radionuclides. Otherwise, other particles such as (conversion) electrons or low-energy photons may produce interfering signals that are in coincidence with the alpha particles and may influence the spectrum analysis (Siiskonen and Pöllänen, 2006). In the case of gas detectors, the detector thickness depends on the type and especially on the pressure of the gas in the detector chamber. Around atmospheric pressure, the range of 10 MeV alpha particles is about 5–10 cm (Fig. 5.7), and so this is the desired detector size.

The very short range of alpha radiation enables the design of detectors, which are able to completely absorb the incident radiation, without requiring large detector volumes. This fact, combined with the large number of charge carriers that are produced, is responsible for the high counting efficiency that is typical of alpha spectrometry.

However, the limited range of alpha radiation is also responsible for the major challenge in alpha particle detection, which is ensuring that alpha particles reach the active volume of the detector without significant losses. This means reduction or elimination of absorption in the source, any absorbent layers between source and detector, and in the detector window. Absorption and self-absorption in the detection process can be reduced by ensuring direct contact between the source and the detector, e.g., by placing the source inside the detector gas chamber or by dissolving the source in a liquid detector, as in the case of liquid scintillation counter (LSC). In the case of silicon detectors, absorption can be reduced by the use of vacuum in the chamber surrounding the detector and the alpha source. In addition, considerable efforts have been made to develop radiochemical procedures for the preparation of "infinitely thin" alpha sources that are almost free of self-absorption (see Section VI).

The interaction of alpha radiation with matter is a statistical process. When monoenergetic alpha particles pass through matter, there is a spread of energies at distances along the beam path, which is called **energy straggling**. The energy distribution of an originally monoenergetic beam becomes wider with penetration depth; however, near the end of the range, the distribution becomes narrower again due to the reduction of the beam energy (see Fig. 5.8).

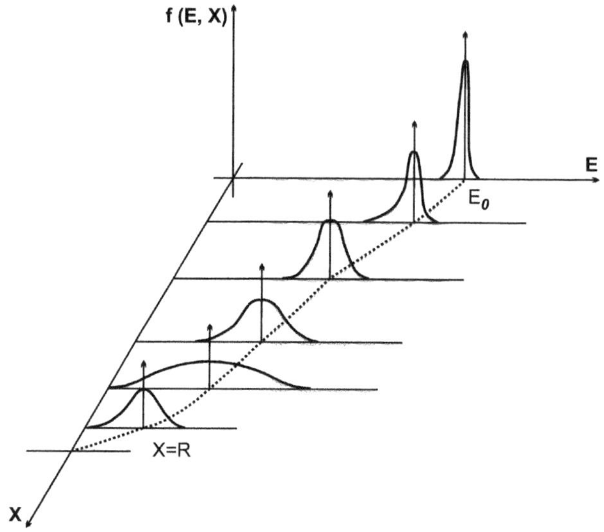

FIGURE 5.8 Schematic drawing of energy straggling of an alpha particle beam (Knoll, 2010). *Reprinted by permission of John Wiley and Sons © 2010.*

Analogously to energy straggling, the fluctuation in the path length of the alpha particles due to the statistical nature of the interactions is known as **range straggling**. Straggling happens in the source, in the media between the source and the detector and in the detector dead layer.

Energy and range straggling are the basic physical processes, which contribute significantly to peak broadening in conventional alpha spectrometry (see the peaks of ^{241}Am in Fig. 5.4). The peak shape in alpha spectrometry is affected by many parameters, including the characteristics of the detector, the absorbers between the detector and the source, and those of the source.

The time required to stop alpha particles in the detector is calculated from its range and average velocity where velocity (v) is a function of the energy (E_α) and the atomic mass of the particle (m_α):

$$v = \sqrt{\frac{2E_\alpha}{m_\alpha}} \tag{5.5}$$

Using typical range values, **stopping times** of alpha particles are a few picoseconds in solids and liquids and a few nanoseconds in gases resulting in the possibility of very fast detector response.

2. Characterization of spectroscopic detectors

To compare various types of detectors, the following parameters are determined:

- energy resolution,
- detection efficiency,

- response time, and
- dead time.

Detector characteristics are not the only important factors that influence the performance of alpha spectrometers. Other major contributors to the overall quality of alpha spectra are other electronic units and the type and quality of the alpha source.

Energy resolution in nuclear spectroscopy is expressed as the FWHM of the peak. In alpha spectrometry, the energy resolution performance of the detector is expressed traditionally as the FWHM in keV at the 5486 keV peak of ^{241}Am.

In alpha spectrometry, the peaks are usually not symmetric (not Gaussian shaped). The asymmetry appears as a low-energy side tailing that can be better characterized by the full width at one-tenth of the peak maximum (FWTM).

The absolute **counting efficiency** (ε_{abs}) in nuclear spectroscopy is defined as *the number of pulses recorded over the number of alpha particles emitted by the source* and it equals

$$\varepsilon_{abs} = \frac{\Omega}{4\pi}\varepsilon_{int} \tag{5.6}$$

where ε_{int} is the intrinsic efficiency of the detector that is defined as the *number of pulses recorded over the number of quanta incident on the detector,* Ω is the solid angle of the detector seen from the actual source position, and $\Omega/4\pi$ is the geometry factor.

A special feature of alpha detectors is that their intrinsic efficiency is practically 100%. This means that all alpha particles that reach the active volume of the detector are counted (although only a part of their energy is registered if the detector thickness is smaller than the range of alpha radiation in the detector material or if the alpha particles are backscattered from the detector). In alpha spectrometry, the determination of the absolute counting efficiency can be limited to the determination of the geometry factor, which can be calculated if the geometry parameters such as source and detector area and source-to-detector distance in coaxial arrangement are known exactly (see Section IV.B for more details.)

Response time is the time interval between the arrival of the alpha particle into the detector and the appearance of the response signal at the detector output. New particles entering the detector during the response time period are either not detected or detected together with another pulse causing a deviating pulse height, i.e., a pileup peak. **Dead time** is the minimum length of time that must separate two events that they can be recorded as two separate pulses.

In semiconductor alpha spectrometry, the response time and the dead time are very short, and high count rates may be tolerated without pileup and dead time corrections. Using gas ionization detectors, the dead time is longer and corrections are more likely to be necessary. The dead time (τ) corrected count rate (i_{cor}) is calculated from the measured count rate (i_{meas}) according to the following equation:

$$i_{cor} = \frac{i_{meas}}{1 - i_{meas}\tau} \tag{5.7}$$

3. Gas ionization detectors

Gas ionization detectors are described in detail in Chapter 2. For alpha spectrometric purposes, ionization chambers and proportional counters can be applied. Geiger–Müller counters are not suitable for measuring the radiation energy because the electron discharge due to the high gas multiplication envelops the entire anode wire independently of the original number of primary electrons produced by the ionizing radiation. Semiconductor detectors have replaced gas detectors in many applications because they are superior in terms of energy resolution, response time, ease of operation, and cost. Gas ionization detectors offer the attractive feature that large alpha sources can be detected, which is often important in the direct analysis of some environmental samples without sample dissolution and chemical separation, such as air samples (Röttger et al., 2001) or the surface of plate glass (Samuelsson, 2011).

Gas ionization detectors usually consist of a gas-filled cylinder that serves as a cathode and an anode wire. The filling gas is usually a noble gas (Ar, He) or a mixture of gases. Alpha particles excite and ionize the gas. In the latter process, positive gas ions and free electrons are created. The average energy needed to form an ion pair is 30–40 eV depending on the type of gas. Therefore, an incident 5 MeV alpha particle creates about 1.5×10^5 ion pairs. In an external electric field, electrostatic forces will move the charges to the electrodes of opposite charge. The mobility of the free electrons is about 1000 times higher than that of the positive gas cations. Therefore, typical collection times for electrons are much shorter (in the range of microseconds) compared with the milliseconds characteristic of cations.

Ionization chambers are operated at low voltage (100–250 V), whereas a somewhat higher voltage (300–600 V) is applied in **proportional counters** where the accelerated electrons generate new charge carriers in collisions with gas molecules. This process is called **gas multiplication**. In proportional counters, the total number of charge carriers is proportional to the primary charge carriers generated by the alpha particle; therefore, the pulse height is also proportional to the particle energy. As a result of gas multiplication (which can be as high as 10^6), the output pulse is much higher than in the case of ionization chambers.

For alpha spectrometry, gas ionization detectors are used in pulse mode. Because of the slow migration of the ions, these detectors are often operated in electron-sensitive mode. By changing the time constant of the electric circuit,

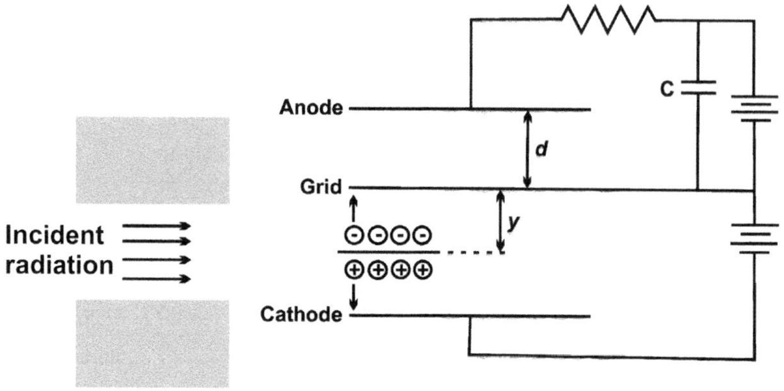

FIGURE 5.9 Operational principle of the gridded ionization chamber according to Knoll (2010). The rise of the pulse results from the drift of the electrons across the grid–anode region *d*. No signal voltage is generated when electrons move across region *y* where *d* and *y* are the distances between the grid–anode and the interaction position–grid. *Reprinted by permission of John Wiley and Sons © 2010.*

the amplitude of the detected pulse reflects only the drift of the electrons. However, the amplitude of the pulse becomes sensitive to the position in the chamber where the alpha particle interaction takes place. This drawback can be overcome by the use of gridded ionization chambers.

In the **Frisch grid detector** (Fig. 5.9), the ionization chamber is divided into two parts by the grid, which is maintained at an intermediate potential between the original electrodes. Alpha particles enter only the grid–cathode compartment through the use of a collimator. Only the upward movement of the electrons in the grid–anode compartment is registered by placing the load resistor in the anode–grid circuit. When the electrons move from the grid to the anode, the grid–anode voltage begins to drop and the signal voltage begins to develop across the resistor. The maximum signal voltage ($V = n\ e^{-}/C$) is reached when the electrons reach the anode.

The grid is maintained at an intermediate voltage between the cathode and anode, and therefore, the grid becomes almost transparent for the electrons. Increasing the grid voltage will cause losses of electrons, and the detected count rate will be reduced (Fig. 5.10).

In the Frisch grid chamber, a fast and position-independent pulse is generated by detecting the electron drift separated from the drift of positive ions.

Gridded ionization chambers have been successfully used for alpha spectrometric analysis of large area sources up to 80 mm radius by Roos and Roos (1999). An absolute counting efficiency close to 40% was achieved, whereas the energy resolution was acceptably high (FWHM was about 30 keV).

Major characteristics of gas ionization detectors:

- Ionization chambers, preferably gridded chambers, operated as pulse chambers have been used to obtain alpha spectra.
- Pulse amplitudes are usually very small because of the lack of gas amplification; consequently, low-noise preamplifiers have to be attached to the ionization chambers.
- Energy resolution of the detector is limited by the relatively high ionization potential (energy needed to create an electron–ion pair) of the gases. Typical resolutions of 35–40 keV have been reported.

FIGURE 5.10 Change in count rate at different grid to anode and anode to cathode voltages (Roos and Roos, 1999).

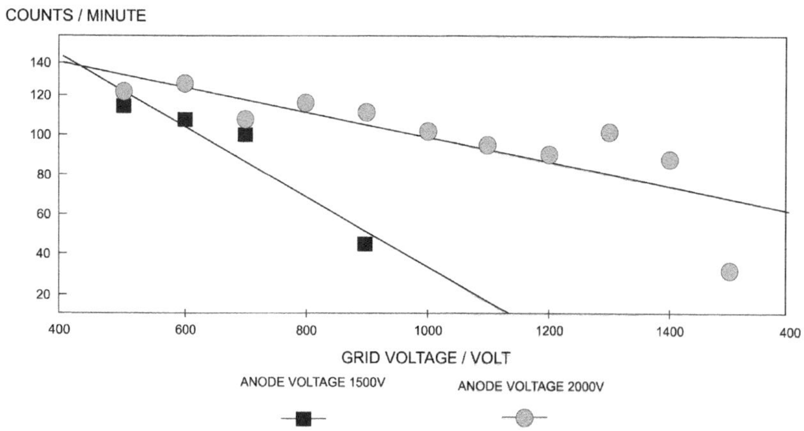

- Response times are relatively slow, electron collection times are in the range of μs, and ion collection times are in the range of ms. In gridded ionization chambers, only the drift of electrons is detected.
- Absolute counting efficiencies are close to 50% if sources are placed inside the chamber.
- Gridded ionization chambers are superior to Si detectors regarding efficiency when large samples (larger than the size of the Si detector) are measured.

4. Semiconductor silicon detectors

Silicon semiconductor detectors are regarded as analogous to gas ionization, as solid-state ionization detectors. The use of a solid medium has the advantage that the dimensions of the detector can be kept small, whereas the counting efficiency is as high as in the case of gases. In alpha spectrometry, this means that a 100% intrinsic efficiency can be reached with a Si detector of 100 μm in thickness. A major advantage of semiconductors is their high (10–20 keV) energy resolution (only the special application cryogenic detectors described below have a higher energy resolution). Their drawbacks include the limitations on size and their high susceptibility to radiation-induced damage, resulting in spectrum degradation.

There are some differences in the pulse generation processes between gas ionization and semiconductor detectors:

- The mobilities of the charge carriers are different. In gas ionization detectors, the mobilities of the electrons are about 1000 times higher than those of the cations, whereas in semiconductor detectors, the mobilities of the electrons are 2–3 times higher than those of the holes.
- The charge collection times are different. The time needed to collect electron–cation pairs is in the range of milliseconds, whereas that for electron–hole pairs is in the range of microseconds.
- The size of the sensitive volume is constant in gas ionization detectors, whereas it is dependent on the applied voltage in semiconductor detectors.

The properties of semiconductor materials, including the description of the band structure, have been discussed in detail in many good monographs (Knoll, 2010; Gilmore, 2008) and are also described in Chapter 4, Volume 1. Here, we focus our discussion on Si detectors, which are the most frequently used detectors in alpha spectrometry.

In Si semiconductor detectors, p-type and n-type Si are brought together in good contact, and a reverse bias (positive voltage to the n-type Si) is applied. At the junction of the two Si layers, an intrinsic layer depleted in any charge carriers is formed. The reverse biased p-n junction, the so-called p-n diode, is used as an alpha particle detector.

The basic interaction between the ionizing particle and the detector takes place in the depleted/intrinsic region. When the alpha particle passes through a semiconductor, **electron–hole pairs** are created along the track. These charge carriers are collected by the electric field, with changes in the electric field being registered as pulses. Pulse amplitudes are small, being in the range of microamperes. Since the height of the pulse is proportional to the energy deposited by the interacting particle, the detector is suitable for spectrometric measurements.

The average energy expended by the ionizing radiation to produce one electron–hole pair in Si is about 3.6 eV (slightly higher than the 3.0 eV in Ge). This value is also somewhat higher than the width of the bandgap in Si (1.12 eV) or the ionization energy because the energy imparted by the alpha particle is not 100% utilized in electron–hole pair production. This value is about an order of magnitude lower than the 30 eV required to create an electron–cation pair in gas ionization detectors. Therefore, the number of charge carriers is increased by about 10 times, and the statistical fluctuation in number is reduced, resulting in a better energy resolution as well as improved signal-to-noise ratio. The ionization energy depends slightly on temperature; in Si, it is 3% greater at liquid nitrogen temperature compared with room temperature.

The electrical resistance of pure Si is high, and therefore, the leakage current across the junction is small. In spectrometric applications, this must not exceed a few nanoamperes. This low current is a major requirement if electric pulses that generate ~ μA current are to be analyzed.

Electrical charges are collected by two ohmic contacts fitted to the opposite faces of the semiconductor material. The ohmic contacts are connected to a detection circuit, and an electric field is applied to collect the charge carriers. Special blocking electrodes are used where the number of charge carriers is reduced during the application of the electric field to reduce the leakage current of the detector.

According to fabrication technology, silicon semiconductor detectors are of three different types: the diffused junction detector, the surface barrier detector (SBD), and the passivated ion–implanted planar Si (PIPSi) detector. Each detector can be regarded as a reverse-biased compensated detector where n- and p-type Si crystals are brought into contact and a depleted/intrinsic (i) layer is formed as the sensitive volume of the detector.

Diffused junction detectors. These detectors, belonging to the first generation of Si diode detectors, are prepared by diffusion from p-type Si. An n-type impurity, usually P, is evaporated on one surface of the crystal and is diffused into the Si crystal to a depth of 0.1–2 μm. A depleted layer is formed under the surface layer that is relatively thick. The incident radiation must pass this "dead layer" before reaching the sensitive volume. The dead layer causes losses in production of charge carriers and a reduction in energy resolution.

Surface barrier detectors. By etching the surface of an n-type Si crystal, a p-type layer (oxide) is formed that is coated by a thin gold layer using vacuum evaporation. SBDs can be produced starting from a p-type Si that is evaporated by aluminum to form an n-type contact. Gold or Al serve as electric contacts and entrance window for the radiation. These detectors are characterized by a very thin dead layer, resulting in high-energy resolution and small losses in charge collection. Due to the very thin entrance window, these detectors are sensitive to mechanical or chemical impacts and are permeable also for photons. They can be operated in light tight chambers that usually also serve as vacuum chambers.

Passivated ion–implanted planar Si detector. At present, these are the detectors most frequently used for alpha spectrometry. N-type Si is used as the starting material for the production of PIPSi detectors (PIPS and ULTRA are the registered trade names of the PIPSi detectors of Canberra and Ortec). Accelerators of about 10 kV are used to implant B or P ions as p-type impurities and an As n-type blocking contact.

PIPSi detectors have the following advantages:

- The implanted Al entrance window is thin and even, allowing high-energy resolution and small losses in charge production.
- The implanted Al window is resistant to mechanical effects; it can be cleaned.
- The leakage current is small.

The flowchart of the production of PIPSi detectors is shown in Fig. 5.11.

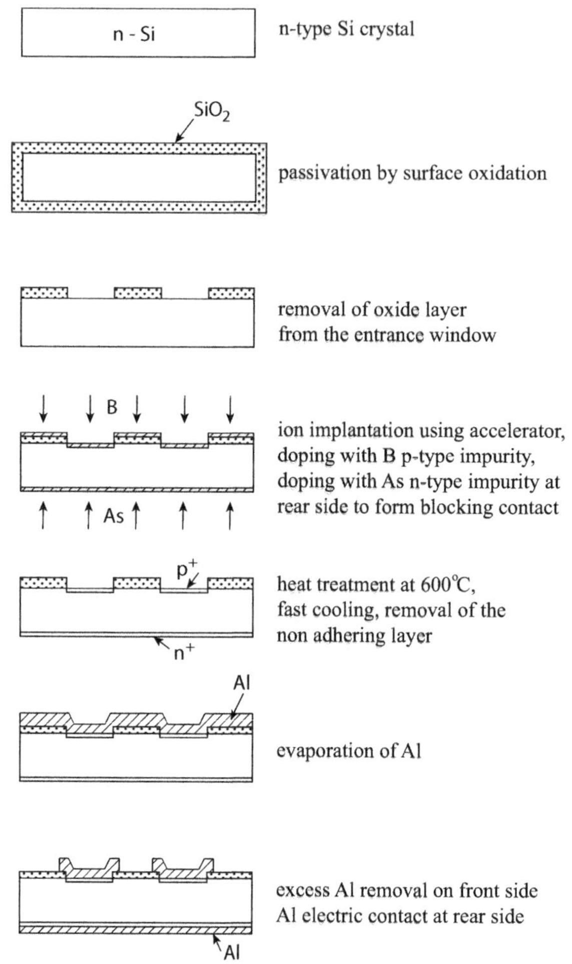

FIGURE 5.11 Steps in PIPSi detector fabrication. PIPSi, passivated ion–implanted planar Si. *Amoudry and Burger (1984), reprinted by permission of Elsevier Ltd. © 1984.*

Silicon detectors, and especially PIPSi detectors, find widespread application in standard alpha spectrometry. Since the 1980s, considerable efforts have gone into improving the detection systems and developing HRAS, which allows the determination of nuclides that emit alpha particles of similar energies.

HRAS is not based on a new detection principle and does not require a new type of instrumentation, but special efforts are made to improve the energy resolution. PIPSi detectors of thin window and small size (20–50 mm^2) are often used. The alpha sources are prepared with special care, typically with vacuum sublimation or electrodeposition to obtain an almost infinitely thin and even sample layer after radiochemical processing. The source to detector distance is kept high (several times the diameter of the detector) to have a small solid angle of measurement (<1% of 4π sr), and so absolute counting efficiencies are typically less than 1% (typically 0.1%–0.2%). High source to detector distance is necessary to minimize the variation of the alpha particle track length in the source and in the dead layer of the detector. Because of the low counting efficiency, a source strength of a few tens of Bq or more are a requisite to avoid a data acquisition time, which is too long. A high vacuum and high stability of the electronic equipment is assured by special technical tools. Temperature control and source collimation might be necessary. Conversion electrons are sometimes deflected by magnetic deflectors to avoid alpha–electron coincidences. The energy resolution of such a system may be even less than 10 keV, which is significantly smaller than the typical values of 30–80 keV for standard alpha spectrometry. HRAS is especially suitable for the analysis of isotope pairs of similar energies such ^{239}Pu and ^{240}Pu, or ^{243}Cm and ^{244}Cm, which are spectroscopically separated.

Major characteristics of Si semiconductor detectors are as follows:

- Surface barrier and preferably passivated implanted planar Si detectors are used to obtain alpha spectra.
- Pulse amplitudes are usually very small (in the range of a few microamperes) but well detectable after amplification.

- The typical leakage current of a PIPSi detector is in the range of nanoamperes.
- Energy resolution of the detector is high due to the low ionization energy needed to create an electron–hole pair. Typical resolution of the detector is 10–20 keV depending on detector performance and size.
- Charge collection times for electron–hole pairs are fast, in the range of microsecond. There is a relatively small difference between the collection times of electrons and holes (a factor of 2–3).
- Absolute counting efficiencies are close to 50% if sources are placed close to the detector inside the chamber.
- Si detectors usually have a low background, resulting in a high counting sensitivity.

Semiconductor Si detectors have proved to have more advantages than gas ionization detectors and have replaced them in most applications.

Research and development trends in Si semiconductor detectors:

Efforts have been made recently to obtain in situ, near real-time alpha spectrometry by combining alpha spectrometry with source preparation on the Si detector. Two techniques have been tested, namely, (1) modifying the Si surface by introducing ion-exchange functional groups and (2) electrodepositing analytes on the metal surface that forms a junction with the semiconductor where the junction can be a rectifying (Schottky contact) or a nonrectifying (Ohmic contact) one.

The first of these techniques was tested by Krupp and Scherer (2018). As a result of a two-step surface modification process, a thiol-functionalized silane was grafted on the surface first, and then sulfonic acid groups were obtained by oxidation, thus obtaining a cation exchange functionalized Si detector. The paper demonstrated the feasibility of measuring alpha particles in solution, and a resolution of about 100 keV was obtained using $^{238}U/^{234}U$ isotopes.

Stika et al. (2018) electrodeposited Th and U on Ni metal plates from molten LiCl-KCl electrolyte. Real-time determination of actinides is of major importance in electrorefining of radioactive wastes. The technique showed promising results.

5. Scintillation detectors

Detection of ionizing radiation by scintillation detectors is based on the emission of light as a result of the interaction of the radiation with the detector material (called a scintillator) followed by collection of light and its conversion into electrical pulses using photomultiplier tubes (PMTs) or photodiodes. Scintillation detection is one of the oldest techniques in the measurement of radioactivity, having had widespread application for the detection of alpha, beta, and gamma radiation in the past. At present, most of the applications are related to gross counting of alpha, beta, and gamma radiation due to the typically high counting efficiency and the low cost of the instrumentation, whereas radiation spectroscopy is limited by the insufficient energy resolution of the scintillators to allow the identification of the radionuclides in an isotope mixture. Scintillation gamma spectrometry was a basic tool till about the 1980s, but it was gradually replaced by semiconductor spectrometry of much better energy resolution.

The operation mechanisms and application fields of LS detectors are described in Chapters 6 and 7 of Volume 1, and those of solid scintillators are summarized in Chapter 9, Volume 1. Here, we summarize the principles of operation and describe the alpha spectrometric application, the so-called high-resolution alpha liquid scintillation spectrometry (α-LSC).

Scintillation detectors include solid and liquid scintillators as well as inorganic and organic scintillators. For all these different types, the basic process used for detection is **fluorescence**, which is the prompt emission of visible radiation (light). Good scintillator materials are characterized by

1. high scintillation efficiency in converting the kinetic energy of the particle to light,
2. linearity of conversion that means that the light yield is proportional to the deposited energy, thus allowing spectrometric analysis,
3. good optical properties, including transparency to the emitted light to allow efficient light collection,
4. fast response time with prompt light emission,
5. high density or high atomic number—this is required only for the detection of gamma radiation.

Inorganic alkali halide crystals, such as NaI(Tl) and CsI(Tl), usually meet the requirements of (1), (2), (3), and (5), but are usually slow in response time. Organic scintillators such as solid plastics and liquid scintillators are usually faster, but the light yield is lower and the requirements of high density and atomic number are not fulfilled.

The main parts of a scintillation detector are the **scintillator**, which converts the radiation energy to visible light photons and the **photoelectron multiplier (PM)** tube containing the photocathode, the multistage electron multiplying section made of a series of electron-multiplying dynodes and an anode for collection of the amplified charge situated in a glass vacuum envelope. Photocathodes are responsible for converting the energy of the photons into electrons. If the light appears as a pulse, the photoelectrons produced will also form a pulse of similar time duration. The number of electrons emitted by the cathode is multiplied on the dynodes where the multiplication factor can be as high as 10^6-10^8 depending on the applied voltage. The electric pulse at the anode of the PM tube can be directly counted without further amplification by a simple counter, a single or MCA. Typical tubes have fast response times. When they are illuminated by a short light pulse, an electron pulse will be produced at the anode with a time width of a few nanoseconds after a delay time of 20–50 ns.

The **sensitivity of the scintillation detector** consisting of a scintillator and a PM tube to radiation can be calculated by the following assumptions: The blue light emitted by the scintillator has a wavelength (λ) of about 400 nm. According to Planck's equation, the energy of the photon is therefore about 3 eV. Taking into account that the conversion efficiency of the scintillator is 5%–20%, then the number of photons generated by a particle of 1 MeV energy is about 15,000–60,000. The efficiency of the photocathode to convert photons into electrons is typically 10%. Therefore, the expected number of electrons at the output of the photocathode is 1500–6000. Altogether, about 200–700 eV energy is required to produce a photoelectron. The corresponding "ionization energy" of gas ionization and semiconductor detectors is one and two orders of magnitude lower, respectively. This difference is the basic reason for the poor energy resolution of scintillators compared with gas ionization and especially to semiconductor detectors.

The mechanism of scintillation in inorganic and organic scintillators is different.

Mechanism of scintillation in inorganic crystals with activators. Inorganic scintillators are insulators characterized by a band structure where the lower band (the valence band) represents the electrons that are bound to lattice sites of the crystals and the upper band (the conduction band) represents the electrons that can migrate throughout the crystal. The gap (the forbidden band) between the two bands is wide (above 5 eV). If energy is deposited in the crystal by radiation, electrons in the valence band can be elevated to the conduction band leaving a hole behind. Addition of small amounts of impurity, called an activator, to the pure crystal will create an intermediate energy level within the gap (a luminescence center) through which the excited electron can deexcite back to the valence band while emitting a visible photon with a reduced energy. This photon will serve as the basis of scintillation detection. Deexcitation processes are typically fast with a half-life of about 100 ns.

Beside the luminescence centers, impurities in the crystal can create electron traps. Deexcitation of the electrons from the electron traps either does not result in visible light emission, or the light emission is delayed causing an "afterglow" appearing in an increased background during the measurement.

Detection of alpha particles by solid scintillators. Among the great variety of inorganic scintillators (e.g., NaI(Tl), CsI(Tl), CsI(Na), CaF_2(Eu), LiI(Eu), bismuth germinate), ZnS(Ag) is the one that has been exclusively used for the detection of alpha particles. It has a high radiation conversion efficiency of about 20%. The maximum wavelength of the emitted photons is 450 nm, close to the maximum sensitivity of the photocathode. The principal decay constant is 0.2 μs. Its major drawback is that it is crystallized as polycrystalline powder, which is not adequate for spectrometric measurements. Usually, a thin layer of 20–30 mg/cm^2 thickness is mounted on a glass surface. A thicker layer of ZnS loses transparency.

Mechanism of scintillation in organic crystals, liquids, and gases. In organic scintillators, excitation and deexcitation processes take place in single molecules independent of their physical state, i.e., solid, liquid, or gaseous state, in marked contrast to inorganic scintillators. Scintillation arises from transitions of electrons between the energy levels of the organic molecule.

The energy spacing between the ground and the excited levels is 3–4 eV for many organic scintillators. In both states, there are several vibrational energy levels where the energy spacing is much smaller (about 0.15 eV), these levels can also be occupied. The molecules at room temperature and in the absence of radiation occupy the ground state. When radiation travels through the scintillator, the deposited energy excites the molecule, and this is followed by fast (ps) deexcitation. Deexcitation to the ground level does not result in light emission. Fluorescent light is emitted in those deexcitation processes where the excited molecule goes back to one of the vibrational levels of the ground state. In organic scintillators, delayed light emission (phosphorescence) is also often observed.

In mixtures of organic molecules, the excitation energy undergoes substantial transfer from one molecule to another. This energy transfer plays an important role in liquid scintillators where often a high concentration of nonscintillating solvent is mixed with a small concentration of an efficient scintillator. The primary energy transfer takes place in the solvent, and the excitation energy is finally emitted as photons by the scintillator molecule. Scintillation cocktails often contain a third organic molecule called a wavelength shifter that can change the wavelength of the emission spectrum so that it is better fitted to the absorption maximum of the photocathode. Scintillator molecules usually contain aromatic rings, and oxygen and nitrogen atoms can be included in the aromatic ring structure.

Detection of alpha particles by LSC. Liquid scintillators are frequently used for the detection of alpha particles. Commercial liquid scintillation counters (LSCs) developed for the detection of low energy beta particles can be used to measure alpha particles without self-absorption losses, giving 100% absolute counting efficiency. This advantage is counterbalanced by the limited resolution of the scintillators. A typical energy resolution using commercial LS system is about 20%–25%, so alpha spectrometry is difficult or almost impossible to achieve.

Detection of low-level alpha radiation by a commercial LSC is difficult because the scintillation yield of the alpha particles is lower by about a factor of 10 than that of beta particles. Therefore, the spectra of the alpha radiation of 4−6 MeV appears in the same region as the spectra of beta particles of 400−600 keV.

Furthermore, the response of organic scintillators to alpha or beta radiation is not completely linear; deviation from linearity is found at low energies. The explanation of this behavior was given by Birks (1964). In organic scintillators such as LSC, only a small fraction of the kinetic energy of the alpha particles is converted to light, and the remainder is dissipated in the matter as heat and lattice vibrations. The high ionization density along the track of the alpha particle causes damages to the organic molecules that leads to quenching (absorption) of the fluorescence light created in the interaction. It is expected that the light output (*dL*/*dx*) is proportional to the specific energy loss of the particle (*dE*/*dx*), if there is no quenching:

$$\frac{dL}{dx} = S\frac{dE}{dx} \tag{5.8}$$

where *S* is the normal scintillation efficiency.

The Birks' formula expresses the deviation from the normal scintillation due to quenching:

$$\frac{dL}{dx} = \frac{S\dfrac{dE}{dx}}{1 + kB\dfrac{dE}{dx}} \tag{5.9}$$

where *kB* is treated as an adjustable parameter to fit experimental data for a given organic scintillator.

The **reduced scintillation efficiency** is a special feature of the interaction of heavy charged particles with organic scintillators and makes alpha spectrometry using commercial LSC very difficult.

To avoid the interference of alpha and beta LS spectra, a technique of **pulse shape discrimination (PSD)** has been introduced. In organic scintillators, fluorescent light is produced in a very rapid "prompt" process as well as in a slower process corresponding to delayed fluorescence. The response times of the processes are in ns and 100 ns ranges, respectively, as shown by the shape of the pulses depicted as pulse intensity versus time in Fig. 5.12. The slow scintillation component is higher for particles of high specific energy loss (*dE*/*dx*), such as alpha particles and heavy charged particles. Based on the differences in the pulse shapes, it is possible to separate pulses due to alpha particles from those originating from beta or gamma radiations. Using special electronic circuits for PSD, alpha and beta spectra are constructed from a single LS spectrum. Using this technique, which is available in modern LS counters, the problems of overlapping alpha and beta spectra due to reduced scintillation efficiency can be eliminated.

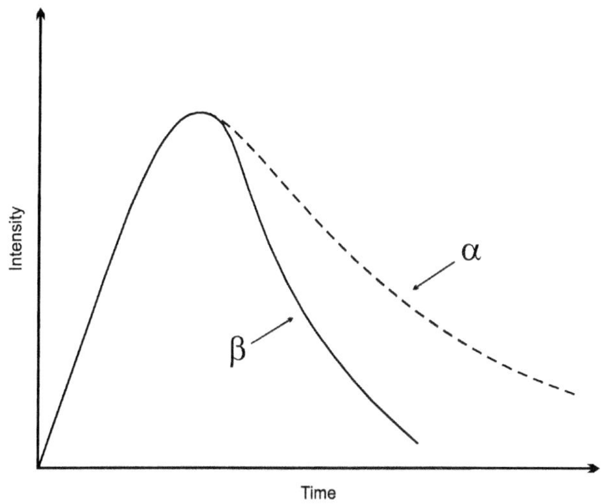

FIGURE 5.12 Typical pulse shapes in liquid scintillation spectrometry. Tailing refers to delayed fluorescence. The contribution of the slow component to the pulse is higher for alpha particles than beta particles.

Alpha spectrometry by high-resolution α-LSC. A so-called high-resolution alpha-LSC characterized by an FWHM of 200−400 keV for alpha particles in the energy range of 4−7 MeV has been developed for the measurement of alpha spectra of separated radioelements such as uranium and thorium. A small volume of extracted sample mixed with an organic scintillator is detected by a single PM tube in a special geometry shown in Fig. 5.13.

The differences in fluorescent yields of alpha and beta particles cause troublesome background interference from beta and gamma radiations in the alpha counting. With the selection of an appropriate scintillator, alpha-produced pulses are about 30 ns longer than beta-produced pulses. Based on this difference, beta-produced pulses are

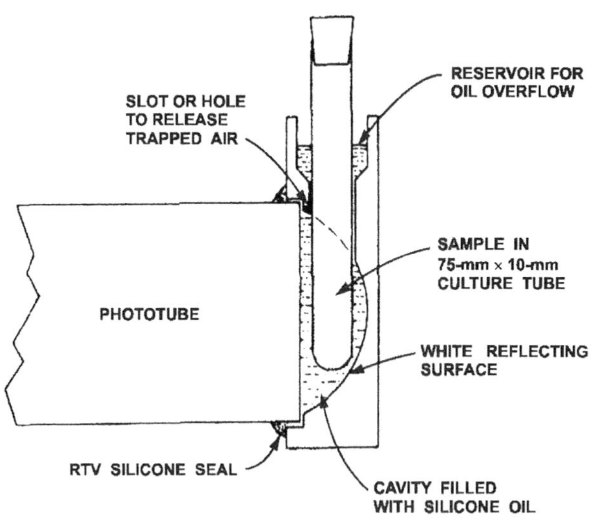

FIGURE 5.13 Schematic drawing of the sample, reflector, and photomultiplier tube arrangement for alpha LSC. *McKlveen and McDowell (1984), reprinted by permission of Elsevier Ltd. © 1984.*

electronically rejected by a PSD. The system is called photon–electron rejecting alpha liquid scintillation (PERALS) spectrometer. The use of PSD is responsible for rejection of the pulses originating from interfering gamma rays and conversion electrons, which are emitted by many of the alpha-decaying nuclides (see Table 5.1 provided earlier in this chapter.). The improved energy resolution of the PERALS system is a consequence of

- the small sample size,
- the efficient geometric arrangement,
- the rejection of electrons and photons by the use of PSD,
- the use of aqueous immiscible scintillator and organic extractant,
- removal of dissolved O_2 by purging the sample with Ar gas.

Eliminating the presence of water and dissolved O_2 helps reduce quench effects that result in highly variable light production and, as a consequence, poor resolution in a conventional LSC. In the PERALS spectrometer—differing from the commercial LSC instruments dedicated for tritium and ^{14}C measurements—a single PM tube is applied. To analyze alpha-emitting nuclides with the relatively low-resolution PERALS spectrometer, selective chemical separations have to be performed. For this purpose, a great variety of special extractive scintillators have been developed that serve as solvent extractants and liquid scintillators at the same time. A list of the available extractive scintillators is given in Chapters 6 and 7 of Volume 1. An alpha spectrum evaluation code has been developed to deconvolute PERALS spectra that will be discussed in Section IV.A.3 (Aupiais, 2004).

Major characteristics of scintillation detectors used for determination of alpha particles:

- Among the great number of activated inorganic scintillators, ZnS(Ag) has been used to detect alpha particles. LS detectors are also used to detect alpha particles. High-resolution alpha LSC is applied for alpha spectrometric measurements.
- Pulse amplitudes are usually directly measurable because amplification is performed by the dynode system of the PM tubes.
- Scintillators are characterized by their light yield or scintillation efficiency, which is usually higher in the case of inorganic than organic scintillators. The total light yield of inorganic scintillators is about 10,000–50,000 photons/MeV.
- The wavelength of the light emitted by organic and inorganic scintillators is in the range of 400–450 nm and 300–550 nm, respectively. This is coupled to the photocathode that has highest sensitivity in the range of 300–600 nm.
- The light yield of organic scintillators is about an order of magnitude lower for alpha particles than for electrons due to quenching caused by radiation-damaged organic molecules.
- Energy resolution of the scintillation detector is poor due to the high energy (a few hundreds of eV) needed to produce a photoelectron. Typical energy resolution of a NaI(Tl) detector for gamma radiation is about 8% –10%. The energy resolution of a commercial LSC for alpha particles is 20%–25%.
- Responses of scintillation detectors are usually fast, in the nanosecond range for organic scintillators and somewhat slower in case of inorganic scintillators (in the ns-μs range). The rise and fall of the light output can be characterized by an FWHM of the light versus time profile, called response time FWHM, analogously to the FWHM of the counts versus energy distribution. The very good time resolution of the scintillators is preserved in pulse processing by PM tubes.
- Absolute counting efficiencies of solid scintillators depend on the counting geometry and the type of radiation to be detected. The counting efficiency of LSC for high-energy beta and alpha particles is 100%; for low energy beta particles such as for ^3H, the efficiency is 30%–40%.
- Scintillation counters generally have high background, increasing the detection limit compared with measurements with gas ionization and semiconductor detectors. The most significant source of noise is the spontaneous emission of thermionic electrons by the photocathode. (Various sources of background and the techniques of background reduction are discussed in detail in Section IV.C.)
- The performance of LSC is strongly dependent on quenching (see Chapter 6, Volume 1), which causes losses in the light output of the scintillator either by hindering the light production (chemical quenching) or by absorbing the produced light (color quenching).

Although semiconductor Si detectors proved to have advantages over scintillation detectors in spectrometric measurements, liquid scintillators are used in various fields, including gross alpha/beta counting; absolute activity determination due to high counting efficiency and the high-resolution alpha LS spectrometry (PERALS) has been successfully used for fast spectrometric determination of some alpha emitters.

6. Cryogenic detectors

A measurement technique has been developed, which is based on the thermal detection of alpha particles using very sensitive detectors, called bolometers. The bolometer is a calorimeter operated at very low temperature, typically below 100 mK. The bolometer has two major parts, an absorber where the energy of the single alpha particle results in a temperature rise and a thermal sensor (thermometer) to measure it. At this very low temperature, the heat capacity of the absorber, which can be a dielectric, a metal, or a semiconductor material, becomes so small that the temperature rise due to the interaction of a single alpha particle becomes detectable (See Chapter 9, Volume 1 for additional information on the operating principle of bolometers).

In the interaction of alpha particles with matter, energy is transferred to electrons by inelastic collisions that induce ionization and excitation. The energy is distributed between the production of charge carriers (in semiconductor materials, electron–hole pairs) and phonons (lattice vibrations, heat). In bolometers, the energy of the phonons is detected. If the electron–hole pairs recombine in the bolometer within the time resolution of the system, their energy is also transferred to phonons that are also detected.

Bolometers have the great advantage that theoretically better energy resolution can be achieved than in the case of Si detectors. Juillard (1999) calculated that the FWHM in a Ge or Si bolometer is in the order of 1 keV for alpha particles of 5 MeV energy. Leblanc et al. (2006) reported 4 keV FWHM for an alpha spectrum measured with a composite copper–germanium bolometer and an external electrodeposited ^{238}Pu source. The detector response was in the range of 10^{-8}–10^{-7} V/keV, and the pulse rise and decay times were about 1 and 5 ms, respectively. Fig. 5.14 shows the schematic drawing of the Cu-Ge bolometer used for the measurements.

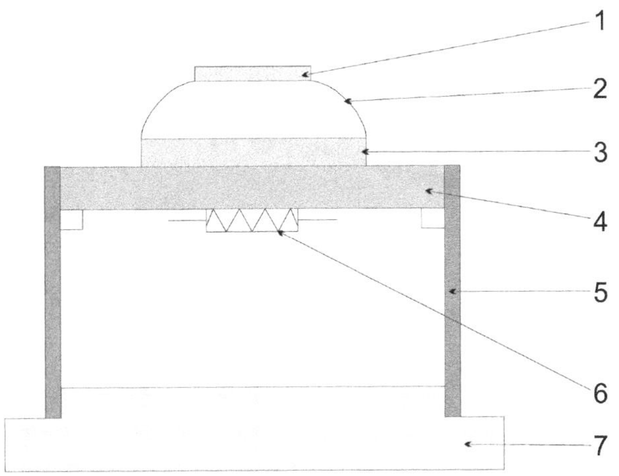

FIGURE 5.14 Bolometer B300 developed by Leblanc et al. (2006). A Cu absorber disk (1) is thermally coupled to a Ge heat diffusing volume (2, 3) coupled to a large thermometer (4). Pure Ge pillars (5) assure a rigid support and the thermal link to the thermal bath (7). The detector frame is fixed on the cold plate of the refrigerator. The bolometer was tested in a 20 mK dilution refrigerator, placed in a Faraday cage and decoupled from ground vibrations. *Reprinted by permission of Elsevier Ltd. © 2006.*

The development of a new type of microcalorimeter also called magnetic spectrometer has been reported. Magnetic calorimeters are particle detectors operated at low temperatures (typically below 100 mK) where the temperature rise due to absorption of the particle in the sensor is transformed into a measurable magnetic flux change in the superconducting quantum interference device (SQUID). Paramagnetic temperature sensors have been used as detectors. Pabinger et al. (2011) used a sputtered Au:Er paramagnetic temperature sensor attached to a superconducting pickup coil and reported 2.83 keV energy resolution for ^{241}Am. Yoon et al. (2012) reported 1.2 keV resolution for the same nuclide.

Hoover et al. (2015) developed a superconducting transition-edge sensor (TES) with the sample encapsulated inside a gold foil that is used to absorb the decay energy (*Q*). The TES is based on a superconducting film, the resistance of which is strongly dependent on temperature. Therefore, it can act as a thermometer. The electric current flowing through the TES changes with temperature, and the current change is measured using the SQUID. Modern refrigerator technology assures the necessary low temperature for the detection of the temperature rises due to single alpha decay events. Energy resolution of about 1 keV has been reported for the *Q*-value spectra.

After further development and optimization, microcalorimeters will probably find application in metrology where extremely HRAS is required for the accurate determination of nuclear data.

B. Electronic units

The basic alpha spectrometric measurement system consists of a detector, electronics to process the signals, a power supply unit, a chamber (light tight chamber in the case of a light-sensitive Si detector and gas-tight chamber in the case of a gas ionization detector) surrounding the detector, a vacuum system to reduce the chamber gas pressure and absorption of alpha radiation, and a software package to analyze the spectra. The signal processing electronics usually contains the preamplifier, the main amplifier, the ADC, and MCA, as illustrated previously in Fig. 5.6.

In semiconductor detectors, the preamplifier is usually a charge sensitive amplifier, whereas in scintillation detectors, it is the PMT. The various types of preamplifiers are described in detail in Chapter 9, Volume 1.

The roles of the main amplifier are (1) to further amplify the signal and (2) to shape the signal from the discharge curve to a signal in pulse shape. In the main amplifier, only the first part with the increase of the discharge curve is used to speed up the signal processing significantly.

The ADC determines the height of the pulse signal supplied from the main amplifier. This pulse height is then converted to a number of channels by the MCA. The MCA counts the number of events per channel and generates the pulse height spectrum.

With the advent of personal computers (PCs) at the beginning of the 1980s, a powerful system of standardized processor, large computing and storage power, and large variety of interfaces became available for spectrometric purposes. At present, there are two types of PC-based MCA systems, PC plug-in cards, and NIM modules under PC control. Spectrum data are stored either in PCs or in standalone storage modules. In both cases, setup and control of the MCA is accomplished via a computer program, and spectrum evaluation software is usually also installed in PCs.

With the increasing development of complex digital electronics, the so-called digital amplifiers for spectroscopy were developed, which are not only amplifiers but also analyzers. These systems digitize the signals from the preamplifier directly and then process them digitally. The great advantage of the digital amplifier is the large signal throughput. As a result, extremely high count rates can be processed almost without interference.

The high voltage supply is adapted to the type of detector and provides a very clean and stable signal. Noise of the high voltage power supply has a detrimental effect on the spectra, resulting in degraded energy resolution or deviations in the amplification of the signals, e.g., in case of scintillation detectors.

A detailed description of the electronic units is given in Chapter 4, Volume 1 on gamma spectrometry and in several good monographs (Knoll, 2010).

IV. Characteristics of the alpha spectrum

A. Peak shape and spectrum analysis

The major characteristics of the alpha particle spectra depend on the type and quality of the detector and the electronic units, the geometric arrangement, and the quality of the alpha source. The operating principles and the major characteristics of the various types of detectors have been discussed above. Energy resolution, pulse shape, response times, and intrinsic efficiency are determined by the detector; other factors will contribute to the overall performance but cannot overcome limitations of the detector. A thick alpha source, a thick absorber layer between the source and the detector, or an uneven detector window will negatively affect the energy resolution, whereas electric noise in the signal processing will increase the background. In the present section, the various factors influencing alpha spectra are discussed first based on the example of semiconductor Si alpha spectrometry, followed by a brief discussion of other detector types.

1. Peak shape and spectrum analysis with Si detectors

a. Peak shape

Typical alpha spectra of various resolution and shape are shown in Fig. 5.15. The sources, which were prepared by membrane filtration, contain different amounts of natural U and so have different thicknesses. The measurements were performed using a PIPSi detector.

Fig. 5.15 shows the effect of self-absorption on the peak shape. The spectrum with the highest resolution was

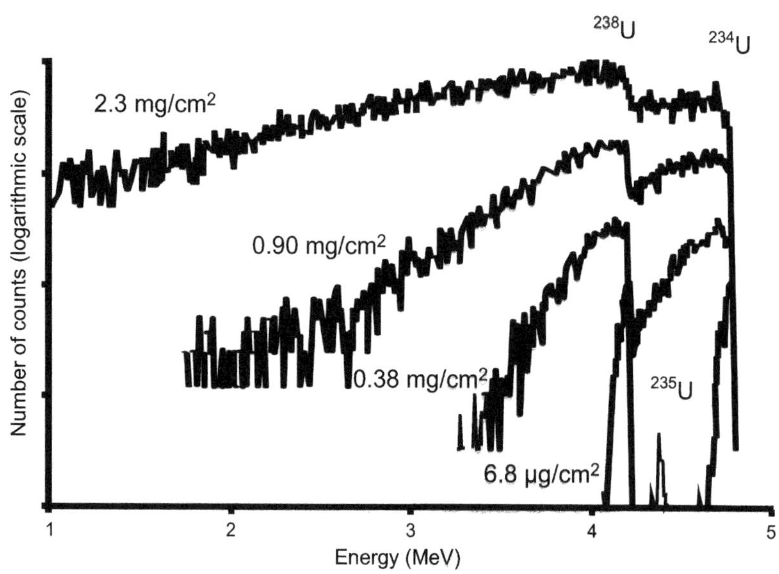

FIGURE 5.15 Alpha spectra of samples composed of natural U with different source thicknesses. The peaks on the left and right correspond to ^{238}U and ^{234}U. Spectra are shifted along the vertical axis to show the differences in peak width (resolution) and tailing. *Pöllänen et al. (2005), reprinted by permission of Elsevier Ltd. © 2005.*

obtained using a typical thin alpha source produced by radiochemical sample processing, similar to the spectrum in Fig. 5.4. This source has a thickness of 6.8 μg/cm^2. By increasing the source thickness to a few mg/cm^2, the self-absorption of alpha radiation is increased, and peaks become wider causing poor energy resolution and the low-energy tailing increases.

Energy resolution (FWHM). Alpha peaks are characterized in routine analytical practice by the FWHM value, which can be easily determined from the spectrum as the peak width in keV at the half of the peak height. The FWHM values of alpha peaks are usually larger than the specified values for the detector. While typical FWHM values of PIPSi detectors are 17–19 keV and 30–37 keV for detectors of 300 and 1200 mm^2 in active area, respectively, those of alpha spectra in routine analytical work may be in the range of 30–80 keV depending on the source quality.

The statistical fluctuation in the number of the electron–hole pairs produced in the detector determines the theoretically achievable energy resolution of an alpha spectrometer. The number of electron–hole pairs varies due to the presence of impurities, due to crystal defects that can trap the charge carriers, and due to possible recombination of the charge carriers. Various secondary processes (such as crystal damage and lattice vibration) may also have an effect. A significant part of the energy deposited in the detector does not create electron–hole pairs but appears as heat and vibrations (phonons). According to theoretical calculations, the limit of the energy resolution in Si detectors (when processes taking place outside the detector such as absorption and self-absorption are neglected) is about 7–10 keV (Steinbauer et al., 1994; Amoudry, 1990). The measured FWHM depends on many other parameters of the detection system, the geometric arrangement of the source and the detector (Fig. 5.16), and the properties of the alpha source itself. Each process causes peak broadening.

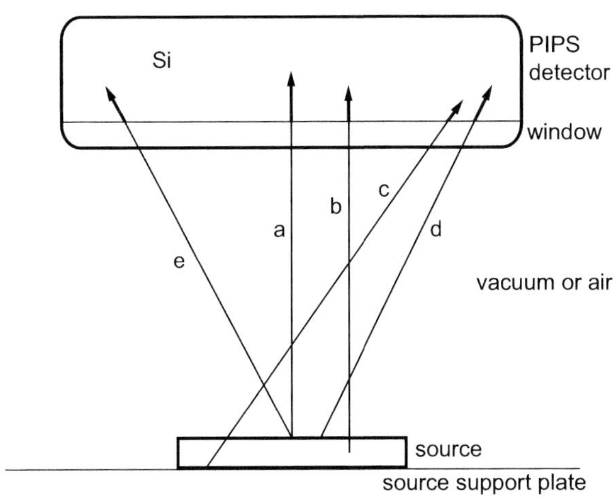

FIGURE 5.16 Paths of alpha particles (a to e) in the source–detector system.

The final FWHM can be calculated from the individual contributions as

$$\mathrm{FWHM}^2 = \mathrm{FWHM}_d^2 + \mathrm{FWHM}_s^2 + \mathrm{FWHM}_a^2 \quad (5.10)$$

where d, s, and a refer to detector, source, and absorber effects, respectively.

The effect of the geometric arrangement on the energy resolution is illustrated in Fig. 5.16.

Alpha particles emitted by the radionuclides embedded in the source material can have different paths before entering the Si detector. They can deposit energy in the source such as particles b and c in Fig. 5.16. All particles (a to e) will deposit energy in the air gap between the source and the detector and in the detector window. The first process represents self-absorption, and the latter ones represent absorption. The role of absorption in air can be neglected if high vacuum is present in the detector chamber, but very often it is desirable to attain a low air pressure to protect the detector from recoil particles (see below). Altogether, the contributions of absorption in the source ("self-absorption"), in the detector dead layer, and in any other absorbent will mutually contribute to the total absorption. Each process causes reduction of the total energy deposited in the Si detector, and each process results in a range and energy straggling of the monoenergetic alpha particles. The lengths of the arrows in the Si material (Fig. 5.16) represent the paths of the alpha particles and the energy deposited by each particle in the active volume of the detector. Variation in these individual energy depositions results in peak broadening.

Absorption and self-absorption have different effects on the alpha spectrum, although from the physics point of view both processes are basically the same. If only absorption has a significant contribution to energy losses (thin source but thick absorber between the source and the detector), then the spectrum is shifted to lower energies because all alpha particles lose part of their energy in interactions outside the detector (Fig. 5.17). It also causes peak broadening due to the variation in energy loss outside of the detector due to varied angles of incidence and to the reduction in number of charge carriers produced in the detector. If only self-absorption has a significant contribution to energy losses (as in the case of thick sources), then the peak is broadened, but the spectrum is not shifted (Fig. 5.18). This is because alpha particles originating from the surface layer of the source can reach the detector without loss of energy.

Absorption and self-absorption depend on the composition, density, homogeneity, and surface roughness of the absorber and the source. All these parameters can be adjusted in the Monte Carlo simulation program known as AASI (https://www.stuk.fi/web/en/services/aasi-program-for-simulating-energy-spectra-in-alpha-spectrometry).

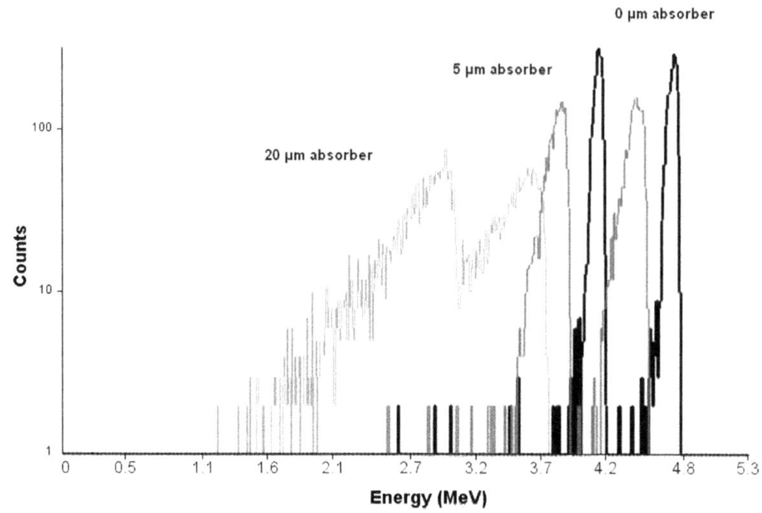

FIGURE 5.17 Simulated alpha spectra of a source containing equal activities of ^{238}U and ^{234}U by assuming different absorber thicknesses (0, 20, and 40 μm). Detector full width at half maximum (here 14 keV) and source thickness (0.4 μm, 6.8 μg/cm^2) do not influence the peak shapes significantly. Spectra were generated by the AASI simulation software (Siiskonen and Pöllänen, 2004).

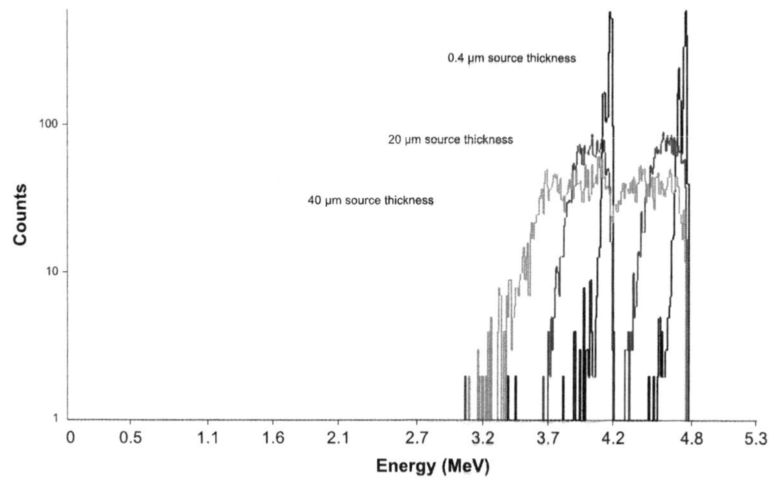

FIGURE 5.18 Simulated alpha spectra of a source containing equal activities of ^{238}U and ^{234}U with varying source thickness (0.4, 20, and 40 μm with source surface densities of 6.4, 320, and 640 μg/cm^2). Detector full width at half maximum and absorber thickness are assumed to be 14 keV and 0 μm (0 μg/cm^2), not influencing the peak shape significantly. Spectra were generated by the AASI simulation software (Siiskonen and Pöllänen, 2004).

The resolution of semiconductor alpha detectors depends on the detector size. The reason is that large detectors cannot be prepared with a completely even thickness of the front face contact and without dead layer inhomogeneity. *Inhomogeneity and surface roughness* in the source, the detector dead layer, or even the absorber will result in increased FWHM values.

The peak resolution depends on the measuring geometry, i.e., the source–detector distance, dimensions of the source and the detector, and their relative positions. When the source is placed parallel to and in line with the detector, peak broadening is observed when the source is moved toward the detector. This phenomenon can be easily interpreted as the increased angle under which the alpha particles can reach the detector, resulting in an increased path in the detector dead layer and, thus, a reduction of energy deposited in the effective volume of the detector (see Fig. 5.16). This effect on FWHM is especially significant in the case of thick detector windows (e.g., in case of DJDs).

Alpha spectra measured with Si detectors are characterized not only by the FWHM but also by the **asymmetric peak shape**. Peaks usually have a Gaussian shape on the high energy side and a low-energy tailing as shown in Figs. 5.15, 5.17 and 5.18.

Due to the complexity of the processes resulting in peak tailing, the FWHM parameter is not enough to describe the peak shape, unlike other spectrometric methods such as gamma ray spectrometry and LSC. In a simplified case, the FWTM parameter is used together with FWHM, but in more detailed studies, the whole peak shape has to be described by peak fitting algorithms. Systematic studies have in recent years revealed the physical processes responsible for the asymmetric shape of alpha peaks.

High-energy tailing originates from pulses that are detected as the sum (coincidence) of the alpha pulses and the pulses due to conversion electrons (or to a lesser extent to low energy gamma radiation) emitted by the same radionuclide if the pulses are detected within the resolution time of the detector. For example, ^{241}Am emits alpha particles and photons with several different energies. Instead of gamma emission, the deexcitation can also occur by the emission of conversion electrons of discrete energies. These relatively low-energy electrons are detected with high efficiency by the Si detector. Thus, pulses on the high energy side of the alpha peak originate basically from the true coincidences of alpha particles and conversion electrons/photons. No true coincidences exist in the case when the alpha decay leads to the ground state of the daughter nuclide. The effects of coincident emissions may be notable if the deexcitation leads to the excited state of the daughter nuclide as pointed out by Siiskonen and Pöllänen (2006).

The origin of **peak broadening and low-energy tailing** is a complex process that can be observed in all types of alpha spectra produced by Si detectors including low- and high-resolution spectra, spectra of isotopes emitting either a single alpha particle or different alpha particles with or without the emission of gamma photons.

Alpha peaks have an asymmetric shape with low-energy tailing in both high- and low-resolution alpha spectrometry, but the physical phenomena responsible for this shape are different. In high-resolution alpha spectrometry, where absorption and self-absorption are reduced to a minimum, the physical processes have been studied in detail by several authors (Steinbauer et al., 1994; L'Hoir, 1984; Bland, 1998). In addition to the (small) variation of the alpha particle track length in the absorbing materials (mainly dead layer of the detector and remaining air in the vacuum chamber), the peak broadening originates from energy straggling in the absorbing material and the electronic noise (e.g., in the detector, the amplifier), which cannot be totally eliminated. This peak broadening can be described by a Gaussian function. The low-energy tailing arises from the processes where the deposited energy of the alpha particles in the Si detector does not result in the production of electron−hole pairs. Such processes are collisions of the alpha particles with the Si nuclei and further interactions of the knocked-on Si ions in the crystal. Stopping of these ions (such as by recoil, heat, and phonon generation) results in pulse height defect that can appear as tailing in the spectrum. Low-energy tailing can be described by exponential functions.

In low-resolution alpha spectrometry, these fundamental processes are overshadowed by other types of interaction of alpha particles usually taking place outside the detector. Low-energy tailing and peak broadening are explained by **absorption and self-absorption** of the alpha particles, such as

- absorption in gases (air) present in the vacuum chamber,
- absorption in the entrance window, and
- self-absorption in the source.

Absorption results in peak shifting, peak broadening, and peak tailing. In the case of self-absorption, there is no peak shifting (as measured at the upper limit of the peak), but there is peak broadening and increased tailing, as discussed above (see Figs. 5.17 and 5.18). Ideally, sources should be infinitely thin to avoid self-absorption. Significant peak deterioration for uranium sources was observed above a source density of 100 μg/cm^2 by Martin and Hancock (2004a), although the significance will depend on the homogeneity of the sources and on the energy difference for the lines of interest.

The processes influencing alpha peak shape in the case of thick sources have been investigated by Monte Carlo simulation (e.g., Ziegler et al., 1996; Pöllänen et al., 2007). Pöllänen found that in thick sources the peak shape distortion is influenced by source thickness, source density, source inhomogeneities, and elemental composition of the source (see Figs. 5.15, 5.17 and 5.18). The AASI software uses Monte Carlo simulations to generate alpha spectra from sources of different characteristics. It was shown by Pöllänen et al. (2005) how the properties of the alpha source (such as inhomogeneities) affect the peak shape. Simulated and measured spectra agreed well when source inhomogeneities were also taken into account.

b. Peak fitting functions

A prerequisite for successful alpha spectrum unfolding is the knowledge of the peak shapes to be used in the analysis. Other relevant points are calibration of the detector (section VII) and how to use decay-related information, i.e., the fitting process peak based or peak family based. Depending on the application, there are other constraints, such as the emission probabilities (Martin Sanchez and Rubio Montero, 1999), and the effects of true coincidences, which may also be taken into account.

Various mathematical models have been proposed to describe the shape of monoenergetic alpha peaks, and these are used for fitting the alpha peaks in spectrum analysis software. The fitting procedure involves

- the selection of the peak shape model that depends on a number of parameters,
- selection of the optimization criteria for fitting the model to the measured data, and
- the choice of the numerical optimization procedure.

As a result of fitting, measured data are compared with calculated values, the goodness of fit is described, and peak areas and peak positions are calculated together with their uncertainties.

To describe the asymmetry of the alpha peaks, **modified Gaussian functions** were used from the end of the 1960s, but these functions failed to describe the peak tailing (García-Toraño et al., 1999). Later, the use of a combination of **Gaussian and exponential functions** was proposed. The analytical function consisting of a Gaussian and a one-sided exponential function was first proposed by L'Hoir (1975), and it has been used for the analysis of alpha peaks in HRAS by several authors (Watzig and Westmeier, 1978; Bortels and Collaers, 1987; Bortels et al., 1995; Martin Sanchez et al., 1996).

Bortels and Collaers (1987) used a Gaussian and one or two exponential functions. In certain cases, a step function had to be subtracted from the spectrum before data processing. In the peak shape model, two random variables (x, y) are considered where x has a normalized Gaussian probability density function $p_x(x)$ with σ resolution:

$$p_x(x) = \frac{1}{\sigma\sqrt{2\pi}}\exp\left(-\frac{x^2}{2\sigma^2}\right) \tag{5.11}$$

and y has a normalized left-sided exponential with parameter τ

$$p_y(y) = \frac{1}{\tau}\exp\left(\frac{y}{\tau}\right)H(-y) \tag{5.12}$$

where $H(-y)$ is the left-sided unit-step function where $H(x) = 1$ for $-\infty < x < 0$, otherwise $H(x) = 0$.

The random variable $u = x + y$ (the energy variable) is characterized by a joint probability density function $p_u(u)$:

$$p_u(u) = \int_{-\infty}^{0} p_y(y)p_x(u-y)dy \tag{5.13}$$

The normalized function after integration is:

$$p_u(u) = \frac{1}{2\tau}\exp\left(\frac{u}{\tau}+\frac{\sigma^2}{2\tau^2}\right)erfc\left[\frac{1}{\sqrt{2}}\left(\frac{u}{\sigma}+\frac{\sigma}{\tau}\right)\right] \tag{5.14}$$

in which *erfc* is the complementary error function. Taking into account the actual peak area A that is different from unity and the position of the peak center μ in the Gaussian function, one obtains the expression of the fitting function $f(u)$ for a single alpha peak:

$$f(u)=A\cdot p_u(u) = \frac{A}{2\tau}\exp\left(\frac{u-\mu}{\tau}+\frac{\sigma^2}{2\tau^2}\right)erfc\left[\frac{1}{\sqrt{2}}\left(\frac{u}{\sigma}+\frac{\sigma}{\tau}\right)\right] \tag{5.15}$$

This fitting function obtained from a Gaussian and one exponential function can describe the asymmetric peak shape relatively well. It has two shape parameters σ and τ and two parameters characterizing the peak area A and the peak position μ. For a measured spectrum composed of several peaks, the sum of fitting functions $F(u)$ is calculated by addition of the individual $f(u)$ functions.

However, it was found that the fitting function (Eq. 5.15) does not describe well the tailing (especially long tailing). Therefore, instead of one exponential, two or three exponential functions or occasionally two exponentials and a constant were included in the fitting functions. It was found useful first to subtract the constant and then use the following fitting function $F(u)$ that comprises a Gaussian function to describe the peak region and two exponential functions to describe the tailing as convolution of two exponential of steeper and a less steep functions. The explicit form of the fitting function is as follows:

$$F(u) = \sum_{i=1}^{m}\frac{A_i}{2}\left[\begin{array}{l}\frac{1-\eta}{\tau_1}\exp\left(\frac{u-\mu_i}{\tau_1}+\frac{\sigma^2}{2\tau_1^2}\right)\\ erfc\left[\frac{1}{\sqrt{2}}\left(\frac{u-\mu_i}{\sigma}+\frac{\sigma}{\tau_1}\right)\right]+\frac{\eta}{\tau_2}\exp\left(\frac{u-\mu_i}{\tau_2}+\frac{\sigma^2}{2\tau_2^2}\right)\\ erfc\left[\frac{1}{\sqrt{2}}\left(\frac{u-\mu_i}{\sigma}+\frac{\sigma}{\tau_2}\right)\right]\end{array}\right] \tag{5.16}$$

This fitting function, consisting of one Gaussian and two exponential functions, has four shape parameters, i.e., σ, τ_1, τ_2, and η, which are the same for all peaks and two parameters, i.e., A_i (peak area) and μ_i (peak position) that are specific for each peak.

The peak fitting model involved in a spectrum analysis software was used successfully for the deconvolution of spectra acquired by HRAS. The method allows even the identification of the strongly overlapping peaks of ^{239}Pu and ^{240}Pu, and it is adequate for the determination of exact emission probabilities of alpha-emitting nuclides. The major drawbacks of the method are the following:

- Relatively high source activities (sometimes 10 Bq) are required even with very long counting times to have acceptable statistics, whereas the energy resolution is also kept excellent (in alpha spectrometry, the requirement toward high resolution is contradicting the requirements toward high efficiency).

- Many fitting parameters are used, which makes the fitting more accurate, but the numerical optimization becomes more complicated.

Other models to fit the alpha spectra have been developed by Koskelo et al. (1996), García-Toraño (1997, 2003), and Lozano et al. (2000). As an example, the results of deconvolution of the spectrum of ^{239}Pu and ^{240}Pu using the ALPACA code developed at the Spanish institute CIEMAT (García-Toraño, 2006) are shown in Fig. 5.19. The spectrum consists of the five strongly overlapping peaks at energies of 5105 keV (^{239}Pu line), 5124 keV (^{240}Pu line), 5144 keV (^{239}Pu line), 5157 keV (^{239}Pu line), and 5168 keV (^{240}Pu line). These isotopes are known as the ones that cannot be determined by routine alpha spectrometry because the five lines lie within a very narrow energy range of 65 keV. (Routinely used alpha spectrometers require about 200 keV peak-to-peak separation to avoid peak overlapping.) Nonetheless, the determination can be done by HRAS using a spectrum deconvolution software.

Bortels et al. (1995) extended the method for the analysis of **alpha spectra of low statistics.** These sources usually have a finite source thickness and a source diameter of a few centimeters; the detector diameter is increased (>100 mm^2), whereas the source to detector distance is reduced increasing the solid angle—all factors resulting in a reduced energy resolution. It was shown that the fitting functions developed for infinitely thin sources with high resolution (<10 keV) cannot be used directly for sources of lower resolution (>10 keV).

It was shown that in low-statistics alpha spectra, there is insufficient information on the spectrum tailing to use the fitting function in Eq. (5.16); therefore, the simpler model using a single exponential function (Eq. 5.15) was used for spectrum deconvolution. Furthermore, it was found that the peak deconvolution is successful only if single isotope spectra are used as constraints when spectra of mixed sources are evaluated. In Pu analysis, pure ^{239}Pu standard was used as a single isotope source. The use of statistical weights in the fitting of low-activity samples has been proposed by García-Toraño (1994).

Bortels et al. (1995) fitted triplets or quadruplets in the common region to the nuclide ^{239}Pu and ^{240}Pu depending on the actual resolution. Besides the activities, the activity ratio of ^{239}Pu/^{240}Pu could be calculated. The deviation from the target value was about 2% when equally large peaks having a separation of about one FWHM are analyzed (Fig. 5.20).

For relatively low statistics alpha spectra from measurements at a relatively high solid angle, ^{239}Pu/^{240}Pu activity ratios in the range of 1.4 down to 0.1 could be determined in the above example. The ^{239,240}Pu peak should be > 100 to obtain reliable values for the ^{239}Pu/^{240}Pu isotopic ratio (Ihantola et al., 2011b).

In low activity samples, several other functions have been used to describe the low-energy side peak tailing, whereas the peak center and the high-energy side of the peak has been described usually by Gaussian function. Fitting models developed by LaMont et al. (1998), Leon et al. (1996), and Mitchell et al. (1998) are much simpler than the models developed for HRAS.

LaMont et al. (1998) also used a modified Gaussian function to model the tailing (See Fig. 5.21). The Gaussian portion of the peak is fitted by Eq. (5.17) when $x > c-t$:

$$f(x) = h \bullet e^{-\frac{(x-c)^2}{w^2}} \tag{5.17}$$

The modified Gaussian portion of the peak tailing is fitted by Eq. (5.18) when $x < c - t$:

$$f(x) = h \bullet e^{-\frac{t(2x-2c+t)^2}{w^2}} \tag{5.18}$$

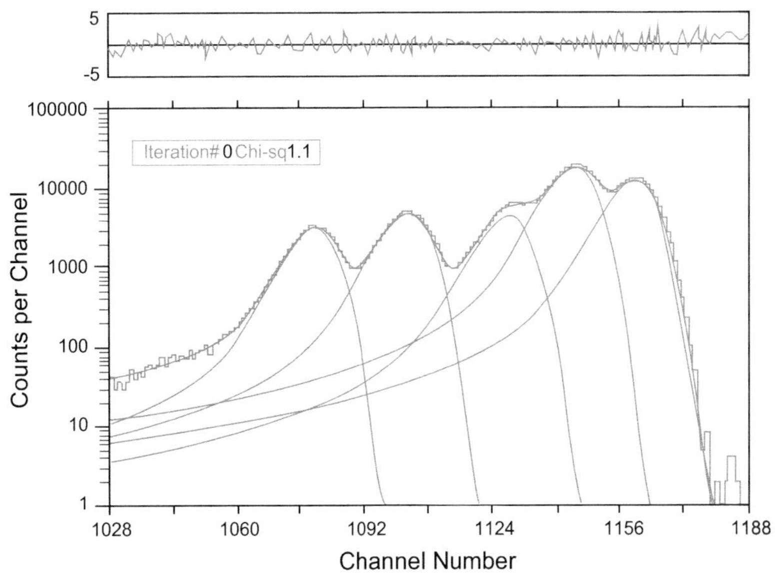

FIGURE 5.19 Graphical presentation of a deconvolution of the multiplet containing the alpha peaks of ^{239}Pu and ^{240}Pu by high-resolution alpha spectrometry. Fitting was done by the ALPACA code. Smooth lines represent the fitted peaks and the sum spectrum, whereas the histogram represents the measured points. The goodness of the fit is shown by plotting the residuals in the upper curve. *García-Toraño (2006), reprinted by permission of Elsevier Ltd. © 2006.*

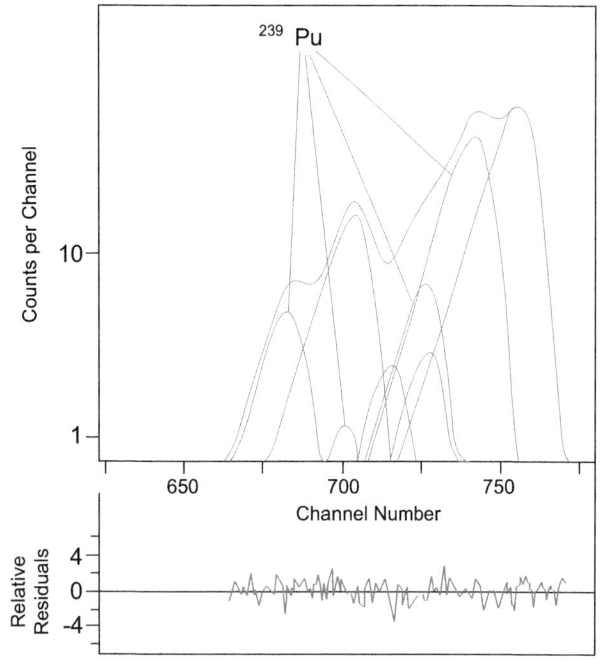

FIGURE 5.20 Fitted spectrum and residuals (relative deviation of measured and fitted values) of a mixed ^{239}Pu and ^{240}Pu source prepared by drop deposition. Peak full width at half maximum is 12.4 keV. Source activity was 17.5 Bq and the activity ratio of $^{239}Pu/^{240}Pu$ was 0.67. Spectra were measured with a 300 mm^2 PIPS detector in a solid angle of about 13% of 4π sr. Quadruplets were used in the fits. (The peaks not labeled in the figure belong to ^{240}Pu.) *Bortels et al. (1995), reprinted by permission of Elsevier Ltd. © 1995.*

where x is the energy variable at which to compute $f(x)$, h is the peak height, c is the center of the peak, t is the distance below c where shape deviates from Gaussian, and $w = \text{FWHM} \cdot 0.60056$.

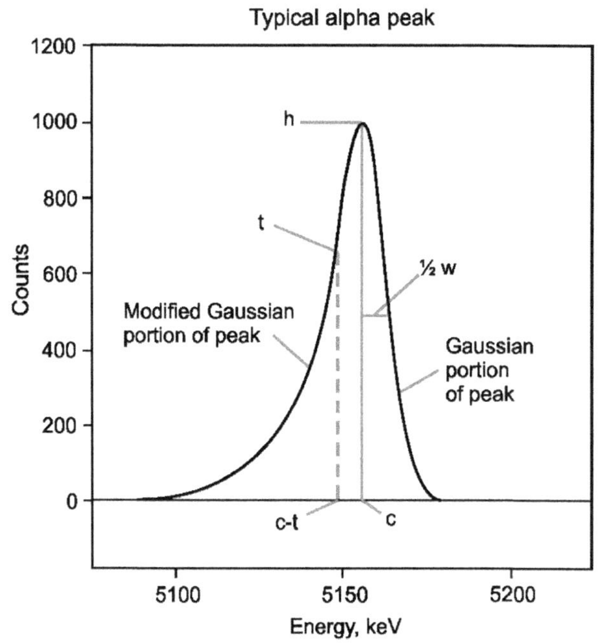

FIGURE 5.21 Alpha peak fitting by a Gaussian and a modified Gaussian function according to LaMont et al. (1998). *Reprinted by permission of Springer-Verlag © 1998.*

The method was used for the determination of $^{239}Pu/^{240}Pu$ activity ratios using a commercially available alpha spectrometer with ULTRA detector of 50 mm^2 in area. Sources were prepared by electrodeposition. Source–detector distance was 1.8 cm, and the resulting resolution was 16 keV FWHM. Peak heights and centroid energies were constrained. An acceptable agreement was found between calculated ^{239}Pu and ^{240}Pu activities and the target values.

MicroSAMPO, a commercially available software originally developed for gamma spectrometry (Aarnio et al., 1987), was used to fit the alpha peaks of low activity samples by Leon et al. (1996) and Mitchell et al. (1998). The fitting model is a modified Gaussian with left- and right-handed exponential tails. The fitting function does not take into account the long-range tailing, but it still was used successfully for deconvolution of ^{239}Pu and ^{240}Pu, as well as ^{243}Cm and ^{244}Cm peaks. The latter issue is especially important because the relatively short-lived Cm isotopes cannot be analyzed by mass spectrometric techniques. To perform spectrum deconvolution of these nuclides in environmental samples, a relatively good energy resolution is required. In the case of the analysis of Cm nuclides, the FWHM value was 13–15 keV, a PIPS detector of 100 mm^2 was used, the source to detector distance was 15 mm, and the sources of about 10 mBq activity were counted for an extremely long time (5×10^6 s).

The **WinALPHA code,** which is freely available from the Physics Section of the IAEA (Capote Noy et al., 2004), uses the following analytical model (García-Toraño, 2003). To describe the peak, a pure Gaussian is used above the peak center, and a second Gaussian with a wider resolution (σ) is used below the peak maximum:

$$f(x) = A \cdot e^{-\frac{(x-c)^2}{2\sigma_{l,r}^2}} \tag{5.19}$$

where $f(x)$ is the value for the channel x, A is the peak amplitude, c is the channel number at peak maximum, and $\sigma_{l,r}$ is the resolution on the left or right side of the peak describing the asymmetry.

The low-energy tailing is described by the following equation:

$$f(x) = B \exp\left[-C\left(\frac{x-c}{\sigma_d}\right)^{0.5}\right] \tag{5.20}$$

where B and C are parameters obtained from the shape parameter n as follows:

$$B = A\frac{\exp(-n^2/2)}{\exp(-2n^2)} \tag{5.21}$$

$$C = 2n^{3/2} \tag{5.22}$$

Thus, the set of shape parameters (n, σ_l, σ_r) are constant within one spectrum, and a fitting function containing relatively few parameters has been obtained. The accurate description of the tailing outside the vicinity of the peak center is not possible.

c. Alpha spectrum analysis software

In the alpha spectrum evaluation process, defined analytical functions containing several shape parameters are fitted to the measured spectrum data. To solve the nonlinear fitting problem, initial parameter values have to be provided, and then optimization procedures are used to solve the nonlinear least squares fitting problem by one of the methods such as the Newton–Raphson, the Levenberg–Marquart, or the simplex method. The goodness of fitting can be characterized by various techniques, e.g., calculating the goodness of fitting (R^2), the chi-square value (χ^2), and/or the fitting residuals. A summary of the mathematical procedures for least squares fitting is given by García-Toraño et al. (1999; García-Toraño 2006).

As a result of fitting, the unknown parameters of the algorithm are determined. Using the fitted function with the known parameters, spectra can be deconvoluted into the individual peaks that contribute to the final measured spectrum. The convoluted spectrum is the simple sum of the individual reconstructed peaks of the given nuclide.

Computerized **spectrum evaluation programs** can be used for the analysis of overlapping peaks of one or more nuclides or also for the analysis of complicated spectra containing several radionuclides. At present, there are several software programs in use, which allow the fitting of alpha spectra with different levels of complexity, from simple peak integration to nonlinear fitting algorithms.

An intercomparison exercise of four commercially available alpha software programs was carried out under the auspices of the IAEA in 1997 (Blaauw et al., 1999; IAEA TECDOC-1104, 1999; García-Toraño et al., 1999). All of these programs are general-purpose, commercial codes, not dedicated for the analysis of high-resolution alpha spectra. **AlphaVision (EG&G Ortec, USA)** uses a Gaussian function with one low-energy exponential function that is fitted to the peaks with the simplex method. The code can be operated in peak search and fit as well as library search and fit mode. The code is operated in selected regions in peak search mode. The **Alps program (Westmeier GmbH, GE)** is based on the fitting of a Gaussian with one or two folded exponentials with a linear least squares algorithm in three passes. By the alpha analysis software of Genie 2000) **(Canberra Industries Inc., USA),** the line shape model is a modified Gaussian function (see Eqs. 5.17 and 5.18) and peak areas can be determined either by summation of singlets or by nonlinear least squares fitting of singlets and multiplets using a modified Marquardt algorithm for fast convergence. Peak search is performed using either a library driven routine or by the generalized second differential method. In **Winner Alpha (Eurisys Mesures, FR)**, the peak areas are determined either by summation of singlets or by nonlinear least squares fitting of singlets or multiplets using a modified Marquardt algorithm for fast convergence. Peak search is performed using either a library-driven routine or by the generalized second differential method. Comparison was done in terms of the determination of peak area, peak position, deconvolution capabilities, uncertainty calculation, nuclide identification in case of large and small peaks, and doublets. All programs had difficulties with the analysis of small peaks. It was clear that the results left room for improvement, e.g., in the uncertainty calculation.

Hurtado et al. (2008) made a comparison of the Genie 2000 and the **WinALPHA programs**. Both are general-purpose codes using a relatively small number of fitting parameters. WinALPHA uses the line shape algorithm described in Eqs. (5.19)–(5.22), and it performs a nonlinear fit of the spectra with the Levenberg–Marquardt optimization method. WinALPHA has two modes of operation: library-driven and automatic peak search modes. The basic properties of Genie 2000) are described in the previous section. The programs were compared in terms of accuracy and precision in the analyses of environmental samples. Both programs provide correct peak areas and peak search results. WinALPHA performs a full uncertainty calculation and a better peak deconvolution than Genie 2000. Nonetheless, Genie 2000 Alpha Analysis and Interactive Peak Fitting software packages were also used for the determination of $^{233}U/^{234}U$ ratios in groundwater samples (Harrison et al., 2016).

The WinALPHA program was successfully used for the determination of the $^{240}Pu/^{239}Pu$ activity ratio in samples from a reprocessing plant and a heavy water reactor by Srinivasan et al. (2005) and Aggarwal and Alamelu (2007), respectively. Electrodeposited Pu sources were measured with Si detectors of 25 mm^2 in area by using HRAS. (The FWHM value was 15 keV in the latter case.) Good agreement was found between measured and targeted activity ratios. Alamelu and Aggarwal (2012) used the WinALPHA program for the determination of the isotopic composition of uranium (^{232}U, ^{233}U, ^{234}U) in irradiated thorium in the presence of ^{228}Th.

Examples of other alpha spectrum evaluation programs are ALFAS by Babelovswsky and Bortels (1993), ALPACA by García-Toraño (1994), FITBOR by Martin Sanchez et al. (1996), Sigma-Plot by Jandel Scientific Inc. (1998), Colegram by Dayras (2002), and ALFITeX by Caro Marroyo et al. (2013, 2014). An improved peak shape fitting was attained by applying a convolution of the Gaussian with more exponential tailing functions (up to 10 on the low- and 4 on the high-energy side) in the BEST algorithm (Pommé and Caro Marroyo, 2015). The method was tested by fitting high-resolution spectra of ^{240}Pu and ^{236}U as well as a thick source of ^{238}U.

While many spectrum unfolding codes have been developed for the analysis of high- and ultrahigh-resolution alpha spectra of thin alpha sources, less attention has been paid to the analysis of thick sources. As discussed above, peak resolution and the peak shape for thick sources are basically determined by self-absorption. **ADAM** (advanced deconvolution of alpha multiplets, Ihantola et al., 2011) is a general-purpose program for alpha spectrometry using Si detectors. The peak shape model in ADAM is convolution of a Gaussian distribution (characterized by FWHM) with low-energy side exponential tail. Up to three exponential functions characterized by the slopes of the exponentials 1/λ, 1/μ, 1/ν (in keV), and their area ratios R1 and R2 are possible. Up to three different peak shapes are possible, and the software can also take into account the coincidences between alpha particles and electrons or low-energy photons. The AASI computer program (Siiskonen and Pöllänen, 2006) provides coincidence information for ADAM to be included in the fitting. Using the ADAM software, the ^{239}Pu/^{240}Pu multiplet in chemically processed samples could be successfully deconvoluted. The method gave good results for the ^{239}Pu/^{240}Pu activity ratio using Si detectors as large as 1200 mm^2 (FWHM 21 keV) (Pöllanen et al., 2011).

Recently, more attention has been paid to the high-energy tailing of peaks, and beside software development, magnet systems have been installed in the spectrometers for the suppression of conversion electrons and their coincidences with alpha particles. By applying HRAS with this technique, the emission probabilities of various alpha particle–emitting radionuclides such as ^{236}U and ^{238}U have been determined with higher accuracy (Marouli et al., 2014; Pomme et al., 2014; Paepen et al., 2014).

2. Peak shape and spectrum analysis with gas ionization detectors

Properties of alpha spectra obtained with gridded ionization chambers are not basically different from the spectra from PIPSi detectors.

Fig. 5.22 shows an example of the asymmetric peak shape with low-energy tailing that is similar to those of the peaks obtained by semiconductor alpha spectrometry. Typical resolutions are somewhat poorer than in the case of PIPSi detectors; the FWHM is not smaller than 30 keV. Spectra obtained with gridded ionization chambers can be evaluated by the same method as for low-resolution alpha spectra using PIPSi detectors.

3. Peak shape and spectrum analysis with liquid scintillation detectors

Peak shapes in LS spectrometry are different from those measured with gas ionization and semiconductor detectors. A typical alpha spectrum of ^{226}Ra and progeny measured with a PERALS spectrometer is shown in Fig. 5.23.

The peaks in the spectrum are symmetric and can be modeled by a Gaussian function. The main reason for the symmetric shape is the lack of absorption and self-absorption in LSC where samples are mixed with the scintillation cocktail. (Absorption and self-absorption are the major reasons for peak broadening and low-energy tailing when "thick" sources are measured with semiconductor detectors.)

LS spectra are characterized by relatively poor resolution even in the case of so-called high-resolution alpha LSC using the PERALS spectrometer. The FWHM is typically a few hundred keV (Fig. 5.24), and the variation of the resolution as a function of energy is more pronounced than in semiconductor alpha spectrometry. This resolution does not allow the detection of the individual alpha lines belonging to many alpha-emitting isotopes. For example, in Fig. 5.23, the alpha radiation of ^{226}Ra at 4.601 and 4.783 MeV appears as a single common peak and cannot be resolved. The Rn and Po isotopes shown in the same spectrum emit monoenergetic alpha particles.

Alpha peaks in LS spectra may also have tailing, but this appears on the high-energy side of the peaks as shown in Fig. 5.24.

This high-energy tailing can be observed in the case of those nuclides that emit several alpha particles of different energies, and the daughter nuclides deexcite to the ground state by internal conversion or the emission of gamma radiation. Internal conversion produces electrons that cause scintillation in the cocktail. Since the scintillation efficiency

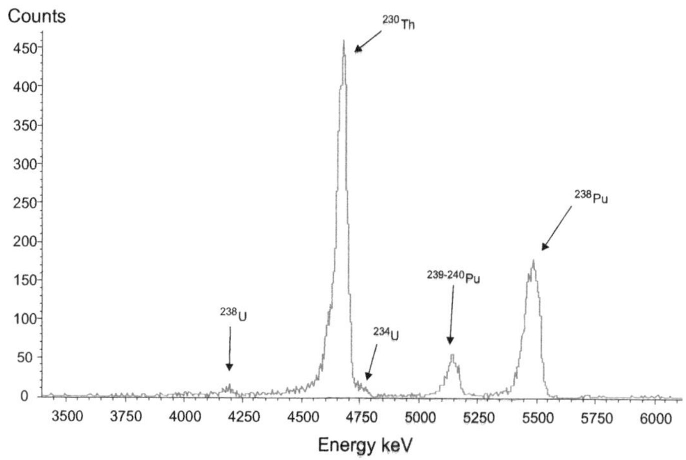

FIGURE 5.22 Schematic drawing of an alpha spectrum of a mixed source measured by a gridded ionization chamber.

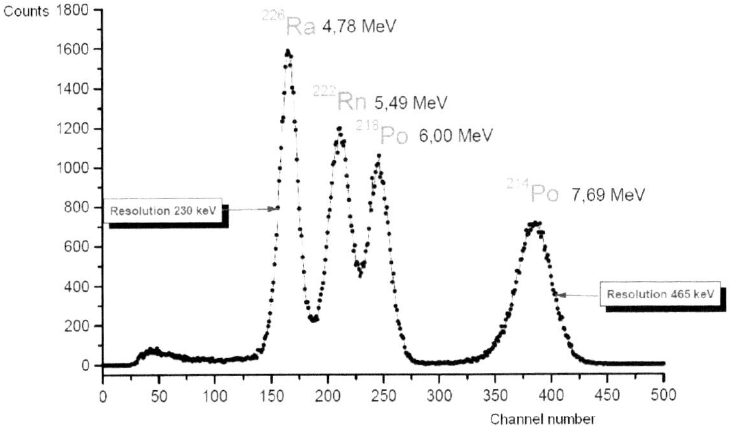

FIGURE 5.23 Schematic drawing of an alpha spectrum of ^{226}Ra and progeny measured by a PERALS spectrometer.

of electrons is about 10 times higher than that of the alpha particles (Section III.A.5), a sum peak resulting from the true coincidence of the alpha particle and conversion electron will appear at a higher energy than the alpha peak itself. This phenomenon was taken into account to better describe the alpha peak shape in LSC by Aupiais (2004), and the spectra successfully deconvoluted by fitting several Gaussian functions to both the direct alpha peaks and the pileup peaks of alpha particles and conversion electrons as shown in Fig. 5.24. The method was successfully used for the determination of U and Th nuclides in water by alpha LSC.

4. Peak shape with cryogenic detectors

The energy resolution of the alpha spectrum using bolometers is extremely high. Measured FWHM values below 5 keV have been reported (Leblanc et al., 2006). The alpha spectrum of a ^{241}Am source is shown in Fig. 5.25.

The superior resolution of alpha spectra using bolometers compared with that of Si detectors is explained by differences in the nature of the interaction of the alpha particle with the detector. In the case of Si detectors, alpha particles have to pass the detector window (the electric contact), whereas there is no detector window in bolometers. Inside the Si detector, there is a relatively large statistical fluctuation in the number of generated and detected electron–hole pairs due to the presence of impurities, crystal defects that can trap the charge carriers, and possible recombination and secondary processes (e.g., crystal damage, lattice vibration). All of these intrinsic processes as well as the instability of the electronic units contribute to degradation in the energy resolution (see Eq. 5.10). Thermal detectors are less sensitive to the fluctuations in the number of electron–hole pairs because the total energy of the phonons (heat, lattice vibrations) is detected. If electron–hole pairs recombine within the time constant of the bolometer, they will be

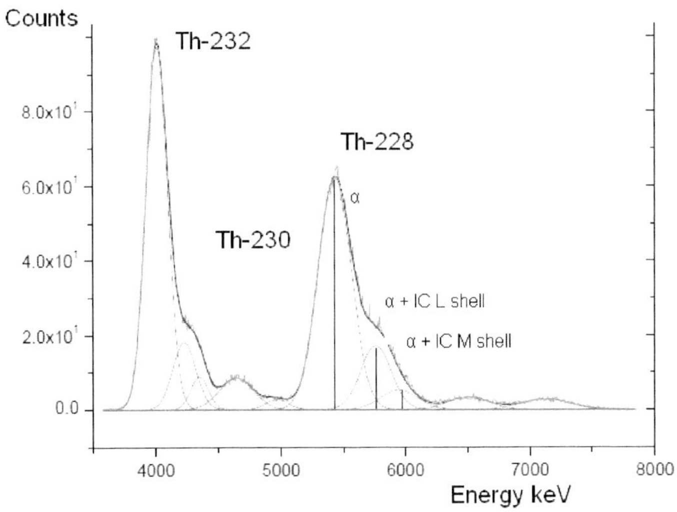

FIGURE 5.24 Schematic drawing of high-resolution alpha LSC (PERALS) spectrum of a mixed source showing spectrum deconvolution. The calculated positions of the α peaks, and the pileup peaks originating from the coincidences of the α peaks and the conversion electrons, are also indicated.

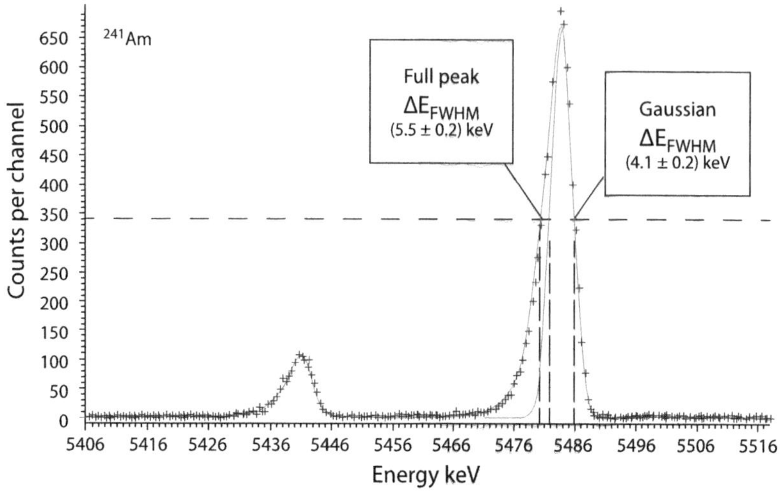

FIGURE 5.25 Alpha spectrum of an external sputtered ^{241}Am source measured with the bolometer B306. Data acquisition time was 12 h. The individual peaks at 5443 and 5486 keV are completely separated. *Leblanc et al. (2006), reprinted by permission of Elsevier Ltd. © 2004.*

detected also as phonons. However, as in the case of Si detectors, energy losses due to secondary emission processes cannot be avoided in bolometers.

According to the calculations of Amoudry (1990) and Steinbauer et al. (1994), the theoretical contribution of the basic processes of detection to the energy resolution of Si detectors is 10.5 and 7.4 keV, respectively, and Juillard (1999) obtained FWHM less than 3 keV for Si and Cu bolometers.

By the development and use of the new types of magnetic microcalorimeters for Q spectrometry, much improved resolution values have been reported. Energy resolutions below 3 keV at the alpha energies of ^{241}Am and Pu isotopes are reported by several research groups (Pabinger et al., 2011; Yoon et al., 2012; Hoover et al., 2015) (See Section A.6.).

B. Counting efficiency

In radiochemical determinations, radioactive tracers are added to the sample, and the activity of the radionuclides present in the sample is determined from the relative number of counts in the analyte and tracer peaks. In this case, a knowledge of the absolute counting efficiency is not necessary to calculate the results (see below). However, it is necessary to determine the chemical recovery, which is an important quality control check. In addition, in cases where the tracer is determined by a separate measurement, the counting efficiency must be known. An example of the latter case is the use of ^{133}Ba as a tracer for Ra determination, where the activity concentration of ^{133}Ba is measured using gamma ray spectrometry.

Since the intrinsic counting efficiency of the Si semiconductor detectors is in practice 100%, the determination of the counting efficiency can be performed by calculating the solid angle presented by the radioactive source to the detector (see Section III.A.2), provided the activity on the source is evenly distributed. A schematic drawing of the source–detector geometrical arrangement is shown in Fig. 5.26.

In the case of a point source placed a distance of H in front of an alpha detector of radius R_D, the following equation can be used to calculate the absolute counting efficiency:

$$\varepsilon_{\text{abs}} = \frac{1}{2}\left(1 - \frac{H}{\sqrt{H^2 + 4_D^2}}\right) \tag{5.23}$$

The solid angle (the absolute counting efficiency) of various source–detector arrangements can be calculated by Monte Carlo computations. Gascon and Munoz (2003) developed two theoretical formulas to calculate the efficiency of extended sources where R_S is the radius of the source and D is given in Fig. 5.26. The simplified geometry formula obtained by an approximation with an infinite series is given below:

$$\varepsilon_{abs} = \frac{1}{2}\left[\frac{R_D^2}{D(D+H)}\right] - \frac{3}{16}\frac{R_{S2}^2 R_D^2 H}{D^5} + \frac{5}{32}\frac{R_{S2}^4 R_D^2 H}{D^9}\left[H^2 - \frac{3}{4}R_D^2\right] \tag{5.24}$$

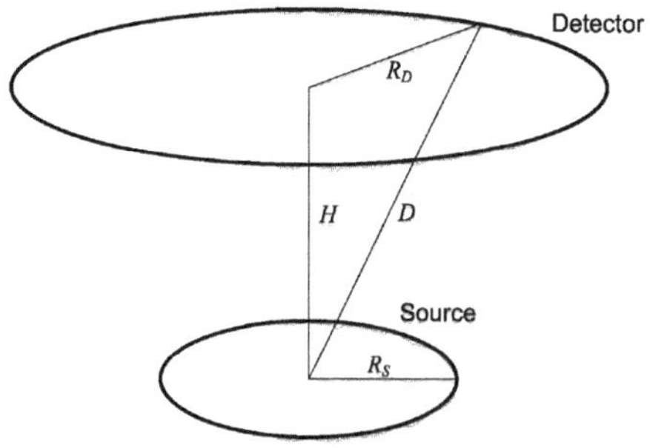

FIGURE 5.26 Geometrical arrangement of the alpha source and the detector used for calculation of the counting efficiency according to Gascon and Munoz (2003).

The calculated efficiencies were compared with each other and with experimentally determined values, and a good agreement was found. Therefore, the use of the above simplified formula can be recommended. Efficiencies are often determined by measurements when standard alpha sources of known activity concentration and the same geometry (same size of the active source area, same distance from detector) are available.

Geometrical detection efficiencies are often calculated by Monte Carlo simulations when the sizes of the source and the detector window and the exact geometric arrangement are known. The AASI simulation software available through the Internet (Siiskonen and Pöllänen, 2004) can calculate the efficiency. Calculated efficiencies as a function of source–detector distance, source diameter, and detector diameter are given in Fig. 5.27.

Typical counting efficiencies for a small source ($R_S < R_D$), close (5–10 mm) to a Si detector of 300–600 mm^2 in area are about 20–40%. The efficiency decreases significantly as the source–detector distance increases.

Calculated efficiencies can be used for a source–detector system if absorption and self-absorption are not high (i.e., not resulting in counting losses) and if radionuclides on the source plate are homogeneously distributed. Counting efficiencies can also be affected (increased/decreased) by backscattering of the alpha particles from the sample holder/detector entrance window.

C. Background and contamination in alpha spectrometry

Semiconductor alpha spectrometers usually have a low background, originating mainly from electronic noise of the instrument. Since alpha pulses have a high energy and electronic noise is usually in the low-energy region, they can be easily discriminated (i.e., separated by pulse height). The typical background of a new spectrometer in the region of a single alpha-emitting isotope can be as low as 1 count per a couple of days (10^{-5}–10^{-6} cps). This low background is responsible for the extremely high sensitivity of alpha spectrometry. **Detection limits** as low as 1 mBq per sample are easily achievable, provided contamination of the detector, the detection chamber, and the source is avoided.

Contamination of the detector can take place when particles from sources are sputtered or splashed onto the detector or the chamber surface. This can happen when the radionuclides do not adhere strongly to the source support. The vacuum can help remove loosely bound particles from the source. This type of contamination can usually be removed by careful cleaning of the detector (PIPSi detectors are washable) or the chamber walls.

Recoil contamination takes place if recoil causes products of alpha decay to sputter from the source onto the chamber walls or (in the worst case) the surface of the detector. If the recoil atoms or their short-lived progeny are themselves alpha emitters, this will raise the background. Recoil atoms, like other accelerated or implanted particles, adhere strongly to the surfaces and are almost impossible to remove. For example, when measuring natural Th, the alpha particle–emitting ^{224}Ra, as the decay/recoil product of ^{228}Th, can contaminate the detector. Since ^{224}Ra has several short-lived progeny, they can all accumulate in the detector window creating a background that contains the following alpha peaks: ^{224}Ra, ^{220}Rn, ^{216}Po, ^{212}Bi, and ^{212}Po. This contamination will "decay" by the half-life of ^{224}Ra (3.7 days). A typical ^{228}Th spectrum is shown in Fig. 5.28. After measurement of the Th source, the background spectrum may contain all these radionuclides except ^{228}Th.

Three techniques have been used to avoid recoil contamination in alpha spectrometry:

- operation of the spectrometer with an air barrier;
- covering the source with a thin even foil (Van der Wijk et al., 1987; Vainblat et al., 2004; Inn et al., 2008); and
- applying a voltage between the source and the detector where the source is negatively biased by a few volts (Sill and Olson, 1970).

The idea behind the first two protective measures is that a properly selected absorber can stop the recoil particles of high mass and low kinetic energy, whereas alpha particles of low mass and high kinetic energy pass through the absorber. An absorber layer of about 12 mg/cm^2 thickness can reduce the recoil contamination by a factor of 1000. Nonetheless, using any type of absorber between the source and detector will increase energy straggling and reduce the energy resolution. It will also slightly shift the spectrum to a lower energy region. Applying a negative bias to the source can retain the positively charged recoil atoms on the source, although a small chamber pressure (at least 12 $\mu g/cm^2$) is still required to prevent recoiling atoms from reaching the detector before they are attracted back to the source (Sill and Olson, 1970).

Alpha spectrometers have to be protected from contamination due to **oil vapors** that can be released into the vacuum system if rotary vacuum pumps are applied to evacuate the alpha chamber. Oil contamination is a real danger if the vacuum pump fails, as oil particles can backstream due to the prevailing vacuum. To avoid oil contamination, a vacuum trap or filter can be installed between the pump and the chamber.

D. Stability of the alpha spectrometer

System stability is very important in alpha spectrometry when long measurements of low activity samples lasting for several days are performed (Pommé and Sibbens, 2004).

Long-term stability is affected by environmental conditions with special respect to the detector junction. Surface barrier detectors fail when the encapsulation of the detector

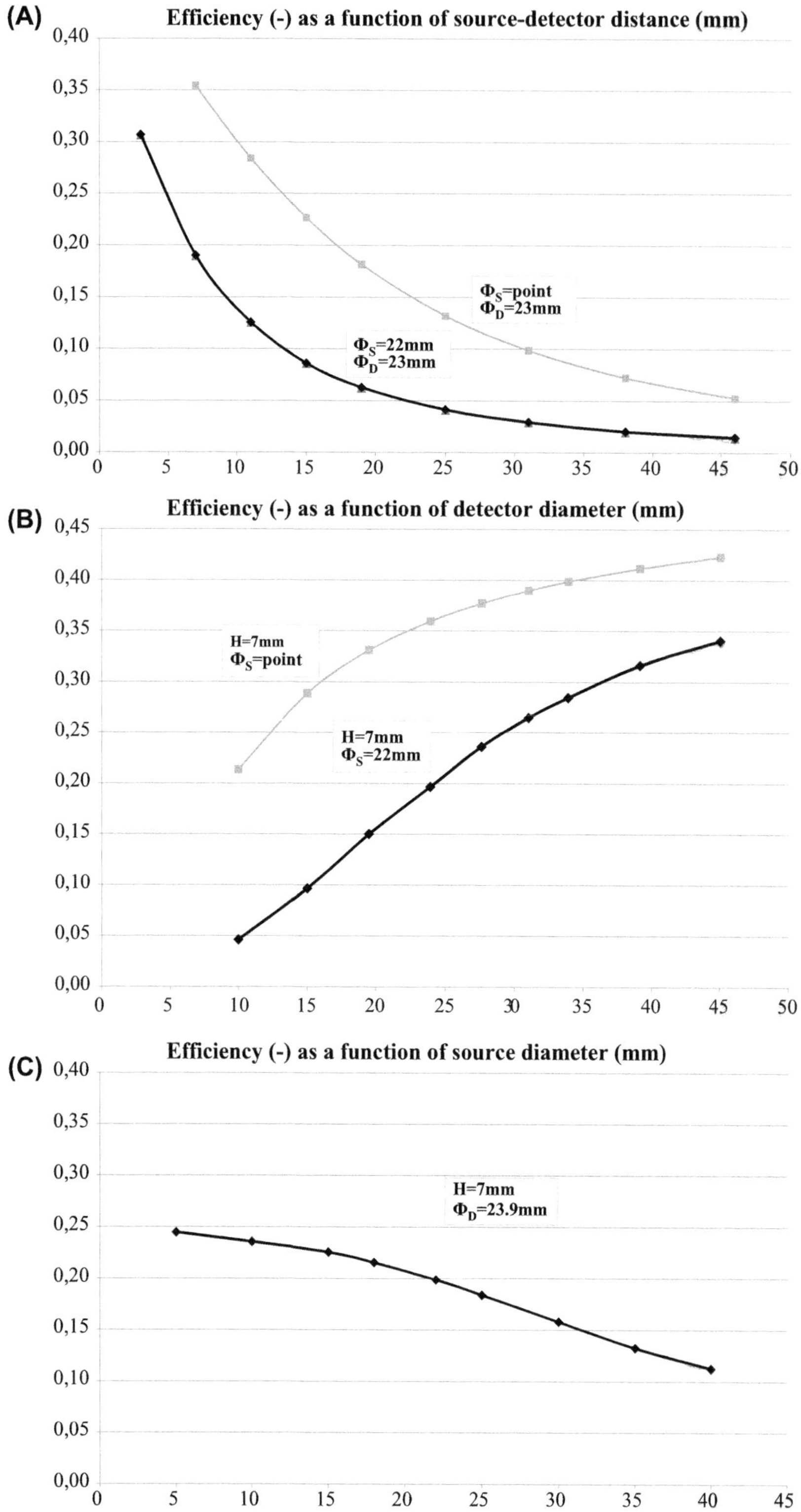

FIGURE 5.27 Counting efficiencies as a function of (A) source–detector distance (H) by assuming point source or extended source of constant diameter and constant detector diameter; (B) detector diameter (Φ_D) by assuming point source or extended source of constant diameter and constant source–detector distance; (C) source diameter (Φ_S) by assuming constant detector diameter and source–detector distance. Efficiencies were calculated by the AASI program according to Siiskonen and Pöllänen (2004) in case of extended sources and according to Eq. (5.23) in case of point sources.

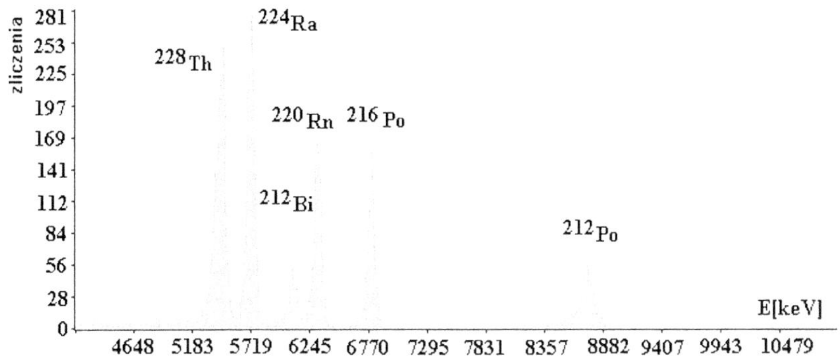

FIGURE 5.28 Alpha spectrum of a ^{228}Th source with ingrown progeny. *From Mietelski (2011).*

is damaged. The junction of the PIPSi detector is implanted in the Si matrix assuring higher intrinsic stability.

Alpha detectors are sensitive to **temperature changes**. The leakage current of a Si diode doubles for every 5.5−7.5°C change. The current change results in a change of the detector voltage, which can affect the overall gain of the preamplifier and cause a spectrum shift.

V. In situ alpha spectrometry with Si detectors

In alpha spectrometry, the chain of sampling, sample processing, data acquisition, spectrum analysis, and reporting and interpreting the results is usually time-consuming and requires appropriate laboratory infrastructure and expertise. However, in some cases, the performance of rapid measurements, while still retaining a fit-for-purpose quality of the spectra, may be of utmost importance; examples include response to a nuclear incident (Dilbeck et al., 2006) and nuclear forensics (Hoffman et al., 2011). In general, the equipment and software can be designed in such a way that makes it possible to shorten sample processing and to simplify processes in the data acquisition and spectrum analysis, allowing the performance of measurements outside of a fixed laboratory and even in the field.

There are several independent approaches to simplifying the abovementioned components of the chain:

- Sampling can be developed so that the samples can be measured as such, i.e., without radiochemical or other sample processing.
- Sample processing can be simplified, enabling a rapid start to the measurements.
- Data acquisition can be made in situ (e.g., at ambient air pressure) without substantial deterioration in the energy resolution or loss of the detection efficiency.
- The processes of spectrum analysis and reporting can be developed and automated in a similar way to those used in gamma ray spectrometry.

Each of these approaches has advantages and disadvantages, which are discussed in the following sections.

A. Sampling and simplified sample processing

In an ideal case, it would be advantageous to analyze the samples by alpha spectrometry without sample processing (Sill, 1995). Unfortunately, this is only rarely possible because of prominent energy absorption of the alpha particles in the source material. Thin, flat, and homogeneous samples with minimum surface roughness are optimal for quantitative alpha spectrometry, but environmental samples seldom have these characteristics. However, there is a group of samples/objects, such as air filters, wipe samples (swipes), wire screens, sheets of solid material, or other smooth surfaces that can be used as such in the measurements.

Applications of continuous aerosol monitors include environmental monitoring outside nuclear facilities or outside hot cells, to give alarm signals in case of a possible release. The monitors are equipped with filters that are counted by an alpha spectrometer operating at ambient air pressure (Moore et al., 1993; Monsanglant-Louvet et al., 2012). Use of an optimized distance between the filter and the detector is important to reduce variation of the alpha particle track length in air and in the filter material. The aim is to minimize the peak tailing and achieve good energy resolution while retaining an acceptable detection efficiency.

Characteristics of the filter are of relevance to the energy resolution and especially peak tailing. The energy resolution achievable with membrane filters is superior to that for glass fiber filters (Fig. 5.29). In the case of glass fiber filters, the radioactive aerosol particles may penetrate the filter, causing notable tailing of the alpha peaks. Optimization is also necessary when a filter material is selected, because the flow rate through membrane filters is usually much smaller than that for glass fiber filters. In addition, collection efficiencies as a function of aerosol particle size are different for various types of filter. The procedures of

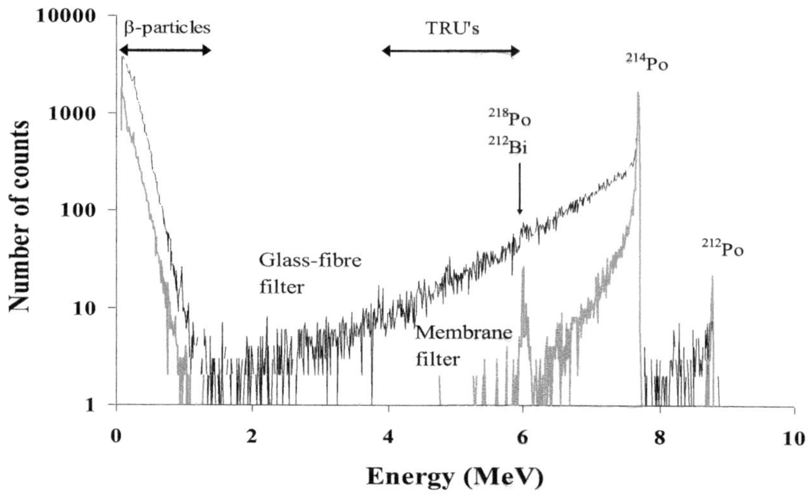

FIGURE 5.29 Alpha particle energy spectra of indoor air radon progeny collected in a glass fibre (top) and membrane filter (bottom) measured in a vacuum (Pöllänen and Siiskonen, 2005). Counts below 1 MeV are originating from beta particles and, to a lesser extent, photons. Most of the transuranium nuclides (TRUs) that might be present in the source in a hypothetical nuclear incident would locate between 4–6 MeV.

wiping and air filtration can be fine-tuned to facilitate measurements in a vacuum by taking into account the requirements needed, especially in the field or other challenging environments.

There are a number of applications in which nondestructive alpha spectrometry is used to detect alpha particle–emitting radon progeny in air or in soil gas (Goliáš et al., 2005). The samples are usually obtained by filtration. Measurement of the activity size distribution of short-lived radon decay progeny is another example for special applications in alpha spectrometry. Porstendörfer (2000) used a multistage impactor equipped with alpha detectors for this purpose. Another approach is to use the diffusion of ultrafine particles from an air stream passing through wire screens; in this case, the alpha detectors either face the screen or a backup filter placed behind the screen (Solomon and Ren, 1992; Solomon, 2001).

Nondestructive alpha spectrometry allows the measurement of alpha spectra without sample processing, which may facilitate subsequent analysis techniques (Fig. 5.30). Sometimes, minor sample processing, such as evaporation of water from the sample (Semkow et al., 2009; Vesterbacka and Pöllänen, 2011; Parsa et al., 2011; Pöllänen and Klemola, 2016; Pöllänen et al., 2016), is necessary to allow direct measurement of the spectra. However, the energy resolution may be relatively poor due to the thickness of the evaporation residue, as evidenced in Fig. 5.31.

Rapid activity determination can be achieved by use of a simplified radioelement separation procedure, for example, separation of all actinides together from the sample (see Section VI). However, this may pose special challenges in the spectrum unfolding because radionuclides of different elements may be present in the source.

B. Data acquisition

The presence of several alpha particle–emitting radionuclides does not necessarily prevent radionuclide identification in the source. For thin sources producing sharp peaks in the spectra, such as presented in Figs. 5.29 and 5.30, the detection efficiency can be determined by the AASI software, which enables activity determination without tracers

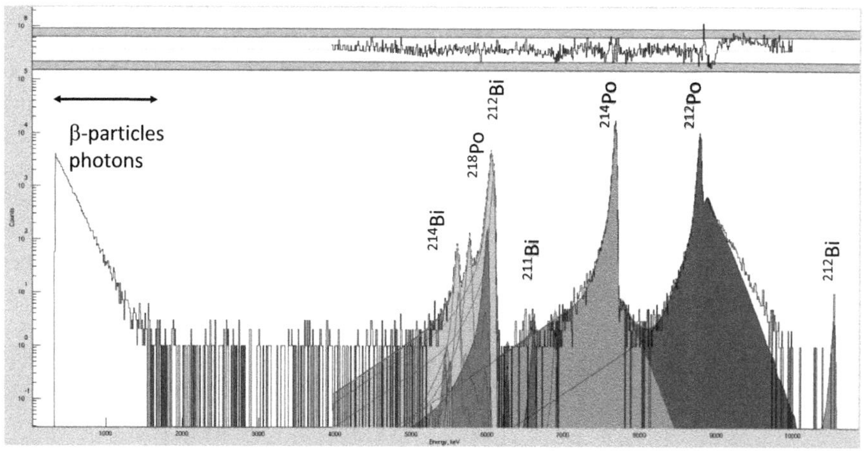

FIGURE 5.30 Unfolded alpha particle energy spectrum (screen snapshot of the ADAM program (Ihantola et al., 2011) measured in a vacuum from a membrane air filter where nuclides of radon progeny are collected. *See Pöllänen and Siiskonen (2014), reprinted by permission of Elsevier Ltd. © 2018, for the details of the spectrum analysis.*

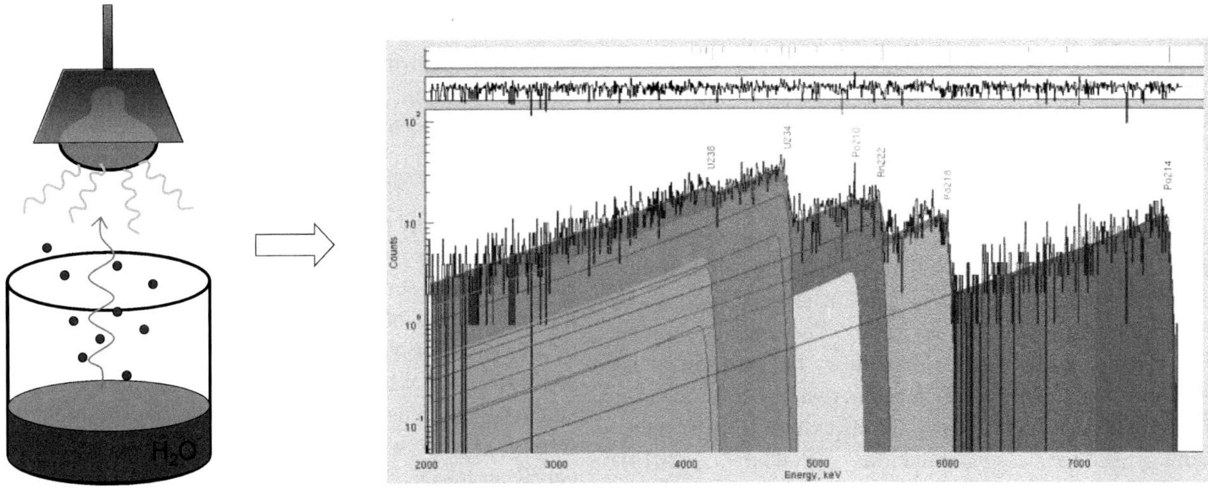

FIGURE 5.31 Schematic representation of the evaporation of drilled well water under an infrared lamp (left) and respective alpha particle energy spectrum measured in a vacuum from the residue containing radon and its progeny (right, screen snapshot of the ADAM program). Stressed alpha peak tailing and poor energy resolution does not prohibit spectrum unfolding.

in the case of in situ spectrometry (Pöllänen et al., 2011), avoiding the necessity of radiochemical processing of the samples. However, for the case of thick sources, although radionuclide identification is still possible, a tracer would need to be present to enable activity determination.

Portable equipment mounted in a mobile platform exist that allow the performance of alpha spectrometry in the same way as in a fixed laboratory (Smolander and Toivonen, 2004; Hoffman et al., 2011). However, in such a laboratory, only simplified sample processing is usually possible. In an optimal case, the energy resolution may be practically as good as in a fixed laboratory (Fig. 5.30; FWHM for the main alpha peaks is 21−23 keV). Simple portable equipment for field measurements have also been developed (Lidström and Tjärnhage, 2001; Martín Sánchez and de la Torre Pérez, 2012).

The data acquisition can also be performed with fairly good energy resolution at ambient air pressure, i.e., without using a vacuum chamber. Except for the thickness of the dead layer, the detectors operating at normal air pressure are basically similar to those operating in a vacuum. The entrance window is covered by aluminum and varnish layers with thickness of approximately 1.5 μm Si equivalent (Pöllänen et al., 2012a,b). These layers are needed to protect the active volume of the detector against daylight, which may cause a large number of background counts. In addition, the layers facilitate decontamination of the detector.

The extra thickness of the dead layer causes worsening of the energy resolution, especially at small source−detector distances. This is because the variation of the alpha particle track length in the dead layer is greater at short distances compared with that at large distances. However, when the data acquisition is performed using large source−detector distances, the detection efficiency will be lower and may be unacceptable for low-activity sources. In addition, at large enough distances at ambient air pressure, the energy loss of the alpha particles in air may become too large to detect alpha particles. For example, if the emission energy of the alpha particles is 5.2 MeV (representing ^{239}Pu and ^{240}Pu), the detected energy at the source−detector distance of 3 cm in air is only around 1 MeV, which is in practice too low an energy to be exploited in detecting alpha particles.

When there is air between the detector and the source, the alpha particles entering the detector on the slant do not contain as much useful information as those entering perpendicularly. Alpha particles α_1 and α_2 in Fig. 5.32 (left) lose substantial and different amounts of energy in air causing wide tails in the alpha spectrum. Their presence does not facilitate spectrum analysis. The variation of the energy loss of alpha particles α_3 and α_4 is much smaller, and

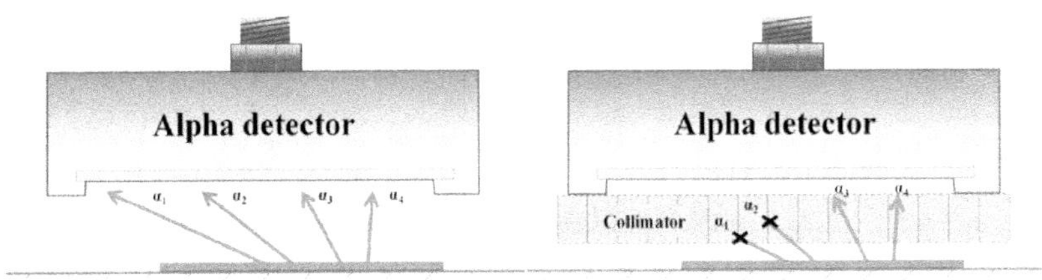

FIGURE 5.32 Detection of alpha particle −emitting radionuclides on a surface below a PIPSi detector at ambient air pressure. The alphas will be detected unless their energy is not fully absorbed in air (figure in the left). A honeycomb collimator placed in front of the detector stops those alphas (α_1 and α_2 in the right), which have an entering angle to the detector too acute to reach the detector. PIPSi, passivated ion−implanted planar Si. *See Pöllänen et al. (2012a,b), reprinted by permission of Elsevier Ltd. © 2018.*

this energy loss only causes a shift of the peak position. This phenomenon can be utilized by using collimation.

Collimation techniques (Park et al., 2015) are widely applied in detecting ionizing radiation, but the question arises how to apply them to alpha spectrometry at ambient air pressure while minimizing loss of detection efficiency. A honeycomb collimator inserted between the source and the detector provides a solution for this problem; the principle of the technique is illustrated in Fig. 5.32 (right). From the efficiency point of view, the cell size of the collimator should not be too small. The smaller the cell size, the better the energy resolution, but the lower the efficiency; the detection efficiency is reduced by a factor of 10 by using a cell size of 3.2 mm, assuming a homogeneous surface contamination (Pöllänen, 2016). Fig. 5.33 shows the improvement in resolution using a honeycomb collimator with a cell diameter of 1.5 mm, demonstrating that it is mainly the alpha particle path length variation in the inactive layers that causes peak widening at small source−detector distances.

A novel equipment was recently developed for alpha spectrometry at ambient air pressure (Pöllänen et al., 2014). The equipment enables alpha spectrometry from smooth and flat sources basically in the same way as for in situ gamma ray spectrometry. Alpha radiation is detected using similar procedures to those used for an alpha counter, but a fixed measurement geometry (including possible alpha particle collimation) is required for radionuclide identification. Because the measured spectra may be extremely complex, sophisticated analysis tools are needed in the activity determination. This may be difficult in the field, but real-time data transmission to a reachback center is included to ensure rapid analysis of the spectra and correct interpretation of the results in a remote location.

C. Spectrum analysis tools

An ideal spectrum unfolding tool should be able to analyze different types of alpha spectra. As shown in Section IV.A, a number of analysis programs have been developed that can be used for radiochemically processed sources, if the resolution is good enough, if the peak shapes can be characterized by an analytical function with appropriate tailing, and if the measurement conditions are stable. These programs may also be used for the spectra obtained by nondestructive alpha spectrometry (see Figs. 5.30 and 5.31). Nevertheless, there are cases for which the presently available spectrum analysis programs are not appropriate. For example, peak shapes in the spectra of Fig. 5.33 deviate from the Gaussian-based shapes, and a good-quality peak fit cannot be obtained.

A method has been developed based on iterative Monte Carlo simulations and subsequent fitting for unfolding of complex alpha spectra. If some characteristics of the source (such as source thickness, particle size, or other relevant properties) are tentatively known, an appropriate alpha peak shape can be obtained by simulations (Pöllänen et al., 2007). This simulated peak shape can be used by a fitting program to perform peak area fitting. The shape simulation with a slightly different input value with subsequent fitting to the measured spectrum is repeated as many times as necessary to obtain acceptable values for the goodness-of-fit indicators (Pöllänen et al., 2009a). An example of the method is presented in Fig. 5.34 in which the peak shapes of the alpha spectrum from particles composed of U are simulated by the AASI program and then fitted by its supplemented version known as AASIFIT.

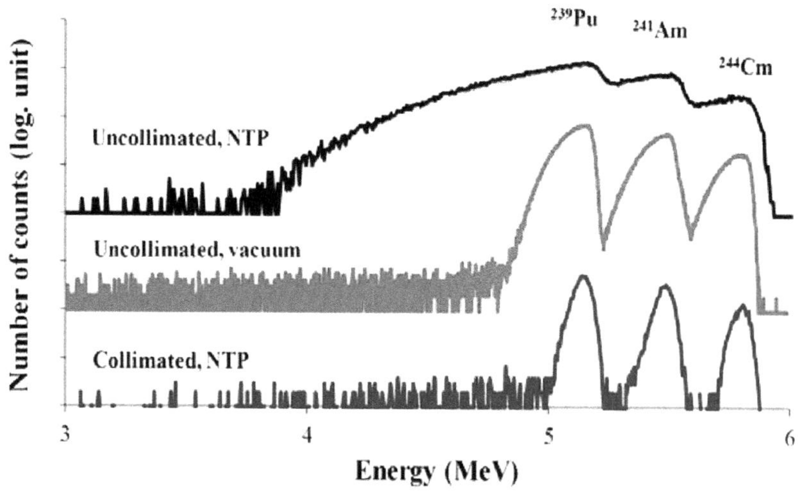

FIGURE 5.33 Alpha particle energy spectra from a thin radiochemically processed source measured at source−detector distance of 9 mm at ambient air conditions, NTP (normal temperature and pressure, spectra at the top and bottom), and in a vacuum (spectrum in the middle). The PIPSi detector used in the measurements was similar to those used in continuous aerosol monitors, having a thick entrance window. The source contains ^{239}Pu, ^{241}Am, and ^{244}Cm, and the energies are calibrated to the nominal emission energy of the alpha particles. Spectrum at the bottom was obtained by using a honeycomb collimator with the cell diameter of 1.5 mm. PIPSi, passivated ion−implanted planar Si. *Pöllänen et al., (2012a,b); reprinted by permission of Elsevier Ltd. © 2018.*

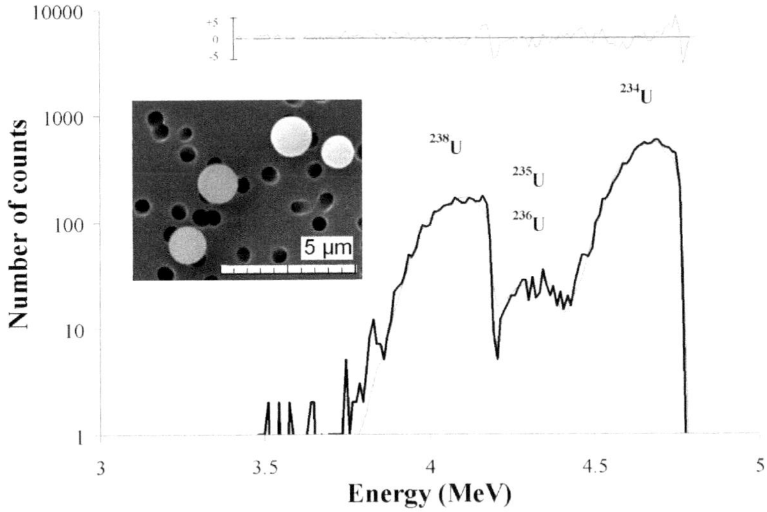

FIGURE 5.34 Measured alpha spectrum (thick line) from spherical U particles (their scanning electron microscopy picture is at the left) and fitted peaks of different U isotopes, and their sum (thin dotted line). The relative residual in terms of standard deviations is presented at the top. *Ranebo et al. (2010), reprinted by permission of Elsevier Ltd. © 2018.*

D. Alpha spectrometry combined with other analysis techniques

Although alpha spectrometry alone is a powerful tool for detecting, identifying, and quantifying alpha particle–emitting radionuclides present in a sample, it can be combined with other analysis methods to get information that is not accessible if the techniques are used independently. In addition to alpha spectrometry, radiochemically processed samples are used for supplementary analysis techniques such as inductively coupled plasma mass spectrometry (Farmer et al., 2003), neutron activation analysis (Montoya et al., 2012), and gamma ray spectrometry (Bem et al., 2002). In combination with X-ray spectrometry, which is an element-based method, it is possible to determine the isotopic ratio of $^{239}Pu/^{240}Pu$ even if the resolution in the alpha spectrum is similar to that presented in Fig. 5.35 (Komura et al., 1984; Arnold, 2006). Parus and Raab (1996) used the combined method of gamma spectrometry with a multigroup analysis (MGA) program (Gunnink et al., 1990) and isotope dilution alpha spectrometry for the determination of plutonium concentration in various nuclear materials.

Alpha spectrometry can also be used in coincidence with gamma ray spectrometry. This technique can be applied, e.g., for checking alpha decay schemes (Martín Sánchez and Caro Marroyo, 2012), analysis of individual radioactive particles (Peräjärvi et al., 2011), or investigations of Rn progeny behavior in filters (Turunen et al., 2011). In connection with event-mode data acquisition (Fig. 5.36), the technique is a powerful tool for determining alpha particle–emitting radionuclides in a sample. Recently, coincidence techniques were also applied for alpha conversion electron measurements (Dion et al., 2016).

VI. Radiochemical processing

To obtain high-energy resolution, "good quality" alpha sources have to be prepared. Prerequisites for good-quality sources are that the deposit be thin and homogeneous with

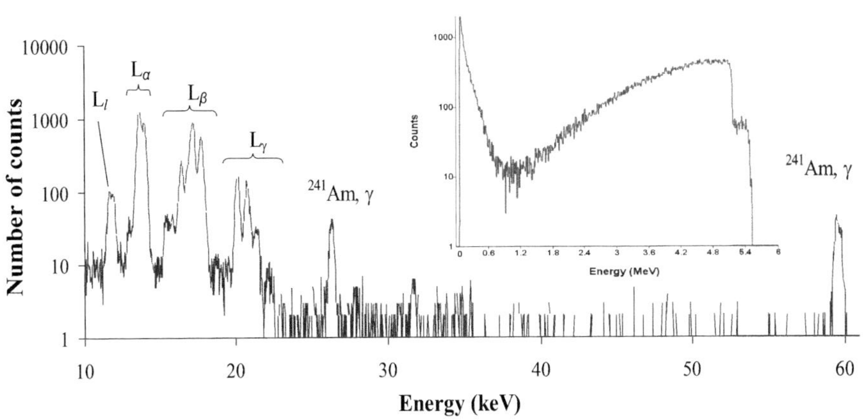

FIGURE 5.35 Alpha (right) and X-ray (left) spectra from a nuclear bomb particle containing Pu isotopes and ^{241}Am. U L-X-ray peaks following the decay of Pu isotopes are denoted in the X-ray spectrum as well as the 26 and 59 keV gamma rays. In the X-ray spectrum, no gamma ray peaks of Pu isotopes are visible. *See Pöllänen et al. (2009b, reprinted by permission of Elsevier Ltd. © 2018) for the details of isotopic ratio calculation.*

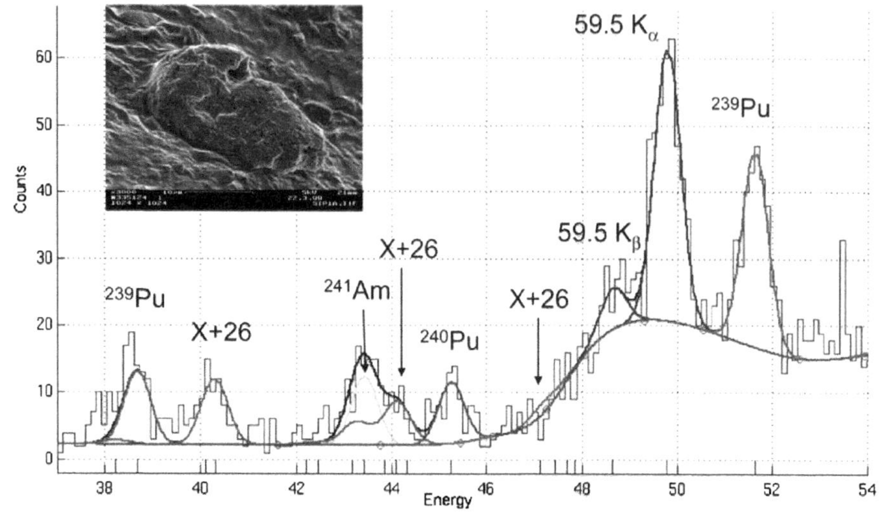

FIGURE 5.36 Part of the gamma ray spectrum from a nuclear bomb particle gated by alpha particles (Fig. 5.35). Inset: scanning electron microscopy image. *Peräjärvi et al. (2009), reprinted by permission of ACS © 2018.*

a smooth flat surface. Since the range of alpha radiation from the alpha-emitting radionuclides is less than 100 μm in typical condensed phases, the source thickness should be desirably much less than this value. Taking into account that the source diameter should not exceed that of the detector, alpha sources should contain less than 10 mg/cm^2 material or less than 50 mg material on a source of 2.5 cm diameter.

To meet this requirement,

- samples are prepared and pretreated by mechanical and physicochemical operations,
- chemical separations are performed to separate the analyte from the matrix components and any interferences of the sample, and
- thin sources are prepared for alpha spectrometry.

The steps involved in standard radiochemical analysis by alpha spectrometry are shown in Fig. 5.37. The full procedure as shown is time-consuming, and recently some of the individual steps have been combined to accelerate the process. Chemical separation has been combined with source preparation by preparing selective PLFs, and efforts have been made to combine chemical separation with deposition and counting by modifying the detector surface with selective ligand.

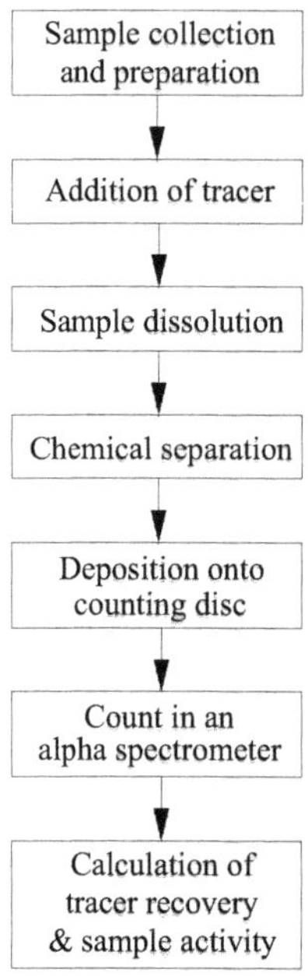

FIGURE 5.37 Steps involved in radiochemical analysis by alpha spectrometry. *Martin and Hancock (2004b), reprinted by permission of Supervising Scientist Division © 2004.*

Sample **preparation and pretreatment** covers all operations that are performed before the chemical separation of the analyte, e.g., drying, homogenization, sieving, destruction of the sample, and preconcentration of the analyte. A short discussion of the methods is given in Section V.A.

Radiochemical processing includes, beside sample preparation and preconcentration, the **chemical separation** of elements using various separation techniques, the most common being precipitation, coprecipitation, distillation, ion exchange, solvent extraction (also called liquid–liquid extraction [LLE]) and extraction chromatography (EC) (also called solid phase extraction), and the source preparation.

In the case of environmental samples, several grams of matrix material are usually removed to obtain thin alpha sources. It is not always necessary to separate all of the alpha-emitting radionuclides from the element to be determined, but there are several nuclide couples whose peaks cannot be resolved by means of standard alpha spectrometry. Examples of well-known **interferences** are the following: ^{238}Pu and ^{241}Am; ^{237}Np, ^{230}Th, and ^{234}U; ^{239}Pu and ^{209}Po. If the radioelements such as Pu and Am, or Np,

Th and U, or Pu and Po, respectively, are chemically separated, the interference issue can be overcome and the originally interfering nuclides can be determined in separate alpha sources. If interferences of the radioisotopes of the same element occur, such as in case of ^{239}Pu and ^{240}Pu, or ^{243}Cm and ^{244}Cm, chemical separations will not solve the problem. In some cases, the ingrowth of short-lived alpha-emitting progeny on the source can be used (e.g., in the case of ^{223}Ra and ^{224}Ra). However, in general, HRAS or other analytical techniques such as mass spectrometry or microcalorimetry have to be introduced in such cases.

Whenever chemical separations are performed, losses of the analytes can appear almost unexpectedly in different parts of the procedure. For example, losses may occur during sample destruction due to volatility or sample splashing. They may also occur during separations due to absorption on the walls of the beakers or on the surfaces of undissolved residues: this is because typically trace or ultratrace amounts of the analytes are processed. To process such small amounts, special methods are used to minimize losses and to determine the **chemical recovery** based on the application of radioactive tracers or chemical carriers.

Tracers are those radionuclides that are added to the sample to quantify the chemical recovery. Either they are not present in the sample or the activity added to the sample is much larger than the amount present in the sample. Tracers are assumed to behave in the same way as the radionuclide(s) to be determined. Tracers are typically but not exclusively the isotopes of the same element as the analyte nuclides. The use of yield tracers was described in detail by Harvey and Lovett (1984). A few examples of frequently used tracers are listed in Table 5.2. The best tracers are relatively long-lived alpha-emitting isotopes that emit alpha radiation of different energy from the analyte nuclides.

Carriers are stable elements showing similar chemical properties to the analyte(s). Unfortunately, many of the alpha-emitting nuclides do not have stable isotopes, but sometimes stable elements belonging to the same group of the Periodic Table can be used as carriers. For example, stable Ba is often used as a carrier for Ra. The mass of the stable carrier before and after separation can be determined by various analytical techniques such as gravimetry, atomic absorption spectrometry, optical emission spectrometry, or mass spectrometry (MS).

As the final step of radiochemical processing, **alpha sources are prepared.** The most frequently used techniques are electrodeposition and microcoprecipitation. Occasionally, drop deposition methods (evaporation) are also used with or without the addition of surface-active agent. The first two techniques have been described in detail in the ASTM Standard Practices C1284-10 and C1163-08, respectively. An overview of alpha source preparation methods is given in Section V.C.

Chemical separations are discussed using as examples the most important alpha-emitting radioelements/radionuclides, including artificial (Pu, Np, Am, Cm) and naturally occurring (U, Pa, Th, Ra, Po) radionuclides in environmental radioactivity measurements in Section V.B. Alpha spectrometry is a measurement method routinely used for the determination of radionuclides of these elements.

To reveal the principle of the radiochemical processes, the basic chemical properties of actinides (Th, Pa, U, Np, Pu, Am, Cm) and Po and Ra relevant for the chemical separations are briefly summarized below. Several monographs and review papers have been published about the radiochemical processing of samples to determine Th (Hyde, 1960), Pa (Kirby, 1959), U (Roberts et al. 1986), Np (Burney and Harbor, 1974), Pu (Coleman, 1965; Cleveland, 1970; Hoffmann, 2001), Am and Cm (Penneman and

TABLE 5.2 Frequently used tracers and analytes in alpha spectrometry.

Analytes	Tracers	Decay mode of tracer
^{239,240}Pu, ^{238}Pu	^{242}Pu, ^{236}Pu	Alpha
^{237}Np	^{239}Np	Beta, gamma
^{241}Am, ^{243}Cm, ^{244}Cm, ^{242}Cm	^{243}Am	Alpha
^{238}U, ^{235}U, ^{234}U	^{232}U, ^{233}U	Alpha
^{232}Th, ^{230}Th, ^{228}Th	^{229}Th	Alpha
^{226}Ra, ^{224}Ra, ^{223}Ra, ^{228}Ra(beta)	^{225}Ra	Beta (alpha particle–emitting progeny)
^{226}Ra, ^{224}Ra	^{133}Ba	Beta, gamma
^{210}Po	^{208}Po, ^{209}Po	Alpha
^{227}Ac	^{225}Ac	Beta (alpha particle–emitting progeny)

^{243}Am and ^{133}Ba are nonisotopic tracers for the determination of Cm and Ra nuclides.

TABLE 5.3 List of Standards and Procedures for the Separation and/or Determination of Th, U, Np, Pu, Am, Cm, Po, and Ra Radionuclides by Alpha Spectrometry.

Standard	Title
ISO 18589-4 (2007)	Measurement of Radioactivity in the Environment. Soil. Part 4. Measurement of Pu Isotopes by α-Spectrometry
ISO 11483 (2005)	Nuclear Fuel Technology—Preparation of Plutonium Sources and Determination of $^{238}Pu/^{239}Pu$ Isotope Ratio by Alpha Spectrometry
ASTM C 1001-05 (2005)	Standard Test Method for Radiochemical Determination of Plutonium in Soil by Alpha Spectroscopy
ASTM D 3865-02 (2002)	Standard Test Method for Pu in Water
ASTM C 1205-07	The Radiochemical Determination of Americium-241 in Soil by Alpha Spectrometry
ASTM C 1475-05 (2010)	Standard Guide for the Determination of ^{237}Np in Soil
ASTM D 3972-09	Standard Test Method for Isotopic Uranium in Water by Radiochemistry
ASTM D 6239-09	Standard Test Method for Uranium in Drinking Water by High-Resolution Alpha-Liquid-Scintillation Spectrometry
ASTM C 1284-10	Standard Practice for Electrodeposition of the Actinides for Alpha Spectrometry
ASTM C 1163-08	Standard Practice for Mounting Actinides for Alpha Spectrometry using Neodymium Fluoride
ASTM D 3084-05	Standard Practice for Alpha-Particle Spectrometry of Water
ASTM C 1561-10	Standard Guide for the Determination of Plutonium and Neptunium in Uranium Hexafluoride and U-rich Matrix by Alpha Spectrometry
IAEA/AQ/11 (2009)	A Procedure for the Rapid Determination of Pu Isotopes and Am-241 in Soil and Sediment Samples by Alpha Spectrometry
IAEA/AQ/12 (2009)	A Procedure for the Determination of Po-210 in Water Samples by Alpha Spectrometry
IAEA/AQ/34 (2014)	A Procedure for the Sequential Determination of Radionuclides in Phosphogypsum: Liquid Scintillation Counting and Alpha Spectrometry for ^{210}Po, ^{210}Pb, ^{226}Ra, Th and U Radioisotopes
IAEA/AQ/37 (2014)	A Procedure for the Sequential Determination of Radionuclides in Environmental Samples: Liquid Scintillation Counting and Alpha Spectrometry for ^{90}Sr, ^{241}Am and Pu Radioisotopes

Keenan, 1960; Warwick et al., 1996), Po (Matthews et al., 2007; Henricsson et al., 2011), and Ra (IAEA, 2010a) using both old and new separation procedures, and good books and book chapters are also available that deal with groups of these radioelements (Bagnall, 1957, 1966; Katz et al., 1986; Wolf, 2006; Nash et al., 2006; Hou and Roos, 2007; Vajda and Kim, 2011; Lehto and Hou, 2011). A list of recently published standards and procedures is given in Table 5.3.

Th, Pa, U, Np, and **Pu** are the first, second, third, fourth, and fifth members of the actinide group, respectively. In aqueous solutions, they exist in various (+3, +4, +5, and +6) oxidation states with the exception of Th, which has only one stable form in solution, Th(IV). The most stable forms of Pa, U, Np, and Pu are Pa(V), U(VI), Np(V), and Pu(IV). In alkaline solutions, U(VII) and Np(VII) are also stable. Because the standard electrode potentials of the various Pu and Np species in acidic solutions are only slightly different, the several various forms can coexist in solutions. Since the oxidation states of Pa, U, Np, and Pu can be changed by the addition of oxidizing or reducing agents, the separation processes can be controlled. According to the actinide theory of Choppin et al. (1997), differences among the various oxidation states of a single actinide are greater than those of different actinides in the same oxidation state.

The formation of complex ions in aqueous solutions with inorganic ions or organic compounds is an important feature of the chemistry of the actinides. The relative complex forming tendency of actinide (An) species is the following:

$$An^{4+}(IV) > An^{3+}(III) \approx An(VI)O_2^{2+} > An(V)O_2^{+}$$

The stability series of the inorganic complexes of tri-, tetra-, and hexavalent actinides are the same:

$$OH^- > carbonate > oxalate > F^- \approx acetate^- > SO_4^{2-} > NO_3^- \approx Cl^- > ClO_4^-$$

There are small differences among the stability constants of the different actinides in the same oxidation state.

Oxidation state adjustment and complex forming conditions have been discussed in many papers, and the conditions of oxidation−reduction reactions as well as stability constants of a great number of complexes have been measured. Formation of the given species is influenced by many parameters (composition of the sample, presence of oxidative/reductive agents, amount and stability of the reagents, kinetics of the process, etc.), and it is an especially challenging job if several actinides are present in the same solution.

Since tetravalent actinides have the highest complexing power, the adjustment of this valence state has formed the crucial part of many separation procedures. In many applications, it would be desirable to keep Th, U, Np, and Pu in the tetravalent oxidation state in the same solution, but this is impossible. Thus, oxidation state adjustment is a kind of compromise among various possibilities.

Np(IV) is easily formed by the addition of various reducing agents, e.g., ferrous ions, iodide ions, to hydrochloric or dilute nitric acid solutions. Strong reducing agents generating Fe^{2+} in situ by the reduction of Fe^{3+}, e.g., hydrazine, hydroxylamine, are adequate for the adjustment of Np(IV) and U(IV) oxidation states. These agents reduce Pu to Pu(III). Trivalent Pu can be oxidized back to tetravalent species by various oxidizing agents. To stabilize the Pu(IV) state without further oxidizing it to Pu(V) or Pu(VI), only a few moderately strong oxidizing agents can be used, e.g., $NaNO_2$ can be added to nitric acid solutions to oxidize Fe^{2+}, and consequently, Pu to Pu(IV) or H_2O_2 can be added to hydrochloric acid solutions, whereas Np still remains tetravalent and U is oxidized back to U(VI). $NaNO_2$ and H_2O_2 are much less effective if Pu in higher oxidation states has to be reduced to the tetravalent form.

Besides tetravalent species, hexavalent ones can often form strong complexes. Hexavalent states of Pu, Np, and U can be obtained using strong oxidizing agents, e.g., $HBrO_3$, $KMnO_4$, $K_2S_2O_8$. Redox conditions of actinides have been studied by many authors, and several reviews are also available.

Am and Cm are the sixth and seventh members of the actinide group. In aqueous solutions, Am can exist in the +3, +4, +5, and +6 oxidation states. Although Am(III) can be oxidized theoretically to Am(IV), that form exists only in basic solutions in the presence of strong complexing agents such as phosphates, but the species is unstable and will be reduced back to the tervalent state easily. Although Am can be oxidized in aqueous solutions to penta- and hexavalent states with a strong oxidizing agent, e.g., Ag catalyzed $K_2S_2O_8$ forming oxo cations $Am(VI)O_2^{2+}$ and Am $(V)O_2^+$, the most stable form is the tervalent one (Osváth et al., 2009). Curium has only exclusively tervalent species in aqueous solutions.

The relative complex forming tendency of Am species depending on the charge density is the same as those of other actinides:

$$Am(VI)O_2^{2+} \sim Am^{3+}(III) > Am(V)O_2^+$$

The lack of the tetravalent Am and Cm species of the highest complex forming power and the chemical similarity of Am to Cm, as well as to the lanthanides, badly influence the separation possibilities. This is the main reason why Am and Cm chemistry is still the most challenging part of actinide analysis.

The stability series of the inorganic complexes of Am(III) and Cm(III) are the same as those of other actinides:

$$OH^-, CO_3^{2-} > F^-, HPO_4^{2-}, SO_4^{2-} > Cl^-, NO_3^- > ClO_4^-$$

Thiocyanate (SCN) complexes of Am and Cm do not have high stability, but because of their relatively high separation factors toward lanthanides, they play an important role in Am and Cm purification.

Ra: Within Group 2, Ca, Sr, Ba, and Ra form a series of closely related elements. All are highly electropositive metals existing in the +2 oxidation state in solution. The chemical properties of these elements vary in a reasonably systematic manner down the group. For example, all form insoluble sulfates, with the solubility product following the series $CaSO_4 > SrSO_4 > BaSO_4 > RaSO_4$. When Ra is coprecipitated with other group 2 sulfates, or with $PbSO_4$, the insolubility of $RaSO_4$ causes Ra to be enriched in the sulfate lattice relative to the primary precipitate cation (Langmuir and Riese, 1985).

The similarity in their chemical properties means that Ca, Sr, and, especially, Ba are very effective carriers for Ra. It also means that separation of Ra from the other group 2 elements is the primary challenge for HRAS of Ra. For samples with a high Ca content, such as bone, separation of Ra from the bulk Ca may present difficulties (Yamamoto et al., 1991). However, for most samples, it is the separation of Ra from Ba, which is the main problem, due to their very similar chemistries.

There are two basic approaches to the difficulty of separation of Ra and Ba. One of these is to separate the group 2 elements from the bulk matrix and then to carry out a final Ra-Ba separation. This can result in a high-resolution source, but it is time-consuming. The alternative approach is to coprecipitate Ra with $BaSO_4$ following the bulk matrix separation and to filter this precipitate for use as the alpha source (e.g., Sill, 1987b). The precipitation and filtration steps are relatively fast, and high recoveries can be expected. However, the poorer peak resolution will generally preclude the use of ^{225}Ra as a tracer. It may also preclude the determination of ^{223}Ra and ^{224}Ra, although the use of microprecipitation techniques may allow this (Nour et al., 2004). In addition, the tracer commonly used, ^{133}Ba, is not an isotope of Ra, and so there is the potential for differences in recoveries between tracer and analyte.

Po: The chemistry of the group 6 elements (O, S, Se, Te, and Po) varies in a more or less regular manner down the series. Due to increased atomic radius, electronegativities decrease down the group, and the elements exhibit increasing metallic character. Hence, whereas O, S, and Se are nonmetals and form covalent bonds, Te shows some metallic character and cationic properties, and Po is essentially a metallic element. There is evidence for the formation of −2, +2, +4, and +6 oxidation states for Po; however, +4 is the most stable state in solution (Figgins, 1961).

The S to Po members of group 6 have *d* orbitals available for use in bonding, and hexacoordinate compounds are most common for Te and Po. Complexes of the form $(PoX_5)^-$ or $PoX_5(H_2O)^-$ are formed in dilute halide solution, whereas $(PoX_6)^{-2}$ is formed in more concentrated acid solutions (X = Cl, Br, or I) (Bagnall, 1983). In the absence of more strongly complexing organic agents, halide complexes can be extracted into some organic solvents.

One important feature of Po is its volatility. This varies with chemical form (Mabuchi, 1963). Po has been shown to be partly volatilized from some samples above 100°C (Martin and Blanchard, 1969). Polonium halides have been reported to be volatile above 150°C, but this may depend on the conditions, as only minimal losses were reported by Svetlik et al. (2010) at 350°C for deposits dried from HCl solution. Po volatility is important for potential losses during sample preparation, digestion, and chemical separation. It also means that Po can migrate from the alpha source in the counting chamber and contaminate the detector. This latter problem seems to be reduced by a period of exposure to air before counting (Bagnall, 1957, 1966).

Po spontaneously autodeposits onto the surfaces of several metals, including Ag, Cu, and Ni from dilute acid solution. Autodeposition onto metal discs (usually Ag) is the most common source preparation technique for Po due to its simplicity and the fact that it results in separation from other metal ions, including other alpha emitters (Smith and Hamilton, 1984; Matthews et al., 2007). This being the case, the preceding chemical separation techniques required are generally less extensive than for other alpha emitters. Some published methods rely only on autodeposition for separation, and for simpler sample matrices, this may be sufficient. However, the presence of interfering elements, especially Fe, can result in lowered recoveries and spectral degradation. Consequently, for general applicability, a separation step is recommended (Lin and Wu, 2009).

A. Sample preparation and pretreatment

The objectives of sample preparation and pretreatment are to obtain a solution that

- is homogenous, is free of any insoluble residues that would interfere during the subsequent chemical processing, and could retain the analyte,
- is free of any organic material that could form complexes with the analyte or the reagents resulting in poor recovery,
- is of relatively small volume to achieve significant volume reduction for large-volume liquid samples,
- contains the analyte preconcentrated usually due to semiselective group separation.

1. Preparation of solid samples

For soil, sediment, and minerals, sample preparation and pretreatment usually mean drying in an oven, homogenizing and sieving (optional), and removing organic material by ashing followed by wet chemical destruction.

a. Sample ashing

In the case of actinide determination, samples are usually ashed at 450−600°C for a couple of hours or days. Elevated temperatures are not desirable because of the formation of refractory particles. In case of volatile elements such as Po, dry ashing should be replaced by wet ashing.

b. Sample dissolution

For sample dissolution, several methods have been used, including acid **leaching** with mineral acids (most frequently with 8 M HNO_3 in the case of actinide determination or aqua regia) and **complete destruction** by a mixture of mineral acids, typically including nitric, hydrochloric, and hydrofluoric acids and hydrogen peroxide. Kenna (2009) studied the leachability of Pu and Np from sediment samples by applying a series of extractants. If hydrofluoric acid is used to volatilize silica as SiF_4, then boric acid is added afterward to the final solubilization to complex any fluorides. Actinides such as Pa and Th often form very refractory materials, e.g., the complete dissolution of Pa oxides requires hydrofluoric, sulfuric, or oxalic acids. **Wet acid decomposition** can be carried out in open systems such as in Teflon beakers on a hot plate or in closed systems such as in a microwave oven or pressure bombs. The higher the pressure and the temperature, the faster is the chemical reaction.

Significantly higher temperatures than the boiling points of the aqueous systems can be attained by **fusion** with a variety of salts, thus speeding up the destruction process. Several good monographs and papers have been published on sample decomposition methods (Bock, 1979; Sill and Sill, 1995; Chiu et al., 1984; Claisse, 1995), which can be applied to the analysis of actinides. Fusion is a very effective way of rapid and complete sample decomposition. It is the best technique to destroy refractory materials, such as Th, Pa, U, and Ra embedded in soil and sediment particles. Unfortunately, large amounts of salts often together with impurities are transferred to the sample solution, which may create problems in the course of subsequent chemical separations. The other disadvantages of fusion are that usually only relatively small samples can be treated and that it is inappropriate for Po determinations due to the large losses that occur at the high fusion temperatures. NaOH, Li borate, and mixed carbonate–borate have been used for fusion in actinide chemistry. Sill et al. (1979) and Sill (1980) proposed a very efficient potassium fluoride/pyrosulfate fusion to treat refractory material. Maxwell and coworkers proposed NaOH as a cheap and efficient chemical for fusion of a variety of samples (up to 100 g soil, 1g concrete or asphalt, 5 kg rice) for the rapid determination of actinides (Maxwell, 2008; Maxwell et al., 2011c, 2013, 2014). They usually used coprecipitation with hydroxide, phosphates, and fluorides to remove the excess NaOH and the interfering elements.

Several standards on sample decomposition are available. Standard ASTM C1342-96 deals with fusion decomposition of solid materials, standard ASTM D1971-95 describes hotplate and convection oven digestion of water and solid samples with nitric and hydrochloric acids, standard ASTM D4309-91 refers to microwave digestion with hydrochloric and nitric acids for the determination of metals and groundwater, and the U.S. EPA Method 3052 (1996) discusses microwave-heated, closed-vessel decomposition of ashes, biological tissues, oils, oil-contaminated soils, sediments, sludges, and soils.

In the case of biological samples, complete destruction is preferred and carried out usually by the combination of dry and wet chemical ashing. Because of the lack of silicates and refractory components, the use of HF can be omitted. Microwave digestion is a preferred way of decomposition of smaller samples. An efficient way of decomposition of organic materials is the Fenton's reaction using hydrogen peroxide and Fe^{2+} catalyst.

2. Preconcentration of liquid samples and sample solutions

a. Preconcentration of actinides

Preconcentration of actinides by coprecipitation is a frequently used method for large volumes of water samples as well as for sample solutions obtained after destruction of solid samples. A well-selected coprecipitant concentrates the analyte in a small volume, removes most of the cationic and anionic interferences, and helps increase the decontamination factor (DF) for the whole radiochemical procedure.

Bismuth phosphate ($BiPO_4$) was the first precipitate applied in large-scale processing of Pu; as $BiPO_4$ carries Pu(III) and Pu(IV), it was used to separate Pu from U and fission products in both technology and analyses (Kimura, 1990). The coprecipitation of Am with $BiPO_4$ was studied by Mathew et al. (1981), who found that below 0.3 M acid concentration, Am is carried with the precipitate.

The most frequently used carriers of actinides are the following: ferric and ferrous hydroxides, lanthanide fluorides and hydroxides, alkaline earth oxalates, phosphates and fluorides, some other phosphates and oxalates, barium sulfate, and manganese dioxide. The coprecipitation behavior of actinides is discussed in detail in the compilation of Katz et al. (1986).

Hydroxide precipitates, e.g., ferrous and ferric hydroxides, lanthanide hydroxides, and calcium phosphate and calcium/magnesium carbonates, are formed in neutral or basic solutions. Hydroxide, phosphate and carbonate precipitates are usually easily dissolved with acids. Ferric hydroxide is known to coprecipitate actinides in all oxidation states, but sometimes recoveries are low, which is why ferrous hydroxide as a reducing agent is preferred. Coprecipitations taking place in acidic media (e.g., fluorides) are usually more selective than those formed in basic solutions (e.g., hydroxides).

Calcium fluoride and especially lanthanide fluorides of low solubility form precipitates in 4–5 M HF solution. Fluoride precipitates can be dissolved as borate complexes and carry only tri- and tetravalent actinides. The separation of actinides by coprecipitations usually takes advantage of the oxidation and reduction cycles.

Barium sulfate carries only tri- and tetravalent actinides. It can be dissolved either with EDTA or with $HClO_4$. The use of EDTA is beneficial if samples have Ca, Mg, and Al content that forms strong complexes and can be separated from actinides during a succeeding lanthanide hydroxide precipitation. Sill and Willis (1966) developed a method for the separation of Np from Ce, Ba, and La based on its coprecipitation with barium sulfate under strict oxidation state control. The oxidation state of Np is adjusted with H_2O_2 to Np(IV), which is coprecipitated with barium sulfate from a sulfuric acid solution containing potassium sulfate. Then, Np is oxidized with chromic acid to Np(VI) and separated from fission products that cannot be oxidized and form precipitate with a next barium sulfate. The Np alpha source is also prepared by coprecipitation with microcrystals of barium sulfate. Sill and Willis (1966) adopted this method for the simultaneous determination of Np, Pu, and U based on a series of barium sulfate coprecipitations. Coprecipitation of Th with $BaSO_4$ may be enhanced by the addition of K^+ to the solution (Burk and Wiles, 1983) or alternatively prevented by use of a La^{3+} holdback carrier (deJong and Wiles, 1984).

Moore (1963) developed a procedure that was later improved by Sill (1984) for the simple separation of Am from many elements, including actinides and lanthanides. Americium was oxidized to its hexavalent state with ammonium peroxydisulfate in a dilute nitric acid solution. The nonoxidizable lanthanides and actinides were than precipitated as fluorides and filtered off. The Am was than reduced and coprecipitated with LaF_3. In the improved method of Sill, the nonoxidized rare earths were coprecipitated with $BaSO_4$ or NdF_3 whereas oxidized Am(VI) was kept in solution. After reduction with H_2O_2, Am(III) was coprecipitated with a small amount of NdF_3 allowing direct detection of the alpha source (discussed subsequently in source preparation). By this method, Am can be separated also from the nonoxidized Cm isotopes.

MnO_2 precipitated from $KMnO_4$ and $MnCl_2$ at pH 8–9 is used for the preconcentration of Am, Pu, and Np from seawater samples at the IAEA's Marine Environmental Laboratory (LaRosa et al., 2005). MnO_2 is dissolved with HNO_3-H_2O_2. Protactinium can be coprecipitated with MnO_2 from siliceous cake after fusion (Dalvi and Verma, 2017).

The pH dependence of the precipitation of metal oxalates was studied by Yamato (1982), who found that calcium oxalate formed at $pH \geq 1.5$ carries Am almost quantitatively. Oxalate precipitates are easily decomposed by ashing or wet chemical oxidation.

The following precipitates have been recently used to preconcentrate actinides from various samples: ferric hydroxide, ferrous hydroxide, calcium oxalate, ferric hydroxide and calcium oxalate, calcium phosphate, calcium fluoride, lanthanum fluoride, manganese dioxide, manganese dioxide, and ferric hydroxide. For the concentration of actinides from a large volume of water or solutions of biological origin, special EC materials of extremely high distribution coefficients have also been developed and applied, e.g., TEVA, TRU, DIPHONIX, and DIPEX resins (see in detail in the next section). Smith et al. (1998) developed a method for the preconcentration of Am and Pu from waste waters using a synthetic water-soluble metal-binding, i.e., phosphonic acid polymer that was removed from the waste by ultrafiltration. Chemical yields over 90% have been achieved.

b. Preconcentration of Po and Ra

Coprecipitation with iron hydroxide (Narita et al., 1989; Jia et al., 2001; Zikovsky, 2002) or MnO_2 (Carvalho, 1997; Peck and Smith, 2000; Martin and Hancock, 2004b; Bojanowski et al., 2005) have commonly been used for the preconcentration of Po and Ra, particularly from bulk water samples. Coprecipitation with $Ca_3(PO_4)_2$ has also been used for Ra and Pb (Nonova et al., 2009). Coprecipitation with MnO_2 is very commonly used for processing seawater and other saline waters (e.g., Towler et al., 1996; Grabowski and Bem, 2010). Direct adsorption from water samples of Ra onto MnO_2-coated discs for alpha spectrometry has also been used (Eikenberg et al., 2004).

Iron tends to interfere with subsequent chemical separation and autodeposition of Po. Consequently, for Po determinations following coprecipitation with iron hydroxide, an Fe removal step, such as solvent extraction with diisopropyl ether, may be required.

Coprecipitation with $PbSO_4$ is commonly used for the preconcentration of Ra and separation from the bulk matrix (Sill, 1987b; Kehagia et al., 2007). Radioisotopes of Pb and Ra have also been coprecipitated with $Pb(NO_3)_2$, with U and Th remaining in solution (Lozano et al., 2010).

Coprecipitation with $BaSO_4$ is also commonly used for preconcentration and separation of Ra, often as an extra step after coprecipitation with $PbSO_4$. Yamamoto et al. (1989) used coprecipitation with barium chromate and performed a subsequent Ra/Ba separation. Commercial salts of Ba may contain traces of Ra, and so the reagent used should be pretested, and the amount used minimized. Use of a Ba salt coprecipitation usually means that a Ra/Ba separation will not be carried out, although it may be possible to do so if only small amounts of Ba are present. Alternatively, LS spectrometry may be performed.

For Po preconcentration, coprecipitation with cobalt–ammonium pyrrolidine diothiocarbamate (Co-APDC) chelate has been used (Gascó et al., 2002; Wildgust et al., 1998). Po(IV) also coprecipitates with $BiPO_4$ at low pH (Hölgye, 2007).

B. Chemical separation

Actinides are amenable to a variety of chemical separation procedures, including precipitation/coprecipitation (see above), LLE, ion exchange chromatography, EC, and the combination of two or more of these methods.

1. Separations by liquid–liquid extraction

Three types of extractants can be distinguished:

i) neutral extractants that form neutral (ion association) complexes with the actinide ions involving the counter-ions, very often nitrates or chlorides and the organic molecule forming chemical bonding via electron donor atoms of the organic molecules (oxygen, phosphoryloxygen, sulfur, nitrogen),
ii) amine extractants where positively charged organic amines form complexes with anions via Coulomb forces, and anions in the complexes can be replaced by anionic actinide complexes in a manner analogous to anion exchange; amine extractants are long-chain

alkyl or aryl quaternary amines or primary, secondary, and tertiary amines that form organic cations with acidic hydrogen,

iii) acidic extractants that form chelate complexes with the actinide cations replacing the H^+ in the organic compound.

Groups (i) and (ii) can be regarded together simply as ion association complexes. Ion association complexes have the advantage of being applied in acidic media. In general, distribution coefficients are increased when acid concentration increases. On the contrary, chelates are used in media of low acidity because H^+ ions are competitors of the cations for the binding sites.

a. Chemical separation of actinides by liquid-liquid extraction

Many of the LLE procedures for actinides have been developed for the large-scale separation of U and Pu for the nuclear industry. Procedures for the separation of minor actinides (Np, Am, Cm) are under investigation. Solvent extraction offers usually moderate selectivity, but the extraction efficiency can be increased by performing the process in a continuous technological system.

Traditional extractants. Originally, the LLE of Pu and U was accomplished with a wide range of *organic acids, ketones, esters, alcohols*, and *ether*; the most well-known being **tributyl phosphate** (TBP) used in the PUREX process. Both Pu and U nitrate complexes have acceptable high (but still relatively low) distribution coefficients for technological purposes. Neptunium follows Pu in the extraction processes whenever the oxidation states are the same, but adjustment often fails. The nitrate complexes are even less retained by TBP (Nash et al., 2006).

Methyl-i-butyl ketone (MIBK) and TBP were used for analytical purposes for the determination of Pu in the 1950s and 1960s. At present, TBP is used in a mixture with carbamoylmethyl-phosphine oxide (CMPO) for EC (discussed subsequently).

While tetra- and hexavalent actinides are readily extracted by organic acids, ketones, esters, alcohols, and ethers, neither of these organic compounds nor tributyl phosphate can extract the trivalent Am. Trivalent Am^{3+} cations do not form stable ion association complexes with simple monofunctional organic compounds; they can be complexed with bifunctional compounds and do not have the affinity to form anions, but they can be extracted/retained by chelates.

Chelate extractants. Among the ketones, **tenoyl-trifluor acetone** (TTA) has quite significant analytical applications. It is a diketone with a sulfur-containing aromatic ring and a fluorine-substituted methyl group that exists mainly in enol form, thus acting as a chelate. TTA has been used for analytical purposes to separate Pu from various samples in the past and recently for the separation and purification of Pu in combination with anion exchangers from marine sediment samples, and extremely high DFs were achieved while Pu recovery was 56% (Donard et al., 2007).

TTA proved to be one of the most selective reagents for the separation of Np(IV) when Pu is kept in the trivalent Pu(III) oxidation state. TTA was used as early as 1947 by Magnusson (Coleman, 1965). Neptunium was reduced and then retained from 1 M HNO_3 or HCl on about 0.5M TTA in xylene, and finally stripped with 8 −10 M HNO_3. TTA has also been used for speciation studies where "TTA extractable" species are regarded as tetravalent ones (Wolf, 2006).

The trivalent Am and Cm cations can form stable chelates with TTA but only at relatively high pH. Sekine et al. (1987) used 0.5 M TTA in xylene to purify Am from a solution at pH $\geq$ 4.

1-Phenyl-3-methyl-4-benzoyl-5-pyrazolone (PMBP) is an N-containing ketone that also forms chelates in the enol form and has gained application in Am separations. It extracts several elements from more acidic media than TTA. Distribution coefficients for Am and the separation factor of lanthanides are slightly higher than in TTA. Jia et al. (1997) extracted Am from a 0.1 M HNO_3 solution into a mixture of 0.05 M PMBP and 0.025 M trioctylphosphine oxide (TOPO) in cyclohexane. Americium was backextracted with 5 M HNO_3. Jia et al. (1998) used PMBP/TOPO extraction to purify Am before preparation of alpha sources. Am was extracted from a DTPA/lactic acid solution directly into 0.05 M PMBP +.025 M TOPO in cyclohexane. Extraction maximum (about 90%) was in the pH range 1−2.

Bis-2-etylhexyl-phosphoric acid (HDEHP) as a chelate shows selectivity for cations of higher charges against those of lower charges; thus, HDEHP is a frequently used extractant in Am and Cm separations and is a good choice for the separation of tetravalent actinides.

The distribution coefficients (D) for Am in HDEHP solution strongly depend on the acidity of the aqueous phase, e.g., lg(D) = 1 for Am extraction by 0.5 M HDEHP in i-octane if the HNO_3 concentration is 0.1 and lg(D) = 4 in 0.01 M HNO_3 solution (Myasoedov et al., 1974). The separation of trivalent actinides from lanthanides is only possible if strong complexants, e.g., lactic acid, DTPA, are added to the aqueous phase. HDEHP has been used for the separation of Am together with Cm and the lanthanides. It was observed that Fe severely interferes with the extraction. Bernabee et al. (1980) extracted Am from 72% $HClO_4$ solution into 15% HDEHP in n-heptane. Americium was backextracted with 4 M HNO_3. Holm et al. (1979) extracted Am from 0.01 to 0.001 M HNO_3 and backextracted it with 4 M HCl.

In the standard test method ASTM C 1205-07 for "The Radiochemical Determination of Americium-241 in Soil by Alpha Spectrometry," ^{241}Am is determined in soil samples up to 10 g. The soil is completely dissolved by the use of pyrosulfate fusion. After the initial separation on barium sulfate, Am together with other actinides and lanthanides is extracted from concentrated $HClO_4$ solution with 15% HDEHP in n-heptane and backextracted with 5 M HNO_3. Americium is separated from other trivalent actinides and lanthanides by oxidation of Am and precipitation of the interferences with LaF_3. The standard test method is based on the procedure developed by Sill and Willis (1966).

In the procedure of Jia et al. (1998), Am was extracted from 0.1 M HNO_3 on a supported KF-HDEHP column, followed by its stripping with 6 M HNO_3. Am was separated from lanthanides on a second KF-HDEHP column using 0.07 M DTPA—1 M lactic acid solution. The average chemical yield for the whole procedure, including Pu and Sr removal and purification of the Am fraction with PMBP/TOPO extraction (see above), was 58%.

Ramebäck and Skillberg (1998) extracted Pu and Np after oxidation with bromate as Pu(IV), Pu(VI), and Np(VI) together with U, Am, and Cm from 0.1 M HNO_3 solution using HDEHP and then stripped Am and Cm with 5 M HNO_3, followed by the backextraction of Np with 1 M HNO_3/hydroxyl amine and Pu(III) with 3 M $HCl/TiCl_3$. Chemical recoveries for Pu, Np, and Am were high, 91%, 82%, and 97%, respectively.

In the recently issued standard test method ASTM D6239-09, uranium is separated from drinking water with an extractive scintillator that contains HDEHP as extractant. Uranium is extracted from a 0.01 M DTPA solution at pH 2.5 and 3.0, whereas extraction of undesired ions is masked by forming strong DTPA complexes. Uranium isotopes are determined by high-resolution alpha liquid scintillation spectrometry.

Ion association complexes. Later attention of scientists was turned to other *organo phosphorous compounds* than phosphates, including phosphonates, phosphinates, and phosphine oxides, which showed higher selectivities for many actinides from acidic media. Beside TBP, TOPO is the most well-known representative of the neutral organophosphorous compounds. Pimpl and Higgy (2000), Ayranov et al. (2005), Lujaniene and Sapolaite (2005) used TOPO as solvent extractant for preconcentration of Pu and Am from acidic leach solutions. By reducing the acidity, TOPO can also extract trivalent actinides. Pu together with Th, U was extracted from 4–6 M HNO_3 to TOPO in cyclohexane. Then, the acidity was reduced, the pH of the solution was adjusted to 1, and Am was also extracted with TOPO and backextracted with 2 M HNO_3. Kalmykov et al. (2004) extracted Pu and Np together from 5 M HCl with TOPO and selectively backextracted Pu as Pu(III) with 1 M HCl and ascorbic acid. Np(IV) remaining in the organic phase was irradiated after evaporation to determine ^{237}Np by neutron activation analysis.

Dibutyl-N,N-diethylcarbamylphosphonate (DDCP) as a bifunctional organophosphorous compound is one of the first representatives of a new group of extractants, which was applied in acidic solutions. DDCP has been used to extract Am (and Cm) from 12 M HNO_3 by Ballestra and Fukai (1983). Lanthanides followed Am in the procedure. Americium was backextracted into 2 M HNO_3. Significant improvement was recorded in the Am recovery (69–79%) for the overall procedure, including coprecipitations, anion exchange separations, and source preparation compared with the procedure using HDEHP. DDCP extraction became the basis of the method for Am determination in seawater at the IAEA's Marine Environmental Laboratory (IAEA, 1989). Actinides were preconcentrated by coprecipitation with MnO_2 and ferric hydroxide; Pu, Np, and Th were separated by anion exchange chromatography; Am was coprecipitated with Ca oxalate and passed through an anion and a cation exchanger. Americium was extracted with DDCP. Finally, Am-lanthanide separation was performed on an anion exchange resin column. Later, DDCP was replaced by a new and more effective bifunctional organophosphorous extractant, the CMPO.

CMPO is **octyl(phenyl)-N,N-di-i-butylcarbamoylmethylphosphine oxide**, a neutral organophosphorous extractant used in conjunction with t-butyl phosphate as a process solvent (Horwitz et al. 1986). The CMPO/TBP extractant was used in the TRUEX (TRansUranium EXtraction) process for the separation of tri-, tetra-, and hexavalent actinides from nitric acid solution, i.e., from the high-level waste (HLW) produced in the PUREX process. The attractive feature of the process is that it allows for the extraction of Am and its separation from tetra- and hexavalent actinides by using different strip solutions. The pitfall of the procedure is that Am is not separated from Cm and lanthanides. The distribution ratios of tri-, tetra-, and hexavalent actinides and nonactinide elements, the mechanism of extraction, were determined in nitric and hydrochloric acids by Horwitz et al. (1987). In the analytical practice, the EC version of the CMPO/TBP extractant called TRU resin has had widespread use (discussed below in the paragraph on Extraction Chromatography).

The higher oxidation states of Am (preferably the complex forming hexavalent state) are not stable in the presence of organic materials.

Amine extractants. Tertiary amines act as strong base anion exchangers and strongly retain the nitrate or chloride complexes of tetravalent Pu and Np, e.g., $[Pu(NO_3)_6]^{2-}$, $[Pu(Cl)_6]^{2-}$ formed in concentrated acidic solutions. **Trioctyl amine** (TOA) is the most frequently used liquid amine extractant. **Alamine 336** also called tricapryl amine is a mixture of n-octyl and n-decyl amines. **Aliquat 336** is a quaternary ammonium salt that contains a mixture of methyl, n-octyl, and n-decyl ammonium salts. The use of amine extractants for Pu analysis has a long tradition, and TOA still has a widespread analytical application, especially as supported liquid extractant (see later under Extraction Chromatography). Jia et al. (1998) separated Pu as Pu(IV) from the leachate of soil samples by extraction with TOA and achieved a Pu recovery of 85%. In the experiment, TOA was fixed to microthene particles. Sill and Willis (1966) used Aliquat 336 to extract Np(IV) from environmental samples. Vance et al. (1998) used amine extraction from 0.25 M H_2SO_4 to separate U from Np. Neptunium was not extracted and could be detected directly from the aqueous phase using ICP-MS.

Chen et al. (2001a) developed a method for the determination of ^{237}Np in soil and sediment samples up to 10 g and seawater up to 100 L using both alpha spectrometry and ICP-MS. Neptunium separation consists of a preconcentration with mixed Fe oxides precipitated in the presence of $K_2S_2O_5$ reducing agent, extraction with TOA from 8−10 M HCl, repeated coprecipitation with $Fe(OH)_2$, and purification of Np(IV) by anion exchange from 8 M HNO_3. An overall recovery of 74%−93% was achieved while DFs for U, Th, and Pu were also high ($>10^4$). The method was extended for the simultaneous determination of $^{239,240}Pu$, ^{241}Am, ^{237}Np, ^{234}U, ^{238}U, ^{228}Th, ^{230}Th, ^{232}Th, ^{99}Tc, and ^{210}Pb-^{210}Po in environmental materials (Chen et al., 2001b). Aliquat 336 was used for the separation of Pa from HCl/HF mixtures by Dinkar et al. (2017). Protactinium was stripped by a mixture of HNO_3/HF.

Alcohol extractant. The extraction of actinides Th, Pa, U, Np, and Am by aliphatic alcohols was investigated by Knight et al. (2016). The selective extraction of Pa from HCl and HNO_3 by 1-octanol was demonstrated.

Trivalent actinides are not well retained by any of the amine extractants from acidic solutions. Amine extractants as well as anion exchangers have found an application in the field of americium chemistry; they have been used for the separation of trivalent actinides from lanthanides.

Separation of Am from lanthanides. For the separation of trivalent actinides from lanthanides, new extraction systems containing a soft donor atom (N, S, Cl) have been proposed based on the experience that actinides form slightly stronger complexes with ligands containing soft donor atoms than lanthanides. The soft donor atom may be present as free ion (SCN^- or Cl^-) or in a water-soluble complexant or in a lipophilic organic extractant. This idea formed the basis for the separations from the early 1940s. Solvent extraction, ion exchange, and EC techniques fit prominently in these processes.

The extractability of tervalent actinides is increased by adding special complexants or high amounts of salts to the aqueous phase. In HLW technology, i.e., in the TALSPEAK (Actinide-Lanthanide Separation by Phosphorous reagent Extraction from Aqueous Komplexes) process, DTPA containing the soft donor N atom and lactic acid were added to the aqueous solution to form stronger complexes with Am than with lanthanides, thus allowing their separation using HDEHP (see above). The procedure of Jia et al. (1998) is based on the same principle.

SCN^- complexant containing the soft donor S and N atoms can help increase the separation factor between Am, Cm, and lanthanides. Moore (1964) first applied quaternary amines to lanthanide/actinide group separations examining the system Aliquat 336/xylene/H_2SO_4-NH_4SCN.

Quaternary amines (0.6 M Alamine 336 in diethylbenzene) are used to selectively extract Am from 11 M LiCl in 0.2 M HCl solution, whereas lanthanides remain in the raffinate. The actinides are subsequently backextracted with 5 M HCl. The separation factor between Am and Eu was 108 (Nash et al., 2006).

Research into the development of new extraction systems to achieve better extractability of Am and higher separation factors for americium−lanthanide separations has been continued.

Comparative evaluation of traditional extractants in LLEs. Acid dependence of the distribution coefficients for all elements, including actinides, were determined by many researchers, and data were compiled in the form of periodic tables. An example of such a data collection was published as the Table of Chromatographers (Lederer, 1992). Among the extractants discussed above, the distribution coefficients (*D*) for tetravalent actinides such as Pu(IV) are increasing in the following order: TBP/7 M HNO_3 < TTA/0.5 M HNO_3 < TOPO/1 M HNO_3 < TOA/4 M HNO_3 < H-DEHP/1 M HNO_3 starting from ~20 for the TBP and ending at ~ 2000 for the HDEHP system (Myasoedov et al., 1974). The retention sequence of the various extractants is the same for other tetravalent species such as Np(IV) and Th. Bigger differences in *D* values exist for different oxidation states. The distribution coefficients (*D*) for trivalent actinides (Am and Cm) are increasing in the following order: TTA/pH ≥ 4 < PMBP/0.1 M HNO_3 < TOPO/0.1 M HNO_3 < HDEHP/0.1 M HNO_3 < DDCP/12 M HNO_3 < CMPO-TBP/2M HNO_3 starting from ~1 for the TTA and ending at ~ 100 for the CMPO system (Myasoedov et al., 1974). For the hexavalent U(VI) species, the following order of *D* has been reported: TTA/1 M HNO_3 < TBP/4M HNO_3 < TOA/4 M HNO_3 < HDEHP/8 M HNO_3 < TOPO/1 M HNO_3 with the highest value of 800 in the latter case (Myasoedov et al., 1974). In the case of ion association complexes, *D* values can be increased by the addition of salting-out agents.

Although in many analytical applications, LLE has been replaced by more efficient chromatographic techniques. Solvent extraction is most appropriately applied to measurements by LS. Extractive scintillators containing solvent extractants discussed above and scintillation cocktails have been developed for the PERALS alpha LS spectrometers (Ensor, 1998; PERALS homepage). In the ETRAC procedure, Pu is extracted by a high-molecular-weight tertiary amine converted to nitrate form. The ALPHAEX system based on HDEHP extractant can be applied for the separation of Am. Uranium can also be extracted with HDEHP containing extractive scintillator (see above). Dazhu et al. (1991) used the extractive scintillator containing TOPO extractant in PPO−naphthalene−toluene scintillator cocktail for the rapid determination of U, Pu, Am, and Cm in the nuclear fuel cycle and in environmental samples. A detailed list of extractive scintillators is found in Chapter 6, Volume 1 of this book.

New extractants. New extractants of high distribution coefficient for trivalent actinides and lanthanides (Nash et al., 2006) are under extensive investigation for the purposes of HLW treatment. **Malonamides** are the non−phosphorous-containing alternatives of CMPO. In the DIAMEX (DIAMide EXtraction) process, dimethyldibutyltetradecylmalonamide (DMDBTDMA) is used for the extraction of Am, Cm, and lanthanides from HLW solutions. In the modified DIAMEX process, a new malonamide, i.e., dimethyldioctylhexylethoxy-malonamide (DMDOHEMA) was used. **Diglycolamides**, i.e., N,N,N′,N′-tetradecyl-3-oxapentanediamide (TODGA) and N,N,N′,N′-tetradecyl-3-oxapentanediamide (TDDGA), were found to have extremely high distribution coefficients for Am nitrate complexes from concentrated nitric acid solutions. TODGA has also been successfully used for the separation of trivalent actinides and lanthanides from HLW process solutions.

For the selective extraction of trivalent actinides from lanthanides, N-donor and S-donor extractants have been developed. With the N-donor molecule bis(5,6-dialkyl-1,2,4-triazine-3-yl)pyridine (**BTP derivative**), high Am/Eu separation factor (>100) from concentrated HNO_3 solution was achieved. The SANEX (Selective ActiNide EXtraction) process is based on the selective separation of trivalent actinides from lanthanides using BTP derivatives. Research into the development of other BTP derivatives and modified SANEX processes is in progress. In the GANEX (Group ActiNide Extraction) process, the majority of U is separated first with a monoamide extractant, and then minor actinides are extracted together by a combined DIAMEX-SANEX process.

The most well-known S-donor compound is bis-(2,4,4-trimethyl)pentyldithiophosphinic acid, also called **Cyanex 301**. Zhu et al. (1996) received a separation factor for the extraction of Am(III)/Eu(III) from HNO_3 solution over 5000. The technological application of Cyanex 301 is in progress.

At present, these extractants exhibit the greatest affinity for trivalent actinides. Crown ethers and calixarene compounds are also under investigation.

Methods have been developed for the separation of Am from Cm for partitioning of radioactive wastes. This is the most difficult separation within spent fuel reprocessing because of the high similarity of the two elements. Theoretically, Am can be oxidized to Am(VI), whereas Cm remains tervalent, and then they could be easily separated, but in practice Am(VI) is very unstable. The Am/Cm separation factor in the DIAMEX process using DMDOHEMA is small (1.6) for effective separation. The highest separation factor for Am/Cm (6−10) has been achieved with a dithiophosphinic acid (bis(chlorphenyl)dithiophosphinic acid), and the separation was in the LUCA (**Lan**thaniden **U**nd **C**urium **A**mericium Trennung) process successfully tested at the Forschungszentrum Jülich (Modolo, 2014).

In analytical practice, anion exchange resins and EC materials are favored over solvent extraction. New EC materials often contain the selective reagents developed for technological purposes (e.g., TEVA, TRU, DGA resins), but unfortunately, many of the extractants of high selectivity (malonamides, BTPs) have not found analytical applications.

b. Chemical separation of Po by liquid−liquid extraction

Po can be extracted into tributyl phosphate from strong HCl solutions (Figgins, 1961). Other organic agents forming complexes include EDTA and diethyldithiocarbamate (DDTC). When complexed with DDTC, Po can be extracted into chlorinated hydrocarbons, such as chloroform from HCl solution over a wide range of acid strengths. This is useful for the separation of Po from Pb and Bi, as Po is extracted from HCl solutions of higher acid concentration than either Bi or Pb (Smithson et al., 1979; Wai and Lo, 1982; Kim et al., 2009). ^{210}Po can be effectively separated from lead samples using a DDTC−toluene extraction with EDTA and citrate as masking agents for the Pb ions (Uesugi et al., 2010).

Other extractants reported in the literature include TOPO in toluene (Jia et al., 2004), di-n-octyl sulfide (Johansson and Skarnemark, 2001), triisooctylamine (TIOA) in xylene solution (Chen et al., 2001c), APDC in MIBK (Shannon and Orren, 1970), "Aliquot 336" (Ibrahim and Whicker, 1987), and trilaurylamine (TLA) into xylenes (Roseberry and Scott, 1985).

2. *Separations by ion exchange*

The most frequently used **cation exchange resins** are the strongly acidic sulfonated resins containing—SO_3^- groups. Less acidic phosphorylated and carboxylated resins are also available. The positively charged counterions can be replaced by any other cations. The ion exchange affinity of the cations increases with increasing charge and decreasing hydrated ion radius. All actinide species, i.e., An^{4+}, An^{3+}, AnO_2^{2+}, AnO_2^+ especially the tetra- and trivalent species, are retained well (high D) at low acidities, and they are removed from the resin by concentrated acids, but no special selectivity against other cations was observed. Ion exchangers are typically used in the form of chromatographic columns assuring much better separation than batch techniques.

Cation exchangers have been used for the preconcentration of actinides from high-volume dilute acidic solutions. Fjeld et al. (2005) used a cation resin column for the preconcentration of actinides from groundwater and liquid radioactive wastes. Cation exchangers have also been used for the chromatographic separation of actinides in ion chromatographs (IC) either alone or together with anion exchangers as mixed-bed resins. IC separation fits well to the purposes of separation of actinides prior to ICP-MS, but it is less adequate for alpha spectrometry due to small sample masses. Pu(III) and Am(III) were successfully separated on a cation exchange column TCC-II by Jernström (2006) using different complex forming eluents and detecting the sources by alpha spectrometry. Samples were loaded from dilute HCl solution, Pu was stripped with dipicolinic acid, and Am eluted with oxalic acid. Jernström (2006) also used mixed-bed ion exchanger CSA5-CGA5 column for IC of Pu and Am.

Separation of Am from lanthanides using cation exchangers. Trivalent Am^{3+} cations are usually retained well (high *D*) at low acidities. Several other elements coadsorb with Am, including lanthanides. As the concentration of HCl increases, the distribution coefficient for Am declines, whereas those for lanthanides is smaller, which is revealed by increasing separation factors.

One of the oldest techniques for the separation of Am from lanthanides is based on cation exchange chromatography using complexing agents, e.g., hydroxycarboxylic acids. Ammonium citrate and ammonium lactate have been used, and α-hydroxy-i-butyric acid (α-HIBA) proved to be the most effective eluent for the selective separation of Am. Choppin and Silva (1956) separated Am from the lanthanides on a jacketed (temperature controlled) cation exchange column using α-HIBA. Aminopolyacetic acids (e.g., EDTA and DTPA) have also been used as eluents. The disadvantages of their use are the low solubility and the slow establishment of the equilibrium.

Ageyev et al. (2005a,b) separated Am and Cm from lanthanides by cation exchange chromatography with gradient elution using 0.16−0.4 M α-HIBA after removal of Pu with anion exchange. Yields of Pu and Am were in the ranges of 60%−80%, 50%−70%, respectively, when 100−200 g of environmental samples exposed to the emissions from the Chernobyl accident were analyzed. A similar technique has been used for the purification of the Am fraction by Berlioz et al. (2005).

In **anion exchangers**, the functional groups are various amines. Strong basic anion exchangers with substituted quaternary amines strongly retain tetra- and hexavalent actinides from concentrated nitric and hydrochloric acid solutions. The selectivity of the anion resin for actinides is high because only a few other cations (e.g., Fe in HCl) are able to form strong anionic complexes in acids. For example, the distribution coefficient (*D*) for Pu(IV) is as high as 10^4. Fig. 5.38 shows the distribution coefficient (*D*) for various elements in nitric and hydrochloric acids.

Uranium and thorium. Thorium can be strongly retained on an anion resin from 8 M HNO_3 whereas U as U(VI) is better retained from 9 M HCl with an acceptable high distribution coefficient (*D*). Uranium can be easily separated from Th with anion resins, but U and Th cannot be retained well from the same solution. These properties have been utilized in the analysis of uranium isotopes in water, urine, and soil samples according to the ASTM standard test methods **D3972-09, C1473-11,** and **C1000-11**, respectively. Uranium is retained on a strongly basic anion exchanger from 8 to 9 M HCl solution after sample destruction and preconcentration with ferrous hydroxide precipitate. Uranium can be stripped from the column with 0.1−1 M HCl. In case of soil samples, Fe is removed from the load solution before separation by extraction of the chloride complexes with di-i-propylether. The retention of Pa from HCl or HF solutions on anion exchange resins is not quantitative (Dalvi and Verma, 2017).

Plutonium. A convenient way to separate Pu selectively from other ions is to load tetravalent Pu(IV) from 8 M HNO_3 on an anion resin and remove the possible interferences, primarily Th, by washing the column with 9 M HCl, because this does not form chloride complexes. Plutonium can be stripping as Pu(III) with 9 M HCl using a reducing agent, e.g., iodide. This is the procedure that became the basis of many methods. Because of the high selectivity of the resin for Pu(IV), very often a single chromatographic separation provides sufficient purity for the analysis. Basically the same procedure was already used in the 1960s (Coleman, 1965) and has been used by many authors recently (Moreno et al.,1997; Qu et al., 1998; Pilviö, 1998; Michel et al., 2008; Komosa and Chibowski, 2002; Solatie, 2002; Hrnecek et al., 2002; Vioque et al., 2002; Berlioz et al. 2005; LaRosa et al., 2005; Ageyev et al., 2005a,b; Giardina et al., 2006; Lee et al., 2000; Jakopic et al., 2007; Michel et al., 2008, 2008).

The standard procedures for Pu analysis by α-spectrometry are usually also based on anion exchange separation. According to the **ASTM C 1001-05 (2005)** standard (Standard Test Method for Radiochemical Determination of Plutonium in Soil by Alpha Spectroscopy), 10−50 g of soil is destructed with a mixture of nitric, hydrofluoric, and hydrochloric acids. Pu is isolated by anion exchange, followed by alpha source preparation. According to the standard **ASTM D3865-02 (2002)** "Test Method for Pu in Water", Pu is coprecipitated with iron as ferric hydroxide, the coprecipitated Pu is dissolved, and the solution is adjusted to 8 M in HNO_3 for anion exchange separation, and then the alpha source is prepared. The recommended procedures of the International Atomic Energy Agency

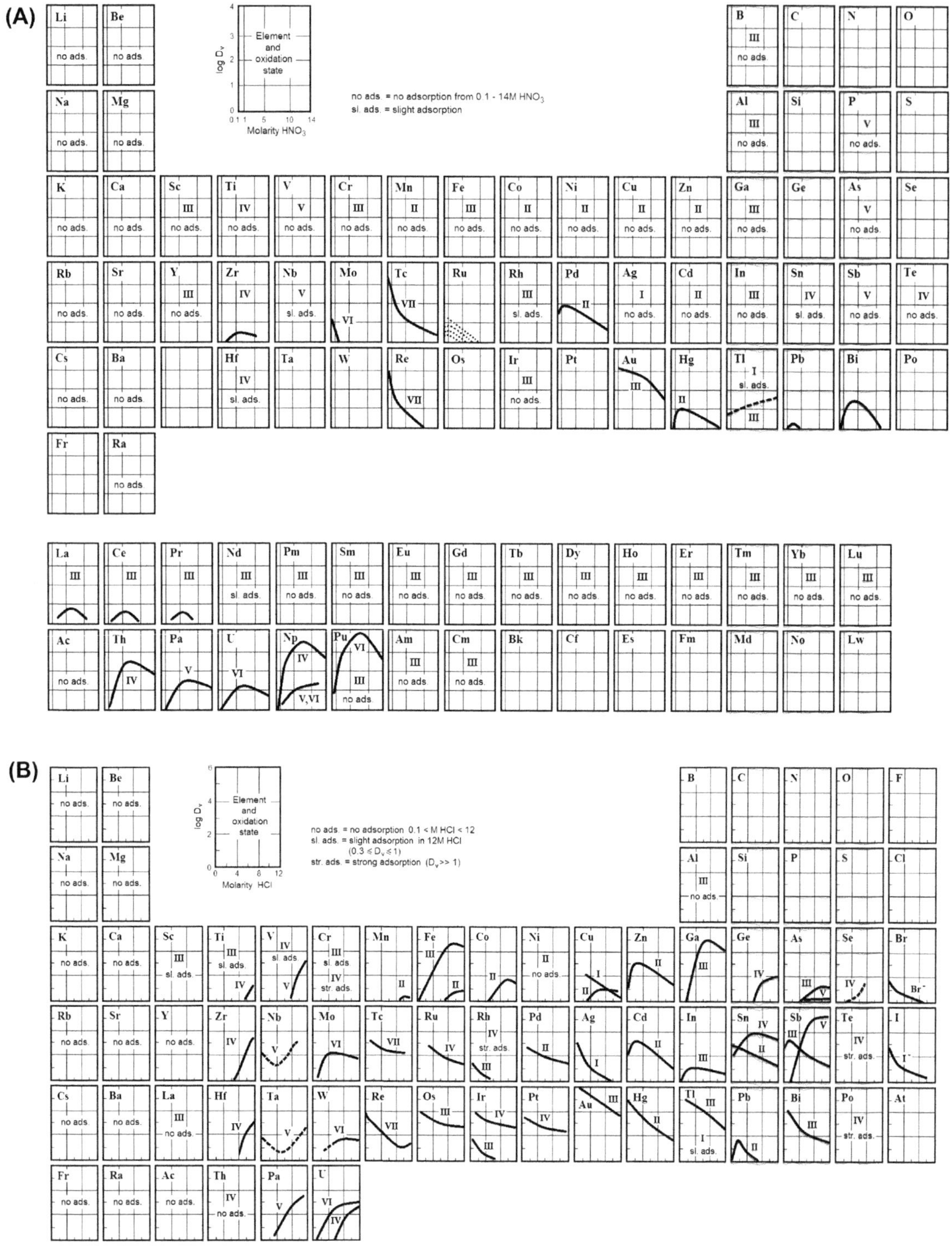

FIGURE 5.38 Distribution coefficient (*D*) for various elements of the periodic chart on a strong basic anion exchange resin in (A) nitric and (B) hydrochloric acids. *Lederer (1992), reprinted by permission of John Wiley and Sons © 1992.*

described in the Technical Report Series 295 from 1989 and in the most recent publication from 2012 (IAEA, 2012) are also based on the separation of Pu by anion exchange chromatography. The latest ISO standard **ISO 18589-4** and **BS 07/30047016 DC** (April 2007) on "Measurement of Radioactivity in the Environment Soil Part 4. Measurement of Pu Isotopes by α-Spectrometry" recommends the optional use of HDEHP extraction, strong basic anion resin chromatography, and EC using TRU resin (see later) for Pu determination.

Many examples exist on the use of anion exchange resins for Pu separation from HCl solution. Samples containing Pu(IV) are loaded from 9 M HCl on the column, and after washing, Pu is stripped using a reducing agent. This procedure has the disadvantage that the anionic ferric chloride complex is also retained by the anion resin. Similar procedures were used from the 1960s up to the present. Lee et al. (2008) studied the oxidation states of Pu in HCl solutions and carried out Pu separation from 9 M HCl solution in the presence of $NaNO_2$. Pu was stripped with 0.36 M HCl +0.01 M HF, and the Pu recovery as a mixture of Pu(IV) and Pu(VI) was >75%.

Neptunium. Anion resins, from both nitric and hydrochloric acid solutions, are used for the selective separation of Np. In reducing media, when Np is turned to the tetravalent form and Pu is reduced to the trivalent state, the separation of Np from Pu is easily accomplished. This is the most common method of selective ^{237}Np separation. Np was retained as Np(IV) together with Th on anion resin from 8 M HNO_3 in reducing media in the procedures of Niese and Niese (1985) and Germian and Pinte (1990). According to the procedures of Koyoma et al. (1998), Michel et al. (1999), Moreno et al. (1997), and La Rosa et al. (2005), Np(IV), Pu(IV), and Th were retained also from 8 M HNO_3 followed by the selective stripping of the actinides. The tetravalent oxidation states were adjusted by applying reduction/oxidation cycles with or without the addition of $NaNO_2$ or H_2O_2. Np(IV) and Pu(IV) together with U can be retained on anion resin from 9–10 M HCl load solution. Then, Pu is stripped selectively with 9–10 M HCl/0.1M NH_4I followed by stripping of Np with more dilute 1–4 M HCl. In the procedure of Rosner et al. (1993), fractionation of Np between the wash and the Np strip solutions was observed. Sumiya et al. (1994) and, more recently, Joe et al. (2007) successfully used similar procedures. To achieve higher DFs, the anion exchange procedures are sometimes repeated (Beasley et al., 1998), or additional purification steps using TTA, TEVA, TOPO extraction (see later) are included. Anion exchange resin has been used to separate Np from U from 4 M acetic acid solution when Np is released while U is retained (Winkel et al., 1971).

If Np and Pu are to be detected in the same fraction, they can be loaded as tetravalent species and stripped also together (Kim et al., 1992; Qiao et al., 2011).

Americium. Am(III) does not form strong complexes with nitrate or chloride ions and is not well retained on anion resins. In concentrated HCl solutions (>11 M), there is an increase in the retention of Am on anion exchangers, whereas lanthanides are adsorbed less. However, the difference is not as high as in the case of cation exchangers. Thus, the separation of Am from lanthanides is less successful under these condition.

Separation of Am from lanthanides using anion exchangers. The presence of alcohol in the media enhances the anion exchange of Am especially in nitric and sulfuric acids. Guseva et al. (1976) measured distribution coefficients as high as 5000 from 0.5 M HNO_3 in 95% methanol. Lighter lanthanides also have high distribution coefficients, whereas heavy lanthanides do not. In a HCl-methanol system, the anionic complexes of Am together with the heavy lanthanides are well retained by the resin, whereas light lanthanides are less retained. Both the HNO_3—methanol and the HCl—methanol systems, and preferably, the combination of the two systems, can be used for Am–lanthanide separation. A mixture of 0.5–1.0 M HNO_3 and 90–96% methanol was reported as the load and wash solutions by many authors (Holm and Fukai, 1976; Ballestra and Fukai, 1983, etc.) followed by stripping of the best retained Am with a mixture of 0.5–1.5 M HNO_3 and 70–86% methanol (Jernström, 2006) or simply with HCl (Livens and Singleton, 1989) or HNO_3 (Bains and Warwick, 1993). Methanol has been replaced with ethanol with similarly good results.

Kraus et al. (1955) found that many elements have higher distribution coefficients in LiCl solutions than in HCl. The sorption of Am on anion exchangers increases with increase in LiCl concentration, whereas the sorption of lanthanides varies only slightly favoring Am–lanthanide separation. The separation is further improved if alcohol is added to the LiCl solution, e.g., a good separation was achieved with 8 M LiCl and 40% methanol on Dowex 1×8 anion exchange resin by Guseva et al. (1976).

The group separation of trivalent actinides and lanthanides can be carried out with concentrated (>8 M) NH_4SCN solution as eluent (Surls and Choppin, 1957). Lanthanides are less strongly retained on an anion column in the SCN form than Am. The Am can be stripped with 4 M HCl. Bojanowski et al. (2005) and Lovett et al. (1990) applied this method successfully for the analysis of marine environmental samples.

Guseva and Tikhomirova (1968) showed that the best separation can be achieved with 1–2 M NH_4SCN solution containing 50–70% methanol, ethanol, or propanol. This procedure became the basis of many methods where Am–lanthanide separation was necessary. Holm et al. (1979) used this procedure for the final purification of the Am fraction from large amounts of environmental samples (200 L seawater, 100 g sediment, 500 g biological material) after the removal of the interfering components. The final residue containing Am and lanthanides was dissolved in a few milliliters of 1 M HNO_3–93% methanol and was loaded on an anion exchange resin column. The column was washed with dilute HCl–NH_4SCN–methanol solution to remove lanthanides, and Am together with Cm was stripped using 1.5 M HCl–80% methanol. The overall chemical recovery varied between 40% and 80%. A similar procedure was followed by Yamato (1982) to determine Pu and Am from 50 g of soil. In the final purification of Am, 1 M HNO_3–93% methanol, 0.1 M HCl–0.5 M NH_4SCN–80% methanol, and 1.5 M HCl–86% methanol solutions were used as load, wash, and strip solutions, respectively, and 75%–92% chemical recoveries for Am were reported for the whole combined procedure.

The procedure has been successfully applied by many laboratories for Am–lanthanide separation. It is the recommended procedure in the laboratories of the IAEA, both in Monaco and in Seibersdorf (LaRosa et al., 1989, 2005). It has been used by Jakopic et al. (2007) for the analysis of environmental samples.

Polonium. Ion exchange resins are less commonly used nowadays for Po separations than methods based on LLE or EC. Polonium is strongly adsorbed onto anion exchange resin across a wide range of acid concentrations. Although this can be used for effective separation, the approach requires large elution volumes (Reischmann et al., 1984; Skwarzec et al., 2001). Cation exchange resins have been used to adsorb Po from dilute acid solutions, with elution by 7M HCl (Reischmann et al., 1984).

Radium. Strong basic anion exchange resins do not retain the group 2 elements (Fig. 5.38) but may be used to remove other alpha emitters, major cations, and/or anions, such as sulfate (Gleason, 1980) or chromate (Yamamoto et al., 1989). Hence, use of anion exchange resins may be preparatory to a final cleanup step for Ra.

Cation exchange resins may be used for preconcentration of Ra and separation from the bulk matrix for sample digests or aqueous samples. The digest is taken up in dilute HCl or HNO_3 for loading onto the resin and washed with dilute acid before eluting Ra with higher concentrations of these acids (e.g., Volpe et al., 1991; Lawrie et al., 2000; Crespo, 2000). However, to effectively separate Ra from Ba and Sr on a cation exchange resin, it is generally necessary to use complexing agents.

Gleason (1980) developed a separation method using cyclohexane-diaminetetraacetic acid (CyDTA), which has been a basis for several methods reported since then (Yamamoto et al., 1989; Rodriguez-Alvarez and Sanchez, 1995; Jia et al., 2006). Hancock and Martin (1991) used EDTA and ammonium acetate on 50WX12 cation exchange resin before washing and elution of Ra. Volpe et al. (1991) also used EDTA solution with cation exchange resin columns.

3. Separations by extraction chromatography

In EC, also called solid-state extraction and reversed-phase partition chromatography, liquid extractants are adsorbed onto the surface of inert solid support material. The theory of EC was described more than 30 years ago in the book of Braun and Gherseni (1975). EC has the following advantages compared with solvent extraction and ion exchange:

- the extraction process takes place in the thin surface layer allowing good contact of the reagents and fast exchange kinetics compared with ion exchangers,
- as a result of the chromatographic technique, more effective separation is achieved than in the batch technique in which LLE is performed,
- less reagents and chemicals are used than in ion exchange or LLC;
- less hazardous waste is produced, and
- the whole process is more economic.

Attempts have been made to replace solvent extraction by EC from the end of the 1960s.

a. Extraction chromatography of actinides

The chelating extractant, **HDEHP** showing high distribution coefficient for Am from dilute acid solutions, has been prepared as a solid EC material on various supports to serve as stationary phase. HDEHP was prepared on hydrophobic support by Horwitz and Bloomquist (1975). Jia et al. (1997) used microporous polythene support. The 0.01 M HNO_3 load solution containing Am was passed through the column and washed with 0.1 M HNO_3, and Am was stripped with 0.07 M DTPA–1 M lactic acid. Pu and Po were separated before using supported TOA columns. Jia et al. (1998) determined Pu, Sr, and Am in a sequential separation procedure. First Pu and Sr were separated by EC, after which Am was concentrated on an HDEHP column. Americium was eluted with a 0.07 M DTPA–1 M lactic acid solution to separate it from lanthanides. The Am fraction was purified by an extraction with PMBP/TOPO. The chemical recovery for Am was 58%, and the source was free from Fe, Po, and lanthanide contamination. Desideri et al. (2002) determined actinides in depleted uranium. After EC separation of the actinides using supported TOA columns, the Am fraction was cleaned using a supported HDEHP column (pH 2.4). Arginelli et al. (2005) prepared HDEHP on polythene support and loaded the pretreated urine sample from dilute acidic solution (pH 2) on the column; Am was stripped with 4 M HCl. Supported HDEHP has become commercially available as Ln resin produced by Eichrom Co. McAlister and Horwitz (2007) characterized the Ln resin by measuring the capacity factors k' of various elements (see Fig. 5.39).

Gleisberg and Köhler (2002) separated Np from uranyl nitrate solution using Ln resin column. Np and Pu species were retained from 10 M HCl load while U entered into the effluent.

Meanwhile, the interest for Am separation by HDEHP has declined due to the development of more efficient EC materials (see TRU resin later).

Supported TOPO on microthene (microporous polyethylene) and TOA on icorene (microporous polyethylene) were prepared by Jia et al. (1998) from the 1970s.

Supported TOPO on Kieselgur was used by Afsar and Schüttelkopt (1988). Americium was loaded onto the column from 0.1 M HNO_3, washed with 0.1 M HNO_3, and

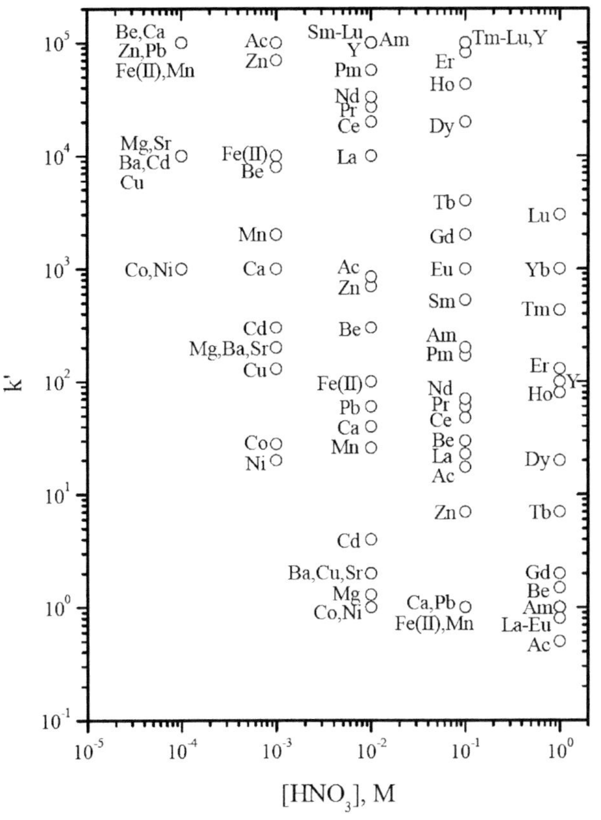

FIGURE 5.39 Uptake of various elements by Ln resin from HNO_3 solutions. *McAlister and Horwitz (2007), reprinted by permission of Eichrom Technologies LLC © 2007.*

stripped with 8 M HNO_3. The supported TOPO was applied for the separation and purification of U and Pu in safeguard samples (ISO 15366-1 and ISO 15366-2). Solatie et al. (2001) investigated the separation/purification methods for the determination of ^{232}U and ^{236}Pu in spent nuclear fuels by combining TOPO, anion exchange resin, and UTEVA EC.

Delle Site et al. (1973) separated actinides from urine samples on microthene-supported TOPO loaded from 4−6 M HNO_3. Actinides were stripped consecutively with 0.3 M H_2SO_4 (Th), 6 M HCl−0.2 M HF (Pa), 1 M HF (U), 6 M HCl + Cl_2 (Np), and 6 M HCl + 0.1 M HI (Pu). Americium was separated on a microthene-supported HDEHP column. Acceptable high recoveries (>70%) were obtained with the exception of Am even in the presence of DTPA complexing agent.

EC using supported TOPO was successfully applied for Pu determination by Cozzella et al. (2002).

Supported liquid anion exchangers were used for the separation of Np from environmental samples in the procedure of Yan-Qin et al. (2001). Neptunium was retained on a Teflon-coated TOA from 2 M HNO_3 and stripped with hot 0.02 M oxalic acid/0.16 M HNO_3. High recovery (>80%) and high DF against U (>10^4) were obtained. In the most recent work of Desideri et al. (2011), Am, Cm, Pu, Np, and U were separated from small volumes of water samples and determined by alpha spectrometry. Samples were loaded onto macroporous PE-supported TOA from 9 M HCl solution, and Am, Pu, Np, and U were sequentially stripped. Recoveries higher than 74% were reported for each actinide.

Horwitz et al. (1966) reported a method that can separate Am from Cm on a supported (hydrophobic diatomaceous earth) Aliquat 336 column using 8 M $LiNO_3$−0.01 M HNO_3 for loading and 3.5 M $LiNO_3$−0.01 M HNO_3 solution to separate Am from Cm. More than 99% of Am and Cm were recovered in a radiochemically pure state (<0.1% impurity). The method was also applied to separate milligram quantities of Am and Cm (Horwitz et al., 1967).

Giardina et al. (2006) used microthene-supported TOPO, icorene-supported TOA, anion exchange resin, and TRU resin (see later) for the separation of Pu from urine samples followed by alpha spectrometric determination. Methods were compared with the standard procedure based on anion exchange chromatography. All methods gave good Pu recoveries (>80%), and DFs were acceptably high for naturally occurring radionuclides. Smaller DFs were obtained by TRU resin (discussed subsequently).

From the 1990s, a family of EC materials has been developed for the separation of actinides by Horwitz and colleagues at the Argonne National Laboratory, USA, and later, these materials became commercially available from EiChrom Co. and Triskem SAS. The EC materials have been carefully characterized by measuring the physical and chemical properties, distribution coefficients, extraction kinetics, and elution behavior and by developing separation procedures for special purposes. If a single resin is not capable of the separation of all the desired radionuclides because of limited selectivity, the use of tandem column arrangements is recommended, in which the effluent of one column serves as the load solution for the subsequent one. Because kinetics of the extraction is fast, the columns can be operated at higher flow rates than gravity flow; for this purpose, vacuum boxes have been proposed.

TEVA resin (Horwitz et al., 1995) dedicated to the separation of TEtraVAlent actinides is a supported quaternary amine-based (Aliquat 336) liquid anion exchanger where the support material is Amberchrom CG-71 ms. TEVA as an analog of strong basic anion exchangers is expected to be an excellent chromatographic material for the retention of Pu(IV) and Np(IV). The resin capacity factors (k') that are directly proportional to the distribution coefficients for Pu(IV) and Np(IV) in 6 M HNO_3 are 3×10^4 and 4×10^3, respectively, and are superior to other actinides. This is the reason why the TEVA resin has received widespread application in Pu and Np analysis since the first report of Horwitz in 1995 (see in Fig. 5.40).

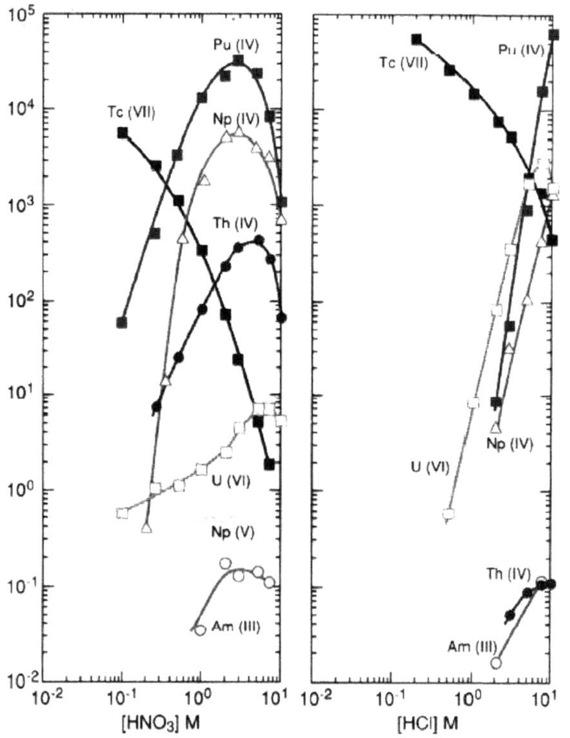

FIGURE 5.40 Capacity factors k' of various actinides on TEVA resin from HNO_3 and HCl solutions. *www.eichrom.com, reprinted by permission of Eichrom Technologies LLC©2012.*

Uranium and thorium: Capacity factors for Th and U(VI) on TEVA resin from HNO_3 solutions as shown in Fig. 5.40 are relatively low, and the selectivity for U and Th in concentrated HCl solutions is limited.

Plutonium: Varga et al. (2007c) used TEVA resin for the separation of Pu from environmental samples of 1–5 g after a CaF_2 coprecipitation. $NaNO_2$ was used to adjust the oxidation state of Pu(IV). The sample was loaded in 3 M HNO_3 and eluted with 6 M HCl, and Pu was stripped with 0.1 M HNO_3/0.1 M HF. The Pu recovery for the whole procedure, including alpha source preparation, was 72%–93%, and the DF for U was $>10^5$, which was also sufficient for ICP-MS measurement. Chamizo et al. (2008) determined Pu in environmental samples without preconcentration using TEVA chromatography. Pu(IV) was loaded on the column in 3 M HNO_3/$Al(NO_3)_3$, followed by washing with 6 M HCl and stripping with 0.5 M HCl. Recovery of Pu (>60%) and DF for U(>100) was somewhat poorer than in the previous method, probably due to the lack of a selective preconcentration. Several authors used TEVA successfully for the separation of Pu for ICP-MS measurements (see the review of Kim et al., 2004). The standard procedures for the determination of actinides in water (ACW13VBS, ACW16, ACW17) developed by EiChrom Technologies, Inc (available from www.eichrom.com) are based on the use of different column sets where Pu is separated by TEVA EC.

Neptunium: For the individual determination of ^{237}Np, samples are usually loaded on TEVA column from 2 to 3 M HNO_3 after reduction of Np to Np(IV), whereas Pu is in the nonretainable Pu(III) oxidation state. Neptunium can be stripped with dilute HNO_3–HF mixture. High Np yields were obtained in the procedures of Anton et al. (2006), Ayranov et al. (2005) and Maxwell and Jones (1998) analyzing soil, sediment, and nuclear materials, respectively.

If Np and Pu are to be determined together, the sample has to be loaded on TEVA after adjusting the Np(IV) and Pu(IV) oxidation states (see earlier). Kenna et al. (2002, 2009) obtained high yields for both elements analyzing sediment samples, whereas Np recoveries varied between 19% and 54% in the procedure of LaRosa et al. (2008). In the rapid separation procedure for the determination of ^{237}Np and Pu isotopes in large soil samples, Maxwell et al. (2011b) also found that Np recoveries were reduced when the matrix content increased (sample masses were higher than 30 g). Plutonium recoveries were high (>82%) and independent of matrix effect. In the standard procedure for the determination of Th and Np in water (ACW08) developed by EiChrom Technologies (1995), quantitative recoveries have been attained using the simple TEVA procedure.

The standard procedure for the determination of ^{237}Np in soil (ASTM C 1475-05) is based on the use of TEVA column. The Np(IV) oxidation state is adjusted using ferrous sulfamate and $NaNO_2$. Sample is loaded from 2.5 M HNO_3/0.5 M $Al(NO_3)_3$, and Np is stripped with 0.02 M HNO_3/0.02 M HF.

Americium: Am(III) is only slightly retained on TEVA resin from nitric or hydrochloric acid solutions. The maximum of the distribution coefficient is 0.1. However, TEVA resin, analog to anion exchange resins, can be used for the separation of Am from lanthanides, which was performed by loading the sample in a 2 M NH_4SCN–0.1 M formic acid solution followed by the elution of lanthanides with 1 M NH_4SCN–0.1 M HCOOH and stripping Am (together with Cm) with 0.25–2 M HCl (Horwitz et al., 1995). Good separation was achieved. A pitfall in the procedure is the same as in case of the classical Am–lanthanide separation using anion exchanger, i.e., it is difficult to dissolve larger sample amounts in a small volume of load solution. The method is applied after the separation of Am from most of the sample components. The same method was applied by Berne (1995) in the standard procedure of the Environmental Measurement Laboratory (EML), USA, for the determination of Am in soil after leaching the actinides, removal of Pu and Th with anion exchange resin, concentrating Am from the effluent with Ca oxalate coprecipitation and separating Am from most of the interferences on a TRU column (see later). The procedure was applied for the purification of the Am fraction by Michel et al. (2008) and Maxwell (2008). Jernström (2006) separated Pu(III) and Am(III) on TEVA resin, whereby plutonium was adsorbed quantitatively (98.5%−99.1%) by the resin when loaded and washed with a mixture of 2 M ammonium thiocyanate in 0.1 M formic acid. Quantitative elution of Pu(III) from the resin was achieved with 0.25 M HCl. Sorption of Am(III) on TEVA resin was quantitative in the same mixture, and americium was eluted from the resin with 0.25 M HCl.

UTEVA resin (Horwitz, 1992) is a dipentyl-pentyl phosphonate impregnated Amberlite XAD-7 EC material that was designed to separate U and TEtraVAlent actinides. Because the resin contains a phosphonate compound, actinides are retained at higher distribution coefficients than on phosphates, e.g., TBP. The k' capacity factor for Pu(IV), Np(IV), U(VI), and Th(IV) are about 10^3, 4×10^2, 4×10^2, and 2×10^2 respectively, in 8 M HNO_3, which assures acceptable retention and easy removal of the tetra- and also the hexavalent actinides (see the capacity factors of actinides on UTEVA from HNO_3 and HCl solutions in Fig. 5.41).

Apostolidis et al. (1998) developed a procedure for the determination of Pu and U from reprocessing solution using a single UTEVA column. The oxidation state of Pu was adjusted with H_2O_2, and the sample in 4M HNO_3 was loaded on the column, and Pu was subsequently eluted with dilute HNO_3 containing hydroxyl amine and ascorbic acid as reducing agents. Finally, U was stripped with bioxalate solution. A high recovery (>95%) was achieved for Pu with a DF for U of about 10^4.

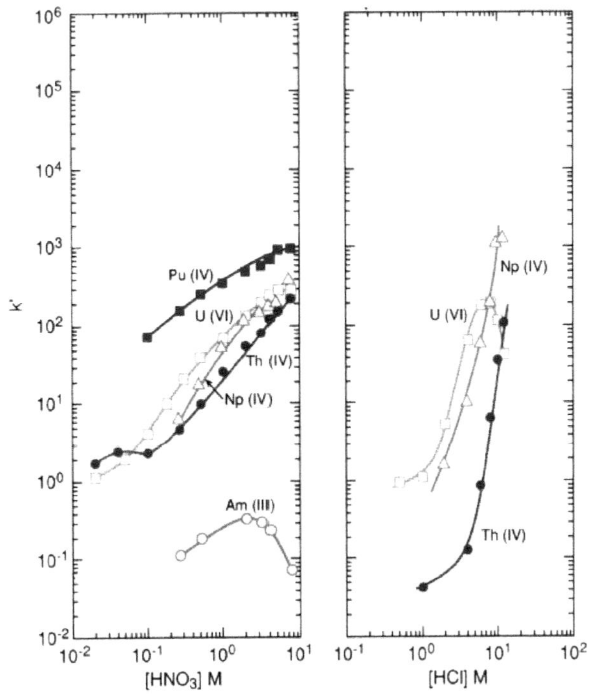

FIGURE 5.41 Capacity factors k' of various actinides on UTEVA resin from HNO_3 and HCl solutions. *www.eichrom.com, reprinted by permission of Eichrom Technologies LLC©2012.*

Pu and Np were determined by UTEVA EC in a combined procedure with the analysis of Th and U from environmental and nuclear waste samples by Vajda et al. (2003) after preconcentration of the actinides in reduced form with ferrous hydroxide. The sample solution was loaded in 8 M HNO_3, which contained ferric nitrate as salting-out agent. Pu was stripped with 9 M HCl/0.1 M NH_4I, as Pu(III), Th, and Np were stripped together with 4 M HCl, and U was eluted finally with dilute acid. Pu recoveries varied in the range of 69%−92%, and the DF was high enough to obtain Pu alpha sources without interferences. Osvath et al. (2009, 2010) improved the procedure by using a strong oxidizing agent to form Np(VI) and Pu(VI), which are retained by the resin together with U, and they purified the Np−Th strip solution by EC separation on a second UTEVA column from 9 M HCl solution. In the modified procedure, Pu and Np recoveries were 65%−82% and 66%−93%, respectively, and the Np source could be analyzed by ICP-MS, whereas Pu and U sources were measured by alpha spectrometry.

Guerin et al. (2010) determined the k' values of Np(VI) on various EC materials, including TEVA, UTEVA, TRU, and DGA from HNO_3 and HCl solutions. The Np oxidation state was adjusted with $KBrO_3$, $KMnO_4$, $K_2S_2O_8$, Ag_2O, and O_3. A simple method using two UTEVA columns was used to detect Np in soil and sediment samples. Np recovery was 70%.

In the rapid procedure of Ohtsuka et al. (2006), UTEVA is used in an ion chromatograph directly connected to ICP-MS. The authors claim that the complete analysis of an environmental sample (1 g), including destruction by fusion, is performed within 60 min. Morgenstern et al. (2002) used UTEVA for the separation of U, Pu, Np, and Am for the determination of U and Pu in safeguarding nuclear materials.

TRU resin (Horwitz et al., 1993; Berne, 1995) is a chromatographic material comprised of a tri-n-butyl phosphate (TBP) solution of the bifunctional organophosphorous extractant octyl(phenyl)-N,N-di(iso-butyl)-carbamoyl-methyl-phosphine oxide (CMPO) adsorbed on an inert polymeric substrate (Amberlite XAD-7). It was designed to separate all transuranium nuclides, including trivalent Am and Cm besides the tetra- and hexavalent species. Extremely high capacity factors were measured on TRU resin in concentrated nitric acid solutions, k' is higher than 10^5, 5×10^4, 10^4, and 10^3 for Pu (IV), Np(IV),Th, and U(VI), respectively, at HNO_3 concentrations above 1 M, and for Am, the k' is about 100. Unfortunately, Fe^{3+} present in the sample is also retained and competes for actinide-bonding sites, especially for those occupied by the less strongly retained Am (Cm). Reduction of Fe to Fe^{2+} helps decrease the interference. The capacity factors of actinides on TRU from HNO_3 and HCl solutions are shown in Fig. 5.42.

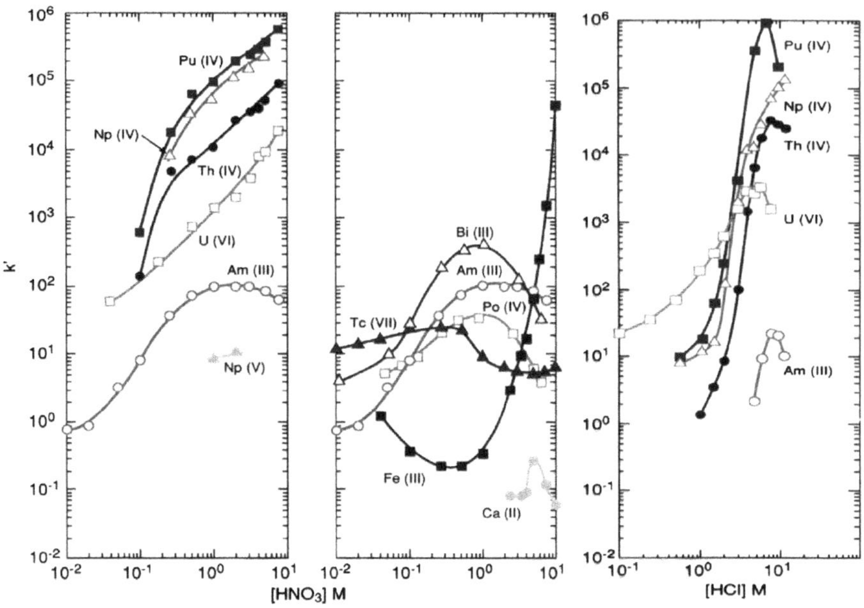

FIGURE 5.42 Capacity factors k' of various actinides on TRU resin from HNO_3 and HCl solutions. TRU, transuranium nuclide. *www.eichrom.com, reprinted by permission of Eichrom Technologies LLC©2012.*

TRU resin has recently become a popular tool for the separation of Am, but it can be used also for the retention of other actinides either for preconcentration or also for analytical separations. It seems to be an ideal material for the simultaneous separation of all actinides. TRU resin replaced almost exclusively all the methods used for concentration of Am in the past, i.e., chelating extractants and cation exchangers. Recently, several new molecules have been synthesized and have proven to be superior to TRU material regarding the distribution coefficient for Am or even separation factors, but the development of analytical procedures has a time-lag and applications are scarce, yet. Due to the moderately high distribution coefficient for Am on TRU and the various interferences from trivalent ions, e.g., Fe and lanthanides, and from matrix components, the separation procedures cannot avoid adequate preconcentration of Am.

Berne (1995) determined Am, Pu, and U in air filter and water samples. The sample was loaded in 8 M HNO_3/ $NaNO_2$. The Am was eluted with 0.025 M HNO_3, and Pu and U together with bioxalate. Because of cross-contaminations, it was recommended to separate Pu first on an anion exchange resin and use the TRU resin for the purification of Am only. This procedure was followed by many laboratories later, including the Environmental Measurements Laboratory, at the IAEA Laboratories at Seibersdorf and Monaco (Moreno et al., 1997, 1998; IAEA, 2012; LaRosa et al., 1989, 2001, 2005).

Horwitz et al. (1993) recommended a procedure for the sequential separation of actinides using a single TRU column. Actinides were loaded in 2 M HNO_3 after reduction when Pu was present as Pu(III). The Am was eluted with 4 M HCl, and Pu was eluted with 4 M HCl/0.1 M hydroquinone, followed by stripping of Th, Np, and U with 1.5 M HCl, 1 M HCl/0.03 M oxalic acid, and 0.1 M ammonium bioxalate, respectively. This procedure was slightly modified and applied for the rapid determination of Am, Pu, Th, and U in small environmental samples (1g) by Vajda et al. (2009). After $LiBO_2$ fusion and CaF_2 coprecipitation, the oxidation state was adjusted with $NaNO_2$ turning Pu to Pu(IV). The sample was loaded in 2 M HNO_2. Am was stripped with 4 M HCl. Pu was eluted after fast on-column reduction with 4 M HCl/0.1 M $TiCl_3$. High recoveries were obtained both for Am (96%) and for Pu (85%), and no traces of contamination were detected in the Am and Pu alpha sources. The behavior of Np on TRU resin was also tested; however, a good separation from U was not attained. A sequential separation procedure for the determination of the actinides in power plant effluents was developed by Spry et al. (2000). In the 3 M HNO_3 load solution, Pu was present as Pu(III), and it was oxidized after retention on-column with $NaNO_2$. Am was stripped as above, Th was eluted with 4 M HCl/0.01 M HF, and Pu with 0.1 M bioxalate. Am and Pu recoveries were acceptably high (52% and 56%, respectively). The robust procedure for the determination of plutonium and americium in seawater developed by Sidhu (2003) is based on MnO_2/ $Fe(OH)_3$ preconcentration of the actinides and their separation using a TRU column. Pu as trivalent ion was loaded together with other actinides from 3 M HNO_3, and it was stripped after Am with 4 M HCl/0.02 M $TiCl_3$. High recoveries were achieved for both nuclides (Pu: 78%, Am: 85%). Similar procedures were applied by Olahova et al. (2005) and Gogorova and Matel (2007).

In the ASTM standard guide C1561-10 for the determination of Pu and Np in uranium hexafluoride and U-rich matrix, two TRU columns are used. U is separated by retention on the first column from a 0.1 M oxalic acid−2 M HNO_3 solution, and then other actinides are retained on the second column from 3 M HNO_3. Americium, Th, and Pu are selectively stripped with 4 M HCl, 1.5 M HCl, and 1 M HCl−0.1 M oxalic acid, respectively.

Separations using UTEVA-TRU tandem resin columns: Several multistage EC methods were proposed for the simultaneous separation of actinides, but the one that became the most frequently used is based on the tandem column set of UTEVA-TRU. Thakkar (2001, 2002) described the detailed procedure in 2001, as the rapid sequential separation of U, Th, Pu, and Am. Actinides were preconcentrated with calcium phosphate (pH = 9) and then dissolved in 3 M HNO_3/1 M $Al(NO_3)_3$. Pu was reduced to Pu(III) with ferrous sulfamate and ascorbic acid. The sample was loaded on the tandem column where UTEVA was the first column and TRU was the second. After loading and washing, the columns were separated. Th and U were removed from the UTEVA column eluting them subsequently with 5 M HCl/0.05 M oxalic acid, respectively. On the TRU column, Pu was oxidized to Pu(IV) with 2 M HNO_3/0.1 M $NaNO_2$, and then Am and Pu were stripped after each other with 4 M HCl and 0.1 M ammonium bioxalate, respectively. This procedure resulted in high recoveries for Pu (>80%) and Am (>72%) as well as for U (>79%), and also relatively high DFs (in the Pu fraction DF for U > 104). This simple and fast procedure was frequently used with minor changes in different laboratories (LaRosa et al., 2001; Varga et al., 2007a, 2007b; Jakopic et al., 2007 and Mellado et al., 2001).

DIPEX resin (Horwitz et al., 1997), also called *actinide* resin, is a neutral, bifunctional organophosphorous extractant, consisting of bis(2-ethylhexyl)methanediphosphonic acid supported on an inert polymeric substrate. It has extraordinary affinity for actinides from dilute acids. The capacity factor for Pu(IV) and Np(IV) from 0.1 M HCl solution is about 2×10^7 and 10^7, respectively, and the same for Am, the traditionally least extractable actinide is even one order of magnitude higher (2×10^8). Actinides together with the extractant can be recovered from the resin by alcohol stripping and wet oxidation. These properties can be exploited in the preconcentration of actinides from large volume samples or complex matrices. The direct analytical application of the resin may be limited by the incomplete removal of the actinides from the resin. Burnett et al. (1997) used DIPEX resin for the preconcentration of actinides from soil solutions and large-volume water samples. Actinides from soil samples after NaOH fusion, dissolution, and ferrous hydroxide scavenge were concentrated on a DIPEX column. Isopropanol was used to solubilize the extractant that retained the actinides followed by the oxidation of the organic component. Basic calcium phosphate was used to scavenge the actinides repeatedly, and the final separation was performed on a UTEVA-TRU tandem column setup as described below. Recoveries of Pu, Am, and U varied in a wide range (31%−91%, 22%−99%, 60%−85%, respectively). *Actinide* resin was used for the preconcentration of Pu and Am from human tissue followed by the separation of Pu with anion exchange resin by Qu et al. (1998). The overall recovery for Pu was high (>80%) assuring the adequacy of the preconcentration, as well. An interesting application of *actinide* resin for the determination of Pu has been proposed by Labrecque et al. (2013). In the cloud point extraction, the selective ligand extracting Pu is enclosed in a micelle shielding that was formed from ionic and nonionic surfactants in the presence of bromine. The method represents a green alternative to conventional LLE. The *actinide* resin was used for the extraction of ^{231}Pa from concentrated HCl solution in the presence of its daughter products (Sankhe et al., 2017).

DIPHONIX resin (Chiarizia et al., 2011) can be regarded as the chelating ion exchange resin analog of the DIPEX EC material. It contains geminally substituted diphosphonic acid groups chemically bound to a styrenic-based polymer matrix. It also contains strongly hydrophilic sulfonic acid groups in the same polymer network. Also, extremely high capacity factors were observed for the retention of all actinides, close to those of DIPEX from dilute acid, but several transition metals (e.g., Cr, Ni, Cu, Zn) are also retained. The removal of the actinides is more problematic due to the massive resin structure. Strong complexing agents can remove the actinides that have to be decomposed or the resin itself has to be destroyed. Analytical procedures have been proposed by the authors, but because of the difficulties in removing the actinides from the resin, they have not been applied in practice. Rosskopfova and Holkova (2005) reported the use of DIPHONIX resin for the preconcentration of actinides from large soil samples followed by their separation using different eluents, i.e., Am was stripped with 6 M HNO_3 and Pu with 4 M HCl/0.015 M $TiCl_3$. Recoveries for Am were high (70%, 78%) but varied for Pu (37%, 93%).

DGA resin (Horwitz et al., 2005) consists of N,N,N′,N′-tetraoctyldiglycolamide adsorbed onto Amberchrom CG-71 resin. DGA is known for its high selectivity for trivalent actinides and lanthanides from HLW technology. Like the DIPEX resin, DGA retains Am strongly; in concentrated HNO_3 solutions (>1M), the capacity factor of Am is higher than 10^4. High capacity factors were found also for Pu ($>10^3$) at high HNO_3 concentrations, but they were not significantly smaller at low acidities. DGA resin can be used advantageously for the separation and purification of Am without losses from various matrices, e.g., soil and water.

The capacity factors of actinides on DGA resin are shown in Fig. 5.43.

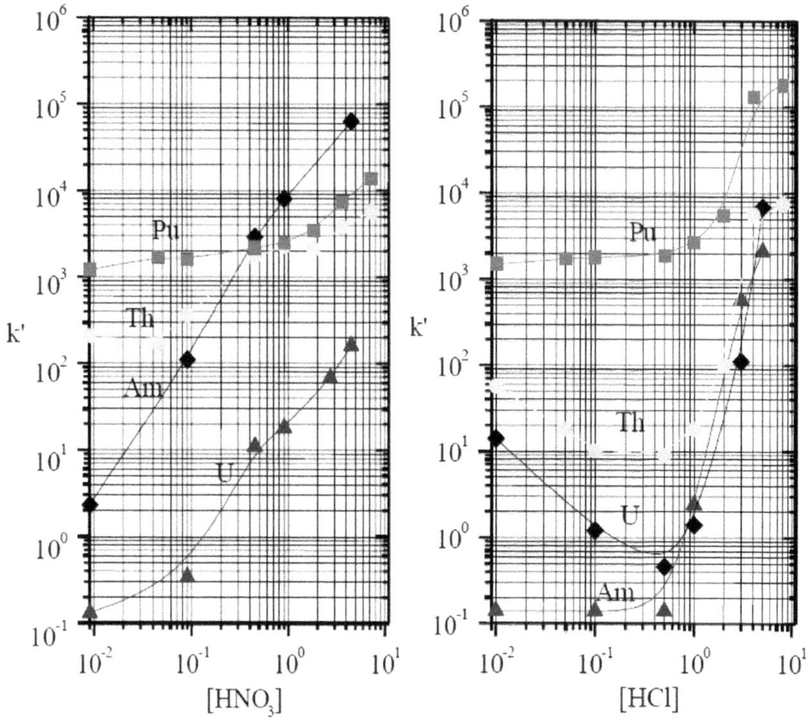

FIGURE 5.43 Capacity factors k' of various actinides on DGA resin from HNO_3 and HCl solutions. *www.eichrom.com, reprinted by permission of Eichrom Technologies LLC©2012.*

In many applications, DGA resin has been used for the separation of Am (together with lanthanides). Maxwell and Culligan (2006) and Maxwell (2008) developed a very promising rapid method for the determination of actinides in large soil samples up to 200 g. Samples are fused with NaOH, and actinides are preconcentrated with $Ce(OH)_3$ and CeF_3 coprecipitation. Stacked TEVA–TRU–DGA cartridges are loaded with 8 M HNO_3. TEVA serves for the separation of Th and Pu(IV), whereas Am (and Cm) are separated with the TRU–DGA multistage column. Americium originally retained on TRU was eluted with 4 M HCl and transferred to the DGA column. Americium was stripped from DGA with dilute (0.25 M) HCl solution. Am and Pu recoveries for the whole procedure were high, varying between 76% and 98% and between 61% and 92%, respectively. The method was adopted for the analysis of water samples (Maxwell, 2006) up to 1 L and extended for the determination of Sr radionuclides using Sr resin. The procedure combined the advantages offered by the rapid and effective destruction with NaOH fusion, the selective preconcentration of actinides with $Ce(OH)_3$ and CeF_3, and the EC separation using highly specific resins, i.e., TEVA for Pu and TRU-DGA for Am.

Maxwell et al. (2011a) developed a rapid simplified method for the determination of Pu and Np in water samples using a stacked column set of TEVA and DGA resins. In the combined procedure, Pu(IV) and Np(IV) were retained on a TEVA column first, and then Pu was reduced and retained as Pu(III) on the DGA column together with U. From the DGA, U was eluted with 0.1 M HNO_3, and Pu(III) was stripped with 0.02 M HCl/0.005 M HF/0.0005 M Ti^{3+}. High recoveries and DFs for U (higher than 10^6 and 10^4 in case of Pu and Np, respectively) were reported. The resulting source quality met the requirements of both alpha spectrometry and ICP-MS.

Since the DGA resin became commercially available, many analytical applications have been reported, but most of them apply DGA resin only for Am separation (Dulanska et al., 2015; Gorietti et al., 2017; Yoon et al., 2018; Luo et al., 2018; Wang et al., 2018). There are a few other examples for the retention of various actinides on DGA.

Groska et al. (2016) developed a procedure for the simultaneous separation and determination of all actinides (U, Th, Np, Pu, Am, Cm isotopes) from low-level waste samples. A single DGA resin column is used to retain actinides from a reductive media. Individual actinides were stripped after adjusting the oxidation states on-column and the concentration of the acids and the oxalic acid complexes. Acceptable high recoveries and decontamination factors were obtained in the alpha spectra.

A new type of DGA resin where DGA is covalently bound to silica has been prepared and tested for the separation of Th, Pa, and U by Hopkins et al. (2018). Th and Pa were separated from U. By using a sulfur-based DGA derivative, the diglycothioamide (DGTA) Pa could be selectively separated from Th and U. The radiochemical purity of ^{231}Pa is an important issue in targeted alpha therapy (Mastren et al., 2018).

1-Octanol impregnated resin, commercialized as TK400 resin (TRISKEM SAS), proved to be a selective EC material for the retention of Pa from HCl solution. High Pa recoveries have been reported by Knight et al. (2016b) for the separation from Np, by Jerome et al. (2018). Protactinium was successfully separated from Th and U in marine sediments with TK400 resin after a preliminary removal of matrix components using an anion exchange resin (Süfke et al., 2017).

b. Extraction Chromatography of Radium and polonium

Sr resin is an EC material particularly useful for separation of group 2 elements, with an especially high affinity for Sr. In nitric acid solutions, uptake of Ra is lower than Pb, Sr, and Ba but higher than Ca. Radium and Ba uptake are highest at approximately 2–3M HNO_3, whereas Sr uptake increases monotonically with HNO_3 concentration. Lawrie et al. (2000) used 3 M HNO_3 on Sr Spec resin as a final cleanup step to separate Ra and Ba prior to alpha spectrometry. Although the main reported applications to date have been for separations preparatory to mass spectrometry (e.g., Yokoyama and Nakamura, 2004), the material shows promise for expanding the application to alpha spectrometry for Ra isotope analysis.

Vajda et al. (1997) developed a method for the separation of Po using Sr Spec resin. The digest is loaded on the column in 2 M HCl. Bismuth and several other matrix components, including Fe, are eluted with the 2 M HCl wash, whereas Po and Pb are retained. Po may then be eluted with 6 M HNO_3, and Pb subsequently eluted with 6 M HCl. The column may be reused one or two times after the first extraction (Vreček et al., 2004; Kim et al., 2009).

3M Empore Radium Rad Disks are a filter-like membrane on which a Ra ion-selective crown ether is embedded. They are especially useful for preconcentration and separation of Ra from water samples and other aqueous solutions. The disks are commonly analyzed using gamma spectrometry, as little further sample preparation is required. However, the Ra may be eluted (for example, with alkaline EDTA) and a source for alpha spectrometry prepared (Ďurecová, 1997; Purkl and Eisenhauer, 2003).

c. Combined procedures

Combined procedures are often used for chemical separations of a single element to have high decontamination factors for one element or for the simultaneous determination of several elements. It is a rare case when an element can be separated adequately for alpha spectrometry by a single procedure. There is a huge variety of combined procedures that cannot be discussed in the frame of this chapter (see the reviews and books referred to at the beginning of this section). The following combinations have been reported recently (Vajda and Kim, 2011):

anion exchange resin—anion exchange resin,
anion exchange resin—TEVA resin,
anion exchange resin—UTEVA resin,
anion exchange resin—TRU resin (—TEVA resin),
anion exchange resin—cation exchange resin (ion chromatography),
anion exchange resin—TEVA-DGA,
LLE with TOPO—UTEVA-TRU,
LLE with TOPO—TRU-TEVA,
LLE with TOPO—TEVA—TRU—TEVA,
LLE with TOPO—HDEHP,
UTEVA—TRU,
UTEVA—DGA,
TEVA—TRU,
TEVA—UTEVA,
TEVA—DGA,
TEVA—UTEVA—TRU,
TEVA—TRU—UTEVA,
TEVA—TRU—DGA,
Sr resin—TEVA—UTEVA (IAEA, 2014a,b).

Remark: Other chromatographic materials have been reported also to be efficient for actinide analysis, but because of lack of information about their composition, they have not been discussed in this review (e.g., molecular recognition materials such as AnaLig-Pu, product of IBC Advanced Technology).

C. Alpha source preparation

Following chemical separation from the sample matrix, the analytes must be prepared as a thin, flat, uniform source suitable for alpha spectrometric analysis. The basic characteristics of a high-quality source are that the alpha particles lose only a small and consistent amount of energy while exiting the source (minimizing energy straggling), the geometry of the source—detector arrangement results in a high counting efficiency, this efficiency is consistent between sources, the deposit is robust, the deposited activity is relatively homogeneous across the deposit, and the source is easy to handle and store.

There are three main methods of source preparation: evaporation, coprecipitation with a microcrystalline precipitate, and electrodeposition. In the ASTM standard D3084—05 on "Standard Practice for Alpha-Particle Spectrometry of Water," all three methods are described.

The **evaporation technique** involves depositing the solution onto a stainless steel or platinum disk or a porcelain disc and evaporating the solution in such a way that a uniformly covered surface is obtained. This method is quick and ensures a very high recovery. A difficulty is ensuring that the source deposit is uniform, and for that purpose, a wetting or spreading agent is usually used (Lally and Glover, 1984). A further disadvantage is that deposits do not always adhere tenaciously to the substrate.

Microcoprecipitation of actinides on a rare earth fluoride, often NdF_3, followed by filtration on a smooth surface membrane filter of small pore size (0.1−0.2 μm), produces a good-quality alpha source (ASTM Test Method C 1163, Sill, 1987a). For Ra, the most commonly used coprecipitation method for source preparation is with $BaSO_4$ as described by Sill (1987b). Coprecipitation and filtration is more rapid than electrodeposition; however, a disadvantage is that the resulting filter is flexible and may require adhesion to a backing to prevent curling.

In **electrodeposition**, metallic radionuclides are electrochemically plated with high yield from an electrolyte solution onto a metal cathode surface. Most commonly, a polished stainless steel cathode planchet is used, although a metal-coated plastic film may be used for specialized applications, such as coincident photon and alpha spectrometry (Johnston et al., 1991). The anode is typically Pt wire or gauze. The anode is often rotated to stir the solution, although bubbling by gas formed at the anode may provide sufficient mixing. A variety of electrolytes have been used to close the electric circuit, including sulfate, oxalate, and acetate buffers.

In most cases, metallic elements are deposited with high yield in a relatively nonselective manner, with the exception of group 1 and 2 elements. For the widely used method of Hallstadius (1983), deposition yields were reported to be between 90% and 99% for Cm, Am, Pu, Np, U, Th, Pb, and Po and <1% for Ra.

An aqueous/alcohol electrolyte solution (e.g., 1 mL 0.1M HNO_3 and 9 mL propanol) may be used to electrodeposit Ra (Smith and Mercer, 1970; Hancock and Martin, 1991). Alternatively, Ra may be deposited from an aqueous solution of ammonium acetate and nitric acid, or ammonium oxalate and HCl, usually with the addition of a small amount of Pt in solution (Roman, 1984; Orlandini et al., 1991; Hamilton et al., 1994).

Electrodeposition is often performed at a constant current density (0.5−1 A/cm^2) and low voltage (<10 V) for about half an hour to 2 hours. A standard procedure for electroplating actinides from a solution of ammonium sulfate was published in 2000 and reapproved in 2005 (ASTM C1284-00, 2000b). For alpha spectrometric purposes, the last two procedures are routinely used.

Polonium spontaneously autodeposits onto the surfaces of several metals, including Ag, Cu, and Ni from dilute acid solution. **Autodeposition** onto metal discs (usually Ag) is the most common source preparation technique for Po due to its simplicity and the fact that it results in separation from other metal ions, including other alpha emitters (Smith and Hamilton, 1984; Matthews et al., 2007). Electrodeposition can also be used for Po source preparation (Uesugi et al., 2010).

A review of electrodeposition methods for the preparation of alpha sources has been prepared by Crespo (2012). Evaluation of procedures for the electrodeposition of uranium, plutonium, and americium has been performed by Oh et al. (2014).

Ultrathin Am, Cm, and Th sources may be prepared by the adsorption of insoluble compounds onto a substrate (Wyllie and Lowenthal, 1984; Martin and Hancock, 2004b). The alpha peak resolution for these sources is very high. However, the recoveries obtained are relatively low, and although useful for the preparation of high-quality test sources, the technique has not been used for routine analysis. Energy resolution of alpha peaks has been improved by the unique technique of ion implantation using ICP-MS device for the preparation of implanted alpha source (Dion et al., 2015). Separation of isotopes by mass with a quadrupole mass spectrometer assures not only high-quality (thin and even) alpha source but also an effective purification.

Whichever method is used for source preparation, perfect homogeneity of the deposited activity across the source cannot be expected. With careful design of the deposition system, a reasonably even and consistent deposit can be attained (Klemenčič and Benedik, 2010). Nevertheless, on a small scale, it can be expected that the activity and source thickness will show some "graininess," which will mean that the effect of the deposit on peak resolution will be greater than that expected purely from the deposited mass per unit area (Martin and Hancock, 2004a).

For the rapid preparation of alpha sources, a new technique has been introduced, the **ligand film technique**. Selective extractive ligands are immobilized in a support that is counted directly after extraction of the alpha-emitting nuclide by alpha spectrometry. The technique combines radiochemical separation with alpha source preparation.

For the rapid extraction of Pu and Am from water and urine, a PLF was prepared from Dipex and polystyrene (Gonzales and Peterson, 2009). Recoveries were poor, but FWHM values were as low as 20 keV. Rim et al. (2013) prepared an HDEHP-based polymer (polystyrene) ligand film for the extraction of Pu, but its distribution in the film was uneven. A Klaui-type tripodal oxygen donor ligand film was prepared by spin-casting solutions onto glass substrate for the rapid analysis of Pu by Hanson et al. (2014). High recoveries and low FWHM values (33 keV) were obtained. Paul et al. (2014, 2015) prepared self-supported and membrane-supported polymer-containing phosphate and sulfonic acid groups for the extraction of Pu and U. Acceptable high recoveries were obtained for both isotopes using the bifunctional polymer layer grafted on poly(ethersulfone) membrane. Ultrathin films of quaternary amine anion exchange polymers on glass and silicon were prepared by dip coating by Mannion et al. (2016)

and successfully used for Pu extraction. Recently, ultrafiltration membranes were grafted with uranium-selective polymer and used for analysis of U isotopes. The great number of publications on PLFs reflects the need for rapid determination of alpha-emitting nuclides in liquid samples.

VII. Determination of activity and recovery

To determine alpha particle–emitting radionuclides conventionally, the following procedure is performed: alpha sources are prepared by radiochemical separation; the alpha spectrometer is calibrated; sample, background, and blank sources are counted with the alpha spectrometer for adequate counting times; the spectra are analyzed (i.e., peaks are identified and peak areas are calculated); and the activities of the identified nuclides are calculated. In this chapter, the major steps of activity determination are discussed using examples from standard semiconductor alpha spectrometry. The same method can be used in the case of gas ionization detectors without significant changes. High-resolution alpha spectra and high-resolution LS spectra are usually evaluated by using computer programs that have been briefly discussed above.

A. Calibration

Calibration of the instrument comprises energy, resolution, and efficiency calibration that may be established from the same source, most likely a calibration source (CS) containing known quantities of at least two radionuclides emitting alpha radiation. The CS should have activities such that instrumental dead time resulting in lost counts can be avoided. The limitation of approximately 1000 Bq is a typical value. The CS is placed in the detector chamber at a distance from the source that is larger than the detector diameter to reduce the geometric effect on FWHM value, and then the counting chamber is evacuated and the system bias is adjusted to the factory given value. The source is counted during a measurement period that produces a relatively low counting uncertainty. A typical alpha spectrum of a CS is shown in Fig. 5.44.

Energy calibration is carried out by plotting the known energies (E) of the peaks belonging to highest alpha abundance against the peak location channel numbers (ch) referring to the peak center position. Usually, a straight line is fitted to the points as

$$E = b + m \cdot ch \tag{5.25}$$

where m is the slope and b is the intercept of the line. Because the energy response of the detector is reasonably linear, the measurement of two points is sufficient for routine analytical purposes. However, modern spectrometry software packages perform calculations based on the measurement of several peaks and may include cubic or quadratic terms to account for minor nonlinearity.

Calibration of the energy resolution is performed by measuring the peak width at channels corresponding to the half of the maximum counts and expressing the width in energy units (see Fig. 5.44). The stability of FWHM values is an important indicator of good system performance.

In principle, **efficiency calibration** can be performed according to the following equation from the measurement

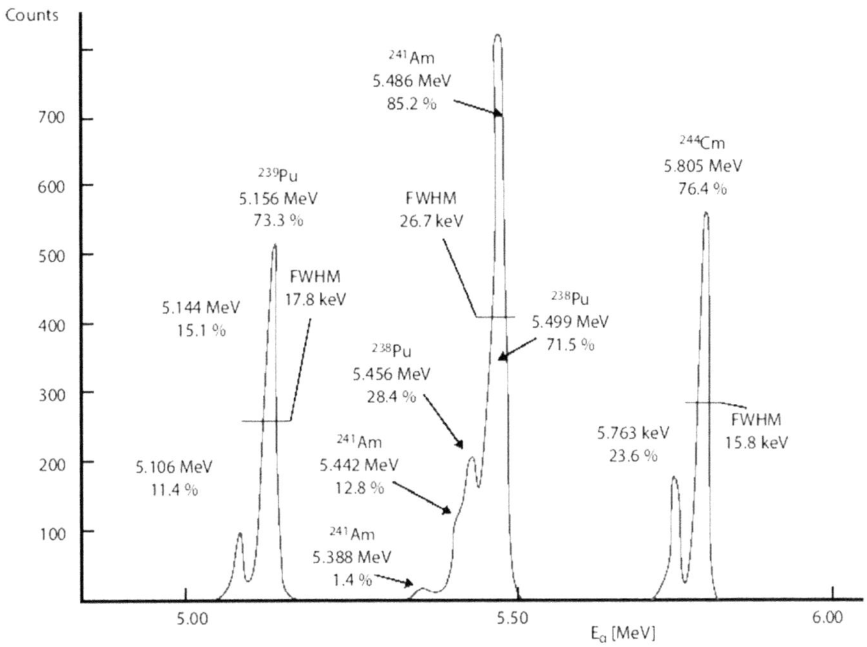

FIGURE 5.44 Alpha spectrum of a commercially available calibration source containing ^{239}Pu, ^{241}Am, and ^{244}Cm. The source was prepared by vacuum evaporation and counted with a PIPSi detector with 300 mm^2 surface area at a distance of 5 cm from the detector. PIPSi, passivated ion–implanted planar Si.

of the same CS, if the calibration standard is measured in the same geometry as the sample source:

$$\varepsilon = \frac{r_{CS} - r_{BG}}{A_{CS} \cdot I_\alpha \cdot D} \tag{5.26}$$

where ε is the counting efficiency, r_{CS} is the gross count rate of the calibration source (cps), r_{BG} is the background count rate (cps), A_{CS} is the activity of the calibration source at the reference date (Bq), I_α is the alpha abundance, and

$$D = e^{-\lambda_{CS} \cdot t_{CS}}$$

is the decay factor for the calibration source from the reference date to the measurement date, where λ_{CS} is the decay constant of the calibration source (s^{-1}) and t_{CS} is the time from the reference date of the calibration source to the measurement date (s).

The law of uncertainty propagation for a single measurement according to the equation above or preferably the standard deviation of multiple measurements can be used to express the standard uncertainty of ε.

In practice, calibration sources rarely have the same geometry as those for the sources prepared from the sample for analysis. Therefore, calibration sources of the required geometry have to be prepared from standard solutions. These sources are used to determine the counting efficiency as described above.

Since the intrinsic detection efficiency of a PIPSi detector has practically a value of 1, the activity determination is possible without using tracers in the sample processing (Pöllänen et al., 2011). If the processed sample is thin, i.e., there is no alpha particle self-absorption, the geometric detection efficiency can be used in the activity determination. Geometric detection efficiency can be calculated by using appropriate computer programs such as AASI (Siiskonen and Pöllänen, 2005).

The efficiency of a solid-state alpha spectrometry detector is not a function of energy over the typical energy range of emitted alpha particles due to the high stopping power of alpha particles in Si. Therefore, the average of the efficiencies of multiple alpha peaks can be calculated to obtain a more accurate efficiency value.

When a sample is analyzed by using an alpha particle–emitting tracer as an internal standard, the efficiency does not affect the result of analysis because the count rates of the analytes are directly related to the count rate of the tracer. Therefore, an exact knowledge of the efficiency is only necessary to calculate the chemical recovery. However, when a sample source is analyzed without an alpha-emitting tracer, then the efficiency is used to calculate the activity of the analyte.

The standard practice of calibration of alpha spectrometers is described in ASTM D7282-06 standard.

B. Measurement procedure

The samples to be analyzed are placed in the vacuum chamber. The selection of the distance of the source from the detector is always a kind of optimization between efficiency and resolution, and it depends on the source activity. The chamber is pumped until a vacuum of about 15 mbar is achieved, if it is desirable to protect the detector from the recoil contamination or of about 0.01 mbar and if high resolution is of major concern. Then the detector voltage is set to the value recommended by the detector manufacturer. Counting times are selected according to the activity of the source and the required counting uncertainty. Low-activity environmental samples are typically counted for several days.

C. Alpha spectrum evaluation

1. Principle of evaluation

The activity of the analyte is determined from the known activity of the tracer that was added to the sample and the ratio of the peak areas belonging to the analyte and the tracer. The method is called **isotope dilution alpha spectrometry** and has the advantage that the calculated activity is inherently corrected for chemical losses because losses of the analyte are taken to be equal to those of the tracer. This is a very realistic assumption whenever the tracer and the analyte are isotopes of the same chemical element in the same chemical form, but the procedure should be used with care if this is not the case.

2. Nuclide identification

In isotope dilution alpha spectrometry, chemical separation is designed to favor one element or a group of elements. Therefore, only a limited number of isotopes are expected in the alpha source, and identification of the nuclides simply means comparing the peak energy in the spectrum with the alpha energies of the expected nuclides and possible contaminants, which could be present due to failure of separation. For example, after chemical separation of Pu from an environmental sample, one can expect the peaks of ^{238}Pu, the common peak of ^{239}Pu and ^{240}Pu, and the peak of the tracer ^{242}Pu or ^{236}Pu (Fig. 5.45).

A very important quality control procedure in alpha spectrometry involves checking the spectrum for the presence of alpha peaks from elements other than the one of interest. This may be due to the failure of the chemical procedure to achieve high separation efficiency due to matrix effects or the presence in the sample of a relatively high activity of the contaminating radionuclide. As examples, possible interferences in the Pu spectrum are Th isotopes, if anion exchange separation of Pu is performed from nitric acid solution, or traces of U isotopes in the Pu source,

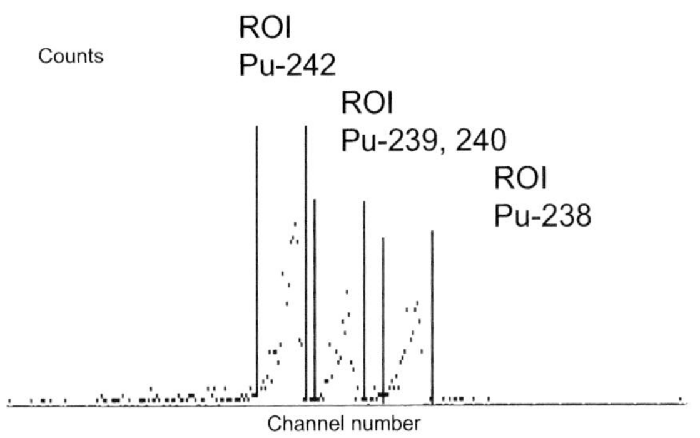

FIGURE 5.45 Setting regions of interest of equal width for the evaluation of the well-separated peaks of Pu nuclides.

if anion exchange separation of Pu has been performed from hydrochloric acid solution.

3. Peak area determination for nonoverlapping peaks

The number of counts in the peak corresponding to the analyte are integrated, and the counts of the tracer nuclide in the channel region of the same width are also integrated. Blank samples are also processed and analyzed to determine the background contribution to the counts in the regions of interest (ROIs).

4. Peak area determination with correction for overlapping peaks

In the case of overlapping peaks, there are basically two different approaches for the peak area determination. The areas can be determined by using computer programs presented in Section IV.A. Accurate results can be obtained provided that correct peak shapes are used in the peak area fitting. Another option for the area determination is to use methods described below. The advantage is that no sophisticated computer programs are necessary.

Due to the asymmetric character and significant low-energy tailing of alpha peaks, the low-energy peaks are sitting on the tails of all the peaks of higher energies. This tailing has to be corrected for, and this correction is especially important, if the peaks are close to each other or if the count rate in the higher energy peak is much greater than that in the lower energy peak.

In Fig. 5.46, partially overlapping peaks of ^{239,240}Pu and ^{238}Pu are shown according to the ISO 11,483:2005 standard. The exact activity ratio of ^{238}Pu/^{239,240}Pu can be determined after tailing correction. There are two standardized methods for the determination of peak area of overlapping peaks, the so-called geometric progression decrease (GPD) method and the exponential decrease (ED) method that have been described in detail in the standard.

The basic idea in the GPD method is that the counts in the tail of a peak decrease in geometric progression. It is assumed that the shapes of the different peaks in the same spectrum are the same, therefore, in the flat portion of tail, peak tailing contributions are proportional to the corresponding peak areas, and the tails are the sum of all single peaks of higher energies.

According to Fig. 5.46, four regions of interest (A, B, C, D) of equal channel numbers are selected. The high-energy end of ROI A is selected as the center energy of the second peak plus the half of its FWHM. The low-energy end of ROI A is fixed so that the region is not extending into the region of the first peak. The high-energy end of ROI B is selected similarly to ROI A as the center energy of the first peak plus the half of its FWHM. The width of ROI B is the same as that of ROI A. ROI C and ROI D are selected so

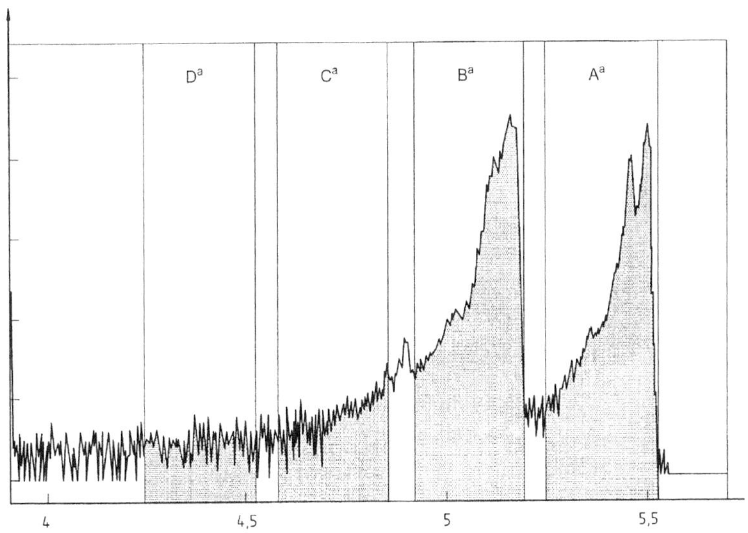

FIGURE 5.46 Tailing correction for partially overlapping peaks according to standard ISO 11483(2005).

that the difference between the two neighboring peaks and the width of each ROI are the same. The constant ROI width is selected so that the ROIs should not extend to the regions of any other peaks. The sum of the counts in ROIs A, B, C, and D are calculated and identified as a, b, c, and d, respectively. The corrected ratio of the peak areas (R_{cor}) is calculated according to the geometric progression decrease as follows:

$$R_{cor} = \frac{a}{b_{corrected}} = \frac{a}{b - \dfrac{ac}{b + \dfrac{ad}{c}}} \tag{5.27}$$

Thus, the values of a and the tailing corrected $b_{corrected}$ are obtained for the calculation of the exact activity or activity concentration. The complicated uncertainty calculation is described in the ISO standard.

The basic idea in the ED method is that the decrease of counts in the tailing region can be described by an exponential function. The ROIs A and B are selected as described above. An exponential function is fitted to the points in the region between ROI A and ROI B:

$$Y_i = A' \cdot e^{B' \cdot X_i} \tag{5.28}$$

where Y_i and X_i are the number of counts and the number of channels belonging to the measured spectrum data point, and A' and B' are the fitted parameters of the exponential function.

Then the area under the assumed tailing curve in ROI B is calculated as the sum of counts ΣY_i using the parameters of the fitted curve. The corrected area of the first peak is obtained as $b - \Sigma Y_i$, and the corrected ratio of the peak areas (R_{cor}) is calculated as

$$R_{\text{cor}} = \frac{a}{b - \sum\limits_{i\_in:ROI\_B} \Sigma Y_i} \tag{5.29}$$

The uncertainty calculation of the corrected peak area is given in the ISO standard.

Both methods of tailing correction have limitations in applications: Spectra of relatively high energy resolution and high counts are required, and the peaks have to be relatively well separated so that the regions can be selected without interferences from other peaks. Both methods have been successfully used for the accurate determination of $^{238}Pu/^{239,240}Pu$ isotope ratios in nuclear fuel samples by alpha spectrometry.

5. Calculation of results

The calculation of results comprises calculation of the activity per unit mass (or per other quantity in question) of the radionuclide and its standard uncertainty. The latter can be occasionally replaced by the calculation of the standard deviation of repeated measurements. Additionally, chemical recoveries and decision thresholds as well as detection limits should be calculated.

The ratio of the net counts of the peaks belonging to the analyte and the tracer (n/n_T) is used to calculate the activity concentration of the analyte in the sample on the date of chemical separation (a_1) taking into account the activity concentration of the tracer solution (C_T), the masses of the sample (m_s) and the tracer solution (m_T) used, and the decay of the tracer between its calibration date and counting (f_T). The decay of the analyte between separation and counting can be neglected, if the nuclides are long lived. The method of calculation of the results has been described by Kanisch (2004).

The activity concentration of the analyte (a_1, Bq kg^{-1}, dry) at the separation time can be calculated by the following formula:

$$a_1 = \frac{C_T \cdot n}{m_s \cdot n_T} \cdot m_T \cdot f_T \tag{5.30}$$

where the decay correction factor is calculated as follows:

$$f_T = \exp(-\lambda_T \cdot t_T) \tag{5.31}$$

where λ_T is the decay constant of the tracer (years^{-1}) and t_T is the time interval between the reference date of the tracer and the separation time (years).

Finally, the activity concentration of the analyte in the sample is calculated back to the date of sample collection (a_0).

$$a_0 = a_1 \exp(\lambda_A t_1) \tag{5.32}$$

where λ_A is the decay constant of the analyte (years^{-1}) and t_1 is the time interval between the sampling date and the separation time (year).

The detection efficiency (ε) is not needed directly for the calculation of the activity concentration in isotope dilution alpha spectrometry. However, it allows one to calculate the chemical recovery (R) from the activity of the tracer added and the net counts in the corresponding peak:

$$R = \frac{n_T}{C_T \cdot m_T \cdot \varepsilon \cdot t_m} \cdot \exp(\lambda_T \cdot t_T) \tag{5.33}$$

The detection efficiency is determined from the activity of a calibration source and the net count rate measured in the same geometry as the sample (see Eq. 5.26).

Calculation of the activities of all analyte isotopes (e.g., ^{238}Pu and $^{239,240}Pu$) is performed in the same way replacing the relevant analyte data in the equation and using the same tracer data (e.g., ^{242}Pu).

6. Calculation of the combined standard uncertainty

Activities in the source and **activities** per unit mass of the analytes in the sample are determined according to Eq. (5.30).

Calculation of the **combined standard uncertainty** can be performed according to the ISO Guide to the Expression of Uncertainty in Measurement (1995) or according to the IAEA (2004) publication on Quantifying Uncertainty in Nuclear Analytical Measurements. First, the individual uncertainties of the input parameters are calculated or estimated, and then the uncertainty of the activity concentration $u(a_0)$ is calculated according to the law of uncertainty propagation.

a. Individual uncertainty components

Uncertainty in weighing of sample mass. The uncertainty associated with the mass of sample is estimated using the data from the calibration certificate of the analytical balance and the manufacturer's recommendations on the uncertainty estimation of the balance. If it is assumed that the balance with 0.1 mg of uncertainty is used for weighing the sample and the linearity distribution is assumed to show a rectangular distribution, the standard uncertainty in weighing of the sample mass can be calculated by the following formula:

$$\frac{0.1}{\sqrt{3}} = 0.058\,mg \tag{5.34}$$

The contribution for the linearity has to be accounted for twice, once for the tare and once for the gross mass (tare + sample), leading to an uncertainty $u(m_s)$ of

$$u(m_s) = \sqrt{2 \times (0.058)^2} = 0.082\,mg \tag{5.35}$$

Uncertainty in the mass of the tracer solution is calculated similarly.

Uncertainty of the activity concentration of the tracer solution. The uncertainty of the certified concentration of the tracer solution on a calibration, $u(C_T)$, is provided by the producer.

Uncertainty of net count rates of analyte and tracer. Since

$$u(n) = \sqrt{n}, \text{ and } r = \frac{n}{t_m} \text{ therefore}$$

$$u(r)^2 = \frac{n}{t_m^2} = \frac{r}{t_m} \tag{5.36}$$

Thus, the uncertainty of count rates for analyte and analogously for tracer in alpha spectra can be calculated according to the equation below:

$$u(r_g)^2 = \frac{n_g}{t_m^2} = \frac{r_g}{t_m} \tag{5.37}$$

$$u(r_{gT})^2 = \frac{n_{gT}}{t_m^2} = \frac{r_{gT}}{t_m} \tag{5.38}$$

In the case of background, the number of counts is usually small, and the $u(n) = \sqrt{n}$ equation is not valid; therefore, background uncertainties are calculated from the average and standard deviation of several long-time background measurements according to

$$\overline{r_0} = \frac{\sum_i r_{0i}}{n_b} \tag{5.39}$$

where $\overline{r_0}$ is the average of n_b measurements each resulting in a r_{0i} background count rate,

$$u(r_0) = \frac{\sum_i \left(r_{0i} - \overline{r_0}\right)^2}{n_b - 1} \tag{5.40}$$

where $u(r_0)$ is the standard deviations of the n_b repeated background measurements.

The average value for $\overline{r_{0T}}$ and the standard deviation $u(r_{0T})$ are derived similarly.

1. The uncertainties for decay correction factor, $u(f_T)$, can be calculated as follows:

$$u(f_T) = f_T \bullet t_T \bullet u(\lambda_T) \tag{5.41}$$

where uncertainties of the decay constants are obtained from nuclear tables.

The uncertainty of measuring time is neglected: $u(t) =$ 0.

b. Combined uncertainty

Combined uncertainty in the determination of the massic activity of the analyte (a_1) is calculated according to the general formula:

$$u(a_1) = \sqrt{\sum_i \left(\frac{\partial a_i}{\partial x_i}\right)^2 u(x_i)^2} \tag{5.42}$$

where x_i are the individual parameters of the result a_1. In the basic Eq. (5.30), net counts are replaced by net count rates expressed as the difference of gross and background count rates to take into account that counting times of sample and background are usually different. The basic equation for the calculation of the result is therefore

$$a_1 = \frac{C_T \bullet (r_g - r_0)}{m_s \bullet (r_{gT} - r_{0T})} \bullet m_T \bullet f_T \tag{5.43}$$

and a_1 is dependent on the parameters $a_1 = f(r_g, r_0, r_T, r_{0T}, C_T, m_s, m_T, f_T)$. Thus, partial derivatives of a_1 are calculated for each parameter as follows:

$$\frac{\partial a_1}{\partial r_g} = \frac{a_1}{(r_g - r_0)} \tag{5.44}$$

$$\frac{\partial a_1}{\partial r_0} = -\frac{a_1}{(r_g - r_0)} \tag{5.45}$$

$$\frac{\partial a_1}{\partial r_{gT}} = -\frac{a_1}{(r_{gT} - r_{0T})} \quad (5.46)$$

$$\frac{\partial a_1}{\partial r_{0T}} = \frac{a_1}{(r_{gT} - r_{0T})} \quad (5.47)$$

$$\frac{\partial a_1}{\partial C_T} = \frac{a_1}{C_T} \quad (5.48)$$

$$\frac{\partial a_1}{\partial m_s} = -\frac{a_1}{m_s} \quad (5.49)$$

$$\frac{\partial a_1}{\partial m_T} = -\frac{a_1}{m_T} \quad (5.50)$$

$$\frac{\partial a_1}{\partial f_T} = -\frac{a_1}{f_T} \quad (5.51)$$

where

$$\frac{\partial a_1}{\partial \lambda_T} = -t_T a_1 \text{ and } \frac{\partial a_1}{\partial t_T} = 0 \quad (5.52)$$

The uncertainty of the analyte concentration on the separation time $u(a_1)$ can be calculated as follows:

$$u(a_1)^2 = \left(\frac{a_1}{r_g - r_0}\right)^2 \bullet u(r_g)^2 + \left(-\frac{a_1}{r_g - r_0}\right)^2 \bullet u(r_0)^2 + \left(\frac{a_1}{r_{gT} - r_{0T}}\right)^2 \bullet u(r_{gT})^2 + \left(\frac{a_1}{r_{gT} - r_{0T}}\right)^2 \bullet u(r_{0T})^2 + \left(\frac{a_1}{C_T}\right)^2 \bullet u(C_T)^2 + \left(-\frac{a_1}{m_s}\right)^2 \bullet u(m_s)^2 + \left(\frac{a_1}{m_T}\right)^2 \bullet u(m_T)^2 + (-t_T a_1)^2 \bullet u(\lambda_T)^2 \quad (5.53)$$

and $u(r_g)$ and $u(r_{gT})$ can be replaced by Eqs. (5.37) and (5.38) to get the final expression for the measurement uncertainty:

$$u(a_1)^2 = \left(\frac{a_1}{r_g - r_0}\right)^2 \bullet \frac{r_g}{t_m} + \left(-\frac{a_1}{r_g - r_0}\right)^2 \bullet u(r_0)^2 + \left(\frac{a_1}{r_{gT} - r_{0T}}\right)^2 \frac{r_{gT}}{t_m} \bullet + \left(\frac{a_1}{r_{gT} - r_{0T}}\right)^2 \bullet u(r_{0T})^2 + \left(\frac{a_1}{C_T}\right)^2 \bullet u(C_T)^2 + \left(-\frac{a_1}{m_s}\right)^2 \bullet u(m_s)^2 + \left(\frac{a_1}{m_T}\right)^2 \bullet u(m_T)^2 + (-t_{T_1} a_1)^2 \bullet u(\lambda_T)^2 \quad (5.54)$$

Combined uncertainty in the determination of the analyte on the sampling date, $u(a_0)$ can be derived from the uncertainty on the separation time $u(a_1)$ by differentiating Eq. (5.32). Each partial derivative can be calculated as follows:

$$\frac{\partial a_0}{\partial a_1} = \frac{a_0}{a_1} \quad (5.55)$$

$$\frac{\partial a_0}{\partial \lambda_A} = a_0 t_1 \quad (5.56)$$

Therefore, the combined relative uncertainty of the basic activity of the analyte (Bq kg^{-1}, dry) on the sampling date ($\mu(a_0)$) can be calculated by the formula

$$\left(\frac{u(a_0)}{a_0}\right)^2 = \left(\frac{u(a_1)}{a_1}\right)^2 \bullet + t_1^2 \bullet u(\lambda_A)^2 \quad (5.57)$$

Calculation of the relative uncertainty of the chemical recovery is made based on Eq. (5.33) that can be expressed with count rates instead of counts:

$$R = \frac{r_{gT} - r_{0T}}{C_T \bullet m_T \bullet \varepsilon} \bullet \exp(\lambda_T \bullet t_T) \quad (5.58)$$

$$u_{rel}(R_{cal})^2 = u_{rel}(r_T)^2 + u_{rel}(C_T)^2 + u_{rel}(\varepsilon)^2 + u_{rel}(m_P)^2 + u_{rel}(f_T)^2 \quad (5.59)$$

Since $u(r_T)^2 = \frac{n_{gT}}{t_m^2} + \frac{n_{0T}}{t_0^2} = \frac{r_{gT}}{t_m} + \frac{r_{0T}}{t_0}$ and (5.60)

$$u_{rel}(f_T)^2 = t_T^2 \bullet u(\lambda_T)^2 \quad (5.61)$$

the following equation is obtained:

$$u_{rel}(R_{cal})^2 = \left(\frac{r_{gT}}{t_m} + \frac{r_{0T}}{t_0}\right) \cdot \left(\frac{1}{r_{gT} - r_{0T}}\right)^2 + u_{rel}(C_T)^2 + u_{rel}(\varepsilon)^2 + u_{rel}(m_P)^2 + t_T^2 u(\lambda_T)^2 \quad (5.62)$$

Similarly, the relative uncertainty of the counting efficiency is calculated from Eq. (5.26):

$$u_{rel}(\varepsilon)^2 = \left(\frac{r_{CS}}{t_{CS}} + \frac{r_{BG}}{t_{BG}}\right) \cdot \left(\frac{1}{r_{CS} - r_{BG}}\right)^2 + u_{rel}(A_{CS})^2 + u_{rel}(I)^2 + t_{CS}^2 u(\lambda_{CS})^2 \quad (5.63)$$

where the indexes *CS* and *BG* refer to the calibration source and the background in the peak of the calibration source, A_{CS} is the known activity of the calibration source, I is its alpha decay probability/abundance, t_{CS} is the time elapsed from source calibration to measurement, and λ_{CS} is the decay constant of the calibration source. Holmes (2004) proposed the spreadsheet approach for the calculation of combined uncertainty in alpha spectrometry, which has the advantage of avoiding having to solve complicated differential equations and identifying how much individual uncertainty components contribute to the combined uncertainty.

Kim et al. (2008b) identified and quantified all uncertainty sources, which can appear in the measurement of ^{210}Pb and ^{210}Po by liquid scintillation counting and alpha particle spectrometry, including correction to the sampling date. The paper also presented how the combined uncertainty in the calculation of ^{210}Po activity on the sampling date depends on the ratio of ^{210}Po/^{210}Pb specific activities on the sampling date (or the reference date in the particular case of analysis of a reference material).

7. Calculation of the decision threshold and the detection limit

a. Decision threshold

In accordance with ISO 11929 (2010), the decision threshold, a^*, is obtained from the following equation:

$$a* = k_{1-\alpha} u(0) \quad (5.64)$$

where $k_{1-\alpha} = 1.65$ is often chosen as default, and $u(0)$ can be calculated from Eqs. (5.54) and (5.57) by substituting the value of r_0 background instead of r_g, the measured value of the analyte. The decision threshold is obtained as activity concentration expressed in Bq kg^{-1}.

b. Detection limit

In accordance with ISO 11929 (2010), the detection limit, $a^{\#}$, is calculated by iteration according to the following equation:

$$a^{\#} = k_{1-\alpha} u(0) + k_{1-\beta} u(a^{\#}) \quad (5.65)$$

Assuming $\alpha = \beta$, then $k_{1-\alpha} = k_{1-\beta} = k$. The detection limit is obtained as activity concentration expressed in Bq kg^{-1} (IAEA, 2017).

VIII. Quality control

General requirements for the competence of testing and calibration laboratories are summarized in the ISO/IEC 17025 (2017) standard. Here, we discuss specific issues of QC of alpha spectrometers and validation of alpha spectrometric methods.

A. Quality control for alpha spectrometers

Instrument performance is regularly monitored by counting check sources. The same alpha source containing at least two alpha emitting radionuclides can be used to check the energy, the FWHM and the efficiency calibration of the detector/alpha spectrometer. The check source is counted under fixed conditions similar to those used for samples. The parameters characterizing the stability of the system are the **peak location, the FWHM, and the counting efficiency** of a selected radionuclide present in the check source. The measured values are plotted in a control chart as a function of counting date/time. It is desirable to set the control limits for mean chart and range chart according to ISO standard (ISO 7870-2, 2013). The coverage factors for the control limits of mean chart and range chart are decided based on the observations in subgroups of size. When measured values are out of the warning level range, the instrument has to be checked or repaired, and it may be that the detector has to be replaced. If the measured values are out of the control limit range, measurements are suspended until the cause analysis for outliers in the control charts has been completed. The frequency of control measurements is regulated in the Quality Manual of the laboratory, but it is desirable to perform QC measurements before each series of analysis. Additionally, **background measurements** are performed on a regular basis to check detector contamination.

As an example, the efficiency control chart of an alpha spectrometer is shown in Fig. 5.47 together with the average value of the measurements and the warning and control levels that were established by the first seven data points.

B. Validation of the procedure

Methods that have been developed at the laboratory, which are nonstandard methods or amplifications and modifications of standard methods, have to be validated to

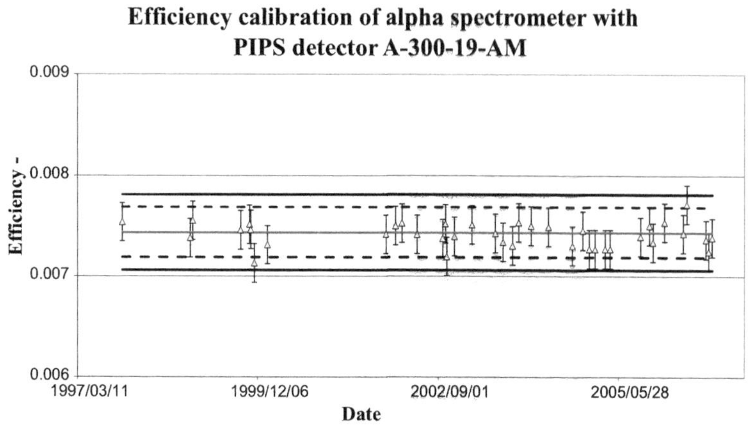

FIGURE 5.47 Control chart of an alpha spectrometer showing the efficiency data and their uncertainty (1σ) as well as the average values, warning levels (±2σ deviation from the average) and control levels (±3σ deviation from the average).

demonstrate that they fit the purposes of application. According to the ISO/IEC 17025 (2005) standard, the techniques used for the determination of the performance of a method are one of, or a combination of, the following:

1. calibration using reference standards or reference materials,
2. comparison of results achieved with other methods,
3. interlaboratory comparisons (intercomparison exercises or proficiency tests),
4. systematic assessment of the factors influencing the result,
5. assessment of the uncertainty of the result based on scientific understanding of the theoretical principles of the method and practical experience.

Whenever standards and reference materials with known concentrations of the analytes are available, it is recommended that measurements be carried out using these materials. Reference materials are provided for a great variety of alpha-emitting nuclides by the IAEA, IRMM, and NIST, among others. Results obtained by alpha spectrometry can often be compared with those obtained by other techniques such as mass spectrometry, gamma spectrometry, and liquid scintillation counting. An increasing number of interlaboratory comparisons have been organized by national and international institutes, and network laboratories (e.g., IAEA, 2010b).

Although systematic assessment of the factors influencing the result (point 4 above) and evaluation of the uncertainty components (point 5) are extremely important, method validation should not be based exclusively on one of these validation techniques. It is desirable to apply—whenever it is possible—the combination of the validation techniques (points 1–3). For example, the recommended procedure of the IAEA for the simultaneous determination of Pu and Am nuclides in environmental samples was validated by preparing the uncertainty budget, by the measurement of six different reference materials by the proposed procedure, and by an interlaboratory comparison with two participating expert laboratories. In the course of the validation procedure, the performance parameters of the method including accuracy, uncertainty, detection limit, range of results, selectivity, linearity, repeatability and reproducibility, robustness against external influences, or cross-sensitivity against interferences have to be assessed (IAEA, 2012). It is recommended to follow the EURACHEM guide (EURACHEM Guide, 2014) for the method validation, which describes the validation procedure in detail.

IX. Conclusions

In recent years, the position of alpha spectrometry as the technique of choice when reliable isotopic analysis of very low activity levels is required has been challenged somewhat by improvements in other radioanalytical techniques, especially mass spectrometry. This is particularly the case for very long-lived radionuclides such as ^{238}U and ^{235}U. However, for shorter-lived alpha emitters such as ^{210}Po, the sensitivity of alpha spectrometry remains unchallenged due to a combination of low background and high detection efficiency.

The need for sample digestion and lengthy chemical separations remains the main drawback for standard alpha spectrometry analysis. The use of microwave digestion and the development of methods for fusion of solid samples or coprecipitation from water are increasingly common approaches for sample preparation and digestion. The development of methods utilizing EC is a major step forward for chemical separation, particularly for actinides. Some progress has also been made in automated chemical separations.

With increasing requirements on laboratories for traceability of results and for accreditation, the existence of standard, validated methods and of proficiency tests is becoming more important. Several methods have been published by ASTM, IAEA, and ISO, partially fulfilling this need. However, the large matrix of analytes and sample types, which can be analyzed by alpha spectrometry, means that a complete and up-to-date set of such methods is unlikely to materialize, and laboratories will continue to develop methods based on their specific needs. A wide variety of proficiency tests and reference materials would also be desirable for a similar reason.

Recently, in situ alpha spectrometry and use of other analytical methods in combination with alpha spectrometry have been under investigation. Coincidence techniques and list mode data acquisition are among the novel methods applied in alpha spectrometry. Special attention has also been paid to the development of spectrum analysis programs. The use of spectrum deconvolution software in conjunction with high- and medium-resolution alpha spectrometry is a promising tool to analyze overlapping peaks and alpha sources of "poor quality." Developments in spectrum evaluation can help replace or at least simplify chemical sample processing and the analysis of those nuclides whose spectra could not be evaluated with conventional alpha spectrometry.

Despite the challenges, alpha spectrometry remains important as a workhorse for many applications, as an essential tool in a number of important specialist applications and as a reliable and high-sensitivity technique for research and other investigative work.

Terms and definitions, symbols, and abbreviations

$a^{\#}$ detection limit (Bq kg^{-1})

a^{*} decision threshold (Bq kg^{-1})

a_0 the activity concentration of the analyte on the sampling date (Bq kg^{-1})

a_1 the activity concentration of the analyte at the separation time (Bq kg^{-1})

C_T the certified concentration of the tracer on the reference date (Bq g^{-1})

f_T the correction factor for the decay of the tracer during elapsed time (days) between the reference date of tracer and the beginning of the measurement

m_s the mass of sample (L)

m_T the mass of the tracer taken (g)

***n* and n_T** net counts in ROIs (region of interest) of analyte and tracer (counts), respectively,

n_0, and n_{0T} blank or background counts in ROIs (region of interest) of analyte and tracer (counts), respectively,

n_g and n_{gT} gross counts in ROIs (region of interest) of analyte and tracer (counts), respectively,

R the chemical recovery of the analyte

***r* and r_T** net count rates in ROIs (regions of interest) of analyte and tracer (counts), respectively,

r_0, and r_{0T} blank or background count rates in ROIs (regions of interest) of analyte and tracer (counts), respectively,

r_g and r_{gT} gross count rates in ROIs (regions of interest) of analyte and tracer (counts), respectively,

t_0 the counting time of blank (s)

t_1 the time interval between the sampling date and the separation time (year)

t_m the sample counting time (s)

t_T the time interval between the reference date of the tracer and the separation time (years)

$u(C_T)$ the uncertainty of the certified concentration of the tracer solution according to the calibration (Bq g^{-1})

$u(f_T)$ the uncertainty of correction factor for decay of the tracer during elapsed time (years) between the reference date of tracer and the separation time

$u(m_s)$ the uncertainty of mass of sample (kg)

$u(m_T)$ the uncertainty of mass of the tracer (g)

$u(n)$, $u(n_T)$ the uncertainty of the net counts in the ROIs of the analyte and the tracer, respectively

$u(\lambda_A)$ the uncertainty of the decay constant of the analyte ($year^{-1}$)

$u_c(a_0)$ combined uncertainty of the activity concentration of the analyte on sampling date (Bq kg^{-1})

λ_A the decay constant of the analyte ($years^{-1}$)

λ_T the decay constant of the tracer ($years^{-1}$)

References

Aarnio, P.A., Routti, G.T., Sandberg, J.V., 1987. MicroSAMPO: personnel computer based advanced gamma spectrum analysis system. In: Proc. American Nuclear Society Topical Conference on Methods and Applications of Radioanalytical Chemistry", Kona, Hawaii, 5-10 April, 14 pp.

Afsar, M., Schüttelkopf, H., 1988. Determination of Am-241, Curium-242 and Curium-244 in Environmental Samples, Kernforschungszentrum. Karlsruhe G.m.b.H., KfK 4346.

Ageyev, V.A., Odintsov, O.O., Sajeniouk, A.D., 2005a. Routine radiochemical method for the determination of ^{90}Sr, ^{238}Pu, $^{239+240}Pu$, ^{241}Am, and ^{244}Cm in environmental samples. J. Radioanal. Nucl. Chem. 264, 337–342.

Ageyev, V.A., Odintsov, O.O., Sajeniouk, A.D., 2005b. Radiochemical Method of Simultaneous Determination of ^{90}Sr, ^{238}Pu, $^{239+240}Pu$, ^{241}Am, and ^{244}Cm in Environmental Samples. NKS-140 Seminar, Tartu, Estonia, 43.

Aggarwal, S.A., 2016. Alpha-particle spectrometry for the determination of alpha emitting isotopes in nuclear, environmental and biological samples: past, present and future. Anal. Methods 8, 5353–5371.

Aggarwal, S.K., Alamelu, D., 2007. High resolution alpha-particle spectrometry for the determination of $^{240}Pu/^{239}Pu$ atom ratio in Pu samples from PHWR. J. Alloys Compounds 444–445, 656–659.

Alamelu, D., Aggarwal, S.K., 2012. Determination of uranium isotopes in irradiated thorium dioxide by alpha spectrometry using WinALPHA for deconvolution of complex spectra. J. Radioanal. Nucl. Chem. 294, 405–408.

Amoudry, F., 1990. L'analyse Par Spectrometrie Alpha. INSTN Ref. From Leblanc 2006.

Amoudry, F., Burger, P., 1984. Determination of the $^{239}Pu/^{240}Pu$ isotopic ratio by high resolution alpha spectrometry. Nucl. Instrum. Methods Phys. Res. 223, 360–367.

Anton, M.P., Espinosa, A., Aragon, A., 2006. Radiochemical determination of ^{237}Np in soil samples contaminated with weapon grade plutonium. Czechoslovak J. Phys. 56, D241–D246.

Apostolidis, C., Molinet, R., Richir, P., Ougier, M., Mayer, K., 1998. Development and validation of a simple, rapid and robust method for the chemical separation of uranium and plutonium. Radiochim. Acta 83, 21–25.

Arginelli, D., Montalto, M., Bortoluzzi, S., Nocente, M., Bonardi, M., Groppi, F., 2005. Radio-analytical determination of americium in human urines by extraction chromatography and high resolution alpha spectrometry. J. Radioanal. Nucl. Chem. 263 (2), 275.

Arnold, D., 2006. Improved determination of plutonium content and isotopic ratios in low activity samples by α-particle and underground L X-ray measurement. Appl. Radiat. Isot. 64, 1137–1140.

Aupiais, J., 2004. Deconvolution of alpha liquid scintillation spectra for quantitative analysis of actinides in water samples. Radiochim. Acta 92, 125–132.

Ayranov, M., Kraehenbuehl, U., Sahli, H., Röllin, S., Burger, M., 2005. Radiochemical separation of actinides from environmental samples for determination with DF-ICP-MS and alpha spectrometry. Radiochim. Acta 93, 249–257.

Babelowsky, T., Bortels, G., 1993. ALFA: a program for accurate analysis of complex alpha-particle spectra on a PC. Appl. Radiat. Isot. 44 (10–11), 1349.

Bagnall, K.W., 1957. Chemistry of the Rare Radioelements: Polonium–Actinium. Butterworths Scientific Publications, London.

Bagnall, K.W., 1966. The Chemistry of Selenium, Tellurium and Polonium. Elsevier Publishing Company, Amsterdam.

Bagnall, K.W., 1983. The chemistry of polonium. Radiochim. Acta 32, 153–161.

Bains, M.E.D.P., Warwick, E., 1993. The separation of actinides from lanthanides by anion exchange in methanol/hydrogen chloride medium, and its application to routine separation. Sci. Total Environ. 130/131, 437.

Ballestra, S., Fukai, R., 1983. Actinide-partitioning from high-level waste (HLW). Talanta 30, 45.

Baskaran, M., 2011. Po-210 and Pb-210 as atmospheric tracers and global atmospheric Pb-210 fallout: a review. J. Environ. Radioact. 102, 500–513.

Beasley, T.M., Kelley, J.M., Maiti, T.C., Bond, L.A., 1998. $^{237}Np/^{239}Pu$ atom ratios in integrated global fallout: a reassessment of the production of ^{237}Np. J. Environ. Radioact. 38, 133–146.

Bem, H., Bem, E., Krzemińska, M., Ostrowska, M., 2002. Determination of radioactivity in air filters by alpha and gamma spectrometry. Nukleonika 47, 87–91.

Berlioz, A.N., Sajenouk, A.D., Tryshyn, V.V., 2005. A technique for the determination of ^{90}Sr, plutonium and americium in environmental samples. J. Radioanal. Nucl. Chem. 263 (2), 307.

Bernabee, R.P., Percival, D.E., Hindman, F.D., 1980. Liquid-liquid extraction separation and determination of Pu and. Am. Anal. Chem. 52 (14), 2351.

Berne, A., 1995. Use of EiChrom TRU Resin in the Determination of Americium, Plutonium and Uranium in Air Filter and Water Samples. EML-575 Procedure for "Americium in Soil", p. 1.

Birks, J.B., 1964. The Theory and Practice of Scintillation Counting. Pergamon Pres, Oxford.

Blaauw, M., Garcia-Torano, E., Woods, S., Fazinic, S., 1999. The 1997 IAEA intercomparison of commercially available PC-based software for alpha-particle spectrometry. Nucl. Instrum. Methods Phys. Res., Sect. A 428, 317–329.

Bland, C.J., 1998. Choosing fitting functions to describe peak tails in alpha-particle spectrometry. Appl. Rad. Isotopes 49 (9–11), 1225–1229.

Bock, R.A., 1979. A Handbook of Decomposition Methods in Analytical Chemistry. Int. Textbook Co., Glasgow.

Bojanowski, R., Holm, E., Whitehead, N.E., 1987. Determination of ^{227}Ac in environmental samples by ion-exchange and alpha spectrometry. J. Radioanal. Nucl. Chem. 115, 23–37.

Bojanowski, R., Radecki, Z., Burns, K., 2005. Determination of radium and uranium isotopes in natural waters by sorption on hydrous manganese dioxide followed by alpha-spectrometry. J. Radioanal. Nucl. Chem. 264, 437–443.

Bortels, G., Collaer, P., 1987. Analytical function for fitting peaks in alpha-particle spectra from Si detectors. Appl. Radiat. Isot. 38 (10), 831–837.

Bortels, G., Hurtgen, C., Santry, D., 1995. Nuclide analysis on low-statistics alpha-particle spectra: an experimental verification for Pu isotopes. Appl. Radiat. Isot. 46 (11), 1135–1144.

Braun, T., Ghersení, G., 1975. Extraction Chromatography. Akadémiai kiadó, Budapest.

Burk, R.C., Wiles, D.R., 1983. Coprecipitation of ^{234}Th with $BaSO_4$ II. The effect of potassium ions. J. Radioanal. Chem. 79, 185–196.

Burnett, W.C., Corbett, D.R., Schultz, M., Horwitz, E.P., Chiarizia, R., Dietz, M., Thakkar, A., Fern, M., 1997. Pre-concentration of actinide elements from soils and large volume water samples using extraction chromatography. J. Radioanal. Nucl. Chem. 226, 121–127.

Burney, G.A., Harbour, R.M., 1974. Radiochemistry of neptunium. In: National Academy of Sciences-National Research Council, Nuclear Science Series NAS-NS-3060.

Capote Noy, R., Garcia-Torano, E., Mainegra, E., Lopez, E., 2004. The WinALPHA code for the analysis of alpha-particle spectra. Nucl. Instrum. Methods Phys. Res. A 525, 522–528.

Caro Marroyo, B., Martin Sanchez, A., Jurado Vargas, M., 2013. ALFITeX: a new code for the deconvolution of complex alpha-particle spectra. J. Radioanal. Nucl. Chem. 296, 1247–1252.

Caro Marroyo, B., Martin Sanchez, A., Jurado Vargas, M., 2014. Improvements to alpha-particle spectrometry techniques. Appl. Radiat. Isot. 87, 328–330.

Carvalho, F.P., 1997. Distribution, cycling and mean residence time of ^{226}Ra, ^{210}Pb and ^{210}Po in the Targus estuary. Sci. Total Environ. 196, 151–161.

Chamizo, E., Jiminez-Ramos, M.C., Wacker, L., Vioque, I., Calleja, A., Garcia-Leon, M., Garcia-Tenorio, R., 2008. Isolation of Pu isotopes from environmental samples using ion chromatography for accelerator mass spectrometry and alpha spectrometry. Anal. Chim. Acta 606, 239–245.

Chen, Q.J., Dahlgaard, H., Nielsen, S.P., Aarkrog, A., Christensen, I., Jensen, A., 2001a. Determination of ^{237}Np in sediment and sea water. J. Radioanal. Nucl. Chem. 249, 527–533.

Chen, Q., Aarkrog, A., Nielsen, S.P., Dahlgaard, H., Lind, B., Kolstad, A.K., Yu, Y., 2001b. Procedures for Determination of $^{239,240}Pu$, ^{241}Am, ^{237}Np, $^{234,238}U$, $^{228,230,232}Th$, ^{99}Tc, and ^{210}Pb-^{210}Po in Environmental Materials. Riso-R-1263(EN). *Report of the Riso National Laboratory* Roskilde.

Chen, Q., Hou, X., Dahlgaard, H., Nielsen, S.P., Aarkrog, A., 2001c. A rapid method for the separation of ^{210}Po from ^{210}Pb by TIOA extraction. J. Radioanal. Nucl. Chem. 249, 587–593.

Chiarizia, R., Horwitz, E.P., Alexandratos, S.D., Gula, M.J., 2011. DIPHONIX resin: a review of its properties and applications. Sep. Sci. Technol. (in press).

Chiu, N.W., Dean, J.R., Sill, C.W., 1984. Techniques of Sample Attack Used in Soil and Mineral Analysis. Atomic Energy Control Board, Ottawa, Canada. Research Report.

Choppin, G.R., Silva, R.J., 1956. Separation of the lanthanides by ion exchange with α-hydroxi-isobutyric acid. J. Radioanal. Nucl. Chem. 3, 153.

Choppin, G.C., Bond, A.H., Hromadka, P.M., 1997. Redox speciation of plutonium. J. Radioanal. Nucl. Chem. 219, 203–210.

Claisse, F., 1995. Glass Disks and Solutions by Fusion in Borates for Users of Claisse Fluxers. Corporation Scientific Claisse Inc.: Sainte-Foy, Quebec, Canada (Doc. PUB951115).

Cleveland, J.M., 1970. The Chemistry of Plutonium. Gordon and Breach, New York.

Coleman, G.H., 1965. The Radiochemistry of Plutonium. National Academy of Sciences National Research Council, NAS-NS 3058, USA.

Cozzella, M.L., Gianquillo, G., Pettirossi, R., Nelms, S., 2002. Determination of plutonium by ICP-MS in environmental samples of Casaccia site: a comparison with alpha spectroscopy. In: European IRPA Congress 2002 Florence (Italy) 8-11 Oct.

Crespo, M.T., 2000. On the determination of ^{226}Ra in environmental and geological samples by α–spectrometry using ^{225}Ra as yield tracer. Appl. Radiat. Isot. 53, 109–114.

Crespo, M.T., 2012. A review of electrodeposition methods for the preparation of alpha-radiation Sources. Appl. Radiat. Isot. 70, 210–215.

Dalvi, A.A., Verma, R., 2017. Radiochemical separation of ^{231}Pa from siliceous cake prior to its determination with gamma ray spectrometry. Radiochim. Acta 105 (10), 831–839.

Dayras, F., 2002. Analysis of ^{239}Pu, ^{244}Cm and ^{243}Am alpha spectra using the unfolding code Colegram without prior use of a nuclear data library. Nucl. Instrum. Methods Phys. Res. A 490, 492–504.

Dazhu, Y., Yongjun, Z., Möbius, S., 1991. Rapid method for alpha counting with extractive scintillator and pulse shape analysis. J. Radioanal. Nucl. Chem. 147 (1), 177.

DDEP recommended data, updated 20.10.2017, "http://www.nucleide.org/DDEP_WG/DDEPdata".

deJong, I.G., Wiles, D.R., 1984. Determination of ^{226}Ra and ^{230}Th: an improved procedure. J. Radioanal. Nucl. Chem. 82, 309–318.

Delle Site, A., Santori, G., Testa, C., 1973. The rapid determination of the transuranium elements by extraction chromatography in urines containing DTPA. In: Proc. Regional Conf. on Radiat. Prot., Jerusalem, 532-527.

Desideri, D., Mali, M.A., Rosalli, C., Testa, C., Boulyga, S.F., Becker, J.S., 2002. Determination of ^{236}U and transuranium elements in depleted uranium ammunition by α-spectrometry and ICP-MS. Anal. Bioanal. Chem. 374, 1091.

Desideri, D., Feduzi, L., Assunta Meli, M., Roselli, C., 2011. Sequential determination of Am, Cm, Pu, Np and U by extraction chromatography. Microchem. J. 97, 264–268.

Dinkar, A.K., Singh, S.K., Dhami, P.S., Gandhi, P.M., Verma, R., Reddy, A.V.R., 2017. Extraction of protactinium from acid media using Aliquat 336. Separation Science Technology 52 (8), 1359.

Dilbeck, G.A., Taylor, B., Leitch, J., Silverstone, M., Moore, B., Honsa, P., 2006. A novel technique for the rapid identification of alpha emitters released during a radiological incident. Health Phys. 91, 311–317.

Dion, M.P., Miller, B.W., Warren, G.A., 2016. Alpha and conversion electron spectroscopy of ^{238,239}Pu and ^{241}Am and alpha-conversion electron coincidence measurements. Nucl. Instrum. Methods Phys. Res. 830, 6–12.

Dion, M.P., Liezers, M., Farmer III, O.T., Miller, B.W., Morley, S., Barinaga, C., Eiden, G., 2015. J. Radioanal. Nucl. Chem. 303, 877–884.

Donard, O.F.X., Bruneau, F., Moldovan, M., Garraud, H., Epov, V.N., Boust, D., 2007. Multi-isotopic determination of Pu (^{239}Pu, ^{240}Pu, ^{241}Pu and ^{242}Pu) in marine sediments using sector field inductively coupled plasma mass spectrometry. Anal. Chim. Acta 587, 170–179.

Dulanská, S., Bilohuščin, J., Remenec, B., Galanda, D., Mátel, L., 2015. Determination of ^{239}Pu, ^{241}Am and ^{90}Sr in urine using pre-filter material and combined sorbents AnaLig® Pu-02, AnaLig® Sr-01, DGA® Resin. J. Radioanal. Nucl. Chem. 304 (1), 127–132.

Ďurecová, A., 1997. Contribution to the simultaneous determination of ^{228}Ra and ^{226}Ra by using 3M's EMPORE^{TM} Radium Rad Disks. J. *Radioanal. Nucl. Chem.* 223, 225–228.

Eichrom Industries, Inc, 1995. Analytical Procedures: ACW08.

Eikenberg, J., Bajo, S., Beer, H., Hitz, J., Ruethi, M., Zumsteg, I., Letessier, P., 2004. Fast methods for determination of anthropogenic actinides and U/Th-series isotopes in aqueous samples. Appl. Radiat. Isot. 61, 101–106.

Ensor, D.D., 1998. Separation and Analytical Chemistry of the Actinides. Report of Tennessee Technological University 493BH0017-3C.

EURACHEM Guide, 2014. The Fitness for Purpose of Analytical Methods-A Laboratory Guide to Method Validation and Related Topics, second ed., ISBN 978-91-87461-59-0.

Farmer, D.E., Steed, A.C., Sobus, J., Stetzenbach, K., Lindley, K., Hodge, V.F., 2003. Rapid identification and analysis of airborne plutonium using a combination of alpha spectroscopy and inductively coupled plasma mass spectrometry. Health Phys. 85, 457–465.

Figgins, P.E., 1961. The Radiochemistry of Polonium. National Academy of Sciences (USA), Nuclear Science Series NAS-NS 3037, Washington, D.C.

Fjeld, R.A., DeVol, T.A., Leyba, J.D., Paulenova, A., 2005. Measurement of radionuclides using ion chromatography and on-line radiation detection. J. Radioanal. Nucl. Chem. 263, 635–640.

García-Toraño, E., 1994. Fitting of low-statistics peaks in alpha-particle spectra. Nucl. Instrum. Methods Phys. Res., Sect. A 339 (1–2), 122–126.

García-Toraño, E., 1997. Proc. of Int. Symposium on "Advances in Alpha-, Beta- and Gamma-Spectrometry". Documentos CIEMAT, pp. 17–23.

García-Toraño, E., 2003. A model shape for the analysis of alpha particle spectra. Nucl. Instrum. Methods Phys. Res., Sect. A 498 (1–3), 289.

García-Toraño, E., 2006. Current status of alpha-particle spectrometry. Appl. Radiat. Isot. 64, 1273–1280.

García-Toraño, E., Crespo, M.T., Woods, S., Blaauw, M., Fazinic, S., 1999. The 1997 IAEA test spectra for alpha-particle spectrometry. Nucl. Instrum. Methods Phys. Res., Sect. A 428, 330–335.

Gascó, C., Antón, M.P., Delfanti, R., González, A.M., Meral, J., Papucci, C., 2002. Variation of the activity concentrations and fluxes of natural (^{210}Po, ^{210}Pb) and anthropogenic (^{239,240}Pu, ^{137}Cs) radionuclides in the Strait of Gibraltar (Spain). J. Environ. Radioact. 62, 241–262.

Gascon, J.L., Munoz, A., 2003. Optimization of the parameters affecting the solid state detector efficiency in alpha-spectrometry. J. Radioanal. Nucl. Chem. 257 (2), 371–374.

Germian, P., Pinte, G., 1990. Neptunium-237 in the marine environment, determination in animal and plant species in the English channel: biological indicators and trophic relationships. J. Radioanal. Nucl. Chem. Art. 138, 49–61.

Giardina, I., Andreocci, L., Bazzarri, S., Mancini, S., Battisti, P., 2006. A comparison of different radiochemical methods applicable for the determination of plutonium isotopes in urine via alpha spectrometry. 15th Radiochemical Conference Marianske Lazne (Czech Republic) 23-28 Apr 2006, *Czech*. J. Phys. 56, D265–D270.

Gilmore, G.R., 2008. Practical Gamma-Ray Spectrometry. J. Wiley and Son, Chichester, 387 pp.

Gleason, G., 1980. An improved ion exchange procedure for the separation of barium from radium. In: Lyon, W.S. (Ed.), Radioelement Analysis Progress and Problems. Ann Arbor Science Co., pp. 47–50

Gleisberg, B., Köhler, M., 2002. Rapid determination of low ^{237}Np activities in uranyl nitrate solutions using extraction chromatography and gamma-ray spectrometry. J. Radioanal. Nucl. Chem. 254, 59–63.

Gogorova, S., Matel, L., 2007. Minulost a Sucasne Trendy Jadrovej Chemie. Omega Info.

Goliáš, V., Pittauerová, D., Procházka, R., Třískala, Z., 2005. Field alpha-spectroscopy of radon (^{222}Rn) and actinon (^{219}Rn) progeny in soil gas: locating a radon source. J. Radioanal. Nucl. Chem. 266, 461–470.

Gonzales, E.R., Peterson, D.S., 2009. Rapid radiochemical sample preparation for alpha spectrometry using polymer ligand films. J. Radioanal. Nucl. Chem. 282, 543–547.

Gorietti, D., Giardina, I., Arginelli, D., Battisti, P., 2017. Determination of plutonium, americium and curium isotopes in radioactive metal wastes deriving from nuclear decommissioning. J. Radioanal. Nucl. Chem. 314, 1785–1792.

Grabowski, P., Bem, H., 2010. Determination of ^{210}Po and uranium in high salinity water samples. J. Radioanal. Nucl. Chem. 2, 455–460.

Groska, J., Vajda, N., Molnar, Z., Bokori, E., Szeredy, P., Zagyvai, M., 2016. Determination of actinides in radioactive waste after separation on a single DGA resin column. J. Radioanal. Nucl. Chem. 309 (3), 1145–1158.

Guerin, N., Langevin, M.A., Nadeau, K., Labrecque, C., Gagne, A., Lanviere, D., 2010. Determination of neptunium in environmental samples by extraction chromatography after valence adjustment. Appl. Radiat. Isot. 68, 2132–2139.

Gunnink, R., Ruhter, W.D., 1990. MGA: A Gamma-Ray Spectrum Analysis Code for Determining Plutonium Isotopic Abundances, A Guide to Using MGA, vol. 2. Lawrence Livermore National Laboratory report, UCRL-LR-103220.

Guseva, L.I., Tikhomirova, G.S., 1968. Radiokhimiya 10, 246 refered by Myasoedov et al. (1974).

Guseva, L.I., Lebedev, I.A., Myasoedov, B.F., Tikhomirova, G.S., 1976. Isolation of trans-plutonium elements on ionites in aqueous/alcohol madia. J. Radioanal. Nucl. Chem. Suppl. 55.

Hallstadius, L., 1983. A method for the electrodeposition of actinides. Nucl. Instrum. Methods Phys. Res. 223, 266–267.

Hamilton, T.F., McRae, V.M., Smith, J.D., 1994. Radium isotope determination by alpha-spectrometry after electrodeposition from solution with added platinum. J. Radioanal. Nucl. Chem. 177, 365–371.

Hancock, G.J., Martin, P., 1991. Determination of Ra in environmental samples by α-particle spectrometry. Appl. Radiat. Isot. 42, 63–69.

Hancock, G.J., Webster, I.T., Ford, P.W., Moore, W.S., 2000. Using Ra isotopes to examine transport processes controlling benthic fluxes into a shallow estuarine lagoon. Geochim. Cosmochim. Acta 64, 3685–3699.

Hanson, S.K., Mueller, A.H., Oldham Jr., W.J., 2014. Klaui ligand films for rapid plutonium analysis by alpha spectrometry. Anal. Chem. 86, 1153–1159.

Harrison, J.J., Payne, T.E., Wilsher, K.L., Thiruvoth, S., Child, D.P., Johansen, M.P., Hotchkis, M.A.C., 2016. Measurement of ^{233}U/^{234}U ratios in contaminated groundwater using alpha spectrometry. J. Environ. Radioact. 151, 537–541.

Henricsson, F., Ranebo, Y., Holm, E., Roos, P., 2011. Aspects on the analysis of ^{210}Po. J. Environ. Radioact. 102, 415–419.

Harvey, B.R., Lovett, M.B., 1984. The use of yield tracers for the determination of alpha-emitting actinides int he marine environment. Nucl. Instrum. Methods Phys. Res. 223, 224–234.

Hoffman, D.C., 2001. Advances in Plutonium Chemistry 1967-2000. American Nuclear Society, USA.

Hoffman, I., Ungar, K., Bean, M., Pöllänen, R., Ihantola, S., Toivonen, H., Karhunen, T., Pelikan, A., 2011. Direct alpha analysis for forensic samples (DAAFS): techniques, applications, and results. In: Proceedings – Third European IRPA Congress, 14–18 June 2010. Nordic Society for Radiation Protection, Helsinki, Finland. Helsinki, 2011 [PDF publication]. Web version online 2010 June 16. [P11-05, 6 pp.].

Holm, E., Ballestra, S., Fukai, R., 1979. A method for ion-exchange separation of low levels of americium in environmental materials. Talanta 26, 791.

Holm, E., Fukai, R., 1976. Determination of americium and curium by using ion-exchange in nitric acid-methanol medium for environmental analysis. Talanta 23, 853.

Holmes, L., 2004. Alpha-spectrometric analysis of environmental samples-Spread sheet approach. In: Quantifying Uncertainty in Nuclear Analytical Measurements. IAEA-TECDOC-1401, International Atomic Energy Agency, Vienna, pp. 141–148.

Homepage of PERALS (2008). www.perals.com. https://nats-usa.com/product/perals-spectrometer-2/.

Hoover, A.S., Bond, E.M., Croce, M.P., Holesinger, T.G., Kunde, G.J., Rabin, M.W., Wolfsberg, L.E., Bennett, D.A., Hays-Wehle, J.P., Schmidt, D.R., Swetz, D., Ullom, J.N., 2015. Measurement oft he ^{240}Pu/^{239}Pu mass ratio using a transition-edge-sensor microcalorimeter for total decay energy spectroscopy. Anal. Chem. 87, 3996–4000.

Hopkins, P.D., Mastren, T., Florek, J., Copping, R., Brugh, M., John, K.D., Nortier, M.F., Birnbaum, E.R., Kleitz, F., Fassbender, M.E., 2018. Dalton Trans. 47, 5189–51395.

Horwitz, E.P., Bloomquist, C.A.A., 1975. Chemical separations for super-heavy element searches in irradiated uranium targets. J. Inorg. Nucl. Chem. 37 (1), 425–434.

Horwitz, E.P., 1992. Separation and preconcentration of uranium from acidic media by extraction chromatography. Anal. Chim. Acta 266, 25–37.

Horwitz, E.P., Orlandini, K.A., Bloomquist, C.A.A., 1966. The separation of Am and Cm by extraction chromatography using a high molecular weight quaternary ammonium nitrate. Inorg. Nucl. Chem. Lett. 2, 87.

Horwitz, E.P., Bloomquist, C.A.A., Orlandini, K.A., Henderson, D.J., 1967. The separation of Milligram Quantities of Am and Cm by extraction chromatography. Radiochim. Acta 8, 127.

Horwitz, E.P., Martin, K.A., Diamond, H., Kaplan, L., 1986. Extraction of Am from nitric acid by carbamoyl-phosphoryl extractants: the influence of substituents on the selectivity of Am over Fe and selected fission. Solv. Extr. Ion Exch. 4 (3), 449.

Horwitz, E.P., Diamond, H., Martin, K.A., 1987. The extraction of selected actinides in the (III), (IV) and (VI) oxidation states from hydrochloric acid by OπD(iB) CMPO: the TRUEX-Chloride Process. Solv. Extr. Ion Exch. 5 (3), 447–470.

Horwitz, E.P., Dietz, M., Chiarizia, R., Dietz, M.L., Diamond, H., 1993. Separation and preconcentration of actinides from acidic media by extraction chromatography. Anal. Chim. Acta 284, 361–372.

Horwitz, E.P., Dietz, M., Chiarizia, R., Diamond, H., Maxwell, S.L., Nelson, M.R., 1995. Separation and preconcentration of actinides by extraction chromatography using a supported liquid anion exchanger: application to the characterization of high-level nuclear waste solutions. Anal. Chim. Acta 310, 63–78.

Horwitz, E.P., Chiarizia, R., Dietz, M.L., 1997. DIPEX: a new extraction chromatographic material for the separation and preconcentration of actinides from aqueous solution. React. Funct. Polym. 33, 25–36.

Horwitz, E.P., McAlister, D.R., Bond, A.H., Barrans Jr., R.E., 2005. Novel extraction chromatographic resins based on tetraalkyldiglycolamides: characterization and potential applications. Solv. Extr. Ion Exch. 23, 319–344.

Hou, X., Roos, P., 2007. Critical Comparison of Radiometric and Mass Spectrometric Methods for the Determination of Radionuclides in Environmental, Biological and Nuclear Waste Samples. Radiation Research Department, Riso, pp. 1–87. www.risoe.dk/rispubl/art/2007_333.pdf.

Hölgye, Z., 2007. Coprecipitation of polonium with bismuth phosphate. J. Radioanal. Nucl. Chem. 274, 647–649.

Hrnecek, E., Heras, L.A., Betti, M., 2002. Application of microprecipitation on membrane filters for alpha spectrometry, liquid scintillation counting and mass spectrometric determination of plutonium isotopes. Radiochim. Acta 90, 721–725.

Hurtado, S., Jiménez-Ramos, M.C., Villa, M., Vioque, I., Manjón, G., Garcia-Tenorio, R., 2008. Numerical analysis of alpha spectra using two different codes. Appl. Radiat. Isot. 66, 808–812.

Hyde, E.K., 1960. The Radiochemistry of Thorium. National Academy of Sciences-National Research Council, Nuclear Science Series NAS-NS-3004.

IAEA, 1989. Measurement of Radionuclides in Food and Environment. A Guidebook. In: International Atomic Energy Agency, Technical Report Series No. 295, Vienna.

IAEA, 2004. Quantifying Uncertainty in Nuclear Analytical Measurements. IAEA-TECDOC-1401, International Atomic Energy Agency, Vienna.

IAEA, 2010a. Analytical Methodology for the Determination of Radium Isotopes in Environmental Samples. IAEA/AQ/19, International Atomic Energy Agency, Vienna.

IAEA, 2010b. Worldwide Open Proficiency Test: Determination of Naturally Occurring Radionuclides in Phosphogypsum and Water. IAEA/AQ/18, International Atomic Energy Agency, Vienna.

IAEA, 2012. A Combined Procedure for Determination of Plutonium Isotopes, 2^{41A}m and 9^{0S}r in Environmental Samples. IAEA/AQ/37, International Atomic Energy Agency, Vienna.

IAEA, 2014a. A Procedure for the Sequential Determination of Radionuclides in Phosphogypsum: Liquid Scintillation Counting and Alpha Spectrometry for ^{210}Po, ^{210}Pb, ^{226}Ra, Th and U Radioisotopes. IAEA/AQ/34. International Atomic Energy Agency, Vienna.

IAEA, 2014b. A Procedure for the Sequential Determination of Radionuclides in Environmental Samples: Liquid Scintillation Counting and Alpha Spectrometry for ^{90}Sr, ^{241}Am and Pu Radioisotopes. IAEA/AQ/37, International Atomic Energy Agency, Vienna.

IAEA, 2017. Determination and Interpretation of Characteristic Limits for Radioactivity Measurements. International Atomic Energy Agency, Vienna. IAEA/AQ/48 (2017).

IAEA TECDOC-1104, 1999. Intercomparison of Alpha-Particle Spectrometry Software Packages. International Atomic Energy Agency, Vienna.

Ibrahim, S.A., Whicker, F.W., 1987. Plant accumulation and plant/soil concentration ratios of ^{210}Pb and ^{210}Po at various sites within a uranium mining and milling operation. Environ. Exp. Bot. 27, 203–213.

Ihantola, S., Pelikan, A., Pöllänen, R., Toivonen, H., 2011. Advanced alpha spectrum analysis based on the fitting and covariance analysis of dependent variables. Nucl. Instrum. Methods Phys. Res. 656, 55–60.

Ihantola, S., Pöllänen, R., Toivonen, H., Pelikan, A., 2011b. Determining the plutonium isotope ratio using alpha spectroscopy. In: Proceedings – 33rd ESARDA Annual Meeting. Symposium on Safeguards and Nuclear Material Management. 2011 May 16–20; Budapest, Hungary.

Inn, K.G.W., Hall, E., Woodward, J.T.I.V., Stewart, B., Pöllänen, R., Selvig, L., Turner, S., Outola, I., Nour, S., Kurosaki, H., LaRosa, J., Schultz, M., Lin, Z., Yu, Z., McMahon, C., 2008. Use of thin collodion films to prevent recoil-ion contamination of alpha-spectrometry detectors. J. Radioanal. Nucl. Chem. 276, 385–390.

ISO Guide to the Expression of Uncertainty in Measurement, 1995.

ISO 11483, 2005. Nuclear Fuel Technology - Preparation of Plutonium Sources and Determina-Tion of ^{238}Pu/^{239}Pu Isotope Ratio by Alpha Spectrometry.

ISO 11929, 2010. Determination of the Characteristic Limits (Decision Threshold, Detection Limit and Limits of the Confidence Interval) for Measurements of Ionizing Radiation – Fundamentals and Application.

ISO 15366-1, 2014. Nuclear Fuel Technology-Chemical Separation and Purification of Uranium and Plutonium in Nitric Acid Solutions for Isotopic and Isotopic Dilution Analysis by Solvent Extraction Chromatography. Part 1: Samples Containing Plutonium in the Microgram Range and Uranium in the Milligram Range.

ISO 15366-2, 2014. Nucelar Fuel Technology-Chemical Separation and Purification of Uranium and Plutonium in Nitric Acid Solutions for Isotopic and Isotopic Dilution Analysis by Solvent Extraction Chromatography. Part 2: Samples Containing Plutonium and Uranium in the Nanogram Range and below.

ISO/IEC 17025, 2017. General Requirements for the Competence of Testing and Calibration Laboratories.

Jakopic, R., Tavcar, P., Benedik, L., 2007. Sequential determination of Pu and Am radio-isotopes in environmental samples; a comparison of two separation procedures. Appl. Radiat. Isot. 65, 504–511.

Jernström, J., 2006. Development of Analytical Techniques for Studies on Dispersion of Actinides in the Environment and Characterization of Environmental Particles. Academic Dissertation, Helsinki.

Jerome, S.M., Collins, S.M., Happel, S., Ivanov, P., Russell, B.C., 2018. Isolation and purification of protactinium-231. Appl. Radiat. Isot. 134, 18–22.

Jia, G., Desideri, D., Guerra, F., Meli, M.A., Testa, C., 1997. Determination of Plutonium and Americium in moss and lichen samples. J. Radioanal. Nucl. Chem. 220 (1), 15.

Jia, G., Testa, C., Desideri, D., Guerra, F., Roselli, C., 1998. Sequential separation and determination of plutonium, americium-241 and strontium-90 in soils and sediments. J. Radioanal. Nucl. Chem. 230, 21–28.

Jia, G., Belli, M., Blasi, M., Marchetti, A., Rosamilia, S., Sansone, U., 2001. Determination of ^{210}Pb and ^{210}Po in mineral and biological samples. J. Radioanal. Nucl. Chem. 247, 491–499.

Jia, G., Torri, G., Petruzzi, M., 2004. Distribution coefficients of polonium between 5% TOPO in toluene and aqueous hydrochloric and nitric acids. Appl. Radiat. Isot. 61, 279–282.

Jia, G., Torri, G., Innocenzi, P., Ocone, R., Di Lullo, A., 2006. Determination of radium isotopes in mineral and environmental water samples by alpha-spectrometry. J. Radioanal. Nucl. Chem. 267, 505–514.

Johansson, M., Skarnemark, G., 2001. Extraction of polonium from aqueous lactic acid solutions using dioctyl sulphide, Cyanex 272, Cyanex 301 or Cyanex 302 in toluene. J. Radioanal. Nucl. Chem. 250, 473–476.

Joe, K., Song, B.C., Kim, Y.B., Han, S.H., Jeon, Y.S., Jung, E.C., Jee, K.Y.(, 2007. Determination of the transuranic elements inventory in high burnup PWR spent fuel samples by alpha spectrometry. Nucl. Eng. Technol. 39, 673–682.

Johnston, P.N., Moroney, J.R., Burns, P.A., 1991. Preparation of radionuclide "sources" for coincident high-resolution spectrometry with low-energy photons and electrons or alpha-particles. Appl. Radiat. Isot. 42, 245–249.

Juillard, A., 1999. Resolution dans des bolometres equips de couches minces d'isolant d'Anderson pour des evenements impliquant des reculs de noyaux (Thesis). University Paris XI Orsay. Ref. from Leblanc 2006.

Kalmykov, S.N., Aliev, R.A., Sapozhnikov, D.Y., Sapozhnikov, Y.A., Afinegonov, A.M., 2004. Determination of Np-237 by radiochemical neutron activation analysis combined with extraction chromatography. Appl. Radiat. Isot. 60, 595–599.

Kanisch, G., 2004. Alpha-spectrometric analysis of environmental samples. In: Quantifying Uncertainty in Nuclear Analytical Measurements. IAEA-TECDOC-1401, International Atomic Energy Agency, Vienna, pp. 127–139.

Katz, J.J., Seaborg, G.T., Morss, L.R., 1986. The Chemistry of the Actinide Elements. Chapman and Hall.

Kehagia, K., Potiriadis, C., Bratakos, S., Koukouliou, V., Drikos, G., 2007. Determination of ^{226}Ra in urine samples by alpha spectrometry. Radiat. Prot. Dosim. 127, 293–296.

Kenna, T.C., 2002. Determination of plutonium isotopes and neptunium-237 in environmental samples by inductively coupled plasma mass spectrometry with total sample. J. Anal. At. Spectrom. 17, 1471–1479.

Kenna, T.C., 2009. Using sequential extraction techniques to assess the partitioning of plutonium and neptunium-237 from multiple sources in sediments from Ob River (Siberia). J. Environ. Radioact. 100, 547–557.

Kim, C.K., Morita, S., Seki, R., Takaku, Y., Ikeda, N., Assinder, D.J., 1992. Distribution and behaviour of ^{99}Tc, ^{237}Np, ^{239,240}Pu, and ^{241}Am in the coastal and estuarine sediments of the Irish sea. J. Radioanal. Nucl. Chem. Art 156, 201–213.

Kim, C.S., Kim, C.K., Lee, K.J., 2004. Simultaneous analysis of ^{237}Np and Pu isotopes in environmental samples by ICP-SF-MS coupled with automated sequential injection system. J. Anal. At. Spectrom. 19, 743–750.

Kim, C.-K., Kim, C.-S., Sansone, U., Martin, P., 2008. Development and application of an on-line sequential injection system for the separation of Pu, ^{210}Po and ^{210}Pb from environ-mental samples. Appl. Radiat. Isot. 66, 223–230.

Kim, C.-K., Martin, P., Fajgelj, A., 2008b. Quantification of measurement uncertainty in the sequential determination of ^{210}Pb and ^{210}Po by liquid scintillation counting and alpha-particle spectrometry. Accred Qual. Assur. 13, 691–702.

Kim, C.-K., Lee, M.-H., Martin, P., 2009. Method validation of a procedure for determination of ^{210}Po in water using DDTC solvent extraction and Sr resin. J. Radioanal. Nucl. Chem. 279, 639–646.

Kimura, T., 1990. Simultaneous determination of neptunium, plutonium, americium and curium using coprecipitation with bismuth phosphate. J. Radioanal. Nucl. Chem. Art. 139, 297–305.

Kirby, H.W., 1959. The Radiochemistry of Protactinium. National Academy of Sciences, National Research Council, Nuclear Science Series NAS-NS 3016.

Klemenčič, H., Benedik, L., 2010. Alpha-spectrometric thin source preparation with emphasis on homogeneity. Appl. Radiat. Isot. 68, 1247–1251.

Knight, A.W., Eitrheim, E.S., Nelson, A.W., Peterson, M., McAlister, D., Forbes, T.Z., Schultz, M.K., 2016. Trace-level extraction behaviour of actinide elements by aliphatic alcohol extractants in mineral acids: insight into the trace solution chemistry of protactinium. Solv. Extr. Ion Exch. 34 (6), 509–521.

Knight, A.W., Nelson, A.W., Eitrheim, E.S., Forbes, T.Z., Schultz, M.K., 2016b. A chromatographic separation of neptunium and protactinium using 1-octanol impregnated onto a solid support. J. Radioanal. Nucl. Chem. 307, 59–67.

Knoll, G.F., 2010. Radiation Detection and Measurement, fourth ed. J. Wiley and Sons, New York. 830 pp.

Komosa, A., Chibowski, S., 2002. Determination of Pu in ground-level air aerosols collected on Petrianov filters. J. Radioanal. Nucl. Chem. 251, 113–117.

Komura, K., Sakanoue, M., Yamamoto, M., 1984. Determination of ^{240}Pu/^{239}Pu ratio in environmental samples based on the measurement of Lx/alpha-ray activity ratio. Health Phys. 46, 1213–1219.

Koskelo, M.J., Burnett, W.C., Cable, P.H., 1996. An advanced analysis program for alpha spectrometry. Radioact. Radiochem. 7, 18–27.

Koyoma, S., Otsuka, Y., Osaka, M., Morozumi, K., Konno, K., Kajitani, M., Mitsugashira, T., 1998. Analysis of minor actinides in mixed oxide fuel irradiated in fast reactor, I; Determination of neptunium-237. J. Nucl. Sci. Technol. 35, 406–410.

Kraus, K.A., Nelson, F., Clough, F.B., Carlston, R.C., 1955. Anion exchange studies. XVI. Adsorption from lithium chloride solutions. J. Am. Chem. Soc. 77, 1391.

Krupp, D., Scherer, U., 2018. Prototype development of ion exchangeing alpha detectors. Nucl. Instrum. Methods Phys. Res. 897, 120–128.

Labrecque, C., Whitty-Leveille, L., Lariviere, D., 2013. Cloud point extraction of pluronium in environmental matrices coupled to ICPMS and α spectrometry in highly acidic conditions. Anal. Chem. 85, 10549–10555.

Lally, A.E., Glover, K.M., 1984. Source preparation in alpha spectrometry. Nucl. Instrum. Methods Phys. Res. 223, 259–265.

Lawrie, W.C., Desmond, J.A., Spence, D., Anderson, S., Edmonson, C., 2000. Determination of radium-226 in environmental and personal monitoring samples. Appl. Radiat. Isot. 53, 133–137.

L'Hoir, 1975. These 3eme Cycle. Universite de Paris. Ref. Bland, 1998.

L'Hoir, 1984. Study of the asymmetric response of silicon surface barrier detectors to MeV light ions. Nucl. Instrum. Methods 223, 336. Ref. Bland, 1998.

LaMont, S.P., Glover, S.E., Filby, R.H., 1998. Determination of plutonium-240/239 ratios in low activity samples using high resolution alpha-spectrometry. J. Radioanal. Nucl. Chem. 234 (1–2), 195–199.

Langmir, D., Riese, A.C., 1985. The thermodynamic properties of radium. Geochim. Cosmochim. Acta 49, 1593–1601.

LaRosa, J.J., Burnett, W., Lee, S.H., Levy, I., Gastaud, J., Povinec, P.P., 1989. Separation of Actinides, Cesium and Strontium from Marine Samples Using Extraction Chromatography and Sorbents. IAEA Technical Report Series TRS-295.

LaRosa, J.J., Burnett, W.C., Lee, S.H., Levy, I., Gastaud, J., Povinec, P.P., 2001. Separation of actinides, cesium and strontium from marine samples using extraction chromatography and sorbents. J. Radioanal. Nucl. Chem. 248, 765–770.

LaRosa, J.J., Gastaud, J., Lagan, L., Lee, S.H., Levy-Polomo, I., Povinec, P.P., Wyse, E., 2005. Recent developments in analysis of transuranics (Np, Pu, Am) in seawater. J. Radioanal. Nucl. Chem. 263, 427–436.

La Rosa, J., Outola, I., Crawford, E., Nour, S., Kurosaki, H., Inn, K., 2008. Radiochemical measurement of ^{237}Np in a solution of mixed radionuclides: experiences in chemical separation and alpha-spectrometry. J. Radioanal. Nucl. Chem. 277, 11–18.

Leblanc, E., Coron, N., Leblanc, J., de Marcillac, P., Bouchard, J., Plagnard, J., 2006. High-energy resolution alpha spectrometry using cryogenic detectors. Appl. Radiat. Isot. 64, 1281–1286.

Lederer, M., 1992. The Periodic Table of Chromatographers. J. Wiley and Sons, Chichester.

Lee, Y.K., Bakhtiar, S.N., Akbarzadeh, M., Lee, J.S., 2000. Sequential isotopic determination of strontium, thorium, plutonium, uranium, and americium in bioassay samples. J. Radioanal. Nucl. Chem. 243 (2), 525.

Lee, M.H., Kim, J.Y., Kim, W.H., Yung, E.C., Jee, K.Y., 2008. Investigation of the oxidation states of plutonium isotopes in a hydrochloric acid solution. Appl. Radiat. Isot. 66, 1975–1979.

Lehto, Y., Hou, X., 2011. Chemistry and Analysis of Radionuclides. Wiley-VCH, New York.

Leon, V., Mitchell, P.J., Condren, O.M., Moran, M., Vives, I., Batlle, J., Sanchez-Cabeza, J.A., 1996. Determination of the ^{240}Pu/^{239}Pu atom ratio in low activity environmental samples by alpha spectrometry and spectral deconvolution. Nucl. Instrum. Methods Phys. Res., Sect. A 369 (2–3), 597. Ref. Garcia-Torano, 2006.

Lin, Z., Wu, Z., 2009. Analysis of polonium-210 in food products and bioassay samples by isotope-dilution alpha spectrometry. Appl. Radiat. Isot. 67, 907–912.

Lidström, K., Tjärnhage, Å., 2001. A method for field measurements using alpha-spectrometry. J. Radioanal. Nucl. Chem. 248, 555–560.

Livens, F.R., Singleton, D.L., 1989. Evaluation of methods for the radiometric determination of americium-241 in environmental samples. Analyst 114, 1097.

Loveland, W., Morrissey, D.J., Seaborg, G.T., 2006. Modern Nuclear Chemistry. J. Wiley and Sons, Hoboken, 671 pp.

Lovett, M.B., Boggis, S.J., Blowers, P., 1990. *MAFF-AEPAM—7*; Ministry of Agriculture Fisheries and Food Directorate of Fisheries Research: Lowestoft.

Lozano, J.C., Madruga, S., Fernández, F., 2000. A function using cubic splines for the analysis of alpha-particle spectra from silicon detectors. Nucl. Instrum. Methods Phys. Res., Sect. A 449, 356–365.

Lozano, J.C., Vera Tomé, F., Blanco Rodríguez, P., Prieto, C., 2010. A sequential method for the determination of ^{210}Pb, ^{226}Ra, and uranium and thorium isotopes by LSC and alpha-spectrometry. Appl. Radiat. Isot. 68, 828–831.

Lujaniene, G., Sapolaite, J., 2005. Determination of cesium, americium and plutonium isotopes in the environmental samples. J. Nordic Nucl. Saf. Res. NKS-140 Seminar, Tartu, Estonia 113–115.

Luo, M., Xing, S., Yang, Y., Song, L., Ma, Y., Wang, Y., Dai, X., Happel, S., 2018. Sequential analysis of actinides in large-size soil and sediment samples with total sample dissolution. J. Environ. Radioact. 187, 73–80.

Mabuchi, H., 1963. On the volatility of some polonium compounds. J. Inorg. Nucl. Chem. 25, 657–660.

Magill, J., Galy, J., 2005. Radioactivity, Radionuclides, Radiation. Springer-Verlag Berlin Heidelberg and European Communities, Karlsruhe, p. 62.

Manickam, E., Sdraulig, S., O'Brien, R., 2010. An improved and rapid radiochemical method for the determination of polonium-210 in urine. Aust. J. Chem. 63, 38–46.

Mannion, J.M., Locklair, W.D., Powell, B.A., Husson, S.M., 2016. Alpha spectroscopy substrates based on thin polymer films. J. Radioanal. Nucl. Chem. 307, 2339–2345.

Marouli, M., Pomme, S., Jobbágy, V., Van Ammel, R., Paepen, J., Stroh, H., Benedik, L., 2014. Alpha-particle emission probabilities of ^{236}U obtained by alpha spectrometry. Appl. Radiat. Isot. 87, 292–296.

Martin, A., Blanchard, R.L., 1969. The thermal volatilization of caesium-137, polonium-210 and lead-210 from in vivo labeled samples. Analyst 94, 441–446.

Martin, P., Hancock, G.J., 2004a. Peak resolution and tailing in alpha-particle spectrometry for environmental samples. Appl. Radiat. Isot. 61, 161–165.

Martin, P., Hancock, G.J., 2004b. Routine analysis of naturally occurring radionuclides in environmental samples by alpha-particle spectrometry. In: Supervising Scientist Report 180, Supervising Scientist for the Alligator Rivers Region. AGPS, Canberra.

Martin-Sanchez, A., Rubio Montero, P., Vera Tome, F., 1996. FITBOR: a new program for the analysis of complex alpha spectra. Nucl. Instrum. Methods Phys. Res., Sect. A 369, 593–596.

Martin Sanchez, A., Rubio Montero, P., 1999. Simplifying data fitting using branching ratios as constraints in alpha spectrometry. Nucl. Instrum. Methods Phys. Res. 420, 481–488.

Martín Sánchez, A., de la Torre Pérez, J., 2012. Portable alpha spectrometer. Appl. Radiat. Isot. 70, 2267–2269.

Martín Sánchez, A., Caro Marroyo, B., 2012. Assembly of an alpha-gamma coincidence measuring device for checking alpha decay schemes. Appl. Radiat. Isot. 70, 2263–2266.

Mastren, T., Stein, B.W., Parker, T.G., Radchenko, V., Copping, R., Owens, A., Wyant, L.E., Brugh, M., Kozimor, S.A., Nortier, M.,F., Birnbaum, E.R., John, K.D., Fassbender, M.E., 2018. Anal. Chem. 90 (11), 7012–7017.

Mathew, E., Matkar, V.M., Pillai, K.C., 1981. Determination of Pu, Am and Cm in environmental materials. J. Radioanal. Chem. 62, 267.

Matthews, K.M., Kim, C.-K., Martin, P., 2007. Determination of ^{210}Po in environmental materials: a review of analytical methodology. Appl. Radiat. Isot. 65, 267–279.

Maxwell, S.L., 2006. Rapid column extraction method for actinides and Sr-89/90 in water samples. J. Radioanal. Nucl. Chem. 267, 537–543.

Maxwell, S.L., 2008. Rapid method for determination of plutonium, americium and curium in large soil samples. J. Radioanal. Nucl. Chem. 275, 395–402.

Maxwell, S.L., Jones, V.D., 1998. Rapid Separation Methods to Characterize Actinides and Metallic Impurities in Plutonium Scrap Materials at SRS, WSRC-MS-98-00122. DOE Contract. No. DE-AC09-96SR18500.

Maxwell, S.L., Culligan, B.K., 2006. Rapid column extraction method for actinides in soil. J. Radioanal. Nucl. Chem. 270, 699–704.

Maxwell, S.L., Culligan, B.A., Jones, V.D., Nichols, S.T., Noyes, G.W., 2011a. Rapid determination of ^{237}Np and Pu isotopes in water by inductively-coupled plasma mass spectrometry and alpha spectrometry. J. Radioanal. Nucl. Chem. 287, 223–230.

Maxwell, S.L., Culligan, B.K., Noyes, G.W., 2011b. Rapid separation method for ^{237}Np and Pu isotopes in large soil samples. Appl. Radiat. Isot. 69, 917–923.

Maxwell, S.L., Culligan, B.K., Kelsey-Wall, A., Shaw, P.J., 2011c. Rapid radiochemical method for determination of actinides in emergency concrete and brick samples. Anal. Chim. Acta 701 (1), 112–118.

Maxwell, S.L., Culligan, B.K., Hutchinson, J.B., 2013. Rapid fusion method for determination of plutonium isotopes in large rice samples. J. Radioanal. Nucl. Chem. 298, 1367–1374.

Maxwell, S.L., Culligan, B.K., Hutchinson, J.B., 2014. Rapid determination of actinides and in asphalt samples. J. Radioanal. Nucl. Chem. 299 (3), 1891–1901.

McAlister, D.R., Horwitz, E.P., 2007. Characterization of extraction of chromatographic materials containing bis(2-ethyl-1-hexyl)phosphoric acid, 2-ethyl-1-hexyl (2-ethyl-1-hexyl) phosphonic acid, and bis(2,4,4-trimethyl-1-Pentyl)phosphinic acid. Solv. Extr. Ion Exch. 25 (6), 757–769.

McKlveen, J.W., McDowell, W.J., 1984. Liquid scintillation alpha spectrometry techniques. Nucl. Instrum. Methods Phys. Res. 223, 372–376.

Mellado, J., Llaurado, M., Rauret, G., 2001. Determination of Pu, Am, U, Th and Sr in marine sediment by extraction chromatography. Anal. Chim. Acta 443, 81–90.

Michel, H., Barci-Funel, G., Dalmasso, J., Ardisson, G., 1999. One step ion exchange process for the radiochemical separation of americium, plutonium and neptunium in sediments. J. Radioanal. Nucl. Chem. 240, 467–470.

Michel, H., Levent, D., Barci, V., Barci-Funel, G., Hurel, C., 2008. Soil and sediment sample analysis for the sequential determination of natural anthropogenic radionuclides. Talanta 74, 1527–1533.

Mietelski, J.W., 2011. Personal Communication.

Mitchell, P.I., Holm, E., Leon Vintro, L., Condren, O.M., Roos, P., 1998. Determination of the ^{243}Cm/^{244}Cm ratio by alpha spectrometry and spectral deconvolution in environmental samples exposed to discharges from the nuclear fuel cycle. Appl. Radiat. Isot. 49, 1283–1288.

Modolo, G., 2014. Untersuchungen zur Abtrennung, Konversion und Transmutation von langlebigen Radionukliden. Ein Beitrag zur fortschrittlichen Entsorgung von hochradioaktiven Abfallen. Schriften des Forschungszentrums Jülich, Reihe Energie und Umwelt 235.

Monsanglant-Louvet, C., Liatimi, N., Gensdarmes, F., 2012. Performance assessment on continuous air monitors under real operating conditions. IEEE Trans. Nucl. Sci. 59, 1414–1420.

Montoya, E.H., Mendoza, P.A., Bedregal, P.S., 2012. A combined method of neutron activation analysis and radiometric measurements for ^{234}U and ^{238}U determination in soil samples of low uranium concentration. J. Radioanal. Nucl. Chem. 291, 175–178.

Moore, F.L., 1963. Separation of americium from other elements. Application to the purification and radiochemical determination of americium. Anal. Chem. 35, 715.

Moore, F.L., 1964. New approach to separation of trivalent actinide elements and lanthanide elements. Selective liquid-liquid extraction with tricaprylmethylammonium thiocyanate. Anal. Chem. 36, 2158.

Moore, M.E., McFarland, A.R., Rodgers, J.C., 1993. Factors that affect alpha particle detection in continuous air monitor applications. Health Phys. 65, 69–81.

Morgenstern, A., Apostolidis, C., Carlos-Marquez, R., Mayer, K., Molinet, R., 2002. Single-column extraction chromatographic separation of U, Pu, Np and. Am. Radiochim. Acta 90, 81–85.

Moreno, J., Vajda, N., Danesi, P.R., LaRosa, J.J., Zeiller, E., Sinojmeri, M., 1997. Combined procedure for determination of ^{90}Sr, ^{241}Am and Pu radionuclides in soil samples. J. Radioanal. Nucl. Chem. 226, 279–284.

Moreno, J., LaRosa, J.J., Danesi, P.R., Burns, K., Vajda, N., Sinojmeri, M., 1998. Determination of 241Pu by LSC in combined procedure for Pu radionuclides, 241Am and 90Sr analysis in environmental samples. Radiact. Radiochem. 9, 35.

Myasoedov, B.S., Guseva, L.I., Lebedev, L.A., Milyukova, M.S., Chnutova, M.K., 1974. Analytical Chemistry of Trans-plutonium Elements. John Wiley and Sons, Jerusalem.

Narita, H., Harada, K., Burnett, W.C., Tsunogai, S., McCabe, W.J., 1989. Determination of ^{210}Pb, ^{210}Bi and ^{210}Po in natural waters and other materials by electrochemical separation. Talanta 36, 925–929.

Nash, K.L., Madic, C., Mathur, J.N., Lacquement, J., 2006. Actinide separation science and technology. In: Morss, L.R., Edelstein, N.M., Fuger, J. (Eds.), The Chemistry of Actinide and Transactinide Elements. Springer, Dordrecht.

Niese, U., Niese, S., 1985. Determination of actinide nuclides in water samples from the pri-mary circuit of a research reactor. J. Radioanal. Nucl. Chem. Art. 91, 17–24.

Nonova, T., Tpsjeva, Z., Kies, A., 2009. Radioactive radium and lead isotopes determinations in biota samples. J. Radioanal. Nucl. Chem. 282, 507–510.

NUDAT 2, 2009. NuDat Database Version 2.5. http://www.nndc.bnl.gov/nudat2.

NUDAT 2, 2019. Brookhaven National Laboratory. Interactive Chart of Nuclides. NuDat 2.7. 2019. https://nndc.bnl.gov/nudat2/. Accessed Jan 2019.

Nour, S., El-Sharkawy, A., Burnett, W.C., Horwitz, E.P., 2004. Radium-228 determination of natural waters via concentration on manganese dioxide and separation using Diphonix ion exchange resin. Appl. Radiat. Isot. 61, 1173–1178.

Oh, J.-S., Warwick, P.E., Croudace, I.W., Lee, S.-H., 2014. Evaluation of three electrodeposition procedures for uranium, plutonium and americium. Appl. Radiat. Isot. 87, 233–237.

Ohtsuka, Y., Takaku, Y., Nishimura, K., Kimura, J., Hisamatsu, S., Inaba, J., 2006. Rapid method for the analysis of plutonium isotopes in a soil sample within 60 min. Anal. Sci. 22, 309–311.

Olahova, K., Matel, L., Rosskopfova, O., 2005. Metody Zakoncentrovania Radionuklidov Z Velkoobjemovych Vzoriek. *NKS-140 Seminar*. Tartu, Estonia, pp. 116–120.

Orlandini, K.A., Gaffney, J.S., Marley, N.A., 1991. An improved technique for the rapid assay of radium isotopes in water. Radiochim. Acta 55, 205–207.

Osváth, S., Vajda, N., Molnár, Z., 2009. Development of a complex method for the determination of actinoides. J. Radioanal. Nucl. Chem. 281, 461–465.

Osváth, S., Vajda, N., Molnár, Z., Széles, É., Stefánka, Z., 2010. Determination of ^{237}Np, ^{93}Zr and other long-lived radionuclides in medium and low level radioactive waste samples. J. Radioanal. Nucl. Chem. 286, 675–680.

Pabinger, A., Pies, C., Porst, J.-P., Schafer, S., Fleischmann, A., Gastaldo, L., Enns, C., Jang, Y.S., Kim, I.-H., Kim, M.S., 2011. Development of cryogenic alpha spectrometers using metallic magnetic calorimeters. Nucl. Instrum. Methods Phys. Res. A 652 (1), 299–301.

Paepen, J., Dirican, A., Marouli, M., Pomme, S., Van Ammel, R., Stroh, H., 2014. A magnet system for the suppression of conversion electrons in alpha spectrometry. Appl. Radiat. Isot. 87, 320–324.

Park, S., Kwak, S.-W., Kang, H.-B., 2015. High resolution alpha particle spectrometry through collimation. Nucl. Instrum. Methods Phys. Res. 784, 470–473.

Parsa, B., Henitz, J.B., Carter, J.A., 2011. Rapid screening and analysis of alpha- and gamma-emitting radionuclides in liquids using a single sample preparation procedure. Health Phys. 100, 152–159.

Parus, J.L., Raab, W., 1996. Determination of plutonium in nuclear materials with combination of alpha and gamma spectrometry. Nucl. Instrum. Methods Phys. Res. 369, 588–592.

Paul, S., Pandey, A.K., Kumar, P., Kaity, S., Aggarwal, S.K., 2014. Tailorod Bifunctional Polymer for Plutonium Monitoring. Anal. Chem. 86, 6254–6261.

Paul, S., Pandey, A.K., Shah, R.V., Aggarwal, S.K., 2015. Chemically selective polymer substrate based direct isotope dilution alpha spectrometry of Pu. Anal. Chim. Acta 878, 54–62.

Peck, G.A., Smith, J.D., 2000. Determination of ^{210}Po and ^{210}Pb in rainwater using measurement of ^{210}Po and ^{210}Bi. Anal. Chim. Acta 422, 113–120.

Penneman, R.A., Keenan, T.K., 1960. The Radio-Chemistry of Americium and Curium. National Academy of Sciences, National Research Council, NAS-NS-3006, USA.

Peräjärvi, K., Hakala, J., Jokinen, A., Moore, I.D., Penttilä, H., Pöllänen, R., Saastamoinen, A., Toivonen, H., Turunen, J., Äystö, J., 2009. Event mode data acquisition for characterization of samples containing radioactive particles. IEEE Trans. Nucl. Sci. 56, 1444–1447.

Peräjärvi, K., Ihantola, S., Pöllänen, R., Toivonen, H., Turunen, J., 2011. Determination of ^{235}U, ^{239}Pu, ^{240}Pu, and ^{241}Am in a nuclear bomb particle using a position-sensitive α-γ coincidence technique. Environ. Sci. Technol. 45, 1528–1533.

Pilviö, R., 1998. Methods for the Determination of Low-Level Actinide Concentrations and Their Behaviour in the Aquatic Environment (Academic dissertation). University of Helsinki.

Pimpl, M., Higgy, R.H., 2000. Improvement of Am and Cm determination in soil samples. J. Radioanal. Nucl. Chem. 248, 537–541.

Pöllänen, R., Siiskonen, T., Vesterbacka, P., 2005. High-resolution alpha spectrometry from thick sources. Radiat. Meas. 39, 565–568.

Pöllänen, R., Siiskonen, T., 2005. Direct high-resolution alpha spectrometry for identifying transuranium elements in outdoor air. In: Proceedings of the International Conference on Monitoring, Assessments and Uncertainties for Nuclear and Radiological Emergency Response. 2005 Nov 21–25; Rio de Janeiro, Brazil.

Pöllänen, R., Siiskonen, T., 2014. Unfolding alpha-particle energy spectrum from a membrane air filter containing radon progeny. Radiat. Meas. 70, 15–20.

Pöllänen, R., Siiskonen, T., Moring, M., Juhanoja, J., 2007. Direct alpha spectrometry for characterizing hot particle properties. Radiat. Meas 42, 1666–1673.

Pöllänen, R., Karhunen, T., Siiskonen, T., Toivonen, H., Pelikan, A., 2009a. Deconvolution of alpha spectra from hot particles. In: Oughton, D., Kashparov, V. (Eds.), Radioactive Particles in the Environment. NATO Science for Peace and Security Series C: Environmental Security. Springer, pp. 209–220.

Pöllänen, R., Ruotsalainen, K., Toivonen, H., 2009b. Determination of ^{239}Pu and ^{240}Pu isotope ratio for a nuclear bomb particle using X-ray spectrometry in conjunction with γ-ray spectrometry and non-destructive α-particle spectrometry. Nucl. Instrum. Methods Phys. Res. 610, 515–521.

Pöllänen, R., Siiskonen, T., Ihantola, S., Toivonen, H., Pelikan, A., Inn, K., La Rosa, J., Bene, B.J., 2011. Activity determination without tracers in high-resolution alpha-particle spectrometry. J. Radioanal. Nucl. Chem. 290, 551–555.

Pöllänen, R., Turunen, J., Karhunen, T., Peräjärvi, K., Siiskonen, T., Wirta, M., Turunen, A., 2014. Novel equipment for in-situ alpha spectrometry with good energy resolution. Health Phys. 109, 601–605.

Pöllänen, R., 2016. Performance of an in-situ alpha spectrometer. Appl. Radiat. Isot. 109, 193–197.

Pöllänen, R., Klemola, S., 2016. Analysis of a liquid sample residue using in situ alpha spectrometry. J. Radioanal. Nucl. Chem. 307, 747–750.

Pöllänen, R., Peräjärvi, K., Siiskonen, T., Turunen, J., 2012a. High-resolution alpha spectrometry at ambient air pressure – towards new applications. Nucl. Instrum. Methods Phys. Res. 694, 173–178.

Pöllänen, R., Siiskonen, T., Ihantola, S., Toivonen, H., Pelikan, A., Inn, K., LaRosa, J., Bene, B.J., 2012b. Determination of $^{239}Pu/^{240}Pu$ isotopic ratio by high-resolution alpha-particle spectrometry using the ADAM program. Appl. Radiat. Isot. 70, 733–739.

Pöllänen, R., Simola, R., Vesterbacka, P., 2016. Rapid alpha spectrometry from liquids doped with ^{209}Po and ^{241}Am using simplified sample processing. Appl. Radiat. Isot. 107, 183–186.

Pommé, S., Caro Marroyo, B., 2015. Improved peak shape fitting in alpha spectra. Appl. Radiat. Isot. 98, 148–153.

Pommé, S., Garcia-Torano, E., Marouli, M., Crespo, M.T., Jobbágy, V., Van Ammel, R., Paepen, J., Stroh, H., 2014. High-resolution alpha-particle spectrometry of ^{238}U. Appl. Radiat. Isot. 87, 315–319.

Pommé, S., Sibbens, G., 2004. Concept for an off-line gain stabilization method. Appl. Radiat. Isot. 60, 151–154.

Porcelli, D., 2008. Investigating groundwater processes using U- and Th-series nuclides. In: Baxter, M.S. (Ed.), Radioactivity in the Environment, vol. 13. Elsevier, Amsterdam, pp. 105–153.

Porstendörfer, J., Zock, C.H., Reineking, A., 2000. Aerosol size distribution of the radon progeny in outdoor air. J. Environ. Radioact. 51, 37–48.

Purkl, S., Eisenhauer, A., 2003. Solid-phase extraction using Empore™ Radium Rad Disks to separate radium from thorium. J. Radioanal. Nucl. Chem. 256, 473–480.

Qiao, J., Hou, X., Roos, P., Miro, M., 2011. High-throughput sequential injection method for simultaneous determination of plutonium and neptunium in environmental solids using macroporous anion-exchange chromatography, followed by inductively coupled plasma mass spectrometric detection. Anal. Chem. 83, 373–381.

Qu, H., Stuit, D., Glover, S.E., Love, S.F., amd Filby, R.H., 1998. Preconcentration of plutonium and americium using the Actinide-CUTM Resin for human tissue analysis. J. Radioanal. Nucl. Chem. 234, 175−181.

Ramebäck, H., Skillberg, M., 1998. Separation of neptunium, plutonium, americium and curium from uranium with di-(2-ethylhexyl)-phosphoric acid (HDEHP) for radiometric and ICP-MS analysis. J. Radioanal. Nucl. Chem. 235, 229−233.

Ranebo, Y., Pöllanen, R., Eriksson, M., Siiskonen, T., Niagolova, N., 2010. Characterization of radioactive particles using non-destructive alpha spectrometry. Appl. Radiat. Isot. 68, 1754−1759.

Reischmann, F.-J., Trautmann, N., Herrmann, G., 1984. Chemistry at low concentrations: polonium at a level of 10^8 to 10^5 atoms. Radiochim. Acta 36, 139−143.

Rim, J.H., Gonzales, E.R., Armenta, C.E., Ünlü, K., Peterson, D.S., 2013. Developing and evaluating di(2-ethylhexyl) orthophosphoric acid (HDEHP) based polymer ligand film (PLF) for plutonium extraction. J. Radioanal. Nucl. Chem. 296, 1099−1103.

Roberts, R.A., Choppin, G.R., Wild, J.F., 1986. The Radiochemistry of Uranium, Neptunium and Plutonium − an Updating. National Academy of Sciences-National Research Council, Nuclear Science Series NAS-NS-3063,1986.

Robbins, J.A., 1978. Geochemical and geophysical applications of radioactive lead. In: Nriagu, J.O. (Ed.), The Biogeochemistry of Lead in the Environment. Part A. Ecological Cycles. Elsevier/North Holland Biomedical Press, Amsterdam, pp. 285−393.

Rodriguez-Alvarez, M.J., Sanchez, F., 1995. Measurement of radium and thorium isotopes in environmental samples by alpha-spectrometry. J. Radioanal. Nucl. Chem. 191, 3−13.

Roman, D., 1984. Electrodeposition of radium on stainless steel from aqueous solutions. Int. J. Appl. Radiat. Isot. 35, 990−992.

Roos, P., Roos, B., 1999. Low-level alpha spectrometry of environmental samples using solid state detectors and gridded ionization chambers. In: Garcia-Torano, E. (Ed.), Status of Alpha-Particle Spectrometry: Instrumentation and Spectral Analysis. IAEA Consultants' Meeting on Development of Alpha Particle Spectrometry Instrumentation, Methods and Applications, Vienna.

Roseberry, L.M., Scott, T.G., 1985. Radiochemical analysis of ^{210}Po in coal gasification samples. J. Radioanal. Nucl. Chem. Lett. 93, 271−278.

Rosner, G., Winkler, R., Yamamoto, M., 1993. Simultaneous radiochemical determination of ^{237}Np and ^{239}Np with ^{235}Np as a tracer, and application to environmental samples. J. Radioanal. Nucl. Chem. Art. 173, 273−281.

Rosskopfova, O., Holkova, D., 2005. Separation of Actinides and Strontium by Extraction Chromatography on Diphonix Resin. NKS-140 Seminar Tartu, Estonia, p. 185.

Röttger, S., Paul, A., Honig, A., Keyser, U., 2001. On-line and medium-level measurements of the radon activity concentration. Nucl. Instrum. Methods Phys. Res., Sect. A 466, 475−481.

Samuelsson, C., 2011. Excerpts from the history of alpha recoils. J. Environ. Radioact. 102, 531−533.

Sankhe, R.H., Mirashi, N., Chaudhury, S., Naik, H., 2017. Estimation of ^{231}Pa in presence of its daughter products by extraction chromatography. J.Radioanal. Nucl. Chem. 311 (1), 681−685.

Sekine, K., Imai, T., Kasai, A., 1987. Liquid-liquid extraction separation and sequential determination of Pu and Am in environmental samples by alpha spectrometry. Talanta 34 (6), 567.

Semkow, T.M., Khan, A.J., Haines, D.K., Bari, A., 2009. Rapid alpha spectroscopy of evaporated liquid residues for emergency response. Health Phys. 96, 432−441.

Shannon, L.V., Orren, M.J., 1970. A rapid method for the determination of polonium-210 and lead-210 in sea water. Anal. Chim. Acta 52, 166−169.

Sidhu, R.S., 2003. A robust procedure for the determination of plutonium and americium in seawater. J. Radioanal. Nucl. Chem. 256, 501−504.

Siiskonen, T., Pöllänen, R., 2004. Simulation of alpha particle spectra from aerosol samples. Appl. Radiat. Isot. 60, 947−953.

Siiskonen, T., Pöllänen, R., 2005. Advanced simulation code for alpha spectrometry. Nucl. Instrum. Methods Phys. Res. 550, 425−434.

Siiskonen, T., Pöllänen, R., 2006. Alpha-electron and alpha-photon coincidences in high-resolution alpha spectrometry. Nucl. Instrum. Methods Phys. Res., Sect. A 558, 437−440.

Sill, C.W., 1980. Determination of gross alpha, plutonium, neptunium, and/or uranium by gross alpha counting on barium sulfate. Anal. Chem. 52, 1452−1459.

Sill, C.W., 1984. Separation of Americium from Tervalent Lanthanides for High-Resolution Alpha Spectrometry. DOE report, contract No. DE-AC07-76IDO1570.

Sill, C.W., Willis, C.P., 1966. Determination of radioisotopes of cerium, barium, lanthanum, and neptunium after separation by barium sulfate. Anal. Chem. 38, 97−102.

Sill, C.W., Olsen, D.G., 1970. Sources and prevention of recoil contamination of solid-state detectors. Anal. Chem. 42, 1596−1607.

Sill, C.W., Hindman, D., Anderson, J.I., 1979. Simultaneous determination of alpha-emitting nuclides of radium through californium in large environmental and biological samples. Anal. Chem. 51, 1307−1314.

Sill, C.W., 1987a. Precipitation of actinides as fluorides or hydroxides for high-resolution alpha spectrometry. Nucl. Chem. Waste Manag. 7, 201−216.

Sill, C.W., 1987b. Determination of radium-226 in ores, nuclear wastes and environmental samples by high-resolution alpha spectrometry. Nucl. Chem. Waste Manag. 7, 239−256.

Sill, C.W., Sill, D.S., 1995. Sample dissolution. Radioact. Radiochem. 6, 8−14.

Sill, W.S., 1995. Rapid monitoring of soil, water, and air dusts by direct large-area alpha spectrometry. Health Phys. 69, 21−33.

Skwarzec, B., Strumińska, D.I., Borylo, A., 2001. The radionuclides ^{234}U, ^{238}U and ^{210}Po in drinking water in Gdańsk agglomeration (Poland). J. Radioanal. Nucl. Chem. 250, 315−318.

Smith, B.F., Gibson, R.R., Jarvinen, G.D., Robison, T.W., Schroeder, N.C., Stalnaker, N.D., 1998. J. Radioanal. Nucl. Chem. 234 (1−2), 225−229.

Smith, J.D., Hamilton, T.F., 1984. Improved technique for recovery and measurement of polonium-210 from environmental materials. Anal. Chim. Acta 160, 69−77.

Smith, K.A., Mercer, E.R., 1970. The determination of ^{226}Ra and ^{228}Ra in soils and plants using ^{225}Ra as a yield tracer. J. Radioanal. Chem. 5, 303−312.

Smithson, G., Muzaffer, F., Petrow, M., 1979. Radiochemical determination of lead-210 in environmental samples resulting from uranium mining-milling operations. In: Radiochemical Procedures for Determination of Selected Members of the Uranium and Thorium Series. Appendix B, CANMET, Ottawa. Report 78-22.

Smolander, P., Toivonen, H., 2004. Mobile in-field measurements in nuclear or radiological threat situations. In: 11th International Congress of the International Radiation Protection Association, 23-28 May 2004, Madrid, Spain. http://www.irpa11.com/new/pdfs/7a24.pdf.

Solatie, D., Carbol, P., Peerani, P., Betti, M., 2001. Investigation of separation/purification methodologies for the determination of 232U and

236Pu in solution of spent nuclear fuels by alpha-spectrometry. Radiochim. Acta 89, 551–556.

Solatie, D., 2002. Development and Comparison of Analytical Methods for the Determination of Uranium and Plutonium in Spent Fuel and Environmental Samples (Academic dissertation). University of Helsinki.

Solomon, S.B., 2001. Field tests of a radon progeny sampler for the determination of effective dose. Sci. Total Environ. 272, 303–313.

Solomon, S.B., Ren, T., 1992. Counting efficiencies for alpha particles emitted from wire screens. Aerosol Sci. Technol. 17, 69–83.

Spry, N., Parry, S., Jerome, S., 2000. The development of a sequential method for the determination of actinides and 90Sr in power station effluent using extraction chromatography. Appl. Radiat. Isot. 53, 163–171.

Srinivasan, S., Seshadri, H., Kumar, T., Rajan, S.K., 2005. Determination of the ^{240}Pu/^{239}Pu ratio by alpha spectrometry using the WinALPHA program. Appl. Radiat. Isot. 64, 591–593.

Steinbauer, E., Bortels, G., Biersack, J.P., Burger, P., Ahmad, I., 1994. A survey of the physical processes which determine the response function of silicon detectors to alpha particles. Nucl. Instrum. Methods Phys. Res., Sect. A 339, 102–108.

Stika, M., Padilla, S., Jarrell, J., Blue, T., Cao, L.R., Simpson, M., 2018. Thin-layer electro-deposition of uranium metal from molten LiCl-KCl. J. Electrochem. Soc. 165 (3), D135–D141.

Sufke, F., Lippold, J., Happel, S., 2017. Improved separation of Pa from Th and U in marine sediment with TK400 resin. Anal. Chem. 90 (2), 1395–1401.

Sumiya, S., Morita, S., Tobita, K., Kurabayashi, M., 1994. Determination of technetium-99 and neptunium-237 in environmental samples by inductively coupled plasma mass spectrometry. J. Radioanal. Nucl. Chem. Art 177, 149–159.

Surbeck, H., 2000. Alpha spectrometry sample preparation using selectively adsorbing thin films. Appl. Radiat. Isot. 53, 97–100.

Surls, J.P., Choppin, G.R., 1957. Ion exchange study of thiocyanate complexes of actinides and lanthanides. J. Radioanal. Nucl. Chem. 4, 62.

Svetlik, I., Belanova, A., Vrskova, M., Hanslik, E., Ivanovova, D., Meresova, J., Tomaskova, L., Novakova, T., 2010. Volatility of ^{210}Po in the gross alpha determination. J. Radioanal. Nucl. Chem. 286, 547–551.

Thakkar, A.H., 2001. A rapid sequential separation of actinides using Eichrom's extraction chromatographic material. J. Radioanal. Nucl. Chem. 248, 453–456.

Thakkar, A.H., 2002. A rapid sequential separation of actinides using Eichrom's extraction chromatographic material. J. Radioanal. Nucl. Chem. 252, 215–218.

Towler, P.H., Smith, J.D., Dixon, D.R., 1996. Magnetic recovery of radium, lead and polonium from seawater samples after preconcentration on a magnetic adsorbent of manganese dioxide coated magnetite. Anal. Chim. Acta 328, 53–59.

Turunen, J., Ihantola, S., Peräjärvi, K., Pöllänen, R., Toivonen, H., Hrnecek, E., 2011. Collection and behaviour of radon progenies on thin Mylar foils. Radiat. Meas. 46, 631–634.

Uesugi, M., Noguchi, M., Yokoyama, A., Nakanishi, T., 2010. Improvements on the method for determining of ^{210}Pb and ^{210}Po in lead. J. Radioanal. Nucl. Chem. 283, 577–584.

Vainblat, N., Pelled, O., German, U., Haquin, G., Tshuva, A., Alfassi, Z.B., 2004. Determination of parameters relevant to alpha spectrometry when employing surface coating. Appl. Radiat. Isot. 61, 307–311.

Vajda, N., LaRosa, J., Zeisler, R., Danesi, P., Kis-Benedek, G., 1997. A novel technique for the simultaneous determination of ^{210}Pb and ^{210}Po using a crown ether. J. Environ. Radioact. 37, 355–372.

Vajda, N., Molnár, Z., Kabai, E., Osvath, S., 2003. Simultaneous determination of long-lived radionuclides in environmental samples. In: Birsen, N., Kadyrzhanov, K.K. (Eds.), "Environmental Protection against Radioactive Pollution". RCS, pp. 133–146.

Vajda, N., Törvényi, A., Kis-Benedek, G., Kim, C.K., Bene, B., Mácsik, Z., 2009. Rapid method for the determination of actinides in soil and sediment samples by alpha spectrometry. Radiochim. Acta 97, 395–401.

Vajda, N., Kim, C.K., 2011. Determination of transuranium isotopes (Pu, Np, Am) by radiometric techniques: a review of analytical methodology. Anal. Chem. (in press).

Vance, D.E., Belt, V.F., Oatts, T.J., Mann, D.K., 1998. Neptunium determination by inductively coupled plasma mass spectrometry. J. Radioanal. Nucl. Chem. Art. 234, 143–146.

Van der Wijk, A., Venema, L., Steendam, S.P., 1987. The use of thin plastic foils in low-level alpha spectrometry. Appl. Radiat. Isot. 38, 1061–1065.

Varga, Z., Stefanka, Z., Suranyi, G., Vajda, N., 2007a. Rapid sequential determination of americium and plutonium in sediment and soil samples by ICP-SFMS and alpha-spectrometry. Radiochim. Acta 95, 81–87.

Varga, Z., Surányi, G., Vajda, N., Stefánka, Z., 2007b. Improved sample preparation method for environmental plutonium analysis by ICP-SFMS and alpha-spectrometry. J. Radioanal. Nucl. Chem. 274, 87–94.

Varga, Z., Surányi, G., Vajda, N., Stefánka, Z., 2007c. Determination of plutonium in environ-mental samples by inductively coupled plasma sector field mass spectrometry and alpha spectrometry. Microchem. J. 85, 39–45.

Vesterbacka, P., Pöllänen, R., 2011. Simple method to determine ^{234}U and ^{238}U in water using alpha spectrometry. In: Proceedings – Third European IRPA Congress, 14–18 June 2010, Helsinki, Finland. Nordic Society for Radiation Protection, Helsinki, 2011. Web version online 2010 June 16. [P12-07, 6 pp.].

Vioque, I., Manjón, G., García-Tenorio, R., El-Daoushy, F., 2002. Determination of alpha-emitting Pu isotopes in environmental samples. Analyst 127, 530–535.

Volpe, A.M., Olivares, J.A., Murrell, M.T., 1991. Determination of radium isotope ratios and abundances in geologic samples by thermal ionization mass spectrometry. Anal. Chem. 63, 913–916.

Vreček, P., Benedik, L., Pihlar, B., 2004. Determination of ^{210}Po and ^{210}Pb in sediment and soil leachates and in biological materials using a Sr-resin column and evaluation of column reuse. Appl. Radiat. Isot. 60, 717–723.

Wai, C.M., Lo, J.M., 1982. Extraction and separation of ^{210}Pb, ^{210}Bi and ^{210}Po by diethyldithiocarbamate. Radiochem. Radioanal. Lett. 50, 293–298.

Wang, Z., Lin, J., Li, S., Guo, Q., Huang, W., Wen, W., Dan, G., Tan, Z., 2018. Rapid method for accurate determination of actinides (U, Th, Pu, and Am) in water samples for emergency response. J. Radioanal. Nucl. Chem. 315, 103–110.

Warwick, P.E., Croudace, I.W., Carpenter, R., 1996. Review of analytical techniques for the determination of americium-241 in soils and sediments. Appl. Radiat. Isot. 47 (7), 627–642.

Watzig, W., Westmeier, W., 1978. ALFUN-A programme for the evaluation of complex spectra. Nucl. Instrum. Methods 153, 517. Ref. Bland, 1998.

Wildgust, M.A., McDonald, P., White, K.N., 1998. Temporal changes of ^{210}Po in temperate coastal waters. Sci. Total Environ. 214, 1–10.

Wiltshire, R.A.P., 1984. Alpha spectrometry –a tool for nuclear data measurements. Nucl. Instrum. Methods Phys. Res. 223, 535–539.

Winkel, P., Corte, F., Hoste, J., 1971. Anion exchange in acetic acid solution. Anal. Chim. Acta 56, 241.

Wolf, S.F., 2006. Trace analysis of actinides in geological, environmental, and biological matrices. In: Morss, L.R., Edelstein, N.M., Fuger, J. (Eds.), The Chemistry of Actinide and Transactinide Elements. Springer, Dordrecht.

Wyllie, H.A., Lowenthal, G.C., 1984. Ultra-thin radioactive sources. Int. J. Appl. Radiat. Isot. 35, 257–258.

Yamamoto, M., Komura, K., Ueno, K., 1989. Determination of low-level ^{226}Ra in environmental water samples by alpha-ray spectrometry. Radiochim. Acta 46, 137–142.

Yamamoto, M., Kawamura, H., Igarashi, Y., Shiraishi, K., Ueno, K., 1991. Measurement of low-level ^{226}Ra in human bone by using ^{225}Ra tracer and alpha-spectrometry: approach without use of Ba carrier. Radiochim. Acta 55, 163–166.

Yamato, A., 1982. An anion exchange method for the determination of ^{241}Am and plutonium in environmental and biological samples. J. Radioanal. Chem. 75 (1–2), 265.

Yan-Qin, J., Jin-Ying, L., Shang-Geng, L., Tao, W., Jun-Ling, L., 2001. Determination of traces of 237Np in environmental samples by ICP-MS after separation using TOA extraction chromatography. Fresenius J. Anal. Chem. 371, 49–53.

Yokoyama, T., Nakamura, E., 2004. Precise analysis of the ^{228}Ra/^{226}Ra isotope ratio for short-lived U-series disequilibria in natural samples by total evaporation thermal ionization mass spectrometry (TE-TIMS). J. Anal. At. Spectrom. 19, 717–727.

Yoon, S., Ha, W.-H., Park, S., Jin, Y-w., Kim, J-m., 2018. Improved procedure for Pu/Am, and U isotope analysis in fecal samples. J. Radioanal. Nucl. Chem. 316, 1151–1155.

Yoon, W.S., Jang, Y.S., Kim, G.B., Kim, K.J., Kim, M.S., Lee, J.S., Lee, K.B., Lee, S.J., Lee, H.J., Yuryev, Y.N., Kim, Y.H., 2012. High energy resolution cryogenic alpha spectrometers using magnetic calorimeters. J. Low Temp. Phys. 167, 280–285.

Ziegler, J.F., Biersack, J.P., Littmark, U., 1996. The Stopping and Range of Ions in Solids. Pergamon Press, New York.

Zikovsky, L., 2002. Determination of alpha-radioactivity in ground water by precipitation of alpha-emitters with sulphates and hydroxides. J. Radioanal. Nucl. Chem. 251, 329–331.

Zhu, Y., Chen, J., Jiao, R., 1996. Extraction of Am(III) and Eu(III) from nitrate solution with purified Cyanex 301. Solv. Extr. Ion Exch. 14, 61.

Dr. Nóra Vajda

Dr. Nóra Vajda is radiochemist and professor of the Hungarian Academy of Sciences. She has been the head of RADANAL Ltd. for the last 10 years and had been working at the Budapest University of Technology and Economics as an associate professor. Her main research area is radiation measurements, especially radiochemical alpha, beta, and gamma ray spectrometry and neutron activation. Other areas of interest are the determination of difficult-to-measure radionuclides in radioactive wastes, technological water of nuclear power plants, and the environment. During her 40-year career, she has published more than 100 scientific papers, was the supervisor of several PhD, MSc, and BSc students, and participated in missions to Member States on peaceful applications of nuclear energy organized by the International Atomic Energy Agency.

Dr. Roy Pöllänen

Dr. Roy Pöllänen is an adjunct professor in the physics department of the University of Helsinki. He works in the Radiation and Nuclear Safety Authority (STUK) of Finland as a senior inspector. His main research area is radiation measurements, especially nondestructive alpha spectrometry and gamma ray spectrometry. Other areas of interest are environmental radiation monitoring and in particular air monitoring, atmospheric transport calculations, determination of radioactive particle characteristics, dose calculations, radiation hazard estimation, radiation detection methods in general, and in-field measurements. Radiation measurements by using unmanned aerial vehicles, emergency preparedness, and organizing training and radiation detection exercises are other topics of interest. During his 30-year career in STUK, Roy Pöllänen has published more than 100 scientific papers and reports. He represents STUK in the International Committee for Radionuclide Metrology, ICRM, and he is active in various (inter)national research projects.

Paul Martin

Dr. Paul Martin has 35 years of experience in radioactivity measurements, with a special emphasis on the determination of the natural series radionuclides in environmental samples by alpha and gamma spectrometry. Particular areas of his professional activity have been uranium mining–related environmental and radiological impact studies, behavior of radionuclides in the tropical environment, radon and radon progeny in the atmosphere, airborne gamma spectrometry as a site assessment tool, sample analysis for nuclear safeguards, and reference material production and interlaboratory comparison exercises for analytical quality assurance. From 2012 to 2017, Paul was an associate editor for the Journal of Environmental Radioactivity.

Chang-Kyu Kim

Chang-Kyu Kim studied chemistry at the University of Tsukuba (Japan), where he received a PhD degree in 1990. From 1990 to 1998, Mr. Kim worked as a principal researcher for the Environmental Radiation Assessment Unit, Korea Institute of Nuclear Safety, Korea. Mr. Kim was in charge as project manager for the project on the development of advanced analytical techniques for ultratrace-level long-lived radionuclides in environmental samples using high-resolution ICP-MS. From 2004 to 2010, Mr. Kim worked as a radiochemist specialist at the Terrestrial Environment Laboratory, IAEA Environment Laboratories, Seibersdorf, Austria. From 2011 to 2012, Mr. Kim worked as a senior radiochemist for the Forensic Scientific Services, Health Physics, Queensland Health, Australia. He has more than 30 years of working experience in the field of radioanalytical chemistry, and he has published more than 150 papers and reports. Since 2013, Mr. Kim has worked as a nuclear safeguards inspector and quality manager at the On-Site Laboratory team, Office of Safeguards Analytical Services, IAEA.

Chapter 6

Liquid scintillation analysis: principles and practice*

Michael F. L'Annunziata[1], Alex Tarancón[2], Héctor Bagán[2] and José F. García[2]
[1]The Montague Group, Oceanside, CA, United States; [2]Department of Chemical Engineering and Analytical Chemistry, University of Barcelona, Barcelona, Spain

Chapter outline

* Previous Co-author: Michael J. Kessler, Packard Instrument Co., Meriden, CT, USA.

Handbook of Radioactivity Analysis: Volume 1. https://doi.org/10.1016/B978-0-12-814397-1.00006-6

I. Introduction

Liquid scintillation counting (LSC) or liquid scintillation analysis (LSA) has been a very popular technique for the detection and quantitative measurement of radioactivity since the early 1950s. The technique has been most useful in studies of the life sciences and the environment, and it is also a powerful tool in the chemical and physical sciences. Many of the principles of liquid scintillation analysis overlap into the fields of low-level environmental radioactivity monitoring, the detection of singular decay events in basic nuclear physics, and the measurement of higher levels of radioactivity used in research, radioisotope applications, and nuclear power. However, the techniques and principles used in the LSA of environmental radioactivity per se will not be covered in detail in this chapter. The reader is directed to Chapter 2 of Volume 2 of this book for additional information on the use of LSA for the measurement of either natural levels of radionuclides or low-levels of man-made radionuclides found in the environment.

Applications of liquid scintillation analysis to the measurement of radionuclides used as tracers in research has led to a large number of cutting edge and Nobel prize-winning discoveries in the life sciences over the past 50 years. The LSA technique in scientific research remains one of the most popular experimental tools used for the quantitative analysis of radionuclides. These include principally the analysis of alpha and beta particle-emitting nuclides, but may also include nuclides that emit gamma radiation, as well as atoms that decay with the emission of x radiation, and those that emit Auger and internal-conversion electrons. Included in this chapter are also advances that continue to be made in the application of liquid scintillation to the analysis of neutrons, and the discrimination of neutron and gamma radiation, the analysis of high-energy charged particles, and the detection and measurement of unique events, such as π^+ decay, μ^+ decay, $\beta\beta$ decay, and the detection of neutrinos.

Some advances in the use of liquid scintillation analysis in the standardization of radionuclides will be included in this chapter. A more comprehensive treatment of radionuclide standardization by means of all available radiation detection and measurement methods will be included in Chapter 7, Volume 2 of this book.

The wide popularity of liquid scintillation analysis is a consequence of numerous advantages, which may include high efficiencies of detection, improvements in sample preparation techniques, automation including computer data processing, and the spectrometer capability of liquid scintillation analyzers permitting the simultaneous assay of different radionuclides.

II. Basic theory

A. Scintillation process

The discovery of scintillation in organic compounds was documented in a thesis by Herforth (1948) under the leadership of Hartmut Kallmann, which is related in a historical account by Niese (2003, 1999). In her thesis presented on September 13, 1948, at the Technical University Berlin – Charlottenburg, Herforth reported that aromatic compounds could convert absorbed energy of nuclear radiation into light photons. Herforth coauthored a subsequent publication with Kallmann and Immanuel Broser, another student of Kallmann, on the scintillation process in naphthalene (Broser et al., 1948) as well as a paper on fluorescence from liquid (molten) naphthalene, diphenyl, and phenanthrene as a consequence of interaction with alpha particles, fast electrons, and gamma rays (Herforth and Kallmann, 1948). Both Herforth and Broser were fellow students, who had researched radioactivity counters under Hans Geiger. Herforth's thesis and publications with Kallmann were followed by papers authored by Kallmann (1950) and Reynolds et al. (1950) on liquid scintillation counting that demonstrated certain organic compounds in solution-emitted fluorescent light when bombarded by nuclear radiation. The origin of liquid scintillation analysis as a technique for the quantification of radioactivity is attributed to the original papers by Kallmann and Reynolds in 1950. The fluorescence or emission of photons by organic compounds (fluors) as a result of excitation can be readily converted to a burst of electrons with the use of a photomultiplier tube (PMT), and subsequently measured as an electric pulse.

The technique of liquid scintillation counting involves placing the sample containing the radioactivity into a glass or plastic container, called a scintillation vial, and adding a special scintillation cocktail containing organic fluors dissolved into suitable solvents. Common capacities of scintillation vials that can be accommodated easily in conventional automatic liquid scintillation analyzers vary from 4 to 20 mL; however, microfuge tubes of 0.5–1.5 mL capacity can also be counted directly in a conventional LSA with the use of special microtube holders. Samples may also be analyzed by high-sample-throughput LSA in plastic microplates containing 24, 96, or 384 sample wells per microplate, which accept sample-fluor cocktail volumes in the range of 20–150 μL. High-sample-throughput microplate liquid scintillation counting is described in Section XIII of this chapter.

Both plastic and glass liquid scintillation counting vials have certain advantages and disadvantages in terms of background, solvent permeability, fragility, and transparency, etc. Polyethylene plastic vials are permeable when stored containing fluor solvents such as benzene, toluene, and xylene; however, these vials do not display solvent diffusion when more environmentally safe commercial fluor cocktails are used (e.g., Ultima Gold, Pico-fluor, Optifluor, etc.), which use diisopropylnaphthalene (DIN), pseudocumene, or linear alkylbenzene solvents. The plastic vials are also unbreakable, less expensive, and display lower backgrounds than the glass vials. Glass vials, however, provide the advantage of transparency to visualize the sample and fluor cocktail solution to permit inspection for undesirable properties such as color, residue, or sample inhomogeneity.

The scintillation cocktail may be composed of a solvent such as DIN, or a linear alkylbenzene together with a fluor solute such as 2,5-diphenyloxazole (PPO) dissolved in a concentration of approximately 2–10 g/L. Many liquid fluor cocktails are available commercially, and these are made to be compatible and mixable with radioactive samples dissolved in either organic solvents or aqueous media. When samples are dissolved in aqueous media, three different chemical components are required in the fluor cocktail solution: the organic solvent, organic scintillator, and a surfactant (emulsifier). The choice of solvent, scintillator, and surfactant for the preparation of contemporary fluor cocktails is dictated by the need for efficient energy transfer and light output in the scintillation process even under very high aqueous sample loads exceeding 50% water as well as the need for environmentally safe solutions with low toxicity, high flash point, and low disposal costs. To meet these needs, some commercially available formulations use diisopropylnaphthalene or a linear alkylbenzene solvent. A few of these commercial fluor cocktails were noted in the previous paragraph. The properties and performance of the modern environmentally safe solvents and some of the commercially available cocktails made from these solvents have been reviewed and tested by Feng et al. (2015), Ruano Sánchez et al. (2013), Verrezen et al. (2008), Medeiros et al. (2003), DeVol et al. (1996), Neuman et al. (1991), Thomson (1991), and a comprehensive review with recommendations is provided in Chapter 7, Volume 1 of this book. The findings of these studies, as summarized by Verrezen et al. (2008), are that no specific liquid scintillation cocktail can be identified to fit all radioanalytical needs, and cocktail formulations are designed to compromise between a performance aspect (*e.g.*, high sample load or sample compatibility) at the expense of another (*e.g.*, counting efficiency or stability). Chapter 7 of Volume 1 of this book provides detailed information on the composition and performance of many liquid scintillation fluor cocktails and sample preparation techniques.

Unique scintillation cocktails may be prepared to meet special detection needs, such as the detection of neutrinos, neutron/gamma discrimination, alpha/beta discrimination, and the detection of neutrinos or unique decay events, such as muon-, pion-, and double-beta decay. The characteristics of the liquid scintillation media or cocktails used for these specific detection needs are noted within the chapter.

A sample, which is to be analyzed for its radioactivity or disintegration rate, is placed into the scintillation cocktail to form a homogeneous counting solution. The liquid scintillation process that occurs in a scintillation cocktail is shown in Fig. 6.1. The first step in the process is the interaction of the nuclear radiation or particle under analysis with the solvent molecules of the liquid scintillation cocktail. These solvent molecules, as depicted in Fig. 6.1, are organic in nature and contain at least one aromatic ring. Because the solvent molecules are in greater concentration than the solute fluor molecules in the fluor cocktail, the solvent molecules will absorb the major portion of the nuclear radiation energy emitted by the radionuclides in the cocktail solution. The

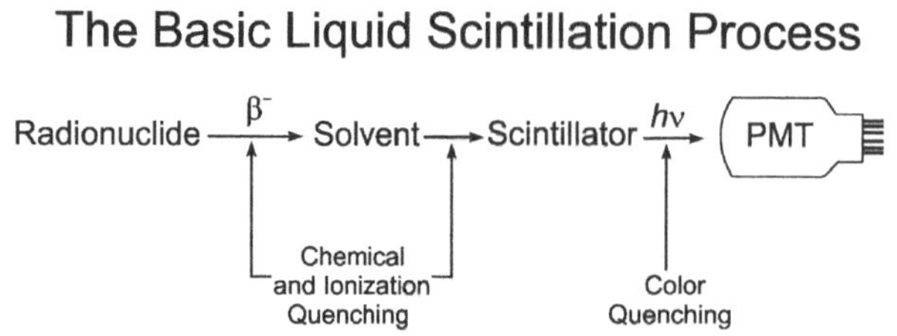

FIGURE 6.1 An illustration of the sequence of events in the basic liquid scintillation process. A radionuclide will dissipate its energy of decay (e.g., β^-–particle energy) in the liquid scintillation cocktail containing solvent and fluor (scintillator). The aromatic solvent absorbs most of the energy of the beta particle. The energy of excitation of the solvent is then transferred to the scintillator (fluor) molecules, which upon deexcitation emit photons of visible light. The intensity of the light photons are detected by a photomultiplier tube (PMT), which converts the light photons into a flow of electrons and further amplifies the current pulse. Points of interference caused by chemical, ionization, and color quenching are also indicated.

result is the formation of activated organic solvent molecules followed by a transfer of energy to the organic scintillator or fluor, as illustrated in Fig. 6.1. Organic scintillators are chosen because they are soluble in the organic solvent, they can easily accept the energy from the activated solvent molecule, and they produce an activated or excited scintillator molecule. These excited scintillator molecules rapidly lose their energy and return to their original ground state by way of a fluorescence mechanism. The energy is released as photon emissions in the wavelength range of 375–430 nm for each radioactive decay process occurring in the fluor cocktail. The wavelength of emission depends on the scintillator dissolved in the fluor cocktail. The photon emission intensity is a function of the energy and the type of nuclear decay.

The scintillation cocktail converts the original nuclear decay energy to photons of light by way of the process shown in Fig. 6.1. The intensity of the photon emission is directly proportional to the original nuclear energy dissipated in the fluor cocktail. The higher the energy absorbed by the fluor cocktail, the greater is the intensity of the photon emission. For example, tritium, which is a low-energy beta-particle emitter ($E_{max} = 18.6$ keV), would produce relatively very low intensity photon emissions for each beta particle absorbed in the fluor cocktail. However, ^{32}P, which is a high-energy beta-particle emitter ($E_{max} = 1710$ keV), would produce a light intensity approximately 100 times greater in the fluor cocktail. Thus, the light intensity emitted by a scintillation fluor cocktail reflects the original nuclear decay energy, and the photon emissions per unit time is proportional to the number of nuclear decays in that time period or, in other words, the sample radioactivity (*e.g.*, disintegrations per minute or DPM).

A liquid scintillation analyzer may also be used to measure the photon emissions produced when radioactive nuclides are adsorbed onto or in close proximity to the surface of a plastic or glass scintillator (solid scintillator) located within a conventional liquid scintillation counting vial or well of a microplate scintillation analyzer. The solid scintillation counting process uses a solid inorganic scintillator (e.g., yttrium silicate) to produce the light emissions, which are quantified by the liquid scintillation counter. The photon emissions are produced directly by the interaction of the decaying nuclear event with the inorganic scintillator. The photon emission intensities produced are proportional to the energies of the radiation emitted from the nuclear decay that are absorbed by the scintillator, similar to that for the liquid scintillation process. This technique, known as scintillation proximity assay (SPA), is used to measure binding reactions, commonly studied in the fields of medicine, biochemistry, and molecular biology, without the need to separate bound from free fractions. It uses glass or plastic solid scintillation microspheres together with a low-energy-emitting isotope-labeled (^{3}H or ^{125}I) ligand. The method is described briefly in Section XIII.B of this chapter and in more detail in Chapter 9, "Solid Scintillation Analysis", in Volume 1 of this book, as it is a solid scintillation technique, which utilizes a liquid scintillation counter.

B. Alpha-, beta-, and gamma-ray interactions in the LSC

The scintillation process and light output are different for the alpha, beta, and gamma decay. These decay processes are described in detail in Chapter 1, Volume 1. Only a brief treatment is provided here. The alpha decay process is illustrated by Eq. (1.154) and Fig. 1.40 of Chapter 1. During the alpha decay process, a helium nucleus, which is composed of two protons and two neutrons, is released with a specific energy (monoenergetic) from the atomic nucleus. The general decay energy range for alpha particles is 2−8 MeV. When alpha decay occurs in a liquid scintillation cocktail, the alpha particles interact with the fluor cocktail to produce photons of light (approximately 1 photon/keV of original decay energy). The photon intensity is converted into an electric pulse of magnitude proportional to the photon intensity via a photomultiplier tube described in Section III of this chapter.

If we compare the linear range (R_{cm}) in centimeters of a 5.5 MeV alpha particle from ^{241}Am in water ($R_{cm} = 0.0048$ cm) to the range of a 0.55 MeV beta particle from ^{10}Be in water ($R_{cm} = 0.178$ cm), we see that the alpha particle travels a much shorter distance, only 2.7 hundredths (0.0048/0.178) that of the beta particle, regardless of the fact that the alpha particle possessed 10 times the energy of the beta particle (See Chapter 1 for calculations of range and energy for alpha and beta particles). The higher charge and mass of the alpha particle compared with the beta particle are responsible for the reduced range of the alpha particle (See Chapter 1) and less efficient excitation energy transfer to solvent and fluor. Alpha particles produce photons of light in the liquid scintillation cocktail at about one-tenth the intensity per unit of particle energy of beta particles (Horrocks, 1974). Therefore, in the case of alpha particles, which are monoenergetic, a single pulse height peak is seen for each alpha decay, at a pulse height equivalent to approximately one-tenth its original nuclear decay energy. A 5-MeV alpha particle, therefore, would be detected by a pulse height equivalent to approximately 500 keV in a liquid scintillation cocktail. Consequently, the pulse heights of alpha particles and beta particles in the same sample often overlap even when the alpha particles emitted from certain radioactive nuclides are of energy about 10 times greater than the E_{max} of beta particles emitted by other radionuclides in the same sample. The overlapping liquid scintillation pulse height spectra of ^{210}Po and $^{90}Sr(^{90}Y)$ in the same sample are illustrated in Fig. 6.2.

The pulse decay times of the photon emissions produced in liquid scintillator from alpha and beta events are also different. A pulse in the scintillation process produced by an alpha particle can be about 35−40 nsec longer than a pulse event produced by a beta particle. Using this characteristic, simultaneous analysis of alphas and betas in the same pulse height energy range can be performed. The discrimination of alpha and beta particles, which produce overlapping pulse height intensities, is described in Section XV of this chapter.

The counting efficiency (i.e., how efficiently the nuclear decay is detected) is approximately 100% for almost all alpha decays using a liquid scintillation cocktail. Because of the unique pulse height spectral characteristics of alpha detection in a liquid scintillation counter (See Fig. 6.2) and their slower pulse decay times, alpha particles can be distinguished easily from most other nuclear decay radiations with the liquid scintillation analyzer.

A very common radionuclide decay process is the production and emission of beta particles. Beta decay can take place by either negatron (β^-) or positron (β^+) emission. The production of a negative beta particle (negatron) is described by Eqs. (1.173) and (1.174) of Chapter 1. During the beta decay process with negatron emission, a neutron is converted to a proton and an electron (negative beta particle) and an antineutrino. The beta particle (negatron) is equivalent to an electron in property, and the antineutrino is a particle of zero charge and near zero mass. The total decay energy that is released in the beta decay process is shared between the beta particle and the antineutrino. This total decay energy is usually expressed as the E_{max}, which is the maximum energy that is released in the decay process. The decay energy is shared between the beta particle and antineutrino, but only the beta particle can be detected directly by the scintillation process. Thus, the

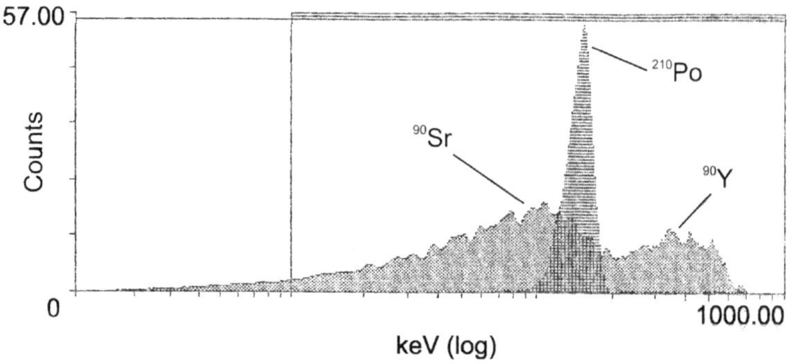

FIGURE 6.2 The overlapping pulse height spectra produced by a mixture of the 5.30 MeV alpha particles of ^{210}Po and the 0.55 MeV energy maximum of the beta particles of ^{90}Sr and 2.28 MeV energy maximum of the beta particles of ^{90}Y in a PerkinElmer 2750 TR liquid scintillation analyzer. The sample contains a mixture of $^{210}Po + {}^{90}Sr(^{90}Y)$ in a scintillation cocktail of 50% water (1:1 mixture of water and Ultima Gold AB fluor cocktail) and displayed a tSIE quench indicating parameter of 277. Notice the relatively sharp peak of the 5.3 MeV alpha-particle pulse height spectrum of ^{210}Po overlapping with the 0.55 MeV energy maximum beta-particle pulse height spectrum of ^{90}Sr and 2.28 MeV maximum beta-particle pulse height spectrum of ^{90}Y. The ^{90}Y daughter is in secular equilibrium with its parent ^{90}Sr. The α-peak of ^{210}Po and the β peaks of $^{90}Sr(^{90}Y)$ appear in a separate α-MCA and β-MCA. The two pulse height spectra of the MCAs are overlapped to demonstrate the overlapping pulse heights produced by the ^{210}Po and ^{90}Sr. (L'Annunziata, M. F. 1997, unpublished work of the writer.)

resultant spectrum for all beta decays starts at zero and goes to the maximum decay energy (E_{max}) as illustrated in Fig. 1.45 of Chapter 1. Approximately 10 photons of light per keV of beta-particle decay energy are produced in the liquid scintillation process. Because of the broad spectrum of beta-particle energies emitted by a given radionuclide sample, beta decays can easily be recognized by this distinct broad spectral pattern as illustrated in Fig. 1.45 of Chapter 1 on the linear energy scale or in Fig. 6.2 of this chapter illustrating the pulse height spectra of $^{90}Sr(^{90}Y)$ on a logarithmic energy scale.

Another type of beta decay produces a positron or positive beta particle. This beta decay process converts a proton to a neutron and a positively charged electron (positron) accompanied by the emission of a neutrino. Positron emission is described by Eqs. (1.202) and (1.203) of Chapter 1. The positron is an antiparticle of an electron; it possesses an opposite charge and a spin in the opposite direction to that of the electron. The total energy released in the positron decay process is shared between the positron and the neutrino. The positron will lose its kinetic energy in matter via ionization. When it comes to a near stop, it comes into contact with an electron, its antiparticle, and is annihilated with the simultaneous production of two gamma-ray photons of 0.51 MeV energy equivalent to the two annihilated electron rest masses. See Chapter 1 for more detailed information on both the negatron and positron decay processes.

The liquid scintillation counting efficiency for beta particles (negatrons or positrons) is dependent on the original energy of the beta decay. For most beta particles with a decay energy above 100 keV, the counting efficiency is 80%–100%, but for lower energy beta decays (e.g., ^{3}H, $E_{max} = 18.6$ keV) the counting efficiency is normally in the range of 10%–60% depending upon the degree of quench in the sample. The phenomenon of quench and its effect on liquid scintillation counting efficiency are described in Sections IV and V of this chapter.

Another common nuclear decay process is gamma-ray emission. In this process, a gamma ray is emitted from the nucleus of the decaying atom. The gamma ray is electromagnetic radiation or, in other respects, a photon particle. The general energy range for gamma rays is 50–1500 keV. Gamma-ray emission often accompanies alpha-emission, beta-emission, or electron capture (EC) decay processes. Bremsstrahlung or X-radiation, which is electromagnetic radiation originating from electron energy transitions, also accompanies the EC decay process. When gamma-emitting radionuclides are detected by the liquid scintillation counter, it is not the gamma ray that is detected to a very significant degree, but rather the alpha particles, beta particles, or atomic electrons (Auger and internal-conversion electrons) that may be produced during decay processes occurring in the liquid scintillation fluor cocktail. Gamma rays from sample radionuclides in the scintillation cocktail can produce Compton electrons, although these interactions are less significant in magnitude in the liquid fluors. In general, electromagnetic radiation makes only a minor contribution to excitation in liquid scintillation fluor compared to charged-particle radiation. For example, if we consider the liquid scintillation analysis of ^{125}I, which decays by electron capture with the emission of gamma rays and daughter X-radiation, liquid scintillation counting efficiencies as high as 85% are reported. However, the excitations in the liquid scintillation fluor are due mainly to the absorption of Auger and internal-conversion electrons and only a minor contribution ($\sim$8%) is the result of X-rays produced during the decay process (L'Annunziata, 1987).

C. Cherenkov photon counting

Beta particles of energy in excess of the threshold energy of 263 keV can be detected and quantified in water using the liquid scintillation analyzer without the use of scintillation fluor cocktail. Other threshold energies exist for media other than water, and these are described in Chapter 6, Volume 2 of this book. The sample is simply placed in a clear liquid solution (often aqueous) and detected by the photons of light produced by the Cherenkov Effect. Charged particles, such as beta particles, that possess sufficient energy can travel at a velocity exceeding the speed of light in transparent media such as water, organic solvents, plastic, and glass. When this occurs, the charged particle will produce Cherenkov photons, which extend from the ultraviolet into the visible wavelengths. The light, which is produced, is of low intensity, and it is detected in the low-energy counting region of a liquid scintillation analyzer with pulse heights equivalent to those produced by beta particles of 0−50 keV. High-energy beta-particle emitters, which emit a significant number of beta particles in water in excess of 263 keV, can be analyzed by counting the Cherenkov photons in the liquid scintillation analyzer without fluor cocktail. Some examples are ^{32}P (E_{max} = 1710 keV), ^{90}Sr(^{90}Y) where the E_{max} of ^{90}Y beta particles is 2280 keV, ^{86}Rb (E_{max} = 1770 keV occurring at an 88% intensity (probability per decay) or 680 keV at a 12% intensity), and ^{89}Sr (E_{max} = 1490 keV). The Cherenkov counting efficiency of these radionuclides in water is in the range of approximately 35%−70% depending on color quench of the sample and E_{max} of the beta-particle emissions. The process of Cherenkov counting is treated in detail in Chapter 6, Volume 2 of this book, and a thorough treatment of the origin, principles, and theory of Cherenkov radiation and the Cherenkov Effect is provided in previous books by the writer (L'Annunziata, 2007, 2016). In general, it is important to remember that when quantifying radionuclides by Cherenkov counting, the counting region should be set to encompass only the low pulse height spectra produced by Cherenkov photons, which is equivalent to the pulse heights produced by low-energy (0−50 keV) beta particles, and no fluor cocktail is required.

III. Liquid scintillation counter (LSC) or analyzer (LSA)

As described previously, the scintillation process involves the conversion of nuclear decay energy into photons of light. Therefore, to quantify the nuclear decay event and to satisfy needs for automation and multiple user programs, a liquid scintillation counter must be able to perform the following functions: (1) it must be able to detect photon emissions that occur in the scintillation vial containing fluor cocktail or solid scintillator (SPA) and be able to determine the number of photon emissions and their intensity; (2) it must be able to hold a large number of scintillation vials (>400) of various sizes (*e.g.*, 20 mL, 8 mL, 7 mL, 4 mL, and microfuge or Eppendorf tubes); (3) it must have the ability to process automatically various types of samples using different counting conditions and counting programs (*e.g.*, single radionuclides, multiple radionuclides, quench corrections, direct DPM, or Cherenkov counting) using programmable counting setups or counting protocols; (4) it must be able to process the data from photon emissions per unit time (e.g., photons per minute) to counts per minute (CPM) and then convert these count rates to actual nuclear disintegration rates or nuclear decay events per unit time (e.g., disintegrations per minute or DPM) using a quench correction method or direct DPM method; and (5) it must perform data analysis and reduction, special computer-managed data application programs, and instrument performance assessment.

The first and most important task of the liquid scintillation analyzer is the detection and quantification of the number of photon emissions and their corresponding intensities. This is accomplished by the heart of the LSC, the light detection, and quantification components. A simple block diagram of a conventional LSC with two photomultipliers is illustrated in Fig. 6.3. Three basic components are found in this part of the LSC, namely, the

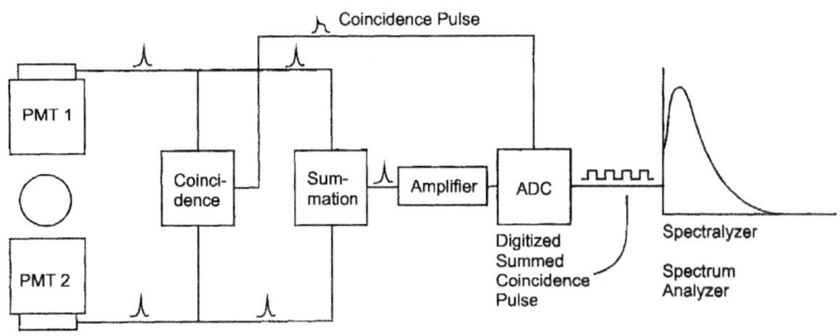

FIGURE 6.3 Schematic diagram of the components of a basic liquid scintillation analyzer. The circle between photomultiplier tubes PMT 1 and PMT2 represents a vial containing the sample and scintillation fluor cocktail. *From Kessler (1989), ©1998−2018, PerkinElmer, Inc. Printed with permission. See also Cook and van der Plicht (2007).*

detector(s), a counting circuit, and a sorting circuit. Liquid scintillation analyzers with three photomultiplier tubes (PMTs) are also available, but yet less common. LSAs operating with three PMTs provide unique methods of activity analysis, and these are discussed in Section IX.C of this chapter.

In order to quantify the radioactivity in the sample, the sample is loaded into the counting chamber using either an up- or downloading elevator mechanism. The downloading mechanism is able to prevent any external light from entering the counting chamber by using a double light seal mechanism. The double light seal is implemented by automatic loading of the sample vial from the sample chamber deck to a holding area, where the sample is sealed from external light. The sample is mechanically moved into the counting chamber, which is below the holding area. Because of this unique downloading mechanism, the photomultiplier tube (PMT) high voltage can remain on at all times and the PMT background stabilized. Once the sample has been loaded into a light-tight chamber, the photons of light emitted from the vial are detected using two photomultiplier tubes (PMT 1 and PMT 2 of Fig. 6.3).

The photomultiplier tubes (PMTs) convert the photons of light emitted from the liquid scintillation vial to electrons when the photons hit a bialkali photocathode located inside the face of the PMT, as illustrated in Fig. 6.4. The photoelectrons produced at the PMT photocathode are amplified through a series of positively charged dynodes, each dynode having an increasing positive voltage along the series. The increasing voltage accelerates the initial photoelectrons produced at the PMT photocathode to yield an avalanche of secondary electrons, resulting in a pulse amplification. A photon, which is produced in the scintillation vial, is thereby converted to a corresponding electronic signal. Because the amount of light produced in the scintillation vial is normally very low ($\sim$10 photons per keV energy absorbed in the liquid fluor cocktail), the PMT must be able to amplify the pulse resulting from secondary electrons produced by a single photon of light by a large factor. This pulse amplification factor or gain is $> 10^6$ for the standard PMT used in the LSC.

As illustrated in the block diagram of Fig. 6.3, two PMTs are used for the measurement of the light intensity from the nuclear decay processes in the sample vial. The two PMTs permit coincidence light detection and coincidence pulse summation required for the LSC to be able to detect low-energy radionuclides such as tritium ($E_{max} = 18.6$ keV) and to distinguish instrument background from true nuclear events. If only a single PMT were used in the LSC, the background level would be approximately 10,000 CPM for a 0–2000 keV counting region. This high background is normally due to the large amplification factor from the PMT. This high background count rate mainly occurs in the 0–10 keV region (thermal and electronic background noise). In the LSC, two PMTs and a coincidence circuit are used to help differentiate background signals from true nuclear decay events in the scintillation vial, which is referred to as coincidence counting. The principle behind coincidence counting is based upon the fact that, when a nuclear decay event occurs in the scintillation vial, light is produced which is isotropic (*i.e.*, is emitted equally in all directions). Since the decay process and resultant scintillation process produce multiphoton events (about 10 photons per keV of nuclear energy dissipated in a liquid scintillation cocktail), light is emitted in all directions from the scintillation vial. The decay

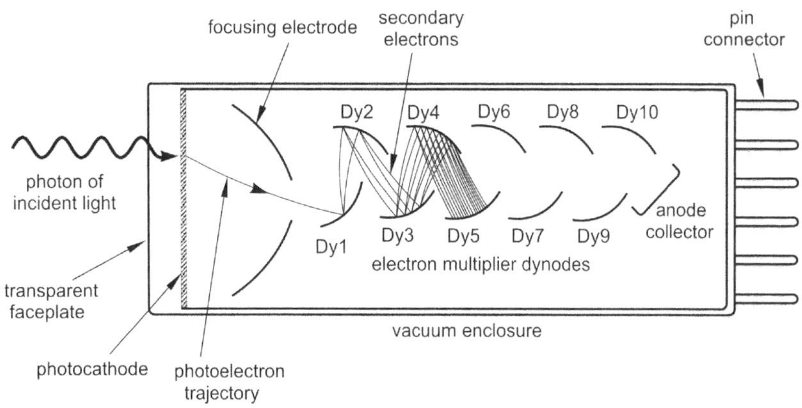

FIGURE 6.4 Schematic of a photomultiplier tube. The drawing is based on schematics by Itaya et al. (2004) and Polyakov (2013). A photon of visible light emitted from a liquid scintillation vial collides with a bialkali photocathode producing a photoelectron. The trajectory of the photoelectron is focused and accelerated toward a positive dynode (Dy1). The impact of the photoelectron onto the first dynode (Dy1) produces secondary electrons that are accelerated toward a second dynode (Dy2) of higher positive voltage. The secondary electrons are accelerated sequentially toward dynodes of increasing potential difference (Dy3 to Dy10), which result in an avalanche of secondary electrons. The avalanche of electrons (not illustrated with dynodes Dy6 to Dy10) are collected at the anode, yielding a pulse amplification $> 10^6$ with a typical rise time of $\sim$2 ns. The rise time is the time for the anode output pulse to rise from 10% to 90% of the peak amplitude. The faceplate is made generally of borosilicate glass with a diameter of $\sim$5 cm.

process and resultant scintillation are very rapid (approximate light decay time is 2–10 ns). Because the scintillation process produces multiphoton events and the events decay rapidly, we can distinguish most background from true nuclear decay in the scintillation vial. If light is produced in the scintillation vial inside the analyzer detection area, it will be emitted in all directions and be detected by the two PMTs in the very short pulse decay time of 2–10 ns. When a signal is detected in both the PMTs within a coincidence resolving time of 18 ns, it is accepted as a true nuclear decay event. If on the other hand, a background event occurs in one of the PMTs or in the electronic circuitry (*e.g.*, thermal or electronic noise), it will produce a single event, which will be detected by only one of the two PMTs in the 18 ns time frame. Such a single event is rejected as occurring external to the sample or, in other words, a background event. By using two PMTs and the coincidence circuit, the instrument background can be reduced from 10,000 CPM with a single PMT to about 30 CPM with two PMTs for a wide-open 0–2000 keV pulse height counting region. The PMT signal that is sent to the coincident circuit is an analog signal with a pulse height that reflects and is proportional to the original nuclear decay energy.

The next part of the detection area, illustrated in Fig. 6.3, is the summation circuit. This circuit has a dual purpose. The first is to reassemble the original two coincident signals into an individual signal with the summed intensity. This helps to optimize the signal-to-noise ratio in the instrument. The second purpose is to compensate for the light intensity variations due to the position of the nuclear decay in the vial that would occur when samples containing color are counted. If only one of the two PMT signals were used in counting a colored sample, the signal height would be dependent on where in the scintillation vial the light was produced. If the light was produced near the edge of the scintillation vial closer to the face of the PMT, a more intense emission of light would be detected by the PMT due to its closer proximity to that edge of the vial. However, with two PMTs and a summed signal, the final pulse height produced by the PMT is not affected by the position of the nuclear decay in the presence of color in the sample counting vial.

Subsequent to pulse summation in the LSC, the signal is further amplified and sent to the analog-to-digital converter (ADC). The ADC converts the signal from an analog signal, which is a pulse with a certain height, to a single number that represents its pulse height or intensity. The digital pulses are finally sorted on the basis of their magnitude or pulse height number. The sorting can be accomplished by one of two methods: pulse height analysis (PHA) or multichannel analysis.

PHA, which is the older of the two methods, utilizes only two discriminators, an upper- and a lower-energy discriminator. An upper-level discriminator is set such that all of the pulses with a certain energy of interest are always lower than this upper level. A lower-level discriminator is also set to reject pulse magnitudes that are below the lower-level discriminator setting thereby reducing background and other counting interferences of low pulse height. When an event is detected, its pulse height is measured, and if it has an intensity lower than the upper discriminator and higher than the lower discriminator, it is accepted as a true nuclear event. If any of the pulses fall outside this range, they are rejected and lost by the counting circuitry. All of the pulses that fall into the accepted range are counted, hence the term liquid scintillation counting (LSC).

The second and more contemporary method of sorting pulses is MCA. The multichannel analyzer is a series of bins or slots, where different pulse height magnitudes are placed once they have been detected. Two types of MCAs are commonly used: linear and logarithmic. The linear MCA provides data with pulse heights calibrated to represent decay energy in keV on a linear scale. For a common 4000 channel linear MCA, each channel may represent approximately 0.5 keV of energy. The logarithmic MCA displays the pulse heights in channels plotted along a logarithmic scale as illustrated in Fig. 6.2.

All of the pulses collected in MCAs are not only counted but analyzed in terms of their number and pulse height; therefore, the liquid scintillation counter is now more often referred to as the liquid scintillation analyzer (LSA). A linear MCA output with a typical beta-particle pulse height spectrum is illustrated in Fig. 6.5.

The second function of the modern LSA is to count automatically various types and sizes of samples in vials containing scintillation fluor cocktail. Most modern LSAs

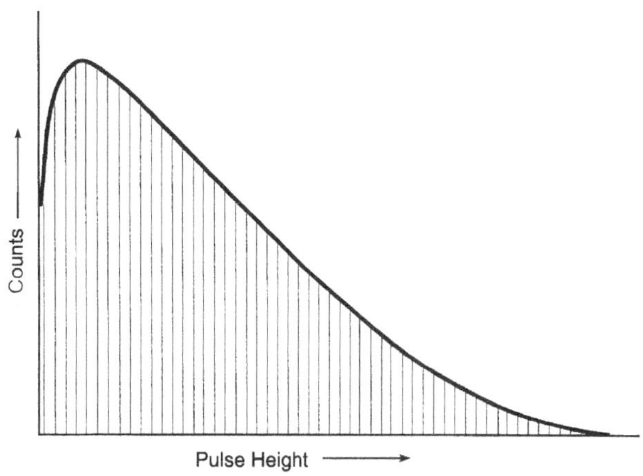

FIGURE 6.5 Illustration of a typical liquid scintillation beta-particle pulse height spectrum collected in the many channels of a multichannel analyzer (MCA). A typical linear MCA will have as many as 4000 channels within which pulses of certain heights are collected and counted. The channels are calibrated over the energy range of 0–2000 keV. *©1998–2018, PerkinElmer, Inc. Printed with permission.*

are cassette based. This means that sample vials are placed in racks holding between 12 and 18 individual scintillation vials or samples. Specific cassettes are available for holding scintillation vials of different volumes up to 20 mL.

Many persons can use the same instrument by establishing counting protocols set up via computer to analyze different radionuclides under different counting conditions and sample sizes. The key functions of the LSA—analyzing sample data by determining sample quench levels, converting count rate (*e.g.*, counts per minute or CPM) to disintegration rate (*e.g.*, disintegrations per minute or DPM) for unknown samples, and automatic monitoring of the performance of the instrument (instrument performance assessment)—will be described in detail further on in this chapter.

The majority of liquid scintillation analyzers in use today contain two photomultipliers, and these are installed with large-capacity sample changers as nonportable equipment in laboratories around the world. Portable and very compact liquid scintillation analyzers with multichannel analysis have been developed, which is very applicable to mobile laboratories (Capogni and De Felice, 2014; Johansson et al., 2014; Kossert et al., 2014a,b,c; Gudjónsson et al., 2009; Theodórsson, 2006; Theodórsson and Jónsson, 2006; Theodórsson et al., 2006; Jónsson et al., 2006). In some portable single-photomultiplier-tube liquid scintillation analyzers, the single PMT measures the scintillation photons from the bottom of the counting vial. Such a small portable liquid scintillation analyzer can find numerous applications in mobile-laboratory on-site monitoring of radionuclides. The application of the single-photomultiplier LSA for the monitoring of environmental tritium and radon has been demonstrated (Theodórsson and Skripkin, 2003; Theodórsson and Gudjónsson, 2003; Gudjónsson et al., 2009). An example of such a portable liquid scintillation analyzer and luminescence counter is the Triathler (Hidex, Inc., Turku, Finland) described by Santiago et al. (2014).

IV. Quench in liquid scintillation counting

In scintillation counting, the sample is either dissolved in a liquid scintillation cocktail or adsorbed onto a solid scintillator in a sample vial or microplate well. In order to quantify the nuclear events as activity in terms of disintegrations per minute (DPM), the LSA counts the number of photon emissions in a preselected time period to provide a count rate (CPM) of the sample. The sample count rate is dependent on how efficiently the nuclear decay events are converted to photons of visible light that are detected and quantified by the LSA. Because the sample solution is always present, it can absorb nuclear decay energy thereby preventing this energy from being absorbed by the solvent and chemical fluor molecules, or the solution can absorb photons of light emitted by the fluor molecules, thus reducing the photon intensity measured by the PMT. This causes the phenomenon called quench. We can define quench as interference in the conversion of nuclear decay energy to photons of light emitted from the sample vial. Quenching can be the result of three phenomena, namely, (1) chemical quenching, (2) ionization quenching, and (3) color quenching. These interference phenomena are discussed in more detail in the subsequent paragraph. Many texts refer only to chemical and color quench phenomena as mechanisms to be considered for the determination of radionuclide detection efficiency with little or no reference to ionization quenching. This may be due to the much smaller effect that ionization quenching has compared to chemical and color quenching in liquid scintillation counting, and the attention to this phenomenon has been demonstrated to be relevant only in the application of liquid scintillation analysis to the standardization of radionuclides (Grau Carles et al., 2004, 2006a,b, Kossert and Grau Carles, 2006, Grau Malonda and Grau Carles, 2008, Bagán et al., 2008, 2009, Bignell et al., 2010a,b, Cassette et al., 2010b, Thiam et al., 2012, van Wyngaardt et al., 2012, Kossert et al., 2014b, and Altzitzoglou and Rožkov, 2016). Radionuclide standardization is carried out only by specialized laboratories such as those found in national institutes of standards. The application of liquid scintillation analysis (LSA) to the standardization of radionuclide samples, as a specialized application of LSA, is discussed in Section IX of this chapter. The points of interference of chemical, ionization, and color quenching in the liquid scintillation process are illustrated in Fig. 6.1.

The three quenching phenomena, which interfere with the liquid scintillation measurement of radionuclides, are described in more detail as follows:

1. *Chemical quenching*. This is the most common quench mechanism and the one that generally can have the most deleterious effect on the liquid scintillation detection efficiency. Chemical quench is caused by the presence of chemical substances in the sample, including the sample solution, that absorbs nuclear decay energy in the scintillation process, thereby obstructing to a certain degree the transfer of nuclear decay energy to the scintillation cocktail solvent. A chemical quenching agent can be thought of as a sponge that absorbs energy before it can produce light in the scintillation process. In addition to reducing the number of photon emissions resulting in a loss of radiation counts, the quenching process can, and often does, decrease the light intensity and thereby reduce the pulse heights as measured by the scintillation process. Chemical quenching occurs to some degree in most liquid scintillation counting samples.

2. *Ionization quenching*. The second mechanism of quench, ionization quenching, as described by Grau Carles et al. (2004), is related to the density of excited solvent molecules in the scintillation fluor cocktail. The response of a liquid scintillator to the interaction of charged particles is not linear with the energy deposited. The nonlinearity of the photon intensity output by the scintillator and the energy deposited by the charged particles increases with the mass of the particle, and the nonlinearity is a function of the linear energy transfer (LET) of the particle. Also, as defined by Bignell et al. (2010a), ionization quenching in scintillators is the reduction of luminosity associated with a high density of excited molecules caused by a large LET to the scintillant. Thus, low-energy beta particles, such as those emitted from ^{3}H, exhibit a greater nonlinearity of light output versus energy deposited than high-energy beta particles or electrons, because the LET of low-energy beta particles or electrons is higher than the LET of high-energy beta particle or electrons (See Tables 1.29 and 1.30 of Chapter 1). The cause of this nonlinearity is ionization quenching, which is a consequence of the density of excited solvent molecules in the scintillation fluor cocktail. Charged particles, which exhibit a high LET, will create a higher number of excited solvent molecules. When the concentration of the excited molecules is high, there is a higher probability of interaction between two excited molecules whereby one molecule can lose its energy of excitation to the other resulting in one molecule becoming superexcited. The superexcited molecule has a high probability of reaching the state of ionization. The overall outcome of this process is two solvent molecules, originally excited by the charged particle (e.g., β-particle or electron), which lose their energy of excitation and thereby do not transfer this energy to the fluor molecules or transfer only a diminished portion of that excitation energy. This results in a loss of excitation energy and a reduction of the photon intensity from the scintillation fluor, which may manifest itself by a reduced count rate or a reduction in the magnitude of the pulse height measured by the photomultiplier from the photon intensity output. Grau Carles et al. (2004) explain that this process is observed clearly in the interaction of α-particles with liquid scintillator. Also, radionuclides decaying by low-energy β transitions, such as ^{3}H, as well as those decaying by electron capture with the emission of Auger or Coster-Kronig electrons, are those that display this ionization quenching, that is, the nonlinearity between the intensity of the photon emissions from the scintillation counting vial and the energy deposited by the β particle or electron. Electron-capture decay results in the emission of X-rays from subsequent atomic electron rearrangements following EC decay, which can produce X-ray Compton electrons and photoionization, all of which are low-energy electrons exhibiting a relative high LET compared to the higher-energy beta particles from most beta-emitting radionuclides. This mechanism of quench is relevant only in the specialized application of liquid scintillation analysis to the standardization of radionuclides (See Section IX of this chapter).

3. *Color quenching*. The phenomenon of color quench occurs when color is visible in the sample that is being counted. The color quench phenomenon normally acts by absorbing photons of light in the scintillation vial before they can be detected and quantified by the PMT. This is similar to what happens when a colored filter is used on a camera to filter out certain wavelengths of light. Chemical quench absorbs nuclear decay energy and color quench absorbs photons of light. Color quenching is often less a problem than chemical quenching, because samples most often can be decolorized easily with minute quantities of bleaching agent or other means (See Chapter 7 of Volume 1 on Sample Preparation Techniques for Liquid Scintillation Counting).

The quenching phenomena reduce the photon intensities emitted from scintillation counting vials and consequently reduce the magnitudes of the pulse heights measured by the liquid scintillation analyzer; in addition, the quenching phenomena can also reduce the count rate of a sample measured by the liquid scintillation analyzer (*i.e.*, counts per minute, CPM). To compensate for the count rate loss due to quench and to determine the sample activity or DPM (nuclear decay rate), it is necessary to know the counting efficiency, defined by the following equation:

$$\%\ \text{efficiency} = \text{CPM}/\text{DPM} \times 100. \tag{6.1}$$

where CPM is the count rate of the sample determined by the LSA, and DPM is the actual disintegration rate of the sample. The relationship between CPM and DPM of the sample varies according to the energy of the nuclear decay at a given degree of quench. The lower the energy of the decay, the greater is the effect of quench on the counting efficiency for beta-emitting radionuclides. This is illustrated in Fig. 6.6, which shows the liquid scintillation pulse height spectra of seven tritium samples ($E_{max} = 18.6$ keV) that were prepared with the same activity (DPM), but with different amounts of 5.0 M HNO_3, which acts as a chemical quenching agent. The liquid scintillation analyzer determined the CPM for each sample by summing the area under the pulse height spectrum of each sample. As illustrated in Fig. 6.6, the least quenched sample is that which contains no HNO_3. The area under the pulse height spectrum of the first sample had 126,287 CPM and the highest pulse heights with a maximum equivalent to approximately

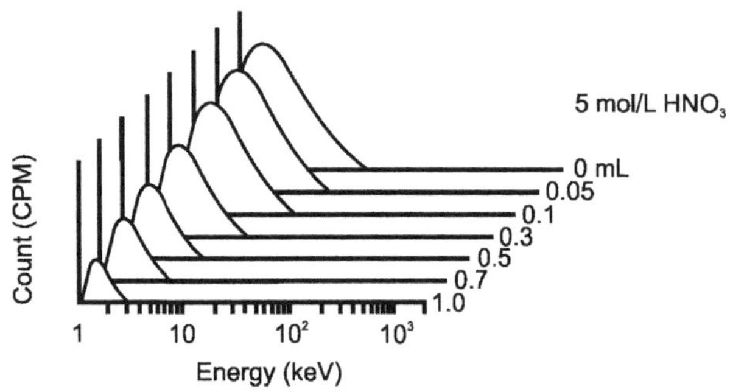

HNO_3 (0.5M) (mL)	^3H (CPM)	^3H (DPM)	^3H (%E)
0.00	126,287	210,000	60.1
0.05	115,834	210,000	55.1
0.10	102,218	210,000	48.7
0.30	61,211	210,000	29.1
0.50	39,846	210,000	19.0
0.70	25,239	210,000	12.0
1.00	16,091	210,000	7.7

FIGURE 6.6 Pulse height spectra of seven samples of ^3H of equal activity containing varying amounts of 0.5M HNO_3 quenching agent. The pulse height spectra are plotted on a logarithmic scale with pulse heights calibrated to equivalence in keV energy. The liquid scintillation counting (detection) efficiencies for each sample are listed as percentages. *©1998–2018, PerkinElmer, Inc. Printed with permission.*

18.6 keV. The counting efficiency for this sample is calculated as 126,287 CPM/210,000 DPM or 60.1%. The second sample is quenched by the added 0.05 mL of HNO_3, and as a result, (1) the CPM is reduced to 115,834, (2) the endpoint or maximum intensity of the pulse height spectrum of this sample is reduced, and (3) the counting efficiency of the sample is reduced to 55.1%. As illustrated in Fig. 6.6, when the sample is quenched more and more, the maximum observed pulse height is reduced further and the CPM collected under the pulse height spectrum is reduced. For example, the last sample listed, which contains the highest amount of quenching agent, gave the lowest count rate of 16,091 CPM and a calculated counting efficiency of only 7.7%. Thus, as the quench increases for tritium, both the maximum pulse height and the total CPM are reduced significantly. Chemical quenching agents, although dilute and small in quantity, can have a significant effect on the counting efficiency of tritium.

On the other hand, an isotope such as carbon-14, which emits beta particles of energy almost 10 times higher ($E_{max} = 156$ keV) than tritium, quenching agents cause a significant reduction in the maximum pulse heights but have a less significant effect on the pulse counts collected than was observed in the case of tritium. Table 6.1 shows the effect of the quenching agent on five samples of carbon-14. The five samples contained the same activity (100,000 DPM), but increasing amounts of quenching agent. The quenching agent is not given here, but a common quenching agent used for these studies is nitromethane over the range of 0–100 μL per 20 mL of fluor cocktail. The endpoint of the pulse height spectra (maximum pulse height expressed in keV) of each sample listed in Table 6.1 changed significantly from sample 1 to sample 5 as chemical quench increased; however, the sample count rates (total counts collected under the pulse height spectra per given period of time) did not change as drastically as for tritium. As can be seen from Table 6.1, pulse height spectral intensity (maximum pulse height) changes as the sample is quenched, but the efficiency or CPM value (area of energy spectrum) changes only slightly. The overall conclusion is that for beta-particle emitters, the lower the energy (E_{max}) of the beta decay, the greater is the effect of quench on the counting efficiency of the radionuclide.

For alpha-emitting radionuclides the phenomenon of quench does not significantly affect the counting efficiency as shown in Fig. 6.7. As the quench of the sample is increased, the monoenergetic alpha peak is simply shifted to lower pulse heights, but the total area under the pulse height spectrum or detection efficiency (equivalent to counting efficiency) is not changed significantly. Also, as illustrated in Fig. 6.7, the alpha-particle pulse height spectrum in the LSA undergoes some peak broadening (reduced resolution) proportional to the level of quench; although this will have no significant effect on detection efficiency. The resolution is determined by the percent full width at half maximum (FWHM), as described in Chapter 9

TABLE 6.1 Effect of quench on carbon-14 counting efficiency in liquid scintillation analysis.

Sample	Maximum pulse height (keV)	CPM	DPM	Efficiency (%)
1	156	95,000	100,000	95.0
2	112	94,500	100,000	94.5
3	71	92,500	100,000	92.5
4	43	90,500	100,000	90.5
5	32	87,000	100,000	87.0

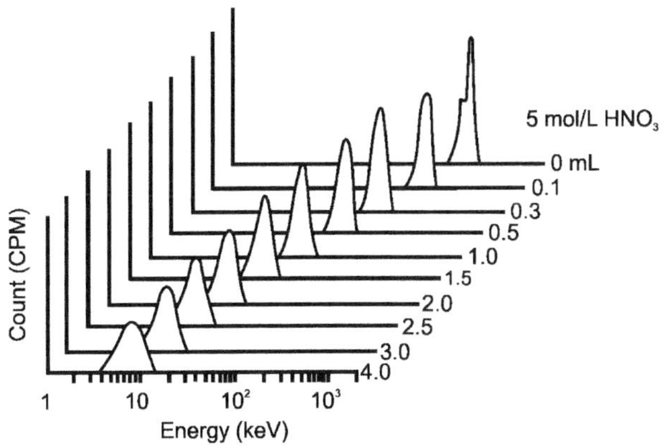

HNO$_3$ (0.5M) (mL)	^{241}Am (CPM)	^{241}Am (DPM)	^{241}Am (%E)	^{241}Am %FWHM
0.00	46,660	46,700	100	42
0.05	46,632	46,700	100	a
0.10	46,692	46,700	100	a
0.30	46,149	46,700	99	47
0.50	46,234	46,700	99	a
0.70	46,371	46,700	99	a
1.00	46,394	46,700	99	a
1.50	46,448	46,700	99	53
2.00	46,256	46,700	99	a
3.00	46,148	46,700	99	a
4.00	46,080	46,700	99	a

[a]% FWHM not calculated

FIGURE 6.7 Liquid scintillation pulse height spectra of various samples of ^{241}Am of equal activity containing varying amounts of 0.5M HNO_3 quenching agent. The pulse height spectra are plotted on a logarithmic scale with pulse height calibrated to equivalence in keV energy. The liquid scintillation counting (detection) efficiencies for each sample are listed as percentages. The alpha peak resolutions are measured as percent full width at half-maximum. *©1998 PerkinElmer, Inc. Printed with permission.*

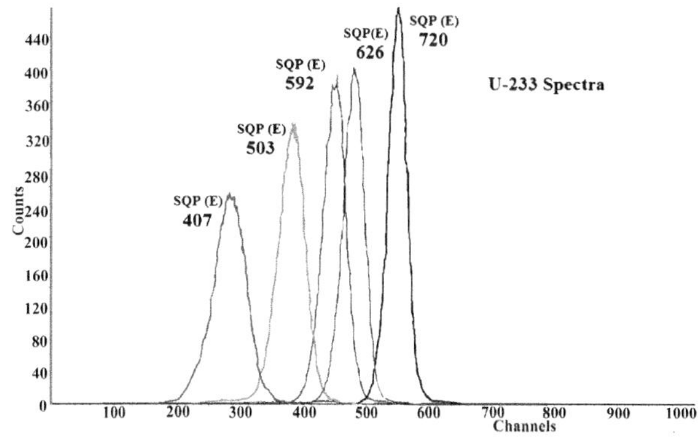

FIGURE 6.8 The shifting of alpha pulse spectra over a wide range of quench from lower levels of quench with a quench indicating parameter SQP(E) = 720 to higher levels of quench at a quench indicating parameter SQP(E) = 407. The broading of the alpha peaks is visible as the quench level increases. The measurement of quench indicating parameters is discussed further on in this chapter. *From Bhade et al. (2017), reprinted with permission from Elsevier © 2017.*

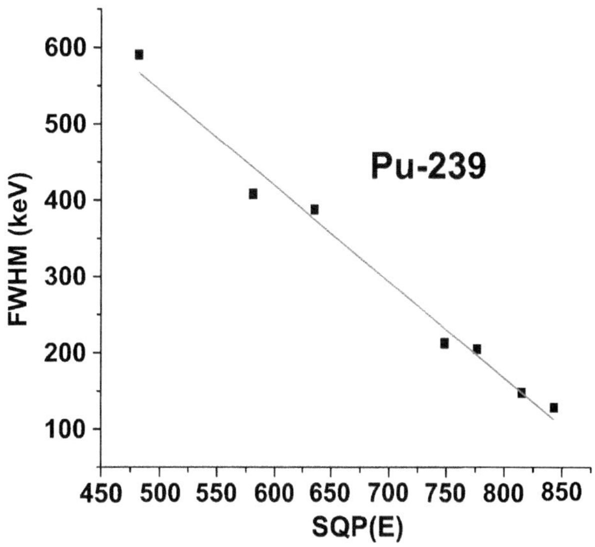

FIGURE 6.9 Influence of quenching on the alpha spectral resolution measured by the full width at half maximum (FWHM). *From Bhade et al. (2017), reprinted with permission from Elsevier © 2017.*

of Volume 1 of this book. Vera Tomé et al. (2002) and Bhade et al. (2017) studied alterations in alpha-peak shape in liquid scintillation with the potential of utilizing LSA for alpha spectrometry. Bhade et al. (2017) measured the liquid scintillation detection efficiencies and alpha spectra resolution of several alpha emitting radionuclides. They demonstrated the shifting of the alpha-particle pulse height spectra to lower level pulse heights and a linear decrease in alpha spectra resolution according to the quench level of the sample, as illustrated in Figs. 6.8 and 6.9, respectively. In addition, they found no significant reduction in the counting efficiencies for alpha particles over a wide range of quench levels. In an early work, ~~by~~ Horrocks (1974a) ~~he~~ reported "alpha particles are counted with 100% efficiency even when the sample is highly quenched." This conclusion was based on the work of several researchers including Horrocks (1974a, 1968, 1966), Seidel and Volf (1972), Joshima and Matsuoka (1970), Scott and Good (1967), Wright et al. (1961), Seliger (1960), Basson and Steyn (1954), who demonstrated detection efficiencies of 98%–100% for alpha emitters over a wide range of quench levels.

For gamma emitting radionuclides, the quenching phenomenon is very similar to that observed with beta-particle interactions in liquid scintillation fluor cocktails (Ishikawa and Takiue, 1973), as Compton-electrons produced by γ-photon interactions in scintillation fluor cocktail will undergo similar quenching effects as beta particles. See Section VI of this chapter for a treatment on the liquid scintillation analysis of gamma-emitting radionuclides.

The effect of quench when solid scintillators are used in a liquid scintillation analyzer is described in Chapter 9 on Solid Scintillation Analysis of Volume 1. When using solid scintillators, the sample is normally placed directly on the solid scintillator and dried or the sample is counted adsorbed onto the solid scintillator as in a scintillation proximity assay (SPA). See Chapter 9 for more detailed information on SPA. When scintillation proximity analysis is undertaken, the sample is in direct intimate contact with the solid scintillator, and therefore no chemical quenching exists for these types of samples. Under these circumstances, color quench can occur with colored samples.

For the Cherenkov counting of samples containing high-energy beta-emitting radionuclides, color quench can occur; however, neither chemical quench nor ionization quench are possible. See Chapter 6, Volume 2 for a detailed treatment of Cherenkov counting with conventional liquid scintillation analyzers.

All chemical substances that either dilute the solvent of the fluor cocktail or compete with it for nuclear decay energy will cause quench. Even dissolved oxygen from the air is a chemical quenching agent (Takiue and Ishikawa, 1974); its effect can be seen in the LSA of weak (low-energy) beta particle-emitting radionuclides such as tritium. More information on chemical quenching agents and their classifications can be obtained in a previous book (L'Annunziata, 1987). As chemical quenching agents in samples generally cannot be avoided and the effect of quench on detection efficiency is significant with many radionuclides, it is important to correct for quench when necessary. This will permit accurate measurement of sample activities in disintegration rate (*e.g.*, DPM).

V. Methods of quench correction in liquid scintillation counting

Some degree of quenching exists in almost all samples, which are quantified by the liquid scintillation counting process. It is consequently important to utilize a method that can be used to correct for quench. The quench correction methods allow us to determine the detection efficiency of a particular sample and convert the count rate (CPM) to the actual number of nuclear decays or disintegration rate (DPM) of a sample. Quench correction can be accomplished by one of the following techniques: (1) an internal standard method, (2) a sample spectrum method, (3) an external standard method, and (4) a direct DPM method. These techniques can be used for quench correction and DPM determination. Each may have distinct advantages for various sample types, instrument model utilized and/or radionuclides to be analyzed. These methods will be discussed subsequently together with explanations of the when and why of using these techniques.

A. Internal standard (IS) method

The internal standard (IS) method is the oldest and most tedious, and it can be the most accurate method if great care is taken in its implementation. The technique involves a series of steps for each sample. The first step is to count each sample and obtain an accurate count rate (CPM) value for each. Then the samples are removed from the liquid scintillation counter, and a known activity (DPM) of a radionuclide standard is added to each sample; hence, the term internal standard is applied to this technique. After the addition of the internal standard and thorough mixing of the standard and sample, the samples are recounted to obtain the CPM of the sample with its internal standard. Once the CPM of the sample and the CPM of the sample with internal standard are obtained, the following equation is applied to determine the counting efficiency of the sample:

$$E = \frac{C_{s+i} - C_s}{D_i}. \tag{6.2}$$

where C_{s+i} is the sample count rate after the addition of the internal standard, C_s is the sample count rate before the addition of the internal standard, and D_i is the disintegration rate of the added aliquot of internal standard. The disintegration rate of the sample, D_s, may then be calculated as follows:

$$D_s = C_s/E \tag{6.3}$$

For example, if the counting efficiency for a given sample was found to be 0.25 according to Eq. (6.2) and the sample count rate was found to be 25,000 CPM, the activity of the sample can be calculated to be 25,000 CPM/ 0.25 = 100,000 DPM.

Several assumptions and restrictions are made for the internal standard method, some of which may be intuitively obvious. These are described as follows: (1) The same radionuclide must be used for the internal standard as the sample radionuclide; for example, a tritium-labeled standard must be used with samples containing tritium. Hendee et al. (1972) showed that [^{3}H]toluene and [^{3}H]hexadecane are good internal standards for organic-compatible fluor cocktails and [^{3}H]water or [^{3}H]hexadecane serve well for aqueous-compatible fluor cocktails when assaying for tritium. The organic standards labeled with ^{14}C are good internal standards for counting efficiency determinations of samples containing ^{14}C. (2) The internal standard added to the sample must have a count rate at least 100 times that of the sample. (3) The addition of the internal standard to the sample must not alter the quench in the sample to any significant degree. (4) The activity (DPM) of the internal standard must be accurately known, as with a National Institute of Standards and Technology (NIST) traceable standard. (Commercially available ^{3}H and ^{14}C standards, as well as mixtures of the two radionuclides, designed specifically for application in the internal standard quench correction technique for either aqueous or organic solutions are available from PerkinElmer, Inc. as well as other standard radionuclide sources. (5) This method of determining sample activities requires accurate sample transfer procedures, which can be tedious when working with many samples and small volumes of internal standard. Dobbs (1965) and Thomas et al. (1965) have investigated syringe dispensing techniques for the addition of internal standards to samples in scintillation counting vials.

This method, if performed properly, can provide accurate activity measurements. Some examples of the internal standard quench correction technique applied to the analysis of ^{3}H may be obtained from the works of Alongi et al. (2003), Sweet et al. (2004), Batistoni et al. (2005, 2007), Klin et al. (2010), Eikenberg et al. (2011), Osman et al. (2016) and Vagner et al. (2016), to the analysis of ^{14}C from the works of Alongi et al. (2003), Martins and Boekel (2003), TenBrook and Tjeerdema (2006), Eikenberg et al. (2011) and Osman et al. (2016), and to the analysis of ^{226}Ra from the work of Godoy et al. (2016). The major disadvantages of this technique are the time and the number of sample-handling steps required for each sample.

B. Sample spectrum characterization methods

Sample spectrum characterization methods of quench correction involve the use of some characteristic of the sample spectrum as a measure of quench in the sample. Some of these methods are described subsequently.

1. Sample channels ratio (SCR)

The sample channels ratio (SCR) method was applied often during the early generations of liquid scintillation counters that were equipped with only the pulse height analysis (PHA) or single channel analyzer for data storage and analysis. Nevertheless, the method is applicable with most commercial liquid scintillation analyzers today. It also remains a useful method for modern liquid scintillation analyzers not equipped with external standards, and the SCR method has other applications described further on in this chapter. The method involves counting the sample in two counting regions defined by lower-level (LL) and upper-level (UL) pulse height discriminator settings. The count rate in each counting region varies according to the level of quench in the sample due to the pulse height spectral shift from higher to lower magnitudes caused by sample quench. An example of the pulse height shift according to quench level is illustrated in Fig. 6.10. As illustrated, a sample that is more highly quenched will produce pulse events of lower magnitude (height) than a sample that is lesser quenched.

The SCR quench correction method requires firstly defining the widths of two counting regions also referred to as counting channels or windows. The lower and upper discriminator levels of one region are selected so as to provide a narrow counting region, which can register pulses of only low magnitude (e.g., Channel 1, 0–300 of Fig. 6.10). The discriminator levels of the second counting region are set to provide a wider counting region, which can register most of the pulses of both low and high magnitude (e.g., Channel 2, 0–700 of Fig. 6.10).

A shift in pulse height due to quenching produces a change in the ratio of the pulses registered (counts) by the two regions. The degree of spectral shift and magnitude of change in the sample channels ratio (SCR), such as CPM_1/CPM_2 or sample count rate in Channel 1 over the sample count rate in Channel 2, are dependent on the severity of quench. Consequently, if a series of quenched standards consisting of scintillation vials each containing the same amount of radioactive standard but increasing amounts of quenching agent were counted, they would show a variation in the channels ratio and counting efficiency, such as that illustrated in Fig. 6.11. The procedures used to prepare sets of quenched standards are described in Section V.D. of this chapter and in Chapter 7 on "Sample Preparation Techniques for LSC" of Volume 1 of this book. The quench correction curve, once prepared for a given

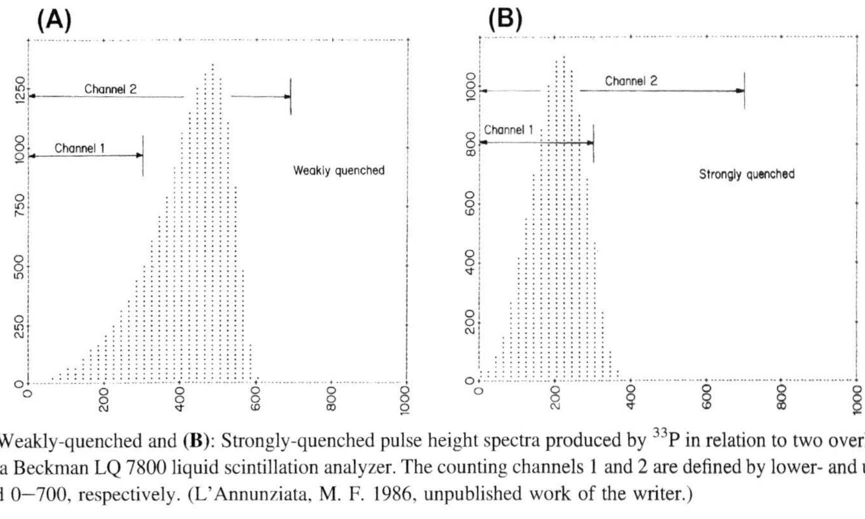

FIGURE 6.10 **(A)**: Weakly-quenched and **(B)**: Strongly-quenched pulse height spectra produced by ^{33}P in relation to two overlapping counting regions (Channels 1 and 2) of a Beckman LQ 7800 liquid scintillation analyzer. The counting channels 1 and 2 are defined by lower- and upper-level discriminator settings of 0–300 and 0–700, respectively. (L'Annunziata, M. F. 1986, unpublished work of the writer.)

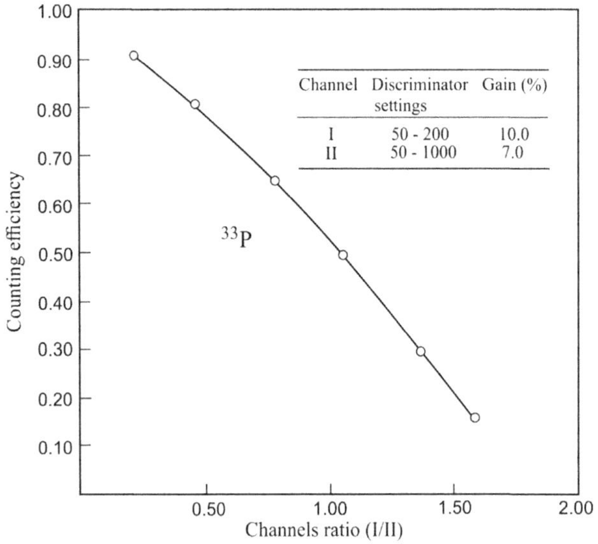

FIGURE 6.11 Typical channels ratio quench correction curve. The channels ratio (I/II) represents the count rate of ^{33}P from Channel I divided by the count rate from channel II. The discriminator and gain settings for each channel are given. The data were obtained from six samples each containing 15 mL of commercial scintillation cocktail (Insta Gel) and 1.0 mL of [^{33}P]orthophosphate of known activity (0.103 μCi equivalent to 3.81 kBq). Each sample contained increasing amounts of CCl_4 as quenching agent. (L'Annunziata, M. F. 1986, unpublished work of the writer.)

radionuclide and fluor cocktail, may be used as a standard curve for determining the counting efficiency of a sample from its channels ratio. The values of counting efficiency for the standard curve are calculated according to

$$E = C_{std}/D_{std} \tag{6.4}$$

where E is the counting or detection efficiency with values between 0 and 1.0, C_{std} is the count rate of the quenched standard in units of counts per minute (CPM) or counts per second (CPS) in the wider Channel 2, and D_{std} is the disintegration rate of the quenched standard in units of disintegrations per minute (DPM) or disintegrations per second (DPS). The activities of the unknown samples are determined from the count rate of the sample in the wider Channel 2 divided by the detection efficiency obtained from the SCR quench correction curve or

$$D_s = C_s/E \tag{6.5}$$

where D_s is the disintegration rate of the sample, C_s is the count rate of the sample in the wide-open channel (i.e., the wider channel from which the detection efficiencies of the quenched standards were determined), and E is the detection efficiency obtained from the SCR quench correction curve, as illustrated in Fig. 6.11.

A more detailed treatment of this method can be found in reviews by L'Annunziata (1984a,b, 1987 and Bagán et al., 2008). The method is less often used with modern liquid scintillation analyzers due to the advent of multichannel analyzers (MCAs) in commercial LSA instrumentation, which utilize sample spectrum quench indicating parameters or external standard quench correction methods. Tarancón et al. (2004) used the SCR quench measurement technique as well as an external standard quench measurement in the evaluation of plastic scintillation beads in a conventional liquid scintillation analyzer. The application of the SCR technique in Cherenkov counting is useful with commercial liquid scintillation analyzers, which are not equipped with an external standard suitable for the production of Cherenkov photons (Todorović et al., 2017). The application of the SCR method to Cherenkov counting is described in detail in Chapter 6 of Volume 2 of this book. The SCR technique is generally not useful with samples of low count rate or high quench, because the counts in one or both of the channels may be so low that a channels ratio becomes meaningless, or long periods of counting time would be required to achieve acceptable levels of statistical accuracy.

A variant of the SCR technique is the Self Constant Channel Ratio (SCCR) method, which is utilized by Hitachi Aloka liquid scintillation analyzers. The SCCR technique utilizes the sample pulse height spectrum to measure a quench indicating parameter. The magnitudes and intensities of the pulse heights produced by the beta particles will depend on the energy of the beta emissions and the quench level of the sample. A multichannel analyzer measures the entire sample pulse height spectrum and the channel equivalent to 75% of the pulse heights measured from a lower level discriminator to 75% of the pulse heights. The pulse height at the 75% cut-off of the sample pulse height spectrum is taken as the quench indicating parameter of a sample. Quench standards are counted and the SCCR quench indicating parameter is measured for each sample. Excellent quench correction curves of % E versus SCCR are obtained, which is evidenced by the strong correlation of a broad range of SCCR values for tritium counting efficiencies ranging from 5% to 60% (Hitachi Aloka, 2014).

2. *Combined internal standard and sample channels ratio (IS-SCR)*

Dahlberg (1982) devised a combination of the IS and SCR methods (IS-SCR), which ameliorates the disadvantages of the two techniques. The high dependence on accurate dispensing of internal standards in the IS technique and the high errors encountered at low count rates in the SCR technique have been eliminated in the combined IS-SCR method for quench correction. In the combined IS-SCR method, the disadvantage of the SCR method at low count rates is avoided by the addition of internal standard

(IS) to low radioactivity samples. The SCR values are then taken for quench correction, instead of calculating the efficiency by the 'classical' IS method of measuring the contribution to count rate by the known amount of standard added. As only an SCR value is required after adding an internal standard, the dependence of the 'classical' IS method on accurate dispensing of standard to sample is also eliminated.

A similar combined IS-SCR technique was devised by McQuarrie and Noujaim (1983) for the counting efficiency determinations of either ^{3}H, ^{14}C, or both nuclides as a mixture. The unique characteristic of this method is the use of ^{67}Ga as the internal standard for either ^{3}H, ^{14}C, or the dual nuclide mixture. The liquid scintillation pulse-height spectrum of ^{67}Ga is characterized by two peaks (Fig. 6.12) corresponding to 8 keV Auger electrons and 93 keV conversion electrons, which are similar in energy to the average beta-particle energy of 5.7 keV for ^{3}H and 49 keV for ^{14}C. A ratio of the measured activity of the two ^{67}Ga peaks is used to reflect the degree of quenching in the sample. The sample is easily recovered after the internal standard ^{67}Ga decays $\left(t_{1/2} = 78 \text{ hours}\right)$ and accurate dispensing of the internal standard to sample is not required, because only the ratio of activity between the two peaks is used to monitor quench.

3. Sample spectrum quench indicating parameters

With the development of the multichannel analyzer (MCA), sample spectrum quench indicating parameters (QIPs) have become more sophisticated, as all of the channels of the MCA can be used simultaneously to measure quench. Examples of quench indicating parameters that measure quench by sample spectrum characterization are the spectral index of the sample (SIS), the spectral quench parameter of the isotope, SQP(I), and the asymmetric quench parameter of the isotope, AQP(I).

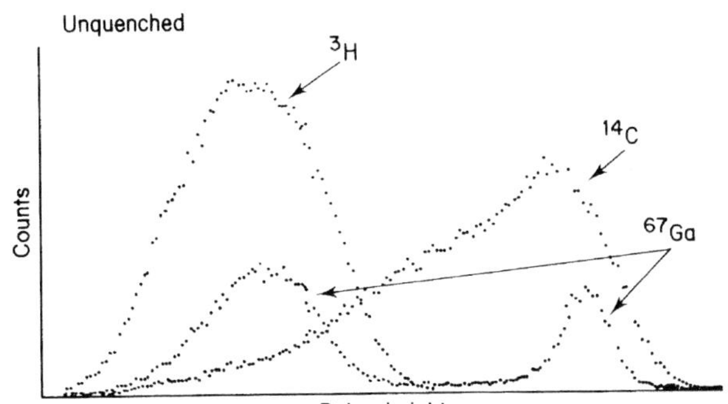

FIGURE 6.12 Liquid scintillation spectra of ^{3}H, ^{14}C, and ^{67}Ga. *From McQuarrie and Noujaim (1983). Reprinted with permission from Faculty of Pharmacy and Pharmaceutical Sciences © 1983 University of Alberta, Edmonton.*

a. Spectral index of the sample (SIS)

The SIS is a measure of the mean pulse height or center of gravity of the sample pulse height spectrum, which is utilized in the Tri-Carb liquid scintillation analyzers of PerkinElmer, Inc. The pulses produced from photon events are linearly amplified, digitized, and stored in a multichannel analyzer to produce a complete sample pulse height spectrum in a region of pulse heights calibrated to represent the energy scale from 0 to 2000 keV. The SIS is a measure of the first moment of the pulse height spectrum proportional to the average energy of the beta spectrum times a factor K or

$$\text{SIS} = K \frac{\sum_{X=L}^{U} X \cdot n(x)}{\sum_{X=L}^{U} n(x)} \tag{6.6}$$

where X is the channel number (see the beta-particle pulse height spectrum with respect to the numerous channels of the MCA in Fig. 6.5), $n(x)$ is the number of counts in Channel X, L, and U are the lowest and uppermost limits of the pulse height spectrum, and K is a factor, which fixes the SIS of unquenched ^{3}H and ^{14}C at 18.6 and 156, respectively, corresponding to the respective maximum beta-particle energies of ^{3}H and ^{14}C in keV. Therefore, the SIS reflects the endpoint or maximum energy of the sample pulse height spectrum as well as the magnitude and shape of the spectrum. From Eq. (6.6) we see that the value of SIS is (1) unitless, (2) always greater than 1.0, (3) becomes smaller in magnitude as quench increases for a given radionuclide, and (4) at a given level of quench, beta-emitters of higher E_{max} will produce higher values of SIS.

An example of count rate (CPM) and quench indicating parameter (SIS) data collected for a series of 10 quenched tritium standards is given in Table 6.2. This data was collected by the liquid scintillation analyzer when the instrument counted each tritium quenched standard to provide a count rate (CPM) for each standard, which is listed in column 2 of Table 6.2. After the count rate of each standard is obtained, the LSA measures the QIP of each standard, in this case SIS, according to Eq. (6.6). The next step required for the preparation of the quench correction curve is the calculation of the percentage counting efficiency for each standard according to Eq. (6.1). The instrument makes this calculation by taking the CPM (column 2) and dividing by the DPM (column 3) of each quenched standard and multiplying by 100 to obtain the percent counting efficiency.

The data of counting efficiency and quench indicating parameter, SIS, listed in Table 6.2 is then taken automatically by the instrument to plot the quench correction curve for tritium illustrated in Fig. 6.13. Another quench correction curve, that for ^{14}C, is also plotted in Fig. 6.13. The ^{14}C

TABLE 6.2 Data collected for the preparation of a ^3H quench correction curve of counting efficiency versus the quench indicating parameter SIS (© 1998–2018 PerkinElmer, Inc. Printed with permission.).

Standard	CPM	DPM	Efficiency (%)[a]	SIS
1	68,000	100,000	68	18.6
2	64,000	100,000	64	16.0
3	58,000	100,000	58	14.8
4	52,000	100,000	52	13.6
5	48,000	100,000	48	12.0
6	38,000	100,000	38	11.0
7	29,000	100,000	29	10.5
8	23,000	100,000	23	9.2
9	18,000	100,000	18	8.5
10	13,000	100,000	13	8.0

[a] *The % efficiency here refers to the % counting efficiency calculated according to Eq. (6.1). For the calculation of sample activities from count rate the decimal equivalent of percent counting efficiency is used (e.g., 0.68 for 68%).*

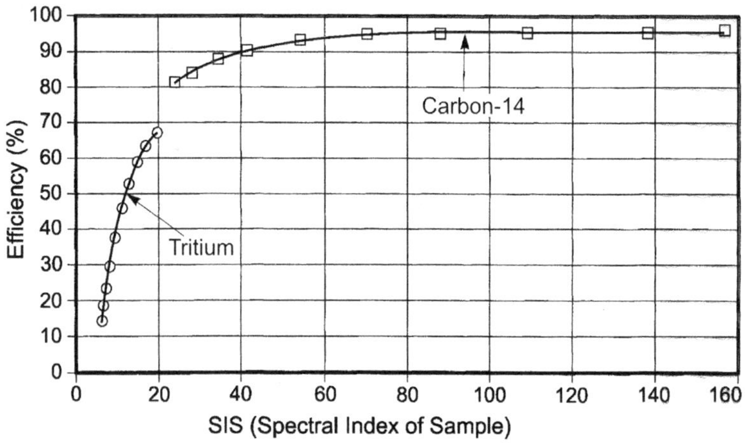

FIGURE 6.13 Quench correction curves for ^3H and ^{14}C based on the quench indicating parameter SIS, a sample pulse height spectrum characterization method. © *1998–2018 PerkinElmer, Inc. Printed with permission.*

quench correction curve was prepared in a fashion similar to the procedure described above for ^3H with the exception that ^{14}C quenched standards were used. Fig. 6.13, therefore, illustrates plots of the quench correction curves for two radionuclides on the same graph. Several observations can be made from these two curves. The first observation is that for ^{14}C, which is a beta-particle-emitting radionuclide of intermediate energy ($E_{max} = 156$ keV), quench has a marked effect on the endpoint or maximum spectral pulse height, as the SIS decreases from 156 to 25). However, the count rate (area under the pulse height spectrum of each standard) or counting efficiency (CPM/DPM) decreases only slightly (0.95–0.83) as illustrated in Fig. 6.13. Therefore, for midrange to higher-energy beta particle-emitting radionuclides, quench does not have as marked effect on the counting efficiency of the sample as on the apparent endpoint of the spectral pulse height. The second observation is related to the tritium quench correction curve. In the case of tritium, both the pulse height spectrum endpoint and the counting efficiency are dramatically reduced as a result of quench. The curve of percent counting efficiency versus SIS is very steep for tritium. This dramatic slope can result in a rather large error in DPM values, if accurate SIS values are not obtained. Also, it is intuitively obvious that the spectrum characterization method of determining the quench indicating parameter is dependent on the counts in the sample. The larger the number of counts, the more accurate is the measurement of sample spectrum quench parameter (*e.g.*, SIS). From these observations, it is clear that the sample spectrum characterization method of determining the QIP should be used only when mid-to high-energy radionuclides are being quantified and when the count rate of the sample is well above background ($> \sim 1000$ CPM). We shall see further on in this chapter that quench indicating parameters derived from an external standard are more versatile and applicable to samples of both low and high activity (See Section IV.C.). However, quench indicating parameters derived from the sample spectrum are particularly useful when external standards cannot be applied such as in color quench correction for Cherenkov counting as demonstrated by L'Annunziata and coworkers (see Noor et al., 1996a). The SIS parameter was used by Vincze et al. (2007) in tests on the applicability of plastic scintillator beads in lieu of scintillation cocktail in a conventional liquid scintillation analyzer. The SIS is also a valuable tool in spectrum unfolding for the analysis of a mixture of two beta-particle-emitting radionuclides (L'Annunziata, 1997b, Lee et al., 2002a,b and Reddy et al., 2009) described further on in this chapter.

Once a quench correction curve is plotted by the liquid scintillation analyzer (LSA) and stored in its memory, it can be applied by the instrument computer to calculate the activity (DPM) of an unknown sample. For example, a sample of ^{3}H of unknown activity, prepared in the same fluor cocktail as the tritium standards used for the preparation of the quench correction curve, is counted and the LSA provides a count rate of 36,000 CPM and a SIS value of 12. A tritium quench curve of percent efficiency versus SIS, as illustrated in Fig. 6.13, is used by the LSA to determine the percent counting efficiency of that unknown sample. The instrument is programmed to read the stored quench curve and obtains the percent counting efficiency of 48% from the curve. The sample activity is calculated by the LSA according to the equation

$$DPM_s = \frac{CPM_{net}}{E} \tag{6.7}$$

Where DPM_S is the sample activity in disintegrations per minute, CPM_{net} is the net count rate (i.e., background-subtracted count rate) of the unknown sample, and E is the counting efficiency obtained from the quench correction curve as a decimal, not as a percent. The value of E should be generally in the range between 0 and 1.0, as the decimal representation of the percent counting efficiency over the range of 0%–100%. Therefore, in this example, the instrument calculates the activity of the unknown sample as 36,000 CPM/0.48 and the resultant value of 75,000 DPM is obtained. The LSA can perform this type of analysis for all samples of unknown activities.

b. Spectral quench parameter of the isotope spectrum or SQP(I)

The spectral quench parameter of the isotope or SQP(I) is also referred to as the mean pulse height of the isotope spectrum (Rundt, 1991; Warrick and Croudace, 2013; Croudace et al., 2016). It is utilized as a sample quench indicating parameter with some liquid scintillation analyzers of Wallac Oy such as the LKB Rackbeta liquid scintillation analyzer. The SQP(I) is measured by the liquid scintillation analyzer in a similar fashion as the previously described quench indicating parameter SIS. As described by Grau Malonda (1999), the SIS provides a value for the center of gravity of the sample pulse height spectrum plotted on a linear pulse height scale, while the SQP(I) is a calculation of the sample pulse height spectrum plotted on a logarithmic scale according to the equation

$$\text{SQP(I)} = \frac{\sum_{i=L}^{U} iN_i}{\sum_{i=L}^{U} N_i} \tag{6.8}$$

where i is the channel number of the MCA (see Fig. 6.5), N_i is the number of counts in channel i, and L and U are lowest and uppermost limits, respectively of the sample pulse height spectrum. By comparing Eqs. (6.6) and (6.8), we can see the close similarities of the sample spectrum quench indicating parameters, SIS and SQP(I).

The sample spectrum quench indicating parameter SQP(I) has the same applications and limitations as SIS previously described. All methods of characterizing the sample spectrum to define the degree of quench in a sample require the use of quenched standards for the preparation of a standard quench correction curve of counting efficiency plotted against the QIP. The procedures used to prepare a standard quench correction curve are described in Section V.D of this chapter.

A relatively new and unique application of the SQP(I) quench indicating parameter to the rapid screening of unknown radionuclides in drinking waters and samples under analysis for emergency response and nuclear forensics are described by Warrick and Croudace (2013) and Croudace et al. (2016). The method involves utilization of the SQP(I) to determine the asymmetry of the peak characteristic of the sample pulse height spectrum, which provides a rapid screening of samples to indicate the presence of multiple radionuclides as well as distinguish between alpha- and beta-emitters. As described by Croudace et al. (2016), a quench parameter SQPI(x) is measured from the sample spectrum and defined in terms of the spectrum channel number below which a percentage area (x) of the spectrum lies. Two values of the SQPI(x) are chosen to define a peak shape factor (PSF), which is the ratio of the SQPI values calculated for 95% of the spectrum and 50% of the spectrum where

$$\text{PSF} = \frac{\text{SQPI (95)}}{\text{SQPI (50)}} \tag{6.9}$$

Croudace et al. (2016) recognize that the deconvolution of alpha and beta emitting radionuclides can be achieved using techniques such as pulse decay discrimination described in Section XV of this chapter; however, they point out that specific pulse processing capability is not available on all commercial liquid scintillation counters and that such discrimination is also quench dependent. The first report of the development of the use of the peak shape factor (PSF) with SQP(I) to discriminate between alpha and beta emitting radionuclides and to screen rapidly samples for the presence of multiple radionuclides was provided by Warrick and Croudace (2013). An example of the potential of this method to screen water samples and other media for radionuclides of interest in emergency response and nuclear forensics is provided by Fig. 6.14, which is a plot of the sample PSF versus its SQP(I) calculated for 95% of the spectrum. Warrick and Croudace (2013) report that the method has been tested on contaminated drinking waters and pipeline scales with radionuclide compositions typical of those that could be encountered in a real case. The method has been demonstrated to be highly effective in the identification and quantification of the radionuclides present, and it is being incorporated into an integrated analytical system for emergency response with a rapid and effective analytical system to monitor the accidental or intentional radionuclide contamination of drinking water supplies. The identification of beta-emitting radionuclides at different quench levels was demonstrated with various beta emitting radionuclide standards with their characteristic beta-particle maximum energies (E_{max}). The radionuclide identification is due to the fact that the sample peak position at SQPI(95) and the external standard quench parameter SQPE varies depending on the beta particle E_{max}. Thus, the ratio of SQPE/SQPI(95) was a function of the radionuclide E_{max} irrespective of a range of sample quench levels as illustrated in Fig. 6.15. The external standard quench parameter SQPE is described in Section V.C.4 of this chapter.

c. Asymmetric quench parameter of the isotope or AQP(I)

The asymmetric quench parameter of the isotope [AQP(I)] is a sample spectrum quench indicating parameter employed with certain microplate scintillation counters of PerkinElmer, Inc (Hughes et al., 2001). Quench correction curves based on AQP(I) provide an improvement over the SQP(I) sample spectrum quench correction curves for low-energy beta-emitters such as tritium. The previously described SQP(I) is a quench indicating parameter that provides an MCA channel number equating the midpoint of the isotope spectrum. When quench occurs in tritium samples, the shift in the MCA channel number according to quench is limited because of the small (low-energy range) of the tritium beta-particle pulse height spectrum. The AQP(I) makes use of two multichannel analyzers, MCA_1 and MCA_2 providing two pulse height spectra of the sample. The MCA_1 produces a pulse height spectrum of the

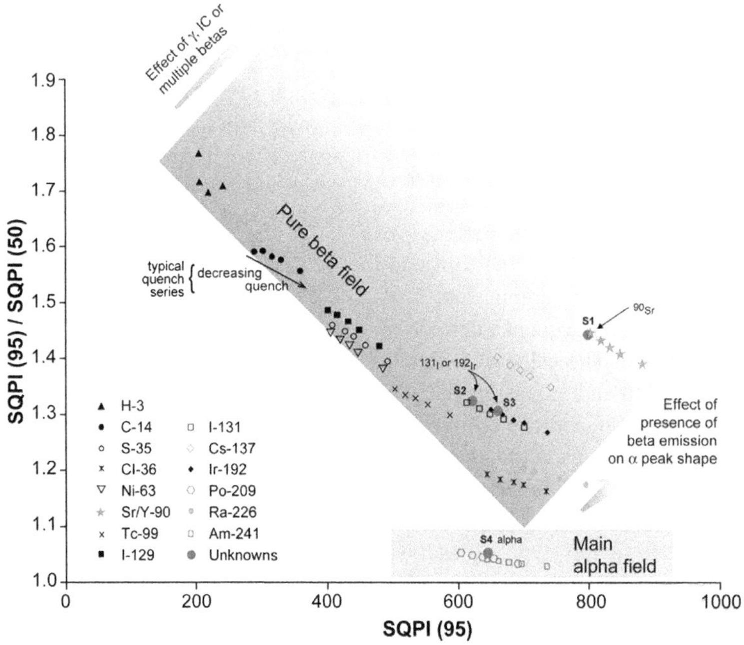

FIGURE 6.14 Identification of radionuclides using a peak shape factor (SQPI(95)/SQPI(50)). S1–S4 represents test samples spiked with an unknown radionuclide that are superimposed on standards data. The radionuclide present in each sample was identified by its position on the plot along with data from the combined internal/external quench ratio. *Reproduced with permission from Warwick et al. (2013), © 2013, American Chemical Society.*

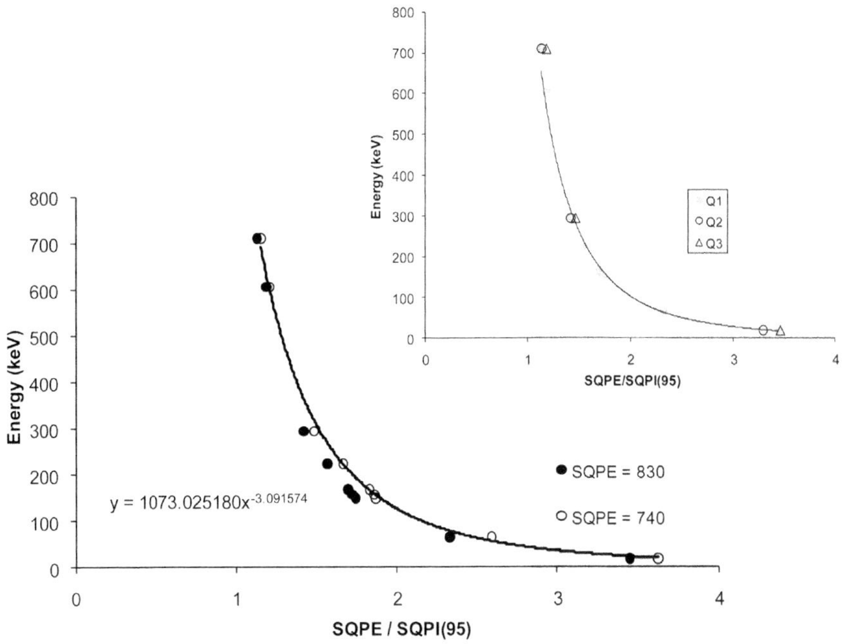

FIGURE 6.15 Relationship between beta E_{max} and SQPE/SQPI(95) over two quench ranges and between three Quantulus LS counters (Q1, Q2, and Q3). *Reproduced with permission from Warwick et al. (2013), © 2013, American Chemical Society.*

sample of beta events detected by the two photomultiplier tubes in coincidence. The MCA_2 records the beta events from only one MCA where pulse events of low height (*e.g.*, below channel number 150) are discarded. The setting of the channel discriminator setting can be adjusted to optimize quench correction curves. The AQP(I) is then calculated from the ratio of the counts in the two MCAs or AQP(I) = MCA_1/MCA_2. Since the pulse height spectrum of MCA_1 is always greater than that of MCA_2, the value of the quench indicating parameter is greater than one, and its value reduces in magnitude as quench increases.

A major advantage of the AQP(I) quench indicating parameter is a very broad range of QIP values over a broad range of quench levels providing more accurate determination of detection efficiencies for low-energy beta-emitters, such as tritium. For example, ^3H samples with counting efficiencies over the range of approximately 2.5% −35% will yield AQP(I) values over the broad range of approximately 5−120, respectively. Therefore, a tritium quench correction curve based on a plot of % counting efficiency of ^3H versus AQP(I) can be used to provide more accurate measurements of counting efficiency as a function of AQP(I).

C. External standard quench indicating parameters

A method of quench correction more popular than the aforementioned sample spectral methods is the measurement of a quench indicating parameter based on an external standard or external radiation source. This method uses an external source of gamma radiation to create a Compton spectrum in the scintillation cocktail as a tool to facilitate the measurement of quench in the sample. The sample with fluor cocktail in its scintillation counting vial is counted first in the absence and subsequently in the presence of the external gamma-ray source. This external gamma-ray source is located within the instrument, and it is positioned into close proximity of the counting vial when needed. The general interaction of the gamma rays with the scintillation vial material and cocktail is mainly via the Compton effect (See Chapter 1). The external gamma radiation produces Compton electrons with a wide spectrum of energies within the scintillation cocktail. The Compton electrons produce a scintillation effect in the fluor cocktail and a characteristic pulse height spectrum. The pulse height spectrum produced by the external standard is used to create a quench indicating parameter (QIP) for the measurement of quench in homogeneous samples in fluor cocktail. A series of radionuclide quenched standards are prepared generally as described in Section V.D, and a quench correction curve of percent counting efficiency versus a QIP created by the external gamma-ray source is plotted. Once the quench correction curve is prepared, the quench levels and activities of experimental samples are determined automatically by the LSA. The instrument determines the count rate of the sample and then uses the external standard gamma-ray source to measure the QIP of the sample. The value of the QIP is then used to obtain the radionuclide counting efficiency from the quench correction curve. The counting efficiency, in turn, is used to convert the sample count rate to disintegration rate according to Eq. (6.7).

Some external standard gamma-ray sources used in liquid scintillation analyzers are ^{133}Ba $(t_{1/2} = 10.6\ y)$, ^{137}Cs $(t_{1/2} = 30\ y)$, ^{152}Eu $(t_{1/2} = 13.2\ y)$, ^{226}Ra $(t_{1/2} = 1559\ y)$, and ^{241}Am $(t_{1/2} = 432\ y)$. The QIPs that are measured against the ^{133}Ba, ^{137}Cs, and ^{152}Eu external standards are defined as the tSIE (transformed spectral index of the external standard), H# (Horrock's number), and SQP(E) (*i.e.*, spectral quench parameter of the external standard), respectively. The various quench indicating parameters (QIPs) are determined differently and with different gamma sources. These will be described subsequently in addition to a quench indicating parameter common to all external standard sources, namely, the external standard channels ratio (ESCR) technique.

1. External standard (source) channels ratio (ESCR)

The external standard channels ratio (ESCR) technique, which may also be referred to as the external source channels ratio technique, for the determination of counting efficiencies is similar to the SCR method described previously. The principal difference is that the channels ratio produced by the external standard Compton pulse height spectrum is utilized rather than that produced by the sample pulse height spectrum.

The ESCR remains a very popular quench indicating parameter. The ESCR technique offers the advantage that the optimum channel widths (i.e., counting regions) and gains for the channels ratio determinations are often factory set to monitor the scintillation events produced by Compton electrons that result from the interaction of gamma rays of the external standard with the scintillation cocktail and vial wall. In the previously described SCR technique, the channel widths and gain settings, which produce the best (most linear) quench correction curve, must be determined experimentally and will differ from radionuclide to radionuclide. Nevertheless, some researchers opt to define the widths of channels for the measurement of the ESCR (Hofmann et al., 1990, Tsroya et al., 2012).

In practice, the ESCR quench correction curve is prepared firstly by counting a series of variably-quenched radionuclide standards in a preselected counting region from which counting efficiency values are obtained. The standards are prepared as described in Section V.D of this chapter. Each standard will contain a known activity (DPM) of a standard of the radionuclide under analysis with different amounts of quenching agent in scintillation fluor cocktail. The % detection efficiency of the radionuclide in each of the quenched standards will differ, and the values are calculated from the count rates of each quenched standard in the preselected counting region and the known activity of the radionuclide in the quenched standards (i.e., %E = CPM/DPM × 100). After each of the standards are counted in a preferred counting region, an additional count is made for each of the variably-quenched standards exposed to the external gamma-ray source; however, the external standard counts are collected in two other preselected or factory set counting channels, and the net count rate in these two channels due to the external standard is computed by subtracting from both channels those pulses or counts contributed by the radionuclide standards. The ratio of the counts due to the Compton spectrum produced by the external standard source in the two counting channels provides a channels ratio (ESCR) for each of the radionuclide standards. The detection efficiency of the radionuclide standards is then plotted against the ESCR (% detection efficiency vs. ESCR) as illustrated in Fig. 6.16. A radionuclide sample of unknown activity prepared in the same scintillation fluor cocktail is counted in the same counting region used to measure the detection efficiencies of the radionuclide standards, and the ESCR for that sample is measured by the liquid scintillation analyzer. From the value of the ESCR of a sample of unknown activity, the detection efficiency of that sample is then found from the relationship (i.e., curve) of % detection efficiency versus ESCR (Feng et al., 2017; Nakayama et al., 2017; Kanisch, 2016; Yang et al., 2006; Liu et al., 2005; Zimmerman and Collé, 1997b; L'Annunziata, 1987; Takiue et al., 1983).

Fig. 6.16 illustrates three quench correction curves with three sets of ^{14}C quenched standards provided by three suppliers, namely Hitachi Aloka, Packard Bioscience (now

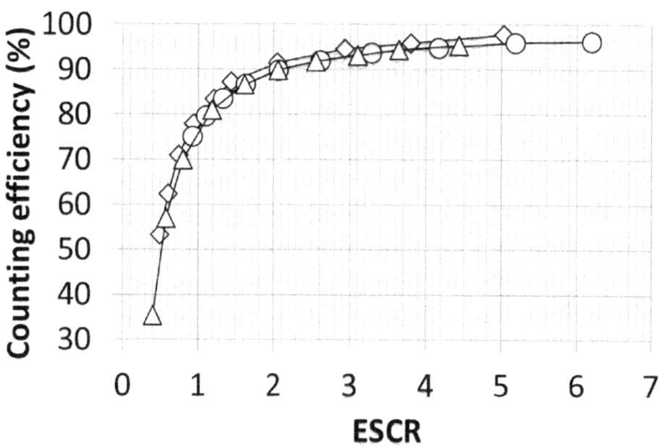

FIGURE 6.16 Quench correction curves based on the external standard channels ratio (ESCR) for three sets of ^{14}C quenched standards measured by an ALOKA LSC-5100 liquid scintillation counter equipped with a 4096-channel pulse-height analyzer. The ^{14}C quenched standards were provided by ALOKA, PACKARD, and PerkinElmer. *From Nakayama, M., Hara, M., Matsuyama, M., Hirokami, K. 2017. Modified integral counting method with various quenched samples for different scintillators.* Radiat. Safety Manag. *16, 1–7; printed with permission from the Japanese Society of Radiation Safety Management.*

a part of PerkinElmer), and PerkinElmer. The three quench correction curves are quite similar; however, they do illustrate differences by a small but not insignificant extent. The quenched standards provided by suppliers are not identical, as reported by Nakayama et al. (2017); they can vary in solvent (e.g., toluene, diisopropylnaphthalene) or quantity and type of scintillation fluor (PPO, POPOP, bis-MSB). Thus, Fig. 6.16 illustrates that for more accurate determinations of sample activities, it is best to analyze samples in the same liquid scintillation cocktail as used for the preparation of quenched standards. The detailed procedure for the preparation of quenched standards is provided in Section V.D of this chapter and Chapter 7, Volume 1. If one prefers not to prepare quenched standards, there remains the possibility of using a direct DPM method of analysis as described by Nakayama et al. (2017) and expounded on in Part V.F of this chapter.

Because the channels ratio in the ESCR method arises from counts produced by an external source, the ratio determination does not suffer from poor statistical accuracy for samples with low count rates, as does the SCR method. The quench correction curves may display a greater difference for color and chemical quenching; although decolization of samples is often an easy task as described in Chapter 7 of Volume 1 of this book. Wigfield and Cousineau (1978) found excellent agreement between counting efficiencies and various combinations of chemical and color quenching. The ESCR method remains a common method of quench correction. It has been used with a wide variety of commercial liquid scintillation analyzers including Wallac LKB, PerkinElmer, and Hitachi Aloka instruments, as reported by Feng et al. (2017) and Al-Haddad et al. (1999) for the measurement of ^{3}H, Nakayama et al. (2017) in the measurement of ^{14}C and ^{35}S, Kanisch (2016) in general environmental radionuclide analysis, van Wyngaardt et al. (2014) in the measurement of ^{99}Tc, Lin et al. (2008), Yang et al. (2006) and Liu et al. (2005) for the measurement of ^{45}Ca, Wells et al. (2001) for the measurement of ^{32}P, and Hofmann et al. (1990) in the measurement of ^{32}Si and ^{32}P.

2. H-number (H#)

The *H*-number (*H#*), as a quench indicating parameter, was first proposed by Horrocks (1976a, 1976b, 1977, 1978a), and it remains a versatile method for quench correction in liquid scintillation analysis with Beckman Instruments (Huang et al. (2016); Collé et al., 2016 and Huang et al. (2016), Hui et al., 2012, Laureano-Perez et al., 2010, Ray et al., 2009, Bonardi et al., 2004, Zimmerman et al., 1998, Zimmerman and Collé, 1997a,b, Sinor et al., 1997, Samuels and Scott, 1995, Newell and Krambeck, 1995, and L'Annunziata, 1987).

The technique involves the irradiation of liquid scintillation counting vials containing radionuclide standards in scintillation cocktail, varying in their degree of quench, with an external radionuclide source $^{137}Cs(^{137m}Ba)$ emitting monoenergetic gamma radiation. The radiation reaching the scintillation vials and samples consists exclusively of 0.662 MeV gamma rays, as the 0.032 MeV X-rays from ^{137m}Ba are absorbed by the source container. Via the Compton effect, the gamma radiation produces a spectrum of Compton scatter electrons of varying energies between zero and an energy maximum (E_{max}) in the scintillation fluor cocktail.

The spectrum of energies of the Compton electrons are constant from sample to sample. However, the scintillation photon intensities and concomitant pulse heights produced by the Compton electrons will vary depending on the amount and type of quenching agent in each sample. The Compton scatter electrons produce a spectrum of pulse events as illustrated in Fig. 6.17. If only those Compton

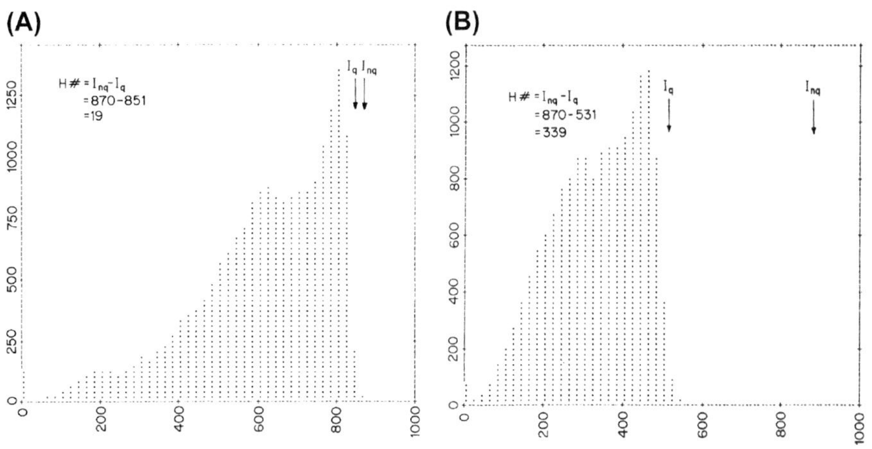

FIGURE 6.17 Effect of quenching on the inflection point (I_q) of the external standard ^{137}Cs Compton edge for (A) a weakly-quenched sample (*H#* = 19) and (B) a strongly-quenched sample (*H#* = 339) in comparison to the inflection point of a nonquenched sample (I_{nq}) factory set at 870 in a Beckman LS 7800 liquid scintillation analyzer (L'Annunziata, M. F. 1986, unpublished work of the writer.).

scatter electrons with energy E_{max} are considered, these would produce pulse heights of maximum magnitude as a peak referred to as the Compton edge. The magnitude of the pulse height spectrum at the Compton edge is maximum for a sample free of quenching agents and saturated with nitrogen gas (nonquenched sample). The Compton edge of quenched samples is encountered at lower pulse heights than that of the nonquenched sample, and the degree of spectral shift is a function of the amount of quench in the sample. A measure of the degree of spectral shift or difference in E_{max} pulse response is called the *H#*, and it is a measure of the amount of quench in a sample.

In practice, the *H#* concept is applied to the quench correction of samples counted with liquid scintillation analyzers with logarithmic pulse height conversion. Such systems convert the initial pulse-height response to the logarithm of the pulse height. Thus, initial pulse responses, which may differ by a factor of 1000, may be handled by a single amplifier and pulse-height analyzer and, as reported by Horrocks (1978b), there is a constant logarithmic difference between response relationships at different quench levels. For example, a 50% reduction in photon yield or an increase in quench by a factor of two represents a constant difference of 0.301 or the logarithm of two between the logarithmic response relationships of different quench levels. This is illustrated in Fig. 6.18.

With logarithmic response relationships, the measured pulse height, *H*, may be defined using the notation of Horrocks as

$$H = a + b \log E \tag{6.10}$$

where *H* is commonly expressed in a discriminator division, *a* is the pulse height response for a 1 keV electron, and *b* is the slope of the energy response curve. For different levels of quench, the slope *b* will remain constant, but the value *a* will differ. For a nonquenched sample, the measured pulse height, H_o, may be defined as

$$H_o = a_o + b \ \log E \tag{6.11}$$

Likewise, for a quenched system the measured pulse height, H_q, may be written as

$$H_q = a_q + b \log E \tag{6.12}$$

The difference between the measured pulse-height responses for electrons of the same energy (*e.g.*, E_{max} from ^{137}Cs Compton edge) in nonquenched and quenched systems is defined as the *H#* or

$$H\# = H_o - H_q \tag{6.13}$$

$$= a_o - a_q + b \log E - b \log E \tag{6.14}$$

$$= a_o - a_q \tag{6.15}$$

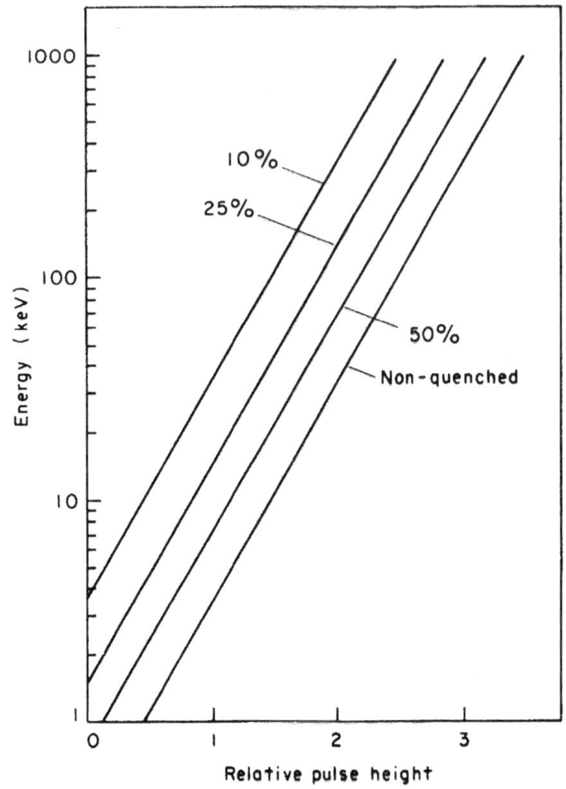

FIGURE 6.18 Relative quench effect on the logarithmic response for different electron energies. Curves are marked to indicate logarithmic response at quench levels of 50, 25, and 10% compared to the pulse-height response from nonquenched scintillation media. Curves marked '25%', '50%,' and 'nonquenched' are separated by a constant value of 0.301 relative pulse height. *From Horrocks (1980). Reprinted with permission from Elsevier © 1980.*

The *H#* is determined by taking the difference between the relative pulse heights at the inflection points of the ^{137}Cs Compton spectra of nonquenched and quenched samples, as illustrated in Fig. 6.17. In the examples presented in Fig. 6.17, *H#* values of 19, and 339 are illustrated. Greater and lesser degrees of quench will result in a corresponding variation in the magnitude of the *H#*. The inflection points of the quenched and nonquenched samples are determined automatically by multichannel analysis while exposing the samples to an external $^{137}Cs(^{137m}Ba)$ source. The multichannel analyzer divides the pulse height scale into narrow channels and accumulates the counts in each channel over a given period of time. A microprocessor then compares the counts in each channel to define the Compton spectra and precisely locate the inflection points.

With certain liquid scintillation spectrometers, for example, those of Beckman Instruments, the inflection point of the external standard Compton edge produced by a nonquenched fluor cocktail is factory set at 870 discriminator units. Quenched samples produce Compton edges with inflection points at lower discriminator levels, and the magnitude of the difference defines the *H#* (see the examples in Fig. 6.17).

In practice, a standard curve is prepared to relate counting efficiency to *H#*. This requires the preparation of a set of standards in liquid scintillation vials containing the same and known activity (DPM) of radionuclide and increasing amounts of quenching agent in scintillation fluor cocktail (see Section V.D for procedures for preparing quenched standards). These standards are then counted in optimal region settings (LL and UL discriminator settings) and the counting efficiency (E = cpm/dpm or cps/Bq) for each quenched standard is plotted against the *H#* as illustrated in Fig. 6.19. The counting efficiency of specific radionuclide samples of unknown activity are determined from their *H #* and the standard curve for that radionuclide. Fig. 6.19 illustrates the magnitude of quench as a function of the E_{max} of the beta-particle emitting nuclide. Quenching agents in the liquid scintillation fluor cocktail have a greater effect on the detection efficiency of beta-emitters of lower E_{max}. The four radionuclides illustrated in Fig. 6.19 emit beta particles with the following values of E_{max} included in parenthesis: ^{3}H (0.018 MeV), ^{14}C (0.15 MeV), ^{33}P (0.25 MeV), and ^{32}P (1.71 MeV). The quench correction curve for ^{3}H in Fig. 6.19 shows a linear relationship between detection or counting efficiency and *H#* over the *H#* range from ~50 to ~200 where most experimental samples would be found. This is confirmed by Colle et al. (2016) as illustrated in Fig. 6.20. A linear quench correction curve may be the most desirable, but it not necessary provided a good correlation between detection efficiency and quench indicating parameter is established.

The *H#* technique offers certain advantages over the 'classical' quench correction methods such as SCR and ESCR. These advantages are (1) any sample can have only one *H#* value, contrary to channels ratio techniques, (2) the *H#* technique results in less variable quench correction curves over a wider range of counting efficiency, (3) if the *H#* of a nonquenched standard is properly calibrated, the *H#* of any given sample would be constant from instrument to instrument, although the counting efficiency may not necessarily be constant.

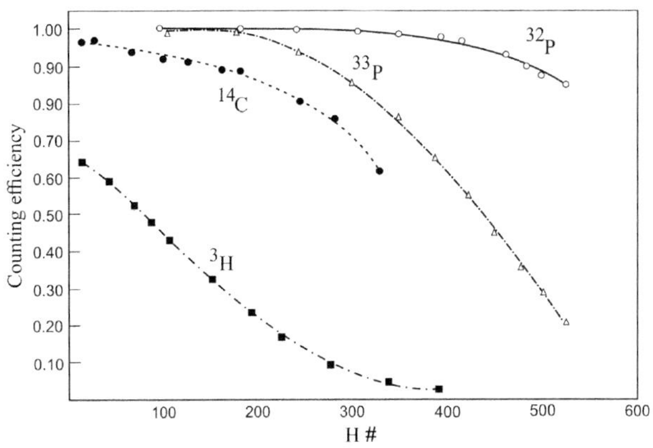

FIGURE 6.19 Quench correction curves for ^{3}H, ^{14}C, ^{33}P, and ^{32}P based on the quench indicating parameter *H#*. The ^{3}H and ^{14}C plots were obtained from commercially obtained quenched standards. Those of ^{33}P and ^{32}P were obtained from standards prepared with Insta Gel scintillation cocktail and increasing amounts of CCl_4 as quenching agent. *From L'Annunziata, M. F. (1986), unpublished work of the writer.*

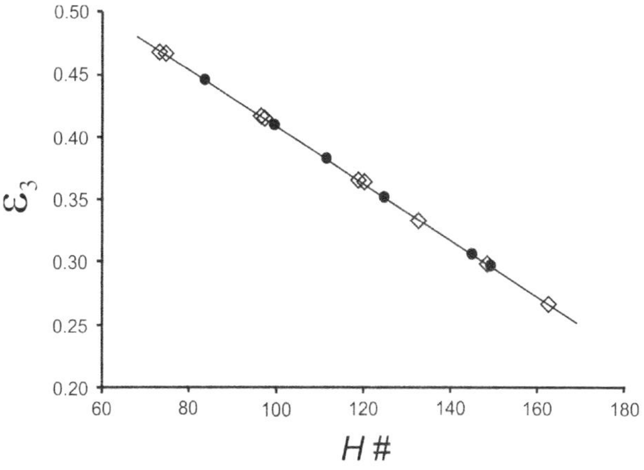

FIGURE 6.20 A typical quench curve in terms of the ^{3}H efficiency ε_3 as a function of the quench indicating parameter H# for a Beckman counter. The open diamonds and closed circles represent data from two ^{3}H sources. *From Collé et al. (2016) reprinted with permission from Elsevier © 2016.*

3. *Relative pulse height (RPH) and external standard pulse (ESP)*

The relative pulse height (RPH) and external standard pulse (ESP) quench correction techniques are similar in concept to the previously described *H#* quench monitor procedure.

In the ESP technique, reported by Laney (1976, 1977) and evaluated by McQuarrie et al. (1980), the liquid scintillation spectrometer measures the degree of quench in a sample by the shift in the average pulse height, P_s, originating from Compton electrons produced by an external ^{133}Ba gamma-ray source as compared to the average pulse height, P_r, produced in a sealed nonquenched reference vial stored in the elevator mechanism of the counter. The shift in the average pulse heights is defined by the ratio

$$\text{ESP} = \frac{P_r - P_\infty}{P_s - P_\infty}, \qquad (6.16)$$

where P_∞ is a correction term corresponding to Compton electrons produced by the external gamma-ray source under 'infinite' quenching. In ESP determinations, the entire external standard pulse-height spectrum of quenched and nonquenched samples is stored in a multichannel analyzer from which the average pulse height is determined.

As described in the case of *H#*, a quench correction curve is prepared relating counting efficiency with ESP, using a set of quenched standards containing the same

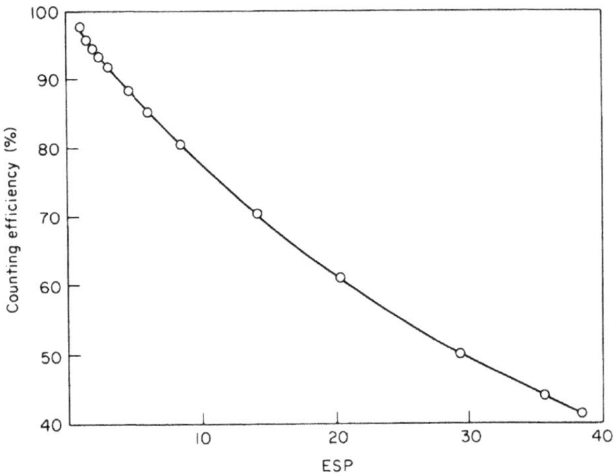

FIGURE 6.21 Variations in counting efficiency with ESP using quenched ^{14}C standards. *Plotted from data of McQuarrie et al. (1980).*

activity of radionuclide and varying amounts of quenching agent in fluor cocktail. A quench correction curve is obtained by plotting counting efficiency against ESP as illustrated in Fig. 6.21. Laney (1976) defined the relative pulse height (RPH) as the reciprocal of the ESP. Both can serve as quench indicating parameters, but ESP produces quench correction curves with more linearity (Grau Malonda, 1999).

4. Spectral quench parameter of the external standard or SQP(E)

The quench indicating parameter SQP(E) is measured with ^{226}Ra or ^{152}Eu (Günther, 1998) as the external standard source in liquid scintillation analyzers of the LKB and Wallac (now PerkinElmer) instruments (Kouru, 1991; Grau Malonda, 1999). A multichannel analyzer with 1024 logarithmic channels is used to determine the position of 99.5% of the endpoint of the external standard Compton spectrum to define the value of the SQP(E). As described by Grau Carles (2006b) the endpoint of the Compton spectrum is a well-defined point for spectrometers with logarithmic amplifiers. The SQP(E) parameter, like all quench indicating parameters (QIPs), is a relative number, and it is a meaningful number when compared to other SQP(E) values. As a QIP in liquid scintillation analysis, SQP(E) remains a robust and widely used parameter to monitor and correct for sample quench levels for the determination of detection efficiencies (Fons-Castells et al., 2017d; Jiang et al., 2017; Kossert et al., 2017a,b; Krištof and Logar, 2017, 2013; Stojkovićet al., 2017a,b,c; 2016, 2015; Collé et al., 2016; Feng et al., 2015, 2013; 2012; Chmeleff et al., 2010; Jörg et al., 2010; Kossert and Grau Carles, 2010, 2008; 2006; Wanke et al., 2010; Kossert et al., 2009a,b; Krajcar Bronić et al., 2009; Varlam et al. 2009; Kossert et al., 2009a,b; Werth and Kuzyakov, 2008; Laureano-Pérez et al., 2007; Tarancón et al., 2007; Grau Carles et al., 2006; Grau Malonda et al., 2006; Hou, 2005; and Hou et al., 2005).

The gamma source for the SQP(E) determination is positioned below the counting vial containing scintillation fluor cocktail. The SQP(E) defines the uppermost channel number (endpoint) that comprises 99.5% of the total counts of the external standard pulse height spectrum. Only a small portion of the endpoint, the remaining 0.5% of the total counts or area under the pulse height spectrum are excluded (Kessler, 1989). As described by Grau Malonda (1999), the SQP(E) for a nonquenched sample is defined as

$$\mathrm{SQP(E)} = P\text{–}400 \tag{6.17}$$

where SQP(E) corresponds to the i value of the equation

$$\sum_{j=1}^{n} N_j \geq (1-r)\sum_{j=400}^{n} N_j > \sum_{j=i+1}^{n} N_j \tag{6.18}$$

where N_j is the total number of external standard counts in channel j, $r = 0.995$, $n =$ the total number of channels $= 1024$, and $\sum_{j=400}^{n} N_j = N_{\mathrm{tot}}$ is the total number of external standard counts or area under the external standard pulse height spectrum above channel 400. The above formula indicates that the first 400 channels of the external standard pulse height spectrum are excluded from the calculations. The 400 channels at the lower end of the pulse height spectrum correspond to approximately 0–20 keV events, and the objective of the exclusion is to reduce that portion of the spectrum that could vary from any 'wall effect' that would occur whenever scintillation fluor solvent penetrates into the plastic wall of the counting vial. The wall effect is enhanced scintillation resulting from organic solvents such as benzene or toluene penetrating into the plastic vial wall. However, this effect generally does not occur to any appreciable extent with the modern solvents based on linear alkylbenzene and diisopropylnaphthalene (See Chapter 7, Volume 1). In the elucidation by Grau Malonda (1999), the value of P is obtained from the equation

$$P = i + \frac{1}{N_i}\left[\sum_{j=i}^{n} N_j - (1-r)N_{\mathrm{tot}}\right] \tag{6.19}$$

where N_i is obtained from the equation

$$N_i = \frac{1}{3}\left[\sum_{j=i-1}^{i+1} N_j\right] \tag{6.20}$$

As required with other quench indicating parameters, it is necessary to count a set of quenched standards all having a known and constant activity of radionuclide but varying levels of quench. From the count rates of each standard and SQP(E) value measured by the liquid scintillation analyzer, a standard curve of counting efficiency for a given

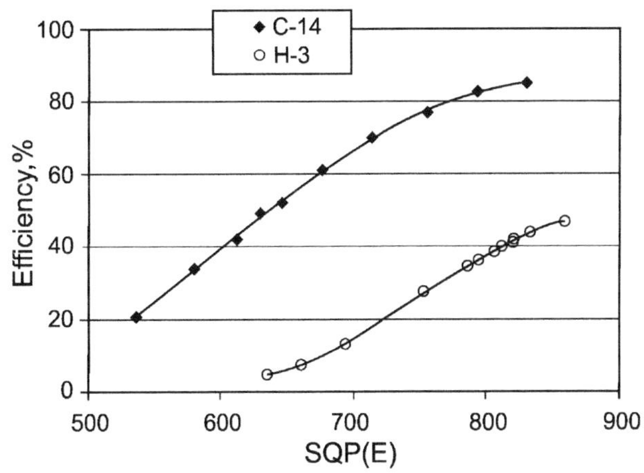

FIGURE 6.22 Quench correction curves for ^3H and ^{14}C with SQP(E) as the quench indicating parameter measured with a Quantulus 1220 liquid scintillation counter. *From Hou (2005), reprinted with permission from Elsevier © 2005.*

radionuclide versus SQP(E) is plotted, such as illustrated in Fig. 6.22 for the radionuclides ^3H and ^{14}C. When a sample of unknown activity of a given radionuclide is analyzed in the liquid scintillation analyzer, the instrument will determine the SQP(E) value of the sample, and subsequently extract the counting efficiency of that radionuclide from the standard curve. It is important that the samples of unknown activity be prepared in the same fluor cocktail formula as the radionuclide standards used to prepare the quench correction curve.

The quench correction curve for ^3H, illustrated in Fig. 6.22, appears to be near linear over the SQP(E) range of ~700−~800, which is a range of quench level in samples generally counted. Collé et al. (2016) determined a similar near linearity in a quench curve, illustrated in Fig. 6.23, determined with two different ^3H radionuclide standards, which can illustrate the robustness of this QIP.

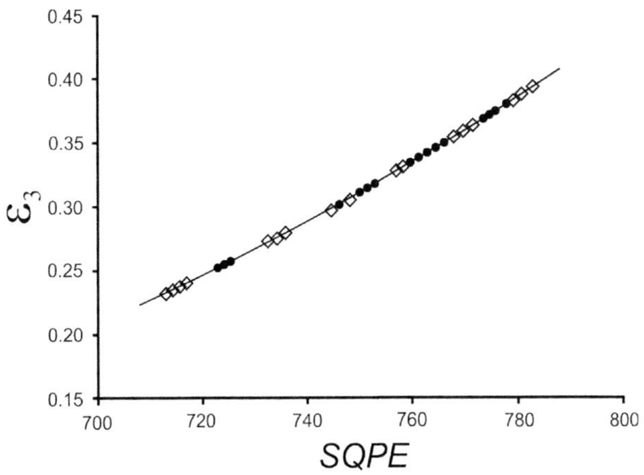

FIGURE 6.23 Typical quench curve in terms of the ^3H efficiency ε_3 as a function of the quench indicating parameter SQPE for a Wallac counter. The open diamonds and closed circles represent data from two ^3H sources. *From Collé et al. (2016) reprinted with permission from Elsevier © 2016.*

Quench correction curves need not be linear provided there is a good correlation between detection efficiency and quench indicating parameter.

5. Transformed spectral index of the external standard (tSIE)

The external standard quench correction methods previously described define specific characteristics of the external standard pulse height spectrum as quench indicating parameters such as, (1) the magnitude of the average pulse heights of the external standard Compton spectrum (*e.g.*, ESP and RPH), (2) the inflection point at the Compton edge (*e.g.*, *H#*), and (3) the endpoint of the external standard pulse height spectrum [*e.g.*, SQP(E)]. Another popular external standard quench indicating parameter was first reported by Everett et al. (1980) and Ring et al. (1980) under the designation of spectral index of the external standard (SIE). The SIE is similar to the SIS previously described with the exception that the SIE characterizes the external standard pulse height spectrum in the same fashion as the SIS characterizes the sample pulse height spectrum (Kessler, 1989). The objective of SIE is to characterize the external standard pulse height spectrum to the extent of quantifying the various features of the pulse height distribution (i.e., pulse height spectral shape) and any changes in these features, which could occur as a result of quench. Features such as the spectral peak, slope at various points of the spectrum, and maximum pulse height will govern the center of gravity of the pulse height spectrum, which will obviously change according to quench level (L'Annunziata, 1987). The SIE is calculated as

$$\mathrm{SIE} = k\frac{\sum_{X=L}^{U} X\cdot n(x)}{\sum_{X=L}^{U} n(x)} \tag{6.21}$$

where k is a factor assigned to provide a maximum value to the SIE of a nonquenched standard, X is the channel number, $n(x)$ is the number of counts or pulse events in channel X, and L and U are lower and upper limits that encompass the pulse height spectrum. The lower limit L is set above zero sufficient to eliminate changes in pulse events of low magnitude produced by the 'wall effect', which could occur if fluor cocktail solvent were to penetrate the plastic wall of the scintillation counting vial. Notice the close similarity of the above equation for SIE to that used to calculate SIS (Eq. 6.6). The values of SIE are unitless, always greater than 1.0, and of magnitude that will vary according to quench (*i.e.*, the higher the quench level in the sample, the lower will be the SIE value.

A further development based on the SIE is the transformed spectral index of the external standard (tSIE) introduced by Packard Instruments and now a product of

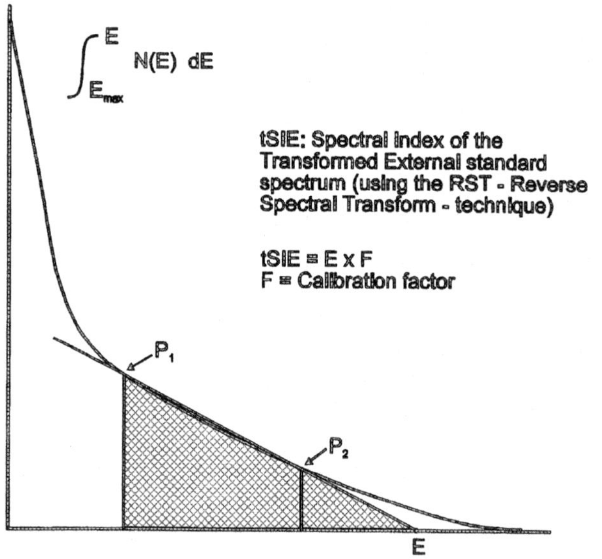

FIGURE 6.24 Transformed liquid scintillation pulse height spectrum of an external ^{133}Ba standard. The tSIE is calculated by the extrapolated value *E* times a calibration factor *F* to provide a quench indicating parameter in the range of zero to 1000. The highest value of tSIE = 1000 is set using an unquenched ^{14}C standard. *©1998–2018 PerkinElmer, Inc. Printed with permission.*

PerkinElmer, Inc. The tSIE method of quench correction uses ^{133}Ba as the external gamma-ray source. The Compton spectrum of this external standard is measured in a MCA, such as the SIE described previously and transformed by performing a reverse back sum on the spectrum to obtain the transformed spectrum as illustrated in Fig. 6.24. From the transformed spectrum, an endpoint energy is determined by a reverse spectral transform (RST) technique using two points on the spectrum and extrapolating to the energy axis (Kessler, 1989). The simplified mathematical expression of the reversed spectral transform is

$$\int_{\widehat{E}_{\max}}^{\widehat{E}} N(\widehat{E})d\widehat{E} = \left(\int_{\widehat{E}_{\max}}^{\widehat{E}_2} N(\widehat{E})d\widehat{E} - \int_{\widehat{E}_{\max}}^{\widehat{E}_1} N(\widehat{E})d\widehat{E} \right) \frac{\widehat{E} - \text{tSIE}}{\widehat{E}_2 - \widehat{E}_1} \tag{6.22}$$

where $\widehat{E}$ is the transformed energy and tSIE is calculated as one of the parameters of the RST function as

$$\text{tSIE} = \widehat{E} - \left(\widehat{E}_2 - \widehat{E}_1\right) \frac{\int_{\widehat{E}_{\max}}^{\widehat{E}} N(\widehat{E})d\widehat{E}}{\int_{\widehat{E}_{\max}}^{\widehat{E}_2} N(\widehat{E})d\widehat{E} - \int_{\widehat{E}_{\max}}^{\widehat{E}_1} N(\widehat{E})d\widehat{E}} \tag{6.23}$$

The final tSIE (extrapolated endpoint times a calibration factor) is calculated on the basis of the tSIE being equal to 1000 for a nonquenched ^{14}C sample, which is used for instrument calibration and normalization. Additional information on the measurement of tSIE and its applications is given by Kessler (1991a,b). The ^{133}Ba gamma-ray source for the measurement of the tSIE is positioned below the sample vial. The positioning of the external standard under the sample produces a quench measurement that compensates for variations of sample volume. The value of tSIE, therefore, can be accurately determined even for small (<1 mL) sample-fluor cocktail mixtures.

The major advantages of using the external standard quench indicating parameter tSIE rather than the QIP based on the sample spectrum (SIS) can be ascertained from Fig. 6.25, which shows quench correction curves for

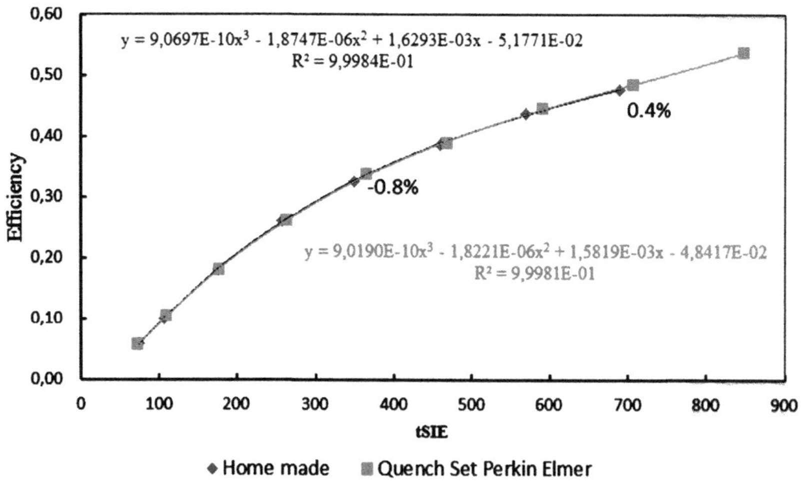

FIGURE 6.25 Quench correction curves for ^{3}H using tSIE as the external standard quench indicating parameter with a PerkinElmer Tri-Carb 3180TR/SL liquid scintillation analyzer. Data points on the curve established with a set of ^{3}H quenched standards provided by PerkinElmer (square red points) and home-made laboratory prepared (blue diamond points) ^{3}H quenched standards in Ultima Gold LLT fluor cocktail in low diffusion polyethylene vials. *From Grahek et al. (2016); reprinted with permission from Elsevier © 2016.*

tritium, ^{14}C. The plots illustrates the dynamic range of the quench indicating parameter <1000 for a nonquenched cocktail mixture to less than 100 for a highly quenched sample.

The quench correction curve illustrated in Fig. 6.25 is a good example of the need to compare the performance of commercially available quenched standards with a set of quenched standards prepared in the laboratory with a specific cocktail and vial type. Commercially available quenched standards may be prepared with scintillation cocktail solvents and fluors different than the cocktails used in the laboratory. The quench correction curves produced by two different sets of quenched standards with varying cocktail and vial type often yield different quench curves as illustrated previously with Fig. 6.16. Grahek et al. (2016) demonstrated in Fig. 6.26 the fortunate equal performance of a commercially available set of quench standards and a set of standards prepared in the laboratory. Such tests are warranted if one wishes to use a commercially obtained set of quenched standards in lieu of preparing quenched standards with fluor cocktail identical to the cocktail to be used for the analysis of samples of unknown activity.

It is important to recall that tSIE, like other external standard QIPs, is radioisotope independent, and it is a function of the quality or quench level of the fluor cocktail. The second advantage of the external standard method over the sample spectrum characterization method for the determination of QIP is that the external standard method is sample count-rate independent. The external standard quench indicating parameter (QIP) does not depend on the count rate of the sample; rather it depends on the counts created by the external gamma-ray source and the resultant Compton electrons produced within the scintillation fluor cocktail. The only disadvantage of the external standard method is that each sample must be counted alone and then counted again with the external standard present. This extra counting step usually requires about 6–120 seconds additional counting time depending on the sample volume and quench level. This disadvantage is of little significance, because sample counting with and without the presence of the external standard and the measurement of the QIP are fully automated.

Examples of quench correction curves prepared in the laboratory with radionuclide standards and scintillation fluor cocktails similar to the fluor cocktails to be used with samples of unknown activity are illustrated in Fig. 6.26. Also, we can note from Fig. 6.26 that, for a given level of quench, the counting efficiencies are higher for radionuclides that emit beta particles of higher energy For example, the maximum detection efficiencies in the scintillation cocktails used by Ahn et al. (2013) illustrated in Fig. 6.26 are 45.3% for ^{3}H and 86.4% for ^{14}C. Also, from the bend and flat parts of the quench correction curve for ^{14}C, we can see that quench has less effect on the counting efficiencies of radionuclides that emit beta particles of higher energy at moderate levels of quench.

The quench indicating parameter tSIE is, like the previously described SQP(E) and ESCR external standard QIPs, a very robust and commonly employed parameter for the measurement of quench and the determination of radionulide detection efficiencies. Numerous examples are found in the literature, and a few covering a wide range of applications are the following: the analysis of ^{3}H and/or ^{14}C (Gomes et al., 2017; Larionova et al., 2017; Grahek et al., 2016; Ahn et al., 2013; Fang et al., 2013; Ishimori et al., 2010; Zhilin et al., 2010; Mingote et al., 2006; Harms and Jerome, 2004), the analysis of ^{10}Be (Chmeleff et al., 2010), ^{18}F (Dziel et al., 2016) ^{55}Fe (Mikelic et al., 2007), ^{32}P, ^{33}P, ^{35}S (Altzitzoglou, 2004), ^{59}Fe (Kossert and Nähle, 2014), ^{64}Cu (Wanke et al., 2010), ^{90}Sr (Maringer et al., 2017; Jäggi and Eikenberg, 2014; Lazare et al., 2009), ^{131}I (Kulkarni et al., 2011a,b), ^{151}Sm (Ramírez-Guinart et al., 2017), ^{210}Pb Laureano-Pérez et al. (2007), ^{241}Pu (Jäggi and Eikenberg, 2014).

Low-energy beta-emitters, such as tritium ($E_{max} = 18.6$ keV), will exhibit a more significant variability in detection efficiency as a function of quench when compared to radionuclides that have higher beta-decay

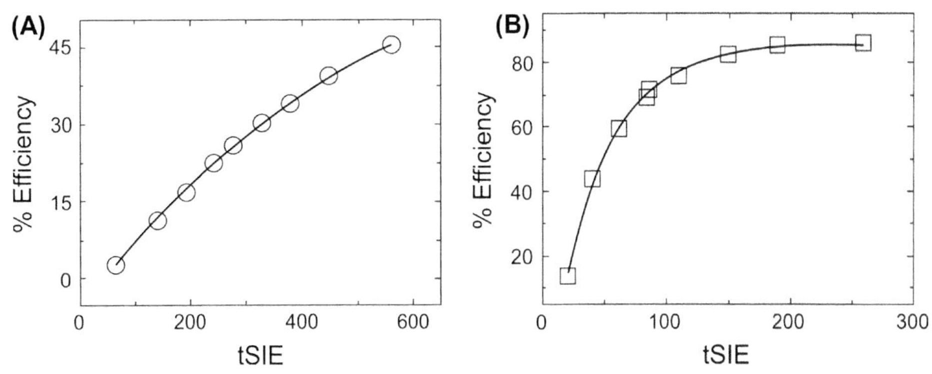

FIGURE 6.26 Quench correct curves for (A) ^{3}H and (B) ^{14}C prepared in the laboratory with ^{3}H and ^{14}C standards. *From Ahn et al. (2013), reprinted with permission from Elsevier © 2013.*

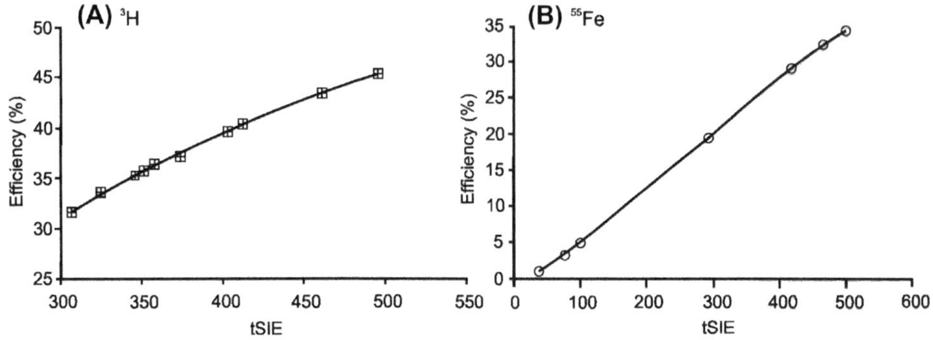

FIGURE 6.27 Quench correction curves of detection efficiency (%) versus the quench indicating parameter tSIE for (A) tritium and (B) for ^{55}Fe. *(A) from Harms and Jerome (2004), reprinted with permission from Elsevier © 2004). (B) from Mikelic et al. (2007), reprinted with permission from Elsevier © 2007.*

energies, such as ^{14}C ($E_{max} = 156$ keV), ^{33}P ($E_{max} = 249$ keV), and ^{32}P ($E_{max} = 1710$ keV). This is illustrated clearly in Figs. 6.19, 6.26 and 6.27. A maximum liquid scintillation detection efficiency for nonquenched tritium (*i.e.*, a sample containing no quenching agent) in modern liquid scintillation analyzers is ~68% (See Table 6.2 and Figs. 6.13 and 6.19), which is achievable with an unquenched tritium standard generally used for instrument calibration. However, in practice, unquenched samples are not encountered, as the sample solution (*e.g.*, H_2O is a quenching agent as well as dissolved oxygen gas in the counting solutions). Thus, typical quench correction curves for tritium are provided by Ahn et al. (2013), Grahek et al. (2016) and Mikelic et al. (2007) illustrated in Figs. 6.25 and 6.26 and 6.27A. The quench correction curve for ^{3}H illustrated in Fig. 6.26 was prepared from a series of tritium quenched standards in counting vials containing the same and known activity of tritium standard and increasing amounts of nitromethane quenching agent in the range of 0−100 μL yielding an optimum detection efficiency of 45.3% for the tritiated water containing no nitromethane. If we consider a radionuclide, such as ^{55}Fe, which emits Auger electrons of energy well below the E_{max} of the tritium beta particles, we will observe yet lower detection efficiencies for samples with similar quench levels as tritium (See Fig. 6.27A and B). The radionuclide ^{55}Fe decays by electron capture (EC) to stable ^{55}Mn with the emission of 5.2 keV Auger *K* electrons and Mn *K*-series X-rays of 5.9 and 6.5 keV (NNDC, 2011). A quench correction curve for ^{55}Fe, illustrated in Fig. 6.27B, reported by Mikelic et al. (2007), was prepared from a series of quench standards of ^{55}Fe in fluor cocktail containing increasing amounts of nitromethane quenching agent. Although quenching agents can have a very significant affect on the liquid scintillation detection efficiencies of low-energy beta-emitting nuclides (*e.g.*, ^{3}H) and on the detection efficiencies of radionuclides that emit Auger electrons (*e.g.*, ^{55}Fe), the quench indicating parameter tSIE as well as the previously described SQP(E) and ESCR are robust QIPs capable of measuring accurately very low detection efficiencies (<10% (E).

6. G-number (G#)

The *G#* is a quench indicating parameter (QIP) first described in detail by Grau Carles and Grau Malonda (1992) as a patent and subsequently reported by Grau Carles et al. (1993a). The method was designed to provide an accurate QIP regardless of the level of quench in a sample, even when the quench level is so high that the counting efficiency of the beta-emitting radionuclide is reduced to less than 1%. The idea behind this approach of Grau Carles and Grau Malonda is based on the use of an external standard which emits considerable quantities of high-energy gamma radiation sufficient to produce appreciable numbers of Compton electrons in the sample scintillation cocktail that have energies above the Cherenkov threshold of 263 keV. The LKB Rack Beta liquid scintillation analyzers are equipped with such a gamma source, namely ^{226}Ra, and the development of this technique was therefore demonstrated with the LKB instrument. The Compton electrons produced by the ^{226}Ra external standard will generate a pulse height spectrum, which is a result of photons emitted by the sample originating from a combination of scintillation and Cherenkov events. As explained by Grau Carles and Grau Malonda (1992), both scintillation and Cherenkov photons are detected simultaneously within the 18-nanosecond time window of the coincidence circuitry of the liquid scintillation analyzer. However, when samples are very highly quenched in the liquid scintillation cocktail, the liquid scintillation diminishes considerably to the point that the pulse height spectrum produced by the ^{226}Ra external standard becomes the result of mostly Cherenkov photons produced by the portion of Compton electrons of energy in excess of 263 keV. In this fashion, regardless of the level of quench, a characteristic of the pulse height spectrum produced by the Cherenkov photons can be measured to provide an accurate QIP even when the

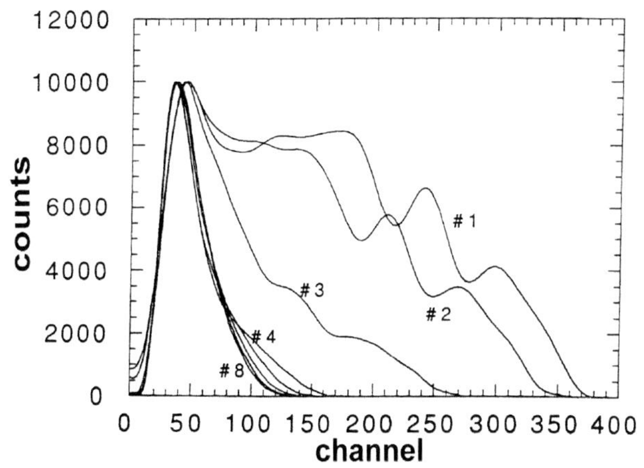

FIGURE 6.28 Compton spectra from ^{226}Ra adjusted and normalized for different quench levels. *From Grau Carles and Grau Malonda (1992), printed with permission.*

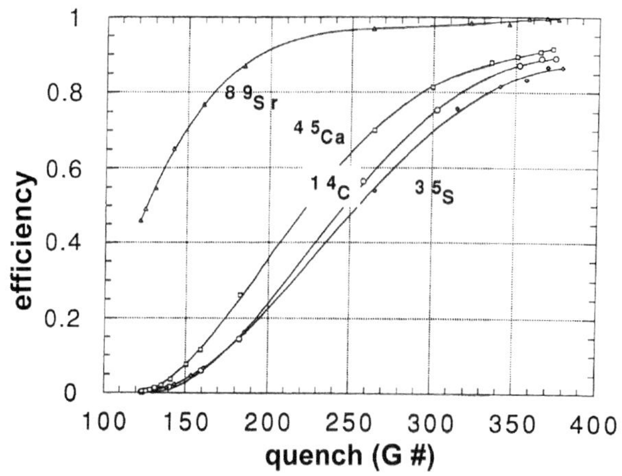

FIGURE 6.29 Calibration or quench correction curves of counting efficiency as a function of the G-number for ^{35}S, ^{14}C, ^{45}Ca, and ^{89}Sr. *From Grau Carles and Grau Malonda (1992), printed with permission.*

scintillation process is quenched to the extent that beta-particle emitting radionuclides are counted at an efficiency of less than 1%.

Like all other methods of quench correction, this technique requires the preparation of a set of quenched standards of the radionuclide of interest. The quenched standards are counted in a suitable counting region defined by lower-level and upper-level discriminator settings. The count rate of each quenched standard is determined in this counting region, and the counting efficiency is calculated. The quenched standards are also exposed to the ^{226}Ra external standard gamma-ray source, and the resulting ^{226}Ra pulse height spectrum is produced in the channel or counting region between 10 and 500. Fig. 6.28 illustrates pulse height spectra produced by eight quenched standards exposed to ^{226}Ra external standard. The spectra illustrate how the pulse heights shift from higher to lower magnitudes (higher channels to lower channels) as quench increases. At very high levels of quench (standards # 4 to # 8 of Fig. 6.28), the pulse height spectra from ^{226}Ra are the consequence of mainly Cherenkov photons produced by the Compton electrons in excess of 263 keV energy.

The G-number is based on the analysis of the displacement of the final part (end-point) of the ^{226}Ra external standard pulse height spectra as a function of quench. According to Grau Carles et al. (1993a), Fourier series are fitted to the pulse height spectra produced by the ^{226}Ra external standard to enable their normalization to the number of counts, y_N, of the first peak in the pulse height spectra due to Cherenkov photons created by the most energetic Compton electrons (i.e., the left-most peaks in Fig. 6.28). The spectral interval within the limits $y_N/10$ and $y_N/50$ is taken from the final part of each external standard pulse height spectrum. A linear relationship in the selected interval is obtained by raising the number of counts y to the power α or

$$y \rightarrow y^{\alpha} \quad 0 < \alpha < 1 \tag{6.24}$$

where α is the value that provides the best regression coefficient to the line

$$y^{\alpha} = ax + b \tag{6.25}$$

For the channel with a number of counts $y = y_N/100$, the G-number is given by

$$G = \frac{(y_N/100)^{\alpha} - b}{a} \tag{6.26}$$

Examples of typical quench correction curves obtained with quenched standards of ^{35}S, ^{14}C, ^{45}Ca and ^{89}Sr are illustrated in Fig. 6.29, where it is clearly illustrated that the G-number serves as an excellent quench indicating parameter over the widest possible range of counting efficiency from the highest detection efficiency to the lowest (<1%). This is demonstrated by the expanded quench correction curve for highly-quenched ^{45}Ca standards illustrated in Fig. 6.30 over the counting efficiency range of <1% to approximately 27%.

There is no documented rationale for the selection of the name 'G-number' for the identification of this quench indicating parameter. However, the writer can only assume that the letter 'G' calls to mind the first letter of the family names of Grau Carles and Grau Malonda, who devised this technique. Consequently, this QIP could likewise be called the 'Grau-number'. There exists a similar corollary in the previous development of the *H*-number by D. L. Horrocks described previously in Section V.C.2 of this chapter. In recognition of its founder, the *H*-number is also referred to as the Horrocks number.

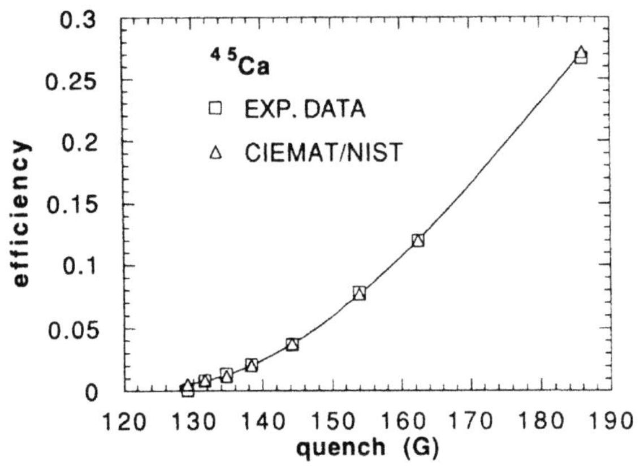

FIGURE 6.30 Experimental and CIEMAT/NIST computed efficiencies for ^{45}Ca. *From Grau Carles et al. (1993a), reprinted with permission from Elsevier ©1993.*

D. Preparation and use of quenched standards and quench correction curves

A quench correction curve is a calibration curve of percent efficiency versus a quench indicating parameter (e.g., ESCR, *H#*, SQP(E), and tSIE). Examples of quench correction curves are found in previous parts of this chapter. The quench correction curve is prepared from a set of quenched standards, which is a series of liquid scintillation counting vials containing a certified standard of a radionuclide in scintillation cocktail, all with the same radionuclide activity (DPM) but different levels of quench. The various levels of quench are achieved by adding small amounts (usually measured in microliters) of a quenching agent such as nitromethane, carbon tetrachloride, or other suitable chemical that will shift the sample and external standard pulse height spectrum from higher to lower magnitudes. Usually about 10 vials are prepared with radionuclide standard with each vial containing the same radionuclide activity in a scintillation cocktail, which is identical to the cocktail that will be used for the activity measurements of the same radionuclide of unknown activity. Among the 10 vials, one will contain no quenching agent and, to the remaining nine vials, an incremental amount of quenching agent is added to each vial. The vial with no quenching agent will be the least quenched among the standards, and the remaining nine vials will yield increasing levels of quench. The general procedure for the preparation of quenched standards is as follows:

1. *Preparation of quenched standards*

There are two methods used for obtaining a set of quenched standards. The first method is to purchase a set of sealed standards for the radionuclide of interest and scintillation cocktail type that one plans to use, and with the quenched standards, prepare a quench correction curve of percent counting efficiency versus a suitable QIP. Quenched standards of ^{3}H and ^{14}C are available commercially, because of their relatively long half-lives. Some suppliers (*e.g.*, PerkinElmer, Inc.) can provide the quenched standards according to specification including variables such as (1) radionuclide standard activity, (2) scintillation cocktail, (3) quenching agent, (4) counting vial size and type, and (5) sample volume. Also, sets of quenched standards for ^{3}H and ^{14}C are sold on the market without customer specifications. If a user is interested in procuring these, it is important that he or she procure the set of quenched standards with vial size, sample volume, and scintillation cocktail compatible with their experimental samples. The second method is to prepare a series of quenched standards in the laboratory and to prepare the quench correction curve from these quenched standards. This method can be the most reliable when the most accurate DPM values are required, because the user can control all aspects of the preparation of the quenched standards to most closely represent the chemistry of his or her experimental samples. A detailed description of the procedure for the preparation of quenched standards and a quench correction curve from the quenched standards is provided next. An outline of the steps is as follows:

a. **Choose the type and size of counting vial and sample volume that will be used.** Counting vials come in various sizes (e.g., 4, 6, 8, 20 mL) and as glass or plastic. Although glass and plastic vials may perform similarly (Elliott, 1984), there can be differences depending on the cocktail used and radionuclide analyzed. Also, background counts can be lower with plastic vials. The size of the vial and cocktail and sample volume can have a significant effect on counting geometry and the quench correction curve could vary significantly according to this variable. The vial size and type, scintillation cocktail-sample volume, and cocktail composition of the quenched standards should be the same as the experimental samples (Collé, 1997a,b).
b. **Choose the scintillation cocktail that will be used.** Commercial scintillation cocktails come in various chemical compositions with differing properties, some miscible with organic solutions and others with aqueous sample solutions (See also Chapter 7, Volume 1). Scintillation cocktails that use solvents such as toluene, xylene, pseudocumene or linear alkylbenzene may be used with samples in organic solvents while cocktails using diisopropylnaphthalene (DIN) or phenylxylylethane (PXE) as solvents may be mixed with aqueous samples. Most accurate results are obtained when the chemistry of the quenched standards is the same as the experimental samples. Detection efficiencies can vary between different scintillation cocktail compositions. Error can be introduced when determining the

activity (e.g., DPM) of an experimental sample mixed in a scintillation cocktail different from that from which the quenched standards were prepared. The error is particularly pronounced in the case of the LSA of low-energy radionuclides such as tritium (Collé, 1997a,b). It is best therefore to prepare quenched standards with the same scintillation cocktail to be used with the experimental samples.

c. **Prepare a stock solution to contain the radionuclide standard of interest of known activity (DPM) in the scintillation cocktail.** The radionuclide standard should have an activity that is accurately known such as one traceable to a primary standard (e.g., NIST traceable standard) or a radionuclide standardized according to a known method of standardization. The standard used must be compatible and thoroughly miscible with the cocktail. The stock solution should be of sufficient volume to prepare more vials than standards that are needed to allow for the possibility of discarding some standard vials for reasons described subsequently. For example, if 10 quenched standards each containing 10 mL of scintillation cocktail are desired, then prepare a stock solution of more than 150 mL of scintillation cocktail containing radionuclide standard to allow the testing of 15 standards with the possibility of discarding 5 as described in step (e) below. The level of the radioactivity in the stock scintillation cocktail should be high enough to require counting the standards for only a short period of time and still achieve good or acceptable counting statistics. Activities of approximately 200,000 DPM per vial of low-energy emitting radionuclide, such as ^{3}H, or approximately 100,000 DPM per vial of higher-energy emitting radionuclide, such as ^{14}C, should be adequate.
d. **Transfer the exact aliquot of radionuclide standard – scintillation cocktail stock solution into each of the vials to be used for the preparation of the quench correction curve.** The stock solution may be added by pipette or gravimetrically to approximately 15 counting vials. Gravimetric addition can be the most accurate. The volume of the radionuclide standard in scintillation cocktail should be identical to the volume of cocktail to be used when counting experimental samples of unknown radionuclide activity.
e. **Count each of the counting vials containing standards and determine whether the count rate (CPM) of each is within acceptable counting statistics.** A counting region that encompasses the entire radionuclide pulse height spectrum can be used or other suitable counting region that would encompass a large part of the pulse-height spectrum and avoiding most background pulse events at the low-energy end of the background spectrum. The standards are of high activity and background counts can be subtracted or ignored if insignificant. Replicate counts of each standard (e.g., count each standard from 5 to 10 times). Obtain a mean count for all of the standards. As an excess number of standards are prepared, any of the standards that deviate more than 2% from the mean count rate can be discarded. This provides us with a set of standards of equal activity in scintillation cocktail.
f. **Select a suitable quenching agent and add increasing amounts of the quenching agent to each standard. If the experimental samples of unknown radionuclide will be counted eventually in a specific volume (e.g., cocktail + 5 mL of water sample) than a blank water sample of the same volume should be added also to the cocktail. Water is a very weak quenching agent.** In this step, the amounts of quenching agent (e.g., nitromethane) do not have to be added with precision. Only the amount of radionuclide standard in each vial, prepared in the previous step, must be exact. With respect to the quenching agent added, it is only important that each vial have increasing amounts of quenching agent so that a quench correction curve of counting efficiency versus QIP can be established over a broad range of quench levels. Nitromethane is a popular quenching agent, because it is a strong quencher and only small increments are required. For example, if the scintillation cocktail uses toluene as a solvent and there are 10 quenched standards of either ^{3}H or ^{14}C, then the following increments of nitromethane can be added to the vials: 0, 1, 5, 10, 18, 25, 35, 45, 55 and 65 μL to provide a broad range of quench levels. However, for a scintillation cocktail miscible with water such as Ultima Gold LLT, larger incremental amounts of nitromethane may be needed for a ^{3}H quench correction curve (e.g., from 0 to 100 μL, as described by Ahn et al., 2013). Notice that the first vial contains no added quenching agent. It would be the least-quenched among the standards. Radionuclide standards that emit beta particles of higher energy than ^{3}H ($E_{max} = 0.018$ MeV) generally require larger increments of quenching agent to provide a quench curve that would display a significant reduction in counting efficiency against a QIP. For example, Ahn. et al. used a range of 0–200 μL of nitromethane as a quenching agent for a ^{14}C ($E_{max} = 0.15$ MeV) quench correction curve with a cocktail containing 10 mL of Permafluor E+ and 10 mL of Carbo-Sorb E. The amount of quenching agent required will depend on the E_{max} of the beta particle spectrum of the radionuclide and the type of fluor cocktail used. Some experimental testing on the part of the user may be required to determine the most appropriate range of quenching agent to be used. For example, Fig. 6.19 shows ^{32}P ($E_{max} = 1.71$ MeV) undergoes little change in counting efficiency over a wide range of quench level.

Quenching agents have less effect on counting efficiency as the E_{max} of the beta-emitter increases, as illustrated in Fig. 6.19, using ^{3}H, ^{14}C, ^{33}P and ^{32}P as examples. Other scintillation cocktails respond differently to nitromethane. For example, cocktails containing DIN or PXE solvent may require larger increments of nitromethane quenching agent. In such cases the following increments of nitromethane for a set of 10 quenched standards may be appropriate: 0, 5, 10, 15, 28, 45, 70, 110, 150 and 230 μL for a ^{3}H or ^{14}C quench correction curve. Other quenching agents are CCl_4, chloroform, acetone, and water, which are not as strong quenchers as nitromethane, also display differing degrees in their relative strength of quench. The amounts of quenching agent required will differ, because not all agents quench equally, e.g., CCl_4 is a stronger quencher than water by two orders of magnitude (Grau Carles, 2006b). Classification of quenching agents according to their quenching power is given by L'Annunziata (1987). The quenching agent used should be soluble and not react with the scintillation cocktail. A color dye can be selected as a quenching agent for a set of quenched standards when color quench is high in the experimental samples.

g. **If in doubt concerning the amounts of quenching agents to add, predetermine this experimentally.** Whenever the amounts of quenching agent required are in doubt, it is easy to predetermine this by adding various amounts of quenching agent to scintillation cocktail in counting vials without radionuclide standard. The vials can then be counted without radionuclide only to determine the external standard quench indicating parameter (QIP) for each vial. Adjustments can be made with more or less quenching agent in order to achieve a wide range of quench levels according to the QIP.
h. **Label the quenched standards by number or letter (e.g., 1 to 10 or A to J) and isotope, date, etc.** The quenched standards can be labeled only on the top of the counting vial cap. A round self-adhesive label may be placed on the cap or the information written on the top of the cap with a fine-tipped indelible pen.
i. **Store the quenched standards in the dark for a day or more for stability against any possible photo- or chemiluminescence.** Photo- and chemiluminescence are possibilities that can occur when either the counting vials are open or when quenching agent is added, respectively. The samples can be counted on a daily basis to observe if there is any reduction in count rate with time as evidence of luminescence. Once stability is confirmed the quenched standards can be used to prepare a quench correction curve.

A procedure alternative to the above for the preparation of a set of quenched standards would be to dispense the desired volume of scintillation cocktail and blank sample medium to a set of counting vials. The volume would depend on the combined volume of sample and scintillation cocktail planned for the experimental counting vials (*e.g.*, 10, 15 or even 20 mL depending on vial size and capacity). The radionuclide standard is then added to each vial in equal amounts using a very precise microliter syringe with an adapter to help assure the addition of the same amounts to each vial. A Hamilton syringe equipped with a Cheney adapter (Hamilton Company, Reno, NV 89,502, USA or CH-7402 Bonaduz, Switzerland) may be suitable. The writer finds it easier and more precise to prepare standards of equal activity by preparing a stock solution of radionuclide standard in scintillation cocktail and dispensing this solution into counting vials as described in steps c and d above.

If stored under refrigeration (5−10°C), sets of quenched standards may be stable for two to 3 years. It is best to keep records of the quench correction curves prepared from a given set of quenched standards from time to time (*e.g.*, monthly basis) to check their stability.

2. Preparation of a quench correction curve

The quench correction curve or plot of counting efficiency versus quench indicating parameter (QIP) is determined with a set of quenched standards as described above, and this curve is stored generally in the memory of the liquid scintillation analyzer. The preparation of the quench correction curve is described as follows:

a. Set up a counting protocol on the LSC to plot a quench correction curve of percent counting efficiency versus a quench indicating parameter. One of the QIPs that uses a sample spectrum characterization method may be used, although an external standard QIP is more often preferred for reasons described in Section V.C of this chapter. Some modern liquid scintillation analyzers store the entire pulse height spectrum of the sample counted onto the memory of the hard disk. With these instruments it is often not necessary to set the counting region defined by lower level (LL) and upper level (UL) discriminator settings. Once the pulse height spectrum of each quenched standard is stored in the instrument, the count rate in any counting region defined by LL and UL discriminator settings can be extracted. However, in older generation instrumentation, it is necessary to first define the counting region according to LL and UL discriminator settings prior to counting the set of quenched standards. The counting efficiency will vary according to counting region settings, and consequently quench correction curves of counting efficiency versus QIP will also differ according to counting region settings.

b. Count the quenched standards at a statistical accuracy of at least 0.5% 2s. Due to high activities of standards in each vial (100,000 to 200,000 DPM) the counting time required to reach the 0.5% 2s statistics may not exceed 5 min per quenched standard.

c. Obtain a plot of the percent counting efficiency of the radionuclide standards versus the QIP. Modern liquid scintillation analyzers will store this data in computer memory. When experimental samples are counted, the instrument should use the QIP measured for the sample and determine the counting efficiency from the correction curve.

3. Use of a quench correction curve

The objective of the quench correction curve (also referred to as a calibration curve) is to determine the counting efficiency of experimental samples of unknown radionuclide activity. From the counting efficiency, the count rate (e.g., CPM) of the sample is converted to radionuclide activity (e.g., DPM) according to Eq. (6.7) described previously. When using a quench correction curve, it is important to keep in mind certain rules, some of which may be intuitively obvious. These are the following:

1. A quench correction curve is good for only one radionuclide, instrument and a specific scintillation cocktail, cocktail-sample volume, and even the vial type (e.g., glass or plastic).
2. The quench correction curve is dependent on counting region defined by lower-level and upper-level pulse height discriminator settings.
3. The quench correction curve is scintillation cocktail dependent (Collé, 1997a,b). It is important to be certain that the scintillation cocktail and volume of sample used for the experimental samples is the same as that used for the preparation of the quench correction curve. The quench correction curve is prepared with standards in fluor cocktail and blank sample without radionuclide. For example, if 5 mL of experimental samples of radionuclide of unknown activity are mixed with 15 mL of fluor cocktail, the quench correction curve should be prepared with 15 mL fluor cocktail containing radionuclide standard mixed with 5 mL of blank water sample in addition to the quenching agent. If a different cocktail is used for experimental samples compared to that used to prepare the quench correction curve, it is necessary to confirm that there is no significant difference between the cocktails in performance for a given radionuclide. Differences in performance of cocktails are more pronounced when analyzing for relatively low-energy beta-emitters such as ^{3}H and ^{14}C.
4. Quench correction curves using an external standard QIP are preferred, because these are independent of sample activity.

E. Combined chemical and color quench correction

When there are significant quantities of chemical or color quench in scintillation cocktail there can be a significant difference in the chemical or color quench correction curve obtained. This is particularly the case when there is a high degree of either chemical or color quench (at high quench levels) and for the liquid scintillation analysis of relatively weak beta-particle emitting radionuclides such as ^{3}H (E_{max} = 0.0186 MeV), ^{63}Ni (E_{max} = 0.066 MeV) and ^{14}C (E_{max} = 0.156 MeV). At high values of SQP(E), which correspond to low levels of quench, Fons-Castells et al. (2017d) observed that chemical and color quench could not be differentiated, as illustrated in Fig. 6.31. However, at high levels of quench corresponding to low SQP(E) values, they found that colored substances in the sample would decrease the counting efficiency greater than samples quenched chemically, i.e., noncolored substances. Under high levels of color, it is best to decolorize the sample before adding fluor cocktail whenever possible, as described in Chapter 7, Volume 1. Fons-Castells et al. (2017d) report high correlations between % detection efficiency and SQP(E) of $R^2 = 0.996$ and 0.999 for the chemical and color quench curves.

Differences in color and chemical quench can be observed regardless of the quench indicating parameter used. The differences in the two curves is based on the two different mechanisms of quench; namely chemical quench that entails the inhibition of energy transfer from cocktail solvent to fluor molecules, and color quench that entails the absorption of light photons emitted by the scintillation cocktail (See Fig. 6.1).

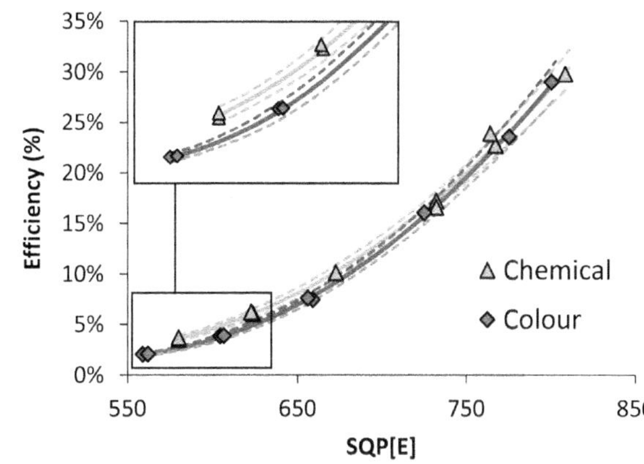

FIGURE 6.31 Tritium efficiency quench curves prepared with $FeCl_3$ and CH_3NO_2 as color and chemical quenching agents, respectively, with an enlargement of low SQP[E] values. *From Fons-Castells et al. (2017d) reprinted with permission from Elsevier © 2017.*

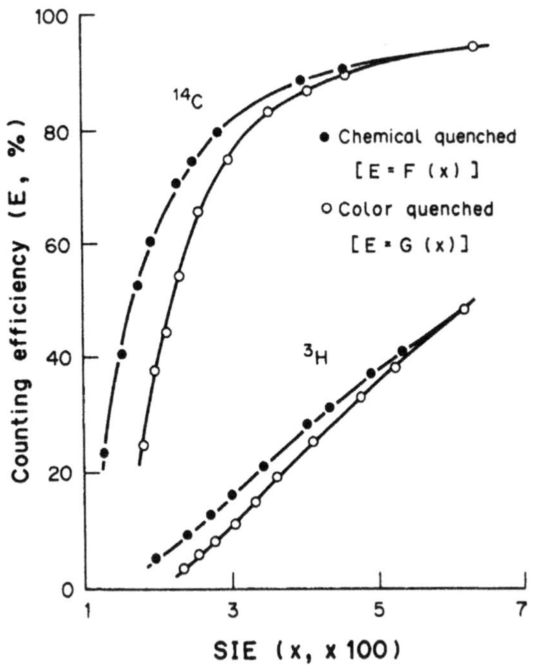

FIGURE 6.32 Chemical and color quench correction curves based on an external standard. *From Takiue et al. (1991a), reprinted with permission from Elsevier © 1991.*

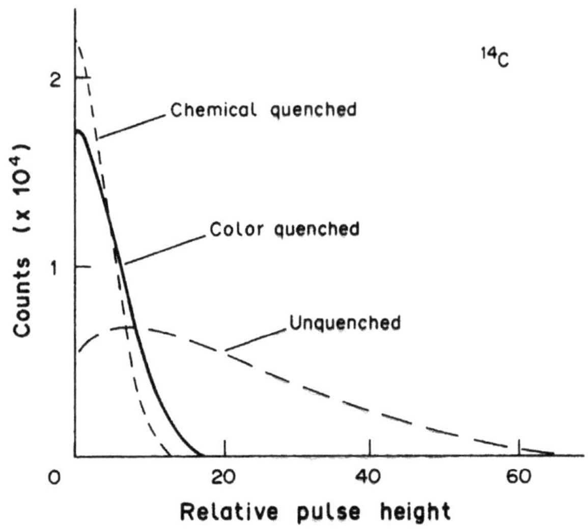

FIGURE 6.33 Liquid scintillation pulse height distributions of chemical- and color-quenched ^{14}C samples. Both samples have identical activity and counting efficiencies. *From Takiue et al. (1991a), reprinted with permission from Elsevier © 1991.*

In older generation liquid scintillation analyzers, the differences between color and chemical quench correction curves may be more pronounced. For example, Takiue et al. (1991a) used an older model Packard 460C liquid scintillation analyzer, which utilized the older SIE (spectral index of the external standard) rather than the tSIE (transformed spectral index of the external standard) described previously in this chapter. With this older quench indicating parameter Takiue et al. (1991a) measured a considerable difference in color and chemical quench correction curves for ^{14}C and ^{3}H illustrated in Fig. 6.32. With these older generation instruments, the liquid scintillation pulse height distribution of a color-quenched sample can be very different from that of a chemical-quenched sample, even if both the samples have the same activity (see Fig. 6.33). Therefore, the pulse height distribution of the external standard can be different for either color- or chemical-quenched samples producing different QIPs and different quench correction curves. The difference can be more significant at high levels of quench as depicted on Figs. 6.31 and 6.32, and it becomes less significant as the beta-particle energy (E_{max}) of the radionuclide increases. The colors (*e.g.*, blue, red, green, yellow, etc.) can have different effects on the degree of color quenching, as the photocathodes of the photomultiplier tubes within the liquid scintillation analyzer (LSA) vary in their degree of sensitivity or response to certain wavelengths. If samples to be counted are colored and highly quenched, it is best to prepare two series of quenched standards, namely, one set of standards that have increasing intensities of coloring agent (color quencher) and another set of standards containing varying amounts of chemical quencher. A comparison of the two quench correction curves derived from these standards would provide evidence for any significant differences that might exist between the two types of quenching. When decolorization (e.g., bleaching) of samples is not possible and samples are highly color quenched, a color-quench correction curve with radionuclide standards may be prepared with a suitable color dye as described in Chapter 7, Volume 1. Certain modern liquid scintillation analyzers are programmed to correct for color quenching (PerkinElmer, Inc.). Bukowski et al. (1992) found no differences between color and chemical quench correction curves for ^{3}H when either picric acid (yellow color quencher) or CCl_4 (chemical quencher) was used as a quenching agent and the H# was measured as the external standard quench indicating parameter (QIP). Also, Effertz *et al.* (1993) demonstrated identical quench correction curves for the measurement of ^{3}H when either nitromethane (chemical quencher) or yellow, red, or green (color quenching agents) were used as color quenchers and tSIE was utilized as the external standard QIP.

The best alternative when color exists in the sample is to decolorize (*e.g.*, sample bleaching or oxidation of organic samples to CO_2 and H_2O, see Chapter 7, Volume 1) before adding fluor cocktail. Decolorizing the sample eliminates the problem of color quench leaving behind only chemical quench, which is present in all samples to be analyzed. The works of Lee et al. (2016), Yuan et al. (2016), Dong et al. (2013), Mui et al. (2013), Norton et al. (2012), Batrakova et al. (2010), Warrick et al. (2010), Williams and Kookana (2010), Sirelkhatim et al. (2008), and Hou et al. (2005) are excellent examples of the use of sample decolorization prior to liquid scintillation analysis of ^{3}H, ^{14}C, ^{63}Ni, and ^{55}Fe (all relatively weak beta-emitters), etc., thereby eliminating color quenching. If decolorization is not possible, some modern liquid scintillation analyzers are equipped with color correction programs or algorithms that will correct for the difference between chemical and color quench.

An example of a practical program for the correction of the difference between color and chemical quench correction curves, whenever these are significant, was formulated by Takiue et al. (1991a). This method entails the preparation of two sets of quenched standards of a given radionuclide. One set of standards is prepared with a color-quenching agent (*e.g.*, bromothymol blue, methyl red, or bromocresol green) that produces minimal chemical quench, and another set of standards is prepared using a chemical-quenching agent (*e.g.*, nitromethane or CCl_4). The sets of quenched standards are used to plot two curves consisting of a color- and chemical-quench correction curve of counting efficiency versus any external standard quench indicating parameter (*e.g.*, ESCR, *H* #, SQP(E), or tSIE, see Fig. 6.32). In addition, the color- and chemical-quenched standards are used to plot a second set of curves consisting of the external standard quench indicating parameter plotted against the external standard counts, referred to as double external standard relation curves (DESR curves; see Fig. 6.34). When an experimental sample, which is

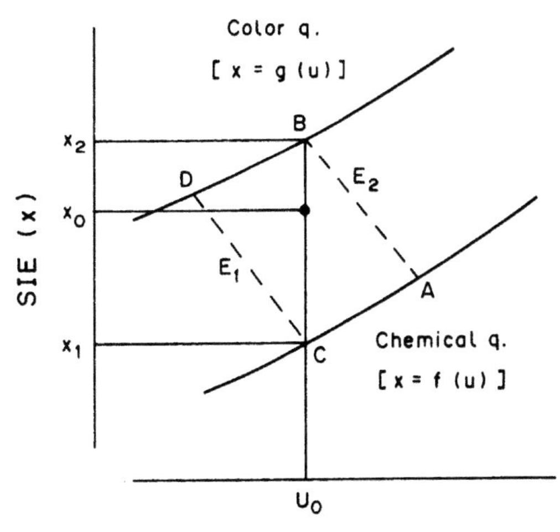

FIGURE 6.34 DESR curves for chemical- and color-quenched radionuclide used for the calculation of the counting efficiency of an experimental combined chemical- and color-quenched sample, where x_o and u_o are the quench indicating parameter and external standard (ES) counts for the experimental combined chemical- and color-quenched sample. *From Takiue et al. (1991a), reprinted with permission from Elsevier © 1991.*

quenched by both chemical and color constituents, is counted, the counting efficiency of the unknown sample has the value between E_1 and E_2, which corresponds to the external standard (ES) counts of u_o of the DESR curves illustrated in Fig. 6.34. Hence, as demonstrated by Takiue et al. (1991a), according to the geometry depicted in Fig. 6.34, the counting efficiency (E_o) is defined as

$$E_o = E_1 + \frac{x_o - x_1}{x_2 - x_1}(E_2 - E_1) \tag{6.27}$$

where x_1 and x_2 are obtained from the chemical and color DESR curves, respectively (Fig. 6.34) where $x_1 = f(u_o)$ and $x_2 = g(u_o)$. The efficiency values E_1 and E_2 are obtained from the chemical and color quench correction curves, respectively (Fig. 6.32) where $E_1 = F(x_1)$ and $E_2 = G(x_2)$. Eq. (6.27) is then written as

$$E_o = F(x_1) + \frac{x_o - f(u_o)}{g(u_o) - f(u_o)}[G(x_2) - F(x_1)]. \tag{6.28}$$

Takiue et al. (1991a) used polynomial curve fitting with the least squares method to define the coefficients of the quench correction and DESR curves. This color correction method is easily applied with the computer application programs of most modern liquid scintillation analyzers. Nevertheless, the problem of combined color and chemical quench in samples is best averted by decolorization of samples prior to the addition of scintillation cocktail.

F. Direct DPM methods

The Direct DPM methods entail the LSA measurement of the disintegration rates of radionuclides, particularly beta-emitters, under various levels of quench without the use of quench correction curves. These methods are described subsequently.

1. *Conventional integral counting method (CICM)*

During the early years of liquid scintillation counting, it was discovered that an extrapolation of integral counting curves to zero discriminator bias could be used to determine the disintegration rates (DPM) of alpha-emitters (Basson and Steyn, 1954) and beta-emitters (Steyn, 1956) without interference from gamma emission. The method applied to alpha-emitters received little attention, because the LSA counting efficiency of alpha-emitters was close to 100% even in these early years of liquid scintillation development. As far as beta-emitters are concerned, this technique, known as integral counting, received some popular attention and applications in the late 1950s and during the 1960s. Further developments made this technique a practical and accurate method for the activity measurement of beta-emitting radionuclides.

The work of Goldstein (1965) demonstrated the broad range of radionuclides that may be analyzed by integral counting as well as the simplicity of the procedure involved. In the development and testing of integral counting, Goldstein (1965) used the first and only commercial LSA available at that time, which was a Packard 314 liquid scintillation spectrometer. The procedure involved three pulse height discriminators labeled AA′, B, and C. The AA′ discriminator was set just above the noise level to reject noise pulses. The C discriminator (upper level discriminator) was turned off or disengaged so that all of the pulses of magnitude above the B discriminator would be registered and counted. The height of the B discriminator was varied in the range of 10–30 V in 5-volt increments. The count rates for a given beta-emitting sample in scintillation cocktail were collected for each setting of the B discriminator. With the B discriminator at its lowest setting the count rate is highest. With each incremental increase in the height of the B discriminator, the count rate diminishes, because fewer and fewer pulses are detected. The resulting plot of count rate on a logarithmic scale versus the B discriminator bias (volts) setting on a linear scale would be linear with negative slope, which could be extrapolated back to zero bias volts. At this point of extrapolation, the count rate (CPM) at zero bias would be the disintegration rate (DPM) of the sample. This extrapolation method is currently referred to as the conventional integral counting method (CICM). It is an effective method for extrapolating to sample DPM for beta-emitters or beta-gamma emitters, which emit beta particles with an $E_{max} > 200$ keV, including ^{147}Pm, ^{45}Ca, ^{99}Tc, ^{36}Cl, ^{204}Tl, ^{89}Sr, ^{90}Sr(^{90}Y), ^{91}Y, ^{32}P, ^{131}I, ^{85}Kr, ^{131m}Xe and ^{60}Co, among others, regardless of quench level. Corrections for quench in the sample scintillation cocktail mixtures are not necessary, because the quench level in the sample affects only the slope of the integral curve, and extrapolation of the curve to zero discriminator bias ends at the same count rate for all quench levels with expected statistical deviations (<2% error). Homma and Murakami (1977) also applied the conventional integral counting method to determine the activity of ^{226}Ra after separating the equilibrated ^{222}Rn in a liquid scintillator. The disintegration rates of ^{222}Rn and its daughters, which include both alpha- and beta-emitters were determined by this method at various quench levels.

The conventional integral counting method for sample DPM determinations generally cannot be applied to the measurement of beta-emitting radionuclides of E_{max} < about 200 keV. However, Homma et al. (1994a) developed the technique into the modified integral counting method (MICM), which can be used to determine the activities of all beta-particle emitters including ^{3}H of very low energy ($E_{max} = 18.6$ keV) and with higher accuracy.

2. Modified integral counting method (MICM)

The modified integral counting method (MICM) was reported by Homma and coworkers (1993a, 1993b), who modified the conventional integral counting method by extrapolating the integral counting curves, not to the zero pulse height as described above for the CICM, but to the zero detection threshold (ZDT) of the liquid scintillation spectrometer, which refers to the average energy required to produce a measurable pulse. They applied the new method to analyze the activity of alpha- and beta-emitters including ^{222}Rn and its daughters as well as the low-energy beta-emitters ^{3}H, ^{14}C, ^{35}S, and ^{45}Ca with 100% detection efficiency. The method is described subsequently in more detail.

The modified integral counting method as was determined by Homma et al. (1994a) is carried out by the following procedure:

a. The first step requires the determination of the zero detection threshold (ZDT) of the particular liquid scintillation analyzer utilized for the analysis. This is carried out by measuring a standardized nonquenched ^{3}H sample according to the integral counting method described earlier. The observed integral count rates of the ^{3}H standard are plotted at several pulse heights and the curve is then extrapolated to the count rate, which is equivalent to the disintegration rate (DPM) of the ^{3}H standard, as illustrated in Fig. 6.35. The keV value (pulse height)

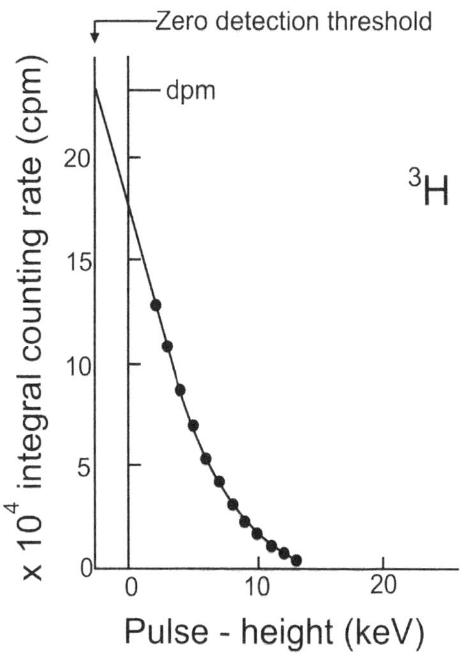

FIGURE 6.35 Determination of the zero detection threshold (ZDT) extrapolating the integral pulse height spectrum of a standardized ^{3}H sample to its disintegration rate, which is indicated by the arrow. *From Homma et al. (1994a); reprinted with permission from Elsevier © 1994.*

at this count rate represents the zero detection threshold (ZDT). The ZDT was found by Homma et al. (1994b) to vary from instrument to instrument over the range of 2.4–3.5 ± 0.2 keV.

b. Once the zero detection threshold is determined for the particular instrument, the absolute disintegration rate of any low-energy beta-emitter (E_{max} < about 200 keV) including ^{3}H as well as high-energy beta-emitter (E_{max} > about 200 keV) is determined by extrapolating the integral pulse height spectrum of the radionuclide of interest to the previously determined zero detection threshold.

Homma et al. (1994a) demonstrated activity measurements by this MICM of ^{3}H and ^{14}C at various quench levels with uncertainties of ±1.3% and 1.4%, respectively. Examples of results obtained from the modified integral counting method applied to the activity determination of ^{35}S and ^{45}Ca are illustrated in Fig. 6.36. As noted by Homma et al. (1994a,b), it is obvious from the plots illustrated that extrapolation of the integral pulse height spectrum to only the zero pulse height leads to an intercept value that is lower than the actual DPM of the radionuclide. However, extrapolation of the integral counting curve to the zero detection threshold leads to the actual disintegration rate of the sample.

The modified integral counting method was reported also by Homma et al. (1993a,b, 1994c) for the determination of ^{222}Rn and its daughters ^{218}Po, ^{214}Pb, ^{214}Bi, and ^{214}Po. Total α and β activity was determined with 100% counting efficiency. The MICM can be applied to the activity measurements of α- and β-emitters as single radionuclide samples or mixtures, and γ-emission does not interfere in most cases (Homma et al., 1994a). Measurements of ^{222}Rn with activity ranges of 0.2–22.9 Bq/L in natural water samples obtained from private wells and springs were carried out by Murase et al. (1998) with the MICM applying 100-minute counting times, which provided activities with an overall uncertainty of 3.1%. An integral counting method was carried out together with improvements in alpha liquid scintillation spectrometry by Yoshikawa et al. (2006) to determine ^{220}Rn/^{222}Rn ratios. For low-activity ratios, they developed a liquid scintillation alpha spectrometry method to measure ^{220}Rn with reduced error. The modified integral counting method has a practical simplicity similar to the efficiency tracing (ET) method described in the following section of this chapter.

Hara et al. (2016) carried out extensive tests on the determination of the optimum quench level for the absolute activity measurements of beta-emitting radionuclides by the modified integral counting method (MICM) and compared the MICM to the CIEMAT/NIST method of radionuclide standardization. They measured a discrepancy of only 1% between the CIEMAT/NIST and MICM with first order function fitting. Nakayama et al. (2017) tested the MICM at various quench levels with ^{14}C and ^{35}S standards using different scintillators including Ultima Gold, Ecoscint, Ultima Gold XR and a toluene-based scintillator. This study demonstrated that the MICM with various quench samples conformed to assayed values.

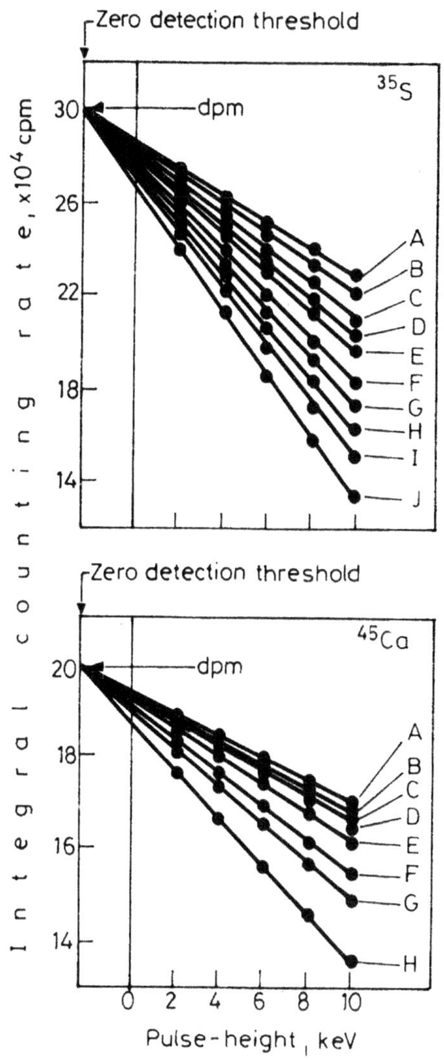

FIGURE 6.36 Extrapolation plots of the integral count rates of quenched ^{35}S and ^{45}Ca to the zero detection threshold for determination of the radionuclide disintegration rates. Letters A, B, C ... denote samples with increasing quench levels. Deviations from actual DPM values were <1% for all plots. *From Homma et al. (1994b) with kind permission from Springer Nature © 1994.*

3. Efficiency tracing (ET) with ^{14}C

Efficiency tracing (ET) with ^{14}C is another practical and simple extrapolation method applied generally to the activity measurements of β-emitting radionuclides with the exception of tritium. This method should not be confused with the CIEMAT/NIST efficiency tracing method described in Section IX.A of this chapter, which is used for radionuclide standardization. The ET method was demonstrated by Takiue and Ishikawa (1978) to provide accurate DPM values for 14 radionuclides. A subsequent study by Ishikawa et al. (1984) showed that the technique provided accurate DPM measurements of 11 additional β and $\beta-\gamma$-emitting radionuclides, namely, ^{14}C, ^{32}P, ^{36}Cl, ^{46}Sc, ^{59}Fe, ^{60}Co, ^{63}Ni, ^{86}Rb, $^{90}Sr(^{90}Y)$, ^{131}I, ^{134}Cs, and ^{147}Pm regardless of quench level. The method was demonstrated by Fujii et al. (1986) in a study whereby the efficiency tracing DPM measurements of ^{14}C, ^{35}S, ^{32}P, ^{36}Cl, ^{45}Ca, and ^{131}I were determined for radionuclides on filter disks in LSA counting vials. The ET method is most often used to determine the activity of single and multiple β and β-γ-emitters; it can be applied also to the activity measurements of mixtures of α- and β-emitters (see Fujii and Takiue, 1988b).

The efficiency tracing (ET) with ^{14}C technique involves the following steps:

a. A ^{14}C nonquenched standard is counted in six separate counting regions, such as, counting regions defined by lower-level discriminators set at 0, 2, 4, 6, 8, and 10, and upper-level discriminators set at the upper limit of the pulse height scale. For example, counting regions, such as, 0–2000, 2–2000, 4–2000, 6–2000, 8–2000 and 10–2000 keV for lower-level (LL) to upper-level (UL) pulse height discriminator settings on a linear keV equivalent scale, may serve as one example of workable counting regions. However, other similar counting regions may be used. Also, equivalent discriminator settings for liquid scintillation analyzers equipped with a logarithmic scale for pulse height analysis may be determined. See L'Annunziata (1997, unpublished work of the writer) and L'Annunziata and coworkers (Noor et al., 1996a).

b. The percent counting efficiency (% E) values of the nonquenched ^{14}C standard in each of the six counting regions are calculated as % E = CPM/DPM × 100.

c. A sample of unknown activity of a β-emitting radionuclide is subsequently counted in the same six regions as the nonquenched ^{14}C standard.

d. The six CPM values of the sample of unknown activity are plotted against the % counting efficiency values of the nonquenched ^{14}C standard.

e. The curve is then extrapolated to 100% counting efficiency, where the CPM of the unknown sample is equal to its DPM. Extrapolation may require a linear or multilinear regression least-squares best fit of the data points and definition of the equation to the line or curve to most accurately determine the point of intersection at 100% counting efficiency. An example of eight efficiency tracing curves for the DPM determination of eight radionuclides from the work of Takiue and Ishikawa (1978) is illustrated in Fig. 6.37, and an example of an efficiency tracing curve for the analysis of ^{35}S from the writer's work is illustrated in Fig. 6.38.

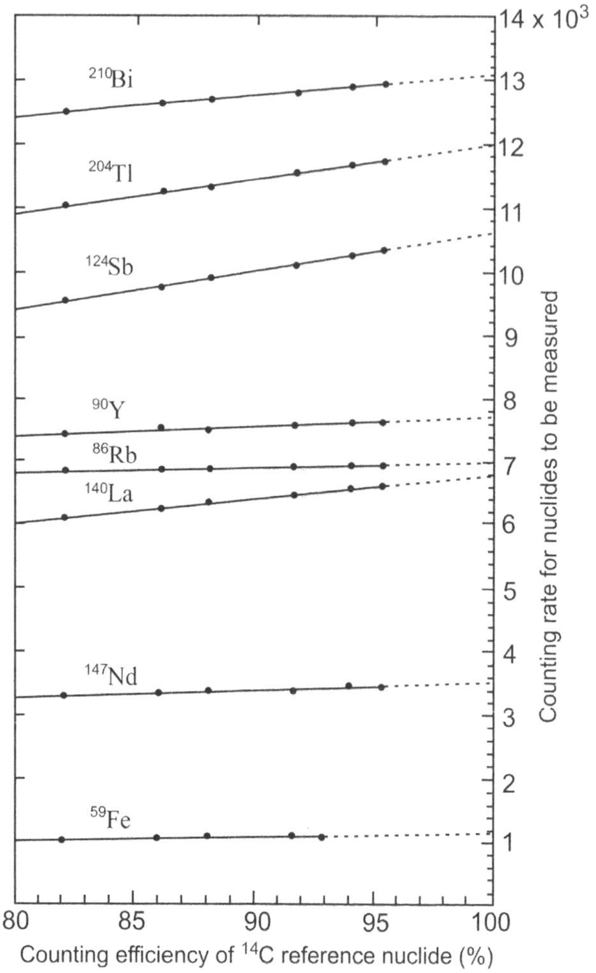

FIGURE 6.37 Efficiency tracing curves of eight radionuclide samples. The dashed portion of the plots are the extrapolated segments to 100% counting efficiency. Extrapolated values indicate the counting rates at 100% counting efficiency or DPM of each nuclide. *From Takiue and Ishikawa (1978), reprinted with permission from Elsevier © 1978.*

Variables such as sample scintillation cocktail volumes over the range of 1–20 mL, composition of the scintillation cocktail, amount or kind of quenching agent, or size and material of the counting vial may affect the slope of the efficiency tracing curve of the sample of unknown activity. At low quench levels the curves are often linear, as illustrated in Fig, 6.37. Linear regression analysis in these cases will define the equation to the line permitting extrapolation of the curve to 100% detection efficiency and DPM of the sample. At higher quench levels the efficiency tracing curves may not be linear, such as illustrated in Fig. 6.38. In these cases, quadratic regression may be needed to fit the data to a second-order polynomial $y = ax^2 + bx + c$, where the independent variable x is the % counting efficiency and

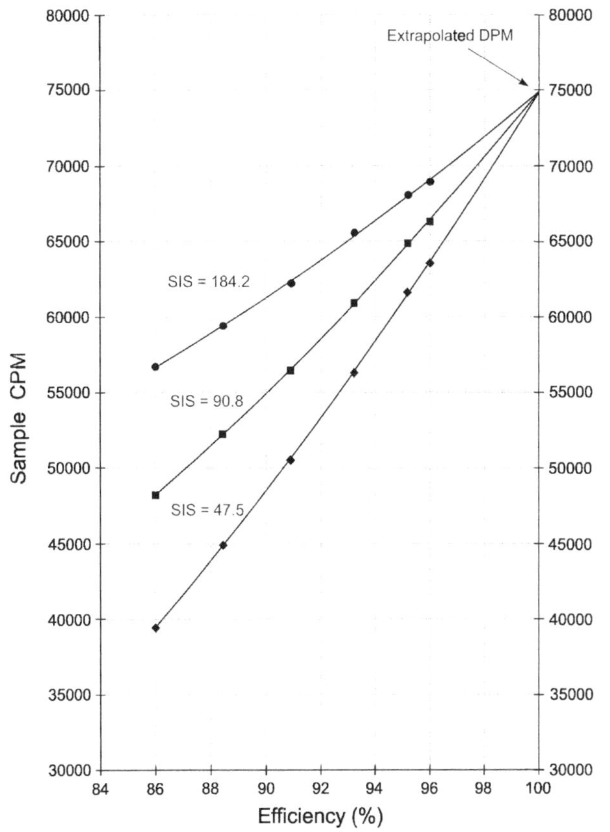

FIGURE 6.38 Efficiency tracing (ET) curves for the activity determinations of three samples of ^{35}S at three levels of quench. The degree of quench is measured by the sample quench indicating parameter SIS where the uppermost curve is that produced by the least-quenched of the three samples. *From the work of L'Annunziata and coworkers, See Noor et al. (1996). Reprinted with permission from Elsevier © 1996.*

the dependent variable y is the sample count rate (CPM). When x of the regression curve is given the value of 100%, the calculated value of y provides the extrapolated sample activity or disintegration rate (DPM). Example of radionuclide activity recoveries that have been reported by this method are given in Table 6.3. At very high levels of sample quench (e.g., SIS < 1000 for ^{86}Rb and SIS < 69 for ^{33}P) error can creep into the dpm calculation, as noted in the table. This is due to the fact that the efficiency tracing curves lose their linearity and even take on a marked concave shape complicating the mathematical extrapolation to 100% efficiency.

An additional attribute of the efficiency tracing technique is the possibility of determining the total DPM of mixtures of β-emitting radionuclides (with the exception of ^{3}H) and α/β-emitters as demonstrated by Fujii and Takiue (1988a,b) and L'Annunziata and coworkers [see Noor et al. (1996a)]. The activity of ^{3}H cannot be measured by this method, as the efficiency tracing curves for ^{3}H are too concave yielding inaccurate extrapolation to 100% detection efficiency. Table 6.4 illustrates the excellent recoveries obtained for total DPM measurements of mixtures of $^{86}Rb + ^{35}S + ^{33}P$. It should be noted, however, that the technique provides the total DPM of the mixture and not the activities of the individual radionuclide components of the mixture.

This direct DPM method is very useful for the determination of activities of radionuclides of relatively short half-life for which NIST-traceable standards are not available commercially. The ET-DPM method is an automatic radionuclide activity analysis option available with some liquid scintillation analyzers (e.g., PerkinElmer, Inc. and Hitachi Aloka LSAs).

The efficiency tracing DPM technique was reviewed by Kessler (1991b). Some practical examples of the application of the efficiency tracing method can be found in the literature including the analysis of ^{222}Rn by Yasuoka et al. (2005), ^{14}C and ^{45}Ca by Dobrin et al. (2006), Takiue et al. (2004), ^{32}P by Liyanage and Yonezawa (2003), and ^{14}C by Wakabayashi et al. (1999). The method was tested by L'Annunziata (1997a) and Edler (2004) for β-emitting radionuclide samples over a wide range of quench and levels of sample activity. Kulkarni et al. (2011a,b) compared the CIEMAT/NIST standardization method to the efficiency tracing (ET) and modified integral counting methods (MICM) and found these to agree within 3.3% of the beta activity. Also, Reddy et al. (2011) have tested both the ET and MICM for activity measurements of beta-emitters, and they confirm that these methods are practical and simple yielding uncertainties down to 1% for activity measurements of beta-emitters with activities in the range of 8.33−100 Bq or higher. On the basis of previous studies, the following conclusions and recommendations are made:

a. The efficiency tracing DPM (ET-DPM) technique is an accurate method for determining the total activity (DPM) of β-emitting radionuclide samples with the exception of ^{3}H.
b. The ET-DPM method can be used to determine the total activity (DPM) of a mixture of β-emitting radionuclides.
c. No quench correction curves are needed, either for chemical or color quench, when using the ET-DPM method.
d. For best results samples should be counted for a duration sufficient to achieve a count rate with a % two sigma standard deviation (i.e., % 2s) of 1% or lower.
e. The ET-DPM method is very useful for the determination of activities (DPM) of radionuclides of short half-life for which quenched standards are not available. When the activity of a nuclide is required within the limits established by a national bureau of standards (e.g., NIST) for what is classified as a primary standard, the CIEMAT/NIST, TDCR, or other liquid scintillation

TABLE 6.3 Calculated recoveries of efficiency tracing (ET) for the determination of activities of ^{86}Rb, ^{35}S, ^{33}P and a composite mixture of the three radionuclides under various levels of quench[a].

Radionuclide	SIS	ET calculated DPM	Recovery (%)[b]
^{86}Rb	1568[c]	43,629[d]	100.0
	1043	43,077[d]	98.7
	83	45,293[d]	103.8
	41	42,563[d]	97.6
	28	42,729[d]	97.9
^{35}S	113.8[c]	74,862[e]	100.0
	82.8	74,087[e]	99.0
	70.6	74,893[e]	100.0
	57.8	75,702[e]	101.1
	50.0	74,822[e]	99.9
	40.3	76,174[e]	101.8
^{33}P	184.2[c]	63,819[f]	100.0
	120.7	64,054[f]	100.4
	90.8	64,370[f]	100.8
	69.5	64,436[f]	100.9
	56.2	65,325[f]	102.4
	47.5	64,868[f]	101.6
^{86}Rb + ^{35}S + ^{33}P	463[c]	176,028[e]	100.0
	358	176,346[e]	100.2
	278	176,643[e]	100.3
	231	175,489[e]	99.7
	198	175,881[e]	99.9
	176	176,836[e]	100.4

[a]*From L'Annunziata and coworkers (See Noor et al., 1996a). Reprinted with permission from Elsevier © 1996.*
[b]*The % recoveries are calculated for those samples containing nitromethane quenching agent on the basis of the efficiency tracing DPM determined for those samples containing no added quenching agent.*
[c]*No quenching agent added.*
[d]*Linear regression extrapolation to 100% counting efficiency, mean correlation coefficient r = 0.985.*
[e]*Linear regression extrapolation to 100% counting efficiency, mean correlation coefficient r = 0.9997.*
[f]*Quadratic regression fitting of data to second order polynomial and extrapolation to 100% efficiency.*

TABLE 6.4 Percent recoveries of calculated activities of five composite mixtures of ^{86}Rb–^{35}S–^{33}P determined by the efficiency tracing (ET) technique.[a]

Sample DPM (in hundreds)^{86}Rb:^{35}S:^{33}P	Total DPM (actual)	Total DPM (ET)	Total DPM recovery (%)
4326:7294:7194	18,814	18,671	99.2
2146:3620:3424	9190	9185	99.9
1042:1794:1550	4386	4408	100.5
3113:5510:4620	13,243	13,237	100.0
432.3:742.6:646.1	1821	1819	99.9

[a]*From L'Annunziata and coworkers (see Noor et al. (1996a), reprinted with permission from Elsevier © 1996.*

method of radionuclide standardization is recommended. These methods are described in Section IX of this chapter and in Chapter 7 of Volume 2.

f. The ET-DPM method may be used to determine the activity of a source radionuclide prior to initiating a tracer experiment with that nuclide. Before beginning an experiment with a radionuclide as a tracer, it is best not to accept blindly the cited activity provided by the radioisotope supplier on the label of the source container. It is best to prepare replicate samples of the radionuclide source in a suitable scintillation cocktail and use the ET-DPM method or other suitable method to confirm the activity (DPM) of the radionuclide source before beginning an experiment with that source.

Liquid scintillation analyzers, such as those manufactured by Hitachi Aloka instruments and Packard (now PerkinElmer) provide an option for the analysis of beta-emitting radionuclides by the efficiency tracing DPM method referred to as ET-DPM, which is a Direct DPM method, because no quench correction curves are required. Some examples in the scientific literature, particularly in the biological sciences, where the Direct DPM methods were used for radioassay are Dulanska et al. (2016), Little et al. (2011), Hawkins et al., 2010, Kobbe et al., 2009, Borthwick et al., 2008, Dianu and Podină, 2007, Mordaunt et al., 2005, Little and Rodríguez, 2005, Breitholtz and Wollenberger, 2003, Eardly et al., 2001, Roberts et al., 2001, and Holst et al., 2000). The efficiency tracing DPM (ET-DPM) method has been demonstrated to be an excellent option when standards of the radionuclide of interest are not available for the preparation of a quench correction curve.

4. Multivariate calibration

The principles of multivariate calibration, including the multivariate methods of multiple linear regression (MLR), principal component regression (PCR), and partial least squares regression (PLSR), among others are described in detail by Varmuza and Filzmoser (2009), Næs (2002), Martens and Næs (1992), Thomas and Haaland (1990), and Beebe and Kowalski (1987). The practical applications of MLR are indisputable, as this statistical method is of widespread use. The authors noted that PCR and PLS is gaining acceptance in chemistry, as the laboratory computer can facilitate data collection and processing required for multivariate calibration. As a statistical mathematical tool, multivariate calibration can be applied to a chemical or physical analysis when more than one measurement is acquired for each sample. Mathematical data matrices are written according to the numbers of samples and variables implicated in an analytical result. As explained by Beebe and Kowalski (1987), PLS is a factor-based modeling procedure in which factors are defined for any linear combination of the variables in the data matrices. The PLS algorithm estimates the factors in these matrices to provide a prediction of the observation on an unknown sample. Information on the mechanics of calibration and prediction using the PLS algorithm can be obtained from Gelaldi and Kowalski (1986) and a rigorous treatment of the methodology from Lorber et al. (1986).

García et al. (1996) applied partial least squares (PLS) as a multivariate calibration to determine (predict) ^{14}C activities in samples over the activity range of 1.48–15.16 DPM per sample-scintillation cocktail mixture. They used a Packard Tri-Carb 2000 CA/LL liquid scintillation analyzer (now PerkinElmer Tri-Carb LSA) and samples with variable quench levels using a quenching agent in the concentration range of 0%–0.6% CCl_4. For the multivariate calibration they applied PLS regression using the PLS-Toolbox package for MATLAB devised by Wise (1992). The factors considered in two models constructed by García et al. (1996) consisted of ^{14}C content, quenching, blank, and spectral variability in one model, with blank omitted from the second model. They obtained slightly improved results omitting the blank (background) among the factors considered. Among 16 samples tested over the activity range of 1.48–15.16 DPM per sample, they obtained predicted activities with a relative error in the range of 0% –5.4% (average = 1.09% relative error). This is the first application of multivariate calibration to activity determinations for single radionuclide samples. The multivariate calibration approach to the analysis of ^{14}C is yet to be applied routinely. However, it offers the advantage of being a time-saving approach to low-level liquid scintillation counting, because background information would not be needed and, therefore, total counting time reduced.

Sakar et al. (2015) performed multivariate regression calibration including multiple linear regression (MLR), principal component regression (PCR) and partial least squares regression (PLSR) algorithms on ^{238}Pu, ^{239}Pu, ^{240}Pu and ^{241}Pu atom% abundances to predict the ^{242}Pu isotopic abundance. They found an excellent correlation between the predicted and standard ^{242}Pu abundances. The standard ^{242}Pu abundances were prepared from 175 Pu samples obtained from pressurized heavy water reactors analyzed during 2008–2010 as referred to as a calibration set (CS), depicted in Fig. 6.39. The method was validated with a test set of 133 Pu samples collected and analyzed during 2011–13.

The multivariate calibration approach has been applied by several researchers to the simultaneous liquid scintillation analysis of mixtures of radionuclides, and these studies are discussed further on in this chapter.

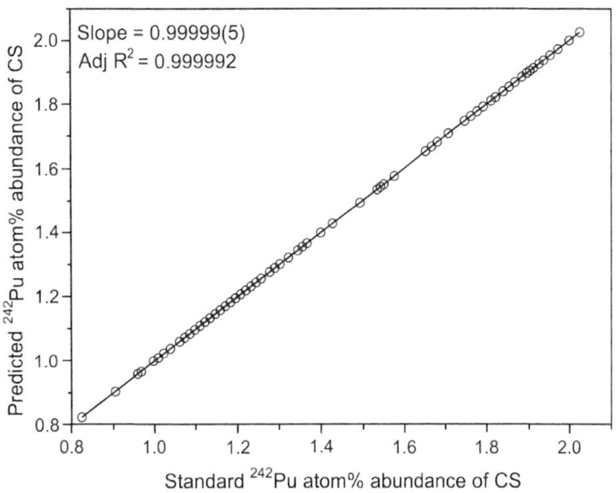

FIGURE 6.39 Predicted ^{242}Pu atom% abundance of CS samples against the standard ^{242}Pu atom% abundance using the MLR calibration model. *From Sarkar et al. (2015) reprinted with permission from Elsevier © 2015.*

VI. Analysis of X-ray, gamma-ray, atomic electron, and positron emitters

Liquid scintillation analysis (LSA) is not restricted to the activity analysis of beta- and alpha-particle emitting nuclides. LSA is also applied to the analysis of nuclides decaying by electron capture (EC) and certain gamma-ray emitters (Grau Malonda, 1999). EC is generally accompanied by the emission of inner bremsstrahlung radiation, X-rays, Auger electrons, and gamma radiation when the daughter nuclide is left in an excited energy state, as illustrated in the following general EC decay process taken from Eq. (1.223) of Chapter1, Volume 1 and presented here again for convenience:

$$e^- + p^+ \rightarrow n + \nu + \text{inner bremsstrahlung} + x - \text{rays} + \text{Auger electrons} + \gamma \qquad (6.29)$$

The electron-capture decay process depicted above, often referred to as *K* capture, is described in detail in Chapter 1, Volume 1. Nuclides emitting gamma rays may also emit internal-conversion (IC) electrons, as described in Section X.C of Chapter 1. Internal conversion occurs when the decay energy is transferred to an atomic electron, and when internal conversion occurs to a significant extent, it results in the emission of atomic electrons of energy equivalent to the nuclear decay energy less the electron binding energy. Thus, when internal conversion occurs to a significant extent, the conversion electrons as well as concurrent events between conversion electron and X-ray emissions contribute significantly to the liquid scintillation detection efficiency of radionuclides decaying by electron capture with gamma emission. The IC electrons as well as X-ray emissions concurrent with IC electron emissions possess higher energies than Auger electrons sufficient to produce pulse heights well above the lower level discriminator setting of liquid scintillation spectrometers.

Tin-113 is an excellent example of a radionuclide that decays exclusively by electron capture with gamma emission commonly used to calibrate gamma spectrometers. The electron capture decay scheme of ^{113}Sn is illustrated in Fig. 6.40, which is from Roteta et al. (2014), and the scheme and decay data are available from Helmer (2002) and online via the National Nuclear Data Center (NNDC, 2018: http://www.nndc.bnl.gov/nudat2/chartNuc.jsp, accessed on January 31, 2018). The ^{113}Sn nuclide decays with a half-life of 115.09 days by exclusively electron capture to ^{113}In by mainly one of two transitions to the metastable state of ^{113m}In at the excited level of 646.8 keV at 2.21% intensity or to a lower excited level of 391.7 keV at 97.79% intensity, and the half-life of the metastable state of ^{113}In is reported by the NNDC to be 99.476 min, which is equivalent to 1.6579 h. Roteta et al. (2014) notes that the gamma transition labeled γ1 of 392 keV in Fig. 6.40 at 84.97% intensity is partially converted with a total conversion coefficient $\alpha_T = 0.540$, which signifies that there are abundant internal-conversion electrons emitted with energy of the gamma transition less

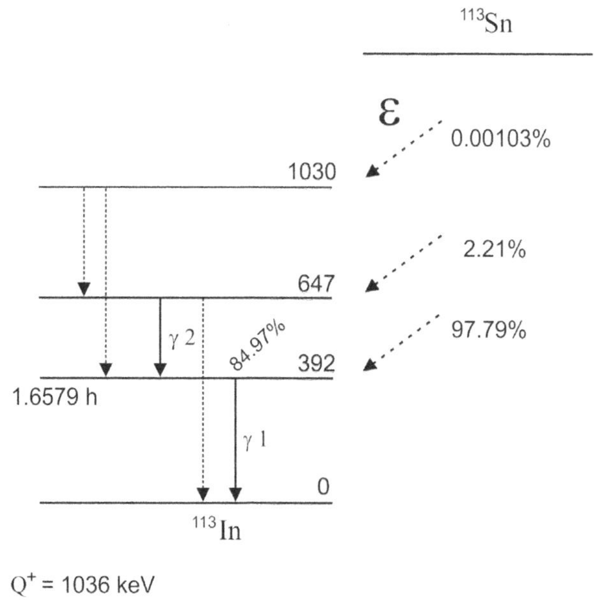

FIGURE 6.40 Decay scheme of ^{113}Sn, which decays exclusively by electron capture denoted by the symbol ε. Data is from Helmer (2002) and the NNDC (http://www.nndc.bnl.gov/nudat2/chartNuc.jsp, accessed January 31, 2018). The energy levels are rounded off and given in keV. Gamma transitions with intensities higher than 0.1% are drawn with solid lines. *From Roteta et al. (2014) reprinted with permission from Elsevier © 2014.*

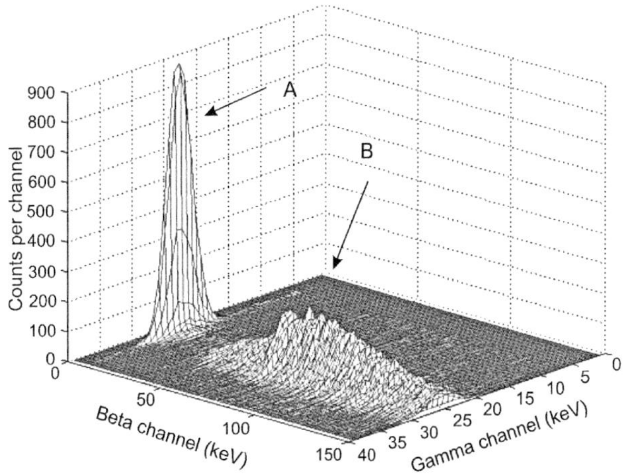

FIGURE 6.41 Three-dimensional coincidence spectrum of ^{113}Sn obtained at CIEMAT with a two-channel digital acquisition system. The spectrum shows two main regions: Region A, which contains coincidences between the low energy Auger electrons (two to four keV) and the K X-rays and Region B, which contains the coincidences between conversion electrons (364–391 keV) and K X-rays. *From Roteta et al. (2014) reprinted with permission from Elsevier © 2014.*

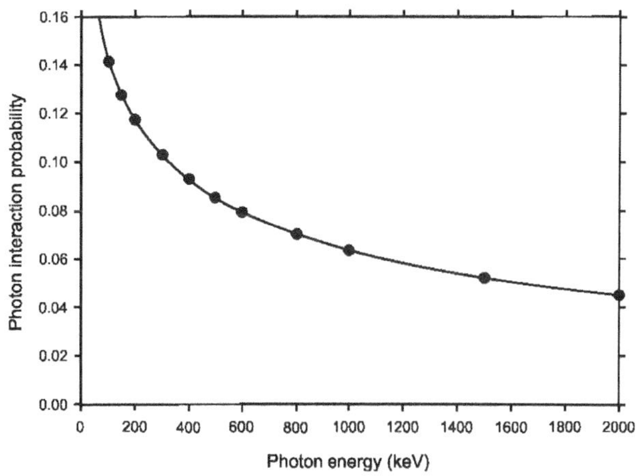

FIGURE 6.42 Simulated photon interaction probability determined at the indicated energies (closed circles) for a counting vial containing 12 mL of *Quicksafe A* liquid scintillator. The solid line is a regression fit to the data. *From Simpson and Morris (2004a), reprinted with permission from Elsevier © 2004.*

the electron binding energy equivalent to 391 keV. The gamma transition labeled γ2 in Fig. 6.40 of 255 keV occurs with an intensity of only 2.1%, and the other transitions from the elevated states of 1030 and 647 keV are minor, and these are illustrated with broken arrows in Fig. 6.40. Roteta et al. (2014) report a total counting efficiency of 0.74 in the liquid scintillation standardization of ^{113}Sn, which would be a result mainly from Auger electron and IC electron interactions with scintillation cocktail together with a smaller fraction the K X-ray emissions, which are illustrated in the 3-dimensional coincidence spectrum of Fig. 6.41.

The electron-capture (EC) decay process often competes with positron emission, as described in Chapter 1, Volume 1. Consequently, some radionuclides that decay by electron capture, may also decay by positron emission. These nuclides are therefore grouped into this section of the chapter, as nuclides that may emit X-rays, gamma rays, atomic electrons (*i.e.*, Auger and IC electrons), and positrons. This broad potential of liquid scintillation analysis was illustrated previously in this chapter (See Fig. 6.12) where the 8 keV Auger electrons and 93 keV conversion electrons emitted from ^{67}Ga produce pulse height spectra with peaks that coincide closely to those produced by ^{3}H and ^{14}C (McQuarrie and Noujaim, 1983). Gallium-67, like the previously described ^{113}Sn, decays exclusively by electron capture. In addition to the emission of Auger and IC electrons, ^{67}Ga displays major gamma-ray emissions at 93, 184, 296, and 388 keV. Altogether, the atomic electrons and X-ray and gamma-ray photons interact with liquid scintillation cocktail, and detection efficiencies between 44% and 76% for ^{67}Ga are reported by Bobin et al. (2007).

Gamma-ray and X-ray photon interactions with liquid scintillation cocktail are a function of the photon energy as depicted in Fig. 6.42. Photon energies of low energy have greater probability of interaction with liquid scintillator, as the linear energy transfer (LET) of the photon is inversely proportional to its energy as illustrated in Table 1.29 of Chapter 1, Volume 1. In general, the lower the X-ray or gamma-ray energy, the higher will be the LET in the scintillation cocktail. As illustrated in Fig. 6.42, the photon interaction probabilities in liquid scintillator are high at photon energies below 100 keV. This is in agreement with the data of Zimmerman (2006a), who demonstrated very high photon interaction probabilities of >0.25 in toluene, > 0.35 in Ultima Gold scintillation cocktail, and >0.5 in Hionic Fluor for photon energies of 20 keV.

Vanadium-49 is another radionuclide that decays solely by pure electron capture. It emits X-rays and Auger electrons of very low energy (<5 keV), and no gamma radiation is emitted. Rodríguez Barquero et al. (1998) and Rodríguez Barquero and Los Arcos (2000) demonstrate that liquid scintillation analysis is the preferred method, because LSA is not affected by self-absorption problems that would otherwise be prevalent with such low-energy electron emitters. They report liquid scintillation counting efficiencies between 8% and 25% for ^{49}V.

In certain cases, the counting efficiencies of radionuclides that decay by electron capture emitting X-rays and Auger electrons may be higher with the liquid scintillation technique than those attainable with the thin-walled NaI(Tl) solid scintillation crystal detector. In the case of a radionuclide, such as ^{49}V emitting only Auger electrons and X-rays of low energy, the liquid scintillation cocktail is the optimum detector, because of its intimate contact with the radionuclide optimizing detection. Also, the liquid scintillation analyzer with its automatic sample changer and computer is more commonly encountered in laboratories than its solid scintillation counterpart. This reflects the driving force behind finding broader ranges of application of liquid scintillation counting.

The interaction of X- and gamma-ray photons with liquid scintillation cocktail (Fig. 6.42) is principally the result of the Compton effect whereby part of the energy of the X- or gamma-ray photon is imparted to orbital electrons. An ejected electron (Compton electron) imparts its energy in material in a fashion similar to that of a beta particle. The absorption of its energy by the liquid scintillation cocktail results in fluorescence with the emission of photons of visible light. In liquid scintillation cocktail, the photoelectric effect usually does not occur over 30 eV. However, the photoelectric interaction is significant and accounted for in the liquid scintillation standardization of radionuclides (particularly low-Z nuclides) decaying by electron capture and emitting low-energy X-rays (Kossert and Grau Carles, 2010, 2008; 2006; Grau Carles, 2006a,b; Grau Carles et al., 2006; Grau Malonda and Grau Carles, 2000). Bransome (1973) reports that the photoelectric effect can be evident at higher gamma-ray energies in the glass vial walls or, in the sample-scintillation cocktail mixture, if the scintillator is loaded with heavy elements.

Cherenkov photons will be produced in liquid scintillators to a significant extent if gamma-ray energies are high enough to produce Compton electrons of sufficient energy to cause the Cherenkov Effect (Grau Carles and Grau Malonda, 1992, and Grau Carles *et al.*, 1993). Radionuclide analysis by Cherenkov counting is discussed in detail in Chapter 6 of Volume 2.

Numerous studies have been undertaken on the liquid scintillation analysis of ^{55}Fe, which decays exclusively by electron capture emitting X-rays and Auger electrons of low energy, 0.6–6.5 keV. Some examples that can be cited are Augeray et al. (2015), Bergeron and Laureano-Pérez (2014), Herranz et al. (2012), Gudelis et al. (2010), Ratel (2008), Razdolescu et al. (2008), Fischer et al. (2007), Mikelic, et al. (2007), Grau Carles and Grau Malonda (2006), Grau Malonda et al. (2006), Grahek and Mačefat (2006, 2004), Warwick and Croudace (2006), Hou et al. (2005), Grau Malonda and Grau Carles (2000), Ceccatelli and De Felice (1999), Günther (1998), Ortiz et al. (1993), Grau Malonda (1982), Cramer et al. (1971), Horrocks (1971), Miller et al. (1969), Cosolito et al. (1968), Eakins and Brown (1966), Perry and Warner (1963), and Dern and Hart (1961a,b). Electron capture decay gives rise to the emission of X-rays, Auger electrons, and internal-conversion electrons, which interact with the liquid scintillation cocktail to cause fluorescence. Researchers have reported counting efficiencies of up to 62% for ^{55}Fe (Huang et al., 2016, Augeray et al., 2015, Bergeron and Laureano-Pérez (2014), Mikelic et al., 2007, Hou et al., 2005, Ortiz et al., 1993, Günther, 1998, and Grau Malonda and Grau Carles, 1999, 2000). A liquid scintillation quench curve with ^{55}Fe detection efficiencies in the range of 3%–35% for ^{55}Fe with tSIE as the quench indicating parameter was illustrated earlier in Fig. 6.27.

In general, the detection efficiencies of radionuclides that decay by 100% electron capture (EC) will be lower than the efficiencies of radionuclides that decay by EC and the alternative β^+ emission or radionuclides that decay via β^- emission, due to the generally higher energies of beta particles in comparison to the lower energies of Auger electrons, X-rays, and internal-conversion electrons resulting from 100% EC decay. For example, a comparison of the liquid scintillation detection efficiencies achievable with a low-energy beta-emitting radionuclide, such as ^{63}Ni (100% β^-, $E_{max} = 66.9$ keV), and a radionuclide decaying by 100% electron capture, such as ^{55}Fe, which emits very low-energy Auger electrons (0.61 keV Auger L and 5.2 keV Auger K and L electrons of 140% and 60% intensities, respectively) and K X-rays (5.9 keV K_α X-rays and 6.5 keV K_β X-rays of 24% and 3% intensities, respectively) is illustrated in Fig. 6.43. Although the liquid scintillation detection efficiencies of 100% EC nuclides may be lower than negatron or positron emitters, they are very adequate for excellent routine activity analysis (Bergeron and Laureano-Pérez (2014)van Wyngaardt et al., 2018, Yuan et al., 2018, Dziel et al., 2016, Augeray et al., 2015, Bergeron and Laureano-Pérez (2014), and Roteta et al., 2014, Herranz et al., 2012).

Another popular radionuclide, which decays exclusively by electron capture, is ^{125}I. The radionuclide is very useful as a tracer in the biological sciences (*e.g.*, Morissette et al., 2018, Xu et al., 2018, Kobayashi et al., 2015, Mochizuki et al., 2016, 2017, Bray et al., 2014, Robinson

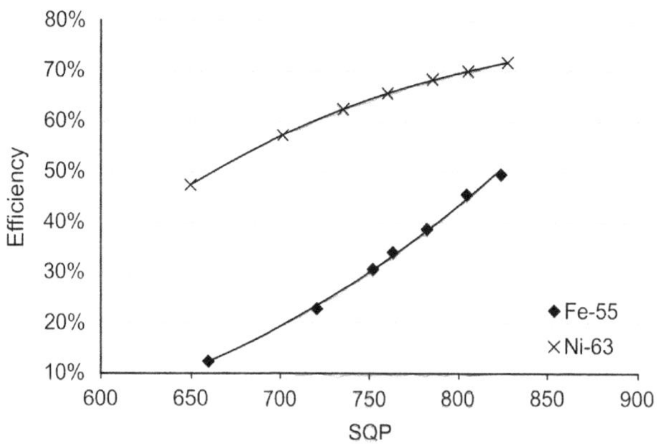

FIGURE 6.43 Quench correction curve for ^{63}Ni (low-energy β^--emitter, $E_{max} = 66.9$ keV) and ^{55}Fe (decaying by 100% EC obtained emitting low-energy Auger electrons and K-X-rays) from the quench indicating parameter (SQP). *From Herranz et al. (2012), reprinted with permission from Elsevier © 2012.*

et al., 2014, Larsson et al., 2001, Malkov et al., 2000 and Teresa et al., 2000) and liquid scintillation is a very convenient and efficient means of analysis. The electron-capture decay results in the emission of 35.5 keV gamma radiation with 6.7% intensity and the emission of 3.7 keV conversion electrons from the K shell with 78% intensity and 30.5–35.4 keV conversion electrons from the L to O shells with 13% intensity; this is accompanied with the emission of 3.7 keV L X-ray photons with 15% intensity and 27 keV K_α and 31 keV K_β X-ray photons of 112% and 20% intensities, respectively, and 3.2 keV and 22.7 KeV Auger electrons with 156% and 20% intensities, respectively (NNDC, 2018 http://www.nndc.bnl.gov/nudat2/chartNuc.jsp, accessed February 1, 2018, Grau Malonda et al., 2006, and Grau Malonda, 1999). Early studies reported a liquid scintillation counting efficiency of 56% for ^{125}I by standard liquid scintillation counting techniques (Rhodes, 1965), and later yet higher counting efficiencies of over 80% were reported by Jordan et al. (1974), Horrocks (1976c), Ring et al. (1980), Chandrasekaran (1981), Kits et al. (1985), and Grau Carles et al. (1994c). Subsequent studies by Grau Malonda and Grau Carles (2000) report counting efficiencies of over 88% for ^{125}I in Insta Gel Plus and Ultima Gold liquid scintillation cocktail. The liquid scintillation detection efficiency is dependent on the quench level, and an example of an ^{125}I quench correction curve taken from the work of Kim et al. (2006) is illustrated in Fig. 6.44.

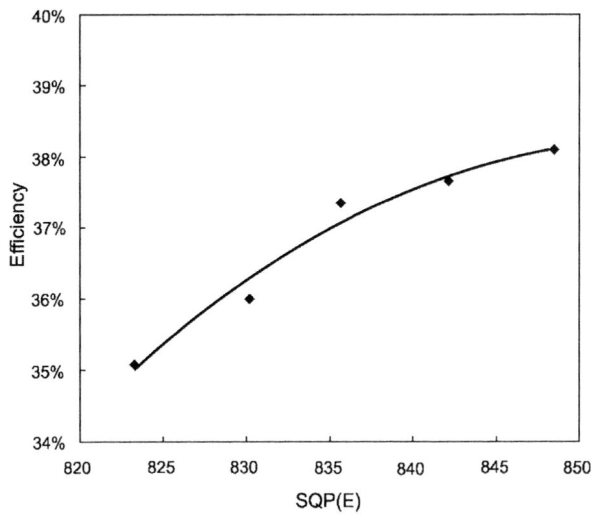

FIGURE 6.44 Quench correction curve for ^{125}I based on the external standard quench indicating parameter, SQP(E). The photon emissions from ^{125}I in scintillation cocktail are due to cocktail interactions with 35 keV gamma-ray emissions in 6.7% of the transitions, internal-conversion electrons in 91% of the transitions and abundant Auger electron and Te K X-ray emissions (NNDC, 2018, http://www.nndc.bnl.gov/nudat2/chartNuc.jsp). *From Kim et al. (2006), reprinted with permission from Radiocarbon, the University of Arizona, © 2006 Arizona Board of Regents on behalf of the University of Arizona.*

As discussed in Chapter 1, either the electron capture decay process or the emission of an internal-conversion electron leaves an orbital electron vacancy. For the case of ^{125}I this vacancy may be filled by electrons from outer shells giving rise to the emission of X-rays of the Te daughter nuclide and Auger electrons. Horrocks (1976c) explained that the electron capture process in ^{125}I involves capture of a *K*-shell electron in 80% of the decay transitions and an *L*-shell in the remaining 20%. The Te L X-ray is 3.5 keV in energy (NNDC, 2018) and totally absorbed by the liquid scintillation cocktail, whereas, the Te *K* X-ray is emitted with 27.7 keV energy and has a high probability of escape. He concluded that the excitations in the liquid scintillation cocktail are due mainly to the absorption of Auger and internal-conversion electrons, and only a minor contribution (about 8% of the fluor excitations) is the result of X-rays produced during the decay process. Liquid scintillation remains a common method of analysis for ^{125}I (Robinson et al., 2014; Tarancón Sanz and Kossert, 2011; Grau Carles and Kossert, 2009; Kossert and Grau Carles, 2008; Ogata, 2007; Grau Malonda et al., 2006; Kossert, 2006; Van Wyngaardt and Simpson, 2006a,b; Pommé et al., 2005). An interesting approach taken by Ogata (2007) was the analysis of ^{125}I in solution within a microtube inserted inside a Bicron (BC-400) plastic scintillator with dimensions suitable to fit within a standard plastic liquid scintillation vial. No liquid scintillator solution is used, there is no fluor waste to contend with, and the sample remains unadulterated and may be reused after counting in a conventional liquid scintillation counter. Ogata (2007) reported a detection efficiency of 4%–8% for the electron-capture nuclide ^{125}I and a detection efficiency of 10%–40% for the high-energy beta-emitting nuclide ^{32}P. The detection efficiency is a function of the sample volume.

Many radionuclides can decay by either electron capture or positron emission, since positron emission can complete with EC as described in detail in Chapter 1, Volume 1. Zinc-65 is a good example of such a radionuclide used in the biological sciences as a tracer (Cresswell et al., 2017; Yamunarani et al., 2016; Huntington et al., 2008 and Wolterbeek *et al.*, 2002). It decays for the most part by electron capture (98.6%), and only 1.4% of the ^{65}Zn nuclide transitions to stable ^{65}Cu occur via β^+ emission with a positron $E_{max} = 325$ keV (NNDC, 2018). Günther (1998) and Sandhya and Subramanian (1998) report liquid scintillation counting efficiencies up to 76%. Positron interaction with scintillation cocktail contributes only a small portion to the overall detection efficiency due to its low intensity of emission. About 50.5% of the transitions occur via electron capture (EC) to the ground state of ^{65}Cu and the remaining 48% by EC with accompanying gamma emission (Günther, 1998). Consequently, the abundant K Auger electrons (7 keV at 47% intensity) and L Auger electrons (1 keV at 127% intensity) and low-energy (1−9 keV) X-ray photons at 40% intensity (NNDC, 2018) that accompany the EC decay process of ^{65}Zn are the emissions that generate significant liquid scintillator excitation and light emission. Liquid scintillation analysis is employed for the standardization of ^{65}Zn (Kossert, 2006; Van Wyngaardt and Simpson, 2006b; Simpson and Morris, 2004a).

Chromium-51 decays by exclusively electron capture and 10% of the excited daughter nuclei simultaneously undergo decay to the ground state with the emission of gamma radiation of 320 keV energy with insignificant emission 0.02% intensity of 319 keV conversion electrons (NNDC, 2018). X-ray and Auger-electron emission, which accompany electron-capture decay, also must be considered among the processes that generate scintillation fluor excitation. Chromium-51 decays with the emission of 5 keV X-rays in 22% of the transitions and 4.4 keV Auger electrons in 66% of the transitions (NNDC, 2018). The double radionuclide tracer ^{59}Fe—^{51}Cr was assayed by Barosi et al. (1980) in red blood-cell kinetic studies. If the double-label is assayed by NaI(Tl) solid scintillation counting of the gamma-ray photopeaks of the two nuclides, optimum counting efficiencies of 15% and 3% are reported for ^{59}Fe and ^{51}Cr, respectively. However, if liquid scintillation counting is used, optimum counting efficiencies of 20% and 15% are reported for the ^{59}Fe and ^{51}Cr double label, respectively. The five-fold increase in the counting efficiency of ^{51}Cr is due mainly to the liquid scintillation cocktail absorption of X-ray and Auger electron energy. Work by Bradshaw et al. (2006) and Kumblad et al. (2005) provide examples of the application of ^{51}Cr as a tracer nuclide, and a 5% detection efficiency for the liquid scintillation analysis of ^{51}Cr was reported (Kumblad et al., 2005).

Positron-emitting nuclides can be assayed by liquid scintillation analysis with a high detection efficiency when positron emission is the principal mode of decay. As described in Chapter 1, Volume 1, positrons have similar interactions, ranges, and stopping powers as negatrons of similar energy; however, in addition positrons will produce annihilation radiation (0.511 keV gamma rays) when the positrons come to rest in the proximity of atomic electrons. The liquid scintillation pulse height spectra produced by positrons are similar to those of negatrons as expected (See Fig. 6.45). Roteta et al. (2006) report a liquid scintillation detection efficiency close to 100% for the positron emissions from ^{18}F. The radionuclide ^{18}F decays to the stable nuclide ^{18}O by positron emission with a positron $E_{max} = 634$ keV in 96.7% of the transitions and by electron capture in the remaining 3.3% of the transitions (NNDC, 2018). Auger electrons from the K shell are

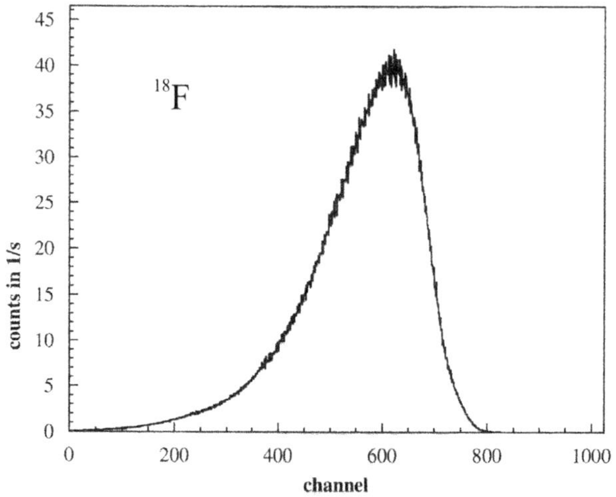

FIGURE 6.45 Pulse height spectrum of a ^{18}F sample measured with a Wallac 1414 Guardian liquid scintillation spectrometer with logarithmic amplifier. Background has been subtracted. *From Schrader et al. (2007), reprinted with permission from Elsevier © 2007.*

emitted with an energy of 0.5 keV at 3% intensity, and 0.5 keV X-rays emissions are insignificant (0.03% intensity). Positron annihilation radiation of 511 keV gamma-ray photons (See Chapter 1, Volume 1) always accompanies positron emission, and, as pointed out by Schrader et al. (2007), the liquid scintillation pulse height spectrum, illustrated in Fig. 6.45, will have some contribution from the secondary Compton electrons of the annihilation radiation. When the E_{max} of the positrons are sufficiently above the 262 keV Cherenkov threshold in water (See Chapter 6 of Volume 2), these may be analyzed by Cherenkov counting (Table 6.5).

Wiebe et al. (1980) and McQuarrie et al. (1981) measured the Compton electron contribution to the count rate resulting from the interactions of 511 keV annihilation gamma rays with the liquid scintillation cocktail. Thus, in addition to positron-scintillation cocktail interactions, they point out that as much as 24% of the observed count rate is due to scintillation cocktail interactions with Compton electrons originating from annihilation radiation. Wiebe et al. (1980) showed that the largest amount of energy that may be deposited in a liquid scintillation cocktail is equivalent to the sum of the highest positron energy and the energy deposited by the annihilation gamma rays or

$$E_{max} = E_{\beta^+} + 2E_{e^-} \tag{6.30a}$$

where E_{max} is the maximum energy deposited by the positron in the scintillation cocktail, $E_{\beta^+}$is the maximum positron energy, and E_{e^-} is the energy of the Compton edge associated with 511 keV annihilation gamma radiation (341 keV). For example, in the case of ^{18}F, $E_{max} = 635\text{ keV} + 2(341\text{ keV}) = 1317\text{ keV}$. Two times

TABLE 6.5 Liquid Scintillation (LS) and Cherenkov Counting Efficiencies of a Few Positron-emitting Nuclides[a].

Nuclide	$E_{\beta^+ max}$	Half-life	LS counting efficiency (%)	Cherenkov counting efficiency in water (%)
^{18}F	0.633 (96.7%)[a]	109.8 m	97[b]	3.7[c]
^{22}Na	0.546 (90.3%) 1.820 (0.05%)	2.60 y	85[d]	unknown
^{34m}Cl[e]	0.501 (0.3%) 1.312 (26%) 2.288 (29%)	32.0 m	100[c]	57[c]
^{34}Cl	4.46 (99.9%)[f]	1.53 s		
^{64}Cu	0.653 (17.6%)[g]	12.70 h	84[h]	unknown
^{68}Ga	0.822 (1.2%) 1.899 (87.7%)[i]	67.7 m	96[j]	47[c]

[a]Energy values are in MeV and the intensities of the decay mode are given alongside in parenthesis (NNDC, 2018).
[b]From Roteta et al. (2006).
[c]Data McQuarrie et al. (1981).
[d]In a restricted LSA counting window in the simultaneous analysis of ^{3}H, ^{14}C, and ^{22}Na. The detection efficiency would be higher in a wider counting window (Miyasawa et al., 1991).
[e]Cl-34m decays by isomeric transition (IT) with 44.6% intensity and β^+ emission with 55.4% intensity (NNDC, 2018).
[f]The counting efficiency is measured from ^{34m}Cl.
[g]Cu-64 decays via β^+ emission with 17.6% intensity, electron capture (EC) with 43.9% intensity and β^- (E_{max} = 0.579 MeV) emission with 38.5% intensity (NNDC, 2018).
[h]From Wanke et al. (2010). The counting efficiency by LSA is due to both the positron and negatron emissions.
[i]88.9% of ^{68}Ga radionuclides decay via positron emission, and the remaining 11.1% via electron capture.
[j]From Roteta et al. (2012).

the energy of the annihilation radiation Compton edge must be accounted for, because a positron annihilates with the simultaneous emission of two gamma rays of 511 keV energy.

Numerous radionuclides that decay by electron capture and/or the competing positron decay mode with the emission of gamma radiation, X-rays, and Auger and internal-conversion electrons may be measured by liquid scintillation. Among these, in order of increasing mass number, are the following: ^{18}F (Nejjari et al., 2008; Roteta et al., 2006 and others cited previously), ^{22}Na (Miyasawa et al., 1991), ^{49}V (Rodriguez Barquero and Los Arcos, 2000 and Rodriguez Barquero et al., 1998), ^{51}Cr (Bradshaw et al., 2006 and others cited previously), ^{54}Mn (Bignell et al., 2010b; Cassette et al., 2006a,b; Grau Malonda and Grau Carles, 1999), ^{55}Fe (Augeray et al., 2015; Bergeron and Laureano-Pérez, 2014 and others cited previously), ^{64}Cu (Wanke et al., 2010), ^{65}Zn (Cresswell et al., 2017 and others cited previously), ^{67}Ga (Bobin et al., 2007 and Zimmerman ey al., 2006a). ^{68}Ga (Oropesa Verdecia et al., 2018; Roteta et al., 2006), ^{85}Sr (Grau Carles et al., 1994c), ^{88}Y (Los Arcos et al., 1991), ^{109}Cd (Wolterbeek and van der Meer, 2002; Grau Carles et al., 1994c), ^{111}In (Grau Carles and Kossert, 2009), ^{113}Sn (Roteta et al. (2014), ^{123}I (Bignell et al., 2010b; Grau Carles and Kossert, 2009), ^{125}I (Morissette et al., 2018; Xu et al., 2018 and others cited previously), ^{133}Ba (Tinker and Smith, 1996), and ^{139}Ce (Van Wyngaardt and Simpson, 2006b).

VII. Common interferences in liquid scintillation counting

The counting interferences most commonly found in liquid scintillation analysis and how each can be recognized and/or corrected to obtain accurate and reproducible DPM values must be considered. Six major counting interferences exist in the scintillation counting of samples: (1) background radiation, (2) quench (color, chemical, and ionization), (3) multiple radionuclides in the same sample, (4) luminescence, (5) static, and (6) wall effect. Each of these interferences will be considered here and in other parts of this chapter with special attention given to their identification, elimination, or means of correcting for any errors that these may generate.

A. Background

Background is defined as counts arising from sources external to the sample, such as cosmic or environmental radiation, and from instrument noise and PMT crosstalk (Jonsson and Theodórsson, 2007). When determining sample count rates (CPM_s) from which sample disintegration rates (DPM_s) will be determined according to procedures described previously in Section V, it is necessary to obtain an accurate measure of the background count rate (CPM_{bkg}) whenever background counts are significant relative to the sample counts. Background count rates are determined by counting a blank counting vial containing the scintillation cocktail plus all other chemical constituents used in the preparation of samples with the exception of the radionuclide of interest. In other words, the blank should have the same matrix (quench level) as the radioactive samples to be analyzed. For example, if the radioactive samples are measured in a sample-scintillation cocktail mixture of 50% water (1:1 water load), the background count rate should be determined in the blank sample 1:1 water-scintillation cocktail mixture. Ideally any other quenching agents that may be present in the sample should also be present in the background blank counting vial. Such a blank can be obtained by preparing a sample containing no radionuclide of interest in a fashion similar to the preparation of the experimental samples (matrix matching). Once a blank is prepared, it must be counted for a sufficient period of time to get an accurate measurement of its count rate. The time required for counting background blanks can be decided by using statistical criteria presented in Chapter 8, Volume 2, and this is discussed in the later part of this chapter concerning measurements of lower limits of detection (LLD).

Once the background count rate is determined, most modern liquid scintillation analyzers store the background pulse height spectrum in computer memory. The background counts for any given counting region of the pulse height spectrum can then be subtracted automatically from the sample count rates to provide a net count rate according to the following equation:

$$CPM_{net} = CPM_s - CPM_{bkg} \tag{6.30b}$$

If the background count rate (CPM_{bkg}) is significant compared to the sample count rate (CPM_s), it is necessary to subtract the background contribution according to the above equation. The net count rate (CPM_{net}) is then used to determine disintegration rates according to Eq. (6.7) described in Section V.B of this chapter. The majority of the background counts are found in the low end of the liquid scintillation pulse height spectrum, such as 0–5.0 keV for instruments that have pulse height spectra calibrated over the energy range of 0–2000 keV.

Methods of reducing background to optimize instrument performance are provided in Section XX of this chapter.

B. Quench

Color, chemical, and ionization quench are described in detail in Section IV of this chapter. In brief, quench affects the scintillation photon intensity and efficiency of detection of radionuclides in the liquid scintillation cocktail. The lower the radiation energy of the radionuclide, the greater is the effect of quench on the counting efficiency of the sample. Four common methods of quench correction are (1) internal standardization, (2) sample spectrum quench correction, (3) external standard quench correction, and (4) Direct DPM methods. The mechanisms of quench and methods of quench correction are treated in detail in Section V of this chapter.

C. Radionuclide mixtures

Multiple radionuclides in samples can present an interference when the pulse height spectra of the two radionuclides overlap. In the case of beta-particle-emitting nuclides, this is due to the fact that all beta-emitting radionuclides produce a continuous spectrum of beta-particle energies from zero to the E_{max}, as illustrated in Fig. 1.45 of Chapter 1, Volume 1. If two beta-emitting radionuclides are present in the same sample (*e.g.*, ^{3}H and ^{14}C of E_{max} 18.6 and 156 keV, respectively), the pulse height spectrum produced by the beta particles of one of the radionuclides (*e.g.*, tritium) produces a broad spectrum of pulse heights created by a broad range of beta-particle energies from zero to the most energetic 18.6 keV beta particle, as illustrated in Fig. 6.46. The magnitude and shape of the spectrum will also depend upon the level of quench in the sample. The second radionuclide ^{14}C, which has a higher decay energy than tritium, emits beta particles of higher energy maximum (E_{max}), produces a pulse height spectrum overlapping that of tritium, and the spectrum of pulse heights produced by the beta particles emitted by ^{14}C extends from zero to a maximum produced by the most energetic ^{14}C beta particles of 156 keV. Fig. 6.46 is a graphic composite spectrum of a sample with both ^{3}H and ^{14}C. In order to quantify the separate radionuclide activities (DPM) of a mixture of two beta-emitting nuclides in a sample, the count rates and counting efficiency of each radionuclide must be determined. Several methods may be employed for

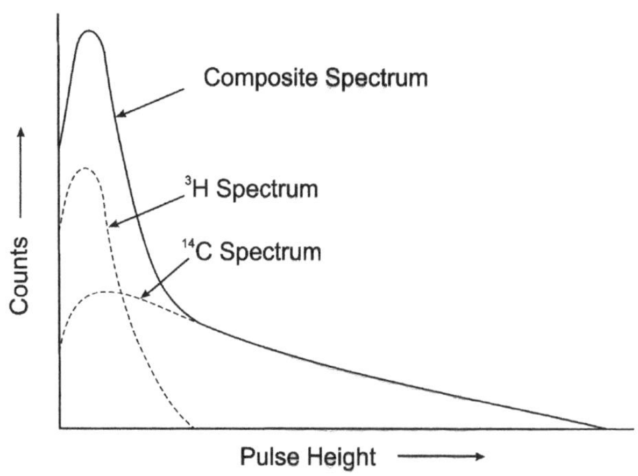

FIGURE 6.46 Overlapping liquid scintillation pulse height spectra of ^{3}H (beta particle $E_{max} = 18.6$ keV) and ^{14}C (beta particle $E_{max} = 156$ keV). The composite spectrum is that produced by the two radionuclides as a mixture. © *1998 PerkinElmer, Inc. Printed with permission.*

the measurement of two, three, or even more beta-emitting radionuclides in a mixture by liquid scintillation analysis. These methods are described in detail in Section VIII of this chapter.

D. Luminescence

Luminescence in liquid scintillation fluor cocktails or in aqueous buffer media refers to the emission of light photons as a consequence of energy absorption and concomitant molecular excitation from origins other than nuclear radiation. Luminescence can be a practical tool in the study of biochemical reactions or an interference in liquid scintillation analysis. This section will provide a treatment of the various types of luminescence encountered in liquid scintillation analysis and recommendations on how to avoid or minimize any interference that some types of luminescence can present.

1. Bioluminescence

Bioluminescence occurs when a biochemical reaction produces photons, which is a desirable reaction when used as a tool to study certain biochemical assays. Some examples are the biochemical reactions catalyzed by avidin-alkaline phosphatase, horseradish peroxidase, β-galactosidase, luciferase, xanthine oxidase, and ATP assays via luciferin-luciferase. The liquid scintillation analyzer can be utilized to count all luminescent events when these are used to study certain biochemical reactions. Under such circumstances, the LSA counting protocol should be set to count the experimental samples in the Single Photon Counting (SPC) mode. In this counting mode, only a single PMT is used without the coincidence counting circuitry. SPC is used because all luminescent events are single photon in nature and would be eliminated by the two PMTs and the coincidence counting circuitry. Therefore, in the SPC mode the coincidence counting circuit is disabled and only one PMT is operational. In addition, because bioluminescence normally produces a higher intensity of light compared to a radioactive event, the high voltage on the PMT is automatically lowered in the SPC mode to prevent saturation of the PMT. The counting region for bioluminescent samples in the liquid scintillation analyzer is generally 0−10 keV for pulse height scales calibrated in keV energy equivalence.

2. Photoluminescence and chemiluminescence

There are primarily two types of luminescence, which can interfere in the assay of radioactive samples in scintillation cocktail. These are photoluminescence and chemiluminescence.

Photoluminescence is the result of the exposure of the sample-scintillation cocktail mixture to ultraviolet light. Photoluminescence is normally single photon in nature and decays in a matter of minutes. Therefore, letting freshly prepared samples in scintillation cocktail remain in the dark of the liquid scintillation analyzer for 10−15 min completely eliminates any photoluminescence that may occur. A precount delay time can also be used with counting protocols of certain liquid scintillation analyzers.

The second type of luminescence, chemiluminescence, is a frequent interference in the liquid scintillation assay of radioactive samples. This is the production of light within the scintillation cocktail due to a chemical reaction. Chemical reactions that cause chemiluminescence often occur when scintillation cocktail is added to the sample solution in the liquid scintillation counting vial. A chemical reaction can occur, for example, when scintillation cocktail is added to a basic sample solution (pH 8−14) or when a chemical substance, such as hydrogen peroxide, is present in the sample. The pH effects and chemical interactions with some component of the scintillation cocktail can cause molecular excitation and light emission. Some types of samples that can produce a considerable chemiluminescence are tissue or cell digests with inorganic bases (*e.g.*, NaOH, KOH, Solvable) or organic bases (*e.g.*, Soluene 350).

During the chemiluminescent reaction, single photons are produced in the scintillation cocktail and, because of their large number, they may bypass the coincidence circuit and be registered as counts together with counts produced by the radionuclides in the sample. The counting of single-photon chemiluminescence events by the coincidence (dual PMT) circuitry can be demonstrated from the equation of Horrocks and Kolb (1981), which is written as

$$N_C = 2\tau_R(N_1N_2) \tag{6.31}$$

where N_C is the coincidence count rate, τ_R is the coincidence resolving time (*e.g.*, 30 ns), and N_1 and N_2 are the single-event count rates from photomultiplier tubes 1 and 2, respectively. As the single-photon events from chemiluminescence increase in frequency, the probability that they produce a coincidence count rate correspondingly increases according to Eq. (6.31). Horrocks and Kolb (1981) calculated the coincidence count rate, N_C, that would be measured by a liquid scintillation analyzer with a resolving time of 30 ns for various single-photon event levels according to Eq. (6.31). For example, they demonstrated that, if photomultiplier tubes 1 and 2 measured the same number of single photon events to be 50,000 (*i.e.*, $N_1 = N_2$), the coincidence count rate, N_C, would be calculated to be

$$N_C = \left[2\left(30\times10^{-9}\right)/60\right](N_1N_2) = 10^{-9}(N_1N_2) = 2.5 \text{ CPM} \tag{6.32}$$

They elaborate further that, if $N_1 = N_2 = 500{,}000$, the coincidence count rate, N_C, would calculate to be 250 CPM, and for $N_1 = N_2 = 5{,}000{,}000$, the coincidence count rate, N_C, would calculate to be 25,000 CPM. Thus, it is obvious from the above calculation that if sample count rates are low, luminescence can be a source of serious error. Luminescence count rates can be very high, depending on the chemical constituents and/or pH of the sample. In any event, luminescence should be identified, if it occurs, and it should be eliminated before counting to avoid error in the activity determinations of experimental samples.

Luminescence, which may occur in sample-scintillation cocktail mixtures, can be detected easily when counting relatively high-energy β-emitting radionuclides ($E_{max} > 70$ keV). For example, Fig. 6.47 illustrates the pulse height spectrum of a luminescent sample, a tritium sample, and a ^{14}C sample. As illustrated in Fig. 6.47, the luminescence spectrum occurs at the very low-energy portion of the pulse height spectrum occurring generally in the 0–6 keV region for pulse height spectra on an energy equivalent scale. Luminescence in a sample can be recognized easily by one of three methods, namely, (1) spectral display, (2) counting region settings, and (3) instrumental detection and measurement. As illustrated in Fig. 6.47, a sample containing appreciable chemiluminescence produces a pulse height spectral peak in the 0–6 keV region on top of (overlapping) the main radioactivity pulse height peak. The figure illustrates the chemiluminescence pulse height spectrum overlapping with ^3H and ^{14}C pulse height spectra. When counting single-radionuclide samples, two counting regions can be used to detect luminescence. For example, when counting ^{14}C the following counting regions may be used: region A: 0–156 keV and region B: 4.0–156 keV. If the counts in these regions are similar, then little if any luminescence would be expected. If, on the other hand, the counts in region A (0–156 keV) are much higher (25%–500%), luminescence may be predicted. In addition, many liquid scintillation analyzers are able to determine the magnitude of luminescence in a sample and calculate percent luminescence on the same page as the CPM and DPM values of each sample. The percent luminescence is calculated by the instrument according to the equation

$$\% \text{ luminescence} = \frac{\text{chance coincidence events}}{\text{true coincidence events}} \times 100 \tag{6.33}$$

When luminescence is detected or even suspected in experimental samples, it can be controlled, corrected for, or even eliminated as discussed subsequently.

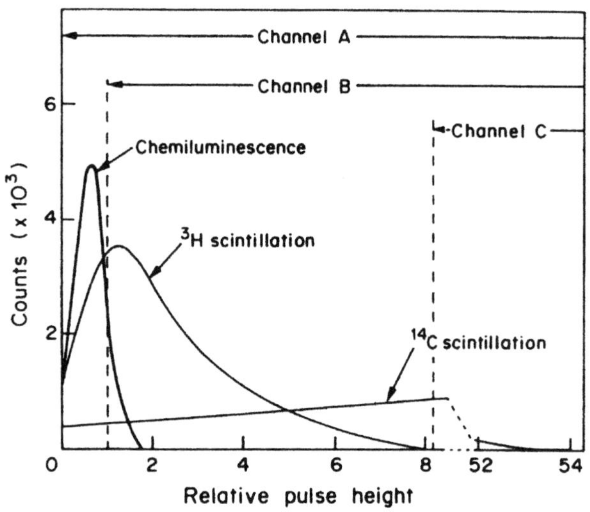

FIGURE 6.47 Pulse height distributions of ^3H and ^{14}C scintillation pulses and of chemiluminescence pulses, and channel settings for the analytical measurement of each activity and chemiluminescence count rate. *From Takiue et al. (1985), reprinted with permission from Elsevier ©1985.*

3. Luminescence control, compensation, and elimination

Once luminescence is recognized as a problem with a particular set of samples, it can be controlled by using proper sample preparation procedures and even eliminated or compensated for by certain liquid scintillation counter controls. The five most common methods of reducing or correcting sample luminescence are (1) dark-adaptation of samples, (2) chemical methods, (3) temperature control, (4) counting region settings, and (5) delayed coincidence counting also referred to as random coincidence counting. A liquid scintillation analyzer equipped with three photomultiplier tubes (*e.g.*, Hidex 300 SL) that can operate in coincidence is reported to avoid sample luminescence (Haaslahti, 2010).

a. Dark-adaptation of samples

The most straightforward method of removing luminescence from liquid scintillation sample-cocktail mixtures is to dark-adapt the samples overnight or until the luminescence decays to the point of not contributing to erroneous sample activities (Warwick et al., 2010; Wiesenberg et al., 2010; Cheng and Wong, 2008; Ren and Mahon, 2007; Warwick and Croudace, 2006; and Groppi et al., 2005a). Photoluminescence generally decays rapidly, and dark adaptation of the samples for an hour or two should suffice for photoluminescence to disappear. The liquid scintillation analysis of samples of low activity require low background counting, and dark adaptation of samples overnight or for more than a day is a practice. Chemiluminescence has a much slower decay rate, and significant levels may persist beyond a day depending on sample temperature (Kessler, 1989). Dark adaptation of samples for 3 days in the liquid scintillation analyzer was reported by Yamada et al. (2004) to suppress chemical luminescence. Some liquid scintillation cocktails are manufactured to be resistant to chemiluminescence (See Part b. Chemical Methods discussed subsequently and Chapter 7, Volume 1). Dark-adaptation of samples exhibiting chemiluminescence, together with other methods described subsequently, may be used to eliminate interference from this phenomenon.

b. Chemical methods

Chemical methods used to avoid or suppress chemiluminescence are reviewed by Peng (1976). Among these, neutralization of alkaline sample solutions with a nonoxidizing acid is recommended, as basic sample solutions are often the main cause of chemiluminescence. This can be accomplished by (1) adding 10 mL of acetic acid to a gallon (3.8 L) of scintillation cocktail, providing this does not alter the performance of the fluor cocktail; (2) neutralizing the basic sample solution before adding the scintillation cocktail to the sample counting vial; or (3) using a special scintillation cocktail designed to reduce or suppress luminescence such as Insta-Fluor, which contains a chemiluminescence inhibitor or Hionic-Fluor, which displays a very fast chemiluminescence decay property (See Chapter 7, Volume 1 for scintillation cocktail properties and characteristics). This normally reduces the amount of chemiluminescence. A problem associated with sample neutralization may occur when counting large macromolecules found in certain biological samples, in which acidification of samples may cause precipitation. Precipitation of sample most often includes precipitation of the radioactive material and, therefore, loss of counting efficiency and incorrect DPM values.

c. Temperature control

Acidification of sample solution, as already described, followed by heating to 40°C is often recommended to reduce chemiluminescence, or acidification may be omitted when it causes precipitation and the sample solution only heated. Heating the sample to 40°C, helps drive the chemiluminescence reaction to its endpoint. This is possible because the reaction is a chemical reaction and every 10°C increase in temperature doubles the reaction rate. Another procedure, alternative to heating, is to cool the reaction using a liquid scintillation analyzer with temperature control. Cooling slows the reaction rather than accelerating its termination. Cooling reduces counts from chemiluminescence. In addition, chemiluminescence can be eliminated altogether by counting-region setting or by delayed-coincidence counting, discussed subsequently in this section.

d. Counting-region settings

This is the recommended method for all radionuclides with the exception of tritium and some nuclides that decay by electron capture (EC) to stable daughter nuclides with the emission of X-rays and Auger electrons. The use of "counting-region settings" to remove pulse events from luminescence is not possible in the measurement of tritium or EC nuclides emitting only X-rays and Auger electrons (*e.g.*, ^{37}Ar, ^{49}V, ^{68}Ge, and ^{131}Cs), because their liquid scintillation pulse height spectra overlap significantly with that of luminescence (See Fig. 6.47 for the case of tritium.). Auger electrons and X-rays following electron capture are of low energy, as noted in Chapter 1, Volume 1, and the liquid scintillation pulse heights from the monochromatic Auger electrons or from X-ray Compton electrons would occur in the same counting region as tritium (Takiue and Ishikawa, 1979a). Nuclides that decay by electron capture with the emission of gamma radiation from unstable daughter nuclei emit also internal-conversion electrons that produce generally pulse heights significantly higher than that of tritium. Consequently, the method of "counting-region settings", which is described in the following paragraph, could be applicable in these cases.

The method of "counting-region settings" to remove erroneous luminescence pulse events from the sample count rate is relatively simple, but not often used. If the counting region for a mid-to high-energy β-emitting radionuclide (*e.g.*, ^{14}C and higher energy β-emitters) is set at approximately 10 keV or above, no luminescence of any kind will be observed in the sample counts and, therefore, no correction will be necessary. The only precaution when performing DPM determination is to set the same counting region for the quench correction curve and for the experimental samples. A counting region for ^{14}C to avoid error from luminescence would be, for example, 10.0−156 keV for lower level and upper level discriminator settings when using instruments that have pulse height discriminator settings calibrated to keV energy equivalence. In the case of tritium, a counting region cannot be set to avoid luminescence, because tritium emits very low energy beta particles (E_{max} = 18.6 keV) producing a pulse height spectrum that greatly overlaps that of luminescence. In the case of tritium, chemiluminescence must be eliminated or measured and subtracted from the sample counts for accurate activity calculations (Takiue et al., 1984). Fig. 6.47 illustrates other discriminator settings to define three counting regions, proposed by Takiue et al. (1985), which permit the simultaneous counting of chemiluminescence, ^{3}H, and ^{14}C, when the DPM analysis of dual-radionuclide samples containing chemiluminescence is required. Equations for calculating the activities (DPM) of the dual-radionuclide samples and the count rates due to chemiluminescence are given by Takiue et al. (1985, 1986). Detailed information on multiple-radionuclide analysis is provided in Section VIII of this chapter.

e. Delayed coincidence counting

Most liquid scintillation analyzers are equipped with delayed coincidence counting, also referred to as random coincidence counting, as a method for the elimination of error resulting from luminescence that can be applied to the liquid scintillation analysis of all radionuclides including tritium (Varlam et al., 2009; Palomo et al., 2007). For this method to work, a delayed coincidence circuit with a 20-ns delay is added to one of the two PMTs. The sample containing the radioactivity is counted simultaneously with and without the delay coincidence circuit enabled. If the coincident circuit is used without the delay or random coincidence circuit, both the luminescence and radioactive decays will be detected. The radioactive decay events are accepted by the coincidence counting circuit, because they are isotropic multiphoton events, and luminescent single-photon events are also accepted by the coincidence counting circuit, because they occur at a high count rate in the sample and are detected within the resolving time of the counting circuit. Now, if the coincident circuit is used with a delay mode added to one of the PMTs, only the single-photon luminescence events will be detected, because of their high count rate of occurrence. Radioactive decay, which produces isotropic multiphoton events, will not be detected by the coincidence circuit due to a 20-ns delay in one of the two PMTs. Finally, the counts collected from the two readings with and without the delay circuit enabled are subtracted channel by channel of the MCA over the pulse height region equivalent to 0–6.0 keV. The resultant spectrum will be a product of the actual nuclear decay events without chemiluminescence. This special delay method is known as the luminescent detection and correction method. It can be applied to all radionuclides independent of radioisotope decay energy. Luminescence detection and correction are available with most state-of-the-art liquid scintillation analyzers (Kessler, 1989); however, caution is recommended in the application of this instrumental method of correction. The user should confirm that the luminescence subtraction is working properly for their samples. For example, Warwick and Croudace (2006) confirmed that the delayed coincidence counting option was not reliable for their particular samples, and they decided that dark-adaptation of the samples was necessary to ensure against erroneously high activities due to chemiluminescence-derived scintillations.

E. Static

Electrostatic discharge is a photon-producing interference in liquid scintillation counting. Static electricity may be generated by friction or pressure between two materials. When nonconductive materials are separated, one material develops a positive and the other a negative charge. Static consists of charged ions, positive or negative, which are atoms electrically out of balance due to the removal or addition of electrons. The intensity of static electricity can be measured as positive or negative voltage on the surface of matter in magnitudes of tens of thousands of volts. The discharge of static electricity is a random event, but the phenomenon commonly occurs with many materials we may come into contact with when preparing samples for counting, including scintillation counting vials. Static electricity is produced easily in low-humidity rooms during the time of the year when dry heat is used to warm laboratories. A common characteristic of static electricity is its stability; it can remain on the surface of scintillation counting vials for relatively long periods of time. When the scintillation vial is placed in the counting chamber and electrostatic discharges occur, it is like an electrical lightning storm occurring in or on the surface of the scintillation vial producing random pulse events. The static charge buildup can have many causes, including shipping, handling, use of plastic gloves, and low humidity in the sample preparation area. Plastic vials tend to build up more of a static charge than glass vials.

There are primarily four methods of reducing or eliminating static from sample vials for liquid scintillation counting: (1) the use of an electrostatic controller, (2) selection of vial type, (3) antistatic wipes, and (4) humidification of the sample preparation and counting area. A brief description of each method is given.

The electrostatic controller is a circular donut-shaped device located in the elevator tube through which the counting vial must pass before it is moved into the counting chamber. In certain instruments it contains eight geometrically located electrodes, which generate a 360-degree field of electrically produced ions. When the counting vial passes through the electrostatic controller, it enters the field of electrically produced counterions, which can neutralize static electricity on the counting vial surface in a matter of 2 s. This process occurs just before the robotic positioning of the vial into the counting chamber located between the two PMTs. Contemporary state-of-the-art liquid scintillation analyzers are equipped with an electrostatic controller (*e.g.*, Mendonça et al., 2006). Although the electrostatic controller offers no guarantee of removing all static from the counting vial surface, there may be no need to take any other steps to control static on the surface of counting vials before placing them in the LSA sample changer.

Further measures may be taken to guard against static charge collection and discharge from counting vial surfaces. One step is to select a type of counting vial that would tend to collect less static electricity. Because plastic tends to hold a static charge more than glass, the use of glass vials helps to reduce the static charge for most samples. The disadvantages of using glass vials is that they are more expensive, more difficult to dispose of, and can yield higher background count rates. The alternative to glass vials is to use special "antistatic" plastic vials (Nakanishi et al., 2009). These vials are manufactured with a special plastic treatment that greatly reduces the amount of static on the vial surfaces compared with standard plastic vials. Yamada et al. (2004) reports that the use of Teflon counting vials with acryl resin caps with the inside wall of the caps lined with aluminum foil could suppress effectively the static electricity. Yoon et al. (2010) took the added precaution of locating the liquid scintillation analyzer 30 cm away from any other electrical devices, ventilation ducts, or air conditioner. Another technique is to wipe each vial with an "antistatic" wipe or with a moist cloth just before placing the vial into the sample changer of the liquid scintillation counter. This readily removes the static charge on the surface of the vial just before counting. The final step that may be taken to reduce static is to increase the humidity in the room where the samples are prepared as well as in the counting area.

F. Wall effect

The 'wall effect' can be classified into two types, namely, (1) the effect of certain traditional organic solvents that penetrate into the walls of plastic counting vials, and (2) a loss of detection efficiency in the liquid scintillation standardization of radionuclides due to the interaction of beta particles with the glass walls of the scintillation counting vial. These two types of 'wall effect' are encountered only in unique modes of liquid scintillation analysis and are thus treated separately as follows:

1. **Wall effect from solvent penetration into plastic counting vial**. When samples are counted in plastic vials with traditional cocktails, the organic scintillator from the cocktail can penetrate the wall of the plastic vial. Traditional cocktails are those made with solvents such as toluene, xylene, and pseudocumene. This can cause a problem when external standard quench correction methods are used, because certain plastic vials with solvent penetration can scintillate causing a distortion of the external standard pulse height spectrum. The result would be inaccurate quench indicating parameters, which would give erroneous counting efficiencies and, as a consequence, error in the DPM measurements of samples. This problem can be overcome easily by always using environmentally safer cocktails. The newer environmentally safer cocktail solvents, such as diisopropylnaphthalene and linear alkylbenzene will not penetrate into the wall of the plastic vial and cause the wall effect. As a general rule, always use glass vials when using traditional scintillation cocktails, and use either plastic or glass vials with the environmentally safer cocktails.
2. **Wall effect in the liquid scintillation standardization of radionuclides**. In the liquid scintillation standardization of radionuclides, discussed further on in this chapter, calculations may include a wall effect, that is, a reduction in detection efficiency due to nuclear decays that occur at the walls of the scintillation vial (Cassette, 2003). In the wall effect it is assumed that a 10-μm layer next to the vial wall will have only a 50% counting efficiency, as electrons interact with the glass envelope of the counting vial (Jaubert and Cassette, 2004; Johansson et al., 2003; Cassette, 2003). The influence of this wall effect is considered negligible in the standardization of high-energy beta-emitting radionuclides [e.g., ^{89}Sr ($E_{max} = 1490$ keV) or ^{90}Y ($E_{max} = 2280$ keV)] by the CIEMAT/NIST or TDCR methods, if the source is unquenched or only moderately quenched (Jaubert and Cassette, 2004; Cassette, 2003). For lower-energy beta-emitting radionuclides and for alpha-emitting nuclides, the wall effect must be corrected for in the standardization of radionuclides (Cassette, 2003). In the liquid scintillation standardization of ^{238}Pu an uncertainty of 0.25% on the efficiency was attributed to the wall effect (Johansson et al., 2003). A similar uncertainty to the detection efficiency of 0.2% for ^{210}Po was attributed to the wall effect (Laureano-Pérez et al., 2007). The significance of the wall effect was demonstrated by Hurtado et al. (2009) by their comparison of the theoretical beta spectrum of ^{90}Y with the spectrum that is produced by the ^{90}Y beta-particle energy deposited in a liquid scintillator (See Fig. 6.48). The comparison was made using a Monte Carlo simulation using GEANT4 code, as a tool for efficiency calculations in radionuclide standardization and stopping power calculations, etc. Fig. 6.48 illustrates how the low-energy portion of the beta spectrum is distorted by the wall effect, as the low-energy beta

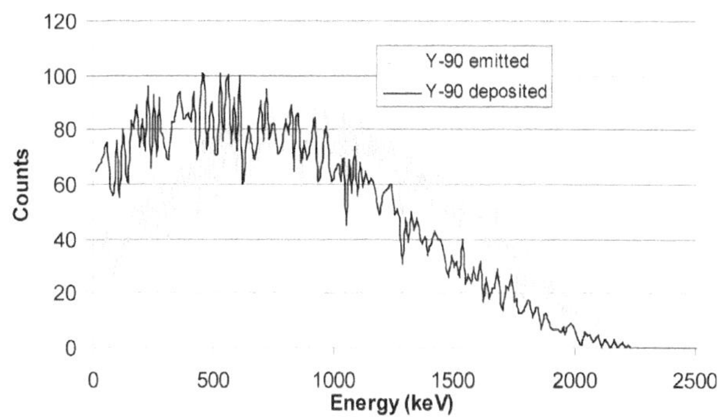

FIGURE 6.48 Theoretical and wall-effect-distorted beta spectra of ^{90}Y. *From Hurtado et al. (2009), Reprinted with permission Radiocarbon, the University of Arizona, © 2009 Arizona Board of Regents on behalf of the University of Arizona).*

particles are absorbed in the walls of the liquid scintillation counting vial. DeVries and Griffen (2008) point out that the high LET of alpha particles in the scintillation cocktail reduce the wall effect, and assigned a liquid scintillation alpha detection efficiency of 100% in the measurement of ^{237}Np.

VIII. Multiple radionuclide analysis

There is a wide-ranging need in the scientific community to analyze multiple radionuclides as mixtures. These include dual-, triple-, and multiple radionuclide mixtures stemming from research in the chemical, biological, nuclear power, and environmental sciences, among others, as reviewed by L'Annunziata (1984b), Bukowski et al. (1992), Takiue et al. (1991b,c, 1992, 1995, 1999), Toribo et al. (1995, 1996, 1997, 1999), Fujii et al. (2000), Kashirin et al. (2000), Nayak (2001) and Reddy et al. (2009). Several methods are available for analyzing the activity of more than one β-emitting radionuclide in the same sample. Because of the broad spectrum of β-particle energies emitted by any given radionuclide anywhere between zero and E_{max}, we always observe a broad pulse height spectrum in the liquid scintillation analyzer from zero to a maximum pulse height. Therefore, all liquid scintillation pulse height spectra from different β-emitting radionuclides overlap to some degree. Because of this spectral overlap and the very broad characteristics of β-particle pulse height spectra, it was traditionally considered feasible to analyze by LSA at most three β-emitting radionuclides in the same sample provided their β-particle energy maxima differed by a factor of three or four (L'Annunziata, 1979, 1984a,b, and 1987). However, advances in LSA have revealed new regionless spectral unfolding and deconvolution methods capable of analyzing several β-emitting radionuclides in the same sample with the aid of computer processing. Even radionuclide mixtures of ^{14}C ($E_{max} = 156$ keV) and ^{35}S ($E_{max} = 167$ keV), which for decades were thought to be impossible to resolve by LSA, because of their similar β-particle energies, can now be identified and quantified as mixtures. A description of the techniques used to resolve and quantify mixtures of β-, and α-emitting radionuclides by LSA is provided in this section.

A. Conventional dual- and triple-radionuclide analysis

The conventional methods described in detail in this section refer, for the most part, to the analysis of two β-emitting radionuclides in a mixture. However, the same principles can apply to the analysis of three beta-emitting radionuclides as a mixture, which is included in this section.

1. Exclusion method

The exclusion method is one of the original methods applied to the analysis of a dual-radionuclide mixture by liquid scintillation analysis. It is described in detail by Okita et al. (1957), Kobayashi and Maudsley (1970) and L'Annunziata (1979). The technique is rarely used today, because of the availability of more efficient methods of dual- or triple-radionuclide analysis. Nevertheless, it is presented briefly here, as the reader will encounter occasional reference to this method in the current literature.

The dual radionuclide mixture of ^{3}H ($E_{max} = 18.6$ keV) and ^{14}C ($E_{max} = 156$ keV) will be taken as an example to describe this method, which requires a relatively large difference in β-particle energies between the two radionuclides. It is recommended generally that the E_{max} of the two radionuclides differ by a factor of 3 or 4. Two counting regions or windows are defined using lower level (LL) and upper level (UL) pulse height discriminators such that in counting region 1, also referred to as counting region A and in this case, we can refer to it as the tritium region (*e.g.*, LL = 0.0 and UL = 18.6 keV), where both the tritium spectrum and the spilldown of the carbon-14 spectrum into the tritium region are found. In this example, the pulse height analyzer is one that utilizes a multichannel analyzer with channel numbers calibrated to approximate keV energy of the ^{3}H and ^{14}C beta particles (See Fig. 6.49). In such a case, a nonquenched sample of ^{3}H will not produce calibrated pulse heights beyond 18.6 keV. In counting region 2, also referred to as region B and in this case the carbon-14 region, pulse height discriminators are set (*e.g.*,

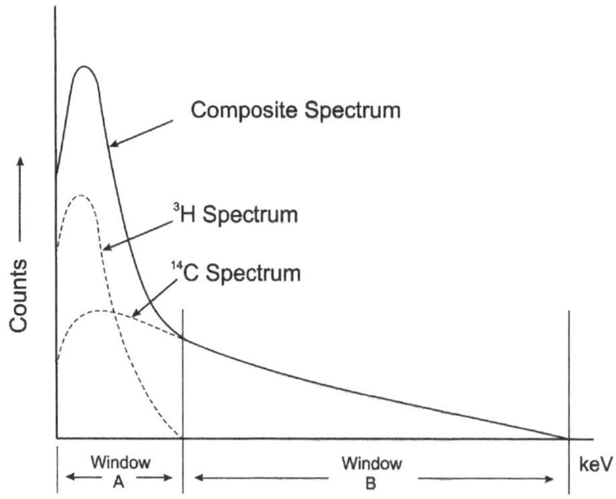

FIGURE 6.49 Typical component and composite liquid scintillation analyzer pulse height spectra produced by approximately equal activities of ^{3}H and ^{14}C. Two counting windows (counting regions) are illustrated, namely, Window A (or Region A), which registers count rates originating from beta emissions of ^{3}H and ^{14}C; whereas, Window B (or Region B) registers a count rate originating only from beta emissions of ^{14}C. *© 1998 PerkinElmer, Inc. Printed with permission.*

LL = 18.6 and UL = 156 keV) to allow only pulses from the carbon-14 spectrum. The name exclusion method is derived from the fact that region 2 excludes all tritium pulses. When counting a mixture of ^{3}H and ^{14}C, count rates (CPM) will be collected in each region, and the following equations are used to calculate the activity or disintegration rate (DPM) for ^{3}H and ^{14}C:

$$H = \frac{N_1 - C_{c_1}}{h_1} \tag{6.34}$$

and

$$C = \frac{N_2}{c_2} \tag{6.35}$$

where H and C are the activities or DPM of ^{3}H and ^{14}C, respectively, in the mixed radionuclide sample, N_1 and N_2 are the net count rates (CPM), *i.e.*, background-subtracted count rates, in regions 1 and 2 of the liquid scintillation analyzer, h_1 and c_1 are the counting efficiencies of ^{3}H and ^{14}C in counting region 1, and c_2 is the counting efficiency of ^{14}C in counting region 2.

From these equations, it is clear that five parameters are needed to calculate the DPM for both the tritium and ^{14}C dual-labeled samples. The net CPM in regions 1 and 2 are determined automatically by the LSC. The three efficiency factors are determined using three quench correction curves, which consist of plots made from a series of tritium and ^{14}C quenched standards. The three curves are constructed by plotting % counting efficiency for tritium and ^{14}C in region 1 and ^{14}C in region 2 versus an external standard quench indicating parameter.

This method is less commonly used today for low-energy beta-emitters, such as ^{3}H$-$^{14}C, because scintillations from beta-particle emissions from both nuclides are highly quenchable, and the counting region of the higher-energy beta-emitter (*e.g.*, ^{14}C in the above example) is restricted to include only pulse events from that nuclide. Under high levels of quench, the detection efficiency of ^{14}C drops significantly, and the pulse events can spilldown excessively from region B to region A. This can yield very low detection efficiencies for the ^{14}C in Region B; although, modern liquid scintillation analyzers equipped with automatic efficiency control (AEC) or automatic quench compensation (AQP) can automatically shift counting region settings according to quench level. With respect to relatively high-energy beta-emitters, such as ^{90}Sr (E_{max} = 546 keV) and ^{90}Y (E_{max} = 2280 keV), which are less easily quenched, the exclusion method is more easily utilized. Eikenberg et al. (2011) report the successful application of this technique for the analysis of ^{3}H and ^{14}C mixtures. They report the use of either internal standards to determine the detection efficiencies of ^{3}H and ^{14}C or quench correction curves. Eikenberg et al. (2011) found the use of internal standards to yield more precise results. (See Section V.A of this chapter for the use of internal standards.) Xiques Castillo et al. (2009, 2010) applied the exclusion method for the analysis of ^{90}Sr(^{90}Y) for clinical applications. They used counting regions defined by channels 5−700 for Region A and 701−1014 for Region B of a Wallac (PerkinElmer) 1029 Rackbeta liquid scintillation analyzer. The upper region (Region B) registers pulse events arising exclusively from ^{90}Y beta particle interactions with liquid scintillator, and these interactions with such a very high-energy beta-emitter are not easily quenched and any spilldown of pulse events from Region B to Region A would be minimal. Another application of the exclusion method is reported by Huntington et al. (2008) for the dual-nuclide liquid scintillation analysis of ^{65}Zn and ^{69m}Zn. A modified version of this method was applied by Feng et al. (2014a) for the analysis of ^{90}Sr and ^{90}Y. They measured the sample vial twice, without and with the addition of a small volume of ^{90}Y. The difference of both measurements was used to establish the activities of both radionuclides, correcting the quenching effect, as long as it is low. They also report that if the activity added is higher than the total initial activity the error in the measurement is lower than 3%. The inclusion method, which is described subsequently, is the most popular method, as it assures, particularly with low-energy beta-emitters, a less variable detection efficiency for both nuclides over a wide range of quench.

2. Inclusion method

In this method the counting regions are set such that there are spillup and spilldown of pulse events in each region from both of the radionuclides when two beta-emitting radionuclides are analyzed as a mixture. When three beta-emitting radionuclides are analyzed, three counting regions are employed.

a. Dual-radionuclide analysis

Again, we will use ^{3}H and ^{14}C as an example of a typical dual mixture, keeping in mind that the procedure presented here and equations used will work for any dual mixture of β-emitting radionuclides provided the β-particle energies (E_{max} values) of the two radionuclides differ significantly. Also, in this discussion we will refer to lower-energy and higher-energy radionuclides to distinguish between nuclides that differ in their beta-decay energies. In this example, ^{3}H (E_{max} = 18.6 keV) is the lower-energy radionuclide and ^{14}C (E_{max} = 156 keV) the higher-energy radionuclide of the mixture.

For this method to work it is necessary that two counting regions (regions A and B) are established by setting the lower level and upper level pulse height discriminators to assure that there will be significant spillup of pulse events from ^{3}H into region B and the unavoidable

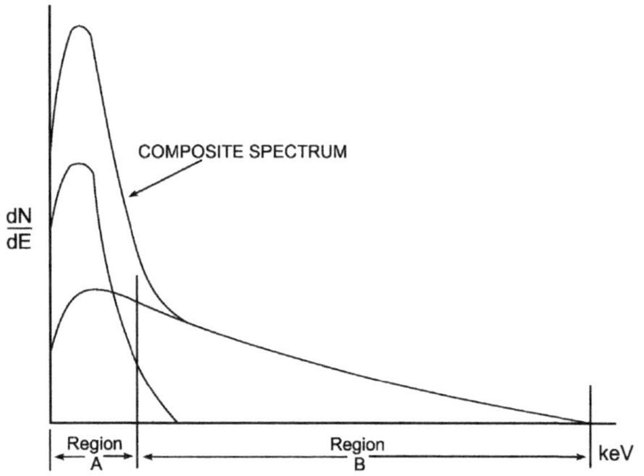

FIGURE 6.50 Typical component and composite pulse height spectra observed for two β-emitting radionuclides (*e.g.*, ^{3}H and ^{14}C) in an approximately 1:1 mixture and having significantly different β-particle energy maxima. The pulse height spectra are plotted as dN/dE, *i.e.*, the number of events or counts per energy channel versus channels of a multichannel analyzer calibrated in keV. Two counting regions are illustrated (Regions A and B also referred to as Windows A and B) for use in the inclusion method, which are set to allow spillup of the lower energy radionuclide from Region A into Region B, whereby counts from both radionuclides appear in both counting regions. © *1998 PerkinElmer, Inc. Printed with permission.*

spilldown of pulse events from ^{14}C into region A. An illustration of pulse height discriminator settings, which establish the counting regions for the analysis of two radionuclides by this inclusion method is given in Fig. 6.50.

Some modern liquid scintillation analyzers are available with preset dual counting regions for the activity analysis of dual mixtures such as $^{3}H-^{14}C$, $^{3}H-^{32}P$ and $^{3}H-^{125}I$. For other radionuclide combinations, it is necessary to establish the LL and UL discriminators settings for the appropriate spillup and spilldown of pulse events from the two radionuclides. The procedure used to establish these regions will be discussed later on in this section. For the case of the $^{3}H-^{14}C$ mixtures, counting region A is normally set by discriminators LL = zero and UL = 12.0 keV, while region B is defined by the discriminator settings LL = 12.0 and UL = 156 keV, when the pulse height spectra are displayed on a linear scale in β-particle energy equivalents (*e.g.*, Fig. 6.50). Some liquid scintillation analyzers simply make use of channel numbers, and the counting regions are defined by the channel numbers (See Fig. 6.56 further on in this section).

After the two counting regions are defined, it is necessary to prepare quench correction curves, which can be used to determine the counting efficiencies of the ^{3}H and ^{14}C (or lower-energy and higher-energy radionuclides) in the two counting regions. Two sets of quench standards are required, one set of ^{3}H and one set of ^{14}C quenched standards. If two other radionuclides need to be analyzed, a set of quenched standards of the lower-energy radionuclide and a set of quenched standards of the higher-energy radionuclide are required. The procedure for preparing a series of quenched standards was given in Section V.D of

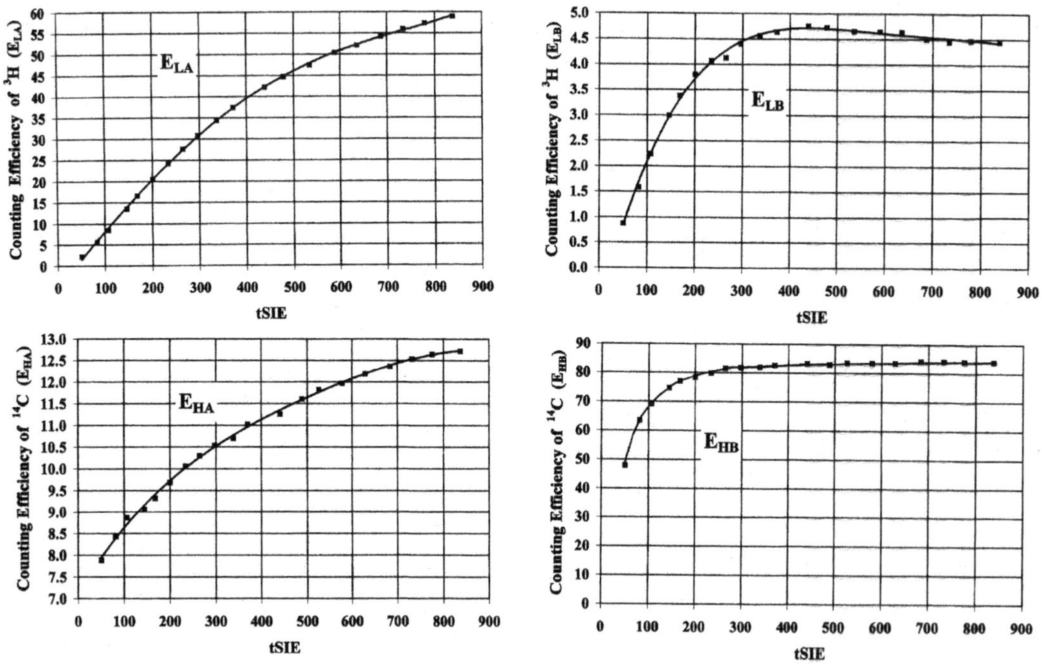

FIGURE 6.51 Quench correction curves of counting efficiency versus tSIE/AEC for ^{3}H and ^{14}C in region A (LL-UL: 0–12.0 keV) and region B (LL-UL: 12.0–156 keV) for the dual radionuclide analysis of $^{3}H-^{14}C$. The notations E_{LA}, E_{LB}, E_{HA}, and E_{HB} are the counting efficiency factors defined in Eqs. (6.36) and (6.37). The quench correction curves were obtained with ^{3}H and ^{14}C quenched standards counted in regions A and B with a PerkinElmer 2770TR/SL liquid scintillation analyzer.

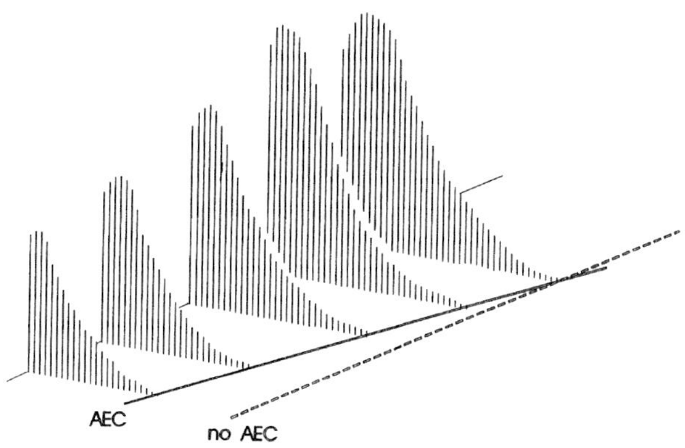

FIGURE 6.52 Illustration of automatic region tracking. Using automatic efficiency control (AEC), the liquid scintillation analyzer automatically moves the upper level discriminator of a counting region from pulse heights of higher magnitude to those of lower magnitude (right to left of the pulse height spectrum) according to the degree of quench in a sample. The pulse height spectra of five samples are illustrated, each at different levels of quench. The samples of higher quench level are those of smallest pulse number and magnitude. © *1998 PerkinElmer, Inc. Printed with permission.*

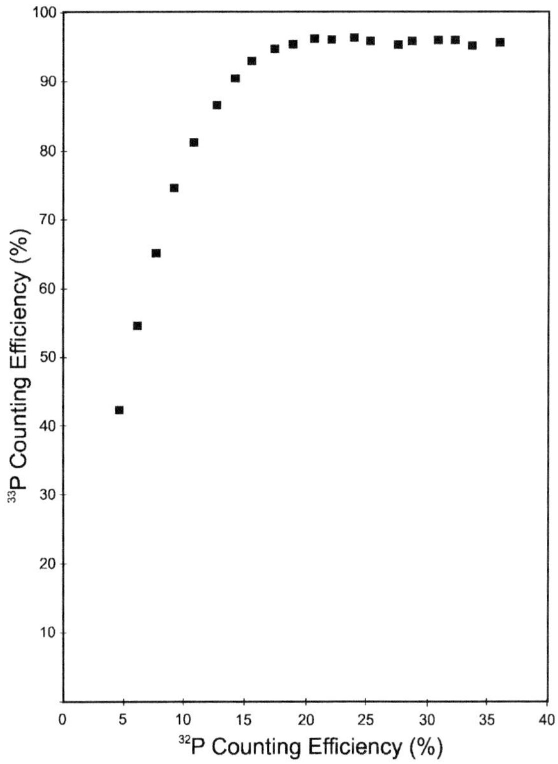

FIGURE 6.53 Effect of region settings on ^{33}P and ^{32}P counting efficiencies over the range of LL-UL, 0−30 to 0−220 keV. The data points represent increasing counting region widths in increments of 10 keV. *From L'Annunziata (1997b), © 1997 PerkinElmer, Inc. Printed with permission.*

this chapter. Each series of quenched standards is counted in regions A and B and, as a result, four quench correction curves are created, as illustrated in Fig. 6.51.

When using quench curve correction, determination of the radionuclide activities becomes more difficult as spillover of ^{14}C into the tritium region (region A) increases with quench or, in other words, when the ratio of ^{14}C to tritium increases in region A. If the number of counts from ^{14}C into the tritium region becomes large, the correction of the tritium counts can result in a small number that is less accurate to measure. Likewise, as quench increases, the spillover of ^{3}H pulse events into region B (^{14}C region) diminishes and can even disappear, which makes the calculations for the ^{3}H and ^{14}C activities invalid. Therefore, to maintain optimal counting conditions, it is necessary to keep the amount of spillover of the ^{14}C pulse events in the tritium region A at a fairly constant level as well as the spillup of tritium events into region B. This is accomplished using an automatic windows tracking method called AEC (automatic efficiency control) or AQC (automatic quench compensation). As the sample is counted, the liquid scintillation analyzer determines the level of quench of the experimental sample using a quench indicating parameter (*e.g.*, *H#*, or tSIE), and the counting regions are adjusted automatically so that the spillover of the ^{14}C into the tritium region (Region A) is kept at a fairly constant level (10%−15%), and the spillup of ^{3}H into the ^{14}C region (Region B) is also preserved. Fig. 6.52 illustrates that the liquid scintillation pulse height spectrum of a radionuclide diminishes as quench increases, and AEC automatically moves a discriminator setting according to the degree of quench in the sample. If dual radionuclide samples of tritium and ^{14}C are counted using quench curves and automatic window tracking methods (*e.g.*, AEC or AQC) are employed, the resultant quench correction curves shown in Fig. 6.51 can be used to determine the counting efficiency of each radionuclide in each counting region. As illustrated, four curves are created, one for each of the two nuclides in each of the two counting regions. The major feature to note is that the amounts of spilldown of ^{14}C into the ^{3}H region and the spillup of ^{3}H into the ^{14}C region are kept constant.

The equations used to calculate the DPM for each radionuclide are derived from the following equations, which describe the count rate in the two counting regions:

$$CPM_A = D_L E_{LA} + D_H E_{HA} \tag{6.36}$$

and

$$CPM_B = D_L E_{LB} + D_H E_{HB} \tag{6.37}$$

where CPM_A and CPM_B are the net (background-subtracted) count rates of a dual radionuclide sample in regions A and B, respectively; D_L and D_H are the disintegration rates (DPM) of the lower-energy radionuclide (*e.g.*, ^{3}H) and higher-energy radionuclide (*e.g.*, ^{14}C), respectively; E_{LA} and E_{LB} are the counting efficiencies of the lower-energy radionuclide in regions A and B, respectively; and E_{HA} and E_{HB} are the counting efficiencies of the higher-energy radionuclide in regions A and B, respectively. The four counting efficiency factors in Eqs. (6.36)

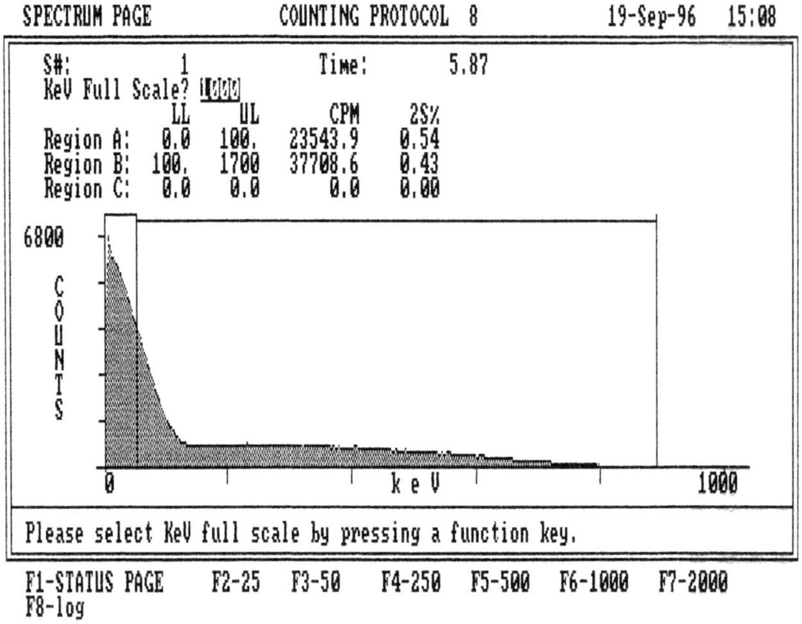

FIGURE 6.54 A composite pulse height spectrum of a ^{33}P—^{32}P dual-radionuclide sample as displayed on the computer screen of a PerkinElmer Tri-Carb 2300 TR liquid scintillation analyzer. Counting regions A and B are set for the dual-radionuclide analysis by the conventional inclusion method. (L'Annunziata, M.F., 1996, unpublished work of the writer.)

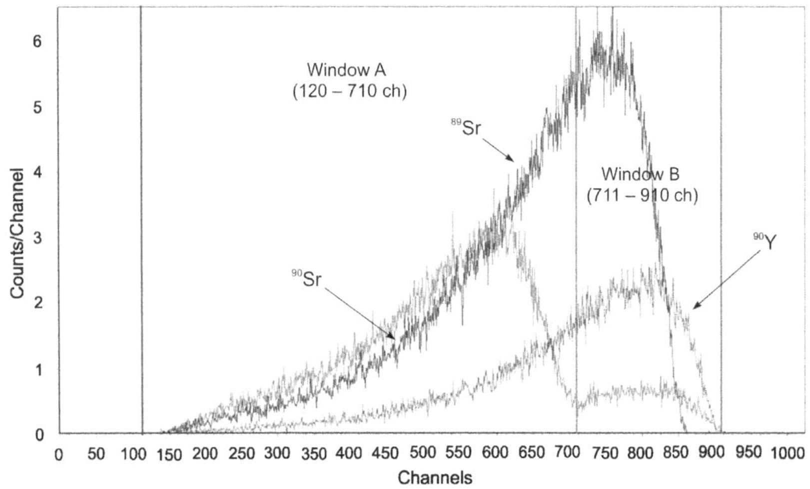

FIGURE 6.55 The liquid scintillation beta pulse height spectra of ^{89}Sr, ^{90}Sr, and ^{90}Y. Two counting regions or windows are illustrated, namely, Window A defined by lower-level (LL) and upper-level (UL) discriminator settings set at channel 120–710, respectively, and Window B with LL and UL discriminator settings at channel 711–910, respectively. A relatively weak and broad peak of pulse events due to ^{90}Y ingrowth with ^{90}Sr parent can be seen in Window B. *From Kim et al. (2009), reprinted with permission from Elsevier © 2009.*

and (6.37) are obtained automatically by the liquid scintillation analyzer from the four quench correction curves (Fig. 6.51) stored in the computer memory of the LSA. Therefore, the preceding two equations still have two unknowns, namely, D_L and D_H, which are solved for simultaneously to obtain

$$D_L = \frac{CPM_A E_{HB} - CPM_B E_{HA}}{E_{LA} E_{HB} - E_{LB} E_{HA}} \tag{6.38}$$

and

$$D_H = \frac{CPM_B E_{LA} - CPM_A E_{LB}}{E_{LA} E_{HB} - E_{LB} E_{HA}} \tag{6.39}$$

For the activity determinations of a dual nuclide mixture, the liquid scintillation analyzer determines the net count rates of the sample in regions A and B, and the quench indicating parameter of the sample. From the value of the quench indicating parameter, the LSA automatically

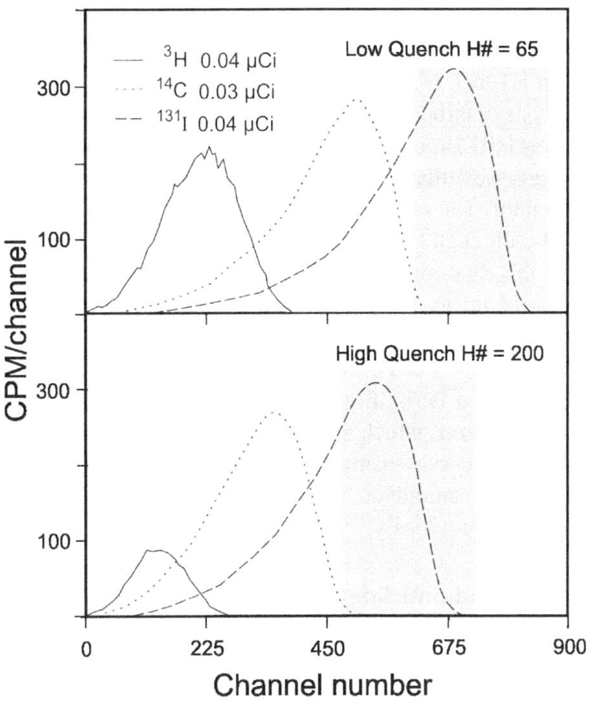

FIGURE 6.56 Liquid scintillation pulse height spectra of ^{3}H, ^{14}C, and ^{131}I at low and high quench levels. As quench increases the pulse height spectra are reduced in intensity (fewer counts/per channel) and pulse height (shift from higher to lower channel numbers). *From Bukowski et al. (1992). Reprinted with permission from Elsevier © 1992.*

extracts the needed four counting efficiency factors from the quench correction curves (*e.g.*, Fig. 6.51) and then automatically calculates the disintegration rates of the two radionuclides in the mixture according to Eqs. (6.38) and (6.39). All of the calculations are performed by the computer of the liquid scintillation analyzer.

Because the discriminator settings defining the counting regions A and B will move automatically according to the amount of quench in the sample, as measured by the QIP, it is important to note that varying the counting region settings changes the quench correction curves (efficiency vs. QIP). This is generally of little concern to the analyst, because most modern liquid scintillation analyzers save, on the hard disk of the computer, the entire pulse height spectra of the quenched radionuclide standards. Consequently, as the counting regions are changed automatically by the instrument according to quench level, so are the resultant new quench correction curves automatically determined by the instrument.

Often it is necessary to analyze radionuclide mixtures for which no preset counting regions have been established in the liquid scintillation analyzer. In such a case, it is necessary to find and select the optimum LL and UL discriminator settings to define counting regions A and B. The procedure used to obtain the proper discriminator settings is as follows, using the ^{33}P—^{32}P dual radionuclide mixture as an example:

1. To find the appropriate LL and UL discriminator settings for the inclusion method for dual-radionuclide activity analysis, we must count firstly a known activity (DPM) of the lower-energy radionuclide (e.g., ^{33}P, $E_{max} = 249$ keV) as a pure radioisotope sample in a wide range of counting regions starting at a LL discriminator setting of zero and progressively increasing the UL discriminator in small increments (e.g., 10 eV). Certain liquid scintillation analyzers (e.g., PerkinElmer Tri-Carb) display the pulse height discriminator settings in units of eV (0—2000 eV); others (e.g., PerkinElmer Quantulus) display the discriminator settings in channel numbers (0—1024). A net count rate (CPM) for each region setting is recorded.
2. This procedure is repeated with a known activity (DPM) of the higher-energy radioisotope (e.g., ^{32}P, $E_{max} = 1700$ keV), which is counted in the same regions selected in the above step 1. The net count rates (CPM) of the higher energy radionuclide in the counting regions are recorded. Both samples of the low-energy radionuclide (e.g., ^{33}P, $E_{max} = 249$ keV) and the higher-energy radionuclide (e.g., ^{32}P, $E_{max} = 1700$ keV) used for this exercise should be at similar and low levels of quench, that is, the lowest level of quench expected for any given unknown mixture.
3. The counting efficiencies of the separate low- and high-energy radionuclides (e.g., ^{33}P and ^{32}P) standards in the various counting regions are then calculated according to the equation %E = (CPM/DPM)(100).
4. The counting efficiencies of the low-energy radionuclide in the individual counting regions are plotted against the counting efficiencies of the high-energy radionuclide in the same counting regions as illustrated in Fig. 6.53.
5. The objective of this exercise is to find discriminator settings at which there will be significant overlap (spillover) of counts from both radionuclides in the two counting regions required for the inclusion method. These conditions are found in the "knee" section of the curve. Hence, a counting region is selected arbitrarily from the knee section of the curve as the counting region A for the dual radionuclide analysis. In the example using the ^{33}P and ^{32}P radionuclide standards, a counting region (LL-UL) of 0—100 keV from the knee of the curve was selected from the eighth data point counting from left to right of Fig. 6.53.
6. Having defined in this way one of the counting regions (region A), we can then proceed to select the discriminator settings for the second counting region (region B). The discriminator settings for region B are defined by selecting its LL discriminator setting to be equivalent to the upper limit of region A (e.g., 100 keV for the ^{33}P—^{32}P double label) and selecting the UL

discriminator setting to encompass all pulses of highest magnitude arising from the higher energy radionuclide. In this example the UL discriminator setting of 1700 keV was selected, because no pulses from this double-radionuclide mixture can reach beyond 1700 keV. The second counting region (region B) for the case of the $^{33}P-^{32}P$ dual radionuclide was therefore defined by the LL and UL discriminator settings of 100−1700 keV.

Fig. 6.54 illustrates the composite pulse height spectrum of a dual-isotope sample of $^{33}P-^{32}P$ with an approximate 1:1 activity ratio as seen on the LSA computer screen. The discriminator settings established for regions A and B, as required for the analysis of these two isotopes in the same sample, are also seen in Fig. 6.54. Data illustrating typical recoveries obtainable by this inclusion method for the analysis of a dual nuclide combination of $^{33}P-^{32}P$ is provided in Table 6.6. The inclusion method has been found to provide excellent recoveries of the $^{3}H-^{14}C$, $^{3}H-^{45}Ca$, $^{14}C-^{36}Cl$, $^{33}P-^{32}P$, $^{55}Fe-^{59}Fe$, $^{90}Sr-^{90}Y$ and other double radionuclide mixtures, for a wide range of activity ratios of the two radionuclides (Hui et al., 2012, Nebelung et al., 2009, Nakanishi et al., 2009, Reddy et al., 2009, Shaffer and Langer, 2007, Rodríguez et al., 2006, Zheng and Bobich, 2004, Zheng et al., 2004, Lee et al., 2002b, Benitez-Nelson and Buesseler, 1998, L'Annunziata, 1984b, 1987, 1997b, Viteri and Kohaut, 1997, and Kessler, 1989). Reddy et al. (2009) made extensive tests of this dual nuclide analysis method for $^{3}H-^{14}C$ combinations at low-activity levels. They obtained minimum detectable activities of 2100 DPM/L for ^{3}H and 1200 DPM/L for ^{14}C with a counting time of 300 min and found the accuracy of results to be within ±10% for ^{3}H:^{14}C activity ratios ranging from 1:1 to 1:18. Fujii and Takiue (2001) report the unique analysis of airborne ^{3}H and ^{14}C in activity concentrations as low as 0.01 Bq/cm^3 by suspension of the radionuclides in a "foggy scintillator" created with an ultrasonic wave generator. The radionuclides could be analyzed as single or dual-radionuclide mixtures. Nebelung et al. (2009) evaluated this dual-radionuclide analysis technique with activity ratios of up to 1/50 for several combinations of nuclides and, including $^{3}H/^{14}C$, $^{14}C/^{60}Co$, $^{14}C/^{90}Sr$, $^{14}C/^{99}Tc$, $^{137}Sc/^{60}Co$, $^{55}Fe/^{3}H$, and $^{99}Tc/^{60}Co$ with good accuracy. They used a PerkinElmer TriCarb 3100 TR liquid scintillation analyzer, which requires the establishment of quench correction curves using the quench indicating parameter, tSIE, with automatic efficiency control (tSIE/AEC), as described previously.

b. Dual-radionuclide analysis with daughter ingrowth

When two beta-emitting radionuclides are analyzed in two counting regions or windows, and there occurs ingrowth of a radionuclide daughter during the analysis, we have essentially the need to analyze simultaneously three radionuclides in the two counting windows. An excellent example is that of the liquid scintillation analysis of ^{89}Sr (E_{max} = 1490 keV) and ^{90}Sr (E_{max} = 546 keV) in two counting windows after the separation of ^{89}Sr and ^{90}Sr from other nuclides in the sample by chemical means, such as ion exchange. When the strontium is separated from other nuclides on an ion-exchange column, activity of the ^{90}Y

TABLE 6.6 DPM Recoveries for $^{33}P-^{32}P$ Dual Nuclide Combinations[a] by the Dual Region Inclusion Method.

Sample[b]	Ratio[c]					% Recovery[d]	
^{33}P:^{32}P	^{33}P:^{32}P	SIS	tSIE	DPM(^{33}P)	DPM(^{32}P)	DPM(^{33}P)	DPM(^{32}P)
1.0 mL: 0.0 mL	1:0	130	470	63,056	188	99.5	0.3
0.0 mL: 1.0 mL	0:1	1039	476	28	63,047	0.0	98.1
0.5 mL: 0.5 mL	1:1	589	472	32,426	32,867	102.2	99.1
1.0 mL: 0.5 mL	2:1	419	444	63,311	32,602	100.0	98.3
0.5 mL: 1.0 mL	1:2	707	445	31,699	65,550	100.1	98.8
1.0 mL: 0.2 mL	5:1	279	457	65,012	13,352	102.6	100.7
0.2 mL: 1.0 mL	1:5	869	460	12,744	65,011	100.6	98.0
1.0 mL: 0.1 mL	10:1	215	464	63,499	6613	100.2	99.7
0.1 mL: 1.0 mL	1:10	936	466	6770	65,194	106.9	98.3

[a]*Activity of ^{33}P = 63,361 DPM/mL, Activity of ^{32}P = 66,328 DPM/mL. Instrumentation: PerkinElmer TriCarb 2300 TR liquid scintillation analyzer.*
[b]*Triplicate samples were counted for a duration of 10 minutes or until a % 2 sigma standard deviation of 1.0% was reached.*
[c]*Ratios listed are only approximate. The exact ratios can be calculated from the isotope activities (DPM/mL) and sample sizes (mL) of each isotope.*
[d]*The standard deviation of the mean of all triplicate measurements ranged from 0.13% to 2.77%.*
From L'Annunziata (1997b) © 1997 PerkinElmer, Inc. Printed with permission.

daughter of ^{90}Sr will appear immediately after the isolation of strontium, and the ^{90}Y activity will increase with time due to daughter nuclide ingrowth until the ^{90}Y activity reaches the activity of the ^{90}Sr parent nuclide, *i.e.*, secular equilibrium (L'Annunziata, 1971 and Fig. 1.144 of Chapter 1). Thus, the liquid scintillation analysis of radiostrontium, *i.e.*, ^{89}Sr and ^{90}Sr, in two counting windows must also include the measurement of the high-energy ^{90}Y ($E_{max} = 2280$ keV) daughter activity, which will produce liquid scintillation events in the same counting windows within which the ^{89}Sr and ^{90}Sr events are counted. Thus, in this case, there is the need to analyze the activities of three beta-emitting radionuclides in two counting regions. The measured activity of the ^{90}Y will be a function of the ^{90}Sr parent nuclide activity and the time in hours since the time of isolation or separation of the strontium from other nuclides (t_0) and the time of measurement or counting (t_1). The recorded time of separation of strontium from other radionuclides and the time of counting will provide the duration in hours during which ^{90}Y ingrowth would have occurred. The calculated ingrowth of ^{90}Y allows for the mathematical correction of the contribution of ^{90}Y to the sample count rates in the two counting windows and permits the final calculated activities of ^{89}Sr and ^{90}Sr.

An excellent example of the dual counting region method with correction for daughter nuclide ingrowth, described in the previous paragraph, was reported by Kim et al. (2009) for the analysis of ^{89}Sr and ^{90}Sr in milk, which is based on the method of Eikenberg et al. (2006). The method of Eikenberg et al. (2006) utilized three counting windows to include the simultaneous analysis of an additional radionuclide, ^{85}Sr, which is used as a tracer nuclide to calculate the chemical recovery of strontium. Kim et al. (2009) isolated strontium including the ^{89}Sr and ^{90}Sr from milk using a column of ion-exchange resin that removed all other radionuclides including the ^{90}Y daughter nuclide of ^{90}Sr. The time of collection of the radiostrontium was taken as t_0, which is the time that the ^{90}Y daughter nuclide of ^{90}Sr is removed and when ingrowth of ^{90}Y with its parent will begin. The isolated radiostrontium was counted in a liquid scintillation analyzer in two counting windows illustrated in Fig. 6.55. When counting a sample of radiostrontium, *i.e.*, ^{89}Sr and ^{90}Sr, isolated from other nuclides in the counting windows, illustrated in Fig. 6.55, Window A will yield a total count rate, which will be sum of count rates from pulse events arising from beta emissions of ^{89}Sr + ^{90}Sr + ingrown ^{90}Y + background events; whereas Window B will yield a summed count rate arising from beta emissions from ^{89}Sr + ingrown ^{90}Y + background events.

Using the notation of Eikenberg et al. (2006), Kim et al. (2009) calculated the net (background-subtracted) count rates in the two counting windows A and B, which are defined as

$$N(t)^A = N(t)^A_{\text{Sr-89}} + N(0)^A_{\text{Sr}-90} + N(t)^A_{\text{Y}-90} \quad (6.40)$$

and

$$N(t)^B = N(t)^B_{\text{Sr-89}} + N(t)^B_{\text{Y}-90} \quad (6.41)$$

where in Eq. (6.40), $N(t)^A$ is the net count rate (CPS) in Window A, $N(t)^A_{\text{Sr-89}}$ is the net count rates in Window A due to ^{89}Sr at the time t of counting (^{89}Sr can undergo appreciable decay from the time of separation from the other nuclides to the time of counting; half-life of ^{89}Sr is 50.5 days), $N(0)^A_{\text{Sr}-90}$ is the count rate due to ^{90}Sr in Window A at the time of counting and the time of separation from other nuclides (^{90}Sr undergoes no observable decay; half-life of ^{90}Sr is 28.8 years), and $N(t)^A_{\text{Y}-90}$is the count rate in Window A due to ingrowth of ^{90}Y at the time t of counting. Also in Eq. (6.41), $N(t)^B$ is the net count rate (CPS) in Window B, $N(t)^B_{\text{Sr-89}}$ is the net count rates in Window B due to ^{89}Sr at the time t of counting, and $N(t)^B_{\text{Y}-90}$ is the count rate in Window B due to ingrowth of ^{90}Y at the time t of counting. According to Eq. (6.7) described earlier in this chapter, it is obvious that the count rate (e.g., CPS) of a radionuclide is the product of the activity or disintegration rate (DPS) and detection efficiency ε of the nuclide, where $0 \leq \varepsilon \leq 1$ for detection efficiencies in the range of 0% and 100%. Thus, Kim et al. (2009) expressed Eqs. (6.40) and (6.41) as

$$N(t)^A = A(t)_{\text{Sr}-89} \cdot \varepsilon^A_{\text{Sr}-89} + A(0)_{\text{Sr}-90} \cdot \varepsilon^A_{\text{Sr}-90} + A(0)_{\text{Sr}-90} \cdot f_1 \cdot \varepsilon^A_{\text{Y}-90} \quad (6.42)$$

and

$$N(t)^B = A(t)_{\text{Sr}-89} \cdot \varepsilon^B_{\text{Sr}-89} + A(0)_{\text{Sr}-90} \cdot f_1 \cdot \varepsilon^B_{\text{Y}-90} \quad (6.43)$$

where $A(t)_{\text{Sr}-89} \cdot \varepsilon^A_{\text{Sr}-89}$is the product of the activity in disintegrations per second (DPS) of ^{89}Sr at the time t of counting and the detection efficiency of ^{89} Sr in counting Window A, $A(0)_{\text{Sr}-90} \cdot \varepsilon^A_{\text{Sr}-90}$is the product of the activity of ^{90}Sr at the time of its separation from ^{90}Y and other radionuclides and the detection efficiency of ^{90}Sr in counting window A, and $A(0)_{\text{Sr}-90} \cdot f_1 \cdot \varepsilon^A_{\text{Y}-90}$ represents the contribution to the count rate in Window B arising from the ingrowth of ^{90}Y from the time t_0 of separation of the ^{90}Sr from yttrium to the time t of counting and

$$f_1 = \left(1 - e^{-\lambda_{\text{Y}-90} t_1}\right) \quad (6.44)$$

where $\lambda_{\text{Y}-90}$ is the decay constant of ^{90}Y and $t_1$is the interval of time from the time of separation of ^{90}Sr from ^{90}Y, t_0, and the time t of counting, i.e., $t_1 = t - t_0$. Thus, the term $A(0)_{\text{Sr}-90} \cdot f_1$ of Eq. (6.43) represents the activity of the ^{90}Y ingrown with the isolated ^{90}Sr, and the activity of the ^{90}Y will grow according to the time interval t_1 to a maximum activity (A_{max}) equivalent to that of isolated ^{90}Sr (L'Annunziata, 1971), or

$$A_{Y-90} = A_{Y-90,\,\max}\left(1 - e^{-\lambda_{Y-90}t_1}\right) \\ = A(0)_{Sr-90}\left(1 - e^{-\lambda_{Y-90}t_1}\right) \quad (6.45)$$

Eq. (6.45) s equivalent to that expressed by Eq. (1.574) of Chapter 1.

From Eqs. (6.42) and (6.43), Kim et al. (2009) arrive at the final expressions for calculating the activities of ^{90}Sr and ^{89}Sr in units of Bq/L when the strontium recovery, r_{sr}, and sample volume, V, are included, which are

$$A(0)_{Sr-90} = \frac{\left(N(t)^A - A(t)_{Sr-89}\cdot \varepsilon^A_{Sr-89}\right)}{\left(\varepsilon^A_{Sr-90} + f_1\cdot \varepsilon^A_{Y-90}\right)\cdot r_{sr}\cdot V} \quad (6.46)$$

$$A(t)_{Sr-89} = \frac{\left(N(t)^B - A(0)_{Sr-90}\cdot f_1\cdot \varepsilon^B_{Y-90}\right)}{\varepsilon^B_{Sr-89}\cdot r_{sr}\cdot V} \quad (6.47)$$

Quench correction curves are not needed in this case to determine the detection efficiencies of ^{90}Sr, ^{89}Sr, and ^{90}Y in the two counting windows, because all samples are prepared uniformly with the same level of quench, and minor variations of quench level have negligible effect on the detection efficiencies of these relatively high-energy beta-emitters. The determination of the detection efficiencies by the use of internal standards of ^{90}Sr, ^{89}Sr, and ^{90}Y, as carried out by Kim et al. (2009) does suffice in this case.

c. Triple-radionuclide analysis

In triple-radionuclide analysis (*e.g.*, a mixture of ^{3}H + ^{14}C + ^{32}P or ^{3}H + ^{14}C + ^{131}I), three counting regions and three sets of quench standards would be required to prepare quench correction curves for each radionuclide in each counting region. Thus, three CPM values, one for each of the three counting regions, are obtained and three equations are solved to calculate the DPM of each of the three radionuclides. For example, Bukowski et al. (1992) describe the liquid scintillation analysis of mixtures of ^{3}H, ^{14}C, and ^{131}I in three counting windows, namely, Window 1 defined by channel numbers 5–300, Window 2 defined by channel numbers 300–550, and Window 3 defined by channel numbers 550–800. The channel numbers (LL and UL discriminator settings) are selected by determining the pulse height spectra of the lowest-quenched sample of each individual radionuclide with the LSA multichannel analyzer, as illustrated in Fig. 6.56. The upper limits of the pulse height spectra of the lowest-quenched samples were selected as the upper-level (UL) discriminator settings for each counting window, as illustrated in Fig. 6.56. Modern liquid scintillation analyzers equipped with automatic efficiency control (AEC) or automatic quench compensation (AQP) will automatically shift the upper-level discriminator setting to lower channel numbers according to the level of quench in the sample, as the uppermost levels of the spectra shift with quench.

When three sets of quenched standards, *i.e.*, quenched standards for each radionuclide are counted in each of the three counting windows, separate quench correction curves of % detection efficiency versus a quench indicating parameter for each radionuclide are obtained in each counting window, as illustrated in Fig. 6.57. In the example given here the detection efficiency of ^{3}H in Windows 2 and 3 are zero, and the detection efficiency of ^{14}C in Window 3 is zero. This is so, because there is no spillover of ^{3}H into Window 2 or 3 as well as no spillover of ^{14}C into Window 3. However, other counting windows can be selected to permit some spillover of pulse events from the ^{3}H into Window 2 or the ^{14}C into Window 3, as the three equations

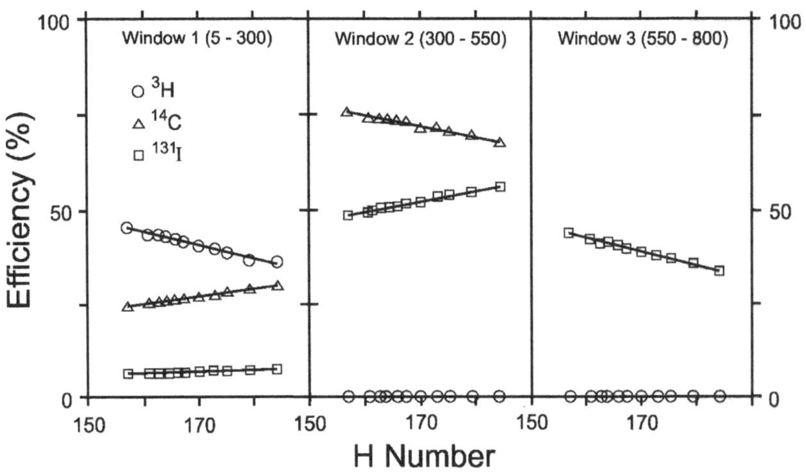

FIGURE 6.57 Quench correction curves of % detection efficiency versus the quench indicating parameter, *H#*, for ^{3}H, ^{14}C, and ^{131}I in three counting windows defined by LL and UL discriminator settings at channel numbers 5–300 for Window 1, 300–550 for Window 2, and 550–800 for Window 3. The detection efficiency of ^{3}H is zero in Windows 2 and 3, and the detection efficiency of ^{14}C is zero in Window 3. *From Bukowski et al. (2009). Reprinted with permission from Elsevier © 1992.*

used to calculate the radionuclide activities can be solved with spillover.

For the analysis of experimental samples, the liquid scintillation analyzer (LSA) will determine the count rate in each counting window, the quench indicating parameter [*e.g.*, H#, tSIE, SQP(E)] of that sample and then determine the nine detection efficiencies for the three radionuclides in each counting window. The LSA will also determine the background count rate in each counting window. With this data collected by the LSA, the following equations can be analyzed according to the notation used by Bukowski et al. (1992) to calculate the activity or disintegration rate of each of the three radionuclides in the sample:

$$\begin{aligned} W_1 &= e_{11}D_1 + e_{12}D_2 + e_{13}D_3 + B_1 \\ W_2 &= e_{21}D_1 + e_{22}D_2 + e_{23}D_3 + B_2 \\ W_3 &= e_{31}D_1 + e_{32}D_2 + e_{33}D_3 + B_3 \end{aligned} \tag{6.48}$$

Bukowski et al. (1992) used a standard Gaussian elimination matrix inversion method to solve the above three equations for D_j, which is the activity (DPM) of radionuclide j, where W_i is the count rate (CPM) in Window i, e_{ij} is the counting efficiency of radionuclide j in Window i, and B_i is the background count rate (CPM) in counting Window i. For the calculation of the activities of three radionuclides utilizing three counting windows according to Eq. (6.48), there is the need for nine counting efficiency values, one for each of the three radionuclides in each counting window (Nebelung et al. (2009). Fig. 6.57 illustrates eight of the nine quench correction curves, the missing ninth curve is that of ^{14}C in Window three; this curve would be similar to that of ^{3}H in Window 3, when there is no spillover of pulse events from ^{14}C into Window 3.

For another excellent example of a triple-radionuclide analysis (^{3}H, ^{22}Na, and ^{36}Cl) of this type and the calculations and quench curves involved, the reader should refer to Schneider and Verbrugge (1993). Nebelung et al. (2009) tested the above method of triple-radionuclide analysis with a PerkinElmer TriCarb 3100 TR liquid scintillation analyzer requiring the preparation of quench correction curves based on the quench indicating parameter tSIE/AEC. They obtained excellent recoveries from the following combination of radionuclides with activity ratios of up to 1/50: $^{55}Fe/^{3}H/^{14}C$, $^{55}Fe/^{14}C/^{99}Tc$, $^{3}H/^{14}C/^{60}Co$, $^{3}H/^{14}C/^{137}Cs$, $^{3}H/^{14}C/^{90}Sr$, and $^{14}C/^{99}Tc/^{60}Co$.

Triple-radionuclide liquid scintillation analysis is utilized for the analysis of radiostrontium, because ^{89}Sr and ^{90}Sr are encountered generally together, and with that comes the unavoidable ^{90}Y daughter nuclide of ^{90}Sr. An exception would be pure $^{90}Sr(^{90}Y)$, which is used in nuclear medicine free of ^{89}Sr as a source of ^{90}Y for clinical applications. The liquid scintillation analysis of ^{89}Sr and

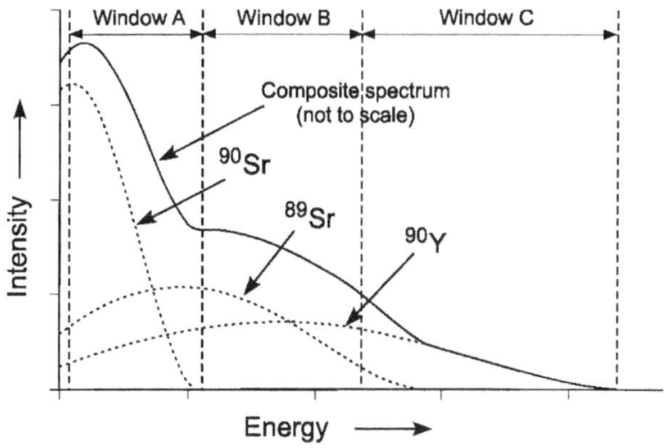

FIGURE 6.58 Approximate relative beta distributions of equal activities of ^{90}Sr, ^{89}Sr, and ^{90}Y and the approximate composite spectrum resulting from a mixture of equal activities the three radionuclides. *Sketched from the work of Chobola et al. (2006). See reference for more precise distributions. (Reprinted with kind permission from Springer Nature © 2006).*

$^{90}Sr(^{90}Y)$ utilizing three counting regions or counting windows was studied in detail by Chobola et al. (2006). They calculated the theoretical beta distributions for ^{90}Sr ($E_{max} = 546$ keV), ^{89}Sr ($E_{max} = 1490$ keV), and ^{90}Y ($E_{max} = 2280$ keV) and compared these with the experimental pulse height spectra produced by the three radionuclides with a PerkinElmer TriCarb 2275 liquid scintillation analyzer. They found the composite pulse height spectrum provided by the LSA to approximate closely to the calculated beta distributions of each radionuclide. Fig. 6.58 illustrates the approximate beta distributions of equal activities of each radionuclide, and a composite spectrum, as depicted by the writer from their work. For the precise spectral distributions see Chobola et al. (2006). They selected three counting windows, as depicted in Fig. 6.58, to be defined by lower-level (LL) and upper-level (UL) discriminator settings of channel 23–200 for Window A, 201–800 for Window B, and 801–4000 for Window C. They counted the radiostrontium after passing the sample to be counted through a strontium-specific ion exchange column (SrSpec, Eichrom Technologies) that would remove all radionuclides other than strontium including the ^{90}Y daughter nuclide of ^{90}Sr.

As depicted in Fig. 6.58, Window A registers pulse events from ^{90}Sr, ^{89}Sr, and ^{90}Y, Window B registers pulse events from ^{89}Sr and ^{90}Y, and Window C registers pulse events from ^{90}Y. Chibola *et al.* (2006) defined three equations for the net (background-subtracted) count rates in each counting window using the matrix

$$\begin{bmatrix} I_A \\ I_B \\ I_C \end{bmatrix} = \begin{bmatrix} A \\ B \\ C \end{bmatrix} \cdot \begin{bmatrix} \eta_{AA} & \eta_{BA} & \eta_{CA} \\ \eta_{AB} & \eta_{BB} & \eta_{CB} \\ \eta_{AC} & \eta_{BC} & \eta_{CC} \end{bmatrix} \tag{6.49}$$

where I_A, I_B, and I_C are the net intensities or net count rates (CPM) obtained in counting Windows A, B, and C, and the sample activities (DPM) of each radionuclide are $A = {}^{90}$Sr activity, $B = {}^{89}$Sr activity, and $C = {}^{90}$Y activity. Within the matrix are the detection efficiencies of each radionuclide in the three counting windows or counting regions where η_{XY}denotes the counting efficiency of radionuclide X in counting region Y. Chobola et al. (2006) solved for activities A, B, and C including the constraint imposed by the ^{90}Y ingrowth. The activity C, which is the activity of ^{90}Y, will depend on the activity of ^{90}Sr in the sample and the time interval in hours from the time of isolation of strontium from the ion exchange column to the time of sample counting, which would govern ^{90}Y ingrowth according to the following:

$$^{90}\mathrm{Y}_{\mathrm{DPM}} = {}^{90}\mathrm{Sr}_{\mathrm{DPM}}\left[1 - e^{-\lambda t}\right] = A\left[1 - e^{-\frac{0.693}{64.1\ \mathrm{h}}(t - t_0)}\right] \quad (6.50)$$

The above Eq. (6.50) is identical to Eq. (6.45) discussed previously in this chapter and that expressed by Eq. (1.574) of Chapter 1, Volume 1. The various detection efficiencies of the three radionuclides required to solve the radionuclide activities, need not be determined by quench correction curves, as the radiostrontium nuclides as well as ^{90}Y are not easily quenched and the samples do not vary significantly in their chemistry or quench level. Internal standards can be used conveniently to determine the detection efficiencies, and internal standards can yield very precise detection efficiencies, if utilized properly, as described in Section V.A of this chapter.

Eikenberg et al. (2006) included a ^{85}Sr yield tracer in the three-window approach to the simultaneous liquid scintillation analysis of ^{89}Sr–^{90}Sr(^{90}Y). Like most analytical preparations for radiostrontium analysis, they separated the radiostrontium from other radionuclides (including ^{90}Y) in the sample by use of cation exchange with a column of SrSpec resin (Eichrom Technologies). Ingrowth of the ^{90}Y daughter nuclide of ^{90}Sr would begin immediately after the radionuclide separation (L'Annunziata, 1971). During the radiostrontium separation process, Eikenberg et al. (2006) allowed for the isolation of the ^{90}Y from the sample for subsequent Cherenkov counting and extrapolation of the ^{90}Y curve to time t_0, the time of separation of the ^{90}Y from the sample when it was in secular equilibrium with the ^{90}Sr.

For liquid scintillation analysis, Eikenberg et al. (2006) selected a three-window counting arrangement in a PerkinElmer TriCarb 2270TR/AB liquid scintillation analyzer, which is equipped with a pulse height analyzer calibrated in keV (See Fig. 6.59). The counting regions were Window A: 0–12 keV, Window B: 12–300 keV, and Window C: 300–1200 keV. Window A would register pulse events mostly from the ^{85}Sr yield tracer. The ^{85}Sr, which decays by electron capture, will produce abundant relatively low-energy Auger and conversion electron interactions with the liquid scintillation fluor cocktail, and there are in addition, X-ray and gamma photon Compton electron

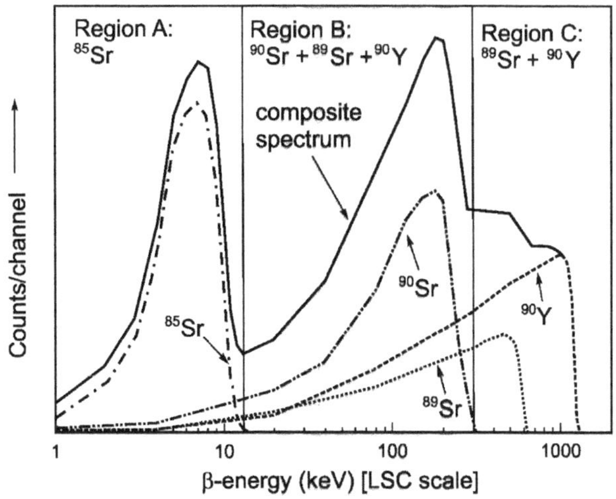

FIGURE 6.59 Approximate LSC beta spectra of ^{85}Sr, ^{90}Sr, ^{89}Sr, and ^{90}Y with the composite spectrum of the four radionuclides. Three counting windows are shown for the more precise measurement of each of the radionuclides. *Sketched from the work of Eikenberg et al. (2006). See reference for more precise distributions (Reprinted with permission of Radiocarbon, University of Arizona, © 2006 Arizona Board of Regents on behalf of the University of Arizona).*

interactions. Window B would register mostly ^{89}Sr and ^{90}Sr beta emissions in addition to ^{90}Y ingrowth.

Eikenberg et al. (2006) expressed the count rates (CPM) in the three counting regions as

$$\begin{aligned} N(t)_m^A &= A(t)_{85}\cdot\varepsilon_{85}^A + A(t)_{89}\cdot\varepsilon_{89}^A + A(0)_{90Sr}\cdot\varepsilon_{90Sr}^A \\ &\quad + A(0)_{90Sr}\cdot f_1\cdot\varepsilon_{90Y}^A \\ N(t)_m^B &= A(t)_{85}\cdot\varepsilon_{85}^B + A(t)_{89}\cdot\varepsilon_{89}^B + A(0)_{90Sr}\cdot\varepsilon_{90Sr}^B \quad (6.51) \\ &\quad + A(0)_{90Sr}\cdot f_1\cdot\varepsilon_{90Y}^B \\ N(t)_m^C &= A(t)_{89}\cdot\varepsilon_{89}^C + A(0)_{90Sr}\cdot f_1\cdot\varepsilon_{90Y}^C \end{aligned}$$

where $N(t)_m^A$, $N(t)_m^B$, and $N(t)_m^C$ are the net (background-subtracted) count rates at time *t(m)*, i.e., time of measurement, in counting regions A, B, and C, respectively, $A(t)_{85}$, $A(t)_{89}$, and $A(0)_{90Sr}$ are the activities (DPM) of the added ^{85}Sr tracer at the time of measurement, the activity of ^{89}Sr at the time of measurement, and the activity of ^{90}Sr at the time of its separation from ^{90}Y on the ion-exchange SrSpec column, respectively, ε_{85}, ε_{89}, ε_{90Sr}, and ε_{90Y} are the detection efficiencies of ^{85}Sr, ^{89}Sr, ^{90}Sr, and ^{90}Y in counting windows A, B, or C, and f_1 is the ^{90}Y ingrowth term described previously (See Eq. 6.44). From the three equations describing the count rates in each counting region (Eq. 6.51), Eikenberg et al. (2006) derived the equation for calculating the activities of ^{85}Sr, ^{89}Sr, and ^{90}Sr. They obtained highly accurate ^{90}Sr activities by collecting the ^{90}Y extracted from the original sample during the ion-exchange separation of radiostrontium from ^{90}Y and other nuclides on the SrSpec exchange column. By Cherenkov counting of the ^{90}Y in a counting window of 0–25 keV, they could follow the ^{90}Y decay over a few half-lives and extrapolate the decay curve back to the time of separation from ^{90}Sr to determine the equal activities of ^{90}Sr and ^{90}Y, which were at secular equilibrium at the time of their separation. (See Section XVII.C and Fig. 1.144 of Chapter 1, Volume 1).

B. Three-over-two fitting and digital overlay technique (DOT)

The three-over-two method and the digital overlay techniques are described by Kouru and Rundt (1991). The three-over-two method, as noted in a previous paper by the writer (Noor et al., 1995), involves the counting of samples containing two radionuclides with three counting regions or windows. The count rate of each radionuclide is determined by solving a set of three equations for the two unknowns. By this technique, three and even more radionuclides in composite samples could be determined by using the number of counting windows greater than the number of radionuclides. This three-over-two fitting method was applied by Kim et al. (2006) to determine the activities of ^{129}I and ^{125}I, which display closely overlapping liquid scintillation pulse height spectra (See Fig. 6.60). Iodine-125 decays by electron capture with the emission of 23 keV Auger electrons (20% intensity), 30 keV conversion electrons (13% intensity), as well as 30 keV X- and 35 keV gamma radiations of the daughter nuclide. Iodine-129 undergoes beta decay with an endpoint energy (E_{max}) of 154 keV and mean beta energy of 40.9 keV, 40 keV gamma, and 30 keV X-radiation (NNDC, 2011). The beta emissions of ^{129}I (average 40 keV) are slightly higher than the Auger and conversion electron emissions of ^{125}I producing an ^{129}I liquid scintillation pulse height spectrum with a peak at higher channel numbers illustrated in Fig. 6.60. Kim et al. (2006) were able to accurately determine the activities of both nuclides in a mixture using the three-over-two fitting method included with the EasyView software (Wallac, PerkinElmer) installed in a PerkinElmer Quantulus 1220 liquid scintillation analyzer.

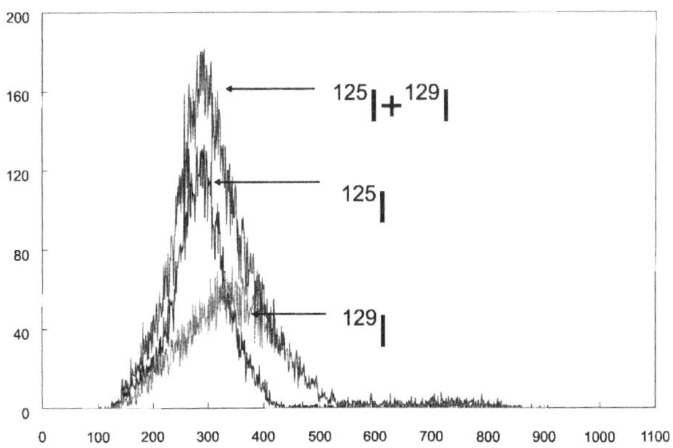

FIGURE 6.60 Pulse height spectra (counts per channel vs. channel number) of ^{125}I and ^{129}I obtained with a PerkinElmer Quantulus 1220 liquid scintillation analyzer. *From Kim et al. (2006), Reprinted with permission of Radiocarbon, University of Arizona, ©2006 Arizona Board of Regents on behalf of the University of Arizona.*

The three-over-two fitting method developed into the digital overlay technique (DOT) whereby, in principle, the number of windows approaches the resolution of the multichannel analyzer of the liquid scintillation counter. This digital overlay technique uses an external standard to measure quench and a specimen overlay to obtain the DPM for single or multiple-radionuclide samples. The shape of the sample spectrum is used to resolve dual or triple radionuclide samples by fitting the spectrum of each component to the measured composite spectrum. The fitting of a reference spectrum to the sample spectrum can give a measure of the level of quench. Spectral fitting is accomplished by the instrument (*e.g.*, Wallac RackBeta (PerkinElmer, Inc.), which maintains a spectrum library that covers a large quench region of both chemical and color quench for the radionuclides. The technique is reviewed and tested by Kouru and Rundt (1991) and described in patents by Rundt and Kouru (1989, 1992). They demonstrate this method as well as the counting region method previously described. The method is applied also in very specific models of Wallac (PerkinElmer, Inc.) liquid scintillation analyzers for quench correction and activity measurements of samples containing only one beta-emitting radionuclide. Kouru (1991), Leppänen and Kukkonen (2006), Wiegand et al. (2007), and Hueber-Becker et al. (2007) report the use of DOT for quench correction and activity analysis of ^{14}C.

C. Full spectrum DPM (FS-DPM)

Full spectrum DPM is a user-friendly method available with PerkinElmer TriCarb liquid scintillation analyzers for the measurement of many dual radionuclide combinations including ^{3}H–^{14}C, ^{3}H–^{32}P, ^{3}H–^{35}S, ^{14}C–^{32}P, ^{33}P–^{32}P, ^{35}S–^{32}P, ^{3}H–^{125}I, ^{125}I–^{131}I, ^{51}Cr–^{14}C, ^{67}Ga-^{68}Ga, ^{55}Fe–^{59}Fe, ^{125}I–^{14}C, ^{59}Fe–^{51}Cr, and ^{89}Sr–^{90}Sr, among others. The analysis protocol for FS-DPM is easy to set up, because no counting regions need to be defined. The full spectrum DPM method utilizes the spectral index of the sample (SIS) of the double radionuclide sample to "unfold" the separate pulse height spectra of the composite spectrum. This is possible because the SIS of the composite spectrum is a function of the individual distributions and the fractional counts of each radionuclide as well as the level of quench in the sample. The direct proportionality of SIS to the radionuclide composition in the sample was demonstrated by L'Annunziata and coworkers (see Noor et al.,

1995, 1996b), who determined the activity ratios of ^{3}H:^{14}C and ^{35}S:^{32}P from only the spectral index of the sample.

The key to spectrum unfolding used in the full spectrum DPM method is the SIS. At a given level of quench, each radionuclide has a defined pulse height-energy distribution and hence a unique SIS value. When two radionuclides are combined into a single sample, the resultant pulse height is the sum of the two individual distributions. The SIS of the total distribution is a function of the SIS of the individual distributions and the fractional counts of each radionuclide. If the SIS_L and the SIS_H are the spectral index values of the low- and high-energy radionuclides of a dual-radionuclide sample, then the SIS of the total distribution SIS_T can be calculated as

$$SIS_T = \frac{(SIS_L)(\sum N_L E + SIS_H)(\sum N_H E)}{\sum N_L E + \sum N_H E} \tag{6.52}$$

where $\sum N_L E$ = accumulated counts from the low-energy radionuclide (*e.g.*, ^{3}H) and $\sum N_H E$ = accumulated counts from the high-energy radionuclide (*e.g.*, ^{14}C).

From Eq. (6.52) the following equations are derived, which define the count rates of the low-energy radionuclide (CPM_L) and the high-energy radionuclide (CPM_H) of a composite sample:

$$CPM_L = \frac{SIS_H - SIS_T}{SIS_H - SIS_L}(CPM_T) \tag{6.53}$$

and

$$CPM_H = \frac{SIS_T - SIS_L}{SIS_H - SIS_L}(CPM_T) \tag{6.54}$$

For a detailed treatment on the derivation of Eqs. (6.53) and (6.54), the reader may refer to Kessler (1989) and van Cauter and Roessler (1991).

To determine the count rates and disintegration rates of unknown dual-radionuclide samples, four quench correction curves must be prepared with two sets of quenched standards. Quenched standards of the low-energy radionuclide and of the high-energy radionuclide are needed. The quenched standards are counted in a regionless environment; that is, no counting region discriminator settings need to be established. The data collected by the liquid scintillation analyzer from the counting of each of the quenched standards are the quench indicating parameters (tSIE), the SIS values of the low- and high-energy radionuclide standards (SIS_L) and (SIS_H), respectively, and the percent counting efficiency of the low- and high-energy radionuclide standards. The LSA then plots automatically the four quench correction curves such as the curves illustrated in Fig. 6.61.

For the analysis of unknown activities of the radionuclide components of a dual-radionuclide sample, the liquid scintillation analyzer will first determine the total count rate (CPM_T) and the tSIE of the sample. From the one value of

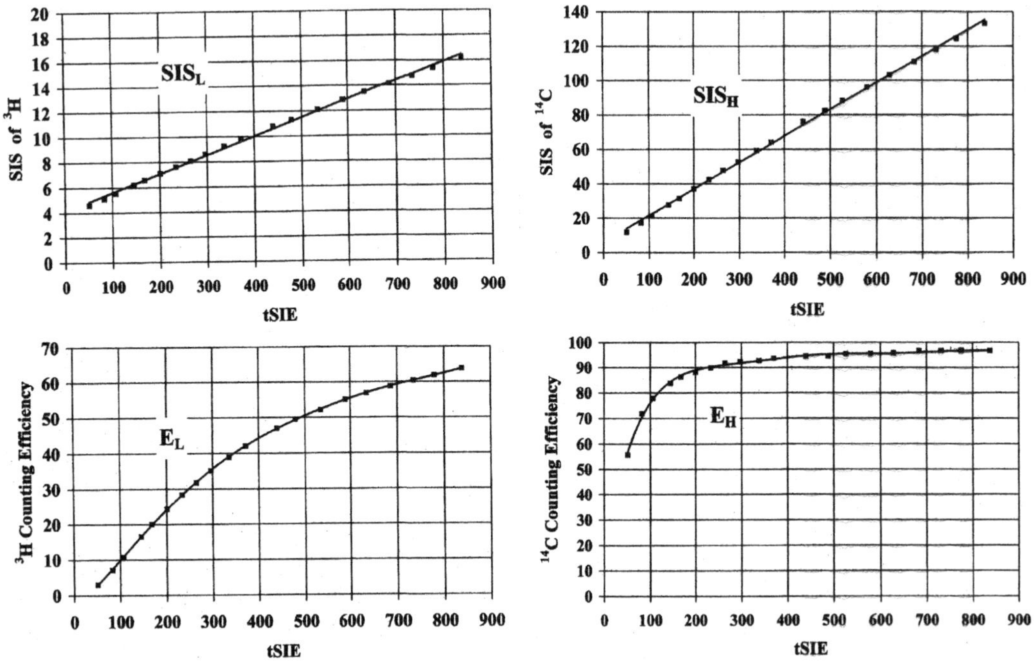

FIGURE 6.61 Quench correction curves for full spectrum DPM (FS-DPM) analysis of ^{3}H and ^{14}C mixtures. From the tSIE of a given sample, the liquid scintillation analyzer obtains from these curves the SIS of the lower- and higher energy radionuclides, SIS_L and SIS_H, respectively, and the percent counting efficiency of the lower- and higher-energy radionuclides, E_L and E_H, respectively. The above curves were obtained from ^{3}H and ^{14}C quenched standards with a PerkinElmer TriCarb 2770TR/SL liquid scintillation analyzer.

tSIE, the instrument extracts automatically the values of SIS_L and SIS_H for the composite sample using two of the quench correction curves (Fig. 6.61), which are maintained in computer memory. The LSA then calculates the count rate values of the low- and high-energy radionuclides CPM_L and CPM_H respectively, according to Eqs. (6.53) and (6.54). The instrument than automatically converts these count rates to the disintegration rates of the low- and high-energy radionuclide DPM_L and DPM_H respectively, according to the equations

$$DPM_L = CPM_L/E_L \tag{6.55}$$

and

$$DPM_H = CPM_H/E_H \tag{6.56}$$

where $E_L$and E_H, the counting efficiencies of the low- and high-energy radionuclides, respectively, are obtained from the respective quench correction curves illustrated in Fig. 6.61.

This DPM method provides accurate DPM values for dual radionuclides when the endpoint energies ($E_{\max}$) of the two radionuclides differ by a factor of 3. The activity ratios can extend over the range of 1:25 and 25:1. For additional information on this technique see Kessler (1989), De Filippis (1991), van Cauter and Roessler (1991), and L'Annunziata (1997b). The method was utilized for the analysis of $^{89}Sr-^{90}Sr$ mixtures by Hong et al. (2001) within 4 h after strontium separation from liquid waste. For low-level counting they report a counting efficiency of 95% for ^{89}Sr and 92% for ^{90}Sr with the full spectrum DPM method and lower limits of detection of 37 mBq/L for ^{90}Sr and 32 mBq/L for ^{89}Sr with a 60-minute counting time. An excellent illustration of the spectral unfolding power of the full spectrum DPM method is provided in Fig. 6.62 using the $^{89}Sr-^{90}Sr$ combination as an example.

Altzitzogluo et al. (1998) and Lee et al. (2002a,b) also report the use of the method for the analysis of ^{90}Sr with its ^{90}Y daughter. Lee et al. (2002b) compared the full spectrum DPM method and inclusion method (Section VIII.A.2) for the $^{90}Sr-^{90}Y$ analysis and found equal performance for both. The full spectrum DPM method is easier to carry out for the less experienced analyst, as no counting regions need to be established. Lee et al. (2002b) tested the FS-DPM method for the analysis of ^{90}Sr in liquid waste, and the method should also be applicable to the analysis of trace activities of ^{90}Sr in samples of ^{90}Y, which are used in clinical applications where ^{90}Sr would be an undesirable contaminant. Fig. 6.63 illustrates the ^{90}Sr and ^{90}Y spectra in the FS-DPM counting mode at various times after the separation of ^{90}Sr. The figure illustrates the potential of this counting mode at any stage of ^{90}Y ingrowth, or for the analysis of traces of ^{90}Sr activity in isolated ^{90}Y.

Excellent recoveries were reported by Lee et al. (2002b) for the determination or ^{90}Sr activities. The measured activities in DPM were within 5% from the expected values with ingrowth of ^{90}Y measured over intervals from 0 to 15 days following the separation of ^{90}Sr. Reddy et al. (2009) made extensive tests of this FS-DPM analysis method for $^{3}H-^{14}C$ combinations at low-activity levels.

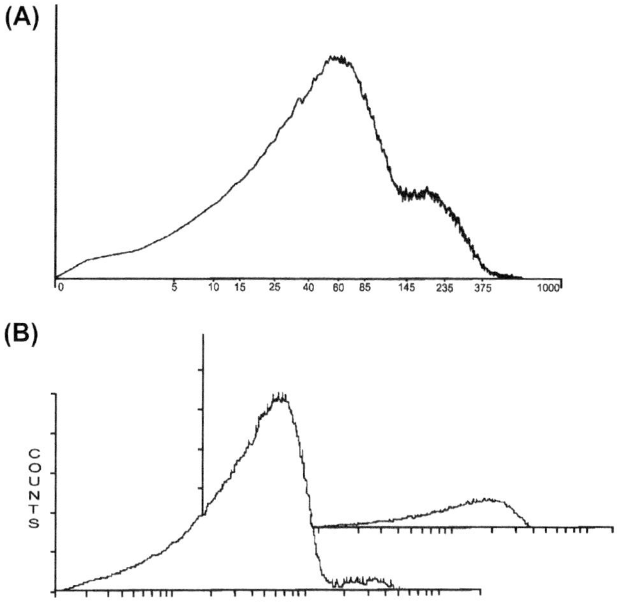

FIGURE 6.62 Unfolded spectrum of ^{90}Sr and ^{89}Sr (B) from mixture spectrum (A) by full spectrum (FS-DPM) method with a PerkinElmer TriCarb 2770 TR liquid scintillation analyzer. The pulse height spectra are plotted as counts versus pulse heights calibrated on a liquid scintillation keV scale. The unfolded spectrum of ^{89}Sr is illustrated in the rear of (B) and the unfolded ^{90}Sr spectrum in the front of (B). A very weak peak of ^{90}Y is seen together with the strong ^{90}Sr spectrum. The ^{90}Y peak was due to the ingrowth of ^{90}Y 2 hours after the separation of ^{90}Y from the sample using a SrSpec ion-exchange column. *From Hong et al. (2001). Reprinted with permission from Elsevier © 2001.*

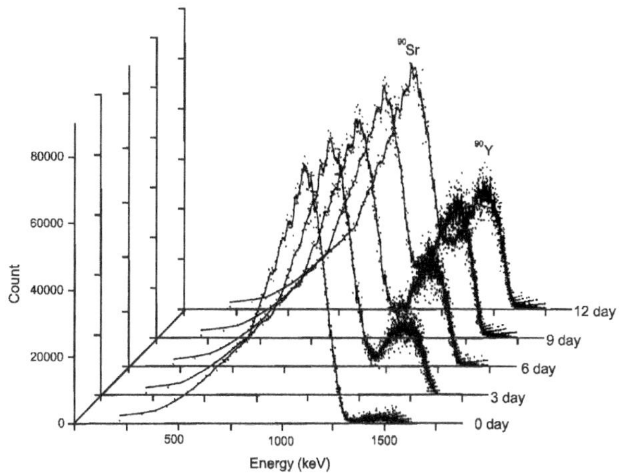

FIGURE 6.63 Spectra of ^{90}Sr and ^{90}Y in FS-DPM counting mode with time after the separation of ^{90}Sr obtained with a PerkinElmer TriCarb 2770 liquid scintillation analyzer. *From Lee et al. (2002b). Reprinted with permission from Elsevier © 2002.*

TABLE 6.7 DPM recoveries for $^{33}P-^{32}P$ dual nuclide combinations[a] by the FS-DPM method.

Sample[b]	Ratio[c]					% Recovery[d]	
^{33}P:^{32}P	^{33}P:^{32}P	SIS	tSIE	DPM(^{33}P)	DPM(^{32}P)	DPM(^{33}P)	DPM(^{32}P)
1.0 mL: 0.0 mL	1:0	130	472	72,167	47	99.5	0.06
0.0 mL: 1.0 mL	0:1	1043	476	28	82,686	0.04	98.2
0.5 mL: 0.5 mL	1:1	612	472	37,184	41,596	102.7	98.6
1.0 mL: 0.5 mL	2:1	441	445	72,493	41,615	100.2	98.7
0.5 mL: 1.0 mL	1:2	729	449	36,400	83,117	100.6	98.5
1.0 mL: 0.2 mL	5:1	292	458	74,487	16,912	102.9	100.2
0.2 mL: 1.0 mL	1:5	885	462	14,521	82,860	100.3	98.2
1.0 mL: 0.1 mL	10:1	234	465	72,319	8584	99.9	101.8
0.1 mL: 1.0 mL	1:10	951	469	7777	82,708	107.4	98.0

[a]*Activity of ^{33}P = 72,373 DPM/mL, Activity of ^{32}P = 84,357 DPM/mL. Instrumentation: PerkinElmer TriCarb 2300 TR liquid scintillation analyzer.*
[b]*Triplicate samples were counted for a duration of 10 minutes or until a % 2 sigma standard deviation of 1.0% was reached.*
[c]*Ratios listed are only approximate. The exact ratios can be calculated from the isotope activities (DPM/mL) and sample sizes (mL) of each isotope.*
[d]*The standard deviation of the mean of all triplicate measurements ranged from 0.085% to 2.97%.*
From L'Annunziata (1997b), Printed with permission. © 1997 PerkinElmer, Inc.

They obtained minimum detectable activities of 2100 DPM/L for ^{3}H and 1200 DPM/L for ^{14}C with a counting time of 300 min and found the accuracy of results to be within ±10% for ^{3}H:^{14}C activity ratios ranging from 1:1 to 1:18. Also, good recoveries of $^{33}P-^{32}P$ combinations were obtained by the writer using the FS-DPM method, and these are illustrated in Table 6.7.

D. Recommendations for multiple radionuclide analysis

The following are a few suggestions to follow when performing DPM determinations for multiple-radionuclide samples by one of the methods previously described: (1) For the preparation of quench correction curves, prepare the sets of quenched standards in the same cocktail, vial size, and total sample volume as the samples to be analyzed. (2) Use of automatic region tracking (AEC or AQC) is recommended for dual- or triple-radionuclide analysis when using the counting-window inclusion methods of DPM analysis to maintain constant spillover of the radionuclides in the required counting regions. (3) DOT or full spectrum DPM methods require an accurate measurement of the sample pulse height spectra by the instrument pulse height analyzer following automatic subtraction of the background. When sample counts are low, long counting times will be needed to obtain well-defined sample pulse height spectra. As noted previously, Reddy et al. (2009) made extensive tests of the FS-DPM as well as the dual-window inclusion method of analysis for $^{3}H-^{14}C$ combinations at low-activity levels. They obtained minimum detectable activities of 2100 DPM/L for ^{3}H and 1200 DPM/L for ^{14}C with a counting time of 300 minutes and found the accuracy of results to be within ±10% for ^{3}H:^{14}C activity ratios ranging from 1:1 to 1:18.

E. Complex spectral analysis

The multiple-radionuclide analysis methods previously discussed are limited to the analysis of not more than three β-emitting radionuclides in the same sample. From the beginning of liquid scintillation analysis in the 1950s up to about 1990, the regionless analysis of more than three β-emitting radionuclides in the same sample was considered to be impracticable or not feasible. The broad pulse height spectra produced by β particles made the task of deconvoluting the pulse height spectra of more than three radionuclides in the same sample appear daunting. However, with the advent of technological advances including applications of multichannel analyzers in liquid scintillation analysis and direct computer processing of LSA data, it has become possible to analyze simultaneously numerous α-, β-, and α-β-emitting radionuclides in a mixture, including nuclides decaying by electron capture emitting relatively low-energy Auger electrons and X-rays. This was evidenced by the early research reported by Takiue et al. (1990b, 1991b,c, 1992, 1995, 1999), Matsui and Takiue (1991), and Fujii et al. (1999, 2000) on the application of the most-probable-value theory to simultaneous multiple-radionuclide (as many as seven) analysis and the work of Grau Carles et al. (1993c), Grau Malonda et al. (1994a,b), and Grau Carles (1996) on the use of spectral deconvolution and interpolation methods to multiple radionuclide (as many as six) activity analysis. These techniques can be applied also to mixtures of α- and β-emitters in one multichannel analyzer by spectral unfolding without α/β discrimination as demonstrated by Grau Carles et al. (1996). Subsequent research advances in liquid scintillation spectral deconvolution for multiple radionuclide analysis were reported by Heilgeist (2000), Kashirin et al. (2000), Verrezen and Hurtgen (2000), Malinovsky et al. (2002a,b), Nebelung (2003), Nebelung and Baraniak (2007), Nebelung et al. (2001, 2009), Ermakov et al. (2006), Altzitzoglou (2008), Remetti and Sessa (2011), Dobrin et al. (2011), Remetti and Franci (2012) and Lee et al. (2017).

It was traditionally believed that ^{14}C ($E_{max} = 156$ keV) and ^{35}S ($E_{max} = 167$ keV) in the same sample could be neither identified nor analyzed by liquid scintillation counting, because of the very close similarities of the β-particle endpoint energies (E_{max}) of these two radionuclides. However, work first reported by Grau Carles and Grau Malonda (1991) demonstrated the accurate analysis of these two radionuclides in a mixture using spectral dilatation-interpolation and least-squares fitting. Activity ratios of ^{14}C/^{35}S were analyzed with an accuracy within about 3%. Grau Carles with Rodríguez Barquero and Grau Malonda (1993a) reported further improvements to this methodology, and Malinovsky et al. (2002a) demonstrated the deconvolution of 1:8 ratios of ^{14}C:^{35}S.

Another advancement is the application of chemometrics including multivariate calibration (MVC) to the deconvolution of α-, β-, and α-β-emitting radionuclides (without the need for α/β discrimination) in the same sample including nuclides decaying by electron capture, as reported by Toribio et al. (1995, 1996, 1997, 1999), Mellado et al. (2005), and Khayatzadeh Mahani et al. (2008, 2009), Mahani et al. (2012) and Fons-Castells et al. (2017a). These techniques were successfully applied to the analysis of α-β-emitting radionuclide mixtures using plastic scintillator microspheres in an LSC (Bagán et al., 2011a). These techniques were also applied as one more step in a radionuclide separation procedure by Fons-Castells (2017b, 2017c).

These advances in the development of techniques for the analysis of several radionuclides in the same sample will be discussed here. The reader is invited to refer to the literature cited for more details on the techniques involved.

1. Most-probable-value theory

Takiue et al. (1990a,b) and Matsui and Takiue (1991) reported the application of the most-probable-value theory as a new technique applied to the simultaneous liquid scintillation analysis of four β-emitting radionuclides in the same sample. This technique was expanded to the simultaneous analysis of six β-emitting radionuclides by Takiue et al. (1991c, 1992) and even seven β- and β-γ-emitting radionuclides (Takiue et al., 1995).

The technique requires only a contemporary liquid scintillation analyzer equipped with a multichannel analyzer, sets of quenched standards for the radionuclides to be measured, and a personal computer for data processing. The approach to this technique, as it was first devised for the analysis of samples containing mixtures of ^{3}H, ^{14}C, ^{32}P, and ^{45}Ca, calls for more counting regions than the number of nuclides to be measured. According to Takiue et al. (1990b) and Matsui and Takiue (1991), the method may be described using the four radionuclide composite sample consisting of ^{3}H, ^{14}C, ^{32}P, and ^{45}Ca, as an example. The count rates of a sample observed in each counting region are defined by the following equations:

$$\begin{aligned} n_1 &= Aa_1 + Bb_1 + Cc_1 + Dd_1 \\ &\vdots \\ n_i &= Aa_i + Bb_i + Cc_i + Dd_i \\ &\vdots \\ n_m &= Aa_m + Bb_m + Cc_m + Dd_m \end{aligned} \tag{6.57}$$

where $n_1, \cdots n_i, \cdots n_m$ are the count rates of a sample in different counting regions ($m >$ number of nuclides in the sample), $A, B, C,$ and D are the activities of ^{3}H, ^{14}C, ^{32}P, and ^{45}Ca, respectively, and a_i, b_i, c_i, and d_i are the respective counting efficiencies in the i th counting region.

The counting efficiencies of a radionuclide in each counting region are determined by means of external standard quench correction curves plotted using sets of quenched standards for each radionuclide. Sets of quenched standards are not commercially available for many radionuclides. These can be prepared in the laboratory in advance by first determining the activities of separate pure β- and β-γ-emitting radionuclides using a direct DPM method, such as efficiency tracing DPM (ET-DPM) as described in Section V.F or a more precise radionuclide standardization according to the techniques described in Section IX of this chapter. Even radionuclides procured commercially should be analyzed to confirm their exact activities by a direct DPM method. Once the DPM values of the separate radionuclide standards are known, sets of quenched standards can be prepared according to the procedure described in Section V.D of this chapter.

From the series of equations listed in Eq. (6.57) and following the derivations of Takiue et al. (1990b) and Matsui and Takiue (1991), the most probable values of $A, B, C,$ and D, that is, the activities of ^{3}H, ^{14}C, ^{32}P, and

^{45}Ca must be determined, and the following equation is derived to search a minimum value (S):

$$S = \sum_{i=1}^{m} w_i\{n_i - (Aa_i + Bb_i + Cc_i + Dd_i)\}^2 \quad (6.58)$$

where w_i is the arithmetic weight of the measurement in the ith counting region, which is calculated by $1/N_i$ where N_i is the total number of counts in the ith counting region.

The most probable value, that is, the activities of each radionuclide $A, B, C,$ and D should satisfy the following condition:

$$\frac{\partial S}{\partial A} = \frac{\partial S}{\partial B} = \frac{\partial S}{\partial C} = \frac{\partial S}{\partial D} = 0. \quad (6.59)$$

The following normal equations are then derived:

$$A\sum w_i a_i^2 + B\sum w_i a_i b_i + C\sum w_i a_i c_i + D\sum w_i a_i d_i = \sum w_i a_i n_i, \quad (6.60)$$

$$A\sum w_i b_i a_i + B\sum w_i b_i^2 + C\sum w_i b_i c_i + D\sum w_i b_i d_i = \sum w_i b_i n_i, \quad (6.61)$$

$$A\sum w_i c_i a_i + B\sum w_i c_i b_i + C\sum w_i c_i^2 + D\sum w_i c_i d_i = \sum w_i c_i n_i, \quad (6.62)$$

$$A\sum w_i d_i a_i + B\sum w_i d_i b_i + C\sum w_i d_i c_i + D\sum w_i d_i^2 = \sum w_i d_i n_i. \quad (6.63)$$

These equations can be solved for radionuclide activities $A, B, C,$ and D, i.e., the disintegration rate (DPM) of ^{3}H, ^{14}C, ^{32}P, and ^{45}Ca, respectively, using the determinant calculated by a personal computer. For example, as given by Matsui and Takiue (1991), the determinant for the calculation of the ^{3}H activity from the above equations is the following:

$$H = \frac{1}{K}\begin{vmatrix} \sum w_i a_i n_i & \sum w_i a_i b_i & \sum w_i a_i c_i & \sum w_i a_i d_i \\ \sum w_i b_i n_i & \sum w_i b_i^2 & \sum w_i b_i c_i & \sum w_i b_i d_i \\ \sum w_i c_i n_i & \sum w_i c_i b_i & \sum w_i c_i^2 & \sum w_i c_i d_i \\ \sum w_i d_i n_i & \sum w_i d_i b_i & \sum w_i d_i c_i & \sum w_i d_i^2 \end{vmatrix} \quad (6.64)$$

where

$$K = \begin{vmatrix} \sum w_i a_i^2 & \sum w_i a_i b_i & \sum w_i a_i c_i & \sum w_i a_i d_i \\ \sum w_i b_i a_i & \sum w_i b_i^2 & \sum w_i b_i c_i & \sum w_i b_i d_i \\ \sum w_i c_i a_i & \sum w_i c_i b_i & \sum w_i c_i^2 & \sum w_i c_i d_i \\ \sum w_i d_i a_i & \sum w_i d_i b_i & \sum w_i d_i c_i & \sum w_i d_i^2 \end{vmatrix} \quad (6.65)$$

For this case six counting regions were used in the multichannel pulse height analyzer for the measurement of the four nuclides. The discriminator settings of counting regions 1, 2, 3, and 4 are set to receive significant pulses from ^{3}H, ^{14}C, ^{45}Ca, and ^{32}P, respectively, with overlapping pulse height distributions. Channels 5 and 6 were set to receive pulses mostly from the medium-energy β-emitting radionuclides ^{14}C and ^{45}Ca, as the use of double channel settings for the medium-energy β-emitters produces more accurate data. The counts of the quenched standards collected in the various counting regions should exceed 10^4 to keep error at a minimum. The mean percent recovery was 2.4% for eight samples containing different proportions of ^{3}H, ^{14}C, ^{45}Ca, and ^{32}P at different quench levels, which represent 32 radionuclide analyses (8 samples $\times$ 4 radionuclides). Matsui and Takiue (1991) modified the technique by using only three counting regions for the analysis of the four radionuclides in a mixture. This required counting the unknown sample at two quench levels determined by the quench indicating parameter tSIE. The sample was counted twice, that is, before and after the addition of quenching agent. Mean recoveries by this modified approach for 28 radionuclide analyses (7 samples $\times$ 4 radionuclides) was 3.6%.

The approach described above can be applied to the simultaneous liquid scintillation analysis of six different β-emitting radionuclides in a mixture as demonstrated by Takiue et al. (1991c, 1992). They demonstrated the activity analysis of ^{3}H–^{63}Ni–^{14}C–^{45}Ca–^{36}Cl–^{32}P by application of the most-probable-value theory. A PerkinElmer Tri-Carb Model 4000 was used, and the samples were counted in 12 counting regions as illustrated in Fig. 6.64. Thus, the measurement of an unknown sample requires 12 observation equations of the general type described in Eq. (6.57). The series of 12 equations are written according to the following:

$$n_i = Aa_i + Bb_i + Cc_i + Dd_i + Ee_i + Ff_i \quad (i = 1 - 12), \quad (6.66)$$

where n_i is the count rate of a sample in the ith counting region. A, B, C, D, E, and F are the activities of the six radionuclides, namely ^{3}H, ^{63}Ni, ^{14}C, ^{45}Ca, ^{36}Cl, and ^{32}P, and

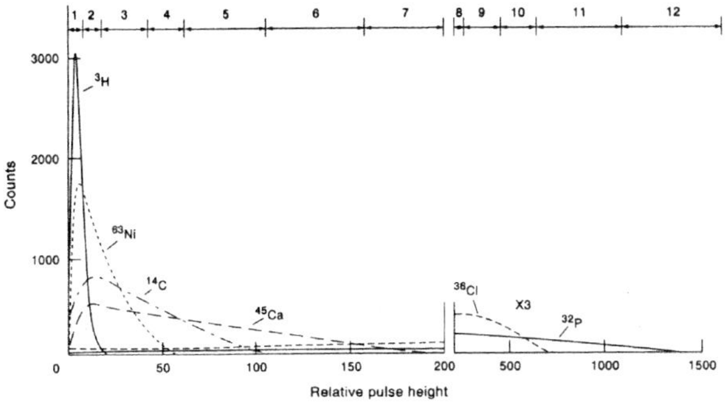

FIGURE 6.64 Liquid scintillation pulse height distributions of six pure beta-emitters and region settings for analytical measurements. *From Takiue et al. (1992), reprinted with permission from Elsevier © 1992.*

a_i, b_i, c_i, d_i, e_i, and f_i are the respective radionuclide counting efficiencies in the *i*th channel. Because six radionuclide activities must be determined simultaneously, a six-by-six matrix is derived and written similarly to the case of a four-by-four matrix (Eq. 6.64) written for the analysis of four radionuclides. The mean recovery for 60 analyses (*i.e.*, 10 samples × 6 radionuclides per sample) was 3.9%.

The technique can be applied to the liquid scintillation analysis of low-level β-γ-emitting radionuclides together with β-emitters as demonstrated by Takiue et al. (1995). In this case, mixtures of the following seven radionuclides were determined by application of the most-probable-value theory: ^{51}Cr—^{3}H—^{125}I—^{14}C—^{45}Ca—^{22}Na—^{32}P. In this case, 14 counting regions were used. The lower limits of detection based on the analysis of 30 samples was calculated as 0.01 Bq/mL for higher energy radionuclides and 0.05 Bq/mL for lower energy radionuclides in the mixtures.

A further development of this technique is its application to the analysis of radionuclide combinations with similar pulse height distributions, such as ^{3}H—^{125}I and ^{3}H—^{51}Cr, regardless of the different decay modes of these radionuclides, that is, β emissions with ^{3}H and electron capture (EC) decay with ^{125}I and ^{51}Cr. Takiue et al. (1991b) demonstrated the successful application of this procedure to the analysis of combinations ^{3}H—^{14}C—^{125}I and ^{3}H—^{14}C—^{51}Cr in a wide range of activity ratios and quench levels.

Subsequent work by Takiue et al. (1999) and Fujii et al. (1999, 2000) included the combined use of liquid scintillation and NaI(Tl) spectrometers to permit the simultaneous determination of the activities of many more nuclides in only one calculation process and, at the same time, enhance the accuracy of the radionuclide activity determinations. This method is referred to as a hybrid radioassay technique, because both liquid and solid scintillation spectrometers are used for a given sample. The NaI(Tl) solid scintillation detector provides additional sensitivity, as it would be particularly sensitive to X- and gamma-ray-photon emitters and high-energy beta-emitters, which produce considerable bremsstrahlung radiation. For example, Takiue et al. (1999)

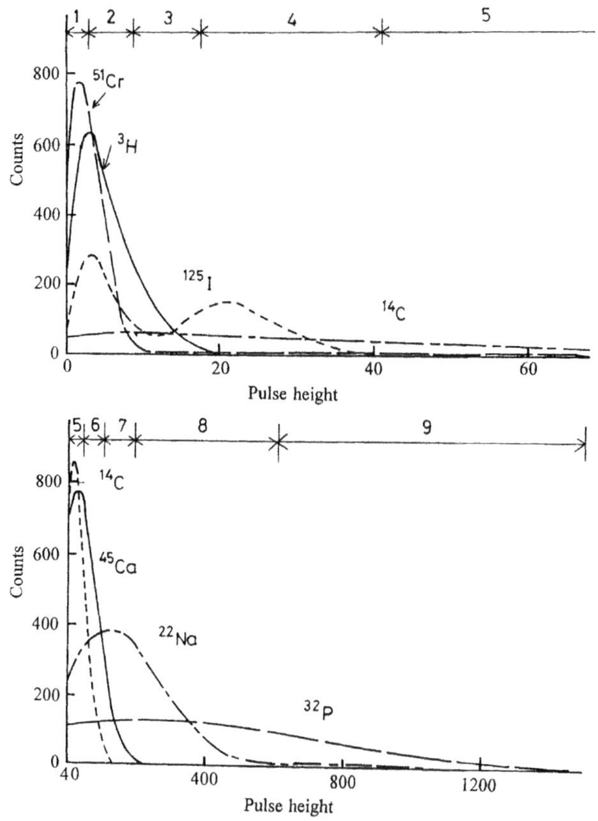

FIGURE 6.65 Liquid scintillation pulse height distributions of seven nuclides and channel settings for analysis. *From Takiue et al. (1999), reprinted with permission from Elsevier © 1999.*

and Fujii et al. (2000) analyzed a mixture of seven radionuclides (*i.e.*, ^{3}H, ^{14}C, ^{22}Na, ^{32}P, ^{45}Ca, ^{51}Cr, and ^{125}I using 12 counting regions defined, as illustrated in Figs. 6.65 and 6.66. The hybrid radioassay technique was applied by Fujii et al. (2000) to the analysis of the seven radionuclides in

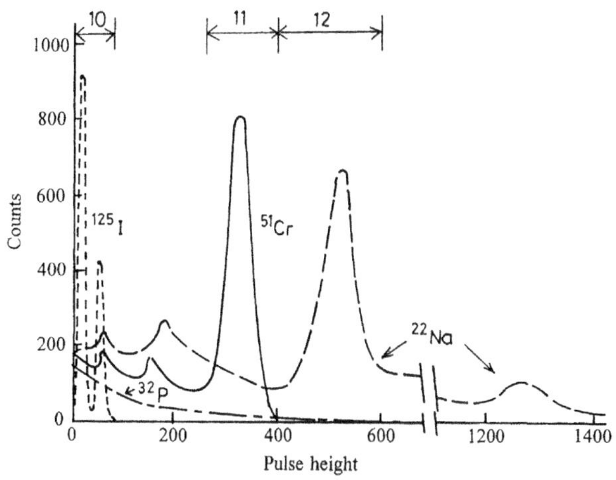

FIGURE 6.66 NaI(Tl) scintillation pulse height distributions of ^{125}I, ^{51}Cr, ^{22}Na, and ^{32}P and channel settings for analysis. *From Takiue et al. (1999), reprinted with permission from Elsevier © 1999.*

waste solutions with a detection limit of approximately 0.005 Bq/mL.

The procedures outlined in this section are carried out easily with any liquid scintillation analyzer equipped with an MCA or, for the hybrid radioassay, data provided by conventional liquid and solid scintillation analyzers without any modification of the equipment and with a personal computer for data processing. If the counting protocols of the LSA allow for only three counting regions with upper- and lower-level discriminator settings, more counting regions can be established by using additional counting protocols. However, this calls for counting the samples in more than one protocol. The liquid scintillation analyzer is classified as a radionuclide spectrometer, because numerous X-ray, β-, and β-γ-emitting radionuclides can be identified and analyzed simultaneously. Beta spectrometry applications of the liquid scintillation analyzer currently play a role of increased importance in the analysis of radionuclides.

2. Spectral fitting, unfolding, and interpolation

Over the period of 1991–1996 a new technique was reported and developed by researchers at CIEMAT, Madrid, which is a powerful spectral unfolding method for the simultaneous activity analysis of numerous β-emitting radionuclides, including β-emitting nuclides of very similar energy maxima (*e.g.*, ^{14}C and ^{35}S) and even some α/β-emitting radionuclides in the same MCA without α/β discrimination. A description of these techniques is provided here. References are cited for additional information.

The procedure involved, as described by Grau Carles (1993), for two types of liquid scintillation analyzers, namely, the LSA that analyzes pulse height spectra on a logarithmic scale and the LSA that uses a linear pulse height scale. As explained by Grau Carles (1993), the method has three key components: spectral fitting, spectrum unfolding, and spectral interpolation. These components to the analysis are described next.

a. Spectral fitting

As noted by Grau Carles and Grau Malonda (1991), pulse height spectra are generally defined in terms of discrete pulse heights or energy values; however, they are histograms. It is therefore necessary to have a continuous mathematical function defining the spectra. For pulse height spectra on a logarithmic scale, the spectral function is obtained by fitting Fourier series to the experimental spectra according to the following function:

$$f_F(\omega) = \begin{cases} a + b\omega + \sum_{k=1}^{N} c_k \sin\dfrac{k\pi\omega}{M} & 0 \le \omega < \omega^* \\ 0 & \omega > \omega^* \end{cases} \tag{6.67}$$

where $\omega = 0$ and $\omega = \omega^* = M$ are the first and the last values of the spectrum and N is the number of harmonics. The coefficients a, b and c_k are:

$$a = y_0$$

$$b = \frac{y_M - y_0}{M}$$

$$c_k = \frac{2}{M}\sum_{j=0}^{M} y'_j \sin\frac{\pi k \omega_j}{M} \tag{6.68}$$

where $y'_j = y_j - (a + b\omega_j)$

$$\omega_j = j = 1, 2, ..., M.$$

For the case of a liquid scintillation analyzer that uses a linear pulse height scale the spectral function is obtained by fitting the Chebyshev series to the experimental spectra determined by the following:

$$f_c(\omega) = \begin{cases} \left[\sum_{k=1}^{N} c_k T_{k-1}(\omega)\right] - \dfrac{1}{2}c_1 & 0 \le \omega < \omega^* \\ 0 & \omega > \omega^* \end{cases} \tag{6.69}$$

where $\omega = 0$ and $\omega = \omega^* = M$ are the first and the last values of the spectrum, $T_k(\omega)$ are the Chebyshev functions defined in the interval $[0, \omega^*]$ and c_k are the coefficients given by

$$c_k = \frac{2}{N}\sum_{j=1}^{N} f(\lambda_j) T_{k-1}(\lambda_j) \tag{6.70}$$

where the values λ_j are the zeros of the function $T_k(\omega)$.

b. Spectrum unfolding

According to descriptions by Grau Carles (1993) and Grau Carles et al. (1993b), the spectral deconvolution method is a simultaneous standardization technique, providing radionuclide activities, based on spectral shape analysis of the component nuclides in the mixture. The spectrum unfolding is based on minimizing the expression:

$$\min\left\{\sum (y_i(X+Y) - ay_i(X) - by_i(Y))^2\right\} \tag{6.71}$$

for a dual-radionuclide mixture, where y_i is the number of counts in channel i for the nuclide in brackets, X and Y are the two radionuclides, and a and b are the parameters obtained from the least-squares fit. This minimum condition, as explained by Grau Carles (1993) and Grau Carles et al. (1994a), can be applied only when all spectra $y_i(X+Y)$, $y_i(X)$, and $y_i(Y)$ have the same quench value, that is, the same quench indicating parameter. Therefore, it is necessary to obtain the spectra $y_i(X)$ and $y_i(Y)$, at the same quench value of the radionuclide mixture. This is achieved by the spectral interpolation described subsequently in the next part of the analysis procedure. The activities in DPM for the two nuclides X and Y are obtained from the following:

$$A(X) = \frac{ay_i(X)}{t\varepsilon(X)} \tag{6.72}$$

and

$$A(Y) = \frac{by_i(Y)}{t\varepsilon(Y)} \tag{6.73}$$

where t is the counting time in minutes and $\varepsilon(X)$ and $\varepsilon(Y)$ are the counting efficiencies of radionuclide X and Y, respectively.

When a mixture of more than two radionuclides is analyzed, Eq. (6.71) is written as

$$\min\left\{\sum_i \left(y_i\left(\sum_j X_j\right) - \sum_j a_j y_i(X_j)\right)^2\right\} \tag{6.74}$$

where X_j are the component nuclides, a_j the coefficients that make the condition minimum, and y_i the number of counts in channel i.

c. Spectral interpolation

Spectra at different quench levels have different end points or maxima. Therefore, as noted earlier, all spectra are interpolated to the same end point and maxima to validate the application of Eq. (6.71) or 6.74 for spectral unfolding of dual or multiple (three or more) radionuclides, respectively. Intepolation of spectra, as explained by Grau Carles et al. (1993c), is carried out in the following steps: (1) a mathematical transformation is found that makes all spectra pass through common maxima, inflection, and endpoints; (2) the transformed spectra are then interpolated channel by channel; and (3) the required spectrum is found by inverse transformation.

As explained by Grau Carles (1993), the spectral function, such as the Chebyshev fitting described previously by Eq. (6.69), is a mathematical manipulation of the spectra that smoothes the spectra and eliminates all superfluous statistical fluctuations. If $f_i(\omega)(i = 1, 2, 3)$ are the spectral functions, these may be divided into two regions, such as

$$f_i(\omega) = \begin{cases} g_i(\omega) & 0 \le \omega < \omega_{i\alpha} \\ h_i(\omega) & \omega_{i\alpha} \le \omega \le \omega_{i\beta} \end{cases} \tag{6.75}$$

where $\omega_{i\alpha}$ is the position of the maximum and $\omega_{i\beta}$ is the endpoint for spectrum i. As explained by Grau Carles (1993), the following mathematical transformation:

$$\omega' = a\omega \tag{6.76}$$

$$a = \frac{\omega_{1\alpha}}{\omega_{i\alpha}} \tag{6.77}$$

allows the transformation of all $g_i(\omega)$ functions, making them pass through the maximum of the spectral function $f_1(\omega)$. In the same way, the transformation

$$\omega' = b\omega + c \tag{6.78}$$

where

$$b = \frac{\omega_{1\alpha} - \omega_{1\beta}}{\omega_{i\alpha} - \omega_{i\beta}} \tag{6.79}$$

$$c = \frac{\omega_{i\alpha}\omega_{1\beta} - \omega_{1\alpha}\omega_{i\beta}}{\omega_{i\alpha} - \omega_{i\beta}} \tag{6.80}$$

takes the maximum and the endpoint of the function $f_1(\omega)$ as common points for each function $h_i(\omega)$.

When the spectral functions are transformed, a channel-by-channel interpolation provides a spectrum y_i^*. As explained by Grau Carles (1993), the following step determines the inverse equation, which transforms the spectrum y_i^* into the spectrum y_i:

Let $f_p^*(\omega)$ be the spectral function for y_i^* distribution and divide the function into the two regions:

$$f_p^*(\omega) = \begin{cases} g_p^*(\omega) & 0 \le \omega < \omega_{1\alpha} \\ h_p^*(\omega) & \omega_{1\alpha} \le \omega \le \omega_{i\beta} \end{cases} \tag{6.81}$$

The positions of the maximum $\omega_{p\alpha}$ and the end point $\omega_{p\beta}$ of the spectrum y_i for Q_p (quench value or quench-indicating parameter) can be found by interpolation in quench correction curves of channel number versus quench indicating parameter, such as one illustrated for ^{14}C in Fig. 6.67. An example of interpolated spectra of ^{14}C and ^{35}S with overlapping spectral maxima, inflection points and endpoints, is illustrated in Fig. 6.68.

Then the inverse transformation

$$\omega' = a'\omega \tag{6.82}$$

where

$$a' = \frac{\omega_{p\alpha}}{\omega_{1\alpha}} \tag{6.83}$$

transforms $g_p^*(\omega)$ into $g_p(\omega)$, and

$$\omega' = b'(\omega) + c' \tag{6.84}$$

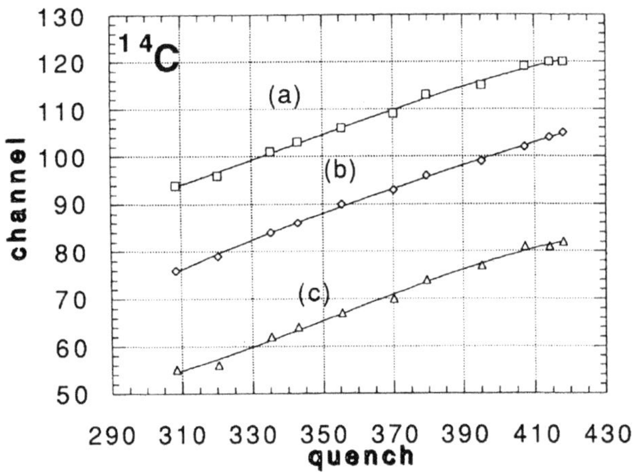

FIGURE 6.67 Fitted curves of ^{14}C local parameters versus quench: (a) end points of the spectra, (b) inflection points, and (c) position of the spectral maxima. The quench indicating parameter is SQP(E). *From Grau Carles et al. (1993a), reprinted with permission from Elsevier © 1993.*

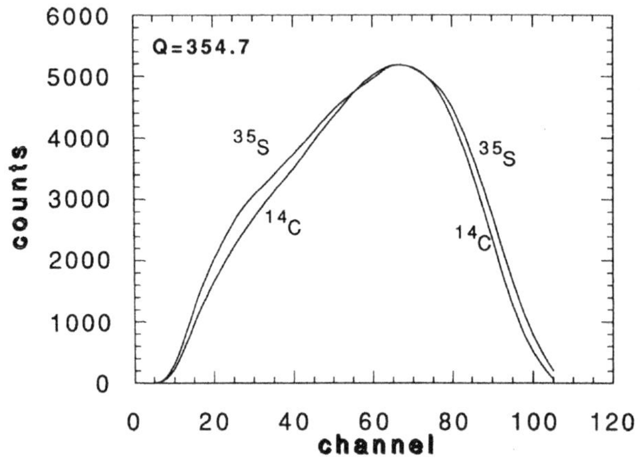

FIGURE 6.68 Computed ^{14}C and ^{35}S spectra for the quench value Q = 354.7, where Q is the SQP(E). *From Grau Carles et al. (1993a), reprinted with permission from Elsevier © 1993.*

where

$$b' = \frac{\omega_{p\alpha} - \omega_{p\beta}}{\omega_{1\alpha} - \omega_{1\beta}} \tag{6.85}$$

$$c' = \frac{\omega_{1\alpha}\omega_{p\beta} - \omega_{p\alpha}\omega_{1\beta}}{\omega_{1\alpha} - \omega_{1\beta}} \tag{6.86}$$

transforms $h_p^*(\omega)$ into $h_p(\omega)$. The functions $g_p(\omega)$ and $h_p(\omega)$ define the spectral function $f_p(\omega)$, the spectrum y_i for the particular quench level Q_p.

The power of this method can be evidenced in the spectral deconvolution of ^{14}C ($E_{max} = 156$ keV) and ^{35}S ($E_{max} = 167$ keV), which have very close β-particle energy maxima with a ratio of the energy maxima of only 1.07. The spectral unfolding of a 2.2:1 mixture of ^{14}C and ^{35}S is illustrated in Fig. 6.69. The difference spectrum illustrated is a result of subtracting the actual ^{14}C—^{35}S pulse height spectrum from the fitted Fourier spectral function $f_F(\omega)$. As

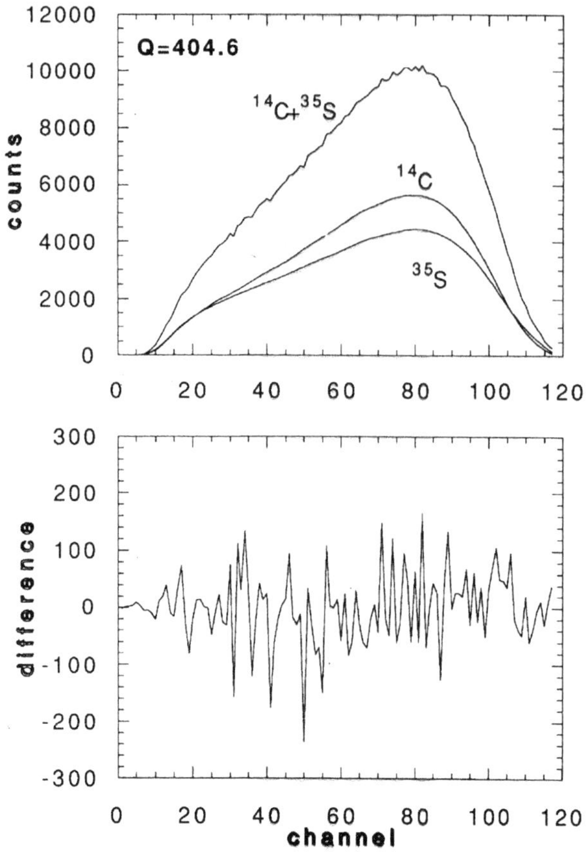

FIGURE 6.69 Spectrum unfolding of a sample containing ^{14}C/^{35}S with an activity ratio of 2.2:1 and the difference spectrum. *From Grau Carles et al. (1993a), reprinted with permission from Elsevier © 1993.*

explained by Grau Carles (1993), the difference spectrum shows that the fluctuations are statistical, that is, Fourier spectral function fits the empirical data well. For the deconvoluted ^{14}C—^{35}S sample spectrum illustrated in Fig. 6.69, Grau Carles et al. (1993b) determined the ^{14}C and ^{35}S activities with a 1.3% and 3.6% discrepancy, respectively, from the true DPM values. The ^{14}C—^{35}S dual mixture was considered traditionally inseparable spectrometrically by liquid scintillation analysis. Another radionuclide pair with very similar pulse height spectral endpoints is ^{3}H—^{55}Fe. Grau Carles et al. (1993c) again demonstrated the potential of this spectrum unfolding method with this nuclide pair. The power of the method is also demonstrated in the analysis of the traditional ^{14}C—^{3}H radionuclide mixtures, as low discrepancies from the true activity values are reported by Grau Carles et al. (1991) at various levels of quench and high-count-rate ratios of ^{14}C—^{3}H up to 100:1. Grau Carles et al. (1993f) demonstrated that this spectral deconvolution method will also provide good recoveries for multiple-radionuclide activity determinations in scintillation cocktails consisting of aqueous gel suspensions, such as aqueous ^{45}Ca—^{35}S in Insta-Gel. Mean recoveries expressed as percent discrepancy between computed and actual radionuclide DPM values for a wide range of ^{45}Ca—^{35}S activity ratios from 10.3:1 to 1:8.1 were 2.3% and 3.1% for ^{45}Ca and ^{35}S, respectively.

In addition to dual-radionuclide mixtures, this spectral unfolding method is applicable to the activity analysis of numerous β-emitting radionuclides. When numerous radionuclides are in the mixture, the previously described Eq. (6.74) is the general expression upon which spectrum unfolding is based. The spectral unfolding and activity analysis of up to six different β-emitting radionuclides in the same mixture are reported by Grau Carles et al. (1993d,e), Grau Malonda et al. (1994a,b) and Grau Carles (1996). Among the radionuclides analyzed by this technique are mixtures of ^{89}Sr, ^{90}Sr, and ^{90}Y even before secular equilibrium was reached between the ^{90}Sr and ^{90}Y. The percent recoveries, that is, percent discrepancy or uncertainty, of the analyzed activities are amazingly low in many circumstances. An example is given in Fig. 6.70, which illustrates the spectral unfolding of a ^{3}H—^{14}C—^{125}I—^{131}I—^{32}P mixture and the computer output with activity analysis and percent uncertainties of the activity determinations for each radionuclide.

The spectrum unfolding technique has been tested by Grau Carles et al. (1996) with a mixture of α- and β-emitting radionuclides consisting of $^{234}Th + ^{234m}Pa + ^{230}Th$. The activities of the α- and β-emitting radionuclides could be determined by the spectrum unfolding method without pulse shape analysis (PSA) for α/β separation. Percent discrepancies for activity determinations of the radionuclides in the mixture by spectrum unfolding were much lower than discrepancies obtained by the traditional PSA method.

The computer program MLOG is described by Grau Carles (1996) for the spectrum unfolding methods previously described. The MLOG program was applied by Grau Carles et al. (1998) to determine the separate activities of the parent-daughter nuclides, ^{125}Sb and ^{125m}Te, and thereby determine the ^{125}Sb to ^{125m}Te beta branching ratio.

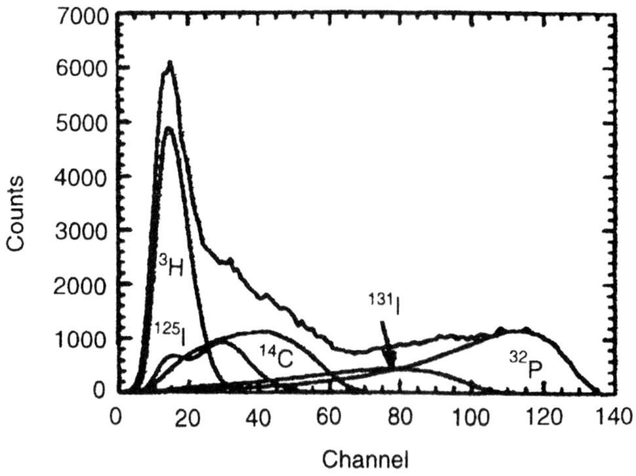

Local Quench Parameters and Efficiency

Nuclide	Quench	Maximum	End	Efficiency
P32	280.8	112.0	136.8	0.9964
I131	280.8	76.6	113.4	0.9620
I125	280.8	29.1	54.1	0.6115
C14	280.8	40.5	71.9	0.8647
H3	280.8	13.7	35.5	0.2129

Nuclide	Counting rate (cpm)
P32	27291.9
I131	12298.1
I125	11669.4
C14	21525.8
H3	28965.3

Nuclide	Activity (dpm)	Uncertainty (%)
P32	27391.1	0.5
I131	12784.0	2.3
I125	19083.2	3.4
C14	24893.0	1.9
H3	136049.3	2.2

FIGURE 6.70 Separated components of a mixture of $^{3}H + ^{14}C + ^{125}I + ^{131}I + ^{32}P$ by spectrum unfolding and the underlying computer output of activity analysis. *From Grau Carles (1996), reprinted with permission from Elsevier © 1996.*

3. Spectral fitting and subtraction

Early work by Heilgeist (2000) applied a spectral-subtraction method to resolve mixed liquid scintillation spectra of ^{85}Sr, ^{89}Sr, and $^{90}Sr/^{90}Y$ with a PerkinElmer Quantulus 1220 liquid scintillation analyzer. Strontium-85 was included in these measurements, as it is commonly used as a tracer nuclide in the determination of chemical recoveries of strontium. The calibration of the liquid scintillation spectrometer and the evaluation of the sample spectrum required the measurement of the pulse height spectra from reference standards of the separate nuclides ^{85}Sr, ^{89}Sr, and $^{90}Sr/^{90}Y$, and isolated ^{90}Y. The reference standards are prepared at the same quench level as the experimental samples.

Pulse height spectra from the separate reference standards permit a determination of the spectral endpoints corresponding to the beta-energy maxima of each radionuclide as illustrated in Fig. 6.71. Window discriminators (pulse height discriminators) are defined, namely, E1, E2, and E3, which are set at the endpoints of the abundant low pulse height events from ^{85}Sr, and the endpoints of the pulse events from ^{90}Sr and ^{89}Sr, respectively. The ^{90}Sr endpoint (E2) was set at Channel 690, and the ^{89}Sr

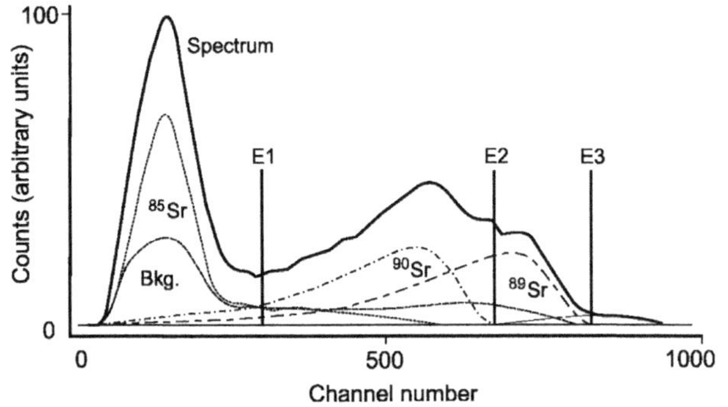

FIGURE 6.71 Spectra deconvolution: ^{85}Sr, ^{89}Sr, and ^{90}Sr in model waste. The solid line is the smoothed sample spectrum, Bkg, is the background spectrum. E1, E2 (channel 690), and E3 (channel 815) are the energy limits of the component spectra. Pulse events arising exclusively from ^{90}Y beta emissions (not identified in the figure) occur beyond E3 in channels 816 to 950. *From Heilgeist (2000). Reprinted with kind permission from Springer Nature © 2000.*

endpoint at channel 815. Pulse events originating from ^{90}Y extend beyond channel 815 (E3) to channel 950. The ^{85}Sr, which decays by electron capture produces a large peak of low pulse heights due to Auger, conversion-electron, and X-ray Compton electron interactions with scintillation cocktail. It also produces a very weak smear of events at higher pulse heights up to discriminator E2 due to gamma-photon Compton interactions.

Background events are automatically subtracted by the liquid scintillation analyzer. A computer evaluation program adjusts the reference spectra to the positions of the measured sample composite spectrum and optimizes the positions of the window discriminators of ^{90}Sr/^{89}Sr at E2 and ^{89}Sr/^{90}Y at E3. The contribution of ^{90}Y to the composite spectrum is calculated first on the basis of its share in channels 816 to 950, which is a counting region free of any pulse events from other radionuclide emissions. Once the contribution of ^{90}Y is subtracted from the composite spectrum, the contribution from ^{89}Sr is determined from its share in the region defined by channels 691 to 815. Finally, the ^{90}Sr share is obtained after the deduction of the ^{90}Y and ^{89}Sr contributions. The contribution of ^{85}Sr due to its overlapping spectrum with ^{90}Sr was eliminated by spectral subtraction using a reference spectrum of ^{85}Sr. Quench effects are avoided by the constant composition of the scintillator solutions. Heilgeist (2000) reported recovery rates of 93.8% and a limit of detection of 0.1 Bq/kg.

Mietelski and Gaca (2002) separated numerically ^{90}Sr and ^{90}Y spectra using a fitting function with a PerkinElmer/Wallac Guardian liquid scintillation analyzer. The analysis could be performed after the separation of strontium on a column of Sr-Spec (Eichrom Industries) resin. There was no need to wait for secular equilibrium to be achieved. The overlapping spectra of ^{90}Sr and ^{90}Y are deconvoluted by the fitting function within 2% error, and the integrals of the ^{90}Sr and ^{90}Y parts of the combined spectrum would provide their count rates.

Nebelung et al. (2009) applied spectral peak functions with a PerkinElmer 1414 Liquid scintillation analyzer to deconvolute dual-, triple-, and multiple- (up to seven) radionuclide combinations for activity determinations. They used the peak fitting module of the graphic program Origin 5.0 (originally Microcal Inc., now OriginLab Corp., https://www.originlab.com/, accessed October 15, 2018), which contains several internal peak fit functions (Nebelung and Baraniak, 2007; Nebelung et al., 2009). Radionuclide standards of each nuclide were used to determine the spectral shape of each nuclide, which was required for the spectral fitting and analysis of multinuclide samples. The results of activity determinations of combinations of only two or three radionuclides by the curve fitting and deconvolution method yielded similar results obtained by the conventional two- and three-window counting method with a PerkinElmer TriCarb 3100 liquid scintillation analyzer. The uncertainty for activity measurements of three or more nuclides in one sample was <10%. An example of the deconvoluted spectra of a mixture of 10 Bq each of ^{3}H/^{14}C/$^{137+}$Cs/^{90}Sr/^{90}Y is illustrated in Fig. 6.72.

A spectral deconvolution method was applied by Altzitzoglou (2008) to the activity analysis of radionuclides from liquid scintillation spectra of two or more radionuclides. The spectral deconvolution method is based firstly on the mathematical expression for the count rate y in each channel i of the LSA multichannel analyzer for the measured spectrum of n radionuclides or

$$y_i = \sum_n c_n \cdot S_{n,i} \quad (6.87)$$

where c_n is the unknown count rate of the radionuclide n in the sample, and $S_{n,i}$ is the content of the corresponding

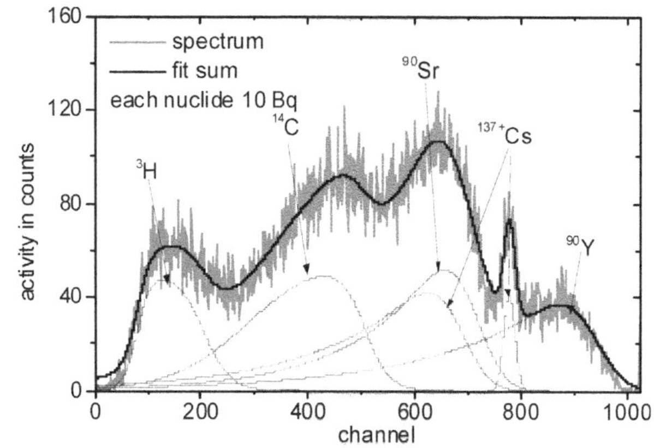

FIGURE 6.72 Liquid scintillation spectrum of a mixture of five radionuclides and the spectra deconvolution measured with a PerkinElmer 1414 liquid scintillation analyzer. *From Nebelung et al. (2009). reprinted with permission from Radiocarbon, University of Arizona, © 2009 Arizona Board of Regents on behalf of the University of Arizona.*

channel i of a pure reference spectrum of radionuclide n. In other words, S_n is the normalized spectrum of radionuclide n, whereby, the total number of counts in the spectrum for radionuclide n would equal unity. Thus, for a complete spectrum of m number of channels, there would be m number of equations similar to Eq. (6.87).

The spectrum deconvolution described by Altzitzoglou (2008) utilizes the shape of the sample spectrum on resolving its components by fitting a pure reference spectrum for each component to the measured composite spectrum. This was performed by a least-squares fitting; whereby, the unknown count rates c_n for each radionuclide n were found by minimizing the sum of the squared differences between the measured (y_i) and predicted (y_i') rates for all channels of the spectra, or

$$\sum_{i}^{m} (y_i - y_i')^2 = \sum_{i}^{m} \left(y_i - \sum_{n} c_n \cdot S_{n,i} \right)^2 \tag{6.88}$$

Altzitzoglou (2008) applied an iterative fitting using the Levenberg–Marquardt algorithm, which is a nonlinear least-squares fitting. The radionuclide activities (DPM) were calculated by c_n/ε_n where c_n is the count rate of nuclide n and ε_n is its detection efficiency. Eqs. (6.87) and (6.88) are valid when the pure and composite spectra have the same quench levels. Altzitzoglou (2008) applied the method to the logarithmic spectra as determined with a PerkinElmer Quantulus 1220 liquid scintillation analyzer, but the method is also applicable to the linear pulse height spectra as provided by the PerkinElmer TriCarb liquid scintillation counters. Fig. 6.73 is an example of a typical measured liquid scintillation composite spectrum of $^{90}Sr(^{90}Y)$ at secular equilibrium and the component spectra

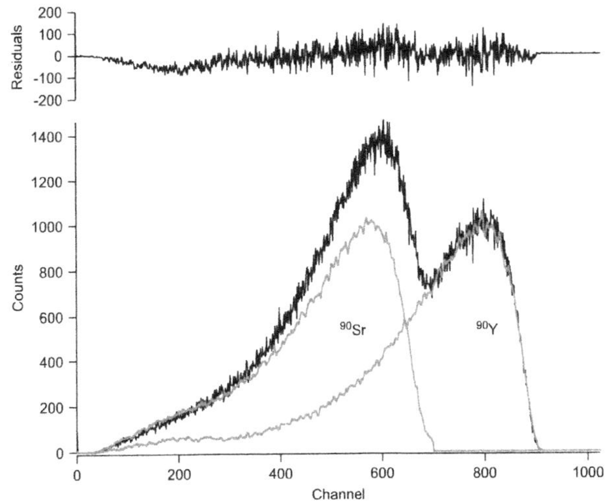

FIGURE 6.73 Typical composite spectrum of $^{90}Sr(^{90}Y)$ at equilibrium as measured with a liquid scintillation analyzer, and the component spectra as calculated by the deconvolution method. The upper spectrum is the plot of the residuals of the fit. *From Altzitzoglou (2008), reprinted with permission from Elsevier © 2008.*

as calculated by the deconvolution method. The deconvolution method was demonstrated to yield satisfactory results with low-activity samples as tested with the measurement of $^{90}Sr(^{90}Y)$ in milk and soil samples. The advantages over traditional methods of multiple radionuclide analysis are no counting windows need to be set, the method is rapid and straightforward and, in the case of $^{90}Sr/^{90}Y$, it is not necessary to attain parent-daughter equilibrium for the activity analysis.

Lee et al. (2017) describe a deconvolution method that uses a spectral data base to obtain the reference pure beta spectra. This deconvolution method can be applied to spectra with the same level of quenching than the mixed total spectrum. The first step of this procedure was the fitting of the count ratio spectra. From the different methods studied, the sum of sine, Eq. (6.89), was the approach that provided the best results.

$$\begin{aligned} Y = {} & a1 \sin(b1E + c1) + a2 \sin(b2E + c2) \\ & + a3 \sin(b3E + c3) + a4 \sin(b4E + c4) \\ & + a5 \sin(b5E + c5) + a6 \sin(b6E + c6) \\ & + a7 \sin(b7E + c7) + a8 \sin(b8E + c8) \end{aligned} \tag{6.89}$$

Next, under the assumption that the count rate at the maximum energy (E_{max}) and 10 channels before ($E_{max\text{-}10}$) is generated by a single radionuclide and that the number of counts per specific energy is proportional to the count ratio value multiplied by the activity, the maximum energy was used to define the radionuclide and the $E_{max\text{-}10}$ to obtain the constant value that multiplies the radionuclide fitted function. Then, the spectrum obtained was substracted from the mixed spectrum and the process was repeated until all nuclides are defined and separated from the multiple beta spectrum. This procedure was applied to the separation of a ^{32}P, ^{90}Y, and ^{106}Rh mixture obtaining relative errors lower than 2%.

A method was developed by Remetti and Sessa (2011) that is capable of deconvoluting complex liquid scintillation pulse height spectra without any preliminary knowledge of the peak shapes of the component radionuclides. They used a PerkinElmer Quantulus 1220 liquid scintillation analyzer with spectral deconvolution by Fourier series fitting with MATLAB software (The MathWorks Inc., version R2009B) according to Eqs. (6.56) and (6.57) described earlier. The calculation algorithm was structured according to the following path:

i. **Identification of the portion of the pulse height spectrum that is due to the radionuclides of highest E_{max}, and the selection of a group of channels for interpolation.**
The radionuclide that possesses the highest energy emission will yield a pulse height spectrum where a limited portion of the upper (highest pulse heights)

part of the spectrum will be free of any overlap of pulse events from lower energy emitting nuclides. Once the channels are identified, which contain pulse events (counts) attributable only to that radionuclide, the algorithm can proceed with the next step, that of the interpolation of its contribution to the entire complex spectrum.

ii. **Interpolation of the high-energy radionuclide contribution with the model "Cubic Spline".**
The model "Cubic Spline", as described by Remetti and Sessa (2011), is a nonparametric interpolation model available among the functions of MATLAB. The interpolation of the high-energy radionuclide contribution to the spectrum is obtained by imposing the condition of passage of the function through the group of points of the above Step 1 and through zero in the first channel.

iii. **Subtraction of the contribution by the high-energy radionuclide to the entire spectrum, and repetition of the procedure (Steps 1 and 2) for the remaining radionuclide contributions to the spectrum.**
Once the shape of the spectrum, attributed to the radionuclide of highest E_{max}, is obtained, it can be subtracted from the entire composite spectrum. This leaves behind a spectrum of the remaining radionuclides of lower E_{max}. Then Steps 1 and 2 are repeated by subtracting singularly, *i.e.*, the pulse height spectrum of one nuclide at a time, until all of the contributing pulse height spectra of each radionuclide are unfolded from the composite spectrum.

iv. **Activity calculation.**
A simple integration of the unfolded pulse height spectra of each radionuclide provides the counts attributable to each radionuclide. The count rate (CPM), of course, will depend on the time needed to achieve the composite pulse height spectrum. The disintegration rate (DPM) of each radionuclide can easily be calculated with a measurement of the detection efficiency, which would be characteristic of the radionuclide emissions, quench level, and counter.

Fig. 6.74 provides an example of the spectral deconvolution obtained by this technique. The discrepancies of the deconvoluted activities of the ^{90}Sr, ^{90}Y, and ^{241}Am from the expected values were reported to be within 5%. This method was implemented in a software called ABCD-Tool with some additional features as the possibility of a manual definition of the number of harmonics and the start and stop channels used for fitting (Remetti and Franci (2012)).

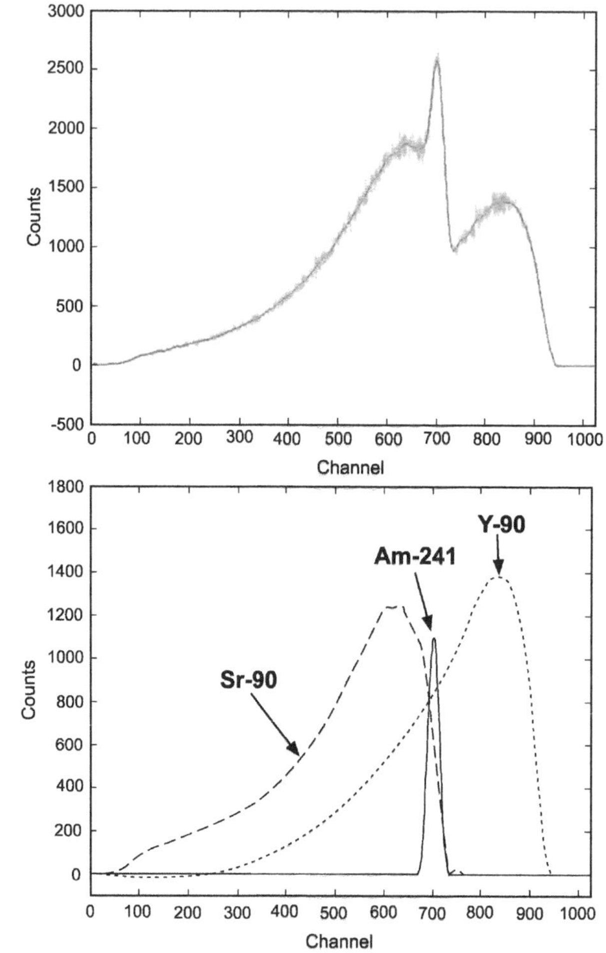

FIGURE 6.74 (Upper graph) Fourier series fitting of a composite spectrum (^{90}Sr + ^{90}Y + ^{241}Am). The continuous line represents the fitted function. (Lower graph) Total unfolding of the shapes of ^{90}Sr, ^{90}Y, and ^{241}Am. *(From Remetti and Sessa, 2011. Reprinted with kind permission from Springer Nature © 2011).*

4. Modeling from spectral library

A different approach to multiple radionuclide analysis involves the establishment of a library of radionuclide spectra constituting different mixtures of radionuclides at varying levels of quench. The spectral library, which is stored as computer data, must be prepared from radionuclide standards in various mixtures and activity ratios including also spectral distortions caused by the wall effect. Kashirin et al. (2000) describe this approach, which they have tested for complex mixtures of up to eight radionuclides including ^{3}H, ^{14}C, ^{60}Co, ^{63}Ni, ^{90}Sr + ^{90}Y, ^{125}I, ^{137}Cs, and ^{241}Am. They applied algorithms described in patents by Belanov et al. (1997, 1998) to minimize deviations between real and model spectra regardless of linear or logarithmic scales used for the presentation of spectra.

Continued research along these lines reported by Malinovsky et al. (2002a,b), Kashirin et al. (2003), and Ermakov et al. (2006) has culminated in the establishment of a software package RadSpecDec for the deconvolution of complex pulse height spectra consisting of combinations of numerous (six or more) alpha-and beta-emitting radionuclides. As described by Ermakov et al. (2006) the deconvolution algorithm is based on the modeling of spectra, which are transformed when convoluted by the superposition of individual reference spectra taken from a previously created radionuclide library of spectra. The spectra library consists of at least 10 spectra at different quench levels. The model spectrum is calculated as a sum of the individual radionuclide spectra with weight coefficients (c_j) that are adjusted during the modeling procedure with the objective that the model spectrum coincides with the sample composite spectrum to a maximum degree.

The model method was proposed by Malinovsky et al. (2002a,b) to be based on the modeling of spectra that are transformed when convoluted into a group by the superpositioning of individual radionuclide spectra taken from the spectral library. As described by Ermakov et al. (2006) the model spectrum is created from a linear combination of normalized spectra of various radionuclides according to the equation

$$M_i = \sum_{i=1}^{j} c_j M_{i,j}(T) \tag{6.90}$$

where M_i is the model spectrum, $M_{i,j}(T)$ is the library spectra of each radionuclide as a function of the quench parameter T, and c_j is the proportion or weight coefficient of each radionuclide specie. The weight coefficients, c_j, of each radionuclide are adjusted to make the model spectrum coincide with the sample spectrum to a maximum degree. Ermakov et al. (2006) describes the solution to the model as generated by a recursive minimization of a function representing the difference between a model spectrum and the real sample composite spectrum.

The modeling is carried out with groups of channels rather than the individual channels. For liquid scintillation spectrometers equipped with a logarithmic analog-to-digital converter (*e.g.*, PerkinElmer Quantulus or Hidex Triathler), a combination of 10 channels form a group. Thus, the composite sample spectrum and the deconvoluted spectra are depicted as plots of Counts versus Group Number or Relative Activity versus Group Number, rather than the conventional Counts versus Channel Number, as illustrated in Fig. 6.75. Liquid scintillation spectrometers equipped with a linear analog-to-digital converter (*e.g.*, PerkinElmer TriCarb) have channel numbers grouped by means of an arithmetic progression. Both logarithmic and linear pulse height spectra are resolved by the modeling technique, whereby the spectral channels become "channel groups". The activity in disintegrations, A_j of the jth radionuclide in the composite spectrum will be a function of the total counts, N, the background-subtracted or net counts, $N - N_{bkg}$, the weight coefficient, c_j, of the jth radionuclide, and the detection efficiency, ε_j, of the jth radionuclide at quench level T or

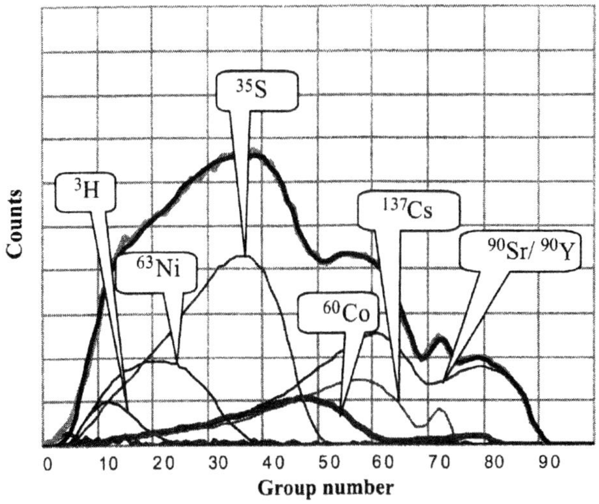

FIGURE 6.75 The spectrum decoding of a mixture (^3H, ^{63}Ni, ^{35}S, ^{60}Co, $^{90}Sr/^{90}Y$, ^{137}Cs) measured with a PerkinElmer Quantulus liquid scintillation analyzer. *From Malinovsky et al. (2002a,b), reprinted with permission from Radiocarbon, University of Arizona, © 2002 Arizona Board of Regents on behalf of the University of Arizona.*

$$A_j = \frac{(N - N_{bkg})c_j}{T\varepsilon_j} = \frac{N_j}{T\varepsilon_j} \tag{6.91}$$

where $T\varepsilon_j$ is the detection efficiency of the jth radionuclide at quench level T. Malinovsky et al. (2002a) and Ermakov *et al.* (2005) report tests on this deconvolution method with a wide variety of sample types and radionuclide combinations including low-level mixtures with good recoveries of the deconvoluted activities compared to the known radionuclide activities.

5. Spectral unfolding by region count ratios

Often the LSA of radionuclide mixtures is complicated by the need to analyze low-energy beta-emitters, such as ^3H, contaminated by much greater activities of high-energy beta impurities. Verrezen and Hurtgen (1996, 2000) describe a multiple-window spectrum unfolding technique for the analysis of low-energy ^3H in the presence of up to 10-fold higher activities of high-energy beta-emitters over a wide range of quench (*e.g.*, tSIE $\geq$ 100). The method is based on the fact that the ratio of the net count rates in two fixed counting regions for a single radionuclide is a constant at a given quench level. Consequently, any spectral shape (*e.g.*, impurity spectrum) can be reconstructed using the established ratio and a pure reference spectrum. On the basis of this hypothesis, Verrezen and Hurtgen (2000) calculate the spectral contribution of the impurity in the low-energy part of the liquid scintillation pulse-height spectrum and correct for the impurity interference on the measured low-energy sample spectrum. The method was demonstrated successfully over a wide range of quenching and activity ratios with mixtures of ^{99}Tc and ^{63}Ni contaminated with ^{60}Co and ^{137}Cs, respectively (Verrezen and Hurtgen, 1996), mixtures of ^{63}Ni/^{3}H, ^{14}C/^{3}H, ^{99}Tc/^{3}H, ^{60}Co/^{99}Tc (Verrezen and Hurtgen, 2000), and mixtures of ^{3}H/^{90}Sr/^{90}Y (Dobrin et al., 2011).

The spectral deconvolution method described here is particularly useful for the analysis of a low-energy beta-emitting nuclide that is contaminated with a higher-energy beta-emitter. Verrezen and Hurtgen (2000) and Dobrin et al. (2011) demonstrate that the method is particularly useful in the analysis of nuclear waste streams of pressurized water reactors where activity ratios of 10^3-10^6 are encountered between impurities and the nuclide of interest. The overlapping character of liquid scintillation pulse height spectra can mask the presence of contaminating radionuclides low in activity relative to the nuclide of interest. The method is based on the hypothesis that the activity ratios of a beta-emitting radionuclide in two counting regions of the liquid scintillation pulse height spectrum is a constant at a constant level of quench regardless of the total activity of the radionuclide. This was demonstrated to be true by Verrezen and Hurtgen (2000) and Dobrin et al. (2011) by counting various activities of a beta-emitting radionuclide in two counting regions at identical quench levels. At a given quench level, the ratio of count rates in two counting regions due to the beta emissions of a radionuclide are a constant.

As described by Verrezen and Hurtgen (2000) the method is based on the following deconvolution algorithm

$$\text{NOI(net, low)} = \text{SMP(net, low)} - \sum_{k=1}^{n} K'_k(Q,\ \text{high})\cdot \text{REF}_k(\text{net, low}) \quad (6.92)$$

where

$$K'_k(Q,\ \text{high}) = \frac{\text{SMP(net, high)}}{\text{REF}_k(\text{net, high})} \quad (6.93)$$

In the above equations SMP(net, *i*) represents the net contribution (calculated as the difference between the gross contribution SMP(gross, *i*) and the background contribution BGD(*i*) of a sample in a specified counting region of interest ROI(*i*) as obtained from the spectral data. NOI(net, *i*) and REF_k(net, *i*) are the net contributions of the nuclide of interest and that of a reference spectrum of an impurity *k* to the beta spectrum in the same region of interest (*i*). The variable *n* refers to the number of radionuclide impurities in the sample. The term *low* represents a region of interest (ROI) in the low-energy range of the spectrum, whereas, the term *high* represents a ROI in the high-energy range, set behind or beyond the endpoint of the spectrum on the nuclide of interest in such a way that NOI(net, high) = 0, or in other terms SMP(net, high) = IMP(net, high). This is illustrated in Fig. 6.76 where the pulse height spectrum of ^{3}H, the nuclide of interest (NOI) is found only in the ROI *low*; whereby, NOI(net, high) = 0 and SMP(net, high) = IMP(net, high).

As described by Verrezen and Hurtgen (2000), Eq. (6.93) relies on the fact that the ratio of the net count rate in two fixed regions of the pulse height spectrum for a single

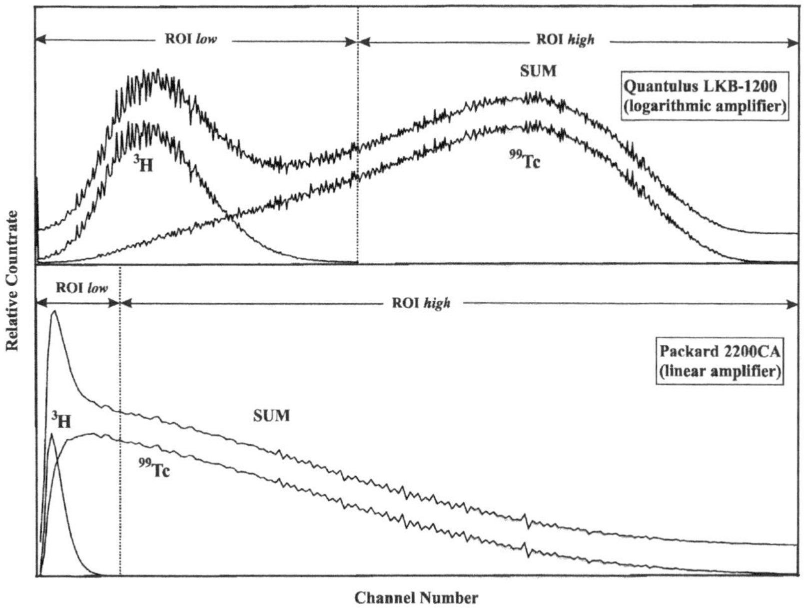

FIGURE 6.76 Effect of contamination of a ^{3}H sample with ^{99}Tc on the spectral shape as measured with a logarithmic amplifier (PerkinElmer/LKB Quantulus 1200, above) and a linear amplifier (PerkinElmer/Packard TriCarb 2200CA, below). Two regions of interest (ROI *low* and ROI *high*) are illustrated defined by a discriminator setting behind or beyond the endpoint of the nuclide of interest (^{3}H) pulse height spectrum. *From Verrezen and Hurtgen (2000), reprinted with permission from Elsevier © 2000.*

radionuclide is a constant for a given level of quenching and that any spectral shape (in this case the spectrum of the impurity, IMP) can be reconstructed using the established ratio and a pure reference spectrum if both are at the same level of quench, that is, if both possess the same quench indicating parameter, or

$$\frac{\mathrm{IMP(net,\ low)}}{\mathrm{IMP(net,\ high)}} = \frac{\mathrm{REF(net,\ low)}}{\mathrm{REF(net,\ high)}} = K(Q) \qquad (6.94)$$

Verrezen and Hurtgen (2000) and Dobrin et al. (2011) used the above equations to recalculate the spectral contribution of the impurity in the low-energy region of the spectrum (ROI, *low*) thereby correcting for the impurity in the measured sample spectrum. The technique is not confined to resolving the interference caused by only one radionuclide impurity. Samples containing n impurities require $n + 2$ pulse height spectra to solve Eq. (6.92), namely, the background spectrum, the sample spectrum and n reference spectra (one for each radionuclide impurity). Thus, the analysis of the contaminated samples by this method requires the preparation of the following counting vials: a background vial, a sample vial, and one internal reference vial for each radionuclide known to be present in the sample. The final mixtures in the counting vials must possess identical chemistry to assure that they all are characterized by the same quench indicating parameter. To assure this was achieved, Dobrin et al. (2011) analyzed the nuclide of interest (NOI) ^{3}H contaminated with the impurity (IMP) ^{90}Sr/^{90}Y by preparing the following counting vials with identical chemistry: (1) a background vial containing 10 mL of distilled water and 10 mL of Ultima Gold XR cocktail; (2) a sample vial containing 1 mL of sample (aqueous), 9 mL of distilled water and 10 mL of Ultima Gold XR cocktail; and (3) reference vials, each containing 1 mL of sample, 1 mL of aqueous reference solution of the appropriate radionuclide present in the sample, 8 mL of distilled water and 10 mL of Ultima Gold XR cocktail. For the analysis of ^{3}H contaminated with the impurity ^{90}Sr/^{90}Y Dobrin et al. (2011) corrected all spectra for the contribution due to background. They subsequently reconstructed the spectrum of the impurity (^{90}Sr/^{90}Y) in the lower energy range using the spectral shape of the reference spectrum and the calculated ratio $K_k'(Q,\ \mathrm{high})$of Eq. (6.93). Lastly, the net spectrum of the nuclide of interest, i.e., ^{3}H, in the lower energy range of the spectrum (ROI *low*) was restored by subtracting the net spectrum of the impurity from the net sample spectrum. If the reference vials were prepared with known activity concentrations of the radionuclides, the detection efficiencies can be obtained for each radionuclide, and the count rates converted to disintegration rates.

6. Multivariate calibration

The application of multivariate calibration for the determination of ^{14}C activities was discussed in Section V.F.4 of this chapter. It has been applied by Toribio et al. (1995, 1996) in the liquid scintillation analysis of mixtures containing the three main α-emitting isotopes of plutonium, ^{238}Pu, ^{239}Pu, and ^{240}Pu. The multivariate methods used were principal component regression (PCR) and partial least squares regression (PLSR). As explained by Toribio et al. (1995), these methods can estimate the concentration of radionuclide of interest without the need to consider the other radionuclides present in the system that affect the experimental signal, provided the isotopes are present in both the unknown samples and the calibration set. The advantage here is that the methods are suitable for the analysis of samples with unknown interferences, thereby avoiding the need of a separation step.

For the PCR and PLSR analysis Toribio et al. (1995) used a program written in FORTRAN 77 run on a personal computer with a DOS operating system described by Tauler et al. (1991), and the programs were validated by comparison with the PLS-Toolbox package for MATLAB written by Wise (1992). The α-emissions of ^{239}Pu and ^{240}Pu have overlapping energies, and the predicted error for the PCR or PLSR separation of these two nuclides was about 35%. These two nuclides are determined together as one component or analyte, $^{240+239}$Pu, which is similar to other established procedures using surface barrier detectors. Samples containing standards and unknowns of low-level activity (3.42–174 DPM) were counted for 5 h and data analyzed by relating known plutonium activities to alpha disintegrations registered in energy channels of a PerkinElmer Tri-Carb CA/LL liquid scintillation analyzer. Relationships were sought between two matrices $\mathbf{R}(M,N)$ and $\mathbf{C}(M,K)$. Matrix $\mathbf{R}(M,N)$ contained the instrumental responses, number of counts at N measuring channels for M calibration samples, and the other matrix $\mathbf{C}(M,K)$ contained the activities of the K analytes in the M samples. The important step in the data treatment in PCR and PLSR models, emphasized by Toribio et al. (1995), is the number of factors chosen as noted previously in Section V.F.4 of this chapter. When the number is chosen correctly, the data compression step filters out noise and useless data without sacrificing significantly the desired information. One approach to the proper selection of factors is to start with a probable number of factors. The analysis is repeated using a different number of factors until the optimum number that fits the data is found. The optimum number of factors in this case was found to be four, all attributable to sample activities. No factor was attributed to background, as predictions did not improve when background was subtracted.

The values obtained for $^{240+239}$Pu and ^{238}Pu activities for samples containing varying amounts of these radionuclides in mixtures over the range of 2.43–174 DPM had errors of about 7% and 10% for the high- and low-activity samples, respectively. Background (blank) subtraction is not required. This approach to radionuclide analysis is still at its early stages of development; however, Toribio et al. (1995) conclude that the multivariate calibration technique provides a detection limit one order of magnitude higher than that obtained with a surface barrier detector. However, the detection limit would be expected to drop with the use of α/β-discrimination and active shielding in LSA. With α/β-discrimination in many liquid scintillation analyzers the β-emitting isotope ^{241}Pu could also be included in the mixture, and activity measurements of mixtures containing the α- and β-emitting isotopes of plutonium in the same sample, as ^{238}Pu + $^{239+240}$Pu + ^{241}Pu, would be possible.

Organic extracting agents are often used to improve the alpha spectral resolution in the liquid scintillation analysis of alpha-emitters. However, subsequent to extraction, alpha peaks are reported to display a continuous shift along their axis with time for periods in excess of 200 h (Toribio et al. (1997). To overcome error due to the instability of alpha spectral peak positions Toribio et al. (1997, 1999) report the application of moving curve fitting (MCF) to the multivariate calibration analysis of mixtures of the four alpha-emitting isotopes ^{242}Pu, $^{239+240}$Pu, and ^{238}Pu with <15% relative errors of predicted values and an activity quantification threshold of 15 DPM.

Mellado et al. (2005) applied multivariate calibration, which constitutes a set of chemometric techniques, to the activity analysis of various combinations, quench levels, and ratios of ^{90}Sr/^{90}Y, ^{99}Tc, ^{63}Ni, ^{137}Cs, and ^{55}Fe within 15% relative error within the range covered by the calibration matrix. Chemometrics including artificial neural network (ANN) and partial least squares (PLS) as complimentary tools are reported by Khayatzadeh Mahani et al. (2008, 2009) for the analysis of ^{226}Ra and uranium as well as thorium and uranium in aqueous samples. Relative errors of prediction for PLS and ANN in synthetic mixtures of ^{226}Ra and uranium were 18.0% and 24.8%, respectively. Also, Mahani et al. (2012) applied PLS for the separation of a mixture of ^{235}U and ^{238}U with prediction relative errors lower than 4%. Fons-Castells et al. (2017a) applied also this tool for the quantification of ^{40}K, ^{60}Co, ^{90}Sr/^{90}Y, ^{134}Cs, ^{137}Cs, U, and ^{241}Am obtaining relative errors of prediction lower than 25%.

Now that the application of plastic scintillator microspheres to LSC are making rapid development (See Section XVI of this chapter), Bagán et al. (2011a) report the use of plastic scintillator microspheres together with multivariate calibration (PLS) to detect and measure the activities of α- and β-emitting radionuclides in mixtures. They quantified mixtures of ^{241}Am, ^{137}Cs, and ^{90}Sr/^{90}Y without the need of pulse height discrimination with errors under 10% in most cases. Measurement time is reduced, as the counting of background blanks are not necessary. The use of plastic scintillator microspheres also obviates the need for liquid scintillation cocktail and thus reduces the disposal costs that occur with mixed radioactive waste.

The PLS tool was also applied as part of a more complex system of radionuclide separation, Fons-Castells et al. (2017b, 2017c) explains a quantification procedure applied to water samples that includes alpha/beta separation to obtain the indicative dose followed by a RAD disk separation to isolate ^{226}Ra, ^{228}Ra, and ^{210}Pb as well as a gamma spectrometry to determine ^{60}Co, ^{134}Cs, ^{137}Cs, and ^{241}Am. With all of this information, the PLS tool was applied to quantify the other radionuclides presents in the original sample (U, ^{40}K, ^{90}Sr/^{90}Y, and $^{239+240}$Pu), as well as the quantification of the isotopes separated by the RAD disk, with a quantification bias below to 15%.

IX. Radionuclide standardization via LSA

Radionuclide standardization is included here as an altogether different specialty of liquid scintillation analysis (LSA), as it pertains to the purest form of measurement of the activity or disintegration rates (DPM) of radionuclides. It includes the methods used for the determination of the absolute activity of radionuclides as reference sources from which other radionuclide activities may be calibrated. The standardization of radionuclides is important in many fields, among which one of the most demanding has been radionuclide standardization for applications and research in therapeutic nuclear medicine (Coursey et al., 1991, 1994; Zimmerman, 2006b). This section will include some of the most popular liquid scintillation analytical methods applied to the standardization of radionuclides. Additional information on these techniques, among others used in radionuclide metrology, is available from Chapter 7 of Volume two entitled "Radionuclide Standardization" by Agustín Grau Malonda.

A. CIEMAT/NIST efficiency tracing

The CIEMAT/NIST method of radionuclide standardization, in its historical context, was first conceived by Agustín Grau Malonda in 1978 with the objective of developing a model/procedure applicable to any liquid scintillation analyzer, any scintillation cocktail, and any radionuclide (Grau Malonda, 1999). As the procedure could not be based solely on theoretical calculations of counting efficiency, Grau Malonda (1982a), in his doctoral dissertation, devised a model that combined theoretical calculations based on the radionuclide to be analyzed and experimental data provided by the liquid scintillation analyzer. Early descriptions of the method applied to the activity analysis of pure beta-particle emitting radionuclides are provided by Grau Malonda (1982a), Grau Malonda and García-Toraño (1982a,b), Grau Malonda et al. (1985), and for radionuclides decaying by electron capture by Grau Malonda (1982b).

Agustín Grau Malonda (1938–)

The name CIEMAT/NIST currently attributed to the method is the result of a collaboration on the method that started in 1984 (Coursey et al., 1986; Grau Malonda and Coursey, 1987) between the Centro de Investigaciones Energéticas Medioambientales y Tecnológicas (CIEMAT) in Madrid (the affiliation of Agustín Grau Malonda) and the National Institute of Standards and Technology (NIST), Gaithersburg, which at that time was called the National Bureau of Standards (the affiliation of Bert M. Coursey). Collaboration between the two institutions was intended to develop the method into one that could be used as a reference method for the standardization of radionuclides by any laboratory in the world with a conventional liquid scintillation analyzer. Other acronyms for the CIEMAT/NIST efficiency tracing method are found in the scientific literature including CNET or C/N efficiency tracing. The method is also referred to as 4πβ efficiency tracing.

The principles of the method and many of the major references to this technique will be presented here. Comprehensive treatments on the procedures and calculations involved are available from books by Grau Malonda (1995, 1999). A paper by García-Toraño *et al.* (1991) also provides a good summary of the method. Excellent reviews on this liquid scintillation radionuclide standardization method as well as other related methods of standardization by liquid scintillation are provided by Grau Malonda (1999), Günther (2002a,b), Zimmerman (2006b), Broda et al., 2007, Collé, 2009, and Grau Carles and Grau Malonda, 2010).

1. Theory and principles (^{3}H as the tracer)

As noted by Grau Malonda (1995, 1999) the original plan, conceived in 1978, was to develop a procedure for the standardization of radionuclides, which is applicable to any commercial liquid scintillation analyzer available in most laboratories, any liquid scintillation cocktail, and any radionuclide. To achieve this goal the method had to be based on a combination of theoretical calculations related to the particular radionuclide under measurement and its emissions as well as theoretical calculations related to a primary standard tracer nuclide (*e.g.*, ^{3}H) and its emissions together with experimental data from the liquid scintillation analyzer which provides information concerning the instrument, scintillation cocktail, and radionuclide (*e.g.*, sample count rate and quench indicating parameter).

The method is centered on the theoretical computer-based calculations of the counting efficiency of the radionuclide to be analyzed and that of a reference primary standard, such as tritium, for different values of figure of merit (*M*) and an experimental quench correction curve obtained from a set of ^{3}H quench standards. The figure of merit (*M*) is a term used in this technique, which is defined as the β-particle energy in keV required to produce one photoelectron by the photocathode of the photomultiplier tube (PMT) or, in other words, one photoelectron that reaches the first dynode of the PMT. The relationship of radionuclide counting efficiency and figure of merit for pure β-emitters and liquid scintillation systems comprised of two photomultiplier tubes in coincidence is defined by

$$\varepsilon = \int_0^{E_{max}} N(E)\left\{1 - \exp\left[\frac{-E\cdot X(E)}{2\cdot M}\right]\right\}^2 dE \qquad (6.95)$$

where E_{max} is the maximum β-particle energy, *N*(*E*) is the theoretical β-particle energy distribution, *X*(*E*) is the correction for ionization quenching and wall losses (wall effect), and *M* is the figure of merit (Grau Malonda, 1982b; Grau Malonda and García-Toraño, 1982a,b; Grau Malonda and Los Arcos, 1983; Grau Malonda et al., 1985; and Coursey et al., 1986, 1989).

The objective of the method is to obtain the counting efficiency of a nuclide under investigation (*e.g.*, ^{14}C) from the counting efficiency of an absolute or primary standard of ^{3}H. Hence, the term 'efficiency tracing' is used in naming the technique. Because the ^{3}H is a primary standard, its efficiency is taken as experimental data, while that of the nuclide under measurement (*e.g.*, ^{14}C) is obtained via a theoretical calculation and experiment. It was essential, therefore, that the calculation model for the counting efficiency, as envisaged by Grau Malonda, be so complete and accurate that the final results of the counting efficiency of the nuclide under investigation would be better than the precision limits of the experiment and instrumentation.

The term 'figure of merit' in this section has a totally different meaning than that used in other parts of this book, where the reader will find reference to figure of merit relating counting efficiency to background in the measurement of low activity samples. In this section, figure of merit refers exclusively to the quantitative yield of the photomultiplier dynode and its relation to counting efficiency. To avoid confusion, the term 'free parameter' or λ

has been adopted instead of figure of merit (M) when relating photomultiplier dynode yield and counting efficiency in this radionuclide standardization technique. Consequently Eq. (6.95) for the expression of the theoretical counting efficiency is often written as

$$\varepsilon = \int_0^{E_{max}} N(E)\left\{1 - \exp\left[\frac{-EQ(E)}{2\lambda}\right]\right\}^2 dE \quad (6.96)$$

where $Q(E)$ is the ionization quenching correction factor or $X(E)$ of Eq. (6.95), and λ is the free parameter or figure of merit M of Eq. (6.95).

The calculation model relies on the decay schemes of the nuclide of interest, and for this purpose, tables of beta particle (i.e., negatron) and positron spectra were published by Grau Malonda and García-Toraño (1978 and 1981a) from which the spectral area over any given energy range may be obtained. Also, tables of calculated counting efficiencies as a function of the free parameter were published for numerous radionuclides decaying by pure negatron emission (Grau Malonda and García-Toraño, 1981b; Grau Malonda et al., 1985), electron capture decay (Grau Malonda, 1982b), positron emission (Grau Malonda and García-Toraño,1982b), beta-gamma-ray emission (García-Toraño and Grau Malonda, 1988), and electron capture decay with gamma-ray emission (Grau Malonda and Fernández, 1985).

As the calculation of the theoretical counting efficiency must account for all beta transitions, Eq. (6.96) is modified accordingly. For example, ^{125}Sb undergoes eight beta-decay transitions to the daughter nuclide ^{125m}Te. Consequently, to account for all beta-decay transitions, Grau Carles et al. (1998) used the following expression, according to Grau Malonda (1995, 1999), for the calculation of ^{125}Sb counting efficiency

$$\varepsilon = \sum_{i=1}^{8} p_i \int_0^{(E_m)_i} N_i(E)\left\{1 - \exp\left[\frac{-EQ(E)}{2\lambda}\right]\right\}^2 dE \quad (6.97)$$

Where p_i, $(E_m)_i$, and $N_i(E)$ are respectively the eight beta-particle transition intensities, end-point energies (E_{max}), and Fermi (beta-particle energy) distributions; E, $Q(E)$ and λ are as previously described. The calculated counting efficiency of ^{125}Sb as a function of the free parameter is therefore, a sum of the calculated efficiencies for each of the eight beta decay transitions.

The calculation of the theoretical counting efficiency can be yet more complex when numerous beta- and gamma transitions are part of the radionuclide decay scheme. The standardization of ^{110m}Ag, described by García-Toraño et al. (2000), is an interesting example. It decays to ^{110}Cd by β emission (98.65%) and isomeric de-excitation by γ emission (1.35%) to ^{110}Ag. The decay scheme comprises more than 12 β branches and 50 gamma rays. García-Toraño et al. (2000) used the CIEMAT/NIST method to standardize samples of ^{110m}Ag, which required the calculation of the Fermi spectra of all β branches, as well as the Compton spectra produced by the γ-ray interactions with the scintillator. A total of 128 decay pathways are possible, and the researchers calculated their probabilities and counting efficiencies to obtain the overall counting efficiency. The expression used for the calculation of the theoretical counting efficiency, in this case, was that described by García-Toraño et al. (1991) for a radionuclide that undergoes β decay followed by a cascade of n γ transitions

$$\varepsilon = \int_0^{E_\beta}\int_0^{E_{\gamma 1}}\cdots\int_0^{E_{\gamma n}} N(E)S_1(E_1)\cdots S_n(E_n)$$

$$\times\left\{1 - \exp\left[\frac{-E_\beta Q(E_\beta) - \sum_{i=1}^{n} E_{\gamma i}Q(E_{\gamma i})}{2\lambda}\right]\right\}^2 dEdE_1\cdots dE_n \quad (6.98)$$

where $N(E)$ and $S_i(E)$ are the β spectrum and Compton spectrum of a given γ transition respectively, $Q(E)$ is the ionization quenching factor, λ is the free parameter or figure of merit, and E_β, $E_{\gamma 1}$, ... $E_{\gamma n}$ are the maximum energies of the β and the energies of the γ transitions, respectively. García-Toraño et al. (2000) factorized the integral expressions of Eq. (6.98) so that single integrals could be solved and the results combined to yield the complete counting efficiency.

Computer programs or codes have been developed for the rapid computation of counting efficiency as a function of the figure of merit (*i.e.*, free parameter) as described by Grau Malonda (1995, 1999) and Grau Malonda et al. (1987). The computer program BETA (García-Toraño and Grau Malonda, 1985), BETA3 (Grau Malonda, 1999), EFFY (García-Toraño and Grau Malonda, 1981, 1985), and PTB code (Physikalisch-Technische Bundesanstalt) code, which is a modified EFFY4 code (Kossert and Schrader, 2004) have been written for computations involving pure beta (negatron or positron) emitters, and the programs EFFY2, EFFY4 (García-Toraño and Grau Malonda, 1981, 1985 and Kossert, 2003), CN2001A (Günther, 2001 and Zimmerman et al., 2002), CN2003 (software available via the Laboratoire National Henri Becquerel, http://www.nucleide.org/ICRM_LSC_WG/icrmsoftware.htm, accessed October 16, 2018) and CN2005 (Günther, 2002a, 2002b), and MICELLE2 (Kossert and Grau Carles, 2010; Altzitzoglou and Rožkov, A., 2016) are available for the computation of counting efficiency as a function of free parameter for complex-decay beta-emitters, and CN2004 code for a positron-emitter (Cessna et al., 2008) and for series of beta and alpha decays Cessna and Zimmerman (2010). The programs calculate the beta spectra using the theory of radioactive beta decay, taking into account the shape factors (*e.g.*, Grau Carles, 1995, 2005; Grau Carles and Kossert, 2006, 2007; Grau Carles et al., 2008; and Kossert et al., 2011a). An updated MICELLE2 code, reported by Kossert and Grau Carles (2010) cited further on in this chapter and the EMI2 (Grau Malonda, 1999) and EMI2X computer programs (Broda et al., 2012) were prepared to facilitate the computation of counting efficiencies of radionuclides with complex beta-decay schemes, such as those occurring in radionuclides decaying by electron capture. The total detection probability is obtained by dividing the spectrum into bands, calculating the partial detection probabilities of each of them, and summing. The average and mean energies in the spectra are also calculated. Because of the different decay processes involved, special programs have been written to compute the counting efficiency as a function of free parameter such as EFYGA for X-ray and gamma-emitters (García-Toraño and Grau Malonda, 1987), EBEGA for beta-gamma-emitters (García-Toraño et al., 1988), and VIASKL (Los Arcos et al., 1987) and VIAS1 (Grau Malonda, 1999) for K-L shell electron-capture nuclides, and EMI (Grau Carles et al., 1994b) and EMI2 (Grau Malonda et al., 1999) for electron capture-gamma and isomeric transition nuclides, and EMIS (Grau Malonda, 1999) for radionuclides decaying by electron capture, electron capture coincident with a gamma transition, and pure isomeric transition, and ADDI (Grau Malonda, 1999) computes the counting efficiency of a radionuclide decaying by electron capture and positron emission. An updated program, EMILIA (Grau Carles, 2006a) is applied to counting efficiency calculations for electron-capture and capture-gamma emitting nuclides. A paper by Los Arcos et al. (1991) provides a detailed treatment of the method applied to multigamma electron-capture radionuclides. The computer programs are available from CIEMAT, Avda. Complutense 22, 28040 Madrid, Spain. Also, the program CIENIST99 written in Visual Basic 5 reported by Günther (2000) combines the CIEMAT/NIST codes BETA, EBEGA, and EMI to accommodate multi-beta/multi-gamma transitions.

The CIEMAT/NIST procedure integrates theoretical calculations with experimental measurements. The free parameter (figure of merit) is an essential component in the theoretical calculation of the counting efficiency, while the quench indicating parameter (QIP) of a particular liquid scintillation analyzer (*e.g.*, tSIE, SQP(E), or H#) is required for the experimental aspect of this method. However, as explained by Grau Malonda and García-Toraño (1982a,b), the figure of merit cannot be obtained directly from experiment and the quenching parameter is not theoretically computable. Therefore, the method makes use of a plot that relates the figure of merit (free parameter) with the quench-indicating parameter. Such a plot is obtained by integrating an experimental quench correction curve of a primary standard, such as ^{3}H (*i.e.*, experimental counting efficiency of ^{3}H vs. a QIP), with a theoretically computed curve of counting efficiency versus figure of merit. The term figure of merit may be abbreviated in the scientific literature as λ, *M*, or FOM.

Tritium is one of the most suitable radionuclides to use as a standard for this method when the analysis of beta-emitters is required. It is readily available as an absolute standard as tritiated water or in organic form, it has a relatively long half-life and, as noted by Coursey et al. (1998), it provides for more sensitive extrapolations to the low-energy portions of the beta-particle spectra than higher-energy standards. Radioactive standards of tritiated water disseminated by the Laboratoire Primaire des Ionizants (LPRI), France, and the National Institute of Standards and Technology (NIST), USA, were intercompared by Zimmerman and Collé (1997a) and demonstrated an apparent mean disagreement between standards of $<0.4\%$ on a relative basis. As a standard for the CIEMAT/NIST method, ^{3}H has the advantage that any uncertainties on the efficiency curve of this radionuclide are reduced for the efficiency calculations of the more energetic β-emitters. See Grau Malonda and García-Toraño (1982a,b) and Grau Malonda (1995, 1999) for calculations demonstrating the uncertainty propagations for tritium.

The objective is to obtain a counting efficiency quench correction curve for any radionuclide, which may decay by β-emission, β/γ with X-ray and γ-photon emissions, EC decay, and isomeric transition. The counting efficiency quench correction curve is then used to determine the absolute activity (DPM) of a radionuclide sample under investigation from its experimental quench indicating parameter.

A most concise and comprehensive explanation of the principles, upon which the CIEMAT/NIST method is based, is provided by Kulkarni et al. (2011a,b) in the following excerpt with the writer's notes in parenthesis (reprinted with permission from Elsevier © 2011):

> *The CIEMAT/NIST method involves the theoretical calculation of detection efficiency of a sample as a function of free parameter, the Figure of Merit (FOM). FOM cannot be measured experimentally and the quench level cannot be calculated theoretically, hence the relationship between FOM and quench indicating parameter is obtained by counting a set of* ^{3}H *quench standards. If the sample is prepared in the same scintillator, with same composition as that of the* ^{3}H *quench standards, then the relationship between FOM and quench indicating parameter, established using the* ^{3}H *nuclide, holds good for any other radioactive sample for that particular liquid scintillation counter. An advantage of using* ^{3}H *as a tracer is that the uncertainty of* ^{3}H *activity has a small effect on the uncertainty of the*

measured sample activity (Grau Malonda and García-Toraño, 1982a). This is due to the fact that ${}^3H$ has (emits) low energy beta particles and (as a consequence) is affected the most due to quenching. This means a slight variation in quenching leads to a large variation in the measured efficiency and hence the FOM. In (the) case of higher energy beta-emitters, the variation of efficiency with quenching is less. Thus, a large variation in the tritium efficiency leads to a very small variation in the sample efficiency and hence less uncertainty.

2. Procedure

The procedure for obtaining the counting efficiency quench correction curve for a nuclide under investigation with ${}^3H$ as the standard is outlined next. The sequence of steps in the procedure is illustrated in Fig. 6.77, and the curves established as a result of each step are illustrated in Fig. 6.78:

a. A set of ${}^3H$ quenched standards is counted in the liquid scintillation analyzer and a quench correction curve of tritium counting efficiency versus the quench-indicating parameter is plotted. See curve (A) of Fig. 6.78.

b. The theoretical counting efficiency of ${}^3H$ (ε_T) as a function of figure of merit is computed. See curve (B) of Fig. 6.78.

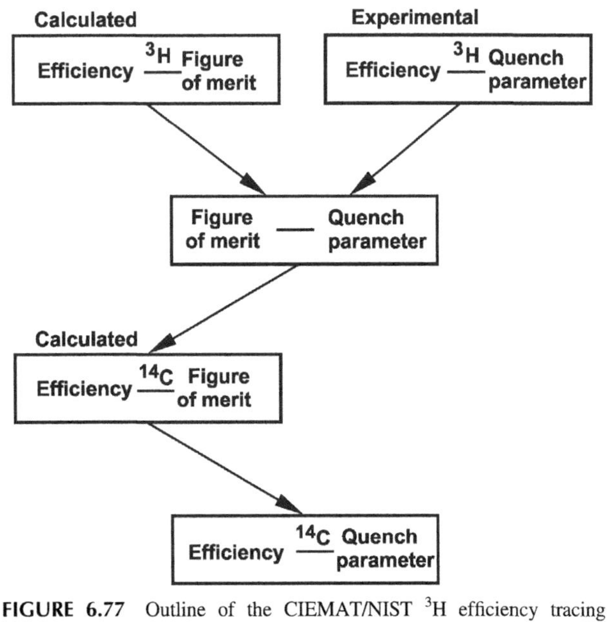

FIGURE 6.77 Outline of the CIEMAT/NIST ${}^3H$ efficiency tracing method of radionuclide standardization. The diagram illustrates the sequence of calculated and experimental relationships, which are defined to yield the quench correction curve of any radionuclide under investigation. Tritium is the standard (or tracer nuclide) in all cases, while ${}^{14}C$, illustrated here as an example, is the radionuclide under investigation. *From Grau Malonda and García-Toraño (1982a), reprinted with permission from Elsevier ©1982.*

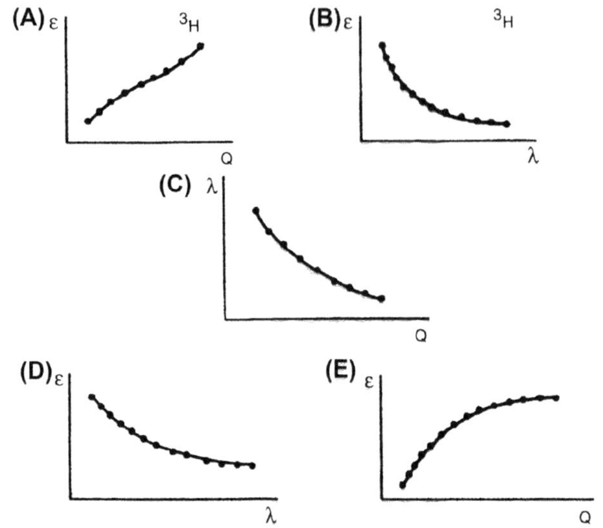

FIGURE 6.78 Calculated and experimental calibration curves obtained in the radionuclide standardization method known as the CIEMAT/NIST ${}^3H$ efficiency tracing method. The curves are described in the sequence by which they are obtained as follows: (A) experimental quench correction curve of ${}^3H$ counting efficiency (ε_T) versus quench indicating parameter (Q); (B) computed ${}^3H$ counting efficiency (ε_T) versus figure of merit or free parameter (λ); (C) "universal" curve of figure of merit (free parameter, λ) versus quench indicating parameter (Q) derived from the above two relationships; (D) computed curve of the counting efficiency of the nuclide under investigation (ε_{nuc}, *e.g.*, ${}^{14}C$) versus the figure of merit or free parameter (λ); and (E) quench correction curve of counting efficiency of the nuclide under investigation (ε_{nuc}, *e.g.*, ${}^{14}C$) versus the instrumental quench indicating parameter (Q) derived from curves (C) and (D). *From Grau Malonda (1995, 1999), reprinted with permission from CIEMAT © 1995–99.*

c. From the preceding two relationships, a curve of the figure of merit as a function of the quench indicating parameter is obtained. Although thse two relationships are determined from ${}^3H$ standards, the figure of merit as a function of quench indicating parameter is independent of any radionuclide. It is a "universal" curve, which may be applied to any radionuclide, that undergoes beta decay. The curve is suitable only for the particular liquid scintillation analyzer and scintillator used. See curve (C) of Fig. 6.78.

d. The theoretical counting efficiency of the nuclide under investigation (ε_{nuc}) as a function of figure of merit is calculated. See curve (D) of Fig. 6.78.

e. Consequently, the relationship between the counting efficiency of the nuclide under investigation (ε_{nuc}) and the quench indicating parameter can be obtained. This plot is obtained from the two preceding relationships, that of the universal curve of figure of merit versus the quench indicating parameter and the computed curve of ε_{nuc} versus figure of merit. See curve (E) of Fig. 6.78.

f. The final curve obtained in step e can be used to determine the counting efficiencies of experimental samples. The experimental sample of the nuclide under investigation is counted and the count rate (CPM) and quench indicating parameter are recorded. From the quench indicating parameter and the quench correction curve of ε_{nuc} versus QIP [Curve (E) of Fig. 6.78], the counting efficiency of the experimental sample ε_{nuc} is obtained.

g. The activity A in units of disintegrations per minute (DPM) of the nuclide under investigation is then calculated as

$$A = \frac{CPM}{\varepsilon_{nuc}} \tag{6.99}$$

The computer calculations of counting efficiency as a function of figure of merit (M) were reported by Grau Malonda et al. (1985) for 35 pure β-emitting radionuclides. These fitted data of counting efficiency as a function of figure of merit are reported as polynomial coefficients to the fifth degree according the equation

$$\ln \varepsilon = A_0 + A_1 \ln M + A_2(\ln M)^2 + \cdots \tag{6.100}$$

The coefficients are listed in Table 6.8. The computer programs for calculating these coefficients and those of many other radionuclides are described in the publications cited in the previous section under Theory and Principles (^{3}H as the Tracer), and the entire programs are available from Grau Malonda (CIEMAT, Avda. Complutense 22, 28,040 Madrid, Spain) or Grau Malonda (1999). The figure of merit of the standard ^{3}H is obtained from the relationship of (ε_T) versus M. The good fit for these data is reported by Grau Malonda et al. (1985) according to the following fifth-degree polynomial:

$$\ln M = B_0 + B_1(\ln \varepsilon) + B_2(\ln \varepsilon)^2 + \cdots, \tag{6.101}$$

where the values of B_i for ^{3}H are reported to be B_0 = 3.11669, B_1 = −0.585619, B_2 = 0.137378, B_3 = −0.148844, B_4 = 0.0540364, and B_5 = −0.00705037. Finally, the quench curve of ε_{nuc} versus a QIP for 34 of the radionuclides listed in Table 6.8 can be obtained from the experimental quench correction curve for ^{3}H (*i.e.*, ε_T vs. QIP) and the data obtained using Eqs. (6.101) and (6.102) and the respective polynomials.

3. Specific examples

An excellent example of the CIEMAT/NIST method using ^{3}H as the primary standard tracer nuclide can be taken from the work of Kulkarni et al. (2011a,b) in the liquid scintillation standardization of the radionuclide ^{131}I. A chronological outline of the procedure is provided as follows:

a. The first step taken by Kulkarni et al. (2011a,b) was to count a set of ^{3}H standards, quenched with nitromethane, with a Packard Tri-Carb liquid scintillation counter (now produced by PerkinElmer). The liquid scintillation counter is equipped with two photomultipliers operating in coincidence and contains a ^{133}Ba external standard providing a quench indicating parameter, known as tSIE, described previously in this chapter. The ^{3}H quenched standards each contained the same known activity (0.9985 ± 0.0035 KBq/g), and same scintillation cocktail with different amounts of quenching agent. From the count rates provided by the liquid scintillation analyzer and known activity of ^{3}H in each counting vial, Kulkarni et al. (2011a,b) were able to plot the relationship of ^{3}H counting efficiency (%E) as a function of the instrumental quench indicating parameter (tSIE) illustrated by curve (A) of Fig. 6.79. The reader may want to compare this curve (A) with the same curve in the previous Fig. 6.78, which was used by Grau Malonda (1995, 1999) to describe the basic principle of the technique.

b. Kulkarni et al. (2011a,b) then proceeded to compute the theoretical ^{3}H counting efficiency using the computer code CN2003 reported by Günther (2002a) and available from the Laboratoire National Henri Bequerel (http://www.nucleide.org/ICRM_LSC_WG/icrmsoftware.htm accessed October 16, 2018). The computed results provided the data for the plotting of the relationship of computed ^{3}H counting efficiency versus figure of merit (FOM), also abbreviated as λ or M (See curve B of Fig. 6.79 and compare that to curve B of Fig. 6.78).

c. From curves A and B, Kulkarni et al. (2011a,b) were able to construct the universal curve of figure of merit (FOM) versus the quench indicating parameter (tSIE) as illustrated by curve C of Fig. 6.79. This may be compared to curve C of Fig. 6.78 in the example provided by Grau Malonda (1995, 1999).

d. The subsequent relationship of the theoretical/computed ^{131}I counting efficiency as a function of figure of merit (FOM) was determined using the computer program CN2003. This is illustrated by curve D of Fig. 6.79. See also the same curve D of Fig. 6.78 provided by Grau Malonda (1996, 1999) as an example in his explanation of the method.

e. From the relationships depicted in curves C and D of Fig. 6.79, Kulkarni et al. (2011a,b) were able to plot finally the relationship of the % counting efficiency of ^{131}I versus the instrumental quench indicating parameter (tSIE) illustrated by curve E of Fig. 6.79. This final relationship of ^{131}I counting efficiency was then used to determine the absolute activity or disintegration rate of samples of ^{131}I in terms of disintegrations per minute (DPM) from the instrumental measurements of the sample count rate in counts per minute (CPM) and the measured quench indicating parameter (tSIE). For example, if the liquid scintillation analyzer measured

TABLE 6.8 Polynomial coefficients for fitting counting efficiency as a function of the figure of merit.

Nuclide	Polynomial coefficients						Maximum disc (%)
	A_0	A_1	A_2	A_3	A_4	A_5	
^{3}H	0.405213 4− 1	−0.519852	−0.238284	−0.542988 − 1	0.233709 − 1	−0.229258−2	0.2
^{10}Be	0.460257 + 1	−0.833326 − 3	−0.14108 − 2	−0.369043−2	0.239038 − 2	−0.675201 − 3	0.07
^{14}C	0.456074 + 1	−0.498280−1	−0.212707−1	0.599262 − 2	−0.114070−1	0.955890 − 3	0.3
^{31}Si	0.460386 + 1	−0.996311−3	−0.651159−3	−0.102653 − 2	0.592699 − 3	−0.174146 − 3	0.02
^{32}Si	0.456478 + 1	−0.405983−1	−0.175982−1	0.923926 − 3	−0.493393−2	−0.792020−4	0.1
^{32}P	0.460414 4 + 1	−0.759601−3	−0.494713−3	−0.795852 − 3	0.460529 − 3	−0.133668−3	0.01
^{33}P	0.457255 4 + 1	−0.319945−1	−0.144253 − 1	−0.128012−2	−0.230966−2	−0.452069 − 3	0.07
^{35}S	0.454759 + 1	−0.580615 − 1	−0.236917−1	0.373332 − 2	−0.928844−2	0.654008 − 3	0.2
^{39}Ar	0.459712 + 1	−0.619246−2	−0.340035−2	−0.367905−2	0.187337−2	−0.608778−3	0.06
^{45}Ca	0.457021 + 1	−0.338821 − 1	−0.151187−1	−0.174736−2	−0.212018 − 2	−0.456534−3	0.06
^{63}Ni	0.441361 + 1	−0.166749	−0.657592−1	−0.176547−1	−0.139047−1	0.280594 − 2	0.07
^{66}Ni	0.455387 + 1	−0.504729−1	−0.212871 − 1	0.588624 − 3	−0.573001−2	0.103022−3	0.2
^{69}Zn	0.459965 + 1	−0.455061−2	−0.239135 − 2	−0.231221−2	0.110588−2	−0.369529−3	0.03
^{79}Se	0.457489 + 1	−0.375947 − 1	−0.161750−1	0.961543−2	−0.109656 − 1	0.656409 − 3	0.3
^{85}Kr	0.459722 + 1	−0.682161−2	−0.340506−2	−0.288481−2	0.123660 − 2	−0.433468−3	0.04
^{87}Rb	0.459557 + 1	−0.777543−2	−0.575931−2	−0.465487−2	0.291824−2	−0.128978−2	0.06
^{89}Sr	0.460272 + 1	−0.237735−2	−0.119424−2	−0.101322−2	0.438239 − 3	−0.154756−3	0.01
^{90}Sr	0.459429 + 1	−0.892563−2	−0.451710 − 2	−0.390183−2	0.172480 − 2	−0.597759−3	0.06
^{90}Y	0.460413 + 1	−0.106759 − 2	−0.558493−3	−0.520378 − 3	0.242267 − 3	−0.808739−4	0.008
^{91}Y	0.460279 + 1	−0.218860 − 2	−0.110022−2	−0.945244 − 3	0.409650 − 3	−0.143794 − 3	0.01
^{99}Tc	0.457235 + 1	−0.312754−1	−0.140836−1	−0.330923−2	−0.891300−3	−0.555615 − 3	0.03
^{106}Ru	0.426331 + 1	−0.279785	−0.120473	−0.527075−1	0.389097 − 2	0.928245 − 3	0.4
^{109}Pd	0.446763 + 1	−0.156553	−0.109538	−0.755600−1	0.902461 − 2	0.629885 − 3	0.6
^{113m}Cd	0.459211 + 1	−0.114637−1	−0.560371−2	−0.371747−2	0.133895−2	−0.579840−3	0.04
^{115}In	0.460051 + 1	−0.234863−2	−0.216760−2	−0.437909 − 2	0.277809 − 2	−0.799807−3	0.08
^{121}Sn	0.458112 + 1	−0.214372−1	−0.100221−1	−0.422650−2	0.702033 − 3	−0.677667−3	0.03

^{123}Sn	0.460229 + 1	−0.283529−2	−0.139369−2	−0.108105 − 2	0.430745 − 3	−0.161408−3	0.02
^{135}Cs	0.457944 + 1	−0.268855 − 1	−0.129479−1	0.900326−3	−0.278706 − 2	−0.546387 − 3	0.1
^{143}Pr	0.459885 + 1	−0.551151−2	−0.275820−2	−0.223641−2	0.930999 − 3	−0.342315−3	0.03
^{185}W	0.458369 + 1	−0.191761 − 1	−0.898716−2	−0.414450 − 2	0.842663 − 3	−0.645690−3	0.03
^{187}Re	0.167732 + 1	−0.165469 + 1	−0.152546	0.391800−1	−0.572250−2	0.368647 − 3	0.01
^{188}W	0.457681 + 1	−0.256581 − 1	−0.117439−1	−0.394376 − 2	0.558416−4	−0.620994 − 3	0.01
^{209}Pb	0.459299 + 1	−0.109936−1	−0.529640−2	−0.334930−2	0.112054−2	−0.511350−3	0.04
^{241}Pu	0.394623 + 1	−0.535624	−0.217935	−0.533980−1	0.206078 − 1	−0.184730−2	0.2
^{249}Bk	0.450771 + 1	−0.938177−1	−0.365791 − 1	0.258343 − 2	−0.141435 − 1	0.176016 − 2	0.3

From Grau Malonda et al. (1985), reprinted with permission from Elsevier © 1985.

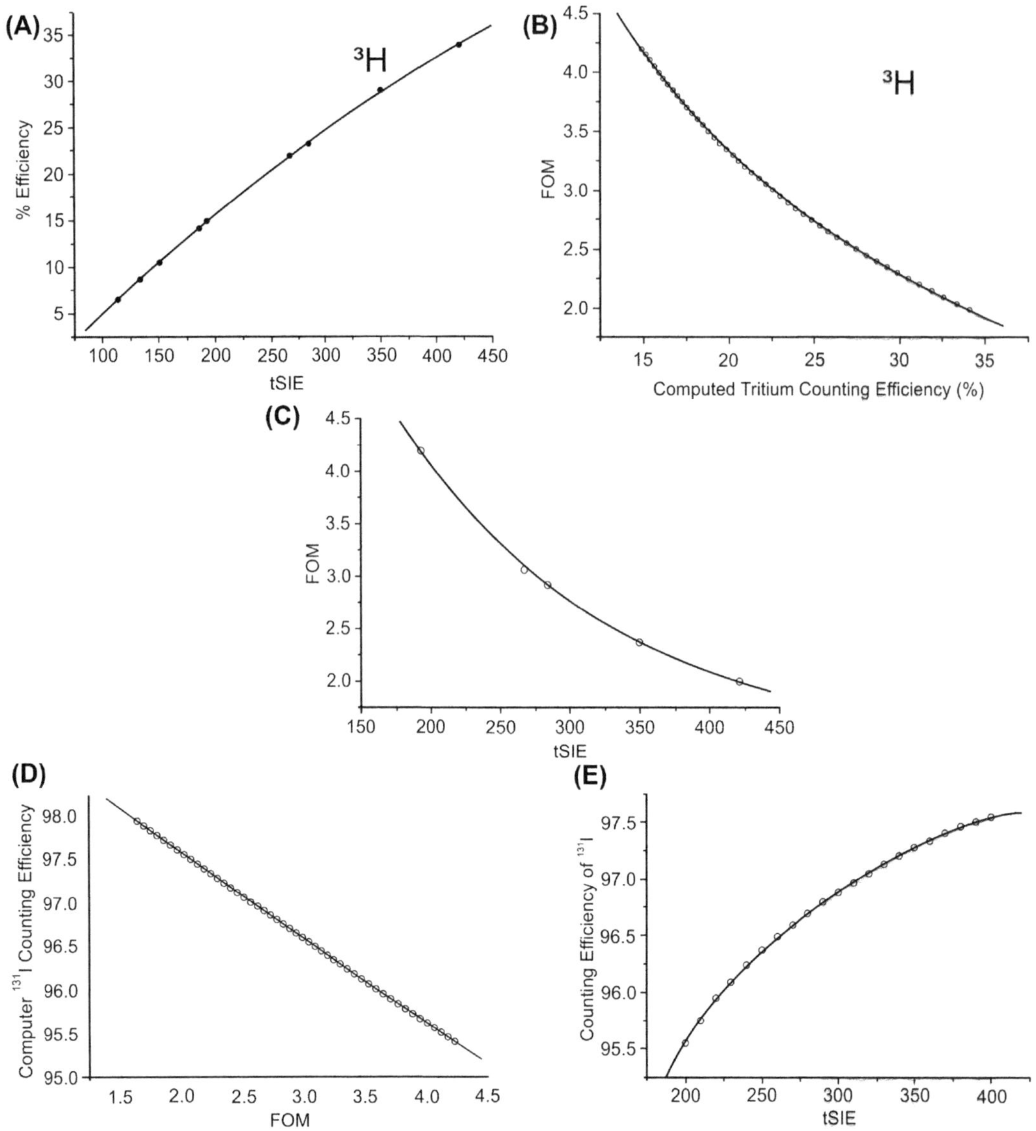

FIGURE 6.79 Set of curves/relationships established for the standardization of ^{131}I by the CIEMAT/NIST ^{3}H efficiency tracing technique yielding ^{131}I activity concentration measurements with a combined relative standard uncertainty of only 0.35% at a coverage factor $k = 1$. To facilitate the narrative of this example, the writer had arranged the curves to correspond to the sequence of curves of Grau Malonda illustrated in the previous Fig. 6.78, The sequence of curves are as follows: (A) Experimentally determined quench correction curve of ^{3}H efficiency versus quench indicating parameter (tSIE), (B) curve of figure of merit (FOM) versus theoretically computed ^{3}H efficiency, (C) universal curve of figure of merit versus tSIE from the previous two curves, (D) computed efficiency of ^{131}I (%) versus figure of merit, and (E) counting efficiency of ^{131}I (%) versus quench indicating parameter (tSIE) from the universal curve C and the computed curve (D) *From Kulkarni et al. (2011a,b) reprinted with permission from Elsevier © 2011.*

a tSIE of 325 of an unknown sample of ^{131}I, the % counting efficiency of 97.13 may be extracted from the relationship depicted in curve e of Fig. 6.79. The measured % counting efficiency (%E = 97.13%) expressed as a decimal (E = 0.9713) would then be used to calculate the disintegration rate (DPM) of the sample from its instrumentally measured count rate (CPM) according to the relation DPM = CPM/E.

The relationships established for this standardization of ^{131}I, depicted by curves A through E of Fig. 6.79, can apply only to the specific liquid scintillation analyzer, cocktail formula, cocktail volume, and counting vials used. If another liquid scintillation analyzer (LSA) is used, even another LSA of the same make and model, the relationships described by curves A through E of Fig. 6.79, would need to be redetermined. Cassette (2016) describes a software package entitled QUENCH, which is freely available for the determination of quenching curves. The computer program provides the best fitting function to experimental data for curves with uncertainties in terms of

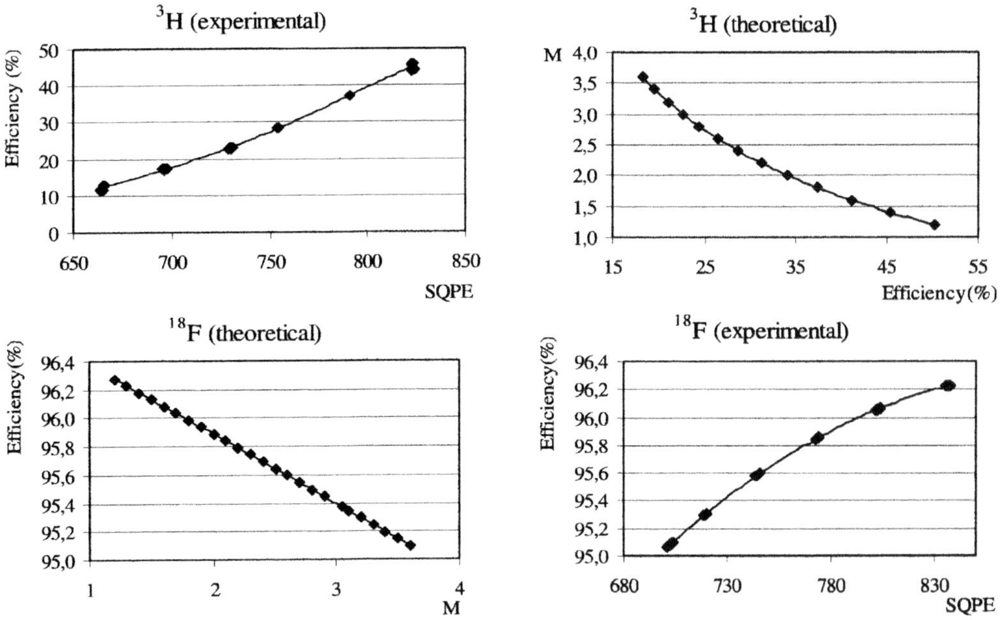

FIGURE 6.80 CIEMAT/NIST method applied to the standardization of ^{18}F. *From Capogni et al. (2006), reprinted with permission © 2006 IOP Publishing Ltd.*

the variables efficiency and quenching. The program provides the user with calculated detection and uncertainty with respect to a measured quench indicating parameter. The QUENCH program and user guide are available on the website of the Laboratoire National Henri Becquerel (LNHB), http://www.lnhb.fr/icrm_lsc_software/, (accessed February 9, 2018).

For the standardization of ^{131}I, Kulkarni et al. (2011a,b) were able to demonstrate that the CIEMAT/NIST method could be used to determine the ^{131}I activities with a combined relative standard uncertainty of 0.35% at a coverage factor of $k = 1$ after summing all sources of uncertainty.

Another illustrative example can be taken from the work of Capogni et al. (2006) on the standardization of [^{18}F]fluorodeoxyglucose, a radiopharmaceutical used in positron emission tomography (PET). The procedure used by Capogni et al. (2006) was the following:

a. The counting efficiency of the tracer nuclide, ^{3}H, as a function of the quench indicating parameter, SQP(E), was determined experimentally with a PerkinElmer "Wallac 1414 WinSpectral" liquid scintillation counter using a set of tritium standards at various quench levels. The plot of % detection efficiency of the tracer nuclide versus SQP(E), i.e., ε_T versus SQP(E) is illustrated in Fig. 6.80 [with the heading ^{3}H (experimental)].

b. The counting efficiency of the tracer nuclide (^{3}H) is calculated as a function of the free parameter (*M*), defined as the effective energy necessary to create one photoelectron at the photocathode. For this calculation, Capogni et al. (2006) used the CN2001 code reported by Günther (2002a). The plot of *M* versus % detection efficiency of the tracer nuclide, i.e., *M* versus ε_T, is illustrated in Fig. 6.80 [with the heading ^{3}H (theoretical)].

c. The independent nuclide relationship between the free parameter (*M*) and the quench indicating parameter SQP(E) was then derived from the functions previously determined in steps (1) and (2), that is, ε_T versus SQP(E) and *M* versus ε_T. This relationship is not illustrated as a plot in Fig. 6.80, but it would be a plot analogous to curves C of Figs. 6.78 and 6.79.

d. The counting efficiency of the nuclide of interest, ε_{nuc}, (in this case ^{18}F) as a function of the free parameter (*M*) was calculated using the CN2001 code. The plot of % detection efficiency of the nuclide of interest (^{18}F) versus the calculated free parameter (*M*), i.e., ε_{nuc} versus *M*, is illustrated in Fig. 6.80 [with the heading ^{18}F (theoretical)].

e. Finally, the counting efficiency of the nuclide of interest (^{18}F) as a function of the quench indicating parameter [SQP(E)], which is the curve illustrated in Fig. 6.80 [with the heading ^{18}F (experimental)], is obtained from the calculated counting efficiency of ε_{nuc} versus *M* of the previous step (d) and the derived function or relationship between the free parameter (*M*) and the quench indicating parameter [SQP(E)], determined in step (c).

f. The count rate (CPM) of a sample of the nuclide of interest (^{18}F) and its quench indicating parameter [(SQP(E)] are measured by the liquid scintillation counter. The detection efficiency of the sample is then obtained directly from the derived curve illustrated in Fig. 6.80, labeled ^{18}F (experimental). The activity of the nuclide of interest A_{nuc} in DPM is then calculated from the sample count rate and its detection efficiency according to the basic relationship $A_{nuc} = CPM/\varepsilon_{nuc}$. In the case of ^{18}F and other short-lived nuclides, correction for decay during counting is required.

The CIEMAT/NIST efficiency tracing with ^{3}H as the tracer nuclide has become a very popular method for radionuclide standardization, as no special equipment is required. A common modern liquid scintillation analyzer and personal computer is all that is needed. The principal advantage of the CIEMAT/NIST efficiency tracing method is that only standards of ^{3}H are required, that is, no standards of the nuclide under investigation are needed. This is particularly relevant when the activities of short-lived nuclides must be determined very accurately and on-site, such as those utilized in nuclear medicine. Also, the calculated uncertainties reported in the activity analysis of a broad range of radionuclides are low. Table 6.9 lists some examples of radionuclides in order of increasing mass number reported in the literature as standardized by the CIEMAT/NIST method with the calculated standard uncertainties of the determinations expressed as a percent. The computer codes used for the calculations of the detection efficiencies against the figure of merit can be obtained from the references cited in the table. The radionuclides and references cited in Table 6.9 are not exhaustive. They serve only as examples of the broad spectrum of beta-decay radionuclides that can be standardized by this technique. See also the chapter entitled Radionuclide Standardization by Agustín Grau Malonda (Chapter 7 of Volume 2) for more additional information on radionuclides standardized by the CIEMAT/NIST method.

4. Sample, cocktail, and spectrometer stability

Liquid scintillation cocktail volume can have a significant effect on the count rate of weak β-emitters, such as ^{63}Ni ($E_{max} = 67$ keV). A study by Zimmerman and Collé (1997b) demonstrated that a 7% variation in the liquid scintillation counting efficiency of ^{63}Ni can be observed over the scintillation cocktail volume range of 1−20 mL. However, they demonstrate that the CIEMAT/NIST method is able nevertheless to trace the observed ^{63}Ni activity to about 0.1%. To assure precise standardizations when employing the CIEMAT/NIST efficiency tracing method some researchers will centrifuge the liquid scintillation counting vial to assure that no significant amount of nuclide and cocktail remain adhered to the inner surface of the counting vial cap (Kulkarni et al., 2011a,b; Kossert et al., 2012; Kossert and Nähle, 2014; Tarancón et al., 2015). Also, to reduce error and uncertainty, a common practice is the gravimetric dispensing of accurately weighed aliquots of radionuclide solutions with a pycnometer (Kulkarni et al., 2011a,b; Van Wyngaardt et al., 2014; Altzitzoglou and Rožkov, 2016; Bergeron et al., 2016) or a very sensitive analytical balance (e.g., Mettler balance) traceable to a national mass balance (Kossert et al., 2012; Kossert et al., 2017a,b).

The physical and chemical stability of samples in scintillation cocktail is a priority. Terlikowska et al. (1998) identified the cause of instability of samples which include (1) degradation of the scintillator or cocktail, (2) adsorption of the nuclide on the walls of the counting vial, (3) settlement of the aqueous phase of the nuclide source, (4) change in the physical characteristics (*e.g.*, size) of the aqueous micelles in cocktail mixtures, and (5) evolution of quenching in the cocktail due to a chemical change in its composition (*e.g.*, an increase of dissolved oxygen in the cocktail with time). A quench indicating parameter measured with an external radiation source may not provide any evidence of a problem in the cases of the above instabilities 2 and 4. Radionuclide samples may be counted repeatedly over a period of days to test for stability.

Bergeron et al. (2016) describe both reversible and irreversible mechanisms of cocktail instability. They provide examples, such as a settling or phase separation processes within the cocktail, which could account for a short-term reversible instability or, on the other hand, the absorption of radionuclide on the vial walls could explain the longer-term irreversible instabilities. Consequently, the stability of cocktails must be studied thoroughly for each nuclide standardized (Fitzgerald et al., 2014; Keightley et al., 2015; Bergeron et al., 2016). It is recommended also that standardizations of any particular nuclide be carried out in accord with previously tested and recommended cocktails and procedures cited in the literature.

Dark adaptation of quenched standards and samples in liquid scintillation counting vials within the stabilized temperature of the liquid scintillation analyzer is also recommended. For the standardization of radionuclides, researchers will leave sample counting vials within the counter for periods of up to 12 h to provide time for the decay of possible luminescence that can arise from laboratory room lighting (Fitzergerald et al., 2014; Altzitzoglou and Rožkov, 2016; Jiang et al., 2017). Fitzergerald et al. (2014) prepared samples for counting under incandescent light rather than fluorescent lighting to reduce possible luminescence and the time needed for dark adaption.

Stable nuclide carrier is used in liquid scintillation radionuclide standardization to avoid error that might occur due to adsorption of radionuclide on the walls of counting vials. For example, the following carriers have been added to nuclide samples for this purpose: 0.25 mL of Sr carrier solution containing 4 mg Sr/g in 1 mol/L HCl in the standardization of ^{89}Sr (Coursey et al. 1998); vials previously

TABLE 6.9 Some radionuclides standardized by the CIEMAT/NIST efficiency tracing method.

Radionuclide	Relative standard uncertainty (%)[a]	References
^{10}Be	0.15[b]	Chmeleff et al. (2010)
^{14}C	0.20[c]	Grau Malonda and García Toraño (1982a), Coursey et al. (1986), Bergeron et al. (2016)
^{18}F	0.30[c]	Grau Malonda and García Toraño (1982a), Capogni et al. (2006), Roteta et al. (2006), Schrader et al. (2007), Cessna et al. (2008), Nedjadi et al. (2010), Fitzgerald et al. (2014)
^{22}Na	0.20[b]	Nähle et al. (2008), Liu et al. (2017)
^{32}P	0.30[c]	Grau Malonda and García Toraño (1982a), Rodríguez Barquero et al. (2004), da Cruz et al. (2004), Cessna (2002), Nähle and Kossert (2011)
^{35}S	0.9	Calhoun et al. (1991), Koskinas et al. (2014)
^{36}Cl	0.3	Grau Malonda and García Toraño (1982a), Rodríguez Barquero et al. (1989), Tarancón et al. (2015)
^{40}K	0.23[c]	Grau Carles and Grau Malonda (1997), Grau Malonda and Grau Carles (2002)
^{41}Ca	0.38[c]	Kossert et al. (2009a,b), Rodríguez Barquero and Los Arcos (1996)
^{44}Sc	0.50[c]	García-Toraño et al. (2016)
^{45}Ca	0.37[c]	Rodríguez et al. (1994), Ortiz et al. (1993), Tarancón Sanz and Kossert (2011)
^{54}Mn	<0.7[c]	Rodríguez et al. (1992), Kossert and Grau Carles (2006), Cassette et al. (2006a,b)
^{55}Fe	0.6	Ortiz et al. (1993), Grau Malonda et al. (2006), Kossert and Grau Carles (2010)
^{59}Fe	0.19[b]	Günther (1994), Kossert and Nähle (2014)
^{60}Co	0.36[c]	Kulkarni et al. (2013), Kossert et al. (2017a,b, in press)
^{63}Ni	0.29[c]	Coursey et al. (1989), Ceccatelli and De Felice (1999), Collé et al. (2008), Nähle and Kossert (2011), Kossert and Mougeot (2015), Tarancón et al. (2015)
^{64}Cu	0.49[b]	Capogni et al. (2008), Wanke et al. (2010), Bakhshandeiar et al. (2011), Amiot et al. (2012)
^{68}Ga	0.51[c]	Roteta et al. (2012)
$^{68}Ge/^{68}Ga$	0.67[c]	Zimmerman et al. (2008), Kulkarni et al. (2017)
^{79}Se	0.48[c]	Bienvenu et al. (2007), Jörg et al. (2010)
^{85}Sr	0.83[c]	Grau Carles et al. (1994c), Broda et al. (2012)
^{87}Rb	0.64[b]	Kossert (2003)
^{89}Sr	0.15[c]	Grau Malonda and García Toraño (1982a), Altzitzoglou et al. (2002), Coursey et al. (1998), Cruz et al. (2002), Nähle and Kossert (2011)
^{90}Sr[d]	0.14[c]	Altzitzoglou et al. (1998), Spasova et al. (2008), Ceccatelli and De Felice (1999)
^{90}Y	0.12[c]	Kossert and Schrader (2004), Zimmerman et al. (2004b), Mo et al. (2005), Schultz et al. (2008), Xiques Castillo et al. (2010), Yamada and Ishii (2012), Ferreira et al. (2016)
^{93m}Nb	0.56[c]	Günther and Schötzig (1992), Ferreira et al. (2016)
^{93}Zr	3.8[c]	Cassette et al. (2010)
^{99}Tc	0.48[c]	Grau Malonda and García Toraño (1982a), Laureano-Perez et al. (2010), van Wyngaardt et al. (2014)
^{99m}Tc	0.41[c]	Grau Malonda and Coursey (1987)

Continued

TABLE 6.9 Some radionuclides standardized by the CIEMAT/NIST efficiency tracing method.—cont'd

Radionuclide	Relative standard uncertainty (%)[a]	References
^{109}Cd	1.5[c]	Grau Carles et al. (1994c), Kossert and Grau Carles (2010)
^{110m}Ag	0.29[c]	García-Toraño et al. (2000)
^{113}Sn	0.60[c]	Roteta et al. (2014)
^{113m}Cd[d]	0.34[b]	Kossert et al. (2011b)
^{124}Sb	0.60[c]	Chauvenet et al. (2010)
$^{125}Sb/^{125m}Te$	<1.50	Grau Carles et al. (1998)
^{125}I	0.50[b]	Grau Carles et al. (1994c), Grau Malonda et al. (2006), Kossert (2006)
^{129}I	0.38[c]	Altzitzoglou and Rožkov (2016)
^{131}I	0.25[c]	Günther (1994), Oropesa Verdecia and Kossert (2009), Kulkarni et al. (2011a,b, 2016)
^{134}Cs	0.19[c]	García-Toraño et al. (2002), Wätjen et al. (2006), Kulkarni et al. (2016)
^{138}Cs	0.74[c]	Wätjen et al. (2006)
^{142}Pr	1.25[c]	Yeltepe and Yücel (2017)
^{151}Sm	0.49[c]	Altzitzoglou and Rožkov (2016)
^{153}Sm	0.19[c]	Schötzig et al. (1999), Dziel et al. (2014)
^{161}Tb	6.1[c]	Jiang et al. (2017, in press)
^{166m}Ho	0.40[c]	Altzitzoglou and Rožkov (2016)
^{170}Tm	0.84[c]	Yeltepe et al. (2017)
^{176}Lu	0.95[b]	Kossert et al. (2013)
^{177}Lu	0.36[b]	Schötzig et al. (2001), Zimmerman et al. (2001), Capogni et al. (2012), Kossert et al. (2012)
^{185}W	3	Grau Malonda and and García Toraño (1982a)
^{186}Re	1.61[c]	Coursey et al. (1991)
$^{188}W/^{188}Re$	0.42[c]	Zimmerman et al. (2002)
^{204}Tl	0.24[c]	Rodríguez Barquero et al. (2004), da Cruz et al. (2004), Hult et al. (2000), Johansson, et al. (2002) and Lee et al. (2004a,b)
^{210}Pb	1.2[c]	Laureano-Perez et al. (2007), Collé and Laureano-Perez (2009)
^{223}Ra[e]	0.23[b]	Cessna and Zimmerman (2010), Keightley et al. (2015), Kossert et al. (2015)
^{227}Ac	0.93[b]	Kossert et al. (2015)
^{229}Th	0.30[c]	Fitzgerald et al. (2010)
^{237}Np	0.46[c]	Günther (2000), Laureano-Perez et al. (2014)
^{241}Pu	1.06[b]	Coursey et al. (1989), Kossert et al. (2011a)

[a]*The lowest standard uncertainty is listed when more than one reference is cited.*
[b]*Square root of the quadratic sum of uncertainty components.*
[c]*Quadratic sum of all uncertainty components at a coverage factor* k = 1 *(See Chapter 7, Vol. II).*
[d]*For low-level activities (~45 Bq/kg) in dry bone ash and in milk.*
[e]*Radium-223 decays with a half-life of 11.4 d (See Appendix A) via a chain of shorter-lived daughter radionuclides yielding a total of five alpha decays and three beta decays. With ^{223}Ra samples in secular equilibrium with daughter nuclides these researchers used the CIEMAT/NIST liquid scintillation efficiency tracing method as part of the primary standardization of ^{223}Ra to measure the detection efficiencies of the three beta decays.*

saturated with a carrier solution containing 100 μg/g of $NiCl_2$ in 0.1 M HCl in the standardization of ^{63}Ni (Terlikowska et al., 1998); 97 μg of inactive $FeCl_3 \cdot 6H_2O$ per gram of solution in the standardization of ^{59}Fe (Kossert and Nähle, 2014); carrier containing approximately 0.06 g of $LuCl_3$ per mL of 1 mol/L HCl in the standardization of ^{177}Lu solution (Zimmerman et al., 2001; Capogni et al., 2012; Kossert et al., 2012); 0.05 mg/g of NaI, 0.02 mg/g of $Na_2S_2O_3$ and 0.02 mg/g of LiOH in the standardization of ^{131}I (Kulkarni et al., 2011a,b); 120 mg $NiCl_2 \cdot H_2O$ per liter of 0.1M HCl in the standardization of ^{63}Ni (Kossert and Mougeout, 2015); 0.5 mg of stable samarium, europium, and terbium carriers in the standardization of ^{161}Tb (Jiang et al., 2017); and 50 μ of $CoCl_2$ per g of solution in the standardization of ^{60}Co (Kossert et al., 2017a,b).

Rodríguez Barquero and Los Arcos (2004) carried out a thorough long-term study that monitored the stability of samples and drift in electronic components of the spectrometer over time intervals that extended to up to 4 years. The tests were carried out with low-energy beta-emitters, namely, ^{3}H, ^{14}C, and ^{63}Ni where detection efficiency could be more susceptible to disturbing changes in sample and cocktail stability or spectrometer response. Their findings were that the CIEMAT/NIST method can compensate for count losses and quench parameter compensation values provided the experimental quench curve of the tracer nuclide, *i.e.*, ε_T versus QIP, is prepared close to the time of nuclide measurements. If this requirement is not met, significant discrepancies of up to 5.6% for ^{3}H, 1.5% for ^{63}Ni, and 0.35% for ^{14}C can be obtained for a 1-year time difference.

5. Cross-efficiency curves

In the preceding treatment on the theory and experimental procedure involved in the CIEMAT/NIST method, it was established that the counting efficiency as a function of the 'figure of merit' (*M*) or 'free parameter' (λ) is found for both the standard (tracer) nuclide (ε_T) and the radionuclide under investigation (ε_{nuc}). Under such circumstances, as noted by Grau Malonda et al. (1985), the ratio of the counting efficiencies ($\varepsilon_T/\varepsilon_{nuc}$) is also known as a function of *M*. Consequently, as the counting efficiency of the tracer nuclide (*e.g.*, ^{3}H) is known, the other can be computed. Tritium is considered, therefore, not only as a standard nuclide, but also a tracer nuclide, because the counting efficiency of the nuclide of interest in found against that of the reference nuclide, ^{3}H in this case. (Most often ^{3}H is used as the tracer for the standardization of beta-emitters, but we will see in the next Section that other tracer nuclides, such as ^{54}Mn, may be used for the standardization of radionuclides that decay by pure electron capture). With this in mind and in view of the curves illustrated in Figs. 6.78 and 6.79, we can see that the curves ε_T versus QIP (obtained from a series of quenched standards of the tracer nuclide, *e.g.*, ^{3}H) and ε_{nuc}versus QIP (obtained from the relationships of ε_T vs. QIP, ε_T vs. *M*, *M* vs. QIP, and ε_{nuc} vs. *M*) can be integrated to provide a curve directly relating the counting efficiency of the nuclide under investigation (ε_{nuc}) and that of the tritium standard or tracer nuclide (ε_T), which is referred to as a cross-efficiency curve. Two such curves of ε_{nuc} versus ε_T are illustrated in Fig. 6.81. With this relationship, the counting efficiency of a given nuclide (ε_{nuc}) at a certain QIP could be determined from the counting efficiency of tracer nuclide (ε_T) at the same QIP.

In the example illustrated in Fig. 6.81, Kulkarni et al. (2017) determined the total activity of ^{68}Ge/^{68}Ga in equilibrium from their combined count rate and total efficiency, the latter of which is obtained by combining the two efficiencies from the cross-efficiency curves. They used a PerkinElmer (formerly Packard) liquid scintillation analyzer, which measures the tSIE as the quench indicating parameter (QIP) of the sample and ^{3}H standards. They calculated the counting efficiency of ^{3}H versus Figure of Merit (FOM), ^{68}Ga versus FOM, and ^{68}Ge versus FOM using the freeware program CN2003 from Günther (2002a). Germanium-68 decays by purely electron capture resulting in the emission of weak conversion and Auger electrons (See Table of Radioactive Isotopes, Appendix A in Volume 2), which accounts for its much lower liquid scintillation detection efficiencies compared to its daughter ^{68}Ga, which decays mostly via positron emission (90% probability) as well as electron capture (10% probability).

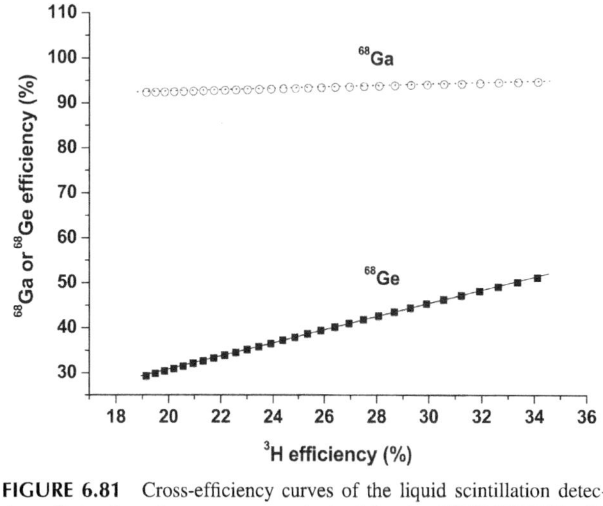

FIGURE 6.81 Cross-efficiency curves of the liquid scintillation detection efficiencies of ε_{nuc} versus ε_T obtained by the CIEMAT/NIST efficiency tracing method for the activity analysis of the parent-daughter radionuclides ^{68}Ge/^{68}Ga. The detection efficiencies of the nuclides to be analyzed, namely, ^{68}Ge and ^{68}Ga, are plotted against the detection efficiency of the tracer nuclide ^{3}H. *From Kulkarni et al. (2017) reprinted with permission from Elsevier © 2017.*

To facilitate the application of the CIEMAT/NIST efficiency tracing method, Günther (1996) proposed the use of 'universal' cross-efficiency curves for the standardization of radionuclides that decay by β- or β-γ emission according to the following procedure:

a. **Calculation of the efficiencies.** Calculating the efficiencies of radionuclides versus the corresponding tritium efficiency (i.e., cross-efficiency curve ε_{nuc} vs. ε_{T}) can be done in a national standards laboratory. The results should be parameterized as coefficients k_i of the polynomial in Eq. (6.102).

$$\varepsilon_{\text{nuc}} = \sum_{i=0}^{n} k_i \varepsilon_{\text{T}}^i \tag{6.102}$$

b. **Verifying the results.** This step would best be done at one of the national standards laboratories, because they have a stock of standard solutions. An experimental comparison of the results with those obtained with other methods of radionuclide standardization would be necessary.
c. **Laboratory procedure.** The user would determine the quench indicating parameter Q of the radionuclide sample automatically with each single measurement, calculating the corresponding tritium efficiency ε_{T} of the individual measurement using his/her own tritium quench curve (ε_{T} vs. Q), and calculating the efficiency ε_{nuc} with this tritium efficiency (ε_{T}) and the coefficients k_i of Eq. (6.102).

The procedure proposed by Günther (1996) is summarized in Fig. 6.82.

An example of a typical cross-efficiency curve (ε_{nuc} vs. ε_{T}) defined by polynomial coefficients k_i for the CIEMAT/NIST activity analysis of ^{161}Tb is illustrated in Fig. 6.83. With the single measurement of the instrumental quench indicating parameter (Q) of a radionuclide sample, the detection efficiency of the tritium tracer (ε_{T}) can be extracted from its quench correction curve (ε_{T} vs. Q). The value of ε_{T} is then used to determine the detection efficiency of the radionuclide sample ε_{nuc} according to Eq. (6.102) defining the best fit to the cross-efficiency curve ε_{nuc} versus ε_{T}.

Cassette and Tartès (2014) warn against considering the cross-efficiency curves as 'universal'. They demonstrated that cross-efficiency curves, particularly in the cases of ^{63}Ni and ^{55}Fe versus ^{3}H, are dependent on the scintillator, on the counter used, and on the nature of the quenching phenomenon (e.g., chemical or color quenching), and thus the cross-efficiency curves cannot be considered as 'universal'. Following the testing of cross-efficiency curves prepared

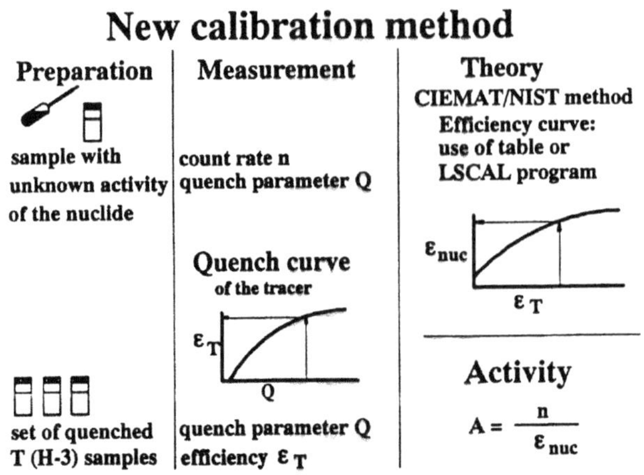

FIGURE 6.82 The procedure is illustrated in three columns from left to right, namely, Sample Preparation, Sample Measurement, and CIEMAT/NIST Theory for the preparation of the cross-efficiency curve (ε_{nuc} vs. ε_{T}). The counting efficiency of the nuclide to be measured (ε_{nuc}) is obtained with the quench parameter (Q) of the measurement, the tritium quench curve (ε_{T} vs. Q), and the tabulated coefficients of a cross-efficiency curve of the nuclide (ε_{nuc} vs. ε_{T}). The LSCAL computer program facilitated the calculation of ε_{nuc} from the equation for a least-squares best fit of the cross-efficiency curve. *From Günther (1996), reprinted with permission from Radiocarbon. University of Arizona © 1996 Arizona Board of Regents on behalf of the University of Arizona.*

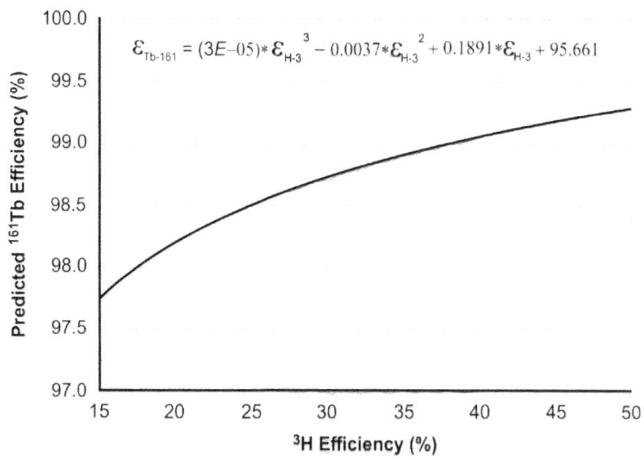

FIGURE 6.83 Cross-efficiency curve (ε_{nuc} vs. ε_{T}) providing the detection efficiency of the radionuclide ^{161}Tb as a function of the detection efficiency of the ^{3}H tracer nuclide calculated by the computer program CN2004. The equation for the best fit of the cross-efficiency curve with polynomial coefficients k_i, illustrated above the curve, facilitates the computation of the value of ε_{nuc} (i.e., detection efficiency of ^{161}Tb) as a function of ε_{T} (detection efficiency of ^{3}H). *From Jiang et al. (2018), in press reprinted with permission from Elsevier © 2018.*

with three commercial liquid scintillation counters (Models Guardian 1414, TriCarb 3170 and Quantulus 1220), three different liquid scintillation cocktails, and chemical- and color-quenched sources, Cassette and Tartès (2014) conclude the following:

a. The quench curves are scintillator dependent, and the degree of this dependence depends on the radionuclide.

b. The quench curves are counter dependent. Although this is obvious for different liquid scintillation counters that provide differing quench indicating parameters (Q) such as tSIE, H#, SQP(E), etc., but it also applies to counters that have the same type of external standard, which produce the same type of quench indicating parameter. For example, the values of the tSIE for a given sample from different TriCarb counters or SQP(E) from different Quantulus counters can differ.
c. The quench curves are dependent on the type of quenching phenomena (e.g., chemical- or color-quenching. The degree of the difference is radionuclide dependent and particularly noteworthy for ^{63}Ni and ^{55}Fe quench curves.
d. The quench curves are dependent on the volume of the source and nature of the counting vial (i.e., clear or diffusive).

To underscore the importance of the above variables when applying the CIEMAT/NIST standardization method, Cassette and Tartès (2014) underscore the need for good laboratory practice, whereby the liquid scintillation counter used for the calibration of a quench correction curve be used also for the nuclide measurements and maintaining the same counting vial, scintillation cocktail, and quenching conditions between calibration and measurement.

If the liquid scintillation counter is equipped with a self-normalization procedure, the repetition of the determination of the tritium quench correction curve need not be carried out except over extended intervals of time. Best results are obtained when the quenched standards of the tracer nuclide (^{3}H) are counted in a time span close to that of counting the nuclide of interest as reported by Rodríguez Barquero and Los Arcos (2004).

Further streamlining of the CIEMAT/NIST efficiency tracing standardization method is described by Günther (1998, 2000, 2002b). As described previously the methodology involves the calculated theoretical efficiency curve of the nuclide under investigation (ε_{nuc}) versus the theoretical efficiency curve of the tracer nuclide (ε_T). The procedure, as portrayed by Günther (1998, 2000), is reduced to the following steps, which are also depicted in Fig. 6.84:

a. The liquid scintillation analyzer is calibrated with a set of quenched standards of a tracer nuclide (usually ^{3}H) by producing an instrument-measured quench curve of counting efficiency, ε_T, versus a quench indicating parameter, Q, [e.g., SQP(E) or tSIE].
b. The count rate (CPM) and quench indicating parameter, Q, are determined for a sample of the nuclide to be measured (i.e., a sample of the nuclide of which the counting efficiency and activity are unknown). From this measured value of the quench indicating parameter of the nuclide under investigation, a fictitious tracer efficiency for this individual measurement is obtained from the tracer calibration curve produced in step 1 (i.e., the value of ε_T from the calibration curve corresponding to the Q of the sample).
c. The efficiency of the nuclide of the sample under measurement is taken from the calculated theoretical curve of efficiency of the nuclide under investigation (ε_{Nuc}) versus the theoretical efficiency curve of the tracer nuclide (ε_T).
d. The activity of the nuclide is then calculated by dividing the sample count rate by its efficiency (i.e., Activity = count rate/ε_{Nuc}.

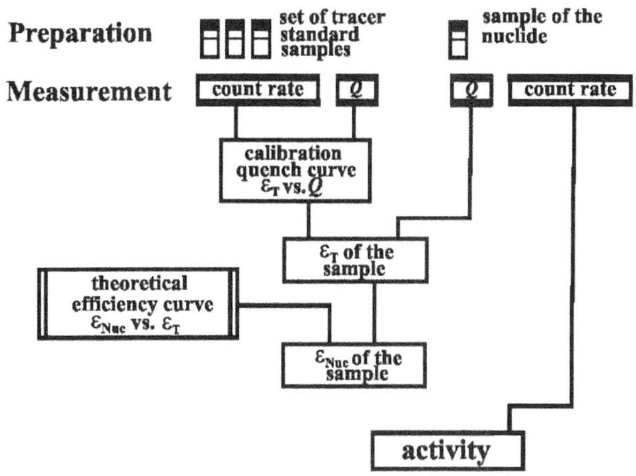

FIGURE 6.84 Principle of the CIEMAT/NIST method. Q is the quench indicating parameter measured, ε_T is efficiency of the tracer, ε_{Nuc} the efficiency of the nuclide to be measured. *From Günther (1998), reprinted with permission from Elsevier © 1998.*

The experimentally determined quench correction curve prepared with quenched standards of the tracer nuclide (*e.g.*, ^{3}H) in step 1 can be useful for an extended period of time (several months or more) with modern liquid scintillation analyzers that are properly cared for in terms of normalization and calibration, and monitored frequently with automated instrument performance assessments (IPAs). The monitoring of instrument stability by regular automated IPAs is discussed further on in this chapter. However, the CIEMAT/NIST method depends highly on the fact that the tracer nuclide and the nuclide under investigation are measured in a cocktail that possesses the same physical and chemical characteristics. Thus, Rodríguez Barquero and Los Arcos (2004) demonstrate that best results are obtained when the quenched standards of the tracer nuclide are measured close in time to the measurement of the sample of the nuclide under investigation, as the physical and chemical characteristics of a sample in cocktail can change in time.

The theoretically calculated counting efficiencies of the nuclide under investigation and the tracer nuclide are carried out via computer programs (codes) already elaborated for specific nuclide decay schemes as previously described in this chapter or available from databases containing the polynomial coefficients for the relationship of the theoretical calculated nuclide efficiency and the tracer efficiency described by the equation.

$$\varepsilon_{\mathrm{Nuc}} = \sum_{i=0}^{n} k_i \varepsilon_{\mathrm{T}}^{i} \qquad (6.103)$$

which is the general equation for the polynomial of the least-squares best fit for the relationship, as described previously (Eq. 6.102) with n number of coefficients. With the establishment of the above two curves or relationships described in Fig. 6.84 (i.e., ε_{T} vs., Q and $\varepsilon_{\mathrm{Nuc}}$ vs. ε_{T}), it becomes a simple matter to determine the counting efficiency of a sample ($\varepsilon_{\mathrm{Nuc}}$) and thereby standardize a radionuclide sample simply from its count rate (CPM) and its quench-indicating parameter (Q).

Broda et al. (2007) provide an excellent and more detailed description of the CIEMAT/NIST method, which is illustrated in Fig. 6.85. The first step, as illustrated in the figure, is the theoretical calculations of the detection efficiency of the tracer nuclide as a function of the free parameter, M, [$\varepsilon_{\mathrm{tracer}}$ (M), *e.g.*, ^{3}H] and of the nuclide under investigation [$\varepsilon_{\mathrm{nuclide}}(M)$] where $M = \lambda$ or FOM of Eq. (6.96). This leads to the establishment of the relationship between the two detection efficiencies, that is, the detection efficiency of the nuclide under investigation as a function of the detection efficiency of the tracer nuclide or $\varepsilon_{\mathrm{nuclide}}(\varepsilon_{\mathrm{tracer}})$, as illustrated in Figs. 6.81–6.84, which is also referred to as the 'efficiency curve'. The next step, as illustrated in Fig. 6.85 of Broda et al. (2007), involves the counting of a set of quench standards of the tracer nuclide (*e.g.*, ^{3}H). The quenched standards, as previously described in this chapter, are prepared with primary standard ^{3}H in a series of counting vials containing scintillation cocktail and increasing amounts quenching agent (See Section V.D). The count rates (CPM) and quench indicating parameters (QIPs) of each vial are recorded. As the tracer nuclide is of a known activity (DPM), the detection efficiency of each vial is calculated (CPM/DPM), and the quench correction curve of the detection efficiency of the tracer nuclide as a function of the quench indicating parameter is obtained, *i.e.*, $\varepsilon_{\mathrm{tracer}}$(QIP) of Fig. 6.85. The next step is the preparation of a sample of the radionuclide under investigation for counting in liquid scintillation cocktail. The radionuclide sample in the liquid scintillation counting vial must possess the sample physical and chemical characteristics (*i.e.*, same cocktail, cocktail volume, and chemical composition as the quench standards of the tracer nuclide. The count rate (CPM) of the sample and its' quench indicating parameter (QIP) are measured. The value of the QIP is then used to determine the efficiency of the tracer nuclide (See curved arrow in Fig. 6.85) from the curve or function $\varepsilon_{\mathrm{tracer}}$(QIP). With the corresponding tracer efficiency thus obtained, the corresponding nuclide efficiency ($\varepsilon_{\mathrm{nuclide}}$) is obtained from the efficiency curve, $\varepsilon_{\mathrm{nuclide}}(\varepsilon_{\mathrm{tracer}})$. Finally, the efficiency of the nuclide under investigation, so determined, is used to convert its count rate (CPM), i.e., $\mathrm{R_{nuclide}}$ of Fig. 6.85, to the activity of the nuclide, according to the equation illustrated in Fig. 6.85, namely, $\mathrm{a_{nuclide}} = \mathrm{R_{nuclide}}/(\varepsilon_{\mathrm{nuclide}} \cdot m)$, where m is the sample mass. The samples are generally added gravimetrically to the liquid scintillation cocktail. Consequently, including the sample mass m in the calculation of the sample activity will yield the sample activity concentration (*e.g.*, DPM/g), which can be converted to units of Bq/g.

This approach has been tested by other researchers with excellent results using ^{3}H as the tracer nuclide. Günther (2002b) reports excellent agreement between the CIEMAT/

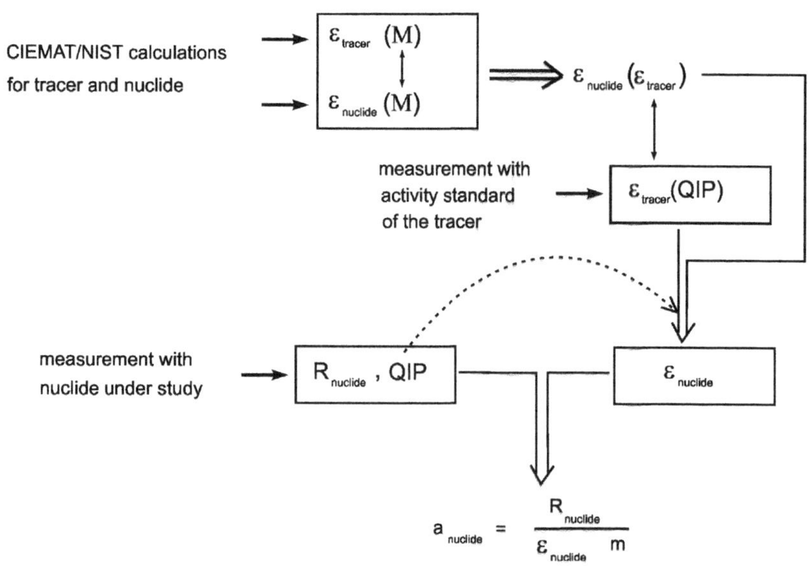

FIGURE 6.85 Illustration of the CIEMAT/NIST method: R denotes the measured count rate, M is the free parameter, QIP is the quench indicating parameter, ε is the counting efficiency, $\mathrm{a_{nuclide}}$ is the nuclide activity concentration, and m is the mass of the solution in an LSC sample. *From Broda et al. (2007). Reprinted with permission © 2007 IOP Publishing, Ltd.*

NIST and $4\pi\beta$-γ standardization methods with the following deviations between the two methods expressed as a percent: ^{134}Cs (<0.3%), ^{153}Sm (<0.2%), ^{169}Er (<0.2%), ^{177}Lu (0.6%), and ^{188}Re (0.2–1.0%). It is a common practice for researchers to employ more than one liquid scintillation standardization method together with the CIEMAT/NIST method in the standardization of a radionuclide, such as the TDCR method or $4\pi\beta-\gamma$ coincidence counting technique. This further demonstrates the robustness of the methods used and reconfirms the standardization determinations.

6. ^{54}Mn as tracer nuclide

All of the previous treatments on the CIEMAT/NIST radionuclide standardization method dealt with the use of standard ^{3}H as the tracer nuclide. Standard tritium has proven to be the best tracer for the standardization of β- or β/γ-emitting nuclides, and it has been the standard of choice since the inception of this technique by Agustín Grau Malonda in 1978. However, for the standardization of the electron-capture decay nuclides ^{55}Fe and ^{65}Zn, Günther (1998) and Bergeron and Laureano-Pérez (2014) have demonstrated ^{54}Mn to be a more robust tracer. Iron-55 decays by pure electron capture and ^{65}Zn decays by electron capture with 98.6% intensity and the remaining 1.4% by positron emission. Günther (1998) demonstrated that the use of standard ^{54}Mn, which decays by pure electron capture, as the tracer nuclide provides radionuclide activities for ^{55}Fe with a total relative uncertainty of only 0.44% whereas the uncertainty when using tritium as the tracer was 1.67%. Likewise, the total uncertainty in the standardization of ^{65}Zn using ^{54}Mn as the tracer nuclide was only 0.44%, while the uncertainty when using ^{3}H as the tracer increased to 0.93%. The lower relative standard uncertainty of ^{54}Mn activity standards compared to ^{3}H standards attributes additional potential to ^{54}Mn as a tracer nuclide, which could even be considered as a promising tracer nuclide for the standardization of ^{3}H by the CIEMAT/NIST method (Kossert and Grau Carles, 2006). In addition, Kossert and Grau Carles (2010) demonstrated that the measurement of ^{55}Fe with ^{54}Mn as the tracer nuclide results in a smaller dependence of the *kB* factor in the model calculation of detection efficiencies. The role of the *kB* value in the computation of counting efficiencies is discussed further on in this chapter. Bergeron and Laureano-Pérez (2014) discovered that long-lived impurities, which can accumulate in ^{54}Mn, can yield unexpectedly high detection efficiencies. Therefore, they recommend the ^{54}Mn sources more than a few years old should be reevaluated for radionuclide impurities prior to use as a tracer.

Calculations of the detection efficiency by the CIEMAT/NIST method require comprehensive information on the energy transferred to the scintillator, and good data on the photon absorption coefficients of X-rays emitted by electron-capture decaying nuclides is necessary. Cassette et al. (2006a,b) report new experimental data on X-ray photon absorption coefficients of liquid scintillators in the energy range of 5–12 keV, which could prove helpful in the standardization of EC nuclides, such as ^{55}Fe. The use of ^{54}Mn as a tracer nuclide for the measurement of nuclides decaying by electron capture is discussed further on in this chapter.

7. Ionization quenching and efficiency calculations (^{3}H or ^{54}Mn as the tracer)

The theoretical calculation of the counting efficiency of the nuclide under investigation and of the tracer nuclide (usually ^{3}H for β and β/γ-emitters or potentially ^{54}Mn for nuclides of low atomic number that decay by electron capture) is carried out using the basic equations Eq. (6.95) or 6.96 of Grau Malonda and coworkers (Grau Malonda, 1982b; Grau Malonda and García-Toraño, 1982a,b; Grau Malonda and Los Arcos, 1983; Grau Malonda et al., 1985), and variations of the equation depending on the decay scheme of the radionuclide under measurement (*e.g.*, Eqs. 6.97 and 6.98). A key factor in the theoretical calculation is the computation of the ionization quench function $Q(E)$, which is included in specific subroutines of the computer codes used to calculate the theoretical detection efficiencies described previously in this chapter. The procedures used to compute the ionization quench function is described by several researchers including Grau Malonda (1999), Grau Malonda and Grau Carles (1999, 2000, 2008), Grau Carles and Grau Malonda (2001, 2006), and García and Grau Malonda (2002), Grau Carles et al. (2004), Kossert and Schrader (2004), Grau Carles et al. (2006), Zimmerman (2006), Grau Malonda et al. (2006), Grau Carles (2006a,b), Kossert and Grau Carles (2006, 2008, 2010), Broda et al. (2007, 2012), Bagán et al. (2008), Bignell et al. (2010a), Broda et al. (2012), Kossert and Nähle (2014), Kossert et al. (2011a,b, 2013, 2014, 2015, 2017a,b), van Wyngaardt et al. (2012, 2014), Tarancón et al. (2015) and Altzitzoglou and Rožkov (2016).

Ionization quenching is defined and discussed in detail in Section IV of this chapter. In brief, as described by Bignell et al. (2010a), ionization quenching in scintillators is a reduction in the luminosity (photon intensity) associated with a high density of excited molecules caused by a large linear energy transfer (LET) to the scintillant. The high density of excited molecules results in a transfer of energy from one excited molecule to another resulting in superexcitation of a molecule and its ionization. Thus, this ionization, caused by the nuclear radiation as it travels through the scintillator solvent, results in a reduction in the amount of energy transferred from the charged particle to the fluor molecules and a consequent reduction in the photon intensity emitted by the scintillation fluor. As described by Grau Malonda (1999), the fluorescence yield or number of photons $L(E)$ produced by an ionizing particle traveling through a liquid scintillator is a nonlinear function of the particle energy (E). The nonlinearity increases with the linear energy transfer (LET) or stopping power of the particle. Therefore, a low-energy electron or beta particle will cause higher ionization quenching than a higher-energy beta particle. Radionuclides decaying by electron capture undergo, as a consequence of the capture of an atomic electron by the nucleus, a cascade of atomic electron transitions resulting in the emission of X-rays, Auger electrons, and Coster-Kronig electrons. The Auger and Koster-Cronig electrons are of low energy (See Chapter 1, Volume 1) with relatively high LET, and the X-rays can produce Compton and photoelectrons, all of which have a high LET relative to the higher-energy beta particles emitted by most beta emitting radionuclides.

According to Birks (1964) the specific fluorescence or number of photons produced per particle path length of travel, dL/dx, is defined by

$$\frac{dL}{dx} = \frac{\eta_0 \frac{dE}{dx}}{1 + kB\frac{dE}{dx}} \tag{6.104}$$

where η_0 is the scintillation efficiency or number of fluorescence photons emitted per unit of absorbed energy, dE/dx is the radiation stopping power (i.e., stopping power of the scintillator for one beta particle of energy E) in units of MeV/cm or MeV cm^2/g (See Chapter 1, Volume 1, Section XIV for the calculation of stopping power) and kB is the ionizing quench parameter or constant with units of cm/MeV or g/MeV cm^2.

In the absence of ionization quench the specific fluorescence would be defined as

$$\frac{dL}{dx} = \eta_0 \frac{dE}{dx} \tag{6.105}$$

and the term $[1 + kB(dE/dx)]^{-1}$ of Eq. (6.104) is the reduction in fluorescence due to ionization quenching. The formula usually used for the ionization quench function in the CIEMAT/NIST calculation of counting efficiency is the formula of Birks (1964).

$$Q(E) = \frac{1}{E} \int_0^{E_{\max}} \frac{dE}{1 + kB(dE/dx)} \tag{6.106}$$

According to Grau Malonda and Grau Carles (1999), the ionization quench function $Q(E)$ is determined by applying the Birks formula (Eq. 6.106) of which the stopping power (dE/dx) is approximated to a polynomial equation. Also, the ionization quench function is evaluated analytically whereby $Q(E)$ is expressed in terms of kB. An empirical equation for $Q(E)$ is expressed by Grau Malonda and Grau Carles (1999) as a quotient of polynomial equations that are fitted for different kB values between 0.001 and 0.020 g/MeV cm^2. They also analyzed the influence of the tracer nuclide (i.e., ^3H or ^{54}Mn) and demonstrated that ^3H remains the recommended tracer for the analysis of nuclides that undergo beta decay, while ^{54}Mn provides a more robust standardization of certain nuclides that decay by electron capture and possess a low atomic number close to that of iron. Also, they note that the calculated formula for $Q(E)$ and kB values must be the same for the tracer nuclide and the radionuclide under measurement. Computer programs BETAKB and EMISKB are provided by Grau Malonda (1999) for the calculation of the theoretical counting efficiencies of radionuclides that undergo beta-decay or electron capture decay, respectively, with subroutines for the computation of the ionization quench function $Q(E)$ for a given Birks parameter kB (See also Los Arcos and Ortiz, 1997). Günther (1998) also demonstrated the improved accuracy of standardization of the electron-capture nuclides ^{55}Fe and ^{65}Zn when ^{54}Mn is used as the tracer. Radionuclide standardization procedures will include the determination and use of the optimum kB factor (i.e., ionization quench parameter) including reports by Kossert and Schrader (2004), Broda et al. (2012), van Wyngaardt et al. (2012, 2014), Kossert and Nähle (2014), Altzitzoglou and Rožkov (2016), Tarancón et al. (2015), and Kossert et al. (2011a,b, 2012, 2014, 2015, 2017a,b).

8. Other factors affecting efficiency calculations

In a review of the CIEMAT/NIST and TDCR methods, Grau Carles and Grau Malonda (2001) presented concepts to improve the robustness of the radionuclide standardization methods. Currently, in all of the models, the counting efficiency is computed on the basis of the photoelectron output of the photomultiplier (PM) photocathode. Although the results of the model calculations agree well with the experimental results, they proposed that the correct calculated theoretical counting efficiency should be based on the final PM anode output rather than on the photocathode or any dynode outputs. As a result, modifications to the theoretical counting efficiency calculations and determinations of the optimum kB parameter were proposed (Grau Carles and Grau Malonda, 2001, García and Grau Malonda, 2002, and Günther (2002a) and are currently implemented. In general, the Visual Basic CN2001 program has been useful for the calculation of theoretical counting efficiencies of the nuclide under measurement versus the efficiency of the ^3H tracer for nuclides decaying by β, β/γ, EC, EC-γ, and for isomers and nuclides with mixed decay schemes. It is available from the Physikalisch-Technische Bundesanstalt (PTB),

Bundesalee 100, 38116 Braunschweig, Germany, and it is based on the CIEMAT programs EFFY (García Toraño and Grau Malonda, 1985), EMI (Grau Carles et al., 1994b), CEGA2 and KB (Los Arcos and Ortiz, 1997). Günther (2002a) summarized that pure beta-emitters present no problem with activities determined with uncertainties within 0.2% and 0.5%. The standardization of some nuclides decaying by isomeric transition (*e.g.*, ^{93m}Nb) are determined with 100% efficiency). Beta/gamma-emitters are determined with as high precision as pure beta-emitters. The EC nuclides do present a problem with higher levels of uncertainty; however, better models such as those proposed by Grau Malonda and Grau Carles (2000), the EMI2 program of Grau Malonda et al. (1999), and the use of ^{54}Mn as a tracer for some EC nuclides were designed to reduce uncertainties in the standardization of electron-capture nuclides.

9. Radionuclides in decay chains

Computed detection efficiencies of radionuclides that undergo solely beta decay to stable nuclides are in the expected range between 0% and 100%, as illustrated in the efficiency curves among examples described earlier in this chapter. The liquid scintillation detection efficiencies of alpha emissions are generally very close to unity (i.e., 100%) due to the significant energy deposited by alpha particles in the scintillation cocktail. For example, in the computed detection efficiency of ^{223}Ra, Keightley et al. (2015) assigned a fixed efficiency of 0.998 ± 0.002 ($k = 1$) for the alpha emissions in the ^{223}Ra decay chain. There is approximately a 0.2% loss of alpha interactions with cocktail due to the wall effect as determined by Cassette (2001); although the wall effect has been estimated to have yet a lower significance in the detection of alpha events, as determined by Fitzgerald and Fourney (2010).

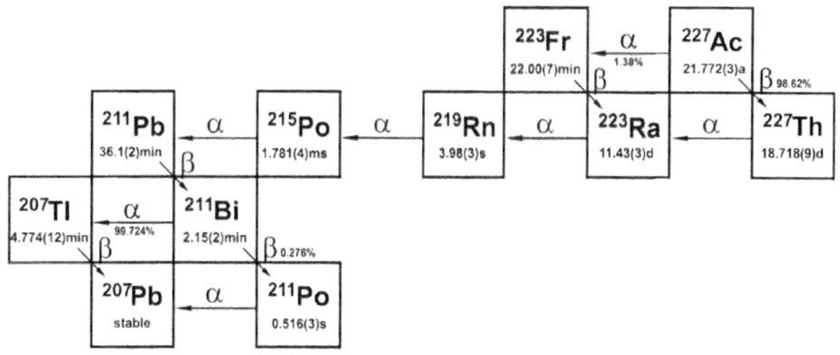

FIGURE 6.86 Radioactive decay chain from ^{227}Ac to stable ^{207}Pb. Within this chain is also the decay series of ^{223}Ra. *From Kossert et al. (2015); reprinted with permission of Elsevier © 2015.*

When alpha emissions are part of a decay chain including beta-decaying radionuclides in equilibrium, such as in the decay schemes of ^{223}Ra and ^{227}Ac, which proceed through a series of progeny comprising a chain of alpha- and beta-decaying radionuclides to the stable ^{207}Pb (See Fig. 6.86), alpha and beta transitions are included in the computer code computations of the CIEMAT/NIST method. The computed detection efficiencies of radionuclides, which decay through such radionuclide decay chains in equilibrium, will include a combination of the detection efficiency of each radionuclide in the decay chain. Thus, the combined or overall computed detection efficiency will exceed 100%, as illustrated in Figs. 6.86 and 6.87. The computation of these detection efficiencies will include, in addition to the unity assigned to the detection efficiency of alpha events and their branching intensities, the branching ratios and energies of beta and gamma transitions taken from nuclear decay data. The computer code used by Kossert et al. (2015) is the MICELLE2 code, which includes all beta and beta/gamma transitions in the decay chain of ^{227}Ac, and that used by Keightley et al. (2015) in the standardization of ^{223}Ra is the CN2004 code. The decay scheme of ^{223}Ra to stable ^{207}Pb, includes essentially the decay of four progeny via alpha emission, namely, ^{223}Ra

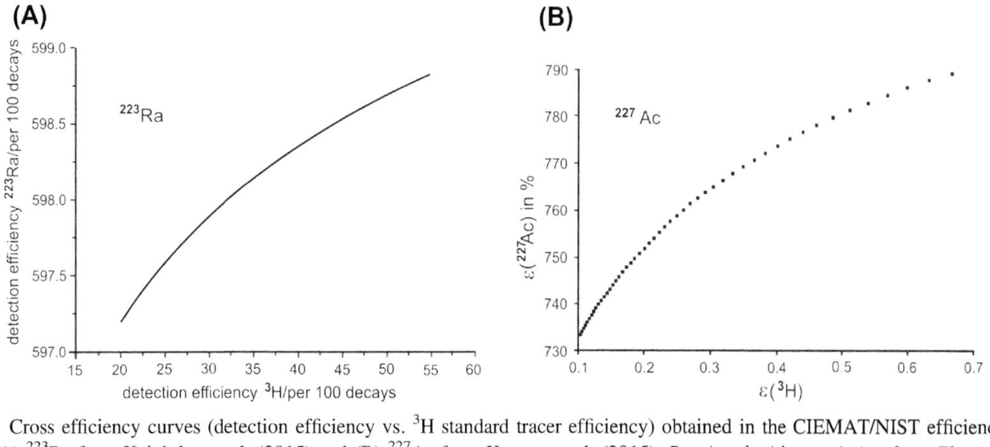

FIGURE 6.87 Cross efficiency curves (detection efficiency vs. ^{3}H standard tracer efficiency) obtained in the CIEMAT/NIST efficiency tracing standardization of (A) ^{223}Ra from Keightley et al. (2015) and (B) ^{227}Ac from Kossert et al. (2015). *Reprinted with permission from Elsevier (c) 2015.*

→ ^{219}Rn → ^{215}Po → ^{211}Pb and ^{211}Bi → ^{207}Tl, and two progeny via beta emission, namely, ^{211}Pb → ^{211}Bi and ^{207}Tl → ^{207}Pb, as illustrated in Fig. 6.86. Thus, the total computed detection efficiency, which includes the four alpha plus two beta transitions should not exceed 600%, as illustrated in Fig. 6.87A. In this case, the detection efficiency computations via the CN2004 computer code, which is an updated version of CN2001 (Keightley et al., 2015), included the four alpha and two beta emissions.

The differing alpha and beta branches in the decay chains of ^{223}Ra and ^{227}Ac serve as excellent examples of two decay chains yielding very different magnitudes of the computed detection efficiencies in the standardization of these two radionuclides. The decay scheme of ^{223}Ra is the shorter of the two schemes, as described in the previous paragraph, with essentially four alpha and two beta transitions yielding a maximum computed detection efficiency of <600% (See Fig. 6.87A). On the other hand, the decay chain of ^{227}Ac yields principally, as illustrated in Fig. 6.86, five alpha emissions, namely, ^{227}Th → ^{223}Ra → ^{219}Rn → ^{215}Po → ^{211}Pb and ^{211}Bi → ^{207}Th. In addition, there are principally three beta decays in the ^{227}Ac decay chain, namely, ^{227}Ac → ^{227}Th, ^{211}Pb → ^{211}Bi, and ^{207}Tl → ^{207}Pb. Two other transitions have very small branches, namely, the alpha emissions of ^{227}Ac with a 1.38% intensity, and the beta transition ^{211}Bi → ^{211}Po of 0.276% intensity. These will contribute little to the computed detection efficiency of ^{227}Ac. Thus, the computed detection efficiency of ^{227}Ac with its essentially five alpha emissions and three beta branches do not exceed 800%, as illustrated in Fig. 8.86B. In this case, the highest computed detection efficiency for ^{227}Ac, which is at its lowest level of quench (highest detection efficiency of ^{3}H tracer) is reported by Kossert et al. (2015) to be <800% as illustrated in Fig. 6.87B.

10. Electron capture radionuclides

The higher levels of uncertainty in the standardization of electron-capture radionuclides by the CIEMAT/NIST method or low-energy beta-emitters by the TDCR method have motivated much research toward improving the efficiency calculations. (The TDCR method is discussed further on in this chapter). It is a model-based method similar to the CIEMAT/NIST method in that both methods involve theoretical calculations of the detection efficiency). With the objective of reducing the uncertainties in the activity determinations, Grau Carles et al. (2004, 2006) and Grau Carles and Grau Malonda (2006) revisited the problem of ionization quenching and photoionization-reduced energy in liquid scintillation counting of electron-capture decaying radionuclides such as ^{55}Fe, ^{51}Cr, and ^{54}Mn. Their studies provided a more detailed simulation of the photoionization process in the atomic rearrangement detection model for these electron-capture nuclides. Further studies were focused on Monte Carlo calculations of spectra and photon interaction probabilities in liquid scintillators (Zimmerman, 2006 and Cassette et al., 2006a,b) for X-ray and gamma-ray emitting nuclides, such as the electron-capture nuclide ^{54}Mn.

Atomic electron rearrangement pathways were modeled and the MOYEN computer program developed (Grau Malonda et al., 2006) for the computation of the energy emissions in LMM Auger transitions upon which the counting efficiency calculations depend strongly. The energy of these transitions was computed for the EC nuclides ^{55}Fe and ^{125}I.

A computer program, EMILIA, that includes the computation of the counting efficiency for the atomic rearrangement detection models, is reported by Grau Carles (2006a), which improved the computed counting efficiency of low-Z electron-capture nuclides under moderate quenching.

Kossert and Grau Carles (2006) utilized EMILIA code to compute the detection efficiency of the low-Z EC nuclides ^{54}Mn, ^{55}Fe, and ^{65}Zn. They also demonstrated that ^{54}Mn could be used as a tracer nuclide for the standardization of ^{55}Fe. They consider a future application would be to standardize ^{3}H by the CIEMAT/NIST method using ^{54}Mn as the tracer nuclide, as the uncertainty of ^{54}Mn activity standards (<± 0.7%) is lower than for ^{3}H standards. The standardization of ^{3}H using ^{54}Mn as a tracer was reported with a relative standard uncertainty of ±1.01% by Nähle et al. (2010). Also, Ratel (2008) underscored the potential of ^{54}Mn as a tracer nuclide instead of ^{3}H to increase the precision of the liquid scintillation standardization of EC nuclides, such as ^{55}Fe.

The computer program, MICELLE, reported by Grau Carles (2007), extends the computation of the liquid scintillation counting efficiency to electron-capture radionuclides of atomic number $30 \leq Z \leq 54$. The program considers all subshells involved in the atomic electron rearrangements following electron capture. MICELLE also simulates samples in the gel phase, as the micelle size of gel scintillators can have an effect on the counting efficiency. The micelle program applied to the standardization of EC radionuclides ^{109}Cd and ^{125}I yielded considerable improvements in the relative standard uncertainties, particularly for ^{109}Cd (±0.5%) when a kB value of 0.0110 cm/MeV was used for the Birks parameter. They report a reduction in the uncertainty in the standardization of ^{125}I, although there was a large dependence of the analysis on the kB value. The uncertainty in the determination of ^{125}I was improved by about 1% (Grau Malonda and Grau Carles, 2008) when models of the simultaneous interactions of electrons in liquid scintillator were included in the efficiency calculation. The simultaneous interaction of electrons produces an overlap of distributions of excited solvent molecules and an increase in the excitation density, which required a modification of the ionization quenching factor in the calculations.

Grau Carles and Grau Malonda (2010) presented a new approach to the problem of elevated uncertainties in the standardization of low-energy beta-emitters, when the TDCR method is applied, and in the standardization of some electron-capture radionuclides, when the CIEMAT/NIST method is applied. This work considers the need to expand the TDCR and CIEMAT/NIST methods by generalizing the equations used in the model and to analyze the influence of ionization and chemical quench on the spectra and counting efficiency. They studied the influence of photomultiplier response in one, two, and three PMT systems. Particularly the effect of electronic noise discriminator level on both experimental spectra and counting efficiency are considered.

In a personal communication of the writer with Grau Malonda (2010), many of the problems with the model arise from the discrepancy between the experimental spectra and the computed one. In other words, when the total efficiency is calculated for beta-emitters, the full spectral area is computed; however, when the experimental radionuclide is counted, only the pulses with pulse heights greater than the noise discriminator level are counted. Thus, for example, the counts for the tritium spectrum turn out to be lower than the computed total spectrum. Low counts yield low efficiency, and a lower efficiency yields a higher figure of merit and consequently a higher figure of merit for the radionuclide to be standardized (*e.g.*, ^{125}I) yielding finally a higher activity determination. Accordingly, an improvement in the accuracy of the liquid scintillation standardization of radionuclides rests with the spectral analysis.

With the objective of making improvements in the spectral analysis of electron capture radionuclides in the CIEMAT/NIST method of radionuclide standardization, Kossert and Grau Carles (2010) report an improved MICELLE2 code for the computation of the detection efficiencies of EC radionuclides to yield a more realistic treatment of photoelectron ejections and consequent atomic electron shell rearrangements yielding a more detailed atomic rearrangement model for the electron capture in the radionuclide of interest. The number of possible electron rearrangements following electron capture increase with the radionuclide atomic number. Although the K shell is the predominate shell for electron capture by the atomic nucleus (See Chapter 1, Volume 1), the MICELLE2 code accounts for probabilities for capture from outer electron shells. Kossert and Grau Carles (2010) note that fractional electron capture probabilities are generated according to the capture of an electron from the K, L_1, L_2, or M shell followed by probabilities for X-ray emission with a fluorescent yield (ω_K) probability or Auger-electron ejection probability ($A_K = 1 - \omega_K$). The MICELLE2 includes specific criteria for the selection of relative probabilities for radiative (X-ray emissions) or nonradiative transitions, namely, Auger and Coster-Kronig electron emissions. As described by Kossert an Grau Carles (2010), the updated MICELLE2 program also contains subroutines that simulate the energy loss of electrons in the liquid scintillation cocktail micelles. The micelles are miniscule and generally invisible spherical volumes of aqueous phase formed within liquid scintillation cocktails that contain emulsifiers to promote the uniform mixing of organic scintillation fluor cocktails with aqueous sample. The energy from electrons or beta particles deposited within the micelles does not produce scintillation light photons and consequently this portion of the particle energy does not contribute to the detection efficiency. In the case of the standardization of ^{55}Fe, Bergamon and Laureano-Perez (2014) report there were no observable reductions in detection efficiency attributable to scintillation cocktail micelles size. Also, they found that the use of ^{54}Mn as a tracer nuclide would reduce the model dependence of the calculation of detection efficiencies provided that sources of ^{54}Mn that are more than a few years old be evaluated for the accumulation of long-lived radionuclide impurities. In a recent study on the standardization of radionuclides that decay by EC or the emission of low-energy beta particles by the TDCR efficiency calculation technique (see next section of this chapter), Nedjadi et al. (2017) computed the ionization quenching function using Birk's law while taking into consideration the energy losses in all of the micelles bisecting the electron pathway through the scintillation cocktail. The computed ionization quenching function was then used to calculate the detection efficiencies for the weak beta-emitters ^{3}H and ^{63}Ni and the EC decaying nuclides ^{54}Mn and ^{55}Fe. They found the effect of micelle size on the calculated detection efficiencies to be small for beta-emitters, but significant for the electron capture nuclides.

There is currently increased interest in the standardization of ^{68}Ga, which has applications in nuclear medicine as a radionuclide useful for positron emission tomography (PET) in medical clinics that do not have a cyclotron for the production of short-lived positron-emitters. Gallium-68 has a half-life of 68.33 min, and it can be milked on-site from a generator containing the ^{68}Ge parent nuclide, which decays with a half-life of 275 days. The beta decay mode of ^{68}Ga includes positron emission at 89.1% abundance and electron capture at the remaining 10.9% abundance. Roteta et al. (2012) have standardized ^{68}Ga by the CIEMAT/NIST method utilizing a Monte Carlo simulation program for the calculation of the detection efficiency. The program simulates the EC and positron decay paths and the consequent cascade of particles and photons for each path. As noted by Roteta et al. (2012), the program includes the atomic electron rearrangements in the K, L, and M shells as well as the deposited energy via the photon interactions with scintillator. The commercial availability of ^{68}Ge/^{68}Ga generators for applications in clinical medicine motivated Kulkarni et al. (2017) to standardize the parent-daughter nuclides. Germanium-68 decays exclusively by electron capture while its daughter nuclide ^{68} Ga day by EC and positron emission. Kulkarni et al. (2017) calculated the efficiencies of ^{68}Ge and ^{68}Ga versus the figure of merit (See Figs. 6.78 and 6.81 illustrated earlier in this chapter) with the freeware computer program CN2003.

11. Applications with plastic scintillation microspheres

Tarancón et al. (2015) applied an adapted version of the MICELLE2 code, which included PENELOPE code to allow for the calculation of the beta-particle energy losses due to varying degrees of particle quenching that can occur with different geometries of plastic scintillation microspheres. Their application of the CIEMAT/NIST method to the metrology of beta-emitters utilizing plastic scintillation microspheres yielded activity calculations for ^{63}Ni, ^{14}C, ^{36}Cl, and ^{90}Sr/^{90}Y with deviations of 8.8%, 1.9%, 1.4%, and 2.1% respectively. A very low-energy beta-emitter, such as ^{63}Ni ($E_{max} = 0.066$ MeV) would be affected obviously by particle quenching to a greater degree than higher-energy beta-emitters.

12. Radionuclide mixtures

The CIEMAT/NIST method has been applied to the standardization of radionuclide mixtures. Early work by Grau Carles (1994) describes computer programs for the deconvolution of ^{89}Sr, ^{90}Sr, and ^{90}Y mixtures with the liquid scintillation standardization of the components by the CIEMAT/NIST method. This was expanded and improved to include the standardization of more complex mixtures of several (up to five) radionuclides (Grau Carles, 1996). Günther (2002c) and Rodríguez Barquero et al. (2004) measured ^{32}P and ^{33}P activities where ^{33}P is an impurity. The relative standard uncertainties for the determination were 0.32% for ^{32}P and 1.6% for ^{33}P impurity. A triple exponential mathematical expression is described by Altsitzoglou (2004) to standardize ^{32}P in the presence of ^{33}P and ^{35}S. Van Wyngaardt and Simpson (2006a) and Van Wyngaardt et al. (2008) applied the liquid scintillation TDCR and CIEMAT/NIST standardization methods to resolved mixtures of ^{14}C and ^{63}Ni and to component mixtures of ^{32}P, ^{33}P, and ^{35}S, respectively. Mixtures of short-lived components were resolved for the CIEMAT/NIST standardization of ^{18}F (Capogni et al., 2006) and ^{64}Cu (Capogni et al., 2008).

B. Secondary standardization by the cross-efficiency method

A secondary standard measurement method that utilizes liquid scintillation counting was reported by Kossert (2006). The method is based on the CIEMAT/NIST method; however, it is distinctly different in that the efficiency curve, i.e., the counting efficiency of the nuclide under investigation as a function of the tritium counting efficiency or $\varepsilon_{nuclide}(\varepsilon_{tracer})$, is not calculated, but rather, it is determined experimentally. The method is illustrated in Fig. 6.88. Notice the similarity between Fig. 6.85, that of

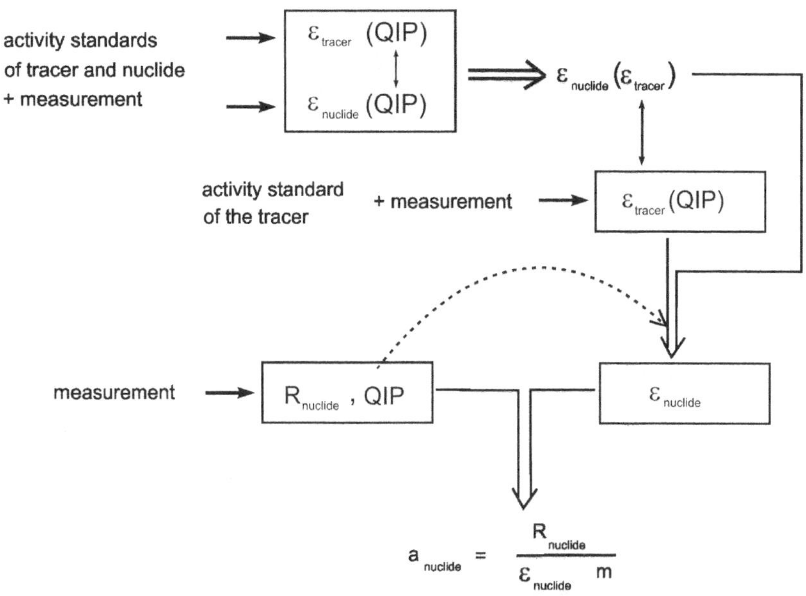

FIGURE 6.88 Secondary standard measuring procedure. R denotes the count rate, QIP the quench indicating parameter, ε the counting efficiency, $a_{nuclide}$ is the solution activity concentration of the nuclide under investigation, and m is the source mass. *From Kossert (2006). Reprinted with permission from Elsevier © 2006.*

the CIEMAT/NIST method, and Fig. 6.88, that of the secondary standardization method. In the secondary standardization method, there is no calculation of figure of merit (*M*); there is instead, the experimental determination of the quench indicating parameter (*QIP*).

This method of secondary standardization of Kossert (2006) is described, in a stepwise fashion, as follows:

a. The first step, as illustrated in Fig. 6.88, involves the measurement of the counting efficiency of the nuclide under investigation, $\varepsilon_{\text{nuclide}}$, as a function of a quench indicating parameter (QIP). To achieve this, a set of quench standards of the nuclide under investigation is prepared. This requires the preparation of a few vials (approx. 10 vials) each containing the scintillation cocktail and a known amount of an activity standard solution. Increasing amounts of quenching agent (e.g., nitromethane) are added to each counting vial, with the exception of the first vial, to provide counting vials with decreasing detection efficiencies (See Section V.D of this chapter). Counting of these standards in a liquid scintillation analyzer will provide the net count rates, (e.g., CPM) and values of a quench indicating parameter (QIP), such as H#, SQP(E), or tSIE, depending on the instrument used and its external standard. From the net count rates and the known activities (DPM) in each vial, the detection efficiencies of each vial are easily calculated (i.e., $\varepsilon_{\text{nuclide}} = \text{CPM/DPM}$). The same procedure is carried out with a set of quenched standards prepared with ^3H, which serves as the tracer nuclide. The counting of the tritium nuclide standards will yield distinct counting efficiencies, $\varepsilon_{\text{tracer}}$, for each vial as well as varying QIP values.

b. The two functions $\varepsilon_{\text{nuclide}}(QIP)$ and $\varepsilon_{\text{tracer}}(QIP)$ obtained from step (a) above are linked by the quench indicating parameter *QIP* to provide a working 'efficiency curve' of $\varepsilon_{\text{nuclide}}(\varepsilon_{\text{tracer}})$ as illustrated in Fig. 6.89 for the case of ^{125}I as an example of a nuclide under investigation. The best fit for the function $\varepsilon_{\text{nuclide}}(\varepsilon_{\text{tracer}})$ for the working 'efficiency curve' is obtained via a fourth-order polynomial, which provides the counting efficiency of the nuclide, $\varepsilon_{\text{nuclide}}$, for a specific counting efficiency of the tritium tracer, $\varepsilon_{\text{tracer}}$, given by the equation

$$\varepsilon_{\text{fit}} = \varepsilon_{\text{nuclide}}(\varepsilon_{\text{tracer}}) = \sum_{i=0}^{4} p_i(\varepsilon_{\text{tracer}})^i \qquad (6.107)$$

Notice the similarity of Eq. (6.107), which uses the experimentally determined QIP as a link between the two detection efficiencies for this secondary standardization method, and Eq. (6.102), which uses the calculated figure of merit (*M*) as the link in the CIEMAT/NIST primary standardization method.

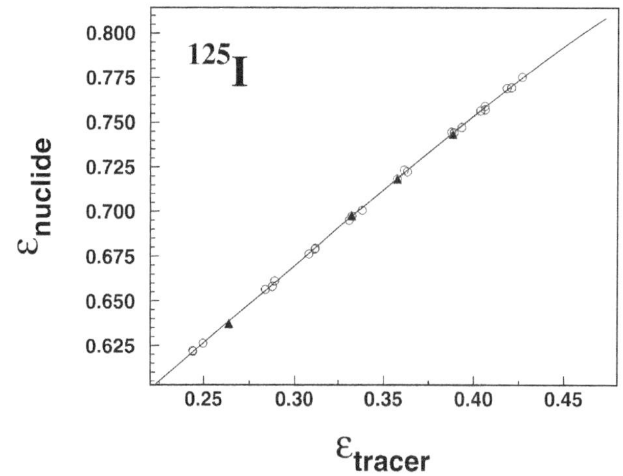

FIGURE 6.89 Cross-efficiency curve of $\varepsilon_{\text{nuclide}}(\varepsilon_{\text{tracer}})$ for ^{125}I in 15 mL of Ultima Gold cocktail plus 1 mL of water. The empty circles represent the determined efficiencies, and the 'efficiency curve' was used to calibrate another solution of ^{125}I, which was also calibrated by coincidence counting. The solid triangles represent the results of efficiencies determined with the second solution of ^{125}I. *From Kossert (2006). Reprinted with permission from Elsevier © 2006.*

c. This following step need not be carried out during the same period of time as the above preparation of the 'efficiency curve'. As pointed out by Kossert (2006), this next step can be performed even years later. This next step involves the preparation of the ^3H 'calibration curve', which is the repeated measurement of the counting efficiency of the tracer nuclide, $\varepsilon_{\text{tracer}}$, expressed as a function of the quench indicating parameter (*QIP*). The newly measured 'calibration curve', $\varepsilon_{\text{tracer}}(QIP)$ is required to account for any instability of the liquid scintillation counter in the event of any changes in time of the instrument response and measured spectra.

d. The sample of the nuclide under investigation is now counted, and the sample count rate, R_{nuclide}, and sample quench indicating parameter (*QIP*) are measured and recorded. From the magnitude of the *QIP*, the corresponding value of the tracer counting efficiency can be obtained from the ^3H 'calibration curve' prepared in Step c. From the known tracer efficiency, the corresponding nuclide efficiency, $\varepsilon_{\text{nuclide}}$, can be obtained from the 'efficiency curve' measured in Step b.

e. With the efficiency of the nuclide under investigation, $\varepsilon_{\text{nuclide}}$, obtained from the 'efficiency curve' and the nuclide count rate, R_{nuclide}, the activity concentration of the nuclide, a_{nuclide}, is calculated, as illustrated in Fig. 6.88, according to the equation $a_{\text{nuclide}} = R_{\text{nuclide}}/(\varepsilon_{\text{nuclide}} \cdot m)$, where m is the mass of the sample counted.

Kossert (2006) and Cassette and Tartès (2014) have tested this secondary standardization method, and certain specifics should be considered, and these are the following:

a. The sample scintillation cocktail composition (i.e., cocktail and sample chemistry) and the sample geometry (i.e., vial size, vial type, and sample-cocktail volume or mass) must be very similar for all measurements.
b. The period between the measurements of the tracer nuclide, and the nuclide under investigation should be short (i.e., within a period of only a few days).
c. The duration between the determination of the 'efficiency curve' and the measurements of the tracer nuclide and the nuclide under investigation is unlimited.
d. The 'calibration curve' of $\varepsilon_{tracer}(QIP)$, e.g., $\varepsilon_{tritium}$ versus SCP(E) or $\varepsilon_{tritium}$ versus tSIE, is easily measured with weighed aliquots of a standard solution of tritiated water added to cocktail in a set of about 10 vials to which increasing amounts of nitromethane are added.
e. The quench curves are counter dependent, and this holds for different counters that have similar external standards and measure the same QIP.
f. The quench curves are dependent on the nature of the quenching, i.e., chemical or color quenching. The differences between the chemical and color quench curves depend on the radionuclide and are more significant in the measurement of ^{55}Fe and ^{63}Ni than in the measurement of ^{3}H.
g. This following point is one of the key advantages of this method: The uncertainty of the tracer activity (e.g., tritium) cancels out, if the aliquots used for the determination of the 'efficiency curve' and the 'calibration curve' stem from the same standard solution.
h. The method can be applied to the activity determinations of radionuclides with low activity concentrations (e.g., 10 Bq/g).

Kossert (2006) tested this secondary standard method against several nuclides standardized by accepted primary standardization methods. This secondary standardization method tested against ^{125}I primary standard yielded an uncertainty of the activity determination lower by only 0.03% of that of the primary standard. Likewise, the activity of ^{65}Zn was measured by this secondary standardization with an uncertainty of only 0.05% lower than that determined by primary standardization. Other nuclides tested by this secondary standardization method were ^{54}Mn, ^{57}Co, ^{67}Ga, ^{85}Sr, ^{88}Y, ^{109}Cd, and ^{111}In. In a subsequent work Kossert et al. (2006) reported the secondary standardization of ^{65}Zn by this method with an uncertainty of 0.4%.

C. Triple-to-double coincidence ratio (TDCR) efficiency calculation technique

The early works of Steyn (1956), Flynn and Glendenin (1959) and Radoszewski (1964) can be given the credit of initiating the possibility of utilizing a liquid scintillation analyzer for the absolute measurement of beta-emitting radionuclides (personal communication of the writer with Ryszard Broda, 2004). The first triple photomultiplier liquid scintillation counting system with an electronic circuit having two different coincident outputs was reported originally by Schwerdtel (1965, 1966a,b); however, the method for the determination of the absolute activities of radionuclides was yet to be developed. An important contribution was made by Kolarov et al. (1970), who proposed a liquid scintillation detection efficiency model used in the current triple-to-double coincidence ratio (TDCR) efficiency calculation technique. Pochwalski (1978) presented the TDCR method as his thesis; he worked many years in Radoszewski's nuclear metrology laboratory in the Institute of Nuclear Research at Świerk, Poland. A triple PMT detector design and the original version of the method of radionuclide standardization now referred to as the TDCR efficiency calculation technique were published in a national report by Pochwalski and Radoszewski (1979) and in the international journals by Pochwalski et al. (1988). Thus, the TDCR efficiency calculation technique was invented in Poland. The integration of theoretical calculations of counting efficiency based on the radiation emissions of the radionuclide under investigation with the experimental data obtained from the LS, somewhat similar to the approach taken in the CIEMAT/NIST efficiency tracing method conceived by Grau Malonda in 1978 described earlier in this chapter, was reported by Broda et al. (1988), Grau Malonda and Coursey (1988), and Grau Carles and Grau Malonda (1989). These papers formed the basis for subsequent research and development in this method, which is described subsequently.

1. Principles

The liquid scintillation analyzer (LSA), in this case, is different from the conventional and more common commercially available equipment, described previously in this chapter. The TDCR efficiency calculation technique requires an LSA equipped with three photomultiplier (3-PM) tubes and two different coincident outputs. Such equipment is available commercially (e.g., Hidex, Turku, Finland), and other 3-PM instruments are designed and manufactured by those laboratories dedicated to radionuclide standardization, such as national institutes of standards and research. Since the publication of the second Edition of the *Handbook of Radioactivity Analysis* in 2003, the national radionuclide metrology laboratories that now apply the TDCR efficiency calculation method for the metrology of radionuclides have become more widespread (Cassette, 2011). A list of several such laboratories in numerous institutions worldwide is provided in Table 6.10.

TABLE 6.10 Some national nuclear metrology laboratories reporting custom-built 3-PM liquid scintillation counting systems.

Nuclear metrology Laboratory[a]	References
Laboratorio de Metrología de Radioisótopos, CNEA, Argentina	Cassette et al. (2018)
Australian Nuclear Science and Technology Organization (ANSTO), Menai, Australia	Qin et al. (2008), Cassette et al. (2018)
European Commission, Joint Research Center, Geel, Belgium	Cassette et al. (2018)
Sophia University "St. Kliment Ohridski", Bulgaria	Mitev et al. (2017)
National Research Council of Canada, Ottawa	Bergeron et al. (2016) and Cassette et al. (2018)
Institute of Atomic Energy, Beijing, China	Wu et al. (2014), Lv et al. (2014)
National Institute of Metrology, Beijing, China	Wang, Z.-Y. (2008), Wu et al. (2015), Cassette et al. (2018)
Centro de Isótopos (CENTIS), Havana, Cuba	Cassette et al. (2018)
Czech Metrology Institute, Prague, Czech Republic	Sochorová et al. (2012)
Laboratoire National Henri Becquerel, Saclay, France	Cassette and Vatin (1992), Cassette and Bouchard (2003), Cassette et al. (2013, 2018)
Physikalisch-Technische Bundesanstalt (PTB), Braunschweig, Germany	Nähle et al. (2010, 2014), Cassette et al. (2018)
Korea Research Institute of Standards and Science (KRISS), Daejeon, Korea	Hwang et al. (2004), Lee et al. (2004a,b), Cassette et al. (2018)
Center for Physical Sciences and Technology, Vilnius, Vilnius, Lithuania	Gudelis et al. (2012, 2017), Cassette et al. (2018)
Instituto Nazionale di Metrologia delle Radiazioni Ionizzanti (INMRI), Rome, Italy	Capogni and Antohe (2014), Capogni and De Felice (2014) and Cassette et al. (2018)
Institute of Atomic Energy, Świerk, Poland	Pochwalski et al. (1988)
National Center for Nuclear Research Radioisotope Center POLATOM	Cassette et al. (2018)
Horia Hulubei National Institute of R&D for Physics and Nuclear Engineering, IFIN-HH, Bucharest, Romania	Razdolescu et al. (2006), Ivan et al. (2008) and Cassette et al. (2018)
D.I. Mendeleyev Institute for Metrology, Russia	Cassette et al. (2018)
National Metrology Institute of South Africa, Cape Town, South Africa	Meyer and Simpson (1990), Simpson et al. (2010)
Centro de Investigaciones Energéticas, Medioambientales y Tecnológicas (CIEMAT), Madrid, Spain	Grau Malonda and Coursey (1988), Grau Carles and Grau Malonda (1989)
Institut de Radiophysique, Lausanne, Switzerland	Nedjadi et al. (2015), Cassette et al. (2018)
National Physical Laboratory, Middlesex, UK	Johansson and Sephton (2010), Johansson et al. (2014) and Cassette et al. (2018)
National Institute of Standards and Technology, (NIST), Gaithersburg, USA	Zimmerman et al. (2004a), Cassette et al. (2018)

[a] *Listed in alphabetical order by country.*

A commercially available liquid scintillation analyzer, the Hidex 300 SL (Hidex Oy, Turku, Finland) manufactured with three photomultiplier tubes, which utilizes the TDCR technique, was released in 2008 (Haaslahti, 2010). The Hidex company also produces a higher capacity (Hidex 600 SL) 3-PM liquid scintillation analyzer. The performance of the Hidex 300 SL is described by Kharitonov et al. (2011). A schematic diagram of the workings of an instrument, referred to as the TDCR liquid scintillation counter, which is applied to the standardization (i.e., direct DPM measurement) of radionuclides, is illustrated in Fig. 6.90, and an illustration of a commercially available three-photomultiplier detector arrangement is illustrated in Fig. 6.91. The optical chamber of the LSA accommodates a

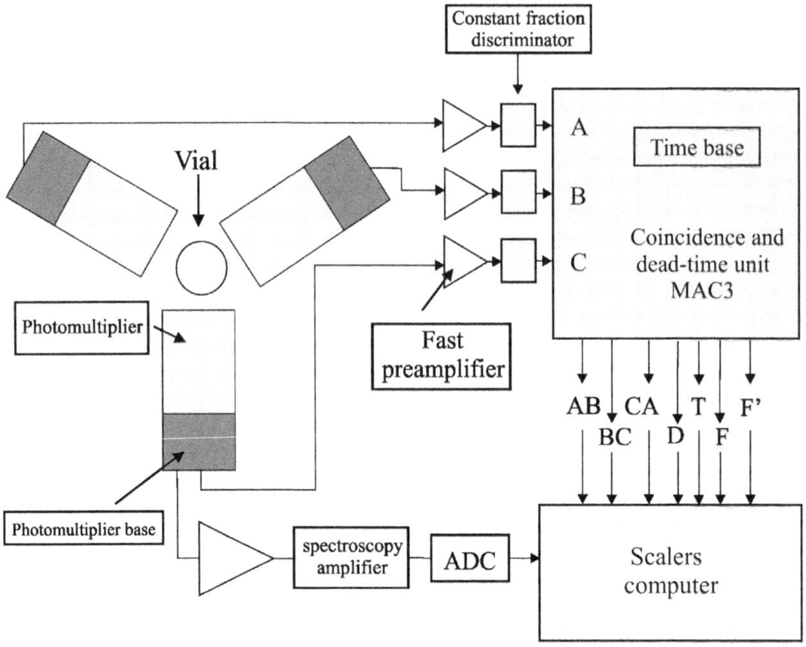

FIGURE 6.90 Diagram of the TDCR system at the Radioisotope Metrology Laboratory, CNEA, Argentina. MAC3 is the electronic coincidence module of Bouchard and Cassette (2000) for the processing of the pulses delivered by the three photomultipliers including double coincidence outputs AB, BC, and CA, the logical sum of the double coincidences D, and the triple coincidences T. *From Arenillas and Cassette (2006). Reprinted with permission from Elsevier © 2006.*

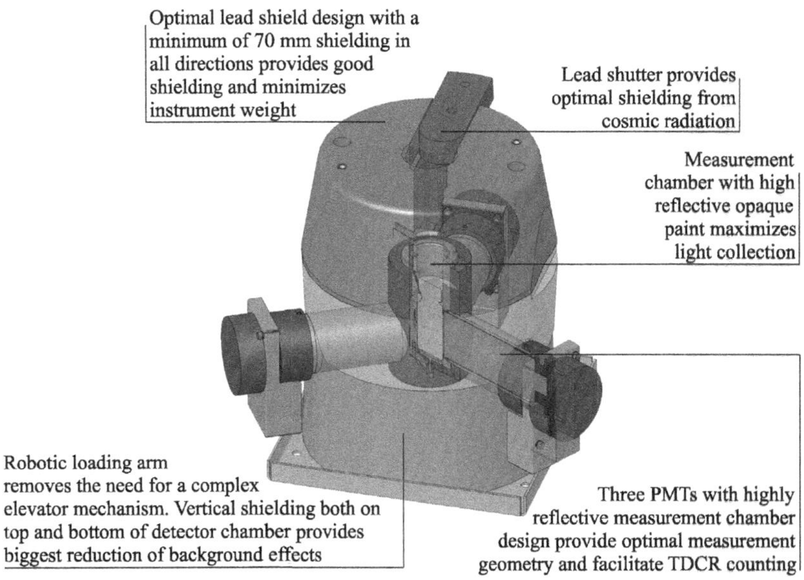

FIGURE 6.91 Hidex 300 SL detector structure. *Courtesy of Hidex, Turku, Finland. Printed with permission.* For the color version of this figure, the reader is referred to the electronic version of the book.

standard 20-mL liquid scintillation counting vial, and facing the vial are three photomultiplier tubes at 120 degree angles to each other. With such an arrangement, dual and triple phototube coincidence outputs are possible as a requirement for the TDCR method. For example, the two coincident outputs used by Pochwalski et al. (1988) from photomultiplier tubes A, B, and C were the count rates from the triple coincidences (ABC), the three double coincidences AB, BC, and AC, and the count rate from the logic sum of the three different double coincidences (AB + BC + CA). Since this initial report of the TDCR method, many developments have been made. Advances following the development of the TDCR radionuclide standardization method are described by Simpson and Meyer (1992 and 1994b), and reviews on the theory and practice of the TDCR method are available from Broda (2003), Lee et al. (2004a,b), Broda et al. (2007), Nähle and Kossert (2011), Cassette (2011), Nähle et al. (2014), and Capogni and Antohe (2014) from which the subsequent treatment and mathematical notation utilized in TDCR are taken.

The triple-to-double coincidence ratio (TDCR) method was developed for the direct determination of the absolute activities of beta- and EC-decaying radionuclides in liquid scintillator cocktail. Like the CIEMAT/NIST method, described previously in this chapter, the TDCR technique combines experimental data with theoretical calculations of the radionuclide detection efficiency. Complete data of the radionuclide decay scheme is used to calculate the counting efficiencies based on a physical and statistical model of the beta-particle, Auger-, and Compton electron interactions with scintillation cocktail. The TDCR method utilizes a physical and statistical model of the distribution of scintillation photons and their detection probabilities in a three-photomultiplier detector system (Broda, 2003). The ionization quenching phenomena, which is the nonlinearity of the scintillator response as a function of the linear energy transfer (LET) or stopping power of the interacting electrons in fluor cocktail, discussed previously in the CIEMAT/NIST method, is also required in the TDCR technique for the theoretical calculations of counting efficiencies.

In the case of a pure β-emitting source on which are viewed simultaneously three normalized photomultiplier tubes the double-tube coincidence count rate N_d and the triple-tube coincidence count rate N_t are defined as

$$N_d = N_0\varepsilon_d \tag{6.108}$$

and

$$N_t = N_0\varepsilon_t \tag{6.109}$$

where N_0 is the disintegration rate of the radionuclide source and ε_d and ε_t are the true double-tube and triple-tube counting efficiencies, respectively. The detection efficiencies are determined without recourse to prior instrumental calibration nor the use of any reference standards. Rather, the detection efficiencies are determined in an absolute manner by combining experimental measurement with theory (Simpson and Morris, 2004b). The detection efficiencies ε_d and ε_t of Eqs. (6.108) and (6.109) are unknown; however, the ratio (R) of the count rates N_t/N_d are determined or measured experimentally with the liquid scintillation counter. For a large number of disintegration rates, the ratio of the experimentally determined triple coincidence count rate to double coincidence count rate, N_t/N_d, converges toward the ratio of counting efficiencies, as described by Broda et al. (2007). Thus, using the notation of Simpson and Morris (2004b) and Lee et al. (2004a,b), the experimentally determined ratio (R) can be written as

$$R = \frac{N_t}{N_d} = \frac{N_0\varepsilon_t}{N_0\varepsilon_d} = \frac{\varepsilon_t}{\varepsilon_d}. \tag{6.110}$$

where N_0 is the radionuclide activity, and the product of activity and detection efficiency (*i.e.*, $N_0\varepsilon_t$ and $N_0\varepsilon_d$) is count rate. As in the previously described CIEMAT/NIST radionuclide standardization method of efficiency tracing (Section IX.A of this chapter), theoretical counting efficiencies are calculated and these are fitted to experimental data. The theoretical counting efficiencies ε_d and ε_t for a pure β-emitter are described, according to a 'free parameter' model by Grau Malonda and Coursey (1988) and Simpson and Meyer (1992a-c) as

$$\varepsilon_2 = \int_0^{E_{\max}} N(E)[1-\exp\{-P\cdot F(E)\cdot E\}]^2 dE, \tag{6.111}$$

and

$$\varepsilon_3 = \int_0^{E_{\max}} N(E)[1-\exp\{-P\cdot F(E)\cdot E\}]^3 dE, \tag{6.112}$$

where ε_2 and ε_3 are the double- and triple-photomultiplier tube efficiencies, respectively, $N(E)$ represents the number of β particles of energy between $E + dE$ of the β-particle spectrum calculated according to the Fermi theory of decay (Fermi spectrum), $E_{\max}$ is the maximum β-particle energy, $F(E)$ is the relative scintillation efficiency Gibson (1968) and P is the figure of merit or free parameter, the only unknown variable of the theoretical efficiency equations. The figure of merit (free parameter) was defined previously in this chapter in Section IX.A concerning the CIEMAT/NIST efficiency tracing technique where a "system" figure of merit M or λ was used, whereby $P = 1/2M$ for a double-phototube system and $P = 1/3M$ for a triple-phototube system (Simpson and Meyer, 1992b), and $M = \lambda$. The theoretical efficiencies, therefore, are a function of a single unknown parameter P. Thus, as described by Simpson and Morris (2004) and Van Wyngaardt and Simpson (2009), the efficiencies can be expressed theoretically as a function of a single floating parameter P, the so-called 'figure-of-merit', which is effectively a measure of the electron emission process and the consequent photon output.

Thus, a single value of P must be found iteratively to fit the experimentally measured count ratio to solve the equation

$$R = \frac{N_t}{N_d} = \frac{\varepsilon_t(P)}{\varepsilon_d(P)} \tag{6.113}$$

The theoretical counting efficiencies as a function of the figure of merit, P, are calculated and the value of P is obtained by mathematical iteration to best fit the experimentally determined triple-to-double count ratio of Eq. (6.113). As noted by Simpson and Meyer (1994b) and Simpson and Morris (2004), the figure of merit P should be single-valued for a given value of R. With the determined value of P for a given experimental sample, the activity of the radionuclide source can then be calculated using Eqs. (6.108) and (6.109) or

$$N_0 = \frac{N_d}{\varepsilon_d} = \frac{N_t}{\varepsilon_t} \tag{6.114}$$

2. Free-parameter model

The calculation of the theoretical efficiencies according to Eqs. (6.111) and (6.112) are similar to Eq. (6.96) previously described in Section IX.A. According to Grau Carles and Grau Malonda (2001), Lee et al. (2004a,b), Nähle and Kossert (2011), and Cassette (2011), the equations applicable to the theoretical counting efficiency calculations for beta-particle emitters and radionuclides that decay by electron capture (EC) in a two- or a three-photomultiplier system are the following:

$$\varepsilon_2 = \int_0^{E_{\max}} S(E)\left(1 - e^{-EQ(E)/2M}\right)^2 dE \tag{6.115}$$

for two photomultipliers working in coincidence, and

$$\varepsilon_T = \int_0^{E_{\max}} S(E)\left(1 - e^{-EQ(E)/3M}\right)^3 dE \tag{6.116}$$

for three photomultipliers in coincidence; whereas, the logic sum of double coincidences in a 3-PM counter is calculated as

$$\varepsilon_D = \int_0^{E_{\max}} S(E)\left[3\left(1 - e^{-EQ(E)/3M}\right)^2 - 2\left(1 - e^{-EQ(E)/3M}\right)^3\right] dE \tag{6.117}$$

Thus, the triple-to-double coincidence ratio (TDCR) is expressed as

$$\frac{N_T}{N_D} = \frac{\varepsilon_T}{\varepsilon_D} = \frac{\int_0^{E_{\max}} S(E)\left(1 - e^{-EQ(E)/3M}\right)^3 dE}{\int_0^{E_{\max}} S(E)\left[3\left(1 - e^{-EQ(E)/3M}\right)^2 - 2\left(1 - e^{-EQ(E)/3M}\right)^3\right] dE} \tag{6.118}$$

where the variables are the same as those defined previously for Eqs. (6.95) and (6.96) in Section IX.A, namely, $E_{\max}$ is the maximum beta-particle energy, $S(E)$ is the normalized energy spectrum transferred to the liquid scintillation cocktail, including energy from beta particles as well as Auger-, Coster-Kronig-, and conversion electrons, and photo- and Compton electrons arising from X-ray- and gamma-ray-interactions in LS cocktail [the term $S(E)$ was previously referred to as $N(E)$ Eqs. 6.95 and 6.96], $Q(E)$ is the ionization quenching correction factor, and M is the free parameter. The free parameter was expressed in Eqs. (6.95) and (6.96) as λ; the expression for the free parameters M or λ depends on the user (*i.e.*, $M = \lambda$). The above calculations are based on a definition of the figure of merit (free parameter) as a computation of the β-particle energy in keV required to produce one photoelectron at the output of the photocathode or the photoelectron yield at the first dynode. The free parameter has units of electrons/keV.

The ionization quench function $Q(E)$ of Eq. (6.118) is determined by applying the Birks formula described previously by Eq. (6.106). The Birks formula expresses the reduction in fluorescence due to ionization quenching $[1 + kB(dE/dx)]^{-1}$ where kB is the ionization quench parameter or constant with units of cm/MeV or g/MeV cm^2, and dE/dx is the radiation stopping power of the scintillator with units of MeV/cm or MeV cm^2/g, as discussed previously with respect to Birks definition of specific fluorescence and ionization quenching according to Eqs. (6.105) and (6.106). As described by Cassette (2011), Broda et al. (2007), and Lee et al. (2004a,b), the left term of Eq. (6.118) (N_T/N_D) is measured experimentally, and the right term ($\varepsilon_T/\varepsilon_D$) is calculated. As the beta-particle energy distribution $N(E)$ for a given radionuclide is known, and the kB ionization quench parameter is an assumed constant, the value of the free parameter must be found whereby the calculated $\varepsilon_T/\varepsilon_D = N_T/N_D$, which is the experimentally measured triple-to-double count ratio.

3. Experimental conditions and efficiency calculations

The practical aspects of the TDCR method are dealt with in detail by Broda et al. (2007). Some of the major points of the TDCR method are presented here.

Simpson and Meyer (1994b) demonstrated that any change in the counting efficiency due to any altered chemical quenching state of the sample will manifest itself in the measured triple and double counting rates, thereby producing a different value of R. They demonstrated, therefore, that the figure of merit P extracted by this method for a given sample includes affects due to chemical quenching so that the calculated sample activity N_0 is independent of chemical quench effects. Using wide ranges of $CHCl_3$ as a quenching agent (18–2287 mg $CHCl_3$ load), they demonstrated no significant quench effect on the calculated activity of ^{14}C. Also, smaller $CHCl_3$ quenching agent loads (13–409 mg) had no affect on the calculated activity of ^{63}Ni, and chloroform loads of up to 60 and 270 μL produced no effect on the calculated activities of ^{99}Tc and ^{3}H, respectively.

The calculation of the counting efficiencies ε_d and ε_t can be carried out with computer codes, such as EFFY, written by García-Toraño and Grau Malonda (1985) previously described in this chapter (see also Bergeron et al., 2016 and Cassette et al., 1998). Simpson and Meyer (1994b) and Simpson and Van Wyngaardt (2006) report the use of a modified EFFY2 program to operate on a personal computer, which calculates the ratio R of the triple-to-double efficiencies as a function of figure of merit. They note that in practice, Eq. (6.113) is solved by incrementally varying the figure of merit, P, in the computer calculation, which in effect reads the counting efficiencies corresponding to the measured ratio R from plots of efficiency curves such as those illustrated in Fig. 6.92. Simpson and Morris (2004) and Simpson and Van Wyngaardt (2006) report the use of EFFY2 in the TDCR standardization of high-energy beta-emitters. Broda et al. (2000) describe a computer program, TDCRB-1, to calculate the radioactive concentration of a solution based on the TDCR method. Also, the DETECSZ program is described by Razdolescu et al. (2002) for the TDCR calculation of the activity concentration of pure beta-emitters. Similar analysis codes, such as TDCRB-02 reported by Broda et al. (2002), are utilized by Nähle et al. (2010), Zimmerman et al. (2010), Bergeron et al. (2012), Gudelis et al. (2012), and Nedjadi et al. (2015).

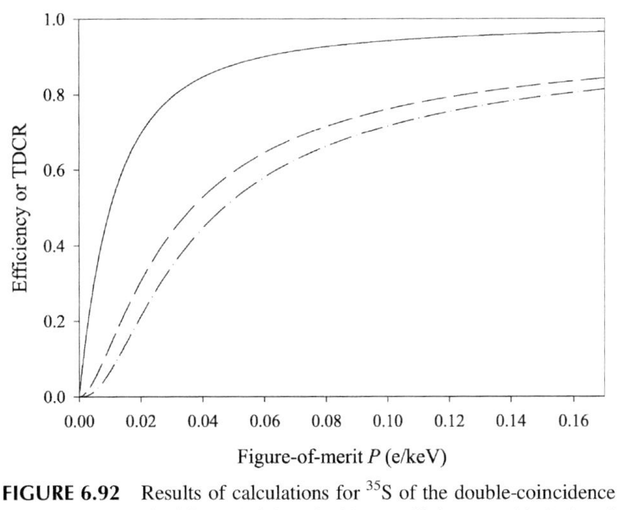

FIGURE 6.92 Results of calculations for ^{35}S of the double-coincidence efficiency ε_d (dashed line). Triple-coincidence efficiency ε_t (dash-dotted line), and the TDCR $\varepsilon_t/\varepsilon_d$ (solid line) are plotted as a function of the figure-of-merit. A value for kB = 0.0100 g/(cm^2 MeV) was adopted in the calculation of the ionization quench function. *From Van Wyngaardt and Simpson (2009). Reprinted with permission from Radiocarbon, University of Arizona © 2009 Arizona Board of Regents on behalf of the University of Arizona, Tucson.*

The program SPECBETA, first reported by Cassette (1992) and Cassette and Vatin (1992) is used for the calculation of the beta spectrum, *i.e.*, $N(E)$ or what is also referred to as $S(E)$, the deposited spectrum in the scintillation cocktail, a required component of the TDCR efficiency calculation. The use of SPECBETA is reported by several researchers including Collé et al. (2008), Zimmerman et al. (2008), Fitzgerald et al. (2010), and Laureano-Perez et al. (2010). Lourenço et al. (2015) calculated beta spectra based on the original home-made code of Grau Malonda (1999).

The calculation of efficiencies of EC nuclides requires data on interaction probabilities of electron-generating X-ray and γ-ray photons with scintillator, and the Monte Carlo code PENELOPE, first reported by Baró et al. (1995) and reviewed and refined by Salvat and Fernandez-Varea (2009) and Salvat et al. (2011), has been used for this purpose (*e.g.*, Cassette et al., 2004, Zimmerman, 2006; and Razdolescu et al., 2008). PENELOPE is noted by Kossert et al. (2014a) as the state-of-the-art simulation tool for the interaction of low-energy radiation in matter. Likewise, the simulation code GEANT4 was developed (Agostinelli et al., 2003; Allison et al., 2006) and utilized to acquire X-ray and gamma-ray energy deposition spectra and interaction probabilities with scintillant for the analysis of EC and beta-gamma-emitting radionuclides (*e.g.*, Bignell et al., 2010a,b, Thiam et al., 2012, and Lourenço et al., 2015).

An updated MICELLE2 code, reported by Kossert and Grau Carles (2010) and the TDEM13 code developed by Broda et al. (2013), were prepared to facilitate the computation of counting efficiencies of radionuclides with complex decay schemes, such as EC nuclides that require a treatment of the ejection of photoelectrons and a detailed model of the subsequent atomic electron rearrangement processes. MICELLE2 is currently a popular code as described by Kossert et al. (2012, 2014b), Jörg et al. (2012), Kossert and Nähle (2014), Fitzgerald et al. (2014), Bergeron et al. (2016) and Colle et al. (2016). A nonradionuclide-specific computer code, TDCR-AR, is described by Rodrigues et al. (2008) for the analysis of data, calculation of detection efficiency, and radionuclide activity in a TDCR system. The TDCR-AR code was satisfactorily tested for the analysis of ^{3}H, ^{14}C, ^{55}Fe, ^{54}Mn, and ^{60}Co. Other simulation codes and programs are available from the literature, and it is common for laboratories to use programs developed 'in-house' for the analysis of data (*e.g.*, Collé et al., 2008). These programs may be obtained by direct contact with the authors of the published results.

Another popular code for the calculation of TDCR detection efficiencies if the TDCR07c program written in Fortran prepared by the Laboratoire National Henri Becquerel, CEA at Saclay (Capogni and De Felice, 2014;

Capogni and Antohe, 2014; Cassette et al., 2016; da Cruz et al., 2014; Mini et al., 2014). As described by Collé et al. (2008), the program derives the detection efficiency for the logical sum of double coincidence from three individual values of triple-to-double coincidence count rate ratios, taking into account the individual quantum efficiency of each photodetector.

The computer programs, such as those described previously in this chapter, may include subroutines for the calculation of the ionization quench function *Q*(*E*) for any given Birks ionizing quench parameter *kB* (See Eq. (6.106) of this chapter.). Cassette et al. (2000) demonstrated that the Birks *kB* factor is independent of the cocktail chemistry and optical design of the TDCR system, the influence of the *kB* factor is negligible for relatively higher-energy nuclides, such as ^{14}C ($E_{max} = 0.155$ MeV) and higher-energy beta-emitters, and the optimum value of the *kB* factor is independent of the high-energy nuclide. Also, for low-energy electron-capture (EC) nuclides decaying toward the fundamental level, such as ^{55}Fe, the counting efficiency is almost independent of the *kB* value, because the energy emission is almost monoenergetic and the nonlinearity of the scintillator (ionization quenching), in this case, has no effect on the light emission (Broda et al. (2007). However, for low-energy beta-emitting nuclides, such as ^{3}H, ^{63}Ni, and ^{241}Pu, the calculated counting efficiency is highly dependent on the *kB* value (Broda et al., 2007 and Kossert *et al.*, 2011). The optimum *kB* value can be determined using different counting conditions described by Broda et al. (2007). The proper fitting of the *kB* ionization quench parameter in the TDCR model, as noted by Broda et al. (2002), is vital to improving the accuracy of this standardization method.

If the quantum efficiencies of the three photomultiplier tubes are different, a set of the following three equations, as described by Arenillas and Cassette (2006), Broda et al. (2007), Van Wyngaardt and Simpson, 2009), Simpson et al. (2010), and Cassette (2011), is solved for the free parameters of each phototube A, B, and C:

$$\frac{\varepsilon_T}{\varepsilon_{AB}} = \frac{\int_0^{E_{max}} S(E)\left(1-e^{-EQ(E)/3M_A}\right)\left(1-e^{-EQ(E)/3M_B}\right)\left(1-e^{-EQ(E)/3M_C}\right)dE}{\int_0^{E_{max}} S(E)\left(1-e^{-EQ(E)/3M_A}\right)\left(1-e^{-EQ(E)/3M_B}\right)dE} \tag{6.119}$$

$$\frac{\varepsilon_T}{\varepsilon_{BC}} = \frac{\int_0^{E_{max}} S(E)\left(1-e^{-EQ(E)/3M_A}\right)\left(1-e^{-EQ(E)/3M_B}\right)\left(1-e^{-EQ(E)/3M_C}\right)dE}{\int_0^{E_{max}} S(E)\left(1-e^{-EQ(E)/3M_B}\right)\left(1-e^{-EQ(E)/3M_C}\right)dE} \tag{6.120}$$

$$\frac{\varepsilon_T}{\varepsilon_{AC}} = \frac{\int_0^{E_{max}} S(E)\left(1-e^{-EQ(E)/3M_A}\right)\left(1-e^{-EQ(E)/3M_B}\right)\left(1-e^{-EQ(E)/3M_C}\right)dE}{\int_0^{E_{max}} S(E)\left(1-e^{-EQ(E)/3M_A}\right)\left(1-e^{-EQ(E)/3M_C}\right)dE} \tag{6.121}$$

Eqs. (6.119) to (6.121) are solved using the multidimensional minimization Downhill Simplex algorithm (Broda et al., 2007), and the function, which must be minimized is

$$\Delta = \left(\frac{\varepsilon_T}{\varepsilon_{AB}} - \frac{T}{AB}\right)^2 + \left(\frac{\varepsilon_T}{\varepsilon_{BC}} - \frac{T}{BC}\right)^2 + \left(\frac{\varepsilon_T}{\varepsilon_{AC}} - \frac{T}{AC}\right)^2 \tag{6.122}$$

where ε_T is the efficiency of the triple coincidence signals (*T*), and ε_{AB}, ε_{BC}, and ε_{AC} are the efficiencies of three double coincidence signals AB, BC, and AC, respectively. The solution gives the free-parameter variables M_A, M_B, and M_C or λ_A, λ_B, and λ_C (*i.e.*, $M = \lambda$) for each *kB* value allowing the counting efficiency of each coincident event (Broda et al. (2007).

Research by Grau Carles and Grau Malonda (1989) and Broda (2003) in the application of the TDCR method to the metrology of nuclides that decay by pure electron capture (EC) demonstrate that, for a free parameter value, at lower ranges of detection efficiency there is a corresponding increase in counting efficiency as a function of the triple to double count ratio; however at higher ranges of detection efficiency there may be as many as three different values of the counting efficiency for different values of the triple to double count rate ratio (See Fig. 7.5 of Chapter 7 of Volume 2 "Radionuclide Standardization" by Agustin Grau Malonda. This different behavior in the metrology of pure EC nuclides is due to the fact that there are different efficiencies corresponding to the discontinuous decay spectra of EC nuclides (Grau Carles and Grau Malonda, 1989), such as K-shell, L-shell, γ, and conversion electron events. A solution to this uncertainty can be resolved by a three-dimensional plot of counting %ε versus TDCR versus λ (i.e., figure of merit or FM) as illustrated by Fig. 7.6 of Chapter 7 of Volume 2 "Radionuclide Standardization". Such a plot illustrates distinct values of efficiency for each triple to double ratio. Also, the observation of three detection efficiencies for certain triple to double ratios occur only at certain levels of quench, and this irregularity can be avoided by controlling the chemical quench level.

4. The TDCR efficiency calculation technique in a nutshell

The previous sections describe in much detail the theory and practice of the TDCR efficiency calculation technique. Additional sources that document the theory, as it applies to radionuclides that undergo beta decay including pure beta-emitters, beta-gamma emitters, and electron capture nuclides, are provided by Simpson and Morris (2004), Simpson and Van Wyngaardt (2006), Van Wyngaardt and Simpson (2009). An excellent summary of the technique is described by van Wyngaardt et al. (2012), and it is provided as follows with the writer's comments in brackets:

In this liquid scintillation counting method, a source is viewed by a three-phototube detection system and the double-coincidence count rate, N_d, and the triple-coincidence count rate, N_t, recorded simultaneously. If N_0 is the disintegration rate, then

$$N_d = N_0 \varepsilon_d \qquad (6.123)$$

and

$$N_t = N_0 \varepsilon_t \qquad (6.124)$$

where ε_d and ε_t are the double- and triple-coincidence detection efficiencies respectively. The efficiencies can be calculated theoretically in terms of a floating parameter, the figure of merit *P*, which gives an indication of the number of electrons at the first dynode of the phototube per keV deposited in the liquid scintillator (Gibson and Gale, 1968). By comparing the ratio of the experimentally measured coincidence count rates with the ratio of the corresponding theoretical efficiencies, *P* can be found by solving Eq. (6.125) iteratively

$$\frac{N_t}{N_d} = \frac{\varepsilon_t}{\varepsilon_d}. \qquad (6.125)$$

If *P* is identical for each phototube/scintillation detector, the theoretical double-and triple-coincidence detection efficiencies are given by

$$\varepsilon_d = \int_0^{E_{max}} S(E)[1 - \exp\{-PQ(E)E\}]^2 dE \qquad (6.126)$$

and

$$\varepsilon_t = \int_0^{E_{max}} S(E)[1 - \exp\{-PQ(E)E\}]^3 dE \qquad (6.127)$$

respectively. The terms comprising Eqs. (6.126) and (6.127) are defined in Section 3 [See Section C, Parts 1 and 2 of this chapter] and various aspects relating to the efficiency calculations are described. Once *P* has been determined, the corresponding efficiencies follow directly and the source activity is given by

$$N_0 = \frac{N_d}{\varepsilon_d} = \frac{N_t}{\varepsilon_t} \qquad (6.128)$$

From van Wyngaardt et al. (2012), and reprinted with permission from Elsevier © 2012.

5. Commercially available 3PM liquid scintillation analyzers

The three-photomultiplier (3-PM) liquid scintillation counters available commercially, are the Hidex 300 SL and the Hidex 600 SL with a larger sample capacity available from Hidex Oy, Turku, Finland. The 3-PM detector structure of the Hidex 300 SL was illustrated previously in Fig. 6.91. The Hidex 300 SL super low-level liquid scintillation analyzer with three PM tubes became available commercially in 2008, and it is gaining in use internationally for the accurate measurement of radionuclides.

As noted in the previous treatment on the TDCR method, there is no need for any external standard to monitor quench to determine sample activities with a liquid scintillation analyzer equipped with three photomultiplier detectors. Jäggi and Eikenberg (2014) compared the performance of the PerkinElmer TriCarb 2200/2550, Quantulus, and Hidex 300 SL liquid scintillation analyzers for the measurement of activity concentrations of both low-energy and intermediate-energy beta-emitters, ^{241}Pu ($E_{max} = 0.021$ MeV) and ^{90}Sr ($E_{max} = 0.546$ MeV) demonstrating that the Hidex 300 SL is an excellent instrument for the routine analysis of pure beta-emitters. Their comparative study provided the results of ^{241}Pu activity concentrations (Bq/kg) determined by four instruments/methods, namely, (i) atom counting by inductively coupled plasma mass spectrometry (ICP-MS), (ii) the TriCarb, (iii) the Quantulus liquid scintillation analyzers that utilize two PM tubes, and (iv) the Hidex 300 SL liquid scintillation analyzer with three PM tubes. For ^{241}Pu beta activity measurements, the total alpha activity of the sample arising from ^{238}Pu, ^{239,240}Pu, and ^{242}Pu had to be determined by α/β discrimination, which can be performed with the TriCarb 2550, Quantulus, and the Hidex 300 SL. The comparative results of the ^{241}Pu activity measurements on five different radioactive RHB slurry samples demonstrated that the relatively new Hidex 300 SL analyzer provided values of ^{241}Pu in agreement with the other standard instrumentation within expanded ($k = 1$) standard uncertainties (Jäggi and Eikenberg, 2014). The acronym RHB refers to the source of the radioactive slurry samples, which is the "*R*ück*h*alte*b*ecken Wald" detention pond at the Paul Scherrer Institut, Switzerland. Strontium-90 activity concentration measured by both the Hidex 300 SL and TriCarb liquid scintillation analyzers were within the expanded ($k = 2$) uncertainties of an IAEA-375 reference soil sample.

Several researchers have put a metrological version (Metro version) Hidex 300 SL TDCR liquid scintillation analyzer to the test in the metrology of radionuclides by the TDCR method by comparing its performance with proven national home-made 3-PM liquid scintillation analyzers. Table 6.11 provides comparative data on radionuclide activity concentrations measured by both the Hidex 300 SL and home-made 3-PM liquid scintillation analyzers. The data illustrates how well the commercially available HIDEX 300 SL stands up to proven national 3-PM scintillation analyzers in accurate radionuclide metrology.

To confirm results obtained by one method of measuring a radionuclide activity concentration,

TABLE 6.11 Comparison of radionuclide activity concentration determinations obtained with the Hidex 300 SL and home-made portable TDCR liquid scintillation analyzers.

Radio nuclide	Activity concentration (kBq/g) and uncertainty (%)[a]				Reference
	Home-made TDCR		Hidex 300 SL[c]		
^{3}H	73.12[b] kBq/g	0.90%	72.59 kBq/g	0.95%	Capogni and Antohe (2014)
^{59}Fe	2027.1[e]	−0.17[d]	2032.5	+0.09[d]	Kossert and Nähle (2014)
^{63}Ni	30.45[b]	0.82	30.63	0.80	Capogni and Antohe (2014)
^{89}Sr	18.720[e]	0.034	18.767	0.051	Kossert et al. (2014c)
^{99}Tc	56.53[b]	0.40	56.56	0.39	Capogni and De Felice (2014)
^{229}Th	9.912[e]	–	9.942	+0.30[f]	Kossert et al. (2014b)

[a]*Relative standard uncertainty (k = 1) unless otherwise specified.*
[b]*ENEA-INMRI home-made portable TDCR, Rome, Italy.*
[c]*Hidex 300 SL "Metro version" (a metrological version).*
[d]*Relative deviation of the mean of activity concentrations derived from results provided by the CIEMAT/NIST method and the standard ENEA TDCR.*
[e]*Physikalisch-Technische Bundesanstalt (PTB) TDCR. Braunschweig, Germany.*

researchers will compare the results provided by a second or even a third method of analysis. For example, Capogni and De Felice (2014) compared their results on the determination of the activity concentrations of ^{63}Ni and ^{99}Tc obtained by three different counters and two different methods of analysis (TDCR and CIEMAT/NIST). Tables 6.12 and 6.13 illustrate measurements made with a national home-made portable 3-PM counter, the Metro version of the Hidex 300 SL 3-PM counter, and the TriCarb 2-PM counter in the measurement of activity concentrations (kBq/g) of ^{63}Ni and ^{99}Tc. To compare instrument performance Capogni and De Felice (2014) utilized the same *kB* factor and photomultiplier coincidence resolving times (t_c) measured in nanoseconds and dead times (t_d) measured in microseconds for each radionuclide analysis. Tables 6.12 and 6.13 illustrate the excellent performance of the Hidex 300 SL against the other TDCR counter and method used for the metrology of ^{63}Ni and ^{99}Tc.

TABLE 6.12 ^{63}Ni activity concentration and uncertainties (k = 1) on reference time (kB = 0.011 cm MeV^{-1}, t_c = 140 ns, and t_{dead} = 50 μs).

LSC system	LSC method	a (kBq g^{-1})	u(a) (kBq g^{-1})	u(a) (%)
Portable TDCR counter	TDCR	30.38	0.21	0.68
Hidex counter[a]	TDCR	30.63	0.25	0.80
Tricarb counter	CIEMAT/NIST	30.38	0.18	0.58

[a]*Metrological version (Metro version).*
From Capogni and De Felice (2014); reprinted with permission from Elsevier © 2014.

TABLE 6.13 ^{99}Tc activity concentration and uncertainties (k = 1) on reference time (kB = 0.011 cm MeV^{-1}, t_c = 120 ns, and t_{dead} = 50 μs).

LSC system	LSC method	a (kBq g^{-1})	u(a) (kBq g^{-1})	u(a) (%)
Portable TDCR counter	TDCR	56.53	0.23	0.40
Hidex counter[a]	TDCR	56.56	0.22	0.39
Tricarb counter	CIEMAT/NIST	56.70	0.28	0.49

[a]*Metrological version (Metro version).*
From Capogni and De Felice (2014); reprinted with permission from Elsevier © 2014.

6. Additional TDCR developments

a. Advances in applications

The TDCR efficiency calculation technique is a proven method for radionuclide standardization. Numerous reports on the standardization of specific radionuclides by the TDCR method have been published. Many of these reports on the metrology of radionuclides will include other methods, such as the CIEMAT/NIST technique to confirm results. Each radionuclide presents a specific challenge with respect to the efficiency calculation due to its characteristic decay scheme, and the reader may refer to the published reports for further information on the specific calculation codes and parameters utilized for each radionuclide. With this in mind, some of the published radionuclide standardizations by the TDCR efficiency calculation method with typical standard uncertainties of the determination in percent are provided in Table 6.14 in order of increasing mass number.

b. Utilization of channel photomultipliers

Channel photomultipliers are very light detectors, which contain a photocathode to convert light photons to electrons, and an electron multiplier consisting of a single narrow and hallow semi-conductive curved channel through which the photoelectrons are accelerated. Secondary electrons are produced by the collisions of the primary electrons on the walls of the channel. These photomultipliers are described in detail in Chapter 9, Volume 1. Ivan et al. (2009) tested channel photomultiplier tubes, in lieu of the traditional photomultiplier equipped with electron multiplier dynodes, in a three-PM detector TDCR liquid scintillation analyzer. They observed significantly lower detection efficiencies; but excellent agreement in activity determinations of ^{3}H, ^{55}Fe, ^{63}Ni, and $^{90}Sr/^{90}Y$ when compared to the conventional counter equipped with the traditional electron dynode multiplier PM tubes. Channel photomultiplier tubes were again utilized by Ivan et al. (2010) in a new TDCR liquid scintillation counter equipped with six photomultiplier tube detectors, which was tested with several beta-emitters and, for comparative measurements, of tritiated water for validation studies. The six photomultipliers were connected in pairs and arranged in a symmetrical pattern about the liquid scintillation vial in the optical chamber of the liquid scintillation counter. The counter can be operated in two modes: as a single system with three CPMs and a double system with six CPMs, connected in pairs in summation mode. The double system with six photomultipliers resulted in a significant improvement in the detection efficiencies over the classical TDCR counter. In light of their reduced weight and size compared to the traditional dynode photomultiplier tubes, channel photomultipliers have been utilized in the construction of portable 3-PM TDCR systems as described by Kossert et al. (2014) and Nähle et al. (2014) and described in the following paragraphs.

TABLE 6.14 Some radionuclides analyzed by the TDCR method.

Radionuclide	Relative standard uncertainty (%)[a]	References
^{3}H	0.44[b]	Razdolescu and Cassette (2004), Zimmerman et al. (2004b), Stanga et al. (2006), Qin et al. (2008), Ivan et al. (2008, 2010), Steele et al. (2009), Mo et al. (2010), and Nähle et al. (2010), Capogni and Antohe (2014), Johansson et al. (2014), Cassette et al. (2016, 2018) and Colle et al. (2016)
^{14}C	0.26[b]	Simpson and Meyer (1994b), Hwang et al. (2004), Qin et al. (2008) and Ivan et al. (2010), Nedjadi et al. (2015) and Bergeron et al. (2016)
^{18}F	0.44[b]	Fitzgerald et al. (2014)
^{32}P	0.5[c]	Jaubert and Cassette (2004)
^{33}P	0.39[b]	Simpson and Morris (2004)
^{36}Cl	0.38[e]	Ivan et al. (2010), Gudelis et al. (2012)
^{41}Ca	1.46[e]	Jörg et al. (2012)
^{45}Ca	0.27[b]	Sochorová et al. (2012), Nedjadi et al. (2015)
^{54}Mn	7.2[d]	Broda and Pochwalski (1993) and Zimmerman et al. (2004b)
^{55}Fe	0.55[b]	Broda and Pochwalski (1993), Ratel (2008), Razdolescu et al. (2008), Johansson and Sephton (2010), Simpson et al. (2010) and Ivan et al. (2008)
^{59}Fe	0.16[e]	Kossert and Nähle (2014)

Continued

TABLE 6.14 Some radionuclides analyzed by the TDCR method.—cont'd

Radionuclide	Relative standard uncertainty (%)[a]	References
^{60}Co	0.20[e]	Kossert et al. (2017a,b, in press)
^{63}Ni	0.30[b]	Simpson and Meyer (1994b), Zimmerman et al. (2004b), Razdolescu et al. (2006), and Collé et al. (2008), Thiam et al. (2012), Capogni and De Felice (2014), Capogne and Antohe (2014), Johansson et al. (2014), and Nedjadi et al. (2015)
$^{68}Ge/^{68}Ga$	0.39[b]	Zimmerman et al. (2008)
^{85}Sr	1.80[b]	Broda et al. (2012)
^{89}Sr	0.41[b]	Razdolescu et al. (2002), Simpson and Van Wyngaardt (2006) and Nähle et al. (2014)
$^{90}Sr/^{90}Y$	1.18[e]	Gudelis et al. (2012) and Mini et al. (2014)
^{90}Y	0.24[b]	Simpson and Van Wyngaardt (2006), Lourenço et al. (2015)
^{99}Tc	0.13[b]	Simpson and Meyer (1994b), Laureano-Perez et al. (2010), Zimmerman et al. (2010), Capogni and De Felice (2014), da Cruz (2014, Johansson et al. (2014) andVan Wyngaardt et al. (2014)
^{103}Pd	1.5[d]	Zimmerman et al. (2004b)
^{113m}Cd	0.39[e]	Kossert et al. (2011b)
^{139}Ce	0.5[b]	Broda et al. (1998)
^{177}Lu	0.31[e]	Kossert et al. (2012)
^{186}Re	0.22[b]	Jaubert (2008)
^{204}Tl	0.62[b]	Simpson and Meyer (1996), Hwang et al. (2002, 2004), Razdolescu and Cassette (2004), Zimmerman et al. (2004b) and Sochorová et al. (2012)
^{229}Th	0.23[e]	Fitzgerald et al. (2010), Kossert et al. (2014b)
^{241}Pu	1.40[b]	Kossert et al. (2011a), van Wyngaardt et al. (2012), and Johansson et al. (2014)

[a] *The lowest standard uncertainty is listed when more than one reference is cited.*
[b] *Quadratic sum of all uncertainty components at a coverage factor* $k = 1$ *(See Chapter 7, Vol. 2).*
[c] *With impurities of up to 10% of* ^{33}P *and* ^{35}S.
[d] *Expanded uncertainty at a coverage factor* $k = 2$ *(See Chapter 7, Vol. 2).*
[e] *Square root of the sum of quadratic standard uncertainty components.*

c. Portable TDCR systems

Miniaturized TDCR systems have been developed under the framework of the project entitled "Metrology for New Generation Nuclear Power Plants" also referred to as the Metrofission Project announced by Johansson et al. (2011) and financed in part by the European Metrology Research Program (EMRP). Within the scope of this project several portable liquid scintillation analyzers equipped with three photomultiplier tubes have been designed as TDCR systems for on-site analysis of radionuclides including low-energy beta-emitters at nuclear power plants or to meet the needs for rapid and accurate radionuclide metrology at facilities where short-lived radionuclides are prepared such as at medical clinics for the diagnosis and treatment of cancer and other infirmities. The development of the portable TDCR systems are described by Cassette et al. (2013), Capogni and Antohe (2014), Capogni and De Felice (2014), Johansson et al. (2014), Kossert et al. (2014b), Nähle et al. (2014), and Mini et al. (2014).

The miniaturization of TDCR systems have included the use of small Hamamatsu PMTs (Capogni and Antohe, 2014; Capogni and De Felice, 2014; Johansson et al., 2014) or light-weight channel photomultipliers (Kossert et al., 2014a,b,c; Nähle et al., 2014) and digital pulse processing as described in detail by Capogni and De Felice (2014) and Mini et al. (2014) consisting of a portable high-speed multichannel digitizer and digital pulse processing software that emulates the MAC3 analog board previously illustrated in Fig. 6.90. The new portable TDCR systems are tested generally against the performance of other instruments (See Table 6.11 previously described) or by comparison of results obtained with radionuclide samples of known activity concentrations as illustrated by Table 6.15. The data provided in Table 6.15 illustrates how

TABLE 6.15 The validation measurement of the new mini-TDCR system at the NPL[a]. Four nuclides were used for the validation and compared against values from previous standardizations (including other methods apart from TDCR).

Radionuclide	Standardized value[b] (kBq g^{-1})	Uncertainty %, $k = 1$	Mini TDCR[c]	Uncertainty %, $k = 1$	Difference (%)
H-3	237.16	2.5	237.28	1.25	0.05
Ni-63	83.95	0.62	83.74	0.56	−0.25
Pu-241	79.74	0.39	78.56	2.60	−1.48
Tc-99	56.70	0.9	56.55	1.14	−0.26

[a]*National Physical Laboratory, Middlesex, UK.*
[b]*Activity concentration determined by other TDCR systems and methods.*
[c]*Activity concentration determined with the mini-TDCR system.*
From Johansson et al. (2014), reprinted with permission from Elsevier © 2014.

a new miniaturized TDCR system (mini-TDCR) performed by comparing the activity concentrations determined by the mini-TDCR system with the known activity concentrations of weak beta emitting nuclides determined by other TDCR systems and methods.

Another illustrative example is the work of Capogni and Antohe (2014), which involved the testing of a new portable TDCR system constructed by the ENEA-INMRI, Rome, Italy. They compared results obtained with a commercial 3-PM liquid scintillation counter, the HIDEX 300 SL "Metro version" to those obtained with the home-made instrument. The HIDEX 300 SL "Metro version" is a metrological version of the HIDEX 300 SL of tested performance at the ENEA-INMRI. Capogni and Antohe (2014) report that the 3-PM liquid scintillation systems provided the ratios of the net count rates for the triple and double coincidences, i.e., *T/AB*, *T/BC* and *T/AC* from which were computed the triple coincidence efficiency (ε_T) and the logical sum of the double coincidence (ε_D). They utilized the TDCR07c software, described previously, and available from the Laboratoire National Henri Becquerel (LNHB) of the CEA, France. They computed the activity concentrations (kBq/g) utilizing the triple coincidences according to

$$A_{C,T} = T/(\varepsilon_T \times m) \quad (6.129)$$

and the activity concentrations from the logical sum of the double coincidences (D) were computed according to

$$A_{C,D} = D/(\varepsilon_D \times m) \quad (6.130)$$

In the analysis of ^{3}H and ^{63}Ni they observed differences of only 0.01% between A_{CT} and $A_{C,D}$ for the two counters, i.e., the HIDEX 300 SL "Metro version" and the portable ENEA counter.

d. TDCR analysis with plastic scintillation microspheres

A development reported by Tarancón Sanz and Kossert (2011) is the application of a free-parameter model to samples containing solid plastic scintillation microspheres and radioactive aqueous solutions. Several beta-emitting radionuclide activities were measured in a TDCR system. The deviations between the determined activity concentrations and the reference values were <2.5%. However, Tarancón et al. (2015) note that the detection efficiency calculation model was highly dependent on the description of the plastic scintillation microspheres geometry, and an additional radionuclide tracing with ^{63}Ni was required to determine the exact scintillation microsphere packing. Thus, Tarancón et al. (2015) applied the CIEMAT/NIST tracing method for the measurement of pure beta-emitters, namely, ^{63}Ni, ^{14}C, ^{36}Cl, and $^{90}Sr/^{90}Y$ by utilizing an adapted MICELLE2 with the inclusion of PENELOPE code capable of measuring energy loss due to particle quenching resulting from varying geometries of heterogeneous mixtures of polystyrene microspheres and water.

D. Compton Efficiency Tracing (CET) method

Another radionuclide metrology method that involves a TDCR counter, referred to as the Compton Efficiency Tracing method, was reported by Cassette and Do (2008). This method utilizes a temporary virtual tracer source created by Compton interaction within the scintillation cocktail. A triple coincidence liquid scintillation counter measures the Compton tracer source after the Compton events are measured and selected by a gamma-ray detector such as a germanium or a cadmium telluride (CdTe) semiconductor detector. Cassette et al. (2010) note that the

CET method can offer the advantage that the tracer source, being a Compton electron source created inside the radionuclide source being measured, would have the identical chemical composition as the source under measurement. Thus, the measurement would have a reduced sensitivity to the *kB* factor and it would be insensitive to the chemical composition and quenching of the source. The CET method is described in Chapter 7 "Radionuclide Standardization", Volume 2.

E. $4\pi\beta-\gamma$ coincidence counting

$4\pi\beta-\gamma$ coincidence counting is a direct method of activity determination, as it is independent of any quench indicating parameters and no reference standards are required for counting efficiency determinations. An additional advantage of this method is that the activity of a nuclide can be determined from experimental counting data only without the need for detector efficiencies.

As may be construed from the name applied to this method, it involves the coincidence detection of two types of radiations from a given radionuclide. The methods can be applied, therefore, to radionuclides that emit in coincidence more than one distinguishable type of radiation. Two types of detectors are required in coincidence to distinguish exclusively two types of radiation emissions, such as a liquid scintillation detector (or gas proportional counter) to measure a beta emission and a solid scintillation (or semiconductor detector) capable of measuring a gamma emission. Because the technique involves, in most cases, the use of a solid scintillation counter as one of the detectors, this method of radionuclide metrology is discussed in Chapter 7 "Radionuclide Standardization" of Volume 2.

X. Neutron/gamma-ray measurement and discrimination

The same year that Herforth (1948) under the leadership of Hartmut Kallmann, documented in her thesis the discovery that beta and gamma radiation produced scintillation in certain organic compounds, P. R. Bell (1948) demonstrated that fast neutrons could be detected in an anthracene scintillator by the proton recoils produced by neutron-proton collisions. Although neutrons have no charge and consequently cannot produce ionization or excitation directly in a scintillation fluor, the essentially equal mass of the neutron and protons (hydrogen atoms) in organic compounds facilitate transfer of energy from the neutron to protons via elastic scattering as described in Chapter 1, Volume 1. Because of the almost equal mass of the neutron and proton, the maximum fraction of the kinetic energy that a fast neutron can lose in a single head-on collision with a proton is calculated as 0.999 or 99.9% (See Table 1.19, Chapter 1, Volume 1). Although, partial energy transfers also occur through less direct neutron-proton collisions, it is clear that fast neutron energy is absorbed more easily by a substance rich in protons (hydrogen atoms) compared to another substance possessing atoms of higher atomic number. The recoil protons with their acquired kinetic energy produce ionization until their energy is dissipated in the medium and they come to a full stop. A liquid scintillation cocktail, rich in protons (high H:C ratio), can be used to measure fast neutron radiation. The energy of the recoil protons dissipated in the scintillation cocktail solvent will excite the solute fluor molecules resulting in the emission of visible light similar to the effect of alpha- and beta-particle interactions discussed previously.

Another radiation type that does not carry charge but does interact in a liquid scintillation fluor cocktail is gamma radiation. Like X-radiation described previously, gamma rays will interact with liquid scintillation cocktail via principally Compton interactions with atomic electrons. The transfer of gamma-ray photon energy to atomic electrons will result in the liberation of Compton electrons. The energy of these electrons will be absorbed by the liquid scintillation solvent in the same way as beta particles described previously. Compton-electron energy absorption will result in fluor excitation and the emission of visible light of intensity proportional to the amount of energy absorbed in the scintillation cocktail.

The two mechanisms of fluor excitation via recoil proton or Compton-electron energy absorption forms the basis for the discrimination of neutron and gamma radiation, respectively, by liquid scintillation. The liquid scintillation discrimination of the two radiation types will be discussed further on in this section.

Only a brief description of the scintillation detection of neutrons and neutron/gamma-ray discrimination will be presented here. More detailed information can be obtained from references specific to the subject (Scherzinger et al., 2017, 2016; Tie et al., 2017; Bourne et al., 2016; Alharbi, 2015; Gamage et al., 2014; Liao and Yang, 2014; Wang et al., 2014; Enqvist et al., 2013; Lawrence et al., 2013; Xiaohui et al., 2013; Zhang et al., 2013; Nakhostin, 2012; Enqvist and Pázsit, 2010; Thomas, 2010; Klein and Neumann, 2002) and from Chapter 9 of Volume 1.

A. Detector characteristics and properties

The liquid scintillation cocktails used for the detection and measurement of neutrons are organic solutions generally high in hydrogen content to enhance the efficiency of trapping the neutron kinetic energy through neutron-proton collisions, which results in the production of recoil protons. Because high gamma-ray backgrounds are often present, the ability of discriminating between incident neutron and gamma radiation through pulse shape discrimination (PSD) becomes important and some liquid scintillation cocktails are most suitable for this purpose. Also, some organic scintillation cocktails are devoid of hydrogen, such as BC-509 or EJ-313 (equivalent to NE-226) or BC-537 or EJ-315 (equivalent to NE-230), which contain deuterated benzene (C_6D_6). Such cocktails have reduced sensitivity to neutrons and can be used for the measurement of gamma radiation in the presence of neutron radiation. Examples of commercially available cocktails for the detection of neutrons and gamma radiation or for neutron/gamma-ray discrimination as well as other applications such as neutrino detection are provided in Table 6.16. These liquid scintillators are used devoid of oxygen gas. Dissolved atmospheric oxygen reduces their light output by about 30%, and eliminates any pulse shape discrimination properties that the cocktails might possess. These cocktails are purged thoroughly with nitrogen gas prior to use.

Liquid scintillation neutron detectors vary in size from small to medium-sized (4−12 cm in diameter by 4−12 cm in length) up to large volume detectors comprising several cubic meters depending on the research objectives (See Table 6.16 and Bourne et al., 2016, Scherzinger et al., 2016, Alharbi, 2015, Enqvist et al., 2013, Zhang et al., 2013, Lavagno et al., 2010, Černý et al., 2004, Vartsky et al., 2003, and Zecher et al., 1997). The detectors are available commercially prefabricated with the liquid scintillation cocktail encapsulated in a sealed inert material (e.g., aluminum, stainless steel, indium, Teflon) of various sizes coated internally with a TiO_2 reflective paint to optimize light output and an optical window of borosilicate glass, which would be linked with the face of a photomultiplier tube. The liquid scintillation cocktail is purged with nitrogen or argon gas before the sealing of the detector to permit n/γ pulse shape discrimination. Liquid scintillation cocktail is also available in bottles or drums for laboratories that need to fabricate detectors according to specific dimensions.

The efficiency and energy resolution of the detector will be governed by its size. Energy is deposited by fast neutrons in the liquid scintillation cocktail by collisions for the most part with hydrogen atoms (n-p collisions) and to a lesser degree with the carbon atoms (See Table 1.19 in Chapter 1, Volume 1), and, as a consequence, recoil proton excitation of cocktail solvent and fluor molecules is predominant. Gamma radiation will deposit energy mainly via

TABLE 6.16 Properties and Applications of Liquid Scintillators Used for Neutron Radiation Measurements or Neutron/Gamma Radiation Discrimination among other Applications.[a]

Scintillator	Properties	Applications
BC-501A[b]	Xylene[e] (>90%)	Yields excellent n/γ pulse shape discrimination (PSD) over a wide energy range up to ∼ 100 MeV (D'Melloe et al. (2007)'Enqvist et al. (2008)'Horwáth et al., 2000, and Nakao et al., 2001, Klein and Neumann, 2002, D'Melloe et al. (2007), Aspinall et al., 2007a, Enqvist et al. (2008), Liu et al., 2009, Yousefi et al., 2009, Braizinha et al., 2010, Lavagno et al., 2010, Bayat et al., 2012, Nakhostin and Walker, 2010, Nakhostin, 2012, Zhang et al., 2013, Gamage et al., 2014, Liao and Yang, 2014, Wang et al., 2014, Alharbi, 2015, Scherzinger et al., 2016, 2017 and Tie et al., 2017).
EJ-301[c]	Aromatic fluors (<10%)[f]	
NE-213[d]	Atomic ratio, H/C: 1.212	
	λ_{max}: 425 nm[g]	
	Light output: 78% anthracene[h]	
	Decay time: 3.2 nsec[i]	
	Density: 0.874 g/cm^3	
	Refractive index, n_D: 1.505	
	Flash point: 26°C	
BC-505	1,2,4-trimethylbenzene[e] (97.5%)	Exhibits high light output and transmission; Fast neutron detectors; Suitable for large volume, including anti-Compton and anti-coincidence shields; High flash point makes it safer to use and transport than xylene or toluene-based scintillators; (Cecil and Scorby, 1992, Hong et al., 2002, Klein and Brooks, 2006, Wurtz et al., 2011, Fukuda et al., 2016 and Cerutti et al., 2017).
EJ-305	Aromatic fluors (<0.5%)	
NE-224	Atomic ratio, H/C: 1.331	
	λ_{max}: 425 nm	
	Light output: 80% anthracene	
	Decay time: 2.5 nsec	
	Density: 0.977 g/cm^3	
	Refractive index, n_D: 1.505	
	Flash point: 48°C	

Continued

TABLE 6.16 Properties and Applications of Liquid Scintillators Used for Neutron Radiation Measurements or Neutron/Gamma Radiation Discrimination among other Applications.[a]—cont'd

Scintillator	Properties	Applications
EJ-309 NE213 with low	Solvent (proprietary): aromatic hydrocarbon (>99%) Organic fluors (<1%) Atomic ratio, H/C: 1.25 λ_{max}: 424 nm Light output: 75% anthracene Decay time: 3.5 nsec[h] Density: 0.96 g/cm^3 Refractive index,n_D: 1.57 Flash point: 144°C	Liquid scintillator developed as an alternative to the more commonly used for pulse shape discrimination (PSD), such as BC-501A, and EJ-301, which contain xylene solvent flash point and high degree of solvent action. EJ-309 exhibits slightly poorer PSD characteristics, but possesses more environmentally favorable chemical properties and higher flash point (Pozzi et al., 2009; Clarke et al., 2009; Enqvist et al., 2010, 2013; Zak et al., 2010; Lawrence et al., 2013; Gamage et al., 2014; Tomanin et al., 2014; Bourne et al., 2016; Lennox et al., 2017 and Bai et al., 2018).
BC-509 EJ-313 NE-226	Hexafluorobenzene[e] Formula: C_6F_6 λ_{max}: 425 nm Light output: 20% anthracene Decay time: 3.1 nsec Density: 1.61 g/cm^3 Refractive index,n_D: 1.38 Boiling point: 80°C Flash point: 10°C	Essentially free of hydrogen atoms (with exception of fluor molecules) only 0.18 atom % of hydrogen atoms; Low sensitivity to fast and moderated neutrons, gamma-ray detection in a fast neutron flux; Neutron insensitivity increased with pulse shape discrimination to enhance gamma-ray detection in n/γ discrimination; Can be used to detect neutrons when it is desireable to avoid thermalizing properties of hydrogen (Cecil and Medley, 1988; Davies et al., 1994; Wolle et al., 1999; Hong et al., 2002, and Hamel et al., 2014).
BC-517H EJ-321H NE-235H	1,2,4-trimethylbenzene (<30%) Mineral oil (>70%) Aromatic fluors (<0.3%) Atomic ratio, H/C: 1.89 λ_{max}: 425 nm Light output: 52% anthracene Mean free path: > 5 m[j] Decay time: 2.5 nsec Density: 0.86 g/cm^3 Refractive index, n_D: 1.476 Flash point: 81°C	Mineral oil-based scintillator where long mean free paths are required and high light output important; Large volume detectors; Compatible with acrylic plastics such as Plexiglas® and Perspex®, as well as many metals and reflective coatings; Pulse shape discrimination capability. Variants of this cocktail exist including BC-717L, BC-517P and BC-517S for large tanks, plastic containers and high light output and equivalents are EJ-321L, EJ-321P, and EJ-321S (Sakai et al., 1992; Arpesella, 1996; Tayloe et al., 2006; Braizinha et al., 2010; Christensen et al., 2014).
BC-519 EJ-325A NE-235C	1,2,4-trimethylbenzene (<40%) Mineral oil (<60%) Fluors (0.5%) Density 0.86 g/cm^3 Atomic ratio, H/C: >NE213 Flash point: >62°C Boiling point: > 204°C	Provides excellent n/γ pulse shape discrimination for neutron energies up to 100 MeV (Horwáth et al., 2000; Chen et al., 2008; Beaulieu et al., 2011; Gozani et et al., 2011; Jandel et al., 2015).

Continued

TABLE 6.16 Properties and Applications of Liquid Scintillators Used for Neutron Radiation Measurements or Neutron/Gamma Radiation Discrimination among other Applications.[a]—cont'd

Scintillator	Properties	Applications
BC-521 EJ-331 NE-323	1,2,4-trimethylbenzene (>85%) Gadolinium (0.5%) Light output: 68% anthracene Atomic ratio, H/C: 1.314 λ_{max}: 424 nm Decay time: 3.6 nsec Density: 0.89 g/cm^3 Refractive index,n_D: 1.50 Flash point: 44°C	Utilized for neutron spectrometry with pulse shape discrimination and neutrino research; Fast neutrons leave a unique signature in Gd-loaded scintillator; They produce a prompt recoil proton pulse and, once thermalized, they are easily captured by Gd, which has an extremely high thermal neutron capture cross section. Neutron capture by Gd is signaled by the emission of three gamma rays totally ~ 8 MeV. Available at 0.25% w/w Gd (EJ-335) with mineral oil to increase hydrogen content and use in large tanks; produced at Gd content of 0%,5% (BC-525) with mineral oil component for increased light transmission, hydrogen content, and higher flash point. Neutron capture by Gd results in a cascade of conversion electrons, X-rays, and γ-rays. BC-525 is used in neutron spectrometry and neutrino research. (Varignon et al., 2006 Banerjee et al., 2007 and Laborie et al., 2012).
BC-523A EJ-339 NE-321A	1,2,4-trimethylbenzene (>30%) Methyl borate (<60%) 1-methylnaphthylene (10%) Light output: 65% anthracene Atomic ratio, H/C: 1.74 Natural ratio$^{10}B/^{11}B$: 0.245 Enriched ratio$^{10}B/^{11}B$: 9.0 λ_{max}: 424 nm Decay time: 3.7 nsec Density: 0.916 g/cm^3 Refractive index,n_D: 1.4 Boiling point: 68.9°C	Available at natural^{10}B isotope abundance or enriched; EJ-339A contains boron loading compound enriched to 90% atom percent in^{10}B and thus contains 4.6%^{10}B by weight; Useful for total neutron absorption spectrometry; fast neutrons produce a prompt recoil proton pulse with initial scatterings in the cocktail; thermalized neutrons may undergo the$^{10}B(n,\alpha)^{7}Li$ capture; the capture pulse is in delayed coincidence with the prompt pulse; Useful in identifying neutron events; Useful for n/γ separation by PSD. (Yen *et al.*, 2000, Vartsky et al., 2003, Jastaniah and Sellin, 2004, Flaska and Pozzi, 2009, Pinoi et al., 2014, and Liao and Yang, 2014).
BC-537 EJ-315 NE-230	Deuterated benzene (>98%) Formula: (C_6D_6) Aromatic fluors (<2%) Light output: 61% anthracene Atomic ratio, D/H: 141:1 Atomic ratio, D/C: 0.99 λ_{max}: 425 nm Blue photons (per 1 MeVee): 12,000[k] Decay time: 1.8 nsec Density: 0.954 g/cm^3 Refractive index,n_D: 1.498 Short decay component: 3.5 nsec[k] Boiling point: 79.1°C Flash point: −11°C (12°F)[k] Mean free path of fission neutrons[k]: 4.96 cm[l]	Useful for n/γ separation as sensitivity to fast neutrons is reduced by deuterium. A nondeuterated version of the scintillator (EJ-315H) is available for comparative research. (Marrone et al., 2002, and Villano et al., 2011, Jaworski et al., 2012, Lawrence et al., 2013, Xiaohui et al., 2012, 2013, Febbraro et al., 2015 and Becchetti et al., 2016, 2017).

Continued

TABLE 6.16 Properties and Applications of Liquid Scintillators Used for Neutron Radiation Measurements or Neutron/Gamma Radiation Discrimination among other Applications.[a]—cont'd

Scintillator	Properties	Applications
EJ-301D	Deuterated xylene[k]	A new and improved deuterated scintillator for fast neutron energy measurements (up uo ~ 20 MeV) without time-of-flight. It is less volatile and less toxic than the benzene-based scintillator detectors. It can provide slightly, more light output and improved pulse shape n/γ discrimination than deuterated benzene scintillators (Becchetti et al., 2016, 2017).
	Formula: (C_8D_{10})	
	Light output: 78% anthracene	
	λ_{max}: 425 nm	
	Blue photons (per 1 MeVee): 9200	
	Atomic ratio, D/C: 120	
	Refractive index, n_D: 1.505	
	Short decay component: 3.2 nsec	
	Flash point: 26°C (79°F)	
	Mean free path of fission neutrons[k]: 4.94 cm[l]	

[a]*Data from ELJEN Technology, 1300 W. Broadway, Sweetwater, TX 79,556, USA, http://www.eljentechnology.com/products/liquid-scintillators, and Saint-Gobain Corp., 750 E. Swedesford Road, Valley Forge, PA 19,482, USA, https://www.crystals.saint-gobain.com/products/liquid-scintillators and Compagnie de Saint-Gobain, Les Miroirs, 18, avenue d'Alsace, 92,400 Courbevoie, France, accessed February 15, 2018.*
[b]*Bicron scintillator (scintillators listed as a group are commercial equivalents).*
[c]*Eljen Technology scintillator (scintillators listed as a group are commercial equivalents).*
[d]*NE Technology scintillator (scintillators listed as a group are commercial equivalents).*
[e]*Solvent.*
[f]*Solute (fluor).*
[g]*Wavelength of fluorescence peak.*
[h]*Percentage of anthracene light output.*
[i]*Short component of de-excitation light.*
[j]*Light transmission.*
[k]*Properties of EJ-301D scintillator are from Becchetti et al. (2016).*
[l]*Mean free path averaged over ^{252}Cf fission-neutron spectrum.*

Compton interactions resulting in Compton-electron excitation of cocktail and fluor. Since both neutrons and gamma rays are generally intermixed, both radiations should be considered. The neutrons will encounter a wide range of collisions with protons including direct head-on collisions and complete transfer of energy in less-direct collisions even just a glancing with a proton resulting in various scattering interactions and only partial energy transfers to protons (See Chapter 1, Volume 1). Likewise, Compton electrons will vary in energy depending on the energy of the incident gamma radiation.

If the liquid scintillation detector is very small or the neutron or γ-ray energies very high, a considerable number of recoil protons or Compton electrons can escape from the detector. This results in an incomplete deposition of either proton-recoil or Compton-electron energy in the cocktail referred to as the 'wall effect' (Bai et al., 2018; Thomas, 2010). When the wall effect is significant, the event (n-p or γ-e) recorded by the photomultiplier tube as a pulse height would be smaller than what would be expected had the recoil proton or Compton electron deposited all its energy in the cocktail. Consequently, the 'response function', which is the pulse height registered per neutron or gamma-ray energy, would be shifted to lower pulse heights. For example, in the analysis of γ-ray photons of energy over the range of 7–20 MeV, Navotny et al. (1997) chose NE-213 scintillation detectors large enough (*e.g.*, 5 cm in diameter by 10 cm in length, 206 cm^3 in volume) taking into account the maximum range of Compton electrons to be detected to avoid a distortion of the pulse height response by wall effects. Likewise, for the measurement of neutron radiation, Hakao *et al.* (2001) demonstrated that for an NE-213 liquid scintillator of 12.7 cm diameter by 12.7 cm long the proton-escape probability ranges from approximately 1% to more than 50% of the recoil protons generated for neutrons over the energy range of 25 MeV to over 300 MeV. The neutron detection efficiencies were determined by Hakao *et al.* (2001) by integrating the pulse height response functions at various calibrated pulse height thresholds. At a lower-limit discriminator setting or pulse height threshold of 1.15 MeV (calibrated with a ^{60}Co source) the neutron detection efficiencies varied from 24% to 10% over the neutron energy range of 23 −132 MeV. Satoh et al. (2001) developed a Monte Carlo code, designated SCINFUL-QMD, for the calculation of neutron detection efficiencies for neutron energies up to 3 GeV in liquid scintillators such as NE-213. Other Monte Carlo and analytical models of neutron scatterings in liquid scintillation cocktail and detector response are provided by Bai et al. (2017), Gohil et al. (2012), Patronis et al. (2007), Pozzi *et al.* (2007), and Enqvist and Pázsit, 2010).

It is important to note here that for heavy particles (*e.g.*, protons and α particles) the light output from the liquid scintillator is nonlinear with energy. Unlike electrons or β particles, which produce a relatively linear light output with energy over the range of 0.100−1.6 MeV (Novotny et al., 1997 and Horwáth et al., 2000), heavy particles exhibit high values of stopping power (dE/dx) in scintillator (See Chapter 1, Volume 1) and thereby produce higher specific ionization and a saturation effect that yields less light per particle energy loss. The nonlinearity of light output per particle energy loss for protons results in recoil proton energies occurring at lower pulse heights in the pulse height spectrum. Consequently, the use of computer codes to unfold neutron and photon energy spectra require accurate measurement of the light output response as a function of energy deposited in the scintillator (Enqvist and Pázsit, 2010; Kornilov et al., 2009; Klein and Neuman, 2002; Neuman *et al.*, 2002; Reginatto et al., 2002; Schmidt et al., 2002). Radionuclide sources with well separated γ-ray energies are used to establish the pulse height scale in terms of electron energies. Gamma-photon sources produce Compton edges in the scintillator where the Compton-electron energies can be used to convert light outputs into electron equivalent energies or MeVee (Tie et al., 2017; Scherzinger et al., 2016; Liao and Yang, 2014; Villano et al., 2011; Hawkes et al., 2010; Zak et al., 2010; Guerrero et al., 2008; Klein and Neumann, 2002; Nakao et al., 2001; Horwáth et al., 2000).

The detector size will govern the number of multiple neutron scatterings. Neutron scattering can result in a broad range of energy transfers to recoil protons, from very little neutron energy when a neutron is simple glanced off a proton (near hit) up to the maximum neutron energy resulting from a head-on collision with a hydrogen atom. Recoil protons from a given neutron after multiple collisions and scatterings will produce photons, which are detected by the photomultiplier tube as they occur within the coincidence time gate of the scintillation analyzer. The sum of the photons produced by the recoil protons from the multiple scatterings of one neutron will produce a pulse height from the photomultiplier proportional to the light output. Due to the summation of photons from multiple scatterings the resultant pulse heights will be higher than what would occur at reduced scattering.

Neutrons are also scattered by carbon atoms in the scintillator; however, carbon recoils do not produce any significant light output in liquid scintillator (Klein and Neumann, 2002). The H:C atomic ratios in liquid scintillator may vary from 1.2 to 1.9 (Table 6.16). The higher the ratio or greater the atom % H in the scintillator, the higher will be the probability of n-p collisions and the production of proton recoils. However, carbon scattering will cause neutron energy loss and consequently a reduction in pulse height response as a function of energy. As illustrated in Table 1.19 of Chapter 1, the maximum fraction of kinetic energy that may be lost by a neutron collision with ^{12}C is 28.6% resulting from a head-on n−^{12}C collision. A neutron may lose, therefore, anywhere between zero and 28.6% of its energy in a single collision with carbon. For example, a scattered neutron that had an initial head-on collision with a ^{12}C atom may then collide with a proton, but the maximum energy of the recoil proton in this case would be 71.4% of the original neutron energy. Consequently, carbon atoms in the scintillator will reduce the pulse height response as a function of neutron energy.

At lower neutron energies, particularly where neutrons may become thermalized in the liquid scintillator, special detectors may be employed, such as BC-523, that contain natural or enriched ^{10}B in the scintillation cocktail (Table 6.16) to enhance detection efficiencies. The incident neutrons are moderated by n-p collisions in the cocktail and then undergo $^{10}B(n,\alpha)^7Li$ capture according to Eq. (1.327) of Chapter 1, Volume 1 and more specifically as follows:

$$ {}^1_0n + {}^{10}_5B \rightarrow \alpha + {}^7_3Li \text{ (g.s.)} + 2.790 \text{ MeV}(6.3\%) \quad (6.131)$$

$$\rightarrow \alpha + {}^7_3Li^* + \gamma \text{ (0.478 MeV)} + 2.312 \text{ MeV}(93.7\%) \quad (6.132)$$

The branching ratios for the two reactions are 6.3% and 93.7%, respectively. The branch described by Eq. (6.131) leaves the ^7Li at the ground state, whereas the other branch (Eq. 7.126) leaves the ^7Li at an excited state, and the α particles plus $^7Li^*$ ions create scintillation light equivalent to light produced by 65-keV electrons or beta particles. Yen *et al.* (2000) report the use of 4 cm diameter detectors of ^{10}B-enriched scintillator joined together in a circular array of 55 detectors. They report detector efficiencies of 95%, 85%, and 71% at neutron energies of 10, 100, and 1000 eV, respectively.

Gd-loaded liquid scintillators, described in Table 6.11, have excellent neutron detection properties, because of the very high thermal neutron capture cross section of Gd (See Table 1.20 of Chapter 1, Volume 1). As described by Banerjee et al. (2007), the mechanism of neutron detection in a Gd-loaded scintillator consists of two stages. Firstly, the fast neutron that enters the liquid scintillator will undergo multiple scatterings primarily with hydrogen atoms in a few tens of nanoseconds producing a prompt signal. After several collisions with atomic nuclei of the scintillator, the neutron slows down and is thermalized, *i.e.*, it reaches the low thermal energy stage, provided the neutron does not escape from the scintillator. The second stage consists in the capture of the thermal neutron by a Gd nucleus, because of its extremely high thermal neutron capture cross section. Neutron capture by Gd results in the emission of three γ-rays with a total energy of ~8 MeV. The detection of the capture gamma rays provides a measure of the total number of neutrons emitted.

Liquid scintillators loaded with ^{6}Li are not available commercially; however, Fisher et al. (2011) and Bass et al. (2013) have easily prepared liquid scintillator loaded with up to 0.40% ^{6}Li capable of fast neutron spectrometry and thermal neutron detection with maximum background discrimination using capture-gated neutron spectroscopy. The detection method is illustrated in Fig. 6.93, which shows a fast neutron entering into a detector containing a ^{6}Li-doped liquid scintillation fluor cocktail. The neutron

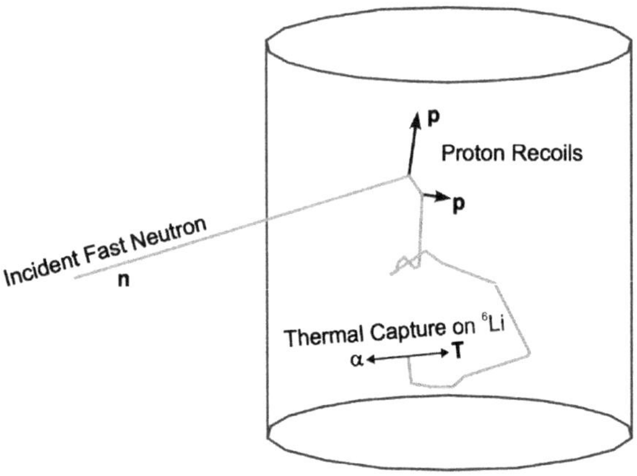

FIGURE 6.93 An illustration of the principle of capture-gated detection. A fast neutron impinges on the ^{6}Li-loaded liquid scintillation detector. It rapidly gives up its energy through nuclear collisions, primarily with protons, in the moderation process. The thermalized neutron diffuses in the medium until it is captured on an atom (e.g., ^{6}Li) with a high capture cross section. *From Fisher et al. (2011); reprinted with permission from Elsevier © 2011.*

scatters from an initial collision with a proton, which on average will recoil with half of the initial neutron energy yielding scintillation light. The neutron continues to scatter off of other protons in the scintillation cocktail until it escapes from the detector or is captured by a ^{6}Li atom after reaching a thermal energy according to the following reaction:

$$^{6}_{3}\mathrm{Li} + ^{1}_{0}\mathrm{n} \rightarrow ^{3}_{1}\mathrm{H}\ (2.05\ \mathrm{MeV}) + ^{4}_{2}\mathrm{He}(2.73\ \mathrm{MeV})\quad Q = 4.78\ \mathrm{MeV} \tag{6.133}$$

As described by Fisher et al. (2011), a large fraction of the fast neutrons entering the detector will lose most of their energy through neutron-proton collisions (i.e., n-p scattering) and most of the scintillation light will originate from these n-p events. A neutron of a few MeV in energy will lose 90% of its energy through n-p scattering in the first 10 ns in the scintillator. Neutron scattering off of carbon atoms can also generate light; however, these are minor when compared to the n-p events. Capture of the thermal neutron will occur in a short time scan of <100 ns depending on the size and geometry of the detector and the concentration of the ^{6}Li, which has a relatively high thermal neutron capture cross section (See Table 1.20 in Chapter 1, Volume 1). The neutron capture reaction with ^{6}Li has a high Q value of 4.78 MeV (Qcalc, https://www.nndc.bnl.gov/qcalc/, accessed Feb 16, 2018). The capture reaction energy is shared between the tritium and alpha particle capture products, and due to their short range, their entire energy is absorbed in the liquid scintillator producing a high scintillation light output. An example of digitized pulse events from a photomultiplier tube resulting from the initial proton recoil from a fast neutron in the liquid scintillator followed by neutron capture by ^{6}Li is illustrated in Fig. 6.94.

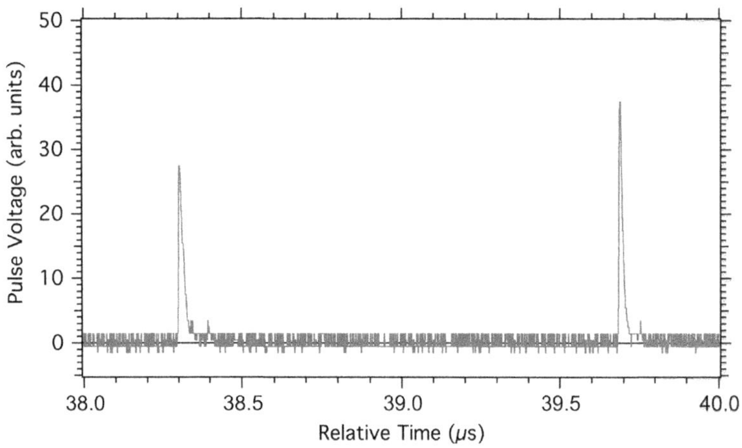

FIGURE 6.94 Plot of the digitized PMT pulse of the detector response to a fast neutron within the ^{6}Li-loaded liquid scintillator. A fast neutron enters the detector and initially scatters off the protons in the scintillator. After thermalizing within the scintillator, it captures on a ^{6}Li, in this example approximately 1.4 μs later. *From Fisher et al. (2011); reprinted with permission from Elsevier © 2011.*

The method offers several advantages over the use of other neutron capture loaded scintillators, such as ^{10}B-loaded scintillators. As described by Bass et al. (2013), these advantages are the following: (i) The products of the $^{6}Li(n,\alpha)^{3}H$ reaction are charged particles and not gamma rays. Their energy is deposited completely in the liquid scintillator. (ii) The *Q*-value of the $^{6}Li(n,\alpha)^{3}H$ reaction (4.78 MeV) is larger than that of the $^{10}B(n,\alpha)^{7}Li$ reaction (2.79 MeV) yielding an energy deposit peak better separated from the noise level. (iii) The preparation and loading of scintillator with enriched $^{6}LiCl$ involves simple chemistry, and at a cost, which is significantly less than commercially available ^{10}B-load scintillators. A disadvantage would be the restricted availability of enriched ^{6}Li for some institutions.

B. Neutron/gamma-ray (n/γ) discrimination

The detection of neutrons is complicated generally with background gamma radiation. The neutron detector can be shielded from radiation background arising from charged particles by use of shielding material between the source and detector or via magnetic field deflection. This leaves only gamma rays, which remain undeflected by magnetic fields and relatively unattenuated by absorbers.

Neutron and γ-ray interactions in a liquid scintillator may be separated by pulse shape discrimination (PSD) according to principles that are similar somewhat to alpha- and beta-particle (*α/β*) discrimination in liquid scintillation described further on in this chapter. The principles behind (n/γ) discrimination rest on different degrees of excitation caused by heavy charged particles versus lighter charged particles in the liquid scintillation cocktail. Neutron interactions produce the relatively heavy recoil protons, whereas γ rays produce much lighter Compton electrons in scintillation cocktail. The heavier recoil protons produce greater specific ionization (higher LET) in cocktail than the Compton electrons and, consequently, proton interactions in cocktail fluors create higher cocktail fluor excited states that take longer for de-excitation to occur. De-excitation is manifested as fluorescence. The fluorescence is analyzed by the electric pulse it produces in the photomultiplier tube (PMT). The shape of the pulse is a function of the specific ionization of the particle (heavy particle vs. lighter particle), and the shape can be defined in two components, namely, a fast (prompt) component or rise time and a slow or delayed component also referred to as the tail. The prompt component of fluorescence has a decay time of only a few nanoseconds, while the delayed component or tail may last a few hundred nanoseconds. The differences in pulse shapes produced by neutron and γ-ray interactions in a liquid scintillator are illustrated in Fig. 6.95. A relatively heavy charged particle, such as a proton arising from neutron interactions, will produce a larger fraction of its

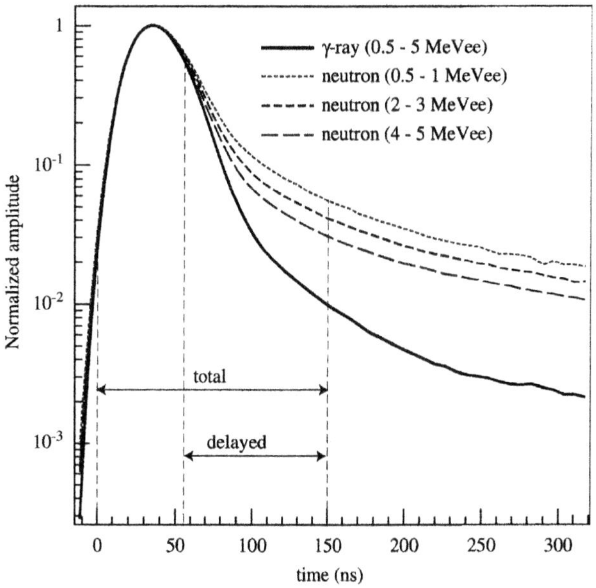

FIGURE 6.95 Neutron and γ-ray true pulse shapes recorded with BC501A liquid scintillator. The shape of the neutron signals shows a dependence on the energy of the recoil nuclei. The emission mechanism exhibits a prompt component, which starts within the first 5 ns and a delayed component, which occurs after a subsequent ~25 ns. *From Guerrero et al. (2008). Reprinted with permission from Elsevier © 2008.*

fluorescence in the slow component, while a much lighter charged particle, such as the Compton electron arising from γ-ray interactions, will cause fluorescence with a much smaller portion of the light in the slow component. Guerrero et al. (2008) demonstrate, with an Am−Be neutron source, that the slower neutrons exhibit a longer delayed component or tail than faster neutrons. This could be explained by the fact that protons of lower energy have a higher LET producing more high-excitation states of scintillator cocktail. The Am−Be source produces neutrons with energies up to ~10 MeV with an average neutron energy of 4 MeV and various gamma lines.

Two of the most common methods of exploiting the phenomena of differing pulse shapes for n/γ discrimination with the use of analogue electronics are (i) the charge-comparison, and (ii) zero-crossing techniques (Liao and Yang, 2014; Nakhostin and Walker, 2010). The charge-comparison method involves the measurement of the integrated charge collected over two different time regions of the pulse to identify neutron and gamma events in the liquid scintillator, and the zero-crossing technique yields n/γ discrimination by a measurement of a zero-crossing time for a suitable fraction of the pulse shape. These techniques will be described subsequently.

1. Digital charge-comparison (CC) method

The method of Heltsley et al. (1988) is commonly used. As described by Liao and Yang (2014), Wang et al. (2014),

Zecher et al. (1997), and Ranucci (1995), this PSD method entails the use of two analog-to-digital converters (ADCs), one that integrates, via a PSD algorithm, the total charge of the pulse, and the other integrates the charge from some fixed time fraction (i.e., portion) of the pulse (*i.e.*, either the first prompt component or the longer tail component). This is illustrated in Fig. 6.96 where A_1 is the total integral charge starting from the very beginning of the pulse to an optimum endpoint where the slow tail reaches background, and A_2 is an integration of the pulse starting from a selected beginning of the tail to the same endpoint referred to as the tail integral. The ratio R of the tail integral to the total integral, as described by Eq. (6.134), is taken to discriminate between the proton and Compton electron events originating from neutron and gamma interactions, respectively in the liquid scintillator.

$$R = A_2/A_1 \tag{6.134}$$

By comparing the magnitude of the charge collected in the fraction of the pulse to the total charge collected in the pulse, one can determine which species created the pulse, proton or electron, and consequently, neutron or γ-ray.

Charge comparison n/γ discrimination techniques that compare the 'short' and 'long', or 'total' and 'delayed', or 'total' and 'tail' components of the scintillation pulse have been utilized by numerous researchers including Bourne et al. (2016), Alharbi (2015), Liao and Yang (2014), Wang et al. (2014), Zhang et al. (2013), Nakhostin (2012), Xiaohui et al. (2012, 2013), Schiffer et al. (2011), Villano et al. (2011), Pozzi et al. (2009), Enqvist et al. (2010), Hawkes et al. (2010), Lavagno et al. (2010), Clarke et al. (2009), Flaska and Pozzi (2009), Hamel et al. (2009a,b), Guerreo *et al.* (2008), D'Mellow et al. (2007), Černý et al. (2004) and Rochman et al. (2004). An example may be taken from the work of Lavagno et al. (2010), who utilized a

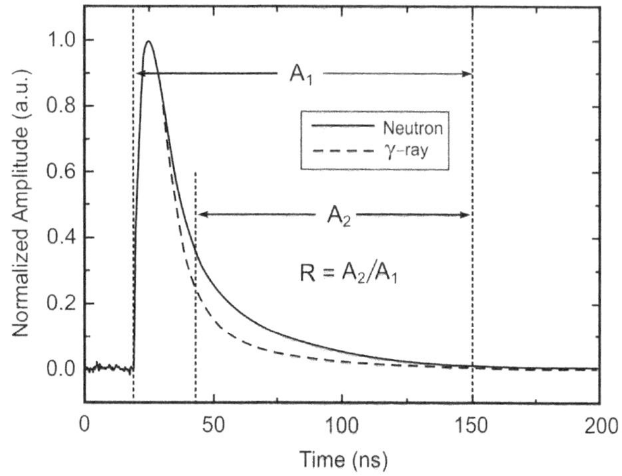

FIGURE 6.96 Pulse events produced by neutron and gamma interaction in the liquid scintillator and regions of charge integration, namely the tail integration A_2 and the total integration, A_1. The ratio R of the two integrations is defined as A_2/A_1. *Adapted from Liao and Yang (2014), Wang et al. (2014), and Alharbi (2015).*

Canberra Nuclear Instrumentation Module (NIM) 2160A Charge-to-Digital Converter (QDC) to exploit the difference in the intensity of the slow or delayed component of the light pulse (See Fig. 6.96) to the space density of the charge carriers produced by the different types of ionizing particles (i.e., protons and electrons), which yields excellent n−γ discrimination. As described by Lavagno et al. (2010), the charge comparison method compares directly the intensity of the slow component to the total light pulse by incident ionizing radiation, *i.e.*, the recoil protons and Compton electrons. A block diagram of the experimental setup and time relation between photomultiplier pulse and gates at the input of the QDCs are illustrated in Fig. 6.97. One QDC at the opening of the gate signal integrates the charge

FIGURE 6.97 NIM electronic block diagram and time relation between photomultiplier pulse and gates at the input of QDCs. *From Lavagno et al. (2010). Reprinted with permission from Elsevier © 2010.*

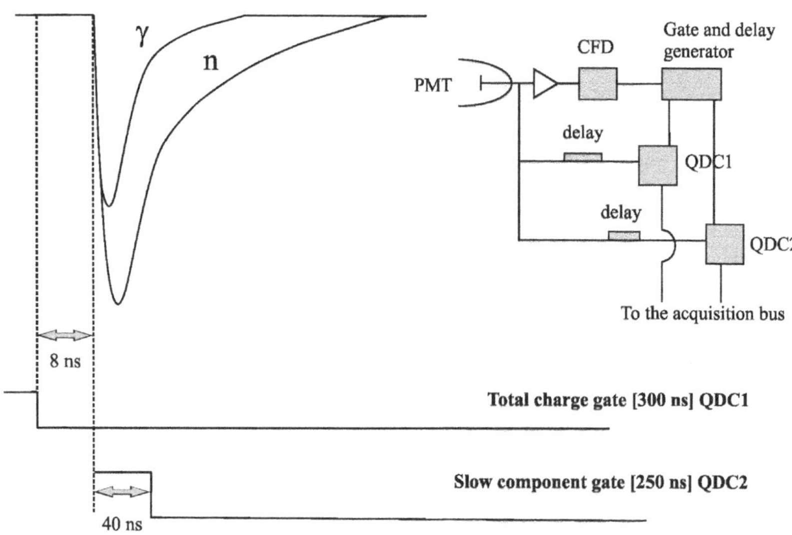

corresponding to the total pulse, and another QDC, triggered by the delay gate signal, collects only the charge from the slow or delayed component. The prompt component of the pulse starts after the first ~5 ns following excitation, and the delayed or slow component in the subsequent 20–25 ns.

The signals arising from γ-ray and neutron interactions may be distinguished by plotting the integrated intensities of the two components of the pulse against each other as illustrated in Fig. 6.98. As described by Guerrero et al. (2008), the integration of the digitized signals is carried out in two different time intervals, and the particle type of each signal is evaluated by a comparison between the two signals. Each signal arising from an electron or proton in the scintillator, as a consequence of a γ-ray or neutron interaction, respectively, is characterized by its time, amplitude, and the two integrals I_{total} and $I_{delayed}$. Fig. 6.98 illustrates well separated signal intensities corresponding to γ-rays and neutrons, which was reported by Geurrero *et al.* (2008) to be obtained from the integration of signals (See Fig. 6.96) over the 0–150 ns interval for I_{total} and from 55 to 150 ns for $I_{delayed}$.

As illustrated previously in Figs. 6.95 and 6.96, the amplitude of the tail section as a result of neutron interactions is higher than that produced by γ-rays. As a consequence, the ratio *R* for neutrons is higher than that for γ-rays (Wang et al., 2014). A histogram for γ-ray and neutron events (i.e., counts) can be constructed as a function of *R*, which with good discrimination, will yield two separate peaks for γ-ray and neutron events, as illustrated in Fig. 6.99. A value of figure of merit (FOM) is calculated according to Eq. (6.135)

$$\mathrm{FOM} = \frac{\mathrm{S}}{\mathrm{FWHM}_{\gamma} + \mathrm{FWHM}_{n}} \tag{6.135}$$

where S is the separation or distance between the gamma and neutron peaks, and FWHMγ and FWHMn are the resolutions of the gamma and neutron peaks, respectively, measured by their full widths at half maximum, as illustrated in Fig. 6.99. The magnitude of the FOM serves to evaluate the degree of n/γ discrimination where optimum discrimination is obtained at highest FOMs.

The ratio of the two integrals, *i.e.*, short integral/long integral is used as the discrimination parameter. The Charge Comparison algorithm described by Hawkes et al. (2010) can determine the optimum starting time for the pulse integration for the short integral by maximizing the FOM, which is calculated over a range of different starting times. A plot, from the work of Hawkes et al. (2010), obtained with the optimized starting time together with Gaussian fits used to measure the peak positions and widths is illustrated in Fig. 6.100.

2. Simplified digital charge-comparison (SDCC) method

The simplified digital charge-comparison (SDCC) method is a relatively new approach to n/γ discrimination, which has been tested thoroughly by Gamage et al. (2011) and Xiaohui et al. (2013). The SDCC method is based on a study by Shippen et al. (2010) demonstrating that the charge-comparison method can be optimized by squaring increments of time that fall in the region of interest of the digital pulse. As described by Xiaohui et al. (2013), the

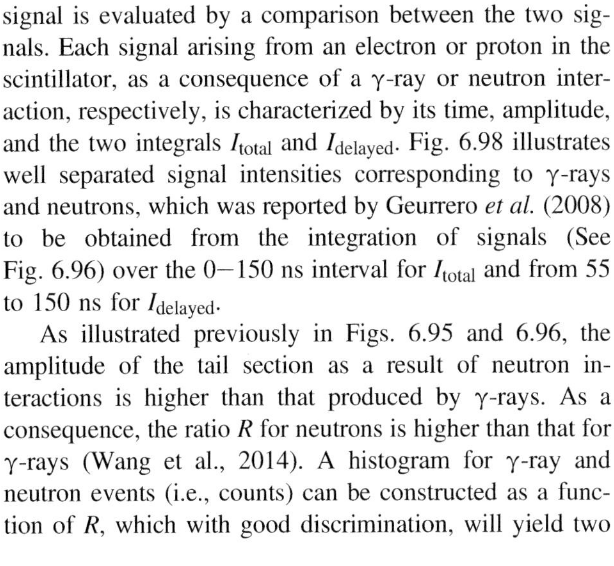

FIGURE 6.98 Distribution of I_{total} and $I_{delayed}$ values in a measurement of an Am/Be source. The two well separated regions correspond to γ-ray and neutron signals. *From Guerrero et al. (2008); reprinted with permission from Elsevier © 2008.*

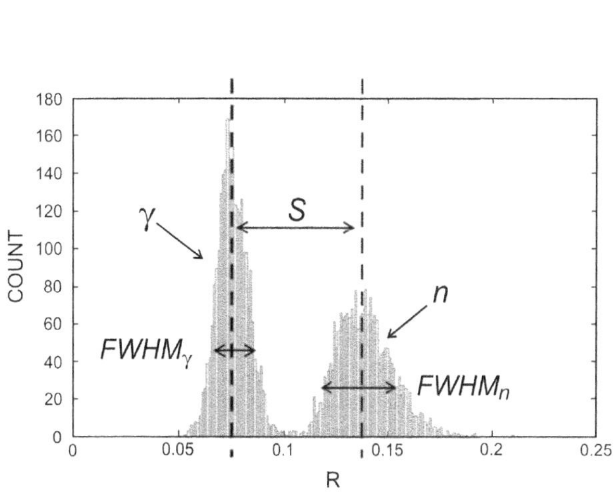

FIGURE 6.99 Histogram of the ratio R. The left peak corresponds to γ-ray and the right peak corresponds to neutrons. *From Wang et al. (2014); reprinted with permission from Elsevier © 2014.*

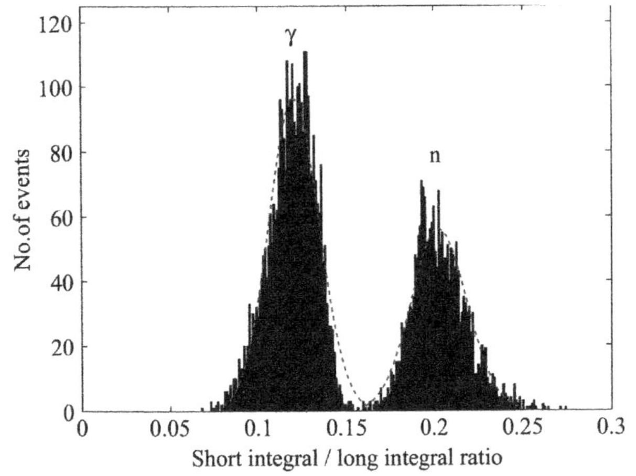

FIGURE 6.100 FOM plot for the optimum value of the short integral starting time in the Charge Comparison Algorithm. The dotted curves show the Gaussian fits used to determine the positions and widths of the peaks. *From Hawkes et al. (2010). Reprinted with permission from Elsevier © 2010.*

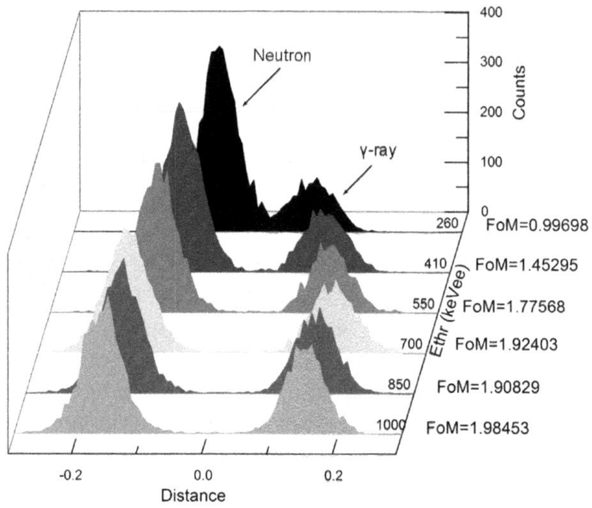

FIGURE 6.101 The results of n/γ discrimination for different energy thresholds (Ethr). Neutron events are illustrated on the left and gamma-ray events on the right. All events were recorded in a detector arrangement containing BC-537 liquid scintillation cocktail. *From Xiaohui et al. (2012); reprinted with permission from Elsevier © 2012.*

SDCC method is based on a discrimination parameter D calculated according to the following:

$$D = \log\left(\sum_{t=a}^{t=b}\left(x_t^2 + x_{t+0.5}^2\right)\right) \qquad (6.136)$$

where x_t and $x_{t+0.5}$ are sample amplitudes at times t and $t+0.5$ and a and b are the start and end times of the pulse region of interest. The optimum time interval for a pulse region of interest would depend on the detector, and Xiaohui et al. (2013) found the values of a = 70 ns and b = 210 ns to be optimum for their apparatus, and the D parameter is always larger for a neutron pulse than for a γ-ray event. Xiaohui et al. (2013) and Gamage et al. (2011) thoroughly tested the CC and SDCC methods as well as the pulse gradient analysis (PGA) method and found excellent n/γ discrimination among the three methods.

Irrespective of the methods utilized for n/γ discrimination, the figure of merit (FOM) or degree of n/γ separation achievable will depend on the energy thresholds applied, which is a preset threshold voltage for the recording of generated pulses arising from the proton recoil and Compton electron events in the liquid scintillator (Binaei Bash et al., 2013; Bayat et al., 2012; Xiaohui et al., 2012; Nakhostin, 2012; Nakhostin and Walker, 2010). The threshold energy scale of a detector system is calibrated by using the Compton edge of standard gamma-ray emitting nuclides sources such as ^{60}Co, ^{137}Cs, and ^{22}Na. For example, ^{22}Na emits a 1275 keV gamma-ray line with a 100% intensity, and it produces a Compton edge at ~ 1060 keV (Binaei Bash et al., 2013; Bayat et al., 2012). Energy calibration using the Compton edge enables the expression of the threshold energies in units of electron-energy equivalents or keVee. Thus, pulse heights that are generated are expressed at neutron and gamma-ray threshold energies in units of keVee, which is an electron energy equivalent unit. Highest figure of merits (FOMs) or optimum separation of neutron and gamma-rays are achieved at high threshold energies, as illustrated in Fig. 6.101. For example, at the threshold energies of 280 and 410 keVee some overlap of neutron and gamma-ray events were recorded by Xiaohui et al. (2012), and at higher threshold energies 550 to 1000 keVee, complete separation is illustrated. Electronic noise is significant at low threshold energies resulting in diminished neutron-gamma discrimination.

As noted in the previous paragraph and illustrated in Fig. 6.101, low FOMs occur at low energy thresholds, that is, there is complete separation of neutron and gamma-rays are not achieved and the two peaks overlap to some degree. Polack et al. (2015) describe an algorithm capable of measuring the neutron and gamma photon misclassification and test as different low-energy thresholds of 40, 60, 90 and 120 keVee.

3. Pulse gradient analysis (PGA)

The digital discrimination of neutrons and γ-rays in mixed radiation fields was described by D'Mellow et al. (2007). This digital discrimination method is referred to as pulse gradient analysis (PGA), which is somewhat distinct from the charge-comparison method described in the previous paragraphs. D'Mellow et al. (2007) note that the principle difference between the pulse produced by a neutron and that produced by a γ-ray is the neutron pulse, which exhibits a slower decay to the baseline, illustrated in

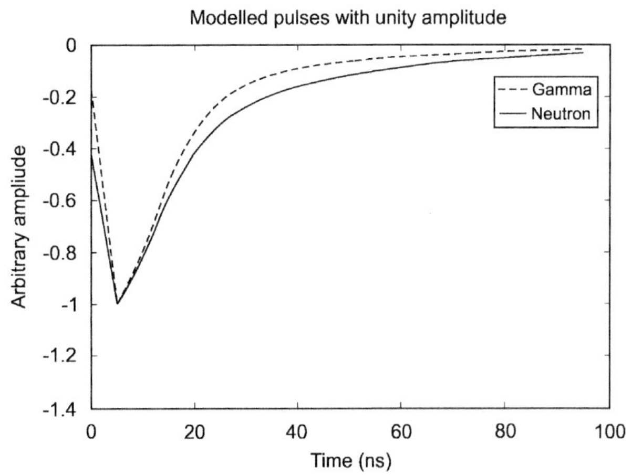

FIGURE 6.102 Plot of amplitude versus time (ns) for model pulses of a neutron and γ-ray in an organic liquid scintillator. When both pulses are attributed the same amplitude, it is clearly seen that the neutron pulse decays more slowly than the γ-ray pulse. *From D'Mellow et al. (2007). Reprinted with permission from Elsevier © 2007.*

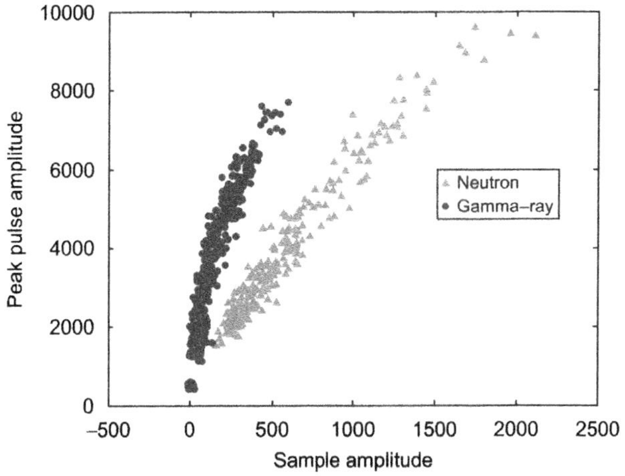

FIGURE 6.103 A scatter plot with data recorded with an Am–Be source. The peak amplitude of the pulse is plotted against the amplitude of the sample for n–γ discrimination by pulse gradient analysis (PGA). *From D'Mellow et al. (2007). Reprinted with permission from Elsevier © 2007.*

Fig. 6.102. This is characterized by a gradient that is shallower on the trailing edge of the pulse. It is this feature that is exploited in the PGA method for n–γ discrimination, *i.e.*, to determine which radiation produced the pulse, neutron or γ-ray.

The PGA method developed by D'Mellow et al. (2007) does not require any measurement of the actual pulse decay rate for the n–γ discrimination. Rather, only a relative measurement of the gradient is found from two suitable samples of the digitized representation of the pulse. These digitized samples are the following: (1) the first sample is the peak magnitude of the pulse; and (2) the second sample measured 20 ns after the first sample. The optimum time for the measurement of the second sample generally occurs 15–25 ns after the peak, and it will depend on the liquid scintillator and PMT properties (D'Mellow, 2006). The second sample, which measures the pulse gradient, is strongly affected by the pulse amplitude, which requires normalizing the pulse to remove this adverse effect. For n–γ discrimination, the amplitude of the pulse is plotted against another parameter proportional to the pulse decay rate. In the PGA algorithm, the pulse decay rate is measured by the amplitude of the second sample. Thus, two distinct groups of events are measured, namely, the peak pulse amplitude and the sample amplitude corresponding to either γ-rays or neutrons. Events labeled as γ-rays had a ratio of peak amplitude to sample greater than 11.41, and those labeled as neutrons for ratios below that value. A scatter plot of γ-ray and neutron events discriminated by the PGA method is illustrated in Fig. 6.103. The optimum ratio of peak amplitude to second sample should be determined for each detection system utilized. Aspinall et al. (2007a) describe the application of algorithms to improve the digital pulse shape discrimination techniques.

Hawkes et al. (2010) developed miniature digitizing electronics with various n–γ discrimination algorithms, which enabled the implementation and testing of the charge-comparison and PGA methods, previously described as well as a model pulse method described by Marrone et al. (2002) and Guerrero et al. (2008). The model pulse method constructs two model pulse shapes, one for neutrons and the other for γ-rays, and the unknown pulses from a mixed neutron/gamma field are characterized according to which of the two models the pulses fit best using a χ^2 criterion. The n/γ methods were implemented with a charge-comparison algorithm and a pulse gradient analysis (PGA) algorithm, respectively, and the model pulse method with a model pulse algorithm. The charge-comparison algorithm was found to have a small but definite figure of merit (FOM) advantage of the other two methods for the neutron fields tested; however, Hawkes et al. (2010) underscore that the FOMs of the other two methods were not much smaller and that each method has its advantages.

Gamage et al. (2011) compared four different algorithms for four different methods of n/γ discrimination including the digital charge-comparison (CC) method, pulse gradient analysis (PGA), a neutron–γ model analysis, and the simplified digital charge collection (SDCC). He evaluated the efficiency of each of these methods in term of the FOM, which is based on the separation of the neutron and gamma events described previously by Eq. (6.135), and the SDCC method yielded the best discrimination performance for the particular experimental factors and algorithms used.

4. Zero-crossing method

The zero-crossing method of n/γ discrimination was first worked out in detail as a result of the pioneering work of

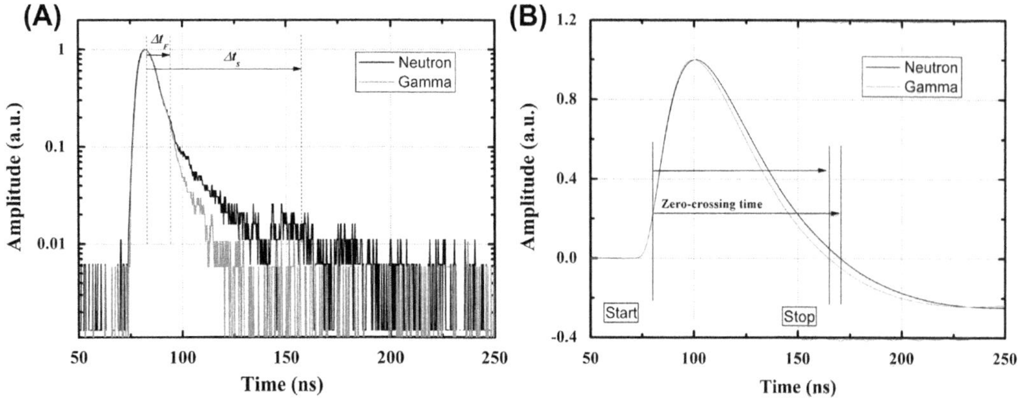

FIGURE 6.104 (A) Typical sample of PMT signals for neutrons and γ-rays. The signals approximately correspond to 500 keVee. The neutron pulse decays more slowly than the γ-ray pulse. (B) The neutron and γ-ray pulses cross the zero line at different times. *From Nakhostin and Walker (2010); reprinted with permission from Elsevier © 2010.*

Roush et al. (1964) at the University of Maryland. It is now one of the most common methods used for neutron-gamma discrimination ~~disintegration~~ in organic liquid scintillation detectors as well as stilbene and plastic scintillation detectors. Neutron-gamma discrimination is achieved by a measurement of the most suitable crossing times of shaped pulses produced by proton recoil and Compton electron events in the scintillator.

Nakhostin and Walker (2010) provide a detailed description of the method together with an algorithm for the digital implementation of the conventional zero-crossing method to achieve excellent neutron-gamma separation with liquid scintillator even at low-energy thresholds. A brief sketch of the zero-crossing method will be provided here in light of the work of Nakhostin and Walker (2010). A photomultiplier tube (PMT) is coupled to a suitable detector (e.g., a cylinder of NE213 liquid scintillator) located at a suitable distance from the neutron-gamma source. The PMT anode signals are digitized with a digital oscilloscope, which are transferred to a personal computer for pulse shaping, as illustrated in Fig. 6.104. As described by Nakhostin and Walker (2010), the pulse shaping process converts the PMT signal to a bipolar pulse, which enables the difference in the decay times of the different PMT signals to be revealed in the zero-crossing time of the shaped signals (See Fig. 6.104B). The zero-crossing time is measured by a constant fraction discriminator (CFD), which determines the start time of a signal and the delay between the start time and the time at which the shaped signal crosses the zero line. As depicted in Fig. 6.104B, pulses from recoil proton events produced by neutron interactions in the liquid scintillator arrive at the zero line after a longer time delay than pulses produced by Compton-electrons from gamma-ray interactions in the scintillator. Neutron-gamma discrimination is based on the differences in zero-crossing times of the signals.

The energy scale of the scintillator performance is expressed in units of keVee (i.e., electron energy equivalents in keV) by measuring the Compton edge produced in the scintillator with a series of gamma-ray sources, such as ^{60}Co, ^{137}Cs, and ^{22}Na. Optimum shaping time constants, which are dependent on the liquid scintillator and PMT used, are determined experimentally with the digital CFD and calculation of the figure of merits (FOMs, according to Eq. (6.135) described previously after fitting Gaussian functions to the neutron and gamma-ray events and calculation of the FWHM for the neutron and gamma-ray peaks. Examples of n/γ discriminations reported by Nakhostin and Walker (2010) are provided in Fig. 6.105. As illustrated in the figure, the performance of this method is reflected in the good n/γ discrimination achieved at the very low energy threshold of 50 keVee, and excellent discrimination at higher energy thresholds. The good performance of the method at very low energy thresholds is a result of the pulse shaping process that removes electronic noise, which is the common cause of poor discrimination at low energy thresholds.

5. Time-of-flight (TOF) spectrometry

Time-of-flight (TOF) spectrometry provides a discrimination between neutrons and gamma-ray photons as well as a measure of the neutron energies. A γ-ray photon travels at the speed of light, whereas neutrons travel slower. For example, a 2.0-MeV neutron travels at only 0.065*c* or 6.5% of the speed of light (See Figs. 1.71 and 1.72 of Chapter 1, Volume 1). Consequently, a γ-ray photon that produces multiple Compton scattering in a scintillation cocktail would create a pulse event in the PMT before a neutron that also undergoes multiple scattering. The n/γ discrimination is enhanced in TOF spectrometry by separating the neutron source from the liquid scintillation detector by a suitable distance that may extend to several meters. As noted by

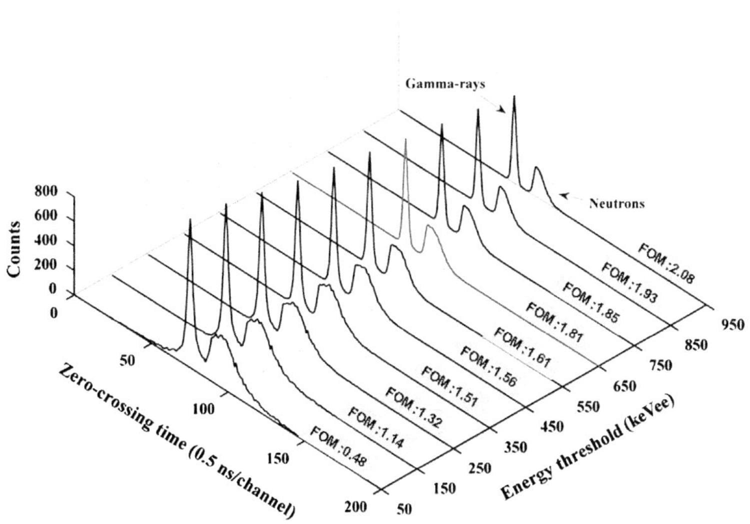

FIGURE 6.105 Neutron-gamma discrimination at different energy thresholds with an NE-213 liquid scintillation detector. *From Nakhostin and Walker (2010); reprinted with permission from Elsevier © 2010.*

Thomas (2010), time-of-flight (TOF) spectrometry requires a measurement of the time taken for the particle (*i.e.*, neutron or γ-ray) to travel a known distance (*e.g.*, the distance from the n−γ source to the detector). Thus, there is a need for two timing signals, a start signal that is triggered when the neutrons and γ-rays are created or when they pass a particular point, and an arrival signal that is triggered when the neutron and gamma rays arrive and are detected by the liquid scintillator.

Thomas (2010) outlines three ways that a start signal can be produced. These are provided below together with some examples added by the writer:

(i) *A pulsed source, such as an accelerator with beam pulsing capabilities.* For example, a proton beam may strike a target material, producing neutrons and γ-rays, situated at a fixed and known distance from the liquid scintillation detector (See Fig. 6.106). Such an experimental arrangement is described by Rochman et al. (2004), Aspinall et al. (2007b), and Liu et al. (2009).

(ii) *The detection of radiation (*e.g.*, γ-ray) that accompanies the formation of a neutron.* Any suitable detector for time-correlated emissions for the start signal from a radiation source may be used. For example, a BaF_2 detector is commonly placed next to a ^{252}Cf source to signal the onset of fission by detecting one of the γ-rays that accompanies the emission of neutrons in ^{252}Cf spontaneous fission (Banerjee et al., 2007; Ronchi et al., 2009). An ionization chamber was used by Flaska and Pozzi (2007) to provide a start signal for a digitizer to determine the time zero for the measurement of neutron pulses, and a plastic scintillator next to a ^{252}Cf source was used by Venkataraman (2008). A liquid scintillator detector is placed a distance away from the source to detect both the neutrons and γ-rays and provide a time-of-flight discrimination. For example, Ronchi et al. (2009) placed a BC-501 liquid scintillator detector 75 cm away from the source. The detected correlated fission events produce a signal in the BC-501 scintillation detector after a time-of-flight delay (t_{TOF}) with respect to the start signal from the BaF_2 detector.

(iii) *A detector used to signal neutron scattering.*

The work of Aspinall et al. (2007a,b) serves as an example of digital time-of-flight (TOF) measurements. Aspinall et al. (2007b) used an accelerator to direct a 2.924 MeV proton beam onto a thin 60 μg/cm^2 LiF target,

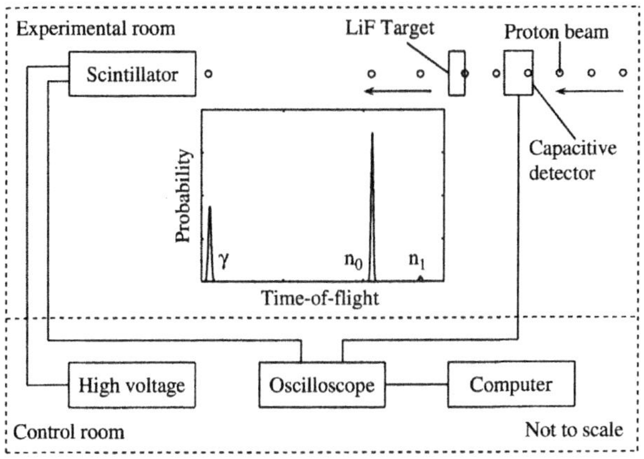

FIGURE 6.106 A schematic diagram of the experimental setup at the Neutron Irradiation Facility, National Physical Laboratory, UK. The experimental arrangement enables the digital time-of-flight discrimination of γ-ray photons and neutrons produced by a proton beam striking a LiF target according to the reaction $^7Li(p,n)^7Be$ with a flight path of 1.614 m between the target and detector. A digital time-of-flight spectrum included in the center of the diagram illustrates the peaks for the intensities of the early arriving γ-ray photons followed by the 1.225 MeV neutrons (strong peak at n_0) and 0.745 MeV neutrons (weak peak at n_1). *From Aspinall et al. (2007b). Reprinted with permission from Elsevier © 2007.*

which produces neutrons via the reaction $^{7}Li(p,n)^{7}Be$ with neutron energies of 1.225 and 0.745 MeV corresponding to transitions to the ground and first excited states, respectively, of the product nucleus. A capacitive-type detector was placed in the proton beam line, as illustrated in Fig. 6.106, at a distance of 1.6 m from the LiF target. In this arrangement, the capacitive detector provided Aspinall et al. (2007b) with the start or proton beam pick-up signal needed to provide timing information for the TOF measurement and information on the proton pulse duration and frequency. Because the beam is monoenergetic and the target very thin, the transit time from the capacitive detector to the LiF target was constant. This provided a constant delay between the beam pick-up signal and the emission of neutrons from the LiF target. Therefore, the beam pick-up signal could be used as the start signal, as well as identify the pulse duration and period of the neutron pulses. The detector containing LS-301 organic liquid scintillator, illustrated in Fig. 6.106, was positioned 1633 mm from the LiF target. The scintillator pulses trigger the arrival or acquisition of γ-rays and neutrons, and the pulses are sent to an oscilloscope and computer to enable n−γ discrimination by pulse gradient analysis and TOF measurements. The acquisition and digital recording of scintillation pulse and corresponding beam pick-up pulse enable all detected events, *i.e.*, γ-ray photons and neutrons, to be sorted in terms of their time of arrival relative to the initial beam pick-up signal (Aspinall et al., 2007b). The discrimination of the γ-ray photons and neutrons as a function of their time-of-flight, graphically depicted in Fig. 6.106, illustrate the early arrival of the γ-rays followed by two well discriminated peaks for neutrons, with a strong peak for neutrons n_0 of 1.225 MeV and a weak peak for neutrons n_1 of 0.745 MeV.

Time-of-flight (TOF) spectrometry is the standard method for testing data obtained from the various n/γ discrimination methods (Lennox et al., 2007). Thus, any improvements or developments into the n/γ pulse shape discrimination (PSD) methods previously described (i.e., the charge-comparison (CC) method, simplified digital charge-comparison (SDCC) method, pulse gradient analysis (PGA) and zero-crossing method) can be tested by TOF spectrometry. For example, Aspinall et al. (2007b) used the TOF spectrometric analysis described above (See Fig. 6.106) to verify the digital discrimination of gamma-ray photons and neutrons by pulse gradient analysis (PGA). The PGA method will provide an n/γ ratio as described previously, and the TOF spectrometry will provide the relative proportions of each event type (i.e., neutrons of certain energies and total gamma-ray photons). The n_{total}/γ from TOF spectrometry can be used to confirm n/γ ratios determined by one of the PSD methods. For example, Table 6.17 provides data from Aspinall et al. (2007b) where the TOF data taken from the time-of-flight spectrum illustrated in Fig. 6.106 confirms the separate PGA discrimination method. The n_{total}/γ from the TOF spectrometry agrees well with the n/γ ratio obtained by PGA analysis. The TOF spectrometry permits the determination of the proportions of each component in the radiation field, i.e., neutrons of various energies and gamma-ray photons. Whereas a pulse shape discrimination (PSD) method, such as PGA will provide only the total proportions of neutrons and gamma-ray photons as illustrated in Table 6.17.

Sunil et al. (2013) utilized time-of-flight to measure neutron energy while implementing the zero-crossing method (described in Section 4 above) to discriminate and separate neutrons from gamma photons. They measured the neutron energy distributions at various angles with respect to a beam of 145 MeV ^{19}F projectiles incident on a thick Al target. The neutron energy distributions were measured with five EJ-301 liquid scintillation detectors situated at a distance of 1.5 m and at 0°, 30 degree, 60 degree, 90 degree and 120 degree angles with respect to the projectile beam.

As underscored by Lennox et al. (2017), time-of-flight experimentation is used as the gold standard for testing data for neutron/gamma pulse shape discrimination (PSD), and as a result, there will be future increasing reliance in time-of-flight data for the algorithm development in PSD. They present a methodology for identifying data sets with minimal contamination of mislabeled neutron and gamma pulses in time-of-flight measurements.

TABLE 6.17 The neutron and γ-ray ratios for the TOF and PGA discrimination methods.

Method	n_0	n_1	n_{total}	γ rays	Scatter	n_{total}/γ
TOF	0.43	0.017	0.45	0.28	0.27	1.61
Error	±0:01	±0:003	±0:01	±0:01	±0:01	±0:07
PGA	–	–	0.59	0.41	–	1.44
Error	–	–	±0:02	±0:01	–	±0:06

From Aspinall et al. (2007b). Reprinted with permission from Elsevier © 2007.

6. General research into n/γ discrimination

Different approaches to the discrimination of n−γ events in an organic scintillator include methods based on the application of artificial neural networks (ANN), which were reported by Liu et al. (2009), Ronchi et al. (2009) and Akkoyun (2013). The methods were verified with mixed neutron/gamma field data assessed by time-of-flight measurements.

Studies by Nakhostin (2012, 2015) demonstrate the performance of recursive algorithms for digital processing of neutron-gamma discrimination in liquid scintillation detectors with normal and low light outputs by the charge-comparison and zero-crossing methods, which remain the most common techniques currently used. Bayat et al. (2012) demonstrate the use of Ultima Gold LLT liquid scintillation as an alternative to the NE-213 and its Bicron equivalent BC-501A scintillator. Zare et al. (2013) carried out Monte Carlo simulations on light transport and experimentation on the timing response of an NE-213 scintillation detector coupled to two PMTs to improve n/γ discrimination. Monte Carlo simulations and experimentation by Binaei Bash et al. (2013) and Ghal-Eh (2011) demonstrate how lightguides and light transport to PMTs can have a significant effect on n/γ discrimination. The use of a Multi-Pixel Photon Counter (MPPC) with detectors based on stilbene and EJ299-33 plastic scintillators reported by Grodzicka-Kobylka et al. (2018) demonstrated excellent capability of MPPC arrays for n/γ discrimination by the zero-crossing method comparable to classical PMTs.

Deuterated benzene- and deuterated xylene-based scintillators have been demonstrated by Becchetti et al. (2017, 2016) and Febbraro et al (2015) to be useful in neutron energy measurements without time-of-flight.

XI. Double beta (ββ) decay detection and measurement

The discovery, principles, and current research on double beta decay are reviewed in Chapter 1, Volume 1. Only a brief treatment of the liquid scintillation techniques employed in the search for neutrinoless double beta decay will be given here.

The first observation of double beta decay was made by Steven R. Elliott, Alan A. Hahn, and Michael K. Moe at the University of California, Irvine in 1987. They measured double beta decay in a sample of ^{82}Se enriched from its natural abundance of 8.7% to an isotope abundance of 97% to provide more source material for the double beta decay as well as reduce contamination from other sources (Eliott et al., 1986, 1987; Moe, 1986; Moe and Rosen, 1989). The decay scheme is written as follows:

$$^{82}_{34}\mathrm{Se} \xrightarrow{t_{1/2}\,=\,9.2\,\times\,10^{19}\ \mathrm{years}} {}^{82}_{36}\mathrm{Kr} + 2\beta^- + 2\,\overline{\nu} \quad (6.137)$$

where the parent nucleus decays to a daughter nucleus of the same mass number A, but with an increase in atomic number Z by two. In general terms, the double beta decay process is expressed in terms of the nuclear atomic and mass numbers as

$$(Z,\ A) \rightarrow (Z+2,\ A) + 2\mathrm{e}^- + 2\overline{\nu} \quad (6.138)$$

Thus, in accord with the beta decay scheme, within the nucleus of the parent nuclide, two neutrons decay to two protons with the simultaneous emission of two beta particles and two antineutrinos. This type of beta decay is abbreviated as 2νββ to signify the simultaneous emission of two neutrinos and two beta particles.

Double beta decay is a rare event manifested by very long half-lives from 10^{18} to 10^{24} years. Because of the rarity of the event, the measurement of double beta decay is extremely difficult requiring highly enriched radionuclide sources and sophisticated coincidence counting instrumentation, which may include very large liquid scintillation detectors as well as other detector types, with the capability of extreme suppression or vetoing of background radiation (Obara, 2017, Gando, 2016, Fukuda et al., 2016, Segui, 2016, Signorelli and Dussoni, 2016, Biller, 2015, Descamps, 2015, Hans et al., 2015, Inoue, 2013, Lefeuvre, 2013, McGrath et al., 2010, Elliott et al., 2006, Gómez et al., 2007, Campbell et al., 2008, Caccianiga and Giammarchi and Fiorini, 2001). The National Nuclear Data Center of the Brookhaven National Laboratory (BNL-99822-2013-CP, http://www.nndc.bnl.gov/bbdecay/list2013.html, accessed February 22, 2018) lists 11 nuclides that are known to undergo 2νββ double beta decay (i.e., the emission of two beta particles and two neutrinos); these are listed in Table 6.18. in order of increasing atomic number. An additional nuclide, ^{130}Ba, listed at the bottom of Table 6.18 has been registered to decay by 2νECEC, which is double Electron Capture with the emission of two neutrinos. The long half-lives of these nuclides ranging from 10^{18} to 10^{24} years underscores the rarity of the double beta decay events, and the need to utilize huge and dedicated detectors for their measurement.

As discussed in Chapter 1, Volume 1, beta decay is not confined to negatron emission. There also exist positron and electron capture (EC) decay processes, which are other forms of beta decay. Research is underway with very large detectors including liquid scintillation detectors for the measurement of $\beta^+\beta^+$, β^+EC, and ECEC decay processes (Belli et al., 2016, Kang et al., 2013; Rukhadze et al., 2012; Andreotti et al., 2011; Kolhinen et al., 2010; Barabash et al., 2007, 2008). The nuclide ^{130}Ba, listed in Table 6.18, is the single nucleus that has been confirmed to decay by 2νECEC.

There is much research underway in numerous laboratories around the world directed toward the search for the detection of neutrinoless double beta decay (0νββ) utilizing liquid scintillation detectors and other detection methods, and only a few of the experiments on this work involving liquid scintillation are cited here, among the citations listed in the previous paragraph (Obara, 2017; Bialek et al., 2016; Fukuda et al., 2016; Gando, 2016; Segui, 2016; Barabash,

TABLE 6.18 Averaged and recommended $T_{1/2}(2\nu)$ values.

Isotope	$T_{1/2}(2\nu)$, yr
^{48}Ca	$4.4^{+0.6}_{-0.5} \times 10^{19}$
^{76}Ge	$1.65^{+0.14}_{-0.12} \times 10^{21}$
^{82}Se	$(0.92 \pm 0.07) \times 10^{20}$
^{96}Zr	$(2.3 \pm 0.2) \times 10^{19}$
^{100}Mo	$(7.1 \pm 0.4) \times 10^{18}$
$^{100}Mo-^{100}Ru(0^+_1)$[a]	$6.7^{+0.5}_{-0.4} \times 10^{20}$
^{116}Cd	$(2.87 \pm 0.13) \times 10^{19}$
^{128}Te	$(2.0 \pm 0.3) \times 10^{24}$
^{130}Te	$(6.9 \pm 1.3) \times 10^{20}$
^{136}Xe	$(2.19 \pm 0.06) \times 10^{21}$
^{150}Nd	$(8.2 \pm 0.9) \times 10^{18}$
$^{150}Nd-^{150}Sm(0^+_1)$[a]	$1.2^{+0.3}_{-0.2} \times 10^{20}$
^{238}U	$(2.0 \pm 0.6) \times 10^{21}$
^{130}Ba, ECEC(2ν)	$\sim 10^{21}$

[a] *$2\gamma\nu\beta\beta$ measurements 0^+_1 for excited states of daughter nuclei.*
From Barabash (2015a,b). Reprinted with permission from Elsevier © 2015.

2015; Maneschg, 2015; Nakamura, 2012; Barabash, 2011; Cremonesi, 2010; Klapdor-Kleingrothaus et al., 2004; Fiorini, 2001; Caccianiga and Giammarchi, 2000).

Beyond the standard model there exists the possibility that the neutrino could be its own antiparticle referred to as a Majorana neutrino. According to this theory, if the electron neutrino possesses a rest mass (now estimated to have a mass upper limit of $< \sim 2.3$ eV/c^2) and is its own antiparticle, a neutrino emitted in one beta decay could be reabsorbed in the second beta decay resulting in a double beta decay in which no neutrinos are emitted. *i.e.*, a neutrinoless double beta decay often abbreviated as 0νββ (See Chapter 1, Volume 1). Patrigani (2016, 2017) reports the upper limit to the rest mass of the electron neutrino to be $m_{\nu_e} < \sim 2.3\ \mathrm{eV}/c^2$ at a 95% confidence level (CL). Future research is aimed at increasing the sensitivity of the mass measurements to ~ the range of ~0.01−0.1 eV (Barabash, 2015). Neutrinoless double beta decay (0νββ) decay has yet to be demonstrated (Barabash, 2015). However, there is now great confidence that 0νββ will be demonstrated, and a renewed effort is underway worldwide toward this objective as a result of work for over a decade that has demonstrated neutrino oscillations, that is, neutrinos can "oscillate" from one flavor into another (*i.e.*, electron-, muon-, and tau-neutrinos) as they travel through space and matter. Oscillation is the changing of neutrino flavors back and forth from one to another, and this could occur only if the neutrino possessed mass. Experimental research on neutrino oscillation is reviewed by Messier (2006), and the Nobel Prize in Physics 2015) was awarded jointly to Takaaki Kajita of the University of Tokyo, Kashiwa, Japan and Arthur B. McDonald of Queen's University, Kingston, Canada for, in the words of the Nobel Committee "the discovery of neutrino oscillations, which show that neutrinos have mass". As a result of the success toward the demonstration of neutrino oscillations, current research in the search for neutrinoless double beta decay is moving forward with great effort, as it would reaffirm the mass of the neutrino, as only a particle with a definite mass can oscillate from one form into another.

A reported observation of 0νββ in ^{76}Ge was made by Klapdor-Kleingrothaus et al. (2004) and confirmed (Klapdor-Kleingrothaus, 2006; Klapdor-Kleingrothaus and Krivosheina, 2006) after data collection over a period of 13 years with high-purity p-type germanium detectors enriched in ^{76}Ge to 86%−88%. Such a detection arrangement is optimum in terms of yielding maximum detection efficiency for a given detection method, because the source (^{76}Ge) is the detector. *i.e.*, source = detector. However, Agostini et al. (2016) reports that a previous claim for evidence of a 0νββ decay signal is strongly disfavored, and the field of research is open again.

Several approaches to the application of liquid scintillation detection and measurement of neutrinoless double-beta decay (0νββ) are underway or in the planning stage. Barabash (2015) provides a review of the present status and prospects for the future for the various double beta decay experiments now planned and underway. The following paragraphs will describe the major projects now underway that involve liquid scintillation detectors.

A. KamLAND-Zen project

The KamLAND-Zen project is aimed at the detection and measurement of neutrinoless double beta decay, which involves the incorporation of the nuclide source into a liquid scintillation detector. Nuclides that undergo double beta decay and are easily loaded into liquid scintillators can be measured by liquid scintillation to observe the neutrinoless double beta decay. A large underground liquid scintillation detector, which is aimed at the detection and measurement of neutrinoless double beta decay in ^{136}Xe is the KamLAND-Zen detector, which is located at the Kamioka Observatory, an underground neutrino detection facility near Toyama, Japan. The acronym KamLAND-Zen signifies KamLAND Zero Neutrino double beta decay experiment. The KamLAND detector was originally designed to detect and measure antineutrinos, which are emitted with abundance from the surrounding commercial nuclear reactors. Thus, the name KamLAND represented Kamioka Liquid Scintillator Antineutrino Detector. The detector was key to providing evidence for neutrino oscillations with solar neutrinos that would diminish the population of electron neutrinos originating from the nuclear reactors. This gave evidence for neutrino mass, as only neutrinos with mass can oscillate from one flavor to another. The KamLAND detector was converted to KamLAND-Zen with the purpose of measureing neutrinoless double beta decay in ^{136}Xe

Double beta decay of ^{136}Xe could occur by either of the following two processes:

$$^{136}_{54}\text{Xe} \rightarrow {}^{136}_{56}\text{Ba} + 2\beta^- + 2\bar{\nu} \quad \}2\nu\beta\beta\} \tag{6.139}$$

or

$$^{136}_{54}\text{Xe} \rightarrow {}^{136}_{56}\text{Ba} + 2\beta^- \quad \}0\nu\beta\beta\} \tag{6.140}$$

The $Q_{\beta\beta}$ value for the ^{136}Xe is 2.479 MeV. In the case of 2νββ, the 2.479 MeV day energy would be shared among four particles, namely, the two beta particles and the two antineutrinos. However, in the case of 0νββ, the 2.479 MeV day energy would be shared only among the two beta particles. Thus, the 2νββ double beta (i.e., two electron) spectrum, for the ^{136}Xe decay would encompass a broad spectrum of energies, as illustrated in Fig. 6.107, because the two beta particles must share the decay energy

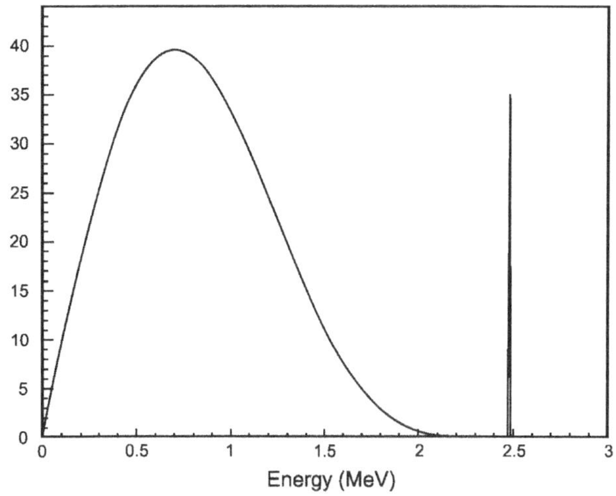

FIGURE 6.107 Double beta decay two-electron spectra for the case of xenon-136. The broad spectrum on the left is that of the double beta energies for the 2νββ decay; and the line spectrum on the right is that of 0νββ decay. The vertical scale is in arbitrary units and the proportion between the neutrinoless line and the $^{136}\text{Xe} \rightarrow {}^{136}\text{Ba} + 2e^- + 2\bar{\nu}$ spectrum is arbitrary as well. The $Q_{\beta\beta}$ value is 2.479 MeV. *From Caccianiga and Giammarchi (2000). Reprinted with permission from Elsevier © 2000.*

with the two antineutrinos. Whereas, 0νββ double beta (i.e., two electron) spectrum would be a line spectrum centered at the decay energy of 2.479 MeV, as illustrated in Fig. 6.107, as the two beta particles are emitted simultaneously and share the entire decay energy. In the search for two neutrino double beta decay (2νββ), two beta particles would be detected simultaneously, and the sum of their energies would always be less than the decay energy. However, evidence for neutrinoless double beta decay (0νββ) would be provided by the simultaneous detection of two beta particles with combined energies equal to the decay energy. The advantage of using liquid scintillation for the detection of double beta decay of ^{136}Xe is that the xenon gas can be dissolved directly into the liquid scintillator permitting the detection of the two beta-particle kinetic energies.

The original KamLAND detector, which was converted to the KamLAND-Zen detector, is illustrated in Figs. 6.107 and 6.108. The KamLAND-Zen is located at 1000 km underground from the top of Mt. Ikenoyama, Gifu Prefecture, Japan. A detailed description of the detector is provided by Asakura et al. (2016), and a brief overview is provided here. The inner balloon of the detector is 3.1 m in diameter made of 25-μm thick transparent nylon containing 300 kg of ^{136}Xe in liquid scintillation cocktail. In the second phase of the project for the measurement of 0νββ decay in ^{136}Xe, the inner balloon will contain 800 kg ^{136}Xe achieved by loading enriched xenon (90.93% ^{136}Xe and 8.89% ^{134}Xe) into liquid scintillator having a combined weight of 13 tons. The liquid scintillator cocktail (Xe-LS) within the inner balloon consists of 2.7 g/L of the fluor PPO

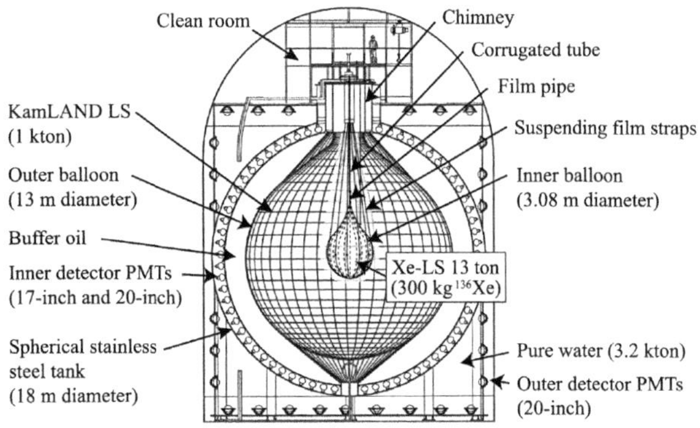

FIGURE 6.108 Schematic diagram of KamLAND-Zen detector. The 300 kg ^{136}Xe in the Xe-loaded liquid scintillator in the inner balloon noted in the figure is currently 383 kg ^{136}Xe, and in the next phase of the experiment, it will be increased to ~800 kg ^{136}Xe (Obara, 2017). For a perspective of the detector dimensions, a figure of a person is illustrated standing in the Clean Room in the upper deck of the detector on the right. *From Asakura et al. (2016); reprinted with permission from Elsevier © 2016.*

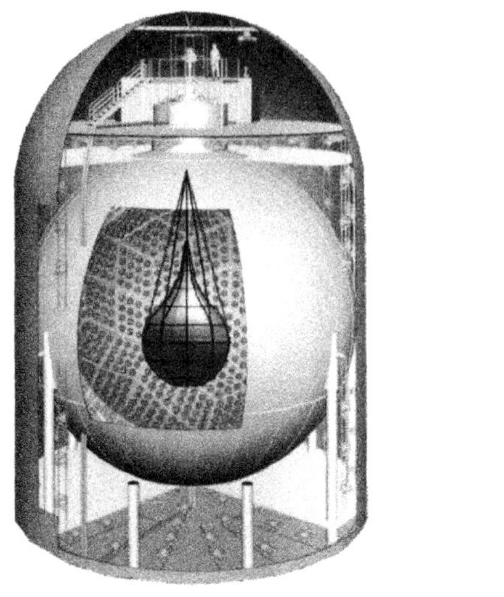

FIGURE 6.109 Cross-sectional illustration of the KamLAND-Zen detector. ^{136}Xe is dissolved into liquid scintillator filled in the mini-balloon. The mini-balloon is suspended from the top. Many of the numerous (1879) photomultiplier tubes (PMTs) can be seen mounted on the interior wall of the spherical stainless-steel tank facing the inner mini-balloon containing the Xe-loaded liquid scintillator. For a perspective of the detector dimensions, a person is illustrated standing at the upper deck to the right. *From Nakamura (2012); reprinted with permission from Elsevier © 2012.*

(2,5-diphenyloxazole) and 2.48% enriched xenon gas dissolved in a solvent consisting of 82% decane and 18% pseudocumene. The 13 tons of Xe-loaded liquid scintillator (Xe-LS) in the inner balloon comprises the detection medium for 2νββ and 0νββ decay of ^{136}Xe. The inner balloon is suspended in 1000 tons of liquid scintillator (KamlAND LS) within an outer balloon measuring 13 m in diameter consisting of 135-μm thick nylon and ethyl vinyl alcohol copolymer. This outer scintillator (KamLAND LS) serves as an active shield for external gamma-radiation as well as a detector for internal radiation from the inner scintillator (Xe-LS) and inner balloon (Asakura et al., 2016). The transparent outer balloon is suspended in 1400 tons of buffer oil contained in a 18 m diameter stainless steel tank. In the inner surface of the stainless-steel tank 1879 photomultiplier tubes (PMTs) are mounted facing the liquid scintillator (See Figs. 6.108 and 6.109). The stainless-steel tank is finally surrounded with 3200 tons of a water Cherenkov outer detector that serves as a cosmic-ray muon veto detector as well as serving to further shield the liquid scintillation detector from external radioactivity.

Due to the rarity of 0νββ relative to 2νββ decay, the main requirements are a high concentration of source material (i.e., ^{136}Xe) within a suitable detector) and the reduction of interfering background radiation to a maximum. Detection of both 2νββ and 0νββ from the initial 112-day and 115-day second phase of KamLAND-Zen measurements are reported by Inoue (2013) and Gando (2016). Efforts are focused toward methods for the reduction of background. Fig. 6.110 illustrates the spectrum acquired by the KamLAND-Zen detector from 2νββ and 0νββ together with spectra from background radiation sources. The spectra from 2νββ ends just under the ^{136}Xe

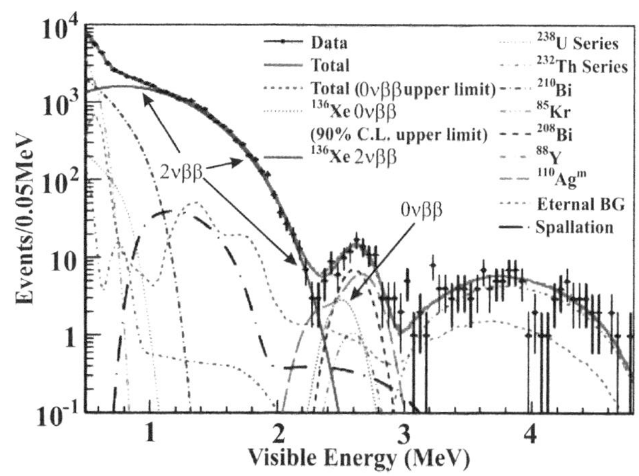

FIGURE 6.110 Observed energy spectrum from 38.6 kg yr of ^{136}Xe exposure from the KamLAND-Zen detector. Best fit spectrum for 2ν2β and backgrounds are shown together with 90% confidence level (C.L.) upper limit for 0ν2β. Arrows pointing to the spectra of the 2ν2β and the 0ν2β decay were added by the writer. The 2ν2β spectrum extends from zero to just under the ^{136}Xe decay energy of 2.479 MeV and the 0ν2β spectrum is a monoenergetic peak (Gaussian) peak at the *Q*-value (2.479 MeV). *From Inoue, 2013; reprinted with permission from Elsevier © 2013.*

decay energy (*Q*-value = 2.479 MeV); whereas that of 0νββ peaks at the *Q*-value. The resolution of the monoenergetic peak for the 0νββ decay is limited by the resolution of the liquid scintillation detector. The interferences from background are apparent from the observed energy spectrum for ^{136}Xe decay illustrated in Fig. 6.110. Asakura et al. (2016) describes in detail the main sources of background, namely, (i) radioactive impurities from and external to the inner balloon, (ii) impurities within the Xe-loaded liquid scintillation fluor cocktail (Xe-LS), and (iii) spallation products generated by cosmic-ray muons. The results illustrated in Fig. 6.110 for the 0νββ are limited by the background peaks from ^{110m}Ag and ^{214}Bi, a daughter nuclide of ^{238}U. Distillation of the liquid scintillator is planned to remove the ^{110m}Ag. The inner balloon was fabricated 100 km from the Fukushima reactor site a few months after the accident, and replacing this inner balloon should reduce radionuclide contamination due to fallout. Obara (2017) recommends replacing the inner balloon material with polyethylene–naphthalate, which has scintillation ability, and it would reduce background from ^{214}Bi from the inner balloon material surface by 99.7%. Gando (2016) notes that obviously it is not possible to eliminate the 2νββ events in the 0νββ signal energy region; however, the energy resolution can be improved from 4.0% to < 2.0% at the Q-value of the ^{136}Xe double beta decay by introducing high quantum efficiency PMTs, light collecting mirrors surrounding the PMTs, and brighter liquid scintillator.

B. SNO+ project

Another detector, which is under development for the liquid scintillation detection and measurement of neutrinoless double beta decay (0νββ) is the SNO+ Project. SNO+ entails the conversion of the Sudbury Neutrino Observatory (SNO) in Canada into a liquid scintillation detector (named SNO+) for double beta decay detection and measurement with ^{130}Te (Bialek et al., 2016; Segui, 2016; Biller, 2015; Descamps, 2015; Kamdin, 2015; Hans et al., 2015). In the previous edition of this book published in 2012, the writer had described the plans for the SNO+ project to measure 0νββ in ^{150}Nd; however, the SNO+ project reevaluated and altered the project in 2012 to measure 0νββ in ^{130}Te in lieu of ^{150}Nd for the following advantages, as described by Segui (2016) and Biller et al. (2015): (i) Te has a high 34% natural abundance of ^{130}Te. Tellerium-130 has the highest natural abundance of any double beta-decay isotope (Hans et al., 2015). (ii) The internal U/Th backgrounds can be actively suppressed via Bi-Po α tagging. (iii) The external gamma radiation can be adequately attenuated in the outer regions of the detector volume. (iv) Tellurium-130 has a much lower rate (factor of ~100) of 2νββ backgrounds relative to ^{150}Nd. (v) Tellurium has no inherent atomic absorption in the optical range. (vi) Natural tellurium is relatively inexpensive.

Double beta decay of Tellerium-130 has a Q-value of 2.527 MeV (QCalc, https://www.nndc.bnl.gov/qcalc/qcalcr.jsp, accessed February 27, 2018). The SNO+ Project will measure the 2νββ and 0νββ decay spectra, which would occur by the following:

$$^{130}_{52}\mathrm{Te} \rightarrow {}^{130}_{54}\mathrm{Xe} + 2\beta^- + 2\bar{\nu} \quad \text{"2νββ"} \tag{6.141}$$

or

$$^{130}_{52}\mathrm{Te} \rightarrow {}^{130}_{54}\mathrm{Xe} + 2\beta^- \quad \text{"0νββ"} \tag{6.142}$$

The SNO+ detector is a large-scale liquid scintillation detector located 2070 m underground (6800 m.w.e) at the Sudbury Neutrino Observatory (SNO) Laboratory in Dudbury, Canada. A cross-sectional illustration of the SNO+ detector is illustrated in Fig. 6.111. The SNO+ detector is the original SNO detector previously used exclusively for the detection and measurement of neutrinos by charged current (CC) and neutral current (NC) reactions with deuterium as well as electron scattering (ES) reactions. The 12 m diameter inner acrylic vessel of the SNO detector was filled with D_2O surrounded by ultrapure H_2O shielding for the previous work on neutrino measurements. The D_2O is replaced with 780 tons of Te-loaded liquid scintillator to create the SNO+ detector for the detection and measurement of neutrinoless double beta in ^{130}Te. The scintillator cocktail consists of 2 g/L of 2,5 diphenyloxazole (PPO) dissolved in linear alkylbenzene (LAB). For the first phase of the experiment, the scintillator will contain 0.3% natural tellurium (34% natural abundance of ^{130}Te) providing ~ 790 kg of ^{130}Te source material

FIGURE 6.111 Cut-away view of the SNO+ detector. The inner 12 m diameter acrylic sphere contains the active volume of Te-loaded liquid scintillation fluor cocktail. A 54% optical coverage is provided by ~9500 photomultiplier tubes (PMTs) placed along the inner walls of a 17 m diameter geodesic structure with interior and exterior shielding by pure water. The PMTs are illustrated as bright dots within the triangular panes of the outer structure. For a perspective of the detector dimensions, a person is illustrated walking toward the upper deck in the cave to the right. *From Kamdin (2015); reprinted with permission from Elsevier © 2015.* For the color version of this figure, the reader is referred to the online version of this book.

providing a high yield of ~10,000 photons/MeV. Higher Te-loading of the scintillator is possible in future upgrades of the detector (Descamps, 2015). It is foreseen that tellurium loading of up to 0.5% in the liquid scintillation will be possible in a future phase of the project, which would provide 1.3 tons of ^{130}Te source material for 0νββ decay detection and measurement. The acrylic vessel containing Te-loaded liquid scintillator is surrounded with 1700 tons of ultrapure water, and an additional 5300 tons are included in the space outside of the detector. The water will provide shielding from background radiation (Segui (2016).

Biller (2015) provides calculated estimated average spectra from 2νββ and 0νββ decay for ^{130}Te as well as the spectra for background radiations with the SNO+ detector. The estimated average spectra with two-full years of data collection are illustrated in Fig. 6.112. Only the tail-end of the ^{130}Te 2νββ spectrum is illustrated, as this is the region of importance in observing the monoenergetic peak for the 0νββ at the *Q*-value of 2.527 MeV. Estimated background radiation spectra are illustrated, which would be due to electron scattering (ES) of ^{8}B solar neutrinos, ^{208}Tl, ^{212}Bi+^{212}Po, ^{214}Bi+^{214}Po, and external radiation sources. Higher Te-loading of liquid scintillator has been demonstrated, and future phases of the experiment with loading ~10 times higher would increase sensitivity. An increase in Te-loading of scintillator of up to 5% has been demonstrated (Segui, 2016), which would provide a total of 13 tons of ^{130}Te source material for future phases of the SNO+ project.

C. EXO-200 project

An alternative approach to the detection of neutrinoless double beta decay in ^{136}Xe is to convert the source into the detector (*i.e.*, source = detector). In other words, rather than add the ^{136}Xe to a liquid scintillation cocktail, the alternative would be to take advantage of the fact that liquid xenon itself undergoes excitation with the emission of scintillation light photons at 175 nm wavelength in the vacuum ultraviolet (VUV) region (See Section XVII of this chapter). Thus, liquid xenon enriched in ^{136}Xe would serve both as the source of the decay emissions and the detector. This technique is currently taken by the Enriched Xenon Observatory (EXO). The first phase of EXO is EXO-200, which utilizes a detector containing 200 kg of liquid xenon enriched in ^{136}Xe to 80%. Construction of the detector was carried out at the SLAC National Accelerator Laboratory of Stanford University, Menlo Park, CA, USA, and it is installed at the Waste Isolation Power Plant (WIPP) near Carlsbad, NM, USA.

The EXO-200 detector consists of a cylindrical chamber that contains the 200 kg of enriched liquid xenon and serves as both a scintillation detector and time projection chamber (TPC) illustrated in Fig. 6.113. The detector is surrounded by a cylindrical wall of VUV Teflon reflectors and two walls of avalanche photodiodes (APDs) for the detection and gain multiplication of the VUV scintillation photons. The conception of a time projection chamber (TPC) originated from the work of David R. Nygren (Nygren, 1974) while working at the Lawrence Berkeley

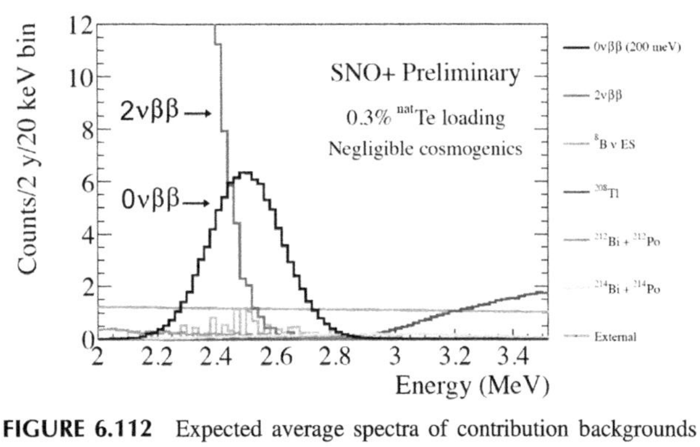

FIGURE 6.112 Expected average spectra of contribution backgrounds for two live years of data compared to a signal corresponding to a 200 meV Majorana neutrino mass, assuming a light level of 200 p.e./MeV (photoelectrons/MeV) and a value for the matrix element of 4. Background radiation spectra are color coded. Arrows added by the writer point to the tail end of the 2νββ spectrum and the peak of the monoenergetic 0νββ spectrum centering at the *Q*-value of 2.527 MeV. Color coding of the background radiation spectra is visible in the electronic version of the book. *From Biller (2015); reprinted with permission from Elsevier © 2015.*

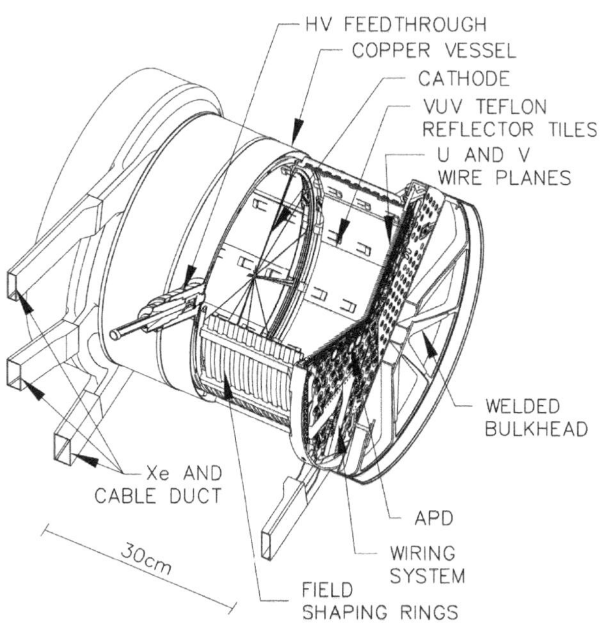

FIGURE 6.113 Drawing of the EXO-200. The thin copper chamber contains ~175 kg of liquid Xenon enriched to 80.6% in the isotope ^{136}Xe. *From Pocar (2013); reprinted with permission from Elsevier © 2013.*

Laboratory of the University of California (L'Annunziata, 2016). The TPC can provide the three-dimensional coordinates of energy deposition in the liquid xenon. The dimensions of the TPC are 40 cm in diameter and 44 cm in length. A complete description of the design, construction and function of the TPC is provided by Auger et al. (2012) and Pocar (2013). As described by Auger et al. (2012), the ionization channel of the TPC includes the cathode and field cage, which defines a region of uniform electric field with two wire planes or grids (U and V) and large avalanche area photodiodes (APDs) that provide a three-dimensional readout of ionization and scintillation in the liquid Xenon. Each end of the TPC cylinder contains two wire grids. The two wire planes or grids consist of a charge collection U wire plane and a shielding V wire plane is located in front of the U wires is biased to provide full electron transparency. The V wires also provide readout in addition to the U wires and the two wire planes are oriented at 60 degrees to each other, which provides a two-dimensional localization of the ionization cloud (See Fig. 6.114). An array of ~250 avalanche photodiodes are in close proximity (6 mm) from the wire grids. The third longitudinal coordinate of an energy deposition is obtained from the time interval between the scintillation signal collected in the APDs and the collection of charge at the wire grids. A set of field-shaping rings lined with VUV Teflon reflector tiles, grades the field and limits the drift region to two cylinders, each of 18.3 cm radius and 19.2 cm length (Pocar, 2013).

When a double beta decay event occurs in the ^{136}Xe, the beta particles will interact with the liquid xenon. The liquid xenon will emit scintillation photons that are detected by the APDs. The simultaneous emission of two beta particles will produce ionization in the liquid Xenon and the electron cloud produced with drift toward the U and V wire grids, as illustrated in Fig. 6.114 for a 0νββ event. The wire grids will register the two coordinates of the event, and the third coordinate will be registered from the liquid scintillation photon emission collected by the APDs.

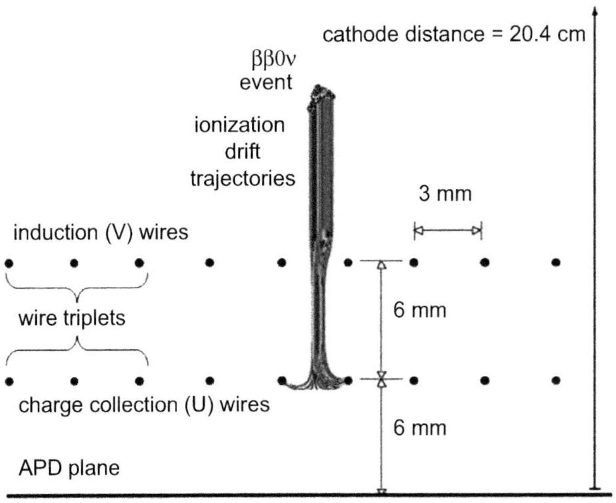

FIGURE 6.114 Geometry of the readout wire planes showing a simulation of an electron cloud drift from a 0νββ event in the liquid Xenon. The electron cloud is colored blue and illustrated with a drift trajectory toward the U and V wire grids. The anode (U) and induction (V) wires, illustrated as colinear, are in reality oriented 60 degrees from each other. *From Auger et al. (2012).* For the color version of the figure, the reader is referred to the online version of this book.

Auger et al. (2012) and Pocar (2013) provide detailed descriptions of the EXO-200, a combined liquid Xenon TPC and scintillation detector. The detector is located underground at a depth of ~1600 m.w.e. in a salt deposit at the Waste Isolation Pilot Plant, near Carlsbad, New Mexico. A low-background cryostat surrounds the detector with 2400 L of high-purity HFE7000 fluid, which provides radiation shielding as well as heat transfer. Additional shielding is provided by 5 cm of low-background copper serving as the cryostat housing encased with further shielding provided by 25 cm of high-purity low-background lead. A clean room housing the detector is surrounded with 50 mm thick plastic scintillation panels, which detect muons traversing the lead shielding with 95.9% efficiency.

In the planning stage of the EXO Collaboration is a larger experiment named nEXO, which would entail a 5-ton detector of liquid Xenon enriched in ^{136}Xe including the possibility of tagging of the ^{136}Ba daughter nuclides, which would provide a complete elimination of background (Dolinski, 2012). Detailed descriptions of the EXO-200 detector and the future prospects of the EXO Collaboration for the detection and measurement of neutrinoless double beta decay in ^{136}Xe are provided in papers by Pocar (2015, 2013), Auger et al. (2012), Dolinski (2012), Neilson et al. (2009), Leonard et al. (2008) and Ueshima et al. (2008).

D. ZICOS project

ZICOS is a new project, which is currently in its initial stages of research and planning (Fukuda, 2016; Fukuda et al., 2016). The acronym ZICOS is used for the ZIrconium Complex in Organic Scintillator neutrinoless double beta decay experiment, which would research the detection and measurement of 0νββ of ^{96}Zr. The ZICOS project has developed a liquid scintillation cocktail that will consist of zirconium organometallic complex in liquid scintillator containing PPO and POPOP as primary and secondary fluors dissolved in anisole solvent. The light yield of the cocktail is comparable to BC505 scintillator with a decay time of 20 ns. The scintillation cocktail has a capacity to contain a Zr concentration of 70 g/L. The isotope ^{96}Zr has a natural abundance of 2.8%; however, it is commercially available enriched to 58.5%, which would yield 128 kg of ^{96}Zr source material in the scintillation detector currently planned.

The ZICOS detector would have a 1.5 m radius or a total volume of 14.1 m^3. The detector would be located within an outer cylindrical tank 5.4 m in diameter and 5 m in height. The tank would be filled with pure water to veto the penetrating muons by the measurement of Cherenkov photons and attenuate the external gamma-ray and neutron radiation. The Q value for the double-beta decay of ^{96}Zr is 3.356 MeV, which is high, and it is expected that the backgrounds from ^{208}Tl would be reduced to one-tenth of the magnitude of the KamLAND-Zen project with the measurement of Cherenkov photons.

XII. Detection and measurement of neutrinos

In a letter addressed to the participants of a Regional Meeting on Radioactivity at Tübingen, Germany, on 4 December 1930, Wolfgang Pauli postulated the existence of an elusive neutral particle of near zero rest mass, which would accompany beta decay. The particle remained elusive and unproven as to its existence for over 25 years. Nevertheless, during that time, the particle remained an essential component in our understanding of beta decay. In 1934, Enrico Fermi coined this particle as the 'neutrino' from the Italian language meaning "little neutral one" in his elaboration of the beta-decay theory (See L'Annunziata, 2016 for a detailed historical account.)

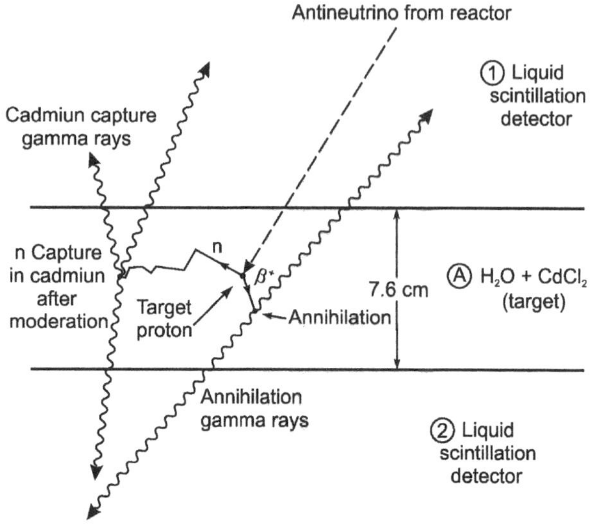

FIGURE 6.115 Detection scheme used by Reines and Cowan for the antineutrino signature signal. An antineutrino is illustrated entering the tank of aqueous $CdCl_2$ solution and striking a target proton. The proton converts to a neutron and positron. The positron annihilates on an electron with the emission of two 0.511 MeV gamma rays in opposite directions detected by the liquid scintillator in tanks above and below the water target tank. The neutron produced by the antineutrino interaction slows down in the water and is captured by a cadmium nucleus, and the resulting gamma rays are detected by the liquid scintillator in the adjacent tanks approximately 10 μs after the positron annihilation. *From Reines (1995), Nobel Lecture with permission from The Nobel Foundation ©1995.*

A. Reines and Cowan reaction

The neutrino remained elusive until 1956, when its existence was demonstrated finally by Nobel Laureate Frederick Reines and Clyde Cowan, Jr (Reines and Cowan, Jr, 1953, 1956, 1957). They confirmed the existence of the neutrino using liquid scintillation detectors to demonstrate inverse beta decay where an antineutrino interacts with a proton to yield a neutron and positron, *i.e.*,

$$\bar{\nu} + p^+ \rightarrow n + e^+ \quad (6.143)$$

Reines and Cowan constructed a liquid scintillation detector next to a 700 MW nuclear reactor at the Savannah River Plant in Aiken, SC, USA, to detect the abundant antineutrinos emitted by the beta decay of fission products. Their detector was composed of a target chamber consisting of 200 L of water containing 40 kg of dissolved $CdCl_2$ sandwiched between two tanks of 1400 L of liquid scintillator schematically illustrated in Fig. 6.115. Each end of the scintillator tanks was viewed by 55 photomultiplier tubes (not illustrated). The water provided target protons for the antineutrinos. As illustrated in Fig. 6.115, the interaction of an antineutrino with a water proton would create a neutron (n) and a positron (β^+). The positron would be annihilated when coming to rest and in contact with an electron, and the resulting annihilation radiation (two 0.51 MeV gamma-ray photons emitted in opposite directions) would be detected in coincidence by the two liquid scintillation detectors above and below the water tank (Fig. 6.115). The neutron produced by the antineutrino interaction would slow down quickly (~10 μsec) in the water and be captured by a cadmium nucleus in the water target chamber. The characteristic multiple gamma rays following the neutron capture would be detected in coincidence by the two liquid scintillation detectors. The antineutrino signature therefore consisted of a delayed coincidence between the prompt scintillation pulses produced by the β^+ annihilation and the scintillation pulses produced microseconds later by the neutron capture in cadmium. With the unique signature for the antineutrino provided by the detector design devised by Reines and Cowan, the high neutron flux produced by the Savannah River reactor ($1.2 \times 10^{13}/cm^2$ sec), and reduced cosmic ray backgrounds from massive shielding 11 m from the reactor and 12 m underground were essential to the success of their experiment. Nevertheless, a detector running time of 100 days over a period of about 1 year was required to provide sufficient conclusive signals from the antineutrino signature.

A more complete account of Pauli's hypothesis of the existence of the neutrino and the work of Reines and Cowan, which culminated in the discovery of the neutrino is provided in a previous book by the writer (L'Annunziata, 2016).

B. Liquid scintillation schemes for neutrino detection and measurement

The detection scheme of Reines and Cowen described above is one among others used today for the liquid scintillation detection and measurement of neutrinos. Neutrinos will interact with matter in various ways including electron scattering (ES), charged current (CC) interactions, and neutral current (NC) interactions. These mechanisms of interaction are described in detail in Chapter 1, Volume 1. Among the schemes available for neutrino detection and measurement by liquid scintillation are the following:

1. Neutrino-electron scattering

The neutrino-electron elastic scattering reaction, $\nu e^- \rightarrow \nu e^-$, is detected in a large mass (>100 tons) of liquid scintillator. Solar neutrinos can be detected via this elastic scattering reaction, such as in the Borexino and SNO+ Collaborations for the measurement of solar neutrinos described subsequently.

The scattered electron can receive any kinetic energy from the neutrino less than or equal to the kinetic energy of the neutrino (Kraus, 2006). A wide spectrum of pulse heights from the different scattering reactions will be produced in the liquid scintillator, but for monoenergetic neutrinos characteristic of certain nuclear reactions inside the sun, a liquid scintillation spectral 'Compton-like' edge can serve as a signature for a ν line energy (Agostini et al., 2017; Bellini, 2016; Davini, 2012; Avanzini, 2011; Alimonti et al., 2002). For example, the solar neutrinos that are a result of the ^{7}Be electron capture reaction in the sun produce a mono-energetic ν line at 0.862 MeV. As described by Agostini et al. (2017), Bellini (2016) Avanzini (2011) in reviews on the Borexino neutrino detector and illustrated in Fig. 6.116, the ^{7}Be$-\nu$ line at 0.862 MeV produces a profile of the recoil electrons in the liquid scintillator with a spectral 'Compton-like edge' at 0.665 MeV. This serves as a spectroscopic signature for the solar ^{7}Be neutrinos. Agostini et al. (2017) explain that the scattering of monoenergetic neutrinos with energy E_ν off electrons leads to recoil electrons with a Compton-like continuous energy spectrum with maximum scattered electron energy defined by

$$T_\nu^{\max} = 2E_\nu^2/(m_e + 2E_\nu) \tag{6.144}$$

where $T_\nu^{\max}$ is the maximum electron recoil energy, E_ν is the neutrino energy, and m_e is the electron mass equivalent to 0.511 MeV. For the monoenergetic 0.862 MeV solar neutrinos from ^{7}Be decay, Eq, 6.144 yields the maximum electron recoil energy of $T_\nu^{\max} = 0.665$ MeV,

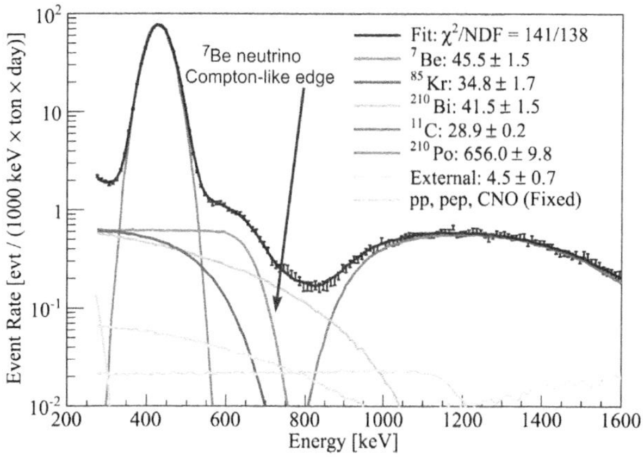

FIGURE 6.116 A Monte Carlo based fit over the energy region 270−1600 keV for the Borexino detector; the event energies are estimated using the number of photons detected by the PMT array. Rate values in the legend are integrated over all energies and are quoted in units of counts/(day·100 ton). The arrow pointing to the ^{7}Be neutrino electron scatter Compton-like edge at 0.665 MeV was added by the writer. Color coding of radiation sources are visible in the electronic version of the book. *From Bellini (2016) and Davini (2012); reprinted with permission from Elsevier © 2016.*

as illustrated in Fig. 6.116, and detailed reviews are provided by Agostini et al. (2017), Bellini (2016), Calaprice (2015), Miramonti (2015), Davini (2012), and Avanzini (2011).

Liquid scintillation detectors designed for the measurement of neutrinos are very large to provide adequate media for background veto as well as massive amounts of liquid scintillator measuring in the hundreds or thousands of tons. A schematic of the Borexino detector is illustrated in Fig. 6.117, which illustrates the immensity of the neutrino detector. The Borexino detector is constructed in the Laboratorio Nazionale del Gran Sasso, Italy, and the

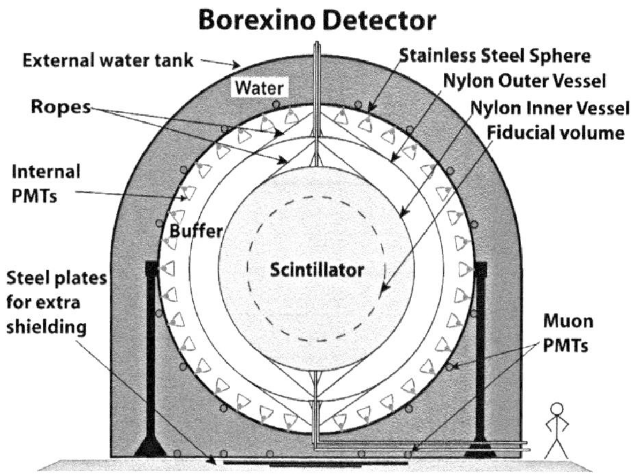

FIGURE 6.117 Schematic view of the Borexino detector. An abstract figure of a person is drawn by the writer on the right side to provide a perspective of the relative size of the detector. *From Agostini et al. (2017); reprinted with permission from Elsevier © 2017.*

name Borexino is derived from the Italian for BOREX experiment (BORon solar neutrino Experiment). The detector, as described by Agostini et al. (2017), is based on the principle of graded shielding, with the inner liquid scintillator core at the center of a set of concentric shells of decreasing radio-purity from inside to outside (see Fig. 6.117). The liquid scintillator detector consists of 278 tons of pseudocumene solvent doped with 1.5 g/L of PPO (2,5-diphenyloxazole) fluor. The liquid scintillator is confined within a thin spherical nylon vessel of 8.5 m diameter surrounded by 2200 photomultipliers tubes (PMTs), defining the so-called Inner Detector (ID). All but 371 PMTs are equipped with aluminum light concentrators designed to increase the light collection efficiency. The detector core is shielded from external radiation by 890 tons of buffer liquid, a solution of pseudocumene and 3–5 g/L of the light quencher dimethylphthalate. The buffer is divided into two volumes by a second nylon vessel with a 11.5 m diameter, preventing inward radon diffusion. A 13.7 m diameter Stainless Steel Sphere encloses the central part of the detector and serves as a support for the PMTs directed toward the scintillator as well as the outer muon PMTs directed toward the outer water detector and shield. An external domed water tank of 18 m diameter and 16.9 m height, filled with ultra-high purity water, serves as a passive shield against neutrons and gamma rays as well as an active muon veto. The Cherenkov light radiated by muons passing through the water is measured by 208 external PMTs defining the so called Outer Detector. The detector is shielded by 3300 tons of radiopure liquids.

2. Reines-Cowan inverse beta decay reaction

The Reines-Cowan reaction described previously is commonly used for the detection and measurement of anti-neutrinos, where fission product beta-decay is abundant emitting electron anti-neutrinos ($\overline{\nu}_e$). The anti-neutrinos are detected by inverse beta decay

$$\overline{\nu}_e + p^+ \rightarrow n + e^+ \tag{6.145}$$

with a threshold energy $Q = 1.80$ MeV (Agostini et al., 2017; Kraus, 2006; Alimonti et al., 2002). The positron (e^+) that is emitted will possess a kinetic energy equal to the neutrino energy (E_ν) – Q, which is deposited in the liquid scintillator. In addition, the positron almost immediately undergoes annihilation in contact with an electron of the scintillation cocktail (e^+e^- annihilation energy = 1.02 MeV). The positron signal created in the liquid scintillator will be the sum of its kinetic energy + 1.02 MeV annihilation energy (Kraus, 2006; Alimonti et al., 2002). Neutrino spectroscopy is therefore possible, as the energy deposited (E_{dep}) and the pulse height produced (PH) in the liquid scintillator by the positron is proportional to the anti-neutrino energy or

$$\begin{aligned} \text{PH} \propto E_{dep} &= E_\nu - 1.80\ \text{MeV} + 1.02\ \text{MeV} \\ &= E_\nu - 0.78\ \text{MeV} \end{aligned} \tag{6.146}$$

An example of the potential of antineutrino energy spectroscopy based on inverse beta decay is provided by Agostini et al. (2017) from data obtained from the Borexino detector provided in Table 6.19. The antineutrino energies presented in the table were correlated with gamma-ray bursts (GRBs) the most energetic events in the universe, which release energy equivalent to 10^{54} erg ($\sim$1 solar mass), under the assumption of isotropic emission of energy (Agostini et al. (2017). The GRBs are detected with a frequency of $\sim$1 event per day from the entire sky.

Even at the anti-neutrino threshold energy where $E_\nu = Q = 1.80$ MeV, the inverse decay reaction produces a liquid scintillation pulse signal at 1.02 MeV, which corresponds to the annihilation radiation. Agostini et al. (2017) and Alimonti et al. (2002) point out that the $\overline{\nu}_e$ tag is made

TABLE 6.19 Borexino confidence level 90% upper limits for fluence of electron antineutrinos from gamma-ray bursts[a].

E_{ν_e} [MeV]	Φ_{ν_e} primary DAQ[b] [cm^{-2}]	Φ_{ν_e} FADC[c] [cm^{-2}]
2	4.36×10^9	4.87×10^9
3	1.13×10^8	2.64×10^8
4	4.00×10^8	2.49×10^8
6	3.81×10^7	5.67×10^7
10	1.50×10^7	3.57×10^7
14	6.96×10^6	1.04×10^7

[a]*From Agostini et al. (2017), The Borexino Collaboration; reprinted with permission from Elsevier © 2017.*
[b]*Primary data acquisition system.*
[c]*Fast analog to digital converters data acquisition system used to cross-check data acquired from a set of 1813 GRBs.*

possible by a delayed coincidence signal produced by a 2.2 MeV γ-ray emitted by the capture of the neutron by a proton. The delayed coincidence pulse occurs ~200 μsec after the emission of a neutron from the inverse beta decay, which is the delay caused by the moderation of the neutron to thermal energies and its capture by a proton of the liquid scintillator. The delayed coincidence of ~200 μs of the positron annihilation signal and the 2.2 MeV gamma-ray signal produced by a neutron capture on a proton serves at the $\bar{\nu}_e$ tag. Only a small number (1%) of the neutron captures occur on ^{12}C producing 4.95 MeV gamma rays. The unique $\bar{\nu}_e$ tag reduces background by a factor of ~100 (Alimonti et al., 2002).

Collaborations have been undertaken to measure reactor neutrino oscillation by locating large Gd-loaded liquid scintillation detectors near and far (~1 km) from the nuclear reactor. Among these collaborations are (i) the Double Chooz experiment (Haser, 2016) located at the Chooz nuclear power plant in France, (ii) Daya Bay Reactor Neutrino Experiment in Shenzhen, China (Beriguete et al., 2014 and Ding et al., 2008), and (iii) the RENO Collaboration (Reactor Experiment for Neutron Oscillation) in the Republic of Korea (Park et al., 2013). Research has been undertaken to improve the quality, purity and performance of Gd-loaded scintillators for large neutrino detectors (Beriguete et al., 2014 and Yeh et al., 2010). Gadolinium has a very high thermal neutron capture cross section (See Table 1.20 of Chapter 1, Volume 1). The delayed signal due to neutron capture in Gd is shorter (~30 μsec) than that produced in conventional liquid scintillator (~200 μsec), and it is identified with the emission of an ~8 MeV gamma ray. The strong gamma-radiation provides a unique signature, for the incident anti-neutrino and thus improved background reduction.

The measurement of the anti-neutrino production by reactors can be used to reveal the fissile composition of reactor fuel thereby providing continuous, nonintrusive, and unattended measurements important to nuclear safeguards (Dai *et al.* (2011).

3. Inverse beta decay (charged current interactions) yielding negatrons and unstable nuclei

In contrast to the detection of anti-neutrinos discussed in the previous Reines-Cowan reaction, neutrinos can be detected by charged current interactions (inverse beta decay) producing negatrons according to the interaction

$$\nu_e + \mathrm{n} \rightarrow \mathrm{p}^+ + e^- \qquad (6.147)$$

where the neutrino interacts with a neutron of a nucleus in the liquid scintillator yielding a proton and a negatron. If the product nucleus is unstable, the energy deposited in the scintillator by the nuclear decay can provide a tag for the neutrino interaction. Several nuclei may be used for neutrino interactions, which would yield an unstable nucleus and suitable tag for the neutrino inverse beta decay. Raghaven (1976) was first to indicate that inverse beta decay on ^{115}In, which has a natural abundance of 95.72%, would make for an efficient solar neutrino detector. Indium-loaded liquid scintillators with up to 15 wt% Indium have been developed for solar neutrino detection (Chang et al., 2011; Motta et al., 2005; Raghaven, 2001; Suzuki et al., 1990; Payne and Booth, 1990). The inverse beta decay reaction of a neutrino with ^{115}In is written as

$$\nu_e + {}^{115}_{49}\mathrm{In} \rightarrow {}^{115}_{50}\mathrm{Sn}^* + e^- \qquad (6.148)$$

where the ^{115}Sn product nucleus is unstable indicated by an asterisk superscript. This reaction has a low threshold energy of 128 keV, which provides the measurement of the incident neutrino energy, which is defined by the equation

$$E_{\nu_e} = E_{e^-} + Q_d \qquad (6.150)$$

where $E_{\nu_e}$is the incident neutrino energy, and E_{e^-} is the beta particle or negatron energy deposited in the liquid scintillator and Q_d is the capture threshold (Pallavicini, 2015; Grieb et al., 2011). In the case of the inverse beta interaction of neutrinos with ^{115}In, the interaction has a low threshold energy of 128 keV (Suzuki et al., 1990), and the negatron energy deposited in the liquid scintillator E_{e^-}would be a function of the neutrino energy according to the equation

$$E_{\nu_e} = E_{e^-} + 128\ \mathrm{keV} \qquad (6.151)$$

The low threshold (128 keV) of the indium interaction provides a high sensitivity of this scintillator to neutrinos from the solar proton-proton fusion reaction (≤430 keV) and to solar neutrinos produced by electron capture (EC) in ^{7}Be (860 keV) as pointed out by Payne and Booth (1990). The ^{115}Sn daughter nucleus decays with a very short half-life (3.2 μsec) by isomeric transition emitting three gamma rays at 101, 116, and 498 keV, which provide a unique signature for the incident neutrino.

On the basis of the above inverse beta interaction of neutrinos with ^{115}In, a large neutrino detector utilizing Indium-loaded liquid scintillator is under development under the scope of the LENS Collaboration. LENS is an acronym referring the experiment named Low Energy Neutrino Spectroscopy, which was initiated at the Virginia Polytechnic and State University (Virginia Tech), Blacksburg, VA, USA, in 2005 where major breakthroughs in detector design, Indium-loaded liquid scintillator development and background rejection were made. Ragu S. Raghaven, who first proposed the utilization of ^{115}In in the detection of solar neutrinos by inverse beta decay in 1976 is on the team of researchers at Virginia Tech. Readers may review papers by Pallavicini (2015) and Grieb et al. (2011) for more information on the development of the LENS detector, which would contain at full capacity up to 125 tons of Indium-loaded liquid scintillator.

Other examples of interactions in liquid scintillator that have potential for the detection and measurement of solar neutrinos by inverse beta decay are the following:

$$\nu_e + {}^{160}_{64}\text{Gd} \rightarrow {}^{160}_{65}\text{Tb}^* + e^- \quad (6.152)$$

and

$$\nu_e + {}^{176}_{70}\text{Yb} \rightarrow {}^{176}_{71}\text{Lu}^* + e^- \quad (6.153)$$

The excited states of $^{160}Tb^*$ and $^{176}Lu^*$ de-excite with very short lifetimes of 75 and 50 nsec, respectively, emitting gamma rays of 63.7 and 72 keV, respectively. Gratta and Wang (1999) point out that the low threshold for these reactions of 240 and 300 keV are well below the maximum solar proton-proton fusion neutrino energy. Also, the very fast (nsec) correlation time between beta-particle energy (e^-) deposition in the scintillator and γ-ray emission by the daughter nuclides provides a powerful signature for the solar neutrino event and background reduction. Lightfoot et al. (2004) report the development of 10 wt% Gd-loaded liquid scintillator for solar neutrino detection as well as neutron measurements. Gadolinium-load scintillator has been produced for some large-scale neutrino detectors such as the RENO neutrino detector (Park et al., 2013) and the Daya Bay reactor neutrino detector (Beriguete et al., 2014). These two detectors measure antineutrinos via the Reines-Cowan inverse beta-decay interaction described above according to $\overline{\nu}_e + p^+ \rightarrow n + e^+$(Eq. 6.145).

4. Neutrino charged current interactions with ^{13}C

The charged current interaction of solar neutrinos with ^{13}C nuclei naturally occurring in liquid scintillator can be used to detect the high-energy solar neutrinos derived from the ^8B reaction in the sun, which yields neutrino energies ≤15 MeV (Ianni et al., 2005). The ^{13}C isotope occurs naturally with an abundance of 1.1%. When very large neutrino detectors are involved, which utilize hundreds or even thousands of tons of liquid scintillator, the mass of ^{13}C in the scintillator will measure in tons. The LENA (Low Energy Neutrino Astronomy) detector for solar neutrino detection and measurement is a next-generation neutrino detector under development as part of an international collaboration. The detector is to be installed at the Center of Underground Physics in the Pyhäsalmi (CUPP) in central Finland. The detector would consist of 50,000 tons of liquid scintillator containing PPO and bis-MSB as primary and secondary fluors dissolved in linear alkylbenzene solvent contained in a tank measuring 100 m in height and 32 m in diameter with 32,000 photomultiplier tubes. The detector design and function is described by Wurm et al. (2015) and Lachenmaier et al. (2010).

The charged current interaction of a neutrino with a ^{13}C nucleus of the liquid scintillator detector with the transition of the unstable ^{13}N product nucleus to the ground state is the following:

$$\nu_e + {}^{13}_{6}\text{C} \rightarrow {}^{13}_{7}\text{N}_{\text{grd}} + e^- \quad (6.154)$$

The eaction threshold is $Q = 2.22$ MeV, and the kinetic energy of the electron or negatron (T_e) can be expressed as $T_e = E_\nu - Q$ neglecting the small recoil energy of the ^{13}N nucleus, which is of the order of a few keV (Möllenberg et al., 2014; Ianni et al., 2005). The pulse height produced by the negatrons can thus provide spectrometry of solar neutrinos derived from the ^8B reaction in the sun, which are of energy in excess of the 2.22 MeV threshold energy. Ianni et al. (2005) point out the advantage of this mode of detection is that the process of ^{13}N decay to its ground state by positron emission according to

$$^{13}_{7}\text{N} \rightarrow {}^{13}_{6}\text{C} + \nu_e + e^+ \quad (6.155)$$

can be used to monitor the solar neutrino interactions with the ^{13}C nuclei, as the visible energy released in the ^{13}N decay is the sum of the positron kinetic energy and the energy released in the e^+e^- annihilation.

C. Collaborations for LS neutrino detection and measurement

There are several international collaborations underway with the objective of measuring the solar neutrino flux and neutrino properties as well as the measurement of neutrinos from other sources by means of large liquid scintillation detectors. The advantage of liquid scintillation is the possibility of preparing hundreds and even tens of thousands of tons of scintillator. Such liquid scintillators provide tons of target material for the elusive neutrino, and the target nuclides can be loaded into the liquid scintillators either enriched or at natural isotope abundance. The geometry of the liquid scintillator detector is easily designed. Even with such large detectors, years of scintillator signal collection and extreme measures of background reduction are often required. Numerous international collaborations are underway in the implementation and development of massive neutrino detectors, and some were described in the previous paragraphs. Reviews describing various detectors, their design and objectives are provided by Agostini et al. (2017), Ianni (2017), Lu (2017), Segui (2016), Calaprice (2015), Jaffe (2015), Pallavicini (2015), Smirnov et al. (2015), Wurm et al. (2015), Oberauer (2012), Wang (2012), Cabrera (2010), and Alimonti et al. (2009).

XIII. Microplate liquid scintillation counting

Many applications use radioactivity as a method of screening numerous compounds or reactions in an effort to find a new therapeutic agent, drug candidate, or to assess a molecular biology function. In these types of applications, many thousands of samples may be counted each week. Because of the numerous samples involved, large amounts of scintillation vials and cocktail are required for traditional liquid scintillation analysis, which add cost to the sample analysis. A large number of samples also creates a large amount of radioactive waste. Consequently, microplate scintillation counting instruments have been developed to reduce waste, increase sample throughput via the simultaneous analysis of several samples with multiple detectors, provide automation capabilities, and reduce the cost of consumable materials and waste disposal.

The microplate scintillation counter uses a microplate sample format, which was developed to provide for the analysis of large numbers of samples. A microplate is a small container with many sample wells. It is about the size of a postal card measuring approximately 12.7 × 8.6 × 1.9 cm (L × W × H). Typical microplates can have numerous wells for holding samples, and the following are examples: (i) 24-well plates (6 × four wells), 96-well plates (12 × eight wells), and 384-well plates (24 × 16 wells). The typical sample capacities for these plates are the following:

- **(i)** The 24-well microplate holds up to 375 and 1500 μL of sample plus scintillation cocktail per well for shallow- and deep-well plates, respectively.
- **(ii)** The 96-well microplate holds 75 and 350 μL of sample per well for shallow- and deep-well plates, respectively.
- **(iii)** The 384-well microplate holds up to 80 μL of sample per well.

It is clear that only relatively small samples can be analyzed in these microplates, but the footprint is small, compact, and easy to automate. As a result of the small sample size and small amount of scintillation cocktail required, the activity analysis of many samples, often encountered in biological research, can be carried out without producing a large amount of radioactive waste. The second major feature of the microplate scintillation counter is the ability to count up to 12 samples simultaneously (*i.e.*, up to 12 detectors) on one microplate.

Only a brief description of the microplate scintillation analyzer is provided here. Additional information on this instrumentation is available in Chapter 9 of Volume 1. The microplate scintillation counter can be of single- or multiple-detector design (up to 12 detectors). Some commercial microplate scintillation counters are the TopCount and MicroBeta counters from PerkinElmer, Inc., Waltham, MA, USA, and the Sense Beta available from HIDEX, Turku, Finland.

A. Detector design and background reduction

The microplate scintillation and luminescence instrument can be of single- or multiple-detector design with up to 12 detectors all packed in close proximity in a very small area. The microplate instrument can count samples directly in microplate wells using one of two mechanisms: single or dual photomultiplier tube (PMT) detectors. In the single PMT detector arrangement, illustrated in Fig. 6.118, the PMT detectors are placed directly above the sample wells; whereas, in the dual PMT arrangement there are two photomultipliers one above and one below each sample well. With the single PMT arrangement, a white reflective and opaque microplate is utilized whereby light photons produced by the radioactivity in the liquid scintillator is reflected back toward the phototube, and a unique method of background reduction, namely, time-resolved liquid scintillation counting (TR-LSC) is utilized. Microplate counters with two PMTs per sample must utilize microplates with clear bottoms to enable the transmission of photons to the second PMT, and coincidence counting is implemented to reduce background noise in addition to TR-LSC. A unique advantage of the dual photomultiplier microplate scintillation analyzers is the option to disable one of the PMTs to permit the utilization of opaque reflective microplates.

The single-PMT time-resolved liquid scintillation counting (TR-LSC) technique uses pulse counting to distinguish between true scintillation pulse events and background noise. This obviates the need for a second PMT per sample and reduces lead shielding requirements. Single-PMT counting with TR-LSC uses scintillators with relatively long decay periods. A scintillator with a long decay constant emits photons, after β-particle excitation,

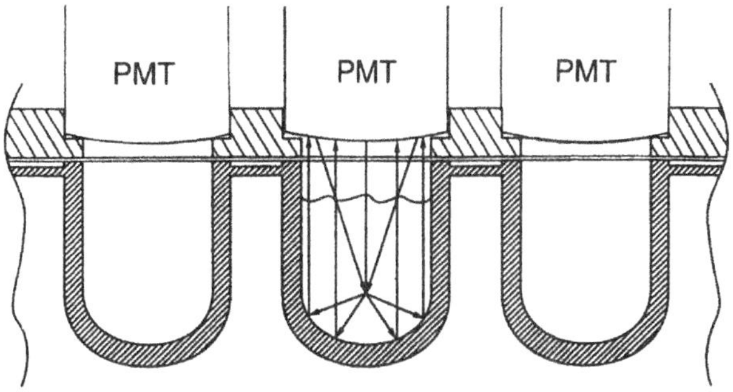

FIGURE 6.118 Single PMT detector design in microplate scintillation counting. A cross section of a microplate segment is illustrated showing three white opaque sample wells and three photomultiplier tubes (PMTs) aligned above each sample well whereby up to 12 PMTs analyze simultaneously as many as 12 samples. © *1998 Perkin Elmer, Inc. Printed with permission.*

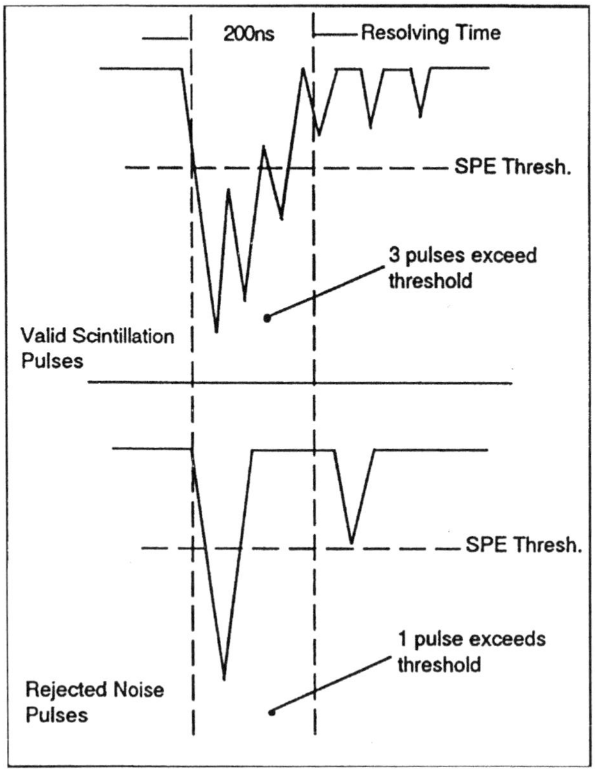

FIGURE 6.119 Time-resolved liquid scintillation counting (TR-LSC) employed in PerkinElmer microplate scintillation and luminescence counters. © *1998 Perkin Elmer, Inc. Printed with permission.*

over a longer period of time. Each scintillation pulse produces a photon packet followed by a series of pulses as illustrated in Fig. 6.119, whereas, PMT noise creates single pulse events. In TR-LSC the characteristics of a pulse are determined over a period of time (*e.g.*, 200 ns) after the initial photon packet is detected. If it is followed by one or more additional pulses within the resolving time period of 200 ns, the pulse is accepted as a true scintillation event. If no additional pulses are detected within the resolving time period, the initial pulse is rejected as background noise. The resolving time circuit to initiate pulse decay discrimination is triggered when a pulse height exceeds the single photon event (SPE) threshold. The number of pulses above the SPE threshold is counted during the resolving time period. Multiple pulses detected in the resolving time are accepted as a valid event, which is further analyzed in the pulse height analyzer of the LSA. If multiple pulses are not detected in the resolving time interval, the triggering pulse is rejected as background noise.

B. Applications

A microplate scintillation counter has numerous applications, and with these are distinct sample preparation methods that are used for radioactivity analysis. All of these applications are intended for small sample volume and high sample throughput analysis. An excellent guide to the use of microplate counters, such as the TopCount and MicroBeta, is provided by Kahl et al. (2012). References to specific applications with microplate counters are included below.

When liquid scintillation counting is required with the microplate analyzer, scintillation cocktail, and sample are added to each well of the microplate. This method is used for counting organic or aqueous samples directly in the microplate wells. An example of this can be taken from enzyme activity studies in which the sample is extracted or filtered directly into the liquid scintillation cocktail and counted. Many examples of analysis of radionuclides, such as ^{3}H, ^{14}C, ^{32}P, ^{35}S, and ^{125}I, in the biosciences can be found in the literature. Among these, a few will be cited here that relate to specific microplate liquid scintillation counters, namely, (i) the TopCount (PerkinElmer, Inc., Waltham, MA, USA) liquid scintillation and luminescence counter (Christiansen, 2018; Hosogi et al., 2018; Kojima et al., 2018; Lozac'h et al., 2018; McCabe et al., 2012; Rauly-Lestienne et al., 2011; He et al., 2011), (ii) the MicroBeta (PerkinElmer, Inc., Waltham, MA, USA) liquid scintillation and luminescence counter (Cunha de Oliveira et al., 2018; Kumar et al., 2018; Ingwersen et al., 2018; McCarthy et al., 2018; Shu et al., 2018; Varaschin et al., 2018), and (iii) the Hidex Chameleon V or Sense Beta (HIDEX, Turku, Finland) liquid scintillation and luminescence counter (Abe et al., 2016; Melkes et al., 2016; Boag et al., 2015; Brun et al., 2014; Ferraz et al., 2014; Matta et al., 2011, 2008; López et al., 2010).

In addition to liquid scintillation analysis, microplate scintillation analyzers are used for the measurement of beta-emitting radionuclides by means of solid scintillation. Solid scintillation counting in microplates involves depositing a sample into the well of a special plastic microplate (*e.g.*, LumaPlate, Scintiplate, Cytostar-T), which contains a solid scintillator at the bottom of the well. The material is dried and then counted in a microplate scintillation counter. The application of scintillating microplates is discussed in more detail in Chapter 9 of this volume.

C. Advantages and disadvantages

The microplate scintillation and luminescence counter has the basic advantage of high sample throughput with up to 12 samples, which can be analyzed simultaneously for radioactivity or luminescence. Hands-off fully automated analysis of thousands of samples is possible, particularly in the simplest case of a microplate analyzer with 32 microplates stacked for counting, each microplate containing 96 samples wells, which provides for 3072 unattended sample analyses. Microplates containing 384 sample wells each with 80 μL capacity can be stacked up to 40 microplates at one time, which provides for over 15,000 samples with unattended analysis. The cost reduction in microplate counting compared with conventional liquid scintillation counting due to the reduced cocktail consumption and consequent reduction in radioactive waste expense and counting vial cost is a definite advantage. Microplates offer also a format that is easy to handle, leaving less chance for error caused by placing samples out of order, as sample wells cannot be mixed up or intermixed. Once samples are prepared on a microplate, they are kept in place on the microplate as a fingerprint. Microplates are easily portable and stored without occupying much space.

A disadvantage of the microplate scintillation analyzer is the small volumes of sample that can be accommodated in the sample wells. A 20-mL liquid scintillation vial can accept approximately 10 mL of aqueous sample mixed with 10 mL of suitable liquid scintillation cocktail. However, the largest sample well of a 24-well microplate has a capacity of only 1.5 mL (1500 μL), of which approximately 0.75 mL at most can be aqueous sample and 0.75 mL suitable cocktail. That is a 13-fold difference in sample size between the two methods of analysis. Therefore, for relatively low sample activities in microplate scintillation counting longer counting times are required.

XIV. PERALS, LS alpha-spectrometry with LAAPDs, and MNPs

Photon-Electron Rejecting Alpha Liquid Scintillation (PERALS) spectrometry was developed to provide an instrument that could measure α-particle radiation with good resolution, high counting efficiency, and low background.

A. PERALS spectrometry

The interaction of alpha particles with liquid scintillation cocktail was described earlier in Section II.B of this chapter. Of particular interest is that alpha particles are generally detected at a very high counting efficiency of about 100% even with high degrees of quench in the scintillation cocktail. However, alpha particles produce only about one-tenth the light intensity (*i.e.* scintillation light output) as beta particles or gamma radiation per unit of radiation energy. A 5-MeV alpha particle will produce a pulse height spectral peak at approximately 500 keV when liquid scintillation pulse heights are calibrated on a scale equivalent to particle energy in keV. Such pulse height energy scales are calibrated with beta particles or electrons, as electrons yield a linear light output per particle energy dissipated in the liquid scintillator. This posed a problem in the liquid scintillation analysis of α-emitters when found in the presence of β-emitters, because their pulse height spectra would overlap even though the energy of the α-particle radiation was 10 times that of the β-particle radiation. For example, the overlapping liquid scintillation pulse height spectra of the α-emitter ^{210}Po ($E = 5.30$ MeV) and the β-emitter ^{90}Sr ($E_{max} = 546$ keV) are illustrated earlier in Fig. 6.2 in Section II.B of this chapter. Therefore, since the early 1970s research has gone into the development of liquid scintillation spectrometers that could reject the pulses produced by β particles and Compton electrons from γ radiation and provide an instrument that could measure α-particle radiation with good resolution, high counting efficiency, and low background. This led to the development of Photon Electron Rejecting Alpha Liquid Scintillation (PERALS) spectrometry. The history of the work that led to the development of PERALS spectrometry, is described by McDowell and McDowell (1991). A description of the instrumentation, its performance, and applications will be provided here.

The development of PERALS spectrometry had to solve two problems of α-particle liquid scintillation spectrometry: (1) the poor energy resolutions obtained with α-emitters in quenching aqueous-accepting scintillation cocktails in conventional liquid scintillation analyzers and (2) the interference caused by overlapping spectra from β- and γ-emitters in the same sample. The problem of resolution of α pulse height spectra was solved by modifying the following elements of the liquid scintillation system as described by McDowell and McDowell (1993): (1) the α-emitters of interest had to be placed into a nonquenching, organophilic complex in a completely organic and highly efficient scintillator; (2) the detector assembly had to be modified with an efficient reflector arrangement and light-coupling optics to transmit the maximum amount of scintillation light to the photocathode of the PMT, and (3) a diffuse reflector had to be used in the detector assembly so that the scintillation light from an α event occurring anywhere in the sample would appear the same to the PMT, because of the nonuniform response of PMTs.

The basic design of the high-resolution α liquid scintillation spectrometer, known as the PERALS spectrometer, is illustrated in Fig. 6.120. The α-emitters are firstly placed into an organic "extractive scintillator" often via a solvent extraction process that enables the radionuclides to pass from the aqueous sample phase into the organic scintillator phase. A culture tube measuring 10 × 75 mm containing 1 mL of the radionuclide sample in the extractive scintillator is placed in the detecting chamber of the PERALS spectrometer (Fig. 6.120). The instrument contains only one photomultiplier tube. Dual photomultiplier tubes are not needed, in contrast to conventional liquid scintillation analyzers, because the thermal electron noise in the phototubes is always well below the pulse heights produced by α-particles (McKlveen, and McDowell, 1984). Light-coupling silicone oil and a white reflecting surface made of barium sulfate and binder are important to meet the light transfer requirements from the sample tube to the

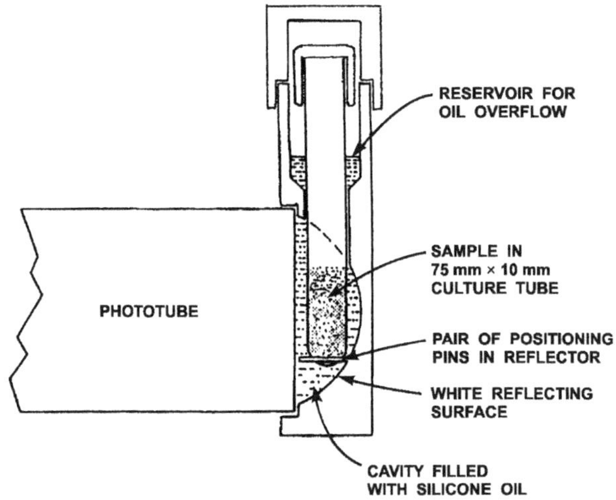

FIGURE 6.120 Cross section of a PERALS spectrometer detector. *From McDowell and McDowell (1993), Reprinted with permission from Radiocarbon, University of Arizona, © 1993 Arizona Board of Regents on behalf of the University of Arizona).*

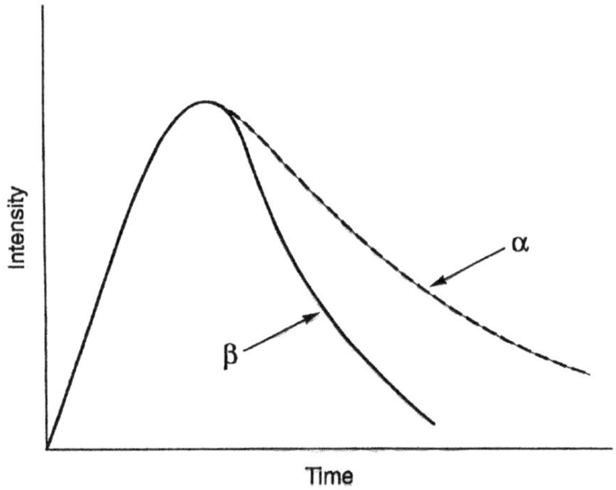

FIGURE 6.121 Alpha and beta pulse decay in a liquid scintillator. The decay time of light produced by the α-particle in certain scintillators is about 35–40 ns longer than that of the faster decaying light produced by the β-particle. *© 1998 PerkinElmer, Inc. Printed with permission.*

photocathode for optimum resolutions of α-particle pulse height spectra.

To provide for clean α-particle pulse height spectra without the interference from β-particle smear, the PERALS spectrometer discriminates the α-particle from β-particle events in the liquid scintillator. The instrument makes use of pulse shape discrimination (PSD), which can take advantage of the longer decay time of light produced by α-particle interactions in scintillation fluor and reject the faster decaying light produced by β-particle or γ-ray interactions. Fig. 6.121 illustrates the light pulse shapes of alpha and beta events in a liquid scintillator. Depending on the scintillator used, the decay times of α-particle interactions can be 30–40 ns longer than β-particle interactions. As explained by Cadieux (1990), each input pulse to the system generates an amplified voltage pulse for pulse height analysis and a pulse shape signal with a voltage proportional to the time length of the incoming pulse. A discriminator generates a gating signal for only the alpha particles after analyzing the pulse shape voltages. The proper discriminator gating signal is set by use of a microcurie γ-ray source outside of the PERALS sample chamber, which generates Compton electrons that simulate β-particle interactions inside the scintillation fluor cocktail, whereby the discriminator can be adjusted to reject pulse events from the Compton electrons. Following pulse shape discrimination, a multichannel analyzer collects and analyzes pulse height spectra from the α-particle interactions.

Technical support, application notes, and procedures in the use of the PERALS spectrometer and extractive scintillators and solvents are available from ORDELA, Inc, Oak Ridge, TN, USA. The company name ORDELA is derived from Oak Ridge DEtector LAboratory (http://www.amale.com/ORDELA/perals/Default.htm, accessed March 9, 2018).

B. Extractive scintillators and solvents for Alpha LS spectrometry

The extractive scintillator or solvent contains an organic extractant (extractive molecule) capable of forming selective complexes with certain alpha-emitters to facilitate their removal from aqueous solutions and transfer into the organic scintillator phase. The organic extractants chosen must be pure, exhibit a minimum of quenching, and be high in the extractive power for the target element(s) as described by McDowell and McDowell (1993) and McDowell (1996). Popular extractive scintillators consist of a combination of scintillator (*e.g.*, PPBO), solvent (*e.g.*, toluene), and an extractant. The extractant is available commercially and may consist of one of the following: di(2-ethyl-hexyl)-phosphoric acid (HDEHP), 1-nonyldecylamine sulfate, tri-*n*-octylamine sulfate (TNOA), tri-octyl-phosphine oxide (TOPO), or similar compound that will serve as a ligand for the extraction of the alpha-emitter from aqueous solution. The extractants are available commercially in pure form or in toluene solution, which provides users the option to select the fluor to add for the preparation of a liquid scintillation cocktail.

Some commercial extractive scintillators used for PERALS spectrometry are produced by ETRAC Laboratories, Inc., Oak Ridge, TN, USA, among which are ALPHAEX (actinide extracting scintillator), THOREX (thorium extracting scintillator), URAEX (uranium extracting scintillator), POLEX (polonium extracting scintillator), RADAEX (radium extracting scintillator), and RADONS (radon extracting scintillator). Extractive

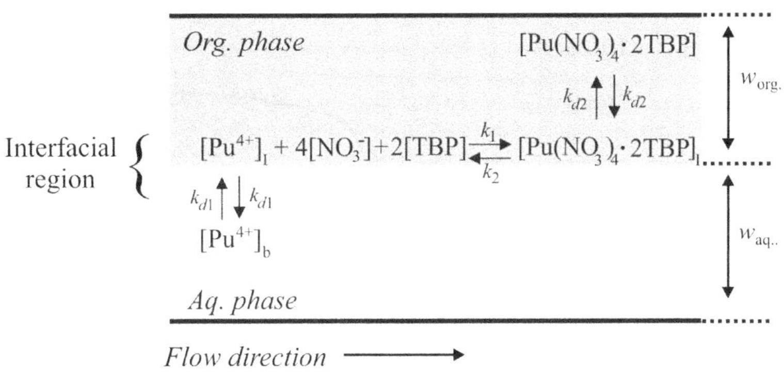

FIGURE 6.122 The extraction model of two-phase flow in the microchannel. The Pu(IV) in the aqueous phase diffuses to the interfacial region. The reaction of Pu(IV) with TBP proceeds at the interface between the two phases and the Pu(IV)-TBP complex is formed. This complex is extracted into the organic phase and diffuses into the organic phase. *From Yamamoto et al. (2015); reprinted with permission from John Wiley and Sons © 2015.*

scintillators generally contain an organic extractant, such as di(2-ethyl-hexyl)-phosphoric acid (HDEHP), 1-nonyldecylamine sulfate, tri-*n*-octylamine sulfate (TNOA), or tri-octyl-phosphine oxide (TOPO).

The extractants serve as complexing agents or ligands that can remove certain radionuclides from aqueous solution (aqueous phase) into an organic phase prior to liquid scintillation alpha spectrometry of the radionuclide. Yamamoto et al. (2015) developed a glass chip microchannel for the measurement of the rates of extraction of alpha-emitting radionuclides from aqueous solution, which provides an excellent means of studying the mechanisms and efficiency or radionuclide extraction. Fig. 6.122 illustrates the method utilized by Yamamoto et al. (2015) to measure the rates of extraction of Pu^{4+} from 3M HNO_3 solution into an organic phase of 30% tri-*n*-butylphosphate (TBP) extractant in *n*-dodecane solvent. The method involves feeding solutions of the organic extractant phase and aqueous Pu^{4+} acid phase simultaneously through a channel of a microchip device measuring 40 μm in width and up to 200 mm length. The two solutions were made to pass through the channel simultaneously at a rate of 5–20 μL/min. The two phases do not mix and travel simultaneously through the channel of 40 μm width, whereby the width of each channel (i.e., the organic extractant phase, $w_{org.}$) and the aqueous Pu^{4+} phase ($w_{aq.}$, See Fig. 6.122) is 20 μm. Upon exiting the channel, each phase is analyzed for Pu^{4+} by PERALS spectrometry, and the extraction rate and optimum contact time between the two phases can be calculated. The technique is useful in the measurement of Pu^{4+} extraction rates and could be applied to other alpha-emitters and organic extractants.

C. Extractive magnetic nanoparticles (MNPs) for Alpha LS spectrometry

O'Hara et al. (2011) demonstrated the potential of magnetic nanoparticles for the extraction of alpha-emitting radionuclides, namely Po, Ra, U, and Am, from chemically unmodified and pH-2 human urine samples. This work demonstrated the use of novel sorbent materials, specifically magnetic nanoparticles: Fe_3O_4 and a composite manganese iron oxide (Mn–Fe–O), which would adhere to alpha-emitting radionuclides to permit the extraction of the radionuclides from urine by collecting the magnetic field induced nanoparticles with a magnet followed by a high-throughput alpha counting of the radionuclides. The method demonstrated a rapid screening method for alpha-emitters in human urine by alpha spectrometry. A subsequent study by O'Hara et al. (2016) demonstrated the ability of the nanoscale amphoteric metal oxides with magnetic attributes, namely commercially available magnetite (Fe_3O_4) and Mn-doped magnetite to extract quantitatively the alpha-emitting radionuclides from aqueous solution over a wide range of pH, as illustrated in Fig. 6.123. This demonstrated the potential of using magnetic nanoparticles for the extraction of alpha-emitting radionuclides from natural waters for their subsequent radionuclide analysis.

The above research led to the development by O'Hara and Addleman (2017) of a new approach to the analysis of the extraction and liquid scintillation analysis of alpha-emitting radionuclides in natural waters. The method involves the utilization of magnetic nanoparticles (MNPs) for the extraction of alpha-emitting radionuclides from aqueous solution, followed by the collection of the MNPs and liquid scintillation analysis. The supermagnetic Fe_3O_4 particles utilized were 27 ± 8 nm in size providing a surface area of 42 m^2/g and a specific magnetization of 81 emu/g. The objective of developing a method for extraction for alpha-emitting radionuclides with magnetic nanoparticles is the finding of a fast method with minimal labor and a reduction in the use of toxic and corrosive reagents. O'Hara and Addleman (2017) demonstrate the efficient extraction of ^{241}Am and ^{210}Po in natural waters with the MNPs with improved minimal detectable activities. They demonstrate that the MNP suspensions can be added directly to liquid scintillation cocktail for LS analysis or dissolved in acid forming a faint colored ferric chloride solution, which is then added to the liquid scintillation cocktail.

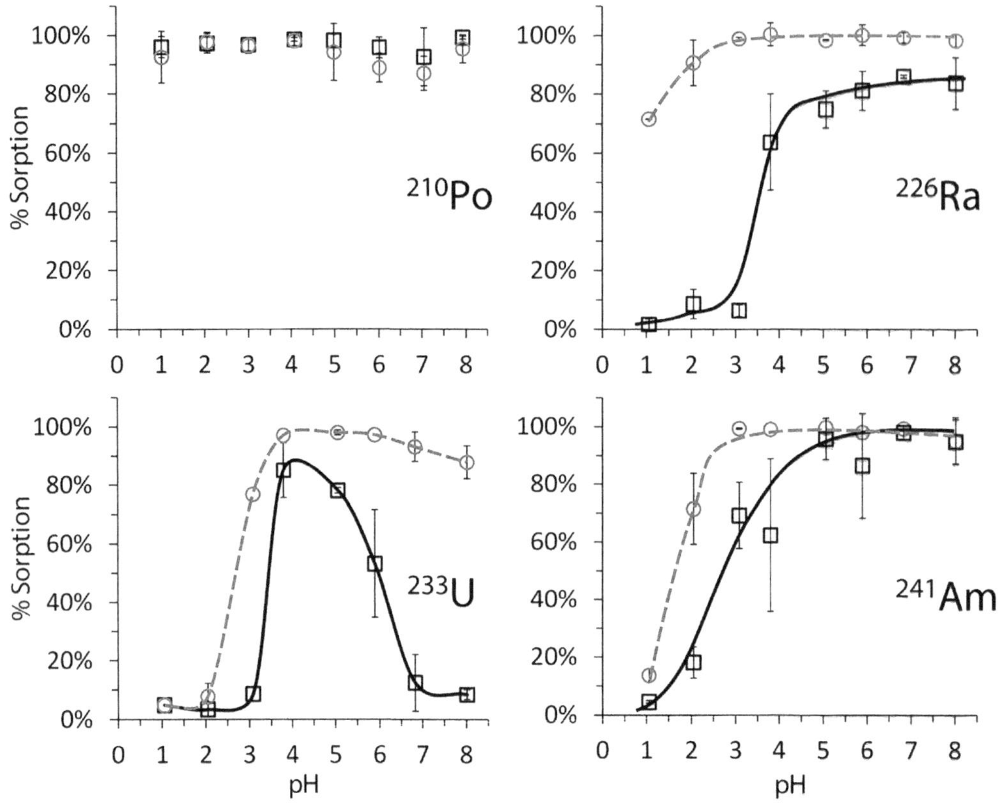

FIGURE 6.123 Sorption of α-emitting radionuclides onto Fe_3O_4 (□) and Mn-doped Fe_3O_4 (O) NPs from Hanford ground water that was pH adjusted between ~1 and the natural pH of ~8. Contact time was 4 h. *From O'Hara et al. (2016); reprinted with permission from The Royal Society of Chemistry © 2016.*

The procedure for extracting ^{241}Am and ^{210}Po from aqueous solutions is outlined in Table 6.20 and illustrated in Fig. 6.124.

The methodology developed is groundbreaking, and the writer foresees the possibility that this methodology can be extended to PERALS spectrometry by the following sequence: (1) alpha-emitting radionuclides are extracted from natural waters by MNPs; (2) the MNPs are then dissolved in acid; (3) the alpha-emitting radionuclides are extracted from the aqueous dissolved MNP-acid solution by organic extractive scintillators; and (4) the organic extractive scintillator is then analyzed by PERALS spectrometry.

D. Applications of PERALS spectrometry

Applications of extractive scintillators to PERALS alpha spectrometry of Pu isotopes are reviewed by Vajda and Kim (2010) and in Chapter 4, Volume 1 on Alpha Spectrometry. PERALS spectrometry has proven to be a valuable tool in the measurement of ^{226}Ra and ^{222}Rn in drinking water with detection limits of 0.006 Bq/L of ^{226}Ra using only 6 mL of sample (Aupiais et al., 1998) and 0.68 Bq/L of ^{222}Rn with the extraction of 1 L of water (Hamanaka et al., 1998). A fast sample preparation method for PERALS analysis of radium in water samples is reported by Aupiais (2005) yielding detection limits of 0.04 Bq/L of ^{226}Ra for 5 mL of sample and 240,000 s counting times. An example of the α-particle discrimination achievable by PERALS with low-activity water samples is provided by Fig. 6.125 illustrating the pulse height spectra of ^{226}Ra, ^{222}Rn, and ^{218}Po with α-particle emissions at 4.78, 5.49, and 6.00 MeV respectively. PERALS spectrometry has demonstrated to be a practical method for the analysis of the actinides with high resolution and at low levels of activity. For example, the following studies can be cited: the analysis of ^{232}U–^{234}U–^{238}U at resolutions of 266, 233, and 200 keV FWHM reported by Dacheux et al. (2000), limits of detection as low as 1 mBq/L with extraction of uranium, thorium, plutonium, americium, and curium in aqueous solutions of up to 250 mL (Dacheux and Aupiais, 1997, 1998). Aupiais (2004a,b) provides alpha spectra deconvolution methods for PERALS analysis of actinide elements in water including the rapid PERALS analysis of uranium activities in water with detection limits of 0.2 μg/kg or 0.003 Bq/kg with a counting time of 240,000 s. Measurements of ^{237}Np by PERALS spectrometry at levels as low as 23.7 and 9.5 pg/L after counting times of 3 and

TABLE 6.20 An outline of the MNP-based analyte collection and analysis method for liquid scintillation analysis of alpha-emitting radionuclides, presented as a supplement to Fig. 6.124. As outlined in the table and illustrated in Fig. 6.124, the LS analysis may be performed by counting the MNP as a suspension or dissolved in solution.

	Step	Description	Footnote
Analyte concentration	A	Obtain sample aliquot	
	B	Analyte sorption onto MNPs	a
	C	Magnetic collection of analyte-bound MNPs	b
LS analysis: MNP suspension	D_1	Decant sample, leaving small volume behind	c
	E_1	Prepare MNP concentrate by resuspending MNPs in solution	d
	F_1	Aliquot MNP concentrate into LS cocktail; perform LS analysis	e
LS analysis: Dissolved MNPs	D_2	Decant sample	
	E_2	Dissolve MNPs in acid	f
	F_2	Aliquot dissolved MNP concentrate into LS cocktail; perform LS analysis	

[a]*MNPs added as a fully suspended concentrate in DI water.*
[b]*Placement location of magnet(s) for efficient MNP collection will depend on sample container dimensions.*
[c]*Alternatively, decant all sample volume and add known volume of water or dilute acid.*
[d]*Requires sonication and vigorous shaking to disrupt particle aggregation and formed during magnetic collection.*
[e]*Abide by MNP/cocktail suspension limits outlines by O'Hara and Addleman (2017).*
[f]*Add a few drops of strong HCl to dissolve MNPs; evaporate to incipient dryness; bring to known volume in water or dilute acid.*
From O'Hara and Addleman (2017); reprinted with permission from The Royal Society of Chemistry © 2017.

10 days, respectively, were demonstrated by Aupiais et al. (1999). Selective extraction of ^{210}Po for PERALS spectrometry by Véronneau et al. (2000) and Aleissa et al. (2006) yielded detection limits bellow 1 mBq/L for 200 mL of solution and 1000 min counting times, and 0.5 L samples and 180 min counting times, respectively. Because of the relatively high spectral resolutions achieved by PERALS spectrometry, the normal Gaussian shape of alpha peaks sometimes display a high-energy tail in addition to the pure Gaussian function. Aupiais and Dacheux (2000) demonstrated that this asymmetry is due to internal conversion. They conclude that for the accurate PERALS analysis of some radionuclides, such as isotopes of actinide elements, it is necessary to account for L- and M-shell internal conversion contributions in the activity

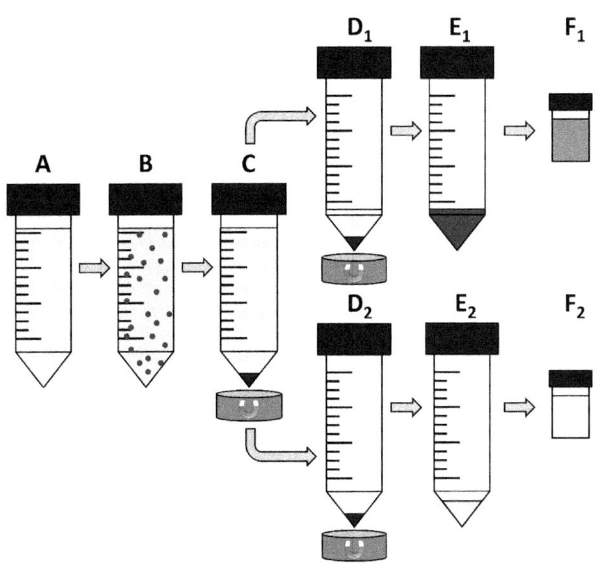

FIGURE 6.124 MNP-based analyte collection and analysis method flow diagram. See Table 6.20 for a description of the steps. *From O'Hara and Addleman (2017); reprinted with permission from The Royal Society of Chemistry © 2017.*

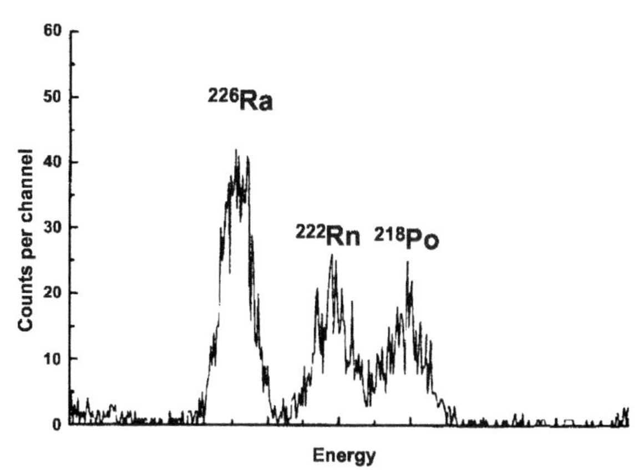

FIGURE 6.125 Radium spectrum of La Bourbouls-Choussy thermal water (3000 ± 200 mBq activity and 88,650 s counting time. *From Aupiais et al. (1998); reprinted with permission © 1998 American Chemical Society.*

measurements. In the spectral deconvolution of actinide elements, Aupiais (2004a) demonstrated the need to account for contributions due to internal conversion.

The PERALS spectrometer can be considered one of the most sensitive liquid scintillation α-spectrometric methods available. As reported by McDowell and McDowell (1993) the PERALS spectrometer rejects 99.99% of ambient and sample β and γ counts from the α spectrum providing ambient radiation backgrounds of 0.001 CPM or less. An alpha pulse height energy resolution of 4.2% FWHM is achievable, which is limited by the quality of contemporary PMTs. The lower limits of detection are estimated at 0.17 mBq (0.01 DPM). The α counting efficiency is 99.7%, as a small 0.3% is lost through α-particle collisions with the sample tube. The use of PERALS instruments in mobile laboratories to diagnose radionuclide contamination for radiological or nuclear events was proposed by Castagnet et al. (2007). A PERALS spectrometer was successfully linked with high purity germanium and bismuth germanate gamma-ray photon detectors by Cadieux et al. (2015) to provide an alpha-gamma coincidence spectrometer for the rapid measurement of actinides in the presence of high beta-gamma fields.

The major disadvantage of PERALS spectrometry is that it is specific for the spectrometric analysis of α-emitters only, that is, this instrument or method cannot be applied to determine α and β activities simultaneously as reviewed by Pates et al. (1996a). Other conventional liquid scintillation analyzers equipped with automatic sample changers are available commercially that provide α/β separation by pulse decay analysis or pulse shape analysis and spectrometric analysis of both α- and β-emitters in the same sample, albeit at a lower α-particle energy resolution than obtained by PERALS spectrometry. However, much improved alpha spectral energy resolutions are possible with conventional LSAs when certain detergent-free and moisture-free extraction cocktails and short optical path (small vials) are used (Aupiais et al., 2003). The simultaneous liquid scintillation measurement of α- and β-emitters in the same sample by conventional liquid scintillation analysis is described in the following Section XV of the chapter.

E. LS alpha-spectrometry with LAAPDs

An improvement in liquid scintillation alpha-spectrometry with energy resolutions of 5% for ^{232}Th (200 keV FWHM) and 3.9% for ^{216}Po (260 keV FWHM) was reported by Reboli et al. (2005) by the use of a silicon Large Area Avalanche Photodiode (LAAPD) in lieu of the standard photomultiplier tube.

Reboli et al. (2005) demonstrate that the LAAPD can be used for alpha liquid scintillation spectrometry and yield better energy resolutions than photomultiplier tubes. The improved energy resolution obtained with the photodiode is reported due to the uniformity of the large spectral response of the APD. The experimental arrangement used by Reboli et al. (2005) for the measurement of the liquid scintillation alpha spectra with the LAAPD is illustrated in Fig. 6.126. Actinides are counted in scintillation cocktails containing fluors in diisopropylnaphthalene isomers (DIN) solvent containing HDEHP (di-(2-ethylhexyl)phosphoric acid) actinide extractant contained within a 10 mm diameter cylindrical fused silica cuvette (see Fig. 6.126). The cuvette containing 400 μL of sample in liquid scintillator is optically coupled to a windowless avalanche photodiode (APD) with silicone fluid or immersion oil. The cuvette is wrapped with Teflon tape with the exception of the face coupled to the APD, which is etched externally to provide uniformity of light collection on the photosensitive surface of the APD. Optimum resolutions of alpha spectra are obtained by cooling of the sample and APD down to a constant temperature, as low as −40°C, and a small cuvette length of 5 mm.

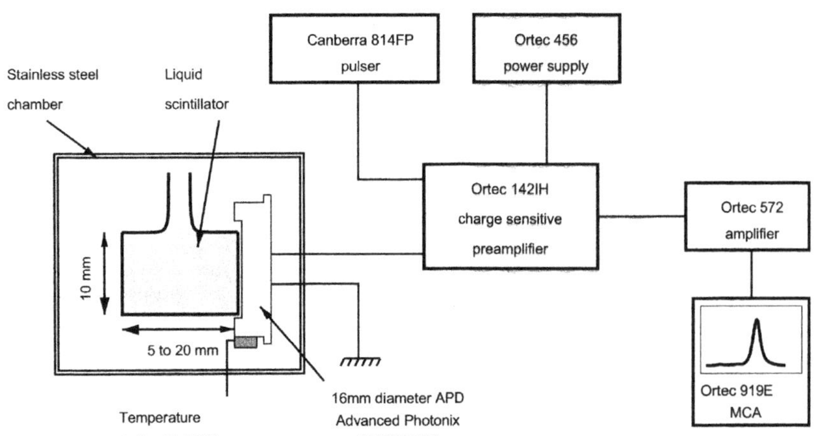

FIGURE 6.126 Experimental arrangement for the determination of liquid scintillation alpha spectra with an avalanche photodiode (APD). *From Reboli et al. (2005); reprinted with permission from Elsevier © 2005.*

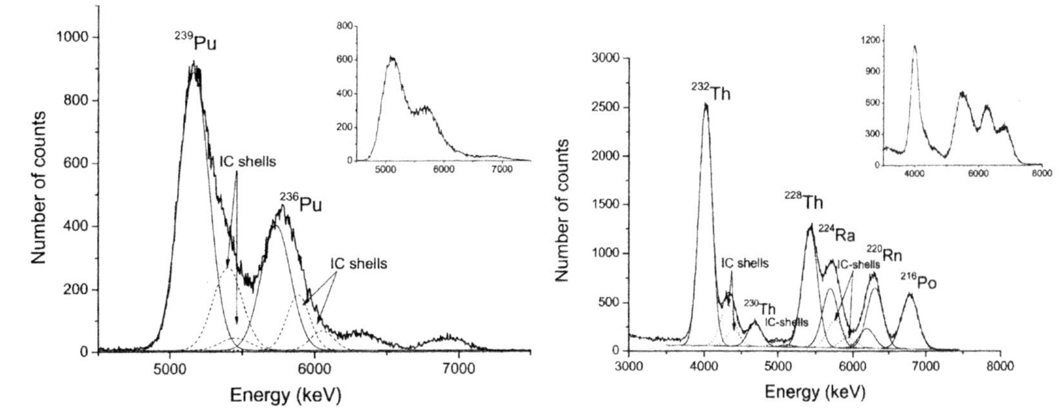

FIGURE 6.127 (Left) Pulse height spectra of ^{239}Pu and ^{236}Pu at −40°C obtained with the liquid scintillation-APD detector. The resolution is equal to ~240 keV for the two isotopes. The ^{236}Pu and ^{239}Pu activities are 4 and 8 Bq, respectively, and the counting time is 17 h. The plot in the upper right shows the ^{239}Pu and ^{236}Pu spectrum measured with the PERALS spectrometer. (Right) Pulse height spectra of ^{232}Th and daughters at −40°C obtained with the liquid scintillation-APD. The resolution is equal to 200 keV for ^{232}Th and 262 keV for ^{216}Po. The ^{232}Th activity is 2 Bq, and the counting time is 6 h. The straight lines represent the main α-particle spectra, and the dashed lines represent the pile-up of the main α-particle + IC shell peaks. The plot in the upper right shows the ^{232}Th and daughters spectrum measured with the PERALS spectrometer. *From Reboli et al. (2005); reprinted with permission from Elsevier © 2005.*

Typical liquid scintillation alpha spectra with the APD detector compared to spectra obtained by PERALS spectrometry are illustrated in Fig. 6.127. Improvement in energy resolution over PERALS is clearly discernible from the spectra. As described by Reboli et al. (2005), the liquid scintillation-APD spectra show a fine structure. The alpha-decay of the plutonium isotopes (^{239}Pu and ^{236}Pu) and the thorium isotopes (^{232}Th, ^{230}Th, and ^{228}Th) do not lead to the ground state and undergo further deexcitation via internal conversion (IC) mainly on L and M shells. The liquid scintillation-APD spectra of Fig. 6.127 offer sufficient energy resolution to illustrate the spectral displacement of the energy peaks of the α-transitions toward higher energy values.

The precise differences in energy resolution between spectra produced by the liquid scintillation-APD and PERALS spectrometers are provided in Table 6.21. The improvement in energy resolution of the liquid scintillation-APD detector is reported by Reboli et al. (2005) to be due to higher quantum efficiency and more uniform active area than that provided by photomultiplier tubes to this date. The silicon photodiodes have a larger spectral response than PMTs, which facilitates the matching of the fluorescence spectra of scintillators with the APD spectral response.

XV. Simultaneous α/β analysis

As described in Sections II.B and XIV of this chapter, α and β pulse height spectra often overlap. The pulse height spectra of the α- and β-emitters ^{210}Po + $^{90}Sr(^{90}Y)$, respectively, illustrated in Fig. 6.2, are a good example of this spectral overlap. In spite of the energies of α and β

TABLE 6.21 Comparison of Energy Resolutions (*R*) for Different α-emitters as Obtained using a Home-Made Liquid Scintillator with an APD as Detector and a Commercial Spectrometer PERALS Equipped with a PMT[a].

Isotope	Energy (keV)	APD	PMT
		R (keV)	*R* (keV)
^{232}Th	4010	200 ± 13	210 ± 13
^{239}Pu	5157	242 ± 13	256 ± 14
^{228}Th	5423	235 ± 11	306 ± 14
^{236}Pu	5768	243 ± 11	293 ± 17
^{216}Po	6785	262 ± 14	448 ± 20

[a]*From Reboli et al. (2005); reprinted with permission from Elsevier © 2005.*

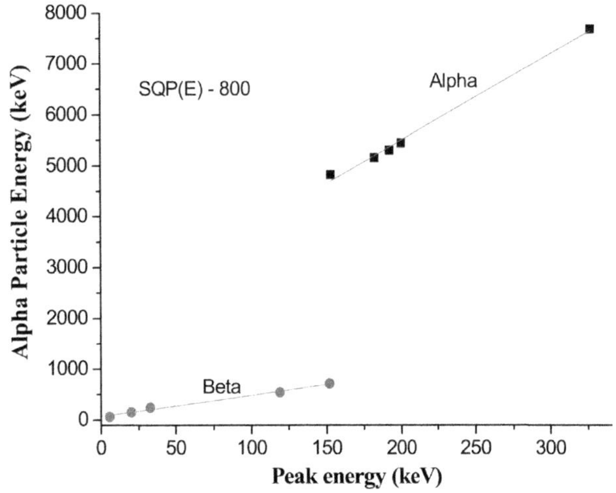

FIGURE 6.128 Alpha and beta particle energies as a function of spectral peak energy. *From Bhade et al. (2017), reprinted with permission from Elsevier © 2017.*

particles differ in one order of magnitude, the energy finally detected by the PMTs is quite similar since α particles have a lower rate of scintillation solvent excitation. This is well illustrated in Fig. 6.128 from Bhade et al. (2017) in which the relation between particle energy and the energy of the peak detected (i.e., the average energy of the spectrum in a Quantulus 1220 detector) for alpha- and beta-emitting radionuclides is shown.

Since α and β discrimination on the basis of the pulse height is not possible, many contemporary liquid scintillation analyzers are equipped with the circuitry capable of analyzing α- and β-emitting radionuclides in the same sample.

A. Detectors

Discrimination between α and β decay events is done by the instruments through pulse decay analysis (PDA) or pulse shape analysis (PSA), first reported by Buchtela et al. (1974) and Thorngate et al. (1974), to differentiate alpha from beta decay events. In pulse decay analysis, a pulse decay time discriminator often referred to as the pulse decay discriminator (PDD), monitors the length of time for the decay of a pulse event originating from photon emission due to fluor cocktail excitation. As described in Section XIV and illustrated in Fig. 6.121, fluor excitation events originating from α-particle interactions have 35–40 ns longer decay lifetimes than events originating from β-particle interactions. This is due to the longer deexcitation and light emission processes in scintillation fluors after α-particle interactions, which exhibit higher linear energy transfer (LET). In PDA, the differences in the lengths of pulse decay events resulting from α- and β-particle interactions can be measured by the charge differences collected at the photomultiplier output at the tail portion of the pulse events. (See also Sections X.B.1 and X.B.2 on the similarity with pulse shape discrimination used to discriminate n/γ interactions with liquid scintillators.) The fluorescence decay resulting from an α-particle interaction, displays a longer tail (Fig. 6.121), and consequently produces a higher charge at the PMT output for the tail portion of the pulse event when compared to a fluorescence decay pulse event produced by a β-particle. The actual difference in decay lifetimes between the α and β events depends on the chemical composition of scintillation fluor cocktail used and particle energies (Pujol and Sanchex-Cabeza, 1997; Pates et al., 1998; Rodriguez Barquero and Grau Carles, 1998).

Instruments that use pulse shape analysis (PSA) to discriminate between α and β events also take advantage of the greater length of the α-produced pulse compared to the β-produced pulse. In PSA, the ratio of the area of the tail of a pulse, beginning at 50 nsec after pulse initiation, is compared to the total pulse area (McDowell, 1966 and Rodriguez Barquero and Grau Carles, 1998), which provides a method of assigning a pulse to a β event (short pulse) or α event (long pulse). This is also referred to as pulse shape discrimination (PSD).

Application of such strategies in commercial liquid scintillation analyzers is combined with data treatment in order to provide the final user with useful data that could lead after optimization of parameters and calibration to the result of alpha and beta decay events in a sample. In Hidex liquid scintillation analyzers (e.g., 300 SL, 600 SL and Triathler) each pulse is classified according to its energy and the pulse length index (PLI). The pulse length index is obtained after the analysis of the pulse after 40 ns of the rise time. As a result, a 2D/3D spectrum is obtained (Eikenberg et al., 2014 and Fig. 6.129). In this case, two parameters can be used for discrimination: PLI-discriminator and energy channels. Therefore, the count rate for alphas is the area above the PLI-discriminator (4 in Fig. 6.129) and the

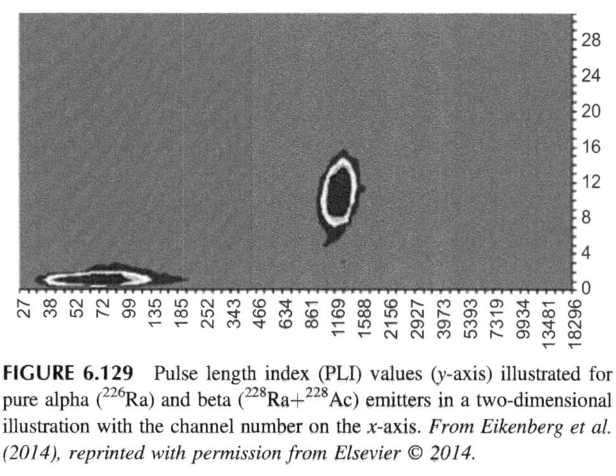

FIGURE 6.129 Pulse length index (PLI) values (*y*-axis) illustrated for pure alpha (^{226}Ra) and beta (^{228}Ra+^{228}Ac) emitters in a two-dimensional illustration with the channel number on the *x*-axis. *From Eikenberg et al. (2014), reprinted with permission from Elsevier © 2014.*

lower and upper energy channels (1000 and 18,269 in Fig. 6.129). The count rate for betas is the area below the PLI-discriminator (4 in Fig. 6.129) and the lower and upper energy channels (1 and 1000 in Fig. 6.129).

In PerkinElmer detectors (e.g., Quantulus and TriCarb series), two spectra are generated. All of the pulse events originating from scintillation photon emissions with a decay time longer than the PDD setting are sent to the α-MCA and those events which have a shorter lifetime are sent to the β-MCA. It is necessary, therefore, to find the optimum PDD setting to get the best separation of α- and β-radionuclide activities into separate MCAs.

B. Establishing the optimum PDD setting

The optimum pulse decay discriminator setting is found by counting a pure α-emitter source and a pure β-emitter source in the liquid scintillation analyzer in the same fluor cocktail, sample composition, sample volume, and type of vial used as the experimental samples to be measured. The optimum PDD setting is affected by sample quench level, the specific quenching agents (chemical and color) in the sample, and the E_{max} of the β-emitter (Pates et al., 1998; DeVol et al., 2007). The β-particle energy is an influencing factor, because the PMT pulse event has a specific length for a given β-particle energy, and the pulse length increases with increasing event energy. Also, as noted by Pates et al. (1998) the delayed component (tail) of pulse events is a function of the amount of π-electron singlet and triplet excitation states in the scintillation solvent produced by the ionizing radiation. Alpha-particles (or heavy charged particles with high LET) will produce more triplet-state excitation than the lighter beta particles. The triplet states take longer to undergo deexcitation fluorescence. However, the solvents as well as quenching agents present in the sample can affect the delayed component of deexcitation. Some components in cocktails can inhibit deexcitation and prolong triplet-state excitation, while other chemicals in the cocktail (*e.g.*, O_2, CCl_4, $CHCl_3$, etc) are electron scavengers, and able therefore, to interact with higher energy states and reduce the delayed component. Moreover, the presence of quenching agents causes a reduction of the number of photons produced and therefore the pulse shape is modified which could affect the final classification of the particle.

When gross α and gross β determinations are needed and the radionuclides in the experimental samples may not be known, one uses an α and β standard of similar energy to that of the α and β radionuclides in the samples (Passo and Cook, 1994) and similar quenching agents to those one might expect in the sample. Different methods for establishing the optimum PSD has been described.

1. Equivalent α and β spillover criteria

This method is the most commonly used and is based on searching the PSD setting in which the spillover of alphas into the beta counting region is equivalent to the spillover of betas into the alpha counting region (Bhade et al., 2010; Zapata et al., 2012; Palomo et al., 2011; Çakal et al., 2015). This PSD setting value usually coincides with the PSD setting with minimum mean spillover. The following calibration steps are required:

1. The pure β-emitter standard dissolved in a suitable liquid scintillation cocktail is placed into the LSA for counting. The chemical composition of this standard should be the same as those of the unknown experimental samples. During this calibration procedure the instrument automatically counts the pure β-emitter standard at various pulse decay discriminator settings and stores the results obtained. In Hidex detectors, pulses are classified according to the its pulse length index and energy in a 3D spectrum, whereas in PerkinElmer detectors, all pulses of duration longer than the PDD setting are sent to the α-multichannel analyzer (α-MCA) and pulses of shorter duration to the β-multichannel analyzer (β-MCA).
2. The pure α-emitter standard dissolved in a suitable cocktail of the same chemistry as the previously described β-emitter (Step 1) is placed into the LSA for counting. Its chemical composition should also be the same as the unknown experimental samples. As in Step 1, the instrument automatically counts the pure α-emitter standard at various pulse decay discriminator settings and stores the results obtained following the same set-up described in Step 1.
3. In PerkinElmer detectors, the percent spillover of the β events into the α-MCA and the percent spillover of the α events into the β-MCA are calculated. A curve of the percent spillover versus the PDD setting as illustrated in Fig. 6.130 is then plotted. In Hidex detectors, the percent spillover of β events into the α-integration region (above PLI) and the percent spillover of the α events into the β-integration region (below PLI) are calculated for one or several α- and β-integration regions (Fig. 6.131).
4. The pulse decay discriminator setting at which the two spillover curves cross is selected as the optimum PDD setting for simultaneous α/β analysis in the same sample (e.g., PDD = 62 of Fig. 6.130 and PLI = 14 of Fig. 6.131). The two spillover curves plotted in this fashion are referred to as crossover plots. At the crossover PDD setting there is minimum spillover of α events into the β-MCA and β events into the α-MCA,

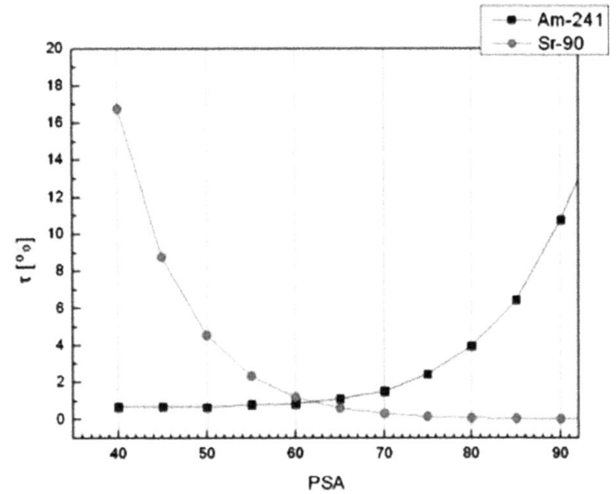

FIGURE 6.130 PSA optimization. *From Todorovíc et al. (2012), reprinted with permission from Elsevier © 2012.*

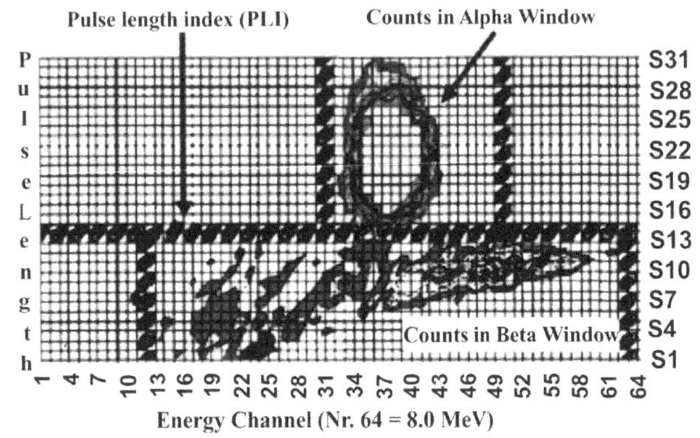

FIGURE 6.131 2-D spectrum of a sample containing natural uranium in a 20 mL vial. *From Wisser et al. (2006), reprinted with permission from Elsevier © 2006.*

when the activities of both α and β events must be determined. However, if only the α-emitter activity is of interest in a sample containing both α- and β-emitters, a higher PDD setting can be selected, well above the crossover setting (e.g., PDD >90 of Fig. 6.130), where the spillover of β events into the α-MCA approaches zero. Alternatively, if only the β-emitter activity is of interest in a sample containing both α- and β-emitters, a lower PDD setting can be selected, well below the crossover setting (e.g., PDD < 40 of Fig. 6.130), where the spillover of α events into the β-MCA is lowest.

2. Inflection point criteria

As described before, the general method for determining the optimum PSA is based on the cross-over plots obtained using an α-emitter and a β-emitter. In the case that a pure α-emitter and a β-emitter is not available, or they are very different from the radionuclides in the real sample, this method cannot be applied without risk of having high errors. For this situation, Feng et al. (2014b) propose a general method to determine the optimum PDD parameter based on the measurement of a sample containing an alpha- and a beta-emitter and determining the PDD parameter value where there is the inflection point of the count rate in the alpha α-MCA. In this point, it is estimated that beta and alpha misclassification are equivalent and compensated. The following calibration steps are required:

1. A solution containing a mixture of an alpha- and a beta-emitter is dissolved in a suitable liquid scintillation cocktail and is placed into the LSA for counting. The chemical composition of this standard should be the same as those of the unknown experimental samples. During this calibration procedure, the instrument automatically counts the sample at various PDD settings and stores the results obtained.
2. A curve of the count rate in the region of interest of the α-MCA versus the PDD setting, as illustrated in Fig. 6.132, is plotted.
3. The data is fitted to the function $y = ax^3 + bx^2 + cx + d$.
4. The inflection point of y (obtained when the second derivative is equal to 0; $x = -b/(3a)$) is the optimum pulse decay discriminator setting (e.g., PDD = 110−130 of Fig. 6.132). At the optimum PDD setting α events spillover is compensated with β events spillover.

3. Plateau criteria

One of the problems usually referred when using the equilibrium spillover criteria is that calibration

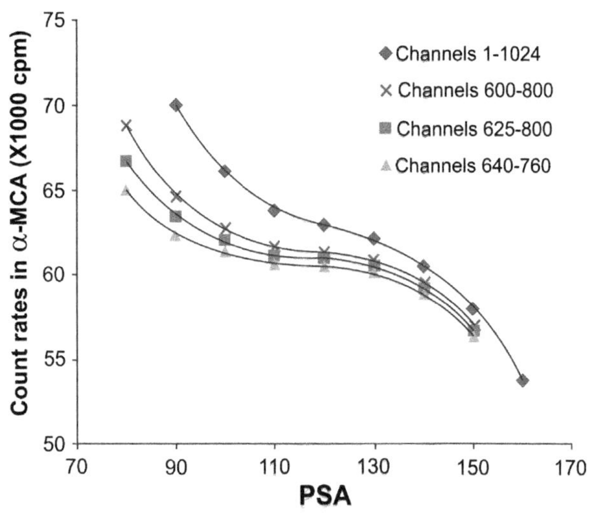

FIGURE 6.132 The effect of ROI on the count rates of Np=Pa at various PSAs. *From Feng et al. (2014b), reprinted with permission from Elsevier © 2014.*

radionuclides do not have the same detection efficiency and/or spillover than the radionuclides on the samples. That usually happens when α and β particles come from ^{226}Ra and daughters. Bhade et al. (2018) propose a method for the analysis radium/radon based on a selection of the optimum PDD parameter as that in which the variability between the detection efficiencies of signals on the α-MCA and on the β-MCA is minimum. In this point, it is estimated that beta and alpha misclassifications are equivalent and compensated. The following calibration steps are required:

1. A ^{226}Ra standard dissolved in a suitable liquid scintillation cocktail is placed into the LSA for counting. The chemical composition of this standard should be the same as that of the unknown experimental samples. During this calibration procedure the instrument automatically counts the ^{226}Ra standard at various pulse decay discriminator settings and stores the results obtained.
2. Region of interest (ROI) is fixed excluding ^{214}Po to avoid errors caused by cancellation of events by the Quantulus guard detector. In the ROI defined the α/β ratio should be 1.5 in a homogeneous liquid scintillator (three alphas ^{226}Ra, ^{222}Rn and ^{218}Po and two betas ^{214}Pb and ^{214}Bi) and one in a homogeneous liquid scintillator (two alphas ^{222}Rn and ^{218}Po and two betas ^{214}Pb and ^{214}Bi).
3. Alpha efficiency and beta efficiency are calculated taking into account all the events into the α-MCA and β-MCA regions of interest respectively. A curve of the efficiencies versus the PDD setting, as illustrated in Fig. 6.133A, is plotted.
4. The optimum pulse decay discriminator setting corresponds to that in which the alpha and beta efficiencies present a plateau (e.g., PDD = 85 of Fig. 6.133A). At the optimum PDD setting α events spillover is compensated with β events spillover and therefore α and β efficiencies can be used. Moreover, it is checked that in this region the α/β ratio has the expected value (Fig. 6.133B).

All of the procedures described above make use of standards without quenching agent added. However, De Vol et al. (2007) and Fons-Castells et al. (2017d) provide a thorough study on the effects of chemical and color quench and radionuclides used to calibrate instrumentation for α/β discrimination. Among these findings, it is important to note that the quenching agent will have a direct effect on total spillover which can change from less than 5% in less quenched samples (i.e., SPQ(E) values between 800 and 850) to 15% or more in highly quenched samples (i.e., SPQ(E) values between 500 and 550) (See Fig. 6.134).

Moreover, both studies show that the effect is dependent of the type of quenching agent. In other words, a liquid scintillation analyzer may determine that two samples in two distinct counting vials have the same level of quench [*i.e.*, same quench indication parameter, tSIE or SQP(E)]; however, if different chemical agents are producing the quench, the spillover can be different between the two samples. Fig. 6.135 illustrates how the α-MCA and β-MCA spectra obtained at equivalent SQP(E) and with the same PDD parameter for chemical and color quenching are very different leading to a different classification. This behavior is also seen in the analysis of a natural uranium sample. In this case, the β/α ratio for color and chemical quenched samples present a very different behavior due to the different misclassification of alpha particles from ^{234}U and ^{238}U (Fig. 6.136). It is important therefore that the matrix of the experimental samples be constant and that the liquid scintillation analyzer be programmed for minimum spillover under the same conditions as the experimental samples.

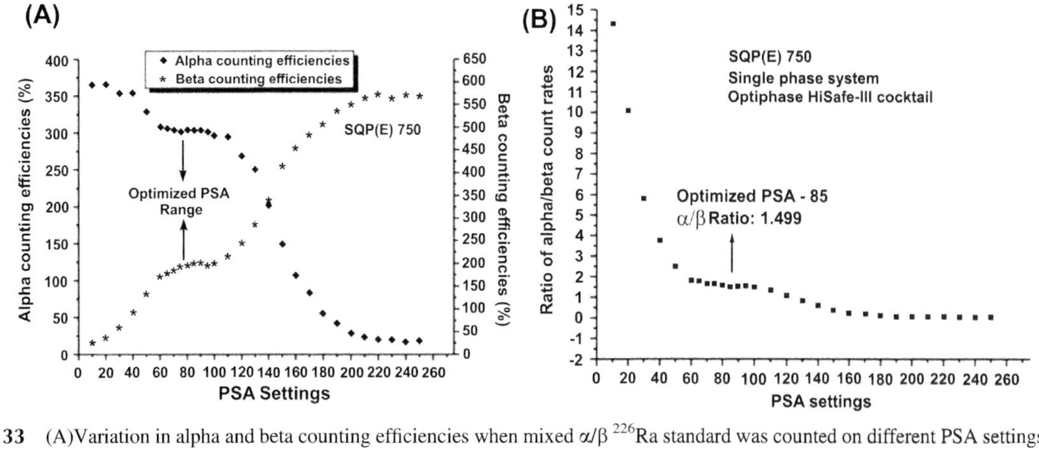

FIGURE 6.133 (A)Variation in alpha and beta counting efficiencies when mixed α/β ^{226}Ra standard was counted on different PSA settings. (B) Ratio of α/β count rates at various PSA levels, as one of the criteria to optimize the PSA level setting. *From Bhade et al. (2018), reprinted with permission from Springer Nature © 2018.*

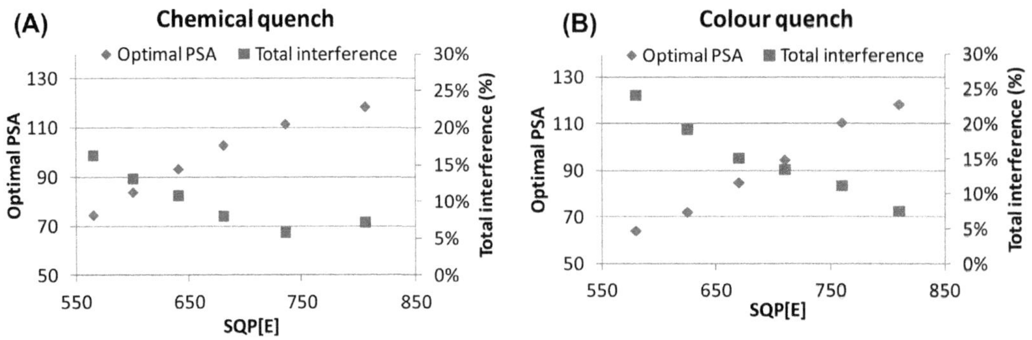

FIGURE 6.134 Optimal PSA and total interference as a function of SQP[E] for chemical (A) and color (B) quenching. *From Fons-Castells et al. (2017d), reprinted with permission from Springer Nature © 2017.*

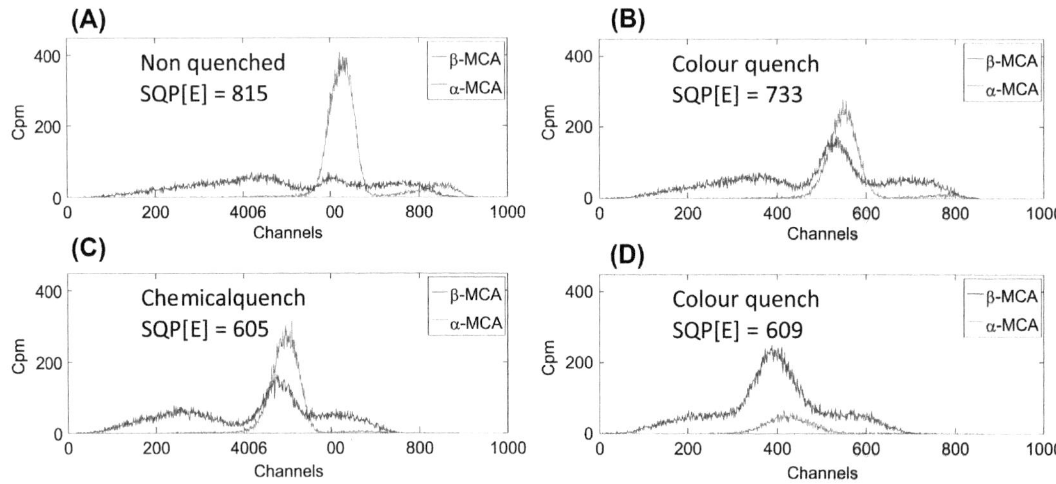

FIGURE 6.135 Spectra in α-MCA and β-MCA of ^{nat}U for (A) nonquenched, (B) moderately color quenched, (C) highly chemical quenched, and (D) highly color quenched. *From Fons-Castells et al. (2017d), reprinted with permission from Springer Nature © 2017.*

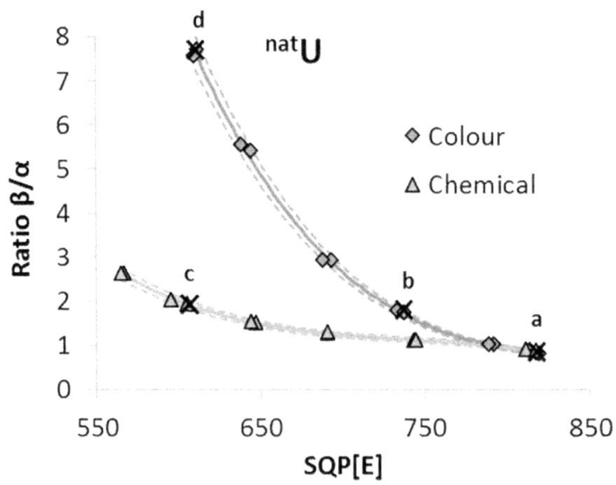

FIGURE 6.136 Ratio between β and α counts as a function of SQP[E] for color and chemical quench in ^{nat}U standards. *From Fons-Castells et al. (2017d), reprinted with permission from Springer Nature © 2017.*

DeVol et al. (2007) also point out that pure alpha-emitters may yield different spillovers than nuclides that emit alpha with gamma cascade events, such as ^{241}Am. Nuclides with gamma cascade events are more likely to be classified by the pulse shape discriminator as alpha events, because of the pulse duration. Gamma cascade events are common with alpha-emitters.

Another important aspect that should be taken into account is the energy of the beta- and alpha-emitters used in the optimum PDD setting. Salonen (2006a) was first to report that alpha spillover is a function also of the energy of the alpha particle. The alpha spillover of lower-energy alpha particles is greater than the spillover of high-energy alpha particles when samples are measured at a constant PDD setting (*i.e.*, constant pulse shape analysis) (See Fig. 6.137). Several combinations of pairs of radionuclides are described in the bibliography for establishing the optimum PDD setting: ^{40}K/^{236}U (Fons-Castells et al., 2017a); ^{90}Sr/^{236}U (Zapata et al., 2012); ^{90}Sr/^{241}Am (Rusconi et al., 2004; Bhade et al., 2010; Çakal et al., 2015; Todorovic et al., 2012; Lin et al., 2012); ^{40}K/^{241}Am (Palomo et al., 2011); ^{36}Cl/^{241}Am (Salonen, 2010); and ^{99}Mo/^{242}Pu (Hou, 2017). In summary, the sample matrix should be as close as possible in chemical and physical properties to the alpha and beta standards used to determine the optimum PDD setting for minimum spillover.

C. α/β spillover corrections and activity calculations

Once the optimum PDD setting is determined, the detection efficiency of the α particles in the α-MCA and that of the β particles in the β-MCA must be measured. This may be best determined by use of internal standardization as described in Section V.A of this chapter. The α-MCA and β-MCA will provide net (background-subtracted) count rates (CPM) for α and β particles, respectively. The net count rates must be corrected for spillover or misclassification and the net count rates converted to α and β activities (disintegration rates). The calculations required for the determination of α and β count rates corrected for spillover are derived by Bhade et al. (2010). The following equations described by Bakir and Bem (1995) provide the sample gross α and β activity concentrations in units of Bq/L taking into account sample volume (*V*) and concentration factor (*C*), for samples concentrated prior to counting:

$$A_\alpha = \frac{I_\beta E_\beta - I_\alpha S_{\beta/\alpha}}{\left(E_\alpha E_\beta - S_{\alpha/\beta} S_{\beta/\alpha}\right) VC60} \tag{6.156}$$

$$A_\beta = \frac{I_\alpha E_\alpha - I_\beta S_{\alpha/\beta}}{\left(E_\alpha E_\beta - S_{\alpha/\beta} S_{\beta/\alpha}\right) VC60} \tag{6.157}$$

where A_α and A_β are the gross α particle and gross β particle activities, respectively, in units of Bq/L, I_α and I_β are the net (background subtracted) count rates in the α and β multichannel analyzers, respectively, $S_{\alpha/\beta}$ and $S_{\beta/\alpha}$are the percent spillover of α events into the β MCA and β events into the α MCA, respectively, $E_\alpha$and E_β are the detection efficiencies for α particles in the α MCA and β particles in the β MCA, respectively, *V* is the volume of analyzed sample in liters, *C* is the concentration factor or the degree of concentration of the sample prior to analysis, and 60 is a conversion factor to change count rate from CPM to CPS.

The backgrounds, count rates, and detection efficiencies in the α MCA and β MCA will depend on the pulse height discriminator settings used in the α and β channels, respectively, and instrumental background-reducing count mode used (*e.g.*, LLCM or NCM) as described in Section XX.C. The optimum PDD setting must be determined at the same pulse height discriminator and background-reducing count mode settings as those used for the experimental sample analysis.

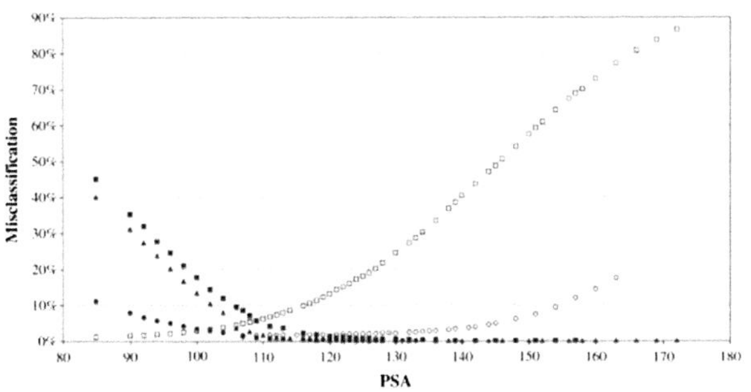

FIGURE 6.137 Alpha and beta misclassification percentages as a function of the PSA value for different alpha- and beta-emitters (● ^{137}Cs, ■ ^{40}K, ▲ ^{90}Y, □ ^{236}U, and ○ ^{241}Am). *From Zapata-García et al. (2012), reprinted with permission from Elsevier © 2012.*

D. Optimizing α/β discrimination in PDA

As described previously, there are several parameters that play a crucial role in α/β discrimination in pulse decay analysis. The following are some guidelines to follow, which can enhance α/β separation performance:

1. Control the chemistry of the sample. The cocktail-sample mixture must be homogeneous and constant for standards and samples. The chemistry of the sample, including the scintillation cocktail and kinds and amounts of quenching agents present, should be the same as those of the α and β standards used to plot the crossover curves for determination of the optimum PDD setting. The sample must be completely solubilized in the liquid scintillation cocktail. Acids are often necessary to keep α-emitter salts in solution in the cocktail. Organic samples and filters are often processed through ashing followed by acid dissolution of salt

TABLE 6.22 Acid Loading Performance of α/β Compatible Liquid Scintillation Cocktails[a].

Cocktail (10 mL)	H_2O (mL)	1 M HCl (mL)	2 M HCl (mL)	1 M HNO_3 (mL)	2 M HNO_3 (mL)	% Misclassification (minimal sample loads)
Ultima Gold	3.50	0.25	0.10	0.70	0.30	0.6
Ultima Gold XR	10.0	2.00	0.90	3.00	1.75	2.2
Ultima Gold AB	10.0	5.50	2.25	3.25	2.25	0.5
Insta-Gel XF + 20% w/v naphthalene	1.5	1.3	1.1	1.4	1.4	0.5

[a]*The loading capacity of the acids are mL acid per 10 mL of cocktail. (© 1998 PerkinElmer, Inc., Printed with permission.)*

residues (Yang and Guo, 1995). Table 6.22 provides some guidelines on the acid loading capacity (mL acid/10 mL cocktail) at room temperature (20–22°C) of certain cocktails. The minimum α/β misclassification (at crossover point from spillover curves) for small sample sizes ($\leq$ 0.5 mL acid) is also provided. Water-soluble paper for smear tests is a sample type that might be used to avoid excessive acids in cocktail (Takiue et al., 1989a), although glass fiber filter, cotton-tipped or polyurethane foam are also demonstrated as a good swipe material for α/β contamination assays (Dai et al., 2012).

2. Minimize the sample quench. The sample volume (e.g., aqueous solution) will dilute the scintillation fluor cocktail and also introduce quenching agents. Moreover, as not all chemical components have the same quenching effect, the optimization of the conditions, such as acid concentration or pH, can lower the quenching effects.
3. Minimum α/β spillover is obtained generally at lowest levels of quench.
4. Use organic acceptor cocktails rather than cocktails that have surfactants or emulsifiers. Table 6.22 lists liquid scintillation cocktails that provide excellent α/β discrimination performance. They all use diisopropylnaphthalene (DIN) as a solvent with the exception of Insta-Gel XF, which has pseudocumene as a solvent. Pseudocumene and the linear alkylbenzenes (LAB) are "fast" solvents, which are less efficient for α/β separation (See Figs. 6.138 and 6.139). Naphthalene is added to Insta-Gel XF in a proportion of 20% w/v to improve α/β separation by acting as an intermediate in the energy transfer process between solvent and fluor increasing the production of π-electron triplet excitation states produced by α-particle interactions (Passo and Cook, 1994). Naphthalene is not required with the Ultima Gold Cocktails, as the DIN solvent serves the same purpose (Thomson, 1991). The DIN solvent-based cocktails, particularly Ultima Gold AB, have been demonstrated to provide excellent α/β separations, while increasing surfactant concentration degrades the cocktail performance for α/β analysis (Pates et al., 1993, 1996b).
5. Use organic extractive scintillators that are selective for the α-emitters of interest. The use of organic (water immiscible) extractive scintillators was described in Section XIV of this chapter. This is recommended when only the α-activity is desired, as after extraction

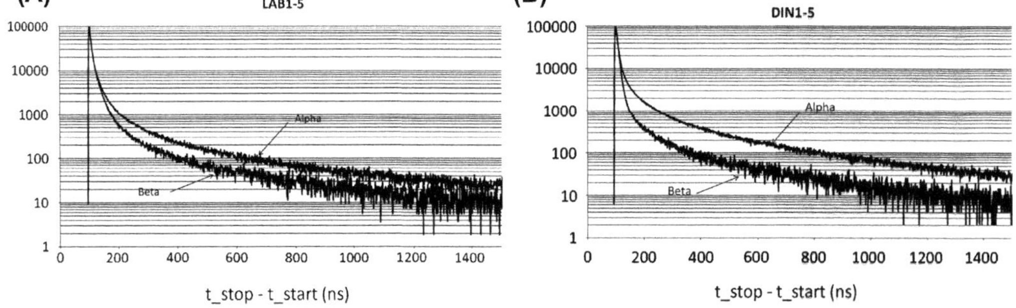

FIGURE 6.138 Raw time histograms of the α and β measurements performed on (A) LAB-based and (B) DIN-based scintillators using the set-up for particle irradiation. The usual display format to normalize the curves to the peak is adopted. *Lombardi et al. (2013), reprinted with permission from Elsevier © 2013.*

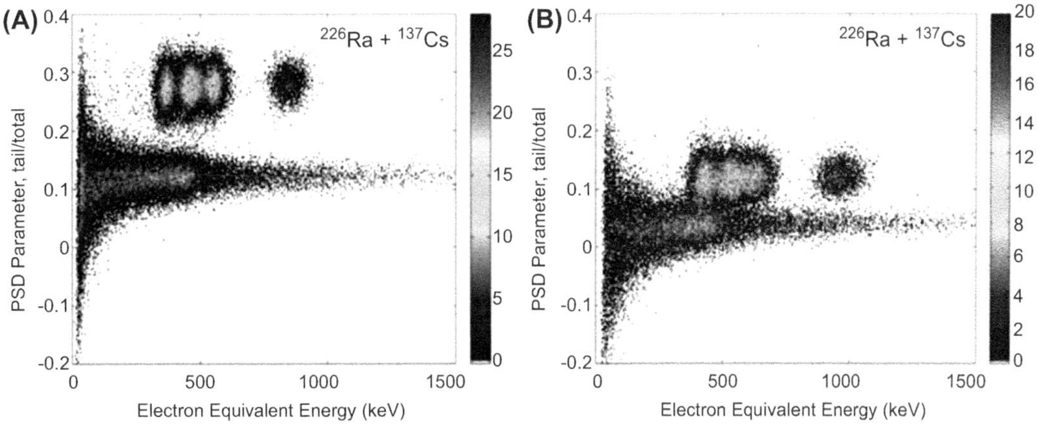

FIGURE 6.139 PSD parameters of commercial cocktails, (A) Ultima-Gold AB and (B) Opti-Fluor, measured with internal ^{226}Ra and external ^{137}Cs sources. *Lee et al. (2012), reprinted with permission from Springer Nature © 2012.*

of the dissolved aqueous sample with the selective organic scintillator, the α-emitters transfer from the aqueous phase to the organic scintillator leaving most of the β-emitters behind in the aqueous phase. Organic extractive scintillators are useful when only α-emitters are of interest, when there is an overwhelming β-emitter activity with α-emitters of interest, and to improve the energy resolution of α pulse height spectra. The work of Yang et al. (1992, 1994) and Aupiais et al. (2003) are recommended as additional resources on this topic.

6. Use specific extractive resins to concentrate the radionuclides of interest. For example, actinide extractive resin, available from Eichrom Technologies, Inc., Darien IL, USA, or Triskem International, Bruz, France, is used to extract elements of high atomic number (e.g., Th, U, Pu, Am, Cm, etc.) from aqueous solutions (Horwitz et al., 1997; Eikenberg et al., 1999). 3M Empore RadDisc, Mapplewood, MN, USA, is used to separate radium isotopes (i.e., ^{228}Ra and ^{226}Ra) from lead isotopes (^{210}Pb) in water samples (Eikenberg et al., 2014; Fons-Castells et al., 2017b). Strong basic anion exchange resin (AG1X4) from Bio-Rad Laboratories, Hércules, CA, USA, was used to separate alpha-emitting radionuclides from ^{99m}Tc eluate to check its purity (Hou, 2017). In all cases, the extraction of the radionuclides of interest together with α/β pulse shape discrimination provide improved limits of detection (Horwitz et al., 1997; Eikenberg et al., 1999).
7. Purging the dissolved oxygen from the cocktail with argon gas will improve α/β separation by reducing quench. This is not commonly performed except in research to obtain unquenched samples, as the additional step is time consuming and the gain in α/β separation is relatively small.
8. Time-resolved pulse decay analysis (TR-PDA) significantly improves α/β separation or, in other words, reduces misclassification of α as β and β as α (Passo, 1996). TR-PDA, available with PerkinElmer liquid scintillation analyzers, is PDA combined with time-resolved background discrimination (TR-LSC), which is described briefly in Section XX.C.5.d of this chapter. This effect, as described by Passo (1996), is due to enhanced α rejection from the β-MCA and associated lowering of the β spillover into the α-MCA due to a change in the shape of the α misclassification curve.

E. Quenching effects in α/β discrimination

As noted previously in Section C, the α/β discrimination is improved by minimizing quench in the sample, which is accomplished by reducing the sample volume as the amount of water and acids among other substances in the sample can cause quench. In this scenario, the best α/β separation performance is obtained by keeping, for the experimental samples, the same quench level (sample chemistry), sample volume, and vial type as for the α- and β standards used to determine the optimum PDD setting. However, if it is not possible to maintain a constant quenching level on the samples, quenching should be taken into account in the activity determinations. A general strategy used for such purpose is to measure the alpha and beta standards at different PDD settings and for a wide range of quench levels. Therefore, activity quantification is done using the optimum PDD setting at the quench level of the sample, taking into account the corresponding alpha and beta detection efficiencies and spillover (Yang, 1996), Çakal et al. (2015), Palomo et al., 2010, Bhade et al. (2010). In spite of quenching minimization, the following are important effects of quench on α/β separation and guidelines to follow in controlling the effects of quench or even taking advantage of these affects, when they occur:

1. Alpha misclassification increases with quench. Once the optimum PDD setting is determined for a given sample quench level as described in Section B above, any

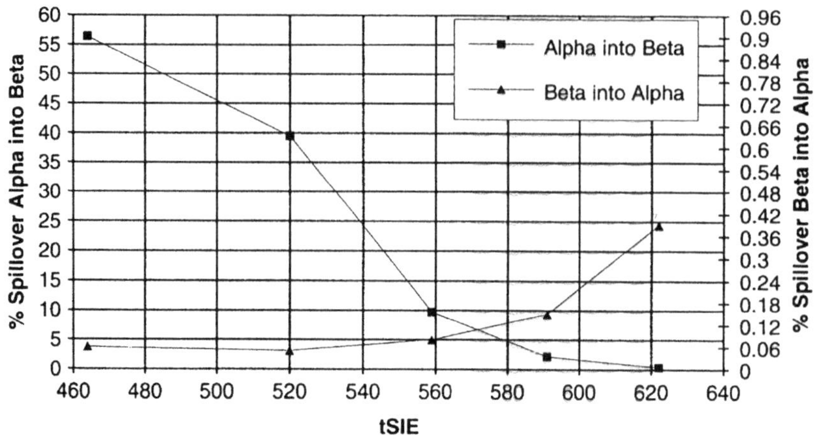

FIGURE 6.140 Percent misclassification curves of alpha and beta events as a function of quench level (tSIE) measured at a constant PDD setting determined at tSIE 620. © *1998 PerkinElmer, Inc. Printed with permission.*

increase in quench level causes an increase in the misclassification of α events into the β-MCA. This can be seen from the alpha into beta curve of Fig. 6.140, which was prepared by measuring firstly the optimum PDD setting for a given value of quench (e.g., tSIE = 620 in this example) and then measuring the percent spillover at various quench levels after adding incremental amounts of quenching agent. The degree of change in misclassification is a function of the type of quenching agent present in the sample (Fig. 6.134, Section XV). Some quenching agents are strong electron scavengers (e.g., O_2, $CHCl_3$, CCL_4), and these are able to interact more easily with π-electron triplet states reducing the delayed component of fluorescence deexcitation (Pates et al., 1998). Therefore, it is important that the chemistry of the α and β standards used to determine the optimum PDD setting and the quenching agents used to model the variation of the optimum PDD setting with the quenching parameter be identical to the chemistry of samples under investigation. Quenching agents used are HNO_3 (Palomo et al, 2010), nitromethane (Bhade et al., 2010), colored compounds, mixture of components usually present in samples (e.g., SIMS solution composed of metallic ions such as Na ($18.3\ g\ L^{-1}$), Fe ($6.0\ g\ L^{-1}$), Al ($5.7\ g\ L^{-1}$), Ni ($2.9\ g\ L^{-1}$) and Nd ($1.5\ g\ L^{-1}$) in 1M HNO_3 in Feng et al. (2017) and dry residue mass (Stojkovic et al., 2017; Çakal et al., 2015).

2. Counting efficiency of α-emitters is expected to remain at 100% (Aburai et al., 1981; Takiue et al., 1989a; Parus et al., 1993); however, loss of α events into the β-MCA occurs with increased quench whereby not all α events are detected. Caution should be taken in assuming 100% detection efficiency, and the detection efficiency should be determined in all cases. Reduced counting efficiencies can also occur with instruments that count afterpulses to discriminate against background as described in Section XX.C.
3. Beta misclassification decreases with an increase in quench as illustrated in Fig. 6.140. However, the percent misclassification and counting efficiency are a function of the E_{max} of the β-particle emitter (Pates et al., 1998). Therefore, to assure accurate measurement of misclassification and optimum PDD setting, it is best to plot a crossover curve using the same β-emitter or one of similar E_{max} as that of the β-emitter expected in the samples to be analyzed.

F. Practical applications of α/β discrimination and analysis

PERALS spectrometry was discussed in the previous Section XIV of this chapter as a method that yields the highest resolution of alpha spectra energy lines by liquid scintillation. However, conventional liquid scintillation analyzers with α/β discrimination and the use of appropriate nonaqueous and nondetergent scintillation cocktails can yield very good and practical energy resolutions. There are numerous applications of the alpha spectrometric analysis of radionuclides, or combined analysis of alpha- and beta-emitting radionuclides with conventional liquid scintillation analyzers equipped with α/β discrimination. Some of these applications will be cited in the following paragraphs.

Aupiais et al. (2003) demonstrated that the energy resolution of a conventional liquid scintillation analyzer (*i.e.*, a PerkinElmer Tri Carb 2770) equipped with α/β discrimination could be enhanced by 200% by using specific non-water miscible cocktails and a short optical path. Under such conditions an energy resolution of about 200 keV for the 4.2 MeV energy line of ^{238}U could be obtained with either the Tri Carb or the PERALS spectrometers. Also, under such conditions the sensitivity of the Tri Carb liquid scintillation analyzer was 40% lower than that of the PERALS spectrometer, because of the higher α/β discrimination of PERALS and its shorter optical path. Figs. 6.141 and 6.142 illustrate the improvement in alpha

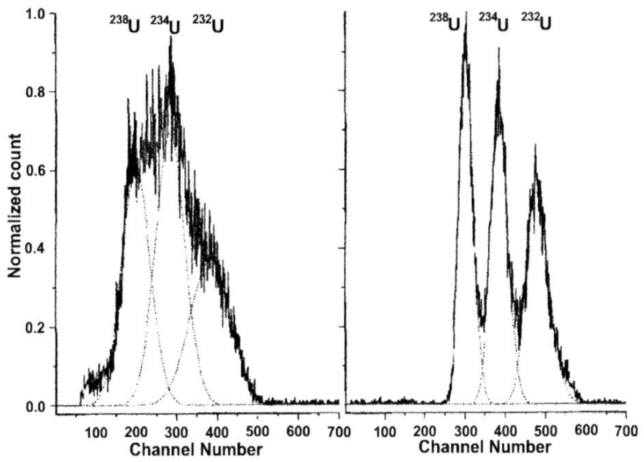

FIGURE 6.141 Uranium isotopes spectra: (left) Ultima Gold AB cocktail in standard PerkinElmer vial ($V = 20$ mL, $\phi = 26$ mm, resolution of E_α of ^{238}U at 4.2 MeV = 509 keV) spectrometer Tri-Carb; (right) Alphaex cocktail in standard culture tube ($V = 1$ mL, $\phi = 10$ mm, resolution of E_α of ^{238}U at 4.2 MeV = 260 keV), spectrometer PERALS. *From Aupiais et al., 2003 © 2003 Oldenbourg Wissenschaftsverlag, München, Reprinted with permission.*

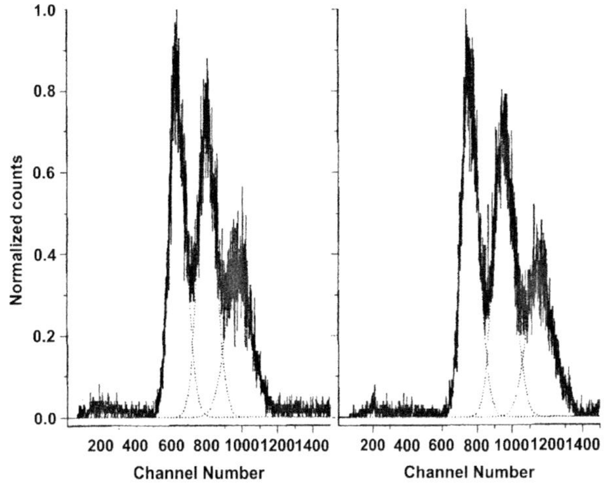

FIGURE 6.142 Improved uranium isotopes spectra: (left) detergent-free Ultima Gold AB cocktail in standard PerkinElmer vial ($V = 5$ mL, $\phi = 26$ mm, resolution of E_α of ^{238}U at 4.2 MeV = 322 keV) spectrometer Tri-Carb; (right) Alphaex cocktail in standard PerkinElmer vial ($V = 5$ mL, $\phi = 26$ mm, resolution of E_α of ^{238}U at 4.2 MeV = 342 keV) spectrometer Tri-Carb. *From Aupiais et al., 2003 © 2003 Oldenbourg Wissenschaftsverlag, München, Reprinted with permission.*

spectral energy resolution obtained with the PerkinElmer Tri Carb LSA compared to the PERALS spectrometer when detergent-free Ultima Gold AB cocktail or Alphaex cocktail (an actinide extracting scintillator specific for alpha measurement) is used. When the sample light path in the Tri Carb LSA is reduced further by employing a 1 mL sample vial ($\phi = 16$ mm), the energy resolution is improved significantly close to that obtained with the PERALS spectrometer. In conclusion, the energy resolution on liquid scintillation alpha spectrometry can be improved by using a lower or shorter optical path and no-water miscible cocktails, such as the toluene-naphthalene based cocktails (*e.g.*, Alphaex) or detergent-free diisoproylnaphthalene (DIN-based) cocktails (*e.g.*, detergent-free Ultima Gold AB). The preparation of the detergent-free Ultima Gold AB cocktail is described by Aupiais et al. (2003) and the Alphaex cocktail is available from ETRAC, Knoxville, TN, USA. Both cocktails contain HDEHP as a nuclide extractant and primary and secondary fluors.

As noted previously, the % spillover of alpha pulse events into the βMCA and beta pulse events into the αMCA will be a function of quench level. In addition, as the quench level of a sample increases, the alpha pulse height spectra will shift from higher to lower pulse heights (*i.e.*, higher to lower channel numbers) and the resolutions of the spectral energy lines will deteriorate with peak broadening (*i.e.*, increase in FWHM). Such an effect is observed in Fig. 6.143, which illustrates the liquid scintillation alpha spectra of ^{226}Ra and its daughter nuclides in equilibrium at various levels of quench (i.e., tSIE = 462, 374, 283, and 218) where the least quenched sample has the highest tSIE. For the alpha spectrometric analysis of ^{226}Ra and isotopes of uranium, Stamoulis et al. (2010) established a spectral deconvolution technique, which involved the correlation of the peak's centroids, corresponding to different radionuclides, to the characteristic energies at different degrees of quenching. Values of the centroids of the peaks and the corresponding FWHMs were used as initial parameters for fitting the spectra with overlapping Gaussians using the Origin analysis program. The derived values of peaks' centroids and FWHMs for each radionuclide were correlated to the quench indicating parameter tSIE.

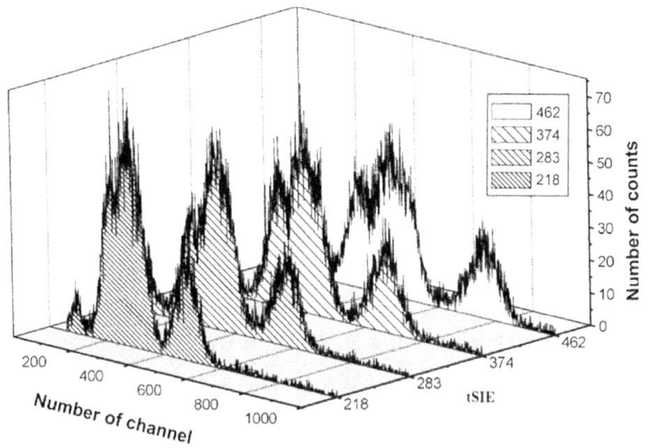

FIGURE 6.143 Alpha spectra of ^{226}Ra and its daughters in equilibrium. These spectra correspond to four degrees of quenching with the tSIE index value varying from 218 (highest quench level) to 482 (lowest quench level). *From Stamoulis et al. (2010). Reprinted with permission from Elsevier © 2010.*

Among the applications of α/β discrimination is the determination of gross alpha and gross beta activities of samples such as drinking water, as governmental regulatory agencies often place limits and regularly monitor groundwater supplies and drinking water. Some example of gross alpha and gross beta activity determinations of water supplies are the works of Bakir and Bem (1995), Warrick and Croudace (2002), Kleinschmidt (2004), Happel et al. (2004), Wong et al. (2005), Wisser et al. (2006), Staffová et al. (2006), Salonen (2006b), Ruberu et al. (2008), Bhade et al. (2010), Forte (2011), Palomo et al. (2011), Lin et al. (2012), Todorovic et al. (2012), Çakal et al. (2015). Gross alpha and gross beta activity measurements have also been applied to the rapid analysis of aerosol filters for environmental monitoring (Sas et al., 2010), urine analysis for terrorism preparedness in the possible event of population exposure (Piraner and Jones, 2009), the assessment of gross alpha- and beta-emitters in nuclear fuel processes (Lara-Robustillo and Rodríguez et al., 2006) and the analysis of gross alpha- and beta-emitters in nasal swabs (Dai et al., 2012).

Other common applications of α/β separation (*i.e.*, discrimination) liquid scintillation spectrometry are the analysis of ^{222}Rn in drinking water (Galán López et al., 2004; Pates and Mullinger, 2007; Salonen, 2010), the spectrometric analysis of isotopes of the ^{238}U and ^{232} Th day series (Bem et al., 2004; Bianchi et al., 2005; Kim et al, 2008, 2017bib_Kim_et_al_2008; Stamoulis et al., 2010; Eikenberg et al., 2014; Stojković et al., 2016; Ozden et al., 2017bib_Kim_et_al_2017; Fons-Castells et al., 2017b), the alpha-emitting nuclides ^{234,238}U, $^{238-240}$Pu, ^{241}Am, ^{244}Cm, and the beta-emitting nuclides ^{90}Sr/^{90}Y and ^{241}Pu (Versilov et al., 2002; Eikenberg et al., 2002), isotopes of Np in nuclear waste (Diodati and Sartori, 2007), analysis of ^{241}Am and ^{243}Am (El Mrabet et al., 2004; Alamelu et al., 2006), analysis of ^{237}Np, $^{238-240}$Pu, and ^{241}Am (Feng and He, 2009), the simultaneous determination of ^{226}Ra, ^{233}U, and ^{237}Np (Nebelung and Baraniak, 2007) and the analysis of alpha-emitter impurities (plutonium and americium) in ^{99m}Tc eluate from ^{99}Mo/^{99m}Tc generator (Hou, 2017).

An interesting approach to α/β discrimination was applied by Groppi et al. (2005b) to the alpha spectrometry of ^{211}At/^{211g}Po nuclides, which are prepared by α-cyclotron irradiation of ^{209}Bi target material according to the reaction ^{209}Bi(α,2n)^{211}At. The ^{211}At product nuclide decays $(t_{1/2} = 7.21\text{ h})$ by alpha emission to ^{207}Bi daughter with the main alpha at 5.87 MeV 41.8% and by EC (58.2%) to the very short-lived ^{211}Po daughter $(t_{1/2} = 0.516\text{ msec})$. The ^{211}Po daughter also decays by alpha emission with the main alpha at 7.45 MeV. The ^{211}At/^{211}Po parent-daughter nuclides at secular equilibrium are used in radiotherapy. The high LET and short range of the alpha particles from both nuclide attributes to these nuclides the power to eradicate tumor cells in the treatment of cancer. At secular equilibrium both nuclides decay at the half-life of the parent, *i.e.*, 7.21 h. Thus, the alpha-emissions remain in the patient long enough to irradiate the cancer cells, and do not remain for any excessive period of time. Alpha/beta discrimination of the ^{211}At/^{211}Po assigns the alpha emissions to the αMCA and pulse events from Auger and conversion electrons to the βMCA. The alpha spectra produced by the ^{211}At/^{211}Po in a Triathler liquid scintillation analyzer (Hidex, Turku, Finland) with α/β pulse shape discrimination is illustrated in Fig. 6.144. The alpha spectra of the two nuclides decay at the same half-life, and the spectra were used to measure quantitatively the activities of the alpha-emitters.

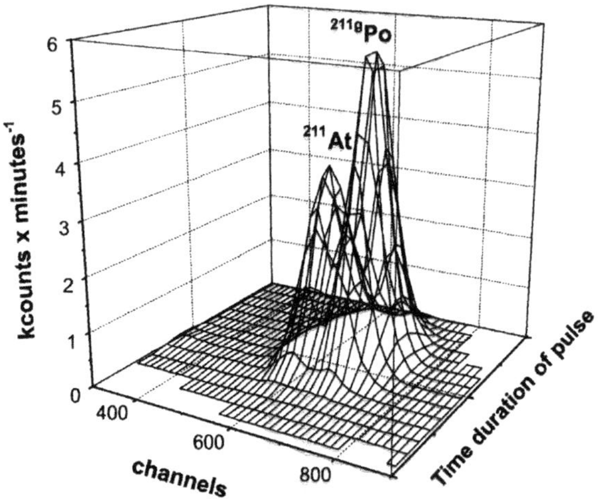

FIGURE 6.144 Liquid scintillation spectrum of ^{211}At/^{210g}Po in extracted organic phase measured with the Triathler (Hidex, Turku, Finland) liquid scintillation counter with pulse shape discrimination as a function of channel number and time duration of pulses. The counts in appropriate windows are used to obtain the decay curves of the two nuclides in secular equilibrium. *From Groppi et al. (2005b). Reprinted with permission from Elsevier © 2005.*

Another approach based on alpha/beta discrimination is the simultaneous analysis of several specific alpha- (^{241}Am, ^{226}Ra, and ^{nat}U) and beta-emitters (^{40}K, ^{60}Co, ^{90}Sr/^{90}Y, and ^{137}Cs) in water samples using partial least squares (PLS) multivariate analysis of the α-MCA and β-MCA spectra (Fons-Castells et al., 2017a). Water samples analyzed were treated following the usual procedure for gross α and gross β: dryness evaporation of 100 mL of sample, redissolution in 10 mL of HCl solution at pH = 1.5, and analysis of an aliquot of 8 mL with 12 mL of Ultima Gold AB in 20 mL PE vials in a Quanutus Detector. PSA discrimination value was set at 100 based on previous calibration done with ^{40}K and ^{236}U. PLS calibration matrix was built using 54 spectra obtained from the analysis of a standard of the six radionuclides, at three activity levels by triplicate in the same way as the samples. ROI for β-MCA and α-MCA were 454–915 and 75–895. Selectivity ratio spectra obtained when applying the PLS model show that predictor channels for ^{241}Am, ^{nat}U, ^{90}Sr/^{90}Y (Fig. 6.145A) do not overlap and good quantification is expected, whereas for ^{40}K, ^{60}Co, and ^{137}Cs there is an overlap that may cause higher interference (Fig. 6.145B).

The PLS model was applied successfully to synthetic samples containing mixtures of radionuclides with deviation lower than 12% for all the isotopes except ^{137}Cs which has higher level of deviation (from 20% to 100%) and with limits of detection lower than 1 Bq/kg for the betas and 0.3 Bq/g for the alphas. It was also seen that the model could be adapted to the sample measured by adding or removing those radionuclides that are expected or not in the sample. In this case the model was adapted for the analysis of two samples from IAEA intercomparisons with relative bias lower than 20% for most of the radionuclides (Tables 6.23 and Table 6.24).

XVI. Plastic scintillators in LSC

Plastic scintillators (PS) are a solid solution of fluorescent solutes in a polymeric aromatic solvent (Bagán et al., 2010). PS behavior as a detector for the analysis of radionuclides in a liquid scintillation counter is very similar to liquid scintillation cocktails, and it was researched by Tarancón et al. (2002a,b, 2003, 2004, 2017). PS composition is equivalent to the basic composition of liquid scintillation cocktails, being PS only differentiated by its solid state. PS may include, as solvent lineal polymers such as polystyrene or polyvinyltoluene, or three-dimensional polymers, by adding a crosslinker as divinylbenzene. In addition to the typical primary and secondary fluorescent solutes as PPO, POPOP, and others. PS may also include secondary solvents like naphthalene or DIN. Plastic scintillators based on lineal polymers are insoluble in aqueous solutions, whereas PS based on three-dimensional polymers are insoluble in aggressive and organic media (Bagán et al., 2014a).

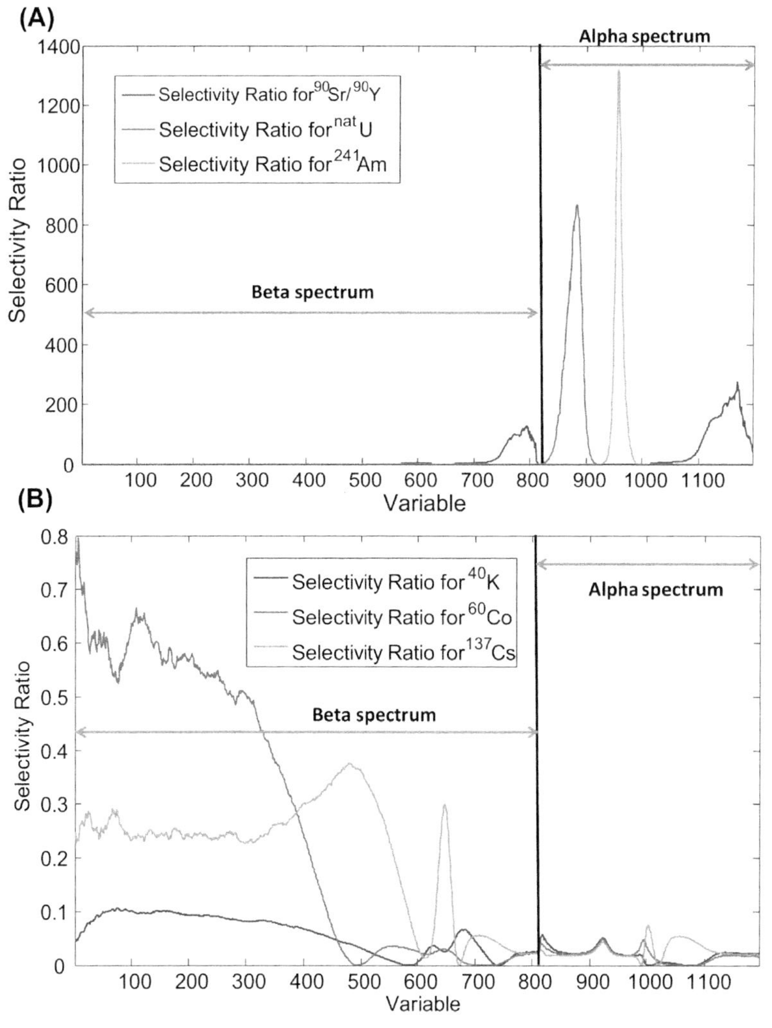

FIGURE 6.145 (A) Spectrum of selectivity ratio for ^{90}Sr/^{90}Y, ^{nat}U, and ^{241}Am. (B) Spectrum of selectivity ratio for ^{40}K, ^{60}Co, and ^{137}Cs. *From Fons-Castells et al. (2017a). Reprinted with permission from Elsevier © 2017.*

TABLE 6.23 Activity determined and reported by the organizer and performance statistics for IAEA-TEL-2015-03 sample 01.

	Organizer		Determined		Performance statistics		
	Activity ($Bq\ kg^{-1}$)	Uncertainty ($Bq\ kg^{-1}$)	Activity ($Bq\ kg^{-1}$)	Uncertainty ($Bq\ kg^{-1}$)	Rel. Bias	Robust SD	z-score
$^{90}Sr/^{90}Y$	29.6	0.8	34.1	3.4	15%	3.0	1.5
^{134}Cs	30.0	0.9	31.9	3.2	6%	1.8	1.1
^{137}Cs	30.1	0.9	32.9	3.3	10%	1.0	2.8

TABLE 6.24 Activity determined and certified by the organizer and relative bias for the quality control material IAEA-TEL-2014-03 sample 03.

	IAEA-TEL-2014-03 sample03		
	Certified ($Bq\ kg^{-1}$)	Determined ($Bq\ kg^{-1}$)	Rel. Bias
$^{90}Sr/^{90}Y$	24.5 ± 0.2	22.60 ± 2.3	−8%
^{134}Cs	26.3 ± 0.2	31.6 ± 3.2	20%
^{137}Cs	19.6 ± 0.1	15.2 ± 1.5	−22%
^{226}Ra	17.9 ± 0.1	15.2 ± 1.5	−15%
^{nat}U[a]	5.48 ± 0.04	6.3 ± 0.6	14%
^{241}Am	20.0 ± 0.1	17.0 ± 1.7	−15%

[a]*Sum of* ^{234}U *and* ^{238}U.

The format of plastic scintillators for radionuclide determination is formulated to favor the proximity between the emitters and detector. Thus, the shape of PS for this application are microspheres (PSm) with a diameter between 10 and 300 μm, pellets measured in mm (Furuta et al., 2017; Furuta, E., Ito, T., 2018), and foils (PSf) with a width between 50 and 100 μm (Merín et al., 2017) (see Fig. 6.146). Among these, the format most used with liquid scintillation spectrometers are PSm.

Plastic scintillator microspheres (PSm) provide several advantages, such as: (i) There are no mixed-waste disposal costs. (ii) Samples can be easily recovered and PSm can be reused innumerable times, if washed properly. (iii) Phase separation is avoided, whereby phase separation causes samples to be unstable during the counting process. This capability opens the possibility of counting samples with very high salt matrices, even those with salt concentrations difficult or impossible to mix suitably with liquid fluor

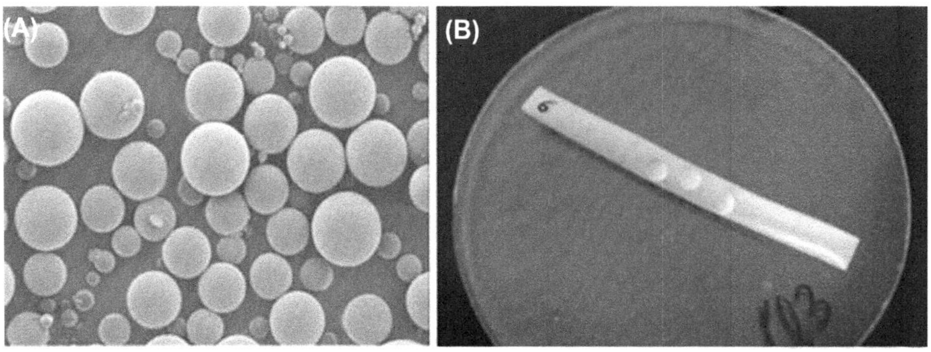

FIGURE 6.146 Images of PSm (Secondary electron microscope image) and PSfoils. *From Tarancón et al. (2017). Reprinted with permission from Springer Nature © 2017.*

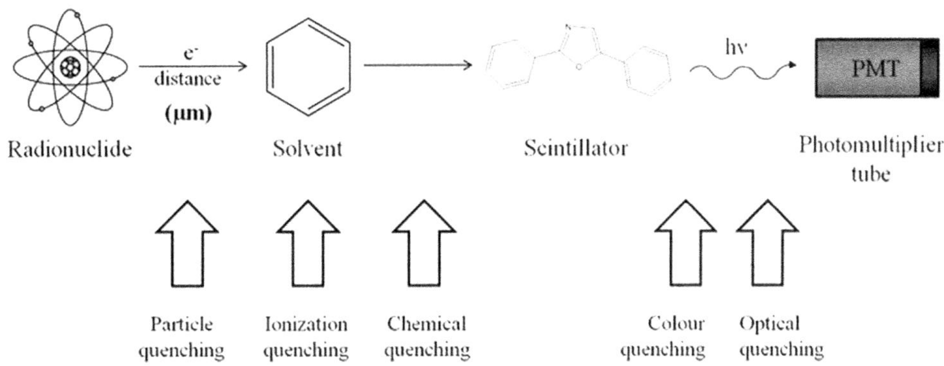

FIGURE 6.147 Scintillation mechanism when using Plastic Scintillator microspheres. *From Tarancón et al. (2017). Reprinted with permission from Springer Nature © 2017.*

cocktails. (iv) PSm can be implemented in selective determinations (PSresins) and even used in continuous measurements as consequence of the PSm solid state. (v) A conventional liquid scintillation counter can be used.

Concerning liquid scintillation cocktails, the use of PSm also exhibit some drawbacks, namely, (i) sample preparation needs an additional step of mixture homogenization, (ii) it causes a decrease in detection efficiency for low energy beta-emitters (e.g., ^{3}H), (iii) the detection process is interfered by two additional quenching phenomena: particle quenching and optical quenching, and (iv) to the current date, the number of suppliers are limited and cost is higher than liquid scintillation cocktails.

In plastic scintillation, sample preparation for radionuclide measurement follows a similar procedure as liquid scintillation. First, the appropriate amount of PSm is added to the counting vial, afterward the sample solution is added, and finally the vial content is homogenized by using a shaking procedure as a Vortex. It is important to pay attention to the relationship between the amount of PSm and the volume of sample in order to avoid an excess of scintillator, which would not be in contact with the sample, or supernatant sample solution, which would not be in contact with the scintillator. The convenient relationship for any kind of PSm diameter is 1.5g of scintillator 0.625 mL of solution. Counting vials are measured in a liquid scintillation counter exactly in the same way as liquid scintillation samples.

The reusability of the PSm after washing in citric acid was demonstrated by (Tarancón et al. (2003).

The generation of a signal as a consequence of the interaction of the emitted beta or alpha particle with the scintillation media is equivalent in plastic and liquid scintillation. When a beta or alpha particle reaches the plastic microspheres, energy is absorbed by the polymeric aromatic compounds proportional to the particle energy deposited, as discussed in Chapter 9, Volume 1. The polymeric aromatic compounds at various excited energy states then transfer this energy to the fluor solutes in the plastic microspheres, which finally emit the photons (See Fig. 6.147). The only differences arise at the beginning and at the end of the process when two additional interferences may happen: particle quenching and optical quenching.

Particle quenching is the loss ~~lost~~ of energy of the beta or alpha particles due to the interaction with the medium during particle trajectory from the point where

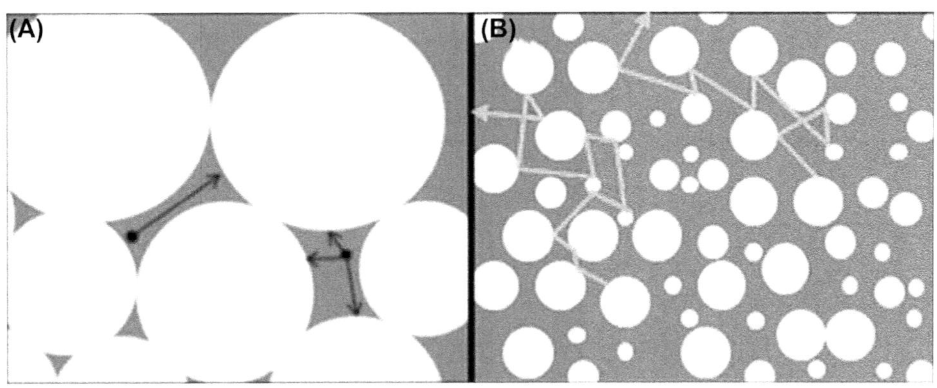

FIGURE 6.148 Scheme of particle (A) and optical (B) quenching processes using plastic scintillator microspheres. *From Tarancón et al. (2017). Reprinted with permission from Springer Nature © 2017.*

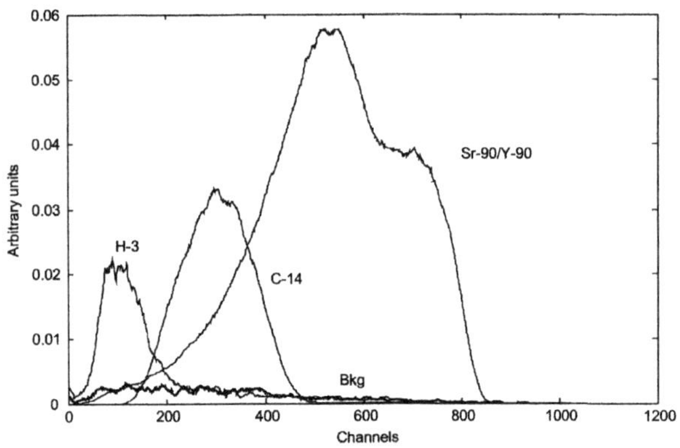

FIGURE 6.149 Pulse height spectra from ^{3}H, ^{14}C, and $^{90}Sr/^{90}Y$ and background spectra produced with plastic scintillator beads with a conventional PerkinElmer Quantulus counter in 7 mL counting vials. *From Tarancón et al. (2004). Reprinted with permission from Elsevier © 2004.*

disintegration happens until reaching the surface of the PSm (Santiago et al., 2011). An increase in PSm diameter will result into a longer particle trajectory to PSm, which will lead to a lower detection efficiency (Fig. 6.148). On the extreme opposite, optical quenching is the loss of photons produced due to the successive changes on the refraction index when photons travel from their emission until leaving the counting vial. The larger the diameter of the PSm, the lower the number of changes in the solution:PSm, therefore the higher is the detection efficiency observed.

These two quenching mechanisms result into opposite outcomes on the detection efficiency. A convenient compromise is achieved when PSm have a diameter in the range of 10–300 μm.

The pulse height spectra produced by beta-emitters in the PSm in a liquid scintillation counter show similarities to those produced by liquid scintillation, as illustrated in Fig. 6.149. Full spectrum detection efficiencies for ^{3}H, ^{14}C, and $^{90}Sr/^{90}Y$ with PSm and aqueous radionuclide solution in 7 mL polyethylene counting vials were reported by Tarancón et al. (2004) to be 0.13% ± 0.04, 17.79% ± 0.54, and 90.02% ± 0.02, respectively, and full spectrum backgrounds were 1.09 ± 0.04, 0.47 ± 0.04, and 1.43 ± 0.07 cpm. The low background count rates are very favorable for this counting method, and optimized counting windows produced remarkably yet much higher figures of merit (E^2/B), where E if the % detection efficiency and B the background count rate. Thus, routine low-level counting (*i.e.*, low activity samples) were demonstrated with relative errors < 5% for medium and high-energy beta-emitters. The low detection efficiency for ^{3}H is understandable, as the low-energy beta particles (E_{max} = 18.6 keV) undergo particle quenching when they travel through the aqueous solution until they reach the plastic microsphere to produce a scintillation pulse event.

Reduction of the PSm diameter produces an increase in the detection efficiency, which becomes relatively more important if the radionuclide considered has lower energy values (see Table 6.25). For ^{3}H a reduction from 145 to 17 μm increases the detection efficiency from 0.81% to 4.51%.

Alpha-particles can also be detected by using PSm with high detection efficiencies. Also, in this case, the values obtained depend on the PSm diameter (see Table 6.26). The alpha-particle spectral distribution as a function of PSm diameter is shown in Fig. 6.150.

Smaller sized plastic scintillator beads (<100 nm), also referred to as nanosuspensions, were tested by Weekley et al. (2004) for the analysis of ^{14}C also to reduce costs of organic liquid scintillation cocktails and cost of mixed organic radioactive waste disposal. They reported a ^{14}C counting efficiency of 45.1%. Zhu and Jay (2007) synthesized fluor-containing nanoparticles (nanosuspensions) from styrene-in-water microemulsions to determine ^{14}C activities in various sample matrices by aqueous LSA. The average diameter of the nanoparticles was ~30 nm. Several α- and β-particle emitting nuclides were tested and the relative counting efficiencies are listed in Table 6.27. The low detection efficiencies of the low-energy β-emitters (*e.g.*, ^{3}H and ^{63}Ni) are considered by Zhu et al. (2006) and Zhu and Jay (2007) to be offset by the positive environmental effects and reduced costs of mixed (organic) waste disposal.

TABLE 6.25 Detection efficiencies of beta-emitters as a function of PSm diameter.

Particle size (μm)	^{3}H (%)	^{14}C (%)	$^{90}Sr/^{90}Y$ (%)
17	4,51	74,3	192,9
32	2,69	68,7	189,8
57	1,73	62,7	190,3
79	1,16	54,5	186,7
145	0,81	45,1	184,8

From Tarancón et al. (2017). Reprinted with permission from Elsevier © 2017.

TABLE 6.26 Detection efficiencies of beta-emitters as a function of PSm diameter.

Particle size (μm)	^{241}Am (%)
17	102,9
32	99,7
57	95,2
79	85,5
145	67,3

From Tarancón et al. (2017). Reprinted with permission from Elsevier © 2017.

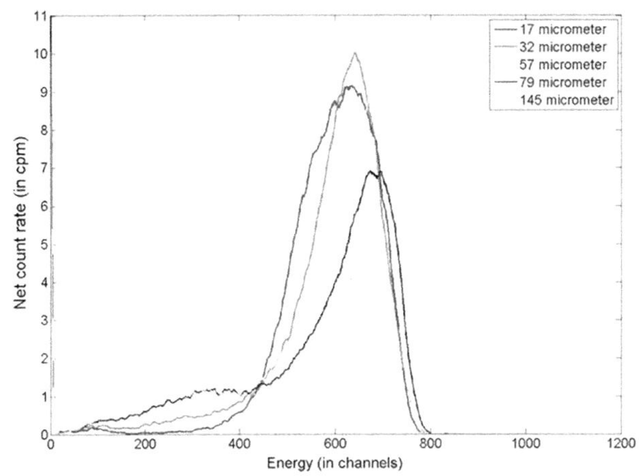

FIGURE 6.150 Spectral distribution of ^{241}Am analyzed by using PSm of different diameters. *From Tarancón et al. (2017). Reprinted with permission from Springer Nature © 2017.*

An infinitesimally small PSm would get the beta- or alpha-emitters embedded within the plastic scintillator closer to the conditions of liquid scintillation measurements. This scenario is almost achieved when the counting solution is evaporated to dryness onto the PSm, resulting in detection efficiencies close to 100% (Bagán et al., 2014b).

Furuta et al. (2009) demonstrated that plastic scintillator of cylindrical shape (capable of fitting in a conventional liquid scintillation counting vial) are a powerful tool for the identification of beta-emitting radionuclides. This is discussed further on in Section XVIII of this chapter.

When PSm are used, detection efficiency calibrations for color and chemical quenching follow the same procedures as in liquid scintillation measurements. A series of standards with increasing quenching effect are measured together with a quenching parameter (e.g., SQP(E)).

For calibrating particle quenching, classical quenching parameters, based on the sample spectrum or on the Compton spectrum generated by a gamma external source, cannot be used since the spectra do not shift as consequence of this quenching effect. However, these parameters can be easily substituted for another quenching parameter such as sample solution density.

Optical quenching, as well as ionization quenching, does not need calibration, because it is constant for all counting samples prepared in the same experimental conditions.

The possibility of applying plastic scintillator microspheres for the standardization of beta-emitting radionuclides using the free-parameter model in lieu of organic liquid scintillation fluor cocktails was demonstrated by Tarancón Sanz and Kossert (2011) and Tarancón et al. (2015). The free-parameter model applied to efficiency tracing and TDCR techniques of radionuclide standardization are discussed in detail in Section IX of this chapter.

TABLE 6.27 Quantification of various α and β-emitting radionuclides by liquid scintillation counting using fluor-containing nanoparticles. Relative counting efficiency is defined as CPM from the nanosuspensions/CPM from conventional organic LSC.

Isotope	Mode of decay (E_{max}) (MeV)	Relative counting efficiency (%)
^{3}H	β^- (0.018)	3.2
^{63}Ni	β^- (0.066)	8.19
^{14}C	β^- (0.156)	45.3
$^{90}Sr/^{90}Y^{a}$	β^- (0.546/2.28)	82.5
^{204}Th	β^- (0.763)	71.3
^{32}P	β^- (1.71)	100
^{241}Am	α (5.49)	96.2
^{239}Pu	α (5.15)	100

[a] ^{90}Sr samples contain progeny isotope ^{90}Y, which is likely why this sample exhibited greater counting efficiency than ^{204}Tl.
From Zhu and Jay (2007). Reprinted with permission from IOP Publishing Ltd. © 2007.

Employing the free-parameter model with the TDCR technique and efficiency tracing with ^{63}Ni, Tarancón Sanz and Kossert (2011) demonstrated the standardization of several beta-emitters with deviations from known reference activities between 0.1% and 2.5%.

The potential of polystyrene PSm as a detector for α/β pulse shape discrimination in conventional liquid scintillation counters was investigated by Bagán et al. (2010). Again, the study was implemented in light of the advantage of avoiding the use of organic liquid scintillation cocktails and the disposal costs of mixed waste. In a comparison of the performance of the PSm to the conventional use of organic fluor cocktails in α/β discrimination, Bagán et al. (2010) demonstrated that the detection efficiency and pulse heights produced by the beta-emitters with plastic scintillator microspheres were lower achieving improved beta discrimination. On the other hand, the analysis of alpha-emitters with the PSm resulted into a poorer alpha discrimination. The α/β discrimination capability of the plastic scintillator is apparently related to the particle energy loss during travel to the plastic microspheres. Bagán et al. (2010) postulate that this energy loss would reduce the excitation of the polystyrene and consequently reduce the number of triplet excitation states key to the pulse shape discrimination of alpha particles from beta particles. The addition of a secondary solvent as naphthalene or di-isopropylnaphtalene translate into a delay on the pulse, especially those proceeding of the alpha-emitters, significatively improving the alpha–beta discrimination capability (see Fig. 6.151).

A new approach to the α/β discrimination taken by Bagán et al. (2014b), which would avoid the energy loss by both alpha and beta particles, is to evaporate the samples to dryness onto the scintillator microspheres.

Yeman et al. (2017) describes the current state-of-the-art of plastic scintillators based on polyvinyltoluene and their capability to neutron signal discrimination in a gamma radiation background.

Another advantage that PSm can offer is the analysis of aqueous samples containing high salt matrices, which can be difficult to mix with conventional liquid scintillation fluor cocktails. Using ^{14}C as a test nuclide in samples coming from studies of oil reservoirs and Detec (UPS-89) polystyrene plastic scintillation beads between 180 and 250 μm diameter provided by Detec (http://detecsciences.com/), Gatineau, QB, Canada or Detec-Europe, F-56,000 Vannes, France, Bagán et al. (2009) demonstrated that samples high in salt content (up to 2.8 M $NaClO_4$) could be analyzed efficiently, and the plastic microspheres can be reused after a simple citric acid wash. The high salt concentrations however, added a new dimension to the detection efficiency calculations, as the increase in sample density with salt concentration would diminish slightly the detection efficiency due to particle quenching. Thus, two quench parameters should be considered, the classical SQP(E) for color quenching correction and the density for particle quenching, as illustrated in Fig. 6.152.

The slight decrease in detection efficiency of the ^{14}C with increased sample density is expected, as the relatively weak beta particles must travel from the solution to the PSm before excitation of plastic and fluor molecules. As two detection efficiency factors are involved, the overall detection efficiency of the radionuclide (E_{unk}) is calculated as

$$E_{unk} = E_C \times E_S \times E_0 \qquad (6.158)$$

where E_C is the relative efficiency caused by color quenching, i.e., the efficiency or the efficiency of the color-quenched sample divided by the efficiency of the standard without quenching agent E_0, and E_S is the relative efficiency caused by the particle quenching (salt concentration), i.e., the efficiency of the salt-quenched standard

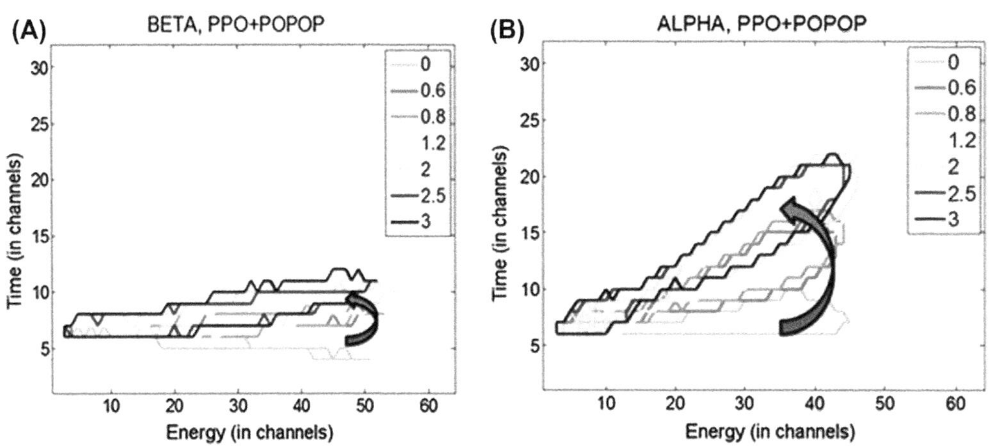

FIGURE 6.151 Energy versus pulse time distribution for a beta-emitter (^{90}Sr/^{90}Y) (A) and an alpha-emitter ^{241}Am (B) as a function of the naphthalene concentration measured in a Triathler detector. *From Tarancón et al. (2017). Reprinted with permission from Springer Nature © 2017.*

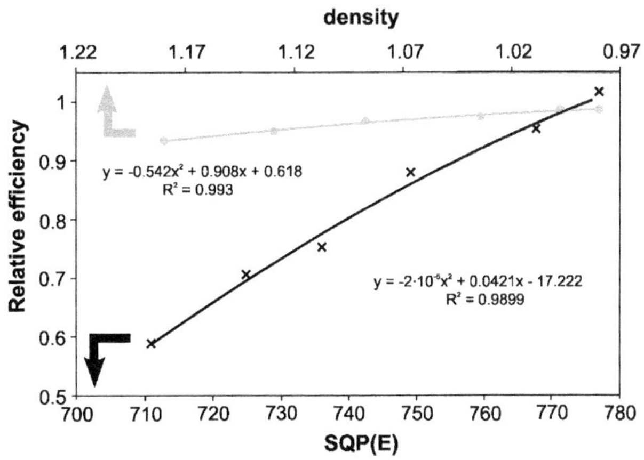

FIGURE 6.152 Relative counting efficiencies of ^{14}C as a function of color quenching (lower curve) measured with an external standard quench indicating parameter [SQP(E)] and particle quenching (salt concentration), *i.e.*, sample density (upper curve). *From Bagán et al. (2009). Printed with permission from Elsevier © 2009.*

divided by the efficiency of the standard without quenching agent, E_0.

Relative errors on the quantification of ^{14}C activity in aqueous samples from the radiotracer studies following this procedure were always under 5%.

Another interesting application of PSm is related to ^{222}Rn. Mitev et al., (2016) and Pelay et al., (2017), studied the absorption behavior of ^{222}Rn in plastic scintillators in comparison to the behavior observed for other polymeric material as polycarbonates. The experimental setup includes the exposition of the PSm produced at the University of Barcelona to a ^{222}Rn atmosphere and their subsequent inclusion in a 20 mL glass vial followed by its measurement in a commercial counter. The results showed that the diffusion length in PSm was around 120 μm whereas the partition coefficients in air were 4.95 ± 0.54 in air and 37.2 ± 1.9 in water. These values are lower than the concentration capabilities exhibited by Makrofol materials (N 112 and DE 26) but the PSm enables the capability of direct detection and does not require any other manipulation. Detection efficiency of the PSm with ^{222}Rn absorbed is 314% (a maximum of 500% is expected due to ^{222}Rn and its progeny) and the resolution of the spectrum distribution depends on PSm diameter and packaging in the measurement vial.

Continuous monitoring of activity in a fluid has been one important goal since the beginning of liquid scintillation techniques. The use of scintillating cocktails mixed with the aqueous samples has been a permanent option with variable success depending on the degree of homogenization of the mixture and always affected by the continuous consumption of reagents and waste production. The use of solid scintillators has appeared historically as an alternative mainly for ^{3}H determination in liquid effluents. Singh and Kadwani (1974) and Noakes et al. (1998) developed a continuous monitoring detector for ^{3}H in air and in aqueous samples with the same common component: a cell filled with plastic scintillator powder. Variations from this approach have continued over the years all taking advantage of the stability of using plastic scintillators as detectors. Bae et al. (2017) describes a specific mobile tritium monitoring detector for a continuous inflow system designed for water samples based on electrolysis and a plastic scintillator detection system. Tarancón et al. (2012) describes a continuous monitoring instrument for beta and alpha-particles quantification in aqueous samples.

Plastic scintillators are solid and this turns into a great advantage for the incorporation to these detector strategies to improve the selectivity of the determinations. Spectra distribution of beta and alpha emissions and the limited resolution and quenching effects on scintillation techniques make a previous separation step obligatory before any specific determination can be made. Plastic scintillating resins (PSr) unify these steps of separation and preparation to the measurement by coating the PSm by an extractive material (see Fig. 6.153).

In practical terms, PS resins are packaged in solid phase extraction cartridges and inserted in the separation experimental set-up as is customary with the commonly used extraction chromatography resins. The only difference is that once the radionuclide of interest is isolated, no further steps of elution are necessary and nor additional preparation for measurement is required due to the fact that PSr cartridges can be disconnected, included directly into a 20 mL vial and introduced into the liquid scintillation counter. No additional scintillator is required, because the PSm core is already included. The main advantages of PSr are the reduction of time and manpower required for sample processing as well as the reduction of reagents and waste. Nevertheless, this technique also has its drawbacks such as

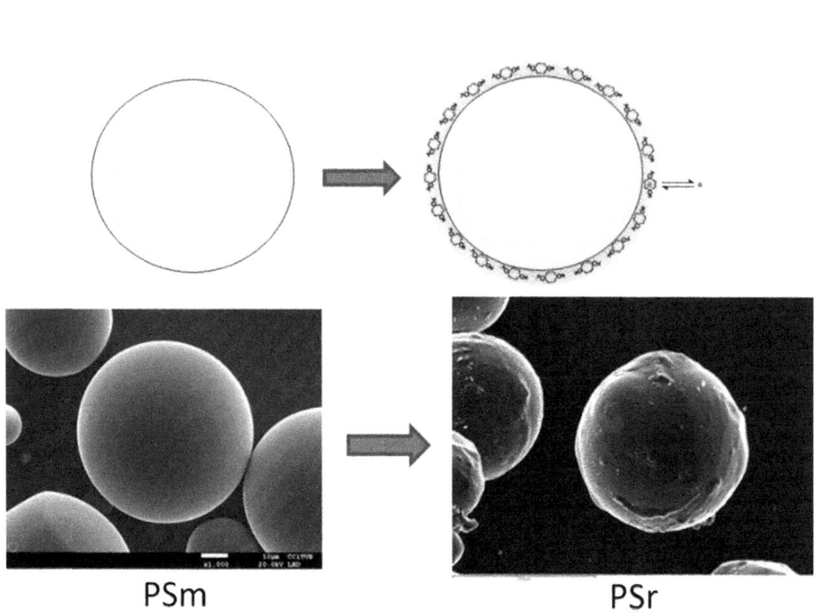

FIGURE 6.153 Scheme and secondary electron microscope images of PSm and PSr produced at the University of Barcelona. *From García et al. (2018).*

the still in progress development of PSr (since only a few are already commercially available) and the potential higher cost specially related to the nonscintillating resins.

There are two main approaches to incorporate the selective extractant onto the plastic scintillator: impregnation or covalent binding. In the first case, extractant is added by deposition, in the second it is bond to the polymer. The first approach is the most widely used, but its stability depends on the solubility of the extractant in the sample solution. The second approach does not have this problem, but often the incorporation of the extractive molecule results in a critical reduction of the scintillation capability. Duval et al. (2016 a,b) describes an extractive scintillating polymer sensor for Uranium determination in ground water. Bliznyuk et al. (2018) describes a more recent approach based on hybrid organic–inorganic beads with the Superlig 620 (SL) material with high affinity to radiostrontium. The application includes SL particles used as core and halloysite clay nanotubes (HNT) modified with a polyvinyltoluene (PVT). The developed SL–HNT–PVT hybrid extractive scintillating material allows real-time detection of low-level concentrations of radiostrontium in water. Bagán et al. (2011b) applied impregnated PSr with a 1M solution of 4,4′(5′)-di-t-butylcyclohexano-18-crown-6 (DtBuCH18C6), to act as an extractant for Sr, in a similar fashion as Sr-specific resins, to extract strontium from solution. The ^{90}Sr is retained by the extractant-impregnated PSr, and the ^{90}Y remains with the rinse solution. The PSr were counted directly in the liquid scintillation analyzer. Drinking water, sea water, and river waters were tested, and a minimum detectable activity of 0.46 Bq/L for a 10 mL sample is reported with sample errors <4% Bagán et al. (2011b).

Sáez-Muñoz et al. (2018) used the same Sr–PSr to establish an emergency procedure to ^{89}Sr/^{90}Sr determinations in milk. For their study, several types of milks were analyzed including cow, sheep and goat, skimmed, semi-skimmed, and in powder form. Total ^{89}Sr+^{90}Sr activity was quantified with relative bias lower than 7% in all cases. The entire analysis, including pretreatment lasted only 5h. Bagan et al. (2012) undertook further studies on radiotracer determinations in oil reservoirs by using a specific PSr for $S^{14}CN^{-}$, which was the active ^{14}C molecule in the radiotracer. The extractant in this case was Aliquat 336. Quantification was achieved with bias lower than 5% and the detection limit for 100 mL of sample was 0.08 Bq L^{-1}.

Barrera et al. (2016) describes a PSr for ^{99}Tc determinations in water and urine using also Aliquat 336. In all samples the quantification bias was lower than 8% and the detection limit for 10 mL of sample was 0.36 Bq L^{-1}. Some other PSr are related to the determination of ^{210}Pb and ^{126}Sn (Lluch et al. 2016).

A potential evolution of the use of PSr can be its incorporation to the protocols followed by the automated separation systems that are being commercially available in the recent years.

XVII. Scintillation in noble liquids

Liquid noble gases, such as helium, neon, argon, and xenon, have properties of high scintillation yield as a medium for the detection of ionizing radiation and nuclear recoil energy. In particular, liquid xenon is considered most suitable for γ-ray detection, because of its high atomic number and short decay constants of the scintillation light ($\tau_1 = 4.3$ ns, $\tau_2 = 22$ ns for short and long components, respectively), which comes close to that of plastic scintillators (Hitachi et al., 1983; Tanaka et al., 2001; Aprile and Doke, 2010). The efficiency of liquid xenon as a scintillator is reported by Aprile et al. (1990) to be similar to that of NaI(Tl) solid scintillation detector, but with a shorter scintillation decay time. Both liquid argon (LAr) and liquid xenon (LXe) are widely studied as efficient liquid scintillation detectors for (i) alpha particles and internal conversion electrons (Aprile et al., 1990, 2005; Miyajima et al., 1992; Tanaka et al., 2001; Ni et al., 2005; Kastens et al., 2009), (ii) α/γ and n/γ pulse height discrimination (Benetti et al., 1993; Peiffer et al., 2008), and (iii) γ radiation and neutron spectrometry from the scintillation efficiency of electron and nuclear recoil events (Arneodo et al., 2000; Akimov et al., 2002; and Aprile et al., 2005). Liquid rare gas detectors are currently being studied for their utility in various fields including (i) nuclear medicine as imaging detectors including positron emission tomography (Gomez-Cadenas et al., 2016, Gallego Manzano et al., 2015; Retière, 2010; Amaudruz et al., 2009; Grignon et al., 2007; Doke et al., 2006; Gallin-Martel et al., and 2006, 2009, 2012), (ii) tomography imaging telescopes (Time Projection Chambers) for research in medium-energy gamma-ray astrophysics (Tomás et al., 2018, Zou et al., 2016; Aprile and Doke, 2010; Cennini et al., 1999, and Aprile et al., 1990), (iii) galactic Dark Matter Search consisting of weakly interacting massive particles (Akerib et al., 2016, Alfonsi et al., 2016, Sorensen et al., 2009, Aprile et al., 2006b, Neves et al., 2005, 2007, and Akimov et al., 2002), (iv) relativistic heavy ion detection (Tanaka et al., 2001), (v) the search for the rare muon decay, *i.e* $\mu^+ \rightarrow e^+ + \gamma$(Ogawa et al., 2017, Ieki, 2016, Adam et al., 2010; Uchiyama, 2010; Iwamoto, 2010; Sawada, 2010; and Nishimura and Natori, 2011), and (vi) the EXO-200 (Enriched Xenon Observatory for 0νββ, *i.e.*, neutrinoless double beta decay) which utilizes a detector containing 200 kg of liquid xenon in the search for neutrinoless double beta decay (Pocar, 2015, 2013; Auger et al., 2012; Dolinski, 2012; Arisaka et al., 2009; Neilson et al., 2009; Leonard et al., 2008; Ueshima et al., 2008). Applications of liquid xenon are numerous in light of the advantageous detector characteristics of high atomic number, scintillation efficiency, and short fluorescence decay times.

The mechanisms of scintillation in liquid argon and liquid xenon are similar and have been investigated thoroughly (Kubota et al., 1978a,b, 1979; Doke, 1981; Doke et al., 1990). As described by Cennini et al. (1999), Mei et al. (2008) and Aprile and Doke (2010), scintillation light in liquid argon (LAr) is produced by ionizing radiation either by direct ionization of an Ar atom followed by excimer formation and de-excitation according to

$$Ar^* + Ar \rightarrow Ar_2^* \rightarrow 2Ar + h\nu(128\ \text{nm}) \tag{6.159}$$

or via ionization, recombination, excimer formation, and de-excitation as

$$Ar^{\oplus} + Ar \rightarrow Ar_2^{\oplus} + e \rightarrow Ar_2^* \rightarrow 2Ar + h\nu(128\ \text{nm}) \tag{6.160}$$

De-excitation from the Ar_2^* to the dissociated ground state 2Ar gives rise to fluorescence displaying fast and slow decay components of about 5 ns and 1 μs, respectively, with relative yields of 23% and 77% at an emission of approximately 128 nm wavelength. The light emission is abundant with approximately one photon emitted per 25 eV of energy dissipated in LAr by ionizing radiation or 40,000 photons per MeV. In the case of liquid xenon (LXe) excitation, light emission is at 175 nm and the lifetimes of the singlet and triplet states of the excited dimer Xe_2^* are 4 and 22 ns, which makes liquid xenon the fastest of the noble liquid scintillators (Hitachi et al., 1983; Aprile et al., 2005).

The wavelength of light emission is in the vacuum ultraviolet (VUV) region. Consequently, glass-faced photomultipliers are used with wavelength shifters. Metal-channel photomultipliers are fabricated by Hamamatsu Photonics to efficiently detect the fast scintillation light from LXe. These PMTs are designed to operate within the LXe to efficiently detect the 175 nm Xe light photons (Giboni et al., 2005). Other photomultipliers with MgF_2 or CaF_2 windows have been used. The addition of liquid xenon in small concentrations ($\leq$100 ppm) into LAr serves as a wavelength shifter from 128 to 173 nm permitting photomultipliers with quartz windows (Cennini et al., 1999). The scintillation mechanism for liquid xenon is similar to that described above for LAr with an average of one photon emitted per 14.7 eV of energy dissipated by ionizing α particles (Tanaka et al., 2001). Avalanche photodiodes (APDs) and silicon photomultipliers are researched as alternatives to the photomultiplier tube for liquid rare gas scintillation detectors (Solovov et al. (2000; Aprile et al., 2006a; and Shagin et al., 2009).

Arneodo et al. (2000) demonstrated the excellent linear response of scintillation fluorescence versus energy deposited in liquid xenon using several γ-ray sources. Typical energy distribution in liquid Xe with ^{57}Co and ^{137}Cs γ-ray sources placed outside the LXe scintillation detector measuring 5.0 and 4.6 cm for the top and bottom diameters and 4.0 cm in height are illustrated in Fig. 6.154. The width of each energy distribution is dominated by a light collection efficiency that varies from 5% to 12% depending on the point of interaction of the γ-ray within

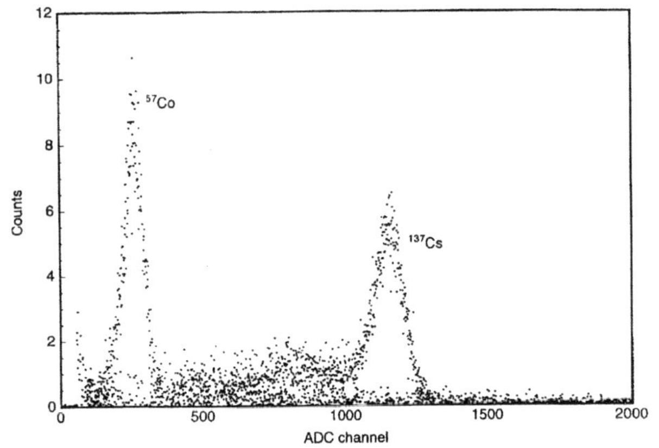

FIGURE 6.154 Liquid xenon scintillation light spectrum of ^{57}Co and ^{137}Cs. *From Arneodo et al. (2000). Printed with permission from Elsevier © 2000.*

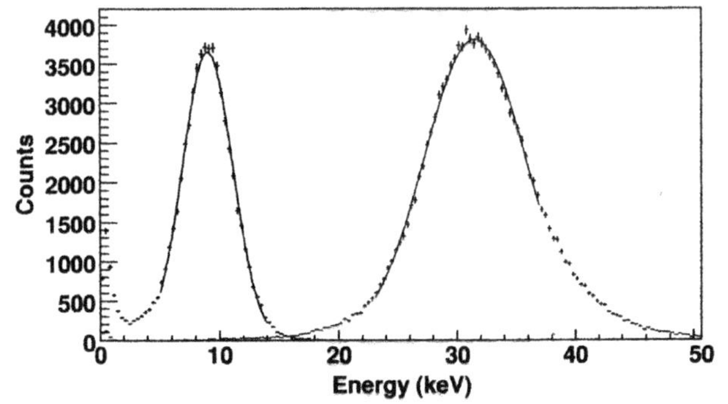

FIGURE 6.156 Activities of energy peaks due to 9.4 and 32.1 keV conversion electrons after doping ^{83m}Kr into the liquid xenon detector. The 9.4 and 32.1 keV peaks have a resolution of 23% and 14%, respectively. *From Kastens et al. (2009). Reprinted with permission © 2009 The American Physical Society.*

the LXe detector volume. The overall detector efficiency was estimated by Arneodo et al. (2000) to be approximately 1.5% for the 175 nm UV photons from the LXe scintillation taking into account the average light collection efficiency of 8%, PMT quantum efficiency of 19%, and photo-extraction potential of LXe of 15 eV.

Calibration of LiXe detectors can be performed from the decay of xenon metastable states, illustrated in Fig. 6.155, produced by fast neutron activation. All of the short-lived metastable states decay within a few days leaving no longtime activity in the detector after the neutron activation, as demonstrated by Ni et al. (2007). The characterization of the response of noble liquid scintillation detectors at low energies can be performed with ^{83m}Kr as a calibration source (Kastens et al., 2009). The ^{83m}Kr decays via isomeric transition with emission of 9.4 and 32.1 keV

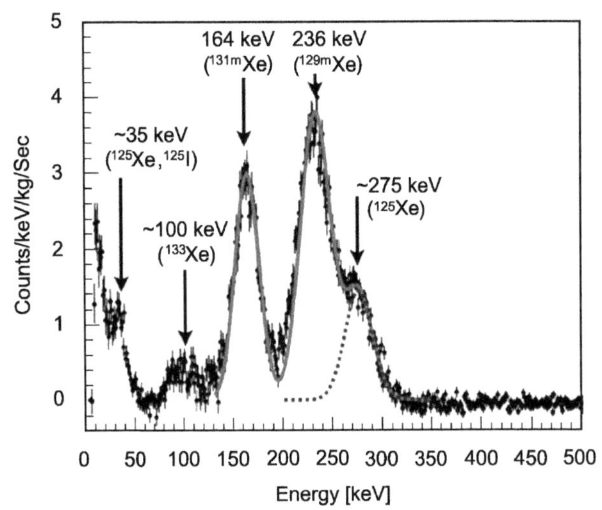

FIGURE 6.155 Background subtracted spectra from activated xenon 12 h after a 5-day neutron activation The energy resolutions for the 164 and 236 keV peaks are 7.0% and 5.1%, respectively. *From Ni et al. (2007). Printed with permission from Elsevier © 2007.*

conversion electrons and with a half-life of 1.83 h. The spectral energy peaks, illustrated in Fig. 6.156, at 9.4 and 32.1 keV are reported to have a 23% and 14% resolution respectively. Such a source with short half-life leaves no long-lived residue in the detector.

XVIII. Radionuclide identification

It is possible to use a liquid scintillation analyzer to identify automatically an unknown pure β-emitting radionuclide sample; once it is identified, the LSA can automatically determine the activity (DPM) of that same sample.

One of the obstacles encountered in identifying β-emitting radionuclides by LSA is that the pulse height spectrum of all β-emitters are broad over the range from 0 keV to the end point energy corresponding to the E_{max} of the β emissions. Also, the pulse height spectrum of a given β-emitter changes in pulse height distribution and endpoint energy according to the level of quench in a sample. Consequently, the pulse height distribution of a higher-energy β-emitter under a high degree of quench could resemble that of a lower-energy β-emitter under a low degree of quench.

By means of a double-ratio technique Takiue et al. (1989b) developed a method for identifying unknown β-emitters. The technique is based on combining the information provided by two quench indicating parameters, tSIE and SIS, which were described previously in this chapter. The tSIE is produced by the external standard, and it provides information defining the level of quench in the sample-scintillation fluor cocktail mixture. The SIS is a sample-spectrum quench indicating parameter, which depends not only on the degree of quench in the sample but also the β-particle energy spectrum. The two quench-indicating parameters together, therefore, provide information that reflects the pulse height distribution and shape of a particular pulse height spectrum at any given quench level. The combined information provided by the two quench indicating parameters is used to identify pure β- or β/γ-emitting radionuclides.

The technique developed by Takiue *et al.*, (1989) requires the preparation of a series of quenched "standards" of all of the pure radionuclides that may need to be identified as unknown samples. The word "standards" is enclosed in quotation marks, because the activity need not be known in the series of quenched standards. The activities of the quenched standards should be high enough to determine accurately the SIS of each of the standards, and they are used only once to prepare a double-ratio curve for a particular radionuclide.

Once the series of quenched standards has been prepared for each radionuclide of interest, the SIS and tSIE of each quenched standard are determined and plotted as illustrated in Fig. 6.157. The family of curves, illustrated in the figure, becomes separated from the lowest curve for ^{3}H to the uppermost curve for ^{32}P according to the maximum energy (E_{max}) and average beta-particle energy (E_{av}) of each radionuclide, because the SIS plotted along the ordinate reflects both E_{max} and E_{av}. The curves, therefore, provide a definition of the magnitudes of the pulse height spectral distributions as a function of quench, which are radionuclide specific.

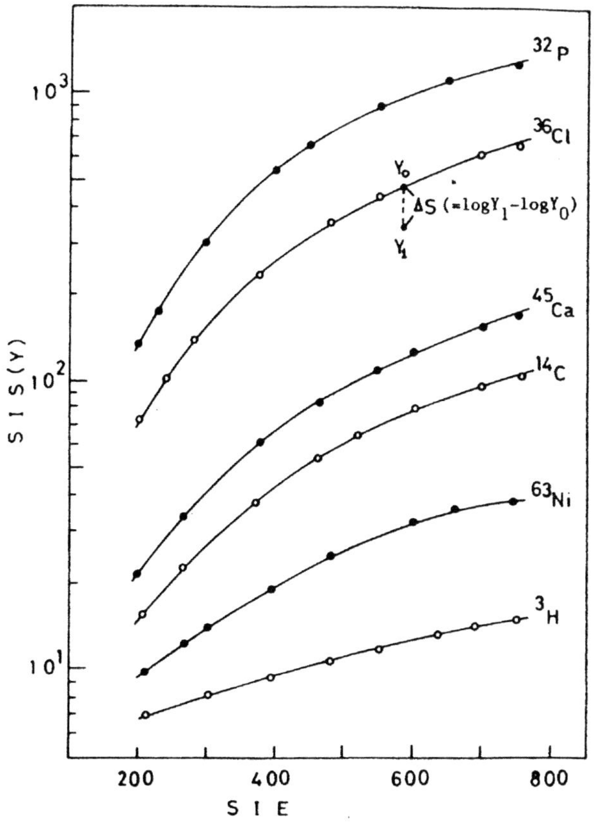

FIGURE 6.157 Double ratio curves for identifying an unknown radionuclide. Each curve is constructed from the SIS and SIE of a series of quenched "standards" of each nuclide. *From Takiue et al. (1989b). Printed with permission from Elsevier © 1989.*

To apply this technique, an unknown sample is monitored for its SIS and tSIE value and then compared with the double-ratio curves, such as those illustrated in Fig. 6.157, that were prepared in advance. From the tSIE and SIS of an unknown sample, the radionuclide can be identified using the double-ratio curves provided the deviation $\Delta S/S$ of the experimental plotted point of the unknown sample is less than ± 0.02, or

$$\frac{\Delta S}{S} = \frac{(\log Y_1 - \log Y_0)}{S} < \pm 0.02 \qquad (6.161)$$

where ΔS is the difference in ordinate ($\log Y_1 - \log Y_0$) between the plotted point Y_1 and the corresponding value Y_0 on the double-ratio curve, and S is the logarithm of the SIS value ($\log Y_0$) on the curve. Takiue et al. (1989b) demonstrated that the activities of the unknown samples can also be determined automatically, if the double-ratio curves are linked by computer program to the quench correction curve of the appropriate radionuclide. In essence, therefore, an unknown pure β-emitter in scintillation fluor cocktail could be placed in an LSA for both automatic identification and activity analysis.

The above procedure was demonstrated to work for several radionuclides with the exception of ^{14}C and ^{35}S. The close similarities of the β-particle energy spectra of ^{14}C ($E_{max} = 156$ keV) and ^{35}S ($E_{max} = 167$ keV) produce overlapping double-ratio curves, which prevents the identification of these two radionuclides in unknown samples. However, Natake et al. (1996) showed that pulse height analysis could be used to identify ^{14}C and ^{35}S in unknown samples. The slight difference in the pulse height distributions could be distinguished using the sample channels ratio (SCR), which is the ratio of two counts in two measurement channels. The optimum channel settings were selected to obtain as great a difference between the two radionuclides as possible for a given quench level. The SCR for a series of quenched "standards" of ^{14}C and ^{35}S are then plotted against the H number in what is referred to as a double quench parameter curve (DQP) curve. The H number is an external standard quench indicating parameter (QIP) described previously in this chapter. Other external standard QIPs, such as tSIE or SQP(E), can also be used in this technique. To apply the technique, the SCR and the H number are determined for the unknown sample and the two QIP values are plotted on the previously prepared DQP curve. The radionuclide is then identified when the plot of that point lies in the neighborhood of either the ^{14}C or ^{35}S DQP curve.

In another approach to identify unknown pure β-emitters Dodson (1996) developed the technique of characterizing the pulse height spectrum according to the following three properties: E_{max}, the isotope center number I and the spectral resolution (FWHM or F). The isotope center number I is defined as

$$I = \frac{\sum(N_i E_i)}{\sum(N_i)} \quad (6.162)$$

where E_i is the energy of the N_i electron. For quenched sets of radionuclide standards, both I and F are found to be linearly dependent on E_{max}. The slopes and intercepts of the linear equations defining I and F as a function of E_{max} over a range of quench levels are tabulated for each radionuclide. The measurement of I, E_{max}, and FWHM of an unknown nuclide's sample spectrum coupled with a relevant algorithm permit the identification of the unknown nuclide within the parameters of the defined reference set.

Alpha-particle emitters were identified by Kaihola (2000) of Wallac Instruments (now PerkinElmer, Inc.) who demonstrated that the external standard quench indicating parameter SQP(E) could be used to correlate the alpha-particle sample pulse height distribution with its emission energy as a function of quench level. Two known energy calibration lines as a function of quench are required to verify the emission energy and identify the radionuclide in liquid scintillation alpha spectroscopy.

Another approach to the identification of pure beta-emitting radionuclides with a conventional liquid scintillation counter was taken by Furuta et al. (2009). They measured the pulse height spectra of six pure beta-emitting radionuclides using plastic scintillator (Saint-Gobain BC-400) that is placed into glass counting vials. Three different instruments were used including Beckman Coulter LS-6000S, LS-6500 and PerkinElmer Tri-Carb 3270TR/SL liquid scintillation analyzers. The radionuclides were dried onto separate plastic scintillator sheets, which were inserted into plastic scintillator cylinders or containers. The plastic scintillator containers were then placed inside a conventional glass LSC vial. After calibrating the x-axis of the pulse height spectra to read 0.5 keV per channel, the channel of the spectrum end-point for ^{3}H, ^{14}C, ^{35}S, ^{45}Ca, and ^{33}P yielded the theoretical maximum beta energy (E_{max}) of each radionuclide, as illustrated in Fig. 6.158. The absolute nuclide decay energy is not measured with the plastic scintillator; however, the plastic scintillator permitted the identification of the nuclides by the end-points and shapes of the spectra. The end-point of the pulse height spectrum of the high-energy ^{32}P ($E_{max} = 1710$ keV) is not illustrated in Fig. 6.158. Furuta et al. (2009) demonstrated that the end-point of this spectrum was reported to reach up to 1612 keV, because the plastic scintillator was not thick enough to absorb all of the beta-particle energy in this case.

A new approach developed to identify and quantify radionuclides in contaminated drinking waters and pipeline deposits was performed by Warwick and Croudace (2013). They measured nine different beta-emitter radionuclides (^{3}H, ^{14}C, ^{35}S, ^{63}Ni, ^{147}Pm, ^{99}Tc, ^{36}Cl, ^{129}I, and ^{131}I) in four different liquid scintillation cocktails (Gold Star, Ultima Gold AB, Ultima Gold XR, and Hisafe 3) using three different Quantultus LS counters. From these measurements they obtained the SQPE parameter as a measurement of the quenching of the sample and the SQPI(95) parameter as a mixed measurement of the quenching and the energy of the radionuclide. The relation of both parameters correlates with the maximum energy of the radionuclide, as illustrated in Fig. 6.14 illustrated previously in this chapter and Fig. 6.159, without influence of the quenching, the instrument or the cocktail used, demonstrating its application for radionuclide identification.

Also, another relationship between SQPI(95) and SQPI(50) was tested in front of the SQPI(95) parameter, as presented in Fig. 6.14. In this case, the factor is influenced

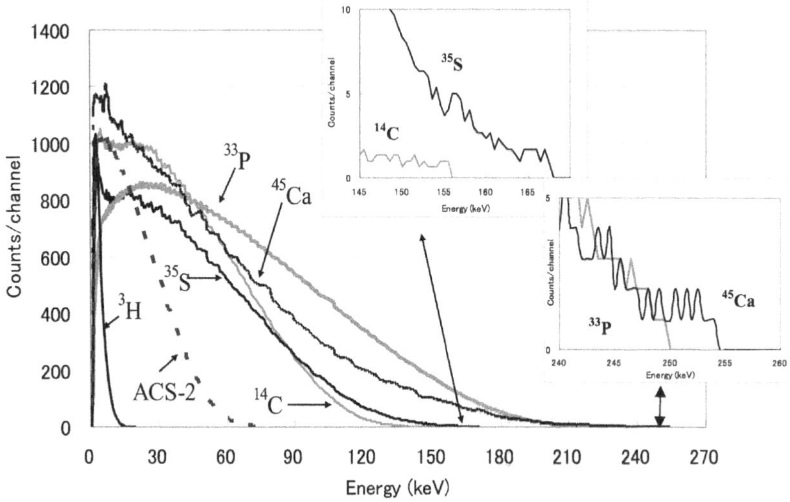

FIGURE 6.158 Spectra of five nuclides measured with plastic scintillator and a quenched spectrum of ^{14}C measured with ACS-2 liquid scintillator (dashed line). *From Furuta et al. (2009), Printed with permission from Radiocarbon, University of Arizona © 2009 Arizona Board of Regents on behalf of the University of Arizona.*

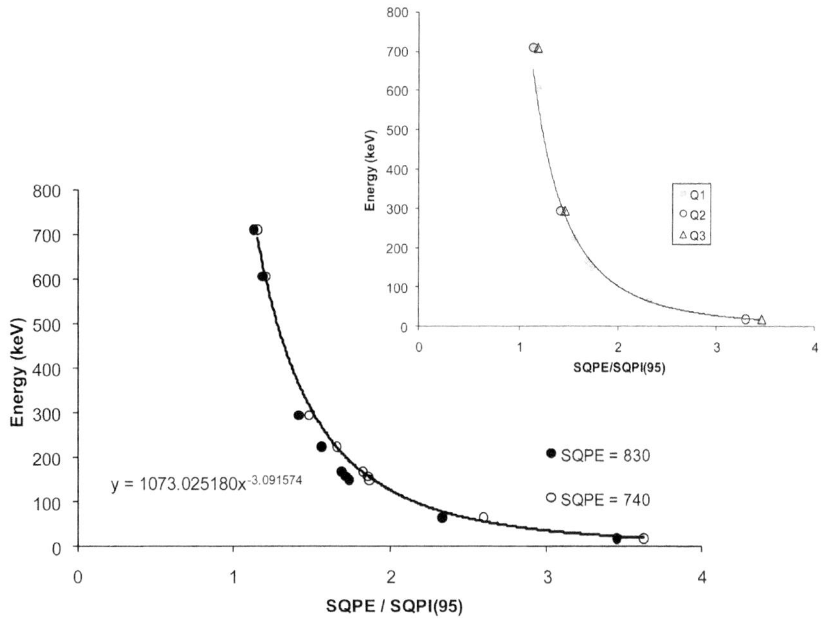

FIGURE 6.159 Application of peak shape factors for radionuclide identification. *From Warwick et al. (2013), reprinted with permission © 2013 American Chemical Society.*

by the quenching. However, radionuclide populations are clearly defined and therefore could be used for identification. In addition, the presence of alpha-emitters could be confirmed but no specific radionuclide identification is possible. This second relation permits the identification of the different radionuclides present but with some limitations, as the discrimination between ^{131}I and ^{192}Ir, that could be solved as a second step with the use of the previous SQPE SQPI relation.

XIX. AIR luminescence counting

An innovative approach to α-particle counting is the use of a liquid scintillation counter to detect luminescence events caused by the α-particle excitation of N_2 in air. The technique was devised by Takiue and Ishikawa (1979b). As explained by these investigators, the alpha particle passing through air dissipates its energy, forming excited nitrogen molecules N_2^*, excited nitrogen molecular ions N_2^{+*}, and excited nitrogen atomic ions N^{+*} as follows with relative energy consumptions as given in parenthesis:

$$N_2 + W_\alpha \longrightarrow \begin{cases} N_2^* & (50\%) \\ N_2^{+*} + e^- & (45\%) \\ N^{+*} + N + e^- & (5\%) \end{cases} \qquad (6.163)$$

where W_α is the α-particle energy dissipated in air, e^- is an electron, and N is the nitrogen atom. The emission spectra are derived from entirely N_2^* and N_2^{+*} bands. Emission wavelengths of the air luminescence are found in the region of 220–520 nm, which is within the range of sensitivity of photomultiplier tubes used in conventional liquid scintillation analyzers. Sand et al. (2014) demonstrate with a calibrated ^{239}Pu source that alpha particles generate 19 ± 3 photons/MeV of alpha-particle energy released in air at normal pressure (22°C and 43% relative humidity), and this relation is linear over the energy range of 0.3 –5.1 MeV.

The oxygen in air (21%) does not contribute to the air luminescence, but rather acts as a quencher by forming a complex with N_2^*, leading to a radiationless deactivation of excited molecular nitrogen. Argon, with a natural concentration in air of only 1%, undergoes luminescence. However, its photon emissions make little contribution to the overall nitrogen luminescence because of the low concentration of argon in air compared to that of nitrogen (78%).

The sample preparation and counting procedure recommended by Takiue and Ishikawa (1979a) is simple and relatively inexpensive. Approximately 0.1 g of a solution of α-particle emitting nuclide is deposited onto a 0.84 mg/cm^2 thick Tetoran film, dried with an infra-red lamp, and inserted into an empty liquid scintillation glass counting vial so that the film and sample remain suspended in the air around the center of the vial.

The sample may be counted as such without further preparation, or the air in the vial may be purged with pure nitrogen as a gas scintillator.

When the nuclides are counted in air, counting efficiencies of 72% and 74% for ^{210}Po and ^{241}Am, respectively, are obtained, and purging with pure nitrogen leads to an increase of 10%–15% in counting efficiency. The method offers several advantages: (1) there is no concern about quenching except for that of oxygen, which remains constant from sample to sample; (2) there is no need to be concerned about the preparation and handling of costly scintillators, as air is the only medium of detection; (3) the technique offers excellent reproducibility from sample to sample because of the absence of variable quenching agents; (4) the spectral emissions of nitrogen overlap well with the spectral response of photomultiplier tubes used in liquid scintillation analyzers, obviating the need for wavelength shifters; and (5) normal air containing 78% nitrogen may be used as the detection medium.

The main disadvantages of the air luminescence method are (1) reduced counting efficiency, as liquid scintillation with fluor cocktail provides 100% counting efficiency for α-emitters even under high levels of quench; (2) there is no α-radionuclide spectroscopy capabilities, since the pulse height spectrum is a peak produced by the air luminescence photons; (3) the limitation of the sample size to about 0.1 mL of radioactive solution that is deposited and dried on the transparent film; and (4) there is no possibility of analyzing β-emitters in the sample, that is, simultaneous α/β discrimination and analysis is not feasible.

These original findings of Takiue and Ishikawa were expanded by Takiue (1980) for the routine analysis of α-emitters on smear (swipe) paper in air without scintillation cocktail. The technique involves swiping a 100 cm^2 contaminated area with a smear paper (25 mm diameter, 30 mg/cm^2) at a pressure of about 0.2 kg/cm^2. The contaminated smear sample, thus obtained, is placed directly in an empty glass scintillation counting vial with the contaminated side of the paper facing upward and measured without adding a liquid scintillation cocktail, as illustrated in Fig. 6.160.

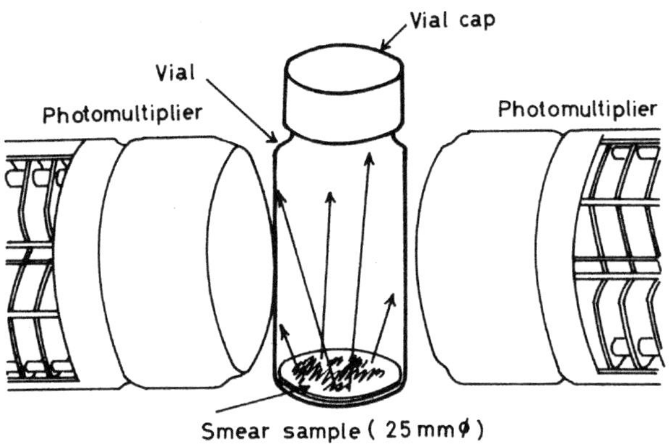

FIGURE 6.160 Measurement geometry for a smear sample in air (contaminated side of paper is facing upward) and detector photomultiplier tubes of a liquid scintillation spectrometer. *From Takiue (1980). Printed with permission from Wolters Kluwer Health/Lippincott, Williams & Wilkins © 1980 Health Physics Society.*

The α-particle counting efficiency is reduced because of the 2π counting geometry with the smear sample at the bottom of the counting vial. Takiue (1980) observed variation in smear removal efficiencies from different contaminated surfaces. The counting efficiencies of ^{210}Po or ^{241}Am as smear samples on paper varied between 10% and 20% from sample to sample. The following advantages of the air luminescence method for monitoring α-emitter radioactive contamination can be cited: (1) sample preparation for α-contamination surface monitoring is simple and inexpensive, because no scintillation cocktail is required or cocktail waste produced; and (2) low backgrounds are achieved ($\sim$9 CPM with the instrumentation used by Takiue), providing a detection limit of 1 x 10^{-7} $\mu Ci/cm^2$ for loose contamination. Disadvantages of the air luminescence method are (1) much lower and more variable counting efficiencies (10%–20%) are obtained in air luminescence analysis of smear samples, compared to a constant 100% counting efficiency obtained by liquid scintillation analysis, and (2) β-emitters cannot be analyzed by air luminescence, that is, simultaneous α/β discrimination and analysis with air luminescence is not possible.

Caution must be taken to assure that there are no significant activities of β particles with energies in excess of 180 keV, which is the threshold energy for the production of Cherenkov radiation by β particles (electrons) in glass. For samples containing high-energy β-emitters, Takiue (1980) demonstrates that a double-region counting technique could be used to discriminate between pulses produced by Cherenkov photons and air luminescence.

Further studies by Homma et al. (1987) demonstrated the possibilities of measuring ^{222}Rn and its daughter nuclides in air in the gaseous phase by air luminescence counting. A counting efficiency of 42% was achieved. The endpoint of the air luminescence spectrum produced by ^{222}Rn is about 18 keV, hence a preset counting region normally used for ^{3}H liquid scintillation analysis, could be used for the air luminescence analysis.

In the liquid scintillation analysis of ^{222}Rn, Murase et al. (1989a) demonstrated that air luminescence produced by ^{222}Rn and its daughters in the gaseous phase above the liquid scintillator can give rise to significant air luminescence counts. Therefore, to avoid error caused by air luminescence in the LSA of ^{222}Rn, currently approved methods require filling the liquid scintillation vial to a total volume of 20 mL with sample and fluor cocktail. For example, the procedure used by the U.S. EPA (EPA/EERF-Manual-78-1) for the analysis of radon in water (see Passo and Cook, 1994) calls for the injection of 10 mL of freshly sampled water into 10 mL of an organic accepting liquid scintillation cocktail (*e.g.*, Opti-Fluor O). The ^{222}Rn partitions into the organic fluor phase, and insignificant gaseous space remains in the standard 20-mL liquid scintillation counting vial for any significant air luminescence interference. To avoid any interference from luminescence by ^{222}Rn alpha particles in the air above the liquid scintillation cocktail, Murase et al. (1999) selected a lower-level discriminator setting of 25 keV. Thus, all pulse events that may originate from air luminescence would be rejected, as these are found in the 0–18 keV region only.

Murase et al. (1989b) demonstrated the air luminescence analysis of ^{210}Po, ^{238}U, and ^{241}Am. The technique involves spotting small amounts (0.1 mL) of aqueous solutions of the α-emitters on the bottom of glass liquid scintillation vials and evaporating the samples to dryness with an infrared lamp. The counting efficiencies obtained were between 33.3 and 33.7 ± 0.23%. The air luminescence pulse height spectra correspond to the region of 0–18 keV for pulse heights calibrated on an energy scale. Therefore, a counting region preset for the analysis of ^{3}H by LSA can be used for the air luminescence counting. Following air luminescence counting, the samples can be recovered from the vials for other studies. The counting efficiencies for the air luminescence determinations are easily determined by subsequent liquid scintillation analysis of samples. These can be carried out at random or for as many vials as desired. Liquid scintillation analysis after air luminescence was carried out by Murase et al. (1989b) after dissolving the radionuclide residue by wetting the bottom of the vial with 0.5 mL H_2O followed with liquid scintillation cocktail. The liquid scintillation analysis of α-emitters is 100%. Therefore, simply the activities (DPM) of samples obtained by liquid scintillation analysis can be used to calculate the air luminescence counting efficiency, as the count rate by LSA is also the disintegration rate. A disadvantage of the air luminescence analysis is the very small sample size of 0.1 mL that can be analyzed. Larger samples might leave too thick a residue at the bottom of the scintillation vial after drying, which would produce samples with significant self-absorption.

Air luminescence counting was used by Homma et al. (1996) to standardize ^{222}Rn samples that are applied to calibrate detectors utilized for the determination of ^{222}Rn in air. The standardization of ^{222}Rn as described by Homma et al. (1996) involves collecting a sample of ^{222}Rn from approximately 20 mL of sample of a 0.1 M HCl solution of 20 kBq $^{226}RaCl_2$ with a syringe and then transferring the ^{222}Rn gas to a standard liquid scintillation counting vial, which was filled with air and closed with a silicon rubber stopper. The ^{222}Rn in the vial is allowed to sit for 3.5 h before measurement to allow the daughters ^{218}Po, ^{214}Pb, ^{214}Bi, and ^{214}Po to reach transient equilibrium with the ^{222}Rn. This sample is referred to as the ^{222}Rn "standard". The ^{226}Ra solution from which the ^{222}Rn is taken does not have to be a standardized solution, because the ^{222}Rn determination is not based on secular equilibrium between ^{226}Ra and ^{222}Rn. It is not even necessary to know how much ^{222}Rn from the ^{226}Ra solution was actually removed. It is necessary only to standardize the ^{222}Rn that is transferred into the rubber stopper-sealed scintillation counting vial. Standardization of the gaseous ^{222}Rn is done by air luminescence counting. The air luminescence counting efficiency of the ^{222}Rn was determined by using the counting efficiency of a standardized ^{210}Po source. The ^{210}Po counting efficiency (E) was determined by calculating $E = \alpha/A$, where α is the air luminescence count rate of a ^{210}Po source and A is the α-particle emission rate of the ^{210}Po source expressed in α particles per second in a solid angle of 2π steradians. Homma et al. (1996) underscore the importance of measuring the air luminescence counting efficiency of ^{222}Rn, as instrument models may vary in optical properties, PMT properties, and circuit properties. Morita-Murase et al. (2001, 2003) have pursued further the application of air luminescence counting to the measurement of ^{222}Rn. They standardized a ^{222}Rn sample collected with a syringe from the air space above a $^{226}RaCl_2$ solution. The ^{222}Rn was injected into a small quartz tube (7 mm in diameter by 40 mm in length) sealed at both ends with silicon rubber stoppers. After 3.5 h, transient equilibrium of ^{222}Rn with its daughters ^{218}Po, ^{214}Pb, ^{214}Bi, and ^{214}Po is reached. The activity of the ^{222}Rn was then determined via analysis of the daughter ^{214}Pb γ-rays with a Ge semiconductor well detector. The small quartz tube with known ^{222}Rn activity was placed into a standard empty liquid scintillation vial, and the vial agitated until the quartz tube was broken liberating the ^{222}Rn into the air of the vial. The ^{222}Rn was counted with a liquid scintillation analyzer to provide the air luminescence counting efficiency of ^{222}Rn for the particular liquid scintillation analyzer used. An air luminescence counting efficiency of 42.6 ± 0.2% for ^{222}Rn was reported. Morita-Murase et al. (2003) report that the low-level count mode (LLCM) of the PerkinElmer TriCarb 3170 TR/SL should not be used for air luminescence counting, because the LLCM rejects pulse events that are not produced in liquid scintillator. Instead, the normal count mode (NCM) may be used with the PerkinElmer Tri-Carb instruments.

XX. Liquid scintillation counter performance

This chapter so far has dealt with the basic principles and procedures of liquid scintillation analysis. The remainder of the chapter will review the methods used to optimize the performance of the LSA. This is particularly important when sensitivity is key to our measurements. In addition, we need methods that can be used to assess the performance of the LSA, commercial or custom made, to insure the reliable and reproducible data from the instrumentation.

A. Instrument normalization and calibration

The first item to be addressed is that of assessing performance and calibration of the LSC to be sure that the activity data (DPMs) are accurate and reproducible. The first area to review is the calibration methods that are used for LSCs. One method is to use a flame sealed standard in scintillation cocktail, usually a ^{14}C source as the calibration normalization source for the LSC. This sealed standard should be a calibrated standard such as a one traceable to a primary standard from a national institute of standards so that its DPM is accurately known and it produces light pulses similar to those found in real samples. Jaubert et al. (2006) report that flame-sealed purely organic liquid scintillation sources can be used as reference sources for periods of more than 10 years, provided the sources are stored at ambient temperature in reduced light.

The calibration procedure of liquid scintillation analyzers varies according to the manufacturer of commercial instrumentation. One method of calibration (*e.g.*, PerkinElmer Tri-Carb LSAs) involves loading an unquenched ^{14}C standard source into the counting chamber, the high voltage for each of the PMTs is set automatically to obtain the optimum ^{14}C efficiency, and the ^{14}C endpoint energy of the standard is fixed at 156 keV with the minimum background. Such a calibration should be performed at least weekly as long as the instrument is being used. Another method of calibration (PerkinElmer Quantulus LSAs) uses a GaAsP LED placed very closed to the PMTs that flashes 60 times per minute and emits a constant light flux. The output from each PMT is analyzed automatically and with the use of a feedback circuit, the high-voltage is modified in order to correct deviations of output from the PMTs. In this way, changes in light detection and electronic parts of the PMT are corrected and stability is guaranteed (McCormac, 1992). This normalization is done automatically every time a sample is measured.

B. Assessing LSA performance

Assessing the performance of the LSA when a new counter is bought or built or on a routine basis is important to have assurance that the instrument is operating within acceptable parameters and to have a standing record over time of the instrument performance. On one hand, when a new counter is introduced in the laboratory, several tests should be performed before starting with routine analysis with two objectives: to assure the correct operation of the instrument and to define the parametric limits for correct and known operation of the counter. These limits will be used later to check the performance of the instrument during routine analysis. On the other hand, to maintain good laboratory practice and satisfy regulatory agencies, it is often necessary to have records of the instrument performance on a routine basis, such as daily, weekly, or monthly, to provide proof that deviations in instrument performance do not have any effect on the analytical results.

1. New commercial counters

When a new commercial counter is introduced in the laboratory, several measurements should be performed to confirm that the counter is working correctly and following the producer specifications. It is also useful to define the conditions in which future measurements are going to be performed since, in agreement with good practices in analysis, it is necessary to measure under known conditions. Basic assessment is done through measurements suggested by the LSA producer. These measurements are focused to reproduce technical specifications of the counter regarding background, counting efficiency for ^{3}H and ^{14}C standards and figure of merit for ^{3}H and ^{14}C in predefined counting windows. Some producers also include recommendations regarding counting efficiency for alpha-emitters or quenching assessment through the quenching indicator of a nonquenched sample.

Since each commercial LSA present their own configuration options, an evaluation of the performance of each option is usually needed. This is more relevant in the case of new models of LSA in which the data or experiences reported in the bibliography by other users is scarce. Wanke et al. (2012) evaluated how different parameters of the metrological version of a new HIDEX 300 SL detector affect the activity determination of several radionuclides though the TDCR method. Results of activity were compared to the data obtained with a well-known and evaluated detector. Parameters evaluated include threshold setting, coincidence time, and linearity. Threshold evaluation (Fig. 6.161) permits an adjustment of the detection to the maximum sensitivity without excessive noise. In the work performed, measurements at a value of 45 resulted in excessive noise counts, and therefore the threshold setting was adjusted to 50. The coincidence time parameter permits the determination of the time interval in which pulses from

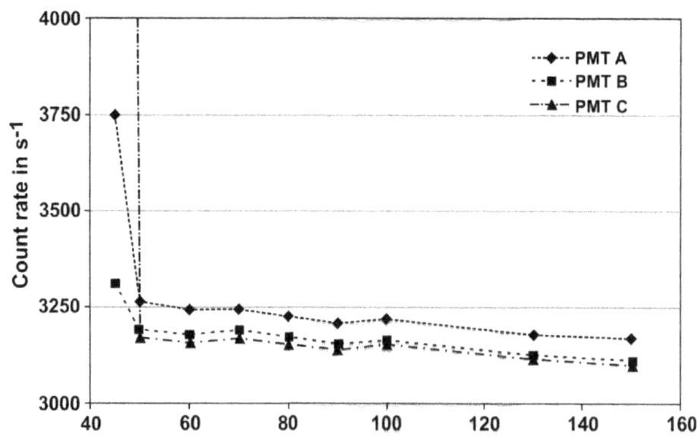

FIGURE 6.161 Single count rates in the PMTs over the threshold setting in the measurement of ^{63}Ni *From Wanke et al. (2012), reprinted with permission from Elsevier © 2012.*

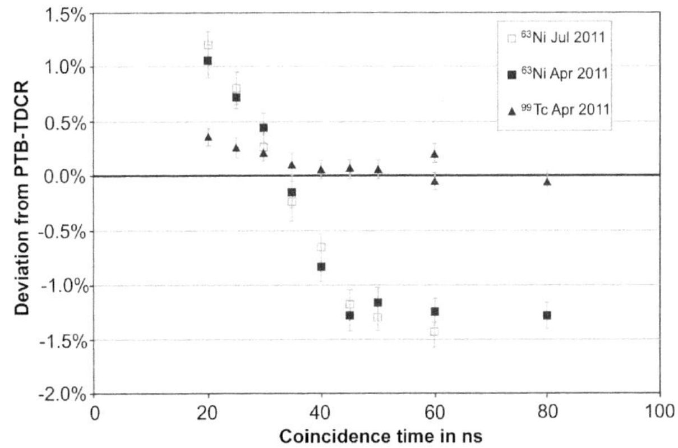

FIGURE 6.162 Deviations in the measurement of ^{63}Ni and ^{99}Tc over the coincidence time setting. The threshold was 50. *From Wanke et al. (2012), reprinted with permission from Elsevier © 2012.*

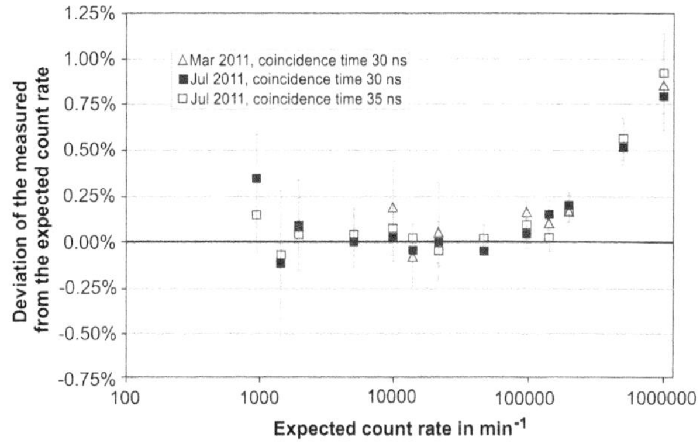

FIGURE 6.163 Linearity measurements—Deviations of the measured count rates from the expected ones in the measurement of ^{241}Am. The bars indicate only statistical uncertainties. *From Wanke et al. (2012), reprinted with permission from Elsevier © 2012.*

each PMT are considered to come from the same disintegration and can be counted as a true signal. A coincident time that is too short can lead to the loss of some scintillations of few photons; whereas a coincident time that is too long can lead to the counting of noise signals. Fig. 6.162 shows that for ^{63}Ni and ^{99}Tc source, a coincidence time of 30 ns was the optimum option. Another feature evaluated by Wanke et al. (2012) was the linearity of the counter. Stability of the counter at high count rates is usually not needed in environmental analysis, but in fields like metrology in which high count rates are measured, an evaluation of the linearity should be performed to define the limits in which highly precise measurements are obtained. The linearity of the HIDEX 300 SL for the analysis of a ^{241}Am standard solution with Ultima Gold^{tm} AB was demonstrated to be good for counts rates below 100,000 cpm; however, deviation regarding the expected count rate starts to be significant at 1,000,000 cpm (Fig. 6.163). Lost of linearity is usually attributed to the dead-time configuration of the counter and how the live-time is calculated.

Final evaluation of this LSA (HIDEX 300 SL) was done by a bilateral comparison performed by two different laboratories (PTB and ENEA) by Kossert et al. (2014c). In this comparison, six LSA instruments were used to determine the activity of LS sources containing ^{90}Sr: four LSAs from the PTB (one noncommercial and three commercial instruments including the new HIDEX 300 SL being one of them); two from ENEA (both commercial including another new HIDEX 300 SL being one of them). The scheme of the comparison is shown in Table 6.28. Results obtained (Fig. 6.164) show that activities were in good agreement, since the uncertainty bars are overlapped and the slightly higher values for both HIDEX 300 SL (0.25% for PTB and 0.32% for ENEA) can be attributed to loss of linearity at the high count rates of the samples (2500–3000 cps).

One key parameter of an LSA that should also be tested is the quenching indicator. Minne et al. (2008) evaluated the stability of the SQP(E) parameter in a Quantulus LSA using a ^{152}Eu gamma source with samples of beta-emitters of different energy with the same quenching level and high activity level (around 1000 Bq). They observed that for $^{90}Sr/^{90}Y$ samples, the quenching indicator changed with the counting time of the external standard. Moreover, for the same level of quenching, samples containing $^{90}Sr/^{90}Y$ or ^{85}Sr presented different quenching indicators (788(9) and 749(0.5)). This behavior was related to the energy of the radionuclides measured. When the spectrum obtained in the measurement of the beta-emitter was at higher energies than the spectrum from the Compton electrons generated by the gamma source, the quenching parameter was defined by the position of the spectrum of the radionuclide. In order to prevent deviations from the calculation of the quenching indicator, Minne et al. (2008) recommend the use of a constant and high (10′ or 15′) counting time for the external standard.

Another strategy also used to assess the performance of a new counter is to measure the same samples with the new counter and an old counter used in routine analysis. With this aim, Jäggi et al. (2014) performed the analysis of ^{241}Pu and ^{90}Sr in several complex samples (waste and soils) in their new commercial detector (Hidex 300 SL) and compared the results with those obtained with their routine counters (PerkinElmer Tricarb 2550, Tricarb 2200 and Quantulus) and other techniques (ICP-MS). Analysis included quenching correction (using TDCR and tSIE) of the counting efficiency since samples presented color. In the case of ^{241}Pu, the results from the analysis of a soil (IAEA-375) and a radioactive slurry sample showed that performance of the new counter was comparable to the routine ones. On the contrary, the analysis of ^{90}Sr in soils yielded a little discrepancy between the detectors. In this case, the authors recommended that some of the features of

TABLE 6.28 Proposed comparison scheme.

No.	Action	Date
1	PTB prepares two sets of LS sources of the radionuclide ^{89}Sr	Week 0
2	The sources will be measured at PTB using established measurement equipment for primary standardization (LS with TDCR and CIEMAT/NIST efficiency tracing)	Week 1
3	The sources will be measured in the Hidex TDCR counter at PTB. At least one measurement will be carried out with defined instrument settings according to an agreed protocol	Week 1
4	One set of sources will be shipped to ENEA	Week 2
5	The sources will be measured in the Hidex TDCR counter at ENEA. At least one measurement will be carried out with defined instrument settings according to an agreed protocol	Week 3
6	ENEA sends sources back to PTB	Week 4
7	Test measurements at PTB of both sets of sources to confirm that sources are stable	Week 5
8	ENEA submits results to PTB	Week 6
9	ENEA also submits file(s) with measured raw data to PTB which then analyzes data of both laboratories with the same software	Week 6
10	Joint report on the comparison including recommendations for other users	After week 10

From Kossert et al. (2014). Reprinted with permission from Elsevier © 2014.

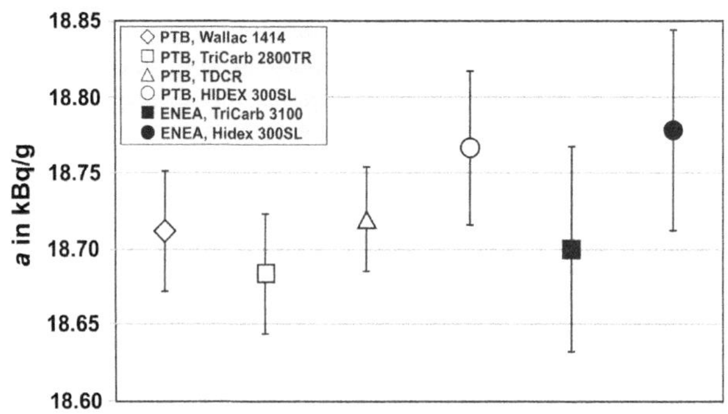

FIGURE 6.164 Linearity measurements—Deviations of the measured count rates from the expected ones in the measurement of ^{241}Am. The bars indicate only statistical uncertainties. *From Kossert et al. (2014), reprinted with permission from Elsevier © 2014.*

the instrument should be checked before it will be used in routine analysis.

2. New custom-made counters

Some specific applications, such as activity measurements by means of the TDCR method with counters with three PMT in metrology, ultra low-level measurements in underground laboratories or on-line continuous analysis, require the use of an LSA with special features. These special characteristics are not implemented in commercial LSAs, and as consequence, custom-made LSAs are built for these applications. Some of these custom-made LSAs take advantage of new instrumental developments, like digital electronics, which facilitate its implementation. In these cases, an optimization process through simulations to know the ideal configuration of the LSA usually precedes the assessment of the correct working of the counter.

One example of the use of simulations in the optimization of the LSA performance, prior to the LSA development, is the Ultra-Low Background Liquid Scintillation Counter (ULB LSC) developed by the Pacific Northwest National Laboratory. Bernacki et al. (2015) reported the evaluation of the optical features of the detection chamber though Monte Carlo simulation using TracePro software. Simulation was done considering a 20 mL vial of ground glass filled with liquid with a refraction index of 1.56 and wavelength emission at 420 nm. Several chamber designs were evaluated (Fig. 6.165) including the classical chamber (vial facing PMT), spherical chambers of different radii and chambers with collinear and noncollinear light collection

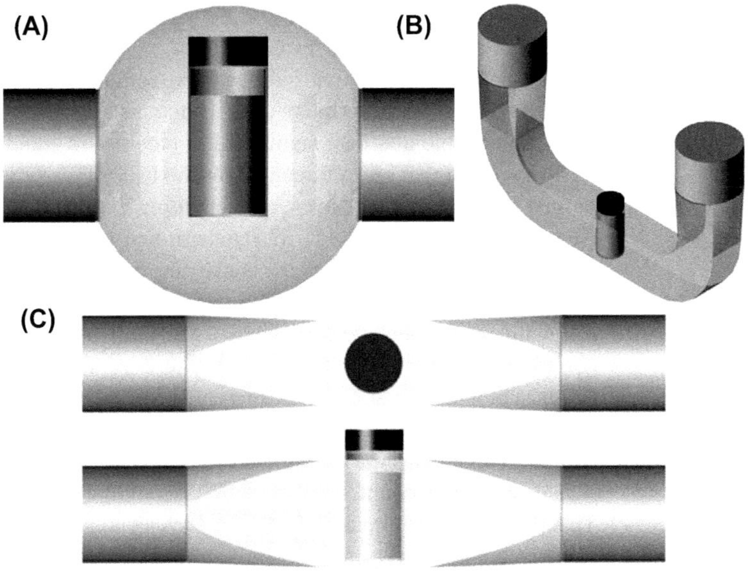

FIGURE 6.165 (A)LSC model with a spherical reflective collection chamber added. The chamber has a radius of 50.8 mm and the assembly is centered on the LSC cocktail column. (B) LSC U-shaped design that can accommodate lead shielding between the PMTs and LSC vial to reduce the effects of NORM background originating in either PMT. (C) Linear hollow cavity reflective light guide for the LSC system with reduced rectangular cross section and protruding vial cap, both top and side views. Rectangular-to-circular light guide sections on both ends mate the light guide to 50.8 mm (45 mm clear aperture) PMT face plates. *From Bernacki et al. (2015), reprinted with permission from OSA publishing © 2015.*

paths. Chambers were evaluated in terms of photon collection efficiency, irradiance map of the PMTs and time-of-flight plot. Results obtained showed that a U-shaped chamber provided the better collection efficiencies whereas TOF response, in spite of the distance between PMT, was acceptable (i.e., lower than 1.5 ns).

For the design of the shielding of the LSA to reduce background, Erchinger et al. (2015) used Monte Carlo simulation with GEANT 4. In a preliminary evaluation of the approach, the gamma-ray flux associated with U/Th/K in the underground laboratory was simulated and measured with a NaI detector. Agreement between simulated and experimental data (Fig. 6.166) showed the validity of the simulation. The final simulation accounted for various measurement parameters including a detection chamber with U-shape geometry and materials of the LSA included: copper, lead of two activities, two PMTs (with its bases), plastic scintillator veto panels, and borate polyethylene teal (Fig. 6.167). They explored the background originated by external gamma sources, lead shield, radioactivity in the IFHC copper, radioactivity in the reflective coating, radioactivity in the PMTs, background from the vial, cross-talk, and neutron background. Total background was estimated to be of 14(1.2) cpd, being cooper shielding (47%), lead shielding (25%), and external gamma rays (14%) the main sources of background.

Regarding the performance assessment of custom-made counters, there are several examples based on counters with three PMTs devoted to activity measurements by the TDCR method. The TDCR method is a primary standardization method of beta-emitting radionuclides (or decaying through more complex decay schemes) and most of the national

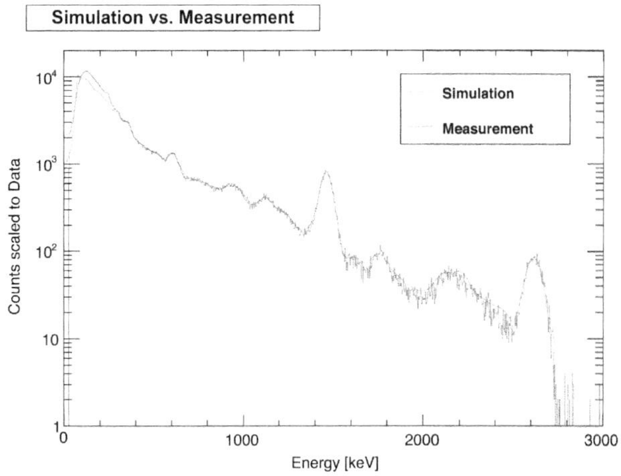

FIGURE 6.166 Comparison of simulated and measured room background for 3 in. × 3 in. NaI detector. *From Erchinger et al. (2015), reprinted with permission from Elsevier © 2015.*

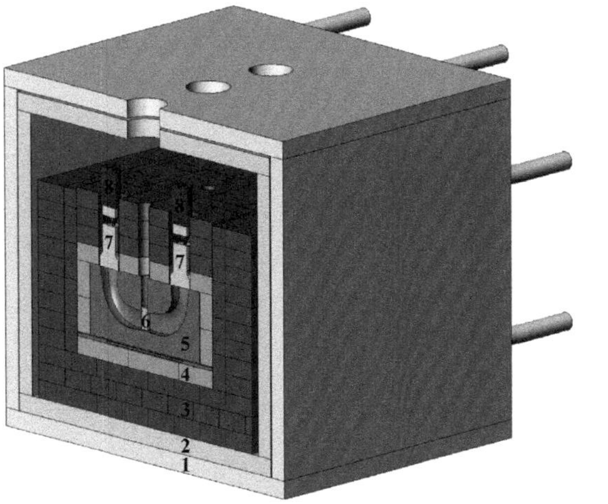

FIGURE 6.167 Cut away view of the shield components. (1) Outer layer of plastic scintillator veto panels (gray). (2) Borated polyethylene (teal). (3) Lead shielding (dark gray). (4) Low background lead shielding (gray). (5) Hollow copper light guide (orange). (6) Liquid scintillation vial (gray). (7) Photomultiplier tubes (PMTs) (light gray). (8) PMT bases (red). See text for additional description. *From Erchinger et al. (2015), reprinted with permission from Elsevier © 2015.*

metrology institutes, and some other laboratories, have developed their own TDCR detector. As a consequence, each TDCR is unique, and a specific performance assessment has to be performed for each one. The differences between detectors lie on the type of PMT used (e.g., classical PMT, channel-PMT, miniature-PMT), reflective material of the counting chamber, electronics (analogical or digital), intended use (portable or non-portable) and applicable for measurements in combination or not with other means of measurement (e.g., 4π(LS)-γ). In spite of the differences, performance assessment is usually based on the same principles, namely, compare the activity determined with the new TDCR with the certified activity of standards measured and/or with the activity determined using a reference detector (TDCR or LSA) from the same institute or from a reference institute. In the case of performance assessment of the electronics, a comparison is usually done regarding the MAC3 unit, the electronics developed at the LNHB (Bouchard et al, 2000). Finally, the instrument performance assessment usually includes an evaluation of parameters such as threshold, dead-time, coincidence resolving time, and linearity. Table 6.29 summarizes the strategies used for the different metrology institutes.

3. Routine instrument performance assessment

A routine Instrument Performance Assessment (IPA) is carried out generally with a set of standards that are traceable to a primary standard from a national institute of standards. These standards can be the same as those used for calibrating the instrument.

The IPA is normally carried out on a regular basis and the results stored and printed in a tabular and/or graphic form with a time and date stamp. Eight parameters should be assessed using a set of three unquenched standards (*e.g.*, purged with argon in a flame-sealed counting vial), namely, (i) a tritium standard, (ii) a ^{14}C standard, and (iii) a background standard, *i.e.* a standard consisting only of scintillation cocktail. The three standards can measure the following instrument performance: (1) tritium efficiency, (2) tritium background, (3) tritium figure of merit, (4) tritium chi-square, (5) ^{14}C efficiency, (6) ^{14}C background, (7) ^{14}C figure of merit, and (8) ^{14}C chi-square. The tritium and ^{14}C efficiencies are determined by the LSA using two sealed standards counted in predefined counting regions. This requires approximately 1 min for each sample, as these samples are generally of sufficiently high activity that only short count times are needed. This test should be performed daily. The backgrounds are determined using a sealed background standard with two preset counting regions for tritium and ^{14}C. This requires approximately 60 min because an accurate CPM value of the background needs to be obtained to assure reproducibility. Background assessment should be performed at least weekly. The ^{3}H figure of merit and ^{14}C figure of merit performance values are determined using the E^2/B calculation, which is the percent counting efficiency squared divided by the background count rate in CPM. This is an excellent indicator of any change in performance for the LSC. It can be calculated using the ^{3}H and ^{14}C standards, and it should be performed daily. The final performance parameters are the ^{3}H chi-square and ^{14}C chi-square. This is a monitor of short-term stability in the LSA and involves counting the ^{3}H and ^{14}C sealed radionuclide standards 20 times for 0.5 min each. The chi-square tests are used to confirm that the count rate measurements vary within the normal and expected deviations; that is, the count rates are neither too consistent nor too variable to be true. The chi-square tests should be performed weekly. By monitoring and recording these eight parameters, Good Laboratory Practices (GLP) can be satisfied.

The instrument performance assessment (IPA) can be either performed manually by the LSC users with the data placed into a spreadsheet or automatically by an LSC that has automatic IPA and instrument calibration. Fig. 6.168 shows a plot of the data obtained with an automatic IPA program on a PerkinElmer Tri-Carb LSA. This type of graph can be displayed, stored in computer memory, and printed for each of the eight IPA parameters described previously. In any given graph for any of the eight parameters, such as the example in Fig. 6.168A, an average line is calculated and the one, two, and three sigma values are provided so that the user can evaluate any trends or

TABLE 6.29 Strategies used for TDCR counter assessment.

References	TDCR detector characteristics	Measurements	Performance assessment
Zimmerman et al. (2004a)	MAC3 unit	^{3}H;^{54}Mn;^{63}Ni;^{90}Y;^{103}Pb;^{204}Tl	• Comparison with certified activity • Comparison with calorimetry measurements • Comparison with CN measurements • Comparison with integral β-counting • Comparison with 4π β-γ counting
Arenillas and Cassette (2006)	EMI 9804A PMTs and MAC3 unit	^{63}Ni	• Comparison with LNHB TDCR system; RC POLATOM TDCR system and CSIR NML TDCR system
Arenillas et al 2006	Ortec 265 PMT and MAC3 unit	^{3}H	• Comparison with activity traceable to LNHB TDCR counter
Ivan et al. (2008)	Channel PMT tubes, TiO_2 reflective paint, and MAC3 unit	^{3}H;^{55}Fe;^{63}Ni;^{90}Sr	• Validation using an LED light pulser. • Comparison with LNHB TDCR system
Nähle et al. (2010)	Burle 8850 PMTs, OP.DI.MA reflective material, and MAC3 unit	^{3}H	• BIPM intercomparison • Comparison with CN measurements in two commercial detectors
Mo et al. (2010)	FPGA based electronics	^{3}H Coincidence resolving time evaluation	• Comparison with certified activity • Comparison with MAC3 unit based values (using the same counter)
Johansson and Sehpthon (2010)	Electron Tubes 9807A, diffusive Scotch tape, NaI(Tl) detector, and MAC3 unit	^{3}H;^{55}Fe	• Comparison with certified activity • CCRI(II) comparison exercises
Halter et al. (2014)	XP2020Q, teflon chamber, MAC3 unit or FPGA, and X-ray tube	2.7; 8.7; 17.3 keV X-ray	• Comparison with ^{3}H activity determined with the same counter and model.
Mini et al. (2014)	Hamamatsu R7600-U200, Digital pulse processing with DT5720 and DT5751 digitalizers	Single photoelectron peak. ^{90}Sr–^{90}Y	• Comparison with activity determined with other TDCR detectors.
Nähle et al. (2014)	Portable TDCR detector, channels photomultipliers, OP.DI.MA reflective material, N978 digitizer, and FPGA.	Threshold adjustment Linearity Dead time Coincidence resolving time ^{3}H;^{45}Ca;^{36}Cl;^{55}Fe;^{63}Ni;^{89}Sr;^{90}Sr–^{90}Y;^{90}Y;^{99}Tc;^{147}Pm;^{229}Th;^{241}Pu	• Comparison with certified activity
Johansson et al. (2014)	Hamamatsu R6095 PMT, cylindrical chamber of aluminum and MAC3 unit. Portable	^{3}H;^{63}Ni;^{99}Tc;^{241}Pu	• CCRI(II) comparison exercise • Bilateral comparison with NIST • Other TDCR detector • Gas counting
Capogni and Antohe (2014).	Hamamatsu R7600-U200, PTFE chamber, digital pulse processing with DT5720 Portable	^{63}Ni;^{99}Tc Pulse shape discrimination	• CCRI(II) comparison exercise samples. • Comparison with CN measurements • Comparison with commercial TDCR counter
Capogni and De Felice (2014).	Hamamatsu Photonics type H1949-51, PTFE chamber, digital pulse processing with DT5720	Single photoelectron peak ^{3}H;^{63}Ni;	• CCRI(II) comparison exercise samples

Continued

TABLE 6.29 Strategies used for TDCR counter assessment.—cont'd

References	TDCR detector characteristics	Measurements	Performance assessment
			• Comparison with commercial TDCR counter
Nedjadi et al. (2015)	Burle 8850 PMTs, extruded duralumin optical chamber painted with TiO_2, and MAC3 unit.	^{45}Ca;^{14}C;^{63}Ni Resolving time	• Comparison with certified activity
Ziemeck et al. (2016)	9214B photomultipliers, NaI(Tl) crystal, aluminum chamber coated with reflective paint, FPGA electronics	Threshold adjustment Linearity Dead time and coincidence resolving time ^{3}H;^{241}Am;^{63}Ni;^{14}C;^{60}Co.	• Comparison with CN • Comparison with other TDCR counter
Mitev et al. (2017)	Hamamatsu R7600-200 square PMTs, PTFE and POM optical chamber, nanoTDCR electronics	Threshold adjustment ^{3}H;^{241}Am;^{63}Ni;^{14}C.	• Comparison with LNHB TDCR system • Comparison with MAC3 unit

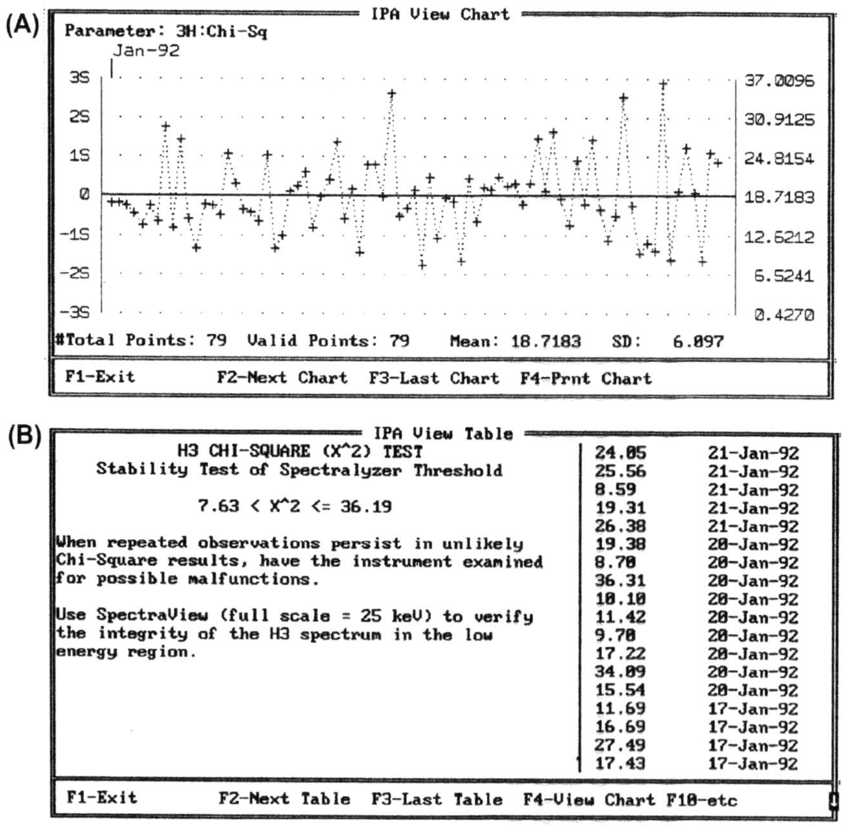

FIGURE 6.168 Chi-square instrument performance assessment (IPA) data displayed on the computer monitor of a liquid scintillation analyzer (a, upper chart) as a chart and (b, lower table) as a table with instructions if measured parameters are obtained too frequently outside of recommended limits. © *1998 PerkinElmer, Inc. Printed with permission.*

outlying points. If an outlying point is observed, this information is printed on the final data output. In addition, a series of suggestions can be provided so that the user can assess the problem further and take preventative action if any parameters are too frequently outside of recommended limits. An example is provided in Fig. 6.168B. Possible problems that might be indicated are counting contamination, PMT failure, dirty PMTs, and electronic failures. Jaubert et al. (2006) and Chałupnik and Mielnikow (2009) provide thorough treatments of the importance of instrument performance assessment or the implementation of a quality assurance system for the monitoring of the long-term stability of LSAs.

C. Optimizing LSC performance

The procedures for determining sample activities in DPM in routine sample analysis are dealt with in detail in previous parts of this chapter. This section will describe methods used to optimize performance particularly for low-count-rate samples when LSA sensitivity is most important. There are several methods of optimizing performance, and these will be described in detail in this section.

1. Counting region optimization

If a radionuclide, such as ^{14}C, is to be analyzed and a low level of detection is required, the counting region, defined by lower level (LL) and upper level (UL) pulse height discriminators, must be optimized for best performance measured by the figure of merit (E^2/B). The figure of merit is calculated as the square of the percent counting efficiency of the radionuclide of interest divided by the background count rate expressed in CPM. Optimizing the counting sensitivity (E^2/B) with spectrum analysis of sample and background is achieved by the following procedure:

1. Count a known activity (DPM) of the radionuclide of interest and store the sample spectrum, as illustrated in Fig. 6.169 (upper plot), in the computer memory. The sample size, sample chemistry, and amount of cocktail used should be similar to those of the experimental samples.
2. Prepare a blank sample (no radionuclide) plus cocktail in a counting vial to measure background. Count the blank sample and store the background spectrum, as illustrated in Fig. 6.169 (lower plot). It is important to prepare the sample and background (blank) counting vials in Steps 1 and 2 with the same cocktail and any other chemical components in identical amounts to provide the same chemistry and the same quench levels in both counting vials. For example, if the experimental samples will have a 50% water load (e.g., 10 mL H_2O + 10 mL cocktail), then the blank background vial should be prepared similarly, but without radioisotope. Ideally, the background counting vial should have the same chemical composition and QIP, that is, the same tSIE, SQP(E) or H#, as the sample vial with the exception that the background vial will not contain any added radionuclide.
3. Start with a wide-open counting region defined with the LL discriminator at 0 keV and the UL discriminator at the sample pulse height spectrum endpoint, and measure the count rates from the sample and background vials in the wide-open counting region. From the count rate data from the sample and background calculate the figure of merit E^2/B.

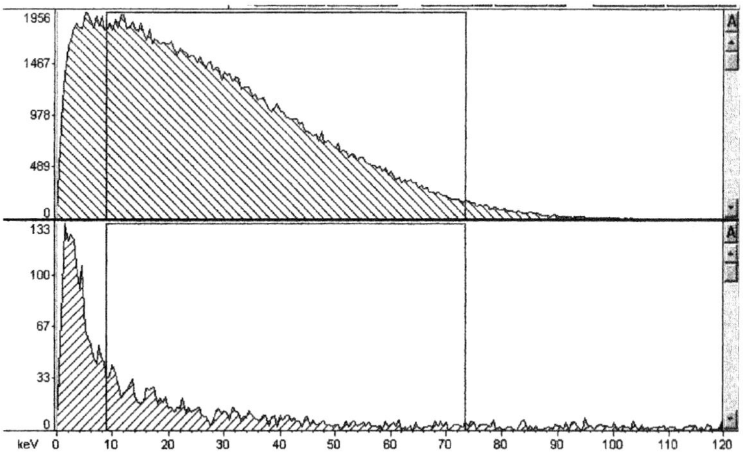

FIGURE 6.169 Pulse height spectra of a ^{14}C sample (above) and a background or blank sample (below). For this particular sample and blank, chemical composition, sample load, cocktail, and vial type, the LL (left) and UL (right) discriminator settings of 9.0–73.5 were found to provide region optimization. In this example, the reduced counting region produced a sacrifice of 18% in the ^{14}C counting efficiency but a reduction of 54% in the background counts. Region optimization is specific to radionuclide, sample-cocktail chemistry or quench level, and vial type and size (L'Annunziata, M.F., 1996, unpublished work of the writer).

4. Repeat Step 3 by increasing the LL discriminator by suitable increments and calculating the E^2/B after each change of the LL discriminator setting.
5. Decrease the UL discriminator by suitable increments and measure the E^2/B for each setting of the LL discriminator already tested. Record the E^2/B for each of the LL-UL counting regions tested.
6. Select the optimum counting region as that region defined by the LL-UL settings that gave the highest E^2/B.

The above procedure can be carried out using a "Replay" option that is included with many liquid scintillation analyzers. This allows the variation of the LL and UL discriminator settings to obtain new count rates from computer-stored pulse height spectra without having to recount the sample or background. Also, certain liquid scintillation analyzers are equipped with SpectraWorks or similar computer program that will (1) process automatically the sample and background spectra, (2) calculate the optimum LL-UL discriminator settings for a counting region, and (3) provide a value of the E^2/B for that optimum region. SpectraWorks will also calculate the lower limits of detection (LLD) or minimal detectable activity (MDA) for a particular radionuclide, counting efficiency, and background obtained. The LLD or MDA is defined further on in this chapter. For more information on SpectraWorks, see Floeckher (1995), which is a reference available on-line.

An example of results obtained with region optimization can be taken from the following case:

A sample of ^{14}C and a background blank were prepared similarly in liquid scintillation counting vials with the exception that the background vial had no radionuclide added. For a wide-open counting region of 0−156 keV defined where LL and UL discriminators are calibrated in keV, the instrument gave a 95% counting efficiency for the ^{14}C, a background count rate of 25.0 CPM, and a calculated E^2/B of 361. After region optimization according to the foregoing procedure, an optimum counting region of 18−102 keV was found, which gave a ^{14}C counting efficiency of only 63.1% but a background of 3.38 CPM, and a calculated E^2/B of 1178. Region optimization caused a reduction by more than 30% in the counting efficiency from 95% to 63.1% but an even greater reduction of more than 10-fold in the background. The counting region of 18−102 keV is the optimum counting region for the particular sample and instrument used in this example. Region optimization is important when sample count rates are low (close to background) and low levels of detection are required.

Another strategy used to determine the optimum counting window consists of the use of the computing power to calculate the figure of merit for all combinations of counting windows (Chapon et al., 2016). This strategy permits, with the aid of 3D representation of the data obtained, to see the regions of the spectrum where figure of merit is higher and simplifies the selection of the best counting window (Fig. 6.170, Figs. 6.171 and Fig. 6.172).

It is important to emphasize here that an optimum counting region so determined is good only for the particular radionuclide, sample size, vial type, cocktail-sample chemistry (*i.e.*, quench level), and liquid scintillation analyzer used. Therefore, we should keep in mind that when samples vary in quench, the optimized counting regions (window) vary as well. If the quench level between samples varies only slightly, one can widen slightly the optimum counting region to accommodate small changes in spectral shifts.

2. Vial size and type

Optimization of performance for the LSC is achieved also by selection of the proper size of vial and vial material (*e.g.*, glass or plastic). The data in Table 6.30 provide ^3H counting efficiencies, backgrounds, and figures of merit (E^2/B) for a standard (wide-open) counting region and an optimized counting region for two counting vial types and

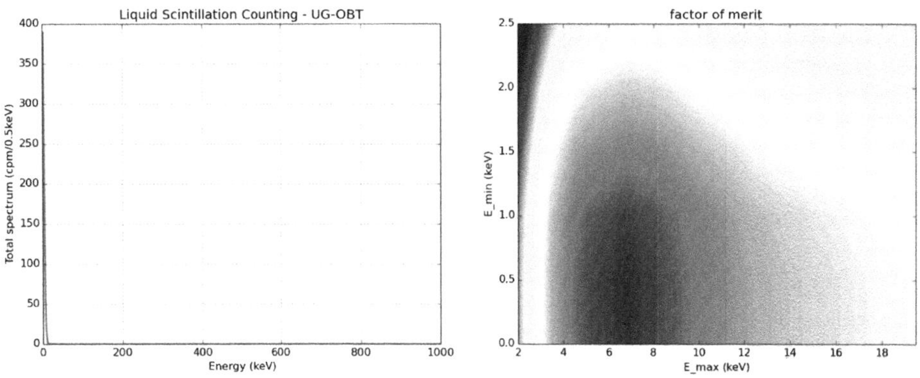

FIGURE 6.170 Organically Bound Tritiated (OBT) solution spectrum and optimization algorithm of the figure of merit for UltimaGold cocktail. *From Chapon et al. (2016) reprinted with permission from Elsevier © 2016 under the terms of the Creative Commons Attribution License (CC BY).*

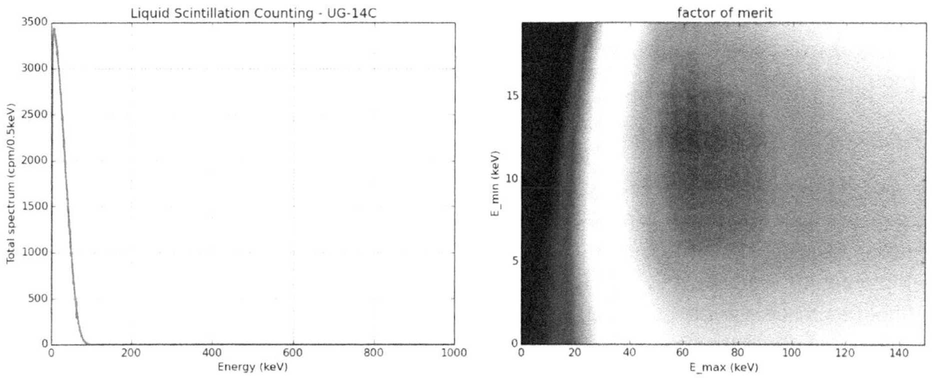

FIGURE 6.171 ^{14}C spectrum and optimization algorithm of the figure of merit for UltimaGold cocktail. *From Chapon et al. (2016) reprinted with permission from Elsevier © 2016 under the terms of the Creative Commons Attribution License (CC BY).*

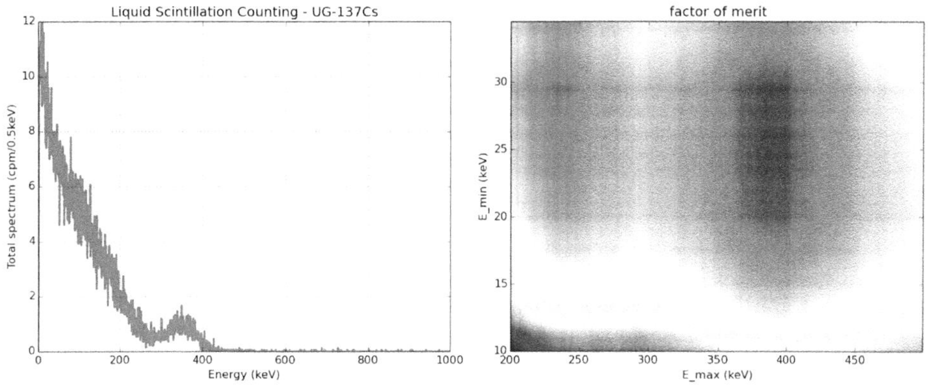

FIGURE 6.172 ^{137}Cs spectrum and optimization algorithm of the figure of merit for UltimaGold cocktail. *From Chapon et al. (2016) reprinted with permission from Elsevier © 2016 under the terms of the Creative Commons Attribution License (CC BY).*

TABLE 6.30 Tritiated water in ultima gold LLT scintillation cocktail in glass and plastic, large and small vials, performance comparison.

			Counting region 0–18.6 keV			Counting region 0.5–5.0 keV		
Vial size (mL)	Vial material	Water (%)	Efficiency[a] (%)	Background (CPM)	E^2/B	Efficiency[a] (%)	Background (CPM)	E^2/B
20	Glass	0.5	30.95	16.87	56.8	28.91	8.04	103.9
20	Plastic	0.5	31.71	10.15	99.1	29.31	3.88	221.1
7	Glass	0.5	29.55	12.81	68.1	27.75	5.92	130.1
7	Plastic	0.5	30.10	7.82	115.8	28.10	2.58	306.1

[a]*% Counting efficiency = (CPM/DPM) (100).*

sizes in a particular liquid scintillation analyzer. The data in this table provide evidence of some well-known generalities that should be considered when optimizing counting performance. First, plastic vials produce a lower background count than the corresponding glass vials. In this example, a 15%–30% improvement in performance (E^2/B) was obtained with plastic vials. Second, small vials for low-volume samples provide a lower background with about the same counting efficiency as large vials. Third, the optimized counting region gives a better performance than the standard counting region. From the example taken in Table 6.30 the total performance was enhanced with small plastic vials over large glass vials in an optimized counting region by almost a factor of 3 (E^2/B 103.9 to 306.1). The only problem with using smaller vials is that the sample capacity is lower, and if maximum sample size is required, the larger vials should be used.

3. Cocktail choice

Choosing the proper liquid scintillation fluor cocktail optimizes counting performance in several aspects, including background, counting efficiency, stability, and loading capacity. When sample counts are low, it is best to choose a cocktail that can hold as much sample as possible and still provide the maximum counting efficiency with a minimum amount of background. Chapter 7, Volume 1 provides a great deal of information that can be helpful in selecting a cocktail as a function of sample chemical properties to optimize counting performance. Two examples of detailed examinations of the performance of several common cocktails were reported by Verrezen et al. (2008) and Komosa and Slepecka (2010). Such cocktail examinations usually take place at the same time as vial types and sizes are tested.

4. Counting time

The error associated with the measurement of sample counts and count rate is a function of counting time. As demonstrated in Chapter 8, Volume 2, the standard deviation of the counts collected and the calculation of the associated count rate are reduced according to the length of the counting time. This is particularly relevant when the sample counts are very close to background, and accurate measurements of sample counts and background counts must be made to distinguish their difference in magnitude within reasonable error limits. Therefore, liquid scintillation counting performance can be optimized by increasing the sample and background counting times. The longer the counting time, the lower is the detection level or minimal detectable activity (MDA). The MDA also referred to as the lower limits of detection (LLD) in units of Bq/L can be calculated by the following equation from Prichard et al. (1992) as described by Passo and Kessler (1993), Passo and Cook (1994), and Biggen et al. (2002):

$$LLD = \frac{L_D}{60EVTX} \tag{6.164}$$

where *LLD* is the lowest activity concentration in Bq/L that yields a net count above background with a 95% probability; the detection limit, L_D, according to Currie (1968) and Hurtgen et al. (2000), is the true net signal that can be detected with a given probability; 60 is a factor for conversion of DPM to Bq (*i.e.*, 60DPM/1Bq); *E* is the fractional detection efficiency (CPM/DPM); *V* is the sample volume in liters, *T* is the count time in minutes; and *X* is any factor that is relevant, such as, decay correction, chemical yield, etc. As described by Passo and Cook (1994), when the background and sample are counted as pairs, that is, at the same counting time, and the background counts (B) collected are > 70, the detection limit, at a 95% confidence level, may be written as

$$L_D = 4.65\sqrt{B} \tag{6.165}$$

Other working expressions of detection limits (L_D) may be used and these are described by Prichard et al. (1992) and Passo and Cook (1994). When the lower limit of detection (LLD) is expressed in the less common units of pCi/L, the conversion factor 60 of Eq. (6.164) is replaced with the factor 2.22 for the conversion of DPM to pCi (i.e., 2.22 DPM/pCi).

To illustrate the application and interpretation of calculated lower limits of detection (*LLD*) and detection limits (L_D) let us take the following example to calculate the *LLD* and, from these calculations, determine how many sample counts must be acquired to reach a calculated *LLD*:

If a particular LSA counting region and cocktail chemistry in the analysis of ^3H yielded a total of 1300 background counts in a 500-minute counting time, that is, $B = 1300$ and background = 2.60 CPM, the counting efficiency for ^3H was 35% or 0.35, and the sample volume analyzed was 10 mL, the lower limit of detection according to Eq. (6.164) would be

$$LLD = \frac{4.65\sqrt{1300 \text{ counts}}}{(60 \text{ DPM/Bq})(0.35)(0.01 \text{ L})(500 \text{ minutes})} = 1.60 \text{ Bq/L} \tag{6.166}$$

In the above calculation the limit of detection is

$$L_D = 4.65\sqrt{1300 \text{ counts}} = 168 \text{ counts} \tag{6.167}$$

Therefore, to achieve the calculated *LLD* of 1.60 Bq/L, a minimum number of sample counts defined by

$$L_D + B = 168 + 1300 = 1468 \text{ counts} \tag{6.168}$$

must be registered with a 10 mL sample in a 500 min counting time to conclude that 1.60 Bq/L are detected with a 95% confidence level.

The denominator of Eq. (6.164) may include additional factors, such as chemical yield, if an extraction of a radionuclide is less than 100% or decay corrections, which is the variable *X* in Eq. (6.164). Detailed treatments of calculations involved in the determination of limits of detection are available from Currie (1968), Prichard et al. (1992), and Passo and Cook (1994). From Eqs. (6.164) and (6.165) and the calculation (6.166), we can see that increasing the counting time and reducing the background will reduce the lower limit of detection or minimal detectable activity. This is illustrated by Fig. 6.173 (Erchinger et al. (2015) where we can see that, by reducing the background count rate, we can achieve a desired *LLD* with a shorter counting time.

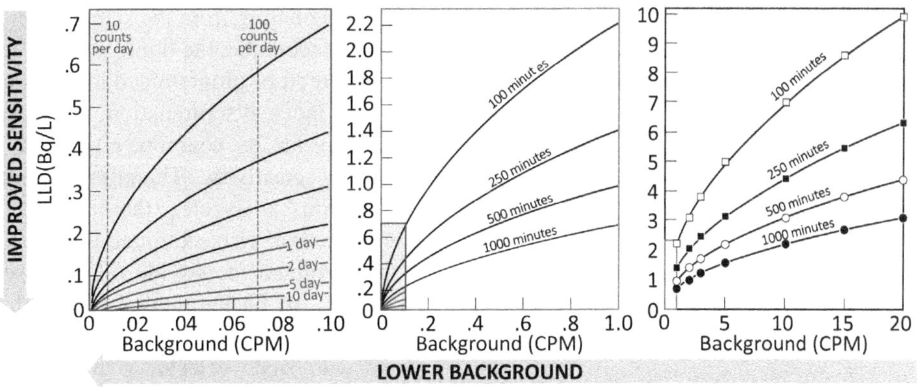

FIGURE 6.173 A "text book" example of the relation between *sensitivity* versus *background and count duration* with extrapolations to a shallow underground, low background liquid scintillation counter design. The "text book" example (right most panel) presents typical sensitivity levels for measurements of a 10 mL tritium sample given typical background rates and count duration and with 35% counting efficiency. The panels to the left progressively reduce the background rate in the proposed instrument and show the relevant sensitivity levels attained for differing count duration. The count duration curves derived from the Currie formulation are provided to guide the eye and may be impractical for some choices of background and count duration. In these figures the background rate is quantified in counts per minute (CPM) and the minimum detectable activity sensitivity is presented as a lower limit of detection (LLD) value. *From Erchinger et al. (2015), reprinted with permission from Elsevier © 2015.*

5. Background reduction

Often the easiest method of optimizing LSC performance is to reduce the background. As described previously in this chapter, region optimization, vial size, and vial material are important variables to control in the reduction of background. This section will discuss five other methods that can be used to reduce background and enhance performance for liquid scintillation analysis.

a. Temperature control

The original purpose of temperature control or cooling during the early years of liquid scintillation counting was to reduce background counts from thermal noise from the PMTs. Modern bialkalie photomultiplier tubes are manufactured to have a low noise even at room temperature. The purpose of temperature control in contemporary liquid scintillation counting is to control and maintain the sample temperature, sometimes required for optimizing sample chemistry even at extremes of external temperatures. The sample chemistry in some cocktails requires lower temperature to accept more sample, and as a result, temperature control is necessary.

b. Underground counting laboratory

This method of background reduction can be used for very low-level counting and this extreme measure has been taken more often in the past for the measurement of very low levels of naturally occurring radionuclides such as ^{3}H in ground water and ^{14}C dating. This involves building a special counting room below the ground to reduce high-energy cosmic radiation that may cause background counts in the liquid scintillation counter. However, modern pulse discrimination electronics, described briefly in Part d following, will reduce instrument backgrounds to levels at which underground laboratories are generally not needed for the conventional measurement of environmental radioactivity. The placement of detectors deep underground is still utilized for the liquid scintillation measurement of rare events, such as double-beta (ββ) decay and the detection and measurement of neutrinos, etc. as described in Sections XI and XII of this chapter.

c. Shielding

Shielding the instrument from external radiation can reduce background. There are two types of shielding, passive and active shielding.

Passive shielding is the "brute force" approach. This may require lead (up to 1000 kg) barriers to reduce background. The extra lead reduces background from environmental gamma photons associated with building materials and instrument construction materials. In addition, this extra lead can reduce the "soft" cosmic muon components of background. The lead shielding is usually lined internally with cadmium and/or copper to absorb any secondary X-rays and cosmic components.

Active shielding can be one of two types, a guard detector with or without anticoincidence circuitry and components. The guard detector is an external detector [*e.g.*, bismuth germanate (BGO)] that surrounds the area between the PMTs and the sample. The BGO has special properties described in detail in Chapter 9, Volume 1. BGO is a solid crystal scintillator with a high density and high atomic number. The high density and atomic number give BGO excellent "stopping power" for the detection of highly penetrating radiation. BGO can be used to discriminate against gamma and muon components of cosmic background. This property of BGO in combination with Time-Resolved LSC (TR-LSC) pulse discrimination electronics (described subsequently) can discriminate more background events.

Another approach to active shielding to reduce background is to manufacture a detector shield above, below, and surrounding the sample vial and PMTs consisting of an external counting chamber of a liquid or solid scintillator, such as mineral oil scintillator, plastic scintillator (*e.g.*, NE110), or even a NaI(Tl) crystal. In addition to the guard material, PMTs and an anti-coincidence circuit are added. The detector guard rejects much of the environmental gamma radiation, as well as soft and hard cosmic radiation.

d. Pulse discrimination electronics

Three different pulse discrimination methods are used, namely, pulse shape analysis (PSA), pulse amplitude comparison (PAC), and time-resolved LSC (TR-LSC).

PSA relies on the fact that a true nuclear decay signal produces a very prompt pulse in the LSC, whereas a background event produces a pulse with a longer decay time. By setting up the instrument to discriminate pulses on the basis of their shape, background can be discriminated from a true nuclear decay signal. This discrimination is specific and must be optimized for each vial material, vial size, and sample-cocktail chemistry.

The second method is PAC, which compares the amplitude of the pulse from each of the two PMTs. If the event occurs in the scintillation vial, photons of approximately the same intensity will be seen by both PMTs, but if the event originates from outside the sample, it will produce pulses in each of the PMTs of different pulse heights. If the PAC deviates significantly from unity, the pulse is discriminated as background.

The third method, TR-LSC, is a patented method of reducing background by discriminating against pulses on the basis of the number of afterpulses that occur following an initial pulse event over a given period of time. These afterpulses are most commonly created by background events. Afterpulses are generally more numerous among nonquenchable background events and less numerous with events originating from the scintillation cocktail. With the TR-LSC technique, the liquid scintillation analyzer can be manufactured or programmed to count afterpulses, known as pulse index discrimination, which can identify background events by counting afterpulses with varying degrees of sensitivity. Therefore, among various liquid scintillation analyzers, the amount of pulse index discrimination for background rejection can be factory set or programmed to provide different sensitivities for the reduction of background (*e.g.*, high-sensitivity count mode, low-level count mode, and ultra-low-level count mode with BGO detector guard). Tables 6.31 and 6.32 illustrate how background can be reduced and performance enhanced significantly using this latter technique of background reduction. The data shown in these two tables might typically be found in a low-level counting laboratory. The samples are placed in the optimum counting geometry, optimum vial size and vial material, optimum cocktail, and counted in the optimum counting region with the TR-LSC method of electronic discrimination at various levels of pulse index discrimination with and without an active guard.

The data of Table 6.31 illustrate that the use of TR-LSC alone and in combination with an active BGO guard detector greatly increases the performance of the LSC for ^{14}C samples. This is also the case for many other radionuclide samples with low count rates. For the case of ^{14}C about 15–20 times better performance is obtained with the ultra-low-level TR-LSC and the BGO guard detector.

The data of Table 6.32 illustrate that the use of TR-LSC alone and in combination with a BGO active guard detector greatly increases the performance of the LSC for tritium samples in general with low count rates. For the case of tritium about a 10-fold better performance is achieved with the use of ultra-low-level TR-LSC in combination with the BGO guard detector.

TABLE 6.31 Radiocarbon Dating Samples as 3.5 mL Benzene with PPO/POPOP Scintillator[a].

Amount of pulse index discrimination	Detector guard	^{14}C efficiency (%)	Background (CPM)	Figure of merit (E^2/B)
None	No	83.45	9.67	720
Normal TR-LSC	No	81.87	7.07	948
High-sensitivity TR-LSC	No	78.51	4.74	1300
Low-level TR-LSC	No	70.10	1.38	3560
Ultra-low-level TR-LSC + BGO Active Guard	Yes	65.01	0.33	12,675

[a]*The optimized counting region 10–102 keV with various levels of TR-LSC pulse index discrimination applied to the pulse events.*

TABLE 6.32 Environmental Tritium Water Samples as 10 mL Water in 10 mL Ultima Gold LLT Scintillation Cocktail[a].

Amount of pulse index discrimination	Detector guard	^{3}H efficiency (%)	Background (CPM)	Figure of merit (E^2/B)
None	No	26.50	18.45	38
Normal TR-LSC	No	26.24	12.75	54
High-sensitivity TR-LSC	No	24.68	9.25	66
Low-level TR-LSC	No	22.59	3.33	153
Ultra-low-level TR-LSC + BGO Active Guard	Yes	20.01	1.00	400

[a]*The optimized counting region is 0.5–5.0 keV with various levels of TR-LSC pulse index discrimination applied to the pulse events.*

For additional information on time-resolved liquid scintillation counting (TR-LSC) see papers by Roessler et al. (1991) and Passo and Roberts (1996).

6. Conclusions

In conclusion, the best ways of optimizing performance of the LSC for low-level samples are to (1) optimize counting geometry, such as optimum sample size, vial size, and vial material; (2) increase counting time; (3) optimize counting region; (4) optimize cocktail selection; and (5) use shielding together with electronic discrimination and an active guard detection method to reduce background.

References

Abe, D., Saito, T., Nogata, Y., 2016. Rosmarinic acid regulates fatty acid and glucose utilization by activating the CaMKK/AMPK pathway in C2C12 myotubes. Fd. Sci. Technol. Res. 22 (6), 779–785.

Aburai, T., Takiue, M., Ishikawa, H., 1981. Quantitative measurement method of α-emitters by using liquid scintillation counter connected with multichannel pulse-height analyzer. Radioisotopes 30, 579–583.

for the MEG Collaboration Adam, J., et al., 2010. A limit for the μ → eγ decay from the MEG experiment. Nucl. Phys. B 834, 1–12.

Agostinelli, S., et al., 2003. Geant4–a simulation toolkit. Nucl. Instrum. Methods Phys. Res. Sect. A. 506, 250–303.

Agostini, M., et al., 2016. Search of neutrinoless double beta decay with the GERDA experiment. Nucl. Part. Phys. Proc. 273–275, 1876–1882.

and the Borexino Collaboration Agostini, M., et al., 2017. Borexino's search for low-energy neutrino and antineutrino signals correlated with gamma-ray bursts. Astropart. Phys. 86, 11–17.

Ahn, H.J., et al., 2013. Application of a wet oxidation method for the quantification of ^{3}H and ^{14}C in low-level radwastes. Appl. Radioat. Isot. 81, 62–66.

Akimov, D., et al., 2002. Measurements of scintillation efficiency and pulse shape for low energy recoils in liquid xenon. Phys. Lett. B 524, 245–251.

Akkoyun, S., 2013. Time-of-flight discrimination between gamma-rays and neutrons by using artificial neural networks. Annals Nucl. Energy 55, 297–301.

Alamelu, D., et al., 2006. Determination of ^{243}Am by pulse shape discrimination liquid scintillation spectrometry. Appl. Radiat. Isot. 64, 579–583.

Aleissa, K.A., Islam, M.S., Alshammari, H., 2006. Measurement of ^{210}Po and ^{210}Pb in groundwater using an alpha liquid scintillation counter, pp. 197–209. In: Chatupnik, S., Schönhofer, F., Noakes, J. (Eds.), LSC 2005, Advances in Liquid Scintillation Spectrometry, Radiocarbon. University of Arizona, Tucson, p. 449.

on behalf of XENON Collaboration Alfonsi, M., 2016. The Xenon dark matter project: from Xenon100 to Xenon1T. Nucl. Part. Phys. Proc. 273–275, 373–377.

Al-Haddad, M.N., Fayoumi, A.H., Abu-Jarad, F.A., 1999. Calibration of a liquid scintillation counter to assess tritium levels in various samples. Nucl. Instrum. Methods Phys. Res. Sect. A 438, 356–361.

Alharbi, T., 2015. Simple algorithms for digital pulse-shape discrimination with liquid scintillation detectors. Radiat. Phys. Chem. 106, 50–55.

for the Borexino Collaboration Alimonti, G., et al., 2002. Science and technology of Borexino: a real-time detector for low energy solar neutrinos. Astropart. Phys. 16, 205–234.

Alimonti, G., et al., 2009. The Borexino detector at the Laboratori Nazionali del Gran Sasso. Nucl. Instrum. Methods Phys. Res. Sect. A 600, 568–593.

Allison, J., et al., 2006. Geant4 developments and applications. IEEE Trans. Nucl. Sci. 53, 270–278.

Alongi, D.M., et al., 2003. The influence of fish cage aquaculture on pelagic carbon flow and water chemistry in tidally dominated mangrove estuaries of peninsular Malaysia. Mar. Environ. Res. 55, 313–333.

Altzitzoglou, T., 2004. Analysis of triple-label samples by liquid scintillation spectrometry. Appl. Radiat. Isot. 60, 487–491.

Altzitzoglou, T., 2008. Radioactivity determination of individual radionuclides in a mixture by liquid scintillation spectra deconvolution. Appl. Radiat. Isot. 66, 1055–1061.

Altzitzoglou, T., Larosa, J.L., Nicholl, C., 1998. Measurement of ^{90}Sr in bone ash. Appl. Radiat. Isot. 49 (9–11), 1313–1317.

Altzitzoglou, T., et al., 2002. Standardization of ^{89}Sr using three different methods. Appl. Radiat. Isot. 56, 447–452.

Altzitzoglou, T., Rožkov, A., 2016. Standardisation of the ^{129}I, ^{151}Sm and ^{166m}Ho activity concentration using the CIEMAT/NIST efficiency tracing method. Appl. Radiat. Isot. 109, 281–285.

Amaudruz, P., et al., 2009. Simultaneous reconstruction of scintillation light and ionization charge produced by 511 keV photons in liquid xenon: potential application to PET. Nucl. Instrum. Methods Phys. Res. Sect. A 607, 668–676.

Amiot, M.N., et al., 2012. Standardization of ^{64}Cu using an improved decay scheme. Nucl. Instrum. Methods Phys. Res. Sect. A 684, 97–104.

Andreotti, E., et al., 2011. Search for β^+/EC double beta decay of ^{120}Te. Astropart. Phys. 34, 643–648.

Aprile, E., Doke, T., 2010. Liquid xenon detectors for particle physics and astrophysics. Rev. Mod. Phys. 82 (3), 2053–2097.

Aprile, E., Mukherjee, R., Suzuki, M., 1990. A study of the scintillation light induced in liquid xenon by electrons and alpha particles. IEEE Trans. Nucl. Sci. 37 (2), 553–558.

Aprile, E., et al., 2005. Scintillation response of liquid xenon to low energy nuclear recoils. Phys. Rev. D 72, 072006.

Aprile, E., et al., 2006a. Detection of liquid xenon scintillation light with a silicon photomultiplier. Nucl. Instrum. Methods Phys. Res. Sect. A 556, 215–218.

Aprile, E., et al., 2006b. Simultaneous measurement of ionization and scintillation from nuclear recoils in liquid xenon for a dark matter experiment. Phys. Rev. Lett. 97, 081302.

Arenillas, P., Cassette, P., 2006. Implementation of the TDCR liquid scintillation method at CNEA-LMR, Argentina. Appl. Radiat. Isot. 64, 1500–1504.

Arisaka, K., et al., 2009. XAX: a multi-ton, multi-target detection system for dark matter, double beta decay and pp solar neutrinos. Astropart. Phys. 31, 63–74.

Arneodo, F., et al., 2000. Scintillation efficiency of nuclear recoil in liquid xenon. Nucl. Instrum. Methods Phys. Res. Sect. A 449, 147–157.

Arpesella, C., 1996. A low background counting facility at Laboratori Nazionali del Gran Sasso. Appl. Radiat. Isot. 47 (9/10), 991–996.

Asakura, K., et al., 2016. Search for double-beta decay of ^{136}Xe to excited states of ^{136}Ba with the KamLAND-Zen experiment. Nucl. Phys. A 946, 171–181.

Aspinall, M.D., et al., 2007a. The empirical characterization of organic liquid scintillation detectors by the normalized average of digitized pulse shapes. Nucl. Instrum. Methods Phys. Res. Sect. A 578, 261–266.

Aspinall, M.D., et al., 2007b. Verification of the digital discrimination of neutrons and γ rays using pulse gradient analysis by digital measurement of time of flight. Nucl. Instrum. Methods Phys. Res. Sect. A 583, 432–438.

Auger, M., et al., 2012. The EXO-200 detector, part I: detector design and construction. J. Instrument. 7, P05010.

Augeray, C., et al., 2015. Development of a protocol to measure iron-55 in solid matrices in the environment. J. Environm. Radioact. 141, 164–173.

Aupiais, J., 2004a. Deconvolution of alpha liquid scintillation spectra for quantitative analysis of actinide elements in water samples. Radiochim. Acta 92, 125–132.

Aupiais, J., 2004b. Rapid determination of uranium activity and concentration in water by alpha liquid scintillation with α/β discrimination. Anal. Chim. Acta 517, 221–228.

Aupiais, J., 2005. Radium measurement in water samples by α-liquid scintillation counting with α/β discrimination. Anal. Chim. Acta 532, 199–207.

Aupiais, J., Dacheux, N., 2000. Understanding the peak asymmetry in alpha liquid scintillation with β/γ discrimination. Radiochim. Acta 88, 391–398.

Aupiais, J., et al., 1998. Determination of ^{226}Ra in mineral drinking waters by α liquid scintillation with rejection of β-γ emitters. Anal. Chem. 70, 2353–2359.

Aupiais, J., et al., 1999. Study of neptunium measurement by alpha liquid scintillation with rejection of β-γ emitters. Anal. Chim. Acta 398, 205–218.

Aupiais, J., Aubert, C., Dacheux, N., 2003. Some rules to improve the energy resolution in alpha liquid scintillation with beta rejection. Radiochim. Acta 91, 63–69.

Avanzini, M.B., 2011. Experimental evidence of electron neutrino oscillations and validation of MSW–LMA model with Borexino. Prog. Part. Nucl. Phys. 66, 412–417.

Bae, J.W., Lee, U.J., Kim, H.R., 2017. Development of continuous inflow tritium measurement in water technology using electrolysis and a plastic scintillator. J. Radioanal. Nucl. Chem. 314 (2), 689–694.

Bagán, H., Tarancón, A., Rauret, G., García, J.F., 2008. Classical vs. evolved quenching parameters and procedures in scintillation measurements: the importance of ionization quenching. Nucl. Instrum. Methods Phys. Res. Sect. A 592, 361–368.

Bagán, H., Hartvig, S., Tarancón, A., Rauret, G., García, J.F., 2009. Plastic vs. liquid scintillation for ^{14}C radiotracers determination in high salt matrices. Anal. Chim. Acta 631, 229–236.

Bagán, H., Tarancón, A., Rauret, G., García, J.F., 2010. Alpha/beta pulse shape discrimination in plastic scintillation using commercial scintillation detectors. Anal. Chim. Acta 670, 11–17.

Bagán, H., Tarancón, A., Rauret, G., García, J.F., 2011a. Mixture quantification using PLS in plastic scintillation measurements. Appl. Radiat. Isot. 69, 898–903.

Bagán, H., Tarancón, A., Rauret, G., García, J.F., 2011b. Radiostrontium separation and measurement in a single step using plastic scintillators plus selective extractants. Application to aqueous sample analysis. Anal. Chim. Acta 686 (1–2), 50–56.

Bagán, H., Tarancón, A., Stavsetrab, L., Rauret, G., García, J.F., 2012. Determination of oil reservoir radiotracer ($S^{14}CN^-$) in a single step using a plastic scintillator extractive resin. Anal. Chim. Acta 736, 30–35.

Bagán, H., Ticoscón, A., Ye, L., García, J.F., 2014a. Crosslinked plastic scintillators: a new detection system for radioactivity measurement in organic and aggressive media. Anal. Chim. Acta 852, 13–19.

Bagán, H., Aboudou, B., Tarancón, A., García, J.F., 2014b. Alpha/beta indices determination in drinking water using plastic scintillation and evaporation to dryness. J. Radioanal. Nucl. Chem. 299, 533–542.

Bakhshandeiar, E., Johansson, L., Pearce, A., Keightley, J., 2011. Standardization of ^{64}Cu using two liquid scintillation methods. In: LSC 2010, Advances in Liquid Scintillation Spectrometry. Radiocarbon. University of Arizona, Tucson.

Bakir, Y.Y., Bem, H., 1995. Application of pulse decay discrimination liquid scintillation counting for routing monitoring of radioactivity in drinking water, pp. 293–299. In: Cook, G.T., Harkness, D.D., MacKenzie, A.B., Miller, B.F., Scott, E.M. (Eds.), Liquid Scintillation Spectrometry 1994, Radiocarbon. University of Arizona, Tucson, p. 396.

Banerjee, K., et al., 2007. Characteristics of Gd-loaded liquid scintillators BC521 and BC525. Nucl. Instrum. Methods Phys. Res. Sect. A 580, 1383–1389.

Barabash, A., 2011. Double beta decay experiments: past and present achievements. Nucl. Phys. B 221, 26–31.

Barabash, A.S., 2015a. Double beta decay experiments: present status and prospects for the future. Phys. Procedia 74, 416–422.

Barabash, A.S., 2015b. Average and recommended half-life values for two-neutrino double beta decay. Nucl. Phys. A 935, 52–64.

Barabash, A.S., Hubert, P., Nachab, A., Umatov, V., 2007. Search for β^{+}EC and ECEC processes in ^{74}Se. Nucl. Phys. A 785 (3–4), 371–380.

Barosi, G., Cazzola, M., Perugini, S., 1980. In: Liquid Scintillation Counting, Recent Applications and Development, vol. II. Academic Press, New York and London, pp. 517–523.

Barrera, J., Tarancón, A., Bagán, H., García, J.F., 2016. A new plastic scintillation resin for single-step separation, concentration and measurement of technetium-99. Anal. Chim. Acta 936, 259–266.

Bass, C.D., et al., 2013. Characterization of a ^{6}Li-loaded liquid organic scintillator for fast neutron spectrometry and thermal neutron detection. Appl. Radiat. Isot. 77, 130–138.

Basson, J.K., Steyn, J., 1954. Absolute alpha standardization with liquid scintillators. Proc. Phys. Soc. 67, 297–298.

Batistoni, P., et al., 2005. International comparison of measuring techniques of tritium production for fusion neutronics experiments. Status and preliminary results. Fision Eng. Design 75–79, 911–915.

Batistoni, P., et al., 2007. Neutronics experiment on a helium cooled pebble bed (HCPB) breeder blanket mock-up. Fision Eng. Design 82, 2095–2104.

Bayat, E., et al., 2012. A comparative study on neutron-gamma discrimination with NE213 and UGLLT scintillators using zero-crossing method. Radiat. Phys. Chem. 81, 217–220.

Becchetti, F.D., et al., 2016. Deuterated-xylene (xylene-d_{10}; EJ301D): a new, improved deuterated liquid scintillator for neutron energy measurements without time-of-flight. Nucl. Instrum. Methods Phys. Res. Sect. A 820, 112–120.

Becchetti, F.D., et al., 2017. Recent developments in deuterated scintillators for neutron measurements at low-energy accelerators. Nucl. Instrum. Methods Phys. Res. Sect. A 874, 72–78.

Beebe, K.R., Kowalski, B.R., 1987. An introduction to multivariate calibration and analysis. Anal. Chem. 59, 1007A–1017A.

Belanov, S.V., et al., 1997. The method of identifying radionuclides with the use of a liquid scintillation counter. RF Patent No 2 (120), 646.

Belanov, S.V., et al., 1998. The method of identifying radionuclides with the use of a liquid scintillation counter. RF Patent No 98 (106), 407.

Bell, P.R., 1948. The use of anthracene as a scintillation counter. Phys. Rev. 73 (11), 1405–1406.

Belli, P., et al., 2016. Search for 2β decay of 106Cd with enriched 106CdWO4 crystal scintillator in coincidence with four HPGe detectors. https://arxiv.org/abs/1603.06363v2.

Bellini, G., 2016. The impact of Borexino on the solar and neutrino physics. Nucl. Phys. B 908, 178–198.

Bem, H., Olszewski, M., Kaczmarek, A., 2004. Concentration of selected natural radionuclides in the thermal groundwater of Uniejów, Poland. Nikleonika 49 (1), 1–5.

Benetti, P., et al., 1993. Detection of energy deposition down to the keV region using liquid xenon scintillation. Nucl. Instrum. Methods Phys. Res. Sect. A 327, 203–206.

Benitez-Nelson, C.R., Buessler, K.O., 1998. Measurement of cosmogenic ^{32}P and ^{33}P activities in rainwater and seawater. Anal. Chem. 70, 64–72.

Bergeron, D.E., Laureano-Pérez, L., 2014. Micelle size effect on Fe-55 liquid scintillation efficiency. Appl. Radiat. Isot. 87, 282–286.

Bergeron, D.E., Fitzgerald, R.P., Zimmerman, B.E., Cessna, J.T., 2012. The effect of impurities on calculated activity in the triple-to-double coincidence ratio liquid scintillation method. Appl. Radiat. Isot. 70, 2170–2175.

Bergeron, D.E., Galea, R., Laureano-Pérez, L., Zimmerman, B.E., 2016. Comparison of ^{14}C liquid scintillation counting at NIST and NRC Canada. Appl. Radiat. Isot. 109, 30–35.

Bernacki, B., Douglas, M., Erchinger, J., Fuller, E., Keillor, M., Morley, S., Mullen, C., Orrell, J., Panisko, M., Warren, G., Wright, M., 2015. Optical design considerations for efficient light collection from liquid scintillation counters. Appl. Opt. 54, 2413–2423.

Bhade, S.P.D., et al., 2010. Standardization of calibration procedures for quantification of gross alpha and gross beta activities using liquid scintillation counter. J. Radioanal. Nucl. Chem. 284, 367–375.

Bhade, S.P.D., et al., 2017. Study of liquid scintillation alpha spectral properties. Applied. Radiat. Isot. 122, 121–126.

Bhade, S.P.D., Reddy, P.J., Anilkumar, S., et al., 2018. Calibration and optimization of alpha-beta separation procedures for determination of radium/radon in single- and two-phase liquid scintillation systems. J. Radioanal. Nucl. Chem. 315, 13.

Bialek, A., et al., 2016. A rope-net support system for the liquid scintillator detector for the SNO+ experiment. Nucl. Instrm. Methods Phys. Res., Sect. A 827, 152–160.

Bianchi, D., Cortese, P., Dellacasa, G., Merlano, I., 2005. Determination of the isotope ratio ^{234}U/^{238}U in liquid scintillation and environmental applications. Nucl. Instrm. Methods Phys. Res., Sect. A 553, 543–549.

Bienvenu, P., Cassette, P., Andreoletti, G., et al., 2007. A new determination of ^{79}Se half-life. Appl. Radiat. Isot. 65, 355–364.

Biggen, C.D., Cook, G.T., MacKenzie, A.B., Pates, J.M., 2002. Time-efficienct method for the determination of ^{210}Pb, ^{210}Bi, and ^{210}Po activities in seawater using liquid scintillation spectrometry. Anal. Chem. 74, 671–677.

Bignell, L.J., Mo, L., Alexiev, D., Hashemi-Nezhad, S.R., 2010a. The effect of multiple γ-ray interactions on ionization quenching corrections in liquid scintillants. Nucl. Instrm. Methods Phys. Res. Sect. A 614, 231–236.

Bignell, L.J., Mo, L., Alexiev, D., Hashemi-Nezhad, S.R., 2010b. Sensitivity and uncertainty analysis of the simulation of ^{123}I and ^{54}Mn gamma and X-ray emissions in a liquid scintillation vial. Appl. Radiat. Isot. 68, 1495–1502.

for the SNO+ Collaboration Biller, S., 2015. SNO+ with tellurium. Phys. Procedia 61, 205–210.

Binaei Bash, Z., Ghal-Eh, N., Bayat, E., Etaati, G.R., 2013. Lightguide influence on timing response of scintillation detectors. Radiat. Phys. Chem. 82, 1–6.

Birks, J.B., 1964. The Theory and Practice of Scintillation Counting. Pergamon Press, Oxford.

Bliznyuk, V.N., Seliman, A.F., Husson, S.M., Lvov, Y.M., DeVol, T.A., 2018. Hybrid nanoparticle–polymer brush composites for detection of low-level radiostrontium in water. Macromol. Mater. Eng. 303 (6), 1700651.

Boag, A.M., et al., 2015. Autoantibodies against cytochrome P450 side-chain cleavage enzyme in dogs (Canis lupus familiaris) affected with hypoadrenocorticism (Addison's disease). PLoS One 10 (11), e0143458. https://doi.org/10.1371/journal.pone.0143458.

Bobin, C., Bouchard, J., Hamon, C., Iroulart, M.G., Plagnard, J., 2007. Standardization of ^{67}Ga using a 4π(LS)$\beta-\gamma$ anti-coincidence system. Appl. Radiat. Isot. 65, 757–763.

Bonardi, M.L., Birattari, C., Groppi, F., et al., 2004. Determination of ^{125}I impurities in [^{123}I]labelled radiopharmaceuticals, by liquid scintillation counting: sensitivity of the method. Nucl. Instrum. Methods Phys. Res. Sect. B 213, 348–350.

Borthwick, L.A., Neal, A., Hobson, L., et al., 2008. The annexin 2-S100A10 complex and its association with TRPV6 is regulated by cAMP/PKA/CnA in airway and gut epithelia. Cell Calcium 44, 147–157.

Bouchard, J., Cassette, P., 2000. MAC3: and electronic module for the processing of pulses delivered by a three photomultiplier liquid scintillation counting system. Appl. Radiat. Isot. 52, 669–672.

Bourne, M.M., et al., 2016. Neutron detection in a high-gamma field using solution-grown stilbene. Nucl. Instrm. Methods Phys. Res. Sect. A 806, 348–355.

Bradshaw, C., Kumblad, L., Fagrell, A., 2006. The use of tracers to evaluate the importance of bioturbation in remobilising contaminants in Baltic sediments. Estuarine Coastal Shelf Sci. 66, 123–134.

Braizinha, B., Esterline, J.H., Karwowski, H.J., Tornow, W., 2010. Determination of the proton and alpha-particle light-response functions for the KamLAND, BC-501A and BC-517H liquid scintillators. Nucl. Instrum. Methods Phys. Res. Sect. A 623 (3), 1046–1049.

Bransome Jr., E.D., 1973. Liquid scintillation counting in nuclear medicine. Semin. Nucl. Med. 3 (4), 389–399.

Bray, L., et al., 2014. A high-affinity, radioiodinatable neuropeptide FF analogue incorporating a photolabile p-(4-hydroxybenzoyl)phenylalanine. Anal. Biochem. 453, 50–54.

Breitholtz, M., Wollenberger, L., 2003. Effects of three PBDEs on development, reproduction and population growth rate of the harpacticoid copepod *Nitocra spinipes*. Aquatic Toxicol 64, 85–96.

Broda, R., 2003. A review of the triple-to-double coincidence ratio (TDCR) method for standardizing radionuclides. Appl. Radiat. Isot. 58, 585–594.

Broda, R., Pochwalski, K., 1993. The ETDCR method of standardizing ^{55}Fe and ^{54}Mn. In: Noakes, J.E., Schönhofer, F., Polach, H.A. (Eds.), Liquid Scintillation Spectrometry 1992, Radiocarbon. University of Arizona, Tucson, pp. 255–260.

Broda, R., Pochwalski, K., Radoszewski, T., 1988. Calculation of liquid-scintillation detector efficiency. Appl. Radiat. Isot. 39, 159–164.

Broda, R., Péron, M.N., Cassette, P., Terlikowska, T., Hainos, D., 1998. Standardization of ^{139}Ce by the liquid scintillation counting using the triple to double coincidence ratio method. Appl. Radiat. Isot. 49 (9–11), 1035–1040.

Broda, R., Cassette, P., Maletka, K., Pochwalski, K., 2000. A simple computing program for application of the TDCR method to standardization of pure-beta emitters. Appl. Radiat. Isot. 52, 673–678.

Broda, R., Maletka, K., Terlikowska, T., Cassette, P., 2002. Study of the influence of the LS-cocktail composition for the standardization of radionuclides using the TDCR model. Appl. Radiat. Isot. 56, 285–289.

Broda, R., Cassette, P., Kossert, K., 2007. Radionuclide metrology using liquid scintillation counting. Metrologia 44, S36–S52. http://iopscience.iop.org/0026-1394/44/4/S06.

Broda, R., Dziel, T., Muklanowicz, A., 2012. Standardization of a ^{85}Sr solution by three methods. Appl. Radiat. Isot. 70, 2222–2226.

Broser, I., Herforth, L., Kallmann, H., Martius, U., 1948. Über den Elementarprozeβ der Lichtanregung von Leuchtstoffen III. Die Anregung des Naphthalins (Weitere Versuche mit dem Leichtmassenzähler). Zeitschrift f. Naturforsch. 3a, 6–15.

Buchtela, K., Tschurlovits, M., Unfried, E., 1974. Eine Methode zur Unterschiedung von α- und β-Strahlen in einem Flüssigszintillationsmessgerät. Int. J. Appl. Radiat. Isot. 25, 551–555.

Brun, A., et al., 2014. High paracellular nutrient absorption in intact bats is associated with high paracellular permeability in perfused intestinal segments. J. Experim. Biol. 217, 3311–3317.

Bukowski, T.R., Moffett, T.C., Revkin, J.H., Ploger, J.D., Bassingthwaighte, J.B., 1992. Triple-label β liquid scintillation counting. Anal. Biochem. 204, 171–180.

Cabrera, A., 2010. The double Chooz detector. Nucl. Instrum. Methods Phys. Res. Sect. A 617, 473–477.

Caccianiga, B., Giammarchi, M.G., 2000. Neutrinoless double beta decay with Xe-136 in Borexino and the Borexino counting test facility. Astropart. Phys. 14, 15–31.

Cadieux, J.R., 1990. Evaluation of a photoelectron-rejecting alpha liquid-scintillation (PERALS) spectrometer for the measurement of alpha-emitting radionuclides. Nucl. Instrum. Methods Phys. Res. Sect. A 299, 119–122.

Cadieux, J.R., Fugate, G.A., King III, G.S., 2015. An alpha–gamma coincidence spectrometer based on the photon–electron rejecting alpha liquid scintillation (PERALS®) system. Nucl. Instrum. Methods Phys. Res. Sect. A 783, 22–27.

Çakal, G., Güven, R., Yücel, H., 2015. An application of LSC method for the measurement of gross alpha and beta activities in spiked water and drinking water samples. Nukleonika 60 (3), 637–642.

Calaprice, F., the Borexino Collaboration, 2015. Solar neutrino results and future opportunities with Borexino. Phys. Procedia 61, 300–309.

Calhoun, J.M., Coursey, B.M., Gray, D., Karam, L., 1991. The standardization of ^{35}S methionine by liquid scintillation efficiency tracing with ^{3}H. In: Ross, H., Noakes, J.E., Spaulding, J.D. (Eds.), Liquid Scintillation Counting and Organic Scintillators. Lewis Publishers, Chelsea, MI 48118, pp. 317–323.

Campbell, D.B., Vetter, K., Henning, R., Lesko, K., Chan, Y.D., et al., 2008. Evaluation of radioactive background rejection in ^{76}Ge neutrino-less double-beta decay experiments using a highly segmented HPGe detector. Nucl. Instrum. Methods Phys. Res. Sect. A 587, 60–67.

Capogni, M., Antohe, A., 2014. Construction and implementation of a TDCR system at ENEA. Appl. Radiat. Isot. 87, 260–264.

Capogni, M., De Felice, P., 2014. A prototype of a portable TDCR system at ENEA. Appl. Radiat. Isot. 93, 45–51.

Capogni, M., De Felice, P., Fazio, A., Simonelli, F., D'Ursi, V., et al., 2006. Development of a primary standard for calibration of [^{18}F]FDG activity measurement systems. J. Phys. Conf. Ser. 41, 506–513. http://iopscience.iop.org/1742-6596/41/1/057.

Capogni, M., De Felice, P., Fazio, A., Latini, F., Abbas, K., 2008. Development of a primary standard for calibration of ^{64}Cu activity measurement systems. Appl. Radiat. Isot. 66, 948–953.

Capogni, M., Cozzella, M.L., De Felice, P., Fazio, A., 2012. Comparison between two absolute methods used for ^{177}Lu activity measurements and its standardization. Appl. Radiat. Isot. 70, 2075–2080.

Cassette, P., 1992. Specbeta programme de calcul du spectre en énergie des électrons émis par des radionucléides émetteurs beta. CEA Technical Note (Saclay: CEA/DAMRI/LPRI).

Cassette, R., 2001. Evaluation of the influence on wall effects on the liquid scintillation counting detection efficiency for the standardization of high-energy beta and alpha radionuclides. In: Moebius, S., Noakes, J.E., Schoenhofer, F. (Eds.), LSC 2001, Advances in Liquid Scintillation Spectrometry, Radiocarbon. University of Arizona, Tucson, p. 456.

Cassette, P., 2003. Evaluation of the influence of wall effects on the liquid scintillation counting detection efficiency for the standardization of high energy beta and alpha radionuclides, pp. 45–55. In: Mobius, S., Noakes, J.E., Schönhofer, J.E. (Eds.), LSC 2001. Advances in Liquid Scintillation Spectrometry, Radiocarbon. University of Arizona, Tucson, p. 456.

Cassette, P., 2011. TDCR in a nutshell. In: LSC 2010, Advances in Liquid Scintillation Spectrometry. Radiocarbon. University of Arizona, Tucson.

Cassette, P., 2016. Quench: a software package for the determination of quenching curves in liquid scintillation counting. Appl. Radiat. Isot. 109, 301–307.

Cassette, P., Vatin, R., 1992. Experimental evaluation of TDCR models for the 3 PM liquid scintillation counter. Nucl. Instrum. Methods Phys. Res. Sect. A 312, 95–99.

Cassette, P., Bouchard, J., 2003. The design of a liquid scintillation counter based on the triple to double coincidence ratio method. Nucl. Instrum. Methods Phys. Res. Sect. A 505, 72–75.

Cassette, R., Tartès, I., 2014. Experimental study of the influence of the counter and scintillator on the universal curves in the cross-efficiency method in LSC. Appl. Radiat. Isot. 87, 32–37.

Cassette, P., et al., 1998. Comparison of activity concentration measurement of ^{63}Ni and ^{55}Fe in the framework of the EUROMET 297 project. Appl. Radiat. Isot. 49, 1403–1410.

Cassette, P., et al., 2000. Analysis of detection-efficiency variation techniques for the implementation of the TDCR method in liquid scintillation counting. Appl. Radiat. Isot. 52, 643–648.

Cassette, P., et al., 2004. Measurement of a ^{103}Pd solution using the TDCR method by LSC. Appl. Radiat. Isot. 60, 439–445.

Cassette, P., et al., 2006a. Comparison of calculated spectra for the interaction of photons in a liquid scintillator. Example of ^{54}Mn 835 keV emission. Appl. Radiat. Isot. 64, 1471–1480.

Cassette, P., et al., 2006b. Measurement of photon absorption coefficients of liquid scintillators in the 5- to 12-keV energy range using a monochromatic x-ray source, pp. 125–133. In: Chałupnik, S., Schönhofer, F., Noakes, J. (Eds.), LSC 2005, Advances in Liquid Scintillation Spectrometry, Radiocarbon. University of Arizona, Tucson, p. 449.

Cassette, P., et al., 2010. Determination of ^{93}Zr decay scheme and half-life. Appl. Radiat. Isot. 68, 122–130.

Cassette, P., et al., 2010b. Study of the influence of the liquid scintillator in the Compton efficiency tracing method. Appl. Radiat. Isot. 68, 1510–1514.

Cassette, P., Capogni, M., Johansson, L., Kossert, K., Nahle, O., et al., 2013. Development of portable liquid scintillation counters for on-site primary measurement of radionuclides using the triple-to-double coincidence ratio method. In: Proc. 3rd Int. Conf. on Advancements in Nuclear Instrumentation, Measurement Methods and Their Applications, ANIMMA 2013 Conference, 23–27 June, Marseille, France.

Cassette, P., Butkus, P., Gudelis, A., Shilnikova, T., 2016. ^{3}H activity comparison between FTMC, VNIIM and LNE-LNHB. Appl. Radiat. Isot. 109, 41–43.

Cassette, P., et al., 2018. Results of the CCRI(II)-S12.H-3 supplementary comparison: comparison of methods for the calculation of the activity and standard uncertainty of a tritiated-water source measured using the LSC-TDCR method. Appl. Radiat. Isot. 134, 257–262.

Castagnet, X., Amabile, J.C., Cazoulat, A., et al., 2007. Diagnosis of internal radionuclide contamination by mobile laboratories. Radiat. Prot. Dosim. 125 (1–4), 469–471.

Ceccatelli, A., De Felice, P., 1999. Standardisation of ^{90}Sr, ^{63}Ni and ^{55}Fe by the $4\pi\beta$ liquid scintillation spectrometry method with ^{3}H-standard efficiency tracing. Appl. Radiat. Isot. 51, 85–92.

Cecil, F.E., Medlet, S.S., 1988. Gamma ray measurements during deuterium and ^{3}He discharges in TFTR. Nucl. Instrum. Methods Phys. Res. Sect. A 271, 628–635.

Cecil, F.E., Scorby, J.C., 1992. Investigation of the neutron detection properties of the fast liquid scintillator BC-505. Rev. Scientific Instrum. 63 (10), 4559. https://doi.org/10.1063/1.1143667.

Cennini, P., Revol, J.P., Rubbia, C., et al., 1999. Detection of scintillation light in coincidence with ionizing tracks in a liquid argon time projection chamber. Nucl. Instrum. Methods Phys. Res. Sect. A 432, 240–248.

Černý, J., Doležal, Z., Ivanov, M.P., et al., 2004. Study of neutron response and $n-\gamma$ discrimination by charge comparison method for small liquid scintillation detector. Nucl. Instrum. Methods Phys. Res. Sect. A 527, 512–518.

Cerutti, F., et al., 2017. Nuclear model developments in FLUKA for present and future applications. EPJ Web Conf. 146, 12005.

Cessna, J.T., 2002. The measurement of activity contained in a ^{32}P stainless-steel stent by destructive assay. Appl. Radiat. Isot. 56, 337–342.

Cessna, J.T., Zimmerman, B.E., 2010. Standardization of radium-223 by liquid scintillation counting. Appl. Radiat. Isot. 68, 1523–1528.

Cessna, J.T., Schultz, M.K., Leslie, T., Bores, N., 2008. Radionuclide calibrator measurements of ^{18}F in a 3 mL plastic syringe. Appl. Radiat. Isot. 66, 988–993.

Chałupnik, S., Mielnikow, A., 2009. Long-term stability of Quantulus LSC counter, pp, 211-218. In: LSC 2008, Advances in Liquid Scintillation Spectrometry Radiocarbon. University of Arizona, Tucson, p. 445.

Chandrasekaran, E.S., 1981. Measurement of iodine-125 by liquid scintillation counting method. Health Phys. 40 (6), 896–898.

Chang, Z., et al., 2011. Indium-loaded liquid scintillator for solar neutrino spectroscopy. Nucl. Phys. B 221, 337.

Chapon, A., Pigrée, G., Putmans, V., Rogel, G., 2016. Optimization of liquid scintillation measurements applied to smears and aqueous samples collected in industrial environments. Results Phy. 6, 50–58.

Chauvenet, B., Bé, M.-M., Amiot, M.-N., et al., 2010. International exercise on ^{124}Sb activity measurements. Appl. Radiat. Isot. 68, 1207–1210.

Chen, Q., et al., 2008. Measurement of the neutron-neutron scattering length using the $\pi^{-}d$ capture reaction. Phys. Rev. C 77 (5), 054002.

Cheng, K.Y., Wong, J.W.C., 2008. Fate of ^{14}C−Pyrene in soil−plant system amended with pig manure compost and Tween 80: a growth chamber study. Bioresource Techn. 99, 8406−8412.

Chobola, R., Mell, P., Daróczi, L., Vincze, A., 2006. Rapid determination of radiostrontium isotopes in samples of NPP origin. J. Radioanal. Nucl. Chem. 267 (2), 297−304.

Chmeleff, J., von Blanckenburg, F., Kossert, K., Jakob, D., 2010. Determination of the ^{10}Be half-life by multicollector ICP-MS and liquid scintillation counting. Nucl. Instrum. Methods Phys. Res. Sect. B 268, 192−199.

Christensen, E., Huber, P., Jaffke, P., Shea, T.E., 2014. Antineutrino monitoring for heavy water reactors. Phys. Rev. Lett. 113, 042503.

Christiansen, C.B., 2018. The impact of short chain fatty acids on GLP-1 and PYY secretion from the isolated perfused rat colon. Amer. J. Physiol.-Gastrointerol. Liver Physiol. PubMed 29494208 https://www.physiology.org/doi/pdf/10.1152/ajpgi.00346.2017.

Clarke, S.D., Flaska, M., Pozzi, S.A., Peerani, P., 2009. Neutron and gamma-ray cross-correlation measurements of plutonium oxide powder. Nucl. Instrum. Methods Phys. Res. Sect. A 604, 618−623.

Collé, R., 1997a. Systematic effects of total cocktail mass (volume) and H_2O fraction on $4\pi\beta$ liquid scintillation spectrometry of ^{3}H. Appl. Radiat. Isot. 48 (6), 815−831.

Collé, R., 1997b. Cocktail mismatch effects in $4\pi\beta$ liquid scintillation spectrometry of ^{3}H: implications based on the systematics of ^{3}H detection efficiency and quench indicating parameter variations with total cocktail mass (volume) and H_2O fraction. Appl. Radiat. Isot. 48 (6), 833−842.

Collé, R., 2009. Radionuclide standardization by primary methods: an overview. J. Radioanal. Nucl. Chem. 280 (2), 265−273.

Collé, R., Laureano-Perez, L., 2009. On the standardization of ^{209}Po and ^{210}Pb, pp. 77−85. In: Eikenberg, J., Jäggi, M., Beer, H., Baehrle, H. (Eds.), LSC 2008, Advances in Liquid Scintillation Spectrometry, Radiocarbon. University of Arizona, Tucson, p. 445.

Collé, R., Zimmerman, B.E., Cassette, P., Laureano-Perez, L., 2008. ^{63}Ni, its half-life and standardization: revisited. Appl. Radiat. Isot. 66, 60−68.

Collé, R., Laureano-Perez, L., Bergeron, D.E., 2016. Comparison of tritiated-water standards by liquid scintillation for calibration of a new Standard Reference MaterialÒ. Appl. Radiat. Isot. 112, 38−49.

Cook, G.T., van der Plicht, J., 2007. Radiocarbon dating, Conventional method, pp. 2899−2911. In: Elias, S.A. (Ed.), Encyclopedia of Quaternary Science. Elsevier, Ltd.

Cosolito, F.J., Cohen, N., Petrow, H.G., 1968. Simultaneous determination of iron-55 and stable iron by liquid scintillation counting. Anal. Chem. 40 (1), 213−215.

Coursey, B.M., Mann, W.B., Grau Malonda, A., et al., 1986. Standardization of carbon-14 by $4\pi\beta$liquid scintillation efficiency tracing with hydrogen-3. Appl. Radiat. Isot. 37, 403−408.

Coursey, B.M., Lucas, L.L., Grau Malonda, A., et al., 1989. The standardization of plutonium-241 and nickel-63. Nucl. Instrum. Methods Phys. Res. Sect. A 279, 603−610.

Coursey, B.M., Cessna, J., Garcia-Toraño, E., et al., 1991. The standardization and decay scheme of Rhenium-186. Appl. Radiat. Isot. 42, 865−869.

Coursey, B.M., Calhoun, J.M., Cessna, J., et al., 1994. Liquid-scintillation counting techniques for the standardization of radionuclides used in therapy. (1994). Nucl. Instrum. Methods Phys. Res. Sect. A 339, 26−30.

Coursey, B.M., Schima, F.J., Golas, D.B., et al., 1998. Measurement standards for strontium-89 for use in bone palliation. Appl. Radiat. Isot. 49 (4), 335−344.

Cramer, C.F., Nicholson, M., Moore, C., et al., 1971. Computer programs for calculation of liquid scintillation data from long-lived and short-lived radionuclides used in single or dual labeling. Int. J. Appl. Radiat. Isot. 22, 17−20.

Cremonesi, O., 2010. Neutrino masses and neutrinoless double beta decay: status and expectations. In: Presented at the "European Strategy for Future Neutrino Physics" Workshop, CERN, October 1−3, 2009, CERN Report No. arXiv: 1002.1437 9 Feb. 2010.

Cresswell, T., et al., 2017. Aquatic live animal radiotracing studies for ecotoxicological applications: Addressing fundamental methodological deficiencies. J. Environm. Radioact. 178−179, 453−460.

Croudace, I.W., Warwick, P.E., Reading, D.G., Russell, B.C., 2016. Recent contributions to the rapid screening of radionuclides in emergency responses and nuclear forensics. Trends Anal. Chem. 85, 120−129.

Cruz, P.A.L., Loureiro, J.S., Bernardes, E.M.O., 2002. Standardization of ^{89}Sr solution from a BIPM intercomparison using a liquid scintillation method. Appl. Radiat. Isot. 56, 457−459.

Cunha de Oliveira, A., et al., 2018. Combined treatment with melatonin and insulin improves glycemic control, white adipose tissue metabolism and reproductive axis of diabetic male rats. Life Sci. (in press).

Currie, L.A., 1968. Limisc for qualitative detection and quantitative determination − application to radiochemistry. Anal. Chem. 40, 586−593.

Dacheux, N., Aupiais, J., 1997. Determination of uranium, thorium, plutonium, americium, and curium ultratraces by photon electron rejecting α liquid scintillation. Anal. Chem. 69, 2275−2282.

Dacheux, N., Aupiais, J., 1998. Determination of low concentrations of americium and curium by photon/electron rejecting alpha liquid scintillation. Anal. Chim. Acta 363, 279−294.

Dacheux, N., Aupiais, J., Courson, O., et al., 2000. Comparison and improvement of the determinations of actinide low activities using several α liquid scintillation spectrometers. Anal. Chem. 72, 3150−3157.

Da Cruz, P.A.L., Iwahara, A., Bernardes, E.M.O., et al., 2004. The absolute standardization of ^{32}P and ^{204}Tl at LNMRI. Appl. Radiat. Isot. 60, 415−418.

da Cruz, P.A.L., da Silva, C.J., Moreira, D.S., et al., 2014. Primary activity standardization of ^{99}Tc by three different absolute methods. Appl. Radiat. Isot. 87, 175−178.

Dahlberg, E., 1982. Quench correction in liquid scintillation counting by a combined internal standard-samples channels ratio technique. Anal. Chem. 54 (12), 2082−2085.

Dai, X., Aaron Liblong, A., Kramer-Tremblay, S., Priest, N., Li, C., 2012. J. Optimisation of nasal swab analysis by liquid scintillation counting. Radiol. Prot. 32, 155−166.

Davies, G.J., Spooner, N.J.C., Davies, J.D., et al., 1994. The scintillation efficiency for calcium and fluorine recoils in CaF_2 and carbon and fluorine recoils in C_6F_6 for dark matter searches. Phys. Letts. B 322, 159−165.

Davini, S., 2012. Recent results of the Borexino experiment. Prog. Part. Nucl. Phys. 67, 553−557.

De Filippis, S., 1991. ^{55}Fe and ^{59}Fe: a qualitative comparison of four methods of liquid scintillation activity analysis. Radioactivity Radiochem 2, 14R&R–21R&R.

Dern, R.J., Hart, W.L., 1961a. Doubly labeled iron. I. Simultaneous liquid scintillation counting of isotopes ^{55}Fe and ^{59}Fe as ferrous perchlorate. J. Lab. Clin. Med. 57, 322–330.

Dern, R.J., Hart, W.L., 1961b. Doubly labeled iron. II. Separation of iron from blood samples and preparation of ferrous perchlorate for liquid scintillation counting. J. Lab. Clin. Med. 57, 460–467.

on behalf of the SNO+ Collaboration Descamps, F., 2015. Neutrino Physics with SNO+. Nucl. Part. Phys. Proc. 265–266, 143–145.

DeVol, T.A., Brown, D.D., Leyba, J.D., et al., 1996. A comparison of four aqueous-miscible liquid scintillation cocktails with an alpha/beta discriminating Wallac 1415 liquid scintillation counter. Health Phys. 70 (1), 41–46.

DeVol, T.A., Theisen, C.D., DiPrete, D.P., 2007. Effect of quench on alpha/beta pulse shape discrimination of liquid scintillation cocktails. Operational Radiat. Safety 92 (Suppl. 2), S105–S111.

DeVries, D., Griffin, H., 2008. X- and γ-ray emissions observed in the decay of ^{237}Np and ^{233}Pa. Appl. Radiat. Isot. 66, 668–675.

Dianu, M., Podină, C., 2007. The safety of environment in final disposal of ultima gold scintillation liquid cocktail used for determination of the radioactive content in various samples at Cernavoda nuclear power plant. Rev. Roumaine Chim. 52 (2), 509–519.

Ding, Y., Zhang, Z., Liu, J., et al., 2008. A new gadolinium-loaded liquid scintillator for reactor neutrino detection. Nucl. Instrum. Methods Phys. Res. Sect. A 584, 238–243.

Diodati, J.M., Sartori, F.M., 2007. ^{239}Np as a tracer of ^{237}Np in effluent samples and low level nuclear waste. J. Radioanal. Nucl. Chem. 272 (1), 11–15.

D'Mellow, B., 2006. Digital Processing in Neutron Spectrometry. Ph.D. Thesis. Lancaster University.

D'Mellow, B., Aspinall, M.D., Mackin, R.O., et al., 2007. Digital discrimination of neutrons and γ-rays in liquid scintillators using pulse gradient analysis. Nucl. Instrum. Methods Phys. Res. Sect. A 578, 191–197.

Dobbs, H.E., 1965. Dispersing Solutions for Liquid Scintillation Counting. Memorandum No. AERE-M1574. UK Atomic Energy Research Establishment, Harwell.

Dobrin, R., Dulama, C., Toma, A., et al., 2006. Study of the performance of efficiency tracing technique on a TriCarb 2100TR liquid scintillation analyzer. In: Proc. of the International Conf. Nucl. Energy for New Europe, Portorož, Slovenia, September 18–21, 2006. http://www.djs.si/port2006.

Dobrin, R.I., Pavelescu, M., Dulama, C.N., et al., 2011. Measurement of multiple labeled samples by beta spectrum unfolding. Romanian J. Phys. 56 (1–2), 93–102.

Dodson, C.L., 1996. Radionuclide identification in liquid scintillation spectrometry, pp. 361–364. In: Cook, G.T., Harkness, D.D., MacKenzie, A.B., Miller, B.F., Scott, E.M. (Eds.), Liquid Scintillation Spectrometry 1994, Radiocarbon. University of Arizona, Tucson.

Doke, T., 1981. Fundamental properties of liquid argon, krypton, and xenon as radiation detector media. Port. Phys. 12 (1–2), 9–48.

Doke, T., Masuda, K., Shibamura, E., 1990. Estimation of absolute photon yields in liquid argon and xenon for relativistic electrons. Nucl Instrum. Methods Sect. A 291, 617–620.

Doke, T., Kikuchi, J., Nishikido, F., 2006. Time-of-flight positron emission tomography using liquid xenon scintillation. Nucl. Instrum. Methods Phys. Res. Sect. A 569, 863–871.

for the EXO Collaboration Dolinski, M.J., 2012. The enriched xenon observatory: EXO-200 and Ba+ tagging. Nucl. Phys. B 229–232, 124–127.

Dong, S.D., Lin, C., Schroeder, M., 2013. Synthesis and evaluation of a new phosphorylated ribavirin prodrug. Antiviral Res. 99, 18–26.

Dulanská, S., Horváthová, B., Remenec, B., Mátel, L., 2016. Determination of Pd in radwaste using NiÒResin. J. Radioanal. Nucl. Chem. 310, 645–650.

Duval, C.E., DeVol, T.A., Wade, E.C., Bliznyuk, V.N., Husson, S.M., 2016a. Stability of polymeric scintillating resins developed for ultratrace level detection of alpha- and beta-emitting radionuclides. J. Radioanal. Nucl. Chem. 310 (2), 583–588.

Duval, C.E., DeVol, T.A., Husson, S.M., 2016b. Extractive scintillating polymer sensors for trace-level detection of uranium in contaminated ground water. Anal. Chim. Acta 947, 1–8.

Dziel, T., Broda, R., Ziemek, et al., 2014. Standardization of Sm-153 solution by absolute methods. Appl. Radiat. Isot. 87, 19–23.

Dziel, T., Listkowska, A., Tymiński, Z., 2016. Standardisation and half-life measurements of ^{111}In. Appl. Radiat. Isot. 109, 345–348.

Eakins, J.D., Brown, D.A., 1966. An improved method for the simultaneous determination of iron-55 and iron-59 in blood by liquid scintillation counting. Int. J. Appl. Radiat. Isot. 17, 391–397.

Eardly, D.F., Carton, M.W., Gallagher, J.M., et al., 2001. Bacterial abundance and activity in deep-sea sediments from the eastern North Atlantic. Prog. Oceanogr. 50, 245–259.

Edler, R., 2004. Efficiency tracing and direct DPM. PerkinElmer (Germany) Applikationsnote 13, 9. http://las.perkinelmer.de/Content/applicationnotes/app_de-efftracingunddirectdpm.pdf.

Eikenberg, J., Zumsteg, I., Rüthi, et al., 1999. Fast radiochemical screening of transuranium radionuclides in urine using actinide extractive resin and low-level α/β LSC. Radioact. Radiochem. 10, 19–30.

Eikenberg, J., Bajo, S., Rüthi, M., et al., 2002. A rapid procedure for determining $^{239+240}$Pu and ^{241}Pu in environmental samples using α/β LSC, pp. 351–362. In: Möbius, S., Noakes, J., Schönhofer, F. (Eds.), LSC 2001, Advances in Liquid Scintillation Spectrometry, Radiocarbon. University of Arizona, Tucson, p. 456.

Eikenberg, J., Beer, H., Rüthi, M., et al., 2006. Precise determination of ^{89}Sr and ^{90}Sr/^{90}Y in various matruces: the LSC 3-window approach, pp. 237–249. In: Chałupnik, S., Schönhofer, F., Noakes, J. (Eds.), LSC 2005, Advances in Liquid Scintillation Spectrometry, Radiocarbon. The University of Arizona, Tucson, p. 449.

Eikenberg, J., Jäggi, M., Rüthi, M., 2011. Determination of H-3 and C-14 in the frame of decommissioning projects at the Paul Scherrer Institute. In: LSC 2010, Advances in Liquid Scintillation Spectrometry. Radiocarbon. University of Arizona, Tucson.

Eikenberg, J., Beer, B., Jäggi, M., 2014. Determination of 210Pb and 226Ra/228Ra in continental water using HIDEX 300SL LS-spectrometer with TDCR efficiency tracing and optimized α/β-discrimination. Appl. Radiat. Isot. 93, 64–69.

Elliott, J.C., 1984. Effect of vial composition and diameter on determination of efficiency, background, and quench curves in liquid scintillation counting. Anal. Chem. 56, 758–761.

Elliott, S.R., Hahn, A.A., Moe, M.K., 1986. Experimental investigation of double-beta decay in ^{82}Se. Phys. Rev. Lett. 56, 2582–2585.

Elliott, S.R., Hahn, A.A., Moe, M.K., 1987. Direct evidence for two-neutrino double-beta decay in ^{82}Se. Phys. Rev. Lett. 59, 2020–2023.

Elliott, S.R., Gehman, V.M., Kazkaz, K., et al., 2006. Pulse shape analysis in segmented detectors as a technique for background reduction in Ge double-beta decay experiments. Nucl. Instrum. Methods Phys. Res. Sect. A 558, 504–510.

El Mrabet, R., Abril, J.M., Manjón, G., et al., 2004. Experimental and modeling study of ^{241}Am uptake by suspended matter in freshwater environment from southern Spain. J. Radioanal. Nucl. Chem. 261 (1), 137–144.

Enqvist, A., Flaska, M., Pozzi, S., 2008. Measurement and simulation of neutron/gamma-ray cross-correlation functions from spontaneous fission. Nucl. Instrum. Methods Phys. Res. Sect. A 595, 426–430.

Enqvist, A., Pázsit, I., 2010. Calculation of the light pulse distributions induced by fast neutrons in organic scintillation detectors. Nucl. Instrum. Methods Phys. Res. Sect. A 618, 266–274.

Enqvist, A., Pozzi, S.A., Flaska, M., et al., 2010. Initial evaluation for a combined neutron and gamma ray multiplicity counter. Nucl. Instrum. Methods Phys. Res. Sect. A 621, 493–497.

Enqvist, A., Lawrence, C.C., Wieger, B.M., Pozzi, S.A., Massey, T.N., 2013. Neutron light output response and resolution functions in EJ-309 liquid scintillation detectors. Nucl. Instrum. Methods Phys. Res. Sect. A 715, 79–86.

Erchinger, J., Aalseth, C.E., Bernacki, B., Douglas, M., Fuller, E., Keillor, M., Morley, S., Mullen, C., Orrell, J., Panisko, M., Warren, G., Williams, R., Wright, M., 2015. Development of a low background liquid scintillation counter for a shallow underground laboratory. Appl. Radiat. Isot. 105, 209–218.

Ermakov, A.I., Malinovsky, S.V., Kashirin, I.A., et al., 2006. Rapid analysis of radionuclide composition (screening) of liquid samples via deconvolution of their LS spectra, pp. 89–98. In: Chałupnik, S., Schönhofer, F., Noakes, J. (Eds.), LSC 2005. Advances in Liquid Scintillation Spectrometry, Radiocarbon. The University of Arizona, Tucson, p. 449.

Everett, L.J., Ring, J.G., Nguyen, D.C., 1980. In: Peng, C.-T., Horrocks, D.L., Alpen, E.L. (Eds.), Liquid Scintillation Counting, Recent Applications and Development, Physical Aspects, vol. I. Academic Press, New York and London, pp. 119–128.

Fang, H., Wang, C., Wang, J., 2013. An innovative distillation device for tritiated water analysis with high decontamination factor. Appl. Radiat. Isot. 81, 272–275.

Feng, X., He, Q., 2009. Simultaneous determination of ^{237}Np, $^{238\text{-}240}$Pu and ^{241}Am in HNO_3 solution by combining extraction, liquid scintillation counting, and α spectrometry. Nucl. Instrum. Methods Phys. Res. Sect. A 609, 165–171.

Feng, X., He, Q., Wang, J., Chen, J., 2012. The long-term stability on basic performances of a diisopropylnaphthalene-based liquid scintillation cocktail. Appl. Radiat. Isot. 70, 1536–1540.

Feng, X., He, Q., Wang, J., Chen, J., 2013. Expanded applications of liquid scintillation counting and several related issues. Energy Proc. 39, 141–150.

Feng, X., He, Q., Wang, J., Chen, J., 2014a. Modified 2-window approach for rapid determination of ^{90}Sr and ^{90}Y at low quench level using liquid scintillation counting. J. Radioanal. Nucl. Chem. 299, 509–516.

Feng, X.-G., He, Q.-G., Wang, J.-C., Chen, J., 2014b. A method for optimum PSA setting in the absence of a pure α or β emitter and its application in the determination of 237Np/233Pa. Applied. Radiat. Isot. 93, 114–119.

Feng, X., He, Q., Wang, J., Chen, J., 2015. The effect of sample stability on the determination of radioactivity for various radionuclides by liquid scintillation counting. Applied. Radiat. Isot. 104, 147–154.

Feng, B., Chen, B., Zhuo, W., Zhang, W., 2017. A new passive sampler for collecting atmospheric tritiated water Vapor. Atmospheric Environ 154, 308–317.

Ferraz, R.P.C., et al., 2014. Cytotoxic properties of the leaf essential oils of *Guatteria blepharophylla* and *Guatteria hispida* (Annonaceae). Flavor Frag. J. 29, 228–232.

Ferreira, K.M., Fenwick, A.J., Arinc, A., et al., 2016. Standardisation of ^{90}Y and determination of calibration factors for ^{90}Y microspheres (resin) for the NPL secondary ionisation chamber and a Capintec CRC-25R. Appl. Radiat. Isot. 109, 226–230.

Fiorini, E., 2001. Double beta decay: The future. Nucl. Phys. B 91, 262–269.

Fischer, A.C., Kroon, J.J., Verburg, T.G., et al., 2007. On the relevance of iron adsorption to container materials in small-volume experiments on iron marine chemistry: ^{55}Fe-aided assessment of capacity, affinity and kinetics. Marine Chem. 107, 533–546.

Fisher, B.M., et al., 2011. Fast neutron detection with ^{6}Li-loaded liquid scintillator. Nucl. Instrum. Methods Phys. Res. Sect. A 646 (1), 126–134.

Fitzgerald, R., Fourney, A.M., 2010. Wall effect for alpha particles in liquid scintillation counting. In: Cassette, P. (Ed.), LSC 2010, Advances in Liquid Scintillation Spectrometry, Radiocarbon. University of Arizona, Tucson, p. 377.

Fitzgerald, R., Collé, R., Laureano-Pérez, et al., 2010. A new primary standardization of ^{229}Th. Appl. Radiat. Isot. 68, 1303–1308.

Fitzgerald, R., Zimmerman, B.E., Bergeron, D.E., et al., 2014. A new NIST primary standardization of ^{18}F. Appl. Radiat. Isot. 85, 77–84.

Flaska, M., Pozzi, S.A., 2007. Identification of shielded neutron sources with the liquid scintillator BC-501A using a digital pulse shape discrimination method. Nucl. Instrum. Methods Phys. Res. Sect. A 577, 654–663.

Flaska, M., Pozzi, S.A., 2009. Digital pulse shape analysis for the capture-gated liquid scintillator BC-523A. Nucl. Instrum. Methods Phys. Res. Sect. A 599, 221–225.

Flynn, K.F., Glendenin, L.E., 1959. Half-life and beta spectrum of Rb-87. Phys. Rev. 116, 744–748.

Floeckher, J., 1995. SpectraWorks™ for windows spectrum analysis software. Cronical Software Notes SC-003. PerkinElmer, Inc., Waltham, MA, USA, p. 4. https://www.perkinelmer.co.uk/lab-solutions/resources/docs/APP_SpectraWorksWINSpectrumAnalysis.pdf.

Fons-Castells, J., Tent-Petrus, J., Llauradó, M., 2017a. Simultaneous determination of specific alpha and beta emitters by LSC-PLS in water samples. J. Environ. Radioact. 166, 195–201.

Fons-Castells, J., Oliva, J., Tent-Petrus, J., Llauradó, M., 2017b. Simultaneous determination of ^{226}Ra, ^{228}Ra and ^{210}Pb in drinking water using 3M Empore™ RAD disk by LSC-PLS. Appl. Radiat. Isotopes 124, 83–89.

Fons-Castells, J., Tent-Petrus, J., Llauradó, M., 2017c. Strategy for the determination of mixtures of alpha and beta emitters in water samples with a combination of rapid methods. J. Radioanal. Nucl. Chem. 314, 797–802.

Fons-Castells, J., Díaz, V., Badia, A., et al., 2017d. Implications of quenching in efficiency, spectrum shape and alpha/beta separation. Appl. Radiat. Isot. 128, 263–269.

Forte, M., 2011. Gross alpha and beta measurements by LSC in waters: advantages, problems and limitations. In: LSC 2010, Advances in Liquid Scintillation Spectrometry. Radiocarbon. University of Arizona, Tucson.

Fujii, H., Takiue, M., 2001. Foggy scintillation counting technique. Appl. Radiat. Isot. 55, 517–520.

Fujii, H., Takiue, M., Ishikawa, H., 1986. Activity determination of disc samples with liquid scintillation efficiency tracing technique. Appl. Radiat. Isot. 37, 1147–1149.

Fujii, H., Matsuno, K., Takiue, M., 1999. Construction of analytical beta ray monitor for liquid waste. Radioisotopes 48, 465–471.

Fujii, H., Matsuno, K., Takiue, M., 2000. Hybrid radioassay of multiple radionuclide mixtures in waste solutions by using liquid and NaI(Tl) scintillation monitors. Health Phys. 79 (3), 294–298.

Fukuda, Y., 2016. ZICOS - New project for neutrinoless double beta decay experiment using zirconium complex in liquid scintillator. J. Phys. Conf. Ser. 718, 062019.

Fukuda, Y., Gunji, T., Moriyama, S., et al., 2016. Development of a liquid scintillator containing a zirconium β-keto ester complex for the ZICOS experiment. Nucl. Part. Phys. Proc. 273–275, 2615–2617.

Furuta, E., Yokota, S., Watanabe, Y., 2009. Identification of beta nuclides measured by using plastic scintillator and liquid scintillation counter, pp. 19–26. In: Eikenberg, J., Jäggi, M., Beer, H., Baehrle, H. (Eds.), LSC 2008, Advances in Liquid Scintillation Spectrometry, Radiocarbon. University of Arizona, Tucson, p. 445.

Furuta, E., Kato, Y., Fujisawa, S., 2017. Measurement of tritium with plastic scintillators in large vials of a low background liquid scintillation counter: an organic waste-less method. J. Radioanal. Nucl. Chem. 314 (2), 701–708.

Furuta, E., Ito, T., 2018. Prototype apparatus for the measurement of tritium in expired air using plastic scintillator pellets. Applied. Radiat. Isot. 132, 151–156.

Galán López, M., Martín Sánchez, A., Gómez Escobar, V., 2004. Application of ultra-low level liquid scintillation to the determination of ^{222}Rn in groundwater. J. Radioanal. Nucl. Chem. 261 (3), 631–636.

Gallego Manzano, L., et al., 2015. XEMIS: A liquid xenon detector for medical imaging. Nucl. Instrum. Methods Phys. Res. Sect. A 787, 89–93.

Gallin-Martel, M.-L., Martin, P., Mayet, F., et al., 2006. Experimental study of a liquid xenon PET prototype module. Nucl. Instrum. Methods Phys. Res. Sect. A 563, 225–228.

Gallin-Martel, M.-L., Gallin-Martel, L., Grondin, Y., et al., 2009. A liquid xenon positron emission tomograph for small animal imaging: First experimental results of a prototype cell. Nucl. Instrum. Methods Phys. Res. Sect. A 599, 275–283.

Gallin-Martel, M.-L., Grondin, Y., Gac, N., et al., 2012. Experimental results and first ^{22}Na source image reconstruction by two prototype modules in coincidence of a liquid xenon positron emission tomograph for small animal imaging. Nucl. Instrum. Methods Phys. Res. Sect. A 682, 66–74.

Gamage, K.A.A., Joyce, M.J., Hawkes, N.P., 2011. A comparison of four different digital algorithms for pulse-shape discrimination in fast scintillators. Nucl. Instrum. Methods Phys. Res. Sect. A 642, 78–83.

Gamage, K.A.A., Joyce, M.J., Cave, F.D., 2014. Sector-shaped fast organic liquid scintillation detectors based neutron coincidence counter. Appl. Radiat. Isot. 92, 1–5.

for the KamLAND-Zen Collaboration Gando, Y., 2016. Latest results from KamLAND-Zen second phase. Nucl. Part. Phys. Proc. 237–275, 1842–1846.

García, G., Grau Malonda, A., 2002. The influence of stopping power on the ionization quench factor. Appl. Radiat. Isot. 56, 295–300.

García, J.F., Izquierdo-Ridorsa, A., Toribio, M., et al., 1996. Classical versus multivariate calibration for a beta emitter (^{14}C) activity determination by liquid scintillation counting. Anal. Chim. Acta 331, 33–41.

García-Toraño, E., Grau Malonda, A., 1981. EFFY, A program to calculate the counting efficiency of beta particles in liquid scintillators. Comput. Phys. Commun. 23, 385–391.

García-Toraño, E., Grau Malonda, A., 1985. EFFY2, A new program to compute the counting efficiency of beta particles in liquid scintillators. Comput. Phys. Commun. 36, 307–312.

García-Toraño, E., Grau Malonda, A., 1987. EFYGA, A Monte Carlo program to compute the interaction probability and the counting efficiency of gamma rays in liquid scintillators. Comput. Phys. Commun. 47, 341–347.

García-Toraño, E., Grau Malonda, A., 1988. Cálculo de la eficiencia de recuentro de nucleidos que experimentan desintegración beta y desexcitación gamma simple. Report CIEMAT 616, Madrid.

García-Toraño, E., Grau Malonda, A., Los Arcos, J.M., 1988. EBEGA-The counting efficiency of a beta-gamma emitter in liquid scintillators. Comput. Phys. Commun. 50, 313–319.

García-Toraño, E., Martin Cassallo, M.T., Rodríguez, L., et al., 1991. On the standardization of beta-gamma-emitting nuclides by liquid scintillation counting. In: Ross, H., Noakes, J.E., Spaulding, J.D. (Eds.), "Liquid Scintillation Counting and Organic Scintillators. Lewis Publishers, Chelsea, MI, pp. 307–316.

García-Toraño, E., Roteta, M., Rodríguez Barquero, L., 2000. Standardization of ^{110m}Ag by liquid scintillation and $4\pi\beta$-γ coincidence counting. Appl. Radiat. Isot. 52, 637–641.

García-Toraño, E., Rodríguez Barquero, L., Roteta, M., 2002. Standardization of ^{134}Cs by three methods. Appl. Radiat. Isot. 56, 211–214.

García-Toraño, E., Peyrés, V., Roteta, M., et al., 2016. Standardisation and precise determination of the half-life of ^{44}Sc. Appl. Radiat. Isot. 109, 314–318.

Geladi, P., Kowalski, B.R., 1986. Partial least-squares regression: a tutorial. Anal. Chim. Acta 185, 1–17.

Ghal-Eh, N., 2011. Light transport contribution to the timing characteristics of scintillation detectors. Radiat. Phys. Chem. 80, 365–368.

Giboni, K.-L., Aprile, E., Majewski, P., et al., 2005. Fast timing measurements of gamma-ray events in liquid xenon. IEEE Trans. Nucl. Sci. 52 (5), 1800–1804.

Gibson, J.A.B., Gale, H.J., 1968. Absolute standardization with liquid scintillation counters. J. Sci. Instrum. 1 (2), 96–106.

Godoy, J.M., et al., 2016. Determination of ^{226}Ra in produced water by liquid scintillation counting. J. Environ. Radioact. 160, 25–27.

Gohil, M., et al., 2012. Measurement and simulation of neutron response function of organic liquid scintillator detector. Nucl. Instrum. Methods Phys. Res. Sect. A 664, 304–309.

Goldstein, G., 1965. Absolute liquid-scintillation counting of beta emitters. Nucleonics 23, 67–69.

Gomes, A.R., Abrantes, J., Libânio, A., et al., 2017. Determination of tritium in water using electrolytic enrichment: methodology improvements. J. Radioanal. Nucl. Chem. 314, 669–674.

Gómez, H., Cebrián, S., Morales, J., et al., 2007. Background reduction and sensitivity for germanium double beta decay experiments. Astroparticle Phys. 28, 435–447.

Gomez-Cadenas, J.J., Benlloch-Rodriguez, J.M., Ferrario, P., 2016. Application of scintillating properties of liquid xenon and silicon photomultiplier technology to medical imaging. Spectrochim. Acta Part B 118, 6–13.

Gozani, T., Stevenson, J., King, M.J., 2011. Neutron threshold activation detectors (TAD) for the detection of fissions. Nucl. Instrum. Methods Phys. Res. Sect. A 652, 334–337.

Grahek, Ž., Mačefat, M.R., 2004. Isolation of iron and strontium from liquid samples and determination of ^{55}Fe and ^{89,90}Sr in liquid radioactive waste. Anal. Chim. Acta 511, 339–348.

Grahek, Ž., Mačefat, M.R., 2006. Extraction chromatographic separation of iron from complex liquid samples and the determination of ^{55}Fe. J. Radioanal. Nucl. Chem. 267 (1), 131–137.

Grahek, Ž., et al., 2016. Measurement of tritium in the Sava and Danube Rivers. J. Environm. Radioact. 162–163, 56–67.

Gratta, G., Wang, Y.F., 1999. Towards low-threshold, real-time solar neutrino detectors. Nucl. Instrum. Methods Phys. Res. Sect. A 438, 317–321.

Grau Carles, A., 1993. A new linear spectrum unfolding method applied to radionuclide mixtures in liquid scintillation spectrometry. Appl. Radiat. Isot. 45, 83–90.

Grau Carles, A., 1994. SRLOG, the simultaneous standardization of ^{90}Sr + ^{90}Y + ^{89}Sr mixtures. Comp. Phys. Commun. 82, 17–22.

Grau Carles, A., 1995. New methods for the determination of β-spectra shapefactor coefficients. Appl. Radiat. Isot. 46, 125–128.

Grau Carles, A., 1996. MLOG, the simultaneous standardization of multinuclide mixtures. Comp. Phys. Commun. 93, 48–52.

Grau Carles, A., 2005. Beta shapefactor determinations by the cutoff energy yield method. Nucl. Instrum. Methods Phys. Res. Sect. A 551, 312–322.

Grau Carles, A., 2006a. EMILIA, the LS counting efficiency for electron-capture and capture-gamma emitters. Comp. Phys. Commun. 174 (1), 35–46.

Grau Carles, A., 2006b. Synergic quenching effects of water and carbon tetrachloride in liquid scintillation gel samples. Appl. Radiat. Isot. 64, 1505–1509.

Grau Carles, A., 2007. MICELLE, the micelle size effect on the LS counting efficiency. Comput. Phys. Commun. 176, 305–317.

Grau Carles, A., Grau Malonda, A., 1989. Electron-capture standardization with a triple phototube system. Anales de Física Ser. B 85, 160–176.

Grau Carles, A., Grau Malonda, A., 1991. A new procedure for multiple isotope analysis in liquid scintillation counting. In: Ross, H., Noakes, J.E., Spaulding, J.D. (Eds.), Liquid Scintillation Counting and Organic Scintillators. Lewis Publishers, Chelsea, MI 48118, pp. 295–306.

Grau Carles, A., Grau Malonda, A., 1992. Precise System for the Determination of the Quench Parameter of Radioactive Samples in Liquid Phase. Patent P. 9202639, December 29, 1992. Registry of Industrial Property, Madrid, Spain.

Grau Carles, A., Grau Malonda, A., 1997. Calibración del ^{40}K por centelleo líquido. In: Determinación del periodo de semidesintegración. Report CIEMAT 831, Madrid.

Grau Carles, A., Grau Malonda, A., 2001. Free parameter, figure of merit and ionization quench in liquid scintillation counting. Appl. Radiat. Isot. 54, 447–454.

Grau Carles, A., Grau Malonda, A., 2006. Computational aspects in modelling the interaction of low-energy X-rays with liquid scintillators. Appl. Radiat. Isot. 64, 1515–1519.

Grau Carles, A., Grau Malonda, A., 2010. Liquid scintillation high resolution spectral analysis. Informes Técnicos Ciemat 1208, 28 (in English), Editorial CIEMAT, Madrid.

Grau Carles, A., Kossert, K., 2006. New advances in the determination of the ^{87}Rb shape factor function. Nucl. Phys. A 767, 248–258.

Grau Carles, A., Kossert, K., 2007. Measurement of the shape-factor functions of the long-lived radionuclides ^{87}Rb, ^{40}K and ^{10}Be. Nucl. Instrum. Methods Phys. Res. Sect. A 572, 760–767.

Grau Carles, A., Kossert, K., 2009. Monte Carlo simulation of Auger-electron spectra. Appl. Radiat. Isot. 67, 192–196.

Grau Carles, A., Martin-Casallo, M.T., Grau Malonda, A., 1991. Spectrum unfolding and double window methods applied to standardization of ^{14}C and ^{3}H mixtures. Nucl. Instrum. Methods Phys. Res. Sect. A 307, 484–490.

Grau Carles, A., Grau Malonda, A., Rodríguez Barquero, L., 1993a. Cherenkov radiation effects on counting efficiency in extremely quenched liquid scintillation samples. Nucl. Instrum. Methods Phys. Res. Sect. A 334, 471–476.

Grau Carles, A., Rodríguez Barquero, L., Grau Malonda, A., 1993b. Standardization of ^{14}C and ^{35}S mixtures. Nucl. Instrum. Methods Phys. Res. Sect. A 335, 234–240.

Grau Carles, A., Rodríguez Barquero, L., Grau Malonda, A., 1993c. A spectrum unfolding method applied to standardization of ^{3}H and ^{55}Fe mixtures. Appl. Radiat. Isot. 44, 581–586.

Grau Carles, A., Rodríguez Barquero, L., Grau Malonda, A., 1993d. Standardization of multi-nuclide mixtures by a new spectrum unfolding method. J. Radioanal. Nucl. Chem. Letters 176, 391–403.

Grau Carles, A., Rodríguez Barquero, L., Grau Malonda, A., 1993e. Simultaneous standardization of ^{90}Sr-^{90}Y and ^{89}Sr mixtures. Appl. Radiat. Isot. 44, 1003–1010.

Grau Carles, A., Rodríguez Barquero, L., Grau Malonda, A., 1993f. Double-label counting of heterogeneous samples, pp. 239–249. In: Noakes, J.E., Schönhofer, F., Polach, H.A. (Eds.), Liquid Scintillation Spectrometry 1992, Radiocarbon. University of Arizona, Tucson.

Grau Carles, A., Rodríguez Barquero, L., Grau Malonda, A., 1994a. Deconvolution of ^{204}Tl/^{36}Cl and ^{147}Pm/^{45}Ca dual mixtures. Nucl. Instrum. Methods Phys. Res. Sect. A 339, 71–77.

Grau Carles, A., Grau Malonda, A., Grau Carles, P., 1994b. EMI, the counting efficiency for electron capture, electron capture-gamma and isomeric transition. Comput. Phys. Commun. 79, 115–123.

Grau Carles, A., Grau Malonda, A., Rodríguez Barquero, L., 1994c. Standardization of ^{125}I, ^{85}Sr and ^{109}Cd by CIEMAT/NIST method. Appl. Radiat. Isot. 45, 461–464.

Grau Carles, A., Grau Malonda, A., Gómez, G., 1996. Standardization of U(X_1 + X_2): The ^{234}Th + ^{234}mPa + ^{230}Th mixture. Nucl. Instrum. Methods Phys. Res. Sect. A 369, 431–436.

Grau Carles, A., Rodriguez Barquero, L., Jimenez De Mingo, A., 1998. ^{125}Sb to ^{125m}Te branching ratio. Appl. Radiat. Isot. 49 (9–11), 1377–1381.

Grau Carles, A., Gunther, E., García, G., Grau Malonda, A., 2004. Ionization quenching in LSC. Appl. Radiat. Isot. 60, 447–451.

Grau Carles, A., Gunther, E., Grau Malonda, A., 2006. The photoionization-reduced energy in LSC. Appl. Radiat. Isot. 64, 43–54.

Grau Carles, A., Kossert, K., Grau Malonda, A., 2008. Determination of the shape factor of ^{90}Sr by means of the cutoff energy yield method. Appl. Radiat. Isot. 66, 1021–1025.

Grau Malonda, A., 1982a. Metrología de la radiactividad beta mediante el recuento por centelleo en fase líquida. Doctoral Dissertation. Universidad Complutense de Madrid, (Spain), 1982, pp. 282; 10131924. http://dissexpress.umi.com/dxweb/results.html?QryTxt=&By=Grau+Malonda&Title=&pubnum=10131924.

Grau Malonda, A., 1982b. Counting efficiency for electron-capturing nuclides in liquid scintillator solutions. Int. J. Appl. Radiat. Isot. 33, 371–375.

Grau Malonda, A., 1995. Modelos de Parámetro Libre en Centelleo Líquido. Editorial CIEMAT, Madrid, ISBN 84-7834-287-7, p. 387.

Grau Malonda, A., 1999. Free Parameter Models in Liquid Scintillation Counting. Editorial CIEMAT, Madrid, p. 416. ISBN: 84-7834-350-4.

Grau Malonda, A., Coursey, B.M., 1987. Standardization of isomeric-transition radionuclides by liquid-scintillation efficiency tracing with hydrogen-3: Application to Technetium-99m. Appl. Radiat. Isot. 38, 695–700.

Grau Malonda, A., Coursey, B.M., 1988. Calculation of beta-particle counting efficiency for liquid-scintillation systems with three phototubes. Appl. Radiat. Isot. 39, 1191–1196.

Grau Malonda, A., Fernández, A., 1985. Cálculo de la eficiencia de detección de nucleidos que se desintegran por captura electrónica y emission gamma. Informe Junta de Energía Nuclear, JEN, 575, Madrid.

Grau Malonda, A., García-Toraño, E., 1978. Espectros beta, I. Espectros simples de negatrons. Informe Junta de Energía Nuclear, JEN, 427, Madrid.

Grau Malonda, A., García-Toraño, E., 1981a. Cálculo de la eficiencia de detección en liquidos centelleadores. I. Nucleidos que se desintegran por emission simple de negatrons. Informe Junta de Energía Nuclear, JEN, 488, Madrid.

Grau Malonda, A., García-Toraño, E., 1981b. Espectros beta, II. Espectros simples de positrons. Informe Junta de Energía Nuclear, JEN, 489, Madrid.

Grau Malonda, A., García-Toraño, E., 1982a. Evaluation of counting efficiency in liquid scintillation counting or pure β-ray emitters. Int. J. Appl. Radiat. Isot. 33, 249–253.

Grau Malonda, A., García-Toraño, E., 1982b. Cálculo de la eficiencia de detección en liquidos centelleadores. II. Nucleidos que se desintegran por emission simple de positrons. Informe Junta de Energía Nuclear, JEN, 518, Madrid.

Grau Malonda, A., Grau Carles, A., 1999. The ionization quench factor in liquid-scintillation counting standardizations. Appl. Radiat. Isot. 51, 183–188.

Grau Malonda, A., Grau Carles, A., 2000. Standardization of electron-capture radionuclides by liquid scintillation counting. Appl. Radiat. Isot. 52, 657–662.

Grau Malonda, A., Grau Carles, A., 2002. Half-life determination of ^{40}K by LSC. Appl. Radiat. Isot. 56, 153–156.

Grau Malonda, A., Grau Carles, A., 2008. The ionization quenching function for coincident electrons. Appl. Radiation Isotopes 1043–1048.

Grau Malonda, A., Los Arcos, J.M., 1983. Un nuevo procedimiento para la Calibración del ^{90}Sr y del ^{90}Y mediante centelleo en fase liquido. An. Fisica B 79, 5–9.

Grau Malonda, A., García-Toraño, E., Los Arcos, J.M., 1985. Liquid-scintillation counting efficiency as a function of the figure of merit for pure beta-particle emitters. Int. J. Appl. Radiat. Isot. 36, 157–158.

Grau Malonda, A., García-Toraño, E., Los Arcos, J.M., 1987. Free parameter codes to compute the counting efficiency in liquid scintillators. Trans. Amer. Nucl. Soc. 55, 55–56.

Grau Malonda, A., Rodriguez Barquero, L., Grau Carles, A., 1994. Radioactivity determination of ^{90}Y, ^{90}Sr and ^{89}Sr mixtures by spectral deconvolution. Nucl. Instrum. Methods Phys. Res. Sect. A 339, 31–37.

Grau Malonda, A., Grau Carles, A., Grau Carles, P., 1994b. EMI, the counting efficiency for electron capture, electron capture-gamma and isomeric transitions. Comput. Phys. Commun. 79, 115–123.

Grau Malonda, A., Grau Carles, A., Grau Carles, P., et al., 1999. EMI2, the counting efficiency for electron capture by a $KL_1L_2L_3M$ model. Comput. Phys. Commun. 123, 114–122.

Grau Malonda, A., Grau Carles, A., Garcia, G., 2006. Mean values of the LMM Auger transition in a KLM model. Appl. Radiat. Isot. 64, 1485–1491.

Grirb, C., et al., 2011. Measuring the Neutrino Luminosity of the Sun with LENS & the MINILENS prototype. Nucl. Phys. B 221, 349.

Grignon, C., Barbet, J., Bardiès, M., et al., 2007. Nuclear medical imaging using β^+ γ coincidence from ^{44}Sc radio-nuclide with liquid xenon as detection medium. Nucl. Instrum. Methods Phys. Res. Sect. A 571, 142–145.

Grodzicka-Kobylka, M., et al., 2018. Study of n-γ discrimination by zero-crossing method with SiPM based scintillation detectors. Nucl. Instrum. Methods Phys. Res. Sect. A 883, 159–165.

Groppi, F., Mainardi, H.S., Martinotti, A., et al., 2005a. The use of liquid scintillation spectrometry as a very sensitive radio analytical tool for the determination of alpha, beta and monchromatic electron emitting impurities in radiopharmaceutical compounds. J. Radioanal. Nucl. Chem. 263 (2), 521–525.

Groppi, F., Bonardi, M.L., Birattari, C., et al., 2005b. Optimization study of α-cyclotron production of At-211/Po-211g for high-LET metabolic radiotherapy purposes. Appl. Radiat. Isot. 63, 621–631.

Gudelis, A., Druteikienė, R., Lukšienė, B., et al., 2010. Assessing deposition levels of ^{55}Fe, ^{60}Co and ^{63}Ni in the Ignalina NPP environment. J. Environ. Radioact. 101, 464–467.

Gudelis, A., Vinčūnas, A., Butkus, P., et al., 2012. Measurements of some radionuclides using a new TDCR system and an ultra low-level conventional LSC counter in CPST, Lithuania. Appl. Radiat. Isot. 70, 2204–2208.

Gudelis, A., Gaigalaitė, L., Butkus, P., 2017. Standardization of tritium with the TDCR instrument. In: LSC2017 Conference, 1–5 May, Copenhagen.

Gudjonsson, G., Theodórsson, P., Sigurdsson, K., et al., 2009. A study of the simple open-vial liquid scintillation method for measurement of radon in air, pp. 361–366. In: Eikenberg, J., Jäggi, M., Beer, H., Baehrle, H. (Eds.), LSC 2008. Advances in Liquid Scintillation Spectrometry, Radiocarbon. The university of Arizona, Tucson, p. 445.

Guerrero, C., Cano-Ott, D., Fernández-Ordóñez, M., et al., 2008. Analysis of the BC501A neutron detector signals using the true pulse shape. Nucl. Instrum. Methods Phys. Res. Sect. A 597, 212–218.

Günther, E.W., 1994. Standardization of ^{59}Fe and ^{131}I by liquid scintillation counting. Nucl. Instrum. Methods Phys. Res. Sect. A 339, 402–407.

Günther, E.W., 1996. A simple method for transferring the tritium calibration of an LSC system to other radionuclides, pp. 373–379. In: Cook, G.T., Harkness, D.D., MacKenzie, A.B., Miller, B.F., Scott, E.M. (Eds.), Liquid Scintillation Spectrometry 1994, Radiocarbon. University of Arizona, Tucson.

Günther, E., 1998. Standardization of the EC nuclides ^{55}Fe and ^{65}Zn with the CIEMAT/NIST LSC tracer method. Appl. Radiat. Isot. 49 (9–11), 1055–1060.

Günther, E., 2000. Standardization of ^{237}Np by the Ciemat/NIST LSC tracer method. Appl. Radiat. Isot. 52, 471–474.

Günther, E., 2001. Computer program CN2001A. Physikalisch-Technische Bundesanstalt, Braunschweig, Germany.

Günther, E., 2002a. What can we expect from the CIEMAT/NIST method? Appl. Radiat. Isot. 56, 357–360.

Günther, E., 2002b. Determination of the activity of radionuclide sources emitting beta and gamma radiation using the CIEMAT/NIST method, pp. 57–63. In: Möbius, S., Noakes, J.E., Schönhofer, F. (Eds.), LSC 2001, Advances in Liquid Scintillation Spectrometry, Radiocarbon. The University of Arizona, Tucson, p. 456.

Günther, E., 2002c. Determination of the ^{32}P activity in angioplastic balloons by LSC. Appl. Radiat. Isot. 56, 291–293.

Günther, E., Schötzig, U., 1992. Activity determination of ^{93m}Nb. Nucl. Instrum. Methods Phys. Res. Sect. A 312, 132–135.

Haaslahti, V., 2010. LSC 2010 – Hidex, Products for Liquid Scintillation Counting. In: Presented at the International Conference "LSC 2010 Advances in Liquid Scintillation Spectrometry" held at the Cité Internationale Universitaire de Paris, France, 6–10 September 2010.

Halter, E., Thiam, c., Bobin, C., Bouchard, J., Chambellan, D., Chauvenet, C., Hamel, M., Rocha, L., Trocmé, M., Woo, R., 2014. First TDCR measurements at low energies using a miniature x-ray tube. Appl. Radiat. Isot. 93, 7–12.

Hamanaka, S., Shizuma, K., Wen, X., et al., 1998. Radon concentration measurement in water by means of α liquid scintillation spectrometry with a PERALS spectrometer. Nucl. Instrum. Methods Phys. Res. Sect. A 410, 314–318.

Hamel, M., Normand, S., Frelin, A.-M., et al., 2009a. Novel 1,8-naphthamides for fast neutron/gamma discrimination in liquid scintillators, pp. 13–18. In: Eokenberg, J., Jäggi, M., Beer, H., Baehrie, H. (Eds.), LSC 2008 , Advances in Liquid Scintillation Spectrometry, Radiocarbon. University of Arizona, Tucson, p. 445.

Hamel, M., Frelin-Labalme, A.-M., Simic, V., et al., 2009b. *N*-(2′,2′-di-*t*-butylphenyl)-4-ethoxy-1,8-naphthalimide: A new fluorophore highly efficient for fast neutrons/gamma-rays discrimination in liquid media. Nucl. Instrum. Methods Phys. Res. Sect. A 602, 425–431.

Hans, S., et al., 2015. Purification of telluric acid for SNO+ neutrinoless double-beta decay search. Nucl. Instrum. Methods Phys. Res. Sect. A 795, 132–139.

Happel, S., Letessier, P., Ensinger, W., et al., 2004. Gross alpha determination in drinking water using a highly specific resin and LSC. Appl. Radiat. Isot. 61, 339–344.

Hara, M., Nakayama, M., Hirokami, K., et al., 2016. Appropriate quenching level in modified integral counting method by liquid scintillation counting. J. Radioanal. Chem. 310, 857–863.

Harms, A.V., Jerome, S.M., 2004. Development of an organically bound tritium standard. Appl. Radiat. Isot. 61, 389–393.

on behalf of the Double Chooz collaboration Haser, J., 2016. Current status of the Double Chooz experiment. Nucl. Part. Phys. Proc. 273–275, 1915–1921.

Hawkes, N.P., Gamage, K.A.A., Taylor, G.C., 2010. Digital approaches to field neutron spectrometry. Radiat. Meas. 45, 1305–1308.

Hawkins, B.T., Sykes, D.B., Miller, D.S., 2010. Rapid, reversible modulation of blood–brain barrier P-glycoprotein transport activity by vascular endothelial growth factor. J. Neurosci. 30 (4), 1417–1425.

He, Z.-H., Zhou, R., He, M.-F., et al., 2011. Anti-angiogenic effect and mechanism of rhein from Rhizoma Rhei. Phytomed 18 (6), 470–478.

Heilgeist, M., 2000. Use of extraction chromatography, ion chromatography and liquid scintillation spectrometry for rapid determination of strontium-89 and strontium-90 in food in cases of increased release of radionuclides. J. Radioanal. Nucl. Chem. 245 (2), 249–254.

Helmer, R.G., 2002. ^{113}Sn. In: Bé, M., et al. (Eds.), Table of Radionuclides. CnEA-ISBN2 727202008. Website. http://www.nucleide.org/DDEP_WG/Nuclides/Sn-113_tables.pdf.

Heltsley, J.H., Brandon, L., Galonsky, A., et al., 1988. Particle identification via pulse-shape discrimination with a charge-integrating ADC. Nucl. Instrum. Methods Phys. Res. Sect. A 263, 441–445.

Hendee, W.R., Ibbott, G.S., Crisha, K.L., 1972. ^{3}H-toluene, ^{3}H-water and ^{3}H-hexadecane as internal standards for toluene- and dioxane-based liquid scintillation cocktails. Int. J. Appl. Radiat. Isot. 23, 90–95.

Herforth, L., 1948. Die Fluoreszensanregung organischer Substanzen mit Alphateilchen, schnellen Elektronen und Gammastrahlen. Thesis presented Sept. 13, 1948. Technical University, Berlin-Charlottenburg.

Herforth, L., Kallmann, H., 1948. Die Fluoreszensanregung von festem und flüssigem Naphthalin, Diphenyl und Phenanthren durch Alphateilchen, schnelle Elektronen und Gammastrahlung. Ann. der Physik 6 (4), 231–245.

Herranz, M., Idoeta, R., Abelairas, A., et al., 2012. Uncertainties in ^{63}Ni and ^{55}Fe determinations using liquid scintillation counting methods. Appl. Radiat. Isot. 70, 1863–1866.

Hitachi Aloka, 2014. Hitachi Aloka AccuFlex LSC-8000 Liquid Scintillation System Instruction Manual, RN1-4695.02. Hitachi Aloka Medical, Ltd., Tokyo, p. 292.

Hitachi, A., Takahashi, T., Funayama, N., et al., 1983. Effect of ionization density on the time dependence of luminescence from liquid argon and xenon. Phys. Rev. B 27 (9), 5279–5285.

Hofmann, H.J., et al., 1990. A new determination of the half-life of ^{32}Si. Nucl. Instrum. Methods Phys. Res. Sect. B 52, 544–551.

Holst, P.B., Christophersen, C., Engvild, K.C., 2000. In vivo incorporation of radioactive ^{36}Cl, a method for monitoring chloro compounds in biological material. J. Chromat. A 903, 267–270.

Homma, Y., Murakami, Y., 1977. Study on the applicability of the integral counting method for the determination of ^{226}Ra in various sample forms using a liquid scintillation counter. J. Radioanal. Chem. 36, 173–184.

Homma, Y., Murase, Y., Takiue, M., 1987. Determination of ^{222}Rn by air luminescence method. J. Radioanal. Nucl. Chem. Letters 119, 457–465.

Homma, Y., Murase, Y., Handa, K., 1993a. Comparison of a modified integral counting method and efficiency tracing method for the determination of ^{222}Rn by liquid scintillation counting, pp. 59–62. In: Noakes, J.E., Schönhofer, F., Polach, F.H.A. (Eds.), Liquid Scintillation Spectrometry 1992, Radiocarbon. University of Arizona, Tucson.

Homma, Y., Murase, Y., Handa, K., 1993b. Determination of atmospheric radioactivity using a membrane filter and liquid scintillation spectrometry, pp. 63–67. In: Noakes, J.E., Schönhofer, F., Polach, H.A. (Eds.), "Liquid Scintillation Spectrometry 1992, Radiocarbon. University of Arizona, Tucson.

Homma, Y., Murase, Y., Handa, K., 1994a. The zero detection threshold of a liquid scintillation spectrometer and its application to liquid scintillation counting. Appl. Radiat. Isot. 45, 341–344.

Homma, Y., Murase, Y., Handa, K., 1994b. Absolute liquid scintillation counting of ^{35}S and ^{45}Ca using a modified integral counting method. J. Radioanal. Nucl. Chem. 187, 367–374.

Homma, Y., Murase, Y., Handa, K., 1994c. A modified integral counting method and efficiency tracing method for measuring ^{222}Rn by liquid scintillation counting. Appl. Radiat. Isot. 45, 699–702.

Homma, Y., Murase, Y., Handa, K., et al., 1996. Rapid calibration of detectors for determining ^{222}Rn using air luminescence counting, pp. 111–116. In: Cook, G.T., Harkness, D.D., MacKenzie, A.B., Miller, B.F., Scott, E.M. (Eds.), Liquid Scintillation Spectrometry 1994, Radiocarbon. University of Arizona, Tucson.

Hong, J., Craig, W.W., Graham, P., et al., 2002. The scintillation efficiency of carbon and hydrogen recoils in an organic liquid scintillator for dark matter searches. Astropart. Phys. 16, 333–338.

Hong, K.H., Cho, Y.H., Lee, M.H., et al., 2001. Simultaneous measurement of ^{89}Sr and ^{90}Sr in aqueous samples by liquid scintillation counting using the spectrum unfolding method. Appl. Radiat. Isot. 54, 299–305.

Horrocks, D.L., 1966. Low-level alpha disintegration rate determinations with a one-multiplier phototube liquid scintillation spectrometer. Int. J. Appl. Radiat. Isot. 17, 441–446.

Horrocks, D.L., 1968, p. 45. In: Horrocks, D.L. (Ed.), Organic Scintillators. Gordon and Breach, New York.

Horrocks, D.L., 1971. Obtaining the possible maximum of 90 percent efficiency for counting of ^{55}Fe in liquid scintillator solutions. Int. J. Appl. Radiat. Isot. 22, 258–260.

Horrocks, D.L., 1974a. Measurement of low levels of normal uranium in water and urine by liquid scintillation alpha counting. Nucl. Instrum. Methods 117, 589–595.

Horrocks, D.L., 1974. Applications of Liquid Scintillation Counting. Academic Press, New York, p. 340.

Horrocks, D.L., 1976a. The mechanism of the liquid scintillation process. In: Noujaim, A.A., Ediss, C., Wiebe, L.I. (Eds.), Liquid Scintillation Science and Technology. Academic Press, New York and London, pp. 1–16.

Horrocks, D.L., 1976b. Absolute disintegration rate determination of beta-emitting radionuclides by the pulse height shift extrapolation method. In: Noujaim, A.A., Ediss, C., Wiebe, L.I. (Eds.), Liquid Scintillation Science and Technology. Academic Press, New York and London, pp. 185–198.

Horrocks, D.L., 1976c. Measurement of ^{125}I by liquid scintillation methods. Nucl. Instrum. Methods 133, 293–301.

Horrocks, D.L., 1977. "The H-Number Concept", Publication No. 1095 NUC-77-IT. Beckman Instruments Inc., Irvine, p. 26.

Horrocks, D.L., 1978a. A new method of quench monitoring in liquid scintillation counting: The H number concept. In: Crook, M.A., Johnson, P. (Eds.), Liquid Scintillation Counting, vol. 5. Heyden, London, pp. 145–168.

Horrocks, D.L., 1978b. A new method of quench monitoring in liquid scintillation counting: The H number concept. J. Radioanal. Chem. 43, 489–521.

Horrocks, D.L., 1980. Effect of quench on the pulse height distribution for tritium-containing samples – high quench levels. In: Peng, C.-T., Horrocks, D.L., Alpen, E.L. (Eds.), Liquid Scintillation Counting, Recent Applications and Development, vol. I. Academic Press, New York and London, pp. 199–210.

Horrocks, D.L., Kolb, A.J., 1981. Instrumental methods for detecting some common problems in liquid scintillation counting. Lab. Pract. 30, 485–487.

Hosogi, J., et al., 2018. An iminium ion metabolite hampers the production of the pharmacologically active metabolite of a multikinase inhibitor KW-2449 in primates: Irreversible inhibition of aldehyde oxidase and covalent binding with endogenous proteins. Biopharm Drug Dispos. 2018 https://doi.org/10.1002/bdd.2123.

Horváth, Á., Ieki, K., Iwata, Y., et al., 2000. Comparison of two liquid scintillators used for neutron detection. Nucl. Instrum. Methods Phys. Res. Sect. A 440, 241–244.

Horwitz, E.P., Chiarizia, R., Dietz, M.L., 1997. DIPEX: A new extraction chromatographic material for the separation and Preconcentration of actinides from aqueous solution. Reactive & Functional Pol. 33, 25–36.

Hou, X., 2005. Rapid analysis of ^{14}C and ^{3}H in graphite and concrete for decommissioning of nuclear reactor. Appl. Radiat. Isot. 62, 871–882.

Hou, X., Frøsig Østergaard, L., Nielsen, S.P., 2005. Determination of ^{63}Ni and ^{55}Fe in nuclear waste samples using radiochemical separation and liquid scintillation counting. Anal. Chim. Acta 535, 297–307.

Hou, X., 2017. Determination of radionuclidic impurities in ^{99m}Tc eluate from ^{99}Mo/^{99m}Tc generator for quality control. J. Radioanal. Nucl. Chem. 314 (2), 659–668.

Huang, B., et al., 2016. Facile synthesis of ^{55}Fe-labeled well-dispersible hematite nanoparticles for bioaccumulation studies in nanotoxicology. Environm. Pollution 213, 801–808.

Hueber-Becker, F., Nohynek, G.J., Dufour, E.K., et al., 2007. Occupational exposure of hairdressers to [^{14}C]-*para*-phenylenediamine-containing oxidative hair dyes: A mass balance study. Fd. Chem. Toxicol. 45, 160–169.

Hughes, K.T., Ireson, J.C., Jones, N.R.A., et al., 2001. Color quench correction in scintillation proximity assays using Paralux Count Mode. Application Note, p. 12. PerkinElmer Life Sciences, Boston, MA.

Hui, S.K., Sharma, M., Bhattacharyya, M.H., 2012. Liquid scintillation based quantitative measurement of dual radioisotopes (^{3}H and ^{45}Ca) in biological samples for bone remodeling studies. Appl. Radiat. Isot. 70, 63–68.

Hult, M., Altzitzoglou, T., Denecke, B., et al., 2000. Standardization of ^{204}Tl at IRMM. Appl. Radiat. Isot. 52, 493–498.

Huntington, C.E., Veum, T.L., Morris, J.S., 2008. Zinc uptake in swine intestinal brush border membrane vesicles using a ^{65}Zn/^{69m}Zn duel isotopre experiment. J. Radioanal. Nucl. Chem. 276 (1), 129–134.

Hurtado, S., Villa, M., Manjon, G., et al., 2009. Monte Carlo simulation of a liquid scintillation counter using GEANT4 code, pp. 155–159. In: Eikenberg, J., Jäggi, M., Beer, H., Baehrle, H. (Eds.), LSC 2008, Advances in Liquid Scintillation Spectrometry, Radiocarbon. University of Arizona, Tucson, p. 445.

Hurtgen, C., Jerome, S., Woods, M., 2000. Revisiting Curie – how low can you go? Appl. Radiat. Isot. 53, 45–50.

Hwang, H.-Y., Park, J.H., Park, T.S., et al., 2002. Development of MCTS technique for 3-PM liquid scintillation counting. Appl. Radiat. Isot. 56, 307–313.

Hwang, H.-Y., Kwak, S.-I., Lee, H.Y., et al., 2004. Development of 3-PM liquid scintillation counting system with geometrical efficiency variation. Appl. Radiat. Isot. 60, 469–473.

Ianni, A., 2017. Solar neutrinos. Prog. Part. Nucl. Phys. 94, 257–281.

Ianni, A., Montanino, D., Villante, F.L., 2005. How to observe ^{8}B solar neutrinos in liquid scintillator detectors. Phys. Lett. B 627, 38–48.

Ieki, K., 2016. Upgrade of the MEG liquid xenon calorimeter with VUV-light sensitive large area SiPMs. Nucl. Instrum. Methods Phys. Res. Sect. A 824, 686–690.

Ingwersen, J., et al., 2018. Nimodipine confers clinical improvement in two models of experimental autoimmune encephalomyelitis. J. Neurochem. https://doi.org/10.1111/jnc.14324 (in press).

Inoue, K., 2013. Results from KamLAND-Zen. Nucl. Phys. B 235–236, 249–254.

Ishikawa, H., Takiue, M., 1973. Liquid scintillation measurement for β-ray emitters followed by γ-rays. Nucl. Instrum. Methods 112, 437–442.

Ishikawa, H., Takiue, M., Aburai, T., 1984. Radioassay by an efficiency tracing technique using a liquid scintillation counter. Int. J. Appl. Radiat. Isot. 35, 463–466.

Ishimori, K., Kameo, Y., Matsue, H., et al., 2010. Carbon-14 analysis in solidified product of non-metallic solid waste by a combination of alkaline fusion and gaseous CO_2 trapping. Appl. Radiat. Isot. 69 (2), 506–510.

Itaya, M., Inagaki, T., Iwata, T., et al., 2004. Development of a new photomultiplier tube with high sensitivity for a wavelength-shifter fiber readout. Nucl. Instrum. Methods Phys. Res. Sect. A 522, 477–486.

Ivan, C., Cassette, P., Sahagia, M., 2008. A new TDCR-LS counter using channel photomultiplier tubes. Appl. Radiat. Isot. 66, 1006–1011.

Ivan, C., Wätjen, A.C., Cassette, P., et al., 2010. Participation in the CCRI(II)-K2.H-3 comparison and study of the new TDCR-LS counter with 6 CPMs. Appl. Radiat. Isot. 68, 1543–1545.

Iwamoto, T., 2010. System overview of liquid xenon calorimeter for the MEG experiment. Nucl. Instrum. Methods Phys. Res. Sect. A 617, 92–95.

Jaffe, D.E., 2015. Future Reactor Neutrino Experiments (RRNOLD). Phys. Procedia 61, 319–322.

Jäggi, M., Eikenberg, J., 2014. Comparison of the Tri Carb and Hidex 300SL technique using measurements of ^{241}Pu and ^{90}Sr on various samples. Appl. Radiat. Isot. 93, 120–125.

Jandel, M., et al., 2015. Capture and fission with dance and neuance. European Phys. J. A 51 (12), 179–187.

Jastaniah, S.D., Sellin, P.J., 2004. Digitial techniques for n/γ pulse shape discrimination and capture-gated neutron spectroscopy using liquid scintillators. Nucl. Instrum. Methods Phys. Res. Sect. A 517, 202–210.

Jaubert, F., 2008. Standardization of a ^{186}Re sodium perrhenate radiochemical solution using the TDCR method in liquid scintillation counting. Appl. Radiat. Isot. 66, 960–964.

Jaubert, F., Cassette, P., 2004. Standardization of a ^{32}P solution containing pure-beta impurities using the TDCR method in liquid scintillation counting. Appl. Radiat. Isot. 60, 601–606.

Jaubert, F., Tartès, I., Cassette, P., 2006. Quality control of liquid scintillation counters. Appl. Radiat. Isot. 64, 1163–1170.

Jaworski, G., et al., 2012. Monte Carlo simulation of a single detector unit for the neutron detector array NEDA. Nucl. Instrum. Methods Phys. Res. Sect. A 673, 64–72.

Jiang, J., Davies, A.V., Britton, R.E., 2017. Measurement of ^{160}Tb and ^{161}Tb in nuclear forensics samples. J. Radioanal. Nucl. Chem. 314, 727–736.

Jiang, J., Davies, A., Arrigo, L., et al., 2018. Analysis of ^{161}Tb by radiochemical separation and liquid scintillation counting. Appl. Radiat. Isot. (in press).

Johansson, L., et al., 2002. Self-absorption correction in standardization of ^{204}Tl. Appl. Radiat. Isot. 56, 199–203.

Johansson, L., et al., 2003. Standardization of ^{238}Pu using four methods of measurement. Nucl. Instrum. Methods Phys. Res. Sect. A 505, 699–706.

Johansson, L.C., Sephton, J.P., 2010. Validation of a new TDCR system at NPL. Appl. Radiat. Isot. 68, 1537–1539.

Johansson, L., Filtz, J.-R., De Felice, P., et al., 2011. Advanced metrology for new generation nuclear power plants. In: Proceedings of the 2nd IMEKO TC 11 International Symposium Metrological Infrastructure, June 15–17th, Cavtat, Dubrovnik Riviera, Croatia.

Johansson, L., et al., 2014. A miniature TDCR system dedicated to in-situ activity assay. Appl. Radiat. Isot. 87, 287–291.

Jónsson, G., Theodórsson, P., Sigurdsson, K., 2006. Auto-radon: A new automatic liquid scintillation system for monitoring radon in water and air, pp. 119–123. In: Chałupnik, S., Schönhofer, F., Noakes, J. (Eds.), LSC 2005. Advances in Liquid Scintillation Spectrometry, Radiocarbon. The University of Arizona, Tucson, p. 449.

Jonsson, G., Theodórsson, P., 2007. Background components of a liquid scintillation counter in the14C window. Radiocarbon 49 (2), 315–323.

Jordan, W.C., Spiehler, V., Haendiges, R., et al., 1974. Evaluation of alternative counting methods for radioimmunoassay of hepatitis-associated antigen (HB Ag). Clin. Chem. 20 (7), 733–737.

Jörg, G., Bühnemann, R., Hollas, S., et al., 2010. Preparation of radiochemically pure ^{79}Se and highly precise determination of its half-life. Appl. Radiat. Isot. 68, 2339–2351.

Jörg, G., Amelin, Y., Kossert, K., et al., 2012. Precise and direct determination of the half-life of ^{41}Ca. Geochim. Cosmochim. Acta 88, 51–65.

Joshima, H., Matsuoka, O., 1970. Pu Special Research Report. National Institute of Radiological sciences NIRS-Pu-5, 53.

Kahl, S.D., Sittampalam, G.S., Weidner, J., 2012. Calculations and instrumentation used for radioligand binding assays. In: Sittampalam, G.S., et al. (Eds.), Assay Guidance Manual [Internet], Eli Lilly & Company and the National Center for Advancing Translational Sciences, Bethesda (MD). https://www.ncbi.nlm.nih.gov/books/NBK91997/.

Kaihola, L., 2000. Radionuclide identification in liquid scintillation alpha-spectroscopy. J. Radioanal. Nucl. Chem. 243 (2), 313–317.

Kallmann, H., 1950. Scintillation counting with solutions. Phys. Rev. 78 (5), 621–622.

for the SNO+ Collaboration Kamdin, K., 2015. Understanding the SNO+ Detector. Phys. Procedia 61, 719–723.

Kang, W.G., et al., 2013. Ultra-low gamma-ray measurement system for neutrinoless double beta decay. Appl. Radiat. Isot. 81, 290–293.

Kanisch, G., 2016. Generalized evaluation of environmental radioactivity measurements with UncertRadio. Part I: Methods without linear unfolding. Appl. Radiat. Isot. 110, 28–41.

Kashirin, I.A., et al., 2000. Liquid scintillation determination of low level components in complex mixtures of radionuclides. Appl. Radiat. Isot. 53, 303–308.

Kashirin, I.A., Ermakov, A.I., Malinovsky, S.V., et al., 2003. LSC: from routine counter to the real spectrometer, pp. 174–184. In: Warwick, P. (Ed.), Environmental Radiochemical Analysis II. Royal society of Chemistry, Cambridge, p. 420.

Kastens, L.W., Cahn, S.B., Manzur, A., et al., 2009. Calibration of a liquid xenon detector with $^{83}Kr^{m}$. Phys. Rev. C 80 (4), 045809. http://prc.aps.org/abstract/PRC/v80/i4/e045809.

Keightley, J., Pearce, A., Fenwick, A., et al., 2015. Standardisation of ^{223}Ra by liquid scintillation counting techniques and comparison with secondary measurements. Appl. Radiat. Isot. 95, 114–121.

Publ. No. 169-3052. In: Kessler, M.J. (Ed.), 1989. Liquid Scintillation Analysis, Science and Technology. PerkinElmer Life and Analytical Sciences, Boston.

Kessler, M.J., 1991a. Applications of quench monitoring using transformed external standard spectrum (tSIE). In: Ross, H., Noakes, J.E., Spaulding, J.D. (Eds.), Liquid Scintillation Counting and Organic Scintillators. Lewis Publishers, Chelsea, MI 48118, pp. 343–364.

Kessler, M.J., 1991b. Absolute activity liquid scintillation counting: an attractive alternative to quench-corrected DPM for higher energy isotopes. In: Ross, H., Noakes, J.E., Spaulding, J.D. (Eds.), Liquid Scintillation Counting and Organic Scintillators. Lewis Publishers, Chelsea, MI 48118, pp. 647–653.

Kharitonov, I., Shilnikova, T., Haaslahti, J., 2011. Uncertainty of a result of tritium activity measurement carried out with counter SL-300 at simplified and conventional processing of measurement results. In: LSC 2010, Advances in Liquid Scintillation Spectrometry. Radiocarbon. University of Arizona, Tucson.

Khayatzadeh Mahani, M., et al., 2008. A new method for simultaneous determination of ^{226}Ra and uranium in aqueous samples by liquid scintillation using chemometrics. J. Radioanal. Nucl. Chem. 275 (2), 427–432.

Khayatzadeh Mahani, M., et al., 2009. Application of chemometrics for simultaneous determination of thorium and uranium in aqueous samples by liquid scintillation, pp. 203–210. In: Eikenberg, J., Jäggi, M., Beer, H., Baehrle, H. (Eds.), LSC 2008. Advances in Liquid Scintillation Spectrometry, Radiocarbon. University of Arizona, Tucson, p. 445.

Kim, C.-K., Martin, P., Fajgelj, A., 2008. Quantification of measurement uncertainty in the sequential determination of ^{210}Pb and ^{210}Po by liquid scintillation counting and alpha-particle spectrometry. Accred Qual. Assur. 13, 691–702.

Kim, C.K., Al-Hamwi, A., Törvényi, A., et al., 2009. Validation of rapid methods for the determination of radiostrontium in milk. Appl. Radiat. Isot. 67, 786–793.

Kim, H., Jung, Y., Ji, Y.-Y., Lim, J.-M., Chung, K.-H., Kang, M.-J., 2017. Validation of a procedure for the analysis of 226Ra in naturally occurring radioactive materials using a liquid scintillation counter. J. Environm. Radioact. 166 (1), 188–194.

Kim, Y.J., et al., 2006. Determination of ^{129}I using liquid scintillation counting, pp. 273–276. In: Chałupnik, S., Schönhofer, F., Noakes, J. (Eds.), LSC 2005. Advances in Liquid Scintillation Spectrometry., Radiocarbon University of Arizona, Tucson, p. 449.

Kits, J., Látalová, M., Látal, F., et al., 1985. Determination of the activity of ^{125}I by liquid scintillation measurement. Int. J. Appl. Radiat. Isot. 36, 320.

Klapdor-Kleingrothaus, H.V., 2006. Lessons after the evidence for $0\nu\beta\beta$ decay. *Phys. Scr.* T 127, 40.

Klapdor-Kleingrothaus, H.V., Krivosheina, I.V., 2006. The evidence for the observation of $0\nu\beta\beta$ decay: The identification of $0\nu\beta\beta$ events from the full spectra. Mod. Phys. Letts. A 21 (20), 1547–1566.

Klapdor-Kleingrothaus, H.V., Krivosheina, I.V., Dietz, A., et al., 2004. Search for neutrinoless double beta decay with enriched ^{76}Ge in Gran Sasso 1990–2003. Phys. Letts. B 586, 198–212.

Klein, H., Neumann, S., 2002. Neutron and photon spectrometry with liquid scintillation detectors in mixed fields. Nucl. Instrum. Methods Phys. Res. Sect. A 476, 132–142.

Klein, H., Brooks, F.D., 2006. Scintillation detectors for fast neutrons. Proceedings of Science. PoS (FNDA2006) 097. http://cds.cern.ch/record/1111862/files/FNDA2006-097.pdf.

Kleinschmidt, R.I., 2004. Gross alpha and beta activity analysis in water—a routine laboratory method using liquid scintillation analysis. Appl. Radiat. Isot. 61, 333–338.

Klin, Y., et al., 2010. Distribution of radiolabeled L-glutamate and D-aspartate from blood into peripheral tissues in naive rats: Significance for brain neuroprotection. Biochem. Biophys. Res. Comm. 399, 694–698.

Kobayashi, Y., Maudsley, D.V., 1970. Practical aspects of double isotope counting. In: Bransome Jr., E.D. (Ed.), Current Status of Liquid Scintillation Counting. Grune and Stratton, New York, pp. 76–78.

Kobayashi, M., et al., 2015. Development of radioiodine-labeled 4-hydroxyphenylcysteamine for specific diagnosis of malignant melanoma. Nucl. Med. Biol. 42, 536–540.

Kobbe, D., Focke, M., Puchta, H., 2009. Purification and characterization of RecQ heliocases of plants, pp. 195–209. In: Abdelhaleem, M.M. (Ed.), Heliocases, Methods and protocols. Humana Press, p. 404.

Kojima, T., et al., 2018. Discovery of 1,2,3,4-tetrahydropyrimido[1,2-a] benzimidazoles as novel class of corticotropin releasing factor 1 receptor antagonists. Bioorgan. Medic. Chem (in press).

Kolarov, V., Le Gallic, Y., Vatin, R., 1970. Mesure absolue directe de l'activité des emetteurs β purs par scintillation liquide. Int. J. Appl. Radiat. Isot. 21, 443–452.

Kolhinen, V.S., et al., 2010. Accurate *Q* value for the ^{74}Se double-electron-capture decay. Phys. Letters B 684 (Issue 1), 17–21.

Kornilov, N.V., Fabry, I., Oberstedt, S., et al., 2009. Total characterization of neutron detectors with a ^{252}Cf source and a new light output determination. Nucl. Instrum. Methods Phys. Res., Sect. A 599, 226–233.

Komosa, A., Ślepecka, K., 2010. Effect of liquid scintillating cocktail volume on ^{3}H and ^{14}C measurement parameters using a Quantulus spectrometer. Nukleonika 55 (2), 155–161.

Koskinas, M.F., Nascimento, T.S., Yamazaki, I.M., et al., 2014. Covariance methodology applied to ^{35}S disintegration rate measurements by the CIEMAT/NIST method. Rad. Phys. Chem. 95, 131–133.

Kossert, K., 2003. Half-life measurements of ^{87}Rb by liquid scintillation counting. Appl. Radiat. Isot. 59, 377–382.

Kossert, K., 2006. A new method for secondary standard measurements with the aid of liquid scintillation counting. Appl. Radiat. Isot. 64, 1459–1464.

Kossert, K., Grau Carles, A., 2006. The LSC efficiency for low-Z electron-capture nuclides. Appl. Radiat. Isot. 64, 1446–1453.

Kossert, K., Grau Carles, A., 2008. Study of a Monte Carlo rearrangement model for the activity of electron-capture nuclides by means of liquid scintillation counting. Appl. Radiat. Isot. 66, 998–1005.

Kossert, K., Grau Carles, A., 2010. Improved method for the calculation of the counting efficiency of electron-capture nuclides in liquid scintillation samples. Appl. Radiat. Isot. 68 (7–8), 1482–1488.

Kossert, K., Nähle, O.J., 2014. Activity determination of ^{59}Fe. Appl. Radiat. Isot. 93, 33–37.

Kossert, K., Mougeot, X., 2015. The importance of the beta spectrum calculation for accurate activity determination of ^{63}Ni by means of liquid scintillation counting. Appl. Radiat. Isot. 101, 40–43.

Kossert, K., Schrader, H., 2004. Activity standardization by liquid scintillation counting and half-life measurements of ^{90}Y. Appl. Radiat. Isot. 60, 741–749.

Kossert, K., Janβen, H., Klein, R., Schneider, M., Schrader, H., 2006. Standardization and nuclear decay data of ^{65}Zn. Appl. Radiat. Isot. 64, 1420–1424.

Kossert, K., Jörg, G., Lierse von Gostomski, C., 2009a. Activity standardization of ^{41}Ca by means of liquid scintillation counting. Radiochim. Acta 97, 1–8.

Kossert, K., et al., 2009b. High-precision measurement of the half-life of ^{147}Sm. Appl. Radiat. Isot. 67, 1702–1706.

Kossert, K., Nähle, O.J., Grau Carles, A., 2011a. Beta shape-factor function and activity determination of ^{241}Pu. Appl. Radiat. Isot. 69, 1246–1250.

Kossert, K., Nähle, O.J., Warwick, P.E., Wershofen, H., Croudace, I.W., 2011b. Activity determination and nuclear decay data of ^{113m}Cd. Appl. Radiat. Isot. 69, 500–505.

Kossert, K., Nähle, O.J., Ott, O., Dersch, R., 2012. Activity determination and nuclear decay data of ^{177}Lu. Appl. Radiat. Isot. 70, 2215–2221.

Kossert, K., Jörg, G., Lierse, v., Gostomski, C., 2013. Experimental half-life determination of ^{176}Lu. Appl. Radiat. Isot. 81, 140–145.

Kossert, K., Cassette, P., Grau Carles, A., Jörg, G., Lierse, V., et al., 2014a. Extension of the TDCR model to compute counting efficiencies for radionuclides with complex decay schemes. Appl. Radiat. Isot. 87, 242–248.

Kossert, K., Nähle, O.J., Janßen, H., 2014b. Activity determination of ^{229}Th by means of liquid scintillation counting. Appl. Radiat. Isot. 87, 274–281.

Kossert, K., Capogni, M., Nähle, O.J., 2014c. Bilateral comparison between PTB and ENEA to check the performance of a commercial TDCR system for activity measurements. Appl. Radiat. Isot. 93, 38–44.

Kossert, K., Bokeloh, K., Dersch, R., Nähle, O., 2015. Activity determination of ^{227}Ac and ^{223}Ra by means of liquid scintillation counting and determination of nuclear decay data. Appl. Radiat. Isot. 95, 143–152.

Kossert, K., Marganiec-Gałązka, J., Mougeot, X., Nähle, O.J., 2017a. Activity determination of ^{60}Co and the importance of its beta spectrum. Appl. Radiat. Isot. (in press).

Kossert, K., Marganiec-Gałązka, J., Nähle, O.J., 2017b. Primary activity standardization of ^{134}Cs. J. Radioanal. Nucl. Chem. 314, 545–553.

Kouru, H., 1991. A new quench curve fitting procedure: fine tuning of a spectrum library, pp. 247–255. In: Ross, H., Noakes, J.E., Spaulding, J.D. (Eds.), Liquid Scintillation Counting and Organic Scintillators. Lewis Publishers, Chelsea, MI 48118.

Kouru, H., Rundt, K., 1991. Multilabel counting using digital overlay technique, pp. 239–246. In: Ross, H., Noakes, J.E., Spaulding, J.D. (Eds.), Liquid Scintillation Counting and Organic Scintillators. Lewis Publishers, Chelsea, MI 48118.

Krajcar Bronić, I., Horvatinčić, N., Barešić, J., Obelić, B., 2009. Measurement of ^{14}C activity by liquid scintillation counting. Appl. Radiat. Isot. 67, 800–804.

for the SNO+ Collaboration Kraus, C., 2006. SNO with liquid scintillator: SNO+. Progr. Part. Nucl. Phys. 57, 150–152.

Krištof, R., Logar, J.K., 2013. Direct LSC method for measurements of biofuels in fuel. Talanta 111, 183–188.

Krištof, R., Logar, J.K., 2017. Direct LSC method for determination of bio-origin by C-14 measurement. J. Radioanal. Nucl. Chem. 314, 715–719.

Kubota, S., Nakamoto, A., Takahashi, T., Hamada, T., Shibamura, E., et al., 1978a. Recombination luminescence in liquid argon and in liquid xenon. Phys. Rev. B 17 (6), 2762–2765.

Kubota, S., Hishida, M., Ruan, J.-Z., 1978b. Evidence for a triplet state of the self-trapped excitation states in liquid argon, krypton and xenon. J. Phys. C 11, 2645–2651.

Kubota, S., Hishida, M., Suzuki, M., Ruan, J.-Z., 1979. Dynamical behavior of free electrons in the recombination process in liquid argon, krypton, and xenon. Phys. Rev. B 20 (8), 3486–3496.

Kulkarni, D.B., Anuradha, R., Reddy, P.J., Joseph, L., 2011a. Standardization of ^{131}I: Implementation of CIEMAT/NIST method at BARC, India. Appl. Radiat. Isot. 69, 1512–1515.

Kulkarni, D.B., et al., 2011b. Comparison of efficiency tracing and zero detection threshold techniques with CIEMAT/NIST standardization method under different quench conditions with liquid scintillation spectrometer. Current Sci 90 (1), 83–87.

Kulkarni, D.B., Anuradha, R., Joseph, L., Tomar, B.S., 2013. Development of liquid scintillation based $4\pi\beta$ (LS)-γ coincidence counting system and demonstration of its performance by standardization of ^{60}Co. Appl. Radiat. Isot. 72, 68–72.

Kulkarni, D.B., Anuradha, R., Joseph, L., Kulkarni, M.S., Tomar, B.S., 2016. Performance demonstration of $4\pi\beta$(LS)-γ coincidence counting system for standardization of radionuclides with complex decay scheme. Appl. Radiat. Isot. 108, 24–29.

Kulkarni, D.B., Joseph, L., Anuradha, R., Kulkarni, M.S., Tomar, B.S., 2017. Standardization of ^{68}Ge-^{68}Ga using $4\pi\beta$(LS)-γ coincidence counting system for activity measurements. Appl. Radiat. Isot. 123, 6–10.

Kumar, M., et al., 2018. Reversed isoniazids: Design, synthesis and evaluation against *Mycobacterium tuberculosis*. Bioorg. Medicin. Chem. 26, 833–844.

Kumblad, L., Bradshaw, C., Gilek, M., 2005. Bioaccumulation of ^{51}Cr, ^{63}Ni and ^{14}C in Baltic Sea benthos. Environ. Poll. 134, 45–56.

Laborie, J.-M., et al., 2012. Measurement of the neutron-induced deuteron breakup reaction cross-section between 5 and 25 MeV. Eur. Phys. J. A 48, 87'.

Lachenmaier, T., von Feilitzsch, F., Göger-Neff, M., Lewkea, T., Marrodán Undagoitia, T., et al., 2010. Physics with the large liquid-scintillator detector LENA. Prog. Part. Nucl. Phys. 64, 381−383.

Laney, B.H., 1976. External standard method of quench correction: advanced techniques. In: Noujaim, A.A., Ediss, C., Wiebe, L.I. (Eds.), Liquid Scintillation Science and Technology. Academic Press, New York and London, pp. 135−152.

Laney, B.H., 1977. Two-parameter pulse height analysis in liquid scintillation. In: Crook, M.A., Johnson, P. (Eds.), Liquid Scintillation Counting, vol. 4. Heyden, London, pp. 74−84.

L'Annunziata, M.F., 1971. Birth of an unique parent-daughter relation: secular equilibrium. J. Chem. Educ. 48, 700−703.

L'Annunziata, M.F., 1979. Radiotracers in Agricultural Chemistry. Academic Press, London, p. 536.

L'Annunziata, M.F., 1984a. The detection and measurement of radionuclides. In: L'Annunziata, M.F., Legg, J.O. (Eds.), Isotopes and Radiation in Agricultural Sciences, vol. 1. Academic Press, London, pp. 141−231.

L'Annunziata, M.F., 1984b. Agricultural biochemistry: reaction mechanisms and pathways in biosynthesis. In: L'Annunziata, M.F., Legg, J.O. (Eds.), Isotopes and Radiation in Agricultural Sciences, vol. 2. Academic Press, London, pp. 105−182.

L'Annunziata, M.F., 1987. Radionuclide Tracers, Their Detection and Measurement. Academic Press, San Diego, p. 505.

L'Annunziata, M.F., 1997a. Efficiency tracing DPM (ET-DPM) and Direct-DPM - Instrument performance data. Tri-Carb LSC Application Note. Packard BioScience Co., Meriden, CT, p. 8.

L'Annunziata, M.F., 1997b. Comparison of conventional and full spectrum DPM (FS-DPM) analysis of ^{33}P-^{32}P double labels - Instrument performance data. Tri-Carb LSC Application Note. PerkinElmer Inc., p. 6. Available via. http://las.perkinelmer.com/Content/applicationnotes/app_tricarbconventnlfsdpm.pdf

L'Annunziata, M.F., 2007. Radioactivity: Introduction and History. Elsevier, Amsterdam and Boston, p. 609.

L'Annunziata, M.F., 2016. Radioactivity: Introduction and History: From the Quantum to Quarks. Elsevier, Amsterdam and Boston, p. 903.

Lara-Robustillo, E., Rodríguez Alcalá, M., 2006. Simultaneous determination of alpha and beta emitters by liquid scintillation counting, pp. 41−49. In: Chałupnik, S., Schönhofer, F., Noakes, J. (Eds.), LSC 2005, Advances in Liquid Scintillation Spectrometry, Radiocarbon. University of Arizona, Tucson, p. 449.

Larionova, N.V., et al., 2017. Plants as indicators of tritium concentration in ground water at the Semipalatinsk test site. J. Environm. Radioact. 177, 218−224.

Larsson, J., Wingårdh, K., Berggård, T., Davies, J.R., Lögdberg, L., et al., 2001. Distribution of iodine-125-labeled α_1-microglobulin in rats after intravenous injection. J. Lab. Clin. Med. 137 (3), 165−175.

Laureano-Perez, L., Collé, R., Fitzgerald, R., Outola, I., Pibida, L., 2007. A liquid-scintillation-based primary standardization of ^{210}Pb. Appl. Radiat. Isot. 65, 1368−1380.

Laureano-Perez, L., Collé, R., Fitzgerald, R., Zimmerman, B.E., Cumberland, L., 2010. Investigation into the standardization of ^{99}Tc. Appl. Radiat. Isot. 68, 1489−1494.

Laureano-Perez, L., Fitzgerald, R., Collé, R., 2014. Standardization of ^{237}Np. Appl. Radiat. Isot. 87, 269−273.

Lavagno, A., Gervino, G., Marino, C., 2010. High efficiency large volume multiparametric neutron detector. Nucl. Instrum. Methods Phys. Res. Sect. A 617, 492−494.

Lawrence, C.C., Enqvist, A., Flaska, M., Pozzi, S.A., Becchetti, F.D., 2013. Comparison of spectrum-unfolding performance of (EJ315) and (EJ309) liquid scintillators on measured ^{252}Cf pulse-height spectra. Nucl. Instrum. Methods Phys. Res. Sect. A 729, 924−929.

Lazare, L., Crestey, C., Bleistein, C., 2009. Measurement of ^{90}Sr in primary coolant of pressurized water reactor. J. Radioanal. Nucl. Chem. 279 (2), 633−638.

Lee, J.B., et al., 2016. A Glu-urea-Lys ligand-conjugated lipid nanoparticle/siRNA system inhibits androgen receptor expression *in vivo*. Molec. Therapy-Nucleic Acids 5, e348. https://doi.org/10.1038/mtna.2016.43.

Lee, J.S., Kim, Y.H., Lee, K.B., et al., 2012. Pulse shaping analysis with LAB-based liquid scintillators. J. Korean Phys. Soc. 60 (3), 554−557.

Lee, K.B., Lee, J.M., Park, T.S., 2004a. Implementation of CIEMAT/NIST LSC efficiency tracing method in KRISS: ^{204}Tl standardization. Appl. Radiat. Isot. 60, 893−897.

Lee, K.B., Lee, J.M., Park, T.S., Hwang, H.Y., 2004b. Implementation of TDCR method in KRISS. Nucl. Instrum. Methods Phys. Res. Sect. A 534, 496−502.

Lee, M.H., Chung, K.H., Lee, C.W., 2002a. Sequential radiochemical separation of ^{90}Sr and ^{99}Tc in aqueous samples and measurement by liquid scintillation counting, pp. 397−403. In: Möbius, S., Noakes, J.E., Schönhofer, F. (Eds.), LSC 2001. Advances in Liquid Scintillation Spectrometry, Radiocarbon. University of Arizona, Tucson, p. 456.

Lee, M.H., Chung, K.H., Choi, G.K., Lee, C.W., 2002b. Measurement of ^{90}Sr in aqueous samples using liquid scintillation counting with full spectrum DPM method. Appl. Radiat. Isot. 57 (2), 257−263.

Lee, U., Bae, J.W., Kim, H.R., 2017. Multiple beta spectrum analysis based on spectrum fitting. J. Radioanal. Nucl. Chem. 314, 617−622.

for the SNO+ Collaboration Lefeuvre, G., 2013. From SNO to SNO+, upgrading a neutrino experiment. Nucl. Instrum. Methods Phys. Res. Sect. A 718, 506−508.

Lennox, K.P., et al., 2017. Assessing and minimizing contamination in time of flight based validation data. Nucl. Instrum. Methods Phys. Res. Sect. A 870, 30−36.

the EXO-200 Collaboration Leonard, D.S., et al., 2008. Systematic study of trace radioactive impurities in candidate construction materials for EXO-200. Nucl. Instrum. Methods Phys. Res. Sect. A 591, 490−509.

Leppänen, M.T., Kukkonen, J.V.K., 2006. Evaluating the role of desorption in bioavailability of sediment-associated contaminants using oligochaetes, semipermeable membrane devices and Tenax extraction. Environ. Poll. 140, 150−163.

Liao, C., Yang, H., 2014. n/γ Pulse shape discrimination comparison of EJ301 and EJ339A liquid scintillation detectors. Annals Nucl. Energy 69, 57−61.

Lightfoot, P.K., Kudryavtsev, V.A., Spooner, N.J.C., Liubarsky, I., et al., 2004. Development of a gadolinium-loaded liquid scintillator for solar neutrino detection and neutron measurements. Nucl. Instrum. Methods Phys. Res. Sect. A 522, 439−446.

Lin, C.-F., Wang, J.-J., Huang, J.-C., Yeh, C.-H., Yuan, M.-C., Chang, B.-J., 2012. Evaluating practicability of an LSC method for routine monitoring gross alpha and beta activities in water samples in Taiwan. Applied. Radiat. Isot. 70 (9), 1981−1984.

Lin, R., Ye, L., Yang, Y., Liao, J., Mo, S., Liu, N., 2008. Calcium antagonistic effects of *Bambusa Rigida* investigated by ^{45}Ca and its protection on myocardial ischemia in rats. Nucl. Sci. Techn. 19, 99–104.

Little, R.J., Rodríguez, C.G., 2005. Modeling of human corticosteroid binding globulin. Use of structure–activity relations in soft steroid binding to refine the structure. Pharmac. Res. 22 (11), 1783–1792.

Little, R., et al., 2011. The transient receptor potential ion channel TRPV6 is expressed at low levels in osteoblasts and has little role in osteoblast calcium uptake. PLoS One 6 (11), E28166. https://doi.org/10.1371/journal.pone.0028166.

Liu, G., Aspinall, M.D., Ma, X., Joyce, M.J., 2009. An investigation of the digital discrimination of neutrons and γ rays with orgainic scintillation detectors using an artificial neural netwrok. Nucl. Instrum. Methods Phys. Res. Sect. A 607, 620–628.

Liu, N., Yang, Y., Mo, S., Liao, J., Jin, J., 2005. Calcium antagonistic effects of Chinese crude drugs: Preliminary investigation and evaluation by ^{45}Ca. Appl. Radiat. Isot. 63, 151–155.

Liu, H.R., Liang, J.C., Yuan, D.Q., 2017. Standardization of Na-22 by the CIEMAT/NIST method. In: LSC2017 Conference, 1–5 May, Copenhagen.

Liyanage, J.A., Yonezawa, C., 2003. A new analytical method for ^{32}P: Liquid scintillation counting with solvent extraction. J. Radioanal. Nucl. Chem. 256 (2), 279–282.

Lluch, E., Barrera, J., Tarancón, A., Bagán, H., García, J.F., 2016. Analysis of ^{210}Pb in water samples with plastic scintillation resins. Anal. Chim. Acta 940, 38–45.

Lombardi, P., Ortica, F., Ranucci, G., Romani, A., 2013. Decay time and pulse shape discrimination of liquid scintillators based on novel solvents. Nucl. Instrum. Methods Phys. Res. Sect. A 701, 133–144.

López, M.L., Vommaro, R., Zalis, M., de Souza, W., Blair, S., et al., 2010. Induction of cell death on *Plasmodium falciparum* asexual blood stages by *Solanum nudum*. Parasit. Int. 59, 217–225.

Lorber, A., Wangen, L.E., Kowalski, B.R., 1986. A theoretical foundation for the PLS algorithm. J. Chemom. 1 (1), 19–31.

Los Arcos, J.M., Ortiz, F., 1997. KB: a code to determine the ionization quench function *Q(E)* as a function of the *kB* parameter. Comp. Phys. Commun. 103, 83–94.

Los Arcos, J.M., Grau Molanda, A., Fernandez, A., 1987. VIASKL: A computer program to evaluate the liquid scintillation counting efficiency and its associated uncertainty for K-L-atomic shell electron-capture nuclides. Comput. Phys. Commun. 44, 209–220.

Los Arcos, J.M., Grau, A., García-Toraño, E., 1991. LSC standardization of multigamma electron-capture radionuclides by the efficiency tracing method. In: Ross, H., Noakes, J.E., Spaulding, J.D. (Eds.), Liquid Scintillation Counting and Organic Scintillators. Lewis Publishers, Chelsea, MI 48118, pp. 611–622.

Lourenço, V., Bobin, C., Chisté, V., Lacour, D., Rigoulay, F., et al., 2015. Primary standardization of SIR-Spheres based on the dissolution of the ^{90}Y-labeled resin microspheres. Appl. Radiat. Isot. 97, 170–176.

Lozac'h, F., et al., 2018. Evaluation of cAMS for ^{14}C microtracer ADME studies: opportunities to change the current drug development paradigm. Bioanalysis. https://doi.org/10.4155/bio-2017-0216 (in press). https://www.ncbi.nlm.nih.gov/pubmed/29451392.

Lu, H., 2017. Status of the JUNO reactor anti-neutrino experiment. Nucl. Part. Phys. Proc. 287–288, 143–146.

Lv, X.-X., Chen, X.-L., Yao, S.-H., 2014. Development of TDCR liquid scintillation system. Atomic Energy Sci. Technol. 48 (11), 2133–2136.

Mahani, M., Ghomi, H.S., Mazloomifar, A., Salimi, B., 2012. Application of multi-way partial least squares calibration for simultaneous determination of radioisotopes by liquid scintillation technique. Nucl. Technol. Radiat. Pro. 27, 125–130.

Malinovsky, S.V., Kashirin, I.A., Ermakov, A.I., Tikhomirov, V.A., Sobolev, A.I., et al., 2002a. New software for analyzing complex spectra obtained with ultra low level liquid scintillation spectrometer 'Quantulus', pp. 119–126. In: Möbius, S., Noakes, J.E., Schönhofer, F. (Eds.), LSC 2001. Advances in Liquid Scintillation Spectrometry, Radiocarbon. University of Arizona, Tucson, p. 456.

Malinovsky, S.V., Kashirin, I.A., Ermakov, A.I., Tikhomirov, V.A., Belanov, S.V., et al., 2002b. Mathematical aspects of decoding complex spectra applied to liquid scintillation counting, pp. 127–135. In: Möbius, S., Noakes, J.E., Schönhofer, F. (Eds.), LSC 2001. Advances in Liquid Scintillation Spectrometry, Radiocarbon. University of Arizona, Tucson, p. 456.

Malkov, V.A., Panyutin, I.G., Neumann, R.D., Zhurkin, V.B., Camerini-Otero, R.D., 2000. Radioprobing of a recA-three-stranded DNA complex with iodine-125: evidence for recognition of homology in the major groove of the target duplex. J. Molec. Biol. 299 (3), 629–640.

Maneschg, W., 2015. Review of neutrinoless double beta decay experiments: Present status and near future. Nucl. Part. Phys. Proc. 260, 188–193.

Maringer, F.J., et al., 2017. Long-term environmental radioactive contamination of Europe due to the Chernobyl accident - Results of the Joint Danube Survey 2013. Appl. Radiat. Isot. 126, 100–105.

Marrone, S., Cano-Ott, D., Colonna, N., Domingo, C., Gramegna, F., et al., 2002. Pulse shape analysis of liquid scintillators for neutron studies. Nucl. Instrum. Methods Phys. Res. Sect. A 490, 299–307.

Martens, H., Næs, T., 1992. Multivariate Calibration. Wiley, p. 419.

Martins, S.I.F.S., van Boekel, M.A.J.S., 2003. Malanoidins extinction coefficient in the glucose/glycine Maillard reaction. Food Chem. 83, 135–142.

Matsui, Y., Takiue, M., 1991. Liquid scintillation radioassay of multi-labeled beta-emitters. Appl. Radiat. Isot. 42, 841–845.

Matta, C., Juhász, T., Szíjgyártó, Z., Kolozsvári, B., Somogyi, C., et al., 2011. PKCdelta is a positive regulator of chondrogenesis in chicken high density micromass cell cultures. Biochim 93, 149–159.

McCabe, M.T., et al., 2012. Mutation of A677 in histone methyltransferase EZH2 in human B-cell lymphoma promoteshypertrimethylation of histone H3 on lysine 27 (H3K27). Proc. Nat. Acad. Sci. USA 109 (8), 2989–2994.

McCarthy, M.T., et al., 2018. Purine nucleotide metabolism regulates expression of the human immune ligand MICA. J. Biol. Chem. (in press) http://www.jbc.org/cgi/doi/10.1074/jbc.M117.809459.

McCormac, F., 1992. Liquid Scintillation Counter Characterization, Optimization and Benzene Purity Correction. Radiocarbon 34 (1), 37–45.

McDowell, W.J., 1996. Recent applications of PERALS® spectrometry, pp. 157–165. In: Cook, G.T., Harkness, D.D., MacKenzie, A.B., Miller, B.F., Scott, E.M. (Eds.), Liquid Scintillation Spectrometry 1994, Radiocarbon. University of Arizona, Tucson.

McDowell, W.J., McDowell, B.L., 1991. Liquid scintillation alpha spectrometry: A method for today and tomorrow. In: Ross, H., Noakes, J.E., Spaulding, J.D. (Eds.), Liquid Scintillation Counting and Organic Scintillators. Lewis Publishers, Chelsea, MI 48118, pp. 105–122.

McDowell, W.J., McDowell, B.L., 1993. The growth of a radioanalytical method: Alpha liquid scintillation spectrometry, pp. 193–200. In: Noakes, J.E., Schönhofer, F., Polach, H.A. (Eds.), Liquid Scintillation Spectrometry 1992, Radiocarbon. University of Arizona, Tucson.

McGrath, J., Fulton, B.R., Joshi, P., Davies, P., Muenstermann, D., et al., 2010. Detecting multi-hit events in a CdZnTe coplanar grid detector using pulse shape analysis: A method for improving background rejection in the COBRA $0\nu\beta\beta$ experiment. Nucl. Instrum. Methods Phys. Res. Sect. A 615, 57–61.

McKlveen, J.W., McDowell, W.J., 1984. Liquid scintillation alpha spectrometry techniques. Nucl. Instrum. Methods Phys. Res. 223 (2–3), 372–376.

McQuarrie, S.A., Wiebe, L.I., Ediss, C., 1980. Observations of the performance of ESP and H# in liquid scintillation counting, pp. 291–299. In: Peng, C.-T., Horrocks, D.L., Alpen, E.L. (Eds.), Liquid Scintillation Counting, Recent Applications and Development. Academic Press, New York and London.

McQuarrie, S.A., Noujaim, A.A., Ediss, C., Wiebe, L.I., 1981. Liquid scintillation counting of positron emitters. ANS 1981 Winter Meeting, San Francisco, CA, November 29-December 3, 1981. Trans. Amer. Nucl. Soc. 39, 27–28.

McQuarrie, S.A., Noujaim, A.A., 1983. ^{67}Ga: A novel internal standard for LSC. In: McQuarrie, S.A., Ediss, C., Wiebe, L.I. (Eds.), Advances in Scintillation Counting. University of Alberta Printing Services, Edmonton, pp. 57–65.

Medeiros, R.B., Godinho, R.O., Mattos, M.F.S.S., 2003. Comparison of the efficacy of biodegradable and non-biodegradable scintillation liquids on the counting of tritium- and [^{14}C]-labeled compounds. Braz. J. Med. Biol. Res. 36 (12), 1733–1739.

Mei, D.-M., Yin, Z.-B., Stonehill, L.C., Hime, A., 2008. A model of nuclear recoil scintillation efficiency in noble liquids. Astropart. Phys. 30, 12–17.

Melkes, B., et al., 2016. Biased μ-opioid receptor agonists diversely regulate lateral mobility and functional coupling of the receptor to its cognate G proteins. Naunyn-Schmiedeberg's Arch. Pharmacol. 389, 1289–1300.

Mellado, J., Tarancón, A., García, J.F., Rauret, G., Warwick, P., 2005. Combination of chemical separation and data treatment for ^{55}Fe, ^{63}Ni, ^{99}Tc, ^{137}Cs and ^{90}Sr/^{90}Y activity determination in radioactive waste by liquid scintillation. Appl. Radiat. Isot. 63, 207–215.

Mendonça, M.L.T.G., Godoy, J.M., da Cruz, R., P., Perez, R.A.R., 2006. Radiocarbon dating of archaeological samples (*sambaqui*) using CO_2 absorption and liquid scintillation spectrometry of low background radiation. J. Environ. Radioact. 88, 205–214.

Merín, R., Tarancón, A., Bagán, H., García, J.F., 2017. Plastic scintillation foils for the measurement of the alpha and beta emitting radionuclides. In: LSC2017 Conference, 1–5 May, Copenhagen.

Messier, M.D., 2006. Review of neutrino oscillation experiments. In: Flavor Phys. and CP Violation Conf., Vancouver 2002. hep-ex/0606013v1 5 Jun 2002.

Meyer, B.R., Simpson, B.R.S., 1990. A direct method for ^{55}Fe activity Measurement. Appl. Radiat. Isot. 41, 375–379.

Mietelski, J.W., Gaca, P., 2002. Measurements of ^{90}Sr and ^{241}Pu in various matrices, pp. 373–378. In: Möbius, S., Noakes, J.E., Schönhofer, F. (Eds.), LSC 2001, Advances in Liquid Scintillation Spectrometry, Radiocarbon. The University of Arizona, Tucson, p. 456.

Mikelic, L., Orescanin, V., Lulic, S., 2007. Determination of ^{55}Fe in waste waters of the Krsko nuclear power plant measured simultaneously by liquid scintillation spectrometer (LSC) and X-ray spectrometer (XRS). Nucl. Instrum. Methods Phys. Res. Sect. B 263, 95–98.

Miller, M., Kereiakes, J.G., Friedman, B.I., 1969. Determination of iron-59 and iron-55 in [blood] plasma using liquid scintillation counting. Int. J. Appl. Radiat. Isot. 20 (2), 133–135.

Mingote, R.M., Barbeira, P.J.S., Rocha, Z., 2006. Methodoly for rapid tritium determination in urine. J. Radioanal. Nucl. Chem. 269 (2), 475–479.

Mini, G., Pepe, F., Tintori, C., Capogni, M., 2014. A full digital approach to the TDCR method. Appl. Radiat. Isot. 87, 166–170.

Minne, E., Heynen, F., Hallez, S., 2008. Possible overestimation of the external standard quench parameter on wallac 1220 quantulus™ with high energetic beta-emitters. J. Radioanal. Nucl. Chem. 278 (1), 39–45.

Miramonti, L., 2015. on behalf of the Borexino Collaboration Geoneutrinos from 1353 days with the Borexino detector. Phys. Procedia 61, 340–344.

Mitev, K., Dimitrova, I., Tarancón, A., Pressyanov, D., Tsankov, D., Boshkova, T., Georgiev, S., Sekalova, R., García, J.F., 2016. Pilot Study of the Application of Plastic Scintillation Microspheres to Rn-222 Detection and Measurement. IEEE Trans. Nucl. Sci. 63 (2), 1209–1217.

Mitev, K., Cassette, P., Jordanov, V., Liu, H.R., Dutsov, C., 2017. Design and performance of a miniature TDCR counting system. J. Radioanal. Nucl. Chem. 314 (2), 583–589.

Miyajima, M., Sasaki, S., Tawara, H., Shibamura, E., 1992. Absolute number of scintillation photons in liquid xenon by alpha-particles. IEEE Trans. Nucl. Sci. 39 (4), 536–540.

Miyazawa, Y., et al., 1991. Application of simultaneous determination of ^{3}H, ^{14}C. and ^{22}Na by liquid scintillation counting to the measurement of cellular ion-transport. Anal. Biochem. 198, 194–199.

Mo, L., Avei, B., James, D., Simpson, B., Van Wyngaardt, et al., 2005. Development of activity standard for ^{90}Y microspheres. Appl. Radiat. Isotopes 63, 193–199.

Mo, L., Bignell, L.J., Steele, T., Alexiev, D., 2010. Activity measurement of ^{3}H using the TDCR method and observation of source stability. Appl. Radiat. Isotopes 68, 1540–1542.

Mochizuki, M., et al., 2016. Discovery of a 7-arylaminobenzimidazole series as novel CRF1 receptor antagonists. Bioorganic Medicinal Chem 24, 4675–4691.

Mochizuki, M., et al., 2017. Discovery of 4-chloro-2-(2,4-dichloro-6-methylphenoxy)-1-methyl-7-(pentan-3-yl)-1H-benzimidazole, a novel CRF1 receptor antagonist. Bioorganic Medicinal Chem 25, 1556–1570.

Moe, M.K., 1986. Double-beta decay in ^{82}Se, ^{128}Te, and ^{130}Te. AIP Conf. Proc. 150, 1012–1016.

Moe, M.K., Rosen, S.P., 1989. Double-beta decay. Scient. Amer. 48–55 (November).

Möllenberg, R., et al., 2014. Detecting the upturn of the solar ^{8}B neutrino spectrum with LENA. Phys. Letts. B 737, 251–255.

Mordaunt, C.J., Gevao, B., Jones, K.C., Semple, K.T., 2005. Formation of non-extractable pesticide residues: observations on compound differences, measurement and regulatory issues. Environm. Poll. 133, 25–34.

Morissette, M., et al., 2018. Membrane cholesterol removal and replenishment affect rat and monkey brain monoamine transporters. Neuropharmacol (in press).

Morita-Murase, Y., Murakami, I., Homma, Y., 2001. The air luminescence count for the rapid determination of ^{222}Rn in a liquid scintillation spectrometer. Chem. Lett. 238–239.

Morita-Murase, Y., Murakami, I., Homma, Y., 2003. Counting efficiency for rapid preparation of known amounts of ^{222}Rn by the air luminescence method. J. Nucl. Radiochem. Sci. 4 (2), 23–26.

Motta, D., Buck, C., Hartmann, F.X., Lasserre, T., Schönert, S., et al., 2005. Prototype scintillator cell for an In-based solar neutrino detector. Nucl. Instrum. Methods Phys. Res. Sect. A 547, 368–388.

Murase, Y., Homma, Y., Takiue, M., 1989a. Effect of air luminescence counts on determination of ^{222}Rn by liquid scintillation counting. Appl. Radiat. Isotopes 40, 295–298.

Murase, Y., Homma, Y., Takiue, M., Aburai, T., 1989b. Determination of air luminescence spectra for alpha-emitters with liquid scintillation spectrometers. Appl. Radiat. Isotopes 40, 291–294.

Murase, Y., Homma, Y., Murakami, I., Handa, K., 1998. Assay of ^{222}Rn in water samples by a modified integral counting method. Appl. Radiat. Isot. 49 (7), 861–865.

Murase, Y., Homma, Y., Murakami, I., Handa, K., Horiuchi, K., 1999. Indoor ^{222}Rn measurements using an activated charcoal detector. Appl. Radiat. Isot. 50, 561–565.

Næs, T., 2002. User Friendly Guide to Multivariate Calibration and Classification. NIR Publications, p. 352.

Nähle, O., Kossert, K., 2011. Comparison of the TDCR method and the CIEMAT/NIST method for the activity determination of beta emitting nuclides. In: LSC 2010, Advances in Liquid Scintillation Spectrometry. Radiocarbon. University of Arizona, Tucson.

Nähle, O., Kossert, K., Klein, R., 2008. Activity standardization of ^{22}Na. Appl. Radiat. Isot. 66, 865–871.

Nähle, O., Kossert, K., Cassette, P., 2010. Activity standardization of ^{3}H with the new TDCR system at PTB. Appl. Radiat. Isot. 68, 1534–1536.

Nähle, O., Zhao, Q., Wanke, C., Weierganz, M., Kossert, K., 2014. A portable TDCR system. Appl. Radiat. Isot. 87, 249–253.

Nakamura, K., for the KamLAND-Zen Collaboration, 2012. Liquid scintillator based experiments in neutrinoless double beta decay. Nucl. Phys. B 229–232, 128–132.

Nakanishi, T., Kusakabe, M., Aono, T., Yamada, M., 2009. Simultaneous measurements of cosmogenic radionuclides ^{32}P, ^{33}P and ^{7}Be in dissolved and particulate forms in the upper ocean. J. Radioanal. Nucl. Chem. 279 (3), 769–776.

Nakao, N., Kurosawa, T., Nakamura, T., Uwamino, Y., 2001. Absolute measurements of the response function of an NE213 organic liquid scintillator for the neutron energy range up to 206 MeV. Nucl. Instrum. Methods Phys. Res. Sect. A 463, 275–287.

Nakayama, M., Hara, M., Matsuyama, M., Hirokami, K., 2017. Modified integral counting method with various quenched samples for different scintillators. Radiat. Safety Manag. 16, 1–7.

Natake, T., Takiue, M., Fujii, H., 1996. Nuclide identification for pure beta-emitting radionuclides with very similar beta end-point energies using a liquid scintillation spectrometer. Nucl. Instrum. Methods Phys. Res. Sect. A 378, 506–510.

Nayak, D., 2001. Multitracer techniques: applications in chemical and life sciences. Appl. Radiat. Isot. 54, 195–202.

Nakhostin, M., 2012. Recursive algorithms for digital implementation of neutron/gamma discrimination in liquid scintillation detectors. Nucl. Instrum. Methods Phys. Res. Sect. A 672, 1–5.

Nakhostin, M., 2015. A comparison of digital zero-crossing and charge-comparison methods for neutron/γ-ray discrimination with liquid scintillation detectors. Nucl. Instrum. Methods Phys. Res. Sect. A 797, 77–82.

Nakhostin, M., Walker, P.M., 2010. Application of digital zero-crossing technique for neutron–gamma discrimination in liquid organic scintillation detectors. Nucl. Instrum. Methods Phys. Res. Sect. A 621, 498–501.

Nebelung, C., 2003. Separation of ^{14}C, ^{234}U and ^{226}Ra in liquid scintillation spectra. Report FZR-373, 53, Rossendorf.

Nebelung, C., Baraniak, L., 2007. Simultaneous determination of ^{226}Ra, ^{233}U and ^{237}Np by liquid scintillation spectrometry. Appl. Radiat. Isot. 65, 209–217.

Nebelung, C., Henniger, J., Mann, G., 2001. Schnelles Freimessverfahren für alpha-aktive Nuklide in Bauschutt durch Direktmessung von groβflächigen dünnen Messpräparaten. Final Report BMBF Project 02 S 7768, Rossendorf.

Nebelung, C., Jähnigen, P., Bernhard, G., 2009. Simultaneous determination of beta nuclides by liquid scintillation spectrometry, pp. 193–201. In: Eikenberg, J., Jäggi, M., Beer, H., Baehrle, H. (Eds.), LSC. 2008. Advances in Liquid Scintillation Spectrometry, Radiocarbon. University of Arizona, Tucson, p. 445.

Nedjadi, Y., Bailat, C., Caffari, Y., Bochud, F., 2010. Standardization of ^{18}F by a coincidence method using full solid angle detectors. Appl. Radiat. Isot. 68, 1309–1313.

Nedjadi, Y., Bailat, C., Caffari, Y., Cassette, P., Bochud, F., 2015. Set-up of a new TDCR counter at IRA-METAS. Appl. Radiat. Isot. 97, 113–117.

Nedjadi, Y., Laedermann, J.-P., Bochud, F., Bailat, C., 2017. On the reverse micelle effect in liquid scintillation counting. Appl. Radiat. Isot. 125, 94–107.

the EXO-200 Collaboration Neilson, R., et al., 2009. Characterization of large area APDs for the EXO-200 detector. Nucl. Instrum. Methods Phys. Res. Sect. A 608, 68–75.

Nejjari, M., Kryza, D., Poncet, G., Roche, C., Perek, N., et al., 2008. In vitro and in vivo studies with [^{18}F]fluorocholine on digestive tumoral cell lines and in an animal model of metastasized endocrine tumor. Nucl. Med. Biol. 35, 123–130.

Neumann, K.E., Roessler, N., ter Wiel, J., 1991. Safe scintillation chemicals for high efficiency, high throughput counting. In: Ross, H., Noakes, J.E., Spaulding, J.D. (Eds.), Liquid Scintillation Counting and Organic Scintillators. Lewis Publishers, Chelsea, MI 48118, pp. 35–41.

Neves, F., Chepel, V., Solovov, V., Pereira, A., Lopes, M.I., et al., 2005. Performance of a chamber for studying the liquid xenon response to γ-rays and nuclear recoils. IEEE Trans. Nucl. Sci. 52 (6), 2793–2800.

Neves, F., Solovov, V., Chepel, V., Lopes, M.I., Pinto de Cinha, J., et al., 2007. Position reconstruction in a liquid xenon scintillation chamber for low-energy nuclear recoils and γ-rays. Nucl. Instrum. Methods Phys. Res. Sect. A 573, 48–52.

Newell, S.Y., Krambeck, C., 1995. Responses of bacterioplankton to tidal inundations of a saltmarish in a flume and adjacent mussel enclosures. J. Exp. Marine Biol. Ecol. 190, 79−95.

Ni, K., Aprile, E., Day, D., Giboni, K.L., Lopes, J.A.M., et al., 2005. Performance of a large area avalanche photodiode in a liquid xenon ionization and scintillation chamber. Nucl. Instrum. Methods Phys. Res. Sect. A 551, 356−363.

Ni, K., Hasty, R., Wongjirad, T.M., Kastens, L., Manzur, A., McKinsey, D.N., 2007. Preparation of neutron-activated xenon for liquid xenon detector calibration. Nucl. Instrum. Methods Phys. Res. Sect. A 582, 569−574.

Niese, S., 1999. The discovery of organic solid and liquid Scintillators by H. Kallmann and L. Herforth 50 years ago. J. Radioanal. Nucl. Chem. 241 (3), 499−501.

Niese, S., 2003. The discovery of the excitation of liquid biphenyl and phenanthrene fluorescence by fast electrons; laudatio on the inauguration of the German Society for Liquid Scintillation Spectrometry (DGFS), pp. 447−450. In: LSC 2001. Advances in Liquid Scintillation Spectrometry. The University of Arizona, Tucson, p. 456.

Nishimura, Y., Natori, H., 2011. Performance of the liquid xenon detector for the MEG experiment. Nucl. Instrum. Methods Phys. Res. Sect. A 628 (1), 376−380.

NNDC, 2018. National Nuclear Data Center, Brookhaven National Laboratory. In: Nuclear Decay Data (NuDat). Decay Radiation Data, NuDat 2.7. http://www.nndc.bnl.gov/nudat2/chartNuc.jsp.

NNDC-BNL, 2011. National Nuclear Data Center, Brookhaven National Laboratory. In: Double Beta ($\beta\beta$) Decay Data. http://www.nndc.bnl.gov/bbdecay/.

Noakes, J.E., Spaulding, J.D., Neary, M.P., 1998. A remotely operated, field-deployable tritium analysis system for surface and ground water measurement. Radiocarbon 40 (1), 183−192.

Noor, A., Kasim, N., L'Annunziata, M.F., 1995. Application of pulse height spectral analysis to double-label counting of ^{35}S-^{32}P. Appl. Radiat. Isot. 46, 791−797.

Noor, A., Zakir, M., Burhanuddin, R., Maming, L'Annunziata, M.F., 1996a. Cerenkov and liquid scintillation analysis of the triple label ^{86}Rb-^{35}S-^{33}P. Appl. Radiat. Isot. 47, 659−668.

Noor, A., Zakir, M., Burhanuddin, R., Kasim, N., Nurr, L.A., Anthony, M., Agung, M., L'Annunziata, M.F., 1996b. Pulse height spectral analysis of ^{3}H:^{14}C ratios. Appl. Radiat. Isot. 47, 767−775.

Norton, G.A., Cline, A.M., Thompson, G.C., 2012. Use of radiocarbon analyses for determining levels of biodiesel in fuel blends − Comparison with ASTM Method D7371 for FAME. Fuel 96, 284−290.

Novotny, T., Büermann, L., Guldbakke, S., Klein, H., 1997. Response of NE213 liquid scintillation detectors to high-energy photons (7 MeV < E_γ < 20 MeV). Nucl. Instrum. Methods Phys. Res. Sect. A 400, 356−366.

Nygren, D.R., 1974. Proposal to Investigate the Feasability of a Novel Concept in Particle Detection. LBL Internal Report, p. 13. http://inspirehep.net/record/1365360/files/Original_TPC_Proposal.pdf.

Obara, S., for the KamLAND-Zen Collaboration, 2017. Status of balloon production for KamLAND-Zen 800 kg phase. Nucl. Instrum. Methods Phys. Res. Sect. A 845, 410−413.

Oberauer, L., 2012. Liquid scintillator detectors. Nucl. Phys. B 229−232, 358−362.

Ogata, Y., 2007. Radioactivity measurement with a plastic scintillation vial. J. Radioanal. Nucl. Chem. 273 (1), 253−256.

O'Hara, et al., 2011. Investigation of magnetic nanoparticles for the rapid extraction and assay of alpha-emitting radionuclides from urine: Demonstration of a novel radiobioassay method. Health Phys. 101 (2), 196−208.

O'Hara, et al., 2016. Magnetic iron oxide and manganese-doped iron oxide nanoparticles for the collection of alpha-emitting radionuclides from aqueous solutions. RSC Adv. 6, 106239−105251.

O'Hara, M.J., Addleman, R.S., 2017. Magnetic iron oxide nanoparticles for the collection and direct measurement of absorbed alpha-emitting radionuclides from environmental waters by liquid scintillation analysis. Anal. Methods 9, 2791−2804.

Okita, G.T., Kabara, J.J., Richardson, F., LeRoy, G.V., 1957. Assaying compounds containing ^{3}H and ^{14}C. Nucleonics 15, 111−114.

Oropesa Verdecia, P., Kossert, K., 2009. Activity standardization of ^{131}I at CENTIS-DMR and PTB within the scope of a bilateral comparison. Appl. Radiat. Isot. 67, 1099−1103.

Oropesa Verdecia, P., et al., 2018. ^{68}Ga activity calibrations for nuclear medicine applications in Cuba. Appl. Radiat. Isot. (in press).

Ortiz, J.F., Los Arcos, J.M., Grau Malonda, A., 1993. CIEMAT/NIST standardization method extended to anode outputs for beta and electron-capture nuclides, pp. 261−267. In: Noakes, J.E., Schönhofer, F., Polach, H.A. (Eds.), Liquid Scintillation Spectrometry 1992, Radiocarbon. University of Arizona, Tucson.

Osman, A.A.A., et al., 2016. Radioecological investigation of ^{3}H, ^{14}C, and ^{129}I in natural waters from Fuhrberger Feld catchment, Northern Germany. J. Environ. Radioact. 165, 243−252.

Ozden, B., Vaasma, T., Kiisk, M., Henry-Tkaczyk, A., 2017. A modified method for the sequential determination of ^{210}Po and ^{210}Pb in Ca-rich material using liquid scintillation counting. J. Radioanal. Nucl. Chem. 311, 365−373.

Pallavicini, M., 2015. Solar neutrinos: experimental review and prospectives. J. Phys. Conf. Ser. 598, 012007.

Palomo, M., Peñalver, A., Aguilar, C., Borrull, F., 2007. Tritium activity levels in environmental water samples from different origins. Appl. Radiat. Isot. 65, 1048−1056.

Palomo, M., Villa, M., Casacuberta, N., Peñalver, A., Borrull, F., Aguilar, C., 2011. Evaluation of different parameters affecting the liquid scintillation spectrometry measurement of gross alpha and beta index in water samples. Applied. Radiat. Isot. 69 (9), 1274−1281.

Park, J.S., et al., 2013. Production and optical properties of Gd-loaded liquid scintillator for the RENO neutrino detector. Nucl. Instrum. Methods Phys. Res. Sect. A 707, 45−53.

Parus, J.L., Raab, W., Radoszewski, T., 1993. Liquid scintillation counting of plutonium and/or americium concentrations, pp. 233−237. In: Noakes, J.E., Schönhofer, F., Polach, H.A. (Eds.), Liquid Scintillation Spectrometry 1992, Radiocarbon. University of Arizona, Tucson.

Passo Jr., C.J., 1996. Time-resolved pulse decay analysis (TR-PDA): A refinement for alpha/beta discrimination, pp. 37−41. In: Cook, G.T., Harkness, D.D., MacKenzie, A.B., Miller, B.F., Scott, E.M. (Eds.), Liquid Scintillation Spectrometry 1994, Radiocarbon. University of Arizona, Tucson.

Passo Jr., C.J., Cook, G.T., 1994. Handbook of Environmental Liquid Scintillation Spectrometry, A Compilation of Theory and Methods. Publ. No. PMC0387. PerkinElmer Life and Analytical Sciences, Boston.

Passo Jr., C.J., Roberts, D.J., 1996. Expanded energy for time-resolved liquid scintillation counting: an enhancement for programmable TR-LSC®, pp. 67−74. In: Cook, G.T., Harkness, D.D.,

MacKenzie, A.B., Miller, B.F., Scott, E.M. (Eds.), Liquid Scintillation Spectrometry 1994, Radiocarbon. University of Arizona, Tucson.

Pates, J.M., Mullinger, N.J., 2007. Determination of ^{222}Rn in fresh water: Development of a robust method of analysis by α/β separation liquid scintillation spectrometry. Appl. Radiat. Isot. 65, 92–103.

Pates, J.M., Cook, G.T., MacKenzie, A.B., Thomson, J., 1993. The development of alpha/beta separation liquid scintillation cocktail for aqueous samples. J. Radioanal. Nucl. Chem. Articles 172, 341–348.

Pates, J.M., Cook, G.T., MacKenzie, A.B., 1996a. Alpha/beta separation liquid scintillation spectrometry: current trends, pp. 267–281. In: Cook, G.T., Harkness, D.D., MacKenzie, A.B., Miller, B.F., Scott, E.M. (Eds.), Liquid Scintillation Spectrometry 1994, Radiocarbon. University of Arizona, Tucson.

Pates, J.M., Cook, G.T., MacKenzie, A.B., Thomson, J., 1996b. The effect of cocktail fluors on pulse shapes and alpha/beta separation liquid scintillation spectrometry, pp. 317–326. In: Cook, G.T., Harkness, D.D., MacKenzie, A.B., Miller, B.F., Scott, E.M. (Eds.), Liquid Scintillation Spectrometry 1994, Radiocarbon. University of Arizona, Tucson.

Pates, J.M., Cook, G.T., MacKenzie, A.B., Passo Jr., C.J., 1998. Implications of beta energy and quench level for alpha/beta liquid scintillation spectrometry calibration. Analyst 123, 2201–2207.

Patronis, N., Kokkoris, M., Giantsoudi, D., Perdikakis, G., Papadopoulos, C.T., Vlastou, R., 2007. Aspects of GEANT4 Monte-Carlo calculations of the BC501A neutron detector. Nucl. Instrum. Methods Phys. Res. Sect. A 578 (1), 351–355.

Payne, A.G.D., Booth, N.E., 1990. Tests of some prototype indium-loaded scintillators for solar neutrino detection. Nucl. Instrum. Methods Phys. Res. Sect. A 288, 632–640.

Peiffer, P., Pollmann, T., Schönert, S., Smolnikov, A., Vasiliev, S., 2008. Pulse shape analysis of scintillation signals from pure and xenon-doped liquid argon for radioactive background identification. J. Instrum. 3, P08007.

Pelay, E., Tarancon, A., Mitev, K., Dutsov, C., Georgiev, S., Tsankov, L., García, J.F., 2017. Synthesis and characterisation of scintillating microspheres made of polystyrene/polycarbonate for 222Rn measurements. J. Radioanal. Nucl. Chem. 314, 637–649.

Peng, C.T., 1976. Chemiluminescence. In: Noujaim, A.A., Ediss, C., Wiebe, L.I. (Eds.), Liquid Scintillation Science and Technology. Academic Press, New York and London, pp. 313–329.

Perry, S.W., Warner, G.T., 1963. A method of sample preparation for the estimation of ^{55}Fe in whole blood by the liquid scintillation counting technique. Int. J. Appl. Radiat. Isot. 14, 397–400.

Pino, F., et al., 2014. Detecting fast and thermal neutrons with a boron loaded liquid scintillator, EJ-339A. Appl. Radiat. Isot. 92, 6–11.

Piraner, O., Jones, R.L., 2009. Urine gross alpha/beta analysis by liquid scintillation counting for terrorism preparedness, pp. 41–46. In: Eikenberg, J., Jäggi, M., Beer, H., Baehrie, H. (Eds.), LSC 2008, Advances in Liquid Scintillation Spectrometry, Radiocarbon. University of Arizona, Tucson, p. 445.

on behalf of the EXO Collaboration Pocar, A., 2013. Neutrinoless double beta decay search with EXO-200. Nucl. Instrum. Methods Phys. Res. Sect. A 718, 462–465.

on behalf of the EXO and nEXO Collaborations Pocar, A., 2015. Searching for neutrino–less double beta decay with EXO–200 and nEXO. Nucl. Part. Phys. Proc. 265–266, 42–44.

Pochwalski, K., 1978. Measurement of the activity of pure β-emitters in liquid scintillator by use of the new triple-to-double coincidence ratio method. Thesis, unpublished. Institute of Nuclear Research, Warsaw (in Polish).

Pochwalski, K., Radoszewski, T., 1979. Disintegration rate determination by liquid scintillation counting using triple to double coincidence ratio (TDCR) method in liquid scintillation counting (IBJ Report INR 1848/OPiDI/E/A).

Pochwalski, K., Broda, R., Radoszewski, T., 1988. Standardization of pure beta emitters by liquid-scintillation counting. Appl. Radiat. Isot. 39, 165–172.

Polyakov, S.V., 2013. Photomultiplier tubes, pp. 69–82. In: Single-Photon Generation and Detection, vol. 45. Elsevier, Inc. https://doi.org/10.1016/B978-0-12-387695-9.00003-2.

Polack, J.K., et al., 2015. An algorithm for charge-integration, pulse-shape discrimination and estimation of neutron/photon misclassification in organic scintillators. Nucl. Instrum. Methods Phys. Res. Sect. A 795, 253–267.

Pommé, S., Altzitzoglou, T., Van Ammel, R., Sibbens, G., 2005. Standardisation of ^{125}I using seven techniques for radioactivity measurement. Nucl. Instrum. Methods Phys. Res. Sect. A 544, 584–592.

Pozzi, S.A., Clarke, S.D., Flaska, M., Peerani, P., 2009. Pulse-height distributions of neutron and gamma rays from plutonium-oxide samples. Nucl. Instrum. Methods Phys. Res. Sect. A 608, 310–315.

Prichard, H.M., Venso, E.A., Dodson, C.L., 1992. Liquid scintillation analysis of ^{222}Rn in water by alpha-beta discrimination. Radioact. Radiochem. 3 (1), 28–36.

Pujol, L., Sanchez-Cabeza, J.-A., 1997. Role of quenching on alpha/beta separation in liquid scintillation counting for several high capacity cocktails. Analyst 122, 383–385.

Qin, M.J., Mo, L., Alexiev, D., Cassette, P., 2008. Construction and implementation of a TDCR system at ANSTO. Appl. Radiat. Isot. 66, 1033–1037.

Radoszewski, T., 1964. An absolute measurement of ^{14}C by the 4π scintillation counter method. Thesis, unpublished (In Polish).

Raghaven, R.S., 1976. Inverse β decay of ^{115}In → ^{115}Sn*: A new possibility for detecting solar neutrinos from the proton-proton reaction. Phys. Rev. Lett. 37 (5), 259–262.

Raghaven, R.S., 2001. pp-Solar neutrino spectroscopy: return of the indium detector. https://arxiv.org/abs/hep-ex/0106054.

Ramírez-Guinart, O., Salaberria, A., Vidal, M., Rigol, A., 2017. Assessing soil properties governing radiosamarium sorption in soils: Can trivalent lanthanides and actinides be considered as analogues? Geoderma 290, 33–39.

Ranucci, G., 1995. An analytical approach to the evaluation of the pulse shape discrimination properties of scintillators. Nucl. Instrum. Meth. Phys. Res. Sect. A 354, 389–399.

Ratel, G., 2008. Analysis of the results of the international comparison of activity measurements of a solution of ^{55}Fe. Appl. Radiat. Isot. 66, 729–732.

Rauly-Lestienne, I., Lestienne, F., Ailhaud, M.-C., et al., 2011. Competitive interaction of 5-HT_{1A} receptors with G-protein subtypes in CHO cells demonstrated by RNA interference. Cell. Sign 23, 58–64.

Ray, B., Simon, J.R., Lahiri, D.K., 2009. Determination of high-affinity choline uptake (HACU) and choline acetyltransferase (ChAT) activity in the same population of cultured cells. Brain Res. 1297, 160–168.

Razdolescu, A.C., Cassette, P., 2004. Standardization of tritiated water and ^{204}Tl by TDCR liquid scintillation counting. Appl. Radiat. Isot. 60, 493–497.

Razdolescu, A.C., Sahagia, M., Cassette, P., Grigorescu, E.L., Luca, A., Ivan, C., 2002. Standardization of ^{89}Sr. Appl. Radiat. Isot. 56, 461–465.

Razdolescu, A.C., Broda, R., Cassette, P., Simpson, B.R.S., Van Wyngaardt, W.M., 2006. The IFIN-HH triple coincidence liquid scintillation counter. Appl. Radiat. Isot. 64, 1510–1514.

Razdolescu, A.C., Cassette, P., Sahagia, M., 2008. Measurement of ^{55}Fe solution activity by LSC-TDCR method. Appl. Radiat. Isot. 66, 750–755.

Reboli, A., Aupiais, J., Mialocq, J.C., 2005. Application of large area avalanche photodiodes for alpha liquid scintillation counting. Nucl. Instrum. Methods Phys. Res. Sect. A 550, 593–602.

Reddy, P.J., Bhade, S.P.D., Narayan, K.K., Narayanan, A., Babu, D.A.R., Sharma, D.N., 2009. Comparative study of different methods for the activity quantification of ^{3}H and ^{14}C radionuclides in dual labeled samples using liquid scintillation analyzer. Appl. Radiat. Isot. 67, 1945–1951.

Reddy, P.J., Bhade, S.P.D., Babu, D.A.R., Sharma, D.N., 2011. Validation of efficiency tracing and zero detection threshold techniques using liquid scintillation analyser TriCarb. Radiat. Protect. Dosimetry 147 (3), 417–422.

Reginatto, M., Goldhagen, P., Neumann, S., 2002. Spectrum unfolding, sensitivity analysis and propagation of uncertainties with the maximum entropy deconvolution code MAXED. Nucl. Instrum. Methods Phys. Res. Sect. A 476, 242–246.

Reines, F., Cowan Jr., C.L., 1953. Detection of the free neutrino. Phys. Rev. Lett. 92, 330.

Reines, F., Cowan Jr., C.L., 1956. The neutrino. Nature 178, 446–449.

Reines, F., Cowan Jr., C.L., 1957. Neutrino physics. Phys. Today 10 (8), 12–18.

Remetti, R., Sessa, A., 2011. Beta spectra deconvolution for liquid scintillation counting. J. Radioanal. Nucl. Chem. 287 (1), 107–111.

Remetti, R., Franci, D., 2012. ABCD-Tool, a software suite for the analysis of α/β spectra from liquid scintillation counting. J. Radioanal. Nucl. Chem. 292, 1115–1122.

Ren, Y.L., Mahon, D., 2007. Evaluation of ^{35}S-residues in grains and grain fractions fumigated with ^{35}S-labelled carbonyl sulfide. J. Stored Prod. Res. 43, 356–361.

Retière, F., 2010. A liquid xenon detector for micro-PET. Nucl. Instrum. Methods Phys. Res. Sect. A 623, 591–593.

Reynolds, G.T., Harrison, F.B., Salvini, G., 1950. Liquid scintillation counters. Phys. Rev. 78 (4), 488.

Rhodes, B.A., 1965. Liquid scintillation counting of radioiodine. Anal. Chem. 37 (8), 995–997.

Ring, J.G., Nguyen, D.C., Everett, L.J., 1980. In: Peng, C.-T., Horrocks, D.L., Alpen, E.L. (Eds.), Liquid Scintillation Counting, Recent Applications and Development, Physical Aspects, vol. I. Academic Press, New York and London, pp. 89–104.

Roberts, D., Moore, H.M., Berges, J., et al., 2001. Sediment distribution, hydrolytic enzyme profiles and bacterial activities in the guts of *Oneirophanta mutabilis, Psychropotes longicauda* and *Pseudostichopus villosus*: what do they tell us about digestive strategies of abyssal holothurians? Prog. Oceanogr. 50, 443–458.

Robinson, M.K., et al., 2014. Albumin adsorption onto surfaces of urine collection and analysis containers. Clin. Chim. Acta 431, 40–45.

Rochman, D., Haight, R.C., O'Donnell, J.M., Devlin, M., Ethvignot, T., Granier, T., 2004. Neutron-induced reaction studies at FIGARO using a spallation source. Nucl. Instrum. Methods Phys. Res. Sect. A 523, 102–115.

Rodrígues, D., Arenillas, P., Capoulat, M.E., Balpardo, C., 2008. General data analysis code for TDCR liquid scintillation counting. Appl. Radiat. Isot. 66, 1049–1054.

Rodríguez, M., Piña, G., Lara, E., 2006. Radiochemical analysis of ^{36}Cl. Czech. J. Phys. 56 (Suppl. 4), D211–D217.

Rodríguez Barquero, L., Grau Carles, A., 1998. The influence of the primary solute on alpha/beta discrimination. Appl. Radiat. Isot. 49 (9–11), 1065–1068.

Rodríguez Barquero, L., Los Arcos, J.M., 1996. ^{41}Ca standardization by the CIEMAT/NIST LSC method. Nucl. Instrum. Methods Phys. Res. Sect. A 369, 353–358.

Rodríguez Barquero, L., Los Arcos, J.M., 2000. Measurement of ^{49}V half-life. Appl. Radiat. Isot. 52, 569–571.

Rodríguez Barquero, L., Los Arcos, J.M., 2004. Compensation by the CIEMAT/NIST-method of long-term effects in LSC measurements of beta emitters. Appl. Radiat. Isot. 61, 1403–1411.

Rodríguez Barquero, L., Grau Malonda, A., Los Arcos Merino, J.M., Suarez Contreras, C., 1989. Preparación y calibración por centelleo liquido de una muestra de Cl-36. An. Fisica B 85, 55–69.

Rodríguez Barquero, L., Garcia-Toraño, E., Los Arcos, J.M., 2004. Standardization of ^{32}P/^{33}P and ^{204}Tl by liquid scintillation counting. Appl. Radiat. Isot. 60, 615–618.

Rodríguez, L., Los Arcos, J.M., Grau, A., 1992. LSC standardization of ^{54}Mn in inorganic and organic samples by the CIEMAT/NIST efficiency tracing method. Nucl. Instrum. Methods Phys. Res. Sect. A 312, 124–131.

Rodríguez, L., Los Arcos, J.M., Grau Malonda, A., Garcia-Toraño, E., 1994. LSC standardization of ^{45}Ca by the CIEMAT/NIST efficiency tracing method. Nucl. Instrum. Methods Phys. Res. Sect. A 339, 6–13.

Rodríguez, L., Los Arcos, J.M., Ortiz, F., Jiménez, A., 1998. ^{49}V standardization by the CIEMAT/NIST LSC method. Appl. Radiat. Isot. 49 (9–11), 1077–1082.

Roessler, N., Valenta, R.J., van Cauter, S., 1991. Time-resolved liquid scintillation counting. In: Ross, H., Noakes, J.E., Spaulding, J.D. (Eds.), Liquid Scintillation Counting and Organic Scintillators. Lewis Publishers, Chelsea, MI 48118, pp. 501–511.

Ronchi, E., Söderström, P.-A., Nyberg, J., et al., 2009. An artificial neural network based neutron–gamma discrimination and pile-up rejection framework for the BC-501 liquid scintillation detector. Nucl. Instrum. Methods Phys. Res. Sect. A 610, 534–539.

Roteta, M., García-Toraño, E., Rodríguez Barquero, L., 2006. Standardization of ^{18}F by coincidence and LSC methods. Appl. Radiat. Isot. 64, 1199–1202.

Roteta, M., Peyres, V., Rodríguez Barquero, L., García-Toraño, E., Arenillas, P., et al., 2012. Standardization of Ga-68 by coincidence measurements, liquid scintillation counting and $4\pi\gamma$ counting. Appl. Radiat. Isot. 70, 2006–2011.

Roteta, M., Peyres, V., García-Toraño, E., 2014. Standardization of Sn-113. Appl. Radiat. Isot. 87, 162–165.

Ruano Sánchez, A.B., et al., 2013. On the use of different scintillation cocktails for determining gross alpha and beta activities in water samples. Applied Radiat. Isot. 81, 175–178.

Roush, M.L., Wilson, M.A., Hornyak, W.F., 1964. Pulse shape discrimination. Nucl. Instrum. Methods 31, 112–124.

Ruberu, S.R., Liu, Y.-G., Perera, S.K., 2008. An improved liquid scintillation counting method for the determination of gross alpha activity in groundwater wells. Health Phys. 95 (4), 397–406.

Rukhadze, N.I., et al., 2012. Experiment TGV-2 – search for double beta decay of ^{106}Cd. Nucl. Phys. B 229–232, 478.

Rundt, K., 1991. The effect of quench curve shape of the solvent and quencher in a liquid scintillation counter. In: Ross, H., Noakes, J.E., Spaulding, J.D. (Eds.), Liquid Scintillation Counting and Organic Scintillators. Lewis Publishers, Chelsea, MI 48118, pp. 257–268.

Rundt, K., Kouru, H., 1989. Liquid scintillation counter for measuring the activity of radioactive samples containing a multiple of radioactive isotopes. US Patent No. 4,918,310.

Rundt, K., Kouru, H., 1992. Apparatus and a method for measuring the activity of radioactive samples containing a multiple of radioactive isotopes. US Patent No. 5,134,294.

Rusconi, R., Azzellino, A., Bellinzona, S., Forte, M., Gallini, R., Sgorbati, G., 2004. Assessment of drinking water radioactivity content by liquid scintillation counting: set up of high sensitivity and emergency procedures. Anal. Bioanal. Chem. 379, 247–253.

Saez-Muñoz, M., Bagan, H., Tarancon, A., Garcia, J.F., Ortiz, J., Martorell, S., 2018. Rapid method for radiostrontium determination in milk in emergency situations using PS resin. J. Radioanal. Nucl. Chem. 315, 543–555.

Sakai, H., et al., 1992. Construction and performance of one- and two-dimensional large position-sensitive liquid and plastic scintillation detectors – an application to a neutron polarimeter. Nucl. Instrum. Methods Phys. Res. Sect. A 320, 479–499.

Salonen, L., 2006a. Alpha spillover depends on alpha energy: A new finding in alpha/beta liquid scintillation spectrometry, pp. 135–148. In: Chałupnik, S., Schönhofer, F., Noakes, J. (Eds.), LSC 2005, Advances in Liquid Scintillation Spectrometry, Radiocarbon. University of Arizona, Tucson, p. 449.

Salonen, L., 2006b. Alpha/Beta liquid scintillation spectrometry in surveying Finnish groundwater samples. Radiochemistry 48 (6), 606–612.

Salonen, L., 2010. Comparison of two direct LS methods for measuring ^{222}Rn in drinking water using α/β liquid scintillation spectrometry. Appl. Radiat. Isot. 68, 1970–1979.

Salvat, F., Fernandez-Varea, J.M., 2009. Overview of physical interaction models for photon and electron transport used in Monte Carlo codes. Metrologia 46, S112–S138.

Salvat, F., Fernandez-Varea, J.M., Sempau, J., 2011. PENELOPE-2011, A Code System for Monte Carlo Simulation of Electron and Photon Transport. Tutorial for PENELOPE (version 2011).

Samuels, E.R., Scott, J.E., 1995. Ca^{+2}-phosphatidylserine-dependent protein kinase C activity in fetal, neonatal and adult rabbit lung and isolated lamellar bodies. Life Sci. 57 (17), 1557–1568.

Sand, J., et al., 2014. Radioluminescence yield of alpha particles in air. New J. Phys. 16, 053022.

Sandhya, D., Subramanian, M.S., 1998. Radiometric determination of trace amounts of zinc using liquid scintillation counting. Talanta 46, 921–926.

Santiago, L., Bagán, H., Tarancón Sanz, A., Rauret, G., García, J.F., 2011. Systematic study of the different quenching phenomena in organic scintillators. In: LSC 2010, Advances in Liquid Scintillation Spectrometry. Radiocarbon. University of Arizona, Tucson.

Santiago, L., Bagán, H., Tarancón, A., García, J.F., 2014. Synthesis of plastic scintillation microspheres: Alpha/beta discrimination. Applied Radiat. Isot. 93, 18–28.

Sarkar, A., et al., 2015. Isotopic correlation for ^{242}Pu composition prediction: Multivariate regresssion approach. Appl. Radiat. Isot. 95, 169–173.

Sas, D., Sládek, P., Janda, J., 2010. Measuring alpha and beta activity of filter and swab samples with LSC. J. Radioanal. Nucl. Chem. 286, 513–517.

Satoh, D., et al., 2001. Study of neutron detection efficiencies for liquid organic scintillator up to 3 GeV. IEEE Trans. Nucl. Sci. 48 (4), 1165–1167.

Sawada, R., 2010. Performance of liquid xenon gamma ray detector for MEG. Nucl. Instrum. Methods Phys. Res. Sect. A 632 (1), 258–260.

Scherzinger, J., et al., 2016. The light-yield response of a NE-213 liquid-scintillator detector measured using 2–6 MeV tagged neutrons. Nucl. Instrum. Methods Phys. Res. Sect. A 840, 121–127.

Scherzinger, J., et al., 2017. A comparison of untagged gamma-ray and tagged-neutron yields from ^{241}AmBe and ^{238}PuBe sources. Applied Radiat. Isot. 127, 98–102.

Schiffer, R.T., Flaska, M., Pozzi, S.A., Carney, S., Wentzloff, D.D., 2011. A scalable FPGA-based digitizing platform for radiation data acquisition. Nucl. Instrum. Methods Phys. Res. Sect. A 652 (1), 491–493.

Schmidt, D., Asselineau, B., Böttger, R., et al., 2002. Characterization of liquid scintillation detectors. Nucl. Instrum. Methods Phys. Res. Sect. A 476, 186–189.

Schneider, E.W., Verbrugge, M.W., 1993. Radiotracer method for simultaneous measurement of cation, anion and water transport through ion-exchange membranes. Appl. Radiat. Isot. 44, 1399–1408.

Schötzig, U., Schönfeld, E., Günther, E., Klein, R., Schrader, H., 1999. Standardization and decay data of ^{153}Sm. Appl. Radiat. Isot. 51, 169–175.

Schötzig, U., Schrader, H., Schönfeld, E., Günther, E., Klein, R., 2001. Standardization and decay data of ^{177}Lu and ^{188}Re. Appl. Radiat. Isot. 55, 89–96.

Schrader, H., Klein, R., Kossert, K., 2007. Standardisation of ^{18}F and ionization chamber calibration for nuclear medicine. Appl. Radiat. Isot. 65, 581–592.

Schultz, M.K., Cessna, J.T., Anderson, T.L., et al., 2008. A performance evaluation of ^{90}Y dose-calibrator measurements in nuclear pharmacies and clinics in the United States. Appl. Radiat. Isot. 66, 252–260.

Schwerdtel, E., 1965. G.I.T. Fachzeitschrift für das Laboratorium 9, 881.

Schwerdtel, E., 1966a. Recent developments in liquid scintillation counting. Kerntechnik 8, 517–520.

Schwerdtel, E., 1966b. Simple method for an exact efficiency determination in liquid scintillation counting of low-energy β-emitters. Atomkernenergie 11, 324–325.

Scott, C.A., Good, M.L., 1967. Use of liquid scintillation counting for the determination of distribution ratios of alpha emitters. J. Inorg. Nucl. Chem. 29, 255–257.

for the SNO+ Collaboration Segui, L., 2016. The SNO+ project. Nucl. Part. Phys. Proc. 273–275, 2654–2656.

Seidel, A., Volf, V., 1972. Rapid determination of some transuranium elements in biological material by liquid scintillation counting. Int. J. Applied Radiat. Isot. 23, 1–4.

Seliger, H.H., 1960. Liquid scintillation counting of α-particles and energy resolution of the liquid scintillator for and β-particles. Int. J. Appl. Radiat. Isot. 8, 29–34.

Shaffer, C.L., Langer, C.S., 2007. Metabolism of $^{14}C/^{3}H$-labeled $GABA_A$ receptor partial agonist in rat, dog and human liver microsomes: Evaluation of a dual-radiolabel strategy. J. Pharmaceut. Biomed. Anal. 43, 1195–1205.

Shagin, P., Gomez, R., Oberlack, U., et al., 2009. Avalanche photodiode for liquid xenon scintillation: Quantum efficiency and gain. J. Instrum. 4, P01005.

Shu, S., et al., 2018. A novel series of 4-methyl substituted pyrazole derivatives as potent glucagon receptor antagonists: Design, synthesis and evaluation of biological activities. Bioorg. Medic. Chem. (in press).

Signorelli, G., Dussoni, S., 2016. Possible usage of Cherenkov photons to reduce the background in a ^{136}Xe neutrino-less double-beta decay experiment. Nucl. Instrum. Methods Phys. Res. Sect. A 824, 658–660.

Simpson, B.R.S., Meyer, B.R., 1992a. The standardization of ^{201}Tl by liquid scintillation coincidence counting. National Accelerator Centre Report. NAC/92-01, PO Box 72, Faure, 7131 South Africa.

Simpson, B.R.S., Meyer, B.R., 1992b. Direct determination of the activity of non-gamma-emitting radionuclides by the TDCR efficiency calculation technique: a review of the present status. Report NAC/92-02, September 1992, National Accelerator Centre, PO Box 72, Faure, 7131 South Africa, p. 18.

Simpson, B.R.S., Meyer, B.R., 1992c. Further investigations of the TDCR efficiency calculation technique for the direct determination of activity. Nucl. Instrum. Methods Phys. Res. Sect. A 312, 90–94.

Simpson, B.R.S., Meyer, B.R., 1994b. Direct activity measurement of pure beta-emitting radionuclides by the TDCR efficiency calculation technique. Nucl. Instrum. Methods Phys. Res. Sect. A 339, 14–20.

Simpson, B.R.S., Meyer, B.R., 1996. Activity measurement of ^{204}Tl by direct liquid scintillation measurements. Nucl. Instrum. Methods Phys. Res. Sect. A 369, 340–343.

Simpson, B.R.S., Morris, W.M., 2004a. Direct activity determination of ^{54}Mn and ^{65}Zn by a non-extrapolation liquid scintillation method. Appl. Radiat. Isot. 60, 475–479.

Simpson, B.R.S., Morris, W.M., 2004b. The standardization of ^{33}P by the TDCR efficiency calculation technique. Appl. Radiat. Isot. 60, 465–468.

Simpson, B.R.S., Van Wyngaardt, W.M., 2006. Activity measurements of the high-energy pure β-emitters ^{89}Sr and ^{90}Y by the TDCR efficiency calculation technique. Appl. Radiat. Isot. 64, 1481–1484.

Simpson, B.R.S., Van Wyngaardt, W.M., Lubbe, J., 2010. Fe-55 activity measurements at the NMISA revisited. Appl. Radiat. Isot. 68, 1529–1533.

Singh, A.N., Kadwani, M.G., 1974. A sensitive detector system for the continuous monitoring of tritium in the air. IEEE Trans. Nucl. Sci. 21 (1), 188–193.

Sinor, J.D., Boeckman, F.A., Aizenman, E., 1997. Intrinsic redox properties of *N*-methyl-D-aspartate receptor can determine the developmental expression of excitotoxicity in rat cortical neurons in vitro. Brain Res. 747, 297–303.

Sirelkhatim, D.A., Sam, A.K., Hassona, R.J., 2008. Distribution of ^{226}Ra–^{210}Pb–^{210}Po in marine biota and surface sediments of the Red Sea, Sudan. J. Environ. Radioact. 99, 1825–1828.

Smirnov, O., et al., 2015. Short distance neutrino Oscillations with BoreXino: SOX. Phys. Procedia 61, 511–517.

Sochorová, J., Auerbach, P., Dutka, Z., 2012. Activity standardization of ^{45}Ca and ^{204}Tl using the new TDCR system at CMI. Appl. Radiat. Isot. 70, 2200–2203.

Sorensen, P., , et al.the XENON10 Collaboration, 2009. The scintillation and ionization yield of liquid xenon for nuclear recoils. Nucl. Instrum. Methods Phys. Res. Sect. A 601, 339–346.

Spasova, Y., Wätjen, U., Altzitzoglou, T., 2008. European measurement comparison of ^{137}Cs, ^{40}K and ^{90}Sr in milk powder. J. Radioanal. Nucl. Chem. 277 (1), 211–215.

Staffová, P., Němec, M., John, J., 2006. Determination of gross alpha and beta activities in water samples by liquid scintillation counting. Czech. J. Phys. 56 (Suppl. D), D299–D305.

Stamoulis, K.C., Ioannides, K.G., Karamanis, D., 2010. Deconvolution of liquid scintillation alpha spectra of mixtures of uranium and radium isotopes. Anal. Chim. Acta 657, 108–115.

Stanga, D., Moreau, I., Cassette, P., 2006. Standardization of tritiaited water by two improved methods. Appl. Radiat. Isot. 64, 1203–1206.

Steele, T., Mo, L., Bignell, L., Smith, M., Alexiev, D., 2009. FASEA: a FPGA acquisition system and software event analysis for liquid scintillation counting. Nucl. Instrum. Methods Phys. Res. Sect. A 609, 217–220.

Steyn, J., 1956. Absolute standardization of beta-emitting isotopes with a liquid scintillator counter. Proc. Phys. Soc. 69, 865–867.

Stojković, I., Tenjović, B., Nikolov, J., Todorović, N., 2015. Radionuclide, scintillation cocktail and chemical/color quench influence on discriminator setting in gross alpha/beta measurements by LSC. J. Environ. Radioact. 144, 41–46.

Stojković, I., Todorović, N., Nikolov, J., Tenjović, B., 2016. PSA discriminator influence on ^{222}Rn efficiency detection in waters by liquid scintillation counting. Appl. Radiat. Isot. 112, 80–88.

Stojković, I., Tenjović, B., Nikolov, J., Todorović, N., 2017a. Possibilities and limitations of color quench correction methods for gross alpha/beta measurements. Appl. Radiat. Isot. 122, 164–173.

Stojković, I., Nikolov, J., Tomić, M., Mićić, R., Todorović, N., 2017b. Biogenic fraction determination in fuels – Optimal parameters survey. Fuel 191, 330–338.

Stojković, I., Todorović, N., Nikolov, J., Tenjović, B., 2017c. Establishment of rapid LSC method for direct alpha/beta measurements in waters. J. Radioanal. Nucl. Chem. 314, 623–627.

Sunil, C., et al., 2013. Thick target neutron yield from 145 MeV $^{19}F+^{27}Al$ system. Nucl. Instrum. Methods Phys. Res. Sect. A 721, 21–25.

Suzuki, Y., Inoue, K., Nagashima, Y., Hashimoto, S., Inagaki, T., 1990. Development of indium-loaded liquid scintillators with long attenuation length. Nucl. Instrum. Methods Phys. Res. Sect. A 293, 615–622.

Sweet, I.R., Cook, D.L., Lernmark, Å., et al., 2004. Systematic screening of potential β-cell imaging agents. Biochem. Biophys. Res. Commun. 314, 976–983.

Takiue, M., 1980. Simple and rapid measurement of α-rays on smear samples using air luminescence. Health Phys. 39 (1), 29–32.

Takiue, M., Ishikawa, H., 1974. Quenching analysis of liquid scintillation. Nucl. Instrum. Methods 118, 51–54.

Takiue, M., Ishikawa, H., 1978. Thermal neutron reaction cross section measurements for fourteen nuclides with a liquid scintillation spectrometer. Nucl. Instrum. Methods 148, 157–161.

Takiue, M., Ishikawa, H., 1979a. Absolute measurement of internal conversion electrons with a liquid scintillation spectrometer. Nucl. Instrum. Methods 164, 343–347.

Takiue, M., Ishikawa, H., 1979b. α-ray measurement due to air luminescence employing a liquid scintillation spectrometer. Nucl. Instrum. Methods 159, 139–143.

Takiue, M., Natake, T., Hayashi, M., 1983. Double ratio technique for determining the type of quenching in liquid scintillation measurement. Int. J. Appl. Radiat. Isot. 34, 1483–1485.

Takiue, M., Hayashi, M., Natake, T., Ishikawa, H., 1984. Elimination of chemiluminescence in liquid scintillation measurement. Nucl. Instrum. Methods Phys. Res. 219, 192–195.

Takiue, M., Natake, T., Hayashi, M., Yoshizawa, Y., 1985. Analytical subtraction of chemiluminescence counts for dual-labeled samples in liquid scintillation measurement. Int. J. Appl. Radiat. Isot. 36, 285–289.

Takiue, M., Natake, T., Fujii, H., Ishikawa, H., 1986. Modification of a dual-label analysis data processing system for chemiluminescence corrections in liquid scintillation counting. Nucl. Instrum. Methods Phys. Res. Sect. A 247, 395–398.

Takiue, M., Fujii, H., Aburai, T., Ishikawa, H., 1989a. Usefulness of water-soluble paper for smear test of low-energy β- and α-emitters using a liquid scintillation technique. Health Phys. 56, 367–371.

Takiue, M., Natake, T., Fujii, H., 1989b. Nuclide identification of β-emitter by a double ratio technique using a liquid scintillation counter. Nucl. Instrum. Methods Phys. Res. Sect. A 274, 345–348.

Takiue, M., Fujii, H., Homma, Y., 1990a. Reliability of the activity determined by the quenching correction method for two groups of emulsion scintillators. Appl. Radiat. Isot. 41, 195–198.

Takiue, M., Matsui, Y., Natake, T., Yoshizawa, Y., 1990b. A new approach to analytical radioassay of multiple beta-labeled samples using a liquid scintillation spectrometer. Nucl. Instrum. Methods Phys. Res. Sect. A 293, 596–600.

Takiue, M., Matsui, Y., Fujii, H., 1991a. Activity determination of simultaneously chemical and color quenched samples using a liquid scintillation counter. Appl. Radiat. Isot. 42 (3), 241–244.

Takiue, M., Matsui, Y., Fujii, H., 1991b. Liquid scintillation radioassay for multiple radionuclide mixtures by the most probable value theory. J. Radioanal. Nucl. Chem. 152, 227–236.

Takiue, M., Fujii, H., Natake, T., Matsui, Y., 1991c. Analytical measurements of multiple beta-emitter mixtures with a liquid scintillation spectrometer. J. Radioanal. Nucl. Chem. Letters 155, 183–193.

Takiue, M., Matsui, Y., Natake, T., Fujii, H., 1992. Nuclide identification of pure-beta emitter mixtures with liquid scintillation spectrometry. Appl. Radiat. Isot. 43, 853–857.

Takiue, M., Natake, T., Fujii, H., 1995. Liquid scintillation radioassay for low-activity beta-emitter mixtures by the method of least squares. J. Radioanal. Nucl. Chem. Letters 200, 247–258.

Takiue, M., Natake, T., Fujii, H., 1999. A hybrid radioassay technique for multiple beta-emitter mixtures using liquid and NaI(Tl) scintillation spectrometers. Appl. Radiat. Isot. 51, 429–434.

Takiue, M., Yoshizawa, Y., Fujii, H., 2004. Cerenkov counting of low-energy beta-emitters using a new ceramic with high refractive index. Appl. Radiat. Isot. 61, 1335–1337.

Tanaka, M., Doke, T., Hitachi, A., et al., 2001. LET dependence of scintillation yields in liquid xenon. Nucl. Instrum. Methods Phys. Res. Sect. A 457, 454–463.

Tarancón, A., García, J.F., Rauret, G., 2002a. Mixed waste reduction in radioactivity determination by using plastic scintillators. Anal. Chim. Acta 463, 125–134.

Tarancón, A., Alonso, E., García, J.F., Rauret, G., 2002b. Comparative study of quenching correction procedures for $^{90}Sr/^{90}Y$ determination by Cerenkov, liquid scintillation and plastic scintillation techniques. Anal. Chim. Acta 471, 135–143.

Tarancón, A., García, J.F., Rauret, G., 2003. Reusability of plastic scintillators used in beta emitter activity determination. Appl. Radiat. Isot. 59, 373–376.

Tarancón, A., García, J.F., Rauret, G., 2004. Determination of beta emitters (^{90}Sr, ^{14}C and ^{3}H) in routine measurements using plastic scintillation beads. Nucl. Instrum. Methods Phys. Res. Sect. A 516, 602–609.

Tarancón, A., García, J.F., Rauret, G., 2007. First approach to radionuclide mixtures quantification by using plastic scintilators. Influence of the diameter of the plastic beads. Anal. Chim. Acta 590, 232–238.

Tarancón Sanz, A., Kossert, K., 2011. Application of a free parameter model to plastic scintillation samples. Nucl. Instrum. Methods Phys. Res. Sect. A 648, 124–131.

Tarancón, A., Pujadas, M., Novella, O., Cros, J., Rauret, G., Garcia, J.F., 2012. A sensor for the continuous monitoring of alpha and beta radionuclides in Ebro Rives. In: ICRM LLRMT. 6th International Conference on Radionuclide Metrology Low-Level Measurement Techniques.

Tarancón, A., Barrera, J., Santiago, L.M., Bagán, H., García, J.F., 2015. Application of the CIEMAT–NIST method to plastic scintillation microspheres. Appl. Radiat. Isot. 98, 13–22.

Tarancón, A., Bagán, H., García, J.F., 2017. Plastic scintillators and related analytical procedures for radionuclide analysis. J. Radioanal. Nucl. Chem. 314 (2), 555–572.

Tauler, R., Izquierdo-Ridorsa, A., Casassas, E., 1991. Comparison of multivariate calibration methods applied to the spectrophotometric study of mixtures of purine and pyrimidine bases. An. Quím. 87, 571–579.

Tayloe, R., et al., 2006. A large-volume detector capable of charged-particle tracking. Nucl. Instrum. Methods Phys. Res. Sect. A 562, 198–206.

TenBrook, P.L., Tjeerdema, R.S., 2006. Biotransformation of clomazone in rice (*Oryza sativa*) and early watergrass (*Echinochloa oryzoides*). Pest. Biochem. Physiol. 85, 38–45.

Teresa, M., Robela, C.P., Camargo, I.M.C., Oliveira, J.E., Bartolini, P., 2000. Single-step purification of recombinant human growth hormone (hGH) directly from bacterial osmotic shock fluids, for the purpose of ^{125}I-hGH preparation. Protein Expression Purif 18 (2), 115–120.

Terlikowska, T., Cassette, P., Péron, M.N., Broda, R., Hainos, D., Tartes, I., Kempisty, T., 1998. Study of the stability of ^{63}Ni sources in Ultima Gold® liquid scintillation cocktail. Appl. Radiat. Isot. 49 (9–11), 1041–1047.

Theodórsson, P., 2006. Determining fundamental parameters of a single-phototube liquid scintillation counter, pp 99 – 107. In: Chałupnik, S., Schönhofer, F., Noakes, J. (Eds.), LSC 2005. Advances in Liquid Scintillation Spectrometry, Radiocarbon. The University of Arizona, Tucson, p. 449.

Theodórsson, P., Gudjonsson, G.I., 2006. Number of cathode electrons per absorbed energy in liquid scintillation counters, derived from resolution of radon spectra, pp 109 – 114. In: Chałupnik, S., Schönhofer, F., Noakes, J. (Eds.), LSC 2005. Advances in Liquid Scintillation Spectrometry, Radiocarbon. The University of Arizona, Tucson, p. 449.

Theodórsson, P., Gudjonsson, G.I., 2003. A simple and sensitive liquid scintillation counting system for continuous monitoring of radon in water, pp. 249–252. In: Möbius, S., Noakes, J.E., Schönhofer, F. (Eds.), LSC 2001. Advances in Liquid Scintillation Spectrometry, Radiocarbon. The University of Arizona, Tucson, p. 456.

Theodórsson, P., Skripkin, V., 2003. Measuring environmental tritium bound in benzene using a single-photomultiplier tube LSC system, pp. 41–44. In: Möbius, S., Noakes, J.E., Schönhofer, F. (Eds.), LSC 2001. Advances in Liquid Scintillation Spectrometry, Radiocarbon. The University of Arizona, Tucson, p. 456.

Theodórsson, P., Sigurdsson, K., Jónsson, G., 2006. A portable, single-phototube liquid scintillation system with automatic sample changing, pp. 115 – 118. In: Chałupnik, S., Schönhofer, F., Noakes, J. (Eds.), LSC 2005. Advances in Liquid Scintillation Spectrometry, Radiocarbon. The University of Arizona, Tucson, p. 449.

Thiam, C., Bobin, C., Chauvenet, B., Bouchard, J., 2012. Application of TDCR-Geant4 modeling to standardization of ^{63}Ni. Appl. Radiat. Isot. 70, 2195–2199.

Thomas, D.J., 2010. Neutron spectrometry. Radiat. Meas. 45, 1178–1185.

Thomas, E.V., Haaland, D.M., 1990. Comparison of multivariate calibration methods for quantitative spectral analysis. Anal. Chem. 62, 1091–1099.

Thomas, R.C., et al., 1965. Dispenser for addition of internal standard in liquid scintillation counting. Anal. Biochem. 13, 358–360.

Thomson, J., 1991. Di-isopropylnaphthalene - A new solvent for liquid scintillation counting. In: Ross, H., Noakes, J.E., Spaulding, J.D. (Eds.), Liquid Scintillation Counting and Organic Scintillators. Lewis Publishers, Chelsea, MI 48118, pp. 19–34.

Thorngate, J.H., McDowell, W.J., Christian, D.J., 1974. An application of pulse shape discrimination to liquid scintillation counting. Health Phys. 27, 123–126.

Tie, H., et al., 2017. Benchmark integral neutron experiments for Fe, Be and C with DT neutron by liquid scintillation detector. Appl. Radiat. Isot. 124, 56–61.

Tinker, R.A., Smith, J.D., 1996. Simultaneous measurement of ^{226}Ra and ^{133}Ba using liquid scintillation counting with pulse shape discrimination. Anal. Chim. Acta 332, 291–297.

Todorović, N., Nikolov, N., Tenjović, B., Bikit, B., Veskovic, M., 2012. Establishment of a method for measurement of gross alpha/beta activities in water from Vojvodina region. Radiat. Meas. 47 (11–12), 1053–1059.

Todorović, V., et al., 2017. ^{90}Sr determination in water samples using Čerenkov radiation. J. Environ. Radioact. 169–170, 197–202.

Tomanin, A., et al., 2014. Characterization of a cubic EJ-309 liquid scintillator detector. Nucl. Instrum. Methods Phys. Res. Sect. A 756, 45–54.

Tomás, A., et al., 2018. Study and mitigation of spurious electron emission from cathodic wires in noble liquid time projection chambers. Astropart. Phys. 103, 49–61.

Toribio, M., García, J.F., Izquierdo-Ridorsa, A., Tauler, R., Rauret, G., 1995. Simultaneous determination of plutonium alpha emitters by liquid scintillation counting using multivariate calibration. Anal. Chim. Acta 310, 297–305.

Toribio, M., García, J.F., Izquierdo-Ridorsa, A., Rauret, G., 1996. Optimization of counting conditions and simultaneous determination of ^{238}Pu, ^{239}Pu and ^{240}Pu by liquid scintillation counting. In: Cook, G.T., Harkness, D.D., MacKenzie, A.B., Miller, B.F., Scott, E.M. (Eds.), Liquid Scintillation Spectrometry 1994, Radiocarbon, pp. 157–165.

Toribio, M., García, J.F., Izquierdo-Ridorsa, A., Rauret, G., 1997. Multivariate calibration and spectrum position correction for simultaneous determination of alpha and beta emitting plutonium isotopes by liquid scintillation. Anal. Chim. Acta 356, 41–50.

Toribio, M., Padró, A., García, J.F., Rauret, G., 1999. Determination of mixtures of alpha emitting isotopes (^{242}Pu, $^{239+240}Pu$, ^{238}Pu) by using liquid scintillation-moving curve fitting. Anal. Chim. Acta 380, 83–92.

Uchiyama, Y., 2010. Gamma ray reconstruction with liquid xenon calorimeter for the MEG experiment. Nucl. Instrum. Methods Phys. Res. Sect. A 617, 118–119.

the XMASS Collaboration Ueshima, K., et al., 2008. Scintillation yield of liquid xenon at room temperature. Nucl. Instrum. Methods Phys. Res. Sect. A 594, 148–154.

Vajda, N., Kim, C.-K., 2010. Determination of Pu isotopes by alpha spectrometry: a review of analytical spectrometry. J. Radioanal. Nucl. Chem. 283, 203–223.

Van Cauter, S., Roessler, N., 1991. Modern techniques for quench correction and DPM determination in windowless liquid scintillation counting: a critical review. In: Ross, H., Noakes, J.E., Spaulding, J.D. (Eds.), Liquid Scintillation Counting and Organic Scintillators. Lewis Publishers, Chelsea, MI 48118, pp. 219–237.

Van Wyngaardt, W.M., Simpson, B.R.S., 2006a. A simple counting technique for measuring mixtures of two pure β-emitting radionuclides. Nucl. Instrum. Methods Phys. Res. Sect. A 564, 339–346.

Van Wyngaardt, W.M., Simpson, B.R.S., 2006b. Absolute activity measurement of the electron-capture-based radionuclides ^{139}Ce, ^{125}I, ^{192}Ir and ^{65}Zn by liquid scintillation coincidence counting. Appl. Radiat. Isot. 64, 1454–1458.

Van Wyngaardt, W.M., Simpson, B.R.S., 2009. Standardization of sulfur-35 by the TDCR efficiency calculation technique, pp. 173–181. In: LSC 2008, Advances in Liquid Scintillation Spectrometry Radiocarbon. University of Arizona, Tucson, p. 445.

Van Wyngaardt, W.M., et al., 2008. Further investigations of a simple counting technique for measuring mixtures of two pure β-emitting radionuclides. Appl. Radiat. Isot. 66, 1012–1020.

van Wyngaardt, W.M., Simpson, B.R.S., van Staden, M.J., Lubbe, J., 2012. Absolute standardization of ^{241}Pu by the TDCR technique and effect of the beta spectral shape. Appl. Radiat. Isot. 70, 2188–2194.

Van Wyngaardt, W.M., van Staden, M.J., Lubbe, J., Simpson, B.R.S., 2014. Standardization of Tc-99 by three liquid scintillation counting methods. Appl. Radiat. Isot. 87, 254–259.

van Wyngaardt, W.M., et al., 2018. Development of the Australian standard for Germanium-68 by two liquid scintillation counting methods. Applied. Rad. Isot. (in press).

Varaschin, R.K., et al., 2018. Histamine H3 receptors decrease dopamine release in the ventral 4 striatum by reducing the activity of striatal cholinergic interneurons. Neurosci (in press). https://doi.org/10.1016/j.neuroscience.2018.01.027.

Varmuza, K., Filmoser, P., 2009. Introduction to Multivariate Statistical Analysis in Chemometrics. CRC Press, p. 336.

Varlam, C., Stefanescu, I., Duliu, O.G., Faurescu, I., Popescu, I., 2009. Applying direct liquid scintillation counting to low level tritium measurement. Appl. Radiat. Isot. 67, 812–816.

Vartsky, D., Goldberg, M.B., Breskin, A., Chechik, R., Guerard, B., Clergeau, J.F., 2003. Large area imaging detector for neutron scattering based on boron-rich liquid scintillator. Nucl. Instrum. Methods Phys. Res. Sect. A 504, 369–373.

Vera Tomé, F., Gómez Escobar, V., Martín Sánchez, A., 2002. Study of the peak shape in alpha spectra measured by liquid scintillation. Nucl. Instrum. Methods Phys. Res. Sect. A 485 (3), 444–452.

Véronneau, C., Aupiais, J., Dacheux, N., 2000. Selective determination of polonium by photon electron rejecting alpha liquid scintillation (PERALS system). Anal. Chim. Acta 415, 229–238.

Verrezen, F., Hurtgen, C., 1996. Radioassay of low-level, low-energy beta activity in multilabeled samples containing high-energy beta impurities using liquid scintillation spectrometry. Radiocarbon 381–389.

Verrezen, F., Hurtgen, C., 2000. A multiple window deconvolution technique for measuring low-energy beta activity in samples contaminated with high-energy beta impurities using liquid scintillation spectrometry. Appl. Radiat. Isot. 53, 289–296.

Verrezen, F., Loots, H., Hurtgen, C., 2008. A performance comparison of nine selected liquid scintillation cocktails. Appl. Radiat. Isot. 66, 1038–1042.

Verzilov, Y., Curtius, H., Fachinger, J., 2002. Determination of alpha/beta radionuclides in concentrated salt brines from fuel leaching by liquid scintillation spectrometry using solvent extraction, pp. 379–388. In: Möbius, S., Noakes, J., Schönhofer, F. (Eds.), LSC 2001, Advances in Liquid Scintillation Spectrometry, Radiocarbon. University of Arizona, Tucson, p. 456.

Villano, A.N., Becchetti, F.D., Kolata, J.J., Ojaruega, M., Roberts, A., 2011. Efficiency measurements of deuterated liquid scintillators using d(d, n)^{3}He coincidence events. Nucl. Instrum. Methods Phys. Res. Sect. A 652 (1), 280–283.

Vincze, Á., Solymosi, J., Kása, I., Sáfrány, Á., 2007. Extractive-scintillating resin produced by radiation polymerization. Radiat. Phys. Chem. 76, 1395–1398.

Viteri, F.E., Kohaut, B.A., 1997. Improvement of the Eakins and Brown method for measuring ^{59}Fe and ^{55}Fe in blood and other iron-containing materials by liquid scintillation counting and sample preparation using microwave digestion and ion-exchange column purification of iron. Anal. Biochem. 244 (1), 116–123.

Wang, Y., 2012. Neutrino detectors: present and future. Phys. Procedia 37, 22–33.

Wang, Z.-Y., 2008. The effect probe into of the measurement radioactivity nuclide for TDCR method by change liquid scintillation sample volume test. Nucl. Electronics Detection Technol. 28 (1), 118–122.

Wang, Z., et al., 2014. Optimization of integration limit in the charge comparison method based on signal shape function. Nucl. Instrum. Methods Phys. Res. Sect. A 760, 5–9.

Wanke, C., Kossert, K., Nähle, O.J., Ott, O., 2010. Activity standardization and decay data of ^{64}Cu. Appl. Radiat. Isot. 68, 1297–1302.

Wanke, C., Kossert, K., Nahle, O.J., 2012. Investigations on TDCR measurements with the HIDEX 300 SL using a free parameter model. Appl. Radiat. Isot. 70, 2176–2183.

Warwick, P.E., Croudace, I.W., 2002. Measurement of gross alpha and beta activities in acid leachates using alpha-beta discriminating liquid scintillation counting, pp. 75–82. In: Möbius, S., Noakes, J., Schönhofer, F. (Eds.), LSC 2001, Advances in Liquid Scintillation Spectrometry, Radiocarbon. University of Arizona, Tucson, p. 456.

Warwick, P.E., Croudace, I.W., 2006. Isolation and quantification of ^{55}Fe and ^{63}Ni in reactor effluents using extraction chromatography and liquid scintillation analysis. Anal. Chim. Acta 567, 277–285.

Warwick, P.E., Croudace, I.W., 2013. Identification and quantification of radionuclides in contaminated drinking waters and pipeline deposits. Anal. Chem. 85, 8166–8172.

Warwick, P.E., Kim, D., Croudace, I.W., Oh, J., 2010. Effective desorption of tritium from diverse solid matrices and its application to routine analysis of decommissioning materials. Anal. Chim. Acta 676, 93–102.

Wakabayashi, G., et al., 1999. Simple measurement of ^{14}C in the environment using a gel suspension method. J. Radioanal. Nucl. Chem. 239 (3), 639–642.

Wätjen, U., Szántó, Z., Altzitzoglou, T., Sibbens, G., Keightley, J., Hult, M., 2006. EC intercomparisons for laboratories monitoring environmental radioactivity. Appl. Radiat. Isot. 64, 1108–1113.

Weekley, J.C., Wuenschel, S., Rosenstiel, P.E., Mumper, R.J., Jay, M., 2004. Aqueous liquid scintillation counting with fluor-containing nanosuspensions. Appl. Radiat. Isot. 60, 887–891.

Wells, J.M., Thomas, J., Boddy, L., 2001. Soil water potential shifts: developmental responses and dependence on phosphorus translocation by the saprotrophic, cord-forming basidiomycete *Phanerochaete velutina*. Mycol. Res. 105 (7), 859–867.

Werth, M., Kuzyakov, Y., 2008. Root-derived carbon in soil respiration and microbial biomass determined by ^{14}C and ^{13}C. Soil Biol. Biochem. 40, 625–637.

Wiebe, L.I., McQuarrie, S.A., Ediss, C., 1980. In: Peng, C.-T., Horrocks, D.L., Alpen, E.L. (Eds.), Liquid Scintillation Counting, Recent Applications and Development, vol. 1. Academic Press, New York and London, pp. 81–87.

Wiegand, C., Pehkonen, S., Akkanen, J., Penttinen, O.-P., Kukkonen, J.V.K., 2007. Bioaccumulation of paraquat by *Lumbriculus variegatus* in the presence of dissolved natural organic matter and impact on energy costs, biotransformation and antioxidative enzymes. Chemosphere 66, 558–566.

Wiesenberg, G.L.B., Gocke, M., Kuzyakov, Y., 2010. Fast incorporation of root-derived lipids and fatty acids into soil – Evidence from a short term $^{14}CO_2$ pulse labelling experiment. Organic Geochem. 41, 1049–1055.

Wigfield, D.C., Cousineau, C.M.E., 1978. Some empirical observations concerning the liquid scintillation counting of coloured solutions. Can. J. Chem. 56, 2173–2177.

Williams, M., Kookana, R., 2010. Isotopic exchangeability as a measure of the available fraction of the human pharmaceutical carbamazepine in river sediment. Sci. Total Environ. 408, 3689–3695.

Wise, B.M., 1992. PLS-Toolbox for use with MATLAB™. Center for Process Analytical Chemistry (CPAC), University of Washington, Seattle.

Wisser, S., Frenzel, E., Dittmer, M., 2006. Innovative procedure for the determination of gross-alpha/gross-beta activities in drinking water. Appl. Radiat. Isot. 64, 368–372.

Wolle, B., Beikert, G., Gadelmeier, F., 1999. Effect of anisotropic D–D fusion neutron emission on counter calibration using activation techniques. Nucl. Instrum. Methods Phys. Res. Sect. A 424, 561–568.

Wolterbeek, H.T., van der Meer, A.J.G.M., 2002. Transport rate of arsenic, cadmium, copper and zinc in *Potamogeton pectinatus* L.: radiotracer experiments with ^{76}As, ^{109,115}Cd, ^{64}Cu and ^{65,69m}Zn. Sci. Total Environ. 287, 13–30.

Wong, C.T., Soliman, V.M., Perera, S.K., 2005. Gross alpha/beta analysis in water by liquid scintillation counting. J. Radioanal. Nucl. Chem. 264 (2), 357–363.

Wright, P.M., et al., 1961. Half-life of samarium-147. Phys. Rev. 123 (1), 205–208.

Wu, Y.L., Liang, J.C., Liu, H.R., et al., 2014. Construction and implementation of a liquid scintillation TDCR system. Acta Metrol. Sin. 35 (1), 83–86.

Wu, Y.L., Liu, H.R., Liang, J.C., et al., 2015. Standardization of tritiated water by the CIEMAT/NIST and TDCR methods. Sci. China Technol. Sci. 58 (3), 559–564.

Wurm, M., et al., 2015. Low-energy neutrino astronomy in LENA. Phys. Procedia 61, 376–383.

Wurtz, W.A., et al., 2011. Measurement of the absolute and differential cross sections for $^{7}Li(\gamma,n_0)^{6}Li$. Phys. Rev. C 84, 044601.

Xiaohui, C., Rong, L., Tonghua, Z., Xinxin, L., 2012. Digital discrimination of neutrons and γ -rays in a deuterated liquid scintillator using the Simplified Digital Charge Collection method. Nucl. Instrum. Methods Phys. Res. Sect. A 694, 11–15.

Xiaohui, C., et al., 2013. Analysis of three digital n/γ discrimination algorithms for liquid scintillation neutron spectrometry. Radiat. Meas. 49, 13–18.

Xiques Castillo, A., Isaac Olivé, K., Casanova González, E., et al., 2009. An adapted purification procedure to improve the quality of ^{90}Y for clinical use. Radiochim. Acta 97, 739–746.

Xiques Castillo, A., Pérez-Malo, M., Isaac-Olivé, K., Mukhallalati, H., Casanova González, E., Torres Berdeguez, M., Cornejo Díaz, N., 2010. Production of large quantities of ^{90}Y by ion-exchange chromatography using an organic resin and a chelating agent. Nucl. Med. Biol. 37 (8), 935–942.

Xu, Y., et al., 2018. Role of angiopoietin-like 3 (ANGPTL3) in regulating plasma level of low-density lipoprotein cholesterol. Atherosclerosis 268, 196–206.

Yamada, T., Ishii, K., 2012. Precise practical measurements of ^{90}Y used in radiopharmaceuticals using re-entrant ionization chambers. Appl. Radiat. Isot. 70, 2246–2250.

Yamada, Y., Yasuike, K., Itoh, M., Kiriyama, N., Komura, K., 2004. Temporal variation of tritium concentration in tree-ring cellulose. J. Radioanal. Nucl. Chem. 261 (1), 9–17.

Yamunarani, R., et al., 2016. Genetic diversity for grain Zn concentration in finger millet genotypes: Potential for improving human Zn nutrition. Crop J. 4, 229–234.

Yang, D., 1996. Calibration and quench correction for alpha liquid scintillation analysis, pp. 339–344. In: Cook, G.T., Harkness, D.D., MacKenzie, A.B., Miller, B.F., Scott, E.M. (Eds.), Liquid Scintillation Spectrometry 1994, Radiocarbon. University of Arizona, Tucson.

Yang, D., Guo, Y., 1995. Determination of alpha radioactivity in vegetable ashes with liquid scintillation analysis. Alpha Beta Application Note, ABA-007. Packard BioScience Company, Meriden, CT 06450, p. 4.

Yang, D., Zhu, Y., Mobius, S., 1992. Rapid method for alpha counting with extractive scintillator and pulse decay analysis. Alpha Beta Application Note, ABA-002. Packard BioScience Company, Meriden, CT 06450, p. 8.

Yang, D., Zhu, Y., Jiao, R., 1994. Determination of Np, Pu and Am in high level radioactive waste with extraction-liquid scintillation counting. J. Radioanal. Nucl. Chem. 183, 245–260.

Yang, Y., et al., 2006. Influence of a chinese crude drug on Ca2+ influx and efflux in rat visceral organs: Investigation and evaluation by ^{45}Ca. Appl. Radiat. Isot. 64, 241–246.

Yasuoka, Y., Ishii, T., Sanada, T., et al., 2005. Measurement of radon concentration in water using direct dpm method of liquid scintillation counter. Intern. Congr. Series 1276, 299–300.

Yeh, M., Cumming, J.B., Hans, S., Hahn, R.L., 2010. Purification of lanthanides for large neutrino detectors: Thorium removal from gadolinium chloride. Nucl. Instrum. Methods Phys. Res. Sect. A 618, 124–130.

Yeltepe, E., Yücel, H., 2017. Standardization of ^{142}Pr activity concentration. Appl. Radiat. Isot. (in press).

Yeltepe, E., Yücel, H., Kossert, K., 2017. Standardization of 170Tm with CIEMAT/NIST efficiency tracing, TDCR and gamma spectrometric methods. In: LSC2017 Conference, 1–5 May, Copenhagen.

Yemam, H.A., Mahl, A., Tinkham, J.S., Greife, U., Sellinger, 2017. A highly soluble p-Terphenyl and fluorene derivatives as efficient dopants in plastic scintillators for sensitive nuclear material setection. Chem. Eur J. 23 (37), 8921–8931.

Yoon, S., Chang, B.-U., Kim, Y., Byun, J.-I., Yun, J.-Y., 2010. Indoor radon distribution of subway stations in a Korean major city. J. Environ. Radioact. 101, 304–308.

Yoshikawa, H., Nakanishi, T., Nakahara, H., 2006. Determination of thoron and radon ratio by liquid scintillation spectrometry. J. Radioanal. Nucl. Chem. 267 (1), 195–203.

Yousefi, S., Lucchese, L., Aspinall, M.D., 2009. Digital discrimination of neutron and gamma-rays in liquid scintillators using wavelets. Nucl. Instrum. Methods Phys. Res. Sect. A 598, 551–555.

Yuan, L., et al., 2016. Investigation of the "true" extraction recovery of analytes from multiple types of tissues and its impact on tissue bioanalysis using two model compounds. Anal. Chim. Acta 945, 57–66.

Yuan, M., Lin, Y., Chu, W., Yeh, C., 2018. Standardization of ^{109}Cd by two methods. Appl. Radiat. Isot. (in press).

Zak, T., Clarke, S.D., Bourne, M.M., Pozzi, S.A., Xu, Y., Downar, T.J., Peerani, P., 2010. Neutron spectroscopy of plutonium oxide using matrix unfolding approach. Nucl. Instrum. Methods Phys. Res. Sect. A 622, 191–195.

Zapata-García, D., Llauradó, M., Rauret, G., 2012. The implications of particle energy and acidic media on gross alpha and gross beta determination using liquid scintillation. Applied. Radiat. Isot. 70 (4), 705–711.

Zare, S., Ghal-Eh, N., Bayat, E., 2013. On timing response improvement of an NE213 scintillator attached to two PMTs. Radiat. Phys. Chem. 90, 6–10.

Zecher, P.D., Galonsky, A., Kruse, J.J., et al., 1997. A large area, position-sensitive neutron detector with neutron/γ-ray discrimination capabilities. Nucl. Instrum. Methods Phys. Res. Sect. A 401, 329–344.

Ziemek, T., Jeczmieniowski, A., Cacko, D., Broda, R., Lech, E., 2016. A new 4π(LS)-γ coincidence counter at NCBJ RC POLATOM with TDCR detector in the beta cannel. Appl. Radiat. Isot. 109, 290–295.

Zhang, C., et al., 2013. Measuring fast neutrons with large liquid scintillation detector for ultra-low background experiments. Nucl. Instrum. Methods Phys. Res. Sect. A 729, 138–146.

Zheng, Q., Bobich, J.A., 2004. ADP-ribosylation factor6 regulates both [^{3}H]-noradrenaline and [^{14}C]-glutamate exocytosis through phosphatidylinositol 4,5-biphosphate. Neurochem. Int. 45, 633–640.

Zheng, Q., McFadden, S.C., Bobich, J.A., 2004. Phosphatidylinositol 4,5-biphosphate promotes both [^{3}H]-noradrenaline and [^{14}C]-glutamate exocytosis from nerve endings. Neurochem. Int. 44, 243–250.

Zhilin, C., Shixiong, X., Heyi, W., Ruimin, C., Guanyin, W., Yinhang, Z., 2010. The effect of vial type and cocktail quantity on tritium measurement in LSC. Appl. Radiat. Isot. 68, 1855–1858.

Zhu, D., Mu, Z., Mooty, C., Kovarik, M., Jay, M., 2006. Suspensions of fluor-containing nanoparticles for quantifying β^--emitting radionuclides in non-hazardous media. J. Pharm. Innov. 1 (1), 76–82.

Zhu, D., Jay, M., 2007. Aqueous polystyrene-fluor nanosuspensions for quantifying α and β^- radiation. Nanotechn 18 (22), 225502. http://iopscience.iop.org/0957-4484/18/22/225502.

Zimmerman, B.E., 2006a. Monte Carlo calculations of spectra and interaction probabilities for photons in liquid scintillators for use in the standardization of radiionuclides. Appl. Radiat. Isot. 64, 1492−1498.

Zimmerman, B.E., 2006b. Radionuclide metrology in the life sciences: Recent advances and future trends. Appl. Radiat. Isot. 64, 1351−1359.

Zimmerman, B.E., Collé, R., 1997a. Comparison of the French and U. S. National ^{3}H (tritiated H_2O) standards by $4\pi\beta$ liquid scintillation spectrometry. Appl. Radiat. Isot. 48, 521−526.

Zimmerman, B.E., Collé, R., 1997b. Cocktail volume effects in $4\pi\beta$ liquid scintillation spectrometry with ^{3}H-standard efficiency tracing for low-energy β-emitting radionuclides. Appl. Radiat. Isot. 48 (3), 365−378.

Zimmerman, B.E., Cessna, J.T., Schima, F.J., 1998. The standardization of the potential bone palliation radiopharmaceutical $^{117m}Sn(+4)$DTPA. Appl. Radiat. Isot. 49 (4), 317−328.

Zimmerman, B.E., Unterweger, M.P., Brodack, J.W., 2001. The standardization of ^{177}Lu by $4\pi\beta$ liquid scintillation spectrometry with ^{3}H-standard efficiency tracing. Appl. Radiat. Isot. 54, 623−631.

Zimmerman, B.E., Cessna, J.T., Unterweger, M.P., 2002. The standardization of $^{188}W/^{188}Re$ by $4\pi\beta$ liquid scintillation spectrometry with the CIEMAT/NIST ^{3}H-standard efficiency tracing method. Appl. Radiat. Isot. 56, 315−320.

Zimmerman, B.E., Collé, R., Cessna, J.T., 2004a. Construction and implementation of the NIST triple-to-double coincidence ratio (TDCR) spectrometer. Appl. Radiat. Isot. 60, 433−438.

Zimmerman, B.E., Cessna, J.T., Millican, M.A., 2004b. Experimental determination of calibration settings for plastic syringes containing solutions of ^{90}Y using commercial radionuclide calibrators. Appl. Radiat. Isot. 60, 511−517.

Zimmerman, B.E., Cessna, J.T., Fitzgerald, R., 2008. Standardization of $^{68}Ge/^{68}Ga$ using three liquid scintillation counting based methods. J. Res. Natl. Inst. Stand. Technol. 113 (5), 265−280.

Zimmerman, B.E., Altzitzoglou, T., Rodrigues, D., Broda, R., Cassette, P., et al., 2010. Comparison of triple-to-double coincidence ratio (TDCR) efficiency calculations and uncertainty assessments for ^{99}Tc. Appl. Radiat. Isot. 68, 1477−1481.

Zou, S., et al., 2016. Test of *Topmetal-II⁻* in liquid nitrogen for cryogenic temperature TPCs. Nucl. Instrum. Methods Phys. Res. Sect. A 830, 275−278.

Further reading

Akerib, D.S., 2016. FPGA-based trigger system for the LUX dark matter experiment. Nucl. Instrum. Methods Phys. Res. Sect. A 818, 57−67.

Alamelu, D., Aggarwal, S.K., 2009. Determination of $^{235}U/^{238}U$ atom ratio in uranium samples using liquid scintillation counting. Talanta 77, 991−994.

Attie, M.R.P., et al., 1998. Absolute disintegration rate measurements of Ga-67. Appl. Radiat. Isot. 49 (9−11), 1175−1177.

Baerg, A.P., 1973. The efficiency extrapolation method in coincidence counting. Nucl. Instrum. Methods 112, 143−150.

Bai, H., 2018. Simulation of the neutron response matrix of an EJ309 liquid scintillator. Nucl. Instrum. Methods Phys. Res. Sect. A 886, 109−118.

Baró, J., Sempau, J., Fernández-Varea, J.M., Salvat, F., 1995. PENELOPE: an algorithm for Monte Carlo simulation of the penetration and energy loss of electrons and positrons in matter. Nucl. Instrum. Methods Phys. Res. Sect. B 100 (1), 31−46.

Batrakova, E.V., 2010. Effects of pluronic and doxorubicin on drug uptake, cellular metabolism, apoptosis and tumor inhibition in animal models of MDR cancers. J. Controlled Rel. 143, 290−301.

Batrakova, E.V., Li, S., Brynskikh, A.M., Sharma, A.K., Li, Y., Boska, M., et al., 2010. Effects of pluronic and doxorubicin on drug uptake, cellular metabolism, apoptosis and tumor inhibition in animal models of MDR cancers. J. Controlled Release 143 (3), 290−301.

Beaulieu, D.R., 2011. Timing resolution of fast neutron and gamma counting with plastic microchannel plates. Nucl. Instrum. Methods Phys. Res. Sect. A 659, 394−398.

Beriguete, W., 2014. Production of a gadolinium-loaded liquid scintillator for the Daya Bay reactor neutrino experiment. Nucl. Instrum. Methods Phys. Res. Sect. A 763, 82−88.

Coursey, B.M., Gibson, J.A.B., Heitzmann, M.W., Leak, J.C., 1984. Standardization of technetium-99 by liquid scintillation counting. Appl. Radiat. Isot. 35, 1103−1112.

Craddock, B.L., Orchiston, E.A., Hinton, H.J., et al., 1999. Dissociation of apoptosis from proliferation, protein kinase B activation, and BAD phosphorylation in interleukin-3-mediated phosphoinositide 3-kinase signaling. J. Biol. Chem. 274 (15), 10633−10640.

Dai, X., Jonkmans, G., Cui, Y., et al., 2010. Gadolinium-loaded liquid scintillator for an anti-neutrino detector. In: LSC 2010, Advances in Liquid Scintillation Spectrometry. Radiocarbon. University of Arizona, Tucson.

Dziel, T., 2016. Radionuclidic purity tests in ^{18}F radiopharmaceuticals production process. Appl. Radiat. Isot. 109, 242−246.

Effertz, R., Neuman, K., Englert, D., 1993. Single photomultiplier technology for scintillation counting in microplates. In: Noakes, J.E., Schoenhofer, F., Polach, H.A. (Eds.), Liquid Scintillation Spectrometry 1992, Radiocarbon, pp. 37−42.

Erchinger, J., Orrell, J., Aalseth, C.E., Bernacki, B., Douglas, M., Fuller, E., Keillor, M., Marianno, C., Morley, S., Mullen, C., Panisko, M., Warren, G., 2017. Background characterization of an ultra-low background liquid scintillation counter. Appl. Radiat. Isot. 126, 168−170.

Febbraro, M., 2015. Deuterated scintillators and their application to neutron spectroscopy. Nucl. Instrum. Methods Phys. Res. Sect. A 784, 184−188.

Fujii, H., Takiue, M., 1988a. Radioassay of dual-labeled samples by sequential Cherenkov counting and liquid scintillation efficiency tracing technique. Nucl. Instrum. Methods Phys. Res., Sect. A 273 (1), 377−380.

Fujii, H., Takiue, M., 1988b. Radioassay of alpha- and beta-emitters by sequential Cherenkov and liquid scintillation countings. Int, J. Radiat. Applic. Instrum. Part A, Appl. Radiat. Isot. 39 (4), 327−330.

Funck, E., Nylandstedt Larsen, A., 1983. The influence from low energy x-rays and Auger electrons on $4\pi\beta$-γ coincidence measurements of electron-capture-decaying nuclides. Int. J. Appl. Radiat. Isot. 34 (3), 565−569.

García, J.F., Tarancón, A., Bagán, H., 2017. Plastic Scintillators and related analytical proposals for radionuclide analysis. In: LSC2017 Conference, 1−5 May, Copenhagen.

Grau Carles, A., 2008. MICELLE, the micelle size effect on the LS counting efficiency. Comp. Phys. Commun. 176, 305−317.

Grau Carles, A., Rodriguez Barquero, L., Jimenez De Mingo, A., 1997. ^{125}Sb to ^{125m}Te branching ratio. In: ICRM '97, 1−12. Biennial Conference, 19−23 May 1997, National Institute of Standards and Technology, Gaithersburg, Maryland, USA.

Grigorescu, L., 1963. Mesure absolue de l'activité des radionucléides par la méthode des coïncidences beta-gamma. In: Corrections des coincidences instrumentals et de temps morts. MESUCORA Congress, Paris.
Grigorescu, L., 1973. Accuracy of coincidence measurements. Nucl. Instrum. Methods 112 (1–2), 151–155.
Grigorescu, E.L., Sahagia, M., Razdolescu, A., et al., 1998. Standardization of ^{110m}Ag and ^{75}Se by the beta-efficiency extrapolation method. Appl. Radiat. Isot. 49 (9–11), 1165–1170.
Hamel, M., 2014. A fluorocarbon plastic scintillator for neutron detection: Proof of concept. Nucl. Instrum. Methods Phys. Res. Sect. A 768, 26–31.
Harwood Jr., H.J., 1995. Protein farnesyltransferase: Measurement of enzymatic activity in a 96-well format using TopCount microplate scintillation counting technology. Anal. Biochem. 226, 268–278.
Hino, Y., Ohgaki, H., 1998. Absolute measurement of ^{192}Ir. Appl. Radiat. Isot. 49 (9–11), 1179–1183.
Hong, J., 2002. The scintillation efficiency of carbon and hydrogen recoils in an organic liquid scintillator for dark matter searches. Astropart. Phys. 16, 333–338.
Houtermans, H., Miguel, M., 1962. 4π-β-γ coincidence counting for the calibration of nuclides with complex decay schemes. Int. J. Appl. Radiat. Isot. 13, 137–142.
Ianni, A., 2011. Neutrino physics with Borexino. Prog. Part. Nucl. Phys. 66, 405–411.
Johansson, L.C., et al., 2004. Primary standardization of ^{204}Tl using the efficiency tracing method. Appl. Radiat. Isot. 60, 347–351.
Mann, W.B., Rytz, A., Spernol, A., McLaughlin, W.L., 1988. Radioactivity Measurements, Principles and Practice. Pergamon Press, Oxford, p. 937.
Mui, B.L., 2013. Influence of polyethylene glycol lipid desorption rates on pharmacokinetics and pharmacodynamics of siRNA lipid nanoparticles. Molec. Therapy-Nucleic Acids 2, e139. https://doi.org/10.1038/mtna.2013.66.
Mui, B.L., Tam, Y.K., Jayaraman, M., Ansell, S.M., Du, X., Tam, Y.Y.C., et al., 2013. Influence of polyethylene glycol lipid desorption rates on pharmacokinetics and pharmacodynamics of siRNA lipid nanoparticles. Mol Ther Nucleic Acids 2, e139.
NCRP, 1985. A Handbook of Radioactivity Measurements Procedures. Report No. 64. National Council on Radiation Protection and Measurements, Washington, D.C, 20014.
Ni, K., Aprile, E., Giboni, K.L., Majewski, P., Yamashita, M., 2006. Gamma ray spectroscopy with scintillation light in liquid xenon. J. Instrum. 1, P09004.
Ogawa, S., on behalf of the KEG II Collaboration, 2017. Liquid xenon calorimeter for MEG II experiment with VUV-sensitive MPPCs. Nucl. Instrum. Methods Phys. Res. Sect. A 845, 528–532.
Park, T.S., Hwang, H.Y., Lee, J.M., 1998. An improved coincidence counting technique for standardization of radionuclides. Appl. Radiat. Isot. 49 (9–11), 1147–1151.
Passo Jr., C.J., Kessler, M.J., 1993. Selectable delay before burst - a novel feature to enhance low-level counting performance, pp. 51–57. In: Noakes, J.E., Schönhofer, F., Polach, H.A. (Eds.), Liquid Scintillation Spectrometry 1992, Radiocarbon. University of Arizona, Tucson.
Pates, J.M., Cook, G.T., MacKenzie, A.B., Passo Jr., C.J., 1996c. Quenching and its effect on alpha/beta separation liquid scintillation spectrometry, pp. 75–85. In: Cook, G.T., Harkness, D.D., MacKenzie, A.B., Miller, B.F., Scott, E.M. (Eds.), Liquid Scintillation Spectrometry 1994, Radiocarbon. University of Arizona, Tucson.
(Particle Datea Group) Patrigani, C., et al., 2016. Neutrino mass, mixing, and oscillations. Chin. Phys. C 40, 100001.
(Particle Datea Group) Patrigani, C., et al., 2017. Neutrino Properties. Chin. Phys. C 40, 100001.
Razdolescu, A.C., Grigorescu, E.L., Sahagia, M., Luca, A., Ivan, C., 2000. Standardization of ^{169}Yb by the $4\pi\beta$-γ method. Appl. Radiat. Isot. 52, 505–507.
Shippen, D.I., Joyce, M.J., 2010. A wavelet packet transform inspired method of neutron-gamma discrimination in liquid organic scintillation detectors. IEEE Trans. Nucl. Sci. 57, 2617–2624.
Simpson, B.R.S., 2002. Radioactivity standardization in South Africa. Appl. Radiat. Isot. 56, 301–305.
Simpson, B.R.S., Meyer, B.R., 1988. A multiple-channel 2- and 3-fold coincidence counting system for radioactivity standardization. Nucl. Instrum. Methods Phys. Res., Sect. A 263, 436–440.
Simpson, B.R.S., Meyer, B.R., 1994a. Standardization and half-life of ^{201}Tl by the $4\pi(x,e)-\gamma$coincidence method with liquid scintillation counting in the -4πchannel. Appl. Radiat. Isot. 45, 669–673.
Simpson, B.R.S., van Oordt, E.J., 1997. Data acquisition program for the NAC radioactivity standards laboratory. NAC Report NAC/97-03, National Accelerator Centre, PO Box 72, Faure, 7131 South Africa.
Suárez, J.A., Rodríguez, M., Espartero, A.G., Piña, G., 2000. Radiochemical analysis of ^{41}Ca and ^{45}Ca. Appl. Radiat. Isot. 52, 407–413.
Tarancón Sanz, A., Bagán, H., Kossert, K., Nähle, O., 2011. Application of a free parameter model to plastic scintillation samples. In: LSC 2010, Advances in Liquid Scintillation Spectrometry. Radiocarbon. University of Arizona, Tucson.
Tsroya, S., Pelled, O., German, U., Katorza, E., Abraham, A., Alfassi, Z.B., 2012. A comparative study of color quenching correction methods for cerenkov counting. Appl. Radiat. Isot. 70 (2), 397–403.
Vagner, I., 2016. Method for organically bound tritium analysis from sediment using a combustion bomb. Appl. Radiat. Isot. 118, 136–139.
Vagner, I., Varlam, C., Faurescu, I., Faurescu, D., Bogdan, D., Bucura, F., 2016. Method for organically bound tritium analysis from sediment using a combustion bomb. Appl. Radiat. Isot. 118, 136–139.
Varignon, C., 2006. A new neutron beam line for (n,xn) reaction studies. Nucl. Instrum. Methods Phys. Res. Sect. B 248, 329–335.
Venkataramanan, S., Gupta, A., Golda, K.S., et al., 2008. A compact pulse shape discriminator module for large neutron detector arrays. Nucl. Instrum. Methods Phys. Res. Sect. A 596, 248–252.
Walton, T.D., 1996. Implementation of a robotics system for high throughput screening utilizing glow luminescence. TopCount Topics, TCA-026. PerkinElmer,Inc., p. 6. available at: http://las.perkinelmer.co.uk/content/applicationnotes/app_topcountimplroboticshts.pdf
Warwick, P.E., Croudace, I.W., 2013. Identification and quantification of radionuclides in contaminated drinking waters and pipeline deposits. Anal. Chem. 85 (17), 8166–8172.
Yan, C.G., 1996. Improvement of accuracy of efficiency extrapolation method in 4π β-γ coincidence counting. Nucl. Instrum. Methods Phys. Res. Sect. A 369, 383–387.
Yi-Fen, Y., Bowman, J.D., Bolton, R.D., Crawford, B.W., et al., 2000. A high-rate ^{10}B-loaded liquid scintillation detector for parity-violation studies in neutron resonances. Nucl. Instrum. Methods Phys. Res. Sect. A 447, 476–489.
Zimmerman, B.E., Cessna, J.T., 1999. The standardization of ^{62}Cu and experimental determinations of dose calibrator settings for generator-produced ^{62}CuPTSM. Appl. Radiat. Isot. 51, 515–526.

Michael F. L'Annunziata

Michael F. L'Annunziata, Ph.D. is the Founding Editor and co-author of the Handbook of Radioactivity Analysis now in its Fourth Edition. His graduate thesis research at the University of Arizona, Tucson, AZ, USA in the 1960s, financed by the then U.S. Atomic Energy Commission, dealt with the analysis of ^{89}Sr and ^{90}Sr in the event of nuclear fallout. During 1970–71, he worked in the chemical industry (Amchem Products, Inc., Ambler, PA, USA) as ^{14}C-tracer chemist. Michael was then appointed Professor at the Postgraduate College, Chapingo, Mexico (1972–75) and Senior Research Scientist and graduate thesis advisor at the Nuclear Center of the National Institute of Nuclear Research (ININ), Mexico City during (1975–77). Michael served as Scientific Officer in the Department of Research and Isotopes and Department of Technical Co-operation of the International Atomic Energy Agency (IAEA) in Vienna, Austria during 1977–1991 and Head of IAEA Fellowships and Training during 1987–91. Michael F. L'Annunziata was first to postulate the soil microbial epimerization of myo-inositol to other inositol stereoisomers as the chemical pathway and origin of the various inositol phosphate stereoisomers in soils (Ph.D. dissertation, 1970, https://dissexpress.proquest.com/dxweb/results.html?QryTxt=&By=L%27Annunziata&Title=&pubnum =) and in 1975 (*SSSA Journal* 39(2), 377–379), and first to demonstrate in 1977, with the use of the radioisotope carbon-14, the soil microbial epimerization of myo-inositol to chiro-inositol as a mechanism involved in the origin of the unique soil inositol phosphate stereoisomers (*SSSA Journal* 41(4), 733–736, https://dl.sciencesocieties.org/publications/sssaj/abstracts/41/4/SS0410040733).
From 1977 to 2007, he served as IAEA Expert in peaceful applications of nuclear energy in over 50 countries of the world in Europe, Asia, Africa, Latin America and the Middle East. His book entitled "Radioactivity: Introduction and History", First Edition, published by Elsevier, was on the Best Sellers List in Physics (*LibraryJournal Academic Newswire* in 2008), and his much expanded second edition entitled *Radioactivity: Introduction and History: From the Quantum to Quarks* published by Elsevier in 2016 (https://www.sciencedirect.com/book/9780444634894/radioactivity) was awarded an Honorary Mention in the 2017 PROSE AWARDS in the category of Chemistry & Physics.

Alex Tarancón

Alex Tarancón holds a PhD (2007) in Chemistry from the University of Barcelona. He is Serra Hunter Associate Professor of the Department of Chemical Engineering and Analytical Chemistry at the University of Barcelona. He has more than 15 years of experience in the development of innovative methodologies for the measurement of radioactivity using plastic scintillators. His experience covers several topics on the development of scintillation instrumentation, optimization of the response of scintillators, studies on the scintillation mechanism, pulse shape discrimination, LS standardization methods and the complete analysis of samples containing beta-emitting radionuclides. His current research is focused on the development of advanced strategies (fast determinations, selective analysis, on-line sensing, in-situ measurements, screening methods, ...) for radionuclide detection based on scintillation. He has been a member of the scientific committee of the two last conferences on Advances in Liquid Scintillation Spectrometry.

Héctor Bagán

Héctor Bagán received his PhD in chemistry at the University of Barcelona in 2011 working on the development of procedures to improve the selectivity in radioactivity measurements. He completed his 2 year post-doctorate at Lund University in the field of polymer synthesis for the separation and purification of biomolecules and proteins. Since then, he has been an adjunct lecturer at the Department of Chemical Engineering and Analytical Chemistry at the University of Barcelona. His research focuses on the use and development of analytical methodology for artworks characterization and also on the development of polymeric materials for the selective characterization of radioactive samples. This last topic includes the development of plastic scintillation resins that unify the separation and measurement preparation in one step, for the radionuclide measurements in environmental and nuclear power plant decommissioning samples.

José F. García

José F. García holds a PhD in Chemistry from the University of Barcelona. He is Associate Professor of the Department of Chemical Engineering and Analytical Chemistry at the University of Barcelona. José F. García has researched radioactivity analysis for the last 30 years. His work has been focused on scintillation analysis and his main contributions in this field are related to the introduction of chemometrics, the development of plastic scintillators and inprovements in the continuous fluids monitoring for different radionuclide applications. These applications have included environmental and decommissioning studies and dating methods development. He has participated in several research projects at the national and international level and technology transfer contracts. José F. García has published more than 70 articles in scientific journals and has co-organized the periodical editions of the International Workshop on Plastic Scintillators.

Chapter 7

Sample preparation techniques for liquid scintillation analysis

James Thomson and Simon Temple
Meridian Biotechnologies Ltd., Chesterfield, United Kingdom

Chapter outline

Handbook of Radioactivity Analysis: Volume 1. https://doi.org/10.1016/B978-0-12-814397-1.00007-8

I. Introduction

It is a sad but very true observation that sample preparation is considered an inconvenient step on the way to obtaining results. Consequently, it is only when the results obtained are different from those expected that the cry "There's something wrong with the cocktail" is heard. While this may or may not be true, the technique of sample preparation, in general, is only rarely considered responsible for the unexpected results. Correct sample preparation in liquid scintillation analysis (LSA) is essential for both accurate and reproducible analyses, and no amount of instrumental sophistication can ever fully compensate for the problems attendant to a badly prepared sample. Good sample preparation guarantees that the sample will be stable during the analysis and therefore provides a solid foundation for accurate results. Sample preparation encompasses a wide variety of methods and techniques and includes dissolution, distillation, extraction, solubilization, digestion, suspension, and combustion. All of these methods of sample preparation hold pitfalls for the unwary, and some expertise is generally required. Common to all sample preparation methods is the liquid scintillation counting (LSC) cocktail, and this is the medium that holds the sample during the analysis process. The LSC cocktail is fundamental to and necessary for the analysis; therefore, correct cocktail selection is a critical step in sample preparation.

II. Liquid scintillation counting cocktail components[1]

To appreciate the significance of correct cocktail selection, it is useful to explain some fundamentals about the components of LSC cocktails and thus gain some insight into not only what components are present but also what properties they impart to the final cocktail. Cocktails can be divided into two main groups: emulsifying cocktails (sometimes called aqueous cocktails) and organic cocktails (sometimes called nonaqueous or lipophilic cocktails). Organic cocktails have only two major components: organic aromatic solvent and scintillators. Emulsifying cocktails have three major components: organic aromatic solvent, emulsifier, and scintillators.

A. Solvents

The traditional or classical aromatic solvents used in LSC are toluene, xylene, and pseudocumene (1,2,4-trimethylbenzene), and these are effective LSC solvents due to the high density of π electrons associated with these solvents. The increase in mono alkyl substitution causes increased electron donation and thus a higher ring electron density; this results in higher counting efficiency:

$$\underset{\text{increasing efficiency}\;\longrightarrow}{\text{benzene} \rightarrow \text{toluene} \rightarrow \text{xylene} \rightarrow \text{pseudocumene}}$$

Although still popular, the use of toluene, xylene, and pseudocumene has decreased and continues to decrease, due to problems with toxicity, flammability, vapor pressure, smell, and permeation through plastics. These solvents have generally been replaced by the newer generation of "safer" solvents:

di – isopropylnaphthalene	phenylxylylethane	dodecylbenzene
(DIN)	(PXE)	(LAB)

These solvents are characterized by high flash point (>145°C), low vapor pressure (<1 mm Hg at 20°C), low toxicity ($LD_{50} > 3000$ mg/kg), low odor, and no permeation through plastics. In addition, both DIN (Thomson, 1985, 1991) and PXE (Wunderley, 1986) exhibit higher counting efficiency than the classical solvents, and the overall order is

benzene → LAB → toluene → xylene → pseudocumene → PXE → DIN
increasing efficiency →

Commercial mixtures (e.g., petroleum distillates) have been used instead of pseudocumene, but these contain many species other than alkyl benzenes and include indane, indene, methyl indane, and ethyl indane. These impurities have a detrimental effect on counting efficiency and background and can be particularly strong chemiluminescers. Indane and indene, even at the 50−100 ppm level, will produce severe chemiluminescence when contacted with alkaline materials such as tissue solubilizers. This effect is so severe that the mixture can turn brown or purple and produce backgrounds $>1 \times 10^6$ CPM (counts per minute).

These petroleum distillates can be purified by,for example, distillation, column chromatography, reactive distillation, solvent extraction, and mild sulfonation, but the final product never really approaches the pure solvent as a completely acceptable LSC solvent. An earlier approach at producing a safer solvent system was to use paraffinic and naphthenic solvents in conjunction with a secondary LSC solvent. The paraffinic and naphthenic solvents have poor scintillation properties, but the addition of this secondary LSC solvent (sometimes referred to as an energy transfer medium) can boost the counting efficiency up to within 80% of the pure LSC solvent. This approach was largely abandoned due to the large-scale availability of the new safer solvents. A summary of the important characteristics of LSC solvents is shown in Table 7.1.

B. Scintillators

The fluor or scintillator is a light transducer that converts nuclear energy into light photons. All LSC cocktails contain at least one and usually two scintillators. The primary scintillator is responsible for the initial or primary energy exchange, but the wavelength of the emitted light does not match the optimum detection wavelength of the photomultiplier tube (PMT). Hence, a secondary scintillator (or wavelength shifter) is added to shift the wavelength of the emitted light to match the PMT. An efficient scintillator should have a high fluorescence quantum efficiency (or high photon yield); a spectrum maximally matched to the response of the photocathode, short fluorescence decay time, a large Stokes' shift, sufficient solubility, and a low sensitivity to quenching agents.

Derivatives of oxazoles, oxadiazoles, phenylenes, p-oligophenylenes, styrylbenzenes, benzoxazoles, benzoxazolyl

TABLE 7.1 Safety characteristics of scintillation solvents.

Solvent	Boiling point	Flash point (°C)	Vapor pressure mmHg@25°C	Classification (international)	Hazards
Toluene	110°C	4°C	28	Flammable	Inhalation Skin absorption Irritating to skin/eyes
Xylene	138°C	25°C	8	Flammable	Inhalation Skin absorption Irritating to skin/eyes
Cumene	152°C	31°C	5	Flammable	Inhalation Irritating to skin/eyes
Pseudocumene	168°C	50°C	2	Flammable	Inhalation Irritating to skin/eyes
LAB	300°C	149°C	<1	Harmless	None (as classified)
PXE	305°C	149°C	<1	Harmless	None (as classified)
DIN	300°C	140°C	<1	Harmless	None (as classified)
Benzyl toluene	290°C	135°C	<1	Harmless	None (as classified)
Diphenyl ethane	290°C	135°C	<1	Harmless	None (as classified)

thiophenes, pyrazolines, and nitriles have scintillation properties and are potential scintillators. The concentration of the scintillator determines the photon yield of a liquid scintillator. At optimum scintillator concentration, the light output is maximum, but beyond this point, self- or concentrationquenching occurs due to the formation of excimers. The probability of excimer formation is greater in scintillators with planar configuration than those with bulky substituent groups offering steric hindrance; for example, PPO (2,5-diphenyloxazole) exhibits a reduced quantum yield value at high concentrations. Substituted polyaryls and p-oligophenylenes currently used as scintillators can be rendered more soluble by substitution of hydrogen with alkyl or alkoxy groups. The alkyl groups tend to enhance the solubility in toluene, whereas the alkoxy groups favor an increased solubility in more polar solvents. Primary fluors are generally substituted fluorescent polyaryls and include the following:

PPO	2,5-Diphenyloxazole
PPD	2,5-Diphenyl-1,3,4-oxadiazole
PBO	2-(4-biphenylyl)-5-phenyloxazole
PBD	2-Phenyl-5-(4-biphenylyl)-1,3,4-oxadiazole
BBD	2,5-Di-(4-biphenylyl)-1,3,4-oxadiazole
Butyl-PBD	2-(4-t-Butylphenyl)-5-(4-biphenylyl)-1,3,4-oxadiazole
BBOT	2,5-Bis-2-(5-t-butyl-benzoxazoyl) thiophene
TP	p-Terphenyl

The most popular and widely used is PPO, and this preference is due to its performance, purity, cost, and availability on a large scale. Butyl-PBD is also occasionally used, but it suffers from the drawback that it yellows when in contact with alkaline species such as tissue solubilizers. The situation with secondary scintillators is similar in that there are many scintillators that are useful but only a few are in common use. The important secondary scintillators include the following:

Bis-MSB	p-bis-(o-methylstyryl)benzene
POPOP	1,4-Bis-2-(5-phenyloxazolyl)benzene
Dimethyl POPOP	1,4-Bis-2-(4-methyl-5-phenyloxazolyl) benzene
NPO	2-(1-Naphthyl)-5-phenyloxazole
NPD	2-(1-Naphthyl)-5-phenyl-1,3,4-oxadiazole
BBO	2,5-Di(4-biphenylyl)oxazole
PBBO	2-(4-Biphenylyl)-6-phenylbenzoxazole

The most popular and widely used is bis-MSB, and this preference is due to its performance, solubility, purity, cost, and availability on a large scale. Dimethyl POPOP is also occasionally used, but it suffers from limited solubility in organic aromatic solvents. Virtually, all commercially available LSC cocktails are based on the combination of PPO and bis-MSB, and this has been the case now for many years.

C. Surfactants

The most efficient energy transfer takes place in aromatic organic solvents. Since the majority of radioisotopes are present in an aqueous environment that is not miscible with the aromatic solvents, detergents are used to bring the aqueous phase into close contact with the organic phase by forming a microemulsion. Such close contact allows efficient energy transfer, which means that radioisotopes in aqueous solutions can be successfully measured both qualitatively and quantitatively. In an LSC cocktail, a microemulsion is a dispersion of very small droplets (radius of approximately 10 nm) of "water-in-oil" and is achieved by the use of detergents. Many detergents have been investigated, but only a few specific types have been successfully used in LSC cocktails. The major groups are nonionics, anionics, cationics, and amphoterics, and within each group, there are certain useful or favored detergents, which are found in many LSC cocktails.

1. Nonionics

This group includes the ethoxylates; one particular type, the alkyl phenol ethoxylate (APE), is found in virtually all current LSC cocktails. However, recent studies have shown that, when APEs biodegrade in the environment, they produce endocrine disrupters. Such endocrine disrupters cause male fish to turn into female fish and produce other sexual dysfunctionality. Consequently, some countries have introduced legislation to ban the drain disposal of products containing APEs. A comprehensive review of this situation is given in Section XIV of this chapter. Until now, only one company has addressed this issue and produced a range of LSC cocktails that are completely APE-free and that company is Meridian Biotechnologies Ltd. Meridian uses alcohol ethoxylates as alternatives to APEs. The ethoxylates are the building blocks upon which cocktail performance is based. Ethoxylates are produced by the vapor- or liquid-phase ethoxylation of the free hydroxyl group of the alkyl phenol in the presence of a catalyst. Such ethoxylates are described as having a specific ethylene oxide or EO length, but this number, which refers to the length of the ethoxylate polymer chain, is in fact an indication based on properties, not on chemical structure. These ethoxylates have a range of polymer chain lengths present in a normal distribution, and the EO length is an indication of the average distribution. Commonly encountered ethoxylates include APEs, alcohol ethoxylates, amine ethoxylates, acid ethoxylates, castor oil ethoxylates, and ester ethoxylates; however, as stated above, only the APEs and now the alcohol ethoxylates are useful in LSC cocktail formulation. These other ethoxylates are not used, because they fail to produce a stable microemulsion, are generally colored, have limited solubility in aromatic solvents, produce unacceptable chemical quench, and have poor counting efficiency properties. Historically, in the 1960s, Patterson and Greene (1965) examined a few nonionic detergents of the alkyl phenol type and concluded that Triton X-100 was the detergent of choice. Unfortunately, factors such as the variability of feedstock and ethoxylation conditions were not taken into account, and this resulted in a severe batch-to-batch variation in the original Triton X-100 product. When used in LSC cocktails, the result was microemulsion instability and chemiluminescence; therefore, this meant that detergent batch selection was necessary. Refinements in production methods significantly improved the quality of the APEs to the extent that standard production material can now be used in LSC cocktails without reverting to batch selection of raw materials. Due to the disposal ban, many manufacturers have dropped APEs from their product offering, but there are still some available. Sourcing acceptable raw material ethoxylate can only be achieved through a lengthy trial-and-error process. When mixed with aromatic LSC solvents, both alcohol ethoxylates and APEs are able to produce microemulsions when aqueous samples are added. The choice of solvent and ethoxylate is critical in that there is a limited range of EO chain lengths, which will work with each individual aromatic solvent, and desirable properties such as clarity of microemulsion or gelling can be chosen by careful EO selection. In addition, altering the chain length of the alkyl group of the ethoxylate affects performance. To extend or enhance performance, other detergents are needed, and these are normally selected from the other groups.

2. Anionics

This group includes, among others, alkyl and alkylaryl sulfonates, alcohol sulfates, alkyl sulfosuccinates, and phosphate esters (PEs). Of these, the sulfosuccinates have been found useful primarily due to their synergistic effect when combined with ethoxylates. Sulfosuccinates, even in relatively small quantities, are able to extend the microemulsion region of ethoxylates without affecting the other phases and have a secondary effect, in which they tend to stabilize the gel region. Various sulfosuccinates have been examined, but the most popular is sodium dioctylsulfosuccinate. This popularity is due to both its availability in a pure, water-white solid form and its ability to extend the microemulsion range of most ethoxylates. Another group of anionics, which are of exceptional benefit, are the PEs. The use of neutralized PEs has been patented (ter Wiel and Hegge, 1983), and the benefits resulting from their use include chemiluminescence resistance and microemulsion formation with difficult sample types. In small quantities, the free acid or neutralized PEs act as hydrotropes and can remove the microemulsion instability often encountered with ethoxylates. When used as the free acid, the disadvantages are that they can react with certain metals and form colored compounds; they will protonate PPO; and they can cause cocktail degradation through reaction with the solvents. In general, the detergent properties of PEs are enhanced by neutralization, but the use of the wrong alkaline material has certain previously unreported drawbacks. If the PEs are neutralized with inorganic alkalis, the salt of the PE has a tendency to precipitate slowly out of the LSC cocktail, especially if the concentration is significant. If ammonia has been used, the presence of an alkaline sample will cause a reaction, resulting in the release of ammonia gas within the vial. Both of these situations are undesirable since they can produce variable quench conditions either in the original cocktail or within the counting vial. Other anionics that are known to have been investigated include sarcosinates, taurates, isothionates, and sulfonates, but little is reported on their benefits. If anionics are used in cocktails either alone or in conjunction with ethoxylates, then care is needed when adding cationics, such as quaternary ammonium hydroxides. There is a potential for reaction between the anionic and cationic, and insoluble compounds can form, either immediately or slowly upon standing. Thus far, the formation of an insoluble product has only been seen when sulfosuccinates are the anionic species.

3. Cationics

This class of detergents carries a positive charge on the hydrophobic portion—a property rendering it substantive to negatively charged surfaces, for example, metals and organic surfaces, in general. As stated previously, the fact that cationics form salts with anionics virtually precludes their use in cocktails since anionics are commonly present. One area that was tentatively researched was the deliberate reaction between free acid PEs and quaternary ammonium hydroxides to see if the products had any beneficial detergent properties. The results were a little inconclusive but suggested that further investigation might be worthwhile. In LSC, cationics, such as the quaternary ammonium hydroxides that are powerful bases, are used to solubilize/dissolve a host of animal fats and tissues. Although these cationics have only a limited use in LSC, they nevertheless affect many of the cocktails in use. Being very basic in nature, they have the potential to initiate chemiluminescence in many cocktails.

4. Amphoterics

This class of detergents has both positive and negative centers in the molecule, and depending on the condition prevailing, cationic, anionic, or nonionic type properties may be exhibited. Examples of molecules peculiar to this class include betaines, imidazolines, and carboxylates. The commercially available products are at present not suitable due to problems with low concentration availability, high color, and poor solubility in aromatic solvents. Although most cocktails contain APEs, it is possible to use a mixed anionic system effectively and one such cocktail was OptiPhase RIA, which was based on sodium dioctyl sulfosuccinate and neutralized PE in pseudocumene solvent.

D. Cocktails

LSC cocktails can be categorized either as "classical" or "safer." The classical cocktails encompass all those based on toluene, xylene, or pseudocumene, whereas the safer cocktails include those based on DIN, PXE, LAB, benzyl toluene, and 2-phenylethane. Other high flash point solvents, such as 2-ethylnaphthalene, C_5-C_{10} alkylbenzenes, hydrogenated terphenyls (Actrel 400), high flash point aromatic fractions (Solvesso 150 and 200), ethylbiphenyl, butylbiphenyl, phenylcyclohexane, phenylnaphthalene, isopropylbiphenyl, diphenylpropane, dixylylethane (Freiberg, 1990), and dicumylethane (Mirsky, 1990), have been investigated but, for various reasons, have not been used commercially in LSC cocktails. The main reasons include purity, scintillation efficiency, cost, and availability on a large scale. Among the safer cocktails, diisopropylnaphthalene (DIN)-based cocktails, such as the Ultima Gold range from PerkinElmer and the ProSafe + range from Meridian Biotechnologies, will give high ^{3}H efficiency, and better color and chemical quench resistance. Other safer cocktails based on DIN include some of the Optiphase Hi-Safe range from PerkinElmer and the AquaSafe and QuickSafe cocktails from Zinsser. Some cocktail manufacturers use PXE as the safer solvent base, and examples include the ScintiSafe range from Fisher and the Ecoscint range from National Diagnostics. Finally, there are many cocktails based on linear alkyl benzene (LAB), and these include Opti-Fluor and Emulsifier Safe from PerkinElmer, Bio-Safe from RPI, Ecolume from ICN, and LumaSafe from Lumac. Most LSC cocktail suppliers feature both types in their ranges, and a full table of the known equivalents is shown in Table 7.2.

III. Dissolution

The primary objective of all sample preparation procedures for LSC is to obtain a homogeneous solution for efficient energy transfer from the sample to the LS cocktail. Aqueous solutions are some of the simplest and are most commonly found in LSA. In general, they provide the environment necessary for many assays and separations, and include the most encountered solvent media for the numerous radioisotopes used in LSC. The main methods for producing aqueous samples are by dissolution, extraction, and distillation. Dissolution simply involves dissolving the sample in water. Extraction can include solid/liquid extraction whereby the sample is extracted from a solid matrix by water, or liquid/liquid extraction, which involves the extraction of the sample from a liquid matrix by water. Distillation involves separation of the aqueous component by evaporation. A variety of LSC cocktails have evolved over the years to accommodate the diverse types, volumes, and concentrations of aqueous samples presented for analysis by LSC and some are considered by Kobayashi and Maudsley (1974a,b), ter Wiel and Hegge (1991), and Peng (1983).

A. Anions

Different types of anions encountered in LSC include chlorides, nitrates, phosphates, acetates, and formates, with sample volumes ranging from less than 100 μL to greater than 10 mL and concentrations varying from less than 10 mM to greater than 2 M. For the purpose of cocktail selection, these aqueous samples can be roughly divided into the following categories:

1. Buffers (e.g., sodium chloride, PBS, and potassium phosphate)
   a. Low ionic strength (less than 0.1 M)
   b. Medium ionic strength (0.1–0.5 M)
   c. High ionic strength (0.5 M to greater than 1 M)
2. Acids (e.g., hydrochloric acid and nitric acid)
3. Alkalis (e.g., sodium hydroxide and potassium hydroxide)
4. Other aqueous sample types (e.g., urea, sucrose, and imidazole)

By using this list of categories, it is now possible to assign cocktails for each category and therefore present a simpler and more comprehensive method of cocktail selection than was previously possible. For each category, cocktails will be recommended based on sample acceptance, performance, and safety.

B. Low ionic strength buffers

Buffers encountered in this group include 0.01 M PBS (phosphate-buffered saline), 50 mM Tris–HCl [tris(hydroxymethyl)aminoethane hydrochloride], 0.1 M NaCl (sodium chloride), 0.01 M Na_2SO_4 (sodium sulfate), etc. Since these aqueous buffers are relatively dilute, there are comparatively few problems. Both di- and trivalent anions such as SO_4 and PO_4 are potentially problematic, due in part to their charge and to their relative size (e.g., chlorides [Cl^-] are much smaller than sulfates [SO_4^{2-}]). These characteristics can impede the formation of a stable

TABLE 7.2 Liquid Scintillation Counting cocktail equivalents (valid as of 08/08/2018—data from websites).

Meridian	MPBio	N.D.	NEN	PerkinElmer	PerkinElmer	Roth	RPI	Zinsser
LSC cocktails for aqueous samples								
ProSafe+	Ecolite+	EcoscintA		Ultima Gold	HiSafe 2			Aquasafe 300+
								QuicksafeA
Gold Star	Universol ES	Ecoscint XR		Ultima Gold XR	HiSafe 3	RotiszintEco-Plus	3a70b	Aquasafe 500+
ProSafe HC+					Supermix			IrgasafePlus
ProSafe FC+	Cytoscint ES	Ecoscint H		Ultima Gold MV				Filtersafe
Gold Star LT[2]				Ultima Gold AB				
Gold Star LT[2]		EcoscintUltra		Ultima Gold LLT				Quicksafe 400
ProSafe LT+								
		Uniscint BD		Opti-Fluor			Bio-Safe II	
	Ecolume	Ecoscintoriginal		Emulsifier Safe			Econo-Safe	
	Universol	Hydrofluor		Insta-Gel Plus			3a70	Quickszint 1
		Liquiscint					Budget-Solve	Quickszint 212
							BioCount	
	Cytoscint		BiofluorPlus					
							Safety-Solve	
	Beta Blend	Monofluor		Pico-Fluor Plus			RiaSolve II	
ProSafe TS+		Bioscint		Hionic-Fluor			Neutralizer	
Gold Star		Soluscint XR						
ProSafe FC+		Filtron-X		Filter-Count				Filtersafe
LSC cocktails for nonaqueous/organic samples								
ProScint Rn/AB	Beta Max ES	Ecoscint O	Mineral Oil Scintillator	Ultima Gold F	BetaplateScint	RotiszintEco	Bio-Safe NA	Quicksafe N
Gold Star O								
				Opti-Fluor O				
RadonCount		Betafluor		Insta-Gel Plus			4a20 and3a20	

Continued

TABLE 7.2 Liquid Scintillation Counting cocktail equivalents (valid as of 08/08/2018—data from websites).—cont'd

Meridian	MPBio	N.D.	NEN	PerkinElmer	PerkinElmer	Roth	RPI	Zinsser
LSC cocktails for flow detectors								
		Monoflow II		Flo-Scint II				QuickszintFlow 302
		Monoflow III		Flo-Scint III				QuickszintFlow 302
				Flo-Scint A				QuickszintFlow 302
		Monoflow IV						
ProFlow G+		EcoscintFlow		Ultima-Flo M				QuicksafeFlow 2
ProFlowP+				Ultima-Flo AP				
Oxidizer reagents								
CarbonCount		Oxosol 306		Carbosorb E				
CarbonTrap		Carbamate-1		Permafluor E+				
		Oxosol C14						Oxysolve C-400
TritiumCount				Monophase-S				Oxysolve-T
Tissue and gel solubilizers								
GoldiSol	NCS solubilizer	Solusol		Soluene-350			TS-2	
AquiGest		Biosol		Solvable				
Other reagents								
Hyamine hydroxide	Hyamine hydroxide			Hyamine hydroxide				Hyamine hydroxide

TABLE 7.3 Cocktail Selection for Low Ionic Strength Samples (based on the use of 10 mL of cocktail).

	Safer cocktail	Classical cocktail
Low sample volume (0–2.5 mL)	Ultima Gold, MV, AB, LLT, Opti-Fluor, Emulsifier Safe (Poly-Fluor), Gold Star, Gold Star LT,[2]ProSafe+ range, and other equivalent cocktails	Insta-Gel Plus (Insta-Gel XF), Pico-Fluor 15, Biofluor Plus, and other equivalent cocktails
High sample volume (>2.5 mL)	Ultima Gold, XR, AB, LLT, Gold Star, Gold Star LT,[2]ProSafe HC+, and other equivalent cocktails	Insta-Gel XF, Pico-Fluor Plus, and other equivalent cocktails

microemulsion and can cause phase instability, especially with high concentrations and large volumes. Surprisingly, similar problems can occur with small volumes, particularly within the range of 0.1–0.5 mL of sample in 10 mL cocktail (1%–5% sample load). The only other area of concern is color quench problems when using certain metallic salts, which are intrinsically colored (e.g., $FeCl_3$[ferric chloride]). Any phase instability problem can usually be resolved by decreasing the sample volume or by increasing the cocktail volume. If the problem persists, then it may be necessary to change to a cocktail that can accept higher strength ionic samples. Color quench problems can be reduced either by diluting the sample with water (if practicable) or by using a cocktail, which is more resistant to color quenching, e.g., any of the Meridian ProSafe+ and Gold Star, PerkinElmer Ultima Gold and OptiPhaseHiSafe cocktails. Of the classical solvent-based cocktails, Insta-Gel Plus (10 mL) can accommodate greater than 2.5 mL of certain sample types and forms a stable gel (usually at greater than 3 mL sample volume), thus making large sample volumes possible. Pico-Fluor 40 (10 mL) can accommodate greater than 2.5 mL of certain samples and remains in a single liquid phase. Cocktails suitable for these samples are shown in Table 7.3.

C. Medium-ionic-strength buffers

Aqueous buffer concentrations encountered in this group range from 0.1 to 0.5 M, and typical buffers are 0.1–0.5 M PBS, 0.15–0.5 M NaCl, 0.25 M ammonium acetate, etc. The cocktails suitable for these sample types are shown in Table 7.4.

The cocktails are designed to overcome any phase instability problems, and therefore, cocktail selection is limited by volume and concentration factors. The Gold Star, ProSafe+, and Ultima Gold families will, in general, give higher quench resistance than the classical cocktails, i.e., higher efficiency at the same sample load. Gold Star, ProSafe+, and Ultima Gold work well with low sample volumes of aqueous buffers up to 0.5 M. Although Insta-Gel Plus will accept small volumes of certain sample types, Pico-Fluor Plus is the recommended classical cocktail for these samples.

D. High-ionic-strength buffers

With high-ionic-strength buffers, the choice of cocktails includes Gold Star, ProSafe HC+, Ultima Gold XR, Optiphase Hi-Safe 3, LumaSafe Plus, Hionic-Fluor, or Pico-Fluor Plus. Certain high-ionic-strength samples can be accommodated in other cocktails; however, the capacity is usually very low (less than 0.5 mL). The only other method of overcoming the problem of low sample acceptance of high-ionic-strength samples is to dilute the buffer sample with water and convert it into a medium-ionic-strength buffer, which simplifies cocktail selection. The sample capacity of selected cocktails for increasing ionic-strength solutions is shown in Table 7.5.

E. Acids

This group includes commonly encountered mineral acids such as hydrochloric acid, nitric acid, sulfuric acid, perchloric acid, orthophosphoric acid, and hydrofluoric acid as well as some aqueous miscible organic acids such as

TABLE 7.4 Cocktail Selection for Medium-Ionic-Strength Samples (based on the use of 10 mL of cocktail).

	Safer cocktail	Classical cocktail
Low sample volume (0–2.5 mL)	Ultima Gold, XR, AB, LLT, Gold Star, Gold Star LT,[2]ProSafe+ range, and other equivalent cocktails	Insta-Gel Plus (Insta-Gel XF), Pico-Fluor Plus, Hionic-Fluor, and other equivalent cocktails
High sample volume (>2.5 mL)	Ultima Gold XR, Gold Star, ProSafe HC+, and other equivalent cocktails	Pico-Fluor Plus, Hionic-Fluor, and other equivalent cocktails

TABLE 7.5 Sample Capacity of Selected Cocktails for Various-Ionic-Strength Buffers (sample capacities are for 10 mL cocktail at 20°C).

Ionic strength	Ultima Gold XR (mL)	Hionic-Fluor (mL)	Ultima Gold (mL)	Gold Star (mL)	ProSafe HC+ (mL)
0.5M NaCl	9.0	1.4	1.5	10.0	6.5
0.75M NaCl	6.5	2.25	0.7	10.0	3.6
1.0M NaCl	5.5	8.5	0.5	10.0	2.5

acetic acid, formic acid, and trichloroacetic acid (TCA). Acids are commonly used as extractants (Thomson and Burns, 1996a), pH modifiers, and solubilizers (Thomson and Burns, 1996b). There are a number of potential problems associated with this particular sample group, and these include quenching, reaction with cocktail components, and chemiluminescence. Strong mineral acids can also cause marked quenching effects primarily due to interaction with the scintillators. This can be overcome by using a cocktail that is known to be compatible with mineral acids or, preferably, by diluting the acid with water prior to the addition of the cocktail. Certain strong mineral acids will react with cocktail components causing both color development and changes in surfactant characteristics. For example, adding even small amounts of concentrated sulfuric acid to a cocktail will result in almost immediate color formation and eventual sulfonation of the surfactants (emulsifiers). This alteration to the surfactants will result in a change or loss of emulsifying properties and lead to phase instability. The color formation is due to sulfonation of minor impurities in the solvent and, in addition to color, significant amounts of chemiluminescence may be produced. Another example involves adding small amounts of concentrated nitric acid, which results in the formation of a yellow/brown color, due to the dissociation of nitric acid and release of NO_2. This problem can be overcome by diluting the acid with water prior to adding it to the cocktail. With some cocktails, the addition of TCA can produce chemiluminescence. Although a rare occurrence, acid-induced chemiluminescence can be avoided by using a cocktail that is resistant to it such as Gold Star LT^2 or Ultima Gold LLT. An overview of suitable cocktails for acids is shown in Table 7.6.

F. Alkalis

This group of samples includes bases such as sodium hydroxide, potassium hydroxide, and ammonium hydroxide. Alkaline samples are produced from applications involving pH modification, cell lysis, CO_2 trapping (Thomson and Burns, 1994), and solubilization. The major problem normally encountered is chemiluminescence, and in general, the amount of chemiluminescence is influenced by both the volume and concentration of alkali added. The standard method of avoiding this problem is to use a cocktail that is known to be resistant to chemiluminescence. Other methods of overcoming the problem include diluting the base with water to reduce the effect, allowing the chemiluminescence to decay in the dark before counting, and neutralizing the base with acid. Prolonged storage of cocktails with alkalis present is not recommended due to the potential for color formation. Where possible, counting should be performed within 1 or 2 days.

G. Other aqueous sample types

This final group covers other aqueous samples/mixtures that are occasionally used in LSC methods and assays.

TABLE 7.6 Cocktail selection for acids (based on the use of 10 mL of cocktail).

Acid	Concentration	Safer cocktail	Classical cocktail
Mineral acids	0–2M	Ultima Gold AB, LLT, Gold Star, Gold Star LT,^{2}ProSafe+, and other equivalent cocktails	Insta-Gel Plus (Insta-Gel XF*), Pico-Fluor Plus, Hionic-Fluor, and other equivalent cocktails
	>2M	Ultima Gold AB, LLT, Gold Star LT^2, and other equivalent cocktails	
Trichloroacetic acid	0%–20%	Ultima Gold LLT, Gold Star LT^2, and other equivalent cocktails.	Hionic-Fluor and other equivalent cocktails

These aqueous mixtures are usually fairly specific for certain types of assays, for example, sucrose gradients in DNA and RNA separation, urea as a denaturing and reducing buffer, and imidazole as a biological buffer. Gold Star, ProSafe+, Ultima Gold, and OptiPhaseHiSafe2 are the cocktails of choice for these sample types. A slightly expanded compilation of one manufacturer's cocktails for different sample types is shown in Table 7.7 (Safer Cocktails), Table 7.8 (Safer Cocktails for Acids), and Table 7.9 (Classical Cocktails). Other available cocktails are suitable for most of these sample types, and the correct selection can be made using the cocktail equivalents table (Table 7.2). This may prove useful in providing a basic guide to cocktail selection. To further help with cocktail selection, sample capacities are presented to help with both the selection and suitability of a cocktail.

H. Selection and suitability of a cocktail based on ionic strength

1. Determine the approximate ionic strength using the sample molarity,

$$\text{Mixture Molarity} = \frac{[(v_a \times m_a) + (v_b \times m_b) + \ldots]}{(v_a + v_b + \ldots)}$$

where v_a is the volume of solution A, m_a is the molarity of solution A, v_b is the volume of solution B, and m_b is the molarity of solution B.

Example: 10.0 mL of 0.2 M NaCl (Sol^n A) added to 2.0 mL of 0.5 M KH_2PO_4 (Sol^n B):

$$\text{Mixture Molarity} = \frac{[(10 \times 0.2) + (2 \times 0.5)]}{(10 + 2)}$$

$$= 0.25 \text{ M (a medium molarity sample)}$$

2. Select appropriate cocktail based on ionic strength.
3. Test for sample/cocktail compatibility.
   a. Dispense 10.0 mL cocktail into a 20 mL glass vial. The use of a glass vial allows a clear view of the mixture.
   b. Add the desired sample volume, cap, and shake thoroughly.
   c. If the mixture is clear, proceed with the stability test.
   d. If the mixture is cloudy or hazy, try increasing the cocktail volume and/or decreasing the sample volume. If the mixture does not clear, select a cocktail that can accept higher-ionic-strength samples such as Gold Star, ProSafe HC+, Ultima Gold XR, OptiPhaseHiSafe 3, or an equivalent cocktail.
   e. If the mixture separates into two distinct phases (such as oil and water) or is milky, select a cocktail that can accept higher-ionic-strength samples.
4. Test for stability: Use a mixture that has passed the sample/cocktail compatibility test. Allow it to stand at the LSC temperature for a minimum of 2 hours or the proposed count time for the sample, if this is greater than 2 hours. If the mixture remains stable, successful counting will be possible.
5. Count.

Note: Once the proper proportions and stability of the sample/cocktail mixture have been established, plastic vials can be considered for routine counting.

IV. Solubilization[2]

In its simplest terms, solubilization is the action of certain chemical reagents on the chemical bonds of a macromolecular structure (such as animal or plant tissue) that affects a structural breakdown (or digestion) into smaller, simpler subunits, which can then be directly dissolved in a liquid scintillation cocktail (Kobayashi and Maudsley, 1974a,b). The tissue sample may be whole, homogenized, macerated, or in some other state of subdivision prior to solubilization. When the digested samples are added to an appropriate liquid scintillation cocktail, they should yield clear, colorless, homogeneous liquids exhibiting a minimum of quench, a minimum of chemiluminescence, and a maximum of counting stability. The chemical reagents used should be capable of rapid and complete digestion with respect to both small and large sample sizes and should not require any complex care or methodology. In addition, the combination of reagents and the method of digestion should allow an accurate determination of the isotopic content with a minimum of systematic error. Solubilizers are predominantly used for the traditional animal metabolism studies and more recently have been increasingly used in cell and tissue culture applications. Another area of growing interest is the fate of radionuclides in the environment; in this field of interest, solubilizers have been found to be an invaluable tool in sample preparation. Fundamentally, there are still only three major classifications of solubilizing reagents, and these are as follows:

1. alkaline systems (e.g., Soluene-350, GoldiSol, and Solvable & AquiGest),
2. acidic systems (e.g., perchloric acid), and
3. other systems (e.g., sodium hypochlorite).

A. Systems

The mode of action of "alkaline systems" is solubilization by hydrolysis, and a wide range of samples including animal tissues, blood, urine, bone tissue, muscle, amino acids, nucleic acids, and proteins can be digested with these

2. ©1998–2019, PerkinElmer Inc. Printed with permission.

TABLE 7.7 Performance of Safer Cocktails with Various Aqueous Buffers at 20°C (based on theuse of 10 mL of cocktail). Dotted lines indicates no or very limited sample capacity.

Sample	Ultima Gold (mL)	Ultima Gold XR (mL)	Ultima Gold AB (mL)	Ultima Gold LLT (mL)	Ultima Gold MV (mL)	Gold Star (mL)	Gold Star LT^2 (mL)	ProSafe+ (mL)
Deionized water	3.2	10.0	10.0	10.0	2.0	10.0	10.0	3.7
0.01M PBS (pH 7.4)	6.5	10.0	8.5	8.0	4.0	10.0	9.5	7.0
0.1M PBS (pH 7.4)	4.0	8.5	<0.2	<0.2	3.0	10.0	5.5	4.2
0.5M PBS (pH 7.4)	0.5	1.25	–	–	<0.5	9.5	<0.3	7.0
0.05M Tris-HCl (pH 7.4)	4.5	10.0	10.0	9.0	2.75	10.0	10.0	4.5
0.15M sodium chloride	6.5	10.0	7.5	6.5	5.0	10.0	9.5	6.5
0.5M sodium chloride	1.5	9.0	8.0	6.0	1.25	10.0	6.0	3.0
1.0M sodium chloride	0.5	5.5	4.5	3.5	0.5	10.0	5.2	1.0
0.04M NaH_2PO_4 (pH 7.4)	8.0	10.0	0.75–8.0	0.7–7.0	2.25	10.0	No data	3.0
0.1M NaH_2PO_4 (pH 4.9)	10.0	10.0	8.0	8.0	7.0	10.0	7.5	7.5
0.2M NaH_2PO_4 (pH 4.9)	3.5	10.0	1.0–6.5	1.0–5.0	2.75	10.0	5.0	4.2
0.25M ammoniumacetate	3.25	8.0	5.0	5.5	3.0	10.0	7.0	6.0
0.1M ammoniumsulfate	3.25	10.0	1.0–7.0	1.0–5.5	2.25	No data	No data	No data
0.1M sodium sulfate	4.25	10.0	–	–	3.25	10.0	–	5.5
0.1M hydrochloric acid	6.5	7.0	10.0	10.0	4.5	10.0	8.5	4.7
10% TCA	3.0	7.0	4.5	4.0	1.5	No data	No data	No data
20% TCA	2.0	5.0	3.0	3.0	0.5	No data	No data	No data
0.1M sodium hydroxide	2.5	10.0	10.0	7.5	1.5	10.0	10.0	2.7
1.0M sodium hydroxide	3.0	1.0	–	–	1.75	10.0	<0.2	2.5
0.1M imidazole (pH 7.4)	10.0	10.0	10.0	10.0	2.0	No data	No data	No data
8M urea	1.0	2.5	2.0	3.5	0.5	No data	No data	No data

TABLE 7.8 Recommended Safer Cocktails for Mineral Acids (based on the use of 10 mL of cocktail). Dotted lines indicates no or very limited sample capacity.

Sample	Ultima Gold (mL)	Ultima Gold XR (mL)	Ultima Gold AB (mL)	Ultima Gold LLT (mL)	Gold Star LT[2] (mL)	ProSafe HC+ (mL)
0.1M HCl	6.5	7.0	10.0	10.0	8.5	10.0
1.0M HCl	0.5	2.5	5.5	5.0	10.0	10.0
2.0M HCl	–	1.0	2.25	3.0	3.5	5.2
5.0M HCl	–	<0.5	2.0	1.5	2.0	2.5
1.0M HNO_3	–	2.5	3.25	3.5	4.5	10.0
2.0M HNO_3	0.5	2.0	2.25	2.5	2.5	9.0
3.0M HNO_3	–	1.0	2.0	2.25	2.5	6.5
1.0M H_2SO_4	–	0.25	6.5	7.0	10.0	3.5
2.0M H_2SO_4	–	–	4.0	4.0	3.5	4.0
1.0M $HClO_4$	2.0	2.0	2.25	2.25	No data	No data
2.0M $HClO_4$	1.5	1.5	2.0	2.5	No data	No data
1.0M H_3PO_4	–	1.5	0.5–10.0	0.5–10.0	10.0	10.0
2.0M H_3PO_4	–	0.5	0.5–4.0	0.5–6.0	4.5	10.0

reagents. With "acidic systems," the sample is oxidized to soluble products by the action of certain strong acids, usually oxidizing acids. Samples such as cartilage, bone, collagen fibers, and dried and hard plant samples can be digested by these reagents. Occasionally, mixed acid reagents and acids with an added oxidizing agent are preferred due to their increased oxidative power. Under "other systems," a number of different reagents can be considered; however, the most useful reagent is sodium hypochlorite whose mode of action is by the process of oxidative bleaching. This is particularly useful when dealing with plant samples, especially those containing chlorophyll, where the sodium hypochlorite effectively prevents color quench in subsequent LSC by bleaching out all of the color present. It is not possible to cover the use of every solubilizer; the intention, therefore, is to focus on the most commonly used solubilizers and their usefulness for LSC applications. This section will identify those sample types that are routinely encountered in solubilization work and will offer helpful hints on sample preparation as well as recommending suitable reagents.

B. Sample preparation methods

The following sample preparation techniques, using the reagents detailed in Table 7.10, were carried out using high-performance glass vials. All ^3H counting efficiencies presented were determined using a Tri-Carb_3100 with 67% absolute ^3H efficiency (sealed argon purged standard) operating at 19°C.

1. Whole tissue

The method of solubilizing whole tissue is relatively straightforward and, apart from color formation with certain tissue types, no major problems should be encountered during sample preparation and counting. Although this section only mentions GoldiSol, Soluene-350, AquiGest, and Solvable, it is also possible in certain cases to use hyamine hydroxide.

2. Muscle (50–200 mg)

The method for processing muscle samples is shown below, and typical results are illustrated in Table 7.11. The choice of LSC cocktail influences the maximum sample size that can be processed.

Procedure:

1. Place selected sample size in a 20-mL glass scintillation vial.
2. Add an appropriate volume of solubilizer (1–2 mL depending on sample size).
3. Heat in an oven or water bath at 50–60°C for the specified time with occasional swirling.
4. Cool to room temperature and add 10 mL of a selected cocktail.
5. Temperature and light adapt for at least 1 hour before counting.

3. Liver

The method for processing liver samples (50–100 mg) is shown below, and typical results obtained are illustrated in

TABLE 7.9 Performance of Classical Cocktails with Various Aqueous Buffers at 20°C (based on the use of 10 of cocktail). Dotted lines indicates no or very limited sample capacity.

Sample	Insta-Gel XF/Plus (mL)	Pico-Fluor 15 (mL)	Pico-Fluor 40 (mL)	Hionic-Fluor (mL)
Deionized water	0–1.7 2.9–10.0	1.6	2.3	1.2
0.01M PBS (pH 7.4)	0.2–1.6 3.1–10.0	2.9	10.0	1.4
0.1M PBS (pH 7.4)	1.0–2.0	<0.25	6.4	1.6
0.5M PBS (pH 7.4)	–	–	2.0	7.0
0.05M Tris-HCl (pH 7.4)	0–1.8 3.0–10.0	2.0	3.0	3.0
0.15M sodium chloride	0–1.8 4.9–10.0	4.0	10.0	1.1
0.5M sodium chloride	0–2.1 3.0–10.0	1.2	3.0	1.4
1.0M sodium chloride	0.4–7.0	<0.3	2.3	8.5
0.04M NaH_2PO_4 (pH 7.4)	0.3–1.9 3.0–10.0	2.0	4.0	1.75
0.1M NaH_2PO_4 (pH 4.9)	0.6–2.0	6.0	10.0	1.75
0.2M NaH_2PO_4 (pH 4.9)	1.0–2.0 4.0–0.0	2.0	7.1	1.75
0.25M ammoniumacetate	0–1.75 3.5–10.0	1.5	5.0	1.75
0.1M ammoniumsulfate	0.5–1.75 3.5–10.0	2.5	8.5	1.75
0.1M sodium sulfate	–	2.5	8.5	1.75
0.1M hydrochloric acid	0–1.5 3.0–10.0	1.8	7.25	1.3
10% TCA	0–2.1 (no gel phase)	1.5	4.0	1.5
20% TCA	0–3.75 (no gel phase)	1.25	3.25	4.5
0.1M sodium hydroxide	0–2.0 3.0–10.0	3.3	2.5	1.2
1.0M sodium hydroxide	–	0.5	4.5	1.2
0.1M imidazole (pH 7.4)	0–1.75 3.0–10.0	3.0	5.5	1.75
8M urea	0–1.5[a]	1.0*	2.5	1.0

[a]*Clears only after extended agitation.*

Table 7.12. As before, the choice of LSC cocktail influences the maximum sample size that can be processed.

Procedure:

1. Place selected sample size in a 20 mL glass scintillation vial.
2. Add 1–2 mL of solubilizer.
3. Heat in an oven or water bath at 50–60°C for the specified time with occasional swirling.
4. Cool to room temperature.
5. Add 0.2 mL of 30% hydrogen peroxide in two aliquots of 0.1 mL, with swirling between additions. Allow any reaction to subside between additions of the hydrogen peroxide.
6. Heat again at 50–60°C for 30 minutes to complete decolorization.
7. Add 10 mL of a selected cocktail, then temperature and light adapt for at least 1 hour before counting.

Notes: Solubilization of liver always results in highly colored samples due to the presence of bilirubin. The above

TABLE 7.10 Characteristics of Solubilizers[a].

Reagent	Type	Concentration	Flash point	Density (g/mL)	Warning
GoldiSol	Alkaline#	~0.5M in IPA	12°C	0.83	Corrosive, flammable
Soluene-350	Alkaline#	~0.5M in toluene	5°C	0.88	Corrosive, flammable
Hyamine hydroxide	Alkaline#	1.0M in toluene	11°C	0.93	Corrosive, flammable, toxic
Solvable and AquiGest	Alkaline	0.4M in water	–	1.02	Corrosive
Perchloric acid	Acidic	70%	–	1.70	Corrosive
Nitric acid	Acidic	68%–70%	–	1.42	Corrosive
Sodium hypochlorite	Other	~7% chlorine	–	1.16	Corrosive
Hydrogen peroxide	Other	30% (100 vols)	–	1.11	Corrosive

[a]*Quaternary ammonium hydroxide type.*

TABLE 7.11 Reagents for solubilization and liquid scintillation counting (LSC) of muscle.

Sample size (mg)	Solubilizer (1.0 mL used)	Digestion time at 50–60°C	Sample appearance	LSC cocktail	^{3}H counting efficiency
50–200	Soluene-350 and GoldiSol	1½–4 hours	Clear	Hionic-Fluor	41%–33%
50–200	Soluene-350 and GoldiSol	1½–4 hours	Clear	Ultima Gold/Gold Star	49%–37%
50–150	Soluene-350 and GoldiSol	1½–4 hours	Clear	Pico-Fluor	42%–36%
50–200	Solvable and AquiGest	2–3½ hours	Clear	Hionic-Fluor	42%–33%
50–200	Solvable and AquiGest	2–3½ hours	Clear	Ultima Gold/Gold Star	48%–41%
50–150	Solvable and AquiGest	2–3½ hours	Clear	Pico-Fluor	39%–34%

TABLE 7.12 Reagents for solubilization and liquid scintillation counting (LSC) of liver.

Sample size (mg)	Solubilizer (1.0 mL used)	Digestion time at 50 –60°C	Sample Appearance[a]	LSC cocktail	^{3}H counting efficiency
100	Soluene-350 andGoldiSol	4 hours	Clear, yellow	Hionic-Fluor	15%
50	Solvable andAquiGest	1 hour	Clear, pale yellow	Ultima Gold	47%
100	Solvable andAquiGest	1½ hours	Clear, pale yellow	Hionic-Fluor	23%

[a]*Appearance after decolorization with hydrogen peroxide.*

work was therefore restricted to a viable sample size (which should ideally not exceed 75 mg). In our experience, AquiGest or Solvable has proved to be better than GoldiSol and Soluene-350 for this particular sample type, mainly due to more rapid solubilization and lower final color.

4. Kidney, heart, sinew, brains, and stomach tissue

The method for processing the above five sample types is shown below, and typical results are illustrated in Table 7.13. As previously stated, the choice of LSC cocktail influences the maximum sample size, which can be processed.

Procedure:

1. Place selected sample size in a 20-mL glass scintillation vial.
2. Add 1–2 mL of solubilizer.
3. Heat in an oven or water bath at 50–60°C for the specified time with occasional swirling.
4. Cool to room temperature.
5. Add 0.2 mL of 30% hydrogen peroxide in two aliquots of 0.1 mL with swirling between additions. Allow any reaction to subside between additions of hydrogen peroxide.
6. Heat again at 50–60°C for 30 minutes to complete decolorization.
7. Add 10 mL of selected cocktail, then temperature and light adapt for 1 hour before counting.

5. Feces

The digestion of feces (Morrison and Franklin, 1978) strongly depends on the type of animal. It is possible to use both GoldiSol & Soluene-350 and AquiGest & Solvable. However, there can be problems with residual color and incomplete digestion due to the presence of cellulose-type material present in feces from species such as rabbit. As an alternative, the use of a sodium hypochlorite solution is recommended. Sodium hypochlorite resolved a problem for one researcher (unpublished work), who was attempting to digest guinea pig feces. Sodium hypochlorite substantially digested this sample rapidly, and isotope recoveries of greater than 98% for ^{3}H

TABLE 7.13 Reagents for solubilization and liquid scintillation counting (LSC) of various tissues.

Sample size	Solubilizer (1.0 mL used)	Digestion time at 50–60°C	Sample Appearance[a]	LSC cocktail	^{3}H counting efficiency
Kidney					
50–100 mg	Soluene-350 and GoldiSol	1½–5 hours	Clear, pale yellow	Hionic-Fluor and Gold Star	41%–34%
50–100 mg	Solvable and AquiGest	1–2 hour	Clear, pale yellow	Hionic-Fluor and Gold Star	40%–38%
Heart					
50–100 mg	Soluene-350 and GoldiSol	2–3 hour	Clear, pale yellow	Hionic-Fluor and Gold Star	40%–38%
50–100 mg	Solvable and AquiGest	1–3 hour	Clear, pale yellow	Hionic-Fluor and Gold Star	40%–38%
Sinew					
50–100 mg	Soluene-350 and GoldiSol	1–4 hour	Clear, pale yellow	Hionic-Fluor and Gold Star	44%–38%
50–100 mg	Solvable and AquiGest	1–2 hour	Clear, pale yellow	Hionic-fluor and Gold Star	42%-39%
Brain					
50–100 mg	Soluene-350 and GoldiSol	1½–2 hour	Clear, pale yellow	Hionic-Fluor and Gold Star	43%–41%
50–100 mg	Solvable and AquiGest	1–2 hour	Clear, pale yellow	Hionic-Fluor and Gold Star	42%–40%
Stomach[b]					
50–100 mg	Soluene-350 and GoldiSol	1½–3 hour	Clear, pale yellow	Hionic-Fluor and Gold Star	41%–39%

[a]*Appearance after decolorization with hydrogen peroxide.*
[b]*Note: It is not possible to digest stomach tissue with Solvable.*

were achieved. This recovery level was confirmed by combustion in a sample oxidizer (PerkinElmer Tri-Carb Sample Oxidizer, Model 307). The solubilization method used for processing this feces sample follows.

Procedure:

1. Weigh 50–150 mg of feces into a 20-mL glass scintillation vial.
2. Add 0.5 mL of sodium hypochlorite solution and cap tightly.
3. Heat in an oven or water bath at 50–55°C for about 30–60 minutes with occasional swirling.
4. Cool to room temperature.
5. Remove the cap and blow out any remaining chlorine using a gentle stream of air or nitrogen.
6. Add 15 mL of Gold Star or Hionic-Fluor and shake to form a clear mixture.
7. Temperature and light adapt for 1 hour before counting.

Note: After digestion, a small amount of white residual matter may remain; however, this should not affect the recovery.

6. *Blood*

The successful preparation of blood samples (see Moore, 1981) for LSC can often be technically difficult, and successful digestion can be largely dependent on the practical experience of the researcher. The source of blood and the correct choice of solubilizer also influence the results of digestion. Consequently, methods are given for both GoldiSol & Soluene-350 and AquiGest & Solvable, and the final choice of method rests with the individual researcher. Some typical results, obtained in an independent laboratory, are shown in Table 7.14.

GoldiSol and Soluene-350 method

Procedure:

1. Add a maximum of 0.4 mL of blood to a glass scintillation vial.
2. Add, while swirling gently, 1.0 mL of a mixture of either GoldiSol or Soluene-350 and isopropyl alcohol (IPA) (1:1 or 1:2 ratio). Ethanol may be substituted for the IPA if desired.
3. Incubate at 60°C for 2 hour. The sample at this stage will be reddish-brown.
4. Cool to room temperature.
5. Add 0.2–0.5 mL of 30% hydrogen peroxide dropwise or in small aliquots. Foaming will occur with Soluene-350 after each addition; hence, gentle agitation is necessary. With GoldiSol, there will only be minimal foaming. Keep swirling the mixture until all foaming subsides and then continue swirling until all of the hydrogen peroxide has been added.
6. Allow the sample to stand for 15–30 minutes at room temperature to complete the reaction.
7. Cap the vial tightly and place in an oven or water bath at 60°C for 30 minutes. The sample at this stage should now have changed to pale yellow.
8. Cool to room temperature, and add 15 mL of Gold Star or Hionic-Fluor.
9. Temperature and light adapt for 1 hour before counting.

AquiGest and Solvable method

Procedure:

1. Add a maximum of 0.5 mL blood to a glass scintillation vial.
2. Add 1.0 mL AquiGest or Solvable.
3. Incubate the sample at 55–60°C for 1 hour. Sample at this stage will be brown/green in appearance.
4. Add 0.1 mL of 0.1 M EDTA-disodium salt solution, which helps reduce foaming when the subsequent hydrogen peroxide is added.
5. Add 0.3–0.5 mL of 30% hydrogen peroxide in 0.1 mL aliquots. Gently agitate between additions to allow reaction foaming to subside.
6. Allow the sample to stand for 15–30 minutes at room temperature to complete the reaction.
7. Cap the vial tightly and place in an oven or water bath at 55–60°C for 1 hour. The color will change from brown/green to pale yellow.
8. Cool to room temperature, and add 15 mL of Gold Star or Pico-Fluor Plus.
9. Temperature and light adapt for 1 hour before counting.

7. *Plant material*

There are two main problems associated with the digestion of plant material: the presence of pigments (especially chlorophyll) and the difficulty of digesting cellulose. Some

TABLE 7.14 Reagents for solubilization and liquid scintillation counting (LSC) of blood.

Sample size	Solubilizer (1.0 mL used)	Sample Appearance[a]	LSC cocktail	^{3}H counting efficiency
0.2–0.4 mL	Soluene-350 and GoldiSol	Clear, pale yellow	Hionic-Fluor/Gold Star	29%–19%
0.25–0.5 mL	Solvable and AquiGest	Clear, pale yellow	Hionic-Fluor/Gold Star	37%–27%
0.5 mL	Solvable and AquiGest	Clear, pale yellow	Pico-Fluor plus	~29%
0.2 mL	Solvable and AquiGest	Clear, pale yellow	Ultima Gold/Gold Star	~44%

[a]*Appearance after decolorization with hydrogen peroxide.*

of the colored pigments can be bleached with hydrogen peroxide, but not all, and therefore, many samples will remain highly colored. The primary problem with cellulose is that it is not soluble in the alkaline solubilizers, and in general, some form of skeletal cellulose material remains after attempted solubilization. To overcome these two problems, a number of different reagent systems have been devised with the overall result that plant material can be processed. Since there is such a wide variety of plant sample types, this section will be confined to describing the solubilization methods together with their associated advantages and drawbacks. GoldiSol and Soluene-350 can be used to solubilize various plant materials, but in general the sample size must be kept small (<50 mg). With such small samples, it is often possible to achieve limited decolorization with hydrogen peroxide; however, color quenching remains a problem and the cellulose is not dissolved. This classifies the use of GoldiSol and Soluene-350 for the digestion of plant material in the "of limited use" category. AquiGest and Solvable are also not ideally suited to solubilize plant materials and suffer from the same drawbacks associated with GoldiSol and Soluene-350. Perchloric acid/nitric acid solution and perchloric acid/hydrogen peroxide solution reagent systems are suitable for digesting samples by the method of "wet oxidation" and are particularly useful for solubilizing samples such as hard and dried plant material. They have also proved useful for the digestion of cartilage, bone, collagen fibers, and even some highly colored samples such as blood and liver. The general methods for these latter two reagents are as described next.

a. Perchloric acid/nitric acid (Wahid et al., 1985)

Procedure:

1. Prepare the solubilizing reagent by adding one volume of 70% perchloric acid to one volume of 70% nitric acid.
2. Where possible, the sample should be oven-dried and then finely cut.
3. Place the prepared sample (up to 200 mg) in a glass scintillation vial fitted with a polycone-lined urea screw cap.
4. Add approximately 0.6 mL of the prepared $HClO_4$/HNO_3 reagent (1:1).
5. Digest the sample in the closed vial at 50–70°C for 1 hour or until an almost colorless solution is obtained.
6. Cool the vial to room temperature, and add 15 mL of Gold Star or Hionic-Fluor.
7. Temperature and light adapt for 1 hour before counting.

b. Perchloric acid/hydrogen peroxide (Sun et al.,1988; Mahin and Lofberg, 1966; Recalcati et al.,1982; Fuchs and De Vries, 1972)

Procedure:

1. Where possible, the sample should be oven-dried and then finely cut.
2. Place the prepared sample (up to 200 mg) in a glass scintillation vial fitted with a polycone-lined urea screw cap.
3. Add 0.2 mL of 70% perchloric acid and swirl gently to completely wet the sample.
4. Add 0.4 mL of 30% hydrogen peroxide and again swirl gently. (This sequence of addition prevents frothing.)
5. Digest the sample in a closed vial at 50–70°C for 1 hour or until an almost colorless solution is obtained.
6. Cool the vial to room temperature, and add 15 mL of Hionic-Fluor.
7. Temperature and light adapt for 1 hour before counting.

Notes (for both methods): With both of these acidic reagent systems, if the isotope label is ^{14}C, there is a potential for loss of the label as radiocarbon dioxide ($^{14}CO_2$) during solubilization. Tritium (^{3}H) losses are virtually prevented due to the formation of $^{3}H_2O$, which condenses inside the vial. Ensure that a polycone insert is used in the cap in place of a foil lined insert, as these aggressive reagents will oxidize the foil (aluminum) insert and may contaminate the digest. Do not heat these mixtures above 90°C; perchloric acid may decompose violently above this temperature. Due to the aggressive nature of these mixtures, the use of gloves is recommended during all handling steps. The perchloric acid/nitric acid method has also been successfully used for the determination of ^{45}Ca and ^{35}S in cartilage and bone (unpublished work).

c. Sodium hypochlorite

It is also possible to use sodium hypochlorite, as reported by Fuchs and De Vries (1972), Porter (1980), and Smith and Lang (1987). Although sodium hypochlorite does not completely dissolve cellulose, it is capable of decolorizing not only chlorophyll but also almost all other pigments found in plant materials. Therefore, provided the radioisotope is not assimilated within the cellulose structure, this reagent should be considered for preparing plant materials for LSC. The advantages of using this reagent for plant solubilization are that it is simple and rapid and does not result in loss of ^{14}C as radiocarbon dioxide gas. In practice, sodium hypochlorite penetrates the plant material and rapidly decolorizes the pigments, leaving behind a white skeleton of cellulose material. The general method is described next.

Procedure:

1. Place the sample (up to 200 mg) in a glass scintillation vial.
2. Add 1.0 mL of sodium hypochlorite solution.
3. Swirl gently until the entire sample has been completely wetted.
4. Cap tightly and place in an oven or water bath at 50–60°C for approximately 1–2 hour.

5. Completeness of digestion is usually indicated by removal of pigmentation and/or the appearance of white skeleton of cellulose.
6. Cool the vial to room temperature, and carefully vent each vial under a fume hood (decolorization is by action of released chlorine and some residual chlorine remains).
7. Blow out any remaining chlorine with a gentle stream of air or nitrogen.
8. Add 15 mL of Gold Star or Hionic-Fluor.
9. Temperature and light adapt for 1 hour before counting.

Note: Sodium hypochlorite is more commonly known as bleach and should have greater than 5% available chlorine if it is to be an effective solubilizer.

8. Electrophoresis gels

Polyacrylamide gel electrophoresis (PAGE) is a technique commonly used for the separation and identification of biological species. Electrophoresis can be best described as the movement of small ions and macromolecules in solution under the influence of an electric field. The rate of migration depends on the size and shape of the molecule, the charge carried, the applied current, and the resistance of the medium. Electrophoresis is carried out in gels cast either in tubes or as slabs. A number of gel materials have been used successfully, including agar, agarose, and polyacrylamide. Agar and agarose gels are prepared by heating the granular material in the appropriate electrolyte buffer, casting the gels, and allowing them to set on cooling. Polyacrylamide gels are made from acrylamide and *N,N*-methylene bisacrylamide (bisacrylamide) mixtures dissolved in electrolyte and polymerized by the addition of chemical catalysts. Cross-linking agents other than bisacrylamide have been used and include DATDA (diallyltartardiamide), BAC (*N,N*-bisacrylylcystamine), and ethylene diacrylate (Choules and Zimm, 1965). The location of the various compounds separated within the gel is determined by staining and by the presence of radioactivity in discrete zones. Sample preparation for analysis by LSC involves either elution of the sample from the gel or complete dissolution of the gel.

a. Gel elution

The complete solution of the gel is not required for satisfactory counting. This is indicated by the observation that when acrylamide gels are treated with a solubilizer, such as GoldiSol or Soluene-350 or their equivalents, the gel swells rapidly and the entrapped macromolecules diffuse out into the solubilizer. The addition of an appropriate cocktail results in a suitable counting medium. During this diffusion process, it is presumed that the macromolecules (proteins) are partially hydrolyzed by the strong organic base (Terman, 1970; Basch, 1968).

After 10−20 hours of digestion with solubilizer at ~45°C of 1 mm gel slices, the undissolved gel can be removed after being counted and no perceptible loss of either ^{3}H or ^{14}C counts or changes in the isotope ratios of the samples is observed (Terman, 1970). Another method of preparing the gel slices for counting is to add the 1−2 mm gel slice to 5 mL water and macerate the slice with the aid of a stirring rod. This mixture is allowed to stand at 50°C for about 2 hours, cooled, and then 10 mL of a suitable cocktail, such as Insta-Gel Plus or equivalent, is added.

b. Gel dissolution

The alternative to elution is the complete dissolution of gels; this can be accomplished using a variety of reagents and these include hydrogen peroxide (Diener and Paetkau, 1972), periodic acid (Spath and Koblet, 1979), 2-mercaptoethanol (Hansen et al., 1980), piperidine (containing EDTA), alkalis, and ammonia. The selection of a suitable reagent depends on the gel, in which the choice of cross-linker dictates the dissolving reagent required.

Acrylamide gels cross-linked with DATDA are dissolved in 0.5 mL of 2% periodic acid in 2 hours at room temperature (Anderson and McClure, 1973), and the digest can be counted in 10 mL of a suitable cocktail such as Insta-Gel Plus, Gold Star, Ultima Gold, OptiPhase HiSafe 2, or equivalents.

Acrylamide gels cross-linked with BAC are dissolved in 2-mercaptoethanol (Hansen et al., 1980). Polyacrylamide gels that are soluble in alkalis can be constructed using ethylene diacrylate as the cross-linker instead of bisacrylamide.

Polyacrylamide gels cross-linked with bisacrylamide can be dissolved by adding the 1−2 mm gel slice to 0.5 mL of 30% (100 vol) hydrogen peroxide and heating at 50°C until the gel dissolves. Some authors use a 1-hour digestion at 55°C (Diener and Paetkau, 1972), overnight at 40°C (Benjamin, 1971), and overnight at 60°C (Dion and Moore, 1972). The hydrogen peroxide also decolorizes the stained gels. When using hydrogen peroxide, cocktail selection is extremely important due to the potential for chemiluminescence. When peroxides are present in an alkaline medium, the chances of chemiluminescence are high. Consequently, after dissolving the gel slice in hydrogen peroxide, it is advisable to add a small amount of 1−2 M HCl to keep the mixture neutral or slightly acidic. In addition, sometimes traces of copper ions have been added to aid the decomposition of the excess hydrogen peroxide, and the presence of these ions intensifies the chemiluminescence in an alkaline medium. Cocktails suitable for this type of sample include Gold Star, Hionic-Fluor, and Aquasol-2 (all acidified with 0.5 mL 2 M HCl per 10 mL cocktail). Certain researchers (Goodman and Matzura, 1971) have criticized the use of hydrogen peroxide and heat on the grounds that during digestion, radioactive carbon dioxide and water may be lost. They proposed the digestion of 1 mm gel slices using 0.25 mL of a mixture containing 1 part of concentrated ammonium hydroxide and 99 parts of 30% hydrogen peroxide at room temperature for 4–8 hours. This method is reported to be free from the danger of loss of radioactivity because of the low-temperature digestion.

Agarose gels can be rapidly and easily dissolved by treatment with sodium hypochlorite solution for about 1 hour at ~45°C. In this instance, cocktail selection is very important, as any residual hypochlorite solution has a tendency to decompose under slightly acidic conditions producing chlorine, and this can lead to chemiluminescence. This chemiluminescence only becomes apparent after standing for >24 hour, and therefore, cocktail selection is important. At present, the recommended cocktail for this application is Hionic-Fluor.

All of the above methods are suitable for the isotopes normally encountered in gel electrophoresis, and in one modification, ^{32}P can be detected and counted in wet gel slices, either alone or covered in buffer, by Cerenkov counting.

As an alternative to elution or dissolution, the gel slices can be combusted using a commercially available oxidizer system. Although only applicable to ^{3}H and ^{14}C, combustion offers a rapid and accurate means of sample preparation free from the chemiluminescence problems associated with most acrylamide gel procedures.

A variety of LSC cocktails, of both high flash point and classical solvent types, are suitable for use with the various solubilizing reagents. The information presented in this section is condensed into a quick reference guide (Table 7.15), and this may prove particularly useful in selecting the most appropriate solubilizer and cocktail for a specific sample type.

V. Combustion

Combustion (sample oxidation) is the complete combustion of a sample using the "open/closed flame combustion technique." This technique is principally used for ^{3}H and ^{14}C. During combustion, the organic portion of the sample is completely converted to water and carbon dioxide, whereas the inorganic portion remains in the particle trap. The respective reactions are as follows:

$$4H + O_2 \rightarrow 2H_2O$$

$$C + O_2 \rightarrow CO_2$$

Since the products of combustion are water and carbon dioxide, problems such as color quenching, self-absorption, and chemiluminescence are eliminated. The two systems operating within a combustion process are (1) water production and incorporation into a cocktail and (2) carbon dioxide production, absorption, and incorporation into a cocktail. On the ^{3}H side, the process is straightforward in that the water formed is incorporated into an emulsifying cocktail such as monophase S. On the ^{14}C side, the basic chemical reaction involves the production of carbon dioxide by the combustion (oxidation) of organic carbon, followed by absorption by an amine. The absorption reaction scheme is that 2 mol of amine reacts with 1 mol of carbon dioxide to form a carbamate. This carbamate is then added to a special cocktail (e.g., CarbonCount or Permafluor E+) for counting. Since the counting quench of the ^{3}H is only water and the quench for ^{14}C is carbamate/CarbonTrap or carbamate/Carbo-Sorb E, the sample size processed can be as much as 2 g for ^{3}H and as much as 1.5 g for ^{14}C-labeled materials.

For low-level ^{3}H counting, multiple combustions are possible with the collection of the combusted material (water) in the same counting vial, and hence, better counting statistics and shorter counting time. There are three principal oxidizer systems in use today, and these are the PerkinElmer Sample Oxidizer, the "Harvey Oxidizer" (also known as R.J. Harvey Biological Oxidizer), and the RaddecPyrolyser. All are able to both combust and separate ^{3}H and ^{14}C quantitatively, but the respective mechanisms by which this is achieved are slightly different. Differences in combustion chamber, plumbing, heating techniques, timing cycles, and reagents account for the main operational differences. Consultation with each manufacturer's product brochures shows how these individual instruments operate. A complete description of the PerkinElmer system, together with details on alternative methods of combustion,is given in *Liquid Scintillation Analysis—Science and Technology* (Kessler, 1989). Combustion can be found in many areas, including biochemistry, metabolic studies, pharmacokinetics, and agrochemical studies. Some typical applications include the determination of ^{3}H and ^{14}C in waste oil and the determination of ^{3}H and ^{14}C in soils. Combustion is the only technique that, prior to counting, can physically separate ^{3}H and ^{14}C, thereby eliminating many of the errors associated with the counting of dual labeled samples.

VI. Comparison of sample oxidation and solubilization techniques[3]

The primary objective of all sample preparation methods is to obtain a stable homogeneous solution suitable for analysis by LSC. There are no absolutes in sample preparation; whichever method produces a sample that lends itself to accurate and reproducible analysis is acceptable. However,

3. ©1998–2019, PerkinElmer Inc. Printed with permission.

TABLE 7.15 Solubilization selection guide.

Sample type	Solubilizer	Maximum sample size	Suitable liquid scintillation counting cocktails
Muscle	Soluene-350 and GoldiSol	150 mg 200 mg	Pico-Fluor Plus, Ultima Gold, Gold Star, or Hionic-Fluor
	Solvable and AquiGest	150 mg 200 mg	Pico-Fluor Plus, Ultima Gold, Gold Star, or Hionic-Fluor
Liver	Soluene-350 and GoldiSol	100 mg	Hionic-Fluor/Gold Star
	Solvable and AquiGest	50 mg 100 mg	Ultima Gold, Gold Star or Hionic-Fluor
Kidney	Soluene-350 and GoldiSol	100 mg	Hionic-Fluor/Gold Star
	Solvable and AquiGest	100 mg	Hionic-Fluor/Gold Star
Sinew	Soluene-350 and GoldiSol	150 mg	Hionic-Fluor/Gold Star
	Solvable and AquiGest	150 mg	Hionic-Fluor/Gold Star
Brain	Soluene-350 and GoldiSol	150 mg	Hionic-Fluor/Gold Star
	Solvable and AquiGest	150 mg	Hionic-Fluor/Gold Star
Stomach	Soluene-350 and GoldiSol	100 mg	Hionic-Fluor/Gold Star
Feces	Soluene-350 and GoldiSol	150 mg	Hionic-Fluor/Gold Star
	Solvable and AquiGest	20 mg	Hionic-Fluor/Gold Star
	Sodium hypochlorite	20 mg	Hionic-Fluor/Gold Star
Blood	Soluene-350 and GoldiSol	0.4 mL	Hionic-Fluor/Gold Star
	Solvable and AquiGest	0.5 mL 0.2 mL	Pico-Fluor Plus, Ultima Gold, Gold Star or Hionic-Fluor
Plant material	Soluene-350 and GoldiSol	<50 mg	Hionic-Fluor/Gold Star
	Solvable and AquiGest	<50 mg	Hionic-Fluor/Gold Star
	$HClO_4/H_2O_2$	200 mg	Hionic-Fluor/Gold Star
	$HClO_4/HNO_3$	200 mg	Hionic-Fluor/Gold Star
	Sodium hypochlorite	200 mg	Hionic-Fluor/Gold Star
Gels			
Bisacrylamide	H_2O_2	1–2 mm slices	Hionic-Fluor + HCl Aquasol-2 + HCl
	Soluene-350 and GoldiSol	1–2 mm slices	Hionic-Fluor
	Solvable and AquiGest	1–2 mm slices	Hionic-Fluor or Ultima Gold
	Water	1–2 mm slices	Insta-Gel XF
DATDA	Periodic acid	1–2 mm slices	Hionic-Fluor or Ultima Gold

there will be occasions when more than one method will be both suitable and available and the selection of either method will depend on other factors. It is precisely this situation that occurs when considering solubilization and sample combustion for sample preparation. Both techniques are routinely used to process samples that are not directly soluble in LSC cocktails. There are numerous sample types that fall into this category, and typical examples include tissue, muscle, kidney, liver, feces, blood, plant material, etc. Many of these samples are encountered in ADME (absorption, distribution, metabolism, and excretion) studies in which the biological behavior and potential toxicological effects of a test substance are investigated.

A. Solubilization

Solubilization is the action of certain chemical reagents on organic materials (such as animal or plant tissue) that affects a structural breakdown (or digestion) into a liquid form that can then be directly dissolved in a liquid scintillation cocktail. Typical solubilizers include organic and inorganic alkalis, which act by the process of alkaline hydrolysis and certain mineral acids that affect solubilization by acidic oxidation. The solubilization process usually involves heating the sample/solubilizer at elevated temperature (40–65°C) for periods ranging from <1 to 24 hour, until a homogeneous mixture is formed. Certain samples that remain colored after solubilization are optionally treated with hydrogen peroxide, and following this, a recommended LSC cocktail is added and the sample is ready for analysis by LSC.

B. What is sample combustion?

The principle of sample oxidation is that the sample is combusted in an oxygen-rich atmosphere, and any hydrogen present is oxidized to water,whereas any carbon is oxidized to carbon dioxide. If tritium is present, then the combustion product will be ^3H_2O, and if ^{14}C is present, then the combustion product will be $^{14}CO_2$. In the PerkinElmer Model 307 Sample Oxidizer, the water is condensed in a cooled coil and then washed into a vial where it is mixed with an appropriate LSC cocktail. The CO_2 is trapped by vapor-phase reaction with an amine, and the resulting product is mixed with an appropriate LSC cocktail. At the end of the combustion cycle, two separate samples (a tritium sample and a ^{14}C sample) are trapped at ambient temperature, thus minimizing cross-contamination. An alternative instrument—the Pyrolyser-Trio (Raddec Ltd., Southampton, UK)—was designed to simultaneously extract $^3H_{total}$ (and ^{14}C) from virtually any type of nuclear and environmental sample. The standard methodology consists of placing weighed samples into silica glass boats that are then inserted into their own silica worktube (1 m long) containing 10 g of platinized alumina catalyst. A laminar flow of air or oxygen is passed over the sample while it is progressively ramped in temperature typically from 50 to 900°C over a period of about 3 h. This controlled oxidation ensures that decomposition is slow and does not lead to any pressure excursions as the sample components decompose and liberate CO_2, hydrocarbons, water, etc. The oxidized combustion products (water/^3H_2O and CO_2) are collected in bubblers from which subsamples are taken for LSC measurement.

C. Advantages and disadvantages

The oxidation process described above can be compared with the steps in a typical solubilization method in Fig. 7.1. As can be seen from Fig. 7.1, there are more steps involved in solubilization as compared with combustion; while this

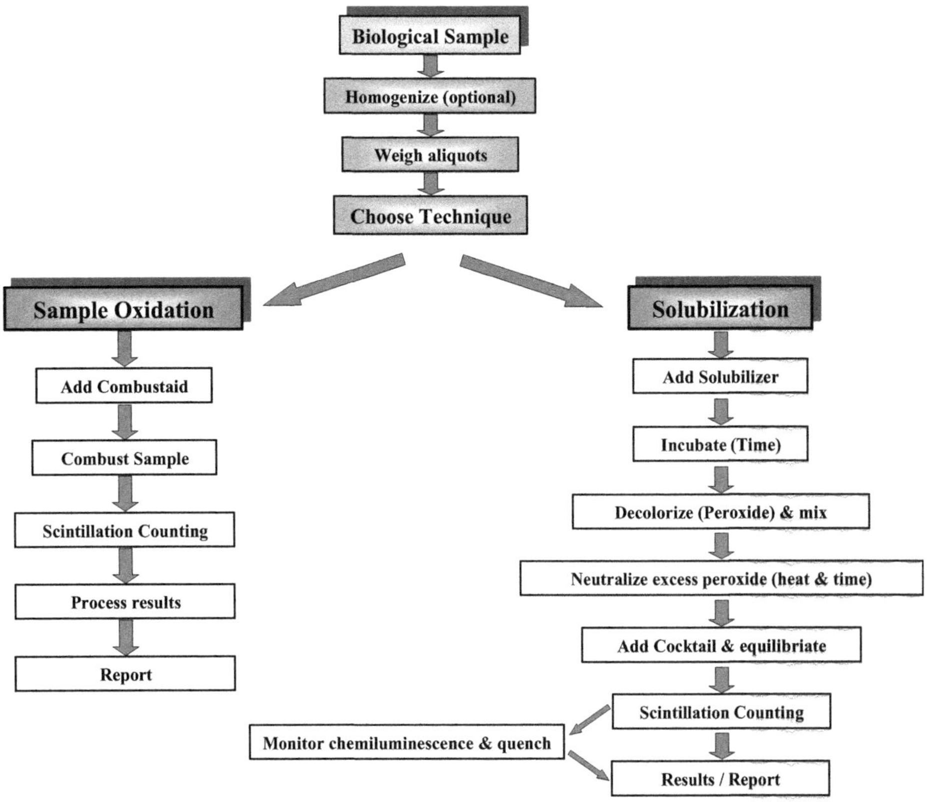

FIGURE 7.1 Sample oxidation and solubilization techniques compared. For the color version of the figure, the reader is referred to the online version of the book.

may seem less attractive, there are other important factors to consider before deciding which methodology is best.

1. Solubilization methods and suitability

The decision to use solubilization as the preferred method of sample preparation depends primarily on the nature of the sample and the number of samples that need to be processed. When the number of samples is low, then solubilization is usually the method of choice. Solubilization is suitable for many organic sample types, but certain of these are prone to problems such as color formation, limited sample size, and time to complete solubilization. These include biological samples such as whole blood, plasma, serum, liver, kidney, fatty tissue, and most plant materials. Many other biological sample types including, muscle, whole tissue, brain, stomach, intestines, nerve cells, cornea, and cartilage can be easily and rapidly processed. There is no major cost involved in setting up to do solubilization. The cost of 1.0 L solubilizer reagent and 10.0 L appropriate LSC cocktail is relatively small, and this is sufficient for 1000 analyses, assuming 1.0 mL solubilizer used with 10.0 mL cocktail. The method is relatively straightforward and is described earlier. To process a sample, simply add the solubilizer and heat at 50−60°C until the sample is dissolved. After solubilization, the sample may be colored and this color can usually be removed or reduced by treatment with hydrogen peroxide. The final step is to add the recommended LSC cocktail, and the sample is ready for counting. Using this method, many samples can be processed simultaneously and then counted sequentially.

a. The typical advantages of solubilization
- Capital outlay is low.
- Homogeneous samples are produced.
- Many samples can be processed simultaneously.
- Color quench is corrected using a quench curve.
- It is suitable for a diversity of isotopes.

b. The disadvantages of solubilization
- Sample sizes are generally $\leq$200 mg.
- Time to complete solubilization can vary from a few hours to many hours.
- Certain samples require modified techniques or longer solubilization times.
- With acidic solubilizers, loss of radioactivity by volatilization may occur.

Solubilization has been in use for many years, and experienced researchers are typically able to optimize the methodologies through attention to detail and judicious selection of reagents and cocktails. Examples of sample types, reagents, and cocktails are shown in Table 7.16.

2. Sample combustion methods and suitability

Since sample combustion is suitable for any organic and even some inorganic samples, the selection of this method is usually governed by the number of samples that need to be processed. When the sample load exceeds 50 per day, then sample oxidation becomes the method of choice for many sample types. In manual mode, the PerkinElmer Model 307 oxidizer can process 100 samples daily, whereas the robotic version can process 240 samples in 8 hours. High-temperature flame combustion at 1300°C enables wet, dry, or freeze-dried samples up to 1.5 g to be processed. For those samples containing dual-label $^{3}H/^{14}C$, the combustion cycle produces single-label ^{3}H and ^{14}C samples in separate vials with no cross-contamination.

a. Advantages of sample combustion
- Sample processing time is rapid.
- Robotic sample processing is possible.
- Sample can be wet, dry, or freeze-dried.
- Any sample containing H and/or C can be combusted.
- It is ideally suited for both single- and dual-label ^{3}H and ^{14}C.
- Sample sizes up to 1.5 g are possible.
- Radioactive recovery is excellent (>97%).
- Memory effect is <0.08%.
- There is no loss of radioactivity by volatilization.
- There is no chemiluminescence interference.
- There is no color quench interference.

b. Disadvantages of sample combustion
- Initial capital investment.
- It is only suitable for ^{3}H and ^{14}C.
- Need a gas supply (oxygen and nitrogen).
- Must be operated in a fume hood.
- Reagents are corrosive and flammable.

The flame combustion technology has proven to be a simple and reliable means of sample preparation and can process a diverse array of samples with a high degree of precision and accuracy. The technique requires a minimal amount of time and sample handling and eliminates any potential interference from color quenching or chemiluminescence. The diversity of samples that can be processed using flame combustion is shown in Table 7.17.

Both solubilization and flame combustion are viable sample preparation techniques for a diverse array of animal and plant tissues. Flame combustion and solubilization procedures each has specific advantages and some disadvantages. Solubilization is more suited to those situations where the sample numbers are low. There can be problems

TABLE 7.16 Sample types that can be processed by solubilization.

Sample type	Solubilizer	Maximumsample size	*Suitable LSC Cocktails*
Muscle	Soluene-350 andGoldiSol	150 mg 200 mg	Pico-Fluor plus Ultima Gold/Hionic-Fluor/Gold Star
	Solvable andAquiGest	150 mg 200 mg	Pico-Fluor plus Ultima Gold/Hionic-Fluor/Gold Star
Liver	Soluene-350 andGoldiSol	100 mg	Hionic-Fluor
	Solvable andAquiGest	50 mg 100 mg	Ultima Gold/Gold Star Hionic-Fluor/Gold Star
Kidney	Soluene-350 andGoldiSol	100 mg	Hionic-Fluor/Gold Star
	Solvable andAquiGest	100 mg	Hionic-fluor/Gold Star
Heart	Soluene-350 andGoldiSol	100 mg	Hionic-Fluor/Gold Star
	Solvable andAquiGest	150 mg	Hionic-Fluor/Gold Star
Sinew	Soluene-350 andGoldiSol	150 mg	Hionic-Fluor/Gold Star
	Solvable andAquiGest	150 mg	Hionic-Fluor/Gold Star
Brains	Soluene-350 andGoldiSol	150 mg	Hionic-Fluor/Gold Star
	Solvable andAquiGest	150 mg	Hionic-Fluor/Gold Star
Stomach	Soluene-350 andGoldiSol	100 mg	Hionic-Fluor/Gold Star
Feces	Hypochlorite	150 mg	Hionic-Fluor/Gold Star
	Soluene-350 andGoldiSol	20 mg	Hionic-Fluor/Gold Star
	Solvable andAquiGest	20 mg	Hionic-Fluor/Gold Star
Blood	Soluene-350 andGoldiSol	0.4 mL	Hionic-Fluor/Gold Star
	Solvable andAquiGest	0.5 mL 0.2 mL	Hionic-Fluor or Pico-Fluor plus Ultima Gold/Gold Star
Plant material	Soluene-350 andGoldiSol	<50 mg	Hionic-Fluor
	Solvable andAquiGest	<50 mg	Hionic-Fluor
	$HClO_4/H_2O_2$	200 mg	Hionic-Fluor
	$HClO_4/HNO_3$	200 mg	Hionic-Fluor
	Hypochlorite	200 mg	Hionic-Fluor

TABLE 7.17 Types that can be processed by flame combustion.

Liver	Whole blood	Bone	Gels
Spleen	Lung	Egg shell	Plastics
Skin	Heart	Plant tissue	Filters
Plasma	Fat	Bacteria	Crude oil
Muscle	Intestines	Insects	TLCs
Kidney	Hair	Glands	Toluene
Brain	Adipose	Water	Synthetic fibers
Feces	Bladder	Urine	Soil

such as color formation, chemiluminescence, and sample size limitations when solubilizing with certain sample types, but by careful attention to detail and experience, these can be overcome. Flame combustion, operated in either manual or robotic format, provides a very powerful tool to rapidly process many different sample types with a high degree of precision and accuracy. The technique requires a minimal amount of time and sample handling and eliminates color quench and chemiluminescence interferences. The only limitation is the initial capital investment and the restriction to ^{3}H and ^{14}C. The PerkinElmer Model 307 manual oxidizer is well suited for sample loads of up to 50 per day.

TABLE 7.18 Trapping capacity of suitable reagents for carbon dioxide.

	mM CO_2 Per mL	mL required for 1 mM CO_2	mL required for 5 mM CO_2	mL required for 10 mM CO_2	Flash point
Sodium hydroxide 0.1M	0.05	20.0	–	–	–
Sodium hydroxide 1.0M	0.50	2.0	10.0	–	–
Hyamine hydroxide 1.0M (in methanol)	0.50	2.0	10.0	–	11°C
Ethanolamine	8.10	0.12	0.62	1.23	93°C
Carbo-Sorb E/CarbonTrap	4.80	0.21	1.04	2.08	27°C

TABLE 7.19 Suitable liquid scintillation counting solutions and capacity for Ethanolamine/CO_2.

Cocktail	Ethanolamine	Methyl cellosolve	mM CO_2trapping capacity
Hionic-Fluor (10.0 mL)	1.0 mL	4.0 mL	8.1 mM
Hionic-Fluor (10.0 mL)	2.0 mL	6.5 mL	16.2 mM
Hionic-Fluor (10.0 mL)	3.0 mL	8.5 mL	24.3 mM

VII. Carbon dioxide trapping and counting[4]

Certain assays result in the generation of discrete gaseous $^{14}CO_2$ samples (not originating from combustion), and these require a modified approach for the trapping of the gas and subsequent LSC (Qureshi et al., 1985; Schadewaldt et al., 1983; Pfeiffer et al., 1981; Riffat et al., 1985). $^{14}CO_2$ gas samples originate from a variety of sources, including $^{14}CO_2$ in expired breath (Bird, 1997), expired by plants, expulsion from blood (Kaczmar and Manet, 1987), and release in enzymatic studies (Sissons, 1976). There are a number of potentially useful reagents available for trapping carbon dioxide, and some of these are shown in Tables 7.18 and 7.19.

A. Sodium hydroxide

Sodium hydroxide absorbs/traps CO_2 by a reaction that produces a sodium carbonate solution. Potassium hydroxide performs in a similar way by forming potassium carbonate. Some of the recommended LSC cocktails for use with this mixture include the following:

- Emulsifier Safe, which accepts up to 2 mL of 0.1 M sodium hydroxide/CO_2 in 10 mL cocktail;
- Opti-Fluor, which accepts up to 5 mL of 0.1 M sodium hydroxide/CO_2 in 10 mL cocktail;
- Hionic-Fluor, which accepts up to 5 mL of 1.0 M sodium hydroxide fully saturated with CO_2 in 10 mL cocktail;
- Ultima-Flo AF, which accepts 10 mL of 0.5 M NaOH/CO_2 or 5 mL of 1.0 M NaOH/CO_2 in 10 mL cocktail; and
- Gold Star, which accepts 5 mL of 1.0 M sodium hydroxide fully saturated with CO_2 in 10 mL cocktail.

Notes: Hionic-Fluor is suitable for use with hydroxide/carbonate solutions due to its sample capacity for concentrated solutions and alkaline pH. Ideally, the final pH should be above 9 to avoid liberation of trapped CO_2. Cocktails containing mixed surfactant systems, such as Ultima Gold, Gold Star, or Ultima Gold XR, can be used; however, counting should be performed the same day, as these cocktails have the potential for slow release of CO_2 on prolonged storage (characterized by dropping CPM levels). Cocktails containing free-acid PEs such as Aqua-Safe should not be used, as their acid nature will cause rapid loss of ^{14}C as $^{14}CO_2$.

B. Hyaminehydroxide

Chemically, hyamine hydroxide performs similarly to sodium and potassium hydroxide, in which it forms hyamine carbonate on reaction with CO_2 (Bird, 1997). Some of the recommended LSC cocktails include Insta-Fluor Plus,

4. ©1998–2019, PerkinElmer, Inc. Printed with permission.

which will accept up to 7.5 mL of hyamine hydroxide saturated with carbon dioxide in 10 mL cocktail, and Emulsifier Safe, which will accept up to 3 mL of hyamine hydroxide saturated with carbon dioxide in 10 mL cocktail, providing a safer system due to the high flash point of this LSC cocktail.

Note: Foaming of hyamine hydroxide used to absorb carbon dioxide expelled in rat breath has been reported. This can be overcome by the addition of one drop of silicone antifoam per 10 mL of hyamine hydroxide. Both of the above cocktails are resistant to any potential chemiluminescence possible with this system.

C. Ethanolamine

Ethanolamine reacts with CO_2 in a rather different way than previously discussed reagents, in that a carbamate is formed as opposed to a carbonate. The main difference between these two species is that a carbamate is more stable under slightly acidic conditions, whereas a carbonate reacts rapidly with acids to release carbon dioxide. Ethanolamine/CO_2 is notoriously difficult to incorporate into LSC cocktails, and thus the recommended LSC solutions may seem a little unusual. The main system known, which is capable of accepting this unusual reagent, is shown in Table 7.20. The recommended LSC cocktail requires the use of a cosolvent, for example, methyl cellosolve, to facilitate the acceptance or mixing of the ethanolamine/CO_2 with cocktail. There may be systems based on other equivalent cocktails, but at present they are unreported.

D. CarbonTrap and Carbo-Sorb E

CarbonTrap and Carbo-Sorb E will react with CO_2 to form a carbamate as well and were developed to work in the PerkinElmer sample oxidizer Model 307. It is, however, possible to use these reagents as carbon dioxide-trapping agents outside of the Model 307. As with the Model 307, the recommended cocktail is either CarbonCount or Permafluor E+, and the following conditions are recommended:

For ratios of CarbonTrap/Carbo-Sorb E to CarbonCount/Permafluor E+ from 1:10 up to 1:1, maximum saturation of carbon dioxide is possible with no phase separation of the resulting carbamate. Should any minor haziness develop, then either add 2–4 mL methanol to the mixture until a clear solution forms or add extra CarbonTrap/Carbo-Sorb E to the mixture until a clear solution forms. This latter method simply dilutes the absorption

TABLE 7.20 Reference table for CO_2 trapping and liquid scintillation counting(LSC).

CO_2 absorber	mM CO_2 per mL	mL required for 1 mM CO_2	mL required for 5 mM CO_2	mL required for 10 mM CO_2	LSCCocktail	Cocktail Volume	mL of Absorber	Maximum CO_2 capacity
NaOH (0.1M)	0.05	20.0	–	–	Emulsifier Safe Opti-Fluor	15.0 mL 14.0 mL	3.0 mL 7.0 mL	0.15 mM 0.35 mM
NaOH (0.5M)	0.25	4.0	–	–	Ultima-Flo AF	10.0 mL	10.0 mL	2.50 mM
NaOH (1.0M)	0.50 0.50	2.0 2.0	10.0 10.0	–	Hionic-Fluor Ultima-Flo AF Gold Star	14.0 mL 14.0 mL 10.0 mL	7.0 mL 7.0 mL 10.0 mL	3.50 mM 3.50 mM 5.00 mM
Hyamine hydroxide (in methanol)	0.50 0.50	2.0 2.0	10.0 10.0	–	Emulsifier Safe Insta-Fluor	15.0 mL 12.0 mL	4.5 mL 9.0 mL	2.25 mM 4.50 mM
Ethanolamine	8.10	0.12	0.62	1.23	Hionic-Fluor/Methyl cellosolve	10.0 mL/ 8.5 mL	3.0 mL	24.3 mM
Carbo-Sorb E/CarbonTrap	4.80	0.21	1.04	2.08	Permafluor E+	10.0 mL	10.0 mL	48.0 mM

capacity below the critical 90% level. The use of Carbon-Trap/Carbo-Sorb E is not recommended in enzymatic, plant, or human studies due to the corrosive nature of the volatile amine present.

This information is condensed into a quick reference table (Table 7.21), and it may prove particularly useful when the total trapping capacity per standard 20-mL LSC vial is required.

TABLE 7.21 Cocktail selection guide for filter counting.

Filter type		Filter-Count	Ultima Gold F	Ultima Gold MV	Soluene-350/ Hionic-Fluor	Filter-Count/Ultima Gold AB
Glass fiber	Dry	√	√	√		√
	Wet	√		√		√
	Dissolved					
Cellulose nitrate	Dry		√	√		
	Wet			√		
	Dissolved	√				√
Cellulose acetate	Dry	√	√	√		√
	Wet	√		√		√
	Dissolved					
Mixed cellulose	Dry		√	√		
Esters	Wet			√		
	Dissolved	√				√
PVC	Dry		√	√		
	Wet			√		
	Dissolved	√				√
Polyacrylonitrile	Dry	√	√	√		√
	Wet	√		√		√
	Dissolved					
Polycarbonate	Dry	√	√	√		√
	Wet	√		√		√
	Dissolved					
Teflon	Dry	√	√	√		√
	Wet	√		√		√
	Dissolved					
Nylon	Dry	√	√	√		√
	Wet	√		√		√
	Dissolved					
PET	Dry	√	√	√	√	√
	Wet	√		√	√	√
	Dissolved					
Normal paper	Dry	√	√	√	√	√
	Wet	√		√	√	√
	Dissolved					

VIII. Biological samples encountered in absorption, distribution, metabolism, and excretion

ADME studies are a vital part of the comprehensive safety evaluation of a new chemical entity. The objective of this study is to investigate the excretion rate and tissue distribution of the chemical entity and involves the collection and analysis of urine and feces, blood, and major internal organs. Total radioactivity level in each sample is assayed, and the recovered dose is compared with the administered dose. The percentage of distribution of the administered dose in urine, feces, and each major internal organ and the half-time for excretion are calculated. Given that in ADME studies all samples are, or are derived from, biological species, the preparation of such samples has always been treated as a special case, because of the difficulties associated with their preparation for LSC. In general, all biological samples contain components not generally found in any other sample types; therefore, it is of benefit to understand the composition of these samples to fully understand the need for the slightly unusual sample preparation methods. The analysis of any biological sample is usually complicated by its complex structure. The components present in biological samples require that either specialist cocktails are needed or, more usually, modified sample preparation techniques are employed. Therefore, it is of benefit to understand the composition of these samples to fully understand the need for specialist cocktails and, in some cases, the slightly unusual sample preparation methods. Within the context of ADME, the samples generally encountered include urine, feces, blood, serum, plasma, milk, and tissue samples (liver, kidney, brain muscle, etc.), and each will be considered in its own right as a sample species.

A. Urine

Urine is formed in the kidneys. Each minute, approximately 600 mL of blood enters each kidney through the renal artery, and as the blood is filtered, water, glucose, vitamins, amino acids, and salts are reabsorbed back into the bloodstream. Normal urine consists of 95% water, and the remaining ingredients include waste products from the breakdown of protein, hormones, electrolytes, pigments, toxins, and any abnormal components. Thus, the main components of urine are (salt) water and waste products. The major waste product from cells in the body is ammonia, and the major waste product from blood is a broken form of heme called bilirubin. In the liver, each of these is converted into a less hazardous form. Ammonia is converted to urea, and bilirubin is degraded to urobilins. Salt, water, and urea are all colorless, but urobilins (which come from degraded pigments) are yellow. Freshly voided urine is generally clear and pale to deep yellow in color. The yellow color is due to the presence of urochrome, a pigment derived from the breakdown of hemoglobin. The more concentrated the urine, the deeper the yellow color. An abnormal color may be due to certain foods or may be due to the presence of bile pigments or blood in the urine. Some drugs and vitamins may also alter the color of urine. Urine is usually slightly acidic (pH around 6) but may vary from 4.5 to 8.0 depending on the acid/base status of the body. A vegetarian diet, prolonged vomiting, and bacterial infection of the urinary tract all cause the urine to become alkaline. From the point of view of sample incorporation into a cocktail, the components present, which are likely to cause problems, are color, protein and protein breakdown products, and divalent/trivalent ions. Individually, these are relatively easy to cope with, but together they present an unusual problem. Color is present to some degree in every urine sample, and the base effect of such color is quenching. The degree of quenching is related to the color present, and therefore, whenever counting urine samples, it is critical to measure sample activities in DPM (disintegrations per minute). In addition, the volume of sample will also contribute to quenching. Since the color of urine samples is infinitely variable and sample volume can also vary, working in DPM is strongly recommended. This will ensure that the use of the correct quench curves will correct for the effect of quench and report the true activity present in the sample.

1. Sample preparation

Protein and protein breakdown products cause an interesting problem, in which, after the addition of urine to certain cocktails, a wispy precipitate forms, usually just below the surface. This can develop either rapidly (within 1 hour) or slowly (over 1 day). This phenomenon is most probably due to protein unfolding in the presence of the detergents present in the cocktail, and such unfolding causes the proteins to precipitate out, albeit slowly. This effect does not usually affect counting reproducibility, but if it is seen as a potential problem, it can be stopped usually by adding 10%−15% of IPA or ethanol. Simply add 1.0 mL IPA to the 10.0 mL cocktail prior to adding the urine sample. Since count times for urine samples in ADME are not excessive, this phenomenon can generally be dismissed unless very low levels are being measured. For low levels of activity, the count times are much longer, and therefore, stability is of major importance.

Divalent and trivalent ions (e.g., sulfate and phosphate) will adversely affect the microemulsion capability of many cocktails. However, the concentration of these ions in urine is usually so low that they should not cause a problem; however, in combination with the other ions present in urine (urea, sodium, potassium, magnesium, calcium, etc.), there can be problems, especially in concentrated urine samples. If the urine does not mix properly with the cocktail (forms a milky mixture), then such a problem has occurred. To resolve this problem either change to a cocktail that is able to accept more concentrated solutes or dilute the urine sample with distilled or deionized water (urine:water is 1:1). Urine from small animals (e.g., rat and mouse) is usually much more concentrated than that from humans and therefore will cause all the problems mentioned above; it is recommended that the sample volume be kept below 0.5 mL per 10 mL cocktail. Specialist cocktail selection as well as dilution with water may be necessary to obtain a stable counting condition.

One final problem that can occur is chemiluminescence, and this is characterized by a dropping CPM with time. If chemiluminescence is suspected, then resolution usually involves either selecting a specialist cocktail or allowing the chemiluminescence reaction to go to completion,that is, allowing the sample/cocktail mixture to stand for a fixed time period before counting. The determination of this period is straightforward, but a nonactive urine sample is required according to the following:

1. Add cocktail to a glass scintillation vial.
2. Add a urine sample that does not contain radioactivity.
3. Place in the counter and arrange for repeat 2 minute counts.
4. Continue repeat counting until the CPM stabilizes.
5. Calculate the total time taken for CPM stabilization, and this becomes the fixed period the sample needs to stand before starting counting.

Provided attention is paid to the above, the preparation and counting of even difficult urine samples can be successfully completed.

B. Whole blood

Because whole blood contains living cells, blood is alive. Whole blood is living tissue circulating through the heart, arteries, veins, and capillaries carrying nourishment, electrolytes, hormones, vitamins, antibodies, heat, and oxygen to the body's tissues. Whole blood contains red blood cells, white blood cells, and platelets suspended in a watery fluid called plasma. Red blood cells and white blood cells are responsible for nourishing and cleansing the body. Since the cells are alive, they too need nourishment. Vitamins and minerals keep the blood healthy. The blood cells have a definite life cycle, just as all living organisms do. Red blood cells, or erythrocytes, carry oxygen throughout the body. A drop of blood, of the size of a pinhead, contains approximately 5 million red blood cells. They are small biconcave disks without a nucleus and get their red color from an iron-containing protein called hemoglobin, which is the iron-containing oxygen-transport metalloprotein in the red cells of the blood in mammals and other animals. The molecule consists of globin, the apoprotein, and four heme groups, an organic molecule with an iron atom.

1. Sample preparation

Because of its complex nature and extreme color, blood should not be added directly to a cocktail. Such an attempt will result in a highly color-quenched and nonhomogeneous mixture that will not lend itself to accurate or reproducible counting. To prepare blood samples for counting, treatment by solubilization (digestion) is required. As described previously, solubilization is the action of certain chemical reagents on organic materials (such as animal tissue) that affects a structural breakdown (or digestion) into a form that can then be directly incorporated into a liquid scintillation cocktail. The source of blood and the correct choice of solubilizer will significantly influence the results of digestion. In general, most of the sample preparation problems occur with blood samples from smaller animals such as rats and mice. In this case, it may be necessary to consider smaller sample volumes, and even then, there can still be color quench problems. An alternative method that may be considered is sample combustion, but this is only applicable if the isotopes of interest are tritium and/or carbon-14. The principle of sample oxidation is that the sample is combusted in an oxygen-rich atmosphere, and any hydrogen present is oxidized to water, whereas any carbon is oxidized to carbon dioxide. If tritium is present, then the combustion product will be ^3H_2O, and if carbon-14 is present, then the combustion product will be $^{14}CO_2$. In commercially available instruments, the water is condensed and then washed into a vial where it is mixed with an appropriate LSC cocktail. The CO_2 is trapped by reaction with an amine, and the resulting product is mixed with a specialist LSC cocktail. At the end of the combustion cycle, two separate samples (a tritium sample and a carbon-14 sample) are produced at ambient temperature, thus minimizing cross-contamination.

The successful preparation of blood samples for LSC counting by solubilization can often be technically difficult, and successful digestion can be largely dependent on the practical experience of the researcher. To process a sample, simply add the solubilizer and heat at 50–60°C until the sample is dissolved. After solubilization, the sample may be colored, and this color can usually be removed or reduced by treatment with hydrogen peroxide. The final step is to add the recommended LSC cocktail, and the sample is ready for counting. Using this method, many samples can be processed simultaneously and then counted sequentially. Currently available on the market are two types of solubilizers: a quaternary ammonium hydroxide in organic solvents and an inorganic alkali in water with added detergents. Examples of the former include Soluene-

350 (PerkinElmer), GoldiSol (Meridian Biotechnologies Ltd.), and ScintiGest (Fisher Scientific). Soluene-350 is based on a mixture of toluene and methanol, and as such is classified as both toxic and flammable. GoldiSol, on the other hand, is based on a mixture of IPA and water, and it is only classified as flammable. ScintiGest is similar to Soluene-350 and is based on toluene and methanol.

The only two versions of an aqueous-based solubilizer are Solvable (PerkinElmer) and AquiGest (Meridian Biotechnologies Ltd.). The respective stepwise processes for solubilization with these different solubilizers are as follows:

a. Quaternary ammonium hydroxide method (GoldiSol and Soluene-350)
   1. Add a maximum of 0.4 mL of blood to a glass scintillation vial.
   2. Add, while swirling gently, 1.0 mL of a mixture of solubilizer and IPA (1:1 or 1:2 ratio). Ethanol may be substituted for the IPA if desired.
   3. Incubate at 55–60°C for 2 h. The sample at this stage will be reddish-brown.
   4. Cool to room temperature.
   5. *If using GoldiSol solubilizer,* the following treatment with H_2O_2 should be followed:
      Up to 0.5 mL of 30% hydrogen peroxide can be added in one addition as almost no foaming occurs. This significantly reduces the sample preparation time.
      If using Soluene-350 solubilizer, the following treatment with H_2O_2 should be followed:
      Add 0.2–0.5 mL of 30% hydrogen peroxide in 100 μL aliquots. Foaming occurs after each addition; therefore, gentle agitation is necessary. Keep swirling the mixture until all foaming subsides, and then repeat the process until all of the hydrogen peroxide has been added.
   6. Allow to stand for 15–30 minutes at room temperature to complete the reaction.
   7. Cap the vial tightly and place in an oven or water bath at 55–60°C for 30 minutes. The samples at this stage should now have changed to slightly yellow. This is important as heating destroys residual hydrogen peroxide, which, if present, can induce unwanted chemiluminescence.
   8. Cool to room temperature, and add 15 mL of recommended LSC cocktail.
   9. Temperature and light adapt for 1 hour before counting.

b. Aqueous-based solubilizer (AquiGestandSolvable)
   1. Add a *maximum* of 0.4 mL blood to a glass scintillation vial.
   2. Add 1.0 mL *solubilizer.*
   3. Incubate the *sample* at 55–60°C for 1 hour. The sample at this stage will be brown/green in appearance.
   4. Add 0.1 mL of 0.1 M EDTA-disodium salt solution, which helps reduce foaming when the subsequent hydrogen peroxide is added. This reagent also complexes iron that has been released during digestion and produces a mixture that is less colored.
   5. Add 0.3–0.5 mL of 30% hydrogen peroxide in 100 μL aliquots. Slight foaming will occur after each addition; therefore, gentle agitation is necessary. Keep swirling the mixture until all foaming subsides, and then repeat the process until all of the hydrogen peroxide has been added.
   6. Allow to stand for 15–30 minutes at room temperature to complete the reaction.
   7. Cap the vial tightly and place in an oven or water bath at 55–60°C for 1 h. This is important as heating destroys residual hydrogen peroxide, which, if present, can induce unwanted chemiluminescence. The color will change from brown/green to pale yellow.
   8. Cool to room temperature, and add 15 mL of recommended LSC cocktail.
   9. Temperature and light adapt for 1 hour before counting.

Both the above methods apply not only to blood but also to any other biological tissue sample, and the only modifications required are concerned with the sample sizes. Since the products of such solubilization are in an alkaline mixture, there is a high degree of probability that chemiluminescence may be present. Whenever a new method is being evaluated, it is strongly recommended that checks are carried out to confirm that the solubilizers and LSC cocktails do not produce unwanted chemiluminescence. This check should also be carried out on new deliveries of LSC cocktail to confirm quality. To determine if there is a potential problem, follow the steps below:

1. Add the appropriate volume of selected cocktail to a glass scintillation vial.
2. Add the appropriate volume of solubilizer.
3. Shake thoroughly to form a stable mixture.
4. Place in the counter,and arrange for repeat 2 minutes counts.
5. If normal background levels are observed, there is no chemiluminescence.
6. If higher than expected CPMs are observed, repeat counting until CPM falls to normal background levels.
7. Calculate the total time taken for CPM stabilization, and this becomes the fixed period that the sample needs to stand before starting counting after the solubilization process.

Chemiluminescence is produced by an interaction between free peroxides and alkaline material. The free peroxides originate from the cocktail and more specifically from the ethoxylate detergents present in the cocktail. Over time, the ethoxylate polymer chain breaks down, and the result is a free peroxide entity. In general, freshly manufactured LSC cocktail is peroxide-free, but with prolonged storage, the level of peroxide will increase, and therefore, attention should be given to the age of the cocktail being used in solubilization studies. Although blood is a complex mixture and solubilization is an involved process, it is possible to successfully process and count blood samples both accurately and reproducibly. As with urine, it is critical that quench curves be used to record activity in DPM due to the variable quench present in solubilized samples.

C. Plasma and serum

Plasma constitutes 55% of blood volume and serves as a transport medium for glucose, lipids, hormones, products of metabolism, carbon dioxide, and oxygen. Plasma is obtained by separating the liquid portion of the blood from the cells. It is a straw-colored clear liquid made up of 90% water and rich in protein. Significant numbers of red and white blood cells as well as platelets are suspended in plasma. It also contains various proteins: albumin (the main protein), fibrinogen, and globulins (including antibodies). Serum is the same as blood plasma except that clotting factors (such as fibrin) have been removed. Serum resembles plasma in composition but lacks the coagulation factors and is obtained by letting a blood specimen clot prior to centrifugation.

1. Sample preparation

Plasma and serum samples can be added directly to LSC cocktails, but consideration needs to be given to the presence of proteins, color, sample volume, sample/cocktail compatibility, and the biological origin of the sample. The high concentration of proteins in plasma and serum usually produces a wispy precipitate, similar to that seen with urine samples. This can develop either rapidly (within 1 h) or slowly (over 1 day). This phenomenon is most probably due to protein unfolding in the presence of the detergents in the cocktail, and such unfolding causes the proteins to precipitate out, albeit slowly. This effect does not usually affect counting reproducibility, but if it is seen as a potential problem, it can be stopped usually by adding 10%−15% of IPA or ethanol. Simply add 1.0 mL or 1.50 mL IPA to 10.0 mL LSC cocktail prior to adding the plasma or serum sample.

Color is present to some degree in plasma and serum samples, and the base effect of such color is quenching. The degree of quenching is related to the color present, and therefore, whenever counting plasma and serum samples, it is important to use quench curves and report sample activities in DPM, thus ensuring that the true activity present in the sample is reported.

The complex nature of plasma and serum signifies that plasma and serum samples do not mix easily with most LSC cocktails; therefore, it is suggested that manufacturers are consulted for recommendations as to which cocktails are most suitable. Whatever cocktail is recommended, it should always be evaluated prior to commencing any work. This should be carried out in glass vials, as it allows close visual inspection of the cocktail/sample mixture. The appearance of such samples will be somewhat different from others encountered in LSC in that they will fluctuate from slightly to highly opalescent. Samples thus prepared should be allowed to stand at the counting temperature for at least 1 hour for confirmation of stability. Provided there is no obvious two-phase separation, they will remain stable. Whatever cocktail is selected, the maximum sample volume will be low; most suitable cocktails will only accept a maximum of 1.0 mL plasma or serum in 10.0 mL cocktail. Samples from small animals are considerably more problematic, and although sample volume is usually 100−250 μL, compatibility with LSC cocktail will be an issue.

Alternative sample preparation methods include sample combustion and solubilization. If solubilization is being used, either of the methods previously detailed under blood can be used, and these are suitable for processing up to 1.0 mL of plasma or serum. Sample combustion is also suitable for processing plasma and serum samples, but it should be remembered that this is only suitable for samples containing tritium and/or carbon-14.

D. Muscle, skin, heart, brains, and stomach

The epithelium is usually the layer of cells that resides "closer" to the outside world. The outermost layer of skin is composed of dead scaly or plate-like epithelial cells. Other examples are the mucous membranes lining the inside of mouths and body cavities. Internally epithelial cells line the insides of the lungs, the gastrointestinal tract, and the reproductive and urinary tracts and make up the exocrine and endocrine glands. The primary functions of epithelial cells include secretion, absorption, and protection. As the name suggests, connective tissue holds everything together. Loose connective tissue holds organs and epithelia in place and contains a variety of proteinaceous fibers, including collagen and elastin. Ligaments and tendons are a form of fibrous connective tissue. In most vertebrate cartilage, it is found primarily in joints, where it provides cushioning. Muscle cells contain contractile filaments that move past each other and change the size of the cell, and there are three general types of muscle. The first two are "striated," they contain filaments of striated muscle, and the third type is "smooth." Smooth muscle differs from striated muscle in that the fibers of the smooth muscles are not arranged regularly, whereas those of striated muscles are arranged regularly. Smooth muscles are used to sustain longer or even near-permanent contraction, whereas the striated muscles are often used for short, burst activities. The most obvious examples of nervous tissue are the cells forming the brain, spinal cord, and peripheral nervous system. The nervous system consists chiefly of two types of cells: neurons and glia. Neurons are the primary cells of the nervous system, and glia are secondary cells involved in nourishment and structural support. The next levels of organization in the body are organs, which are structures that contain at least two different types of tissue functioning together for a common purpose, e.g., heart and stomach.

1. Sample preparation

Since any radioactivity that resides within an organ cannot be accurately measured by LSC, some type of destructive sample preparation is necessary to release or convert the radioactivity into a form more conducive to analysis by LSC. Such sample preparation usually involves solubilization (alkaline hydrolysis), which is the action of certain chemical reagents on organic materials (such as animal tissue) that affects a structural breakdown (or digestion) into a liquid form that can then be directly mixed with a liquid scintillation cocktail. Since we are dealing with solubilization, the recommended reagents are the same as mentioned previously.

The method of solubilizing all of the above is relatively straightforward, and apart from possible color formation with certain tissue types, no major problems should be encountered during sample preparation and LSC. The method can be considered a general method and is as follows:

1. Place selected sample (up to 200 mg) in a 20-mL glass scintillation vial.
2. Add an appropriate volume of solubilizer (1−2 mL depending on sample size).
3. Heat in an oven or water bath at 50−60°C with occasional swirling until the sample is completely dissolved.
4. Cool to room temperature.
5. If color is present, add 100 μL of 30% (100 vol) hydrogen peroxide and allow to stand for 10−15 minutes or until any reaction has subsided. If no color is present proceed to step 7.
6. Heat again at 50−60°C for 30 minutes to complete decolorization.
7. Add 10 mL of a recommended cocktail.
8. Temperature and light adapt for at least 1 hour before counting.

Although all of these samples can be solubilized, it is important to note that there are a few exceptions. There are two particular samples that cannot be solubilized with the aqueous-based solubilizer, and these are arteries/veins and stomach tissue. Both of these sample types are only very slowly or partially solubilized by an aqueous-based solubilizer, and for complete solubilization, the quaternary ammonium hydroxide type must be used.

E. Liver and kidney

The liver plays a major role in metabolism and has a number of functions in the body, including detoxification, glycogen storage, and plasma protein synthesis. The kidney is part of the urinary system and filters wastes (especially urea) from the blood. It then excretes them, along with water, as urine. Renal functions include the excretion of waste material from the bloodstream, secretion of hormones, and maintenance of serum electrolyte, acid−base levels, and osmolality. Both liver and kidney samples are brownish in color and in addition contain blood or blood decomposition products. As a consequence, sample preparation by solubilization will be similar to that used for whole blood, and as with blood, it is possible to use both solubilizer types mentioned previously.

1. Sample preparation

The methods utilized for the preparation of liver and kidney samples are as follows:

a. Aqueous-based solubilizer method
1. Add a maximum of 100 mg liver or kidney to a glass scintillation vial.
2. Add 1.0 mL solubilizer (AquiGest or Solvable).
3. Incubate the sample at 55−60°C for 1 hour.
4. Add 0.1 mL of 0.1M EDTA-disodium salt solution, which helps reduce foaming when the subsequent hydrogen peroxide is added. This reagent also complexes iron that has been released during digestion and produces a mixture that is less colored.
5. Add 0.3−0.5 mL of 30% hydrogen peroxide in 100 μL aliquots. Slight foaming will occur after each addition; therefore, gentle agitation is necessary. Keep swirling the mixture until all foaming subsides, and then repeat the process until all of the hydrogen peroxide has been added.
6. Allow to stand for 15−30 minutes at room temperature to complete the reaction.
7. Cap the vial tightly, and place it in an oven or water bath at 55−60°C for 1 hour. This is important as heating destroys residual hydrogen peroxide, which, if present, can induce unwanted chemiluminescence. The color will change to pale yellow.
8. Cool to room temperature, and add 15 mL of recommended LSC cocktail.
9. Temperature and light adapt for 1 hour before counting.

b. Quaternary ammonium hydroxide method

1. Add a maximum 100 mg liver or kidney to a glass scintillation vial.
2. Add 1.0 mL of selected solubilizer (GoldiSol or Soluene-350).
3. Incubate at 55–60°C for 2 hours.
4. Cool to room temperature.
5. *If using GoldiSol solubilizer*, the following treatment with H_2O_2 should be followed:

   Up to 0.5 mL of 30% hydrogen peroxide can be added in one addition as almost no foaming occurs. This significantly reduces the sample preparation time.

   If using Soluene-350 solubilizer, the following treatment with H_2O_2 should be followed:

   Add 0.2–0.5 mL of 30% hydrogen peroxide in 100 μL aliquots. Foaming occurs after each addition; therefore, gentle agitation is necessary. Keep swirling the mixture until all foaming subsides, and then repeat the process until all of the hydrogen peroxide has been added.
6. Allow to stand for 15–30 minutes at room temperature to complete the reaction.
7. Cap the vial tightly, and place it in an oven or water bath at 55–60°C for 30 minutes. The samples at this stage will now be slightly yellow. This is important as heating destroys residual hydrogen peroxide, which, if present, can induce unwanted chemiluminescence.
8. Cool to room temperature, and add 15 mL of recommended LSC cocktail.
9. Temperature and light adapt for 1 hour before counting.

Solubilization of liver and kidney always results in highly colored samples due to the presence of bilirubin; in addition, such samples from small animals will produce even more color. It is strongly recommended that sample size should not exceed 75 mg for small animal samples. The aqueous-based solubilizer type has proved to be better than the quaternary ammonium hydroxide solubilizer for these particular sample types, mainly due to significantly less color development at the end of the solubilization process. If excessive color still develops, then one way of reducing the interference is to increase the volume of LSC cocktail from 15 to 20 mL. In effect, this is diluting the color, but it is nevertheless a solution to the problem. Due to the presence of color quenching, it is critical that quench curves be used to record sample activity in DPM. Do not report results in CPM, as these will vary from sample to sample.

F. Fatty tissue

Fatty tissue is a kind of body tissue containing stored fat that serves as a source of energy. It also contains adipose tissue and serves to cushion and insulate vital organs. Fatty tissue is made up of fat cells, which are a unique type of cell. White fat cells are large cells that have very little cytoplasm (only 15% cell volume), a small nucleus, and one large fat droplet that makes up 85% of cell volume. Due to the chemical nature of fats, they are difficult to solubilize and only the quaternary ammonium hydroxide type of solubilizers are capable of solubilizing fatty tissue. Even then, the rate of solubilization is slow and may require up to 24 hours to completely solubilize animal fatty tissue.

1. Sample preparation

Fat samples can be completely solubilized, but it is a matter of time, temperature, and sample condition. Recommendations include the following:

1. cutting the fat samples into small pieces, which increases the surface area and hence the rate of reaction;
2. carrying out the solubilization at 60°C or even slightly higher; and
3. adding a small amount of IPA.

Other than the rate of solubilization, the only problem that will occur is color formation. Solubilization of fatty tissue almost invariably results in yellow coloration of the final mixture.

The recommended method is as follows:

1. Add a maximum 100 mg finely divided fat to a glass scintillation vial.
2. Add 1.0 mL of selected solubilizer and optionally 1.0 mL IPA.
3. Incubate at 60°C until the entire fat sample is solubilized (may take 10–15 hours).
4. Cool to room temperature.
5. *If GoldiSol is used as the solubilizer*, the following treatment with H_2O_2 should be followed:

   Up to 0.5 mL of 30%, hydrogen peroxide can be added in one addition, as almost no foaming occurs. This significantly reduces the sample preparation time.

   If Soluene-350 is used as the solubilizer, the following treatment with H_2O_2 should be followed: Add 0.2–0.5 mL of 30% hydrogen peroxide in 100 μL aliquots. Foaming occurs after each addition; therefore, gentle agitation is necessary. Keep swirling the mixture until all foaming subsides, and then repeat the process until all of the hydrogen peroxide has been added.
6. Allow to stand for 15–30 minutes at room temperature to complete the reaction.
7. Cap the vial tightly, and place it in an oven or water bath at 55–60°C for 30 minutes. The samples at this stage will now be slightly yellow. This is important as heating destroys residual hydrogen peroxide,

which, if present, can induce unwanted chemiluminescence.
8. Cool to room temperature, and add 15 mL of recommended LSC cocktail.
9. Temperature and light adapt for 1 hour before counting.

As described previously, due to the presence of color quenching, it is critical that quench curves be used to record activity in DPM. Do not report results in CPM, as these will vary from sample to sample.

G. Feces

Feces are body waste formed of undigested food that has passed through the gastrointestinal system to the colon. Feces are produced and stored in the colon until eliminated. A proportion of feces is inner gut lining. Normally, when blood cells become old, they are trapped and destroyed by the spleen. When this occurs, the hemoglobin must be broken down in the liver to bilirubin to be disposed of. Bilirubin is eventually excreted in the bile and leaves the body in the feces. Its presence is the reason why feces are brown in color. Feces are mostly made of water (about 75%); the rest is made of dead bacteria that helped with food digestion, living bacteria, protein, undigested food residue (known as fiber), waste material from food, cellular linings, fats, salts, and substances released from the intestines (such as mucus) and the liver.

1. Sample preparation

The digestion of feces strongly depends on the type of animal. It is possible to use both the quaternary ammonium hydroxide type and the aqueous-based solubilizers; however, there can be problems with residual color and incomplete digestion due to the presence of cellulose-type material present in feces from species such as rabbit. The method is the same as shown previously for muscle, skin, heart, stomach, brains, and stomach tissue.

An alternative method involves the use of a sodium hypochlorite solution, and this reagent was successfully used to digest guinea pig feces. During this procedure, the sodium hypochlorite solution decomposes and a small amount of free chlorine gas is produced. Care should be taken when opening the vials after solubilization as chlorine gas will be present. It is recommended that this part of the procedure be carried out under an efficient fume hood. With this method, the choice of LSC cocktail is critical. The addition of sodium hypochlorite to almost every cocktail produces an unusual form of chemiluminescence, which manifests itself either immediately or gradually over a 24 hour period. Until now, the only LSC cocktail that is unaffected by this phenomenon is the PerkinElmer cocktail Hionic-Fluor. With this method, isotope recoveries of greater than 98% for tritium were achieved, and these were confirmed by combusting the same sample in a sample oxidizer (PerkinElmer Tri-Carb Sample Oxidizer, Model 307).

The solubilization method used for processing the feces sample is as the following:

1. Weigh 50–150 mg of feces into a 20-mL glass scintillation vial.
2. Add 0.5 mL of sodium hypochlorite solution, and cap tightly.
3. Heat in an oven or water bath at 50–55°C for about 30–60 minutes with occasional swirling. All the brown color of the feces will disappear, and a white fibrous material remains.
4. Cool to room temperature.
5. Remove the cap, and blow out any remaining chlorine in a fume hood using a gentle stream of air or nitrogen.
6. Add 15 mL of Hionic-Fluor or Gold Star cocktail, and shake to form a clear mixture.
7. Temperature and light adapt for 1 hour before counting.

After digestion, a small amount of white residual matter (most probably undigested cellulose) may remain; however, this should not affect the recovery. Sodium hypochlorite is more commonly known as bleach and should have greater than 5% available chlorine if it is to be an effective solubilizer.

H. Homogenates

A homogenate is produced when a tissue sample has been mechanically disrupted. This releases the organelles from the cell. The resultant fluid is known as the homogenate, which is then filtered to remove any complete cells and large pieces of debris. The only problem with direct counting of homogenates is sample/cocktail compatibility. This can be resolved by either using a cocktail, which is compatible with the sample (e.g., Ultima Gold, Gold Star, ProSafe+, or Hionic-Fluor), reducing the sample size (recommended sample volume is 0–2 mL) or increasing the cocktail volume to obtain a clear mixture. Usually, the homogenate is prepared in either water or 7:3 ethanol:water, and the counting is carried out at or slightly below 20°C. With some homogenates, it may be necessary to reduce sample volume considerably or even change to the use of a solubilizer with a cocktail (e.g., GoldiSol and Gold Star or Soluene-350 and Hionic-Fluor). Alternatively, the use of combustion (sample oxidation) is a more than acceptable alternative.

I. Solubilization and combustion

Solubilization is the action of certain chemical reagents on organic materials (such as animal or plant tissue) that affects a structural breakdown (or digestion) into a liquid form that can then be directly dissolved in a liquid scintillation cocktail. Typical solubilizers include organic and inorganic alkalis, which act by the process of alkaline hydrolysis, and certain mineral acids, which affect solubilization by acidic oxidation. The solubilization process usually involves heating the sample/solubilizer at an elevated temperature (40–65°C) for periods ranging from <1 to 24 hour, until a homogeneous mixture is formed. Certain samples that remain colored after solubilization are optionally treated with hydrogen peroxide, and following this, a recommended LSC cocktail is added and the sample is ready for analysis by LSC. The decision to use either solubilization or combustion can be a difficult one. Below is a list of the relative advantages and disadvantages of each technique, and these should help with the decision process. Whichever is chosen, careful attention to sample preparation, reagent, and LSC cocktail selection is critical to successful analysis.

The principle of sample oxidation is that the sample is combusted in an oxygen-rich atmosphere, and any hydrogen present is oxidized to water, whereas any carbon is oxidized to carbon dioxide. If tritium is present, then the combustion product will be ^3H_2O, and if ^{14}C is present, then the combustion product will be $^{14}CO_2$. In the Model 307 Sample Oxidizer (PerkinElmer, Inc.), the water is condensed in a cooled coil and then washed into a vial where it is mixed with an appropriate LSC cocktail. The CO_2 is trapped by vapor-phase reaction with an amine, and the resulting product is mixed with an appropriate LSC cocktail. At the end of the combustion cycle, two separate samples (a tritium sample and a ^{14}C sample) are trapped at ambient temperature, thus minimizing cross-contamination. An alternative instrument—the Pyrolyser-Trio (Raddec Ltd., Southampton, UK)—was designed to simultaneously extract $^3H_{total}$ (and ^{14}C) from virtually any type of nuclear and environmental sample. The standard methodology consists of placing weighed samples into silica glass boats that are then inserted into their own silica worktube (1 m long) containing 10 g of platinized alumina catalyst. A laminar flow of air or oxygen is passed over the sample while it is progressively ramped in temperature typically from 50–900°C over a period of about 3 hours. This controlled oxidation ensures that decomposition is slow and does not lead to any pressure excursions as the sample components decompose and liberate CO_2, hydrocarbons, water, etc. The oxidized combustion products (water/^3H_2O and CO_2) are collected in bubblers from which subsamples are taken for LSC measurement.

The typical advantages of solubilization are as follows:

- Capital outlay is low.
- Homogeneous samples are produced.
- Many samples can be processed simultaneously.
- Color quench is corrected using a quench curve.
- It is suitable for a diversity of isotopes.

The disadvantages of solubilization are as follows:

- Sample sizes are generally ≤200 mg.
- Time to complete solubilization can vary from a few hours to many hours.
- Certain samples require modified techniques or longer solubilization times.
- With acidic solubilizers, loss of radioactivity by volatilization may occur.
- Excess peroxide must be destroyed after decolorization.
- Chemiluminescence may be present.
- If ^3H and ^{14}C are both present in the sample, then dual-label DPM calculation is necessary.

The advantages of sample combustion are as follows:

- Sample processing time is rapid.
- Sample can be wet, dry, or freeze-dried.
- Any sample containing H and/or C can be combusted.
- It is ideally suited for both single- and dual-label ^3H and ^{14}C.
- Sample sizes up to 1.5 g are possible.
- Radioactive recovery is excellent (>97%).
- Memory effect is <0.08%.
- There is no loss of radioactivity by volatilization.
- There is no chemiluminescence interference.
- There is no color quench interference.

The disadvantages of sample combustion are as follows:

- Initial capital investment is required.
- It is only suitable for ^3H and ^{14}C.
- It needs a gas supply (oxygen and nitrogen).
- It must be operated in a fume hood.
- Reagents are corrosive and flammable.

IX. Filter and membrane counting[5]

Filter counting, or solid support counting, as it is sometimes known, is probably best described as heterogeneous counting. The main difference between homogeneous counting and heterogeneous counting is that homogeneous counting relies on 4π geometry, whereas 2π geometry applies to heterogeneous counting. An explanation of the terms 4π and 2π geometry is needed to appreciate the differences between the two counting techniques. In homogeneous counting, the sample is completely "dissolved" in the liquid scintillation cocktail; therefore, the photons of light (scintillations), which are emitted as the end result of the energy transfer process within the liquid scintillation cocktail, are free to radiate in any direction. In

5. ©1998–2019, PerkinElmer, Inc. Printed with permission.

geometric terms, this freedom of radiation is described by a sphere or globe whose surface area is $4\pi r^2$. In heterogeneous counting, this freedom is restricted by the presence of the filter or membrane, which absorbs both the kinetic energy of β-particles and the photons of light passing in one plane; therefore, the emitted light can only occupy the surface area of a hemisphere, which is $2\pi r^2$. Hence, the derivation of the expressions 4π and 2π counting geometries, and these are illustrated in Fig. 7.2.

In essence, filter counting can be a relatively simple technique where the sample is isolated or collected on a filter and usually dried. This filter is placed in a scintillation vial, a small volume of an appropriate LSC cocktail such as Ultima Gold F or ProSafe FC+ is added, and, after ensuring that the filter is completely wetted, the vial is counted.

A. Elution situations

The difficulty in counting on filters or other solid supports is that when the sample is immersed in a cocktail, four situations may develop:

1. The sample may remain bound to the filter or solid support—no elution situation.
2. The sample may be partially eluted by the cocktail—partial elution situation.
3. The sample may have certain solubility in the cocktail—equilibrium situation.
4. The sample may be completely dissolved in the cocktail—complete elution situation.

Of these, the one to avoid is partial elution as the soluble fraction is counted with 4π geometry, whereas the insoluble (filter bound) portion is counted with 2π geometry. This will make measurement under these circumstances irreproducible. However, a partial elution situation may go to equilibrium with time and therefore should not be discounted out of hand. Repeat counting of the sample, over several hours, will determine if an equilibrium situation has been reached, and this is characterized by a constant count rate being obtained over time. If the sample is insoluble (no elution), the efficiency and reproducibility of counting will depend on the magnitude of the β-energy, the nature of the filter or solid support, its orientation in the vial, and the size of the sample molecule. If the sample is completely dissolved or eluted into the cocktail (complete elution), counting considerations will be similar to those of solubilized samples, where a true homogeneous state is obtained. Other factors that affect counting are the presence and composition of the sample precipitate and the amount of sample that becomes soluble in the cocktail (Bransome and Grower, 1970).

B. Sample collection and filters

Self-absorption affects the efficiency of counting on solid supports with low-energy ^3H samples being more susceptible to self-absorption than higher-energy isotope samples such as ^{14}C. The type and amount of sample, thickness, absorption level, and the material of construction of the solid support also influence the self-absorption effect. The order of counting efficiency of solid supports is

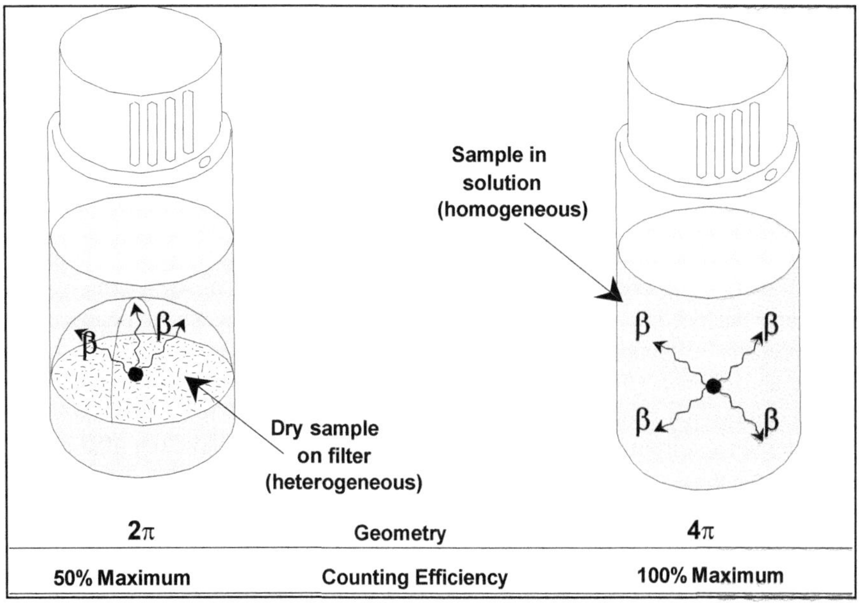

FIGURE 7.2 Sample counting geometries encountered in liquid scintillation analysis.

Glass fiber > cellulose acetate > standard cellulose > chromatographic paper

← Increasing efficiency

(Wang and Jones, 1959; Gill, 1967). This order of efficiency will vary depending on the size of the molecules. Smaller molecules can readily diffuse into amorphous regions of the cellulose fibers, whereas the larger molecules may remain on the surface. Microscopically, glassfiber filters appear as an impermeable virtual network of threads, whereas the paper filters appear as capillary tubes. For glassfiber filters, the efficiency can be markedly different for sample material trapped on the surface as opposed to that embedded in the pores. This is particularly true for low-energy betaparticles from ^{3}H. In some cases, reproducible counting efficiencies can be obtained by the addition of a carrier of known weight (many times more than the sample), which subsequently induces the same amount of self-absorption for each sample (Wang and Willis, 1965). The carrier must be added before the filtration step, and time must be allowed for complete mixing with the real sample. When using chromatographic paper to isolate or collect samples, one should remember that some grades contain a UV enhancer, and this can be eluted into the cocktail, producing unwanted chemiluminescence. Such spurious counts can lead to an overestimation of the activity of the sample. Provided all of these factors and effects are taken into consideration during sample preparation, successful and reproducible counting can be accomplished using this technique.

C. Filter and membrane types

In practice, there are a number of filter types, which can be used to isolate or collect various sample types for LSC. The choice of filter type will depend on both the nature and particle size of the sample; however, glassfiber filters are recommended if at all practical. Other filter types, which have been used, include cellulose nitrate, cellulose acetate, mixed cellulose esters, polyvinyl chloride (PVC), polyacrylonitrile, normal paper, polycarbonate, Teflon, nylon, and polythene terephthalate. The categories of sample types, which can be analyzed by this technique, include precipitates of macromolecules (such as nucleic acids and proteins), aquatic and terrestrial ecosystem samples (such as algae and phytoplankton), and other deposits (such as airborne particulate matter).

D. Sample preparation methods

The different elution situations influence both the choice of the sample preparation method and the recommended liquid scintillation cocktail.

1. No elution

This situation is highly desirable since sample preparation for counting by LSC is both simple and rapid. Sample quench is constant, and a simple CPM mode on the LSC is preferred, as external standard quench correction cannot be used. With constant quench and therefore constant efficiency, the CPM results obtained are as accurate as DPM results obtained through normal 4π homogeneous counting. After collection of the sample on the filter, the filter is dried and placed in the vial. Approximately 2−3 mL of cocktail is added (ensuring that the filter is completely wet), and counting can be carried out immediately. For best counting performance using this method, it is recommended that the filter is completely dried prior to the addition of cocktail. Additionally, knowledge of the solubility characteristics of the sample will aid in the selection of the most appropriate cocktail. In general, the most applicable cocktail for dried filter counting is an organic cocktail such as Ultima Gold F or BetaPlateScint, which provides the highest counting efficiency. Occasionally, it is not practical to completely dry the filters; in these cases, a cocktail such as Ultima Gold MV should be used with the slightly damp filters for highest counting performance. The type of samples routinely counted using this method include precipitates from DNA and RNA studies, phytoplankton from seawater, algae from aquatic environments, as well as samples from enzyme activity assays, cell proliferation, and receptor-binding assays.

Note: A simple method to confirm that a no elution situation exists is to decant the cocktail into another vial and recount the cocktail—the absence of activity confirms that no elution has occurred and that the sample is completely bound to the filter. If necessary, accurate quantitation of the total isotope activity (i.e., DPM) can be carried out by removing the filter and using either solubilization or combustion techniques.

2. Partial elution

As previously mentioned, this situation is the least desirable due to the presence of both 2π and 4π geometry within the counting mixture. Any results from this situation will be inaccurate and cannot be reproduced. It is possible, however, using one or a combination of the following methods, to convert the system from a partial to an equilibrium or a complete elution situation:

1. After sample preparation, shake the contents for a fixed time period and recount. Repeat this procedure until constant CPMs are obtained (i.e., equilibrium situation).
2. Change to a cocktail in which the sample has either zero or complete solubility.
3. Extract the sample from the filter with a suitable solvent prior, and add the appropriate cocktail.

3. Complete elution

The goal with complete elution is to convert from 2π to 4π geometry. Two slightly different sample preparation methods are used, in which either the entire filter is dissolved in an appropriate cocktail or the sample is extracted or eluted from the filter prior to the addition of cocktail. In the first instance, the cocktail of choice is Filter-Count, and filter types, which can be dissolved by this cocktail, include cellulose nitrate, mixed cellulose esters, and PVC. Filter-Count will not dissolve cellulose acetate, glass fiber, normal paper, PTFE, nylon, or phosphocellulose filters. With cellulose acetate and glassfiber filters, a transparent appearance results, whereas the others remain relatively unaltered. Filter-Count will not give color formation with any filter whether soluble or insoluble. The use of Filter-Count is extremely simple, in which sample preparation involves adding cocktail (Filter-Count) to the filter, allowing it to dissolve (with optional heating), and then counting. Cellulose acetate is not dissolved by Soluene-350 or by Filter-Count. This filter type can be dissolved by strong hydrochloric acid, and after dissolution and dilution with water, the LSC cocktails Ultima Gold AB or Gold Star LT^2 are recommended for trouble-free counting. It is important to note that with the exception of cellulose acetate, normal paper, and PET filters, virtually all other filters produce color when used with GoldiSol and Soluene-350. In this variant of the technique, dissolving the filter overcomes the self-absorption problems, saves on drying time (accepts wet or dry filters), and provides reproducible results.

In the second case, as previously described, the sample is extracted or eluted from the filter with a suitable solvent and then counted using the appropriate cocktail.

Note: It is possible to adapt this technique for alpha/beta counting of airborne particulates. Provided the correct filter type is used, the sample filter can be dissolved in Filter-Count, and then Ultima Gold AB can be added (ratio of 2:1 Filter-Count:Ultima Gold AB).

The benefits of such a method include the removal of self-absorption problems (especially important for alphas) and significant time saved on sample preparation (ashing and acid extraction steps are eliminated).

The introduction of the PerkinElmer TopCount microplate scintillation and luminescence counters together with the development of various filter plates offers the ability to count labeled samples in filter plates (24 or 96 samples per plate), minimizing sample preparation steps and increasing sample throughput. A number of publications on the applicability of this assay method are available (PerkinElmer, TopCount Topics 11, 12, and 18). A quick reference table (Table 7.21) can be used as a guide for selecting the correct LSC cocktail type for a particular filter or membrane type.

X. Sample stability troubleshooting

Sample stability troubleshooting is the identification and resolution of problems, which can occur as a result of incorrect sample preparation. These problems become apparent during counting and are sometimes not visible during the sample preparation process. These problems can be found wherever samples are mixed with cocktail, irrespective of whether the sample is solid, dissolved in an organic solvent, or dissolved in an aqueous system.

Fundamentally, this problem becomes apparent when one or more of the following LSC situations occur:

1. unstable count rate (decreasing count rate),
2. unstable count rate (increasing count rate), and
3. reduced counting efficiency (reduced more than expected).

All the three abovementioned problems can usually be resolved by minor changes to sample preparation procedures, and/or by careful selection of the correct cocktail.

A. Decreasing count rate

Potential reasons for a decreasing count rate are a two-phase separation of the sample–cocktail mixture (sometimes described as "milky"appearance), precipitation of the radiolabeled sample, and chemiluminescence. When a two-phase situation occurs and the activity is more soluble in the aqueous phase than the organic phase, then, as the two phases separate the activity, it migrates with the aqueous phase and results in lower CPMs. This can be remedied by increasing the cocktail volume or changing to a cocktail more suited to the sample type. In addition, check that the cocktail is suitable for use at the operating temperature (both sample preparation and counting temperature). Precipitation of the radiolabeled sample can be overcome either by diluting the sample with water or a suitable cosolvent or by changing to a more suitable cocktail. A chemiluminescence problem is usually characterized by higher than expected CPMs that decay with time. Chemiluminescence can be overcome either by changing to a chemiluminescence-resistant cocktail such as Hionic-Fluor, Gold Star, Ultima Gold, or equivalent cocktail or by waiting for the chemiluminescence to decay. Additional suggestions include ensuring that the correct cocktail type is in use (i.e., do not use a lipophilic cocktail with aqueous samples) using a glass vial to check that the sample/cocktail mixture is homogeneous and checking that the sample and cocktail have been thoroughly mixed (shaken) and formed a stable microemulsion. Finally, consult the cocktail phase diagram to verify that the selected cocktail is suitable for use with the sample type.

B. Increasing count rate

Potential reasons for an increasing count rate are a two-phase separation of the sample–cocktail mixture (sometimes described as "milky" appearance), inadequate mixing of sample and cocktail, incomplete elution of the sample from a solid support, and incomplete or slow solubilization of the sample in the cocktail. When a two-phase situation occurs and the activity is more soluble in the organic phase than in the aqueous phase, then, as the two phases separate the activity, it remains in the organic phase and results in increasing CPMs due to the reduction in quench in the organic phase. This can be remedied by increasing the cocktail volume or changing to a cocktail more suited to the sample type. Inadequate mixing of the sample and cocktail, which in essence is the same as the preceding situation, can be easily remedied by ensuring that the mixture is thoroughly shaken. If a sample is incompletely or only slowly eluted from a solid support, increasing CPM will probably result. This can be corrected by modifying the elution conditions to get complete elution. Reagents such as water, alkalisolubilizer, acid, or other solvent may be useful in this instance. In certain cases, provided the appropriate filter material is in use, it may be possible to use a cocktail such as Filter-Count to dissolve the filter and thus separate the sample from the solid support. Where the sample is not completely soluble in the cocktail, then dissolve the sample in a solvent that is more compatible with the cocktail, or add a suitable cosolvent.

C. Reduced counting efficiency

Potential reasons for reduced counting efficiency are either high color quench or high chemical quench in the samples. If the color quench is a result of solubilization, then bleach the sample with hydrogen peroxide, prior to adding cocktail. If the sample is naturally colored (e.g., some metallic salts), dilute with water or use a color-quench-resistant cocktail (e.g., Ultima Gold, Gold Star, ProSafe+ families of cocktail, or equivalent cocktail). Usually, chemical quench results from the use of an inappropriate solvent, which is highly quenching, and therefore, the remedy may simply be to change to a less chemical quenching alternative (e.g., dichloromethane is less quenching than chloroform). The sequence of chemical quench strength is shown in Table 7.22. Other remedies include either increasing cocktail volume or decreasing sample volume. Overall, if the sample preparation method cannot be changed, it may be necessary to increase the count time to obtain better statistical results.

TABLE 7.22 Strength of chemical quenchers.

Solvent	Quench strength
Nitro groups (nitromethane)	Strongest quencher
Sulfides (diethyl sulfide)	↓
Halides (chloroform)	↓
Amines (2-methoxyethylamine)	↓
Ketones (acetone)	↓
Aldehydes (acetaldehyde)	↓
Organic acids (acetic acid)	↓
Esters (ethyl acetate)	↓
(Water)	↓
Alcohols (ethanol)	↓
Ethers (diethyl ether)	↓
Other hydrocarbons (hexane)	Mildest quencher

XI. Swipe assays

All laboratories handling radionuclides are required to conduct radiation safety surveys to maintain their license. These surveys include those necessary to evaluate external exposure to personnel, surface contamination levels, and concentrations of airborne radioactive material in the facility and in effluents from the facility. In Klein et al. (1992), the wisdom of relying on wipe tests for detecting removable radioactivity due to low-energy beta emitters from surfaces has been questioned. Their opinion is that "Without considering the kind and amount of radioactivity present, wipe testing is scientifically misguided and wasteful of resources and personnel ..." They conclude by saying well-reviewed operating procedures and good working practice are the best protection in the work place. It is true that wipe testing may not be very efficient, but it does give some measure of what is removable. For weak beta emitters, especially tritium, there is little choice but to use a wipe test and a liquid scintillation counter to detect its presence. For better or for worse, wipe testing remains an important requirement for all licensees.

A. Wipe media and cocktails

Various wipe media have been evaluated for their ability to gather and release radiolabeled compounds into counting solutions and include filter paper, glassfiber filter, cotton swab, and plastic squares. Virtually, any emulsifying cocktail is suitable for use in this particular area, and the only major prerequisites are that the cocktail has a high tritium counting efficiency and is able to accommodate small volumes of aqueous and alcohol–aqueous samples in a clear microemulsion. Paper filter circles are not soluble in any of the counting solutions used. Glassfiber filters, in general, became translucent when placed into the counting solution, and the faint outline of the glassfiber filter can be seen on the bottom of the vial. The plastic swabs show different solubility patterns depending on the LSC cocktail used. In some cases, the blue plastic stem can dissolve into a series of blue globs that adhere to the sides and bottom of the vial. When shaken, these blobs remained immiscible. Other cocktails can extract the blue color from the plastic tube. The cotton swab can appear dissolved as a clear layer of fluid on top. When shaken, the sample can turn cloudy, again depending on the cocktail used. The thin plastic squares dissolve in most LSC cocktails, but the speed of dissolution depends on the type of cocktail: fast in classical and slow in a safer cocktail. The plastic wipes dissolve rapidly in classical cocktails and appear as droplets either floating on top or laying on bottom of the vial. When shaken, the samples become cloudy. In all cases, when the samples containing the swabs or plastic squares became cloudy on shaking, the count rates did not change when recounted (remained within $\pm 3\%$ of the last count rate).

For quantitation of wiped radioactive contaminants by LSC, it is essential that all the radioactive material be in solution. The contaminants are generally classified as being either water- or organic-soluble material. Organic-soluble material will normally dissolve in the organic solvents contained in the LSC cocktail. To accommodate water-soluble material, an emulsifying cocktail is used with added water. Without water, water-soluble material, such as leucine, cannot be emulsified and will remain undissolved on the solid support, resulting in poor recoveries.

B. Regulatory considerations

Any laboratory using radioactivity would be classed as a restricted area. Therefore, the action level for any $\alpha-\beta$ contamination would be the presence of radioactivity exceeding 20,000 DPM/100 cm^2. Assuming that the wiping procedure picked up at least 1% of the surface contamination (a very conservative estimate), there would be a minimum of 200 DPM deposited on the wipe. For tritium, 100% recovery means that 200 DPM would be recovered. With a counting efficiency of 50%, this translates into 100 net CPM. This amount of activity, being greater than five times background (about 20 CPM), can be detected easily. Even if this level of activity were quenched 50%, the activity of 50 net CPM is still easily detected. In a real situation dealing with removable contamination, it would be reasonable to expect that an amount greater than 1% would be picked up by wiping. With this assumption, it would appear that wiping is not an unreasonable method for detecting and following the removal of radioactive contamination from a surface in any biomedical laboratory.

C. Practical considerations

For most isotopes, it is important to have the sample completely dissolved in the cocktail, especially with weak beta emitters such as ^{3}H. Wipe medium selection is largely up to the individual, but the consensus of opinion is that glassfiber filter is the best overall performer. However, paper is probably the most popular medium because of its availability, ease of use, and low cost. In one institution, waste computer paper is cut to size and used as wipes. Cotton swabs are also easy to use, but recoveries are relatively low for both paper and swabs compared with glass fiber and plastic squares. The plastic squares are principally intended for pseudocumene- and xylene-based cocktails in which they are readily soluble. Safer cocktails based on LAB, phenylxylylethane, or diisopropylnaphthalene solvents slowly dissolve the plastic wipes but form an immiscible second phase. Surprisingly, it appears that the radioactivity is released into the cocktail during this process, based on reported recoveries. Klein et al. (1992) found that wetting the wipe medium greatly increased the amount of radioactive contamination removed. The problem of wetting wipes is that any solvent will increase the possibility of spreading the contamination. The solvent can cause contamination to further penetrate a surface as well as increase the probability of transfer to the hands. If paper is used, wet paper will lose its strength and tend to shred when moderate pressure, which is recommended, is applied during wiping. The contaminated surface should be exposed to solvents only during the decontamination procedure. Furthermore, it would be an added burden to control the amount of water added to each wipe before wiping a surface.

D. General procedure for wipe testing

Most regulatory guides do not contain any detail on how to conduct these surveys, and the protocols are left to the licensee. However, guidelines do suggest an area of 100 cm^2 be monitored and the action level of contamination for weak beta emitters to be 2000 DPM/100 cm^2 in unrestricted areas and 20,000 DPM/100 cm^2 in restricted areas. In an earlier publication, U.S.A.E.C. Regulatory Guide 1.86, the action level for removable $\alpha-\beta$ contamination was set at 1000 DPM/100 cm^2. Before doing any wipes, it is advisable to prepare a map of the area to be monitored. The locations to be wiped should be circled and numbered to identify the wipes after they are completed. Use the same dry wipe medium for all wipes to maintain uniform monitoring conditions. Wipe an area about 100 cm^2 (approximately 4×4 in.) using a moderate amount of pressure. The wipe can be numbered, or it can be placed directly into a numbered counting vial. After the wipe is placed into a vial, counting solution containing $\sim 2\%$ water

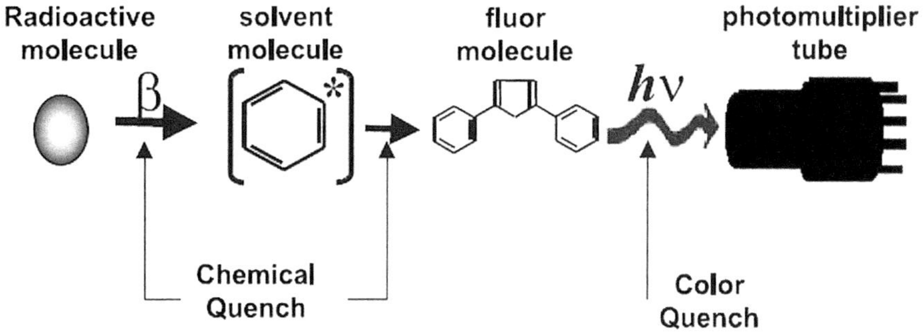

FIGURE 7.3 Quenching in the energy transfer process.

is added. The vial is capped and vortexed a few seconds to promote elution from the wipe into the counting solution. The vial is assayed by LSC. One of the most useful consequences of the presence of multichannel analyzers (MCAs) in liquid scintillation counters is that the spectrum of any sample being counted can be visualized online. An inspection of the spectrum can be very helpful in identifying the nature of the sample being counted. For example, if the contamination is due to a single radionuclide, then the visualization of the expected spectrum would confirm this fact. If the sample contained two radionuclides, then the spectrum would also reflect this fact. The ability to identify the components of a mixed sample depends on the relative content of each component. The spectrum can be visualized in either the log or linear form. In general, the log mode is more helpful in detecting the presence of more than one radionuclide in the sample. Finally, any laboratory handling radioactivity should be staffed with properly trained personnel. This assumption is generally true when laboratories first begin this type of work. However, experience has shown that, as time goes by, people tend to forget or neglect the safeguards designed into approved protocols. This is why wipe tests are necessary. Its purpose is to continually ensure safe working conditions within laboratories handling radionuclides. If contamination is detected, viewing the sample spectrum should be helpful in identifying the radionuclides present. Although the wipe test is qualitative at best, the detection of radioactivity significantly over background levels should be the cause for further investigation.

XII. Preparation and use of quench curves in liquid scintillation counting[6]

Basically, the liquid scintillation process is the conversion of the energy of a radioactive decay event into photons of light. PMTs are used to detect and convert the photons into electrical pulses. Both the sample and the scintillator are dissolved in an aromatic solvent, which allows energy to be transferred. Any factor that reduces the efficiency of the energy transfer or causes the absorption of photons (light) results in quenching in the sample. There are two main types of quench: chemical quench (Gibson, 1980; Birks, 1971) and color quench (Gibson, 1980; Ross, 1965; Ten Haaf, 1975).

A. Chemical quench

Chemical quench occurs during the transfer of energy from the solvent to the scintillator. Any chemical species that is electronegative (electron capturing) will affect the energy transfer process by capturing or stealing the pi electrons associated with the aromatic solvent and thus reduces the availability of pi electrons necessary for efficient energy transfer.

B. Color quench

Color quench is an attenuation of the photons of light. The photons produced are absorbed or scattered by the color in the solution, resulting in reduced light output available for measurement by the PMTs. The steps in the energy transfer process affected by chemical and color quenching are indicated in Fig. 7.3.

The collective effect of quench is a reduction in the number of photons produced and therefore detected CPM. Counting efficiency is affected by the degree of quenching in the sample. To determine absolute sample activity in DPM (absolute activity), it is necessary to measure the level of quench of the samples first and then make the corrections for the measured reduction in counting efficiencies.

C. Measurement of quench

It is possible to measure quench accurately via high-resolution spectral analysis. Quenching manifests itself by a shifting of the energy spectrum toward lower-energy channels in the MCA. On PerkinElmer's Tri-Carb series LSAs, there are two methods of spectral analysis for measuring quench.

6. ©1998–2019, PerkinElmer, Inc. Printed with permission.

The first method is the spectral index of the sample (SIS) that uses the sample isotope spectrum to monitor the quench of the solution. The SIS value decreases as quench increases, reflecting the shift of the spectrum to lower energy.

The second method used to measure quench is the transformed spectral index of the external standard (t-SIE), which is calculated from the Compton spectrum induced in the scintillation cocktail by an external ^{133}Ba gamma source. The source is positioned under the sample vial, causing a Compton spectrum to be produced in the cocktail solution. From a mathematical transformation of this spectrum, the t-SIE value is determined, and t-SIE is a relative value, on a scale from 0 (most quenched) to 1000 (unquenched). The calculated t-SIE value is adjusted to 1000 when the instrument is calibrated. Like SIS, t-SIE decreases as quench increases.

Both SIS and t-SIE are used as quench-indicating parameters (QIPs). t-SIE is independent of the sample isotope and of the activity in the vial and has a large dynamic range. This makes it a very reproducible means of tracking the quench of the cocktail. SIS uses the sample isotope spectrum to track quench; it is most accurate with high-count-rate samples. The range of SIS values reflects the energy range of the isotope. Both can be used as QIPs to create quench curves, although the use of the external standard is preferred for samples containing low activity and is required for multilabeled samples.

D. Quench curve

A quench standard curve is prepared with a series of standards in which the absolute radioactivity (DPM) per vial is constant and the amount of quench increases from vial to vial. A quench curve uses the relationship between counting efficiency and QIP to correct the measured CPM to DPM. When a quench curve is made, the DPM value in each standard is known. Each standard is counted, and the CPM is measured. The counting efficiency is calculated using the following relationship:

$$\frac{\text{CPM} \times 100}{\text{DPM}} = \%\text{Counting Efficiency}$$

At the same time, the QIP is measured for each standard. A correlation is made using the QIP on one axis (X) and the % efficiency on the other axis (Y). A curve is fitted to the standard points. Fig. 7.4 shows the quench curves for ^{3}H and ^{14}C using SIS as the QIP, and Fig. 7.5 shows the quench curves for the same isotopes using t-SIE as the QIP. Once the quench curve is stored in the instrument computer, it can be used for automatic DPM calculations. When unknowns are counted, the sample CPM and the QIP are measured. Using the QIP, the counting efficiency is determined from the quench curve. Sample DPMs are then calculated by applying the appropriate efficiency to the CPM of the sample:

$$\text{DPM} = \frac{\text{CPM}}{\text{Efficiency (expressed as a decimal)}}$$

The standards and unknowns must be counted with the same energy regions. PerkinElmer LSCs with spectra-based libraries (2500 series, 2700 series, and the new 2900 and 3100 series) store the curve in a $0-E_{max}$ window and allow the curve to be recalculated for the windows used in the protocol. For other LSCs (1600, 1900, 2100, 2200, and 2300), the windows used to acquire the quench curve must be used in the actual DPM determination.

1. Preparation of quench curves

a. Method 1

As mentioned, a quench standard curve is prepared with a series of standards in which the absolute radioactivity (DPM) per vial is constant and the amount of quench increases from vial to vial. The quench is increased from vial to vial by the addition of a quenching agent. A quenching agent is any chemical or colored material added to the vial that causes a shift in the standard spectrum to a lower energy and a subsequent decrease in the counting efficiency of the radioactive standard.

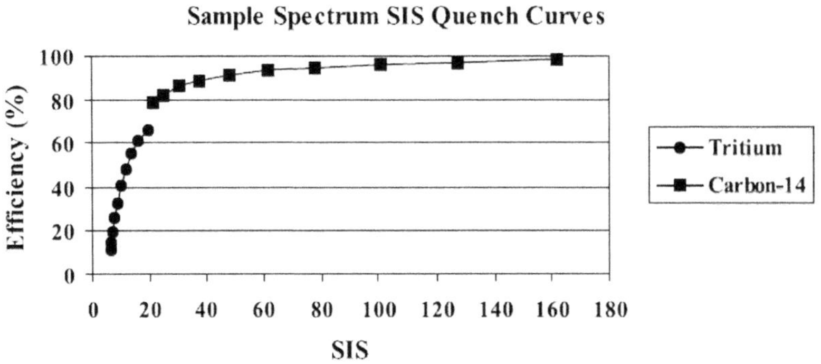

FIGURE 7.4 Quench curves for ^{3}H and ^{14}C using spectral index of the sample (SIS) as the quench-indicating parameter.

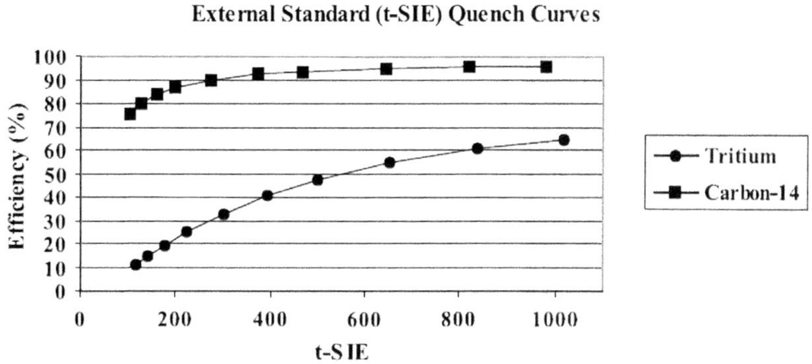

FIGURE 7.5 Quench curves for ^{3}H and ^{14}C using transformed spectral index of the external standard (t-SIE) as the quench-indicating parameter.

Usually, a series of 6–10 quench standards are prepared per radionuclide. This series is sometimes referred to as the quench set. PerkinElmer provides factory-stored quench curves in the instrument for ^{3}H and ^{14}C. On occasion, it is necessary for the investigator to prepare a quench curve, for isotopes other than ^{3}H or ^{14}C, for example, ^{35}S. There are some basic considerations to note before preparing a quench curve, which are the following:

- It is necessary to obtain a calibrated source of radioactivity to use as the source of the activity (DPM). It is essential that a known amount of activity be added per vial. Also, the standard material must be compatible with the cocktail chosen.
- A suitable quenching agent must be chosen. It is desirable to closely approximate the chemical environment in the samples. If samples contain water with various other constituents, add the same material in increasing amounts to the standards. Additional quenching agents that are most often used, and available in the laboratory are carbon tetrachloride (CCl_4), acetone (CH_3CH_3CO), chloroform ($CHCl_3$), and nitromethane (CH_3NO_2).

To prepare the quench standards, perform the following steps:

1. Make a batch of radioactive solution in the chosen cocktail such that the desired DPMs are transferred to each individual vial when dispensing the cocktail. Prepare the standards with a sufficient level of activity, typically 50,000–200,000 DPM per vial, to be able to count the standards with good statistics in a short time. If 10 standards are to be made with 10 mL of cocktail per vial, then 100 + mL of radioactive cocktail solution is required. If 15 mL of cocktail is to be used, then 150 + mL of radioactive cocktail is required. Note: If the unknowns to be counted contain two radioisotopes (both ^{3}H and ^{14}C, for example), then individual standard curves must be prepared for each isotope.
2. Count the individual standards for at least 5 minutes to check for constant activity (CPM). Any sample that deviates more than 2% from the mean should be discarded.
3. Add incremental amounts of the quenching agent to vials 2 ... n (quenching agent is not added to vial 1) to obtain the desired quench range. It may be necessary to predetermine the amounts to add per vial by testing various volumes of quenching agent added to the cocktail only (no radioactivity) and monitor the amount of cocktail quench using t-SIE; otherwise, add the suggested amounts of nitromethane based on the information given in Table 7.23.
4. Count the complete set under the conditions described in the instrument operation manual for storing a quench curve. Practically, we suggest that the standards are counted to a preselected level of statistical accuracy (generally 0.5% 2 s), and this is usually achieved within 5 min per sample with the sealed standards provided.

TABLE 7.23 Volume of nitromethane needed for quench curve.

Quench level	Toluene standards (15 mL)	Ultima Gold standards (15 mL)
A (1)	0 μL	0 μL
B (2)	1 μL	5 μL
C (3)	5 μL	10 μL
D (4)	11 μL	15 μL
E (5)	17 μL	26 μL
F (6)	25 μL	45 μL
G (7)	35 μL	70 μL
H (8)	45 μL	110 μL
I (9)	55 μL	150 μL
J (10)	66 μL	230 μL

5. Once the quench curve(s) are counted and stored, count unknown samples using the stored quench curve(s) to determine the DPM value for each sample.

b. Method 2

Basically, the preparation of a quench curve with any LSC cocktail is relatively straightforward, and the following procedure is given as a guideline. Many researchers use their own methods and equipment, and the procedure is therefore open to modification.

1. Dispense 10.0–15.0 mL of LSC cocktail into 10 high-performance glass vials.
2. Add activity to each vial (200,000 DPM for ^{3}H or 100,000 DPM for ^{14}C).
3. Count all 10 vials to ensure that the same amount of activity is in each vial. A count time of about 5 min per vial will be sufficient. Any sample that deviates more than 2% from the mean should be discarded.
4. Number the vials 1–10 or A–J, and add the suggested amounts of nitromethane based on the information given in Table 7.23.
5. Count the complete set under the conditions described in the instrument operation manual for storing a quench curve. PerkinElmer recommends counting each standard for 30 min or until a preselected level of statistical accuracy (using %2S terminator, generally 0.5%) is reached.

Notes:

1. For dispensing the activity, use a glass-barreled microliter syringe fitted with a Chaney adapter. Such an adapter ensures reproducible dispensing of activity.
2. After preparation, the standards should be stored in the dark, preferably at 5–10°C for best stability.

2. Notes on using the quench curves

1. t-SIE is independent of the sample isotope and of the activity in the vial, and has a large dynamic range. This makes it a very reproducible means of tracking quench in the cocktail.
2. SIS should only be used when there is at least 500 CPM activity in the sample. Remember that SIS uses the sample isotope spectrum to track quench; it is most accurate with high-count-rate samples. For an accurate SIS, a good sample spectrum needs to be acquired.
3. SIS should not be used for low-activity samples since an accurate sample spectrum cannot be acquired.
4. Most customers prefer to purchase quenched standards. For cocktails based on toluene, xylene, pseudocumene, or LAB as the solvent, toluene-quenched standards should be used. For cocktails based on DIN (diisopropylnaphthalene) or PXE (phenylxylylethane) as the solvent, Ultima Gold–quenched standards should be used. If the wrong quenched standard is used, there can be an error in DPM. This error is most pronounced with low-energy isotopes such as tritium (see Tables 7.25–7.28).
5. Be sure that your prepared quench curve covers a wide t-SIE range (i.e., 800–300) to provide accurate DPM results.

3. Color quench

When a small amount of color is present in a sample, there is virtually no difference between chemical and color quenching, and the standard chemical quench curves are suitable. This applies to samples where the t-SIE is in the range of 100–400. However, if a significant amount of color is present in the sample (t-SIE is < 100), it may be necessary to consider preparing a color quench curve. When preparing a color quench curve, the selection of a suitable color quench agent is important. Aqueous soluble food dyes are usually a good choice as they provide both a stable color and a wide range of colors. The color of the sample must match the color used in the quench curve. Do not use a pH indicator since some cocktails contain acidic components, and these will alter the color. Preparing a color quench curve is very similar to the method used for a chemical quench curve, and the only difference is the quench agent. If considerable color is present in the sample, it may be wise to modify or change the sample preparation method to either remove or reduce the level of color. Some techniques that have proved useful are shown in Table 7.24.

Finally, the easiest way to reduce color quench is to either decrease the sample size or increase the cocktail volume, or both.

4. Quench curve errors

The errors that can be present when the wrong quench curve is used with an LSC cocktail are shown in Tables 7.25–7.28. Table 7.25 shows that only Ultima Gold

TABLE 7.24 Techniques for reducing color in certain samples.

Nature of sample	Suggested remedy
Color from sample solubilization	Treat with hydrogen peroxide to bleach out the color
Plant material	Consider sample oxidation
Inorganic matrix	Change to alternate colourless anion

TABLE 7.25 Disintegrations per minute Errors for Various Cocktails versus Ultima Gold Quench Curve (Tritium).

	Ultima Gold	Toluene	Opti-Fluor	Insta-Gel Plus	Pico-Fluor 15
No quench	−0.12%	−1.04%	+6.00%	+2.70%	+4.89%
Low quench	−0.46%	+4.24%	+7.06%	+5.14%	+6.45%
Medium quench	+0.04%	+5.87%	+8.43%	+5.82%	+6.91%
High quench	−0.14%	+10.10%	+14.41%	+10.02%	+11.89%
Highest quench	−0.20%	+13.42%	+18.01%	+13.36%	+13.43%

TABLE 7.26 Disintegrations per minute Errors for Various Cocktails versus Toluene Quench Curve (Tritium).

	Ultima Gold	Toluene	Opti-Fluor	Insta-Gel Plus	Pico-Fluor 15
No quench	−4.10%	−0.49%	−0.57%	−1.58%	−0.32%
Low quench	−5.33%	−0.27%	−0.13%	−1.22%	+0.23%
Medium quench	−6.51%	+0.01%	+0.45%	−0.19%	+0.60%
High quench	−10.39%	−0.01%	+1.21%	−0.79%	−0.49%
Highest quench	−16.16%	−0.70%	+0.11%	+0.56%	−0.21%

TABLE 7.27 Disintegrations per minute Errors for Various Cocktails versus Ultima Gold Quench Curve (Carbon-14).

	Ultima Gold	Toluene	Opti-Fluor	Insta-Gel Plus	Pico-Fluor 15
No quench	+0.06%	−1.37%	+2.51%	+2.25%	+1.96%
Low quench	+0.03%	+0.12%	+2.04%	+0.78%	+0.80%
Medium quench	−0.13%	+0.84%	+1.72%	+1.15%	+1.20%
High quench	+0.02%	+1.30%	+2.51%	+3.11%	+1.71%
Highest quench	−0.63%	+4.52%	+3.81%	+3.59%	+2.77%

TABLE 7.28 Disintegrations per minute Errors for Various Cocktails versus Toluene Quench Curve (Carbon-14).

	Ultima Gold	Toluene	Opti-Fluor	Insta-Gel Plus	Pico-Fluor 15
No quench	+0.45%	+0.42%	+1.71%	+1.90%	+1.96%
Low quench	+0.27%	−0.03%	+0.89%	+1.20%	+0.43%
Medium quench	−0.54%	−0.37%	+1.28%	+0.31%	+0.83%
High quench	−0.81%	+0.01%	+1.86%	+0.77%	+0.81%
Highest quench	−6.51%	+0.33%	+1.04%	+0.49%	+0.88%

cocktails should be used for tritium DPM measurements with the Ultima Gold tritium quench curve. Table 7.26 shows that the Ultima Gold cocktails should not be used for tritium DPM measurements with the toluene tritium quench curve. Table 7.27 shows that only Ultima Gold cocktails should be used for carbon-14 DPM measurements with the Ultima Gold carbon-14 quench curve. Table 7.28 shows that the Ultima Gold cocktails should not be used for carbon-14 DPM measurements with the toluene carbon-14 quench curve.

5. Using a quench curve

Once the quench curve is stored in the instrument computer, it can be used for automatic DPM calculations. When unknowns are counted, the sample CPM and the QIP are measured. Using the QIP (SIS or t-SIE), the counting efficiency is determined from the appropriate quench curve.

For example:

- A tritium sample is analyzed and has the following count rate and QIP:

$$\text{Count rate} = 10,000\ \text{CPM}$$

$$\text{t-SIE} = 400$$

From the quench curve shown in Fig. 7.5, the instrument uses the t-SIE of 400 to determine that the counting efficiency is 42%. Since we now know the CPM and the counting efficiency, it is possible to calculate the DPM as follows:

$$\text{DPM} = \frac{\text{CPM}}{\text{Efficiency (expressed as a decimal)}}$$

$$= \frac{10,000}{0.42} = 23,809\ \text{DPM}$$

A compilation of recommended quench curves for different LSC cocktails is shown in Table 7.29, and this will provide a basic guide to correct quench curve selection. For accurate DPM recovery, it is imperative that the quench curve selected be appropriate for the LSC cocktail being used. PerkinElmer manufactures quench curves using either a safer, high flash point (Ultima Gold) cocktail or a classical solvent (toluene) that is suitable for use with all Perkin–Elmer LSC cocktails. Whenever there is doubt about the appropriateness of a quench curve, researchers should always prepare their own using the same chemistry as that found in the sample.

TABLE 7.29 Recommended quench curves for various cocktails.

Cocktail	Recommended quench curve (^{3}H and ^{14}C)
Ultima Gold	Ultima Gold
Ultima Gold AB	Ultima Gold
Ultima Gold LLT	Ultima Gold
Ultima Gold MV	Ultima Gold
Ultima Gold XR	Ultima Gold
Gold Star/Gold Flow	Ultima Gold
ProSafe+	Ultima Gold
ProSafe HC+	Ultima Gold
ProSafe FC+	Ultima Gold
ProSafe TS+	Ultima Gold
OptiPhaseHiSafe 2	Ultima Gold
OptiPhaseHiSafe 3	Ultima Gold
Ultima Gold F	Ultima Gold
OptiScintHiSafe	Ultima Gold
BetaPlateScint	Ultima Gold
StarScint	Ultima Gold
Opti-Fluor/Opti-Fluor O	Toluene
Emulsifier-Safe	Toluene
Insta-GelPlus	Toluene
Pico-Fluor 15	Toluene
Pico-Fluor Plus	Toluene
Insta-Fluor	Toluene
Hionic-Fluor	Toluene
Filter-Count	Toluene
Carbo-Sorb E/Permafluor E+	Toluene
CarbonTrap/CarbonCount	Toluene
Monophase S	Toluene
Formula 989	Toluene
Aquasol/Aquasol II	Toluene
Aquassure/Biofluor®	Toluene
Atomlight	Toluene
Econofluor-2	Toluene

E. Quench correction using selectable multichannel analyzer

The External Standard Channels Ratio (ESCR) was in frequent use in scintillation counting until the introduction of computer-controlled spectrum analysis methods in the late 1980s when it fell into disuse. ESCR is discussed in Chapter 6 of this book in detail. The method of quench correction used in the recently introduced Hitachi Aloka LSC-8000 is ESCR. The ESCR in the Hitachi Aloka AccuFLEX LSC-8000 liquid scintillation counter uses a

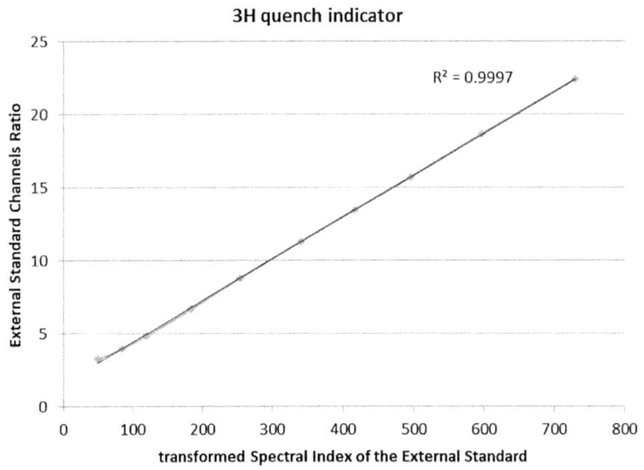

FIGURE 7.6 ^{3}H transformed spectral index of the external standard versus External Standard Channels Ratio. PerkinElmer ^{3}H Standards 283,500 DPM/Std ±3%, 30 min, three cycles, region 0–18.6 keV.DPM, disintegrations per minute.

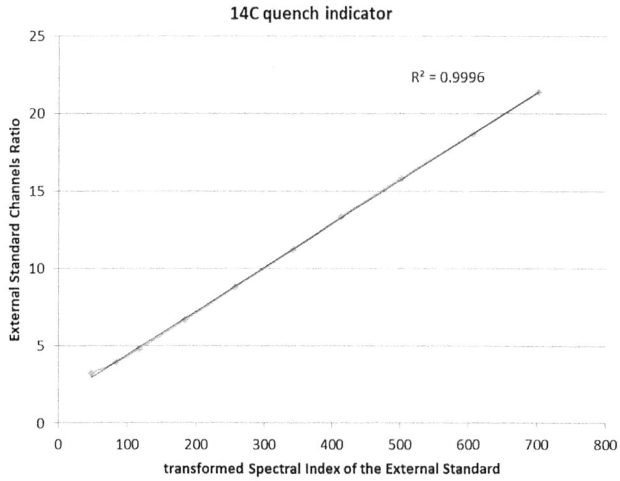

FIGURE 7.7 ^{14}C transformed spectral index of the external standard versus External Standard Channels Ratio. PerkinElmer 14C Standards 119,000 DPM/Std ±2%, 30 minutes, three cycles, region 0–156 keV.DPM, disintegrations per minute.

selectable MCA for calculations using ESCR. Data shown in (Figs. 7.6 and 7.7) show excellent correlation between t-SIE and ESCR. This confirms the validity of ESCR as compared with another more commonly used method of quench correction. Hitachi Aloka LSC-8000 uses a selectable 4000 channel MCA with automatic gain switching and high-resolution spectral analysis.

The t-SIE is graphed against ESCR.

The selectable 4000-channel MCA allows the gain setting to be such that in the 0–20 keV region the increment is in 0.005 keV (100× gain), in the 0–200 keV the increment is 0.05 keV (10 × gain), and in the 0–200 keV it is 0.5 keV (1 × gain). This allows for accurate measurement of the counts in the spectrum. Fig. 7.8 shows the histogram of the spectrum of a ^{3}H standard counted for 10 min and the subsequent increase in data available.

XIII. Environmental sample preparation[7]

The advance of the nuclear industry in all its forms coupled with growing concerns for possible environmental contamination has led to an increased interest in the quantification of radioisotopes in the environment. Radionuclides currently present in the environment originate from a variety of natural sources such as fallout from nuclear weapons testing and in discharges from both nuclear and nonnuclear industries. Natural sources and fallout from nuclear devices provide the main input to terrestrial ecosystems, except for certain radionuclides emanating from nuclear installations.

In the late 1950s and early 1960s, during and immediately after the period of most frequent aboveground nuclear weapons testing, numerous studies were performed to determine the distribution and movement of fallout radionuclides in air, precipitation, agricultural produce, animals, and soils (e.g., L'Annunziata, 1967). Around this time, the nuclear power industry was developing, and consequently, both the diversity and amount of radioactive species in the environment increased. Currently, the majority of the high-level waste from the nuclear industry is stored for ultimate disposal in sites classified as stable, such as deep geological strata. However, as many nuclear facilities are situated in coastal areas, the bulk of low-level radioactive waste is discharged to the sea.

Because of nuclear fallout and discharges from the nuclear industry (including releases from the Chernobyl accident in 1986 and more recently from the Fukushima accident in 2011), certain radionuclides are studied more than others. This is due to either their radiotoxicity, increased presence in the environment, or ease of entry into the food chain (MAFF, 1984; MAFF, 1994; Bocock, 1981; NRPB, 1993; HMIP, 1993; MAFF, 1993; Passo and Cook, 1994); a selection of these is presented in Table 7.30. The separation and isolation of these radionuclides from the complex sample matrices often encountered are presenting researchers with a myriad of problems; however, many of these have been eased by the introduction of novel extraction chromatographic separation technology (Eichrom Industries Inc. Darien, Illinois, USA). By employing this new technology, previously difficult and time-consuming radionuclide separations are completed more effectively and efficiently. In combination with recent advances in LSC technology by the Packard Instrument Company (now PerkinElmer, Inc.), it is now possible to

7. ©1998–2019, PerkinElmer, Inc. Printed with permission.

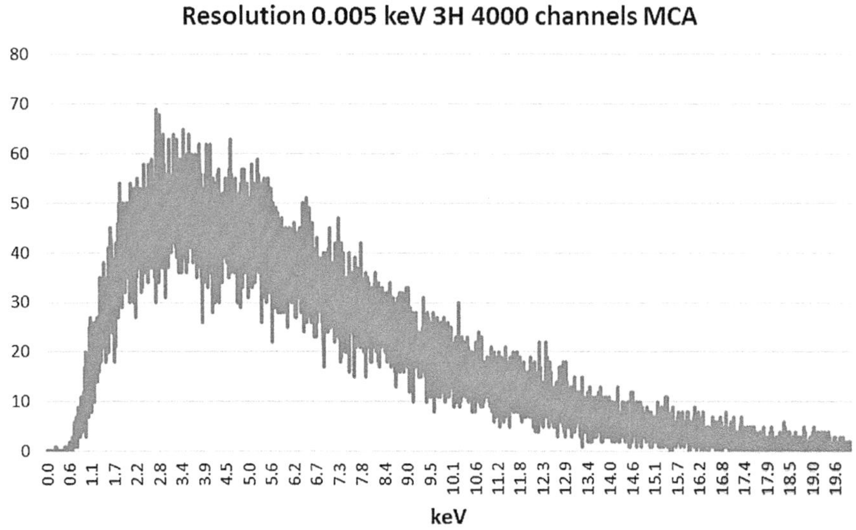

FIGURE 7.8 ^{3}H spectrum with a 0.005 keV resolution in a 4000-channel multichannel analyzer (MCA).

consider LSC as an alternative screening tool to alpha spectrometry and gas flow proportional counting.

A. Extraction chromatographic sample preparation

A range of products that use the technique of extraction chromatography to efficiently preconcentrate and separate radionuclides from a variety of different matrices have been introduced by Eichrom Technologies LLD (Lisle, Illinois, USA) and TrisKem International (Bruz, France). Extraction chromatography combines the power and selectivity of solvent extraction with the ease of use of a chromatographic column. Through careful selection of the extractant, which is adsorbed onto the resin support, a material is produced that is highly specific for a particular radionuclide or group of radionuclides. In the majority of cases, either retention or elution of the radionuclide of interest is achieved by modification of the concentration of the acidic eluent used. Using such technology has resulted in an accurate and reproducible sample preparation method for the determination of radionuclides in environmental samples.

TABLE 7.30 Environmental radionuclides of specific interest[a].

Radionuclide	Environmental location
^{3}H	Milk, crops, groundwater, and precipitation
^{14}C	Milk, crops, animals, and seawater
^{35}S	Milk, crops, animals, and soil/sediments
^{90}Sr	Milk, crops, animals, and soil/sediments
$^{134,137}Cs$	Milk, crops, animals, and soil/sediments
^{238}Pu, $^{239+240}Pu$, ^{241}Pu, ^{241}Am	Milk, crops, animals, and soil/sediments

[a]*© 1998–2011 PerkinElmer, Inc. Printed with permission.*

Eichrom/Triskem has published methods for radionuclide separations using their chromatography technology, and these can be conveniently explained as follows:

ACW, ACS, ACU methods	Refer to actinides in water, soil, and urine
SRW, SRS, SRU methods	Refer to strontium in water, soil, and urine
TCW, TCS methods	Refer to technetium in water and soil
H3W methods	Refer to tritium (^{3}H) in water
OTW methods	Refer to other (e.g., lead in water)

A method for the determination of Cl-36 (and I-129) in environmental and decommissioning samples has been published by TrisKem (TKI-CL). A summary of these methods together with the recommended LSC cocktail for isotopic determination by LSC is presented in Table 7.31. Ultima Gold, Gold Star, and ProSafe+ cocktails are recommended for these applications due to the use of diisopropylnaphthalene (DIN) as the solvent base, which enhances alpha/beta resolution in LSC (Passo and Cook, 1994). The information presented in Table 7.31 shows both the eluent used in the final stripping of the radionuclide from the chromatographic column and the appropriate LSC cocktail, which will accommodate either all or at least a sizable aliquot of the eluent. Table 7.31 should be used in conjunction with Table 7.32, which demonstrates the maximum capacity of each of the LSC cocktails with the each of the eluents.

TABLE 7.31 Compatibility of liquid scintillation countingcocktails with Eichrom eluents[a].

Method No.	Analyte	Emitter	Strip volume (mL)	Eluent	Recommended cocktail
ACS07	$^{234-238}U$	Alpha, gamma	15	1M HCl	AB/XR/LLT/GS/LT[2]
	^{234}Th	Beta, gamma	20	5M HCl/0.05M oxalic acid	AB*/GS
ACW01	$^{234-238}U$	Alpha, gamma	15 (20 + 5)	0.01M HCl	AB/XR/LLT/GS/LT[2]
	^{234}Th	Beta, gamma		9M+6M HCl	AB*/LLT*/GS
ACW03	$^{234-238}U$	Alpha, gamma	15	1M HCl	AB/XR/LLT/GS/LT[2]
	^{241}Pu	Beta, gamma	10 (3 + 20)	0.1M ammonium H oxalate	AB/XR
	^{241}Am	Alpha, gamma		9M+4M HCl	AB*/LLT*/GS
ACW10	$^{234-238}U$	Alpha, gamma	5 (20 + 5)	2.5M HNO_3	AB/XR/LLT/GS/LT[2]
	^{234}Th	Beta, gamma		9M+6M HCl	AB*/GS
OTW01	^{210}Pb	Beta, gamma	20	Water	LLT/LT[2]
SRW01	$^{89,90}Sr$	Beta	10	0.05M HNO_3	AB/XR/GS/LT[2]
TCS01	^{99}Tc	Beta	0.7g (2 mL)	TEVA resin	AB/XR/LLT/GS/LT[2]
TCW01	^{99}Tc	Beta	0.7g (2 mL)	TEVA resin	AB/XR/LLT/GS/LT[2]
ACW04	^{241}Am	Alpha, gamma	15	2M HCl	AB*/LLT*/GS/LT[2]
ACU02	$^{234-238}U$	Alpha, gamma	15	1M HCl	AB/XR/LLT/GS/LT[2]
	^{241}Pu	Beta, gamma	10 (3 + 20)	0.1M ammonium H oxalate	AB/LLT/LT[2]
	^{241}Am	Alpha, gamma		9M+4M HCl	AB*/LLT*/LT[2]
OTW02	Tritium	Beta	25	Nonacidified water sample	LLT/LT[2]
NIW01	^{63}Ni	Beta	15	3M HNO_3	AB*/LLT*/LT[2]

*Indicates limited sample uptake capacity (see Table 7.3 for further details).
[#]Ascorbic acid causes yellowing upon storage (>2 days).
AB = Ultima Gold AB; XR = Ultima Gold XR; LLT = Ultima Gold LLT; GS = Gold Star; LT[2] = Gold Star LT.[2]

It should be remembered that the use of alpha/beta LSC for alpha determination will only provide a gross alpha measurement and is capable of limited alpha—alpha resolution. An example of such an alpha/beta LSC is the Packard Tri-Carb 2770TR/AB (now PerkinElmer), which uses time-resolved pulse decay analysis (TR-PDA) technology to separate the alpha spectrum from the beta spectrum.

An alternative method to acidic stripping of the radionuclide from the column is to elute with IPA. IPA effectively strips the resin coating (containing the radionuclide), and this can be counted with 4π geometry in a suitable LSC cocktail. The information in both these tables is presented to help researchers investigate the use of LSC as an alternative technology for gross alpha determination. There are also methods in which the loaded resin is directly mixed with the cocktail and counted.

B. Aqueous sample preparation

Many of the radioactive species of interest to low-level radioactivity researchers are present in an aqueous medium, usually water. Therefore, any suitable LSC cocktail must not only have a high capacity for water but also be compatible with water from a variety of different sources. These include distilled, deionized, tap, rain, river, and even seawater.

In addition to high sample capacity, other preferred requirements for LSC include a very low background contribution and high counting efficiency. Ideally, the LSC cocktail should also be based on the high flash point, safer solvent DIN. Ultima Gold LLT and Gold Star LT[2] are such cocktails and are primarily designed for the low-level tritium (LLT) monitoring and research sectors.

TABLE 7.32 Sample capacity of liquid scintillation counting cocktails for chromatographic eluents[a].

Eluent	Strip Volume (mL)	UG-AB ml/ 10 mL @ 20°C	UG-LLT ml/ 10 mL @ 20°C	UG-XR ml/ 10 mL @ 20°C	GS ml/ 10 mL@ 20°C	GS-LT[2] ml/ 10 mL @ 20°C
0.01M hydrochloric acid	15	10.0 mL	8.0 mL	10.0 mL	10.0 mL	8.5 mL
0.02M hydrochloric acid	15–20	9.0 mL	7.0 mL	10.0 mL	10.0 mL	8.0 mL
2.0M hydrochloric acid	15	3.5 mL	3.5 mL	1.0 mL		3.5 mL
5.0M hydrochloric acid	15	2.0 mL	1.5 mL	<0.5 mL		
6.0M hydrochloric acid	20	1.0 mL	1.5 mL	<0.5 mL		
4.65M hydrochloric acid (9M + 4M mixture)	3 + 20	1.5 mL	2.0 mL	<0.5 mL		
9.0M hydrochloric acid (concentrated HCl 1.16 S G.)	20	1.0 mL	1.0 mL	<0.2 mL	<0.2 mL	
3M HCl/0.25M ascorbic acid	15	2.0 mL	2.0 mL	0.5 mL	No data	No data
0.05M nitric acid	10	8.0 mL	7.0 mL	9.0 mL	10.0 mL	7.0 mL
3.0M nitric acid	15	2.0 mL	2.25 mL	1.0 mL	3.25 mL	2.25 mL
0.02M HNO_3/0.02M HF	10	8.0 mL	10.0 mL	10.0 mL	No data	No data
0.1M ammonium H oxalate	10	8.0 mL	6.0 mL	9.0 mL	No data	No data
Water	25	10.0 mL	10.0 mL	10.0 mL	10.0 mL	10.0 mL

Additionally, they have other unique performance characteristics that set them apart from currently available cocktails. They meet additional requirements for a low-level counting cocktail such as long-term stability and subambient temperature stability and can accept the important mineral acid species normally encountered in alpha/beta counting applications. A selection of the more important properties is illustrated in Tables 7.33 and 7.34. Table 7.33 shows the performance of Ultima Gold LLT in a Tri-Carb 2770 TR/SL operated in low-level count mode at 15°C. All samples were prepared in duplicate in standard polyethylene vials; background count time used was 500 min per sample. The minimum detectable activity (MDA) was calculated according to Currie(1968).

Table 7.34 shows that Ultima Gold LLT can accept the important mineral acid species normally used in environmental sample preparation methods. In addition to these sample types, Ultima Gold LLT has a high sample capacity for urine and is therefore ideally suited for use in the bioassay of ^{3}H in urine.

As a quick reference guide, the main radionuclides present in the environment and originating from either nuclear devices or the nuclear industry are summarized in Table 7.35.

Environmental sample preparation encompasses a wide variety of techniques, including extraction chromatography, acid extraction, ashing and solvent extraction, and chemical separation.

TABLE 7.33 Performance of Ultima Gold LLT for low-level tritium[a].

Water to cocktail ratio	Cocktail	^{3}H efficiency in 0.4–4.5 keV	Background	E^2/B	E^2V^2/B	MDA Bq L^{-1}
8 mL:12 mL	Ultima Gold LLT	24.6%	1.15 CPM	526	33,680	1.22
10 mL:10 mL	Ultima Gold LLT	21.2%	1.11 CPM	405	40,490	1.11
11 mL: 9 mL	Ultima Gold LLT	18.1%	0.95 CPM	345	41,730	1.06

TABLE 7.34 Sample holding capacity of Ultima Gold LLT with various acids[a].

Temperature°C	1M HCl (mL)	2M HCl (mL)	1M HNO_3 (mL)	2M HNO_3 (mL)	1M H_2SO_4 (mL)	2M H_2SO_4 (mL)	1M H_3PO_4 (mL)	2M H_3PO_4 (mL)
14	4.0	2.0	3.0	2.25	5.5	3.5	4.5	4.0
16	4.0	2.5	3.5	2.25	6.5	4.0	5.0	4.0
18	4.0	2.5	3.5	2.25	7.0	4.0	7.0	4.0
20	5.0	3.0	3.5	2.5	7.0	4.0	4.0	4.0

[a]*© 1998–2019 PerkinElmer, Inc. Printed with permission.*

Advances in liquid scintillation technology, together with new and emerging sample preparation techniques, now enable researchers to consider LSC as an alternative environmental sample radionuclide counting method or as a potentially useful screening tool.

XIV. Waste cocktails—environmental consequences

A. Generation of waste cocktails

Scintillation waste consists of liquid scintillation cocktails (including dissolved and suspended samples) and associated containers such as counting vials. LSC currently accounts for 65% of mixed radioactive, organic waste. Such waste is generated as a result of environmental monitoring, research, teaching, and patient care activities, and all waste that contains or is contaminated with radioactive material requires disposal. This means that if a hazardous chemical or biological hazardous waste contains or is contaminated with radioactive material, it must be disposed of as radioactive waste. However, little attention has been paid to the secondary chemical/biological hazards that may be present.

B. Disposal methods

The components of the LSC cocktail may represent a hazard or a disposal problem in addition to the radioactivity. For many experiments, only a small percentage of the samples counted will have significant radioactivity, so disposal of the LSC cocktail is the predominating issue.

Incineration is a thermal treatment technology that involves the combustion of material through contact with a controlled flame in enclosed chambers.

Chemical oxidation—The process uses UV light and hydrogen peroxide to oxidize organic compounds to carbon dioxide and water.

Steam reforming—Reforming reactions occur in a near-oxygen-free environment under high pressure and temperature and allow an extremely high conversion of organics (wastes) to the primary conversion products of CO_2, H_2O, and H_2.

Bioprocessing is an alternative technology that involves the use of microorganisms, which can metabolize aromatic solvents of the type used in LSC waste. The use of microorganisms serving as biocatalysts in a continuous flow system designed for the processing of stored liquid scintillation waste has demonstrated success in processing stored, "hot" LSC waste. Details of the applicability of bioprocessing to treating mixed waste were presented at the third Biennial Symposium of Mixed Waste in a report entitled "'Bioprocessing' of Mixed Waste: A Success Story." A bioprocess demonstration was conducted on a commercial LSC formulation; the initial results of this technology indicate that hazardous organic solvents, xylene, and pseudocumene have been removed due to processing. The process was applied to two barrels containing LSC waste with a tritium radioactive component and pseudocumene-based hazardous constituent. Further research will determine the applicability of the process to treat mixed waste. Initial results have revealed that bioprocessing can satisfactorily remove certain hazardous constituents from mixed waste. The technology is considered an alternative to incineration and could be considered a more environmentally friendly means of treating this waste.

C. Biodegradability

Biodegradation is the breakdown of organic contaminants by microbial organisms into smaller compounds. The microbial organisms transform the contaminants through metabolic or enzymatic processes. Biodegradation processes vary greatly, but frequently the final product of the degradation is carbon dioxide or methane and water. Biodegradation is a key process in the natural attenuation of contaminants at hazardous waste sites, and it is a microbial process that occurs when all of the nutrients and physical conditions involved are suitable for growth. Temperature is an important variable; keeping a substance frozen can

TABLE 7.35 The main radionuclides originating from nuclear devices and the nuclear industry[a].

Element	Radionuclide	Half-life	Type of emission	Energy (MeV)	Sources
Hydrogen	^{3}H	12.3 years	Beta	0.02	Fallout, nuclear industry
Carbon	^{14}C	5730 years	Beta	0.16	Fallout, nuclear industry
Phosphorus	^{32}P	14.3 days	Beta	1.71	Fallout, nuclear industry
Sulfur	^{35}S	88 days	Beta	0.17	Nuclear industry
Argon	^{41}Ar	1.83 hours	Beta	1.20, 2.49	Nuclear industry
			Gamma	1.29	
Calcium	^{45}Ca	165 days	Beta	0.26	Nuclear industry
Chromium	^{51}Cr	27.8 days	Gamma	0.32	Nuclear industry
			EC	0.752	
Manganese	^{54}Mn	303 days	Gamma	0.84	Fallout, nuclear industry
Iron	^{55}Fe	2.6 years	EC	0.232	Fallout, nuclear industry
	^{59}Fe	45.1 days	Beta	0.273,0.475	Nuclear industry
			Gamma	0.142–1.29	
Cobalt	^{58}Co	71.3 days	Beta	0.474	Nuclear industry
	^{60}Co	5.3 years	Gamma	0.810,0.864	Fallout, nuclear industry
			EC	2.31	
			Beta	0.315,1.49	
			Gamma	1.17,1.33	
Nickel	^{63}Ni	92 years	Beta	0.07	Nuclear industry
Zinc	^{65}Zn	244 days	Beta	0.325	Nuclear industry
			Gamma	1.11	
			EC	1.11	
Arsenic	^{76}As	1.1 days	Beta	0.35–2.96	Nuclear industry
			Gamma	0.51–2.66	
Krypton	^{85}Kr	10.8 years	Beta	0.67	Fallout, nuclear industry
			Gamma	0.14	
Strontium	^{89}Sr	52 days	Beta	1.46	Fallout, nuclear industry
	^{90}Sr	28.1 years	Beta	0.546	Fallout, nuclear industry
Yttrium	^{90}Yr	2.67 d	Beta	2.27	Nuclear industry
	^{91}Yr	58.8 days	Beta	0.33,1.55	Nuclear industry
			Gamma	1.21	
Zirconium	^{95}Zr	65 days	Beta	0.36–1.13	Fallout, nuclear industry
			Gamma	0.236,0.723	
Niobium	^{95}Nb	35.1 days	Beta	0.160	Fallout, nuclear industry
			Gamma	0.766	
Technetium	^{99}Tc	2.12×10^{5} years	Beta	0.29	Fallout, nuclear industry
Ruthenium	^{103}Ru	39.6 days	Beta	0.203–0.90	Fallout, nuclear industry
	^{106}Ru	367 days	Gamma	0.04–0.61	Nuclear industry
			Beta	0.039	
			Gamma	0.512–2.64	
Silver	^{110m}Ag	253 days	Beta	0.087,0.529	Nuclear industry
			Gamma	0.657,0.818	
Antimony	^{124}Sb	60.3 days	Beta	0.06–2.32	Nuclear industry
	^{125}Sb	2.7 years	Gamma	0.044–2.30	Fallout, nuclear industry
			Beta	0.10–0.619	
			Gamma	0.036–0.671	
Tellurium	^{125m}Te	58 days	Gamma	0.035,0.110	Nuclear industry
	^{132}Te	3.25 days	Beta	0.22	
			Gamma	0.049–0.228	
Iodine	^{129}I	1.7×10^{7} years	Beta	0.189	Fallout, nuclear industry
	^{131}I	8.07 days	Gamma	0.040	Fallout, nuclear industry
			Beta	0.257–0.806	
			Gamma	0.080–0.723	

Continued

TABLE 7.35 The main radionuclides originating from nuclear devices and the nuclear industry[a].—cont'd

Element	Radionuclide	Half-life	Type of emission	Energy (MeV)	Sources
Xenon	^{131m}Xe	11.8 days	Gamma	0.164	Nuclear industry
	^{133}Xe	5.27 days	Beta	0.267, 0.347	Nuclear industry
			Gamma	0.080,0.382	
Cesium	^{134}Cs	2.05 years	Beta	0.089,0.410	Nuclear industry
	^{136}Cs	13 days	Gamma	0.475–1.40	Fallout
	^{137}Cs	30.2 years	Beta	0.341, 0.560	Fallout, nuclear industry
			Gamma	0.067, 1.24	
			Beta	0.511, 1.18	
			Gamma	0.662	
Barium	^{140}Ba	12.8 days	Beta	0.47–1.02	Fallout
			Gamma	0.139–0.537	
Lanthanum	^{140}La	1.67 days	Beta	1.25–2.17	Fallout, nuclear industry
			Gamma	0.110–2.55	
Cerium	^{141}Ce	33 days	Beta	0.444, 0.582	Fallout, nuclear industry
	^{144}Ce	285 days	Gamma	0.145	Fallout, nuclear industry
			Beta	0.175–0.309	
			Gamma	0.034–0.134	
Neodymium	^{147}Nd	11.1 days	Beta	0.38, 0.82	Fallout
			Gamma	0.091–0.69	
Promethium	^{147}Pm	2.5 years	Beta	0.23	Nuclear industry
Europium	^{154}Eu	16 years	Beta	0.27–1.86	Nuclear industry
	^{155}Eu	1.81 years	Gamma	0.060–1.60	Nuclear industry
			Beta	0.10–0.25	
			Gamma	0.043–0.105	
Polonium	^{210}Po	138.4 days	Alpha	5.30	Nuclear industry
			Gamma	0.803	
Thorium	^{234}Th	24.1 days	Beta	0.100, 0.191	Nuclear industry
			Gamma	0.030–0.094	
Uranium	^{234}U	2.47×10^5 years	Alpha	4.60, 4.72	Nuclear industry
	^{235}U	7.1×10^8 years	Gamma	0.053–0.580	Nuclear industry
	^{236}U	2.39×10^7 years	Alpha	4.16–4.60	Nuclear industry
	^{238}U	4.51×10^9 years	Gamma	0.074–0.367	Nuclear industry
			Alpha	4.33, 4.44	
			Gamma	0.050	
			Alpha	4.14, 4.15	
			Gamma	0.048	
Neptunium	^{237}Np	2.14×10^6 years	Alpha	4.40–4.87	Nuclear industry
			Gamma	0.020–0.240	
Plutonium	^{238}Pu	86 years	Alpha	5.36, 5.46, 5.50	Fallout, nuclear industry
	^{239}Pu	2.44×10^4 years	Gamma	0.044	Fallout, nuclear industry
	^{240}Pu	6580 years	Alpha	5.01–5.16	Fallout, nuclear industry
	^{241}Pu	13.2 years	Gamma	0.039–0.769	Fallout, nuclear industry
	^{242}Pu	3.79×10^5 yeasr	Alpha	5.02, 5.12, 5.17	Nuclear industry
			Gamma	0.045, 0.104	
			Alpha	4.80–5.05	
			Beta	0.021	
			Alpha	4.86, 4.90	
Americium	^{241}Am	458 years	Alpha	5.39–5.55	Fallout, nuclear industry
			Gamma	0.026–0.060	
Curium	^{242}Cm	163 days	Alpha	5.97–6.11	Nuclear industry
			Gamma	0.044	

[a]*© 1998–2019 PerkinElmer, Inc. Printed with permission.*

prevent biodegradation. Most biodegradation occurs at temperatures between 10 and 35°C. Water is essential for biodegradation, and bacteria and fungi, including yeasts and molds, are the microorganisms responsible for biodegradation. Biodegradability falls into one of the following three categories: primary biodegradation, readily biodegradable, and inherently biodegradable.

Primary biodegradation is more limited in scope and refers to the disappearance of the compound as a result of its biotransformation to another product.

Readily biodegradable refers to specific tests where a specified percentage of the compound on test must biodegrade within a fixed time period.

Inherently biodegradable is similar in specification to readily biodegradable, but there is an initial acclimation period prior to starting the biodegradability testing protocol.

Provided the sample passes one of the following tests, it can be properly described as biodegradable.

1. Testing for biodegradability

The biodegradation rate measures the speed at which microorganisms can break down the test materials. There are several ways in which this can be determined. One of the most widely accepted tests is biological oxygen demand (BOD), which measures the rate of oxygen use of microorganisms feeding on a sample of test material. The higher the rate of oxygen use, the faster the metabolism of the sample by the microorganism. Materials that are nonbiodegradable will return a very low BOD. Samples are usually tested for 28 days, with oxygenuse measurements taken every 5 days. The increase in the rate of oxygen use shows that the microorganisms are acclimatizing to the sample more as a source of carbon in their diets.

There are a variety of methods used to establish biodegradability, and the most commonly used are the Organisation for Economic Cooperation and Development (OECD) methods.

One of the most important screening tests is the MITI-I test (Japanese Ministry of International Trade and Industry), also known as OECD 301C. The MITI-I is a screening test in which the test substance is initially present at 100 mg/L, and the inoculum is 30 mg sludge solids/L. The test measures BOD, and like other OECD ready biodegradability tests, it normally lasts for 28 days. If oxygen demand due to degradation of the test substance reaches or exceeds 60% of theoretical, the test substance is considered readily biodegradable. The MITI inoculum is prepared using a process of feeding a mixture of sludges from various sources for 30 days with peptone only. This standardization reduces the diversity of microorganisms in the sludge and also their ability to acclimate to and degrade various substrates. Apparently, this reduces variability in the results and thereby makes the test of higher utility.

2. Biodegradability test methods

a. OECDTG 301 A ready biodegradability: DOCdie-away test

The biodegradation of organic products, dissolved in water, by chemicoorganotrophic microorganisms using the products as the sole source of carbon and energy is observed. The initial content of organic carbon is 40 mg/L (40 ppm). The test flasks are shaken mechanically throughout the test. The dissolved organic carbon (DOC) left in solution is measured after 3, 7, 14, and 28 days. The corresponding level of biodegradation is calculated and evaluated on the basis of this level.

Pass level is 70%.

b. OECDTG 301 B ready biodegradability: CO_2 evolution test

A chemically defined liquid medium with a carbon content of 5–20 mg/L, essentially free of other organic carbon sources, is spiked with the test material and inoculated with sewage microorganisms. CO_2-free air is bubbled through the solution. The CO_2 released is trapped as $BaCO_3$. After reference to suitable blank controls, the total amount of CO_2 produced is determined for the test period. It is calculated as the percentage of total CO_2 that the test material could have had theoretically.

Pass level is 60%.

c. OECDTG 301 C ready biodegradability: modified MITItest (I)

An automated, closed system device for measuring BOD is used. The samples are stirred continuously. Chemicals to be tested are the sole organic carbon sources and are inoculated with microorganisms in the testing vessels. During the test period, BOD is measured continuously. Biodegradability is calculated on the basis of BOD. Supplemental chemical analysis is undertaken to measure DOC, residual chemicals, etc.

Pass level is 60%.

d. OECDTG 301 D ready biodegradability: closed bottle test

A predetermined amount of the compound is dissolved in a mineral nutrient solution, providing a usual concentration of 2 mg/L active substance. The solution is inoculated with a small number of microorganisms from a fixed population and kept in closed bottles, without any stirring, in the dark at a constant temperature of 20°C ± 1°C. The degradation is followed by oxygen analyses over a 28-day period.

A control with inoculum, but without test material, is run in parallel for the determination of oxygen blanks.

Pass level is 60%.

e. OECDTG 301 E ready biodegradability: modified OECDscreening test

A predetermined amount of the compound is dissolved in a mineral nutrient solution, fortified with a trace element and essential vitamin solution. The concentration reached is 5–40 mg/LDOC. The solution is inoculated with a small number of microorganisms from a mixed population and aerated at 20–25°C in the dark or at least in diffuse light only. The degradation is followed by DOC analysis over a 28-day period. The procedure is checked by means of a standard. A control with inoculation, but without either test material or a standard, is run in parallel.

Pass level is 70%.

f. OECDTG 301 F ready biodegradability: also called the manometric respirometry test

This is the same test as the 301 D but automated to obtain a continuous oxygen measurement. The principal differences are that the test flasks are open, exposed to daylight, and continuously stirred. An advantage is that the test gives data on both primary and ultimate degradation.

Pass level is 60%.

g. OECD 302 A inherent biodegradability: modified semicontinuous activated sludgetest

This test is an adaptation of the semicontinuous activated sludge procedure for assessing LABsulfonate.

>20% indicates inherently biodegradable, and >70% indicates ultimately biodegradable.

h. OECD 302 B inherent biodegradability: Zahn-Wellens/EMPAtest

Modified Zahn-Wellens test (inherent biodegradability):a static test for ultimate degradation.

>20% indicates inherently biodegradable, and >70% indicates ultimately biodegradable.

i. OECD 302 C inherent biodegradability: modified MITItest (II)

Modified MITI test (inherent biodegradability):basically the same test as 301 C.

OECD 304 A inherent biodegradability in soil evaluation of the mineralization rate of a ^{14}C-labeled test chemical.

D. Incineration

Incineration reduces the volume and toxicity of waste materials by converting hazardous organic compounds to non-hazardous products such as water and carbon dioxide. Currently, most of the LSC wastes are sent offsite to licensed facilities where they are combusted. Advantages of the technique are as follows: it is proven and well understood, it results in almost complete reduction of toxic organics if strict monitoring procedures are emphasized, it is a low-treatment technology if operated properly, and it eliminates liquids, producing a waste that is dense and easy to transport. Some of the notable drawbacks of incineration include the relatively high cost, the risks associated with transportation, the potential for release of radioactive materials into the environment, and finally the reintroduction of carbon dioxide straight back into the atmosphere. As an example, if 10 L of LSC cocktail is incinerated, it will produce ~26,000 L of CO_2 (0.05 tonnes of CO_2), which will be injected directly into the atmosphere. There are also costs associated with the disposal of the radioactive ash generated from incineration.

E. Legislation and regulatory information

Almost all currently available LSC cocktails contain nonyl phenol ethoxylates (NPEs). The NPEs are not completely biodegraded and produce metabolites that are known endocrine disrupters. Endocrine disrupters are a group of chemicals (natural, synthetic, industrial chemicals, or by-products), and exposure to which can cause adverse health effects in an intact organism or its offspring or (sub) population by altering the function of the endocrine system. In wildlife, endocrine disrupters have been clearly shown to cause abnormalities and impaired reproductive performance in some species and to be associated with changes in immunity, behavior, and skeletal deformities. In humans, endocrine disrupters have been suggested as being responsible for apparent changes seen in human health patterns over recent decades. These include declining sperm counts in some geographical regions, increased incidences in numbers of male children born with genital malformations, and increased incidences of certain types of cancer that are known to be sensitive to hormones. More controversially, links have been suggested with impairment in neural development and sexual behavior.

As a result of extensive studies, the European Community (EC) directive 2003/53/EC came into force in 2003 and put in place controls to restrict the marketing and use of nonyl phenol (NP) and NPEs. This directive covers all current 27 EC member states (as of 2011 and is the 26th amendment (Directive 89/677/EC) to Directive 76/769/EC). NPs are toxic to fish, aquatic invertebrates, and algae, affect plant growth, and impact upon the reproduction and mortality of terrestrial invertebrates. They can also affect organisms higher up the chain through consumption of lower organisms (particularly for terrestrial organisms).

Canada has also declared a ban on the use of NPEs for certain applications and is in the process of phasing them out completely (Hoponick and Sierra Club, 2005). Japan is also seriously reviewing the issue.

In the United States, the Environmental Protection Agency (EPA) will consider initiating rulemaking under the Toxic Substance Control Act (TSCA) Section 5(B) (4) to add NP and NPEs to the list of chemicals that present or may present an unreasonable risk of injury to health or the environment. The EPA initiated rulemaking in late 2011 to add NP and NPEs to the TRI (Toxic Release Inventory) list. At present, neither NPs nor NPEs are on this EPCRA (Emergency Planning and Community Right-to-Know Act) section 313 list of toxic chemicals. The EPA is supporting and encouraging the elimination of other uses of NP and NPEs, followed by regulatory actions as necessary. The EPA intends to encourage the manufacturers of all NPE-containing direct-release products to move to NPE-free formulations. For example, as the Textile Rental Services Association (TRSA) implements its phase-out commitment through SDSI (Safer Detergents Stewardship Initiative), it will provide an example for other product sectors to follow, especially those with "direct releases" to the environment. The EPA-sponsored CleanGredients database offers a source of ready alternatives—functionally equivalent to NPEs based on performance characteristics and safer, because they meet the EPA Criteria for Safer Surfactants. The EPA will develop an alternative analysis and encourage the elimination of NPE use in other industries that discharge NPEs to water, such as the pulp and paper processing and textile processing sectors, where safer alternatives may be available.

A direct consequence of EC directive 2003/53/EC is that, from 2005, cocktail waste containing NPEs cannot be drain-disposed anywhere within the 27 member states of the EC.

R.E.A.C.H: Liquid scintillation cocktails are sold globally. Their manufacture and subsequent distribution into the supply chain is highly regulated within the chemical industry, and the requirements of each country vary significantly from each other. For example, in the European Union, chemicals are regulated under REACH: **R**egistration, **E**valuation and **A**uthorisation of **Ch**emicals. In the United States of America, it is the TSCA: Toxic Substances Control Act, and in China, chemicals are governed under the Provisions on Environmental Administration of New Chemical Substances. These three regulations differ greatly and highlight the complexity of bringing an LSC cocktail to a global market.

Chemical regulations are wide reaching and cover every aspect of LSC cocktail production starting with importation of raw materials, through formulation, manufacturing, packaging, classification and labeling for supply, finishing with transport and safety documentation that is sent out with the final product. To comply with each of these development stages and with each country's different requirements, an in-depth knowledge is required and expert help, which can be costly. For example, the cost to register a single raw material in Europe is in the region of £250k.

The overall aims of all the chemical regulations are to ultimately ensure that chemicals are safe for human health and the environment and to increase economic efficiency. However, escalating regulatory requirements and subsequent costs have, in some cases, driven up the cost of raw materials to such an extent that they are being withdrawn from the market as they are no longer viable, forcing some of the smaller chemical manufacturers to shut down. Other historic raw materials are being highlighted as being unsafe and have limited time left on the market while safer alternatives are sought.

The raw materials used in LSC cocktails are highly specialized and chosen specifically for the properties they deliver. Surfactants based on APEs that are a key component within several LSC cocktails have been highlighted in Europe as being substances of very high concern. This has recently led to several companies within the EU ceasing production of them, which will vastly reduce the portfolio of LSC cocktails on offer. At the time of printing withdrawal of APEs from the market will reduce 57 currently available LSC cocktails down to only 9 in 2021 when they can no longer be used without permission. Meridian Biotechnologies have proactively developed a range of LSC cocktails based on safer alcohol ethoxylates with equivalent performance as alternatives.

Staying abreast of chemical regulations is critical to enable targeted formulation development of LSC cocktails to keep up with global market demand.

F. Waste cocktails—the way forward

The best way forward is to reformulate the LSC cocktails using alcohol ethoxylates, which are the obvious alternatives. This approach brings both advantages and drawbacks. The significant advantage is that they are readily biodegradable and do not form endocrine disrupters upon degradation. Since the alcohol ethoxylates are aliphatic in nature, they do not contribute to the energy transfer process, and consequently, any LSC cocktail based on alcohol ethoxylates will exhibit slightly lower counting efficiency with respect to lower-energy isotopes such as tritium. For all other higher-energy isotopes, there will be equivalent performance to currently available LSC cocktails. Another consequence of this change to alcohol ethoxylates is that ease of microemulsion formation is partially compromised. The alcohol ethoxylates are slightly inferior to the NPEs with respect to incorporation of aqueous samples, and to overcome this problem, required a significant amount of research and development time.

As of early 2011, the only companies that have overcome these issues and offer NPE-free LSC cocktails are Meridian Biotechnologies Ltd. and PerkinElmer Inc.

Meridian Biotechnologies Ltd. has complete ranges, formulated in a high flash point biodegradable solvent available for both routine LSC and flow scintillation counting. These are called the ProSafe+ and Pro-Flow+ ranges, and both are completely NPE-free, readily biodegradable, and suitable for drain disposal subject to local authority approval. PerkinElmer Inc. has thus far introduced two replacement products called Pico-fluor Plus and Biofluor Plus, and although they are NPE-free, they are based on pseudocumene and therefore totally unsuitable for drain disposal.

The Meridian Biotechnologies Ltd. ProSafe+ and Pro-Flow+ ranges have been tested and been shown to be biodegradable; a further study has shown that, upon biodegradation, the only by-products, apart from water and some CO_2, are simple alcohols and glycols, with no endocrine disrupter species present. At present, many users are still using NPE-based LSC cocktails and accepting the extra costs and inconvenience of disposal by incineration. Change within the LSC community is notoriously slow, but the introduction of new regulations and a greater awareness of environmental consequences are slowly but surely driving a change to NPE-free products.

Acknowledgments

The authors would like to thank PerkinElmer Inc. for their kind permission to use excerpts from their Counting Solutions series of Application Notes. The authors are also grateful that they were able to draw upon material from the work of this chapter on "Sample Preparation for Liquid Scintillation Counting" published in 1998 and 2003 in the *Handbook of Radioactivity Analysis* for the publication of a Chapter 9 "Sample Preparation for Liquid Scintillation Counting" in the book *Radiotracers in Drug Development* by Graham Lappin and Simon Temple published in 2006 by the CRC Press.

References

Anderson, L.E., McClure, W.O., 1973. An improved scintillation cocktail of high solubilising power. Anal. Biochem. 51, 173–179.

Basch, R.S., 1968. An improved method for counting tritium and carbon-14 in acrylamide gels. Anal. Biochem. 26, 184–188.

Benjamin, W.B., 1971. Selective in vitro methylation of rat chromatin associated histine following partial hepatectamy. Nature (London) 234, 18.

Bird, N., 1997. Packard Counter Intelligence Applications, Detection of *Helicobacter pylori* Infection Using A 14C-Urea Breath Test. CIA-001. PerkinElmer Life and Analytical Sciences, Boston.

Birks, J.B., 1971. Liquid scintillator solvents. In: Peng, C.T., Horrocks, D.L., Alpen, E.L. (Eds.), Organic Scintillators and Liquid Scintillation Counting. Academic Press, New York, pp. 3–23.

Bocock, K.L., 1981. Radionuclides in Terrestial Ecosystems, 0904282422.

Bransome Jr., E.D., Grower, M.F., 1970. Liquid scintillation counting of ^{3}H and ^{14}C on solid supports: a warning. Anal. Biochem. 38, 401–408.

Choules, G.L., Zimm, B.H., 1965. An acrylamide gel soluble in scintillation fluid. Anal. Biochem. 13, 336–344.

Currie, L.A., 1968. Limits of qualitative detection and quantitative determination. Anal. Chem. 40 (3), 586–593.

Diener, E., Paetkau, V.H., 1972. Antigen recognition: early surface-receptor phenomena induced by binding of a tritium labelled antigen. Proc. Nat. Acad. Sci. U.S.A 69, 2364–2368.

Dion, A.S., Moore, D.H., 1972. Gel electrophoresis of reverse transcriptase of musine mammary tumour virions. Nature (London) 240, 17.

Freiberg, S., 1990. Scintillation Medium and Method. International Patent WO/91/11735.

Fuchs, A., De Vries, F.W., 1972. A comparison of methods for the preparation of 14C-labeled plant tissues for liquid scintillation counting. Int. J. Appl. Radiat. Isot. 23, 361–369.

Gibson, J.A.B., 1980. Modern techniques for measuring the quench correction in a liquid scintillation counter. In: Peng, C.T., Horrocks, D.L., Alpen, E.L. (Eds.), Liquid Scintillation Counting, Recent Applications and Developments. Academic Press, New York, pp. 153–172.

Gill, D.M., 1967. Liquid scintillation counting of tritiated compounds supported by solid filters. Int. J. Appl. Radiat. Isot. 18, 393–398.

Goodman, D., Matzura, H., 1971. An improved method of counting radioactive acrylamide gels. Anal.Biochem. 42, 481–486.

Hansen, J.N., Pheiffer, P.H., Boehnert, J.A., 1980. Chemical and electrophorestic properties of solubilisable disulfide gels. Anal.Biochem. 105, 192–201.

Hoponick, J., Sierra, C., November 2005. A Safer Alternative Exists to This Toxic Cleaning Agent.

Kaczmar, B.U., Manet, R., 1987. Accurate determination of $^{14}CO_2$ by expulsion from blood. Appl. Radiat. Isot. 38 (7), 577–578.

Kessler, M.J., 1989. Liquid Scintillation Analysis — Science and Technology. Packard Publication, PerkinElmer Life and Analytical Sciences, Boston, pp. 7-1–7-9.

Klein, R.C., Linins, I., Gershey, E.L., 1992. Detecting removable surface contamination. Health Phys. 62, 186–189.

Kobayashi, Y., Maudsley, D.V., 1974a. Biological Applications of Liquid Scintillation Counting. Academic Press, New York, pp. 153–167, pp. 153–167.

Kobayashi, Y., Maudsley, D.V., 1974b. Biological Applications of Liquid Scintillation Counting. Academic Press, New York, pp. 58–68, pp. 58–68.

L'Annunziata, M.F., 1967. The Chelation and Movement of ^{89}Sr and $^{90}Sr(^{90}Y)$ in a Calcareous Soil. The University of Arizona, Tucson. MSc Thesis. https://repository.arizona.edu/handle/10150/318640.

Mahin, D.T., Lofberg, R.T., 1966. A simplified method of sample preparation for determination of tritium, or sulfur-35 in blood or tissue by liquid scintillation counting. Anal.Biochem. 16, 500–509.

Ministry of Agriculture, Fisheries and Food, 1994. Radionuclides in Foods, ISBN 0 11 242975 0. Food Surveillance Paper No. 43.

Mirsky, J., 1990. Liquid Scintillation Medium with a 1,2-dicumylethane solvent. U.S. Patent 5,135,679.

Moore, P.A., 1981. Preparation of whole blood for liquid scintillation counting. Clin. Chem. 27 (4), 609–611.

Morrison, B.J., Franklin, R.A., 1978. A rapid, hygienic method for the preparation of faecal samples for liquid scintillation counting. Anal.-Biochem. 85, 79–85.

National Radiological Protection Board, 1993. Radiation Exposure of the UK Population. Review (NRPB-R263). ISBN 0 85951 364 5HMSO, ISBN 0 85951 364 5.

Passo Jr., C.J., Cook, G.T., 1994. Handbook of Environmental Liquid Scintillation Spectrometry 1994. Packard Instrument Company, Meriden, CT 06450 USA.

Patterson, M.S., Greene, R.C., 1965. Measurement of low energy beta-emitters in aqueous solution counting. Anal. Chem. 37, 854–857.

Peng, C.T., 1983. Sample preparation in liquid scintillation counting. In: McQuarrie, S.A., Ediss, C., Wiebe, L.I. (Eds.), Advances in Scintillation Counting. University of Alberta, Alberta, Canada, pp. 279–306.

PerkinElmer, TopCount Topics #11, Direct Counting of Millipore Multi-Screen Filtration Plates, PerkinElmer Life and Analytical Sciences, Boston.

Pfeiffer, K., Rank, D., Tschurlovits, M., 1981. A method of counting ^{14}C as $CaCO_3$ in a liquid scintillator with improved precision. Int. J. App. Radiat. Isot. 32, 665–667.

Porter, N.G., 1980. A method for bleaching plant tissues prior to scintillation counting. Lab. Pract. 29 (1), 28–29.

Qureshi, R.M., Fritz, P., Dsimmie, R.J., 1985. The use of CO_2 absorbers for the determination of specific ^{14}C activities. Int. J. Appl. Radiat. Isot. 36 (2), 165–170.

Recalcati, L.M., Basso, B., Albergoni, F.G., Radice, M., 1982. On the determination of ^{14}C-labeled photosynthesis products by liquid scintillation counting. Plant Sci. Lett. 27, 21–27.

Riffat, M., Qureshi, R.M., Fritz, P., 1985. ^{14}C dating of hydrological samples using simple procedures. Int. J. Appl. Radiat. Isot. 36 (10), 825.

Ross, H.H., 1965. Colour quench correction in liquid scintillator systems. Anal. Chem. 37 (4), 621–623.

Schadewaldt, P., Forster, M.E.C., Munch, U., Staib, W., 1983. A device for the liberation and determination of $^{14}CO_2$. Anal.Biochem. 132, 400–404.

Sissons, C.H., 1976. Improved technique for accurate and convenient assay of biological reactions liberating $^{14}CO_2$. Anal.Biochem. 70, 454–462.

Smith, I.K., Lang, A.L., 1987. Decolourisation and solubilisation of plant tissue prior to determination of^{3}H, ^{14}C and ^{35}S by liquid scintillation counting. Anal.Biochem. 164, 531–536.

Späth, P.J., Koblet, H., 1979. Properties of SDS-polyacrylamide gels highly cross-linked with N, N′-diallyltartardiamide and the rapid isolation of macromolecules from the gel matrix. Anal.Biochem. 93, 275–285.

Sun, D., Wimmers, L.E., Turgeon, R., 1988. Scintillation counting of 14C-labeled soluble and insoluble compounds in plant tissue. Anal. Biochem. 169, 424–428.

Ten Haaf, F.E.L., 1975. Colour quenching in liquid scintillation counters. In: Crook, M.A., Johnson, P. (Eds.), Liquid Scintillation Counting, vol. 3. Heyden& Son, Ltd., London, pp. 41–43.

ter Wiel, J., Hegge, T.C.J.M., 1983. Mixture for Use in the LSC Analysis Technique. U.S.patent 4,624,799.

ter Wiel, J., Hegge, T.C.J.M., 1991. Advances in scintillation cocktails. In: Ross, H.H., Noakes, J.E., Spalding, J.D. (Eds.), Liquid Scintillation Counting and Organic Scintillators. Lewis Publishers, Chelsea, Michigan, pp. 51–67.

Terman, S., 1970. Relative effect of transcription-level and transcription-level control of protein synthesis during early development of the sea urchin. Proc. Nat. Acad. Sci. U.S.A 65, 985–992.

Thomson, J., 1985. Scintillation Counting Medium and Counting Method. US Patent 4,657,696, April 14th 1987.

Thomson, J., 1991. Di-isopropylnaphthalene-A new solvent for liquid scintillation counting. In: Ross, H.H., Noakes, J.E., Spaulding, J.D. (Eds.), Liquid Scintillation Counting and Organic Scintillators. Lewis Publishers, Chelsea, Michigan, pp. 19–34.

Thomson, J., Burns, D.A., 1994. Packard LSC Counting Solutions, "Radio-Carbon Dioxide ($^{14}CO_2$) Trapping and Counting", CS-001. PerkinElmer Life and Analytical Sciences, Boston.

Thomson, J., Burns, D.A., 1996a. Packard LSC Counting Solutions, "Environmental Sample Preparation for LSC," CS-004. PerkinElmer Life and Analytical Sciences, Boston.

Thomson, J., Burns, D.A., 1996b. Packard LSC Counting Solutions, "LSC Sample Preparation by Solubilization," CS-003. PerkinElmer Life and Analytical Sciences, Boston.

Wahid, P.A., Kamalam, N.V., Sankar, J., 1985. Determination of ^{32}P in wet-digested leaves by Cerenkov counting. Int. J. Appl. Radiat. Isot. 36 (4), 323–324.

Wang, C.H., Jones, D.E., 1959. Liquid scintillation counting of paper chromatograms. Biochem. Biophys. Res. Commun. 1, 203–205.

Wang, C.H., Willis, D.L., 1965. Radiotracer Methodology in Biological Science. Prentice-Hall, Englewood Cliffs, New Jersey, pp. 202–206, pp. 202–206.

Wunderley, S.W., 1986. Composition for Liquid Scintillation Counting. U.S. Patent 4,867,905.

Further reading

Her Majesty's Inspectorate of Pollution. report - Monitoring Programme – Radioactive Substances Report for 1993, HMIP, Cameron House, Lancaster LA1 4XQ.

Ministry of Agriculture, report - Fisheries and Food. Terrestial Radioactivity Monitoring Programme (TRAMP) Report for 1993, Radioactivity in food and agricultural products in England and Wales - (TRAMP/8).

PerkinElmer, TopCount Topics #12, Biological Applications of Microplate Scintillation Counting. PerkinElmer Life and Analytical Sciences, Boston.

PerkinElmer, TopCount Topics #18, Counting Radioisotopic and Luminescent Labels on Filters and Membranes with the FlexiFilter2 Plate. PerkinElmer Life and Analytical Sciences, Boston.

Chapter 8

Radioisotope mass spectrometry

Clemens Walther[1] and Klaus Wendt[2]

[1]*Institute of Radioecology and Radiation Protection, Leibniz University Hannover, Hannover, Germany;* [2]*Institute of Physics, Johannes Gutenberg-University Mainz, Mainz, Germany*

Chapter outline

I. Introduction

In all of its various and widespread areas of analytical applications, mass spectrometry serves as an ultimately sensitive and highly selective method for the determination of major, minor, trace, and even ultratrace components (Bacon et al., 2008; Butler et al., 2018, and earlier issues of both annual reviews). The field of radioisotope determination primarily concerns the quantification of inorganic elemental trace and ultratrace components in lowest contaminations, frequently aside of stable isotopes of the same element. For this purpose, a multitude of highly specialized techniques

Handbook of Radioactivity Analysis: Volume 1. https://doi.org/10.1016/B978-0-12-814397-1.00008-X

have been developed, both for analytical routine applications and for specific experiments in various fields of fundamental or applied research. Applications cover the full range of studies on radioisotope distributions in environmental and technical samples. In addition to research on the distribution of naturally occurring radioactivity and its applications in science and technology, the surveillance of anthropogenic radioactive fallout from weapon tests and the nuclear industry by nuclear safeguards is of major importance (Boulyga et al., 2015). This field has gained strong additional relevance by the public awareness to the two major nuclear accidents, in Chernobyl, Ukraine, in 1986 and Fukushima, Japan, in 2011. Also, we want to point out the increasing use of radioisotopes in medicine, in industrial production, and even applications in forensic studies (Mayer et al., 2015; Wang et al., 2018). In general, the characterization of nuclear materials, recycled products, and by-products, that is, spent nuclear fuel or depleted uranium ammunition, as well as the control and management of nuclear waste, is of highest concern regarding nuclear safety (Betti and de las Heras, 2004; Chartier et al., 1999). Industrial applications require the study of radioisotope contents in glasses, ceramics, or the semiconductor industry (Rohr et al., 1994; Fukuda and Sayama, 1997), and the sensitive characterization of highest-purity substances (Beer and Heumann, 1992; Herzner and Heumann, 1992). Finally, very specific fields of lowest-level radioisotope mass spectrometry concern the analysis of hot particles in the environment (Tamborini and Betti, 2000; Saibu et al., 2018), the surveillance of radiopharmaceuticals for nuclear–medical applications, geochronology, and cosmochemistry. Work at considerably higher radioactivities includes isotope enrichment, purification, and characterization of dedicated radioactive sources.

Obviously, radiometric techniques detecting α-, β-, or γ-radiation are the standard methods for the determination of radionuclides. Nevertheless, with respect to long-lived radionuclides and in particular those dominantly emitting β-radiation, these standard methods fall short of the detection limits required for ultratrace analyses, depending on the decay type, half-life, and numerous sources of background or noise. Trace analytical measurements of β-emitters, for example, $^{90,89}Sr$ or ^{99g}Tc, require time-consuming chemical separations to remove any interfering source of radioactivity. With respect to α-spectroscopy, using conventional surface barrier detectors, almost carrier-free samples are a prerequisite to achieve a reasonable energy resolution for radioisotope assignment. Even then, unambiguous radiometric isotope analysis is often prevented by interferences of very similar α-energies, e.g., for $^{239}Pu/^{240}Pu$.

On the contrary, mass spectrometric techniques directly count individual atoms of the species of interest involving dedicated experimental setups and ionization methods to minimize background and interferences. In this manner, mass spectrometric methods achieve a very good sensitivity together with isotopic as well as isobaric selectivity. In addition, the recording of a full isotopic pattern of an element often provides beneficial information about the origin and composition of a radioisotope contamination. Correspondingly, mass spectrometric data are considered to be superior to radiometric data in quite many analytical cases, and its higher significance in many cases is appreciated accordingly.

The choice of the best suited analytical technique depends on a number of aspects including the following:

- Type and available amount of sample material
- Specific decay characteristics of the radioisotope in question
- Expected overall content of radioisotopes and their composition in the sample
- Possible existence of interferences, such as, molecules, or other radionuclides

An important criterion for the application of radioisotope mass spectrometry for a specific isotope is given by the following reasoning: The number of events from radioactive decay counting during a given measuring time, which should not exceed approximately 1 week, increases proportional to the number of atoms in the sample and to the decay rate (proportional to 1/half-life). The number of counts obtained by mass spectrometry depends on the number of atoms alone and, of course, in both cases radiometric and mass spectrometric methods, on the efficiency of the detection method. Hence, mass spectrometry becomes increasingly advantageous with decreasing decay rate, that is, increasing half-life of the radionuclides, as compiled for a number of important long-lived radioisotopes in Table 8.1. In a conservative assessment, a measuring time of 24 h, an efficiency of 100% and absence of any background events are assumed for an optimum radiometric count rate to compare with the atom number present in the sample. For each of these isotopes, the number of atoms exceeds the possible decay counts by far more than three orders of magnitude.

A compilation of about 40 relevant "environmental radioisotopes" with half-lives beyond about 1 year is given in Fig. 8.1, plotted with their half-life against the proton number of the element and indicating their origin by the plotting symbols. For comparison, the age of both, the universe as well as the earth, were added, indicated as solid (age of the universe) and dashed (age of the earth) horizontal lines.

TABLE 8.1 Some selected long-lived α- and β-emitting radioisotopes.

Radionuclide	Primary decay type	Half-life $t_{1/2}$	Activity (mBq)	Ideal radiometric count rate within 24 h	Atom number for 1 fg sample content
^{3}H	β^-	12.3 a	$4.5 \cdot 10^4$	$3.9 \cdot 10^9$	$2.0 \cdot 10^8$
^{14}C	β^-	5730 a	$2.1 \cdot 10^1$	$1.8 \cdot 10^3$	$4.3 \cdot 10^7$
^{41}Ca	β^-	$1.03 \cdot 10^5$ a	0.38	32	$1.5 \cdot 10^7$
^{90}Sr	β^-	28.5 a	$6.7 \cdot 10^2$	$5.8 \cdot 10^4$	$6.7 \cdot 10^6$
^{99}Tc	β^-	$2.1 \cdot 10^5$ a	0.078	6.7	$6.1 \cdot 10^6$
^{129}I	β^-	$1.57 \cdot 10^7$ a	0.001	0.086	$4.7 \cdot 10^6$
^{239}Pu	α	$2.41 \cdot 10^4$ a	0.27	23	$2.5 \cdot 10^6$

Atom numbers, activities and optimally expected radiometric count rates for a sample content of 1 fg of the species are compared. Note that 10^{-2} mBq, corresponding to about 1 decay/24 h, is the absolute limit for radiometric determination under optimum conditions.

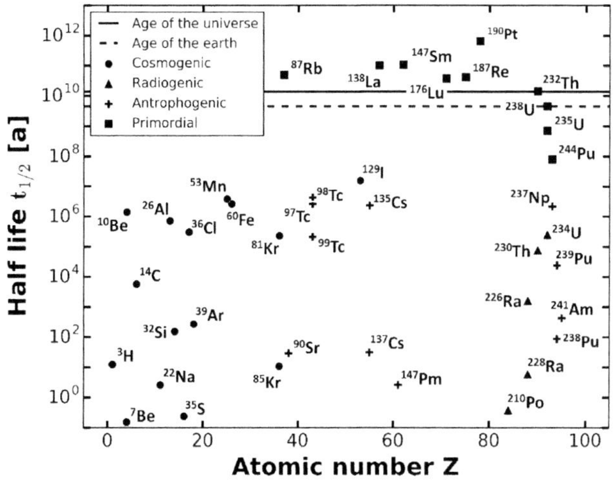

FIGURE 8.1 Relevant "environmental radioisotopes" with half-lives exceeding about 1 year. Half-lives are plotted against the proton number of the element. The origin of individual isotopes is indicated by the plotting symbols.

As noted previously, the sensitive and rapid (in case of an emergency) determination of some long-lived radioisotopes is of highest relevance in areas such as follows:

- Risk assessment
- Low-level surveillance and discrimination of environmental contaminations
- Studies of biological radiation effects and hazards
- Mid- and long-term radioactive waste control
- Fundamental research in geo- and cosmochemistry
- Investigations on the speciation and migration behavior of elements without stable isotopes, such as Pm, Tc, and the actinides

The most prominent mass spectrometric techniques used for these investigations on radioisotopes are based on a limited set of basic instrumentation types. These include the following:

- Thermal ionization mass spectrometry (TIMS) (Platzner, 1997; Aggarwal, 2016)
- Glow discharge mass spectrometry (GDMS) (Betti and de las Heras, 2004)
- Secondary ion mass spectrometry (SIMS) (Adriaens et al., 1992; Betti et al., 1999)
- Inductively coupled plasma mass spectrometry (ICP-MS) (Kim et al., 1991; Crain, 1996; Pröfrock and Prange, 2012) or laser ablation inductively coupled plasma mass spectrometry (LA-ICP-MS) (Becker, 2010)
- Resonance ionization mass spectrometry (RIMS) (Trautmann and Wendt, 2012)
- Accelerator mass spectrometry (AMS) (Fifield, 1999; Kutschera 2005, 2016)

RIMS and AMS are used for high-sensitivity monoelemental ultratrace analyses and determinations of isotope ratios requiring specific isobaric selectivity and very high dynamical range. The other mass spectrometric methods, in general, represent sensitive multielement techniques permitting the determination of absolute concentrations and isotopic abundances of trace and ultratrace elements, which might be somewhat affected by interferences.

The different mass spectrometric methods can be categorized according to the processes used sequentially for evaporation, atomization, and ionization of the sample species. For the analysis of bulk materials, for example, compact solids or powder samples, the initial step of evaporation/atomization could involve different approaches, which include thermal evaporation, for example, from a filament or oven, evaporation and atomization in a plasma discharge, or ablation by impact of laser radiation. In contrast, the atomization may be induced by a concerted particle interaction involving electrons or a well-controlled ion beam for initiating a sputtering process. Subsequent ionization is achieved by different processes among which are the following: as a side effect of this atomization process, for example, directly on a hot surface, by collisional ionization induced by the electrons, ions, or in the plasma, or finally, by resonant or nonresonant single- or multi-photon absorption processes induced by laser light. Simultaneous or quasisimultaneous evaporation and ionization occurs commonly in the ion sources of TIMS, SIMS, ICP-MS, and AMS apparatuses. An incremental procedure, separating evaporation and atomization from ionization of the atomic species in time and/or space, is arranged in TIMS by using multiple filaments, and this occurs in GDMS, LA-ICP-MS, and RIMS.

Ion acceleration, ion beam formation, and guidance into the subsequent mass dispersive system are chosen in a way that is well adapted to the specific properties of the ions formed. Furthermore, possible interference by elemental, molecular, and cluster ions that may occur in large surplus determines the choice of the optimum technique. Both static magnetic sector-field and combinations of electric and magnetic sector-field systems, for example, in Mattauch-Herzog or Nier-Johnson type mass spectrometers (Prohaska et al., 2015), which apply high ion acceleration energies in the range of several keV, are in use. Such a combination of a magnetic mass analyzer with an electrostatic analyzer leads to a so-called double-focusing sector-field mass analyzer. The electrostatic sector establishes ion trajectories by energy only. In contrast, the magnetic sector determines trajectories by both mass and energy, that is, momentum. When these two ion-optical elements are combined, ions are sorted by mass only while the two sectors provide angular focusing. In this way, the total system is considered to be double-focusing. Resulting mass resolutions are sufficient to separate interfering isobaric components and even study elements affected by elemental or molecular isobaric interferences.

For many applications, dynamic ion separation systems such as compact but powerful radiofrequency quadrupole-mass filters (Paul and Raether, 1955; Dawson, 1976), time-of-flight (TOF) systems, or the variety of different ion trap mass spectrometer designs are highly competitive. These most often very compact devices in most cases use ion sources operating near ground potential, which makes them beneficial for a number of routine or specific applications, for example, when continuous sample supply is required.

Ion detection after mass separation is performed by a variety of detector types. Historically, mass spectrometry started using ion-sensitive photographic emulsion plates to record full mass spectra. Unfortunately, these systems were limited in their sensitivity, linearity, and dynamical range. In addition, they required time-consuming photographic development before data analysis. Faraday cups, measuring directly and quantitatively the ion current, became the workhorses of mass spectrometry for long time. For this purpose, the efficient suppression of any emitted secondary electrons, which would falsify a quantitative reading, was mandatory. Today, highly sensitive detectors, most often capable of recording single ions, revolutionized the field with respect to sensitivity and background suppression. The far majority of detector devices in the field are based on the technique of secondary electron generation and multiplication in discrete electrode arrangements, in microstructured channel plates, or in single-channel multipliers (channeltrons). Fully enclosed, more durable, and protected scintillation type detectors with photomultipliers (Daly detectors) are advantageous for some specific tasks in radioisotope detection (Daly, 1960). In addition, pixel-based solid-state detectors are in use, which offer spatial resolution for imaging applications (Jungmann and Heeren, 2013).

A direct analysis of the sample material without any preparatory chemical separation, as practicable in the case of GDMS, SIMS, or LA-ICP-MS, is very favorable but only possible in a few cases and specific samples. In such circumstances, possible inevitable contaminations, which could be introduced during chemical pretreatment, for example, in TIMS, RIMS, or AMS measurements, or which might result from interferences within the required aqueous solutions (ICP-MS), are prevented. However, the quantification in such direct, solid-state mass spectrometric analyses is much more difficult than in chemically pretreated samples, since often there is no suitable standard reference material available in the same form as the sample (Becker et al., 2000). Correspondingly, usual quantification methods such as the addition of internal or external standards or the use of isotope dilution techniques (Muccio and Jackson, 2009; Garcia Alonso and Rodriguez Gonzalez, 2013) might fail. In recent years, an increasing number of so-called hyphenated techniques were developed and used routinely. These techniques combine a (often chemical) preselection step, usually involving chromatographic separation, to produce pure fractions, which afterward are online coupled to the analyzing mass spectrometric methods (Bürger et al., 2007; May et al., 2008).

In the following sections of this chapter, a short introduction into the relevant figures of merit of mass spectrometric techniques, the most prominent approaches used in radioisotope analysis and a set of specific, yet typical applications for the determination of long-lived radioisotopes and related isotope ratio measurements will be presented.

II. Figures of merit

Mass spectrometric techniques are characterized by a set of basic specifications regarding the performance of the instrumental setup with respect to individual aspects of analytical relevance. Also, for radioisotope mass spectrometry, these numbers are associated with the common figures of merit typical for conventional mass spectrometry, that is, the mass resolution, the dynamical range, the selectivity, and finally the sensitivity. The latter is often expressed by the level of detection (LOD). In general, the experimental information, which is extracted from any mass-dispersive spectrometric system, is just a raw mass spectrum. Such a spectrum gives intensities and positions of most often irregularly shaped peaks along a sequence of masses. These data are afterward converted into useful results via blank measurements, the recording of a calibration curve as well as by systematic studies on the technology at hand. The understanding of the specific performance of the setup is mandatory, and the related parameters are common for all different techniques and are defined as follows:

- The basic quantity **mass resolution** is given by the ratio of the mass number divided by the width of the mass peak as

$$R = \frac{m}{\Delta m}.$$

  It determines the location in the mass spectrum and defines the precision, with which a specific mass value can be assigned to a recorded peak. In the analysis of radioisotopes, as for similar studies in inorganic mass spectrometry of atoms and small molecules, a rather low and easily achievable resolution with values beyond 500 is sufficient for the mass dispersive element. This is easily achieved by magnetic sector fields, TOF, or most quadrupole mass filters. Only in the case of high-precision mass determination or in studies on the radioisotope content on large biomolecules, much higher values of mass resolution in the order of 10^5 and more are needed, which are provided by ion trap spectrometers.
- The **dynamical range** of a mass spectrometer is given by the specifications of the detector—or possibly different detector channels—for the maximum permissible count rate, considering dead time corrections, versus the unavoidable background count rate. It is determined experimentally from the calibration curve, which ideally is taken from the lowest detectable concentration up to the maximum permitted count rate. Under optimum conditions, it should confirm a linear response in this full range of up to many orders of magnitude and, for multicollector systems, relate the readings of the different channels.
- The **selectivity** of a mass spectrometric device is loosely related to the resolution but quantitatively specified by the inverse of the abundance sensitivity. The latter is given by the contribution of a chosen interfering peak—not necessarily from a directly neighboring mass—to the desired mass peak, which easily could amount to many orders of magnitude surplus. Due to the usually irregular, asymmetric, and not analytically reproducible mass peak shape, this quantity cannot be predicted theoretically but must be determined experimentally under operation conditions. In addition, it must be clarified if its value is based on a ratio of peak heights or integrated areas, which could well deliver significant variations. According to the experimental conditions, it usually differs toward the lower or higher side of a mass peak and is strongly affected by specific experimental conditions such as residual vacuum pressure, ionization technique, mass analyzer type, and preconditioning of the sample.
- The **sensitivity** is an experimentally defined quantity, which depends on the efficiencies of the individual process steps: sample preparation, evaporation and atomization, ionization, mass spectrometer transmission, and detection. Quantitatively, it is often defined in a relative form, giving a detected number of ions (or charges) versus the sample amount used. The sensitivity is related to the **LOD**. This quantity is the smallest (true) value of the measurand that can be detected with high reliability. The LOD is calculated based on the background or blank measurement. Furthermore, the expected best estimates and uncertainties of the measurand's assumed true value (ion counts) are needed. Using the probability distribution of the measurand's value (e.g., a Gaussian), the probability $(1-\gamma)$ for the true value to lie inside the coverage interval is calculated. Since the width of this Gaussian probability function decreases with increasing number of counts, the LOD decreases. It thus defines the lowest relative contents m/M of the species m under study, which can be reliably determined in the sample of a given size M.
- A final specification is the **precision**, which can be achieved in the measurement of isotope ratios. It is significantly dependent on the dynamical range, which must be covered for the determination of the different trace and reference isotopes, and usually lies in the range of few percent for RIMS or AMS up to better than 0.1% for TIMS.

More details on the figures of merit of analytical techniques are found in textbooks on statistics, analytic chemistry, or specifically mass spectrometry, e.g., Prohaska et al. (2015). For radioisotope mass spectrometry, concentrations commonly in the range of ng/g to pg/g for solids and down to sub-pg/L for aqueous solutions can be detected. Exceptional precision of 0.02% relative standard deviation (RSD) for isotope ratio measurements has been demonstrated using TIMS and ICP-MS (Becker, 2002), whereas RIMS and AMS can detect 10^6-10^7 (fg) and 10^4 atoms (ag) per sample, respectively, for which sample sizes in the μg (RIMS) up to mg (AMS) range are used after chemical preseparation. These values correspond to lowest isotope ratios in the 10^{-15} range and below, but with reduced precision in the low percentage range.

III. Thermal ionization mass spectrometry

A. Principle of surface ionization

TIMS is well suited for very sensitive and, at the same time, precise measurements of isotope ratios, achieving overall detection efficiencies in the region of 0.1% and precision values of even better than 0.01% (Platzner, 1997; Bürger et al., 2015). Usually, it is performed with sector-field mass spectrometers providing acceleration voltages in the order of 3−10 kV. For TIMS measurements, a small volume (∼10 μL) of an aqueous solution is deposited on a filament, that is a thin ribbon made of a refractory metal (Kawai et al., 2001). In many cases, the ion source contains an arrangement of two or even three filaments. The first filament is used for exhaustive and smooth evaporation of the analyte by careful thermal heating and the additional filaments produce the efficient ionization of the evaporated atoms or molecules on an extra hot surface. Ion acceleration toward the mass spectrometer is carried out using an extraction electrode together with a beam-forming electrostatic lens.

The release of atoms or molecules from a hot surface populates final states according to their excitation energies following statistical thermodynamics. As described by the Saha-Langmuir equation, positively ionized states are preferred for species with low ionization potential I_p as compared with the work function Φ of the ionizer material, which determines electron emission or capture at the - usually metallic - surface (Dresser, 1968). The ratio of positive ions to neutrals in thermal equilibrium is expressed by

$$\frac{n_+}{n_0} \propto e^{(\Phi - I_p)/kT}.$$

Negative surface ionization only occurs for those elements exhibiting a high electron affinity E_a, and it is determined by the relation

$$\frac{n_-}{n_0} \propto e^{(E_a - \phi)/kT}.$$

In both cases, T is the temperature of the filament and k is the Boltzmann constant.

In vacuum, all elements with ionization potentials I_p well below 7 eV may be efficiently evaporated and are ionized positively at temperatures around $T \approx 1000{-}2500$ K, respectively (**p-TIMS**). Generally, filaments manufactured of the refractory elements Ta, Re, or W are in use. These exhibit comparably high electron work functions Φ around 5 eV, which slightly depend on the electronic configuration at the surface, together with ultimate temperature stability expressed by the highest melting points of all the elements lying beyond 3000 K. In this way, singly charged positive ions are obtained with high efficiency for alkali and alkaline earth elements, lanthanides, and actinides. Background from molecular hydrocarbons present in the vacuum system is strongly suppressed (Jakopic et al., 2010). For elements with a first ionization potential above 7 eV, special ion emitting reagents such as silica gel and H_3PO_4 and/or H_3BO_3 are used to stimulate the formation of MO_x^+ ions.

Negative ions of only a very few elements, i.e., the halogens, S, Se, Os, Re, and Te, are accessible by thermal ionization with reasonable sensitivity (**n-TIMS**), most often studied in the form of an oxide molecule MO_x^- (Liu et al., 2000). For the determination of precise isotope ratios of Os and Re, specific corrections on the oxygen isotope are required (Liu et al., 1998). Thoriated W or LaB_6 surfaces with low electron work functions $\Phi \approx 2.7$ eV are used frequently. Similarly, when volatilizing refractory metal elements in the form of MO_x molecules, corresponding negatively charged oxide ions are produced by thermal ionization and may be studied. On the other hand, n-TIMS quite often exhibits memory effects, and the stable and reliable operation of the ion source is much more delicate than in the case of p-TIMS. Thus, in general, it is not of practical relevance for radioisotope studies, as discussed in a direct comparison between the pros and cons of both technologies, i.e., p-TIMS and n-TIMS (Heumann et al., 1995). Also, the application of inductively coupled plasma mass spectrometry (see Section VI) on negative ions has been reported, but it is a similarly exotic system (Vickers et al., 1988).

Here, we consequentially limit the discussion of TIMS to the generation and study of positive ions in p-TIMS. A sketch of a p-TIMS system using a 90 degrees sector-field magnet is given in Fig. 8.2. Typical arrangements for single, double, or triple filaments are included schematically in the inlay aside of the source. Before use, filaments are thoroughly cleaned and thermally conditioned by controlled heating sequences in high vacuum, including short excess heating to maximum temperature just before destruction.

Samples for TIMS, containing rare radionuclides, must be particularly well prepared using pure reagents to avoid isobaric contaminants. These controlled conditions reduce errors due to fractionation of the isotope content during the measurement (Johnson and Beard, 1999). The correction for the isotope fractionation occurring in thermal ionization has been investigated in detail (Ramakumar and Fiedler, 1999; Habfast, 1998), whereas isobaric interferences in the sample could significantly

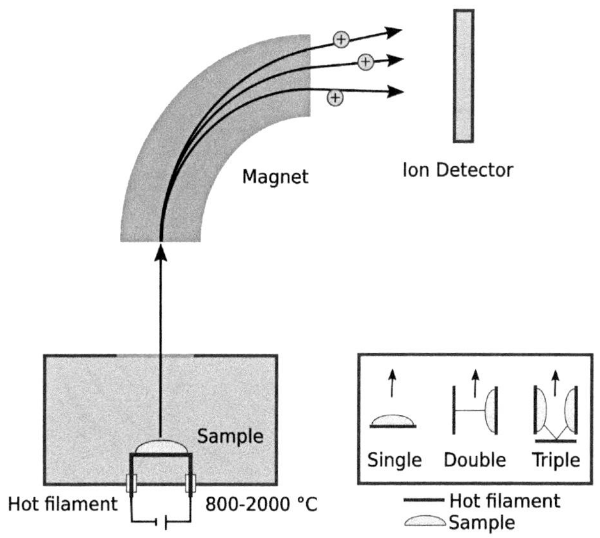

FIGURE 8.2 Schematic view of thermal ionization at a sector-field mass spectrometry. The inlay shows details of a single filament (left), as used, for example, for alkali elements, or of multiple filaments (middle and right) as used for alkaline earth, lanthanides, and actinides.

affect and limit investigations, for example, for measurements of Pu in the presence of a vast surplus of U (Kumar et al., 1995).

B. Applications

High-precision TIMS measurements for the determination of radioisotope contents and isotope ratios in a sample imply a proper quantification. In most cases, this is done via isotope spiking to apply isotope dilution techniques, as pointed out in detail in Fassett and Paulsen (1989), Heumann (1992), Muccio and Jackson (2009) and Garcia Alonso et al., (2013).

Fast magnet switching and multicollector detection, as developed by Wieser and Schwieters (2005), and applied in total evaporation TIMS (Fiedler, 1995), was demonstrated for the determination of ^{187}Re as well as the actinide isotopes ^{238}U and ^{239}Pu, to minimize errors introduced by instabilities in the ion source and the overall operation of the mass spectrometer (Ramakumar and Fiedler, 1999). For measuring radioactive isotopes of alkaline earth elements, that is, radium, preseparation using chromatography is coupled to total evaporation TIMS, achieving detection limits as low as 0.1 pg/g and a precision in isotope ratios better than 1% for ^{226}Ra in silicate rocks (Yokoyama and Nakamura, 2004). Analytics in geochronology or in nuclear industry–related applications comprise high-precision isotope ratio determination with medium to high sensitivity, which is ensured by the good overall efficiency of the technique with correspondingly achievable high statistics (Richter and Goldberg, 2003). For these tasks, a high dynamic range and specifically high stability of the thermal ionization process are required for proper quantification of the element/isotope under investigation versus the reference.

1. *Isotope ratio determination with thermal ionization mass spectrometry*

Geochronology based on ^{39}Ar and ^{40}K decay has been reported exhibiting a reproducibility of 0.03% for measurements on sub-μg samples (Nägler and Villa, 2000). Studies on Li isotopes have included investigations of nucleosynthetic processes in the solar system and the precise measurement of Li in open seawater and several international rock standards (James and Palmer, 2000). The monitoring of accidental Pu intake of workers has been performed by TIMS at different laboratories (Inkret et al., 1998; Ham and Harrison, 2000; LaMont et al., 2005; and Hare et al., 2010). The Institute for Reference Materials and Measurements IRMM at Geel, Belgium, reports on U abundances and isotope ratios in natural U (Richter et al., 1999, 2010) as well as in nuclear materials (Richter and Goldberg 2003). A direct and fast TIMS method was developed to overcome the influence of impurities on the measurement in ^{235}U/^{238}U isotope ratio measurements in nuclear spent fuel safeguards by using refractory metal oxide (Al-Nuzal, 2011). The interlaboratory and other errors in Pb isotope analyses were carefully investigated with a combined TIMS and GDMS study using a ^{207}Pb–^{204}Pb double spike (Thirlwall, 2000). A reliable analytical technique for Mo isotope abundances in fission products (Wieser and De Laeter, 2000) and also in samples for double-beta decay studies (Wieser and De Laeter, 2001) was tested on molybdenites and other high-purity metals and was compared with IUPAP reference materials. A modified thermal ionization mass spectrometer with deceleration components was presented in measurements on ^{234}U/^{238}U standards and on ^{230}Th/^{232}Th samples (Van Calsteren and Schwieters, 1995), delivering an abundance sensitivity in the order of 10^{-8}. In parallel for the isotope ratios of ^{234}U/^{238}U of ~0.7% and ^{230}Th/^{232}Th of 6.1×10^{-6}, a precision of 1.3% was obtained.

2. *High-sensitivity measurements with thermal ionization mass spectrometry*

Interesting work has been reported on TIMS applications regarding high-sensitivity measurements at lowest concentrations, for which any kind of isobaric interference must be reliably suppressed (Taylor et al., 1998). One example is the precise counting of atom numbers for half-life determination of extremely long-lived isotopes, undertaken, for example, in the case of ^{126}Sn (Oberli et al., 1999) and ^{244}Pu (Aggarwal, 2006). Other applications concern radiogenic isotope geology and geochronology, as described in detail in textbooks, such as Dickin (2018). Sensitive ultratrace determinations in the attogram per liter level in seawater have been reported for the actinide isotope ^{231}Pa, demonstrating overall efficiencies exceeding 0.5% (Shen et al., 2003), whereas an overall efficiency of 2.7% for femtogram level studies on Pu isotopes was demonstrated for TIMS, including separation recovery from an initial chromatographic preseparation step (Grate et al., 2011).

IV. Glow discharge mass spectrometry

A. Principle of ionization in a glow discharge

Glow discharge mass spectrometry (GDMS) represents a very powerful and efficient analytical method for direct trace analysis and depth profiling (Harrison, 1988; Betti and de las Heras, 2004; Venzago and Pisonero, 2015). It has found widespread applications in the determination of trace-level distributions at surfaces and within conducting and nonconducting solids or dispersed liquids (Duckworth et al., 1993; Barshick et al., 1993, 2000). It is also frequently applied to the characterization of nuclear samples (Betti, 1996) as well as nuclear forensics (Betti et al., 1999). GDMS has the advantage of almost entirely avoiding sample preparation, as measurements are carried out directly on the initial sample material. This serves as the cathode for a low-pressure argon plasma, burning at a pressure of ca. 1 mbar. The Ar^+ ions are accelerated by the source potential of a few hundred volts and sputter off sample constituents as atoms, ions, molecules, and clusters. The plasma conditions ensure an efficient dissociation and atomic ion formation by Penning and/or electron impact ionization. Positive ions from the glow discharge are emitted via an exit aperture through the anode and are analyzed usually by a double-focusing mass spectrometer, for example, in reverse Nier–Johnson geometry like the commercial VG-9000 instrument (Winsford, UK), as sketched in Fig. 8.3. Such a system provides high transmission (>75%) and a mass resolution of up to $m/\Delta m \approx 10{,}000$. For the handling of highly radioactive nuclear materials, GDMS systems are operated in sealed glove boxes (Betti et al., 1994). Depending on the specific sample material, either a direct current (dc) or radiofrequency (RF) power is used as the ion source. With dc-GDMS in electrically conducting materials, LODs in the ng/g range with a reproducibility of ~10% were reported (Van Straaten et al., 1994; Venzago and Weigert, 1994). Due to electrical charging of the sample surface, the analysis of nonconducting materials is more demanding. Admixtures of nonconducting powders with high-purity metal or graphite powder (De Gent et al., 1995) or alternatively the use of an electron flooding secondary cathode (Schelles et al., 1996) was applied successfully. RF-GDMS permits the analysis of conducting as well as semiconducting or nonconducting materials without major restriction (Marcus, 1996).

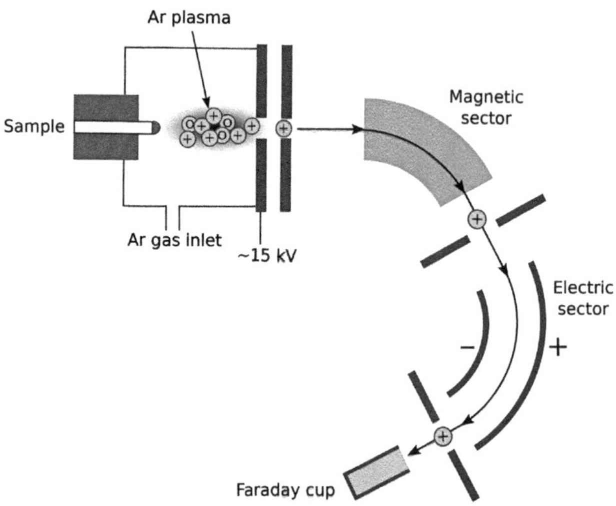

FIGURE 8.3 Schematics of a glow discharge mass spectrometry system in reverse Nier–Johnson geometry. The glow discharge source is shown enlarged and worked out in more detail.

B. Applications

In the fields of nuclear technology, the characterization of nuclear fuels with respect to elements and individual isotopes is highly relevant. This includes high-temperature metallic cladding materials, before and after exposure, nuclear waste, and undeclared nuclear material, often on high to highest activity levels. On the other hand, the direct determination of suspected low-level radiocontamination in environmental samples and hot particles is of major concern. GDMS is successfully applied in both fields.

1. Trace and bulk analysis of nuclear samples

Metallic alloy fuels consisting of UNdZr and UPuZr with uranium content of up to 81% were analyzed semi-quantitatively by GDMS using the signal intensity of the analyte and taking into account the specific element sensitivity of uranium. The relative sensitivity factor (RSF) values were obtained from a comparison with other methods such as ICP-MS and ICP-AES in analyzing uranium metal specimens of different origin. A good agreement between concentration values as determined directly by GDMS and other techniques, which involve specific chemical sample preparation, was found (Betti, 2002, 2005). Further applications of GDMS include the analysis of zirconium alloys, which are used as cladding materials for nuclear fuel elements. Quantitative data on these zircaloy cladding materials were obtained in agreement by applying either matrix specific or RSF values for a uranium metal sample. In this way, GDMS meets the requirements for bulk solids analysis in nuclear systems, providing a full elemental analysis at major, minor, and trace concentration levels with lowest matrix dependency (Robinson and Hall, 1987).

Nonconductive nuclear samples, especially oxide-based compounds such as uranium and plutonium oxide samples, are investigated by GDMS (Betti, 1996; Betti and de las Heras, 2004). Different arrangements were developed for this purpose and are in use, often in combination with each other, such as the following: (1) a secondary cathode is placed directly in front of the (nonconductive) sample surface (Tong and Harrison, 1993); (2) a powder sample is mixed with a conducting host matrix in an appropriate ratio (Winchester and Marcus, 1988); and (3) in a modern approach, a pulsed glow discharge source combined with time-of flight mass spectrometry is used (Ganeev et al., 2017).

Oxygen as a major element in the matrix causes problems due to its release during the discharge process. Introduced into the plasma, it influences the signal by quenching metastable atoms and by forming oxide complexes with the analyte. Tantalum and titanium are thus well suited as host matrix, because these materials significantly reduce the formation of UO^+ and PuO^+ species. Due to its getter capability for oxygen, tantalum is used specifically as a secondary cathode material as well. It was shown that, for several elements, the RSFs strongly depend on the oxygen content in the sample (Betti 1996), which implies a specific matrix reference sample for the quantitative analysis of oxygen-containing samples. A good agreement for plutonium oxide samples using ICP-MS and GDMS with tantalum as secondary cathode was obtained for a number of radioisotopes (De las Heras et al., 2000).

Using a pulsed glow discharge, U contents and $^{235}U/^{238}U$ isotope ratios were studied in Chernobyl microparticles (Ganeev et al., 2017). Routinely, materials used for nuclear fuel production are tested by dc-GDMS on their isotopic composition of major elements and the concentration of trace elements, which might lead to severe radiocontaminations by neutron capture in the reactor. Both, initial raw materials and fuel pellets were studied (De la Heras et al., 2000). Data from a uranium oxide reference sample analysis (Morille, CEA, France) taken by secondary cathode GDMS gave precisions of $\sim 10\%$ RSD or better (Betti, 2002). LODs for GDMS in the low μg/g level are reported for most elements, including stable as well as radioactive isotopes (Zhang et al., 2016).

2. Determination of radioisotopes in the environment

In the safety assessment of repositories for highly radioactive waste, the study of the potential migration behavior of the enclosed radionuclides into the nearby environment is a major focus. In this respect, GDMS is in use for the detection of anthropogenic uranium (Duckworth et al., 1993) as well as other radioisotopes in soil, sediment, and vegetation samples (Betti, 1996). The determination of ^{237}Np in Irish Sea sediment samples is another example for such measurements using secondary electrode GDMS (De las Heras et al., 2002). The influences of host matrix interferences must be considered, for example, in the determination of ^{236}U, which leads to detection limits in the pg/g range. This purely anthropogenic isotope indicates the presence of irradiated uranium in a sample.

3. Determination of isotopic compositions

The precise and accurate measurement of isotope ratios is a further application of GDMS, which is of high relevance, for example, in nuclear forensics. Here, the main interest is the isotopic composition of all types of declared or undeclared nuclear materials with respect to U and Pu isotopes. Direct investigations are carried out by means of GDMS (Barshick et al., 1998), whereas a comparison with TIMS once again showed that GDMS is competitive in terms of precision and accuracy for the relevant elements (Mayer et al., 2005).

4. Depth profiling measurements

Depth profiling by GDMS is in use for the study of the mechanism of corrosion of zircaloy cladding of nuclear fuels by measuring the diffusion of impurities in the ZrO_2 layers (Actis-Dato et al., 2000). In general, the investigation of the diffusion mechanism of trace elements is an important aspect in the characterization of environmental processes. An example is the sorption or desorption of actinides in various host rocks or technical barrier materials. These types of studies, which imply depth information, are important for the conception and planning of nuclear waste repositories, and GDMS is a highly versatile tool for that purpose (Betti and de las Heras, 2004).

V. Secondary ion mass spectrometry

A. Principle of ionization through ion impact

SIMS is a surface imaging technique, mainly applied for mono- and multielement trace analysis of solid materials or thin layers, depth profiling, and mapping of elemental and isotopic distributions with high sensitivity. A beam of primary ions with energy ranging from about 1 to 25 keV is focused onto the sample. The ions transfer their kinetic energy to the atoms of the solid by sputtering neutral particles from the surface and—to a lesser extent—positively or negatively charged secondary ions, from which the name of the technique is derived (Benninghoven et al., 1992).

The emitted secondary ions can be atomic, molecular, or cluster ions. They are separated in a mass spectrometer according to their mass-to-charge ratio and are measured with a sensitive ion detection system. To cover a wide dynamic range, often both an electron multiplier and a Faraday cup are installed as sensitive ion detection systems. For a given element, the yield of secondary ions depends on its ionization potential and on the physicochemical properties of the specimen such as its chemical composition, thickness, and even surface quality. Hence, SIMS efficiencies strongly depend on element and sample composition, making it hard to obtain absolute concentrations. Therefore, SIMS is primarily used for element imaging by measuring relative quantities such as isotope ratios or for depth profiling of trace elements in the form of relative abundances.

There are two operational regimes in SIMS, namely, *dynamic SIMS*, which uses a high primary ion current density for LbL erosion and depth profiling studies, and *static SIMS*, which operates at a low primary ion current. *Dynamic SIMS* is restricted to elemental and isotopic analysis but allows the three-dimensional characterization of the sample composition by scanning the primary ion beam over the surface and combining this imaging with depth profiling. Furthermore, the sample is ablated, that is, destroyed in the course of the measurement. Imaging can be performed either via the so-called microscope or the microprobe mode. In the microscope mode, the sample is illuminated by a broad beam of primary ions and the produced secondary ions are accelerated by an electrostatic field and focused with an electrostatic lens. Here, the lateral resolution is about 0.4 μm and determined by the ion optics. For microprobe investigations, a tightly focused illuminating beam is used and an immersion lens acts as the collecting optical system. In this mode, the resolution is defined by the beam diameter (Chabala et al., 1995); 50 nm and below is achieved by standard commercial instruments (e.g., CAMECA nano-SIMS, https://www.cameca.com/, accessed November 29, 2018)).

In addition to element and isotope analysis, *static SIMS* (Van Vaeck et al., 1999; Adriaens et al., 1999) enables the detection of structural ions from organic compounds with an information depth limited to the uppermost surface layer, that is, speciation and organic analysis are feasible. The very low sample consumption makes static SIMS a quasi-nondestructive method. However, depth profiling is possible only by using an additional sputtering ion beam.

With both types of SIMS, primary ions are produced in so-called "ion guns" by either electron or plasma ionization of both, reactive or inert gases (e.g., O_2^+, Ar^+), by surface ionization of alkalines (Cs^+), or by liquid metal ion field emission (Ga^+). Depending on the energy of the ions and the nature of the sample, the primary ions penetrate only fractions of a nanometer up to about 10 nm into the solid. The primary ion's kinetic energy is transferred to the atoms of the sample in a cascade process producing the secondary atoms and ions only in a thin layer. This makes SIMS a surface-sensitive technique with depth resolution in atomic dimensions.

For many sample materials, the majority of sputtered entities remains neutral and hence is lost to MS detection, which most often limits the sensitivity for trace analysis determination. By use of pulsed lasers, these neutral atoms are postionized in a technique called secondary neutral mass spectrometry (SNMS). For many years, neutral species were ionized in a nonselective process by high-energy ultraviolet photons (Hayashi et al., 1991; Kaesdorf et al., 1990). However, ionization can also be performed in a multistep resonant process (see the Section VII) by using up to three well-tuned photon energies of high-repetition-rate tunable pulsed lasers. Based on an early study by Erdmann et al. (2008, 2009), the multistep resonant process was further developed (Franzmann et al., 2017) and is now applied at several laboratories (Franzmann et al., 2018; Savina et al., 2017; Stephan et al., 2016).

In most modern instruments, the secondary ions are analyzed by single focusing magnetic sector fields or using radiofrequency quadrupoles for compactness. Time-of-flight mass spectrometers allow imaging of a wide range of mass-to-charge ratios simultaneously. A few highly sophisticated SIMS instruments use a high-resolution double-focusing mass separation system. These were developed for resolution of interferences from molecular or cluster ions with respect to atomic ions. A schematic representation of a secondary ion mass spectrometer equipped with sector-field analysis is provided in Fig. 8.4.

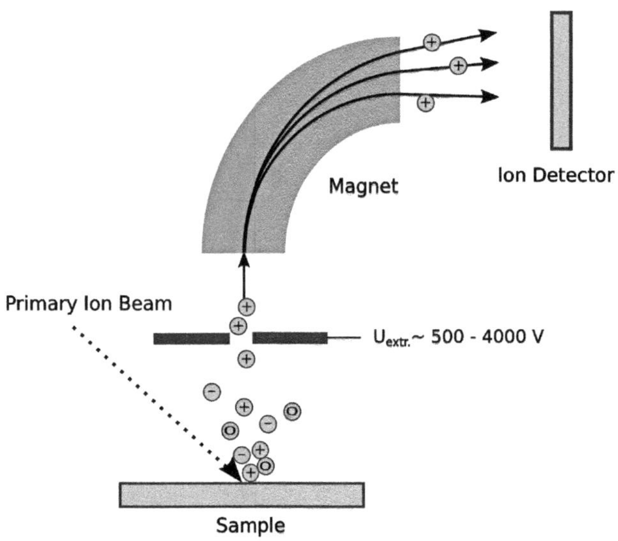

FIGURE 8.4 Schematic representation of the secondary ion mass spectrometry principle.

With a double-focusing mass spectrometer or, more commonly, TOF analyzers, resolutions exceeding $m/\Delta m = 10{,}000$ are routinely achieved. A detection limit for SIMS in the ng/g to pg/g range can be achieved (Tamborini et al., 1998; Tamborini and Betti, 2000). While element imaging and relative abundance measurements are easily carried out, quantification of the analytical results obtained by SIMS is a challenging task. The ion count rates must be corrected with relative sensitivity coefficients (RSCs), which vary, among others, as a function of the primary ion beam, the physical and chemical properties of the element to be investigated, the composition of the matrix, and the charge of the secondary ions. Many of these parameters have to be determined by using standard reference materials. However, successful use of SIMS for the quantitative analysis of partly oxidized uranium surfaces was demonstrated (Kuruc et al., 2011).

B. Applications

SIMS has many applications in the field of radioisotope analysis. Due to the good lateral resolution of the primary ion beam, SIMS is successfully applied for particle identification down to micrometer size and for isotope analyses in astrophysics, for example, of presolar dust grains (Savina et al., 2003). The characterization of environmental samples, possibly contaminated by radioisotopes, studies for nuclear medical applications using, that is, ^{99}Tc, ^{89}Sr, and ^{76}Br as radiotracers (Fourre et al., 1997), investigations on surface compositions, depth profile measurements, and material science studies are examples for the wide range use of SIMS. In the following paragraphs, some important applications are exemplified without any claim to be exhaustive.

1. Radioecology

SIMS has been applied to the identification of single uranium particles and for the determination of their isotopic composition (Tamborini et al., 1998). This is an important issue in environmental monitoring and radioprotection. Such particles can be released from nuclear installations either during regular operation or in the case of accidents. Due to their small size, they are transported as aerosols and are recovered in environmental samples or collected, for example, in air filters. The isotopic composition serves as a fingerprint characteristic of the origin of the particles. From certain isotope ratios, such as ^{241}Pu/^{241}Am, the age of the particle can be calculated (Jernstrom et al., 2006). Furthermore, the properties of these particles allow conclusions on the course of the accident itself as, for example, in the case of micrometer-sized fuel debris released from the Chernobyl nuclear power plant in 1986 (Salbu et al., 2001). To understand weathering mechanisms, UO_2F_2 particles were stored in environmental samples, and the alterations were investigated by SIMS (Kips and Kristo, 2009).

Many military actions and accidents led to the release of U- and Pu-bearing material, spanning the range from depleted uranium ammunition (Danesi et al., 2003a,b; Lind et al., 2009) to highly enriched weapons grade materials (Lind et al., 2008). One of the most prominent cases is the Thule accident. By means of SIMS, particles extracted from marine sediments were traced back to this origin (Nielsen et al., 2009). Plutonium released from the underground nuclear weapon tests in the Nevada test site were shown to be much more mobile than expected, which is ascribed to colloid-mediated transport (Kersting et al., 1999). At the same time, large amounts of actinides were released from the Mayak reprocessing plants during several accidents in the 1950s and 1960s. Speciation of these radionuclides and evidence of colloid-mediated transport were provided by using SIMS, X-ray absorption fine structure, and X-ray photoelectron spectroscopy as described by Kalmykov et al. (2007). SIMS applied in combination with electron probe microanalysis and glow discharge mass spectrometry allowed the characterization of environmental samples contaminated with traces of radioisotopes (Lefèvre et al., 1996). Here, the main emphasis was put on the radionuclides ^{127}Cs, ^{129}I, and ^{238}U, localized in soil and grass samples.

2. Safeguards and nonproliferation control

SIMS was applied to the characterization of plutonium and highly enriched uranium particles with a diameter of $\sim 10\ \mu m$ in nuclear forensic studies (Betti et al., 1999; Tamborini and Betti, 2000). The isotopic composition of these particles was measured with an accuracy and precision of 0.5%. Furthermore, trace elements were determined in the particles by SIMS, and their composition revealed the history of the sample. SIMS was also applied to the identification of plutonium particles and their characterization according to the isotope ratios, especially with respect to the ratio ^{239}Pu/^{240}Pu. Other examples are the isotopic determination of uranium and plutonium in glass microparticles (Simons, 1986) and in clay microspheres (Stoffels et al., 1994). The high efficiency of large geometry SIMS for safeguard purposes was reported by Ranebo et al. (2009). The isotopic composition of laboratory synthesized monodisperse uranium oxide particles could be verified with SIMS (Erdmann et al., 2000; Ranebo et al., 2010). The dependence of sputtered molecular ions on the stoichiometry of the uranium particle (Hocking et al., 2012) provides forensic evidence for chemical processing of a sample, even if the isotope ratio was not changed (Schuler, 2010).

3. Cosmochemistry

The chemical analysis of meteorites and the isotope analyses of presolar dust grains (Stephan et al., 2018; Trappitsch et al., 2018) are two examples for the application of SIMS in astrophysics. Results from such analysis make it possible to detect the origins and pathways through the interstellar medium and thus give new insights into the cosmic nucleosynthesis (Diehl and Hillebrandt, 2002). For the isotope analyses of dust grains consisting mainly of SiC, nanoscale secondary ion mass spectrometry (nanoSIMS) has enabled investigations with spatial resolution in the range of 100 nm. Calcium–aluminum-rich inclusions (CAIs) from the Allende Meteorite were investigated by means of SIMS to get information on the $^{11}B/^{10}B$ and $^{9}Be/^{11}B$ compositions. The $^{10}B/^{11}B$ values from various spots show ^{10}B excesses, which are correlated with the beryllium to boron ratio in a manner indicative of the in situ decay of the radioactive isotope ^{10}Be ($T_{1/2} = 1.6 \times 10^6$ a). ^{10}Be is produced only by nuclear spallation reactions and not by stellar nucleosynthesis, so its existence in early solar system material (CAI) attests to intense irradiation processes in the solar nebula (McKeegan et al., 2000). The technique also has been useful in the investigation of indigenous hydrocarbons in extraterrestrial material, possible essential precursors to the formation of life on earth (Stephan, 2001).

4. Geosciences

In the geosciences, radionuclides are used for dating purposes and for the investigation of erosion rates. Recent SIMS developments in geoscience applications include (1) the lowering of the total sampling mass down to the 10 fg range, (2) the enhancement of spatial resolution to better than 100 nm, and (3) a reduction in the uncertainties on major element isotope ratios to well below $\pm 0.2‰$ (Linge et al., 2017; Wiedenbeck, 2008).

5. Trace analysis

The application of the SIMS technique for the assessment of trace concentrations of uranium and plutonium in urine samples was explored (Amaral et al., 1997). Concentrations in the order of 10^{-10} g/L of ^{238}U and 10^{-11} g/L of ^{239}Pu can be determined with this method. The accurate quantification of a number of elements among them Th and U in complex matrices is another example (Ottolini and Oberti, 2000). Among the physical methods for in vitro analytical imaging in the microscopic range, SIMS plays an important role because it images the isotopes of quite a number of elements with good detection limits and an excellent spatial resolution (Thellier et al., 2001).

6. Radiochemistry and material sciences

Within the context of the disposal of radioactive waste, radionuclide migration and mobilization and retention mechanisms need to be understood. In a combined scanning electron microscopy and SIMS study, Ra in contact with already existing barite in aqueous solution was shown to undergo recrystallization into a $Ba_{1-x}Ra_xSO_4$ solid solution (Klinkenberg et al., 2014). While considerably slower than coprecipitation, as used for Ra removal from naturally occurring radioactive materials, recrystallization forms long-term stable phases, which is desirable for the safe disposal of long-lived radioactive waste. The investigation of uranium solids for use as fuels was investigated in many studies. An additional application is the proof of uranium passivation by carbon implantation (Nelson et al., 2006).

VI. Inductively coupled plasma mass spectrometry

A. Principle and instrumentation

Inductively coupled plasmas (ICPs) were introduced into analytical applications in the early 1960s, when powerful radiofrequency generators became available (Reed, 1961). The technique, in general, is based on a flame-like electrical discharge that operates at ambient air pressure. It has revolutionized the practice of elemental and isotope ratio analyses, in particular by the use of argon ICPs. In addition to applications in atomic emission spectrometry (AES) and atomic fluorescence spectrometry, ICPs were established as universal and most efficient ion sources for mass spectrometry in 1980 (Houk et al., 1980), combining the processes of vaporization, atomization, excitation, and complete ionization of a vaporized sample within the hot plasma. A comprehensive presentation of ICP-MS is found in various textbooks and review articles (see, e.g., Montaser, 1998; Ammann, 2007; Beauchemin, 2010). Only a brief introduction to the general principle and the relevant instrumentation is provided here. The plasma torch is formed by a stream of high-pressure argon gas of typically 8–20 L/min, flowing through an assembly of usually three concentric quartz tubes. The analyte is introduced through the central bore after nebulization of initially liquid solutions as aerosol particles or droplets. Most common for ICP-MS is a torch diameter of 18 mm. The torch is encircled at the top by a cooled induction coil, called the load coil, connected to a free-running or crystal-controlled radiofrequency (RF) generator. Most ICP-based mass spectrometers use an RF of either 27 or 40 MHz with typical output powers in the kilowatt range. The magnetic field generated by the RF current through the load coil induces a strong electrical current in the argon gas stream. A plasma is instantaneously formed and afterward self-sustained, when the Ar gas is seeded with energetic electrons from a high-voltage discharge or from a solid-state piezoelectric transducer. The largest current flow occurs in the periphery of the plasma, which gives the ICP a distinctive annular doughnut-type structure and ensures the efficient introduction of the sample material through the central channel into the plasma. Due to highest gas temperature in the range of up to 8000 K and still higher electron temperatures up to 10,000 K in the plasma, the Ar-ICP is capable of ionizing a wide range of elements, in particular entirely all metals. As a prerequisite, these conditions together with the comparably long plasma-to-sample interaction time in the range of some milliseconds lead to complete vaporization and atomization of the sample aerosols. ICP-MS thus permits a simultaneous multielement determination with lowest LODs.

ICP-MS exhibits a number of special attributes, which are among the following: As a result of the axial sample inlet arrangement into the core of the ICP discharge, the sample aerosols do not undergo any kind of interactions with the surrounding hot walls or a possibly reactive atmosphere. On the contrary, the argon gas and the plasma effectively isolate the analyte's pathway. This enclosure of the analyte ionic species provides optimum conditions for transfer from the ICP on ambient air pressure into the evacuated mass spectrometer system. For this purpose, a sampling cone orifice together with, typically, a two-stage interface, formed by a sampler and a skimmer aperture, is positioned in front of the torch on its central axis. The sampler provides the first differential pumping aperture of approximately 1 mm resulting in a pressure of typically 5 mbar downstream. Behind the sampler, a concentric shock wave is formed. This structure surrounds the so-called zone of silence and terminates in a perpendicular Mach disc of the gas stream. The skimmer orifice is located in the zone of silence for optimal extraction of the analyte ions into the mass spectrometer. The velocity of the particles passing through the MS interface is around 2.5×10^5 cm/s resulting in a mass-dependent axial kinetic energy in the range 0.5−10 eV for ions of 6−240 u. Final acceleration and ion beam formation is achieved by the acceleration potential and an electrostatic lens system. The longitudinal distances of the arrangement are chosen in a way to exhibit maximum analyte signal together with minimal background in the subsequent high vacuum mass spectrometer section. Fig. 8.5 sketches the principal components of an ICP-MS arrangement, illustrating the plasma torch and the interface system in some detail. As the plasma properties and analytical characteristics of the AR-ICP are not significantly affected by a variation in the sample composition, a wide range of elements contained in major, minor, trace, and ultratrace concentration levels in a variety of liquid or easily vaporable matrices can be studied.

Most spectroscopic studies of the ions produced by an ICP discharge are conducted using compact quadrupole mass spectrometers, which offer the advantages of easy use by operating the ion source near ground potential and low cost. Nevertheless, there are some major drawbacks: a poor peak shape, as compared with magnetic double-focusing mass analyzers, and an intrinsically low mass resolution in the range of 0.1 u. This is insufficient to separate analyte ions from isobaric interferences generated by the ICP ion source. These may arise from molecular species formed, for example, by a reaction of sample, plasma gas, and matrix or solvent constituents. Examples are the interferences of $^{40}Ar^{16}O$ with ^{56}Fe, or $^{35}Cl^{16}O$ with ^{51}V or numerous hydrides, which might generate signals at the same nominal masses as the analyte.

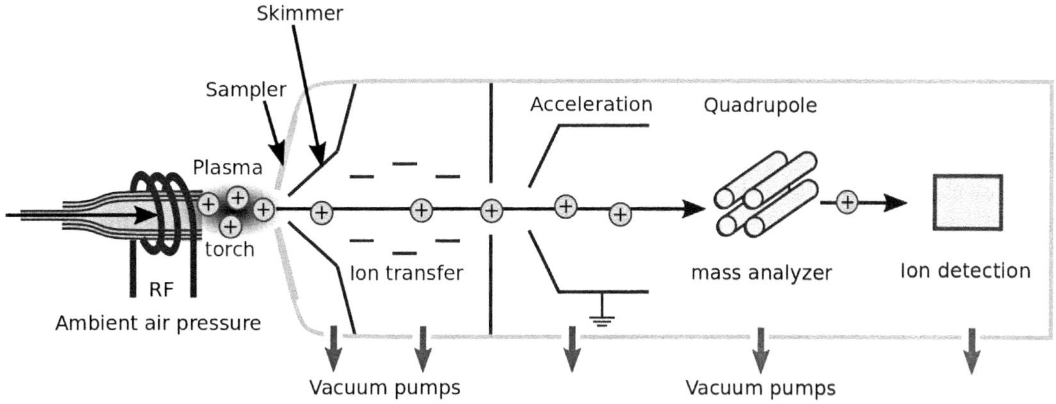

FIGURE 8.5 Principal components of a quadrupole inductively coupled plasma mass spectrometry (ICP-MS) system. Ions are created in the ICP plasma torch and transferred into the quadrupole mass spectrometer through the MS interface. This involves sampler and skimmer apertures and the ion transfer lens arranged inside a differentially pumped vacuum section. After separation according to the mass-to-charge ratio, indicated here by a quadrupole mass filter, ions are recorded at the detector.

To overcome the drawbacks described above, various refinements of the basic ICP-technique have been developed which include

- the use of non-Ar or mixed gas plasmas (Montaser, 1998, Balcaen et al., 2007),
- electrothermal vaporization (Becker and Hirner, 1994),
- laser ablation (Gray, 1985; Richner et al., 1994; Becker et al., 1998),
- matrix separation by chromatography and preconcentrations within hyphenated techniques (Ebdon et al., 1993),
- shielded torches (Gray, 1986; Uchida and Ito, 1994),
- cold plasma operation (Jiang et al., 1988; Tanner, 1995),
- the introduction of collision or reaction cells (Feldmann et al., 1999 a,b; Tanner et al., 2002),
- the use of tandem mass spectrometry (Balcaen et al., 2015), or
- the use of multiple ion collectors (Yang, 2008).

Using a single ion detector, optimum figures of merit in isotope ratio measurements by ICP-MS were achieved by the introduction of a collision cell interface. This device serves to dissociate or neutralize disturbing argon gas–based molecular ions from the plasma torch and to reduce the kinetic energy of the ions. Its application results in higher ion transmission, improved sensitivity, and better precision of isotope ratio measurements compared with conventional ICP-MS (Boulyga and Becker, 2002a). A further significant instrumental improvement for isotope ratio measurements by sector-field ICP-MS was the implementation of multiple ion collectors. In this way, a precision of isotope ratios of up to 0.002% was demonstrated (Becker, 2002).

Apart from all the modifications to ICP-MS, the most effective approach for the improvement of ICP-MS analyses is the use of a high-resolution sector-field mass analyzer (Becker and Dietze, 1999b). This configuration allows for the proper separation and interpretation of spectral interferences in elemental analyses and generally permits the operation of the plasma in its normal "hot" mode in which the level of oxides and hydrides is much higher as in the cold, but inefficient plasma operation. Even though experimentally much more extensive, the use of a high-resolution system is the preferred method for investigating low levels of ultratrace components, that is, radioisotopes in complicated matrices. They also serve best for precision measurements of isotope ratios by ICP-MS. Less common mass analyzers used in ICP-MS include ion cyclotron resonance mass spectrometers, ion trap mass spectrometers, for example, the Orbi trap, as well as time-of-flight mass spectrometers (Olesik, 2000).

B. Sample introduction

Regarding ICP-MS, the introduction of the sample into the plasma plays a key role in the production of ions and interfering species as discussed in numerous publications (Greenfield and Montaser, 1992; Montaser, 1998). Generally, the sample can be introduced in gaseous, liquid, and solid form. Here, we restrict ourselves to liquids and solids and those techniques that have been applied in the analysis of long-lived radionuclides.

1. Nebulization

Liquids are dispersed into fine aerosols before being introduced into the ICP by using different nebulizers: pneumatic (PNs), ultrasonic (USNs), or thermospray nebulizers are in use. Although a large variety of individual sample introduction devices are commercially available (Sharp, 1988), they all are formed on two basic designs: in the concentric type, the sample solution passes through a capillary surrounded by a high-velocity gas stream, which flows parallel to the gas stream axis. Alternatively, the cross-flow type has a liquid-carrying capillary set at a right angle to the tube carrying the high-velocity gas stream. In both configurations, a pressure differential created across the sample draws the solution through the capillary. The analyte transport efficiency of commonly used PNs reaches a few percent, limited by nebulizer clogging due to excessively concentrated solutions. In ultrasonic nebulization, the sample solution is applied onto the surface of a piezoelectric transducer driven by an ultrasonic generator at a frequency of several 100 kHz. The total production rate of droplets generated by the USN is significantly higher than that obtained with PNs, and desolvation is essential to remove excess solvent and prevent plasma cooling. Analyte transport efficiencies approaching 20% are reached for USNs operating at an uptake rate of 1 mL/min. This value can be further optimized if the spray chamber is heated or if the solution uptake rate is reduced to a few microliters per minute. For example, analyte transport efficiencies close to 100% are achieved with a microflow USN at solution uptake rates of 5–20 μL/min (Montaser, 1998; Beauchemin, 2010). Direct injection nebulizers introduce 100% of the aerosol into the plasma, thereby increasing the solvent load. Most recent versions of these include microconcentric pneumatic nebulizers positioned directly inside the ICP torch. Further developments include a specific high-efficiency nebulizer or a direct injection high-efficiency nebulizer (DIHEN), as reviewed in textbooks (Montaser, 1998; Prohaska et al., 2015).

2. Hyphenated systems

The combination of liquid chromatography (LC) with ICP-MS is a straightforward approach, which provides attractive opportunities for element speciation. In most cases, a direct transfer of the eluate from the column through a nebulizer into the torch (Ebdon et al., 1993) provides the interfacing of LC to the ICP. Similarly, capillary electrophoresis, being another attractive separation technique with excellent speciation capabilities, is hyphenated to ICP-MS. This was reported first in 1995 using a conventional concentric nebulizer and a conical spray chamber as an interface to the ICP. LODs in the low pg/g range and precision in the range of 1%–5% RSD were reported (Olesik et al., 1995).

3. Laser ablation

Laser ablation is the most versatile sampling technique used for solid materials in ICP spectrometry. Laser ablation was first introduced into ICP-MS in 1985 (Gray, 1985). When sufficient energy in the form of a focused laser beam is directed onto a solid sample, material from the surface is sputtered and vaporized. The plume of vapor and particulate matter is transported within an argon carrier gas to the ICP plasma torch for atomization and ionization, as sketched in Fig. 8.6.

The specific strength of the combination of LA-ICP-MS is found in the ability to directly study diverse materials ranging from conducting and nonconducting inorganic and organic compounds available as solids or as powders (Becker et al., 1998). In addition to conventional bulk analysis, the focusing characteristics of lasers permit sampling of small areas, such that localized microanalysis and spatially resolved studies are feasible (Krachler et al., 2018).

C. Applications to radionuclides

A number of reviews summarize recent applications of ICP-MS in environmental sciences, including radioecology (Pröfrock and Prange, 2012) and with a focus on uranium analysis (Tissot and Dauphas, 2015; Bu et al., 2017).

1. Methodical developments on isotope ratios

The early application of a double-focusing sector-field ICP-MS for ultratrace and isotopic analysis of uranium in radioactive waste samples was performed in high-purity water, showing a detection limit in the low fg/mL range (Kerl et al., 1997). Under optimum experimental conditions, a precision in $^{235}U/^{238}U$ isotope ratio determinations was 0.07% for ratios around unity. For the isotopic standard U-020 ($^{235}U/^{238}U = 0.0208$), a precision of 0.23% RSD at the 100 pg/mL level using ultrasonic nebulization was achieved. With $^{234}U/^{238}U$ ratios as low as 10^{-5}, data obtained by ICP-MS and α-spectrometry were shown to agree well. Nuclide analysis in a tantalum target, which was used in a spallation source by irradiation with 800 MeV protons, is another early example. Concentration of radionuclides within a highly radioactive matrix in the range from 1 ng g^{-1} to 50 μg g^{-1} was performed to verify the theoretical predictions for spallation yields. Double-focusing sector-field ICP-MS was used in combination with liquid–liquid extraction from the tantalum matrix to remove the ^{182}Ta activity, and results were compared with quadrupole ICP-MS. Comparison of the ICP-MS results to radiometric data on some nuclides, obtained by neutron activation analysis, yielded reasonable agreement (Becker and Dietze, 1997; Becker and Dietze, 1998). Precise long-term measurements of uranium and thorium isotope ratios

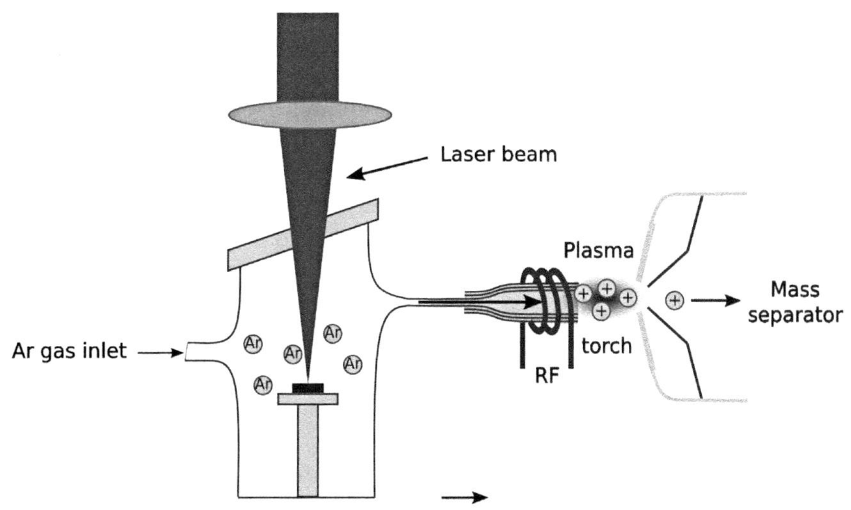

FIGURE 8.6 Principle of the laser ablation inductively coupled plasma (ICP) coupling. By the interaction of a focused laser beam with a solid surface, sample material is ablated and vaporized, and the ablated material is transported using argon as the carrier gas into the ICP.

were carried out in standard 1 μg/L solutions, using a conventional ICP-MS (Becker and Dietze, 1999a). Uranium isotope ratios of $^{235}U/^{238}U = 1$, 0.02, and 0.00725 were determined, using different approaches for nebulization: either a cross-flow nebulizer (CFN) at a solution uptake of 1 mL/min or a low-flow microconcentric nebulizer (MCN) at a solution uptake of 0.2 mL/min over 20 h. For an isotope ratio $^{235}U/^{238}U \approx 1$, RSDs of $\leq 0.05\%$ using both, CFN and MCN, were achieved. Additional isotope ratio measurements using a DIHEN at a solution uptake of 0.1 mL/min on the same uranium solution yielded a RSD of 0.06–0.08%. The sensitivity of the solution introduction for uranium, thorium, and plutonium was increased significantly compared with CFN and MCN. In parallel, the solution uptake could be reduced to 1 μL/min in DIHEN-ICP-MS. Isotope ratio measurements at an ultralow concentration level (e.g., determination of $^{240}Pu/^{239}Pu$ isotope ratio in a 10 ng/L Pu waste solution) were carried out for the characterization of radioactive waste and environmental samples. Recently, Pu isotope ratio studies were reported in different environmental and biomedical samples using ICP-MS (Evans et al., 2003). In a study utilizing IAEA standard soils, it has been shown that a uranium surplus of about six orders of magnitude prohibits any suitable application of ICP-MS (Xu et al., 2014). On the other hand, Pu isotope ratio determinations for the isotopes $^{239,240,241}Pu$ in marine sediments near the Fukushima Dai-ichi nuclear power plant were successfully achieved after uranium decontamination by about six orders of magnitude (Bu et al., 2014).

Systematic investigations were undertaken on the capability of double-focusing sector-field ICP-MS coupled with a plasma-shielded torch using different nebulizers. A total amount of analyte between 0.4 and 10 pg was used for a number of long-lived radionuclides (^{226}Ra, ^{230}Th, ^{237}Np, ^{238}U, ^{239}Pu, and ^{241}Am; each available at concentrations of 1 ng/L in aqueous solution) with different nebulizers (Becker and Dietze, 1999b). Detection limits in the sub-pg/L range and precisions ranging from 1% to 2% RSD for a 1 ng/L concentration level were reported. At lowest solution uptake rates (down to 1 μL/min), uranium solutions were analyzed by DIHEN-ICP-MS using the double-focusing sector-field instrument, which exhibits significantly higher transmission and sensitivity than the conventional quadrupole ICP-MS. Using the shielded torch arrangement, a precision of Pu isotopic analysis of 0.2%, 2%, and 14% RSD for 1000, 100, and 10 pg/L was obtained.

The performance of a specific ICP-MS system with a hexapole collision cell (HEX-ICP-QMS) was studied to obtain figures of merits and to prepare for real applications (Boulyga and Becker, 2001). A significant improvement in sensitivities and detection limits was achieved in particular for those ions affected by interferences with argon or argon-based molecular ions due to the removal of Ar^+ and ArX^+ ions (with X = H, C, N, O, Ar). Investigation on heavier elements generally profited due to an enhanced ion transmission by the hexapole ion guide. Subsequent applications of HEX-ICP-MS were focused on the determination of uranium in soil samples from the Chernobyl area (Boulyga and Becker, 2002b). A limit of the $^{236}U/^{238}U$ ratio determination in 10 μg/L uranium solution of 3×10^{-7} was measured, corresponding to a detection limit of 3 pg/L. The precision of uranium isotope ratio measurements in 10 μg/L standard solutions was 0.13% ($^{235}U/^{238}U$) and 0.33% ($^{236}U/^{238}U$).

A very specific application of ICP-MS has developed in the separation of individual radioisotopes for the production of ultrapure targets for fundamental research and calibration purposes (Liezers et al., 2015). Both actinide isotopes ^{238}Pu and ^{241}Am could be purified with a signature of more than a factor of 100, and subsequently, background-free α-spectrometry at the intrinsic resolution of the detector could be realized. The limitation remaining is the low efficiency of the process (Dion et al., 2015).

2. Radioecology

Soil samples from the former Semipalatinsk nuclear test site (SNTS) in Eastern Kazakhstan were investigated using double-focusing ICP-MS to determine the contamination level and isotope ratios of plutonium (Yamamoto et al., 1996). After dissolution of the samples, the Pu fraction was purified by anion exchange chromatography, evaporated to dryness, and dissolved in nitric acid for introduction into the ICP-MS. Isotope ratios were determined using an ultrasonic nebulizer. High transmission measurements were carried out at relatively low resolution of $m/\Delta m \approx 400$ (by opening slits), and isotope ratios were determined at a level of about 0.1 pg/g solution. Low $^{240}Pu/^{239}Pu$ ratios of 0.036(1) and 0.067(1) were found at different test sites and $^{242}Pu/^{239}Pu$ ratios of $4.8(4) \times 10^{-5}$ and $2.9(5) \times 10^{-4}$, respectively. These values are significantly different to the values commonly accepted for global fallout from the atmospheric nuclear weapons tests, but they represent the situation after a Pu weapon explosion similar to data from the Nagasaki area. Recently, this work was extended to the study of altogether 32 human tissue and bone samples collected by autopsy from individuals that lived near the SNTS. A significant difference in the $^{240}Pu/^{239}Pu$ isotope ratios for tissue samples and for the global fallout value indicated that weapons-grade plutonium was incorporated into the human tissues, especially lungs (Yamamoto et al., 2008).

3. Treatment and storage of nuclear waste

ICP-MS was also used for the characterization of long-term storage and geological isolation by vitrification of highly radioactive waste, that is, defense-related material, at the Defense Waste Processing Facility (DWPF), Savannah River Site in South Carolina, United States. Safety considerations require the verification of repository criteria for 24 radioisotopes, including major fission products, activation products, actinides, and their daughter nuclei that grow in the waste. Based on a routine determination of four isotopes, that is, ^{90}Sr, ^{137}Cs, ^{238}U, and ^{238}Pu, it was found that the concentrations of the other radiocontaminations in the final glass product can be predicted. For that, a thorough characterization of the high-level waste tanks and a knowledge of the concentrations of the predominant nonradioactive components in the vitrification process are sufficient (Kinard et al., 1997).

The capability of LA-ICP-MS was systematically investigated for measurements of long-lived radionuclides, for example, in nonconducting concrete matrices that are common in nuclear waste packaging. The LODs of long-lived radionuclides were determined in the low pg/g range using a quadrupole LA-ICP-MS or a double-focusing sector-field LA-ICP-MS. For both systems, different calibration procedures, including the correction of analytical results with experimentally determined RSCs and the use of calibration curves or solution calibration by coupling LA-ICP-MS with an ultrasonic nebulizer, were compared and applied to the determination of long-lived radionuclides, that is, Th and U in different solid samples (Becker et al., 2000). The application of dedicated hyphenated systems, that is, ion chromatography (IC) attached on-line to ICP-MS (IC-ICP-MS), for the determination of the quality achieved in separation of fission products and actinides and the determination of trace elements in the different nuclear waste solutions, is reviewed (Betti, 1997). The coupling of IC and a low-resolution ICP-MS has been found to be effective in eliminating isobaric interferences in the determination of, for example, ^{90}Sr, ^{137}Cs, lanthanides, and actinides (Barrero Moreno et al., 1996). This online coupling has been installed inside a glove box for the determination of the complete inventory of fission products and actinides in spent nuclear fuel and high-level liquid waste samples. With this system, for example, for the radioactive isotopes of ^{134,135,137}Cs, isobaric interferences by stable isotopes of Ba were eliminated giving a detection limit of 16 pg/g for total Cs with a precision of 2.5% at a concentration level of 10^{-7} (Barrero Moreno et al., 1999). The accuracy of the ratio ^{134}Cs/^{137}Cs was found to be 2.5%. The same system was applied successfully to the determination of actinides in aqueous leachate solutions from uranium oxide (Solatie et al., 2000). The results were compared with those of classical radiometric methods and determinations by α-, γ-spectrometry, and liquid scintillation counting (LSC). The comparison confirmed that IC-ICP-MS is a powerful method for the detection of long-lived radionuclides. IC-ICP-MS was found to be the most convenient choice being much less time-consuming than the radiometric methods with respect to sample preparation (α-spectrometry) as well as counting time (γ-spectrometry and LSC).

4. Application to Chernobyl and Fukushima samples

Numerous specific studies on environmental contamination by different radionuclides using different ICP-MS systems were published after the occurrence of the two major reactor accidents at Chernobyl in 1986 and Fukushima in 2011. Concentrations of the isotopes ^{239}Pu and ^{240}Pu and their isotope ratios were determined by ICP-MS in soil samples collected from three forest sites of the Chernobyl 30-km exclusion zone. Overall, ^{239,240}Pu activity levels of 1 up to above 10 kBq/m^2 were measured, whereas the isotope ratio of ^{240}Pu/^{239}Pu $\approx$ 0.4, which is well above the value for nuclear fallout of 0.18, confirmed the origin of the contamination and the high burn-up grade of the nuclear fuel (Muramatsu et al., 2000). The highly mobile fission fragment ^{99}Tc was analyzed in soil and in plant leaf samples similarly taken from the 30 km zone around Chernobyl. Strongly varying ^{99}Tc activity levels in the order of 1–15 Bq/kg in soil and 0.2–6 Bq/kg in plants were measured, and soil-to-plant transfer factors of similar magnitude such as for radiocesium have been extracted (Uchida et al., 2000).

To study contaminations around Fukushima, radioisotope ratios of ^{129}I/^{127}I in iodine (Ohno et al., 2013) and ^{134}Cs/^{137}Cs and ^{135}Cs/^{137}Cs in cesium (Ohno and Muramatsu, 2014) were analyzed with ICP-MS in collected rainwater from the vicinity within and around the 30 km exclusion zone. To suppress fully isobaric interferences from Xe, Ba, and molecular ions in both studies, tandem mass spectrometry at a triple quadrupole system equipped with an octopole reaction cell was used. For iodine, the measured isotope ratios in the order of ^{129}I/^{127}I $\approx 10^{-7}$ are in good agreement with independent AMS studies, whereas the determined cesium ratios were verified with radiometric data from a germanium semiconductor detector. The Cs results differ from global fall-out values and from that of Chernobyl, indicating that measurements of rainwater samples can be used as a radiocesium tracer in the environment. Hyphenated ICP-MS-based techniques were applied to study the role of organics for iodine transport in Japan after the Fukushima accident (Xu et al., 2016). Uranium and plutonium isotope ratios in environmental samples collected close to the Fukushima Dai-ichi Nuclear Power Plant (FDNPP) were determined by a number of groups (Zheng et al., 2012a, 2012b; Yang et al., 2017). A general overview of mass spectrometric activities in the assessment of the radiocontamination around the Fukushima accident site is found in Bu et al. (2018).

VII. Resonance ionization mass spectrometry

A. Principle and requirements for the laser light sources

The different mass spectrometric methods discussed so far, in general, apply to conventional ionization processes. Most of these show strong variations in the ionization efficiency depending on the element, which is expressed, for example, by the RSF factor in GDMS, and might even exhibit isotope fractionation. Thus, these techniques are generally limited in their sensitivity as well as in the significance of their data by the occurrence of isobaric interferences. In addition, due to finite abundance sensitivity values, cross-contaminations, present in the form of peak wings from the neighboring masses, could obscure the mass spectra and limit the precision of results on overall contents and isotope ratios. In contrast, RIMS, using a stepwise resonant excitation and ionization process of specific atoms of the analyte by well-tuned laser light, achieves highest elemental and, if needed, even may contribute additional isotopic selectivity on top of the mass separation. It is a dedicated single-element determination technique and not suitable for simultaneous multielement analysis in one sample; however, it can be adapted to the majority of elements by appropriate tuning of the lasers. Due to the high cross sections for resonant optical processes, that is, the individual excitation steps, and the availability of well-controllable high-power tunable lasers, RIMS offers outstanding properties. These include

- complete suppression of atomic and molecular isobaric interferences,
- very good overall sensitivity in the fg range down to $\sim 10^6$ atoms per sample,
- feasibility of ultrahigh isotope selectivity by taking advantage of the isotope shift in the atomic transitions in addition to the abundance sensitivity of the mass spectrometer, and
- universal applicability to the far majority of elements as a single-element determination technique.

The principle of a typical three-step resonant excitation process leading to final ionization is shown in Fig. 8.7. The initial step is a preferably complete conversion of the sample material into individual atoms within a hot furnace or in free vacuum above a heated filament. Starting from the atomic ground state (or from a thermally populated low-lying state), the sample atoms are excited stepwise to a high-lying excited state along strong optical resonance lines. As illustrated in the figure, usually at least two resonant photon absorption steps are required. Either a real or a virtual (indicated as dotted line) intermediate state can be involved. Finally, the highly excited atoms are ionized

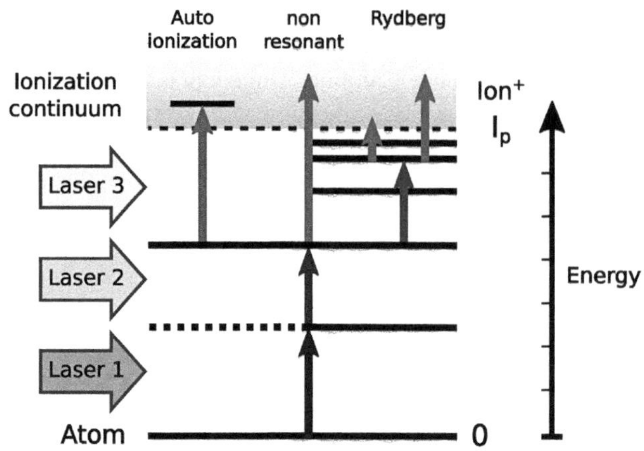

FIGURE 8.7 Principle of a three-step resonant excitation of an atom by light of tuneable lasers. Inefficient nonresonant ionization (center) can be circumvented by addressing autoionizing levels (left) or high-lying bound Rydberg states. For the efficient ionization from the highly excited atomic state besides photon absorption from a laser beam, additional processes may be interjected as discussed in the text.

by absorption of another laser photon or by an alternative process as discussed below.

The probability for induction of one of the individual excitation and ionization steps, W_{opt}, determines the final ionization rate. It is given by the simple formula

$$W_{opt} = \sigma_{res} \cdot n_{photons} \cdot t_{int},$$

where σ_{res} is the absorption cross section, $n_{photons}$ is the number of photons per cm^2 per second, and t_{int} is the interaction time. In resonance, the cross section σ_{res} for a strong optical transition between two bound atomic states is in the order of $\lambda^2/2\pi$, where λ is the wavelength of the transition, corresponding to about 10^{-9} cm^2 (or $\sim 10^{15}$ b). Total laser intensities, measured as photon flux, of 10^{15}–10^{18} photons/($cm^2 \cdot s$) are easily available from continuous wave lasers or during pulses of pulsed laser systems. In a cloud of free atoms, that is, an atomic vapor, thermally expanding above a filament or within a heated atomizer furnace, interaction times in the order of few microseconds to milliseconds are common. Correspondingly, saturation of the optical excitation with maximum transition rate and an efficiency near 100% in each individual step toward ionization are easily realized in such a sequential process. This makes it one of the most efficient ionization processes for atoms (Hurst et al., 1979).

A bottleneck exists in the ionization step into the ionization continuum. Nonresonantly raising the electron energy beyond the ionization limit (I_P) is very inefficient with low cross sections in the order of 10^{-17} cm^2 and requires very high laser power. Resonant population of an autoionizing level or, alternatively, a high-lying bound Rydberg state in the final excitation step can circumvent this shortcoming. An atom in an autoionizing level immediately decays into a positively charged ion under emission of an electron, whereas an atom in the energetically high-lying Rydberg state may be ionized efficiently through various processes, which include the influence of an external electric field, absorption of an additional laser photon, or a far-infrared absorption process. Even making use of collisions or the omni-present black blackbody radiation is possible (Letokhov, 1987).

Power and spectral bandwidth, given both by the spectral power density of the laser radiation, is a crucial factor that determines the efficiency and selectivity of the resonant ionization process. The intrinsic, so-called natural linewidth of the optical transition, Δν, which is the inverse of the spontaneous decay rate A_{12}, determines the laser intensity needed for saturation and affects the isotopic selectivity. The latter is achieved through resolution of the isotope shift in the optical excitation process. Therefore, experimental conditions must properly be chosen to reduce or entirely prevent broadening effects such as Doppler broadening, induced by the velocity distribution in the atomic ensemble, pressure broadening, induced by collisions in the atomic vapor, or transit time broadening, caused by the limited interaction time between laser (pulse) and atom.

The extremely high elemental selectivity of RIMS is a consequence of the uniqueness of excitation schemes for individual elements. This is due to the fact that the density of levels in an atom, which are accessible to electric dipole transitions, is of the order of ~1 per eV (for low-lying levels) up to few 100 per eV for higher excited levels around principal quantum number n = 20. In the contrary, the typical natural linewidth of an optical transition is around 10^{-7} eV. As linewidths of a usual pulsed tunable laser are below 10^{-4} eV and for continuous wave lasers even below 10^{-8} eV, the probability of accidentally matching a transition of an unwanted atomic species is negligible, especially in any two- or three-step excitation process. Correspondingly, highest suppression of isobaric interferences or contributions from isotopes of neighboring elements is achieved, and the only contribution from other elements or molecules may originate from parasitic processes. These include thermal ionization on the surfaces of the atomic beam source or laser-induced nonresonant ionization processes, primarily observed on molecules and when high-power blue to ultraviolet light is used. Since its first proposal and demonstration in 1972 (Ambartzumian and Letokhov, 1972), RIMS has been widely used in fundamental research as well as in analytical applications (Letokhov, 1987; Hurst and Payne, 1988; Payne et al., 1994). Today, prominent applications of RIMS include

- preparatory and fundamental studies in atomic and high-precision physics,
- ultratrace analysis of radioisotopes,
- ionization of exotic radioisotopes at all the major online facilities worldwide and, finally,
- a wide field of nuclear physics investigations.

Due to the broad availability of tunable light from powerful laser sources, presently about two-thirds of the elements of the Periodic Table are accessible to resonance ionization, as shown in Fig. 8.8. One great outcome in the field of fundamental atomic physics research with the aid of RIMS is the determination of the fundamental quantity of the ionization potential for those elements without stable isotopes, which in most cases were not precisely measured before or were just predicted from systematics or theory. These values are of particular relevance in (1) our understanding and interpretation of chemical behavior and are basic requirements in current quantum chemical calculations on chemical reactions of the most often radiotoxic trace elements, (2) the prediction of properties of heavy actinides up to the superheavy elements as well as (3) the optimization of laser based ultra-trace analytical methods.

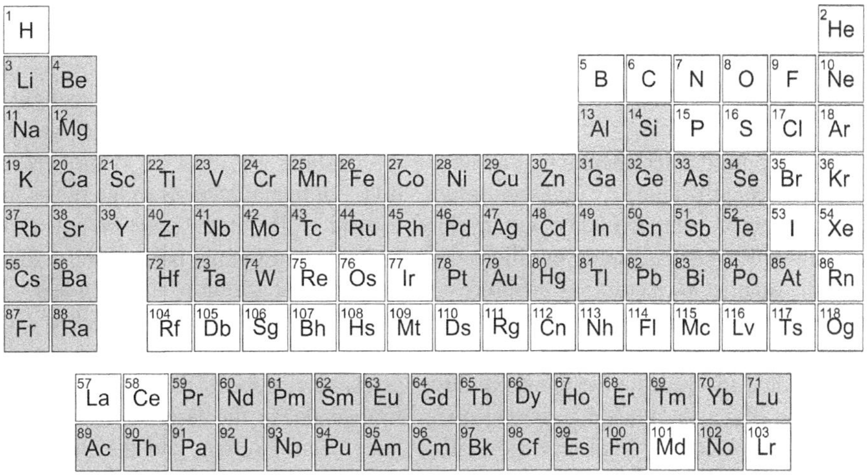

FIGURE 8.8 Periodic Table, elements accessible by resonance ionization mass spectrometry are indicated in gray color.

Data were taken on technetium (Mattolat et al., 2010), astatine (Rothe et al., 2013), and 11 of the 14 actinide elements up to einsteinium (Erdmann et al., 1998; Peterson et al., 1998, Rossnagel et al., 2012). For most of these cases, data were taken on samples in the pg range ($\sim 10^{12}$ atoms). A further application concerns the high precision measurement of atomic and nuclear masses, which is carried out at online facilities using Penning-type ion trap or multireflectron mass spectrometers (Dilling et al., 2018).

In the field of analytical applications, RIMS is well established for the fast and sensitive ultratrace determination of long-lived α- and β-emitters, providing LODs in the fg range (10^6-10^7 atoms) and, in addition, providing the isotopic composition as valuable information on the source of a contamination (Kluge et al., 1994; Trautmann et al., 2004; Trautmann and Wendt, 2012). Beyond that, the ultratrace determination of long-lived radioisotopes is also applied in biomedical tracer investigations involving highest isotopic and/or elemental selectivity (Ham and Harrison, 2000; Bushaw et al., 2001; Müller et al., 2001; Hennessy et al., 2005; Denk et al., 2006) and in cosmochemistry (Geppert et al., 2005).

B. Resonance ionization mass spectrometry systems and applications

1. Elemental-selective resonance ionization mass spectrometry using pulsed lasers

For studies involving high sensitivity but only moderate demands on isotopic selectivity, which is accomplished by the mass dispersion of the mass spectrometer, pulsed lasers are regularly in use for RIMS. In this way, saturation of all optical excitations and the ionization step and corresponding high efficiency in the ionization process can be ensured. The resulting pulsed generation of ions is most easily combined with mass separation in a time-of-flight mass spectrometer, where the laser pulse delivers the start signal and a complete mass spectrum is recorded for each laser pulse. The temporal and spatial overlap of the evaporated atoms from the sample and the laser pulses needs to be adapted properly:

- For low-repetition-rate pulsed-laser operation at around 10−200 Hz, sample volatilization by pulsed laser desorption is favorable to prevent significant duty cycle losses of up to some orders of magnitude in efficiency (Krönert et al., 1991; Maul et al., 2006). As the process is furthermore strongly affected by the pulse-to-pulse stability of power and spatial profile of the desorption laser, it is not well suited for quantitative ultratrace determinations. With the advent of reliable laser systems, it was almost entirely replaced by high-repetition-rate pulsed-laser RIMS.
- For high-repetition-rate pulsed-laser operation in the 10 kHz range, duty cycle losses during laser pulse pauses are negligible due to the relatively slow movement of atoms with thermal energies, even for continuous evaporation of the sample from an electrothermally heated oven or filament (Grüning et al., 2001). In combination with a TOF-MS, an easy-to-handle and reliable analytical apparatus is formed, which just requires advanced data acquisition to manage the high throughput with refreshment of entire mass spectra in the kilohertz range. Alternatively, high-repetition-rate pulsed-laser RIMS is performed in combination with well-synchronized sample volatilization by ion beam sputtering or laser ablation. This approach not only gives complete temporal overlap of atomization to the laser pulses but even ensures a direct and spatially resolved analysis of surfaces and particles avoiding chemical preparation steps. This technique of SNMS was already discussed in the section on SIMS of this chapter together with some prominent applications in radioisotope determination.

Although not commercially available, analytical systems based on the concept of pulsed laser RIMS are routinely in use for the ultratrace determination of actinides, primarily plutonium, in environmental and biological samples (Passler et al., 1997; Trautmann et al., 2004; Kratz, 2015). After an initial chemical preseparation step extracting predominantly the plutonium content from the sample, plutonium hydroxide is formed and electrolytically deposited on a tantalum filament, which subsequently is covered with a thin titanium layer (~ 1 μm) by sputtering technique. This so-called sandwich filament is inserted into the TOF-MS and heated to about 1000 °C. During this procedure, plutonium hydroxide is converted into the oxide, which immediately releases free atoms through the reducing titanium layer. These atoms are ionized near the filament surface by a three-step three-color laser excitation/ionization process. The ions formed are accelerated in the two-stage electric field of the TOF mass selection, and the mass selectively is recorded by a multichannel plate detector at the end of the time-of-flight drift tube, which is operated either in linear mode or as a reflectron by using an electrostatic ion reflector to further improve mass resolution. For quantification, an internal standard tracer isotope - usually ^{244}Pu in the case of environmental samples - is added to the sample prior to the chemical treatment. Independent monitoring of the chemical yield is carried out by α-spectroscopy on ^{236}Pu added as second tracer. Fig. 8.9 shows the setup with details of the filament, the two suppression electrodes preventing electrons and surface ions from entering the laser ionization region, the transversal laser beam overlap, and the different components of the TOF-MS. The latter include two acceleration grids for

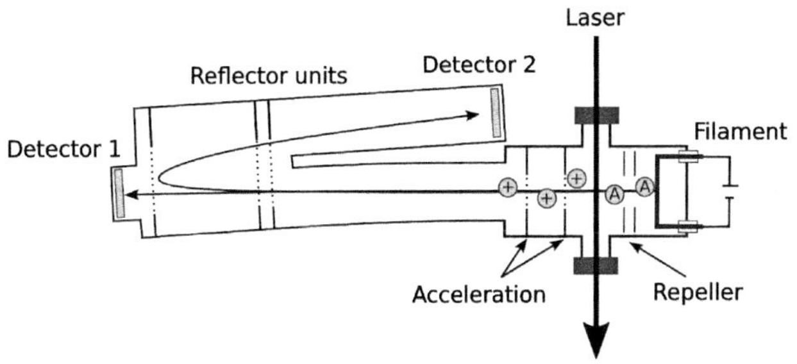

FIGURE 8.9 Sketch of a typical setup for pulsed-laser resonance ionization mass spectrometry with time-of-flight mass analysis, used, for example, for sensitive and isotope-selective determination of plutonium and other actinides.

double-stage acceleration, the reflector unit, and the two detectors used either in linear mode or in reflectron geometry.

A similar technique is also used for ultratrace determination of ^{99}Tc, while an application to minor actinides is under development, for example, for Np, where process efficiencies of 0.3% were reported (Raeder et al., 2011). For the determination of ^{99}Tc in soil, a resonant laser excitation scheme was developed (Wendt et al., 2012) and the rare isotope ^{97}Tc was used as a tracer for quantification. A LOD for ^{99}Tc determination of 0.5 fg was determined, which compares favorably with the competing techniques TIMS and AMS (Schönberg et al., 2017).

A way to overcome limitations inherent in the filament-based sample evaporation, that is, incomplete temporal and spatial overlap of the expanding atom vapor cloud in front of the filament with the beams of the pulsed lasers, is the "laser ion source" technique. By means of this technique, the sample atoms are confined inside a narrow heated furnace cavity of typically ~3 mm diameter and >30 mm length, wherein they are resonantly ionized by the laser beams entering through the ion exit hole. The sample atoms are evaporated from a heated reservoir, enter the atomizer cavity at the rear end, and are "stored" inside the cavity by multiple wall collisions before escape. Once ions are generated by the laser radiation or spurious effects, these are guided along the cavity by the field leakage of the acceleration voltage and efficiently extracted toward the subsequent ion beam/mass separator system. The laser ion source is used for analytical applications in combination with a high-transmission radiofrequency quadrupole filter or a magnetic sector-field instrument. In this case, a significantly higher efficiency in the order of 1% and beyond has been demonstrated for technetium and plutonium determination as compared with TOF-RIMS, but spectra are obscured by unspecific background contributions, which brings the LOD to similar values as discussed above (Raeder et al., 2012; Wendt et al., 2012).

Due to the high ionization efficiency and isobaric selectivity demonstrated, hot cavity-type laser ion sources today are standard at the different isotope separator online (ISOL)-type facilities for the artificial production of short-lived radioactive isotopes (Barzakh et al., 1997; Marsh, 2014). Similarly, also at radioactive ion beam facilities, which use the in-flight fragment separator technique (Blumenfeld et al., 2013), resonance ionization is used in gas cells and jets. In this case, the highly energetic products from nuclear reactions are thermalized and neutralized in a high-pressure noble gas for selective resonant laser ionization carried out before or during flushing out by a gas jet toward the mass spectrometer (Kudryavtsev et al., 2002). A combination of a gas-filled ion guide with a laser ion source enabled the first analysis on fermium (Backe et al., 1997; Sewtz et al., 2003) and served as a prototype for combined laser spectroscopic and mass spectrometric investigations on the heaviest actinides toward the transactinides (Laatiaoui et al., 2016). At a pressure of about 35 mbar, the storage time within the interaction volume illuminated by the laser light is about 40 ms, limited by diffusion and thermal convection in the buffer gas. Yet, at 200 Hz repetition rate, a tunable high-power laser system is well suited for efficient resonant ionization. The ions are guided by electric fields to a nozzle and flushed out of the cell by a gas jet. By means of skimmers, a quadrupole ion guide, and mass filter structures, separation from the buffer gas and mass selection is achieved with highest overall sensitivity.

To date, hundreds of radioisotopes of more than 50 different elements have been ionized at the different online facilities by laser resonance ionization, and these were studied after mass separation in high transmission magnetic sector-field spectrometers with respect to their atomic and nuclear properties, that is, mass, size, and shape of the nucleus (Fedosseev et al., 2012). For mainly all those elements, the ionization efficiencies of laser ion sources are generally in the order of 20%–50% and thus more than comparable with conventional ion sources (e.g., surface or plasma ion sources) (Fedosseev et al., 2017). The additional specific advantage of the laser ion source, that is, the drastically improved selectivity in suppression of isobars and neighboring masses, is only limited by interfering of nonspecific ionization processes—primarily thermal ionization at the cavity walls. To achieve the highest isobaric selectivity by complete prevention of these effects, the so-called laser ion source trap has been developed, which combines mass spectrometric, ion trap, and laser spectroscopic aspects (Blaum et al., 2003; Schwellnus et al., 2008). In this approach, the laser ionization takes place just when the atoms have effused out of the hot ion source cavity within a quadrupole structure. Spurious ions are repelled and driven back into the hot source, whereas laser ions are produced inside the radiofrequency quadrupole trap structure, where they are accumulated and possibly cooled and bunched within buffer gas, for optimum mass resolution and background reduction (Fink et al., 2015). The enrichment and purification of rare radioisotopes for specific applications in fundamental research is a further upcoming application, where mass spectrometry profits from the high-efficiency selectivity and reliability of laser resonance ionization. One particular example is the implantation of ^{163}Ho ions into microscopic detector pixels for the determination of the neutrino mass (Dorrer et al., 2018).

2. Highest isotopic selectivity using continuous wave lasers

There are a number of analytical challenges, where exceptionally high isotopic selectivities are required. This is the case, for example, in the determination of a rare radioisotope at ultratrace level aside of a vast surplus of neighboring stable isotopes of the same element. The latter could be numerous orders of magnitude higher in number than the isotope of interest and, in addition, could even be accompanied by additional molecular interferences. Isotopic selectivities in the range well above 10^9 are, in general, not achievable by conventional inorganic mass spectrometry due to the omnipresence of elements and the ubiquitous formation of small molecules such as hydrides or oxides (Heumann et al., 1995). Using narrow-bandwidth laser radiation, as delivered by continuous wave lasers, in a suitable experimental geometry, high-resolution (HR) RIMS contributes a remarkable additional isotopic selectivity of up to multiple orders of magnitude already within the ionization process. By taking advantage of the isotope shift in well-chosen atomic transitions, this optical selectivity enhances the abundance sensitivity of the mass spectrometer, reduces molecular interferences, and leads to extremely high isotopic selectivities, which can be estimated to reach up to 10^{15}.

Sensitive laser mass spectrometric investigations directed toward achieving such high isotope selectivities have been carried out on a long list of ultrarare radioisotopes. Early studies concerned the radioisotopes ^{81}Kr and ^{85}Kr of the noble gas krypton, which either are of relevance for radioisotope dating or are of environmental concern as a major fission product release after a possible nuclear reactor disaster or from nuclear reprocessing, respectively. RIMS determination techniques for noble gases use fully encapsulated vacuum chambers and refrigeration or recirculation of the sample material to obtain extremely low LODs in the low fg (10^{-15} g) to ag (10^{-18} g) region and even beyond (Payne et al., 1994; Crowther et al., 2008; Iwata et al., 2010; Strashnov et al., 2011). Due to these outstanding specifications, unusual applications of this technology, for example, for the detection of exotic species like dark matter are envisaged (Iwata et al., 2015).

Regarding the surveillance of the use of nuclear power and possibly associated lowest-level contaminations in the environment, dedicated RIMS techniques have been developed for a number of fission products, which are radiometrically not easily accessible. These include the alkaline isotopes ^{135}Cs, which is one of the only seven long-lived fission products existing, and ^{137}Cs. Both isotopes were measured at a surplus of about six orders of magnitude of stable ^{133}Cs. Results showed good agreement with TIMS studies (Pibida et al., 2004). Significantly higher isotopic selectivity is required for investigations on the alkaline earth isotopes ^{41}Ca, ^{89}Sr, and ^{90}Sr, which we shall discuss in detail below, as well as for analytical studies of ^{210}Pb and ^{236}U. The determination of traces of the long-lived lead isotope in the range of few fg per g of sample material from human hair was investigated as an indicator for accumulated radon exposure (Bushaw, 1992).

The artificial isotope ^{236}U is frequently used to distinguish anthropogenic material, stemming from the nuclear fuel cycle, from natural uranium. For the sensitive surveillance of ^{236}U content in environmental samples, an isotopic selectivity of a minimum of 9 up to 14 orders of magnitude is needed (Schumann et al., 2005). For most of these studies, a quite similar compact analytical setup is used, which applies three-step optical resonance ionization induced by small diode lasers and ionization at a compact quadrupole mass filter. This apparatus will be described in the example of ^{41}Ca determinations as follows: For selective ultratrace determination of the long-lived radioisotope ^{41}Ca ($T_{1/2} = 1.04 \times 10^5$ a) aside a huge surplus of stable calcium, an excellent overall isotope selectivity $> 10^{12}$ has been demonstrated at an LOD of well below 10^6 atoms ^{41}Ca (Müller et al., 2001). This enables various applications, for example, using ^{41}Ca as a biomedical tracer isotope for (1) in vivo studies of human calcium kinetics, (2) cosmochemical investigations on the ^{41}Ca content in meteorites (together with related studies on the production cross section in nuclear reactions), and finally, (3) the determination of ^{41}Ca activity in reactor concrete and other materials for an integrated neutron dosimetry. For these studies, ^{41}Ca abundances of 10^{-9} to 10^{-12} relative to the predominant stable isotope ^{40}Ca (96.94%) were addressed. Due to the much lower estimated ambient ^{41}Ca concentrations of only about 10^{-15}, radioisotope dating by ^{41}Ca, which would open up a new time window around 10^6 years before present, is not in reach by HR-RIMS. In addition to the sophisticated laser system, the setup for ultratrace determination of ^{41}Ca consists of a vacuum recipient with a graphite furnace for efficient atomization of the sample and a quadrupole mass filter for mass-selective detection of the ions ionized via the selective laser excitation process. The calcium atomic beam is crossed perpendicularly by the laser beams at a minimum distance in front of the orifice to ensure highest overall efficiency. For maximum isotope selectivity, a three-step three-color excitation scheme is used to populate a high-lying Rydberg state, from where efficient nonresonant ionization is induced by a 10.6-μm CO_2 laser. All three resonant excitation steps are driven by narrow-bandwidth radiation from compact extended cavity diode lasers, which can be tuned well onto the respective Ca resonances and kept well controlled in resonance. For quantification of the ^{41}Ca content, the low abundance isotope ^{43}Ca (0.134%) is used as a reference, which significantly reduces the required dynamic range by three orders of magnitude. In samples of concrete from the shield of a nuclear research reactor, very low relative abundances in the order of a few 10^{-10} corresponding to contents of only ~20 mBq/g, were determined by quantifying the total calcium content via X-ray fluorescence spectroscopy (Müller et al., 2000). In the frame of biomedical and nutritional studies in connection with osteoporosis prevention and therapy, human urine samples were measured by this technique with somewhat equally low abundances (Denk et al., 2006).

In strontium, a desired fast and sensitive determination of the radiotoxic isotopes ^{89}Sr and ^{90}Sr in air samples, for example, after a nuclear accident, would require an overall selectivity of Sr $\geq 10^{11}$ (Monz et al., 1993). For instance, about 10^8 atoms of ^{90}Sr and ^{89}Sr need to be detected among some 10^{18} atoms of stable strontium, mainly ^{88}Sr, assumed for a collection of 1000 m^3 of ambient air. Due to a very small optical isotope shift in Sr, this requirement is not met by the triple-resonance diode laser-based RIMS on a thermal atomic beam, as discussed previously for the ^{41}Ca determination (Bushaw et al., 2001). In contrast, the complex setup of resonance ionization mass spectrometry on a fast atomic beam has been adapted and applied. Sample atoms are initially ionized in a conventional thermal ion source and undergo sector-field mass spectrometry at an ion beam energy of about 30 keV. After neutralization in a cesium vapor cell, resonance ionization is carried out in collinear geometry between laser and atomic beam, which is applied on the resulting fast neutral atomic beam by utilizing the strong Doppler shift for optical isotope selection. Highly excited Rydberg atoms are generated, selectively field-ionized, and counted by a channeltron detector after passing an electrostatic energy filter for further background suppression. With this approach, an overall efficiency of 1×10^{-5} and a selectivity for ^{90}Sr/^{88}Sr of well above 10^{10} have been demonstrated and ^{90}Sr measurements within a variety of environmental and technical samples were reported (Wendt et al., 1997). The LOD for ^{90}Sr has been determined to be 3×10^6 atoms, corresponding to an activity of 1.5 mBq per sample.

VIII. Accelerator mass spectrometry

A. Principle

AMS is a modern and sophisticated mass spectrometric technique, specialized to provide highest absolute sensitivity and selectivity with respect to isobaric and isotopic contaminations. Individual ions are detected after acceleration to energies in the hundred keV to few MeV range. Any interferences from atomic or molecular species are efficiently suppressed in the different stages of the mass spectrometric measuring process. Thus, AMS permits the determination of ultrarare radioisotopes with abundances far below 10^{-9} of a dominant neighboring isotope and surpasses the usual selectivity limits in isotope ratio measurements of conventional mass spectrometric techniques by far (Purser et al., 1981). AMS data provide relative isotope abundances but do not give absolute concentrations of an element or isotope. Hence, AMS in most cases is combined with an absolute concentration measurement of at least one stable isotope by, for example, ICP-MS. The development of AMS has been boosted primarily in connection with the determination of the cosmogenic radioisotope ^{14}C. The numerous applications include (1) radioisotope dating, (2) atmospheric and oceanographic circulation studies, (3) determination of anthropogenic radioactive contamination levels, particularly as a consequence of the atmospheric nuclear weapons tests, and (4) biomedical applications, for example, in studies on cancer prevention by using ^{14}C labeled compounds (Hellborg and Skog, 2008). Many of these fields have only become possible as a result of the development of AMS, which has partly replaced the formerly used technique of low-level radiometric counting (Libby, 1955; García-León, 2018) and also neutron activation analysis (Szidat et al., 2000) due to its higher sensitivity, smaller sample size requirement, and faster response. Radiocarbon dating, carried out on ^{14}C isotopic levels as low as 10^{-15}, is the broadest field of application of AMS with thousands of samples per year. Further applications on many other radioisotopes have been developed, which contribute to the value and acceptance of the technique (Michel, 1999).

Following a suggestion by Oeschger and coworkers (Williams et al., 1969), which was triggered by a first, very early mass spectrometric experiment at an accelerator by Alvarez and Cornog (1939), it was realized in the late 1970s that isobaric contaminations can often be eliminated very efficiently by performing mass spectrometry with high-energy (~MeV) ion beams (Müller, 1977; Nelson et al., 1977; Bennett et al., 1977). In this regime, a variety of experimental techniques can be applied and quantitative, and highly selective counting of the species of interest with unrivalled background suppression is enabled, including the following:

- Already in the ion source of the mass spectrometer, unwanted atomic isobars can be suppressed by using negative ions in the first stage of a tandem accelerator. Many interfering elements do not form such anions. For example, $^{14}N^-$, the only abundant atomic isobar of $^{14}C^-$, is not stable and consequently is neither produced nor accelerated.
- A second technique to suppress interfering ions directly inside the source is laser-induced photodetachment. If the wanted species' electron affinity exceeds that of the unwanted species, the latter can be neutralized by photodetachment, and hence it is removed from the beam. At the Vienna Environmental Research Accelerator at University of Vienna, this isobar suppression was demonstrated successfully for ^{63}Cu (Forstner et al., 2015), ^{93}Zr, and ^{93}Nb (Martschini et al., 2015).
- Furthermore, chemical reactions in gas-filled cells might prove a very universal tool for removal of interfering ions in the future (Eliades et al., 2015).
- Molecular isobars can be efficiently removed by passing the accelerated beam through a thin foil or a gaseous target inducing a so-called stripping process. Here, negative atomic ions are efficiently converted into positive ones by stripping off some electrons. Due to the high energy transfer from the collisions, molecules are effectively dissociated, removing an additional source of interfering ions.
- Finally, versatile and discriminatory ion detection techniques, for example, elemental selective energy-loss measurements in ionization chambers, are applied on high-energy ions (~MeV) to identify the species via its mass and proton number and further reduce the influence of isobaric interferences.

Examples for such interferences in the case of the $^{14}C^+$ ion are $^{28}Si^{2+}$, $^{14}N^+$, or $^{13}CH^+$. Other limiting effects include background effects originating from gas kinetics, wall scattering, and charge exchange collisions. Nevertheless, these contributions can be strongly suppressed by optimizing the experimental conditions. A typical AMS system is based on a tandem accelerator with a terminal voltage of 2.5−10 MV, usually involving a beam line length of several tens of meters. On the side of the ion source, the so-called low-energy side, intense beams of negative atomic or molecular ions of the species to be analyzed are produced. This is accomplished usually by sputtering, applying Cs^+ ion bombardment with about 10 keV energy, focused onto the target of the analyte. The latter is introduced in the form of about 10 mg of material, fixed within a target wheel or a sample exchange mechanism with more than 60 positions, which enables rapid changes between analytical sample, blank, and standard. The importance of a proper chemical pretreatment of the sample, including removal of extraneous material, dissolution, and, when appropriate, addition of carrier, separation of the element of interest, and conversion into the most suitable form for the sputtering process, must be emphasized (Tuniz et al., 1998; Lopez-Lora et al., 2018).

A minor drawback of the AMS technique is the fact that rather large sample sizes around 10 mg are required for the initial ionization process. Negative target ions from the sputtering process are accelerated toward a first low-energy (typically ~50 keV) mass separator for an initial mass selection. By varying the electric potential applied at the magnet chamber, a rapid change from one mass to another is possible, avoiding complications that arise from the scanning of the magnetic field, for example, hysteresis or slow switching speed. Subsequently, acceleration to the high terminal voltage of several megavolts takes place within the first half of the tandem accelerator. On the high positive potential of the terminal, the ions undergo a stripping process in a thin foil (~5 μg/cm^2) or a gas target, and a spectrum of positive charge states is populated. Now positively charged, these ions are accelerated a second time, this time from the terminal voltage down to ground; for example, in the case of carbon, typically charge states between C^{2+} and C^{7+} are populated, depending on the ion velocity. For the case of the C^{5+} ion, this leads to a total beam energy of 15 MeV on the high energy side, when a typical 2.5 MV tandem accelerator is used. All molecular ions are dissociated during the stripping process, and remaining fragments have significantly altered mass-to-charge ratio and can easily be separated afterward. For this purpose, the high-energy ion beam undergoes another mass selective step in a second sector field magnet as well as charge state and energy filtering within an electrostatic deflector before reaching the detectors. Due to the high energy of the incoming ions, some kind of selectivity via species identification with respect to charge number Z and mass A is possible by measuring energy and specific energy loss in a sensitive semiconductor telescope or an energy-sensitive ionization chamber. A typical layout of an AMS machine, which shows these individual components, is given in Fig. 8.10.

Generally, AMS determines the relative abundance of the rare radioisotope under study by comparing its count rate with the ion current of an abundant stable (or reference) isotope of the same element. The measured quantity is therefore, in most cases, an isotope ratio, which might cover many orders of magnitude. In most applications, a high accuracy in the order of $\leq 1\%$ for this value is required. Such precise isotope ratio measurements with unrivalled dynamical range of up to 15 orders of magnitude are accomplished experimentally by accelerating both isotopes alternately in a fast-cyclic mode. While the transmitted atoms of the rare ultratrace isotope are counted

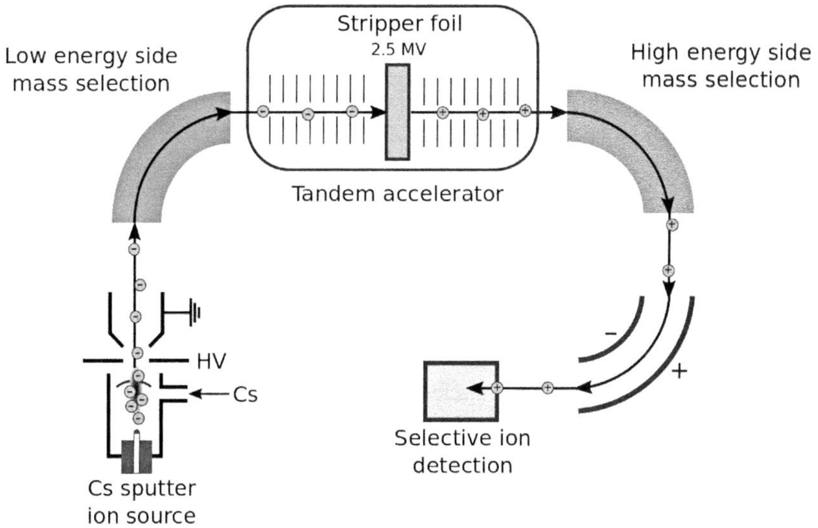

FIGURE 8.10 Schematics of the primary components of a typical accelerator mass spectrometry machine. Details are provided in the text.

individually as ion counts in a two-dimensional Z versus A plane, the reference isotope, which typically is up to a factor of 10^{12} more abundant, is measured as electrical current using a Faraday cup detector for this intense beam. Proper calibration and stability of both detectors during the measurement is mandatory. Hence, each analytical measurement is flanked by measurements on blinds and calibration samples for background surveillance and proper quantification of results. For the heavier isotopes, the energy resolution of semiconductor detectors or ionization chambers is not sufficient for a proper species identification and reasonable suppression of isobars and/or neighboring isotopes. A better background suppression is possible by combining the energy determination in the detector with a velocity analysis of the incoming ions by an additional time-of-flight measurement (Coffin and Engelstein, 1984). In a different approach, a gas-filled magnet in front of the detector leads to a spatial separation of projectiles with different proton number Z and hence significantly increases the selectivity with respect to those isobars, which have not been filtered out sufficiently during the stripping process (Paul, 1990). For some isotopes, the process of fully stripping the ions after gaining sufficient high energy is favorable for complete background suppression, but this can only be realized at large accelerators.

During the last decade, enormous effort can be observed within the AMS community to reduce size and cost of AMS systems. By incorporating well-matched technical features such as gas strippers and state-of-the-art silicon detectors, very compact low-voltage (300 kV – 1 MV) AMS machines have been developed (Synal et al., 2000). By now, these systems have demonstrated their broad range of applicability not only for ^{14}C measurements but also in the analysis of other radioisotopes such as ^{3}H, ^{26}Al, ^{41}Ca, ^{129}I, and Pu isotopes with selectivities close to that of a standard large-frame AMS machine (Suter et al., 2000).

The technique of AMS has currently become a standard method, applied for investigations on a variety of long-lived radioisotopes. Commonly utilized isotopes in routine operation are ^{10}Be, ^{14}C, ^{26}Al, ^{36}Cl, ^{41}Ca, and ^{129}I. Many other isotopes, namely ^{3}H, ^{3}He, ^{7}Be, ^{22,24}Na, ^{32}Si, ^{39}Ar, ^{44}Ti, ^{53}Mn, ^{55,60}Fe, ^{59,63}Ni, ^{79}Se, ^{81}Kr, ^{90}Sr, ^{93}Zr, ^{93}Mo, ^{99}Tc, ^{107}Pd, ^{151}Sm, ^{205}Pb, ^{236}U, ^{237}Np, and $^{238-244}$Pu have also been investigated by AMS during recent years. Most of the radioisotopes mentioned in both categories have natural abundances in the range of 10^{-9} down to 10^{-18}, as illustrated in Fig. 8.11. Here some isotopes are missing, as their natural abundance is either not known precisely or it varies across the earth, due to an inhomogeneous distribution of natural or anthropogenic

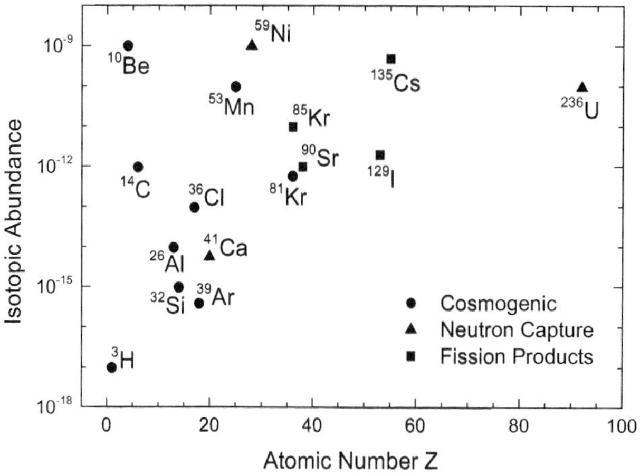

FIGURE 8.11 Natural radioisotopes with abundances below 10^{-9} together with their process of origin.

sources. The isotopes shown in Fig. 8.11 are produced by cosmic radiation, neutron capture, or as fission product. More details about AMS and a compilation of the different AMS versions for individual radioisotopes can be found in the textbook on AMS (Tuniz et al., 1998) and a number of review articles (Litherland et al., 1987; Elmore and Philips, 1987; Hotchkis et al., 2000; Woelfli, 1987; Vogel et al., 1995; Fifield, 1999; Kutschera et al., 2000; Synal, 2013). The application of AMS in various fields is increasing rapidly since the late 1970s (Kutschera, 2016). Since the year 2012, there were approximately 110 uniquely dedicated AMS facilities in use around the world, whereas quite some more nuclear physics accelerator facilities have been adapted to perform AMS measurements, often also on a routine basis (Kutschera, 2012). Whereas some facilities measure the full spectrum of "AMS nuclides," many others focus on a subset such as the lighter ones up to ^{129}I (Rugel et al., 2016) or the actinides (Hotchkis et al., 2018).

B. Applications of accelerator mass spectrometry

1. Radioisotope dating in archeology and other applications of the isotope ^{14}C

AMS has become best known due to its application to the analysis of the isotope ^{14}C, where thousands of samples per year are investigated worldwide. Carbon-14 is produced by the neutron component of cosmic rays from ^{14}N in the upper atmosphere with a rate of about 2 atoms/cm^2/s. Through the formation of CO_2, it is very homogeneously distributed worldwide and is easily assimilated by living organisms. With a half-life of $T_{1/2} = 5730$ years and an average abundance in living organisms of about 10^{-12}, the determination of the remaining ^{14}C in fossils not only gives access to a dating range of about 40,000 years before present but also opens up a broad field of further investigations, for example, circulation studies in the atmosphere or the oceans, authentication of minerals, plants, and even human artifacts such as precious artworks. Most important is the fact that a well-adapted continuous calibration curve of the ^{14}C content in the biosphere, dating back to 7000 BC can be established from counting tree rings. Outstanding applications for ^{14}C radiodating by AMS include the dating of the Ice Man from the Tyrolean Alps (3120–3350 BC), the Shroud of Turin (1260–1390 AD), as well as numerous approaches of dating ice cores from the artic regions or glaciers. Novel hyphenated techniques are continuously developed. At the University of Bern, a continuous flow gas interface was coupled to a low-energy AMS of TANDY type for online measurements of CO_2 gas, for instance, produced by organic/elemental carbon analyzers (Agrios et al., 2017).

2. Accelerator mass spectrometry applications in geo- and cosmoscience

Numerous additional studies in the field of geosciences currently are carried out with the AMS technique with respect to radioisotopes other than ^{14}C. The cosmogenic production mechanism and the short residence time of ^{10}Be in the atmosphere make it an ideal cosmic ray and hence solar activity monitor, which is preserved in arctic ice for hundreds of years and serves as a paleoclimatic archive. Together with data on a few other long-lived radionuclides, for example, ^{26}Al (Auer et al., 2009) and ^{36}Cl both measured by AMS, further information such as the strength of the earth's magnetic field, the mode of the atmospheric mixing, and precipitation patterns are extracted. In this way, for example, the 11-year solar cycle as well as severe changes during the end of the last glacial period, 40 ka back, have been obtained. From investigations of the same three isotopes, the age of the Arizona Meteor crater has been determined to about 49,000 years (Vogel et al., 1995). As a different AMS application in geosciences, ocean currents were probed by detecting actinides and long-lived fission products released from reprocessing plants (Christl et al., 2015). Abundance of ^{60}Fe in sea sediments yielded findings on the history of supernova explosions in the vicinity of our solar system during the past 10 Mio. years (Wallner et al., 2016).

The production mechanisms for long-lived radioisotopes are significantly different in space than on the earth's surface, due to the different composition of cosmic radiation (Michel, 1999). Thus, AMS analyses in supposed extraterrestrial material like micrometeorites, which exhibit isotopic anomalies from the well-known terrestrial abundances, serve as the most stringent test for its authenticity. In addition, if data of at least some of the different routine AMS isotopes, that is, ^{10}Be, ^{14}C, ^{26}Al, ^{36}Cl, ^{41}Ca, and ^{129}I, are combined, these measurements provide valuable information, for example, on the terrestrial age of the material, the exposure age, and the geometrical conditions of exposure (Herzog, 1994). Recent reviews summarize the use of AMS and other techniques in the field of geo- and cosmosciences (Linge et al., 2017; Wiedenbeck et al., 2012, 2014).

3. Noble gas analysis

Due to their chemical inertness, noble gases form an ideal tracer for dating ground- and surface waters, for performing related circulation studies or for the analysis of gaseous inclusions in ice cores. Since noble gases generally do not form stable negative ions, long-lived noble gas isotopes, for example, ^{39}Ar or ^{81}Kr, cannot be analyzed by standard AMS facilities. Thus, W. Kutschera and coworkers used an electron cyclotron resonance source instead, to produce positive ions and a high-energy cyclotron accelerator to accelerate them. To separate ^{81}Kr from its abundant isobar ^{81}Br, they applied the full-stripping technique, accelerating the ions up to a 4-GeV beam energy and removing all electrons of the ions (Collon et al., 1997). Once fully stripped, $^{81}Kr^{+36}$ can be well separated from the maximum charge stage of $^{81}Br^{+35}$ with $Z = 35$. With a detection efficiency of $\sim 1 \times 10^{-5}$, a 10% precision measurement required 0.5 cc STP of a krypton sample (Collon et al., 2000). For groundwater dating, this amount had been extracted from 16 tons of groundwater. In a closely related development, ^{39}Ar has been detected at an isotopic abundance level of 1×10^{-16} with an efficiency of 1×10^{-3} using a high-energy ($\sim$200 MeV) linear accelerator (Collon et al., 2004). Since the charge number of ^{39}Ar ($Z = 18$) is lower than that of its abundant isobar, ^{39}K ($Z = 19$), the full stripping technique cannot be used here. Instead, ^{39}Ar has been separated from ^{39}K using the standard technique of a gas-filled magnet, which exploits the fact that the energy loss of ^{39}K is significantly different from that of ^{39}Ar.

4. Accelerator mass spectrometry in life sciences

Research with tracer isotopes has become enormously relevant for various kinds of in vivo studies concerning influences of nutrition or deficiencies as well as analyzing the cause and prevention of numerous diseases. In most cases, no suitable enriched stable isotopes for such studies are available, and radioisotopes must be chosen. However, radionuclide concentrations as high as required by radiometric detection techniques would cause a significant radiation dose to the test person. Furthermore, such high doses could affect the proper physiological response itself. Hence, AMS with its high sensitivity and selectivity has become a valuable system for analyzing biomedical samples (Jackson et al., 2001). Again, ^{14}C is the key isotope, used for labeling numerous biomedically relevant compounds, while their influence can be studied by AMS measurements on their content in bodily excretions or blood (Vogel, 2000). Other radioisotopes used in this field are ^{10}Be, ^{26}Al, ^{41}Ca, ^{79}Se, and ^{129}I. Aluminum was considered an essential trace element, while, only in the late 1990s, its toxicity has been recognized. A number of AMS studies with ^{26}Al on gastrointestinal and intravenous absorption have been carried out, which indicated pathways and established exchange rates for Al crossing the blood–brain barrier (Tuniz et al., 1998). ^{41}Ca has become extremely versatile in the measurement of nutritional and other effects on calcium kinematics and the prevention of diseases such as osteoporosis (Freeman et al., 1997). The high demand on determining the capacity for these life science applications in the field of biomedical and nutritional research has led to the recent commissioning of dedicated commercial AMS instruments (Barker and Garner, 1999). Additional biomedical applications comprise the uptake and retention of Pu in living organisms (Fifield, 2000).

AMS is used for so-called retrospective dosimetry. The release of ^{131}I from nuclear installations can cause significant dose to the human thyroid. Due to its short half-life of only 8 days, screening is often not performed fast enough, as was the case in the Chernobyl accident and also partly in Fukushima. However, measuring the long-lived fission product ^{129}I in environmental compartments permits a determination of the thyroid doses with rather high accuracy (Michel et al., 2005, 2015).

AMS-based measurements gained new insight into the function of the human brain. Contrasting up to then assumption of little or no renewal of so-called hippocampal cells in the adult human brain, Spalding et al. (2013) proved an annual turnover of almost 2% of the cells by measuring the concentration of nuclear bomb test–derived ^{14}C in genomic DNA.

5. Accelerator mass spectrometry measurements on long-lived radionuclides in the environment

Most of the anthropogenic long-lived radioisotopes in the environment arise from the nuclear fuel cycle or nuclear explosions. They are produced either by neutron capture (e.g., actinides) or as fission products (e.g., ^{90}Sr, ^{99}Tc, ^{135}Cs). The determination of low-level contaminations of these isotopes is of major importance for nuclear safeguards; however, it is often complicated by the necessity to distinguish these contributions from the naturally existing levels. A few high-transmission, high-voltage AMS instruments are specialized for measuring these radioisotopes (Hotchkis et al., 2000). Overall efficiencies of up to 10^{-4} enabling measurements with as few as 10^6 atoms of the radioisotope of interest have been demonstrated in favorable cases with typical selectivities above 10^{10} (Fifield, 2008). A recent review summarizes selected radioecological studies on ^{14}C, ^{36}Cl, ^{129}I, ^{236}U, and Pu (García-León, 2018). Further AMS investigations in this field have been reported for ^{90}Sr (Paul et al., 1997), ^{99}Tc (Fifield, 2000; Bergquist et al., 2000),^{129}I (Daraoui et al., 2016) (Daraoui et al., 2012; Michel et al., 2012), ^{135}Cs (Lachner et al., 2015), ^{206}Pb (Paul et al., 2000), ^{236}U (Zhao et al., 1994, 1997; Berkovits et al., 2000), and Pu isotopes (Fifield et al., 1996; Wallner et al., 2000; McAninch et al., 2000; Hotchkis et al., 2000; Lachner et al., 2010). Plutonium and Americium transport and scavenging by snow was investigated by Gückel et al. (2017). The isotope abundance of ^{236}U in depleted uranium particles was used as an indicator that the material originates from the enrichment of reprocessed uranium (Danesi et al., 2003a, 2003b). Several of the investigations on Pu and ^{236}U release from the

Fukushima Daiichi accident used AMS (Schneider et al., 2013, 2017). Applications of AMS in the nuclear field include trace detection of activation products in shielding material to be decommissioned (Warwick et al., 2009) and highly sensitive and selective detection of Am and Cm (Christl et al., 2014). In a field experiment, colloid-mediated transport of ^{233}U, ^{237}Np, ^{242}Pu, and ^{243}Am was monitored at ultratrace concentrations (Quinto et al., 2017). AMS was also used in the search for naturally occurring superheavy elements (Dellinger et al., 2011) and in the precise measurement of the half-lives of long-lived nuclides (Vockenhuber et al., 2004).

References

Actis-Dato, L.O., Aldave de las Heras, L., Betti, M., Toscano, E.H., Miserque, F., Gouder, T., 2000. Investigation of mechanisms of corrosion due to diffusion of impurities by direct current glow discharge mass spectrometry depth profiling. J. Anal. At. Spectrom. 15, 1479–1484.

Adriaens, A.G., Fassett, J.D., Kelly, W.R., Simons, D.S., Adams, F.C., 1992. Determination of uranium and thorium concentrations in soils: comparison of isotope dilution a secondary ion mass spectrometry and isotope dilution-thermal ionization mass spectrometry. Anal. Chem. 64, 2945–2950.

Adriaens, A., van Vaeck, L., Adams, F., 1999. Static secondary ion mass spectrometry (S-SIMS) Part 2: material science applications. Mass Spectrom. Rev. 18, 48–81.

Aggarwal, S.K., 2006. Precise and accurate determination of alpha decay half-life of 244Pu by relative activity method using thermal ionization mass spectrometry and alpha spectrometry. Radiochim. Acta 94, 397–401.

Aggarwal, S.K., 2016. Thermal ionisation mass spectrometry (TIMS) in nuclear science and technology – a review. Anal. Methods 8, 942–957.

Agrios, K., Salazar, G., Szidat, S., 2017. A continuous-flow gas interface of a thermal/optical analyzer with C-14 AMS for source apportionment of atmospheric aerosols. Radiocarbon 59, 921–932.

Al-Nuzal, S.M.D., 2011. Fast evaluation of $^{235}U/^{238}U$ ratio in nuclear spent fuel safe guard by thermal ionization mass spectrometry by using refractory metal oxides. J. Radioanal. Nucl. Chem. 288, 313–317.

Alvarez, L.W., Cornog, R., 1939. ^{3}He in helium. Phys. Rev. 56, 379.

Amaral, A., Galle, P., Cossonnet, C., Franck, D., Pihet, P., Carrier, M., Stephan, O., 1997. Perspectives of uranium and plutonium analysis in urine samples by secondary ion mass spectrometry. J. Radioanal. Nucl. Chem. 226, 41–45.

Ambartzumian, R.V., Letokhov, V.S., 1972. Selective two-step (STS) photoionization of atoms and photodissociation of molecules by laser radiation. Appl. Opt. 11, 354–358.

Ammann, A.A., 2007. Inductively coupled plasma mass spectrometry (ICP MS): a versatile tool. J. Mass Spectrom. 42, 419–427.

Auer, M., Wagenbach, D., Wild, E.M., Wallner, A., Priller, A., Miller, H., Schlosser, C., Kutschera, W., 2009. Cosmogenic ^{26}Al in the atmosphere and the prospect of a $^{26}Al/^{10}Be$ chronometer to date old ice. Earth Planet. Sci. Lett. 287, 453–462.

Backe, H., Eberhardt, K., Feldmann, R., Hies, M., Kunz, H., Lauth, W., Martin, R., Schöpe, H., Schwamb, P., Sewtz, M., Thörle, P., Trautmann, N., Zauner, S., 1997. A compact apparatus for mass selective resonance ionization spectroscopy in a buffer gas cell. Nucl. Instrum. Meth. B 126, 406–410.

Bacon, J.R., Linge, K.L., Parrish, R.R., Van Vaeck, L., 2008. Atomic spectrometry update. Atomic mass spectrometry. J. Anal. At. Spectrom. 23, 1130–1162.

Balcaen, L.I.L., De Samber, B., De Wolf, K., Cuyckens, F., Vanhaecke, F., 2007. Hyphenation of reverse-phase HPLC and ICP-MS for metabolite profiling-application to a novel antituberculosis compound as a case study. Anal. Bioanal. Chem. 389 (3), 777–786.

Barker, J., Garner, R.C., 1999. Biomedical applications of accelerator mass spectrometry/isotope measurements at the level of the atom. Rapid Commun. Mass Spectrom. 13, 285–293.

Barrero Moreno, J.M., Garcia Alonso, J.I., Arbore, P., Nicolaou, G., Koch, L., 1996. Characterization of spent nuclear fuels by ion chromatography–inductively coupled plasma mass spectrometry. J. Anal. At. Spectrom. 11, 929–935.

Barrero Moreno, J.M., Betti, M., Nicolaou, G., 1999. Determination of caesium and its isotopic composition in nuclear samples using isotope dilution-ion chromatography-inductively coupled plasma mass spectrometry. J. Anal. At. Spectrom. 14, 875–879.

Barshick, C.M., Duckworth, D.C., Smith, D.H., 1993. Analysis of solution residues by glow discharge mass spectrometry. J. Am. Soc. Mass Spectrom. 4, 47–53.

Barshick, C.M., Goodner, K.L., Clifford, H.W., Eyler, J.R., 1998. Application of glow discharge Fourier-transform ion cyclotron resonance mass spectrometry to isotope ratio measurements. Int. J. Mass Spectrom. 178, 73–79.

Barshick, C.M., Duckworth, D.C., Smith, D.H., 2000. Inorganic Mass Spectrometry. Marcel Dekker Inc., New York.

Barzakh, A.E., Denisov, V.P., Fedorov, D.V., Orlov, S.Y., Seliverstov, M.D., 1997. A mass-separator laser ion source. Nucl. Instrum. Meth. **B** 126, 85–87.

Beauchemin, D., 2010. Inductively coupled plasma mass spectrometry. Anal. Chem. 82 (12), 4786–4810.

Becker, J.S., 2002. State-of-the-art and progress in precise and accurate isotope ratio measurements by ICP-MS and LA-ICP-MS. J. Anal. At. Spectrom. 17, 1172–1185.

Becker, J.S., 2010. Determination of long-lived radionuclides in nuclear, environmental and clinical samples by ICP-MS and LA-ICP-MS. In: Gross, M.L., Caprioli, R.M. (Eds.), Encyclopedia of Mass Spectrometry, Elemental, Isotope Ratio Mass Spectrometry. Elsevier, New York, pp. 200–211.

Becker, J.S., Dietze, H.-J., 1997. Double-focusing sector field inductively coupled plasma mass spectrometry for highly sensitive multi-element and isotopic analysis. J. Anal. At. Spectrom. 12, 881–889.

Becker, J.S., Dietze, H.-J., 1998. Determination of long-lived radionuclides by double-focusing sector field ICP mass spectrometry. Adv. Mass Spectrom. 14, 681–689.

Becker, J.S., Dietze, H.-J., 1999a. Precise isotope ratio measurements for uranium, thorium, and plutonium by quadrupole-based inductively coupled plasma mass spectrometry. Fresenius J. Anal. Chem. 364, 482–488.

Becker, J.S., Dietze, H.-J., 1999b. Application of double-focusing sector field ICP mass spectrometry with shielded torch using different nebulizers for ultratrace and precise isotope analysis of long-lived radionuclides. J. Anal. At. Spectrom. 14, 1493–1500.

Becker, S., Hirner, A.V., 1994. Coupling of inductively coupled plasma mass spectrometry (ICP-MS) with electrothermal vaporisation (ETV). Fresenius J. Anal. Chem. 350, 260–263.

Becker, J.S., Gastei, M., Tenzler, D., Dietze, H.J., 1998. LA-ICP-MS for trace analysis of long-lived radionuclides in solid non-conducting radioactive waste samples. Adv. Mass Spectrom. 14, 687.

Becker, J.S., Pickhardt, C., Dietze, H.-J., 2000. Laser ablation inductively coupled plasma mass spectrometry for the trace, ultratrace and isotope analysis of long-lived radionuclides in solid samples. Int. J. Mass Spectrom. 202, 283–297.

Beer, B., Heumann, K.G., 1992. Trace analyses of U, Th and other heavy metals in high purity aluminum with isotope dilution mass spectrometry. Fresenius J. Anal. Chem. 343, 741–745.

Bennett, C.L., Beukens, R.P., Clover, M.R., Gove, H.E., Liebert, R.B., Litherland, A.E., Purser, K.H., Sondheim, W.E., 1977. Radiocarbon dating using electrostatic accelerators: negative ions provide the key. Science 198, 508–510.

Benninghoven, A., Janssen, K.T.F., Tümpner, J., Werner, H.W., 1992. Secondary Ion Mass Spectrometry, SIMS VIII. John Wiley & Sons, New York.

Bergquist, B.A., Marchetti, A.A., Martinelli, R.E., McAninch, J.E., Nimz, G.J., Proctor, I.D., Southon, J.R., Vogel, J.S., 2000. Technetium measurements by accelerator mass spectrometry at LLNL. Nucl. Instrum. Meth. B 172, 328–332.

Berkovits, D., Feldstein, H., Ghelberg, S., Hershkowitz, A., Navon, E., Paul, M., 2000. ^{236}U in uranium minerals and standards. Nucl. Instrum. Meth. **B** 172, 372–376.

Betti, M., 1996. Use of a direct current glow discharge mass spectrometer for the chemical characterization of samples of nuclear concern. J. Anal. At. Spectrom. 11, 855–860.

Betti, M., 1997. Use of ion chromatography (ICP/MS) for the determination of fission products and actinides in nuclear applications. J. Chromatogr. A 789, 369–379.

Betti, M., 2002. Analysis of samples of nuclear concern with glow discharge atomic spectrometry. In: Marcus, R.K., Broekaert, J.A.C. (Eds.), Glow Discharge Plasmas in Analytical Chemistry. John Wiley & Sons, New York, pp. 273–290.

Betti, M., 2005. Isotope ratio measurements by secondary ion mass spectrometry (SIMS) and glow discharge (GDMS). Int. J. Mass Spectrom. 242, 169–182.

Betti, M., de las Heras, L.A., 2004. Glow discharge spectrometry for the characterization of nuclear and radioactively contaminated environmental samples. Spectrochim. Acta B 59, 1359–1376.

Betti, M., Rasmussen, G., Hiernaut, T., Koch, L., Milton, D.M.P., Hutton, R.C., 1994. Adaption of a glow discharge mass spectrometer in a glove-box for the analysis of nuclear materials. J. Anal. At. Spectrom. 9, 385–391.

Betti, M., Tamborini, G., Koch, L., 1999. Use of secondary ion mass spectrometry in nuclear forensic analysis for the characterization of plutonium and highly enriched uranium particles. Anal. Chem. 71, 2616–2622.

Blaum, K., Geppert, C., Kluge, H.-J., Mukherjee, M., Schwarz, S., Wendt, K., 2003. A novel scheme for a highly selective laser ion source. Nucl. Instrum. Methods Phys. Res. B 204, 331–335.

Blumenfeld, Y., Nilsson, T., Van Duppen, P., 2013. Facilities and methods for radioactive ion beam production. Phys. Scr. T **152**, 014023.

Boulyga, S.F., Becker, J.S., 2001a. ICP–MS with hexapole collision cell for isotope ratio measurements of Ca, Fe, and Se. Fresenius J. Anal. Chem. 370, 618–623.

Boulyga, S.F., Becker, J.S., 2001b. Determination of uranium isotopic composition and ^{236}U content of soil samples and hot particles using inductively coupled plasma mass spectrometry. Fresenius' J. Anal. Chem. 370, 612–617.

Boulyga, S.F., Becker, J.S., 2002a. Improvement of abundance sensitivity in quadrupole based ICP-MS with hexapole collision cell. J. Anal. At. Spectrom. 17, 1202–1206.

Boulyga, S.F., Becker, J.S., 2002b. Isotopic analysis of uranium and plutonium using ICP-MS and estimation of burn-up of spent uranium in contaminated environmental samples. J. Anal. At. Spectrom. 17, 1143–1147.

Boulyga, S.F., Konegger-Kappel, S., Richter, S., Sangely, S., 2015. Mass spectrometric analysis for nuclear safeguards. J. Anal. At. Spectrom. 30, 1469–1489.

Bu, W., Zheng, J., Guo, Q., Aono, T., Tazoe, H., Tagami, K., Uchida, S., Yamada, M., 2014. A method of measurement of ^{239}Pu, ^{240}Pu, ^{241}Pu in high U content marine sediments by sector field ICP–MS and its application to Fukushima sediment samples. Environ. Sci. Technol. 48, 534–541.

Bu, W.T., Zheng, J., Ketterer, M.E., Hu, S., Uchida, S., Wang, X.L., 2017. Development and application of mass spectrometric techniques for ultra-trace determination of U-236 in environmental samples-A review. Anal. Chim. Acta 995, 1–20.

Bu, W., Ni, Y., Steinhauser, G., Zheng, W., Zheng, J., Furuta, N., 2018. The role of mass spectrometry in radioactive contamination assessment after the Fukushima nuclear accident. J. Anal. At. Spectrom. 33, 519–546.

Bürger, S., Banik, N.L., Buda, R.A., Kratz, J.V., Kuczewski, B., Trautmann, N., 2007. Speciation of the oxidation states of plutonium in aqueous solutions by UV/Vis spectroscopy, CE-ICP-MS and CE-RIMS. Radiochim. Acta 95, 433–438.

Bürger, S., Vogl, J., Kloetzli, U., Nunes, L., Lavelle, M., 2015. Thermal ionisation mass spectrometry. In: Prohaska, T., Irrgeher, J., Zitek, A., Jakubowski, N. (Eds.), Sector Field Mass Spectrometry for Elemental and Isotopic Analysis. Royal Society of Chemistry, Cambridge.

Bushaw, B.A., 1992. Attogram measurements of rare isotopes by cw resonance ionization mass spectrometry. Inst. Phys. Conf. Ser. 128, 31–36.

Bushaw, B.A., Nörtershäuser, W., Müller, P., Wendt, K., 2001. Diode-laser-based resonance ionization mass spectrometry of the long-lived radionuclide ^{41}Ca with $< 10^{-12}$ sensitivity. J. Radioanal. Nucl. Chem. 247, 351–356.

Butler, O.T., Cairns, W.R.L., Cook, J.M., Davidson, C.M., Mertz-Kraus, R., 2018. Atomic spectrometry update. J. Anal. At. Spectrom. 33, 8–56.

Chabala, J.M., Soni, K.K., Li, J., Gavrilov, K.L., Levi-Setti, R., 1995. High-resolution chemical imaging with scanning ion probe SIMS. Int. J. Mass Spectrom. Ion Processes 143, 191–212.

Chartier, F., Aubert, M., Pilier, M., 1999. Determination of Am and Cm in spent nuclear fuels by isotope dilution inductively coupled plasma mass spectrometry and isotope dilution thermal ionization mass spectrometry after separation by high-performance liquid chromatography. Fresenius J. Anal. Chem. 364, 320–327.

Christl, M., Dai, X., Lachner, J., Kramer-Tremblay, S., Synal, H.-A., 2014. Low energy AMS of americium and curium. Nucl. Instrum. Meth. B 331, 225–232.

Christl, M., Casacuberta, N., Lachner, J., Maxeiner, S., Vockenhuber, C., Hans-Arno-Synal, Goroncy, I., Herrmann, J., Daraoui, A., Walther, C., Michel, R., 2015. Status of ^{236}U analyses at ETH Zurich and the distribution of 236U and 129I in the north sea in 2009. Nucl. Instrum. Meth. **B** 361, 510–516.

Coffin, J.P., Engelstein, P., 1984. Time-of-flight systems for heavy ions. In: Bromley, D.A. (Ed.), Treatise on Heavy-Ion Science. Plenum Press, New York.

Collon, P., Antaya, T., Davids, B., Fauerbach, M., Harkewicz, R., Hellstrom, M., Kutschera, W., Morrissey, D., Pardo, R., Paul, M., Sherrill, B., Steiner, M., 1997. Measurement of Kr-81 in the atmosphere. Nucl. Instrum. Methods Phys. Res. B123, 122–127.

Collon, P., Kutschera, W., Loosli, H.H., Lehmann, B.E., Purtschert, R., Love, A., Sampson, L., Anthony, D., Cole, D., Davids, B., Morrissey, D.J., Sherrill, B.M., Steiner, M., Pardo, R.C., Paul, M., 2000. Kr-81 in the Great Artesian Basin, Australia: a new method for dating very old groundwater. Earth Planet. Sci. Lett. 182, 103–113.

Collon, P., Bichler, M., Caggiano, J., Dewayne Cecil, L., El Masri, Y., Golser, R., Jiang, C.L., Heinz, A., Henderson, D., Kutschera, W., Lehmann, B., Leleux, P., Loosli, H.H., Pardo, R.C., Paul, M., Schlosser, P., Scott, R.H., Smethie Jr., W.M., Vondrasek, R., 2004. Development of an AMS method to study oceanic circulation characteristics using cosmogenic ^{39}Ar. Nucl. Instrum. Meth. **B** 223–224, 428–434.

Crain, J.S., 1996. Applications of inductively coupled plasma mass spectrometry in environmental radiochemistry. Atom. Spectrosc. Perspect. 11, 30–39.

Crowther, S.A., Mohapatra, R.K., Turner, G., Blagburn, D.J., Kehm, K., Gilmour, J.D., 2008. Characteristics and applications of RELAX, an ultrasensitive resonance ionization mass spectrometer for xenon. J. Anal. At. Spectrom. 23, 938–947.

Daly, N.R., 1960. Scintillation type mass spectrometer ion detector. Rev. Sci. Instrum. 31, 264–267.

Danesi, P.R., Bleise, A., Burkart, W., Cabianca, T., Campbell, M.J., Makarewicz, M., Moreno, J., Tuniz, C., Hotchkis, M., 2003a. Isotopic composition and origin of uranium and plutonium in selected soil samples collected in Kosovo. J. Environ. Radioact. 64, 121–131.

Danesi, P.R., Markowicz, A., Chinea-Cano, E., Burkart, W., Salbu, B., Donohue, D., Ruedenauer, F., Hedberg, M., Vogt, S., Zahradnik, P., Ciurapinski, A., 2003b. Depleted uranium particles in selected Kosovo samples. J. Environ. Radioact. 64, 143–154.

Daraoui, A., Michel, R., Gorny, M., Jakob, D., Sachse, R., Synal, H.A., Alfimov, V., 2012. Iodine-129, iodine-127 and caesium-137 in the environment: soils from Germany and Chile. J. Environ. Radioact. 112, 8–22.

Daraoui, A., Tosch, L., Gorny, M., Michel, R., Goroncy, I., Herrmann, J., Nies, H., Synal, H.A., Alfimov, V., Walther, C., 2016. Iodine-129, iodine-127 and cesium-137 in seawater from the north sea and the Baltic sea. J. Environ. Radioact. 162–163, 289–299.

Dawson, P.H. (Ed.), 1976. Quadrupole Mass Spectrometry and its Applications. Elsevier, Amsterdam reissued (1995) by AIP Press, Woodbury, N.Y.

De Gent, S., van Griecken, R., Hang, W., Harrison, W.W., 1995. Comparison between direct current and radiofrequency glow discharge mass spectrometry for the analysis of oxide-based samples. J. Anal. At. Spectrom. 10, 689–695.

De las Heras, L.A., Bocci, F., Betti, M., Actis Dato, L.O., 2000. Comparison between the use of direct current glow discharge mass spectrometry and inductively coupled plasma quadrupole mass spectrometry for the analysis of trace elements in nuclear samples. Fresenius J. Anal. Chem. 368, 95–102.

De las Heras, A.L., Hrnecek, E., Bildstein, O., Betti, M., 2002. Neptunium determination by dc glow discharge mass spectrometry (dc-GDMS) in Irish Sea Sediment samples. J. Anal. At. Spectrom. 17, 1011–1014.

Dellinger, F., Forstner, O., Golser, R., Priller, A., Steier, P., Wallner, A., Winkler, G., Kutschera, W., 2011. Ultrasensitive search for long-lived superheavy nuclides in the mass range A=288 to A=300 in natural Pt, Pb and Bi. Phys. Rev. C 83 (065806), 1–13.

Denk, E., Hillegonds, D., Vogel, J., Synal, A., Geppert, C., Wendt, K., Fattinger, K., Hennessy, C., Berglund, M., Hurrell, M.B., Walczyk, T., 2006. Labeling the human skeleton with ^{41}Ca to assess changes in bone calcium metabolism. Anal. Bioanal. Chem. 386, 1587–1602.

Dickin, A.P., 2018. Radiogenic Isotope Geology. Cambridge University Press, Cambridge.

Diehl, R., Hillebrandt, W., 2002. Astronomy with radioactivity. Phys. J. 1, 47–53.

Dilling, J., Blaum, K., Brodeur, M., Eliseev, S., 2018. Penning-trap mass measurements in atomic and nuclear physics. Annu. Rev. Nucl. Part Sci. 68, 45–74.

Dion, M.P., Liezers, M., Farmer III, O.T., Miller, B.W., Shannon Morley, S., Barinaga, C., Eiden, G., 2015. Improving alpha spectrometry energy resolution by ion implantation with ICP-MS. J. Radioanal. Nucl. Chem. 303, 877–884.

Dorrer, H., Chrysalidis, K., Day Goodacre, T., Düllmann, C.E., Eberhardt, K., Enss, C., Gastaldo, L., Haas, J.R., Harding, J., Hassel, C., Johnston, K., Kieck, T., Köster, U., Marsh, B., Mokry, C., Rothe, S., Runke, J., Schneider, F., Stora, T., Türler, A., Wendt, K., 2018. Production, isolation and characterization of radiochemically pure 163Ho samples for the ECHo-project. Radiochim. Acta 106, 535–547.

Dresser, M.J., 1968. The Saha-Langmuir equation and its application. J. Appl. Phys. 39, 338–339.

Duckworth, D.C., Barshick, C.M., Smith, D.H., 1993. Analysis of soils by glow discharge mass spectrometry. J. Anal. At. Spectrom. 8, 875–879.

Ebdon, L., Fisher, A., Handley, H., Jones, P., 1993. Determination of trace metals in concentrated brines using inductively coupled plasma mass spectrometry on-line preconcentration and matrix elimination with flow injection. J. Anal. At. Spectrom. 8, 979–981.

Eliades, J.A., Zhao, X.L., Litherland, A.E., Kieser, W.E., 2015. Negative ion-gas reaction studies using ion guides and accelerator mass spectrometry I: SrF3-, YF3-, ZrF3-, YF4- and ZrF5- in NO2. Nucl. Instrum. Meth. B 361, 294–299.

Elmore, D., Philips, F.M., 1987. Accelerator mass spectrometry for measurement of long-lived radioisotopes. Science **236**, 543–550.

Erdmann, N., Nunnemann, M., Eberhardt, K., Herrmann, G., Huber, G., Köhler, S., Kratz, J.V., Passler, G., Peterson, J.R., Trautmann, N., Waldek, A., 1998. Determination of the first ionization potential of nine actinide elements by resonance ionization mass spectroscopy (RIMS). J. Alloys Comp. 271–273, 837–840.

Erdmann, N., Betti, M., Stetzer, O., Tamborini, G., Kratz, J.V., Trautmann, N., van Geel, J., 2000. Production of monodisperse uranium oxide particles and their characterization by scanning electron microscopy and secondary ion mass spectrometry. Spectrochim. Acta **B** 55, 1565–1575.

Erdmann, N., Passler, G., Trautmann, N., Wendt, K., 2008. Resonance ionization mass spectrometry for trace analysis of long-lived radionuclides. Radioact. Environ. 11, 331–354.

Erdmann, N., Kratz, J.-V., Trautmann, N., Passler, G., 2009. Resonance ionization mass spectrometry of ion beam sputtered neutrals for element- and isotope-selective analysis of plutonium in micro-particles. Anal. Bioanal. Chem. 395, 1911–1918.

Evans, P., Elahi, S., Lee, R., Fairman, B., 2003. A rapid and accurate method for the determination of plutonium in food using magnetic sector ICP-MS with an ultra-sonic nebulizer and ion chromatography. J. Environ. Monit. 5, 175–179.

Fassett, J.D., Paulsen, P.J., 1989. Isotope dilution mass spectrometry for accurate elemental analysis. Anal. Chem. 61, 643A–649A.

Fedosseev, V., Chrysalidis, K., Day Goodacre, D., Marsh, B., Rothe, S., Seiffert, C., Wendt, K., 2017. Ion beam production and study of radioactive isotopes with the laser ion source at ISOLDE. J. Phys. G 44, 084006.

Fedosseev, V.N., Kudryavtsev, Y., Mishin, V.S., 2012. Resonance laser ionization of atoms for nuclear physics. Phys. Scr. 85, 058104.

Feldmann, I., Jakubowski, N., Stuewer, D., 1999a. Application of a hexapole collision and reaction cell in ICP-MS Part I:Instrumental aspects and operational optimization. Fresenius J. Anal. Chem. 365, 415–421.

Feldmann, I., Jakubowski, N., Thomas, C., Stuewer, D., 1999b. Application of a hexapole collision and reaction cell in ICP-MS Part II: analytical figures of merit and first applications. Fresenius J. Anal. Chem. 365, 422–428.

Fiedler, R., 1995. Total evaporation measurements: experience with multi-collector instruments and a thermal ionization quadrupole mass spectrometer. J. Mass Spectrom. 146, 91–97.

Fifield, L.K., 1999. Accelerator mass spectrometry and its applications. Rep. Prog. Phys. 62, 1223–1274.

Fifield, L.K., 2000. Advances in accelerator mass spectrometry. Nucl. Instrum. Meth. B 172, 134–143.

Fifield, L.K., 2008. Accelerator mass spectrometry of long-lived heavy radionuclides. Radioact. Environ. 11, 263–293.

Fifield, L.K., Cresswell, R.G., di Tada, M.L., Ophel, T.R., Day, J.P., Clacher, A.P., King, S.J., Priest, N.D., 1996. Accelerator mass spectrometry of plutonium isotopes. Nucl. Instrum. Meth. B 117, 295–303.

Fink, D.A., Richter, S.D., Blaum, K., Catherall, R., Crepieux, B., Fedosseev, V.N., Gottberg, A., Kron, T., Marsh, B.A., Mattolat, C., Raeder, S., Rossel, R.E., Rothe, S., Schwellnus, F., Seliverstov, M.D., Sjödin, M., Stora, T., Suominen, P., Wendt, K.D.A., 2015. On-line implementation and first operation of the laser ion source and trap at ISOLDE/CERN. Nucl. Instrum. Meth. B 344, 83–95.

Forstner, O., Andersson, P., Hanstorp, D., Lahner, J., Martschini, M., Pitters, J., Priller, A., Steier, P., Golser, R., 2015. The ILIAS project for selective isobar suppression by laser photodetachment. Nucl. Instrum. Meth. B 361, 217–221.

Fourre, C., Clerc, J., Fragu, P., 1997. Contribution of mass resolution to secondary ion mass spectrometry microscopy imaging in biological microanalysis. J. Anal. At. Spectrom. 12, 1105–1110.

Franzmann, M., Bosco, H., Walther, C., Wendt, K., 2017. A new resonant Laser-SNMS system for environmental ultra-trace analysis: installation and optimization. Int. J. Mass Spectrom. 423, 27–32.

Franzmann, M., Bosco, H., Hamann, L., Walther, C., Wendt, K., 2018. Resonant Laser-SNMS for spatially resolved and element selective ultra-trace analysis of radionuclides. J. Anal. At. Spectrom. 33, 730–737.

Freeman, S.P.H.T., King, J.C., Vieira, N.E., Woodhouse, L.R., Yergey, A.L., 1997. Human calcium metabolism including bone resorption measured with a Ca-41 tracer. Nucl. Instr. Meth B 123, 266–270.

Fukuda, M., Sayama, Y., 1997. Determination of traces of uranium and thorium in (Ba, Sr)TiO3 ferroelectrics by inductively coupled plasma mass spectrometry. Fresenius J. Anal. Chem. 357, 647–651.

Ganeev, A., Gubal, A., Korotetski, A., Bogdanova, O., Burakov, B., Titova, A., Solovyev, N., Ivanenko, N., Drobyshev, E., Iakovleva, E., Sillanpää, M., 2017. Direct isotope analysis of Chernobyl microparticles using time-of-flight mass spectrometry with pulsed glow discharge. Microchem. J. 132, 286–292.

Garcia Alonso, J.I., Rodriguez Gonzalez, P., 2013. Isotope Dilution Mass Spectrometry. Royal Society of Chemistry, Cambridge.

García-León, M., 2018. Accelerator mass spectrometry (AMS) in radioecology. J. Environ. Radioact. 186, 116–123.

Geppert, Ch., Müller, P., Wendt, K., Schnabel, Ch., Synal, H.-A., 2005. Intercomparison measurements between accelerator and laser based mass spectrometry for ultra-trace determination of ^{41}Ca in the 10^{-11} to 10^{-10} isotopic range. Nucl. Instr. and Meth. in Phys. Res. B 229, 519–526.

Grate, J.W., O'Hara, M.J., Farawila, A.F., Douglas, M., Haney, M.M., Petersen, S.L., Maiti, T.C., Aardahl, C.L., 2011. Extraction chromatographic methods in the sample preparation sequence for thermal ionization mass spectrometric analysis of plutonium isotopes. Anal. Chem. 83, 9086–9091.

Gray, A.L., 1985. Solid sample introduction by laser ablation for inductively coupled plasma source mass spectrometry. Analyst 110, 551–556.

Gray, A.L., 1986. Influence of load coil geometry on oxides and doubly charged ion response in inductively coupled plasma mass spectrometry. J. Anal. At. Spectrom. 1, 247–249.

Greenfield, S., Montaser, A., 1992. Common RF generators, torches, and sample introduction systems. In: Montaser, A., Golightly, D.W. (Eds.), Inductively Coupled Plasmas in Analytical Atomic Spectrometry, second ed. VCH, New York.

Grüning, C., Huber, G., Kratz, J.V., Passler, G., Trautmann, N., Waldek, A., Wendt, K., 2001. Resonance ionization mass spectrometry for ultratrace analysis of plutonium with a new solid state laser system. Int. J. Mass Spectrom. 235, 171–178.

Guckel, K., Shinonaga, T., Christl, M., Tschiersch, J., 2017. Scavenged Pu-239, Pu-240, and Am-241 from snowfalls in the atmosphere settling on Mt. Zugspitze in 2014. Sci. Rep.

Habfast, K., 1998. Fractionation correction and multiple collectors in thermal ionization isotope ratio mass spectrometry. Int. J. Mass Spectrom. 176, 133–148.

Ham, G.J., Harrison, J.D., 2000. The gastrointestinal absorption and urinary excretion of plutonium in male volunteers. Radiat. Prot. Dosim. 87, 267–272.

Hare, D., Tolmachev, S., James, A., Bishop, D., Austin, C., Fryer, F., Doble, P., 2010. Elemental bio-imaging of thorium, uranium, and plutonium in tissues from occupationally exposed former nuclear workers. Anal. Chem. 82, 3176–3182.

Harrison, H.W., 1988. Glow discharge mass spectrometry. In: Adams, F., Gijbels, R., van Griecken, R. (Eds.), Inorganic Mass Spectrometry, Chemical Analysis, vol. 95. John Wiley & Sons, New York, pp. 85–123.

Hayashi, S., Hashiguchi, Y., Suzuki, K., Ohtsubo, T., McIntosh, B.J., 1991. Quantitiative trace analysis by nonresonant laser post-ionization. Surf. Interface Anal. 17, 773–778.

Hellborg, S.R., Skog, G., 2008. Accelerator mass spectrometry. Mass Spectrom. Rev. 27, 398–427.

Hennessy, C., Berglund, M., Ostermann, M., Walczykc, T., Synald, H.-A., Geppert, C., Wendt, K., Taylor, P.D.P., 2005. Certification of a ^{41}Ca dose material for use in human studies (IRMM-3703) and a corresponding set of isotope reference materials for ^{41}Ca measurements (IRMM-3701). Nucl. Instrum. Meth. B 229, 281–292.

Herzner, P., Heumann, K.G., 1992. Ultratrace analysis of U, Th, Ca and selected heavy metals in high-purity refractory metals with isotope dilution mass spectrometry. Mikrochim. Acta 106, 127–135.

Herzog, G.F., 1994. Applications of accelerator mass spectrometry in extraterrestrial materials. Nucl. Instrum. Meth. B 92, 492–499.

Heumann, K.G., 1992. Isotope dilution mass spectrometry (IDMS) of the elements. Mass Spectrom. Rev. 11, 41–67.

Heumann, K.G., Eisenhut, S., Gallus, S., Hebeda, E.H., Nusko, R., Vengosh, A., Walczyk, T., 1995. Recent developments in thermal ionization mass spectrometric techniques for isotope analysis – a review. Analyst 120, 1291–1299.

Hocking, H.E., Burggraf, L.W., Duan, X.F., Gardella, J.A., Yatzor, B.P., Schuler, W.A., 2012. Composition of uranium oxide particles related to TOF-SIMS ion distributions. Surf. Interface Anal. 45, 545–548.

Hotchkis, M., Fink, D., Tuniz, C., Vogt, S., 2000. Accelerator mass spectrometry analyses of environmental radionuclides: sensitivity, precision and standardisation. Appl. Radiat. Isot. 53, 31–37.

Hotchkis, M.A.C., Child, D.P., Froehlich, M.B., Wallner, A., Wilcken, K., Williams, M., 2018. Actinides AMS on the VEGA accelerator. Nucl. Instrum. Methods Phys. Res. Sect. B Beam Interact. Mater. Atoms.

Houk, R.S., Fassel, V.A., Flesch, G.D., Svec, H.J., Gray, A.L., Taylor, C.E., 1980. Inductively coupled argon plasma as an ion source for mass spectrometric determination of trace elements. Anal. Chem. 52, 2283–2289.

Hurst, G.S., Payne, M.G., Kramer, S.D., Young, J.P., 1979. Resonance ionization spectroscopy and one-atom detection. Rev. Mod. Phys. 51, 767–810.

Hurst, G.S., Payne, M.G., 1988. Principles and Applications of Resonance Ionization Spectroscopy. Hilger Publications, Bristol.

Inkret, W.C., Efurd, D.W., Miller, G., Roskop, D.J., Benjamin, T.M., 1998. Applications of thermal ionization mass spectrometry to the detection of ^{239}Pu and ^{240}Pu intakes. Int. J. Mass Spectrom. 178, 113–120.

Iwata, Y., Ito, C., Harano, H., Aoyama, T., 2010. Improvement of the resonance ionization mass spectrometer performance for precise isotope analysis of krypton and xenon at the ppt level in argon. Int. J. Mass Spectrom. 296, 15–20.

Iwata, Y., Sekiya, H., Itoa, C., 2015. Ultrasensitive resonance ionization mass spectrometer for evaluating krypton contamination in xenon dark matter detectors. Nucl. Instrum. Meth. A 797, 64–69.

Jakopic, R., Richter, S., Kuehn, H., Aregbe, Y., 2010. Determination of ^{240}Pu/^{239}Pu, ^{241}Pu/^{239}Pu and ^{242}Pu/^{239}Pu isotope ratios in environmental reference materials and samples from Chernobyl by thermal ionization mass spectrometry (TIMS) and filament carburization. J. Anal. At. Spectrom. 25, 815–821.

Jackson, G.S., Weaver, C., Elmore, D., 2001. Use of accelerator mass spectrometry for studies in nutrition. Nutr. Res. Rev. 14, 317–334.

James, R.H., Palmer, M.R., 2000. The Li isotope composition of international rock standards. Chem. Geol. 166, 319–326.

Jernstrom, J., Eriksson, M., Simon, R., Tamborini, G., Bildstein, O., Marquez, R.C., Kehl, S.R., Hamilton, T.F., Ranebo, Y., Betti, M., 2006. Characterization and source term assessments of radioactive particles from Marshall Islands using non-destructive analytical techniques. Spectrochim. Acta B 61, 971–979.

Jiang, S.-J., Houk, R.S., Stevens, M.A., 1988. Alleviation of overlap interferences for determination of potassium isotope ratios by inductively coupled plasma mass spectrometry. Anal. Chem. 60, 1217–1220.

Johnson, C.M., Beard, B.L., 1999. Correction of instrumentally produced mass fractionation during isotopic analysis of Fe by thermal mass spectrometry. Int. J. Mass Spectrom. 193, 87–99.

Jungmann, J.H., Heeren, R.M.A., 2013. Detection systems for mass spectrometry imaging: a perspective on novel developments with a focus on active pixel detectors. Rapid Commun. Mass Spectrom. 27, 1–23.

Kaesdorf, S., Schröder, H., Kompa, K.L., 1990. Aspects of quantification of off-resonance laser ionization for SNMS. Vacuum 41, 1669–1670.

Kalmykov, S.N., Kriventsov, V.V., Teterin, Y.A., Novikov, A.P., 2007. Plutonium and neptunium speciation bound to hydrous ferric oxide colloids. Compt. Rendus Chem. 10, 1060–1066.

Kawai, Y., Nomura, M., Murata, H., Susuki, T., Fujii, Y., 2001. Surface ionization of alkaline earth iodides in double filament system. Int. J. Mass Spectrom. 206, 1–5.

Kerl, W., Becker, J.S., Dietze, H.-J., Dannecker, W., 1997. Isotopic and ultratrace analysis of uranium by double-focusing sector field ICP mass spectrometry. Fresenius J. Anal. Chem. 359, 407–409.

Kersting, A.B., Efurd, D.W., Finnegan, D.L., Rokop, D.J., Smith, D.K., Thompson, J.L., 1999. Migration of plutonium in ground water at the Nevada test site. Nature 397, 56–59.

Kim, C.K., Seki, R., Morita, S., Yamasaki, S., Tsumura, A., Takaku, Y., Igarashi, Y., Yamamoto, M., 1991. Application of a high resolution inductively coupled plasma mass spectrometer to the measurement of long-lived radionuclides. J. Anal. At. Spectrom. 6, 205–209.

Kinard, W.F., Bibler, N.E., Coleman, C.J., Dewberry, R.A., 1997. Radiochemical analyses for the defense waste processing facility startup at the Savannah River Site. J. Radioanal. Nucl. Chem. 219, 197–201.

Kips, R.S., Kristo, M.J., 2009. Investigation of chemical changes in uranium oxyfluoride particles using secondary ion mass spectrometry. J. Radioanal. Nucl. Chem. 282, 1031–1035.

Klinkenberg, M., Brandt, F., Breuer, U., Bosbach, D., 2014. Uptake of Ra during the recrystallization of barite: a microscopic and time of flight-secondary ion mass spectrometry study. Environ. Sci. Technol. 48, 6620–6627.

Kluge, H.-J., Bushaw, B.A., Passler, G., Wendt, K., Trautmann, N., 1994. Resonance ionization spectroscopy for trace analysis and fundamental research. Fresenius J. Anal. Chem. 350, 323–329.

Krachler, M., Varga, Z., Nicholl, A., Wallenius, M., Mayer, K., 2018. Spatial distribution of uranium isotopes in solid nuclear materials using laser ablation multi-collector ICP-MS. Microchem. J. 140, 24–30.

Kratz, J.-V., 2015. Ultratrace analysis of long-lived radionuclides by resonance ionization mass spectrometry (RIMS). Radioanal. Nucl. Chem. 303, 1361–1366.

Krönert, U., Becker, S., Bollen, G., Gerber, M., Hilberath, T., Kluge, H.-J., Passler, G., the ISOLDE Collaboration, 1991. On-line laser spectroscopy by resonance ionization of laser-desorbed, refractory elements. Nucl. Instrum. Meth. A 300, 522–537.

Kudryavtsev, Y., Bruyneel, B., Franchoo, S., Huyse, M., Gentens, J., Kruglov, K., Mueller, W.F., Prasad, N.V.S.V., Raabe, R., Reusen, I., Van den Bergh, P., Van Duppen, P., Van Roosbroeck, J., Vermeeren, L., Weissman, L., 2002. The Leuven isotope separator on-line laser ion source. Nucl. Phys. A701, 465C–469C.

Kumar, S., Saxena, M.K., Shah, P.M., Jain, H.C., 1995. Investigations for isobaric interferences of ^{238}Pu at ^{238}U during thermal ionization mass spectrometry of uranium and plutonium from the same filament loading. Int. J. Mass Spectrom. 178, 113–120.

Kuruc, J., Harvan, D., Galanda, D., Matel, L., Jerigova, M., Velic, D., 2011. Alpha spectrometry and secondary ion mass spectrometry of electrodeposited uranium films. J. Radioanal. Nucl. Chem. 289, 611–615.

Kutschera, W., 2005. Progress in isotope analysis at ultratrace level by AMS. Int. J. Mass Spectrom. 242, 145–160.

Kutschera, W., 2012. https://vera2.rad.univie.ac.at/share/WWW_Exchange/Miscellaneaous/AMS%20facilities%20of%20the%20world.pdf.

Kutschera, W., 2016. Accelerator mass spectrometry: state of the art and perspectives. Adv. Phys. X 1, 570–595.

Accelerator mass spectrometry. In: Kutschera, W., Golser, R., Priller, A., Strohmaier, B. (Eds.), Nucl. Instrum. Meth. B 172, 1-977 and earlier

Proceedings of the Series of International Conference on Accelerator Mass Spectrometry, all printed in Nucl. Instr. Meth. B.

Laatiaoui, M., Lauth, W., Backe, H., Block, M., Ackermann, D., Cheal, B., Chhetri, P., Düllmann, C.E., van Duppen, P., Even, J., Ferrer, R., Giacoppo, F., Götz, S., Heßberger, F.P., Huyse, M., Kaleja, O., Khuyagbaatar, J., Kunz, P., Lautenschläger, F., Mistry, A.K., Raeder, S., Ramirez, E.M., Walther, T., Calvin Wraith, C., Yakushev, A., 2016. Atom-at-a-time laser resonance ionization spectroscopy of nobelium. Nature 538, 495–498.

Lachner, J., Christl, M., Bisinger, T., Michel, R., Synal, H.-A., 2010. Isotopic signature of plutonium at Bikini atoll. Appl. Radiat. Isot. 68, 979–983.

Lachner, J., Kasberger, M., Martschini, M., Priller, A., Steier, P., Golser, R., 2015. Developments towards detection of Cs-135 at VERA. Nucl. Instrum. Meth. B 361, 440–444.

LaMont, S.P., Shick, C.R., Cable-Dunlap, P., Fauth, D.J., LaBone, T.R., 2005. Plutonium determination in bioassay samples using radiochemical thermal ionization mass spectrometry. J. Radioanal. Nucl. Chem. 263, 477–481.

Lefèvre, O., Betti, M., Koch, L., Walker, C.T., 1996. EPMA and mass spectrometry of soil and grass containing radioactivity from the nuclear accident at Chernobyl. Mikrochim. Acta 13, 399–408.

Letokhov, V.S., 1987. In: Laser Photoionization Spectroscopy. Academic Press, Orlando.

Libby, W.F., 1955. In: Radiocarbon Dating. University of Chicago Press, Chicago.

Liezers, M., Farmer, O.T., Dion, M.P., Thomas, M.L., Ei, G.C., 2015. The production of ultra-high purity single isotopes or tailored isotope mixtures by ICP-MS M. Int. J. Mass Spectrom. 376, 58–64.

Lind, O.C., Nygren, U., Sidhu, R.S., Roos, P., Pöllänen, R., Ranebo, Y., Salbu, B., 2008. Overview of sources of radioactive particles of Nordic relevance as well as a short description of available particle characterisation techniques. NKS 180, 1–24. ISBN:978-87-7893-246-4.

Lind, O.C., Salbu, B., Janssens, K., Proost, K., Danesi, P.R., 2009. Characterisation of DU particles from Kosovo and Kuwait. In: Oughton, D.H.K.V. (Ed.), Radioact. Particles in the Environment, pp. 57–67.

Linge, K.L., Bédard, L.P., Bugoi, R., Enzweiler, J., Jochum, K.P., Kilian, R., Liu, J., Marin-Carbonne, J., Merchel, S., Munnik, F., Morales, L.F.G., Rollion-Bard, C., Souders, A.K., Sylvester, P.J., Weis, U., 2017. GGR biennial critical review: analytical developments since 2014. Geostand. Geoanal. Res. 41, 493–562.

Litherland, A.E., Allen, K.W., and Hall, E.T., Eds. 1987. Proc. Royal Discussion Meeting. Phil. Trans. Roy. Soc. A323, 1–173

Liu, Y., Huang, M., Masuda, A., Masao, I., 1998. High-precision determination of Os and Re isotope ratios by in situ oxygen isotope rate correction using negative thermal ionization mass spectrometry. Int. J. Mass Spectrom. 173, 163–175.

Liu, Y., Masuda, M., Inoue, M., 2000. Measurement of isotopes of light rare earth elements in the form of oxide ions: a new development in thermal ionization mass spectrometry. Anal. Chem. 72, 3001–3005.

Lopez-Lora, M., Chamizo, E., Villa-Alfageme, M., Hurtado-Bermudez, S., Casacuberta, N., Garcia-Leon, M., 2018. Isolation of U-236 and Pu-239,Pu-240 from seawater samples and its determination by accelerator mass spectrometry. Talanta 178, 202–210.

Marcus, R.K., 1996. Radiofrequency powered glow discharges: opportunities and challenges. J. Anal. At. Spectrom. 11, 821–828.

Marsh, B., 2014. Resonance ionization laser ion sources for on-line isotope separators. Rev. Sci. Instrum. 85, 02B923.

Martschini, M., Buchriegler, J., Collon, P., Kutschera, W., Lachner, J., Lu, W.T., Priller, A., Steier, P., Golser, R., 2015. Isobar separation of Zr-93 and Nb-93 at 24 MeV with a new multi-anode ionization chamber. Nucl. Instrum. Methods Phys. Res. Sect. B Beam Interact. Mater. Atoms 361, 201–206.

Mattolat, C., Gottwald, T., Raeder, S., Rothe, S., Schwellnus, F., Wendt, K., Thörle-Pospiech, P., Trautmann, N., 2010. Determination of the first ionization potential of technetium. Phys. Rev. A 81, 052513.

Maul, J., Strachnov, I., Eberhardt, K., Karpuk, S., Passler, G., Trautmann, N., Wendt, K., Huber, G., 2006. Spatially resolved ultra-trace analysis of elements combining resonance ionization with a MALDI-TOF spectrometer. Anal. Bioanal. Chem. 386, 109–118.

May, C.C., Worsfold, P.J., Keith-Roach, M.J., 2008. Analytical techniques for speciation analysis of aqueous long-lived radionuclides in environmental matrices. Trends Anal. Chem. 27, 160–168.

Mayer, K., Wallenius, M., Ray, I., 2005. Nuclear forensics - a methodology providing clues on the origin of illicitly trafficked nuclear materials. Analyst 130, 433–441.

Mayer, K., Wallenius, M., Lützenkirchen, K., Horta, J., Nicholl, A., Rasmussen, G., van Belle, P., Varga, Z., Buda, R., Erdmann, N., Kratz, J.-V., Trautmann, N., Fifield, L.K., Tims, S.G., Fröhlich, M.B., Steier, P., 2015. Uranium from German nuclear power projects of the 1940s — a nuclear forensic investigation. Angew. Chem. Int. Ed. 54 (45), 13452–13456.

McAninch, J.E., Hamilton, T.F., Brown, T.A., Jokela, T.A., Knezovich, J.P., Ognibene, T.J., Proctor, I.D., Roberts, M.L., Sideras-Haddad, E., Southon, J.R., Vogel, J.S., 2000. Plutonium measurements by accelerator mass spectrometry at LLNL. Nucl. Instrum. Meth. B 172, 711–716.

McKeegan, K.D., Chaussidon, M., Robert, F., 2000. Incorporation of short-lived ^{10}Be in a calcium-aluminium-rich inclusion from the Allende meteorite. Science 289, 1334–1337.

Michel, R., 1999. Long-lived radionuclides as tracers in terrestrial and extraterrestrial matter. Radiochim. Acta 87, 47–73.

Michel, R., Handl, J., Ernst, T., Botsch, W., Szidat, S., Schmidt, A., Jakob, D., Beltz, D., Romantschuk, L.D., Synal, H.A., Schnabel, C., Lopez-Gutierrez, J.M., 2005. Iodine-129 in soils from Northern Ukraine and the retrospective dosimetry of the iodine-131 exposure after the Chernobyl accident. Sci. Total Environ. 340, 35–55.

Michel, R., Daraoui, A., Gorny, M., Jakob, D., Sachse, R., Tosch, L., Nies, H., Goroncy, I., Herrmann, J., Synal, H.A., Stocker, M., Alfimov, V., 2012. Iodine-129 and iodine-127 in European seawaters and in precipitation from Northern Germany. Sci. Total Environ. 419, 151–169.

Michel, R., Daraoui, A., Gorny, M., Jakob, D., Sachse, R., Romantschuk, L.D., Alfimov, V., Synal, H.A., 2015. Retrospective dosimetry of Iodine-131 exposures using Iodine-129 and Caesium-137 inventories in soils — a critical evaluation of the consequences of the Chernobyl accident in parts of Northern Ukraine. J. Environ. Radioact. 150, 20–35.

Montaser, A., 1998. Inductively Coupled Plasma Mass Spectrometry. Wiley-VCH, New York.

Monz, L., Hohmann, R., Kluge, H.-J., Kunze, S., Lantzsch, J., Otten, E.W., Passler, G., Senne, P., Stenner, J., Stratmann, K., Wendt, K., Zimmer, K., Herrmann, G., Trautmann, N., Walter, K., 1993. Fast, low-level detection of strontium-90 and strontium-89 in environmental samples by collinear resonance ionization spectroscopy. Spectrochim. Acta 48B, 1655–1671.

Muccio, Z., Jackson, J.P., 2009. Isotope ratio mass spectrometry. Analyst 134, 213–222.

Müller, R.A., 1977. Radioisotope dating with a cyclotron. Science 196, 489–494.

Müller, P., Blaum, K., Bushaw, B.A., Diel, S., Geppert, C., Nähler, A., Nörtershäuser, W., Trautmann, N., Wendt, K., 2000. Trace detection of ^{41}Ca in nuclear reactor concrete by diode-laser-based resonance ionization mass spectrometry. Radiochim. Acta 88, 487–493.

Müller, P., Bushaw, B.A., Blaum, K., Diel, S., Geppert, C., Nähler, A., Trautmann, N., Nörtershauser, W., Wendt, K., 2001. ^{41}Ca ultratrace determination with isotopic selectivity $> 10^{12}$ by diode-laser-based RIMS. Fresenius J. Anal. Chem. 370, 508–512.

Muramatsu, Y., Rühm, W., Yoshida, S., Tagami, K., Uchida, S., Wirth, E., 2000. Concentrations of ^{239}Pu and ^{240}Pu and their isotopic ratios determined by ICP-MS in soils collected from the Chernobyl 30-km zone. Environ. Sci. Technol. 34, 2913–2917.

Nägler, T.F., Villa, I.M., 2000. In pursuit of the ^{40K} branching ratios: K-Ca and 3^{9A}r-4^{0A}r dating of gem silicates. Chem. Geol. **169**, 5–16.

Nelson, D.E., Korteling, R.G., Stott, W.R., 1977. Carbon-14: direct detection at natural concentrations. Science 198, 507–508.

Nelson, A.J., Felter, T.E., Wu, K.J., Evans, C., Ferreira, J.L., Sickhaus, W.J., McLean, W., 2006. Uranium passivation by C+ implantation: a photoemission and secondary ion mass spectrometry study. Surf. Sci. 600, 1319–1325.

Nielsen, S.P., Roos, P., Dahlgaard, H., Olsen, S.K., Jernstrom, J., Eriksson, M., 2009. Thule expedition 2003 – studies on radioactive contamination and particles. In: Oughton, D.H.K.V. (Ed.), Radioactive Particles in the Environment, pp. 93–109.

Oberli, F., Gartenmann, P., Meier, M., Kutschera, W., Suter, M., Winkler, G., 1999. The half-life of ^{126}Sn refined by thermal ionization mass spectrometry measurements. Int. J. Mass Spectrom. 184, 145–152.

Ohno, T., Muramatsu, Y., 2014. Determination of radioactive cesium isotope ratios by triple quadrupole ICP-MS and its application to rainwater following the Fukushima Daiichi Nuclear Power Plant accident. J. Anal. At. Spectrom. 29, 347–351.

Ohno, T., Muramatsu, Y., Shikamori, Y., Toyama, C., Okabea, N., Matsuzakic, H., 2013. Determination of ultratrace 129I in soil samples by Triple Quadrupole ICP-MS and its application to Fukushima soil samples. J. Anal. At. Spectrom. 28, 1283–1287.

Olesik, J.W., 2000. Inductively coupled plasma mass spectrometry. In: Barshick, C.M., Duckworth, D.C., Smith, D.H. (Eds.), Inorganic Mass Spectrometry, Fundamentals and Applications. Marcel Decker, New York.

Olesik, J.W., Kinzer, J.A., Olesik, S.V., 1995. Capillary electrophoresis inductively coupled plasma spectrometry for rapid elemental speciation. Anal. Chem. 67, 1–12.

Ottolini, L., Oberti, R., 2000. Accurate quantification of H, Li, Be, B, F, Ba, REE, Y, Th, and U in complex matrixes: a combined approach based on SIMS and single-crystal structure refinement. Anal. Chem. 72, 3731–3738.

Passler, G., Erdmann, N., Hasse, H.-U., Herrmann, G., Huber, G., Köhler, S., Kratz, J.V., Mansel, A., Nunnemann, M., Trautmann, N., Waldek, A., 1997. Application of laser mass spectrometry for trace analysis of plutonium and technetium. Kerntechnik 62, 85–90.

Paul, M., 1990. Separation of isobars with a gas-filled magnet. Nucl. Instrum. Meth. B 52, 315–321.

Paul, W., Raether, M., 1955. Das Elektrische Massenfilter. Z. Phys. 140, 262–273.

Paul, M., Berkovits, D., Cecil, L.D., Feldstein, H., Hershkovitz, A., Kashiv, Y., Vogt, S., 1997. Environmental ^{90}Sr measurements. Nucl. Instrum. Meth. B 123, 394–399.

Paul, M., Berkovits, D., Ahmad, I., Borasi, F., Caggiano, J., Davids, C.N., Greene, J.P., Harss, B., Heinz, A., Henderson, D.J., Henning, W., Jiang, C.L., Pardo, R.C., Rehm, K.E., Rejoub, R., Seweryniak, D., Sonzogni, A., Uusitalo, J., Vondrasek, R., 2000. AMS of heavy elements with an ECR ion source and the ATLAS linear accelerator. Nucl. Instrum. Meth. B 172, 688–892.

Payne, M.G., Deng, L., Thonnard, N., 1994. Applications of resonance ionization mass spectrometry. Rev. Sci. Instrum. 65, 2433–2459.

Peterson, J.R., Erdmann, N., Nunnemann, M., Eberhardt, K., Huber, G., Kratz, J.V., Passler, G., Stetzer, O., Thörle, P., Trautmann, N., Waldek, A., 1998. Determination of the first ionization potential of einsteinium by resonance ionization mass spectroscopy (RIMS). J. Alloys Comp. 271–273, 876–878.

Pibida, L., McMahon, C.A., Bushaw, B.A., 2004. Laser resonance ionization mass spectrometry measurements of cesium in nuclear burn-up and sediment samples. Appl. Radiat. Isot. 60, 567–570.

Platzner, I.T., 1997. Modern isotope ratio mass spectrometry. In: Chemical Analysis, vol. 145. John Wiley & Sons, Chicester, pp. 1–530.

Pröfrock, D., Prange, A., 2012. Inductively coupled plasma-mass spectrometry (ICP-ms) for quantitative analysis in environmental and life sciences: a review of challenges, solutions, and trends. Appl. Spectrosc. 66, 843–868.

Prohaska, T., Irrgeher, J., Zitek, A., Jakubowski, N., 2015. Sector Field Mass Spectrometry for Elemental and Isotopic Analysis. Royal Society of Chemistry, Cambridge.

Purser, K.H., Williams, P., Litherland, A.E., Stein, J.D., Storms, H.A., Gove, H.E., Stevens, C.M., 1981. Isotopic ratio measurement at abundance sensitivities greater than 1: 10^{15} : a comparison between mass spectrometry at keV and MeV energies. Nucl. Instr. Meth. 186, 487–498.

Quinto, F., Blechschmidt, I., Perez, C.G., Geckeis, H., Geyer, F., Golser, R., Huber, F., Lagos, M., Lanyon, B., Plaschke, M., Steier, P., Schafer, T., 2017. Multiactinide analysis with accelerator mass spectrometry for ultratrace determination in small samples: application to an in situ radionuclide tracer test within the colloid formation and migration experiment at the Grimsel test site (Switzerland). Anal. Chem. 89, 7182–7189.

Raeder, S., Stobener, N., Gottwald, T., Passler, G., Reich, T., Trautmann, N., Wendt, K., 2011. Determination of a three-step excitation and ionization scheme for resonance ionization and ultratrace analysis of Np-237. Spectrochim. Acta B 66, 242–247.

Raeder, S., Hakimi, A., Stöbener, N., Trautmann, N., Wendt, K., 2012. Detection of plutonium isotopes at lowest quantities using in-source resonance ionization mass spectrometry. Anal. Bioanal. Chem. 404, 2163–2172.

Ramakumar, K.L., Fiedler, R., 1999. Calibration procedures for a multicollector mass spectrometer for cup efficiency, detector amplifier linearity, and isotope fractionation to evaluate the accuracy in the total evaporation method. Int. J. Mass Spectrom. 184, 109–118.

Ranebo, Y., Hedberg, P.M.L., Whitehouse, M.J., Ingeneri, K., Littmann, S., 2009. Improved isotopic SIMS measurements of uranium particles for nuclear safeguard purposes. J. Anal. At. Spectrom. 24, 277–287.

Ranebo, Y., Niagolova, N., Erdmann, N., Eriksson, M., Tamborini, G., Betti, M., 2010. Production and characterization of monodisperse plutonium, uranium, and mixed uranium-plutonium particles for nuclear safeguard applications. Anal. Chem. 82, 4055–4062.

Reed, T.B., 1961. Induction-coupled plasma torch. J. Appl. Phys. 32, 821–824.

Richner, P., Evans, D., Wahrenberger, C., Dietrich, V., 1994. Applications of laser ablation and electrothermal vaporization as sample introduction techniques for ICP-MS. Fresenius J. Anal. Chem. 350, 235–241.

Richter, S., Goldberg, S.A., 2003. Improved techniques for high accuracy isotope ratio measurements of nuclear materials using thermal ionization mass spectrometry. Int. J. Mass Spectrom. 229, 18–197.

Richter, S., Alfonso, A., De Bolle, W., Wellun, R., Taylor, P.D.P., 1999. Isotopic "fingerprints" of natural uranium ore samples. Int. J. Mass Spectrom. 193, 9–14.

Richter, S., Eykens, R., Kühn, H., Aregbe, Y., Verbruggen, A., Weyer, S., 2010. New average values for the n(^{238}U)/n(^{235}U) isotope ratios of natural uranium standards. Int. J. Mass Spectrom. 295, 9–97.

Robinson, K., Hall, E.F.H., 1987. Glow discharge mass spectrometry for nuclear materials. J. Metal 39, 14–16.

Rohr, U., Meckel, L., Ortner, H.M., 1994. Ultratrace analysis of uranium and thorium in glass; Part 1: ICP-MS, classical photometry and chelate-GC. Fresenius J. Anal. Chem. 348, 356–363.

Rossnagel, J., Raeder, S., Hakimi, A., Ferrer, R., Trautmann, N., Wendt, K., 2012. Determination of the first ionization potential of actinium. Phys. Rev. A 85 (1), 012525.

Rothe, S., Andreyev, A.N., Antalic, S., Barzakh, A.E., Borschevsky, A., Capponi, L., Cocolios, T.E., De Witte, H., Elseviers, J., Fedorov, D.V., Fedosseev, V.N., Fink, D., Ghys, L., Huyse, M., Kudryavtsev, Y., Radulov, D., Rajabali, M.M., Rapisarda, E., Van den Bergh, P., Van Duppen, P., Eliav, E., Fritzsche, F., Imai, N., Kaldor, U., Köster, U., Lane, J., Lassen, L., Liberati, J., Lynch, K.M., Marsh, B.A., Nishio, K., Wakabayashi, Y., Pauwels, D., Pershina, V., Popescu, L., Procter, L.J., Raeder, S., Rossel, R.E., Sandhu, K., Seliverstov, M.D., Sjödin, A.M., Venhart, M., Wendt, K., 2013. Measurement of the first ionization potential of astatine by laser ionization spectroscopy. Nat. Commun. 4, 1835.

Rugel, G., Pavetich, S., Akhmadaliev, S., Baez, S.M.E., Scharf, A., Ziegenrucker, R., Merchel, S., 2016. The first four years of the AMS-facility DREAMS: status and developments for more accurate radionuclide data. Nucl. Instrum. Meth. B 370, 94–100.

Salbu, B., Krekling, T., Lind, O.C., Oughton, D.H., Drakopoulos, M., Simionovici, A., Snigireva, I., Snigirev, A., Weitkamp, T., Adams, F., Janssens, K., Kashparov, V.A., 2001. High energy X-ray microscopy for characterisation of fuel particles. Nucl. Instrum. Meth. A 467, 1249–1252.

Salbu, B., Kashparov, V., Lind, O.C., Garcia-Tenorio, R., Johansen, M.P., Child, D.P., Roos, P., Sancho, C., 2018. Challenges associated with the behaviour of radioactive particles in the environment. J. Environ. Radioact. 186, 101–115.

Savina, M.R., Pellin, M.J., Tripa, C.E., Veryovkin, I.V., Calaway, W.F., Davis, A.M., 2003. Analyzing individual presolar grains with CHARISMA. Geochim. Cosmochim. Acta 67, 3215–3225.

Savina, M.R., Isselhardt, B.H., Kucher, A., Trappitsch, R., King, B.V., David Ruddle, D., Raja Gopal||, R., Ian Hutcheon, I., 2017. High useful yield and isotopic analysis of uranium by resonance ionization mass spectrometry. Anal. Chem. 89, 6224–6231.

Schelles, W., de Gent, S., Maes, K., van Grieken, R., 1996. The use of a secondary cathode to analyse solid non-conducting samples with direct current glow discharge mass spectrometry: potential and restrictions. Fresenius J. Anal. Chem. 355, 858–860.

Schneider, S., Walther, C., Bister, S., Schauer, V., Christl, M., Synal, H.-A., Shozugawa, K., Steinhauser, G., 2013. Plutonium release from Fukushima Daiichi fosters the need for more detailed investigations. Nat. Sci. Rep. 3 (2988), 1–5.

Schneider, S., Bister, S., Christl, M., Hori, M., Shozugawa, K., Synal, H.-A., Steinhauser, G., Walther, C., 2017. Radionuclide pollution inside the Fukushima Daiichi exclusion zone, part 2: forensic search for the "Forgotten" contaminants Uranium-236 and plutonium. Appl. Geochem. 85, 194–200.

Schönberg, P., Mokry, C., Runke, J., Schönenbach, D., Stöbener, N., Thörle-Pospiech, P., Trautmann, N., Reich, T., 2017. Application of resonance ionization mass spectrometry for ultratrace analysis of technetium. Anal. Chem. 89, 9077–9082.

Schuler, W.A., 2010. Nuclear forensics: measurements of uranium oxides using time-of-flight secondary mass spectrometry (TOF-SIMS): department of Engineering Physics. In: Master of Science in Combating Weapons of Mass Destruction. Air Force Institute of Technology, Wright-Patterson Air Force Base, Ohio.

Schumann, P.G., Wendt, K.D.A., Bushaw, B.A., 2005. High-resolution triple-resonance autoionization of uranium isotopes. Spectrochim. Acta B 60, 1402–1411.

Schwellnus, F., Blaum, K., Geppert, C., Gottwald, T., Kluge, H.J., Mattolat, C., Nörtershäuser, W., Wies, K., Wendt, K., 2008. The laser ion source and trap (LIST) – a highly selective ion source. Nucl. Instrum. Meth. B 266, 4383–4386.

Sewtz, M., Backe, H., Dretzke, A., Kube, G., Lauth, W., Schwamb, P., Eberhardt, K., Grüning, C., Thörle, P., Trautmann, N., Kunz, P., Lassen, J., Passler, G., Dong, C.Z., Fritzsche, S., Haire, R.G., 2003. First observation of atomic levels for the element fermium ($Z = 100$). Phys. Rev. Lett. 90, 163002.

Sharp, B.L., 1988. Pneumatic nebulizers and spray chambers for inductively coupled plasma spectrometry, A Review: Part 1. Nebulizers. J. Anal. At. Spectrom. 3, 613–652.

Shen, C.C., H.C., Edwards, R.L., Moran, S.B., Edmonds, H.N., Hoff, J.A., Thomas, R.B., 2003. Measurement of attogram quantities of ^{231}Pa in dissolved and particulate fractions of seawater by isotope dilution thermal ionization mass spectroscopy. Anal. Chem. 75, 1075–1079.

Simons, D.S., 1986. Single particle standards for isotopic measurements of uranium by secondary ion mass spectrometry. J. Trace Microprobe Tech. 4, 185–195.

Solatie, D., Carbol, P., Betti, M., Bocci, F., Hiernaut, T., Rondinella, V.V., Cobos, J., 2000. Ion chromatography inductively coupled plasma mass spectrometry (IC-ICP-MS) and radiometric techniques for the determination of actinides in aqueous leachate solutions from uranium oxide. Fresenius J. Anal. Chem. 368, 88–94.

Spalding, K.L., Bergmann, O., Alkass, K., Bernard, S., Salehpour, M., Huttner, H.B., Bostrom, E., Westerlund, I., Vial, C., Buchholz, B.A., Possnert, G., Mash, D.C., Druid, H., Frisen, J., 2013. Dynamics of hippocampal neurogenesis in adult humans. Cell 153, 1219–1227.

Stephan, T., 2001. TOF-SIMS in cosmochemistry. Planet. Space Sci. 49, 859–906.

Stephan, T., Trappitsch, R., Davis, A.M., Pellin, M.J., Rost, D., Savina, M.R., Yokochi, R., Liu, N., 2016. CHILI – the Chicago Instrument for Laser Ionization – a new tool for isotope measurements in cosmochemistry. Int. J. Mass Spectrom. 407, 1–15.

Stephan, T., Trappitsch, R., Davis, A.M., Pellin, M.J., Rost, D., Savina, M.R., Jadhav, M., Kelly, C.H., Gyngard, F., Hoppe, P., Dauphas, N., 2018. Strontium and barium isotopes in presolar silicon carbide grains measured with CHILI-two types of X grains. Geochim. Cosmochim. Acta 221, 109–126.

Stoffels, J.J., Briant, J.K., Simons, D.S., 1994. A particulate isotopic standard of uranium and plutonium in an aluminosilicate matrix. J. Am. Soc. Mass Spectrom. 5, 852–858.

Strashnov, I., Blagburn, D.J., Gilmour, J.D., 2011. A resonance ionization time of flight mass spectrometer with a cryogenic sample concentrator for isotopic analysis of krypton from extraterrestrial samples. J. Anal. At. Spectrom. 26, 1763–1772.

Suter, M., Jacob, S.W.A., Synal, H.-A., 2000. Tandem AMS at sub-MeV energies – status and prospects. Nucl. Instrum. Meth. B 172, 144–151.

Synal, H.-A., 2013. Developments in accelerator mass spectrometry. Int. J. Mass Spectrom. 349–350, 192–202.

Synal, H.-A., Jacob, S., Suter, M., 2000. The PSI/ETH small radiocarbon dating system. Nucl. Instrum. Meth. B 172, 1–7.

Szidat, S., Schmidt, A., Handl, J., Jakob, D., Michel, R., Synal, H.A., Schnabel, C., Suter, M., Lopez-Gutierrez, J.M., 2000. RNAA and AMS of iodine-129 in environmental materials —comparison of analytical methods and quality assurance. Kerntechnik 65, 160—167.

Tamborini, G., Betti, M., 2000. Characterisation of radioactive particles by SIMS. Mikrochim. Acta 132, 411–417.

Tamborini, G., Betti, M., Forcina, V., Hiernaut, T., Giovannone, B., Koch, L., 1998. Application of secondary ion mass spectrometry to the identification of single particles of uranium and their isotopic measurement. Spectrochim. Acta B53, 1289–1302.

Tanner, S.D., 1995. Characterization of ionization and matrix suppression in inductively coupled "cold" plasma mass spectrometry. J. Anal. At. Spectrom. 10, 905–921.

Tanner, S.D., Baranov, V.I., Bandura, D.R., 2002. Reaction cells and collision cells for ICP-MS: a tutorial review. Spectrochim. Acta B 57, 1361–1452.

Taylor, R.N., Croudace, I.W., Warwick, P.E., Dee, S.J., 1998. Precise and rapid determination of $^{238}U/^{235}U$ and uranium concentration in soil samples using thermal ionization mass spectrometry. Chem. Geol. 144, 73–80.

Thellier, M., Derue, C., Tafforeau, M., Le Sceller, L., Verdus, M.-C., Massiot, P., Ripoll, C., 2001. Physical methods for in vitro analytical imaging in the microscopic range in biology, using radioactive or stable isotopes. J. Trace Microprobe Tech. 19, 143–162.

Thirlwall, M.F., 2000. Inter-laboratory and other errors in Pb isotope analyses investigated using a ^{207}Pb–^{204}Pb double spike. Chem. Geol. 163, 299–322.

Tissot, F.L.H., Dauphas, N., 2015. Uranium isotopic compositions of the crust and ocean: age corrections, U budget and global extent of modern anoxia. Geochim. Cosmochim. Acta 167, 113–143.

Tong, S.L., Harrison, W.W., 1993. Glow discharge mass spectrometric analysis of non-conducting materials. Spectrochim. Acta B48, 1237–1245.

Trappitsch, R., Stephan, T., Savina, M.R., Davi, A.M., Pellin, M.J., Rost, D., Gyngard, F., Gallino, R., Bisterzo, S., Cristallo, S., Dauphas, N., 2018. Simultaneous iron and nickel isotopic analyses of presolar silicon carbide grains. Geochim. Cosmochim. Acta 221, 87–108.

Trautmann, N., Wendt, K., 2012. Fast chemical separations and laser mass spectrometry – tools for nuclear research. Radiochim. Acta 100, 675–815.

Trautmann, N., Passler, G., Wendt, K., 2004. Ultratrace analysis and isotope ratio measurements of long-lived radioisotopes by resonance ionization mass spectrometry (RIMS). Anal. Bioanal. Chem. 378, 34–355.

Tuniz, C., Bird, J.R., Fink, D., Herzog, G.F., 1998. Accelerator Mass Spectrometry. CRC Press, LLC, Boca Raton.

Uchida, H., Ito, T., 1994. Comparative study of 27.12 and 40.68 MHz inductively coupled argon plasmas for mass spectrometry on the basis of analytical characteristic distributions. J. Anal. At. Spectrom. 9, 1001–1006.

Uchida, S., Tagamia, K., Rühm, W., Steiner, M., Wirth, E., 2000. Separation of Tc-99 in soil and plant samples collected around the Chernobyl reactor using a Tc-selective chromatographic resin and determination of the nuclide by ICP-MS. Appl. Rad. Istotop 53, 69–73.

Van Calsteren, P., Schwieters, J.B., 1995. Performance of a thermal ionization mass spectrometer with a deceleration lens system and post-deceleration detector selection. J. Mass Spectrom. 146, 119–129.

Van Straaten, M., Swenters, K., Gijbels, R., Verlinden, J., Adriaenssens, E., 1994. Analysis of platinum powder by glow discharge mass spectrometry. J. Anal. At. Spectrom. 9, 1389–1397.

Van Vaeck, L., Adriaens, A., Gijbels, R., 1999. Static secondary ion mass spectrometry (S-SIMS) Part 1: methodology and structural interpretation. Mass Spectrom. Rev. 18, 1–47.

Venzago, C., Pisonero, J., 2015. Glow discharge mass spectrometry. In: Prohaska, T., Irrgeher, J., Zitek, A., Jakubowski, N. (Eds.), Sector Field Mass Spectrometry for Elemental and Isotopic Analysis. Royal Society of Chemistry, Cambridge.

Venzago, C., Weigert, M., 1994. Application of the glow discharge mass spectrometry (GDMS) for the multielement trace and ultratrace analysis of sputtering targets. Fresenius J. Anal. Chem. 350, 303–309.

Vickers, G.H., Wilson, D.A., Hieftje, G.M., 1988. Detection of negative ions by inductively coupled plasma mass spectrometry. Anal. Chem. 60, 1808–1812.

Vockenhuber, C., Oberli, F., Bichler, M., Ahmad, I., Quitté, G., Meier, M., Halliday, A.N., Lee, D.-C., Kutschera, W., Steier, P., Gehrke, R.J., Helmer, R.G., 2004. New half-life measurement of Hf-182: improved chronometer for the early solar system. Phys. Rev. Lett. 93, 172501.

Vogel, J.S., 2000. Accelerator mass spectrometry for human biochemistry: the practice and the potential. Nucl. Instrum. Meth. B 172, 884–891.

Vogel, J.S., Turteltaub, K.W., Nelson, D.E., 1995. Accelerator mass spectrometry. Anal. Chem. 367A, 353–359.

Wallner, C., Faestermann, T., Gerstmann, U., Hillebrandt, W., Knie, K., Korschinek, G., Lierse, C., Pomar, C., Rugel, G., 2000. Development of a very sensitive AMS method for the detection of supernova-produced longliving actinide nuclei in terrestrial archives. Nucl. Instrum. Meth. B 172, 333–337.

Wallner, A., Feige, J., Kinoshita, N., Paul, M., Fifield, L.K., Golser, R., Honda, M., Linnemann, U., Matsuzaki, H., Merchel, S., Rugel, G., Tims, S.G., Steier, P., Yamagata, T., Winkler, S.R., 2016. Recent near-Earth supernovae probed by global deposition of interstellar radioactive Fe-60. Nature 532, 69–72.

Wang, F., Zhang, Y., Zhao, Y.G., Guo, D.F., Xie, S.K., Tan, J., Li, J.Y., Lu, J., 2018. Application of laser ionization mass spectrometry for measurement of uranium isotope ratio in nuclear forensics and nuclear safeguards. Meas. Sci. Technol. 29 (9).

Warwick, P.E., Croudace, I.W., Hillegonds, D.J., 2009. Effective determination of the long-lived nuclide ^{41}Ca in nuclear reactor bioshield concretes: comparison of liquid scintillation counting and accelerator mass spectrometry. Anal. Chem. 81, 1001−1906.

Wendt, K., Bhowmick, G.K., Bushaw, B.A., Herrmann, G., Kratz, J.V., Lantzsch, J., Müller, P., Nörtershäuser, W., Otten, E.W., Schwalbach, R., Seibert, U.-A., Trautmann, N., Waldek, A., 1997. Rapid trace analysis of $^{89,90}Sr$ in environmental samples by collinear laser resonance ionization mass spectrometry. Radiochim. Acta 79, 183−190.

Wendt, K., Geppert, C., Mattolat, C., Passler, G., Raeder, S., Schwellnus, F., Wies, K., Trautmann, N., 2012. Progress of ultra trace determination of technetium using laser resonance ionization mass spectrometry. Anal. Bioanal. Chem. 404, 2173−2176.

Wiedenbeck, M., 2008. GGR biennial review: advances in geo-SIMS during 2006-2007. Geostand. Geoanal. Res. 32, 489−494.

Wiedenbeck, M., Bugoi, R., Duke, M.J.M., Dunai, T., Enzweiler, J., Horan, M., Jochum, K.P., Linge, K., Košler, J., Merchel, S., Morales, L.F.G., Nasdala, L., Stalder, R., Sylvester, P., Weis, U., Zoubir, A., 2012. GGR biennial critical review: analytical developments since 2010. Geostand. Geoanal. Res. 36, 337−398.

Wiedenbeck, M., Bédard, L.P., Bugoi, R., Horan, M., Linge, K., Merchel, S., Morales, L.F.G., Savard, D., Souders, A.K., Sylvester, P., 2014. GGR Biennial critical review: analytical developments since 2012. Geostand. Geoanal. Res. 38, 467−512.

Wieser, M.E., De Laeter, J.R., 2000. Thermal ionization mass spectrometry of Mo isotopes. Int. J. Mass Spectrom. 197, 253−261.

Wieser, M.E., De Laeter, J.R., 2001. Evidence of the double β decay of zirconium-96 measured in $1.8x10^9$ year-old zircons. Phys. Rev. C 64, 024308.

Wieser, E.M., Schwieters, J.B., 2005. The development of multiple collector mass spectrometry for isotope ratio measurements. Int. J. Mass Spectrom. 242, 97−115.

Williams, P.M., Oeschger, H., Kinney, P., 1969. Natural radiocarbon activity of the dissolved organic carbon in the north-east pacific ocean. Nature 224, 256−258.

Winchester, M.R., Marcus, R.K., 1988. Glow discharge sputter atomization for atomic absorption analysis of nonconducting powder samples. Appl. Spectrosc. 42, 941−944.

Woelfli, W., 1987. Advances in accelerator mass spectrometry. Nucl. Instrum. Meth. B 29, 1−13.

Xu, Y., Qiao, J., Hou, X., Pan, S., Roos, P., 2014. Determination of plutonium isotopes (^{238}Pu, ^{239}Pu, ^{240}Pu, ^{241}Pu) in environmental samples using radiochemical separation combined with radiometric and mass spectrometric measurements. Talanta 119, 590−595.

Xu, C., Zhang, S., Sugiyama, Y., Ohte, N., Ho, Y.-F., Fujitake, N., Kaplan, D.I., Yeager, C.M., Schwehr, K., Santschi, P.H., 2016. Role of natural organic matter on iodine and $^{239,240}Pu$ distribution and mobility in environmental samples from the northwestern Fukushima Prefecture, *Japan*. J. Environ. Radioact. 153, 156−166.

Yamamoto, M., Tsumura, A., Katayama, Y., Tsukatani, T., 1996. Plutonium isotopic composition in soil from the former Semipalatinsk nuclear test site. Radiochim. Acta 72, 209−215.

Yamamoto, M., Oikawa, S., Sakaguchi, A., Tomita, J., Hoshi, M., Apsalikov, K.N., 2008. Determination of 240Pu/239/Pu isotopic ratios in human tissues collected from areas around the Semipalatinsk nuclear test site by sector filed high resolution ICP-MS. Health Phys. 95, 291−299.

Yang, L., 2008. Accurate and precise determination of isotopic ratios by MC ICP MS: a review. Mass Spectrom. Rev. 28, 990−1011.

Yang, G.S., Tazoe, H., Hayano, K., Okayama, K., Yamada, M., 2017. Isotopic compositions of U-236, Pu-239, and Pu-240 in soil contaminated by the Fukushima Daiichi nuclear power plant accident. Sci. Rep. 7, 13619.

Yokoyama, T., Nakamura, E., 2004. Precise analysis of the $^{228}Ra/^{226}Ra$ isotope ratio for short-lived U-series disequilibria in natural samples by total evaporation thermal ionization mass spectrometry (TE-TIMS). J. Anal. At. Spectrom. 19, 717−727.

Zhang, J., Zhou, T., Tang, Y., Cui, Y., Li, J., 2016. Determination of relative sensitivity factors of elements in high purity copper by doping-melting and doping-pressed methods using glow discharge mass spectrometry. Anal. At. Spectrom. 31, 2182−2191.

Zhao, X.L., Nadeau, M.J., Kilius, L.R., Litherland, A.E., 1994. The first detection of naturally - occurring ^{236}U with accelerator mass spectrometry. Nucl. Instrum. Meth. B 92, 249−253.

Zhao, X.L., Kilius, L.R., Litherland, A.E., Beasley, T., 1997. AMS measurement of environmental U-236: preliminary results and perspectives. Nucl. Instrum. Meth. B 126, 297−300.

Zheng, J., Aono, T., Uchida, S., Zhang, J., Honda, M.C., 2012a. Distribution of Pu isotopes in marine sediments in the Pacific 30 km off Fukushima after the Fukushima Daiichi nuclear power plant accident. Geochem. J. 46, 361−369.

Zheng, J., Tagami, K., Watanabe, Y., Uchida, S., Aono, T., Ishii, N., Yoshida, S., Kubota, Y., Fuma, S., Ihara, S., et al., 2012b. Isotopic evidence of plutonium release into the environment from the Fukushima DNPP accident. Sci. Rep. 2, 304.

Clemens Walther

Clemens Walther studied Physics (thesis on metal cluster physics using laser spectroscopy in an ion trap) at Mainz University and University of Seattle, WA. He completed his habilitation in Nuclear Chemistry in Mainz in 2008, where he also served as lecturer. From 1999 to 2011, he was employed at the Karlsruhe Institute of Technology, Institute for Nuclear Waste Disposal, serving as group leader "Actinide Nanoparticle Analysis" and deputy department head of "Actinide Speciation" where he headed a group developing and applying mass spectrometric and laser spectroscopic techniques. He is currently full professor, head of the Institute of Radioecology and Radiation Protection, and dean at the Leibniz University Hannover. CW is member of the German Commission of Radiation Protection (and head of the subgroup "Radioecology"), former chair of the mass spectrometry section of the German Physical Society (DPG), chair of the NRC section of the German Chemical Society (GDCh), and chair of the European Network on Nuclear and Radiochemistry Education and Training.

Klaus D.A. Wendt

Klaus Wendt studied physics at Johannes Gutenberg University of Mainz, where he got attracted to laser physics and quantum optics. After receiving his diploma in 1981, he went on to CERN to complete a PhD on the study of nuclear structure using laser spectroscopy and mass spectrometry at the radioactive ion beam facility ISOLDE. After graduation, he stayed as CERN Fellow up to 1986. He then returned back to Mainz University, and the Chernobyl nuclear disaster and its consequences motivated him to establish a research group working on ultratrace isotope determination using dedicated laser mass spectrometric techniques. Today, he is full professor and group leader in the field of laser resonance ionization spectroscopy and its applications. In addition, he strongly addresses aspects of teacher education and physics didactics.

Chapter 9

Solid scintillation analysis

Michael F. L'Annunziata
The Montague Group, Oceanside, CA, United States

Chapter outline

Handbook of Radioactivity Analysis: Volume 1. https://doi.org/10.1016/B978-0-12-814397-1.00009-1

I. Introduction

Solid scintillation is a process whereby the absorption of energy from ionizing radiation (e.g., alpha particles, beta particles, heavier charged particles, and X- or gamma rays) by certain crystalline inorganic or organic materials results in the emission of photons of light from the solid absorbing material. Neutron radiation, which is nonionizing, can produce indirectly the scintillation effect in certain solid scintillators, which may be used specifically for the detection and measurement of neutrons. The effect of neutrons in scintillators is generally the result of ionization and scintillation caused by (n,p) or (n,α) nuclear reactions within the scintillator material, which will be described further on in this chapter. Even neutrinos, which are nonionizing and, in addition, of very low rest mass, cause detectable interactions in scintillators, via neutrino-electron scattering, neutral current excitation of nuclei, and inverse beta decay. These processes are discussed in Chapter 6, Volume 1, and Chapter 6, Volume 2, and expounded further in this chapter.

An excellent account of the discovery of solid scintillation is given by Kolar and den Hollander (2004). Henri Becquerel, who was awarded the Nobel Prize in Physics for the discovery of radioactivity as Professor at the Ecole Polytechnic in Paris (An historical account is provided by the writer, L'Annunziata, 2007, 2016), was aware of the phosphorescent properties of crystalline zinc sulfide, referred to as "Sidot's blende", as related in his Nobel Lecture (Becquerel, 1903). "Sidot's blende" was named after Théodore Sidot of France, who established a procedure for the preparation of crystalline zinc sulfide. In historical accounts by Fricke (2001) and Kolar and den Hollander (2004), a pioneer in research on radioactivity by the name of Friedrich Oskar Giesel (1852–1927), while an employee of the Quinine Factory Braunschweig Buchler & Co. in Braunschweig, Germany, synthesized his own "Sidot's blende" by heating zinc and sulfur in a copper vessel. The crystalline zinc sulfide thus contained traces of copper, which Friedrich Giesel mixed with gelatin and spread the mixture over glass plates to manufacture phosphorescent screens. The "Sidot's blende" or zinc blende is reported to have a hexagonal (wurzite) crystalline structure with traces of copper ions ($ZnS:Cu^{2+}$), which is responsible for its yellow-green luminescence (Bol et al., 2002; Kolar and den Hollander, 2004). In a biography by Fricke (2001), Friedrich Giesel was under the impression that the light emitted by the ZnS screens was not continuous but appeared to flicker when exposed to radioactivity, and on December 9, 1902, he wrote to Julius Elster (1854–1920) to encourage him to research this phenomenon. Thus, Julius Elster teamed up with his friend Hans Geitel (1855–1923), both of whom were working in Wolfenbüttel, Germany, to scientifically investigate the luminescence. Shortly thereafter, the phenomenon of solid scintillation was reported by Julius Elster and Hans Geitel in both German and English scientific journals (Elster and Geitel, 1903a,b). With the aid of a magnifying glass they were able to see the individual flashes of light emitted by a ZnS screen placed in front of a sample emitting alpha particles.

Also in 1903, Sir William Crookes (1832–1919) in London studied the scintillations systematically, and he developed the apparatus for counting the light flashes (Crookes, 1903a,b). It consisted of a brass tube fitted with a lens at one end and a ZnS screen at the other. The position of the radioactive source, radium, from the ZnS screen could be adjusted by means of a thumbscrew. The closer the radium was adjusted to the screen, the more numerous and frequent were the flashes of light. He named the apparatus the Spinthariscope from the Greek "spintharis" meaning spark. The instrument was used by Rutherford (1919, 1920) and his students Hans Geiger and Ernest Marsden. They were able to visibly count the individual scintillations produced by alpha particles as the particles bombarded the ZnS screen. It was a crucial instrument in their research with alpha particles, leading to Rutherford's discovery of the atomic nucleus described in Chapter 1, Volume 1.

The first report of scintillation in organic compounds was documented much later in a thesis by Herforth (1948), who studied under the leadership of Hartmut Kallmann. An historical account of this discovery is related by Niese (1999, 2003). In her thesis presented on September 13, 1948, at the Technical University Berlin—Charlottenburg, Herforth reported that certain crystalline organic compounds could convert absorbed energy of alpha and beta particles. The discovery of this solid scintillation phenomenon in organic crystals led to liquid scintillation analysis through the work of Kallmann (1950) and Reynolds et al. (1950), which is covered in Chapter 6, Volume 1. The advances made in the solid scintillation analysis of nuclear radiation detection since these early discoveries are astounding. These may be best compared with advances made in development from the abacus to the modern electronic computer.

This chapter focuses mainly on solid scintillation analysis with inorganic crystalline materials. Certain solid inorganic crystalline materials have a high "stopping power" for X- and gamma radiation because of the high densities and high effective atomic numbers (Z_{ef}) that these inorganic crystals possess. Therefore, many inorganic scintillators are used throughout many fields of basic and applied research and applications (Diehl et al., 2017; Glodo et al., 2017; Moszyński et al., 2016, 2003; van Eijk, 2014, 2001; Ghelman et al., 2011; Novotny, 2005; Nikl et al., 2006; Higgens et al., 2006, 2008; van Eijk et al., 1994) and in almost all medical diagnostic imaging instrumentation (Ghabrial et al., 2018; Kertzscher and Beddar, 2018;

Germano et al., 2016; Lecoq, 2016; Slomka et al., 2016; Llosá, 2015; Deprez et al., 2011; Heller and Zanzonico, 2011; Ninos et al., 2010; Bieberle et al., 2007; Korzhik et al., 2007; Kandarakis et al., 2005; Ricard, 2004; van Eijk, 2003) for the detection and measurement of penetrating X- and gamma radiation. Depending on detector types and properties, solid inorganic scintillators are also applied to the detection and measurement of particulate radiation, such as alpha particles, beta particles, and neutrons. Neutrons are detected indirectly through nuclear reactions in crystal detectors. Special applications with solid scintillation detectors used for the analysis of alpha, beta, and neutron radiation will be discussed in this chapter. Applications of solid scintillation analysis are broad. We can find applications in the measurement of radioactivity and radionuclides in almost every field of science, including physics, chemistry, industry, space exploration, and the clinical and biochemical fields. This chapter will cover the basic principles and current practice of solid scintillation for the analysis of nuclear radiation and radionuclides.

II. Principles of solid scintillation

A. Inorganic crystal scintillators and their properties

Chapter 6 of Volume 1 dealt with the scintillation phenomenon that occurs when radiation energy is absorbed by certain organic materials (fluors) such as anthracene, PTP, and PPO dissolved in suitable organic solvents. The interaction of radiation with the organic solvents and the transfer of that energy to the organic fluors occur at the molecular level; that is, the organic molecules in a ground energy state absorb radiation energy and are elevated to an excited energy state. On returning to its ground state, the molecule emits absorbed energy as photons of visible radiation. The detection and analysis of electromagnetic radiation (i.e., X- and gamma-ray photons) by these organic materials, either solid or liquid, generally are not efficient due to the high penetrating power of the electromagnetic radiation. The relatively low density of organic materials (e.g., anthracene, 1.3 g/cm^3 and PTP, 1.2 g/cm^3) and their low effective atomic number (Z_{ef}) diminishes the "stopping power" of these scintillators for X- and gamma radiation.

The higher density (σ) and higher effective atomic number (Z_{ef}) of certain inorganic crystalline materials such as NaI (3.7 g/cm^3), CsI (4.5 g/cm^3), and bismuth germanate ($Bi_4Ge_3O_{12}$), also known as BGO (7.1 g/cm^3), give these solids a greater gamma-ray stopping power, that is, a greater ability to absorb energy of impinging gamma radiation. In addition to the higher density and atomic number offered by these inorganic crystalline scintillators, there are other basic properties of scintillators in general that are desirable for applications involving gamma and neutron radiation measurements. The desirable properties of a scintillator, which would constitute an "ideal" scintillator, are reviewed in detail by Derenzo et al. (2003) and Melcher (2005) and listed in Table 9.1. These properties of the ideal scintillator are the following:

- ***High detection efficiency.*** A scintillator of high density (ρ) and high effective atomic number (Z_{eff}) will provide the optimum stopping power for gamma radiation, whereas a scintillator enriched in isotopes with high neutron-capture cross-section (e.g., ^{6}Li, ^{10}B or ^{157}Gd) would yield improved neutron detection.
- ***Low cost of manufacture.*** The cost of manufacture is governed by difficulties of crystal growth and their production into various shapes and sizes.
- ***Chemical stability and resistance to thermal and mechanical shock.*** Many inorganic crystal detectors are hygroscopic and must be protected from the atmosphere, and must be handled with care to avoid fracturing to preserve their optical properties. Resistance to environmental effects and high mechanical strength are desirable properties.
- ***High count rate capability and coincidence timing.*** Short decay times and minimal afterglow would permit higher count rates. Afterglow, which is caused by relatively long-lived excitation states, is the emission of light after radiation excitation in the crystal has terminated. Strong afterglow can cause pulse pileup at high count rate applications and reduced energy resolution.
- ***Good spatial resolution and signal-to-noise ratio.*** A high light yield (i.e., high photon emission per radiation excitation energy deposited in the crystal scintillator or photons/MeV) with low noise will yield best spatial resolution.
- ***Good energy resolution.*** A light yield that is both high and proportional to the radiation energy deposited in the crystal will provide the best energy resolution.
- ***Suitable emission wavelength.*** The optimum wavelength of spectral emission of the crystal scintillator is the wavelength most efficiently detected by the photodetector, be it the photocathode of the photomultiplier tube (PMT) or by the photodiode (PD) detector.
- ***Radiation resistance.*** This refers to resistance to damage by high levels of ionizing radiation particularly in the field of high energy physics.
- ***Good light transmission from scintillator to photodetector.*** Optimization of light transmission from the scintillator to the photodetector is achieved by matching as close as possible the index of refraction of the scintillator at its maximum wavelength of emission to the index of refraction of the optical coupling of the photodetector, such as the PMT or PD window.

Crystal scintillators will not possess all of the optimum properties, and a tradeoff is made in selecting a scintillator

TABLE 9.1 Properties of the ideal scintillator.

Performance criteria	Scintillator properties	Notes
Detection efficiency	High density and Z for gamma-rays; or high cross-section for neutrons	$\rho > 7\ g/cm^3$, $Z > 60$; ^{6}Li, ^{10}B, ^{157}Gd
Practical manufacturing at low cost	Low cost raw materials; practical growth technique; congruent melting; acceptable melting point.; no phase transitions	Usually Czochralski or Bridgman growth techniques are the most practical
Chemical, thermal, and mechanical stability	Inert/non-hygroscopic; light yield insensitive to temp.; low/uniform thermal expansion; no cleavage; shock resistant	Facilitates fabrication of practical detectors
High count-rate capability and coincidence timing	Short decay time; low afterglow	T < 10 ns Afterglow <0.1% at 100
Good spatial resolution; good signal-to-noise	High light yield	Definition of "high" varies widely according to appl.
Good energy resolution	High light yield and proportional response to energy of excitation	
Suitable emission wavelength	Near 400 nm for use with PMTs; Near 600 nm for use with PD/APDs	
Radiation resistant	Insensitivity of optical properties to ionizing radiation and light	Considerable success has been achieved in improving radiation hardness
Good transmission of light to photodetector	Low index of refraction	$n \sim 1.5$ is usually considered to be optimal in order to match PMT windows

From Melcher (2005); reprinted with permission from Elsevier © 2005.

according to its properties and the particular application for which it is needed. Some of the scintillators possess one of the properties listed in Table 9.1, that are at the optimum or at what is considered currently to be the theoretical limit of that property. Some scintillators that exhibit an optimum or near optimum value for one of the qualifying properties are listed in Table 9.2.

There are basic requirements of inorganic crystal scintillators, which are principally a fast response or decay time for light output (10−100 ns) and high light yield (≥20,000 photons per MeV of gamma-ray or particle energy absorbed), and many of these scintillators are described in review papers, by Gundacker et al. (2018), Yanagida (2016, 2013), Ahmed (2015), Ghelman et al. (2011), Higgens et al. (2006, 2008), Nikl et al. (2006), Novotny (2005), van Eijk et al. (1994), and van Eijk (1993, 1997a,b, 2001). As characterized by van Eijk (1997a,b, 2001), these inorganic scintillators can be placed into two groups. In the first group are included NaI:Tl, CsI:Tl, CsI:Na, BGO, and $CdWO_4$ with light yields of $\geq 10^4$ photons per MeV and decay times of >200 ns. The second group includes BaF_2, CsF, CeF_3, and CsI with faster decay times (~1−30 ns), but lower light yields (a few thousand photons per MeV). The GSO:Ce scintillator (i.e., Gd_2SiO_5:Ce) falls in between these two groups with a 60 ns day time and 8000 photons per MeV light output (see Table 9.3). The use of a colon in the nomenclature of these detectors separates the molecular formula from the atomic activation cites or luminescence centers of the detector, respectively. Parentheses are also used. For example, thallium activated sodium iodide may be represented as NaI:Tl, NaI:Tl^+, or NaI(Tl). The mechanisms of activation and luminescence in inorganic scintillators are described further on in this chapter. To improve solid scintillation performance for the measurement of gamma radiation in many applications, researchers are searching actively for a scintillator that meets all of the basic requirements of fast response, high light yield, high density, and high effective *Z*.

Table 9.3 lists the properties of several commercially available and commonly used inorganic crystalline scintillators. Many other crystal scintillators are available, and some of these scintillators and their specific applications will be described further on in this chapter. Scintillator research is directed toward the development of more efficient scintillators with faster response, higher light yields and energy resolution, and low production costs. Examples are the high energy resolution of YAP:Ce, $LaCl_3$:Ce, $LaBr_3$:Ce, and SrI_2:Eu^{2+} which show a good proportionality of the light output in a large range of gamma-ray energies (Moszyński, 2010; Neal et al., 2011; Quarati

TABLE 9.2 Examples of performance currently achieved by scintillators.

Property	Current examples	Current value	Factors affecting the theoretical limit	Possible future materials
Light output	NaI (at 77 K) $LaBr_3$:Ce CsI:Tl	120,000 ph/MeV 61,000 ph/MeV 52,000 ph/MeV	Energy to produce an e−h pair, efficient energy transfer to luminescence center, high quantum efficiency	$LnPO_4$:Ln; $K_3Ln(PO_4)_2$:Ln $LnBO_4$:Ln $LnTaO_4$:Ln near band edge semi- conductors (e.g., CuI, ZnO:Ga)
Energy resolution	$LaBr_3$:Ce	2.6%	Good photon statistics, linear energy response	Mixed rare earth halides
Decay time	BaF_2 (core-valence luminescence) LaF_3:Nd (electric dipole emission)	0.6 ns 6 ns	Oscillator strength, wave-length, index of refraction	New core-valence compounds; Pr^{3+}, Nd^{3+}, and Yb^{3+} doped compounds; near band edge semi-conductors (e.g., CuI, ZnO:Ga)
Density	Lu_2O_3:Eu (ceramic)	9.4 g/cm^3	Crystal structure	Ln, Tl, Pb, Bi compounds
Effective atomic number	$PbWO_4$	76	Atomic species	Ln, Tl, Pb, Bi compounds
	$Bi_4Ge_3O_{12}$	75		
Cost	$PbWO_4$	~$2/cm^3	Raw materials, growth pro-cess, equipment investment	Ceramics

From Melcher (2005); reprinted with permission from Elsevier © 2005.

TABLE 9.3 Survey of characteristics of selected single crystal inorganic scintillators.

Crystal	Density (g/cm^3)	Light yield (phot/MeV)[a]	Total scint. efficiency (% BGO)[b]	Emission maximum (nm)	Dominant scint. decay time (ns)
CsI:Tl	4.51	66,000	n.m.[c]	550	800
NaI:Tl	3.67	41,000	n.m.	410	230
$LaBr_3$:Ce	5.3	61,000	n.m.	358	35
BaF_2 (only fast comp.)		1500	n.m.	180–220	0.8
$Bi_4Ge_3O_{12}$	7.1	8600	100	480	300
$PbWO_4$	8.28	300	3–5	410	3
$YAlO_3$:Ce	5.6	21,000	n.m.	360	25
$LuAlO_3$:Ce	8.34	12,000	n.m.	365	18
$Y_3Al_5O_{12}$:Ce	4.56	24,000	500–700	550	90
$Lu_3Al_5O_{12}$:Ce	6.67	12,500	500–700	530	55
Gd_2SiO_5:Ce	6.7	8000	600–700	420	60
Lu_2SiO_5:Ce	7.4	26,000	1500–1800	390	30
$Y_3Al_5O_{12}$:Pr	4.56	13,000	1000–1200	318	17
$Lu_3Al_5O_{12}$:Pr	6.67	13,000	1200–1500	308	20
Lu_2SiO_5:Pr	7.4	2200	800	247	26

Continued

TABLE 9.3 Survey of characteristics of selected single crystal inorganic scintillators.—cont'd

Crystal	Density (g/cm^3)	Light yield (phot/MeV)[a]	Total scint. efficiency (% BGO)[b]	Emission maximum (nm)	Dominant scint. decay time (ns)
$Y_{0.7}Yb_{0.3}AlO_3$	6.4	n.m.	10	350	0.9
ZnO—fast UV em. only	5.61	n.m.	>10	385	<0.8

[a]*Light yield is the number of generated UV/visible photons within a certain time gate (typically 1 μs), after the high-energy photon absorption.*
[b]*Total scintillation efficiency is given by all of the generated UV/visible light, which is usually obtained by an absolute comparison of radio-luminescence spectra against a "standard scintillator sample".*
[c]*Not measured.*
From Nikl et al. (2006); reprinted with permission from Elsevier © 2006.

et al., 2011). For example, the europium-activated alkaline earth halide $SrI_2:Eu^{2+}$ is reported by Neal et al. (2011) to exhibit a light yield of 81,400 photons/MeV with an associated energy resolution of 3.7% (FWHM for 662 keV gamma rays) with measured decay times from 1.1 to 2.0 μs. Many industrial, research, and medical applications have different demands for detector characteristics and efficiencies. Nevertheless, the basic properties of fast response (10–100 ns), high light yield (>25,000 photons/MeV), high density, and high effective atomic number (Z_{ef}) are required for many applications. Several inorganic crystal scintillators are available commercially and under research as efficient detectors for positron emission tomography (PET) and other nuclear medical imaging applications in which stopping power, speed, and energy resolution are crucial (Ghabrial et al., 2018; Kertzscher and Beddar, 2018; Germano et al., 2016; Lecoq, 2016; Slomka et al., 2016; Llosá, 2015; Deprez et al., 2011; Ninos et al., 2010; Korzhik et al., 2007; Valais et al., 2007; Ricard, 2004; van Eijk, 2003; Korzhik and Lecoq, 2001). Some of these scintillators together with their properties are listed in Table 9.4.

TABLE 9.4 Inorganic scintillators for PET applications with comparable or higher light yields and shorter decay times than BGO.

Material	$\rho(g/cm^3)$	Z_{eff}/photo-abs. coeff. At 511 keV (cm^{-1})	LY[a] (ph/MeV)	$\lambda_{s,c}$(nm)	$\tau_{s,c}$(ns)	Status of technology
$Bi_3Ge_4O_{12}$ (BGO)	7.13	75.2/0.37	8200	505	300	Under production
LuF_3:Ce	8.3	61.1/0.31	8000	310	23 + slow	Samples
Lu_2S_3:Ce	6.2	66.7/0.24	28,000	592	32	Samples
$Lu_3Al_5O_{12}Sc$ (LuAG)	6.7	62.9/0.21	22,500	270	610	Samples
$LuAlO_3Ce$ (LuAP)	8.34	64.9/0.29	11,400	365	17 + slow	Samples
$Lu_{0.7}Y_{0.3}AlO_3$:Ce (LYAP)	7.1	60/0.21	13,000	375	23 + slow	Under production
Lu_2SiO_5:Ce (LSO)	7.4	66/0.28	27,000	420	40	Under production
$(Lu-Y)_2SiO_5$:Ce (LYSO)	7.1	60/0.20	32,000	420	41	Under production
$(Lu-Gd)_2SiO_5$:Ce (LGSO)	6.5	59/0.18	17,000	430	65	Under production
Gd_2SiO_5:Ce (GSO)	6.7	57/0.175	12,500	430	60,600	Under production
$Lu_2Si_2O_7$:Ce (LPS)	6.23	64.4/0.21	30,000	380	30	Samples
$LuBO_3$:Ce	7.4	64.5/0.28	26,000	410	39	Samples
$LaBr_3$:Ce	5.29	47/0.065	63,000	380	17–35	Under production

[a]*LY = Light Yield, λ = emission maximum, τ = decay time.*
From Korzhik et al. (2007); reprinted with permission from Elsevier © 2007.

The inorganic atoms in the solid crystalline state are usually "doped" with varying concentrations (mol %) of an element impurity such as thallium, cerium, or other elements, which act as centers of activation or luminescence centers in the crystal. For example, one of the oldest and most commonly used crystal detector of this type is NaI:Tl, which is sodium iodide activated with approximately 1.3×10^{-3} mol fraction of thallium (Mundy and Rothman, 1983) referred to as a thallium-activated sodium iodide detector. The concentration of element impurity, that creates the luminescence centers in the crystal detector, will affect the light output or pulse heights. For example, Hamada et al. (2001) demonstrated that increasing the Tl^+ concentration from 10^{-3} to 10^{-2} mol in CsI:Tl detectors increased the pulse heights originating from heavy particle (alpha) excitation compared with those of electron excitation. The optimum concentration of dopant is determined experimentally. For example, Balamurugan et al. (2006) determined that a 0.06 mol% of Tl doped CsI crystal yielded optimum gamma-energy resolution of the detector and increasing the Tl^+ doping concentration decreased the hardiness of the crystal. Another example can be taken from the work of Fukabori et al. (2011a) where Ce^{3+} doping of $LiYF_4$ single crystals was tested over the range of 0.1–10 mol% of Ce content. They found that energy resolutions of the crystal for ^{137}Cs gamma quanta would decrease at Ce concentrations that would exceed 2 mol%. Co-doping of inorganic crystal scintillators with a second dopant can affect also the efficiency of the scintillator. The addition of Eu^{2+} codopant to thallium doped CsI (i.e., CsI:Tl) crystal reduces the intensity of the scintillator afterglow by two orders of magnitude in the time range of 10 μs–100 ms (Nagarkar et al., 2005; Ovechkina et al., 2007). Strong afterglow is a deleterious effect, as it can cause pulse pileup in high count rate applications, reduce energy resolution in radionuclide imaging, and produce reconstruction artifacts in computed tomography (CM) (Ovechkina et al., 2007). Europium-doped polycrystalline lutetium oxide scintillator at 0.5 mol% Eu^{3+} was co-doped with promethium over the range of 0–0.5 mol% Pr^{3+} by Jie et al. (2011) to give a crystal detector abbreviated as Lu_2O_3:(Eu,Pr) whereby the added promethium reduced the crystal afterglow. Neal et al. (2011) improved the performance of europium-activated (or europium-doped) strontium iodide crystals, $SrI_2:Eu^{2+}$ with the divalent europium at 3%–6% by substituting 30 atom% of the strontium ions with Mg^{2+}, Ca^{2+} or Ba^{2+}. They achieved improved light yields of 81,400 photons/MeV with an associated energy resolution of 3.7% (FWHM for 662 keV gamma rays) and fast decay times of 1.0–2.0 μs. Several examples of ion-activated inorganic scintillators are listed in Tables 9.2–9.4, where a colon separates the element acting as the center of activation from the detector formula. The thallium (Tl^+) or other such ions (e.g., Ce^{3+}, Eu^{2+} or Pr^{3+}, among others) serve as activation centers in the inorganic crystals, where deexcitation of absorbed energy from interacting nuclear radiation and subsequent emission of visible light occur. Other scintillators do not require doping with element impurities to improve the scintillation efficiency. Some of these scintillators are also described in Tables 9.2–9.4.

The mechanisms of the conversion of nuclear radiation energy to photons of visible light in such crystal detectors are described subsequently.

B. Scintillation mechanisms in inorganic crystals

The mechanisms of scintillation in inorganic crystal detectors have been studied by Pejchal et al. (2018), Sun et al. (2018), Liu et al. (2017), Drozdowski et al. (2016), Nikl et al. (2016, 2013), Sahi et al. (2016), Yang et al. (2016), Bartosiewicz et al. (2015), Bizarri (2010), Alexandrov et al. (2008), Pidol et al. (2005), Weber (2002, 2004), Belsky et al. (2001), Hamada et al. (2001), Rodnyi (2001), Zazubovich (2001), Saoudi et al. (1999), Wojtowicz et al. (1994b, 1995, 1996), Lempicki (1995), Lempicki et al. (1993), Heath et al. (1979), Kabler (1975), and Dietrich et al. (1973), among others. Bizarri (2010) and Lempicki (1995) outline the scintillation process in three stages: (1) creation of electron-hole pairs by an absorbed gamma- or X-ray photon, (2) energy transfer after electron-hole pair propagation to a luminescent center, and (3) light emission at a luminescent center. The treatment that follows will refer at times to the interaction of electromagnetic gamma or X-radiation with the crystal detector. However, the same basic principles apply to any type of radiation, that produces ionization and electron-hole pairs in matter, such as electrons, protons, alpha particles and other heavy ions.

As described in Chapter 1, Volume 1, the principal mechanisms of the interaction of gamma- or X-ray photons with matter are the Compton effect, photoelectric effect, and pair production. In each of these mechanisms of interaction with scintillating crystals, photon energy is absorbed by the crystal. As a result of the initial interaction of each gamma photon in the crystal, an electron will be produced, either a Compton electron, photoelectron, or an electron from pair production. The Compton and photoelectrons are the result of ionization producing electron-hole pairs. A Compton scatter photon will travel on and produce further ionization and additional electron-hole pairs in the crystal detector until its energy is absorbed or until it exits the crystal detector. Electron-hole pair production is, therefore, the first step in the absorption of gamma or X-radiation in crystal scintillators.

Let us now look at the mechanisms whereby the electron-hole pairs produced in the crystal scintillators can result in the emission of visible light photons from the crystal lattice. To explain this process we should consider first that there are two types of crystal scintillator detectors. Lempicki (1995) and Weber (2002, 2004) classify the detectors into the following two types:

1. **Intrinsic (i.e., Self-activated)**. These crystals do not contain any doped elements as centers of activation. Examples are BGO ($Bi_4Ge_3O_{12}$), $PbWO_4$, $CdWO_4$, BaF_2, and CsI, among others listed in Tables 9.2–9.5. In these cases, luminescence may be considered a property of the lattice itself.
2. **Extrinsic (i.e., Activated)**. These crystal detectors generally are doped with a trace amount of element, such as Tl^+, Ce^{3+}, Pr^{3+}, which are introduced into the crystal lattice to serve as centers of activation and luminescence. Examples of this type are NaI:Tl, CsI:Tl, $LuAlO_3$:Ce (LuAP), and Lu_2SiO_5:Ce (LSO), among others listed in Tables 9.2–9.5.

The luminescence resulting from nuclear radiation energy deposition and ionization in the intrinsic or self-activated crystal detectors may involve electron–hole (e–h) recombination, self-trapped and defect-trapped exciton luminescence, charge transfer transitions (metal-to-ligand, ligand-to-metal) within the crystal molecule, or core–valence band transitions (Weber, 2004). The emission transitions, listed in Table 9.5, have characteristic emission wavelengths and decay times, which are described in detail by Weber (2004). Excitation and luminescence in the extrinsic (activated) crystals occurs at a site of the dopant ion (e.g., Tl^+, Ce^{3+}, Eu^{2+}), which undergoes excitation following the capture of ionization holes and electrons (Weber, 2004; Bizarri, 2010, See also Fig. 9.1).

One of the oldest crystal detectors still in extensive use today since its discovery by Nobel Laureate Robert Hofstadter in 1948, while at Princeton University (Hofstadter, 1948, 1950), is thallium-activated sodium iodide [NaI(Tl) or NaI:Tl]. We can confine our explanation of scintillation to this crystal, because of its widespread use and the extensive study devoted to this detector, and the general mechanisms apply to other activated inorganic scintillators.

TABLE 9.5 Scintillator materials and their radiative transitions.

Scintillator material	Emission transition
Intrinsic (self-activated)	
CeF_3	5d–4f
$Bi_4Ge_3O_{12}$ (BGO)	6p–6s
CsI, BaF_2, LaF_3	Self-trapped exciton
$CdWO_4$, $YTaO_4$	Charge transfer
CsF, BaF_2, $KMgF_3$	Core–valence band
Extrinsic (activated)	
Gd_2O_3:Eu^{3+}, Gd_2O_2S:Pr^{3+} or Tb^{3+}	4f–4f
CaF_2:Eu^{2+}	5d–4f
Lu_2SiO_5:Ce^{3+}, LaF_3:Nd^{3+}	5d–4f
NaI:Tl^+, CsI:Tl^+	6p–6s
CsI:Na^+, CdS:Te^{2-}	Excitonic
ZnS:Ag, CuI, PbI_2	Donar–acceptor pair

From Weber (2004); reprinted with permission from Elsevier © 2004.

In brief, the principal mechanisms of photon emission subsequent to electron-hole production in the NaI(Tl) crystal occur in the following two sequences: (1) electron and hole trapping processes and (2) radiative recombination mechanisms (Lempicki, 1995; Weber, 2004; Alexandrov

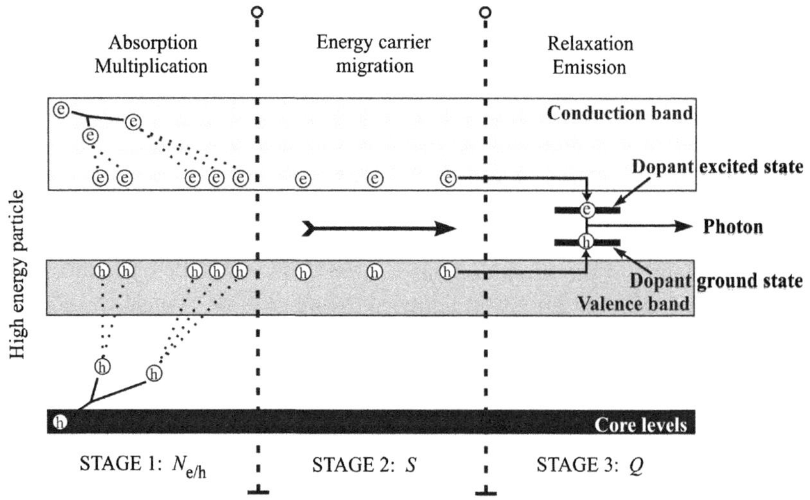

FIGURE 9.1 Basic scheme of various stages involved in scintillation. *From Bizarri (2010); reprinted with permission from Elsevier © 2010.*

et al., 2008; Bizarri, 2010). Predominate electron-hole trapping processes are explained as follows:

The Tl is normally found in the crystal lattice as Tl^+ ions. Electrons, e^-, are trapped by Tl^+ to form Tl^0 as described by

$$e^- + Tl^+ \rightarrow Tl^0 \quad (9.1)$$

and holes (h^+) are trapped by Tl^+ to form Tl^{++} according to

$$h^+ + Tl^+ \rightarrow Tl^{+2} \quad (9.2)$$

The predominate radiative recombination mechanisms then come into play to include the interaction of trapped electrons with free holes to form activated thallium or $(Tl^+)^*$ followed by the emission of visible light upon thallium deactivation according to

$$h^+ + Tl^0 \rightarrow (Tl^+)^* \rightarrow Tl^+ + h\nu(335,\ 420\ \text{nm}) \quad (9.3)$$

and the interaction of trapped holes with free electrons (e^-) to produce activated thallium or $(Tl^+)^*$ followed by the emission of visible light upon thallium deactivation as illustrated by

$$e^- + Tl^{+2} \rightarrow (Tl^+)^* \rightarrow Tl^+ + h\nu(335,\ 420\ \text{nm}) \quad (9.4)$$

The luminescence at dopant activation centers is illustrated in Fig. 9.1 where the scintillation scheme is depicted to occur in three stages, as described by Bizarri (2010). Firstly, the high-energy particle (HEP) deposits energy in the scintillator crystal forming N number of electron–hole pairs ($N_{e/h}$) in Stage 1. The electron–hole pairs travel to luminescent centers depicted in Stage 2, where the major constraint is the transport or transfer efficiency S of the crystal of the e–h pair energy to the luminescence centers. Finally the Stage 3 occurs when the dopant ions act as the centers for electron–hole trapping and dopant ion excitation and deexcitation with photon emission, governed by the luminescence efficiency, Q, of the scintillation crystal. The overall light yield (LY) in units of photons emitted per MeV of energy deposited in the crystal by the nuclear radiation is a function of the transport efficiency of the e–h pair energy (S) and the luminescence efficiency (Q) of the crystal detector, as described by Eq. (9.8) further on in this chapter.

Although of lower density than other available inorganic scintillator crystal detectors, the NaI(TI) crystal in combination with low-noise photomultipliers remains among the most commonly used today for X- and gamma-ray photon measurements as well as other particulate radiation and as a promising tool for the detection of the nuclear recoil (NR) events from weakly interacting massive particles (WIMPs) interactions (Bayat et al., 2018; Gütlein et al., 2017; Kasban et al., 2017; Oliván et al., 2017; Varley et al., 2017; Kim et al., 2015; Amaré et al., 2014; Bezuidenhout, 2013; Cuesta et al., 2013; Parker et al., 2013; Caciolli et al., 2012; Peeples and Gardner, 2012; Ianakiev et al., 2009; Alexandron et al., 2008; Hell et al., 2000). The popularity of this detector is due partly to the development of polycrystalline NaI(Tl) by Harshaw/Filtrol Partnership (Solon, OH) under the trade name Polyscin, which is prepared by the recrystallization of single-crystal ingots under heat and pressure and other more modern techniques of producing large ingots of NaI(Tl) detector material (Taranyuk et al., 2011). The ingots are more durable than single-crystal material and can be fabricated in a wide range of sizes and shapes to accommodate specific applications without any cost to scintillation performance. Mechanical and thermal shock inflicted on single-crystal NaI(Tl) can cause fractures, which can cleave among many planes and propagate across the entire crystal, resulting in an overall reduction in light output efficiency and pulse height resolution. Fractures in the single-crystal ingots, however, usually remain confined to localized areas of the crystal with no overall interference with light collection efficiency and pulse height resolution.

Other important factors that have contributed to the popularity of NaI(TI) are the relative short decay time of this detector and its high light output or relative scintillation conversion efficiency. The decay time is the time necessary for deactivation. Shorter decay times permit greater resolution of succeeding deactivation events, referred to as resolving time, in the crystal detector. The resolving time of an NaI(Tl) crystal detector is reported to be around 10 ns (Seo et al., 2010).

Optical emission in the scintillation process is governed by an exponential decay with decay constants varying with detector type (see Table 9.3). Heath (1983) explained the effect of variable optical emission decay on the basis of detector afterglow. That is, although most excited states of a scintillator or phosphor may have the same lifetime, phosphors are characterized by the presence of long-lived states referred to as afterglow. Grabmaier (1984) defines afterglow as the emission of light from the scintillator after the cessation of radiation excitation. This is a property of all scintillation crystal phosphors, and NaI(Tl) and BGO display this property to a minimal degree. The high density and atomic number of BGO give it a greater stopping power than NaI(Tl) for electromagnetic radiation. This is of particular importance in improving detection efficiencies of high-energy gamma radiation. Therefore, BGO has become a very popular detector. The properties of BGO detectors and their applications are provided in Tables 9.3 and 9.4, and they are discussed further on in this chapter.

Research has produced scintillation crystal detectors with higher density, higher light yields, and faster responses than BGO. Some of these fast solid scintillation detectors are the Ce^{3+}-activated crystals found in Tables 9.3 and 9.4 among others activated by Pr^{3+} and

Eu^{2+}. The properties and the scintillation processes in these materials are described by Khan et al. (2018), Kim et al. (2016), Sun et al. (2015), Gundiah et al. (2014), Samulon et al. (2014), Derenzo et al. (2011), Neal et al. (2011), Bourret-Courchesne et al. (2009, 2010), Mares et al. (2005), Belsky et al. (2001), Korzhik and Lecoq (2001), Rodnyl (2001), van Loef et al. (2001), Pauwels et al. (2000), Saoudi et al. (1999), Wojtowicz et al. (1994a, 1995, 1996), and Lempicki et al. (1993). The scintillation mechanism at sites of Ce^{3+} in crystal detectors is similar to that of Tl^{+} described previously. Following the production of electron-hole pairs by the impinging gamma radiation, a two-step scintillation process involves first the trapping of holes by Ce^{3+} according to

$$Ce^{3+} + h^{+} \rightarrow Ce^{4+} \tag{9.5}$$

and the interaction of the trapped holes with conduction band electrons yielding excited Ce^{3+} ions $(Ce^{3+})^*$ as follows

$$Ce^{4+} + e^{-} \rightarrow \left(Ce^{3+}\right)^{*} \tag{9.6}$$

The excited cerium ion $(Ce^{3+})^*$ falls to a stable ground state with the emission of a scintillation photon, $h\nu$, according to

$$\left(Ce^{3+}\right)^{*} \rightarrow Ce^{3+} + h\nu \tag{9.7}$$

where h^{+} and e^{-} represent a valence band hole and conduction band electron, respectively (see Fig. 9.1). Wojtowicz et al. (1994a,b) demonstrated that the high hole-capture cross section of the Ce^{3+} ions and good transport properties of electrons are responsible for the high scintillation light yield of these Ce-doped crystals, which are properties yielding high light yields with other dopants, such as Eu^{2+}.

The scintillation processes previously described with Tl^{+} and Ce^{3+} activation centers can be described similarly with other ions doped in a crystal lattice to serve as centers of luminescence. Bressi et al. (2001) modeled the scintillation process in the following steps: (1) absorption of radiation, (2) production of an electron-hole pair, (3) thermalization of the electron and hole, (4) energy transfer to a luminescence or activation center, and (5) emission, which is similar to that described by Bizarri (2010) and depicted in Fig. 9.1. Bressi et al. (2001) define the light yield (LY) in units of photons/MeV as

$$\mathrm{LY} = 4.35 \times 10^{5} \frac{\beta SQ}{E_g[\mathrm{eV}]} \frac{\text{photons}}{\text{MeV}} \tag{9.8}$$

where β is the conversion efficiency (Step 1), E_g is the band-gap energy (Step 2), that is, the energy difference between the valence and conduction band of the detector, S is the transport or transfer efficiency of the e-h pair/energy to the luminescence center (Steps 3 and 4), and Q is the luminescence efficiency (Step 5). A similar definition for photon production or light yield is provided by Rodnyi et al. (1995) and van Eijk (2001). The light yield could be controlled by the selection of a scintillation detector with a small band gap (Kurudirek et al., 2018; Wang et al., 2018; Kowalski et al., 2017; Shendrik et al., 2017; Yanagida et al., 2017; Auffray et al., 2016; Wu et al., 2016; Shwetha et al., 2015; Ninos et al., 2010; Hawrami et al. 2008; van Loef et al., 2005). Band-gap energies may vary from <1 eV to several eV. Values are provided by van Eijk (2001) for several types of detectors. Bizarri (2010) points out, that for an efficient Q, the ground state of the dopant ion should be above the valence band, and the excited state of the dopant below the conduction band (see Fig. 9.1).

The relative light output of all scintillation crystal detectors is a function of temperature, and the effect of temperature varies significantly from one detector type to another over the temperature range -100 to $+140°C$. These temperature effects are generally less significant for NaI(Tl) detectors in normal room measurements of radionuclides, because this detector displays a rather stable light output response at $\pm20°C$ at either side of room temperature (Heath et al., 1979; Sabharwal et al., 1982; Ianakiev et al., 2009). In space exploration and borehole logging, where more extremes in temperature are experienced, temperature effects on the light output of scintillation crystal detectors are very significant. NaI(Tl) is often the detector of choice for in-situ gamma spectroscopy including borehole logging, because of its high light output (40,000 photons/MeV), high effective atomic number (Z_{eff}), and high scintillation efficiency (Asfahani, 1999; Rudin et al., 2001). However, Rudin et al. (2001) demonstrate that CaF_2(Eu) may be a suitable replacement for NaI(Tl) particularly as groundwater probes and at liquid waste sites. The CaF_2(Eu) detector has a lower light yield (24,000 photons/MeV), but it is nonhygroscopic and more fracture resistant. The low effective atomic number of CaF_2(Eu) renders it less suitable for gamma-ray spectroscopy at energies above a few hundred keV.

C. Conversion of detector scintillations to voltage pulses

The utilization of the solid scintillation phenomenon for the analysis of nuclear radiation requires the conversion of the visible photon emissions of the inorganic scintillator crystal to voltage pulses. The voltage pulses can be measured to determine their magnitude and number. As in the case of liquid scintillation analysis, a PMT with a photosensitive cathode is used in many instruments to convert a given photon emission to photoelectrons, which are amplified by a series of dynodes to an avalanche of electrons (see Fig. 6.4, Chapter 6, Volume 1). Other types of photomultipliers can also do this job, and these are described in Section III.B of this chapter. The magnitude of the resulting pulse height produced by the PMT is proportional to the photon intensity emitted by the crystal scintillator. The photon intensities depend, in turn, on the amount of nuclear radiation energy absorbed by the scintillator. In Table 9.3 it is noted that NaI(Tl) produces around 40 photons per keV absorbed. Therefore, in summary, we can state that a given pulse height in volts produced by the PMT (within the resolving time of the PMT, e.g., 5 ns) is proportional to the scintillation photon intensity, and the scintillation photon intensity is, in turn, proportional to the energy of the nuclear radiation absorbed by the crystal detector. The following equation can be written to describe this process, which is the basis of the principle behind solid or liquid scintillation analysis of the energies of nuclear radiation:

$$\text{Pulse height(volts)} \propto \text{scintillation photon intensity} \propto \text{gamma-ray energy absorbed} \tag{9.9}$$

The absorption of a monoenergetic beam of gamma radiation by the crystal detector is governed by the exponential attenuation Eq. (1.442) (Chapter 1, Volume 1), $I = I_0 e^{-\mu x}$, and the attenuation coefficient, μ, is a sum of the independent coefficients due to the Compton, photoelectric, and pair production effects in the crystal, defined by Eq. (1.454) (Chapter 1). The magnitude of the attenuation coefficients for these radiation absorption processes in a given absorber material is a function of the incident gamma-ray energy. Attenuation coefficients due to the Compton, photoelectric, and pair production processes as well as the total attenuation coefficient (sum of these three processes) in NaI as a function of gamma-ray energy are plotted in Fig. 1.127 of Chapter 1, Volume 1.

Because gamma rays, originating either from the sample or from annihilation radiation, and X-radiation, produced in the crystal and detector shielding, may impart any fraction of their energy in Compton interactions, a wide spectrum of photon intensities of visible light are emitted from the crystal detector. Consequently, a spectrum of pulse magnitudes in volts is produced as an output from the PMT. Pulses low in magnitude are also produced by thermal noise, that is, thermionic emission of electrons from the cathode of the PMT. All pulses produced may be amplified and then registered as counts over a period of time to provide a count rate, and the pulse events may be analyzed by a multichannel analyzer (MCA) to determine the spectrum of pulse heights produced and gamma-ray energies absorbed by the crystal detector. The principles of operation and applications of the PMT as well as other photomultipliers and the MCA as components of the solid scintillation analyzer are discussed further on in this chapter and other parts of this book.

III. Solid scintillation analyzer

From the information provided up to this point of the chapter we might conclude that inorganic crystal scintillators are used most often for the measurement of electromagnetic X- and gamma radiation. This is the case; however, these detectors are also commonly used to measure other ionizing radiations such as electrons, heavy ions and elementary particles as well as non-ionizing radiations such as neutrons and neutrinos. The latter are detected indirectly via the reactions they produce within the inorganic crystal scintillators. This section of the chapter will focus mostly on the analysis of electromagnetic X- and gamma radiation. Special applications of solid scintillation to the measurement of charged-particle radiation and the neutrally charged neutrons and neutrinos are covered separately in this chapter.

Automated modern state-of-the-art solid scintillation analyzers are computer controlled to enable the automatic hands-off analysis of many samples, sample data processing, instrument calibration and control, and programmed instrument performance assessment (IPA). Linked to the computer are the hardware components, including the sample changer, solid scintillation detector, or even several detectors in one instrument to enable the high-throughput simultaneous analysis of multiple samples, and the electronic components required to convert scintillation photons to voltage pulses, and pulse height MCAs required to correlate these voltage pulses with radiation energies and to quantify their intensities. The basic components of the solid scintillation analyzer will be discussed subsequently.

A. Scintillation crystal detectors

The chemical structure, properties, and performance of a wide range of inorganic crystal scintillation detectors are provided in Tables 9.2–9.4 of this chapter. The properties of many other inorganic scintillators will be discussed further on in this chapter.

Scintillation crystal detectors may be employed in various shapes or geometries. Among these are (1) the planar detector, (2) the well-type detector, and (3) the through-hole detector. The crystalline scintillation detectors are discussed in this section of the chapter. Yet other detector formats exist including the noncrystalline plastic or glass type. There are commercially available 24- or 96-well plastic microplate solid scintillators and microsphere plastic or glass scintillators for high-throughput scintillation counting and scintillation proximity assay (SPA). Also, there are plastic and glass fiber-optic scintillators. These, as well as other scintillator detector types, are discussed separately further on in this chapter.

1. Planar detector

The planar solid scintillation detector usually consists of an inorganic crystal that is cylindrical in shape, and NaI(Tl) is the most commonly employed detector material. The dimensions will vary as described subsequently. One flat surface of the cylindrical crystal detector is exposed to the radiation source and the other end of the crystal is connected to a PMT by means of a light pipe and light-transmitting grease. The PMT converts the individual scintillation events into measurable voltage pulses as described later in this section. The crystal surfaces that are not in contact with the PMT are covered with a protective coating (e.g., aluminum ~ 0.5 mm thick), which blocks out external light and protects the scintillation crystal detector from atmospheric moisture (particularly hygroscopic crystal detectors) and from physical damage. The detector coating, if sufficiently thick, may also serve as a shield against alpha and low-energy beta radiation. The inner part of the coating is often whitened or covered with a reflector material (e.g., MgO) in order to reflect photons of visible light toward the PMT.

The surface of a crystal of this type can receive, at best, only 50% of the radiation emitted from a given sample, since the crystal can be exposed to a maximum of 180 degrees of a possible 360 degrees of radiation emission. This counting geometry is often referred to as 2π geometry. This is not a very efficient counting geometry, because of the very limited exposure the detector can have to the sample radiation, as this type of detector can "see" only 50% of the sample at best. Detection efficiencies with this counting geometry are relatively low compared with other detector geometries. Consequently, planar detectors are generally not used for the analysis of samples, which are handled and prepared in the laboratory, as more efficient detector geometries are available for sample analysis. The planar detector is used in other applications such as the monitoring of surface radioactivity, area monitors, space exploration, and imaging detectors used in medical diagnosis, such as single-photon emission computed tomography (SPECT) and PET, which require the detection of gamma-ray photons from only one direction (Knoll et al., 2016; Phillips et al., 2015; Retière et al., 2012).

The crystal detector is most commonly 75 × 75 mm in size. Smaller crystals are used, but they provide lower gamma-ray detection efficiencies, because the probabilities become greater that a gamma ray may pass through the crystal detector without dissipating its energy within. Belle et al. (1974) demonstrated that oversized crystals are characterized by an enhanced background without increased sensitivity. They concluded that a 70 × 70 mm (~3 × 3″) NaI(Tl) crystal is best with gamma-ray energies up to 0.4 MeV, whereas a 150 × 150 mm (~6 × 6″) crystal is recommended for gamma-ray energies between 0.4 and 3.0 MeV.

2. Well-detector

The well-type detector consists of a crystal detector with a cavity drilled at the center of the crystal to form a well within which a sample may be placed in a suitable sample tube or other container for radiation analysis and counting, as illustrated in Fig. 9.43 further on in this chapter. For example, a cavity of 15 mm diameter (~0.6″ dia.) may be drilled into the center of a 75-mm-diameter (3″ dia.) cylindrical crystal. The crystal surfaces not in contact with the PMT are covered with a thin aluminum coat for the reasons given previously in Section III.A.1. The sample to be counted is enclosed within a plastic tube and placed within this cavity, which is referred to as the detector well. This detector geometry was designed in the early years of solid scintillation analysis and used by the writer in the early 1960s. Manufacturers of this type of detector utilize a very thin aluminum coat (0.25−0.5 mm) to achieve the highest counting efficiency possible. The detector becomes vulnerable to damage if the sample tube is dropped into the well. Contamination of the well cavity or coating over long-term use is also a concern.

The well-type detector was envisaged with the objective of surrounding the sample with detector as much as possible. Optimum counting geometry is obtained when a sample is only a point source and that source is at the very center and completely surrounded with a radiation detector. Such an arrangement enables the detector to receive all of the possible radiation emissions, which may emanate in all directions (360 degrees) from a given sample. This is referred to as a 4π counting geometry. Obviously, such a counting geometry with a well-detector is not possible. The well-type detector was designed to try to approach, as closely as possible, a 4π counting geometry.

The well-detector remains a commonly used counting geometry. In this counting geometry there is a higher degree of variability in detection efficiency as a function of sample size or volume. For example, if the sample is a point source at the bottom of a sample tube and the tube is inserted into a well-detector, the sample is as far as possible from the well opening. As the sample size or volume is increased, the surface of the sample in the tube approaches the well entrance. Consequently, as the sample volume is increased, the detection efficiency will vary. It may increase and then diminish, as illustrated in Fig. 9.2 for ^{65}Zn; or the detection efficiency may remain stable for small volumes of sample and then drop as the sample volume is increased, as in the case of ^{22}Na (Fig. 9.2). The different response of the well-type detector to counting geometry is due to the

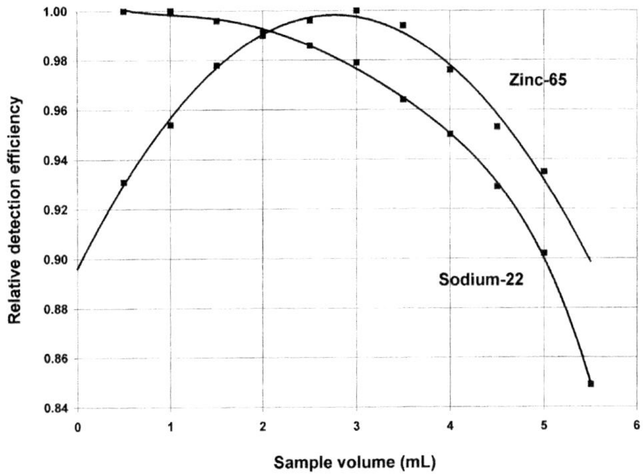

FIGURE 9.2 Changes in detection efficiency of a 6 × 6″ (150 × 150 mm) NaI(Tl) well-type detector for the analysis of ^{22}Na and ^{65}Zn as a function of sample volume. *From L'Annunziata, M.F. (1965) unpublished work of the writer for ^{22}Na and Garcia Burciaga (1976) and Garcia Burciaga et al. (1978) for ^{65}Zn.*

energies of the photon or charged-particle emissions and the volume of crystal detector exposed to radiation emissions from a sample. As the sample volume increases beyond a certain point, there is increasing probability of radiation escape without penetrating or interacting with the crystal detector. The two radionuclide samples, illustrated in Fig. 9.2 serve as examples. If we compare the gamma-ray emissions of ^{22}Na and ^{65}Zn, it is possible to explain the different sample geometry effects on detection efficiencies in the well-type detector. The decay schemes of the two radionuclides are given in Chapter 1, Volume 1 (see Eqs. (1.224)−(1.227) of Chapter 1, Volume 1). Sodium-22 emits a relatively high-energy gamma of 1.275 MeV with 100% intensity and 0.511-MeV annihilation gamma with 180% intensity (90% of the ^{22}Na radionuclides decay by positron emission). On the other hand, ^{65}Zn emits a lower energy gamma ray (1.115 MeV) with an intensity of 50% and a lower abundance of 0.511-MeV annihilation gamma (3.0% arising from the 1.5% of the ^{65}Zn nuclides decaying by positron emission). Most of the ^{65}Zn nuclides (98.5%) decay by electron capture (EC), which give rise to a large number of low-energy X-rays of the copper daughter nuclide. See Table A.1 of the Appendix for data on radiation types, energies, and radiation intensities of many of the radionuclides. The effect of sample volume on the detection efficiency of the well-type detector varies significantly with radionuclide radiation emissions and energies.

El-Khatib et al. (2016), Pommé (2012, 2009) and Pommé et al. (2009) have elaborated analytical models and software for the calculation of detection efficiencies of well-type detectors for photons, electrons, and positrons for a radioactive source at an arbitrary position relative to the detector and sample well. These calculations are used in the standardization of radionuclides. In the routine measurement of radionuclide activities, the optimum volumes for measuring the activities of radionuclides can be determined by simple measurements of count rates as a function of sample volume such as illustrated in Fig. 9.2. Ramadhan and Abdullah (2018) carried out thorough studies on the effects of gamma-ray photon energy and sample geometry on the gamma-ray full energy peak absolute efficiency. They used the following equation for the calculation of gamma-ray energy peak efficiencies (ε) of various radionuclides:

$$\varepsilon = \frac{N}{A \times t \times f_\gamma} \times 100\ \% \tag{9.10}$$

where N is the net peak area for each photopeak (see Fig. 9.3), A is the radionuclide activity and f_γ is the probability of a gamma-ray line for a particular radionuclide.

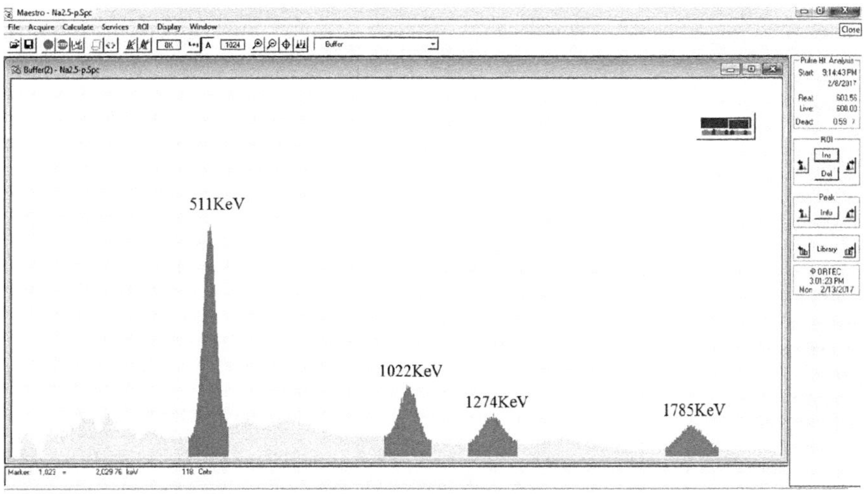

FIGURE 9.3 Gamma-ray spectrum of Na-22 obtained with a 3 × 3 inch (7.6 × 7.6 cm) NaI(Tl) well-type scintillation detector. *From Ramadhan and Abdullah (2018); reprinted with permission from the Korean Nuclear Society © 2018.*

Various photopeaks are found in the Na-22 gamma spectrum illustrated in Fig. 9.3. The photopeak at 511 keV arises from positron annihilation photons and that at 1022 keV arise from the coincident detection of two 511 keV photons. The photopeak at 1274 keV is that of the gamma-ray photon emission of Na-22, and the photopeak at 1785 Kev is a result of the detection in coincidence of a 1274 keV gamma-ray photon and 511 keV annihilation photon.

The distance of the source from the bottom of the sample well will affect the photopeak detection efficiency, and this effect is more prevalent for photopeaks of the lower gamma-ray photon energy lines, as depicted in Fig. 9.4. Gamma-ray photons of lower energy have a greater probability of absorption and energy loss in the scintillation detector; wheras, those of higher energy have a greater probability of traveling through the detector with less energy loss. When measured at a specific distance from the bottom of the sample well, the relationship of the detection efficiency of a point source as a function of gamma-ray energy line is illustrated in Fig. 9.5. The analysis of samples do not generally involve point sources; however, the Figure illustrates the higher probability of gamma-ray photon energy absorption in the scintillator for gamma-ray photons of lower energy.

For consistant results in the routine measurement of radionuclide activities, sample volumes should be maintained constant to assure constant sample counting geometry. For all radionuclides measured including the low-energy radionuclides, such as, ^{57}Co, with gamma energies of 0.122 MeV (87% intensity) and 0.014 MeV (9% intensity) and Fe daughter K X-rays (~55% intensity), and ^{125}I, with gamma energy of 0.035 MeV (7% intensity) and Te daughter K X-rays (138% intensity), sample volumes should be maintained constant and the optimum volume determined, which may be under 1 mL for the 15-mm detector well dimension in the 75-mm diameter cylindrical crystal cited previously or corrections can be made for changes in detection efficiency as a function of sample volume. Hunter et al. (1975) demonstrated significant

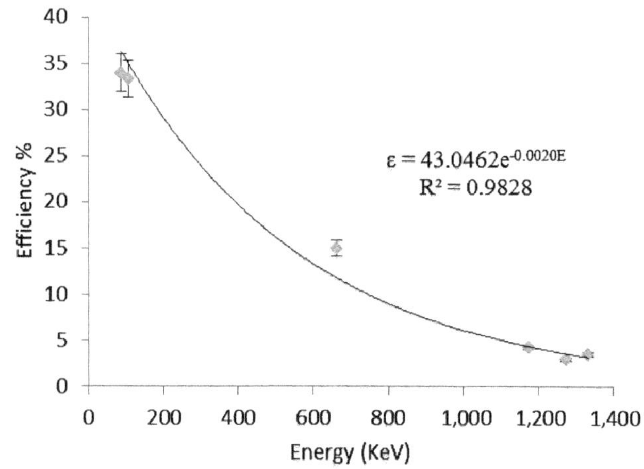

FIGURE 9.5 Detection efficiency of a 3 × 3 inch NaI(Tl) well-type scintillation detector as a function of gamma-ray energy with a source placed at a 2.5 cm height above the bottom of the sample well. *From Ramadhan and Abdullah (2018); reprinted with permission from the Korean Nuclear Society © 2018.*

decreases in detection efficiency of these radionuclides with the well-type detector with sample volumes over 1 mL. In many cases where detector dimensions will vary, the user should plot the detection efficiencies against sample volume for the particular radionuclide. Corrections for the effect of sample geometry on detection efficiency can then be made. Automatic gamma analyzers, which can analyze up to 1000 samples automatically, with well-type detectors are available commercially with up to 10 detectors to provide for the simultaneous analysis of 10 samples in tubes of variable dimension.

3. Through-hole detector

The through-hole solid scintillation detector consists of a cylindrical crystal with a hole machined all the way through the center of the crystal. Fig. 9.6 illustrates four crystal detectors of the through-hole type with a sample tube in the central cavity of each detector. Such a detector geometry

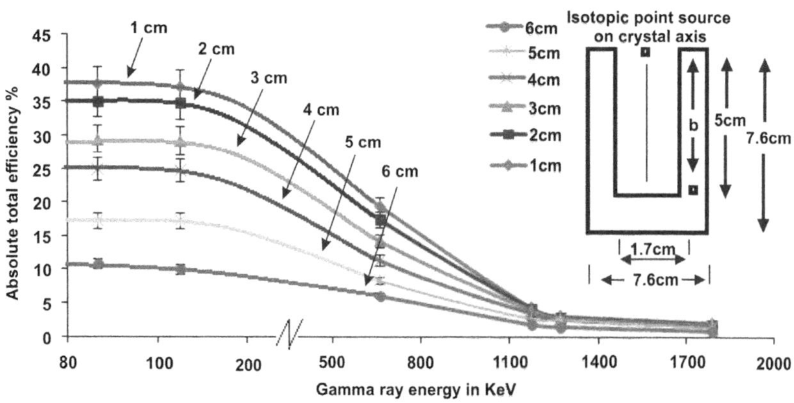

FIGURE 9.4 Experimental absolute efficiency measured with point gamma-ray sources for gamma-ray energy lines at 86.5, 105, 662, 1274 and 1785 KeV with a 3 × 3 inch NaI(Tl) well-type scintillation detector as a function of the source location (b) from the bottom of the well. Arrows were added by the writer to point to curves derived from sources at various distances from the well bottom. For the color version of this figure, the reader is referred to the electronic version of the book. *From Ramadhan and Abdullah (2018); reprinted with permission from the Korean Nuclear Society © 2018.*

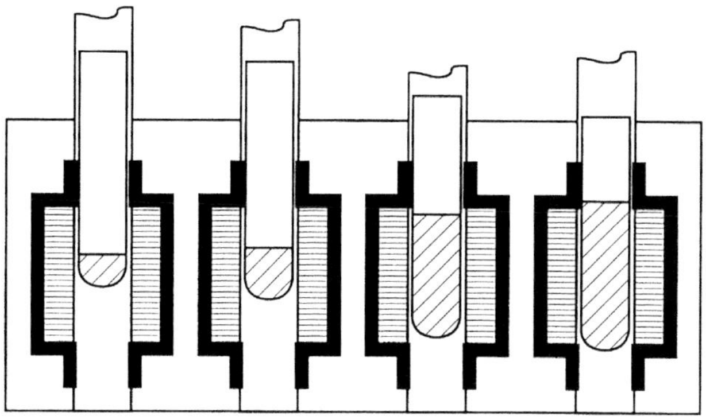

FIGURE 9.6 Four through-hole solid scintillation crystal detectors with sample tubes containing different volumes of sample. The samples are illustrated in four positions centered in the detectors automatically by instrument counting protocols. *©1998 PerkinElmer, Inc. Printed with permission.*

permits the automatic loading of samples from the bottom and automatic instrument adjustment of the sample elevation to position the sample at the very center of the crystal detector. The four samples illustrated in Fig. 9.6 are of different volume, but positioned automatically by the instrument to provide optimum detection efficiency and minimal variation of detection efficiency as a function of volume.

Automatic gamma analyzers with through-hole detectors have been produced commercially with up to 10 detectors to provide for the simultaneous analysis of 10 samples in tubes of variable dimension up to a maximum size of 14 × 100 mm for a 1.5-inch (3.8 cm) diameter crystal. Other models of automatic gamma analyzers can accommodate sample tubes up to 10 × 100 mm in size for crystal detectors 2 or 3 inches (5 cm or 7.6 cm) in diameter. The larger crystal detectors provide higher detection efficiencies, particularly desirable for the higher energy gamma-emitting radionuclides. These gamma analyzers have automatic sample changers, which can accommodate up to 1000 samples for hands-off analysis.

B. Photomultipliers

A photomultiplier is an electronic device used to measure low intensities of visible radiation by converting the photons, which strike the light-sensitive surface of the photomultiplier, into a photocurrent (e.g., flow of electrons). The magnitude of the electric pulse produced by the photomultiplier is a function of the intensity of the light that strikes the photosensitive surface of the photomultiplier. The most common type of photomultiplier used with commercial automatic laboratory gamma analysis instrumentation is the dynode photomultiplier, more commonly known as the PMT. Other types of photomultipliers may be used where resistance to magnetic fields are required or where miniature devices are needed for field applications, space exploration, and so on. Various types of photomultipliers are described in this section.

1. Dynode photomultiplier or PMT

The dynode photomultiplier or PMT is placed in direct contact with the solid scintillation crystal via a plastic light pipe and clear translucent grease to ensure optimum light transmission from the scintillation crystal to the PMT. The PMT is constructed of an outer glass surface with an inner photocathode (see Chapter 6, Volume 1, Fig. 6.4). The photocathode consists of a photosensitive substance (e.g., bialkali alloy, Sb−K−Cs) that produces photoelectrons when bombarded with photons of visible light. The photoelectrons produced at the photocathode are then accelerated toward a positively charged dynode in the PMT. The dynode is a metal plate given a positive potential, which emits secondary electrons upon impact with accelerated electrons. The acceleration of the photoelectrons and the impact of these on the dynode produce multiple secondary electrons. A series of several additional dynodes are contained in the PMT, and each subsequent dynode is given a higher positive potential to further accelerate and produce a larger number of secondary electrons. The final outcome is an avalanche of electrons at the last dynode. This final avalanche of secondary electrons is collected as a current pulse, which is of a magnitude that can be handled by the electronic circuitry and further amplified, counted, and analyzed for its pulse height.

From the description of the dynode photomultiplier or PMT provided in the previous paragraph, it is easy to see why this electronic component is also referred to as an electron multiplier phototube. Therefore, from the preceding account, we can state that the PMT has a twofold purpose: (1) to convert any given scintillation of visible light emission from the scintillation detector into a current pulse of secondary electrons and (2) to amplify the current pulse to a magnitude that can be handled by the counting and pulse height analysis circuitry associated with the radiation analyzer. It is important to note at this point that the number of photoelectrons produced at the PMT photocathode and the magnitude of the final current pulse collected after the series of dynode amplifications are a function of the light intensity, which is in turn a function of the radiation energy absorbed by the scintillation detector. As a consequence, the current gain from the PMT will depend on the initial scintillation photon intensity, and it will vary within the range 10^6 to 10^9. This range of current gain provides a source of variable pulse heights, which may be analyzed by other components of the radiation analyzer, such as the pulse height analyzer and multichannel analyzer (MCA) discussed further on in this chapter.

2. Hybrid photomultiplier tube

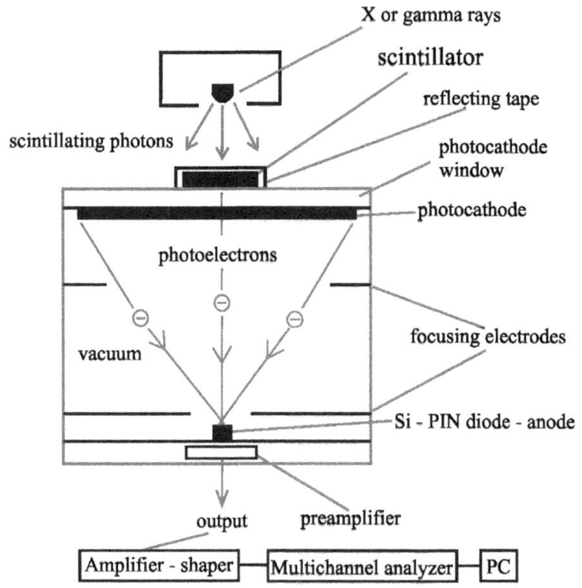

FIGURE 9.7 Scheme of the hybrid photomultiplier (HPMT); and an experimental setup for scintillator response measurement with the HPMT. *From Mares and D'Ambrosio (2007); reprinted with permission from Elsevier © 2007.*

The hybrid photomultiplier tube (HPMT) is constructed as a vacuum tube with a photocathode face as in the classical PMT. The HPMT differs from the classical PMT in that it does not contain dynodes for electron amplification, found in the classical PMT; rather, the HPMT uses high accelerating voltages (10–20 kV) and a high focusing voltage to accelerate the photoelectrons produced at the photocathode onto a small anode consisting of a Si p-i-n diode or avalanche photodiode (PD) only a few mm^2 in area, as illustrated in Fig. 9.7. The gain or electron amplification achieved by the HPMT is a consequence of both the high voltage applied to the tube for electron acceleration and the energy needed to create one electron–hole (e–h) pair in the p-i-n diode (3.6 eV). Thus, ~ 280 (e–h) pairs are produced for a 1 kV accelerating voltage; and a typical gain from a 12 kV accelerating voltage would be ~3000 due to some energy loss in the diode entrance window (Mares and D'Ambrosio, 2007; Burger et al., 2008). Amplifications of 5000 are reported with the silicon diode anode and an accelerating field of 20 kV; and a yet higher gain of 10^5 with an avalanche PD anode would allow simpler readout electronic circuits (Iijima, 2011; Korpar, 2011 and Renker, 2004, 2007).

The hybrid photomultiplier can be equipped with a pixelated silicon PD anode capable of high resolution imaging as required in medical imaging. Depending on the pixel size, single photon detection is possible (D'Ambrosio et al., 2006; Burger et al., 2008; Anton et al., 2009). A hybrid photomultiplier concept employing the pixelated semiconductor detector Timepix for the detection of photoelectrons and resultant imaging is described by Llopart et al. (2007), Anton et al. (2009), and Medipix Collaboration (2011). Timepix comprises 256 × 256 pixels with a pixel pitch of 55 μm electronically connected to a 300 μm thick silicon layer developed by the MEDIPIX, which is a worldwide collaboration of institutes producing and evaluating semiconductor pixel detectors for tracking applications in high energy physics including medical imaging. Canberra Industries has coupled pixelated PDs, illustrated in Fig. 9.8, which can be mounted in an HPMT. In the vacuum tube the photoelectrons produced at the photocathode are accelerated by the electric field whereby the electrons penetrate into the depleted region of the silicon diode (Burger et al., 2008). High resolution permitting single photon counting is possible subject to pixel size.

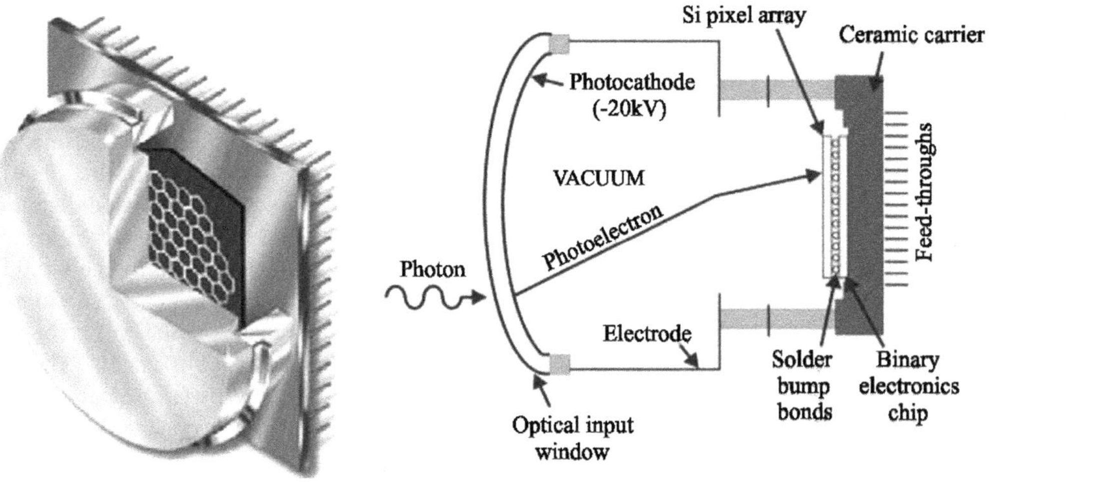

FIGURE 9.8 (Left) Artistic view of an array of hexagonal pixels of a pixelated PD and (Right) An HPMT with Si pixelated detector. *(Left) From Burger et al. (2008); reprinted with permission from Elsevier © 2008; (Right) From Fontanelli (2009); reprinted with permission from Elsevier © 2009.*

A pixelated hybrid photomultiplier is employed in the Ring-Imaging Cherenkov (RICH) detector at the CERN Large Hadron Collider (LHC). The hybrid photomultiplier, illustrated in Fig. 9.8 converts the Cherenkov photons to photoelectrons in a multialkali photocathode; and the photoelectrons are accelerated and focused by an electrostatic field of ~20 kV onto a reverse–biased silicon pixel detector, which is bump-bonded onto a 40 MHz read-out chip (Fontanelli, 2009). The kinetic energy of each accelerated photoelectron is dissipated in the silicon with the creation of ~5000 electron–hole pairs, estimated as 20 kV/3.6 eV/e–h pair with some energy lost in the diode entrance window. The RICH detectors, described by Fontanelli (2009), utilize 484 pixelated hybrid photomultipliers covering a total surface area of ~3.3 m^2; and the granularity of the pixels is 2.5×2.5 mm^2 resulting in approximately half a million pixels. RICH detectors are described in Chapter 6, Volume 2.

Advance designs of the HPMT with Si diode, also referred to as the hybrid Si photomultiplier (SiPM) and vacuum silicon photomultiplier tube (VSiPMT) have been designed and are described by Barbarino et al. (2018, 2013a,b), Barbato et al. (2017), Berneking et al. (2018), Suda (2014), Ko et al. (2013), Sanfilippo et al. (2013). The first 1-inch industrial VSiPMT is illustrated in Fig. 9.9. The VSiPMT, as described by Barbarino et al. (2018) is a silicon photomultimpler (SiPM) intended to replace the classical dynode PMT with unprecedented performance capabilities, including negligible power consumption, single-photon emission resolution, low voltage–based stabilization, and good timing performance. The 1-inch SiPM contains a pixelated PD with an array of 54 SiPMs 3×3 mm^2.

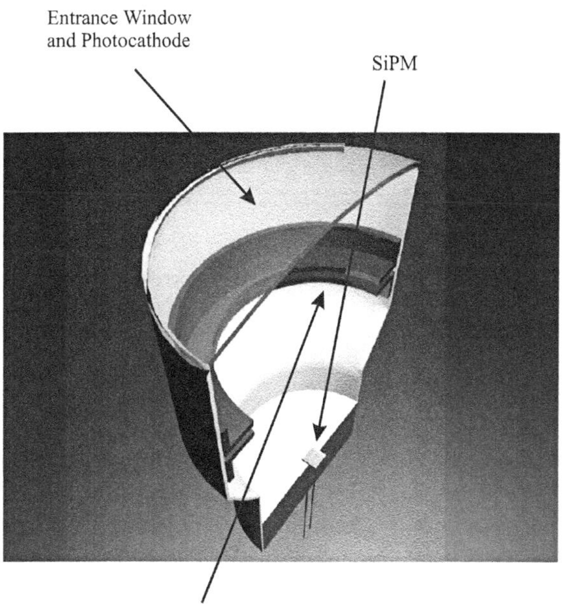

FIGURE 9.9 A schematic cutaway of the VSiPMT showing the interior composition of the device. On the top there is the light entrance window, then a photocathode for the conversion of photons to electrons (yellow). In the middle there is a focusing ring (dark gray) producing an electric field which accelerates and focuses the photoelectrons on the SiPM surface (yellow square at the bottom). On the bottom the SiPM acts in this configuration as a (photo)electron detector and (photo)electron current amplifier. Everything is assembled into a hermetically sealed container. Lettering was added by the writer. For the color version of this figure, the reader is referred to the electronic version of the book. *From Barbarino et al. (2018); reprinted with permission from Elsevier © 2018.*

3. Microchannel plate photomultiplier

Microchannel plate photomultipliers (MCP-PMs) can have advantages over the conventional multidynode PMTs described previously, particularly when the size of the photomultiplier is critical (Fraser, 2002; Lees and Fraser, 2002; Nappi, 2009; Wetstein, 2011), and when low susceptibility to magnetic fields (<2 T) is a requirement (Barnyakov et al., 2010; Baturin et al., 2006; Iijima, 2009; Kosev et al., 2010; Lehmann et al., 2008). A microchannel plate (MCP) consists of closely packed and fused micro lead glass capillary tubes (~10 μm in diameter and ~400 μm in length) with inner walls containing a material that can easily yield secondary electrons (Iijima, 2011). A photocathode is placed at the entrance of the fused capillary tubes and an anode collector at the exit to produce an MCP-PM. Certain MCPs are manufactured as radioactivity detectors and imagers (Lapington et al., 2009; Lees and Fraser, 2002; Lees and Hales, 2001; Lees et al., 1997; Price and Fraser, 2001). These are referred to as MCP detectors, which are manufactured to detect directly nuclear radiation (e.g., beta particles and X-rays) when the entrance to the MCP is made of a nickel-based electrode, which may be coated also with an alkali halide (e.g., CsI) capable of producing electrons directly from the impinging nuclear radiation.

Light photons emitted from a scintillation crystal detector are first absorbed by the photocathode, yielding photoelectrons in the process. The photoelectrons are then accelerated across the ends of MCP capillary tubes by a voltage potential applied across the channels. As the photoelectrons travel through the microchannels, secondary electrons are produced via impact or collisions with the inner walls of the channels. These secondary electrons, in turn, produce other secondary electrons through further collisions with the capillary walls. The result is a cascade or avalanche of electrons. An illustration of the production of secondary electrons and consequent electron amplification in a microchannel is illustrated in Fig. 9.10. For clarity, only a few secondary electrons are illustrated; the avalanche of electrons or output charge cloud produced at the end collector plate is not shown. Various MCP-PM designs can provide electron multiplications in the range

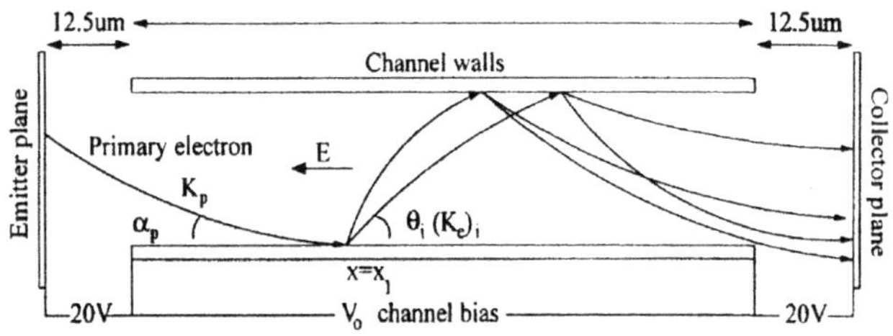

FIGURE 9.10 Signal amplification in a microchannel. K_p and α_p are the energy and grazing angle for the incident primary electron initiating the avalanche. $\theta_i(i = 1, 2\ldots$ in the schematic shown here) and $(K_e)_i$ are the emission angles and energies of the emitted secondaries. *From Price and Fraser (2001); reprinted with permission from Elsevier © 2001.*

of 10^5-10^6 (Pfaffinger et al., 2019; Hirose et al., 2015, 2014; Matsuoka, 2014; Kim et al., 2012a,b; Iijima, 2009, 2011; Kosev et al., 2010; Nappi, 2009); and gains of almost 10^7 have been reached with MCPs of 6 μm pore size (Lehman et al., 2008). MCP-PMs are very compact in size relative to a classical PMT. The MCP-PMs vary in dimensions with diameters of 25–100 mm, channel pore (channel) sizes of 3–25 μm, and are 0.5–1.5 mm thick (Korpar et al., 2008; Kosev et al., 2010; Lapington et al., 2009; Lees and Fraser, 2002; Lees and Hales, 2001; Lehmann et al., 2008; Leutz, 1995; Nappi, 2009).

Dual MCP-PMs are constructed to augment electron amplification; and such an arrangement is illustrated in Fig. 9.11. An incident photon of light is converted to a photoelectron at the photocathode via the photoelectric effect. The photoelectron is accelerated across a potential gap toward a dual set of MCPs. Approximately a 1 kV potential is applied across each plate causing a photoelectron entering a pore to accelerate and strike the pore walls and thus create an avalanche of secondary electrons (Wetstein, 2011). The amplified pulse of secondary electrons exits from a pore of the MCPs; and the electrons are collected on the anode and the signal digitized. Time resolutions of the dual microplate photomultiplier, reported by Brook et al. (2019), Pfaffinger et al. (2019), Harnew et al. (2016) and Wetstein (2011), are ~ 10 ps, which are under development for application in time-of-flight RICH detectors described in Chapter 6, Volume 2. The relative dimensions of the MCP PMT with 4 × 4 anodes yielding excellent position resolution and imaging capabilities to the photomultiplier is illustrated in Fig. 9.12; and an example of its application in the detection of Cherenkov photons in a time-of-propagation (TOP) counter for particle identification (PID) is illustrated in Fig. 9.12. TOP counters are

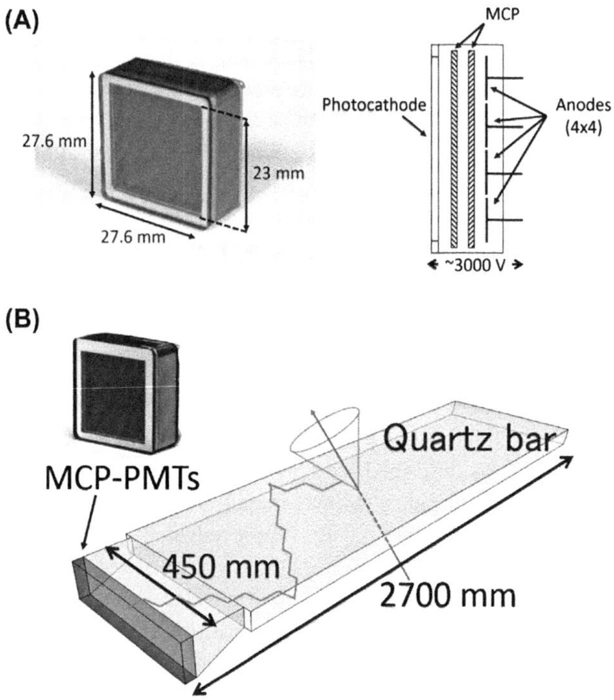

FIGURE 9.12 (A) Photograph of the MCP-PMT with dimensions and its cross-section view.; and (B) schematic drawing of a TOP counter illustrating the path of an incoming particle, the production of a Cherenkov photon cone, and the propagation of the Cherenkov photon along the quartz bar and it final detection by an MCP-PMT. *(A) From Hirose et al. (2015); reprinted with permission from Elsevier © 2015; (B) Hirose (2014); reprinted with permission from Elsevier © 2014.*

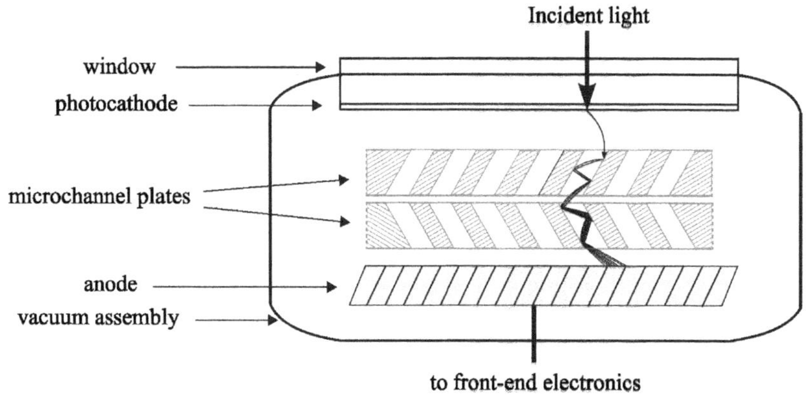

FIGURE 9.11 The structure of a generic MCP-PM tube. *From Wetstein (2011); reprinted with permission from Elsevier © 2011.*

described in detail in the chapter on Cherenkov Counting (Chapter 6, Volume 2).

The MCP-PM offers certain advantages over the classical PMT; among these are the following: (1) small size (0.5−1.5 mm thick); (2) low background noise; (3) capable of detecting single photons with pore diameters < 10 μm; (4) operational in magnetic field up to 2 T with appropriate orientation; (5) improved timing resolutions (~10−20 ps) suitable for time-of-flight RICH imaging; and (6) high quantum efficiency (QE) with use of appropriate photocathode (e.g., GaAsP) of ~40% at 500 nm and extending up to 700 nm (Brook et al., 2019; Harnew et al., 2016; Hirose et al., 2015, 2014; Wetstein, 2011; Iijima, 2009; Lapington et al., 2009; Nappi, 2009). The principal disadvantages of MCP-PMs compared with conventional PMTs are (1) their relatively high cost of fabrication, (2) lower gains or electron magnifications, (3) magnetic fields (>2 T) prevent electrons from hitting the MCP walls and, therefore, stop electron multiplication, (4) they exhibit a lower photon detection efficiency, and (5) they have a reduced lifetime (Nappi, 2009). The feasibility of MCP-PMs for the coincidence detection of 0.511 keV annihilation gamma rays in PET has been tested by Kim et al. (2012a,b, 2010a). Their initial findings resulted in an energy resolution of ~11% at 511 keV, a coincidence timing resolution of ~323 ps, and a coincidence detection efficiency of ~40%.

MCP-PMs led to the development of MCP detectors. MCP detectors were developed originally for their use in photon counting in astronomical X-ray telescopes (Weisskopf et al., 1995). MCP-PMs are applied as imagers (e.g., beta autoradiography) for commonly used radioisotopes in the biological sciences and other radiation sources, including neutron, gamma, and particle beam imaging (Tremsin et al., 2018; Wiggins et al., 2017, 2018; Ackermann et al., 2016; Wang et al., 2015a,b; Cazalas et al., 2014; Lapington, 2012). For radionuclide imaging, these detectors contain a nickel-based electrode at the face or entrance to the microchannel, which may be also coated with an alkali halide (e.g., CsI) as a photocathode. Beta particles originating from a radionuclide (e.g., ^3H, ^{14}C, ^{35}S, ^{32}P) or gamma- or X-ray radiation from certain radionuclides (e.g., ^{125}I) are capable of producing primary electrons upon colliding with the microchannel photocathode. Low-energy beta-emitting radionuclides (e.g., ^3H) are measured by the MCP in a windowless fashion where the beta-particles (electrons) interact directly with the MCP to produce the avalanche of electrons along the microchannel (Lees et al., 2005). The primary electrons collide with the walls of the microchannels resulting in an electron multiplication or avalanche as described in the previous paragraph for the MCP-PM. MCPs of this type may be only 1.5 mm thick with channel diameters ranging from 6 to 12.5 μm. The microchannels are bundled side by side to form a relatively large area detector with dimensions of 100 × 100 mm (Lees and Hales, 2001). This forms a radioactivity imaging device, as each microchannel acts as an individual detector. Spatial resolutions of ~70 μm full width half maximum are reported (Lees et al., 1997, 2005) for the large area 100 × 100 mm detector. The detectors have proven to be useful devices for the imaging of radionuclides in the biological sciences (Lees and Fraser, 2002; Lees et al., 1997, 1998, 1999, 2005). MCP detectors have also been employed as neutron imaging devices when the MCPs are doped with a suitable converter, which have high interaction cross-sections for low-energy neutrons, such as ^{10}B or ^{157}Gd (Tremsin et al., 2018, 2008; 2005; Wiggins et al., 2018; Wang et al., 2015a,b; Cazalas et al., 2014; Siegmund et al., 2007).

4. Channel photomultiplier

The channel photomultiplier (CPM) is a light detector, which contains a photocathode to convert light photons to electrons, and an electron multiplier consisting of a single narrow and hollow semiconductive curved channel through which the photoelectrons are accelerated. A diagram of the CPM is illustrated in Fig. 9.13. The CPM is very distinct from the MCP-PM, as the CPM consists only of a single curved channel rather than a more complex sheet of numerous and straight microchannels. Thus, the CPM, as an individual unit, cannot provide spatial resolution of radiation sources. Light photons produce photoelectrons at the photocathode. The photoelectrons are accelerated under vacuum in an electric field along a narrow curved semiconductive channel. Collisions of the electrons on the walls of the channel create secondary electrons, which terminate in an avalanche of electrons collected at the anode with a gain exceeding 10^8 (Ivan et al., 2008). The tortuous path of the curved channel increases the number of electron collisions with the glass wall of the channel and the consequent multiplication effect of the electron avalanche.

The CPM is produced by several commercial suppliers; and some specifications are the following: bialkali photocathode, cerium-activated yttrium aluminum perovskite [YAP(Ce)] scintillator faceplate, 10−19 mm diameter faceplate, 300−650 nm spectral sensitivity, 22.2% QE at 410 nm, and a single photoelectron gain of 3.3×10^8 at

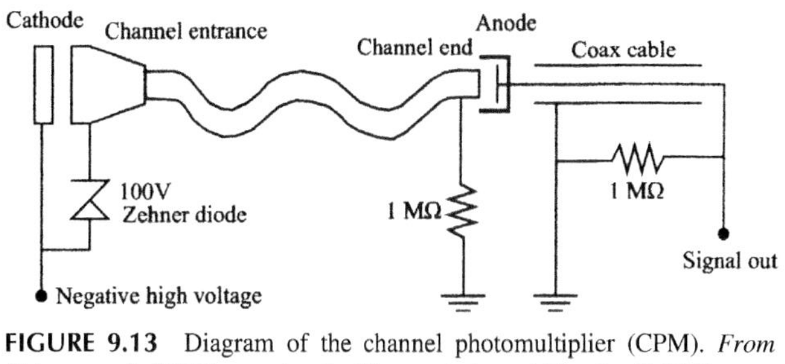

FIGURE 9.13 Diagram of the channel photomultiplier (CPM). *From Grimm et al. (2003); reprinted with permission of Elsevier © 2003.*

3.0 kV (Grimm et al., 2003; Ivan et al., 2010). The YAP(Ce) faceplate avoids considerable light loss due to the large difference in index of refraction (n of YAP = 1.95) and a typical borosilicate glass ($n = 1.49$) or UV-glass window (Grimm et al., 2003). CPM tubes were successfully utilized by Ivan et al. (2008, 2010) for the standardization of radionuclides in a TDCR liquid scintillation counter containing three and six photomultipliers in lieu of the classical dynode PMTs. As described by Ivan et al. (2008) the CPM operates in a saturated mode, so a photoelectron peak is a signature of any number of photoelectrons created at the same time at the photocathode. Thus, unlike the classical dynode PMT, the pulse amplitude is not the function of the number of photons reaching the photocathode. This feature prohibits the use of the CPM for spectroscopy, which is a useful but unnecessary feature of scintillation counters (Ivan et al., 2008).

5. Semiconductor photomultipliers

There is much ongoing research into the development of photodetectors that are good substitutes for the relatively large albeit very efficient PMT in applications of radioactivity analysis particularly in situations that demand small size and low weight such as in the fields of nuclear medicine and space exploration, and resistance to magnetic fields often required in studies of high-energy physics. In particular, imaging technologies, such as PET and computed tomography (e.g., SPECT), are advancing rapidly placing a high demand for efficient and small photomultiplier detectors, that can provide high resolution images. Research is underway at improving the performance of semiconductor photomultipliers (PDs) for this purpose. Although PDs have yet to serve as general replacements for the commonly used dynode PMTs in the general laboratory analysis of radionuclides, PDs are more widely utilized in many fields of radiation detection and measurement due to their very compact size, less bias to operate, resistance to magnetic fields, and excellent X- and γ-photon energy resolutions achievable with matching crystal scintillators (Kim et al., 2010b). For example, PDs with wafer dimensions are used as substitutes for PMTs in high-energy physics experiments where intense magnetic fields and space limitations preclude the use of PMTs (Lee et al., 2019; Legou et al., 2019; Whitaker et al., 2018; Li et al., 2017; Ohno et al., 2016; Cooper et al., 2012; Monteiro et al., 2012; Nelson et al., 2012; Brown and Cockerill, 2011; Nakamura et al., 2011; Hobson, 2009; Renker, 2009; Badalà et al., 2008; Beznosko et al., 2006; Fernandes et al., 2003). In the field of nuclear medicine imaging, in particular, PET and SPECT, there is increased demand to improve on image resolution from photon emissions, which only miniscule PDs can provide. Research in the improvement of PD performance in conjunction with scintillation detectors in nuclear medicine imaging of X- and gamma radiation with SPECT and PET is intense (Samson et al., 2018; Herzog and Lerche, 2016; Kang and Choi, 2016; Schaart et al., 2016; Slomka et al., 2014; Rato Mendes et al., 2013; Shimazoe et al., 2013; Hu et al., 2012; Schneider et al., 2012; Beltrame et al., 2011a,b; Kato et al., 2011; Moadel, 2011; Yoshino et al., 2011; Bolle et al., 2010; Nishikido et al., 2010; Bérard et al., 2009; Braem et al., 2009; Kataoka et al., 2009). In space exploration, the characteristics of low weight, small volume, and low power consumption generally exclude the use of the bulky PMTs. Compact X- and gamma-ray spectrometers consisting of scintillators resistant to intense cosmic radiation coupled to small wafer PDs are in use on planetary probes and orbital vehicles. Research on the discovery of new PDs and their applications in astrophysics and planetary exploration is ongoing (Ohno et al., 2016; Boll et al., 2015; Kasahara et al., 2012; Monteiro et al., 2012; Anderhub et al., 2011a,b,c; Leroy et al., 2010; Kataoka et al., 2010a,b; Braun et al., 2009; Koteswara Rao et al., 2009; Labanti et al., 2009; Otte et al., 2009).

a. p-i-n photodiodes

The classical dynode PMT can be substituted, at least for certain applications, by the solid-state light amplifier (Orphan et al., 1978). As described by Heath et al. (1979), the light amplifier consists of a conventional photocathode coupled to a silicon diode. The photoelectrons produced at the photocathode of the device are accelerated at a potential of 15 keV and focused on to the reverse-biased p-i-n silicon junction diode. Because about 3.6 eV is required to produce an electron-hole pair in silicon, the light amplifier produces a net single-stage gain of about 4000 (i.e., 15,000 eV/3.6 eV). A subsequent slight gain produces an amplification equivalent to that of a PMT. A photocathode need not be used in conjunction with the p-i-n semiconductor diode, because photons with energies not much in excess of 1 eV can produce electron-hole (e−h) pairs directly in a semiconductor diode with nearly 100% efficiency (Groom, 1983, 1984). These devices are called p-i-n diode detectors, when particles are detected directly without the aid of a scintillator crystal. For example, PIN-diode detector arrays were used by Laitinen et al. (2002) directly without a scintillator to determine the depth profile of ^{31}Si radiotracer (β^-, $E_{max} = 1.5$ MeV) in amorphous ceramic. Such detectors were used by Wauters et al. (2009) to accurately measure the beta-spectra of ^{60}Co with a 500 μm-thick Si PIN PD with a sensitive area of 81 mm^2 exposed directly to the β-particle emissions of a ^{60}Co source in an experimental arrangement that would prevent beta-particle backscatter and under vacuum (10^{-4} torr) to prevent absorption by air molecules. The β spectrum is illustrated in Fig. 9.14. The endpoint energy of 318.1 keV of the principle beta branch is shown in the Figure. Cobalt-60 decays by two beta

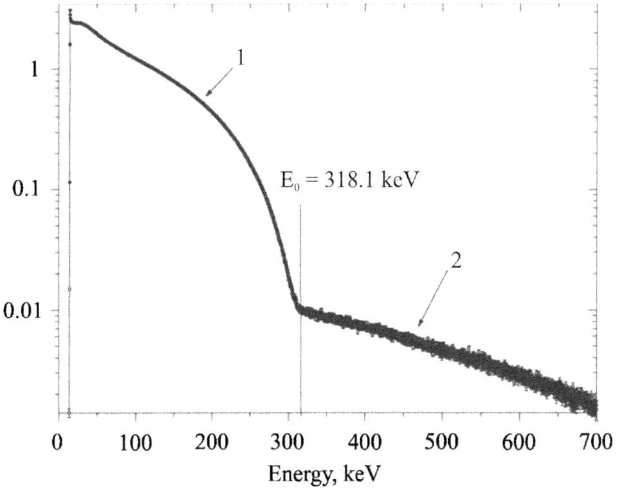

FIGURE 9.14 ^{60}Co β spectrum obtained with a "scattering free" source and the Si PIN PD. The regions below and above the endpoint energy of 318.1 keV are labeled as 1 and 2, respectively. The Si PIN detector energy resolution is 7 keV. *From Wauters et al. (2009); reprinted with permission from Elsevier © 2009.*

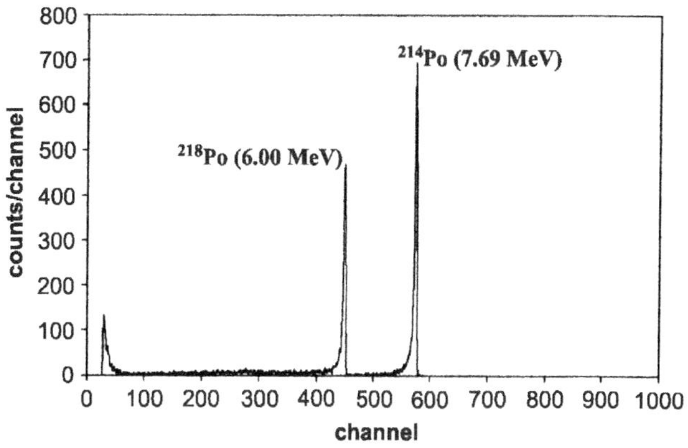

FIGURE 9.15 Pulse height spectra of alpha particles from radon daughter products with an inverse bias voltage of −9V applied to the PIN photodiode. *From Martín-Martín et al. (2006); reprinted with permission from Elsevier © 2006.*

branches of the following decay energies and intensities: 318.2 keV (99.9%) and 1491.4 keV (0.1%), as reported in the database of the National Nuclear Data Center, Brookhaven National Laboratory (NuDat 2.7, 2018). The higher decay energy is of insignificant intensity to interfere with the determination of the principle beta decay spectrum. The region of the spectrum beyond the 318.1 keV endpoint energy, illustrated in Fig. 9.14 as region 2, is attributed to Compton scattering of the 1173 and 1332 keV γ lines, which follow the ^{60}Co decay (Wauters et al., 2009).

The direct measurement of the energies of α-particles by the PIN PD can yield highly resolved energy lines, as the PIN PD absorbs directly the α-particle energy instead of measuring the photons of visible light from a crystal scintillator. Martín-Martín et al. (2006) devised a method of measuring ^{222}Rn in air with an 18 × 18 mm PIN PD. Air was passed at a given flow rate through a light-tight chamber containing the PIN PD; and the daughter nuclides of the ^{222}Rn were collected electrostatically on the surface of the PD. The close proximity of the daughter nuclides on the PD surface permitted their energy dissipation directly into the PIN PD. The highly resolved α spectra of the daughter nuclides ^{218}Po and ^{214}Po are illustrated in Fig. 9.15. Barlas et al. (2013) devised a hemispherical collection chamber to force the electrostatic collection of alpha particles to the surface of a windowless PIN PD resulting in highly resolved alpha spectra of the ^{218}Po and ^{214}Po radon progeny.

Although certain applications may call for the measurement of nuclear radiation directly with the PIN PD, most applications employ the efficient conversion of the radiation energy into photons of visible light by crystal scintillators, which are subsequently measured with the PIN PD coupled to the scintillator. When a PIN diode is coupled to a scintillation detector they are more often referred to as PIN PDs rather than PIN diode detectors. The photodiodes (PDs) are exposed directly to photons of visible light emitted by the crystal scintillation detector. The e-h pairs produced in the depletion layer are collected to provide what is referred to as a photocurrent signal. The principle of the PD is described by Lorenz (1983). In brief, the PDs, made of high-resistivity silicon (>1 kΩ cm), have a very thin p-type diffusion front layer and an n-type diffusion layer on the back of the wafer. A reverse bias voltage of a few tens of volts generates a depletion layer over nearly the entire depth of a 200-μm-thick diode.

A common PD similar to that used by Lorenz is the Hamamatsu p-i-n-PD or PIN PD. Electron-hole pairs are produced in these diodes when photons are absorbed by the diodes as illustrated in Fig. 9.16. The electron-hole pairs

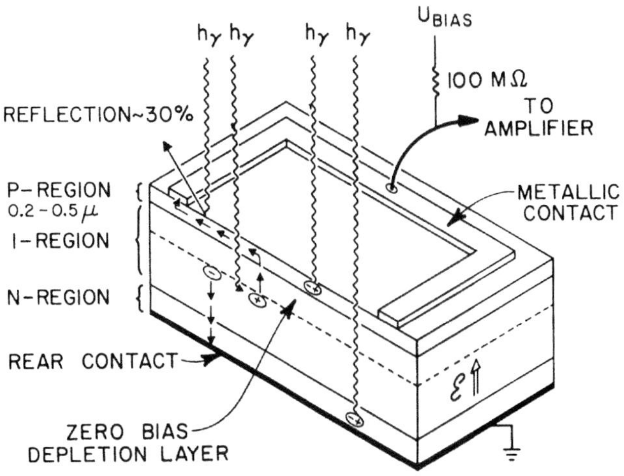

FIGURE 9.16 Cross-section of a p-i-n semiconductor photodiode illustrating photon interactions with different components of the diode including the migration of electron-hole pairs formed in the depletion layer of the diode. *From Lorenz (1983).*

formed in the depletion layer are separated by the electric field; and a photocurrent is generated when the electrons and holes migrate to the n- and p-regions, respectively. The e−h pairs formed in the p and n regions recombine and produce no charge effects (see Fig. 9.16). Surface reflection of photons, illustrated in Fig. 9.16, accounts for a reduced QE. Silicon PDs, at optimum photon wavelengths, exhibit QEs in the range of 70%−80% (Renker, 2004). In Si, 3.62 eV of energy on the average must be deposited in the PD to create an e−h pair. This would translate into ~2.76×10^5 e−h pairs (i.e., 10^6 eV/3.62 eV/e−h pair) produced for each MeV of energy deposited in the Si PD (Yang et al., 2011). Thus, a 59 keV γ photon from ^{241}Am, fully absorbed in the depletion layer of a Si PIN PD, would produce ~16.3×10^3 e−h pairs (Otte et al., 2006). The pulse height is dependent on the number of electron-hole pairs produced and, in turn, dependent on the intensity of photons emitted from the solid scintillation detector. See Chapter 4, Volume 1 of this book for additional reading on the mechanisms of interaction of radiation with semiconductor materials.

Meisner et al. (1995) report the application of two side-mounted PIN PDs as photomultipliers for a cylindrical (15 × 61 mm) CsI(Tl) scintillation detector. The detector was used successfully in a soil probe as a rugged and magnetic field-insensitive means of identifying gamma ray-emitting ^{137}Cs and ^{60}Co contaminants at soil depths of 20 m. Schotanus (1995) of SCIONIX Holland BV demonstrated the energy resolutions achieved with the coupling of a small 1-cm^3 CsI(Tl) crystal to a 10 × 10 mm^2 silicon PIN PD. The resolutions reported were 8.7% and 4.7% FWHM for 511- and 1275-keV photopeaks of ^{22}Na, respectively. For 662- and 1332-keV photopeaks of ^{137}Cs and ^{60}Co sources, energy resolutions obtained from the coupled miniature crystal and PD were 7.4% (662 keV) and 4.4% (1332 keV). Typical pulse height spectra obtained from a CsI(Tl) activated crystal with an entrance area of 25 × 25 mm^2 and 10.0 mm thickness coupled to a Si-PIN PD 300 μm thick and active area of 18 × 18 mm^2 is illustrated in Fig. 9.17. Energy resolutions of ~3.6% for ~5 MeV α particles and ~7.8% for 1.33 MeV γ rays are reported (Gupta et al., 2011). Comparable resolutions for γ-ray energy lines were measured by Kim et al. (2009, 2010b) with a CsI(Tl) scintillator and PIN PD. An energy resolution of 2.7% was obtained for the 662 keV γ ray from a ^{137}Cs source coupled to a LaBr$_3$ crystal scintillator Moszyński et al. (2009).

PMTs are sensitive to weak magnetic fields and can be shielded to only a few hundred gauss, whereas PDs are considered insensitive to magnetic fields. The insensitivity to magnetic fields has been placed at a theoretical limit of 15 T (Renker, 2004). PMTs also occupy much more space than PDs, and applications employing many crystal detectors and their individual PMTs are bulky and awkward, particularly when shielding is also required. PD wafers are small and can be placed in direct intimate contact with the crystal detector.

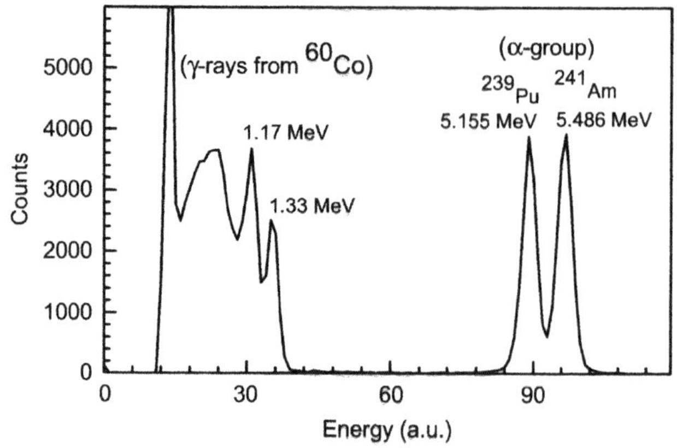

FIGURE 9.17 Pulse height spectra for α particles from ^{241}Am−^{239}Pu and γ rays from ^{60}Co obtained with a CsI(Tl) crystal scintillator detector coupled to a Si-PIN PD. The broad peak below 1.17 MeV is due to Compton scattering of γ rays of 1.17 and 1.33 MeV. *From Gupta et al. (2011); reprinted with permission from Elsevier © 2011.*

Silicon PDs can display high electronic noise due to capacitance and dark leakage current. As explained by Holl et al. (1995), the PD noise can be reduced by introducing long filter time constants unfit for high-count-rate applications. In addition, the PDs are highly sensitive to charged particles that may pass the depletion layer and produce false signals of magnitude similar to that of expected signals from the incident photon radiation. With a PIN PD readout coupled to a CsI(Tl) scintillator, Mikhailov and Panteleev (2001) observed a long-lasting "light noise" originating from the crystal similar to a crystal afterglow. This noise together with PD leakage current are contributing factors that define the energy resolution of the CsI(Tl)-PD gamma detector. Reduced temperature can improve energy resolution. They report a resolution of $<5.5\%$ FWHM for the 1275 keV γ line of ^{22}Na at 5°C.

An efficient handheld portable neutron detector consisting of a ^{6}LiI(Eu) crystal scintillator enriched to 96% ^{6}Li coupled with a light guide to two silicon PD detectors sandwiched on both sides of the crystal scintillator is described by Yang et al. (2011). Such an arrangement enabled a coincident double PD readout to permit the discrimination of gamma-ray background from neutron-capture events in the detector. It is very compact, as the scintillator is only 3 mm thick and 1 in. (2.54 cm) diag.; and the PDs measured 10 × 10 mm in active area with a 0.3 mm depletion layer thickness. The high enrichment of ^{6}Li makes the crystal essentially opaque to thermal neutrons with an intrinsic thermal neutron detection efficiency of 99.3%. The small thickness of the scintillator minimizes γ-energy dissipation while maximizing thermal neutron stopping power. The ^{6}Li(n,α)^{3}H reaction has a high Q-value of

4783 keV (QCalc, 2018), which attributes the light yield of a neutron capture signal equivalent to a 3–4 MeV gamma photon making discrimination against gamma necessary (Yang et al., 2011). Pulse-shape n/γ discrimination is possible; but this would require more cumbersome electronics for a handheld portable device. In lieu of pulse-shape discrimination, gamma-ray interference was reduced to natural background by employing a coincidence output from the two Si PD detectors coupled to both sides of the scintillator. The schematics of the coincidence electronics is illustrated in Fig. 9.67 much further on in this chapter. Neutron capture events produce significant coincidence signals in both PDs; whereas, with a ^{137}Cs gamma source, it was demonstrated by Yang et al. (2011) that the probability of an interfering γ-ray interaction producing a detectable pulse in both PDs was very low (1.69×10^{-6}%).

Another novel design of a sandwiched scintillator crystal between two Si PIN PDs was devised by Kah et al. (2011). A CsI(Tl) crystal was sandwiched between two silicon photostrip sensors whereby the p+ strip arrays of each sensor were oriented orthogonally to each other. Such a configuration provided the two-dimensional position information and depth of interaction of the radiation particle in the crystal scintillator by measuring signal ratios between the two photostrip sensors.

Lioliou and Barnett (2016) used GaAs p-i-n PDs to measure directly ^{55}Fe X-ray and ^{63}Ni beta spectra with p-i-n PDs. Highly resolved K_α and K_β X-ray spectra are reported by Singh et al. (2017) with a Si p-i-n diode detector. The X-rays were produced by bombardment of 25 keV electrons on Ti and Cu targets. A resolution of 165 eV at 5.9 keV is reported.

Pixelated Si PIN PDs coupled to inorganic crystal scintillators can provide gamma-camera imaging in nuclear medicine applications (Kang and Choi, 2016; Schaart et al., 2016; Huh et al., 2015; Michail et al., 2010, 2011; Park et al., 2009; Safavi-Naeini et al., 2008; Kim et al., 2006). The pixelated Si PIN PDs coupled to scintillator crystal segments provide imaging information due to the different origins of nuclear events from the emitted γ-ray photons within a patient administered an appropriate radiopharmaceutical or imaging provided by the varying intensities of X-ray photons traversing the patient.

b. Avalanche photodiodes

Advances continue to be made in the development of avalanche photodiodes (APDs) for scintillator readout capable of accepting high-count-rate applications for large multichannel systems used in PET, studies of HEP physics, and general radiation detection and measurement. Multidynode PMTs are the classical readouts and are commonly used for scintillation detectors in many fields; however, many advances have made the APD photomultiplier a preferred substitute to the PMT in many applications.

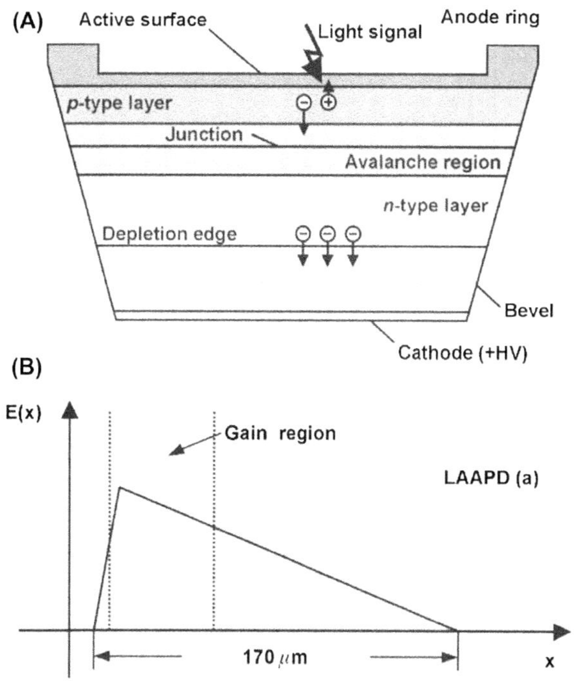

FIGURE 9.18 (A) Schematic cross-section of a Large Area Avalanche Photodiode (LAAPD) up to 16 mm diameter, and (B) electric field profile according to Advanced Photonics, Inc. *From Moszyński et al. (2003); reprinted with permission from Elsevier © 2003.*

As described by Holl et al. (1995) and Moszyński at al. (2002a), light photons produce electron-hole pairs as they penetrate the silicon of the APD, illustrated in Fig. 9.18. The depth of photon penetration is dependent on the wavelength and governed by their attenuation coefficients. The e–h pairs combine instantaneously in the very thin p+ front layer, while electrons in the depleted volume drift toward the n+ layer under the applied electric field potential. When the electrons enter the high-field region of the pn transition, impact ionization occurs. Impact ionization refers to collisions of electrons with other electrons, whereby more e–h pairs are produced and an avalanche multiplication of e–h pairs occurs. Only electrons are able to produce the e–h pairs at a certain field strength. Therefore, the initial e–h pairs produced by the incident photon, which penetrates only a shallow depth, result in an avalanche of electrons (pulse amplification). A relatively wide depletion layer extends into the n-type region resulting in a low dark current. The broad gain region of the electric field profile (Fig. 9.18) permits operation at high gain with low excess noise.

The advantages of the APD over the conventional PD described in the previous section are much lower noise for fast readout and lower interference (by a factor of 10–100) from charged particles that traverse the APD. Moszyński et al. (2002a) outline the advantages and limitations of the APD compared to the conventional dynode PMT. Among

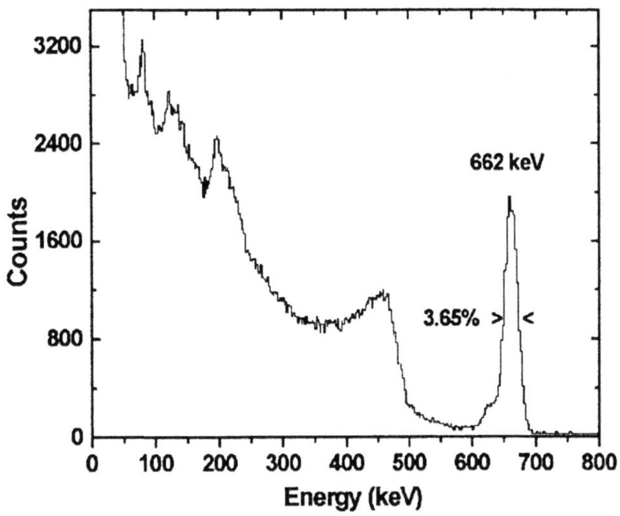

FIGURE 9.19 Pulse height spectrum of ^{137}Cs measured with $LaCl_3(Ce^{3+})$ (Ø 8 × 5 mm) coupled to an APD no. 70-73-510 from Advanced Photonix, Inc. The energy resolution (FWHM) at 662 keV is 3.65 ± 0.05%. *From Allier et al. (2002), reprinted with permission from Elsevier ©2002.*

the advantages are high QE, good spatial uniformity, insensitivity to magnetic fields, compactness, low power consumption, rugged, and flexible geometrical configuration. Some limitations to performance are lower gain, higher noise contribution, and susceptibility to radiation damage.

For the measurement of gamma radiation, the APD is used in conjunction with a solid scintillator whereby the photons from the scintillator are detected by the APD. The APD coupled to an appropriate solid scintillator can yield energy resolutions for X- and γ-ray photons superior to those provided by the conventional PMT (Grodzicka-Kobylka et al., 2017; Silva et al., 2015; Sasano et al., 2013; Cooper et al., 2012; Sato et al., 2006; Yatsu et al., 2006; Kataoka et al., 2005; Ikagawa et al., 2003, 2005; Kaputsa et al., 2003; Moszyński et al., 2002a, 2003; Allier et al., 2002; Balcerzyk et al., 2002). An energy resolution of 3.65% FWHM for the 662 keV gamma ray from ^{137}Cs measured with a small (Ø 8 × 5 mm) $LaCl_3(Ce^{3+})$ scintillator coupled to an APD serves as an example (Fig. 9.19) of the application of an APD in lieu of a PMT.

BGO scintillator has favorable properties for gamma-ray photon measurements, including high density and a high photon absorption coefficient at 0.511 MeV, which is the positron annihilation photon energy of interest in PET. Grodzicka-Kobylka et al. (2017) demonstrated the excellent gamma-ray spectra obtained with various scintillators, including BGO, LSO, and CsI(Tl) coupled to APD cells. Examples of gamma-ray photon spectra of gamma emissions from ^{157}Eu, ^{22}Na, and ^{137}Cs obtained with APDs coupled to a BGO scintillator crystal detector are illustrated in Fig. 9.20.

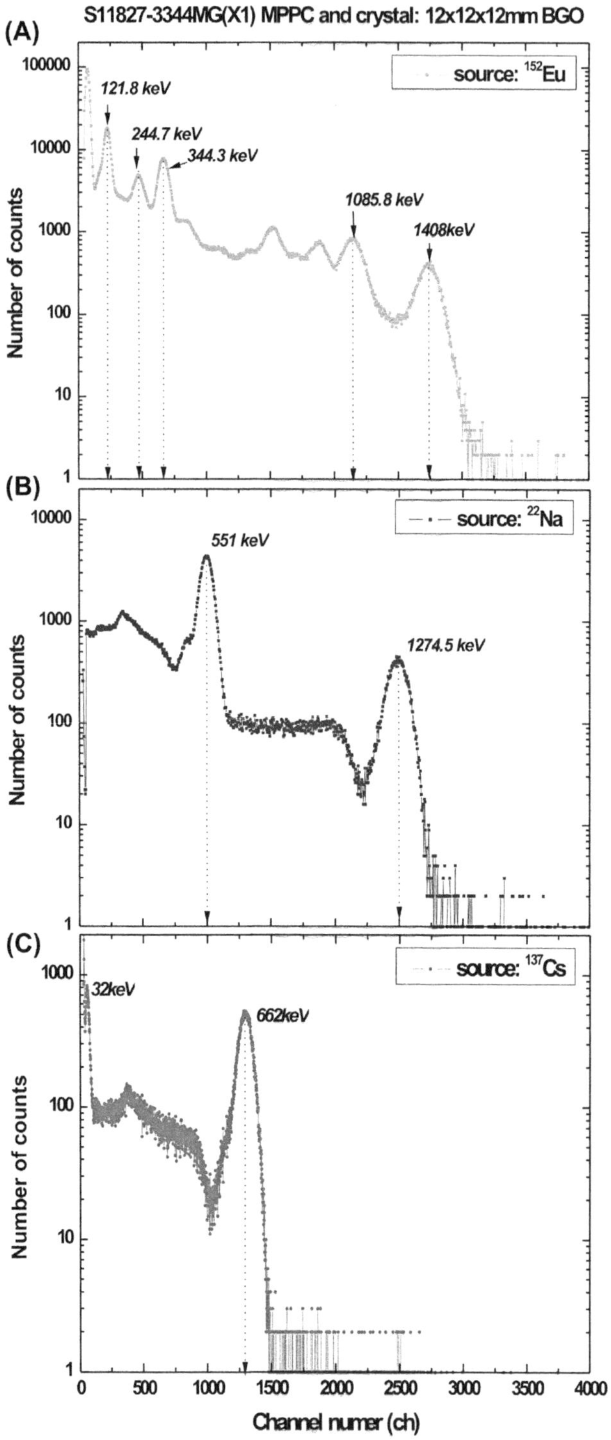

FIGURE 9.20 The energy spectra measured with a 12 × 12 × 12 mm BGO crystal coupled to 4 × 4 inch Hamamatsu SiPM (silicon photomultiplier) consisting of arrays of APDs. Radioactive sources: (A) ^{152}Eu, (B) ^{22}Na, (C), and ^{137}Cs. *From Grodzicka-Kobylka et al. (2017); reprinted with permission from Elsevier © 2017.*

The APD can be used directly as a radiation detector without the use of a scintillation detector such as in the measurement of β-emitting radionuclides (Kriss and Hamby, 2004) and spectra of low-energy X-rays (Yatsu et al., 2006; Kataoaka et al., 2005; Fernandes et al., 2003; Ikagawa et al., 2003; Kaputsa et al., 2003; Moszyński et al., 2000a,b; 2002a). Typical X-ray photon and β-particle spectra obtained from direct measurement with an APD are illustrated in Fig. 9.21. The β-particle spectra of ^{14}C, ^{36}Cl, $^{90}Sr/^{90}Y$, and ^{99}Tc were measured with a 16-mm diameter LAAPD by Kriss and Hamby (2004). They observed that the active thickness of the LAAPD is only $\sim$15 μm; and thus, most of the β-particles of the higher-energy emitters (e.g., ^{36}Cl and $^{90}Sr/^{90}Y$) pass through the active thickness, depositing some energy behind. These nuclides exhibit the same endpoint on the MCA). However, they report that the lower-energy radionuclides ^{14}C and ^{99}Tc retained their characteristic spectral shapes. Thus, the LAAPD can function as a general β-particle counter; however, its usefulness as a spectroscopy tool would be limited to very low energy beta emitters (Kriss and Hamby, 2004). The X-ray spectrum of ^{55}Fe, illustrated in Fig. 9.21 was measured by Kataoka et al. (2005) with a 3-mm diameter Hamamatsu APD at $-20°C$. The effect of temperature on resolution of the 5.8 keV X-ray line of ^{55}Fe at -20 degrees, 0 and 20 °C was studied by Yatsu et al. (2006). The resolution (% FWHM) was optimum at $-20°C$; and the % FWHM increased by only $\sim$1% at 0°C and by another $\sim$2% increase at 20 °C.

Due to the penetration power of higher-energy γ-rays, the use of the LAAPD for the measurement of most γ-emitting radionuclides is performed with the coupling of the APD and a scintillation crystal detector. For example, the experimental arrangements for measuring directly the soft 5.4 keV X-rays from ^{54}Mn with an LAAPD is illustrated in Fig. 9.22A; and the indirect measurement of the 835 keV γ-rays from ^{54}Mn by measurement of the visible-light photons from an inorganic crystal scintillator by the LAAPD is illustrated in Fig. 9.22B. The LAAPD detects the crystal scintillator light output at $\sim$ 520 nm produced by the γ-ray interactions (Fernandes et al., 2003); and the amplitude of the LAAPD output is proportional to the scintillation light intensity and the γ-ray photon energy absorbed by the crystal scintillator.

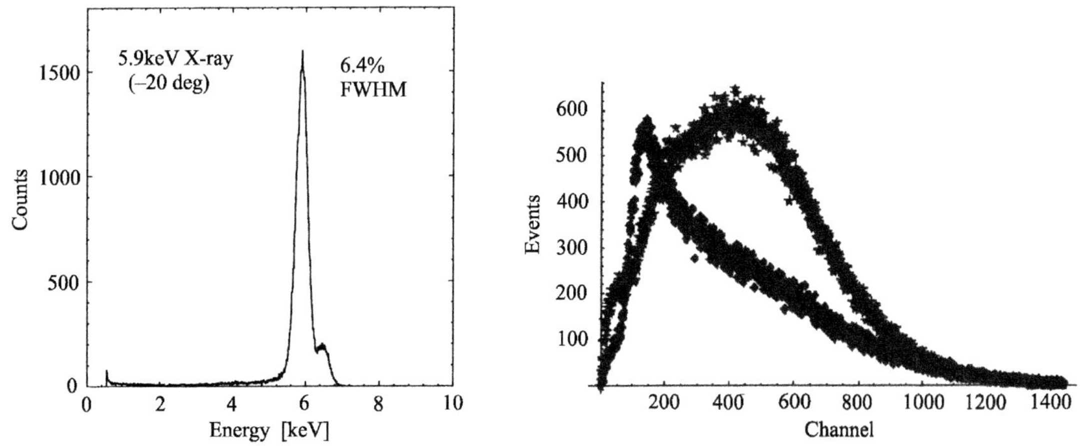

FIGURE 9.21 (Left) Energy spectrum of 5.9 keV X-rays at $-20°C$ from an ^{55}Fe source measured directly with a 3-mm diameter "reach through" type APD. The energy threshold is $\sim$0.5 keV. The K-shell peaks of the Mn K_α and K_β (small peak at the right) are marginally resolved in the line profile. The resolution (FWHM) of the 5.9 keV peak was ΔE $\sim$379 eV or 6.4%; (Right) Beta-particle pulse height spectra of ^{14}C (stars) and ^{99}Tc (diamonds) measured directly with a 16-mm diameter LAAPD. *(Left) From Kataoka et al. (2005); reprinted with permission from Elsevier © 2005. (Right) From Kriss and Hamby (2004); reprinted with permission from Elsevier © 2004.*

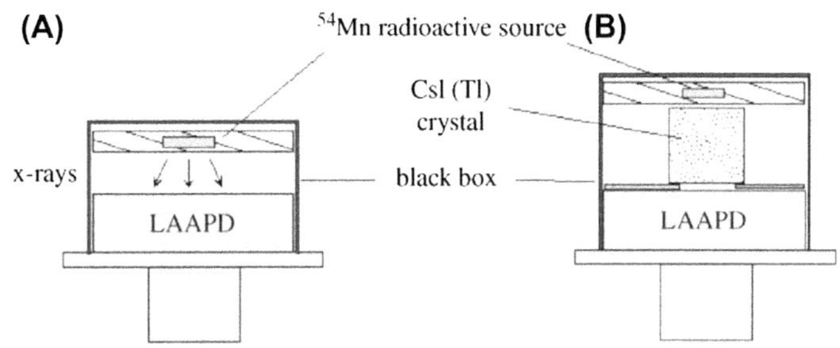

FIGURE 9.22 Source/detector arrangements for LAAPD detection of (A) X-rays and (B) visible-light photons. *From Fernandes et al. (2003); reprinted with permission from Elsevier © 2003.*

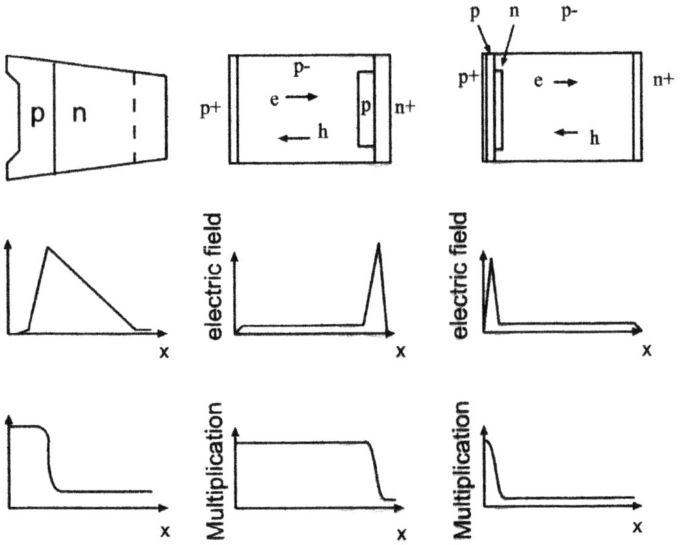

FIGURE 9.23 Internal structures of three different types of APDs (i) beveled-edge, (ii) reach through, and (iii) reverse type (from left to right) The electric field profiles and gain profiles in each type of APD are shown in the middle and bottom rows. *From Kataoka et al. (2005); reprinted with permission from Elsevier © 2005.*

The spectrometric measurement and analysis of X-ray and α-emitting radionuclides by the APD measurement of the liquid scintillation photon spectral intensities from liquid Xe (Fernandes et al., 2003; Ni et al., 2005) and from liquid scintillation fluor cocktails (Reboli et al., 2005) is also possible. This is discussed in Chapter 6, Volume 1.

APDs are commercially available as three types, namely (i) "beveled edge", (ii) "reach through", and (iii) "reverse type". These are illustrated in Fig. 9.23 and described by Kataoka et al. (2005) and Yatsu et al. (2006). The three types of APDs are the following:

i. The "beveled-edge" type This type of APD is a traditional p+n junction in which the n-type resistivity is selected to require a very high breakdown voltage, which is ~2 kV (Kataoka et al., 2005). The beveled-edge APDs have advantages of high gain and a low excess noise factor due to the broad multiplication region (Yatsu et al., 2006). Examples are described and tested by Moszyński et al. (2001, 2002c, 2003). An energy resolution of 4.9% for the 662 keV line from a ^{137}Cs source was observed with a CsI(Tl) crystal scintillator coupled to a beveled-edge LAAPD (Moszyński et al., 2001). Other examples reported by Moszyński et al. (2003) are 550 eV for the 5.9 keV X-ray energy line from an ^{55}Fe source is reported from the direct interaction of the X-rays with a beveled-edge LAAPD; and a 4.3% energy resolution for 662 keV gamma rays with the LAAPD coupled to a YAP:Ce crystal detector. The broad gain region enables the operation of this device at high gain with low excess noise and excellent gain uniformity (Moszyński et al., 2003).

ii. The "reach through" type The reach-through type of APD contains an avalanche region consisting of a p-type and n-type semiconductor layers; the depletion layer consists of a relatively wide drift region of low field (~2 V/μm) and a relatively narrow region (25–30 V/μm) for impact ionization (Kataoka et al., 2005). The depletion region is separated into two regions: (i) a thick drift region, comprising >90% of the APD volume, where photons and charged-particles are absorbed, and (ii) a thin multiplying region where electrons are multiplied (Yatsu et al., 2006). The reach-through APD offers an advantage of direct X-ray detection (i.e., no scintillator required) due to the think depletion layer (≥100 μm) in front of the amplification region; and this APD is sensitive to weak scintillation photons from γ-ray interactions with scintillators at a high QE of ~80% at ~500 nm of scintillation photon wavelength (Tanaka et al., 2007). Optimum energy resolutions (FWHM) at +20°C with the APD coupled to a $4 \times 4 \times 4\ mm^3$ CsI(Tl) scintillator reported by Tanaka et al. (2007) for 5.9 keV X-rays, 32 keV X-rays, and 662 keV γ-rays were 15%, 6.6%, and 7.6%, respectively; and X-ray and γ-ray signals can be discriminated using pulse shape discrimination (PSD) applied to the APD output. An X-ray and γ-ray imaging spectrometer over the energy range of 1 keV to 1 MeV with a $5 \times 5\ mm^2$ reach-through APD coupled to a CsI(Tl) scintillator is reported by Nakamori et al. (2010).

The thin-surface dead layer (~340 Å) and a thick depletion layer (140 μm) of the reach-through APD permit the measurement of low-energy electrons from 3 keV up to 100 keV with good linearity and excellent energy resolution of 4 keV at 100 keV electrons (i.e., 4%); and measurements may be made over a wide range of temperatures (Ogasawara et al., 2008, 2009). The measurement of heavier ions is also possible. Ogasawara et al. (2010) demonstrate the pulse height distribution of the reach-

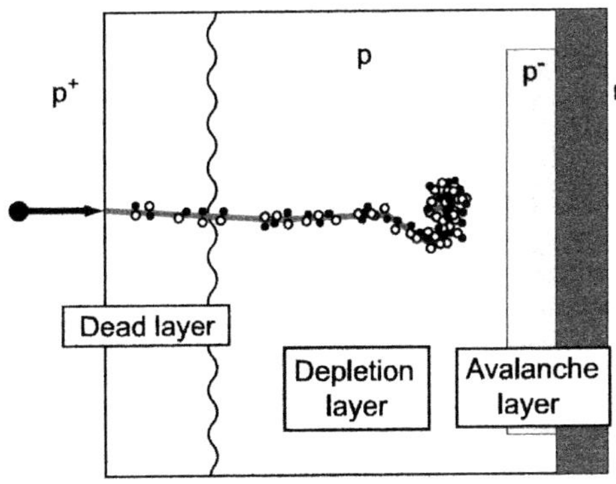

FIGURE 9.24 Schematic structure of a reach-through APD with an ion trajectory. Electrons of the e−h pairs drift in the direction of the avalanche layer due to the electric field in the depletion layer. Electron-hole (e−h) pairs are represented as dots and circles along the ion trajectory. *From Ogasawara et al. (2010); reprinted with permission from Elsevier © 2010.*

through APDs with active layers or depletion-layer thickness of ~150 μm show a significant peak for energies >2.5 keV for protons and 3.4 keV for helium ions with a noise level of ~300 electrons. Energy resolutions are 15% for 58 keV protons, 16% for 56 keV helium ions, 43% for 55 keV nitrogen ions, and 46% for 55 keV argon ions. A schematic illustration of the reach-through APD and ion trajectory is illustrated in Fig. 9.24. The p+ or dead layer of the APD was measured by Ogasawara et al. (2010) to be 340 Å thin. The dead layer is followed by the depletion layer (~150 μm thick), which provides an appreciable depth for absorption of high-energy particles. Electron-hole (e−h) pairs are formed throughout the particle or ion trajectory; and the electrons drift toward the avalanche layer where they undergo multiplication.

iii. The "reverse" type The reverse-type of APD is designed to be coupled to scintillators, as the multiplication region is very close to the detector surface (~5 μm) where absorption of photons of visible light from scintillators is effective. The peak emission wavelength of many inorganic crystal scintillators is at ~ 550 nm or less, and thus most photons of visible light are absorbed in the first 1−3 μm of the depletion layer and electron generation undergoes full multiplication (Kataoka et al., 2005). The reverse-type of APD exhibits very low noise. This is due to the multiplication region being just behind the detector surface and the dark current undergoes only hole multiplication and not amplified (Kataoka et al., 2005; Yatsu et al., 2006). Advantages of the reverse-type APD, summarized by Yoshino et al. (2011), are the efficient detection of weak scintillation light signals due to the narrow high field amplification region close to the front end, a high QE of ≥80% over the wavelength range of 500−830 nm, operation at a low bias voltage (~400 V for an avalanche gain of 100), and excellent noise characteristics.

Reverse-type APDs, coupled to inorganic scintillators, have been developed for high-resolution X-ray and γ-ray spectrometers in satellite space research (Arimoto et al., 2018; Saito et al., 2013; Ikagawa et al., 2005; Kataoka et al., 2005, 2010a). A typical energy resolution for the 662 keV ^{137}Cs γ-ray absorption peak with a 5 × 5 × 5 mm^3 BGO crystal scintillator is 8.3% (Sato et al., 2007). Pixelated reverse-type APDs are developed for high resolution X-ray imaging and PET (Kataoka et al., 2009, 2010b; Yoshino et al., 2011). Coupled to appropriate crystal scintillators they report energy resolutions of 7% FWHM for 662 keV γ-rays and a coincidence timing resolution of 4 ns (FWHM) for 511 keV annihilation photons, which is a vital measurement in PET. Because the APDs are not affected by high magnetic fields, these can be applied in detectors providing combined PET and MRI imaging.

c. Geiger-mode avalanche photodiode (GM-APD) — silicon photomultiplier (SiPM)

An APD is operated in the Geiger-mode when the applied bias voltage is raised above the avalanche breakdown point — the threshold voltage for Geiger discharge; and the avalanche process is stopped by momentarily reducing the electric field in the amplification region by means of an active circuit or a quench resistor (Korpar, 2011). This process introduces dead time when the PD recovers to its operating voltage. The Geiger-mode avalanche photodiode (GM-APD) was devised to obtain single-photon sensitivity; however, dark counts, dead time, and recovery time following a breakdown permitted PD areas of only ~100 μm (Renker, 2009). Such a device is very small; and PDs of larger area of this type are subdivided into many PD cells connected in parallel via an individual limiting resistor (Renker, 2009). Simultaneous signals from the numerous subdivided PD cells add up to a multiple of a single cell resulting in a gain of up to 10^7. These devices are reviewed by Huh et al. (2015), Pellion et al. (2015), Berra et al. (2013), Barbarino et al. (2012), Korpar (2011) and Renker (2009). An array of small GM-APDs, which operate in parallel, is illustrated in Fig. 9.25; and because such an array serves as a photomultiplier, the device is very often referred to as a SiPM, which is currently the name most commonly applied to this device. Other names, that have been applied to this photomultiplier are the multi-pixel photon counter or MPPC (Huh et al., 2015; Kato et al., 2011), the micropixel avalanche PD or MAPD (Sadygov et al., 2006, 2009a,b), the solid-state photomultiplier or SSPM (Chen et al., 2010), and the avalanche microchannel PD or AMPD (Britvitch et al., 2007).

The operation of the SiPM is described by Korpar (2011), which is basically as follows: A single incoming photon, such as that arising from an inorganic crystal

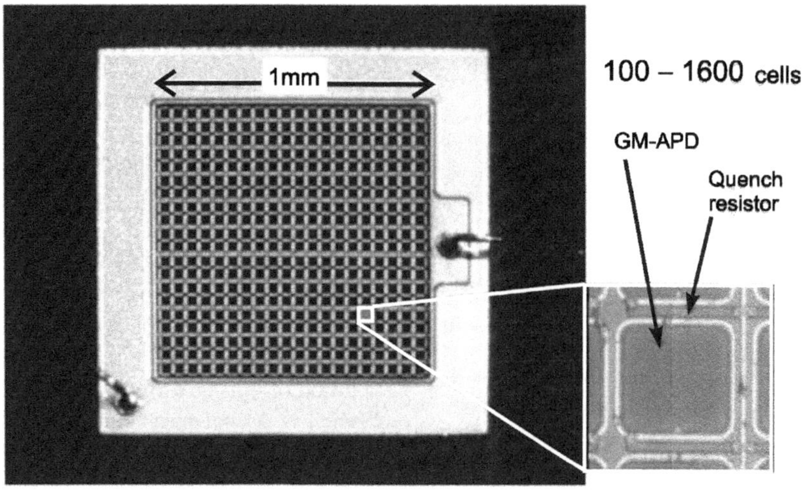

FIGURE 9.25 Silicon photomultiplier (SiPM) consisting of an array of Geiger-mode avalanche photodiode (GM-APD) cells with a magnified or close-up image of one of the GM-APD cells. *From Korpar (2011); reprinted with permission from Elsevier © 2011.*

scintillator or a photon of Cherenkov light, produces an electron-hole pair in one of the GM-APD microcells. This initiates the avalanche process; and the microcell discharges to the breakdown voltage, which stops the avalanche. The microcell then recharges to the bias potential. A signal is produced by this discharge/recharge process; and as all quenching resisters are connected to a common terminal, simultaneous signals from various microcells of the array add up to an integer multiple of the single cell signal.

The summing of signals from the various APDs of the SiPM array produces a final signal proportional to the photon intensity. The amount of X-ray or γ-ray energy deposited in an inorganic crystal scintillator will govern the photon intensity emitted by the scintillator and, in turn, govern the signal output from the SiPM. The process of this solid-state scintillation measurement of nuclear radiation is depicted in Fig. 9.26. The figure illustrates a low-energy (i.e., longer wavelength) γ-ray interacting with a crystal scintillator; and, on the right, a γ-ray of higher energy (i.e., shorter wavelength) depositing energy in another scintillator. The visible light emitted from the scintillator on the left is of lower intensity than that produced on the right. Both scintillators are coupled with a SiPM array depicted just below each scintillator. The scintillator emitting the brighter or more intense light emission on the right triggers a larger number of GM-APD cells of the SiPM array; and the corresponding amplitude of the pulse event registered by the SiPM array on the right is larger than that on the left.

The field of application, which is currently the most demanding of the SiPM, is nuclear medicine. Research is intensely focused on the production of the next generation of nuclear medicine X-ray and γ-ray imaging detectors, which utilize the SiPM in lieu of the classical dynode photomultiplier tube, i.e., PMT (Legou et al., 2019; Samson et al., 2018; Herzog and Lerche, 2016; Kang and Choi, 2016; Schaart et al., 2016; Huh et al., 2015; Silva et al., 2015; Rato Mendes et al., 2013; Shimazoe et al., 2013; Hu et al., 2012; Schneider et al., 2012; Beltrame et al., 2011a,b; Kato et al., 2011; Lee et al., 2011; Powolny et al., 2011; Roncali and Cherry, 2011; Verheyden et al., 2011; Zorzi et al., 2011; Bolle et al., 2010; Efthimiou et al., 2010;

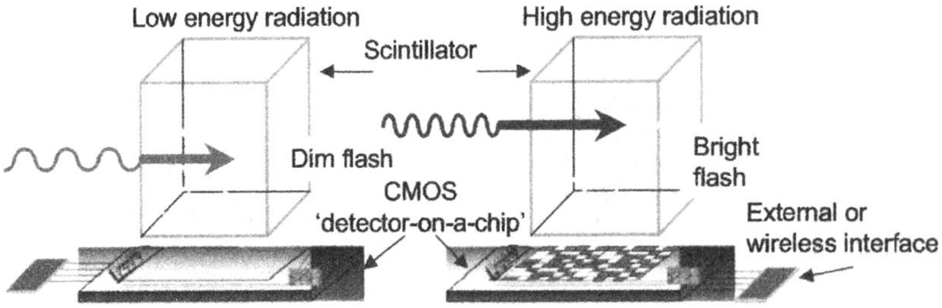

FIGURE 9.26 Concept of the solid-state scintillation detection with a silicon photomultiplier (SiPM). The crystal scintillator converts an X-ray, γ-ray, or neutron event into an optical pulse (i.e., photon emission of given intensity). The optical pulse is detected by the SiPM array of GM-APDs; and the number of GM-APDs that are triggered by the optical pulse is proportional to the amplitude of the optical pulse (i.e., proportional to the photon intensity emitted by the scintillator). CMOS is an acronym for Complementary Metal-Oxide-Semiconductor, which is a technology for constructing integrated circuits used in microprocessors. *From Christian et al. (2010); reprinted with permission from Elsevier © 2010.*

Pestotnik et al., 2010; Persson et al., 2009). Imaging techniques in which the SiPM may be utilized are X-ray CT, SPECT, and PET. The SiPM offers numerous advantages including high gain, low noise, low voltage, excellent energy resolution and timing properties, and insensitivity to magnetic fields. Excellent energy resolution for the 511 keV annihilation γ-ray detected in PET with the SiPM coupled to an appropriate crystal scintillator is reported by Legou et al. (2019), Huh et al. (2015), Silva et al. (2015), and Efthimiou et al. (2010), among others. The insensitivity of the SiPM to magnetic fields offers the added advantage of the development of the emerging hybrid PET/MRI imaging systems (Mannheim et al., 2018; Ferda, 2017; Herzog and Lerche, 2016; Rato Mendes et al., 2013; Roncali and Cherry, 2011; Efthimiou et al., 2010).

There is a temperature dependence when making energy calibrations of crystal scintillators with gamma emitters. A temperature-dependent light yield of the crystal scintillator results in a gain shift of the detector system. Fig. 9.27 illustrates the temperature dependence with a CsI(Tl) scintillator coupled to either a PMT or APD as photosensor readouts for the crystal scintillator. Silva et al. (2015) utilized pulse shape analysis to classify the pulse shape with respect to the internal temperature of the crystal detector to compensate for the resultant gain shift; and they report that gain shifts due to temperature variations can be corrected to a precision of better than 1% with either a PMT or APD as photosensor. Compensation for temperature dependence is particularly relevant when making coincident detection of the 0.511 MeV gamma photons emitted in opposite directions required in PET in nuclear imaging for medical diagnosis and treatment monitoring. Huh et al. (2015) developed a compensation method for the stabilization of the temperature-dependent performance of Geiger-mode APD sensor arrays coupled to scintillator crystals in PET. Fig. 9.28 illustrates the PET detector arrangement with associated electronics for temperature compensation. The positron annihilation gamma emissions are detected by two arrays of 4 $\times$ 4 channels of LYSO scintillator crystal detectors facing each other and a positron emitter. An array of Geiger-mode APDs are coupled to the crystal scintillator as MPPCs. The bias voltage of each APD is controlled by the DC power supply; and preamplifier gains are controlled by a digital potentiometer with microprocessor (MICOM). Huh et al. (2015) demonstrated that the photopeak variation can be stabilized by adjusting the preamplifier gain within a specified bias voltage range for the development of PET systems with Geiger-mode APD arrays.

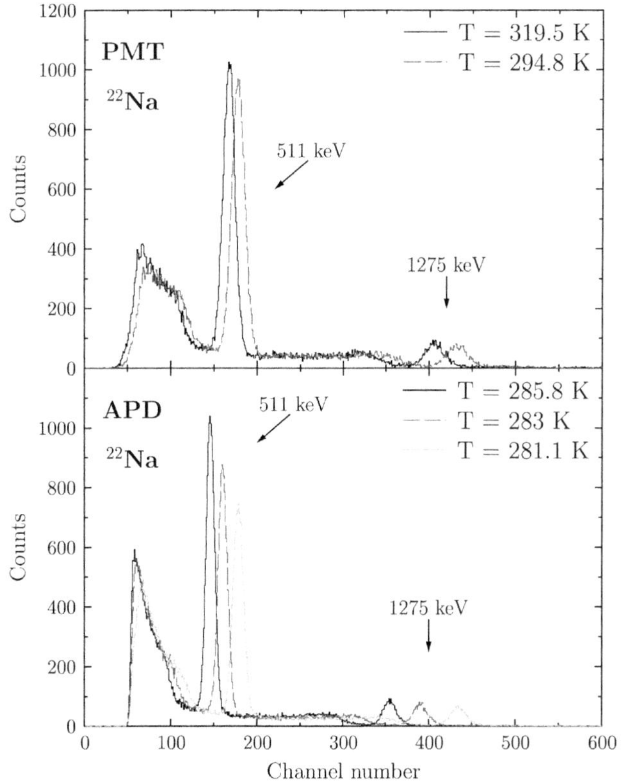

FIGURE 9.27 Upper frame: Energy spectra of ^{22}Na measured at different temperatures using as detector a CsI(Tl) crystal coupled to a PMT sensor. The spectra are not calibrated to illustrate the temperature gain dependence. Lower frame: Energy spectra measured at different temperatures using as detector a CsI(Tl) crystal coupled to an APD sensor. Shifts in the 0.511 MeV positron annihilation gamma peak and the 1.275 MeV gamma emission of ^{22}Na are clearly observed. *From Silva et al. (2015); reprinted with permission from Elsevier © 2015.*

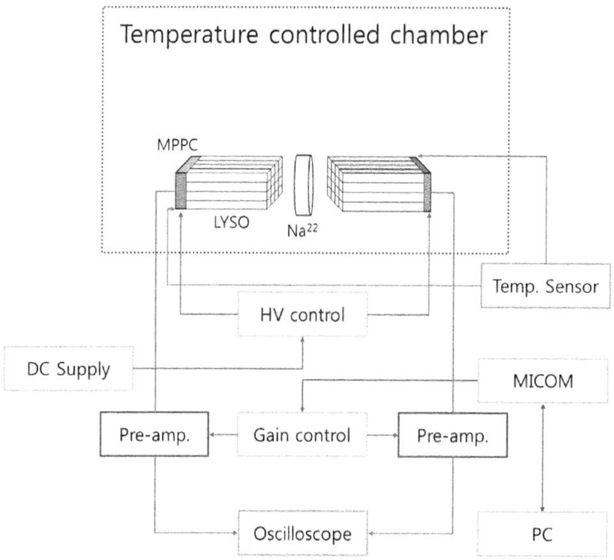

FIGURE 9.28 Experimental setup for the measurement of temperature-dependent performance of PET detectors using Geiger-mode avalanche photodiodes (APDs) as photosensors as multi-pixel photon counters (MPPCs) coupled to the LYSO crystal scintillator arrays facing the ^{22}Na positron emitting source in opposite directions for the detection of 0.511 MeV positron annihilation gamma-rays. *From Huh et al. (2015); reprinted with permission from Elsevier © 2015.*

The SiPM offers promise as a photodetector when coupled to appropriate crystal scintillators in radionuclide and radiation metrology. Research in the development of inorganic crystal scintillator detectors coupled to the SiPM are under development (Czyz et al., 2018; Morishita et al., 2018; Park et al., 2017; Ginzburg et al., 2015, 2011; Bellini et al., 2010; McClish et al., 2007).

In the field of astroparticle physics, energetic phenomena, and rare events are measured with detectors of very large volume, which may involve the measurement of scintillation or Cherenkov photons. Currently very large dynode PMTs are used for this purpose, as described in Chapter 6, Volume 1 and Chapter 6, Volume 2. The vacuum silicon photomultiplier tube (VSiPMT) shows promise as a potential substitute for the classical PMT, for a photodetector that can view and detect large volumes with high gain and single-photon counting capability (Barbarino et al., 2009, 2018, 2013a,b; Barbato et al., 2017; Berneking et al., 2018; Suda, 2014; Ko et al., 2013; Sanfilippo et al., 2013; Russo et al., 2010, 2009). The VSiPMT is a large hemispherical vacuum tube capable of viewing large active areas with a photocathode face. The photoelectrons emitted from the photocathode are then accelerated toward an SiPM (see Fig. 9.9 illustrated earlier in this chapter). The SiPM offers several advantages such as small size, low cost, high gain, absence of external voltage divider, and no passive power consumption (Russo et al., 2010).

The SiPM is capable of measuring the low-intensity levels of Cherenkov photons; and the employment of the SiPM in RICH detectors has been demonstrated (Balossino et al., 2017; Barnyakov et al., 2016; Contalbrigo et al., 2014; Tahirović et al., 2015; Gamal et al., 2011; Korpar et al., 2010, 2009). The insensitivity of the SiPM to magnetic fields is a key advantage, as subatomic particles identified by RICH detectors often utilize magnetic fields to determine particle momentum, as described in the chapter on Cherenkov Counting (Chapter 6, Volume 2).

C. Pulse height discriminators

Subsequent to amplification and pulse shaping, pulse height discriminators come into play. A pulse height discriminator is an electronic gate that permits pulses of a certain magnitude to continue on in the circuit to be registered and counted. In general, there are two discriminators, which are used to identify the pulses, which may continue on to be registered. These are the lower level (LL) and the upper level (UL) discriminators. The LL discriminator can be used to eliminate electronic noise from the photomultiplier. The settings of both the LL and UL discriminators can be used to define a counting region or counting window. The heights of the LL and UL discriminators may be input manually when setting up a computer-controlled counting protocol. According to the settings of the LL and UL discriminators, the instrument circuitry will allow only pulses of magnitude (volts) higher than the LL discriminator and lower than the UL discriminator to continue on to be registered or counted by the scalar. Although the amplifier output pulse heights are measured in volts, the LL and UL discriminator settings may be calibrated in other units, such as, kiloelectron volts (keV), which allows correlation between amplifier output pulse heights from the crystal scintillator detector and the actual gamma-ray energies absorbed by the detector.

D. Single-channel analyzer

The correlation between the gamma-ray energy dissipation within the crystal scintillation detector and pulse height amplitude output from the PMT or PD, preamplifier, and linear amplifier are used to measure the gamma-ray energies emitted by a given radionuclide or mixture of radionuclides in a sample. For this purpose a single-channel analyzer (SCA) may be employed.

An SCA contains two electronic pulse height discriminators described in the previous section of this chapter, namely the LL and UL discriminators. The output pulses from the amplifier, that have heights or amplitudes below the LL discriminator setting or above the UL discriminator setting, are rejected. Only pulses that have magnitudes higher than the LL discriminator setting and lower than the UL discriminator setting, referred to as the discriminator window, pass the discriminator circuit and go on to be registered by a scalar circuit. If we expose a solid scintillation detector to a sample containing gamma ray-emitting radionuclides, we may use the LL and UL discriminators to analyze the pulse heights produced and consequently the energies of the gamma ray photons emitted by the sample. For example, if we begin by arbitrarily setting the LL discriminator at its lowest possible pulse height, namely 0.0 keV, and the UL discriminator at 1.0 keV, the discriminator circuit will allow only pulses of magnitude between 0.0 and 1.0 keV equivalents to go on to be registered and counted by a scalar circuit. A time interval may be selected to give the instrument a specific time to register pulses passing the discriminator circuit between 0.0 and 1.0 keV. The pulses collected by the scalar circuit in the allotted time interval are summed. The total counts collected in the 0.0−1.0 keV counting region (window) are registered; these counts may also be reported as a count rate (e.g., counts per minute, CPM), if the total counts collected are divided by the time interval. The procedure as described will produce one data value, the total number of counts or count rate in the one counting region (channel) defined by the discriminator settings of 0.0−1.0. We may repeat this counting procedure after changing the LL and UL discriminator settings to 1.0 and 2.0 keV, respectively. This will provide a second value of counts or count rate

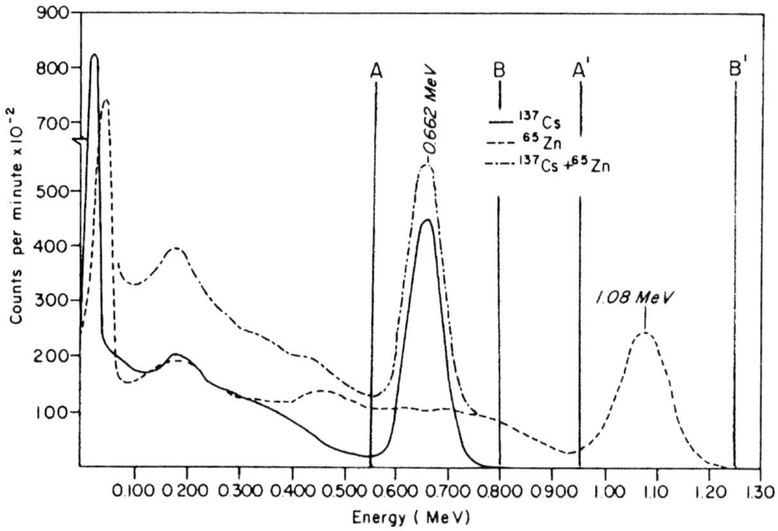

FIGURE 9.29 Gamma-ray spectra of ^{137}Cs and ^{65}Zn as separate radiation sources and as a mixture measured with a NaI(Tl) detector. The lines A and B represent LL and UL discriminator settings for a counting region which receives all the pulses arising from the photopeak of 0.662-MeV gamma rays from the ^{137}Cs sample in secular equilibrium with its daughter radionuclide ^{137m}Ba. Ihe lines A′ and B′ represent LL and UL discriminator settings for another counting region, which receives pulses from the photopeak of the 1.08-MeV gamma rays of ^{65}Zn. *From L'Annunziata, M. F. (1965), unpublished work of the writer.*

collected of pulses in the energy range 1.0–2.0 keV. The procedure may be continued as described by collecting counts in the regions 2.0–3.0, 3.0–4.0, 4.0–5.0, and so on up to a maximum, for example, 1999–2000 keV. The counts or count rate collected in each channel selected are then plotted against the window height or channel height, which is the LL discriminator setting in keV of each selected channel. This plot will produce a spectrum of gamma-ray photon energies absorbed by the scintillation crystal detector and consequently a spectrum of the gamma-ray energies emitted by a given sample. Fig. 9.29 provides an example of gamma-ray spectra plotted using an SCA according to the procedure described. The use of two discriminators in this fashion is referred to as single-channel gamma-ray analysis. The difference between LL and UL in keV is referred to as the window width or channel width, and the LL discriminator setting for a given channel setting is referred to as the window height or channel height. Manual variation of the channel height while maintaining the channel width constant permits scanning of the entire spectrum of pulse heights produced by a given gamma radiation source. The procedure is simple but tedious and time consuming for the measurement of γ-ray spectra. On the other hand, the SCA is a simple, relatively inexpensive, and a popular component used to analyze radionuclide activities by counting the scintillation pulse events within a given energy region defined by the LL and UL discriminator settings (Zhao et al., 2013; Zafrir et al., 2011; Kumara et al., 2010; Toyohara et al., 2006; Vrinda Devi et al., 2010; Cendre et al., 2005; Kraus et al., 2005; Dupasquier et al., 2004). For example, the SCA may be used to conveniently count all pulse events arising from the photopeak of a ^{137}Cs source in the counting window defined by LL and UL discriminator settings A and B (i.e., 550–800 keV) of Fig. 9.29 for the particular scintillation analyzer used in this example. A wide counting window defined by LL and UL discriminator settings between zero and 800 keV would encompass all events arising from β-particle, X-ray, and γ-ray interactions in the scintillator crystal from ^{137}Cs/^{137m}Ba decay. The LL discriminator is generally set somewhere above zero to discriminate against background noise. In such a case, the precise setting of the LL discriminator above zero, would be determined according to the background count rate. Background count rates due to photomultiplier noise occur at the lower end of the pulse height spectrum.

Scintillation analyzers may also be equipped with an MCA, which provides automatically the gamma-ray spectrum of a sample. The MCA is described subsequently.

E. Multichannel analyzer

With the SCA previously discussed, one can count at one time only the pulses received by a given channel (counting window or region) after selection of the channel position (window height) and channel width. Numerous channel positions corresponding to definite pulse height ranges must be tediously chosen in order to construct energy spectra similar to those illustrated in Fig. 9.29. For example, approximately 40 channel heights of 80 keV width were manually selected to produce the gamma-ray spectrum of ^{137}Cs/^{137m}Ba in Fig. 9.29. At each of these channel positions the count rate was determined for a 1-minute time interval. A minimum of 40 minutes was thus necessary to scan the ^{137}Cs gamma-ray spectrum, exclusive of the time involved in changing the channel heights. If a narrower channel width is selected (e.g., 20 keV rather than 80 keV), as many as 160 count rate determinations in the same number of counting channels would be required.

On the other hand, the MCA permits the simultaneous accumulation of pulses in as many as several thousands of energy levels (channels) in a semiconductor core memory. The stored spectra of pulses in the core memory may be viewed simultaneously in real time on the computer monitor of a solid scintillation analyzer and saved on disk or printed whenever desired.

The following is a description of the basic principle of operation of a MCA. Gamma rays from a sample dissipate energy in the crystal scintillator, e.g., NaI(Tl), CsI(Tl), or other scintillator detector. A wide spectrum of photon intensities will be emitted by the scintillator according to the amount of energy dissipated in the crystal by each γ-ray interaction in the crystal according to photoelectric and Compton scattering interactions among other mechanisms of gamma-ray energy dissipation. The PMT or PD will then convert the photon intensities to voltage pulses of various magnitudes as a function of the various photon intensities emitted by the scintillator. As previously described in this chapter, pulses of different magnitude arise from different energy quanta absorbed by the solid scintillation crystal detector. The pulse events of various magnitudes produced by the photomultiplier undergo amplification and pulse shaping before processing by the MCA, as depicted in the pulse amplification and detection circuit illustrated in Fig. 9.30. Multichannel analysis involves processing the pulses onto an analog-to-digital converter (ADC), where each pulse amplitude in volts is converted into an equivalent time interval by the production of an equivalent number of pulses from a periodic oscillator. An address register then counts the number of pulses produced by the periodic oscillator, which gives a quantified measure of the amplitude of each input signal. This quantification identifies the channel among many thousands within which the pulse is to be stored in the core memory. In most MCAs, counts in excess of 10^9 can be stored in each channel. The contents of the core memory are viewed on the computer monitor as a spectrum of pulse heights while the sample is being counted, as illustrated in Fig. 9.30; and, if desired, the spectrum is saved on disk or printed.

MCAs are available commercially with thousands of channels. Scintillation analyzers are equipped with a computer-controlled MCA, which can provide automated sample processing and spectral analysis over the gamma-ray energy range (e.g., 0–2000 keV). An example of gamma spectra produced by an automatic modern solid scintillation analyzer is illustrated in Fig. 9.31. The simultaneous accumulation and readout of pulses in many thousands of channels corresponding to pulses of defined amplitudes are very timesaving, and the hands-off construction of the gamma spectra is effortless (Neal et al., 2011; Quarati et al., 2011; Bourret-Courchesne et al., 2009, 2010; Limkitjaroenporn et al., 2010).

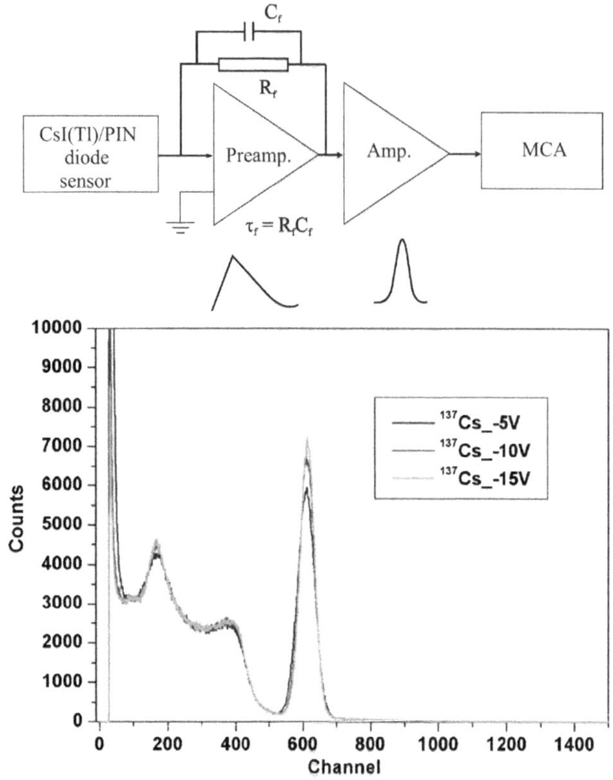

FIGURE 9.30 (Top) Pulse amplification and detection circuit for the multichannel analysis of pulse events produced by a CsI(Tl) scintillator coupled to a Si-PIN photodiode photomultiplier. Preamp and AMP are the preamplifier and shaping amplifier; and C_f and R_f are the feedback capacitor and resistor, respectively. (Bottom) Typical gamma spectra provided by the multichannel analyzer (MCA) at three bias voltages to the Si-PIN photodiode. *From Kim et al. (2009); reprinted with permission from Elsevier © 2009.*

Research is underway in the development of digital multichannel analyzers (DMCAs) based on a field programmable gate array (FPGA), as described by Garcia-Duran et al. (2019), Becker and Farsoni (2014), and Zeng (2012). As described by Zeng (2012), the signals from the scintillation detector and photo detector (PMT or APD) are processed by the FPGA, which utilizes the signals acquired by the analog-to-digital converter to perform the calculus, identify spectral peaks, determine the threshold, and plot the energy spectrum utilizing a personal computer for energy spectrum processing. As underscored by Garcia-Duran et al. (2019), the MCA embedded into an FPGA is a virtual instrument for the handling and visualization of the pulse height spectrum. An example of the output of the multichannel analysis obtained with a commercial ORTEC MCA and that from the MCA embedded into an FPGA is illustrated in Fig. 9.32. The two pulse height spectra illustrated in Fig. 9.32 are similar in terms of resolution of the 0.662 MeV photopeak and the appearance of the Compton shoulder; however, as underscored by Garcia-Duran, the embedded MCA offers the advantage of

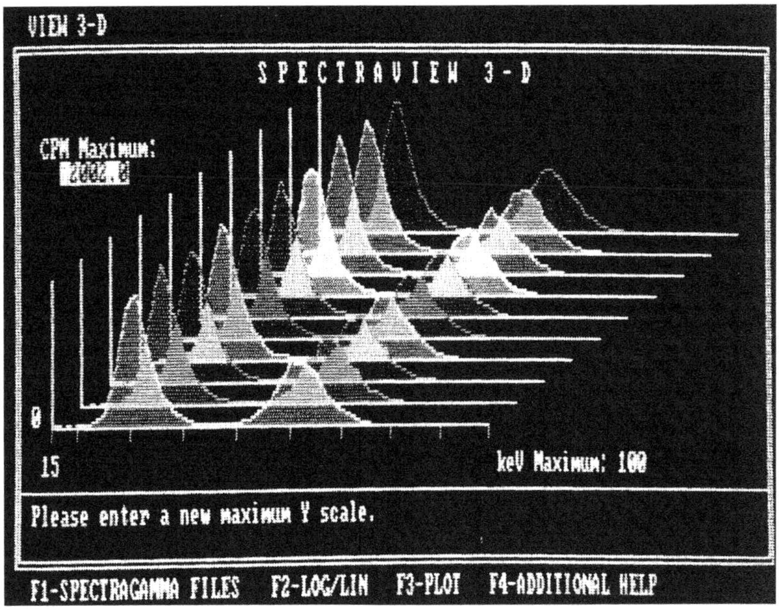

FIGURE 9.31 Multichannel gamma spectra of ^{125}I over the energy range 15–100 keV as seen on the computer monitor of a solid scintillation gamma analyzer. A gamma spectrum of a given sample is determined by MCA and plotted and analyzed in real time. © *1998 PerkinElmer, Inc. Printed with permission.*

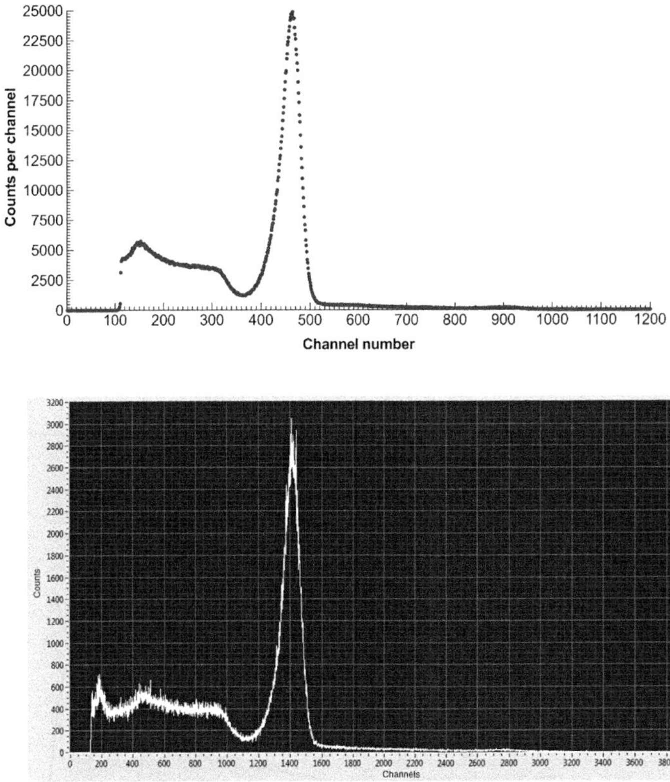

FIGURE 9.32 Upper panel: The pulse height spectrum of ^{137}Cs from a NaI(Tl) scintillator obtained with an MCA from ORTEC. Lower panel: The pulse height spectrum of ^{137}Cs obtained with the embedded MCA. The differences in channel numbers of the two spectra are due to the embedded MCA scaling; and the differences in counts of the two spectra are due to counting time. *From Garcia-Duran et al. (2019); reprinted with permission from Elsevier © 2019.*

optimization of hardware use related to the programmable logic device, that is, only 9% of the hardware available in the FPGA is used. This provides the possibility to add other components to improve the MCA or develop a network involving several detectors. With this objective in mind, Becker and Farsoni (2014) developed an FPGA-based miniature gamma-ray spectrometer, which may be accessed wirelessly by mobile phone providing a device with applications in emergency response, Battlefield assessment and personal dosimetry. They report excellent resolution of 5.9% FWHM for the 662 keV photopeak of ^{137}Cs gamma spectrum and 54.3% FWHM for the 32 keV X-ray peak from the Ba K X-rays of the ^{137}Cs daughter nuclide utilizing CsI(Tl) scintillator coupled to a SiPM photodiode.

F. Other components

The previous paragraphs in this section have provided details of the physical crystal scintillator detector, the scintillation process, the photodetector, and the electronic components needed to convert X- or γ-ray photon emissions into voltage pulses that may be analyzed in terms of number (counts) and magnitude (pulse height or keV energy) and the data converted into count rate (CPM) or gamma spectra. There are other components of the modern solid scintillation analyzer that are vital to the automatic processing and analysis of samples. These components include (1) computer system control for the hands-off analysis of samples and processing of data; (2) a sample changer for the automatic unattended analysis of up to 1000 samples tubes; (3) automatic and adjustable counting regions for single-, dual-, and multiple-radionuclide measurements; (4) many (up to 60) different analysis protocols that can be set by users for the automatic analysis of samples; (5) automatic protocol-adjustable sample positions in detectors for optimum sample counting geometry (for through-hole detectors); (6) live real-time spectral display with an MCA; (7) computer, hard disk, and printer for the processing, storage, and recording of data; (8) computer-controlled automatic IPA, detector normalization, decay corrections, background reductions, spillover calculations, and radionuclide activity (DPM) calculations; and (9) computer-controlled special application programs for instrumental analysis of samples, such as single- and multiple-site receptor binding assays, radioactive waste tracking, radioimmunoassay data processing, chromium release assays, hepatitis screening, and traditional and radioallergosorbent test (RAST) allergy screening, among other assays. These components, among others, that are characteristic of automatic gamma analyzer systems are discussed in Section V of this chapter.

IV. Concepts and principles of solid scintillation analysis

A. Gamma-ray spectra

The spectrum of pulse heights, being directly proportional to the gamma-ray energies dissipated in the crystal detector, provides a measure of the gamma-ray energy spectrum of the radiation source. Such energy spectra are illustrated in Figs. 9.3, 9.17, 9.19, 9.20, 9.27 and 9.29−9.32 in previous parts of this chapter. The gamma-ray spectrum is simply a plot of the relative abundances of the pulses versus the pulse heights calibrated in gamma ray energies in units of keV or MeV. Two spectral peaks of Fig. 9.29 are enclosed within the A and B and A′ and B′ discriminators, which represent the LL and UL discriminator settings of two counting regions. These two peaks are referred to as photopeaks because they are the result of the interaction of 0.662 MeV gamma rays from $^{137}Cs/^{137m}Ba$ or 1.08 MeV gamma rays from ^{65}Zn with the crystal detector via the photoelectric effect or a combination of Compton effects resulting in the total dissipation of the gamma-ray energy in the crystal.

Gamma-ray spectra are very useful for the identification of radionuclides. For example, the photopeak is usually the highest energy peak. An exception is the "doubles peak," which can occur at a higher energy, as it is due to the coincidence detection of two gamma-ray photons (see Section IV.C of this chapter). The photopeak gives a measure of the gamma-ray energy, which is a characteristic of each radionuclide, and may be used to identify unknown radionuclides.

Gamma-ray spectra may also be characterized by several other peaks such as those illustrated in Fig. 9.33. These are often not well defined because of the Compton

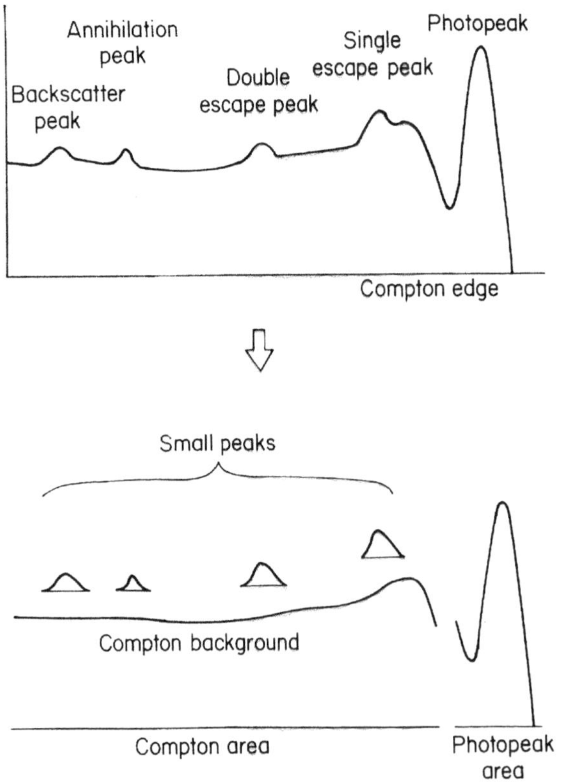

FIGURE 9.33 Segmentation of a spectral curve showing the case of a typical spectrum for gamma rays of about 2 MeV. *From Tominaga (1983); reprinted with permission from Elsevier © 1983.*

smear or background produced by a wide spectrum of Compton electron energies. The Compton edge and backscatter peaks are characteristic of most gamma-ray spectra. Compton electrons of maximum energy produce the Compton edge. A portion of the radiation, which may pass through the detector without interaction, may be scattered at 180 degrees by the detector shielding back into the detector. Absorption of the backscattered radiation by the crystal detector produces the backscatter peak.

Three other peaks illustrated in Fig. 9.33 may also be observed in gamma spectra. These are the annihilation peak, double-escape peak, and single-escape peak. These peaks are the result of pair production in the crystal detector and cannot be observed in the spectra of gamma rays of energy below the theoretical threshold of 1.02 MeV. In practice, significant numbers of pair production events are confined to high-energy (>1.5 MeV) gamma ray—photon interactions. The annihilation peak is located at an energy value of 0.51 MeV; and it is the result of the absorption of one of the two 0.51-MeV gamma-ray photons following positron annihilation accompanying pair production. When one of the two annihilation gamma rays escapes from the detector without interaction, the single-escape peak is produced at an energy level of $E_\gamma - 0.51$ MeV, where E_γ is the gamma-ray energy characterized by the photopeak. Both of the annihilation gamma rays may escape the detector without interaction and produce the double escape peak at the energy level of $E_\gamma - 1.02$ MeV.

Gamma-ray spectrometry may also be applied to the assay of more than one gamma ray—emitting nuclide in a mixture. If these nuclides emit gamma rays of characteristic energies where photopeaks of the gamma spectra are well separated, they may be assayed separately by setting the counting regions or windows directly over the photopeak of each nuclide. For example, Fig. 9.29 illustrates the combined spectrum of a mixture of ^{137}Cs and ^{65}Zn and discriminator settings that encompass the photopeak of each nuclide. The ^{65}Zn activity can be determined with LL and UL discriminator settings A′ and B′ providing a 0.950—1.25 MeV counting window; whereas the 0.550—0.800 MeV counting window defined by LL and UL discriminator settings A and B provides pulse contributions from both nuclides. The contribution due to ^{65}Zn in this window, which may be determined with a pure ^{65}Zn standard of known activity, must be subtracted from the total number of counts in this counting region to provide the net counts due to the nuclide ^{137}Cs. The determination of the gamma spectrum of a sample and the analysis of the gamma-emitting nuclides in a given sample are most effectively carried out with a MCA. The use and mode of operation of a MCA are discussed in Section III.E of this chapter.

B. Counting and detector efficiencies

Solid scintillation gamma counting fortunately does not have the disadvantages of quenching normally associated with the liquid scintillation counting technique. The sample is physically separate from the crystal scintillator, which prevents any possible physical or chemical interference with the gamma energy transfer and scintillation processes. Nevertheless, the sample composition and the counting geometry, namely the sample's shape, density, and volume as well as its distance from the scintillation detector, can have a marked effect on the count rate or the counting efficiency of a particular sample. Also, in the literature it is possible to encounter several terms related to counting and detection efficiency, which can be confusing. These terms and the factors that affect counting and detection efficiency will be discussed in more detail in this section.

1. Counting efficiency

The counting efficiency is defined as the ratio of the number of counts collected in a given period of time (count rate) to the disintegration rate or activity of the source. This is the detection efficiency that is used generally for routine measurements of radionuclide activities, such as in applications involving radionuclide tracers. It is calculated as

$$E = \mathrm{cpm_{std}}/\mathrm{dpm_{std}} \tag{9.11}$$

where $\mathrm{cpm_{std}}$ is the count rate in counts per minute of a standard source measured by the analyzer and $\mathrm{dpm_{std}}$ is the disintegration rate in disintegrations per minute or activity of the radiation standard source. The value of the counting efficiency, also expressed as a percentage, is a characteristic of a particular detector, the counting region selected as defined by the LL and UL discriminator settings, the radionuclide, and the counting geometry. The counting geometry will include the sample, its composition (e.g., density governing self-absorption), geometry (e.g., point source, shape, or volume and distance from the detector), and the combined geometry of the sample, sample tube or container, and detector.

The counting efficiency of a single-nuclide sample is easily determined from the count rate of an internal radionuclide standard of known disintegration rate. By internal radionuclide standard it is meant a radionuclide of known activity that is added to the sample. The internal standard is counted with counting conditions and geometry similar to those of the sample when it is counted alone without the internal standard. The procedure involves first counting the sample without the standard. This is followed by adding the internal standard to the sample and counting a second time. The counting efficiency of the sample can be calculated as follows:

$$E = \frac{\text{cpm}_{s+i} - \text{cpm}_s}{\text{dpm}_i} \quad (9.12)$$

where cpm_s is the count rate of the sample, cpm_{s+i} is the count rate of the sample after the addition of the internal standard, and dpm_i is the disintegration rate or activity of the internal standard. This method of determining counting efficiencies is valid when the following conditions are met: (1) The volume and composition of the internal standard have no effect on the counting geometry and counting efficiency of the sample; (2) The internal standard is counted with the same instrumental parameters (e.g., LL and UL discriminator settings) as the sample; and (3) The standard used is the same nuclide as that of the sample of unknown activity.

Once known, the counting efficiency for a particular sample can be used to convert the count rate of the sample to disintegration rate (activity) according to the equation

$$\text{dpm}_s = \text{cpm}_s/E \quad (9.13)$$

For certain radionuclides the disintegration rates of unknown sample activities can be determined directly by the "doubles-peak" method, which does not require any measure of counting efficiency. This method is described in Section IV.C of this chapter.

2. Detector efficiency

Detector efficiencies are often used to compare and evaluate different types of solid scintillator crystals and counting geometries. Detector efficiency may be referred to as the full-energy peak efficiency, photopeak efficiency, or photoelectric efficiency. These terms are used to define detector efficiency, and these will be discussed subsequently.

a. Full-energy peak efficiency

The full-energy peak efficiency or photopeak efficiency is defined as the ratio of the number of gamma-ray photons detected in the photopeak to the number of gamma rays of the photopeak energy emitted by the source (Scafè et al., 2011; Anil Kumar et al., 2009a,b; Abbas, 2007), which may be written as described by Anil Kumar et al. (2009b) as

$$\varepsilon = \frac{\text{no. gammas detected under the photopeak}}{\text{no. of gammas emitted by the source}} \quad (9.14)$$

The full-energy peak efficiency can be measured according to the equation

$$\varepsilon = \frac{N}{AP_\gamma} \quad (9.15)$$

where N is the gamma count rate in the full energy peak, A is the activity of the source during the measurement time in Bq, and P_γ is the gamma emission probability (Abbas, 2001a, 2007). The same equation can be expressed in terms of the total number or counts of the gamma-rays detected in the photopeak rather then count rate as follows

$$\varepsilon = \frac{N_{\text{peak}}}{AtP_\gamma} C_i \quad (9.16)$$

where N_{peak} is the number of net counts in the photopeak, A is the radionuclide activity, t is the time or duration of the measurement, P_γ is the emission probability, and C_i is the correction factor due to dead time (Anil Kumar et al., 2009a,b; Hurtado et al., 2004).

The value of the full-energy peak efficiency of a given source is a function of the distance from the radiation source to the scintillator detector, the path length that the photon traverses within the active volume of the gamma detector, and the solid angle subtended by the front surface of the crystal detector (Abbas and Noureddeen, 2011; Abbas, 2006; Moss et al., 1984). It can be used to evaluate the properties of the solid crystal detector to fully stop or absorb impinging gamma radiation by the photoelectric effect or a series of Compton interactions resulting in complete absorption of the gamma-ray energy. Evans (1980) compared the full-energy peak efficiencies of BGO and NaI(Tl) detectors of the same dimensions with radionuclides that span a wide range of gamma energies. The full energy peak efficiency is inversely proportional to the gamma-ray energy (Evans, 1980; El-Khatib et al., 2017). The work of Evans (1980) indicated that the photopeak resolution of BGO was only about half as good as that obtained with equivalent NaI(Tl) detectors; however, the higher photopeak efficiency of BGO and the near absence of Compton continua (Compton smear) make BGO the detector of choice when a high-resolution detector is not needed. Both density (g/cm^3) of the scintillator and effective atomic number (Z_{eff}) are important properties that can govern the gamma radiation stopping power and photopeak efficiency. For a given optimized detection geometry and given γ-energy, the photopeak efficiency will be a function of the density and effective atomic number of the scintillator, as the probability of the photoelectron effect rises with Z (Valentini and Vianden, 2010). For example, the photopeak or full-energy peak efficiency of BGO ($\rho = 7.1$ g/cm^3, $Z_{\text{eff}} = 75$) is greater than that of NaI(Tl) scintillator ($\rho = 3.7$ g/cm^3, $Z_{\text{eff}} = 51$) for a given γ-energy and detector size and geometry (Evans, 1980). Also, Valentini and Vianden (2010) demonstrated a considerably higher photopeak efficiency of LSO scintillator ($\rho = 7.4$ g/cm^3, $Z_{\text{eff}} = 65.8$) over that of BaF_2 ($\rho = 4.89$ g/cm^3, $Z_{\text{eff}} = 52.2$).

b. Total or absolute efficiency

The total or absolute efficiency, which should not be confused with the counting efficiency previously discussed (Eq. 9.11), is defined as the probability that the incident gamma rays will interact with the crystal detector via at least one of the predominate mechanisms of gamma-ray energy absorption, namely the photoelectric effect, Compton effect, or pair production. It is defined as the ratio of the number of gamma-ray photons detected by the scintillator to the number of gamma rays emitted by the source (in all directions). This detector efficiency may be written, as described by Anil Kumar et al. (2009b), according to the expression

$$\varepsilon_{abs} = \frac{\text{total no. of gammas detected}}{\text{no. of gammas emitted by source}} \tag{9.17}$$

The total efficiency is different from the counting efficiency defined earlier in the chapter by Eq. (9.11) because the counting efficiency calculated according to Eq. (9.11) is calculated on the basis of (i) the possible detection of other particles from the source and (ii) the disintegration rate (dpm) of a source; whereas the absolute efficiency is calculated on the basis of the number of gamma rays emitted per radionuclide disintegration. Because gamma emission is often characteristic of the daughter nuclide in various decay processes, gamma emission may occur via several energy states and may not occur with each disintegration of a given radionuclide, as described in Chapter 1, Volume 1. According to the radionuclide, the gamma rays emitted per disintegration may vary in number as well as probability of emission.

The relationship between photopeak efficiency and total efficiency can be expressed by peak-to-total ratio (photofraction) and expressed as

$$P/T = \frac{\varepsilon_P}{\varepsilon_T} \tag{9.18}$$

where P/T is the peak-to-total ratio, and ε_P and ε_T are photopeak or full-energy peak efficiency and the total efficiency, respectively (Abbas, 2001b).

The magnitudes of the photoelectric efficiency (or photopeak efficiency) and total detector efficiency diminish as the photon energy increases. The relationship between photon energy and detector efficiency for a detector of given geometry is nearly linear when plotted on a logarithmic scale (Sudarshan et al., 1992; Suzuki, 1982; Brinkman and Veenboer, 1979). As illustrated in Fig. 9.34, the photoelectric and total detection efficiencies decrease as the photon energy increases. Thus, nuclides that emit high-energy gamma radiation (e.g., ^{24}Na, 1.38 and 2.76 MeV) are detected with lower efficiencies than nuclides that emit lower energy gamma radiation (e.g., ^{54}Mn, 0.83 MeV). The inverse relation between gamma energy and detector

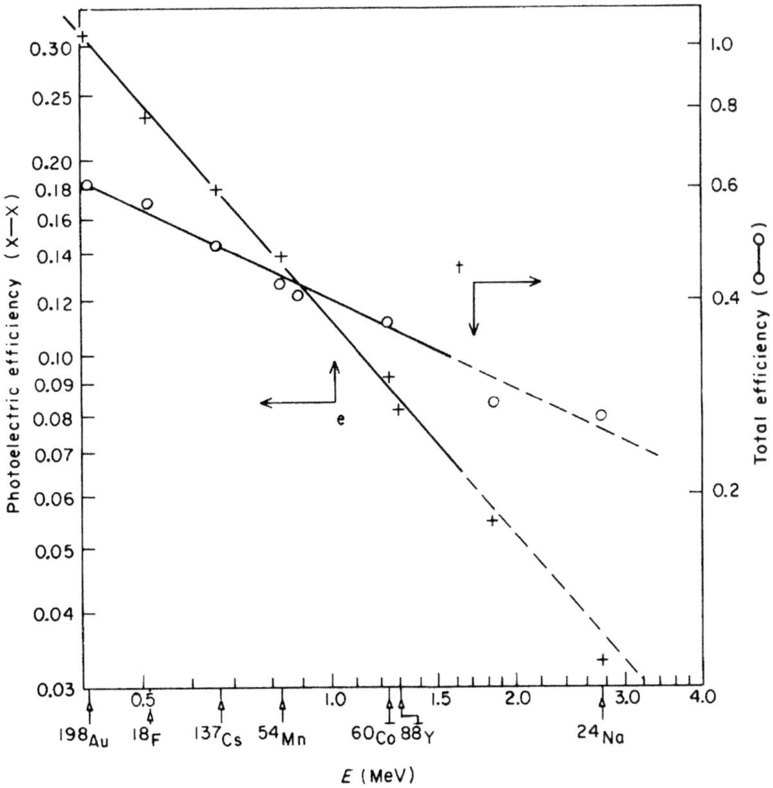

FIGURE 9.34 Photoelectric, *e*, and total efficiency, *t*, curves for samples placed on the bottom of the well of a 60 × 60 mm NaI(Tl) crystal. Diameter of well is 17 mm, depth of well is 38 mm. *From Brinkman and Veenboer (1979); reprinted with permission from Elsevier © 1979.*

efficiency can be extrapolated to counting efficiency. Such comparisons can be made, of course, when there is negligible self-absorption among the nuclide samples. The relationship between counting efficiency and photon energy is explained by the fact that most of the photons of low energy (<0.100 MeV) hitting the detector are absorbed due to high absorption coefficients (Sudarshan et al., 1992; Holmberg et al., 1972). As the photon energy increases, the absorption coefficients decrease or, in other words, the stopping power of the detector crystal decreases.

For a given radionuclide source, the total or absolute detection efficiency is dependent on the source and detector counting geometry (Abbas and Noureddeen, 2011; Enqvist et al., 2010; Yalcin et al., 2007); and it can be a measure for evaluating and comparing various scintillator materials (Ayas-Maierhafer and DeVol, 2007).

C. Sum-peak activity determinations

The activity or disintegration rate of some radionuclides may be determined directly from the X- or gamma-ray spectra, without the use of standards, if they decay by the emission of two photons in coincidence (i.e., X- or gamma-ray photons). The technique, known as the sum-peak or doubles-peak method, was developed following the pioneering work of Brinkman and Alten (1963, 1965), Brinkman et al. (1963a,b, 1965), and Eldridge and Crowther (1964). The method has been applied often to the activity analysis of ^{60}Co and ^{125}I (Marsoem et al., 2014; Sahagia et al., 2008; Capogni et al., 2006; da Silva et al., 2006; Kim et al., 2003; Miyahara et al., 2000; Pommé et al., 2005; Yuan and Hwang, 2000; Kawano and Ebihara, 1990, 1991, 1992). In the case of ^{60}Co, the nuclide decays by β emission with the concomitant emission of two gamma rays from the daughter nuclide ^{60}Ni in coincidence, namely, 1.17 and 1.33 MeV gamma rays of equal intensity. The radionuclide ^{125}I decays exclusively by EC followed by a 35 keV gamma-ray emission of the daughter nuclide ^{125}Te. Following EC the K X-rays of 27.4 and 31 keV are emitted in coincidence with the 35 keV gamma-ray photons of the ^{125}Te daughter.

Based on the pioneering work of Brinkman et al. (1963a,b), the disintegration rate of a nuclide source emitting two gamma rays in coincidence or an X-ray and gamma ray in coincidence, such as the cases for ^{60}Co or ^{125}I, can be determined by the sum-peak equation

$$N_0 = T + \frac{A_1 A_2}{A_{12}} \tag{9.19}$$

where T is the total count rate of the entire spectrum extrapolated to zero energy (or the area under the total spectrum), A_1, A_2, and A_{12} are the count rates of the photopeaks (or areas under these peaks) of γ_1, γ_2, and the sum-peak, respectively. The photopeaks of γ_1 and γ_2 are also referred to as the singles peaks, and these may consist of two separate photopeaks of two gamma rays of different energy as for the case of ^{60}Co or, they may consist of a gamma-ray and X-ray peak or two X-ray peaks as in the case of ^{125}I. The sum-peak is also called the doubles-peak, as it is the result of coincidence summing of two photons. High-purity germanium (HPGe) semiconductor detectors yielding well-resolved singles and doubles peaks have been applied to the sum-peak method of activity analysis of the following radionuclides: ^{22}Na (Ogata et al., 2016, 2015; de Oliveira et al., 2012), ^{57}Co (Novković et al., 2012; Iwahara et al., 2009), ^{60}Co (Ogata et al., 2016, 2015; Marsoem et al., 2014; Bikit et al., 2009a; Vidmar et al., 2009; Kim et al., 2003; Miyahara et al., 2000), ^{65}Zn (Oliveira et al., 2012), ^{125}I (Marsoem et al., 2014), ^{133}Ba (Nađđerđ et al., 2013; Bikit et al., 2009a,b; Novković et al., 2009), ^{134}Cs (Ogata et al., 2018a; b, 2016), ^{139}Ce (Korun and Martinčič, 1993), and ^{152}Eu (Ferreira Filho et al., 2018; Nemes et al., 2011). The classic NaI(Tl) crystal scintillation detector can be used to measure the activities of ^{60}C and ^{125}I by photon energy spectral lines with poorer resolution than that provided by semiconductor detectors; and Eq. (9.19) serves as the basis for working equations used in these experimental activity determinations, subsequently described.

An example of a typical gamma spectrum of ^{60}Co measured with a NaI(Tl) detector is illustrated in Fig. 9.35. Kawano and Ebihara (1990, 1991, 1992) measured the activity N_0 of ^{60}Co according to Eq. (9.19) and the measured gamma spectrum according to the following calculated spectral areas:

$$T = I_T - B_T \tag{9.20}$$

$$A_1 = A_2 = \frac{(I_{1+2} + B_s)}{(2 + \alpha)} \tag{9.21}$$

$$A_{12} = I_{12} - \frac{1}{2}(\mathrm{Ch}_5 - \mathrm{Ch}_4 + 1)(C5 + C4) \tag{9.22}$$

where I_T and B_T are the area and background, respectively, under the entire spectrum; I_{1+2} is the total area between channel 1 (Ch1) and channel 2 (Ch2) of the spectrum illustrated in Fig. 9.35, and B_S is the background for the same area; α is the ratio of the area of the Compton scattering of γ_2 under the photopeak of γ_1 to A_2; I_{12} is the total area between channel 4 (Ch4) and channel 5 (Ch5), and $C4$ and $C5$ are the counts in channels Ch4 and Ch5, respectively.

The sum-peak method for the activity analysis of ^{125}I has been used by several researchers (Marsoem et al., 2014; Sahagia et al., 2008; Capogni et al., 2006; da Silva et al., 2006; Pommé et al., 2005; Wang et al., 2001; Yuan and Hwang, 2000; De Felice and Myteberi, 1995; Dias and Koskinas, 1995; Martin and Taylor, 1992). The case of ^{125}I

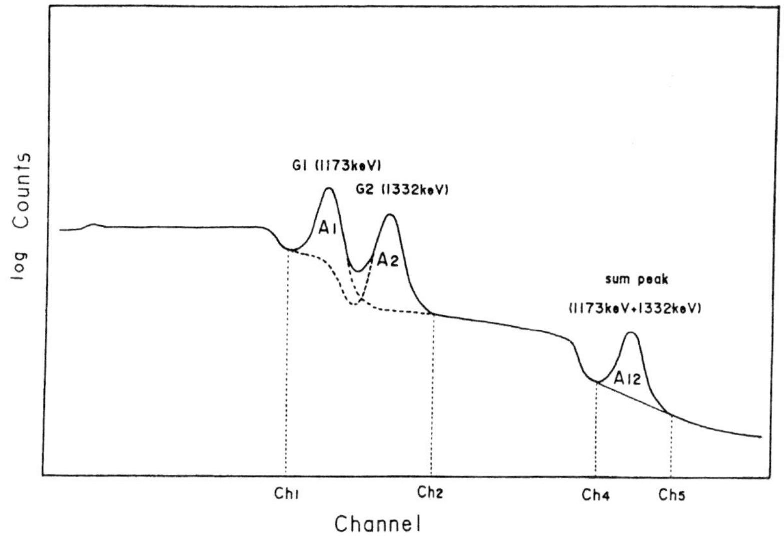

FIGURE 9.35 Typical gamma spectrum of ^{60}Co as measured with a NaI(Tl) crystal. The photopeaks are identified as follows: A_1, photopeak of γ_1 (G1); A_2, photopeak of γ_2 (G2); A_{12}, sum peak. *From Kawano and Ebihara (1990, 1992); reprinted with permission from Elsevier © 1990.*

is treated differently than the previous example, as it decays by EC into the 35.5 keV excited state of ^{125}Te with the emission of a 27.4- and 31-keV X-rays in coincidence with the 35-keV gamma-ray photon emitted by the excited state of the ^{125}Te daughter (6.67%) or by emission of conversion electrons (CEs) (93.33%), as illustrated in Fig. 9.36. The photopeak at the lower end of the spectrum occurring at ~28.5 keV is due to the unresolved single-photon detection of the X-rays and the 35-keV gamma rays, while the coincidence photopeak (sum peak) occurring at ~56.8 keV is the result of coincident summing of two X-rays or an X-ray and a 35-keV gamma ray. The singles peaks for the two X-rays and the one gamma ray remain unresolved under one peak at ~28.5 keV. Only with the use of a high-resolution semiconductor detector can the singles peaks for each of the photon emissions be observed (Wang et al., 2001).

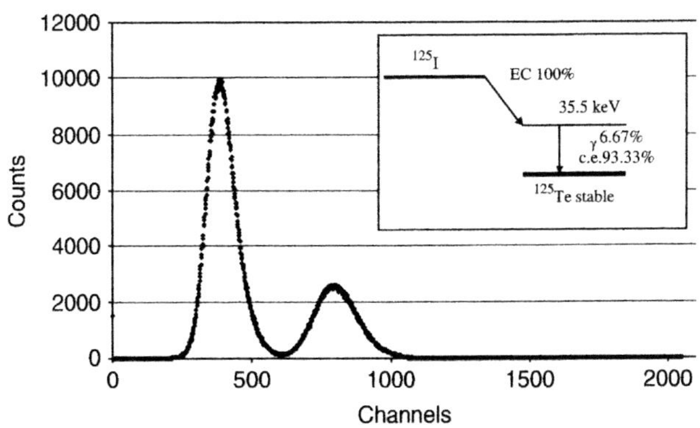

FIGURE 9.36 Typical ^{125}I spectrum taken from a NaI(Tl) well-type detector. *From Capogni et al. (2006); reprinted with permission from Elsevier © 2006.*

For the case of the unresolved singles peaks of ^{125}I, as illustrated in Fig. 9.36, Capogni et al. (2006) define the count rates under the two photopeaks of Fig. 9.36 as

$$A_1 = \varepsilon N_0 P_1(1 - \varepsilon P_2) + \varepsilon N_0 P_2(1 - \varepsilon P_1) \tag{9.23}$$

$$A_{11} = \varepsilon^2 P_1 P_2 N_0 \tag{9.24}$$

where A_1 is the count rate under peak 1 (the more intense peak at the lower end of the spectrum (see Fig. 9.36) corresponding to the count rate of the unresolved K X- and gamma-ray singles peak area, A_{11} is the count rate under the coincident event peak (the lesser intense peak at the higher-energy end of the spectrum (see Fig. 9.36), ε is the detection efficiency for 30 keV photons, as it is assumed that the detection efficiency would be constant for all photons between 27.2 and 35.5 keV, N_0 is the decay rate or activity of the ^{125}I, P_1 is the K X-ray emission probability per EC decay, i.e., $P_1 = P_K\omega_K = 0.6990 \pm 0.0056$ where P_K is the K shell EC probability and ω_K is the fluorescence yield of K X-ray, and P_2 is the sum of the K X-ray emission probability for internal conversion events plus the 35 keV gamma-ray emission probability; i.e., $P_2 = (1 + \alpha_K\omega_K)/(1 + \alpha) = 0.7624 \pm 0.0075$ where α and α_K are total and K shell conversion coefficients (Capogni et al., 2006; Wang et al., 2001; Yuan and Hwang, 2000; Dias and Koskinas, 1995; Martin and Taylor, 1992).

From Eqs. (9.23) and (9.24) Capogni et al. (2006) derive the following equation of Eldridge and Crowrther (1964) as

$$N_0 = \frac{P_1 P_2}{(P_1 + P_2)^2} \frac{(A_1 + 2A_{11})^2}{A_{11}} \tag{9.25}$$

De Felice and Myteberi (1995) demonstrated the application of the sum-peak method to the absolute activity measurement of ^{125}I with an overall uncertainty of 0.3% (1σ). Capogni et al. (2006) report an experimental standard uncertainty of 0.14% for various instrumental setups used with two NaI(Tl) crystal detectors of different volumes. Marsoem et al. (2014) report total relative standard uncertainties of 0.39% and 0.36% for the standardization of ^{125}I and ^{60}Co by the sum-peak method. Also, Dias and Koskinas (1995) and Yuan and Hwang (2000) found excellent agreement between the sum-peak method and other methods of radionuclide standardization of ^{125}I.

In the case of ^{22}Na sum peak activity measurements with a NaI(Tl) detector, de Oliveira et al. (2012) reported combined uncertainties in the analysis to be only 0.15% and 0.11% depending on the detector geometry of $< 4\pi$ or 4π, respectively. A counting geometry of $< 4\pi$ is that of a single well-type NaI(Tl) detector; and the 4π detector geometry was achieved by placing a planer NaI(Tl) detector directly over the well-type NaI(Tl) detector thereby completely surrounding the sample with detector. A typical gamma-ray spectrum of the ^{22}Na with the NaI(Tl) detector in the 4π geometry is illustrated in Fig. 9.37. Sodium-22 decays principally by positron emission with 90% intensity and EC with 10% intensity followed by gamma-photon emission at 1274 keV with 100% intensity. The various peaks in the gamma spectrum of ^{22}Na are identified by in Fig. 9.37 as the annihilation photopeak at 511 keV, the coincident detection of two annihilation photons at 1022 keV, the gamma photopeak at 1274 keV, the coincident detection of the 1274 keV gamma-photon and annihilation photon at 1785 keV and the coincident detection of two annihilation photons and a 1274 keV gamma photon producing a peak at 2296 keV. The basic equations used for the sum-peak activity measurement of ^{22}Na are described by de Oliveira et al. (2012), namely,

$$N_0 = N_T + \frac{A_1 A_2}{A_{12}} \tag{9.26}$$

where $N_T = T$ of Eq. (9.19) described previously and N_T is the number of rate interactions under the total spectrum, A_1 and A_2 are the count rates of two photopeaks, and A_{12} is the sum-peak count rate resulting from the coincident interactions of the two radiation photons represented by peaks A_1 and A_2. Eq. (9.26) is expressed by de Oliveira et al. (2012) as

$$N_0 = N_T + R \tag{9.27}$$

where, namely,

$$R = \frac{A_{1274}A_{1022}}{A_{2296}} = \frac{A_{1274}A_{511}}{A_{1785}} \tag{9.28}$$

where A_{511} = the count rate under the 511 keV annihilation photopeak, A_{1022} = the count rate under the 511 + 511 keV sum peak, A_{1274} = count arte under the 1274 gamma emission photopeak, A_{1785} = the count rate under the 511 + 1274 keV sum peak, and A_{2296} = count rate under the 511 + 511+1274 sum peak. A third ratio of count rates under the singles and sum peaks may be used for the ratio R, which is the following:

$$R = \frac{A_{511}^2}{A_{1022}} \tag{9.29}$$

where A_{511}^2 = the square of the count rate under the 511 keV annihilation photon peak, and A_{1022} = the count rate under the 511 + 511 keV sum peak. de Oliveira et al. (2012) note that the R value defined by Eq. (9.29) can be used in the above Eq. (9.27) only in the case of 4π counting geometry where the detector completely surrounds the sample. This is clear in light of the positron annihilation phenomenon which results in the emission of two 511 keV gamma photons in opposite (180 degrees) direction. When using a well counter, which is $<4\pi$ counting geometry, a coincidence factor W must be included in Eq. (9.27) or

$$N_0 = W\left(N_T + \frac{A_{511}A_{511}}{A_{1022}}\right) \tag{9.30}$$

where $W = 1$ in 4π counting geometry. The reader may refer to de Oliveira et al. (2012) for the working equation in the calculation of W when the detector provides $<4\pi$ counting geometry.

Taking advantage of the high resolution and efficiency for X-rays and low-energy gamma rays provided by a well-type HPGe semiconductor detector, Wang et al. (2001) was

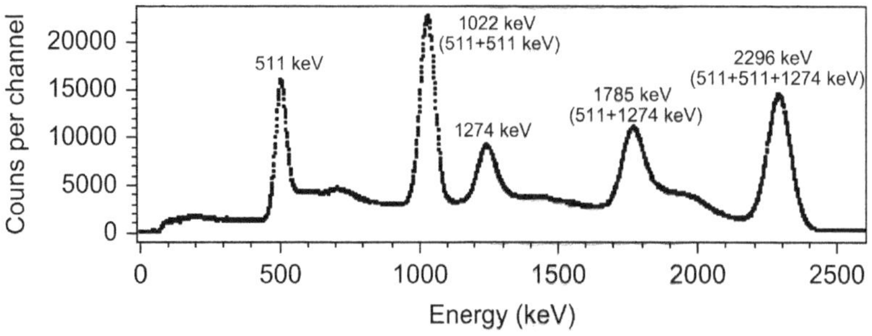

FIGURE 9.37 Gamma spectrum of ^{22}Na obtained by a well-type NaI(Tl) detector covered by another NaI(Tl) detector in 4π counting geometry. *From de Oliveira et al. (2012); reprinted with permission from Elsevier © 2012.*

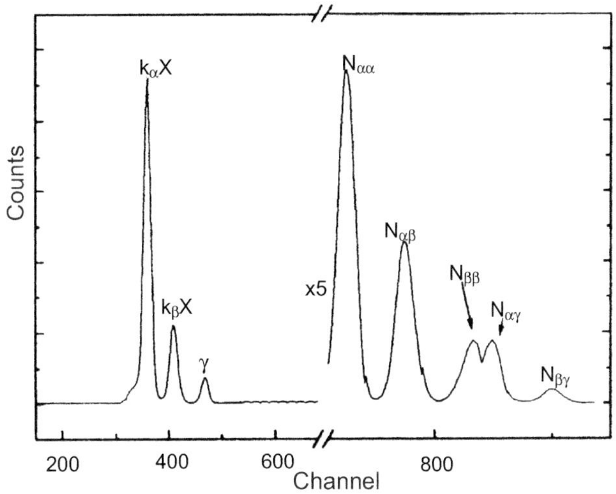

FIGURE 9.38 Spectrum of ^{125}I obtained with a well-type HPGe semiconductor detector. The singles peaks are identified as K_αX, K_βX, and γ, while the doubles or sum peaks are denoted as $N_{\alpha\alpha}$, $N_{\alpha\beta}$, $N_{\beta\beta}$, $N_{\alpha\gamma}$, and $N_{\beta\gamma}$. *From Wang et al. (2001), reprinted with permission from Elsevier © 2001.*

able to define the count rates of the individual resolved single peaks for two K X-rays and the gamma ray as well as the sum peaks for the ^{125}I spectrum illustrated in Fig. 9.38. The spectrum illustrates the three singles peaks for the two K X-ray emissions (K_αX and K_βX) from the EC decay of ^{125}I as well as the 35-keV gamma emission line. The five sum peaks for all probabilities of coincidence summing between the two K X-rays and the one gamma ray are also discernible. From the peak areas and the following equation derived by Wang et al. (2001), the ^{125}I activity was calculated:

$$N_0 = T + \frac{N_\gamma N_{\alpha\beta}}{N_{\beta\gamma}(2N_{\alpha\alpha} + N_{\alpha\beta})}\left[N_\alpha + N_\beta - N_\gamma\left(\frac{N_{\alpha\beta}}{2N_{\beta\gamma}} + \frac{N_6}{N_{\beta\gamma}} - \frac{N_{\alpha\alpha}}{N_{\alpha\beta}}\right)\right] \tag{9.31}$$

where N_α, N_β, and N_γ are the singles peaks for the 27.3 keV K_α X-ray, 31.0 K_β X-ray, and 35.49 keV γ-ray, respectively, $N_{\alpha\alpha}$, $N_{\alpha\beta}$, $N_{\beta\beta}$, $N_{\alpha\gamma}$, and $N_{\beta\gamma}$ are sum peaks for all possible combinations of coincidence detection of the two X-ray emissions or one of the X-ray emissions with the gamma ray, and N_6 represents the unresolved sum peak of K_β X-rays ($E_{\beta\beta} = 62.0$ keV) and sum peak of K_α X-rays and γ-rays ($E_{\alpha\gamma} = 62.8$ keV), which could not be separated and therefore treated as one peak, $N_6 = N_{\beta\beta} + N_{\alpha\gamma}$. With this method, Wang et al. (2001) determined the ^{125}I activity with 0.6% uncertainty.

Nemes et al. (2016) demonstrate the use of pileup peaks for corrections of accidental coincidences and Monte Carlo techniques for angular correlation, and lower-level discriminator corrections, the source activity can be measured by the sum-peak method with an accuracy and reproducibility below 1%. Further studies by Nemes et al. (2018) show, that at high count rates, random coincidence corrections must be considered and provide an equation for the sum-peak activity calculations that include these corrections.

In Chapter 1, Volume 1, examples were provided of radionuclides that exhibited different electron-capture decay rates as a function of the chemical environment of the radionuclide. Because EC involves extranuclear atomic electrons, which can be affected by chemical environments, the decay rates of EC radionuclides can be affected. Singh et al. (2001) utilized this phenomenon and the sum-peak method (gamma-gamma summing) to correlate variations in nuclear precession frequencies with electric field gradient components in different chemical environments. They measured the variations in the ^{75}Se sum-peak intensity at the 400 keV line due to the simultaneous detection of the 121 + 279 keV gamma rays relative to the singles peak intensities at 121 and 279 keV.

D. Modified sum-peak activity determinations

Ogata et al. (2015, 2016) developed a modified sum-peak method, which simplifies activity determinations requiring only the singles peak and sum-peak count rates and permitting the activity determinations of a radionuclide within a mixture of radionuclides. Ogata et al. (2016) confirmed the technique with known activities of ^{60}Co, ^{22}Na, and ^{134}Cs. They used HPGe detectors for gamma-ray spectrometry; however, a solid scintillation spectrometer could be used if the singles and sum peaks can be accurately defined and measured. As noted by Ogata et al. (2016), the conventional sum-peak method requires the total count rate (See N_T of Eqs. (9.26) and (9.27) decribed previously); and when the sample contains radionuclides other than that of interest, the total activity of the radionuclide of interest cannot be easily determined. In contrast to this, the modified sum-peak method requires only the count rates of the singles peak(s) and that of the sum peak.

Ogata et al. (2016) break down the basic equation that defines the sum-peak method as follows:

$$N_0 = R + T = \left[\frac{N_1 N_2}{N_{12}} + N_t\right]\overline{W}(0) = \frac{N_1 N_2}{N_{12}}\overline{W}(0) + N_t\overline{W}(0) \tag{9.32}$$

where N_0 is the sample activity, as described previously, R is the ratio of count rates in the singles and sum peaks, T is the total count rate under the entire spectrum, N_1 and N_2 are the count rates of the photopeaks for the single gamma-ray photons, N_{12} is the sum-peak count rate, N_t is the total count rate under the spectrum, and $\overline{W}(0)$ is the angular correlation of successive gamma-ray emissions (Evans, 1955; Frauen felder and Steffen, 1965; Visinelli, 2014). The angular correlation of 1.111 was used by Ogata et al.

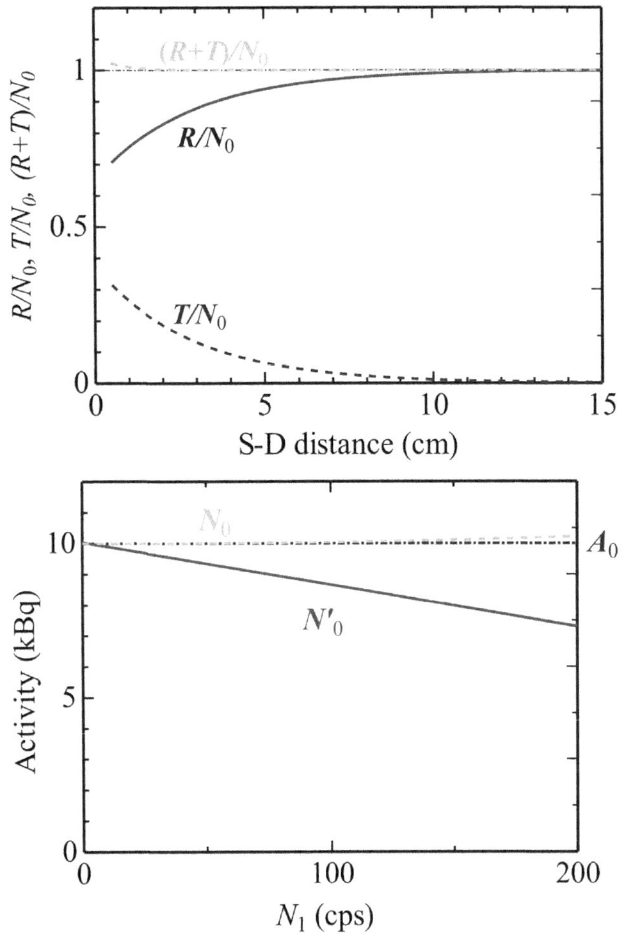

FIGURE 9.39 Semitheoretical plots estimated by the typical efficiencies for HPGe detector of 40% relative efficiency (RE), where (A) represents the partition of the activity R, T, and $(R+T)$ as a function of S–D distance, and (B) represents the dependences of N_0 and N_0' on N_1, where A_0 is assumed to be 10 kBq of ^{60}Co. *From Ogata et al. (2016); reprinted with permission from Elsevier © 2016.*

(2016). As noted by Agata et al. (2016), the value of T in Eq. (9.32) decreases almost proportionally to the total efficiency, which is inversely proportional to the square of the source-to-detector (S–D) distance; however, the value of R is not proportional to the efficiency because the efficiency cancels in the ratio term N_1N_2/N_{12}. The activity N_0 is a constant value if there is negligible decay during the assay; and thus, R becomes a dominant variable with increasing source-to-detector distance, as illustrated in Fig. 9.39A. in light of the fact that R/N_0 becomes dominant with increasing distance from the source, Ogata et al. (2016) modified Eq. (9.32) by removing N_t, to provide the following solution to the sample activity:

$$N_0' = \frac{N_1N_2}{N_{12}}\overline{W}(0) = R \tag{9.33}$$

As depicted in Fig. 9.39A, N_0' approaches N_0 with increasing sample-to-detector distance where the curve R/N_0 approaches 1.0 at infinite distance. Thus, by taking several measurements at various source-to-detector distances, the value of N_0 is obtained at the extrapolated value of N_0' at infinite distance. Ogata et al. (2016), simplified the method yet further, by plotting N_0' as a function of one of the singles peak count rates, N_1, as a function of sample-to-detector (S–D) distance to yield the following relation:

$$N_0 = N_0'(N_1),\ N_1 \to 0 \tag{9.34}$$

In the case of ^{60}Co taken as an example, the source activity A_0 is 10 kBq determined by the conventional sum-peak method as N_0, Ogata et al. (2016) illustrate the dependence of N_0 and N_0' on N_1 as plotted in Fig. 9.39B where N_0' can be extrapolated to A_0 with decreasing count rate of N_1. An added advantage of this method is that the removal of N_t from the equation of the sum-peak method permits the activity measurements of a radionuclide containing other radionuclides in the sample. This modified sum-peak method was verified with activity determination of ^{60}Co, ^{22}Na, and ^{134}Cs.

E. Self-absorption

The counting efficiencies of samples can be affected significantly by self-absorption, particularly when the radionuclide emits low-energy X- or gamma radiation. Self-absorption may be defined as the attenuation of radiation by the sample emitting the radiation. This absorbed radiation does not reach the detector and, if appreciable, counting efficiencies can be markedly reduced. Self-absorption is a function of the sample composition and geometry, namely its effective atomic number, density, thickness, and the X-ray or gamma-ray photon energy (Lopes et al., 2019; Carrazana González et al., 2016; Długosz-Lisiecka and Ziomek, 2015; Dziri et al., 2014; Landsberger et al., 2013). Photons of low energy such as the 28- and 32-keV X-rays and 35-keV gamma rays emitted from ^{125}I are much more susceptible to self-absorption than the higher energy gamma ray photons of, for example, ^{60}Co (1.17 and 1.33 MeV).

The X- and gamma rays of ^{125}I can be attenuated significantly before escaping from a typical 1-cm-diameter sample tube (Malcolme-Lawes and Massey, 1980), which is a common sample geometry with crystal scintillator well detectors. The effect of salt concentration and ions of different atomic mass absorption coefficients on the solid scintillation counting efficiency of ^{125}I is illustrated in Fig. 9.40. Pommé et al. (2005) emphasize the importance of maintaining constant counting geometry and self-absorption in the analysis of ^{125}I.

Rieppo et al. (1975) and Rieppo (1976a,b) have studied extensively the relationship of counting efficiency and

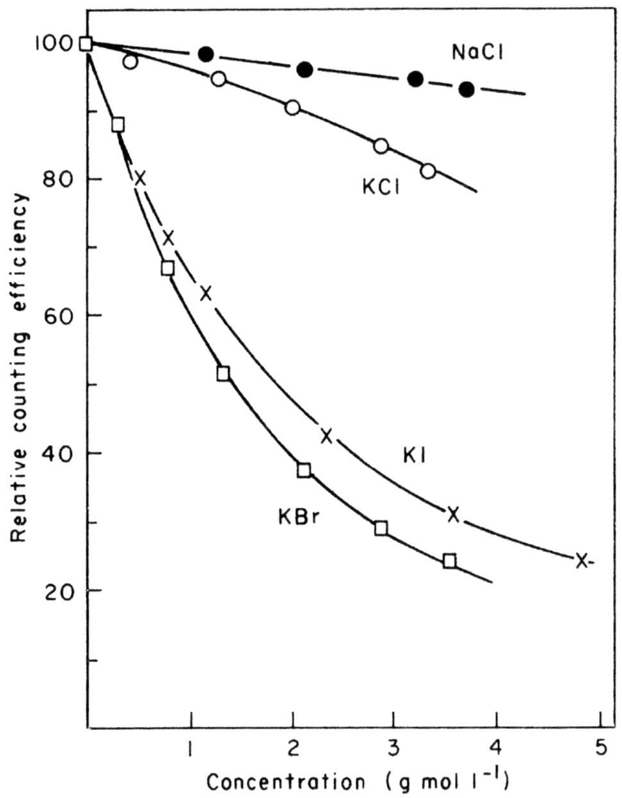

FIGURE 9.40 Variation of counting efficiency for ^{125}I (relative to efficiency in water = 100) with concentration of materials in aqueous samples with a NaI(Tl) well detector. Sample volumes are 5 mL and bottle diameter is 1.3 cm. *From Malcolme-Lawes and Massey (1980); reprinted with permission from Elsevier © 1980.*

sample density (e.g., water, bone, and aluminum) for low-energy photons in the energy range 0.010–0.200 MeV for samples with a wide range of counting geometries. Self-absorption can be significant in the solid scintillation analysis of low-energy X- or gamma-ray photons (<400 keV); and it is a function of the thickness and density of the samples (Lopes et al., 2019; Huy et al., 2013; Elane Streets, 1994) because of the relatively poor penetrating power of low-energy photons. Abbas (2006) and Carrazana González (2016) provide analytical formulae for the efficiency calibration of gamma detectors by determining the attenuation of the gamma photons by the source with calculations of the photon path length through the source material. A Monte Carlo GEANT simulation tool, which automatically accounts for sample self-absorption, is described by Ashrafi et al. (2006) for the modeling of a NaI(Tl) scintillator detector. In a review of radiometric methods, Hou and Roos (2008) emphasize the importance of chemical separation of radionuclides prior to the measurement of X- or gamma-emitting radionuclides to minimize self-absorption.

F. Counting geometry

The counting geometry is the geometric relationship between the sample and detector, including the sample and detector size and shape, the distance between the sample and detector, sample thickness, and angle of orientation of the detector around the sample (e.g., 180 degrees or 2π for a flat sample and planar detector or $>>2\pi$ geometry for a well detector or through-hole detector). The effect of sample geometry on the detection and counting efficiencies for various detector types including the planar, well-type, and through-hole detectors is discussed in Section III.A of this chapter.

G. Resolution

The resolution of a solid scintillation detector is a measure of the "sharpness" of the X-ray or gamma-ray energy peaks of the pulse height spectra. The sharper or narrower these peaks, the greater the resolution and the ability of the crystal detector to distinguish and measure the photon energy lines of one or more radionuclides in a sample. The resolution is expressed most often as the width of the photon peak at half-maximum peak height. The width of the photopeak in keV at the half-maximum height is expressed in terms of the percentage of the photon energy or peak center and is referred to as % full width at half-maximum (% FWHM). For example, the resolution of the photopeak at the 662 keV gamma energy line of ^{137}Cs illustrated previously in this chapter in Fig. 9.19 is shown to be 3.65%; and it is calculated as follows:

$$\% \text{ FWHM} = \frac{674.1 \text{ keV} - 649.9 \text{ keV}}{662.0 \text{ keV}} \times 100 = 3.65\% \tag{9.35}$$

where 674.1 and 649.9 keV in the numerator are the energy values obtained from the energy axis taken at both sides of the half-maximum peak height (position of the two arrows of Fig. 9.19 and 662 keV in the denominator is the magnitude of the gamma-energy line.

The resolutions of scintillators are energy dependent; and experimental values have been reported by many researchers. Resolutions experimentally measured with a 38 × 38 mm BGO detector as a function of gamma-ray energy were measured by Evans (1980). When comparing resolutions for photopeaks of different energies, a high % FWHM for a low-energy photon peak could mean a better resolution than a lower % FWHM for a high-energy photon. For example, a 17% FWHM resolution for the 514-keV gamma peak of ^{137}Cs represents an energy resolution of 87 keV (i.e., 0.17 × 514 keV), whereas a 10% FWHM resolution for the 1275-keV gamma peak of ^{22}Na was calculated by Evans (1980), i.e., an energy resolution of 127.5 keV. Evans (1980) pointed out that the resolution of the BGO detector is only about half as good as that which can be obtained with equivalent NaI(Tl) crystals. However, the higher efficiency of BGO and the near absence of Compton continua make this detector the choice when a high-resolution detector is not needed.

While energy resolution is an important property in X-ray and gamma-ray photon spectroscopy, other important properties in the selection of desirable characteristics of inorganic crystal scintillators are high density and effective atomic number, high light output (i.e., luminosity), short decay time without afterglow, convenient emission wavelength, mechanical ruggedness, radiation hardiness, and low cost (Moses, 2002). Measured energy resolution at 662 keV of numerous scintillators as a function of light output or luminosity is illustrated in Fig. 9.41. All of the ideal properties of a scintillator may not be found with just one scintillator. Often it is a tradeoff between the most desired properties against other desirable ones. For example, the relatively high density and high effective atomic number (Z_{eff}) of BGO ($\rho = 7.13$ g/cm^3), GSO ($\rho = 6.71$ g/cm^3), and LSO ($\rho = 7.4$ g/cm^3) provide these scintillators with a relatively high stopping power; however, they are not as efficient in terms of energy resolution and light output as other scintillators of lower density and lower effective atomic number such as $LaCl_3$:Ce ($\rho = 3.8$ g/cm^3) and $LaBr_3$:Ce ($\rho = 5.3$ g/cm^3) as depicted in Fig. 9.41.

A scintillation crystal detector can become damaged when the crystal undergoes fracture via thermal or mechanical shock or when it absorbs atmospheric moisture, which can occur with the hygroscopic NaI(Tl) crystals, as well as other hygroscopic crystal detectors, when the protective cover is damaged. When damage occurs the energy transfer scintillation processes and light transmission of the crystal are impaired and the energy resolution will drop. The photopeaks may become broadened and/or transformed. Modern gamma scintillation analyzers are equipped with a MCA and automatic calibration programs to monitor the integrity of the photopeak resolution with calibrated NIST-traceable gamma-emitting standards.

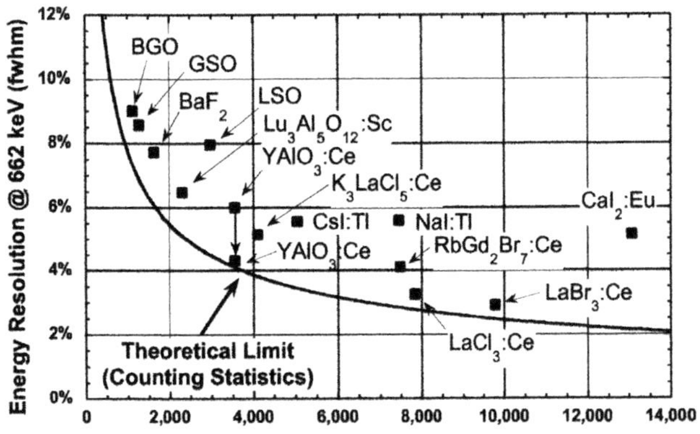

FIGURE 9.41 Measured energy resolution of several scintillators for 662 keV gamma rays as a function of their light output (expressed as the number of photoelectrons observed with a photomultiplier tube). The solid curve indicates the theoretical lower limit placed by counting statistics. *From Moses (2002) and Dorenbos (2002); reprinted with permission from Elsevier © 2002.*

H. Background

Background includes counts collected by the scintillation detector produced by radiation events arising from sources outside the sample under analysis. The sources of background radiation are cosmic rays, natural radionuclides in the sample tube and neighboring materials (environment), detector contamination, and radioactivity from samples that may be in close proximity to the sample under analysis. The latter source of background should be considered particularly in automatic gamma-counting systems where multiple detectors arrays and multiple sample-changing cassettes are all in close proximity and when high-energy gamma-emitting radionuclides are present in sample tubes.

Background count rates should always be determined for the particular counting region of interest defined by the LL and UL discriminator settings. The background for a particular counting protocol may be determined by using a blank sample, that is, a sample tube or microplate well containing the volume of reagents used without any radioactive sample. This is similar to a reagent blank. When background count rates are significantly high relative to the sample count rates, background counts must be determined and subtracted from the sample counts. Under these circumstances, sample counts and background counts should be collected over a sufficient period of time to get a measure of the sample and background count rates with a low %2S (% 2 sigma standard deviation).

The measurement of background also serves another purpose, that of testing the stability of the instrument detector systems and associated electronics. This fits within a program of IPA, which is dealt with in Chapter 6, Volume 1. The stability of the background count rates serves as an index of the stability of the counting system.

Modern solid scintillation analyzers are equipped with an MCA, which can be programmed to collect the background counts according to pulse height over the entire spectrum of pulse heights for a predetermined time period. The background for any counting region of interest defined by LL and UL discriminator settings is easily obtained from the entire pulse height spectrum. The instrument can be set to sum the pulse heights in those channels for any region of interest.

V. Automated solid scintillation analyzers

Automated solid scintillation analyzers are referred to here as high-throughput computer-based multiple-detector instruments capable of analyzing simultaneously as many as 10 or 12 samples with automated IPA, multiple-user programming, "walk-away" sample analysis, and "hands-off" data processing of up to 1000 samples with potential links to computer mainframe and local area network (LAN) assay automation and data management. Two types of automated scintillation analyzers will be described in this section: (1) automated gamma analyzers, which utilize up to 10 NaI(Tl) detectors for the simultaneous measurement of gamma- and X-ray-emitting radionuclides in sample tubes, and (2) microplate scintillation analyzers, which utilize 24-, 96-, or 384-well microplates for the solid or liquid scintillation analysis of beta particle- and atomic electron-emitting radionuclides.

A. Automated gamma analysis

An automated gamma analyzer should possess all the components needed to provide a complete unattended analysis of relatively large numbers of samples containing X-ray, gamma-ray, or positron-emitting radionuclides. Among the main components of such a system are (1) a large-capacity sample changer capable of holding and moving large numbers of sample tubes (up to 1000) to and from detectors, (2) multiple detectors for the simultaneous analysis of up to 10 samples when a high sample throughput is required, (3) automatic computer-controlled system with numerous (hundreds) of independent programmable assay protocols, (4) sample number and sample cassette identification via binary code, which can be linked to a LAN data processing and management system through a programmable computer interface, (5) automatic isotope decay correction according to date and time, (6) live gamma spectrum displays, (7) automatic detector normalization and detector cross-talk measurement for multiple-detector instruments, (8) correction for radionuclide interference between two counting regions via automatic spillover calculations, (9) automatic background correction in all counting regions, (10) multiple simultaneous counting regions, either preset for certain radionuclides, such as ^{125}I, ^{57}Co, ^{60}Co, ^{75}Se, ^{51}Cr, ^{131}I, ^{59}Fe, and ^{22}Na, or defined arbitrarily with LL and UL discriminator settings over a wide spectral energy region (e.g., 15–2000 keV), (11) automatic IPA, and (12) special applications software for sample data processing and calculations.

1. Multiple detector design

In an early part of this chapter, a description of the various types of detector geometries was given. Two types of detector geometries for automated gamma counters have been developed. One detector design is the through-hole detector described previously in this chapter (see Fig. 9.6). Another design of automated gamma counters, which provides increased detection efficiencies, is one equipped with as many as 10 well-type inorganic crystal detectors equipped with an automatic sample changer with analysis capacity of up to 1000 samples, such as the PerkinElmer Wizard Automatic Gamma Counter.

Automated gamma counters utilizing a well-type detector geometry are currently the most common commercial detector models utilized, such as the following, among others:

(i) the PerkinElmer Wizard automatic gamma counters; and a few examples are described by Lodge et al. (2015), Watanabe et al. (2016), Wingelhofer et al. (2016), Chen et al. (2017), Kim et al. (2017), Monteiro et al. (2017, 2018), Sims-Williams et al. (2017), Wang et al. (2017), Dal Maso et al. (2018), Hadj Rabia et al. (2018), Li et al. (2018), Makino et al. (2018), O'Hara et al. (2018a,b), and Okamura et al. (2018),

(ii) the Hidex AMG automatic gamma counter (Hellberg et al., 2017; Westrøm et al., 2017; Muuri et al., 2018), or

(iii) the Hitachi Aloka ARC automatic gamma counter (Yamamoto et al., 2017; Oshima et al., 2017a,b; Zhu et al., 2018a,b).

Automatic gamma counters are commonly used to analyze the activities of positron-emitting radionuclides utilized in PET as evidenced by the numerous citations provided in the previous paragraph. The 0.51 MeV gamma-ray resulting from positron annihilation and the 1.02 MeV gamma-ray line from coincident detection of the two simultaneous 0.51 MeV emissions from positron annihilation are efficiently detected with the well-type NaI(Tl) detector. Lodge et al. (2015) evaluated the performance of a PerkinElmer Wizard automatic well-type gamma counter for the activity analysis of positron emitting radionuclides, which is equipped with NaI(Tl) crystal detector measuring 75 mm in diameter and 80 mm height with a center well measuring 33 mm in diameter and 60 mm deep hole. Radioactive samples are automatically transferred from a sample test-tube rack to the inside of the well detector. The well-type detector design provides improved detection efficiencies over the through-hole detector, as the crystal

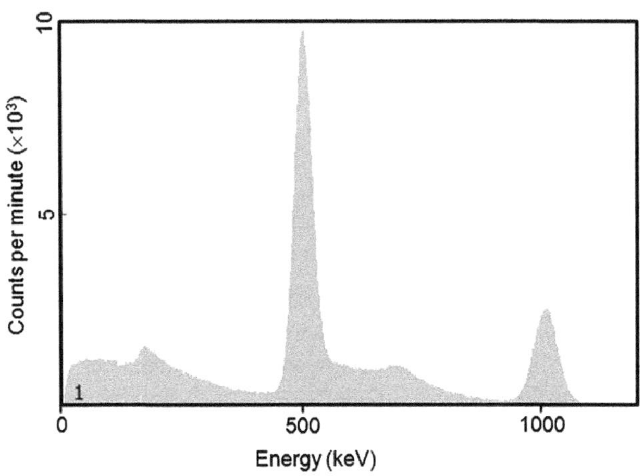

FIGURE 9.42 ^{18}F positron annihilation gamma energy spectrum measured with the PerkinElmer 2480 Wizard² automatic gamma counter. With the sample located in a sample tube inside the NaI(Tl) detector well a photopeak corresponding to detection of a single 511 keV positron annihilation gamma-ray photon and a coincidence sum peak at 1.022 keV corresponding to the coincident detection of the two 511 annihilation photons. *From Lodge et al. (2015).*

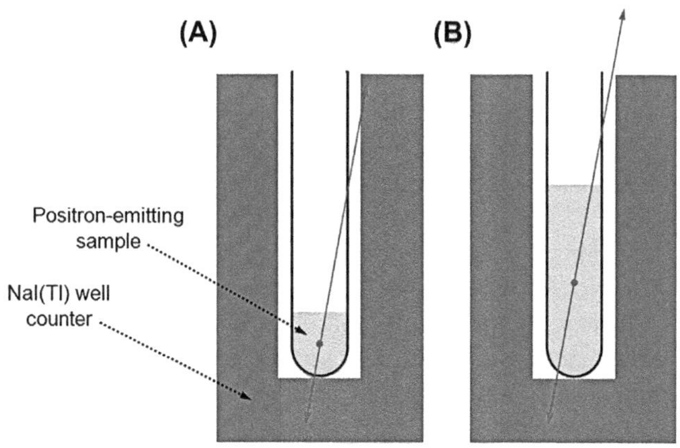

FIGURE 9.43 Sample volume effect. (A) An example pair of annihilation photons contributing to the coincidence sum peak of a positron-emitting radionuclide. (B) As the sample volume increases, there is an increasing likelihood that photons will escape from the hole at the top of the well. Counts lost from the coincidence sum peak due to reduced geometric efficiency become single-photon events. This mechanism tends to increase counts in the 511-keV peak, partially offsetting the loss of counts due to reduced geometric efficiency. The sample volume effect will occur also with gamma-emitting radionuclides. *From Lodge et al. (2015).*

detector surrounds the sample with the exception of the sample well opening.

A typical positron annihilation gamma spectrum from a positron emitting radionuclide is illustrated in Fig. 9.42. The spectrum will show a photopeak corresponding to a single 0.511 MeV positron annihilation gamma ray and a smaller coincidence sum peak at 1.022 MeV corresponding to the simultaneous detection of the two 0.511 MeV gamma-ray photons that are emitted in opposite directions following positron annihilation.

The sample volume will affect the counting geometry. In well-type detectors, there is an increased possibility of gamma-ray photon escape without interaction with the detector as the sample volume increases. This is illustrated in Fig. 9.43 where the detection of annihilation gamma-ray photons resulting from positron-emitting radionuclides has a greater possibility of escaping detection through the well opening, as the sample volume increases. This effect of sample volume on detection efficiency will occur with all gamma emitting radionuclides in addition to nuclides that produce annihilation gamma-ray photons from positron emission. It is therefore necessary to determine the effect of sample volume on detection efficiency, which will differ for each detector and radionuclide; and a plot of detection efficiency versus sample volume will provide a measure of sample geometry effects. Lodge et al. (2015) studied sample volume effects on the detection efficiency of positron and gamma-ray emitting radionuclides as a function of window width illustrated in Fig. 9.44. As illustrated in Fig. 9.44, the volume effect is minimal when measuring a positron-emitting radionuclide (e.g., ^{11}C or ^{18}F) in the 511 keV photopeak region (409–613 keV) because the escape of one of an annihilation photon with reduce the coincidence sum peak at 1.022 keV; but the same positron annihilation may be registered by the detector in the 511 keV photopeak region. For gamma-emitting radionuclides (e.g., ^{125}I, ^{99m}Tc, and ^{51}Cr), we do not have the benefit of coincident annihilation photons; and an escape of gamma-ray photons from the hole of the detector well should result in a reduction of detection efficiency. The counting of positron emitters in a wide-open counting region or window, which would encompass the 511 keV photopeak and the coincidence 1.022 keV sum peak, shows a more significant effect of sample volume on detection efficiency, as illustrated in Fig. 9.44, due to the escape of annihilation photons from the well opening.

The proximity of the multiple detectors to each other and to samples in the sample changer combined with the penetrating power or range of high-energy gamma radiation emitted by some radionuclides could give rise to possible error from cross-talk. We can define cross-talk as the registration of radiation events by a detector when those events originate from samples other than the sample in the detector.

There are two types of detector cross-talk, namely (1) detector-to-detector cross-talk, which occurs when a detector registers one or more events produced by radiation emanating from a sample located in another adjacent detector, and (2) deck-to-detector cross-talk, which occurs when a detector registers events arising from radiation emitted by samples on the sample changer deck. In multidetector automatic gamma analysis instrumentation the gamma radiation from one or more samples may be of

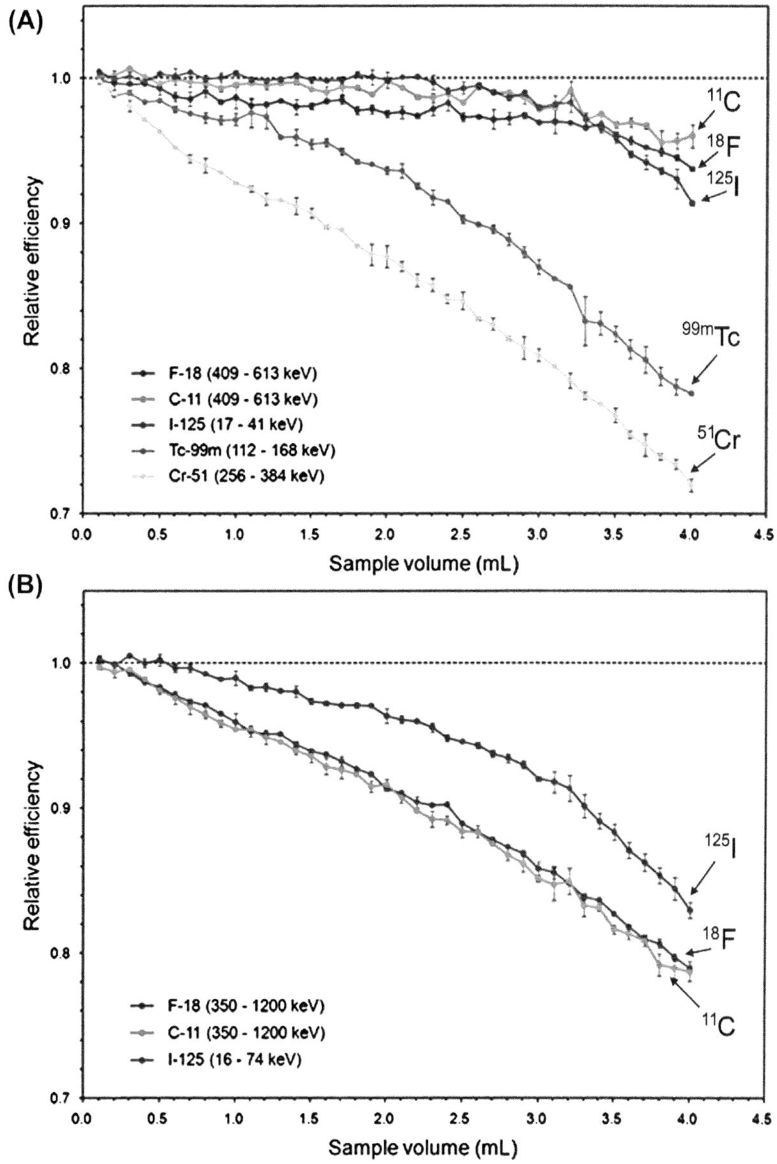

FIGURE 9.44 Relative efficiency as a function of sample volume for five different isotopes: ^{18}F, ^{11}C, ^{125}I, ^{99m}Tc, and ^{51}Cr with the PerkinElmer 2480 Wizard^2 automatic gamma counter. Energy windows are indicated in the Figure. (A) Data acquired in a window centered on the lower photopeak where detection efficiencies for a sample of 4 mL relative to those at 0.1 mL are 0.96, 0.94, 0.91, 0.78, and 0.72 for ^{11}C, ^{18}F, ^{125}I, ^{99m}Tc, and ^{51}Cr, respectively; and (B) data acquired in a wide window encompassing both the lower photopeak and the coincidence sum peak. *From Lodge et al. (2015).*

sufficient energy and intensity to reach neighboring detectors and register events in the neighboring detectors.

Detector-to-detector cross-talk can be calculated by determining the count rate (CPM) from a detector containing a blank sample and dividing that count rate by the disintegration rate (DPM) of a sample in an adjacent detector. The ratio is expressed as a percentage. Also, conveyor-to-detector cross-talk could arise from samples that are conveyed by the instrument robotics from the sample deck adjacent to a detector. Conveyor-to-detector cross-talk is also measured with a detector containing a blank sample and dividing that count rate by the disintegration rate of the closest sample in the sample deck, as described by Lodge et al. (2015). When cross-talk is suspected, the degree of cross-talk should be measured; and sample count rates corrected for cross-talk. Modern automatic gamma analyzers can employ a mathematical matrix calculation to correct for detector cross-talk; and they are equipped with sufficient lead shielding to reduce cross-talk.

Typical detection efficiencies and cross-talk for gamma-emitting radionuclides with a modern automatic gamma counter with a NaI(Tl) detector measuring 50 mm in height and 32 mm in diameter are provided in Table 9.6.

TABLE 9.6 Detection efficiencies, energy resolution, and crosstalk for various gamma-emitting radionuclides with the PerkinElmer 2470 Wizard[2] automatic gamma counter.[a]

			Cross-talk[c]	
Radionuclide	Detection efficiency[b]	Energy resolution	Detector-to-detector	Conveyor-to-detector
^{125}I	78%	<30%	Negligible	Negligible
^{129}I	58	<30	Negligible	Negligible
^{51}Cr	3	<14	<0.5%	Negligible
^{137}Cs	26	<12	<4	<0.12%
^{58}Co	3.5	<8	<5	<0.2

[a]*Data from PerkinElmer, http://www.perkinelmer.com/lab-solutions/resources/docs/SPC_01_WIZARD_2470_Final.pdf.*
[b]*Calculated as CPM/DPM × 100 for a window of 15–1000 keV.*
[c]*Crosstalk is a function of gamma-ray photon energy and intensity.*

2. Multiuser automatic gamma activity analysis

Automatic gamma counters are computer based. The entire operation of the gamma analyzer, including the sample changing, counting, single- and multiple-radionuclide spectral analysis, count rate statistical analysis, isotope decay corrections, activity analysis, and data processing, as well as instrument calibration and performance assessment, is controlled and completed by computer. It is necessary only to place the sample tubes in the appropriate sample cassettes and to input the basic information that the computer needs to carry out the automatic hands-off gamma analysis. Some of the major features of a typical automatic gamma analyzer will be outlined briefly here.

The modern automatic gamma analyzer has no toggle switches, buttons, or dials for the operator to set up and carry out an instrumental gamma analysis program. The one and only switch on the instrument is the main power switch, which is kept in the "On" position at all times. The only means of controlling the instrument is via the computer keyboard. Up to 999 user-definable analysis protocols may be available depending on detector model, and a user will select and name or identify one of the analysis protocols and input the necessary information into the instrument computer to define how the instrument will analyze his or her samples. Therefore, an analysis protocol is simply a customized procedure edited by the user that the instrument will implement to count and analyze a particular set of samples. The user will first identify and edit a protocol. The computer keyboard procedure for editing a counting protocol should be user friendly because the instrument should provide on the computer screen the available options for the user on what may be imputed as instructions to the computer. Editing an analysis protocol on the computer screen involves (1) naming the protocol, which helps the user identify that protocol for future applications and also informs others that the protocol belongs to a particular application or user, (2) defining the count conditions and parameters, which control the counting of samples, and (3) defining the data reduction parameters, which can include assay-specific data calculations, curve fitting, and data printout. Automatic gamma analysis of as many as 1000 samples and the simultaneous counting of 10 samples in 10 well-type NaI(Tl) detectors are possible in commercially available automatic gamma counters, such as the PerkinElmer Wizard gamma counters.

3. Multiple gamma-emitting nuclide analysis

There are two options for measuring multiple gamma-emitting radionuclides in mixtures, namely the solid scintillation crystal detector or the germanium semiconductor detector. The solid inorganic crystal scintillation detectors provide higher detection efficiencies than the semiconductor detector. The higher detection efficiencies of the crystal scintillators require shorter counting times for lower activity samples, but the semiconductor detector yields much improved energy peak resolutions permitting the more facile quantification of complex mixtures particularly when the photopeaks of different nuclides in the mixture are close (L'Annunziata, 1984, 1987; Roig et al., 1999). The differences between the detection efficiencies and resolutions achieved by the crystal scintillators (NaI(Tl) and $LaBr_3$) and semiconductor (HPGe) detectors were measured by Bender et al. (2015) and illustrated in Fig. 9.45. The differences in energy resolution and detection efficiency between the solid scintillator and semiconductor detector are underscored by Roig et al. (1999), who reported a detection efficiency and photopeak resolution of 22% and 60 keV FWHM at 661.7 keV of ^{137}Cs, respectively, for a 3 inch NaI(Tl) detector, whereas the lower 1.7% detection efficiency but much improved

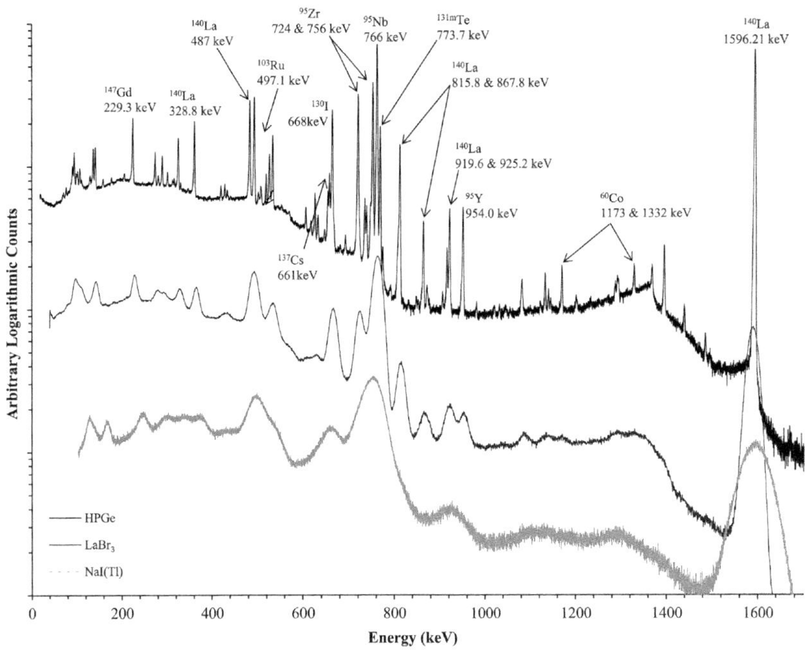

FIGURE 9.45 Comparison of gamma-ray spectra from a nuclear fuel element measured using each detector (NaI(Tl), $LaBr_3$ and HPGe) which highlights the difference in energy resolution. *From Bender et al. (2015); reprinted with permission from Elsevier © 2015.*

1.4 keV resolution for the 661.7 keV photopeak of ^{137}Cs was obtained with the germanium semiconductor detector. The high detection efficiencies of inorganic crystal scintillation detectors makes these detectors very desirable for the analysis of single radionuclides or, of multiple radionuclides when one of the following conditions can be met: (1) photopeaks of these have energies separated enough to count the areas under the photopeaks of each radionuclide of the mixture, or (2) where corrections for any overlap or spillover of activity from interfering nuclides may be calculated, or (3) the complex spectrum of multiple radionuclides may be mathematically unfolded to provide the intensities of the photopeaks produced by the various radionuclides in the mixture.

Simultaneous analysis of more than one gamma-emitting nuclide in the same sample with the NaI(Tl) scintillation analyzer is common. It is necessary only that the photopeaks of the nuclides be separated enough so that counting regions can be defined to encompass the photopeaks of each nuclide. This will maximize the counts collected for each nuclide in each region, although there can be overlap of counts from the other radionuclide in each counting region. The following sections provide examples of the simplest case of two nuclides in a mixture and a more complicated case of several nuclides in the same sample.

a. Dual-nuclide analysis

The gamma spectrum of two nuclides in the same sample was seen previously in Fig. 9.29 for the case of a mixture of ^{137}Cs and ^{65}Zn. In this case two counting regions defined by setting the LL and UL energy or pulse height discriminator settings of A and B (550 and 800 keV) for the ^{137}Cs counting region and A′ and B′ (950 and 1250 keV) for the ^{65}Zn counting region, respectively. The counting regions encompass the photopeaks of each radionuclide; however, the region set for counting the lower energy gamma emitter ^{137}Cs also contains counts arising from Compton smear of the higher energy nuclide ^{65}Zn. This is known as spilldown because counts from the high-energy gamma emitter "spill down" into the counting region of the lower-energy gamma emitter. It is necessary, therefore, to calculate the percentage of the counts in the ^{137}Cs region that are due to ^{65}Zn to correct for the spilldown. Often, the photopeaks of both radionuclides of a mixture have counts arising from each nuclide. We would, therefore, also have to correct for spillup, which is the percentage of counts that "spill up" into the counting region of the higher-energy gamma emitter from the lower-energy gamma emitter of the mixture.

The spilldown and spillup corrections that must be made can be illustrated easily by using the dual nuclide $^{125}I-^{57}Co$ as an example. This is a dual radionuclide tracer used for the common folic acid-vitamin B_{12} assays in serum samples (Jing et al., 2015; Hampel et al., 2014; Aléisso et al., 2011; Kerkeni et al., 2009; Lim et al., 2008; Lin and Yin, 2009; Maioli et al., 2009). Two counting regions are fixed, namely region A defined by setting the LL-UL discriminators of 15–75 keV for the ^{125}I counting region (35-keV photopeak) and region B defined by the LL and UL discriminator settings of 75–165 keV for the ^{57}Co counting region (122-keV main photopeak). The spectra of the two radionuclides overlap. The CPM of ^{125}I in the mixture is obtained from the count rate in region A after subtracting the counts in that region due to spilldown from the higher-energy gamma emitter ^{57}Co. Also, the CPM of ^{57}Co in the mixture is determined from the count rate in region B after subtracting the counts in that region due to the spillup from the lower-energy nuclide ^{125}I.

The spilldown and spillup correction factors are determined by counting the pure samples of each nuclide in each counting region. If regions A and B are the counting regions for the lower and higher-energy nuclides, respectively, we can determine the spilldown and spillup correction factors as follows: First, a pure sample of the higher-energy nuclide ^{57}Co is taken and counted in both regions and the spilldown correction factor calculated as

$$\text{Spilldown} = (\text{CPM region A} / \text{CPM region B})(100) \tag{9.36}$$

The spilldown calculation gives the percentage of counts in the counting region of the lower-energy nuclide generated by the higher-energy radionuclide. Then a pure sample of the lower-energy radionuclide is taken and counted in the two counting regions. From the count rates in these regions the spillup correction factor is calculated as

$$\text{spillup} = (\text{CPM region B} / \text{CPM region A})(100) \tag{9.37}$$

The spillup calculation gives the percentage of counts in the counting region of the higher-energy nuclide generated by the lower-energy nuclide.

Once the percent spilldown and spillup correction factors are determined, the values are imputed into the counting protocol, and the automatic gamma analyzer will make the calculations required for the corrected and final count values for the two radionuclides in the mixture. The following equations form the basis upon which the count rates of the two nuclides in the mixture are calculated:

$$y_1 = x_1 + s_2x_2 \tag{9.38}$$

$$y_2 = x_2 + s_1x_1 \tag{9.39}$$

where y_1 and y_2 are the total number of counts from a dual-nuclide mixture in photopeak counting regions 1 and 2 (regions A and B), $x_1$and $x_2$are the unknown number of counts of nuclides 1 and 2 in their photopeak regions, and s_1 and s_2 are constants, which are the spillover values for nuclides 1 and 2, respectively. Typical values for the spilldown and spillup for the ^{125}I–^{57}Co dual nuclide with the PerkinElmer Cobra automatic gamma analyzer are 3.00% or 0.03 for the spilldown value (s_2) of nuclide 2 (^{57}Co) into the counting region of ^{125}I and 0.80% or 0.008 for the spillup value (s_1) of nuclide 1 (^{125}I) into the counting region of ^{57}Co. However, the magnitude of spilldown of ^{57}Co into the ^{125}I region is reportedly reduced from the 3.0% to <1.0% when utilizing the preset counting regions utilized by the PerkinElmer 2480 Wizard[2] automatic gamma counter (http://www.perkinelmer.com/CMS Resources/Images/46-129502SPC_02_WIZARD_2480_Final.pdf). Eqs. (9.38) and (9.39) are solved simultaneously for the two unknown values of $x_1$and x_2, which are the corrected count rate values of nuclides 1 and 2.

The availability of ^{125}I-labeled human serum albumin has renewed interest is using this isotope tracer together with ^{51}Cr-labeled red blood cells to measure simultaneously plasma volume (PV) and red cell volume (RCV), as reported by Gómez Perales (2015). The photopeaks of ^{125}I and ^{51}Cr are at 18 and 320 keV, respectively, and the two NaI(Tl) detector photopeaks overlap significantly. Gómez Perales (2015) describes the count rate determinations of the dual radionuclide tracers by the counting of the two radionuclides in their photopeak windows or gates as previously determined by Wood and Levitt (1965). The method is based on determining the proportions of the two radionuclide counts that appear in each counting window or gate (i.e., gates 1 and 2 with discriminators set to encompass the major photopeaks of each radionuclide) and then solving two equations with two unknowns to resolve the count rates of the two radionuclides as follows:

$$K_1I + K_2C = A \quad (\text{gate } 1) \tag{9.40}$$

and

$$K_3I + K_4C = B \quad (\text{gate } 2) \tag{9.41}$$

where I = total count (i.e., combined count) of ^{125}I in gates 1 and 2, C = the total count of ^{51}Cr in both gates, A = the observed count in gate 1 and B = the observed count in gate 2, and K_1 and K_2 are the proportions of the activity of ^{125}I and ^{51}Cr in the total observed count A in gate 1, and K_3 and K_4 are the proportions of the activity of ^{125}I and ^{51}Cr in the total observed count B in gate 2. The coefficients K_1 through K_4 are determined by counting samples of pure ^{125}I and pure ^{51}Cr in the two gates as follows:

K_1 = amount of ^{125}I counted in gate 1/amount of ^{125}I counted in gate 1 and gate 2.
K_2 = amount of ^{51}Cr counted in gate 1/amount of ^{51}Cr counted in gate 1 and gate 2.
K_3 = amount of ^{125}I counted in gate 2/amount of ^{125}I counted in gate 1 and gate 2.
K_4 = amount of ^{51}Cr counted in gate 2/amount of ^{51}Cr counted in gate 1 and gate 2.

The two Eqs. (9.40) and (9.41) containing two unknowns are solved for I and C, namely the total counts for ^{125}I and ^{51}Cr in both gates, to give

$$I = (K_4A - K_2B)/(K_1K_4 - K_2K_3) \tag{9.42}$$

and

$$C = (K_1B - K_3A)/(K_1K_4 - K_2K_3) \tag{9.43}$$

Other dual-radionuclides combinations that are measured by solid scintillation gamma spectrometry are ^{99}Tc$-^{131}$I as renal tubular tracers (Lipowska et al., 2017), ^{125}I$-^{131}$I (Vaidyanathan et al., 2018, 2012, 2011; Choi et al., 2014; Zalutsky et al., 2012; Chen et al., 2011; Jia et al., 2011; Medine et al., 2010; Miyamoto et al., 2010; Wilbur et al., 2010; Yeşilagâç et al., 2011), and ^{57}Co$-^{58}$Co used in the classical Schilling test to diagnose deficiencies in vitamin B_{12} absorption (Chen and Chan, 2011; Nay et al., 2011; Malizia et al., 2010; Mazokopakis et al., 2009; Tzellos et al., 2009). The technique of calculating percent spilldown and percent spillup of the overlapping photopeaks is also required in these dual-radionuclide measurements.

b. Spectrum unfolding of multiple radionuclide spectra

The simplest case of the solid scintillation gamma analysis of two nuclides in a mixture is described in the previous section. The solid scintillation analysis of multiple-nuclide mixtures (more than two radionuclides) gives rise to a more complex pulse height spectrum where the counting regions, which are selected to surround the photopeaks of each nuclide of the mixture, will register counts from the nuclide of interest as well as counts from other nuclides spilling up and spilling down into the counting regions. An example of such a case is illustrated in Fig. 9.46, which illustrates the gamma spectrum of a mixture of six radionuclides and six counting regions (regions A through F) selected to encompass the photopeaks of each nuclide. In the fashion of Eqs. (9.38)–(9.41), six equations can be written to describe the counts collected in each of the six counting regions, as illustrated in Fig. 9.46, including the correction factors for the spillup and spilldown of counts from radionuclides in each counting region. The six equations need to be solved simultaneously. To facilitate the analysis of multiple radionuclides of this type, computer software such as PC-GERDA V2.9 can be employed with the automatic gamma analyzer for data processing with corrections for background, radionuclide decay, and spillover. The application of PC-GERDA software for the measurement of regional myocardial blood flow, mechanisms and consequences of vascular remodeling, and coronary vasoregulation, among other applications by solid scintillation gamma spectrometry of radionuclide mixtures is described in detail in the literature (Stabley et al., 2013; Broisat et al., 2011; Dominguez et al., 2010; Hatada et al., 2006; Kingma et al., 2006; Krause et al., 2004; Riou et al., 2003; Ruiz et al., 2002). Since 2013, the scientific literature ceased to make direct reference to the use of PC-GERDA software for the analysis of multiple radionuclides by solid scintillation gamma-ray spectrometry. This is probably due to the fact that PerkinElmer automatic gamma counters, such as the 2480 Wizard[2], is equipped with the software for the

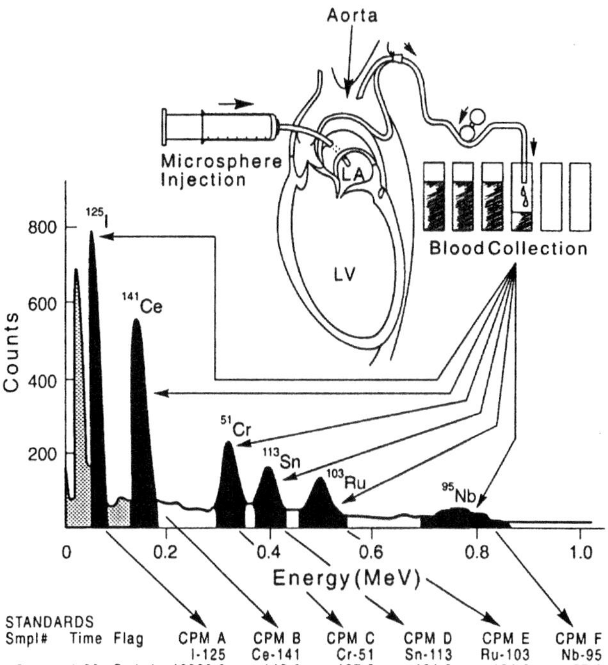

STANDARDS

Smpl#	Time	Flag	CPM A I-125	CPM B Ce-141	CPM C Cr-51	CPM D Sn-113	CPM E Ru-103	CPM F Nb-95
2	1.00	S 1-1	46980.0	142.0	127.0	101.0	124.0	57.0
3	1.00	S 2-1	11564.0	853275.	634.0	134.0	124.0	90.0
4	1.00	S 3-1	12977.0	26220.0	333936.	704.0	239.0	84.0
5	1.00	S 4-1	25232.0	40865.0	20562.0	401218.	738.0	599.0
6	1.00	S 5-1	10882.0	18717.0	8122.0	4309.0	151486.	116.0
7	1.00	S 6-1	28155.0	55642.0	62314.0	53181.0	56565.0	42689.0

FIGURE 9.46 Six photopeaks from a composite spectrum of ^{125}I, ^{141}Ce, ^{51}Cr, ^{113}Sn, ^{103}Ru, and ^{95}Nb and count rates (CPM) obtained from six counting regions after spillover corrections were applied using PCGERDA software for calculating regional blood flow with radionuclide-labeled microspheres. *© 1998–2010 PerkinElmer, Inc. Printed with permission.*

automatic analysis of up to six different radionuclides with automatic spillup and spilldown calculations (http://www.perkinelmer.com/CMSResources/Images/46-129502SPC_02_WIZARD_2480_Final.pdf).

The poor resolution of inorganic crystal scintillation detectors may be overcome with the use of multivariate calibration methods (Roig et al., 1999). Multivariate calibration techniques were discussed in Chapter 6, Volume 1. Among these methods are multivariate regression (MLR), partial least squares regression (PLS), and principal component analysis (PCA), which can be used to determine amounts of radionuclides of interest although other radionuclides are present that affect the experimental signal provided the isotopes are present in both the unknown samples and a calibration set (Coble et al., 2017; Hellesen et al., 2017; Alharbi, 2016; Liu et al., 2016; Dayman et al., 2014; Orton et al., 2011, 2012; Toribo et al., 1995; Martin and Naes, 1989). The advantage is that these methods have proven successful for the analysis of samples with unknown interferences. The MLR and PLS methods were tested by Roig et al. (1999) to determine activities of ^{85}Sr, ^{134}Cs, and ^{110m}Ag in a mixture, which produced solid scintillation spectra with overlapping photopeaks and Compton smear illustrated in Fig. 9.47. They used the

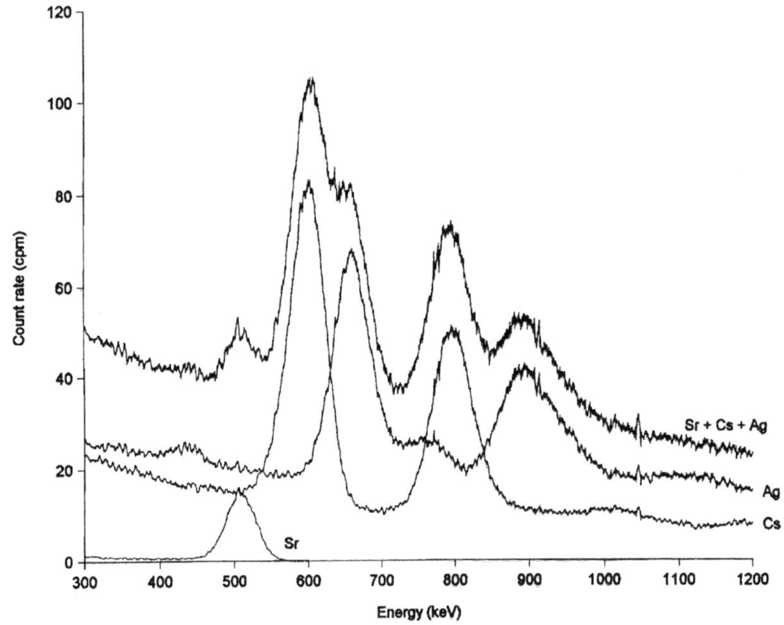

FIGURE 9.47 Smoothed spectra obtained with a NaI(Tl) solid scintillation detector (PerkinElmer/Packard Auto-Gamma 5530) of single solutions of ^{85}Sr, ^{134}Cs, and ^{110m}Ag and a sample with the three radionuclides (upper spectrum). Radionuclide activities (when present): 100 Bq of ^{85}Sr; 1390 Bq of ^{134}Cs; 1220 Bq of ^{110m}Ag. *From Roig et al. (1999); reprinted with permission from Elsevier © 1999.*

multivariate calibration, MATLAB (The Mathworks, Inc., Natick, Massachusetts, https://www.mathworks.com/) and the PLS-Toolbox for MATLAB written by Wise (1992). They tested the influence of different calibration sets on the prediction of radionuclide activities; and the prediction errors to samples without interferences for the aforementioned mixture were demonstrated to be <4%, which are comparable to errors obtainable with germanium detectors.

Spectrum unfolding methods have proven to be a powerful and practical tool in the measurement of radionuclides with poorly resolved spectra masked with a combination of background, photopeak spillover, and Compton smear. Other spectrum unfolding methods for the solid scintillation analysis of gamma-ray-emitting nuclides are described by Baré et al. (2010), Baré and Tondeur (2011), Mertens et al. (2007), Muravsky et al. (1998), Nguyen et al. (1996), Rahman et al. (2009), Skipper and Hangartner (2002), Sükösd et al. (1995) and Suzuki et al. (1993).

A water-tight housing for a NaI(Tl) detector was designed by van Put et al. (2004), Mertens et al. (2007), Baré et al. (2010), and Baré and Tondeur (2011) for the monitoring of gamma-emitting radionuclides in large bodies of water, such as lakes and oceans, at depths of up to 50 m. Their technique relies on the calculation of response functions of the detector by the Monte Carlo code MNCP4c2 (LANL, 2003) for gamma energies between 15 and 3000 keV; and an applications of the UMG 3.3 package (PTB, 2004) for unfolding the experimental spectra according to the response matrix built by the MNCP code (Baré et al., 2010, and Baré and Tondeur (2011); Mertens et al., 2007; van Put et al., 2004) simplified the application of the Monte Carlo code by establishing an isotropic source-detector geometry, that is, the source being a large body of water is considered to be an infinite, homogeneous, and constant density medium surrounding the water-tight NaI(Tl) detector and electronic console. Fig. 9.48 illustrates an experimental gamma spectrum obtained by Baré and Tondeur (2011) with 10 hours of count data collection with the NaI(Tl) detector assembly immersed in a 20 m^3 tank of simulated sea water containing 1 Bq/L of ^{139}Ce and ^{113}Sn, 12 Bq/L of ^{40}K by addition of KCl, plus 0.10 Bq/L of ^{137}Cs and an unspecified amount of ^{133}Ba. The only visible structures of the gamma spectrum obtained were the main ^{133}Ba peak at 356 keV and the ^{40}K peak at 1460 keV (see upper curve of Fig. 9.48). The unfolded spectrum was obtained by performing 6000 mathematical iterations using the GRAVEL iterative algorithm GRV_MC33 (see lower curve of Fig. 9.48). Baré and Tondeur (2011) demonstrate the power of the spectral unfolding method by a reduction of the minimal detectable activity (MDA) by a factor of 10 for the monitoring of environmental radioactivity in large bodies of water.

For laboratory and field measurements of soil samples Chiozzi et al. (2000a,b) used an EG&G Ortec NaI(Tl) detector and MCA for the measurement of radionuclides of uranium, thorium, and potassium. Samples of 0.65–0.85 kg were measured in polyethylene Marinelli beakers. The analysis was based on solid scintillation gamma spectra of the environmental samples, which provided adequately separated photopeaks for ^{214}Bi (^{238}U series), ^{208}Tl (^{232}Th series), and ^{40}K. Takiue et al. (1999) and

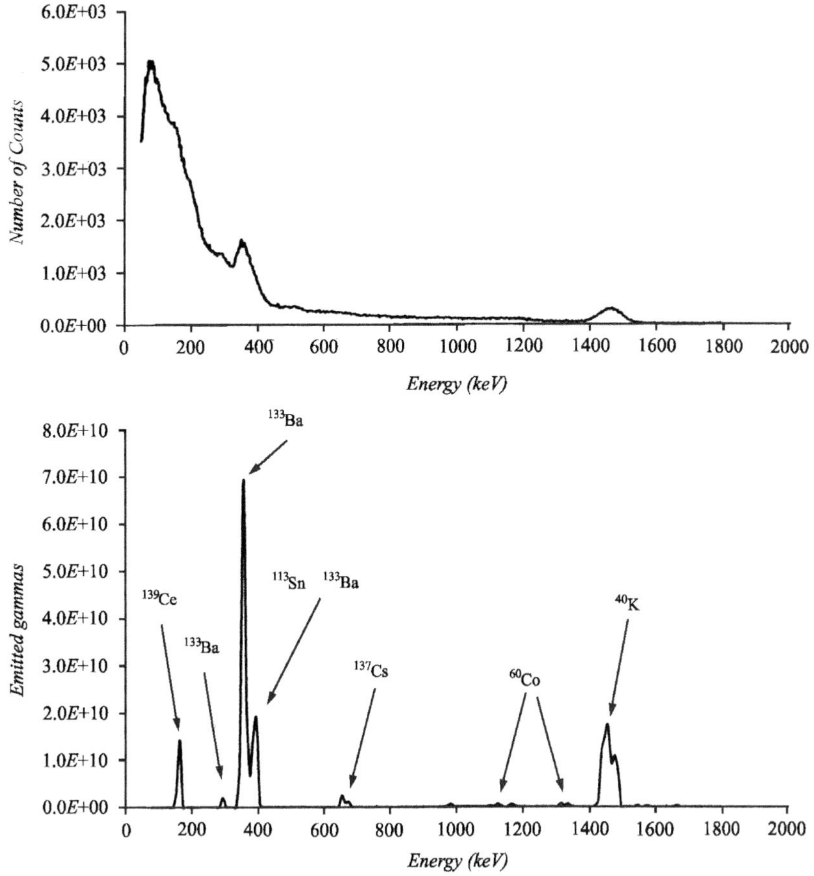

FIGURE 9.48 Top: Spectrum of 0.10 Bq/L of ^{137}Cs together with ^{40}K, ^{139}Cs, ^{133}Ba and ^{113}Sn obtained by a water-tight NaI(Tl) detector immersed in simulated sea water; Bottom: The unfolded spectrum. *From Baré and Tondeur (2011); reprinted with permission from Elsevier © 2011.*

Fujii et al. (1999, 2000) used the combined spectral information from a NaI(Tl) solid scintillation detector and liquid scintillation analyzer to measure seven beta-, gamma-, and X-ray-emitting nuclides in a mixture. The method is described in detail in Chapter 6, Volume 1.

B. Microplate scintillation analysis

The automated scintillation analyzers with the highest throughput today are the microplate counters, which can analyze all beta particle- and atomic electron-emitting nuclides by liquid or solid scintillation counting as well as carry out nonradioactive luminescence counting. These counters are referred to as microplate scintillation and luminescence counters. In this chapter, we will limit our treatment of this counter to only the solid scintillation counting of radionuclides. The use of these instruments to analyze radionuclides by liquid scintillation and Cherenkov counting are provided in Chapters 6, Volume 1 and Chapter 6, Volume 2, respectively.

1. Solid scintillation counting in microplates

Some commonly used microplate scintillation counters are the TopCount Microplate Scintillation and Luminescence Counter and the MicroBeta Plate Counter, both of which are products of PerkinElmer, Inc., Waltham, MA, USA, and the Hidex Sense Beta Plus Microplate Reader produced by Hidex, Turku, Finland. The TopCount microplate scintillation counter utilizes 2, 6, or 12 PMTs for the simultaneous counting of 2, 6, or 12 samples. The MicroBeta has 1, 2, 6, or 12 PMT detectors. The samples are counted in 24-, 96-, or 384-well microplates, which are stacked in the counter's microplate changing mechanism. The instrument moves the microplates one at a time for the simultaneous counting of up to 12 samples in separate microplate sample wells. Fig. 9.49 illustrates the location of a PMT relative to one of the microplate sample wells of a PerkinElmer TopCount scintillation counter. The highly reflective surface of the microplate well assures reflection of light photons from scintillation events toward the PMT.

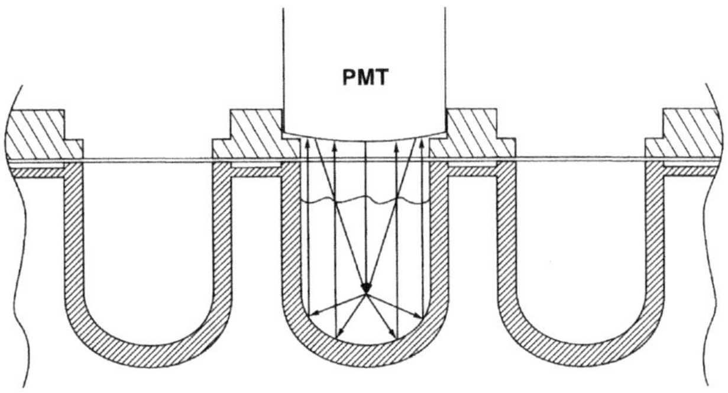

FIGURE 9.49 Diagram illustrating the position of a PMT and sample in a microplate well. In TopCount high-throughput automatic scintillation analysis, up to 12 PMTs are positioned simultaneously above 12 sample wells. In SPA aqueous liquid medium is in the sample well, which has a white reflective surface. In solid scintillation analysis the sample is dry and adhered to the surface of a LumaPlate plastic scintillator-coated well. *©1998 PerkinElmer, Inc. Printed with permission.*

Although only one PMT per sample is employed, the backgrounds are low due to a patented method of Time-Resolved Liquid Scintillation Counting (TR-LSC), which can distinguish between true scintillation events and background noise (US patent 4,651,006). The theory of TR-LSC for background reduction in TopCount liquid and solid scintillation counting is described in Chapter 6, Volume 1 and in a paper published by PerkinElmer (1991a). In the solid scintillation analysis of beta particle- and atomic electron-emitting nuclides the samples are analyzed in the dry state in the microplate wells as discussed further on in this section.

PerkinElmer MicroBeta Plate Counters are equipped with dual PMTs, which are situated above and below the sample well, with the utilization of microplates with transparent bottoms. The dual PMTs yield higher detection efficiencies with background rejection by coincidence counting. TopCount instruments utilize only one PMT above the sample well and TR-LSC for background rejection.

The 24-, 96-, and 384-well microplates contain wells oriented in a 6 × 4, 12 × 8, and 24 × 16 fashion along the length and width of each microplate. The maximum usable well volumes are 1500, 350, and 80 μL for the 24-, 96-, and 384-well microplates, respectively. Up to 40 microplates may be stacked in the counter microplate changing mechanism. If we consider the 384-well microplate, the stacking of 40 microplates in the counter would allow the unattended counting of 15,360 samples. The counter is computer controlled and fully automated for complete hands-off sample analysis.

Solid scintillation counting in microplates involves depositing a sample into the well of a special plastic scintillator microplate (e.g., LumaPlate, FlashPlate, ScintiPlate, and Cytostar-T). The microplate wells contain a scintillator embedded in clear plastic in one of two possible formats, namely, (1) a clear plastic scintillator is coated as a thin layer along the entire inside wall of a white opaque plastic microplate-well to be used with a microplate counter having only one PMT detector per sample well (e.g., TopCount counters) or (2) the scintillator is embedded into the microplate-well with a transparent bottom to be used with a microplate counter equipped with two PMT detectors per sample well (e.g., MicroBeta counters).

The material to be assayed can be dried and then counted in the microplate scintillation counter. In addition, solid scintillation microplate counting is used for the purpose of the off-line radioassay of high-performance liquid chromatography (HPLC) effluent by fraction collecting into the wells of a scintillation microplate. This is followed by a rapid drying of the collected fractions in the microplate wells and subsequent counting in a microplate scintillation counter. The method is referred to as the combination of HPLC and microplate scintillation counting (MSC) and abbreviated as HPLC-MSC or more simply abbreviated as LC-MSC. The effluent from an HPLC column can be split so that a fraction of the effluent (e.g., 75%) is collected into the wells of a microplate and the remaining HPLC effluent (e.g., 25%) goes on to a mass spectrometer for the structural identification of organic compounds as these are separated off the HPLC column. The combination of HPLC with MSC and mass spectrometry (MS) is abbreviated HPLC-MSC-MS. The radioactivity measured in the various fractions by the microplate counter provides a quantitative measure of the organic compounds in the HPLC effluent (Glaenzel et al., 2018; Gu et al., 2018; Zhuo and Zheng, 2018; Jensen et al., 2017; Dymond et al., 2016; Lacy et al., 2015; Tolonen and Pelkonen, 2015; Koppen et al., 2014; Tong et al., 2012; Barros et al., 2011; Kiffe et al., 2003, 2007; Dear et al., 2006, 2008; Bruin et al., 2006; Zhu et al., 2005a,b; Nedderman et al., 2004).

The basic sequence of combining HPLC with MSC and MS is illustrated in Fig. 9.50. The figure illustrates two possible approaches. One method shown in Fig. 9.50A would be to split the HPLC effluent so that a larger portion of the effluent (e.g., 75%) is directed to a fraction collector for the collection of fractions into the wells of a deep well scintillating microplate; and a smaller remaining portion of the effluent (e.g., 25%) is directed to a mass spectrometer for the analysis of the organic compounds separated on the liquid chromatogram. The on-line spectroscopic analysis of organic compounds in HPLC effluents is described by L'Annunziata and Nellis (2001a,b). The HPLC fractions collected in the wells of the scintillating microplates are vacuum dried; and the radioactivity measured with a microplate counter. A plot of the radioactivity versus the time in minutes for the collection of HPLC fraction will provide an HPLC radioprofile with peaks representing the radioisotope-labeled compounds eluted from the liquid

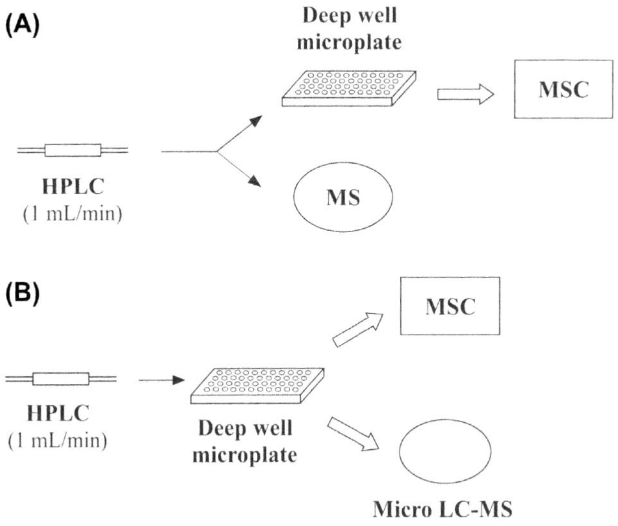

FIGURE 9.50 Setups of the combination HPLC-MSC with mass spectrometry. System A is the setup for simultaneous radioactivity profiling and metabolite identification. The HPLC eluent is split into both MSC and MS. System B is the setup for tandem analysis of metabolite profile by HPLC-MSC and metabolite characterization by micro LC-MS. *From Zhu et al. (2005a); reprinted with permission from Elsevier © 2005.*

chromatography column, as illustrated in Fig. 9.52. The count rates under each peak will provide a measure of the relative abundance of isotope labeled compound eluted based on the level of radioactivity applied to the column assuming a constant detection efficiency. The mass spectra determined in line with the radioactivity peaks will provide evidence for the identity of the compound eluted from the HPLC column. Another approach is illustrated in Fig. 9.50B where the HPLC effluent is not split; rather, it is collected as fractions into a scintillating microplate and submitted to tandem analysis by MSC and micro liquid chromatography MS (micro LC-MS).

The splitting of the effluent of a liquid chromatography column to allow a fraction of the effluent to go on to a mass spectrometer for organic molecular structure determination and another fraction to go on to be collected into the wells of scintillating microplates is illustrated in Fig. 9.51. From the effluent splitter the fraction going on to the scintillating microplate wells is controlled by a fraction collector, which moves stacked microplates and a syringe plunger in *X*, *Y*, and *Z* directions for the collection of the column effluent fractions into separate microplate wells. Typical fraction collectors are the Collect Pal (Leap Technologies, Inc., Morrisville, NC, USA, https://www.leaptec.com/) and the Gilson FC 204 Fraction Collector (Gilson, Inc., Middleton, WI, USA, http://www.gilson.com/en/default.aspx). After fraction collection the microplates are dried under vacuum and then counted in a microplate scintillation analyzer. The coupling of liquid chromatography with microplate scintillation counting (LC-MSC) offers an advantage over flow scintillation analysis (FSA) in the detection and measurement of radionuclide-labeled compounds as they are eluted from the chromatography column. The advantage of LC-MSC over FSA is an increased sensitivity because the counting time of the microplate scintillation analyzer can be controlled by the user; whereas the residence time of flow scintillation analyzers is relatively short at the flow rates commonly utilized in liquid chromatography. However, FSA offers the advantage of rapid real-time measurement of radioactivity in the effluent column (see Chapter 10, Volume 2 on Flow Scintillation Analysis). By this technique the count rates (CPM) or even the activities (DPM) of small amounts of ^{3}H- or ^{14}C-labelled compounds separated by liquid chromatography can be determined. The process of drying the fractions in the wells of the plastic scintillator microplates places the ^{3}H- or ^{14}C-labelled compounds into intimate contact with the scintillator-embedded plastic

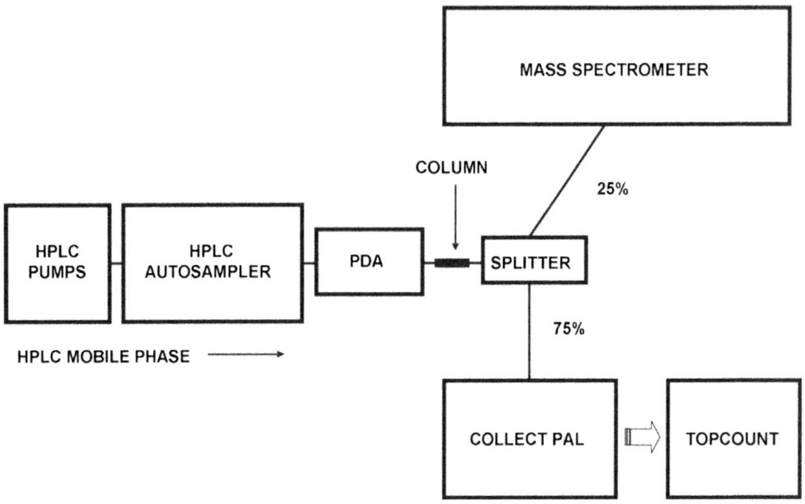

FIGURE 9.51 Typical system configuration for the HPLC radioprofiling with simultaneous mass spectrometry analysis. A photodiode array (PDA) is a UV/Vis detector generally utilized in liquid chromatography systems; and the effluent splitter in this case is shown to direct 25% of the effluent on to the mass spectrometer and the remaining 75% on to the Collect Pal fraction collector. From the fraction collector the samples are dried over night and then counted with a microplate scintillation analyzer, e.g., TopCount. *From Barros et al. (2011); reprinted with permission from Elsevier © 2011.*

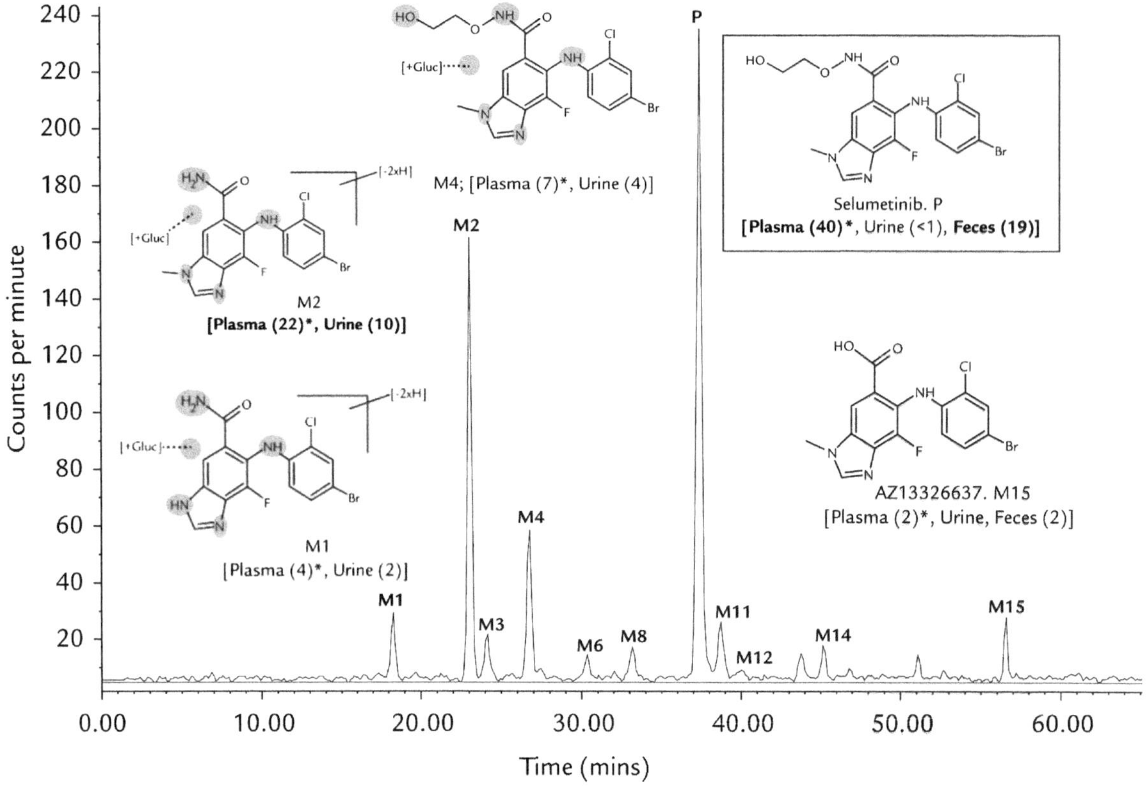

FIGURE 9.52 A radioprofile constructed from data of radioactivity measured in fractions of an HPLC effluent of the carbon-14 labeled drug Selumetinib (most intense peak marked P) and its metabolites in human plasma collected in the wells of scintillating microplates (ScintiPlates) after a single 75-mg oral dose administered to a representative subject. Square brackets indicate matrices in which metabolite was detected (only the radiochromatogram of human plasma extract is illustrated). Values in parentheses indicate mean percentage of dose excreted as each metabolite. *Plasma values indicate percentage chromatogram only. Bold type denotes >10% dose excreted as this metabolite in this matrix or >10% of the radioactivity in the plasma chromatogram. All of the drug metabolites of the numbered peaks were identified; but not illustrated in the figure. *From Dymond et al. (2016); reprinted with permission from Elsevier © 2016.*

coating of the microplate well. One should keep in mind that the drying process can result in loss of volatile organic compounds; and self-absorption of low-energy beta particles can occur in dried samples. Nevertheless, the LC-MSC technique has proven to be a valuable tool in studies involving the chromatographic separation and measurement of radioisotope-labeled organic compounds. A typical liquid radiochromatogram illustrating the radioactivity peaks of ^{14}C-labelled organic compounds eluted from a liquid chromatography column and collected in 96-well scintillating microplates (ScintiPlates) and the identification of eluted ^{14}C-labeled compounds by MS is illustrated in Fig. 9.52. The HPLC postcolumn effluent was split (4:1) with the majority of the flow directed to the fraction collector into the ScintiPlate wells; and the remainder of the flow directed to a mass spectrometer for organic structural analysis (Dymond et al., 2016).

In certain circumstances the solid scintillation counting technique in plastic scintillator microplates can replace conventional liquid scintillation analysis in counting vials when nonvolatile forms of the radionuclides and small samples need to be analyzed. MSC offers the following advantages: (1) the samples are safer to handle because of their solid nonvolatile state; (2) waste disposal costs are reduced due to the small volumes of the plastic microplates and the solid state of the waste; (3) samples may be recovered from the wells after counting for further studies, if they are stable in the dry state; (4) once in the wells, the sample positions cannot be changed mistakenly by laboratory personnel, which ensures positive sample identification; and (5) sample throughput is increased because as many as 12 samples in wells can be analyzed simultaneously.

The procedure involves simply applying the samples to the wells of a plastic-scintillator microplate (e.g., LumaPlate). The plates are dried in a fume hood overnight, or a heat lamp, hair dryer, or a centrifugal evaporator may be used to facilitate sample drying. Prior to counting, the microplate is sealed with cover film (e.g., TopSeal) to facilitate handling and to prevent contamination of the samples and counter. Börnsen (2000) and PerkinElmer (1996) provide detailed information on sample handling and performance for solid scintillation counting with the LumaPlate.

The TopCount solid scintillation counting efficiencies (%) reported by Börnsen (2000) for various radionuclide-labeled compounds deposited in separate wells of the 96-well LumaPlate are ^{3}H (49%), ^{14}C (85%), ^{32}P (87%), ^{51}Cr (24%), and ^{125}I (75%) with backgrounds of only 8–9 CPM. The low backgrounds provide high figures of merit ($FOM = E^2/B$). The 24-well LumaPlate will provide significantly higher counting efficiencies, for example, ^{3}H (55%), ^{14}C (95%), ^{32}P (93%), ^{51}Cr (48%), and ^{125}I (83%), but with higher backgrounds of about 19–20 CPM. The FlashPlate is used generally for scintillation proximity assays (SPAs) discussed further on in this chapter. Some examples of the solid scintillation analysis of various radionuclides with the ScintiPlate, LumaPlate, and Cytostar T microplates are found in the following works: ^{3}H (Omori et al., 2015), ^{14}C (Dymond et al., 2016; Koppen et al., 2014), ^{18}F (Tang et al., 2017), ^{32}P (Crosbie et al., 2013, 2012), ^{51}Cr (Aguilar et al., 2017; Ando et al., 2017; Gravelle et al., 2014), and ^{125}I (Hallinger et al., 2017).

Because the scintillation phenomenon occurs in the solid scintillation detector (wall of the microplate well) separate from the dried sample, there is no chemical quench. However, if the sample residues along the walls of the microplate wells are colored, it is possible to have quenching from the absorption of the scintillation photons from color. The effect of color quenching can be determined with colored agents as described subsequently in Section V.B.2.f of this chapter and in a report by PerkinElmer (1996). A quench correction curve of counting efficiency (E) versus a quench-indicating parameter (QIP) (e.g., transformed spectral index of the sample, tSIS) is prepared with color-quenched radionuclide standards. The quench correction curve is, in turn, used by the instrument to convert count rates (CPM) to activity (DPM) according to $DPM = CPM/E$.

The close proximity of the wells and PMTs could contribute to well-to-well cross-talk, which can be defined as pulse events registered by one PMT arising from nuclear radiation originating in a neighboring well. The LumaPlate structural material is opaque to light, which prevents optical well-to-well cross-talk. High-energy beta radiation and low-energy gamma rays could theoretically pass through the walls of one well and produce a scintillation event in the walls of another. Cross-talk expressed as percentage is defined and calculated as described in Section V.A.1 of this chapter. Cross-talk with the TopCount plate counters for the 96-well PicoPlates are reported by Börnsen (2000) to be about 1% or less for radionuclides emitting penetrating radiation such as ^{32}P and ^{125}I; insignificant for ^{14}C and ^{51}Cr, and undetectable for ^{3}H.

The plastic scintillator microplates serve as excellent platforms for many SPAs such as immunoassays, receptor binding assays, enzyme activity, and other bioassays. These techniques carried out with microplate solid scintillation counting are described in the following paragraphs.

2. Scintillation proximity assay

SPA is a technology for the analysis of binding reactions, commonly studied in the medical and biochemical sciences, which circumvents the need to separate bound from free fractions. Glass or plastic solid scintillator microspheres or scintillating microplates are used in this assay together with an isotope-labeled ligand. With the use of modern liquid handling equipment and an automatic scintillation counter for samples in a microplate format, hundreds of samples may be prepared and analyzed in a single day, because traditional processes for the separation of bound and free fractions are not required. This section provides a brief treatment of SPA. Additional sources of general information on SPA are available from PerkinElmer (2018), Wildey et al. (2017), Xia et al. (2016), Sadena and Sadena (2011), Heuvel (2010), Alcami and Viejo-Borbolla (2009), Glickman et al. (2008), Flaumenhaft (2007), Minor (2006), Schnurr et al. (2007), de Jong et al. (2005), Jones et al. (2003), Park et al. (1997), Cook (1996), and PerkinElmer (1991b, 1997). Additional detailed information and procedures may be obtained from references cited in the following paragraphs.

a. Basic principles

SPA makes use of the weak (low-energy) radiation emitted by certain radionuclides, such as the very low-energy beta-particles emitted by ^{3}H ($E_{max} = 18.6$ keV) or the weak Auger electrons emitted by ^{125}I ($E_{max} = 30$ keV). The radiations emitted by these radionuclides have very short ranges of travel in water. For example, the maximum range of ^{3}H beta particles in water is only ~ 25 μm (see Table 1.12 in Chapter 1). The Auger electrons of ^{125}I, which are of slightly higher energy, travel only a few micrometers further in water. Therefore, if these radionuclides are in aqueous solution their radiations will not be detectable unless they are in intimate or very close contact with a scintillating compound. SPA, therefore, makes use of yttrium silicate scintillating glass microspheres, polyvinyltoluene (PVT) scintillator microspheres, or scintillating sample wells of microplates, which emit visible light when the beta particles or Auger electrons from ^{3}H or ^{125}I are able to penetrate the scintillating glass or plastic. Most of the material in this section will focus on scintillator microspheres; but the principles apply also to scintillating sample wells of microplates. Because of the short ranges of travel of ^{3}H and ^{125}I radiations in water, their penetration into microsphere scintillators can happen only when the ^{3}H or ^{125}I nuclides are in intimate or very close contact with the microspheres. If the microsphere scintillators are coated

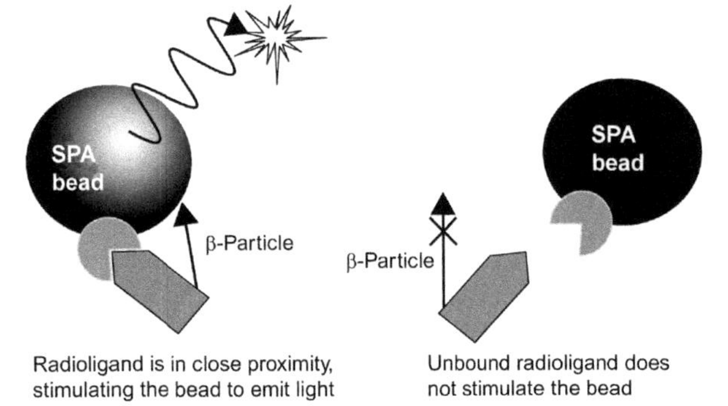

FIGURE 9.53 Illustration of the principle of scintillation proximity assay (SPA). The scintillation proximity assay is based on the emission of light as a result of energy transfer from the radioactive decay of the radioisotope-labeled ligand (radioligand) to the microsphere that contains a scintillant (SPA bead). The emission of light only proceeds, if the radioisotope-labeled ligand and receptor are in close proximity (~10 μm) illustrated on the left of the Figure. Otherwise the energy of the radioligand is absorbed by the buffer illustrated on the right of the figure. *From Wildey et al. (2017); reprinted with permission from Elsevier © 2017.*

with an antibody or receptor molecule and the ^{3}H or ^{125}I resides as a label on an antigen or ligand molecule, a binding reaction can occur between the antibody and antigen or receptor and ligand molecules, as illustrated in Fig. 9.53. The close proximity of the ^{3}H or ^{125}I through the binding reactions enables the beta-particle or Auger-electron radiations to penetrate the microsphere scintillator with concomitant isotropic emission of photons of visible light by the scintillator microspheres. A scintillation counter can detect the photons of light emitted by any aqueous medium (e.g., buffer solution) containing a mixture of (1) microsphere scintillator with surrounding antibody or receptor molecule, (2) free radioisotope-labeled antigen or ligand molecules, and (3) bound radioisotope-labeled antibody-antigen or bound radioisotope-labeled receptor-ligand molecules on scintillator microspheres. The count rate (CPM) of the photon emissions will serve as a measure of the bound radioisotope-labeled fraction, because radiations emitted from labeled molecules not bound close to the microsphere surface dissipate their energy in the aqueous medium and are not detected.

SPAs are not limited to applications with radionuclides emitting very low-energy particulate radiations, such as ^{3}H or ^{125}I. However, the use of radionuclides emitting higher-energy beta-particle radiation, such as ^{14}C ($E_{max} = 155$ keV), ^{35}S ($E_{max} = 167$ keV), and ^{33}P ($E_{max} = 249$ keV) is possible (Donnier-Maréchal et al., 2017; Eldeeb et al., 2017; Erdmann et al., 2017; Aas et al., 2011; Chapman et al., 2011; Ojo et al., 2011; Vakalopoulos et al., 2011; Zaghdane et al., 2011; Herdemann et al., 2010; Seimandi et al., 2010; Zhang et al., 2010; Gaster, 2009; Grobler et al., 2009). Because of the relative high energies of the beta particles emitted by ^{14}C, ^{35}S, or ^{33}P and the consequent relative longer lengths of travel of these particles in water (see ^{14}C beta-particle maximum range in water in Table 1.12 of Chapter 1), it is necessary to minimize scintillation proximity bead contact with free radionuclide label in the solution medium. This is accomplished by either (1) letting the scintillation proximity beads settle for several hours or overnight, (2) microplate centrifugation, or (3) suspension (flotation) of the scintillation proximity beads over 7 M cesium chloride.

The suspension or flotation technique noted in the previous paragraph is described by Jeffery et al. (2002) and illustrated in Fig. 9.54, which also exemplifies the use of

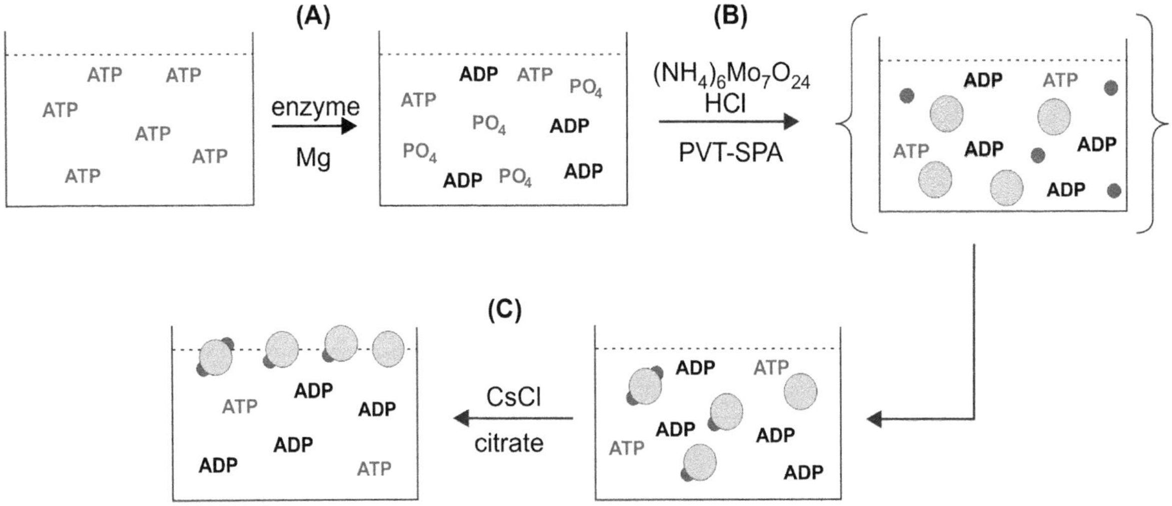

FIGURE 9.54 Schematic representation of the scintillation proximity ATPase assay. (A) The ATPase catalyzes the hydrolysis of ^{33}P-ATP to generate radioactive inorganic phosphate $[^{33}P]PO_4^{3-}$. (B) Addition of polyvinyltoluene scintillation proximity beads (PVT-SPA beads, large gray circles) suspended in a solution of ammonium molybdate and HCl. The $[^{33}P]$phosphomolybdate anion $(^{33}PMo_6O_{40})^{3-}$ (small dark circles or small red circles in colored version of Figure) forms and then absorbs onto the surface of SPA beads (larger gray circles). (C) Addition of CsCl to float the beads prior to radioactivity analysis in a microplate counter *From Jeffery et al. (2002); reprinted with permission from Elsevier © 2002.*

SPA in the study of enzyme activity. In this example, ATPase catalyses the hydrolysis of ^{33}P-labeled ATP to generate radioactive inorganic phosphate. Scintillation proximity beads suspended in a solution of ammonium molybdate are added. The ammonium molybdate forms an anion with the ^{33}P-phosphate, and the phosphomolybdate anion then absorbs onto the SPA beads. The addition of 3.5 M CsCl solution causes the SPA beads to float to the surface of the suspension separating the beads with the bound ^{33}P-phosphate from the unbound ^{33}P-ATP in the solution. The amount of light emitted by the beads provides a measure of the ^{33}P-phosphate adhered to the SPA beads. The technique was demonstrated as useful for kinetic studies of the enzyme activity.

b. Immunoassay applications

In the case of the binding of a ^{3}H- or ^{125}I-labeled antigen to an antibody adhered to an SPA bead (see the general illustration in Fig. 9.53), the following interactions adapted from Watson (1996) can be written:

$$\begin{array}{lc} \text{Step 1:} & \text{Ag}^* + \text{Ab—SPA bead} \longleftrightarrow \text{Ag}^*\text{—Ab—SPA bead} \\ & + \\ \text{Step 2:} & \text{Ag (unlabeled antigen)} \\ & \uparrow\downarrow \\ & \text{Ag—Ab—SPA bead (unlabeled complex)} \end{array} \tag{9.44}$$

where "Ag*" denotes the radioisotope-labeled antigen, "Ab–SPA bead" represents the antibody adhered to an SPA bead, and "Ag*–Ab–SPA bead" the complex between the radioisotope-labeled antigen and antibody on the SPA bead, which reach equilibrium when mixed together (Step 1). Immunoassays are quantitative analytical methods for measuring the amount of unlabeled antigen "Ag" in a sample based on the competitive binding of the unlabeled antigen and labeled antigen for the antibody. In other words, if we add unlabeled antigen to Eq. (9.44) (Step 2) in the form of either a standard solution or an unknown sample, the amount of labeled antigen bound to the antibody on the SPA bead will be reduced by the formation of unlabeled "Ag-antibody-SPA bead" complex (abbreviated Ag-Ab-SPA bead in Eq. 9.44), and the amount of radioactivity detected by the microsphere will diminish corresponding to the amount of unlabeled antigen added to the system. This is also referred to as competitive binding between the labeled antigen (Ag*) and unlabeled antigen (Ag) for the antibody (Ab). Given a limited concentration of antibody molecules attached to microspheres, adding unlabeled antigen to the system will reduce the number of labeled antigen molecules associated with the microsphere. Consequently the rate of scintillation photon emission and radioactivity measured by the scintillation detector will be reduced correspondingly. If we compare this reduction in scintillation caused by unlabeled antigens in an experimental unknown sample to the same effects produced by known concentrations of unlabeled antigens used as standards, we can determine the concentration of the antigens in our unknown sample. With the preparation of an appropriate standard curve, the count rate of an unknown sample can be used to determine the amount or concentration of unlabeled antigen or ligand in the sample. An example of such a standard curve is illustrated in Fig. 9.55, which was prepared from a commercial SPA kit. The standard curve, also known as a Scatchard plot, is named after George Scatchard, who carried out many of the pioneering studies of macromolecule-ligand interactions. The ordinate gives the ratio of the radioactivity or quench-corrected count rate of the antibody-labeled antigen complex (B) to the radioactivity or quench-corrected count rate (B_0) of the antibody-

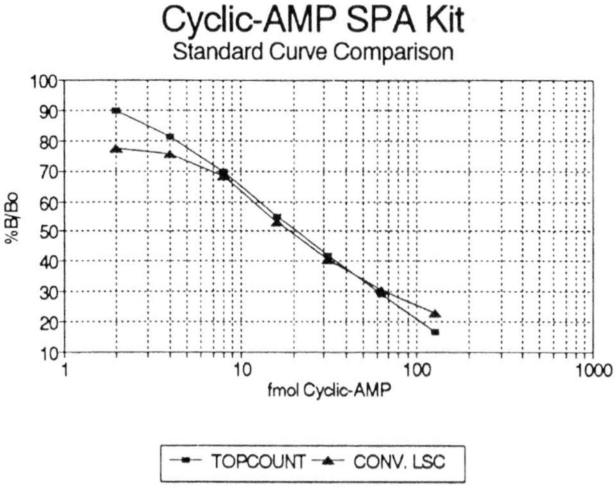

FIGURE 9.55 Standard curves (Scatchard plots) for the scintillation proximity assay (SPA) of cyclic AMP determined with a TopCount microplate scintillation counter and a conventional liquid scintillation counter using a commercially available SPA kit. *From PerkinElmer (1991a,b); printed with permission © 1991PerkinElmer, Inc.*

labeled antigen complex in the absence of any unlabeled antigen. In other words, the nomenclature B/B_0 arises from the definitions of $B =$ the antibody-bound radioactivity and $B_0 =$ the antibody-bound radioactivity at zero dose of unlabeled antigen. The amount (e.g., fmol) or concentration (e.g., nM) of unlabeled antigen is plotted on the abscissa. The ratio B/B_0 may also be expressed as a percentage. It is obvious that, at zero dose of unlabeled antigen, $B = B_0$, and therefore, the ratio is unity or 100% when no unlabeled antigen is added to the medium. The ordinate may also be expressed in other units, which define the distribution of the radioisotope label, such as the ratio of the bound to the free fraction. The curve, illustrated in Fig. 9.55, shows the similarity in standard curves obtained with two types of detector systems, namely TopCount or MicroBeta microplate scintillation counters and conventional liquid scintillation counting. Both detector systems may be used for SPA. The microplate scintillation counter is preferred for high-throughput assays because the microplate scintillation counter provides the possibility of assaying up to 12 samples simultaneously, as discussed previously in this chapter.

The standard curve, illustrated in Fig. 9.55, is prepared from a given quantity of microspheres containing adsorbed antibody (20−200 μg), a given activity of ^{3}H- or ^{125}I-labeled antigen, and various amounts of unlabeled antigen. Adequate buffer, which may contain a low concentration (e.g., 0.1%) of bovine serum albumin (BSA), detergent, and 10%−12% glycerol are also added in a total volume of about 200 μL. The BSA serves as a nonspecific protein which, together with the detergent, helps reduce any nonspecific binding of the isotope-labeled antigen with the microspheres. The glycerol prevents settling of the microspheres, keeping them in suspension. The total sample volume is, therefore, very small and generally does not exceed 400 μL.

The small sample volumes involved in SPA (<400 μL) and the large numbers of samples that must be analyzed by most laboratories make SPAs of samples prepared in microplate formats most favorable. Samples may be prepared in sample microplates equipped with 24-, 96-, and 384-sample-well formats. The plates measure only 12.8 × 5.6 × 1.9 cm (L × W × H). Automatic microplate scintillation analyzers are equipped to analyze up to 12 samples in microplates simultaneously. The sample wells are organized in three formats, namely, 24-well plates (6 × 4 wells), 96-well plates (12 × 8 wells), and 384-well plates (24 × 16 wells), which accommodate the simultaneous analysis of up to 6 to 12 samples, respectively, by an equal number of PMTs in a microplate scintillation counter. Some applications of SPA as a tool in radioimmunoassay are reported by Meena et al. (2017), Wildey et al. (2017), Wheelan et al. (2015), Wang et al. (2013), Jackson et al. (2011), Shen et al. (2011) and Zhao et al. (2011).

c. Receptor binding assays

A receptor binding assay provides information on the relative competitiveness of ligands for binding sites on receptor molecules. The scintillation proximity technique with microsphere scintillator beads may be used to implement receptor binding assays. The technique involves mixing microsphere scintillator beads with solubilized or membrane-bound receptors such as membrane protein. The high affinity of the membrane protein for the microspheres is the basis of this assay. Once the receptor molecules are immobilized on the microspheres, a radioisotope-labeled ligand is added. The isotope-labeled ligand then binds to the receptor molecules on the microsphere beads [Scintillator coated microplates (e.g., FlashPlate or ScintiPlate described further on in this section of the chapter) may also be used in lieu of scintillating microspheres (Wildey et al., 2017; de Jong et al., 2005)]. The binding of the isotope-labeled ligand with the receptor molecule brings the isotope-labeled ligand into close proximity to the microsphere scintillator beads, which will excite the scintillator. The scintillation photons emitted from the microspheres can be assayed with the microplate scintillation counter. If, however, we add another competing ligand or drug (generally referred to as an analyte), which is not labeled with radioisotope, it will displace the isotope-labeled ligand on the receptor molecules according to its competitive affinity for the receptor and its concentration. The displacement of isotope-labeled ligand from the receptor into the medium solution and away from the scintillator microspheres will result in a reduction of the measured radioactivity. Different ligands over a range of ligand concentrations can be tested in this fashion against a given receptor. A Scatchard plot can then be drawn for each ligand as illustrated in Fig. 9.56 over the entire range of unlabeled competitive ligand concentrations tested where $B =$ the bound ligand radioactivity and $B_0 =$ the bound ligand radioactivity at zero dose of unlabeled competitive ligand. We can determine from the Scatchard plots which ligands or drugs tested have greater or lesser affinity for the receptor. The SPA Scatchard plots illustrated in Fig. 9.56 from the work of Nose et al. (2009) depict the competitive binding of various agonists estrogen receptor α ligand-binding domain (hERα-LBD), which contains a molecular pocket able to capture a hormone-like chemical to build up its activation conformations. The work of Nose et al. (2009) illustrated in Fig. 9.56 demonstrates that the 17β-estradiol binds more strongly to the receptor molecule than 4-(1-adamantyl)phenol; and 4-*tert*-octylphenol and 4-*n*-octylphenol, which bind increasingly less strongly to the receptor molecule. In receptor binding assays the term IC_{50} and its value are often used to compare the competitive binding of antagonists for the receptor molecule. The IC_{50} is calculated as the concentration required to inhibit 50% of

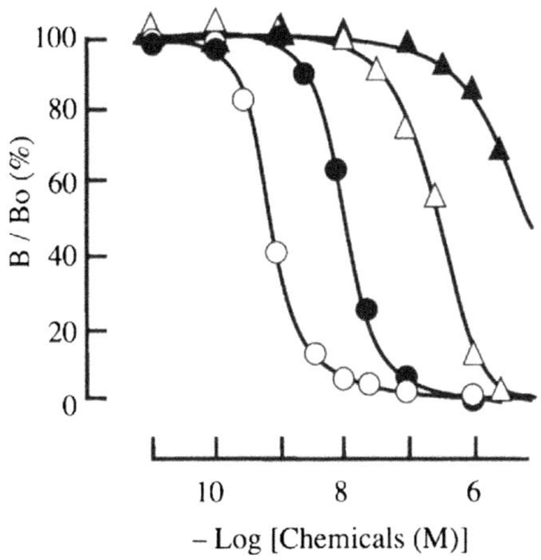

FIGURE 9.56 Receptor binding assay using purified human estrogen receptor α ligand-binding domain (hERα-LBD receptor) and [^{3}H]17β-estradiol radioligand over a range of four analyte concentrations (M) of 1×10^{-11} to 1×10^{-4} nM of 17β-estradiol (open circle), 4-(1-adamantyl) phenol (closed circle), 4-*tert*-octylphenol (open triangle), and 4-*n*-octylphenol (closed triangle). *From Nose et al. (2009); reprinted with permission from Elsevier © 2009.*

the labeled ligand binding to its receptor. The relative IC_{50} values for the four antagonists illustrated in Fig. 9.56 from left to right are 1.40, 82.3, 884 nM (nanomolar), and N.D (not determined, see the broken curve in Fig. 9.56), as reported by Nose et al. (2009).

Temperature can have a significant effect on SPAs. The work of Blair et al. (1995) demonstrates the significant differences in IC_{50} values obtained over the temperature range 15−27°C. It can be important, therefore, to control assay temperature by using a scintillation counter, which has a temperature control. Reviews on the principles and applications of receptor binding assays is given by Wildey et al. (2017), Cook et al. (2006), Powell and Price-Jones (2006), and de Jong et al. (2005). There are numerous examples of the application of SPA to receptor binding studies in the current literature; and a select few are the following: Luethi et al. (2018), Flanagan (2016), Meleza et al. (2016), Wang and Ma (2015), Zeng and Xu (2015), Dechavanne et al. (2011), Schepmann et al. (2011), Tawa et al. (2011), Stanley et al. (2010), Nose et al. (2009).

d. Enzyme assays

Enzyme assays can be included among the list of applications of SPA. Enzyme assays can be of two types: (1) the enzyme activity toward a substrate is determined by the hydrolytic action of the enzyme (e.g., protease or nuclease) on an isotope-labeled substrate attached to the SPA microspheres, resulting in the release of radioactivity from the scintillation microspheres; and (2) the enzyme activity toward a substrate is determined by the polymerizing action of the enzyme (e.g., polymerase or transferase) on a donor isotope-labeled substrate in the free medium, which becomes attached to the SPA microspheres, resulting in an increase in radioactivity detection from the scintillation proximity beads.

An illustrative example of a scintillation proximity enzyme assay is that of an ATPase assay reported by Jeffery et al. (2002) illustrated previously in Fig. 9.54. In this case, the enzyme catalyzes a reaction on a substrate, whereby the product of the reaction (^{33}P-inorganic phosphate) becomes attached to the SPA bead. As enzyme activity increases, an increase in radioactivity is observed from the SPA beads. Consequently enzyme kinetics and the effects of inhibitors on enzyme activity can be easily observed without the need of tedious procedures and separation steps. For example, Fig. 9.57 illustrates the ATPase reaction rates as a function of substrate (ATP) concentration and inhibitor ATP-γ-S concentration as measured with SPA. This serves as only one example of many enzyme activity measurements that can be carried out without the tedious preliminary preparations such as HPLC, electrophoreses, or centrifugation.

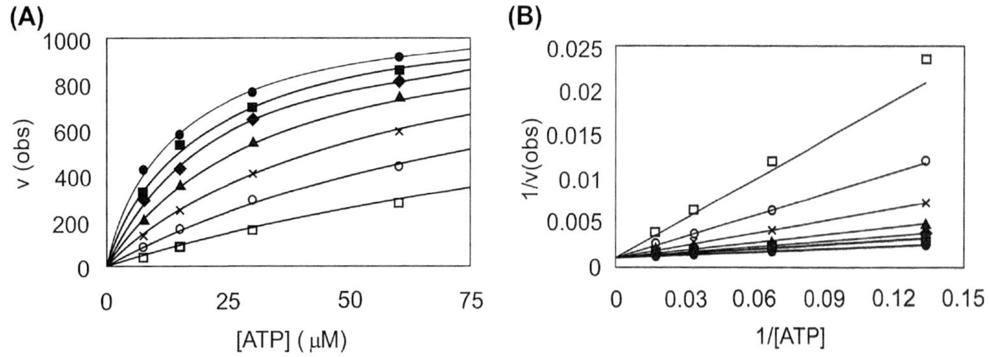

FIGURE 9.57 Observed reaction rates (ATPase activity) of HPV6 E1 (a E1 helicase from human papillomavirus) via SPA using [γ-^{33}P]ATP substrate. (A) The observed rates (in pM inorganic phosphate, P_i, produced per s) for substrate concentrations ranging from 7.5 to 60 μM at inhibitor (ATP-γ-S) concentrations of 0 (●), 3.1 μM (■), 6.3 μM (◆), 12.5 μM (▲), 25 μM (*), 50 μM (○), and 100 μM (□). (B) The same data points plotted in double reciprocal format. *From Jeffery et al. (2002); reprinted with permission from Elsevier © 2002.*

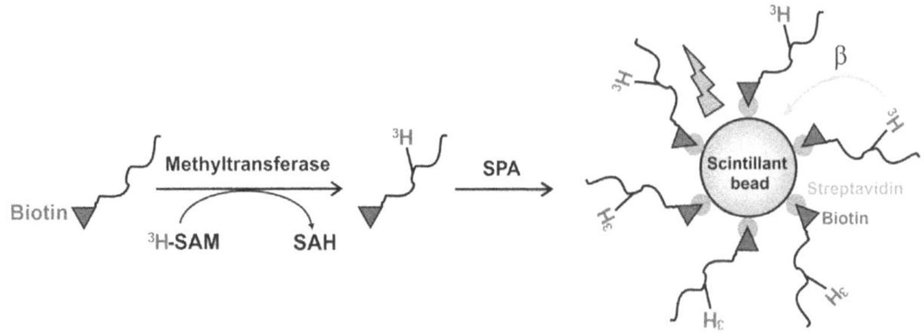

FIGURE 9.58 Bead-based scintillation proximity assay for methyltransferase enzyme activity. *From Zeng and Xu (2015); reprinted with permission from Elsevier © 2015.*

Another example of enzyme assay with scintillating microspheres is described by Zeng and Wu (2015) in the measurement of methyltransferase enzyme activity analysis. The assay is illustrated in Fig. 9.58. The in vitro methylation assay catalyzed by the methyltransferase enzyme proceeds with a combination of substrate, methyltransferase, and a universal methyl donor labeled with ^{3}H (i.e., ^{3}H-SAM or *S*-adenosyl-L-methionine). The methyl group is transferred enzymatically to a nitrogen atom of lysine or arginine side chains of the protein substrate, which produces the ^{3}H-labelled methylated substrate and the SAH (*S*-adenosyl-homocysteine) substrate. Streptavidin-coated microscopic scintillating beads capture the ^{3}H-labelled biotinylated substrates subsequent to the methylation reaction. This brings the ^{3}H radionuclides in close proximity to the scintillation beads sufficient for the beads to absorb β-particles from ^{3}H decay exciting the scintillator with the emission of light detected by the microplate scintillation counter. Unreacted ^{3}H-SAM substrate remains in solution too distant from scintillation beads to produce any scintillation photons. This avoids any experimental need to filter off any unreacted substrate.

Numerous enzyme assays using SPA technology with scintillating microspheres can be found in the literature, among which the following are excellent examples: Flanagan (2016), Wang and Ma (2015), Zeng and Xu (2015), Kumar et al. (2014). Stevenson et al. (2013), Ojo et al. (2011), Scott et al. (2011), Sun et al. (2011), Thirstrup et al. (2011), and Zhang et al. (2010, 2011).

f. SPA with scintillating microplates

Microsphere scintillators are not the only means of carrying out SPA. It is also possible to use specially prepared 96- and 384-well polystyrene microplates, which contain plastic scintillator-coated wells to which are bonded the antibody or receptor molecules. An example of such a microplate is known as the FlashPlate. A diagram illustrating a FlashPlate microplate well with a radioisotope-labeled ligand (radioligand) in solution and another radioligand bound to a receptor molecule attached to the surface of the scintillant-coated well is shown in Fig. 9.59. The receptor molecules (capture molecules) are coated on the plastic scintillator in each of the 96 or 384 wells of the microplate. Therefore, no scintillator beads are used. Radionuclide-labeled ligand is added and allowed to incubate with the receptor molecules. Only radionuclide labeled ligand bound to a receptor molecule on the surface of the plastic scintillator is in sufficiently close proximity to the plastic scintillator to produce a scintillation effect in the scintillator and the concomitant emission of light. Radioligand in the solution will be too far away from the capture molecules coated on the well surface and scintillator to produce any significant scintillation effect. The addition of unlabeled competitive binding ligand will reduce the number of radionuclide-labeled compounds on the capture molecules and thereby reduce the count rates detected by the microplate scintillation analyzer. The polystyrene-based scintillator on a microplate format provides the platform for nonseparation assays using low- and moderate-energy beta particle-emitting nuclides, such as ^{3}H ($E_{max} = 18.6$ keV), ^{14}C ($E_{max} = 155$ keV), ^{35}S ($E_{max} = 167$ keV), ^{33}P ($E_{max} = 249$ keV) and the atomic-electron emitter ^{125}I ($E_{max} = 30$ keV).

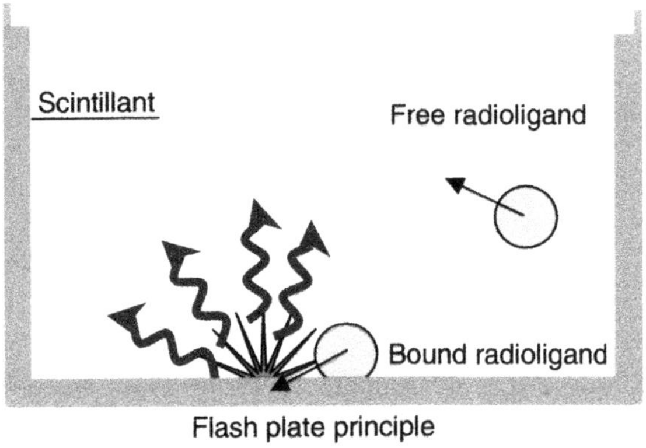

FIGURE 9.59 Diagram illustrating the principle of FlashPlate technology. The scintillant is embedded in a clear plastic coating on the inner wall of the microplate-well. Radioligand bound to the surface of the assay-well leads to the emission of light. Free radioligand, in solution and not proximal to the scintillant, causes only minor background emission. *From Schnurr et al. (2007); reprinted with permission from Elsevier © 2007.*

An example of the application of the FlashPlate to enzyme assays is provided by Zeng and Xu (2015)

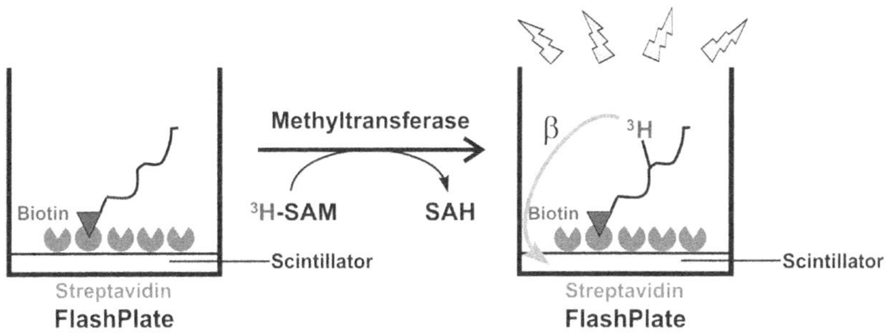

FIGURE 9.60 A FlashPlate-based in vitro scintillation proximity assay for methyltransferase activity. *From Zeng and Xu (2015); reprinted with permission from Elsevier © 2015.*

according to the illustration provided in Fig. 9.60. The figure illustrates a procedure for in vitro methylation assay with methyltransferase enzyme utilizing a FlashPlate-based SPA. In this case, a Streptavidin FlashPlate is used whereby biotinylated histone peptide or protein substrates are coated onto the wells of the Streptavidin FlashPlate. The methylation reaction is then initiated with the addition of methyltransferase and ^{3}H—*S*-adenosyl-L-methionine (^{3}H-SAM), which is a universal methyl donor labeled with ^{3}H. The methyl group is transferred enzymatically to a nitrogen atom of lysine or arginine side chains of the protein substrate, which produces the ^{3}H-labelled methylated substrate and the SAH (*S*-adenosyl-homocysteine) substrate. The ^{3}H label on the biotinylated histone peptide binds strongly to the Streptavidin; and the proximity of the ^{3}H label to the plastic scintillator coated onto the FlashPlate well produces scintillation photons from the beta decay, which are counted by a suitable microplate counter. The radiolabeled ^{3}H-SAM in solution in the sample well of the FlashPlate remains too distant from the scintillator coating to produce any scintillation events.

There are numerous literature references to the applications of FlashPlates in 96-well or 384-well microplate formats. A few examples will be cited here as applied to specific assays, including

(i) immunoassays (Witkin et al., 2016; Lauritzen et al., 2014; McNamara et al., 2012; Moore, 2012a,b; Møller et al., 2011; Shen et al., 2011; Xiao et al., 2008; Hertel et al., 2007),

(ii) receptor binding assays (Song et al., 2018; Flanagan, 2016; Wang and Ma, 2015; Knudsen et al., 2012; Harndahl et al., 2011; Shen et al., 2011; Matsushima et al., 2010; Bednarek et al., 2008), and

(iii) enzyme assays (Prause et al., 2018; Yan et al., 2017; Pagire et al., 2015; Rutter et al., 2014; Yu et al., 2014; Oka et al., 2013; Dechavanne et al., 2011; Kharel et al., 2011; Thirstrup et al., 2011; Reynisson et al., 2009; Miesel et al., 2007), among other specific assays used in molecular biology studies.

g. Color quench correction

The main advantages of SPA are (1) the circumvention of separation steps usually linked to immunoassays, receptor binding assays, and enzyme activity assays, among other assays carried out in the medical and biochemical sciences; (2) the reduced costs per analysis arising from a reduction in reagent volumes consumed and the waste accumulated; and (3) the consequent high-throughput analysis possible when SPA is linked to automatic liquid handling and microplate sample counting techniques. The first advantage of the SPA technique listed above can give rise to colored samples because purification steps in the assay procedure are avoided. Color in the assayed samples can absorb the visible photons of light emitted by the microsphere scintillation beads or scintillator microplates before this light can reach the photomultipliers. Absorption of the light emissions will cause a reduction of the light intensities reaching the PMTs and consequently a reduction in the pulse heights registered by the scintillation counter. A reduction in count rate is generally the result because many reduced pulse heights fall below the detectable levels set by a LL pulse height discriminator setting. The phenomenon of reduced pulse heights and concomitant reduction in count rate of samples due to color of the sample is known as color quenching. It is important, therefore, that sample assays be corrected for color quench. We should keep in mind that chemical quench does not occur in SPA because this is a solid scintillation technique in which the beta particles from ^{3}H, ^{14}C, ^{35}S or ^{33}P radionuclides or the Auger electrons from ^{125}I enter into the solid scintillator to cause scintillation where no chemical quenching agents can be found.

The correction of quenched sample count rates (CPM) requires a measurement of the detection efficiency of each sample via a quench correction curve of counting efficiency or relative efficiency versus a suitable QIP such as that provided by an external standard (e.g., tSIE). However, fortunately the SPA technique does not require absolute activity measurements and the count rate of the sample

(CPM) will suffice provided the sample count rates are corrected for color quench. In reports by Hughes et al. (2001), Jones et al. (1997), Neumann et al. (1994), and PerkinElmer (2001, 1997), it is demonstrated that the quench-corrected count rate (QC-CPM) produced by a microplate scintillation counter is simple and provides accurate results for most SPAs. For example, the PerkinElmer TopCount and MicroBeta microplate scintillation counters can count simultaneously up to 12 wells of a microplate and, at the same time, determine the color quench of each sample from the sample pulse height spectrum. The QIP measured is the tSIS determined by the TopCount or spectral quench parameter [SQP(I)] or asymmetric quench parameter [AQP(I)] measured with the MicroBeta. A thorough treatment of QIPs and their use are provided in Chapter 6 of Volume 1. These QIPs are excellent for monitoring the color quench level of samples, because they are not affected by the color of the sample; that is, all color quench correction curves will be identical regardless of the color of the sample. A quench correction curve can be prepared using standard (e.g., ^{3}H-PVT, ^{125}I-PVT or other isotope) SPA beads according to the procedure described in detail in the references cited earlier in this paragraph. Of course, the standard isotope-labeled beads must contain the same radioisotope that is used for the SPAs. Fig. 9.61 illustrates the constant relationship between count rate and tSIS one can expect regardless of sample color. The count rate (CPM) on the ordinate of this curve can be converted to read percent relative efficiency by placing the highest count rate (corresponding to the highest tSIS value) at 100% with all other count rates converted to a percentage relative to the highest count rate. In practice, a researcher will determine the average CPM of an unquenched radionuclide-bound microsphere scintillator sample (colorless sample) and then input this value into a counting protocol of the instrument to prepare a quench correction curve. Several quenched standards are prepared by adding the same amount of radionuclide-bound microspheres to sample wells including increasing amounts of color-quenching agent. The instrument will then count each sample (unquenched and quenched) and plot a quench curve of percent counting efficiency versus the QIP [e.g., tSIS, SQP(I), or AQP(I)]. Once the curve is prepared, it is stored in the memory of the microplate scintillation analyzer protocol so that each experimental sample that is counted can be corrected for color quench according to the QIP of the sample. For example, if an experimental sample gave a QIP value corresponding to a relative efficiency of 25% from a standard curve, and the count rate of the sample $(CPM)_s$ was 5250, the instrument would calculate a quench corrected count rate (QC-CPM) as

$$QC - CPM = (CPM)_s/E = 5250/0.25 = 21,000 \quad (9.45)$$

Hughes et al. (2001) demonstrated the application of this color quench correction technique to SPAs.

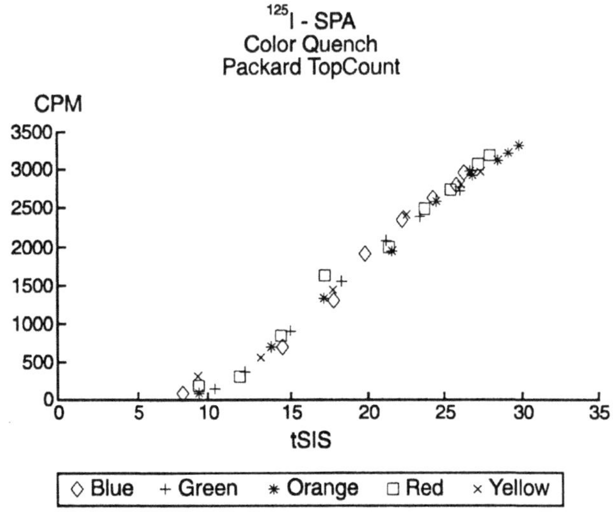

FIGURE 9.61 Count rates versus the quench-indicating parameter tSIS observed with ^{125}I-labeled SPA beads on TopCount microplate scintillation counter with various concentrations of five different colors of dye. Similar curves are obtained with PerkinElmer MicroBeta microplate scintillation counters using the quench-indicating parameter [SQP(I)] or [AQP(I)] (see Hughes et al., 2001). *From Neumann et al. (1994), printed with permission; © 1994 PerkinElmer Inc.*

VI. Detection of neutrons

The neutron-induced nuclear reactions that occur in inorganic scintillator detectors as well as in scintillators of other types (e.g., organic), and scintillators in plastic or glass media depend on the energy of the neutron. Fast neutrons are detected in organic and plastic scintillator converters via proton recoil following elastic scattering (Shan et al., 2016; Meshkian, 2015; Li et al., 2014) or via converter material containing ^{232}Th or ^{238}U (Yamane et al., 1998, 1999), which undergo fast-neutron induced fission yielding fission products detectable with inorganic scintillators. Also, fast neutrons are detected by exploiting neutron capture reactions, such as the ^{35}Cl(n,p)^{35}S and ^{35}Cl(n,α)^{32}P reactions with Cl in crystal scintillator (Giaz et al., 2016). Reviews on scintillation detectors for fast neutrons are provided by Ryzhikov et al. (2018, 2017), Giaz et al. (2016), Tarasenko et al. (2013), and Klein and Brooks (2006). Thermal neutrons, on the other hand, are detected on the basis of the high neutron reaction capture cross-sections with particular nuclides, such as ^{6}Li, ^{10}B, ^{155}Gd, ^{157}Gd, ^{235}U (Finocchiaro et al., 2018; Lo Meo et al., 2017; Köhli et al., 2016; Mendicino et al., 2015; Boatner et al., 2012; Oed, 2004; van Eijk, 2004; van Eijk et al., 2004).

A. Inorganic neutron scintillators

Due to their low velocity and neutral charge, thermal neutrons are detected by means of their nuclear reaction with target nuclei. These reactions, first discovered by Enrico Fermi as described in Chapter 1, Volume 1, result in the emission of charged particles (e.g., protons, alpha particles, conversion electrons, or fission fragments) or gamma radiation. The ionization or scintillator excitation produced by these reaction products are the basis for the detection and measurement of thermal neutrons. Commonly used target nuclei for thermal neutron detection are ^3He, ^6Li, ^{10}B, ^{235}U, and ^{157}Gd due to their high thermal-neutron reaction cross-sections. The nuclear reactions of thermal neutrons with commonly used target nuclei for neutron detection are listed in Table 9.7. As described by Oed (2004) all of the reactions listed in Table 9.7, with the exception of that of ^{157}Gd, are fission processes with the two fission fragments ejected in opposite directions. The energies of the fission fragments are listed in Table 9.7.

Thermal neutron detection with ^3He listed in Table 9.7 involves gaseous ionization detectors; and these will not be treated in this chapter. Also, the current shortage of ^3He gas led to rigorous research and development of alternative neutron-capturing materials; and ^6Li and ^{10}B have become the current major nuclides of research for the development of thermal neutron detectors (Finocchiaro et al., 2018; Lo Meo et al., 2017; Köhli et al., 2016; Mendicino et al., 2015; Boatner et al., 2012; Ianakiev et al., 2011). The natural abundance of ^6Li and ^{10}B are 7.5% and 19.9%, respectively. Also, natural Gd consists of several isotopes, among which ^{155}Gd (14.8% natural abundance) and ^{157}Gd (15.6% natural abundance) have very high thermal neutron capture cross-sections at low thermal neutron energies (see Fig. 9.62); although Gd detectors have some disadvantages discussed further on in this section. Thus, inorganic Gd crystalline scintillators (Pan et al., 2018b; Zaman et al., 2017; Trtik and Lehmann, 2015; Zhuravleva et al., 2011; Sidletskiy et al., 2010; Kawamura et al., 2007) as well as Gd-loaded plastic scintillators (Watanabe et al., 2015; Ovechkina et al., 2009) also receive due attention in the development of thermal neutron detector materials.

The reactions abbreviated in Table 9.7 are described in more detail by van Eijk et al. (2004) for the key target nuclei utilized in the scintillation measurement of thermal neutrons, i.e., ^6Li, ^{10}B, ^{155}Gd, and ^{157}Gd according to the following reactions where the natural abundance (%) is given alongside each target isotope, the neutron-capture cross-sections are expressed as σ (b = barns) at a neutron wavelength of 1 Å, and the relevant branching ratios (BRs) are given:

$$n + {}^6Li(7.5\%) \rightarrow {}^3H(2.75\ MeV) + {}^4He(2.05\ MeV),\quad \sigma = 520\ b \tag{9.46}$$

$$\begin{aligned} n + {}^{10}B(20\%) &\rightarrow {}^7Li(1.0\ MeV) + {}^4He(1.8\ MeV)BR = 7\% \\ &\rightarrow {}^7Li(0.83\ MeV) + {}^4He(1.47\ MeV) \\ &\quad + \gamma(0.48\ MeV)BR = 93\% \\ &\quad \sigma = 2100\ b \end{aligned} \tag{9.47}$$

$$\begin{aligned} n + {}^{155}Gd(15\%) &\rightarrow {}^{156}Gd + \gamma\text{'s} + \text{conversion electrons} \\ &\quad + \text{X-rays}(39\ keV - 199\ keV;\ BR \sim 80\%), \\ &\quad \sigma = 17,000\ b \end{aligned} \tag{9.48}$$

$$\begin{aligned} n + {}^{157}Gd(16\%) &\rightarrow {}^{158}Gd + \gamma\text{'s} + \text{conversion electrons} \\ &\quad + \text{X-rays}(29\ keV - 182\ keV;\ BR \sim 85\%), \\ &\quad \sigma = 70,000\ b \end{aligned} \tag{9.49}$$

About 80% of the neutron captures in Gd occur in ^{157}Gd and 18% of the neutron captures occur in ^{155}Gd (Reeder, 1994b). As described in Chapter 1, Volume 1, the

TABLE 9.7 Commonly used isotopes for thermal neutron detection, reaction products and their kinetic energies.

	Reaction	Light fragment (l.fr)	Energy (MeV)	Heavy fragment (h.fr.)	Energy (MeV)
	n (^3He, p) ^3H	p	0.57	^3H	0.19
	n (^6Li, α) ^3H	^3H	2.74	α	2.05
93%	n (^{10}B, α) $^7Li + \gamma$	α	1.47	^7Li	0.83
7%	n (^{10}B, α)	α	1.77	^7Li	1.01
	n (^{235}U, l. fr.) h. fr.	l. fr.	≤80	h. fr.	≤60
	n (^{157}Gd, Gd) e^-	Conversion electron	0.07–0.182		

From Oed (2004); reprinted with permission from Elsevier © 2004.

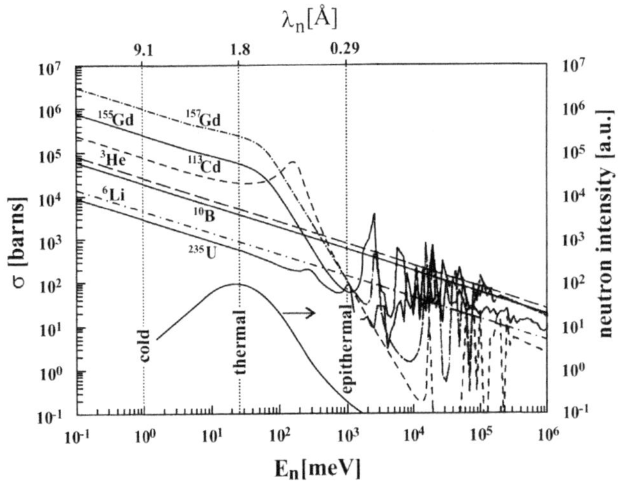

FIGURE 9.62 Neutron-capture cross-sections of some isotopes important for neutron detectors as a function of neutron energy. The lowest curve shows a typical moderated neutron spectrum of a nuclear reactor (right y-axis). *From van Eijk (2004); reprinted with permission from Elsevier © 2004.*

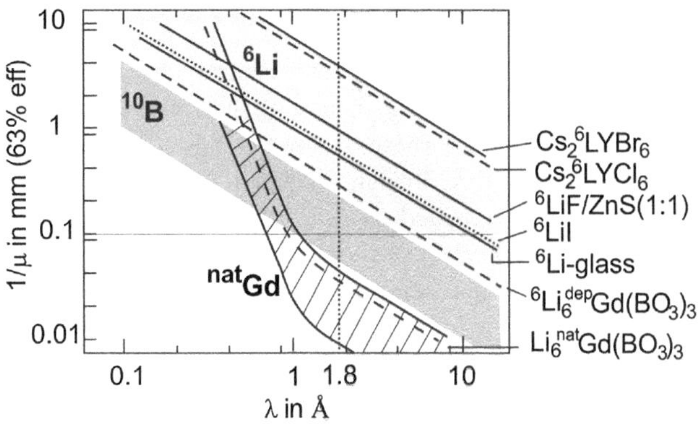

FIGURE 9.63 Absorption lengths of typical inorganic scintillator compounds based on ^{6}Li, ^{10}B, and ^{nat}Gd as a function of neutron wavelength. The absorption lengths are positioned into three zones for the ^{6}Li, ^{10}B, and ^{nat}Gd compounds; and data for some traditional and new neutron scintillators are shown. Li_6 (without superscript) and ^{dep}Gd represent natural Li and depleted Gd, respectively. *From von Eijk et al. (2004); reprinted with permission from Elsevier © 2004.*

neutron-capture cross-sections are a function of the neutron energy. The cross-section (σ) as a function of the neutron energy for the above reactions 9.46–9.49 were plotted by van Eijk (2004), which is illustrated in Fig. 9.62. While ^{157}Gd has the highest neutron-capture cross-section for cold neutrons and the low-energy fraction of thermal neutrons, its neutron-capture cross-section drops rapidly at neutron energies > ~ 0.050 eV; and ^{6}Li and ^{10}B have the highest neutron-capture cross-sections at neutron energies >1 eV, as illustrated in Fig. 9.62.

The linear absorption coefficients μ_n, which has units of cm^{-1} (see Chapter 1, Volume 1), were calculated by van Eijk (2004) as a function of the neutron-capture cross-sections and the weight fraction of the neutron-sensitive nucleus in various scintillators containing 100% enriched ^{6}Li, ^{10}B, and natural and depleted Gd. The absorption lengths or inverse of the absorption coefficients, $1/\mu_n$ (~63% absorption) were plotted by van Eijk et al. (2004) as a function of neutron wavelength, which is illustrated in Fig. 9.63. In the thermal region, the neutron absorption lengths are only of the order of tens of μm for natural Gd (^{nat}Gd) scintillators, 0.1 mm for ^{10}B scintillators, and 1 mm for ^{6}Li scintillators. The very short absorption length of thermal neutrons in Gd scintillator compounds was reported by Reeder (1994a,b), who studied thermal neutron scintillation in gadolinium orthosilicate, Gd_2SiO_5 (GSO), and noted that thick GSO crystals are not required, as thermal neutrons will penetrate only the first 11 μm from the surface of the GSO crystal. The small size requirement of this detector for thermal neutrons is an advantage in applications in which detector size is a limiting factor. In addition, GSO is reported to have a light output of about 20% of that of NaI(Tl) and a short decay time constant of 60 ns (Table 9.4) suitable for measuring arrival times of neutrons from a pulsed source at high count rates (Ishibashi et al., 1998; Tokanai et al., 2000).

The thermal neutron capture product nuclides ^{156}Gd and ^{158}Gd are themselves stable nuclides; however, the binding energies of the neutron capture processes leave the nuclides in an excited state. Deexcitation of ^{156}Gd and ^{158}Gd results in a release of a cascade of gamma rays (up to 2 MeV), X-rays, and conversion electrons (Tokanai et al., 2000). Knitel (1998) noted that the thermal neutron capture reactions of gadolinium yield high probabilities of internal conversion. The internal conversion electrons have short ranges in the scintillator yielding fluorescence close to the point of neutron interaction. The 29-keV conversion electrons, because of their charge, have a range of only about 2 μm in the GSO and, therefore, these do not escape the crystal and produce excitation and luminescence. The 44-keV X-rays produced have a range of about 75 μm in GSO. Reeder (1994b) explains that about half of these escape from the GSO crystal without detection because the thermal neutron capture occurs near the crystal surface. Also, the probability for the production of only γ-rays is ~20% (van Eijk et al., 2004). In addition, the gamma radiation produced has a very high probability of escape from the scintillation detector (i.e., essentially all of the gamma radiation escapes as reported by Knitel (1998) and van Eijk et al. (2004)). The gamma radiation escape would be due again to the fact that the thermal neutron capture occurs near the crystal surface, the small size of the detector, and the greater penetrating power of the gamma radiation. Knitel (1998) provides an excellent example of the relevance of high probability for internal conversion. He provides the example of ^{113}Cd, which has a very high thermal neutron capture, but a relatively low probability of internal conversion, only 4% that of ^{155}Gd and ^{157}Gd.

For the detection of 6 Å cold neutrons, Tokanai et al. (2000) characterized a 0.5 mm-thick GSO:Ce scintillator activated with a Ce concentration on 0.5 mol % coupled to a PMT. They obtained a pulse height spectra with clear 31 and 81 keV peaks corresponding to conversion electrons or X-ray singles peaks and their coincidence doubles peaks, respectively. A position resolution of 1.3 mm FWHM of a two-dimensional neutron image was measured. Tokanai et al. (2000) concluded that with a GSO scintillator as thin as 20 μm, more than 90% of 80 keV gamma rays and 44 keV X-rays, induced by neutron absorption by Gd, would escape the GSO scintillator, and the main component of the pulse height spectrum would be the 31 keV X-ray conversion electron peak. This is ideal for the reduction of background from gamma- and X radiation and, in addition, the spatial resolution would be improved because the range of 31 keV conversion electrons in GSO is only 2 μm. The neutron position resolution in such a detector system would be limited only by the light imaging system.

While thermal neutron detectors based on natural gadolinium have specific applications, van Eijk et al. (2004) point out that Gd scintillators are less attractive because of several factors including the small amount of energy deposited in the Gd detector, the ~80% limit on its detection efficiency (i.e. ~ 20% is lost as undetected gamma radiation), and the absence of the possibility of neutron-gamma discrimination, among other factors. Thus, research and development are focused on scintillator compounds with ^{6}Li as the target nucleus as well as some compounds containing ^{10}B target nuclei (Mahl et al., 2018; Giaz et al., 2016; Celentano et al., 2015; Ianakiev et al., 2015; Lee et al., 2013). An advantage of ^{6}Li and ^{10}B over Gd, as target nuclei in scintillation detectors for the detection of neutrons, is that the reaction products are relatively heavy nuclei, which obviously have very short ranges in the scintillator detectors with total energy absorption. The energies of the product nuclei are of the order of magnitude of a few MeV (see Eqs. 9.46−9.49); and their ranges in a scintillator generally are ≤40 μm (van Eijk et al., 2004).

Thermal neutron scintillators along with their properties and limitations are listed in Table 9.8. Compounds are synthesized with enriched ^{6}Li at 95 atom % or higher to enhance thermal neutron capture. The column of mean free path (mm) of Table 9.8 is also referred to as the neutron absorption length (mm); and these values are calculated at a neutron wavelength of 1.8 Å. The mean free path is calculated as the inverse of the absorption coefficient, which is a function of the neutron energy or wavelength. The term α/β ratio in the last column of Table 9.8 refers to the ratio of the number of photons per MeV produced by the heavy charged particles (e.g., ^{3}H and ^{4}He product nuclides from thermal neutron capture by ^{6}Li) and the number of photons produced by electrons from gamma-ray interactions via photoelectric, Compton, and pair-production interactions. A high α/β ratio is preferred to facilitate discrimination between neutron radiation and background gamma.

Research and development in thermal neutron scintillators include the following, among other detector designs:

(i) ^{6}Li-glass scintillator (Ianakiev et al., 2015; Rich et al., 2015; Fukabori et al., 2011a), ^{6}Li-loaded plastic scintillator (Balmer et al., 2015),
(ii) ^{6}Li metal foil sandwiched together with scintillator and light pipe (Ianakiev et al., 2011, 2015),
(iii) ^{6}Li-enriched crystal scintillators, such as, Cs_2LiYCl_6:Ce (Qin et al., 2018; Giaz et al., 2016),
(iv) LiAlO2 (Pejchal et al., 2011), $^{6}LiCaAlF_6$ (Iwanowska et al., 2011), $LiSrAlF_6$ (Yanagida et al., 2011),
(v) mixed-alkali rare-earth double phosphate [e.g., $CsLi_2Lu(PO_4)_2$] scintillator (Neal et al., 2007),
(vi) the measurement of ultracold neutrons (UCNs) with ^{6}Li-glass scintillator (Jamieson and Rebenitsch, 2015; Ban et al., 2009), and

TABLE 9.8 Properties of state-of-the-art thermal neutron scintillators.

Host material	Dopant	Density (g/cm^3)	Mean free path at thermal energy (mm)	Decay time(s) (ns)	Limitations
^{6}Li glass	Ce	2.5	0.52	75	Poor light output; low α/β ratio
LiI	Eu	4.1	0.54	1400	Gamma sensitize; hygroscopic
LiF/ZnS	Ag	2.6	0.8	>1000	Poor efficiency; slow
$LiBaF_3$	Ce,K,Rb	5.3	7.9	1/34/2100	Long mean free path; UV window needed; intrinsic radioactivity
Cs_2LiYBr_6	Ce	4.1	3.7	89/2500	Large mean free path; hygroscopic; neutrons absorbed in Cl and Cs

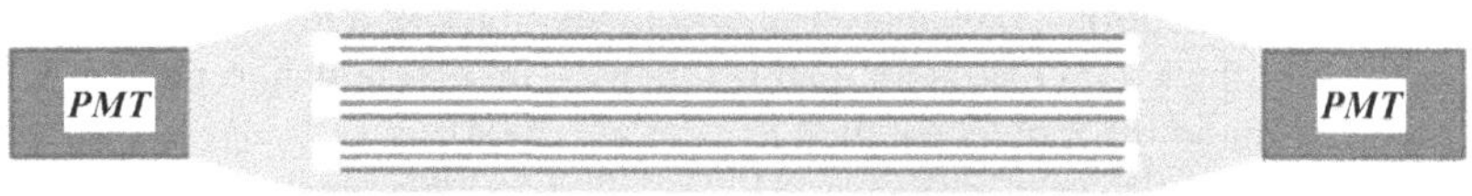

FIGURE 9.64 Lithium foil sandwich detector concept. The ^{6}Li sheets are shown in red, the PVD scintillator film is shown in blue, and the polymethyl methacrylate (PMMA) light-guide strips and fishtail light guide are shown in yellow. The electronic version of this text will display the color coding of the elements of the sandwiched neutron detector. *From Ianakiev et al. (2011); reprinted with permission from Elsevier © 2011.*

(vii) the suppression of gamma-radiation background with the design of a portable double photodiode coincident readout system using a 96% enriched ^{6}LiI(Eu) neutron detector (Yang et al., 2011).

The use of thin ^{6}Li foil sandwiched between scintillator and strips of light guide is an interesting neutron detector design reported by Ianakiev et al. (2011) to reduce interference from gamma radiation. The detector design is illustrated in Fig. 9.64. The thermal neutrons are captured by the ^{6}Li metal foil, the charged particles produced (primarily Triton) are detected by a physical vapor deposition (PVD) scintillator, and the light guide sandwiched with the scintillator transfers the light photons from the scintillator to PMT tubes on either side of the sandwiched detector. Gamma energy discrimination is improved by low-energy deposition in the thin scintillator film.

With the objective of designing a detector with properties similar to ^{3}He tubes embedded in a High Density Polyethylene (HDPE) moderator, Ianakiev et al. (2015) designed a ^{6}Li glass scintillator as a thermal neutron detector due to the attractive properties of ^{6}Li glass scintillator, which include excellent light output with a single peak neutron response (~7000 photons/neutron), very good thermal neutron macroscopic cross-section (15 cm^{-1}), fast light decay time (~70 ns), and the flexibility of glass to be fabricated into any desirable shape or size. The only drawback of using ^{6}Li glass scintillator is its gamma sensitivity, which Ianakiev et al. (2015) were able to reduce by embedding small ^{6}Li glass scintillator particles attached to a glass support structure embedded in an organic light-guide medium consisting of a mineral oil, which matches the indices of refraction and minimizes the scintillation light output at 396 nm. They demonstrate with various neutron and gamma-ray sources very good separations of neutron and gamma distributions (see Fig. 9.65) with normalized plateau characteristics similar to ^{3}He detectors.

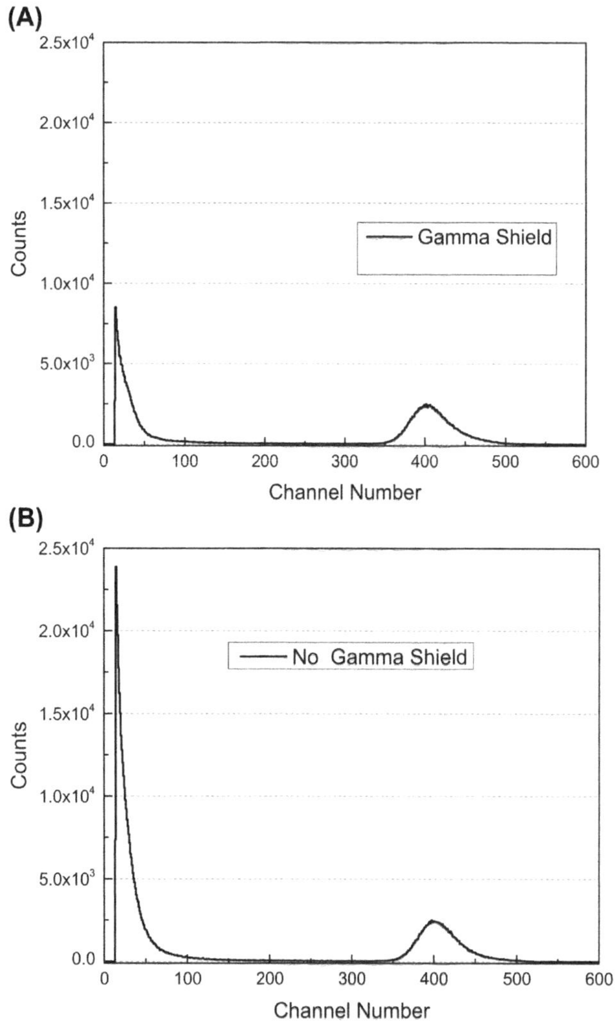

FIGURE 9.65 ^{252}Cf pulse height spectra: (A) with lead gamma shield and (B) without lead shield. The spectra are normalized to the neutron peak to show the difference in the gamma pulse-height distribution. *From Ianakiev et al. (2015); reprinted with permission from Elsevier © 2015.*

Another sandwich detector design was reported by Ban et al. (2009) for the detection of ultra-cold neutrons (UCNs), which is illustrated in Fig. 9.66. The detector consists of a ^{6}Li-depleted glass scintillator sandwiched together with a ^{6}Li-enriched glass scintillator. Both scintillators are of 100 μm thickness. The principle of the UCN detection design is depicted in Fig, 9.66. A UCN crosses the ^{6}Li-depleted scintillator and then reaches the ^{6}Li-enriched scintillator. A neutron capture occurs close to the entrance surface of the ^{6}Li-enriched glass scintillator, the escaping Triton, which has a range of 37 μm deposits its energy in the upper scintillator; whereas, the alpha particle deposits its energy in the first 6 μm of the lower scintillator. The differences in the ^{6}Li densities of the depleted and enriched scintillators are 6.5×10^{17} and 2.2×10^{22} cm^{-3}, respectively. Thus, the UCN absorption length, based on the ^{6}Li neutron capture reaction for UCN velocities ranging from 3 to 10 m/s, are determined to vary

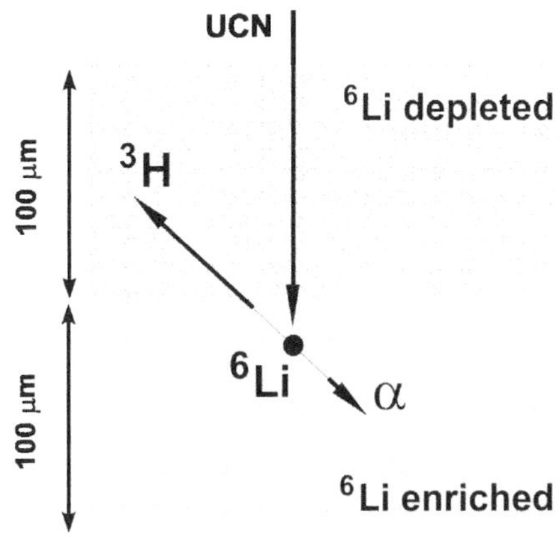

FIGURE 9.66 Scheme of an Ultra-Cold Neutron (UCN) detector. *From Ban et al. (2009); reprinted with permission from Elsevier © 2009.*

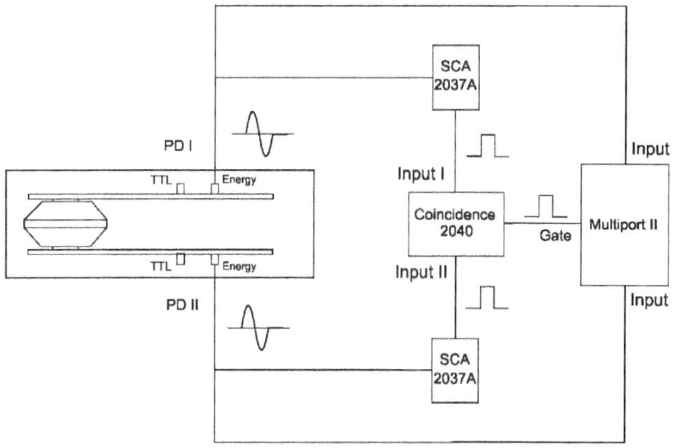

FIGURE 9.67 Coincidence electronics schematic of a portable handheld neutron detector. *From Yang et al. (2011); reprinted with permission from Elsevier © 2011.*

between 2.4×10^4 and 7.7×10^4 μm for the ^{6}Li-depleted scintillator and between 0.6 and 2.2 μm for the ^{6}Li-enriched scintillator (Ban et al., 2009). Thus, the upper ^{6}Li-depleted scintillator is practically transparent to the UCN; whereas, all UCN are stopped in the lower ^{6}Li-enriched scintillator. The events occurring at the edge or interface between the two scintillators lead to full energy deposition and a greater discrimination between the UCN and gamma background contribution (Ban et al., 2009). The two sides of this type of detector are very similar to the microscopic level, as both sides consist of cerium-doped lithium glass, one side depleted in ^{6}Li and the other side enriched in ^{6}Li. It is important therefore to determine which side of the glass should face toward the UCN source. Jamieson and Rebenitsch (2015) report a method of determining, which side of the glass is enriched with ^{6}Li with the use of a Si surface barrier detector, which sees an excess of events of α and Triton energies of neutron capture when the enriched side is facing the Si surface barrier detector and the detector is exposed to an AmBe and ^{252}Cf neutron sources.

A coincidence photodiode (PD) detector design with a 96% enriched ^{6}LiI(Eu) neutron detector as a portable handheld neutron-sensitive device capable of discriminating against background gamma radiation was reported by Yang et al. (2011). The detector utilizes a coincident readout by two PDs attached to two sides of a ^{6}LiI(Eu) crystal scintillator. The output signal is accepted only when both PDs register a coincident output signal; and the probability for a gamma-ray to generate a significant pulse in the active volumes of both PDs simultaneously is very low. Yang et al. (2011) demonstrated the method to effectively suppress gamma-radiation background in the neutron counting window to natural background levels. A gamma-radiation background of 0.034 CPS is reported with three ^{137}Cs sources having a total activity of 128.1 μCi in contact with the detector. Such low gamma-ray backgrounds would be equivalent to a probability of 1.69×10^{-6}% for a 662 keV gamma-ray from the ^{137}Cs impinging on the detector to create a false coincident event. Pulse-shape discrimination could be used to discriminate against gamma-radiation; however, that would increase the size and weight of the hand-held device. To avoid this Yang et al. (2011) modified the conventional detector design by placing two PDs on either side of a ^{6}LiI(Eu) neutron scintillator and combined this with a coincidence circuit, illustrated in Fig. 9.67, that would accept relevant neutron events in the scintillator. The high intrinsic thermal neutron detection efficiency of 99.83% together with the coincidence circuitry provide the hand-held detector with high gamma-discrimination capability, which is needed generally in scenarios involving the inspection of special nuclear materials.

The detector and its associated electronics, as described by Yang et al. (2011), is very compact, as it is all housed within an aluminum enclosure measuring externally only $89 \times 36 \times 30$ mm. The detector consists of a ^{6}LiI(Eu) crystal of 1 in. (2.5 cm) diag. by 3 mm thick coupled with a light guide on each side with two PDs on both sides of the crystal, which are labeled in the schematic of Fig. 9.67 as PD I and PD II. The PDs have a compact 10×10 mm active area with a 0.3 mm depletion layer thickness. The readout circuit has a built-in preamplifier and shaping amplifier. The outputs are one analog energy signal and one TTL (transistor–transistor logic) timing signal. The energy signal has a bipolar shape with 3 μs shaping time. The TTL is generated at the zero crossing point of the bipolar energy pulse, which is fed from the front end electronics to an SCA (2037A) and an input on the Multiport II multichannel analyzer (Canberra MCA). The gate on the MCA is only open when the coincidence output is active, which ensures that only the coincident outputs of both PDs are registered in the MCA.

From Fig. 9.62, illustrated previously in this chapter, it can be seen that the neutron capture cross-section of the ^{10}B target nucleus is much greater than that of ^{6}Li throughout the cold, thermal and epithermal neutron energy regions. We would expect, therefore, that replacing ^{6}Li with ^{10}B in scintillator compounds would improve neutron detection efficiency. However, ^{10}B scintillator compounds as neutron converters result in decreased light yields when compared to ^{6}Li scintillator compounds, which is due to the reduced nuclear reaction energy of ^{10}B and to the degradation in the transparency of the ^{10}B compounds (Kojima et al., 2004). The Q value of the $n(^{6}Li,\alpha)^{3}H$ reaction is 4.78 MeV; whereas, the Q value of the $n(^{10}B, \alpha)^{7}Li$ is only 2.79 MeV. Research into the development of efficient ^{10}B scintillators is ongoing (Mahl et al., 2018; Pan et al., 2018; Yin et al., 2018; Celentano et al., 2015). Kojima et al. (2004) indicate that the reduced light yields can be overcome somewhat by placing the scintillator (e.g., ZnS:Ag) and ^{10}B compound (e.g., $^{10}B_2O_3$) as close as possible so that the scintillator can receive efficiently the energy of the α- and ^{7}Li-particle. Also increasing the transparency of the scintillator is important. Ishii et al. (2005) studied Ce-doped oxide glass containing B and Li. The light output of the $B_2O_3-Li_2O:CeO_2$ glass by neutron irradiation was 9.2% of Li-glass; however, the low density, high neutron absorption and short decay time (30−45 ns) provide this glass scintillator with the properties useful to neutron detection. Fujimoto et al. (2011) have synthesized Ce-doped $Ba_3(BO_3)_2$ crystals as neutron scintillators, which exhibit short decay times ($\sim$36 ns); however, light yield is still a limiting property, which can be increased with improvements in crystalline quality.

B. Solid organic neutron scintillators

Organic scintillators are rich in protons and thus effective moderators of fast neutrons. Proton recoil resulting from elastic scattering of fast neutrons in solid organic scintillators can result in the transfer of most if not all of the neutron energy to protons depending on the scintillator dimensions. A fast neutron of a few MeV can lose 90% of its energy via collisions with protons during the first 10 ns in the scintillator (Fisher et al., 2011). As described in Chapter 1, Volume 1, the maximum energy, that may be transferred by a fast neutron to a recoil proton via a head-on collision, would be equivalent to the neutron energy. The recoil protons, which are charged particles, can then deposit their energy in the organic scintillator; and the scintillation photon intensities measured with a PMT or PD. Pulse shape discrimination (PSD), which is discussed further on in this chapter, may be used to discriminate between pulse events produced by the recoil protons and those produced by electrons, which result from background gamma-ray photon interactions in the organic scintillator.

Various forms of organic scintillators serve as fast neutron detectors including:

(i) organic crystalline materials such as stilbene or *p*-terphenyl (Becchetti et al., 2018; Carman et al., 2018; Grodzicka-Kobylka et al., 2018; Bourne et al., 2016; Sardet et al., 2015; Yanagida et al., 2015; Tarasenko et al., 2013; Robinson et al., 2011; Mertens et al., 2010; Hull et al., 2009; Budakovsky et al., 2007),

(ii) plastic scintillators available commercially from suppliers or prepared in the laboratory, such as polystyrene-based plastic containing primary and secondary fluors (e.g., 1% PPO and 0.1% POPOP, among others (Roy et al., 2018; Yu et al., 2017; Hartman et al., 2015; Lawrence et al., 2014a,b; Stevanato et al., 2011; Ovechkina et al., 2009), and

(iii) plastic scintillators doped with target nuclei having high neutron capture cross-sections, such as ^{6}Li, ^{10}B, or $^{155,157}Gd$ (Dumazert et al., 2018, 2016; Mahl et al., 2018; Balmer et al., 2015; Pawełczak et al., 2014; Zaitseva et al., 2013). This type of plastic neutron detector offers the possibility of capture-gated neutron detection for discrimination against undesirable background events.

Stilbene crystal slabs may be solution-grown up to 10 cm in thickness, as illustrated in Fig. 9.68; and organic polycrystalline detectors of diameters up to 25 cm can be made by pressing crystalline powder of p-terphenyl doped with diphenylbutadiene molecules or stilbene crystals from

FIGURE 9.68 A deuterated stilbene-d_{12} crystal grown in the shape of a hexagon. The initial seed is still seen through the optically transparent material that does not show any visible defects. The crystal mass is 38 g with a height of 35 mm. The crystal bottom has prints of the square marks lined on the platform for the growth rate measurements. *From Carman et al. (2018); reprinted with permission from Elsevier © 2018.*

a melt (Carman et al., 2018; Zaitseva et al., 2015; Budakovsky et al., 2007). Currently the most well-investigated organic crystal scintillator for the measurement of fast neutrons, in a n/γ mixed field with good n/γ PSD, is stilbene (Hull et al., 2009). The crystalline stilbene may be prepared at proton natural abundance or deuterated to facilitate PSD (Carman et al., 2018). The stilbene scintillators may be manufactured as single crystals grown from a melt or from stilbene crystalline powder and polymerized in a compound transparent to scintillation photons (Budakovsky et al., 2007; Robinson et al., 2011).

The response of stilbene scintillator as a single crystal measuring 30 mm in diameter × 30 mm in length to fast neutrons from a ^{239}Pu–Be source was compared to that of stilbene scintillators prepared from stilbene crystalline powder by Budakovsky et al. (2007). Three scintillators measuring 25 mm diameter × 20 mm length were prepared with stilbene powder having three different stilbene powder particle-size dimensions $L < 0.5$ mm, $L = 1 \ldots 2$ mm, and $L > 2.5$ mm. The three stilbene crystalline powder fractions were immersed into a glue compound that yielded a silicone scintillator matrix transparent to scintillation photons. The response of the various stilbene scintillators to fast neutrons from a ^{239}Pu–Be source is illustrated in Fig. 9.69.

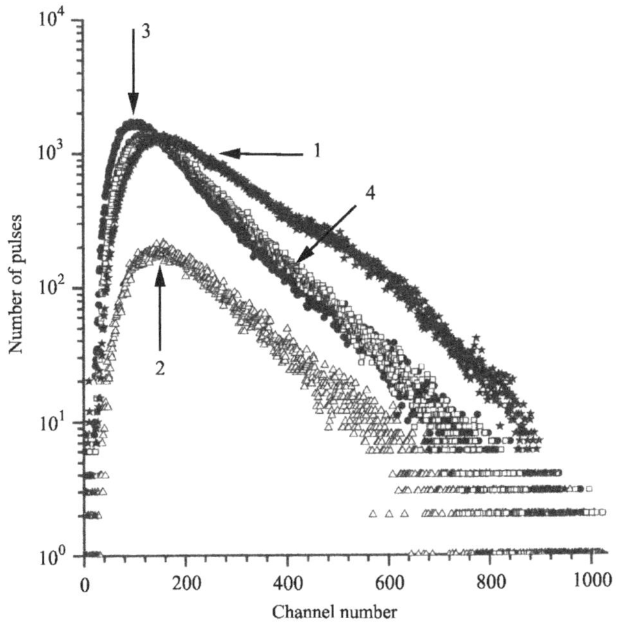

FIGURE 9.69 Recoil proton spectra of ^{139}Pu–Be source of fast neutrons for a reference stilbene scintillator consisting of a single crystal (curve 1) and for the composite scintillators prepared on the base of stilbene crystalline powder in silicone matrix. The fractions of stilbene powder are $L < 0.5$ mm (curve 2), $L = 1 \ldots 2$ mm (curve 3), and $L > 2.5$ mm (curve 4). The relative efficiencies for the detection of the fast neutrons are 0.12, 0.83, and 0.78 for the composite scintillators yielding curves 2, 3, and 4, respectively, compared to the reference scintillator consisting of a single crystal (curve 1). *From Budakovsky et al. (2007); reprinted with permission from Elsevier © 2007.*

Budakovsky et al. (2007) used pulse-shape n/γ discrimination to obtain the proton-recoil spectra from the ^{239}Pu–Be neutrons net of the accompanying gamma radiation. The recoil proton pulse height spectra cover a broad range of energies, as the neutrons can undergo numerous scatterings in the organic crystal yielding proton recoil energies varying between zero and the neutron energy. The stilbene reference detector consisting of a single crystal yields the highest neutron detection efficiency and higher light yield than the composite stilbene neutron detectors prepared from crystal powders. The composite scintillator made from very fine-meshed stilbene crystalline powder (<0.5 mm) yields the lowest neutron detection efficiency. This may be due to proton scattering that can occur close to the surface of the small crystalline powder resulting in the emission of the recoil proton after depositing only a fraction of its energy in the crystal. The range of a 2 MeV recoil proton in the stilbene is only ~0.07 mm (Budakovsky et al., 2007); however, the elastic scattering of fast neutrons in the small crystal powder (<0.5 mm) at depths of <0.07 mm can result in the emission and loss of the recoil proton from the scintillator.

The current limited supply of ^{3}He has sparked a concerted research effort to find suitable alternatives for neutron detectors; and the performance of these new detectors is sometimes compared to that of ^{3}He. Robinson et al. (2011) used a Monte Carlo N-Particle (MCNP) radiation transport code to simulate the performance of stilbene crystals of various dimensions as fast neutron detectors compared to that of a ^{3}He detector embedded in a moderator, which is commonly used as a radiation portal monitor. The intrinsic detection efficiency of each of the neutron detectors, which is defined as the number of neutrons counted per neutron incident on the detector surface, was used to evaluate detector performance. The simulated response of stilbene crystal detectors, based on perfect discrimination between neutron and gamma radiation events, relative to that of the moderated ^{3}He detector system is illustrated in Fig. 9.70. All detectors were attributed the same counting geometry and same solid angle so that the collection efficiency (neutrons incident on the detector per emitted neutron were the same for all detectors). As seen from Fig. 9.70, optimum neutron detection is a function of stilbene crystal thickness; and the efficiency of the moderated ^{3}He detector system is that of the horizontal line. To achieve equivalent intrinsic efficiency to the moderated ^{3}He system, while maintaining the same detector area, require stilbene thicknesses in excess of 2 cm, given contemporary pulse-shape discrimination thresholds (Robinson et al., 2011).

Organic scintillators including plastic scintillators loaded with target nuclei having high thermal neutron capture cross-sections (e.g., ^{10}B or ^{6}Li) can serve as excellent detectors of fast neutrons via the principle known

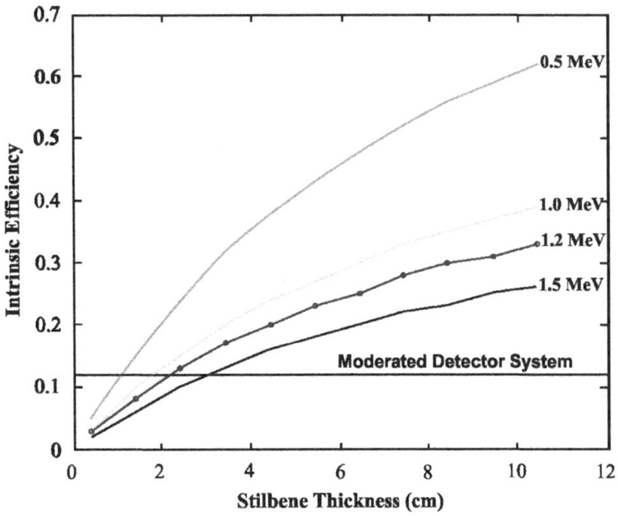

FIGURE 9.70 Simulated intrinsic efficiencies of stilbene crystal detectors as a function of thickness for several PSD thresholds (labeled by neutron energy) and that of a moderated ^{3}He detector system. *From Robinson et al. (2011); reprinted with permission from Elsevier © 2011.*

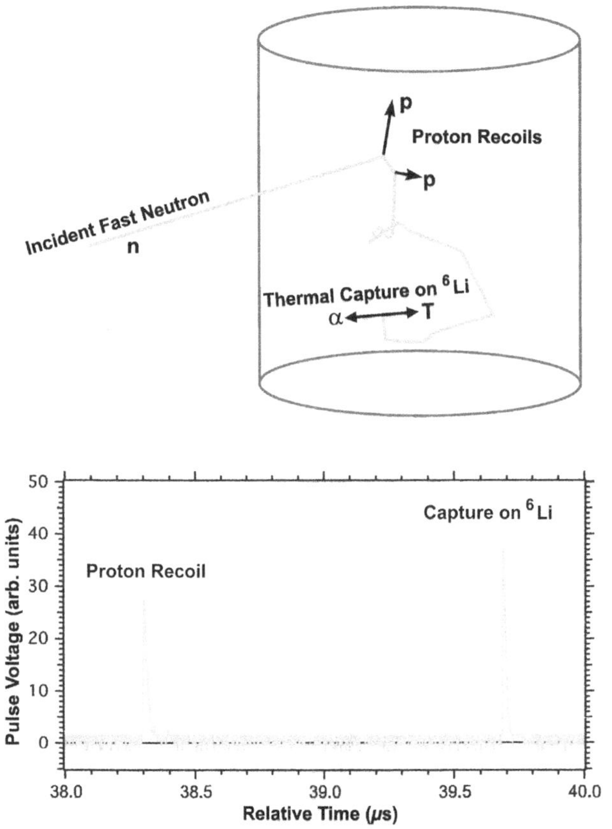

FIGURE 9.71 (Upper) An illustration of the principle of capture-gated detection. A fast neutron impinges on the detector. It rapidly gives up its energy primarily with protons. The thermalized neutron diffuses in the medium until it is captured by a nucleus with a high capture cross-section. (Lower) Plot of the PMT pulse of the detector response to a fast neutron within a scintillator following fast neutron scatter with proton recoil and neutron capture on ^{6}Li. Time difference between the two events is ~1.4 μs. *From Fisher et al. (2011); reprinted with permission from Elsevier © 2011.*

as **capture-gated neutron detection**. The organic scintillator has a high density of protons, which can moderate efficiently the energy of the fast neutrons, as described in Chapter 1, Volume 1. The principle of capture-gated neutron detection is illustrated in Fig. 9.71. As described by Fisher et al. (2011), a fast neutron enters the organic scintillator; and elastic scattering causes a proton recoil with approximately half (on average) of the neutron energy, producing scintillation light. The neutron then scatters off another proton producing a second recoil proton (see Fig. 9.71); and the neutron continues to scatter until it leaves the detector volume or is captured generally at thermal energies by a target nucleus, such as ^{6}Li or ^{10}B, which have a high thermal neutron capture cross-section. The major portion of the detected scintillation light will come from the recoil protons; and a neutron of a few MeV energy will lose 90% of its energy in the organic scintillator via neutron scattering with proton nuclei during the first 10 ns in the scintillator (Fisher et al., 2011). The thermalized neutron will diffuse in the scintillator until it is captured by a nucleus particularly that of an isotope that was introduced into the crystal during its manufacture, because of its high thermal neutron capture cross-section, such as ^{6}Li or ^{10}B. The time to capture of the fast neutron from the time of its entry into the scintillator is ~100 ns depending on the size and geometry of the scintillator and the isotope and it's concentration introduced into the scintillator as a capture target nucleus. The delayed-coincidence signature (i.e., the fast proton recoil scintillation at ~10 ns and the delayed capture scintillator at ~100 ns) indicates that the fast neutron dissipated essentially all of its energy within the scintillator, which provides a method for the rejection of uncorrelated background events together with PSD (Fisher et al., 2011; Wilhelm et al., 2017).

Britvich et al. (2005) recommends ^{10}B isotope in natural boron as a target isotope in molded polystyrene scintillators in lieu of ^{6}Li target nuclei. The ^{10}B has a natural abundance of 19.9%, which can be suitable for fast neutron measurements in plastic scintillators; and it avoids the expensive procedures of its enrichment. Boron-10 also has a higher neutron-capture cross-section than ^{6}Li (see Fig. 9.62); although ^{6}Li is often used as a target nucleus at very high isotope enrichments exceeding 95%. Quarenta et al. (2011) report the synthesis and testing of boron-doped polysiloxane scintillators with boron concentrations ranging from 4 wt.% up to 8 wt.% with excellent sensitivity to fast and thermal neutrons measurement. Holm et al. (2014) designed a capture-gated neutron spectrometer with boron loaded plastic scintillator with 5% weight with excellent neutron and gamma photon discrimination with one out of 10^5 photons misclassified. The capture-gated neutron detection in ^{10}B-doped scintillator is illustrated in Fig. 9.72. With a ^{10}B-loaded liquid rather than a solid scintillator, Hunt et al. (2016) provide an excellent example of a detailed analysis of PSD in capture-gated neutron spectrometry. Fig. 9.72 illustrates two possibilities for the interaction of incident neutrons in the ^{10}B-doped scintillation. In the interaction (a) illustrated, the incident neutron interacts with protons resulting in proton recoil and neutron attenuation and finally capture by ^{10}B at the thermal neutron energy state. The neutron capture by ^{10}B results in the production of an α-particle and ^7Li and the concomitant emission of a gamma-ray photon from the ^7Li nuclide. The moderation of the neutron via proton recoil produces a moderation peak; and the α-particle produces the capture peak, illustrated in the lower panel of Fig. 9.72. Partial moderation of a neutron via neutron scatter and proton recoil with the eventual escape of the neutron from the scintillator depicted in the interaction of the incident neutron (b) does not result in neutron capture within the scintillator and thus only a moderation waveform is produced, illustrated in the waveform (b). Hunt et al. (2016) explain that noise can produce a second pulse in the tail of a neutron scattering peak as illustrated in the waveform (c) illustrated in Fig. 9.72. With the double-peak signature, illustrated in waveform (a) of Fig. 9.72, full energy deposition of the neutron occur; and the incident neutron energy can be estimated from the pulse height of the first moderation peak. Hunt et al. (2016) use PSD to identify fast neutrons from thermal neutrons as well as γ-ray photons and noise.

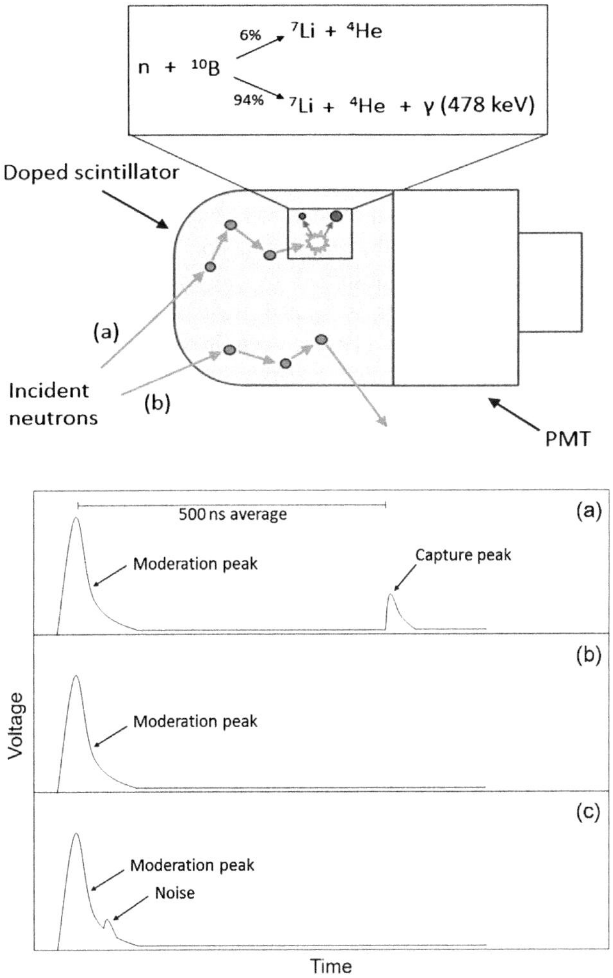

FIGURE 9.72 Upper Panel: Neutron-induced events in a ^{10}B doped liquid scintillator. (a) Full energy deposition of fast neutron energy, giving rise to the waveform shown below plot (a). (b) Partial energy deposition of fast neutron energy, giving rise to the waveform shown below in plot (b). Lower Panel: Schematic waveforms of a ^{10}B-doped liquid scintillator. (a) From full energy deposition of an incident fast neutron. The first peak is caused by the moderation of the fast neutron and carries the information about the incident energy, while the second peak is caused by the moderated neutron undergoing a $^{10}B(n,\alpha)^7Li$ reaction. The mean time between moderation pulse and capture pulse is 500 ns. (b) From an incident γ-ray or partial energy deposition of an incident fast neutron, i.e., no $^{10}B(n, \alpha)^7Li$ reaction has occurred. (c) Noise on the tail of a moderation peak. *From Hunt et al. (2016); reprinted with permission from Elsevier © 2016.*

The isotopes ^{155}Gd and ^{157}Gd occur with natural abundances of 15% and 16%, respectively; and ^{157}Gd has the highest thermal-neutron capture cross-section among all isotopes commonly used as target nuclei for thermal neutron energies < 0.05 eV (see Fig. 9.62). Ovechkina et al. (2009) synthesized and tested a Gd-loaded polystyrene scintillator. The plastic scintillator was prepared from liquid styrene mixed with concentrations of 1% 2,5-diphenyloxazole (PPO) as primary fluor and 0.1% 4-bis(5-phenyl-2-oxazole) abbreviated POPOP as secondary fluor; and gadolinium isopropoxide, $Gd[OCH(CH_3)_2]_3$, was added as the neutron-sensitive dopant in concentrations up to 6.4%. Polymerization of the mixture was accomplished at 125−140°C in an argon atmosphere into plastic scintillator having a 14 mm diameter and up to 13 mm thickness. A 13 mm thick scintillator containing 0.5% gadolinium by weight detected 46% of incident thermal neutrons.

A unique approach to the design of a large-volume Gd-loaded plastic scintillator is described by Pawełczak et al. (2011). The addition of Gd into large-volume plastic scintillators can cause opacity and significantly reduce the light attenuation length. Thus, Pawełczak et al. (2011) devised a capture-gated Gd-loaded plastic scintillator, NSTAR (Neutron Sandwich Transmuter/Activation-γ Radiator), consisting of Gd foil sandwiched between layers of plastic scintillator (Saint Gobain BC-408) as illustrated in Fig. 9.73. Separating physically the Gd target nuclei from the scintillator avoids the opacity problem associated with the target nuclei loaded directly in the scintillator and permits the manufacture of large volume detectors. The

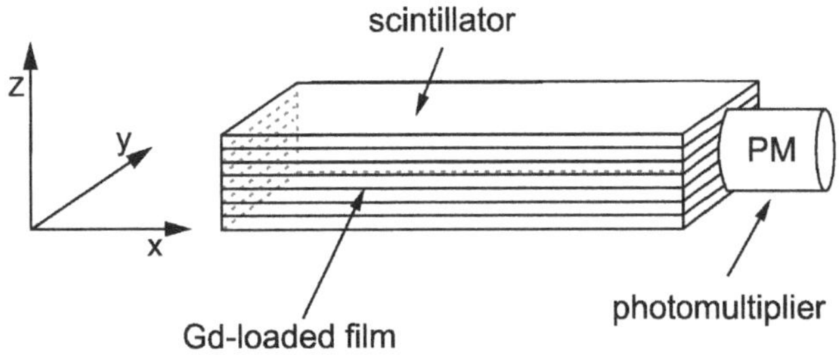

FIGURE 9.73 Schematic view of an NSTAR capture-gated Gd-loaded neutron prototype detector. *From Pawełczak et al. (2011); reprinted with permission from Elsevier © 2011.*

detector consisted of 12 × 20 × (50 or 100) cm^3 stacks of plastic scintillator slabs alternating with thin Gd converter films viewed by fast photomultipliers. The principle of the capture-gated measurement of fast neutrons in the NSTAR, as described by Pawełczak et al. (2011), is as follows: Fast neutrons are rapidly moderated in the plastic scintillator where they produce a "prompt" scintillation light output signal. The thermalized neutrons diffuse slowly through the detector until captured by the nuclei in the second detector component (i.e., Gd-loaded film, See Fig. 9.73), the $n_{th}-\gamma$ converter, which gives rise to a delayed output signal. The ^{nat}Gd film, which contains natural gadolinium isotopes, in particular, ^{155}Gd and ^{157}Gd at 16% and 15% natural abundance respectively, having very high thermal-neutron capture cross-sections, emit a cascade of two to three gamma rays (see Eqs. 9.48 and 9.49) following the neutron capture process corresponding to neutron binding energies of 8.46 MeV (^{155}Gd) and 7.87 MeV (^{157}Gd), which are detected with the plastic scintillator component of NSTAR via Compton scattering and are absorbed with an increasing efficiency with active detector volume to yield a delayed photopeak at ~ 8 MeV to signal the total absorption of capture γ-rays (Pawełczak et al., 2011).

C. Neutron detectors with scintillating and optical fibers

Fiber-optic arrays can be used as detectors of nuclear radiation including the detection and measurement of neutrons. There are two types of basic detector arrangements when using scintillating fibers for the detection of nuclear radiation including neutron measurements. These are (i) a scintillating fiber (i.e., one that contains fluor molecules) absorbs energy of the impinging nuclear radiation and the scintillation photons propagate along the fiber to a photodetector and (ii) a scintillation detector (e.g., plastic or inorganic scintillator) is coupled to an optical fiber, which propagates the scintillation photons on to a photodetector. These are described in the following sections.

1. Scintillating fibers arrays

Plastic scintillating fibers are reported to provide better photon yields, longer light attenuation, shorter decay times, and longer nuclear reaction and radiation lengths than glass fiber (Leutz, 1995). As reported by Leutz (1995), the plastic fibers are manufactured with a scintillating plastic core several micrometers up to a few millimeters in diameter, and the cores are surrounded by a cladding layer ($\leq 5\ \mu m$) of lower refractive index than the scintillating core (see Fig. 9.74). The scintillating core may consist of polystyrene (PS) plastic polymer and the cladding of polymethyl methacrylate (PMMA). When the cladding is of lower refractive index than the scintillating core (i.e., $n_g < n_c$) the fiber traps a fraction t of the scintillating light, as illustrated in Fig. 9.74 according to the equation

$$t = 1 - (n_2 / n_1) \tag{9.50}$$

where n_2 and n_1 are the index of refraction of the cladding material and that of the scintillating core, respectively. Any untrapped light is absorbed by an extramural absorber (EMA), if the fiber is coated with a black layer. As explained by Leutz (1995), even in the absence of an EMA the untrapped light escapes at an angle, which prevents it from being trapped again by a neighboring fiber; and it becomes lost after about 50 mm of fiber length. Therefore, the untrapped light will not contribute to appreciable cross-talk between neighboring scintillating fibers.

The fiber diameter and type of scintillator used in the plastic core material govern the spatial resolution because proper selection of scintillating fluor in the core minimizes cross-talk between neighboring fibers. Leutz (1995) points out that the ionizing nuclear radiation will first cause excitation of the polystyrene. Because the polystyrene fluorescence yield is poor, an aromatic scintillator must be included in the polystyrene core polymer. Both primary and secondary scintillators have been used, such as *p*-terphenyl (PTP) and POPOP, respectively; however, POPOP has an overlapping absorption and emission spectrum. A single scintillator concentration of >0.015 M fraction of 1-phenyl-3-mesityl-2-pyrazoline (PMP) as a one-component fluor in the polystyrene scintillating plastic core can be used as it has a large Stokes shift, that is, practically no overlap of its absorption and emission spectra. Leutz (1995) indicates that cross-talk between scintillating fibers is avoided by using only one fluor component, which displays a large Stokes shift in the polystyrene core, such as PMP (Güsten and Mirsky, 1991). This is explained by the fact that light emitted by the first fluor (dopant) can escape from fibers of small diameter, excite the wavelength shifter (second fluor or dopant) in neighboring fibers, and cause undesirable cross-talk when the absorption and emission spectra of the two dopants overlap.

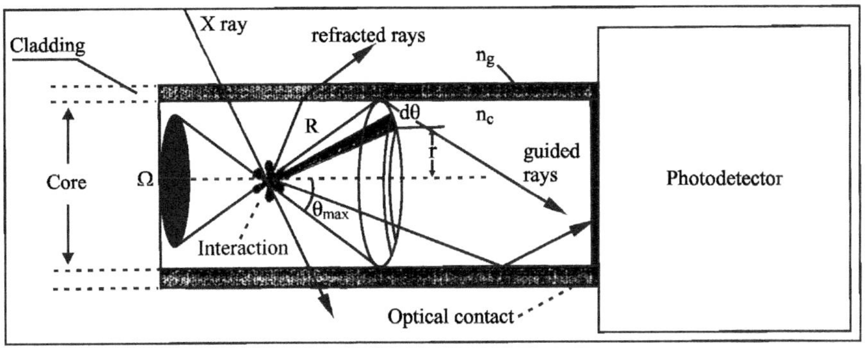

FIGURE 9.74 Schematic diagram of the optical propagation in the plastic scintillating fiber. *From Ikhlef et al. (2000); reprinted with permission from Elsevier ©2000.*

Barbagallo et al. (2011) used an array of scintillating fibers in the design of a thermal-neutron detector suitable for measurements in the decommissioning of storage and radioactive waste or the mapping of flux in the out-of-core regions of fission reactors. The detector designed by Barbagallo et al. (2011) is illustrated in Fig. 9.75, which operates on the following principle:

- a thermal neutron impinges on a layer of ^{6}LiF enriched to 96% in ^{6}Li;
- the thermal neutron is captured by the ^{6}Li triggering the reaction ^{6}Li(n,t)^{4}He;
- the triton and the α-particle are emitted back to back; and each hit the scintillator in opposite directions;
- the scintillation photons produced by the triton or α-particle propagate along a scintillating fiber (9 cm in length) to be detected by a corresponding SiPM at each end of the scintillating fiber bundle.

The neutron converter consisted of a 3 μm thick and 10 mm diameter film of ^{6}LiF deposited by evaporation on a glass substrate. The converter was assembled in contact with the middle point of a bundle of nine plastic scintillating fibers (0.3 mm in diameter). The bundle (9 cm long) is arranged flat onto the converter; and at each end the bundle of 3 × 3 fibers is squeezed into a square array with dimensions of 0.9 × 0.9 mm, which fits comfortably with optical grease onto the sensitive area of 1 × 1 mm of each SiPM. This experimental arrangement enabled left-right coincidence counting to suppress spurious counts, which could arise from SiPM dark noise and cross-talk between adjacent SiPMs.

Gamma background radiation was determined with a detector insensitive to neutrons, i.e., without the ^{6}Li neutron converter; and the difference of the pulse events of the two detectors would yield the spectra due the neutron flux as measured by the alpha-particle and tritium reaction products of neutron capture, as illustrated in Fig. 9.76. The gamma-ray rejection is good, as the gamma-ray efficiency is reported to be 10^{-4}; and the neutron/gamma rate was

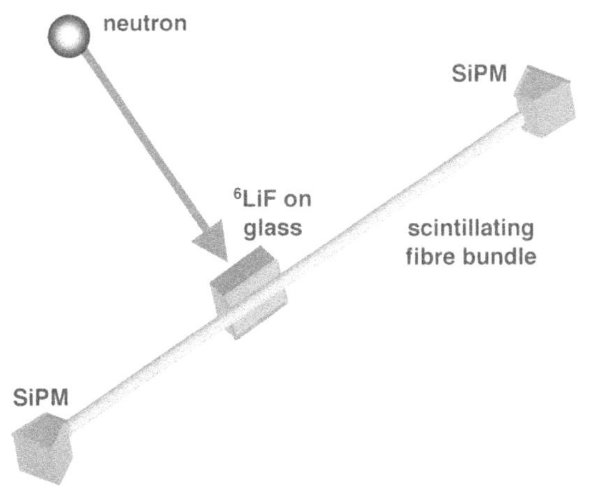

FIGURE 9.75 Sketch of the operating principle of the thermal neutron detector, based on neutron capture by ^{6}Li and detection of either the alpha-particle or tritium produced by means of a scintillating fiber bundle and two SiPM photosensors. *From Barbagallo et al. (2011); reprinted with permission from Elsevier © 2011.*

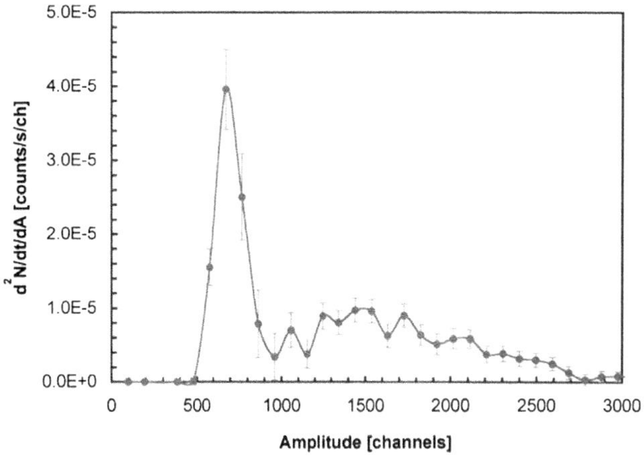

FIGURE 9.76 Difference between two spectra produced by detectors with and without ^{6}LiF neutron converter. Two bumps in the spectrum are due to the detection of α-particles (lower channels, 500–1000) and tritium (higher amplitude channels, 1000–3000). The pulse amplitudes due to tritium appear at the higher channels since the lighter and more energetic tritium on the average produces more light in the scintillation fibers. *From Barbagallo et al. (2011); reprinted with permission from Elsevier @ 2011.*

~2%. Barbagallo et al. (2011) propose to improve the detector by depositing the ^{6}LiF neutron converter directly onto the scintillating fiber bundle and sandwiching or stacking two or more bundles.

A thermal neutron detector with a spatial resolution of ~1 micron is proposed by Song et al. (2016). They utilized a Monte Carlo toolkit Geant4 to design a detector assembly composed of cerium-doped lithium glass scintillator microfibers of 1 μm diameter. The location of neutron absorption is reconstructed from the charged particle products (α-particle and tritium) of neutron capture on ^{6}Li, as described by Eq. (9.46) earlier in this chapter. The detector design includes surrounding each scintillating fiber with air-filled glass capillaries with the same diameter of the fiber to suppress cross-talk of scintillation light from fiber to fiber and thereby improve on spatial resolution, as illustrated in Fig. 9.77. The front end of the detector is coated with a thin optical reflector to maximize the light collection efficiency at the output end. The scintillation light produced by neutron absorption is magnified and recorded by an intensified CCD camera (ICCD) illustrated in Fig. 9.77.

Scintillating fiber arrays are applied to various applications including:

(i) the real-time measurement of gross beta radiation in drinking water (Lv et al., 2018),

(ii) the coupling of scintillating fiber arrays to silicon photodiode (Si-PD) or SiPM photo sensors as hodoscope detectors in beam monitoring where their fast response and insensitivity to magnetic fields make them ideal particle trackers where timing information is required (Damyanova and Bravar, 2017; Blasco et al., 2015), and

(iii) the construction of trackers associated with the LHCb detector of the Large Hadron Collider (LHC) (Kirn on behalf of the LHCb collaboration, 2017; Tobin on behalf of the LHCb Upgrade Scintillating Fiber Tracker group, 2016; Leverington et al., 2013).

The basic principle of the scintillating fiber (SciFi) particle tracker designed for the LHCb detector located at the LHC is illustrated in Fig. 9.78. As described by Kirn on behalf of the LHCb collaboration (2017), the SciFi detector consists of scintillating fiber mats having a length of 5 m and a width of 0.53 m consisting of layers of honeycomb and carbo fiber–reinforced polymer (CFRP) glued to both sides of the scintillating fiber mats to provide rigidity to the module. The plastic scintillating fibers are 0.25 mm in diameter. An epoxy glue loaded with 25% TiO_2 reflective material minimized optical cross-talk between adjacent fibers while at the same time established the fibers as a composite unit. Fig. 9.78 illustrates how a charged particle traveling through the honeycomb of scintillating fibers

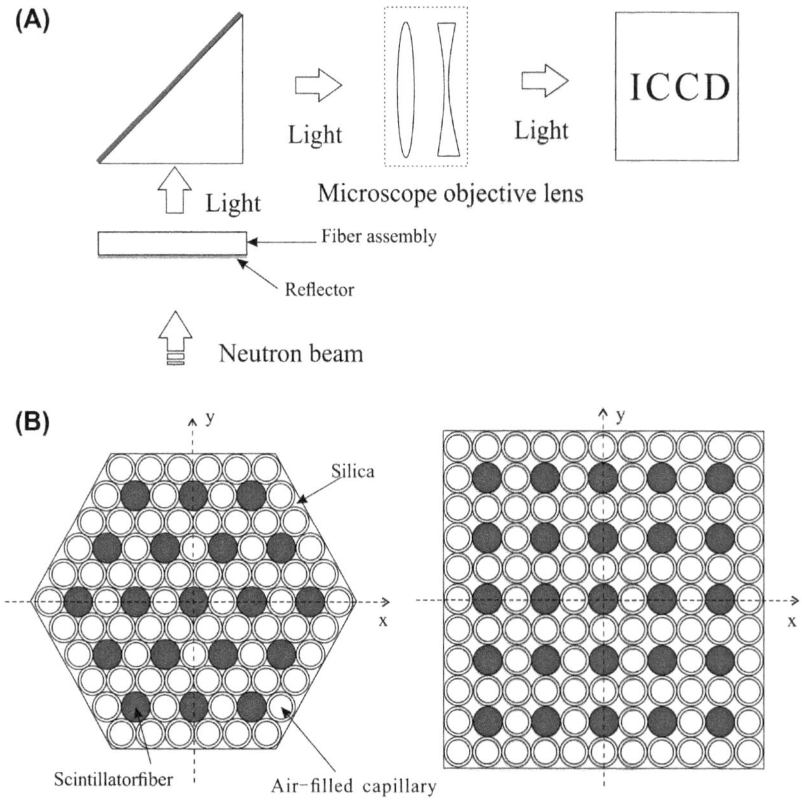

FIGURE 9.77 (A) The schematic layout (not to scale) of the neutron detector system. (B) Both hexagonal and square microfiber assemblies are viewed in the neutron beam direction. *From Song et al. (2016); reprinted with permission from Elsevier © 2016.*

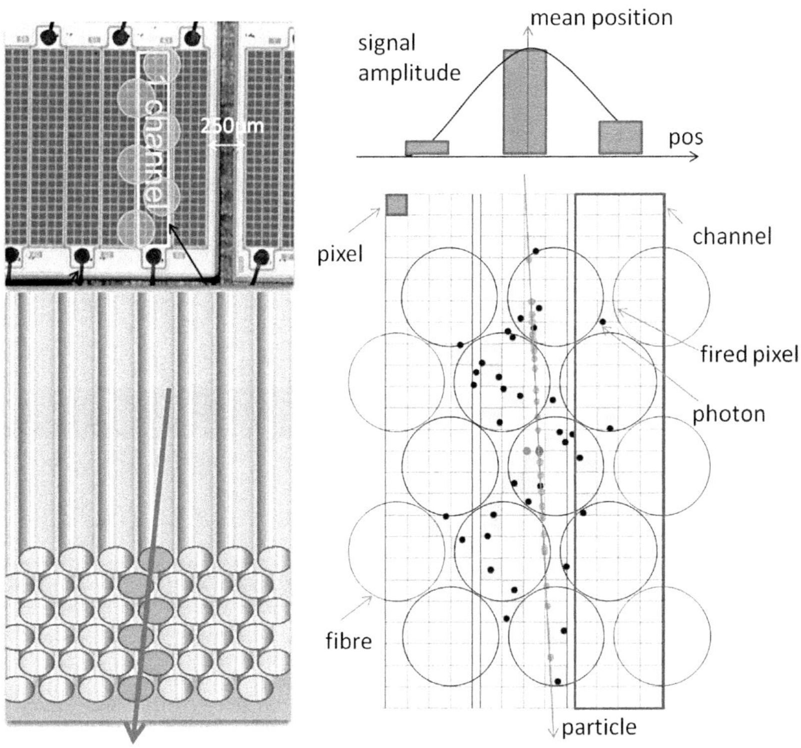

FIGURE 9.78 A sketch of the basic principle of the SciFi (scintillating fiber) tracker. A charge particle passing through the fiber produces scintillation light which travels to the end and is detected in a SiPM array. The black dots indicate photons arriving at the SiPM, the yellow squares indicate the pixels that fire. (For interpretation of the references to color in this figure legend, the reader is referred to the web version of this chapter). *From Kirn on behalf of the LHCb collaboration (2017); reprinted with permission from Elsevier © 2017.*

initiates excitation and scintillation with emissions of photons captured by the SiPM pixel array, which provide a signal amplitude corresponding the energy absorbed by the scintillating fibers.

2. Optical fiber—neutron detector arrays

Plastic or quartz optical fiber of lengths extending to several meters may be coupled to a scintillation detector with or without neutron converter, in lieu of the scintillating fibers described in the previous paragraphs. Arrangements with optical fiber may include a small ($\sim$1.0 mm dia.) organic or inorganic scintillator surrounded with a neutron converter (e.g., ^{6}LiF enriched to $>95\%$ ^{6}Li) coupled to a desirable length of plastic or quartz optical fiber. The measurement of gamma background is possible with the use of the scintillator detector without the neutron converter. Such detector arrangements are illustrated in Fig. 9.79.

To improve the detection of photons emitted by the scintillator a short section ($\sim$1 cm) of wavelength shifting fiber (WSF) may be coupled between the scintillator and optical fiber. The WSF absorbs the scintillation photons and then emits photons at a longer wavelength, which may be more efficiently detected by the photomultiplier. A neutron detector arrangement with WSF and optical fiber is illustrated in Fig. 9.80. The measurement principle of this optical fiber—neutron detector arrangement, as described by Yagi et al. (2011) is as follows: (1) A converter material, such as ^{6}LiF, generates charged particles by reacting with neutrons; (2) scintillation light is emitted from a scintillator, such as ZnS(Ag), after the charged particles have collided with it; (3) the scintillation light is absorbed by the WSF; and the WSF then emits light with a longer wavelength than the original light; and (4) the light inside the WSF is then transmitted to the tip of the optical fiber; and finally travels through the optical fiber until it reaches the PMT.

Commonly utilized converter materials for thermal neutrons, which emit fission fragments upon thermal neutron capture, discussed previously, are enriched ^{6}Li and ^{10}B. Also fissile ^{235}U and ^{239}Pu, which undergo nuclear fission yielding fission fragments, may also be utilized as converter isotopes. On the other hand, common converter materials for fast neutrons are the fissionable radionuclides ^{232}Th (Haas et al., 2011) and ^{238}U, which undergo fission upon reaction with fast neutrons ($>$1 MeV). With neutron converter materials in mind, a detector was developed at Nagoya University (Mori et al., 1994), which consists of an optical fiber with its tip covered with a mixture of a neutron converter and ZnS(Ag) scintillator. Along these lines, Yamane et al. (1998, 1999) designed and tested small detectors for thermal and fast neutron flux measurements in a research reactor. They chose ^{6}Li as converter for the analysis of thermal neutron flux and ^{232}Th as converter for fast neutron flux measurements. The scintillator selected was ZnS(Ag), which was mixed with a converter and glued with

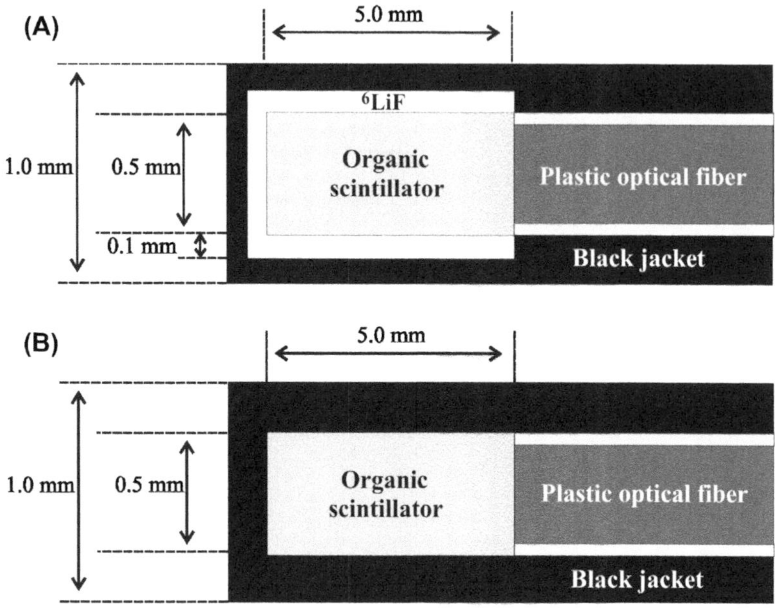

FIGURE 9.79 Cross-section of the fiber-optic (A) neutron sensor and (B) gamma sensor. The detector arrangement utilizes ^{6}LiF and BCF-20 (Saint Gobain) organic scintillator. A thin cladding surrounding the plastic optical fiber is depicted as a lighter color or hue between the plastic optical fiber and black jacket. *From Jang et al. (2011); reprinted with permission from Elsevier © 2011.*

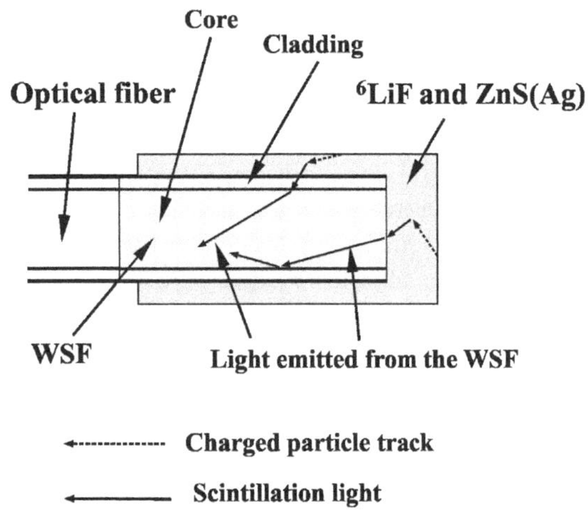

FIGURE 9.80 Transmission mechanism of an optical fiber detector with a WSF and surrounding ^{6}LiF neutron converter and ZnS(Ag) scintillator. The WSF and its surrounding neutron converter and scintillator are ~1 cm in length. The optical fiber is only 1 mm in diameter; and its length is measured in meters. An optical grease connects the WSF and optical fiber. Charged particle tracks in the scintillator are illustrated with broken arrows; and the scintillation light in the scintillator and WSF is depicted with solid arrows. *From Yagi et al. (2011); reprinted with permission from Elsevier © 2011.*

epoxy to the tip of glass optical fiber (2 mm diameter quartz). Glass optical fiber was selected because it is more resistant to radiation damage than plastic fiber. A 0.5 mm thick aluminum cap covers the neutron detector tip, which both serves to protect the scintillator and converter material and reflect light toward the optical fiber. A schematic view of the detector and measurement layout is provided in Fig. 9.81. The electric signals from the PMT are sent through a preamplifier and amplifier (AMP), and MCA for pulse height measurements. An SCA and multichannel scalar were also applied to the measurements.

The advantages of this type of detector for neutron flux measurements demonstrated by Yamane et al. (1999) are the following: (i) The detector is slim (2 mm dia.) and capable of fitting into narrow measurement positions such as between reactor fuel plates. (ii) Due to the small size of the detector, it has reduced spatial dependence of the neutron flux providing good flux measurement resolution. (iii) Small changes or fluctuations in the neutron flux at certain known positions in the reactor core, such as dips caused by spacers, can be measured. (iv) The neutron current or gradient of the flux can be measured. (v) A neutron flux mapping can be made over large distances of the reactor core is a short period of time. This is a major advantage as conventional measurements of the axial flux profile of a reactor core using cobalt wire activation measurements may take several hours, while the use of a glass fiber detector would take only a few minutes for the same measurement. Yagi et al. (2011) increased the sensitivity of such a small neutron detector by using a WSF between the ^{6}LiF neutron converter/ZnS(Ag) scintillator and the optical fiber. The sensitivity of the detector for thermal neutrons increased significantly with the WSF. They utilized the detector with a WSF of up to 20 cm in length attached to the optical fiber for the measurement of reaction rate distributions in a reactor core.

Jang et al. (2011) underscore the advantages of optical fiber in the detection and transmission of scintillation light signals, among which are long-distance signal transmission, real-time measurement, and no disturbances from

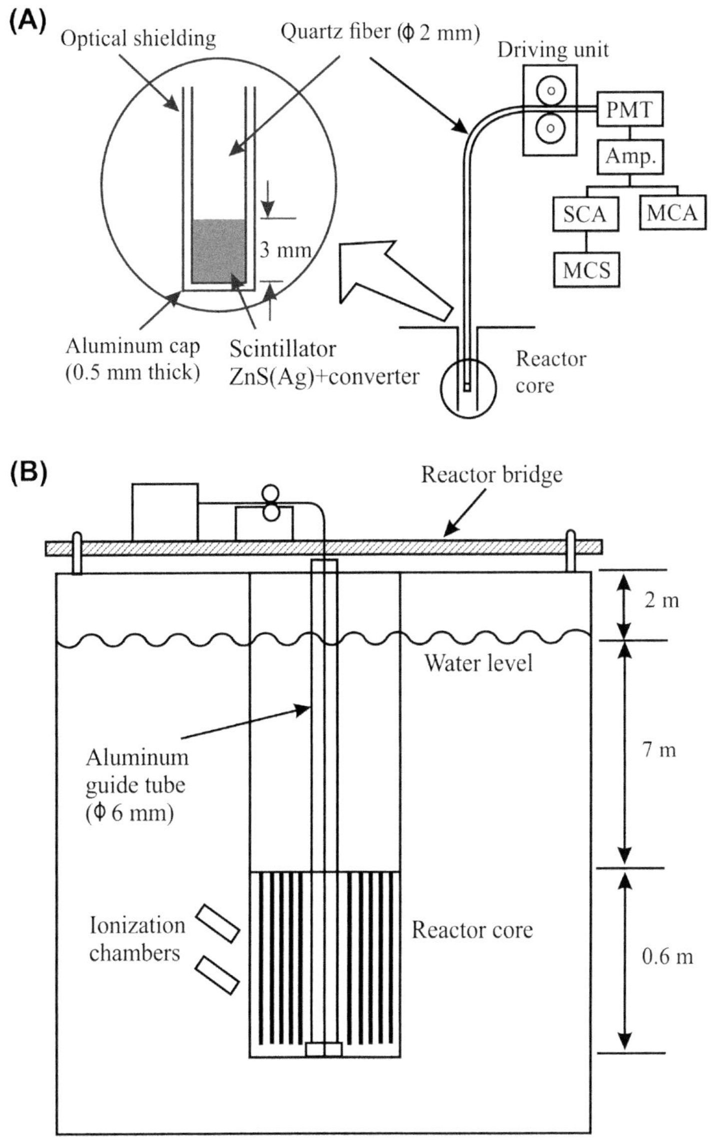

FIGURE 9.81 Schematic view of (A) the quartz fiber tip and the measurement arrangement, and (B) layout of the measurement of neutron flux in a research reactor. *From Yamane et al. (1999); reprinted with permission from Elsevier © 1999.*

temperature, pressure, and electromagnetic waves. As gamma-radiation background is often present in the measurement of neutron radiation, they devised an integrated fiber-optic neutron-, gamma-, and Cherenkov sensors using a cable containing three fiber optic sensors, which are described as follows: (1) thermal neutrons were measured with the ^{6}LiF converter coated onto organic scintillator at the tip of a 0.5 mm diameter optical fiber, as illustrated in Fig. 9.79, (2) the gamma-ray background was measured with organic scintillator without the neutron converter at the tip of an optical fiber, as illustrated in Fig. 9.79 and (3) background from high-energy particles, which produce Cherenkov radiation, was determined with a fiber optic cable containing neither an organic scintillator detector nor neutron converter. The three fiber optic sensors housed in a light tight jacket as an integrated fiber optic sensor are illustrated in Fig. 9.82.

Large area neutron-sensitive scintillators for neutron diffractometers and reflectometers were designed by McKnight et al. (2008) and Nakamura et al. (2009). These contain ZnS/^{6}LiF as the scintillator and neutron converter material, which surround numerous WSFs, which can lead to specific PMTs. The detector designed by Nakamura et al. (2009) measured 32 mm long and 70 mm wide. It exhibited a spatial resolution of 0.5 mm, a detection efficiency of 35% for 1 Å neutrons, and a gamma sensitivity of less than 10^{-7}.

A common application of the coupling of scintillation detectors with optical fiber is the coupling of scintillators sensitive to neutrons with a wide variety of wavelength

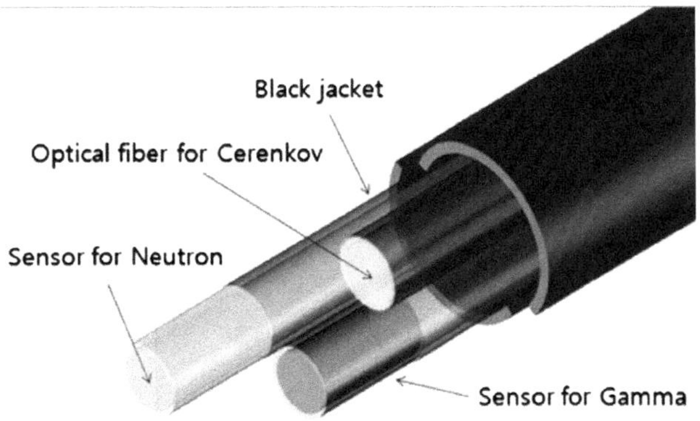

FIGURE 9.82 Structure of the integrated fiber optic sensors. *From Jang et al. (2011); reprinted with permission from Elsevier © 2011.*

shifting optical fiber arrays such as the following: (i) the use of 0.5 cm thick wafers of Europium-activated lithium calcium aluminum fluoride (LiCAF) crystals mated to SiPMs to measure neutron flux (Ford et al., 2018), (ii) $ZnS/^6LiF$ scintillating layers with embedded WSF with SiPM readout (Mosset et al., 2014), (iii) a scintillating screen composed of $ZnS/^6LiF$ coupled to WSFs of 100 μm diameter oriented in the x- and y-directions with 34 × 34 μm pixel size output yielding sub 100 μm spatial resolution (Nakamura et al., 2014a), (iv) large area (256 × 256 mm^2) neutron-sensitive $ZnS/^{10}B_2O_3$ screens coupled to cross-arranged (x- and y-directions) WSF arrays for application in a neutron diffractometer (Kawasaki et al., 2014), and (v) a two-dimensional high-spatial-resolution scintillating-based neutron detector consisting of a $ZnS/^{10}B_2O_3$ scintillating screen coupled to wavelength-shifting fibers oriented in x- and y-directions to provide a position-sensitive neutron detector (Nakamura et al., 2015).

A schematic view of the high-spatial-resolution neutron detector designed by Nakamura et al. (2015) is illustrated in Fig. 9.83. The neutron-sensitive spatial resolution detector covers a detection area of 320 × 320 mm consisting of $ZnS/^{10}B_2O_3$ scintillator screens, which are placed on both sides of 1 mm diameter wavelength shifting (WLS) fibers. The WLS fibers are situated at a 2.5 mm pitch in the x and y directions; and the crossed WLS fibers are in direct contact (cross-talk) with each other. Overall, the detector contains 256 WLS fibers comprised of 128 fibers in two layers oriented in the x and y directions. As noted by Nakamura et al. (2015), a neutron incident on the scintillator screen is absorbed and captured by the nuclear reaction $^{10}B(n,\alpha)^7Li$; and the secondary α-particle and ^7Li ion deposit energy in the ZnS scintillator. The ZnS phosphor has a peak photon emission at 450 nm with ∼100,000 photons emitted per neutron absorbed. About one-half of the photons escape the screen and enter the WSFs, which reemits the light at a longer wavelength of 500 nm. The light propagates along the affected fibers to the PMTs at the end of each fiber. The PMTs and detector electronics are operated in the photon counting method to ensure a high detection efficiency. The photoelectrons (p.e.) detected at each fiber are counted for 2 μs when a trigger signal opens a gate; and a scintillation event from a fiber is recorded as a neutron event when the total number of photoelectrons exceeds a threshold level. The position of the incident neutron is calculated based on an MPC (Model Predictive Control) method, which utilizes algorithms similar to those implemented with Anger cameras.

VII. Scintillation in plastic media

Plastic scintillators may be considered as solid solutions of one or more organic scintillation fluors (fluorescent dopants) in translucent plastic. The solid plastic medium plays a role similar to that of the liquid solvent in liquid scintillation counting (Chapter 6, Volume 1). Many types of plastic scintillators are in use today mainly for the measurement of low penetrating radiation, such as particulate radiation, X-rays, and low-energy gamma rays. Plastic scintillators have also become popular for the measurement of neutrons via neutron-induced nuclear reactions in the plastic media, as described in the previous paragraphs. Two main types of plastic scintillator designs exist: (1) the integral plastic scintillators, which are constructed of one whole piece of plastic polymer containing scintillator varying in size from microspheres of 100−1000 μm in diameter to large detectors measuring a meter or more in diameter, and (2) fiber-optic arrays of plastic scintillator consisting of as many as several thousand plastic scintillating fibers each 25−60 μm in diameter as individual light pipes bundled together into only a few mm diameter.

A. The scintillation process in plastic

Plastic scintillators are classified into binary, ternary, or even higher order scintillator solutions. A binary scintillator consists of two molecular species, namely the solvent *X* (plastic or organic liquid) and the solute *Y* (organic fluor or dopant). The solvent (plastic) medium, which is the bulk material, absorbs the major part of the energy dissipated by incident radiation (see Brookes, 1979). This absorbed energy, that is, excitation energy of the solvent molecules, is then transferred to the solute molecules, which is represented as follows:

$$X^* + Y \rightarrow X + Y^* \tag{9.51}$$

where X^* and Y^* represent excited states of solvent *X* and solute *Y*, respectively. Decay of Y^* from its excited state to the ground state results in fluorescence or the emission of visible light.

The addition of a wavelength shifter or secondary solute, *Z*, to the medium constitutes a ternary system. In such a system, the energy transfer process is extended to a third

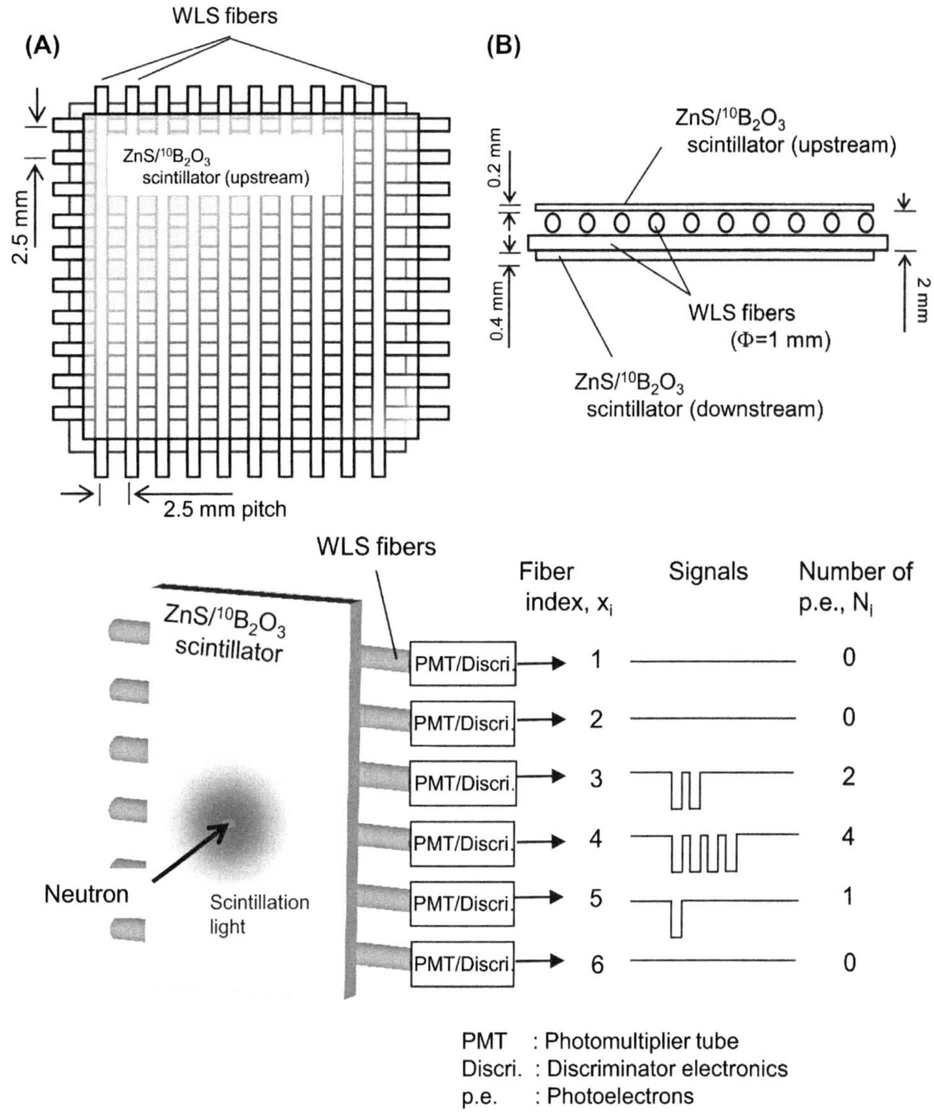

FIGURE 9.83 Upper Panel: Schematic views of the neutron-detecting head: (A) top view and (B) side view. Only 10 fiber channels for each direction are shown. Lower Panel: Schematic diagram of the scintillator and wavelength-shifting (WLS) fiber detector (only the WLS-fibers array in the x direction is presented for simplicity). In this case, the incident position of the neutron is calculated as ~3.9(=(3 × 2 + 4 × 4 + 5 × 1)/(2 + 4+1)) with the MPC method. *From Nakamura et al. (2015); reprinted with permission from Elsevier © 2015.*

molecular species, namely the secondary solute. In other words, the excited primary solute Y^* transfers its energy of excitation both radiatively and nonradiatively to the secondary solute, Z, as follows:

$$Y^* + Z \rightarrow Y + Z^*. \tag{9.52}$$

Decay of Z^* occurs with the emission of visible radiation of longer wavelength than that emitted by Y^*. Secondary solutes are used primarily to provide fluorescent radiation with a maximum emission wavelength that may be better matched to the sensitivity of the PMT or PD. A unitary scintillator is one that consists of a single pure molecular species such as the organic crystalline phosphor (e.g., anthracene). Unitary scintillators, described previously in this chapter, are not discussed here, because plastic scintillators consist of only the binary or higher order scintillators.

Polystyrene-based plastic scintillators have become popular, as polystyrene (PS) is easily drawn into fibers used in scintillator fiber-optic arrays. In a study of polystyrene-based scintillators, Bross (1990) reviews the scintillation process and its efficiency. As explained by Bross (1990), solutions of polystyrene with an efficient fluorescent dopant can give a 100% quantum yield, which is defined as the probability that an excited state in the polymer leads to an emission of a photon by the dopant (fluor). The primary dopant referred to as Y in Eqs. (9.51) and (9.52), which couples directly to the primary scintillation of the solvent, is of higher concentration (~1%). The secondary dopant (Z of Eq. 9.52), which is added to the polystyrene polymer in low concentration (~0.01%), shifts the mean fluorescent wavelength of the primary dopant from the deep blue 350–390 nm further into the red.

Bross (1990) underscores the ideal conditions for fluorescence in polystyrene scintillators. These are (1) a wavelength shifter with emission and absorption spectra that do not overlap, such as exhibited by compounds that undergo intramolecular proton transfer (IPT compounds) upon excitation; (2) a primary dopant, with an absorption spectrum that overlaps the fluorescence emission distribution of polystyrene; and (3) a fluorescence spectrum of the primary dopant with emission peaks longer than 500 nm, which obviates the need for a secondary dopant. The primary dopants 3-hydroxyflavone (3HF), 2-(2-hydroxyphenyl)-benzothiazole (HBT), *p*-terphenyl (PTP), and 1-phenyl-3-mesityl-2-pyrazoline (PMP), among others, have been tested as favorable dopants in polystyrene and poly(vinyltoluene) polymers (Rahmanifard et al., 2018; Güsten and Mirsky, 1991; Bross, 1990; Destruel et al., 1989; Zorn et al., 1988). Reviews on the development, composition, and applications of plastic scintillators and dopants utilized in plastic polymers are provided by Rahmanifard et al. (2018) and (L'Annunziata, 1987).

B. Applications of plastic scintillators

The applications of plastic scintillator in the detection and measurement or radionuclides are numerous. Many applications are treated in this book, including the use of plastic scintillator beads for the measurement of beta, beta-gamma, and alpha-emitting radionuclides (see Chapter 6, Volume 1), high sample-throughput scintillation analysis with plastic scintillator microplates (see Section V.B of this chapter), and the use of plastic scintillators loaded with a neutron converter as capture-gated neutron detectors (Section VI.B of this chapter).

Plastic scintillators are useful in the detection of fast neutrons through collisions of the neutrons with protons in the scintillator. Plastic scintillators are appropriate because of the relatively high hydrocarbon polymer proton (hydrogen) content of plastics. Inorganic scintillator detectors described earlier in this chapter contain no protons (hydrogen atoms) in their molecular structures.

The collisions of fast neutrons with protons in plastic scintillators result in transfer of energy to the proton with recoil of the proton (see Fig. 1.78, Chapter 1) with subsequent conversion of the proton recoil energy to light in the scintillator. The light is then subsequently converted to an electric pulse by a photomultiplier in intimate contact with the plastic scintillator, as illustrated in Fig. 9.84.

VIII. n/γ pulse shape discrimination

Often it is necessary to measure neutron radiation in fields of high gamma-ray intensity. Effective discrimination between neutron and gamma radiation is possible with certain inorganic and organic scintillators that display significant differences in their scintillation pulse shapes analyzed by their decay time. Many scintillators display significantly different pulse shapes produced by two groups of particles, namely, (i) heavy charged particles (i.e., p, d, t, or α), which are products of neutron interactions with scintillators (e.g., neutron scattering or capture reactions) and (ii) electrons, which are products of γ-ray interactions with scintillators (e.g., Compton electrons). For example, under gamma irradiation $LiBaF_3$:Ce yields a very fast (~1 ns) light output due to core-valence (CV) transitions or luminescence followed by an additional light output with a short lifetime of about 35 ns arising from Ce^{3+} excitation. Both of these light outputs arising from gamma interactions are considered fast, and both of these fast components are absent under neutron irradiation. Both neutron and gamma irradiation produce light output with a relatively long component with a lifetime of about 2.5 μs due to self-trapped-excitation (STE) luminescence. The n/γ radiation can be discriminated, therefore, on the basis of the presence or absence of the short (ns) lifetime components in the scintillation pulse events.

Reeder and Bowyer (2001, 2002) used a data acquisition system consisting of a multi-input charge-to-digital converter (QDC). The scintillator is coupled to a PMT, and the signal from the PMT is set to two separate inputs to the QDC. Each input is gated to record the amount of charge in two portions of the pulse. A "short" gate of about 30 ns records the fast components due to the CV and Ce^{3+} fluorescence arising from gamma interactions, and the "total" gate of about 1.6 μs records both the short and long components due to STE excitations as a consequence of both gamma and neutron interactions. A two-dimensional

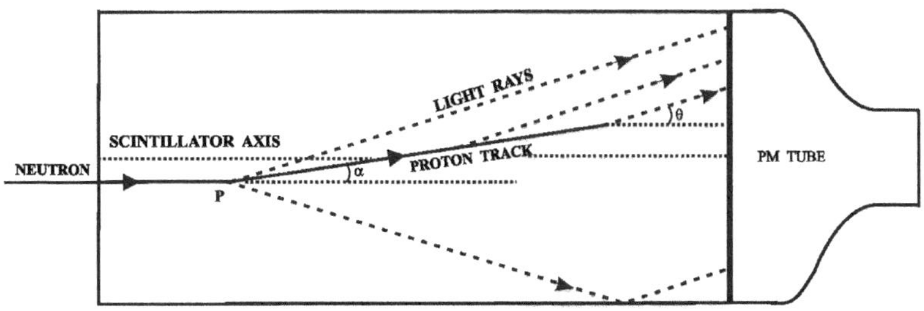

FIGURE 9.84 Schematic of a neutron path in a plastic scintillation neutron detector illustrating a recoil proton track and the emitted scintillation light within the scintillator. The symbols θ and α represent the angles of light emission and recoil proton track, respectively. *From De et al. (1963); reprinted with permission © IEEE.*

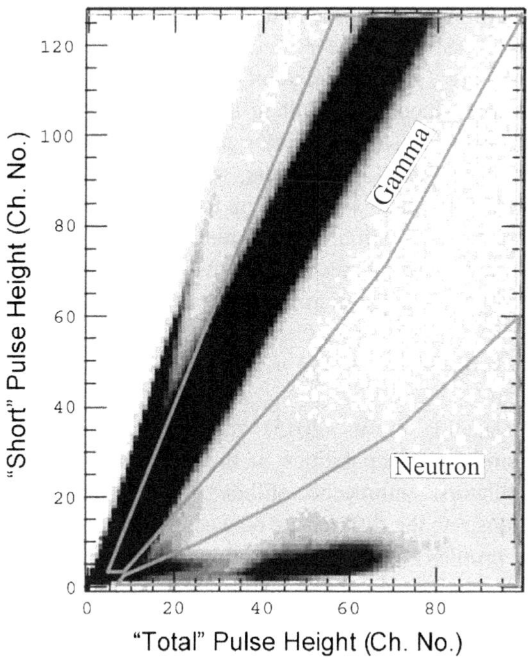

FIGURE 9.85 Two-dimensional array of pulse heights from "short" to pulse heights from "total" gate for ^{252}Cf at 11.55 cm from crystal. Data were obtained with a Cd shield around the scintillator and a 1.27-cm thick Pb absorber between the source and scintillator. Outlined regions define events attributed to neutrons or to gammas. *From Reeder and Bower (2002), reprinted with permission from Elsevier © 2002.*

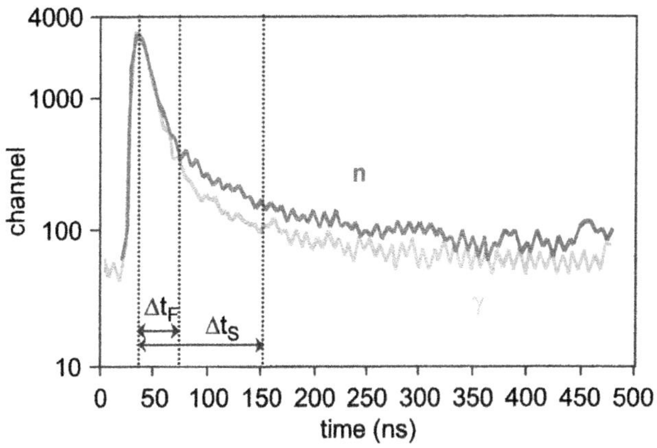

FIGURE 9.86 Neutron and γ-ray pulses recorded with 12-bit 200 MSamples/s transient recorder card; time intervals for charge comparison method analysis are shown. *From Kaschuck and Esposito (2005); reprinted with permission from Elsevier © 2005.*

array of the "short" signal versus the "total" signal demonstrates the discrimination of n/γ events, as is illustrated in Fig. 9.85. This technique of n/γ discrimination is referred to as the charge-comparison technique, because a fractional charge collected (e.g., a fast signal) is compared to the total charge collected being a slow or long signal.

Kaschuck and Esposito (2005) describe the performance of a n/γ digital pulse shape discrimination (DPSD) system by means of a commercial 12-bit 200 MSample/s transient recorder card with three organic scintillators, namely, stilbene, NE 213, and anthracene, as analog electronic pulse-shape discrimination (PSD) modules, which do not allow data processing and are limited in count rate capability (up to ~ 200 kHz). They used the charge-comparison method to obtain the simultaneous n/γ discrimination and pulse height analysis with the organic scintillators demonstrated to perform with count rates in excess of 1 MHz (Esposito et al., 2010; Kaschuck and Esposito, 2005). The charge comparison method to discriminate between the n and γ events, as described by Kaschuck and Esposito (2005) is the following:

If two equal amplitude n- and γ-pulses are considered, the neutron pulse has a larger slow component (see Fig. 9.86). Each pulse is integrated numerically for two different time intervals (Δt_F and Δt_S) corresponding to the fast and slow components. The value of the ratio $R = Q_S/Q_F$ (Q_F and Q_S are the charges in the two intervals Δt_F and Δt_S) provides the indication whether a n- or a γ-event has taken place. Two integrating systems were selected $\Delta t_1 = 25$ ns and $\Delta t_2 = 120$ ns; and the integration was performed starting at the time at the peak value of each pulse. After the end of the data processing, both the pulse height spectra for neutron and γ-rays as well as the total spectrum are provided simultaneously.

By recording events from neutron and γ-ray radionuclide sources Kaschuck and Esposito (2005) plotted the two-dimensional intensity distributions Q_F versus Q_S for stilbene scintillator illustrated in Fig. 9.87 using the DPSD data processing software. An Am−Be source was used with the source shielded in ~15 cm-thick polyethylene shielding and the n/γ separations at different electron energies are illustrated in Fig. 9.87B.

Jhingan et al. (2008) describe how the charge comparison technique for n/γ discrimination, which has been implemented over the past 2 decades, can be implemented electronically in more than one fashion. The conventional method is to compare the fractional or slow charge in the tail (see Fig. 9.86 above) with the full or total charge of the PMT anode pulse fed to a charge integrating analog digital converter (ADC) or charge digital converter (QDC). Jhingan et al. (2008) devised an alternative electronic method of comparing the shaped-dynode pulse, fed to a peak-sensing ADC, with the fractional slow charge integrated in the QDC, which reduces the number of gate and delay generators employed in the conventional method.

Watanabe et al. (2002) report a unique application of plastic scintillator sandwiched together with pure lucite plastic to separate high-energy neutron and gamma radiation. The detector consists of 10 layers of plastic scintillator plates alternately sandwiched between 10 Lucite plastic plates. The Lucite plastic plates are pure, that is, they do not

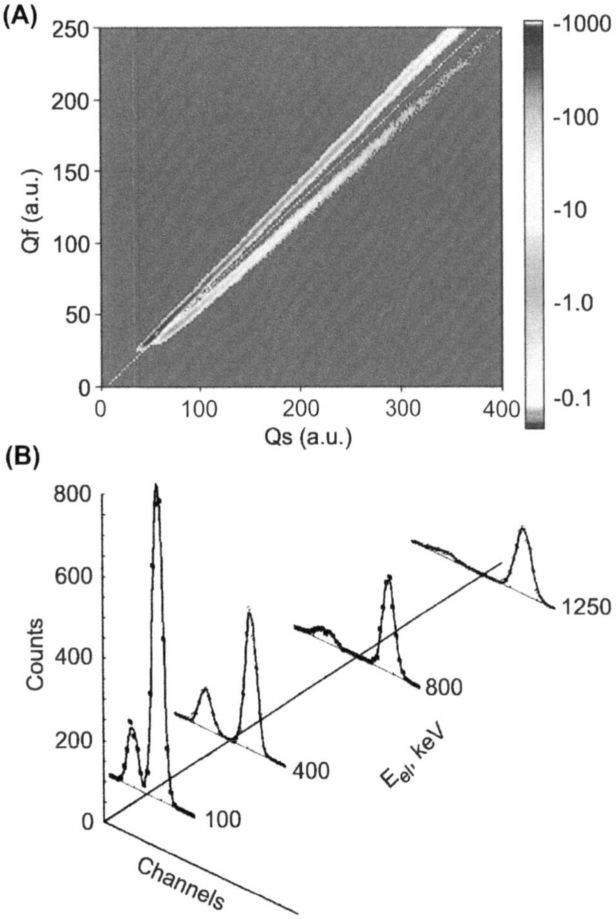

FIGURE 9.87 Stilbene (HV = −1400 V) and Am−Be source: (A) Two-dimensional intensity distribution in Q_F−Q_S space. The color scheme of intensities can be observed with the electronic version of the figure. (B) n/γ separation profiles at various electron energies (neutron peaks are on the left) obtained with 12-bit 200 MSamples/s sampling. *From Kaschuck and Esposito (2005); reprinted with permission from Elsevier © 2005.*

contain scintillator and consequently act as Cherenkov radiators. The plates measure 100 × 103 × 3.7 mm (width × length × thickness). When sandwiched together and coupled to two photomultipliers at two ends they form a n/γ discriminator, referred to as a scintillator-Lucite sandwich detector (SLSD). The principle behind the SLSD is that high-energy gamma radiation will form e^+ and e^- in the detector, which will result in the emission of scintillation light in the plastic scintillator sheets and Cherenkov photons in the pure Lucite sheets; while the neutron radiation will not produce any light in the pure Lucite, but will produce scintillation in the plastic scintillator through proton and carbon recoil from n−p and n−C collisions. Effective n/γ separation is demonstrated in the energy region of 20 MeV to 12 GeV with a neutron detection efficiency of 7%−10% and a γ-misclassification probability of $<10^{-2}$.

Yanagida et al. (2015) studied the neutron-gamma discrimination capabilities of the pure solid-state organic scintillators, anthracene, stilbene, and p-terphenyl. They demonstrate the differences between the scintillation decay time profiles of each scintillator under neutron and gamma-ray excitation, illustrated in Fig. 9.88; and the discrimination achieved between neutron and gamma-ray events when plotting the events against the slow to fast components of the scintillation decay time profiles of each scintillator, as illustrated in Fig. 9.89. Anthracene produced the best n/γ discrimination as measured by the figure of merit (FOM) defined as

$$\mathrm{FOM} = \frac{S}{\mathrm{FWHM_{neutron}} + \mathrm{FWHM_{gamma}}} \tag{9.53}$$

where S is the distance between the neutron and gamma-ray peaks and $\mathrm{FWHM_{neutron}}$ and $\mathrm{FWHM_{gamma}}$ are the full width at half maximum of the neutron and gamma-ray peaks, respectively. The FOMs calculated by Yanagida et al. (2015) were 1.53, 1.10, and 1.26 for anthracene, stilbene, and p-terphenyl, respectively.

Pulse shape n/γ discrimination is not limited to solid organic scintillators; but it is also carried out efficiently with liquid scintillators as described in Chapter 6, Volume 1. Jhingan et al. (2008) used two detectors, a plastic scintillator to provide a start signal and a second liquid scintillator detector at a 30-cm distance to establish time-of-

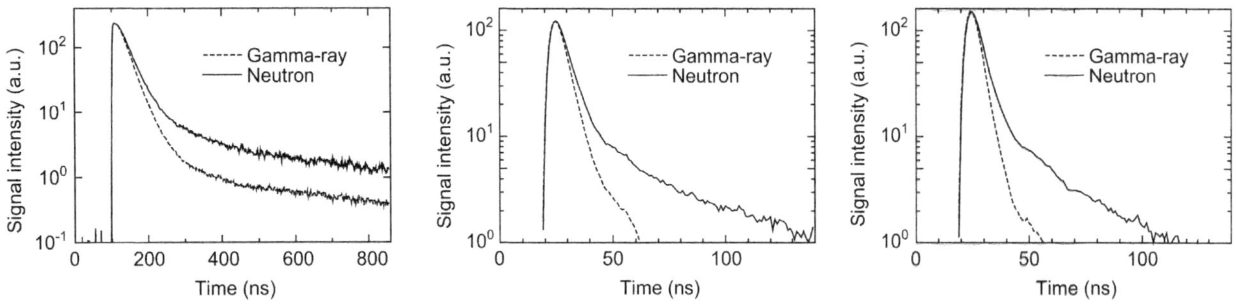

FIGURE 9.88 Scintillation decay time profiles under neutron and gamma-ray excitation of anthracene (left), stilbene (center), and p-terphenyl (right) by event selection. *From Yanagida et al. (2015); reprinted with permission from Elsevier © 2015.*

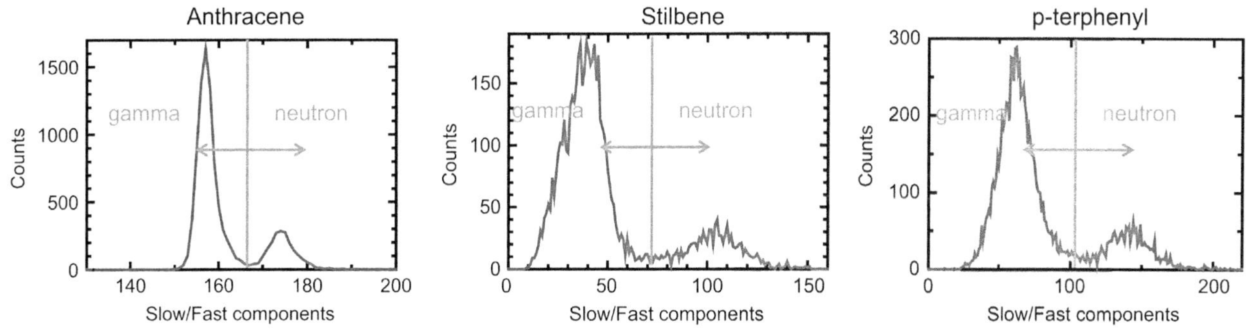

FIGURE 9.89 Histograms of neutron and gamma-ray events of anthracene (left), stilbene (center), and p-terphenyl (right), plotted on the ratio of slow/fast components. The fast/slow integrated time windows of anthracene, stilbene, and p-terphenyl were 20–40/40–150 ns, 20–40/40–150 ns, and 20–30/30–150 ns, respectively. *From Yanagida et al. (2015); reprinted with permission from Elsevier © 2015.*

flight (TOF) discrimination in addition to PSD. TOF measurements for n/γ discrimination are described in Chapter 6, Volume 1 and Chapter 6, Volume 2. Effective n/γ PSD has been demonstrated with several new inorganic and organic scintillators including ^{6}Li–boron–phosphorus glass scintillator and ^{6}Li-loaded plastic scintillators (Katagiri et al., 2004), $Cs_2LiYCl_6:Ce^{3+}$ (Bessiere et al., 2005), 9,10-diphenylanthracene (van Loef et al., 2011), and Ce:LiCaAlF$_6$ (Yamazaki et al., 2011) and well as ^{6}Li-containing organic scintillators (Bertrand et al., 2015; Zaitseva et al., 2013). Liao and Yang (2015) describe a frequency gradient analysis (FGA), as a new method of neutron-gamma discrimination developed by Liu et al. (2010). The FGA method, as described by Liao and Yang (2015), exploits the difference between the Fourier transforms of the neutron and gamma pulse.

Reviews of recent developments and state of the art of neutron/gamma PSD in plastic scintillator is provided by Zaitseva et al. (2018), Bertrand et al. (2015) and Liao and Yang (2015).

IX. Bonner sphere neutron spectrometry

The methods by which one determines the energy spectra of neutron radiation have an interesting similarity to that by which we determine gamma-ray energy spectra. Both neutrons and gamma-ray photons are neutral particles. Therefore, their energies can only be determined indirectly by the energies that they transfer to charged particles. Neutron energies are measured by the energies they transfer to recoil protons via n–p scattering; and gamma-ray energies are measured by the transfer of gamma-ray energy to atomic electrons via Compton scatter or the photoelectric effect. Unfortunately, neutron spectra are not characterized with a specific photopeak or Compton edge; rather, there is a broad proton-recoil pulse height distribution due to the large range of neutron scatter angles and proton recoil energies that result.

Various techniques may be used to measure the energy of neutrons; and these are reviewed by Thomas (2010) including (i) time-of-flight, (ii) nuclear reaction methods [i.e., $^{3}He(n,p)^{3}H$] where the neutron energy is shared between the proton and Triton products, and the neutron energy measured by the sum of the proton and Triton energies, (iii) hydrogen-filled proton-recoil spectrometers, and (iv) Bonner sphere spectrometers (BSSs). Most of these methods are beyond the scope of this chapter; however, BSSs may utilize a solid crystal scintillator; and these will be discussed here. For other methods, the reader is referred to reviews on neutron spectrometry by Thomas (2010, 2004), Brooks and Klein (2002) and Klein et al. (2002).

The Bonner sphere neutron spectrometer was first designed by Richard L. Bramblet, Ronald I. Ewing, and Tom W. Bonner (Bramblett et al., 1960) at Rice University with a ^{6}LiI(Eu) scintillator mounted on a polystyrene light pipe as coupling to a PMT. The detector is named after Tom W. Bonner (1910–61), who is known for his work on the development of instrumentation and methods for studies of neutron and nuclear physics. The Bonner sphere has developed into a popular instrument to measure neutron fields up to energies of ~10 MeV in various environments including the nuclear industry (Roberts et al., 2017), boron neutron capture therapy (BNCT) and accelerators associated with hadron therapy in nuclear medicine (Masuda et al., 2017; Tamaki et al., 2017, 2015; Bedogni et al., 2012), nuclear reactor beam port measurements (Mazrou et al., 2016; Vega-Carrillo et al., 2014), around a linear accelerators (Lagares et al., 2016; Mohammadi et al., 2015; Benites-Rengifo et al., 2014; Barquero et al., 2005), around HEP accelerators (Mares et al., 2017; Agosteo et al., 2012), cosmic-ray shower neutrons (Lee and Sheu, 2017; Vega-Carrillo and Manzanares-Acuña, 2004), neutron spectra associated with cyclotrons for the production of PET radionuclides (Benavente et al., 2015; Vega-Carrillo, 2001), and neutron field measurements associated with a nuclear fusion reactor (Hu et al., 2018) among other applications.

The first BSS consisted of a small 4 mm diameter × 4 mm thick ^{6}LiI(Eu) scintillation crystal containing 96.1% ^{6}Li and many modern BSS still utilize the small ^{6}LiI(Eu) detector (Bedogni et al., 2010, 2018; Lacerda et al., 2017; Vega-Carrillo et al., 2014; Mirzajani et al., 2013; Ortiz-Rodríguez et al., 2013; Agosteo et al., 2012; Marek and Viererbl, 2011, 2008; Kim et al., 2010; Howell et al., 2009), Mazrou et al. (2008), and Burgett et al. (2008); although other thermal neutron detectors, such as ^{3}He based on the thermal neutron capture reaction ^{3}He(n,p)^{3}H are used (Masuda et al., 2017; Roberts et al., 2017; Lagares et al., 2016; Mazrou et al., 2016; Tamaki et al., 2015; Bedogni et al., 2014a, 2012; Rühm et al., 2014; Birattari et al., 2010; Méndez-Villafañe et al., 2010; Pioch et al., 2010) and passive detectors, such as ^{197}Au or ^{55}Mn activation foil material (Ueda et al., 2015; Domingo et al., 2010; Burgett et al., 2008; Fernandez et al., 2007), as well as spectrometers utilizing dysprosium activation foils based on the reaction ^{164}Dy(n,γ)^{165}Dy (Bedogni et al., 2013, 2014b; 2015; Gómez-Ros et al., 2012).

The spectrometer is calibrated with various polyethylene spherical moderators of differing diameters to provide an isotropic response to neutrons. The moderator spheres are manufactured to dimensions measured to the exact inch or half-inch. The units used to define Bonner sphere moderator dimensions since the onset was in inches and the dimension in inches in this particular case has remained popular, although centimeters are used by some just as well. A schematic of the apparatus is illustrated in Fig. 9.90. Approximately 10−12 polyethylene moderator spheres of different diameter are used to calibrate the spectrometer. The sphere dimensions are precisely defined and may range from 2 to 18 inches in diameter. Each sphere, of course, has a hole to accommodate the detector, which is situated at the center of the sphere.

Each polyethylene moderator sphere of various diameters are constructed with a cylindrical hole drilled at the bottom leading to the center of the sphere to accommodate either a ^{6}LiI(Eu) or ^{3}He thermal neutron detector. Fast neutrons are thermalized according to the size or diameter of the polyethylene sphere, as described by Reginatto (2009), and illustrated in Fig. 9.91. Neutrons that are thermalized within the polyethylene sphere and captured by the ^{6}Li according to the reaction ^{6}Li(n,α)^{3}H produce a scintillation event within the LiI(Eu) crystal scintillation detector and a pulse event in the PMT.

Each polyethylene sphere with its thermal neutron detector combination has a different energy response to neutrons; and as the size of the moderator is increased, the peak of the response function shifts to a higher energy. Measurements of thermal neutrons without a sphere (bare detector) are also made. In addition to measuring thermal neutrons without a polyethylene sphere, i.e., the "bare detector", measurements may be made with the thermal neutron detector covered with cadmium, which is referred to as the "Cd-covered bare detector" (Reginatto, 2009). The

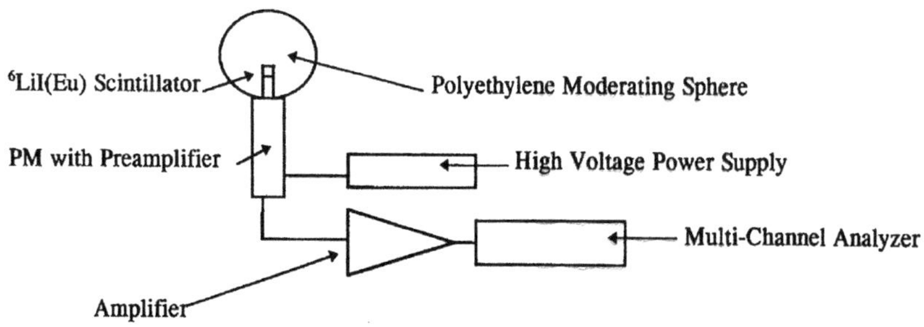

FIGURE 9.90 Schematic of the Bonner sphere neutron spectrometer. *From Haney et al. (1999); reprinted with permission from Elsevier © 1999.*

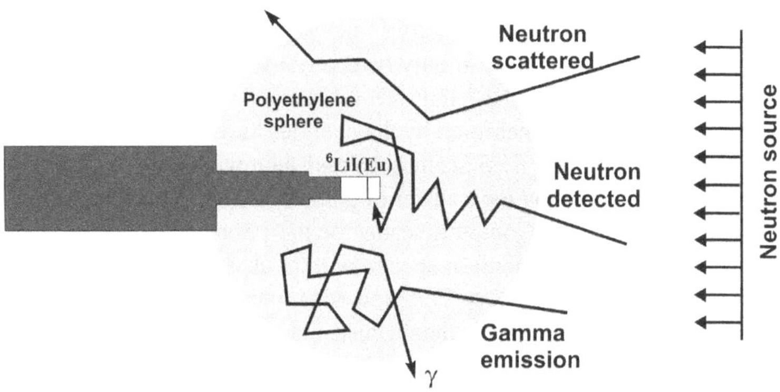

FIGURE 9.91 Bonner sphere system for a ^{6}LiI(Eu) neutron radiation detector. *From Ortiz-Rodriguez et al. (2013); reprinted with permission © 2013 Manuel Ortiz-Rodríguez et al., licensee InTech.*

multiple n−p scatterings that occur in the polyethylene sphere result in neutron thermalization before they hit the detector as illustrated in Fig. 9.91. Reginatto (2009) underscores that a BSS with a wide range of polyethylene spheres of various diameters is suitable for measurements of neutron energies in the range from thermal to ~ 20 MeV; and for measurements of higher neutron energies, it is necessary to use an extended-range Bonner sphere spectrometer (ERBSS) which includes, in addition to the standard sphere of a BSS, a few modified spheres with shells of high-Z metals embedded in them. Modified polyethylene spheres with concentric shells of Cu, Pb, and W provide response to neutrons energies >100 MeV (Lee and Sheu, 2017, 2015; Mares et al., 2017; Wang et al., 2015a,b; Esposito et al., 2010; Howell et al., 2010, 2009 Reginatto, 2009; Wiegel et al., 2009; Burgett et al., 2008).

The size and shape of the ^{6}LiI(Eu) crystal and its high ^{6}Li enrichment is chosen to provide good gamma-ray discrimination and high efficiency for thermal neutrons, respectively. Approximately 80% of the thermal neutrons are absorbed in the surface 1.0 mm of the ^{6}LiI(Eu) crystal, while gamma- and fast neutron interactions would be proportional to the volume of the detector. The small 4 × 4 mm detector provides a large surface to volume ratio to reduce detector response to fast neutrons and gamma rays. Also, the small dimensions of the ^{6}LiI(Eu) detector permit the use of small polyethylene spheres (e.g., 50−60 mm in diameter) for the detection of low-energy neutrons. Either the ^{6}LiI(Eu) crystal or the SP9 spherical ^{3}He proportional counter of 32 mm diameter operated at 2 atm pressure (Centronic Ltd., UK) can be used as thermal neutron sensors. The ^{3}He proportional counter is sensitive to thermal neutrons reacting according to the ^{3}He(n,p)^{3}H reaction; and the ^{6}Li(Eu) scintillator is also sensitive to thermal neutrons according to the reaction ^{6}Li(n,α)^{3}H. The detector has superior neutron fluence response and gamma-ray discrimination. The use of pressurized ^{3}He detectors, which are highly sensitive to thermal neutrons, has been used commonly in the past; however, Bedogni et al. (2018) reports that due to the supply restrictions and high cost or ^{3}He, the scientific community is leaning toward cheaper detectors with acceptable sensitivity. With this in mind Bedogni et al. (2018) tested a Large ^{6}LiI(Eu) Bonner Sphere Spectrometer (LL-BSS) measuring 11 × 3 mm, which is significantly larger than the standard 4 × 4 mm ^{6}LiI(Eu) detector. It's sensitivity is lower than the 32 mm diameter pressurized (200 kPa) SP9 spherical ^{3}He detector; however, the large ^{6}LiI(Eu) detector responds 50% better than the smaller 05NH1 pressurized (8 kPa) cylindrical (10 × 9 mm) ^{3}He detector.

The spherical spectrometer is calibrated as a function of sphere diameter and neutron energy over a broad range of neutron energies (e.g., 10^{-7} − >1000 MeV). The response of small moderator spheres will show an increase, peak out, and then decreases rapidly as the neutron energy is increased. This is what would be expected, because when the moderator is small low energy neutrons will have a good probability of reaching the detector; but as the neutron energy increases, small moderators become increasingly less efficient in thermalizing neutrons. The response of larger spheres is low for the low-energy neutrons, then increases according to neutron energy, reaches a maximum, and finally decreases at very high neutron energies. The low response of large-sphere moderators to low−energy neutrons is due to the capture of the thermalized neutrons by hydrogen of the polyethylene moderator. The path-length of travel of neutrons after thermalization is longer in the larger spheres, which increases their probability of capture before reaching the ^{6}LiI(Eu) or ^{3}He detector. The high-energy neutrons, therefore, have increased chances of becoming thermalized in the larger polyethylene spheres and yet reach the detector. At very high neutron energies (>10 MeV) probabilities for thermalization in the larger polyethylene spheres diminishes and the response drops.

A key to the BSS is its calibration in terms of its response as a function of neutron energy. Bedogni et al. (2018) defines the response function of a given sphere as the counts in the central detector, per unit incident neutron fluence, when the sphere is uniformly irradiated. The response functions for different sphere sizes over a wide range of neutron energies are calculated; and Monte Carlo methods are used (Bedogni et al., 2018; Ghal-Eh et al., 2017; Lacerda et al., 2017; Rühm et al., 2014; Mirzajani et al., 2013; Mazrou et al., 2010; Howell et al., 2009; Vega-Carillo et al., 2008, 2009). The plot of Fig. 9.92 illustrates that the peak of the response function varies according to neutron energy and sphere size, and the response peak moves to higher energies as the sphere size increases. In a review by Thomas and Alevra (2002) they describe the sphere readings according to the following:

$$M_i = \int R_i(E)\phi(E)dE \tag{9.54}$$

where M_i is the reading for sphere i, $R_i(E)$ is the response function of sphere i exposed to a neutron field with spectral fluence $\phi(E)$. Good numerical methods are available that provide approximations to $R_i(E)$ supported with measurements from monoenergetic neutron sources. From measurements taken by the Bonner sphere set in an unknown neutron field, the extraction of $\phi(E)$ from the above calibration data permits the unfolding of the energy spectrum. As elucidated by Thomas and Alevra (2002) and Thomas (2010), neutron spectra is represented as an array where ϕ_j is the neutron fluence in group j extending over the energy range of E_j *to* E_{j+1} and a measured reading is given by

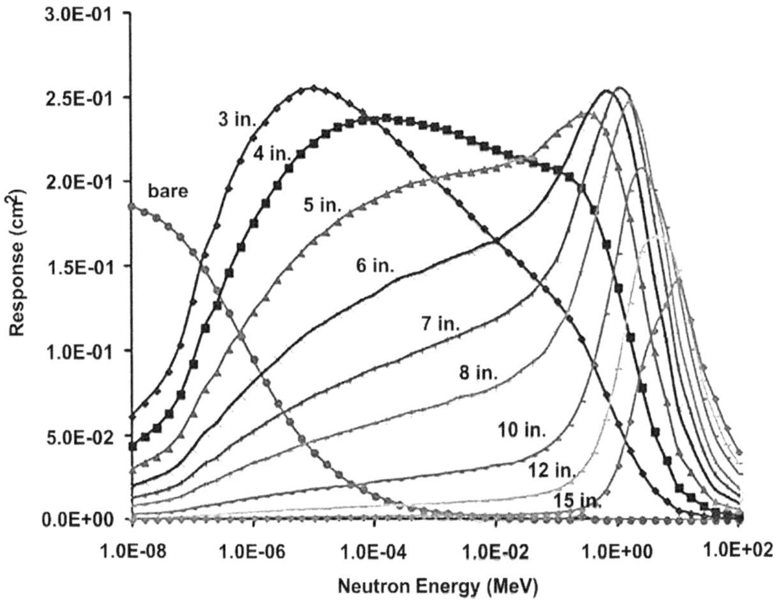

FIGURE 9.92 The calculated response functions of a ^{6}LiI(Eu) BSS with the bare detector and with polyethylene spheres ranging in diameter from 3 inches to 15 inches. *From Mirzajani et al. (2013); reprinted with permission from Elsevier © 2013.*

$$M_i = \sum_{j=1}^{n} R_{ij}\phi_j \tag{9.55}$$

where R_{ij} represents the response function $R_i(E)$ averaged over group j. Eq. (9.54) provides a set of m linear equations, one for each sphere; and if $m \geq n$, the equations can be solved by least squares (Zaidins et al., 1978) to provide values of the spectral fluence ϕ_j. However, since the number of spheres is usually small ($m \approx 10$), the solution provides a poor representation of the neutron spectrum (Thomas and Alevra, 2002), and additional a priori information combined with spectral unfolding techniques are used. Many spectral unfolding codes have been developed; and the methods are reviewed by Roberts et al., (2017), Mazrou et al. (2016), Bedogni et al. (2014c), Thomas (2010) and Matzke (2002).

With the use of additional measurements with a "bare detector" (i.e., no polyethylene moderator sphere) and moderator spheres containing shells of high-Z metals providing ERBSS, a response function over a broader range of neutron energies can be obtained.

Chiles et al. (1990) developed a combined neutron and gamma-ray spectrometer using a cerium-activated ^{6}Li glass scintillator coupled to a Bicron BC501 liquid scintillator, which can discriminate between thermal neutrons, high-energy neutrons, and gamma radiation. The ^{6}Li glass scintillator is sensitive to the thermal neutrons and emits light with a time constant of ~60 ns, while light emitted in the liquid scintillator by proton recoil from fast neutrons is emitted with a decay constant of ~30 ns, and the time constant for scintillations occurring from gamma-scattered Compton electrons is only ~3.7 ns. The three different light decay constants make possible the electronic discrimination of pulses arising from the three different radiations. This type of detector was fabricated to facilitate neutron monitoring and surveillance in nuclear facilities and weapons storage, where a possibility of fission excursion exists.

Astronauts living and working in orbital space are exposed to various type of radiation including neutrons, protons, gamma radiation, heavier charged particles, etc. A unique neutron spectrometer capable of measuring the neutron spectrum in mixed neutron-gamma and neutron-proton radiation fields was developed by Taniguchi et al. (2001). The spectrometer utilizes a pair of ^{6}Li and ^{7}Li glass scintillators in a spherical polyethylene multi-moderator spectrometer similar to the Bonner sphere. A schematic view of the spectrometer is provided in Fig. 9.93. The spectrometer is designed to discriminate neutron from other particles, such as protons, which are dominate components of space radiation, by subtracting the light outputs of ^{7}Li glass scintillator from light produced by ^{6}Li glass scintillator. NE912 and NE913 glass-scintillator detectors are employed, which measure 2.54 cm long × 2.54 cm diameter. The NE912 glass scintillator is doped with 7.7 wt% of 95% ^{6}Li-enriched lithium, whereas the NE913 glass scintillator contains 8.3 wt% of 99.99 ^{7}Li-enriched lithium. The ^{6}Li has a high sensitivity to thermal neutrons through the ^{6}Li(n,α)^{3}H reaction discussed earlier in this chapter, and the ^{7}Li has a low sensitivity to neutrons. The scintillators are coupled to photomultipliers and the output pulse events are amplified and analyzed by a multichannel pulse height analyzer. Taniguchi et al. (2001) calculated the response functions to neutrons for the various polyethylene

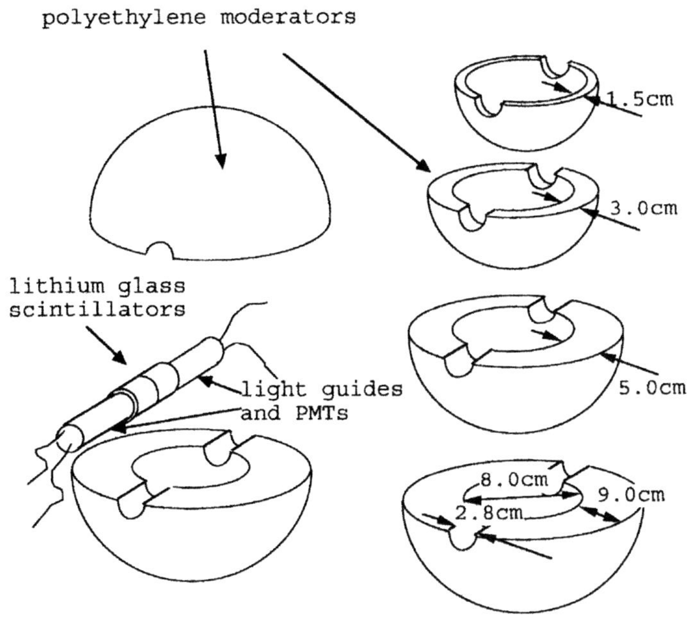

FIGURE 9.93 Schematic view of the multi-moderator spectrometer with a pair of ^{6}Li and ^{7}Li glass scintillators. *From Taniguchi et al. (2001); reprinted with permission from Elsevier © 2001.*

moderator thicknesses and neutron energies (0.25–22 MeV). The detector demonstrated clear separation of neutron from gamma-ray events, when the neutron to gamma-ray flux was >0.3, and neutron from proton events could be clearly discriminated.

X. Lucas cell

The Lucas cell is a popular instrument and among the most sensitive for the measurement of radon from various parts of the environment including soil gas (Sethy et al., 2014a; Cosma et al., 2013; Künze et al., 2012, 2013; Al-Hilal and Al-Ali, 2010; Mujahid and Hussain, 2010; Prasad et al., 2008), water with adequate water bubbling techniques (Shivakumara et al., 2014; Khan et al., 2009; Cosma et al., 2008; Singh et al., 2008; Machaj and Bartak, 2004), and in the general indoor environment of buildings and homes (Cosma et al., 2013; Frutos Vázquez et al., 2011; Hassan et al., 2011; Maharana et al., 2010; Al-Masri and Shwiekani, 2008; Prasad et al., 2008; Quindos Poncela et al., 2005; Abbady et al., 2004).

The Lucas cell is a solid scintillation chamber designed for the specific purpose of measuring the alpha-particle emissions of environmental levels of ^{222}Rn. It was first designed by Lucas (1957) at the Argonne National Laboratory, Lemont. Illinois; and the basic components of the cell have not changed since. A typical Lucas cell is illustrated in Fig. 9.94. The cell consists of a vessel of ~100 mL volume or larger and made of Kovar metal with a narrow brass collar at the top as the only orifice and a clear silica window at the flat bottom (face). The inner wall of the cell is coated with a thin layer (~20–50 mg/cm^2) of

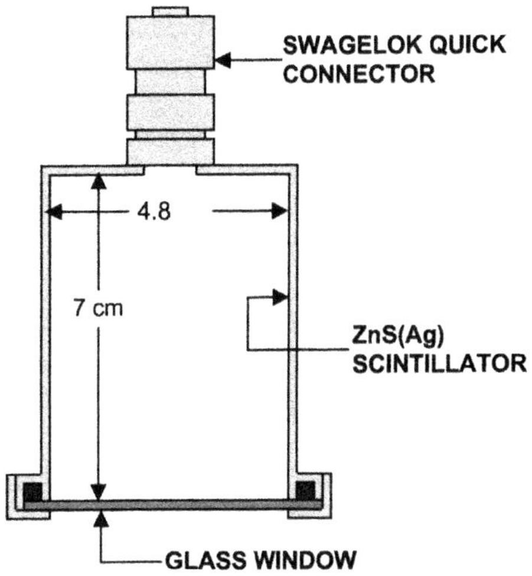

FIGURE 9.94 Schematic diagram of a Lucas cell. *From Eappen et al. (2008); reprinted with permission from Elsevier © 2008.*

ZnS(Ag) scintillator. The inner surface of the flat window at the face of the Lucas cell is not coated with scintillator; it is covered with a transparent coat of tin oxide to maintain electrical conductivity and prevent any accumulation of radon or its daughters on the window. Environmental ^{222}Rn is first collected by a variety of methods, which may include grab sampling of soil gas into a large 120 mL gas tight syringe, as described by Cosma et al. (2013), Künze et al. (2012, 2013) with injection into an evacuated Lucas cell (Sethy et al., 2014b) or air sampling pump (El-Abnoudy and Hassan, 2016) including the continuous sampling of air into collection chambers (Chu et al., 2016) for the measurement of time dependence on radon concentration (Tommasino, 2005) or bubbling the radon in water samples into evacuated Lucas cells (Shivakumara et al., 2014). To measure radon activity, the clear window of the Lucas cell is placed in contact with the face of a conventional PMT. The alpha particles of radon and its daughters in the Lucas cell interact with the scintillator; and the scintillation photons from these interactions are detected by the PMT and counted.

Radon (^{222}Rn) is the dominant source of human exposure to ionizing radiation worldwide (Tommisino, 2005). It is a colorless and odorless noble gas. Radon-222 is produced as a product of radium (^{226}Ra) in the uranium-238 decay series described in Chapter 1, Volume 1. Radon occurs in soil gas and groundwater from which it can seep into dwellings through building foundations and household bath and kitchen water. Added to the hazard are the four immediate radioactive decay products of radon-222, which achieve secular equilibrium with radon-222, illustrated in Fig. 9.95. Tommisino (2005) underscores that the alpha-particle emissions from the decay products exhibit a high linear energy transfer to human tissue. Continued exposure to excessive levels can impart damage to the human lungs increasing cancer risk, which is the underlying motivation for the analysis, monitoring, and remediation of radon in our environment.

Semkow et al. (1994) have determined the detection efficiency of the Lucas cell for ^{222}Rn and the three alpha particle emissions from ^{222}Rn and its daughters ^{218}Po and ^{214}Po. Cosma et al. (2013) reports a detection efficiency of 2.2 counts sec^{-1} to 1 Bq of radon activity. This corresponds to 2.2. counts for three alpha-particle emissions or a detection efficiency of 2.2/3.0 = 0.73 when the ^{222}Rn and its daughter nuclides are in secular equilibrium within the Lucas cell. Shivakumara et al. (2014) also reported a detection efficiency of 0.74. Tommasino (2005) reports the efficiency of the Lucas cell to vary between 70% and 80%. It is necessary to allow at least 3 hours for the ^{222}Rn and its short-lived decay products ^{218}Po, ^{214}Pb, ^{214}Bi, and ^{214}Po to reach secular equilibrium with the longer-lived ^{222}Rn before counting with the Lucas cell (El-Abnoudy and Hassan, 2016; Shivakumara et al., 2014; Künze et al., 2012). The ingrowth of the short-lived daughter nuclides with the ^{222}Rn immediately after radon sampling is illustrated in Fig. 9.96, based on the ingrowth equations provided by Robley D. Evans (1907–95) in his classic paper on radon daughters (Evans, 1969).

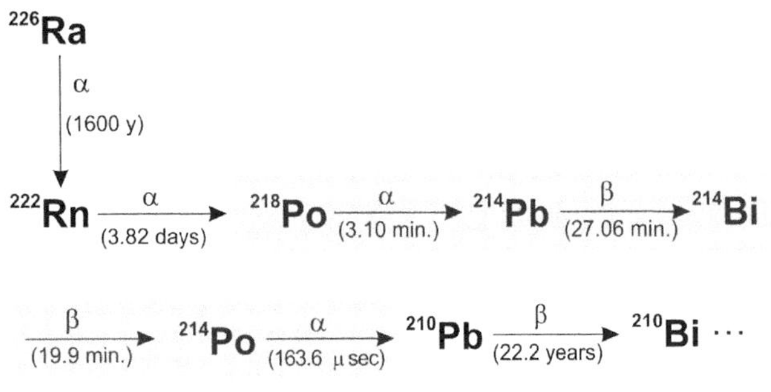

FIGURE 9.95 Radium-226 subseries illustrating the radon-222 short-lived decay products (shaded area) that achieve secular equilibrium with the radon-222.

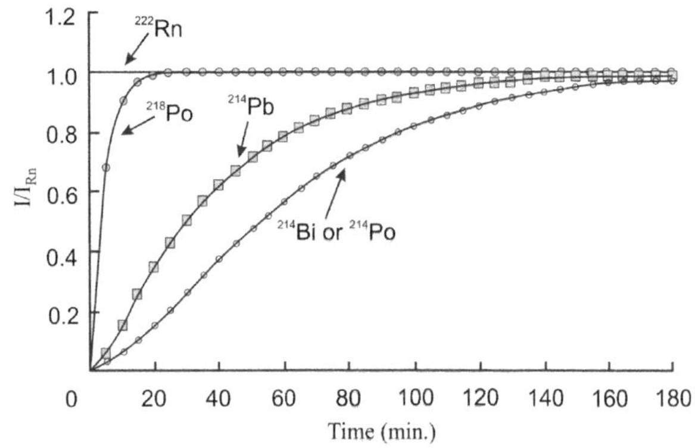

FIGURE 9.96 Ingrowth of daughter nuclides ^{218}Po, ^{214}Pb, ^{214}Bi, and ^{214}Po with newly isolated ^{222}Rn plotted as I/I_{Rn}, which is the ratio of the radionuclide intensity to that of the parent ^{222}Rn, against time in minutes. The activities of ^{214}Bi and ^{214}Po are equal, as noted by Robley D. Evans, due to the very short 163.5 μs half-life of ^{214}Po. The radionuclide symbols are added by the writer. *From El-Abnoudy and Hassan (2016); reprinted with permission from Elsevier © 2016.*

The Lucas cell can also be used for the measurement of ^{226}Ra in water via the measurement of its emanating ^{222}Rn daughter after secular equilibrium is reached. The detailed procedure is provided by Peters et al. (1993) and Shivakumara et al. (2014). In brief, the principle of the method described by Peters et al. (1993) entails the concentration and separation of the radium-226 in water by coprecipitation with barium sulfate. The concentrated sample is redissolved and sealed in a purged flask (bubbler). After 21 days to allow for ingrowth of the radon-222 daughter, the gaseous ^{222}Rn is transferred to a Lucas cell using helium as a carrier gas and then counted as described. The detection limit is reported to be about 0.05 pCi per sample when counted for 1000 minutes, depending on the individual background of the Lucas cell used.

To improve the detection efficiency for a large volume Lucas cell and reduce background counts, Sakamoto and Takahara (2001) designed a scintillation cell with an inside diameter of 7.5 cm and a length of 30 cm constructed of a vinyl chloride tube with acrylic resin windows at each end. Condensing lenses were also place at each end of the tube to aid in the detection of scintillation light by PMTs placed at each end. The inside wall of the tube was coated with ZnS(Ag) scintillator adhered onto aluminum sheet. The two PMTs provided for coincidence counting to reduce background. Pulse height analysis provides a spectrum of the alpha peaks, whereby proper selection of a counting window with discriminator settings would further facilitate background reduction. The average counting efficiency for radon and its progenies was reported to be 0.55 with a background count rate of 0.015 cps. The lower limit of detection (LLD) for a 2-h counting time was less than 10 Bq/m^3. Quindos-Poncela et al. (2003) devised a modified Lucas cell whereby the inside walls of the cell were lined with ZnS(Ag) coated Mylar; and devised a removable transparent bottom to the cell. These modifications enabled the opening and removal of used ZnS(Ag) coated Mylar and its replacement with new ZnS(Ag) coated Mylar whenever contamination or background buildup would occur. This reduces the cost of recoating the inside surfaces of the cell with new scintillator; and the cell is rendered reusable at any time.

Eappen et al. (2007) used the Lucas cell for the on-line measurement of thoron (^{220}Rn) by modifying the method used to measure radon (^{222}Rn). In radon analysis, the α-counting with the Lucas cell is performed after a delay period varying from a few hours to a few days; however, in the thoron measurements the α-counting was initiated immediately after sampling due to the short half-life of thoron (55.6 seconds). Eappen et al. (2008) further developed the technique to permit the simultaneous analysis of radon and thoron with the Lucas cell. The method utilizes two arbitrary counting intervals from zero time with respect to the sampling and the formulation of simultaneous linear equations with two unknown variables, the radon and thoron activities in the Lucas cell.

XI. PHOSWICH detectors

A phoswich detector consists of two or more scintillation detectors optically coupled as a phosphor sandwich, from which the scintillation light output is viewed by a single PMT. This unique detector arrangement is designated as a PHOSWICH, which is the acronym for PHOSphor sandWICH. Research is directed to the development of phoswich detectors as practical instruments for the simultaneous measurement and discrimination of alpha, beta, gamma, and neutron radiation.

The principle behind phoswich detectors is the different properties of the interaction between scintillators and different types of radiation. In particular, scintillators will differ in their propensity to interact or absorb radiation depending on the type and even the energy of the radiation as well as display differing light output decay times with concomitant different pulse shapes. Consequently placing multiple scintillators of different types layered on top of each other as a sandwich detector and coupling the combined scintillators to one photomultiplier (e.g., PMT) will constitute a detector capable of measuring simultaneously several types of radiation.

The selection of the scintillators, their dimensions, and their arrangement as a sandwich will depend on the types of radiation to be analyzed. For example, a phoswich detector designed to measure alpha/beta/gamma and neutron radiation will consist of scintillators sandwiched in a fashion so that the alpha particles are first absorbed by a thin scintillator sensitive to alpha particles, followed by a second scintillator that may be thicker but capable of absorbing the more penetrating beta particles, and another scintillator sensitive to the more penetrating gamma radiation and finally a fourth scintillator for the measurement of neutron radiation. Also, the scintillators that constitute the sandwich are carefully selected taking into account their differing scintillation light decay times, which can further facilitate pulse shape analysis for radiation discrimination.

A. Simultaneous counting of α-, β-, and γ-rays or α-, $\beta(\gamma)$-rays, and neutrons

There is a need for detectors that can measure simultaneously radiations of several types particularly in the different processes and nuclear safety management required in the nuclear fuel cycle. Much research and development had gone initially into the development of phoswich detectors for the simultaneous measurement and discrimination of α-, β-, and γ-rays Usuda (1992); Usuda and Abe (1994); and Usuda et al. (1994a,b). An example of such a phoswich detector is the coupled three-detector array of ZnS(Ag), NE102A, and BGO scintillators designed by Usuda et al. (1994a) for the simultaneous counting of α, β, and γ rays. The phoswich consisted of a thin (10 mg/cm^2) ZnS(Ag) scintillator sandwiched with a thicker (5 mm) NE102 plastic scintillator, which was sandwiched with a third 5 mm thick BGO scintillator. The α particles penetrate only the ZnS(Ag) scintillator, while the β particles and γ rays continue on to interact with the NE102A plastic scintillator, where β and soft γ interactions occur, and finally the hard γ rays continue on to interact with the BGO scintillator. This phoswich detector was tested for the discrimination of radiation types from a mixture of several sources such as ^{244}Cm for α particles, $^{90}Sr(^{90}Y)$ for β particles, ^{241}Am for soft γ rays as well as α particles, and ^{137}Cs and ^{60}Co for β and γ counting. Usuda et al. (1994a,b) demonstrated that ZnS(Ag) is insensitive to β and γ rays with a pulse rise time the slowest among the scintillators, and the NE102A plastic scintillator has the highest sensitivity to β particles (low-Z scintillator), the fastest pulse rise time among the scintillators used, and a relatively narrow peak width (FWHM). The third scintillator, BGO, has a rise time intermediate between those of ZnS(Ag) and NE102A and a high sensitivity to γ rays (high Z and density). The BGO may be replaced with an NaI(Tl) scintillator for the third phoswich for radiation monitoring purposes, as the latter provides lower backgrounds in this triple-phoswich detector.

Usuda (1995) noted that, in the radiation monitoring of high burn-up spent fuels, significant neutron emission must be considered in addition to other forms of radiation. He devised a modification of the triple-phoswich detector previously described to include a ^{6}Li glass scintillator (NS8), which has a high reaction cross-section for thermal neutrons. A triple-phoswich detector assembly consisting of ZnS(Ag), NE102A plastic, and ^{6}Li glass scintillators permits the simultaneous counting of α, $\beta(\gamma)$ rays and thermal neutrons. The ^{6}Li glass scintillator used was cerium-activated 7.5% lithium silicate glass containing 95.6% enriched ^{6}Li (NS8 Nikon scintillator). The detector was further developed by Usuda et al. (1997, 1998a,b) and Yasuda et al. (2001) with the careful selection of scintillator detectors according to their pulse decay rise times, light outputs, and emission wavelengths with the employment of optical filters to further enhance the discrimination of pulse events from each scintillator according to pulse shape analysis. The phoswich detectors were developed to measure simultaneously α-, $\beta(\gamma)$-rays, and fast neutrons (n_f) and thermal neutrons (n_{th}). A typical arrangement of a phoswich detector with these capabilities is illustrated in Fig. 9.97. Alpha particles are fully absorbed by the thin (10 mg/cm^2) ZnS(Ag) scintillator. Beta particles of low energy (<20 keV) interact slightly with the ZnS(Ag), but the higher-energy beta particles continue on to be fully absorbed by the 10 mm thick anthracene scintillator. The fast neutrons interact with the anthracene via proton recoil while the thermal neutrons undergo neutron capture in the 5 mm thick NS8 ^{6}Li glass scintillator enriched in ^{6}Li to 95.6%. The phoswiches are protected from ambient light with Al-coated Mylar film (0.25 mg/cm^2). The phoswiches can also be placed into radioactive solutions if protected with Au-coated Mylar film (0.9 mg/cm^2). The output signals from the PMT have specific rise times and pulse heights that are analyzed by PSD. The rise-time distributions of α-, $\beta(\gamma)$-rays, n_f and n_{th} from ^{244}Cm, ^{137}Cs, ^{252}Cf and $^{252}Cf + ^{244}Cm$ reported by Usuda et al. (1997, 1998a,b) with the ZnS(Ag)/anthracene/NS8 phoswich are illustrated in Fig. 9.98. The acronym FOM in the figure refers to crystal figure of merit. The FOMs are separated into different categories, to reflect the ability of the scintillator to discriminate pulse events from different radiation sources. For example, Usuda et al. (1997) classified FOMs into $\beta(\gamma)/\alpha$, $\beta(\gamma)/n_f$, n_f/n_{th}, and n_f/α. Yasuda et al. (2001) used a more simple dual detector phoswich of ZnS(Ag)/^{6}Li glass or ZnS(Ag)/anthracene and a three-dimensional analysis of pulse event number, height, and rise time to fully separate β/γ, n_f, and α events.

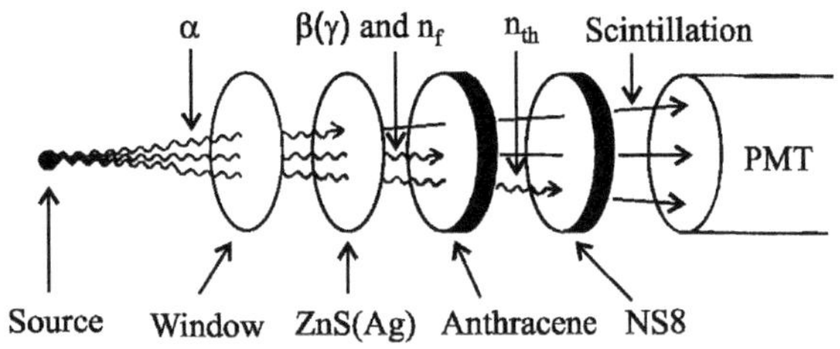

FIGURE 9.97 Arrangement of a ZnS(Ag)/anthracene/NS8 phoswich for simultaneous counting of α-, $\beta(\gamma)$-rays and neutrons. *From Usuda et al. (1997); reprinted with permission from Elsevier © 1997.*

Researchers at the University of Missouri−Columbia designed a triple crystal phoswich detector for the simultaneous analysis of alpha-, beta-, and gamma radiation (Childress and Miller, 2002; White and Miller, 1999), and they have evaluated each crystal for mischaracterizations of beta- and gamma radiation. Their detector design consisted of a sandwich of the following three detectors provided in the order of radiation penetration: (1) a 10 mg/cm^2 thick (0.002445 cm) ZnS:Ag for the detection of alpha particles, (2) a 0.254 cm thick CaF_2:Eu crystal for the detection of beta particles and some low-energy gamma radiation, and

FIGURE 9.98 Pulse shape spectra of α, β particles (including γ rays), thermal and fission neutrons observed with ZnS(Ag)/anthracene/NS8 phoswich. *From Usuda et al. (1998a); reprinted with permission from Elsevier © 1998.*

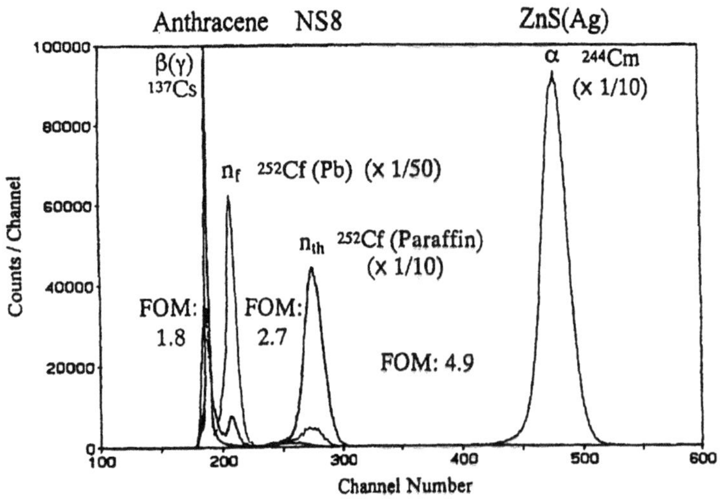

(3) a 2.54 cm thick NaI:Tl crystal for the measurement of gamma radiation. They used MCNP (Monte Carlo N-particle) simulation analysis (Briesmeister, 2000) capable of simulating electron, photon, and neutron interactions in detectors of simple geometry including spheres, cylinders, planes, cones, and ellipsoids. The application of Monte Carlo computer simulations to model interactions in detectors has been demonstrated to yield low errors (<5%) when compared to collected data (Bronson and Wang, 1996; Kamboj and Kahn, 1996; Ródenas et al., 2000). Analysis by Childress and Miller (2002) demonstrates that the most probable beta-particle energy loss in the first scintillator ZnS:Ag is around 20 keV due to the short path length that electrons can traverse in ZnS:Ag. The Monte Carlo code showed that the ZnS:Ag attenuation of gamma rays was restricted to gamma photons below 50 keV. The CaF_2:Eu preferentially interacts with beta particles. The Monte Carlo simulations found that the phoswich design has minimum inherent energy limits of 250 keV E_{max} for beta particles and 50 keV for gamma rays. The 2.54 cm thick NaI:Tl crystal yielded intrinsic gamma efficiency ranges from a maximum of 80% for 100 keV to 26% for 2 MeV photons. Mischaracterization of gamma events in the CaF_2:Eu crystal can be calculated and corrected.

Yamamoto and Ishibashi (2015) developed a three-layer phoswich as an alpha-beta-gamma imaging detector, which may be used remotely for the imaging and analysis of radiation at sites of nuclear power plant accidents such as the Fukushima Daiichi accident. In such situations nuclear fission products abound with emissions of beta and gamma radiation and some nuclear fuels contain plutonium giving rise to alpha particle emissions. The imaging detector was comprised of a three-layer phoswich coupled to a positive photomultiplier tube (PSPMT) illustrated in Fig. 9.99. The face of the detector consisted of a thin plastic scintillator with a very short decay time of ~5 ns for the measurement and imaging of alpha particle sources. The second layer of the detector was made of thin Gd_2SiO_3 (GSO) scintillator with 1.5 mol % Ce with a decay time of 35 ns for the detector of mostly beta particles. The third layer consisted of a thin layer of GSO scintillator with a high concentration of Ce (0.4 mol %), which yielded a decay time of 70 ns. With PSD, the images produced by the alpha, beta, and gamma radiations were separated on the basis of the Anger Principle utilizing 8 × 8 anode signals from the PSPMT with resolutions of 1.25 mm FWHM for alpha and beta particles and 2 mm FWHM for gamma photons. The pulse shape spectra from alpha and beta particle and gamma photons are illustrated in Fig. 9.100. The peak to the right of the figure is from the first layer of plastic scintillator of the phoswich arising from mainly the alpha-particles of ^{241}Am, the middle peak is due mainly to the beta-particle radiation of ^{90}Sr/^{90}Y; and the peak to the left is due to gamma photons from ^{137}Cs.

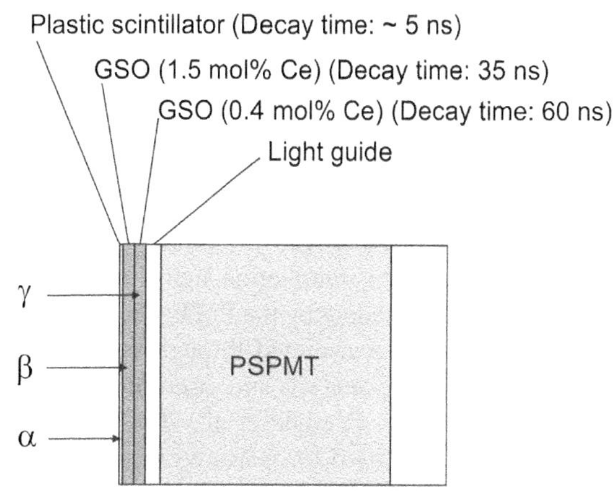

FIGURE 9.99 Schematic drawing of alpha-beta-gamma imaging detector. *From Yamamoto and Ishibashi (2015); reprinted with permission from Elsevier © 2015.*

B. Remote glass-fiber-coupled phoswiches

Studies by Yasuda et al. (2000a) found that a YAP scintillator ($YAlO_3$:Ce) has a fast rise time and sharp peak (small FWHM) with only 10 mg/cm^2 thickness in a

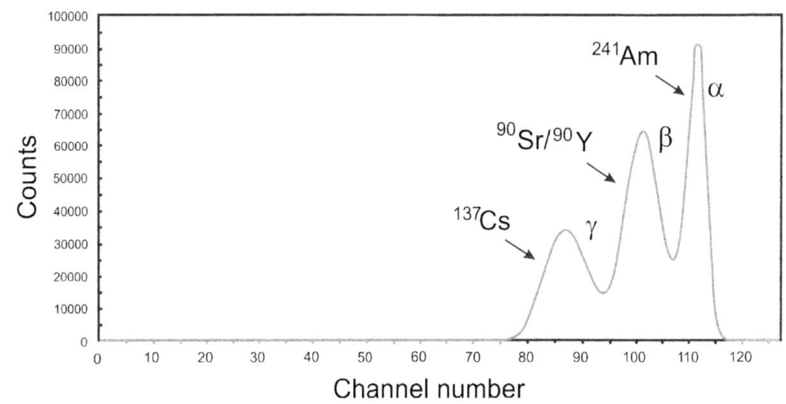

FIGURE 9.100 Pulse shape spectra for alpha and beta particles and gamma photons in the three-layer phosphwich detector. The peak to the right is from the first layer of thin plastic scintillator mainly due to alpha particles from ^{241}Am. The middle peak is from the second layer of GSO scintillator with 1.5 mol % Ce due to mainly ^{90}Sr/^{90}Y; and the peak to the left is from the third layer of GSO scintillator with 0.4 mol % Ce mainly from ^{137}Cs gamma photons. *From Yamamoto and Ishibashi (2015); reprinted with permission from Elsevier © 2015.*

phoswich detector coupled to a 5 mm thick YAG ($Y_3Al_5O_{12}$:Ce) scintillator, which was sufficient to accept high count rates and clearly distinguish alpha particles from beta and gamma rays. A new type of phoswich detector was developed by coupling a ZnS(Ag)/NE102A phoswich to the PMT via a quartz optical fiber. A NE172 wavelength shifter was positioned between the phoswich and the optical fiber for optimum scintillation light transmission via the optical fiber and readout by the PMT. The light from the phoswich detector is transmitted through the optical fiber for remote pulse shape analysis and discrimination of alpha and beta(gamma) rays (Yasuda et al., 2000b). This phoswich detector can be used to measure radioactivity in liquids as a dip-type inline monitor or to measure radioactivity in narrow or isolated spaces such as glove boxes or hot cells.

C. Low-level counters

Phoswich detectors are also employed in low-level counting of radionuclides in the environment in the presence of interfering γ radiation. For example, Wang et al. (1994) utilized a double CaF_2(Eu)-NaI(Tl) phoswich central (β,γ) detector in an anticoincidence low-background detector assembly for the measurement of β-emitting $^{89}Sr + ^{90}Sr(^{90}Y)$ activities in waste effluent of nuclear power plants in the presence of γ-emitting radionuclides. The window of the CaF_2(Eu) crystal used in the phoswich assembly was coated with aluminized Mylar; and the crystal was very thin (0.0635 mm) compared with the coupled NaI(Tl) detector (6.36 cm thick). This allowed the low-atomic-number CaF_2(Eu) to perform as an effective β detector with negligible interaction of γ radiation. A separate detector guard consisting of a surrounding NaI(Tl) crystal operated in anticoincidence mode to reduce backgrounds in a high-background laboratory from 5880 $\pm$ 229 cps without shielding to 1.0 $\pm$ 0.1 cpm with shielding and anticoincidence counting. An MDA for ^{90}Sr of 20 pCi/L is reported.

Rajabi Moghadam et al. (2015) developed a combination of a fast and slow scintillator for the simultaneous spectroscopy of alpha particles and gamma photons for the analysis of environmental solid samples. They chose a BC400 plastic scintillator with 2.4 ns decay time sandwiched with a CsI(Tl) scintillator with a fast decay time of 0.68 μs. The stark differences in time responses of the sandwiched scintillators, pulses arising from only one decay are easily discriminated against those obtained with both decay components by means of pulse shape analysis. A 50 μ BC400 scintillator was demonstrated sufficient to completely stop and absorb the entire energy of 5485.56 keV alpha particles; and 2.5 cm thick CsI(Tl) was sufficient to capture the energy of 100 keV gamma photons, which demonstrated the suitability of this detector for gamma-ray photons. The Minimum Detectable Activity (MDA) of 62.37 $\pm$ 5 mBq for alpha particles and 372.08 $\pm$ 10 mBq for gamma-rays is reported. The higher MDA for gamma photons is due to higher backgrounds from the thicker CsI(Tl) detector.

D. Simultaneous counting of n/γ/p fields

Neutron spectrometers that can adequately discriminate between neutrons and charged particles and measure equivalent doses to the human body are needed in space exploration and inside large human spacecraft such as the International Space Station. In these environments high-energy charged particles in space produce high-energy neutrons by their interaction with the spacecraft structural materials. With this objective in mind, Takada et al. (2002) developed a phoswich detector for neutron spectrometry in a mixed field of neutrons and charged particles. They coupled two organic scintillators of different light-output decay times, which could measure high-energy neutrons up to 130 MeV.

The phoswich neutron detector consists of a 133 mm diameter by 133 mm long NE213 organic liquid scintillator surrounded by a 15 mm thick NE115 plastic scintillator. A glass cell encapsulates the liquid scintillator to protect the optically coupled NE115 plastic scintillator. The phoswich is coupled to a single PMT via a light guide. The NE213 scintillator was selected by Takada et al. (2002) for the measurement of high-energy neutrons because of its ability to discriminate between gamma rays and neutrons. The light-output decay times of this scintillator are 3.7 ns for gamma-ray induced scintillation and 30 ns for neutron-induced scintillation. The surrounding NE115 scintillator provides for charged-particle detection because it has a much longer 225 ns day time.

The light decay time constants between the two scintillators employed in this phoswich detector was used to discriminate pulses of the three different particle species (n/γ/p). Fig. 9.101 illustrates how the light output signals are produced in the scintillators via interactions of the three particle species, and how PSD of the light output are used to characterize the signals from the detector. In brief, the Compton electron scattered by a gamma ray dissipates its energy only in the NE213 scintillator, and its signal displays a fast component of 3.7 ns (Fig. 9.101A). The neutron is detected via proton recoil, illustrated in Fig. 9.101B, that occurs only in the NE213 scintillator; and it has a slower decay time of 30 ns compared to the gamma-ray. The proton dissipates its energy in both scintillators (NE115 + NE213), and its signal is much longer as it becomes the sum of the fast and slow components with a decay time of 225 ns for the slow component.

The three particles were detected by Takada et al. (2002) using PSD based on the standard Computer

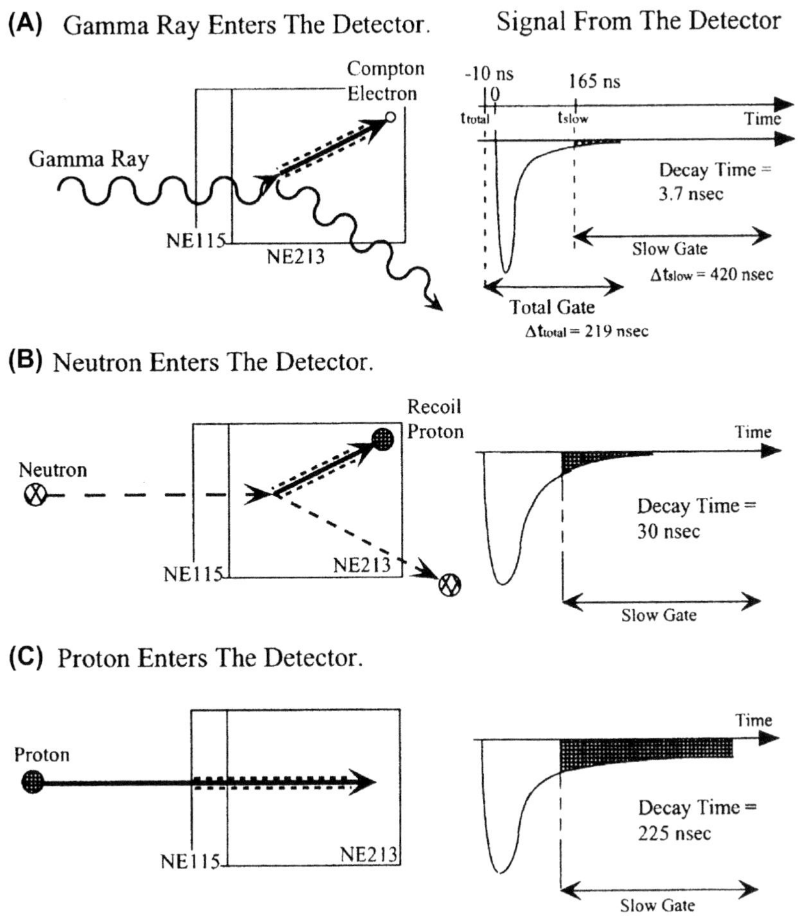

FIGURE 9.101 Schematic models of signals produced by the NE115/NE213 phoswich detector. (A), (B), and (C) are the cases of a gamma ray, neutron, or proton entering into the detector, respectively. The top-right sketch is the time relation between the pulse from the detector and the gate inputs to two ADCs. Two light outputs (total and slow components) are obtained by integrating the gate signal with the time intervals. *From Takada et al. (2002); reprinted with permission from Elsevier © 2002.*

Automated Measurement and Control (CAMAC) charge-integration ADCs (Lecroy 2249w, Lecroy Corp., Chestnut Ridge, NY). Analog-to-digital converters (ADCs) can be of two kinds, namely, an ADC that measures either charge or voltage and produces a digital number proportional to the magnitude of the input signal. Takada et al. (2002) developed energy-response functions of the phoswich detector to photons, neutrons, and protons. From the response functions, the energy spectra of the gamma photons, neutrons, and protons could be obtained through spectrum unfolding. Research in this direction was reported previously by Takada et al. (2001a,b).

Bismuth- and lithium-load transparent plastic (polyvinyltoluene, PVT) scintillators were fabricated and tested by Cherepy et al. (2015) for gamma photon and neutron discrimination. A phoswich detector consisting of bismuth-loaded PVT plastic coated with ZnS(^{6}Li) paint provided simultaneous gamma and neutron measurements via PSD with an FOM of 3.8 together with gamma spectroscopy with energy resolution of 15% at 662 keV. The FOM is a measure of the separation of the gamma-ray and neutron peaks, which is measured by the difference between the centroid of the two peaks divided by the sum of the FWHM of each peak described further on in this section and defined by Eq. (9.53) earlier in this chapter.

A portable phoswich detector was designed by Hartman and Barzilov (2017) capable of differentiating fast and thermal neutrons and gamma photons. The phoswich consisted of ^{6}Li-loaded glass (5 cm dia. and 1 mm thick disc) enriched with 1.58×10^{22} ^{6}Li atoms/cm^3 and EJ-299-33A plastic scintillators coupled to a single PMT. The ^{6}Li-glass scintillator has a high sensitivity to thermal and epithermal neutrons while it exhibits a low sensitivity to fast neutrons. A key factor in the design of the phoswich is the difference in the decay times of the two scintillators. The thermal neutron decay time in the ^{6}Li-glass is ~60 ns; whereas, the plastic scintillator has a scintillation time constant of 4 ns for gamma photons and ~30 ns for neutrons via proton recoil, which enabled excellent discrimination between neutrons and gamma photons. It was designed as a highly portable handheld instrument measuring 9 × 25 × 17 cm with a liquid crystal display touchscreen and a WiFi module for wireless communication.

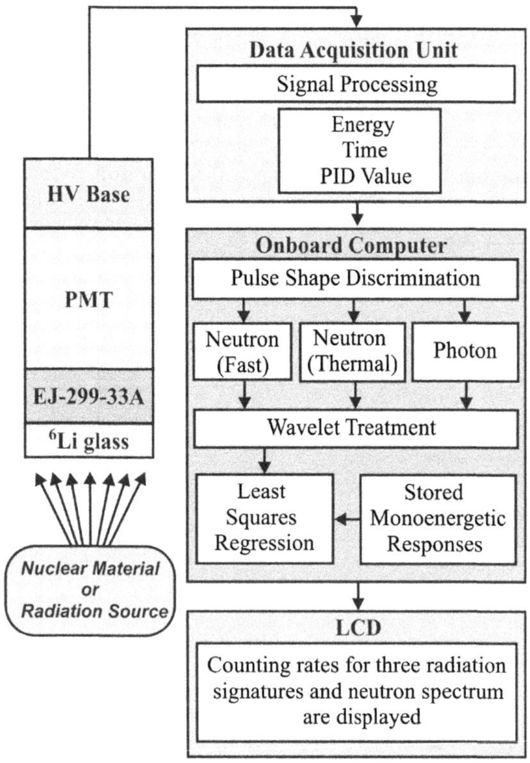

FIGURE 9.102 Scheme of operation of the handheld portable phoswich detector for neutron-gamma discrimination. *From Hartman and Barzilov (2017).*

In brief, the portable neutron-gamma phoswich detector of Hartman and Barzilov (2017) processes a gamma-neutron source as follows: The PMT anode signals created by scintillation photons go on to a data acquisition (DAQ) chip (See Fig. 9.102) which processes the data assigning an energy, time, and PID value to each signal. The PID value is the ratio of the integral under the tail of a signal (slow component) to the integral of the front part of the pulse (fast component), as illustrated in Fig. 9.103. The PID values go on to a computer chip within the handheld device for PSD identifying the signals separate categories for fast neutrons, thermal neutrons, or gamma photons. The segregated waveforms then undergo a wavelet treatment plotting the shape of the response with data smoothing, which facilitates spectral unfolding and least squares regression for the decomposition of a polyenergetic neutron detector response as the sum of monoenergetic neutron responses. The slower scintillation decay time of the ^{6}Li-glass scintillator provided a measurement of the thermal neutron count rate. Hartman and Barzilov (2017) measured an FOM of 1.3 for the separate gamma photons and neutrons by the plastic scintillator, as illustrated in Fig. 9.104. The FOM is calculated as the difference between the centroid of the two peaks divided by the sum of the FWHM of the two peaks.

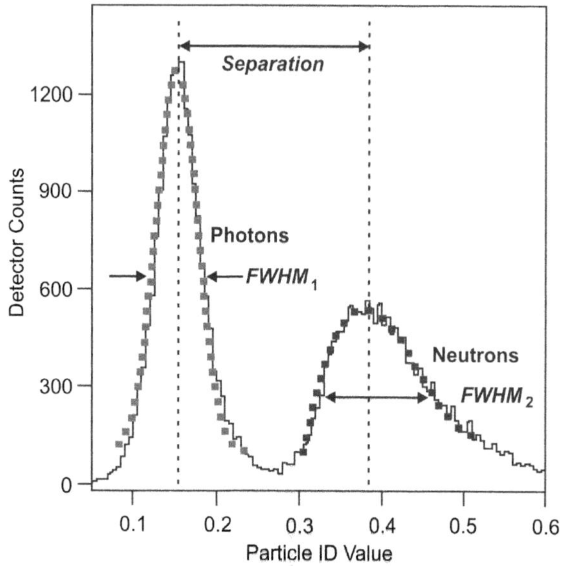

FIGURE 9.104 Neutron/gamma photon figure of merit. *From Hartman and Barzilov (2017).*

FIGURE 9.103 Digitized PMT neutron and gamma photon signal waveforms for the EJ-299-33A plastic scintillator. *From Hartman and Barzilov (2017).*

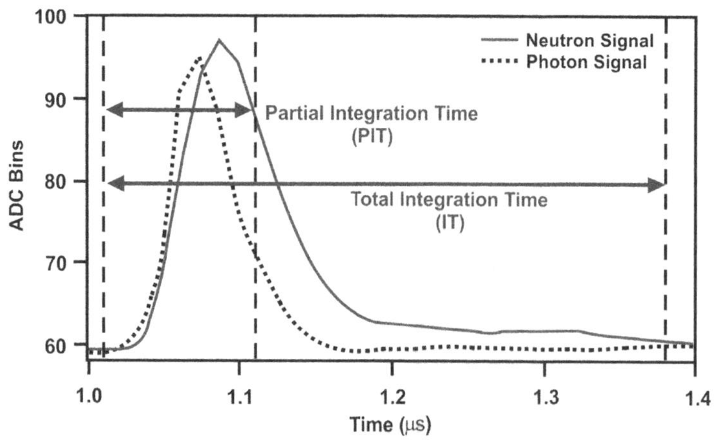

Huang et al. (2018) developed a gamma and neutron phoswich detector as a spectroscopic personal radiation detector (SPRD) in the form of a pocket-sized device for applications in homeland security related applications. The phoswich consists of a $\Phi 25 \times 25$ mm NaI(Tl) for gamma photon detection and a $\Phi 25 \times 3$ mm LiI(Eu) crystal for neutron detection. Pulse shape and pulse amplitude discrimination were used to separate gamma photon and neutron radiation. Energy resolutions for 661.6 keV gamma photons in NaI(Tl) and thermal neutrons in LiI(Eu) were measured as $7.0 \pm 0.2\%$ and $11.2 \pm 0.2\%$ respectively at a selected PSD threshold; and an FOM for gamma-neutron separation as a function of the PSD metric was measured as 4.4 ± 0.2.

E. Neutron spectrometry

In the measurement of neutron energy by the recoil-proton method, undesired events in the scintillator are those where the recoil proton does not deposit all of its energy in the scintillator, but rather escapes via scatter out of the scintillator. To eliminate those protons that deposit only part of their energy in the scintillator and thereby optimize energy resolution, Watanabe et al. (2008) devised a phoswich detector consisting of a plastic scintillator (NE 102A) with a fast decay constant (24 ns) surrounding a NaI(Tl) scintillator with a slow decay constant (230 ns). The large difference in decay constants of the two scintillators permits the use of PSD to discriminate between a pulse event that occurs in only one scintillator from an event that occurs in both scintillators. The principle of discrimination between recoil protons produced by a proton radiator (e.g., polyethylene) in the path of a neutron beam is illustrated in Fig. 9.105. The phoswich detector illustrated in the left-hand portion of Fig. 9.105 has a trapezoid body of NaI(Tl) inorganic crystal scintillator 362.5 mm in length surrounded with a NE102A plastic scintillator of 10 mm thickness with the exception of the detector face where the particles are incident. The length of the detector is sufficient to fully stop 400-MeV incident protons. Scintillation light

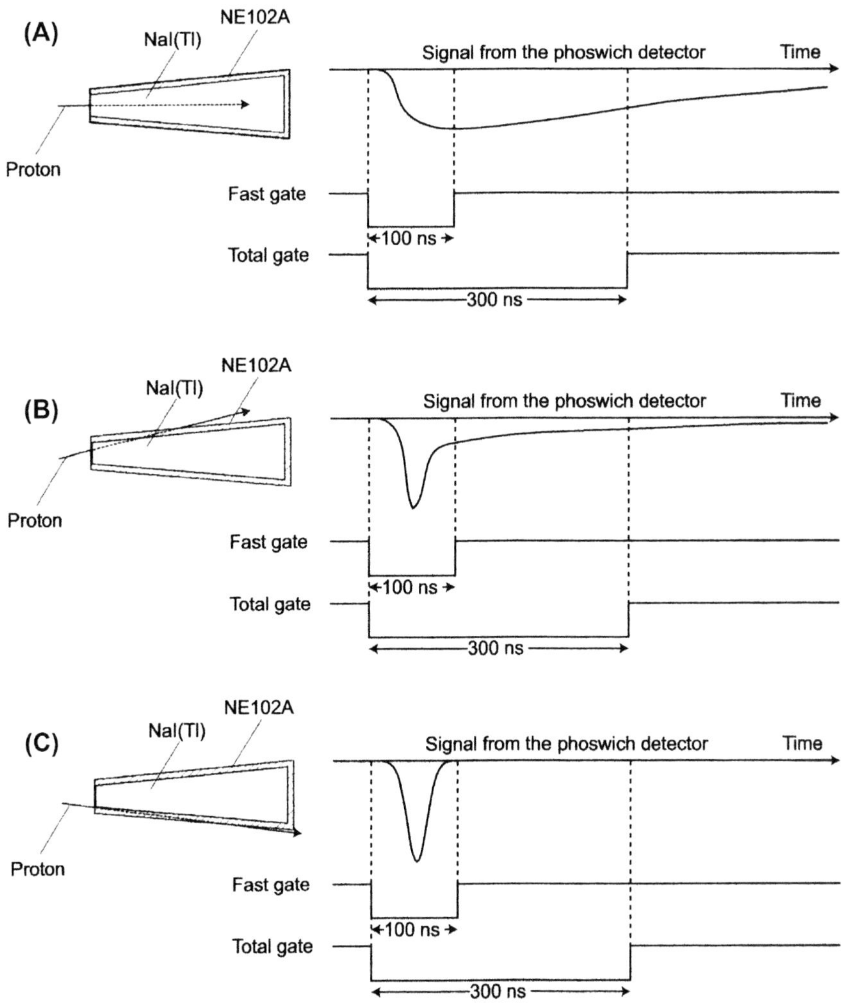

FIGURE 9.105 Schematic description of the principle of discriminating recoil protons with the phoswich detector: (A) recoil-proton deposits its energy in the NaI(Tl) crystal; (B) recoil-proton deposits its energy in both scintillators; and (C) recoil proton deposits its energy in the plastic scintillator. *From Watanabe et al. (2008); reprinted with permission from Elsevier © 2008.*

from the phoswich detector is transmitted onto a single PMT via a light guide at the rear end of the detector, which is not illustrated. Watanabe et al. (2008) took advantage of the fast decay time of the plastic scintillator (a few ns) and the slow decay time of the NaI(Tl) scintillator (230 ns) and utilized two different gate-charge integrations, illustrated schematically in Fig. 9.105, to discriminate between proton events occurring only in the NaI(Tl) scintillator against events occurring in both scintillators. Fig. 9.105A illustrates the case where a proton deposits all of its energy in the NaI(Tl) scintillator whereby the output signal has a relatively small fraction in the fast component compared to the total scintillation light. On the other hand, when a recoil proton passes through both the NaI(Tl) and plastic scintillators, the resultant output signal has a larger fast component, as illustrated in Fig. 9.105B. When a recoil proton losses energy only in the plastic scintillator, the output signal consists only of a fast component, as illustrated in Fig. 9.105C. Watanabe et al. (2008) set charge integration to be made over time durations defined by the fast gate and the total gate, as illustrated in Fig. 9.105. The fast gate was set to encompass the scintillation peak of the fast scintillator (100 ns); and the total gate covers a large portion of the long tail of the NaI(Tl) scintillator output signal.

The phoswich detector was calibrated for proton deposited energy by establishing the relationship between the deposited energy and the ADC channel for the NaI(Tl) scintillator. The calibration procedure, which made use of the Particle and Heavy Ion Transport code System (PHITS) described by Iwase et al. (2002), is elaborated by Watanabe et al. (2008) as follows: The energy deposited by the recoil proton was first calculated by PHITS, at the peak position generated by the H(n,p) reaction for an incident neutron energy. Then the calculated deposition energy was related to the ADC channel of the full-energy peak in the response function that was measured for the same incident neutron energy. The calibration demonstrated a linear dependence of the ADC channel on the deposition energy over the proton-recoil deposited energy range of 40−400 MeV. Examples of phoswich detector response functions for incident neutron energies of 195−205 and 245−255 MeV are illustrated in Fig. 9.106.

F. Simultaneous beta- and gamma spectroscopy

Farsoni and Hamby (2005) carried out initial studies on the β-particle energy depositions at different depths of a triple-layer phoswich detector consisting of BC-400/CaF_2:Eu/BC-444 scintillators with decay constants of 2.4, 940, and 264 ns, respectively. The scintillators had thicknesses sufficient to completely stop 0.1 MeV electrons in the first layer, 1.0 MeV electrons in the second layer, and 2.5 MeV electrons in the third layer. Using Monte Carlo N-Particle (MCNP) analysis, it was demonstrated that the traditional rise time measurement technique could not provide appropriate and accurate PSD with this type of phoswich detector. These results guided future work by Farsoni and Hamby (2007) into the development of a triple-layer phoswich detector capable of the simultaneous discrimination and spectroscopy of beta- and gamma-ray particles. The phoswich detector contained three scintillators layered as illustrated in Fig. 9.107. The first layer after the aluminized Mylar window is a thin plastic scintillator (BC-400) with very fast light decay characteristics (2.4 ns); and the second layer is an inorganic crystal scintillator (CaF_2:Eu), which displays a very slow light decay signal (900 ns). The effective atomic numbers (Z_{eff}) are low for the first two scintillators; and thus both layers are appropriate for β-

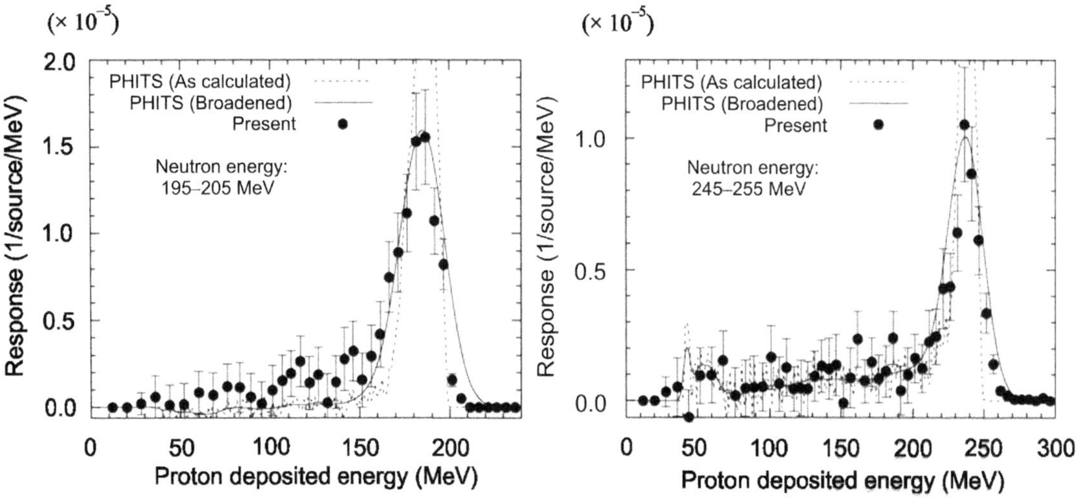

FIGURE 9.106 Experimental response functions of the phoswich detector with the calculation by PHITS. *From Watanabe et al. (2008); reprinted with permission from Elsevier © 2008.*

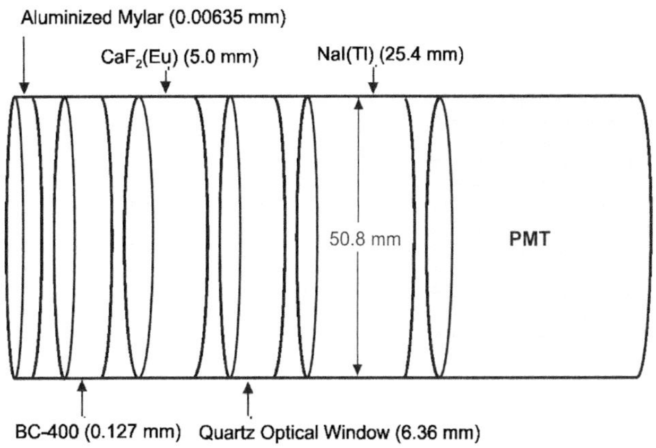

FIGURE 9.107 Schematic diagram of a triple-layer phoswich detector for the simultaneous discrimination and spectroscopy of β-particles and γ-rays. Layers of the detector are separated for clarity; but the real design has no spaces. *From Farsoni and Hamby (2007); reprinted with permission from Elsevier © 2007.*

particle detection with a thickness sufficient to stop electrons with energies up to 3.2 MeV. Beta particles are accepted, if they deposit energy in one or both of these two scintillators. The third scintillator of the phoswich is layered under or after a 6.36 mm thick quartz optical window. It consists of the inorganic crystalline scintillator NaI(Tl), which displays moderately slow scintillation light decay characteristics (230 ns). This last scintillator is intended to measure only γ-rays. The combined thickness of the previous two scintillators and the quartz optical window prevents the incursion of electrons (i.e., β-particles) with energies $\leq$ 6.7 MeV. Thus, any commonly encountered β-particles cannot reach the NaI(Tl) scintillator.

Simultaneous beta/gamma spectroscopy was demonstrated with the phoswich detector devised by Farsoni and Hamby (2007); and an example of typical spectra obtained for ^{14}C and ^{137}Cs sources are illustrated in Fig. 9.108. The system utilizes a customized digital pulse processor and characterization logarithm. The MCNP software package was used to establish response characteristics of the phoswich detector to electrons and photons. When a mixed beta/gamma field is analyzed, the detector generates pulses from which the type of radiation and its energy can be distinguished. The energy maximum (E_{max}) of the ^{14}C β-particles of 156 keV illustrated in Fig. 9.108 corresponds to the theoretical value; and the γ-ray spectrum of ^{137}Cs at 662 keV has an energy resolution of ~6.7% (Farsoni and Hamby, 2007).

A phoswich detector composed of two detectors consisting of an inorganic scintillator (calcium chlorborate) together with a scintillating plastic was devised by Arsaev and Kladov (2007) for beta spectrometry. The phoswich detector was able to discriminate backgrounds by a factor of 9.3 lower than that of a single detector based on plastic scintillator and pulse heights were linearly proportional to the beta-particle energy.

The measurement of radioactive xenon in the atmosphere is one of several methods that are used to monitor nuclear weapons testing. Farsoni et al. (2013) underscore that the measurement of the four xenon radioisotopes listed in Table 9.9 together with their relative concentration ratios in the atmosphere will enable the International Monitoring

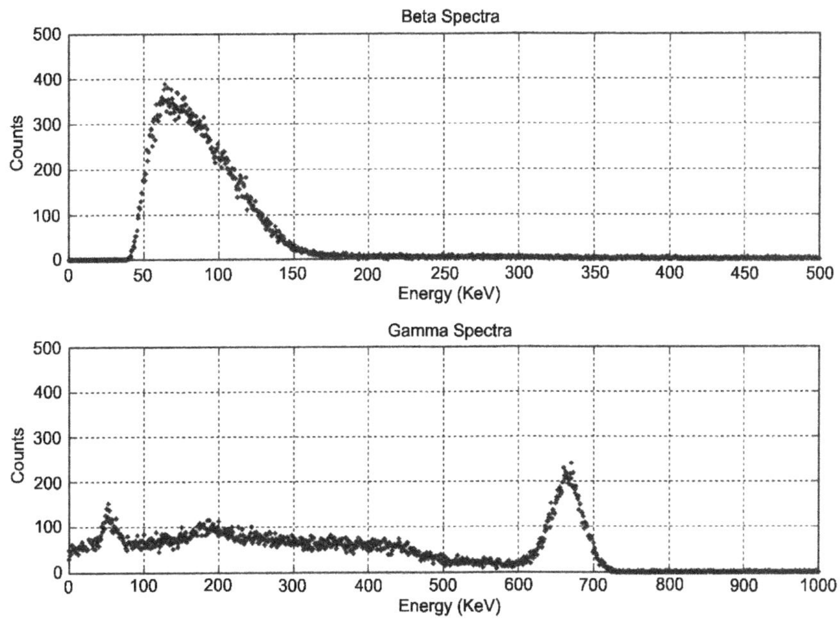

FIGURE 9.108 Simultaneous beta and gamma-ray energy deposition spectra from ^{14}C and ^{137}Cs obtained with a triple-layer phoswich detector. *From Farsoni and Hamby (2007); reprinted with permission from Elsevier © 2007.*

TABLE 9.9 Half-lives and characteristic energies for the decay of ^{131m}Xe, ^{133m}Xe, ^{133}Xe, and ^{135}Xe.

Radionuclide	^{131m}Xe	^{133m}Xe	^{133}Xe	^{135}Xe
Half-life	11.93 d	2.19 d	5.25 d	9.14 h
Gamma-rays (keV)	163.9	233.2	81.0	250.0
Gamma-ray abundance (%)	1.96	10.3	37.0	90.0
X-ray, K-shell (keV)	30	30	31	31
X-ray abundance (%)	54.1	56.3	48.9	5.2
Beta, maximum energy (keV)	–	–	346	905
Beta abundance (%)	–	–	99	97
CE, K-shell (keV)	129	199	45	214
CE abundance (%)	60.7	63.1	54.1	5.7

From Farsoni et al. (2013); reprinted with permission from Elsevier © 2013.

System (IMS) to monitor nuclear weapon testing worldwide. Research is directed toward improving a highly efficient and sensitive method of measuring the xenon radioisotopes by beta-gamma coincidence, which involves measuring simultaneously the beta-particles or conversion electrons and the gamma-rays or X-rays emitted by the xenon radioisotopes.

For the simultaneous beta-gamma analysis, small amounts of xenon must be extracted from large volumes of air; and to circumvent the use of separate monitoring stations and separate detectors to analyze low-levels of beta and gamma emissions from radioisotopes of xenon, the development of a single phoswich detector for this purpose has been researched (Yongchun et al., 2017; Farsoni et al., 2013; Warburton et al., 2011; Hennig et al., 2009; de la Fuente et al., 2008; de Celis et al., 2007; Henning et al., 2007; Ely et al., 2005; Ringbom et al., 2003; McIntyre et al., 2001). A phoswich detector composed of ZnS(Ag), CaF_2(Eu), and NaI(Tl) scintillators was coupled to a single PMT for $\alpha/\beta/\gamma$ recognition and a fast electronic card for PSD analysis (de la Fuente et al., 2008). The phoswich detector could be operated in coincidence mode to identify a composed signal produced by the simultaneous detection of α/β particles and X-ray/γ particles. The coincidence detection technique can enable the monitoring of low levels of gaseous fission products produced by underground nuclear explosions. Warburton et al. (2011) reviews phoswich detector designs, which employ a BC404 plastic scintillator to detect the beta-particles and a CsI(Tl) or NaI(Tl) scintillator to detect the gamma photons in coincidence with the beta-particles, referred to as "PhosWatch", which can achieve radioxenon detection limits required for nuclear explosion monitoring. They also research the application of an atomic layer deposition of Al_2O_3 as a coating to prevent the "memory effect" arising from the diffusion of Xe into the BC404 scintillator. Yongchun et al. (2017) research the absorption-desorption behavior of xenon gases on BC-404 scintillator, as the phoswich detector is required with sufficient sensitivity for its utilization in verifying the Comprehensive Nuclear Test Ban (CTBT) treaty with the measurement of radioxenon.

Farsoni et al. (2013) designed a unique phoswich for the beta-gamma coincidence measurement of xenon radioisotopes applicable to the monitoring of nuclear weapons testing. The phoswich is designed with three scintillation layers illustrated in Fig. 9.109. The beta-gamma coincidence events from radioxenon are identified when a coincidence energy absorption is measured in the first BC-400

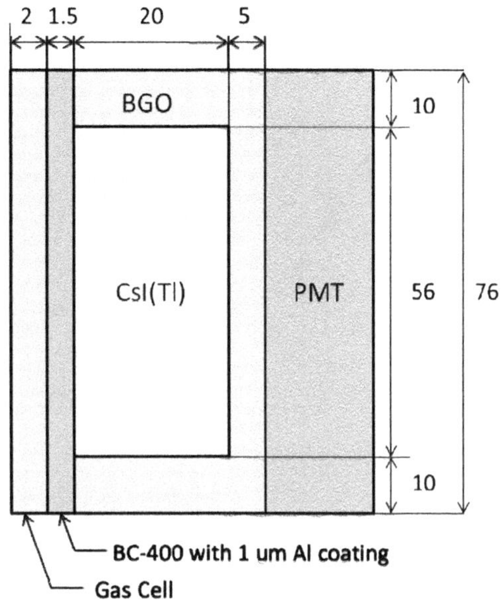

FIGURE 9.109 Schematic diagram of the three-layered phoswich detector for beta-gamma coincidence measurements of atmospheric radioxenon. *From Farsoni et al. (2013); reprinted with permission from Elsevier © 2013.*

plastic scintillator and the second CsI(Tl) crystal scintillator. A BGO scintillator layer surrounds the CsI(Tl) crystal to identify and reject Compton scattering events in the gamma photon detection layer. When a gamma photon triggers both the CsI(Tl) and BGO layers, the event is rejected, which provides a Compton suppression capability to reduce the Compton continuum in the gamma photon energy spectrum measurements. A 1 μm aluminum coating on the face of the BC-400 scintillator blocks the diffusion of Xe into the scintillator. The decay times of the BC-400 and CsI(Tl) scintillators are 2.4 and ~1000 ns, which facilitates the beta-gamma PSD. The detection system implements an online real-time digital processing firmware implemented in a Field-Programmable Gate Array (FPGA) device as well as a PC-based digital pulse processing algorithm. By means of the FPGA algorithm structure the system is triggered, pulse shapes are discriminated, Compton events are rejected, beta-gamma coincidence events are identified, the beta and gamma energies are measured, and the energy histograms are updated. An example of the real-time beta-gamma coincidence analysis and discrimination of beta and gamma events for ^{135}Xe and ^{133}Xe is provided by Figs. 9.110 and 9.111. The beta-spectrum of ^{135}Xe in Fig. 9.110 extends slightly beyond the 850 keV E_{max} of the beta-particle emissions due to the coincident detection of conversions electrons at 45 keV.

G. Other phoswich detectors

Phoswich detectors are under development for a wide range of applications not mentioned previously in this chapter among which include the detection and measurement of heavy ions in the region of Z »10 (Carbone et al., 2018; Muralithar et al., 2013), for the discrimination and measurement of neutrons, photons, and protons in the showers of cosmic radiation (Kole et al., 2015; Lawrence et al., 2014a,b; Takada et al., 2010) and in high-energy nuclear reaction experiments (Cherepy et al., 2015; Nácher et al., 2015; Tengblad et al., 2013), the measurement of fast (1−10 MeV) solar neutrons in the inner heliosphere (McKibben et al., 2005), and for X- and low-energy γ-ray astronomy at high altitudes (Mazumdar et al., 2010).

There is a current impetus in research on the development of phoswich detectors for the coincident measurement of annihilation radiation in PET as a tool in the medical field as an advanced imaging technique for cancer diagnosis and research (Kobayashi et al., 2017a,b; Olaya Dávila et al., 2017; Shimazoe et al., 2017; Yamamoto et al., 2016, 2014). As described by Olaya Dávila et al. (2017), an image produced by a PET is based on the external coincidence detection of two 0.511 MeV gamma-ray photons emitted at opposite directions from their point of origin; and the validation of an annihilation event requires a coincidence

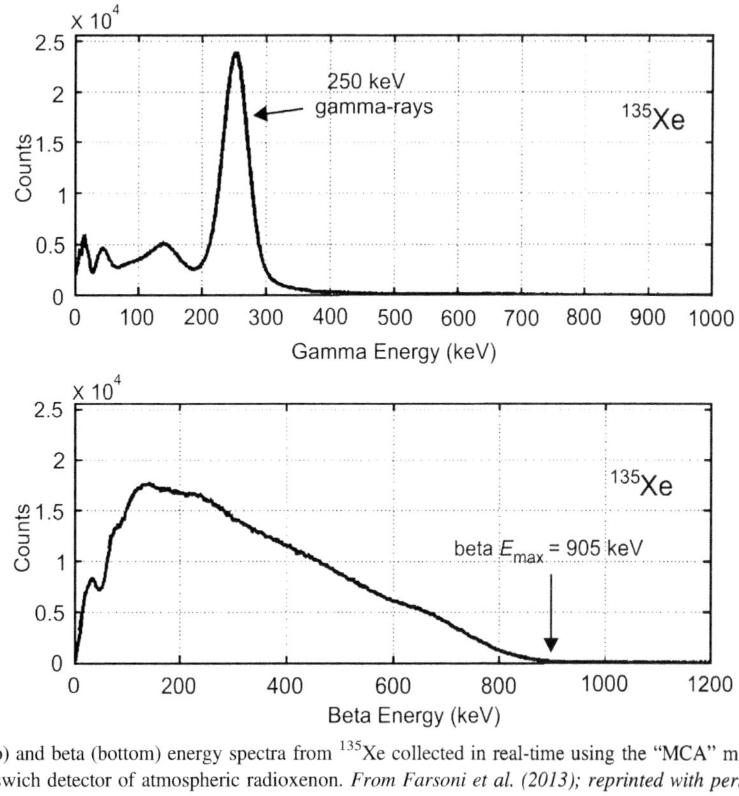

FIGURE 9.110 Gamma (top) and beta (bottom) energy spectra from ^{135}Xe collected in real-time using the "MCA" mode of the Field-Programmable Gate Array device of the phoswich detector of atmospheric radioxenon. *From Farsoni et al. (2013); reprinted with permission from Elsevier © 2013.*

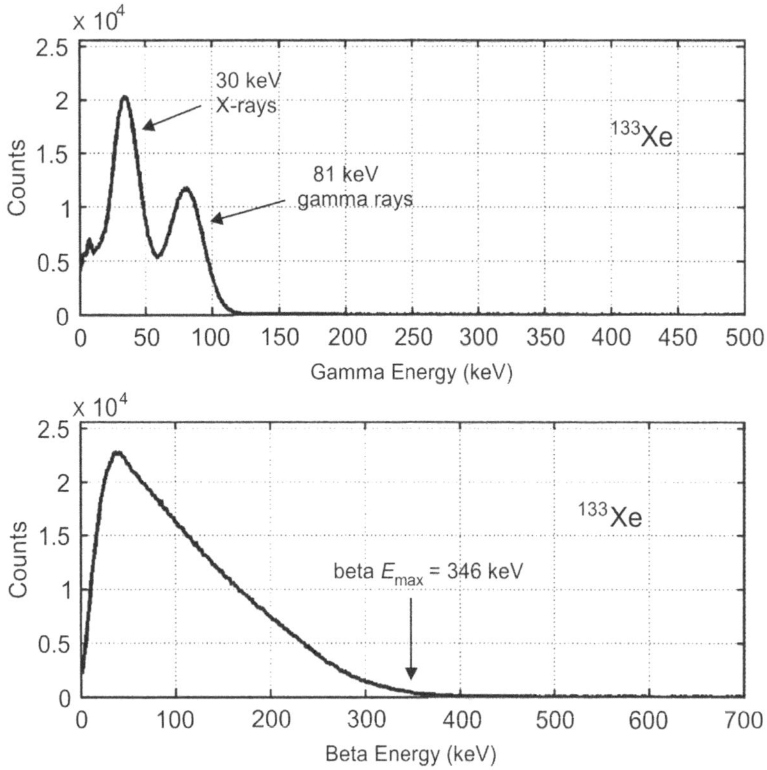

FIGURE 9.111 Gamma (top) and beta (bottom) energy spectra from ^{133}Xe collected in real-time using the "MCA" mode of the Field-Programmable Gate Array device of the phoswich detector of atmospheric radioxenon. *From Farsoni et al. (2013); reprinted with permission from Elsevier © 2013.*

measurement in the order of 12 ns of difference between the opposite detectors in a PET system, as illustrated in Fig. 9.112. Coincident events are registered and an image constructed according to the points of origin of the positron-negatron annihilations. Research is underway to improve the spatial resolution of the images produced in PET by utilizing phoswich detectors. High-resolution PET can be achieved by the determination of the depth-of-interaction (DOI), which improves the spatial resolution in peripheral areas of the field of view by reducing the parallax error for the 511 keV annihilation gamma photons (Shimazoe et al., 2017). Depth-of-interaction measurements can be achieved with phoswich detectors, which have different decay times and PSD for scintillation crystal identification in the PET electronics (Kobayashi et al., 2017a,b; Olaya Dávila et al., 2017; Shimazoe et al., 2017; Yamamoto et al., 2016).

H. Analytical expressions

The analytical expressions for the total and full energy peak efficiencies of phoswich detectors are provided by Abbas (2010), Abbas and Noureddeen (2011), and Yalcin et al. (2009). These expressions involve accurate analytical calculations of the path length, that the photon traverses in the active volume of the detector and the geometrical solid

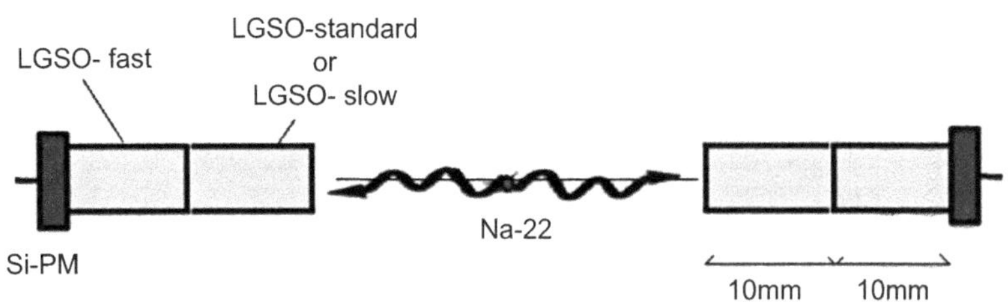

FIGURE 9.112 Schematic drawing of a phoswich detector arrangement in a PET instrument. A positron emitted by the Na-22 undergoes annihilation emitting two 511 keV gamma photons in opposite directions, which are detected at opposite ends in a 12 ns coincidence window. The ^{22}Na of 2.6 y half-life is used only for research into PET experimental design. Very short-lived positron emitters are utilized in nuclear medicine imaging such as ^{11}C, ^{13}N, ^{15}O or ^{18}F, among others. The detectors consist of two sandwiched Ce-doped $Li_{1.8}Gd_{0.2}SiO_5$ (LGSO) scintillators with different Ce concentrations yielding scintillating crystals with fast and slow pulse decay times. The scintillators are coupled to highly sensitive silicon photomultipliers. *From Yamamoto et al. (2016); reprinted with permission from Elsevier © 2016.*

angle subtended by the source to the detector at the point of entrance. Monte Carlo and experimental methods are described by these researchers.

XII. Neutrino interactions

As discussed in Chapter 1, Volume 1, the neutrino is an elusive particle of zero charge and very low mass. The detection and measurement of neutrinos is traditionally carried out with very large masses of liquid scintillators (see Chapter 6, Volume 1), which can provide tons of target material.

Solid scintillators including inorganic crystal scintillators have yet been found to be as useful compared to liquid scintillators, for the detection and measurement of neutrinos because of the large mass of target material needed and the difficulty of fabricating inorganic scintillator crystals to very large size and mass. Nevertheless, a large 500 kg CsI(Tl) detector is under development for the detection and measurement of nuclear reactor antineutrinos near the core of the Kuo-sheng Nuclear Power Station in Taiwan. Nuclear power reactors are an abundant source of electron antineutrinos ($\overline{\nu}_e$) in the low-energy (MeV) range. The work is a collaboration mainly between China and other countries, referred to as the TEXONO Collaboration, which is described in detail by Wong (2015), Zhu et al. (2006), Liu et al. (2002), Lai et al. (2001), Li et al. (2001) and Wong and Li (1999, 2000). More than one detector may be utilized inside an inner target volume; and one of the detectors is an array of CsI(Tl) crystal scintillators illustrated in Fig. 9.113. The detector, consists of about 187 kg of CsI(Tl) crystals (Unique Crystals, Beijing) arranged in a 17 × 15 matrix with each crystal module weighing ~ 2 kg. The individual CsI(Tl) crystal modules are hexagonal in shape with a length of 40 cm and side measuring 2 cm. PMTs are situated at both ends of each module for scintillation light readout; whereby, the sum of the two PMT signals ($Q_{tot} = Q_1 + Q_2$) can provide the energy of the event and the difference of the two signals can serve as a measure of the longitudinal position of the event. The PMT signals are received by amplifiers and shapers, and then digitized by a Flash Analog Digital Converter (FADC). Pulse shape analysis is possible to discriminate between pulse events produced by heavy charged particles, such as alpha particles, and those produced by electrons from γ events in the crystal.

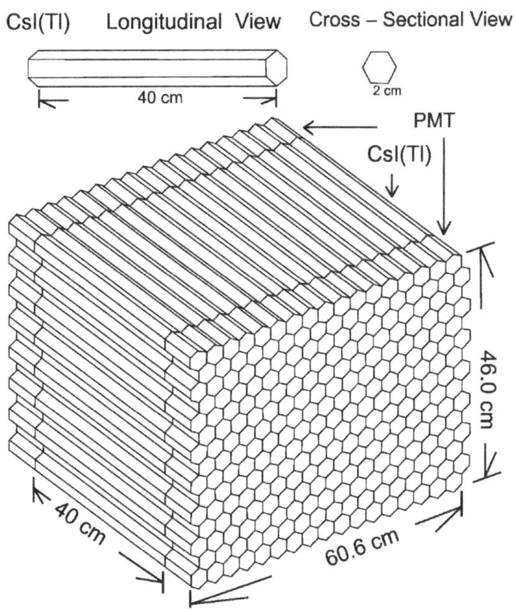

FIGURE 9.113 Schematic of the CsI(Tl) target configuration. *From Wong (2015), and Zhu et al. (2006); reprinted with permission from Elsevier © 2006.*

The compact and massive CsI(Tl) detector enables sufficient shielding to be built around it including an outer plastic scintillator cosmic-ray veto, 15 cm of lead and 5 cm of steel to shield against ambient background radioactivity, 25 cm of boron-loaded polyethylene to absorb cosmic-ray induced neutrons, and a dry-nitrogen outer atmosphere sealed in plastic to prevent radon gas from diffusing into the target region. Wong (2015) and Liu et al. (2001) describe advantages of the CsI(Tl) detector when compared to the conventional liquid scintillation detector, in addition to its compactness described above. These are the following:

(1) a large γ-ray photon attenuation of the crystal detector due to its high Z nuclei and density, particularly in the low energy range below 500 keV. For example, the attenuation length for a 100 keV γ ray is only 0.12 cm in the CsI(Tl) detector; whereas, in a liquid scintillator it would be 6.7 cm. Thus, at this low γ-ray energy, 10 cm of CsI(Tl) detector would have the same attenuation power as 5.6 m of liquid scintillator.
(2) a characteristic detector response, that is, the energy resolution of the CsI(Tl) detector is better than a typical liquid or plastic scintillator; and the low-energy γ-ray photons in the crystal scintillator would be fully stopped.

The major event that would be measured in the CsI(Tl) detector are neutrino−electron scattering, which is described in Chapter 1, Volume 1, Section X.G and Chapter 6, Volume 1, Section XII.B. The electron antineutrinos originating from the decay of reactor fission products interact with electrons in the CsI(Tl) detector via the elastic scattering reaction

$$\overline{\nu}_e + e^- \rightarrow \overline{\nu}_e + e^- \quad (9.56)$$

due to charged current (CC) and neutral (NC) weak current interactions (Deniz, 2012). The scattered electron can receive any kinetic energy from the neutrino less than or equal to the kinetic energy of the neutrino (Kraus, 2006). Experimentally, the recoil energies of the electrons from the neutrino interactions can be measured in the scintillator; and experimental searches for the neutrino magnetic moment will focus on the reduction in the threshold for the recoil electron energy (Li et al., 2001).

Other electron antineutrino interactions that can occur with the CsI(Tl) detector are inverse beta decay with ^{133}Cs and ^{127}I. Inverse beta decay is discussed in detail in Chapter 1, Volume 1, which is described by the following:

$$\bar{\nu}_e + p^+ \rightarrow n + \beta^+ \quad (9.57)$$

Inverse beta decay results in a reduction in atomic number of the isotope be it ^{133}Cs or ^{127}I according to the following:

$$\bar{\nu}_e + (A,Z) \rightarrow (A,Z-1)^* + \beta^+ \quad (9.58)$$

The positron yields a prompt signature with the emission of 511 keV annihilation photons. Characteristic gamma lines from the daughter nuclides are additional signatures. Li et al. (2001) note that with CsI(Tl) as the active target, the gamma lines with M1 transitions would be 81 and 160 keV for ^{133}Cs and 58, 202, and 418 keV for ^{127}I. These would be observed in the pulse height spectra from the CsI(Tl) detector.

Bressi et al. (2001) and Antonini et al. (2001) proposed the use of Yb-doped $Y_3Al_5O_{12}$ (YAG:Yb) scintillator for the detection of low-energy neutrinos. Subsequent research support the potential of Yb-doped YAG as well as other Yb-doped crystal scintillators for neutrino measurements (Yoshida et al., 2018; Zhu et al., 2018a,b; Chékir-Mzali et al., 2016; Nikl et al., 2013; Ferhi et al., 2012; Ricci et al., 2011 Belogurov et al., 2003, 2004; Chipaux et al., 2002). The electron neutrino reacts with the ^{176}Yb as follows:

$$\nu_e + {}^{176}\text{Yb} \rightarrow {}^{176}\text{Lu}^* + e^- \quad (Q = 301\ \text{keV}) \quad (9.59)$$

$$^{176}\text{Lu}^* \rightarrow {}^{176}\text{Lu} + \gamma \quad (E_\gamma = 72\ \text{keV}) \quad (9.60)$$

The asterisk in Eqs. (9.59) and (9.60) denote an excited state of ^{176}Lu. The signature for this neutrino event is a prompt electron and delayed gamma signal, and to provide discrimination against background noise, a delayed coincidence within 50 ns is required as well as high light yield. The charge transfer luminescence of Yb^{3+} has a great potential to be used for fast scintillator applications. Scintillators having fast response can be a strong tool for applications requiring detections with a high count rate (Yoshida et al., 2018). Also, Zhu et al. (2018a,b) notes that the charge-transfer luminescence of Yb^{3+} in Yb-doped YAG crystals can have a decay time of <4 ns; and that such an ultrafast scintillation property of the Yb:YAG crystal has attracted broad attention in many research areas including applications in solar neutrino detection. The neutrino reaction with ^{176}Yb differs from that of reactions described by Eqs. (9.57) and (9.58), as the neutrino reacts with a neutron resulting in an increase in atomic number, as described in Chapter 1, Volume 1 according to the following:

$$\nu_e + n \rightarrow p^+ + e^- \quad (9.61)$$

$$\nu_e + (A,Z) \rightarrow (A,Z+1) + e^- \quad (9.62)$$

Tests on YAG:Tb crystals with as much as 15% Yb have been conducted by Bressi et al. (2001) and Antonini et al. (2001). They demonstrated a fluorescence time of approximately 10 ns and a light yield of 8000 photons/MeV, which would be suitable for neutrino detection.

Other potential inorganic scintillators for the detection and measurement of neutrinos are the Indium-based crystal scintillators (Lasserre, 2013; Hsiao and Chang, 2011; Borisevich et al., 2005). Inverse beta decay with ^{115}In was first proposed by Raghaven (1976) for the detection and measurement of neutrinos. The inverse beta decay in nuclei of ^{115}In would occur in indium-based crystal scintillators as follows:

$$\nu_e + {}^{115}_{49}\text{In} \rightarrow {}^{115}_{50}\text{Sn}^* + e^- \quad (9.63)$$

The superscript asterisk in Eq. (9.63) denotes an excited state of ^{115}Sn, which decays to the ground state with a lifetime of 4.76 μs and emits two γ-ray photons of 116 and 497 keV. This provides a triple-coincidence signal, namely the emitted electron together with the two γ-ray photons (Fukuda et al., 2010).

XIII. Double beta (ββ) decay measurements

The discovery, principles, and current research on double beta decay are reviewed in Chapter 1, Volume 1. A treatment of the liquid scintillation techniques and semiconductor detectors employed in the search for neutrinoless double beta decay is provided in Chapters 6 and 4, Volume 1. Only a brief treatment of the methods applied in the use of solid inorganic crystal scintillators to the measurement of double beta decay and the search for neutrinoless double beta decay will be given here.

Double beta decay was predicted by Nobel Laureate Maria Goeppert-Mayer (1935) at Johns Hopkins University. The first observation of double beta decay was made by Steven R. Elliott, Alan A. Hahn, and Michael K. Moe at the University of California, Irvine in 1987. They measured double beta decay in a sample of ^{82}Se enriched from its natural abundance of 8.7% to an isotope abundance of 97% to provide more source material for the double beta decay as well as reduce contamination from other sources (Elliott et al., 1986, 1987; Moe, 1986; Moe and Rosen, 1989). The decay scheme is written as follows:

$$^{82}_{34}\mathrm{Se} \xrightarrow{t_{1/2}\,=\,9.2\times10^{19}\ \mathrm{years}} {}^{82}_{36}\mathrm{Kr} + 2\beta^- + 2\,\overline{\nu} \qquad (9.64)$$

where the parent nucleus decays to a daughter nucleus of the same mass number A, but with an increase in atomic number Z by two. In general, the double beta decay process is expressed in terms of the nuclear atomic and mass numbers as

$$(A,Z) \rightarrow (A,Z+2) + 2\mathrm{e}^- + 2\overline{\nu} \qquad (9.65)$$

Thus, in accord with the beta decay scheme, within the nucleus of the parent nuclide, two neutrons decay to two protons with the simultaneous emission of two beta particles and two antineutrinos. This type of beta decay is abbreviated as $2\nu\beta\beta$ or $2\nu2\beta$ to signify the simultaneous emission of two neutrinos and two beta particles.

Double beta decay is a rare event manifested by very long half-lives from 10^{18} to 10^{24} years. Because of the rarity of the event, the measurement of double beta decay is extremely difficult requiring a large mass of nuclide often achieved with highly enriched radionuclide sources and the use of instrumentation with the capability of extreme suppression or vetoing of background radiation. The National Nuclear Data Center of the Brookhaven National Laboratory currently lists 12 nuclides that are known to undergo double beta decay, and for which decay data are well documented; these are in order of increasing atomic order $^{48}_{20}$Ca, $^{76}_{32}$Ge, $^{82}_{34}$Se, $^{96}_{40}$Zr, $^{100}_{42}$Mo, $^{116}_{48}$Cd, $^{128}_{52}$Te, $^{130}_{52}$Te, $^{136}_{54}$Xe, $^{130}_{56}$Ba, $^{150}_{60}$Nd, and $^{238}_{92}$U (NNDC-BNL, 2018). As discussed in Chapter 1, Volume 1, beta decay is not confined to negatron emission. Other modes of beta decay also exist, which include positron emission and EC decay processes. Research is underway with inorganic crystalline scintillation detectors including low-temperature scintillating bolometers for the measurement of $\beta^+\beta^+$, β^+/EC, and double EC (i.e., ECEC) decay processes (Barabash, 2015; Kang et al., 2013; Krutyak et al., 2013; Belli et al., 2008, 2009; Kim et al., 2007; Danevich et al., 2003).

There is much research underway in numerous laboratories around the world directed toward the potential application of solid inorganic crystal scintillation detectors among other detection methods (see also Chapter 4, Volume 1 and Chapter 6, Volume 1) for the detection and measurement of neutrinoless double beta decay, and only a few of the reports on this work are cited here. Among these are research projects into the measurement of double beta decay of the following nuclides with crystal scintillation detectors including:

(i) ^{106}Cd and ^{116}Cd with $CdWO_4$ crystal scintillator (Sisti et al., 2016; Belli et al., 2010; Bardelli et al., 2006; Danevich et al., 2003, 2005a, 2006),

(ii) ^{48}Ca with CaF_2 and $CaWO_4$ crystal scintillators (Iida et al., 2016; Sisti et al., 2016; Maneschg, 2015; Umehara et al., 2015; Yoshida et al., 2005, 2009; Ogawa et al., 2003, 2004),

(iii) ^{64}Zn, ^{180}W and ^{186}W with $ZnWO_4$ crystal scintillator (Danevich et al., 2014; Krutyak et al., 2013; Belli et al., 2008, 2009),

(iv) ^{100}Mo with $CaMoO_4$ crystal scintillator (Jiang et al., 2018; Pan et al., 2018a; Park et al., 2016; Sisti et al., 2016; Maneschg, 2015; Kang et al., 2013; Annenkov et al., 2008),

(v) ^{150}Nd with yttrium–aluminum garnet doped with neodymium, abbreviated as YAG:Nd (Danevich et al., 2005b),

(vi) ^{82}Se and ^{100}Mo with ZnSe and $ZnMoO_4$ crystals (Maneschg, 2015), and

(vii) ^{160}Gd with Gd_2SiO_5(GSO):Ce scintillating crystals (Kobayashi et al., 2012).

Beyond the Standard Model there exists the possibility that the neutrino could be its own antiparticle, referred to as a Majorana neutrino. According to this theory, if the neutrino possesses a rest mass (now estimated to have a mass upper limit of 2.3 eV/c^2) and is its own antiparticle, a neutrino emitted in one beta decay could be reabsorbed in the second beta decay resulting in a double beta decay in which no neutrinos are emitted. i.e., a neutrinoless double beta decay often abbreviated as $0\nu\beta\beta$ or $0\nu2\beta$ (See Chapter 1, Volume 1 for a more detailed treatment). Current investigation into the search for neutrinoless double beta decay is important, because it would affirm the mass of the neutrino, as only a particle with a definite rest mass can oscillate from one form into another. The Nobel Prize in Physics 2015 was awarded jointly to Takaaki Kajita and Arthur B. McDonald for the discovery of neutrino oscillations, which provides evidence that neutrinos have mass. The current search and measurement of neutrinoless double beta decay would allow the measurement of the effective neutrino mass and define the absolute neutrino mass scale as described by Maneschg (2015).

In the $2\nu\beta\beta$ process, the two beta particles that are emitted simultaneously share the total decay energy, $Q_{\beta\beta}$, with the two neutrinos. Thus, the decay energy is shared between four particles, namely, the two neutrinos and two beta particles. In the $0\nu\beta\beta$ process, the two beta particles emitted share the entire decay energy, as there would be no neutrinos emitted with which to share the decay energy. The general beta-particle energy spectra for $2\nu\beta\beta$ and $0\nu\beta\beta$ decay are illustrated in Fig. 9.114. The two beta-particle spectra are very distinct. In the $2\nu\beta\beta$ decay process, the numbers of beta particles (N) vary over a wide range of energy plotted as the sum of the two beta-particle energies divided by the decay energy, i.e., $\sum\left(E_{\beta_1}+E_{\beta_2}\right)/Q_{\beta\beta}$. The beta-particle energy spectrum for the $2\nu\beta\beta$ decay would

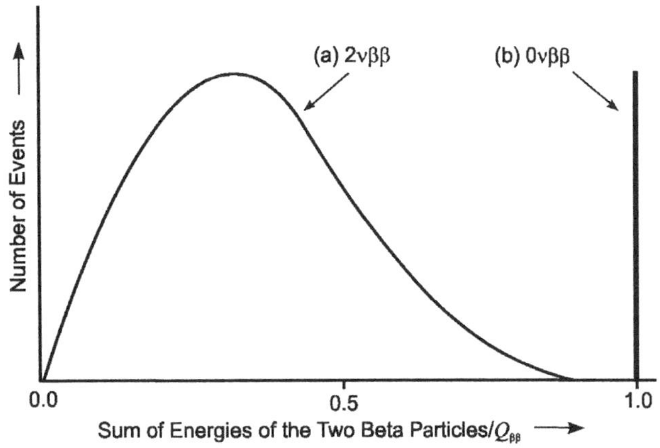

FIGURE 9.114 Beta-particle energy spectra for (A) two-neutrino double-beta decay ($2\nu\beta\beta$), and (B) neutrinoless double-beta decay ($0\nu\beta\beta$). In the $2\nu\beta\beta$ decay process, the sum of the two beta particle energies divided by the decay energy, $\sum(E_{\beta_1}+E_{\beta_2})/Q_{\beta\beta}$, displays a continuous spectrum and the maximum will never reach 1.0; whereas, the spectrum for $0\nu\beta\beta$ would display a single line at $\sum(E_{\beta_1}+E_{\beta_2})/Q_{\beta\beta}=1$. *From L'Annunziata (2016); reprinted with permission from Elsevier © 2016.*

encompass the continuous energy range with a spectral shape similar to that of the single beta-particle emission, because they share the decay energy with neutrinos. From Fig. 9.114 it can be seen, that the endpoint of the continuous beta-particle spectrum of the $2\nu\beta\beta$ decay does not reach the value of 1.0. This is due to the fact that the two beta particles emitted in the $2\nu\beta\beta$ decay process must share decay energy with two neutrinos, and thus the sum of the two beta particle energies can never equal the decay energy, $Q_{\beta\beta}$. However, the beta-particle energy spectrum for the $0\nu\beta\beta$ decay, where neutrinos do not escape with any of the decay energy, would display a single energy line for the sum of the two beta particle energies, and this line would occur at $\sum(E_{\beta_1}+E_{\beta_2})/Q_{\beta\beta}=1$. Current research is underway in the search for the sum of the two beta particles to equal the decay energy, $Q_{\beta\beta}$, or where $\sum(E_{\beta_1}+E_{\beta_2})/Q_{\beta\beta}=1$, which would confirm $0\nu\beta\beta$ decay.

An example may be taken from the case of double beta decay of ^{116}Cd, where $2\nu\beta\beta$ decay has been measured with $^{116}CdWO_4$ crystal scintillator; and the $0\nu\beta\beta$ decay is yet to be confirmed (Sisti et al., 2016; Arnaboldi et al., 2010; Gironi et al., 2009; Bardelli et al., 2006; Danevich et al., 2003, 2005a,b, 2006). The $2\nu\beta\beta$ decay of ^{116}Cd can be written as

$$^{116}_{48}\mathrm{Cd} \xrightarrow{t_{1/2}=2.9\times10^{19}\ \mathrm{years}} {}^{116}_{50}\mathrm{Sn} + 2\beta^- + 2\bar{\nu} \qquad (9.66)$$

where the mass number remains constant at 116; and the atomic number increases from 48 to 50. The beta decay scheme of ^{116}Cd; and the simulated response functions of a $^{116}CdWO_4$ crystal scintillator in $2\nu\beta\beta$ and $0\nu\beta\beta$ decay of the ^{116}Cd are illustrated in Fig. 9.115. The double beta decay energy ($Q_{\beta\beta}$) of the ^{116}Cd of 2813.49 keV, given in Fig. 9.115, is calculated by the difference in parent and daughter nuclide atomic mass units ignoring the negligible difference in the binding energies of the electrons of the two nuclides (see Eq. (1.247) of Chapter 1, Volume 1) or

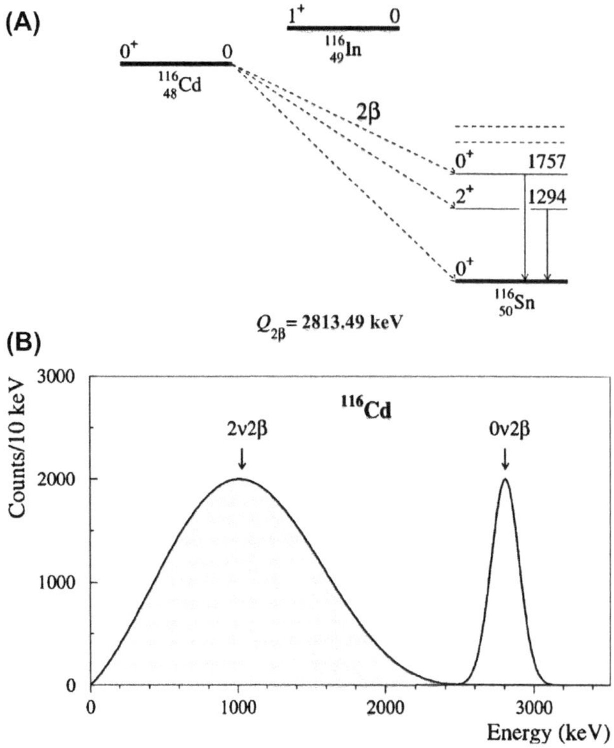

FIGURE 9.115 (A) The scheme of ^{116}Cd–^{116}In–^{116}Sn triplet and (B) Simulated response functions of the $^{116}CdWO_4$ crystal scintillators to the 2ν and 0ν modes of the 2β decay of ^{116}Cd. *From Danevich et al. (2003); reprinted with permission from Elsevier © 2003.*

$$\begin{aligned} Q_{\beta\beta} &= (M_{^{116}\mathrm{Cd}} - M_{^{116}\mathrm{Sn}})c^2 \\ &= (115.90476323\mathrm{u} - 115.90174282\mathrm{u})(931494\ \mathrm{keV/u}) \\ &= 2813.49\ \mathrm{keV} \end{aligned} \qquad (9.67)$$

where $M_{^{116}\mathrm{Cd}}$ and $M_{^{116}\mathrm{Sn}}$ are the atomic masses of the parent and daughter nuclides ^{116}Cd and ^{116}Sn, respectively, in atomic mass units; and where $1\mathrm{u} = 931.494\ \mathrm{MeV}/c^2$ (see Eqs. (1.10) and (1.11) of Chapter 1, Volume 1). Atomic Mass Units are available from IOP Science of which the latest table was published in 2017 (Chinese Physics C 41, No. 3 (2017) 030003, http://iopscience.iop.org/article/10.1088/1674-1137/41/3/030003).

The ^{116}Cd decays by double beta ($\beta\beta$) directly to ^{116}Sn, as illustrated in Fig. 9.115A. There is no transition of ^{116}Cd via a single beta decay to ^{116}In because the mass and thus energy level of the ^{116}In is higher than the ^{116}Cd; and this transition is thus forbidden. The simulated response functions of a ^{116}Cd-enriched $CdWO_4$ crystal scintillator is illustrated in Fig. 9.115B. The $2\nu\beta\beta$ decay would produce a broad peak covering a wide range of energies, as this peak represents a broad spectrum of energies that the two β-particles share with two neutrinos (i.e., four particles share the 2813.49 keV decay energy, two β-particles and two neutrinos); whereas, the neutrinoless double beta decay ($0\nu\beta\beta$) would be manifested as a single peak at 2813.49 keV, as illustrated in Fig. 9.115B, because two β-particles would be emitted simultaneously without neutrino emission; and the sum of their energies would be equivalent to the decay energy.

To take the example of ^{116}Cd further, the calculated $\beta\beta$ decay energy and simulated $^{116}CdWO_4$ crystal scintillator response, discussed in the previous paragraph, was put to experiment by Danevich et al. (2006, 2005a, 2003). These researchers utilized four cadmium tungstate $^{116}CdWO_4$ scintillators enriched to 83% ^{116}Cd providing a total mass of 330 g of detector material. The $^{116}CdWO_4$ detector was surrounded with an active shield consisting of 15 $CdWO_4$ crystals of large volume providing a mass of 20.6 kg. An additional outer active shield was made of plastic scintillator measuring 40 × 40 × 95 cm providing a complete 4π active shield of the main $^{116}CdWO_4$ detector. Additional outer passive shielding with high purity copper, lead, and polyethylene layers were utilized. For each event in the $^{116}CdWO_4$ detector, signal amplification, arrival time and pulse shape were recorded. With the combined active and passive shielding and pulse shape analysis, the background count rate in the energy region 2.5–3.2 MeV was reduced to 0.037(10) counts/(yr·kg·keV), which was the lowest background achieved with crystal scintillators. Low backgrounds in the 2.5–3.2 MeV energy region was important because the calculated $Q_{\beta\beta}$ is 2.813 MeV (see Eq. 9.67). The measured spectra from the $^{116}CdWO_4$ detector are illustrated in Fig. 9.116, among which is the clearly defined energy spectrum from the $2\nu\beta\beta$ decay of ^{116}Cd (curve a).

The $0\nu2\beta$ decay peak is not observed from the experimental data illustrated in Fig. 9.116; however, the $2\nu2\beta$ decay peak (curve a) is clearly defined; and Danevich et al. (2003, 2005a) illustrate the experimentally determined value of the double-beta decay energy $Q_{\beta\beta} = 2808 \pm 43$ keV (see Inset of Fig. 9.116). The experimentally determined double-beta decay energy is very close to the theoretically computed value ($Q_{\beta\beta} = 2813.49$ keV) as computed above in Eq. (9.67); and calculated with QCalc (2018), which is the on-line computer and data base of the National Nuclear Data Center, Brookhaven National Laboratory (https://www.nndc.bnl.gov/qcalc/).

From the inset of Fig. 9.116, we can see from the experimental spectrum that the peak corresponding to $0\nu\beta\beta$ decay should be measured unequivocally at $Q_{\beta\beta}$, which is

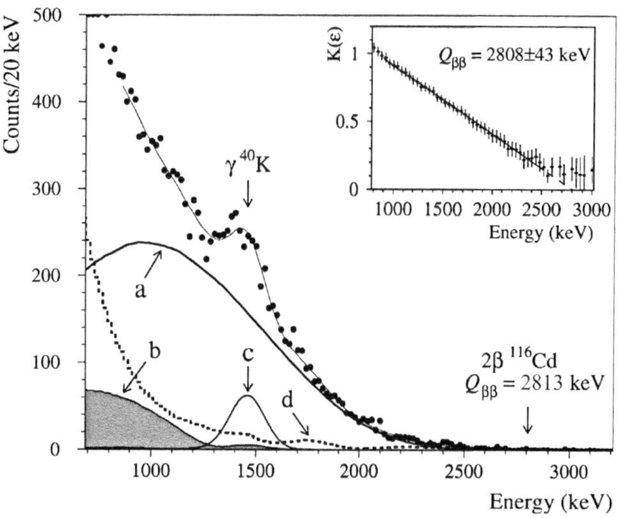

FIGURE 9.116 Spectra of $\gamma(\beta)$ events measured with $^{116}CdWO_4$ detector during 12,649 hours. Also shown are the most important model components: (A) the $2\nu2\beta$ spectrum of ^{116}Cd; (B) external γ background caused by radioactive contamination of the PMTs; ^{40}K contamination of the (C) nonenriched and (D) enriched $CdWO_4$ scintillators. Solid line represents the fit of the data in the 860–2700 keV energy interval. (Inset) The $2\nu2\beta$ decay Kurie plot and its fit. *From Danevich et al. (2003, 2005a); reprinted with permission from Elsevier © 2005.*

also where the tail of the $2\nu\beta\beta$ spectrum will appear. Thus, a sufficient number of counts from $0\nu\beta\beta$ decay must be measured over very long periods of measurement. This is an extremely challenging task that requires low backgrounds and large masses of double-beta decaying nuclide. In view of the difficulty in meeting the challenge of measuring $0\nu\beta\beta$ decay, Orio (2016) underscores the following criteria: *"In order to have chances of observing the $0\nu\beta\beta$ peak, it is mandatory to have a consistent detector mass (at least of the order of 100 – 1000 kg) to maximize the number of $\beta\beta$ emitters, and a very low background in the region of interest (ROI) to minimize the spurious events. Another fundamental requirement to fulfill is a good energy resolution, to discriminate the $0\nu\beta\beta$ peak from the right tail of the $2\nu\beta\beta$ distribution. The difficulty in observing the $0\nu\beta\beta$ peak can be appreciated by the inset of* Fig. 9.117."

Examples of the application of inorganic crystal scintillators as well as other detector types utilized in the search for $0\nu\beta\beta$ decay are listed in Table 9.10. The table also lists the eight principle nuclide candidates to the measurement of $0\nu\beta\beta$, the Q value for the double beta decay ($Q_{\beta\beta}$), the natural abundance of the nuclide of interest, the world production of the elements with natural isotope abundance of the nuclide of interest, and the acronym for the research project or collaboration, detector type, and fiducial $\beta\beta$ mass. Enrichment of the nuclide of interest whenever possible is obviously an advantage. Also, the greater the fiducial mass of the $\beta\beta$ nuclide, the greater will be the potential for the measurement of $0\nu\beta\beta$ decay.

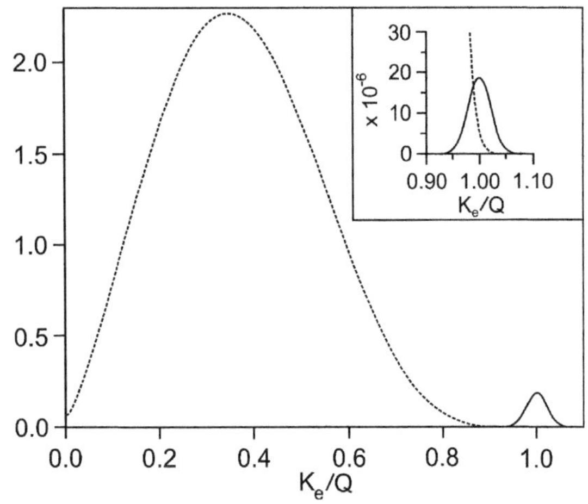

FIGURE 9.117 $2\nu\beta\beta$ and $0\nu\beta\beta$ two electrons sum energy spectra. $2\nu\beta\beta$ is a continuum between 0 and $Q_{\beta\beta}$, while $0\nu\beta\beta$ is a peak centered at $Q_{\beta\beta}$. In the inset is shown that a good energy resolution is needed to disentangle the two spectra. The y-axis scale and the $0\nu\beta\beta$ relative peak height are arbitrary. *From Orio (2016).*

Inorganic crystal scintillators offer, in certain circumstances, advantages over other detector types or they may be the only alternative for the measurement of $\beta\beta$ decay on the basis of the following:

- With inorganic crystal scintillators several nuclides of interest that undergo $\beta\beta$ decay can be incorporated directly into the molecular structure of crystal scintillators, and often at an atom-percent enrichment. In other words, the source = the detector. Examples are (1) the measurement of ^{116}Cd decay with $^{116}CdWO_4$ crystal enriched in ^{116}Cd to 83% (Sisti et al., 2016; Belli et al., 2010; Bardelli et al., 2006; Danevich et al., 2003, 2005a, 2006); (2) the measurement of ^{106}Cd decay with $^{106}CdWO_4$ crystal scintillator enriched in ^{106}Cd to 66.4% (Belli et al., 2010); (3) the search for double beta decay processes in ^{64}Zn with $ZnWO_4$ crystal scintillator at the ^{64}Zn natural abundance of 48.6% (Danevich et al., 2014; Krutyak et al., 2013; Belli et al., 2008, 2009); (4) the development of $CaMoO_4$ crystal scintillators for double beta decay measurements of ^{100}Mo (Jiang et al., 2018; Pan et al., 2018; Park et al., 2016; Sisti et al., 2016; Maneschg, 2015; Kang et al., 2013; Annenkov et al., 2008); (5) the development of yttrium−aluminum garnet doped with Nd (YAG:Nd) scintillator to measure double beta decay of ^{150}Nd (Danevich et al., 2005b); (6) the use of $^{48}CaWO_4$ to measure ^{48}Ca double beta decay (Zdesenko et al., 2005); (7) The search for double beta decay processes in ^{64}Zn, ^{70}Zn, ^{180}W and ^{186}W with $ZnWO_4$ crystal scintillators (Belli et al., 2009) and $ZnMoO_4$ (Maneschg, 2015); and (8) The measurement of ^{48}Ca $2\nu\beta\beta$ decay and the search for $0\nu\beta\beta$ decay with $^{48}CaF_2$ crystal scintillator at a 2% enrichment of ^{48}Ca (Suzuki, 2010; Ogawa et al., 2003, 2004; Yoshida et al., 2005, 2009).
- Measurements of radionuclide decay where source = detector with inorganic crystal scintillators can provide the potential for a large mass of source and target material in a small or minimal volume of detector. For example, one of the largest projects ongoing in the search of $0\nu\beta\beta$ decay with a crystal scintillator is the CANDLES (CAlcium fluoride for the study of Neutrinos and Dark matter by Low Energy Spectrometer) system utilizing $^{48}CaF_2$ crystal scintillator in Osaka, Japan. A 305 kg prototype detector is undergoing experimentation (Iida et al., 2016; Umehara et al., 2015). The project will increase the target mass to 3.4 tons and eventually 30 tons with 2% enrichment (Suzuki, 2010).

Bonvicini et al. (2014) lists advantages and disadvantages of certain scintillation detectors over other detector types in the search for $0\nu\beta\beta$ decay. In brief, these advantages are listed as follows:

1. Ge diode detectors and bolometers have excellent energy resolutions; and they offer solutions for background reduction. However, Ge diodes and bolometers are limited in the achieveable mass increase, which is due to cost in the case of Ge diode detectors and technical issues in the case of bolometers.
2. Large-scale organic scintillation detectors have no apparent limitation in terms of mass scalability in low background environment; however, these detectors yield poor energy resolution.
3. Xenon liquid- or gas-based time projection chambers must make a trade off between source mass and energy resolution; although they provide powerful background reduction concepts.
4. The ideal is to include, in one single instrumental setup, all of the critical factors for the measurement of $0\nu\beta\beta$, namely (i) high energy resolution, (ii) powerful background reduction methods, and (iii) source mass scalability. Large arrays of inorganic scintillators may achieve this end.
5. Inorganic scintillating crystals offer the following advantages: (i) They can be grown with a high level of purity with the source isotope within the crystal structure often at a high isotopic enrichment. The high purity achievable with inorganic crystal scintillators can yield excellent scintillation properties needed for a high energy resolution. (ii) Several $0\nu\beta\beta$ candidate nuclei are contained in a wide range of inorganic scintillators, which can provide flexibility on the choice of scintillation detector. (iii) Quenching factors (i.e., differences in light yield for an α-particle or nuclear recoil compared to that of an β-particle or γ-photon) and PSD remove all alpha background from the energy region around

TABLE 9.10 The main candidate isotopes for the search for $0\nu\beta\beta$ decay.

Isotope	$Q_{\beta\beta}$ (keV)[a]	Natural Abundance (%)[b]	Production 2013 (tons)[c]	Experiment[c]	Source/detector	Fiducial $\beta\beta$ mass (kg)[c]
^{48}Ca	4268.1	0.187	–	CANDLES[d]	CaF_2 crystal scintillator	6→>40
^{76}Ge	2039.1	7.73	155	GERDA[e]	Ge diodes at 86%–88% ^{76}Ge	15 → 30
				Majorana demonstrator[c]	Enriched ^{76}Ge in vacuum cryostats	25
^{82}Se	2997.9	8.73	>2.3 × 10³	SuperNEMO[f]	^{82}Se track calorimeter	7 → 100
				LUCIFER[g]/CUPID-0[h]	$Zn^{82}Se$ scintillating bolometer	–
^{100}Mo	3034.4	9.82	2.7 × 10⁵	MOON[i]	^{100}Mo film + NaI(Tl) scintillator	–
				AMoRE[j]	$^{40}Ca^{100}MoO_4$ crystal scintillator	100
				LUMINEU[k]	99.5% ^{100}Mo scintillating bolometer	7 → 338
^{116}Cd	2813.5	7.49	2.2 × 10⁴	COBRA[l]	CdZnTe semiconductor	–
^{130}Te	2527.5	34.08	>95	CUORE[m]	TeO_2 crystal bolometer	10 → 200
				SNO+[n]	Te-doped liquid scintillator	160–270
^{136}Xe	2457.8	8.85	3–4	EXO[o]	81% ^{136}Xe liquid Xe time projection chamber	80
				KamLAND-Zen[p]	^{136}Xe-loaded liquid scintillator	150 → 260–340
				NEXT[q]	^{136}Xe gas xenon TPC	9 → 90
^{150}Nd	3371.4	5.64	~1.7 × 10⁴	DCBA[r]	Drift chamber/Magnetic tracker	–

[a]*Data from National Nuclear Data Center, Brookhaven National Lab (Qcalc, https://www.nndc.bnl.gov/qcalc/).*
[b]*Data from National Nuclear Data Center, Chart of Nuclides, https://www.nndc.bnl.gov/chart/chartNuc.jsp.*
[c]*Data from Maneschg (2015); reprinted with permission from Elsevier © 2015.*
[d]*Acronym for CAlcium fluoride for the study of Neutrinos and Dark matters by Low Energy Spectrometer.*
[e]*Acronym for GERmanium Detector Array.*
[f]*Acronym for Neutrino Ettore Majorana Observatory.*
[g]*Acronym for Low-background Underground Cryogenic Installation For Elusive Rates.*
[h]*Acronym for CUORE Upgrade with Particle Identification.*
[i]*Acronym for Mo Observatory of Neutrinos.*
[j]*Acronym for Advanced Mo-based Rare process Experiment.*
[k]*Acronym for Luminescent Underground Molybdenum Investigation for NEUtrino mass and nature (Poda, 2016; Tenconi, 2015).*
[l]*Acronym for Cadmium Zinc Telluride 0-Neutrino Double-Beta Research Apparatus.*
[m]*Acronym for Cryogenic Underground Observatory for Rare Events.*
[n]*Acronym for Sudbury Neutrino Observatory (upgraded to $0\nu\beta\beta$ decay search).*
[o]*Acronym for Enriched Xenon Observatory.*
[p]*Acronym for KAMioka Liquid scintillator Anti-Neutrino Detector-ZEro Neutrino double-beta decay experiment.*
[q]*Acronym for Neutrino Experiment with a Xenon TPC.*
[r]*Drift Chamber Beta-ray Analyzer.*

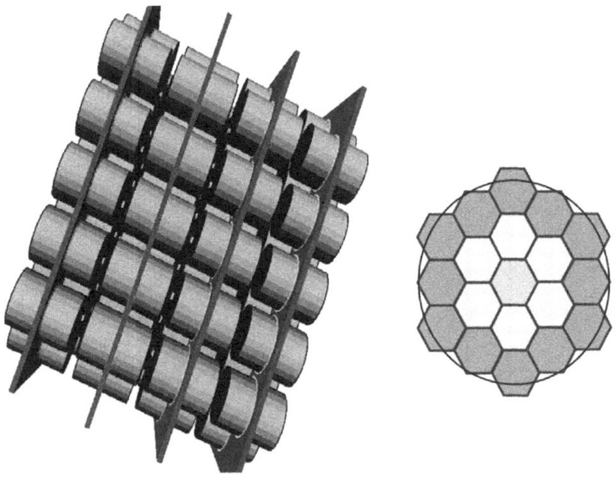

FIGURE 9.118 A prototype array of high-performance scintillation crystal detectors for $0\nu\beta\beta$ searches. Left: A side view of an array of 108 $^{40}Ca^{100}MoO_4$ crystal scintillators, Right: Front view of 19 stacked scintillator detectors where a silicon drift detector as photodetector is coupled to each circular surface of each detector. The calcium can be deleted in the isotope ^{48}Ca and enriched in ^{100}Mo. The $CaMoO_4$ scintillator has a very high molybdenum content of 48% in mass. *From Bonvicini et al. (2014).*

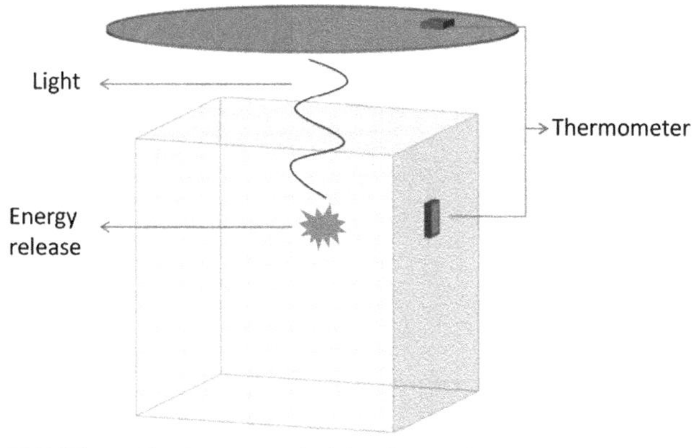

FIGURE 9.119 Operating principle of a scintillating bolometer. The energy of an incident particle is absorbed by the scintillating crystal. The release of energy inside the scintillating crystal follows two channels: light production and thermal excitation. The heat produced in the process is read out by temperature sensors (e.g., thermistors) attached to the crystal scintillator and to the light detector. The energy absorbed by the scintillator and the light intensity are converted into heat. *From Gironi (2010); reprinted with permission from Elsevier © 2010.*

$Q_{\beta\beta}$, which are powerful tools for background reduction. (iv) Large arrays of inorganic scintillators can be assembled without much technical complications in a relatively low-cost experimental setup. For example, inorganic scintillation crystals are easily shaped as separate units and stacked together to form a large detector with an optimum mass of $\beta\beta$ source nuclide, as illustrated in Fig. 9.118.

General reviews on the methods used for the measurement of double beta decay and the search for neutrinoless double beta decay are available from Henning (2016), Maneschg (2015) and Ejiri (2010).

XIV. Scintillating bolometers

A scintillating bolometer consists of a crystal scintillator connected to a thermistor, which will measure the temperature increase in the scintillator caused by energy absorbed from the incident radiation, and connected to a second bolometer (e.g., thin film of ~100 μm of Ge), which will measure the energy of the scintillation photon by the increase in temperature caused by the absorbed light. Scintillating bolometers are used to discriminate and measure nuclear particles; and they are promising tools in the search for neutrinoless double beta ($0\nu\beta\beta$) decay and for Weakly Interacting Massive Particles (WIMPs) of Dark Matter.

A. Operating principle

The operating principle of a scintillating bolometer is illustrated in Fig. 9.119. Current scintillating bolometers may be referred to as containing dual bolometers because they contain the crystal scintillating bolometer, which absorbs most of the energy of the incident particle with that energy measured as heat increase, while in turn, a second optical bolometer absorbs emitted scintillation light, which is also measured by the temperature increase produced by the absorbed light.

The possibility of the simultaneous measurement of the heat and light signals as a new tool for particle identification was proposed by González-Mestres and Perret-Gallix (1988). The very different scintillation yield of different particles (e.g., β/γ particles, α particles, and recoiling nuclei) could be used to identify the type of particle, whereas the heat signal would serve as a measure of the deposited energy with high resolution (Ortigoza et al., 2011). The β/γ particles are grouped together here because γ photons produce Compton electrons in the crystal scintillator and these electrons deposit their energy like β particles. The work of Alessandrello et al. (1992) demonstrated the first simultaneous detection of scintillation light and heat signals at low temperature (20 mK) with a CaF(Eu) scintillating bolometer. The light intensity was measured with a Si PD. Bobin et al. (1997) advanced the technology further with the simultaneous measurement of heat and light with a CaF_2(Eu) target bolometer optically coupled to a second infrared bolometer for the measurement of the scintillation light intensity. Alessandrello et al. (1998) demonstrated the strong suppression of background with the scintillating bolometer, and thus the potential of the scintillating bolometer for the measurement of double beta ($\beta\beta$) decay.

A schematic diagram of a current scintillating dual bolometer (i.e., scintillating crystal bolometer with optical

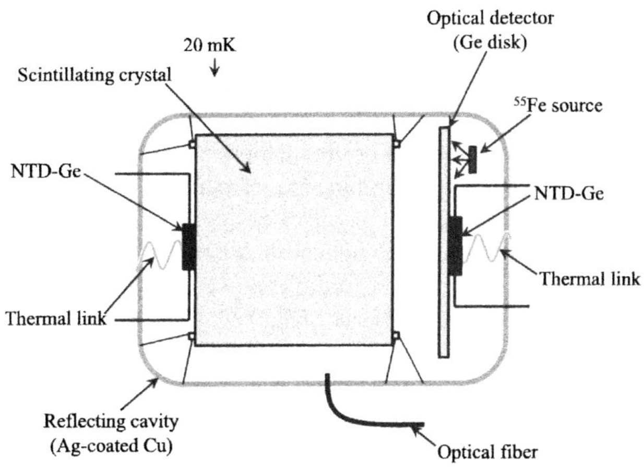

FIGURE 9.120 Schematic view of a dual bolometer configuration. The ^{55}Fe source, is not a necessary component. It is included here as an inner ^{55}Fe X-ray source placed opposite the Ge optical bolometer to calibrate the energy absorbed on the optical detector. *From Ortigoza et al. (2013, 2011); reprinted with permission from Elsevier © 2011.*

bolometer) is illustrated in Fig. 9.120. The principle of operation is described in detail by several researchers (Artusa et al., 2014a,b; Arnaboldi et al., 2011a,b; Ortigoza et al., 2011; Cebrián et al., 2005; Coron et al., 2004, 2008, 2009). In brief, the principle of operation, in accord with the schematic illustrated in Fig. 9.120, is as follows: A scintillating crystal (e.g., ~50 g BGO) is located within a reflecting cavity of Ag-coated Cu. The crystal scintillator is connected to a Neutron Transmutation Doped Ge (NTD-Ge) thermistor, which measures the heat signal from the scintillator resulting from the absorption of an incident particle. A second bolometer, the optical detector, consists of a thin (~100 μm) Ge disk, which is connected to a second NTD-Ge thermistor, receives the heat signal following the absorption of scintillation light from the crystal scintillator. The dual bolometers (i.e., scintillating crystal and optical detector) are thermally linked to a mixing chamber of a dilution refrigerator at 20 mK. A plastic optical fiber permits the periodic feeding of light pulses ($\lambda = 820$ nm) to the bolometers to monitor bolometer stability and correction for gain drifts. An inner ^{55}Fe X-ray source may be placed on the back side of the Ge optical bolometer to calibrate the absorbed energy. When an incident particle deposits its energy in the scintillation crystal, the major portion of that energy will be converted to heat; and the remaining energy is emitted as scintillation light, which is likewise converted to heat in the optical detector. The heat signals from the scintillator crystal and Ge optical sensors are received by their respective NTD-Ge thermistors.

The energy-absorbing scintillator and Ge-optical detector must be coupled to a constant temperature source by means of a thermal link. As described by Arnaboldi et al. (2011a,b), the temperature variation induced by an energy release (E) in the absorber is inversely proportional to the heat capacitance (C) of the bolometer, which is written as

$$\Delta T = \frac{E}{C} \tag{9.68}$$

The accumulated heat flows to the heat sink through the thermal link; and the absorber returns to the base temperature with a time constant $\tau = C/G$, where G is the thermal conductance of the link:

$$\Delta T(t) = \frac{E}{C}e^{-t/\tau} \tag{9.69}$$

To obtain a measurable temperature rise, the heat capacity of the absorber must be very small. Thus, bolometers must be operated at cryogenic temperatures of the order of 10−100 mK (Arnaboldi et al., 2011a,b).

The pulse amplitude of the signal from the scintillator bolometer is registered as a Heat Signal, and the pulse amplitude from the optical bolometer is registered as a Light Signal, as illustrated in Fig. 9.121. The intensities of the light signals or scintillation yields (i.e., light outputs) will differ in accord with the type of particle absorbed in the scintillator at a given energy. The light/heat amplitude ratios of the following orders of magnitude for a $CaWO_4$ crystal scintillator have been measured: 1, 2.5, and 10 for neutron-induced nuclear recoil events, α events, and β/γ events, respectively (Cébrian et al., 2003, 2005); and different scintillation yields by the different particle events in the scintillator serve as the basis for the discrimination of particles, as discussed in the subsequent paragraphs.

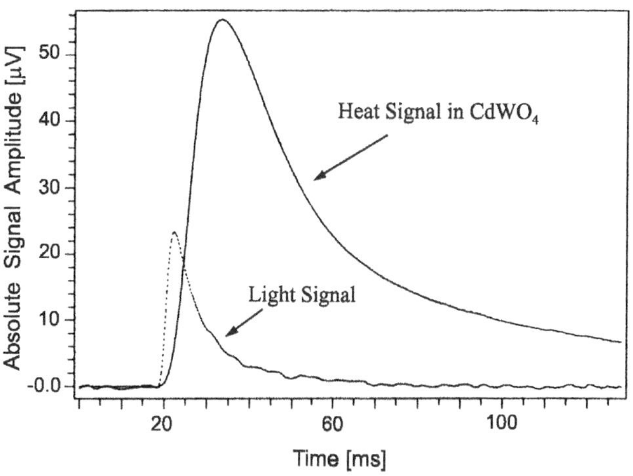

FIGURE 9.121 Signal due to the energy released (2615 keV) by a γ-ray in a $CdWO_4$ crystal scintillator (solid line) and light detected simultaneously by the Ge light detector faced toward the $CdWO_4$ crystal (dashed line). *From Gironi et al. (2009); reprinted with permission from Elsevier © 2009.*

B. Search for neutrinoless double beta ($0\nu\beta\beta$) decay

The theory of neutrinoless double beta ($0\nu\beta\beta$) decay is discussed in detail in Chapter 1, Volume 1 and techniques used in the search for $0\nu\beta\beta$ decay are provided in Chapters 4 and 6 of Volume 1, and previously in Section XIII of this chapter. Scintillating bolometers promise to be an important tool in the detection of $0\nu\beta\beta$ decay.

Double beta ($\beta\beta$) decay with the simultaneous emission of two neutrinos, abbreviated $2\nu\beta\beta$ decay, exhibits a continuous beta spectrum as illustrated in Fig. 1.59 of Chapter 1, Volume 1 and Figs. 9.114, 9.115 and 9.117 of this chapter. The continuous $2\nu\beta\beta$ decay has been demonstrated in liquid scintillation measurements to present a significant background interference in the observation of the peak, which would be due to $0\nu\beta\beta$ decay (see Fig. 6.112 of Chapter 6, Volume 1). However, Gironi (2010) underscores that with the excellent resolutions provided by scintillating bolometers the continuous spectrum of $2\nu\beta\beta$ would seldom present a significant source of background. Research efforts are underway in several countries to develop the scintillating bolometer for the ongoing worldwide search for $0\nu\beta\beta$ decay (Arnaboldi et al., 2010, 2011a,b; Barinova et al., 2010; Gironi, 2010; Gironi et al., 2009; Pirro et al., 2006).

Artusa et al. (2014a,b) provide a detailed description of the varying light and heat yields of particles in scintillating bolometers, which forms the basis for particle discrimination; and the following is a brief explanation based on their account: The ratio of light and heat signals from the scintillating bolometer is a function of particle mass and charge. The thermal response of a bolometer will not vary to a high degree with particle type (~7% with TeO crystals and ~10%–20% for scintillating crystals); however, the light yield from the scintillator (i.e., fraction of particle energy emitted as photons of light) will differ significantly with ionization energy density. Thus, the light emission produced by β-particles and γ-photons will be similar; whereas, the light emission induced by α-particles and neutrons will differ from that produced by the β- or γ-particles. Thus, in the search for $0\nu\beta\beta$ with scintillating crystals containing a double beta decaying nuclide, the energy deposited by the two β-particles from the double beta decay can be identified from an α signal; and only beta-particles and gamma photons contribute to background events.

The varying scintillation yields of particle-events of different types, namely neutron recoil events, α events, and β/γ events in the scintillating bolometer crystal absorber permits the discrimination of these particle events, and consequently a means of significantly reducing background. The β/γ events are classified together, because γ-rays interact with scintillator by the production of Compton electrons, which interact with scintillator similar to β-particles. Obviously then γ-ray and β-particle interactions are indistinguishable; and any γ-ray interference as background must be eliminated in the search for $0\nu\beta\beta$ decay. Background reduction or elimination is carried out mainly by means of the discrimination of particles according to their different scintillation yields or light outputs. The pulse amplitude of the Ge optical bolometer (Light Signal) measured in coincidence with the pulse amplitude of the scintillator bolometer (Heat Signal) are plotted against each other (Light Pulse Amplitude vs. Heat Pulse Amplitude) to provide a scatter plot whereby α events and nuclear recoil events can be discriminated against β/γ events, as illustrated in Fig. 9.122. Further discrimination of background α events from β/γ events is achievable through PSD of the heat signals (Arnaboldi et al., 2011a,b).

One of the major sources of interfering background in the search for $0\nu\beta\beta$ decay is γ-rays because events due to γ-rays are indistinguishable from β-particle events. However, Pirro et al. (2006) and Arnaboldi et al. (2011a,b) point out that the highest interfering γ line is the 2615 keV line of ^{208}Tl with the exception of the extremely rare high-energy γ-rays from ^{214}Bi; and that the problem of interfering γ-ray backgrounds would be avoided by confining the search for $0\nu\beta\beta$ decay to nuclides with $\beta\beta$ decay energies ($Q_{\beta\beta}$) above 2615 keV. Nuclides with double beta decay energies above the 2615 keV line are listed in Table 9.11.

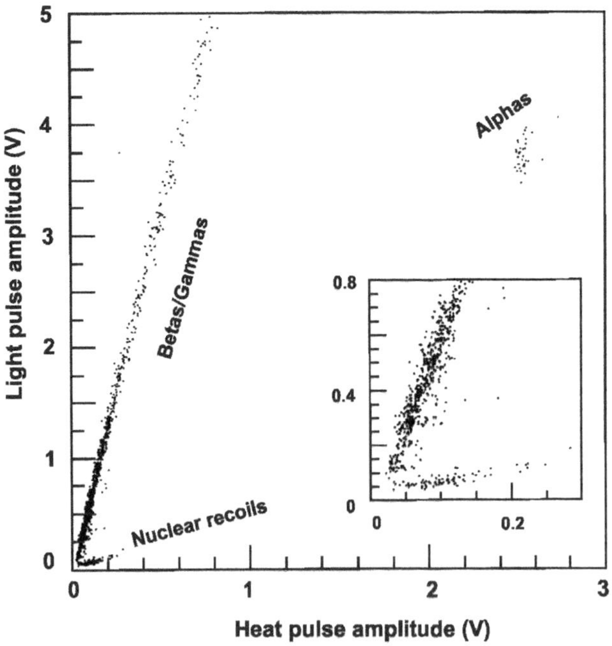

FIGURE 9.122 Light versus heat signals plot obtained at Orsay under ^{252}Cf neutron source irradiation. The inset shows a detail of the low-energy region. *From Cebrián et al. (2003); reprinted with permission from Elsevier © 2003.*

TABLE 9.11 Double beta decay isotopes with endpoint energies above the ^{208}Tl line.

Isotope	$Q_{\beta\beta}$ (MeV)	Natural abundance (%)
^{116}Cd	2.80	7.5
^{82}Se	3.00	9.2
^{100}Mo	3.03	9.6
^{96}Zr	3.35	2.8
^{150}Nd	3.37	5.6
^{48}Ca	4.27	0.19

From Arnaboldi et al. (2011b); reprinted with permission from Elsevier © 2011.

Pirro et al. (2006) demonstrate that by confining the search for $0\nu\beta\beta$ decay to nuclides that have decay energies above the 2615 keV γ-line of ^{208}Tl will altogether eliminate background events. They demonstrate that, in a 417 hours live-time background measurement with a $CdWO_4$ bolometer, the background above 2615 keV drops to zero using the optical detector to discriminate the background α events. The discrimination of α- and γ-events is illustrated by the scatter plot of Fig. 9.123; and an arrow in Fig. 9.123 points out the location in the scatter plot where the search for $0\nu\beta\beta$ decay events (abbreviated 0v−DBD in Fig. 9.123) of ^{116}Cd should concentrate ($Q_{\beta\beta} = 2805$ keV). This is also in the region where no background events from either α- or γ-events are observed. The small size of the Ge optical detector ($\leq$1 mm thickness) does not permit the detector calibration with external sources; and thus, a ^{55}Fe calibration X-ray source can be placed on the back side of the optical detector, as illustrated previously in Fig. 9.120 (Gironi et al., 2009; Ortigoza et al., 2011).

Crystal and scintillating bolometers are currently under development in the search for $0\nu\beta\beta$ decay with the isotopes ^{82}Se, ^{100}Mo, ^{116}Cd, and ^{130}Te in the following crystal chemical formula: **$Zn^{82}Se$** (Dafinei et al., 2017; Artusa et al., 2016; Gironi, 2016; Orio, 2016; Beeman et al., 2015; Dafinei, 2014; Tenconi et al., 2014), **$Zn^{100}MoO_4$** (Armengaud et al., 2017; Poda, 2016; Chernyak et al., 2015, 2013; Tenconi, 2015; Barabash et al., 2014; Shlegel et al., 2014; Beeman et al., 2012a, 2012b), **$Li_2^{100}MoO_4$** (Buşe et al., 2018; Chen et al., 2018; Armengaud et al., 2017; Velázquez et al., 2017; Bekker et al., 2016), **$^{116}CdWO_4$** (Barabash et al., 2016; Arnaboldi et al., 2010; Gironi et al., 2009), and **$^{130}TeO_2$** (Casali, 2017; Artusa et al., 2015a,b; 2014a,b; Gironi, 2013; Beeman et al., 2012c).

Artusa et al. (2014a) reviewed the strategy and potential for future scintillating bolometers in light of the properties of very promising scintillating crystals containing ^{82}Se, ^{116}Cd, ^{100}Mo, and ^{130}Te $\beta\beta$ candidate isotopes. For comparative purposes, Table 9.12 lists the total number of the $\beta\beta$ isotopes in towers of individual bolameters comprising 988 crystals of a given dimension. The rarity of a $0\nu\beta\beta$ event requires a detector with large numbers of $\beta\beta$ candidate isotope and thus a large number of scintillating bolometers to provide an adequate mass of source material to permit the detection and measurement of the $0\nu\beta\beta$ decay. Thus numerous (up to ~ 1000) single bolometers must be coupled into towers that are enclosed within the cryostat and radiation shielded barriers of the detector. Artusa et al. (2015a,b, 2014a,b) describe the CUORE (Cryogenic Underground Observatory for Rare Events) towers that will encompass 988 bolometers containing either $^{130}TeO_2$ or

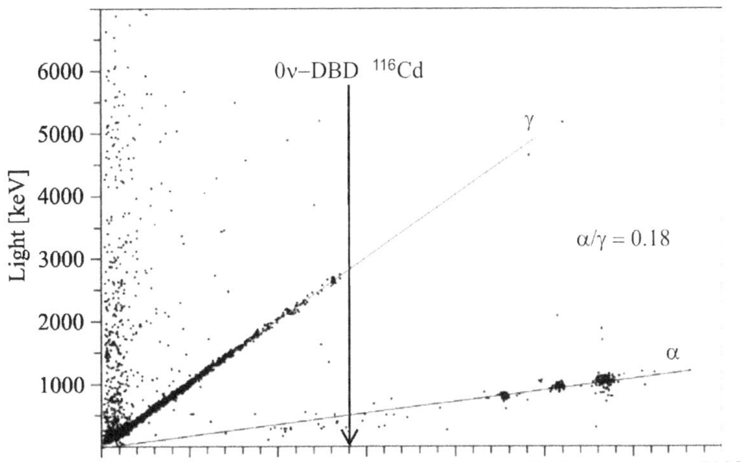

FIGURE 9.123 Scatter plot of light versus heat obtained in a 417 hours live-time background measurement. The energy scale of the light detector is normalized to the one of the $CdWO_4$. The points in the 0−300 keV (heat) region are due to pileup. The α-induced background is clearly discriminated. *From Pirro et al. (2006); reprinted with permission from Elsevier © 2006.*

TABLE 9.12 Characteristics of different $\beta\beta$ candidates in the search for $0\nu\beta\beta$ with scintillation crystal bolameters. Listed are the $\beta\beta$ candidate isotope, crystal containing the isotope, the total mass of 988 crystal arrays in individual bolameters with crystals measuring 5 × 5 × 5 cm^3, the total number of $\beta\beta$ candidates ($N_{\beta\beta}$), the number of decays in 5 years ($N_{0\nu\beta\beta}$) for the most and least favorable values of $F_N^{0\nu}$, the nuclear Figure of merit[b], for $|m_{ee}|$ = 50 meV and $|m_{ee}|$ = 10 meV with a 90% isotope enrichment in the $\beta\beta$ emitting isotope.[a]

Isotope	Crystal	Mass (kg)	$N_{\beta\beta}$	$N_{0\nu\beta\beta}^{50meV}$ (cnts)	$N_{0\nu\beta\beta}^{10meV}$ (cnts)
^{82}Se	ZnSe	664	2.4 × 10^{27}	10–85	0.4–3.4
^{116}Cd	$CdWO_4$	985	1.5 × 10^{27}	13–44	0.5–1.8
^{100}Mo	$ZnMoO_4$	740	1.3 × 10^{27}	12–99	0.5–4
^{130}Te	TeO_2	751	2.4 × 10^{27}	13–89	0.5–3.6

[a]*From Artusa et al. (2014a), Eur. Phys. J. C, 74 10 (2014) 3096, DOI: https://doi.org/10.1140/epjc/s10052-014-3096-8.*

[b]*The nuclear figure of merit is a variable in the following relation:* $1/T_{1/2}^{0\nu} = \left(|m_{ee}|^2/m_e^2\right)F_N^{0\nu}$ *where* $T_{1/2}^{0\nu}$ *is the 0νββ decay half-life,* $|m_{ee}|$ *is the effective Majorana neutrino mass, and* m_e *is the electron mass.*

Zn^{82}Se crystals. The CUORE detector has 19 towers with each tower containing 13 floors with 4 bolometer modules on each floor. An example of a single bolometer containing a 0.51 kg $^{116}CdWO_4$ scintillating crystal is illustrated in Fig. 9.124; that of a tower of 30 single scintillating bolometers each with its individual crystal and light detector is illustrated in Fig. 9.125. Gironi (2016) reports that the tower illustrated in Fig. 9.125 for the LUCIFER (Low-background Underground Cryogenic Installation For Elusive Rates) project would contain 30 bolometers each with 430 g crystals of Zn^{82}Se at a 96.5% isotope enrichment providing ~5 × 10^{25} ^{82}Se nuclei.

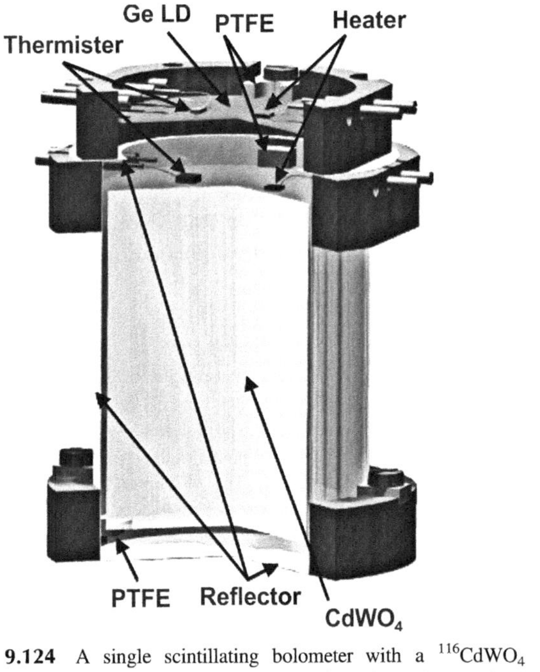

FIGURE 9.124 A single scintillating bolometer with a $^{116}CdWO_4$ crystal. It is held together with four L-shaped Teflon pieces (PTFE) fixed to two cylindrical Cu frames. The frames are held together with two Cu columns (not visible from the chosen perspective). The crystal is surrounded by a 43 mm reflecting sheet. The light detector (Ge LD) consists of a 1 mm thick pure Ge crystal absorber. The surface of the Ge LD facing the crystal is darkened with a 600 Å layer of SiO_2 to increase light absorption. A resister of ~300 kΩ is attached to each absorber and acts as a heater to stabilize the gain of the bolometer. *From Arnaboldi et al. (2010); reprinted with permission from Elsevier © 2010.*

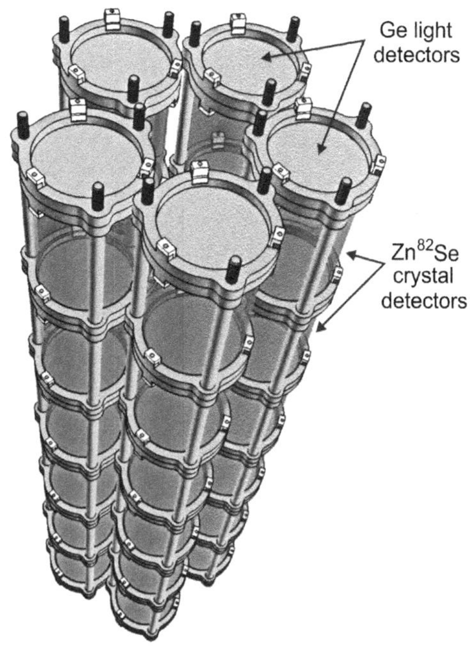

FIGURE 9.125 Five towers of Zn^{82}Se crystal bolometers for the LUCIFER project in the search for $0\nu\beta\beta$ decay. Each tower contains six floors of scintillating bolometers. Each bolometer contains a single Zn^{82}Se crystal and its individual Ge light detector. The 30 crystals each weighing 0.43 kg of Zn^{82}Se at a ^{82}Se isotope enrichment to 96.5% yielding a total of ~5 × 10^{25} $\beta\beta$ ^{82}Se nuclei. *From Gironi (2016); reprinted with permission from Elsevier © 2016.*

C. Search for weakly interacting massive particles

There is a concerted effort worldwide to determine the nature of dark matter in the universe of which a component is considered to be a neutral WIMP. Galactic halo WIMPs could be detected by their weak interaction with nuclei in scintillating bolometers with the production of low-energy recoils. The nuclear recoil (NR) energies from WIMP interactions would be of the order of tens of keV at low event rates, i.e., below tenths of events per kilogram per day (Cebrián et al., 2003). The ability to detect the NR events would depend on the elimination of interfering background due to β/γ events. Numerous laboratories are researching the use of scintillating bolometers as a promising tool for the detection of the NR events from WIMP interactions due to the excellent resolutions and discrimination power against β/γ events (Angloher et al., 2016a,b, 2014, 2005; Eitel, 2015; Armengaud, 2012, 2013; Beeman et al., 2013; Coron et al., 2013, 2004, 2008, 2009; Ortigoza et al., 2013, 2011; Schmidt et al., 2013; Arnaboldi et al., 2011a,b; Cebrián et al., 2003, 2005). Among these research efforts are (i) The Rare Objects Search with Bolometers UndergrounD (ROSEBUD) Collaboration, which is a joint effort between the Institut d'Astrophysique Spatiale (IAS, Orsay, France) and the University of Zaragoza (UZ, Zaragoza, Spain). This project is underway in an ultralow-level radioactivity environment in the Canfranc Underground Laboratory (2450 m.w.e) in northeastern Spain, (ii) the Cryogenic Rare Event Search with Superconducting Thermometers (CRESST) Collaboration including universities in Germany, the United Kingdom and Italy at the Gran Sasso underground laboratory in Italy, (iii) the Expérience pour Detector Les WIMPs En Site Souterrain (EDELWEISS) project at the Modane Underground Laboratory in France, and (iv) The European Underground Rare Event Calorimeter Array (EURECA), which will combine researchers on the CRESST and EDELWEISS experiments at the Modane Underground Laboratory.

The discrimination of β/γ events from NR events with the scintillating bolometer is based on the higher light output for a given energy that a β/γ event produces compared to that of a heavy charged particle, such as a NR produced by neutrons or WIMPs. The evaluation of the ability of a scintillating bolometer to discriminate nuclear recoils from β/γ events is based on the "quenching factor" of the event. The "quenching factor" is defined as the reduction factor for the light output of a NR event relative to an electron/photon event of the same energy (Angloher et al., 2005). Quenching factors in the range of 10 to 8.2 were measured by Angloher et al. (2005) for recoil energies in the energy range of 10−150 keV with a bolometer equipped with a $CaWO_4$ crystal detector. A quenching factor of 10 was measured by Cebrián et al. (2003, 2005) for nuclear recoils. The ratio of 1:10 for the weak light output from NR events to the stronger light output from β/γ events offers the foundation for the discrimination of β/γ background events from NR events.

By exposing a 50 g undoped sapphire (Al_2O_3) crystal in a scintillating bolometer equipped with a Ge optical bolometer (i.e., light detector) to sources of ^{241}Am, ^{57}Co, and ^{252}Cf, Ortigoza et al. (2013) demonstrated the excellent discrimination among NR events from the ^{252}Cf source and the electron/gamma events in the 100 keV energy range of interest, which would be the energy range for the observation of WIMP recoil events. An example is the scatter plot of the light pulse amplitude measured by the Ge bolometer light detector against the energy in keV deposited as heat in the scintillating bolometer illustrated in Fig. 9.126. The energy scale is calibrated with the γ-lines of ^{241}Am and ^{57}Co sources. Dark matter search experiments with scintillating crystal bolometers, as noted in previous sections, utilize the differences in light intensity produced by β/γ events in the scintillating bolometer and that produced by NR events. Ortigoza et al. (2013) measured the ratios of the scintillating signals (i.e., light signals or light intensities) produced by β/γ events to those produced by NR events, expressed as β/γ:NR ratios, in BGO and Al_2O_3 scintillating bolometers. β/γ:NR ratios in the energy region between 0—100 keV were of the order of magnitude of 8 and 17 for the BGO and Al_2O_3 scintillating bolometers, respectively. The magnitude of these ratios provide good discrimination from β/γ background events from WIMP NR events in the low-energy region where WIMP recoil events would be detected.

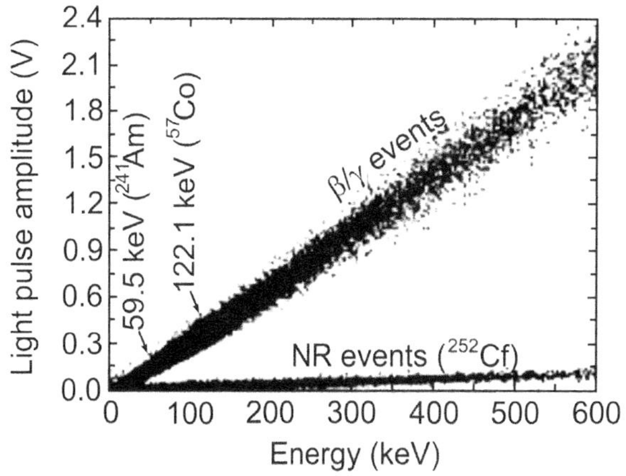

FIGURE 9.126 Light verses heat scatter plot obtained with a Al_2O_3 scintillating bolometer when irradiated with neutrons from a ^{252}Cf source and photons from ^{241}Am and ^{57}Co. The heat channel was calibrated in energy with the gamma lines of the ^{241}Am and ^{57}Co. *From Ortigoza et al. (2013); reprinted with permission from Elsevier © 2013.*

References

Aas, V., Hessvik, N.P., Wettergreen, M., Hvammen, A.W., Hallén, S., Thoresen, G.H., Rustan, A.C., 2011. Chronic hyperglycemia reduces substrate oxidation and impairs metabolic switching of human myotubes. Biochim. Biophys. Acta 1812, 94–105.

Abbady, A., Abbady, A.G.E., Michel, R., 2004. Indoor radon measurement with the Lucas cell technique. Appl. Radiat. Isot. 61, 1469–1475.

Abbas, M.I., 2001a. HPGe detector photopeak efficiency calculation including self-absorption and coincidence corrections for Marinelli beaker sources using compact analytical expressions. Appl. Radiat. Isot. 54, 761–768.

Abbas, M.I., 2001b. Analytical formulae for well-type NaI(Tl) and HPGe detectors efficiency computation. *Appl. Radiat. Isot.* 55, 245–252.

Abbas, M.I., 2006. Validation of analytical formulae for the efficiency calibration of gamma detectors used in laboratory and in-situ measurements. Appl. Radiat. Isot. 64, 1661–1664.

Abbas, M.I., 2007. Direct mathematical method for calculating full-energy peak efficiency and coincidence corrections of HPGe detectors for extended sources. Nucl. Instrum. Methods Phys. Res., Sect. B 256, 554–557.

Abbas, M.I., 2010. A new analytical method to calibrate cylindrical phoswich and LaBr3(Ce) scintillation detectors. Nucl. Instrum. Methods Phys. Res., Sect. A 621, 413–418.

Abbas, M.I., Noureddeen, S., 2011. Analytical expression to calculate total and full-energy peak efficiencies for cylindrical phoswich and lanthanum bromide scintillation detectors. Radiat. Meas. 46, 440–445.

Ackermann, U., Eschbaumer, S., Bergmaier, A., et al., 2016. Position and time resolution measurements with a microchannel plate image intensifier: a comparison of monolithic and pixelated $CeBr_3$ scintillators. Nucl. Instrum. Methods Phys. Res., Sect. A 823, 56–64.

Agosteo, S., et al., 2012. Characterization of extended range Bonner Sphere Spectrometers in the CERF high-energy broad neutron field at CERN. Nucl. Instrum. Methods Phys. Res., Sect. A 694, 55–68.

Aguilar, O.A., et al., 2017. A viral immunoevasin controls innate immunity by targeting the prototypical natural killer cell receptor family. Cell 169, 58–71.

Ahmed, S.N., 2015. Chapter 6. Scintillation detectors and photodetectors, pp. 331–433. In: Physics and Engineering in Radiation Detection, second ed. Elsevier, p. 784.

Alcami, A., Viejo-Borbolla, A., 2009. Chapter 8. Identification and characterization of virus-encoded chemokine binding proteins, pp. 173–191. In: Abelson, J.N., Simon, M.I. (Eds.), Methods in Enzymology, vol. 460. Elsevier, Amsterdam.

Alessandrello, A., et al., 1992. Development of a thermal scintillating detector for double beta decay of 48Ca. Nucl. Phys. B 28A, 233–235.

Alessandrello, A., et al., 1998. A scintillating bolometer for experiments on double beta decay. Phys. Lett. B 420, 109–113.

Aléssio, A.C.M., Santos, C.X.C., Debbas, V., et al., 2011. Evaluation of mild hyperhomocysteinemia during the development of atherosclerosis in apolipoprotein E-deficient and normal mice. Exp. Mol. Pathol. 90, 45–50.

Alexandrov, B.S., Ianakiev, K.D., Littlewood, P.B., 2008. Branching transport model of NaI(Tl) alkali-halide scintillator. Nucl. Instrum. Methods Phys. Res., Sect. A 586, 432–438.

Alharbi, T., 2016. Principal Component Analysis for pulse-shape discrimination of scintillation radiation detectors. Nucl. Instrum. Methods Phys. Res., Sect. A 806, 240–243.

Al-Hilal, M., Al-Ali, A., 2010. The role of soil gas radon survey in exploring unknown subsurface faults at Afamia B dam, Syria. Radiat. Meas. 45, 219–224.

Allier, C.P., van Loef, E.V.D., Dorenbos, P., Hollander, R.W., van Eijk, C.W.E., Krämer, K.W., Güdel, H.U., 2002. Readout of a $LaCl_3(Ce^{3+})$ scintillation crystal with a large area avalanche photodiode. Nucl. Instrum. Methods Phys. Res., Sect. A 485, 547–550.

Al-Masri, M.S., Shwiekani, R., 2008. Radon gas distribution in natural gas processing facilities and workplace air environment. J. Environ. Radioact. 99, 574–580.

Amaré, J., Cebrián, S., Cuesta, C., 2014. Preliminary results of ANAIS-25. Nucl. Instrum. Methods Phys. Res., Sect. A 742, 187–190.

Anderhub, H., Backes, M., Biland, A., Boller, A., Braun, I., Bretz, T., Commichau, S., Commichau, V., Dorner, D., Gendotti, A., et al., 2011a. A G-APD based camera for imaging atmospheric Cherenkov telescopes. Nucl. Instrum. Methods Phys. Res., Sect. A 628, 107–110.

Anderhub, H., Backes, M., Biland, A., Boller, A., Braun, I., Bretz, T., Commichau, S., Commichau, V., Dorner, D., Gendotti, A., et al., 2011b. Results of the prototype camera for FACT. Nucl. Instrum. Methods Phys. Res., Sect. A 639, 55–57.

Anderhub, H., Backes, M., Biland, A., Boller, A., Braun, I., Bretz, T., Commichau, S., Commichau, V., Domke, M., Dorner, D., et al., 2011c. FACT—the first Cherenkov telescope using a G-APD camera for TeV gamma-ray astronomy. Nucl. Instrum. Methods Phys. Res., Sect. A 639, 58–61.

Ando, M., et al., 2017. Potent therapeutic activity against peritoneal dissemination and malignant ascites by the novel anti-folate receptor alpha antibody KHK2805. Trans. Oncol. 10 (5), 707–718.

Angloher, G., Bucci, C., Christ, P., Cozzini, C., von Feilitzsch, F., Hauff, D., Henry, S., Jagemann, T., Jochum, J., Kraus, H., et al., 2005. Limits on WIMP dark matter using scintillating CaWO4 cryogenic detectors with active background suppression. Astropart. Phys. 23, 325–339.

The EURECA Collaboration Angloher, G., et al., 2014. EURECA conceptual design report. Phys. Dark Univ. 3, 41–74.

Angloher, G., et al., 2016a. Results on light dark matter particles with a low-threshold CRESST-II detector. Eur. Phys. J. C 76, 25.

Angloher, G., Dafinei, I., Gektin, A., et al., 2016b. A CsI low-temperature detector for dark matter search. Astropart. Phys. 84, 70–77.

Anil Kumar, G., Mazumdar, I., Gothe, D.A., 2009a. Efficiency calibration and simulation of a LaBr3(Ce) detector in close-geometry. Nucl. Instrum. Methods Phys. Res., Sect. A 609, 183–186.

Anil Kumar, G., Mazumdar, I., Gothe, D.A., 2009b. Experimental measurements and GEANT4 simulations for a comparative study of efficiencies of LaBr3:Ce, NaI(Tl), and BaF2. Nucl. Instrum. Methods Phys. Res., Sect. A 610, 522–529.

Annenkov, A.N., Buzanov, O.A., Danevich, F.A., Georgadze, A.S., Kim, S.K., Kim, H.J., Kim, Y.D., Kobychev, V.V., Kornoukhov, V.,N., Korzhik, M., et al., 2008. Development of CaMoO4 crystal scintillators for a double beta decay experiment with 100Mo. Nucl. Instrum. Methods Phys. Res., Sect. A 584, 334–345.

Anton, G., Gebert, U., Michel, T., Rügheimer, T.K., 2009. A hybrid photodetector using the Timepix semiconductor assembly for photoelectron detection. Nucl. Instrum. Methods Phys. Res., Sect. A 602, 205–208.

Antonini, P., Bressi, G., Carugno, G., Iannuzzi, D., 2001. Scintillation properties of YAG:Yb crystals. Nucl. Instrum. Methods Phys. Res., Sect. A 460, 469–471.

Arimoto, M., Harita, S., Sugita, S., et al., 2018. Development of a 32-channel ASIC for an X-ray APD detector onboard the ISS. Nucl. Instrum. Methods Phys. Res., Sect. A 882, 138–147.

Armengaud, E., et al., The EDELWEISS Collaboration, 2013. Background studies for the EDELWEISS dark matter experiment. Astropart. Phys. 47, 1–9.

Armengaud, E., Augier, C., Barabash, A.S., et al., 2017. Development of ^{100}Mo-containing scintillating bolometers for a high-sensitivity neutrinoless double-beta decay search. Eur. Phys. J. C 77, 785.

Armengaud, E., 2012. Direct detection of WIMPs. Comptes Rendus Phys. 13 (6–7), 730–739.

Arnaboldi, C., Beeman, J.W., Cremonesi, O., Gironi, L., Pavan, M., Pessina, G., Pirro, S., Previtali, E., 2010. CdWO4 scintillating bolometer for Double Beta Decay: light and heat anticorrelation, light yield and quenching factors. Astropart. Phys. 34, 143–150.

Arnaboldi, C., Capelli, S., Cremonesi, O., Gironi, L., Pavan, M., Pessina, G., Pirro, S., 2011a. Characterization of ZnSe scintillating bolometers for double beta decay. Astropart. Phys. 34, 344–353.

Arnaboldi, C., Brofferio, C., Cremonesi, O., Gironi, L., Pavan, M., Pessina, G., Pirro, S., Previtali, E., 2011b. A novel technique of particle identification with bolometric detectors. Astropart. Phys. 34, 797–804.

Arsaev, M.I., Kladov, A.V., 2007. A phoswich detector for β-ray spectrometry. Instrum. Exp. Tech. 50 (1), 104–107.

Artusa, D.R., et al., 2014a. Exploring the neutrinoless double beta decay in the inverted neutrino hierarchy with bolometric detectors. Eur. Phys. J. C 73, 3096.

Artusa, D.R., et al., 2014b. CUORE Collaboration, Initial performance of the CUORE-0 experiment. Eur. Phys. J. C 74 (8), 2956.

Artusa, D.R., et al., 2015a. CUORE and beyond: bolometric techniques to explore inverted neutrino mass hierarchy. Phys. Proc. 61, 241–250.

Artusa, D.R., et al., 2015b. Searching for neutrinoless double-beta decay of ^{130}Te with CUORE. Adv. High Energy Phys. 2015 (2015), 879871.

Artusa, D.R., et al., 2016. First array of enriched $Zn^{82}Se$ bolometers to search for double beta decay. Eur. Phys. J. C 76, 364.

Asfahani, J., 1999. Optimization of spectrometric gamma-gamma probe configuration using very low radioactivity sources for lead and zinc grade determination in borehole logging. Appl. Radiat. Isot. 51, 449–459.

Ashrafi, S., Anvarian, S., Sobhanian, S., 2006. Monte Carlo modeling of a NaI(Tl) scintillator. J. Radioanal. Nucl. Chem. 269 (1), 95–98.

Auffray, E., Korjik, M., Lucchini, M.T., 2016. Free carrier absorption in self-activated $PbWO_4$ and Ce-doped $Y_3(Al_{0.25}Ga_{0.75})_3O_{12}$ and $Gd_3Al_2Ga_3O_{12}$ garnet scintillators. Opt. Mater. 58, 461–465.

Ayaz-Maierhafer, B., DeVol, T.A., 2007. Determination of absolute detection efficiencies for detectors of interest in homeland security. Nucl. Instrum. Methods Phys. Res., Sect. A 579, 410–413.

Badalà, A., Blanco, F., La Rocca, P., Librizzi, F., Pappalardo, G.S., Pulvirenti, A., Riggi, F., Vernet, R., Awes, T.C., Muller, H., Tupikin, N., 2008. Characterization of avalanche photodiodes (APDs) for the electromagnetic calorimeter in the ALICE experiment. Nucl. Instrum. Methods Phys. Res., Sect. A 596, 122–125.

Balamurugan, N., Arulchakkaravarthi, A., Selvakumar, S., Lenin, M., Rakesh, K., Muralithar, S., Sivaji, K., Ramasamy, P., 2006. Growth and characterization of undoped and thallium doped cesium iodide single crystals. J. Cryst. Growth 286, 294–299.

Balcerzyk, M., Klamra, W., Moszyński, M., Kapusta, M., Szawlowski, M., 2002. Energy resolution and light yield non-proportionality of ZnSe:Te scintillator studied by large area avalanche photodiodes and photomultipliers. Nucl. Instrum. Methods Phys. Res., Sect. A 482, 720–727.

Balmer, M.J.I., Gamage, K.A.A., Taylor, G.C., 2015. Comparative analysis of pulse shape discrimination methods in a ^{6}Li loaded plastic scintillator. Nucl. Instrum. Methods Phys. Res., Sect. A 788, 146–153.

Balossino, I., Barion, L., Contalbrigo, M., et al., 2017. Cherenkov light imaging tests with state-of-the-art solid state photon counter for the CLAS12 RICH detector. Nucl. Instrum. Methods Phys. Res., Sect. A 876, 89–92.

Ban, G., Bodek, K., Lefort, T., Naviliat-Cuncic, O., Pierre, E., Plonka, C., Rogel, G., 2009. UCN detection with 6Li-doped glass scintillators. Nucl. Instrum. Methods Phys. Res., Sect. A 611, 280–283.

Barabash, A.S., 2015. Double beta decay experiments: present status and prospects for the future. Phys. Proc. 74, 416–422.

Barabash, A.S., Chernyak, D.M., Danevich, F.A., et al., 2014. Enriched $Zn^{100}MoO_4$ scintillating bolometers to search for $0\nu2\beta$ decay of ^{100}Mo with the LUMINEU experiment. Eur. Phys. J. C 74, 3133.

Barabash, A.S., Belli, P., Bernabei, R., et al., 2016. Improvement of radiopurity level of enriched $^{116}CdWO_4$ and $ZnWO_4$ crystal scintillators by recrystallization. Nucl. Instrum. Methods Phys. Res., Sect. A 833, 77–81.

Barbagallo, M., Cosentino, L., Greco, G., Guardo, G., Montereali, R.M., Pappalardo, A., Scirè, C., Scirè, S., Vincenti, M.A., Finocchiaro, P., 2011. A thermal neutron mini-detector with SiPM and scintillating fibers. Nucl. Instrum. Methods Phys. Res., Sect. A 652, 355–358.

Barbarino, G., de Asmundis, R., De Rosa, G., Fiorillo, G., Russo, S., 2009. A new design for an high gain vacuum photomultiplier: the silicon PMT used as amplification stage. Nucl. Phys. B Proc. Suppl. 197 (1), 52–56.

Barbarino, G., de Asmundis, R., De Rosa, G., Mollo, C.M., Russo, S., Vivolo, D., 2012. High gain hybrid photomultipliers based on solid state p-n junctions in Geiger mode and their use in astroparticle physics. Phys. Proc. 37, 703–708.

Barbarino, G., Barbato, F.C.T., Campajola, L., de Asmundis, R., De Rosa, G., Mollo, C.M., Vivolo, D., 2013a. Vacuum silicon photomultipliers: recent developments. Nucl. Instrum. Methods Phys. Res., Sect. A 718, 582–583.

Barbarino, G., de Asmundis, R., De Rosa, G., Mollo, C.M., Vivolo, D., 2013b. VSiPMT for underwater neutrino telescopes. Nucl. Instrum. Methods Phys. Res., Sect. A 725, 162–165.

Barbarino, G., Barbato, F.C.T., Mollo, C.M., Nocerino, E., Vivolo, D., Fukasawa, A., 2018. Another step towards photodetector innovation: the first 1-inch industrial VSiPMT. Astropart. Phys. 101, 27–35.

Barbato, F.C.T., Barbarino, G., Campajola, L., Di Capua, F., Mollo, C.M., Valentini, A., Vivolo, D., 2017. R & D of a pioneering system for a high resolution photodetector: the VSiPMT. Nucl. Instrum. Methods Phys. Res., Sect. A 876, 48–49.

Bardelli, L., Bini, M., Bizzeti, P.G., Carraresi, L., Danevich, F.A., Fazzini, T.F., Grinyov, B.V., Ivannikova, N.V., Kobychev, V.V., Kropivyansky, B.N., 2006. Further study of CdWO4 crystal scintillators as detectors for high sensitivity 2β experiments: scintillation properties and pulse-shape discrimination. Nucl. Instrum. Methods Phys. Res., Sect. A 569, 743−753.

Barlas, E., Bayrak, A., Emirhan, E., Haciomeroglu, S., Ozben, C.S., 2013. Spatial distribution of ^{214}Po ions in the electrostatic collection. Appl. Radiat. Isot. 80, 23−26.

Bartosiewicz, K., Babin, V., Kamada, K., Yoshikawa, A., Nikl, M., 2015. Energy migration processes in undoped and Ce-doped multicomponent garnet single crystal scintillators. J. Lumin. 166, 117−122.

Baré, J., Tondeur, F., 2011. Gamma spectrum unfolding for a NaI monitor of radioactivity in aquatic systems: experimental evaluations of the minimal detectable activity. Appl. Radiat. Isot. 69, 1121−1124.

Baré, J., Evrard, M., Mertens, C., De Lellis, C., Tondeur, F., 2010. Gamma spectrum unfolding for a NaI monitor of radioactivity in aquatic systems. Appl. Radiat. Isot. 68, 836−838.

Barinova, O.P., Danevich, F.A., Degoda, V.Y., Kirsanova, S.V., Kudovbenko, V.M., Pirro, S., Tretyak, V.I., 2010. First test of Li2MoO4 crystal as a cryogenic scintillating bolometer. Nucl. Instrum. Methods Phys. Res., Sect. A 613, 54−57.

Barnyakov, A.Y., Barnyakov, M.Y., Beloborodov, K.I., Bobrovnikov, V.S., Buzykaev, A.R., Danilyuk, A.F., Golubev, V.B., Kirillov, V.L., Kononov, S.A., Kravchenko, E.A., Onuchin, A.P., Martin, K.A., Serednyakov, S.I., Vesenev, V.M., 2010. Particle identification aerogel counter with $n = 1.13$ for π/K separation. Nucl. Instrum. Methods Phys. Res., Sect. A 623, 336−338.

Barnyakov, M.Y., Frach, T., Kononov, S.A., Kuyanov, I.A., Prisekin, V.G., 2016. Radiation hardness test of the philips digital photon counter with proton beam. Nucl. Instrum. Methods Phys. Res., Sect. A 824, 83−84.

Barquero, R., Mendez, R., Vega-Carrillo, H.R., Iñiguez, M.P., Edwards, T.M., 2005. Neutron spectrum and dosimetric features around a 18 MV linac accelerator. Health Phys. 88, 48−58.

Barros Jr., A., Ly, V.T., Chando, T.J., Ruan, Q., Donenfeld, S.L., Holub, D.P., Christopher, L.J., 2011. Development and evaluation of a multiple-plate fraction collector for sample processing: application to radioprofiling in drug metabolism studies. J. Pharm. Biomed. Anal. 54, 979−986.

Baturin, V., Burkert, V., Kim, W., Majewsky, S., Park, K., Popov, V., Smith, E.S., Son, D., Stepanyan, S.S., Zorn, C., 2006. Nucl. Instrum. Methods Phys. Res., Sect. A 562, 327−337.

Bayat, E., Afarideh, H., Abbasi Davani, F., Ghal−Eh, N., 2018. Noise reduction of PGNAA spectrum using empirical mode decomposition technique. Radiat. Phys. Chem. 149, 49−53.

Becchetti, F.D., Torres-Isea, R.O., Di Fulvio, A., et al., 2018. Deuterated stilbene (stilbene-d_{12}): an improved detector for fast neutrons. Nucl. Instrum. Methods Phys. Res., Sect. A (in press).

Becker, E.M., Farsoni, A.T., 2014. Wireless, low-cost, FPGA-based miniature gamma ray spectrometer. Nucl. Instrum. Methods Phys. Res., Sect. A 761, 99−104.

Becquerel, H., 1903. On radioactivity, a new property of matter. In: Nobel Lectures, Physics 1901−1921. Elsevier Publishing Company, Amsterdam (1967).

Bednarek, M.A., MacNeil, T., Tang, R., Fong, T.M., Cabello, M.A., Maroto, M., Teran, A., 2008. Cyclic analogs of α-melanocyte-stimulating hormone (αMSH) with high agonist potency and selectivity at human melanocortin receptor 1b. Peptides 29, 1010−1017.

Bedogni, R., Esposito, A., Chiti, M., 2008. Determination of workplace neutron spectra at a high energy hadron accelerator using active and passive Bonner sphere spectrometers. Radiat. Meas. 43, 1113−1117.

Bedogni, R., Domingo, C., Esposito, A., Chiti, M., Garca-Fust, M.J., Lovestam, G., 2010. Testing Bonner sphere spectrometers in the JRC-IRMM mono-energetic neutron beams. Nucl. Instrum. Methods Phys. Res., Sect. A 620, 391−396.

Bedogni, R., et al., 2012. Measurement of neutron spectra generated by a 62 AMeV carbon-ion beam on a PMMA phantom using extended range Bonner sphere spectrometers. Nucl. Instrum. Methods Phys. Res., Sect. A 681, 110−115.

Bedogni, R., et al., 2013. Testing a newly developed single-sphere neutron spectrometer in reference monochromatic fields from 147 keV to 14.8 MeV. Nucl. Instrum. Methods Phys. Res., Sect. A 714, 110−114.

Bedogni, R., et al., 2014a. Spectrometry of 50 and 100 MeV quasi monochromatic neutron fields with extended range Bonner spheres. Nucl. Instrum. Methods Phys. Res., Sect. A 746, 59−63.

Bedogni, R., et al., 2014b. First test of SP^2: a novel active neutron spectrometer condensing the functionality of Bonner spheres in a single moderator. Nucl. Instrum. Methods Phys. Res., Sect. A 767, 159−162.

Bedogni, R., et al., 2014c. Investigation of the neutron spectrum of americium−beryllium sources by Bonner sphere spectrometry. Nucl. Instrum. Methods Phys. Res., Sect. A 763, 547−552.

Bedogni, R., et al., 2015. Experimental test of a newly developed single-moderator, multi-detector, directional neutron spectrometer in reference monochromatic fields from 144 keV to 16.5 MeV. Nucl. Instrum. Methods Phys. Res., Sect. A 782, 35−39.

Bedogni, R., Pola, A., Costa, M., Monti, V., Thomas, D.J., 2018. A Bonner Sphere Spectrometer based on a large ^{6}LiI(Eu) scintillator: calibration in reference monoenergetic fields. Nucl. Instrum. Methods Phys. Res., Sect. A 897, 18−21.

Beeman, J.E., Bellini, F., Benetti, P., et al., 2015. Double-beta decay investigation with highly pure enriched ^{82}Se for the LUCIFER experiment. Eur. Phys. J. C 75, 591.

Beeman, J.W., Bellini, F., Capelli, S., et al., 2012a. $ZnMoO_4$: a promising bolometer for neutrinoless double beta decay searches. Astropart. Phys. 35, 813−820.

Beeman, J.W., Danevich, F.A., Degoda, V.Y., 2012b. A next-generation neutrinoless double beta decay experiment based on $ZnMoO_4$ scintillating bolometers. Phys. Lett. B 710, 318−323.

Beeman, J.W., Bellini, F., Cardani, L., et al., 2012c. Discrimination of α and β/γ interactions in a TeO_2 bolometer. Astropart. Phys. 35, 558−562.

Beeman, J.W., Gentils, A., Giuliani, A., et al., 2013. Effect of SiO_2 coating in bolometric Ge light detectors for rare event searches. Nucl. Instrum. Methods Phys. Res., Sect. A 709, 22−28.

Bekker, T.B., Coron, N., Danevich, F.A., 2016. Aboveground test of an advanced Li_2MoO_4 scintillating bolometer to search for neutrinoless double beta decay of ^{100}Mo. Astropart. Phys. 72, 38−45.

Belle, Y.S., Zlobin, L.I., Moroz, G.I., Yushkevich, G.F., 1974. Selection of an optimum detector size for radioisotope analysis. Spectrom. Meth. Anal. Radioakt. Zagryaz. Pochv. Aerozol., Dokl. Vses. Soveshch. 11−14.

Belli, P., Bernabei, R., Cappella, F., Cerulli, R., Dai, C.J., Danevich, F.A., Grinyov, B.V., Incicchitti, A., Kobychev, V.V., Nagornaya, L.L., et al., 2008. Search for 2β processes in 64Zn with the help of ZnWO4 crystal scintillator. Phys. Lett. B 658, 193−197.

Belli, P., Bernabei, R., Cappella, F., Cerulli, R., Danevich, F.A., Grinyov, B.V., Incicchitti, A., Kobychev, V.V., Mokina, V.M., Nagorny, S.S., et al., 2009. Search for double beta decay of zinc and tungsten with low background ZnWO4 crystal scintillators. Nucl. Phys. A 826, 256–273.

Belli, P., Bernabei, R., Boiko, R.S., Brudanin, V.B., Bukilic, N., Cerulli, R., Chernyak, D.M., Danevich, F.A., d'Angelo, S., Degoda, V.Y., et al., 2010. Nucl. Instrum. Methods Phys. Res., Sect. A 615, 301–306.

Bellini, V., Capogni, M., Febbraro, V., Finocchiaro, P., 2010. Implementation of a new low cost detector for low intensity light pulses produced by a radioactive gamma-source. Appl. Radiat. Isot. 68, 1320–1323.

Belogurov, S., Bressi, G., Carugno, G., Moszyński, M., Czarnacki, W., Kapusta, M., Szawlowski, M., 2003. Characterization of Yb:YAG and Yb:YAP scintillators by means of LAAPD at temperature around 100 K. Nucl. Instrum. Methods Phys. Res., Sect. A 496, 385–389.

Belogurov, S., Bressi, G., Carugno, G., Grishkin, Y., 2004. Properties of Yb-doped scintillators: YAG, YAP, LuAG. Nucl. Instrum. Methods Phys. Res., Sect. A 516, 58–67.

Belsky, A.N., Auffray, E., Lecoq, P., Dujardin, C., Garnier, N., Canibano, H., Pedrini, C., Petrosyan, A.G., 2001. Progress in the development of $LuAlO_3$-based scintillators. IEEE Trans. Nucl. Sci. 48 (4), 1095–1100.

Beltrame, P., Bolle, E., Braem, A., Casella, C., Chesi, E., Clinthorne, N., De Leo, R., Dissertori, G., Djambazov, L., Fanti, V., et al., 2011a. Demonstration of an Axial PET concept for brain and small animal imaging. Nucl. Instrum. Methods Phys. Res., Sect. A 628, 426–429.

Beltrame, P., Bolle, E., Braem, A., Casella, C., Chesi, E., Clinthorne, N., Cochran, E., De Leo, R., Dissertori, G., Djambazov, G., et al., 2011b. Construction and tests of demonstrator modules for a 3-D axial PET system for brain or small animal imaging. Nucl. Instrum. Methods Phys. Res., Sect. A 636, S226–S230.

Benavente, J.A., Vega-Carrillo, H.R., Lacerda, M.A.S., Fonseca, T.C.F., Faria, F.P., da Silva, T.A., 2015. Neutron spectra due ^{13}N production in a PET cyclotron. Appl. Radiat. Isot. 99, 20–24.

Bender, S., Heidrich, B., Ünlü, K., 2015. Compton suppressed $LaBr_3$ detection system for use in nondestructive spent fuel assay. Nucl. Instrum. Methods Phys. Res., Sect. A 784, 474–481.

Benites-Rengifo, J.L., Vega-Carrillo, H.R., Velazquez-Fernandez, J., 2014. Photoneutron spectrum measured with Bonner sphere spectrometer in planetary method mode. Appl. Radiat. Isot. 83, 256–259.

Bérard, P., Bergeron, M., Pepin, C.M., Cadorette, J., Tétrault, M.-A., Viscogliosi, N., Fontaine, R., Dautet, H., Davies, M., Deschamps, P., Lecomte, R., 2009. Development of a 64-channel APD detector module with individual pixel readout for submillimetre spatial resolution in PET. Nucl. Instrum. Methods Phys. Res., Sect. A 610, 20–23.

Berneking, A., Gola, A., Ferri, A., Finster, F., Rucatti, D., et al., 2018. A new PET detector concept for compact preclinical high-resolution hybrid MR-PET. Nucl. Instrum. Methods Phys. Res., Sect. A 888, 44–52.

Berra, A., Bolognini, D., Bonfanti, S., et al., 2013. SiPM based readout system for $PbWO_4$ crystals. Nucl. Instrum. Methods Phys. Res., Sect. A 718, 63–65.

Bessiere, A., Dorenbos, P., Van Eijk, C.W.E., Krämer, K.W., Güdel, H.U., 2005. Luminescence and scintillation properties of Cs2LiYCl6:Ce3+ for γ and neutron detection. Nucl. Instrum. Methods Phys. Res., Sect. A 537, 242–246.

Beznosko, D., Blazey, G., Dyshkant, A., Rykalin, V., Schellpffer, J., Zutshi, V., 2006. Modular design for narrow scintillating cells with MRS photodiodes in strong magnetic field for ILC detector. Nucl. Instrum. Methods Phys. Res., Sect. A 564, 178–184.

Bertrand, G.H.V., Hamel, M., Normand, S., Sguerra, F., 2015. Pulse shape discrimination between (fast or thermal) neutrons and gamma rays with plastic scintillators: State of the art. Nucl. Instrum. Methods Phys. Res. Sect. A 776, 114–128.

Bezuidenhout, J., 2013. Measuring naturally occurring uranium in soil and minerals by analysing the 352 keV gamma-ray peak of ^{214}Pb using a NaI(Tl)-detector. Appl. Radiat. Isot. 80, 1–6.

Bieberle, A., Kronenberg, J., Schleicher, E., Hampel, U., 2007. Design of a high-resolution gamma-ray detector module for tomography applications. Nucl. Instrum. Methods Phys. Res., Sect. A 572, 668–675.

Bikit, I., Nemes, T., Mrda, D., 2009a. Simple method for absolute activity measurement of 60Co source. Nucl. Instrum. Methods Phys. Res., Sect. A 603, 333–336.

Bikit, I., Nemes, T., Mrda, D., Jovančević, N., 2009b. On the absolute source activity measurement with a single detector: the 133Ba case. Nucl. Instrum. Methods Phys. Res., Sect. A 612, 103–111.

Birattari, C., Dimovasili, E., Mitaroff, A., Silari, M., 2010. A Bonner Sphere Spectrometer with extended response matrix. Nucl. Instrum. Methods Phys. Res., Sect. A 620, 260–269.

Bizarri, G., 2010. Scintillation mechanisms of inorganic materials: from crystal characteristics to scintillation properties. J. Cryst. Growth 312, 1213–1215.

Blair, J., Smith, J.T., Wood, E., Picardo, M., Skinner, R., Cook, N., 1995. The Effect of Temperature on the GAP.SH2/phosphotyrosine Peptide Interaction; Analysis Using Scintillation Proximity Assay (Poster). *2nd European Conference On High Throughput Screening*, Budapest, May. (Copy available from Amersham International plc., Whitechurch, Cardiff CF4 7YT, UK.).

Blasco, J.M., Sanchis, E., Granero, D., et al., 2015. Study and simulation of the read-out electronics design for a high-resolution plastic scintillating fiber based hodoscope. Nucl. Instrum. Methods Phys. Res., Sect. A 784, 232–235.

Boatner, L.A., Neal, J.S., Blackston, M.A., 2012. Dual-chamber/dual-anode proportional counter incorporating an intervening thin-foil solid neutron converter. Nucl. Instrum. Methods Phys. Res., Sect. A 693, 244–252.

Bobin, C., Berkes, I., Hadjout, J.P., Coron, N., Leblanc, J., De Marcillac, P., 1997. Alpha/gamma discrimination with a CaF2(Eu) target bolometer optically coupled to a composite infrared bolometer. Nucl. Instrum. Methods Phys. Res., Sect. A 386, 453–457.

Bol, A.A., Ferwerda, J., Bergwerff, J.A., Meijerink, A., 2002. Luminescence of nanocrystalline $ZnS:Cu^{2+}$. J. Lumin. 99, 325–334.

Boll, N.J., Salazar, D., Stelter, C.J., Landis, G.A., Colozza, A.J., 2015. Venus high temperature atmospheric dropsonde and extreme-environment seismometer (HADES). Acta Austron. 111, 146–159.

Bolle, E., Braem, A., Casella, C., Chesi, E., Clinthorne, N., Cochran, E., De Leo, R., Dissertori, G., Djambazov, G., Fanti, V., et al., 2010. The AX-PET project: demonstration of a high resolution axial 3D PET. Nucl. Instrum. Methods Phys. Res., Sect. A 617, 214–216.

Bonvicini, V., Capelli, S., Cremonesi, O., et al., 2014. A flexible scintillation light apparatus for rare event searches. Eur. Phys. J. C 74, 3151. https://doi.org/10.1140/epjc/s10052-014-3151-5.

Borisevich, A.E., Korzhik, M.V., Drobychev, G.Y., Cavaignac, J.-F., Chipaux, R., 2005. New class of indium-containing room temperature inorganic scintillators. Nucl. Instrum. Methods Phys. Res., Sect. A 537, 228–231.

Börnsen, O., 2000. Using the TopCount™ microplate scintillation and luminescence counter and deep-well LumaPlate™ microplates in combination with micro separation techniques for metabolic studies. Application Note TopCount AN004-TC, PerkinElmer. http://www.perkinelmer.co.jp/Portals/0/resource/products_ls/ls_cocktails/pdf/AN004-TC.pdf.

Bourne, M.M., Clarke, S.D., Adamowicz, N., et al., 2016. Neutron detection in a high-gamma field using solution-grown stilbene. Nucl. Instrum. Methods Phys. Res., Sect. A 806, 348–355.

Bourret-Courchesne, E.D., Bizarri, G., Borade, R., Yan, Z., Hanrahan, S.M., Gundiah, G., Chaudhry, A., Canning, A., Derenzo, S.E., 2009. Eu^{2}+-doped Ba2CsI5, a new high-performance scintillator. Nucl. Instrum. Methods Phys. Res., Sect. A 612, 138–142.

Bourret-Courchesne, E.D., Bizarri, G., Hanrahan, S.M., Gundiah, G., Yan, Z., Derenzo, S.E., 2010. BaBrI:Eu2+, a new bright scintillator. Nucl. Instrum. Methods Phys. Res., Sect. A 613, 95–97.

Braem, A., Joram, C., Séguinot, J., Dissertori, G., Djambazov, L., Lustermann, W., Nessi-Tedaldi, F., Pauss, F., Schinzel, D., Solevi, P., et al., 2009. AX-PET: a novel PET detector concept with full 3D reconstruction. Nucl. Instrum. Methods Phys. Res., Sect. A 610, 192–195.

Braun, I., Commichau, S.C., Rissi, M., Backes, M., Biland, A., Bretz, T., Britvitch, I., Commichau, V., von Gunten, H., Hildebrand, D., et al., 2009. First Avalanche-photodiode camera test (FACT): a novel camera using G-APDs for the observation of very high-energy γ-rays with Cherenkov telescopes. Nucl. Instrum. Methods Phys. Res., Sect. A 610, 400–403.

Bramblett, R.L., Ewing, R.I., Bonner, T.W., 1960. A new type of neutron spectrometer. Nucl. Instrum. Methods 9, 1–12.

Bressi, G., Carugno, G., Conti, E., Del Noce, C., Iannuzzi, D., 2001. New prospects in scintillation crystals. Nucl. Instrum. Methods Phys. Res., Sect. A 461, 361–364.

Briesmeister, J., 2000. MCNP–A General Monte Carlo N-Particle Transport Code. Los Alamos National Laboratory Manual, Los Alamos, NM.

Brinkman, G.A., Veenboer, J.T., 1979. Calibration of well-type thallium-activated sodium iodide detectors. Int. J. Appl. Radiat. Isot. 30, 171–176.

Brinkman, G.A., Aten Jr., A.H.W., 1963. Absolute standardization with a NaI(Tl) crystal—III: calibration of β+-emitters. Int. J. Appl. Radiat. Isot. 14, 503–510.

Brinkman, G.A., Aten Jr., A.H.W., 1965. Absolute standardization with a NaI(Tl) crystal—V: calibration of isotopes with complex decay schemes. Int. J. Appl. Radiat. Isot. 16, 177–181.

Brinkman, G.A., Aten Jr., A.H.W., Veenboer, J.T., 1963a. Absolute standardization with a NaI(Tl) crystal—I: calibration by means of a single nuclide. Int. J. Appl. Radiat. Isot. 14, 153–157.

Brinkman, G.A., Aten Jr., A.H.W., Veenboer, J.T., 1963b. Absolute standardization with a NaI(Tl) crystal—II: determination of the total efficiency. Int. J. Appl. Radiat. Isot. 14, 433–437.

Brinkman, G.A., Aten Jr., A.H.W., Veenboer, J.T., 1965. Absolute standardization with a NaI(Tl) crystal—IV: determination of the photoelectric efficiency. Int. J. Appl. Radiat. Isot. 16, 15–18.

Britvich, G.I., Vasil'chenko, V.G., Gilitsky, Y.V., Chubenko, A.P., Kushnirenko, A.E., Mamidzhanyan, E.A., Pavluchenko, V.P., et al., 2005. A neutron detector on the basis of a boron-containing plastic scintillator. Nucl. Instrum. Methods Phys. Res., Sect. A 550, 343–358.

Britvitch, I., Lorenz, E., Olshevski, A., Renker, D., Sadygov, Z., Scheuermann, R., Stoykov, A., Werner, A., Zheleznykh, I., 2007. Development of scintillation detectors based on avalanche microchannel photodiodes. Nucl. Instrum. Methods Phys. Res., Sect. A 571, 317–320.

Broisat, A., et al., 2011. Myocardial uptake of 7′-(Z)-[123I]Iodorotenone during vasodilator stress in dogs with critical coronary stenoses. Circ. Cariovasc. Imaging 4 (6), 685–692. https://doi.org/10.1161/CIRCIMAGING.110.961763.

Bronson, F.L., Wang, L., 1996. Validation of the MNCP Monte Carlo code for germanium detector gamma efficiency calibrations. In: Proc. Waste Management 1996 Congress, February 28, Tucson,AZ, USA.

Brookes, F.D., 1979. Development of organic scintillators. Nucl. Instrum. Methods 162, 477–505.

Brook, N.H., Castillo García, L., Conneely, T.M., et al., 2019. Testbeam studies of a TORCH prototype detector. Nucl. Instrum. Methods Phys. Res., Sect. A (in press).

Brooks, F.D., Klein, H., 2002. Neutron spectrometry - historical review and present status. Nucl. Instrum. Methods Phys. Res., Sect. A 476, 1–11.

Bross, A.D., 1990. Development of intrinsic IPT scintillator. Nucl. Instrum. Methods Phys. Res., Sect. A 295, 315–322.

Brown, R.M., Cockerill, D.J.A., 2011. Electromagnetic calorimetry. Nucl. Instrum. Methods Phys. Res., Sect. A (in press).

Bruin, G.J., Waldmeier, F., Boernsen, K.O., Pfaar, U., Gross, G., Zollinger, M., 2006. A microplate solid scintillation counter as a radioactivity detector for high performance liquid chromatography in drug metabolism: validation and applications. J. Chromatogr. A 1133, 184–194.

Budakovsky, S.V., Galunov, N.Z., Grinyov, B.V., Karavaeva, N.L., Kyung Kim, J., Kim, Y.-K., Pogorelova, N.V., Tarasenko, O.A., 2007. Stilbene crystalline powder in polymer base as a new fast neutron detector. Radiat. Meas. 42, 565–568.

Burger, P., Keters, M., Evrard, O., Van Buul, L., 2008. Industrial silicon detectors, advancements in planar technology. Nucl. Instrum. Methods Phys. Res., Sect. A 591, 1–5.

Burgett, E., Hertel, N., Howell, R., 2008. Measurements of high energy neutron spectra with a bonner sphere extension (BSE) measurement system. Trans. Am. Nucl. Soc. 99, 564–565.

Buşe, G., Giuliani, A., de Marcillac, P., et al., 2018. First scintillating bolometer tests of a CLYMENE R&D on Li_2MoO_4 scintillators towards a large-scale double-beta decay experiment. Nucl. Instrum. Methods Phys. Res., Sect. A 891, 87–91.

Caciolli, A., Baldoncini, M., Bezzon, G.P., et al., 2012. A new FSA approach for in situ γ ray spectroscopy. Sci. Total Environ. 414, 639–645.

Capogni, M., Ceccatelli, A., De Felice, P., Fazio, A., 2006. Random-summing correction and pile-up rejection in the sum-peak method. Appl. Radiat. Isot. 64, 1229–1233.

Carrazana González, J., Vargas, M.J., Castillo, R.G., 2016. Influence of self-absorption corrections in the quantification of ^{210}Pb and ^{241}Am for sediment dating. Radiat. Phys. Chem. 127, 92–96.

Carbone, D., Finocchiaro, P., Agodi, C., 2018. Mini-phoswich and SiPM for heavy ion detection. Nucl. Instrum. Methods Phys. Res., Sect. A (in press).

Carman, M.L., Glenn, A.M., Mabe, A.N., et al., 2018. Solution growth of a deuterated trans-stilbene crystal for fast neutron detection. J. Cryst. Growth 498, 51–55.

Casali, N., 2017. Model for the Cherenkov light emission of TeO_2 cryogenic calorimeters. Astropart. Phys. 91, 44–50.

Cazalas, E., Trivelpiece, C., Jovanovic, I., 2014. Spatial resolution and efficiency of microchannel plate detectors with neutron converter films. Nucl. Instrum. Methods Phys. Res., Sect. A 767, 421–425.

Cebrián, S., Coron, N., Dambier, G., De Marcillac, P., García, E., Irastorza, I.G., Leblanc, J., Morales, A., Morales, J., Ortiz de Solórzano, A., Puimedón, J., Sarsa, M.L., Villar, J.A., 2003. First underground light versus heat discrimination for dark matter search. Phys. Lett. B 563, 48–52.

Cebrián, S., Coron, N., Dambier, G., de Marcillac, P., García, E., Irastorza, I.G., Leblanc, J., Morales, A., Morales, J., Ortiz de Solórzano, A., Puimedón, J., Sarsa, M.L., Villar, J.A., Villar, J.A., 2005. ROSEBUD-II. Light-heat discrimination with scintillating bolometers underground. Nucl. Phys. B 138, 519–521.

Celentano, G., Vannozzi, A., Mancini, A., 2015. ^{10}B enriched film deposited by e-beam technique on Al_2O_3 substrate for high efficiency thermal neutron detector. Surf. Coat. Technol. 265, 160–165.

Cendre, E., Kaftandjian, V., Lecomte, G., Kjaer, K., 2005. Complementarity of a photon-counting system and radioscopy for inspection of cast aluminium components. NDT Int. 38, 239–250.

Chan, J., Chan, H.Y.F., 2011. Usefulness of thyrogastric immune features as predictors of pernicious anaemia that lacks intrinsic factor antibody. Int. J. Lab. Hematol (in press).

Chapman, K.L., Vaswani, D., Hendry, N., Langmead, C.J., Kew, J.N.C., Watson, J.M., 2011. The muscarinic M4 receptor is the functionally predominant subtype in rat and mouse striatum as demonstrated using [35S] GTPγS binding. Eur. J. Pharmacol. 652, 1–6.

Chékir-Mzali, J., Horchani-Naifer, K., Férid, M., 2016. Structural, characterization and spectroscopic properties of $NH_4YbP_2O_7$ micropowders. Optik 127, 5622–5630.

Chen, L., Guo, G., Liu, T., Guo, L., Zhu, R., 2011. Radiochemotherapy of hepatocarcinoma via lentivirus–mediated transfer of human sodium iodide symporter gene and herpes simplex virus thymidine kinase gene. Nucl. Med. Biol. (in press).

Chen, P., Jiang, L., Chen, Y., et al., 2018. Bridgman growth and luminescence properties of Li2MoO4 single crystal. Mater. Lett. 215, 225–228.

Chen, X.J., Johnson, E.B., Staples, C.J., Chapman, E., Alberghini, G., Christiana, J.F., 2010. Optical and noise performance of CMOS solid-state photomultipliers. Proc. SPIE-Int. Soc. Opt. Eng. 7781. Art. No. 77810F.

Chen, S., et al., 2017. Effect of methoxy group position on biological properties of ^{18}F–labeled benzyl triphenylphosphonium cations. Nucl. Med. Biol. 49, 16–23.

Cherepy, N.J., et al., 2015. Bismuth- and lithium-loaded plastic scintillators for gamma and neutron detection. Nucl. Instrum. Methods Phys. Res., Sect. A 778, 126–132.

Chernyak, D.M., Danevich, F.A., Degoda, V.Y., 2013. Optical, luminescence and thermal properties of radiopure ZnMoO4 crystals used in scintillating bolometers for double beta decay search. Nucl. Instrum. Methods Phys. Res., Sect. A 729, 856–863.

Chernyak, D.M., Danevich, F.A., Degoda, V.Y., 2015. Effect of tungsten doping on $ZnMoO_4$ scintillating bolometer performance. Opt. Mater. 49, 67–74.

Childress, N.L., Miller, W.H., 2002. MCNP analysis and optimization of a triple crystal phoswich detector. Nucl. Instrum. Methods Phys. Res., Sect. A 490, 263–270.

Chiles, M.M., Bauer, M.L., McElhaney, S.A., 1990. Multi-energy neutron detector for counting thermal neutrons, high-energy neutrons, and gamma photons separately. IEEE Trans. Nucl. Sci. 37, 1348–1350.

Chiozzi, P., De Felice, P., Fazio, A., Pasquale, V., Verdoya, M., 2000a. Laboratory application of NaI(Tl) γ-ray spectrometry to studies of natural radioactivity in geophysics. Appl. Radiat. Isot. 53, 127–132.

Chiozzi, P., Pasquale, V., Verdoya, M., De Felice, P., 2000b. Practical applicability of field γ-ray scintillation spectrometry in geophysical surveys. Appl. Radiat. Isot. 53, 215–220.

Chipaux, R., Cribier, M., Dujardin, C., Garnier, N., Guerassimova, N., Mallet, J., Meyer, J.-P., Pédrini, C., Petrosyan, A.G., 2002. Ytterbium-based scintillators, a new class of inorganic scintillators for solar neutrino spectroscopy. Nucl. Instrum. Methods Phys. Res., Sect. A 486, 228–233.

Choi, J., et al., 2014. N-Succinimidyl guanidinomethyl iodobenzoate protein radiohalogenation agents: influence of isomeric substitution on radiolabeling and target cell residualization. Nucl. Med. Biol. 41, 802–812.

Christian, J.F., Stapels, C.J., Johnson, E.B., McClish, M., Dokhale, P., Shah, K.S., Mukhopadhyay, S., Chapman, E., Augustine, F.L., 2010. Advances in CMOS solid-state photomultipliers for scintillation detector applications. Nucl. Instrum. Methods Phys. Res., Sect. A 624, 449–458.

Chu, M.C., et al., 2016. The radon monitoring system in Daya Bay reactor neutrino experiment. Nucl. Instrum. Methods Phys. Res., Sect. A 808, 156–164.

Coble, J., Orton, C., Schwantes, J., 2017. Multivariate analysis of gamma spectra to characterize used nuclear fuel. Nucl. Instrum. Methods Phys. Res., Sect. A 850, 18–24.

Contalbrigo, M., Baltzell, N., Benmokhtar, F., et al., 2014. The large-area hybrid-optics CLAS12 RICH detector: tests of innovative components. Nucl. Instrum. Methods Phys. Res., Sect. A 766, 22–27.

Cook, N.D., 1996. Scintillation proximity assay: a versatile high throughput screening technology. Drug Discov. Today 1, 287–294.

Cook, J.R., Graves, R., Price-Jones, M.J., Berry, J.A., Hughes, K.T., 2006. Chapter 11. Scintillation proximity assay (SPA) receptor binding assays, pp. 141–158. In: Minor, L.K. (Ed.), Handbook of Assay Development in Drug Discovery. CRC Press, Boca Raton, FL.

Cooper, R.I., Alarcon, R., Bales, M.J., et al., 2012. A gamma-and X-ray detector for cryogenic, high magnetic field applications. Nucl. Instrum. Methods Phys. Res., Sect. A 691, 64–71.

Coron, N., Dambier, G., Leblanc, E., Leblanc, J., De Marcillac, P., Moalic, J.-P., 2004. Scintillating and particle discrimination properties of selected crystals for low-temperature bolometers: from LiF to BGO. Nucl. Instrum. Methods Phys. Res., Sect. A 520, 159–162.

Coron, N., García, E., Gironnet, J., Leblanc, J., de Marcillac, P., Martínez, M., Ortigoza, Y., Pobes, C., Puimedón, J., Redon, T., Sarsa, M.L., Torres, L., Villar, J.A., 2008. Thermal relative efficiency factor for recoiling 206Pb nuclei in a sapphire bolometer. *Phys. Lett.* B 659, 113–118.

Coron, N., García, E., Gironnet, J., Leblanc, J., de Marcillac, P., Martínez, M., Ortigoza, Y., Ortiz de Solórzano, A., Pobes, C., Puimedón, J., Redon, T., Sarsa, M.L., Torres, L., Villar, J.A., 2009. A BGO scintillating bolometer as dark matter detector prototype. Opt. Mater. 31, 1393–1397.

Coron, N., Cuesta, C., García, E., et al., 2013. Study of parylene-coated NaI(Tl) at low temperatures for bolometric applications. Astropart. Phys. 47, 31–37.

Cosma, C., Moldovan, M., Dicu, T., Kovacs, T., 2008. Radon in water from transylvania (Romania). Radiat. Meas. 43 (8), 1423–1428.

Cosma, C., Cucoş-Dinu, A., Papp, B., Begy, R., Sainz, C., 2013. Soil and building material as main sources of indoor radon in Băiţa-Ştei radon prone area (Romania). J. Environ. Radioact. 116, 174–179.

Crookes, W., 1903a. The emanation of radium. *Proc. R. Soc.* (London) A71, 405–408.

Crookes, W., 1903b. Certain properties of the emanation of radium. Chem. News 87, 241.

Crosbie, P.A.J., et al., 2012. Elevated N3-methylpurine-DNA glycosylase DNA repair activity is associated with lung cancer. Mut. Res./Fund. Molec. Mech. Mutagen. 732, 43–46.

Crosbie, P.A.J., et al., 2013. Topographical study of O^6-alkylguanine DNA alkyltransferase repair activity and N7-methylguanine levels in resected lung tissue. Chem. Biol. Interact. 204, 98–104.

Cuesta, C., Oliván, M.A., Amaré, J., et al., 2013. Slow scintillation time constants in NaI(Tl) for different interacting particles. Opt. Mater. 36, 316–320.

Czyz, S.A., Farsoni, A.T., Ranjbar, L., 2018. A prototype detection system for atmospheric monitoring of xenon radioisotopes. Nucl. Instrum. Methods Phys. Res., Sect. A 884, 64–69.

Dafinei, I., 2014. The LUCIFER project and production issues for crystals needed in rare events physics experiments. J. Cryst. Growth 393, 13–17.

Dafinei, I., Nagorny, S., Pirro, S., et al., 2017. Production of ^{82}Se enriched Zinc Selenide (ZnSe) crystals for the study of neutrinoless double beta decay. J. Cryst. Growth 475, 158–170.

Dal Maso, E., et al., 2018. Characterization of signalling and regulation of common calcitonin receptor splice variants and polymorphisms. Biochem. Pharmacol. 148, 111–129.

D'Ambrosio, C., de Notaristefani, F., Hull, G., Orsolini Cencelli, V., Pani, R., 2006. Study of LaCl3:Ce light yield proportionality with a hybrid photomultiplier tube. Nucl. Instrum. Methods Phys. Res., Sect. A 556, 187–191.

Damyanova, A., Bravar, A., 2017. Scintillating fiber detectors for precise time and position measurements read out with Si-PMs. Nucl. Instrum. Methods Phys. Res., Sect. A 784, 232–235.

Danevich, F.A., Georgadze, A.S., Kobychev, V.V., Nikolaiko, A.S., Ponkratenko, O.A., Tretyak, V.I., Zdesenko, S.Y., Zdesenko, Y.G., Bizzeti, P.G., Fazzini, T.F., Maurenzig, P.R., 2003. Two-neutrino 2β decay of 116Cd and new half-life limits on 2β decay of 180W and 186W. Nucl. Phys. A 717, 129–145.

Danevich, F.A., Bizzeti, P.G., Fazzini, T.F., et al., 2005a. Double β decay of 116Cd. Final results of the Solotvina experiment and CAMEO project. Nucl. Phys. B 138, 230–232.

Danevich, F.A., Kobychev, V.V., Nagorny, S.S., Tretyak, V.I., 2005b. YAG:Nd crystals as possible detector to search for 2β and α decay of neodymium. Nucl. Instrum. Methods Phys. Res., Sect. A 541, 583–589.

Danevich, F.A., Georgadze, A.S., Kobychev, V.V., et al., 2006. Application of PbWO4 crystal scintillators in experiment to search for 2β decay of 116Cd. Nucl. Instrum. Methods Phys. Res., Sect. A 556, 259–265.

Danevich, F.A., Kobychev, R.V., Kobychev, V.V., et al., 2014. Optimization of light collection from crystal scintillators for cryogenic experiments. Nucl. Instrum. Methods Phys. Res., Sect. A 744, 41–47.

da Silva, M.A.L., Poledna, R., Iwahara, A., da Silva, C.J., Delgado, J.U., Lopes, R.T., 2006. Standardization and decay data determinations of 125I, 54Mn and 203Hg. Appl. Radiat. Isot. 64, 1440–1445.

Dayman, K.J., Coble, J.B., Orton, C.R., Schwantes, J.M., 2014. Characterization of used nuclear fuel with multivariate analysis for process monitoring. Nucl. Instrum. Methods Phys. Res., Sect. A 735, 624–632.

De, A., Dasgupta, S.S., Sen, D., 1993. Time dispersion in large plastic scintillation neutron detectors. IEEE Trans. Nucl. Sci. 40 (5), 1329–1332.

Dear, G.J., Patel, N., Kelly, P.J., Webber, L., Yung, M., 2006. TopCount coupled to ultra-performance liquid chromatography for the profiling of radiolabeled drug metabolites in complex biological samples. J. Chromatogr. B 844, 96–103.

Dear, G.J., Patel, N., Weightman, A., Pirard, H., Talvitie, M., 2008. Utilizing a −100 °C microplate CCD imager, yttrium silicate coated 384 microplates and ultraperformance liquid chromatography for improved profiling of radiolabeled drug metabolites in complex biological samples. J. Chromatogr. B 868, 49–57/.

de Celis, B., de la Fuente, R., Williart, A., de Celis Alonso, B., 2007. Coincidence measurements in α/β/γ spectrometry with phoswich detectors using digital pulse shape discrimination analysis. Nucl. Instrum. Methods Phys. Res., Sect. A 580, 206–209.

Dechavanne, V., Barrillat, N., Borlat, F., Hermant, A., Magnenat, L., Paquet, M., Antonsson, B., Chevalet, L., 2011. A high-throughput protein refolding screen in 96-well format combined with design of experiments to optimize the refolding conditions. Protein Expr. Purif. 75, 192–203.

De Felice, P., Myteberi, X., 1995. Standardization of ^{125}I by the sum-peak method and the results of a bilateral comparison between ENEA (Italy) and INP (Albania). J. Radioanal. Nucl. Chem. Lett. 200, 109–118.

de Jong, L.A.A., Uges, D.R.A., Franke, J.P., Bischoff, R., 2005. Receptor-ligand binding assays: technologies and applications. J. Chromatogr. B 829, 1–25.

de la Fuente, R., de Celis, B., del Canto, V., Lumbreras, J.M., de Celis Alonso, B., Martín-Martín, A., Gutierrez-Villanueva, J.L., 2008. Low level radioactivity measurements with phoswich detectors using coincident techniques and digital pulse processing analysis. J. Environ. Radioact. 99, 1553–1557.

for the TEXONO Collaboration Deniz, M., 2012. Measurement of $\overline{\nu_e}$– e– scattering cross-section with a CsI(Tl) scintillating crystal array at the Kuo-Sheng Nuclear Power Reactor. Nucl. Phys. B 229–232, 498.

de Oliveira, E.M., Iwahara, A., Poledna, R., 2012. Use of sum-peak and coincidence counting methods for activity standardization of ^{22}Na. Nucl. Instrum. Methods Phys. Res., Sect. A 687, 69–74.

Deprez, K., Van Holen, R., Vandenberghe, S., Staelens, S., 2011. Design of a high resolution scintillator based SPECT detector (SPECTatress). Nucl. Instrum. Methods Phys. Res., Sect. A (in press).

Derenzo, S.E., Weber, M.J., Bourret-Courchesne, E., Klintenberg, M.K., 2003. The quest for the ideal inorganic scintillator. Nucl. Instrum. Methods Phys. Res., Sect. A 505, 111–117.

Derenzo, S., Bizarri, G., Borade, R., Bourret-Courchesne, E., Boutchko, R., Canning, A., Chaudhry, A., Eagleman, Y., Gundiah, G., Hanrahan, S., Janecek, M., Weber, M., 2011. New scintillators discovered by high-throughput screening. Nucl. Instrum. Methods Phys. Res., Sect. A (in press).

Destruel, P., Taufer, M., D'Ambrosio, C., Da Via, C., Fabre, J.P., Kirkby, J., Leutz, H., 1989. A new plastic scintillator with large Stokes shift. Nucl. Instrum. Methods Phys. Res., Sect. A 276, 69–77.

Dias, M.S., Koskinas, M.F., 1995. Accidental summing correction in ^{125}I activity determination by the sum-peak method. Appl. Radiat. Isot. 46, 945–948.

for the PANA Collaboration Diehl, S., et al., 2017. Measurement and optimization of the light collection uniformity in strongly tapered PWO crystals of the PANDA detector. Nucl. Instrum. Methods Phys. Res., Sect. A 857, 1–6.

Dietrich, H.B., Purdy, A.E., Murray, R.B., Williams, R.T., 1973. Kinetics of self-trapped holes in alkali halide crystals. Experiments in thallium-activated sodium iodide and thallium-activated potassium iodide. Phys. Rev. B 8, 5894–5901.

Długosz-Lisiecka, M., Ziomek, M., 2015. Direct determination of radionuclides in building materials with self-absorption correction for the 63 and 186 keV γ-energy lines. J. Environ. Radioact. 150, 44–48.

Domingo, C., García-Fusté, M.J., Morales, E., Amgarou, K., Terrón, J.A., Roselló, J., Brualla, L., Nuñez, L., Colmenares, R., Gómez, F., Hartmann, G.H., Sánchez-Doblado, F., Fernández, F., 2010. Neutron spectrometry and determination of neutron ambient dose equivalents in different LINAC radiotherapy rooms. Radiat. Meas. 45, 1391–1397.

Dominguez II, J.M., Prisby, R.D., Muller-Delp, J.M., Allen, M.R., Delp, M.D., 2010. Increased nitric oxide-mediated vasodilation of bone resistance arteries is associated with increased trabecular bone volume after endurance training in rats. Bone 46, 813–819.

Donnier-Maréchal, M., et al., 2017. Synthesis and pharmacological evaluation of benzamide derivatives as potent and selective sigma-1 protein ligands. Eur. J. Med. Chem. 138, 964–978.

Dorenbos, P., 2002. Light output and energy resolution of Ce3+-doped scintillators. Nucl. Instrum. Methods Phys. Res., Sect. A 486, 208–213.

Drozdowski, W., Brylew, K., Witkowski, M.E., et al., 2016. Effect of Lu-to-Y ratio and Mo coactivation on scintillation properties of LuYAG:Pr and LuAG:Pr,Mo crystals. Opt. Mater. 59, 107–114.

Dumazert, J., Coulon, R., Carrel, F., 2016. Sensitive and transportable gadolinium-core plastic scintillator sphere for neutron detection and counting. Nucl. Instrum. Methods Phys. Res., Sect. A 828, 181–190.

Dumazert, J., Coulon, R., Lecomte, Q., Bertrand, G.H.V., Hamel, M., 2018. Gadolinium for neutron detection in current nuclear instrumentation research: a review. Nucl. Instrum. Methods Phys. Res., Sect. A 882, 53–68.

Dupasquier, A., Kögel, G., Somoza, A., 2004. Studies of light alloys by positron annihilation techniques. Acta Mater. 52, 4707–4726.

Dymond, A.W., et al., 2016. Metabolism, excretion, and pharmacokinetics of Selumetinib, an MEK1/2 inhibitor, in healthy adult male subjects. Clin. Ther. 38 (11), 2447–2458.

Dziri, S., Nachab, A., Nourreddine, A., Sellam, A., Pape, A., 2014. Elemental composition effects on self-absorption for photons below 100 keV in gamma-ray spectrometry. Nucl. Instrum. Methods Phys. Res., Sect. B 330, 1–6.

Eappen, K.P., Sapra, B.K., Mayya, Y.S., 2007. A novel methodology for online measurement of thoron using Lucas scintillation cell. Nucl. Instrum. Methods Phys. Res., Sect. A 572, 922–925.

Eappen, K.P., Nair, R.N., Mayya, Y.S., 2008. Simultaneous measurement of radon and thoron using Lucas scintillation cell. Radiat. Meas. 43, 91–97.

Efthimiou, N., Argyropoulos, G., Panayiotakis, G., Georgiou, M., Loudos, G., 2010. Initial results on SiPM performance for use in medical imaging. In: 2010 Int. Conf. on Imaging Systems and Techniques, IST 2010 – Proc, pp. 256–260 (Art. No. 5548492).

for the EDELWEISS collaboration Eitel, K., 2015. The EDELWEISS dark matter search: status and perspectives. Phys. Proc. 61, 61–66.

Ejiri, H., 2010. Double beta decays and neutrino nuclear responses. Prog. Part. Nucl. Phys. 64, 249–257.

El-Abnoudy, A.A., Hassan, S.F., 2016. Air renewal times and ventilation rate calculations for underground workings using radioactive measurement. Int. J. Mining Sci. Technol. 26, 843–850.

Elane Streets, W., 1994. Development of self-absorption coefficients for the determination of gamma-emitting radionuclides in environmental and mixed waste samples. Nucl. Instrum. Methods Phys. Res., Sect. A 353, 702–705.

Eldeeb, K., Leone-Kabler, S., Howlett, A.C., 2017. Chapter 1: mouse neuroblastoma CB1 cannabinoid receptor-stimulated [^{35}S]GTPγS binding: total and antibody-targeted Gα protein-specific scintillation proximity assays, pp. 1–21. In: Methods in Enzymology, vol. 593. Elsevier, Amsterdam.

Eldridge, J.S., Crowther, P., 1964. Absolute determination of ^{125}I in clinical applications. Nucleonics 22 (6), 56–59.

El-Khatib, A.M., Badawi, M.S., Thabet, A.A., et al., 2016. Well-type NaI(Tl) detector efficiency using analytical technique and ANGLE 4 software based on radioactive point sources located out the well cavity. Chin. J. Phys. 54, 338–346.

El-Khatib, A.M., Salem, B.A., Badawi, M.S., et al., 2017. Full-Energy peak efficiency of an NaI(Tl) detector with coincidence summing correction showing the effect of the source-to-detector distance. Chin. J. Phys. 55, 478–489.

Elliott, S.R., Hahn, A.A., Moe, M.K., 1986. Experimental investigation of double-beta decay in ^{82}Se. Phys. Rev. Lett. 56, 2582–2585.

Elliott, S.R., Hahn, A.A., Moe, M.K., 1987. Direct evidence for two-neutrino double-beta decay in ^{82}Se. Phys. Rev. Lett. 59, 2020–2023.

Elster, J., Geitel, H., 1903a. Über die durch radioactive Emanation erregte scintillierende Phosphoreszenz der Sidot-Blende. Phys. Z. Jahrgang 4, 439–440.

Elster, J., Geitel, H., 1903b. On the scintillation phosphorescence of Sidot's blende caused by radio-active emanation. Chem. News 88, 37.

Ely, J.H., Aalseth, C.E., McIntyre, J.I., 2005. Novel beta-gamma coincidence measurements using phoswich detectors. J. Radioanal. Nucl. Chem. 263 (1), 245–250.

Enqvist, A., Pozzi, S.A., Flaska, M., Pázsit, I., 2010. Initial evaluation for a combined neutron and gamma ray multiplicity counter. Nucl. Instrum. Methods Phys. Res., Sect. A 621, 493–497.

Erdmann, F., et al., 2017. Augmented reversible photoswitching of drug-target interaction through "surface borrowing". Biochem. Pharmacol. 125, 84–92.

Esposito, A., Bedogni, R., Domingo, C., García, M.J., Amgarou, K., 2010. Measurements of leakage neutron spectra from a high-energy accumulation ring using extended range bonner sphere spectrometers. Radiat. Meas. 45, 1522–1525.

Evans Jr., A.E., 1980. Gamma-ray response of a 38-mm bismuth germanate scintillator. IEEE Trans. Nucl. Sci. NS-27, 172–175.

Evans, R.D., 1969. Engineer's guide to the elementary behavior of radon daughters. Health Phys. 17, 229–252.

Evans, R.D., 1955. The Atomic Nucleus. McGraw-Hill, Inc., p. 972

Farsoni, A.T., Hamby, D.M., 2005. MCNP analysis of a multilayer phoswich detector for β-particle dosimetry and spectroscopy. Nucl. Instrum. Methods Phys. Res., Sect. A 555, 225–230.

Farsoni, A.T., Hamby, D.M., 2007. A system for simultaneous beta and gamma spectroscopy. Nucl. Instrum. Methods Phys. Res., Sect. A 578, 528–536.

Farsoni, A.T., Alemayehu, B., Alhawsawi, A., Becker, E.M., 2013. Real-time pulse-shape discrimination and beta–gamma coincidence detection in field-programmable gate array. Nucl. Instrum. Methods Phys. Res., Sect. A 712, 75–82.

Ferda, J., 2017. Hybrid imaging with PET/MRI: ready for clinical routine? Eur. J. Radiol. 94, A1–A2.

Ferhi, M., Horchani-Naifer, K., Hraiech, S., Férid, M., Guyot, Y., Boulon, G., 2012. Near infrared and charge transfer luminescence of trivalent ytterbium in $KLa(PO_3)_4$ powders optics. Commun. Now. 285, 2874–2878.

Fernandes, L.M.P., Antognini, A., Boucher, M., Conde, C.A.N., Huot, O., Knowles, P., Kottmann, F., Ludhova, L., Mulhauser, F., Pohl, R., Schaller, L.A., dos Santos, J.M.F., Taqqu, D., Veloso, J.F.C.A., 2003. Behaviour of large-area avalanche photodiodes under intense magnetic fields for VUV- visible- and X-ray photon detection. Nucl. Instrum. Methods Phys. Res., Sect. A 498, 362–368.

Fernández, F., Bouassoule, T., Amgarou, K., Domingo, C., Garcia, M.J., Lacoste, V., Gressier, V.V., Muller, H.H., 2007. Monte Carlo calculations and validation of a gold foil-based Bonner sphere system. Radiat. Prot. Dosim. 126, 366–370.

Ferreira Filho, A.L., da Silva, R., da Cruz, P.A.L., et al., 2018. Application of the sum-peak method to activity standardizations of ^{152}Eu sources in LNMRI (BR). Appl. Radiat. Isot. 134, 340–350.

Finocchiaro, P., Cosentino, L., Lo Meo, S., Nolte, R., Radeck, D., 2018. Absolute efficiency calibration of 6LiF-based solid state thermal neutron detectors. Nucl. Instrum. Methods Phys. Res., Sect. A 885, 86–90.

Fisher, B.M., Abdurashitov, J.N., Coakley, K.J., Gavrin, V.N., Gilliam, D.M., Nico, J.S., Shikhin, A.A., et al., 2011. Fast neutron detection with 6Li-loaded liquid scintillator. Nucl. Instrum. Methods Phys. Res., Sect. A 646, 126–134.

Flanagan, C.A., 2016. Chapter 10. GPCR-radioligand binding assays, pp. 191–215. In: Methods in Cell Biology, vol. 132. Elsevier, Amsterdam.

Flaumenhaft, R., 2007. Chapter 3.07. Chemical biology, pp. 129–149. In: Taylor, J.B., Triggle, D.J. (Eds.), Comprehensive Medicinal Chemistry II. Elsevier, Amsterdam.

the LHCb RICH Collaboration Fontanelli, F., 2009. The pixel hybrid photon detector of the LHCb RICH. Nucl. Phys. B 197, 292–295.

Ford, M.A., O'Day, B.E., McClory, J.W., Sharma, M.K., Danagoulian, A., 2018. Evaluation of Eu:LiCAF for neutron detection utilizing SiPMs and portable electronics. Nucl. Instrum. Methods Phys. Res., Sect. A 908, 110–116.

Fraser, G.W., 2002. The ion detection efficiency of microchannel plates. Int. J. Mass Spectrom. 215, 13–30.

Fricke, R.G.A., 2001. Friedrich Oskar Giesel, Pioner der Radioactivitätsforschung, Opfer seiner Wissenschaft. AF Verlag, Wolfenbüttel, p. 179.

Frauenfelder, H., Steffen, R.M., 1965. Alpha-Beta- and Gamma-ray Spectroscopy, first ed. North-Holland Publishing Co., Amsterdam, p. 1033.

Frutos Vázquez, B., Olaya Adán, M., Quindós Poncela, L.S., Sainz Fernandez, C., Fuente Merino, I., 2011. Experimental study of effectiveness of four radon mitigation solutions, based on underground depressurization, tested in prototype housing built in a high radon area in Spain. J. Environ. Radioact. 102, 378–385.

Fujii, H., Matsuno, K., Takiue, M., 1999. Construction of analytical beta ray monitor for liquid waste. Radioisotopes 48, 465–471.

Fujii, H., Matsuno, K., Takiue, M., 2000. Hybrid radioassay of multiple radionuclide mixtures in waste solutions by using liquid and NaI(Tl) scintillation monitors. Health Phys. 79 (3), 294–298.

Fujimoto, Y., Yanagida, T., Tanaka, H., Yokota, Y., Kawaguti, N., Fukuda, K., Totsuka, D., et al., 2011. Growth and characterization of Ce-doped Ca3(BO3)2 crystals for neutron scintillator. J. Cryst. Growth 318, 784–787.

Fukabori, A., Yokota, Y., Yanagida, T., Moretti, F., Chani, V., Kawaguchi, N., Kamada, K., Yoshikawa, A., 2011a. Effect of Ce doping on scintillation characteristics of LiYF4 single crystals for γ-ray detection. Nucl. Instrum. Methods Phys. Res., Sect. A 631, 68–72.

Fukuda, Y., Izawa, T., Koshio, Y., Moriyama, S., Namba, T., Shiozawa, M., 2010. InP solid state detector for measurement of low energy solar neutrinos. Nucl. Instrum. Methods Phys. Res., Sect. A 623, 460–462.

Ghal-Eh, N., Kalaei, M., Mohammadi, A., Vega-Carrillo, H.R., 2017. Replacement of Bonner spheres with polyethylene cylinders for the unfolding of an ^{241}Am–Be neutron energy spectrum. Appl. Radiat. Isot. 128, 292–296.

Gamal, A., Paul, B., Michael, C., Roland, H., Johann, M., Herbert, O., Ken, S., 2011. Application of Geiger-mode photosensors in Cherenkov detectors. Nucl. Instrum. Methods Phys. Res., Sect. A 639, 107–110.

Garcia Burciaga, G., 1976. The Use of ^{65}Zn in the Study of the Organic Compounds of Zinc in Chicken Manure and Its Transformations Upon Incorporation into a Sandy Soil (M.Sc. thesis). Colegio de Postgraduados, Chapingo, Mexico.

Garcia Burciaga, G., L'Annunziata, M.F., Ortega, M.L., Alvarado, R., 1978. Chemistry of ^{65}Zn-labelled chicken manure in the soil. In: Proceedings IAEA International Symposium, Dec. 11-15, Colombo, Sri Lanka, IAEA-SM-235/43, pp. 393–406.

Garcia-Duran, A., Hernandez-Davila, V.M., Vega-Carrillo, H.R., et al., 2019. FPGA embedded multichannel analyzer. Appl. Radiat. Isot. (in press).

Gaster, M., 2009. Reduced lipid oxidation in myotubes established from obese and type 2 diabetic subjects. Biochem. Biophys. Res. Commun. 382, 766–770.

Germano, G., Berman, D.S., Slomka, P., 2016. Technical aspects of cardiac PET imaging and recent advances. Cardiol. Clin. 34, 13–23.

Ghabrial, A., Franklin, D., Zaidi, H., 2018. A Monte Carlo simulation study of the impact of novel scintillation crystals on performance characteristics of PET scanners. Phys. Med. 50, 37–45.

Ghelman, M., Paperno, E., Ginsburg, D., Mazor, T., Cohen, Y., Osovizky, A., 2011. Sub-milliwatt spectroscopic personal radiation device based on a silicon photomultiplier. Nucl. Instrum. Methods Phys. Res., Sect. A (in press).

Giaz, A., Blasi, N., Boiano, C., et al., 2016. Fast neutron measurements with 7Li and 6Li enriched CLYC scintillators. Nucl. Instrum. Methods Phys. Res., Sect. A 825, 51–61.

Ginzburg, D., Kopeika, N., Paran, J., Cohen-Zada, I., Ghelman, M., Pushkarsky, V., Marcus, E., Manor, A., Mazor, T., Kadmon, Y., Cohen, Y., Osovizky, A., 2011. Optimizing the design of a silicon photomultiplier-based radiation detector. Nucl. Instrum. Methods Phys. Res., Sect. A (in press).

Ginzburg, D., Knafo, Y., Manor, A., 2015. Personal radiation detector at a high technology readiness level that satisfies DARPA's SN-13-47 and SIGMA program requirements. Nucl. Instrum. Methods Phys. Res., Sect. A 784, 438–447.

Gironi, L., Arnaboldi, C., Capelli, S., Cremonesi, O., Pessina, G., Pirro, S., Pavan, M., 2009. CdWO4 bolometers for double beta decay search. Opt. Mater. 31, 1388–1392.

Gironi, L., 2010. Scintillating bolometers for double beta decay search. Nucl. Instrum. Methods Phys. Res., Sect. A 617, 478–481.

Gironi, L., 2013. Surface induced background in CUORE. Nucl. Instrum. Methods Phys. Res., Sect. A 718, 497–498.

Gironi, L., 2016. A scintillating bolometer array for double beta decay studies: the LUCIFER experiment. Nucl. Instrum. Methods Phys. Res., Sect. A 824, 620–621.

Glaenzel, U., et al., 2018. Metabolism and disposition of Siponimod, a novel selective S1P1/S1P5 agonist, in healthy volunteers and *in-vitro* identification of human cytochrome P450 enzymes involved in its oxidative metabolism. Drug Metab. Dispos. https://doi.org/10.1124/dmd.117.079574.

Glickman, J.F., Schmid, A., Ferrand, S., 2008. Review: scintillation proximity assays in high-throughput screening. Assay Drug Dev. Technol. 6 (3), 433–455. https://doi.org/10.1089/adt.2008.135.

Glodo, J., Wang, Y., Shawgo, R., Brecher, C., Hawrami, R.H., Tower, J., Shah, K.S., 2017. New developments in scintillators for security applications. Phys. Proc. 90, 285–290.

Goeppert-Mayer, M., 1935. Double-beta disintegration. Phys. Rev. 48, 512–516.

Gómez Perales, J.L., 2015. Blood volume analysis by radioisotopic dilution techniques: state of the art. Appl. Radiat. Isot. 96, 71–82.

Gómez-Ros, J.M., et al., 2012. Designing an extended energy range single-sphere multi-detector neutron spectrometer. Nucl. Instrum. Methods Phys. Res., Sect. A 677, 4–9.

González-Mestres, L., Perret-Gallix, D., 1988. Neutrinos, dark matter and low temperature detectors. In: González-Mestres, L., Perret-Gallix, D. (Eds.), Proceedings of the 2nd European Workshop on Low Temperature Detectors for Neutrinos and Dark Matter. Editions Frontieres, Gif-sur-Yvette, p. 504.

Grabmaier, B.C., 1984. Crystal scintillators. IEEE Trans. Nucl. Sci. NS-31, 372–376.

Gravelle, P., et al., 2014. Cell growth in aggregates determines gene expression, proliferation, survival, chemoresistance, and sensitivity to immune effectors in follicular lymphoma. Am. J. Pathol. 184 (1), 282–295.

Grimm, O., Kim, G.N., Lee, M.W., Röser, U., Viertel, G.M., von Gunten, H., 2003. A channel photomultiplier with a scintillator faceplate. Nucl. Instrum. Methods Phys. Res., Sect. A 513, 644–646.

Grobler, J.A., Stillmock, K.A., Hazuda, D.J., 2009. Scintillation proximity assays for mechanistic and pharmacological analyses of HIV-1 integration. Methods 47, 249–253.

Grodzicka-Kobylka, M., Szczesniak, T., Moszyński, M., 2017. Comparison of SensL and Hamamatsu 4×4 channel SiPM arrays in gamma spectrometry with scintillators. Nucl. Instrum. Methods Phys. Res., Sect. A 856, 53–64.

Grodzicka-Kobylka, M., Szczesniak, T., Moszyński, M., et al., 2018. Study of n-γ discrimination by zero-crossing method with SiPM based scintillation detectors. Nucl. Instrum. Methods Phys. Res., Sect. A 883, 159–165.

Groom, D. E. (1983). Silicon photodiode detection of bismuth germinate scintillation light. pp. 256–267. *In* "Proceedings of the International Workshop on Bismuth Germanate." Princeton University, November 1982 (C.Newman Holmes, Ed.) CONF-821160, DE83 011369, U.S. National Science Foundation and U.S. Dept. of Energy, Washington, DC.

Groom, D.E., 1984. Silicon photodiode detection of bismuth germanate scintillation light. Nucl. Instrum. Methods Phys. Res. 219, 141–148.

Gu, C., et al., 2018. Late-occurring and long-circulating metabolites of $GABA_{A\alpha2,3}$ receptor modulator AZD7325 involving metabolic cyclization and aromatization: relevance to MIST analysis and application for patient compliance. Drug Metab. Dispos. https://doi.org/10.1124/dmd.117.078873.

Gundacker, S., Turtos, R.M., Auffray, E., Lecoq, P., 2018. Precise rise and decay time measurements of inorganic scintillators by means of X-ray and 511 keV excitation. Nucl. Instrum. Methods Phys. Res., Sect. A 891, 42–52.

Gundiah, G., Brennan, K., Yan, Z., et al., 2014. Structure and scintillation properties of Ce^{3+}-activated $Cs_2NaLaCl_6$, Cs_3LaCl_6, $Cs_2NaLaBr_6$, Cs_3LaBr_6, Cs_2NaLaI_6 and Cs_3LaI_6. J. Lumin. 149, 374–384.

Gupta, Y.K., Biswas, D.C., Roy, P., Nayak, B.K., Thomas, R.G., Inkar, A.L., Vind, R.P., John, B., Saxena, A., Choudhury, R.K., 2011. Pulse height and timing characteristics of CsI(Tl)-Si(PIN) detector for γ-rays and fission fragments. Nucl. Instrum. Methods Phys. Res., Sect. A 629, 149–153.

Güsten, H., Mirsky, J., 1991. PMP, a novel scintillation solute with a large Stokes's shift, pp. 1–7. In: Ross, H., Noakes, J.E., Spaulding, J.D. (Eds.), Liquid Scintillation Counting and Organic Scintillators. Lewis Publishers, Chelsea, MI.

Gütlein, A., Angloher, G., Gotti, C., et al., 2017. The COSINUS project: development of new NaI-based cryogenic detectors for direct dark matter search. Nucl. Instrum. Methods Phys. Res., Sect. A 845, 359–362.

Haas, D.A., Bliss, M., Bowyer, S.M., Kephart, J.D., Schweiger, M.J., Smith, L.E., 2011. Actinide-loaded glass scintillators for fast neutron detection. Nucl. Instrum. Methods Phys. Res., Sect. A (in press).

Hadj Rabia, S., Benmoussa, F., Benzaid, A., Baz, A., 2018. Hydatidosis: preparation and evaluation of radiolabeled antigens and antibodies. Exp. Parasitol. 187, 67–74.

Hallinger, D.R., et al., 2017. Development of a screening approach to detect thyroid disrupting chemicals that inhibit the human sodium iodide symporter (NIS). Toxicol. In Vitro 40, 66–78.

Hamada, M.M., Costa, F.E., Pereira, M.C.C., Kubota, S., 2001. Dependence of scintillation characteristics in the CsI(Tl) crystal on Tl^{+} concentrations under electron and alpha particle excitations. IEEE Trans. Nucl. Sci. 48 (4), 1148–1153.

Hampel, D., et al., 2014. Competitive chemiluminescent enzyme immunoassay for vitamin B12 analysis in human milk. Food Chem. 153, 60–65.

Haney, J.H., Barnhart, T.E., Zaidins, C.S., 1999. Extraction of neutral spectral information from Bonner-sphere data. Nucl. Instrum. Methods Phys. Res., Sect. A 431, 551–555.

Harndahl, M., Rasmussen, M., Roder, G., Buus, S., 2011. Real-time, high-throughput measurements of peptide-MHC-I dissociation using a scintillation proximity assay. J. Immunol. Methods (in press).

Harnew, N., Brook, N., Castillo García, L., et al., 2016. The TORCHtime-of-flight detector. Nucl. Instrum. Methods Phys. Res., Sect. A 824, 106–110.

Hartman, J., Barzilov, A., 2017. Study of pulse shape discrimination for a neutron phoswich detector. EPJ Web Conf. 153, 07012. https://doi.org/10.1051/epjconf/201715307012.

Hartman, J., Barzilov, A., Peters, E.E., Yates, S.W., 2015. Measurements of response functions of EJ-299-33A plastic scintillator for fast neutrons. Nucl. Instrum. Methods Phys. Res., Sect. A 804, 137–143.

Hassan, N.M., Ishikawa, T., Hosoda, M., Iwaoka, K., Sorimachi, A., Sahoo, S.K., Janik, M., Kranrod, C., Yonehara, H., Fukushi, M., Tokonami, S., 2011. The effect of water content on the radon emanation coefficient for some building materials used in Japan. Radiat. Meas. 46, 232–237.

Hatada, K., Ruiz, M., Riou, L.M., Lima, R.L., Goode, A.R., Watson, D.D., Beller, G.A., Glover, D.K., 2006. Organ biodistribution and myocardial uptake, washout, and redistribution kinetics of Tc-99m N-DBODC5 when injected during vasodilator stress in canine models of coronary stenoses. J. Nucl. Cardiol. 13 (6), 779–790.

Hawrami, R., Batra, A.K., Aggarwal, M.D., Roy, U.N., Groza, M., Cui, Y., Burger, A., Cherepy, N., Niedermayr, T., Payne, S.A., 2008. New scintillator materials (K2CeBr5 and Cs2CeBr5). J. Cryst. Growth 310, 2099–2102.

Heath, R.L., 1983. Status of photon counting using solid scintillation. In: McQuarrier, S.A., Ediss, C., Wiebe, L.I. (Eds.), Advances in Scintillation Counting. University of Alberta, Edmonton, pp. 156–175.

Heath, R.L., Hofstadter, R., Hughes, E.B., 1979. Inorganic scintillators, a review of techniques and applications. Nucl. Instrum. Methods 162, 431–456.

Hell, E., Knüpfer, W., Mattern, D., 2000. The evolution of scintillating medical detectors. Nucl. Instrum. Methods Phys. Res., Sect. A 454, 40–48.

Hellberg, D., et al., 2017. Effects of atorvastatin and diet interventions on atherosclerotic plaque inflammation and [^{18}F]FDG uptake in Ldlr$^{-/}$ $^{-}$Apob$^{100/100}$ mice. Atherosclerosis 263, 369–376.

Heller, S., Zanzonico, P., 2011. Nuclear probes and intraoperative gamma cameras. Semin. Nucl. Med. 41, 166–181.

Hellesen, C., Grape, S., Jansson, P., Svärd, S.J., Lindel, M.A., Andersson, P., 2017. Nuclear spent fuel parameter determination using multivariate analysis of fission product gamma spectra. Ann. Nucl. Energy 110, 886–895.

Hennig, W., Tan, H., Fallu-Labruyere, A., Warburton, W.K., McIntyre, J.I., Gleyzer, A., 2007. A phoswich well detector for radioxenon monitoring. Nucl. Instrum. Methods Phys. Res., Sect. A 579, 431–436.

Herdemann, M., Heit, I., Bosch, F.-U., Quintini, G., Scheipers, C., Weber, A., 2010. Identification of potent ITK inhibitors through focused compound library design including structural information. Bioorg. Med. Chem. Lett 20, 6998–7003.

Hennig, W., Warburton, W.K., Fallu-Labruyere, A., et al., 2009. Development of phoswich detector system for radioxenon monitoring. J. Radioanal. Nucl. Chem. 282, 681.

Henning, R., 2016. Current status of neutrinoless double-beta decay searches. Rev. Phys. 1, 29–35.

Hereforth, L., September 13, 1948. Die Fluoreszensanregung organisher Substanzen mit Alphateilchen, schnellen Elektronen und Gammastrahlen (Thesis). Technical University, Berlin-Charlottenburg.

Hertel, P., Didriksen, M., Pouzet, B., Brennum, L.T., Søby, K.K., Larsen, A.K., Christoffersen, C.T., et al., 2007. Lu 35-138 ((+)-(S)-3-{1-[2-(1-acetyl-2,3-dihydro-1H-indol-3-yl)ethyl]-3,6-dihydro-2H-pyridin-4-yl}-6-chloro-1H-indole), a dopamine D4 receptor antagonist and serotonin reuptake inhibitor: characterisation of its in vitro profile and pre-clinical antipsychotic potential. Eur. J. Pharmacol. 573, 148–160.

Herzog, H., Lerche, C., 2016. Advances in clinical PET/MRI instrumentation. Pet. Clin. 11, 95–103.

Heuvel, J.P.V., 2010. Chapter 2.03. Receptor theory and the ligand-macromolecule complex, pp. 27–50. In: McQueen, C.A. (Ed.), Comprehensive Toxicology, second ed. Elsevier, Amsterdam.

Higgins, W.M., Glodo, J., van Loef, E., Klugerman, M., Gupta, T., Cirignano, L., Wong, P., Shah, K.S., 2006. Bridgman growth of LaBr3:Ce and LaCl3:Ce crystals for high-resolution gamma-ray spectrometers. J. Cryst. Growth 287, 239–242.

Higgins, W.M., Churilov, A., van Loef, E., Glodo, J., Squillante, M., Shah, K., 2008. Crystal growth of large diameter LaBr3:Ce and CeBr3. J. Cryst. Growth 310, 2085–2089.

for the Belle II PID Group Hirose, S., 2014. Performance of the MCP-PMT for the BelleII TOP counter in a magnetic field. Nucl. Instrum. Methods Phys. Res., Sect. A 766, 163–166.

Hirose, S., Iijima, T., Inami, K., et al., 2015. Development of the micro-channel plate photomultiplier for the Belle II time-of-propagation counter. Nucl. Instrum. Methods Phys. Res., Sect. A 787, 293–296.

Hobson, P.R., 2009. Avalanche photodiodes and vacuum phototriodes for the electromagnetic calorimeter of the CMS experiment at the Large Hadron Collider. Nucl. Instrum. Methods Phys. Res., Sect. A 604, 193–195.

Hofstadter, R., 1948. Alkali halide scintillation counters. Phys. Rev. 74, 100–101.

Hofstadter, R., 1950. Properties of scintillation materials. Nucleonics 6 (5), 70–72.

Holl, I., Lorenz, E., Natkaniez, D., Renker, D., Schmelz, C., Schwartz, B., 1995. Some studies of avalanche photodiode readout of fast scintillators. IEEE Trans. Nucl. Sci. 42, 351–356.

Holm, P., Peräjärvi, K., Ristkari, S., Siiskonen, T., Toivonen, H., 2014. A capture-gated neutron spectrometer for characterization of neutron sources and their shields. Nucl. Instrum. Methods Phys. Res., Sect. A 751, 48–54.

Holmberg, P., Rieppo, R., Passi, P., 1972. Calculated efficiency values for well-type thallium-doped sodium iodide detectors. Int. J. Appl. Radiat. Isot. 23, 115–120.

Hou, X., Roos, P., 2008. Critical comparison of radiometric and mass spectrometric methods for the determination of radionuclides in environmental, biological and nuclear waste samples. Anal. Chim. Acta 608, 105–139.

Howell, R.M., Burgett, E., Hertel, N.E., Kry, S.F., Wang, Z., Salehpour, M., 2009. Measurement of high-energy neutron spectra with a bonner sphere extension system. Nucl. Technol. 168 (2), 333–339.

Howell, R.M., Burgett, E.A., Wiegel, B., Hertel, N.E., 2010. Calibration of a Bonner sphere extension (BSE) for high-energy neutron spectrometry. Radiat. Meas. 45, 1233–1237.

Hsiao, Y.-J., Chang, S.-C., 2011. Photoluminescent properties of LiInO2 nanocrystals. Mater. Lett. 65 (19–20), 2920–2922.

Hu, W., Choi, Y., Hong, K.J., 2012. Free-running ADC- and FPGA-based signal processing method for brain PET using GAPD arrays, 664, 370−375.

Hu, Z., et al., 2018. Neutron field measurement at the experimental advanced superconducting Tokamak using a bonner sphere spectrometer. Nucl. Instrum. Methods Phys. Res., Sect. A 895, 100−106.

Huang, T., Fu, Q., Yuan, C., Lin, S., 2018. A gamma and neutron phoswich read out with SiPM for SPRD. Nucl. Instrum. Methods Phys. Res., Sect. A 881, 48−52.

Hughes, K.T., Ireson, J.C., Jones, N.R.A., Kivelä, P., 2001. Colour Quench Correction in Scintillation Proximity Assays Using ParaLux Count Mode. Application Notes, pp. 12. PerkinElmer, Inc., Waltham, MA, USA. https://shop.perkinelmer.com/Content/applicationnotes/app_microbetaclrqchcrtnscintproxassays.pdf.

Huh, Y., Choi, Y., Jung, J.H., Jung, J., 2015. A method to stabilize the temperature dependent performance of G-APD arrays. Nucl. Instrum. Methods Phys. Res., Sect. A 772, 83−88.

Hull, G., Zaitseva, N.P., Cherepy, N.J., Newby, J.R., Stoeffl, W., Payne, S.A., 2009. New organic crystals for pulse shape discrimination. IEEE Trans. Nucl. Sci. 56 (3), 899−903.

Hunt, S., Iliadis, C., Longland, R., 2016. Characterization of a ^{10}B-doped liquid scintillator as a capture-gated neutron spectrometer. Nucl. Instrum. Methods Phys. Res., Sect. A 811, 108−114.

Hunter, D., Dratz, A.F., Rohrer, R.H., Coberly, J.C., 1975. Potential errors in radioassay of ^{125}I. J. Nucl. Med. 16, 952−954.

Hurtado, S., García-León, M., García-Tenorio, R., 2004. GEANT4 code for simulation of a germanium gamma-ray detector and its application to efficiency calibration. Nucl. Instrum. Methods Phys. Res., Sect. A 518, 764−774.

Huy, N.Q., Binh, D.Q., An, V.X., Loan, T.T.H., Can, N.T., 2013. Self-absorption correction in determining the ^{238}U activity of soil samples via 63.3 keV gamma ray using MCNP5 code. Appl. Radiat. Isot. 71, 11−20.

Ianakiev, K.D., Alexandrov, B.S., Littlewood, P.B., Browne, M.C., 2009. Temperature behavior of NaI(Tl) scintillation detectors. Nucl. Instrum. Methods Phys. Res., Sect. A 607, 432−438.

Ianakiev, K.D., Swinhoe, M.T., Favalli, A., Chung, K., MacArthur, D.W., 2011. ^{6}Li foil scintillation sandwich thermal neutron detector. Nucl. Instrum. Methods Phys. Res., Sect. A 652, 417−420.

Ianakiev, K.D., Hehlen, M.P., Swinhoe, M.T., 2015. Neutron detector based on Particles of ^{6}Li glass scintillator dispersed in organic lightguide matrix. Nucl. Instrum. Methods Phys. Res., Sect. A 784, 189−193.

Iida, T., Kishimoto, T., Nomachi, M., 2016. The CANDLES experiment for the study of Ca-48 double beta decay. Nucl. Part. Phys. Proc. 273−275, 2633−2635.

Iijima, T., 2009. Development of RICH counters towards the KEKB/Belle upgrade. Instrum. Methods Phys. Res., Sect. A 598, 138−142.

Iijima, T., 2011. Status and perspectives of vacuum-based photon detectors. Instrum. Methods Phys. Res., Sect. A 639, 137−143.

Ikagawa, T., Kataoka, J., Yatsu, Y., Kawai, N., Mori, K., Kamae, T., Tajima, H., Mizuno, T., Fukazawa, Y., Ishikawa, Y., Kawabata, N., Inutsuka, T., 2003. Performance of large-area avalanche photodiode for low-energy X-rays and γ-rays scintillation detection. Nucl. Instrum. Methods Phys. Res., Sect. A 515, 671−679.

Ikagawa, T., Kataoka, J., Yatsu, Y., Saito, T., Kuramoto, Y., Kawai, N., Kokubun, M., Kamae, T., Ishikawa, Y., Kawabata, N., 2005. Study of large area Hamamatsu avalanche photodiode in a γ-ray scintillation detector. Nucl. Instrum. Methods Phys. Res., Sect. A 538, 640−650.

Ikhlef, A., Skowronek, M., Beddar, A.S., 2000. X-ray imaging and detection using plastic scintillating fibers. Nucl. Instrum. Methods Phys. Res., Sect. A 442, 428−432.

Ishibashi, H., Kurashige, K., Kurata, Y., Susa, K., Kobayashi, M., Tanaka, M., Hara, K., Ishii, M., 1998. Scintillation performance of large Ce-doped Gd_2SiO_5 (GSO) single crystal. IEEE Trans. Nucl. Sci. 45, 518−521.

Ishii, M., Kuwano, Y., Asai, T., Asaba, S., Kawamura, M., Senguttuvan, N., Hayashi, T., et al., 2005. Boron based oxide scintillation glass for neutron detection. Nucl. Instrum. Methods Phys. Res., Sect. A 537, 282−285.

Ivan, C., Cassette, P., Sahagia, M., 2008. A new TDCR-LS Counter using Channel photomultiplier tubes. Appl. Radiat. Isot. 66, 1006−1011.

Ivan, C., Wätjen, A.C., Cassette, P., Sahagia, M., Antohe, A., Grigorescu, E.L., 2010. Participation in the CCRI(II)-K2.H-3 comparison and study of the new TDCR-LS counter with 6 CPMs. Appl. Radiat. Isot. 68, 1543−1545.

Iwahara, A., Poledna, R., da Silva, C.J., Tauhata, L., 2009. Primary activity standardization of 57Co by sum-peak method. Appl. Radiat. Isot. 67, 1887−1891.

Iwanowska, J., Swiderski, L., Moszynski, M., Yanagida, T., Yokota, Y., Yoshikawa, A., Fukuda, K., Kawaguchi, N., Ishizu, S., 2011. Thermal neutron detection with Ce3+ doped LiCaAlF6 single crystals. Nucl. Instrum. Methods Phys. Res., Sect. A (in press).

Iwase, H., Niita, K., Nakamura, T., 2002. Development of general-purpose particle and heavy ion transport Monte Carlo code. J. Nucl. Sci. Technol. 39 (11), 1142−1151.

Jackson, S.J., et al., 2011. Manganese-enhanced magnetic resonance imaging (MEMRI) of rat brain after systemic administration of MnCl2: hippocampal signal enhancement without disruption of hippocampus-dependent behavior. Behav. Brain Res. 216, 293−300.

Jamieson, B., Rebenitsch, L.A., 2015. Determining the ^{6}Li doped side of a glass scintillator for ultra cold neutrons. Nucl. Instrum. Methods Phys. Res., Sect. A 790, 6−9.

Jang, K.W., Lee, B.S., Moon, J.H., 2011. Development and characterization of the integrated fiber-optic radiation sensor for the simultaneous detection of neutrons and gamma rays. Appl. Radiat. Isot. 69, 711−715.

Jeffery, J.A., Sharom, J.R., Fazekas, M., Rudd, P., Welchner, E., Thauvette, L., White, P.W., 2002. An ATPase assay using scintillation proximity beads for high-throughput screening or kinetic analysis. Anal. Biochem. 304, 55−62.

Jensen, L., et al., 2017. Absorption, metabolism and excretion of the GLP-1 analogue semaglutide in humans and nonclinical species. Eur. J. Pharm. Sci. 104, 31−41.

Jhingan, A., Singh, H., Singh, R.P., Golda, K.S., Sugathan, P., Mandal, S., Bhowmik, R.K., 2008. Simple ways of n-γ discrimination using charge comparison technique. Nucl. Instrum. Methods Phys. Res., Sect. A 585, 165−171.

Jia, Z.-Y., Deng, H.-F., Huang, R., Yang, Y.-Y., Yang, X.-C., Qi, Z.-Z., Ou, X.-H., 2011. In vitro and in vivo studies of adenovirus-mediated human norepinephrine transporter gene transduction to hepatocellular carcinoma. Cancer Gene Ther. 18 (3), 196−205.

Jiang, l., Wang, Z., Chen, H., et al., 2018. Thermal annealing effects on the luminescence and scintillation properties of $CaMoO_4$ single crystal grown by Bridgman method. J. Alloy. Comp. 734, 179−187.

Jie, X., Lianyun, D., Ying, S., Jianjun, X., Huajun, Q., Guoxin, S., 2011. The influence of Pr3+ co-doping on the luminescent properties of Lu2O3:5 mol% Eu films. J. Lumin. 131, 970−974.

Jing, M., et al., 2015. Effects of zinc deficiency and zinc supplementation on homocysteine levels and related enzyme expression in rats. J. Trace Elem. Med. Biol. 30, 77−82.

Jones, N.R.A., Kivelä, P., Hughes, K.T., Ireson, J.C., 1997. Color quench correction in scintillation proximity assays using ParaLux Count Mode. J. Biomol. Screen 2 (3), 179−182. http://journals.sagepub.com/doi/pdf/10.1177/108705719700200308.

Jones, S.A., Parks, D.J., Kliewer, S.A., 2003. Cell-free ligand binding assays for nuclear receptors, pp. 53−71. In: Russell, D.W., Mangelsdorf, D.J. (Eds.), Methods in Enzymology, vol. 364. Elsevier, Amsterdam.

Kah, D.H., Bae, J.B., Hyun, H.J., Kim, H.J., Kim, H.O., Park, H., 2011. Fabrication and performance test of a silicon photo-strip detector coupled with a crystal scintillator. Nucl. Instrum. Methods Phys. Res., Sect. A 628, 256−259.

Kalber, M.N., 1975. Proceedings of the NATO Advanced Study Institute on Radiation Damage in Material. Corsica, 1973. Hordhof International Publishing.

Kallman, H., 1950. Scintillation counting with solutions. Phts. Rev. 78 (5), 621−622.

Kamboj, S., Kahn, B., 1996. Evaluation of Monte Carlo simulation of photon counting efficiency for germanium detectors. Health Phys. 70, 512−529.

Kandarakis, I., Cavouras, D., Sianoudis, I., Nikolopoulos, D., Episkopakis, A., Linardatos, D., Margetis, D., Nirgianaki, E., Roussou, M., Melissaropoulos, P., Kalivas, N., Kalatzis, I., Kourkoutas, K., Dimitropoulos, N., Louizi, A., Nomicos, C., Panayiotakis, G., 2005. On the response of Y3Al5O12: Ce (YAG: Ce) powder scintillating screens to medical imaging X-rays. Nucl. Instrum. Methods Phys. Res., Sect. A 538, 615−630.

Kang, J., Choi, Y., 2016. Simulation study of PET detector configuration with thick light guide and GAPD array having large-area microcells for high effective quantum efficiency. Comput. Methods Progr. Biomed. 131, 79−87.

Kang, W.G., Choi, J.H., Jeon, E.J., et al., 2013. Ultra-low gamma-ray measurement system for neutrinoless double beta decay. Appl. Radiat. Isot. 81, 290−293.

Kapusta, M., Crespo, P., Wolski, D., Moszyński, M., Enghardt, W., 2003. Hamamatsu S8550 APD arrays for high-resolution scintillator matrices readout. Nucl. Instrum. Methods Phys. Res., Sect. A 504, 139−142.

Kasahara, S., Takashima, T., Hirahara, M., 2012. Variability of the minimum detectable energy of an APD as an electron detector. Nucl. Instrum. Methods Phys. Res., Sect. A 664, 282−288.

Kasban, H., Ali, E.H., Arafa, H., 2017. Diagnosing plant pipeline system performance using radiotracer techniques. Nucl. Eng. Technol. 49, 196−208.

Kaschuck, Y., Esposito, B., 2005. Neutron/γ-ray digital pulse shape discrimination with organic scintillators. Nucl. Instrum. Methods Phys. Res., Sect. A 551, 420−428.

Katagiri, M., Sakasai, K., Matsubayashi, M., Kojima, T., 2004. Neutron/γ-ray discrimination characteristics of novel neutron scintillators. Nucl. Instrum. Methods Phys. Res., Sect. A 529, 317−320.

Kataoka, J., Saito, T., Kuramoto, Y., Ikagawa, T., Yatsu, Y., Kotoku, J., Arimoto, M., Kawai, N., Ishikawa, Y., Kawabata, N., 2005. Recent progress of avalanche photodiodes in high-resolution X-rays and γ-rays detection. Nucl. Instrum. Methods Phys. Res., Sect. A 541, 398−404.

Kataoka, J., Koizumi, M., Tanaka, S., Ishibashi, H., Nakamori, T., Kawai, N., Ikeda, H., Ishikawa, Y., Kawabata, N., Matsunaga, Y., Kishimoto, S., Kubo, H., 2009. Development of large-area, reverse-type APD-arrays for high-resolution medical imaging. Nucl. Instrum. Methods Phys. Res., Sect. A 604, 323−326.

Kataoka, J., Toizumi, T., Nakamori, T., Yatsu, Y., Tsubuku, Y., Kuramoto, Y., Enomoto, T., Usui, R., Kawai, N., Ashida, H., et al., 2010a. In-orbit performance of avalanche photodiode as radiation detector on board the picosatellite Cute-1.7+APD II. J. Geophys. Res. 115 (5) https://doi.org/10.1029/2009JA014699. Art. No.: A05204.

Kataoka, J., Matsuda, H., Nishikido, F., Koizumi, M., Ikeda, H., Yoshino, M., Miura, T., Tanaka, S., Ishikawa, Y., Kawabata, N., et al., 2010b. Development of an APD-based PET module and preliminary resolution performance of an experimental prototype gantry. IEEE Trans. Nucl. Sci. 57 (5), 2448−2454.

Kato, T., Kataoka, J., Nakamori, T., Miura, T., Matsuda, H., Sato, K., Ishikawa, Y., Yamamura, K., Kawabata, N., Ikeda, H., Sato, G., Kamada, K., 2011. Development of a large-area monolithic 4×4 MPPC array for a future PET scanner employing pixelized Ce:LYSO and Pr:LuAG crystals. Nucl. Instrum. Methods Phys. Res., Sect. A 638, 83−91.

Kawamura, S., Kaneko, J.H., Higuchi, M., Fujita, F., Homma, A., Haruna, J., Saeki, S., et al., 2007. Investigation of Ce-doped Gd2Si2O7 as a scintillator material. Nucl. Instrum. Methods Phys. Res., Sect. A 583, 356−359.

Kawano, T., Ebihara, H., 1990. Determination of disintegration rates of a ^{60}Co point source and volume sources by the sum-peak method. Int. J. Radiat. Appl. Instrum. A Appl. Radiat. Isot. 41, 163−167.

Kawano, T., Ebihara, H., 1991. Measurement of disintegration rates of small ^{60}Co sources in lead containers by the sum-peak method. Appl. Radiat. Isot. 42, 1165−1168.

Kawano, T., Ebihara, H., 1992. Error estimation for the sum-peak method by use of two ^{60}Co point sources in place of extended sources. Appl. Radiat. Isot. 43, 705−711.

Kawasaki, T., Nakamura, T., Toh, K., et al., 2014. Detector system of the *SENJU* single-crystal time-of-flight neutron diffractometer at J-PARC/MLF. Nucl. Instrum. Methods Phys. Res., Sect. A 735, 444−451.

Kerkeni, M., Letaief, A., Achour, A., Miled, A., Trivin, F., Maaroufi, K., 2009. Hyperhomocysteinemia, paraoxonase concentration and cardiovascular complications in Tunisian patients with nondiabetic renal disease. Clin. Biochem. 42, 777−782.

Kertzscher, G., Beddar, S., 2018. Novel inorganic scintillation detector system for real-time treatment verification during brachytherapy. Phys. Med. 51, 73.

Khan, F., Ali, N., Khan, E.U., Khattak, N.U., Khan, K., 2009. Radon monitoring in water sources of Balakot and Mansehra cities lying on a geological fault line. Radiat. Prot. Dosim. 138 (2), 174−179.

Khan, A., Rooh, G., Kim, H.J., Kim, S., 2018. Ce^{3+}-activated Tl_2GdCl_5: novel halide scintillator for X-ray and γ-ray detection. J. Alloy. Comp. 741, 878−882.

Kharel, Y., Mathews, T.P., Kennedy, A.J., Houck, J.D., MacDonald, T.L., Lynch, K.R., 2011. A rapid assay for assessment of sphingosine kinase inhibitors and substrates. Anal. Biochem. 411, 230−235.

Kiffe, M., Jehle, A., Ruembeli, R., 2003. Combination of high-performance liquid chromatography and microplate scintillation counting for crop and animal metabolism studies: a comparison with classical on-line and thin-layer chromatography radioactivity detection. Anal. Chem. 75 (4), 723–730.

Kiffe, M., Nufer, R., Trunzer, M., Graf, D., 2007. Cytostar-T plates—A valid alternative for microplate scintillation counting of low radioactivity in combination with high-performance liquid chromatography in drug metabolism studies? J. Chromatogr., A 1157, 65–72.

Kim, I.J., Park, C.S., Choi, H.D., 2003. Absolute calibration of 60Co by using sum-peak method and an HPGe detector. Appl. Radiat. Isot. 58, 227–233.

Kim, H.J., Kim, S.C., Kim, S.K., et al., 2007. Searches for the β+/EC decays of 64Zn and 112Sn, and the ββ decay transitions of 124Sn to the excited states of 124Te. Nucl. Phys. A 793, 171–177.

Kim, H.S., Ha, J.H., Park, S.H., Cho, S.Y., Kim, Y.K., 2009. Fabrication and performance characteristics of a CsI(Tl)/PIN diode radiation sensor for industrial applications. Appl. Radiat. Isot. 67, 1463–1465.

Kim, B.H., Kim, S.I., Kim, J.S., Kim, J.L., Lee, J.I., 2010. Response of several neutron measuring devices under the fractional change of thermal and fast neutrons. Radiat. Meas. 45, 1541–1543.

Kim, H., Frisch, H., Chen, C.-T., Genat, J.-F., Tang, F., Moses, W.W., Choong, W.S., Kao, C.-M., 2010a. A design of a PET detector using micro-channel plate photomultipliers with transmission-line readout. Nucl. Instrum. Methods Phys. Res., Sect. A 622, 628–636.

Kim, H.S., Park, S.H., Ha, J.H., Lee, D.-H., Cho, S.Y., 2010b. Characteristics of a fabricated PIN photodiode for a matching with a CsI(Tl) scintillator. IEEE Trans. Nucl. Sci. 57 (3), 1382–1385.

Kim, K.H., Kang, D.-W., Yuk, S., 2006. Design of pixellated scintillator on PIN-type photodiode for digital mammography. Nucl. Instrum. Methods Phys. Res., Sect. A 567, 364–367.

Kim, H., Chen, C.-T., Frisch, H., Tang, F., Kao, C.-M., 2012a. A prototype TOF PET detector module using a micro-channel plate photomultiplier tube with waveform sampling. Nucl. Instrum. Methods Phys. Res., Sect. A 662, 26–32.

Kim, H., Chen, C.-T., Frisch, H., Tang, F., Kao, C.-M., 2012b. An application of micro-channel plate photomultiplier tube to positron emission tomography. Phys. Proc. 37, 1480–1487.

Kim, C., Kim, H., Kim, J., et al., 2015. Replacement of a photomultiplier tube in a 2-inch thallium-doped sodium iodide gamma spectrometer with silicon photomultipliers and a light guide. Nucl. Eng. Technol. 47, 479–485.

Kim, H.J., Rooh, G., Park, H., Kim, S., 2016. Investigations of scintillation characterization of Ce-activated $Tl_2LiGdBr_6$ single crystal. Radiat. Meas. 90, 279–281.

Kim, M.H., et al., 2017. A novel Tc-99m and fluorescence labeled peptide as a multimodal imaging agent for targeting angiogenesis in a murine hindlimb ischemia model. Appl. Radiat. Isot. 121, 22–27.

Kingma Jr., J.G., Vincent, C., Rouleau, J.R., Kingma, I., 2006. Influence of acute renal failure on coronary vasoregulation in dogs. J. Am. Soc. Nephrol. 17, 1316–1324.

on behalf of the LHCb collaboration Kirn, T., 2017. SciFi – a large scintillating fibre tracker for LHCb. Nucl. Instrum. Methods Phys. Res., Sect. A 845, 481–485.

Klein, H., Brooks, F.D., 2006. Scintillation detectors for fast neutrons. In: International Workshop on Fast neutron Detectors, University of Cape Town, South Africa, April 3-6, 2006 (FNDA2006)097. http://pos.sissa.it/archive/conferences/025/097/FNDA2006_097.pdf.

Proceedings of the International workshop on neutron field spectrometry in science, technology and radiation protection, pisa, Italy, June 4–8, 2000. In: Klein, H., Thomas, D., Menzel, H.G., Curzio, G., d'Errico, F. (Eds.), Nucl. Instrum. Methods Phys. Res., Sect. A 476 (1–2), 1–536.

Knitel, M.J., 1998. New Inorganic Scintillators and Storage Phosphors for Detection of Thermal Neutrons. Delft University Press, Delft, The Netherlands, p. 180.

Knoll, P., Mirzaei, S., Rahmim, A., 2016. Scatter correction for planar and SPECT imaging with factor analysis. Phys. Med. 32, 222–250.

Knudsen, L.B., et al., 2012. Functional importance of GLP-1 receptor species and expression levels in cell lines. Regul. Pept. 175, 21–29.

Ko, G.B., Yoon, H.S., Kwon, S.I., 2013. Development of a front-end analog circuit for multi-channel SiPM readout and performance verification for various PET detector designs. Nucl. Instrum. Methods Phys. Res., Sect. A 703, 38–44.

Kobayashi, M., Tamagawa, Y., Tomita, S., et al., 2012. Significantly different pulse shapes for γ- and α-rays in $Gd_3Al_2Ga_3O_{12}$:Ce^{3+} scintillating crystals. Nucl. Instrum. Methods Phys. Res., Sect. A 694, 91–94.

Kobayashi, T., Yamamoto, S., Okumura, S., Yeom, J.-Y., Kamada, K., Yoshikawa, A., 2017a. Basic performance of Mg co-doped new scintillator used for TOF-DOI-PET systems. Nucl. Instrum. Methods Phys. Res., Sect. A 842, 14–19.

Kobayashi, T., Yamamoto, S., Yeom, J.-Y., Kamada, K., Yoshikawa, A., 2017b. Development of high resolution phoswich depth-of-interaction block detectors utilizing Mg co-doped new scintillators. Nucl. Instrum. Methods Phys. Res., Sect. A 875, 111–118.

Köhli, M., Allmendinger, F., Häußler, W., et al., 2016. Efficiency and spatial resolution of the CASCADE thermal neutron detector. Nucl. Instrum. Methods Phys. Res., Sect. A 828, 242–249.

Kojima, T., Katagiri, M., Tsutsui, N., Imai, K., Matsubayashi, M., Sakasai, K., 2004. Neutron scintillators with high detection efficiency. Nucl. Instrum. Methods Phys. Res., Sect. A 529, 325–328.

Kolar, Z.I., den Hollander, W., 2004. 2003: a centennial of spinthariscope and scintillation counting. Appl. Radiat. Isot. 61, 261–266.

Kole, M., Chauvin, M., Fukazawa, Y., 2015. PoGOLino: a scintillator-based balloon-borne neutron detector. Nucl. Instrum. Methods Phys. Res., Sect. A 770, 68–75.

Koppen, V., Jones, R., Bockx, M., Cuyckens, F., 2014. High volume injections of biological samples for sensitive metabolite profiling and quantitation. J. Chromatogr. A 1372, 102–109.

Korpar, S., 2011. Status and perspectives of solid state photon detectors. Nucl. Instrum. Methods Phys. Res., Sect. A 639, 88–93.

Korpar, S., Križan, P., Pestotnik, R., Stanovnik, A., 2008. Timing and cross-talk properties of BURLE multi-channel MCP PMTs. Nucl. Instrum. Methods Phys. Res., Sect. A 595, 160–172.

Korpar, S., Adachi, I., Chagani, H., Dolenec, R., Hara, K., Iijima, T., Križan, P., Nishida, S., Pestotnik, R., Stanovnik, A., 2009. Measurements of Cherenkov photons with silicon photomultipliers. Nucl. Phys. B 197, 283–287.

Korpar, S., Chagani, H., Dolenec, R., Križan, P., Pestotnik, R., Stanovnik, A., 2010. A module of silicon photo-multipliers for detection of Cherenkov radiation. Nucl. Instrum. Methods Phys. Res., Sect. A 613, 195–199.

Korun, M., Martinčič, R., 1993. Coincidence summing in gamma and x-ray spectrometry. Nucl. Instrum. Methods Phys. Res., Sect. A 325, 478–484.

Korzhik, M., Lecoq, P., 2001. Search of new scintillation materials for nuclear medicine applications. IEEE Trans. Nucl. Sci. 48 (3), 628–631.

Korzhik, M., Fedorov, A., Annenkov, A., Borissevitch, A., Dossovitski, A., Missevitch, O., Lecoq, P., 2007. Development of scintillation materials for PET scanners. Nucl. Instrum. Methods Phys. Res., Sect. A 571, 122–125.

Kosev, K., Butterling, M., Anwand, W., Cowan, T., Hartmann, A., Heidel, K., Jungmann, M., Krause-Rehberg, R., Massarczyk, R., Schilling, K.D., Schwengner, R., Wagner, A., 2010. Evaluation of a microchannel-plate PMT as a potential timing detector suitable for positron lifetime measurements. Nucl. Instrum. Methods Phys. Res., Sect. A 624.

Koteswara Rao, V., Agrawal, P.C., Sreekumar, P., Thyagarajan, K., 2009. The scientific objectives of the ASTROSAT mission of ISRO. Acta Astronaut. 65, 6–17.

Kowalski, Z., Kaczmarek, S.M., Berkowski, M., Głowacki, M., Zhydachevskii, Y.A., Suchocki, A., 2017. Growth and optical properties of $ZnWO_4$ single crystals pure and doped with Ca and Eu. J. Cryst. Growth 457, 117–121.

for the SNO+ Collaboration Kraus, C., 2006. SNO with liquid scintillator: SNO+. Prog. Part. Nucl. Phys. 57, 150–152.

Kraus, H., Mikhailik, V.B., Wahl, D., 2005. Multiple photon counting coincidence (MPCC) technique for scintillator characterisation and its application to studies of CaWO4 and ZnWO4 scintillators. Nucl. Instrum. Methods Phys. Res., Sect. A 553, 522–534.

Krause, S.M., Clayton, F.C., Williams Jr., D.L., 2004. Myocardial protection with the non-selective endothelin receptor antagonist L-753,037 following acute coronary artery occlusion in the dog. J. Cardiovasc. Pharmacol. 43 (2), 214–221.

Kriss, A.A., Hamby, D.M., 2004. Beta spectroscopy with a large-area avalanche photodiode module and a plastic scintillator. Nucl. Instrum. Methods Phys. Res., Sect. A 525, 553–559.

Krutyak, N.R., Mikhailin, V.V., Vasil'ev, A.N., et al., 2013. The features of energy transfer to the emission centers in $ZnWO_4$ and $ZnWO_4$:Mo. J. Lumin. 144, 105–111.

Kumar, V.P., Basavannacharya, C., de Sousa, S.M., 2014. A microplate assay for the coupled transglycosylase–transpeptidase activity of the penicillin binding proteins; a vancomycin-neutralizing tripeptide combination prevents penicillin inhibition of peptidoglycan synthesis. Biochem. Biophys. Res. Commun. 450, 347–352.

Kumara, W.A.S., Halvorsen, B.M., Melaaen, M.C., 2010. Single-beam gamma densitometry measurements of oil–water flow in horizontal and slightly inclined pipes. Int. J. Multiph. Flow 36, 467–480.

Künze, N., Koroleva, M., Reuther, C.-D., 2012. ^{222}Rn activity in soil gas across selected fault segments in the Cantabrian Mountains, NW Spain. Radiat. Meas. 47, 389–399.

Künze, N., Koroleva, M., Reuther, C.-D., 2013. Soil gas ^{222}Rn concentration in northern Germany and its relationship with geological subsurface structures. J. Environ. Radioact. 115, 83–96.

Kurudirek, S.V., Menkara, H., Klein, B.D.B., Hertel, N.E., Summers, C.J., 2018. Effect of annealing temperature on the photoluminescence and scintillation properties of ZnO nanorods. Nucl. Instrum. Methods Phys. Res., Sect. A 877, 80–86.

Labanti, C., Marisaldi, M., Fuschino, F., Galli, M., Argan, A., Bulgarelli, A., Di Cocco, G., Gianotti, F., Tavani, M., Trifoglio, M., 2009. Design and construction of the Mini-Calorimeter of the AGILE satellite. Nucl. Instrum. Methods Phys. Res., Sect. A 598, 470–479.

Lacerda, M.A.S., Méndez-Villafañe, R., Lorente, A., Ibañez, S., Gallego, E., Vega-Carrillo, H.R., 2017. A new response matrix for a ^{6}LiI scintillator BSS system. Nucl. Instrum. Methods Phys. Res., Sect. A 868, 122–128.

Lacy, S., Hsu, B., Miles, D., Aftab, D., Wang, R., Nguyen, L., 2015. Metabolism and disposition of Cabozantinib in healthy male volunteers and pharmacologic characterization of its major metabolites. Drug Metab. Dispos. 43, 1190–1207.

Lagares, J.I., et al., 2016. Neutron spectra around a tandem linear accelerator in the generation of ^{18}F with a bonner sphere spectrometer. Appl. Radiat. Isot. 114, 154–158.

The TEXONO Collaboration Lai, W.P., Cheng, K.C., Li, H.B., Sheng, H.Y., Zhuang, B.A., Chang, C.Y., Chen, C.P., Chen, Y.P., Hsu, H.C., Li, J., et al., 2001. The electronics and data acquisition systems of a CsI(Tl) scintillating crystal detector for low energy neutrino experiment. Nucl. Instrum. Methods Phys. Res., Sect. A 465, 550–565.

Laitinen, P., Tiourine, G., Touboltsev, V., Räisänen, J., 2002. Detection system for depth profiling of radiotracers. Nucl. Instrum. Methods Phys. Res., Sect. B 190, 183–185.

Landsberger, S., Brabec, C., Canion, B., et al., 2013. Determination of ^{226}Ra, ^{228}Ra and ^{210}Pb in NORM products from oil and gas exploration: problems in activity underestimation due to the presence of metals and self-absorption of photons. J. Environ. Radioact. 125, 23–26.

LANL, 2003. MCNP5 Monte-Carlo N-particle Transport Code Version 5, LA-CP-03-0245. Los Alamos, NM, USA.

L'Annunziata, M.F., 1984. The detection and measurement of radionuclides. In: L'Annunziata, M.F., Legg, J.O. (Eds.), Isotopes and Radiation in Agricultural Sciences, vol. 1, pp. 141–231.

L'Annunziata, M.F., 1987. Radionuclide Tracers: Their Detection and Measurement. Academic Press, London.

L'Annunziata, M.F., Nellis, S.W., 2001a. Flow Scintillation Analyzer (FSA) Interfaced with the HPLC and Nuclear Magnetic Resonance (NMR) Spectrometer. A State-of-the-Art Application of the Radiomatic FSA. FSA Application Note. FSA-004, pp. 7. PerkinElmer, Inc., Waltham, MA, USA. http://www.perkinelmer.com/CMSResources/Images/44-73061APP_FSAInterfacedHPLC.pdf.

L'Annunziata, M.F., Nellis, S.W., 2001b. Metabolism Studies with On-Line HPLC and Mass Spectrometry (MS) Interfaced with the Flow Scintillation Analyzer (FSA). FSA Application Note. FSA-005, pp. 7. PerkinElmer Inc., Waltham, MA, USA. http://www.perkinelmer.com/CMSResources/Images/44-73062APP_FSAMetabolismStudies.pdf.

L'Annunziata, M.F., 2007. Radioactivity: Introduction and History. Elsevier, Amsterdam, p. 609.

L'Annunziata, M.F., 2016. Radioactivity: Introduction and History: From the Quantum to Quarks. Elsevier, Amsterdam, p. 902.

Lapington, J.S., 2012. High speed imaging using a capacitive division technique. Nucl. Instrum. Methods Phys. Res., Sect. A 695, 410–414.

Lapington, J.S., Fraser, G.W., Miller, G.M., Ashton, T.J.R., Jarron, P., Despeisse, M., Powolny, F., Howorth, J., Milnes, J., 2009. A multichannel high time resolution detector for high content imaging. Nucl. Instrum. Methods Phys. Res., Sect. A 610, 123–127.

Lasserre, T., 2013. Testing the reactor and gallium anomalies with intense (anti)neutrino emitters. Nucl. Phys. B 235–236, 214–219.

Lauritzen, K.H., et al., 2014. Lactate receptor sites link neurotransmission, neurovascular coupling, and brain energy metabolism. Cerebr. Cortex 24, 2784–2795.

Lawrence, C.C., Febbraro, M., Massey, T.N., et al., 2014a. Neutron response characterization for an EJ299-33 plastic scintillation detector. Nucl. Instrum. Methods Phys. Res., Sect. A 759, 16–22.

Lawrence, D.J., Feldman, W.C., Gold, R.E., Goldsten, J.O., McNutt, R.L., 2014b. The neutron, gamma-ray, X-ray spectrometer (NGXS): a compact instrument for making combined measurements of neutrons, gamma-rays, and X-rays. Acta Astronaut. 93, 524–529.

Lecoq, P., 2016. Development of new scintillators for medical applications. Nucl. Instrum. Methods Phys. Res., Sect. A 809, 130–139.

Lee, K.W., Sheu, R.J., 2015. Spectral correction factors for conventional neutron dosemeters used in high-energy neutron environments. Radiat. Prot. Dosim. 164 (3), 210–218.

Lee, K.W., Sheu, R.J., 2017. Comparing two measurements of the same cosmic-ray neutron spectrum using standard Bonner spheres and high-sensitivity Bonner cylinders. Radiat. Prot. Dosim. 177 (4), 450–457.

Lee, C., Soo Kim, Y., Suk Sul, W., Kim, H., Han Shin, S., Cho, G., 2011. Feasibility study on TOF-PET with fill factor improved SiPMs. Nucl. Instrum. Methods Phys. Res., Sect. A (in press).

Lee, H.Y., Taddeucci, T.N., Haight, R.C., 2013. Li-glass detector response study with a ^{252}Cf source for low-energy prompt fission neutrons. Nucl. Instrum. Methods Phys. Res., Sect. A 703, 213–219.

Lee, S.C., Jeon, H.B., Kang, K.H., Kim, B.B., Park, H., 2019. Study of silicon photodiode performance for X-ray detector in cargo system. Nucl. Instrum. Methods Phys. Res., Sect. A (in press).

Lees, J.E., Fraser, G.W., 2002. Efficiency enhancements for MCP-based beta autoradiography imaging. Nucl. Instrum. Methods Phys. Res., Sect. A 477, 239–243.

Lees, J.E., Hales, J.M., 2001. Imaging and quantitative analysis of tritium-labelled cells in lymphocyte proliferation assays using microchannel plate detectors originally developed for x-ray astronomy. J. Immunol. Methods 247, 95–102.

Lees, J.E., Fraser, G.W., Dinsdale, D., 1997. Direct beta autoradiography using microchannel plate (MCP) detectors. Nucl. Instrum. Methods Phys. Res., Sect. A 392, 349–353.

Lees, J.E., Fraser, G.W., Carthew, P., 1998. Microchannel plate detectors for ^{14}C autoradiography. IEEE Trans. Nucl. Sci. 45, 1288–1292.

Lees, J.E., Pearson, J.F., Fraser, G.W., Hales, J.M., Richards, P.G., 1999. An MCP-based system for beta autoradiography. IEEE Trans. Nucl. Sci. 46, 636–638.

Lees, J.E., Bassford, D., Whitford, C., Blake, O., 2005. Novel electronics for MCP imaging detectors. Nucl. Instrum. Methods Phys. Res., Sect. A 537, 621–625.

Legou, P., Perez, P., Sacquin, Y., et al., 2019. A compact fast low energy gamma ray detector insensitive to high magnetic field. Nucl. Instrum. Methods Phys. Res., Sect. A (in press).

Lehmann, A., Britting, A., Eyrich, W., Pizzolotto, C., Teufel, A., Düren, M., Föhl, K., Hoek, M., Lu, S., Schepers, G., Seitz, B., Sfienti, C., 2008. Performance studies of microchannel plate PMTs in high magnetic fields. Nucl. Instrum. Methods Phys. Res., Sect. A 595, 173–176.

Lempicki, A., 1995. The physics of inorganic scintillators. J. Appl. Spectrosc. 62, 209–231.

Lempicki, A., Berman, E., Wojtowicz, A.J., Balcerzyk, M., Boatner, L.A., 1993. Cerium-doped orthophosphates: new promising scintillators. IEEE Trans. Nucl. Sci. 40, 384–387.

Leroy, C., Rancoita, P.-G., Barone, M., Gaddi, A., Price, L., Ruchti, R. (Eds.), 2010. Astroparticle, Particle and Space Physics, Detectors and Medical Physics Applications – Vol. 5 Proceedings of the 11th Conference. World Scientific Publishing, Hackemsack, NJ, 948 pp.

Leutz, H., 1995. Scintillating fibers. Nucl. Instrum. Methods Phys. Res., Sect. A 364, 422–448.

Leverington, B., Rosellini, R., Anelli, M., Campana, P., 2013. Particle trackers based on 1 mm and 0.7 mm scintillating fibers readout by a multi-anode photomultiplier. Nucl. Instrum. Methods Phys. Res., Sect. A 718, 509–512.

Li, H.B., Liu, Y., Chang, C.C., Chang, C.Y., Chao, J.H., Chen, C.P., Chen, T.Y., et al., 2001. A CsI(Tl) scintillating crystal detector for the studies of low-energy neutrino interactions. Nucl. Instrum. Methods Phys. Res., Sect. A 459, 93–107.

Li, H., Wu, Y., Cao, C., Huo, H., Tang, B., 2014. Design optimization, manufacture and response measurements for fast-neutron radiography converters made of scintillator and wavelength-shifting fibers. Nucl. Instrum. Methods Phys. Res., Sect. A 762, 64–69.

Li, Q., et al., 2018. Tumor uptake of pegylated diabodies: balancing systemic clearance and vascular transport. J. Control. Release 279, 126–135.

Liao, C., Yang, H., 2015. Pulse shape discrimination using EJ-299-33 plastic scintillator coupled with a Silicon Photomultiplier array. Nucl. Instrum. Methods Phys. Res. Sect. A 789, 150–157.

Lim, H.-J., Choi, Y.-M., Choue, R., 2008. Dietary intervention with emphasis on folate intake reduces serum lipids but not plasma homocysteine levels in hyperlipidemic patients. Nutr. Res. 28, 767–774.

Limkitjaroenporn, P., Kaewkhao, J., Limsuwan, P., Chewpraditkul, W., 2010. Nonproportionality of electron response using CCT: plastic scintillator. Appl. Radiat. Isot. 68, 1780–1784.

Lin, C.-C., Yin, M.-C., 2009. Vitamins B depletion, lower iron status and decreased antioxidative defense in patients with chronic hepatitis C treated by pegylated interferon alfa and ribavirin. Clin. Nutr. 28, 34–38.

Lioliou, G., Barnett, A.M., 2016. Gallium Arsenide detectors for X-ray and electron (beta particle) spectroscopy. Nucl. Instrum. Methods Phys. Res., Sect. A 836, 37–45.

Lipowska, M., et al., 2017. Monoanionic ^{99m}Tc-tricarbonyl-amino-polycarboxylate complexes with uncharged pendant groups: radiosynthesis and evaluation as potential renal tubular tracers. Nucl. Med. Biol. 47, 48–55.

Liu, Y., Chen, C.P., Li, H.B., Tang, C.H., Chang, C.Y., Hou, L., Lai, W.P., Li, J., Lin, S.T., Luo, C.S., The TEXONO Collaboration, 2002. Studies of prototype CsI(Tl) crystal scintillators for low-energy neutrino experiments. Nucl. Instrum. Methods Phys. Res., Sect. A 482, 125–143.

Liu, G., et al., 2010. A digital method for the discrimination of neutrons and γ rays with organic scintillation detectors using frequency gradient analysis. IEEE Trans. Nucl. Sci. 57 (3), 1682–1691.

Liu, Y., et al., 2016. Determination of the conventional true value of gamma-ray air kerma in a minitype reference radiation. Appl. Radiat. Isot. 118, 238–245.

Liu, S., Mares, J.A., Babin, V., et al., 2017. Effect of reducing Lu^{3+} content on the fabrication and scintillation properties of non-stoichiometric $Lu_{3-x}Al_5O_{12}$:Ce ceramics. Opt. Mater. 63, 179–184.

Llopart, X., Ballabriga, R., Campbell, M., Tlustos, L., Wong, W., 2007. Timepix, a 65k programmable pixel readout chip for arrival time, energy and/or photon counting measurements. Nucl. Instrum. Methods Phys. Res., Sect. A 581, 485–494.

Llosá, G., 2015. Recent developments in photodetection for medical applications. Nucl. Instrum. Methods Phys. Res., Sect. A 787, 353–357.

Lodge, M.A., Holt, D.P., Kinahan, P.E., Wong, D.F., Wahl, R.L., 2015. Performance assessment of a NaI(Tl) gamma counter for PET applications with methods for improved quantitative accuracy and greater standardization. EJNMMI Phys. 2 (11), 1−13. https://doi.org/10.1186/s40658-015-0114-3.

Lo Meo, S., Cosentino, L., Mazzone, A., 2017. Study of silicon + ^{6}LiF thermal neutron detectors: GEANT4 simulations versus real data. Nucl. Instrum. Methods Phys. Res., Sect. A 866, 48−57.

Lopes, J.M., Medeiros, M.P.C., Garcêz, R.W.D., et al., 2019. Comparison of simulated and experimental values of self-absorption correction factors for a fast and credible adjust in efficiency curve of gamma Spectroscopy. Appl. Radiat. Isot. (in press).

Lorenz, E., November 1982. Photodiode readout and related problems, pp. 229−255. In: Newman Holmes, C. (Ed.), Proceedings of the International Workshop on Bismuth Germanate. Princeton University. CONF-821160, DE83 011369, U.S. National Science Foundation and U.S. Dept. of Energy, Washington, DC.

Lucas, H.F., 1957. Improved low-level alpha-scintillation counter for radon. Rev. Sci. Instrum. 28, 680−683.

Luethi, D., et al., 2018. Monoamine receptor interaction profiles of 4-thio-substituted phenethylamines (2C-T drugs). Neuropharmacology 134, 141−148.

Lv, W.-H., Yi, H.-C., Liu, T.Q., et al., 2018. Gross beta determination in drinking water using scintillating fiber array detector. Appl. Radiat. Isot. 137, 161−166.

Machaj, B., Bartak, J., 2004. Fast measurement of radon concentration in water with Lucas cell. Nukleonika 49 (1), 29−31.

Maharana, M., Eappen, K.P., Sengupta, D., 2010. Radon emanometric technique for 226Ra estimation. J. Radioanal. Nucl. Chem. 285 (3), 469−474.

Mahl, A., Yeman, H.A., Fernando, R., Koubek, J.T., Sellinger, A., Greife, U., 2018. ^{10}B enriched plastic scintillators for application in thermal neutron detection. Nucl. Instrum. Methods Phys. Res., Sect. A 880, 1−5.

Maioli, C., Bestetti, A., Mauri, A., Pozzato, C., Paroni, R., 2009. Removal of radioisotopes in solution and bactericidal/bacteriostatic sterilising power in activated carbon and metal silver filters. Environ. Toxicol. Pharmacol. 27, 49−53.

Makino, A., et al., 2018. PET probe detecting non-small cell lung cancer susceptible to epidermal growth factor receptor tyrosine kinase inhibitor therapy. Bioorg. Med. Chem. 26, 1609−1613.

Malcolme-Lawes, D.J., Massey, S., 1980. The variation of γ-counting efficiency for iodine-125 with sample composition. Int. J. Appl. Radiat. Isot. 31, 155−158.

Malizia, R.W., Baumann, B.M., Chansky, M.E., Kirchhoff, M.A., 2010. Ambulatory dysfunction due to unrecognized pernicious anemia. J. Emerg. Med. 38 (3), 302−307.

Maneschg, W., 2015. Review of neutrinoless double beta decay experiments: present status and near future. Nucl. Part. Phys. Proc. 260, 188−193.

Mannheim, J.G., Schmid, A.M., Schwenck, J., et al., 2018. PET/MRI hybrid systems. Semin. Nucl. Med. 48, 332−347.

Marek, M., Viererbl, L., 2011. Bonner sphere spectrometer for characterization of BNCT beam. Appl. Radiat. Isot. (in press).

Mares, J.A., D'Ambrosio, C., 2007. Hybrid photomultipliers − their properties and application in scintillation studies. Opt. Mater. 30, 22−25.

Mares, J.A., Nikl, M., Beitlerova, A., Solovieva, N., D'Ambrosio, C., Blazek, K., Maly, P., Nejezchleb, K., Fabeni, P., Pazzi, G.P., 2005. $Ce^{3}+$-doped scintillators: status and properties of (Y,Lu) aluminium perovskites and garnets. Nucl. Instrum. Methods Phys. Res., Sect. A 537, 271−275.

Mares, V., et al., 2017. Neutron spectrometry and dosimetry in 100 and 300 MeV quasi-mono-energetic neutron field at RCNP, Osaka University, Japan. EPJ Web Conf. 153, Art. No. 08020.

Marsoem, P., Wurdiyanto, G., Candra, H., Saibatulham, H., 2014. Direct measurement of ^{60}Co and ^{125}I activity by the sum-peak method in PTKMR-BATAN. Appl. Radit. Isot. 87, 200−202.

Martens, H., Naes, T., 1989. Multivariate Calibration. Wiley, New York.

Martin, R.H., Taylor, J.G.V., 1992. The standardization of ^{125}I: a comparison of three methods. Nucl. Instrum. Methods Phys. Res., Sect. A 312, 64−66.

Martín-Martín, A., Gutiérrez-Villanueva, J.L., Muñoz, J.M., García-Talavera, M., Adamiec, G., Íñiguez, M.P., 2006. Radon measurements with a PIN photodiode. Appl. Radiat. Isot. 64, 1287−1290.

Masuda, A., et al., 2017. Neutron spectral fluence measurements using a Bonner sphere spectrometer in the development of the iBNCT accelerator-based neutron source. Appl. Radiat. Isot. 127, 47−51.

Matsuoka, K., 2014. For the Belle II PID group development and production of the MCP-PMT for the Belle II TOP counter. Nucl. Instrum. Methods Phys. Res., Sect. A 766, 148−151.

Matsushima, A., et al., 2010. Bisphenol AF is a full agonist for the estrogen receptor ERα but a highly specific antagonist for ERβ. Environ. Health Perspect. 118 (9), 1267−1272.

Matzke, M., 2002. Propagation of uncertainties in unfolding procedures. Nucl. Instrum. Methods Phys. Res., Sect. A 476, 230−241.

Mazokopakis, E.E., Karefilakis, C.M., Giannopoulou, C., 2009. Is Schilling test necessary for the etiological diagnosis of B12 insufficiency? Hellenic J. Nucl. Med. 12 (1), 84−86.

Mazrou, H., Sidahmed, T., Idiri, Z., Lounis-Mokrani, Z., Bedek, S., Allab, M., 2008. Characterization of the CRNA Bonner sphere spectrometer based on 6LiI scintillator exposed to an 241Am-Be neutron source. Radiat. Meas. 43, 1095−1099.

Mazrou, H., Idiri, Z., Sidahmed, T., Allab, M., 2010. MCNP5 evaluation of a response matrix of a Bonner Sphere Spectrometer with a high efficiency 6LiI (Eu) detector from 0.01 eV to 20 MeV neutrons. J. Radioanal. Nucl. Chem. 284 (2), 253−263.

Mazrou, H., Nedjar, A., Seguini, T., 2016. Neutron spectrum measurements at a radial beam port of the NUR research reactor using a Bonner spheres spectrometer. Appl. Radiat. Isot. 114, 63−70.

Mazumdar, I., Kumar, G.A., Gothe, D.A., Manchanda, R.K., 2010. A LaBr3:Ce-NaI(Tl) Phoswich for x- and low energy γ-ray astronomy. Nucl. Instrum. Methods Phys. Res., Sect. A 623, 995−998.

McClish, M., Dokhale, P., Christian, J., Stapels, C., Shah, K.S., 2007. Characterization and scintillation studies of a solid-state photomultiplier. Nucl. Instrum. Methods Phys. Res., Sect. A 572, 1065−1070.

McIntyre, J.I., Abel, K.H., Bowyer, T.W., et al., 2001. Measurements of ambient radioxenon levels using the automated radioxenon sampler/analyzer (ARSA). J. Radioanal. Nucl. Chem. 248, 629−635.

McKibben, R.B., Connell, J.J., MacRi, J.R., McConnell, M.L., Ryan, J.M., Flückiger, E.O., Moser, M.R., Brown, J.C., McKinnon, A.L., 2005. Applications of a phoswich-based detector for fast ($\sim$1−10 MeV) solar neutrons for missions to the inner heliosphere. Adv. Space Res. 36, 1432−1438.

McKnight, T.K., Czirr, J.B., Littrell, K., Campbell, B.J., 2008. The flexible embedded-fiber neutron detector. Nucl. Instrum. Methods Phys. Res., Sect. A 586, 246−250.

McNamara, A., et al., 2012. Preclinical efficacy of THRX-200495, a dual pharmacology muscarinic receptor antagonist and β_2-adrenoceptor agonist (MABA). Pulm. Pharmacol. Ther. 25, 357–363.

Medine, I.E., Unak, P., Sakarya, S., Toksöz, F., 2010. Enzymatic synthesis of uracil glucuronide, labeling with 125/131I, and in vitro evaluation on adenocarcinoma cells. Cancer Biother. Radiopharm. 25 (3), 335–344.

Medipix Collaboration (2011). http://medipix.web.cern.ch/medipix/.

Meena, R., Supriya, C., Pratap Reddy, K., Sreenivasula Reddy, P., 2017. Altered spermatogenesis, steroidogenesis and suppressed fertility in adult male rats exposed to genistein, a non-steroidal phytoestrogen during embryonic development. Food Chem. Toxicol. 99, 70–77.

Meisner, J.E., Nicaise, W.F., Stromswold, D.C., 1995. CsI(Tl) with photodiodes for identifying subsurface radionuclide contamination. IEEE Trans. Nucl. Sci. 42, 288–291.

Melcher, C.L., 2005. Perspectives on the future development of new scintillators. Nucl. Instrum. Methods Phys. Res., Sect. A 537, 6–14.

Meleza, C., et al., 2016. Development of a scintillation proximity binding assay for high throughput screening of hematopoietic prostaglandin D_2 synthase. Anal. Biochem. 511, 17–23.

Méndez-Villafañe, R., Los Arcos Merino, J.M., Díaz, E.G., Lorente Fillol, A., 2010. Determination of the emission rate of an Am–Be neutron source with a Bonner sphere spectrometer. Radiat. Meas. 45, 1271–1275.

Mendicino, R., Boscardin, M., Carturan, S., 2015. Characterization of 3D and planar Si diodes with different neutron converter materials. Nucl. Instrum. Methods Phys. Res., Sect. A 796, 23–28.

Mertens, C., De Lellis, C., Van Put, P., Tondeur, F., 2007. MCNP simulation and spectrum unfolding for an NaI monitor of radioactivity in aquatic systems. Nucl. Instrum. Methods Phys. Res., Sect. A 580, 118–122.

Mertens, C., De Lellis, C., Tondeur, F., 2010. Neutron-photon discrimination and spectrum unfolding with a stilbene detector. Appl. Radiat. Isot. 68, 957–960.

Meshkian, M., 2015. Monte Carlo simulation of a fast neutron counter for use in neutron radiography. Nucl. Instrum. Methods Phys. Res., Sect. A 788, 73–78.

Michail, C.M., Spyropoulou, V.B., Fountos, G.P., Kalyvas, N.E., Mytafidis, A.K., Valais, I.G., Kandarakis, I.S., Panayiotakis, G.S., 2010. Imaging performance of a high resolution CMOS sensor under mammographic and radiographic conditions. In: IEEE Intern. Conf. on Imaging Systems and Techniques, IST 2010 – Proc. 2010, pp. 152–155. Article No. 5548544.

Michail, C.M., Spyropoulou, V.A., Fountos, G.P., Kalyvas, N.I., Valais, I.G., Kandarakis, I.S., Panayiotakis, G.S., 2011. Experimental and theoretical evaluation of a high resolution CMOS based detector under X-ray imaging conditions. IEEE Trans. Nucl. Sci. 58 (1), 314–322.

Miesel, L., Kravec, C., Xin, A.-T., McMonagle, P., Ma, S., Pichardo, J., Feld, B., Barrabee, E., Palermo, R., 2007. A high-throughput assay for the adenylation reaction of bacterial DNA ligase. Anal. Biochem. 366, 9–17.

Mikhailov, M.A., Panteleev, L., 2001. Noise contributions to energy resolutions in CsI(Tl) scintillation detector with PIN photodiode readout. Nucl. Instrum. Methods Phys. Res., Sect. A 463, 288–292.

Minor, L.K. (Ed.), 2006. Handbook of Assay Development in Drug Discovery. CRC Press, Boca Raton, FL, p. 488.

Mirzajani, N., Ciolini, R., Curzio, G., 2013. Analysis of the application of the shadow cone technique for the determination of neutron spectrum with Bonner sphere spectrometer. Nucl. Instrum. Methods Phys. Res., Sect. A 722, 24–28.

Miyahara, H., Narita, N., Tomita, K., Katoh, Y., Mori, C., Momose, T., Shinohara, K., 2000. Activity measurement by using γ-ray sum peak method considering intake radioactivity and its application. Radioisotopes 49, 253–259.

Miyamoto, R., Akizawa, H., Nishikawa, T., Uehara, T., Azuma, Y., Nakase, I., Futaki, S., Hanaoka, H., Iida, Y., Endo, K., Arano, Y., 2010. Enhanced target-specific accumulation of radiolabeled antibodies by conjugating arginine-rich peptides as anchoring molecules. Bioconjug. Chem. 21 (11), 203102037.

Moadel, R.M., 2011. Breast cancer imaging devices. Semin. Nucl. Med. 41 (3), 229–241.

Mohammadi, N., Miri-Halimabad, H., Rafat-Motavlli, L., Akbari, F., 2015. Neutron spectrometry and determination of neutron contamination around the 15 MV siemens primus LINAC. J. Radioanal. Nucl. Chem. 304, 1001–1008.

Moe, M.K., 1986. Double-beta decay in ^{82}Se, ^{128}Te, and ^{130}Te. AIP Conf. Proc. 150, 1012–1016.

Moe, M.K., Rosen, S.P., 1989. Double-beta decay. Sci. Am. 48–55.

Møller, C.L., Raun, K., Jacobsen, M.L., Pedersen, T.Å., Holst, B., Conde-Frieboes, K.W., Wulff, B.S., 2011. Characterization of murine melanocortin receptors mediating adipocyte lipolysis and examination of signalling pathways involved. Mol. Cell. Endocrinol. (in press).

Monteiro, C.M.B., Fernandes, L.M.P., Veloso, J.F.C.A., Oliveira, C.A.B., dos Santos, J.M.F., 2012. Secondary scintillation yield from GEM and THGEM gaseous electron multipliers for direct dark matter search. Phys. Lett. B 714, 18–23.

Monteiro, L.O.F., et al., 2017. Technetium-99 m radiolabeled paclitaxel as an imaging probe for breast cancer in vivo. Biomed. Pharmacother. 89, 146–151.

Monteiro, L.O.F., et al., 2018. Paclitaxel-loaded folate-coated long circulating and pH-sensitive liposomes as a potential drug delivery system: a biodistribution study. Biomed. Pharmacother. 97, 489–495.

Moore, B.C., 2012a. Gonadotropin-induced changes in oviductal mRNA expression levels of sex steroid hormone receptors and activin-related signaling factors in the alligator. Gen. Comp. Endocrinol. 175, 251–258.

Moore, B.C., 2012b. Gene–environment interactions: the potential role of contaminants in somatic growth and the development of the reproductive system of the American alligator. Mol. Cell. Endocrinol. 354, 111–120.

Mori, C., Osada, T., Yanagida, K., Aoyama, T., Uritani, A., Miyahara, H., Yamane, Y., Kobayashi, K., Ichihara, C., Shiroya, S., 1994. Simple and quick measurement of neutron flux distribution by using an optical fiber with scintillator. J. Nucl. Sci. Technol. 31, 248–249.

Morishita, Y., Yamamoto, S., Izaki, K., et al., 2018. Optimization of thickness of GAGG scintillator for detecting an alpha particle emitter in a field of high beta and gamma background. Radiat. Meas. 112, 1–5.

Moses, W.W., 2002. Current trends in scintillator detectors and materials. Nucl. Instrum. Methods Phys. Res., Sect. A 487, 123–128.

Moss, C.E., Tissinger, E.W., Hamm, M.E., 1984. Efficiency of 7.62 cm bismuth germanate scintillators. Nucl. Instrum. Methods Phys. Res. 221, 378–384.

Mosset, J.B., Stoykov, A., Greuter, U., et al., 2014. Evaluation of two thermal neutron detection units consisting of ZnS/^{6}LiF scintillating layers with embedded WLS fibers read out with a SiPM. Nucl. Instrum. Methods Phys. Res., Sect. A 764, 299–304.

Moszyński, M., 2003. Inorganic scintillation detectors in γ-ray spectrometry. Nucl. Instrum. Methods Phys. Res., Sect. A 505, 101–110.

Moszyński, M., 2010. Energy resolution and non-proportionality of scintillation detectors – new observations. Radiat. Meas. 45, 372–376.

Moszyński, M., Szawlowski, M., Kapusta, M., Balcerzyk, M., Wolski, D., 2000a. Large area avalanche photodiodes in X-rays and light detection. IEEE Trans. Nucl. Sci. 47 (4), 1297–1302.

Moszyński, M., Kapusta, M., Balcerzyk, M., Szawlowski, M., Wolski, D., 2000b. Large area avalanche photodiodes in x-ray and scintillation detection. Nucl. Instrum. Methods Phys. Res., Sect. A 442, 230–237.

Moszyński, M., Kapusta, M., Balcerzyk, M., Szawlowski, M., Wolski, D., Wegrzecka, I., Wegrzecki, M., 2001. Comparative study of avalanche photodiodes with different structures in scintillation detection. IEEE Trans. Nucl. Sci. 48 (4), 1205–1210.

Moszyński, M., Szawlowski, M., Kapusta, M., Balcerzyk, M., 2002a. Large area avalanche photodiodes in scintillation and x-ray detection. Nucl. Instrum. Methods Phys. Res., Sect. A 485, 504–521.

Moszyński, M., Czarnacki, W., Szawlowski, M., Zhou, B.L., Kapusta, M., Wolski, D., Schotanus, P., 2002c. Performance of large-area avalanche photodiodes at liquid nitrogen temperature. IEEE Trans. Nucl. Sci. 49 (3), 971–976.

Moszyński, M., Szawlowski, M., Kapusta, M., Balcerzyk, M., 2003. Avalanche photodiodes in scintillation detection. Nucl. Instrum. Methods Phys. Res., Sect. A 497, 226–233.

Moszyński, M., Plettner, C., Nassalski, A., Szczesniak, T., Swiderski, L., Syntfeld-Kazuch, A., Czarnacki, W., Pausch, G., Stein, J., Niculae, A., Soltau, H., 2009. A comparative study of silicon drift detectors with photomultipliers, avalanche photodiodes and PIN photodiodes in gamma spectrometry eith LaBr3 crystals. IEEE Trans. Nucl. Sci. 56 (3), 1006–1011.

Moszyński, M., Syntfeld-Każuch, A., Swiderski, L., Grodzicka, M., Iwanowska, J., Sibczyński, P., Szcześniak, T., 2016. Energy resolution of scintillation detectors. Nucl. Instrum. Methods Phys. Res., Sect. A 805, 25–35.

Mujahid, S.A., Hussain, S., 2010. Natural radioactivity in soil in the Baluchistan province of Pakistan. Radiat. Prot. Dosim. 140 (4), 333–339.

Mundy, J.N., Rothman, S.J., 1983. Radiation detection and pulse counting. Methods Exp. Phys. 21, 50–54.

Muralithar, S., Mukherjee, B., Singh, R.P., et al., 2013. A charged particle detector array for detection of light charged particles from nuclear reactions. Nucl. Instrum. Methods Phys. Res., Sect. A 729, 849–855.

Muravsky, V.A., Tolstov, S.A., Kholmetskii, A.L., 1998. Comparison of the least squares and the maximum likelihood estimators for gamma-spectrometry. Nucl. Instrum. Methods Phys. Res., Sect. B 145, 573–577.

Muuri, E., et al., 2018. The in-diffusion of ^{133}Ba in granitic rock cubes from the Olkiluoto and Grimsel in-situ test sites. Appl. Geochem. 92, 188–195.

Nácher, E., Mårtensson, M., Tengblad, o., et al., 2015. Proton response of CEPA4: a novel $LaBr_3(Ce)$–$LaCl_3(Ce)$ phoswich array for high-energy gamma and proton spectroscopy. Nucl. Instrum. Methods Phys. Res., Sect. A 769, 105–111.

Nađđerđ, L.J., Davidović, M.D., Davidović, D.M., et al., 2013. A possible improvement of the determination of ^{133}Ba activity and detection efficiency by the sum-peak method,by inclusion of the previously neglected transitions. Nucl. Instrum. Methods Phys. Res., Sect. A 698, 60–65.

Nagarkar, V.V., Ovechkina, E.E., Miller, S.R., Gaysinskiy, V., Brecher, C., Lempicki, A., 2005. Luminescence properties of CsI:Tl crystals doped with Eu. Met. Funct. Mater. 12, 645–652.

Nakamori, T., Enomoto, T., Toizumi, T., Yatsu, Y., Kawai, N., Kataoka, J., Ishikawa, Y., Kawai, T., Kawabata, N., Matsunaga, Y., 2010. Development of x-ray/gamma-ray imaging spectrometer with reach-through APD arrays. AIP Conf. Proc. 1279, 400–402.

Nakamura, T., Schooneveld, E.M., Rhodes, N.J., Katagiri, M., Toh, K., Sakasai, K., Soyama, K., 2009. A half-millimetre spatial resolution fibre-coded linear position-sensitive scintillator detector with wavelength-shifting fibre read-out for neutron detection. Nucl. Instrum. Methods Phys. Res., Sect. A 606, 675–680.

Nakamura, M., Imai, K., Hirose, M., Matsumoto, H., Tosaki, M., Ohsawa, D., Makino, S., 2011. Heavy-ion microbeam system for cell irradiation at Kyoto University. Nucl. Instrum. Methods Phys. Res., Sect. B (in press).

Nakamura, T., Toh, K., Kawasaki, T., et al., 2014. A scintillator-based detector with sub-100-μm spatial resolution comprising a fibre-optic taper with wavelength-shifting fibre readout for time-of-flight neutron imaging. Nucl. Instrum. Methods Phys. Res., Sect. A 737, 176–183.

Nakamura, T., Toh, K., Kawasaki, T., et al., 2015. A two-dimensional scintillation-based neutron detector with wavelength-shifting fibers and incorporating an interpolation method. Nucl. Instrum. Methods Phys. Res., Sect. A 784, 202–207.

Nappi, E., 2009. Advances in the photodetection technologies for Cherenkov imaging applications. Nucl. Instrum. Methods Phys. Res., Sect. A 604, 190–192.

Nay, J.W., Aaron, V.D., Gunderman, R.B., 2011. Using radiology to teach physiology. J. Am. Coll. Radiol. 8 (2), 117–123.

Neal, J.S., Boatner, L.A., Spurrier, M., Szupryczynski, P., Melcher, C.L., 2007. Cerium-doped mixed-alkali rare-earth double-phosphate scintillators for thermal neutron detection. Nucl. Instrum. Methods Phys. Res., Sect. A 579, 19–22.

Neal, J.S., Boatner, L.A., Ramey, J.O., Wisniewski, D., Kolopus, J.A., Cherepy, N.J., Payne, S.A., 2011. The characterization of Eu2+-doped mixed alkaline-earth iodide scintillator crystals. Nucl. Instrum. Methods Phys. Res., Sect. A 643, 75–78.

Nedderman, A.N.R., Savage, M.E., White, K.L., Walker, D.K., 2004. The use of 96-well Scintiplates to facilitate definitive metabolism studies for drug candidates. J. Pharm. Biomed. Anal. 34, 607–617.

Nelson, K.A., Geuther, J.A., Neihart, J.L., et al., 2012. Nuclear reactor pulse calibration using a CdZnTe electro-optic radiation detector. Appl. Radiat. Isot. 70, 1118–1120.

Nemes, T., Mrda, D., Bikit, I., 2011. Absolute Activity Measurement of 152Eu Sources With a Single Detector. Nucl. Instrum. Methods Phys. Res., Sect. A (in press).

Nemes, T., Mrda, D., Bikit, I., Grujic, S., 2016. An improved formula for determination of the activity via the sum-peak method. Appl. Radiat. Isot. 114, 173–178.

Nemes, T., Mrda, D., Bikit, I., 2018. Generalized equation for the sum-peak method, accounting for random coincidences. Nucl. Instrum. Methods Phys. Res., Sect. A 898, 11–14.

Neumann, K., Kolb, A., Englert, D., Hughes, K.T., Jessop, R.A., Harris, A., 1994. Quench Correction in Scintillation Proximity Assays. TopCount Topics, TCA-019. PerkinElmer, Ins., Waltham, MA, USA. http://www.perkinelmer.co.jp/Portals/0/resource/products_ls/ls_cocktails/pdf/tca-019.pdf.

Nguyen, H.V., Campbell, J.M., Couchell, G.P., Li, S., Pullen, D.J., Schier, W.A., Seabury, E.H., Tipnis, S.V., 1996. Programs in C for parameterizing measured 5" × 5" NaI gamma response functions and unfolding of continuous gamma spectra. Comput. Phys. Commun. 93, 303–321.

Ni, K., Aprile, E., Day, D., Giboni, K.L., Lopes, J.A.M., Majewski, P., Yamashita, M., 2005. Performance of a large area avalanche photodiode in a liquid xenon ionization and scintillation chamber. Nucl. Instrum. Methods Phys. Res., Sect. A 551, 356–363.

Niese, S., 1999. The discovery of organic solid and liquid scintillators by H. Kallman and L. Hereforth 50 years ago. J. Radioanal. Nucl. Chem. 241 (3), 499–501.

Niese, S., 2003. The discovery of the excitation of liquid biphenyl and phenanthrene fluorescence by fast electrons; laudation on the inauguration of the German Society for Liquid Scintillation Spectrometry (DGFS), pp. 447–450. In: Möbius, S., Noakes, J.E., Schönhofer, F. (Eds.), LSC, 2001. Advances in Liquid Scintillation Spectrometry. The University of Arizona, Tucson, pp. 456.

Nikl, M., Yoshikawa, A., Vedda, A., Fukuda, T., 2006. Development of novel scintillator crystals. J. Cryst. Growth 292, 416–421.

Nikl, M., Yoshikawa, A., Kamada, K., Nejezchleb, K., Stanek, C.R., Mares, J.A., Blazek, K., 2013. Development of LuAG-based scintillator crystals – A review. Prog. Cryst. Growth Charact. Mater. 59, 47–72.

Nikl, M., Babin, V., Mares, J.A., et al., 2016. The role of cerium variable charge state in the luminescence and scintillation mechanism in complex oxide scintillators: The effect of air annealing. J. Lumin. 169, 539–543.

Ninos, K., Cavouras, D., Fountos, G., Kandarakis, I., 2010. The effect of scintillator response on signal difference to noise ratio in X-ray medical imaging. Nucl. Instrum. Methods Phys. Res., Sect. A 622, 246–255.

Nishikido, F., Inadama, N., Oda, I., Shibuya, K., Yoshida, E., Yamaya, T., Kitamura, K., Murayama, H., 2010. Four-layer depth-of-interaction PET detector for high resolution PET using a multi-pixel S8550 avalanche photodiode. Nucl. Instrum. Methods Phys. Res., Sect. A 621, 570–575.

NNDC-BNL, 2018. Double Beta ($\beta\beta$) Decay Data. National Nuclear Data Center, Brookhaven National Laboratory. http://www.nndc.bnl.gov/bbdecay/.

Nose, T., Tokunaga, T., Shimohigashi, Y., 2009. Exploration of endocrine-disrupting chemicals on estrogen receptor α by the agonist/antagonist differential-docking screening (AADS) method: 4-(1-adamantyl)phenol as a potent endocrine disruptor candidate. Toxicol. Lett. 191, 33–39.

Novković, D., Nađđerđ, L., Đurašević, M., Vukanac, I., Kandić, A., Milošević, Z., 2009. The direct measurement of 133Ba activity by the sum-peak method. Nucl. Instrum. Methods Phys. Res., Sect. A 608, 116–120.

Novković, D., Kandić, A., Vukanac, I., Đurašević, M., Milošević, Z., 2012. The direct measurement of ^{57}Co activity by the sum-peak method. Appl. Radiat. Isot. 70, 2154–2156.

Novotny, R., 2005. Inorganic scintillators—a basic material for instrumentation in physics. Nucl. Instrum. Methods Phys. Res., Sect. A 537, 1–5.

Nudat 2.7, 2018. National Nuclear Data Center, Broohaven Nuclear LaboratoryLong Island, NY, USA. http://www.nndc.bnl.gov/nudat2/chartNuc.jsp.

Oed, A., 2004. Detectors for thermal neutrons. Nucl. Instrum. Methods Phys. Res., Sect. A 525, 62–68.

Ogasawara, K., Hirahara, M., Miyake, W., Kasahara, S., Takashima, T., Asamura, K., Saito, Y., Mukai, T., 2008. High-resolution detection of 100 keV electrons using avalanche photodiodes. Nucl. Instrum. Methods Phys. Res., Sect. A 594, 50–55.

Ogasawara, K., Livi, S., McComas, D.J., 2009. Temperature dependence of the thin dead layer avalanche photodiode for low energy electron measurements. Nucl. Instrum. Methods Phys. Res., Sect. A 611, 93–98.

Ogasawara, K., Livi, S.A., Grotheer, E., McComas, D.J., 2010. Thin dead-layer avalanche photodiodes enable low-energy ion measurements. Nucl. Instrum. Methods Phys. Res., Sect. A 614, 271–277.

Ogata, Y., Miyahara, H., Ishigure, N., Ishihara, M., Nishio, M., Yamamoto, S., 2015. Development of a modified sum-peak method for activity determination of some gamma emitters. Nucl. Instrum. Methods Phys. Res., Sect. A 775, 34–40.

Ogata, Y., Miyahara, H., Ishihara, M., Ishigure, N., Yamamoto, S., Kojima, S., 2016. Development of the modified sum-peak method and its application. Appl. Radiat. Isot. 109, 354–357.

Ogata, Y., Itadzu, H., Kojima, S., 2018a. Determination of ^{134}Cs activity with the sum-peak method using a well-type germanium detector. J. Radioanal. Nucl. Chem. 316 (3), 1145–1149.

Ogata, Y., Itadzu, H., Kojima, S., 2018b. Estimation of ^{134}Cs activity by a new approximation referred to the sum-peak method via a well-type Ge detector. Appl. Radiat. Isot. 134, 172–176.

Ogawa, I., Hazama, R., Ajimura, S., Matsuoka, K., Kudomi, N., Kume, K., Ohsumi, H., Fushimi, K., Suzuki, N., Nitta, T., 2003. Double beta decay study of 48Ca by CaF2 scintillator. Nucl. Phys. A 721, 525c–528c.

Ogawa, I., Hazama, R., Miyawaki, H., Shiomi, S., Suzuki, N., Ishikawa, Y., Kunitomi, G., Tanaka, Y., Itamura, M., Matsuoka, K., et al., 2004. Search for neutrino-less double beta decay of 48Ca by CaF2 scintillator. Nucl. Phys. A 730, 215–223.

O'Hara, M., Murray, N.J., Carter, J.C., Kellogg, C.M., Link, J.M., 2018a. Hydroxamate column-based purification of zirconium-89 (^{89}Zr) using an automated fluidic platform. Appl. Radiat. Isot. 132, 85–94.

O'Hara, M., Murray, N.J., Carter, J.C., Morrison, S.S., 2018b. Optimized anion exchange column isolation of zirconium-89 (^{89}Zr)from yttrium cyclotron target: Method development and implementation on an automated fluidic platform. J. Chromatogr. A 1545, 48–58.

Ohno, M., Kawano, T., Edahiro, i., et al., 2016. Development and verification of signal processing system of avalanche photo diode for the active shields onboard *ASTRO-H*. Nucl. Instrum. Methods Phys. Res., Sect. A 831, 410–414.

Ojo, K.K., Arakaki, T.L., Napuli, A.J., Inampudi, K.K., Keyloun, K.R., Zhang, L., Hol, W.G.J., et al., 2011. Structure determination of glycogen synthase kinase-3 from Leishmania major and comparative inhibitor structure-activity relationships with Trypanosoma brucei GSK-3. Mol. Biochem. Parisitol. 176, 98–108.

Oka, Y., et al., 2013. Discovery of N-{5-[3-(3-hydroxypiperidin-1-yl)-1,2,4-oxadiazol-5-yl]-4-methyl-1,3-thiazol-2-yl}acetamide (TASP0415914) as an orally potent phosphoinositide 3-kinase γ inhibitor for the treatment of inflammatory diseases. Bioorg. Med. Chem. 21, 7578−7583.

Okamura, T., Okada, M., Kikuchi, T., Wakizaka, H., Zhang, M.-R., 2018. Kinetics and metabolism of apocynin in the mouse brain assessed with positron-emission tomography. Phytomedicine 38, 84−89.

Oliván, M.A., Amaré, J., Cebrián, S., et al., 2017. Light yield determination in large sodium iodide detectors applied in the search for dark matter. Astropart. Phys. 93, 86−95.

Oliveira, E.M., Iwahara, A., Poledna, R., 2012. Standardization of ^{65}Zn by sum-peak method. Appl. Radiat. Isot. 70, 2087−2090.

Omori, Y., et al., 2015. Analgesic effect of GT-0198, a structurally novel glycine transporter 2 inhibitor, in a mouse model of neuropathic pain. J. Pharmacol. Sci. 127, 377−381.

Orio, F., 2016. Current status and perspectives of the LUCIFER experiment. Nucl. Part. Phys. Proc. 273−275, 1795−1800.

Orphan, V., Polichar, R., Ginaven, R., 1978. A solid-state photomultiplier tube for improved gamma counting techniques. Trans. Am. Nucl. Soc. 28, 119−121.

Ortigoza, Y., Coron, N., Cuesta, C., García, E., Ginestra, C., Gironnet, J., De Marcillac, P., Martínez, M., Pobes, C., Puimedón, J., Redon, T., Sarsa, M.L., Torres, L., Villar, J.A., 2011. Energy partition in sapphire and BGO scintillating bolometers. Astropart. Phys. 34, 603−607.

Ortigoza, Y., Torres, L., Coron, N., et al., 2013. Light Relative Efficiency Factors for ions in BGO and Al_2O_3 at 20 mK. Astropart. Phys. 50−52, 11−17.

Ortiz-Rodríguez, J.M., del Rosario Martinez-Blanco, M., Cervantes Viramontes, J.M., Vega-Carrillo, H.R., 2013. Chapter 4: Robust design of artificial neural networks methodology in neutron spectrometry, pp. 28.10.5772/3409. In: Suzuki, K. (Ed.), Artificial Neural Networks − Architectures and Applications, first ed. Publisher: INTECH. https://doi.org/10.5772/3409. https://doi.org/10.5772/51274.

Orton, C.R., Fraga, C.G., Christensen, R.N., Schwantes, J.M., 2011. Proof of concept simulations of the Multi-Isotope Process monitor: An online, nondestructive, near-real-time safeguards monitor for nuclear fuel reprocessing facilities. Nucl. Instrum. Methods Phys. Res., Sect. A 629, 209−219.

Orton, C.R., Fraga, C.G., Christensen, R.N., Schwantes, J.M., 2012. Proof of concept experiments of the multi-isotope process monitor: An online, nondestructive, near real-time monitor for spent nuclear fuel reprocessing facilities. Nucl. Instrum. Methods Phys. Res., Sect. A 672, 38−45.

Oshima, N., et al., 2017a. ^{111}In-DTPA-D-Phe^{-1}-Asp^{0}-D-Phe^{1}-octreotide exhibits higher tumor accumulation and lower renal radioactivity than ^{111}In-DTPA-D-Phe^{1}-octreotide. Nucl. Med. Biol. 54, 18−26.

Oshima, N., et al., 2017b. Redesign of negatively charged ^{111}In-DTPA-octreotide derivative to reduce renal radioactivity. Nucl. Med. Biol. 48, 16−25.

Otte, A.N., Hose, J., Mirzoyan, R., Romaszkiewicz, A., Teshima, M., Thea, A., 2006. A measurement of the photon detection efficiency of silicon photomultipliers. Nucl. Instrum. Methods Phys. Res., Sect. A 567, 360−363.

Otte, A.N., Britvitch, I., Biland, A., Goebel, F., Lorenz, E., Pauss, F., Renker, D., Röser, U., Schweizer, T., 2009. Detection of Cherenkov light from air showers with Geigermode-APDs. Nucl. Instrum. Methods Phys. Res., Sect. A 610, 415−418.

Ovechkina, E.E., Gaysinskiy, V., Miller, S.R., Brecher, C., Lempicki, A., Nagarkar, V.V., 2007. Multiple doping of CsI:Tl crystals and its effect on afterglow. Radiat. Meas. 42, 541−544.

Ovechkina, L., Riley, K., Miller, S., Bell, Z., Nagarkar, V., 2009. Gadolinium loaded plastic scintillators for high efficiency neutron detection. Phys. Proc. 2, 161−170.

Olaya Dávila, H., Martínez Ovalle, S.A., Pérez, H., Castro, H., 2017. Determination of spatial resolution of positron emission tomograph of clear PET-XPAD3/CT system. Univ. J. Phys. Appl. 11 (4), 97−101.

Pagire, S.H., et al., 2015. Discovery and optimization of adamantane carboxylic acid derivatives as potent diacylglycerol acyltransferase 1 inhibitors for the potential treatment of obesity and diabetes. Eur. J. Med. Chem. 101, 716−735.

Pan, S., Zhang, J., Pan, J., 2018. Thermal expansion, luminescence, and scintillation properties of $CaMoO_4$ crystals grown by the vertical Bridgman method. J. Cryst. Growth (in press).

Pan, S., Zhang, J., Pan, J., et al., 2018b. Optimized crystal growth and luminescence properties of Ce^{3+} ions doped $Li_6Gd(BO_3)_3$, $Li_6Y(BO_3)_3$ and their mixed crystals. J. Alloy. Comp. 751, 129−137.

Park, S.-W., Yuk, S., Park, J.-B., Yi, Y., 2009. A design for a linear array PIN photodiode for use in a Computed mammo-Tomography (CmT) system. Nucl. Instrum. Methods Phys. Res., Sect. A 610, 210−214.

Park, Y.-W., Garyantes, T., Cummings, R.T., Carter-Allen, K., 1997. Optimization of ^{33}P Scintillation Proximity Assays Using Cesium Chloride Bead Suspension. TopCount Topics, TCA-030. PerkinElmer, Inc., Waltham, MA, USA. http://www.perkinelmer.com/CMSResources/Images/44-73160APP_TopCountOpt33PScintProx.pdf.

Park, H.-K., et al., 2016. The AMoRE: Search for Neutrinoless Double Beta Decay in 100Mo. Nucl. Part. Phys. Proc. 273−275, 2530−2632.

Park, H.M., Park, K.H., Kang, S.W., Joo, K.S., 2017. Feasibility of in situ beta ray measurements in underwater environment. J. Environ. Radioact. 175−176, 120−125.

Parker, A.J., Boxall, C., Joyce, M.J., Schotanus, P., 2013. A thallium-doped sodium iodide well counter for radioactive tracer applications with naturally-abundant ^{40}K. Nucl. Instrum. Methods Phys. Res., Sect. A 722, 5−10.

Pauwels, D., Le Masson, N., Viana, B., Kahn-Harari, A., van Loef, E.V.D., Dorenbos, P., van Eijk, C.W.E., 2000. A novel inorganic scintillator. IEEE Trans. Nucl. Sci. 47 (6), 1787−1790.

Pawełczak, I.A., Tõke, J., Henry, E., Quinlan, M., Singh, H., Schröder, W.U., 2011. NSTAR—A capture gated plastic neutron detector. Nucl. Instrum. Methods Phys. Res., Sect. A 629, 230−238.

Pawełczak, I.A., Glenn, A.M., Martinez, H.P., et al., 2014. Boron-loaded plastic scintillator with neutron-γ pulse shape discrimination capability. Nucl. Instrum. Methods Phys. Res., Sect. A 751, 62−69.

Peeples, J.L., Gardner, R.P., 2012. Monte Carlo simulation of the nonlinear full peak energy responses for gamma-ray scintillation detectors. Appl. Radiat. Isot. 70, 1058−1062.

Pejchal, J., Fujimoto, Y., Chani, V., Moretti, F., Yanagida, T., Nikl, M., Yokota, Y., et al., 2011. Crystal growth and luminescence properties of Ti-doped $LiAlO_2$ for neutron scintillator. J. Cryst. Growth 318, 828−832.

Pejchal, J., Babin, V., Kucerkova, R., et al., 2018. Luminescence processes in Ti-doped $LiAlO_2$ single crystals for neutron scintillators. J. Lumin. 201, 231−244.

Pellion, D., Jradi, K., Brochard, N., Prêle, D., Ginhac, D., 2015. Single-photon avalanche diodes (SPAD) in CMOS 0.35 μm technology. Nucl. Instrum. Methods Phys. Res., Sect. A 787, 380–385.

PerkinElmer, 1991a. Theory of TopCount Operation. TopCount Topics, TCA-003. PerkinElmer, Inc., Waltham, MA, USA. http://www.perkinelmer.co.jp/Portals/0/resource/products_ls/ls_cocktails/pdf/tca-003.pdf.

PerkinElmer, 1991b. Scintillation Proximity Assay on the TopCount Microplate Scintillation Counter. TopCount Topics, TCA-004. PerkinElmer, Inc., Waltham, MA, USA. http://www.perkinelmer.co.jp/Portals/0/resource/products_ls/ls_cocktails/pdf/tca-004.pdf.

PerkinElmer, 1996. Solid Scintillation Counting. TopCount Topics, TCA-002. PerkinElmer, Inc., Waltham, MA, USA. http://www.perkinelmer.co.jp/Portals/0/resource/products_ls/ls_cocktails/pdf/tca-002.pdf.

PerkinElmer, 1997. High Efficiency Count Mode for SPA and Cytostar-T™ assays. TopCount Topics, TCA-029. PerkinElmer, Inc., Waltham, MA, USA. http://www.perkinelmer.co.jp/Portals/0/resource/products_ls/ls_cocktails/pdf/tca-029.pdf.

PerkinElmer, 2001. Scintillation Proximity Assay on the TopCount Microplate Scintillation Counter. TopCount Topics TCA-004. PerkinElmer, Inc., Waltham, MA, USA. http://www.perkinelmer.co.jp/Portals/0/resource/products_ls/ls_cocktails/pdf/tca-004.pdf.

PerkinElmer, 2018. SPA Ligand Binding Assays. http://www.perkinelmer.com/lab-products-and-services/application-support-knowledgebase/radiometric/spa-ligand-binding.html.

Persson, A., Khaplanov, A., Cederwall, B., Bohm, C., 2009. A prototype detector module for combined PET/CT or combined photon counting/standard CT based on SiPM technology. IEEE Nucl. Sci. Symp. Conf. Rec 3503–3507 (Art. No. 5401800).

Pestotnik, R., Korpar, S., Chagani, H., Dolenec, R., Križan, P., Stanovnik, A., 2010. Silicon photo-multipliers as photon detectors for PET. Nucl. Instrum. Methods Phys. Res., Sect. A 623, 594–596.

Peters, R.J., Bates, B.B., Knab, D., 1993. Radon-226 in water-radon emanation. pp. ER180-1–ER180-18. In: Tamura, L.S. (Ed.), Health and Environmental Chemistry: Analytical Techniques, Data Management, and Quality Assurance, vol. II. Los Alamos National Laboratory Manual. LA-10300-M.

Pfaffinger, M., Böhm, M., Britting, A., et al., 2019. Recent results with lifetime enhanced microchannel-plate photomultipliers. Nucl. Instrum. Methods Phys. Res., Sect. A (in press).

Phillips, J.J., Straiton, J., Staff, R.T., 2015. Planar and SPECT ventilation/perfusion imaging and computed tomography for the diagnosis of pulmonary embolism: A systematic review and meta-analysis of the literature, and cost and dose comparison. Eur. J. Radiol. 84, 1392–1400.

Pidol, L., Viana, B., Kahn-Harari, A., Bessière, A., Dorenbos, P., 2005. Luminescence properties and scintillation mechanisms of Ce3+-, Pr3+- and Nd3+-doped lutetium pyrosilicate. Nucl. Instrum. Methods Phys. Res., Sect. A 537, 125–129.

Pioch, C., Mares, V., Rühm, W., 2010. Influence of Bonner sphere response functions above 20 MeV on unfolded neutron spectra and doses. Radiat. Meas. 45, 1263–1267.

Pirro, S., Arnaboldi, C., Beeman, J.W., Pessina, G., 2006. Development of bolometric light detectors for double beta decay searches. Nucl. Instrum. Methods Phys. Res., Sect. A 559, 361–363.

for the LUMINEU and the EDELWEISS Collaborations Poda, D.V., 2016. Scintillating bolometers based on $ZnMoO_4$ and $Zn^{100}MoO_4$ crystals to search for 0v2β decay of ^{100}Mo (LUMINEU project): first tests at the Modane Underground Laboratory. Nucl. Part. Phys. Proc. 273–275, 1801–1806.

Pommé, S., 2009. Detection efficiency calculation for photons, electrons and positrons in a well detector. Part I: Analytical model. Nucl. Instrum. Methods Phys. Res., Sect. A 604, 584–591.

Pommé, S., 2012. STEFFY—software to calculate nuclide-specific total counting efficiency in well-type γ-ray detectors. Appl. Radiat. Isot. 70, 2010–2074.

Pommé, S., Altzitzoglou, T., Van Ammeland, R., Sibbens, G., 2005. Standardisation of 125I using seven techniques for radioactivity measurement. Nucl. Instrum. Methods Phys. Res., Sect. A 544, 584–592.

Pommé, S., Sibbens, G., Vidmar, T., Camps, J., Peyres, V., 2009. Detection efficiency calculation for photons, electrons and positrons in a well detector. Part II: Analytical model versus simulations. Nucl. Instrum. Methods Phys. Res., Sect. A 606, 501–507.

Powell, D., Price-Jones, M.J., 2006. Chapter 14. Nuclear receptor scintillation proximity assay, pp. 183–192. In: Minor, L.K. (Ed.), Handbook of Assay Development in Drug Discovery. CRC Press, Boca Raton, FL.

Powolny, F., Auffray, E., Brunner, S.E., Garutti, E., Goettlich, M., Hillemanns, H., Jarron, P., Lecoq, P., Meyer, T., Schultz-Coulon, H.C., Shen, W., Williams, M.C.S., 2011. Time-based readout of a silicon photomultiplier (SiPM) for time of flight positron emission tomography (TOF-PET). IEEE Trans. Nucl. Sci. (in press).

Prasad, G., Prasad, Y., Gusain, G.S., Ramola, R.C., 2008. Measurement of radon and thoron levels in soil, water and indoor atmosphere of Budhakedar in Garhwal Himalaya, India. Radiat. Meas. 43, S375–S379.

Prause, M., et al., 2018. Synthesis, in vitro and in vivo evaluation of ^{18}F-fluoronorimatinib as radiotracer for Imatinib-sensitive gastrointestinal stromal tumors. Nucl. Med. Biol. 57, 1–11.

Price, G.J., Fraser, G.W., 2001. Calculation of the output charge cloud from a microchannel plate. Nucl. Instrum. Methods Phys. Res., Sect. A 474, 188–196.

PTB, 2004. NEA-1665 UMG 3.3 Analysis of Data Measured with Spectrometers Using Unfolding Techniques. http://www.oecd-nea.org/tools/abstract/detail/nea-1665.

QCalc, 2018. Q-value Calculator. National Nuclear Data Center, Broohkhaven National Laboratory, Long Island, NY, USA. http://www.nndc.bnl.gov/qcalc/.

Qin, J., Lai, C., Lu, X., et al., 2018. Characterization of a ^{6}Li enriched Cs_2LiYCl_6:Ce scintillator and its application as a -ray detector. Nucl. Instrum. Methods Phys. Res., Sect. A 886, 55–60.

Quaranta, A., Carturan, S., Marchi, T., Buffa, M., Degerlier, M., Cinausero, M., Guastalla, G., et al., 2011. Doped polysiloxane scintillators for thermal neutrons detection. J. Non-Cryst. Solids 357, 1921–1925.

Quarati, F.G.A., Owens, A., Dorenbos, P., de Haas, J.T.M., Benzoni, G., Blasi, N., Boiano, C., Brambilla, S., Camera, F., Alba, R., Bellia, G., Maiolino, C., Santonocito, D., Ahmed, M., Brown, N., Stave, S., Weller, H.R., Wu, Y.K., 2011. High energy gamma-ray spectroscopy with $LaBr_3$ scintillation detectors. Nucl. Instrum. Methods Phys. Res., Sect. A 629, 157–169.

Quindos-Poncela, L.S., Fernandez, P.L., Sainz, C., Arteche, J., Arozamena, J.G., George, A.C., 2003. An improved scintillation cell for radon measurements. Nucl. Instrum. Methods Phys. Res., Sect. A 512, 606–609.

Quindos Poncela, L.S., Fernandez, P.L., Gomez Arozamena, J., Sainz Fernandez, C., 2005. A method for measuring effective radon diffusion coefficients in radon barriers by using modified Lucas cells. Radiat. Meas. 39, 87–89.

Raghaven, R.S., 1976. Inverse β decay of $^{115}In \rightarrow {}^{115}Sn^*$: a new possibility for detecting solar neutrinos from the proton-proton reaction. Phys. Rev. Lett. 37, 259–262.

Rahman, M.S., Cho, G., Kang, B.-S., 2009. Deconvolution of gamma-ray spectra obtained with NaI(Tl) detector in a water tank. Radiat. Prot. Dosim. 135 (3), 201–210.

Rahmanifard, R., Katebi, F., Zahedi, A.R., Gholipour-Peyvandi, R., 2018. Synthesis and development of a vinyltoluene-based plastic scintillator. J. Lumin. 194, 456–460.

Rajabi Moghadam, S., Feghhi, S.A.H., Safari, M.J., 2015. A phoswich detector for simultaneous alpha–gamma spectroscopy. Nucl. Instrum. Methods Phys. Res., Sect. A 799, 59–63.

Ramadhan, R.A., Abdullah, K.M.-S., 2018. Background reduction by Cu/Pb shielding and efficiency study of NaI(Tl) detector. Nucl. Eng. Technol. 50, 462–469.

Rato Mendes, P., Cuerdo, R., Sarasola, I., 2013. A detector insert based on continuous scintillators for hybrid MR–PET imaging of the human brain. Nucl. Instrum. Methods Phys. Res., Sect. A 702, 80–82.

Reboli, A., Aupiais, J., Mialocq, J.C., 2005. Application of large area avalanche photodiodes for alpha liquid scintillation counting. Nucl. Instrum. Methods Phys. Res., Sect. A 550, 593–602.

Reeder, P.L., 1994a. Neutron detection using GSO scintillator. Nucl. Instrum. Methods Phys. Res., Sect. A 340, 371–378.

Reeder, P.L., 1994b. Thin GSO scintillator for neutron detection. Nucl. Instrum. Methods Phys. Res., Sect. A 353, 134–136.

Reeder, P.L., Bowyer, S.M., 2001. Fast neutron and alpha detection using $LiBaF_3$ scintillator. IEEE Trans. Nucl. Sci. 48 (3), 351–355.

Reeder, P.L., Bowyer, S.M., 2002. Calibration of $LiBaF_3$:Ce scintillator for fission spectrum neutrons. Nucl. Instrum. Methods Phys. Res., Sect. A 484, 469–485.

Reginatto, M., 2009. What can we learn about the spectrum of high-energy stray neutron fields from Bonner sphere measurements? Radiat. Meas. 44, 692–699.

Renker, D., 2004. Photosensors. Nucl. Instrum. Methods Phys. Res., Sect. A 527, 15–20.

Renker, D., 2007. New trends on photodetectors. Nucl. Instrum. Methods Phys. Res., Sect. A 571, 1–6.

Renker, D., 2009. New developments on photosensors for particle physics. Nucl. Instrum. Methods Phys. Res., Sect. A 598, 207–212.

Retière, F., Amaudruz, P.-A., Bishop, D., et al., 2012. Development of a low cost planar detector for positron emission tomography. Nucl. Instrum. Methods Phys. Res., Sect. A 695, 139–142.

Reynisson, J., Court, W., O'Neill, C., Day, J., Patterson, L., McDonald, E., Workman, P., Katan, M., Eccles, S.A., 2009. The identification of novel PLC-γ inhibitors using virtual high throughput screening. Bioorg. Med. Chem. 17, 3169–3176.

Reynolds, G.T., Harrison, F.B., Salvini, B., 1950. Liquid scintillation counters. Phys. Rev. 78 (4), 488.

Ricard, M., 2004. Imaging of gamma emitters using scintillation cameras. Nucl. Instrum. Methods Phys. Res., Sect. A 527, 124–129.

Ricci, P.C., Casu, A., Chiriu, D., Corpino, C., Carbonaro, C.M., Marceddu, M., Salis, M., Anedda, A., 2011. Optical properties of pure Ytterbium Aluminium perovskites. Opt. Mater. 33, 1000–1003.

Rich, G.C., Kazkaz, K., Martinez, H.P., Gushue, T., 2015. Fabrication and characterization of a lithium-glass-based composite neutron detector. Nucl. Instrum. Methods Phys. Res., Sect. A 794, 15–24.

Rieppo, R., Blomster, K., Holmberg, P., 1975. Calculated 10–150 keV gamma-ray efficiency values for well-type sodium iodide detectors. Int. J. Appl. Radiat. Isot. 26, 558–561.

Rieppo, R., 1976a. Calculated 10–150 keV γ-ray efficiency values of sodium iodide detectors for a 4-pi-detector geometry. Intercomparison with the respective values of well- and face-type sodium iodide detectors. Int. J. Appl. Radiat. Isot. 27, 457–459.

Rieppo, R., 1976b. Efficiency values of sodium iodide well-type detectors in the γ-region 10–150 keV for point and needle-shaped source geometries. Int. J. Appl. Radiat. Isot. 27, 453–456.

Ringbom, A., Larson, T., Axelson, A., Elmgren, K., Johansson, C., 2003. SAUNA—a system for automatic sampling, processing and analysis of radioactive xenon. Nucl. Instrum. Methods Phys. Res. 508, 542–553.

Riou, L.M., Ghezzi, C., Vanzetto, G., Broisat, A., Mathieu, J.-P., Bontron, R., Pasqualini, R., Fagret, D., 2003. Verapamil does not inhibit 99mTcN-NOET uptake in situ in normal or ischemic canine myocardium. J. Nucl. Med. 44 (6), 981–987.

Roberts, N.J., Thomas, D.J., Visser, T.P.P., 2017. Improved Bonner sphere neutron spectrometry measurements for the nuclear industry. Radiat. Phys. Chem. 140, 319–321.

Robinson, S.M., Runkle, R.C., Newby, R.J., 2011. A comparison of performance between organic scintillation crystals and moderated 3He-based detectors for fission neutron detection. Nucl. Instrum. Methods Phys. Res., Sect. A 652, 404–407.

Ródenas, J., Martinavarro, A., Rius, V., 2000. Validation of the MCNP code for the simulation of Ge-detector calibration. Nucl. Instrum. Methods Phys. Res., Sect. A 450, 88–97.

Rodnyi, P.A., 2001. Progress in fast scintillators. Radiat. Meas. 33, 605–614.

Rodnyi, P.A., Dorenbos, P., van Eijk, C.W.E., 1995. Energy loss in inorganic scintillators. Phys. Status Solidi B 187, 15–29.

Roig, M., de Juan, A., García, J.F., Toribo, M., Vidal, M., Rauret, G., 1999. Determination of a mixture of gamma-emitting radionuclides using solid scintillation detectors and multivariate calibration. Anal. Chim. Acta 379, 121–133.

Roncali, E., Cherry, S.R., 2011. Application of silicon photomultipliers to positron emission tomography. Ann. Biomed. Eng. 39 (4), 1358–1377.

Roy, P., Banerjee, K., Saha, A.K., et al., 2018. Detailed investigation on the possibility of using EJ-299-33A plastic scintillator for fast neutron spectroscopy in large scale experiments. Nucl. Instrum. Methods Phys. Res., Sect. A 901, 198–202.

Rudin, M.J., Richardson, W.M., Dumont, P.G., Johnson, W.H., 2001. In situ measurement of transuranics using a calcium fluoride scintillation detection system. J. Radioanal. Nucl. Chem. 248 (2), 445–448.

Rühm, W., et al., 2014. Comparison of Bonner sphere responses calculated by different Monte Carlo codes at energies between 1 MeV and 1 GeV e Potential impact on neutron dosimetry at energies higher than 20 MeV. Radiat. Meas. 67, 24–34.

Ruiz, M., Takehana, K., Petruzella, F.D., Watson, D.D., Beller, G.A., Glover, D.K., 2002. Arbutamine stress perfusion imaging in dogs with critical coronary artery stenoses: 99mTc-sestamibi versus 201T1. J. Nucl. Med. 43 (5), 664–670.

Russo, S., Barbarino, G., De Asmundis, R., De Rosa, G., 2009. The Vacuum Silicon Photomultiplier Tube (VSiPMT): A new concept of photon detector. First feasibility results. IEEE Nucl. Sci. Symp. Conf. Rec. 1966–1969. Art. No. 5402144.

Russo, S., Barbarino, G., De Asmundis, R., De Rosa, G., 2010. The Vacuum Silicon Photomultiplier Tube (VSiPMT): A new version of a hybrid photon detector. Nucl. Instrum. Methods Phys. Res., Sect. A 623, 291–293.

Rutherford, E., 1919. Collision of α-particles with light atoms. *Nature* (London) 103, 415–418.

Rutherford, E., 1920. Nuclear constitution of atoms. Proc. Royal Soc. (London) 97A, 374–401.

Rutter, A.R., et al., 2014. GSK356278, a potent, selective, brain-penetrant phosphodiesterase 4 inhibitor that demonstrates anxiolytic and cognition-enhancing effects without inducing side effects in preclinical species. J. Pharmacol. Exp. Ther. 350, 153–163.

Ryzhikov, V.D., Naydenov, S.V., Onyshchenko, G.M., et al., 2018. Multilayer fast neutron detectors based on composite heavy-oxide scintillators for detection of illegal nuclear materials. Nucl. Instrum. Methods Phys. Res., Sect. A 903, 287–296.

Ryzhikov, V.D., Naydenov, S.V., Piven, L.A., 2017. Fast neutron detectors and portal monitors based on solid-state heavy-oxide scintillators. Radiat. Meas. 105, 17–25.

Sabharwal, S.C., Ghosh, B., Phiske, M.R., Navalkar, M.P., 1982. High temperature performance characteristics of a NaI(Tl) detector. Nucl. Instrum. Methods 195, 613–616.

Sadena, A., Sadena, N., 2011. Chapter 4. Biosensors involved in drug discovery, pp. 61–94. In: Handbook of Biosensors and Biosensor Kinetics. Elsevier, Amsterdam.

Sadygov, Z., Olshevski, A., Chirikov, I., Zheleznykh, I., Novikov, A., 2006. Three advanced designs of micro-pixel avalanche photodiodes: Their present status, maximum possibilities and limitations. Nucl. Instrum. Methods Phys. Res., Sect. A 567, 70–73.

Sadygov, Z., Zerrouk, A.F., Ariffin, A., Khorev, S., Sasam, J., Zhezher, V., Anphimov, N., Dovlatov, A., Musaev, M., Muxtarov, R., Safarov, N., 2009a. Performance of new Micro-pixel Avalanche Photodiodes from Zecotek Photonics. Nucl. Instrum. Methods Phys. Res. Sect. A 610 (1), 381–383.

Sadygov, Z., Zerrouk, A.F., Ariffin, A., Khorev, S., Sasam, J., Bokova, T., Olshevski, A., Anfimov, N., Chalishev, V., Pilyar, N., Troitskaya, M., Dovlatov, A., Musaev, M., Muxtarov, R., 2009b. Spatial distribution of photo-sensitivity in new micro-pixel avalanche photodiodes: Assembly of 64-element arrays. Nucl. Instrum. Methods Phys. Res. Sect. A 610 (1), 390–392.

Safavi-Naeini, M., Lerch, M.L.F., Petasecca, M., Pignatel, G.U., Reinhard, M., Rosenfeld, A.B., 2008. Evaluation of pixellated, back-sided planar photodetectors for high-resolution imaging instrumentation. Nucl. Instrum. Methods Phys. Res., Sect. A 589, 259–267.

Sahagia, M., Ivan, C., Grigorescu, E.L., Razdolescu, A.C., 2008. Standardization of ^{125}I by the coincidence method and practical applications. Appl. Radiat. Isot. 66 (6-7), 895–899.

Sahi, S., Wang, Z., Luo, J., et al., 2016. Investigation of luminescence mechanism in $La_{0.2}Y_{1.8}O_3$ scintillator. J. Lumin. 173, 99–104.

Saito, T., Nakamori, T., Yoshino, M., et al., 2013. Development of high performance Avalanche photodiodes and dedicated analog systems for HXI/SGD detectors onboard the Astro-H mission. Nucl. Instrum. Methods Phys. Res., Sect. A 699, 230–234.

Sakamoto, S., Takakura, H., 2001. Detection efficiency improvement of a large volume scintillator. J. Radioanal. Nucl. Chem. 248 (2), 345–351.

Samson, A., Thibaudeau, C., Bouchard, J., et al., 2018. A fully automated and scalable timing probe-based method for time alignment of the LabPET II scanners. Nucl. Instrum. Methods Phys. Res., Sect. A 889, 1–6.

Samulon, E.C., Gundiah, G., Gascón, M., et al., 2014. Luminescence and scintillation properties of Ce^{3+}-activated $Cs_2NaGdCl_6$, Cs_3GdCl_6, $Cs_2NaGdBr_6$ and Cs_3GdBr_6. J. Lumin. 153, 64–72.

Sanfilippo, D., Fallica, P.G., Carbone, B., et al., 2013. Timing properties measurements of STMicroelectronics silicon photomultipliers for PET scanners. Nucl. Instrum. Methods Phys. Res., Sect. A 702, 70–72.

Saoudi, A., Pepin, C., Houdi, D., Lecomte, R., 1999. Scintillation light emission studies of LSO scintillators. IEEE Trans. Nucl. Sci. 46 (6), 1925–1928.

Sardet, A., Varignon, C., Laurent, B., Granier, T., Oberstedt, A., 2015. p-Terphenyl: an alternative to liquid scintillators for neutron detection. Nucl. Instrum. Methods Phys. Res., Sect. A 792, 74–80.

Sasano, M., Nishioka, H., Okuyama, S., 2013. Geometry dependence of the light collection efficiency of BGO crystal scintillators readout by avalanche photodiodes. Nucl. Instrum. Methods Phys. Res., Sect. A 715, 105–111.

Sato, R., Kataoka, J., Kanai, Y., Ishikawa, Y., Kawabata, N., Ikagawa, T., Saito, T., Kuramoto, Y., Kawai, N., 2006. Development of 2 cm-square Hamamatsu avalanche photodiodes for high-resolution x-rays and γ-rays detection. Nucl. Instrum. Methods Phys. Res., Sect. A 556, 535–542.

Sato, M., Yanagida, T., Yoshikawa, A., Yatsu, Y., Kataoka, J., Ishikawa, Y., Saito, F., 2007. Reverse-type avalanche photodiode for scintillation detection and intrinsic energy resolution of BGO and GSO:Ce. IEEE Nucl. Sci. Symp. Conf. Rec. 3, 2023–2032. Art. No. 4436550.

Scafè, R., Auer, P., Bennati, P., La Porta, L., Pisacane, F., Cinti, M.N., Pellegrini, R., De Vincentis, G., Conte, G., Pani, R., 2011. Production of radioactive phantoms using a standard inkjet printer and the public domain multi-printing code GENIA. Phys. Med. (in press).

Schaart, D.R., Charbon, E., Frach, T., Schulz, V., 2016. Advances in digital SiPMs and their application in biomedical imaging. Nucl. Instrum. Methods Phys. Res., Sect. A 809, 31–52.

Schepmann, D., Lehmkuhl, K., Brune, S., Wünsch, B., 2011. Expression of σ receptors of human urinary bladder tumor cells (RT-4 cells) and development of a competitive receptor binding assay for the determination of ligand affinity to human σ2 receptors. J. Pharm. Biomed. Anal. 55, 1136–1141.

Schmidt, B., Armengaud, E., Augier, C., et al., 2013. Muon-induced background in the EDELWEISS dark matter search. Astropart. Phys. 44, 28–39.

Schneider, F.R., Mann, A.B., Konorov, I., et al., 2012. A Positron Emission Tomograph based on LSO-APD modules with a sampling ADC read-out system for a students' advanced laboratory course. Z. Med. Phys. 22, 143–149.

Schnurr, B., Ahrens, T., Regenass, U., 2007. Chapter 3.27. Optical assays in drug discovery, pp. 577–598. In: Taylor, J.B., Triggle, D.J. (Eds.), Comprehensive Medicinal Chemistry II, Vol. 3 Drug Discovery Technologies. Elsevier, Amsterdam.

Schotanus, P., 1995. Miniature radiation detection instruments. Radiat. Meas. 24, 331–335.

Scott, C.W., Tian, G., Yu, X.H., Paschetto, K.A., Wilkins, D.E., Meury, L., Cao, C.Q., Varnes, J., Edwards, P.D., 2011. Biochemical characterization and in vitro activity of AZ513, a noncovalent, reversible, and noncompetitive inhibitor of fatty acid amide hydrolase. Eur. J. Pharmacol. (in press).

Seimandi, M., Bockaert, J., Marin, P., 2010. Techniques for studying inverse agonist activity of antidepressants at recombinant nonedited 5-HT2C-INI receptor and native neuronal 5-HT2C receptors, pp. 61–79. In: Abelson, J.N., Simon, M.I. (Eds.), Methods of Enzymology, vol. 485. Elsevier, Amsterdam.

Semkow, T.M., Parekh, P.P., Schwenker, C.D., Dansereau, R., Webber, J.S., 1994. Efficiency of the Lucas scintillation cell. Nucl. Instrum. Methods Phys. Res., Sect. A 353, 515–518.

Seo, H., Kim, C.H., Park, J.H., Kim, J.K., Lee, J.H., Lee, C.S., Lee, J.S., 2010. Development of double-scattering-type Compton camera with double-sided silicon strip detectors and NaI(Tl) scintillation detector. Nucl. Instrum. Methods Phys. Res., Sect. A 615, 333–339.

Sethy, N.K., et al., 2014a. Assessment of naturally occurring radioactive materials in the surface soil of uranium mining area of Jharkhand, India. J. Geochem. Explor. 142, 29–35.

Sethy, N.K., Jha, V.N., Ravi, P.M., Tripathi, R.M., 2014b. A simple method for calibration of Lucas scintillation cell counting system for measurement of ^{226}Ra and ^{222}Rn. J. Radiat. Res. Appl. Sci. 7, 472–477.

Shan, Q., Chu, S., Ling, Y., Cai, P., Jia, W., 2016. Designing a new type of neutron detector for neutron and gamma-ray discrimination via GEANT4. Appl. Radiat. Isot. 110, 200–204.

Shen, F., Smith, J.A.M., Chang, R., Bourdet, D.L., Tsuruda, P.R., Obedencio, G.P., Beattie, D.T., 2011. 5-HT4 receptor agonist mediated enhancement of cognitive function in vivo and amyloid precursor protein processing in vitro: A pharmacodynamic and pharmacokinetic assessment. Neuropharmacology 61, 69–79.

Shendrik, R., Shalaev, A.A., Myasnikova, A.S., et al., 2017. Optical and structural properties of Eu^{2+} doped BaBrI and BaClI crystals. J. Lumin. 192, 653–660.

Shivakumara, B.C., Chandrashekara, M.S., Kavitha, E., Paramesh, L., 2014. Studies on ^{226}Ra and ^{222}Rn concentration in drinking water of Mandya region, Karnataka State, India. J. Radiat. Res. Appl. Sci. 7, 491–498.

Shimazoe, K., Orita, T., Nakamura, Y., Takahashi, H., 2013. Time over threshold based multi-channel LuAG-APD PET detector, 731, 109–113.

Shimazoe, K., Koyama, A., Takahashi, H., et al., 2017. Color sensitive silicon photomultipliers with micro-cell level encoding for DOI PET detectors. Nucl. Instrum. Methods Phys. Res., Sect. A 873, 12–15.

Shlegel, V.N., Berge, L., Boiko, R.S., et al., 2014. Purification of molybdenum oxide, growth and characterization of medium size zinc molybdate crystals for the LUMINEU program. EPJ Web Conf. 65, 03001.

Shwetha, G., Kanchana, V., Vaitheeswaran, G., 2015. Optical properties of orthovanadates, and periodates studied from first principles theory. Mater. Chem. Phys. 163, 376–386.

Sidletskiy, O., Baumer, V., Gerasymov, I., Grinyov, B., Katrunov, K., Starzhinsky, N., Tarasenko, O., et al., 2010. Gadolinium pyrosilicate single crystals for gamma ray and thermal neutron monitoring. Radiat. Meas. 45, 365–368.

Siegmund, O.H., Vallerga, J.V., Tremsin, A.S., Mcphate, J., Feller, B., 2007. High spatial resolution neutron sensing microchannel plate detectors. Nucl. Instrum. Methods Phys. Res., Sect. A 576, 178–182.

Silva, J., Fiori, E., Isaak, J., et al., 2015. Temperature gain correction for CsI(Tl) detection systems based on digital pulse shape analysis. Nucl. Instrum. Methods Phys. Res., Sect. A 776, 98–106.

Sims-Williams, H., et al., 2017. Deep brain stimulation of the periaqueductal gray releases endogenous opioids in humans. Neuroimage 146, 833–842.

Singh, K., Kawaldeep, Sahota, H.S., 2001. Study of nuclear quadrupole interactions in different environments of decaying atoms of 75Se by sum peak method. Appl. Radiat. Isot. 54, 261–267.

Singh, J., Singh, H., Singh, S., Bajwa, B.S., 2008. Estimation of uranium and radon concentration in some drinking water samples. Radiat. Meas. 43, S523–S526.

Singh, B., Kumar, S., Prajapati, S., Singh, B.K., Llovet, X., Shanker, R., 2017. Measurement of angular distributions of K x-ray intensity of Ti and Cu thick targets following impact of 10–25 keV electrons. J. Electron Spectrosc. Relat. Phenom. 216, 17–22.

Sisti, M., Baldazzi, G., Bonvicini, v., et al., 2016. FLARES: A flexible scintillation light apparatus for rare event searches. Nucl. Instrum. Methods Phys. Res., Sect. A 824, 661–664.

Slomka, P.J., Berman, D.S., Germano, G., 2014. New cardiac cameras: Single-photon emission CT and PET. Semin. Nucl. Med. 44, 232–251.

Slomka, P.J., Pan, T., Germano, G., 2016. Recent Advances and Future Progress in PET Instrumentation. Semin. Nucl. Med. 46 (1), 5–19.

Song, Y., Conner, J., Zhang, Z., Hayward, J.P., 2016. Monte Carlo simulation of a very high resolution thermal neutron detector composed of glass scintillator microfibers. Appl. Radiat. Isot. 108, 100–107.

Song, Y., et al., 2018. Discovery of potent DOT1L inhibitors by AlphaLISA based high throughput screening assay. Bioorg. Med. Chem. 26, 1751–1758.

Skipper, J.A., Hangartner, T.N., 2002. Deblurring of x-ray spectra acquired with a NaI-photomultiplier detector by constrained least-squares deconvolution. Med. Phys. 29 (5), 787–796.

Stabley, J.N., et al., 2013. Chronic skeletal unloading of the rat femur: Mechanisms and functional consequences of vascular remodeling. Bone 57, 355–360.

Stanley, N.J., Hutchinson, M.R., Kvist, T., Nielsen, B., Mathiesen, J.M., Bräuner-Osborne, H., Avery, T.D., et al., 2010. A new metabotropic glutamate receptor agonist with in vivo anti-allodynic activity. Bioorg. Med. Chem. 18, 6089–6098.

Stevanato, L., Fabris, D., Hao, X., Lunardon, M., Moretto, S., Nebbia, G., Pesente, S., Sajo-Bohus, L., Viesti, G., 2011. Light output of EJ228 scintillation neutron detectors. Appl. Radiat. Isot. 69, 369–372.

Stevenson, G.I., et al., 2013. Solid-phase synthesis of Biotin-S-Farnesyl-L-Cysteine, a surrogate substrate for isoprenylcysteine carboxylmethyltransferase (ICMT). Bioorg. Med. Chem. Lett. 23, 5671–5673.

Suda, Y., 2014. Performance evaluation of the large-aperture Hybrid Photo-Detector. Nucl. Instrum. Methods Phys. Res., Sect. A 766, 185–188.

Sudarshan, M., Joseph, J., Singh, R., 1992. Full energy peak efficiency of NaI(Tl) gamma detectors and its analytical and semi-empirical representations. J. Phys. D Appl. Phys. 25, 1561–1567.

Sükösd, C., Galster, W., Licot, I., Simonart, M.P., 1995. Spectrum unfolding in high energy gamma-ray detection with scintillation detectors. Nucl. Instrum. Methods Phys. Res., Sect. A 355, 552–558.

Sun, D., Wang, Z., Caille, S., Degraffenreid, M., Gonzalez-Lopez De Turiso, F., Hungate, R., Jaen, J.C., et al., 2011. Synthesis and optimization of novel 4,4-disubstituted cyclohexylbenzamide derivatives as potent 11β-HSD1 inhibitors. Bioorg. Med. Chem. Lett 21, 405–410.

Sun, X.-Y., Gao, P., Wu, S., et al., 2015. Luminescent properties and energy transfer of Ce^{3+}-activated $Li_2O-B_2O_3-Gd_2O_3$ scintillating glasses under VUV–UV and X-ray citation. Nucl. Instrum. Methods Phys. Res., Sect. B 350, 36–40.

Sun, X.-Y., Han, T.-T., Wu, D.-L., et al., 2018. Investigation on luminescence properties of Dy^{3+}-, Eu^{3+}-doped, and Eu^{3+}/Dy^{3+}-codoped $SrGd_2O_4$ phosphors. J. Lumin. 204, 89–94.

Suzuki, S., 1982. Detection efficiency of NaI(Tl) crystals and loss of position resolution caused by photon interactions in the crystals in γ-cameras. Int. J. Appl. Radiat. Isot. 33, 411–414.

Suzuki, Y., 2010. Review of instrumentation for observational particle physics. Nucl. Instrum. Methods Phys. Res., Sect. A 623, 57–62.

Suzuki, T., Goda, K., Suzuki, N., 1993. Analysis of intensity of two closed gamma ray spectra with NaI scintillator. J. Nucl. Sci. Technol. 30, 1071–1074.

Tahirović, E., Korpar, S., Križan, P., Pestotnik, R., 2015. Characterization of the Hamamatsu MPPC S11834 as photon sensor for RICH. Nucl. Instrum. Methods Phys. Res., Sect. A 787, 203–206.

Takada, M., Taniguchi, S., Nakamura, T., Nakao, N., Uwamino, Y., Shibata, T., Fujitaka, K., 2001a. Neutron spectrometry in a mixed field of neutrons and protons with a phoswich neutron detector. Part I: response functions for photons and neutrons of the phoswich neutron detector. Nucl. Instrum. Methods Phys. Res., Sect. A 465, 498–511.

Takada, M., Taniguchi, S., Nakamura, T., Fujitaka, K., 2001b. Neutron spectrometry in a mixed field of neutrons and protons with a phoswich neutron detector. Part II: application of the phoswich neutron detector to neutron spectrum measurements. Nucl. Instrum. Methods Phys. Res., Sect. A 465, 512–524.

Takada, M., Taniguchi, S., Nakamura, T., Nakao, N., Uwamino, Y., Shibata, T., Fujitaka, K., 2002. Characteristics of a phoswich detector to measure the neutron spectrum in a mixed field of neutrons and charged particles. Nucl. Instrum. Methods Phys. Res., Sect. A 476, 332–336.

Takada, M., Yajima, K., Yasuda, H., Sato, T., Nakamura, T., 2010. Neutron, photon and proton energy spectra at high altitude measured using a phoswich-type neutron detector. Radiat. Meas. 45, 1297–1300.

Takiue, M., Natake, T., Fujii, H., 1999. A hybrid radioassay technique for multiple beta-emitter mixtures using liquid and NaI(Tl) scintillation spectrometers. Appl. Radiat. Isot. 51, 429–434.

Tamaki, S., Sato, F., Murata, I., 2015. A feasibility design study on a neutron spectrometer for BNCT with liquid moderator. Appl. Radiat. Isot. 106, 41–44.

Tamaki, S., Sato, F., Murata, I., 2017. Study on a liquid-moderator-based neutron spectrometer for BNCT—Development and experimental test of the prototype spectrometer. Nucl. Instrum. Methods Phys. Res., Sect. A 870, 90–96.

Tanaka, S., Kataoka, J., Kanai, Y., Yatsu, Y., Arimoto, M., Koizumi, M., Kawai, N., Ishikawa, Y., Kawai, S., Kawabata, N., 2007. Development of wideband x-ray and gamma-ray spectrometer using transmission-type, large-area APD. Nucl. Instrum. Methods Phys. Res., Sect. A 582, 562–568.

Tang, T., et al., 2017. Preparation and evaluation of L- and D-5-[^{18}F]fluorotryptophan as PET imaging probes for indoleamine and tryptophan 2,3-dioxygenases. Nucl. Med. Biol. 51, 10–17.

Taniguchi, S., Takada, M., Nakamura, T., 2001. Development of multimoderator neutron spectrometer using a pair of 6Li and 7Li glass scintillators. Nucl. Instrum. Methods Phys. Res., Sect. A 460, 368–373.

Taranyuk, V., Gektin, A., Kisil, I., Kolesnikov, A., 2011. NaI(Tl) and CsI(Tl) scintillation crystal growth by skull method. J. Cryst. Growth 318, 820–822.

Tarasenko, O., Galunov, N., Karavaeva, N., et al., 2013. Stilbene composite scintillators as detectors of fast neutrons emitted by a ^{252}Cf source. Radiat. Meas. 58, 61–65.

Tawa, P., et al., 2011. High-throughput scintillation proximity assay for stearoyl-CoA desaturase-1. J. Biomol. Screen 16, 506–517.

for the LUMINEU collaboration Tenconi, M., 2015. LUMINEU: a pilot scintillating bolometer experiment for neutrinoless double beta decay search. Phys. Proc. 61, 782–786.

Tenconi, M., Giuliani, A., Nones, C., et al., 2014. Characterization of a ZnSe scintillating bolometer prototype for neutrinoless double beta decay search. EPJ Web Conf. 65, 04005.

Tengblad, O., Nilsson, T., Nácher, E., et al., 2013. $LaBr_3$(Ce):$LaCl_3$(Ce) Phoswich with pulse shape analysis for high energy gamma-ray and proton identification. Nucl. Instrum. Methods Phys. Res., Sect. A 704, 19–26.

Thirstrup, K., et al., 2011. Endogenous 2-oxoglutarate levels impact potencies of competitive HIF prolyl hydroxylase inhibitors. Pharmacol. Res. 64, 268–273.

Thomas, D.J., 2004. Neutron spectrometry for radiation protection. Radiat. Prot. Dosimetry 110 (1–4), 141–149.

Thomas, D.J., 2010. Neutron spectrometry. Radiat. Meas. 45, 1178–1185.

Thomas, D.J., Alevra, A.V., 2002. Bonner sphere spectrometers – a critical review. Nucl. Instrum. Methods Phys. Rev., Sect. A 476, 12–20.

Tobin, M., 2016. on behalf of the LHCb Upgrade Scintillating Fibre Tracker group LHCb Upgrade: Scintillating Fibre Tracker. Nucl. Instrum. Methods Phys. Rev., Sect. A 824, 148–151.

Tokanai, F., Morimoto, K., Oku, T., Ino, T., Suzuki, J., et al., 2000. Cold neutron imaging with a GSO scintillator. Nucl. Instrum. Methods Phys. Res., Sect. A 452, 266–272.

Tolonen, A., Pelkonen, O., 2015. Analytical challenges for conducting rapid metabolism characterization for QIVIVE. Toxicology 332, 20–29.

Tominaga, S., 1983. Generation of precise scintillator response curves by an interpolation technique. Nucl. Instrum. Methods Phys. Res. 215, 231–233.

Tommasino, L., 2005. Radon, pp. 32–44. In: Radiochemical Methods. Reference Module in Chemistry, Molecular Sciences and Chemical Engineering. Encyclopedia of Analytical Science, second ed. Elsevier, Amsterdam.

Tong, W., Chowdhury, S., Wrona, M., Bateman, K., 2012. Coupling of UHPLC with fast fraction collection–microplate scintillation counting and MS for radiolabeled metabolite profiling. Bioanalysis 4 (11), 1299–1309.

Toribo, M., García, J.F., Izquierdo-Ridorosa, A., Tauler, R., Rauret, G., 1995. Simultaneous determination of plutonium alpha emitters by liquid scintillation counting using multivariate calibration. Anal. Chim. Acta 310, 297–305.

Toyohara, J., Hayashi, A., Gogami, A., Fujibayashi, Y., 2006. Alkyl-fluorinated thymidine derivatives for imaging cell proliferation: II. Synthesis and evaluation of N3-(2-[18F]fluoroethyl)-thymidine. Nucl. Med. Biol. 33, 765–772.

Tremsin, A.S., Feller, W.B., Downing, R.G., 2005. Efficiency optimization of microchannel plate (MCP) neutron imaging detectors. I. Square channels with 10B doping. Nucl. Instrum. Methods Phys. Res., Sect. A 539, 278–311.

Tremsin, A.S., Vallerga, J.V., McPhate, J.B., Siegmund, O.H.W., Feller, W.B., Crow, L., Cooper, R.G., 2008. On the possibility to image thermal and cold neutron with sub-15 μm spatial resolution. Nucl. Instrum. Methods Phys. Res., Sect. A 592, 374–384.

Tremsin, A.S., Craft, A.E., Bourke, M.A.M., et al., 2018. Digital neutron and gamma-ray radiography in high radiation environments with an MCP/Timepix detector. Nucl. Instrum. Methods Phys. Res., Sect. A 902, 110–116.

Trtik, P., Lehmann, E.H., 2015. Isotopically-enriched gadolinium-157 oxysulfide scintillator screens for the high-resolution neutron imaging. Nucl. Instrum. Methods Phys. Res., Sect. A 788, 67–70.

Tzellos, T.G., Tahmatzidis, D.K., Lallas, A., Apostolidou, K., Goulis, D.G., 2009. Pernicious anemia in a patient with Type 1 diabetes mellitus and alopecia areata universalis. J. Diab. Complic. 23 (6), 434–437.

Ueda, H., Tanaka, H., Sakurai, Y., 2015. Reprint of The improvement of the energy resolution in epi-thermal neutron region of Bonner sphere using boric acid water solution moderator. Appl. Radiat. Isot. 106, 107–110.

Umehara, S., Kishimoto, T., Nomachi, M., 2015. Search for neutrino-less double beta decay with CANDLES. Phys. Proc. 61, 283–288.

Usuda, S., 1992. Development of ZnS(Ag)/NE102A and ZnS(Ag)/stilbene phoswich detectors for simultaneous α and β(γ) counting. J. Nucl. Sci. Technol. 29 (9), 927–929.

Usuda, S., 1995. Simultaneous counting of α, $\beta(\gamma)$-rays and thermal neutrons with phoswich detectors consisting of ZnS(Ag). ^{6}Li-glass and/or NE102A scintillators. Nucl. Instrum. Methods Phys. Res., Sect. A 356, 334–338.

Usuda, S., Abe, H., 1994. Phoswich detectors for flow monitoring of actinide solutions with simultaneous α and $\beta(\gamma)$ counting. J. Nucl. Sci. Technol. 31 (1), 73–79.

Usuda, S., Abe, H., Mihara, A., 1994a. Phoswich detectors combining doubly or triply ZnS(Ag), NE102A, BGO and/or NaI(Tl) scintillators for simultaneous counting of α, β and γ rays. Nucl. Instrum. Methods Phys. Res., Sect. A 340, 540–545.

Usuda, S., Abe, H., Mihara, A., 1994b. Simultaneous counting α, β and γ rays with phoswich detectors. J. Alloys Compd. 213/214, 437–439.

Usuda, S., Sakurai, S., Yasuda, K., 1997. Phoswich detectors for simultaneous counting of α-, $\beta(\gamma)$-rays and neutrons. Nucl. Instrum. Methods Phys. Res., Sect. A 388, 193–198.

Usuda, S., Yasuda, K., Sakurai, S., 1998a. Development of phoswich detectors for simultaneous counting of alpha particles and other radiations (emitted from actinides). Appl. Radiat. Isot. 49, 1131–1134.

Usuda, S., Yasuda, K., Sakurai, S., 1998b. Simultaneous counting of radiation emitted from actinides with improved phoswich detectors by applying an optical filter. J. Alloys Compd. 271–273, 58–61.

Vaidyanathan, G., et al., 2011. Radioiodinated O6-benzylguanine derivatives containing an azido function. Nucl. Med. Biol. 38 (1), 77–92.

Vaidyanathan, G., et al., 2012. SIB-DOTA: A trifunctional prosthetic group potentially amenable for multi-modal labeling that enhances tumor uptake of internalizing monoclonal antibodies. Bioorg. Med. Chem. 20, 6929–6939.

Vaidyanathan, G., et al., 2018. Brush border enzyme-cleavable linkers: Evaluation for reducing renal uptake of radiolabeled prostate-specific membrane antigen inhibitors. Nucl. Med. Biol. (in press).

Vakalopoulos, A., Schmeck, C., Thutewohl, M., Li, V., Bischoff, H., Lustig, K., Weber, O., Paulsen, H., Elias, H., 2011. Chromanol derivatives - A novel class of CETP inhibitors. Bioorg. Med. Chem. Lett 21, 488–491.

Valais, I., David, S., Michail, C., Nikolopoulos, D., Liaparinos, P., Cavouras, D., Kandarakis, I., Panayiotakis, G.S., 2007. Comparative study of luminescence properties of LuYAP:Ce and LYSO:Ce single-crystal scintillators for use in medical imaging. Nucl. Instrum. Methods Phys. Res., Sect. A. 580, 614–616.

Valentini, R., Vianden, R., 2010. PAC studies with LSO scintillation crystals. Nucl. Instrum. Methods Phys. Res., Sect. A. 623, 1002–1008.

van Eijk, C.W.E., 1993. Research and development of scintillation crystals and glasses. In: De Notaristefani, F., Lecoq, P., Schneegans, M. (Eds.), Heavy Scintillators for Scientific and Industrial Applications, Proceedings of the "Crystal 2000" International Workshop, Sept. 22-26, 1992, Chamonix, France. Editions Frontieres, Gif-sur-Yvette Cedex, France.

van Eijk, C.W.E., 1997a. Development of inorganic scintillators. Nucl. Instrum. Methods Phys. Res., Sect. A. 392, 285–290.

van Eijk, C.W.E., 1997b. In: Win, Z.W., Li, P.J., Feng, X.Q., Xue, Z.L. (Eds.), Proceedings of the 4th International Conference on Inorganic Scintillator Applications., pp. 5–12. Shanghai, China.

van Eijk, C.W.E., 2001. Inorganic-scintillator development. Nucl. Instrum. Methods Phys. Res., Sect. A 460, 1–14.

van Eijk, C.W.E., 2003. Inorganic scintillators in medical imaging detectors. Nucl. Instrum. Methods Phys. Res., Sect. A 509, 17–25.

van Eijk, C.W.E., 2004. Inorganic scintillators for thermal neutron detection. Radiat. Meas. 38, 337–342.

van Eijk, C.W.E., 2014. Chapter 8.16, scintillator-based detectors, pp. 249–270. In: Comprehensive Biomedical Physics, vol. 8. Elsevier. https://doi.org/10.1016/B978-0-444-53632-7.00617-1.

van Eijk, C.W.E., Andriessen, J., Dorenbos, P., Visser, R., 1994. Ce^{3+} doped inorganic scintillators. Nucl. Instrum. Methods Phys. Res., Sect. A 348, 546–550.

Van Eijk, C.W.E., Bessière, A., Dorenbos, P., 2004. Inorganic thermal-neutron scintillators. Nucl. Instrum. Methods Phys. Res., Sect. A 529, 260–267.

van Loef, E.V.D., Dorenbos, P., van Eijk, C.W.E., Krämer, K., Güdel, H.U., 2001. High-energy-resolution scintillator: Ce^{3+} activated $LaBr_3$. Appl. Phys. Lett. 79 (10), 1573–1575.

van Loef, E.V.D., Dorenbos, P., van Eijk, C.W.E., Krämer, K.W., Güdel, H.U., 2005. Scintillation properties of K2LaX5:Ce3+ (X=Cl, Br, I). Nucl. Instrum. Methods Phys. Res., Sect. A 537, 232–236.

van Loef, E.V., Glodo, J., Shirwadkar, U., Zaitseva, N., Shah, K.S., 2011. Solution growth and scintillation properties of novel organic neutron detectors. Nucl. Instrum. Methods Phys. Res., Sect. A (in press).

Van Put, P., Debauche, A., De Lellis, C., Adam, V., 2004. Performance level of an autonomous system of continuous monitoring of radioactivity in seawater. J. Environ. Radioact. 72, 177–186.

Varley, A., Tyler, A., Dowdall, M., Bondar, Y., Zabrotski, V., 2017. An *in situ* method for the high resolution mapping of ^{137}Cs and estimation of vertical depth penetration in a highly contaminated environment. Sci. Total Environ. 605–606, 957–966.

Vega-Carrillo, H.R., 2001. Neutron energy spectra inside a PET cyclotron vault room. Nucl. Instrum. Methods Phys. Res., Sect. A 463, 375–386.

Vega-Carrillo, H.R., Manzanares-Acuña, E., 2004. Background neutron spectrum at 2420 m above sea level. Nucl. Instrum. Methods Phys. Res., Sect. A 524, 146–151.

Vega-Carrillo, H.R., Donaire, I., Gallego, E., Manzanares-Acuña, E., Lorente, A., Iñiguez, M.P., Martin-Martin, A., Gutierrez-Villanueva, J.L., 2008. Calculation of response matrix of a BSS with 6LiI scintillator. Rev. Mex. Fis. 54 (1), 57–62.

Vega-Carrillo, H.R., Gallego, E., Lorente, A., 2009. Response matrix calculation of a bonner sphere spectrometer based on A 6LiI(Eu) scintillator. Nucl. Technol. 168 (2), 359–363.

Vega-Carrillo, H.R., Hernández-Dávila, V.M., Aguilar, F., Paredes, L., Rivera, T., 2014. Neutron spectra at two beam ports of a TRIGA Mark III reactor loaded with HEU fuel. Appl. Radiat. Isot. 83, 252–255.

Velázquez, M., Veber, P., Moutatouia, M., 2017. Exploratory growth in the Li_2MoO_4-MoO_3 system for the next crystal generation of heat-scintillation cryogenic bolometers. Solid State Sci. 65, 41–51.

Verheyden, R., Chagani, H., Dolenec, R., Pestotnik, R., Križan, P., Korpar, S., Stanovnik, A., 2011. Performance study of silicon photomultipliers as photon detectors for PET. Nucl. Instrum. Methods Phys. Res., Sect. A 628, 381–384.

Vidmar, T., Kossert, K., Nähle, O.J., Ott, O., 2009. Application of the sum-peak method to activity standardizations of extended 60Co sources. Appl. Radiat. Isot. 67, 160–163.

Visinelli, L., 2014. Gamma-Gamma Angular Correlation in Ni^{60}. https://www.researchgate.net/publication/259967436_Gamma-Gamma_Angular_Correlation_in_Ni60.

Vrinda Devi, K.V., Soreng, T., Mukherjee, D., Panakkal, J.P., Kamath, H.S., 2010. Non destructive determination of PuO2 content in MOX fuel pins for fast reactors using Passive Gamma Scanning. J. Nucl. Mater. 399, 122–127.

Wang, Y., Ma, H., 2015. Protein kinase profiling assays: a technology review. Drug Discov. Today Technol. 18, 1–8.

Wang, C.-F., Lee, J.-H., Chiou, H.-J., 1994. Rapid determination of Sr-89/Sr-90 in radwaste by low-level background beta counting system. Appl. Radiat. Isot. 45, 251–256.

Wang, Z., Zhang, X., Chang, Y., Liu, D., 2001. The determination of ^{125}I activity using sum-peak method with a well-type HPGe-detector-based spectrometer. Nucl. Instrum. Methods Phys. Res., Sect. A 459, 475–481.

Wang, Z., et al., 2013. The selectivity and potency of the new PDE5 inhibitor TPN729MA. J. Sex. Med. 10, 2790–2797.

Wang, P.-F., et al., 2015a. Simulation of the response functions of an extended range neutron multisphere spectrometer using FLUKA. Chin. Phys. C 39 (7), 076001.

Wang, Y., et al., 2015b. Readout for a large area neutron sensitive microchannel plate detector. Nucl. Instrum. Methods Phys. Res., Sect. A 784, 226–231.

Wang, Y., et al., 2017. Detection of melanoma metastases with PET—comparison of ^{18}F-5-FPN with ^{18}F–FDG. Nucl. Med. Biol. 50, 33–38.

Wang, Z.X., Pei, H., Tao, X.M., et al., 2018. Enhanced scintillation of $Ba_3In(B_3O_6)_3$ based on nitrogen doping. J. Solid State Chem. 258, 351–357.

Warburton, W.K., Asztalos, S.J., Hennig, W., 2011. α-Al_2O_3 Diffusion barriers to eliminate the radioxenon memory effect in phoswich detectors. In: 2011 Monitoring Research Review: Ground-Based Nuclear Explosion Monitoring Technologies. http://www.xia.com/Papers/Xe_Memory.PDF.

Watanabe, H., Abe, K., Harada, E., Inoue, S., Inagaki, T., et al., 2002. Scintillator-Lucite sandwich detector for n/γ separation in the GeV energy region. Nucl. Instrum. Methods Phys. Res., Sect. A 484, 118–128.

Watanabe, T., Arakawa, H., Kajimoto, T., Iwamoto, Y., Satoh, D., Kunieda, S., Noda, S., Shigyo, N., Ishibashi, K., Nakamura, T., Haight, R.C., 2008. Performance of a phoswich detector composed of an inner NaI(Tl) crystal and surrounding NE102A plastic scintillator for neutron spectrometry. Nucl. Instrum. Methods Phys. Res., Sect. A 587, 20–28.

Watanabe, M., Katsumata, M., Ono, H., et al., 2015. First performance test of newly developed plastic scintillator for radiation detection. Nucl. Instrum. Methods Phys. Res., Sect. A 770, 197–202.

Watanabe, H., et al., 2016. Synthesis and evaluation of copper-64 labeled benzofuran derivatives targeting β-amyloid aggregates. Bioorg. Med. Chem. 24, 3618–3623.

Watson, J., 1996. In *Vitro* Measurement of the Second Messenger cAMP: RIA vs FlashPlate™. FlashNews No. FN001. NEN Life Sciences Products, Boston, MA.

Wauters, F., Kraev, I., Zákoucký, D., Beck, M., Golovko, V.V., Kozlov, V.Y., Phalet, T., Tandecki, M., Traykov, E., Van Gorp, S., Severijns, N., 2009. A GEANT4 Monte-Carlo simulation code for precision β spectroscopy. Nucl. Instrum. Methods Phys. Res., Sect. A 609, 156–164.

Weber, M.J., 2002. Inorganic scintillators: today and tomorrow. J. Lumin. 100, 35–45.

Weber, M.J., 2004. Scintillation: mechanisms and new crystals. Nucl. Instrum. Methods Phys. Res., Sect. A 527, 9–14.

Weisskopf, M.C., Odell, S.L., Elsner, R.F., Van Speybroeck, L.P., 1995. Advanced x-ray astrophysical observatory AXAF – an overview. Proc. SPIE-Int. Soc. Opt. Eng. 2515, 312.

Westrøm, S., et al., 2017. Preparation of ^{212}Pb-labeled monoclonal antibody using a novel ^{224}Ra-based generator solution. Nucl. Med. Biol. 51, 1–9.

on behalf of the LAPPD collaboration Wetstein, M., 2011. Development of sub-nanosecond, high gain structures for time-of-flight ring imaging in large area detectors. Nucl. Instrum. Methods Phys. Res., Sect. A 639, 148–150.

Wheelan, N., et al., 2015. Short-term inhibition of 11b-hydroxysteroid dehydrogenase type 1 reversibly improves spatial memory but persistently impairs contextual fear memory in aged mice. Neuropharmacology 91, 71–76.

Whitaker, M.D.C., Lioliou, G., Barnett, A.M., 2018. $Al_{0.2}Ga_{0.8}As$ 2 × 2 square pixel X-ray photodiode array. Nucl. Instrum. Methods Phys. Res., Sect. A 899, 106–114.

White, T.L., Miller, W.H., 1999. A triple-crystal phoswich detector with digital pulse shape discrimination for alpha/beta/gamma spectroscopy. Nucl. Instrum. Methods Phys. Res., Sect. A 422, 144–147.

Wiegel, B., Agosteo, S., Bedogni, R., Caresana, M., Esposito, A., Fehrenbacher, G., Ferrarini, M., Hohmann, E., Hranitzky, C., Kasper, A., et al., 2009. Intercomparison of radiation protection devices in a high-energy stray neutron field, Part II: Bonner sphere spectrometry. Radiat. Meas. 44, 660–672.

Wiggins, B.B., deSouza, Z.O., Vadas, J., Alexander, A., Hudan, S., deSouza, R.T., 2017. Achieving high spatial resolution using a microchannel plate detector with an economic and scalable approach. Nucl. Instrum. Methods Phys. Res., Sect. A 872, 144–149. Nucl. Instrum. Methods Phys. Res., Sect. A 891, 53-57.

Wiggins, B.B., Vadas, J., Bancroft, D., et al., 2018. An efficient and cost-effective microchannel plate detector for slow neutron radiography. Nucl. Instr. Med. Phy. Res. Sec. A.

Wilbur, D.S., Chyan, M.-K., Hamlin, D.K., Perry, M.A., 2010. Preparation and in vivo evaluation of radioiodinated closo-decarbonate(2-) derivatives to identify structural components that provide low retention in tissues. Nucl. Med. Biol. 37 (2), 167–178.

Wildey, M.J., Haunso, A., Tudor, M., Webb, M., Connick, J.H., 2017. High-throughput screening, pp. 149–195. In: Annual Reports in Medicinal Chemistry, vol. 50. Elsevier, Amsterdam.

Wilhelm, K., Nattress, J., Jovanovic, I., 2017. Development and operation of a ^{6}LiF:ZnS(Ag)—scintillating plastic capture-gated detector. Nucl. Instrum. Methods Phys. Res., Sect. A 842, 54–61.

Wingelhofer, B., et al., 2016. Preloading with L-BPA, L-tyrosine and L-DOPA enhances the uptake of [^{18}F]FBPA in human and mouse tumour cell lines. Appl. Radiat. Isot. 118, 67–72.

Wise, B.M., 1992. PLS-Toolbox for Use with MATLAB™. Center for Process Analytical Chemistry (CPAC), University of Washington, Seattle.

Witkin, J.F., et al., 2016. Preclinical findings predicting efficacy and side-effect profile of Y2940094, an antagonist of nociceptin receptors. Pharm. Res. Per. 4 (6), e00275.

Wojtowicz, A.J., Lempicki, A., Wisniewski, D., Boatner, L.A., 1994a. Cerium-doped orthophosphate scintillators. Mater. Res. Soc. Symp. Proc. 123–129. Conf. Record 348 (Scintillators and Phosphor Materials).

Wojtowicz, A.J., Wisniewski, D., Lempicki, A., Boatner, L.A., 1994b. Scintillation mechanisms in rare earth orthophosphates. In: Conferance Record EURODIM94.

Wojtowicz, A.J., Lempicki, A., Wisniewski, D., Balcerzyk, M., Brecher, C., 1995. The role of charge transfer states in Ln^{+3}-activated scintillators. In: Trendler, R.C. (Ed.), IEEE Conference Record, Nuclear Science Symposium and Medical Imaging Conference Oct. 30–Nov 5, 1994, vol. 1, pp. 134–138.

Wojtowicz, A.J., Lempicki, A., Wisniewski, D., Balcerzyk, M., Brecher, C., 1996. The carrier capture and recombination processes in Ln^{+3}-activated scintillators. IEEE Trans. Nucl. Sci. 43, 2168–2173.

Wong, H.T., Li, J., 1999. A pilot experiment with reactor neutrinos in Taiwan. Nucl. Phys. B 77, 177–181.

Wong, H.T.-K., 2015. Taiwan EXperiment On NeutrinO – History, Status and Prospects. The Universe 3 No. 4, 22-37. https://arxiv.org/abs/1608.00306.

Wong, H.T., Li, J., 2000. Scintillation crystal detector for low energy neutrino physics. Mod. Phys. Lett. A 15 (33), 2011–2016.

Wood, C.A., Levitt, S.H., 1965. Simultaneous red cell mass and plasma volume determinations using ^{51}Cr tagged red cells and ^{125}I labeled albumin. J. Nucl. Med. 6, 433–440.

Wu, D., Prange, M.P., Gao, F., Kerisit, S., 2016. First-principles search for efficient activators for LaI_3. J. Lumin. 176, 227–234.

Xia, L., De Vries, H., IJzerman, A.P., Heitman, L.H., 2016. Scintillation proximity assay (SPA) as a new approach to determine a ligand's kinetic profile. A case in point for the adenosine A1 receptor. Purinergic Signal. 12, 115–126.

Xiao, J.C., Jewell, J.P., Lin, L.S., Hagmann, W.K., Fong, T.M., Shen, C.-P., 2008. Similar in vitro pharmacology of human cannabinoid CB1 receptor variants expressed in CHO cells. Brain Res. 1238, 36–43.

Yagi, T., Misawa, T., Pyeon, C.H., Shiroya, S., 2011. A small high sensitivity neutron detector using a wavelength shifting fiber. Appl. Radiat. Isot. 69, 176–179.

Yalcin, S., Gurler, O., Kaynak, G., Gundogdu, O., 2007. Calculation of total counting efficiency of a NaI(Tl) detector by hybrid Monte-Carlo method for point and disk sources. Appl. Radiat. Isot. 65, 1179–1186.

Yalcin, S., Gurler, O., Gundogdu, O., Kaynak, G., 2009. Monte Carlo simulation of gamma-ray total counting efficiency for a Phoswich detector. Radiat. Meas. 44, 80–85.

Yamamoto, S., Kobayashi, T., Yeom, J.-Y., et al., 2014. Development of GAGG depth-of-interaction (DOI) block detectors based on pulse shape analysis. Nucl. Instrum. Methods Phys. Res., Sect. A 767, 289–295.

Yamamoto, S., Ishibashi, H., 2015. Development of a three-layer phoswich alpha–beta–gamma imaging detector. Nucl. Instrum. Methods Phys. Res., Sect. A 785, 129–134.

Yamamoto, S., Kobayashi, T., Okumura, S., Yeom, J.-Y., 2016. Timing performance measurements of Si-PM-based LGSO phoswich detectors. Nucl. Instrum. Methods Phys. Res., Sect. A 821, 101–108.

Yamamoto, K., et al., 2017. Radioimmunoassay of relaxin-like gonad-stimulating peptide in the starfish *Patiria (=Asterina) pectinifera*. Gen. Comp. Endrocrinol. 243, 84–88.

Yamane, Y., Lindén, P., Karlsson, J.K.-H., Pázsit, I., 1998. Measurement of 14.1 MeV neutrons with a Th-scintillator optical fiber detector. Nucl. Instrum. Methods Phys. Res., Sect. A 416, 371–380.

Yamane, Y., Uritani, A., Misawa, T., Karlsson, J.K.-H., Pázsit, I., 1999. Measurement of the thermal and fast neutron flux in a research reactor with a Li and Th loaded optical fiber detector. Nucl. Instrum. Methods Phys. Res., Sect. A 432, 403–409.

Yamazaki, A., Watanabe, K., Uritani, A., Iguchi, T., Kawaguchi, N., Yanagida, T., Fujimoto, Y., Yokota, Y., Kamada, K., Fukuda, K., Suyama, T., Yoshikawa, A., 2011. Neutron-gamma discrimination based on pulse shape discrimination in a Ce:LiCaAlF6 scintillator. Nucl. Instrum. Methods Phys. Res., Sect. A (in press).

Yanagida, T., 2013. Study of rare-earth-doped scintillators. Opt. Mater. 35, 1987–1992.

Yanagida, T., 2016. Ionizing radiation induced emission: Scintillation and storage-type luminescence. J. Lumin. 169, 544–548.

Yan, J., et al., 2017. Discovery of a low-systemic-exposure DGAT-1 inhibitor with a picolinoylpyrrolidine-2-carboxylic acid moiety. Bioorg. Med. Chem. 25, 4701–4714.

Yanagida, T., Kawaguchi, N., Fujimoto, Y., Yokota, Y., Yamazaki, A., Watanabe, K., Kamada, K., Yoshikawa, A., Chani, V., 2011. Ce and Eu-doped LiSrAlF6 scintillators for neutron detectors. Radiat. Meas. (in press).

Yanagida, T., Watanabe, K., Fujimoto, Y., 2015. Comparative study of neutron and gamma-ray pulse shape discrimination of anthracene, stilbene, and p-terphenyl. Nucl. Instrum. Methods Phys. Res., Sect. A 784, 111–114.

Yanagida, T., Koshimizu, M., Kawano, N., Okada, G., Kawaguchi, N., 2017. Optical and scintillation properties of $ScAlMgO_4$ crystal grown by the floating zone method. Mater. Res. Bull. 95, 409–413.

Yang, H., Menaa, N., Bronson, F., Kastner, M., Venkataraman, R., Mueller, W.F., 2011. Evaluation of a LiI(Eu) neutron detector with coincident double photodiode readout. Nucl. Instrum. Methods Phys. Res., Sect. A 652, 364–369.

Yang, B., Xu, J., Zhang, Y., et al., 2016. Improvement and luminescent mechanism of $Bi_4Si_3O_{12}$ scintillation crystals by Dy^{3+} doping. Nucl. Instrum. Methods Phys. Res., Sect. A 807, 1–4.

Yasuda, K., Usada, S., Gunji, H., 2000a. Properties of a YAP powder scintillator as alpha-ray detector. Appl. Radiat. Isot. 52, 365–368.

Yasuda, K., Usada, S., Gunji, H., 2000b. Development of scintillation-light-transmission type phoswich detector for simultaneous alpha- and beta (gamma)-ray counting. IEEE Trans. Nucl. Sci. 47 (4), 1337–1340.

Yasuda, K., Usada, S., Gunji, H., 2001. Simultaneous alpha, beta/gamma, and neutron counting with phoswich detectors by using a dual-parameter technique. IEEE Trans. Nucl. Sci. 48 (4), 1162–1164.

Yatsu, Y., Kuramoto, Y., Kataoka, J., Kotoku, J., Saito, T., Ikagawa, T., Sato, R., Kawai, N., Kishimoto, S., Mori, K., Kamae, T., Ishikawa, Y., Kawabata, N., 2006. Study of avalanche photodiodes for soft X-ray detection below 20 keV. Nucl. Instrum. Methods Phys. Res., Sect. A 564, 134–143.

Yeşilagâç, R., Ünak, P., Medine, E.T., Içhedef, Ç.A., Ertay, T., Müftüler, F.Z.B., 2011. Enzymatic synthesis of 125/131I labeled 8-hydroxyquinoline glucuronide and in vitro/in vivo evaluation of biological influence. Appl. Radiat. Isot. 69, 299–307.

Yin, J., Zhang, J., Wang, J., et al., 2018. Growth and scintillation properties of $Ce:Li_6Y(BO_3)_3$ crystal enriched with ^{10}B isotopes. Radiat. Meas. 113, 20–24.

Yongchun, X., et al., 2017. Studies on adsorption-desorption of xenon on surface of BC-404 plastic scintillator based on soaking method. Nucl. Instrum. Methods Phys. Res., Sect. A 847, 99–103.

Yoshida, S., Kishimoto, T., Ogawa, I., Hazama, R., Umehara, S., Matsuoka, K., Yokoyama, D., Mukaida, K., Ichihara, K., Tatewaki, Y., 2005. CANDLES project for double beta decay of 48Ca. Nucl. Phys. B 138, 214–216.

Yoshida, S., Kishimoto, T., Ogawa, I., Hazama, R., Umehara, S., Matsuoka, K., Yokoyama, D., Ichihara, K., Tatewaki, Y., 2009. Ultraviolet wavelength shift for undoped CaF2 scintillation detector by two phase of liquid scintillator system in CANDLES. Nucl. Instrum. Methods Phys. Res., Sect. A 601, 282–293.

Yoshida, Y., Shinozaki, K., Igashira, T., et al., 2018. Characterizations of Pr-doped Yb3Al5O12 single crystals for scintillator applications. Solid State Sci. 78, 1–6.

Yoshino, M., Kataoka, J., Nakamori, T., Matsuda, H., Miura, T., Katou, T., Ishikawa, Y., Kawabata, N., Matsunaga, Y., Kamada, K., Usuki, Y., Yoshikawa, A., Yanagida, T., 2011. The development and performance of UV-enhanced APD-arrays for high resolution PET imaging coupled with pixelized Pr:LuAG crystal. Nucl. Instrum. Methods Phys. Res., Sect. A (in press).

Yu, W., et al., 2014. A scintillation proximity assay for histone demethylases. Anal. Biochem. 463, 54–60.

Yu, L., Terashima, S., Ong, H.J., et al., 2017. Multi-layer plastic scintillation detector for intermediate- and high-energy neutrons with n-γ discrimination capability. Nucl. Instrum. Methods Phys. Res., Sect. A 866, 118–128.

Yuan, M.-C., Hwang, W.-S., 2000. The absolute counting of ^{125}I. Appl. Radiat. Isot. 52, 523–526.

Zafrir, H., Haquin, G., Malik, U., Barbosa, S.M., Piatibratova, O., Steinitz, G., 2011. Gamma versus alpha sensors for Rn-222 long-term monitoring in geological environments. Radiat. Meas. (in press).

Zaghdane, H., Boyd, M., Colucci, J., Simard, D., Berthelette, C., Leblanc, Y., Wang, Z., et al., 2011. New indole amide derivatives as potent CRTH2 receptor antagonists. Bioorg. Med. Chem. Lett 21, 3471–3474.

Zaidins, C.S., Martin, J.B., Edwards, F.M., 1978. A least-squares technique for extracting neutron spectra from Bonner sphere data. Med. Phys. 5 (1), 42–47.

Zaitseva, N., Glenn, A., Martinez, H.P., et al., 2013. Pulse shape discrimination with lithium-containing organic scintillators. Nucl. Instrum. Methods Phys. Res., Sect. A 729, 747–754.

Zaitseva, N., Glenn, A., Carman, L., et al., 2015. Scintillation properties of solution-grown *trans*-stilbene single crystals. Nucl. Instrum. Methods Phys. Res., Sect. A 789, 8–15.

Zaitseva, N.P., Glenna, A.M., Mabe, A.N., Carman, M.L., Hurlbut, C.R., Inman, J.W., Payne, S.A., 2018. Recent developments in plastic scintillators with pulse shape discrimination. Nucl. Instrum. Methods Phys. Res. Sect. A 889, 97–104.

Zalutsky, M.R., et al., 2012. Radioimmunotargeting of malignant glioma by monoclonal antibody D2C7 reactive against both wild-type and variant III mutant epidermal growth factor receptors. Nucl. Med. Biol. 39, 23–34.

Zaman, F., Rooh, G., Srisittipokakun, N., et al., 2017. Scintillation and luminescence characteristics of Ce^{3+} doped in $Li_2O-Gd_2O_3-BaO-B_2O_3$ scintillating glasses. Radiat. Phys. Chem. 130, 158–163.

Zazubovich, S., 2001. Physics of halide scintillators. Radiat. Meas. 33, 699–704.

Zdesenko, Y., Avignone, F.T., Brudanin, V.B., Danevich, F.A., Kobychev, V.V., Kropivyansky, B.N., Nagorny, S.S., Tretyak, V.I., Vylov, T., 2005. CARVEL experiment with 48CaWO4 crystal scintillators for the double β decay study of 48Ca. Astropart. Phys. 23, 249–263.

Zeng, W., 2012. The design of digital multi-channel analyzer based on FPGA. Energia Procedia 39, 428–433.

Zeng, H., Xu, W., 2015. Chapter 16, enzymatic assays of histone methyltransferase enzymes, pp. 333–360. In: Zheng, Y.G. (Ed.), Epigenetic Technological Applications. Elsevier, Amsterdam.

Zhang, J., Amin, A.G., Hölemann, A., Seeberger, P.H., Chatterjee, D., 2010. Development of a plate-based scintillation proximity assay for the mycobacterial AftB enzyme involved in cell wall arabinan biosynthesis. Bioorg. Med. Chem. 18, 7121–7131.

Zhang, Z., Lu, X., Xu, J., Rothfuss, J., Mach, R.H., Tu, Z., 2011. Synthesis and in Vitro Evaluation of New Analogues as Inhibitors for Phosphodiesterase 10A. Eur. J. Med. Chem. (in press).

Zhao, W., et al., 2011. Chicken type II collagen induced immune tolerance of mesenteric lymph node lymphocytes by enhancing beta2-adrenergic receptor desensitization in rats with collagen-induced arthritis. Int. Immunopharmacol. 11, 12–18.

Zhao, Y., Bi, Q., Hu, R., 2013. Recognition and measurement in the flow pattern and void fraction of gas-liquid two-phase flow in vertical upward pipes using the gamma densitometer. Appl. Thermal. Eng. 60, 398–410.

Zhu, M., Zhang, D., Skiles, G.L., 2005a. Chapter 9. Quantification and structural elucidation of low quantities of radiolabeled metabolites using microplate scintillation counting (MSC) techniques in conjunction with LC-MS, pp. 195–223. In: Chowdhury, S.W. (Ed.), Identification and Quantification of Drugs, Metabolites and Metabolizing Enzymes by LC-MS. Elsevier, Amsterdam.

Zhu, M., Zhao, W., Vazquez, N., Mitroka, J.G., 2005b. Analysis of low level radioactive metabolites in biological fluids using high-performance liquid chromatography with microplate scintillation counting: Method validation and application. J. Pharm. Biomed. Anal. 39, 233–245.

Zhu, M., Qi, H., Pan, M., et al., 2018a. Growth and luminescent properties of Yb:YAG and Ca co-doped Yb:YAG ultrafast scintillation crystals. J. Cryst. Growth 490, 51–55.

The Texono Collaboration Zhu, Y.F., et al., 2006. Measurement of the intrinsic radiopurity of 137Cs/235U/238U/232Th in CsI(Tl) crystal scintillators. Nucl. Instrum. Methods Phys. Res., Sect. A 557, 490–500.

Zhu, W.J., et al., 2018b. Development of radioiodine labeled acetaminophen for specific, high-contrast imaging of malignant melanoma. Nucl. Med. Biol. 59, 16–21.

Zhuo, X., Zheng, X., 2018. Drug metabolism: metabolite isolation and identification. In: Reedijk, J. (Ed.), Reference Module in Chemistry, Molecular Sciences and Chemical Engineering, third ed. Elsevier, Amsterdam, p. 108.

Zhuravleva, M., Yang, K., Green, A., Melcher, C.L., 2011. Crystal growth and scintillation properties of Ce3+-doped $KGd2Cl_7$. J. Cryst. Growth 318, 796–799.

Zorn, C., Bowen, M., Majewski, S., Walker, J., Wojcik, R., Hurlbut, C., Moser, W., 1988. Pilot study of new radiation-resistant plastic scintillators doped with 3-hydroxyflavone. Nucl. Instrum. Methods Phys. Res., Sect. A 273, 108–116.

Zorzi, N., Melchiorri, M., Piazza, A., Piemonte, C., Tarolli, A., 2011. Development of large-area silicon photomultiplier detectors for PET applications at FBK. Nucl. Instrum. Methods Phys. Res., Sect. A 636 (Suppl. 1), S201–S213.

Further reading

Armengaud, E., 2012. Direct detection of WIMPs. Comptes Rendus Phys. 13, 730–739.

Audi, G., Wapstra, A.H., 1995. The 1995 update to the atomic mass evaluation. Nucl. Phys. A 595 (4), 409–480.

Briesmeister, J., 2002. MCNP–A General Monte Carlo N-particle Transport Code. Version 4C. Los Alamos National Laboratory Publication LA-13709-M, Los Alamos, NM.

Buchan, K.W., Sumner, M.J., Watts, I.S., 1993. Human placental membranes contain predominantly ET_B receptors. J. Cardiovasc. Pharmacol. 22 (Suppl. 8), S136–S139.

On behalf of the CLAS 12 RICH Group Contalbrigo, M., 2015. Tests of innovative photon detectors and integrated electronics for the large-area CLAS12 ring-imaging Cherenkov detector. Nucl. Instrum. Methods Phys. Res., Sect. A 787, 224–228.

Entine, G., 1990. Large area silicon avalanche photodiodes for scintillation detectors. Nucl. Instrum. Methods Phys. Res., Sect. A 288, 137–139.

Fukabori, A., Yanagida, T., Chani, V., Moretti, F., Pejchal, J., Yokota, Y., Kawaguchi, N., et al., 2011b. Optical and scintillation properties of Pr-doped Li-glass for neutron detection in inertial confinement fusion process. J. Non-Cryst. Solids 357, 910–914.

Han, M.-J., Bummer, P.M., Jay, M., 1998. Solid scintillation proximity membranes. II. Use in wipe test assays for radioactive contamination. J. Membr. Sci. 148, 223–232.

Han, M.-J., Lee, K.W., Seo, B.-K., 2003. Preparation and characterization of a double-layered porous film to assay for surface radioactive contamination. J. Membr. Sci. 223, 59–67.

Highes, K.T., Ireson, J.C., Jones, N.R.A., Kivalä, P., 2001. Colour Quench Correction in Scintillation Proximity Assays Using ParaLux Count Mode. Application Notes. PerkinElmer Life and Analytical Sciences, Boston, p. 12.

Holden, N.E., 1997. Table of the isotopes. In: Lide, D.R. (Ed.), Handbook of Chemictry and Physics 1997. CRC Press, Boca Raton.

Kang, J.W., 2016. Association of serum 25-hydroxyvitamin D with serum IgE levels in Korean adults. Auris Nasus Larynx 43, 84–88.

Kubo, N., Mabuchi, M., Katoh, C., Arai, H., Morita, K., Tsukamoto, E., Morita, Y., Tamaki, N., 2002. Validation of left ventricular function from gated single photon computed emission tomography by using a scintillator-photodiode camera: a dynamic myocardial phantom study. Nucl. Med. Commun. 23 (7), 639-664, 522-534.3.

Lees, J.E., Richards, P.G., 1999. Rapid, high-sensitivity imaging of radiolabeled gels with microchannel plate detectors. Electrophoresis 20, 2139–2143.

Li, X., Lockhart, C., Lewellen, T.K., Miyaoka, R.S., 2011. Study of PET detector performance with varying SiPM parameters and readout schemes. IEEE Trans. Nucl. Sci. (in press).

Litton Airtron SYNOPTICS, Technical Brief, 1997. Ce:LuAP Scintillator Crystals. Litton Airtron SYNOPTICS, Synthetic Crystals and Optical Products. Charlotte, NC.

Moszyński, M., Czarnacki, W., Kapusta, M., Szawlowski, M., Klamra, W., Schotanus, P., 2002b. Energy resolution and light yield non-proportionality of pure NaI scintillator studied with large area avalanche photodiodes at liquid nitrogen temperatures. Nucl. Instrum. Methods Phys. Res., Sect. A 486, 13–17.

Nakamura, T., Kawasaki, T., Hosoya, T., et al., 2012. A large-area two-dimensional scintillator detector with a wavelength-shifting fibre readout for a time-of-flight single-crystal neutron diffractometer. Nucl. Instrum. Methods Phys. Res., Sect. A 686, 64–70.

Parikh, S., Matulis, J., 2010. Vitamin B12 deficiency associated with metformin. Endocrinololgist 20 (1), 38–40.

Persson, A., Cederwall, B., 2011. Development of a multimodality sensor for spectral photon counting CT, standard CT and PET. *Nucl. Instrum.* Methods Phys. Res., Sect. A (in press).

Price, E.W., 2016. A comparative evaluation of the chelators H4octapa and CHX-A″-DTPA with the therapeutic radiometal ^{90}Y. Nucl. Med. Biol. 43, 566–576.

Pruszynski, M., et al., 2015. D-amino acid peptide residualizing agents bearing N-hydroxysuccinimido- and maleimido-functional groups and their application for trastuzumab radioiodination. Nucl. Med. Biol. 42, 19–27.

Renker, D., 2006. Geiger-mode avalanche photodiodes, history, properties and problems. Nucl. Instrum. Methods Phys. Res., Sect. A 567, 48–56.

Schieber, M., Zuck, A., Braiman, M., Nissenbaum, J., Turchetta, R., Dulinski, W., Husson, D., Riester, J.L., 1997. Novel mercuric iodide polycrystalline nuclear particle counters. IEEE Trans. Nucl. Sci. 44 (6), 2571–2575.

Schieber, M., Zuck, A., Braiman, M., Nissenbaum, J., Zuck, A., Turchetta, R., Dulinski, W., Husson, D., Riester, J.L., 1998a. Evaluation of mercuric iodide ceramic semiconductor detectors. Nucl. Phys. B 61, 321–329.

Schieber, M., Zuck, A., Braiman, M., Melekhov, J., Nissenbaum, J., Turchetta, R., Dulinski, W., Husson, D., Riester, J.L., 1998b. Radiation-hard polycrystalline mercuric iodide semiconductor particle counters. Nucl. Instrum. Methods Phys. Res., Sect. A 410, 107–110.

Schotanus, P., 1996. SCIONIX Scintillation Detectors. SCIONIX Holland B.V., Bunnik, The Netherlands, p. 48.

Shih, B.-B., 2017. SPECT imaging evaluation of ^{111}indium-chelated cetuximab for diagnosing EGFR-positive tumor in an HCT-15-induced colorectal xenograft. J. Chin. Med. Assoc. 80, 766–773.

van Eijk, C.W.E., Dorenbos, P., van Loef, E.V.D., Krämer, F., Güdel, H.U., 2001. Energy resolution of some new inorganic-scintillator gamma-ray detectors. Radiat. Meas. 33, 521–525.

Wong, H.T., Li, J., 2000. Scintillation crystal detector for low energy neutrino physics. Mod. Phys. Lett. A 15 (33), 2011–2016.

Wunderly, S.W., 1989. Solid scintillation counting: a new technique for measuring radiolabeled compounds. Appl. Radiat. Isot. 40, 569–573.

Wunderly, S.W., 1993. Simultaneous measurement of alpha and beta emissions on ready cap. In: Noakes, J.E., Schönhofer, F., Polach, H.A. (Eds.), Liquid Scintillation Spectrometry 1992. Radiocarbon, Tucson, AZ, pp. 217–223.

Yang, D., 1993a. Alpha liquid scintillation analysis: some recent developments and applications. In: Noakes, J.E., Schönhofer, F., Polach, H.A. (Eds.), Liquid Scintillation Spectrometry 1992. Radiocarbon, Tucson, AZ, p. 201.

Yang, D., 1993b. Alpha-counting with a solid scintillator. J. Radioanal. Nucl. Chem. Lett. 175, 393–400.

Yang, D., 1993c. Alpha-particle counting with solid scintillator cap. Radioact. Radiochem. 4, 8–13.

Zimmermann, H., Müller, B., Hammer, A., Herzog, K., Seegebrecht, P., 2002. Large-area lateral P-I-N photodiode on SOI. IEEE Trans. Nucl. Sci. 49 (2), 334–336.

Michael F. L'Annunziata

Michael F. L'Annunziata, Ph.D. is the Founding Editor and co-author of the Handbook of Radioactivity Analysis now in its Fourth Edition. His graduate thesis research at the University of Arizona, Tucson, AZ, USA in the 1960s, financed by the then U.S. Atomic Energy Commission, dealt with the analysis of ^{89}Sr and ^{90}Sr in the event of nuclear fallout. During 1970–71, he worked in the chemical industry (Amchem Products, Inc., Ambler, PA, USA) as ^{14}C-tracer chemist. Michael was then appointed Professor at the Postgraduate College, Chapingo, Mexico (1972–75) and Senior Research Scientist and graduate thesis advisor at the Nuclear Center of the National Institute of Nuclear Research (ININ), Mexico City during (1975–77). Michael served as Scientific Officer in the Department of Research and Isotopes and Department of Technical Co-operation of the International Atomic Energy Agency (IAEA) in Vienna, Austria during 1977–91 and Head of IAEA Fellowships and Training during 1987–91. Michael F. L'Annunziata was first to postulate the soil microbial epimerization of myo-inositol to other inositol stereoisomers as the chemical pathway and origin of the various inositol phosphate stereoisomers in soils (Ph.D. dissertation, 1970, https://dissexpress.proquest.com/dxweb/results.html?QryTxt=&By=L%27Annunziata&Title=&pubnum=) and in 1975 (*SSSA Journal* 39(2), 377–379), and first to demonstrate in 1977, with the use of the radioisotope carbon-14, the soil microbial epimerization of myo-inositol to chiro-inositol as a mechanism involved in the origin of the unique soil inositol phosphate stereoisomers (*SSSA Journal* 41(4), 733–736, https://dl.sciencesocieties.org/publications/sssaj/abstracts/41/4/SS0410040733). From 1977 to 2007, he served as IAEA Expert in peaceful applications of nuclear energy in over 50 countries of the world in Europe, Asia, Africa, Latin America and the Middle East. His book entitled "Radioactivity: Introduction and History", First Edition, published by Elsevier, was on the Best Sellers List in Physics (*LibraryJournal Academic Newswire* in 2008); and his much expanded Second Edition entitled "Radioactivity: Introduction and History, From the Quantum to Quarks" published by Elsevier in 2016 (https://www.sciencedirect.com/book/9780444634894/radioactivity) was awarded an Honorary Mention in the 2017 PROSE AWARDS in the category of Chemistry & Physics.

Index

'*Note*: Page numbers followed by "f" indicate figures and "t" indicate tables.'

B

C

H

I

M

Q

R

S

T

9780128143971